P9-CQP-671

Original photographs by Betty M. Barnes

New art work by Susan Heller

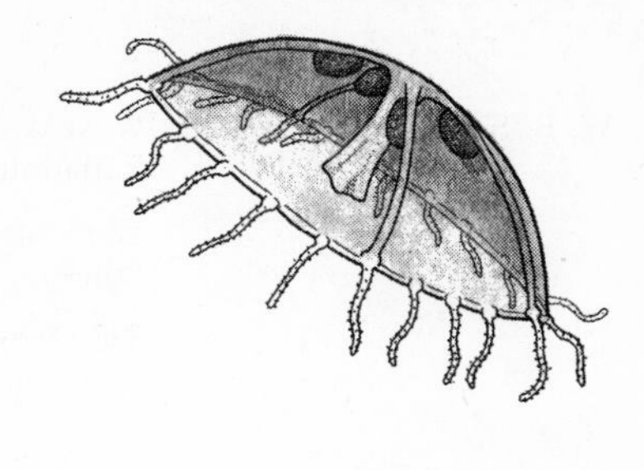

Robert D. Barnes, Ph.D.

Professor of Biology
Gettysburg College

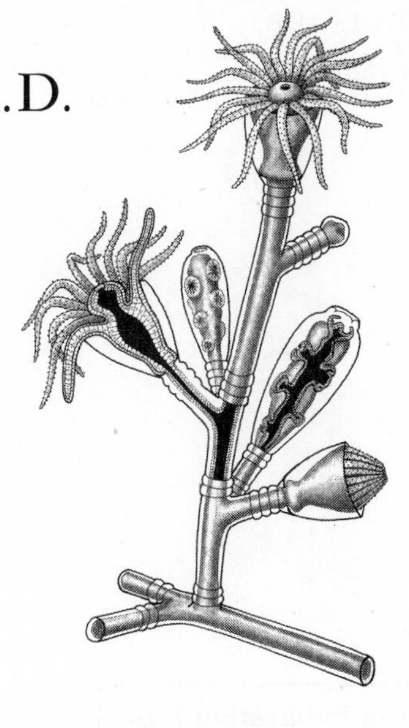

INVERTEBRATE ZOOLOGY

THIRD EDITION

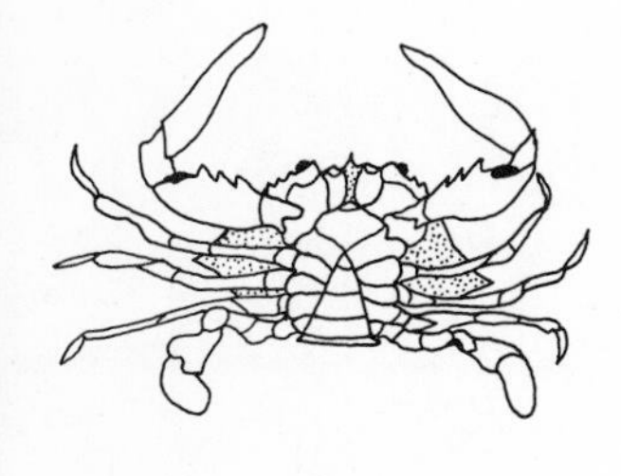

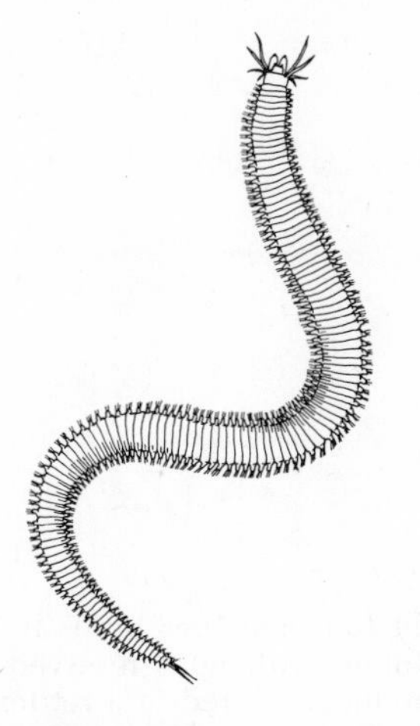

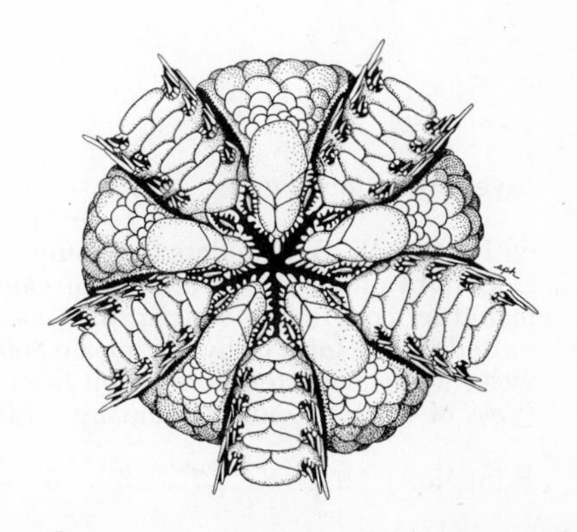

W. B. SAUNDERS COMPANY

Philadelphia, London, Toronto 1974

W. B. Saunders Company: West Washington Square
Philadelphia, PA 19105

12 Dyott Street
London, WC1A 1DB

833 Oxford Street
Toronto, Ontario M8Z 5T9, Canada

Library of Congress Cataloging in Publication Data

Barnes, Robert D.

Invertebrate zoology.

Includes bibliographies.

1. Invertebrates. I. Title.

QL362.B27 1974 592 73-89172

ISBN 0-7216-1562-7

Invertebrate Zoology ISBN 0-7216-1562-7

Print No: 9 8 7 6 5 4 3 2 1

PREFACE TO THE THIRD EDITION

The emphasis of this revision was placed on the major free-living invertebrate groups that constitute the core of most courses in Invertebrate Zoology. Although updating and correction of errors were important concerns, much attention was devoted to improving the coverage and presentation. A considerable part of some chapters, especially those on flatworms, mollusks, and lophophorates, was rewritten.

I am increasingly convinced that evolution and adaptive biology are the most useful themes for facilitating an understanding and retention of the tremendous diversity presented by invertebrates. Considerable new material of this sort has been added and such topics as locomotion, feeding, and gas exchange, which most strikingly reflect different modes of existence, have been given more attention than some others. The conventional sequence of topics has sometimes been altered where I felt that other arrangements would enhance an overall understanding. Thus, for example, "gas exchange and water circulation" is an early topic in the section on gastropods, and the discussion of the evolution of feeding precedes that on adaptive diversity of bivalves.

Although a particular evolutionary hypothesis has frequently been followed in order to develop a context or framework for understanding, opposing viewpoints have always been indicated, and I have increasingly taken a neutral stand in many controversies on phylogeny.

The literature on invertebrates—books, reviews, and research papers—continues to grow at an almost overwhelming rate. The transfer of important material from that accumulation into a textbook such as this one is increasingly difficult, not only because of the dispersed volume involved, but also because of the need to keep the length of the text within reasonable limits. For this edition, considerable anatomical and embryological descriptive details were eliminated to make room for new material. The increase of approximately 100 pages over the previous edition results largely from new illustrations.

Courses in invertebrate zoology vary greatly in the emphasis of their coverage. This text, like most others, is designed to accommodate that variation. Therefore, more material is presented than a student should be expected to learn or remember. I continue to urge that this text be used to augment the instructor's course rather than be treated as a course in itself.

The illustrations received an extensive overhauling and have been greatly enriched and improved. Some older figures have been deleted, and some have been redone. Many new figures have been added. Largely responsible for this new appearance of the book are Susan Heller and my wife, Betty M. Barnes. Susan Heller, who is not only a fine artist but a knowledgeable invertebrate zoologist, provided the new drawings, including the composite figures of invertebrate inhabitants of *Sargassum* and the surfaces of coralline stones. My wife provided many fine original photographs of a great variety of material.

This text has undergone considerable evolution since the first edition appeared in 1963. Although the many changes certainly reflect the expansion of literature over the past ten years, they also reflect the author's own continual study and more sophisticated knowledge of invertebrates. The new edition further reflects the interest and generosity of many readers and friends who have provided a wealth of helpful comments, suggestions, and information. To all I am much indebted.

I am especially grateful to those persons who read sections of the manuscript for this edition. They gave valuable advice and information and corrected errors. They are not responsible for errors that may still remain. The chapter on protozoans was reviewed by R. Barclay McGhee, University of Georgia; sponges by Klaus Ruetzler, Smithsonian Institution; cnidarians by Charles E. Cutress, University of Puerto Rico; turbellarians by Reinhard Rieger, University of North Carolina; gastrotrichs by William D. Hummon, Ohio University; nematodes by W. Duane Hope, Smithsonian Institution; polychaetes by Marian H. Pettibone, Smithsonian Institution; leeches by Roy T. Sawyer, College of Charleston; mollusks by Ruth D. Turner, Armelie Scheltema, Robert B. Bullock, Elaine Hoagland, and Carol Jones, all at Harvard University; arachnids by Herbert W. Levi, Harvard University; arthropod compound eyes by Jerome J. Wolken, Carnegie-Mellon University; insects by Pedro Barbosa, University of Massachussetts; non-decapod crustaceans by Thomas E. Bowman, Smithsonian Institution; and decapod crustaceans by Dorothy E. Bliss, American Museum of Natural History. The larger parasitic invertebrate groups were reviewed by Sherman S. Hendrix, Gettysburg College.

I would also like to thank Carol Ann Gray, June Fox, Philip Price, Esther Clapsaddle, Katherine Barnes, and Christine Tougas for helping with many tasks involved in the preparation of the manuscript.

As always, the help and encouragement provided by those associated with the W. B. Saunders Co. did much to reduce the burden of the task. To my editors, Richard Lampert and Jay Freedman, I am especially grateful.

ROBERT D. BARNES

CONTENTS

Chapter 20

THE LESSER DEUTEROSTOMES

Chapter 21

Chapter 1

INTRODUCTION

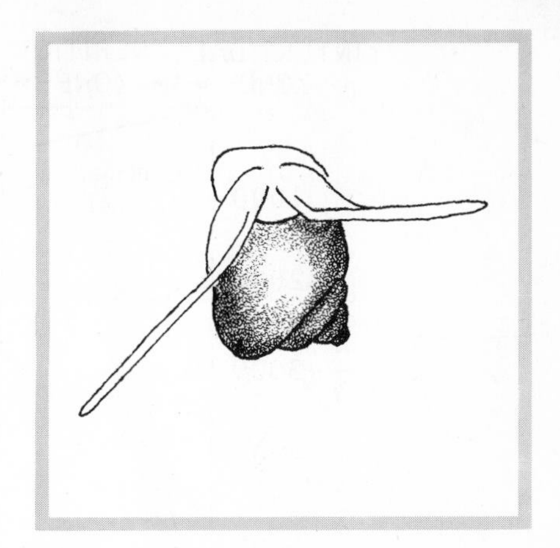

There are over a million described species of animals. Of this number about 5 per cent possess a backbone and are known as vertebrates. All others, comprising the greater part of the Animal Kingdom, are invertebrates. These animals are the subject of this book.

Division of the Animal Kingdom into vertebrates and invertebrates is artificial and reflects human bias in favor of man's own relatives. One characteristic of a single subphylum of animals is used as the basis for separation of the entire Animal Kingdom into two groups. One could just as logically divide animals into mollusks and non-mollusks or arthropods and non-arthropods. The latter classification could be supported at least from the standpoint of numbers, since approximately 85 per cent of all animals are arthropods.

The artificiality of the invertebrate concept is especially apparent when one considers the vast and heterogeneous assemblage of groups that are lumped together in this category. There is not a single positive characteristic that invertebrates hold in common. The range in size, in structural diversity, and in adaptations to different modes of existence is enormous. Some invertebrates have common phylogenetic origins; others are only remotely related. Some are much more closely related to the vertebrates than to other invertebrate groups.

Quite obviously, invertebrate zoology cannot be considered a special field of zoology, certainly not in the same sense as protozoology or entomology. A field that embraces all biological aspects—morphology, physiology, embryology, and ecology—of 95 per cent of the Animal Kingdom represents no distinct area of zoology itself. For the same reason, no zoologist can truly be called an invertebrate zoologist. He is a protozoologist, a malacologist, an acarologist, or is concerned with some aspect of physiology, embryology, or ecology of one or more animal groups. Beyond such limited areas the number and diversity of invertebrates are too great to permit much more than a good general knowledge of the major groups.

The Marine Environment. The Animal Kingdom is generally believed to have originated in Archeozoic oceans long before the first fossil record. Every major phylum of animals has at least some marine representatives; some groups, such as coelenterates and echinoderms, are largely or entirely marine. From the ancestral marine environment, different groups of animals have invaded fresh water; some have moved onto land.

Compared to fresh water and to land, the sea is a relatively uniform environment, but life is not uniformly distributed throughout the depth and breadth of the world's oceans, which cover approximately 71 per cent of the earth's surface. The margins of the continents gradually slope seaward in the form of underwater shelves to depths of 150 to 200 m. and then slope steeply to depths of 3000 m. or more. Before reaching the ocean floor, the continental slope is interrupted by a terrace or more gradual incline, formed by the continental rise (Fig. 1–1). The floor of the ocean basins, called the abyssal plain, ranges from 3000 to 5000 meters in depth and may be marked by such features as sea mounts, ridges, and trenches. Widths of the different continental shelves vary considerably. The edge of the western Atlantic shelf is some 75 miles from the shore, but along the Pacific coast of North America the continental shelf is very narrow. The widest shelf occurs off the coast of Siberia and ex-

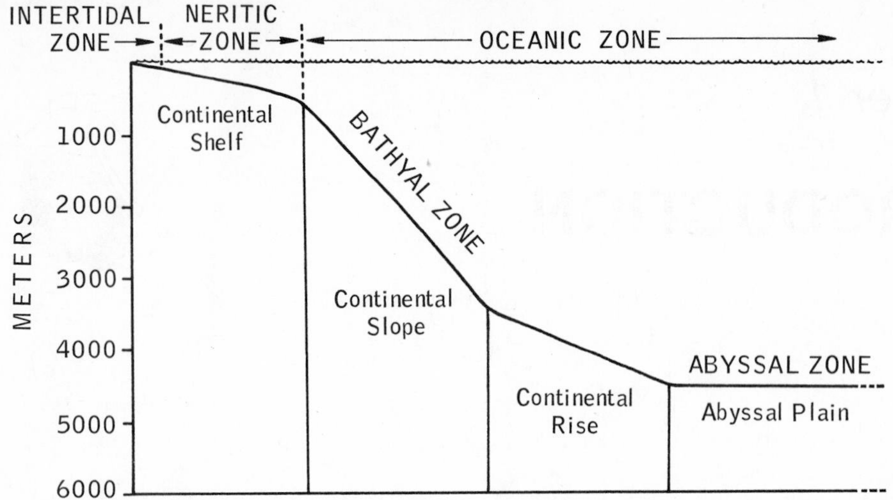

Figure 1–1. A diagrammatic cross section through an ocean basin, showing the principal horizontal and vertical zones.

tends as much as 800 miles into the Arctic Ocean.

Waters over the continental shelves comprise the *neritic zone,* and those beyond the shelf comprise the *oceanic zone* (Fig. 1–1). The edge of the sea, which rises and falls with the tide, is the *intertidal (littoral) zone.* The region above is the *supratidal (supralittoral)* and that below is the *subtidal (sublittoral).* The continental slopes form the *bathyal zone,* the abyssal plains form the *abyssal zone,* and the trenches form the *hadal zone.*

Vertical distribution of marine organisms is largely controlled by the depths of light penetration. Light sufficient for photosynthesis to exceed respiration penetrates to only a short distance below the surface or to depths as great as 200 m., depending upon the turbidity of the water. Below this upper *euphotic zone* is a transition zone, where some photosynthesis can occur but the production rate is less than the loss through respiration. From the transition zone down to the ocean floor, total darkness prevails. This region constitutes the *aphotic zone.* The animals that are permanent inhabitants of the aphotic and transition zones are carnivorous, suspension, or detritus feeders and depend ultimately on the photosynthetic activity of the microsopic algae in the upper, lighted regions. The aphotic zone animals are sometimes red, purple, or black.

The suspended or swimming animals of ocean waters constitute the *pelagic fauna,* and those that live on the bottom compose the *benthic fauna.* Bottom dwellers may live on the surface (*epifauna*) or beneath the surface (*infauna*) of the ocean floor, and usually strikingly reflect the character of the substratum, that is, whether it is a hard bottom of coral or rock, or a soft bottom of sand or mud. Many animals are adapted for living in the spaces between sand grains and compose what is commonly referred to as the *interstitial fauna* or *meiofauna.* This group includes representatives of virtually every major phylum of animals, and a number of previously unknown groups of animals have been discovered here in recent years. Pelagic and benthic animals are found in all of the horizontal zones. For example, one can refer to neritic pelagic animals or to the infauna of the abyssal zone.

The neritic, or coastal, waters support a greater population of marine life than do those of the open ocean (oceanic zone). The abundance results from the supply of nitrates, phosphates, and other nutrients dumped into coastal waters by rivers and streams or brought to the surface by upwellings and turbulence. These substances are required by photosynthetic organisms, the producers which form the base of the food chain for animal life. Oceanic waters that have a low productivity, such as the Gulf Stream and the Sargasso Sea, are clear and blue. The low concentration of plankton allows light to penetrate to a considerable depth, and the blue wavelengths are reflected from the water molecules. Sea water that is rich in plankton is green. Plankton and organic detritus reflect yellow wavelengths, which, combined with the blue wavelengths reflected by the water molecules, produce a green color.

In tropical waters of many areas the population levels are lower than in temperate and cold oceans. However, tropical and subtropical seas commonly contain a greater number of species than do temperate waters. One of the world's richest marine faunas in numbers of species occurs in the Indonesian region. The islands of this region—Borneo, Sumatra, the Celebes, and others—represent all that remains of the great, sunken land mass that once connected Australia with southern Asia. The surrounding seas, from the Gulf of Siam to the Arafura Sea between New Guinea and Australia, are very shallow and form the center of the rich Indo-Pacific fauna.

At the other extreme the oceans' abyssal plains, which are perpetually dark and icy cold, support a comparatively small fauna, both in number of species and in number of individuals.

Freshwater and Estuarine Environments. The lakes of the world also exhibit a horizontal and vertical zonation but their smaller size, shallower depth, and freshwater content make them ecologically different in many ways from oceans. The margin of a lake, where light can reach the bottom, is called the *littoral zone*. Within the littoral zone the upper lighted layer of water, equivalent to the euphotic zone of the sea, is in lakes termed the *limnetic zone*. Within and below the limnetic zone, the waters and bottom of the lake belong to the *profundal zone*.

Temperature is a primary factor controlling the environment of lakes. In contrast to salt water, which becomes increasingly dense at decreasing temperatures, fresh water reaches its greatest density at 4° C.; and thus when lakes in temperate parts of the world are warmed during spring and summer, the warm water stays at the surface while the heavier, colder water remains at the bottom. Little circulation occurs between the upper and lower levels, so that not only is the bottom zone dark, but it is also relatively stagnant from lack of oxygen and supports only a limited fauna. With the advent of cold weather, water of the upper stratum becomes heavier and sinks, resulting in a general turnover between the surface and the bottom. Conditions are stabilized again in the winter but with a reversed temperature stratification, for now the lighter, colder water in the form of ice floats at the surface, and the warmer (4° C.), heavier water is at the bottom. In the spring, following the melting of the winter ice, there is another turnover as in the fall.

Tropical lakes either have a single winter turnover or exhibit a highly stable condition, with little vertical circulation.

The junction of freshwater rivers and streams with the sea is not abrupt. Rather, the two environments grade into one another, creating the estuarine environment, characterized by brackish water, i.e., salinities considerably below the 35 per cent typical of the open sea. The estuarine environment embraces river mouths and surrounding deltas, coastal marshes, small embayments, and the finger-like extensions of the sea that probe the coast or margins of sounds. It is usually affected by tides, from which the word estuary (*aestus*, tide) is derived. The majority of marine animals are osmoconformers and stenohaline and cannot survive greatly reduced salinities. The lower and fluctuating salinities of estuaries thus restrict the estuarine fauna to those euryhaline marine invaders and few freshwater species that can tolerate these conditions. The fauna also contains some animals which have become especially adapted for estuarine conditions and are found nowhere else.

In the tropics a characteristic community of estuarine environments, as well as of more saline areas where waters are quiet, is mangrove. Mangroves are species of small trees that can tolerate saline conditions. They occupy the intertidal zone and commonly possess prop roots or special aerial roots (pneumatophores) which project above the water's surface. The most highly developed mangrove communities are found in the Indo-Pacific, where numerous species form a number of zones extending seaward. Such mangroves may occupy vast coastal areas and are virtually impenetrable. Red mangrove, *Rhizophora mangle*, which possesses long prop roots extending straight downward from the limbs, is the common mangrove of tropical America (Fig. 1–2). Mangroves trap sediment and contribute to land building. They create a habitat that is occupied by many animals and other plants.

Plankton. Both oceans and fresh-water lakes contain a large assemblage of microscopic organisms that are free-swimming or suspended in the water. These organisms comprise the plankton and include both plants (phytoplankton) and animals (zooplankton). Although many planktonic organisms are capable of locomotion, they are too small to move independently of currents.

Figure 1–2. A mangrove at low tide. This is red mangrove, *Rhizophora mangle.* Note the bolsters of algae surrounding the prop roots. (By Betty M. Barnes.)

Phytoplankton is composed of enormous numbers of diatoms and other microscopic algae.

Marine zooplankton includes representatives from virtually every group of animals, either as adults or as developmental stages. Some species (holoplankton) spend their entire lives in the plankton; the larvae of others (meroplankton) enter and leave the plankton at different points in the course of their development. About 70 per cent of the bottom-dwelling marine invertebrates have a planktonic larval life. The animal constituents of fresh-water plankton are more limited in number. Plankton, especially marine plankton, is of primary importance in the aquatic food chain. The photosynthetic phytoplankton—chiefly diatoms, dinoflagellates, and minute flagellates—form the primary trophic level and serve as food for larger animals. As would be expected, plankton attains its greatest density in the upper, lighted zone; and in productive waters planktonic organisms may occur in such enormous numbers that the water appears turbid.

Phylogeny. In the subsequent chapters, the evolutionary histories of the various phyla are explored. Their evolution is frequently used as a basis for understanding the adaptive diversity within the phylum or class. The final chapter in the book examines the different theories and schemes which have been proposed to explain the phylogenetic, or evolutionary, relationships *between* phyla. In discussing phylogenetic relationships, it is convenient to use such terms as *primitive, advanced, lower, higher,* and *specialized.* Unfortunately, these terms are not always understood by students and tend to create the erroneous impression that evolution has proceeded from one group to another toward some ideal state or goal of perfection. Such terms as *primitive* and *advanced* are relative and are significant primarily in discussing evolution within a particular group of animals. For example, *primitive species* are those that possess many, or the greatest number of, characteristics believed to have been possessed by the ancestral stock from which the living members of the group arose. *Advanced species* are those that have changed considerably compared with the primitive condition, usually as a result of different environmental situations or the assumption of a different mode of existence. *Specialized* usually refers to characteristics of species that are especially adapted to a particular ecological niche. The terms *specialized* and *advanced,* however, should not be thought of as meaning *more perfect* or *better,* because the environmental conditions for which a particular, specialized species is adapted may not prevail in primitive forms. Moreover, while certain species possess some primitive characteristics, these same species frequently are specialized in other respects.

The terms *primitive* and *advanced* can be particularly misleading when comparing different phyla, because usually only one, or a few, characteristics are being referred to. For example, since multicellular animals evolved from single-celled forms, the protozoans are, in respect to this characteristic, primitive as compared to multicellular phyla. However, in other ways protozoans are not necessarily primitive as compared to metazoans, for they have undergone at the unicellular level a great evolutionary development, leading to an intracellular specialization unequaled by the cells of metazoan animals.

The terms *higher* and *lower* usually refer to the levels at which species or groups have stemmed from certain main lines of evolution. Thus, sponges and coelenterates are often referred to as *lower* phyla, since they are believed to have originated near the base of the Animal Kingdom phylogenetic tree. This does not imply that sponges and coelenterates are primitive in all respects; for they, like all other groups of animals, have followed certain independent lines of specialization. Furthermore, this does not necessarily imply that higher groups have evolved *directly* through sponges and coelenterates.

No treatment of invertebrates is complete, nor can there be a proper understanding of invertebrate phylogeny, without some consideration of fossil forms and extinct groups. As much invertebrate paleontology has been included as space permits. From time to time, reference will be made to different geological eras and periods. For the student who has only a slight background in geology, the geological time table of Figure 1–3 may be of some value for later reference.

Bibliographies. The bibliography at the end of each chapter is not intended to be a selection of titles recommended for further reading. The literature on invertebrates is enormous, as one would expect, considering the vast area of biology it covers. Most of this literature consists of research papers scattered through a great number of biological journals published throughout the world over the last 80 years. Obviously, any list of such papers that might be given here would have to be so limited in number and so arbitrary a selection that its value would be highly questionable. The bibliographies in this text therefore are of a different nature. They consist of two categories. One category comprises the literature cited in the text; the other category, which is of considerably more importance to the student, consists of important reference works on the group of in-

ERAS	PERIODS		MOUNTAIN-MAKING EPISODES	LIFE	YEARS AGO
CENOZOIC	QUATERNARY	RECENT	ALPINE-CASCADIAN	AGE OF MAN	15,000
		PLEISTOCENE			1,500,000
	TERTIARY	PLIOCENE		AGE OF MAMMALS AND ANGIOSPERMS	
		MIOCENE			30,000,000
		OLIGOCENE			
		EOCENE	HIMALAYAN		
MESOZOIC	CRETACEOUS		LARAMIDE	AGE OF REPTILES AND GYMNOSPERMS	
	JURASSIC		SIERRA NEVADA		
	TRIASSIC				180,000,000
PALEOZOIC	"CARBONI-FEROUS"	PERMIAN	APPALACHIAN		225,000,000
		PENNSYLVANIAN	HERCYNIAN	AGE OF FISHES AND PTERIDOFHYTES	
		MISSISSIPPIAN			
	DEVONIAN				
	SILURIAN		CALEDONIAN-ACADIAN		370,000,000
	ORDOVICIAN			AGE OF INVERTEBRATES AND THALLOPHYTES	
	CAMBRIAN				
PROTEROZOIC	"PRE-CAMBRIAN"		GRAND CANYON YOUNGER LAURENTIANS		500,000,000
ARCHEOZOIC			OLDER LAURENTIANS		2,500,000,000?

Figure 1–3. Subdivisions of geologic time. (From Darrah, W. C., 1960: Principles of Paleobotany. 2nd Ed. Roland Press, New York.)

vertebrates with which the chapter deals. Those reference citations in which the coverage is not clearly indicated by the title have been provided a brief annotation. Many of these works contain the extensive bibliographies that the student should consult for additional information on particular topics.

A listing of journals containing papers on invertebrate zoology would serve little purpose because of the large number involved, and because so few are devoted to invertebrates alone. A few hours spent in a good biology library with a bibliography from one of the reference works described above would be of much greater value in initiating a student into the nature of journals of biology.

Bibliography

MULTIVOLUME WORKS COVERING INVERTEBRATE GROUPS

Bronn, H. G. (Ed.), 1866– : Klassen und Ordnungen des Tierreichs. C. F. Winter, Leipzig and Heidelberg. (Many volumes; series still incomplete.)

Grassé, P. (Ed.), 1948– : Traite de Zoologie. Masson et Cie, Paris. (Covers entire Animal Kingdom; still incomplete.)

Hyman, L. H., 1940– : The Invertebrates. Six volumes. McGraw-Hill Co., N.Y. (Volumes on annelids and arthropods not yet completed.)

Kaestner, A., 1967–1970. Invertebrate Zoology. Three volumes. Interscience Publishers, N. Y. (Completed, although lophophorates and echinoderms are not included.)

Moore, R. C. (Ed.), 1952– : Treatise on Invertebrate Paleontology. Geological Society of America and University of Kansas Press. (A detailed treatment of fossil invertebrates. Many volumes, but series is still incomplete.)

ONE-VOLUME GENERAL WORKS ON INVERTEBRATES

Barrington, E. J. W., 1967: Invertebrate Structure and Function. Houghton Mifflin Co., Boston. 549 pp. (A textbook organized from the standpoint of morphology and physiology.)

Gardiner, M., 1972: The Biology of Invertebrates. McGraw-Hill Co., N. Y. 945 pp. (A textbook organized from the standpoint of morphology and physiology.)

Hickman, C. P., 1973: Biology of the Invertebrates. 2nd Edition. C. V. Mosby Co., Saint Louis. 757 pp.

Marshall, A. J., and Williams, W. D., 1972: Textbook of Zoology: Invertebrates. American Elsevier Publishing Co., N. Y. 874 pp.

Meglitsch, P. A., 1972: Invertebrate Zoology. 2nd Edition. Oxford University Press, N. Y. 834 pp.

WORKS ON MORPHOLOGY, PHYSIOLOGY, OR ECOLOGY OF INVERTEBRATES

Beklemishev, W. N., 1969: Principles of Comparative Anatomy of Invertebrates. Two volumes. University of Chicago Press, Chicago.

Bücherl, W., and Buckley, E. E. (Eds.), 1971: Venomous Animals and Their Venoms. Vol. 3. Venomous Invertebrates. Academic Press, N. Y. 560 pp.

Bullock, T. H., and Horridge, G. A., 1965: Structure and Function of the Nervous System of Invertebrates. Two volumes. W. H. Freeman, San Francisco.

Eltringham, S. K., 1971: Life in Mud and Sand. Crane, Russak, and Co., N.Y. 218 pp. (An ecology of marine mud and sand habitats.)

Florkin, M., and Scheer, B. J. (Eds.), 1967–1972: Chemical Zoology. Seven volumes to date. Academic Press, N. Y. (A compilation of articles covering various aspects of biochemistry and physiology of animal groups.)

Giese, A. C., and Pearse, J. S., 1973: Reproduction of Marine Invertebrates. Vol. 1. Acoelomate Metazoans. Academic Press, N. Y.

Halstead, B. W., 1965: Poisonous and Venomous Marine Animals of the World. Vol. I. Invertebrates. U. S. Government Printing Office, Washington, D.C. 994 pp.

Kume, M., and Dan, K., 1968: Invertebrate Embryology. Clearinghouse for Federal Scientific and Technical Information, Springfield, Va.

Laverack, M. S., 1968: On the receptors of marine invertebrates. Oceanogr. Mar. Biol., Ann. Rev., *6*: 249–324.

MacGinitie, G. E., and MacGinitie, N., 1968: Natural History of Marine Animals. 2nd Edition. McGraw-Hill Co., N. Y. 523 pp.

Mileikovsky, S. A., 1971: Types of larval development in marine bottom invertebrates, their distribution and ecological significance: a re-evaluation. Mar. Biol., *10*:193–213.

Newell, R. C., 1970: Biology of Intertidal Animals. American Elsevier Publishing Co., N. Y. 555 pp.

Nicol, J. A. C., 1960: Biology of Marine Animals. Interscience Publishers, N. Y. 707 pp.

Prosser, C. L. (Ed.), 1973: Comparative Animal Physiology. 3rd Edition. W. B. Saunders Co., Philadelphia. 966 pp.

Russell, F. E., 1965: Marine toxins and venomous and poisonous animals. Adv. Mar. Biol., *3*:256–384.

Schaller, F., 1968: Soil Animals. Univ. Michigan Press. 114 pp.

Stephenson, T. A., and Stephenson, A., 1972: Life Between Tidemarks on Rocky Shores. W. H. Freeman, San Francisco. 425 pp. (Ecology of the intertidal zone of rocky shores. Systematic coverage of specific regions of the world.)

Swedmark, B., 1964: The interstitial fauna of marine sand. Biol. Rev., *39*:1–42.

Tombes, A. S., 1970: An Introduction to Invertebrate Endocrinology. Academic Press, N. Y. 217 pp.

Trueman, E. R., and Ansell, A. D., 1969: The mechanisms of burrowing in soft substrata by marine animals. Oceanogr. Mar. Biol., Ann. Rev., *7*:315–366.

Vernberg, W. B., and Vernberg, F. J., 1972: Environmental Physiology of Marine Animals. Springer-Verlag, N. Y. 346 pp.

Wells, M. J., 1965: Learning by marine invertebrates. Adv. Mar. Biol., *3*:1–62.

GENERAL REFERENCES FOR IDENTIFICATION

Edmondson, W. T., Ward, H. B., and Whipple, G. C. (Eds.), 1959: Freshwater Biology. 2nd Edition. John Wiley and Sons, N.Y. 1248 pp. (Guide to identification of freshwater organisms.)

Gosner, K. L., 1971: Guide to Identification of Marine and Estuarine Invertebrates. Interscience Publishers, N.Y. 693 pp. (Descriptions of the important criteria and keys for identification of marine invertebrates from Cape Hatteras to New England.)

Light, S. F., Smith, R. I., Pitelka, F. A., Abbott, D. P., and Weesner, F. M., 1967: Intertidal Invertebrates of the Central California Coast. University of California Press, Berkeley. 446 pp.

Pennak, R. W., 1953: Freshwater Invertebrates of the United States. Ronald Press, N.Y. 768 pp.

Smith, R. I. (Ed.), 1964: Keys to Marine Invertebrates of the Woods Hole Region. Contribution No. 11, Systematics-Ecology Program, Mar. Biol. Lab., Woods Hole, Mass.

LABORATORY GUIDES

Bullough, W. S., 1958: Practical Invertebrate Anatomy. Macmillan Co., N.Y. 483 pp.

Dales, R. P. (Ed.), 1970: Practical Invertebrate Zoology. University of Washington Press, Seattle. 356 pp.

Freeman, W. H., and Bracegirdle, B., 1971: An Atlas of Invertebrate Structure. Heinemann Educational Books, London. 129 pp.

Sherman, I. W., and Sherman, V. G., 1970: The Invertebrates: Function and Form. Macmillan Co., N.Y. 304 pp.

Welsh, J. H., Smith, R. I., and Kammer, A. E., 1968: Laboratory Exercises in Invertebrate Physiology. 3rd Edition. Burgess Publishing Co., Minneapolis. 219 pp.

Woolley, T. A., 1963: Laboratory Directions for Invertebrate Zoology. Burgess Publishing Co., Minneapolis. 140 pp.

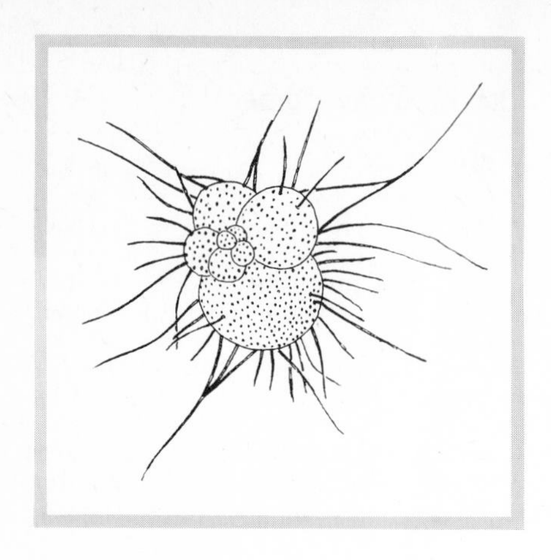

Chapter 2

THE PROTOZOANS

The protozoans are a heterogeneous assemblage of some 50,000 microscopic single-celled organisms possessing typical cellular structures. Classically, this assemblage has been treated as a single phylum within the Animal Kingdom—the phylum Protozoa—and such an approach is still utilized by the majority of contemporary protozoologists.

Despite its usage, the concept of the Protozoa as an animal phylum poses several problems. First, the unicellular level of organization is the only characteristic by which the phylum can be described; in all other respects the phylum displays extreme diversity. Protozoans exhibit all types of symmetry, a great range of structural complexity, and adaptations for all types of environmental conditions. As organisms, the protozoans have remained at the unicellular level, but have evolved along numerous lines through the specialization of parts of the protoplasm. That is, specialization has occurred through the evolution of organelles. Protozoan evolution parallels that of multicellular animals, in which specialization has occurred through the differentiation of cells within a multicellular body. Thus, simplicity and complexity in protozoans are reflected in the number and nature of their organelles, in the same way that simplicity and complexity in multicellular animals are reflected in the development of tissues and organ systems.

A phylum, as is true of any taxon, should be a monophyletic group, its members having evolved from some not too remote common ancestor. The protozoans cannot be considered closely monophyletic, and herein lies the second major problem presented by the concept Protozoa. Virtually all motile unicellular organisms have been thrown into a single phylum, with little regard to evolutionary relationships. It is among the flagellate protozoans that this acquisition by protozoologists produces the greatest aberration. Most of the free-living flagellates are plant-like autotrophic organisms that, when assembled together, clearly constitute a hodgepodge collection of largely unrelated forms. The majority represent algal organisms that are placed within a number of different phyla by the phycologists. Their affinity with other members of the algal phyla is clearly evident in many cases. For example, certain green flagellate "protozoans" are much more closely related to multicellular green algae, many of which produce motile zoospores, than they are to other flagellates.

Several solutions have been proposed to the difficulties inherent in the traditional concept of the phylum Protozoa. One solution has been to abandon the Protozoa as a phylum and to raise each of the protozoan classes to phylum rank. This rectifies the absence of unifying characters for the assemblage when treated as a single entity. It does not resolve the flagellate problem, although this could be accomplished by restricting the flagellate group to animal-like forms.

Such a restriction, plus removal of the protozoans from the Animal Kingdom, is found in the classification of living organisms into a number of kingdoms, a concept that has found favor among many biologists. Of the various schemes that have been proposed (Whittaker, 1969; Dodson, 1971), some place the blue-green algae and the bacteria, which possess poorly defined nuclei and few organelles, in the Kingdom Monera. The autotrophic multicellular organisms which form embryos constitute the Plant Kingdom, the Metaphyta. The holozoic multicellular embryo-forming organisms make up the Animal

Kingdom, the Metazoa. All of the remaining organisms form the Kingdom Protista. They possess typical (eukaryote) cells and may be unicellular or multicellular, but the gametes are always produced in single cells rather than in multicellular organs. No embryo is formed from the zygote. With this definition, the protozoan classes become phyla of the Kingdom Protista, along with the algal and fungal groups. The autotrophic flagellate "protozoans" are separated from the heterotrophic flagellates and placed among the different algal protistans.

Since the classical concept is still used by most protozoologists and in protozoology texts, the following discussion will consider the protozoans as representing a single phylum. The classification of the phylum will follow that published by the Society of Protozoologists in 1964.

Protozoans occur wherever moisture is present—in the sea, in all types of fresh water, and in soil. There are commensal, mutualistic, and many parasitic species. In fact, the sporozoans are entirely parasitic.

Although most protozoans occur as solitary individuals, there are numerous colonial forms. Some colonial forms, such as *Volvox*, attain such a degree of cellular interdependence that they approach a true multicellular level of structure (Fig. 2–2*C*). Both solitary and colonial species may be either free-moving or sessile.

The great majority of protozoans are microscopic. *Anaplasma*, a blood parasite, is so small that it occupies only 1/6th to 1/10th of a red blood cell. At the other extreme, the fresh-water ciliate, *Spirostomum*, may reach a length of 3.0 mm. and be seen with the naked eye. The colonies of some colonial species are even larger. *Nummulites*, a fossil foraminiferan of the early Cenozoic era, was 19 cm. across, probably the largest size that has been attained by any protozoan.

Protozoan Organelles and General Physiology. The protozoan cell possesses all typical cellular structures and carries out all basic cellular processes. But a protozoan represents more than the equivalent of a single cell in a multicellular animal. Rather, a protozoan, although unicellular, must be recognized as being a complete organism, carrying out all functions found in any multicellular animal. Many of these functions take place by means of organelles, specialized parts of the protoplasm. A protozoan cell can attain an extreme complexity. Sonneborn (1950) makes the statement that *Paramecium*—commonly used as an example of a "simple" animal—is "far more complicated in morphology and physiology than any cell of the body of man."

The protozoan body is usually bounded only by the cell membrane, which possesses the typical three-ply sheet ultrastructure of cells in general. The rigidity or flexibility of the protozoan body is largely dependent upon the nature of the underlying cytoplasm. This cortical cytoplasm, called ectoplasm, is usually gelatinous, in contrast to the more fluid, internal cytoplasm, called endoplasm. Ectoplasm and endoplasm are merely different colloidal states of protoplasm and are reversible. Nonliving external coverings or shells occur in many different groups. Such coverings may be simple gelatinous or cellulose envelopes; or they may be distinct shells, composed of various inorganic and organic materials, or sometimes foreign particles cemented together.

The nucleus is most commonly vesicular, containing considerable nucleoplasm and one or more nucleoli, sometimes called endosomes; or the nucleus may be densely packed. The locomotor organelles of protozoans may be flagella, cilia, or flowing extensions of the body called pseudopodia. Since the type of locomotor organelle is important in the classification of the phylum, discussion of the structure of these organelles is deferred until later.

Characteristic of many protozoans is an organelle called the contractile vacuole, or water expulsion vesicle (Fig. 2–14*B*). Contractile vacuoles are water-balancing structures, acting as pumps to remove excess water from the cytoplasm. These usually spherical vacuoles periodically collapse, releasing their fluid contents to the outside. The vacuole develops initially from the confluence of small droplets. The membrane of the vacuole, as appears to be true of all intracellular membranes, has the same ultrastructure as that of the cell membrane. The protoplasm surrounding the contractile vacuole membrane is filled with tubules and vesicles. The contents of these structures appear to dump into the main vacuole during the filling stages of its cycle (Fig. 2–43). One to several contractile vacuoles may be present within the animal, and the position and structure vary within different groups. Contractile vacuoles are most commonly encountered in freshwater protozoans with cytoplasm hypertonic to the aqueous environment. However, contractile vacuoles are also present in some marine groups and in certain parasites.

All types of nutrition occur in protozoans. Some are autotrophic and others are saprozoic; many are holozoic, and digestion occurs intracellularly within food vacuoles. Intracellular digestion has been most studied in amebas and ciliates. The food vacuoles undergo definite changes in hydrogen ion concentration (pH) and in size during the course of digestion. Following ingestion, the vacuole contents become increasingly acid and smaller. After the initial acid phase, enzymes in an alkaline medium pass from the cytoplasm into the vacuole, and the vacuole increases in size and becomes alkaline. The vacuole contents are then digested, the products absorbed, and the undigestible remnants egested to the outside. Roth (1960) studied the various phases of intracellular digestion with electron microscopy. During the phase when the vacuole is receiving alkaline fluid and enlarging, the vacuole membrane is greatly folded and protruded toward the interior of the cavity. During the actual digestion and absorption phase, the outer or cytoplasmic side of the vacuole membrane is surrounded by minute vesicles. These vesicles arise by evagination and pinching off of the vacuole membrane (pinocytosis) and probably represent the transfer of material from the vacuole to the cytoplasm. These observations probably reflect the association of the food vacuole with the lysosome system of the cell, an association which is also characteristic of intracellular digestion in multicellular animals.

Gas exchange occurs by the diffusion of oxygen across the cell membrane. Protozoans that live in water where there is active decomposition of organic matter, or live in the digestive tract of other animals, can exist with little or no oxygen present. Some protozoans are facultative anaerobes, utilizing oxygen when present but also capable of anaerobic respiration. Changing availability of food supply and of oxygen associated with decay typically results in a distinct succession of populations of protozoan species (Bick, 1973).

Metabolic wastes are diffused to the outside of the organism. Ammonia is the principal nitrogenous waste, and the amount eliminated varies directly with the amount of protein consumed.

Reproduction. The protozoan reproductive processes and life cycles are varied, and a formidable terminology has been created to describe peculiarities of different groups. Only a few of the more common terms are described here.

Asexual reproduction occurs in most protozoans and is the only known mode of reproduction in some species. Division of the animal into two or more daughter cells is called fission. When this process results in two similar daughter cells, it is termed binary fission; when one daughter cell is much smaller than the other, the process is called budding. In some protozoans multiple fission, or schizogony, is the rule. In schizogony, after a varying number of nuclear divisions, the cell divides into a number of daughter cells. With few exceptions, asexual reproduction involves some replication of missing organelles following fission.

Sexual reproduction may involve fusion (syngamy) of identical gametes (called isogametes), or gametes that differ in size and structure. The latter, called anisogametes, range from types that differ only slightly in size to well differentiated sperm and eggs. Meiosis commonly occurs in the formation of gametes, but in many flagellate protozoans meiosis is post-zygotic, that is, occurs following the formation of the zygote as in most algae. In ciliate protozoans there is no formation of distinct gametes; instead two animals adhere together in a process called conjugation, and they exchange nuclei. Each migrating nucleus fuses with a stationary nucleus in the opposite conjugant to form a zygote nucleus (synkaryon). Less common is a process called autogamy, in which two nuclei, each representing a gamete, fuse to form a zygote, all within a single individual.

Encystment is characteristic of the life cycle of many protozoans, including the majority of freshwater species. In forming a cyst, the protozoan secretes a thickened envelope about itself and becomes inactive. Depending on the species, the protective cyst is resistant to desiccation or low temperatures, and encystment enables the animal to pass through unfavorable environmental conditions. The simplest life cycle includes only two phases; an active phase and a protective, encysted phase. However, the more complex life cycles are often characterized by encysted zygotes or by formation of special reproductive cysts, in which fission, gametogenesis, or other reproductive processes take place.

Protozoans may be dispersed long distances in either the motile or encysted stages. Water currents, wind, and mud and debris on the bodies of water birds and other animals are common agents of dispersion.

SUBPHYLUM SARCOMASTIGOPHORA

The flagellate and ameboid protozoans are included within the subphylum Sarcomastigophora. With few exceptions the nuclei are of one type, i.e., monomorphic, although more than one may be present. Spores are not formed, and sexual reproduction, when present, involves syngamy.

Superclass Mastigophora

The superclass Mastigophora (or Flagellata) includes those protozoans that possess flagella as adult locomotor organelles; it is generally considered to be the most primitive of the major protozoan groups.

Mastigophorans are conveniently divided into the phytoflagellates and the zooflagellates.

The phytoflagellates (class Phytomastigophorea) usually bear one or two flagella, and typically possess chromoplasts, usually called chromatophores,* which are bodies containing the pigments necessary in photosynthesis (Fig. 2–1A and B). These protozoans are thus plant-like and usually holophytic. In fact, phycologists treat most species in this division as algae. The phytoflagellate division contains most of the free-living members of the class and includes such common forms as *Euglena, Chlamydomonas, Volvox,* and *Peranema.*

The zooflagellates (class Zoomastigophorea) possess one to many flagella, lack chromoplasts, and are either holozoic or saprozoic. Some are free-living, but the majority of species are commensal, symbiotic, or parasitic in other animals, particularly

*Although *chromatophore* is the more correct term, *chromoplast* has been used here to avoid confusion with the special pigment cells of multicellular animals, which are also called chromatophores.

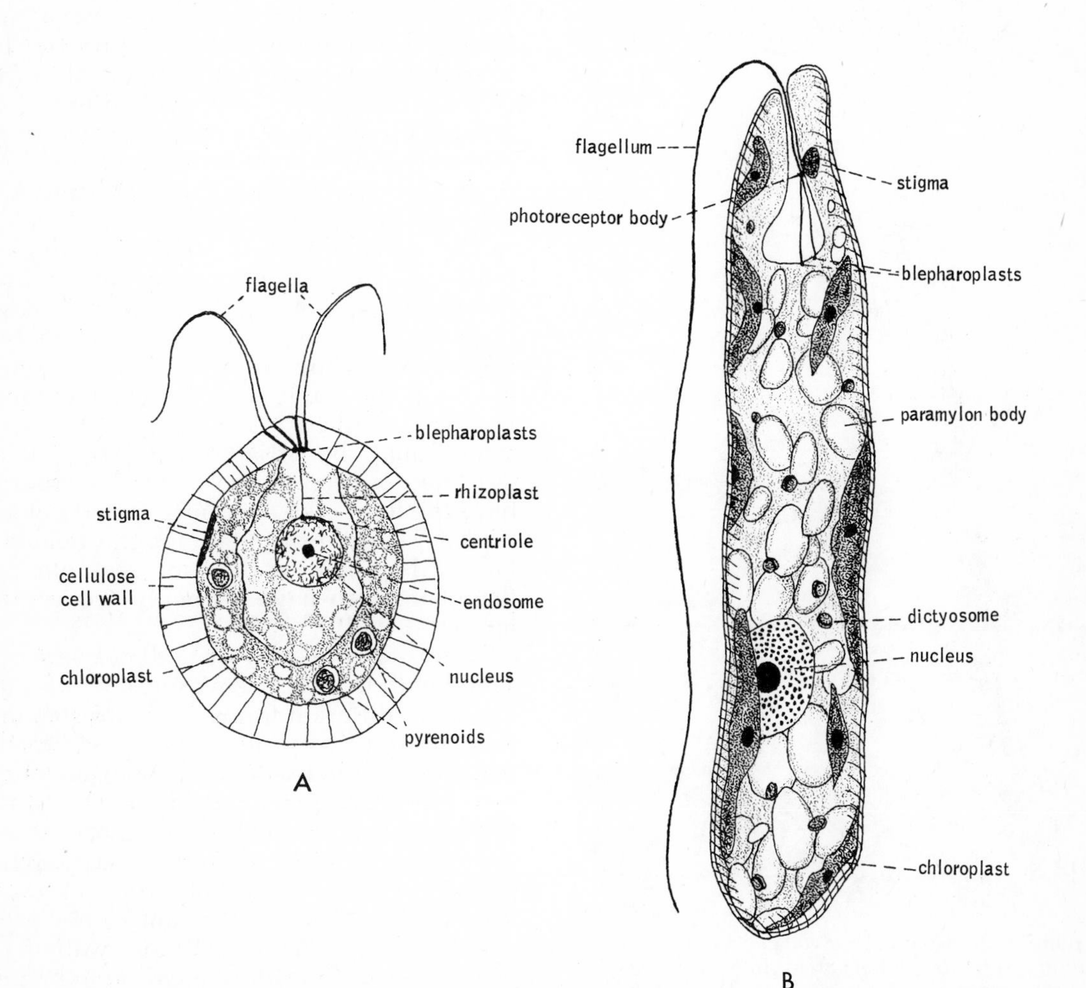

Figure 2–1. *A. Haematococcus*, a phytoflagellate with a single cup-shaped chloroplast. (After Elliot from Hyman.) *B. Euglena gracilis.* (After Hollande.)

arthropods and vertebrates. Many groups have become highly specialized. It is generally agreed that this division does not represent a closely related phylogenetic unit. The zooflagellates have probably evolved from a number of different holophytic groups through the loss of pigments.

Form and Structure. Most flagellates possess distinct anterior and posterior ends, although almost any plan of symmetry occurs. Numerous species of colonial flagellates exist, and in the phytoflagellate order Volvocida, which contains many freshwater representatives, colonial organization reaches a remarkably high level. These colonies exist in the form of flattened curved plates or hollow spheres. A gelatinous envelope surrounds the individual cell or surrounds the entire colony as a sheath. *Gonium* consists of 4 to 16 cells in a quadrangular plate (Fig. 2–2*A*). Colonies of *Pandorina* (Fig. 2–2*B*) and *Eudorina* have spherical or ovoid shapes with internal cavities of varying sizes and consist of 4 to 64 cells. The largest volvocid colonies, which occur in species of *Volvox* (Fig. 2–2*C*) and *Pleodorina,* are large, hollow spheres. Each of these volvocid colonies possesses a synchronized flagellar beat, and swims with a definite anterior pole directed forward. In *Pleodorina* and *Volvox*, a degree of cellular specialization even exists. Some of these cells are strictly somatic, while others are reproductive. The reproductive cells are larger and located away from the anterior pole.

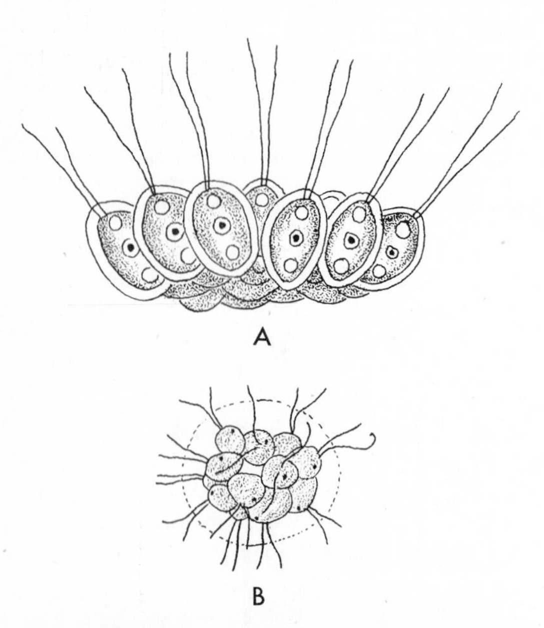

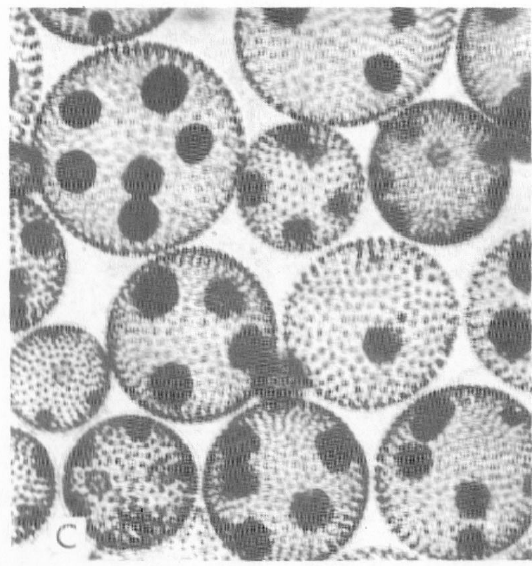

Figure 2–2. Colonial volvocids. A. Side view of *Gonium pectorale.* (After Stein from Pavillard.) *B. Pandorina morum.* (After Smith from Hall.) *C. Volvox.* Note daughter colonies within parent colonies. (Courtesy of General Biological Supply House, Inc.)

The choanoflagellates, a group of freshwater zooflagellates that contains a number of interesting colonial forms, are peculiar in having a cylindrical collar around the base of each flagellum (Fig. 2–3*A*). Most of these colonies are sessile and attached to the substratum directly or by a stalk. In *Proterospongia,* however, the colony consists of a gelatinous mass into which the flagellated collar cells are imbedded.

The cell membrane may form the only body covering, as in euglenids, some dinoflagellates, and most zooflagellates; it may permit some change in shape, such as bulging and bending or even ameboid movement. In the euglenids, the cell membrane is often modified into a thickened pellicle, which may be striated (Fig. 2–1*B*). Under the electron microscope the striations have the form of deep grooves lying between ridges. The ridges contain longitudinal tubular fibrils. Some of the more complex zooflagellates also have a deeply grooved surface (Fig. 2–10).

Many of the phytoflagellates are provided with nonliving coverings, which take a variety of different forms. The chrysomonads may have a gelatinous covering; or they may be provided with a siliceous or calcium carbonate skeleton in the form of scales or plates, to which long spines are often attached (Fig. 2–3*B*). All Volvocida, which include *Volvox, Pandorina, Haematococcus* (Fig. 2–1*A*), *Polytomella,* and other green, plantlike forms, have a cellulose wall of variable thickness outside the cell membrane.

One of the most interesting groups of phytoflagellates possessing a nonliving

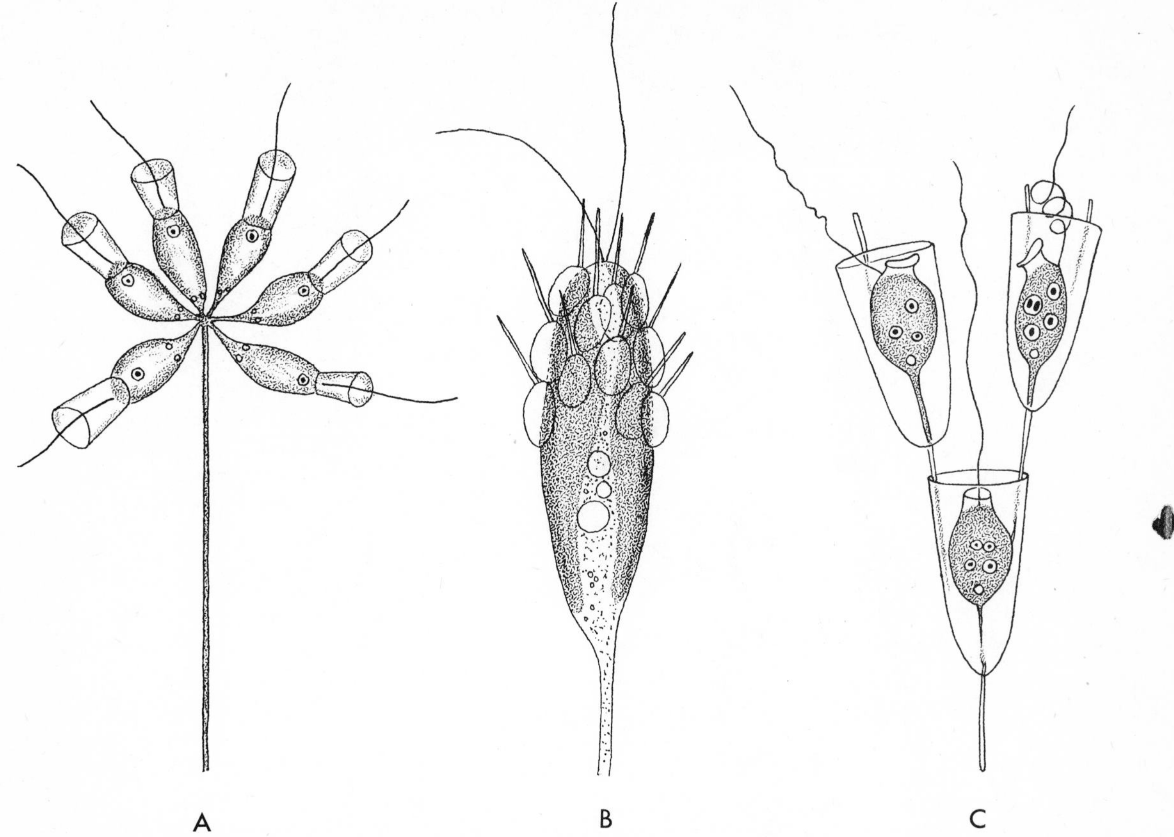

Figure 2–3. *A.* Colony of the choanoflagellate, *Codosiga botrytis.* (After Stein from Manwell.) *B. Synura splendida* with scales and spines. (After Korschikov from Hollande.) *C. Poteriodendron.* A stalked, colonial genus of the loricate zooflagellate order Bicosoecida. (After Stein from Kudo.)

covering is the marine and freshwater Dinoflagellida. The primitive dinoflagellate body is roughly ovoid, but asymmetrical (Fig. 2–4*A*). Typical dinoflagellates possess two flagella. One is attached a short distance behind the middle of the body; it is directed posteriorly and lies in a longitudinal groove called the sulcus. The other flagellum is transverse and is located in a groove that either rings the body or forms a spiral of several turns. The transverse groove is called the girdle, if it is a simple ring, or the annulus, if it is spiraled.

Dinoflagellates can be divided into unarmored and armored types. The unarmored forms are naked or enclosed in a simple, continuous, cellulose envelope. The common freshwater and marine genus, *Gymnodinium*, is an unarmored type (Fig. 2–4*A*). Armored dinoflagellates have either a cellulose covering consisting of two valves or a covering made of cellulose plates cemented together (Fig. 2–4*B*). Frequently the armor is sculptured, and often long projections or winglike extensions protrude from the body, creating very bizarre shapes. For example, some species of *Ceratium*, a common marine and freshwater genus, possess an anterior horn and two, long posterior horns, which curve forward, giving the body an anchor-like appearance (Fig. 2–4*E*). *Histiophysis* is shaped somewhat like a jug (Fig. 2–4*C*), and *Ornithocercus* has beautiful sail-like wings (Fig. 2–4*D*). Another interesting genus is *Polykrikos*, a colonial group, in which 2, 4, 8, or 16 individuals are permanently united (Fig. 2–4*G*). The large and aberrant *Noctiluca* (Fig. 2–4*F*), and smaller species of some common genera, are luminescent and are the principal contributors to planktonic bioluminescence. When present in large numbers they may produce a striking effect in a quiet sea at night.

Dinoflagellates occur in countless numbers in marine plankton, and a number of forms are abundant in fresh water. Marine species of the genera *Gymnodinium* and *Gonyaulax* are responsible for outbreaks of the so-called "red tides" off the coasts of New England, Florida, California, Europe, and elsewhere. Under ideal environmental conditions, and perhaps with the presence of a growth-

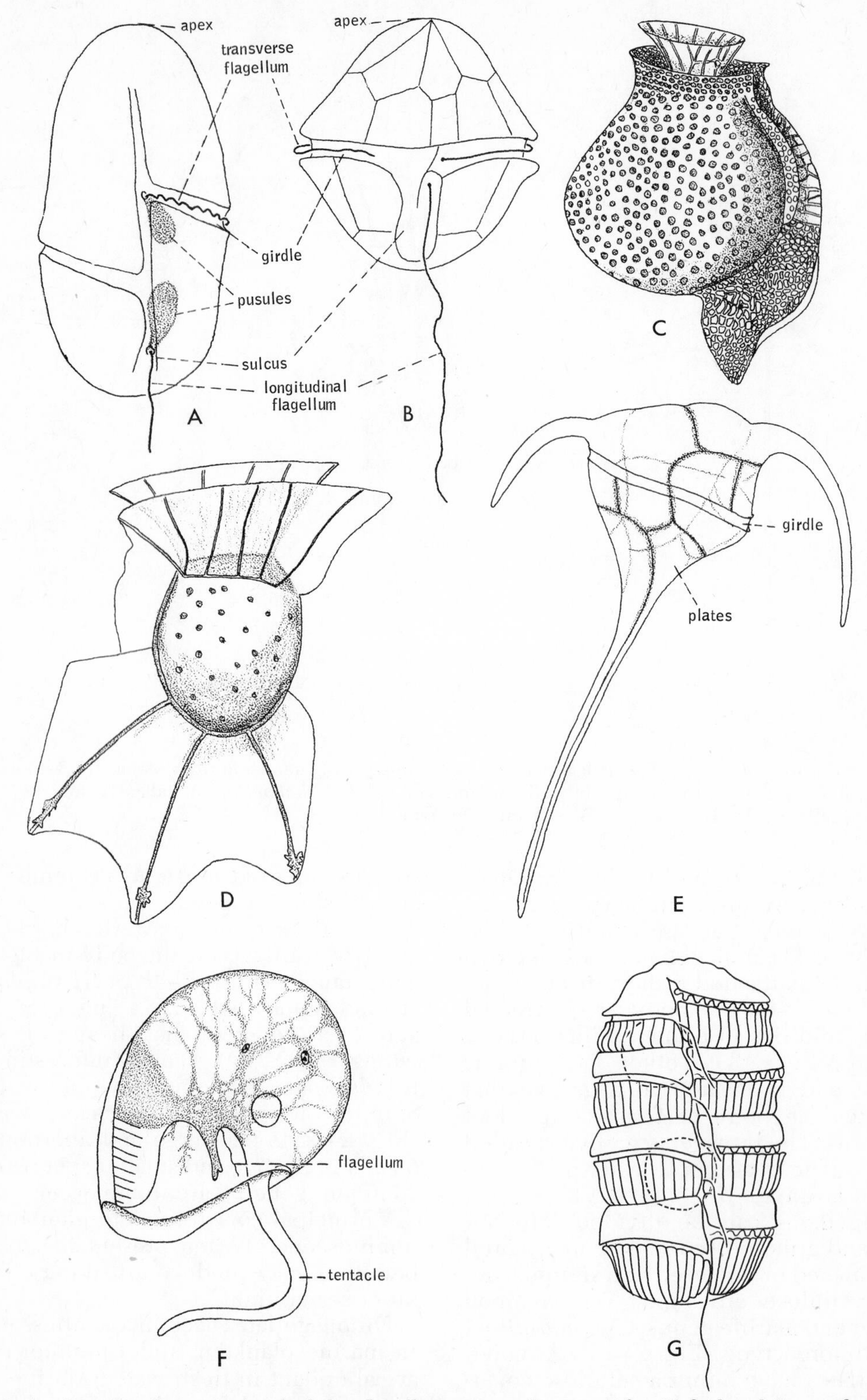

Figure 2–4. Dinoflagellates. *A.* A naked dinoflagellate, *Gymnodinium.* (After Kofoid and Swezy from Hall.) *B.* A freshwater armored dinoflagellate, *Glenodinium cinctum.* (After Pennak.) *C. Histiophysis rugosa.* Flagella are hidden. *D. Ornithocercus.* (*C* and *D*, after Kofoid and Skogsberg from Chatton.) *E. Ceratium.* (After Jörgenson from Hyman.) *F. Noctiluca.* An aberrant bioluminescent dinoflagellate. Only one of the small flagella is visible within the "oral" depression. (After Robin from Kudo.) *G. Polykrikos kofoidi.* A colonial dinoflagellate composed of four individuals. (After Kofoid and Swezy from Kudo.)

promoting substance, populations of certain species increase to enormous numbers. Riegel and others (1949), describing an outbreak of "red tide" in Monterey Bay, California, reported that the dinoflagellate *Gonyaulax* reached a density of 20 to 40 million organisms per cubic centimeter of sea water. As a result, the water was colored red in daylight and was brightly luminescent at night. However, "red tides" are not always red. The water may be colored yellow or brown. Concentrations of certain toxic metabolic substances reach such high levels that other marine life may be killed. The 1972 red tides off the coasts of New England and Florida killed thousands of birds, fish, and other animals and wreaked havoc with the shellfish industry because of the danger of eating clams and oysters which had fed upon the dinoflagellates.

Among the zooflagellates, a nonliving vase-like encasement, or lorica, is characteristic of the order Bicosoecida (Fig. 2–3*C*). Species of this largely freshwater group may be attached or free-swimming, solitary or colonial.

The greatest complexity among flagellates in organelle structure is attained by certain mutualistic and parasitic zooflagellates (Fig. 2–13). They will be discussed in more detail in connection with nutrition.

Locomotion. Locomotion of flagellates is typically by means of flagella. The phytoflagellates usually have one or two flagella; the zooflagellates, one to many. When two or more flagella occur, they may be of equal or unequal length, and one may be leading and one trailing, as in *Peranema* (Fig. 2–5*A*) and the dinoflagellates. Flagella are characteristically attached at the anterior end, except in dinoflagellates and some zooflagellates. In the cryptomonads (Fig. 2–5*B*) and the euglenids (Figs. 2–1*B* and 2–5*A*), the flagella originate from an anterior pit that, when deep and vase-like, is known as a cytopharynx or reservoir.

Electron microscopy has revealed that the structures of flagella and cilia are fundamentally similar throughout the Animal and Plant Kingdoms. A single flagellum (or cilium) is constructed very much like a cable. Two central microtubules form a core, which is in turn encircled by nine outer microtubules. The outer microtubules are actually double (Fig. 2–6*A*). The entire bundle is en-

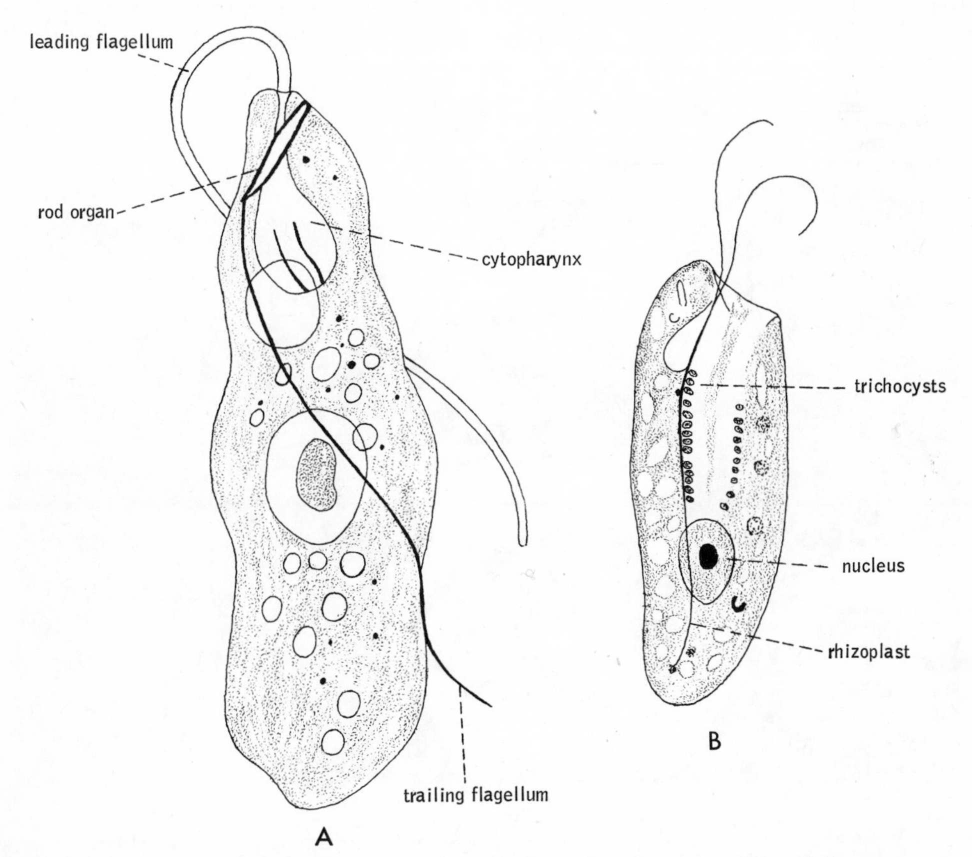

Figure 2–5. *A. Peranema trichophorum,* a holozoic euglenoid. (After Chen.) *B. Chilomonas paramecium,* a cryptomonad. (After Prowazek from Hyman.)

closed within a sheath that is continuous with the cell membrane. The flagellum always arises from a basal body, or kinetosome, that lies just beneath the surface. Like centrioles, the kinetosomes have an ultrastructure somewhat like the flagellum or cilium, except that the central fibrils are absent and the nine fibrils in the outer circle are in triplets, two of the three being continuous with the doublets of the flagellum. The kinetosome may act as a centriole; if not, a fibrillar connection called a rhizoplast runs between the kinetosome and the centriole or nucleus (Figs. 2–1*A* and 2–5*B*). Associated with the flagellar apparatus of some mutualistic zooflagellates is an intracellular filament or rod, called the axostyle (Fig. 2–13*C*). The axostyle usually connects anteriorly with the kinetosome and runs through the length of the body. The complex of flagellum, kinetosome, and other structures is called a mastigont, or, when the complex is associated with the nucleus, a karyomastigont.

The flagella of many species of phytoflagellates bear delicate lateral fibrils, called mastigonemes, which cannot be seen with the ordinary light microscope (Fig. 2–6*B*). Manton (1959) believes that the mastigonemes arise from two of the nine enveloping fibers of the flagellum. Not uncommonly, only one of the two flagella possesses mastigonemes, the trailing or less active flagellum being simple. Mastigonemes probably function to increase the surface area of the flagellum.

Flagellar propulsion in most mastigophorans essentially follows the same principle as that of a propeller, the flagellum undergoing spiral undulations which either push or pull. In some species the undulatory waves pass from base to tip and drive the organism in the opposite direction (Fig. 2–7*A*). This type of propulsion is found in *Chlamydomonas* and the dinoflagellates (Fig. 2–7*D*). In other species, such as some trypanosomes, euglenas, *Peranema*, and mastigamoebas, the undulations pass from tip to base and pull the organism (Fig. 2–7*B* and *C*). In its progression, the body commonly describes a spiral course and, at the same time, rotates.

An explanation of flagellar (or ciliary) movement in relation to the ultrastructure of the organelle is still to be provided. That such a relationship exists seems obvious from the uniformity of flagellar structure among all organisms except the monerans. The prevalent view holds that motion results from con-

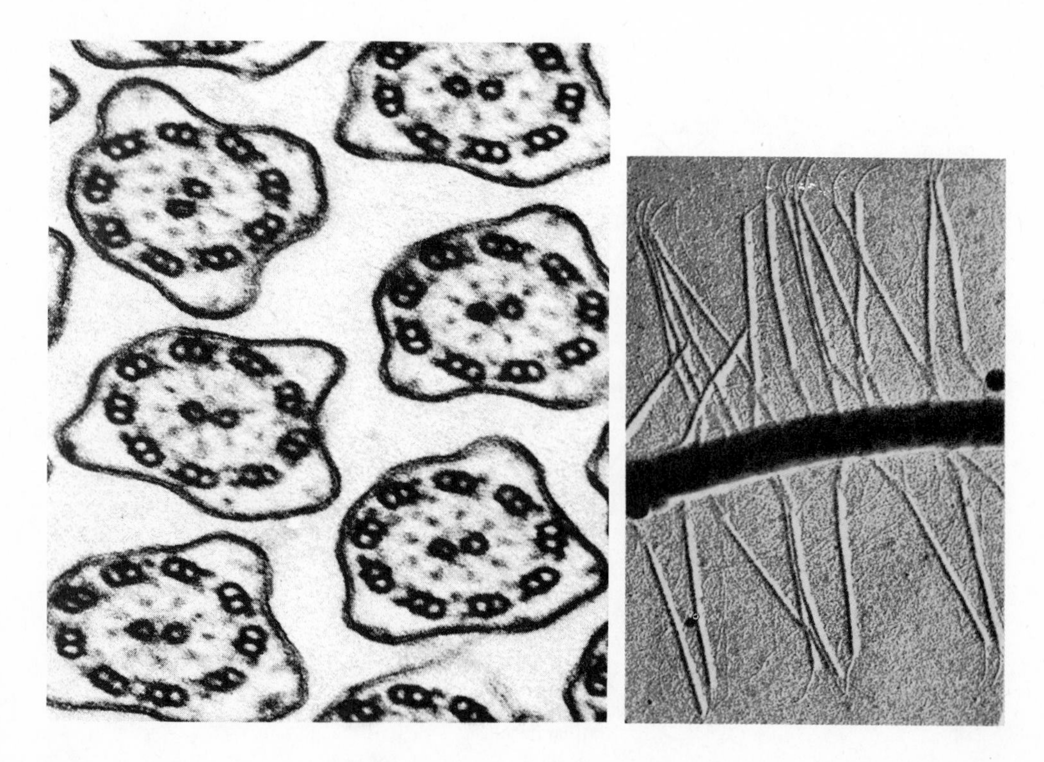

Figure 2–6. *A*. Electron photomicrograph of cross sections of some flagella of a zooflagellate. The two central fibrils are surrounded by nine double fibrils. The outer bounding membrane is continuous with the surface body membrane. (Courtesy of I. R. Gibbons.) *B*. Mastigonemes of flagellum of *Ochromonas*. (Courtesy of D. R. Pitelka.)

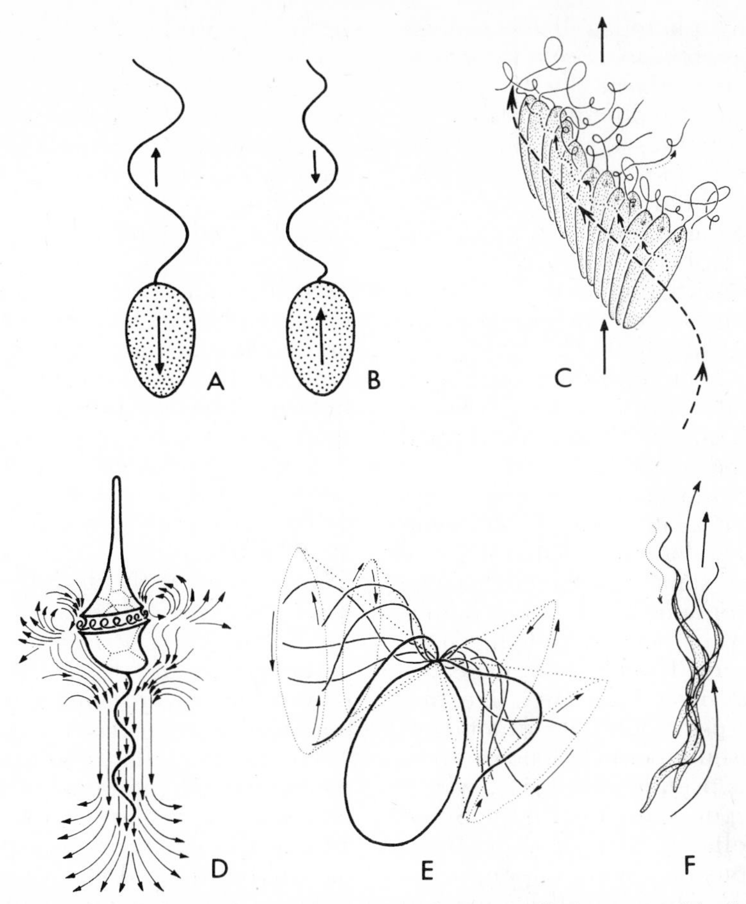

Figure 2–7. Flagellary locomotion. *A.* Pushing force (like a boat propeller) generated by base to tip undulations of the flagellum. *B.* Pulling force (like an airplane propeller) generated by tip to base undulations of the flagellum. *C.* Movement in *Euglena viridis.* Actual path indicated by dashed arrows. *D.* Locomotion in the dinoflagellate *Ceratium.* Arrows indicate the water currents generated by the transverse and posterior flagella. *E.* Locomotion in the phytoflagellate *Polytomella.* Arrows indicate the direction of flagellar beat. *F.* Locomotion of the blood parasite *Trypanosoma.* Dotted arrow indicates movement of undulating membrane; fine solid arrow, the actual path of movement. (From Jahn, T. L., and Bovee, E. C., 1967: Motile behavior of Protozoa. *In* Chen, T. (Ed.), Research in Protozoology. Pergamon Press, N.Y., pp. 41–200.)

traction of the peripheral fibers of the organelle. The subject has been reviewed by Kinosita and Murakami (1967).

Mastigophorans which have thin, flexible pellicles are often capable of ameboid movement; and some forms, such as chrysomonads, may cast off their flagella and assume an ameboid type of locomotion entirely.

Nutrition. Phytoflagellates are primarily holophytic; but some are saprozoic or holozoic, or combine any two of these three modes of nutrition. Colored species possess chromoplasts, which contain chlorophyll and any of a number of xanthophylls. When the chlorophyll is not masked by other pigments, a flagellate appears green in color, as in the phytomonads and euglenids. If the xanthophylls dominate, the color is red, orange, yellow, or brown, as in the dinoflagellates.

Flagellate nutrition has provoked considerable interest, and the literature on the subject has become enormous. Culturing has been the most frequent method of approach in these studies. A species is grown in a prepared medium, and population growth rate is used as an index of the utilization of particular medium constituents. Such studies indicate that many flagellates cannot be classified as strict holophytes or saprophytes (the term *saprophytic,* used with reference to plant-like forms, is considered here as synonymous to *saprozoic*) and that intermediate conditions between these two types of nutrition exist. Some species, such as *Euglena gracilis,* are strictly photoautotrophic and can synthesize organic compounds from inorganic sources. But many phytoflagellates require either an organic carbon source or an organic nitrogen source for synthesis.

Although all colored phytoflagellates are capable of photosynthesis, a number are also facultative saprophytes. For example, *Haematococcus* is holophytic in light and saprophytic in the dark (Fig. 2–1*A*). *Euglena* also becomes saprophytic in the absence of light and even loses its pigments. A number of chrysomonads and some dinoflagellates are both holozoic and holophytic.

There have also been numerous studies on vitamin and metal requirements in flagellates. A review of some of these nutritional studies is presented in Kidder (1967).

Holozoic nutrition occurs in colorless phytoflagellates and in zooflagellates, but some colored phytoflagellates, such as the chrysomonads, and some dinoflagellates also display holozoic tendencies. The mechanics of ingestion and the processes of digestion are still very obscure. The chrysomonads engulf food particles in a food cup, like that in many amebas (page 31). The dinoflagellates also utilize pseudopodia in food capture. *Ceratium* forms a pseudopodial net, which extends through pores in the theca. Some marine dinoflagellates also possess two vacuoles, called pusules, which open to the outside (Fig. 2–4*A*). These pusules appear to *draw in* fluid and perhaps are used to ingest particulate material.

The mode of ingestion in *Peranema,* a colorless freshwater euglenid, has been studied in some detail. The anterior of this euglenid contains two parallel rods, making up a so-called rod organ, located adjacent to the reservoir (Fig. 2–5*A*). The anterior of the rod organ terminates at the cytostome, which is just below the outer opening of the canal leading from the reservoir. *Peranema* feeds on a wide variety of living organisms, including *Euglena;* and the cytostome can be greatly distended to permit engulfment of large prey. In feeding, the rod organ is protruded and used as an anchor to pull in prey, which is swallowed whole; or the rod organ can cut into the victim, and the contents are then sucked out (Fig. 2–8). Following ingestion, the prey is digested within the food vacuoles.

Among zooflagellates, the choanoflagellates could be described as filter feeders. The collar surrounding the flagellum possesses an ultrastructure of parallel contractile fibrils. The beating of the flagellum sweeps water from the outside through the collar (Fig. 2–9*A* and *B*). Particles are trapped on the outer surface of the collar mesh and then moved downward into a receiving vacuole behind the collar base.

The zooflagellates contain most of the parasitic mastigophorans; but there are some dinoflagellates that are ectoparasites on the gills of fish or on marine annelids and crustaceans, or are endoparasites in the digestive tracts of different invertebrates. The euglenoids *Euglenomorpha* and *Phacus* are gut parasites of tadpoles. Parasitic zooflagellates occur largely in arthropods and vertebrates and often have complex life cycles involving two hosts. *Leishmania* and *Trypanosoma* are the most notable parasites in the group.

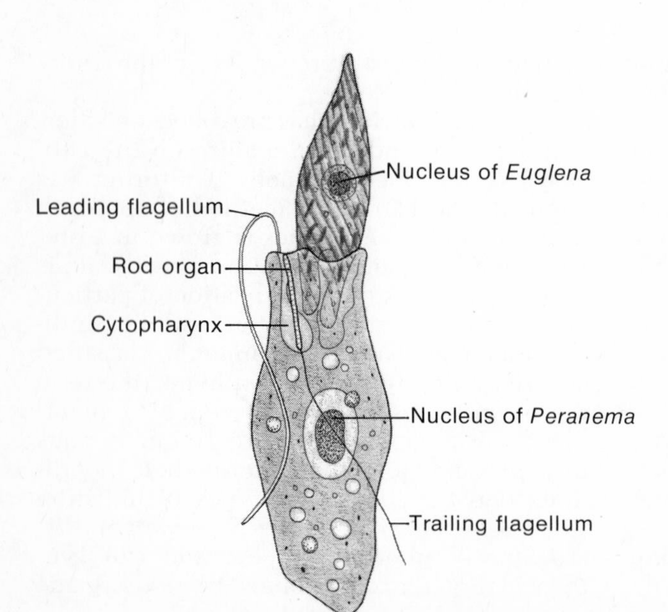

Figure 2–8. A *Peranema* swallowing a *Euglena.* (After Chen.)

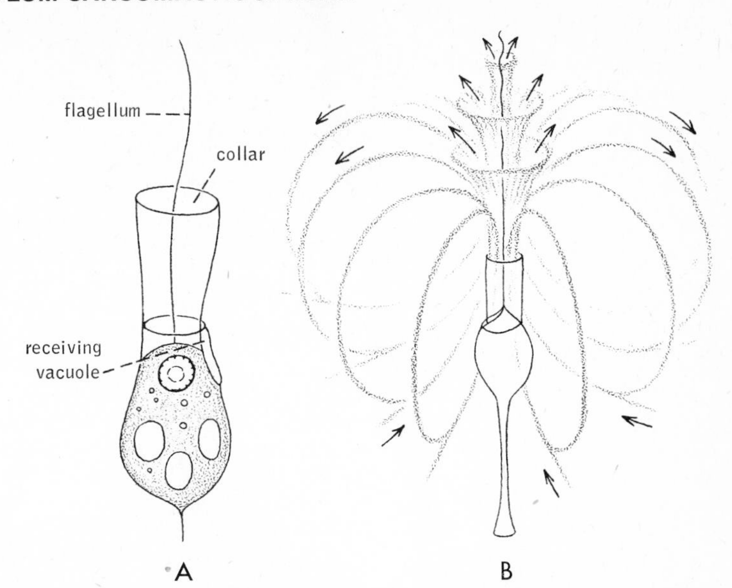

Figure 2–9. *A*. The choanoflagellate *Codosiga*. *B*. Water currents produced by the beating of the flagellum of *Codosiga*. (After Dogiel.)

Parasitic zooflagellates are not all limited to animal host. A few attack certain families of flowering plants.

There are many commensal and symbiotic flagellates. Among the most interesting are the intestinal symbionts of cockroaches, wood-eating cockroaches, and termites. These flagellates comprise members of several orders, particularly the Hypermastigida, and many are extremely complex (Fig. 2–10*A*). The hypermastigids are nearly all multi-flagellate with a saclike or elongated body, usually bearing an anterior rostrum and cap. The kinetosome complex from which the flagella arise is called the mastigont. The ultrastructure of *Trichonympha*, which possesses thousands of flagella and virtually defies description, has been studied by Pitelka and Schooley (1958), Gibbons and Grimstone (1960), and Grimstone (1959, 1961) (Fig. 2–10*A* and *B*). However, there is also a hypermastigid in termites which has no flagella but is moved by attached bacteria (spirochaetes).

In some termites and wood-eating cockroaches, the host is dependent upon its flagellate fauna for the digestion of wood. According to Cleveland (1925, 1928, 1934, and 1960), the wood consumed by the roach or termite is ingested by the flagellates, and the products of digestion are also utilized by the insect. The ingestion of the wood particles occurs at the posterior end of the flagellate by pseudopodial engulfment. The termite host loses its fauna with each molt; but by licking other individuals, by rectal feeding, or by eating cysts passed in feces (in the case of roaches), a new fauna is obtained. In wood-eating cockroaches, the life cycles of the flagellates are closely correlated with the production of molting hormones by the late nymphal insect.

Another commensal group, the opalinids, occurs in the gut of fish, frogs, and toads. These leaf-like, mouthless protozoans are of interest because they are uniformly covered by longitudinal rows of cilia rather than flagella. However, they possess monomorphic nuclei and their reproduction is like that of flagellates. The opalinids are currently placed in a separate superclass, the Opalinata, within the Sarcomastigophora. They are thus raised to equal rank with the flagellate and the ameboid members of the subphylum.

Phytoflagellates store reserve foods such as oils or fats; or they may store carbohydrates in the form of typical plant starch (as in phytomonads), paramylum (in euglenids [Fig. 2–1*B*]), or leucosin. In zooflagellates, glycogen is the usual reserve food product.

Water Balance, Excretion, and Sensory Organelles. Little information is available on water balance and excretion in flagellates, although contractile vacuoles (usually one per organism) occur in freshwater species and some trypanosomatid parasites. Euglenids have one or two contractile vacuoles, which open into a reservoir (Fig. 2–5*A*).

Eye spots, or stigmata, composed of granules of a carotin pigment called hematochrome, occur in phytomonads, green euglenids, and dinoflagellates. In *Euglena* the stigma is composed of a cup of pigment that shades a light-sensitive material, or photoreceptor body, located near the base of the flagellum (Fig. 2–1*B*). *Euglena* is positively

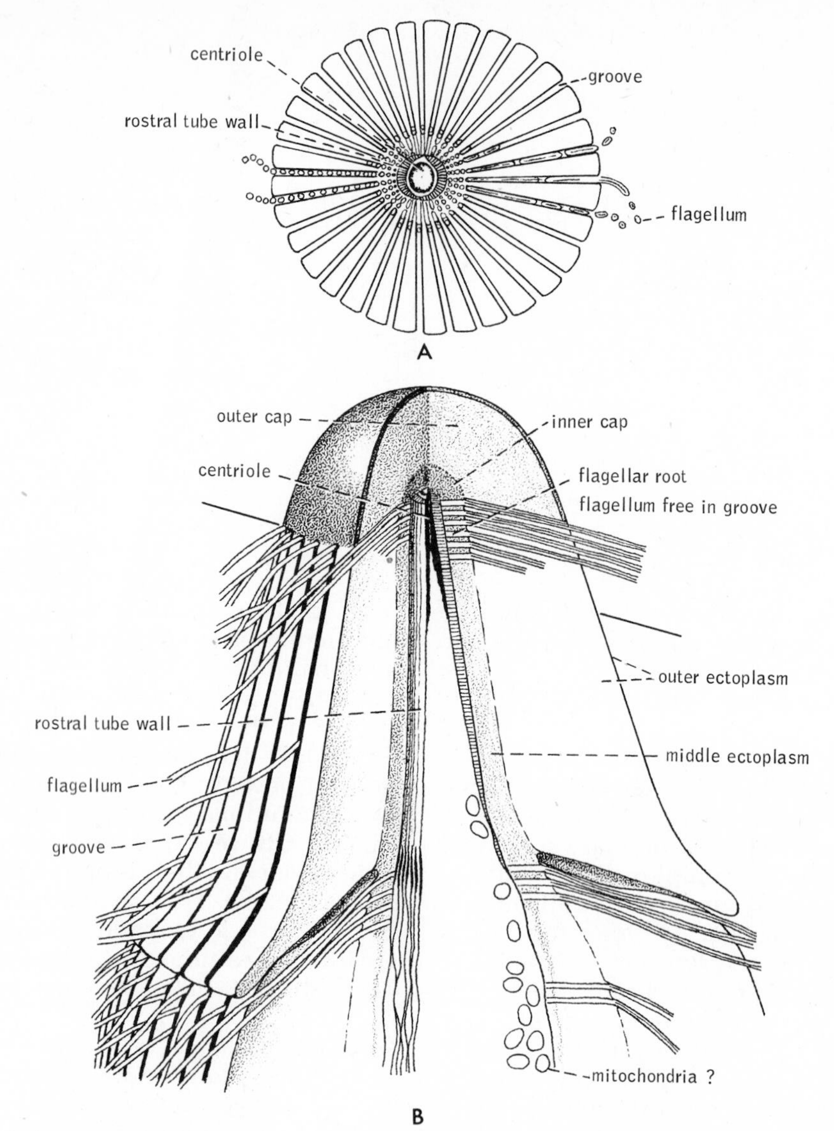

Figure 2–10. Ultrastructure of anterior end of *Trichonympha*. *A*. The numerous flagella emerge from deep longitudinal grooves on the body surface. *B*. Cross section through rostrum. (After Pitelka and Schooley.)

phototactic; and since the pigment shield permits photoreception from one direction only, the animal is able to orient and move in the direction of the light source. In some dinoflagellates the eye spot is even provided with a lens. In addition to acting as a locomotor organelle, the flagellum probably plays an accessory role as a sensory organelle. This appears to be true for *Peranema*, in which flagellar contact with prey elicits a general excitatory response.

Reproduction and Life Cycles. In the majority of flagellates asexual reproduction occurs by binary fission, and most commonly the animal divides longitudinally. Division is thus said to be symmetrogenic, that is, producing mirror-image daughter cells (Fig. 2–11). In multiflagellate species the flagella are divided between the daughter cells. In those species with few flagella, the one or several flagella may duplicate prior to cell dvision; or they may be equally apportioned to each daughter cell, resorbed and formed anew in each daughter cell, or even unequally apportioned. The same may apply to other organelles. Thus, division in many flagellates is not perfectly symmetrogenic.

In the dinoflagellates, division is oblique. In armored species the two fission products regenerate the missing plates; but in a few species, such as *Glenodinium*, division occurs inside the original parental envelope, which then ruptures to allow two naked daughter cells to escape. Incomplete fission occurs in some dinoflagellates, such as species of *Ceratium* and *Gymnodinium*, and results in a chain of individuals. Multiple fission is typical of some dinoflagellates.

Asexual reproduction in the colonial volvocids can be complex. In *Volvox*, for example, any one of a certain few cells at the posterior of the colony may undergo fission to form a daughter colony. As fission proceeds, the daughter cells become arranged in the form of a hollow ball, called a plakea, in which the flagellated ends of the cells are directed toward the interior. The plakea then everts, or turns inside out, through an opening in one side of the sphere. Following eversion, the hollow spherical form is reassumed, but now the flagellated ends of the cells are directed to the exterior in the normal position. The daughter colonies usually escape by a rupturing of the parent wall.

Sexual reproduction in the flagellates has rarely been observed except in the Volvocida, the symbiotic Hypermastigida, and the Oxymonatida. The Volvocida display all gradations of sexual reproduction, from isogamy to highly developed heterogamy. Some species of *Chlamydomonas* form isogametes that are but miniatures of the adults. Other species show the beginnings of sex differentiation by having gametes that differ just slightly in size. In *Platydorina*, heterogamy is well developed, but the large macrogametes still retain flagella and are free-swimming. Finally, in *Volvox* true eggs and sperm develop from special reproductive cells at the posterior of the colony. The egg is stationary and is fertilized within the parent colony. Volvocid colonies may be either monoecious or dioecious. Meiosis in the volvocids is postzygotic rather than prezygotic.

The life cycles of phytoflagellates and free-living zooflagellates are usually simple. Non-flagellated (palmella) stages are characteristic of most well-known phytoflagellates. In the palmella stage the organism loses its flagellum and becomes a ball-like, nonmotile, usually floating, relatively undifferentiated cell, located inside the original parental envelope when such an envelope is present (Fig. 2–12*A*). Fission often follows so that the palmella consists of a cluster of cells (Fig. 2–12*B*). Many species, such as some dinoflagellates, may remain in the palmella stage for a long time, making it the dominant stage in the life cycle. The symbiotic zooxanthellae, the yellow-brown unicellular "algae" that occur in some invertebrates, are actually the palmella stage of certain dinoflagellates.

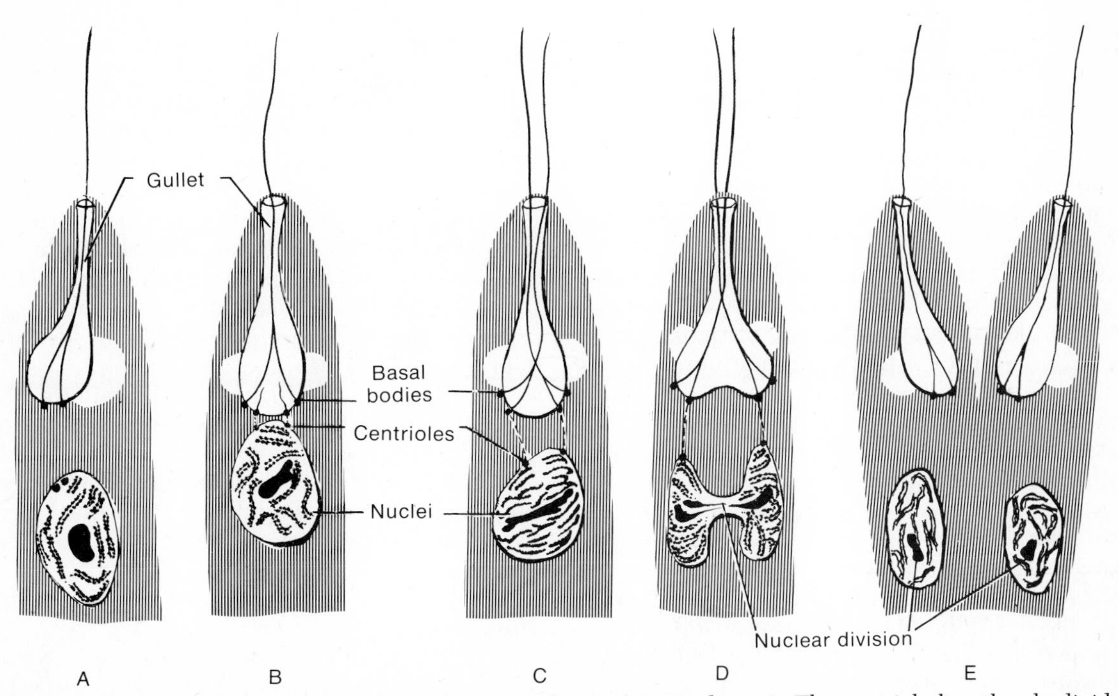

Figure 2–11. Asexual reproduction (symmetrogenic division) in *Euglena*. *A*. The centriole has already divided. *B*. Each centriole produces a new basal body and flagellum. The nucleus is in prophase and the contractile vacuole is double. *C*. The old pair of flagellar roots separate and fuse with the new roots. *D*. Mitosis proceeds and the gullet begins to divide. *E*. Anterior end dividing following duplication of organelles. (Redrawn from Ratcliffe, 1927.)

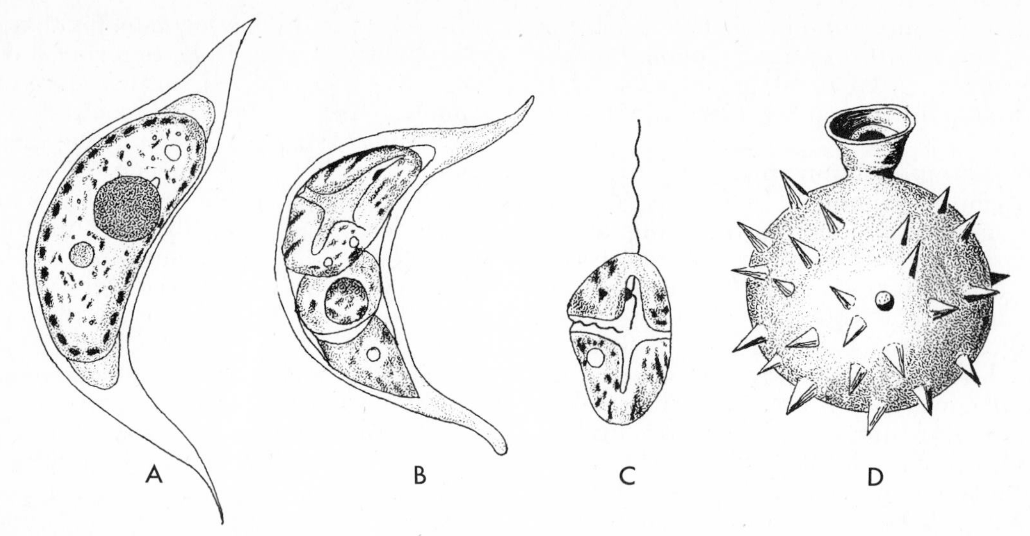

Figure 2–12. *A* to *C*. Stages in the life cycle of the dinoflagellate, *Cystodinium steinii*. *A*. Palmella stage. *B*. Fission, or spore formation, within cyst. *C*. Dinospore, or flagellate, stage. (After Klebs from Chatton.) *D*. Siliceous cyst of the chrysomonad, *Ochromonas fragilis*. (After Conrad from Hollande.)

Distinctive cysts are formed in a number of different phytoflagellate groups. Cysts of chrysomonads (Fig. 2–12*D*) typically have siliceous walls; cysts of dinoflagellates (Fig. 2–12*A*) are often crescent-shaped. In *Volvox* the zygote secretes a thick, tuberculate cyst wall. After disintegration of the parent colony, the zygote sinks to the bottom, where it may remain in the encysted stage for many months.

SYSTEMATIC RÉSUMÉ OF THE SUPERCLASS MASTIGOPHORA

One to many flagella present. Asexual reproduction by binary, more or less symmetrogenic, fission. Autotrophic or heterotrophic or both.

Class Phytomastigophorea. Mostly free-living, plantlike flagellates with or without chromoplasts, and usually one or two flagella.

Order Chrysomonadida. Small flagellates with yellow or brown chromoplasts and one to three flagella. Siliceous cysts. Marine and freshwater. *Chromulina, Ochromonas, Synura.*

Order Silicoflagellida. Flagellum single or absent and chromoplasts brown. Internal siliceous skeleton. Marine. *Dictyocha.* Known mostly from fossil forms.

Order Coccolithophorida. Tiny marine flagellates covered by calcareous platelets—coccoliths. Two flagella and yellow to brown chromoplasts. No endogenous siliceous cysts. *Coccolithus, Rhabdosphaera.*

Order Heterochlorida. Two unequal flagella and yellow-green chromoplasts. Siliceous cysts. *Heterochloris, Myxochloris.*

Order Cryptomonadida. Compressed, biflagellate, with an anterior depression or reservoir. Two chromoplastids, yellow to brown or colorless. Marine and freshwater. *Chilomonas* is a common colorless genus in polluted water (Fig. 2–5*B*); species of this order are commonly used as a food organism in *Paramecium* cultures.

Order Dinoflagellida. An equatorial and a posterior longitudinal flagellum located in grooves. Body either naked or covered by cellulose plates or valves, or by a cellulose membrane. Brown or yellow chromoplasts and stigma usually present. Largely marine; some parasites. Includes the marine genera *Gonyaulax, Noctiluca, Histiophysis, Ornithocercus;* the marine and freshwater genera *Glenodinium, Gymnodinium, Ceratium, Oödinium.*

Order Ebriida. Biflagellate, with no chromoplasts; internal siliceous skeleton. Mainly fossil. *Ebria.*

Order Euglenida. Elongated, green or colorless flagellates with one or two flagella arising from an anterior recess. Stigma present in colored forms. Primarily freshwater. *Euglena, Phacus, Peranema, Rhabdomonas.*

Order Chloromonadida. Small, dorsoventrally flattened flagellates with numerous green chromoplasts. Two flagella, one trailing. *Gonyostomum.*

Order Volvocida. Body with

green, usually single cup-shaped chromoplast, stigma, and often two to four apical flagella per cell. Many colonial species. Largely freshwater forms. *Chlamydomonas, Polytomella, Haematococcus, Gonium, Pandorina, Platydorina, Eudorina, Pleodorina, Volvox.*

Class Zoomastigophorea. Flagellates with neither chromoplasts nor leucoplasts. One to many flagella, in most cases with basal granule complex. Many commensals, symbionts, and parasites.

Order Choanoflagellida. Freshwater flagellates, with a single flagellum surrounded by a collar. Sessile, sometimes stalked; solitary or colonial. *Codosiga, Proterospongia.*

Order Bicosoecida. Largely freshwater flagellates encased within a lorica. Two flagella, one free, the other attaching posterior end of body to shell. *Salpingoeca, Poteriodendron.*

Order Rhizomastigida. Ameboid forms, with one to many flagella. Chiefly freshwater. *Mastigamoeba, Dimorpha.*

Order Kinetoplastida. Basically one, but up to four, flagella. Mostly parasitic. *Bodo, Leishmania, Trypanosoma.*

Order Retortamonadida. Gut

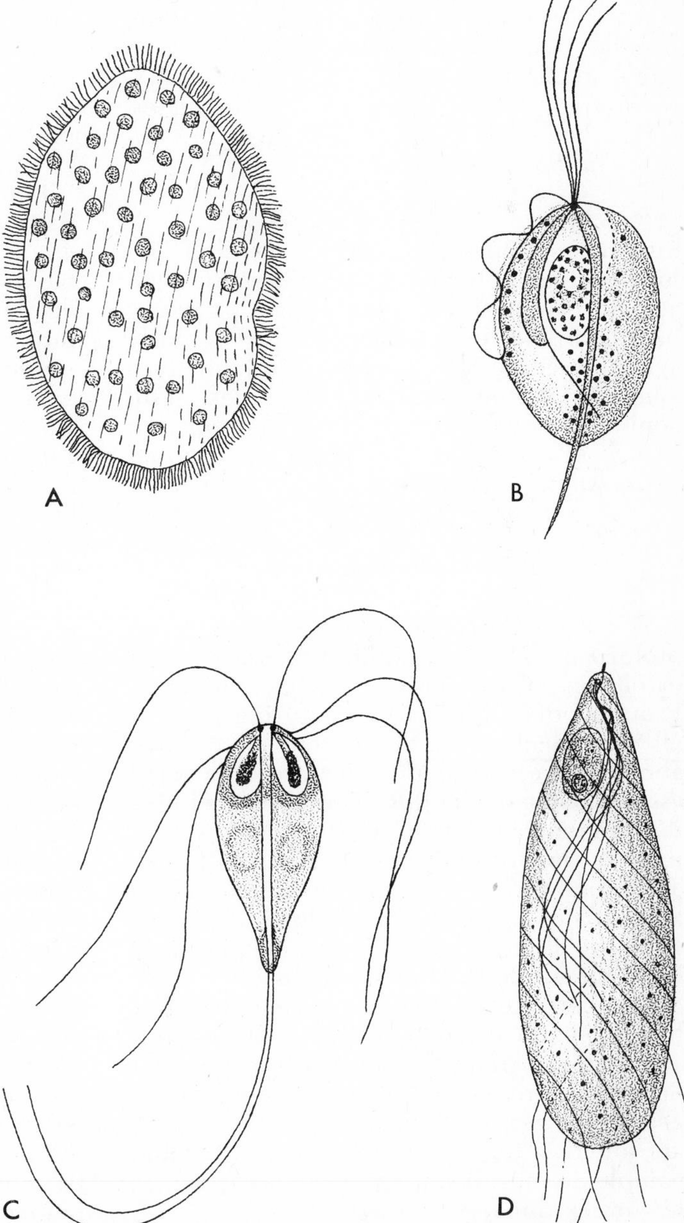

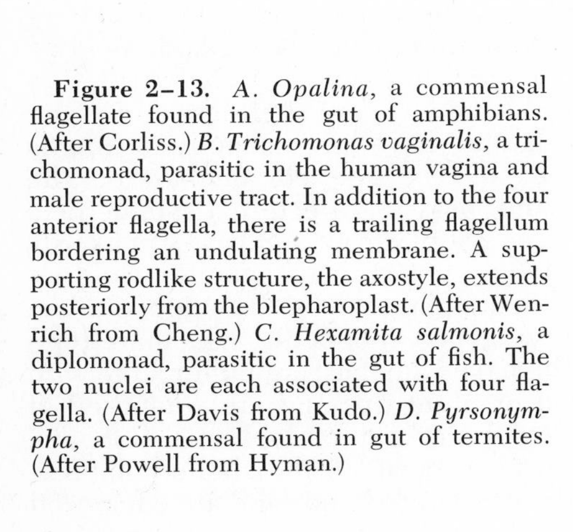

Figure 2–13. *A. Opalina,* a commensal flagellate found in the gut of amphibians. (After Corliss.) *B. Trichomonas vaginalis,* a trichomonad, parasitic in the human vagina and male reproductive tract. In addition to the four anterior flagella, there is a trailing flagellum bordering an undulating membrane. A supporting rodlike structure, the axostyle, extends posteriorly from the blepharoplast. (After Wenrich from Cheng.) *C. Hexamita salmonis,* a diplomonad, parasitic in the gut of fish. The two nuclei are each associated with four flagella. (After Davis from Kudo.) *D. Pyrsonympha,* a commensal found in gut of termites. (After Powell from Hyman.)

parasites of insects or vertebrates, with one to four flagella. One flagellum associated with ventrally located cytostome. *Chilomastix.*

Order Diplomonadida. Bilaterally symmetrical flagellates, with two nuclei, each nucleus associated with four flagella. Mostly parasites. *Hexamita* (Fig. 2–13*C*), *Giardia.*

Order Oxymonadida. Symbiotic flagellates having one to many nuclei, each nucleus associated with four flagella, some of which are turned posteriorly and adhere to body surface. *Oxymonas, Pyrsonympha.*

Order Trichomonadida. Symbiotic flagellates. Four to six flagella, one of which is trailing. *Trichomonas* (Fig. 2–13*B*).

Order Hypermastigida. Many flagella, with kinetosomes arranged in a circle, plate, or longitudinal or spiral rows. Symbionts in gut of termites, cockroaches, and wood roaches. *Lophomonas, Trichonympha, Barbulanympha.*

Superclass Opalinata. Body covered by longitudinal, oblique rows of cilia rising from anterior subterminal rows. Two or many monomorphic nuclei. Binary fission generally symmetrogenic. Sexual reproduction involves syngamy with flagellated gametes. Gut commensals of anurans; less commonly of fishes, salamanders, and reptiles. *Opalina, Zelleriella.*

Superclass Sarcodina

The superclass Sarcodina contains those protozoans in which adults possess flowing extensions of the body called pseudopodia. Pseudopodia are used for capturing prey in all Sarcodina; and in those groups that move about actively, pseudopodia are also used as locomotor organelles. The class includes the familiar amebas as well as many other marine, freshwater, and terrestrial forms. The slime molds are sometimes included in the Sarcodina, but in the following discussion, the slime molds will be considered to be fungi and left to the mycologists.

The Sarcodina either are asymmetrical or have a spherical symmetry. They possess relatively few organelles, and in this respect the Sarcodina are perhaps the simplest protozoans. However, skeletal structures, which are found in the majority of species, reach a complexity and beauty that is surpassed by few other animals.

There is considerable evidence that a close phylogenetic relationship exists between the Sarcodina and the Mastigophora. As already mentioned, many mastigophorans undergo ameboid phases, in which the flagellum is lost and movement is achieved by pseudopodia. Some Sarcodina have flagellated stages. The close relationship between the two classes is beautifully illustrated by a small number of interesting protozoans, such as *Mastigamoeba* (Fig. 2–14*A*), which have been placed in the order Rhizomastigida. These animals look like typical amebas, but a long flagellum is also present. The Rhizomastigida have arbitrarily been included as an order of the Mastigophora; but they could just as logically be considered Sarcodina. The presence of flagellated gametes among many Sarcodina and the tendency of many flagellates to lose their flagella during some phase of the life cycle would seem to indicate that perhaps the Mastigophora is the ancestral group.

Form and Structure in the Orders of Sarcodina. The Sarcodina contains four groups—the amebas, the foraminiferans, the heliozoans, and the radiolarians—each displaying relatively distinct form and structure.

Amebas. The amebas may be naked or enclosed in a shell. The naked amebas, which include the genera *Amoeba* (Fig. 2–14*B*) and *Pelomyxa*, are shell-less and asymmetrical, and have a constantly changing body shape. The shelled amebas, on the other hand, possess an external symmetrical shell. The cytoplasm in amebas is markedly divided into an external ectoplasm and an internal endoplasm. The pseudopodia are one of two types: lobopodia, which are typical of the naked amebas, are rather large with rounded or blunt tips and are composed of both ectoplasm and endoplasm; filopodia, which usually occur in the shelled amebas, tend to have pointed ends and are composed of ectoplasm only. However, environmental conditions can produce considerable variation in pseudopodial structure.

In shelled amebas, the shell is either secreted by the cytoplasm, in which case it is either chitinoid or siliceous, or it is composed of foreign materials imbedded in a cementing matrix. The ameba is attached by protoplasmic strands to the inner wall of the shell. Shelled amebas always have a large opening through which the pseudopodia or body can be protruded. The shape is therefore often similar to that of a vase or a helmet. In *Arcella* (Fig. 2–15*A* and *B*), one

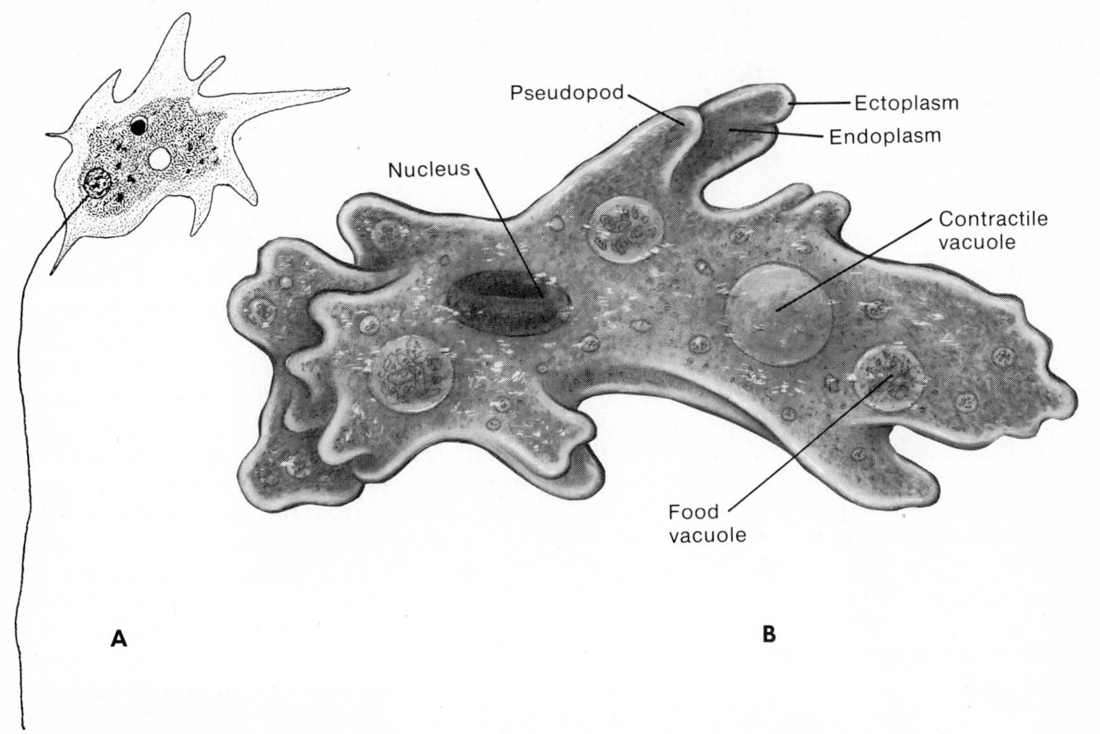

Figure 2–14. *A. Mastigamoeba.* (After Calkins, modified from Hyman.) *B. Amoeba.*

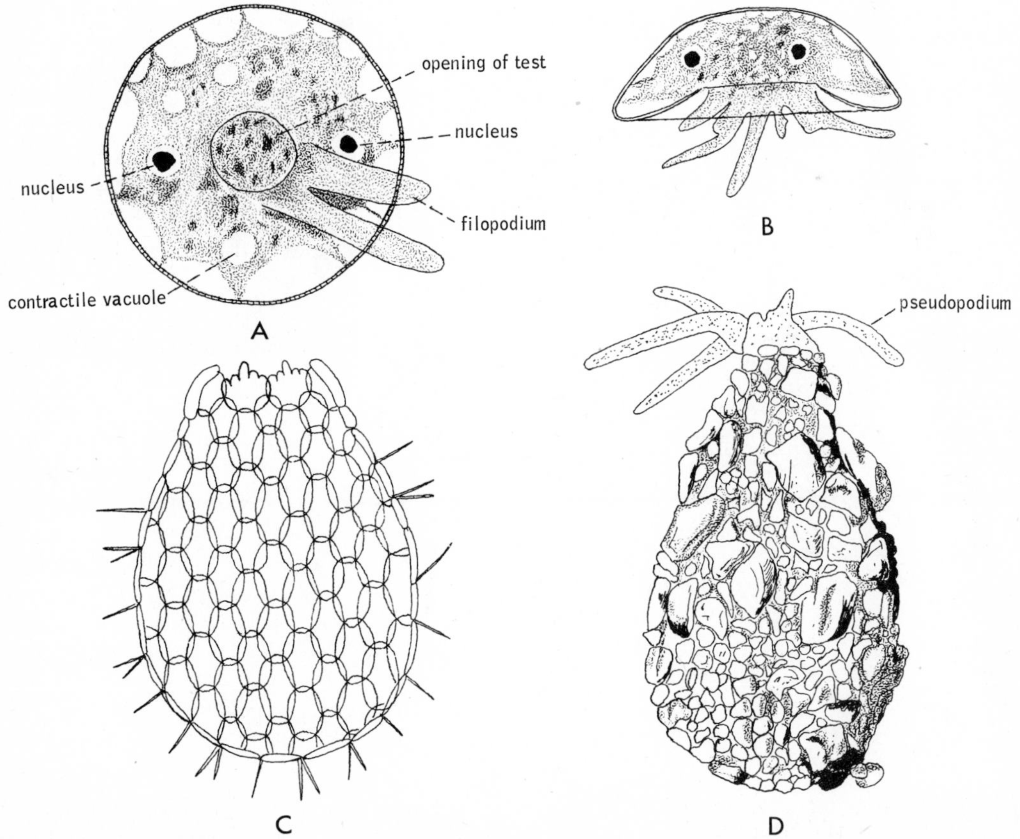

Figure 2–15. Shelled amebas. *A. Arcella vulgaris,* apical view; *B.* Side view. (*A* and *B* after Deflandre.) Test of *Euglypha strigosa,* composed of siliceous scales and spines. (After Wailes from Deflandre.) *D. Difflugia oblonga* with test of mineral particles. (After Deflandre.)

of the most common freshwater amebas, the brown or straw-colored keratin shell has the shape of a flattened dome with the aperture in the middle of the underside. When viewed from above, the tips of the filopods can be seen projecting just beyond the periphery of the shell. In *Euglypha* the secreted shell is constructed of round siliceous, often spiny scales (Fig. 2–15*C*). *Difflugia* has a shell composed of mineral particles that are ingested by the animal and imbedded in a secreted matrix (Fig. 2–15*D*). The shape of the shell is somewhat like that of an egg with an aperture located at one end. The shell of *Difflugia* and related forms is usually colorless but may be tinted red, blue, or violet from iron, manganese, or cobalt in the sand grains.

Foraminiferans. The order Foraminiferida is primarily marine; its members exist in great numbers in the sea. The pseudopodia are threadlike, branched, and interconnected and are called reticulopodia (Fig. 2–16*B*). Foraminiferans secrete a shell that is most commonly composed of calcium carbonate plus small amounts of other inorganic compounds, such as silica and magnesium sulfate. The structure of the shell is quite different from that of the shelled amebas. Some species live within a single-chambered shell and are said to be unilocular; but most forams are multilocular, having many-chambered shells. Multilocular forms begin life in a single chamber, the proloculum; but as the animal increases in size, the protoplasm overflows through a large opening in the first chamber and secretes another compartment.

This process is continuous throughout the life of the animal and results in the formation of a series of many chambers, each larger than the preceding one. Since the addition of new chambers follows a symmetrical pattern, the shells of multilocular forams have a distinct shape and arrangement of chambers. In some forms the chambers are

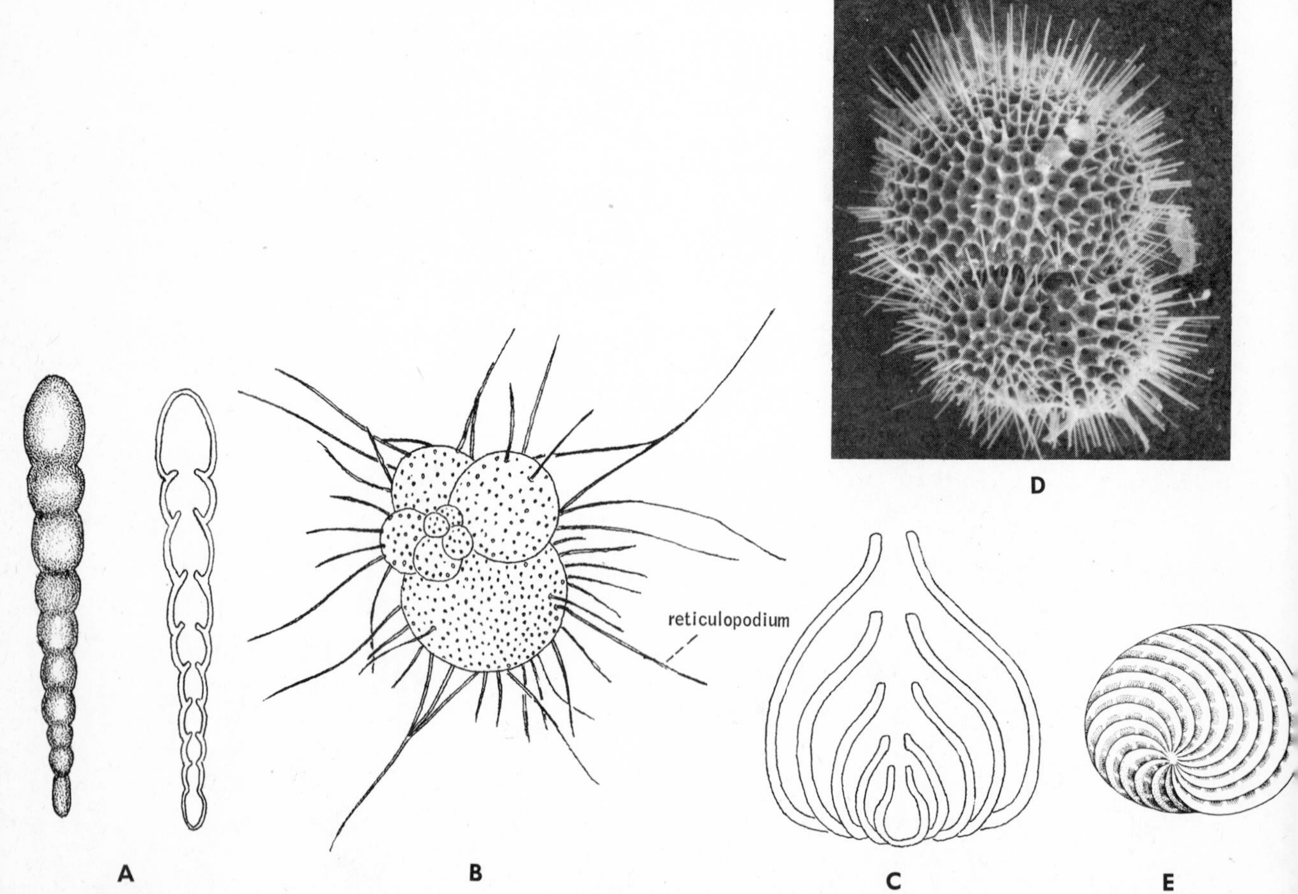

Figure 2–16. Foraminiferans. *A*. Shell of *Rheophax nodulosa* (entire, and in section). (After Brady from Calvez.) *B*. Living *Globigerina*. (After Hyman.) *C*. Shell of an ellipsoidinid foraminiferan (in section). (After Hyman.) *D*. *Globigerinoides sacculifer*, a tropical planktonic foram, in which the test bears spines. (By A. W. H. Bé, 1968: Science, *161*:881–884. Copyright 1968 by the American Association for the Advancement of Science.) *E*. *Archaias* sp., a common benthic foram of shallow tropical waters.

arranged in a straight line, so that the shell looks like a string of successively larger beads (Fig. 2–16*A*). In others, each new chamber nearly encloses the previous chamber, resulting in a multichambered shell, somewhat like an onion in structure (Fig. 2–16*C*). Not infrequently new chambers are added in a spiral manner so that the shell is like that of some snails (Fig. 2–16*E*).

All the chambers in multilocular species open successively into another chamber; the entire shell is filled with protoplasm that is continuous from one chamber to the next. Unlike the shelled amebas, in which pseudopodia extend through one large opening in the shell, every chamber in a foram shell is usually pierced by many small openings. It is through these perforations that the reticulopods project.

It should be realized that multilocular forams are not colonies but represent single individuals. Many are visible to the naked eye and a few attain several centimeters in diameter; for example, the so-called Mermaid's pennies of Australia.

Most forams are benthic, but species of *Globigerina* and related genera are common planktonic forms. The chambers of these multicolored species are spherical but arranged in a somewhat spiral manner (Fig. 2–16*B* and *D*). Planktonic forams have more delicate shells than do benthic species; and cold-water (denser-water) planktonic forams have smaller and less porous shells than do species of tropical oceans. Thus, the distribution of fossil planktonic species is an important indicator of climates in past geological ages.

A few forams are sessile. *Homotrema* forms large, red, calcareous tubercles about the size of a wart on the underside of coral heads (Fig. 10–25). The pink sands of the beaches of Bermuda result from the large number of pieces of *Homotrema* tests.

Heliozoans. Members of this group of spherical Sarcodina are commonly known as sun animalcules. They occur primarily in fresh water and may be either free or stalked. The fine, needle-like pseudopodia, called axopodia, radiate from the surface of the body (Fig. 2–17). Each axopod contains a central axial rod, which is covered with a granular, adhesive ectoplasm. Although the axial rod has a supporting function, it is not a permanent skeleton but is a protoplasmic bundle of fibers or tubes, which can shorten, bend, or even "melt" into typical cytoplasm. Axopods are not used in locomotion but function only for capturing food.

The body of a heliozoan consists of two parts (Figs. 2–17*A* and 2–18*C*). There is an outer cytoplasmic sphere, called the cortex, which is often greatly vacuolated. The inner part of the body, or medulla, is composed of dense endoplasm, containing one to many nuclei and the bases of the axial rods. The axial rods all may be attached to the membrane of a single central nucleus, as in *Actinophrys;* they may be attached to the membranes of numerous nuclei, as in *Camptonema* (Fig. 2–18*A*); or they may have

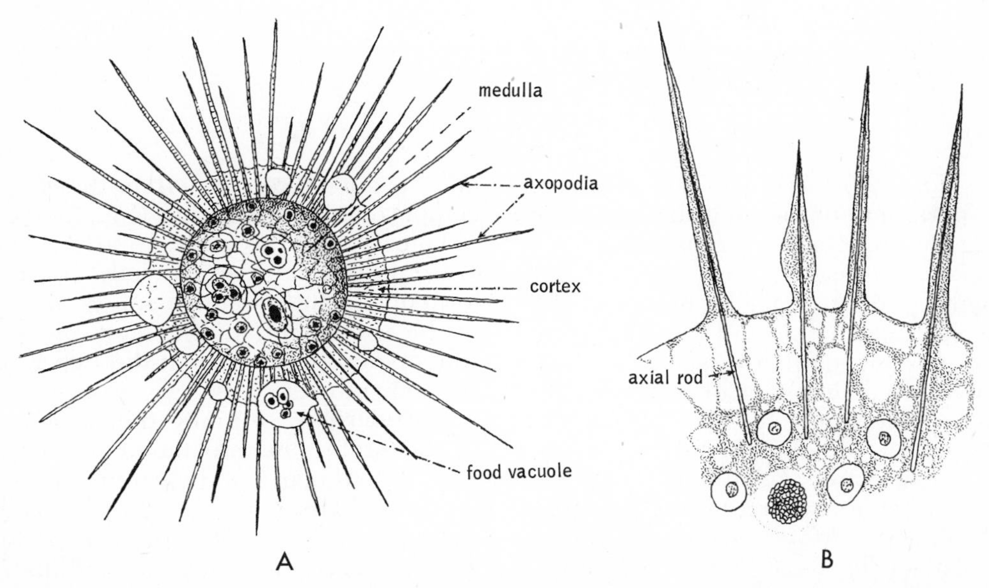

Figure 2–17. A multinucleate heliozoan, *Actinosphaerium eichorni*. *A*. Entire animal. (After Doflein from Trégouboff.) *B*. Section of cortex with adjacent axopodia. (After Bütschli from Trégouboff.)

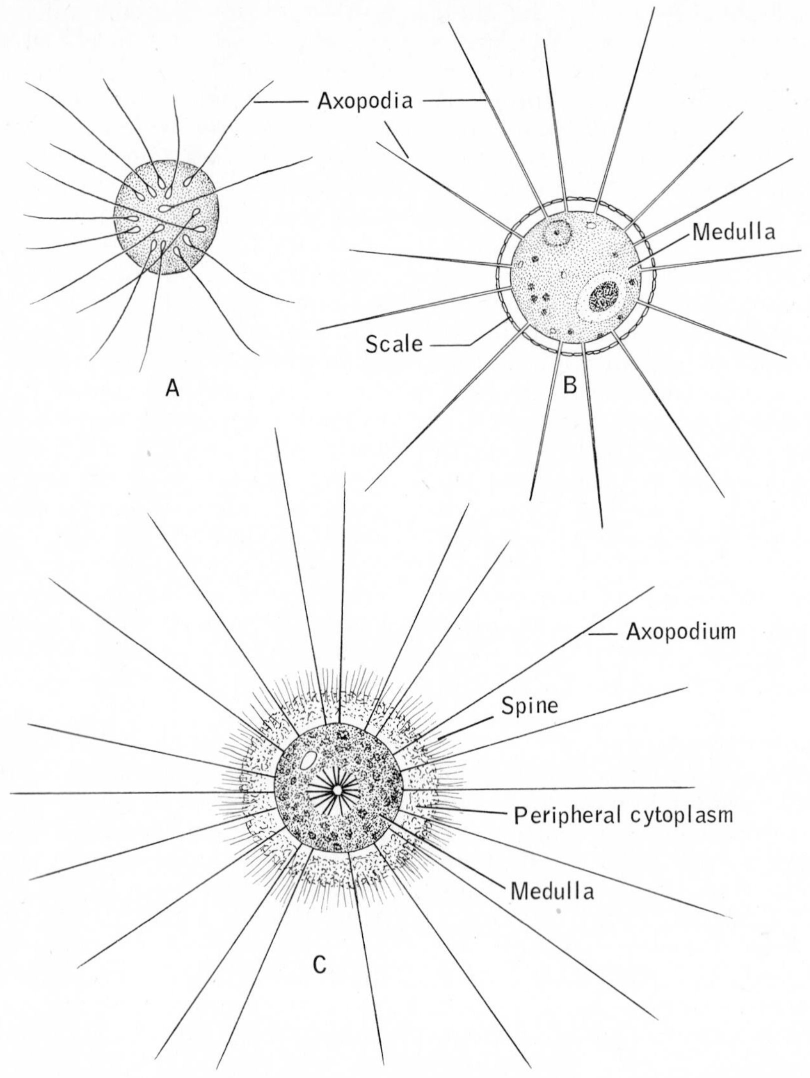

Figure 2–18. Heliozoans. *A. Comptonema nutans* with axopodia attached to nuclear membranes. (After Schaudinn from Trégouboff.) *B. Pinaciophora fluviatilis* with skeleton of scales. *C. Heterophrys myriopoda* with skeleton in form of spines. (*B* and *C* after Penard from Hall.)

no connection with the nuclei and originate from the periphery of the medulla, as in the common multinucleate *Actinosphaerium* (Fig. 2–17*A*).

Although heliozoans may be naked, skeletons are not uncommon and may be composed of foreign particles, such as sand grains or living diatoms, or of separate siliceous pieces secreted by the animal. In either case the skeletal components are imbedded in an outer gelatinous covering. When the skeleton is secreted, the siliceous pieces assume a great variety of forms, such as plates (Fig. 2–18*B*), spheres, tubes, or needles. These siliceous pieces may be arranged either tangentially to the body in one or more layers or, when the skeleton is composed of long needles (Fig. 2–18*C*), may radiate like the axopods. In a few cases the skeleton is composed of tectin and is in the form of a beautiful lattice sphere. Regardless of the nature and arrangement of the skeleton, openings are present through which the axopods project.

Radiolarians. Among the most beautiful of the protozoans are members of the subclasses Radiolaria and Acantharia. They are entirely marine and primarily planktonic. Radiolarians are relatively large protozoans; some species are several millimeters in diameter, and some colonial forms attain a diameter of several centimeters. Like heliozoans, the bodies of radiolarians are usually spherical and divided into an inner and outer part (Fig. 2–19). The inner region, which contains one to many nuclei, is bounded by a central capsule. In the subclass Radiolaria, in which the central capsule is highly developed, the chitinoid capsule membrane is perforated by openings, which may be evenly distributed (Fig. 2–20*A*) or may be restricted to one to three regions (called pore fields) of the membrane. The perforations allow the cytoplasm of the central capsule (or intracapsular cytoplasm) to be continuous with the cytoplasm of the outer division of the body. This extracapsular cytoplasm forms a broad cortex, called the calymma, which surrounds the central capsule.

In many species the calymma contains large numbers of symbiotic dinoflagellates (zooxanthellae) in a palmella stage as well as many vacuoles, which give the cytoplasm a frothy appearance (Fig. 2–19).

The pseudopodia are filopods, reticulopods, or axopods, and radiate from the surface of the body. They originate from the central capsule or just outside of it and extend through the calymma as dense cytoplasm.

A skeleton is almost always present in radiolarians and is usually siliceous, but in the subclass Acantharia it contains strontium sulfate. Two types of skeletal arrangements occur. One type has a radiating structure, in which the skeleton is composed of long spines or needles that radiate out from the center of the central capsule and extend beyond the outer surface of the body (Fig. 2–19). The points where the skeletal rods leave the body surface are surrounded by contractile fibrils (Acantharia). The action of these fibrils allows the spines to be moved and can cause the entire calymma to be contracted. The second type of skeleton is constructed in the form of a lattice sphere (Fig. 2–20*B*). There may be any number of such spheres arranged concentrically outside and inside the body. Furthermore, the lattice network is often sculptured and ornamented with barbs and spines of varying size and number, and in some species even branched projections are present. Finally, the lattice skeleton may not be spherical, and skeletons of many different bizarre patterns may occur (Fig. 2–21).

In colonial radiolarians there are many central capsules embedded in a common

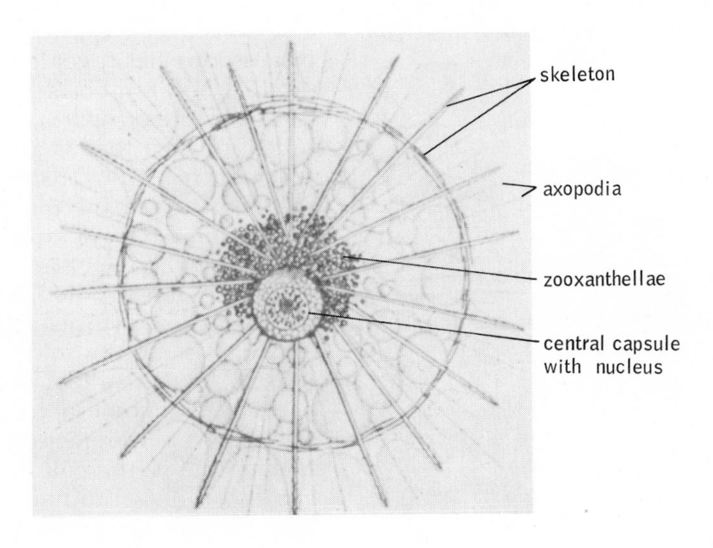

Figure 2–19. *Aulacantha scolymantha*, a radiolarian with a skeleton of radiating pieces. (After Kuhn from Hanson, E. D., 1967: *In* Florkin, M., and Scheer, B. J. (Eds.), Chemical Zoology. Academic Press, N.Y., p. 399.)

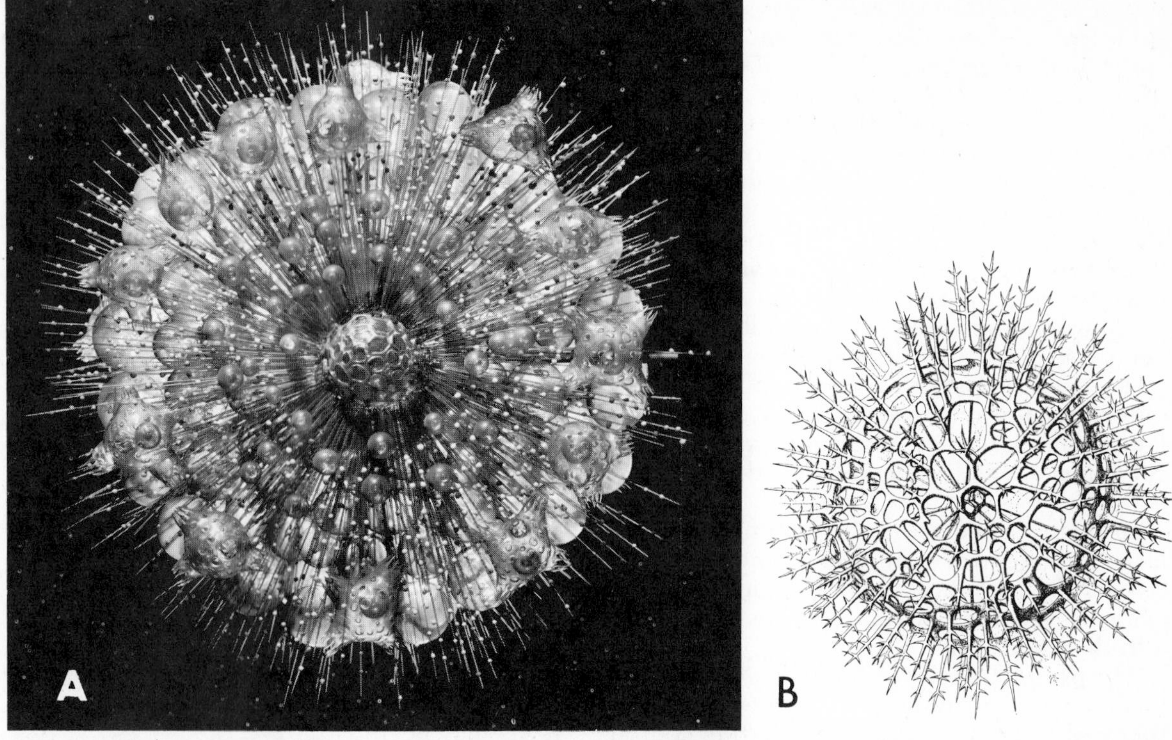

Figure 2–20. *A*. Glass model of a radiolarian, *Trypanosphaera regina*. (Courtesy of the American Museum of Natural History.) *B*. The internal, siliceous skeleton of a radiolarian. (Courtesy of E. Giltsch, Jena.)

mass. Each capsule may be surrounded by a skeleton, as for example in *Sphaerozoum*, where the central capsules are enveloped by loose spicules.

Locomotion. Flowing ameboid movement is limited to those Sarcodina that possess lobopods or filopods and is best developed in the naked amebas. In these animals, locomotion may involve a single large lobopod or several small ones, but in either case only the tips of the pseudopods are in contact with the substratum. Electron microscopy has revealed a filamentous fringe on the outer mobile plasma membrane of some species. Such a fringe may represent mucoproteins that facilitate adhesion to the substratum. It has also been suggested that the highly plastic plasma membrane possesses some type of rapid self-replicating system. The theory of ameboid movement accepted by most zoologists at the present time assumes that cytoplasmic flow results from changes in the colloidal state of the protoplasm. As a result of some initial stimulus, ectoplasm at some point on the body surface undergoes a liquefaction and becomes endoplasm; that is, the colloidal state shifts from a condition of gel to one of sol. As a result of this change, internal pressure causes the cytoplasm to flow out at this point, forming a pseudopodium.

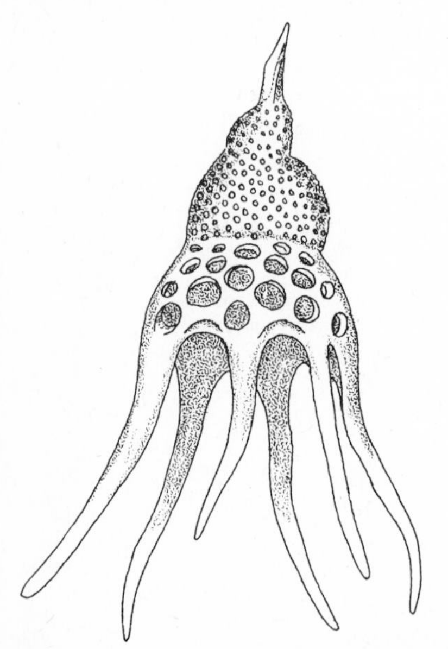

Figure 2–21. A nonspherical lattice skeleton of a radiolarian of the suborder Nasselarina. (After Haeckel from Hyman.)

In the interior of the pseudopodium, the fluid endoplasm flows forward along the line of progression; around the periphery, endoplasm is converted to ectoplasm, thus building up and extending the sides of the pseudopodium like a sleeve. At the posterior of the body, ectoplasm is assumed to be undergoing conversion to endoplasm.

Some investigators suggest that on the molecular level ameboid movement is actually a type of slow contraction, similar in many ways to muscle contraction. One theory postulates that protein chains in the endoplasm undergo contraction at the anterior end. Here the endoplasm is converted to ectoplasm, and in ectoplasm (the gel state) the protein chains are in the contracted state. The protein chains become extended at the posterior end where the ectoplasm is liquefied during its conversion to endoplasm. Thus, according to this front-contraction theory, the body of the animal is literally pulled forward by contraction at the anterior end.

Recent studies by Allen and others (1971) have shown that strong suction applied to one pseudopodium does not change the direction of flow in others. This indicates that ameboid streaming, at least in the *Chaos-Amoeba* group, does not result from high internal positive pressures generated at the posterior end and provides additional support to the "front-contraction theory."

A different type of ameboid flow has been demonstrated in the ectoplasmic reticulopods of *Allogromia*, considered to be either a foraminiferan or a shelled ameba. *Allogromia* extends a rigid reticulopod net out for some distance from the body. Particles adhering to the surface could readily be followed with the light microscope, and were found to move up one side and down the other reticulopod. Thus, in each pseudopodium there exist two opposing cytoplasmic streams. A kind of shearing or sliding effect of opposing cytoplasmic units or bundles has been suggested. A similar type of shearing force movement may also operate in axopods and filopods. Ameboid movement has been reviewed by Jahn and Bovee (1967).

Although there are some foraminiferans that are pelagic or cling to a surface film, the majority are bottom dwellers and creep slowly over the substratum. The body is pulled or dragged along by the reticulopods.

Radiolarians and unattached heliozoans are adapted for a planktonic existence, and their pseudopodia are food-capturing rather than locomotor organelles. However, radiolarians are able to float and even move vertically in the water by extending or contracting the calymma, axopods, and skeletal pieces, by increasing or decreasing the vacuolated condition of the calymma, by changing the density of the calymma cytoplasm, and by the presence of endoplasmic oil droplets.

Nutrition. The Sarcodina are entirely holozoic with the exception of some parasitic species. Their food consists of all types of small organisms—bacteria, algae, diatoms, protozoans, and even small multicellular animals such as rotifers and nematodes. The prey is captured and engulfed by means of the pseudopodia.

In the amebas, pseudopodia extend around the prey in a cuplike fashion, eventually enveloping it completely with cytoplasm. The enclosing of the captured organism by cytoplasm results in the formation of a food vacuole within the ameba (Fig. 2–14*B*). Commonly, the pseudopodia are not in intimate contact with the prey during engulfment, and a considerable amount of water is enclosed within the food vacuole along with the captured organism.

Engulfment also may involve complete contact with the surface of the prey, and the resulting vacuole is then completely filled by food. Death of the prey takes from three to sixty minutes and results primarily from lack of oxygen. Engulfment in shelled amebas follows essentially the same pattern as in naked forms, although in species with filopods little water is engulfed with the food. The vacuole is then moved through the aperture of the shell into the interior of the body.

In foraminiferans, heliozoans, and radiolarians the numerous radiating pseudopodia (chiefly reticulopods or axopods) act primarily as traps in the capture of prey. Any organism that comes in contact with the pseudopodia becomes fastened to the granular, adhesive surface of these organelles. A granular mucoid film is especially evident on the surface of foraminiferan reticulopods and quickly coats the surface of captured prey. This film apparently contains proteolytic secretions, which aid in paralyzing the prey and initiate digestion even during capture. In all three groups the food is eventually enclosed in a food vacuole and drawn toward the interior of the body. The axial rods of heliozoans may contract, drawing the prey into the ectoplasmic cortex; or the axopods may liquefy and surround the food, forming a vacuole at the site of capture. The vacuole will then be moved inward.

Digestion occurs in the cortex of heliozoans and the calymma of radiolarians. Where an enveloping skeletal sphere is present, the food passes through the openings for the pseudopodia. In foraminiferans food is digested outside of the shell and only the products of digestion pass to the interior.

Protease, amylase, lipase, and cellulase have been detected in different species. Egestion can take place at any point on the surface of the body, and in the actively moving amebas, wastes are usually emitted at the posterior, as the animal crawls about. However, in radiolarians waste products are retained and accumulated in a mass within the interior of the body.

Some naked amebas are parasitic. With the exception of *Hydramoeba* and a few species that are parasitic during immature stages in algae, the majority are endoparasites in the digestive tracts of annelids, insects, and vertebrates. Several species occur in the human intestine, but only *Entamoeba histolytica*, which is responsible for amebic dysentery, is ordinarily pathogenic. The life cycle of these intestinal amebas is direct, and the parasites are usually transmitted from the digestive tract of one host to that of another by means of cysts that are passed in feces.

Water Balance. Contractile vacuoles (Fig. 2–14*B*) occur in freshwater species but are absent in marine Sarcodina. There may be one to many contractile vacuoles, and the expulsion of fluid can occur anywhere on the body surface. In amebas the single or few contractile vacuoles are carried about in the flowing endoplasm.

Reproduction and Life Cycles. Asexual reproduction in most amebas, heliozoans, and radiolarians is by binary fission (Fig. 2–22*A*). In amebas with a soft shell, the shell divides into two parts, and each daughter cell forms a new half. When the shell is dense and continuous, such as in *Arcella*, a mass of protoplasm extrudes from the opening prior to division; this extruded mass secretes a new shell (Fig. 2–22*B*). The double-shelled animal now divides.

Division in the radiolarians is somewhat similar to that in the shelled amebas. Either

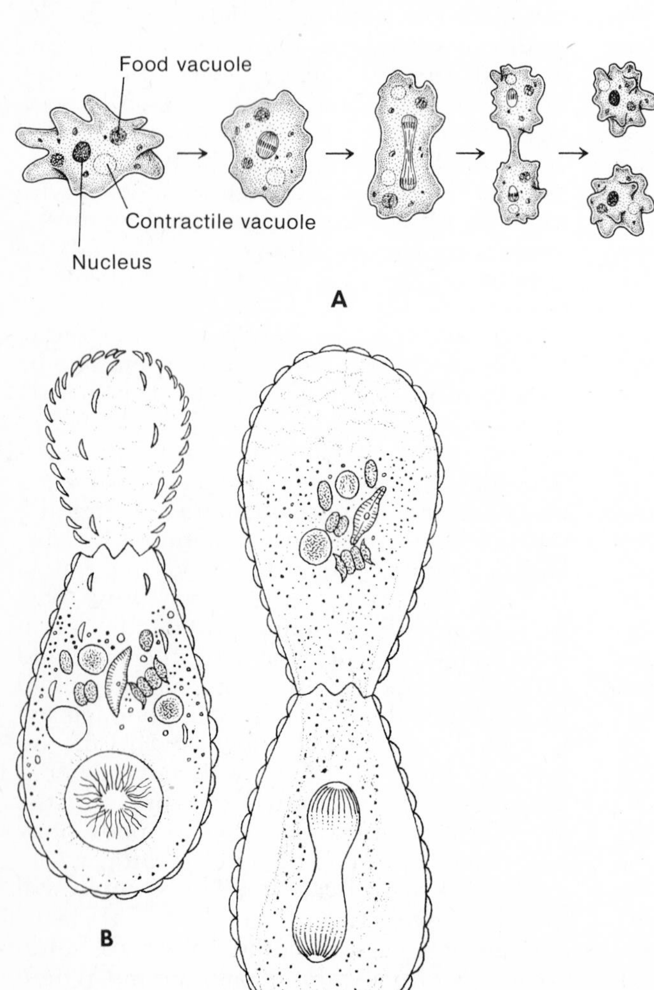

Figure 2–22. *A*. Fission in a naked ameba. *B* and *C*, two stages in the division of *Euglypha*, a shelled ameba: *B*, formation of skeletal plates on protoplasmic mass protruding from aperture; *C*, division of nucleus. (*B* and *C* after Sevajakov from Dogiel.)

the skeleton itself divides, and each daughter cell forms the lacking half; or one offspring receives the skeleton, and the other secretes a new one (Fig. 2–23).

Multiple fission is common in multinucleated amebas and heliozoans. In certain shelled amebas the parent animal sporulates a large number of little, naked amebas, each of which produces a new shell in the process of growth.

Sexual reproduction occurs in Sarcodina, although it is more frequently encountered and better known in some groups than in others. Sexual reproduction has rarely been observed among the amebas; but hologamy—the fusion of two individuals, each acting as a gamete—has been reported in some species.

Among the heliozoans, sexual reproduction is known in *Actinosphaerium* and *Actinophrys*. In *Actinophrys*, which is naked, the axopods are withdrawn and a cyst is formed. Within the cyst the animal divides into two daughter cells, each of which then undergoes two maturation divisions. Only the nuclei are involved in the maturation divisions, and the chromosome number is reduced from 44 to 22. Following each division, the contents of one nucleus are extruded as a sort of polar body. The two gametic nuclei now fuse to form a single, diploid zygote nucleus. A similar although somewhat more complicated process occurs in the multinucleate *Actinosphaerium*.

Radiolarians are known to form flagellated isogametes, following extensive nuclear divisions; fusion occurs in the sea water, and the zygote develops directly into an adult form.

Reproduction in the foraminiferans is complex but relatively uniform throughout the order and involves a definite alternation of asexual and sexual generations. Each species of foraminiferans is dimorphic, and in most multilocular species the two types of individuals differ in the size of the proloculum. One type of individual, known as a schizont, reproduces asexually and has a shell with a small proloculus, called a microspheric shell. The other type, which has a megalospheric shell, is known as a gamont and reproduces sexually.

An example of a multilocular foraminiferan life cycle is illustrated in Figure 2–23. The schizont becomes multinucleate and gives rise to a large number of small amebas, supposedly haploid, each containing a single nucleus. These offspring secrete a megalospheric shell and become gamonts. When mature, the gamonts produce a large number of biflagellated anisogametes, pairs of which fuse to form a zygote. The zygote gives rise to a schizont with a microspheric shell, thus completing the cycle. A similar life cycle occurs in monolocular species, but the shells of the schizonts and the gamonts are not distinguishable.

Distribution of the Sarcodina. The naked amebas are found in both fresh and salt water and in soil to a depth of several feet. The shelled amebas and the heliozoans are primarily freshwater groups; the foraminiferans are largely marine; the radiolarians are exclusively marine. With the exception of the radiolarians and some foraminiferans, which are planktonic, the majority of Sarcodina are bottom dwellers or crawl about over the surfaces of submerged vegetation or other objects. Most sessile forms are heliozoans.

The planktonic radiolarians display a distinct vertical stratification from the ocean surface down to 4600 m. depths. Some groups occur only at very great depths; others are limited to the upper, lighted zone; some migrate between the surface layer and lower depths, depending on the season. For example, representatives of the orders Nassellaria and Spumellaria occur in great numbers in the surface waters of the Mediterranean during winter months; but as the temperature of the water rises in the spring, they migrate down to approximately the 360 m. level.

Although most foraminiferans are benthic, a number of genera, such as *Globigerina* (Fig. 2–16*B*) and *Orbulina*, are adapted for a planktonic existence and occur in enormous numbers. Their shells are thin and often bear long spines to increase floatation. They are found in marine waters throughout the world.

The great numbers in which planktonic foraminiferans and radiolarians occur are indicated by the fact that their shells, sinking to the bottom at death, form a primary constituent of many ocean bottom sediments. Where they compose 30 per cent or more of the sediment, it is called a foram or radiolarian ooze. However, at depths below 4000 meters the great pressure tends to dissolve foram shells.

Many Sarcodina harbor symbiotic algae or phytoflagellates in a palmella stage. As mentioned earlier, symbiotic dinoflagellates (zooxanthellae) are common in radiolarians. Zooxanthellae as well as zoochlorellae (unicellular green algae) also occur in many foramini-

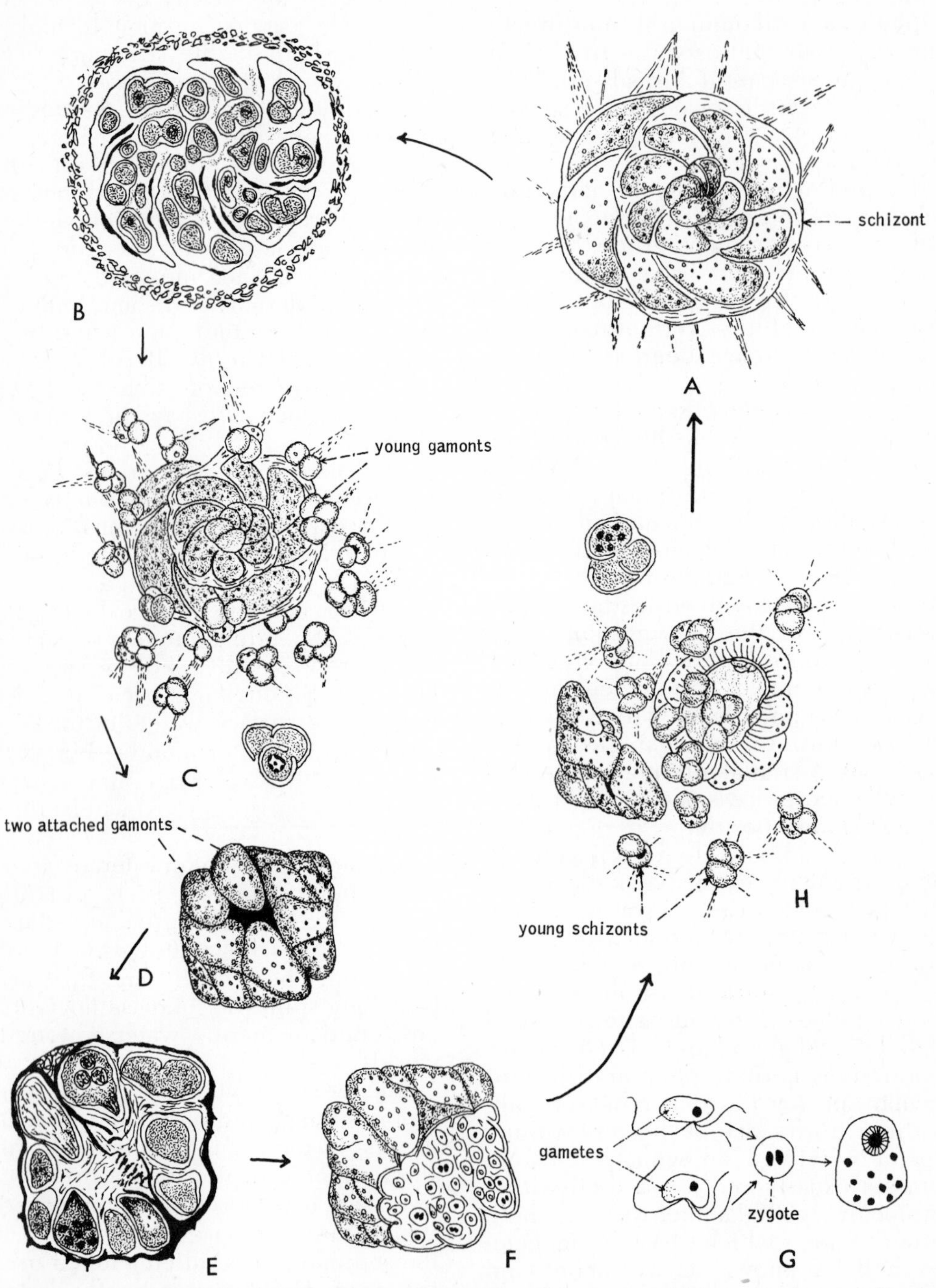

Figure 2–23. Life cycle of the foraminiferan, *Discorbis patelliformis*. *A*. Schizont. *B*. Asexual development of gamonts within chambers of schizont. *C*. Liberation of young gamonts from parent schizont. *D–G*. Formation and fusion of gametes within two attached gamonts. *H*. Separation of attached gamonts accompanied by liberation of young, schizonts. (After Myers from Calvez.)

ferans. A few amebas, such as *Amoeba stigmata,* contain zoochlorellae. The physiological importance of such symbiotic relationships is still uncertain.

The Sarcodina are the only class of protozoans that has an extensive fossil record. Fossil forms are, of course, restricted to those with skeletons or shells—the shelled amebas, the foraminiferans, and the radiolarians. The fossil record of the shelled amebas is relatively brief and recent. The group appears as fossils only in the Cenozoic era and consists of forms that are virtually identical with those living today. However, the foraminiferans and radiolarians have a long and abundant fossil record. In fact, the radiolarians are among the oldest known fossils and are said to have been found in Pre-Cambrian rocks. The foraminiferans first appeared in the Cambrian period, and from the late Paleozoic era on, there is an abundant fossil record. Extensive accumulations of foram shells occurred during the Mesozoic and early Cenozoic eras and contributed to the formation of great limestone and chalk deposits in different parts of the world.

SYSTEMATIC RÉSUMÉ OF SUPERCLASS SARCODINA

Protozoans with pseudopodia as feeding and locomotor organelles; flagella, when present, only in developmental stages. Little development of cortical organelles. Skeletons of various forms and composition characteristic of some groups.

Class Rhizopodea. Lobopodia, filopodia, or reticulopodia used for locomotion and feeding.

Subclass Lobosia. Pseudopodia usually lobopods.

Order Amoebida. Naked. Largely freshwater, some marine; many parasites. *Pelomyxa, Amoeba, Entamoeba, Hydramoeba.*

Order Arcellinida. Body enclosed in a shell or test with an aperture through which the pseudopodia protrude. Free-living, largely in fresh water. *Arcella, Difflugia, Centropyxis.*

Subclass Filosia. Marine and freshwater Sarcodina, with tapering and branching filopodia. Naked or with a test having a single aperture. *Penardia* (naked), *Allogromia.*

Subclass Granuloreticulosia. Delicate and granular reticulopodia.

Order Athalamida. Naked. *Biomyxa.*

Order Foraminiferida. Chiefly marine species with typically multichambered shells. Shells may be membranous or composed of chitin or foreign particles, but most commonly the shells are calcareous. *Globigerina, Orbulina, Discorbis, Spirillina, Nummulites.*

Class Piroplasmea. Small, round, rod-shaped or ameboid parasites in vertebrate red blood cells. Formerly included with the sporozoans, but spores are absent. *Babesia,* which is transmitted by ticks, is cause of cattle-tick fever.

Class Actinopodea. Primarily floating or sessile Sarcodina with pseudopodia radiating from a spherical body.

Subclass Radiolaria. Perforated chitinoid central capsule, siliceous skeleton; strontium sulfate reported in one group. Reticulopodia, filopodia, or sometimes axopodia. Marine. *Collozoum, Sphaerozoum, Aulocantha.*

Subclass Acantharia. Imperforate nonchitinoid central capsule without pores; anisotropic skeleton of strontium sulfate; axopodia. Marine. *Acanthometra.*

Subclass Heliozoia. Without central capsule. Naked, or if skeleton present, of siliceous scales and spines. Axopodia or filopodia. Primarily in fresh water. *Actinophrys, Actinosphaerium, Camptonema.*

Subclass Proteomyxidia. Largely marine and freshwater parasites of algae and higher plants. Filopodia and reticulopodia in some species. *Vampyrella.*

THE SPOROZOANS: SUBPHYLA SPOROZOA AND CNIDOSPORA

The sporozoans are parasitic protozoans which were formerly united within one class, the Sporozoa, because of the presence of sporelike infective stages in some members of both groups. The presence of flagellated gametes, the ability to move by gliding or body flexions, and the possession of pseudopodia as feeding organelles in different groups suggest a relationship to the flagellates and to the sarcodines; but the assemblage is so heterogeneous that it has long been recognized as an artificial group. In an attempt to rectify the polyphyletic nature of the old group, the former class has now been divided into the Sporozoa and the Cnidospora and each raised to the rank of subphylum. The term sporozoan is still used as a common name for both groups.

The subphylum Sporozoa contains the most familiar of the sporozoan parasites—the gregarines and the coccidians. The coccidians infect the intestinal or blood cells, and include the parasites causing human malarias and the coccidioses of domesticated animals. Species of Sporozoa are widespread and parasitize vertebrates and most invertebrate phyla.

The life cycle is complex. It usually involves asexual and sexual generations and sometimes two hosts. A generalized life cycle might be described as beginning with the infection by a sporozoite, which is taken into the host. The sporozoite develops into a trophozoite, which may move by gliding or flexion of the body. The trophozoite undergoes multiple fission, called schizogony. The resulting daughter cells, or merozoites, continue the infection of the host. Schizogony, at least in well-known sporozoans, is repeated a number of times. Eventually, in what may involve a maturation, some merozoites give rise to gamonts, a process called gamogony. The gamonts ripen to gametes, which may be similar or different, with the male cell commonly flagellated. Gametes unite (zygosis). The zygote now undergoes sporogony, which often involves the elaboration of a spore envelope and, invariably, division of the zygote into sporozoites, in some kinds into thousands of sporozoites. The life cycle can thus be divided into three phases: schizogony, an asexual multiplication of the parasite following infection of the host; gamogony, the development of gametes; and sporogony, also a multiplication and an infective stage, typically a spore (Fig. 2–24). Although this pattern of the life cycle is more or less basic for the class, there are innumerable variations.

The gregarines attain the largest size among the sporozoans. These members of the Sporozoa are parasites of invertebrates, especially annelids and insects. Intracellular parasitic species are only a few microns long, but those which inhabit the body or gut cavities of the host may reach 10 millimeters in length. The body of a gregarine trophozoite is elongate and may be divided into an anterior and posterior portion separated by a membrane (Fig. 2–25). The anterior part sometimes possesses hooks, a sucker or suckers, or simple filament or knob for anchoring the parasite into the host's cells (Fig. 2–26). The host becomes infected through ingesting spores containing sporozoites of the parasite (Fig. 2–27). Depending upon the species, the liberated gregarine sporozoites either remain in the gut of the host or penetrate the gut wall to reach other areas of the body. In some species schizogony is present, but in many, including the common *Gregarina blattarum* of cockroaches or related forms in meal worms or termites, there is no schizogony. These gregarines are thus unusual among protozoans in exhibiting only sexual reproduction. Whether or not there is schizogony, the trophozoite eventually becomes a sexual form called a gamont. Two gamonts unite, a process called syzygy, and then form a common cyst. Formation of gametes and syngamy occur within the cyst, after which each zygote undergoes sporogony to form sporozoites enclosed within a spore. The spore is liberated from the host in the feces.

The coccidians are intracellular parasites of vertebrates and invertebrates. They attack the intestinal epithelium, blood, or other cells of the host. In many gut parasites the zygote, enclosed within a cyst wall, is liberated in the feces of the host (Fig. 2–28). Sporogony takes place within the cyst, each zygote producing a spore with a number of sporozoites characteristic of its genus. The infective spores are ingested by another host and the sporozoites are liberated in the gut. Host cells are invaded. Schizogony occurs, increasing the number of infections. Some of these form gametes; the microgametes are small and flagellated, and they move to large stationary macrogametes within the host's cells.

Species of the coccidian genera *Eimeria* and *Isospora* infect the intestinal epithelium of the host and include many economically important pathogenic forms. Coccidioses from species of *Eimeria* in chickens, turkeys, and calves are especially destructive (Fig. 2–24). *Isospora*, usually characteristic of fishes, occurs in man but the resulting coccidiosis is not serious.

Two hosts are commonly involved in the life cycle. In *Aggregata eberthi*, the parasite infects the gut cells of a crab and a cuttlefish, but schizogony occurs in the crab, and gamogony and sporogony take place in the cuttlefish. The latter becomes parasitized by the merozoites when feeding on infected crabs. Crabs are infected by ingested spores from the cuttlefish.

In many coccidians, vertebrate red blood cells are the site of schizogony, and gamogony and sporogony occur in a blood-sucking invertebrate, such as a leech, mite, or one of

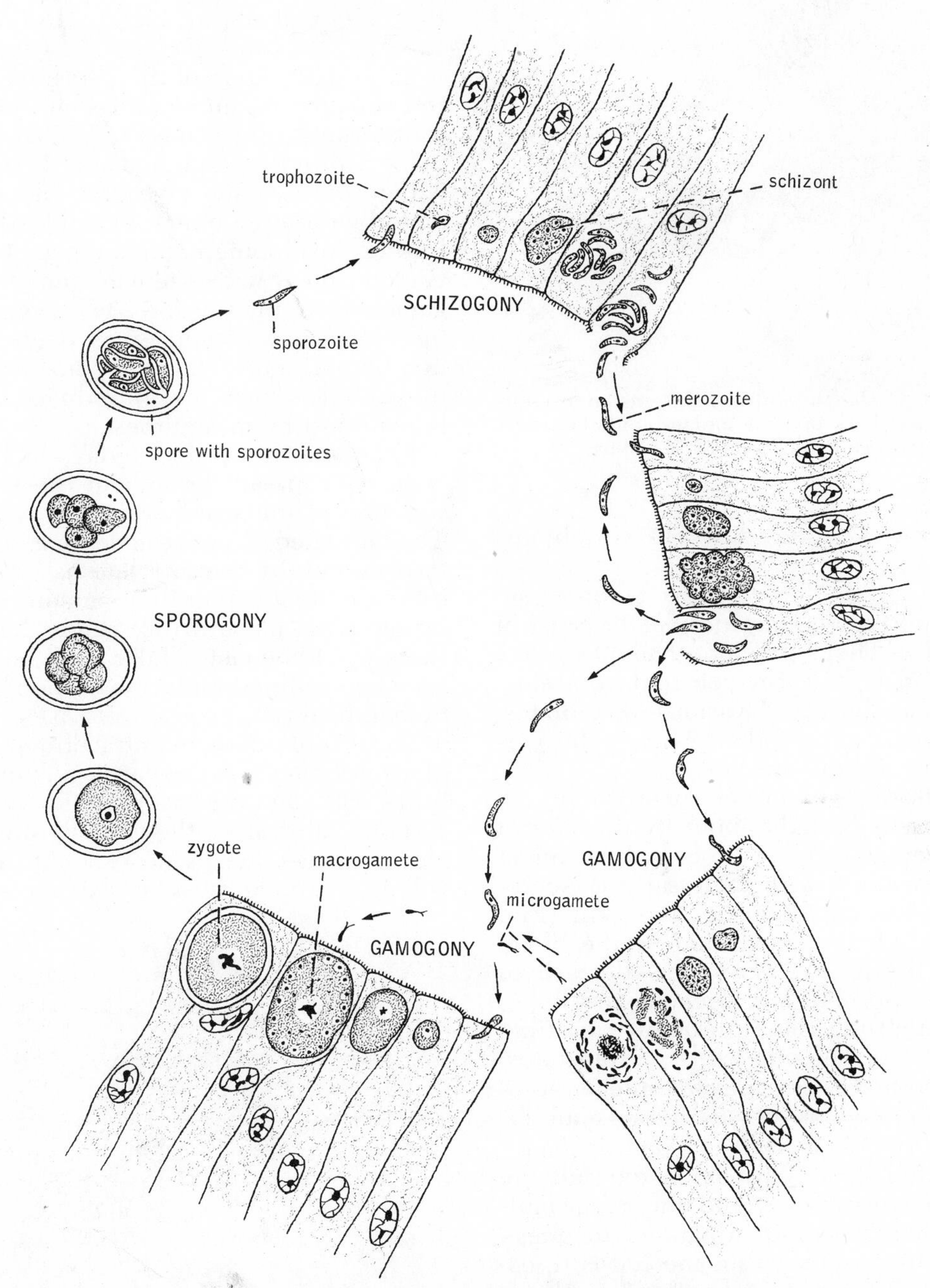

Figure 2–24. Life cycle of an eimeriid coccidian. (After Morgan and Hawkins from Noble and Noble.)

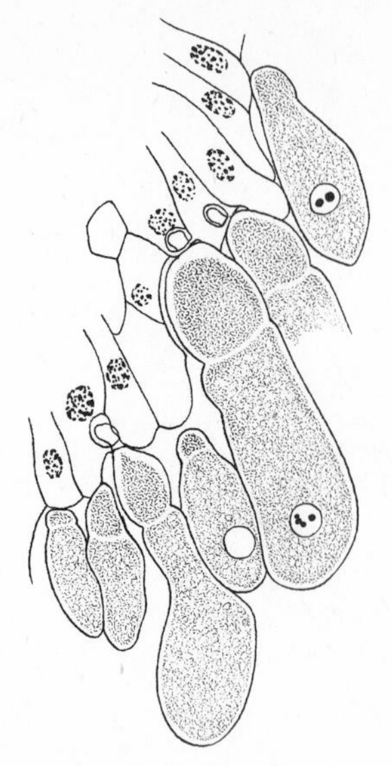

Figure 2–25. Trophozoites of the gregarine *Gregarina garnhami*, attacking the midgut epithelium of a locust. (After Canning from Noble and Noble.)

various insects. Many different vertebrates may be hosts.

To this group of coccidians belongs the much studied genus *Plasmodium*, the cause of the malarias. There are more than 50 species of *Plasmodium*, all of which require a mosquito as one host and various vertebrates, mostly birds, for the other. Man is the vertebrate host for four species.

The introduction of the parasite into a human host is brought about by the bite of certain species of mosquitoes, which inject sporozoites along with their salivary secretions into the capillaries of the skin (Fig. 2–29). The parasite is carried by the blood stream to the liver, where it invades a liver cell. Here further development results in asexual reproduction through multiple fission. These daughter cells (cryptozoites) invade other liver cells and continue to reproduce. After a week or so there is an invasion of red blood cells by parasites produced in the liver. Within the red cell the parasite increases in size and undergoes multiple fission. These individuals (merozoites), produced by fission within the red cell, escape and invade other red cells. The liberation and re-invasion does not occur continually but occurs more or less synchronously from all infected red blood cells. The timing of the event depends upon the period of time required to complete the developmental cycle within the host's cells. The release causes chills and fever, the typical symptoms of malaria. A specific periodicity is characteristic of different species. For example, in *Plasmodium vivax* 48 hours are required to complete the development of erythrocytic stages, i.e., the period from invasion of the red cell to release of daughter cells. The chills and fever produced by this species thus recur at approximately 48-hour intervals.

Eventually some of the parasites invading red cells do not undergo fission but become transformed into gametes. The gametes remain within the red blood cell. If such a cell is ingested by a mosquito, the gamete is liberated within the new host's gut. After some further development, a male gamete (microgamete) fuses with a female gamete (macrogamete) to form a zygote. The zygote enters the stomach wall and gives rise to a large number of spore stages (sporozoites). It is these stages that are introduced into the human host by mosquitoes.

Although in decline today, malaria was once widespread throughout the world and was one of the worst scourges of mankind. The untreated disease can be long-lasting and terribly debilitating. As late as 1957 it was estimated that 250 million persons were suffering from malaria, of which 2.5 million cases would be fatal. Malaria has been a continual and important factor in the shaping of human history.

The toxoplasmids are intracellular parasites of vertebrates, particularly mammals and birds. Although the toxoplasmids may infect a variety of tissues, they chiefly attack muscle or connective tissue. The life cycle involves a single host and is abbreviated.

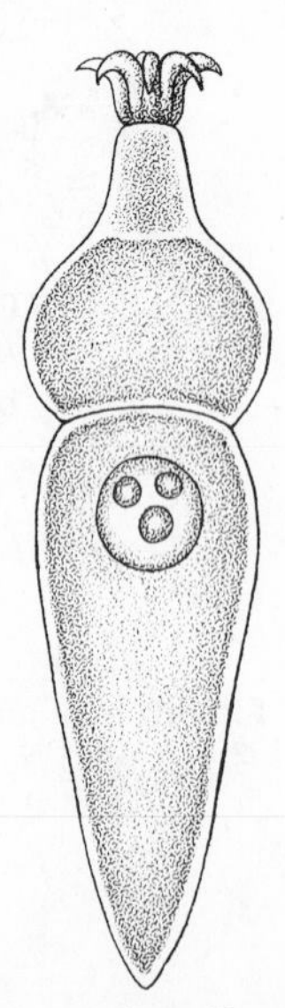

Figure 2–26. A gregarine, *Corycella armata*, parasitic in water beetles. (After Grell from Noble and Noble.)

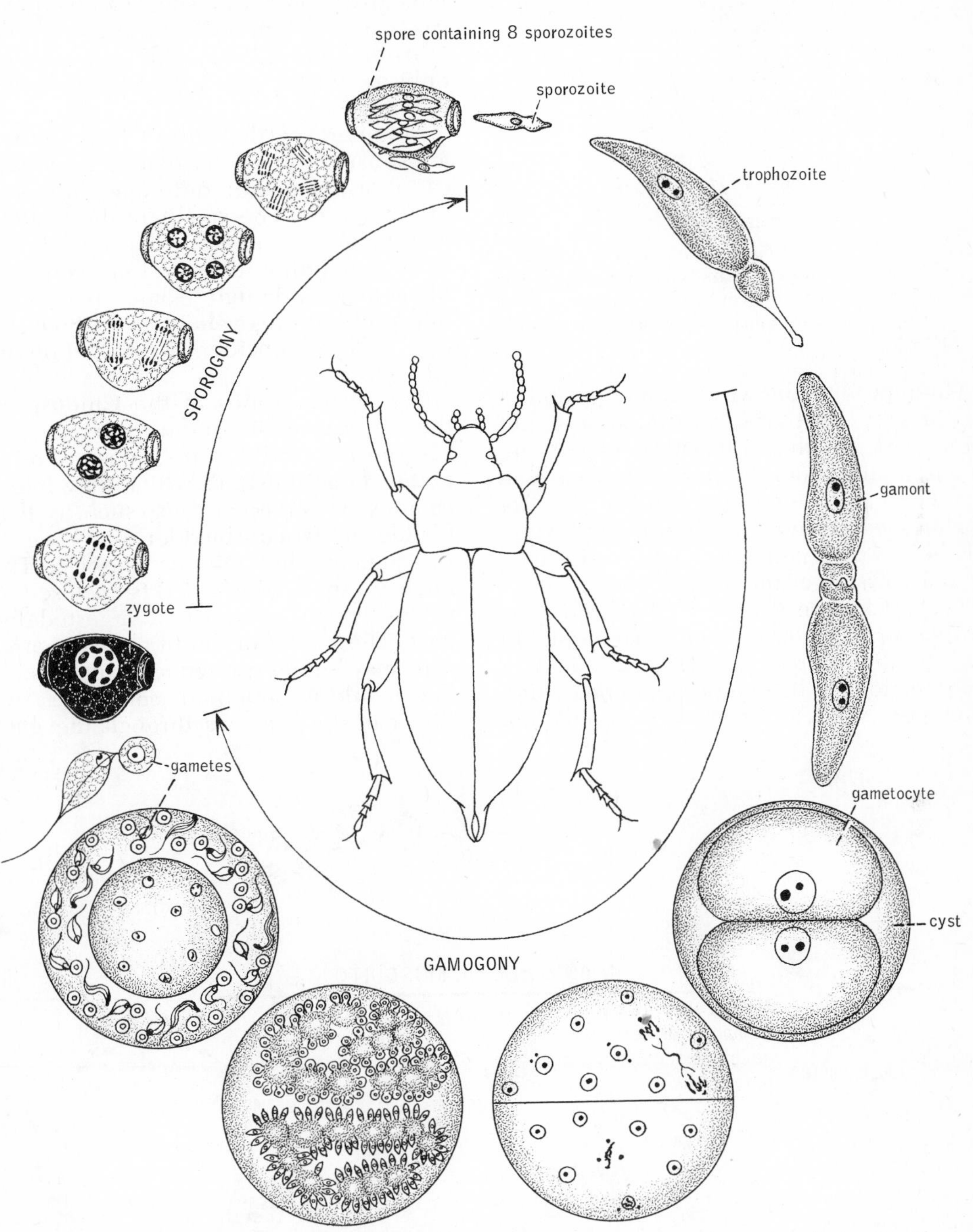

Figure 2–27. Life cycle of a gregarine, *Stylocephalus longicollis*, an intestinal parasite of a beetle. There is no schizogony in this species. (After Grell from Noble and Noble.)

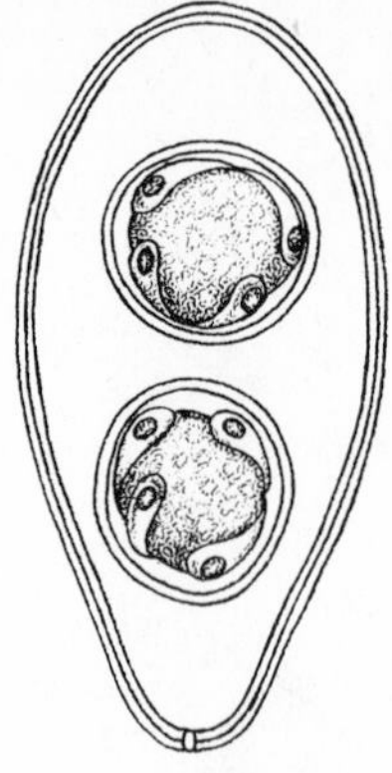

Figure 2–28. Cyst of *Isospora hominis*, a coccidian parasite in the intestine of man. The cyst, which has been passed in the feces, contains two spores. Each spore, in turn, contains four sporozoites. (After Noble and Noble.)

Sexual reproduction was unknown, and it was thought that there were no spores; but recent work has shown that these stages are the coccidial form of *Isospora*. Transmission is by a cyst containing trophozoites. The trophozoites can move by gliding or flexion, and the parasite multiplies within the host by binary fission or budding.

The best known genus of toxoplasmids is *Toxoplasma*, which is very common in many birds and mammals (Fig. 2–30). Human infection has been reported to be as high as 33 per cent of the world's population, making this organism perhaps one of the most prevalent human parasites. A great variety of tissues may be attacked, and the symptoms of human toxoplasmosis depend, in large part, on the tissues involved. Severe cases are not common. At present, congenital transmission through the placenta is the only means by which the parasite is known to be passed.

Members of the subphylum Cnidospora are separated from those of the Sporozoa because of the marked difference in the nature of the spores. The spores of the Cnidospora possess one or more peculiar polar capsules, each containing one to four coiled polar filaments. Although some parasitologists interpret the life cycle to be without gamogony, others consider all three phases to take place.

The major groups of the Cnidospora are the myxosporidians and the microsporidians. The myxosporidians are parasites of vertebrates, largely fish, in which they infect the cavities of certain organs, such as the gall bladder or urinary bladder, or tissues of the integument, the gills, or muscles. The binucleate spore (Fig. 2–31) is enclosed in two valves which open when ingested by the host. The nuclei of the liberated spore fuse. This may be interpreted as sexual fusion and the resulting ameboid cell as a zygote. The parasite burrows through the gut wall

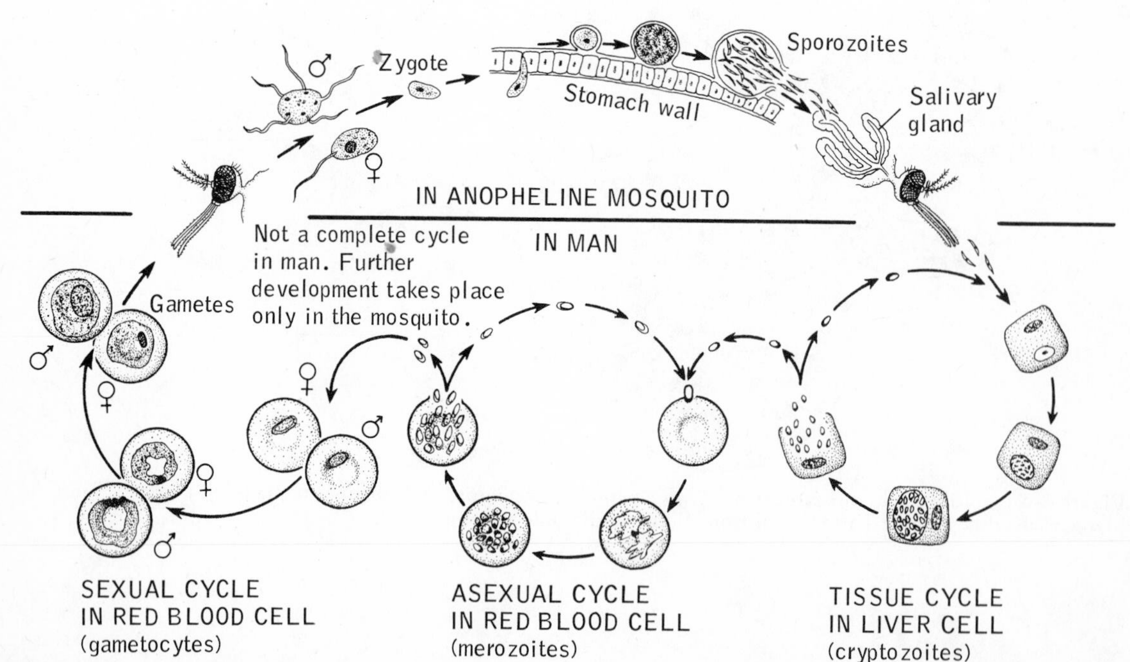

Figure 2–29. The life cycles of *Plasmodium* in a mosquito and in man. (Redrawn and modified from Blacklock and Southwell.)

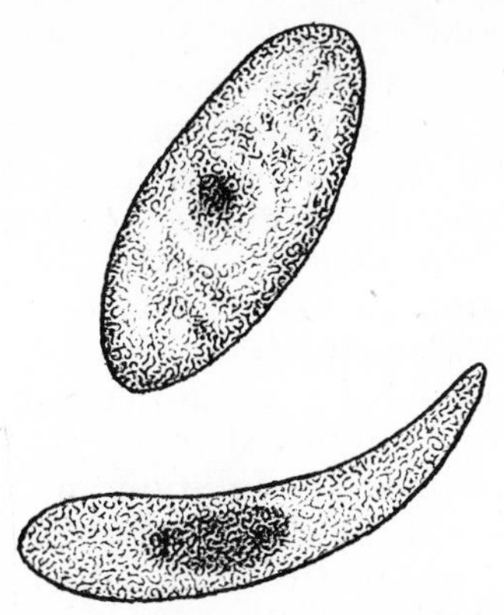

Figure 2–30. Sporozoites of *Toxoplasma gondii*, a toxoplasmid sporozoan attacking various tissues in man and other mammals and birds. (After Noble and Noble.)

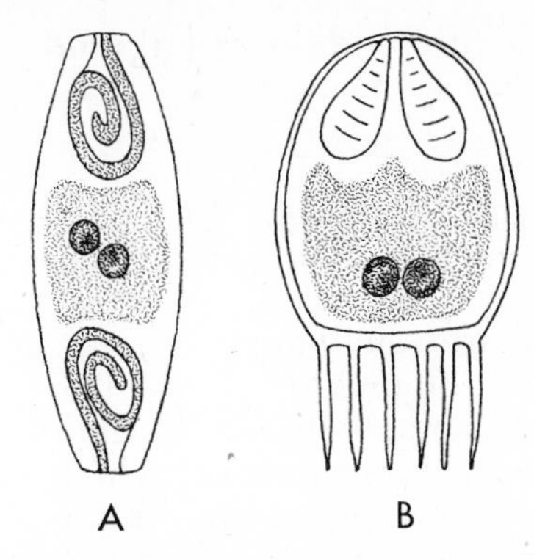

Figure 2–31. Binucleate myxosporidian spores, each with two polar capsules. *A.* Spore of *Sphaeromyxa*. *B.* Spore of *Mitraspora*. Both parasites of fish. (After Noble and Noble.)

and migrates to particular organs. There it undergoes schizogony within a host cell, increasing the infection. Sporogony finally takes place, giving rise to cysts with one to two or more spores. The polar capsules arise from nuclei during sporogony.

The microsporidians, as the name implies, are very small sporozoans, with spores sometimes no larger than bacteria. Only a single polar filament is present. Microsporidians are intracellular parasites and the principal hosts are fish, in which they infect the skin and muscles, and arthropods, in which they infect the gut. As parasites of the honeybee and silkworm, they are of economic importance. One of the early studies of sporozoans was that of Pasteur in 1870 on *Nosema bombycis* of silkworms.

SYSTEMATIC RÉSUMÉ OF THE SPOROZOANS

Subphylum Sporozoa. Spores usually present but lacking polar filaments. All species parasitic.

Class Telosporea. Reproduction sexual and asexual.

Subclass Gregarinia. Invertebrate parasites, in which the mature trophozoites are large and occur in the host's gut and body cavities. *Gregarina, Monocystis* (common parasite of earthworm's seminal receptacles).

Subclass Coccidia. Mature trophozoites small and intracellular. *Eimeria, Isospora, Aggregata, Plasmodium.*

Class Toxoplasmea. Spores absent and only asexual reproduction present. *Toxoplasma.*

Class Haplosporea. Only asexual reproduction occurs, but spores present. A small group of parasites of invertebrates and a few vertebrates.

Subphylum Cnidospora. Spores with polar filaments present. All species parasitic.

Class Myxosporidea. Spores developed from several nuclei and enclosed in two or three valves. *Ceratomyxa.*

Class Microsporidea. Spores developed from one nucleus and enclosed in a single valve. *Nosema, Glugea.*

SUBPHYLUM CILIOPHORA; CLASS CILIATEA

The subphylum Ciliophora has only one class, the Ciliatea. This is the largest and the most homogeneous of the protozoan classes. Some 6000 species have been described, and many groups are still not well known.

All possess cilia or compound ciliary structures as locomotor or food-acquiring organelles at some time in the life cycle. Also present is an infraciliary system, composed of basal granules, or kinetosomes, below the cell surface and interconnected by longitudinal fibrils. Such an infraciliary system may be present at all stages in the life cycle even with marked reduction in surface ciliation. Most ciliates possess a cell mouth, or cytostome. In contrast to the other protozoan classes, ciliates are characterized by the presence of two types of nuclei; one vegetative (the macronucleus) and the other reproductive (the micronucleus). Fission is transverse, and sexual reproduction never involves the formation of free gametes.

Ciliates are widely distributed in both fresh and marine waters. There are also many ecto- and endocommensals and some parasites.

Form and Structure. The body shape is usually constant and in general is asymmetrical. Although the majority of ciliates are solitary and free-swimming, there are both sessile and colonial forms. The bodies of tintinnids and some heterotrichs and peritrichs are housed within a lorica, which is either secreted or composed of foreign material cemented together. In the peritrichs the lorica is attached to the substratum, but in some heterotrichs and most tintinnids the lorica is carried about by the animal (Fig. 2–32*B*).

The ciliate body is typically covered by a complex, living pellicle, usually containing a number of different organelles. The pellicular system has been studied in detail in a number of genera, including *Paramecium, Tetrahymena, Colpidium,* and *Glaucoma.* There is an outer limiting surface membrane, which is continuous with the membrane surrounding the cilia. Beneath the outer membrane are closely packed vesicles, or alveoli, which are moderately to greatly flattened (Figs. 2–33 and 2–34). The outer and inner membranes bounding a flattened alveolus would thus form a middle and inner membrane of the ciliate pellicle. Between adjacent alveoli emerge the cilia and mucigenic or other bodies. Two rows of alveoli lie between each longitudinal row of cilia (Fig. 2–34). Beneath the alveoli are located the infraciliary system, the kinetosomes, and parallel fibrils. The alveoli contribute to the stability of the pellicle and perhaps limit the permeability of the cell surface (Pitelka, 1970).

The pellicle of the familiar *Paramecium* has inflated kidney-shaped alveoli (Fig. 2–34). The inflated condition and the shape of the alveoli produce a polygonal space about the one or two cilia which arise between them. Alternating with the alveoli are bottle-shaped organelles, the trichocysts, which form a second, deeper compact layer of the pellicular system.

Trichocysts are peculiar, rodlike or oval organelles characteristic of some ciliates. They are oriented at right angles to the body surface and may have a general distribution or may be limited to certain regions of the body. There are different kinds of trichocysts. The explosive trichocysts of *Paramecium,* for example, can be discharged to the exterior in only a few milliseconds. The discharged trichocyst consists of a long, striated, threadlike shaft surmounted by a barb, which is shaped somewhat like a golf tee (Fig. 2–35). The shaft is not evident in the undischarged state and is probably formed in the process of discharge.

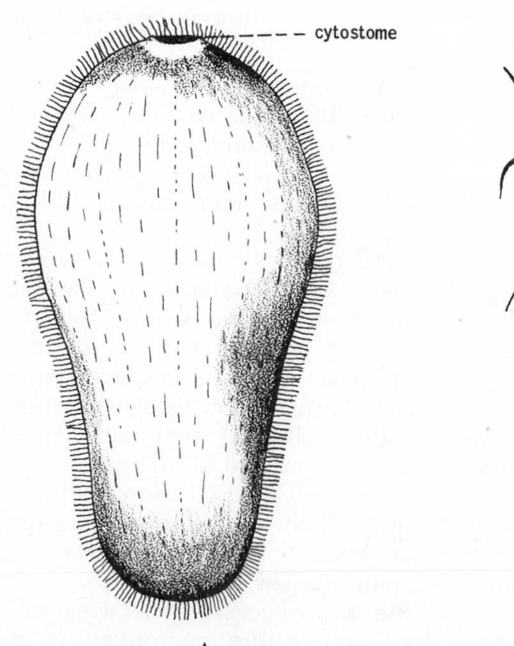

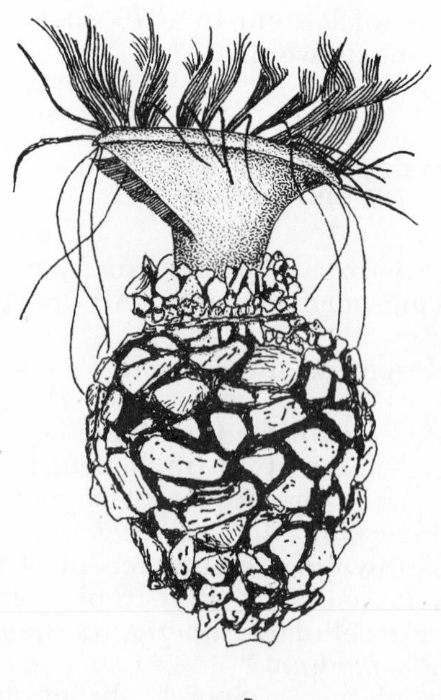

Figure 2–32. *A. Prorodon,* a primitive ciliate. *B. Tintinnopsis,* a marine ciliate with lorica, or test, composed of foreign particles. Note conspicuous membranelles and tentacle-like organelles interspersed between them. (After Fauré-Fremiet from Corliss.)

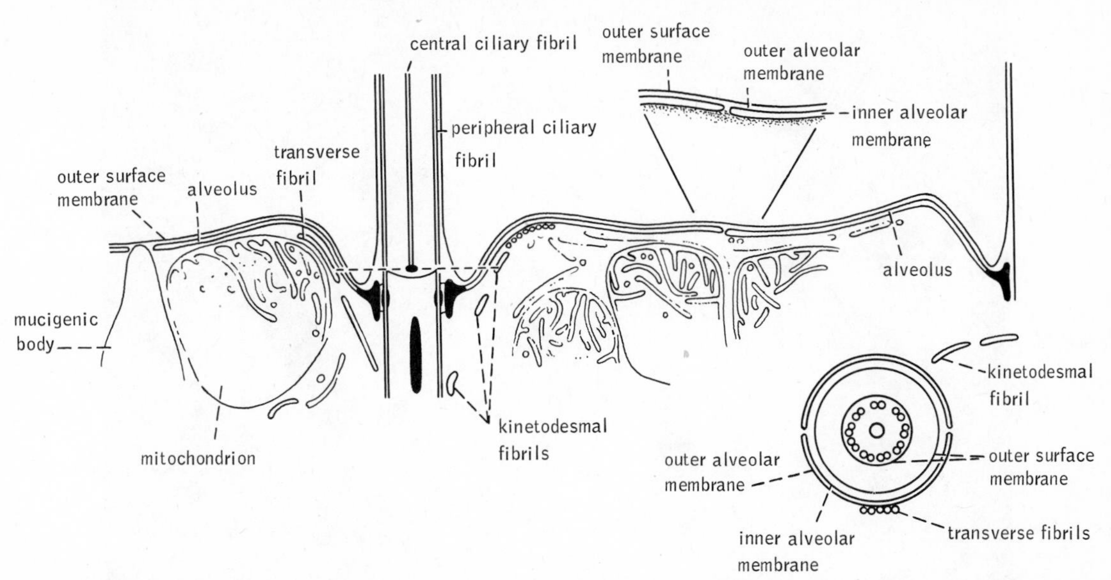

Figure 2–33. Section through cilium and pellicle of *Colpidium.* Note that alveoli are greatly flattened and their inner and outer membranes fused at base of cilium. At top right is an enlarged view of surface and alveolar membranes. At lower right is a cross section of a cilium and surrounding pellicle taken at the level indicated by the dashed line. Note the circle of nine doubled peripheral ciliary fibrils. (After Pitelka.)

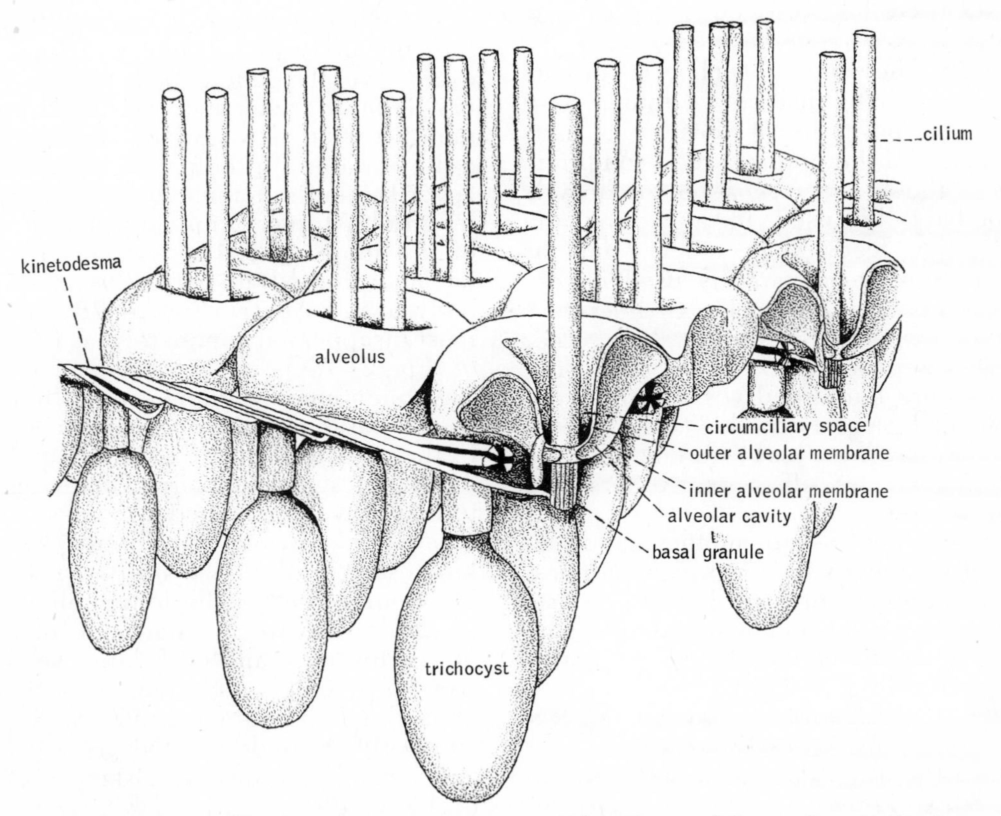

Figure 2–34. Pellicular system in *Paramecium.* (After Ehret and Powers from Corliss.)

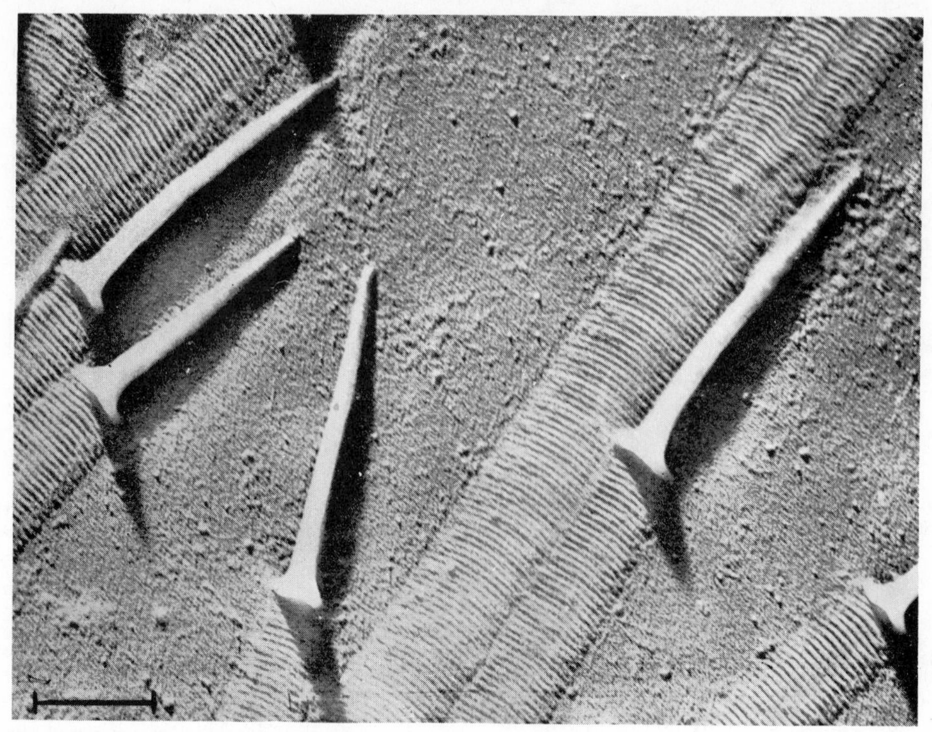

Figure 2–35. Electron micrograph of discharged trichocysts of *Paramecium*. Note golf-tee-shaped barb and long, striated shaft. (By Jacus and Hall, 1946: Biol. Bull., *91*:141.)

In *Dileptus anser* the trichocysts are apparently turned inside out at discharge (Fig. 2–36*A* and *B*). The expelled trichocyst consists of a bulbous base that tapers into a long thread. The positions of these two parts are reversed in the undischarged condition. In some ciliates the discharged trichocysts may perhaps be used in anchoring the animal when feeding. In others, such as *Dileptus*, they are apparently used for defense or for capturing prey. The trichocysts of *D. anser* are reported to paralyze other small protozoans. In *Dileptus gigas* the trichocysts are fluid-filled vesicles, the contents of which can be expelled (Fig. 2–36*C*). Such trichocysts, called toxicysts, have a definite paralyzing effect on rotifers and other protozoans.

Mucigenic bodies are another group of pellicular organelles found in many ciliates. They are arranged in rows like trichocysts and discharge a mucoid material that may function in the formation of cysts or protective coverings (Fig. 2–33).

Cilia have the same structure as flagella; they differ from flagella chiefly in that they are generally more numerous and are considerably shorter. Compound ciliary organelles, evolved from the adhesion of varying numbers of individual cilia, are of common occurrence and will be described later.

The ciliature can be conveniently divided into the body (or somatic) ciliature, which occurs over the general body surface, and the oral ciliature, which is associated with the mouth region. Distribution of body cilia is quite variable. In the primitive groups, cilia cover the entire animal and are arranged in longitudinal rows (Fig. 2–32*A*); but in many of the more specialized groups they have become limited to certain regions of the body, as in *Euplotes* (Fig. 2–37*B*), or have almost disappeared completely, as in *Vorticella* (Fig. 2–37*A*).

As mentioned earlier, each cilium arises from a basal granule, or kinetosome, located in the alveolar layer (Figs. 2–34 and 2–38). The kinetosomes that form a particular longitudinal row are connected by means of fine, striated fibrils, called kinetodesma. The kinetosomes plus the fibrils of that row make up a kinety. The longitudinal bundle of fibrils runs to the right side of the row of kinetosomes, and each kinetosome gives rise to one kinetodesmos (fibril), which joins the longitudinal bundle and extends anteriorly. A single kinetodesmos is tapered and extends for varying distances as a part of the bundle.

A kinety system is apparently characteristic of all ciliates. Even groups such as the Suctorida, which possess cilia only during developmental stages, retain the kinety system in the adult. It is commonly stated that the system coordinates ciliary beat, but convincing evidence for such a function is still lacking. The kineties probably do contribute to the anchorage of the cilium, and may be important in the spacing of cilia in morphogenesis.

Locomotion. The ciliates are the fastest moving of the protozoans, some reaching a speed of 2 mm. per second. In its beat each cilium performs an effective and a recovery stroke. During the effective stroke the cilium is outstretched and moves from a forward to a backward position. In the recovery stroke the cilium is bent over to the right against the body (when viewed from above and looking anteriorly) and is brought back to the forward position in a counterclockwise movement. The recovery position offers less water resistance and is somewhat analogous to feathering an oar. A cilium moves in three different planes in the course of a complete cycle of beat, and the positions have been captured and recorded in scanning electron micrographs of freeze-dried *Paramecium* (Tamm, 1972).

The beating of the surface cilia is synchronized and waves of ciliary beat progress down the length of the body from anterior to posterior. The direction of the waves is slightly oblique, which causes the ciliate to swim in a spiral course and at the same time to rotate on its longitudinal axis.

The ciliary beat can be reversed, and the animal can move backward. This backward movement is associated with the so-called avoiding reaction. In *Paramecium*, for example, when the animal comes in contact with some undesirable substance or object, the ciliary beat is reversed (Fig. 2–39). It moves backward a short distance, turns slightly clockwise or counterclockwise, and moves forward again. If unfavorable conditions are still encountered, the avoiding reaction is repeated. Detection of external stimuli is probably through the cilia; and although perhaps all of the cilia can act as sensory receptors in this respect, there are certain long, stiff cilia that play no role in movement and are probably entirely sensory. The direction and intensity of the beat is controlled by levels of free intracellular calcium ions (Eckert, 1972).

The highly specialized hypotrichs, such as *Urostyla*, *Stylonychia*, and *Euplotes* (Fig. 2–37*B*), have greatly modified body cilia. The body has become differentiated into distinct dorsal and ventral surfaces, and cilia have

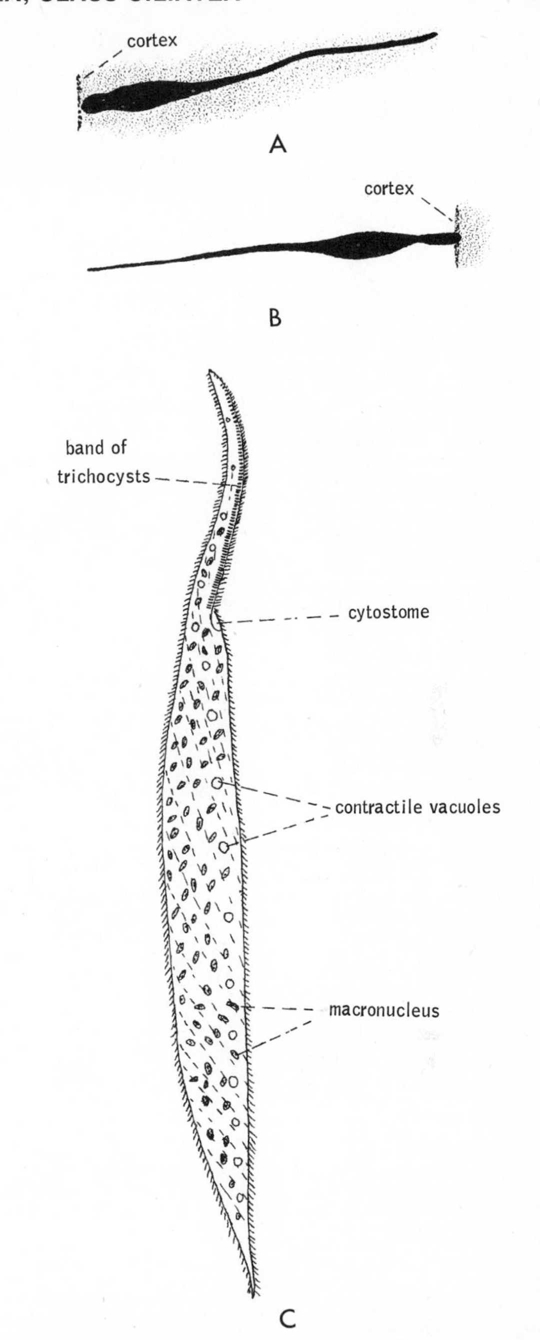

Figure 2–36. *A* and *B*. Trichocyst of *Dileptus anser*. *A*. Before discharge. *B*. After discharge. (After Hayes from Hall.) *C*. *Dileptus gigas*. (After Hyman.)

largely disappeared except on certain areas of the ventral surface. Here the cilia occur as a number of tufts (called cirri), arranged in rows as in *Urostyla*, or in a small number of groups as in *Euplotes*. Even more remarkable is the fact that any single cirrus can be moved in any direction independently of another cirrus.

A peculiar method of locomotion occurs

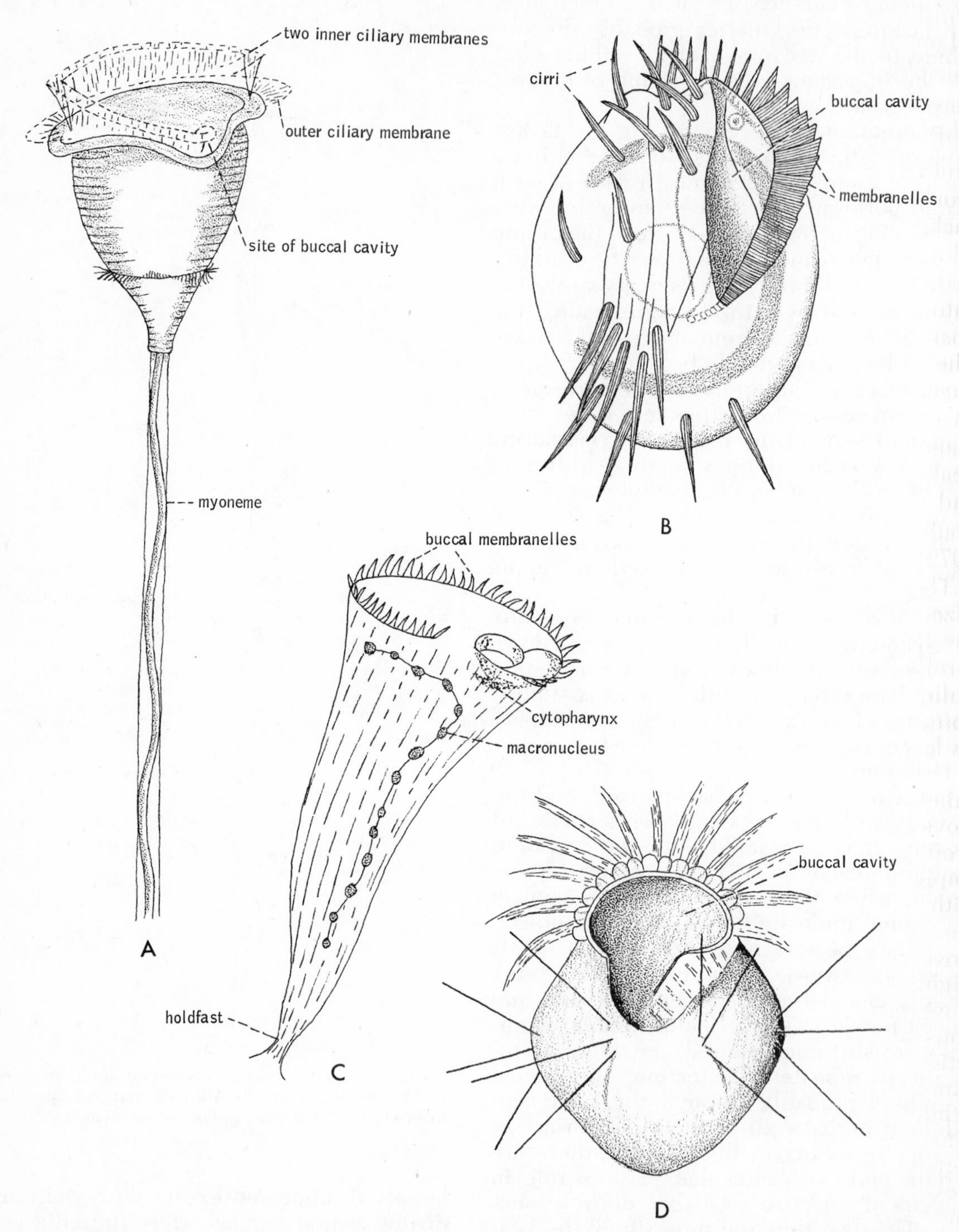

Figure 2–37. *A. Vorticella. B. Euplotes.* (After Pierson from Kudo.) *C. Stentor.* (After Tartar from Manwell.) *D. Halteria.* Note girdle of long, bristle-like cilia, which produce a popping type of locomotion. (After Corliss.)

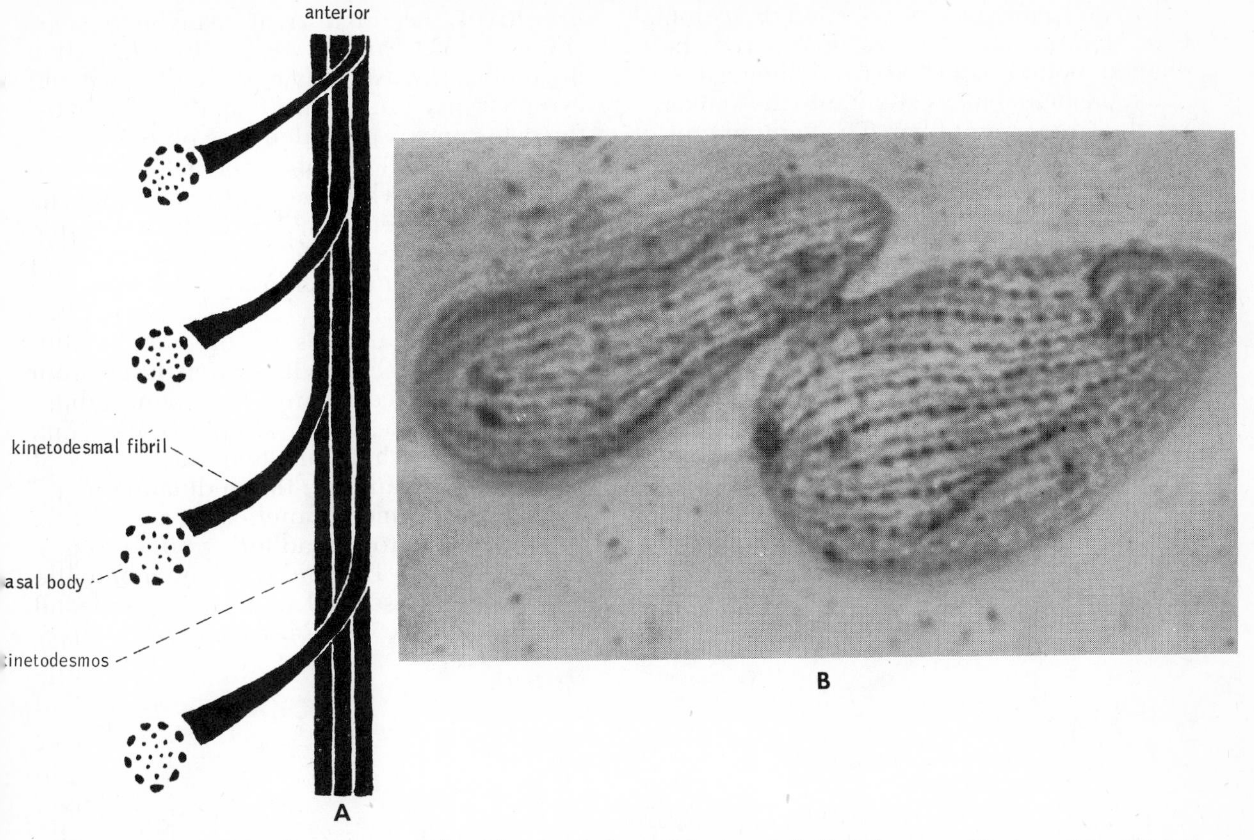

Figure 2–38. *A.* A ciliate kinety. (After Metz *et al.* from Grimstone.) *B. Tetrahymena.* Kinety system appears as dark beaded longitudinal lines. Fainter intervening line represents line of adjoining alveoli. To right side of the anterior buccal cavity can be seen the membranelles. The undulating membrane appears as a curved line on left margin of buccal cavity.

in the little spirotrich, *Halteria* (Fig. 2–37*D*). A girdle of approximately 11 long bristles, probably representing modified cilia, surrounds the body and replaces other ciliation, except that associated with food acquisition. The animal swims in a popping or bouncing manner as a result of sudden movements of the bristles.

Some ciliates, especially sessile forms, can undergo contractile movements, either shortening the stalk by which the body is attached as in *Vorticella,* or shortening the entire body as in *Stentor.* Contraction is brought about by contractile fibrils, or "myonemes," that lie in the pellicle. These fibrils, which can be very numerous as in *Stentor,* parallel the longitudinal axis of the animal. In *Spirostomum* the contractile microfilaments are arranged as bundles located beneath longitudinally spiral ciliated grooves.

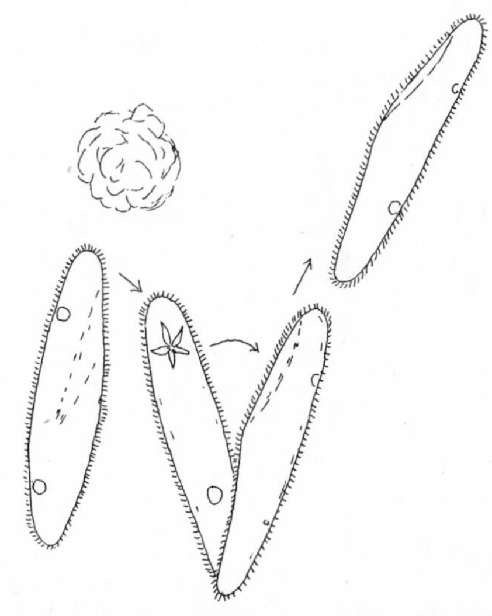

Figure 2–39. The avoiding reaction of *Paramecium.* (After Hyman.)

In *Vorticella* (Fig. 2–37*A*) and the colonial *Carchesium,* both of which have bell-shaped bodies attached by a long slender stalk, the myonemes extend into the stalk as a single, large, spiral fiber. The contractions of this spiral myoneme, which functions very much like a coiled spring, produce the familiar popping movements that are so characteristic of *Vorticella* and related genera.

Nutrition. The free-living ciliates are almost entirely holozoic. Typically a distinct mouth, or cytostome, is present, although it has been secondarily lost in some groups such as the suctorians and the parasitic astomatids. In primitive groups the mouth is located anteriorly (Fig. 2–32*A*), but in most ciliates it has been displaced posteriorly to varying degrees. The mouth opens into a canal or passageway called the cytopharynx, which extends into the endoplasm. The cytopharynx is devoid of cilia, and at its terminal end the food vacuoles are formed. Primitively, as in a great many holotrichs (Figs. 2–32*A* and 2–40*A*), the ingestive organelles consists only of the cytostome and cytopharynx; but in the majority of ciliates the cytostome is preceded by a preoral chamber. The preoral chamber may take the form of a vestibule, which varies from a slight depression to a deep funnel, with the cytostome at its base (Fig. 2–40*B*). The vestibule is clothed with simple cilia derived from the somatic ciliature.

In the higher ciliates the preoral chamber is typically a buccal cavity, which differs from a vestibule by containing compound ciliary organelles instead of simple cilia (Fig. 2–40*C* to *F*). There are two basic types of such ciliary organelles: the undulating membrane and the membranelle. An undulating membrane is a row of adhering cilia forming a sheet (Fig. 2–41*A* and *B*). A membranelle is derived from several short rows of cilia, all of which adhere to form a more-or-less rec-

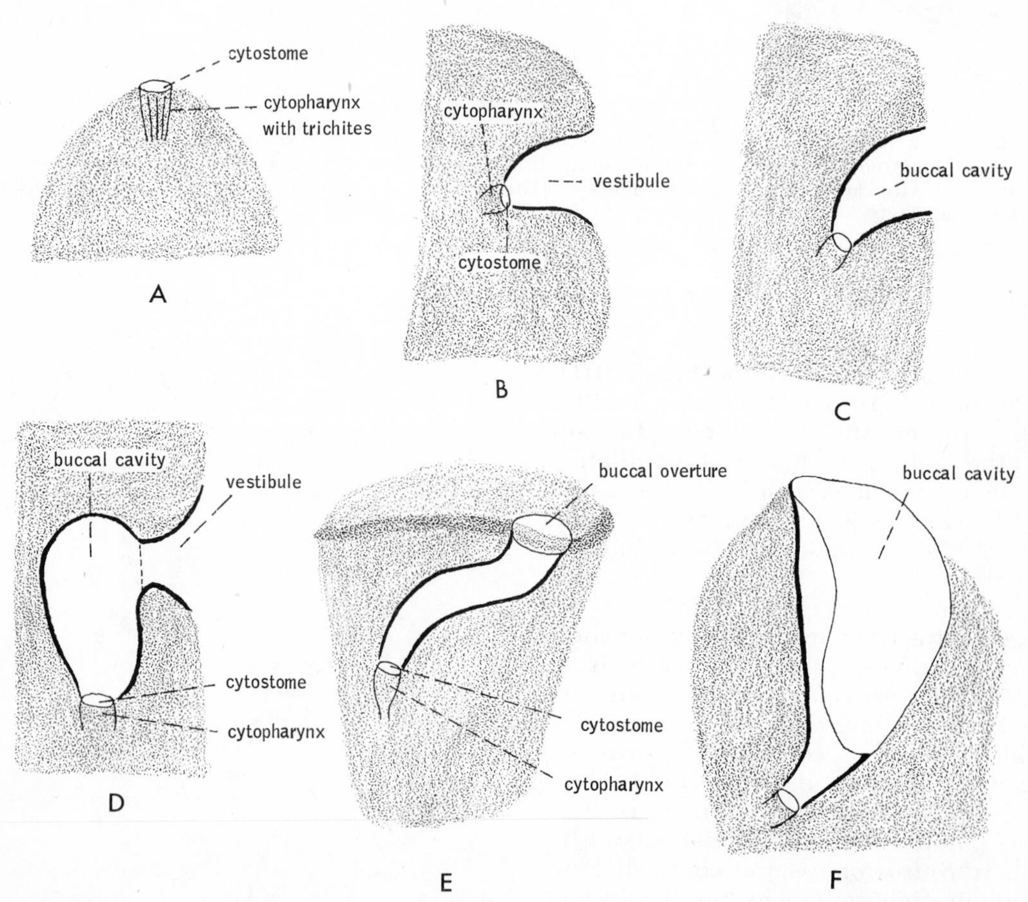

Figure 2–40. Oral areas of various ciliates. *A.* In rhabdophorine gymnostomes; such as, *Coleps, Prorodon,* and *Didinium. B.* In a trichostome with vestibule displaced from anterior end. *C.* In a tetrahymenine hymenostome, such as *Tetrahymena. D.* In a peniculine hymenostome, such as *Paramecium. E.* In a peritrich, such as *Vorticella. F.* In a hypotrich, such as *Euplotes.* (After Corliss.)

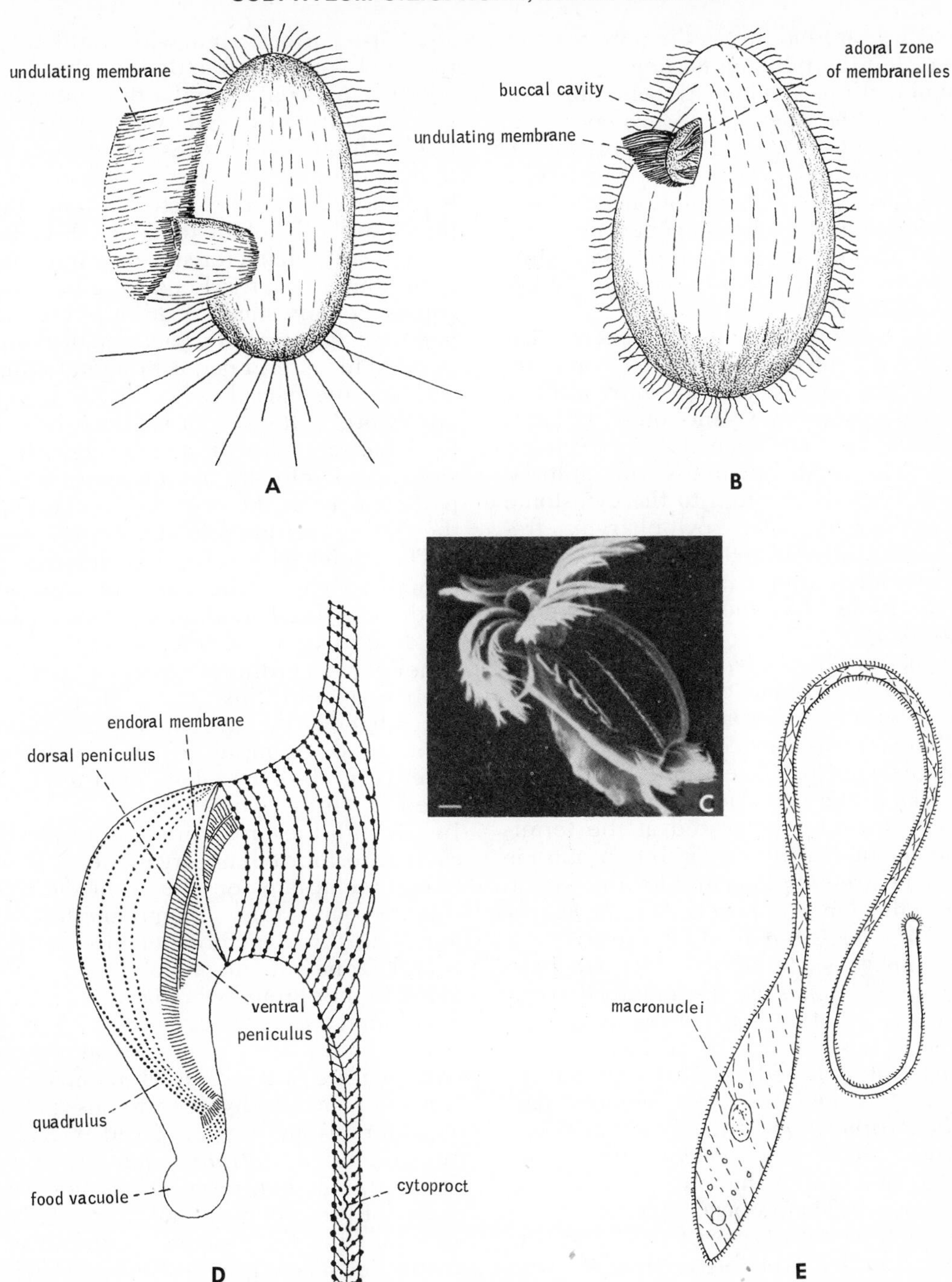

Figure 2–41. *A. Pleuronema.* (After Noland from Corliss.) *B. Tetrahymena.* (After Corliss.) *C.* Scanning electron photomicrograph of *Uronychia*, a marine ciliate, showing the highly developed membranelles. (By Small, E. B., and Marszalek, D. S., 1969: Science, *163*:1064–1065. Copyright 1969, American Association for the Advancement of Science. *D.* Buccal organelles of *Paramecium.* (After Yusa from Manwell.) *E. Lacrymaria.* (After Conn from Hyman.)

tangular plate and typically occur in a series (Figs. 2–32*B;* 2–37*C;* and 2–41*B*). Fusion of cilia tends to restrict the beat to one plane, in contrast to the spiral pattern of a solitary cilium.

The term peristome, which is commonly encountered in the literature, is synonymous with buccal cavity. In members of a number of orders the buccal organelles project from the buccal cavity; or, as in the Hypotrichida (Fig. 2–37*B*), the buccal cavity is somewhat shallow so that the organelles occupy a flattened area around the oral region. Such an area is called the peristomial field.

Although primarily ingestive in function, in some ciliates such as *Pleuronema* and *Uronychia* (Fig. 2–41) the ciliary organelles are very large and project from the buccal cavity to serve also in locomotion. The buccal cavity may in turn be preceded by a vestibule. This is true of the peniculine hymenostomes, such as *Paramecium* (Figs. 2–40*D* and 2–41*D*).

The free-swimming holozoic species display two types of feeding habits. Some are raptorial, and attack and devour rotifers, gastrotrichs, protozoans, and other ciliates. Others have become specialized for ciliary feeding. The oral apparatus of raptorial ciliates is typically limited to the cytostome and cytopharynx. The cytopharynx frequently has walls strengthened by bundles of microtubules (nemadesma), which surround the lumen like the staves of a barrel (Fig. 2–40*A*).

The mouth in *Coleps* and *Didinium* (Fig. 2–42*A*) can open to a great diameter, almost as wide as the diameter of the body itself; these forms can consume very large prey. Some species have their anterior extended as a proboscis. The proboscis may be short with the mouth located at the terminal end as in *Didinium*, or the proboscis may be an extremely long and flexible snout. Although the long proboscis of *Lacrymaria* (Fig. 2–41*E*) bears the mouth at the terminal end, in many such ciliates (for example *Loxodes* and *Trachelius*) the mouth is a slit or circle located at the base of the proboscis.

Didinium has perhaps been most studied of all the raptorial feeders. This little barrel-shaped ciliate feeds on other ciliates, particularly *Paramecium* (Fig. 2–42*A*). When *Didinium* attacks a *Paramecium*, the proboscis attaches to the prey through the terminal mouth. The proboscis apparently adheres only to certain types of pellicles; it is this limitation that probably restricts the diet of *Didinium*.

An interesting group of raptorial ciliates is the aberrant subclass Suctoria, formerly considered a separate class. Free-living suctorians are all sessile and are attached to the substratum directly or by means of a stalk (Fig. 2–42). Cilia are present only in the immature stages. The body is commonly somewhat globular or cone-shaped and differs from that of other ciliates in bearing tentacles (Fig. 2–42). The tentacles may be knobbed at the tip; they are occasionally shaped like long spines (Fig. 2–45*B*). Each tentacle consists of an inner, somewhat rigid tube surrounded by a contractile sheath. The prey adheres on contact with the tentacles and is apparently paralyzed by an adhesive secretion. Through a process that is still not understood, the captured protozoan gradually is sucked into the suctorian body through the central tube of the tentacles (Fig. 2–42*C*).

Typically characteristic of ciliary feeders is the buccal cavity. Food for ciliary feeders consists of any small, dead or living, organic particles, particularly bacteria that are suspended in water. Food is brought to the body and into the buccal cavity by the compound ciliary organelles. From the buccal cavity the food particles are driven through the cytostome and into the cytopharynx. When the particles reach the end of the cytopharynx, they collect within a food vacuole.

The holotrich order, Hymenostomatida, contains some of the most primitive ciliary feeders, at least judging from the nature of the buccal cavity and associated ciliary organelles. According to Corliss (1961), it is probably from this group that the more specialized forms have evolved. *Tetrahymena* is a good example of such a primitive type (Fig. 2–41*B*). The cytostome is located a little posteriorly and ventrally. Just within the broad opening to the buccal cavity are four ciliary organelles—an undulating membrane on the right side of the chamber and three membranelles on the left. The three membranelles constitute an adoral zone of membranelles, which in many higher groups of ciliates is much more developed and extensive.

The four buccal ciliary organelles are a basic feature of the order Hymenostomatida. However, many hymenostomes possess a considerably more specialized buccal apparatus than does *Tetrahymena*.

In *Paramecium*, perhaps the most familiar genus of the order, an oral groove along the side of the body leads posteriorly to a vestibule, located about midway back from the anterior end. The vestibule, buccal cavity, and cytopharynx together form a large curved funnel (Figs. 2–40*D* and 2–41*D*). The undulating membrane, here called the endoral membrane, runs transversely along the right wall and marks the junction of the vestibule and buccal cavity. The three membranelles are also modified. Two, called peniculi, are greatly lengthened and thus tend to be more similar to an undulating membrane in function than to the more typical membranelle.

The peniculi run down the left wall of the

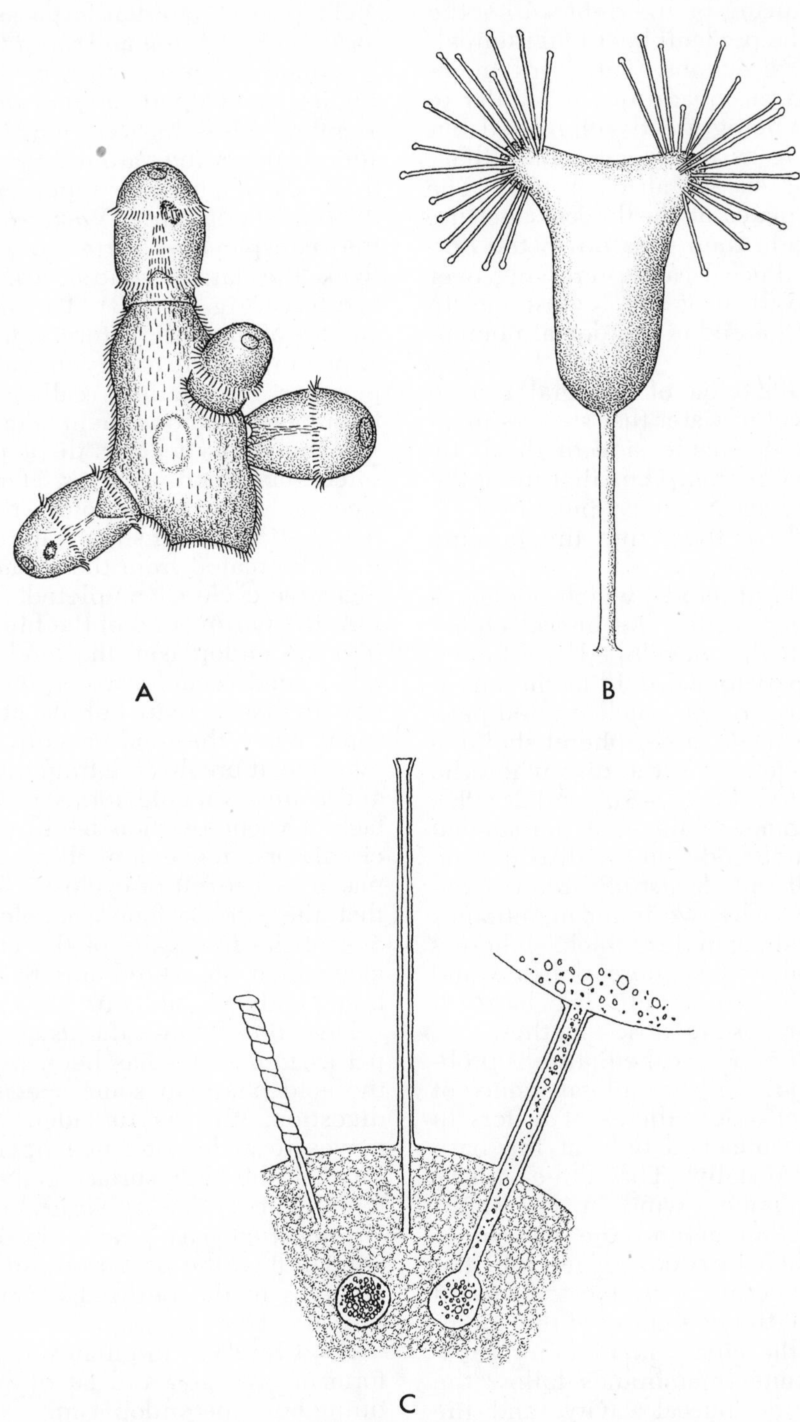

Figure 2–42. *A. Didinium*, a raptorial ciliate, attacking a *Paramecium*. (After Mast from Dogiel.) *B. Acineta*, a suctorian. (After Calkins from Hyman.) *C.* Tentacles of the suctorian, *Tokophrya lemnarum*, retracted, extended, and during feeding. (After Noble from Manwell.)

buccal cavity. The ventral peniculus stops at the cytostome, but the longer dorsal peniculus crosses over to the right wall at the cytostome and terminates on the right wall of the cytopharynx. The peniculi beat longitudinally in opposite directions. The third membranelle, called the quadrulus, is similar to the peniculi in being greatly elongated but differs in having its four component ciliary rows separated except at each end. The quadrulus originates at the buccal aperture near the peniculi, then runs down the dorsal wall of the buccal cavity, crossing over to the right wall near the cytostome to terminate near the end of the dorsal peniculus.

In feeding, the cilia of the oral groove produce a current of water that sweeps in an arclike manner down the side of the body and over the oral region. The ciliature of the vestibule and buccal cavity pull in food particles and drive them into the forming food vacuole.

In the order Peritrichida, which possesses little or no somatic cilia, the buccal ciliary organelles are highly developed and form a large disclike peristomial field at the apical end of the animal. In the much studied peritrich genus *Vorticella,* a peripheral shelflike projection can close over the disc when the animal is retracted (Figs. 2–37A and 2–40*E*). The ciliary organelles lie in a peristomial groove between the edge of the disc and the peripheral shelf and consist of three ribbonlike membranes. The two inner membranes are three basal granules thick, whereas the outer membrane is more delicate and projects laterally like a shelf. The bases of all three membranes are close together.

Although peritrich buccal ciliature is probably homologous to the adoral zone of membranelles of other ciliates, it differs in having the cilia attached only at the base and separated distally. The three ciliary bands, or membranes, wind in a counterclockwise direction around the margin of the disc and then turn downward into the funnel-shaped buccal cavity, located toward the outer side of the beginning of the spiral. In this region the ciliary membranes separate; the two inner membranes follow the inner wall of the buccal cavity, and the outer membrane follow the outer wall.

In feeding, food particles are passed along the outer membrane, driven by the undulations of the two inner membranes. Within the buccal cavity water flows downward between the inner membranes and the outer membrane and then flows outward between the inner wall of the cavity and the inner membranes.

Ciliates of the subclass Spirotricha, which includes such familiar forms as *Stentor, Halteria, Spirostomum* and *Euplotes,* are typically ciliary feeders. They usually possess a highly developed adoral zone of many membranelles. In *Stentor* (Fig. 2–37*C*) the adoral zone winds around the apical pole of the body in a manner superficially similar to the buccal ciliature of *Vorticella.* In *Euplotes* the conspicuous series of membranelles flank the large, ventral, triangular buccal aperture (Fig. 2–37*B*). The loricate marine ciliates of the order Tintinnida have a crown of pectinate (or feathery) membranelles with peculiar organelles, called tentaculoids, interspersed among them (Fig. 2–32*B*).

Many ciliary feeders display considerable selection of food particles. *Stentor* and *Paramecium,* for example, often reject nonnutritive particles; or if such material is ingested, it is eliminated from the vacuole before the digestive cycle is completed.

At the narrow end of the funnel that opens into the endoplasm, the food particles pass into a food vacuole, which forms at the funnel tip like a soap bubble at the end of a pipe. When the food vacuole reaches a certain size, it breaks free from the cytopharynx, and a new vacuole forms in its place. Detached vacuoles then begin a more or less circulatory movement through the endoplasm as a result of cyclosis. It is significant that the site of food vacuole formation is one of the few parts of the ciliate cell surface which is covered only by the cell membrane and lacks alveoli.

Digestion follows the usual pattern, and a pH as low as 1.4 has been reported during the acid phase in some species. Following digestion, the waste-laden food vacuole moves to a definite anal opening, or cytoproct, at the body surface and expels its contents. The cytoproct varies in position. In *Paramecium* it is located on the side of the animal, near the posterior end (Fig. 2–41*D*), whereas in the peritrichs it opens into the buccal cavity.

Food reserves in ciliates are stored in the form of glycogen and fat droplets scattered throughout the endoplasm.

There are relatively few parasitic ciliates, although there are many ecto- and endocommensals. One interesting parasite, the apostomatid *Foettingeria,* is remarkable in having a life cycle that requires two hosts. This little marine ciliate lives inside the gastrovascular cavity of sea anemones. The motile young leave the anemone and encyst on small crustaceans. When such infected crustaceans are eaten by anemones, the ciliates emerge from their cysts.

Some suctorians are endoparasites. Hosts include fishes and certain ciliates. *Sphaerophyra,* for example, lives within the endoplasm of *Stentor,* and *Endosphaera* is parasitic within the body of the peritrich, *Telotrochidium.*

Space limitations permit mention of only a few of the many interesting commensal species. *Kerona,* a little crawling hypotrich, and *Trichodina,* a mobile peritrich, are ectocommensals on the surface of hydras. Other free-swimming peritrichs (all of which have a girdle of cilia about the posterior end of the body and move with the posterior end forward) occur on the body surfaces of freshwater planarians, tadpoles, sponges, and other animals. The holotrich *Balantidium coli* is an endocommensal in the intestine of pigs and is passed by means of cysts in the feces. This ciliate has occasionally been found in man, where in conjunction with bacteria it erodes pits in the intestinal mucosa and produces pathogenic symptoms.

Among the most interesting endocommensals are the highly specialized Entodiniomorphida (Fig. 2–47A). These spirotrichs live as harmless commensals in the digestive tract of many different hoofed mammals. Like the flagellate symbionts of termites and roaches, some of them ingest and break down the cellulose of the vegetation eaten by their hosts. The products of digestion are utilized by the host, but unlike termites and roaches, the host is not dependent upon them. The ectoplasm of these ciliates is very thick, restricting the endoplasm to a little internal saclike area. The cytopharynx opens into this cavity, and a canal leads away to the cytoproct. Thus, these ciliates possess a "complete digestive tract" on a cytoplasmic level of structure.

A few ciliates display symbiotic relationships with algae. The most notable of these is *Paramecium bursaria,* in which the endoplasm is filled with green zoochlorellae.

Water Balance. Contractile vacuoles are found in both marine and freshwater species, but especially in the latter. In the primitive species a single vacuole is located near the posterior, but many species possess more than one vacuole (Fig. 2–36C). In *Paramecium* one vacuole is located at the posterior and one at the anterior of the body (Fig. 2–39). The vacuoles are always associated with the innermost region of the ectoplasm and empty through a distinct canal that penetrates the pellicle. Thus, the position of the contractile vacuole is always fixed in ciliates. In most ciliates a system of radiating collecting canals is located around the vacuole and is particularly evident when the vacuole is small and filling up (Fig. 2–43).

In *Paramecium* and others, there are many collecting canals completely surrounding the vacuole, but in some genera, such as *Stentor* and *Bursaria,* there is only one large collecting canal. It is doubtful if the vacuole itself is a permanent organelle; rather, it probably re-forms after each expulsion by the fusion of small droplets delivered by the collecting

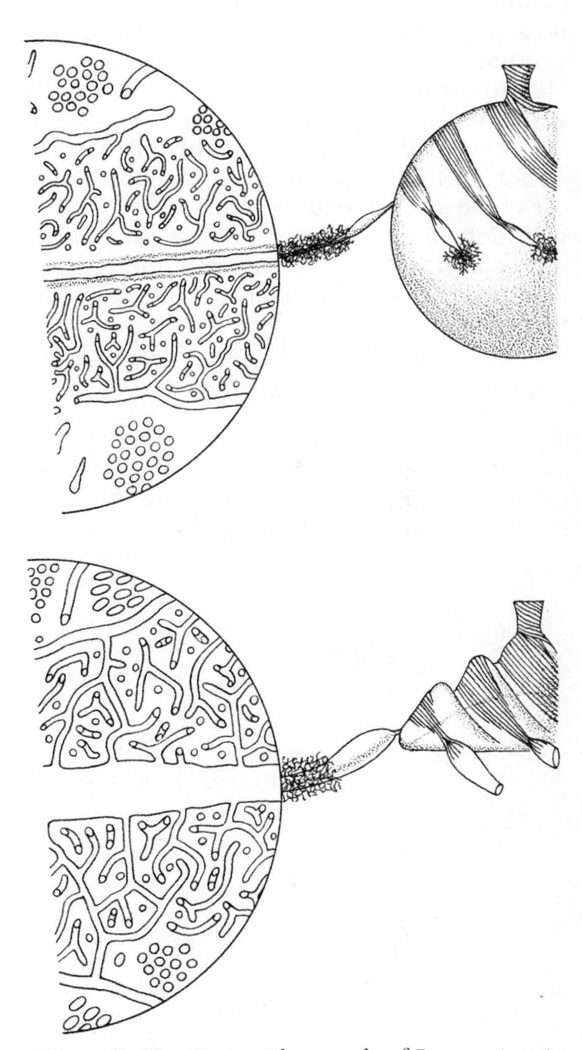

Figure 2–43. Contractile vacuole of *Paramecium* including a radial canal and surrounding cytoplasm. Top figure shows dilated vacuole and exit through pellicle for expulsion of fluid. Bottom figure shows contracted vacuole and enlarged radial canal, into which cytoplasmic tubules are emptying. (After Schneider from Pitelka.)

canals. When there is more than one vacuole present, they pulsate at different rates depending on their position. For example, in *Paramecium* the posterior vacuole pulsates faster than the anterior vacuole because of the large amount of water being delivered into the posterior region by the cytopharynx. Although contractile vacuoles may be present in marine species, the rate of pulsation is considerably slower than that in freshwater species; they are probably removing ingested water.

Reproduction. Ciliates differ from all other animals in possessing two distinct types of nuclei—a usually large macronucleus and one or more small micronuclei. The micronuclei are small, rounded bodies and vary in number from one to as many as 80, depending on the species. The micronuclei are responsible for genetic exchange and nuclear reorganization, and also give rise to the macronuclei. The macronucleus is sometimes called the vegetative nucleus, since it is not critical in sexual reproduction. However, the macronucleus is essential for normal metabolism, for mitotic division, and for the control of cellular differentiation, and is responsible for the genic control of the phenotype.

Usually only one macronucleus is present, but it may assume a variety of shapes (Fig. 2–44). The large macronucleus of *Paramecium* is somewhat oval or bean-shaped and is located just anterior to the middle of the body. In *Stentor* and *Spirostomum* the macronucleus is long and shaped like a string of beads. Not infrequently the macronucleus is in the form of a long rod bent in different configurations, such as a **C** in *Euplotes* or a horseshoe in *Vorticella.* The macronucleus is highly polyploid, the chromosomes having undergone repeated duplication following the micronuclear origin of the macronucleus.

Asexual Reproduction. Asexual reproduction is always by means of binary fission, which is typically transverse. More accurately, fission is described as being homothetogenic, with the division plane cutting across the kineties—the longitudinal rows of cilia or basal granules (Fig. 2–45*A*). This is in contrast to the symmetrogenic fission of flagellates, in which the plane of division (longitudinal) cuts between the rows of basal granules. Mitotic spindles are typically formed by each of the micronuclei. The behavior of macronuclei is more variable. A spindle does not form, and division is usual-

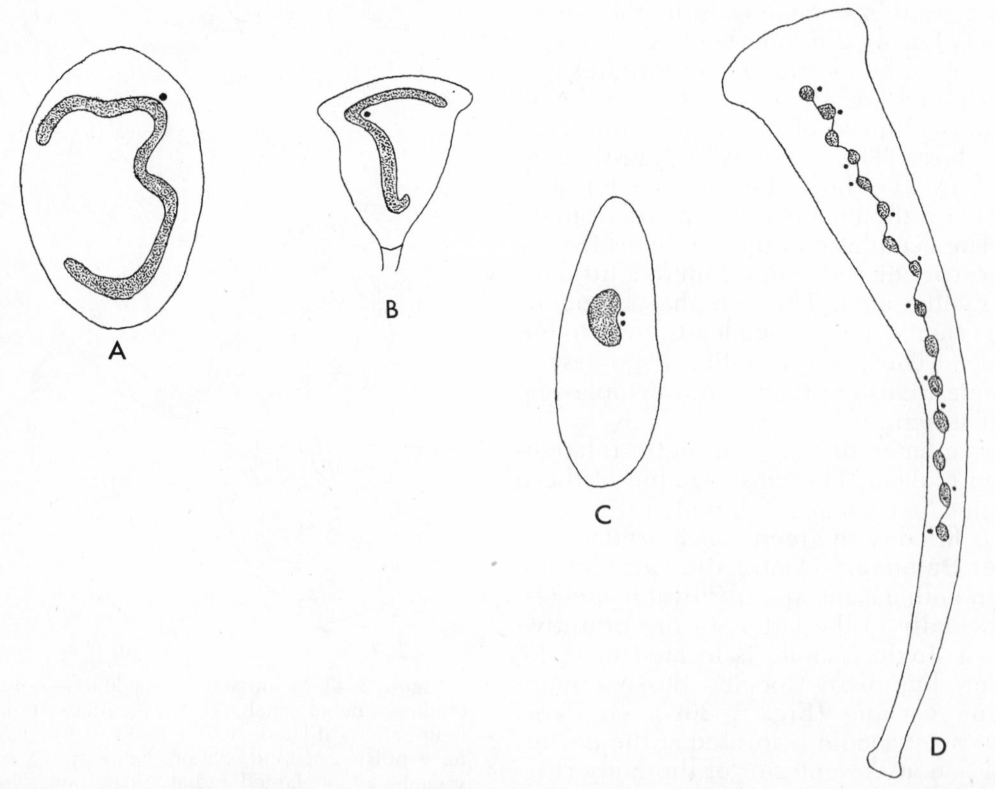

Figure 2–44. Macronuclei (in gray) of various ciliates (micronuclei, in black). *A. Euplotes. B. Vorticella. C. Paramecium. D. Stentor.* (After Corliss.)

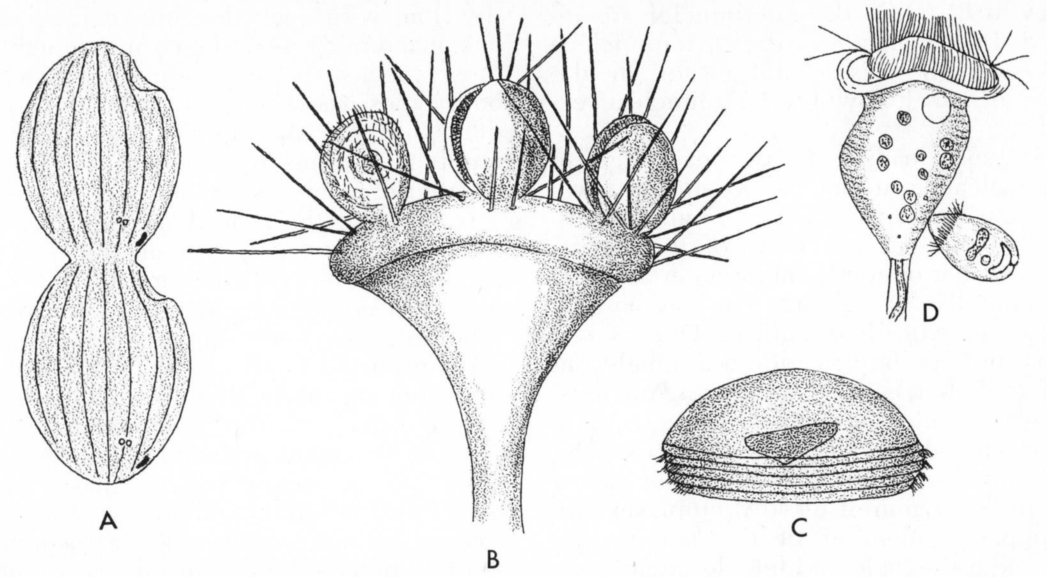

Figure 2–45. *A.* Homothetogenic type of fission, in which the plane of division cuts across the kineties. (After Corliss.) *B.* Suctorian *Ephelota* with external buds. (After Noble from Hyman.) *C.* Detached bud of *Dendrocometes.* (After Pestel from Hyman.) *D.* Conjugation in *Vorticella.* Note the small nonsessile microconjugant. (After Kent from Hyman.)

ly accomplished by constriction. When a number of macronuclei are present, they may first combine as a single body before dividing. The same is true of some forms with beaded or elongated macronuclei. In *Stentor,* for example, the macronuclear chain condenses into a single mass and then divides. In still other ciliates the macronucleus degenerates and arises anew from micronuclei.

Modified fission in the form of budding occurs in some ciliate groups, notably the Suctoria. In most members of this subclass the parent body buds off a varying number of daughter cells from the outer surface (Fig. 2–45*B*); or there is an internal cavity or brood chamber, and the buds form internally from the chamber wall. In contrast to the sessile adults, which lack cilia, the daughter cells, or buds, are provided with several circlets of cilia and are free-swimming (Fig. 2–45*C*). Following a few hours of free existence, the "larva" attaches and assumes the characteristics of the sessile adults.

Although there are no centrioles, the kinetosomes of many ciliates, like the basal granules of flagellates, divide at the time of fission. Furthermore, the kinetosomes play a primary role in the re-formation of organelles. It has been found that all of the organelles can be re-formed providing the cell contains a piece of macronucleus and some kinetosomes. In the more primitive ciliates, in which the cilia have a general distribution over the body surface, the kinetosomes have equal potentials in the re-formation of organelles. However, in the specialized ciliates there is a corresponding specialization of the kinetosomes; only certain ones are involved in the re-formation of new cellular structures during fission. For example, in hypotrichs such as *Euplotes,* all of the organelles are resorbed at the time of fission; and certain of the kinetosomes on the ventral side of the animal divide to form a special group that is organized in a definite field or pattern. These special "germinal" kinetosomes then migrate to different parts of the body, where they form all of the surface organelles—cirri, peristome, cytopharynx, and other structures. A review of ciliate differentiation has been provided by Hanson (1967).

Encystment is typical of most ciliates; however, some, including *Paramecium* and the chonotrichs, are believed never to encyst and are passed only in an active state from one body of water to another.

Sexual Reproduction. An exchange of nuclear material by conjugation is involved in sexual reproduction. By apparently random contact in the course of swimming, two sexually compatible members of a particular species adhere in the oral or buccal region of the body. Adhesion probably results from the secretion of a sticky substance by the cilia. Following the initial attachment, there is a fusion of protoplasm in the region of contact. Two such fused ciliates are called conjugants; attachment lasts for several hours. During this period a reorganization and exchange of nuclear material occurs (Fig.

2–46*A* to *F*). Only the micronuclei are involved in conjugation; the macronucleus breaks up and disappears either in the course of, or following, micronuclear exchange.

The steps leading to the exchange of micronuclear material between the two conjugants are fairly constant in all species. After two meiotic divisions of the micronuclei, all but one of them degenerate. This one then divides, giving two micronuclei that are genetically identical. One is stationary and can be considered a female nucleus; the other, called a wandering nucleus, will migrate into the opposite conjugant and can be considered the male nucleus. The wandering nucleus in each conjugant moves through the region of fused protoplasm into the opposite member of the conjugating pair. There the male and female nuclei fuse with one another to form a "zygote" nucleus, or synkaryon.

Shortly after nuclear fusion the two animals separate, and each is now called an exconjugant. After separation, there follows in each exconjugant a varying number of nuclear divisions, leading to the reconstitution of the normal nuclear condition characteristic of the species. This reconstitution usually, but not always, involves a certain number of cytosomal divisions. For example, in some forms where there is but a single macronucleus and a single micronucleus in the adult, the synkaryon divides once. One of the daughter nuclei forms a micronucleus; the other forms the macronucleus. Thus, the normal nuclear condition is restored without any cytosomal divisions.

However, in *Paramecium caudatum* (Fig. 2–46*G* to *N*), which also possesses a single nucleus of each type, the synkaryon divides three times, producing eight nuclei. Four become micronuclei and four become macronuclei. Three of the micronuclei are then resorbed. The animal now undergoes two cytosomal divisions, during the course of which each of the four resulting daughter cells receives one macronucleus. The single micronucleus in each daughter cell undergoes mitosis at each cytosomal division. Restoration of the normal nuclear state in *Paramecium aurelia,* which possesses one macronucleus and two micronuclei, occurs still differently. Here the synkaryon divides twice to form two micronuclei and two macronuclei. The two micronuclei then each divide again to produce a total of four micronuclei and two macronuclei. The exconjugant then undergoes one cytosomal division, with each daughter cell receiving one macronucleus and two micronuclei. In those species that have numerous nuclei of both types, there is no cytosomal division; the synkaryon merely divides a sufficient number of times to produce the requisite number of macronuclei and micronuclei.

In some of the more specialized ciliates, the conjugants are a little smaller than nonconjugating individuals; or the two members of a conjugating pair are of strikingly different sizes. Such macro- and microconjugants occur in *Vorticella* (Fig. 2–45*D*) and represent an adaptation for conjugation in sessile species. The macroconjugant, or "female," remains attached; while the small bell of the microconjugant, or "male," breaks free from its stalk and swims about. On contact with an attached macroconjugant the two bells adhere, and nuclear exchange occurs. However, there is no separation after conjugation, and the little "male" conjugant degenerates. In the Suctoria, conjugation takes place between two attached individuals that happen to be located side by side. This is also true in the chonotrichs, ectocommensal ciliates attached to crustaceans.

The frequency of conjugation is extremely variable. Some species have rarely been observed to undergo the sexual phenomenon of conjugation; others conjugate every few days or weeks. In the chonotrichs, conjugation is linked with the molting of their crustacean hosts; here the "male" leaves its own pedestal and may be completely engulfed by the female.

Conjugation brings about a shuffling of hereditary characteristics, just as is true of sexual reproduction in other animals; and in some ciliates this nuclear reorganization has a rejuvenating effect and is necessary for continued asexual fission. For example, it has been shown that some species of *Paramecium* can pass through only approximately 350 continuous asexual generations. If nuclear reorganization does not occur, the asexual line (or clone) will die out. However, there are many species in which fission can occur indefinitely and conjugation is an unnecessary adjunct for asexual fission. Some early workers cultured so-called "deathless" clones of *Paramecium,* in which fission supposedly continued indefinitely without any intervening conjugation.

Recent protozoologists have demonstrated that although conjugation may not occur in such "deathless" clones, another type of nuclear reorganization called autogamy takes place and has the same effect on fission as

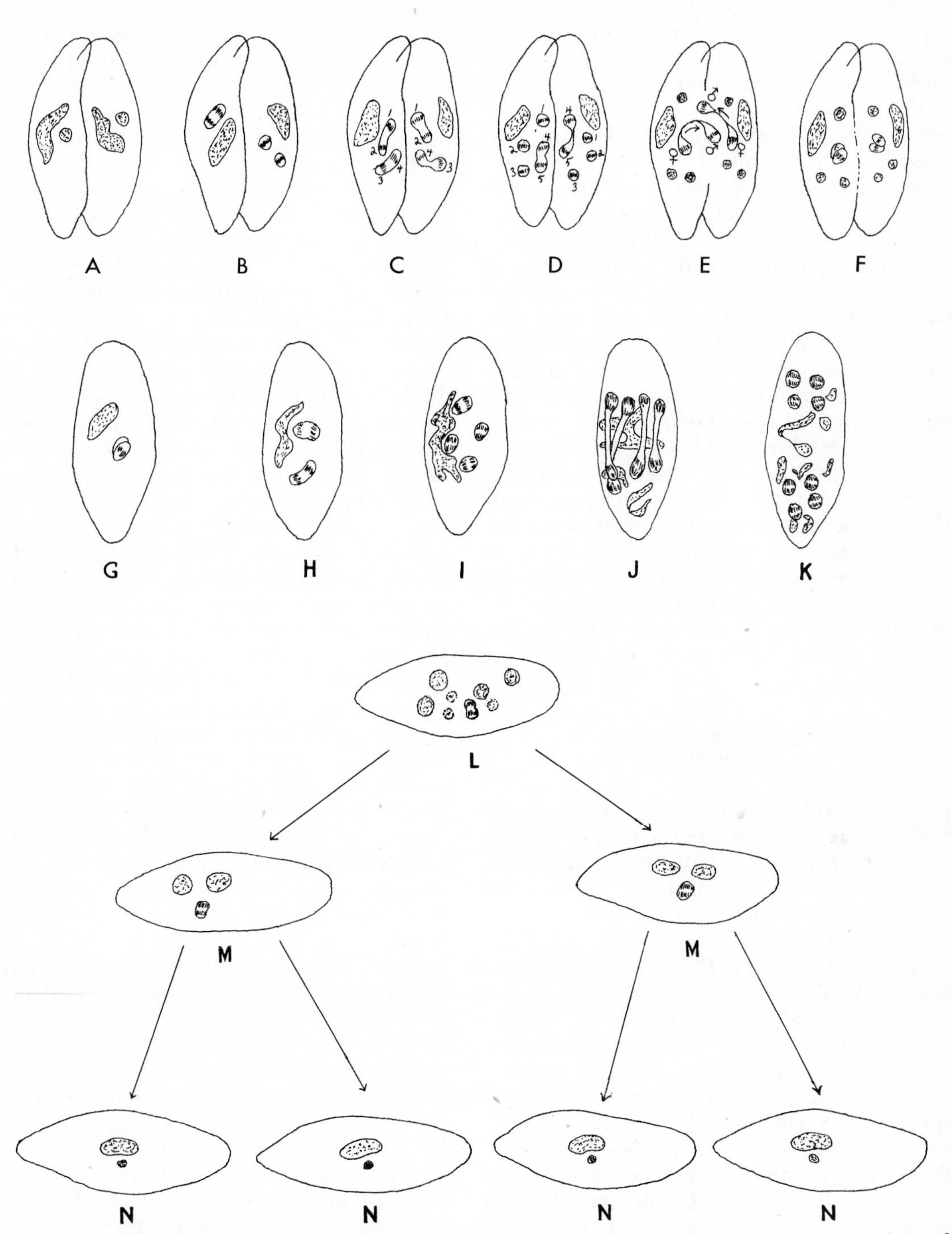

Figure 2–46. Sexual reproduction in *Paramecium caudatum*. *A* to *F*. Conjugation. *B* to *D*. Micronuclei undergo three divisions, the first two of which are meiotic. *E*. "Male" micronuclei are exchanged. *F*. They fuse with the stationary micronucleus of the opposite conjugant. *G*. Exconjugant with macronucleus and synkaryon micronucleus; other micronuclei have been resorbed. *H* to *K*. Three divisions of synkaryon, forming eight micronuclei; old macronucleus is resorbed. *L*. Four micronuclei form macronuclei; three are resorbed. *L* and *M*. Remaining micronucleus divides twice in course of two cytosomal divisions. Resulting daughter cells each receive one of the four macronuclei in *L*. *N*. Normal nuclear condition restored. (Modified after Calkins from Wichterman.)

does conjugation. Autogamy involves the same nuclear behavior as does conjugation, but there is no conjugation and no exchange of micronuclear material between two individuals. The macronucleus degenerates and the micronucleus divides a number of times to form eight or more nuclei. Two of these nuclei fuse to form a synkaryon; the others degenerate and disappear. The synkaryon then divides to form a new micronucleus and macronuclei as occurs in conjugation. Nuclear reorganization, either in the form of conjugation or in the form of autogamy, is not necessary for continued fission in all ciliates. Some species can reproduce asexually indefinitely.

Definite mating types have been shown to exist in species of *Paramecium, Tetrahymena, Euplotes, Stylonychia,* and a few other ciliates. For example, there are a number of varieties, or "syngens," of *Paramecium caudatum* and *P. aurelia,* each with two mating types. Conjugation is always restricted to a member of the opposite mating type and does not occur between members of the same type, apparently owing to a failure of the cytoplasm to adhere. The mating types are hereditary.

SYSTEMATIC RÉSUMÉ OF CLASS CILIATEA

Subclass Holotrichia. With simple and uniform body cilia. Buccal ciliature either absent or, if present, usually inconspicuous.

Order Gymnostomatida. Chiefly large ciliates with no oral ciliature. Cytostome opening directly to outside. *Coleps, Lacrymaria, Prorodon, Dileptus, Didinium, Actinobolina, Nassula.*

Order Trichostomatida. With vestibular but no buccal ciliature. *Colpoda, Tillina, Balantidium.*

Order Chonotrichida. Vase-shaped ciliates lacking body cilia. A "funnel" at the free end of the body bears vestibular cilia. Chiefly marine and ectocommensal on crustaceans. *Spirochona, Lobochona, Chilodochona.*

Order Apostomatida. Body with spirally arranged ciliation. Cytostome mid-ventral and in the vicinity of a peculiar rosette. Marine parasites or commensals with complex life cycles usually involving two hosts, one of which is commonly a crustacean. *Foettingeria, Hyalophysa.*

Order Astomatida. Commensals or endoparasites living chiefly in the gut and coelom of oligochaete worms. Cytostome absent; body ciliation uniform. *Anoplophrya, Hoplitophrya.*

Order Hymenostomatida. Small ciliates having a uniform body ciliation but possessing a buccal cavity. Buccal ciliature consists of an undulating membrane and an adoral zone of membranelles. *Colpidium, Tetrahymena, Blepharostoma, Paramecium.*

Order Thigmotrichida. A small group of marine and freshwater ciliates found in association with bivalve mollusks. Anterior end of body bears a tuft of thigmotactic cilia. *Thigmophrya, Boveria.*

Subclass Peritrichia.

Order Peritrichida. Adult usually lacks body cilia, but the apical end of the body typically bears a conspicuous buccal ciliature. Mostly attached stalked ciliates. *Vorticella, Carchesium, Zoothamnium, Lagenophrys, Trichodina.*

Subclass Suctoria.

Order Suctorida. Sessile, stalked ciliates with the distal end bearing few to many tentacles. Adult stage completely devoid of any ciliature. *Acineta, Podophrya, Sphaerophyra, Dendrocometes, Ephelota.*

Subclass Spirotrichia. With generally reduced body cilia and well-developed, conspicuous buccal ciliature.

Order Heterotrichida. With uniform body cilia or body encased in a lorica and body cilia absent. *Bursaria, Stentor, Blepharisma, Spirostomum, Folliculina.*

Order Oligotrichida. Small ciliates with body cilia reduced or absent. Conspicuous buccal membranelles, commonly extending around apical end of body. *Halteria.*

Order Tintinnida. Loricate, mostly free-swimming ciliates with conspicuous oral membranelles when extended. Chiefly marine. *Codonella, Tintinnus, Favella.*

Order Entodiniomorphida. Endocommensal ciliates in the digestive tract of herbivorous mammals. Body cilia reduced or absent. Prominent buccal ciliature, often in separate anterior clumps. Posterior end may be drawn out into spines. *Entodinium, Cycloposthium, Elephantophilus.*

Order Odontostomatida. A small group of laterally compressed and wedge-shaped ciliates with carapace and reduced body and buccal cilia. *Saprodinium.* (Fig. 2–47*B*).

Order Hypotrichida. Dorsoventrally flattened ciliates in which the body cilia are restricted to fused tufts of cilia, or

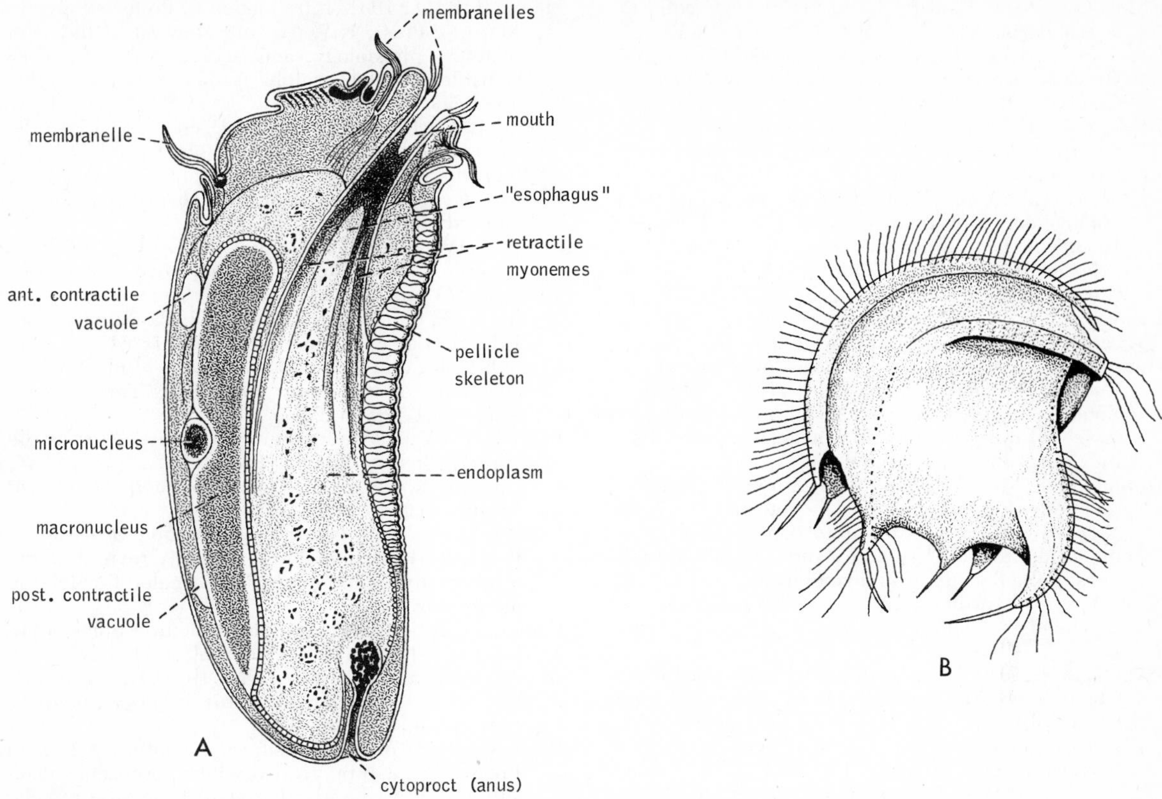

Figure 2–47. A. The entodiniomorph, *Diplodinium ecaudatum,* a commensal in the rumen of cattle. (After Manwell.) *B. Saprodinium dentatum,* an odontostome ciliate. (After Kahl from Corliss.)

cirri, located on the ventral side of the body. *Euplotes, Uronychia, Kerona, Stylonychia, Urostyla, Oxytricha.*

Bibliography

The Journal of Protozoology is devoted entirely to the publication of research on protozoans.

Allen, R. D., Francis, D., and Zek, R., 1971: Direct test of the positive pressure gradient theory of pseudopod extension and retraction in amoebae. Science, *174*:1237–1240. (See also Science (1972), *177*:636–638 for additional contributions to the debate on the nature of ameboid movement.)

Bick, H., 1972: Ciliate Protozoa. An illustrated guide to the species used as biological indicators in freshwater biology. Amer. Public Health Assoc., Washington, D.C. 198 pp.

Bick, H., 1973: Population dynamics of Protozoa associated with the decay of organic materials in fresh water. Amer. Zool. *13*:149–160.

Chen, Y. T., 1950: Investigation of the biology of *Peranema trichophorum.* Quart. J. Micr. Sci., *91*: 279–308.

Cheng, T. C., 1964: The Biology of Animal Parasites. W. B. Saunders, Philadelphia. (A parasitology text with a good account of the parasitic protozoans.)

Cleveland, L. R., 1925: Method by which *Trichonympha* ingests solid particles of wood. Biol. Bull., *48*:282–287.

Cleveland, L. R., 1928: Symbiosis between termites and their intestinal Protozoa. Biol. Bull., *54*:231–237.

Cleveland, L. R., 1960: The centrioles of *Trichonympha* from termites and their function in reproduction. J. Protozool., *7*:326–341.

Cleveland, L. R., Hall, S. R., Saunders, E. P., and Collier, J., 1934: The wood-feeding roach *Cryptocercus,* its Protozoa, and the symbiosis between Protozoa and the roach. Amer. Acad. Sci., Mem. 17, pp. 185–342.

Corliss, J. O., 1959: An illustrated key to the higher groups of the ciliate Protozoa, with definition of terms. J. Protozool., *6*:265–281.

Corliss, J. O., 1961: The Ciliated Protozoa: Characterization, Classification, and Guide to the Literature. Pergamon Press, N.Y. (A definition of the class Ciliatea and its orders and families. This volume contains a very extensive bibliography of the literature on ciliates.)

Corliss, J. O., 1973: Protozoa. *In* Gray, P. (Ed.), Encyclopedia of Microscopy and Microtechnique. Van Nostrand Reinhold Co., N.Y.

Dodson, E. O., 1971: The kingdoms of organisms. Syst. Zool., *20*:265–281.

Dogiel, V. A., 1965: General Protozoology, 2nd Edition. Oxford University Press, N.Y. (A good introductory protozoology text. Emphasis on biology of Protozoa rather than taxonomy.)

Eckert, R., 1972: Bioelectric control of ciliary activity. Science, *176*:473–481.

Ehret, C. F., and Powers, E. L., 1959: The cell surface of *Paramecium.* Int. Rev. Cytol., *8*:97–133.

Gibbons, I. R., and Grimstone, A. V., 1960: On flagellar structure in certain flagellates. J. Biophys. Biochem. Cytol., 7:697–716.

Giese, A. C., 1973: *Blepharisma:* The Biology of a Light-Sensitive Protozoan. Stanford University Press, Stanford, Calif. 366 pp.

Grassé, P. (Ed.), 1952: Traité de Zoologie, Vol. 1, pt.1, Phylogénie; Protozoaires: Généralites, Flagellés. 1953: Vol. 1, pt. 2, Protozoaires: Rhizopodes, Actinopodes, Sporozaires, Cnidosporidies. Masson et Cie, Paris.

Grimstone, A. V., 1961: Fine structure and morphogenesis in Protozoa. Biol. Rev., *36*:97–150.

Hanson, E. D., 1967: Protozoan development. *In* Florkin, M., and Scheer, B. J. (Eds.), Chemical Zoology. Academic Press, N.Y., pp. 395–539. (A good review of many aspects of protozoan reproduction, including ciliate differentiation.)

Honigberg, B. M. (Chairman), 1964: A revised classification of the Phylum Protozoa. J. Protozool., *11*:7–20. (A classification of the protozoans proposed by the Committee on Taxonomy and Taxonomic Problems of the Society of Protozoologists.)

Hutner, S., and Lwoff, A. (Eds.), 1955: Physiology and Biochemistry of Protozoa. Vol. 2. Academic Press, N.Y.

Hyman, L. H., 1940: The Invertebrates: Protozoa through Ctenophora. McGraw-Hill Book Co., N.Y., pp. 44–232. (A concise general treatment of the Protozoa, although now out-of-date in many areas. The chapter *Retrospect* in Vol. 5 (1959) of this series surveys investigation on protozoans from 1938 to 1958.)

Jahn, T. L., and Bovee, E. C., 1967: Motile behavior of Protozoa. *In* Chen, T. (Ed.), Research in Protozoology. Pergamon Press, N.Y., pp. 41–200. (A good review of protozoan locomotion, although already some aspects are out-of-date.)

Jeon, K. W. (Ed.), 1973: The Biology of Amoeba. Academic Press, N.Y.

Kidder, G. W. (Ed.), 1967: Protozoa. *In* Florkin, M., and Scheer, B. J. (Eds.), Chemical Zoology, Vol. 1. Academic Press, N.Y., 912 pp. (This work covers many general aspects of the biology of protozoans in addition to biochemistry.)

Kinosita, H., and Murakami, A., 1967: Control of ciliary motion. Physiol. Rev. *47*:53–82.

Kudo, R. R., 1954: Protozoology. 4th Edition. Charles C Thomas, Springfield, Ill. (A general account of the protozoans with particular emphasis on the taxonomy of the group.)

Lackey, J. B., Deflandre, G., and Noland, L. E., 1959: Zooflagellates; Rhizopoda and Actinopoda; and Ciliophora. *In* Edmondson, W. T., Ward, H. B., and Whipple, G. C. (Eds.), Freshwater Biology. 2nd Edition. John Wiley and Sons, N.Y. pp. 190–297.

Manton, I., 1959: Electron microscopical observations on a very small flagellate: the problem of *Chromulina pusilla* Butcher. J. Mar. Biol. Assoc. U. K., *38*: 319–333.

Manwell, R. D., 1961: Introduction to Protozoology. St. Martin's Press, N.Y. (An introduction to the morphology, physiology, and ecology of protozoans. Considerably less emphasis on taxonomy in this text than in that by Kudo.)

Moore, R. C. (Ed.), 1964 and 1954: Treatise on Invertebrate Paleontology. Protista. Vols. C and D. Geological Society of America and University of Kansas Press. (These volumes cover fossil forams and radiolarians.)

Murray, J. W., 1973: Distribution and Ecology of Living Benthic Foraminiferids. Crane, Russak and Co., N.Y. 274 pp.

Noble, E. R., and Noble, G. A., 1972: Parasitology. 3rd Edition. Lea & Febiger, Philadelphia. 617 pp.

Noland, L. E., and Finley, H. E., 1931: Studies on the taxonomy of the genus *Vorticella.* Trans. Amer. Micr. Soc., *50*:81–123.

Pennak, R. W., 1953: Freshwater Invertebrates of the United States. Ronald Press, N.Y., pp. 20–76. (Figures and keys to the common genera of freshwater protozoans.)

Pitelka, D. R., 1963: Electron-Microscopic Structure of Protozoa. Pergamon Press, N.Y. (A review of the ultrastructure of protozoans as revealed by electron microscopy.)

Pitelka, D. R., 1970: Ciliate ultrastructure: Some problems in cell biology. J. Protozool., *17*(1):1–10.

Pitelka, D. R., and Schooley, C. N., 1958: The fine structure of the flagellar apparatus in *Trichonympha.* J. Morphol., *102*:199–246.

Read, C. P., 1970: Parasitism and Symbiosis. Ronald Press, N.Y., 316 pp. (A text which approaches parasitology from the standpoint of host-parasite relationships rather than from a survey of parasitic groups.)

Riegel, B., Stangers, D. W., Wikholm, D. M., Mold, J. D., and Sommer, H., 1949: Paralytic shellfish poison. V. The primary source of the poison, the marine plankton organism, *Gonyaulax caterella.* J. Biol. Chem., *177*:7–11.

Roth, L. E., 1960: Electron microscopy of pinocytosis and food vacuoles in *Pelomyxa.* J. Protozool., 7:176–185.

Sonneborn, T. M., 1950: *Paramecium* in modern biology. BioS., *21*:31–43.

Sonneborn, T. M., 1957: Breeding systems, reproductive methods, and species problems in Protozoa. *In* May, E. (Ed.), The Species Problem. A.A.A.S. Pub., Washington, D.C., pp. 155–324.

Tamm, S. L., 1972. A scanning electron microscope study. The Journal of Cell Biology, *55*:250–255.

Tartar, V., 1961: The Biology of *Stentor.* Pergamon Press, N.Y. (A detailed treatment of many aspects of the biology of this genus of ciliates.)

Whittaker, R. H., 1969: New concepts of kingdoms of organisms. Science, *163*:150–159.

Wichterman, R., 1953: The Biology of *Paramecium.* Blakiston, N.Y.

Yusa, A., 1957: The morphology and morphogenesis of the buccal organelles in *Paramecium* with particular reference to their systematic significance. J. Protozool., *4*:128–142.

Chapter 3

INTRODUCTION TO THE METAZOANS

Metazoans are multicellular, motile, heterotrophic organisms which develop from embryos. The gametes never form within unicellular structures but rather are produced within multicellular sex organs or at least within surrounding somatic cells. According to the protistan concept, this is the definition of the Animal Kingdom. If the classical view is retained, the definition differentiates the subkingdom Metazoa from the subkingdom Protozoa of the Animal Kingdom. In any case, the metazoans comprise almost all of what are generally considered to be animals. The diversity is enormous. There are twenty-nine phyla, depending upon who is counting, and only one, the Chordata, contains animals which do not belong to the invertebrate assemblage.

For the student who is making the first serious attempt to study invertebrates in some depth, the task may seem overwhelming. Each group has certain structural peculiarities, a special anatomical terminology, and a distinct classification. All of these factors tend to magnify the differences between groups. At the same time the very real danger is also created that the student may lose sight of functional and structural similarities that result from similar modes of existence and similar environmental conditions, as well as of homologies arising from close evolutionary relationships.

This danger perhaps may be lessened if a few basic biological principles are kept in mind. All animals must meet the same problems of existence—the procurement of food and oxygen, maintenance of water balance, removal of metabolic wastes, and perpetuation of the species. The body structure necessary to meet these problems is, in large part, correlated with three factors: the type of environment in which the animal lives, the size of the animal, and the mode of existence of the animal.

Of the three major environments—salt water, fresh water, and land—the marine environment is generally the most stable. Wave action, tides, and vertical and horizontal ocean currents produce a continual mixing of sea water and ensure a medium in which the concentration of dissolved gases and salts fluctuates relatively little. The buoyancy of sea water reduces the problem of support. It is therefore not surprising that the largest invertebrates have always been marine. Since sea water is isotonic to the tissue fluids of most marine animals, there is little difficulty in maintaining water balance.

The buoyancy and uniformity of sea water provide an ideal medium for animal reproduction. Eggs in sea water can be shed and fertilized, and can undergo development as floating embryos, with little danger of desiccation and of salt imbalance, or of being swept away by rapid currents into less favorable environments. Larvae are particularly characteristic of marine animals. Larval stages provide a means of wide dispersal of the species, and feeding larvae can obtain food material for completion of their development without the necessity for large amounts of yolk material within the egg.

Fresh water is a much less constant medium than sea water. Streams vary greatly in turbidity, velocity, and volume, not only along their course, but also from time to time as a result of droughts or heavy rains. Small ponds and lakes fluctuate in oxygen content, turbidity, and water volume. In large lakes the environment changes radically with increasing depth.

Like salt water, fresh water is buoyant and

aids in support. The low salt concentration, however, creates some difficulty in maintaining water balance. Since the body of the animal contains a higher salt concentration than that of the external environment, there is a tendency for water to diffuse inward. The animal thus has the problem of getting rid of excess water. As a consequence, freshwater animals usually have some mechanism for pumping water out of the body while holding on to the salts; i.e., they must osmoregulate. In general, the eggs of freshwater animals are either retained by the parent or attached to the bottom of the stream or lake, rather than being free-floating as is often true of marine animals. Larval stages are usually absent. Floating eggs and free-swimming larvae are too easily swept away by currents. Since development is usually direct, the eggs typically contain considerable amounts of yolk.

Nitrogenous waste of aquatic animals, both marine and freshwater, is usually excreted as ammonia. Ammonia is very soluble and toxic and requires considerable water for its removal; but since there is no danger of water loss in aquatic animals, the excretion of ammonia presents no difficulty.

Terrestrial animals live in the harshest environment. The supporting buoyancy of water is absent. Most critical, however, is the problem of water loss by evaporation. The solution of this problem has been a primary factor in the evolution of many adaptations for life on land. The integument of terrestrial animals presents a better barrier between the inner and the outer environment than that of aquatic animals. Respiratory surfaces, which must be moist, have developed in the interior of the body to prevent drying out. Nitrogenous wastes are commonly excreted as urea or uric acid, which require less water for removal than does ammonia. Fertilization must be internal; the eggs are usually enclosed in a protective envelope or deposited in a moist environment. Except for insects and a few other arthropods, development is direct; and the eggs are usually endowed with large amounts of yolk. Terrestrial animals which are not well adapted to withstand desiccation are either nocturnal or are restricted to humid or moist habitats.

A second factor that is correlated with the nature of animal structure is the size of the animal. As the body increases in size, the ratio of surface area to volume decreases, for volume increases by the cube, and surface area by the square, of body length. In small animals the surface area is sufficiently great in comparison to the body volume that exchange of gases and waste can be carried out efficiently by diffusion through the general body surface. Also, internal transport can take place by diffusion alone.

However, as the body increases in size, distances become too great for internal transport to take place by diffusion alone; and more efficient transportation mechanisms are necessary. In larger animals this has led to the development of coelomic and blood-vascular circulatory systems. Also, through folding and coiling, the surface area of internal organs and portions of the external surface may be increased to facilitate secretion, absorption, gas exchange, and other processes.

A third factor that is related to the nature of animal structure is the mode of existence. Free-moving animals are generally bilaterally symmetrical. The nervous system and sense organs are then concentrated at the anterior end of the body, since this is the part that first comes in contact with the environment. Such an animal is said to be cephalized.

Attached, or sessile, animals are radially symmetrical or tend toward a radial symmetry, in which the body consists of a central axis around which similar parts are symmetrically arranged. Radial symmetry is an advantage for sessile existence, since it allows the animal to meet the challenges of its environment from all directions. Skeletons, envelopes, or tubes are commonly present for support or for protection against motile predators.

The feeding mechanisms of animals are usually correlated with the mode of existence. Animals which can actively swim or crawl are frequently raptorial. The slower movers may be scavengers or herbivores. Those which inhabit bottom sediments are often deposit feeders. Deposit feeders consume the substratum in which they live. The organic detritus, or deposit material, is digested away from the sand grains and other inorganic substances, and the latter is egested and may be ejected as castings. Deposit feeders may apply the mouth against the substratum and ingest material directly; or they may be indirect deposit feeders, utilizing special appendages to gather in deposit material some distance from the mouth.

Sessile animals may feed on passing prey, ingest deposit material, or subsist on organic detritus or microscopic plants and animals suspended in the surrounding water. The last type of nutrition involves suspension feeding, in which the animal traps or collects organisms or detritus suspended in the surrounding water. The term *suspension feeding* replaces the ambiguous term *ciliary feeding,* which may or may not be synonymous with suspension feeding (Jørgensen, 1966). Some suspension feeding animals utilize suspended material that simply adheres to the general body surface. The collected particles are then transported by ciliary currents to the mouth. More often the animal is adapted for filtering a water current for suspended particles. The water current may be produced by the animal or may be a natural current. The term suspension feeding will be used in the ensuing chapters only when filtering is not employed; otherwise the animal will be called a filter feeder. It should be kept in mind, however, that the distinction is not always clearcut.

METAZOAN ORGANIZATION

Symmetry. Two metazoan phyla, the Cnidaria and the Ctenophora, display a fundamental radial symmetry and are commonly grouped together as the Radiata. The cnidarians and the ctenophores are linked not only by a radial symmetry, but by a similar body plan. They both possess a digestive cavity with a mouth as the only major opening to the exterior, and the cells lying between the digestive cavity and the outer body surface are not highly organized. Many of these animals are attached to the substratum; others are free-swimming, but the movement is principally vertical, and the majority are at the mercy of water currents.

The remainder of the metazoans are bilaterally symmetrical and are grouped together as the Bilateria. The bilateral symmetry is correlated with the locomotor movement displayed by these animals. One end is directed forward, and a constant surface is oriented toward the substratum. A few of the Bilateria are radial, such as the echinoderms, or display a tendency toward a radial symmetry, but such a radial symmetry is associated with a secondary assumption of an attached mode of existence.

Metamerism. At least twice in the history of the bilateral animals the condition of metamerism evolved—in the annelid worms and in the chordates. A metameric animal is one in which the body is divided into a linear series of similar parts, segments or metameres. Not included among the segments are the head, or acron, bearing the brain and sense organs, or the terminal pygidium on which the anus opens (Fig. 3–1). New segments arise in front of the pygidium; thus the oldest segments always lie just behind the head. Certain longitudinal structures, such as the gut and principal blood vessels and nerves, extend the length of the body, passing through successive segments; other structures are repeated in each segment, reflecting the basic metameric organization of the body. However, metamerism is fundamentally a mesodermal phenomenon. It is the body wall musculature and sometimes the coelom which are the primary segmental divisions. The mesodermal segmentation, in turn, imposed a corresponding metamerism on the associated "supply system"—nerves, blood vessels, and excretory organs.

Wherever metamerism has evolved, there has been a tendency for the condition to undergo secondary reduction in forms which

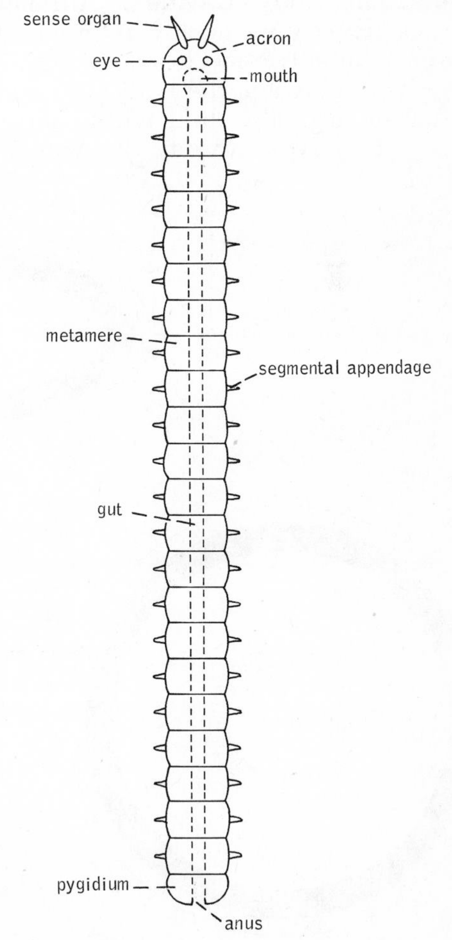

Figure 3–1. Diagram of a metameric animal, such as an annelid worm.

become adapted for new modes of existence. Some indication of the primitiveness of an animal can be determined by the degree of segmentation it displays. Reduction in segmentation can result from loss of segments, fusion of segments, or differentiation of segments so that they are no longer similar. For example, if the segments bear appendages, the appendages may become specialized for different functions. Such differentiation has resulted in the use of the term *serial homology*. Serial homology refers to structures which, though different, segmentally have the same embryonic and evolutionary origin. Thus, the claws of a crab are serially homologous to the terminal pair of abdominal flippers. They both arose originally from similar segmental appendages.

The definition of metamerism described here is one accepted by most zoologists. But there are some who would use a broader definition. Groups which might fall within a less exclusive interpretation of metamerism will be indicated as they are later encountered.

Metazoan Body Cavities. Although a body cavity can mean any internal space, the term generally refers to a large fluid-filled cavity lying between the body wall and the internal organs. The fluid within the body cavity may serve a variety of functions. It often functions as a hydrostatic skeleton, and it may serve as a circulatory medium. The space may be utilized as a temporary site for the accumulation of excess fluids and waste, and may be the site for maturation of eggs and sperm. It provides an area for the enlargement (and increase in surface area) of internal organs.

The radiate metazoans (and the flatworms and proboscis worms among the bilateral phyla), lacking a body cavity, have a solid type of body construction and are said to be acoelomate. In the flatworms and the proboscis worms—the acoelomate Bilateria—the space between the gut and the body wall is filled with a network of cells called mesenchyme (Fig. 3–2A).

Two types of body cavities are found among the remaining bilateral metazoans. One group of phyla possesses what is called a pseudocoel—an unfortunate name, since it implies that there is something false about the body cavity. The pseudocoel is derived from the blastocoel of the embryo; that is, it represents a persistent blastocoel. The internal organs are actually free within the space, for there is no peritoneum bounding the cavity (Fig. 3–2*B*). The pseudocoelomate Bilateria include such familiar animals as rotifers and roundworms.

All other metazoans possess a body cavity

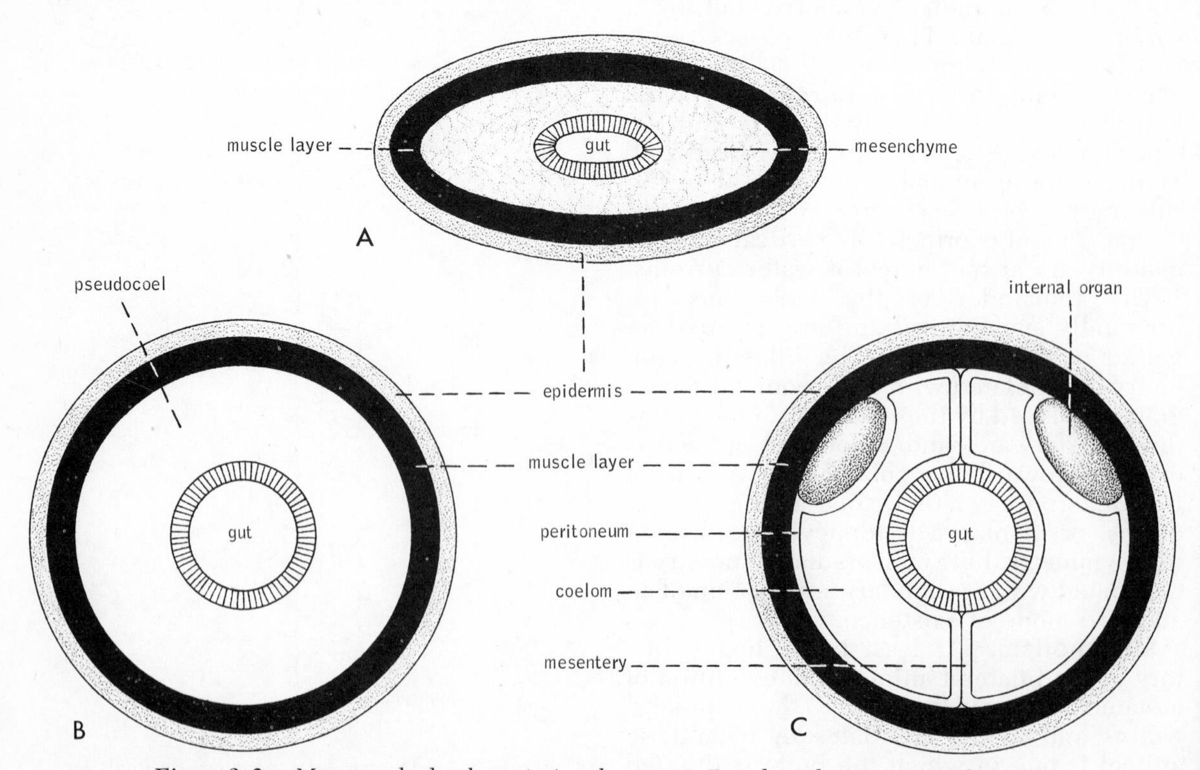

Figure 3–2. Metazoan body plans. *A*. Acoelomate. *B*. Pseudocoelomate. *C*. Coelomate.

called a coelom, or if a body cavity is absent, it clearly represents a secondary loss. A coelom has a different embryogeny than does a pseudocoel. A coelom arises as a cavity within the embryonic mesoderm. The mesoderm provides the cellular lining to the cavity. The lining is called the peritoneum. None of the internal organs are actually free within the coelom. They may bulge into it, but all are bounded by peritoneum. The internal organs are thus all located behind peritoneum; i.e., they are retroperitoneal (Fig. 3–2*C*).

The various theories dealing with evolution of the coelom and metamerism and the relationship of these theories to invertebrate phylogeny will be discussed in Chapter 21.

Metazoan Organ Systems. Specialization among metazoans has taken place through the differentiation of cells to form tissues and organs, which are specialized for various functions. As indicated earlier, there is a close correlation between the evolution of metazoan organs and the size of the animal. As animals become larger, internal distances become too great for transport by simple diffusion, and blood-vascular systems for internal transport make their appearance. In like manner, the surface area becomes proportionately less for the increasing volume of the animal. General body surface is insufficient for gas exchange and the removal of nitrogenous waste. These two problems are rectified with specialized gas exchange and excretory organs.

Of the various metazoan organs, some introductory comments need be made only with regard to excretory organs. Excretory organs are found only among the bilateral phyla, and in general they are tubular structures, although they vary in location and number. They function to concentrate waste from blood or coelomic fluid, and filtration, secretion, and selective reabsorption are usually the processes by which concentration takes place. The degree of importance of these three processes varies, depending upon the importance of the organ in water and salt balance. This dependence is, of course, correlated with the animal's environment.

The most common type of excretory organ in the Animal Kingdom is a simple or branching tubular structure called a nephridium, which opens to the exterior through a nephridiopore. There are several types of nephridia. The nephridial tubules of all acoelomates, all pseudocoelomates, and some coelomates have blind inner ends and are called protonephridia. The protonephridia are branched, and at the inner end of each tubule is located a long single flagellum, or a tuft of cilia. The flagellum is carried by a terminal cell called a solenocyte (Fig. 10–39*C*). The tuft of cilia constitutes a flame bulb. One cell may form a single flame bulb or give rise to many bulbs (Fig. 7–19*A* and *B*).

The second type of nephridium is called a metanephridium. The metanephridium is an unbranched tubule, which is open at the inner end through a funnel called a nephrostome. Metanephridia are found only in coelomate animals; the nephrostome drains the coelom (Fig. 10–40).

METAZOAN EMBRYOGENY

The embryonic development of bilateral metazoans provides the principal basis for the division of these phyla into two rather well-defined groups. One division embraces flatworms, annelids, mollusks, and arthropods, as well as a number of smaller allied phyla. These groups constitute the Protostomia. The other division includes echinoderms, chordates, and several smaller phyla; the members of this division are known as the Deuterostomia.

The protostomes and deuterostomes each display a basic plan of development that is characteristic and distinct from that of the other. This is not to say that all members of each group have identical patterns of development. There are many examples of modification and deviation in every phylum, largely through changes in the distribution and the amount of yolk material. But each line does display certain characteristic features.

Types of Cleavage and Embryonic Development

Determinate and Indeterminate Cleavage. The embryonic fate of blastomeres is not determined until relatively late in the embryogeny of deuterostomes. For example, areas of undifferentiated cells, such as the primitive streak of birds and mammals and the lips of the blastopore in amphibians, persist for some time after gastrulation. This late establishment of the embryonic fate of cells allows the phenomenon of twinning to occur. If a starfish egg is allowed to cleave to the four-cell stage and the cells (blastomeres) are then separated, each cell is capable of forming a complete gastrula and then a larva. This formation of blastomeres in the embryo with unfixed fates is known as *indeterminate cleavage* and is characteristic of deuterostomes.

The flatworm–annelid–mollusk division (the protostomes) displays a very early fixation of the fate of embryonic cells. If a marine annelid egg is allowed to undergo two cleavage divisions and the resulting four blastomeres are separated, each blastomere will develop into only a fixed quarter of the gastrula and larva. Thus, each cell has a predetermined and fixed fate which cannot be altered, even if the cell is moved from its original position in the embryo. The formation of blastomeres with fixed embryonic contributions is known as *determinate cleavage* and is characteristic of development in protostomes.

Spiral Cleavage. Determinate versus indeterminate cleavage is not the only difference between development in protostomes and that in deuterostomes. The two groups also differ strikingly in the pattern of cleavage. In echinoderms and chordates cleavage is said to be radial. The axes of early cleavage spindles are either parallel or at right angles to the polar axis (the axis between animal and vegetal poles). The resulting blastomeres are thus always situated directly above or below one another (Fig. 3–3).

This arrangement is rare in protostomes. Cleavage is total, but the axes of the cleavage spindles are oblique to the polar axis, rather than at right angles or parallel. This position results in the blastomeres having a spiral arrangement, any one cell being located between the two blastomeres above or below it (Fig. 3–4). Thus, one tier or set of cells alternates in position with the next tier. This cleavage pattern is characteristic of protostomes and is known as spiral cleavage.

Since spiral cleavage is determinate and since the fate of the various blastomeres is quite similar throughout the different protostome phyla, it has been found profitable to designate each blastomere in order to trace the lineage of the cells. Such a system of designation was improvised by the American embryologist E. B. Wilson (1892) and has now become the standard system for describing cleavage of the spiral type. An explanation of this system is incorporated in the subsequent discussion on spiral cleavage.

Invertebrates with spiral cleavage usually have moderately telolecithal eggs (having the yolk concentrated toward one pole), so that cleavage is total but not equal. The first two cleavage planes are vertical and divide the egg into four equal blastomeres, designated as macromeres *A*, *B*, *C*, and *D*. The third cleavage plane is horizontal but is shifted toward the animal pole so that the upper set of four blastomeres or micromeres is considerably smaller than those below (Fig. 3–4*A* and *B*). Since the spindle axes are oriented obliquely to the polar axis, the four micromeres sit in the angles formed by the contiguity of the four macromeres. This first set of four micromeres is known as the first quartet and is labeled *1a*, *1b*, *1c*, and *1d*, depending on which of the four macromeres they were derived from. The four macromeres continue to be designated with capital letters plus the numerical prefix 1. They thus are designated *1A*, *1B*, *1C*, and *1D*.

The four macromeres now give off a second set of four micromeres (Fig. 3–4*C* and *D*). If the cleavage of the first quartet was clockwise, then the second quartet issues forth in a counterclockwise direction, again to lie in the angles formed by the first quartet above and the macromeres below. The second quartet is designated *2a*, *2b*, *2c*, and *2d*, and the macromeres become *2A*, *2B*, *2C*, and *2D*.

The macromeres form still a third quartet (*3a*, *3b*, *3c*, and *3d*) (Fig. 3–4*E* and *F*) and a fourth quartet (*4a*, *4b*, *4c*, and *4d*) (Fig. 3–5*B*). The four macromeres, which may now actually be smaller than the cells of the fourth quartet, are designated *4A*, *4B*, *4C*,

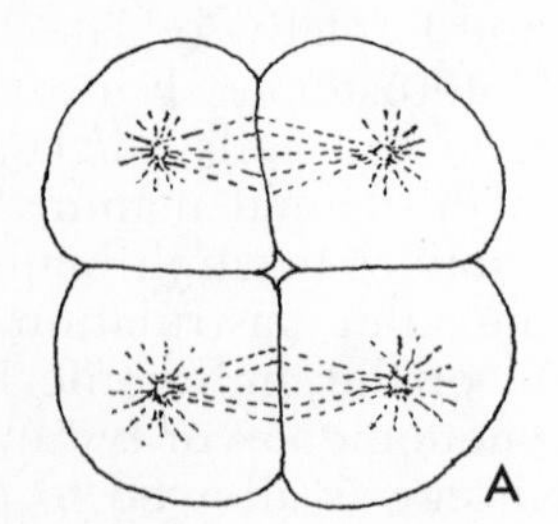

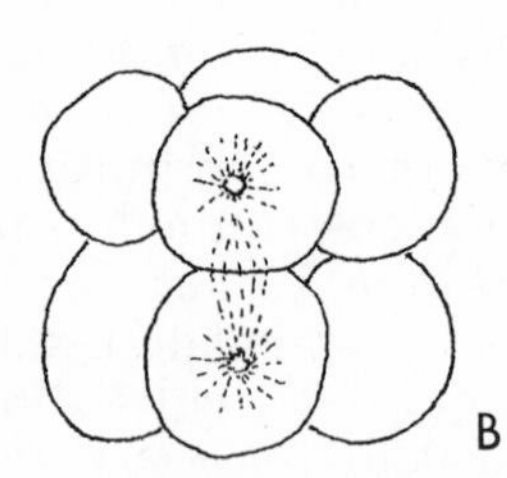

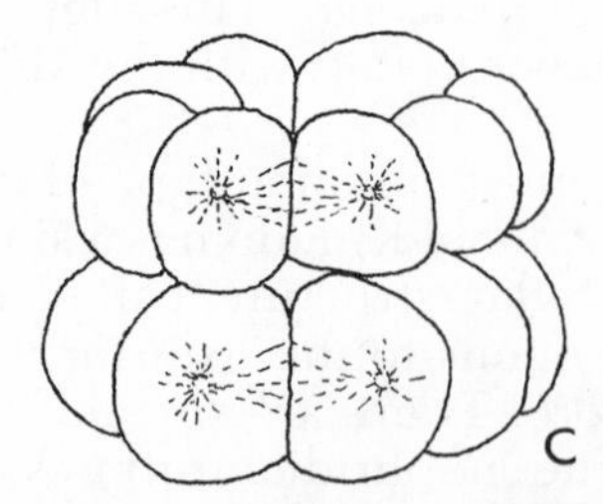

Figure 3–3. Radial cleavage in the sea cucumber, *Synapta*. *A*. Polar view. *B* and *C*. Lateral views. (After Selenka from Balinski.)

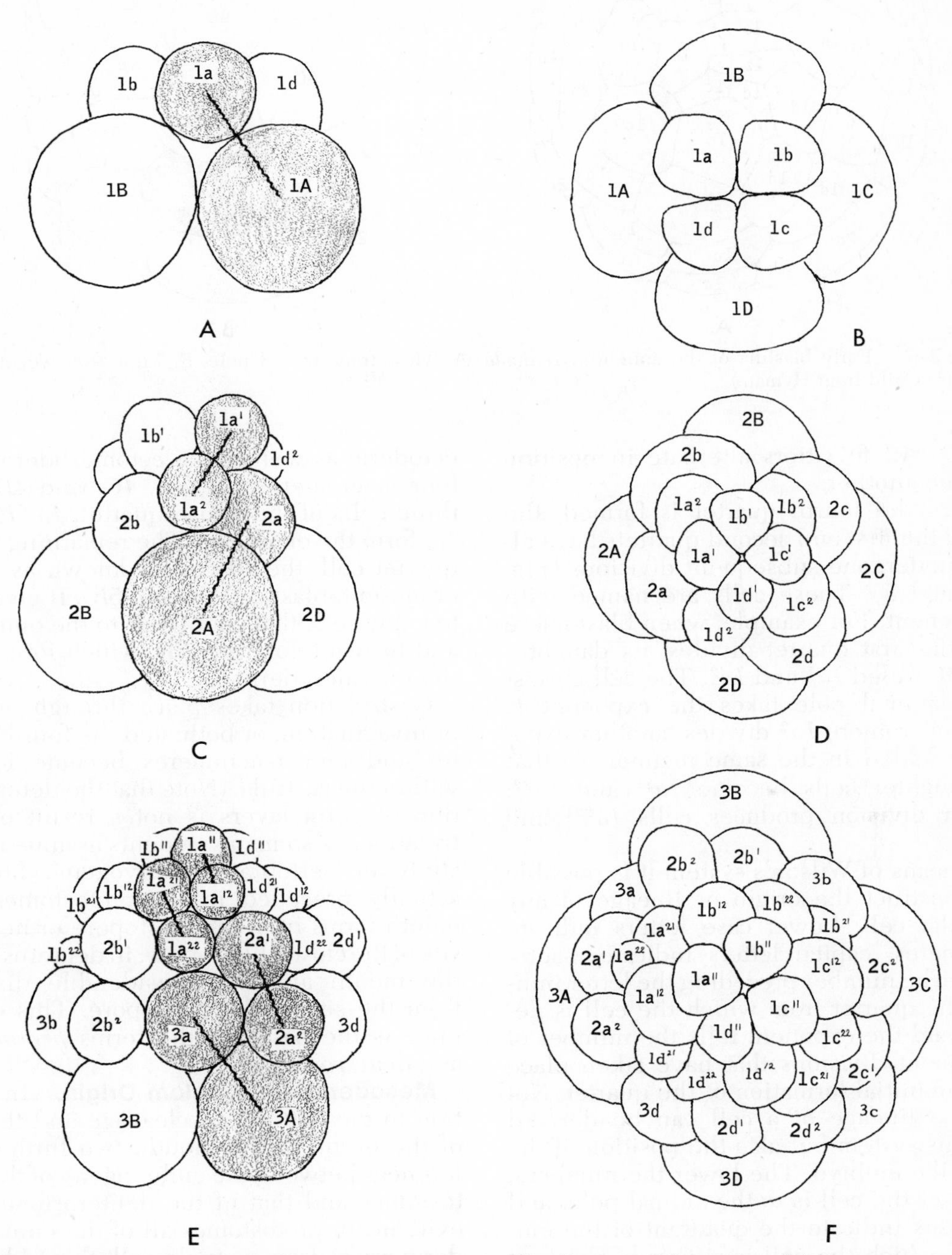

Figure 3–4. *A, C,* and *E.* Spiral cleavage (lateral views). All cells derived from original *A* blastomere are shaded. Wavy black lines indicate orientation of spindles at each cleavage. (After Villee, Walker, and Smith.) *B, D,* and *F.* Same stages viewed from animal pole. (After Hyman.)

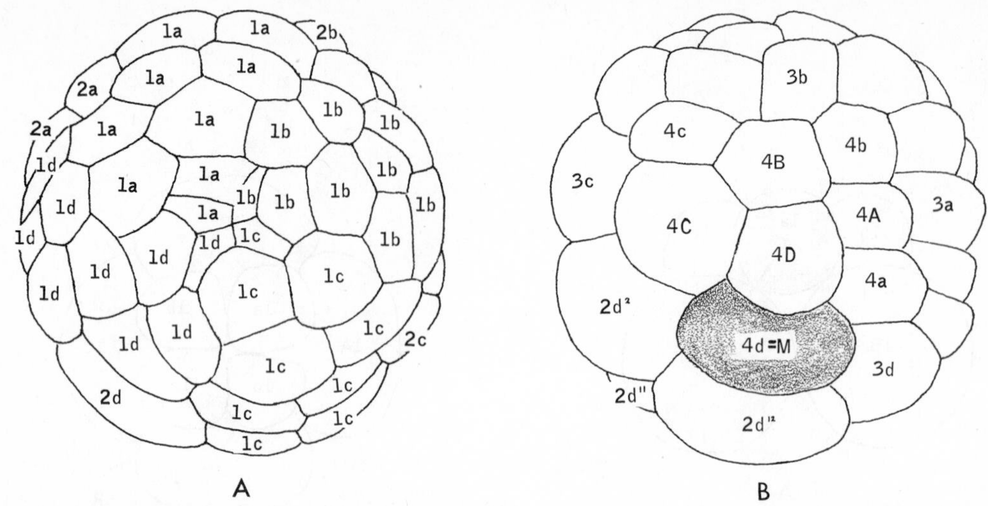

Figure 3–5. Early blastula of the annelid, *Arenicola*. *A*. View from animal pole. *B*. View from vegetal pole (Both after Child from Hyman.)

and *4D*. All five tiers alternate in position with one another.

Before the fourth quartet is formed, the cells of the first and second quartets have already undergone subsequent divisions (Fig. 3–4*E* and *F*). These cells are named with an exponent. For example, when blastomere *1a* of the first quartet divides, its daughter cells are called $1a^{1}$ and $1a^{2}$. The cell closest to the animal pole takes the exponent *1*. When micromere $1a^{2}$ divides, another exponent is added in the same manner, so that the daughter cells become $1a^{21}$ and $1a^{22}$; another division produces cells $1a^{221}$ and $1a^{222}$.

By means of Wilson's system it is possible to reconstruct the origin or lineage of any particular cell. Lower case letters indicate micromeres; capital letters indicate macromeres. The number preceding the letter indicates the quartet from which the cell is derived, and the exponent tells the number of subsequent divisions that have taken place after the initial formation of the quartet. Not only the lineage of a cell can be derived from this system but also the position of the cell in the embryo. The lower the numbers, the closer the cell is to the animal pole, and the letters indicate the quadrant of the embryo in which the cell is situated. The four macromeres are always located at the vegetal pole.

With the formation of the fourth quartet the embryo has become a blastula (Fig. 3–5), which may be either hollow or solid, depending on the amount of yolk present. Also, the germ layers have become fixed. The first three quartets of micromeres form all of the ectoderm as well as the ectomesoderm. The four macromeres, *4A*, *4B*, *4C*, and *4D*, and three cells of the fourth quartet, *4a*, *4b*, and *4c*, form the entoderm. The remaining fourth quartet cell, the *4d* cell, is known as the *M* or mesentoblast cell (Fig. 3–5*B*). It gives rise to a few cells that contribute to the entoderm and to two teloblast cells, which form all of the entomesoderm.

Gastrulation takes place through epiboly or invagination, or both; and the fourth quartet and four macromeres become located within the gastrula. (Note that the determination of germ layers is not a result of gastrulation, as so many students assume from a study of vertebrate embryogeny, but has actually preceded it.) In protostomes, the mouth forms from the blastopore or near the site of the closed blastopore; in deuterostomes the mouth arises a considerable distance from the site of the blastopore. This difference is the basis for the terms *protostome* and *deuterostome*.

Mesoderm and Coelom Origin. In addition to the patterns of cleavage and the site of the origin of the mouth, two further differences between the embryogeny of the protostomes and that of the deuterostomes are evident. In protostomes all of the entomesoderm arises from a single cell, the *4d* blastomere. The *4d*, or *M* cell (mesentoblast cell) gives rise to two teloblasts, or primordial mesoderm cells. Located originally at the posterior end of the animal in front of the site of the future anus, each cell proliferates to form a mass of mesodermal cells located one on either side of the body (Fig. 3–6). In metameric protostomes the mesodermal

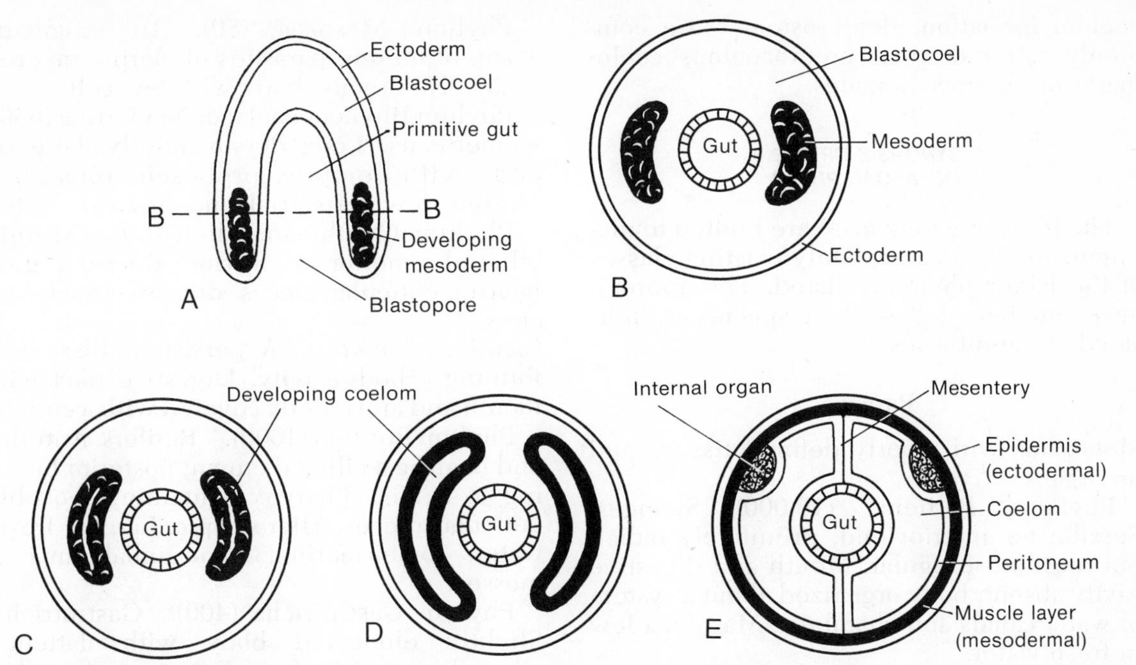

Figure 3–6. Origin of the coelom by schizocoely. *A.* Frontal section of a late gastrula. *B.* Cross section of gastrula at level indicated in *A. C.* Coelom developing as a split (schizocoel) in the mesodermal mass. *D.* Further development of the mesoderm and coelom. Note that the old blastocoel is gradually obliterated. *E.* Cross section through an adult coelomate protostome. Muscle layer develops from mesoderm. Internal organs lie behind peritoneum (retroperitoneal), which lines coelom.

masses form a linear series of segmented blocks of cells. In any event, a split forms within each mesodermal mass, and the resulting cavity enlarges to form the coelom. This mode of coelom formation is termed schizocoely and coelomate protostomes are therefore often referred to as schizocoelous coelomates or schizocoelomates.

Deuterostomes on the other hand, have an entirely different method of mesoderm and coelom formation. The mesoderm primitively arises by a process called enterocoelic pouching, in which the wall of the archenteron evaginates to form pouches. The pouches separate from the archenteron, either as a pair or as a series of lateral pairs in the case of metameric deuterostomes (Fig. 3–7). The cavity of the evagination and later pouch becomes the coelom and the wall becomes the mesoderm. Because of this mode of

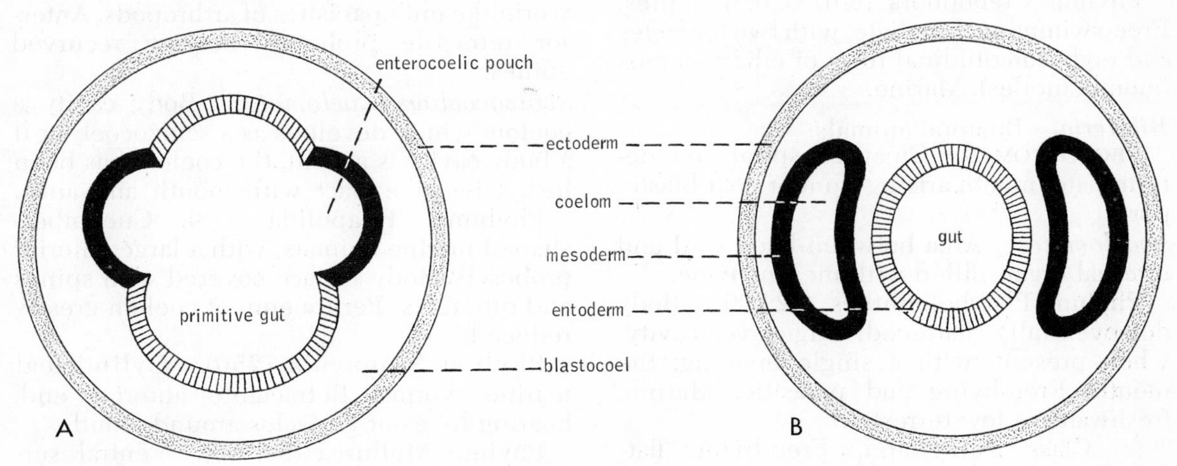

Figure 3–7. Mesoderm and coelom formation by enterocoelic pouching. *A.* Lateral evaginations of wall of primitive gut. *B.* Separation of pouches to form coelom, mesoderm, and definitive gut entoderm.

coelom formation, deuterostomes are commonly referred to as enterocoelous coelomates or enterocoelomates.

SYNOPSIS OF THE METAZOAN PHYLA

The following diagnoses are limited to distinguishing characters. Only existing classes of the larger phyla are listed. The approximate number of described species is indicated in parentheses.

Parazoa

Metazoans with poorly defined tissues and no organs.

Phylum Porifera (10,000). Sponges. Sessile; no anterior end; primitively radial, but most are irregular. Mouth and digestive cavity absent; body organized about a system of water canals and chambers. Marine, a few in fresh water.

Eumetazoa

Metazoans with organs, mouth, and digestive cavity.

Radiata. Tentaculate radiate animals with few organs. Digestive cavity, with mouth the principal opening to the exterior.

Phylum Coelenterata, or Cnidaria (8900). Free-swimming or sessile, with tentacles surrounding mouth. Specialized cells bearing stinging organoids, called nematocysts. Solitary or colonial. Marine, few in fresh water.

Class Hydrozoa. Hydra, hydroids.
Class Scyphozoa. Jellyfish.
Class Anthozoa. Sea anemones, corals.

Phylum Ctenophora (90). Comb jellies. Free-swimming; biradiate, with two tentacles and eight longitudinal rows of ciliary combs (membranelles). Marine.

Bilateria. Bilateral animals.

PROTOSTOMES. Cleavage spiral and determinate; mouth arising from or near blastopore.

Acoelomates. Area between body wall and internal organs filled with mesenchyme.

Phylum Platyhelminthes (12,700). Body dorsoventrally flattened; digestive cavity, when present, with a single opening, the mouth. Free-living and parasitic. Marine, freshwater, a few terrestrial.

Class Turbellaria. Free-living flatworms.
Class Trematoda. Flukes.
Class Cestoda. Tapeworms.

Phylum Mesozoa (50). An enigmatic group of minute parasites of marine invertebrates. No organs; body with few cells.

Phylum Rhynchocoela, or Nemertina (650). Nemerteans. Long dorsoventrally flattened body with complex proboscis apparatus. Marine, few terrestrial and in fresh water.

Phylum Gnathostomulida (80). Minute ciliated acoelomate worms. Buccal region bearing cuticular pieces; digestive tract with anus.

Pseudocoelomates. A persistent blastocoel forming a body cavity. Digestive tract with mouth and anus. Body covered with a cuticle.

Phylum Rotifera (1500). Rotifers. Anterior end bearing a ciliated crown; posterior tapering to a foot. Pharynx containing movable cuticular pieces. Microscopic. Largely freshwater, some marine, some inhabitants of mosses.

Phylum Gastrotricha (400). Gastrotrichs. Slightly elongated body with flattened ciliated ventral surface. Few to many adhesive tubes present; cuticle commonly ornamented. Microscopic. Marine and freshwater.

Phylum Kinorhyncha (75). Slightly elongated body. Cuticle segmented and bearing recurved spines. Spiny anterior end retractile. Less than 1 mm. in length. Marine.

Phylum Nematoda (10,000). Roundworms. Slender cylindrical worms with tapered anterior and posterior ends. Cuticle often ornamented. Free-living and parasitic. Free-living species usually only a few millimeters or less in length. Marine, freshwater, and in soil.

Phylum Nematomorpha (230). Hairworms. Extremely long threadlike bodies. Adults free-living in damp soil, in fresh water and a few marine. Juveniles parasitic.

Phylum Acanthocephala (500). Small wormlike endoparasites of arthropods. Anterior retractile proboscis bearing recurved spines.

Schizocoelous Coelomates. Body cavity a coelom which develops as a schizocoel; or if a body cavity is absent, the coelom has been lost. Digestive tract with mouth and anus.

Phylum Priapulida (8). Cucumber-shaped marine animals, with a large anterior proboscis. Body surface covered with spines and tubercles. Peritoneum of coelom greatly reduced.

Phylum Sipuncula (250). Cylindrical marine worms. Retractable anterior end, bearing lobes or tentacles around mouth.

Phylum Mollusca (80,000). Ventral surface modified in the form of a muscular foot, having various shapes; dorsal and lateral surfaces of body modified as a shell-secreting mantle, although shell may be reduced or absent. Marine, freshwater, and terrestrial.

Class Monoplacophora.

Class Polyplacophora. Chitons.

Class Aplacophora.

Class Gastropoda. Snails, welks, conchs, slugs.

Class Bivalvia, or Pelecypoda. Bivalve mollusks.

Class Scaphopoda. Tusk or tooth shells.

Class Cephalopoda. Squids, cuttlefish, octopods.

Phylum Echiura (60). Cylindrical marine worms, with flattened, anterior, nonretractile proboscis. Trunk with a large pair of ventral setae.

Phylum Annelida (8700). Segmented worms. Body wormlike and metameric. A large longitudinal ventral nerve cord. Marine, freshwater and terrestrial.

Class Polychaeta. Marine annelids.

Class Oligochaeta. Freshwater annelids and earthworms.

Class Hirudinea. Leeches.

Phylum Pogonophora (80). Marine deep water animals, with a long body housed within a chitinous tube. Anterior end of body bearing one to many long tentacles. Digestive tract absent.

Phylum Tardigrada (180). Water bears. Microscopic segmented animals. Short cylindrical body bearing four pairs of stubby legs terminating in claws. Freshwater and terrestrial in lichens and mosses; few marine.

Phylum Onychophora (65). Terrestrial, segmented, wormlike animals, with an anterior pair of antennae and many pairs of short conical legs terminating in claws. Body covered by a thin cuticle.

Phylum Arthropoda (923,000). Body metameric with jointed appendages and encased within a chitinous exoskeleton. Coelom vestigial. Marine, freshwater, terrestrial.

Class Merostomata. Horseshoe crabs.

Class Arachnida. Scorpions, spiders, mites.

Class Pycnogonida. Sea spiders.

Class Crustacea. Water fleas, shrimp, lobsters, crayfish, crabs.

Class Insecta. Insects.

Class Diplopoda. Millipedes.

Class Pauropoda. Pauropods.

Class Symphyla. Symphylans.

Class Chilopoda. Centipedes.

Phylum Pentastomida (70). Wormlike endoparasites of vertebrates. Anterior end of body with two pairs of leglike projections terminating in claws and a median snout-like projection bearing the mouth.

Lophophorate coelomates. With a crown of hollow tentacles (a lophophore) surrounding or partially surrounding mouth.

Phylum Phoronida (15). Marine, wormlike animals with the body housed within a chitinous tube.

Phylum Bryozoa (4000). Bryozoans. Colonial, sessile; the body housed within a gelatinous or more commonly a chitinous or chitinous and calcareous exoskeleton. Mostly marine, few freshwater.

Phylum Entoprocta (60). Body attached by a stalk. Mouth and anus surrounded by the tentacular crown. Mostly marine.

Phylum Brachiopoda (260). Lamp shells. Body attached by a stalk and enclosed within two unequal dorsoventrally oriented calcareous shells.

DEUTEROSTOMES, OR ENTEROCOELOUS COELOMATES. Cleavage radial and indeterminate; mouth arising some distance away (anteriorly) from blastopore. Mesoderm and coelom develop primitively by enterocoelic pouching of the primitive gut.

Phylum Chaetognatha (50). Marine planktonic animals with dart-shaped bodies bearing fins. Anterior end with grasping spines flanking a ventral preoral chamber.

Phylum Echinodermata (5300). Secondarily pentamerous radial symmetry. Most existing forms free moving. Body wall containing calcareous ossicles usually bearing projecting spines. A part of the coelom modified into a system of water canals with external projections used in feeding and locomotion. All marine.

Class Crinoidea. Sea lilies and feather stars.

Class Stelleroidea.

Subclass Asteroidea. Sea stars.

Subclass Ophiuroidea. Brittle stars and basket stars.

Class Echinoidea. Sea urchins, sand dollars, and heart urchins.

Class Holothuroidea. Sea cucumbers.

Phylum Hemichordata (80). Acorn worms. Body divided into proboscis, collar, and trunk. Anterior part of trunk perforated with varying number of pairs of pharyngeal clefts. Marine.

Class Enteropneusta. Acorn worms.

Class Pterobranchia. Pterobranchs.

Phylum Chordata (39,000). Pharyngeal clefts, notochord, and dorsal hollow nerve cord present at some time in life history. Marine, fresh-water, and terrestrial.

Subphylum Urochordata (1600). Sea

squirts, or tunicates. Sessile, nonmetameric invertebrate chordates enclosed with a cellulose tunic. Notochord and nerve cord present only in larva. Solitary and colonial. Marine.

Class Ascidiacea. Sea squirts, or sessile tunicates.

Class Thaliacea. Free-swimming urochordates.

Class Larvacea. Planktonic urochordates.

Subphylum Cephalochordata (25). *Amphioxus*. Metameric fishlike invertebrate chordates.

Subphylum Vertebrata (37,790). The vertebrates. Metameric. Trunk supported by a series of cartilaginous or bony skeletal pieces (vertebrae) surrounding or replacing notochord in adult.

Bibliography

Hyman, L. H., 1951: The Invertebrates: Platyhelminthes and Rhynchocoela. Vol. 2. McGraw-Hill, N.Y., pp. 10–14.

Jorgensen, C. B., 1966: Biology of Suspension Feeding. Pergamon Press, N.Y. (A detailed account of all aspects of this type of feeding mechanism.)

Chapter 4

THE SPONGES

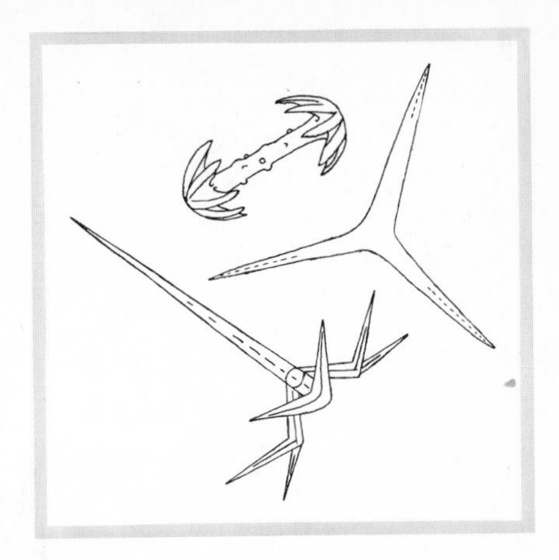

Sponges, which constitute the phylum Porifera, are the most primitive of the multicellular animals. Neither true tissues nor organs are present, and the cells display a considerable degree of independence. All members of the phylum are sessile and exhibit little detectable movement. This combination of characteristics convinced Aristotle, Pliny, and other ancient naturalists that sponges were plants. In fact it was not until 1765, when internal water currents were first observed, that the animal nature of sponges became clearly established.

Except for some 150 freshwater species, sponges are marine animals. They abound in all seas, wherever there are rocks, shells, submerged timbers, or coral to provide a suitable substratum. Some species even live on soft sand or mud bottoms. Most sponges prefer relatively shallow water; but some groups, including most glass sponges, live in deep water.

Sponge Structure. Sponges vary greatly in size. Certain calcareous sponges are approximately the size of a bean, while the large loggerhead sponges would more than fill a large wash tub. Some are radially symmetrical, but the vast majority are irregular and exhibit massive, erect, encrusting, or branching growth patterns (Figs. 4–1, 4–2, and 4–7). The type of growth pattern displayed is influenced by the nature and inclination of the substratum, by availability of space, and by the velocity and type of water

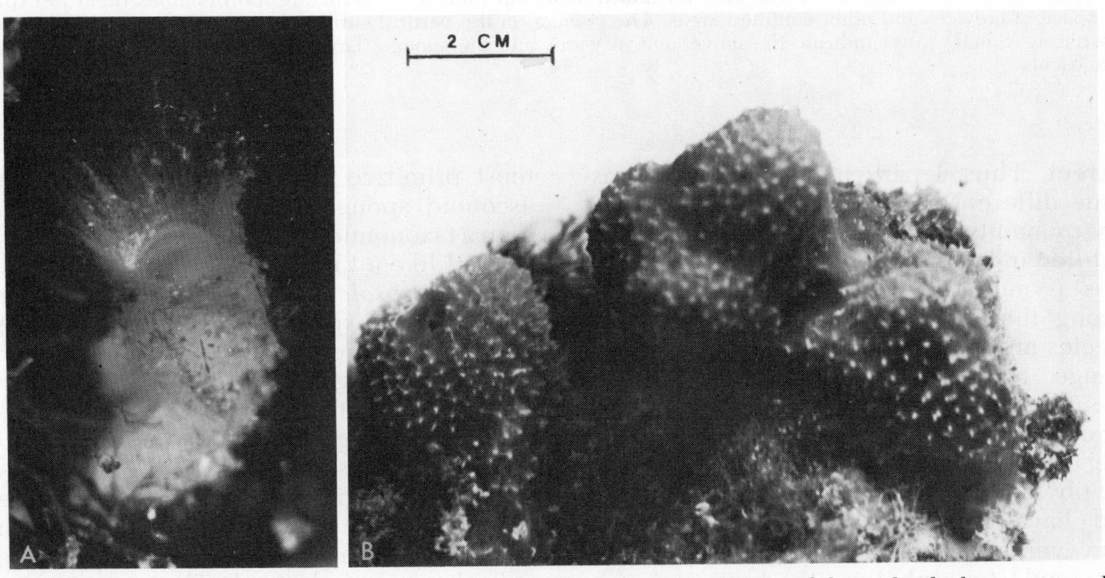

Figure 4–1. *A.* A small calcareous syconoid sponge. The vase-shaped body of this individual is no more than 5 mm. in length. Long spicules fringe the large osculum. *B. Dysidea etheria*, a West Indian leuconoid sponge, which is pale blue in color. One osculum is visible at left. (Both by Betty M. Barnes.)

Figure 4–2. Relationship of sponge form to utilization of substratum. The two massive sponges at the right on top of rock require an exposed surface, but their elevated form enables them to utilize water well above the substratum and their attachment area is a relatively small part of the total body surface area. The encrusting sponges below the rock utilize much of their surface area for attachment, but their low encrusting form enables them to exploit the space of crevices and other confined areas. The sponge on the vertical surface at left utilizes space *within* the substratum. Small arrows indicate the movement of water into the sponge; large arrows indicate the exit of water from oscula.

current. Thus a particular species may assume different appearances under different environmental situations; this variation has resulted in some taxonomic confusion. Drabness is more the exception than the rule among the Porifera; most of the common species are brightly colored. Green, yellow, orange, red, and purple sponges are frequently encountered.

Some sponges are radially symmetrical (Figs. 4–1A and 4–7), but many members of the phylum have lost the ancestral symmetry and have become irregular (Fig. 4–1*B*). However, since many radial species display the simplest morphology, the basic structure and histology of sponges can be more easily understood by beginning with these forms.

Porifera that possess the simplest and the most primitive type of structure are called asconoid sponges, a structural term rather than a taxonomic one. The asconoid sponge is shaped like a tube and is always small (Fig. 4–3). *Leucosolenia,* which is the most common asconoid sponge along the northern Atlantic coast, rarely exceeds 10 cm. in height. Asconoid sponges of this type are not usually solitary but are composed of clusters of tubes fused together along their long axes or at their bases.

The surface of an asconoid sponge is perforated by many small openings, called incurrent pores, from which the name Porifera (pore-bearer) is derived. These pores open into the interior cavity, the atrium (spongocoel). The latter in turn opens to the outside through the osculum, a large opening at the top of the tube. A constant stream of water passes through the incurrent pores into the spongocoel and out the osculum.

The body wall is relatively simple. The outer surface is covered by flattened polygonal cells, the pinacocytes. The margins of pinacocytes can be contracted or withdrawn so that the entire animal may increase or decrease slightly in size. The pores are guarded by a type of cell, called a porocyte, which is shaped like a short tube and extends from the external surface to the spongocoel. The bore, or lumen, of the tube forms the incurrent pore, or ostium; and the outer end of the cell can be closed or opened by contraction. A porocyte is derived from a pinacocyte through the formation of an intracellular perforation.

Beneath the pinacoderm lies a layer called the mesohyl, which consists of a gelatinous protein matrix, containing skeletal material and amoeboid cells.

The skeleton is relatively complex and provides a supporting framework for the living cells of the animal. (To avoid repetition, the discussion presented here on the sponge skeleton applies to the phylum in general, not just to the asconoid sponges.) The skeleton may be composed of calcareous spicules, siliceous spicules, protein spongin fibers (Fig. 4–4), or a combination of the last two. The spicules exist in a variety of forms and are important in the identification and classification of species. An extensive nomenclature has developed through the use of these structures in sponge taxonomy.

Monaxon spicules are shaped like needles or rods and may be curved or straight (Fig. 4–5*A* to *D*). The ends are pointed, knobbed, or hooked. Tetraxons are four-pronged spicules (Fig. 4–5*E* and *F*), but some types

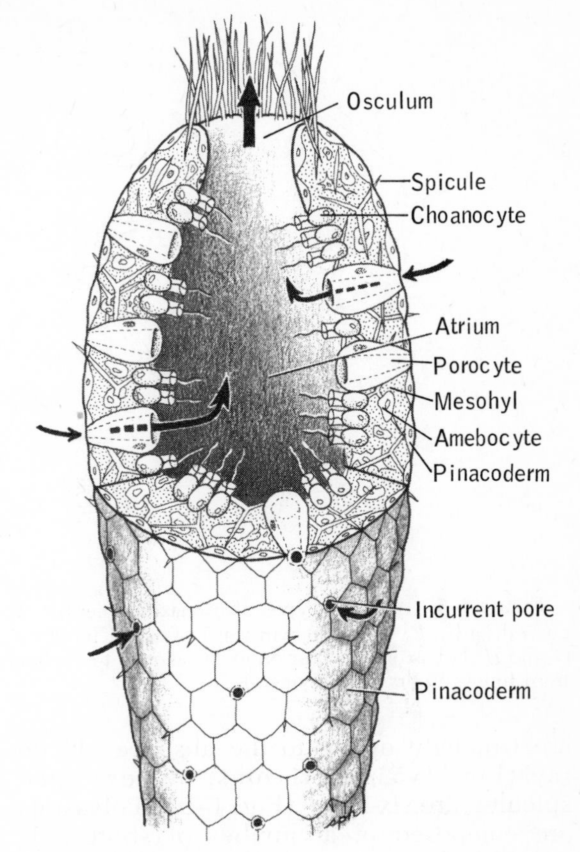

Figure 4–3. A partially sectioned asconoid sponge. (Modified from Buchsbaum.)

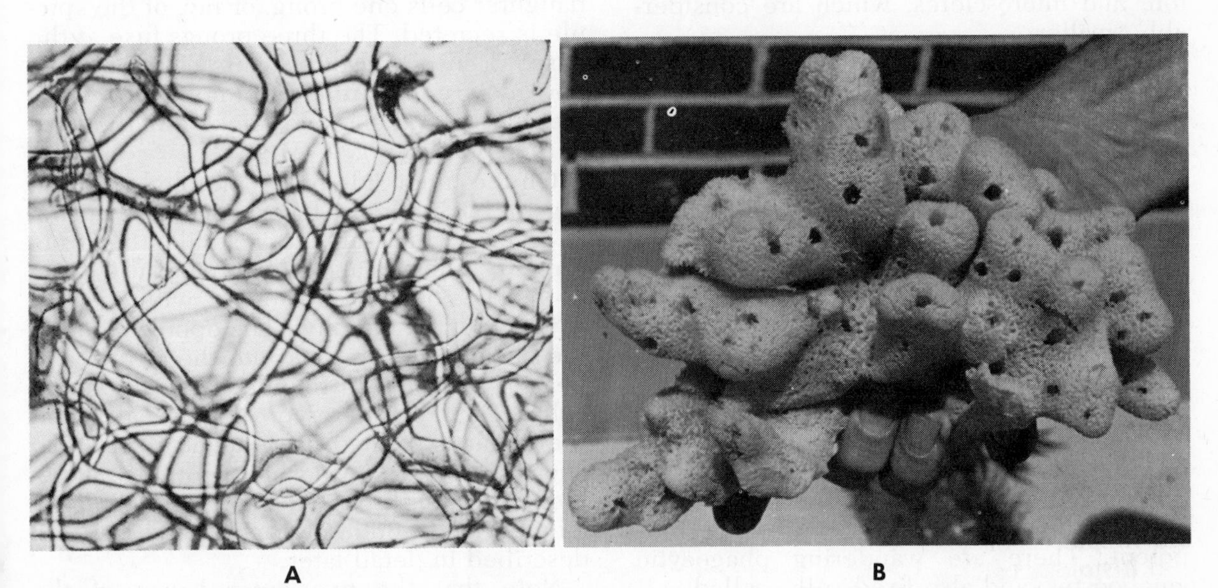

Figure 4–4. *A*. Photomicrograph of spongin fibers (they appear translucent). (Courtesy of the General Biological Supply House, Inc.) *B*. The spongin skeleton of a commercial sponge (*Spongia*) from the Mediterranean. The large openings are oscula. (By Betty M. Barnes.)

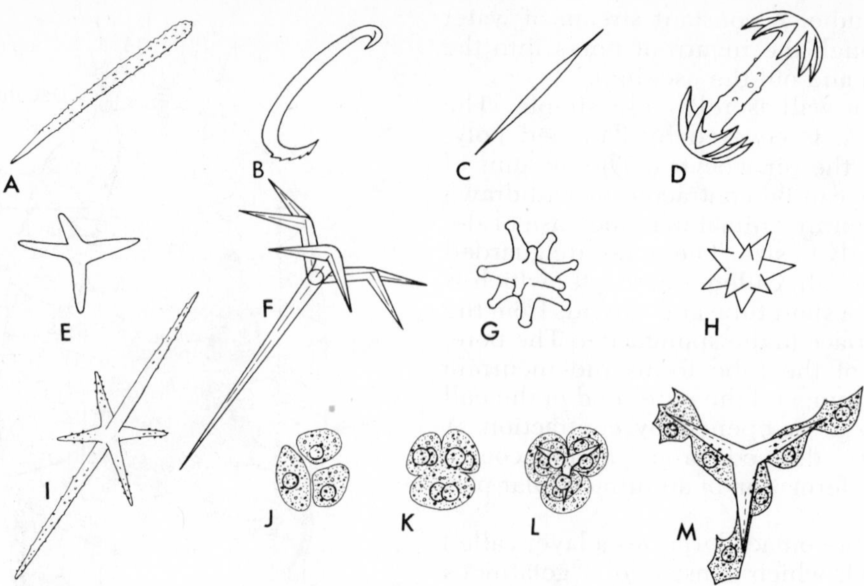

Figure 4–5. *A* to *D*. Types of monaxon spicules. *A* and *C*. Straight. *B*. With recurved ends. *D*. Ends with recurved barbs. *E*. Tetraxon with rays of equal length. *F*. Tetraxon with one long ray and three short, forked rays. *G* and *H*. Polyaxons. *I*. Triaxon or hexaxon. *J* to *M*. Secretion of a calcareous triradiate spicule. (*J–M* after Minchin from Jones; others from Hyman.)

are triradiate owing to the absence of one ray (Fig. 4–5*M*). Triaxons, or hexactinal spicules, are six-rayed (Fig. 4–5*I*). Polyaxons are composed of a number of short rods radiating from a common center (Fig. 4–5*G* and *H*). They may be grouped so they appear like a bur, like a star, or something like a child's jack. Two additional terms are commonly encountered and can apply to any of the spicule types just described: megascleres, which are larger spicules forming the chief supporting elements in the skeleton, and microscleres, which are considerably smaller.

The skeleton is located primarily in the mesohyl, but spicules frequently project through the pinacoderm. The arrangement of spicules within the mesohyl, although sometimes haphazard, often presents a definite and symmetrical pattern. Some of the spicules may even be fused to form a distinct lattice. The substance composing the spongin fibers is a fibrous protein related to keratin and collagen.

Ameboid cells in the mesohyl include a number of types. Large cells with blunt pseudopodia and large nuclei are called archaeocytes. Archaeocytes are capable of forming other types of cells that are needed. Such cells, which can transform into other types within an animal, are said to be totipotent. There are wandering phagocytic amebocytes, and also fixed cells, called collencytes, that are anchored by long cytoplasmic strands.

The skeleton, whether composed of spicules or spongin fibers, is secreted by amebocytes called sclerocytes. Several cells are usually involved in the secretion of a single spicule in the calcareous sponges, and the process is relatively complex. A three-pronged spicule, for example, has its beginnings in the sclerocytes derived from an amebocyte, called a scleroblast. The three sclerocytes partially fuse to form a trio of cells (Fig. 4–5*J* to *M*). Each member of the trio then divides, and between each pair of daughter cells one prong, or ray, of the spicule is secreted. The three prongs fuse at the base. Each of the three pairs of sclerocytes now moves outward along a ray, one cell secreting the end and one thickening the base of the spicule (Fig. 4–5*M*).

On the inner side of the mesohyl, and lining the atrium, is a layer of cells, the choanocytes, which are very similar in structure to choanoflagellate protozoans (Fig. 4–9). The choanocyte is ovoid, with one end adjacent to the mesohyl. The opposite end of the choanocyte projects into the atrium and bears a flagellum surrounded by a contractile collar. Electron microscopy has revealed that the collar is formed of adjacent fibrils. The choanocytes are responsible for the movement of water through the sponge and for obtaining food. Both of these processes are described in detail later.

Note that the two outer layers of the sponge body are properly called pinacoderm and mesohyl. Although epidermis and mesenchyme are terms that are often used, the layers of the sponge body are not like the epidermis and mesenchyme of other metazoans.

The primitive asconoid structure just described imposes very definite size limitations. The rate of water flow may be slow, because the large atrium contains too much water to be pushed out of the osculum rapidly. As the sponge increases in size, the problem of water movement is intensified. An increase in the volume of the atrium is not accompanied by a sufficient increase in surface area of the choanocyte layer to rectify the problem. Thus, asconoid sponges are always small.

The problems of water flow and surface area have been overcome during the evolution of sponges by the folding of the body wall and the reduction of the atrium. The folding increases the surface area of the choanocyte layer, and the reduction of the atrium lessens the volume of water that must be circulated. The net result of these changes is a greatly increased and more efficient water flow through the body. A much greater size now becomes possible, although the primitive radial symmetry is commonly lost.

Living sponges display various stages in the changes just described. These changes are also recapitulated in the embryogeny of many species. Sponges that exhibit the first stages of body wall folding are called syconoid sponges. Syconoid sponges include the

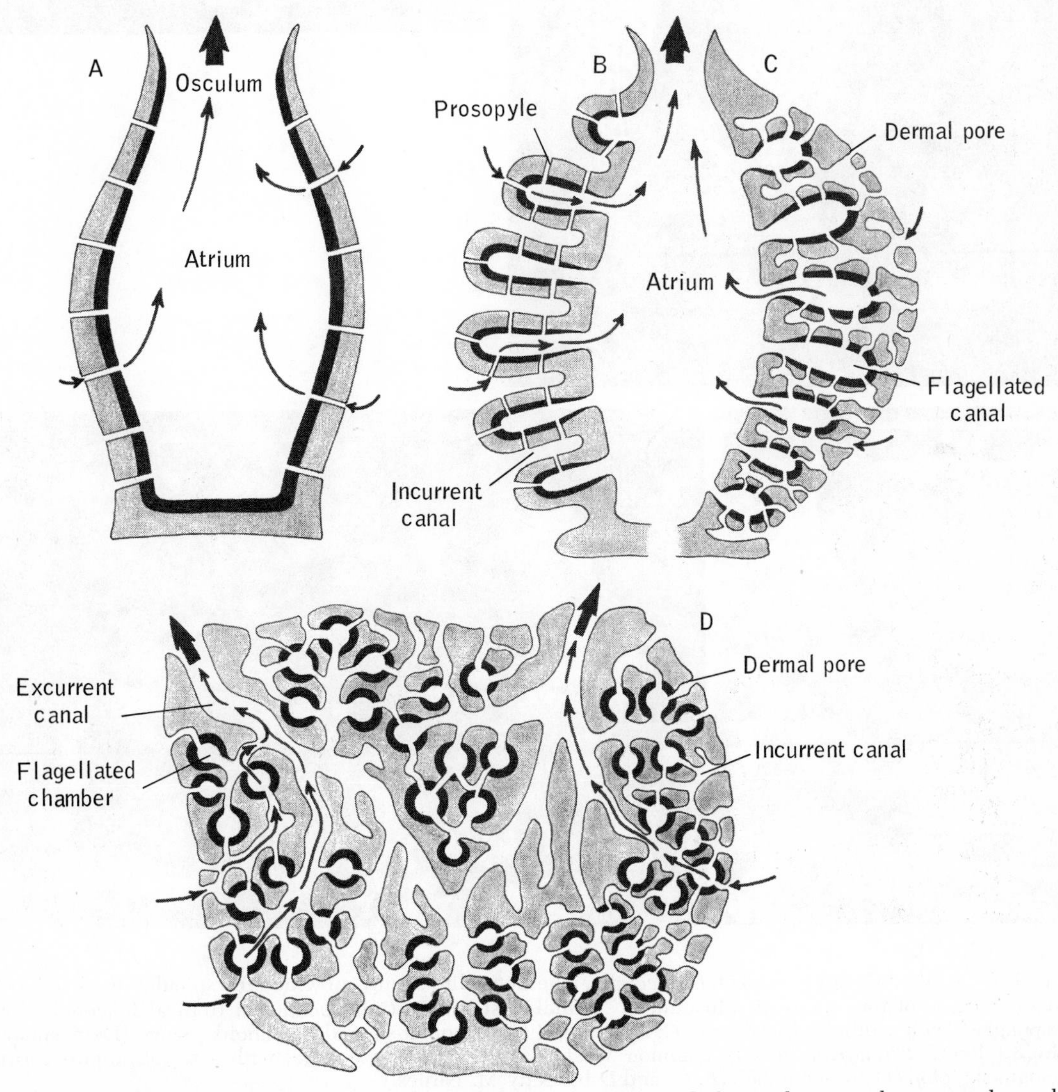

Figure 4–6. Morphological types of sponges. (Pinacoderm and mesohyl in pale gray; choanocyte layer, black.) *A*. Asconoid type. *B*. Syconoid type. *C*. More specialized syconoid type, in which entrance to incurrent canals has been partially filled with pinacoderm and mesohyl. *D*. Leuconoid type. (All modified from Hyman.)

Figure 4–7. *A.* Venus's-flower-basket, *Euplectella*, a hexactinellid sponge in which the spicules are fused to form a lattice. (Courtesy of the American Museum of Natural History.) *B. Callyspongia*, a tropical leuconoid sponge (Demospongiae) with a tubular body form. *C. Poterion*, a large, goblet-shaped leuconoid sponge (Demospongiae). *D.* "Chicken-liver," *Chondrilla*, a very common West Indian encrusting sponge, with a tough, almost cartilage-like spongin skeleton (Demospongiae). (*B, C,* and *D* by Betty M. Barnes.)

well known genus *Sycon* (=*Scypha*, the *Grantia* of supply houses). In syconoid structure, the body wall has become "folded" horizontally, forming finger-like processes (Fig. 4–6*B*). This development produces external pockets, extending inward from the outside, and evaginations, extending outward from the atrium. The two pockets produced by a fold do not meet, but by-pass each other and are blind. To visualize the syconoid condition, one might imagine pushing the fingers of one hand outward from the atrium side and pushing the fingers of the other hand inward from the outer side so that they would pass each other rather than meet.

In this more advanced type of sponge, the choanocytes no longer line the spongocoel but are now confined to the evaginations, which are called flagellated or radial canals. The corresponding invaginations from the epidermal side are known as incurrent canals and are lined by pinacocytes. The two canals are connected by openings called prosopyles, which are equivalent to the pores of asconoid sponges. Water now flows through the incurrent canals, the prosopyles, the flagellated canals, and the atrium, and flows out the osculum.

A slightly more specialized stage of the syconoid structure develops with the plugging of the open ends of the incurrent canals with pinacocytes and mesohyle (Fig. 4–6*C*). Openings remain to permit entrance of water into incurrent canals. Despite the folding of the body wall, syconoid sponges still retain a radial symmetry.

The highest degree of folding takes place in the leuconoid sponge (Fig. 4–6*D*). The flagellated canals have undergone folding, or evagination, to form small, rounded, flagellated chambers; and the atrium has usually disappeared except for water channels leading to the osculum. Water enters the sponge through the dermal pores and passes into subdermal spaces. The spaces lead into branching incurrent channels, which eventually open into the flagellated chambers through prosopyles. Water leaves the chamber through an apopyle and courses through excurrent channels, which become progressively larger as they are joined by other excurrent channels. A single large channel eventually opens to the outside through the osculum.

Although porocytes have been found in the calcareous leuconoid sponges, they are believed to be absent in most of the complex species (Demospongiae and Hexactinellida).

Most sponges are built on the leuconoid plan, which is evidence of the efficiency of this type of structure. Leuconoid sponges are composed of a mass of flagellated chambers and water canals, and may attain a considerable size. They are commonly irregular, although there are vase-shaped and tubular forms in which the excurrent canals empty into a large central chamber (Fig. 4–7). Rather than a single osculum, there may be many. Whether a large leuconoid sponge with many oscula represents a colony or a single individual has evoked considerable debate. A case could be made for either claim.

PHYSIOLOGY

The physiology of a sponge is largely dependent on the current of water flowing through the body. The water brings in oxygen and food and removes waste. Even sperm and eggs are moved in and out by the water currents. The volume of water pumped by a sponge is remarkable. A specimen of *Leuconia (Leucandra)*, a leuconoid sponge 10 cm. in height and 1 cm. in diameter, has roughly 2,250,000 flagellated chambers and pumps 22.5 liters of water per day through its body. The flow through the osculum is 8.5 cm. per second. Regulating the size of the osculum controls the rate of flow, even stopping it altogether. In the Demospongiae, control of the osculum size is facilitated by a special type of epidermal cell called a myocyte, which displays some similarities to the smooth muscle cell in shape and contractility. The myocytes are located primarily around the osculum, where they are arranged like a band of circular muscle.

Reiswig's (1971) studies on tropical Demospongiae demonstrated that some exhibit a diurnal rhythm in the propulsion of water through their bodies; others exhibit an erratic endogenous water flow. External conditions, such as turbulent water caused by storms, may halt water flow regardless of internal conditions.

The current is produced by the beating of the choanocyte flagella, but there is neither coordination nor synchrony of the choanocytes in a particular chamber. The choanocytes are turned toward the apopyle, and each flagellum beats in a spiral manner from its base to its tip (Fig. 4–8). As a result, water is sucked into the flagellated chamber through the small prosopyles located between the bases of the choanocytes. It is then

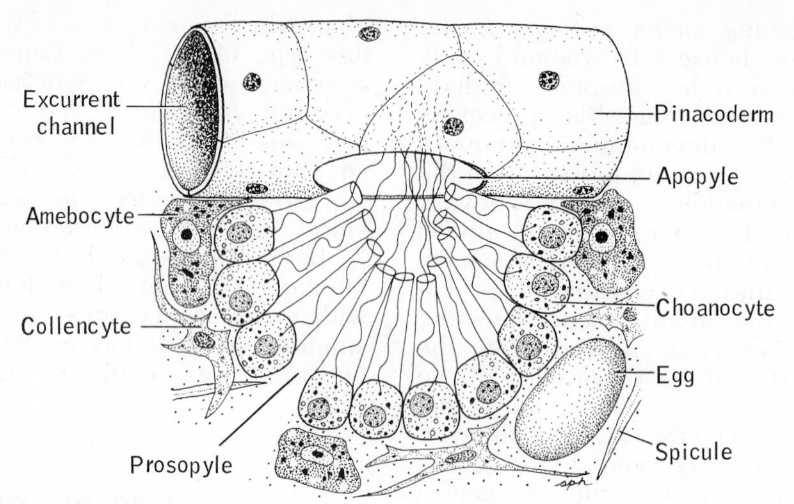

Figure 4–8. Section through flagellated chamber of freshwater sponge, *Ephydatia*. (Modified after Kilian.)

driven to the center of the chamber and out the larger apopyle into an excurrent canal.

Sponges feed on extremely fine particulate material. Studies on three species of Jamaican sponges (Reiswig, 1971) have demonstrated that 80 per cent of the filterable organic matter consumed by sponges is of a size level below that which can be resolved with ordinary microscopy. The other 20 per cent consists of bacteria, dinoflagellates, and other fine plankton.

Food particles are apparently selected largely on the basis of size and are screened in the course of their passage into the flagellated chambers. Only particles smaller than a certain size can enter the dermal pores or pass through the prosopyles. Screening is also provided by protoplasmic strands stretched across the excurrent canals. Food particles are finally filtered by the choanocytes. The trapping of particles by these cells probably results from water passing through the fibrillar mesh composing the collar (Fig. 4–9).

Reiswig believes that the larger particles (5 to 50 μ) are phagocytized by cells lining the inhalant pathways. Particles of bacterial size and below (less than 1 μ) are probably removed and engulfed by the choanocytes. The larger particles, at least, are digested intracellularly as in protozoans. The amebocytes probably act as storage centers for food reserves.

In tropical waters, at least, there is about seven times more available carbon in the unresolved fraction than at the planktonic level. The sponges' ability to utilize this food source undoubtedly accounts for their long success as sessile animals, especially in tropical waters.

Adult sponges are incapable of any locomotion, although some species can contract or alter the body shape to some degree.

Egested waste and nitrogenous waste (largely ammonia) leave the body in the water currents. Gas exchange occurs by simple diffusion between the flowing water and the cells in the sponge along the course of water flow.

No nervous system has been found to exist in sponges, and reactions are local and independent. Several French investigators (Pavans de Ceccatty, 1955) claim to have found bipolar nerve cells in the mesohyl of sponges, but at present there is no physiological evidence that these cells actually have such a function. The problem has been reviewed by Jones (1962), and Bullock and Horridge (1965).

THE CLASSES OF SPONGES

Approximately 10,000 species of sponges have been described at present, and are placed within four classes based on the nature of the skeleton.

Class Calcarea, or Calcispongiae. Members of this class, known as calcareous sponges, are distinctive in having spicules composed of calcium carbonate. In the other two classes the spicules are always siliceous. The spicules of calcareous sponges are monaxons or three- or four-pronged types; they are usually separate. All three grades of structure—asconoid, syconoid, and leuconoid—are encountered. Many Calcarea are drab, although brilliant yellow, red, and

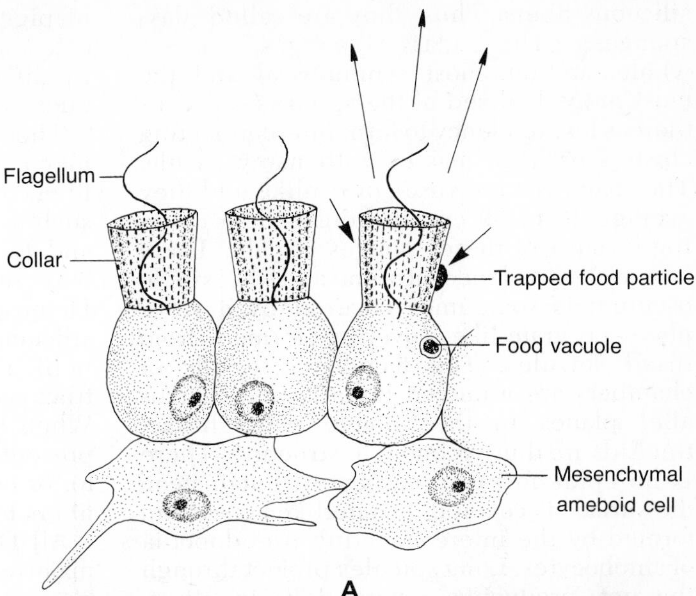

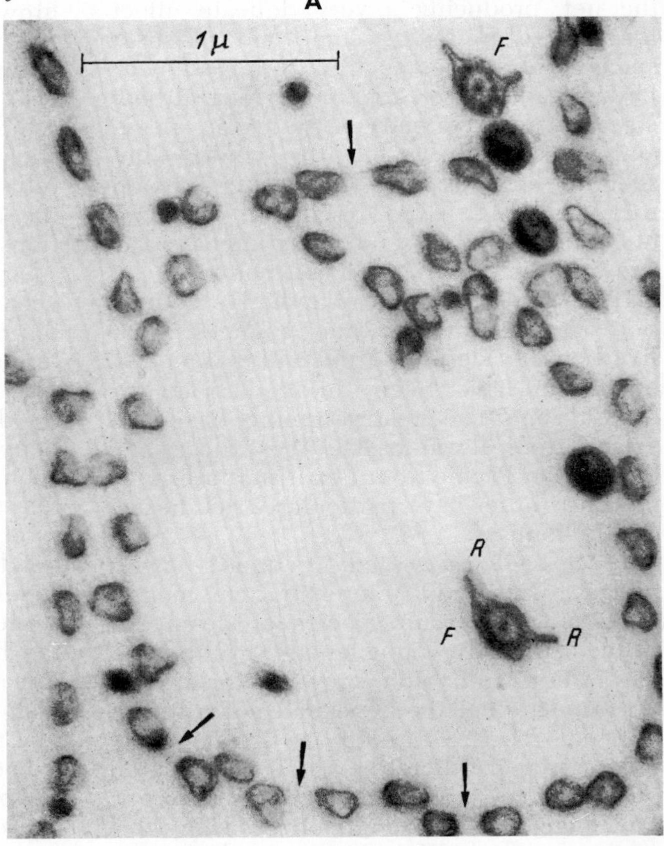

Figure 4–9. *A*. Section of flagellated layer and underlying mesenchyme, showing three choanocytes, or collar cells. Arrows indicate direction of water current. *B*. Photomicrograph of a section taken through collar of a sponge choanocyte. Larger lenticular body within circle represents section through flagellum. (Photograph by E. J. Fjerdingstad, 1961: *In* Zeitschrift für Zellforschung und mikroskopische Anatomie, 53:639–644.)

lavender species are known. They are not as large as species of other classes; most are less than 10 cm. in height. Species of calcareous sponges exist throughout the oceans of the world, but most are restricted to relatively shallow coastal waters. Genera such as *Leucosolenia* and *Sycon* are common examples used in biology courses.

Class Hexactinellida, or Hyalospongiae. Representatives of this class are commonly known as glass sponges. The name Hexactinellida is derived from the fact that the spicules are always of the triaxon, or six-pointed, type. Furthermore, some of the spicules often are fused to form a skeleton that may be lattice-like and built of long, siliceous fibers. Thus, they are called glass sponges. The glass sponges, as a whole, are the most symmetrical and the most individualized of the sponges—that is, there is less tendency to form interconnecting clusters or large masses with many oscula. The shape is cup-, vase-, or urnlike and they average 10 to 30 cm. in height. The coloring in most of these sponges is pale. There is a well-developed atrium, and the single osculum is sometimes covered by a sieve plate—a grate-like covering formed from fused spicules. The flagellated canals or chambers are arranged radially and in parallel planes in the sponge wall; hexactinellids are thus syconoid in structure. There is no epidermis of pinacocytes, but instead the surface is covered by a netlike syncytium formed by the interconnecting pseudopodia of amebocytes. Long spicules project through the net, producing a very delicate effect. Lattice-like skeletons composed of fused spicules in species such as Venus's-flower-basket (*Euplectella*) retain the general body structure and symmetry of the living sponge and are very beautiful; the white, filmy skeleton looks as if it were fashioned from rock wool (Fig. 4–7A). Basal tufts of spicule fibers implanted in sand or sediments adapt many species for living in soft bottoms.

In contrast to the Calcarea, the Hexactinellida are chiefly deepwater sponges. Most live between depths of 450 and 900 meters, but some have been dredged from the abyssal zone. The greatest number of species exist in tropical waters. The West Indies and the eastern Pacific from Japan south through the East Indies have particularly rich hexactinellid faunas.

Species of *Euplectella,* Venus's-flower-basket, display an interesting commensal relation with certain species of shrimp. A young male and a young female shrimp enter the atrium and, after growth, are unable to escape through the sieve plate covering the osculum. Their entire life is spent in the sponge prison, where they feed on plankton brought in by the sponge's water currents. The sponge with its imprisoned shrimp formerly was used in Japan as a wedding present, symbolizing the idea "till death us do part."

Class Demospongiae. This class contains the greatest number of sponge species and includes most of the common and familiar North American sponges. The majority are marine and range in distribution from shallow water to great depths.

Coloration is frequently brilliant because of pigment granules located in the amebocytes. Different species are characterized by different colors, and a complete array of hues is encountered.

The skeleton of this class is variable. It may consist of siliceous spicules or spongin fibers or a combination of both. A few genera, such as *Oscarella,* lack any sort of skeleton, and Lévi (1957) believes that these genera may represent the most primitive of the Demospongiae. Those Demospongiae with siliceous skeletons differ from the Hexactinellida in that their larger spicules are never triaxons but are monaxons or tetraxons. When both spongin fibers and spicules are present, the spicules are usually connected to, or completely imbedded in, the spongin fibers to form a skeletal network.

All Demospongiae are leuconoid, and the majority are irregular; but all types of growth patterns are displayed. Some are encrusting; others have an upright branching habit or form irregular mounds. There are also species, such as *Poterion* (Fig. 4–7*C*), which are goblet- or urn-shaped, and others, such as *Callyspongia* (Fig. 4–7*B*), which are tubular. The great variation in the shapes of the Demospongiae reflects, in part, adaptations to substratum and to the limitations of space. Large upright forms can exploit vertical space and use only a small part of their surface area for attachment. Encrusting forms, although they require more surface area for attachment, can utilize very confined habitats, such as crevices and spaces beneath stones (Fig. 4–2). The largest sponges are members of this class; some of the tropical loggerhead sponges (*Spheciospongia*) form masses of over a meter in height and diameter.

Several families of Demospongiae deserve mention. The boring sponges, composing the family Clionidae, are able to bore into calcareous structures such as coral and mollusk shells (Fig. 4–2). Channels are formed that the body of the sponge fills. At the surface the sponge body projects from the channel openings as small papillae. These papillae represent either clusters of ostia opening into an incurrent canal or an osculum. Excavation, which is begun by the larva, is brought about by the removal of chips of calcium carbonate by special amebocytes. The amebocyte begins the process by etching out the margins of the chip by means of some secreted substance (Cobb, 1969). The chip is then undercut in the same manner, the amebocyte enveloping the chip in the process. Eventually, the chip is freed and in some

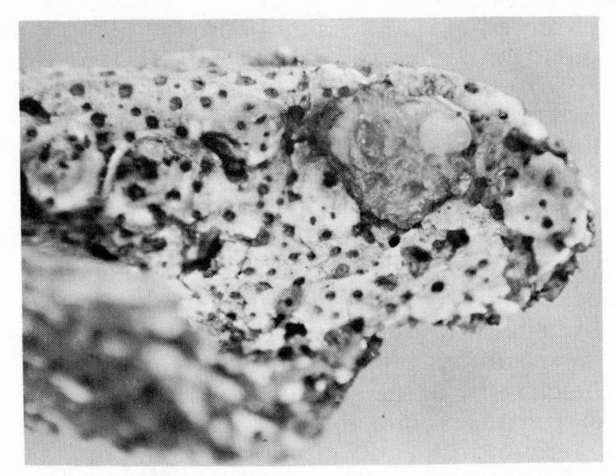

Figure 4–10. Remains of clam shell that has been riddled with boring sponge.

way is transported through the mesohyl to an excurrent water canal. *Cliona celata*, a common boring sponge that lives in shallow water along the Atlantic coast, inhabits old mollusk shells. The bright sulfur yellow of the sponge is visible where the bored channels reach the surface of the shell. *Cliona lampa* of the Caribbean is red, and it commonly overgrows the surface of the coral or coralline rock which it has penetrated as a thin encrusting sheet. Boring sponges are important agents in the decomposition of shell and coral (Fig. 4–10).

Members of two families of sponges occur in fresh water, but the family Spongillidae contains the majority of freshwater species. The Spongillidae are world-wide in distribution and live in lakes, streams, and ponds where the water is not turbid. They have an encrusting growth pattern, and some are green in color owing to the presence of symbiotic zoochlorellae in the amebocytes. The algae are brought in by water currents and are transferred from the choanocytes to the amebocytes.

The family Spongiidae contains the common bath sponges. The skeleton is composed only of spongin fibers. *Spongia* and *Hippospongia*, the two genera of commercial value, are gathered from important sponge fishing grounds in the Gulf of Mexico, the Caribbean, and the Mediterranean. The sponges are gathered by divers, and the living tissue is allowed to decompose in water. The remaining undecomposed skeleton of anastomosing spongin fibers is then washed and bleached (Fig. 4–4*B*). The colored block "sponges" seen on store counters are a synthetic product.

Class Sclerospongiae. Hartman and Goreau (1970) have proposed a fourth class of sponges, the Sclerospongiae, in which they place six species of coralline sponges found on Jamaican fore-reef slopes. These leuconoid sponges differ from other sponges in having an internal skeleton of siliceous spicules and spongin fibers and an outer encasement of calcium carbonate. The numerous oscula are raised on the calcareous skeletal mass and have a star-like configuration from the converging excurrent canals.

REPRODUCTION

Sponges reproduce asexually by budding or by a variety of processes that involve formation and release of an aggregate of essential cells, particularly amebocytes. Spongillid sponges, as well as some marine forms, have aggregates called gemmules (Fig. 4–11). In freshwater sponges a mass of food-filled archaeocytes becomes surrounded by other amebocytes that deposit a hard covering. Spicules are also added, so that a thick resistant shell is formed. Gemmule formation takes place primarily in the fall; a large number of these bodies are formed by each sponge. With the onset of winter, the parent sponge disintegrates. The gemmules are able to withstand freezing and drying and thus are able to carry the species through the winter. In spring the interior cells emerge from the shell through an opening called a micropyle. These cells develop into an adult sponge.

Some marine sponges form gemmules with a spongin coat reinforced by spicules. Others form gemmules without a spongin coat.

Considering the relative independence of sponge cells, it is not surprising that many of these animals should have remarkable powers of regeneration. Some sponges con-

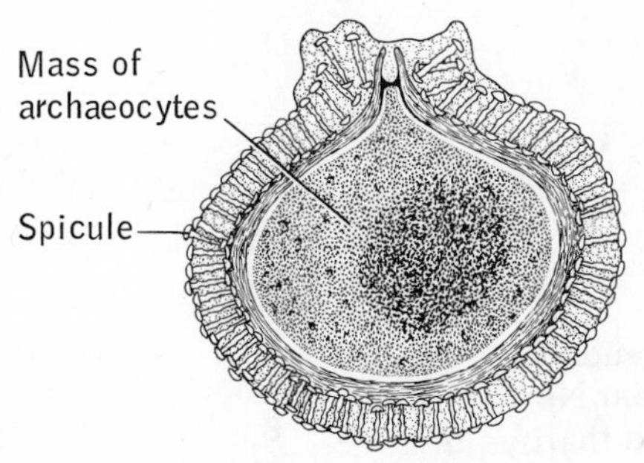

Figure 4–11. Section through gemmule of freshwater sponge. (After Evans from Hyman.)

strict near the ends of branches, causing the branch ends to fall off and regenerate into new individuals. Regeneration is employed in the propagation of commercial sponges in overfished areas off the Florida coast. Pieces of sponge called "cuttings" are attached to cement blocks and dumped into the water. Regeneration and several years' growth produce a sponge of marketable size.

The classic experiment demonstrating the regenerative ability of sponges begins with the forcing of living sponge tissue through a silk mesh. The separated cells reorganize after a short period of time and form themselves into several new sponges. Humphreys' (1970) studies have shown that calcium and magnesium ions plus some cell surface macromolecule are necessary for reaggregation. Whether successful reaggregation will occur only with dissociated cells from the same species is still being debated. However, if a sponge is sliced and a piece from another sponge of the same species is inserted into the wound, host and graft will grow together within a short period of time. The host will reject a graft belonging to a different species.

There appear to be some limits to regenerative ability. Fragments of the calcareous leuconoid sponge *Leucosolenia* must be larger than 0.4 mm. and must contain some choanocytes in order to regenerate.

The sexual reproduction and the embryogeny of sponges display a number of peculiar features. Both hermaphroditic and dioecious sponge species exist, although most are hermaphroditic, producing eggs and sperm at different times. The sperm and eggs arise from amebocytes, and sperm have also been reported to be transformed choanocytes. In any case, the germinal cells develop by engulfing surrounding nurse cells.

Sperm leave the sponge by means of water currents and enter other sponges in the same manner. Certain tropical sponges have been observed to release their sperm suddenly in great milky clouds (Fig. 4–13). Reiswig (1970) reported that a sperm cloud extended two to three meters above the bottom and induced other sponges to release their sperm. Sudden sperm release may be characteristic of most sponges.

After a sperm has reached a flagellated chamber, it enters a choanocyte or an amebocyte. These cells act as carriers and transport the sperm to the egg. After the carrier with its sperm has reached an egg (which would be close by in the surrounding mesohyl), the carrier fuses with the egg and transfers the sperm to it. Fertilization thus occurs in situ.

Embryogeny. The embryogeny of the calcareous sponges is the most familiar. The subsequent description is based largely on this group. The fertilized egg begins its development within the mesohyl of the parent. Cleavage is total and equal; and a flattened, 16-cell embryo, consisting of two tiers of eight cells each, is formed. The tier of eight cells resting against the choanocyte layer of the parent is destined to form the pinacoderm, and the other eight-celled tier forms choanocytes. The future choanocyte tier undergoes rapid mitosis, while the pre-pinacoderm tier remains undivided. Between the two layers a blastocoel develops. The net result of development up to this point is a blastula. One side is composed of many small, elongated micromeres; the other side is composed of the original eight pre-pinacoderm cells called macromeres. Flagella develop on the blastocoel ends of the micromeres, and a mouthlike opening extending from the blastocoel to the outside forms between the macromeres. Through this opening the blastula engulfs surrounding amebocytes for nutrition.

A curious process called inversion then occurs. It is strikingly like the process of inversion in the development of volvocid flagellates. The entire embryo turns inside out through the mouth. One hemisphere of

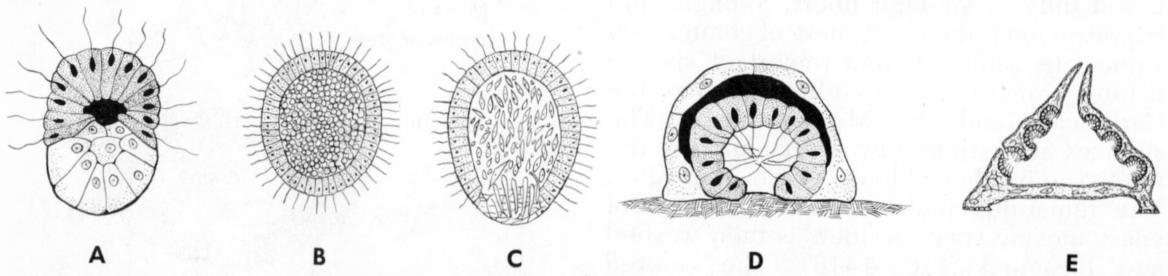

Figure 4–12. Sponge larvae and postlarval development. *A.* An amphiblastula larva. *B* and *C.* Parenchymula larvae. *D.* Gastrulation of a amphiblastula larva following settling. (*C* and *D* redrawn from Hyman.) *E.* Postlarval rhagon stage. (After Sollas from Hyman.)

Figure 4–13. Sperm release from a specimen of the tubular West Indian sponge, *Verongia archeri.* Sponge is about 1.5 m. long. (By Reiswig, H. M., 1970: Science, *170*:538–539.)

the embryo is still composed of eight macromeres, and the other hemisphere of micromeres, but the flagella are now on the exterior. The embryo at this point has reached a stage called the amphiblastula larva (Fig. 4–12*A*), characteristic of many calcareous sponges.

The amphiblastula breaks out of the mesohyl and is carried to the outside by the water currents of the parent sponge. After a short free-swimming existence the amphiblastula settles and attaches, and gastrulation occurs either by epiboly or invagination, or by both. But the process is unusual, in that in calcareous sponges the macromeres overgrow the micromeres (Fig. 4–12*D*), whereas in other metazoans it is the macromeres that typically become internal. The macromeres in sponges give rise to the pinacoderm, and the micromeres to the choanocytes; and both layers produce the amebocytes of the mesohyl.

A hollow amphiblastula larva develops only in calcareous sponges and a few primitive Demospongiae, such as *Oscarella.* A parenchymula larva is characteristic of most Demospongiae, the Hexactinellida, and even some Calcarea. A parenchymula larva is a stereoblastula (Fig. 4–12*B* and *C*). It is similar to the amphiblastula in being flagellated, but the exterior flagellated cells are more extensive and cover all the outer surface, except often the posterior pole. The interior of the larva is filled with archeocytes. The parenchymula larva breaks out of the mesohyl of the parent and has a brief free-swimming existence before settling. Gastrulation now takes place. After settling, the exterior flagellated cells of the parenchymula move inward and form flagellated chambers.

In the freshwater sponges the larva has passed the gastrula stage when it emerges from the parent, for a few internal flagellated chambers are already present. After attachment, the external flagellated cells are phagocytized by amebocytes and take no part in the formation of the choanocytes.

In all the calcareous sponges, development to the larval stage takes place within the body of the parent. However, among the Demospongiae, there are species that incubate the embryos and others that liberate fertilized eggs. Development in the latter occurs in the sea water.

In many of the calcareous sponges that possess a leuconoid structure, the final stages of development after attachment of the larva recapitulate the more primitive asconoid and syconoid structures. For example, a leuconoid sponge undergoes developmental stages resembling the asconoid and syconoid structures before finally reaching the adult leuconoid condition. In other leuconoid sponges, especially the Demospongiae, the leuconoid condition is attained more directly. The first stage is known as a rhagon (Fig. 4–12*E*). It resembles either the asconoid or the synocoid structure except that the walls are quite thick. The leuconoid plan develops directly from the rhagon stage by means of the formation of channels and flagellated chambers in the thick body wall.

Marine sponges may live a number of years, but those in temperate regions are usually dormant in the winter. *Microciona* in Long Island Sound, for example, passes the winter in a reduced state, lacking flagellated chambers and other parts of the water canal system (Simpson, 1968). With increase in water temperature, the sponge redevelops the adult functional condition.

THE PHYLOGENETIC POSITION OF SPONGES

Sponges certainly arose prior to the Paleozoic era, and there have been a number of claims of Pre-Cambrian fossils; but none of these claims have been clearly established. However, beginning with the Cambrian period and extending to the present, the fossil record of sponges is abundant. Early Paleozoic reefs were composed of calcareous blue-green algae (stromatolites) and two now-extinct groups of calcareous sponge-like animals (archaeocyathids and stromatoporoids). There are also some flints which are composed entirely of fused sponge spicules.

The first Calcarea known were from the Devonian period, and during this same period there was an especially great development of glass sponges (Hexactinellida). This class underwent another period of development during the Mesozoic era.

The evolutionary origin of sponges poses a number of interesting problems. The absence of organs and the low level of cellular differentiation and interdependence in sponges certainly seem to be primitive characteristics. Also a body structure built around a water canal system and lacking distinct anterior and posterior ends is found in no other groups of animals. All these features would suggest that sponges are phylogenetically remote from other metazoans. Some zoologists believe that sponges have evolved from a different and separate group of protozoans than did the other metazoans. The principal evidence supporting this theory is the similarity of sponge choanocytes to the choanoflagellates. The chief objection to this theory is that the larval cells of sponges show little similarity to choanoflagellates. The external cells of the larva are flagellated, but they are not choanocytes, and only when they become internal do the flagellated cells transform into collar cells. Also, collar cells are not peculiar to choanoflagellates and sponges. Although not identical, they have been found to occur in widely divergent metazoan animals, such as certain echinoderm larvae, in the oviduct of some sea cucumbers, and in certain corals.

A second and more attractive theory is that sponges arose from a simple, hollow, free-swimming, colonial flagellate, perhaps the same stock that gave rise to the ancestors of the other metazoans. The larvae of sponges are quite similar to such flagellate colonies. The mesohyl perhaps originated from an ingression of cells into the interior, such as occurs in the embryogeny of some sponges. It can be hypothesized that the water canal system developed in conjunction with the assumption of a sessile existence and that the flagellated, exterior cells moved into the interior to become choanocytes. The relegation of flagellated cells to the interior is recapitulated in the embryogeny of living sponges.

Although the origin of sponges is uncertain, there can be little doubt that sponges diverged early from the main line of metazoan evolution and have given rise to no other members of the Animal Kingdom. A considerable number of facts would indicate that sponges are a dead-end phylum: No mouth and no digestive cavity exist; the entire body structure is built around a unique water-canal system; the outer body layer is poorly developed; the embryogeny of sponges displays several peculiarities found in no other group of metazoans. Because of their isolated phylogenetic position, the sponges have often been separated from the other multicellular animals (Eumetazoa) and placed in a separate subkingdom, the Parazoa.

Bibliography

Brien, P., 1968: The sponges, or Porifera. *In* Florkin, M., and Scheer, B. J. (Eds.), Chemical Zoology, Vol. 2. Academic Press, N. Y., pp. 1–30.

Bullock, T. H., and Horridge, G. A., 1965: Structure and Function of the Nervous System of Invertebrates, Vol. 1. W. H. Freeman, San Francisco, pp. 450–453.

Cobb, W. R., 1969: Penetration of calcium carbonate substrates by the boring sponge, *Cliona*. American Zool., *9*:783–790.

Fry, W. G. (Ed.), 1970: The Biology of Porifera. Academic Press, N. Y., 512 pp. (A collection of papers presented at a symposium of the Zoological Society of London.)

Goreau, T. F., and Hartman, W. D., 1963: Control of coral reefs by boring sponges. *In* Sognnaes, R. F. (Ed.), Mechanisms of Hard Tissue Destruction. American Association for the Advancement of Science, Washington, D.C., Vol. 75, pp. 25–54.

Gosner, K. L., 1971: Guide to Identification of Marine and Estuarine Invertebrates: Cape Hatteras to the Bay of Fundy. Wiley-Interscience, N. Y., pp. 51–66.

Hartman, W. T., and Goreau, T. F., 1970: Jamaican coralline sponges: their morphology, ecology, and fossil relatives. *In* Fry, W. G. (Ed.), The Biology of Porifera. Academic Press, N. Y., pp. 205–240.

Humphreys, T., 1970: Biochemical analysis of sponge cell aggregation. *In* Fry, W. G. (Ed.), The Biology of Porifera. Academic Press, N. Y., pp. 324–333.

Hyman, L. H., 1940: The Invertebrates: Protozoa through Ctenophora, Vol. 1. McGraw-Hill, N. Y., pp. 284–364. (The chapter called *Retrospect* in the fifth volume (1959) of this series summarizes investigations on sponges from 1938 to 1958.)

Jewell, M., 1959: Porifera. *In* Edmonson, W. T., Ward, H. B., and Whipple, G. C. (Eds.), Freshwater Biology. 2nd Edition. John Wiley and Sons, N. Y., pp. 298–312. (An illustrated key to the common genera and species of freshwater sponges in the United States.)

Jones, W. C., 1962: Is there a nervous system in sponges? Biol. Rev., *37*:1–50.

Jørgensen, C. B., 1966: The Biology of Suspension Feeding. Pergamon Press, N. Y.

Kaestner, A., 1967: Invertebrate Zoology, Vol. 1. Wiley-Interscience, N. Y., pp. 23–42.

Kilian, E. F., 1952: Wasserströmung und Nahrungsaufnahme beim Süsswasserschwamm *Ephydatia fluviatilis*. Z. vergl. Physiol., *34*:407–447. ("Water flow and food uptake in the freshwater sponge *Ephydatia fluviatilis*.")

Lévi, C., 1957: Ontogeny and systematics in sponges. Syst. Zool., *6*:174–183.

Lévi, C., 1963: Gastrulation and larval phylogeny in sponges. *In* Dougherty, E. (Ed.), The Lower Metazoa. University of California Press, Berkeley, pp. 375–382.

Light, S. F., Smith, R. I., Pitelka, F. A., Abbott, D. P., and Weesner, F. M., 1967: Intertidal Invertebrates of the Central California Coast. University of California Press, Berkeley, pp. 6–21.

Moore, R. C. (Ed.), 1955: Treatise on Invertebrate Paleontology. Archaeocyatha, Porifera. Vol. E. Geological Society of America and University of Kansas Press.

Newell, N. D., 1972: The evolution of reefs. Scientific American, *226*(6):54–65.

Pavans de Ceccatty, M., 1955: Le système nerveux des éponges calcaires et silicieuses. Ann. Sci. Nat. Zool., ser. 11, vol. 17.

Pennak, R. W., 1953: Freshwater Invertebrates of the United States. Ronald Press, N. Y., pp. 77–79. (Chapter on freshwater Spongillidae includes a general account of the family and a key to the North American species.)

Penny, J. T., and Racer, A. A., 1968: Comprehensive revision of a worldwide collection of freshwater sponges (Porifera: Spongillidae). U. S. National Museum Bull., *272*:1–183.

Reiswig, H. M., 1970: Porifera: sudden sperm release by tropical Demospongiae. Science, *170*:538–539.

Reiswig, H. M., 1971: In situ pumping activities of tropical Demospongiae. Mar. Biol., *9*(1):38–50.

Reiswig, H. M., 1971: Particle feeding in natural populations of three marine demosponges. Biol. Bull., *141*(3):568–591.

Simpson, T. L., 1968: The biology of the marine sponge *Microciona prolifera*. J. Exp. Biol. Ecol., *2*: 252–277.

Tuzet, O., 1963: The phylogeny of sponges. *In* Dougherty, E. (Ed.), The Lower Metazoa. University of California Press, Berkeley, pp. 129–148.

Van Weel, P. B., 1948: On the physiology of the tropical freshwater sponge *Spongilla proliferans:* ingestion, digestion, and excretion. Physiol. Comp. Oecol., *1*:110–126.

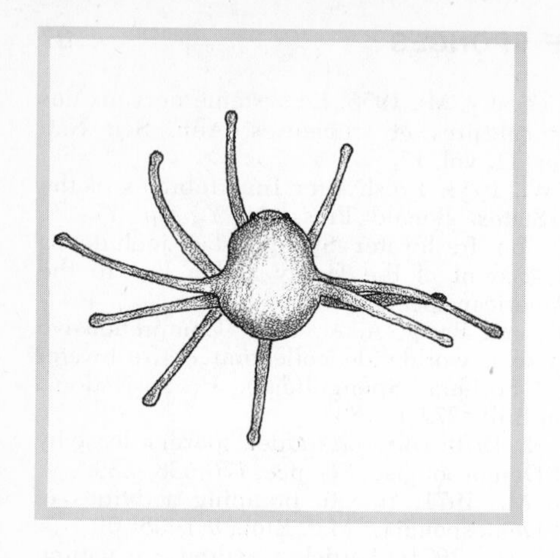

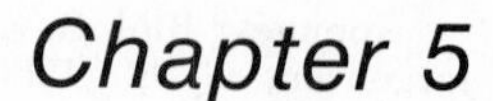

Chapter 5

THE CNIDARIANS

The phylum Cnidaria, or Coelenterata, includes the familiar hydras, jellyfish, sea anemones, and corals. The brilliant coloring of many species, combined with a radial symmetry, often creates a beauty that is surpassed by few other animals. The radial symmetry is commonly considered justification for uniting the cnidarians and the related ctenophores within a division of phyla of the Animal Kingdom called the Radiata. As currently used, the older but familiar name "coelenterate" refers to the cnidarians and ctenophores collectively, and is thus synonymous with Radiata as used here.

The cnidarians possess two basic metazoan structural features. There is an internal space for digestion, called in cnidarians a gastrovascular cavity (Fig. 5–1). This cavity lies along the polar axis of the animal and opens to the outside at one end to form a mouth. The presence of a mouth and digestive cavity permits the use of a much greater range of food sizes than is possible in the protozoans and sponges. In cnidarians a circle of tentacles, representing extensions of the body wall, surrounds the mouth to aid in the capture and ingestion of food.

The cnidarian body wall consists of three basic layers (Fig. 5–1): an outer layer of epidermis, an inner layer of cells lining the gastrovascular cavity, and between these a layer called mesoglea. The mesoglea ranges from a thin, noncellular membrane to a thick, fibrous, jellylike, mucoid material with or without wandering amebocytes. Histologically, the cnidarians have remained rather primitive, although they anticipate some of the specializations that are found in higher metazoans. A considerable number of different cell types compose the epidermis and gastrodermis, but there is only a limited degree of organ development.

Although all cnidarians are basically tentaculate and radially symmetrical, two different structural types are encountered within the phylum. One type, which is sessile, is known as a polyp. The other form is free-swimming and is called a medusa. Typically

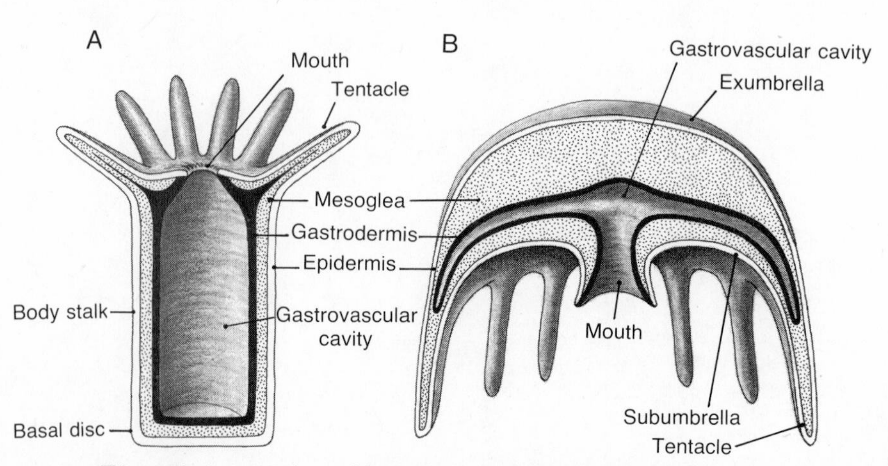

Figure 5–1. *A.* Polypoid body form. *B.* Medusoid body form.

the body of a polyp is a tube or cylinder, in which the oral end, bearing the mouth and tentacles, is directed upward, and the opposite, or aboral, end is attached (Fig. 5–1*A*).

The medusoid body resembles a bell or an umbrella, with the convex side upward and the mouth located in the center of the concave undersurface (Fig. 5–1*B*). The tentacles hang down from the margin of the bell. In contrast to the polypoid mesoglea (middle layer), which is more or less thin, the medusoid mesoglea is extremely thick and constitutes the bulk of the animal. Because of this mass of jelly-like mesogleal material, these cnidarian forms are commonly known as jellyfish. Some cnidarians exhibit only the polypoid form, some only the medusoid form, while others pass through both in their life cycle.

Except for the hydras and a few other freshwater hydrozoans, cnidarians are marine. Most are inhabitants of shallow water; sessile forms abound on rocky coasts or on coral formations in tropical waters. The phylum is composed of approximately 9000 living species, and a rich fossil record dates from the Cambrian period.

CNIDARIAN HISTOLOGY AND PHYSIOLOGY: HYDRAS

For the sake of simplicity, the following introduction to the histology and physiology of cnidarians is based primarily on the familiar hydras, but the greater part of this discussion also applies to the other cnidarians; the more important exceptions will be described in the survey of the three classes later in the chapter.

Hydras are cylindrical, solitary polyps that range from a few millimeters to 1 cm. or more in length (Fig. 5–2). However, the diameter seldom exceeds 1 mm. The aboral end of the cylindrical body stalk forms a basal disc, by means of which the animal attaches to the substratum. The oral end contains a mound, or cone, called the hypostome, with the mouth at the top. Around the base of the cone is a circle of about six tentacles.

The Epidermis. The epidermis is composed of five principal types of cells.

Epithelio-Muscle Cells. The most important type, in terms of body covering, is the epithelio-muscle cell (Figs. 5–3 and 5–4*A*). These cells are somewhat columnar in shape, with the base resting against the mesoglea and the slightly expanded distal end forming most of the epidermal surface. However, unlike true columnar epithelium, epithelio-muscle cells possess two, three, or more basal extensions each containing a contractile fiber or myoneme. The basal extensions are oriented parallel to the axis of the body stalk and tentacles and thus form a cylindrical layer of contractile fibers. Although this layer corresponds to a layer of

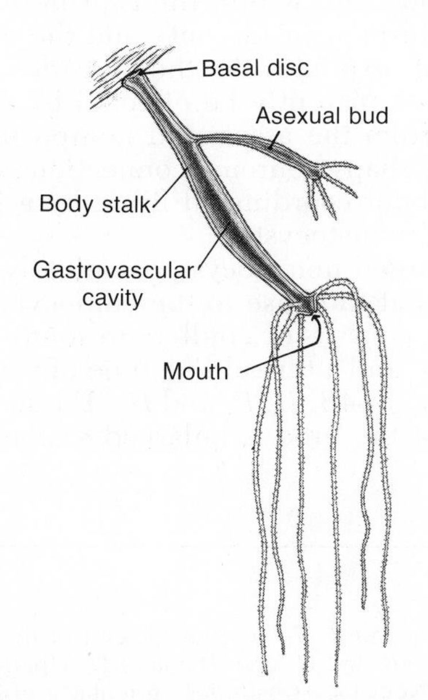

Figure 5–2. *A.* Section through a hydra with an asexual bud. (After Hyman.) *B.* Photograph of a living hydra. Note extended tentacles and the warlike batteries of nematocysts that appear along the tentacles. (By Charles Walcott.)

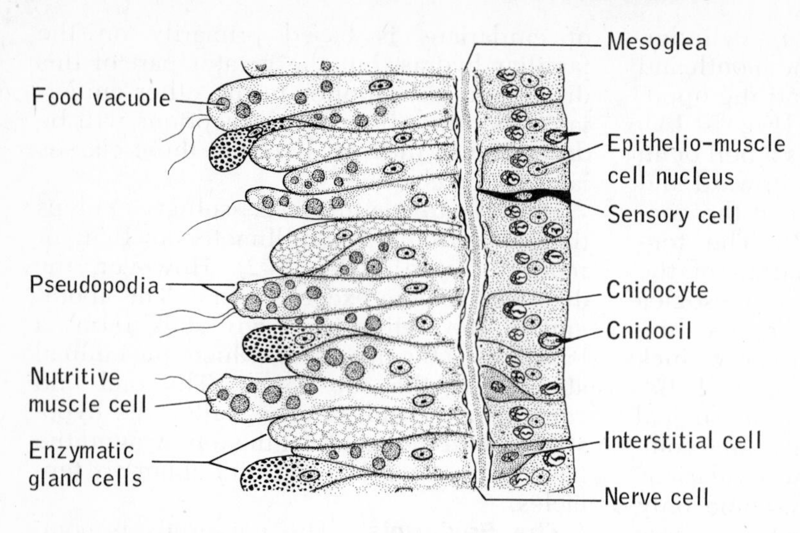

Figure 5–3. Body wall of hydra (longitudinal section).

longitudinal muscle, it is not composed of true muscle cells.

Interstitial Cells. Located beneath the epidermal surface and wedged between the epithelio-muscle cells are small rounded cells with relatively large nuclei. These are known as interstitial cells (Fig. 5–3). The interstitial cells, acting as the germinal or formative cells of the animal, give rise to the sperm and eggs as well as to any other type of cells.

Cnidocytes. A third cell type, the cnidocytes, are located throughout the epidermis, and are lodged between or invaginated within epithelio-muscle cells. They are especially abundant on the tentacles. These specialized cells, which are unique to and characteristic of all cnidarians, contain stinging structures called nematocysts. A cnidocyte is a rounded or ovoid cell with a basal nucleus (Fig. 5–4*G*). One end of the cell contains a short, stiff, bristle-like process called a cnidocil, which has an ultrastructure similar to that of a flagellum and which is exposed to the surface. The cnidocil, which is actually motile in anthozoans, terminates in a kinetosome, or basal body. The interior of the cell is filled by a capsule containing a coiled, pleated tube (Fig. 5–4*G*), and the end of the capsule that is directed toward the outside is covered by a cap or lid. Supporting rods run the length of the cnidocyte, and the base is anchored to the lateral extensions of an epithelio-muscle cell, or in some cnidarians to more than one epithelio-muscle cell. The base of the cnidocyte may also be associated with a neuron terminal.

The nematocysts are discharged from the cnidocyte and are used for anchorage, for defense, and for the capture of prey. The discharge mechanism apparently involves a change in the permeability of the capsule wall. Under the combined influence of mechanical and chemical stimuli, which are initially received and conducted by the cnidocil, the lid of the nematocyst opens. Hydrostatic pressure within the capsule everts the tube (turns it inside out), and the entire nematocyst explodes to the outside. Discharge can apparently be effected by nerve impulses from the associated neuron terminal, and perhaps neuronal connections serve to bring about coordinated firing by a large number of nematocysts.

A discharged nematocyst, completely free or adhered at the base to the cnidocyte and epidermis, consists of a bulb representing the old capsule and a thread-like tube of varying length (Fig. 5–4*B*, *D*, *F*, and *H*). Frequently the base of the tube is enlarged so that the

Figure 5–4. *A.* Epithelio-muscle cell of hydra. (After Gelei from Hyman.) *B.* Volvent nematocyst of hydra. *C.* Discharged volvent and penetrant nematocysts on a crustacean bristle. (*B* and *C* after Hyman.) *D.* Open-tubed nematocyst of the hydroid, *Eudendrium.* (After Weill from Hyman.) *E.* Nematocyst batteries on tentacle of hydra. *F.* Penetrant nematocyst of hydra. *G.* Diagram of a cnidocyte from the hydrozoan jellyfish *Gonionemus.* (From Westfall, J., 1970: Z. Zellforsch., *110*:457–470.) *H.* Spineless nematocyst with open tube of hydra. (*E* to *H* after Hyman.) *I.* Sensory nerve cell of a sea anemone. (After the Hertwigs from Hyman.)

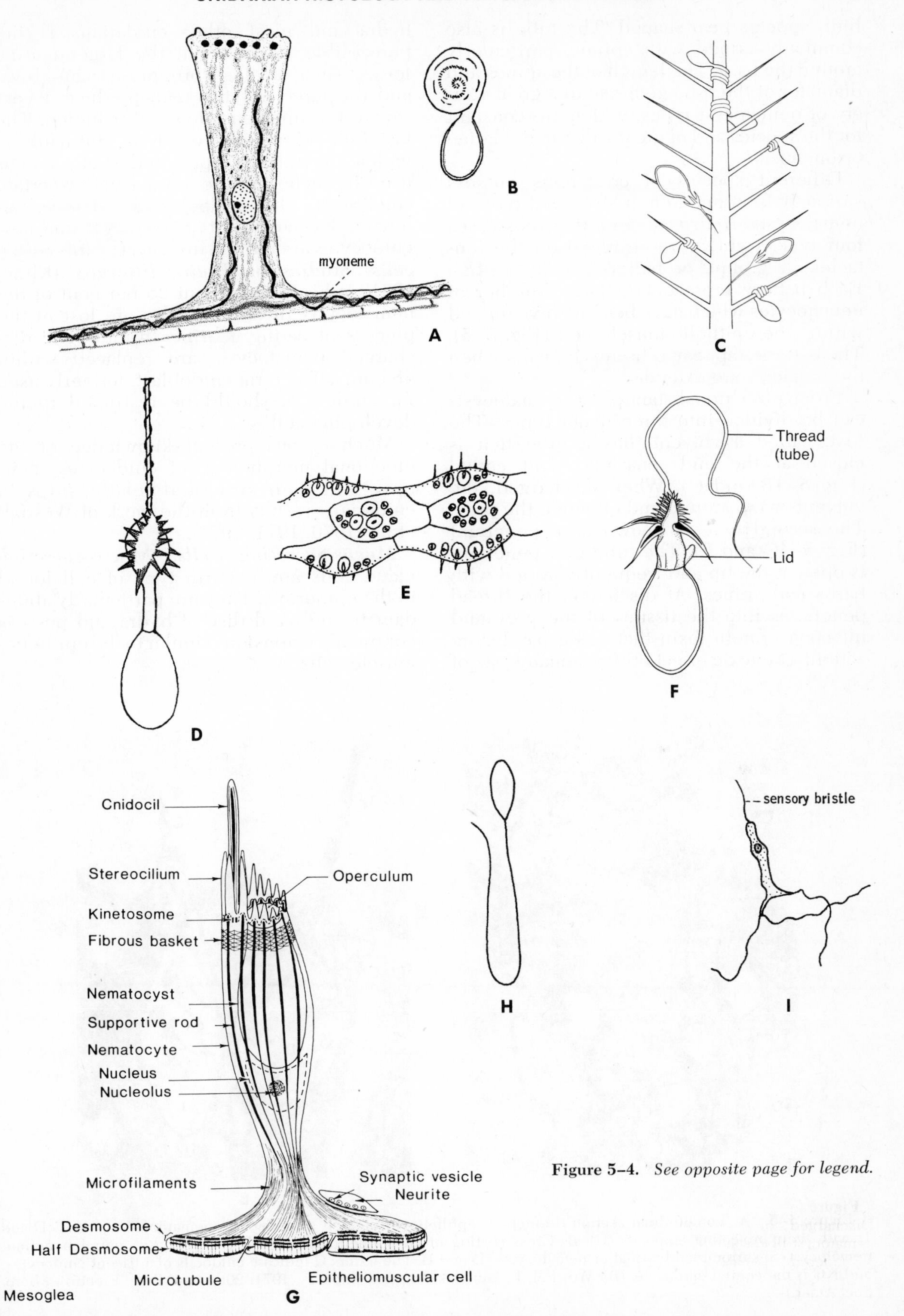

Figure 5–4. *See opposite page for legend.*

bulb appears pear-shaped. The tube is also commonly armed with spines, particularly around the base. The length of the spines and diameter of the tube give rise to a great variety of nematocyst types, which are constant for the species and of great value to cnidarian taxonomists.

Different species of cnidarians possess variously from one to four structural types of nematocysts. *Hydra,* for example, possesses four types, which are arranged on the tentacles in groups or batteries (Fig. 5–4*E*). Each battery represents a large number of nematocytes which have become invaginated within one epithelio-muscle cell (Fig. 5–5). The batteries appear as bumps or warts when the tentacles are extended.

From a functional standpoint, nematocysts can be divided into three major types. The first, called a volvent, has a tube that is closed at the end, unarmed, and coiled (Fig. 5–4*B* and *C*). When discharged, the volvents wrap around and entangle the prey. The second type is known as a penetrant (Fig. 5–4*C* and *F*). The tube of a penetrant is open at the tip and frequently armed with barbs and spines. At discharge, the thread penetrates into the tissues of the prey and injects a protein toxin that has a paralyzing action. The toxic effect of the nematocysts of hydra and most other cnidarians is not perceptible to man; but the larger marine forms, such as the Portuguese man-of-war and the large jellyfish, can produce a very severe burning sensation and irritation. The last type of nematocyst is a glutinant, in which the tube is open and sticky and is used in anchoring the animal under certain conditions. The cnidocyte degenerates following discharge of its nematocyst, and new cnidocytes are formed from nearby interstitial cells. Studies on *Hydra littoralis* (Kline, 1961) indicate that about 25 per cent of the nematocysts in the tentacles are lost in the process of eating a brine shrimp. The discharged nematocysts are replaced within 48 hours. The term cnidoblast, formerly used for cnidocyte, should be restricted to the developing cell.

Much of our present knowledge of the functional morphology of cnidocytes, neurons, and their structural relationships in cnidarians results from the work of Westfall et al. (1970, 1971, 1973).

Mucus-Secreting Cells. Mucus-secreting gland cells are a fourth type of cell found in the epidermis. They are particularly abundant in the basal disc of hydra and possess contractile extensions similar to the epithelio-muscle cells.

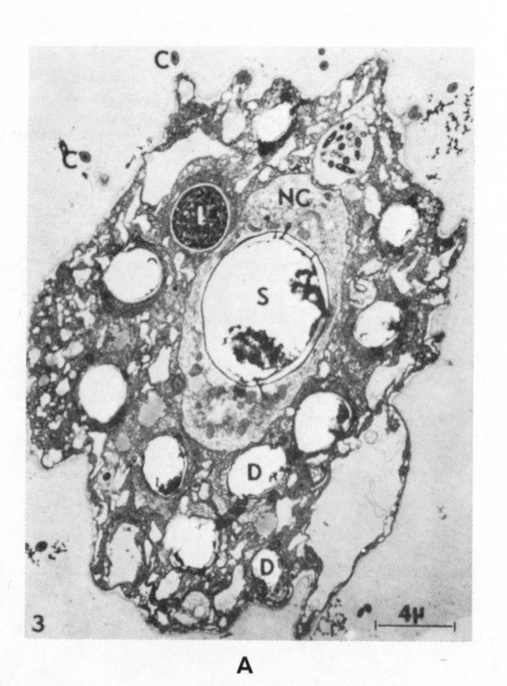

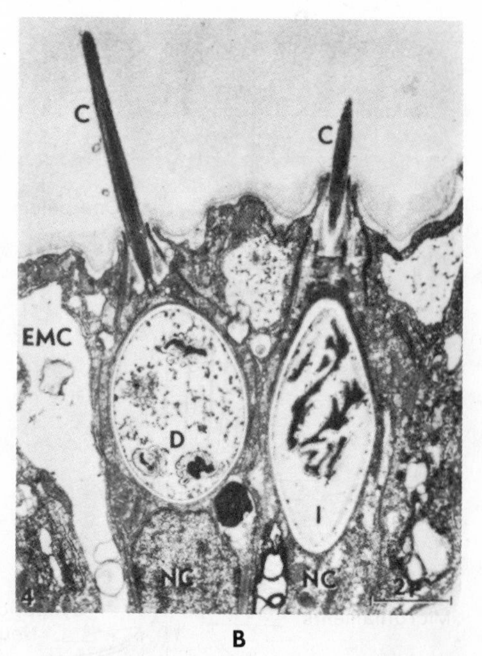

Figure 5–5. *A.* Longitudinal section through an epitheliomuscle cell showing two nematocyst capsules (D and I) with their projecting cnidocils (C). *B.* Cross section of an epitheliomuscle cell containing a central penetrant nematocyst (S) surrounded by smaller nematocysts (D and I). The letters C indicate cnidocils of adjacent cnidocytes, and NC is the central cnidocyte. (By Westfall, J., Yamataka, S., and Enos, P., 1971: 20th Ann. Proc. Electron Micro. Soc. Amer.)

Sensory and Nerve Cells. The remaining two types of cells are the sensory cells and the nerve cells. Sensory cells are elongated cells oriented at right angles to the epidermal surface (Figs. 5–3 and 5–4*I*). The base of each cell gives rise to a number of neuron processes, and the distal end terminates in a sphere or a sensory bristle. Sensory cells are particularly abundant on the tentacles and, like cnidocytes, may be invaginated within epithelio-muscle cells.

The nerve cells, which are more or less similar to the multipolar neurons of other animals, are located at the base of the epidermis next to the mesoglea and are oriented parallel to it (Fig. 5–3). They will be described in the discussion of the nervous system.

The Gastrodermis. The histology of the gastrodermis, or inner layer of the body wall, is somewhat similar to that of the epidermis (Fig. 5–3). However, the cells corresponding to epithelio-muscle cells in the epidermis are called nutritive-muscle cells in the gastrodermis. The two types are similar in shape, but the nutritive-muscle cells are usually flagellated; and the basal contractile extensions, which develop to the highest degree in the hypostome and tentacle bases, are more delicate than those of the epidermis. Furthermore, the contractile fibers of the gastrodermis are oriented at right angles to the long axis of the body stalk and thus form a circular muscle layer.

Interspersed among the nutritive-muscle cells are enzymatic-gland cells. These are wedge-shaped, flagellated cells with their tapered ends directed toward the mesoglea. Enzymatic-gland cells do not exhibit the basal contractile processes.

The remaining cells composing the gastrodermis are identical to those found in the epidermis. Mucus-secreting gland cells are abundant around the mouth. Nerve cells also exist, but in far fewer numbers. Nematocysts are lacking in the gastrodermis of hydras but are present in this layer in many other cnidarians, although restricted in distribution. In some species of hydra the gastrodermal cells contain symbiotic zoochlorellae. This alga, which is apparently *Chlorella*, gives the hydra a bright green color.

Locomotion. The body stalk and tentacles of hydra can extend, contract, or bend to one side or the other. It is usually stated that the epidermal contractile fibers, which are longitudinal, act antagonistically to the circular gastrodermal fibers. However, in hydra the gastrodermal fibers are so poorly developed in most parts of the body that movement is due almost entirely to the contractions of the epidermal fibers. Fluid within the gastrovascular cavity plays an important role as a hydraulic skeleton. By taking in water through the mouth as a result of the beating of the gastrodermal flagella, a relaxed hydra may stretch out to a length of 20 mm. Contraction of the epidermal fibers can shorten the same animal to 0.5 mm.

Although hydras are essentially sessile animals, it is possible for them to shift locations. The tentacles and oral end may bend over and touch the substratum while the basal disc is simultaneously released. The animal can then somersault or move like an inch worm. Adhesive glutinant nematocytes aid in anchoring the tentacles during each movement. Another common method of movement is by floating. The basal disc detaches and secretes a gas bubble that carries the animal to the surface. It may float about upside down for some time.

A characteristic feature of the behavior pattern of hydras is the periodic contraction bursts which draw the animal into a tight ball. The contractions, followed by extension, occur every five to ten minutes in daylight (and at a much lower frequency at night). McCullough (1963) suggests that these bursts provide for regular intermittent sampling of the environment around the hydra.

Nutrition. Hydras, like most cnidarians, are carnivorous and feed mainly on small crustaceans. Contact with the tentacles brings about a discharge of nematocysts, which entangle and then paralyze the prey. The tentacles then pull the captured organism toward the mouth, which opens to receive it. The retraction and bending inward of the tentacles and the opening of the mouth is a reflex response initiated by reduced glutathione liberated by the prey. Mucus secretions aid in swallowing, and the mouth can be greatly distended. Eventually the prey arrives at the gastrovascular cavity. Enzymatic-gland cells then discharge trypsin-like enzymes that begin the digestion of proteins, and the tissues of the prey are gradually reduced to a soupy broth. The beating of the flagella of the gastrodermal cells aids in mixing. From studies made on other cnidarians, it is probable that during this phase of digestion the proteins are hydrolyzed only as far as polypeptides.

Subsequent to this initial extracellular phase, digestion continues intracellularly. The nutritive-muscle cells produce pseudopodia that engulf small fragments of tissue. Continued digestion of proteins and the digestion of fats occur within food vacuoles in the nutritive-muscle cells. The food vacuoles in cnidarians undergo the acid and alkaline phases characteristic of protozoans. Products of digestion are circulated by cellular diffusion; fats and glycogen are the chief storage products of excess food. Undigestible materials are ejected from the mouth on contraction of the body.

Gas Exchange and Excretion. There are no special organs for gas exchange and excretion. Gas exchange occurs through the general body surface. Nitrogenous wastes are largely in the form of ammonia, which also diffuses through the general body surface.

The Nervous System. The nervous system is primitive. The nerve cells are arranged in an irregular nerve net, or plexus, located beneath the epidermis and are particularly concentrated around the mouth. Although it was formerly thought that hydras lacked synapses, the work of Westfall et al. (1971) has revealed that synaptic junctions exist between all neurons and between neurons and effectors—muscle fibers and cnidocytes. Some synapses are symmetrical; that is, both terminals secrete a transmitter substance and an impulse can be initiated in either direction across the synapse. Other synapses are asymmetrical, permitting transmission only in one direction. Neurons serving symmetrical synapses obviously transmit impulses in both directions and are thus directionally nonpolarized, in contrast to the neurons of higher animals. Cnidarian neurons are also unusual in possessing some synaptic junctions along the course of the process rather than only at the terminal; i.e., the terminal of one process may synapse with the process of another neuron at some point along its course. The firing of the nerve net results primarily from the summation of the sensory cells involved, and the degree to which the response is local or general depends upon the strength of the stimulus.

The association of nerve cells to form conducting chains between receptor and effector shows all degrees of complexity. The neurons contain two, three, or more processes. These processes may terminate in muscle fibers, in sensory cells, or in the processes of other ganglion cells. Some neurons have even been shown to possess two branches, each serving a *different kind of effector* (cnidocyte and muscle fiber). The length of the processes varies, and they may give rise to motor and sensory branches. Further, some ganglion cells have only motor processes; some have only sensory processes; and some act as interneurons, having processes that connect with other ganglion cells. This variation may be significant in representing the beginnings of the sensory, motor, and interneuron systems found in higher metazoans.

A double nerve net system in the same body layer is very common in cnidarians other than hydras. One nerve net acts as a diffuse slow conducting system of multipolar neurons; the other is a rapid, through-conducting system of bipolar neurons. Either may be quite localized in extent. Facilitation is important in transmission but is confined to the neuroeffector junction in the through systems.

Reproduction. Hydras reproduce asexually by budding; in fact, this is the usual means of reproduction during the warmer months of the year. A bud develops as a simple evagination of the body wall and contains an extension of the gastrovascular cavity. The mouth and tentacles form at the distal end. Eventually the bud detaches from the parent and becomes an independent hydra.

Considering the facility for asexual reproduction, it is not surprising that hydras, like many cnidarians, have considerable powers of regeneration. A classic experiment is that of Trembley, performed first in 1744. By means of a knotted thread inserted through the basal disc and drawn out through the mouth, Trembley turned a hydra wrong side out. After a short period the gastrodermal cells reoriented themselves on the inner side of the mesoglea, and the epidermal cells migrated to the outside. As in sponges, dissociate cells will aggregate and from the larger aggregates an entire animal will develop. Although interstitial cells can contribute to regeneration, the process is not dependent upon these cells; rather, dedifferentiated epidermal and gastrodermal cells are the principal source of regenerate material. Entire animals have been produced from gastrodermis, and from epidermis in other cnidarians.

A polarity or gradient of dominance exists from oral to aboral end. If the body stalk of a hydra is severed into several sections, each will regenerate into a new individual. Furthermore, the original polarity is retained so that the tentacles always form on the end of the section that was closest to the oral end of the intact animal, and a basal disc forms at the other end. A piece of oral end grafted onto the middle of the body stalk will induce the formation of another oral end. The oral-aboral gradient is also reflected in the rate of regeneration, an aboral piece regenerating more slowly than a more oral piece. The many studies of cnidarian regeneration and the role of bioelectric fields and other factors in explaining dominance and control are described in Rose's (1970) book on regeneration.

Mitotic activity occurs throughout the stalk, but the rate of production of new cells is especially high in the region just below the hypostome. Since the tips of the tentacles, basal disc, and buds are sites of cell death or cell loss (bud), there is a gradual migration of cells down the stalk and out along the tentacles (Fig. 5–6). Within a period of several weeks, all of the cells in the body of an individual are replaced. Hydras never grow old.

Sexual reproduction in hydras occurs chiefly in the fall, because the eggs are a means by which the species survives the winter. The majority of hydras are dioecious. As in all cnidarians, the germ cells originate for the most part from interstitial cells, which aggregate to form ovaries or testes (Fig.

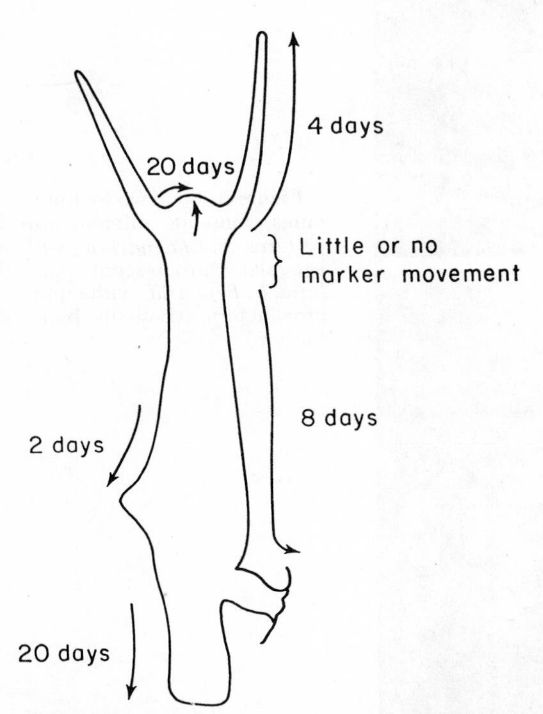

Figure 5–6. Tissue movement in hydra as a result of mitotic activity. Arrows indicate direction of movement and numbers indicate days required for cells to move along path of arrow. (From Campbell, R. D., 1967: Jour. Morph., *121*(1):19–28.)

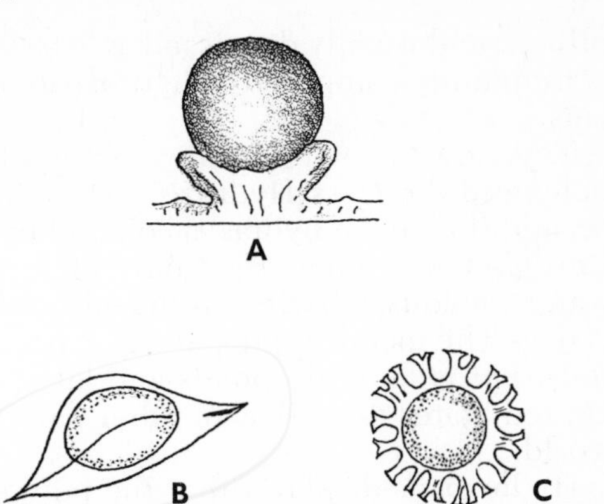

Figure 5–7. *A.* Extruded egg of a hydra. *B.* Embryonated egg surrounded by chitinous shell of *Pelmatohydra*. *C.* Chitinous shell and embryonated egg of *Hydra littoralis*. (After Hyman.)

5–17*B*). In general, the testes are located in the epidermis of the upper half of the stalk and the ovaries in the lower half. This difference in location is true particularly of the few hermaphroditic forms.

A single egg is produced in each ovary, while the other interstitial cells of the ovarian aggregate merely serve as food in the egg's formation. As the egg enlarges, a rupture occurs in the overlying epidermis, exposing the egg (Fig. 5–7*A*). The testis is a conical swelling with a nipple through which the sperm escape. Sperm liberated from the testes into the surrounding water penetrate the exposed surface of the egg, which is thus fertilized in situ.

The egg then undergoes cleavage and simultaneously becomes covered by a chitinous shell (Fig. 5–7*B* and *C*). When shell formation is complete, the encapsulated embryo drops off the parent and remains in its protective casing through the winter. With the advent of spring the shell softens, and a young hydra emerges. Since each individual may bear several ovaries or testes, a number of eggs may be produced each season. Sexual reproduction has been included here to complete the description of hydra, but the pattern described for hydra is not typical of most marine cnidarians.

CLASS HYDROZOA

The class Hydrozoa contains a large number of common cnidarians, but because of their small size and relative inconspicuousness, the layman is largely unaware of their existence. A considerable part of the marine growth attached to rocks, shells, and wharf pilings, and usually dismissed as "sea weed," is frequently composed of hydrozoan cnidarians.

The few known freshwater cnidarians belong to the class Hydrozoa. They include (in addition to the hydras) a colonial hydroid, *Cordylophora lacustris,* and some freshwater medusae with reduced polypoid stages. The medusoid species *Craspedacusta sowerbyi* is found in ponds and lakes in the United States as well as in other parts of the world.

Hydrozoans display either the polypoid or medusoid structure, and some species pass through both forms in their life cycle. Three characteristics unite the members of this class. The mesoglea is never cellular; the gastrodermis lacks nematocysts; and the gonads are epidermal, or if gastrodermal, the eggs and sperm are shed directly to the outside and not into the gastrovascular cavity.

Hydroid Structure. Although some hydrozoans display only the medusoid form, most species possess a polypoid stage in their life cycle. Some forms, such as the hydras, exist as solitary polyps, but the vast majority are colonial. In hydras, buds form on the stalk as simple evaginations of the body wall. The distal end of the bud forms a mouth and a circle of tentacles; then the whole bud drops off to form a new individual. In the development of colonial forms, the buds remain attached; these in turn produce buds so that each polyp is connected to the

Figure 5–8. Two hydroid colonies, showing different growth patterns. *A. Eudendrium,* with an irregular arborescent growth form. *B. Pennaria,* with a pinnate growth form. (Both by Betty M. Barnes.)

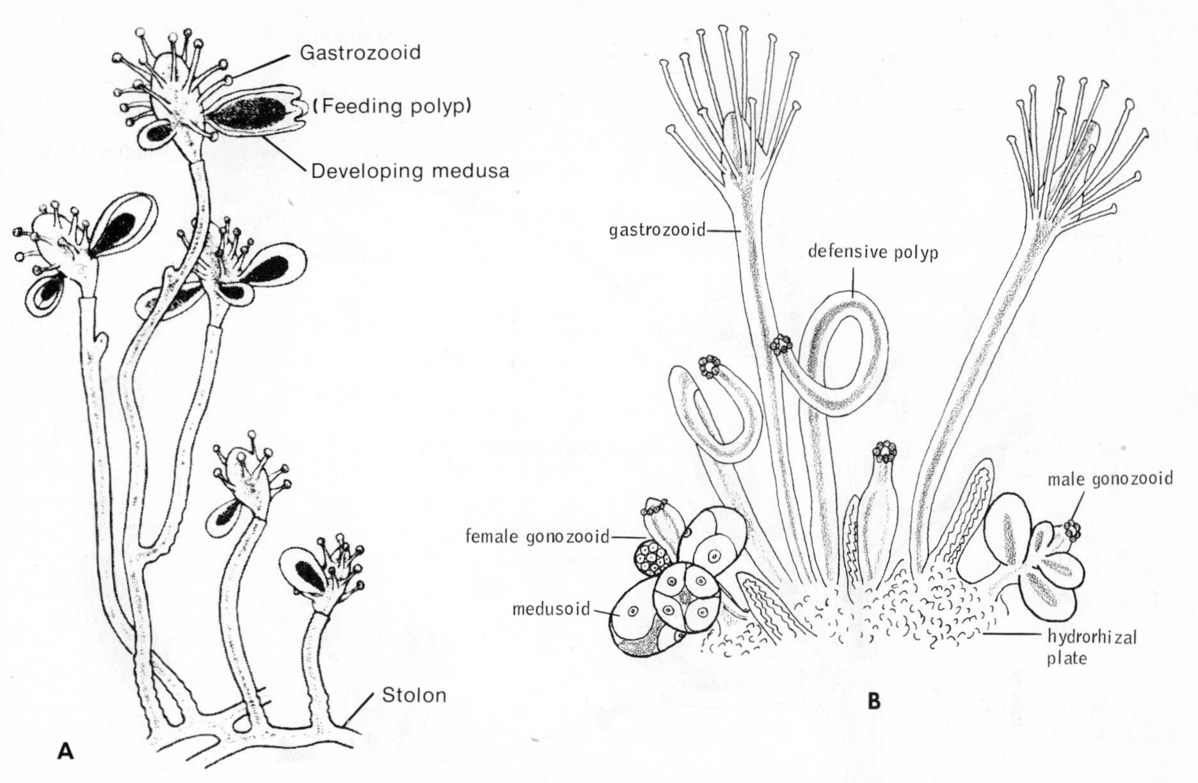

Figure 5-9. A. *Coryne*, a dimorphic hydroid in which medusoids are formed directly on gastrozooids. (Modified after Naumov.) B. *Hydractinia*, a tetramorphic hydroid. Polyps arise separately from mat of stolons, or hydrorhizae. Skeleton limited to covering of stolons, and fusion of adjacent skeletons forms a plate. (After Hyman.)

others. Such a collection of polyps is known as a hydroid colony (Figs. 5-8, 5-9, and 5-10). The three layers in a hydroid colony—epidermis, mesoglea, and gastrodermis—and the gastrovascular cavities are all continuous; it is difficult to tell where one individual begins and another ends.

In describing hydrozoan individuals or colonies, it is convenient to use the term *hydranth*, which refers to the oral end of the polyp bearing the mouth and tentacles, and the term *hydrocaulus*, which refers to the stalk of the polyp.

In most species of hydrozoans the colony is anchored by a horizontal root-like stolon, called the hydrorhiza, which creeps over the substratum (Fig. 5-10). From the stolon arise single upright polyps or branches of polyps. The branches develop through two different growth and budding patterns. In monopodial growth, the first polyp continues to elongate as a result of a growth zone at the distal end of the stalk, or hydrocaulus (Fig. 5-11A). It may even lose its hydranth. As the primary hydrocaulus elongates, it also gives rise to new polyps by lateral budding. These secondary polyps may also elongate and produce buds, and the resulting colony may contain many individuals. In a colony which develops by monopodial growth, the main branch or axis of the colony is formed by the hydrocaulus of the oldest polyp; and its hydranth, if still preserved, occupies the tip of the branch. The same is true of the secondary branches.

The colonies of other species of hydrozoans develop by sympodial growth. The primary polyp does not continue to elongate, but produces one or more lateral polyps by budding (Fig. 5-11*B*). The secondary polyps also do not continue to elongate, but undergo lateral budding. In such a colony resulting from sympodial growth, the main branch or axis of the colony represents the combined hydrocauli of many polyps, and the youngest polyps are located at the tips of the branches.

Whether the colony develops by sympodial or monopodial growth, the form of the colony is quite variable. It may be arborescent (treelike) as in *Eudendrium* (Fig. 5-8*A*), or pinnate as in *Pennaria* and others (Fig. 5-8*B*). The mode of colony growth (i.e., monopodial or sympodial) is not always readily apparent from the appearance of the colony.

Most hydroid colonies are only a few

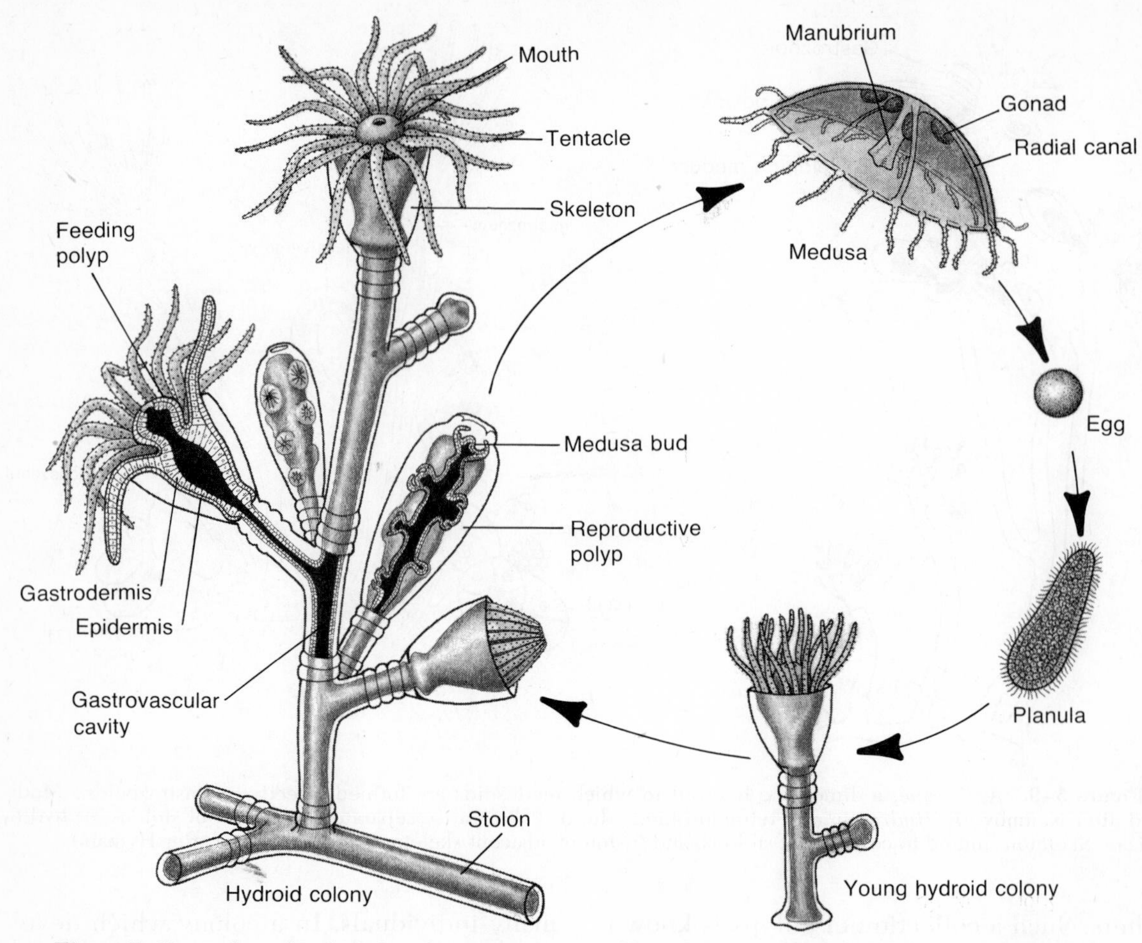

Figure 5–10. Life cycle of *Obelia*, showing structure of hydroid colony. (Adapted from various sources.)

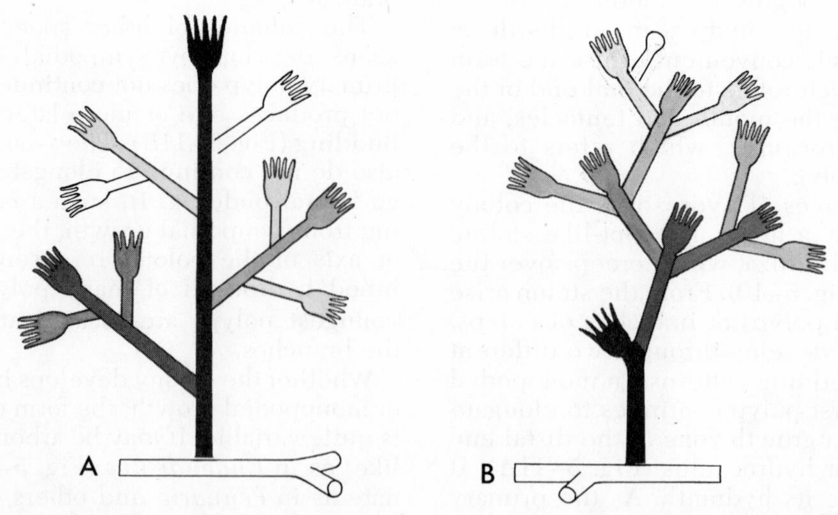

Figure 5–11. Hydroid growth patterns. Oldest polyps shown in black; youngest polyps are clear. *A*. Monopodial growth. *B*. Sympodial growth.

inches or less in length; individual polyps are approximately the size of the oral end of hydras. The coloration, which may be pink or orange, is usually not very striking because of the small size. The largest hydrozoan polyps are solitary species belonging to the genera *Corymorpha*, which may reach a length of 6 to 9 cm., and the giant deep sea *Branchiocerianthus*, which may reach a length of over two meters.

Correlated with the increased size resulting from a colonial organization, most hydroids are, at least in part, surrounded by a supporting, nonliving, chitinous envelope secreted by the epidermis. Such a cylinder is known as the perisarc (Fig. 5–10), and the living tissue that it surrounds is called the coenosarc. The perisarc may be confined to the hydrocaulus (Fig. 5–9*A*), but often it continues upward to enclose the hydranth itself in a casing known as the hydrotheca, as in *Obelia* and *Campanularia*. The hydrotheca may be bell-like and open (Fig. 5–10), or its opening may be covered by a lid of from one to several pieces, as in *Abietinaria* and *Sertularia*. The lid opens when the polyp is extended and feeding, and closes when the polyp contracts. Hydroids with a hydrotheca surrounding the polyp proper are said to be thecate; those without the hydrotheca, athecate.

The hydras and the solitary polyps of other hydrozoans have no skeleton, nor is a skeleton very extensive in some colonial species in which all polyps arise from hydrorhizae. For example, in *Hydractinia echinata*, which lives on snail shells inhabited by hermit crabs, the closely placed naked polyps arise directly from hydrorhizae. Only the hydrorhizae are provided with skeletons, which are fused together to form a more or less continuous plate anchoring the colony to the shell (Fig. 5–9*B*).

Polymorphism is another characteristic of hydroids which is correlated with their colonial organization. All hydroid colonies are at least dimorphic; that is, the colony consists of at least two structurally and functionally different types of individuals. The most numerous and conspicuous type of individual is the nutritive, or feeding, polyp, called a gastrozooid or trophozooid (Figs. 5–9 and 5–10). The feeding polyp looks like a short hydra with a distal mouth, a hypostome, and tentacles.

The gastrozooids capture and ingest prey, and thus provide nutrition for the colony. Most hydrozoans feed on any zooplankton which is small enough to be handled by the gastrozooids. A colony of *Hydractinia echinata* has been reported by Christensen (1967) to consume 47 nematodes, 6 copepods, and 18 arm joints of a brittlestar in 24 hours. Extracellular digestion takes place in the gastrozooid itself; the partially digested broth then passes into the common gastrovascular cavity of the colony, where intracellular digestion occurs. Circulation is probably facilitated by rhythmic pulsations and contraction waves which have been observed in many hydroids.

In most species the gastrozooids also fulfill the defensive functions of the colony, but in some hydroids there are special defensive polyps. The defensive polyps assume a variety of forms but are frequently club-shaped structures, well supplied with nematocysts and adhesive cells (Fig. 5–9*B*). Defensive polyps commonly are located around the gastrozooid.

All hydroids possess reproductive individuals as part of the colony. The reproductive individuals, called gonophores or medusoids, are buds that develop into gamete-producing medusae. Medusoids assume a variety of shapes and locations. They may arise from the hydrocaulus, from the hydrorhiza, from the gastrozooid stalk, or frequently from the body of the gastrozooid itself. The latter development takes place in such forms as *Coryne* (Fig. 5–9*A*) and *Tubularia* (Fig. 5–12).

In *Obelia* and *Campanularia*, the medusoids are restricted to special modified polyps called gonozooids. A common type of gonozooid, called a gonangium, consists of a central stalk, the blastostyle, on which the medusoid buds develop (Fig. 5–10). The perisarc extension around the blastostyle is called the gonotheca. It is usually vase-shaped with a constricted opening at the top. The formation of a medusoid bud is asexual. Through growth, each bud may develop into a complete medusa, which swims away as a tiny jellyfish.

An unusual group of hydroids consists of the members of the genera *Porpita*, *Velella* (Fig. 5–13), and *Porpema*. These are colonial pelagic floating species, which may reach the size of saucers. Each colony consists of a highly modified gastrozooid, in which the body is greatly flattened and the aboral surface has developed a float. Gonozooids bearing gonophores hang down between the central mouth of the gastrozooid and the marginal tentacles.

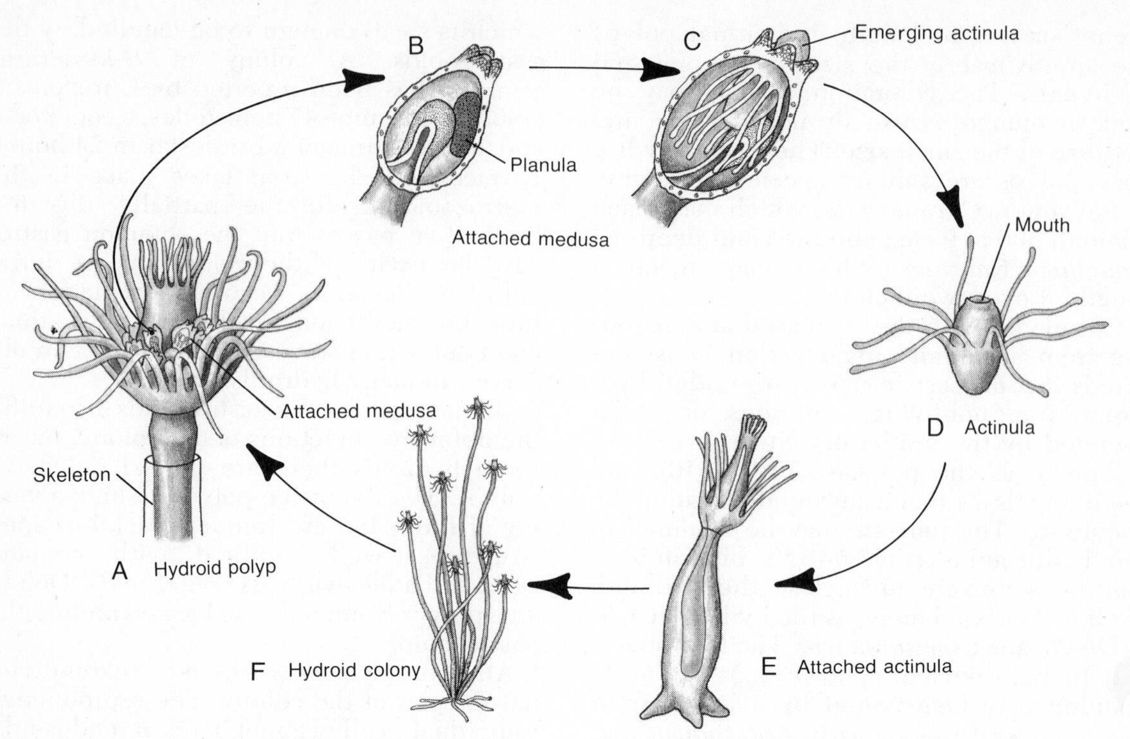

Figure 5–12. Life cycle of *Tubularia*. *A*. Skeleton is restricted to stalk of polyp; medusae are formed on gastrozooid and remain attached. *B*. Egg develops into planula within attached parent medusa. *C*. An actinula larva is released from medusa (*D*) and eventually settles to the bottom and develops into a new hydroid colony (*E* and *F*). (After Allman from Bayer and Owre.)

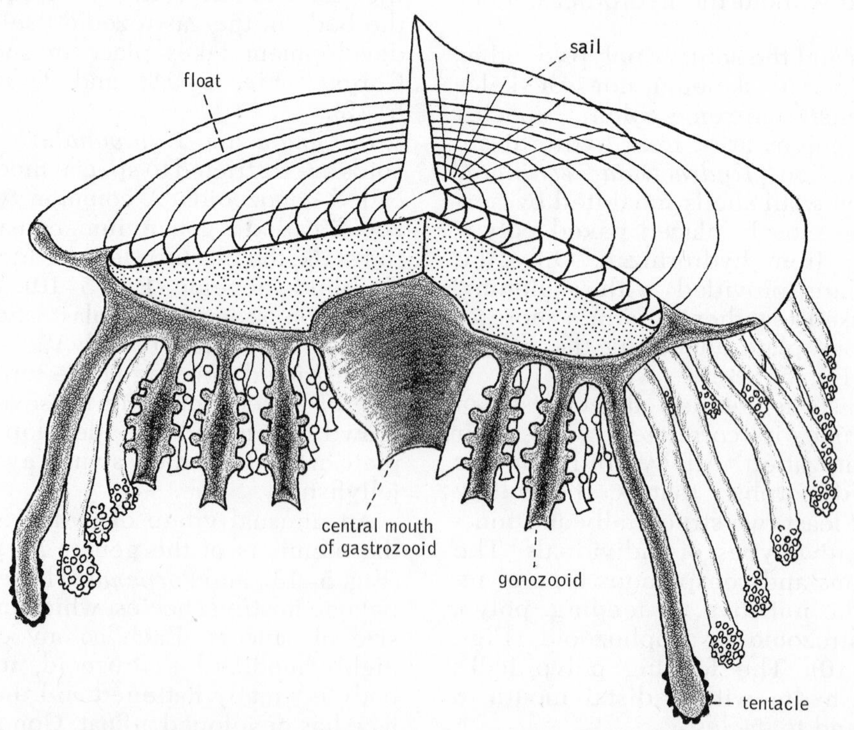

Figure 5–13. The pelagic hydroid *Velella*, with section removed. (After Delage and Herouard redrawn from Buchsbaum.)

Medusoid Structure. Unlike the medusae of the Scyphozoa, hydroid medusae are usually small, ranging from 0.5 cm. to over 6 cm. in diameter (Figs. 5–14 and 5–15). The upper surface of the bell is called the exumbrella, and the lower surface, the subumbrella. The epidermal cells of the exumbrella are not elongated but are flattened like squamous epithelium. The margin of the bell projects inward to form a shelf called a velum (characteristic of most hydromedusae). The tentacles that hang down from the margin of the bell are basically four in number, but there are species with one, two, or more than four tentacles; and, like the tentacles of polyps, they are richly supplied with nematocysts.

The mouth opens at the end of a tubelike extension called the manubrium, which hangs down from the center of the subumbrella. The manubrium also possesses nematocysts and is often lobed or frilled. The gastrovascular cavity is considerably more complex than the simple polypoid sac or tube. The cavity consists of a series of canals, arranged to resemble the hub, spokes, and rim of a wheel. The mouth leads into a central stomach, from which typically extend four radial canals. The radial canals join with a ring canal running around the margin of the unbrella. A swelling known as a tentacular bulb is located at the junction of each radial canal with the ring canal. The manubrium, stomach, and canals are all lined by gastrodermis.

As in all medusoid forms, the mesoglea is extremely thick and gelatinous and constitutes the bulk of the animal. The mesoglea of the hydromedusa is devoid of cells but does contain fibers that are probably secreted by both the epidermis and the gastrodermis. A thin, noncellular membrane separates the mesoglea from the epidermis, and another bounds the gastrodermis.

Medusae are also carnivorous, and feed upon all sorts of animals, including tiny fish, which come in contact with the tentacles. The processes of their nutrition are essentially the same as in the polyp. Although food particles are distributed throughout the gastrovascular cavity, most intracellular digestion takes place in the manubrium, in the stomach, and in tentacular bulbs.

The muscular system of the medusa is somewhat more specialized than in the polyp. The gastrodermal cells lack contractile extension, and the muscular system is thus restricted to the epidermal layer. Furthermore, the muscular system is best developed around the bell margin and subumbrella surface, where the fibers form a radial and circular system. Some of the epithelio-muscle cells of the velum have their contractile extensions oriented to form a powerful, circular band of fibers, which are striated. The contractions of the muscular system, particularly of the circular fibers, produce rhythmic pulsations of the bell, which drive water out from beneath the subumbrella. The velum reduces the subumbrella aperture and thus increases the force of the water jet. Swimming movements are largely vertical, and following a series of pulsations which drive the animal upward, it slowly sinks. Although medusae are isotonic with seawater, some buoyancy is gained by differences in ion content. The mesoglea, for example, has a much lower concentration of sulfate ions than does sea water. Horizontal movement is largely dependent upon water currents. There are a few species of hydromedusae that crawl about over the bottom on their tentacles.

The nervous system of the medusa (Fig. 5–15*E*) is also more highly specialized than that of the polyp. In the margin of the bell, the epidermal nerve cells are usually organized and concentrated into two nerve rings, one above and one below the attachment of the velum. These nerve rings, which represent one of the highest levels of nervous organization in cnidarians, connect with fibers innervating the tentacles, the musculature, and the sense organs. Fibers also interconnect the two rings. The lower ring is the center of rhythmic pulsation; i.e., it contains the pacemakers. Pulsation will continue in the bell as long as any portion of the ring is intact. It is with the lower ring that the statocysts (described later) are connected.

The bell margin is liberally supplied with general sensory cells, and also contains two types of true sense organs—light sensitive ocelli and statocysts. The ocelli consist of patches of pigment and photoreceptor cells organized within either a flat disc or a pit. The ocelli are typically located on the outer side of the tentacular bulbs. Some medusae are negatively phototactic and descend to deeper water during the day; others are attracted to light. Many, however, show no phototaxis.

Statocysts are located between the tentacles or associated with the tentacular bulb at the tentacle base (Fig. 5–15*D*). They may be in the form of either pits or closed vesicles, but in both cases the walls contain sensory cells with bristles projecting into the lumen. Attached to the bristles are from a

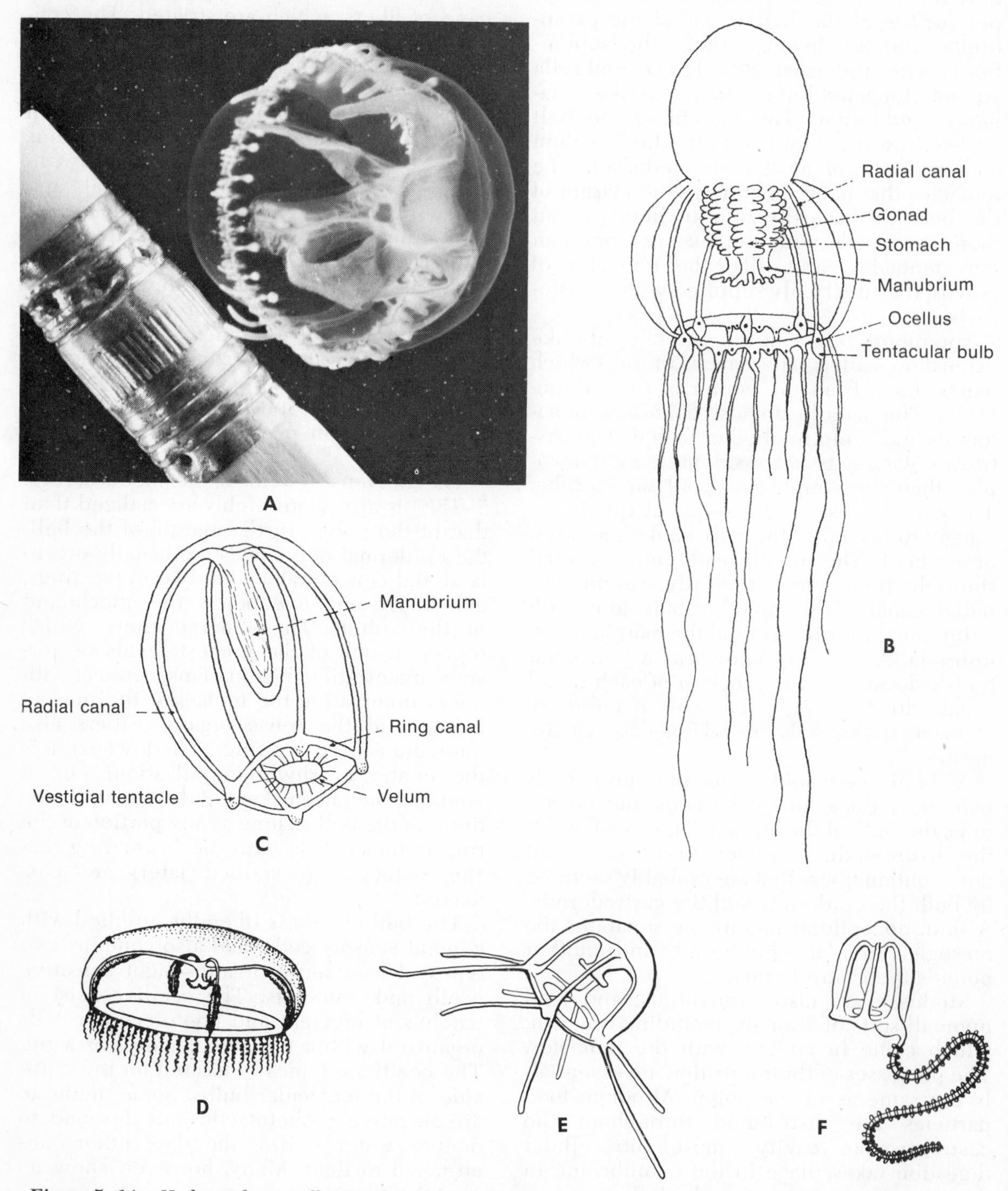

Figure 5-14. Hydromedusae. All are only a few centimeters or less in diameter. *A.* High-speed photograph of the freshwater medusa *Craspedacusta.* (By C. F. Lytle.) *B. Leuckartia.* (After Hyman.) *C. Pennaria.* (After Mayer from Hyman.) *D. Cuspidella. E. Bougainvillia. F. Corymorphia.* (*D–F* after Mayer.)

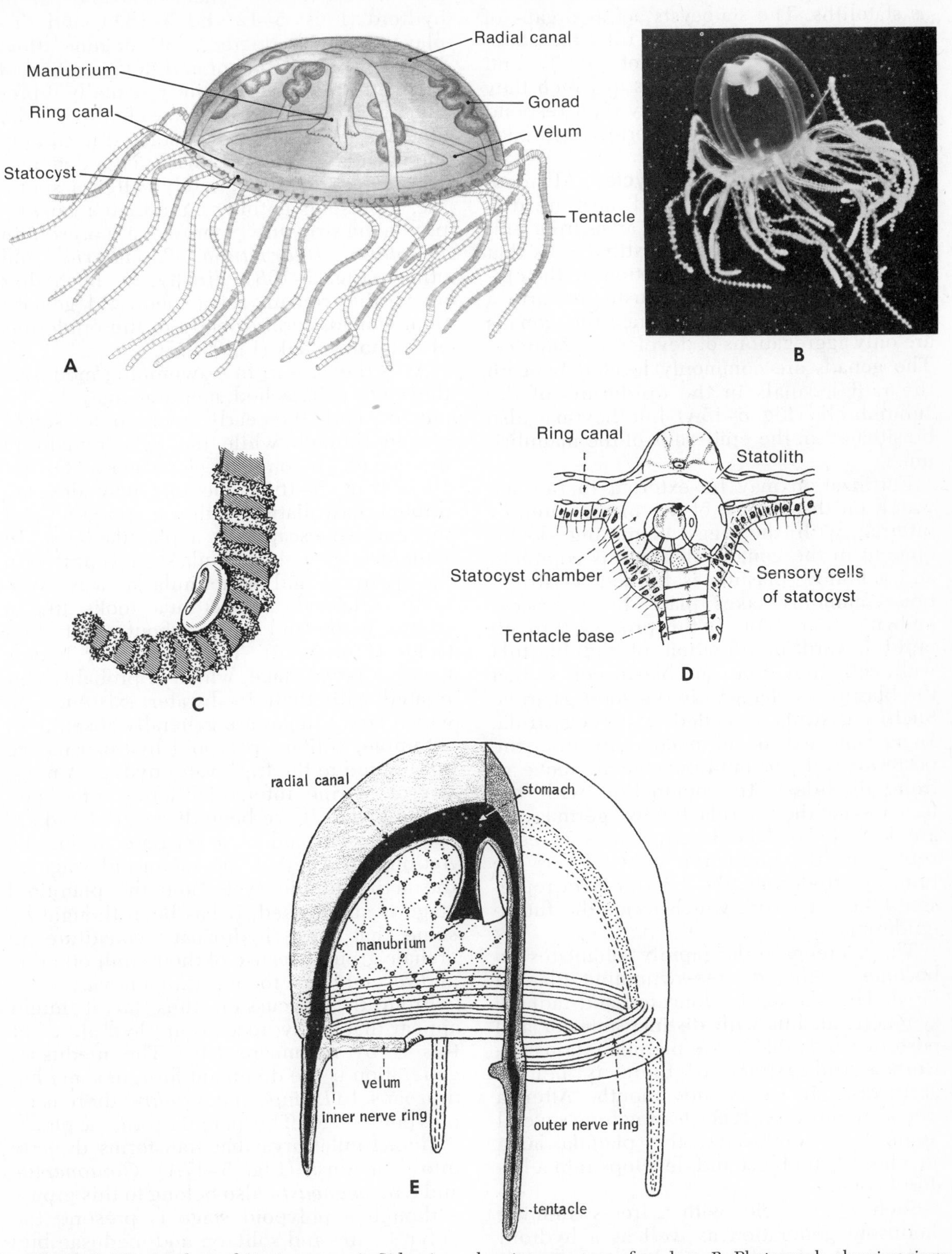

Figure 5-15. Medusa of *Gonionemus*. *A*. Side view, showing structure of medusa. *B*. Photograph, showing rings of nematocysts on tentacles. *C*. Portion of a tentacle. *D*. Statocyst at base of a tentacle of a hydrozoan medusa. *E*. Nervous system of a hydromedusa. (*A* redrawn from Mayer; *B* courtesy of D. P. Wilson; *C* after Hyman; *E* after Bütschli from Kaestner.)

few to many calcareous concretions known as statoliths. The statocysts act as organs of equilibrium. When the bell tilts, the statoliths respond to the pull of gravity and stimulate the sensory bristles to which they are attached. The animal may then respond by muscular contractions to bring itself back into a horizontal position.

Reproduction and Life Cycle. All medusae reproduce sexually, and most are dioecious. The eggs and sperm arise from epidermal or gastrodermal interstitial cells that have migrated to specific locations in the epidermis. Here these cells cluster to form a gonad. However, as in hydras, the gonads are only aggregations of developing gametes. The gonads are commonly located beneath the radial canals in the epidermis of the subumbrella (Fig. 5–15*A*), but they may also be situated in the epidermis of the manubrium.

Fertilization may be external in the sea water, on the surface of the manubrium, or internal with the eggs beginning development in the gonad. Cleavage is complete and a hollow blastula is formed. Gastrulation commonly takes place by a process known as ingression. In this process, through rapid inward proliferation of the blastula wall, cells move into the blastocoel, so that the blastula is changed into a solid gastrula. Such a gastrula is called a stereogastrula. Ingression may be unipolar (only from the posterior wall) or multipolar (cells move in from all sides). In conjunction with the formation of the gastrula, the two germ layers are laid down. The interior mass of cells represents the entoderm, which forms the future gastrodermis; the exterior layer represents the ectoderm, which forms the future epidermis.

The stereogastrula rapidly elongates to become a ciliated, free-swimming planula larva. The planula is elongated and radially symmetrical, but with distinct anterior and posterior ends. Since the planula retains the stereogastrula structure, there is neither gastrovascular cavity nor mouth. After a free-swimming existence lasting from several hours to several days, the planula larva attaches to an object and develops into a hydroid colony.

Such a life cycle, with a free-swimming medusoid generation as well as a hydroid stage, is displayed by *Obelia* (Fig. 5–10), *Pennaria*, *Syncoryne*, and other species of hydrozoans.

The majority of hydroids, such as *Tubularia*, *Sertularia*, and *Plumularia*, do not produce a free-swimming medusa. Instead, the medusa remains attached to the parent hydroid (Figs. 5–12 and 5–16*A*), and displays various degrees of degeneration. Despite the attachment and degeneration of the medusa, it still remains a sexually reproducing individual. In some hydroids the attached medusa has degenerated until only the gonadal tissue remains. Such a degenerate medusa, inappropriately called a sporosac, represents nothing more than a gamete-producing structure. Sporosacs are present in *Tubularia*, *Hydractinia*, *Plumularia*, and others (Fig. 5–16*B*). Finally, in the hydras no vestige remains of a medusoid generation. Gonads form directly in the epidermis of the polyp stalk (Fig. 5–17*C*).

As is the case in free-swimming medusae, the eggs of attached medusae may be retained, and the early embryonic stages passed through while the eggs remain in the gonad. In some species such as *Orthopyris* (Fig. 5–16*B*), the egg may develop through gastrulation within a sporosac; and the embryo escapes as a planula larva. In *Tubularia* even the planula stage is passed in the sporosac, and an actinula larva is eventually released. An actinula looks like a stubby hydra and creeps about on its tentacles (Fig. 5–12). The freshwater hydras have no larval stage, which is probably correlated with their freshwater existence, in which larval stages are generally absent.

Simple, solitary polypoid hydrozoans are not limited to the freshwater hydras. A number of marine interstitial forms, unrelated to the hydras, have been discovered and are sometimes placed in a separate order, the Actinulida. There is no medusoid stage in the actinulid life cycle, and the planuloid stage is suppressed. It has been thought by some that these hydrozoans constitute an archaic group because of their adult ciliation and resemblance to an actinula larva.

From the discussion thus far it might appear that all hydrozoans are hydroids, but this is by no means true. The medusoid generation is the dominant form in some hydrozoans. In *Liriope* and *Aglaura*, there is no polypoid stage. The planula forms a planktonic actinula larva that transforms directly into a medusa (Fig. 5–17*A*). *Gonionemus* and *Craspedacusta* also belong to this group. Although a polypoid stage is present, the polyp is tiny and solitary; and medusae bud off from the sides (Fig. 5–17*B*).

The appearance of hydroid colonies and hydromedusae is usually highly seasonal and, in temperate regions at least, is closely correlated with water temperatures. Under suitable environmental conditions a colony

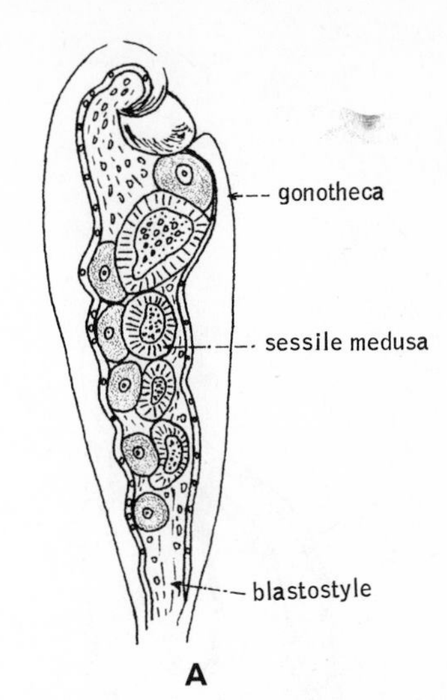

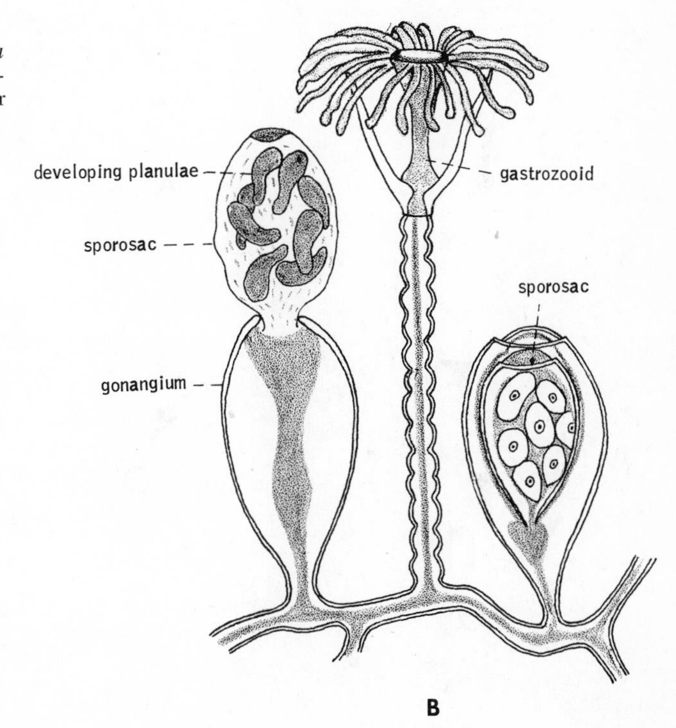

Figure 5–16. *A*. Gonangium of *Campanularia* with sessile medusae. (After Hyman.) *B. Orthopyxis* with uneverted and everted sporosacs. (After Nutting from Hyman.)

will continue to grow, but the life span of the polyps may be limited. For example, after seven days the tissues of hydranths of species of *Campanularia* are resorbed and their materials utilized in developing polyps. Species of *Tubularia* automize (i.e., self-amputate) hydranths on which old attached medusae are empty of sperm or actinulae (the hatching stage in this genus). Hydrozoans that have free medusae may liberate enormous numbers of them in a short period of time. A seven-centimeter colony of *Bougainvillia* produced 4450 medusae over a three-day period; as a result of such production, hydromedusae may often make a dramatic appearance in the plankton. The life span of a medusa, which includes a growth period, ranges from a few days to many months.

Hydroids have been the subject of numerous regeneration studies, and many of the results are similar to those described on page 94 for hydras. Hydroid regeneration studies are reviewed in Rose (1970).

Hydrozoan Evolution. The evolutionary significance of "metagenesis" in hydrozoans is an intriguing problem. Which came first, the polyp or the medusa? In 1886, W. K. Brooks worked out a theory of cnidarian evolution that is still supported by many

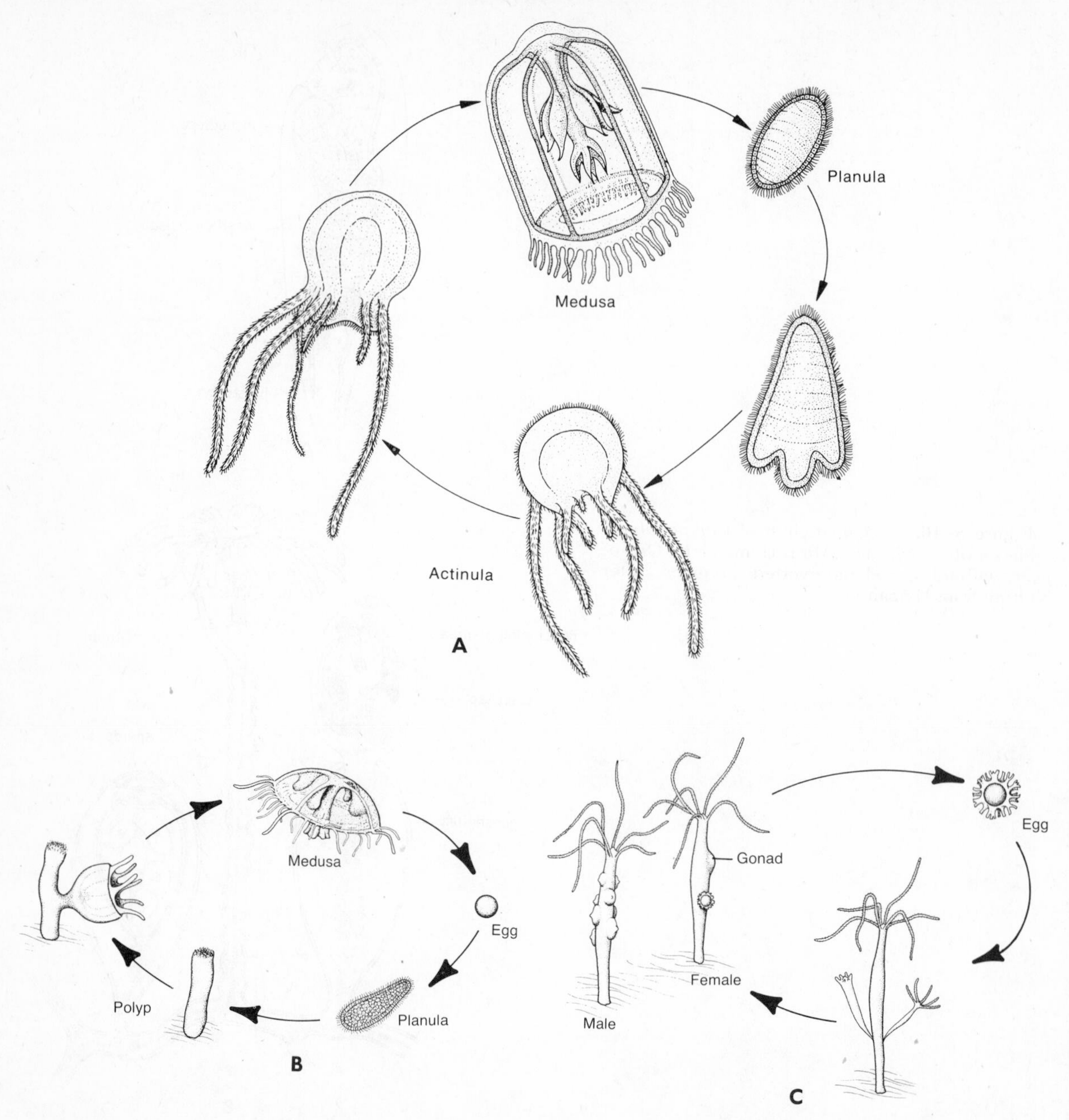

Figure 5–17. Some hydrozoan life cycles. *A. Aglaura*, a hydrozoan which has no polypoid stage. Planula larva develops into an actinula, which develops directly into a medusa. (From Bayer and Owre.) *B. Craspedacusta* (*Gonionemus* would be similar.) Polyp is solitary. *C. Hydra*, a freshwater hydrozoan in which medusoid stage has disappeared and planula larva is suppressed.

zoologists today. According to Brooks' theory, the ancestral cnidarian form was medusoid. The tendency among hydrozoans has been to suppress the medusoid stage so that, in such forms as hydra, the medusa has completely disappeared. The polyp, on the other hand, represents an evolutionary retention and development of the larval condition.

The evolutionary sequence may have occurred in the following manner. The ancestral cnidarians were medusoid, and development led through a planktonic planula larva and a later planktonic actinula larva. The actinula, in which the cnidarian characters first make their appearance, developed into the adult medusa. In some groups of such ancestral medusoid cnidarians the actinula took up an attached benthic existence to give rise to a polyp. Such an attached condition could have been an adaptation to exploit a new food supply, to extend larval life, or to facilitate asexual reproduction of additional polyps by budding. Hydromedusae, rather than arising by the direct transformation of the actinula-polyp, were now produced by budding. The life cycle of such a species would have been: medusa→planula→polyp→medua.

The medusa, which primitively was free, became retained and then gradually suppressed until it disappeared completely. Living hydrozoans display life cycles that illustrate different stages in this evolutionary sequence (see bottom of this page).

The hydroid colony evolved through the retention of budded polyps. Correlated with the evolution of colonial organization was the development of skeletons and polymorphism, but these events probably occurred independently in different evolutionary lines. Medusoid suppression was probably also an event that was repeated numerous times, and carried to various degrees, throughout the Hydrozoa. For example, it is unlikely that the hydras evolved from a colonial hydroid; they are probably derived from a line of hydrozoans, similar to *Gonionemus* and *Craspedacusta,* in which polyps were always solitary but originally budded off free-swimming medusae.

Specialized Hydrozoan Orders. Before the conclusion of the discussion on Hydrozoa, a few specialized orders must be described briefly.

Siphonophora. Members of the order Siphonophora, which includes the familiar *Physalia* (the Portuguese man-of-war (Fig. 5–18)), exist as large pelagic colonies composed of modified polypoid and medusoid individuals, the majority displaying a remarkable degree of polymorphism. A conspicuous feature of many species is a large gas-filled sac that acts as a float for the colony. The float of the Portuguese man-of-war may attain a length of 30 cm. Some species (but not *Physalia*) can regulate the gas content of the float so that the colony can sink below the surface during stormy weather. Some siphonophores, such as *Nanomia,* live permanently below the surface and migrate great distances vertically. During the migration, gas (which is over 90 per cent carbon monoxide) is secreted into or released from the floats. The Pacific *Nanomia bijuga* has been reported to migrate as much as 300 meters in less than an hour. Locomotion is also brought about by swimming bells, or nectophores, which are modified pulsating medusae. Swimming bells are highly developed in the colonies of *Nectalia* and *Stephalia* (Fig. 5–19).

Feeding is carried on by the polypoid members of the colony. Each nutritive polyp has only one long tentacle, which in *Physalia* may hang down several meters below the float (Fig. 5–18A). The tentacles of *Physalia* are strewn with nematocysts, and an accidental encounter with a large colony can be a painful and even a dangerous experience for a swimmer. The toxin from the nematocysts can produce such extreme pain that death may result from drowning. *Physalia* feeds on surface fish, such as mackerel and flying fish, which on colliding with the tentacles are paralyzed by nematocysts and then drawn up to the mouths of the feeding polyps. The products of digestions are shared by all members of the colony through intercommunicating gastrovascular cavities.

Siphonophoran colonies develop through a process of budding. The larval polyp, which develops from the egg, forms the float (if present) and gives rise to budding zones. From these budding zones the various members of the colony develop by primary or secondary budding, and all remain attached together. The larval body thus functions as the main attachment point for the mature colony. There are also attached reproductive medusae.

Siphonophorans are largely tropical and semitropical, but numbers of *Physalia* are

---

Medusa→Egg→Planula→Actinula→Medusa

Medusa→Egg→Planula→Attached Actinula→Medusa

Medusa→Egg→Planula→Polyp→Medusa Bud→Medusa

Attached Medusa→Planula→Polyp→Attached Medusa

Degenerate Attached Medusa→Planula→Polyp→Degenerate Attached Medusa

Polyp→Planula→Polyp

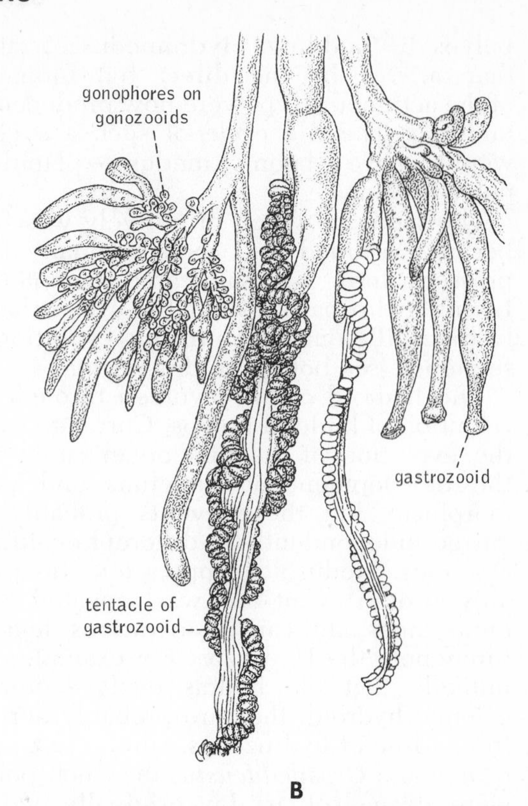

A B

Figure 5–18. *A. Physalia*, the Portuguese man-of-war, a siphonophoran. (Courtesy of the New York Zoological Society.) *B.* Part of Portuguese man-of-war colony. (After Lane.)

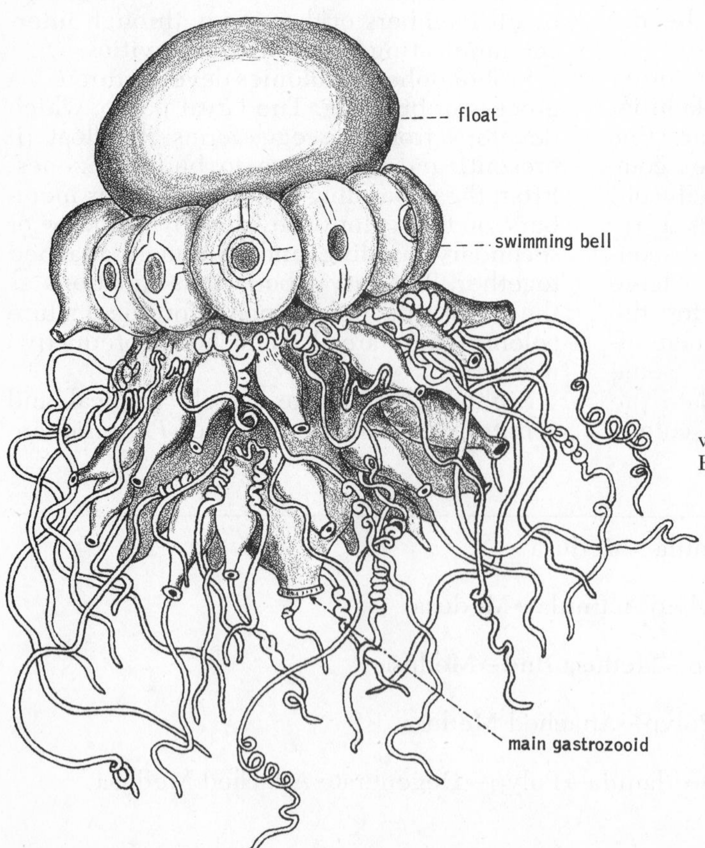

Figure 5–19. *Stephalia*, a siphonophoran with swimming bells. (After Haeckel from Hyman.)

A

B

Figure 5–20. Two coralline hydrozoans. *A. Millepora* of the order Milleporina. (Courtesy of the Encyclopaedia Britannica.) *B. Hydrallmania* of the order Stylasterina. (Photograph by W. J. Rees.)

often seen on the north Atlantic coast, especially following storms that have blown them in from the Gulf Stream.

Milleporina and Stylasterina. The two small groups of hydrozoans sometimes placed in separate orders Milleporina (*Millepora*) and Stylasterina (*Stylaster*) secrete an external calcareous skeleton (Fig. 5–20). Both are colonial polypoid hydrozoans with either an encrusting or an upright growth form. Both may attain considerable size and are often brightly colored. Large numbers of defensive polyps are present and can produce a sting. For this reason the Milleporina are sometimes called fire coral or stinging coral. Most members of these orders are tropical or semitropical and are common components of coral reefs. *Millepora* is very conspicuous on West Indian reefs.

SYSTEMATIC RÉSUMÉ OF CLASS HYDROZOA

Order Trachylina. Medusoid hydrozoans lacking a polypoid stage. Medusa develops directly from an actinula. This order contains perhaps the most primitive members of the class. *Liriope, Aglaura.*

Order Hydroida. Hydrozoans with a well-developed polypoid generation. Medusoid stage present or absent. The majority of hydrozoans belong to this order.

Suborder Limnomedusae. Hydrozoans possessing small solitary polyps and free medusae. *Gonionemus;* the freshwater *Craspedacusta.*

Suborder Anthomedusae. Skeletal covering, when present, does not surround hydranth (athecate). *Tubularia, Pennaria, Syncoryne, Eudendrium, Hydractinia, Millepora* (fire coral), *Stylaster, Branchioceriantus*, the freshwater hydras.

Suborder Leptomedusa. Hydranth surrounded by a skeleton (thecate). *Obelia, Campanularia, Abietinaria, Sertularia, Plumularia, Aglaophenia, Polyorchis.*

Suborder Chondrophora. Pelagic, polymorphic, polypoid colonies. *Velella, Porpita.*

Order Actinulida. Tiny solitary hydrozoans resembling an actinula larva. No medusoid stage present. Interstitial inhabitants. *Halammohydra, Otohydra.*

Order Siphonophora. Pelagic hydrozoan colonies of polypoid and medusoid individuals. Colonies with float or large swimming bells. Largely in warm seas. *Physalia* (Portuguese man-of-war), *Stephalia, Nectalia.*

CLASS SCYPHOZOA

Scyphozoans are the cnidarians most frequently referred to as jellyfish. In this class the medusa (Fig. 5–21*A*) is the dominant and conspicuous individual in the life cycle; the polypoid form is restricted to a small larval stage. In addition, scyphozoan medusae are generally larger than hydromedusae. The majority of scyphozoan medusae have a bell diameter ranging from 2 to 40, cm.; some species are even larger. The bell of *Cyanea capillata* may reach two meters in diameter. Coloration is often striking; the gonads and other internal structures, which may be deep orange, pink, or other colors, are visible through the colorless or more delicately tinted bell.

The some 200 described species of scyphozoans live in all seas from the Arctic to tropical oceans. Although there are a number of deep sea forms, the majority inhabit coastal waters and are often a nuisance on bathing beaches. Their large size and their nematocysts make them unpleasant and sometimes dangerous swimming companions. The so-called stinging nettles, which inhabit the latter Atlantic coast in large numbers during the latter part of the summer, are members of this class. The sea wasps, including certain species such as *Chironex fleckeri* (Fig. 5–22) of the small tropical order Cubomedusae, produce such virulent stings that they are extremely dangerous, more so than the notorious hydrozoan Portuguese man-of-war. Some 50 fatalities have been recorded from Australian coasts. Death, if it occurs, takes place three to twenty minutes after stinging. Lesions from non-fatal stings may be very severe and slow to heal.

Although these cnidarians are typically free-swimming animals, one order, the Stauromedusae, is sessile (Fig. 5–21*B*). In this group, the exumbrellar surface is drawn out into a stalk by which the animal attaches to algae and other objects. Members of this group are thus somewhat polypoid in general structure.

In general, scyphozoan medusae are similar to hydromedusae. The bell varies in shape from a shallow saucer to a deep helmet, and the margin is typically scalloped to form lobes called lappets (Fig. 5–25). A velum is absent except in the Cubomedusae. The manubrium is drawn out into four or eight often frilly, oral arms, which aid in the capture and ingestion of prey. The tentacles around the bell margin vary from four to many but are absent in rhizostomes (Fig. 5–23).

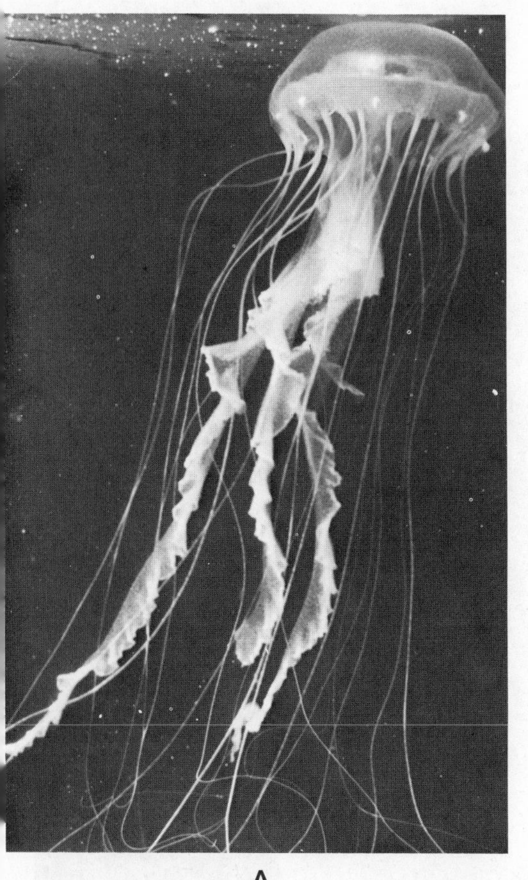

A B

Figure 5–21. *A.* The sea nettle, *Chrysaora quinquecirrha*, a common scyphozoan along the Atlantic coast. (Courtesy of William H. Amos.) *B.* Two species of sessile scyphomedusae attached to an alga. The specimen on the left, *Haliclystus auricula*, has a diameter of less than 1 in. The specimen on the right is *Craterolophus convolvulus*. (Courtesy of D. P. Wilson, *In* Buchsbaum, R. M., and Milne, L. J., 1961: The Lower Animals. Chanticleer Press, New York.)

In *Aurelia*, the type most often studied in introductory courses, the tentacles are small and form a short fringe around the margin (Fig. 5–24); but in other species the tentacles are much longer (Fig. 5–21*A*). The tentacles, the manubrium, and frequently the umbrella surface possess nematocysts.

The mesoglea of the scyphozoan is similar to that in the hydromedusa, being thick, gelatinous, and fibrous; but unlike that of the hydrozoan, it contains wandering ameboid cells, and thus is a true cellular layer. The cells of the scyphozoan mesoglea appear to originate from the epidermis; therefore, if the term *mesoderm* is used, the layer must be considered ectomesodermal in origin and not the true entomesoderm characteristic of higher metazoans.

The scyphozoan muscular system is like that of the hydromedusa. Locomotion is brought about by a band of powerful circular fibers (the coronal muscle) that surrounds the subumbrella margin; the fibers compose true muscle cells located within the mesoglea. Swimming pulsations are similar to those of hydromedusae. Also, the mesoglea of at least some species functions to provide a neutral buoyancy for the animal by maintaining an ion content that is different from that of sea water. Although most scyphomedusae are moved horizontally by currents or waves, some fast swimming tropical Cubomedusae can control the direction of their movement.

The more primitive orders of Scyphozoa do not display the gastrovascular canal system seen in the hydromedusa. The mouth opens through the manubrium into a central stomach, from which extend four gastric pouches (Fig. 5–26). Between the pouches are septa, each of which contains an opening to aid in water circulation. Thus, all four pouches are in direct communication with each other. The margin of the septum, which

A

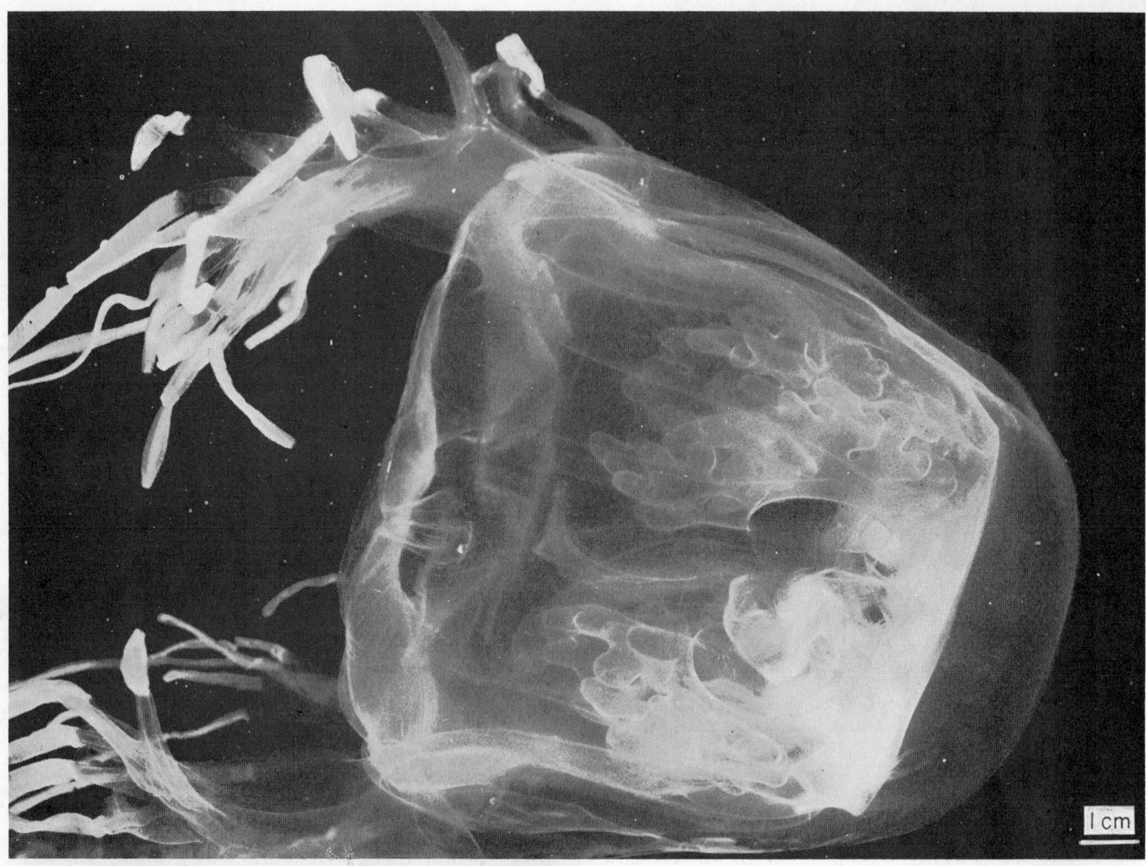

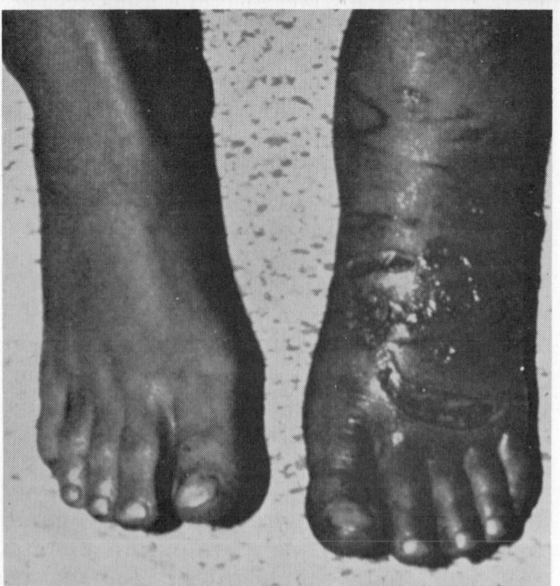

B

Figure 5–22. *A*. The Indo-Pacific sea wasp, *Chironex fleckeri*, a dangerous scyphozoan of the order Cubomedusae. The tentacles have been cut to permit handling of the specimen. *B*. Five-day old lesions produced by a small sting of this sea wasp. (By J. H. Barnes. *In* W. J. Rees (Ed.), 1966. The Cnidaria and Their Evolution. Zoological Society of London.)

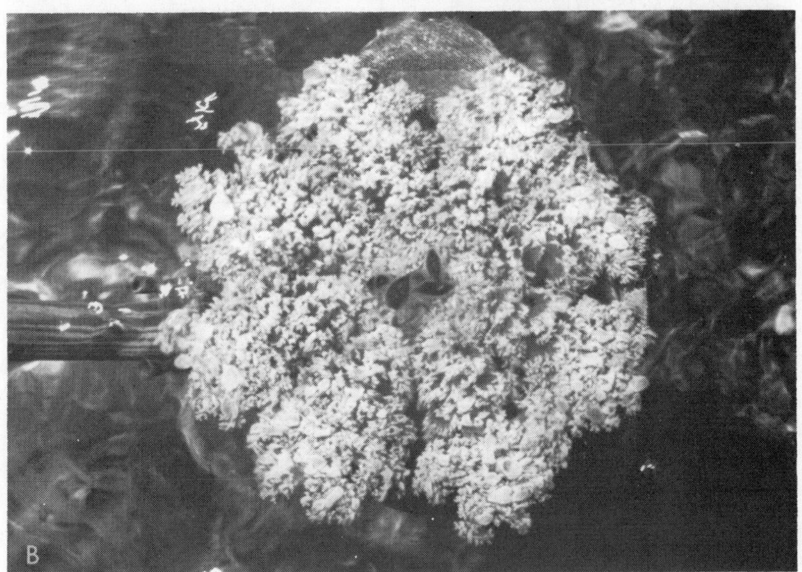

Figure 5–23. *Cassiopeia,* a common tropical genus of rhizostome scyphozoans, which live upside down on the bottom in quiet shallow water. *A.* Specimens resting on bottom, viewed through water. *B.* A single specimen about 21 cm. in diameter, which has been lifted to the surface. (Both by Betty M. Barnes.)

faces the central portion of the stomach, bears a large number of threads, or filaments, containing nematocysts and gland cells.

Medusae of the order Semaeostomeae, which contains the more familiar North Atlantic jellyfish such as *Chrysaora* and *Aurelia,* have gastric pouches and septa in the larval stage only; as adults they possess a system of radial canals which result from a fusion of the upper and lower layers of gastrodermis at the periphery of the stomach. The canal system can be extensive (Fig. 5–24). Four canals called perradials extend from the remaining central stomach. These correspond in position to the oral arms of the manubrium, and divide the animal into quadrants. Between the perradial canals extend interradial canals, and between the interradials and perradials are the adradial canals. The canals are frequently branched as in *Aurelia;* and a ring canal may be absent, but not in *Aurelia.* Although septa are absent, filaments are present and are attached interradially to the periphery of the stomach floor.

Adult scyphozoans feed on all types of small animals, and several eat ctenophores or other medusae. Not many scyphozoans feed upon fish; in fact, larval fish of a number of species use certain species of scyphozoans for protection. As the medusa sinks slowly downward, prey is captured on contact with the tentacles or oral arms of the manubrium. Some species, including *Aurelia,* are actually

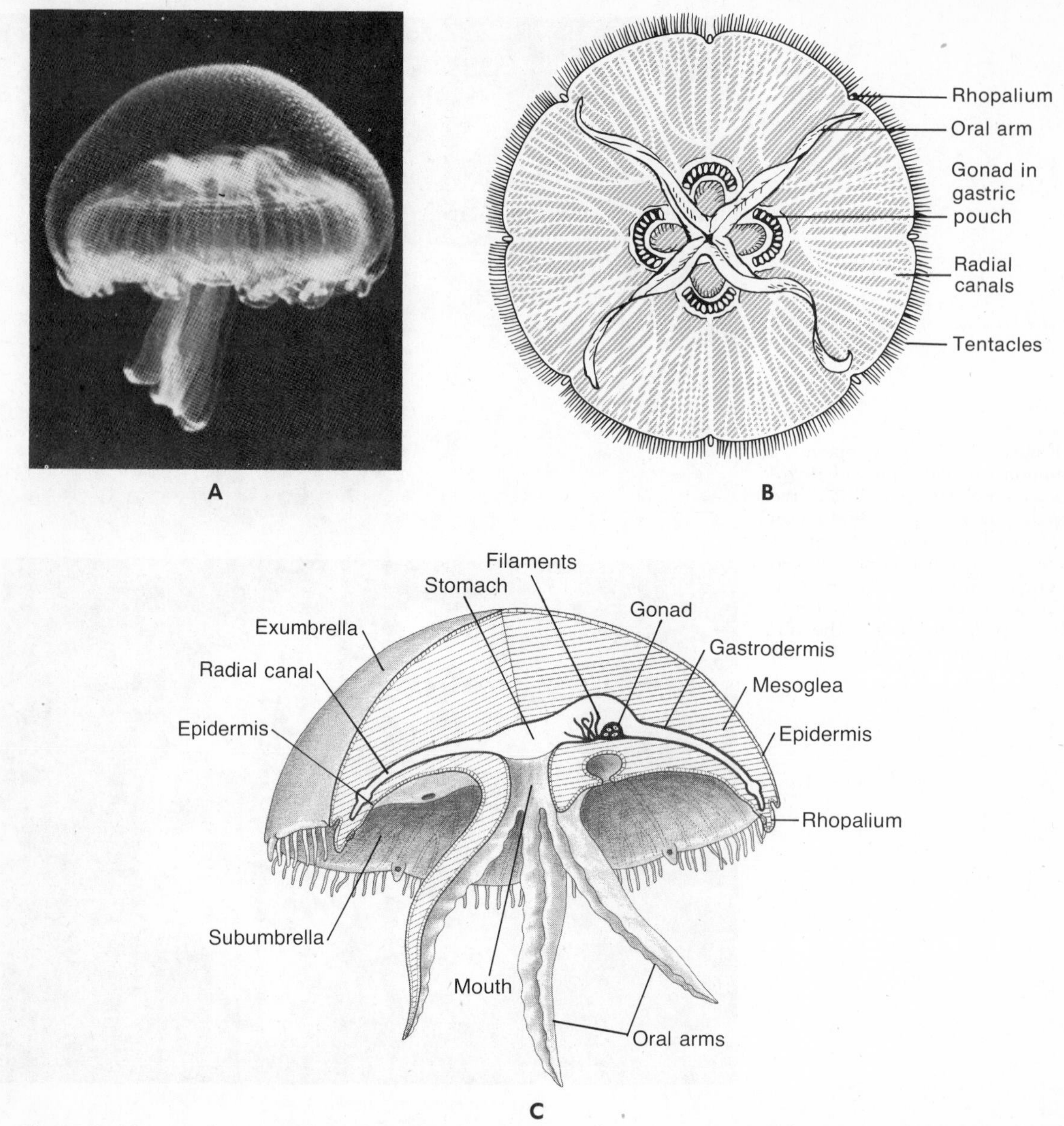

Figure 5–24. *Aurelia,* a scyphozoan medusa. *A.* Young specimen with bell contraction. *B.* Ventral view; *C,* side view in section. (*A,* courtesy of D. P. Wilson.)

suspension feeders. As *Aurelia* sinks, plankton becomes entrapped in mucus on the subumbrellar surface, which is flagellated in these forms. Cilia then sweep the food to the bell margin, where it is scraped off by the oral arms. Ciliated grooves on the oral arms carry the food to the mouth and stomach. *Cassiopeia,* a common jellyfish of Florida and the West Indies, rest upside down on the bottom in the quiet shallow water of mangrove embayments (Fig. 5–23). This genus, like other members of the order Rhizostomeae, possess many small secondary mouths that open into the stomach by way of canals in the oral arms. Small animals trapped on the surface of the frilly oral arms are carried into the mouths within mucous cords. However, *Cassiopeia* possesses symbiotic algae (zooxanthella) in its tissues, and in adequate light it can survive entirely on the products of the algal photosynthesis.

Digestion is essentially as described in hydras. The gastric filaments are the source of extracellular enzymes, and the gastrodermal nematocysts are probably used to quell prey that is still active.

The nervous system is of the nerve-net type and is synaptic. Nerve rings occur only in the orders Coronatae and Cubomedusa. In other scyphozoans the pulsation control is centered in marginal concentrations of neurons. Each such concentration is situated in a little, club-shaped structure called a rhopalium; these are located around the bell margin between lappets and number from four to sixteen. Each rhopalium is flanked by a pair of small specialized lappets called rhopalial lappets and is covered by a hood (Fig. 5–25).

Two sensory pits, a statocyst, and sometimes an ocellus are borne by each rhopalium. The statocyst is located at the tip of the club, and the sensory pits at the base. One sensory pit is located on the outer (exumbrellar) side, and one on the inner side. The sensory pits are merely concentrations of the same type of general sensory cells that are distributed over the body surface. When ocelli are present, as in *Aurelia* (Fig. 5–25*B*), they may be simple pits, containing pigment and photoreceptor cells; but all cubomedusans have complex eyes containing a lens and a retina-like arrangement of sensory cells. Many scyphozoans display distinct phototaxis. They come to the surface of the water during cloudy weather and at twilight, but move downward in bright sunlight and at night.

With few exceptions, scyphozoans are dioecious, and the gonads are located in the gastrodermis, in contrast to the usual epidermal gonad in hydrozoans. In septate groups with gastric pouches, the eight gonads are located on both sides of the four septa (Fig. 5–26). In semaeostome medusae, which lack septa, four horseshoe-shaped gonads lie on the floor of the stomach periphery (Fig. 5–24). When mature, the eggs or

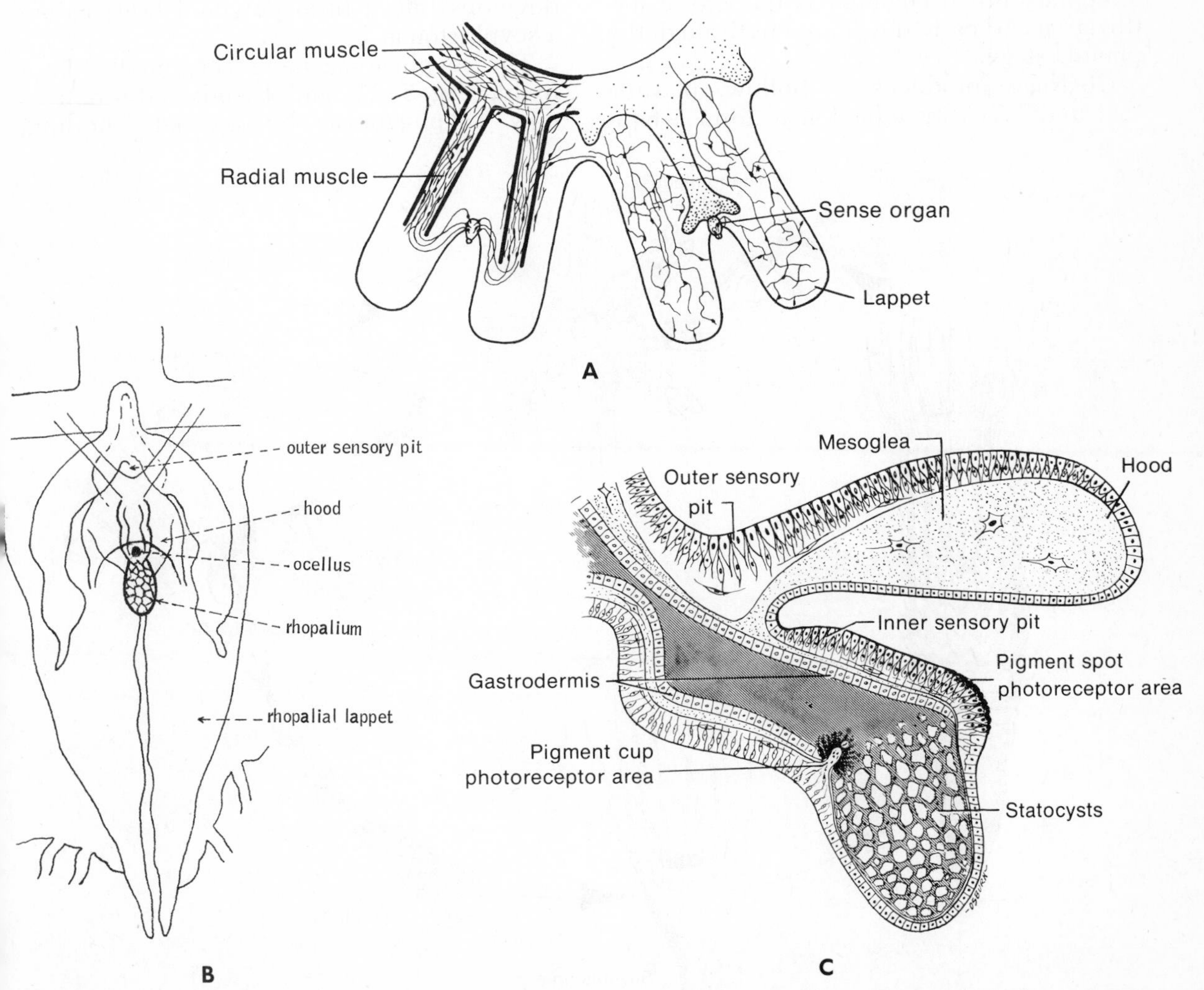

Figure 5–25. Marginal sense organs of schyphozoans. *A.* Diagram of a section of bell margin showing nerve net, lappets, and position of sense organ, or rhopalium. (After Horridge.) *B.* Rhopalium of *Aurelia* (aboral view). (After Hyman.) *C.* Section through a rhopalium showing the hood and the various sensory areas. (Modified from Hyman after Schewiakoff.)

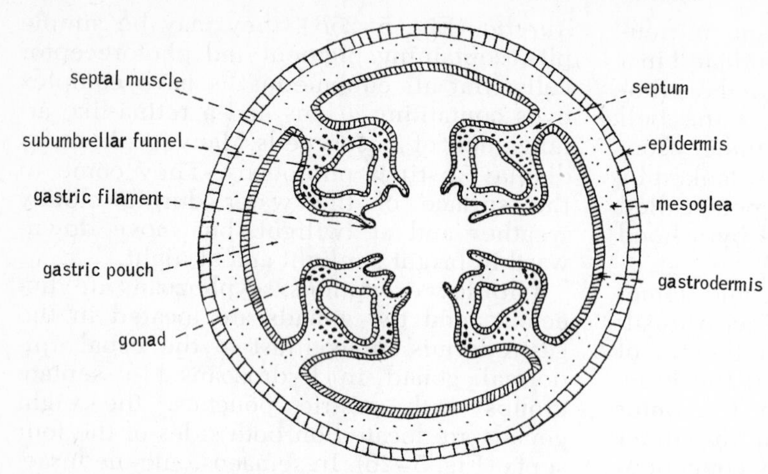

Figure 5–26. Section through a scyphozoan with gastric pouches. (After Hyman.)

sperm break into the gastrovascular cavity and pass out through the mouth. In some semaeostomes, including *Aurelia*, the eggs become lodged in pits on the oral arms. This temporary brood chamber is the site of fertilization and early development through the planula stage.

Cleavage produces a hollow blastula that undergoes invagination to form a typical planula larva (Fig. 5–27). After a brief free-swimming existence, the planula settles to the bottom and becomes attached by its anterior end. The attached planula then develops into a little polypoid larva called a scyphistoma.

The scyphistoma looks very much like a hydra (Fig. 5–27), and it feeds and produces new scyphistomae by asexual budding,

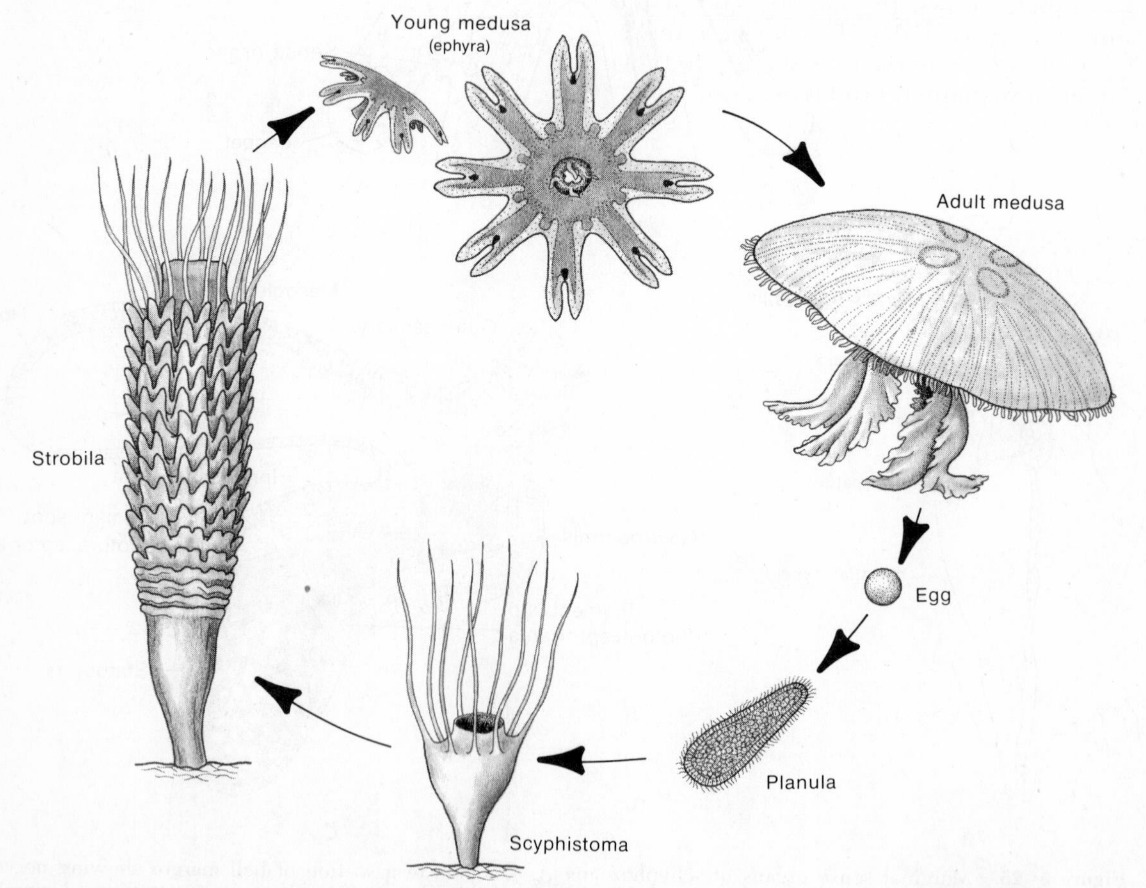

Figure 5–27. Life cycle of the scyphozoan *Aurelia*. The polypoid stage is a larva and produces medusae by transverse budding.

either directly from the midcolumn wall or from stolons (in *Aurelia*). At certain periods of the year young medusae are formed. In some species (Cubomedusae) the scyphistoma transforms directly into a small medusa (Fig. 5–28). In others, medusa formation is accomplished by transverse fission of the oral end of the scyphistoma, a process called strobilization. Buds may form one at a time (monostrobilization) or many simultaneously (polystrobilization) so that the immature medusae, called ephyrae, are stacked up like saucers at the oral end of the body stalk (Fig. 5–27). As formation of the ephyrae is completed, they break away from the oral end of the scyphistoma one by one.

After strobilization, the scyphistoma may resume its polypoid existence until the following year, when formation of ephyrae is repeated. A scyphistoma may live for one or several years.

The ephyra is almost microscopic, has a deeply incised bell margin, and has incompletely developed adult structures (Fig. 5–27). Ephyrae feed largely on small crustaceans, which are caught on the lappets and then wiped across the mouth and manubrium. An ephyra that has formed in the winter takes six months to two years to transform into a sexually reproducing adult medusa.

Modifications of the life cycle which avoid the necessity of a substratum are known. Some pelagic scyphomedusae, such as species of *Pelagia, Atolla,* and *Periphylla,* produce one adult from a single egg. In species of *Chrysaora* and *Cyanea* the larva is retained on the parent in cysts. Highly specialized cysts and scyphistomae occur in gastric brood chambers in the deep sea *Stygiomedusa.*

The larval nature of the scyphozoan polyp and its direct transformation into a medusa in some species would seem to support the belief that the medusoid form is primitive in cnidarians and that the polypoid form evolved as a larval adaptation. There are zoologists, such as Werner (1970), who take the opposite view and consider the scyphistoma to recapitulate the ancestral cnidarian condition.

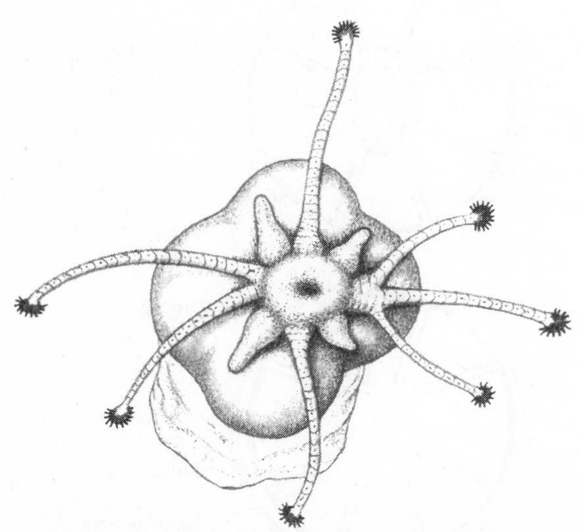

Figure 5–28. Scyphistome larva of the cubomedusan *Tripedalia,* which develops directly into a single ephyra, in contrast to the many ephyrae produced by such forms as *Aurelia* (polystrobilization). (After Werner, Cutress, and Studebaker.)

SYSTEMATIC RÉSUMÉ OF CLASS SCYPHOZOA

Order Stauromedusae, or Lucernariida. Sessile scyphozoans attached by a stalk on the aboral side of the trumpet-shaped body. Chiefly in cold littoral waters. *Haliclystus, Craterolophus, Lucernaria.*

Order Cubomedusae, or Carybdeida. Scyphomedusae with bells having four flattened sides and simple margins. Tropical and subtropical oceans. *Carybdea, Chiropsalmus, Chironex.*

Order Coronatae. Bell of medusa with a deep groove or constriction, the coronal groove, extending around the exumbrella. Many deep sea species. *Periphylla, Nausithoe, Linuche, Atolla.*

Order Semaeostomae. Scyphomedusae with bowl-shaped or saucer-shaped bells having scalloped margins. Gastrovascular cavity with radial canals or channels extending from central stomach to bell margin. Occur throughout the oceans of the world. *Cyanea, Pelagia, Aurelia, Chrysaora, Stygiomedusa.*

Order Rhizostomae. Bell of medusa lacking tentacles. Oral arms of manubrium branched and bearing deep folds into which food is passed. Folds, or "secondary mouths," lead into arm canals. Arm canals of manubrium pass into stomach. Original mouth lost through fusion of oral arm. Tropical and subtropical shallow water scyphozoans. *Cassiopeia, Rhizostoma, Mastigias, Stomalophus.*

CLASS ANTHOZOA

Anthozoans are either solitary or colonial polypoid coelenterates in which the medusoid stage is completely absent. Many familiar forms, such as the sea anemones, corals, sea fans, and sea pansies, are members of this class. This is the largest of the cnidarian classes and contains over 6000 species.

Although the anthozoans are polypoid, they differ considerably from hydrozoan polyps. The mouth leads into a tubular pharynx that extends more than half way into the gastrovascular cavity (Fig. 5–29). The pharynx is derived from invaginated ectoderm. The gastrovascular cavity is divided by longitudinal mesenteries, or septa, into radiating compartments, and the edges of the mesentery bear nematocysts. The gonads, as in the scyphozoans, are gastrodermal, and the fibrous mesoglea contains cells. The nematocysts, in contrast to those of hydrozoans and scyphozoans, do not possess an operculum, or lid. Rather, the anthozoan nematocyst has a three-part tip that folds back on expulsion.

In order to simplify the survey of this class, which is somewhat heterogeneous, the sea anemones, the stony corals, and the octocorallian corals are dealt with separately.

Sea Anemones. Sea anemones are solitary polyps and are considerably larger and heavier than the polyps of hydrozoans (Fig. 5–30). Most sea anemones range from 1.5 cm. to 5.0 cm. in length, with a diameter varying from the size of a dime to that of a half dollar; but there are species of *Goniactinia, Edwardsia,* and *Nematostella* which are as small as 4 mm. in diameter, and others which attain great size. Specimens of *Tealia columbiana* on the North Pacific coast of the United States and *Stoichactis* on the Great Barrier Reef of Australia may have a diameter of more than a meter at the oral end. Sea anemones are often brightly colored. They may be white, green, blue, orange, or red, or a combination of colors. Some are truly spectacular.

Sea anemones inhabit coastal waters throughout the world, but are particularly abundant in tropical oceans. They commonly live attached to rocks, shells, and submerged timbers; a few forms burrow in mud or sand. There are also some which live attached to jellyfish and sea jellies, and a number of species are commensal on the shells of hermit crabs. One species of crab transfers its anemone (*Calliactis*) when it moves to a new shell. An even more intimate relationship exists between the European hermit crab *Eupagurus prideauxi* and the sea anemone *Adamsia palliata*, for the anemone so envelopes the shell and extends the margin of the shell opening with its basal disc that the crab never has to seek a larger house (see review by Ross, 1967). The sea anemone provides some protection and camouflage for the crab, and perhaps in turn

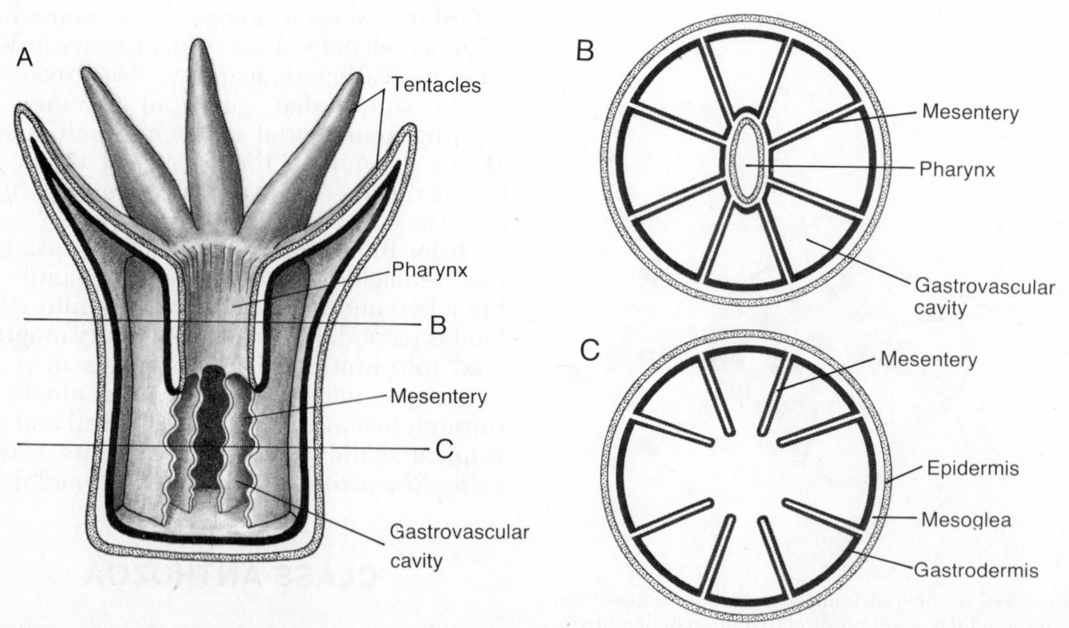

Figure 5–29. Structure of an anthozoan polyp. *A.* Longitudinal section. *B.* Cross section taken at level of pharynx. *C.* Cross section taken below level of pharynx.

Figure 5–30. Sea anemones. *A* and *B*. *Anthopleura,* a genus of common sea anemones found in shallow water along the Pacific coast of the United States. The column of the partially closed specimen in *B* is covered with shell fragments, which adhere to epidermal tubercles. (*A*, courtesy of Turtox News.) *C*. A West Indian species of *Actinia.* (*B* and *C* by Betty M. Barnes.)

obtains some food caught by the crab, or has the advantage of new food sources as it is moved about. Certain crabs attach anemones to their claws, which they then extend toward an intruder as a means of defense.

The major part of the sea anemone body is formed by a heavy column (Fig. 5–31). At the aboral end of the column there is a flattened pedal disc for attachment. At the oral end, the column flares slightly to form the oral disc, which bears eight to several hundred hollow tentacles. In the center of the oral disc is a slit-shaped mouth, bearing at one or both ends a ciliated groove called a siphonoglyph. The groove provides for the circulation of water into the gastrovascular cavity. The current of water perhaps functions to maintain the internal fluid, or hydrostatic, skeleton against which the muscular system can act. Also, the flow of water into the gastrovascular cavity may provide additional opportunity for the exchange of gases through the gastrodermal surface.

The slit-shaped mouth and siphonoglyphs

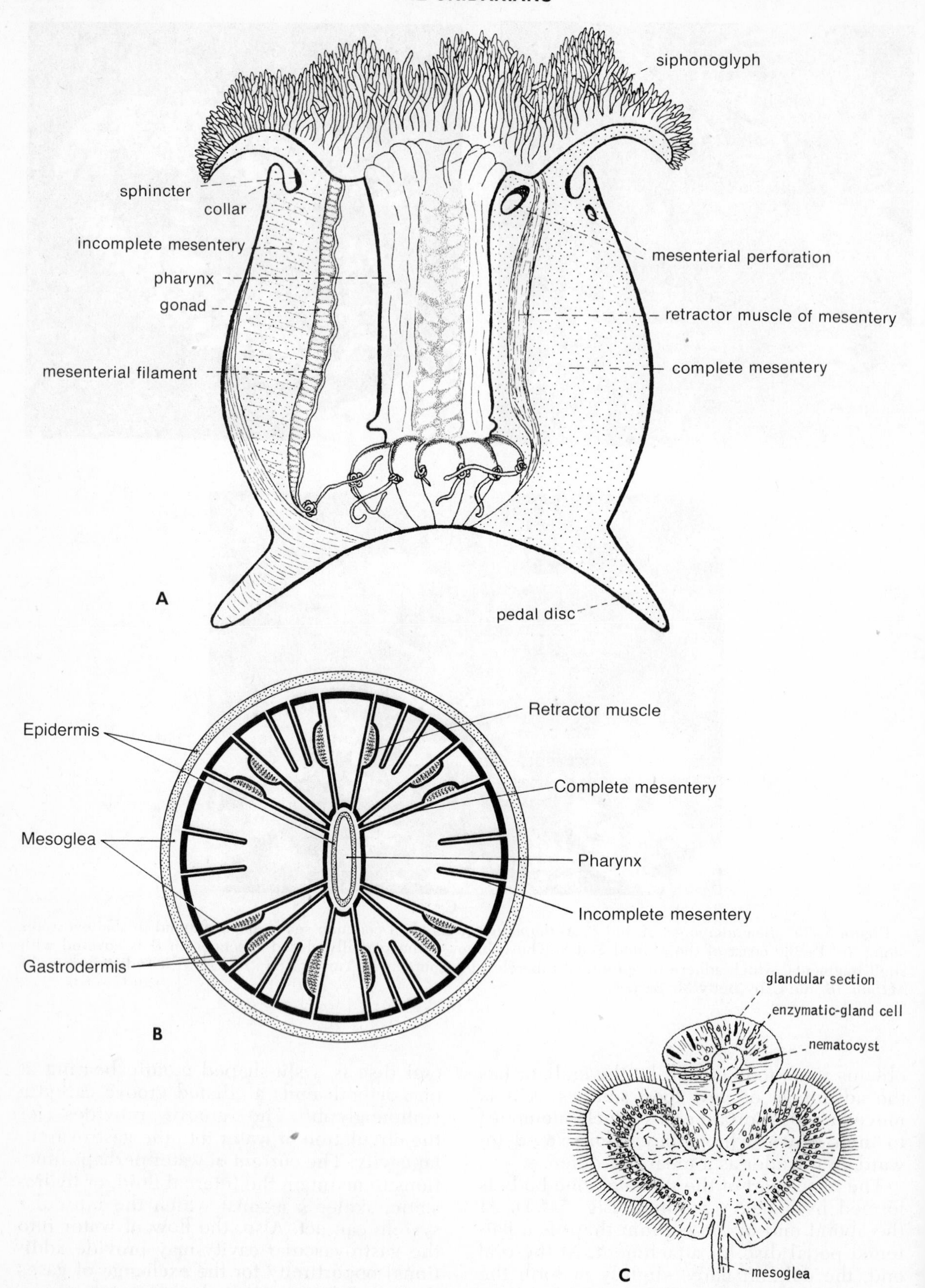

Figure 5-31. Structure of a sea anemone. *A*. Longitudinal section. *B*. Cross section at the level of the pharynx. *C*. Mesenterial filament of a sea anemone. (*A* and *C* after Hyman.)

impose a biradial or bilateral symmetry on sea anemones. They may be considered biradial when there are two siphonoglyphs, and bilateral when there is only one.

When a sea anemone contracts, the upper surface of the column is pulled over and covers the oral disc. In many sea anemones, including the familiar *Metridium* of the Atlantic coast, the column bears a circular fold at its junction with the oral disc. This fold is known as a collar, and it covers the oral surface on contraction of the animal (Fig. 5–31*A*).

The mouth leads into a flattened pharynx, which extends approximately two-thirds of the way into the column (Fig. 5–31). The wall of the pharynx, being derived from a true stomodeum, contains the same layers as the body wall. The inner side is covered by a ciliated epidermis, and the outer by gastrodermis. Between the two is a layer of mesoglea. The pharynx is kept closed and flat by the water pressure in the gastrovascular cavity. The siphonoglyph is kept open by an especially thick mesoglea and very heavy epidermal cells.

As in all anthozoans, the gastrovascular cavity of the sea anemone is partitioned by longitudinal, radiating mesenteries. In the sea anemones, there are usually two types of mesenteries, called complete and incomplete. Complete mesenteries are connected to the body wall on one side and to the wall of the pharynx on the other (Fig. 5–31*B*). Incomplete mesenteries are connected only to the body wall and extend only part way into the gastrovascular cavity. Each mesentery on one side has a mate of the same age on the opposite side, the two together forming a couple. The couples at each end of the tapered pharynx are called directives. The mesenteries usually occur in multiples of 12. When only 12 mesenteries are present, such as the primitive *Halcampoides*, they are complete and are called the primary cycle. The addition of a cycle of 12 secondary couples located between the primary ones brings the total to 24. A tertiary cycle brings the number to 48, and so on. The first two cycles may be composed of complete mesenteries; or only the first may be complete, with subsequent cycles incomplete and successively smaller. Many exceptions to the numerical symmetry just described exist. Moreover, asexual reproduction produces considerable irregularity, particularly in *Metridium*.

In the upper part of the pharyngeal region, the mesenteries are pierced by openings that facilitate water circulation (Fig. 5–31*A*). Below the pharynx, the complete mesenteries have free margins and recurve toward the body wall.

Histologically, each mesentery consists of two layers of gastrodermis separated by a layer of mesoglea. Both the gastrodermis and mesoglea are continuous with their corresponding layers in the body wall, and also in the pharynx when the mesenteries are complete.

The free edge of the mesentery is trilobed and is called a mesenterial filament (Fig. 5–31*C*). The mesenterial filament is longer than the mesentery and thus tends to be somewhat convoluted. The filament's lateral lobes are composed of ciliated cells and aid in water circulation. The middle lobe consists of nematocysts and enzymatic-gland cells. The middle lobe is present on all mesenteries below the pharynx. In some sea anemones, including *Metridium*, the middle lobe continues at the base of the mesentery as a thread called an acontium, which projects into the gastrovascular cavity.

Sea anemone histology is more or less typical. The epidermis may be ciliated, and in some species is covered with a cuticle. The mesoglea is much thicker than that of hydrozoan polyps and contains a large number of fibers, as well as amebocytes. The cnidocil of the cnidocyte is an actual flagellum and is motile.

Sea anemones feed on various invertebrates, and large species can capture fish. The prey is paralyzed by nematocysts, caught by the tentacles, and carried to the mouth. The mouth is opened by radial muscles in the mesenteries, and the prey is swallowed. The complete feeding response of tentacle contraction, mouth opening, and swallowing is activated by asparagine and reduced glutathione liberated by the prey.

The middle lobe of the mesenterial filaments, which may be in contact with the swallowed food, produce the enzymes for extracellular digestion. Not only proteins but also fats are digested extracellularly. These filaments also appear to be the principal site of intracellular digestion and absorption.

Some large sea anemones with short tentacles, such as *Stoichactis* and *Radianthus*, are actually suspension feeders. Planktonic organisms are trapped on the surface of the column and tentacles, where mucus and nematocysts undoubtedly play a role in their capture. Cilia on the surface of the column beat toward the oral disc, where similar ciliary current sweep the food to the tips of the tentacles. The tentacles then bend down and deposit the food in the mouth.

In the Red Sea and Indo-Pacific region, little fish of the genus *Amphiprion* live commensally among the tentacles of large sea anemones. The surface mucus of the fish apparently raises the threshold of nematocyst discharge, making it possible for the fish to live in an otherwise lethal habitat. The anemone provides protection and some food remains; the fish in turn protects the anemone from some predators, removes necrotic tissue, and by its swimming and ventilating movements reduces fouling of the anemone by sediment of various sorts (Mariscal, 1970).

Symbiotic zooxanthellae are found in the tissues of many sea anemones. The relationship is similar to that described for corals (p. 131).

The muscular system in sea anemones is much more specialized than that in the other two classes of cnidarians. The longitudinal, epidermal fibers of the column and pharynx have disappeared except in primitive species. They are present, however, in the tentacles and oral disc. Thus, the muscular system is largely gastrodermal. Bundles of longitudinal fibers in the mesenteries form retractor muscles for shortening the column (Fig. 5–31). Circular muscle fibers in the columnar gastrodermis are well developed. At the junction of the column and the oral disc, the circular fibers form a distinct sphincter in many species for covering the oral disc upon retraction of the tentacles (Fig. 5–31).

The presence of complete mesenteries may facilitate the function of the internal hydraulic skeleton by limiting the maximum extent of the diameter of the column when the retractor muscles contract. Although sea anemones are essentially sessile animals, many species are able to change locations. Movement may be accomplished by slow gliding on the pedal disc, by crawling on the side, or by walking on the tentacles. Burrowing sea anemones slowly plant the body column into sand or mud by peristaltic contractions which change the column diameter. The North American and European *Stomphia coccinea* attempts to escape certain predators by detaching and swimming. The swimming motion results from bending movements of the body. The direction is random. Some other sea anemones swim by a lashing movement of the tentacles. Members of the family Minyadidae are pelagic and float upside down by means of a gas bubble enclosed within the folded pedal disc.

The nervous system exhibits the typical cnidarian pattern and is synaptic. No specialized sense organs are present.

Asexual reproduction is not uncommon in sea anemones. One method is by pedal laceration, in which parts of the pedal disc are left behind as the animal moves. In some instances the disc puts out lobes that pinch off. These detached portions then regenerate into small sea anemones. Many sea anemones reproduce asexually by longitudinal fission, and a few species do so by transverse fission.

Sea anemones may be hermaphroditic or dioecious. The gonads are located in the gastrodermis of all or certain of the mesenteries in the form of longitudinal band-like cushions behind the mesenterial filament (Fig. 5–31*A*). In hermaphroditic species, eggs and sperm are produced at different times from the same gonads. This is known as protandry and is common among invertebrates.

The eggs may be fertilized in the gastrovascular cavity, with development taking place in the mesenterial chambers; or fertilization may occur outside the body in the sea water. Cleavage is either equal or unequal, and a coeloblastula is formed in both cases. The blastula undergoes gastrulation by ingression or invagination to form a typical planula larva. A pharynx forms as a stomodeal invagination. Mesenteries develop from the column wall and grow toward the pharynx. There are still no tentacles, and the young sea anemone lives as a ciliated ball, unattached and free-swimming (Fig. 5–32). With further development, the "larva" settles, attaches, and forms tentacles.

Stony, or Scleractinian, Corals. Closely related to sea anemones are the stony, or scleractinian, corals (also called madreporarian corals). Unlike the sea anemone, the stony coral produces a calcium carbonate skeleton. Some corals are solitary with polyps reaching 25 cm. in diameter, but the majority are colonial with very small polyps averaging 1 to 3 mm. in diameter. However, the entire colony can become very large. Coral polyps are very similar in structure to sea anemones but do not possess siphonoglyphs (Figs. 5–33 and 5–36*D*). The mesenterial filaments contain only one glandular lobe with nematocysts. These filaments project into the central portion of the gastrovascular cavity and may even project from the mouth when feeding upon large prey.

Expansion and feeding typically occur at night, and corals employ the same methods of obtaining food as sea anemones—both raptorial and suspension feeding. Debris is

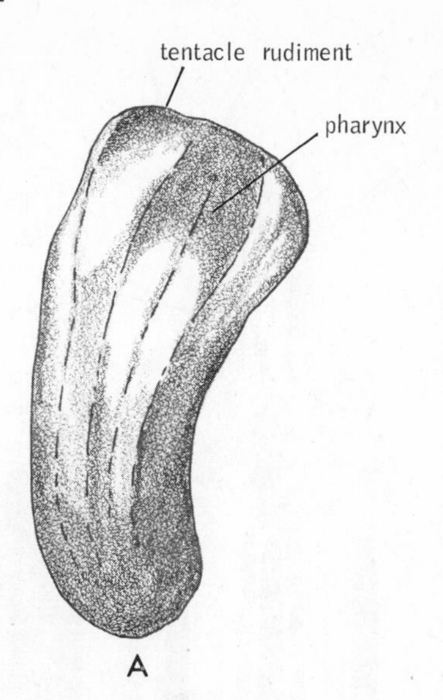

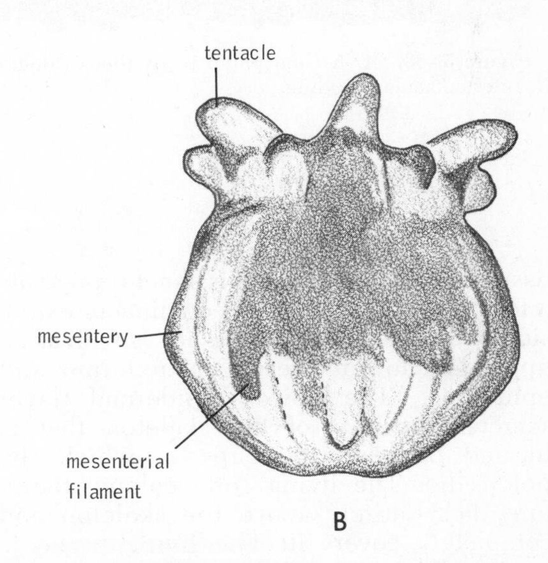

Figure 5-32. Developing sea anemone, *Actinia equina*. *A*. A stage in which mesenteries and pharynx have formed but tentacles not yet visible. *B*. A later stage showing primary tentacles. (After Chia, F.-S. and Rostron, M., 1970: Jour. Marine Biol. Assoc. U.K., *50*:253-264.)

removed by special downward ciliated currents on the column.

The skeleton is composed of calcium carbonate crystals and is secreted by the epidermis of the lower half of the column as well as by the basal disc. This secreting process produces a skeletal cup, within which the polyp is immovably fixed. The cup is termed the calyx; the surrounding walls of the cup, the theca; and the floor of the cup, the tabla. The floor contains thin radiating calcareous septa (Figs. 5-33*A* and 5-34). Each scleroseptum projects upward into the base of the polyp, folding the basal layers and inserting them between a pair of mesenteries. As long as a colony is alive, calcium carbonate is deposited beneath the living tissues.

In addition to providing a substratum on which the living polyp can attach, the skelton (and especially the sclerosepta) also serves for protection. When contracted, the polyps project little above the skeletal platform and are difficult for most fish and other predators to remove.

The polyps of colonial corals are all interconnected, but the attachment is lateral rather than aboral as in hydroids. The column wall folds outward above the skeletal cup and connects with similar folds of adjacent polyps. Thus, all the members of the colony are connected by a horizontal sheet of

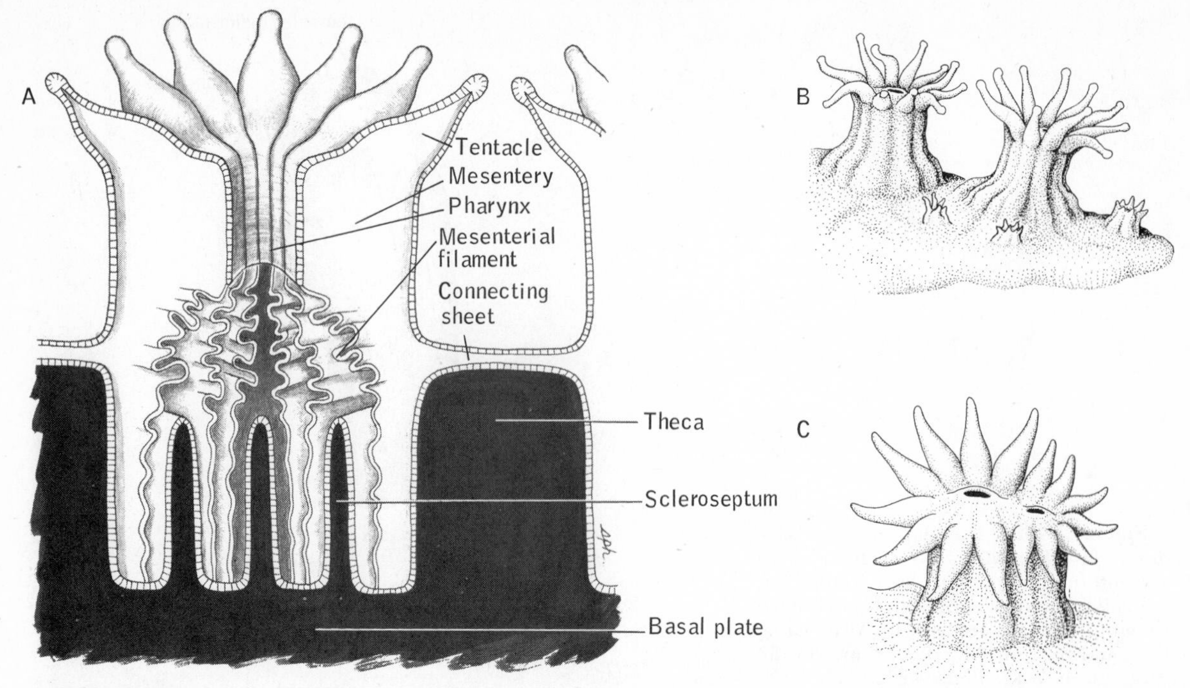

Figure 5–33. *A*. A coral polyp in its theca (longitudinal section). (After Hyman.) *B*. Extratentacular budding. *C*. Intertentacular budding.

tissue (Fig. 5–33). Since this sheet represents a fold of the body wall, it contains an extension of the gastrovascular cavity as well as an upper and lower layer of gastrodermis and epidermis. The lower epidermal layer secretes the part of the skeleton that is located between the cups in which the polyps lie. The living coral colony, therefore, lies entirely above the skeleton and completely covers it. The living tissue is commonly straw-colored because of the presence of symbiotic zooxanthellae.

The skeletal configurations of various species of corals are due in part to the growth pattern of the colony and in part to the arrangement of polyps in the colony (Figs. 5–34 and 5–36). Some species form flat or rounded skeletal masses, while others have an upright and branching growth form. When the polyps are well separated from each other, the coral skeleton has a pitted appearance as in the eyed coral, *Oculina*, *Astrangia*, one of the corals living along the North Atlantic coast, and the reef coral *Monastrea* (Fig. 5–34*C*). The polyps of brain coral are arranged in rows (Figs. 5–34*A* and *B*; 5–35; and 5–36*A*). The rows are well separated, but the polyps comprising a row are so close together that their thecae are confluent. As a result, the skeleton of the colony has the appearance of a human brain, containing troughs or valleys separated by skeletal ridges.

The coral colony expands in size by the budding of new polyps. Depending on the species, budding may be extratentacular, in which the buds arise from the base of old polyps (Fig. 5–33*B*), or intratentacular, in which the buds arise from the oral discs of old polyps. In the latter case, the oral disc of the parent lengthens in one direction (Fig. 5–33*C*). Gradually, the oral disc constricts and the separation extends down the length of the column to form two new polyps. The budding process is accompanied by simultaneous changes in the deposition of the underlying sclerosepta. As might be expected, brain corals arise by intertentacular budding, in which the oral discs and columns never constrict after new mouths are formed. Thus, a row of polyps in brain coral share a common oral disc bearing many mouths (Fig. 5–34*A*).

Sexual reproduction is similar to that in the sea anemones, and there are both dioecious and hermaphroditic species. The single polyp,

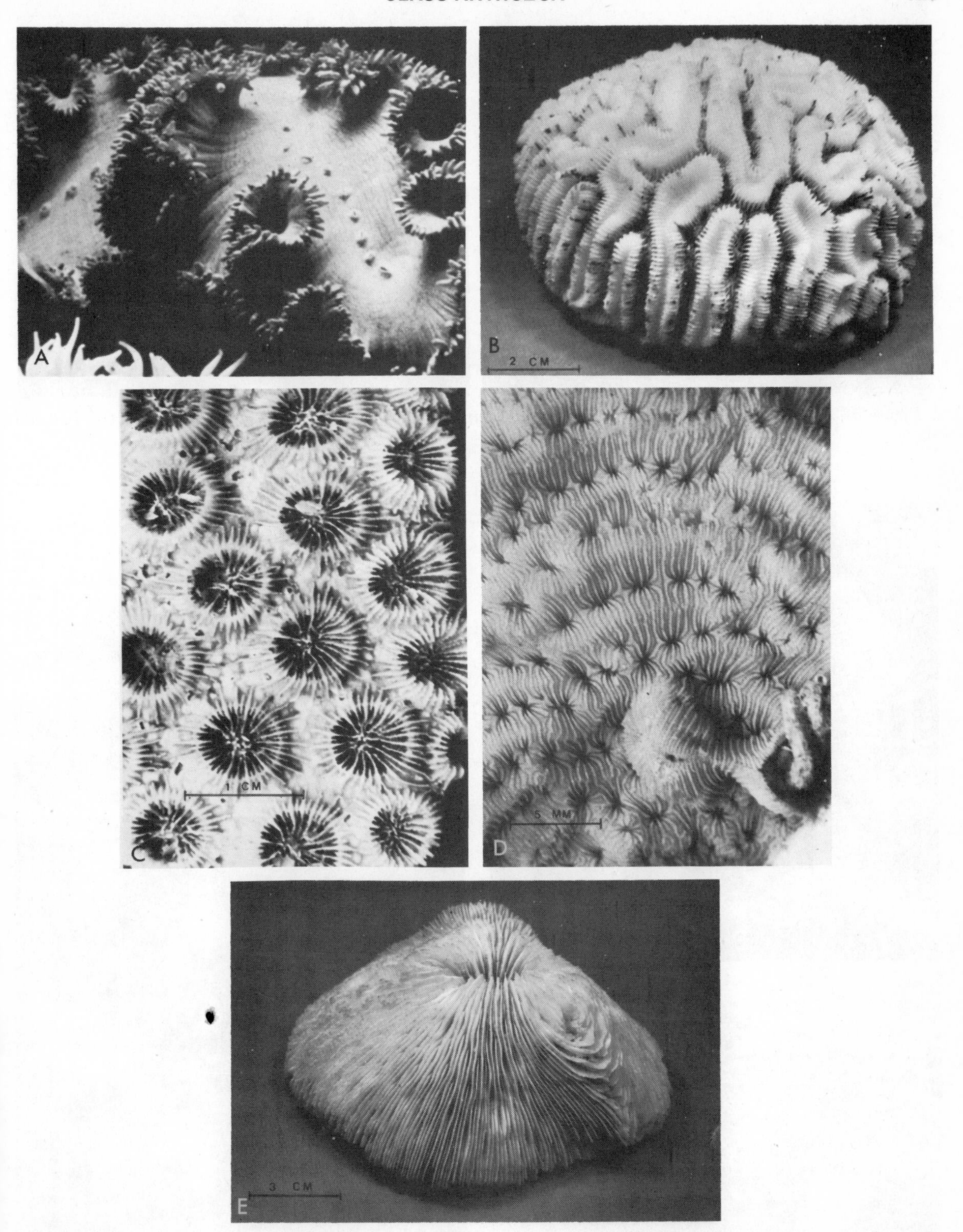

Figure 5–34. *A.* Oral surface of a living brain coral. Note the row of mouths. (By Catala, R. L., 1964: Carnival under the sea. R. Sicard, Paris. ©R. L. Catala.) *B.* Skeleton of the brain coral *Diploria.* Note the arrangement of sclerosepta. *C.* Skeleton of *Monastrea* with large distinct cups, in which the polyps are located in life. *D.* Skeleton of lettuce coral, *Agaricia,* in which the polyps are arranged in rows. *E.* Skeleton of *Fungia,* a solitary coral of the Indo-Pacific. The skeleton of this very large polyp is limited to sclerosepta projecting from the tabla. There is no wall, hence no cup. (*B–E* by Betty M. Barnes.)

Figure 5–35. A living specimen of the brain coral *Diploria* during the day (*A*) and at night (*B*). (By Betty M. Barnes.)

Figure 5–36. Scleractinian, or stony, corals. *A.* Brain coral on the Great Barrier Reef. (Courtesy of Fritz Goro.) *B.* Staghorn coral from the Great Barrier Reef. (By Allen Keast, courtesy of Buchsbaum, R. M., and Milne, L. J., 1961: The Lower Animals. Chanticleer Press, New York.) *C.* Knobbed and lettuce coral on a Bahamian reef. (By John Storr.) *D.* Cup corals. (By D. P. Wilson.)

which is produced by sexual reproduction, attaches and (by asexual budding) becomes the parent of all other members of the colony.

Octocorallian Corals. Sea anemones and corals, because of their structural similarities, are grouped in the subclass Zoantharia. The remaining anthozoans, including many common marine forms such as sea pens, sea pansies, sea fans, whip corals, and pipe corals, form the subclass Octocorallia, or Alcyonaria. The Octocorallia are similar to the Zoantharia in general structure but possess a number of different and distinctive features. Octocorallians always possess eight tentacles, and these are pinnate—that is, they possess side branches as does a feather (Figs. 5–37A and 5–38). Correlated with the number of the tentacles, there are always eight complete mesenteries. Only one siphonoglyph is present.

The octocorallians are colonial cnidarians, and the polyps are rather small, similar to those of stony corals. The polyps (Fig. 5–38) of an octocorallian colony are connected by a mass of tissue called coenenchyme. This consists of a thick mass of mesoglea, which is perforated by gastrodermal tubes that are continuous with the gastrovascular cavities of the polyps. The surface of the entire fleshy mass is covered by a layer of epidermis,

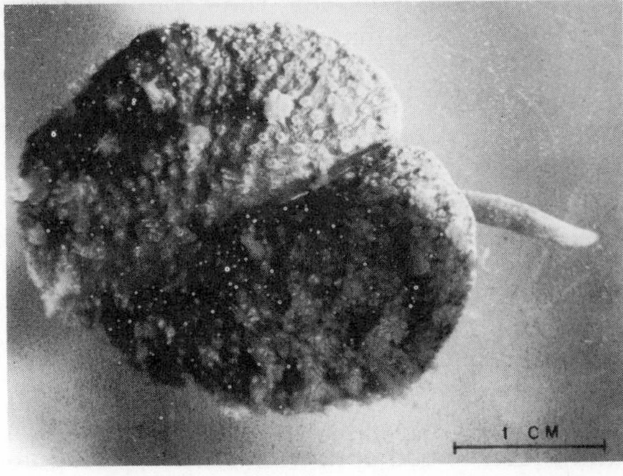

Figure 5–37. Octocorallians. *A.* Polyps of a sea pen. (Courtesy of Turtox News.) *B.* Sea fan. (Courtesy of the American Museum of Natural History.) *C.* Sea rods, the most common gorgonian growth form. (Courtesy of T. Parkinson.) *D.* Skeletal tubes of organ pipe coral. *E.* Sea pansy, *Renilla*, a soft bottom octocorallian. The flat fleshy colony, which is anchored into the substratum by a short stalk, represents the expanded body of a primary polyp. On the surface of this primary polyp arise a large number of secondary polyps. (*E* by Betty M. Barnes.)

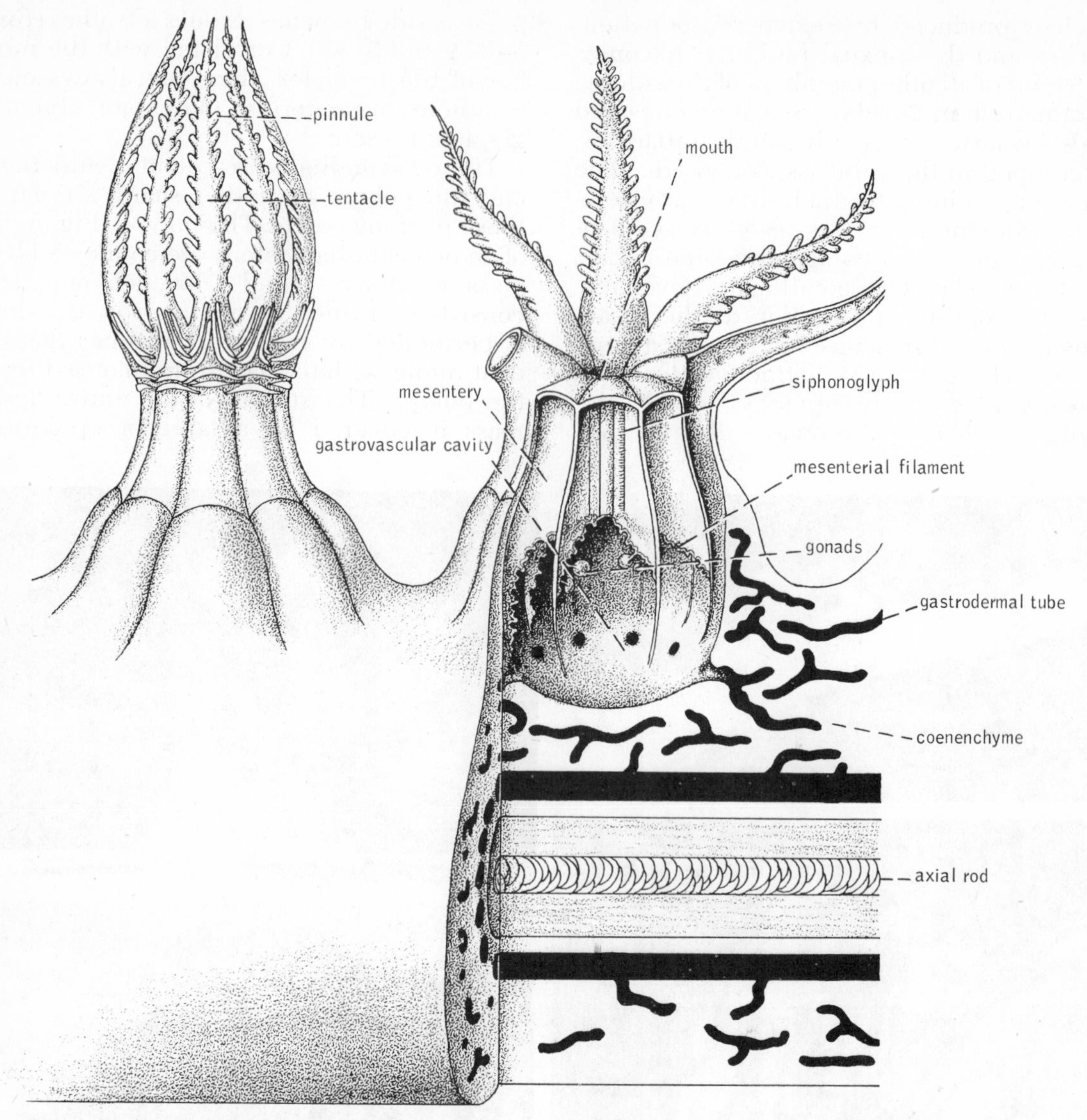

Figure 5–38. Structure of a gorgonian coral of the anthozoan subclass Octocorallia. (After R. C. Moore.)

which joins the epidermis of the polyp column. Only the upper portion of the polyp projects above the coenenchyme.

The amebocytes of the coenenchyme secrete skeletal material that supports the colony. Thus, the skeleton of the Octocorallia is internal and is an integral part of the tissue. This arrangement is in sharp contrast to that of the stony corals, in which the skeleton is entirely external. The octocorallian skeleton may be composed of separate or fused calcareous spicules, or of a horny material.

Among the most familiar of the octocorallians are the gorgonian, or horny, corals (order Gorgonacea) (Fig. 5–37), which include the whip corals, sea feathers, sea fans, and precious red coral (*Corallium*). Gorgonians are common and conspicuous members of reef faunas (Fig. 5–39*B*). The body of most gorgonian corals contains a central axial rod composed of an organic substance called gorgonin. Around the axis is a cylinder of coenenchyme and polyps (Fig. 5–38). The coenenchyme contains embedded calcareous ossicles or spicules of different shapes and colors. It is the color of the calcareous skeletal components that accounts for the yellow, orange, or lavender color of some species. The yellow-brown color of many reef species results from the presence of symbiotic zooxanthellae. Precious red coral, *Corallium*, which is found primarily in the Mediterranean and off the coast of Japan, is used in the making of jewelry. In this genus the central organic gorgonin is replaced by a solid axis of fused red calcareous spicules. The colonies of most gorgonian corals are erect

Figure 5–39. *A*. Exposed coral on the Great Barrier Reef off the coast of northeast Australia. Note different growth forms of various species present. (Courtesy of the Australian News and Information Bureau.) *B*. Photograph of the reef at Bermuda. The large hemisphere at the left is a specimen of the brain coral *Diploria*. The many plant-like organisms are gorgonian corals. (By C. Gebelein.)

branching rods and are thus rather plant-like (Fig. 5–37*C*). Whip coral consists of slightly branched, long cylindrical filaments about the diameter of drinking straws. Sea feathers are more highly branched, as the name implies, and sea fans are composed of filaments oriented in one plane, connected by cross bars to form a lattice (Fig. 5–37*B*). In general, the design of gorgonian corals requires a small amount of surface area for attachment but provides a great surface area for feeding, as a consequence of the branching vertical development. The flexible skeleton permits bending in currents.

The tropical Indo-Pacific organ pipe coral, *Tubipora,* which belongs to another group of octocorallians, is differently organized (Fig. 5–37*D*). Here long parallel polyps are encased in calcareous tubes of fused spicules. The tubes are connected at intervals by transverse calcareous plates, or platforms.

The sea pens and sea pansies (order Pennatulacea) are inhabitants of soft bottoms and are quite different from the coral-like members of the Octocorallia. There is a large primary polyp with a stemlike base, which is anchored in the sand. The body is fleshy, although the coenenchyme contains minute calcareous spicules, or ossicles. The upper part of this primary polyp gives rise to secondary dimorphic polyps. The more typical and conspicuous of these secondary polyps are termed autozooids. In the sea pens, the primary polyp is elongate and cylindrical. In the sea pansy, *Renilla* (Fig. 5–37*E*), which is common along the Atlantic coast and the southern California coast of the United States, the primary polyp is leaflike, with the secondary polyps limited to the upper surface. There are deep sea species. *Umbellula,* which lives on the Atlantic abyssal plain, looks like a pinwheel (secondary polyps) mounted at the end of a long stick (primary polyp) stuck in the bottom (Fig. 5–40).

Figure 5–40. *Umbellula,* a strange deep sea octocorallian related to sea pansies. Photograph was taken at a depth of over 5000 meters some 350 miles off the west coast of Africa. Stalk is estimated to be about a meter in length. (By Walter Jahn, U.S. Naval Oceanographic Service.)

Members of the order Alcyonacea are known as soft corals or leather corals. These strange octocorallians possess soft fleshy or leathery colonies that may reach a large size, one meter in diameter in the case of *Sarcophyton.* The colonies are irregular in shape, with lobate or finger-like projections. Separate calcareous spicules are embedded in the coenenchyme.

Fossil Corals, Reef Corals, and Coral Reefs. The cnidarians are represented in the fossil record largely by the corals. The more than 6000 described fossil species belong to three principal groups: the existing order Scleractinia; the wholly extinct order Rugosa; and the subclass Tabulata. The rugose corals were similar to scleractinian corals but were largely solitary and hornshaped with a ridged surface, and had conspicuous major and minor sclerosepta (Fig. 5–41*A*). The tabulate corals were colonial and housed in external skeletal tubes connected together in a variety of ways (Fig. 5–41*B* and *C*). The base of the tube rested on a transverse platform, called a tabula, but sclerosepta were absent or poorly developed.

The rugose and tabulate corals were inhabitants of the Paleozoic seas where they contributed to the formation of great reefs, especially during the Silurian period. The remains of a great Silurian reef stretches as a belt from Ontario and northern Michigan south by way of Milwaukee and Chicago into Ohio. The formation is 50 miles wide in some places in Indiana.

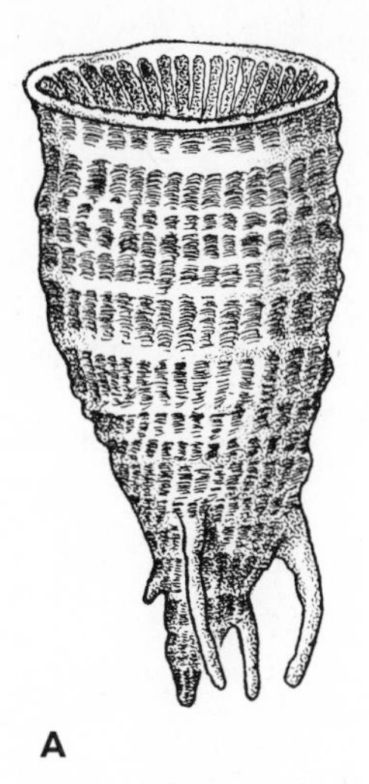

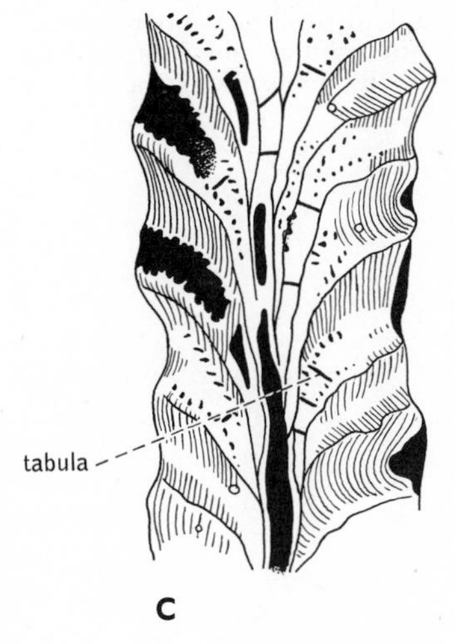

Figure 5–41. *A.* A rugose coral. (After Orbigny.) *B.* A fossil tabulate coral. (See also Fig. 19–62). (By Betty M. Barnes.) *C.* Longitudinal section of a tabulate coral. (After Rominger.)

Scleractinian corals replaced the rugose and tabulate corals at the end of the Paleozoic. They live in temperate, arctic, and antarctic waters, but the richest faunas inhabit tropical seas. It is in warm oceans that the reef-forming, or hermatypic, corals are found. In the West Indies there are some 40 reef-building species; there are over 200 species contributing to the Great Barrier Reef of Australia.

Many species of corals need relatively shallow water and do not live at depths below the light penetration level. This is true of all of the reef-building species, most of which have a vertical distribution of 90 meters or less below the water surface (Fig. 5–42). This vertical restriction is imposed by the symbiotic zooxanthellae (dinoflagellates in the palmella state) that live in the tissues of the corals. The zooxanthellae require light energy for photosynthesis, and the corals cannot exist without the zooxanthellae. Only ahermatypic corals which lack zooxanthellae are found in deep cold waters.

Our knowledge of the physiological relationship between hermatypic corals and their symbiotic algae has grown considerably in recent years. The nutritive needs of the coral are supplied in part by the planktonic animals upon which it feeds, and in part by its algal symbionts. A significant portion of the carbon fixed by the algae in photosynthesis is passed to the coral, largely in the form of glycerol, but including glucose and alanine. The food caught by the coral probably supplies both coral and algae with nitrogen and phosphorus, which is then cycled back and forth between the two (Lewis and Smith, 1971). Although the nutritive dependence of the coral on the algae has been demonstrated experimentally (Franzis-

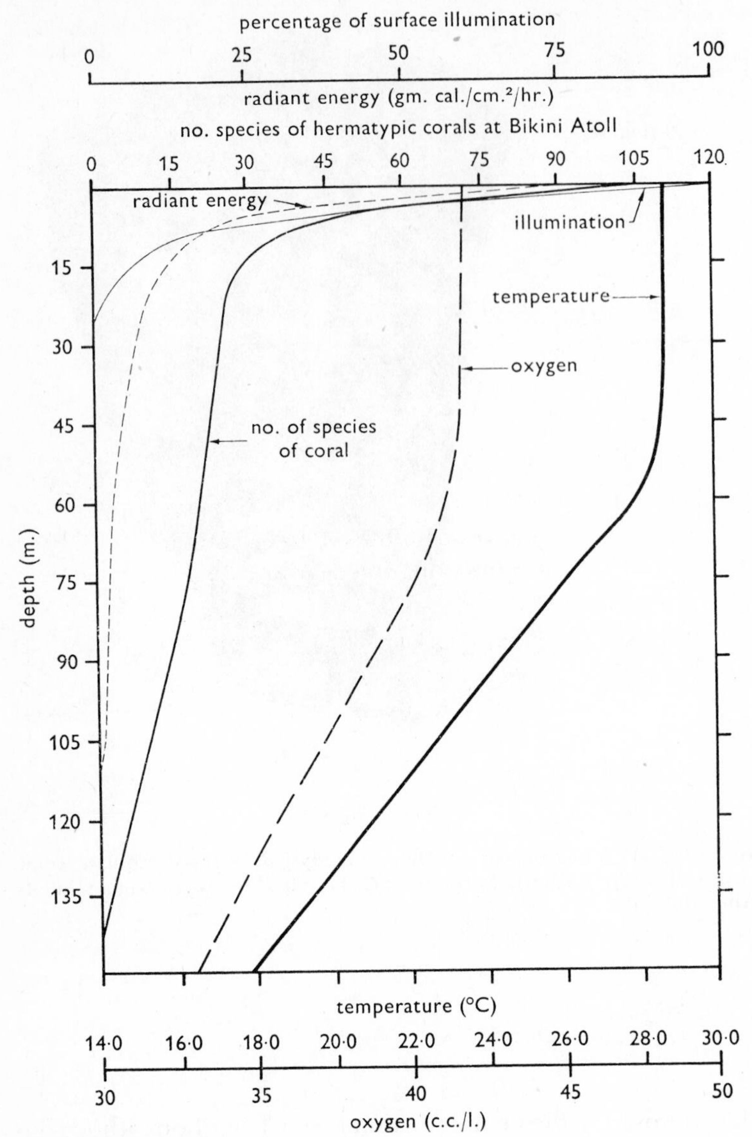

Figure 5–42. Relationship between the vertical distribution of hermatypic corals at Bikini Atoll and environmental factors. (After Wells and Motoda from Stoddart, D. R., 1969: Biological Review *44*:433–498.)

ket, 1970), the degree of dependence probably varies from species to species and perhaps with locality. Mid-ocean Pacific reef corals surrounded by relatively barren water may be especially dependent upon their algal symbionts.

The symbiosis also facilitates the deposition of the coral skeleton, for corals that are deprived of their algae or kept in the dark deposit calcium carbonate at a much slower rate than under normal conditions. Algal photosynthesis may increase calcium carbonate production by removing carbon dioxide and driving the following reaction to the right:

$$Ca(HCO_3)_2 \rightleftharpoons CaCO_3\downarrow + H_2CO_3 \rightleftharpoons H_2O + CO_2$$

Dome-shaped or encrusting coral colonies deposit about 1 to 2 cm. of $CaCO_3$ per year, and branching species may add as much as 10 cm. to the ends of the branches. An entire reef has been estimated to accumulate about 2.5 cm. per year in calcareous deposits, but this rate could fluctuate greatly depending upon conditions. For example, a species of the brain coral *Diploria* grows six times faster in areas of the West Indies than it does in the more northerly latitude of Bermuda, where the annual radial accumulation is 3 to 4 mm.

Although scleractinian corals are the principal contributors to reef structure, additions are made by hydrozoan corals, calcareous algae, bivalves, and other organisms. Gorgonian corals are abundant in reef communities and possess symbiotic zooxanthellae, but their skeletons are not important contributors to reef structure.

The coralline rock that underlies the reef surface is typically a conglomerate, and its formation is a more complex process than the simple accumulation of layers of calcium carbonate deposited by corals and other reef organisms. The large masses of skeletal material produced by corals are eventually penetrated and broken up by boring organisms, such as sponges and bivalves. The resulting large pieces of rubble constitute the "bricks" of coralline rock. Encrusting depositors and small calcareous debris and sediment fill the intervening spaces to compose the mortar. Gradually at lower levels it all becomes cemented together as a solid mass of coralline conglomerate (Scoffin, 1972).

The structural reef contains a great variety of habitats—deep slopes, grottos, crevices, holes, shallow pools, and wave-swept channels—which support a reef community of hundreds of species of animals. The entire association is probably an autotrophic system; that is, the reef is maintained by the radiant energy transformed by its photosynthetic members (Stoddart, 1969; Odum, 1971).

The majority of the existing reefs and reef-based islands in the world today developed during the Cenozoic era and are most widespread in the Indian Ocean, southeast Asia, and South Pacific areas, and in the West Indies. They constitute the contracted remains of much more widely distributed reef formations of the early Cenozoic.

The formation of coral islands, atolls, and fringing reef have usually resulted from the subsidence of the underlying geological formation upon which the reef rests. The many common features of zonation that such reefs exhibit can be illustrated by the Pacific atoll of Eniwetok, a broken island ring surrounding an interior lagoon approximately 20 miles across (Fig. 5–43). The outer, or seaward, side of the reef slopes steeply, and on the slope live the great massive corals, which are major contributors to the reef structure. Above the slope there is a terrace, a fore-reef slope, and a ridge-like reef front. The terrace and the fore-reef slope may be densely populated with coralline algae (but not in the West Indian reefs). The reef front, which is dominated by encrusting corals, is cut through by channels formed by the surging surf. Behind the reef front is an extensive reef flat consisting of several zones and often containing a rich and varied fauna. Here, in quiet pools, are found some of the most delicate corals. The inner side of the reef flat is flanked by a vegetated emergent zone that slopes gently into the interior lagoon.

Core drillings taken on Eniwetok Atoll provide evidence for reconstructing the geologic history of the reef. The drillings extend downward through 1283 m. (almost a mile) of coral rock before striking a basaltic base. This base marks the top of a volcanic cone that rises 3.2 km. (almost two miles) from the ocean bottom. Analysis of the core samples obtained from the drillings reveals strata containing recrystallized sediments and fossils of land plants. The oldest and deepest sediments date from the beginning of the Cenozoic era.

At the dawn of the Cenozoic era, Eniwetok was a volcanic sea mount whose peak emerged above the ocean as an island. Around the perimeter of the island developed a coral reef, called a fringing reef. Gradually during the course of the Cenozoic era, the emergent volcanic cone subsided. The fringing reef continued to grow upward, keeping pace with subsidence and forming a vertical wall resting on the volcanic basaltic base. Within the coral walls the present lagoon of the atoll filled the area left behind by the submerged cone. The extent of submergence can be measured by the 1283 m. height of the reef revealed by drillings. But submergence was not even. The Cenozoic era was punctuated by fluctuations in sea level caused in part by periods of glaciation. When the sea level fell, the reef was an emergent island ring on which vegetation developed, erosion occurred, and the level of most active coral growth shifted. These changes are recorded in the pollen and fossil vegetation embedded in the coral walls of the existing atoll and the submergent erosion terraces which ring its outer perimeter. With rise in sea level, the scene changed to a wave-washed reef ring. The last fluctuation in sea level was a 2 m. fall, which resulted in the emergent atoll of the present Eniwetok.

Many examples of fringing reefs and atolls are found in areas where reef-forming corals are present. They vary in degree of development, and the nature of the reef or associated island may exhibit some differences from the history described for Eniwetok. Bermuda, which is also resting upon a sunken volcano, is only 171 m. above basaltic

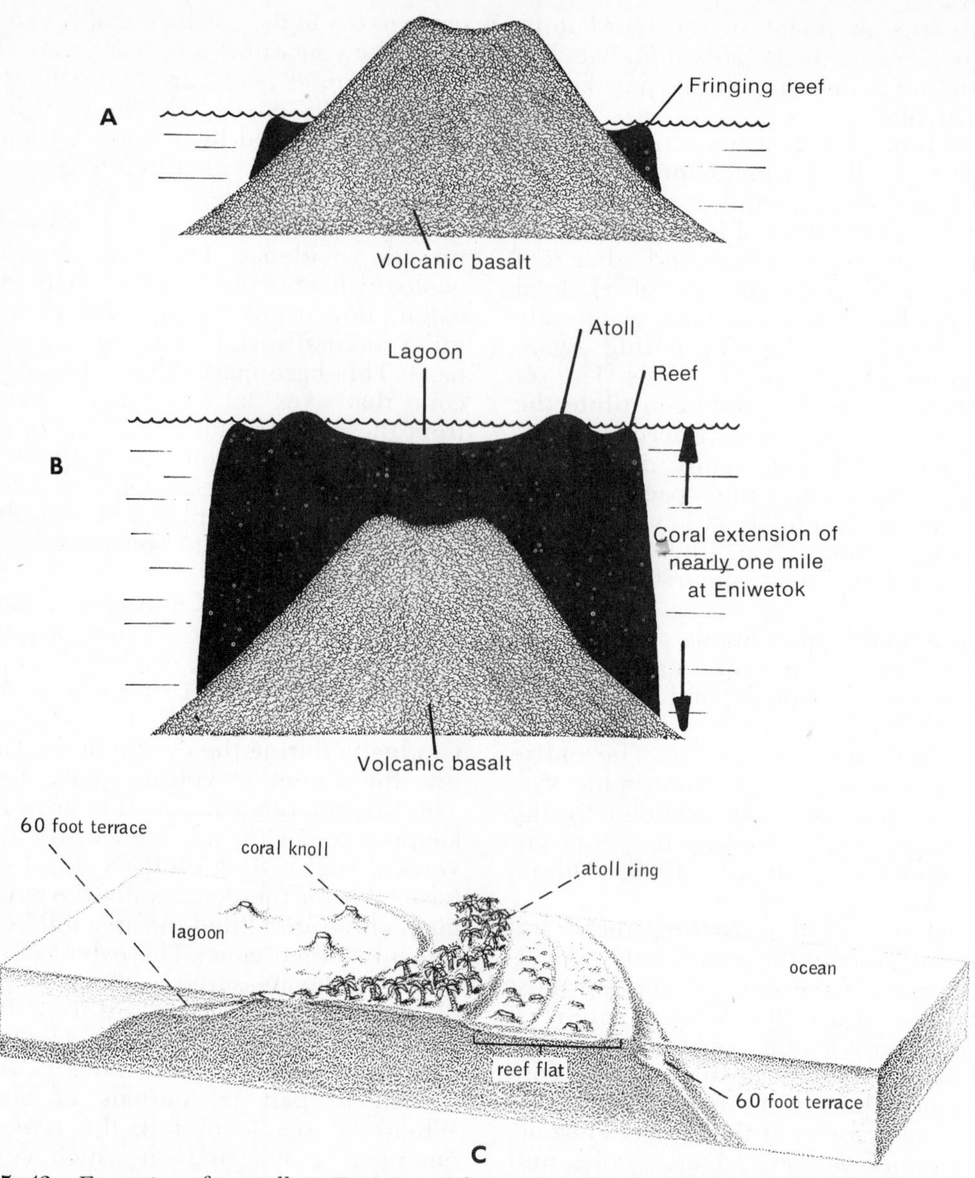

Figure 5–43. Formation of an atoll. *A.* Fringing reef around an emergent volcano. *B.* Continuous deposition of coral as volcanic cone subsides leads to the formation of a great coralline cap; emergent part of cap is atoll. *C.* Section through part of atoll. (After Ladd.)

rock. Drillings on the Bahamas pass through an upper 161 m. of surface limestone and then through 4488 m. of dolomite, the bottom of which dates to the Cretaceous. Large submarine caverns with stalactites attest to the sea level fluctuations in the Bahamas.

Barrier reefs have developed along certain coastlines, the most notable example being the Great Barrier Reef of Australia. This great reef extends for 1200 miles along the east coast of that continent, and the channel lagoon which runs between the reef and mainland ranges in width from 10 to 150 miles. A core drilling on Heron Island, near the southern end of the reef, passes through 154 m. of limestone before striking quartz sandstone.

The biology of reef corals and the ecology of reefs have received much attention in recent years, and there is now a large body of literature devoted to the subject. Those interested in additional information should refer to the extensive reviews by Yonge (1968) and by Stoddart (1969), or to the briefer summary by Odum (1971).

SYSTEMATIC RÉSUMÉ OF CLASS ANTHOZOA

Subclass Octocorallia, or Alcyonaria. Polyp with eight pinnate tentacles. Almost entirely colonial.

Order Stolonifera. No coenenchymal mass; polyps arising from a creeping mat or stolon. Skeleton of calcareous tubes or separate calcareous spicules, or skeleton lacking. Tropical and temperate oceans in shallow water. *Tubipora* (organ-pipe coral), *Clavularia*.

Order Telestacea. Lateral polyps on simple or branched stems. Skeleton of calcareous spicules. *Telesto*.

Order Alcyonacea. Soft corals. Coenenchyme forming a rubbery mass and colony having a massive mushroom shape, or a variously lobate growth form. Skeleton of separate calcareous spicules. Largely tropical. *Alcyonium*, *Gersemia*.

Order Coenothecalia. Contains only the Indo-Pacific blue coral, *Heliopora*, having a massive calcareous skeleton.

Order Gorgonacea. Horny corals or gorgonian corals. Common tropical and subtropical octocorallian cnidarians having a largely upright, plantlike growth form and a skeleton of a horny organic material. Separate or fused calcareous spicules may also be present. *Gorgonia* (sea fan), *Leptogorgia* (sea whip), *Corallium* (precious red coral).

Order Pennatulacea. Sea pens. Colony having a fleshy, flattened or elongate body, or rachis. Skeleton of calcareous spicules.

Figure 5–44. *A*. *Palythoa*, an anemone-like anthozoan belonging to the order Zoanthidea. In the tropics, species of *Palythoa* commonly carpet rocks in shallow water. (By Betty M. Barnes.) *B*. *Cerianthus*, a large burrowing anthozoan. The animal secretes a soft mucous tube, into which the body can be retracted. The members of this order (Ceriantharia) are similar to sea anemones in size and general appearance. (Courtesy of Buchsbaum, R. M. and Milne, L. J., 1961: The Lower Animals. Chanticleer Press, N.Y.)

Stylatula, Veretillum, Renilla (sea pansy), *Umbellula.*

Subclass Zoantharia. Polyps with more than eight tentacles, and tentacles rarely pinnate. Solitary or colonial.

Order Zoanthidea. Small anemone-like anthozoans having one siphonoglyph and no skeleton. Solitary or colonial. *Palythoa* (Fig. 5–44). Some, such as *Epizoanthus* and *Parazoanthus*, epizoic on other invertebrates.

Order Actiniaria. Sea anemones. Solitary anthozoans with no skeleton, with mesenteries in hexamerous cycles, and usually with two siphonoglyphs. *Halcampoides, Edwardsia, Metridium, Epiactis, Stoichactis.*

Order Scleractinia, or Madreporaria. Stony corals. Mostly colonial anthozoans secreting a heavy external calcareous skeleton. Sclerosepta arranged in hexamerous cycles. *Fungia, Acropora, Porites, Astrangia, Oculina.* Many fossil species.

Order Rugosa, or Tetracoralla. An extinct order of mostly solitary corals possessing a system of major and minor radiating sclerosepta. Cambrian to Permian. *Zaphrentis.*

Order Corallimorpharia. Tentacles radially arranged. Resemble true corals but lack skeletons. *Corynactis.*

Order Ceriantharia. Anemone-like anthozoans with greatly elongate bodies adapted for living within secreted mucous tubes buried in sand or mud. One siphonoglyph; mesenteries all complete. *Cerianthus* (Fig. 5–44).

Order Antipatharia. Black or horny corals. Gorgonian-like species with upright, plantlike colonies. Polyps arranged around an axial skeleton composed of a black thorny material and bearing thorns. Largely in deep water in tropics. *Antipathes.*

Subclass Tabulata. Extinct colonial anthozoans with heavy calcareous skeletal tubes containing horizontal platforms, or tabulae, on which the polyps rested. Sclerosepta absent or poorly developed. *Favosites, Halysites.*

CNIDARIAN PHYLOGENY

The evolutionary relationships of the major cnidarian groups is still subject to much debate. The arguments center around the basic premises of Brooks' theory. Was the ancestral cnidarian most like a medusa or a polyp? Among the many recent and contemporary zoologists who support Brooks' view, Hyman (1940), Hand (1959, 1966), Rees (1966), and Swedmark and Teissier (1966) have written most extensively on the subject. All view the polyp as a secondarily derived form representing a development of the larval stage. Hand considers the anthozoans as being derived from the polyp of the scyphozoans. Rees, Swedmark, and Teissier view the actinula larva of hydrozoans as a recapitulation of the ancestral cnidarian from which a medusoid form arose.

In contrast to those who propose the medusa as being closest to the ancestral cnidarian, there are numerous zoologists who reject Brooks' thesis and contend that the cnidarian arose from a polypoid ancestral form. The views of many of these zoologists are related to other evolutionary theories and will be discussed in the final chapter of this volume. However, Thiel (1966) and Werner (1970), who are primarily concerned with cnidarian phylogeny, hold that the polypoid form is most primitive and that the scyphozoans and anthozoans evolved from a common polypoid ancestor. Thiel contends that the medusa arose independently in both the hydrozoans and the scyphozoans as a means of dispersal.

Bibliography

Bayer, F. M., 1961: The shallow-water Octocorallia of the West Indian Region. Martinys Nijoff, Hague.

Bayer, F. M., and Owre, H. B., 1968: The Free-Living Lower Invertebrates. Macmillan Co., N. Y., pp. 25–123.

Bouillon, J., 1968: Introduction to Coelenterates. *In* Florkin, M., and Scheer, B. J. (Eds.), Chemical Zoology, Vol. 2. Academic Press, N. Y., pp. 81–148.

Carlgren, O., 1949: A survey of the Ptychodactiaria, Corallimorpharia and Actiniaria. Kungl. Svenska Vetensk. Handl., ser. 4, vol. 1, no. 1. (A monograph on the sea anemones of the world; includes a list of all known species and keys to genera.)

Chapman, D. M., 1966: Evolution of the Scyphistoma. (See Rees, The Cnidaria and Their Evolution.)

Christensen, H. E., 1967: Ecology of *Hydractinia echinata.* I. Feeding biology. Ophelia, *4*:245.

Franzisket, L., 1970: The atrophy of hermatypic reef corals maintained in darkness and their subsequent regeneration in light. Inter. Rev. Gesamten Hydrobiol., *55*(1):1–12.

Fraser, C., 1954: Hydroids of the Atlantic Coast of North America. University of Toronto Press. (A monograph on the American North Atlantic hydroids. Keys, figures, and extensive bibliography included.)

Gosner, K. L., 1971: Guide to Identification of Marine and Estuarine Invertebrates: Cape Hatteras to the Bay of Fundy. Wiley-Interscience, N. Y., pp. 66–161.

Hand, C., 1959: On the origin and phylogeny of the coelenterates. Syst. Zool., *8*(4):191–202.

Hand, C., 1966: On the evolution of the Actinaria. (See Rees, The Cnidaria and Their Evolution.)

Hyman, L. H., 1940: The Invertebrates: Protozoa through Ctenophora, Vol. 1. McGraw-Hill, N. Y., pp. 365–361. (An excellent general account of the cnidarians; includes an extensive bibliography. The chapter *Retrospect* (1959) in the fifth volume of this series summarizes investigations on cnidarians from 1938 to 1958 and also discusses current ideas regarding invertebrate phylogeny.)

Hyman, L. H., 1959: Coelenterata. *In* Edmondson, W. T., Ward, H. B., and Whipple, G. C. (Eds.), Freshwater Biology. 2nd Edition. John Wiley and Sons, N. Y., pp. 313–344. (Illustrated key to the North American freshwater cnidarians.)

Kline, E. S., 1961: Chemistry of the nematocyst capsule and toxin of *Hydra littoralis*. (See Lenhoff and Loomis, pp. 153–168.)

Kaestner, A., 1967: Invertebrate Zoology. Vol. 1. Wiley-Interscience, N. Y., pp. 43–143.

Kramp, P. L., 1961: Synopsis of the medusae of the world. Mar. Biol. Assoc. U. K., *40*:1–469. (Brief descriptions and distributions of all known hydromedusae and scyphomedusae. Exhaustive bibliography.)

Lenhoff, H. M., and Loomis, W. F. (Eds.), 1961: The Biology of Hydra and Some Other Coelenterates. University of Miami Press, Coral Gables, Florida. (A collection of papers presented at a symposium on the physiology and ultrastructure of hydra and other cnidarians.)

Lenhoff, H. M. (Ed.), 1971: Experimental Coelenterate Biology. University Press of Hawaii, Honolulu. 288 pp. (A collection of papers on symbiosis, reef ecology, calcification, and other topics.)

Lewis, D. H., and Smith, D. C., 1971: The autotrophic nutrition of symbiotic marine coelenterates with special reference to hermatypic corals. Proc. Roy. Soc. London, ser. B, *178*:111–129.

Light, S. F., Smith, R. I., Pitelka, F. A., Abbott, D. P., and Weesner, F. M., 1967: Intertidal Invertebrates of the Central California Coast. University of California Press, Berkeley. Coelenterates, pp. 21–47.

Lytle, C., 1961: Patterns of budding in the freshwater hydroid, *Craspedacusta*. (See Lenhoff and Loomis.)

Mariscal, R. W., 1970: Nature of symbiosis between Indo-Pacific anemone fishes and sea anemones. Mar. Biol., *6*:58.

McCullough, C. B., 1963: The contraction burst pacemaker system in hydras. Proc. XVI Intern. Congr. Zool., *1*:25.

Moore, R. C. (Ed.), 1956: Treatise on Invertebrate Paleontology. Coelenterata. Vol. F. Geological Society of America and University of Kansas Press.

Newell, N. D., 1972: The evolution of reefs. Scientific American, *226*(6):54–65. (A good account of the changing nature of the world's reefs since the Precambrian.)

Odum, E. P., 1971: Fundamentals of Ecology. W. B. Saunders Co., Philadelphia. (A summary of coral reef biology and ecology is given on pp. 344–349.)

Pennak, R. W., 1953: Freshwater Invertebrates of the United States. Ronald Press, N. Y., pp. 98–113. (A brief treatment of the freshwater hydrozoans with a key to the North American species.)

Rees, W. J. (Ed.), 1966: The Cnidaria and Their Evolution. Academic Press, N. Y. (This volume represents a collection of papers presented at a symposium held by the Zoological Society of London in 1965.)

Rees, W. J., 1966: The evolution of the Hydrozoa. (See Rees, The Cnidaria and Their Evolution.)

Rose, S. M., 1970: Regeneration. Appleton-Century-Crofts, N. Y. 264 pp. (An entire chapter is devoted to studies of cnidarian regeneration.)

Ross, D. M., 1967: Behavioral and ecological relationships between sea anemones and other invertebrates. Ann. Rev. Mar. Biol. Oceanogr., *5*:291–316.

Russell, F. S., 1954–1970: Medusae of the British Isles. Vol. I (1954), Vol. II (1970). Cambridge University Press, London.

Scoffin, T. P., 1972: Fossilization of Bermuda patch reefs. Science, *178*:1280–1282.

Smith, F. G. W., 1971: Atlantic Reef Corals. University of Miami Press, Coral Gables, Florida. 164 pp. (Handbook for identification of the Atlantic reef corals.)

Stoddart, D. R., 1969: Ecology and morphology of recent coral reefs. Biol. Rev., *44*:433–498.

Swedmark, B., and Teissier, G., 1966: The Actinulida and their evolutionary significance. (See Rees, The Cnidaria and Their Evolution.)

Thiel, H., 1966: The evolution of the Scyphozoa. (See Rees, The Cnidaria and Their Evolution.)

Vaughn, T. W., and Wells, J. W., 1943: Revision of the suborders, families, and genera of the Scleractinia. Special Paper, Geological Society of America, *44*:1–363.

Werner, B., 1970: Neue Beiträge zur Evolution der Scyphozoa und Cnidaria. *In* First International Symposium on Zoophylogeny. Salamanca, 1969.

Westfall, J. A., 1970: The nematocyte complex in a hydromedusan, *Gonionemus vertens*. Z. Zellforsch., *110*:457–470.

Westfall, J. A., 1973: Ultrastructural evidence for neuromuscular systems in coelenterates. Amer. Zool., *13*:237–246.

Westfall, J. A., Yamataka, S., and Enos, P. D., 1971: Ultrastructural evidence of polarized synapses in the nerve net of *Hydra*. J. Cell Biol., *51*:318–323.

Westfall, J. A., Yamataka, S., and Enos, P. D., 1971: Scanning and transmission microscopy of nematocyst batteries in epitheliomuscular cells of *Hydra*. 29th Ann. Proc. Electron Microscopy Soc. Amer., Boston, Mass., edited by C. J. Arceneaux.

Yonge, C. M., 1963: The biology of coral reefs. Adv. Mar. Biol., *1*:209–261.

Yonge, C. M., 1968: Living corals. Proc. Roy. Soc. London, ser. B, *169*:329–344.

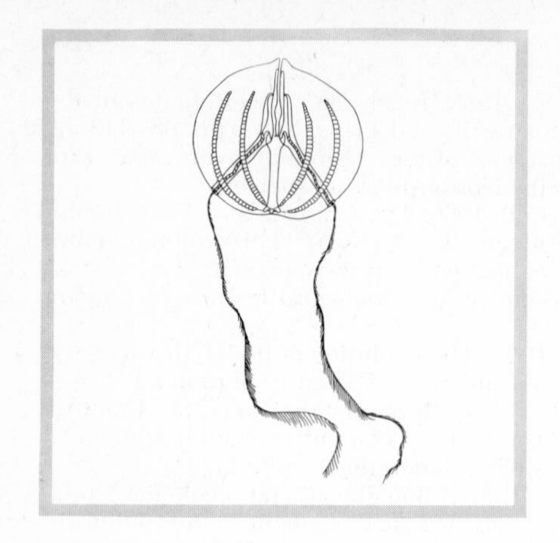

Chapter 6

THE CTENOPHORES

The Ctenophora is a small phylum of marine animals that are commonly known as sea walnuts or comb jellies. The phylum contains approximately 90 species, some of which are abundant in coastal waters. *Pleurobrachia* and *Mnemiopsis* are common genera along the north Atlantic coast.

Ctenophores are believed to be an offshoot from the ancestral medusoid cnidarian. This relationship to the cnidarians is exemplified in a number of ctenophore characteristics. Ctenophores are radially symmetrical, and the general body plan is somewhat similar to that of a medusa. The gastrovascular cavity is in the form of a canal system, and the thick body layer is comparable to the cnidarian mesoglea. On the other hand, ctenophores have undergone considerable specialization and display a number of innovations that indicate a sharp divergence from the cnidarian line. These differences are made clear in the subsequent description of the phylum.

The more primitive or generalized ctenophores, such as *Pleurobrachia* (Fig. 6–1) and *Mertensia*, are spherical or ovoid in shape and range in size from that of a pea to that of a golf ball. They are usually transparent, but various structures such as the tentacles and comb rows may be tinged with white, orange, or purple.

The body wall is composed of an outer epidermis of syncytial, cuboidal, or columnar epithelium. Sensory cells and often mucous gland cells are present, but there are no nematocysts except in a single species. Beneath the epidermis lies a thick layer homologous to the mesoglea of cnidarians. Like the cnidarian mesoglea, this layer is composed of a jelly-like material strewn with fibers and amebocytes, and can be considered a type of loose mesenchyme. The mesenchyme of ctenophores contains true muscle cells, which are smooth and arranged as an anastomosing network.

The spherical body can be divided into two hemispheres (Figs. 6–1 and 6–2). The mouth, on the lower side, forms the oral pole; the diametrically opposite point on the body bears an apical organ and marks the aboral pole.

The body is further divided into equal sections by eight ciliated bands. These bands, called comb rows, are characteristic of ctenophores and are the structures from which the name of the phylum is derived. Each band extends about four-fifths of the distance from the aboral pole to the oral end of the body and is made up of short transverse plates of long, fused cilia called combs. The combs are arranged in succession one behind the other, to form a comb row.

The combs provide the locomotor power in ctenophores, although lobate forms can also swim by contractions of the lobes. The ciliary beat functions in waves beginning at the aboral end of the row. The effective sweep of each comb is toward the aboral pole so that the animal is driven with the mouth, or oral end, forward. The animal usually maintains a vertical position with the mouth directed upward, but can swim downward and can right itself when tilted by water currents. The beat can be temporarily reversed when some object is encountered. The nervous system and apical organ control the synchrony and coordination of the ciliary waves.

From each side of the aboral hemisphere is suspended a long, branched tentacle. Unlike

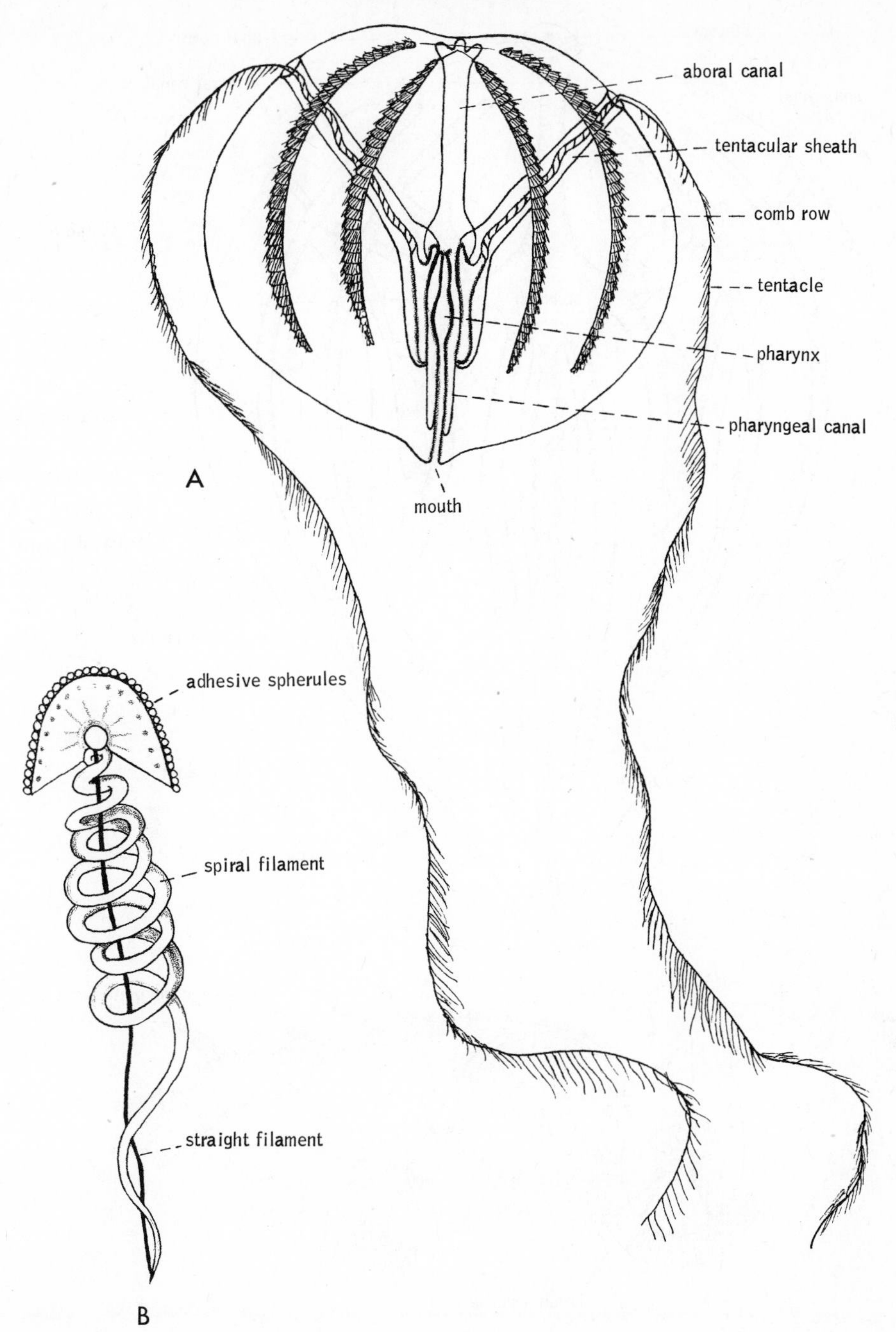

Figure 6–1. A. *Pleurobrachia.* (After Hyman.) B. Colloblast. (After Koma from Hyman.)

the tentacles of cnidarians, those of ctenophores are not attached to the surface of the sphere, but emerge from the bottom of a deep, ciliated epidermal canal called the tentacular sheath, or pouch. There are two pouch openings, which are located between comb rows on opposite sides of the body, each approximately at a 45 degree angle from the aboral pole. Although the comb rows are radially arranged, the two tentacles actually impose a biradial rather than a true radial symmetry on ctenophores.

Each tentacle consists of a mesenchymal core covered by epidermis. The muscle cells of the mesenchyme are frequently arranged in bundles so that the tentacles are very contractile. The tentacular epidermis, although lacking nematocysts, possesses peculiar adhesive cells called colloblasts. A colloblast cell is composed of a hemispherical head

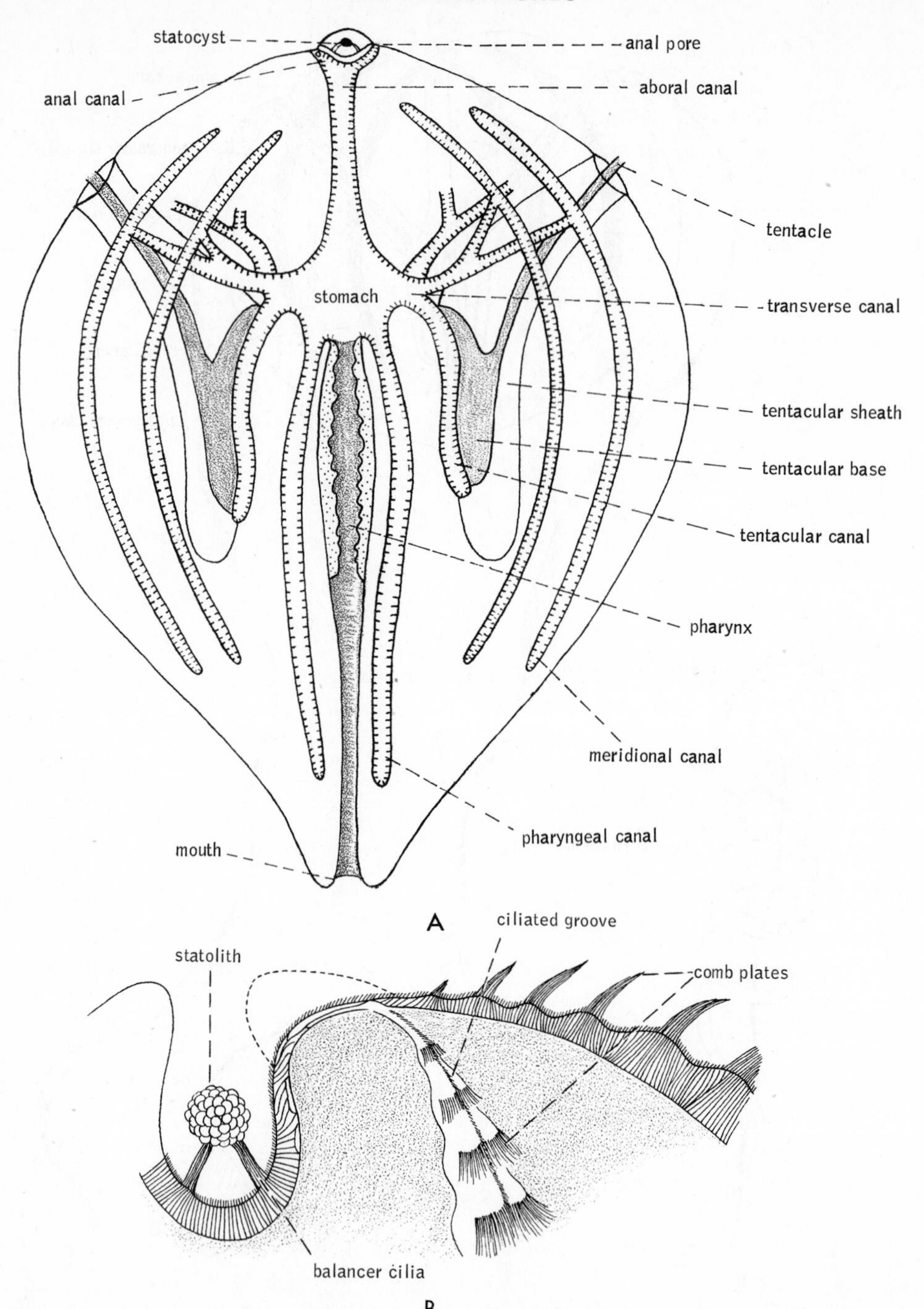

Figure 6–2. *A.* Digestive system of a cydippid ctenophore. (After Hyman.) *B.* Part of anterior end of a ctenophore showing statocyst, ciliated groove, and comb plates. (After Horridge.)

located on the surface of the epidermis (Fig. 6–1*B*). The hemispherical head is connected to the mesenchymal core of the tentacle by means of a straight filament, around which is coiled a contractile spiral filament. The surface head of the colloblast secretes a sticky material used in catching prey. One ctenophore, *Euchlora rubra*, is known to possess nematocysts. This rare species lacks colloblasts, and the nematocysts are arranged in longitudinal tracts on the unbranched tentacles. The existence of nematocysts on this ctenophore is additional evidence of the cnidarian origin of the Ctenophora.

The digestive system is composed of an elaborate biradial system of canals which arise from a central stomach (Fig. 6–2A). The mouth leads into a long tubular pharynx, which extends along the polar axis to the stomach. The digestive system, excluding the pharynx, is lined by a gastrodermal epithelium. In the canals the lining is always thicker on the side toward the polar axis (i.e., the outer side) and is thin and ciliated on the inner side.

The ctenophores are carnivorous, feeding on small planktonic animals. For example, much of the diet of *Pleurobrachia* and *Belinopsis* is composed of copepods. The food is caught on the colloblasts of the extended tentacles and then wiped into the mouth. Digestion begins extracellularly in the pharynx, and the resulting broth then passes into the stomach and canal system, where digestion is completed intracellularly. The cells on the outer, ciliated side of the canals are probably limited to providing circulation of fluid within the system. Indigestible wastes are passed out through the anal pores and mouth.

Ctenophores are noted for their luminescence, which is probably characteristic of the phylum as a whole. Light-production takes place in the walls of the meridional canals, so that externally the light appears to emanate from the comb rows.

The nervous system of ctenophores is a subepidermal nerve network which is particularly well developed beneath the comb rows. The only sense organ is an apical organ containing a statolith (Fig. 6–2*B*). The statolith lies in a deep pit resting on four tufts of balancer cilia. From each tuft of balancer cilia extends a forked ciliated groove. Each fork runs to a comb row and extends through them. When the animal is tilted, the pressure exerted by the statolith on the respective balancer cilia is received by these sensory structures, and signals are transmitted by way of the ciliated grooves to the corresponding comb rows. The rate of beating of the comb rows is altered and the ctenophore rights itself. Severing of a ciliated groove from the apical organ does not halt the beating of the comb rows, but the beating becomes independent and no longer coordinated with that of the other rows.

All members of the phylum are hermaphroditic. The gonads are in the form of two bands located in the thickened wall of each meridional canal. One band is an ovary and the other a testis. The eggs and sperm are usually shed to the exterior through the mouth, and fertilization takes place in the sea water, except in the few species which brood their eggs.

Ctenophore embryogeny anticipates the embryogeny of the protostomes in many respects. Only a very brief description is given here. Cleavage is total and determinate. Eventually a solid blastula is formed that is composed of a large number of micromeres overlying a smaller number of macromeres. Gastrulation takes place by means of epiboly and invagination—that is, the micromeres, which will form the ectoderm, grow downward and enclose the macromeres, which are simultaneously invaginating into the interior to become the entoderm. The mesenchyme appears to derive from the ectoderm like the mesenchyme of cnidarians, and therefore must be considered an ectomesoderm.

The gastrula soon develops into a free-swimming larva. This larva, called a cydippid larva, very closely resembles the adult of ctenophores that have the more ovoid or spherical body structure described previously. The flattened species of ctenophores also possess a spherical cydippid larva that undergoes a more extensive transformation to attain the adult structure. This general existence of a spherical larva in ctenophores seems to substantiate the belief that the primitive shape was spherical or ovoid.

A tendency in the evolution of ctenophores has been for the body to become expanded or lengthened along the tentacular meridian and flattened or compressed along the opposite meridian. Living ctenophores illustrate various degrees of this modification. Such genera as *Pleurobrachia* and *Mertensia* have retained the primitive or ovoid structures and are not markedly flattened.

Mnemiopsis, a common genus along the north Atlantic coast, is moderately flattened (Fig. 6–3*A*). Moreover, as further specialization has developed, the middle of the body has become constricted along the tentacular plane, leaving the expanded outer portions in the form of large lobes. The resulting shape is somewhat similar to that of a clam. In the lobate ctenophores, tentacles are short, lack sheaths, and have moved to a position near the mouth. As a result of this change in the position and size of the tentacles, feeding is somewhat modified from that described previously, although plankton are still utilized as food.

Velamen (Fig. 6–3*B*) and *Cestum*, a genus known as Venus's girdle, have become so expanded and flattened that they look like transparent celluloid belts. One species of *Cestum* reaches a length of over one meter.

They swim not only by means of the comb rows but also by muscular undulations.

Another group of ctenophores are the peculiar and highly aberrant genera that compose the order Platyctenea (Fig. 6–4). In contrast to such ctenophores as *Cestum*, all members of this order are greatly flattened dorsoventrally (i.e., aborally-orally), and most have taken up a creeping or even a sessile mode of existence.

Because of the resemblance of these ctenophores to polyclad flatworms, some past zoologists advocated the theory that the flatworm-annelid-arthropod line of invertebrates arose from the ctenophores. However, the resemblance to flatworms is undoubtedly

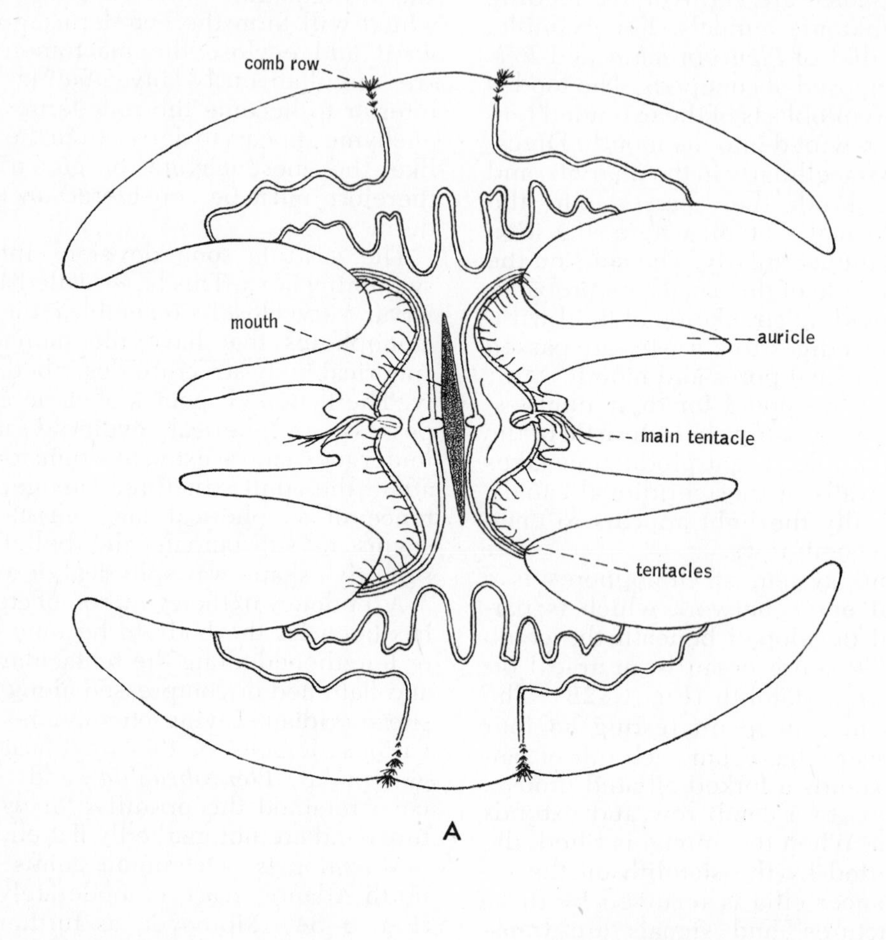

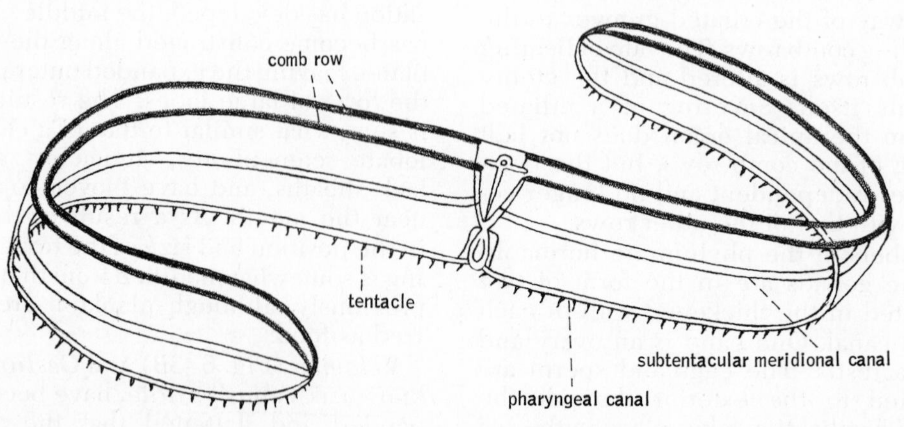

Figure 6–3. A. *Mnemiopsis* (oral view). B. *Velamen*. (Both after Mayer from Hyman.)

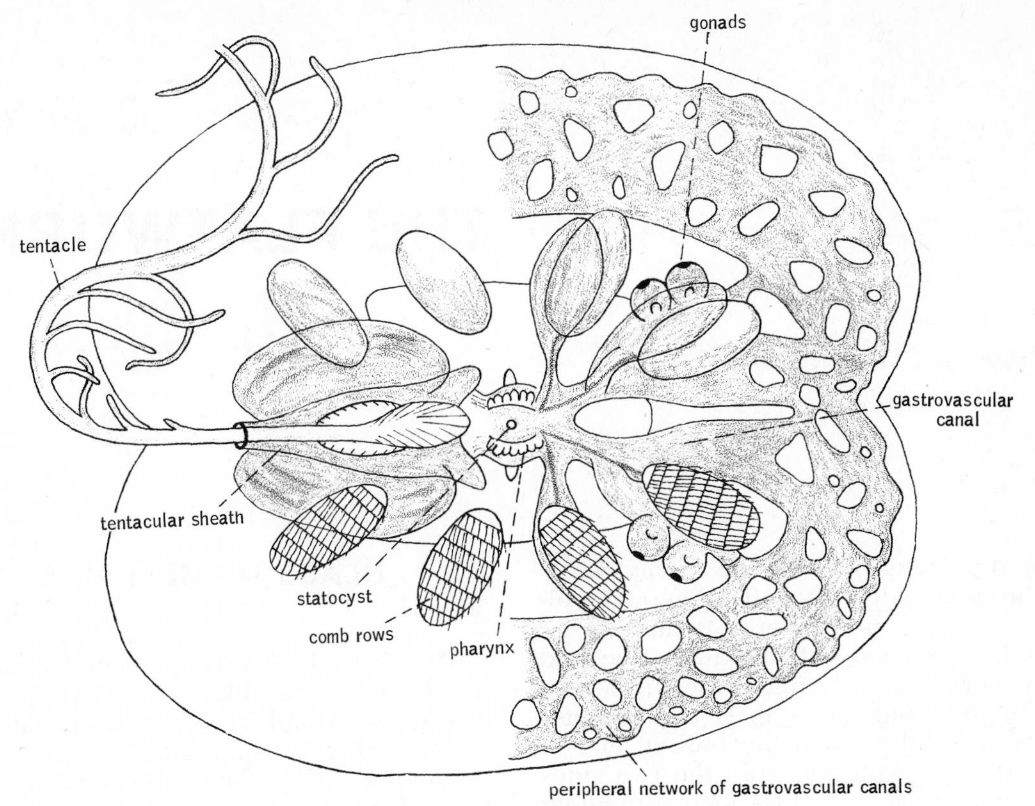

Figure 6–4. *Ctenoplana.* (After Komai from Hyman.)

superficial. These ctenophores possess, in most cases, all the typical structures displayed by other members of the phylum.

SYSTEMATIC RÉSUMÉ OF PHYLUM CTENOPHORA

Class Tentaculata. Ctenophores with tentacles.

Order Cydippida. Body rounded or oval; tentacles branched and retractile into pouches. *Mertensia, Pleurobrachia.*

Order Lobata. Body moderately compressed, with two large oral lobes to either side of tentacular plane. Tentacles not in pouches. *Mnemiopsis.*

Order Cestida. Body ribbon-shaped as a result of great compression along tentacular plane. The two principal tentacles and two of the four comb rows reduced. *Cestum, Velamen.*

Order Platyctenea. Body greatly flattened as a result of a reduction in the oral-aboral axis. Adapted for creeping. Comb rows reduced or absent in adult. *Ctenoplana, Coeloplana.*

Class Nuda. Ctenophores without tentacles.

Order Beroida. Body conical. Pharynx very large. *Beroë.*

Bibliography

Bayer, F. M., and Owre, H. B., 1968: The Free-living Lower Invertebrates. Macmillan Co., N. Y., pp. 124–143.

Gosner, K. L., 1971: Guide to Identification of Marine and Estuarine Invertebrates: Cape Hatteras to the Bay of Fundy. Wiley-Interscience, N. Y. Ctenophora, pp. 161–166.

Horridge, G. A., 1965: Relations between nerves and cilia in ctenophores. Amer. Zool., 5:357–375.

Hyman, L. H., 1940: The Invertebrates: Protozoa through Ctenophora. Vol. 1, McGraw-Hill, N. Y., pp. 662–696. (An old but still valuable general account of the Ctenophora; excellent figures and extensive bibliography. The Chapter *Retrospect* in Vol. 5 (1959) of this series surveys investigations on ctenophores from 1938–1958.)

Kaestner, A., 1967: Invertebrate Zoology. Vol. I. Wiley-Interscience, N. Y., pp. 155–153.

Light, S. F., Smith, R. I., Pitelka, F. A., Abbott, D. P., and Weesner, F. M., 1967: Intertidal Invertebrates of the Central California Coast. University of California Press, Berkeley. Ctenophora, pp. 49–55.

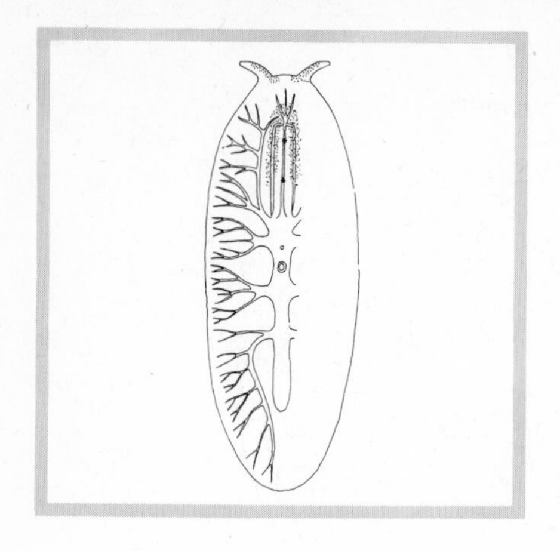

Chapter 7

THE FLATWORMS

The remaining phyla of the Animal Kingdom are bilaterally symmetrical and are collectively called the Bilateria. Bilateral symmetry is correlated with motility. The part of the body first meeting the environment (the anterior end) bears most of the sense organs and differs from the trailing end of the body (the posterior end); the two sides of the body, each contacting the environment in the same way, are identical, but the upper and lower surfaces differ. Radial symmetry may exist, as in the echinoderms, but is always of secondary origin. Of all the bilateral phyla, many zoologists agree that the free-living flatworms of the phylum Platyhelminthes are the most primitive. These animals display in many ways the transition to the more complex structure of higher Bilateria.

The phylum Platyhelminthes embraces three classes of worms. Two of these are entirely parasitic. The Trematoda comprise flukes, and the Cestoda, tapeworms. The third class, the Turbellaria, are free-living and are certainly the ancestors of the two parasitic classes. The members of the phylum are dorsoventrally flattened—accounting for the name flatworm—and display a solid, or acoelomate, grade of structure, in which there is no body cavity and mesenchyme fills the space between the internal organs and body wall. The mouth is the only opening to the digestive tract, where a digestive tract is present. Protonephridia are the osmoregulatory organs, and the reproductive system is hermaphroditic. Flatworms are unusual in possessing biflagellate sperm, and the flagellar ultrastructure of most species is unique in displaying a 9-1 instead of a 9-2 pattern of microtubules.

CLASS TURBELLARIA

The Turbellaria vary in shape from ovoid to elongate and, like the other flatworm classes, are usually dorso-ventrally flattened. In general, the larger the worm, the more pronounced is the flattened condition. Head projections are present in some species. These may be in the form of short tentacles (Fig. 7–1*B*), which vary in number and position, or in the form of lateral projections of the head called auricles (Fig. 7–1*D*). The auricles are frequently found in freshwater planarians. Coloration is mostly in shades of black, brown, and gray, although some groups display brightly colored patterns. A few species are green, owing to the presence of symbiotic algae. Turbellarians range in size from microscopic forms to species that are more than 60 cm. long, although most are less than 10 mm. in length.

Turbellarians are primarily aquatic, and the great majority are marine. Although there are a few pelagic species, most of them are bottom dwellers that live in sand or mud, under stones and shells, or on sea weed. Many species are common constituents of the interstitial fauna. Freshwater forms, such as the common laboratory planarian, *Dugesia* (Fig. 7–1*D*), live in lakes, ponds, streams, and springs, where they occupy bottom habitats. Some species have become terrestrial (Fig. 7–1*F*), but these are confined to very humid areas and hide beneath logs and leaf mold during the day, emerging only at night to feed. The land planarians are the giants of the Turbellaria, some reaching 60 cm. or more in length. They are largely tropical, but a few, such as the North American

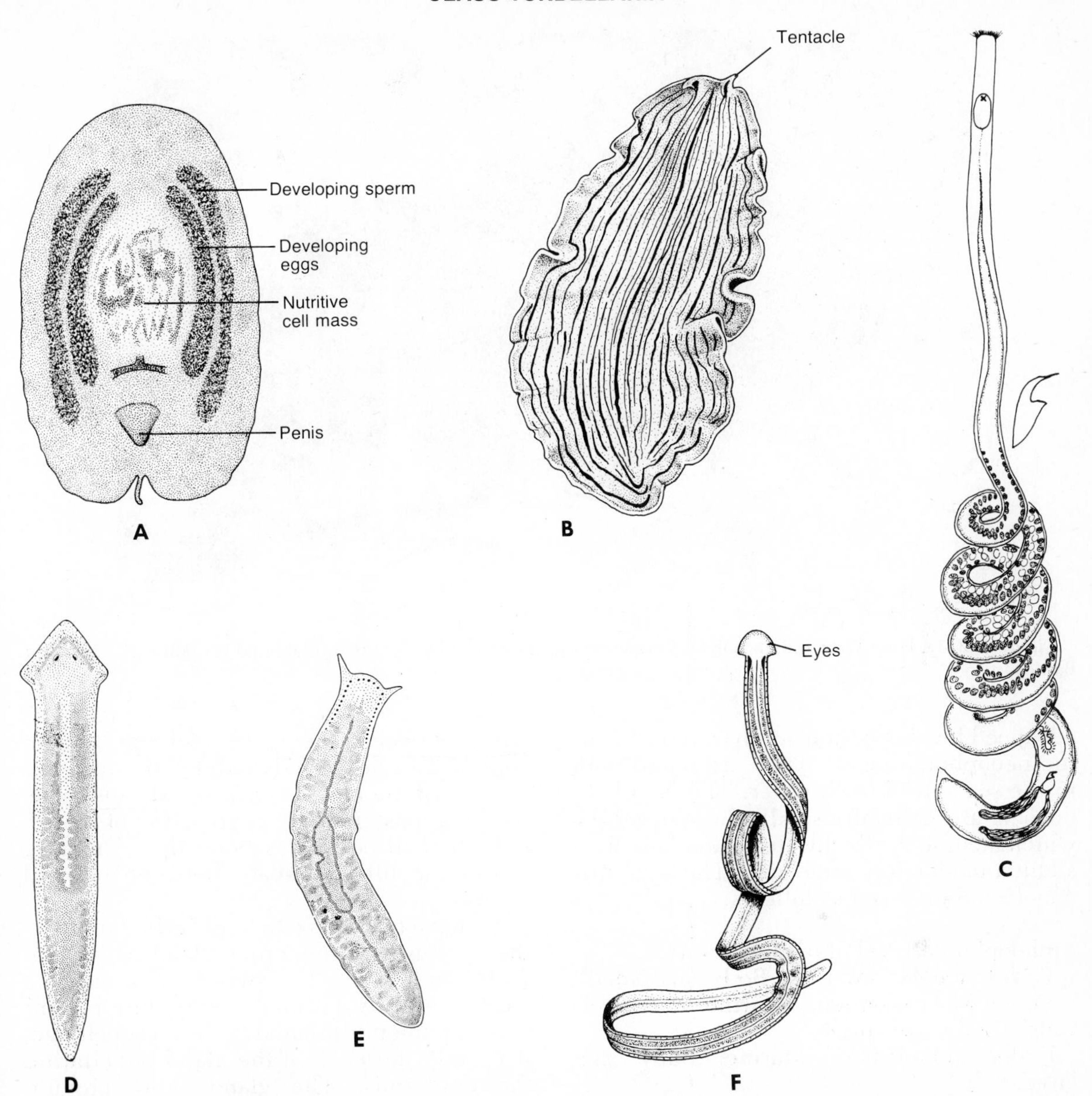

Figure 7–1. Turbellarian flatworms. *A. Polychoerus*, an acoel. *B. Prostheceraeus*, a polyclad. *C. Nematoplana*, an interstitial member of the suborder Proseriata. (From Rieger, R., and Ott, J., 1971: Vie et Milieu, Supplement 22.) *D. Dugesia*, a common freshwater planarian. *E. Polycelis*, a freshwater planarian. (*D* and *E* after Steinmann and Bresslau.) *F. Bipalium kewense*, a cosmopolitan land planarian found in Florida, Louisiana, California, and in greenhouses throughout the United States. (After Hyman.)

Bipalium adventitium, live in temperate regions. Some 3000 species of turbellarians have been described.

Despite their similarity in appearance, turbellarians exhibit considerable internal complexity, and the class is composed of a relatively large number of diverse groups. Those orders which must be referred to in the following discussion can perhaps be more easily kept in mind if they are grouped and designated at the outset. The members of the orders Acoela, Macrostomida, and Polycladida exhibit a more primitive level of organization (often referred to as the archoophoran level). All three are marine, but the Macrostomida are also represented in freshwater habitats. The Acoela and Macrostomida are small to microscopic in size; the Polycladida are large, some species reaching five centimeters or more in length (Fig. 7–2). The Neorhabdocoela and Tricladida are among those turbellarians that exhibit a more

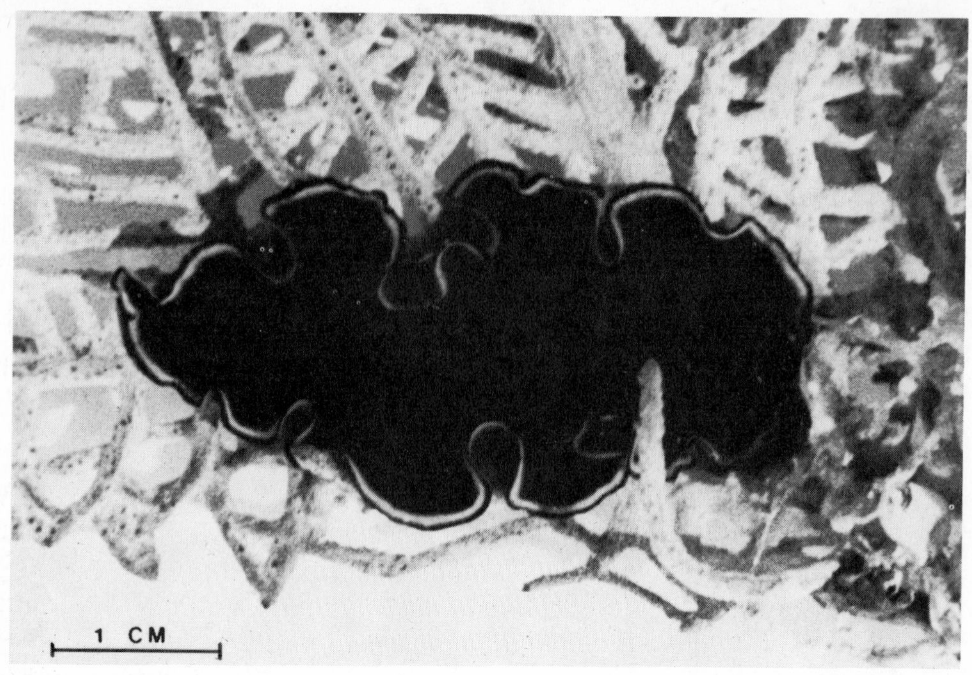

Figure 7–2. A large West Indian polyclad crawling on a sea fan. The light marginal band is orange. (By Betty M. Barnes.)

advanced level of organization (referred to as the neoophoran level). They are found both in the sea and in fresh water. The Neorhabdocoela are generally small; the Tricladida, which include the freshwater planarians, may attain a moderately large size. These groups may be summarized as follows:

Archoophoran level of organization*:
Order Acoela. Marine. Body size small.
Order Macrostomida. Marine and freshwater. Body size small.
Order Polycladida. Marine. Body size large.
Neoophoran level of organization:
Suborder Tricladida (order Seriata). Marine and freshwater. Moderately large.
Order Neorhabdocoela. Marine and freshwater. Small.

Body Wall. The body of turbellarians is covered by a ciliated epidermis (Fig. 7–3). In acoels the epidermis tends to be syncytial, and in triclads cilia are confined to the ventral surface (Fig. 7–4). Beneath the epidermis there is a muscle layer composed of several layers—an outer circular and an inner longitudinal layer, with diagonal fibers lying between them. In addition, dorsoventral muscle strands extend through the interior of the body (Fig. 7–4). Mesenchyme tissue, a loose net-like aggregation of irregularly shaped cells, lies beneath the muscle layers and fills the space between internal organs.

A characteristic feature of turbellarians is the presence of numerous gland cells (Fig. 7–3*A* and 7–4). The gland cells may be located entirely within the epidermis, but are more commonly situated in the mesenchyme, with only the neck of the gland penetrating the epidermis. The glands may provide mucus for adhesion, for covering the substratum in movement, or for enveloping prey. Gland cells are frequently grouped together. Such an anterior aggregation, called a frontal gland, is characteristic of many turbellarians and is believed to be a primitive turbellarian feature (Figs. 7–5, 7–6, and 7–8*B*). Other groupings are also encountered. Some turbellarians have glands at the caudal end; some have them arranged as a marginal ring around the body. In *Bdelloura*, which lives as a commensal on the book gills of the Atlantic horseshoe crab, the caudal part of this adhesive zone is greatly increased to form an adhesive plate (Fig. 7–7). The sticky secretion produced by the glands, plus the fact that the necks of the gland cells can be projected as little papillae, enable the ani-

*This synopsis is greatly abbreviated. See Systematic Résumé on page 163 for complete classification.

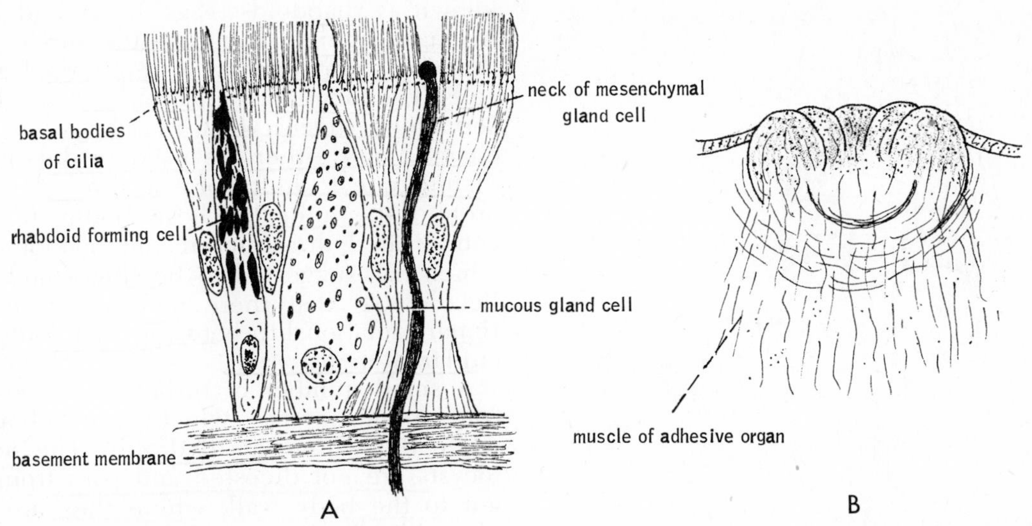

Figure 7–3. *A*. Section through epidermis of a polyclad. (After Bock from Hyman.) *B*. Glandulomuscular adhesive organ of the freshwater planarian, *Procotyla*. (After Hyman.)

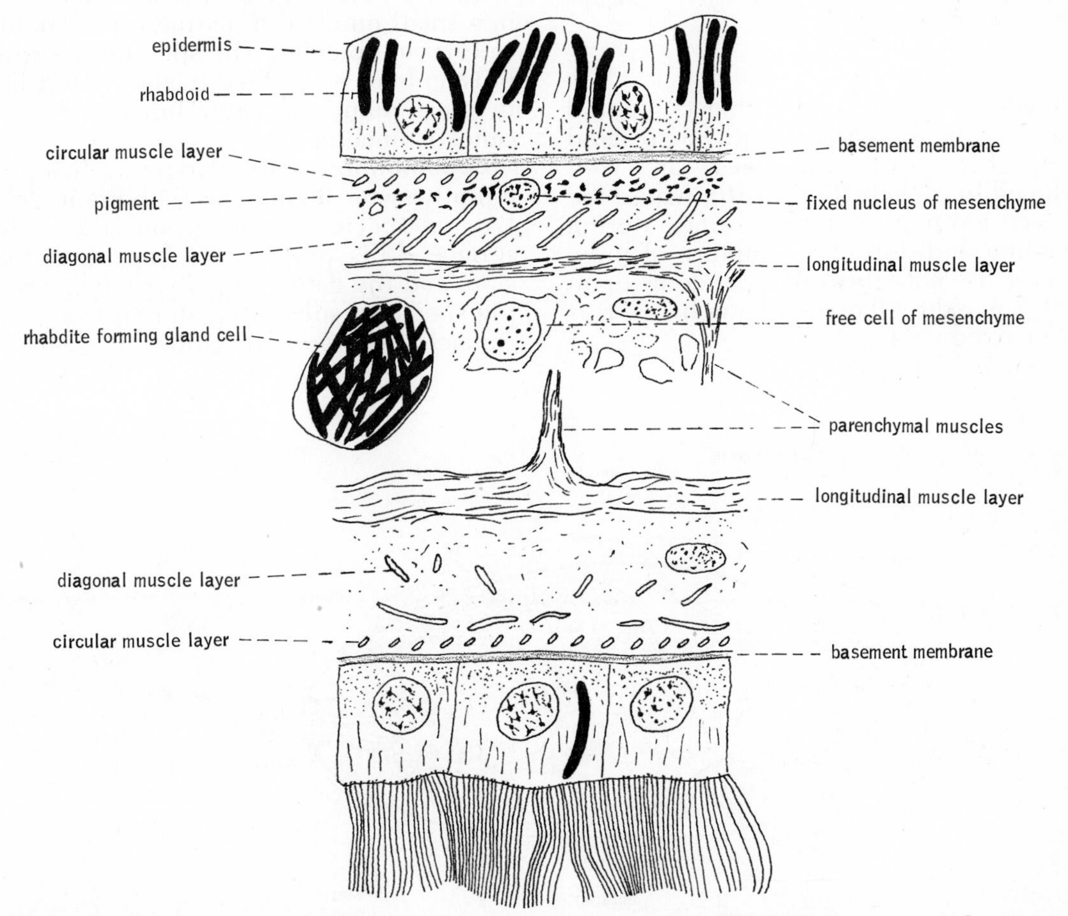

Figure 7–4. Dorsal and ventral body walls of a freshwater planarian (longitudinal section). (After Hyman.)

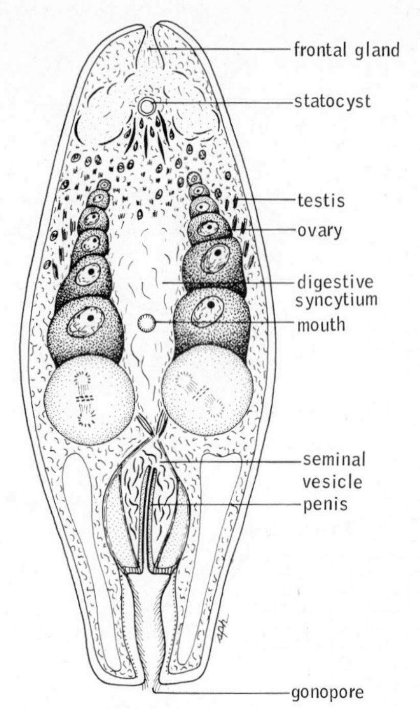

Figure 7-5. *Pseudaphanostoma brevicaudatum,* an acoel flatworm. (After Dörjes.)

mal to adhere very tightly to objects. Ventral adhesive organs—glands with associated muscle fibers about the neck—are well developed in triclads (Fig. 7-3*B*), where the secretions aid in gripping the surface during locomotion and during the capture of prey. Adhesion is not always dependent upon glands. Some turbellarians adhere by the tips of specialized cilia.

Typical of almost all Turbellaria are numerous rod-shaped epidermal bodies known as rhabdoids (Figs. 7-3*A* and 7-4). Arranged at right angles to the surface, the rhabdoids are secreted by epidermal gland cells. The function is uncertain, although observations suggest that rhabdoids are discharged and either are used in defense or disintegrate to help form a slimy covering around the animal. These bodies have a complicated ultrastructure and development, which has suggested to Reisinger and Kelbetz (1964) that there may be an evolutionary relationship between rhabdoids and cnidarian nematocysts.

A small number of turbellarians possess true nematocysts, which are obtained when the worm eats hydroids. The nematocysts are not digested and pass from the gut to the body wall, where they are employed by the worm in defense. Sea slugs, a group of gastropod mollusks, utilize nematocysts in the same way.

Another unexpected structure encountered in a small number of marine turbellarians is calcareous spicules. The spicules are simple rods and are embedded in the mesenchyme in great numbers. Their functional significance is uncertain.

Locomotion. Turbellarians that are minute swim about within bottom debris. Larger species creep upon the ventral surface. Cilia provide the force for propulsion, and the flattened body shape of creeping species results in greater surface area in contact with the substratum. Muscular un-

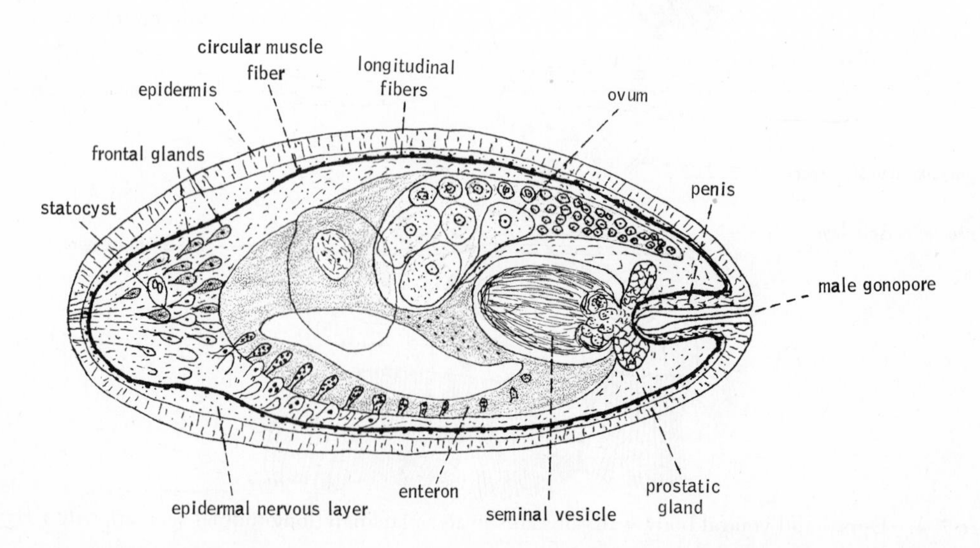

Figure 7-6. Sagittal section of the acoel, *Nemertoderma.* This acoel is unusual in having a digestive cavity. (After Westblad from Hyman.)

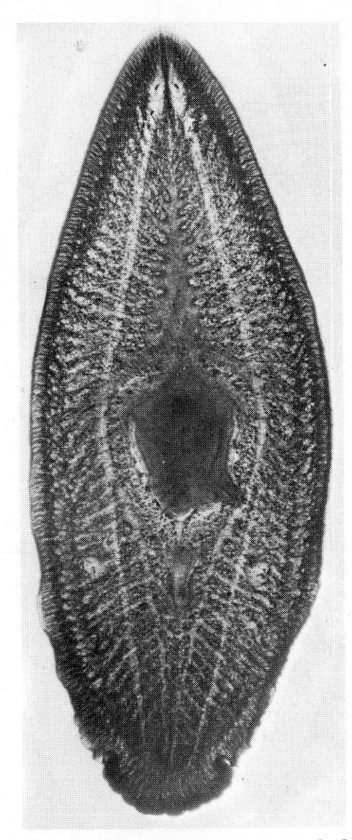

Figure 7–7. *Bdelloura*, a marine triclad commensal that lives on gills of horseshoe crabs. The large central dark mass is the pharynx, in front of which can be seen the anterior branch of the intestine. The two light longitudinal streaks passing to either side of the pharynx are the ventral nerve cords. The pair of ganglia from which they arise, lying beneath the eyes, can be seen at the anterior end. (Courtesy of Ward's Natural Science Establishment.)

dulations are important in the locomotion of flatworms larger than 3 mm., which include the planarians. Transverse waves of contraction sweep the length of the body, raising and lowering the ventral surface. This mode of movement is highly developed in snails and will be described in the discussion of that group (p. 338). All flatworms, minute and large, depend upon muscle contractions to turn and twist the body.

The interstitial fauna contains many minute turbellarians. Those which live in the intertidal or swash zone, where the spaces between sand grains are not continually filled with water, are among the most specialized. Adhesive glands and elongate bodies are common adaptations for gaining traction in moving or maintaining position between sand grains (Fig. 7–1*C*).

Nutrition. The digestive cavity, or intestine, of turbellarians is a blind sac, with the mouth serving for both ingestion and egestion. The wall of the intestine is single-layered and is composed of phagocytic and gland cells. Primitively, as in the Macrostomida, the intestine is ciliated (Fig. 7–8*A*).

The form of the intestine is in part related to the size of the worm. Small turbellarians, such as the macrostomids and neorhabdocoels, have intestines which are simple sacs. The members of the order Acoela lack an intestinal cavity, and the intestinal cells form a solid syncytial mass (Fig. 7–8*B*). It is to the lack of a gut cavity that the name Acoela refers.

The larger turbellarians have intestines with lateral diverticula, which increase the surface area for digestion and absorption and compensate for the absence of an internal transport system. In polyclads the intestine consists of a central tube, from which a great many lateral branches arise (Figs. 7–9*A* and 7–10*A*). These in turn are subdivided, and may anastomose with other branches. The members of the Tricladida, to which the planarians belong, have an intestine composed of three principal branches—one anterior and two posterior lateral (Figs. 7–11 and 7–12*C*). Each of these principal branches, in turn, has many lateral diverticula. The three branches join in the middle of the body, anterior to the mouth and pharynx. The names Polycladida and Tricladida refer to the branching intestine of these groups of turbellarians.

The mouth is primitively located on the midventral surface, but may be situated anywhere along the midventral line. The connection between the mouth and the intestine shows increasing complexity within the class. This region, called the pharynx, is a simple ciliated tube (simple pharynx) in the small primitive Acoela, Macrostomida, and Catenulida (*Stenostomum*) (Figs. 7–8 and 7–13). In other turbellarians the pharynx has become a more complex ingestive organ as a consequence of folding and development of the muscle layers. A folded, or plicate, pharynx is characteristic of the polyclads and triclads, which are large turbellarians with branched intestines. The folded condition is believed to have evolved from a simple pharynx and has resulted in a muscular pharyngeal tube lying within a pharyngeal cavity (Figs. 7–11, 7–12*C*, and 7–13*B*). The free end of the tube is projected out of the mouth during feeding (Fig. 7–12*D*). The

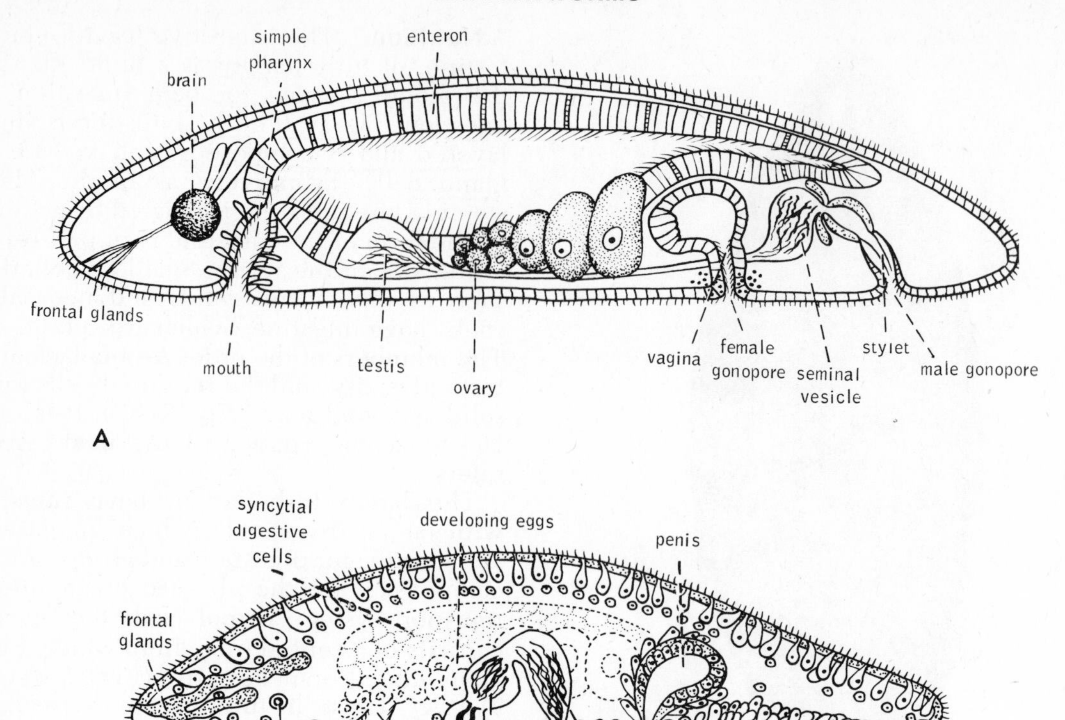

Figure 7–8. Semidiagrammatic sagittal sections of two turbellarians. *A.* The freshwater *Macrostomum*, a member of the order Macrostomida. (After Ax.) *B.* The marine acoel *Convoluta.* (After Westblad.)

pharynx may project backward (Figs. 7–11 and 7–12), as in the common freshwater planarians; or the pharynx may be attached posteriorly and extend forward (Fig. 7–9*A*). In many polyclads, it hangs down from the roof of the pharyngeal cavity like a circular ruffled curtain (Figs. 7–9*B* and 7–10*A*).

A bulbous pharynx, encountered in such orders as the Neorhabdocoela, is believed to be derived from the plicate condition through reduction of the outer pharyngeal cavity and separation by a special muscular septum from the mesenchyme of the body (Figs. 7–13*C* and 7–14). The muscular bulb, which characterizes this type of pharynx, can be protruded from the mouth in many species (Fig. 7–12*B*).

Turbellarians are largely carnivorous and prey on various invertebrates that are small enough to be captured, as well as on the dead bodies of animals that sink to the bottom. Tiny crustaceans—water fleas, copepods, amphipods—are common prey; but there are marine species which feed on sessile animals such as bryozoans and little tunicates, and the polyclad *Stylochus frontalis* feeds on living oysters, eating bits at a time.

Many turbellarians capture living prey by wrapping themselves around it, entangling it in slime, and pinning it to the substratum by means of the adhesive organs. They then ingest the prey by swallowing it whole or by a sucking or pumping action. A few species are known to stab the prey with their penis, which terminates in a hardened stylet and projects from the mouth; and the kalyptorhynch rhabdocoels, as well as some other forms, possess an anterior raptorial proboscis which may be provided with hooks (Fig. 7–1*G*). The proboscis of these species is not connected with the mouth and gut.

Prey is swallowed whole by those turbellarians with a simple pharynx, by those with a protrusible bulbous pharynx, and even by the polyclads that have a plicate pharynx. In the triclads, the pharyngeal tube is extended from the mouth and inserted either into the body of the prey or into carrion. The exo-

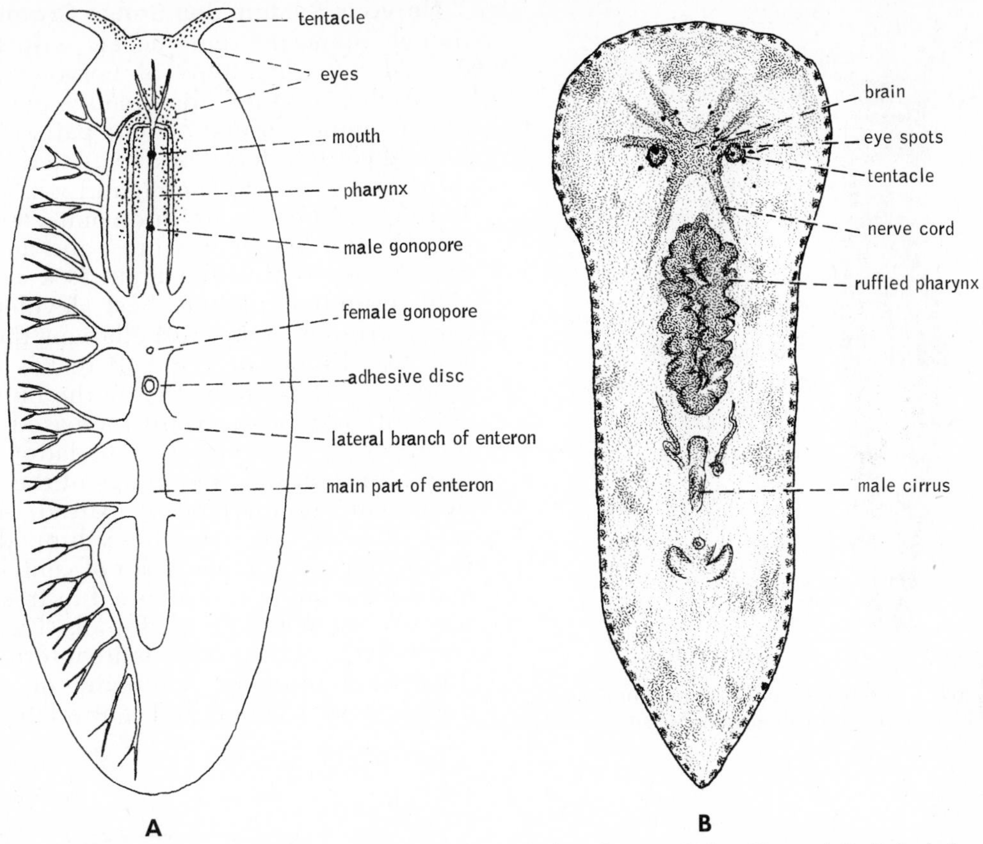

Figure 7–9. *A*. Digestive system of a polyclad with a tubular pharynx. (After Hyman.) *B*. Polyclad with dorsal tentacles and ruffled pharynx.

skeleton of crustaceans is penetrated at thin points, such as the articulations between segments. The penetration by the pharynx is aided by proteolytic enzymes produced by glands in the pharynx. The contents are then pumped into the enteron by peristaltic action (Fig. 7–15). A European land planarian of the genus *Orthodemus* utilizes the same mechanism in feeding on slugs and earthworms. Ingestion by a sucking or pumping action is also employed by forms that have a nonprotrusible bulbous pharynx.

In studies on the acoel *Convoluta paradoxa*, Jennings (1957) found that small prey is captured and is engulfed by the internal mass of digestive cells, which is partially everted through the mouth. Larger prey is pressed into the mouth and swallowed. After ingestion, the prey passes into the internal mass of nutritive cells.

Digestion is first extracellular. Disintegration of the ingested food is initiated by pharyngeal enzymes, and additional proteolytic enzymes are supplied by gland cells of the intestine. The resulting food fragments are then engulfed by the phagocytic cells, in which digestion is completed intracellularly. Digestion in acoels occurs in essentially the same manner as in other turbellarians, and in some cases a temporary digestive cavity may form around the food.

Freshwater planarians are able to with-

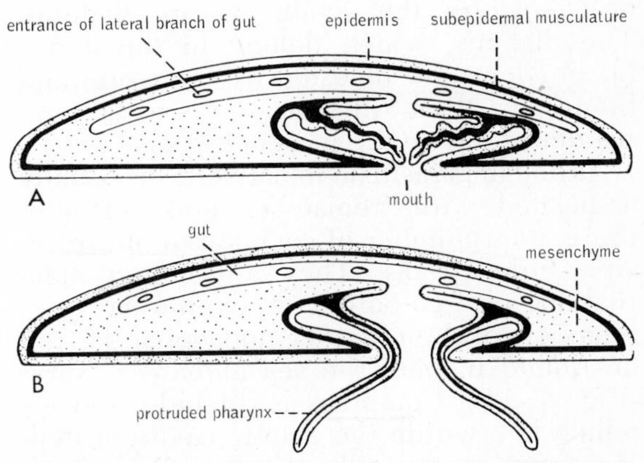

Figure 7–10. The polyclad *Leptoplana*, showing retracted and protruded ruffled, plicate pharynx (sagittal section). (After Jennings.)

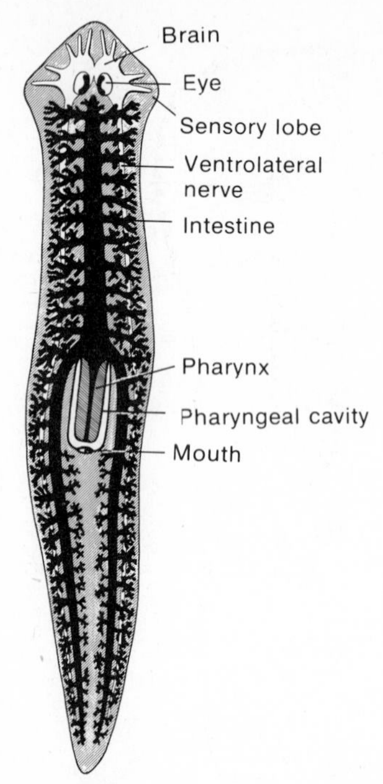

Figure 7–11. Dorsal view of the planarian *Dugesia*, a triclad, showing the three-branched digestive system. (After Hyman.)

stand prolonged experimental starvation. In extreme cases, they utilize part of the enteron and all of the tissue of the mesenchyme and reproductive system. In fact, the body volume may be reduced to as little as 1/300 of the original.

Some species of *Convoluta* harbor green zoochlorellae (Volvocales) within their mesenchyme. These worms may occur in such enormous numbers in certain regions on the European coast that they form green blotches in shallow water. For one species of *Convoluta* the symbionts are diatoms. The diatoms, which belong to the genus *Licmophora*, are ingested by the worm and then leave their siliceous shells to take up residence within the mesenchyme.

Although parasitism in flatworms is usually associated with trematodes and cestodes, there are a number of commensal and parasitic turbellarians. They are largely neorhabdocoels (suborder Dalyellioida) and members of the order Temnocephalida, and are found in both the sea and fresh water (Fig. 7–14). Commensals include species which live within the mantle cavity of mollusks and on the gills of crustaceans. Parasitic forms inhabit the gut of mollusks and the body cavity of echinoderms.

Nervous System and Sense Organs. Although there are some acoels with an epidermal nerve-net type of nervous system, the primitive turbellarian plan is perhaps an arrangement of three or four pairs of longitudinal cords—dorsal or dorso-lateral, lateral or marginal, ventro-lateral, and ventral (Fig. 7–16*A*). The cords are interconnected along their length by commissures, and anteriorly by a brain. Within the brain is a statocyst. This primitive plan is best developed in acoels. In most other turbellarians the statocyst has disappeared and there has been a tendency toward reduction in the number of pairs of cords and toward increased prominence of the ventral pair. This latter feature may represent the beginnings of the ventral nerve cord (embryonically paired) so characteristic of the annelid-arthropod line. Reduction to a dominant pair of ventral cords has been more or less attained in freshwater triclads and most neorhabdocoels (Figs. 7–11 and 7–16*C*). Marine triclads are intermediate, possessing three or four pairs of cords—dorsal, ventral, lateral, and ventro-lateral. The

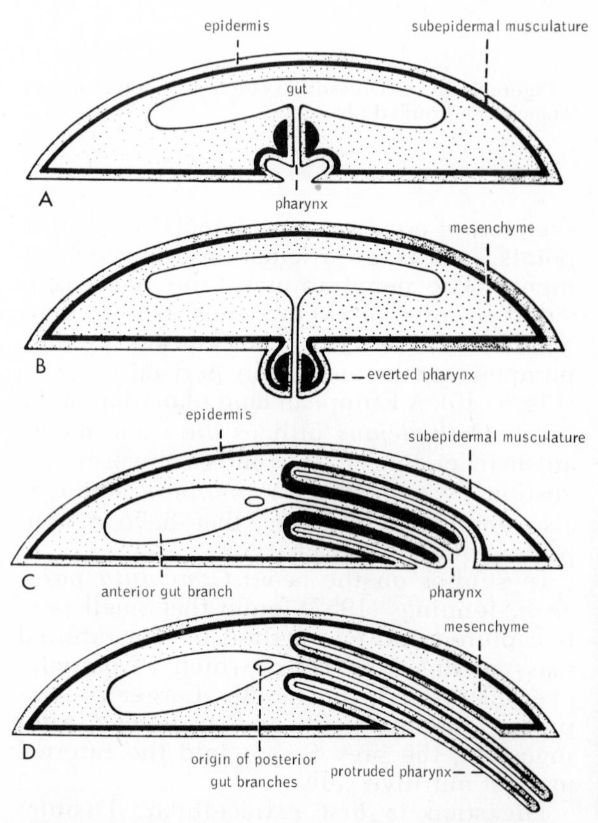

Figure 7–12. *A* and *B*. The rhabdocoel, *Mesostoma*, showing retracted and bulbous pharynx (sagittal section). *C* and *D*. The fresh-water triclad, *Polycelis*, showing retracted and protruded tubular plicate pharynx (sagittal section). (All after Jennings.)

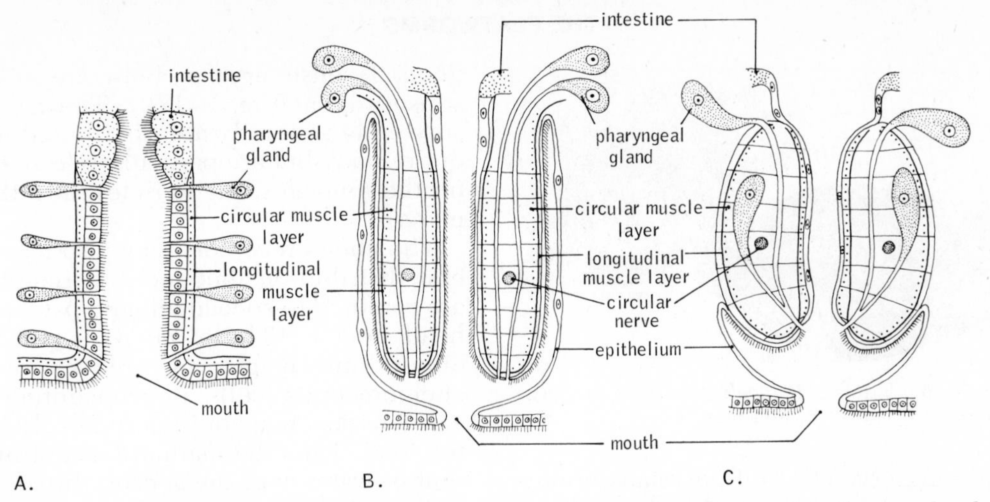

Figure 7–13. The three types of turbellarian pharynx. *A.* Simple pharynx, found in the Acoela, the Macrostomida, and the Catenulida. *B.* Plicate pharynx, found in the polyclads and planarians. *C.* Bulbous pharynx, found in the rhabdocoels. (Modified after Ax.)

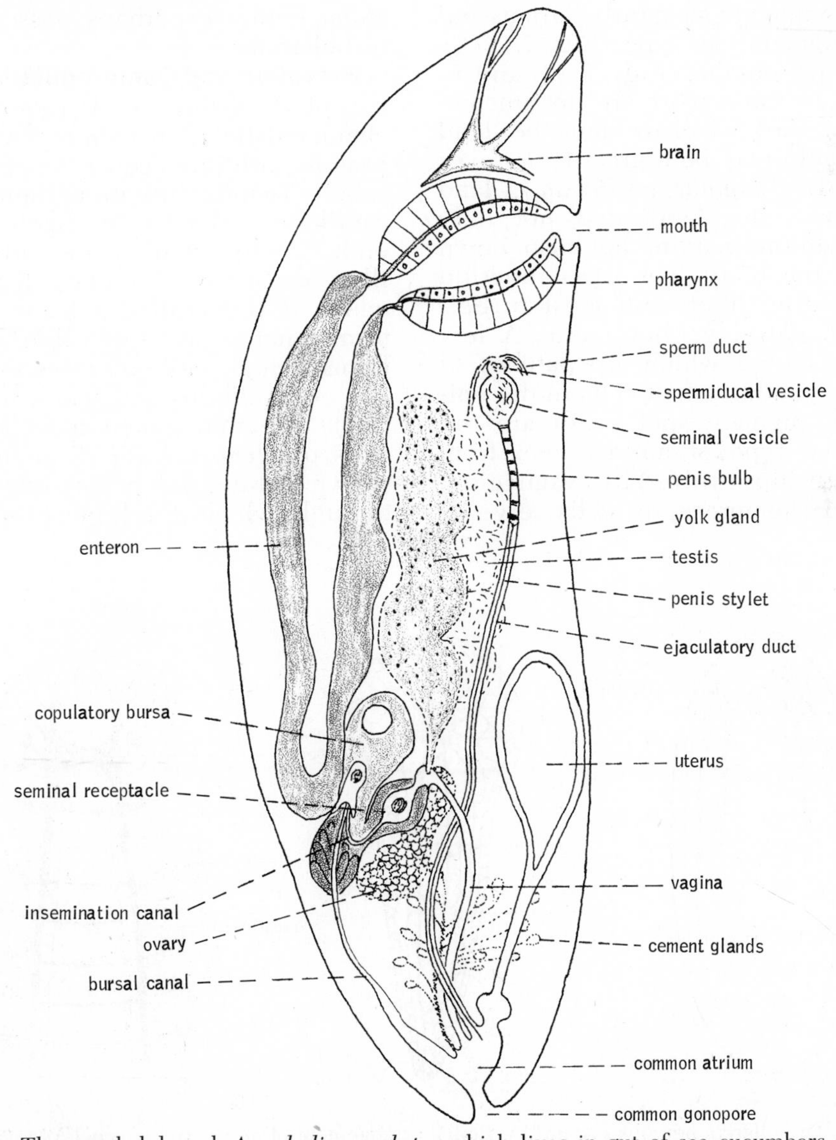

Figure 7–14. The neorhabdocoel, *Anoplodiera voluta,* which lives in gut of sea cucumbers (sagittal section). (After Westblad from Hyman.)

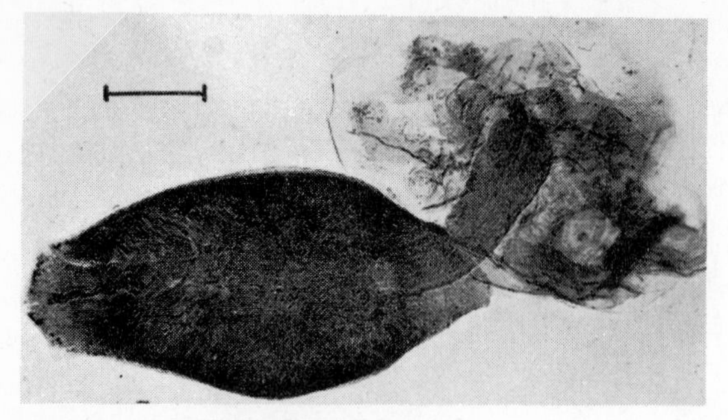

Figure 7–15. Triclad. *Polycelis*, feeding on large *Daphnia*. (Courtesy of J. B. Jennings.)

longitudinal nerve cords of polyclads undergo continuous branching so that the nervous system takes the form of a complex network (Fig. 7–16*B*).

Eyes are common in most turbellarians and are of the pigment-cup type (Fig. 7–17*A*). Two is the usual number (Figs. 7–1*D* and 7–18), but two or three pairs are not uncommon; and in the polyclads and the land planarians, there may be a great many eyes (Fig. 7–9*A*). In *Geoplana mexicana*, a land planarian, eyes are distributed not only along the anterior margin, but also down the sides of the body. The eyes function only in detecting light, and most turbellarians are negatively phototactic. A few acoels possess eyes which are patches of photoreceptors and pigment cells in the epidermis. Such pigment spot ocelli are the most primitive type of animal eye, from which pigment cups evolved by evagination.

Tactile and chemoreceptor cells serve as general sense organs. Both are cells with sensory cilia (Fig. 7–17*B*). The tactile sensory cells generally are distributed over the entire body but are particularly concentrated on the tentacles, the auricles, and the body margins.

The chemoreceptors may also have a general distribution, but more frequently they are confined to special ciliated patches in the head region. These ciliated patches are either sunken pits or grooves in which chemoreceptor cells are concentrated. The sense organs play an important role in locating food. The cilia maintain a continual current of water over the sensory bristles.

Special rheoceptor cells—sensory cells that orient the animal to water currents—are found in *Mesostoma* and *Bothromesostoma* and are perhaps possessed by most turbellarians.

Excretion and Osmoregulation. With the exception of the acoels, protonephridia are characteristic of flatworms. A turbellarian protonephridium consists of a branched tubule terminating in a number of blind capillaries, which bear flagella at the inner end. Usually, such a flame bulb and its capillary consist of a single cell (Fig. 7–19*B*), which is then called a flame cell; but the term is not always applicable, since in some forms a single cell gives rise to a large number of capillaries and flame bulbs (Fig. 7–19*A*). Electron microscopy of the flame cells of *Stenostomum* (Kummel, 1962) has shown that the wall is pierced with openings through which fluid must enter the cell

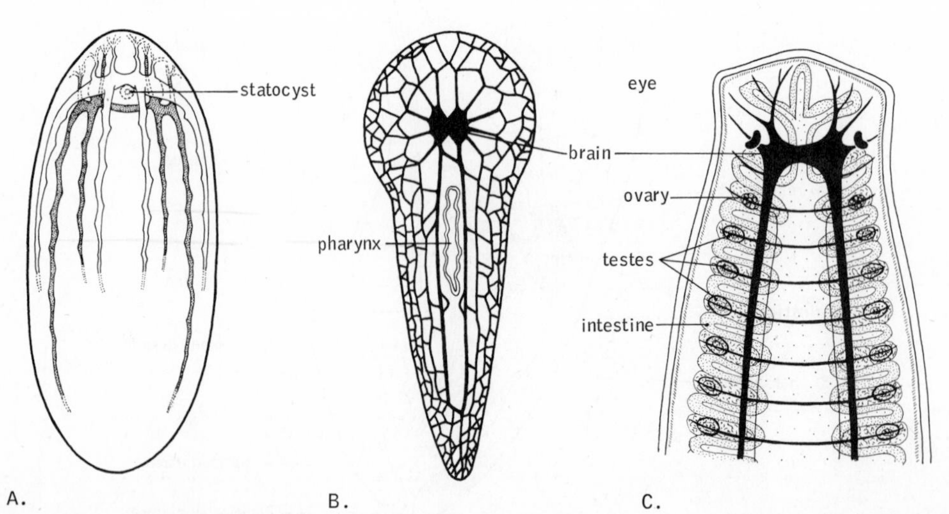

Figure 7–16. Turbellarian nervous systems. *A*. Radial arrangement of nerves in the acoel *Anaperus*. (After Westblad.) *B*. Ventral submuscular nerve net of polyclad, *Gnesioceros*. *C*. Anterior end of the nervous system of the planarian *Procerodes*. (Modified after Lang.)

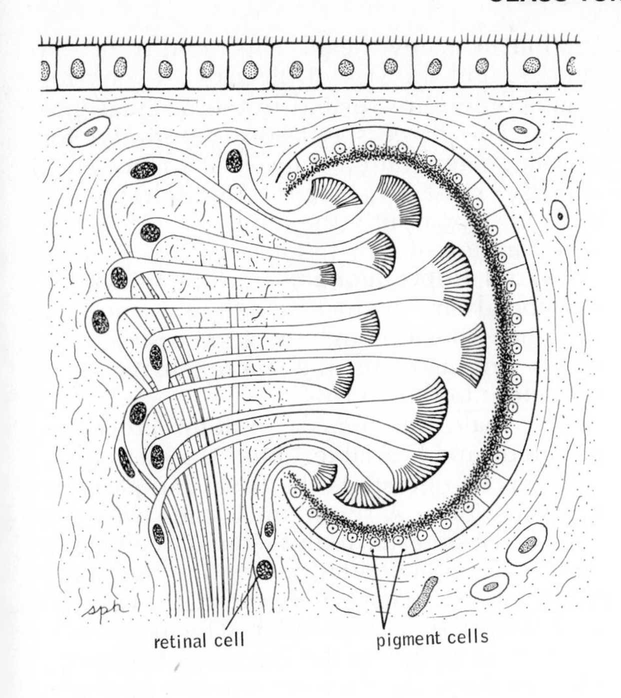

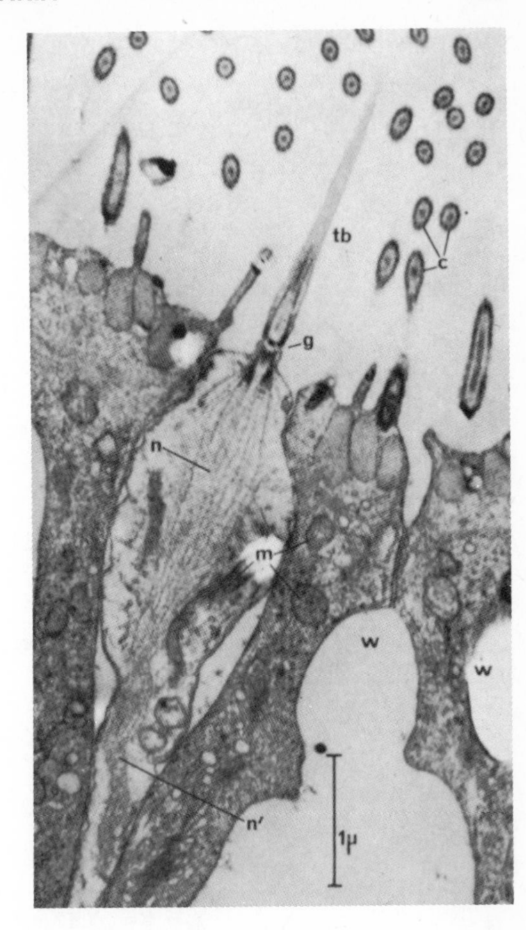

Figure 7–17. *A*. Inverse pigment-cup ocellus of the planarian *Dugesia*. (After Hesse.) *B*. Electron photomicrograph of a sensory cilium (tb) of the marine *Bothrioplana* (order Seriata). The letter c designates the ordinary cilia. (From Reisinger, E., 1968: Z. zool. Syst. Evolutionsforsch., *6*:1–55.)

lumen (Fig. 7–19*B*). The flagella of the flame bulbs drive the fluid down the connecting tubules.

The number of protonephridia and the position of the nephridiopores are variable, but they are typically paired. As many as four pairs occur in the triclads, and they often form an anastomosing network with many nephridiopores (Fig. 7–19*D*).

It is very probable that turbellarian protonephridia are osmoregulatory rather than excretory in function. The size and shape of the animal would present no problems for efficient removal of nitrogenous waste (ammonia) through the general body surface. Moreover, protonephridia are best developed in freshwater species, whereas the protonephridial system is reduced or absent in marine turbellarians, which have no problem in ridding the body of excess water.

Asexual Reproduction and Regeneration. Some turbellarians reproduce asexually by means of fission. For example, in the genera *Catenula*, *Stenostomum*, and *Microstomum*, all of which are composed of small individuals, fission is transverse; but the individuals remain attached, so that chains are formed (Fig. 7–20*D*). The individuals composing such chains are known as zooids. When a zooid attains a fairly complete degree of development, it detaches from the chain as an independent individual.

Some freshwater planarians, such as *Dugesia*, also reproduce by transverse fission, but no chains of zooids are formed and regeneration occurs after separation. The fission plane usually forms behind the pharynx, and separation appears to be dependent upon locomotion. The posterior of the worm adheres to the substratum, while the anterior half continues to move forward until the worm snaps in two. Each half then regenerates the missing parts to form two new small worms.

A few species of freshwater planarians, such as members of the genus *Phagocata*, and some land planarians fragment rather than undergo transverse fission. In *Phagocata*

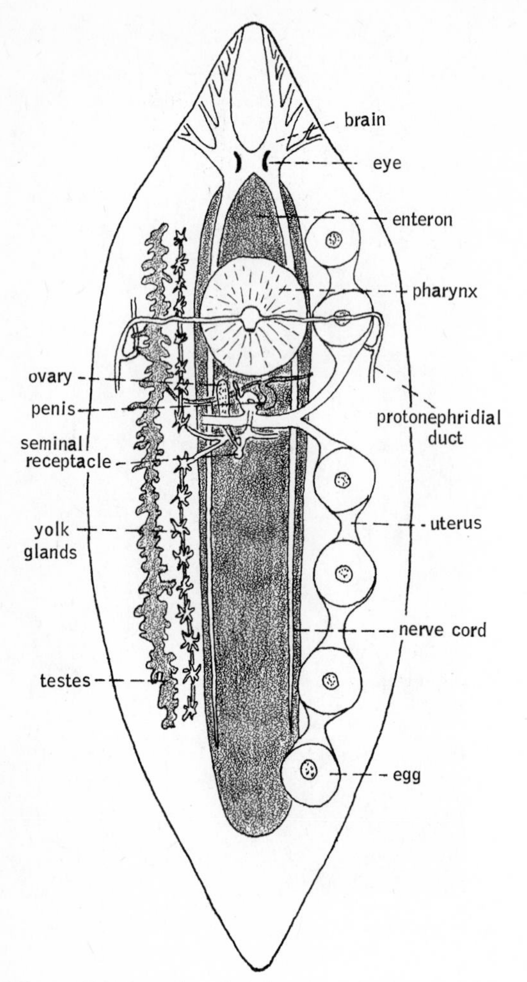

Figure 7–18. The rhabdocoel, *Mesostoma ehrenbergii* (ventral view).

each piece forms a cyst in which regeneration takes place and from which a small worm emerges.

Most flatworms have considerable powers of regeneration, and their regenerative ability has been investigated by numerous workers; Rose (1970) reviews the more important studies. Only a few aspects can be mentioned here. A distinct physiological gradient exists in flatworms, so that the body is polarized, the anterior representing one pole and the posterior representing the other. Regeneration is correlated with this polarity. For example, an excised piece retains its polarity, with the cut surface toward the anterior producing a new head and the posterior surface producing a new tail. One region suppresses the regeneration of the same region at another level of the body. For example, head extract added to the culture medium of a headless worm inhibits regeneration of a new head. The rapidity of the process and the size of pieces necessary for regeneration are related to the original gradient. The process is fastest anteriorly; for normal regeneration, pieces cut more posteriorly must be larger. Dedifferentiated cells are apparently the chief source for the regeneration of new tissue.

Sexual Reproduction. Except in acoels, the gonads are distinct from the surrounding mesenchyme, although the germ cells apparently originate in the mesenchyme and migrate into the gonads.

The male and female systems are complicated and variable, but *Macrostomum* will serve to illustrate the basic and perhaps primitive plan. A sperm duct leads from the single pair of testes to the seminal receptacle (Fig. 7–8A). The latter passes into the penis bulb, which bears glands (prostatic vesicle) and is armed with a stylet. The penis lies within the male genital canal, which opens through the male gonopore onto the posterior ventral surface. An oviduct leads from the pair of ovaries to the bursa, a sperm storage center, and the latter communicates with the vagina and female gonopore located in front of the male opening. Cement glands surround the vagina.

In other turbellarians there may be more than one pair of testes present. The penis is generally muscular but does not always bear a stylet. In some neorhabdocoels there is an eversible copulatory organ called a cirrus. There are also a number of turbellarians, including some polyclads, which display the peculiar condition of having multiple male parts, such as prostatic glands, seminal vesicles, and penises (Fig. 7–21). However, in at least some of these worms the multiple penis bulbs and stylets function in defense rather than in reproduction.

In the female system, as in the case of the male system, there may be one to numerous pairs of ovaries, but with only one pair of oviducts. In some turbellarians the ovaries are like those of other animals, and produce eggs in which yolk material is an integral part of the cytoplasm (Fig. 7–8). In many turbellarians, however, the ovary is of a more specialized form. A division of labor has evolved, and part of the ovary, called the vitellarium (or sometimes yolk gland), has become specialized for the production of yolk cells (modified eggs); while part of the ovary, called the germinarium, has become specialized for the production of yolkless eggs. The germinarium and vitellarium may

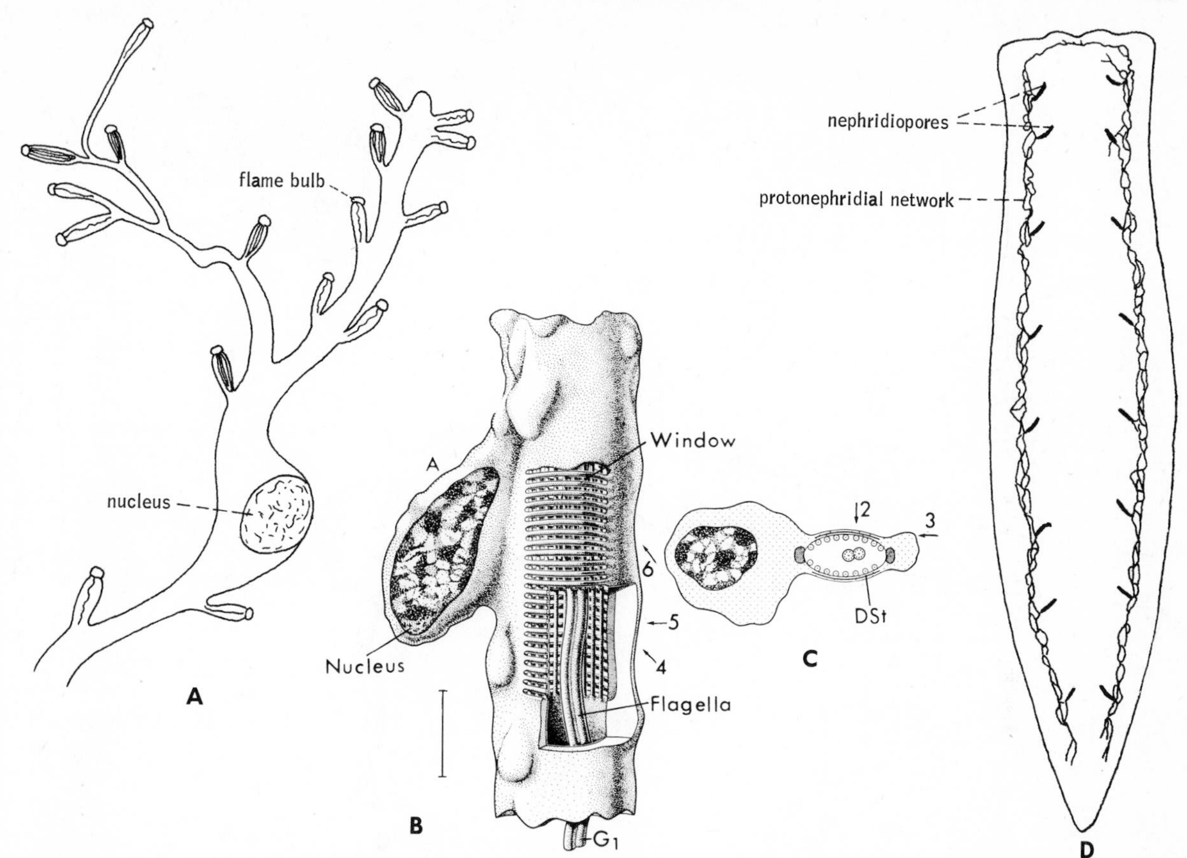

Figure 7–19. *A.* Part of *Mesostoma* protonephridium. (After Reisinger from Hyman.) *B* and *C.* Flame cell, or bulb, of *Stenostomum: B,* lateral view of cell; *C,* cross section at level of nucleus. (*B* and *C* from Kummel, G., 1962: Z. Zellforsch., 57:172–201.) *D.* Excretory system of triclad, *Dendrocoelum lacteum.* (After Wilhelmi from Hyman.)

be united, or they may be completely separated, with a special duct bringing the yolk cells to the oviduct (Figs. 7–14 and 7–18). In either case, the egg, after being released from the germinarium, becomes surrounded by a number of yolk cells and the entire mass is deposited.

The division of the ovaries into two types is the basis for separating the Turbellaria into the two levels of organization, the Archoophora and the Neoophora. The more primitive Archoophora, in which there are no yolk glands and the eggs contain yolk as in other animals, include the acoels, the macrostomids, the catenulids, and the polyclads. The Neoophora, in which the female system contains yolk glands and the eggs are accompanied by yolk cells, include the neorhabdocoels and the triclads. The two other classes of flatworms, the Trematoda and the Cestoda, also possess specialized ovaries with vitellaria.

In addition to the bursa, the oviducts may dilate to provide still another storage center, a seminal receptacle (Figs. 7–14 and 7–18). The sperm are stored in the seminal receptacle until fertilization. The seminal receptacle may receive sperm from the copulatory bursa or, if a bursa is absent, directly after copulation.

In some species, such as in *Bdelloura,* the bursa is not connected to the atrium but has a separate opening and vagina for receiving the male penis. Special canals for transferring the sperm connect the bursa with the oviducts. Such a condition—where two female openings are present, one for the reception of the male penis and sperm, and one for the passage of eggs to the exterior—is not restricted to the flatworms but is encountered among other invertebrates.

Another modification sometimes present in the female system is a temporary storage center, or uterus, for ripe eggs. The uterus may be a blind sac (Fig. 7–18), as in some neorhabdocoels; or it may be merely a dilated part of the oviduct, as in the polyclads. However, most Turbellaria lack uteri, because only a few eggs are laid at a time.

In acoels the female system is less well

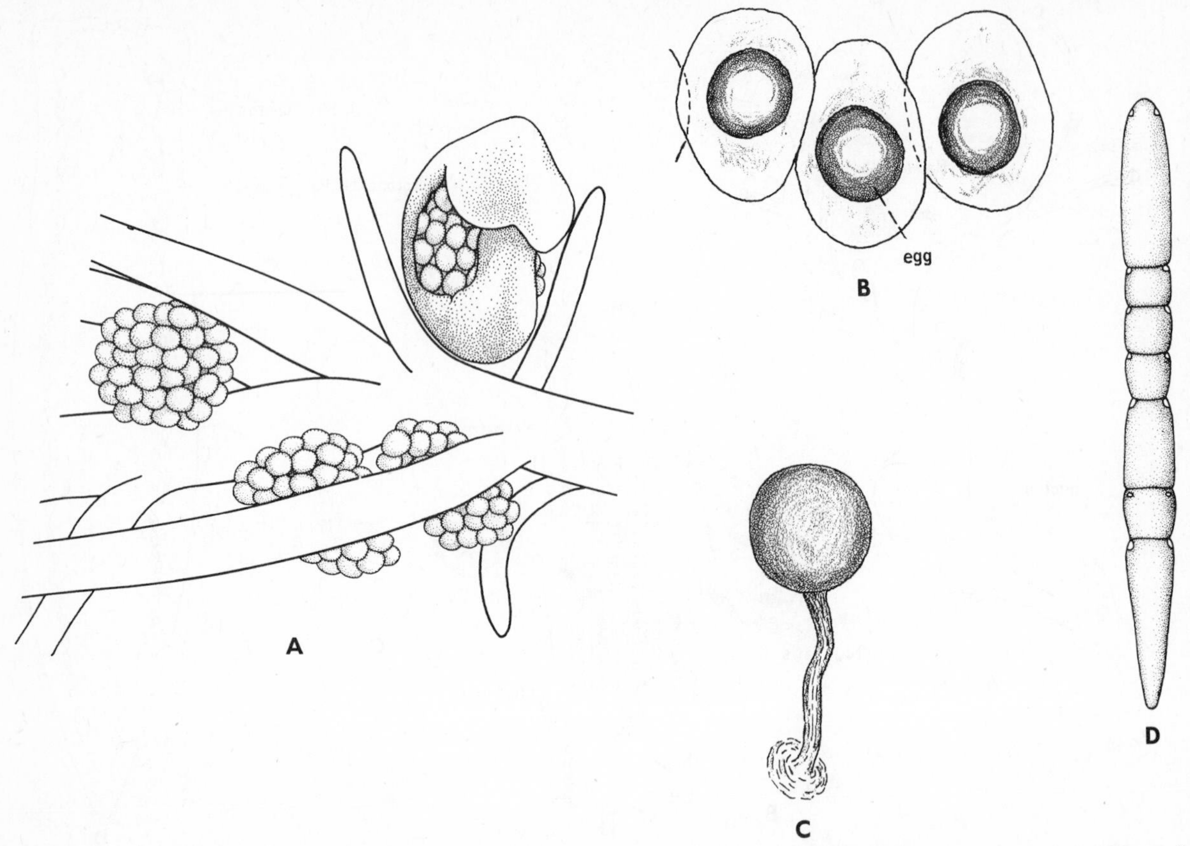

Figure 7–20. *A*. Egg masses of *Convoluta*. (From Apelt, G., 1969: Mar. Biol., *4*(4):279.) *B*. Section of egg ribbon of the polyclad, *Stylochus*. (After Kato from Hyman.) *C*. Stalked cocoon of a fresh-water planarian. (Modified from Pennak.) *D*. Chain of zooids in *Stenostomum*. (After Child from Hyman.)

developed than in other turbellarians. Some possess no female ducts at all, not even a gonopore. In others, there are no oviducts, but a short blind vagina for the reception of the male penis leads from a female gonopore (Fig. 7–8*B*). It is believed by some zoologists that this condition in acoels is a reduction from the more developed condition described for *Macrostomum*. Others consider it primitive. Separate male and female gonopores are characteristic of most Macrostomida, the Acoela, and the Polycladida, and this is probably the primitive turbellarian condition; but many turbellarians, including the common planarian, possess a single gonopore and genital atrium into which both male and female systems open.

Sperm transfer in turbellarians involves copulation and is usually reciprocal. Hypodermic impregnation is common and is especially characteristic of forms that possess a penis stylet, or of acoels that lack female ducts. The penis is rammed deep into the body wall of the copulating partner, depositing sperm into the mesenchyme (Figs. 7–22*A–C*). The sperm then migrate to the ovaries.

In other turbellarians, copulation involves the insertion of the penis into the female gonopore or common gonopore of the other (Fig. 7–22*E*). During copulation the worms orient themselves in a variety of ways with the ventral surfaces around the genital region pressed together and elevated (Fig. 7–22*D*).

Self-fertilization is uncommon in flatworms, as is true of most hermaphroditic animals. In turbellarians, self-fertilization is perhaps precluded, in part, because movement of the sperm mass out of the male system and production of seminal fluid are dependent on ejaculation, which occurs only during copulation.

Turbellarians which possess oviducts but lack yolk glands release only a small number of eggs. For example, species of the acoel genera *Convoluta* and *Archaphanostoma* lay from one to thirty eggs. The marine polyclads lay their eggs in gelatinous strings (Fig. 7–20), the adhesive jelly being produced by glands in the atrium. On reaching the exterior, the surface of each egg or of several eggs throws off a shell or capsule, which becomes hardened (Fig. 7–20*B*). The eggs require up to one week to hatch, and an individual may lay a number of egg masses. Acoels which have no gonoducts release their eggs through the pharynx or by rupture of the body wall.

Neorhabdocoels and triclads possess yolk glands, and, as a result, egg production is somewhat modified. As the fertilized eggs pass through the oviduct, they are accompanied by yolk cells produced by the yolk glands. On reaching the atrium, one or several eggs, along with many yolk cells, are enclosed by a hard capsule produced by the yolk cells themselves. The glands of the atrium produce a cementing secretion that helps the capsules adhere to the substratum. In many forms, including some freshwater triclads such as the common *Dugesia*, the attachment of the capsules is by means of a stalk (Fig. 7–20*C*), so that the brownish capsules resemble little balloons stuck to rocks and other debris. One worm can produce a number of capsules.

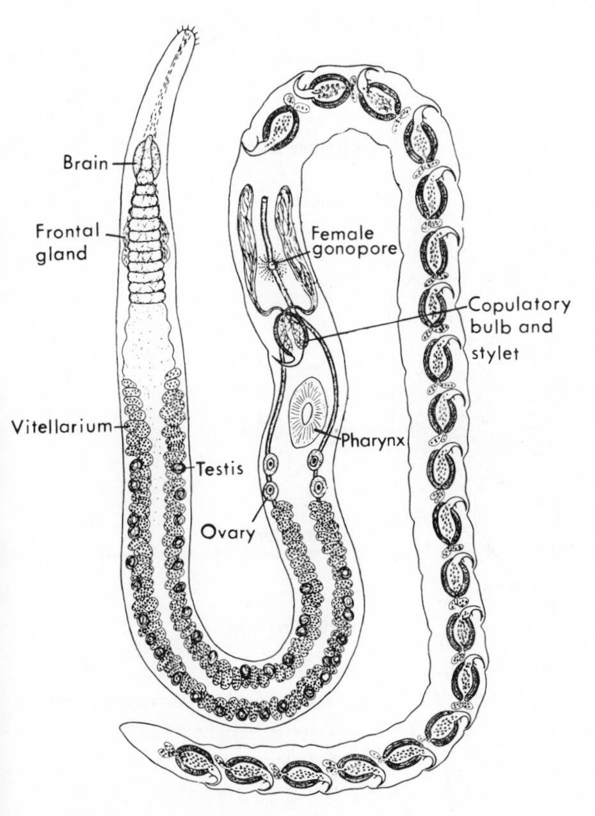

Figure 7–21. *Polystyliphora* (suborder Proseriata), a turbellarian with multiple copulatory bulbs and stylets. Since only the first receives sperm, the others may serve a defensive function. (From Ax, P., 1958: Zool. Anz. Suppl., *21*:227–249.)

Some freshwater turbellarians often produce two types of eggs—summer eggs, which are enclosed in a thin capsule and hatch in a relatively short period, and winter eggs, which have a thicker and more resistant capsule. Winter eggs remain dormant during the winter, resisting freezing and drying, and hatch in the spring with the rise in water temperature. In some neorhabdocoels, such as *Mesostoma*, the dormant eggs are liberated upon the death of the parent. Parthenogenesis is characteristic of some turbellarians, and there are some parthenogenic species in which males are unknown.

Reproduction may be controlled by day length and temperature. Freshwater planarians, for example, which are almost all inhabitants of temperate regions, reproduce asexually by fusion during the summer, and sexually during the fall under the stimulus of shorter day lengths and lower temperatures.

As we have seen, the eggs of archoophoran turbellarians—acoels, polyclads, and macrostomid rhabdocoels—are entolecithal; i.e., the yolk is an integral part of the egg cytoplasm. Early development in these groups is similar to that in other protostome phyla. In acoels, cleavage is spiral but somewhat different from the pattern described in Chapter 3. The egg undergoes a single vertical cleavage rather than two. As a result, there are only two cells, *A* and *B*, to form the micromeres. The micromeres are produced in duets rather than in quartets, although the typical four sets are formed (Fig. 7–23). The gastrula is solid (stereogastrula), and this level of development is perhaps retained by the adult. The blastopore closes, and the mouth forms as a new opening near the same site by means of a stomodeal invagination. The adult digestive tract never progresses beyond this point. The nervous system is derived from the first duet, and the remaining external micromeres form the epidermis. There is no free-swimming larval stage. The acoel *Archaphanostoma agile* completes embryonic development in two to five days and reaches sexual maturity in about 23 days after hatching.

Development in the Polycladida, Macrostomida, and Catenulida conforms much more closely to the typical pattern of spiral

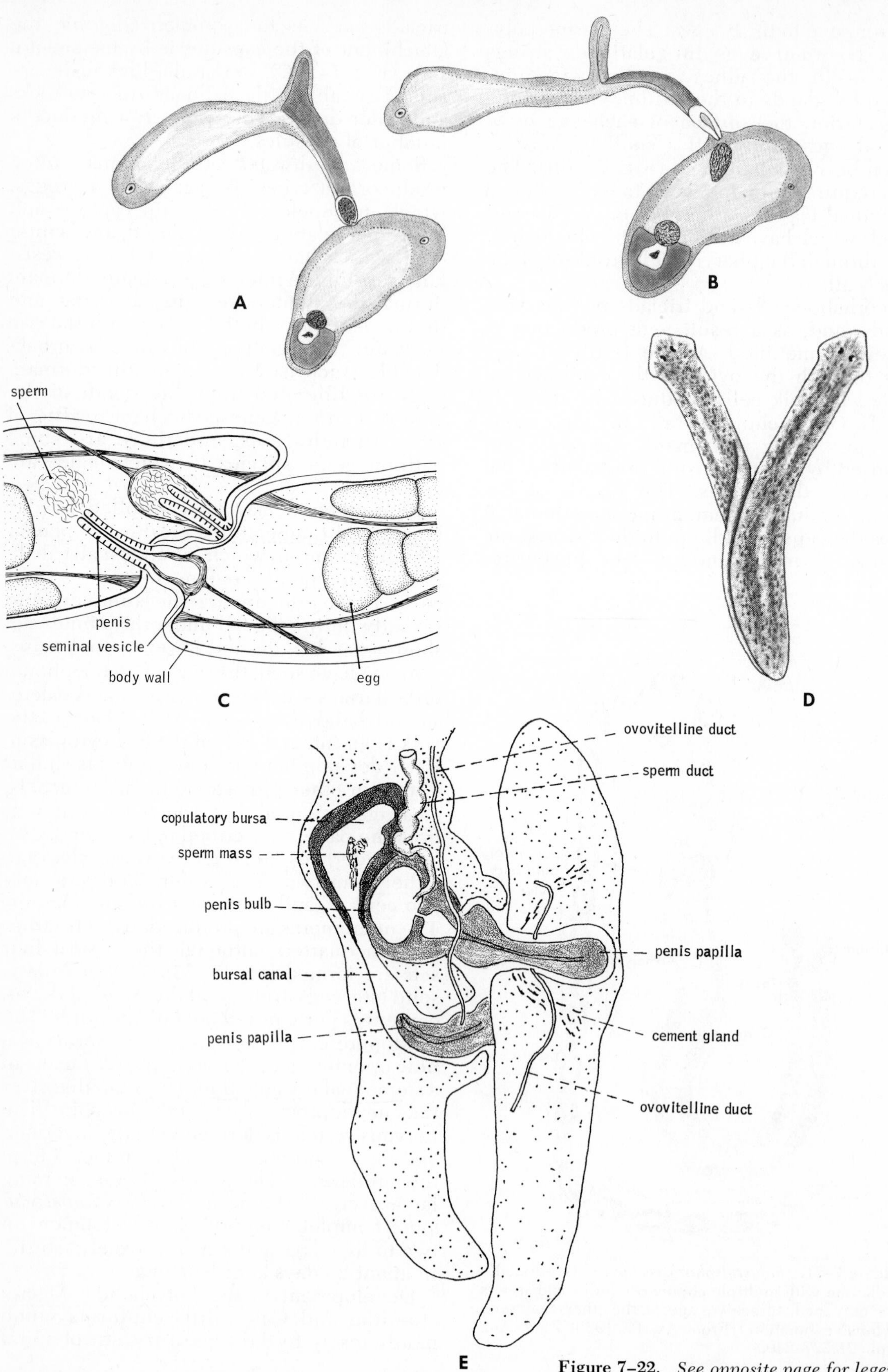

Figure 7–22. *See opposite page for legend.*

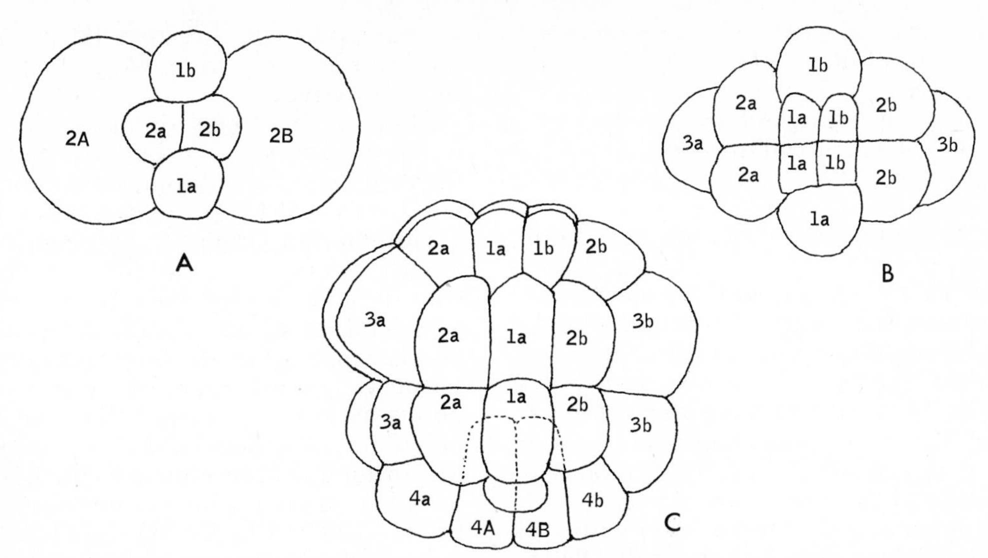

Figure 7–23. Cleavage in the acoel, *Polychoerus*. *A* and *B*. Views from the animal pole. *C*. Lateral view. (After Gardiner from Hyman.)

cleavage described in Chapter 3. Gastrulation is by epiboly and produces a stereogastrula. The mouth and pharynx form from a stomodeal invagination near the original site of the blastopore. This invagination connects with the enteron, which has formed from an entodermal mass and become hollow (Fig. 7–24*A*).

In most polyclads there is no larva, but in some there is a free-swimming stage called a Muller's larva (Fig. 7–24*B*). Eight arms or lobes, which are directed posteriorly and bear long cilia, form as extensions of the body. The ciliary tuft of the frontal gland projects forward. The larva swims about for a few days and then settles to the bottom as

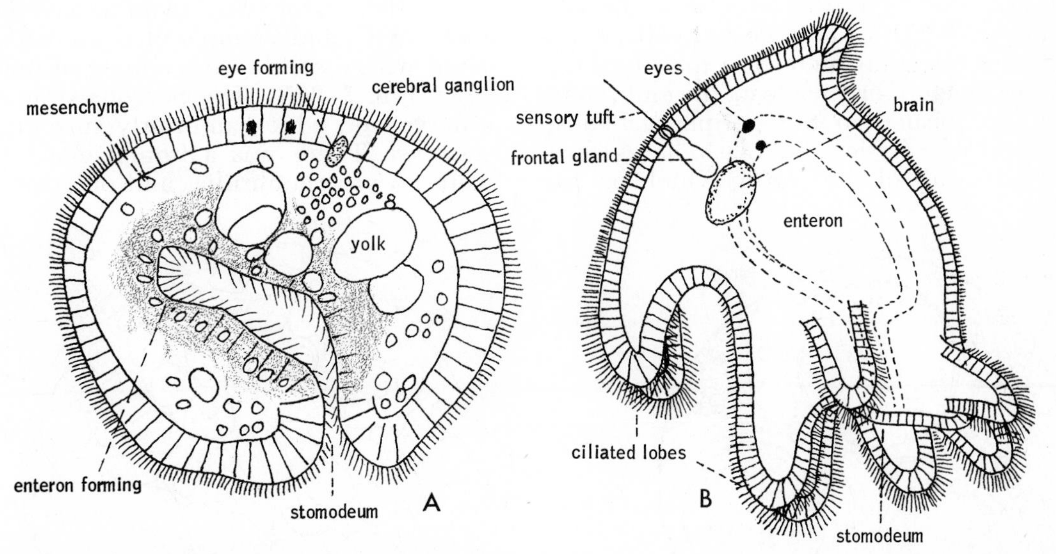

Figure 7–24. *A*. Late gastrula of the polyclad, *Hoploplana*. (After Surface from Hyman.) *B*. A Müller's larva (lateral view). (After Kato from Hyman.)

Figure 7–22. Copulation in turbellarians. *A* and *B*. Hypodermic impregnation in the acoel *Pseudaphanostoma*. (From Spelt, G., 1969: Mar. Biol., *4*:278.) *C*. Detailed section of hypodermic impregnation between two individuals of the acoel *Archaphanostoma agile*. (After Apelt.) *D*. Copulating planarians. *E*. Section through a pair of copulating *Dugesia*. (After Burr from Hyman.)

a young worm. Feeding probably does not commence until after the larval stage has been passed.

Some species of *Stylochus* pass through a similar larval stage called Götte's larva. It differs from a Muller's larva in having only four ciliated lobes.

The neoophorans—triclads and many rhabdocoels, which have yolk glands and whose yolk is separate and outside the egg (i.e., ectolecithal eggs)—have undoubtedly evolved from forms that had spiral cleavage, but in most species the presence of external yolk cells has so altered the pattern of development that it no longer bears any resemblance to the ancestral type. The initial blastomeres divide and separate to form an embryonic mass of cells intermingled with yolk cells. In triclads some of the cells flatten and form a membrane enclosing the remaining blastomeres and part of the yolk. The blastomeres and yolk surrounded by this membrane are called the embryonic mass. The remainder of the yolk forms a syncytium surrounding the membrane and the embryonic mass. Several such masses are enclosed within a single capsule (Fig. 7–25*A*). Some of the blastomeres in such an embryonic mass now collect at one point within the membrane and form a temporary pharynx and intestine, which ingest the surrounding yolk syncytium into the interior of the embryo (Fig. 7–25*B*). The ingested yolk so distends the temporary intestine that the blastomeres that formerly made up the embryonic mass are pushed to the periphery. These now peripherally-located blastomeres form the body wall of the worm, which on the ventral side proliferates three anterior-posterior masses, from which most of the body is derived.

There is no free-swimming larval stage in these orders of Turbellaria that exhibit the ectolecithal condition; the young worms emerge from the capsule in a few weeks.

Turbellarian Origins. Although most zoologists would agree that the neoophorans—triclads and neorhabdocoels—represent an advanced level of organization, and that acoels, macrostomids, and catenulids are primitive groups, there are different viewpoints as to the nature of the ancestral turbellarians. A widely held view considers the Macrostomida to be closest to the base of the turbellarian phylogenetic tree. As elaborated by Ax (1963) (Fig. 7–26*A*), such a "macrostomid" ancestor possessed a simple pharynx and a ciliated intestine without diverticula, a radial arrangement of longitudinal nerve cords, statocyst, frontal gland, a pair of protonephridia, an hermaphroditic system like that described for *Macrostomum*, and entolecithal eggs with spiral cleavage. The Acoela and Catenulida, although considered primitive, are believed to have diverged further from the ancestral form than have the Macrostomida.

Another view holds that the Acoela are the most primitive turbellarians. Such an acoeloid ancestor is believed to have lacked a gut cavity, and a simple pharynx communicated with a solid interior mass of nutritive cells (Fig. 7–26*B*). The nervous system was a nerve net, which anteriorly surrounded a statocyst. There was a frontal gland; there were no protonephridia; and only the male

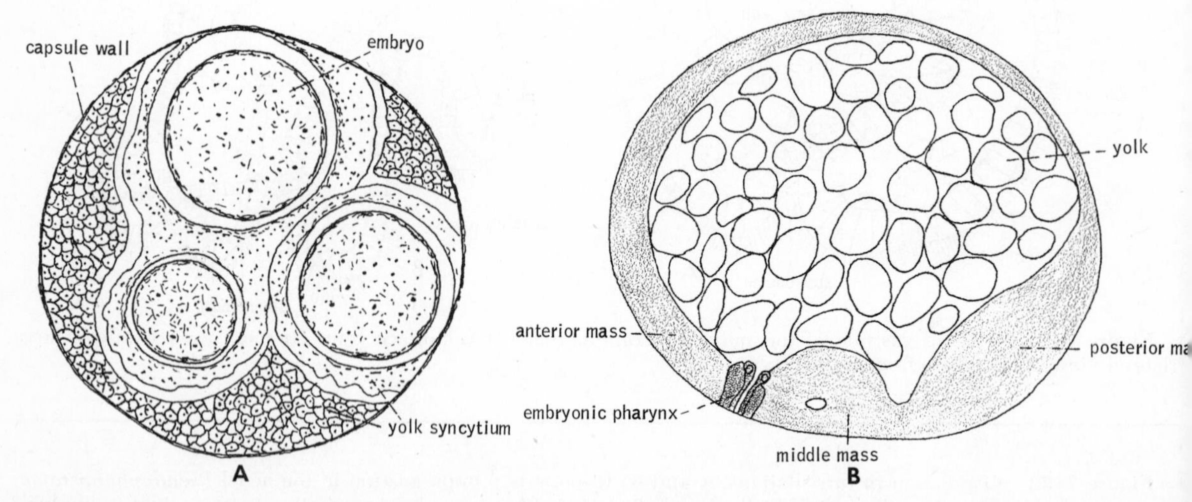

Figure 7–25. *A*. Section through a triclad capsule, showing three embryos imbedded in yolk syncytium. (After Metschnikoff from Hyman.) *B*. Triclad embryo after ingestion of yolk and formation of three embryonic masses. (After Fulinski from Hyman.)

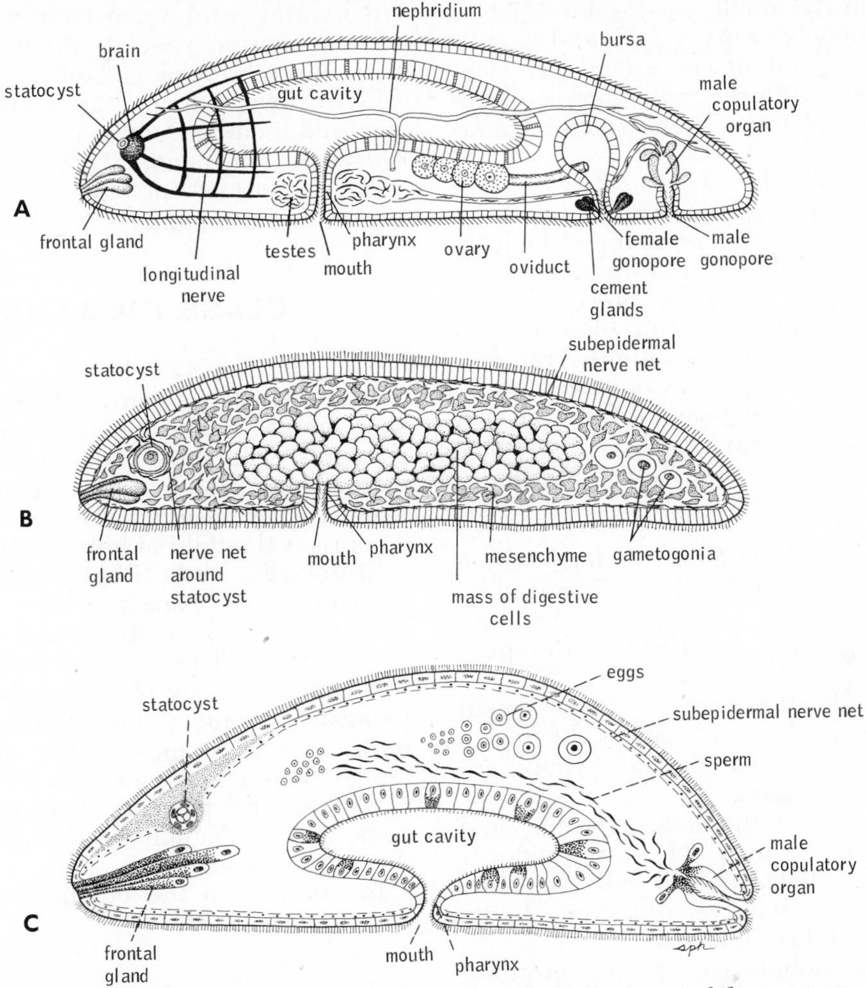

Figure 7–26. Hypothetical turbellarian ancestors (sagittal sections). *A.* Ancestral form suggested by Ax (1963). Similar to members of the existing order Macrostomida. *B.* Ancestral acoeloid flatworm. *C.* Ancestral form according to Karling (1973).

reproductive system possessed ducts and a gonopore.

Most recently, Karling (1973) has postulated a turbellarian ancestor which is somewhat intermediate between the acoeloid and macrostomidan forms (Fig. 7–26*C*). According to Karling, the digestive system was saclike as in the Macrostomida. But, as in acoels, there were perhaps no nephridia, and the nervous system was confined to an epidermal nerve net. No female ducts were present, and sperm transfer would have been by hypodermic impregnation.

SYSTEMATIC RÉSUMÉ OF THE CLASS TURBELLARIA

Archoophoran turbellarians. Orders which reflect a more primitive level of organization. Yolk glands absent; eggs entolecithal; cleavage spiral.

Order Acoela. Small marine flatworms, usually measuring less than 2 mm. in length. Mouth and sometimes a simple pharynx present, but no digestive cavity. Protonephridia absent. No distinct gonads, the gametes originating directly from the mesenchyme. Oviducts absent. *Amphiscolops, Anaperus, Afronta, Polychoerus, Convoluta, Archaphanostoma, Pseudaphanostoma, Nemertoderma.* A few species are commensal within the intestine of various echinoderms.

Order Macrostomida. Small marine and freshwater species having a simple sac-like ciliated intestine, a simple pharynx, and one pair of lateral nerve cords. *Macrostomum, Microstomum.*

Order Catenulida. This aberrant group

contains mostly small freshwater species having a simple pharynx and ciliated sac-like intestine, two pairs of nerve cords, and unpaired gonads, with the male gonopore dorsal above the pharynx. *Stenostomum, Catenula.*

Order Polycladida. Marine flatworms of moderate size, averaging from 3 to 20 mm. in length, with a greatly flattened and more or less oval shape. A pair of anterior marginal or dorsal tentacles may be present. Many are brightly colored. Intestine elongate and centrally located, with many highly branched diverticula. Plicate pharynx either an anteriorly directed tube or pendant from the roof of the pharyngeal cavity. Eyes numerous. *Gnesioceros, Leptoplana, Notoplana, Stylochus, Prostheceraeus.*

Neoophoran turbellarians. Orders which reflect an advanced level of organization. Yolk glands present, eggs ectolecithal, and development greatly modified from the spiral pattern.

Order Prolecithophora. Usually small marine and freshwater species having a plicate or bulbous pharynx and a simple intestine. *Plagiostomum.*

Order Lecithoepitheliata. Marine and freshwater species in which ovary produces both egg and follicle-like yolk cells. Intestine simple, four pairs of nerve cords, penis in the form of a stylet. *Prorhynchus.*

Order Neorhabdocoela. A large group of small marine and freshwater turbellarians, having a bulbous pharynx, a simple intestine, and one pair of nerve cords. Suborder Typhloplanoida contains largely freshwater, free-living species; *Mesostoma, Bothromesostoma.* Suborder Dalyellioida contains marine and freshwater species commensal and parasitic on and within snails, clams, sea urchins, and sea cucumbers; *Anoplodiera.* Suborder Kalyptorhynchia contains mostly marine species; *Gyratrix, Gnathorhynchus.*

Order Temnocephalida. Commensal and parasitic turbellarians on crustaceans, mollusks, and turtles. Posterior ventral surface provided with an adhesive disc, and anterior margin bears finger-like projections.

Order Seriata. Turbellarians with folded pharynx and lateral gut diverticula. The members of the suborder *Proseriata*, in which the gut is not branched, include many marine interstitial forms, as well as some freshwater species; *Otoplana, Nematoplana.* The suborder Tricladida, in which the gut is three-branched, contains relatively large marine, freshwater, and terrestrial turbellarians. Among marine species, *Bdelloura* is commensal on the book gills of horseshoe crabs. The freshwater species are known as planarians and include *Dendrocoelum, Procerodes, Dugesia, Phagocata, Polycelis, Procotyla.* Land planarians include *Bipalium, Orthodemus* and *Geoplana.*

CLASS TREMATODA

The flatworms belonging to the class Trematoda are known as flukes and represent one of the major groups of metazoan parasites. The flukes, along with the tapeworms and roundworms, include the vast majority of parasitic worms, and their study comprises the field of helminthology.

Structure and Physiology. Although trematodes are parasitic, their structure is more like that of turbellarians than is that of the third flatworm class, the parasitic Cestoda. The body of trematodes is oval to elongate and ranges in length from less than 1 millimeter to 7 meters, but most are no longer than a few centimeters. Ventral and oral adhesive organs are characteristic of the class, and the mouth is typically located at the anterior end.

In contrast to the ciliated epidermis of

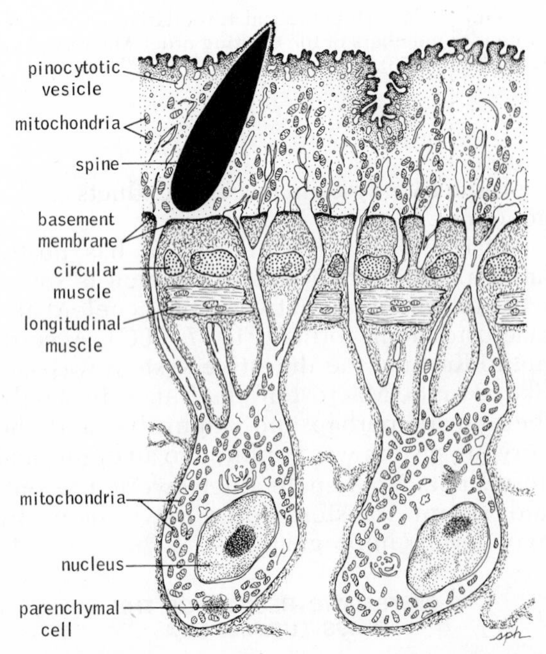

Figure 7–27. Section through the integument of the sheep liver fluke, *Fasciola hepatica.* (After Threadgold.)

turbellarians, the body of trematodes is covered by a non-ciliated cytoplasmic syncytium, the tegument, overlying consecutive layers of circular, longitudinal, and diagonal muscle. The syncytium represents extensions of cells that are located in the mesenchyme (Fig. 7–27).

The mouth leads into a muscular pharynx that pumps into the digestive tract the cells and cell fragments, mucus, tissue fluids, or blood of the host on which the parasite feeds. The pharynx passes into a short esophagus. From the esophagus the digestive tract gives rise to one or—more commonly—two intestinal ceca which extend posteriorly along the length of the body (Fig. 7–28A). The ceca are generally simple cyclindrical tubes but are branched in some species. The physiology of nutrition is still incompletely understood, but secretive and absorptive cells

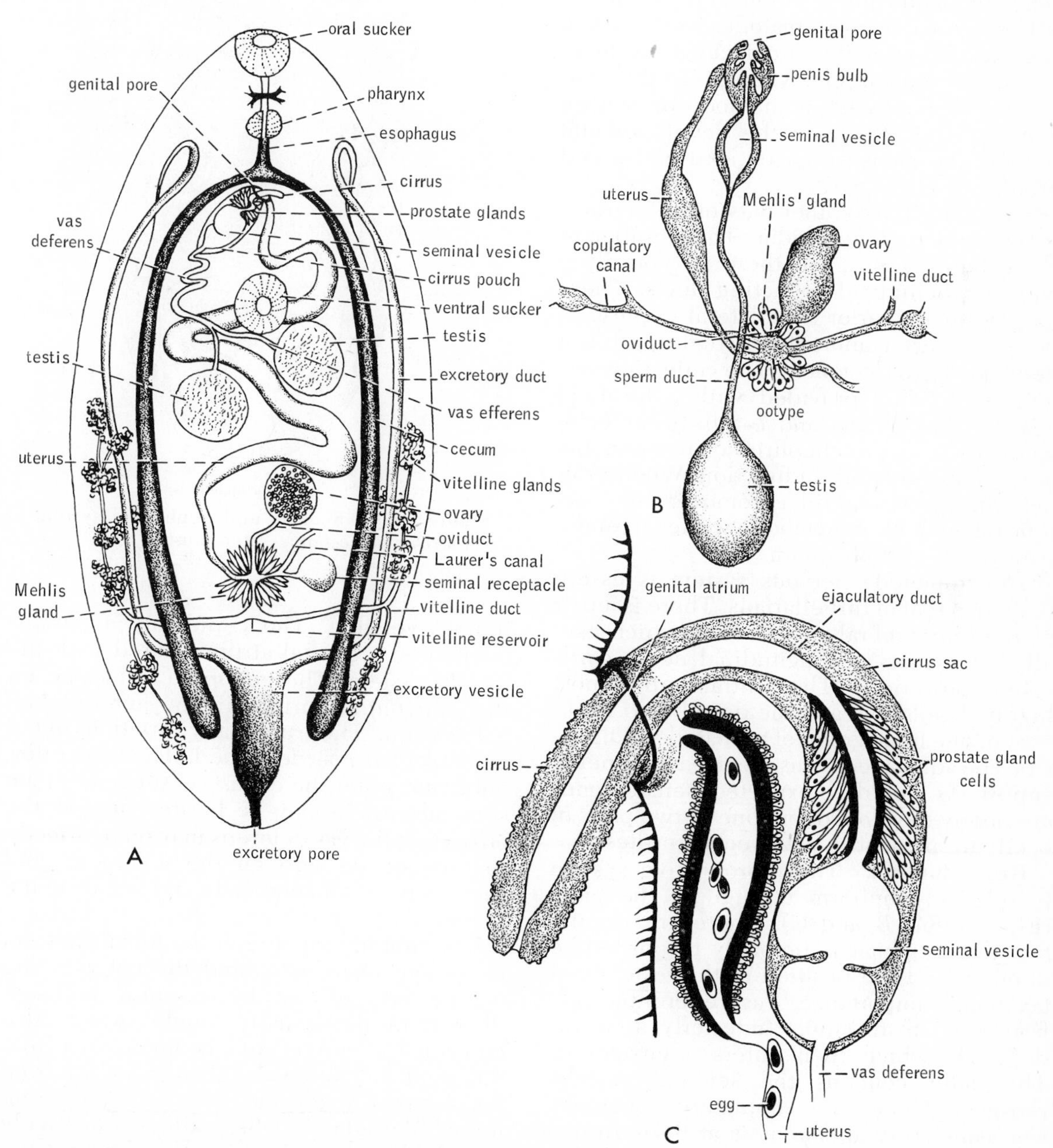

Figure 7–28. *A.* Structure of a generalized trematode. (After Chandler and Read.) *B.* Reproductive systems of a monogenetic trematode. (After Cheng.) *C.* Cirrus protruding from gonopore. (After Noble and Noble.)

have been reported so that digestion is apparently extracellular in part.

The tegument plays a vital role in the physiology of trematodes. It provides protection, especially against the host's enzymes in gut-inhabiting species. Nitrogenous wastes are passed to the exterior through the tegument, and it is the site of gas exchange. In endoparasites there is some amino acid absorption by the tegument. The protein synthesis involved in trematode egg production and in larval reproduction places especially heavy demands on the amino acid supply.

The ectoparasitic trematodes are aerobic, but the endoparasites are facultative anaerobes. The amount of oxygen utilized in respiration depends upon the location within the host and also upon the developmental stage of the parasite. Lactic acid is the end product of glycolysis.

Trematodes, like other flatworms, are provided with protonephridia. The number of flame bulbs varies, but there is typically a pair of longitudinal collecting ducts. There may be two anterior dorsolateral nephridiopores or, very commonly, a single posterior nephridiopore (Digenea). The collecting system is typically provided with a terminal vesicle (Figs. 7–28*A* and 7–29). In the ectoparasites, the protonephridia are probably only osmoregulatory in function. Whether or not the protonephridia remove any significant amount of metabolic wastes in endoparasites is still uncertain.

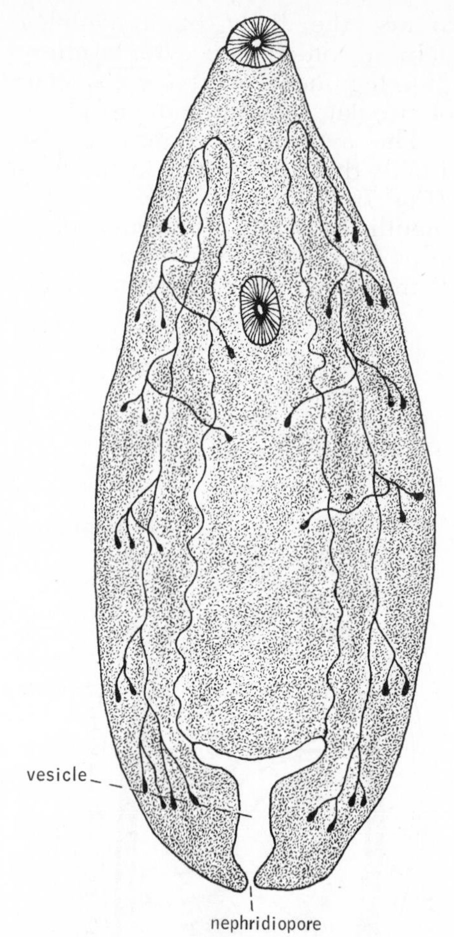

Figure 7–29. Suckers and nephridial system of *Opisthorchis pedicellata*. (After Faust.)

The trematode nervous system is essentially like that of turbellarians. There is a pair of anterior cerebral ganglia from which usually three pairs of longitudinal nerve cords extend posteriorly. The ventral cord is most highly developed, and the dorsal cord is absent in one large group of trematodes (Digenea). The adhesive organs receive a rich nerve supply. As would be expected, sense organs are poorly developed, but one or two pairs of ocelli are present in many ectoparasites.

Reproduction. The reproductive system is relatively uniform throughout the class (Fig. 7–28*A*, *B* and *C*). There are usually two testes, which is probably the primitive number, and the position of the testes is of taxonomic importance. Vasa efferentia, one from each testis, unite anteriorly as a vas deferens, which then enters a cirrus sac. The latter contains the seminal vesicle, prostate glands, and copulatory apparatus. The copulatory apparatus is at the terminal end of the male system, and is called a cirrus if it is an eversible structure or a penis if not eversible. The copulatory apparatus opens into a genital atrium shared with the female system. The gonopore is located on the midventral surface in the anterior half of the worm. There are many variations of the general plan just described. The vasa efferentia may enter the cirrus sac separately; the seminal vesicle may be located outside the cirrus sac; the vas deferens may open directly to the outside through the gonopore; and there may be separate male and female gonopores.

The central structure of the female system is a small chamber called the ootype. The ootype receives eggs by way of a short oviduct from the usually single ovary. Also entering the ootype are a common duct from the right and left yolk glands, a duct from the seminal receptacle, and secretions from unicellular gland cells, called collectively the Mehlis' gland. Leaving the ootype is the uterus, which runs anteriorly to the genital atrium. The atrium has a muscular terminal end that facilitates the expulsion of encapsulated eggs. Two other structures sometimes form a part of the female system; there may be a vitelline reservoir, which is an evagination of the vitelline duct, and there is sometimes present one or two vaginae (Monogenea). When present, the vaginae open separately to the exterior on the dorsal, lateral, or ventral surface. A part of the length may also be modified as a seminal receptacle. An important variation in this basic plan of the female system occurs in the digenetic trematodes, where the vitelline duct and vagina or seminal receptacle join the oviduct prior to entering the ootype.

Sperm, on leaving the testes, are stored in the seminal vesicle. Copulation is mutual and cross fertilization is the general rule, although self-fertilization does occur. During copulation the cirrus or penis of the male system of one worm is inserted into the uterine or vaginal opening of the other worm and sperm are ejaculated. The prostate gland provides semen for sperm survival. Sperm travel down the uterus or the vaginal canal to be stored in the seminal receptacle.

The egg when released from the ovary is fertilized either en route to the ootype or within the ootype. The egg is ectolecithal, as are those of many turbellarians. The vitelline glands supply yolk material for the egg and also a material which hardens around the egg to form the shell. The encapsulated eggs pass through the uterus to be expelled to the exterior. The function of the Mehlis' gland is uncertain, but its secretions may provide lubrication for the passage of eggs through the uterus. Compared to the free-living turbellarians, the number of eggs produced by trematodes is enormous. Cheng (1964) cites an output figure of 10,000 to 100,000 times that of turbellarians.

The trematode life cycle involves one to several hosts. The primary host, that of the adult parasite, is almost always a vertebrate. Fish are by far the principal victims of trematode parasitism. Mammals, which by no means escape trematode infestation, are the hosts for a relatively small number of trematodes as compared to other vertebrate groups.

In order to discuss trematode life cycles, it is necessary to consider the three trematode orders separately:

Order Monogenea. The monogenetic trematodes derive their name from the fact that there is but a single host in the life cycle. They are also distinguished from the digenetic trematodes in the possession of a large posterior adhesive organ called an opisthaptor. The opisthaptor is a complex muscular organ which attaches the parasite to the host and typically bears a number of suckers plus sclerotized pieces called hooks, anchors, and bars (Fig. 7–30*A* and *B*). The structures of the opisthaptor are variously arranged and modified in different species. In addition to the opisthaptor, there is usually a pair of anterior suckers flanking the mouth (Fig. 7–30*A*).

The monogeneid trematodes are largely parasitic on marine and freshwater fish, but amphibians, reptiles, and invertebrates also serve as hosts. The majority are ectoparasites, but some have migrated into body chambers with external openings, such as the mouth, gills, chambers, and urogenital tract. A few are even found in the coelom.

The shell of the elongate egg is provided with a lid and usually one or two threads which function to attach the egg to the host (Fig. 7–31*A*). Early development is highly modified, as is true of the ectolecithal development of turbellarians. The eggs, on hatching, release a free-swimming ciliated larva called an onchomiracidium that enables the parasite to reach a new host (Fig. 7–31*B*). The precocious development of the adhesive organ enables the larva to attach. Space permits only a few specific life cycles to be described, and it should be realized that there are many variations.

Polystoma integerrimum is found in the bladder of frogs and toads and is an example of remarkable synchronization of the life cycle with that of the amphibian host (Fig. 7–32). The eggs are shed when the frog or toad returns to the water to breed. The onchomiracidium attaches to the gills of the tadpoles. When the tadpoles metamorphose, the parasite leaves the gill chamber and passes down the gut into the bladder.

Benedenia melleni is parasitic on the epidermis and eyes of a variety of marine fish. The parasite can cause blindness and serious damage to the host's integument. The life cycle follows the pattern already described. *Dactylogyrus elegans*, which parasitizes the gills of carp, produces a summer egg that develops rapidly and a winter egg that lays over in the bottom mud to hatch in the spring.

Species of the genus *Gyrodactylus* are common ectoparasites on the gills and body surface of marine and freshwater fish and

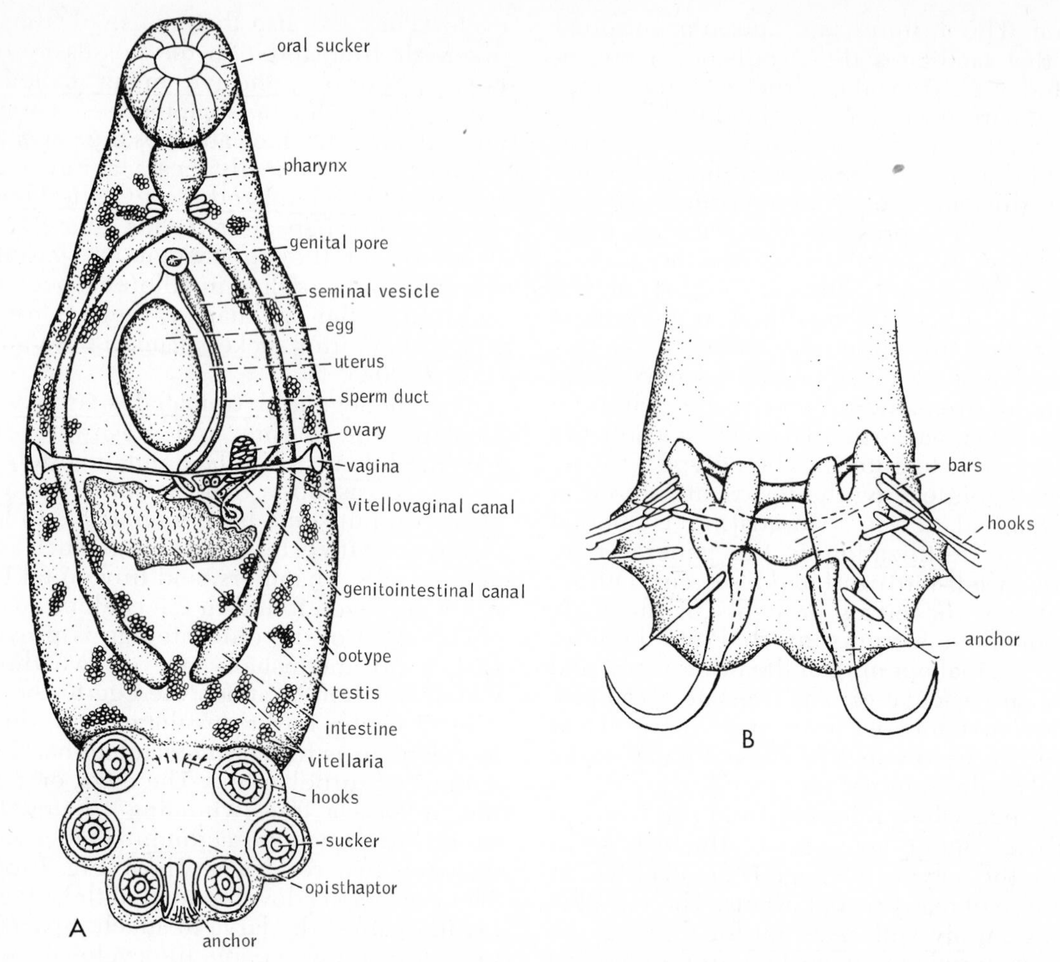

Figure 7–30. *A. Polystomoidella oblongum,* a monogenetic trematode parasitic in the urinary bladder of turtles. (After Cable.) *B.* Opisthaptor of *Dactylogyrus.* (After Mueller and Van Cleave.)

frogs (Fig. 7–33) and may be a serious pest of fish hatcheries. Intrauterine development of the larva takes place; moreover, a secondary larva develops within the first, a third within the second, and possibly even a fourth. After the first larva has been released and attaches to a new host, the second larva containing its progeny is released from the first.

Order Digenea. The order Digenea contains the greatest number of parasitic flatworms. Over 6000 species have been described to date, and new descriptions are continually being published. There are many species that cause parasitic disease in man and domesticated animals.

In contrast to the monogenetic trematodes, the life cycles of the digenetic trematodes involve two to four hosts. The host for the adult is the primary host, and the one to three hosts for the numerous larval stages are termed intermediate hosts. The adhesive organs are typically two large suckers (Fig. 7–35*A*). One sucker, called the oral sucker, is located around the mouth. The other sucker, the acetabulum, is located ventrally in the middle or posterior end of the body.

All the digenetic trematodes are endoparasites. The primary hosts include all groups of vertebrates, and virtually any organ system may be infected. The intermediate hosts are largely invertebrates, of which snails are perhaps the most important.

The life cycle is complex and will be introduced with a generalized scheme followed by more specific examples.

The egg is enclosed within an oval shell with a lid and is deposited into water. A ciliated, free-swimming miracidium hatches from the egg (Fig. 7–34*A*). This first larval stage, on contacting a mollusk, penetrates the epidermis of the intermediate host, and makes its way to the host's gut, ending up in the digestive gland.

Within the digestive gland, the miracidium, which lost its cilia on entrance into the host, develops into a second larval stage, called a sporocyst (Fig. 7–34*B*). Inside the hollow sporocyst, germinal cells give rise to a number of embryonic masses. Each mass develops into another larval stage, called a redia or daughter sporocyst, which is also a chambered form (Fig. 7–34*C*). Germinal cells within the redia again develop into a number of larvae called cercariae (Fig. 7–34 *D* and *E*).

The cercaria, a fourth larval stage in this scheme, possesses a digestive tract, suckers, and tail. The cercaria leaves the host and is free-swimming. If it comes in contact with a second intermediate host, an invertebrate (commonly an arthropod) or a vertebrate, it penetrates the host and encysts. The encysted stage is called a metacercaria (Fig. 7–34*F*). If the host of the metacercaria is eaten by the final vertebrate host, the metacercaria is released, migrates, and develops into the adult form within a characteristic location in the host.

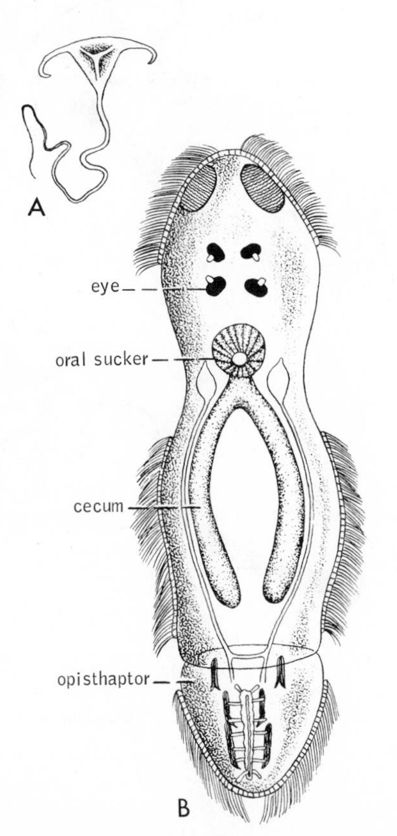

Figure 7–31. Egg (*A*) and onchomiracidium (*B*) of *Benedenia melleni*, a monogenetic trematode parasitic on fish. (Both after Jahn and Kuhn.)

A great many trematodes infect the gut or gut derivatives of their primary host. Lungs, bile ducts, pancreatic ducts, and intestines are common sites. The life cycle of *Echinostoma revolutum*, which infects the intestine of a variety of birds and mammals, including cats, dogs, and man, is a good illustration of the generalized life cycle just described. The eggs pass out with the host's feces and hatch in water. The free-swimming miracidium can penetrate and give rise to a sporocyst in species of three different genera of snails. The sporocyst develops into a first and second redial generation. From the redia, cercariae leave the snail and penetrate other snails, where they form metacercariae. If the second intermediate host is ingested by the primary host, the worm develops into an adult. The Chinese liver fluke, *Opisthorchis (=Clonorchis) sinensis*, and the sheep liver fluke, *Fasciola hepatica*, also have typical life cycles and are usually described in introductory biology and zoology courses. The intermediate hosts for the Chinese liver fluke are a snail and a fish (Fig. 7–35). The sheep liver fluke possesses the typical larval stages, but a snail is the only intermediate host. The metacercaria encysts on vegetation along the edges of ponds and streams.

The members of the family Bucephalidae are small parasites (less than 1 mm. in length) that are found in the gut of carnivorous freshwater and marine fish (Fig. 7–36*A*). The eggs are passed into the water with the host's feces. On hatching, the free-swimming miracidium enters a bivalve mollusk, such as one of the common freshwater clams (*Unio* or *Anodonta*) or oysters. Within the host's digestive gland a sporocyst is formed. The redia stage is skipped and cercariae develop directly from the sporocyst. The free-swimming cercariae attach to small fish, where they encyst beneath the scales or fins as metacercariae. If the second intermediate host is fed upon by certain larger carnivorous fish, the metacercaria is released and develops into adulthood within the host's gut.

The strigeoid trematodes are a large group of several families characterized by having the body divided into distinct anterior and posterior portions. The primary hosts are largely mammals (foxes, cats, dogs, dolphins, whales) that feed on aquatic animals. Some strigeoids are parasites of reptiles. The strigeoid genus *Alaria* (Fig. 7–36*B*), which occurs in the intestines of carnivorous mam-

(Text continued on page 173.)

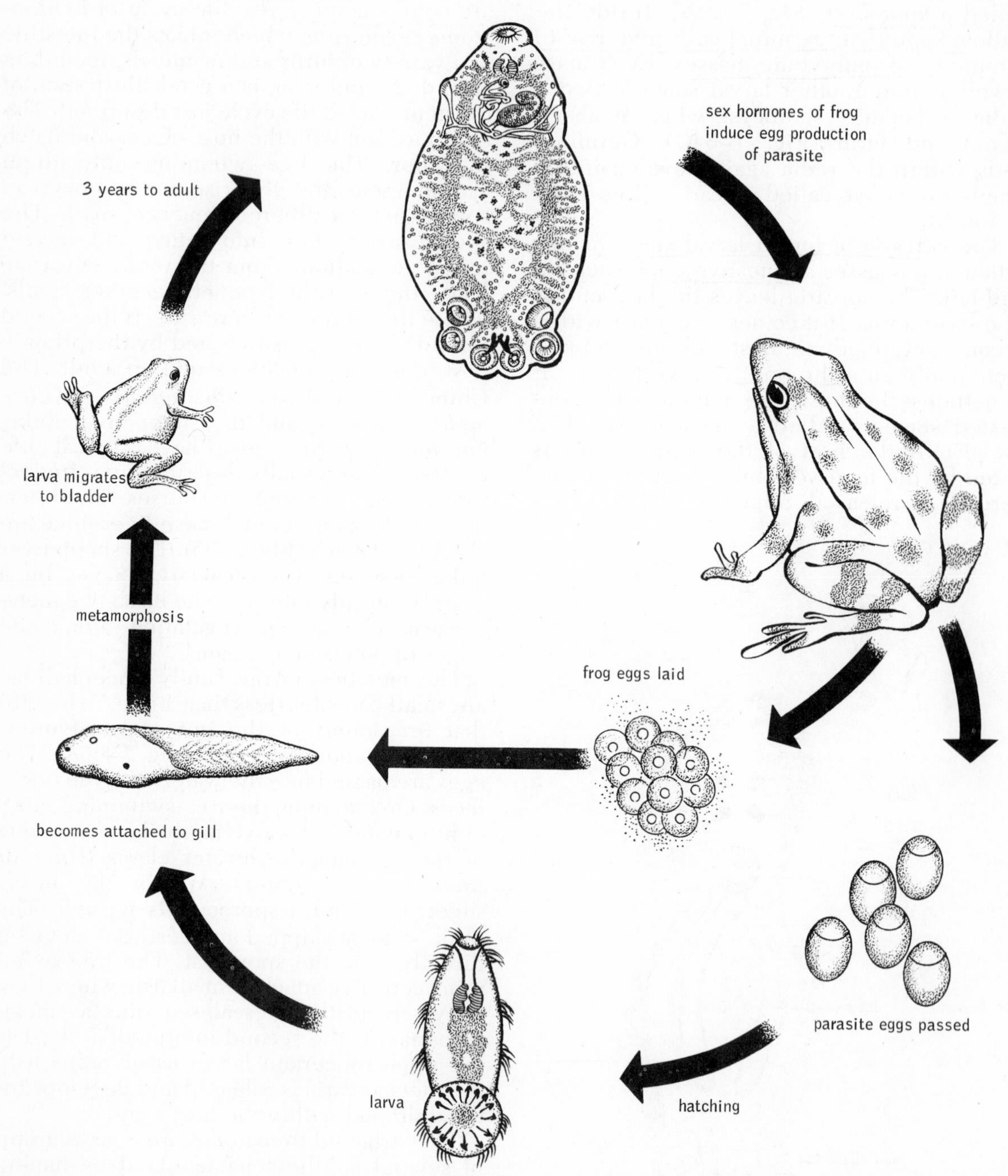

Figure 7–32. Life cycle of *Polystoma integerrimum,* a monogenetic trematode found in the bladder of frogs. (Modified from Williams and Miretski.)

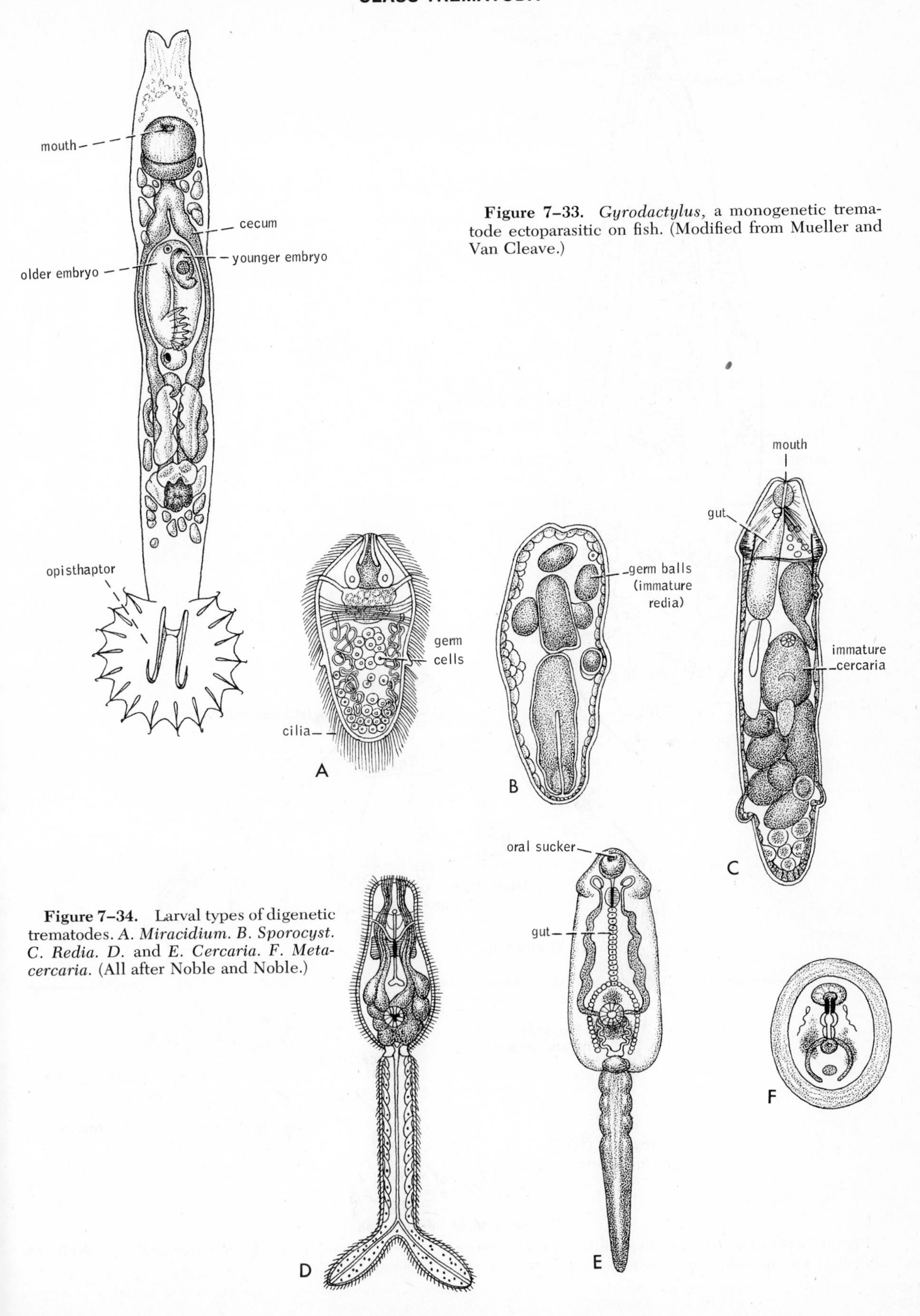

Figure 7–33. *Gyrodactylus*, a monogenetic trematode ectoparasitic on fish. (Modified from Mueller and Van Cleave.)

Figure 7–34. Larval types of digenetic trematodes. *A. Miracidium. B. Sporocyst. C. Redia. D.* and *E. Cercaria. F. Metacercaria.* (All after Noble and Noble.)

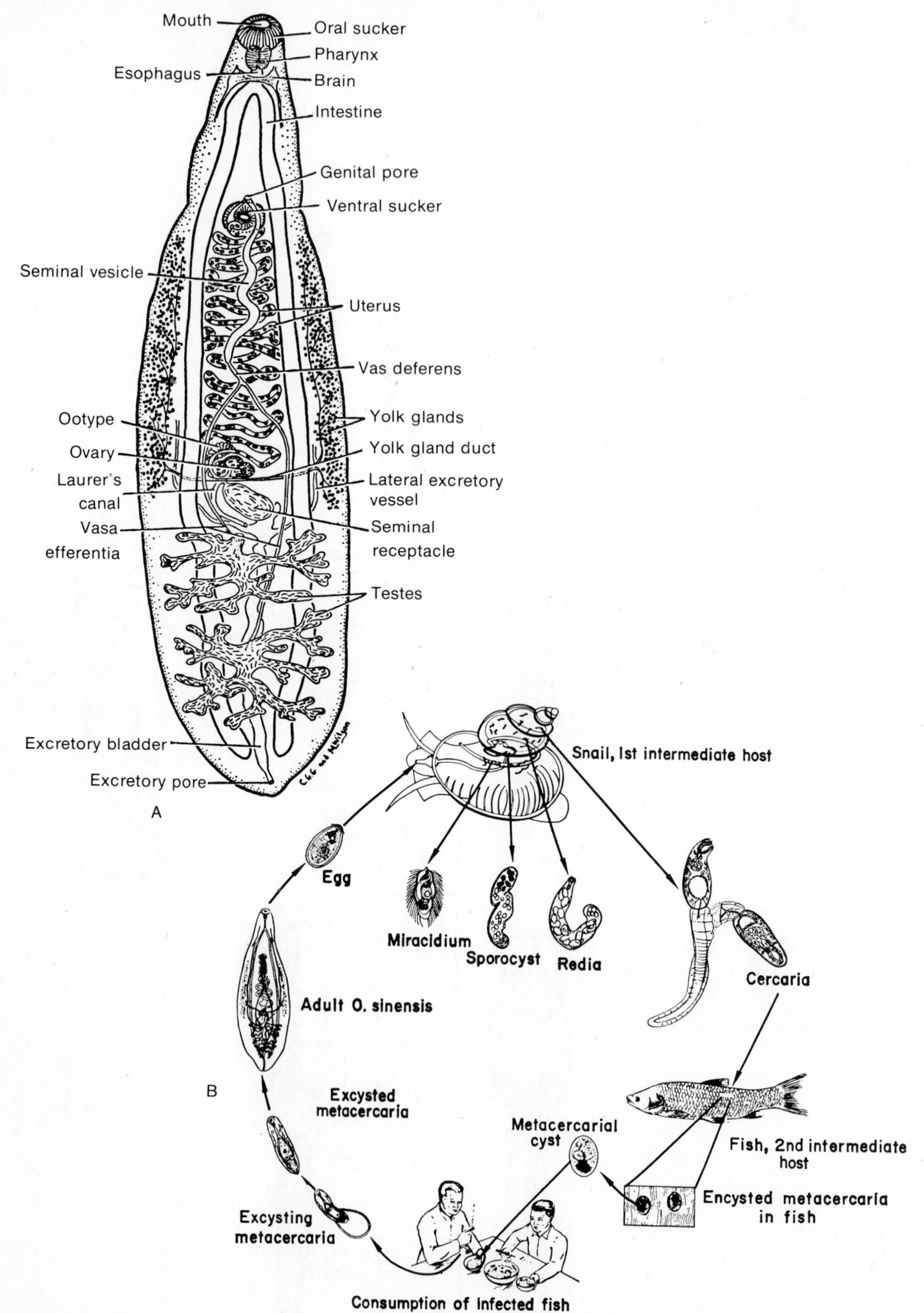

Figure 7–35. The Chinese liver fluke, *Opisthorchis sinensis*. *A*, Dorsal view of adult worm; *B*, life cycle. (*A* after Brown from Noble and Noble; *B* after Yoshimura from Noble and Noble.)

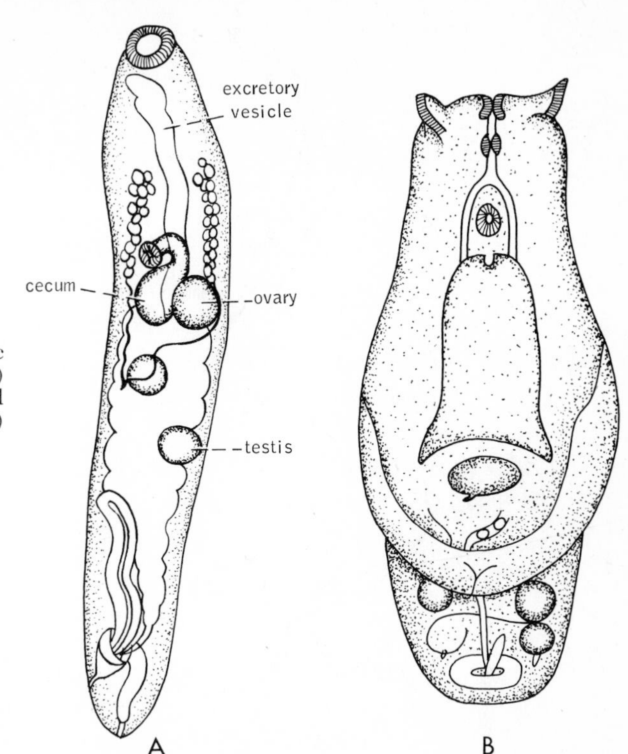

Figure 7–36. *A. Bucephalopsis ovata,* a digenetic trematode found in the gut of fish. (After Yamaguti.) *B. Alaria alata,* a strigeoid digenetic trematode found in the intestine of certain carnivores. (After Baylis.)

mals, is unusual in having three intermediate hosts. From the first host, a snail, the cercariae enter such aquatic vertebrates as tadpoles and frogs. Only partial development to the metacercaria occurs. The complete metacercaria stage is attained when the second intermediate host is eaten by a mouse, rat, or some other mammal. The metacercaria reaches adulthood when an infected third host is eaten by the large carnivorous primary host.

There are a number of families of trematodes which inhabit the blood of their host, but certainly the best known blood flukes belong to the family Schistosomatidae, which contains species producing the human disease schistosomiasis, or bilharziasis. *Schistosoma mansoni,* one of several species parasitic in man, occurs in Africa and tropical areas of the New World (Fig. 7–37). The adult, like other species of schistosomes, inhabits the intestinal veins. The members of this family, in contrast to most other flatworms, are dioecious. The male is 6 to 10 mm. in length and 0.5 mm. in diameter. A ventral groove extends most of the length of the male and into this groove fits the longer but more slender female. Eggs are deposited into the blood and are carried to the large intestine. Here they penetrate the wall, pass into the intestinal lumen, and leave with the host's feces. If the feces are deposited in water, the eggs hatch and the miracidium escapes. The miracidium penetrates a snail, and a sporocyst develops in the digestive gland of the snail. As is true of many other trematodes, the miracidium does not exhibit a great host specificity. Seventeen species of pulmonate snails belonging to three genera serve as intermediate hosts. The sporocyst gives rise to the cercaria without an intermediate redia stage. The cercaria leaves the snail host and, on contact with human skin, penetrates the integument. Penetration involves enzymes and muscular boring movements. On penetration of the skin the cercariae are carried by the blood stream first to the lungs, then to the liver, and finally to the mesenteric veins. During this period the cercaria gradually transforms into an adult.

The two other species of *Schistosoma* attacking man are the oriental *S. japonicum,* which has a life cycle similar to that of *Schistosoma mansoni,* and *S. haematobium* of North Africa. The eggs of *Schistosoma haematobium* leave the primary host by way of the bladder and urine. Schistosomiasis is seriously debilitating and can be lethal. With malaria and hookworm, schistosomiasis represents one of the three greatest parasitic scourges of mankind. The percentage of the population infected in endemic areas is

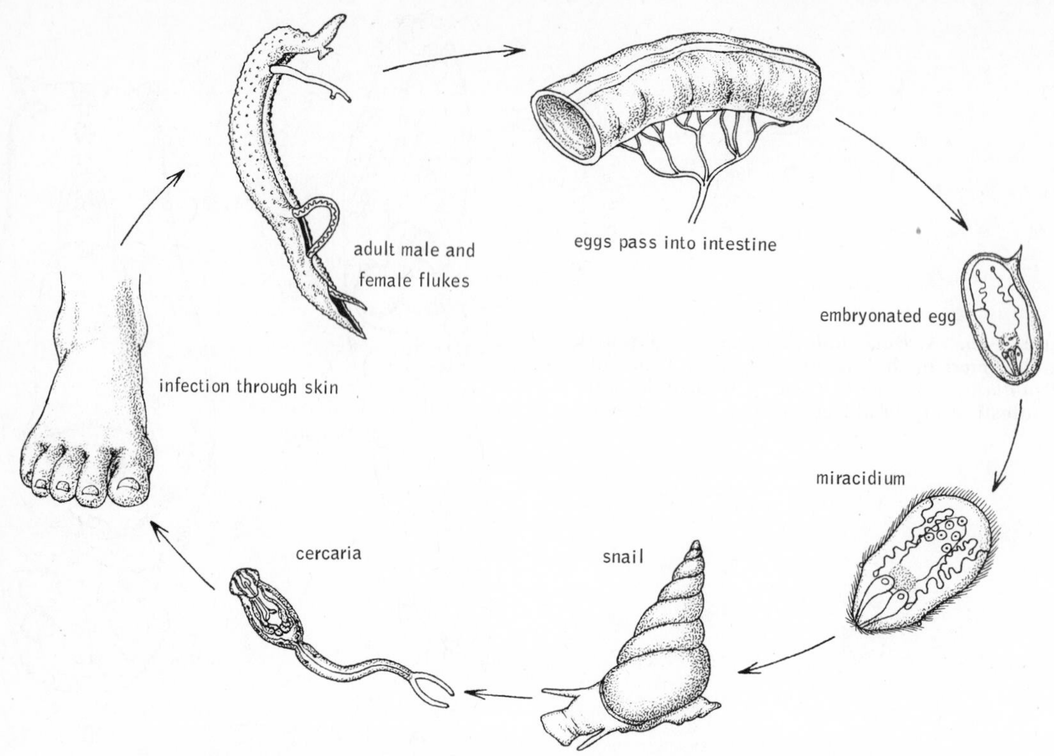

Figure 7–37. Life cycle of *Schistosoma mansoni.* (Modified from several authors.)

enormous. Egg penetration through the intestinal wall and bladder, aberrant lodgment of eggs in various organs, and the developmental stages of the larvae in the lung and liver can result in inflammation, necrosis, or fibrosis depending on the degree of infection.

Other members of the Schistosomatidae infect various birds and mammals, including domestic species. "Swimmer's itch" is an irritation produced by the incomplete penetration into human skin by cercariae of blood flukes of birds.

Order Aspidobothrea. The order Aspi-

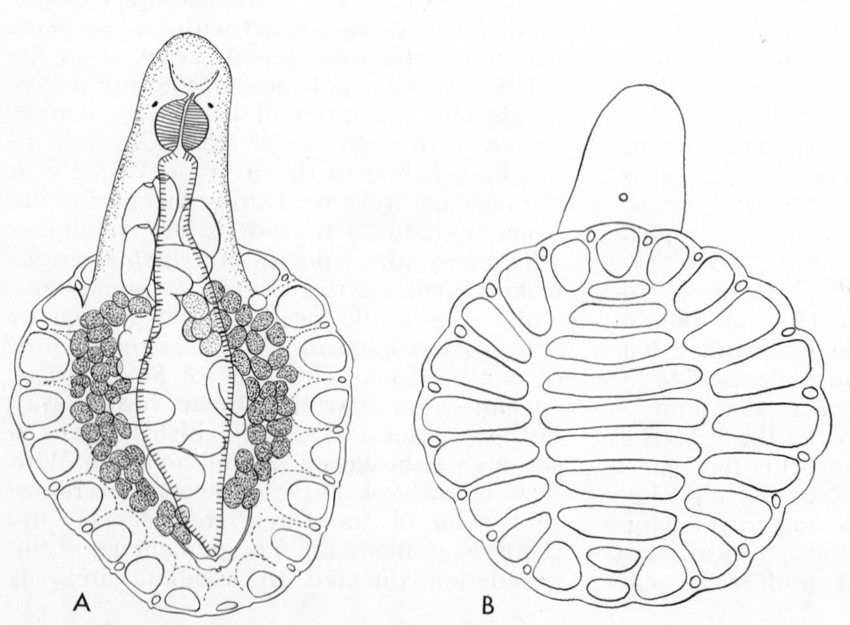

Figure 7–38. *Cotylaspis insignis*, an aspidobothrean trematode parasitic in freshwater mussels. *A.* Dorsal view. *B.* Ventral view showing large ventral sucker with subdivisions. (After Hendrix and Short.)

dobothrea includes a small group of trematodes which show similarities to both the Monogenea and Digenea. The distinguishing feature is the adhesive organ, which is a sucker covering the entire ventral surface (Fig. 7–38*A* and *B*). Usually the sucker is subdivided by ridges. The digestive tract contains only one intestinal caecum. The reproductive system is essentially like that of the digenetic trematodes, but there is typically only one testis in the male system.

The aspidobothreans are mostly endoparasites occurring in the gut of fish and reptiles, and in the pericardial and renal cavities of mollusks. Their life cycles involve one or two hosts.

Evolution of the Trematodes

Among the free-living turbellarian flatworms, the rhabdocoels may well have been the ancestors of the trematodes. Although most rhabdocoels are free-living, there are some commensal and parasitic groups. Hyman (1951) considers the dalyellioid rhabdocoels of particular interest from a standpoint of trematode evolution. The rhabdocoels are ectocommensal (largely) on echinoderms and mollusks, the latter being an important host for trematodes. The reproductive systems of both groups are similar, and the trematode redia is somewhat like that of a dalyellioid rhabdocoel. It is quite possible that the monogenetic and digenetic trematodes had independent origins. Hyman suggests that mollusks could have been the original hosts of digenetic trematodes, and with the evolution of vertebrates, they invaded this new group. They retained their connection with the mollusks, which became intermediate hosts.

CLASS CESTODA (CESTOIDEA)

The Cestoda is the most highly specialized of the flatworm classes. All are endoparasites, and the body is covered by a tegument, as in trematodes. The cestodes differ, however, from the members of the other two classes in the complete absence of a digestive tract. The class is divided into two subclasses, the Cestodaria and the Eucestoda. The subclass Cestodaria is a small group showing certain similarities to the trematodes and will be considered briefly after the subclass Eucestoda.

Subclass Eucestoda

Structures and Physiology. The great majority of cestodes belong to the subclass Eucestoda and are known as tapeworms. The body of the adult tapeworm is quite unlike that of the other flatworms. There is an anterior region, called the scolex, which is adapted for adhering to the host (Fig. 7–39). Behind the scolex is a narrow neck region which gives rise to the third part of the body, called the strobila. The strobila consists of linearly arranged individual sections called proglottids, and composes the greater part of the worm. Tapeworms are generally long, and some species reach lengths of 40 ft. The

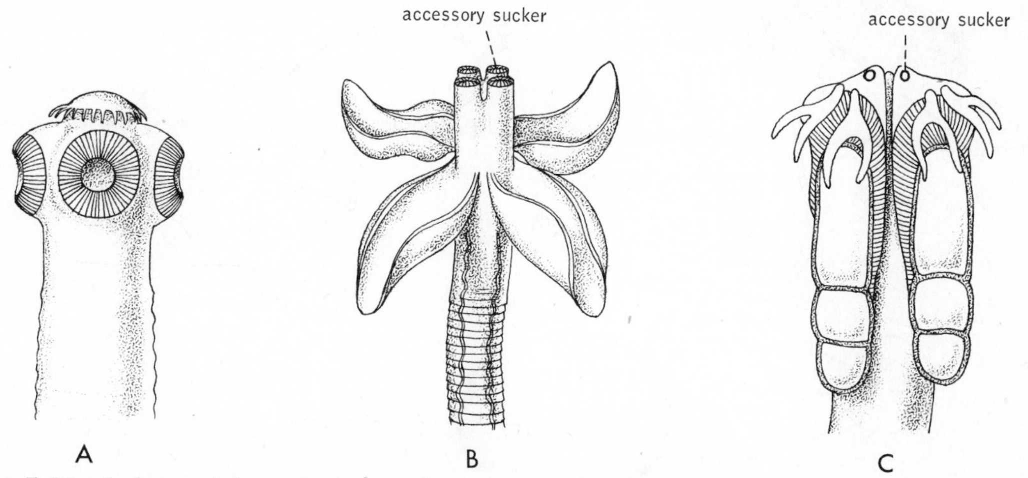

Figure 7–39. Scoleces of three tapeworms, showing the four principal suckers, or sucker-like adhesive structures, the small accessory suckers, and the hooks. *A. Taenia.* (After Southwell.) *B. Myzophyllobothrium.* (After Shipley and Hornell.) *C. Acanthobothrium.* (After Southwell.)

scolex, neck, and strobila are regarded as comprising the body of a single individual and not a colony, as has been suggested in the past.

Compared to the mature proglottids, the scolex is relatively small. It generally has the form of a more or less four-sided knob and is provided with suckers or suckers and hooks for adhering to the host. Although there are generally four large suckers arranged around the sides of the scolex, the scolex is often a more complicated structure than that of the familiar *Taenia* (Fig. 7–39*A*). The main suckers may be leaflike or ruffled and there may be terminal accessory suckers in place of or in addition to hooks (Fig. 7–39*C*).

The neck is a short region behind the scolex, which produces the proglottids by means of transverse constrictions (Fig. 7–39*B*). The youngest proglottids are thus at the anterior end and increase in size and maturity toward the posterior end of the strobila. The strobila is sometimes considered to be metameric, depending merely upon how one wishes to define metamerism. As metamerism was defined in Chapter 3, the zone of formation of new segments is at the posterior end, which is opposite that of tapeworms.

Extending through the chain of proglottids are the nervous system and protonephridial system. An anterior nerve mass lies in the scolex, and two lateral longitudinal cords extend posteriorly through the strobila (Fig. 7–40). There may also be a dorsal and ventral pair of cords and, quite commonly, accessory lateral cords. Ring commissures connect the longitudinal cords in each proglottid.

Flame cells and tubules in the mesenchyme drain into four peripheral longitudinal collecting canals, two of which are dorsolateral and two ventrolateral (Fig. 7–40). The longitudinal canals interconnect in the scolex, and the ventral canals are usually connected by a transverse canal in the posterior end of each proglottid. After proglottids have begun to shed, the collecting ducts open to the exterior through the last proglottid.

The complex body wall of cestodes is illustrated in Figure 7–41. The muscle layer of tapeworms consists of the same circular

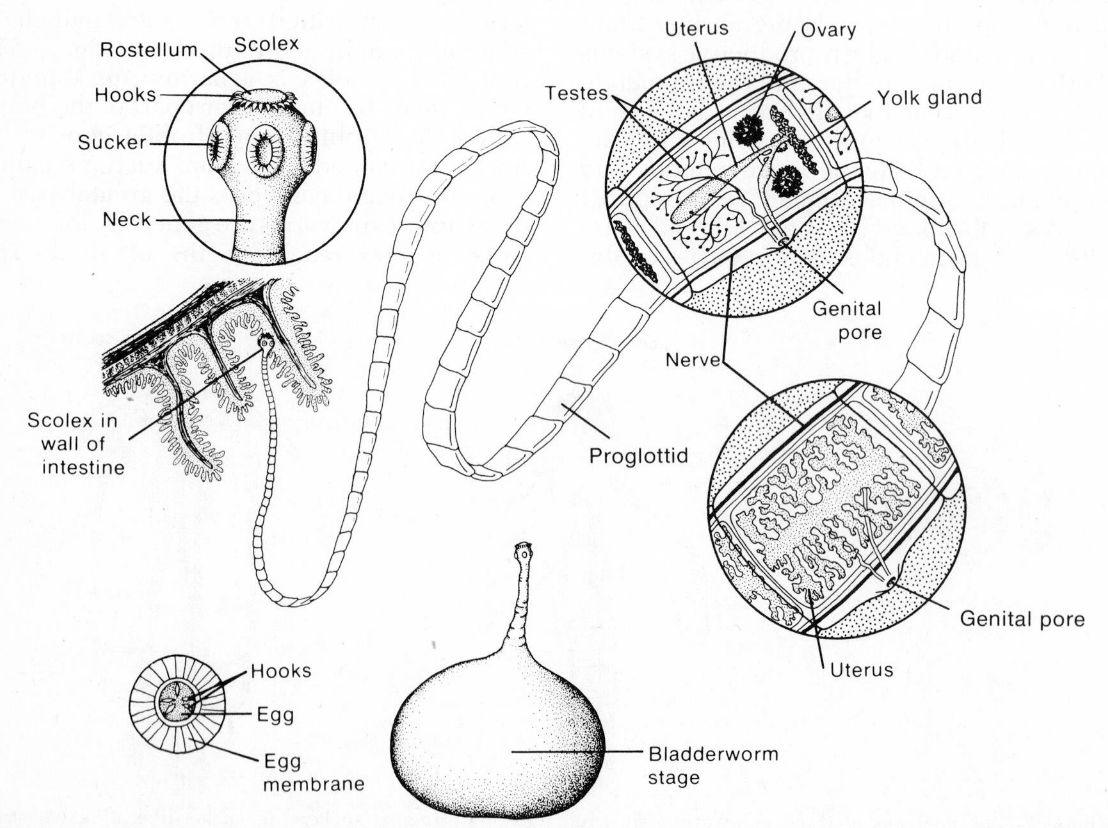

Figure 7–40. The pork tapeworm, *Taenia solium.* Insets show the head and an immature and a mature section of the body. (From Villee, C. A., 1972: Biology, 6th Ed. W. B. Saunders Co.)

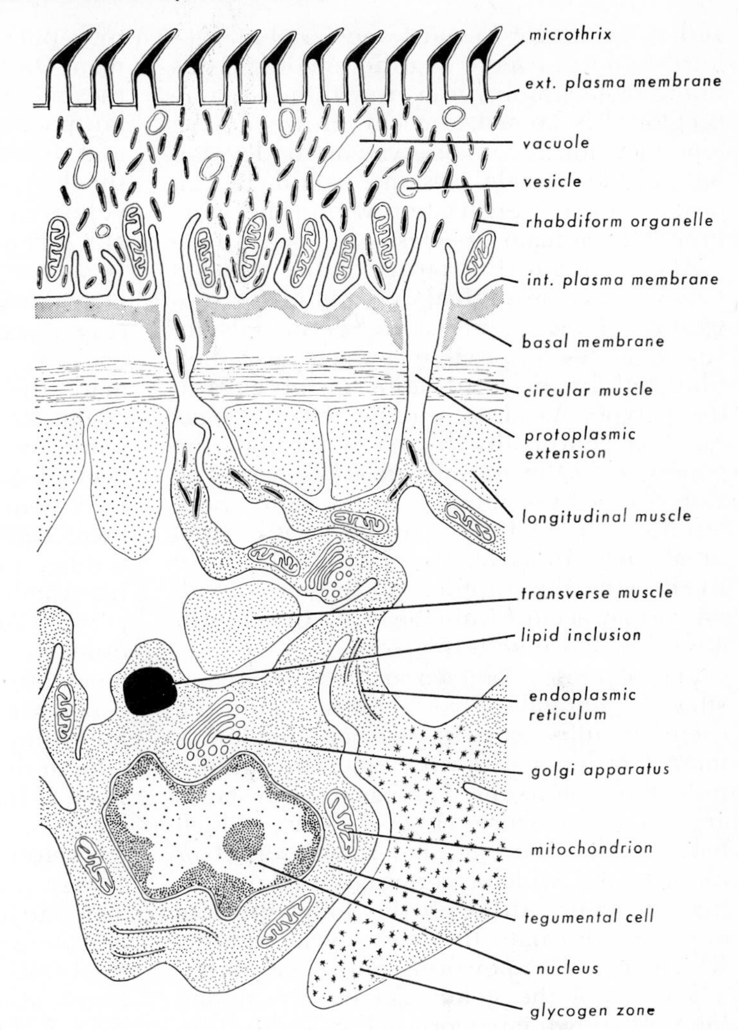

Figure 7–41. Section through the tegument of the tapeworm *Caryophyllaeus*. Other than the folding of the surface membrane, note the similarity to the trematode tegument shown in Figure 7–27. (After Beguin from Smyth, J. D., 1969: The Physiology of Cestodes. W. H. Freeman and Co., San Francisco.)

and longitudinal layers, but in addition there is a secondary mesenchymal musculature of longitudinal, transverse, and dorsoventral fibers. The secondary mesenchymal musculature encloses the interior mesenchyme.

The tapeworm tegument plays the same vital role in the physiology of tapeworms that it does in trematodes. The surface of the outer syncytial cytoplasm is thrown into folds, the microtriches, which increase the surface area (Fig. 7–41). The tegument assumes additional importance in the absorption of food, since tapeworms are devoid of a digestive system. Anaerobic metabolism apparently predominates in tapeworms, but is not their exclusive mode of metabolism.

Reproduction. A complete reproductive system occurs within each proglottid and makes up a major part of each of these body sections. As shown in Figure 7–40, the tapeworm reproductive system is basically like that of digenetic trematodes. There are usually a common male and female atrium and gonopore. The presence of a vaginal canal extending between the atrium and the ootype does differ from the condition found in many trematodes. The canal is enlarged as a seminal receptacle. The uterus leaves the ootype and functions solely in egg transport and storage. In some tapeworms, in which the eggs are continually being produced, the uterus runs as a canal to the gonopore. In many other tapeworms, the uterus is a blind sac which acts to store eggs.

Cross-fertilization is the rule in tapeworms

and is necessary in most species, but self-fertilization between two different proglottids in the same strobila or even within the same proglottid is known to occur. However, the tendency for the male system to develop before the female system would be an obstacle to self-fertilization within the same proglottid in many species.

At copulation, the cirrus of the male is everted into the vaginal opening of the proglottid of an adjacent worm. Sperm cells are stored in the seminal receptacle and then liberated for the fertilization of the eggs in the ootype. As already indicated, the eggs may be continually liberated through the gonopore via the uterine canal or they may be stored in a blind uterus. In the latter case, terminal proglottids packed with eggs break away from the strobila. The eggs are freed with the rupture of the proglottid, which may occur within the host's intestine or after they leave with the feces.

Life Cycles. Tapeworms are endoparasites in the gut of vertebrates. Their life cycles require one to two or sometimes more intermediate hosts, which are arthropods and vertebrates. The basic larval stages are an oncosphere (Fig. 7–42*A*), which hatches from the egg, and a cysticercus or plerocercoid, which is terminal and develops into an adult. Although the following few examples illustrate the basic cycle patterns which occur in tapeworms, variations exist.

Species of the genus *Taenia* are among the best known tapeworms (Fig. 7–40). *Taenia solium* is a parasite of man and reaches a length in the human intestine of up to 10 ft. The proglottids are passed in the host's feces and the eggs do not hatch unless eaten by a pig or one of several other hosts, even man. On hatching, the oncospheres bore into the intestinal wall (Fig. 7–42*A*), where they are picked up by the blood-vascular system and transported to striated muscle. Here the larva leaves the blood stream and develops into a cysticercus larva. The cysticercus, sometimes called a bladder worm, is an oval stage about 10 mm. in length with the scolex invaginated into the interior (Fig. 7–42*B*). If raw pork or pork insufficiently cooked to kill the cysticercus is ingested by man, the cysticercus is freed, the scolex evaginates, and the larva develops into an adult worm. Species of *Taenia pisiformis* occur in cats and dogs, with rabbits as the intermediate hosts. The beef tapeworm of man, *Taeniarhynchus saginatus*, belongs to a related genus, and bovine animals are the intermediate hosts. Oribatid mites are important hosts for other members of the same order (Cyclophyllidea), which contains many tapeworms of mammals.

Dibothriocephalus latus, one of the fish tapeworms, is widely distributed and parasitic in the gut of many carnivores, including man. If the eggs are deposited with feces in water, a ciliated free-swimming oncosphere (coracidium) hatches after an approximately 10-day developmental period (Fig. 7–43*B*). The larva is ingested by certain copepod crustaceans, where it penetrates the intestinal wall and develops within the hemocoel into a six-hooked larva, called a procercoid (Fig. 7–43*D*). When ingested by a variety of freshwater and marine fish, the procercoid, like the oncosphere, penetrates the host's gut and eventually reaches the striated muscles of the fish to develop into a plerocercoid

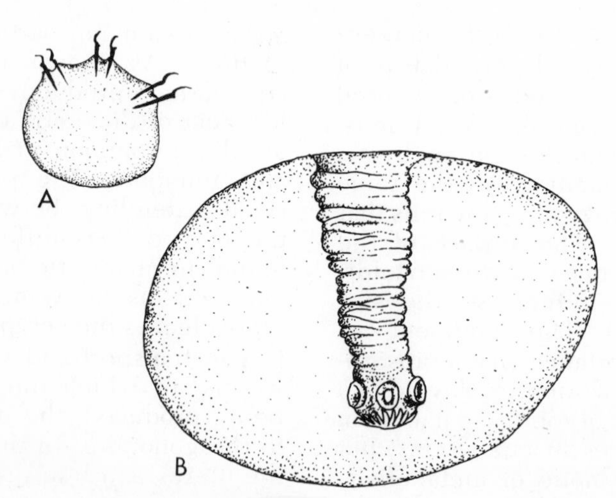

Figure 7–42. Larvae of *Taenia solium*. *A*. Oncosphere with hooklets. (After Blanchard.) *B*. Cysticercus, with scolex invaginated into interior of body. (After Chandler and Read.)

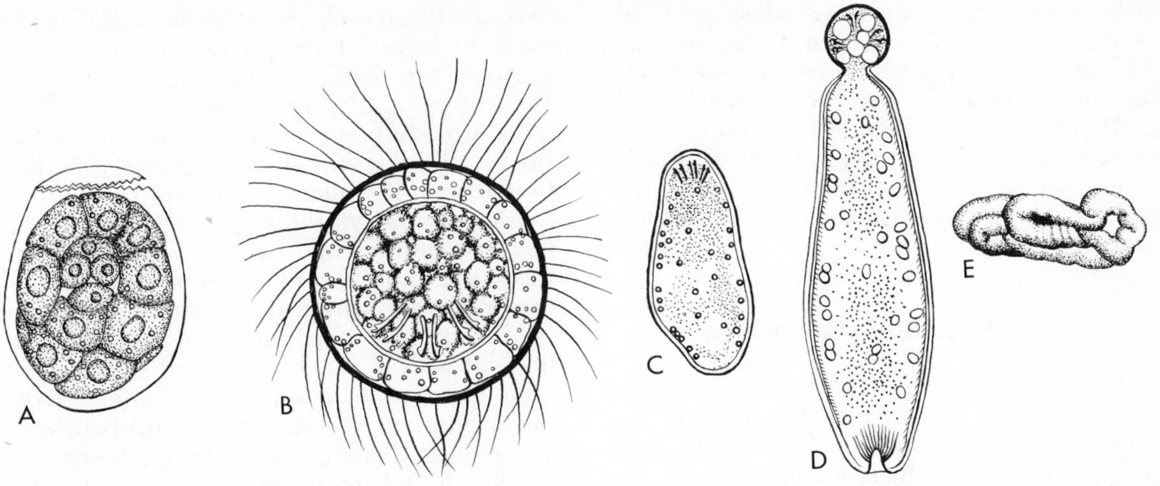

Figure 7–43. Some stages in the life cycles of the fish tapeworm, *Dibothriocephalus latus*. *A*. Egg with embryo. (After Schavinsland.) *B*. Ciliated oncosphere. (After Vergeer.) *C*. Early procercoid from a copepod, the intermediate host. *D*. Mature procercoid. (*C* and *D* after Brumpt.) *E*. Plerocercoid in fish. (After Vergeer.)

larva. The plerocercoid, which looks like a tapeworm without a strobila, develops into an adult tapeworm when ingested by a primary host.

Another group with a ciliated free-swimming oncosphere are members of the order Proteocephala. Water snakes (*Natrix*) are the primary hosts for the proteocephalan, *Ophiotaenia perspicua*. Eggs leave in the host's feces and are ingested by a variety of species of copepods which can act as intermediate hosts for the procercoids. Tadpoles, which ingest the copepods, are the intermediate hosts for the plerocercoid. The plerocercoid remains within the amphibian through metamorphosis. If the frog is ingested by the primary host, the tapeworm completes its life cycle.

Subclass Cestodaria

The subclass Cestodaria is a small group of cestodes which show some similarities to trematodes. They lack a scolex and strobila, and the body contains only one hermaphroditic reproductive system. Trematode-like suckers are sometimes present (Fig. 7–44). However, the absence of a digestive system and the presence of a larva which is similar to the larva of many tapeworms would place them with the cestodes.

The cestodarians are intestinal and coelomic parasites of elasmobranchs and primitive teleost fish. The intermediate hosts are invertebrates.

Origin of the Cestodes

The similarities between the Cestodaria and the Trematoda are generally considered to be superficial and do not indicate a trematode origin for the tapeworms. Most parasitolo-

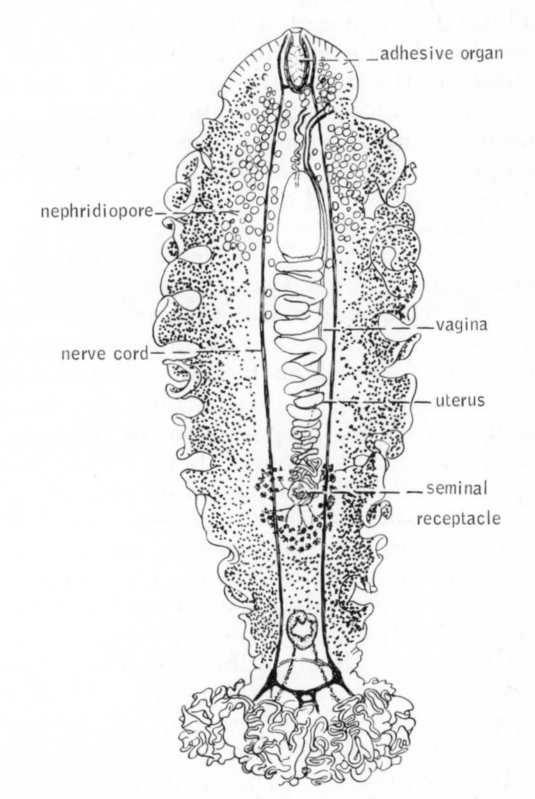

Figure 7–44. *Gyrocotyle fimbriata*, a cestodarian cestode in fish. (After Lynch.)

gists agree that the cestodes probably evolved independently from the trematodes. Although the rhabdocoels are believed to be the most probable ancestors, there is no particular rhabdocoel group that makes an especially good candidate for the position. The position of the cestodarians is an enigma and their relationship to the eucestodes may be very remote.

Host-Parasite Relationships

An extensive discussion of host-parasite relationships is not possible in a book of this scope, and the student who is especially interested in the subject should refer to one of the parasitology texts cited in the bibliography of this chapter. A few general principles, however, should be outlined following a discussion of two such major groups of metazoan parasites.

The trematodes and cestodes illustrate most of the adaptations of animals for a parasitic existence: (1) Adhesive organs for attachment to the exterior of the host or to the wall of interior cavities are usually present. (2) Sense organs are reduced, especially in endoparasites. (3) There are various modifications in nutrition. Such modifications may include development of storage areas for ingested food in the case of ectoparasites, adaptations of the ingestive organs, or direct absorption of food through the body wall. (4) There is an increase in the reproductive capabilities of the parasite through greater egg production, and often asexual reproductive stages that are universal in the life history of the parasite. (5) There are commonly larval stages which permit passage of the parasite from one host to another. The utilization of a single host may be a primitive condition, from which forms having multiple hosts evolved, or a single host may have resulted from the dropping out of intermediate hosts. Undoubtedly, both conditions have evolved in the history of parasitism.

The parasite may have no noticeable effect on the host, or there may be marked effects from a variety of factors. The parasite utilizes food materials that would normally be available to the host. This deprivation has relatively little effect on the host unless the infestation is very great. More serious is the damage to the walls of the organ to which the parasite is attached, or obstruction of the cavity in which the parasite is living. For example, in heavy infestations flukes and tapeworms may completely occlude the gut or associated ducts.

Egg and larval penetration and encystment of larval stages in various organs are perhaps the causes of the most serious effects of parasitism in the host. Such effects have already been mentioned in the case of flukes. Similar larval pathogenicity is particularly well illustrated by the tapeworms, *Echinococcus granulosus* and *E. multilocularis*, of canines and cats. The intermediate hosts are rabbits and rodents, but humans can ingest eggs from dogs. The oncosphere, on hatching in the gut, penetrates the intestinal wall to reach the blood stream. The larva can encyst in a variety of sites—liver, lungs, kidneys, heart, brain, and elsewhere. The encysted larva, called a hydatid cyst, may reach a length of 10 cm. and can cause the death of the host.

The liberation of toxins by the adult or larval parasite can produce deleterious effects on the host and may also be a cause of pathogenicity.

PHYLUM MESOZOA

The Mesozoa are a small group of some 50 species of minute parasitic animals of very simple structure but having complex life cycles. The hosts of all are marine invertebrates. Members of the order Orthonectida have been found in flatworms, proboscis worms, polychaetes, bivalve mollusks, brittle stars, and other groups. The dioecious adults are unattached within the host, and the tiny wormlike body consists of an outer layer of ciliated cells enclosing an internal mass of either sperm or egg cells. Adult males and females are released from the host simultaneously, and sperm from the males penetrate the body of the female and fertilize the eggs (Fig. 7–45*A* and *B*). Cleavage leads to a ciliated larva which infects the tissue spaces of a new host (Fig. 7–45*C*). Within the host the larva loses its cilia, becomes syncytial, and gives rise to agametes (Fig. 7–45*D*). The agametes in turn become mature males and females.

The mesozoans comprising the order Dicyemida live attached within the nephridial cavity of squids, cuttlefish, and octopods. Although this order was known before the orthonectids, knowledge of its complete life cycle is still incomplete. The adult dicyemid is called a nematogen and is 1 to 7 mm. long. The body consists of a long central axial cell surrounded by a single layer of ciliated cells. The anterior cells are used for attachment. The population of individuals within the hosts, especially young cephalo-

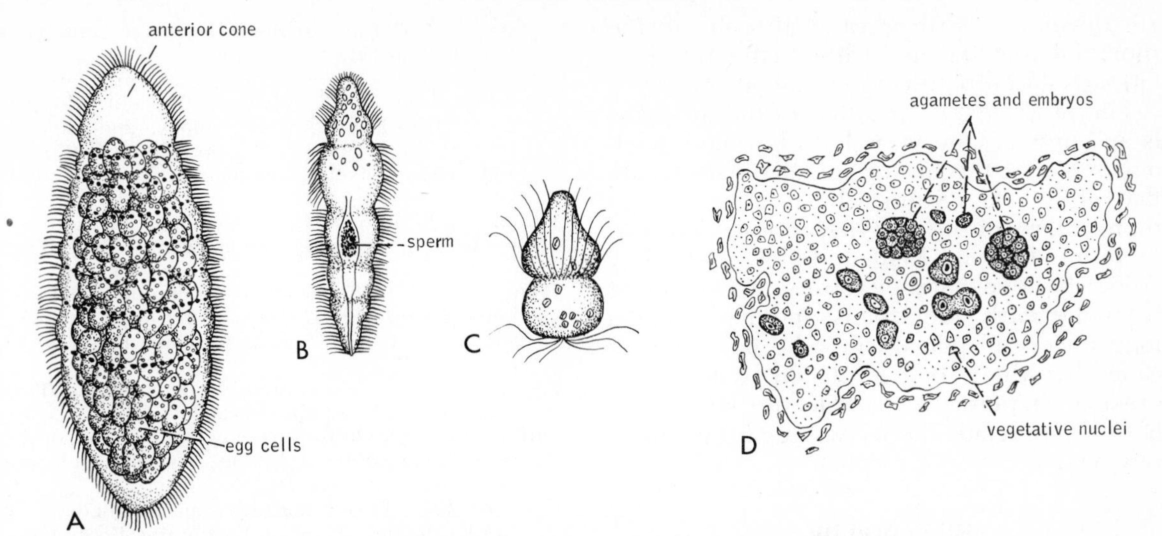

Figure 7–45. Stages in the life cycle of orthonectid mesozoans. *A* and *B*. Mature female and male of the clam parasite, *Rhopalura granosa*. (Both after Atkins.) *C*. Ciliated larva of same species. (After Atkins.) *D*. Male plasmodium of another species of *Rhopalura*. (After Caullery and Mesnil.)

pods, increases through the production of young having the same form as the parent. Such young are called vermiform larvae and are formed *within* the axial cell of the parent (Fig. 7–46*A*).

In mature cephalopods the axial cell of the parasite gives rise to another type of larva, called infusoriform larvae. Individuals producing this type of larvae are similar in form to the nematogens but are called rhombogens. The infusoriform larvae produced by the rhombogens are small with posteriorly directed cilia (Fig. 7–46*B*). They are passed out with the host's urine and, although their fate is unknown, it is believed that some bottom-dwelling invertebrate is required as an intermediate host to complete the life cycle.

A peculiar feature of the life cycle of the dicyemids is the fact that the infusoriform larvae develop from sperm and eggs produced in what McConnaughey (1963) considers to be highly reduced hermaphroditic individuals, called infusorigens. The infusorigens and subsequent infusoriform larvae

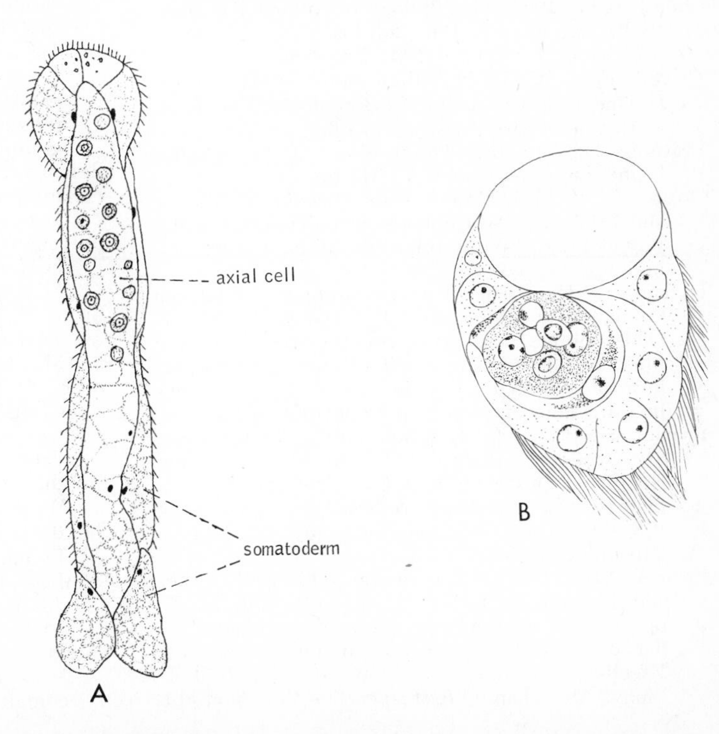

Figure 7–46. Stages in the life cycle of dicyemid mesozoans. *A*. Nematogen of *Pseudicyema*, parasitic in the cuttlefish *Sepia*. (After Whitman.) *B*. Infusoriform larva of *Dicyema*, a parasite of *Octopus*. (After Short.)

all develop within the axial cell of the rhombogen, which according to this interpretation is a highly developed larval stage.

The phylogenetic position of the Mesozoa is an enigma. One view held by many zoologists is that the mesozoans are degenerate flatworms; in fact, Stunkard (1954) would make them a class of the Platyhelminthes. Others, including Hyman (1940) and McConnaughey (1963), believe that they represent an offshoot from the early metazoans and have had a long history of parasitism. The only significance to their being placed within the present chapter is to enable the student to better appreciate the views regarding their phylogeny.

Bibliography

Apelt, G., 1969: Fortpflanzungsbiologie, Entwicklungszyklen und vergleichende Fruehenwicklung acoeler Turbellarien. Mar. Biol., *4*(4):267–325. (Reproductive biology, developmental cycles and comparative early development of acoel turbellarians.)

Ax, P., 1963: Relationships and phylogeny of the Turbellaria. *In* Dougherty, E. C. (Ed.), The Lower Metazoa. University of California Press, Berkeley, pp. 191–224.

Ax, P., and Apelt, G., 1965: Die "Zooxanthellen" von *Convoluta convoluta* (Turbellaria, Acoela) entstehen aus Diatomeen. Die Naturwissenschaften, *52*(15): 444–446.

Ax, P., and Borkett, H., 1968: Organisation und Fortpflanzung von *Macrostomum romanicum* (Turbellaria, Macrostomida). Zool. Anz. Suppl. *32*:344–347.

Bayer, F. M., and Owre, H. B., 1968: The Free-Living Lower Invertebrates. Macmillan Co., N. Y., pp. 144–177.

Cheng, T. C., 1964: The Biology of Animal Parasites. W. B. Saunders Co., Philadelphia.

de Beauchamp, P., Caullery, M., Euzet, L., Grassé, P., and Joyeux, C., 1961: Plathelminthes et Mésozaires. *In* Grassé, P. (Ed.), Traité de Zoologie, Vol. 4, pt. 1, Masson et Cie, Paris, pp. 1–729.

Erasmus, D. A., 1973: The Biology of Trematodes. Crane, Russak & Co., N. Y., 312 pp.

Gosner, K. L., 1971: Guide to Identification of Marine and Estuarine Invertebrates: Cape Hatteras to the Bay of Fundy. Wiley-Interscience, N. Y. Platyhelminthes, pp. 166–174.

Hyman, L. H., 1940: The Invertebrates: Protozoa through Ctenophora. Vol. 1. McGraw-Hill, N. Y. (This volume contains a concise account of the Mesozoa, although it is out of date. The chapter *Retrospect* in Vol. 5 of this series (1959) surveys investigations on mesozoans from 1938 to 1958.)

Hyman, L. H., 1950: Platyhelminthes: *Polychoerus, Stenostomum, Bdelloura, Dugesia, Procotyla,* and *Holoplana. In* Brown, F. A. (Ed.), Selected Invertebrate Types. John Wiley & Sons, N. Y., pp. 141–158. (Directions for laboratory study of representative turbellarians.)

Hyman, L. H., 1951: The Invertebrates: Platyhelminthes and Rhynchocoela. Vol. 2. McGraw-Hill, N. Y., pp. 52–219. (Authoritative general account of the flatworms. All aspects of the phylum are covered. Excellent illustrations and an extensive bibliography. The chapter *Retrospect* in Vol. 5 of this series (1959) summarizes research on Turbellaria from 1950 to 1958.)

Hyman, L. H., and Jones, E. R., 1959: Turbellaria. *In* Edmondson, W. T., Ward, H. B., and Whipple, G. C. (Eds.), Freshwater Biology. 2nd Edition. John Wiley & Sons, N. Y., pp. 323–365. (Key to and figures of the common freshwater turbellarians.)

Jennings, J. B., 1957: Studies on feeding, digestion, and food storage in free-living flatworms. Biol. Bull. *112*:63–80.

Jennings, J. B., 1968: Nutrition and digestion in Platyhelminthes. *In* Florkin, M., and Scheer, B. J. (Eds.), Chemical Zoology. Vol. 2. Academic Press, N. Y., pp. 305–327.

Kaestner, A., 1967: Invertebrate Zoology. Vol. 1. Wiley-Interscience, N. Y., pp. 154–200.

Karling, T., 1973: On the anatomy and affinities of the turbellarian orders. Libbie Hyman Memorial Symposium, McGraw-Hill, N. Y.

Kenk, R., 1972: Freshwater planarians (Turbellaria) of North America. Biota of Freshwater Ecosystems, Identification Manual No. 1. Environmental Protection Agency, Washington, D.C. 81 pp.

Kümmel, G., 1962: Zwei neue Formen von Cyrtocyten, Vergleich der bisher bekannten Cyrtocyten und Erörtenrung des Begriffes "Zelltype." Z. Zellforsch., *57*:172–201. (Two new forms of cyrtocytes, comparison to previously known cyrtocytes and discussion of concepts of "cell type.")

Light, S. F., Smith, R. I., Pitelka, F. A., Abbott, D. P., and Weesner, F. M., 1967: Intertidal Invertebrates of the Central California Coast. University of California Press, Berkeley. Platyhelminthes, pp. 55–60.

McConnaughey, B. H., 1963: The Mesozoa. *In* Dougherty, E. C. (Ed.), The Lower Metazoa. University of California Press, Berkeley, pp. 151–168.

Noble, E. R., and Noble, G. A., 1972: Parasitology. 3rd Edition. Lea & Febiger, Philadelphia. 617 pp.

Pennak, R. W., 1953: Freshwater Invertebrates of the United States. Ronald Press, N. Y., pp. 114–141. (An introductory work on the taxonomy of freshwater turbellarian genera. Key, figures, and bibliography included, as well as an introductory section on the biology of the group.)

Read, C. P., 1970: Parasitism and Symbiology. Ronald Press, N. Y. 316 pp. (A text which approaches parasitology from the standpoint of host-parasite relationships rather than from a survey of parasitic groups.)

Reisinger, E., 1968: *Xenoprorhynchus* ein Modellfall für progressiven Funktionswechsel. Z. zool. Syst. Evolutionsforsch., *6*:1–55.

Reisinger, E., and Kelbetz, S., 1964: Feinbau und Entlandungs-mechanismus der Rhabditen. Z. Wiss. Mikroskopie und mikr. Technik, *65*(8):472–508.

Riser, N., and Morse, P. (eds.), 1974: Biology of the Turbellaria. Libbie Hyman Memorial Symposium. McGraw-Hill Co., N.Y. In press.

Rose, S. M., 1970: Regeneration. Appleton-Century-Crofts, N. Y. (The chapter on regeneration in worms provides a good introduction to studies on flatworm regeneration.)

Smyth, J. D., 1966: The Physiology of Trematodes. W. H. Freeman, San Francisco.

Smyth, J. D., 1969: The Physiology of Cestodes. W. H. Freeman, San Francisco. 279 pp.

Stunkard, H. W., 1954: The life history and systematic relations of the Mesozoa. Quart. Rev. Biol., *29*:230–244.

Stunkard, H. W., 1972: Clarification of taxonomy in Mesozoa. Syst. Zool., *21*(2):210–214. (This paper contains a good literature review.)

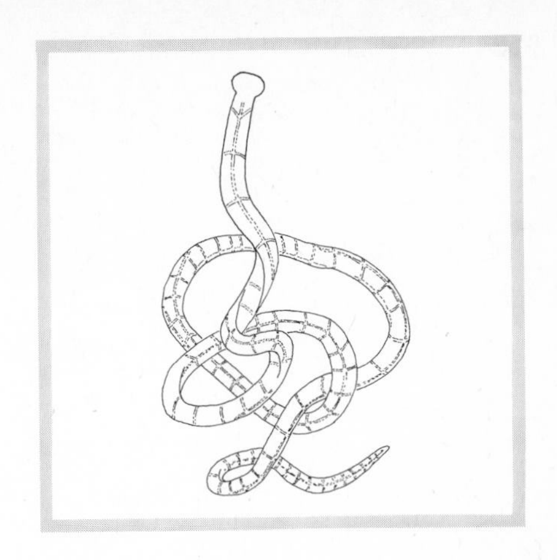

Chapter 8

THE NEMERTEANS

The phylum Rhynchocoela, or Nemertinea, comprises a group of about 650 species of elongated and often flattened worms. They are commonly known as proboscis worms because of the presence of a remarkable proboscis apparatus used in capturing food. Most of the nemerteans are marine and are bottom dwellers. They live in shallow water, beneath shells and stones, or in algae, or burrow in mud and sand. Some species form semipermanent burrows lined with mucus or even distinct tubes. Nemerteans are more abundant in temperate seas than in the tropics, and many species inhabit the Atlantic and Pacific littoral waters of the United States. There is a single genus of freshwater species; and one genus containing terrestrial forms is confined to the tropics and subtropics. A few commensals live with sea squirts, sea anemones, and mollusks, but no parasitic forms are known.

Evolution. Phylogenetically, nemerteans appear to be an offshoot from the free-living flatworms. An acoelomate plan of construction, in which a solid mass of mesenchyme fills the space between the body wall and the intestine, strongly suggests a flatworm origin. Moreover, both flatworms and nemerteans possess protonephridia, a ciliated epidermis, and similar nervous systems and sense organs.

However, despite their ancestry, nemerteans are more highly organized than flatworms, although they are probably not on the direct line of protostome evolution leading to the coelomates. The nervous system displays two large lateral nerve cords and a middorsal nerve cord, and the protonephridial system has assumed an excretory function. In addition, there is a simple circulatory system and a separate exit (or anus) for the digestive tract.

External Structure. In external appearance, nemerteans resemble flatworms but tend to be somewhat larger and more elongated. Although there are many small species, those of some genera, such as the common *Cerebratulus* and *Lineus* (Fig. 8–1*A*) of the Atlantic coast, are ribbon-shaped and reach a length of 2 m. No distinct head is evident, but the anterior is often provided with side lobes that make it resemble a spatula or a flattened heart (Fig. 8–1*B*). In some species the head is lanceolate. Most nemerteans are pale; some have brightly colored hues or patterns of yellow, orange, red, and green.

Body Wall and Locomotion. The body wall is like that of flatworms but is more highly organized (Figs. 8–2 and 8–4*A*). An outer layer of ciliated columnar epithelium rests upon a dermis of connective tissue. Unicellular mucus-secreting glands are scattered throughout the epidermis or are present as clusters of cells, with each cluster having a common duct to the outside. Beneath the dermis is a well-developed body-wall musculature, composed of circular and longitudinal smooth muscle in two or three layers. Mesenchymal tissue occupies the space between the digestive tract and the muscle layers; bands of muscle extend through the mesenchyme, connecting the dorsal and ventral body walls.

Most nemerteans move, like flatworms, by gliding over the substratum on a trail of slime. Depending upon the size of the worm, cilia and muscular contractions account for

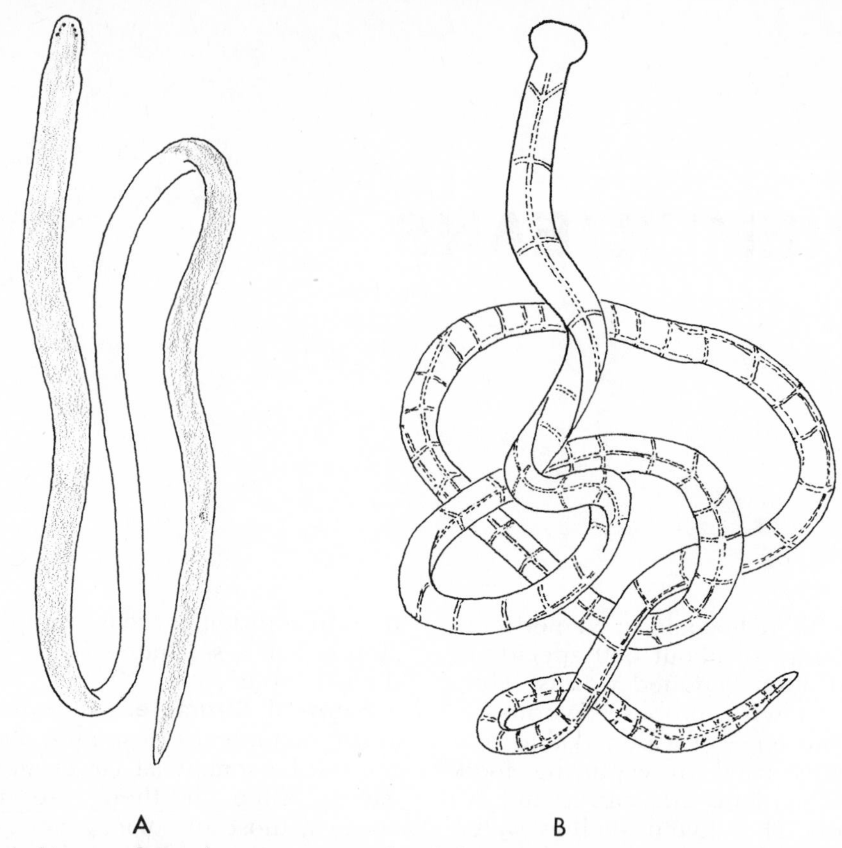

Figure 8–1. *A. Lineus ruber*, a heteronemertean found throughout northern latitudes. (After Hyman.) *B*. The paleonemertean, *Tubulanus capistratus*, from the Pacific coast. (After Coe from Hyman.)

locomotion. Some species, particularly certain pelagic forms, are able to swim and use lateral muscular undulations exclusively.

Nutrition and Digestive System. The most characteristic feature of the phylum is the proboscis apparatus. Although a proboscis, when present, is usually associated with the digestive tract in invertebrates, the proboscis of nemerteans is not connected to the digestive tract except secondarily in some species. Moreover, the embryogeny of the proboscis is entirely independent from that of the gut. The opening of the proboscis apparatus is through a pore at the anterior tip of the worm (Figs. 8–3 and 8–4*A*). This part leads into a short canal known as the rhynchodaeum, which extends back to the level of the brain. The histology of the rhynchodaeum wall, as well as the remainder of the proboscis apparatus, is similar to that of the body wall, because the entire structure develops as an ectodermal invagination. Rhabdites are present in the proboscis wall of some species, providing further evidence of a flatworm origin for the phylum.

The lumen of the rhynchodaeum is continuous with that of the proboscis proper (Fig. 8–4*A*). The latter consists of a long tube, often coiled, lying free in a fluid-filled cavity called the rhynchocoel. The posterior of the proboscis is blind and is attached to the back of the rhynchocoel by a retractor muscle. Because the rhynchocoel is of mesodermal origin and is lined by mesothelium, it could be considered a coelom, although it is probably not homologous with the coelom of higher invertebrates.

In two of the nemertean orders, the paleonemerteans and heteronemerteans, the proboscis is a simple tube (Fig. 8–3*A*), but in the hoplonemerteans the proboscis has become more specialized and is armed with a heavy barb called a stylet, which is set in the proboscis wall (Figs. 8–3*B* and 8–4*B*). Where the stylet is attached, about two-thirds the distance back from the anterior of the animal, the proboscis wall is greatly swollen, particularly on one side, virtually occluding the lumen of the tube. This swelling, known as the diaphragm, bears the bulbous base of the stylet in a socket on its anterior face. In addition to the main stylet, accessory or reserve stylets are present for replacing the main stylet when necessary.

The proboscis is used for defense and for the capture of prey. It is literally shot out of the body, everting in the process. Because the posterior of the proboscis is attached to the back of the rhynchocoel by the retractor muscle, the proboscis can never be completely everted. In the armed nemerteans the proboscis is only projected as far as the stylet, which then occupies its tip (Fig. 8–4C). The force for eversion is provided by muscular pressure on the fluid in the rhynchocoel, and retraction is effected by the posterior retractor muscle. The proboscis coils around the prey and glandular secretions from the proboscis wall aid in holding it. The stylet of armed nemerteans stabs the prey repeatedly and also probably delivers toxic secretions.

Perhaps nemerteans evolved from those rhabdocoel flatworms that possess a short glandular anterior proboscis (Hyman, 1951).

The development of an anus in the nemertean digestive system marks a significant change from the digestive system of flatworms. Ingestion and egestion can thus take place simultaneously, a much more efficient arrangement and one that is present in almost all higher invertebrates. The mouth is ventral and located anteriorly at the level of the brain (Figs. 8–3A; 8–4A; and 8–5). It opens into a foregut consisting of a buccal cavity, an esophagus, and a glandular stomach (Figs. 8–3A and 8–5). The prominence of these three regions varies. The foregut opens into a long intestine, extending the length of the body posteriorly to the anus. The intestine is provided with lateral diverticula (Fig. 8–5), and in some species it extends anteriorly beyond the junction with the foregut as a cecum. The intestinal wall is composed of elongated ciliated cells and is richly provided with gland cells.

In many hoplonemerteans, the mouth has disappeared and the esophagus opens into the rhynchodaeum. In all other rhynchocoels, the digestive system is completely separate from the proboscis apparatus.

Nemerteans are entirely carnivorous and feed primarily on annelids, although they also eat other small living or dead invertebrates, such as mollusks and crustaceans. The prey, after capture by the proboscis, is

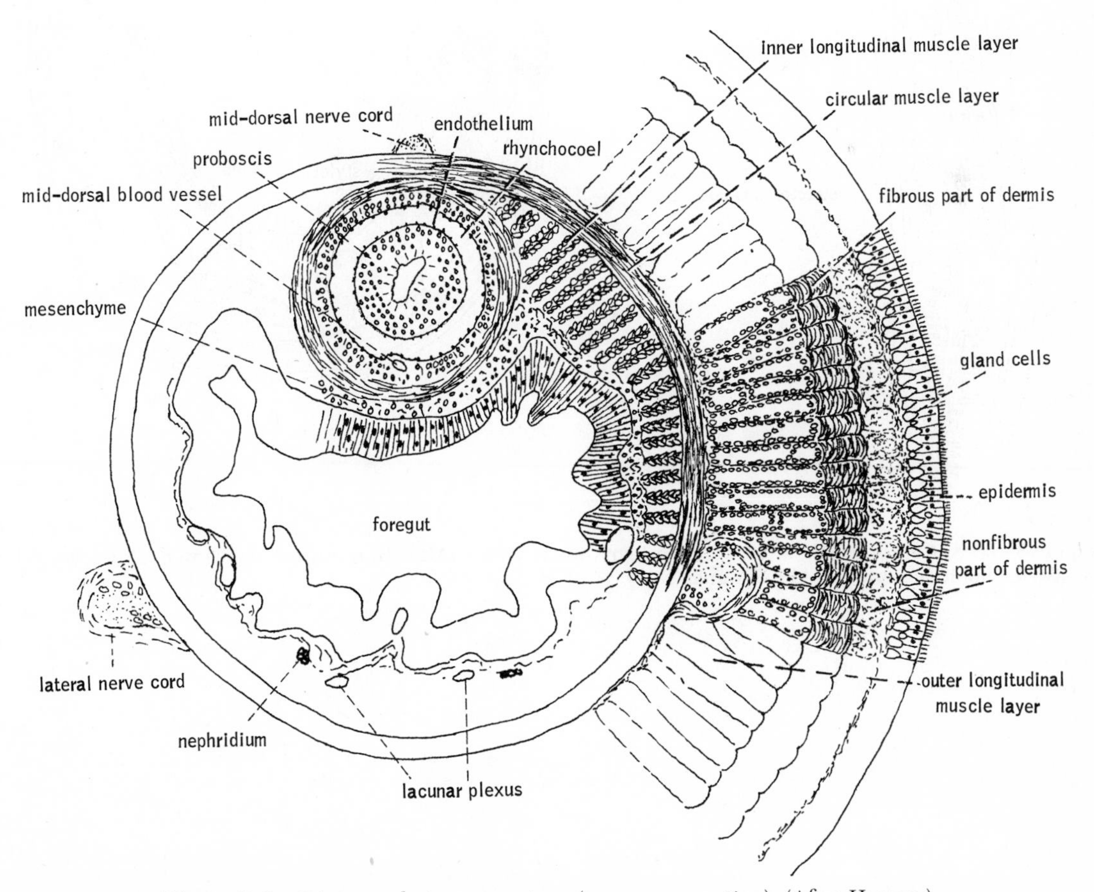

Figure 8–2. *Lineus*, a heteronemertean (transverse section). (After Hyman.)

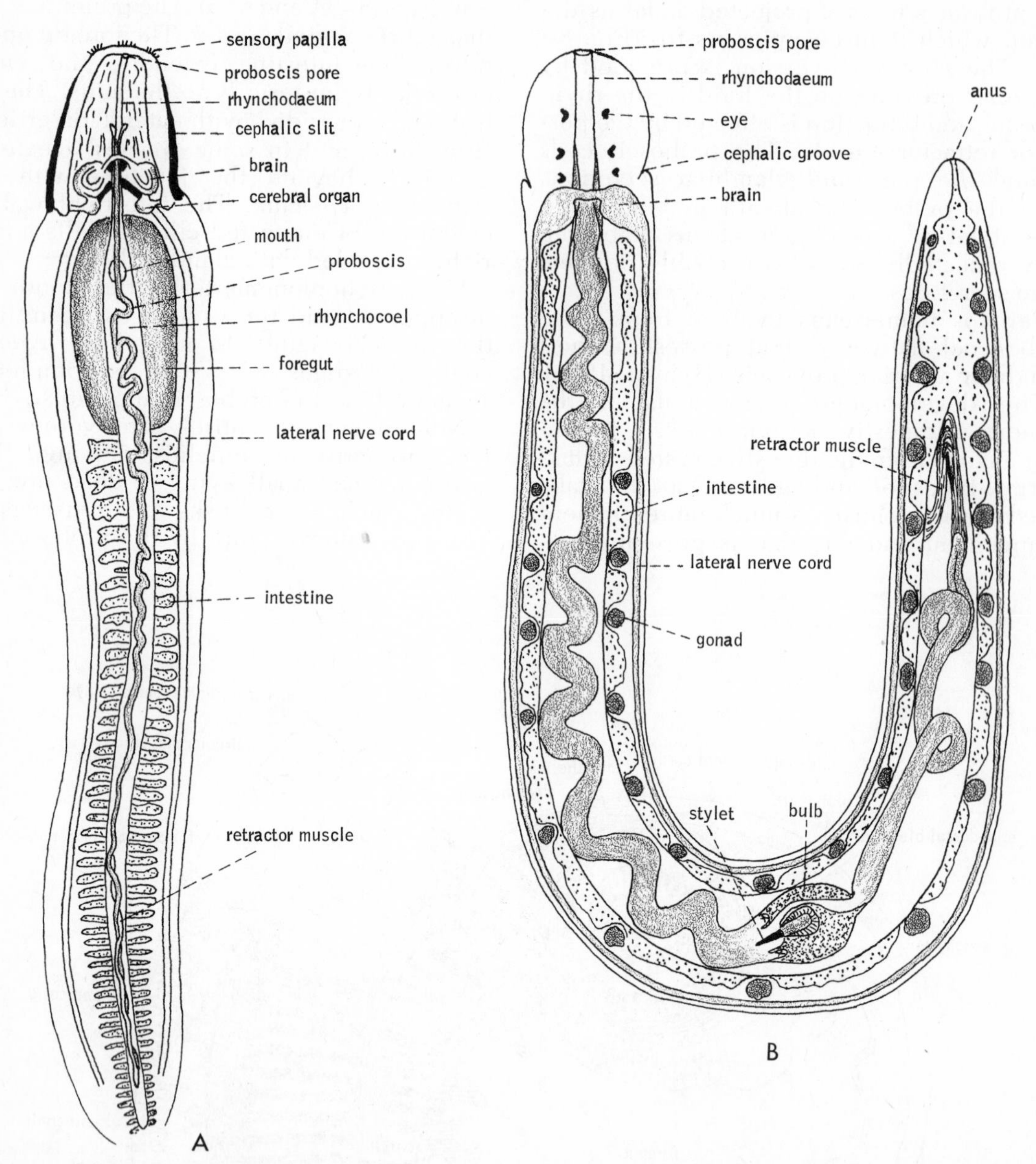

Figure 8–3. *A. Cerebratulus*, a heteronemertean (dorsal view). (After Bürger from Hyman.) *B. Prostoma rubrum*, a freshwater hoplonemertean. (After Pennak.)

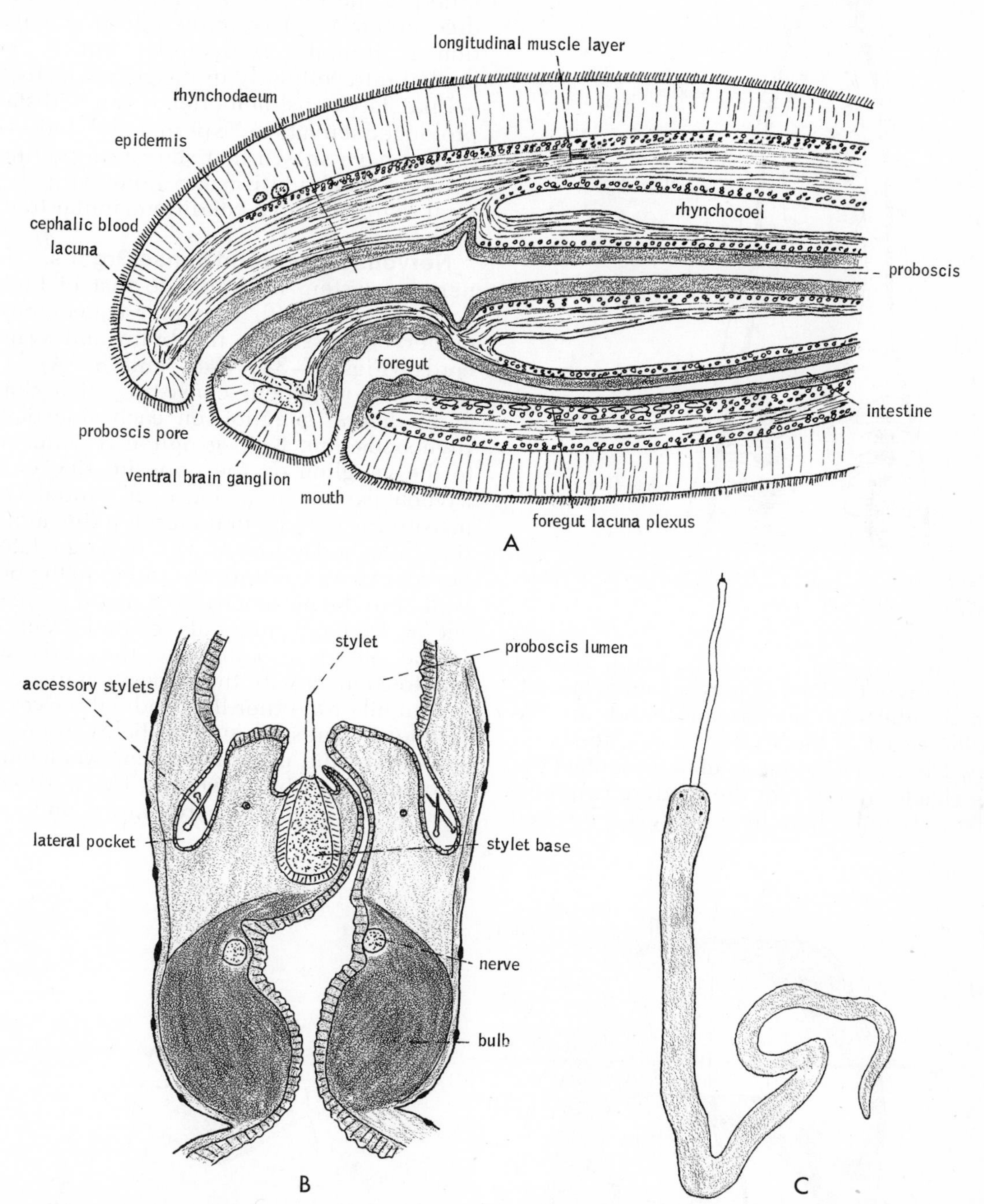

Figure 8–4. *A.* Anterior end of the paleonemertean, *Procarina* (Sagittal section). (Modified after Nawitzki from Hyman.) *B.* Stylet apparatus of the freshwater nemertean, *Prostoma.* (Modified after Böhmig from Hyman.) *C.* Armed nemertean with protruded proboscis.

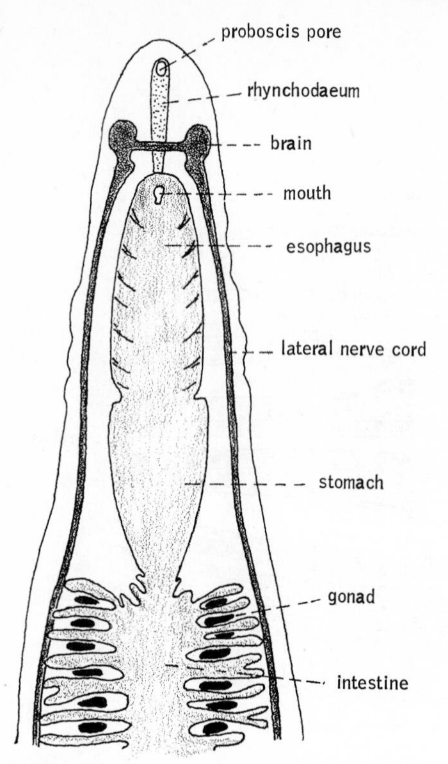

Figure 8–5. Digestive tract of *Carinoma*, a paleonemertean. (After Coe.)

swallowed whole by rapidly being sucked into the mouth. *Paranemertes peregrina*, an armed intertidal nemertean found on the Pacific Coast of the United States, feeds on polychaetes. The nemertean must contact the polychaete in order for the feeding response to be initiated. Then the proboscis is everted and wraps around and stabs the prey several times with the stylet (Roe, 1970).

Jennings (1960) found that in *Lineus*, an unarmed nemertean, the prey is probably killed in the foregut by acid secretions and digestion takes place in the intestine. Digestion is initially extracellular but is concluded intracellularly in phagocytic cells.

Like flatworms, nemerteans can withstand prolonged starvation. Experimental studies on *Lineus* have shown that starvation produces extreme reduction in size, along with structural regression to a condition similar to that of the larva.

Nervous System and Sense Organs. The nervous system is similar to that of higher flatworms but with a greater concentration of nervous tissue in the brain and ventral cords (Figs. 8–3*A*; 8–5; and 8–6*A*). The brain is four-lobed, consisting of a dorsal and a lateral ganglion on each side of the rhynchodaeum. Each ganglion is connected to the ganglion on the opposite side of the rhynchodaeum by a dorsal or ventral commissure, so that the brain forms a ring around the rhynchodaeum. A pair of large lateral nerves extends posteriorly, either in the body wall or in the mesenchyme. Lateral branches set at frequent intervals extend from the lateral nerves to innervate the body wall and to connect with the opposite nerve cord.

A number of other longitudinal nerves are frequently present and probably represent vestiges of the primitive radial arrangement of cords found in flatworms. The most common of these minor cords are a middorsal

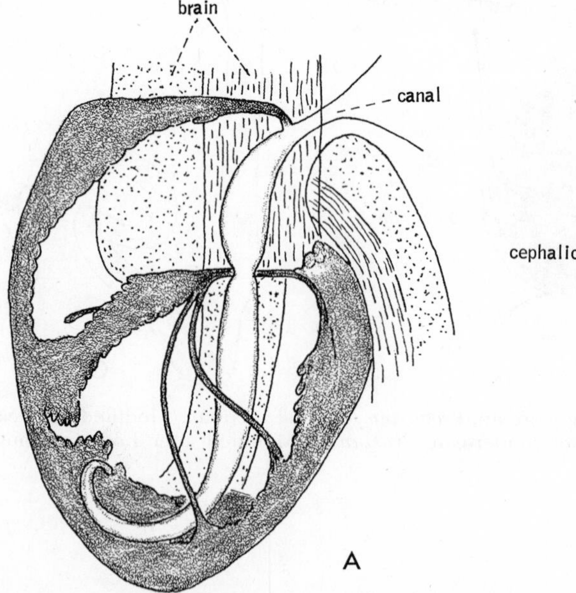

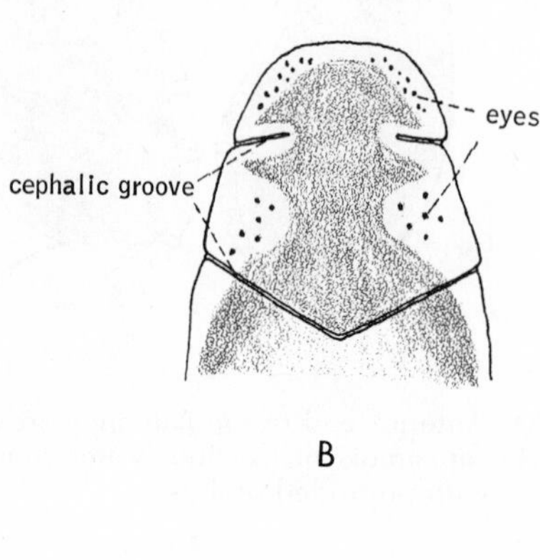

Figure 8–6. *A*. Cerebral organ of *Cerebratulus*. Shaded areas represent gland cells; white areas are filled with ganglion cells. *B*. Head of the hoplonemertean, *Amphiporus angulatus*. (*A* and *B* after Hyman.)

nerve (Fig. 8–2) that extends from the dorsal commissure posteriorly the length of the body, and a pair of esophageal nerves that arise from the ventral ganglia and extend posteriorly to innervate the foregut. Lateral nerves from the lateral cord frequently connect with this middorsal nerve.

The sense organs of nemerteans are also similar to those of flatworms. Tactile sensory cells are located in the epidermis, particularly at the anterior of the animal. Two to six eyes (Fig. 8–3*B*) are frequently present, located on the dorsal surface anterior to the brain; but there may be as many as several hundred eyes, arranged in rows or groups on each side of the anterior end (Fig. 8–6*B*). The eyes are inverse pigment cups like those of flatworms. Nemerteans avoid light and are active only at night.

Also located at the anterior of many nemerteans are two types of sense organs, believed to be chemoreceptors. The first type of sense organ consists of ciliated cephalic grooves or slits (Fig. 8–6*B*), which are glandless epidermal areas richly supplied with ganglion cells. These organs probably have both a tactile and chemoreceptive function.

The second type of sense organ, characteristic of nemerteans and found in most members of the phylum, is the cerebral organs (Fig. 8–6*A*). These consist of a pair of dorsal ciliated tubes or canals arising as invaginations of the epidermis. The inner ends of the canals contain gland and nerve cells associated with the brain. The external openings of the canals are in the cephalic slits, or grooves, or in a pair of pits over the brain area. Water currents in the canals appear to be activated in the presence of food, indicating a chemoreceptive function. *Prostoma,* on which observations of the cerebral organs have been made, is able to locate very small fragments of food in still water up to 17 cm. away. Some investigators have postulated an endocrine function for the cerebral organs because of their association with the brain and because of the abundance of gland cells present.

Circulation and Excretion. Nemerteans are the only acoelomates to possess a true circulatory system. The system is closed—that is, the blood never bathes the tissues directly but is confined to vessels or spaces with distinct walls. In the simplest form (Fig.

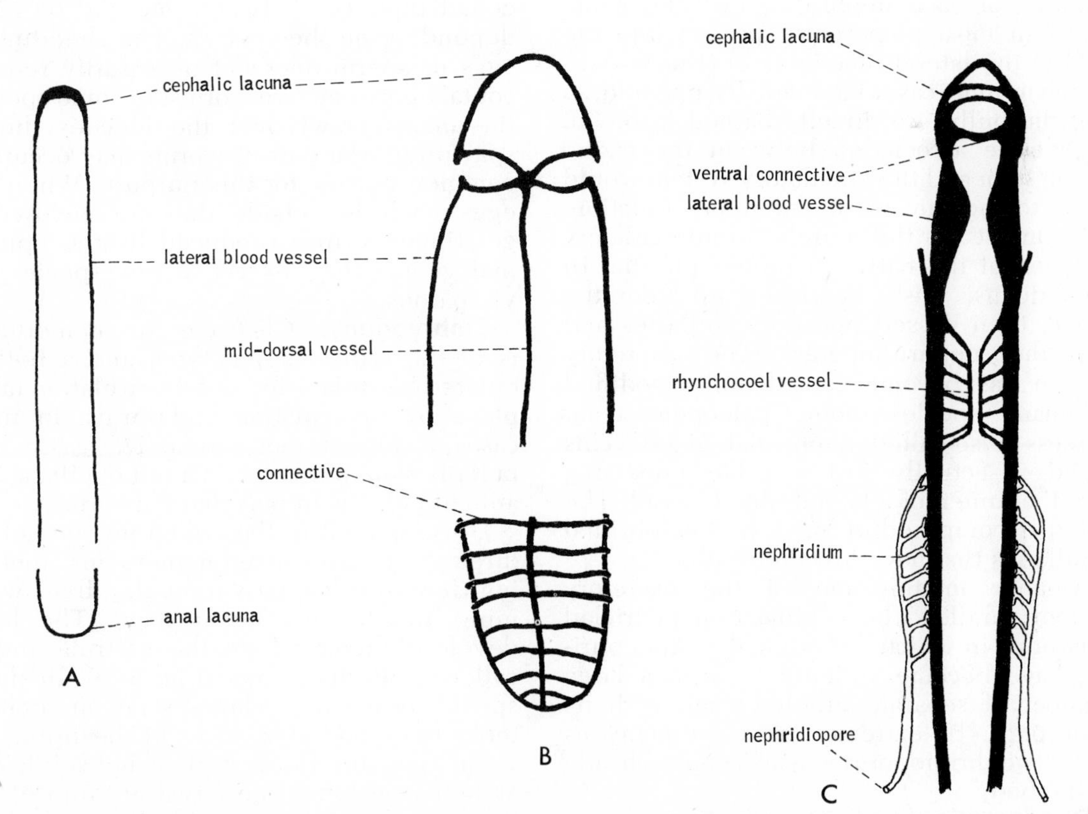

Figure 8–7. Circulatory systems of various nemerteans. *A. Cephalothrix. B. Amphiporus.* (*A* and *B* after Oudemans from Hyman.) *C. Tubulanus.* (After Bürger from Hyman.)

8–7*A*), there are only two vessels, which are located one on each side of the gut. These are connected anteriorly by a cephalic lacuna and posteriorly by an anal lacuna. The lacunae are mesenchymal spaces but are lined by a thin membrane. In many nemerteans considerable elaboration on this basic plan has taken place. One of the commonest additions to the basic plan is a middorsal vessel (Fig. 8–7*B*). This vessel arises from a transverse connection between the proboscis apparatus and the gut and eventually connects to the anal lacuna. In the region of the intestine, transverse vessels serve as connections between the lateral and middorsal vessels.

The large vessels are contractile, but blood flow is very irregular and does not follow a definite circuit. The blood may flow forward and backward in the longitudinal vessels.

Nemertean blood is usually colorless, but in many species it contains corpuscles bearing yellow, red, orange, or green pigments of uncertain function. In addition to the corpuscles, the blood also contains amebocytes.

The excretory system consists primitively of one pair of protonephridia. A nephridiopore is located on each side of the foregut, and a tubule extends anteriorly from the opening of each nephridiopore. The flame bulbs in most nemerteans project into the wall of the lateral blood vessel (Fig. 8–7*C*), which in some cases has even disappeared, so that the bulbs are directly bathed in blood. This close association between the excretory system and the circulatory system would seem to indicate a true excretory function, in contrast to the purely osmoregulatory function of the flatworm protonephridia. In nemerteans, waste is picked up from the blood, then passed into the capillaries and out the nephridiopores. The excretory system has become considerably modified in many species. Some paleonemerteans possess a so-called nephridial gland. This gland is actually just a ridge projecting into the lumen of a lateral blood vessel. The gland is composed of many protonephridial capillaries that have lost their bulbs.

Another modification of the excretory system parallels the modification in triclad flatworms in which the original protonephridia have become split up to form a large number of separate tubules, each with its own pore. In extreme cases, as many as 35,000 nephridiopores may exist on each side of the body.

Regeneration and Reproduction. Nemerteans, especially the larger species, display a marked tendency to fragment when irritated. The collection of large intact specimens is therefore usually difficult to accomplish. Very frequently the proboscis becomes detached when everted. The proboscis soon regenerates, but when the body has fragmented only the anterior end is capable of regenerating the posterior part. Some species, including members of the genus *Lineus*, reproduce asexually by fragmentation, and even posterior sections of the body are capable of regeneration.

The majority of nemerteans are dioecious, hermaphroditism existing chiefly among the freshwater and terrestrial hoplonemerteans. The reproductive system is very simple. The gametes develop from mesenchymal cells that aggregate and become surrounded by a thin-walled sac to form a gonad. A considerable number of such gonads usually form a regular row on each side of the body in the mesenchyme. When intestinal diverticula are present, the gonads are spaced between them (Fig. 8–5).

After maturation of the gametes, a short duct grows from the gonad to the outside, allowing the gametes to escape. The eggs are apparently squeezed out of the body by muscular contractions of the body wall. Each gonad produces between one and 50 eggs, depending on the species. The shedding of eggs or sperm does not necessarily require contact between two worms; in some species the males crawl over the females during spawning, or a pair of worms may occupy a common burrow for this purpose. When the eggs reach the outside, they are enclosed in gelatinous strings produced by the epidermal glands (Fig. 8–8*A*). A few species are viviparous.

Embryogeny. Cleavage in nemerteans is of the typical spiral type, and a hollow ciliated blastula is formed. Gastrulation takes place by invagination and forms, in most cases, a solid stereogastrula. The gastrula is radially symmetrical with a tuft of cilia at the animal pole, as in polyclad flatworms.

Development in the armed nemerteans is direct, but many heteronemerteans, including *Cerebratulus*, pass through a free-swimming and feeding larval stage. The larva develops directly from the gastrula and is called a pilidium larva (Fig. 8–9). In those species possessing a larval stage, an archenteron is formed as a result of the initial invagination, and the gastrula is not solid. The archenteron forms the larval gut and opens through the persistent blastopore, which forms the larval mouth. After a free-swim-

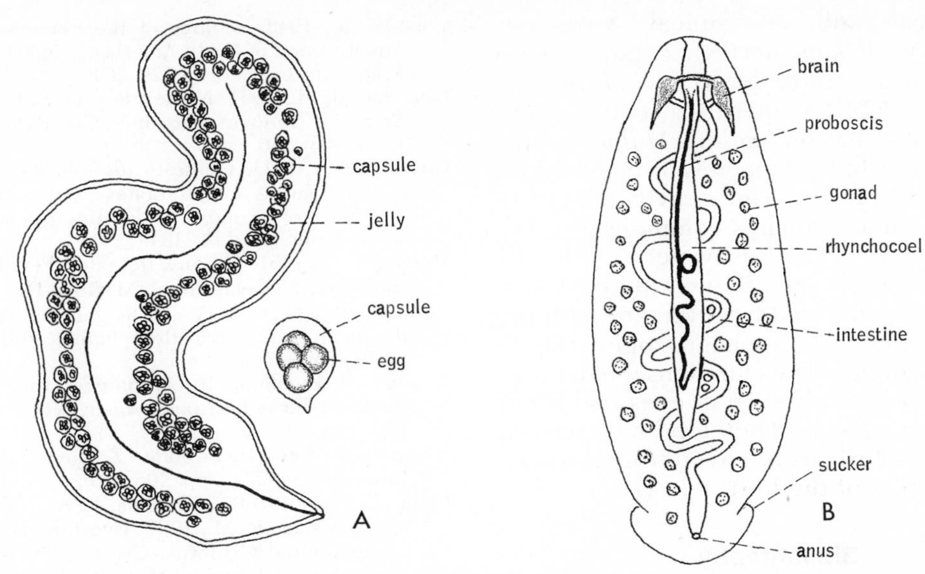

Figure 8–8. *A.* Egg string of *Lineus ruber.* (After Schmidt from Hyman.) *B. Malacobdella grossa,* a commensal in the mantle cavity of marine clams. Proboscis and esophagus open into common chamber. (After Coe.)

ming existence, the pilidium undergoes a rather complicated metamorphosis into a young worm.

The heteronemertean *Lineus* forms a spherical ciliated larva called a Desor's larva. However, the Desor's larva remains within the egg envelope and should really be considered only a modification of the pilidium, which contains no oral lobes and sensory plate.

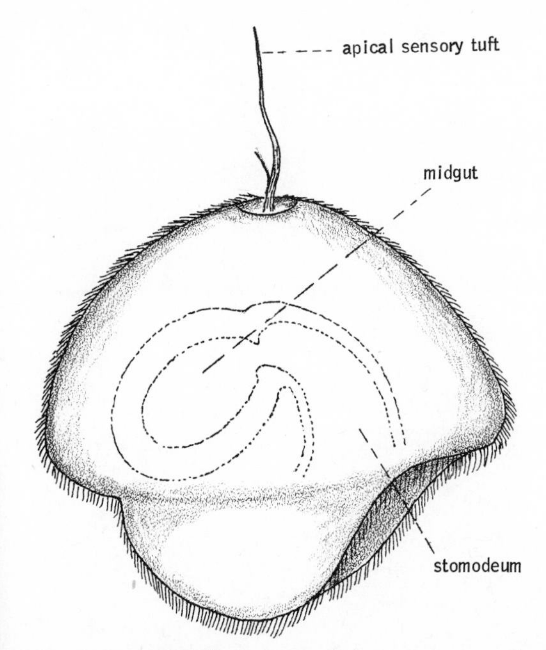

Figure 8–9. Pilidium larva of the nemertean, *Cerebratulus marginatus.* (After Coe.)

SYSTEMATIC RÉSUMÉ OF THE PHYLUM RHYNCHOCOELA

Class Anopla. Contains the two principal orders of unarmed nemerteans. Mouth located posterior to brain.

Order Paleonemertini. The most primitive members of the phylum belong to this order. All are neritic forms. The musculature of the body wall is either two-layered or three-layered, with the longitudinal layer located between two circular layers. The circulatory system is usually limited to two lateral vessels without middorsal or transverse vessels. Eyes and cerebral organs are often absent. *Tubulanus, Carinina, Carinoma,* and *Cephalothrix.*

Order Heteronemertini. The body-wall musculature is three-layered, with a circular layer between two longitudinal layers. Cerebral organs, middorsal blood vessels, and lateral intestinal diverticula are present. *Cerebratulus* and *Lineus.*

Class Enopla. Contains two orders, including the armed nemerteans. Mouth located anterior to brain or gut. Opens into rhynchodeum.

Order Hoplonemertini. All of the armed nemerteans belong to this order. The intestine contains lateral diverticula and a cecum; a dorsal blood vessel is present. This order contains several interesting ecological groups, such as the floating or swimming pelagic nemerteans, which live in the open ocean; the freshwater nemerteans (*Prostoma*);

the tropical and subtropical terrestrial nemerteans (*Geonemertes*); and many marine shallow water forms (*Amphiporus*).

Order Bdellonemertini. This small order contains but one genus, *Malacobdella* (Fig. 8–8*B*), with four species, three of which are commensal in the mantle cavity of marine clams and one of which is commensal in the mantle cavity of a freshwater snail. Although the proboscis is unarmed, it has probably been derived from the armed type. The proboscis and esophagus open into a common chamber. The commensal existence is reflected in the absence of eyes and other special sense organs. They feed on bacteria, algae, and protozoans removed from the ventilating current of the host.

Bibliography

Bayer, F. M., and Owre, H. B., 1968: The Free-Living Lower Invertebrates. Macmillan Co., N. Y., pp. 178–204.

Coe, W. R., 1943: Biology of the nemerteans of the Atlantic coast of North America. Trans. Connecticut Acad. Arts and Sci., *35*:129–328.

Gontcharoff, M., 1961: Nemertiens. *In* P. Grassé (Ed.): Traité de Zoologie. Masson et Cie, Paris, Vol. 4, pt. 1, pp. 783–886.

Gosner, K. L., 1971: Guide to Identification of Marine and Estuarine Invertebrates: Cape Hatteras to the Bay of Fundy. Wiley-Interscience, N. Y. Nemerteans, pp. 174–184.

Hyman, L. H., 1951: The Invertebrates: Platyhelminthes and Rhynchocoela. Vol. 2. McGraw-Hill, N. Y., pp. 459–531. (A very complete general treatment of the nemerteans. Excellent figures and extensive bibliography.)

Jennings, J. B., 1960: Observations on the nutrition of the Rhynchocoelan *Lineus ruber*. Biol. Bull., *119*(2): 189–196.

Kaestner, A., 1967: Invertebrate Zoology, Vol. I. Wiley-Interscience, N. Y., pp. 206–216.

Light, S. F., Smith, R. I., Pitelka, F. A., Abbott, D. P., and Weesner, F. M., 1967: Intertidal Invertebrates of the Central California Coast. University of California Press, Berkeley, Nemerteans, pp. 60–63.

Roe, P., 1970: The nutrition of *Paranemertes peregrina*. I. Studies on food and feeding behavior. Biol. Bull., *139*:80–91.

Chapter 9

THE PSEUDOCOELOMATES

Among the bilateral animals, only flatworms and nemerteans have a solid (or acoelomate) type of body construction, in which the space between the digestive tract and the muscle layers of the body wall is filled with mesenchyme. All other bilateral animals have a fluid-filled space called a body cavity located between the body wall and the internal organs; or if a body cavity is absent, such groups are at least clearly derived from forms with a body cavity. The development of a body cavity in the higher invertebrates was probably associated with a number of different factors. The fluid within a body cavity can function as a hydraulic skeleton, and can also aid in the translocation of gases, food materials, and waste materials. Also, with a body cavity, there is more space for the accommodation of coiled or looped internal organs.

The vast majority of animals exhibit a coelomate type of body structure; i.e., they possess a coelom as a body cavity. A smaller number possess a pseudocoel as a body cavity and are called pseudocoelomates. It is the pseudocoelomate phyla which are the subject of this chapter.

A pseudocoel, as explained in Chapter 3, represents a persistent embryonic blastocoel. The mesodermal derivatives are, for the most part, located at the outer side of the cavity, and the internal organs are actually free within the pseudocoel; i.e., they do not lie behind peritoneum (Fig. 3–2). The pseudocoelomate phyla are structurally very diverse and their phylogenetic interrelationships are obscure. Quite possibly the entire assemblage may not constitute a monophyletic unit. It is generally assumed that the pseudocoelomate groups comprise a side branch stemming from the acoelomates or some line of coelomates. But it should be emphasized that a pseudocoel is not a stage in the evolution of a true coelem; rather, it is an independent evolutionary acquisition. Thus, from an evolutionary standpoint, the pseudocoelomates represent a terminal group of phyla.

Considering the small size of most pseudocoelomates in comparison with the majority of coelomates, it is probable that the principal functions of the pseudocoel are those of circulation and the maintenance of internal turgor pressure.

The Aschelminths

The largest number of pseudocoelomate animals, including such familiar groups as the rotifers, the gastrotrichs, and the nematodes, are often placed within a single phylum, the Aschelminthes. Most members of this assemblage are rather elongate and often somewhat cylindrical, especially the wormlike nematodes. Although the anterior of the body is cephalized to a varying degree and bears the mouth and certain sense organs, there is no well-formed head. The body is typically covered with a distinct collagen cuticle, which in some groups has developed into a skeletal encasement. Surface ciliation is restricted or absent. Adhesive glands in varying numbers are characteristic of many species and often open to the outside of the body by projecting cuticular tubes. The digestive tract is usually a complete tube with a mouth and anus, and a specialized pharyngeal region is almost always present. Since most species are aquatic and very small, no gas exchange organs or circulatory system are present. However, protonephridia often exist. A peculiar feature of many members is the numerical constancy of the cells, or the nuclei, that compose the various organs, a condition known as eutely. The numbers are characteristic for each species but vary from one species to another. The condition is perhaps associated with the extremely small size characteristic of many members of the group. Commonly, mitosis ceases following embryonic development, and growth continues through increase in the size of the cells. Most aschelminths are dioecious, and cleavage is strongly determinate.

Despite these similarities among the aschelminths, they are extremely diverse in other respects, and the group lacks any well-defined, distinctive unifying characters. There is thus considerable justification for questioning how closely related the various aschelminths are to one another and whether the assemblage actually warrants the rank of phylum. Although the concept of the Aschelminthes as a phylum has received some acceptance, there are many zoologists who prefer to use the name aschelminth as a convenient term of reference without taxonomic rank and to treat the different divisions as separate phyla. Such an approach will be used in this chapter. However, for those who prefer the alternative view, each of the following aschelminth phyla can be considered a class.

PHYLUM GASTROTRICHA

The gastrotrichs are a small phylum of some 400 marine and freshwater aschelminths that inhabit the interstitial spaces of bottom sediments and superficial detritus, the surfaces of submerged plants and animals, and the pseudoplanktonic spaces between submerged plants. The phylum is divided into two orders: the order Macrodasyida is composed of marine and brackish water species, and the order Chaetonotida contains all of the freshwater species as well as some marine forms. Although not as abundant as rotifers, many chaetonotids are common animals of ponds, streams, and lakes. They are also not uncommon marine animals. Hummon (1971) reported 23 species of macrodasyids and 19 species of chaetonotids from the intertidal zone of New England beaches.

External Structure. Like rotifers, most gastrotrichs are microscopic in size. They range in length from 50 to 1000 microns though members of some species may approach 4000 microns. The bottle-shaped or strap-shaped body is usually flattened ventrally, and the posterior end is sometimes forked (Figs. 9–1 and 9–2).

The locomotor cilia are always restricted to the ventral surface of the trunk and head. The entire ventral surface may be ciliated or, as in *Chaetonotus* (Fig. 9–3*A*), the cilia may be arranged in two ventral longitudinal bands that run the length of the body. In other gastrotrichs, the cilia are present in patches arranged in two longitudinal bands or in a number of transverse rows (Fig. 9–3*B*). In a few species, cilia are present only on the ventral surface of the head. The head ciliation is variable and frequently specialized. Commonly the cilia are arranged in tufts, as in *Chaetonotus* where four ciliated tufts exist, two on either side of the head. Frequently some of the cilia are modified into bristles, which may or may not be motile and are probably sensory in function. Locomotion takes place by a gliding motion over the substratum. The ventral ciliation of gastrotrichs suggests that these animals may be the most primitive of the aschelminth pseudocoelomates.

The highly specialized cuticle is a distinguishing feature of this phylum. Often the cuticle is modified in the form of scales, which may be ornamented with one or more spines (Fig. 9–2). Adjacent scales may be separated or fused, or more typically may overlap like shingles.

Adhesive tubes (Fig. 9–1) are present. The number and position of the adhesive tubes vary, but they are commonly arranged in anterior, lateral, and posterior series, with as many as 250 present. In *Chaetonotus* (Fig. 9–1), the forked ends at the posterior of the body carry two adhesive tubes. Some gastrotrichs move in a leechlike fashion or may be temporarily sessile, the adhesive tubes providing for attachment to the substratum.

Internal Structure. The body wall has a syncytial epidermis and bands of longitudinal and circular muscle fibers. The pseudocoel in most gastrotrichs consists of only a small, slitlike space between the body wall and the gut.

The terminal mouth opens directly into the pharynx or, as in some chaetonotids, opens into a small protrusible buccal capsule which is lined with cuticle and may bear ridges or projecting teeth. The pharynx, which is in no way similar to the mastax of rotifers, is an

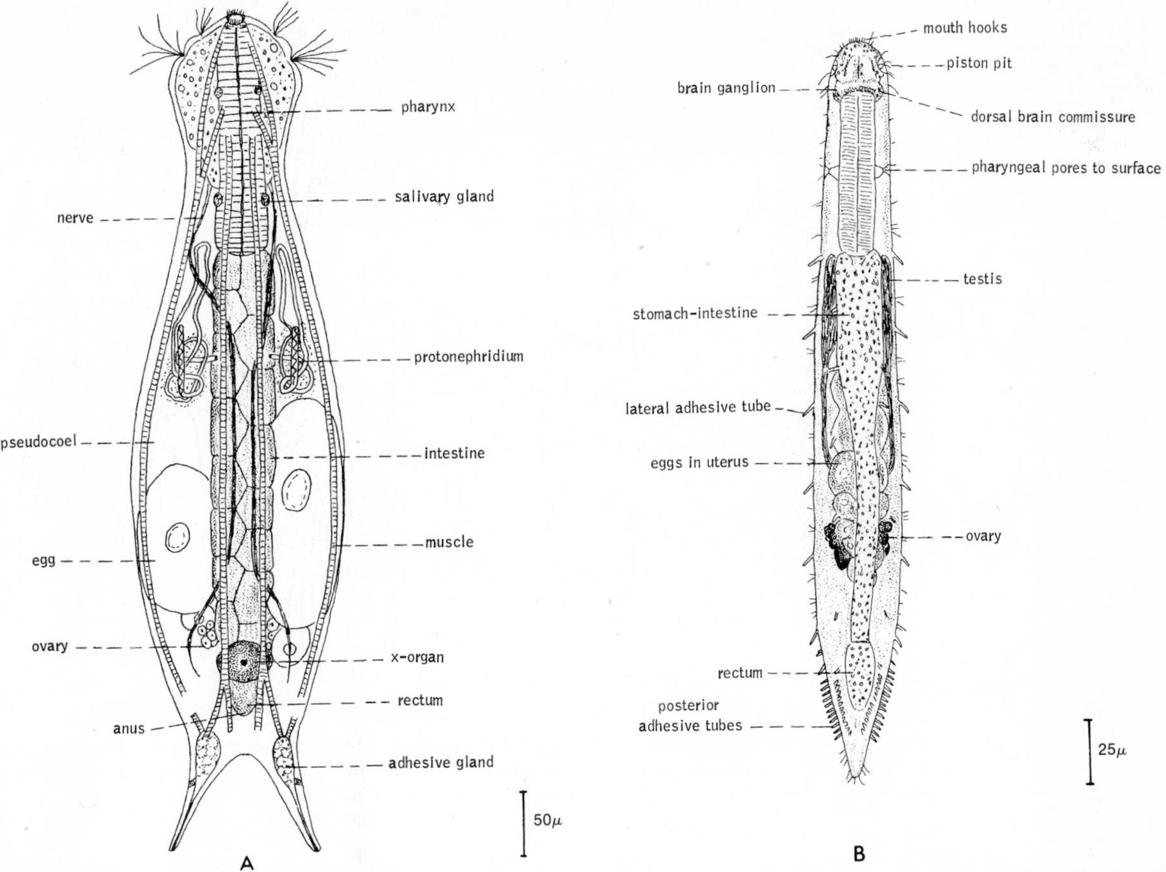

Figure 9–1. *A*. A chaetonotid gastrotrich (ventral view). (Modified after Zelinka from Pennak.) *B*. A macrodasyid gastrotrich (*Macrodasys*). (After Hyman.)

elongated tube, which may contain one to four bulbous swellings (Figs. 9–1 and 9–2*A*). In macrodasyids, the pharynx opens to the exterior through a pair of pores. The pharyngeal wall is composed of columnar epithelium, radial muscle fibers, and some gland cells; the lumen is triangular and lined with cuticle. Posteriorly, the pharynx opens into a cylindrical, tapered stomach-intestinal region, the walls of which are composed of a single layer of cuboidal or columnar epithelium. Gland cells are also present. The anus is located near the terminal end of the trunk.

Gastrotrichs feed on small dead or living organic particles, such as bacteria, diatoms, and small protozoans, all of which are sucked into the mouth by the pumping action of the pharynx or driven in by buccal cilia. In the marine macrodasyids, the paired pharyngeal pores permit the release of excess water ingested with the prey. Digestion takes place in the stomach-intestinal region and is very rapid.

Protonephridia, typically a single pair, are best developed in freshwater gastrotrichs and probably function as osmoregulatory organs. The long, coiled nephridial tubules open through separate nephridiopores on the ventrolateral surface near the middle of the trunk (Fig. 9–1).

The brain is composed of a ganglionic mass (Fig. 9–1*B*) on each side of the anterior of the pharynx; the two masses are connected dorsally by a commissure. Each brain mass gives rise to a ganglionated cord that extends the length of the body. The sense organs are represented by the ciliated tufts and bristles on the head, as well as sensory bristles located over the general body surface. Many gastrotrichs possess ciliated pits located on both sides of the head. In a few species, the bottoms of these pits can be projected as tentacles. Eye spots, which are

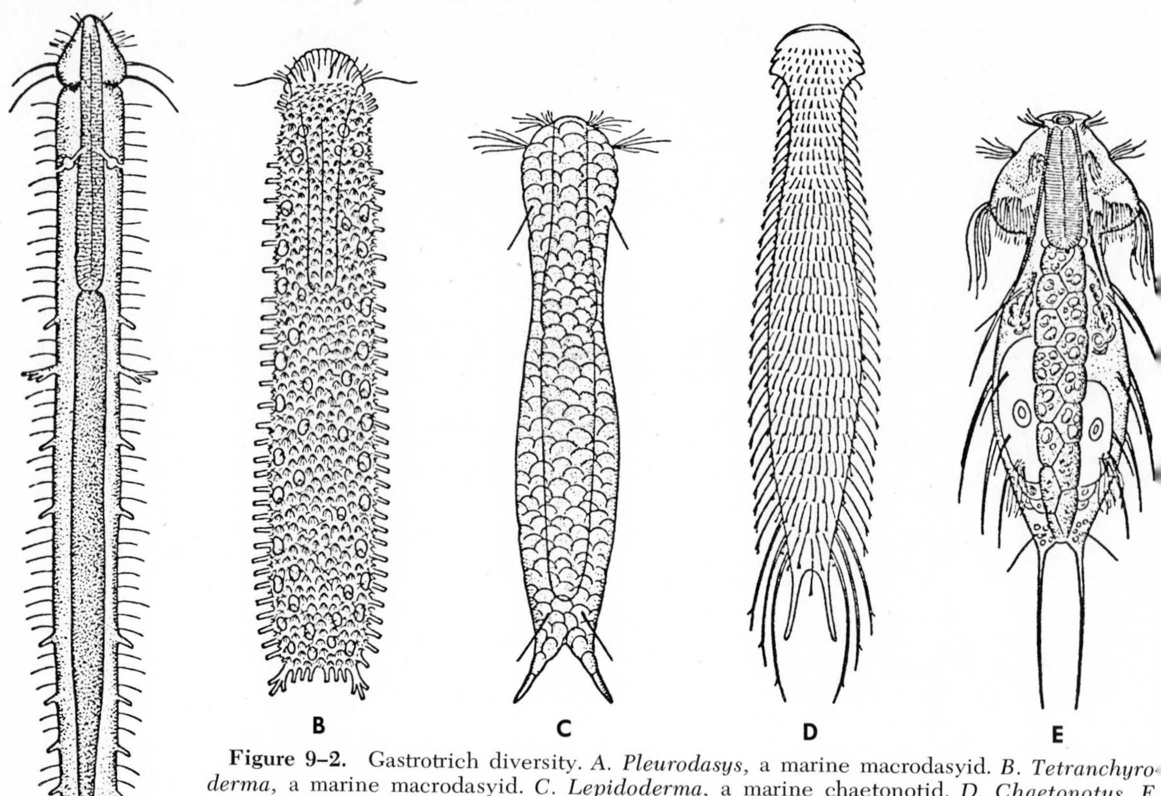

Figure 9–2. Gastrotrich diversity. *A. Pleurodasys,* a marine macrodasyid. *B. Tetranchyroderma,* a marine macrodasyid. *C. Lepidoderma,* a marine chaetonotid. *D. Chaetonotus.* *E Dasydytes,* a chaetonotid. (All from Grassé, P., 1965: Traité de Zoologie.)

clusters of pigment granules within certain brain cells, are present in a few species.

Reproduction. In contrast to other aschelminths, gastrotrichs are hermaphroditic, although the male system of many chaetonotids has become so degenerate that functionally all individuals are females. The one or two ovaries, which are simply clusters of germinal cells, are located in the posterior of the trunk. On leaving the ovary, the eggs pass into a rather poorly defined space called a uterus located just in front of the ovaries; or the eggs pass into a central compartment of the pseudocoel, as in the macrodasyids. The oviduct, which is present only in macrodasyids, is single and contains an anterior seminal receptacle and a posterior copulatory bursa. The oviduct opens through the anus or through a separate female gonopore in front of the anus.

Although chaetonotids lack an oviduct, the eggs pass through a saclike structure called the x-organ (Fig. 9–1), which opens onto the ventral surface of the body. The male system contains one or two testes (Fig. 9–1*B*), each with a sperm duct. The sperm ducts open to the outside separately or in common through pores located ventrally, either anteriorly in the trunk or near the

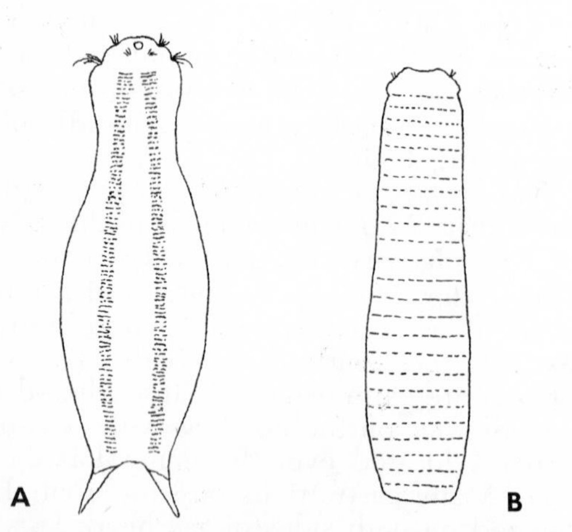

Figure 9–3. *A.* Ventral ciliation in *Chaetonotus.* *B.* Ventral ciliation in *Thaumastoderma.* (After Hyman.)

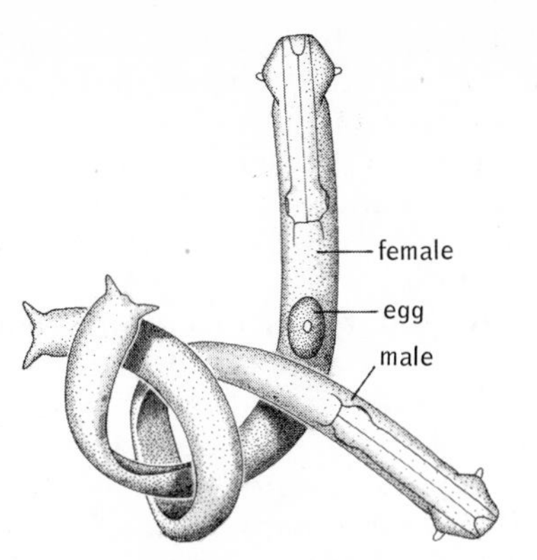

Figure 9–4. Copulation in the marine gastrotrich *Turbanella cornuta.* (After Teuchert.)

female gonopore. In some gastrotrichs, such as *Macrodasys, Urodasys,* and *Turbanella,* the terminal end of the sperm duct is modified to serve as a penis, and sperm transfer is direct (Fig. 9–4). In others, such as *Dactylopodola,* sperm are transferred as spermatophores, which are attached by the male to the right side of the posterior half of the female. In macrodasyids the egg leaves the body by rupture of the body wall.

Laboratory studies on *Lepidodermella* indicate a maximum life of about 40 days, during the first fourth of which a female lays four or five eggs. However, the life expectancy is probably far shorter under natural conditions.

In freshwater chaetonotids, which are mostly parthenogenetic, two types of eggs are produced and attached to the substratum. One type is the dormant egg, like those of rotifers; the other type hatches in one to four days. Cleavage is bilateral but determinate. Young gastrotrichs have most of the adult structures on hatching and reach sexual maturity in about three days. The growth stages of *Turbanella* are illustrated in Figure 9–5.

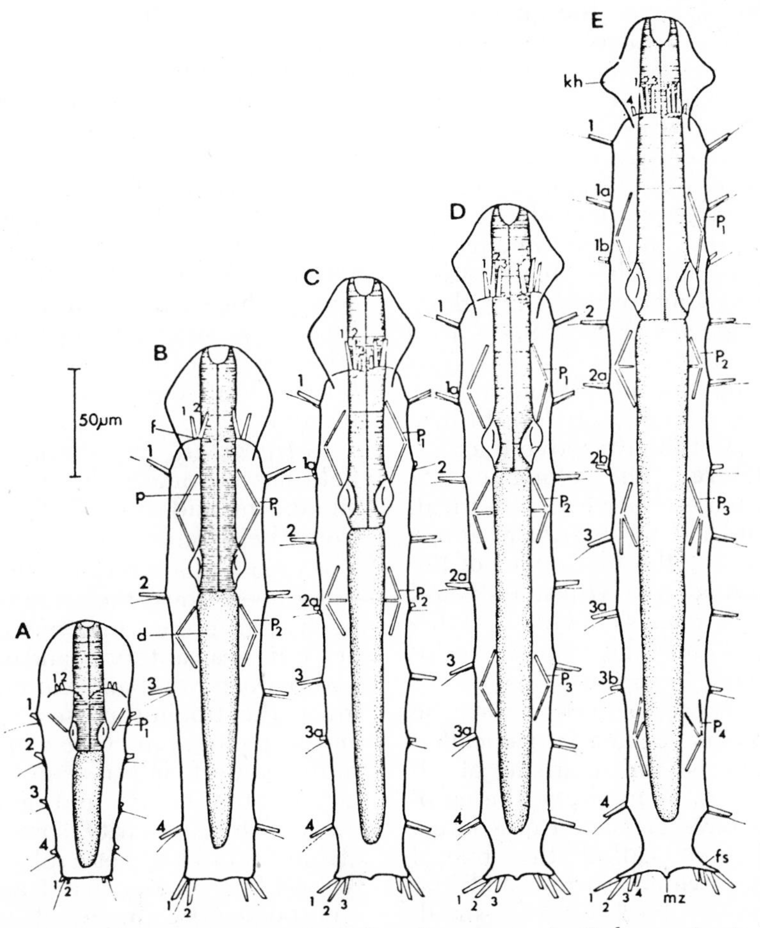

Figure 9–5. Growth in the marine gastrotrich *Turbanella cornuta. A.* Embryo prior to hatching. *B-D.* Post-hatching juvenile stages. *E.* Individual at sexual maturity. p, pharynx; p_1-p_4, protonephridia; d, intestine; 1-4, adhesive tubes. (From Teuchert, G., 1968: Z. Morph. Tiere, *63*:343–418.)

SYSTEMATIC RÉSUMÉ OF PHYLUM GASTROTRICHA

Order Macrodasyida. Marine and brackish water gastrotrichs. Body usually strap-shaped. Adhesive tubes located on anterior, posterior, and sides of body. Pharyngeal pores present. Hermaphroditic. *Macrodasys, Urodasys, Turbanella, Dactylopodola.*

Order Chaetonotida. Freshwater and marine gastrotrichs. Body usually bottle-shaped. Adhesive tubes usually restricted to the posterior end. Pharyngeal pores absent. Most species are parthenogenetic females. *Chaetonotus, Lepidodermella.*

PHYLUM ROTIFERA

The phylum Rotifera contains the very common and abundant freshwater animals commonly known as rotifers, or wheel animalcules. The name is derived from the presence of a ciliated crown, or corona, which when beating, gives the appearance of a rotating wheel in many species (Fig. 9–6*A*). Although some marine species exist and some species live in mosses, the majority inhabit fresh water. Approximately 1500 species have been described.

Some species may reach 3 mm. in length, but most rotifers are approximately the size of ciliate protozoans and are among the smallest metazoan animals. Most rotifers are solitary free-moving animals, but there are some sessile species (Fig. 9–9), as well as some colonial forms (Fig. 9–13*A*). The body is usually transparent, although some rotifers appear green, orange, red, or brown, owing to coloration of the digestive tract.

External Structure. The elongated or saccular body, which is relatively cylindrical, can be divided into a short anterior region, a large trunk composing the major part of the body, and a terminal foot (Figs. 9–6 and 9–11*B*). The body is always covered by a distinct cuticle, which may be ringed, sculptured, or ornamented in various ways.

The broad or narrowed anterior forms the head region and bears a ciliated organ called the corona, which is characteristic of all members of the class. Primitively, the corona is believed to have consisted of a large, ventral, ciliated area called the buccal field (Fig. 9–7*A*), which surrounded the mouth. If rotifers evolved from a small, ciliated, creeping ancestor as is generally believed, then perhaps the buccal field represented a vestige of the ancestral ventral ciliation. From the buccal field, cilia extended around the anterior margins of the head to form a crownlike ring called the circumapical band. The area in the interior of the ring, which is devoid of cilia, is called the apical field.

The different types of coronas characteristic of different groups of rotifers are believed to have evolved from this basic structural plan. During the derivation of these different types, various parts of the buccal field and the circumapical band either have been lost or have become more highly developed. Not infrequently certain cilia have become modified to form cirri, membranelles, or bristles. A few of the more common types of coronal ciliation are described here.

In *Euchlanis,* the original buccal field and circumapical band have been retained except for a portion posterior to the mouth (Fig. 9–8). In this area, only a vestige remains. The cilia along the margin of the buccal field have become modified into stiff cirri, and the cilia of the circumapical band on the lateral sides of the head form large tufts. When the ciliary tuft is located on a projection of the body wall, the projection and tuft collectively are called the auricle. The buccal field in *Collotheca* and related forms is modified into a funnel, and ciliation is reduced (Fig. 9–9). The edges of the buccal field have become expanded, forming a varying number of lobes bearing bristles, or setae, which may be arranged in bundles or tufts. The setae are perhaps derived from cilia. As a result of this expansion of the buccal field, the mouth lies at the apex of the funnel.

In *Hexarthra* and *Testudinella* the buccal field has become reduced and relatively insignificant, and the corona is derived entirely from the circumapical band (Fig. 9–10*B*). The cilia in the interior of the band are greatly reduced; thus, the circumapical band has been transformed into two circlets of modified cilia—an anterior band called the trochus and a posterior band called the cingulum. Since the trochus passes above the mouth and the cingulum below the mouth, these rotifers superficially resemble the trochophore larvae of annelids and mollusks. On the basis of this similarity, some zoologists have claimed a phylogenetic relationship between the rotifers on one hand and the mollusks and the annelids on the other. But if the two ciliated girdles of these rotifers are actually derived from a single circumapical band,

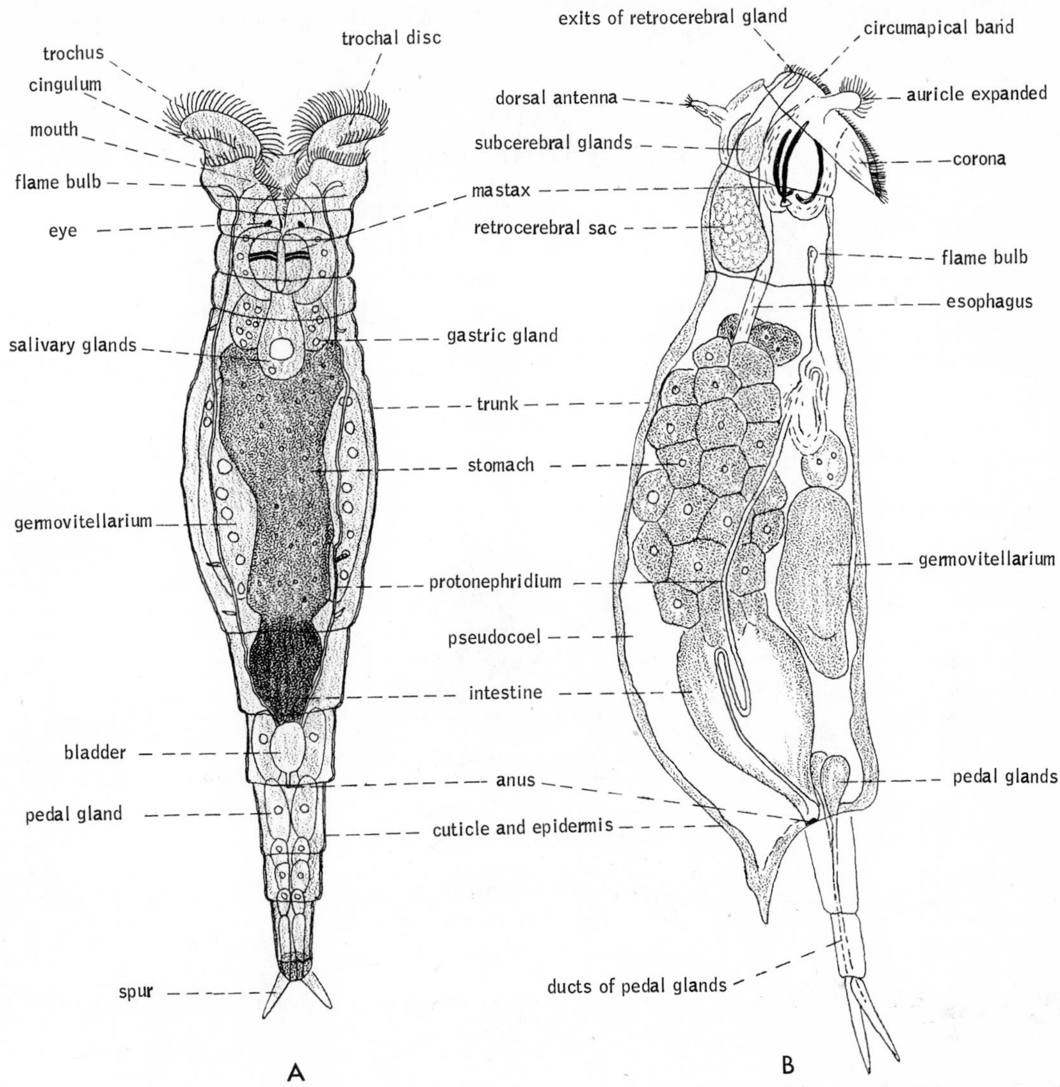

Figure 9–6. A. *Philodina roseola* (ventral view). (After Hickernell from Hyman.) B. *Notommata copeus* (lateral view). (After Hyman.)

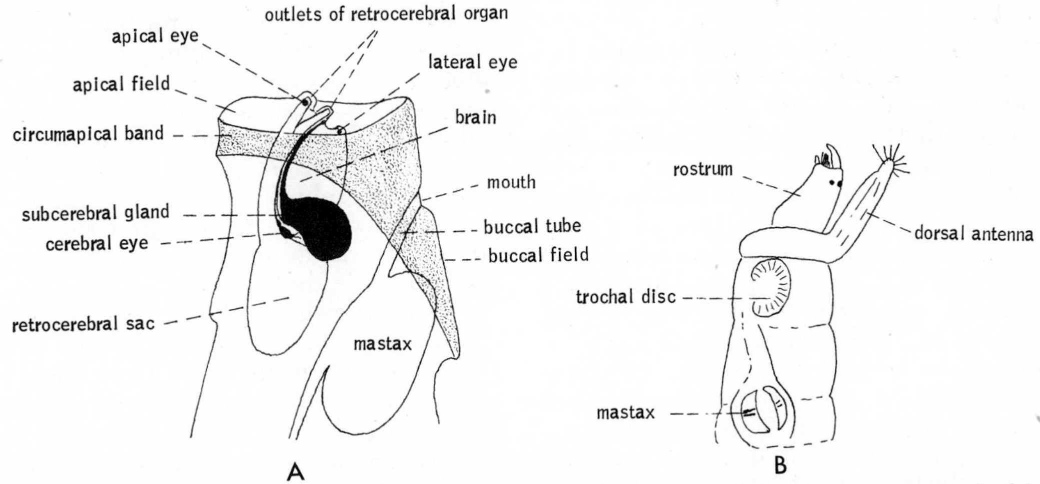

Figure 9–7. A. Primitive corona (lateral view). (After Beauchamp from Hyman.) B. Anterior end of bdelloid rotifer with trochal disc retracted (lateral view). (After Hyman.)

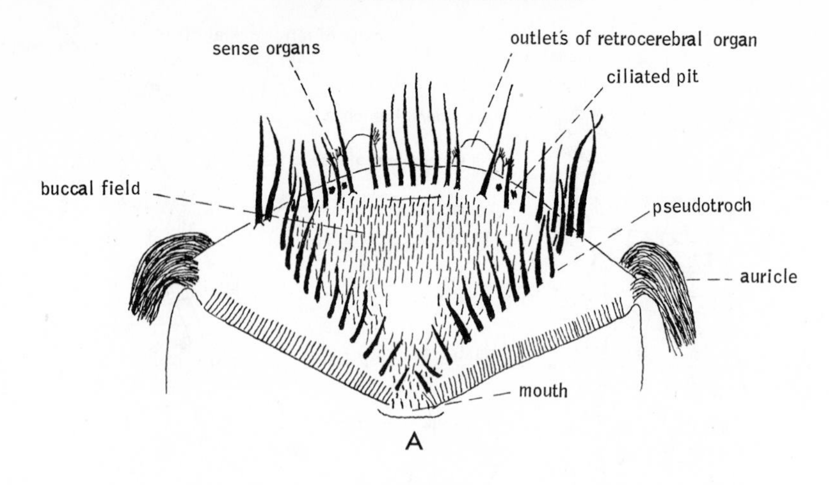

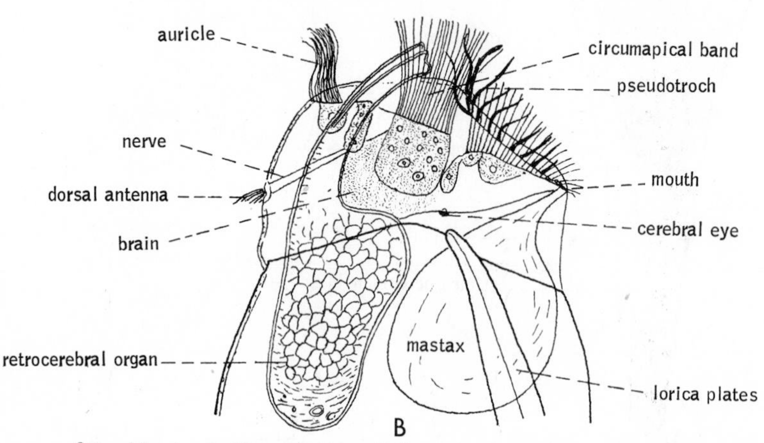

Figure 9–8. Corona of *Euchlanis*. *A*. Ventral view. (After Stoszberg from Hyman.) *B*. Lateral view. (After Beauchamp from Hyman.)

then resemblance to the trochophore larva is entirely superficial; and these rotifers are specialized rather than primitive members of the class.

The corona of bdelloid rotifers, which include many common freshwater species, represents a still further modification of these two ciliated bands (Fig. 9–6*A*). The anterior circlet of cilia (the trochus) is raised up on a pedestal and divided into two discs called trochal discs. The posterior circlet passes around the base of the pedestals and runs beneath the mouth. In living species, the beating membranelles of the trochal discs resemble two rotating wheels at the anterior of the body, and it is from this type of corona that the names rotifer and wheel animalcule are derived. The trochal discs are used both in swimming and in feeding, and the pedestals can be retracted when the discs are not functioning.

The head in bdelloid and notommatid rotifers carries a middorsal projection called the rostrum (Fig. 9–7*B*). This little projection bears cilia and sensory bristles at its tip. Other head structures in rotifers include the eye, which varies in number and location, and the retrocerebral organ. A typical retrocerebral organ consists of a pair of lateral subcerebral glands, located above and behind the brain (Figs. 9–7*A* and 9–8*B*) and a median retrocerebral sac, which is drained by a single duct. Prior to reaching the exterior, the duct forks to open through two pores on the apical field. The ducts of the glands accompany the ducts of the retrocerebral sac.

The elongate, or saccular, trunk composes the major part of the body. The cuticle is frequently much thickened to form a conspicuous encasement, called a lorica (Fig. 9–11). The lorica may be divided into distinct plates or ringlike sections and is usually ornamented with ridges or spines. The spines may be quite long and in some rotifers are

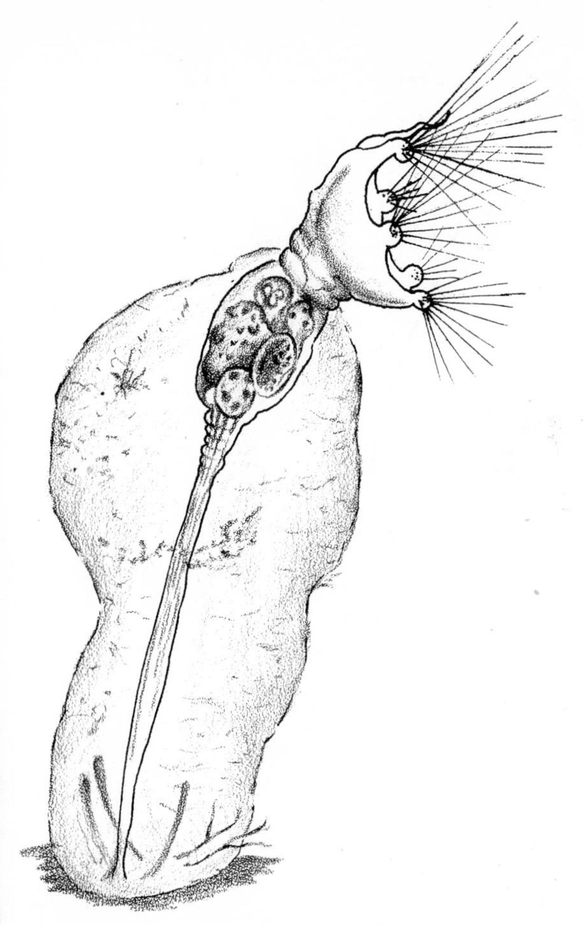

Figure 9–9. *Floscularia cornuta*, a sessile rotifer with funnel-shaped buccal field. Trunk and stalk surrounded by gelatinous mass. Female. (After Hudson.)

movable. In *Polyarthra*, the spines are flattened to form long skipping blades that are located in two clusters of six each on either side of the body (Fig. 9–10*A*).

Typically, the trunk bears three small projections called antennae. Two of the antennae are lateral and are located on each side of the trunk toward the posterior of the body; the other, which is sometimes paired, is located more anteriorly on the dorsal surface. The antennae frequently bear sensory bristles at their tips.

The terminal portion of the body, or foot, is considerably narrower than the trunk region (Figs. 9–6 and 9–11*B*). The cuticle is frequently ringed; and in many bdelloids (Fig. 9–6*A*), the resulting segments or joints of the foot are able to telescope into similar larger joints of the trunk. Even the head may be retracted in this manner. The end of the foot usually bears one to four projections called toes. In both the crawling and sessile rotifers, the foot is used as an attachment organ; in these groups the foot contains two to thirty glandular syncytial masses called pedal glands (Fig. 9–12*A*). These pedal glands produce an adhesive substance that is carried to the exterior by means of ducts opening onto the toes or other parts of the foot.

In a number of common bdelloids, such as *Philodina* (Fig. 9–6*A*) and *Rotaria*, the pedal glands open onto the ends of two long and diverging conical spurs located near

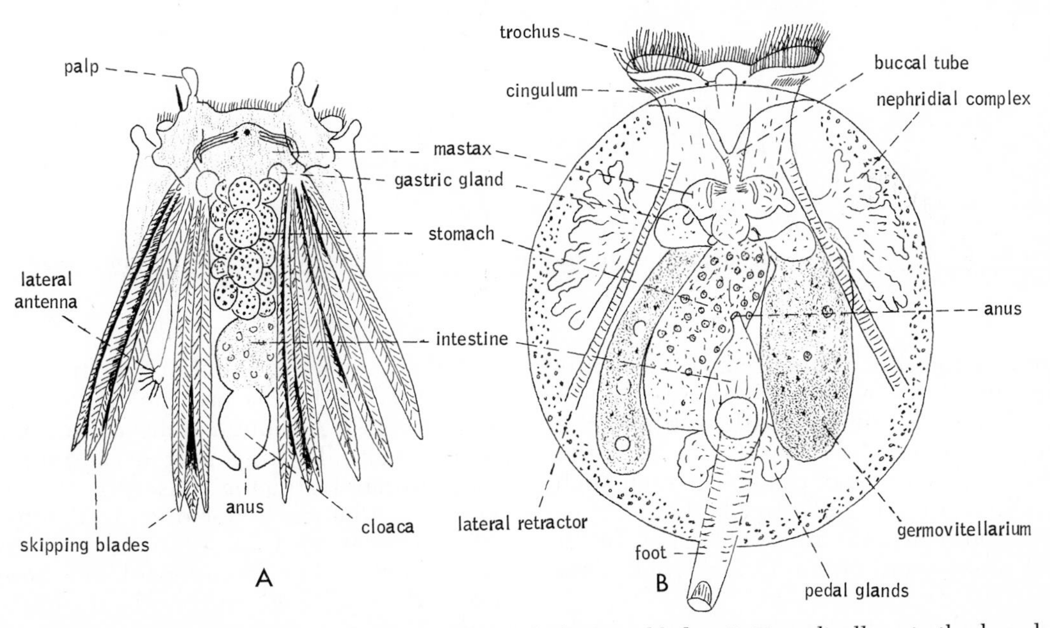

Figure 9–10. *A. Polyarthra trigla*, a pelagic rotifer with skipping blades. *B. Testudinella*, a turtle-shaped swimming rotifer. (Both after Hyman.)

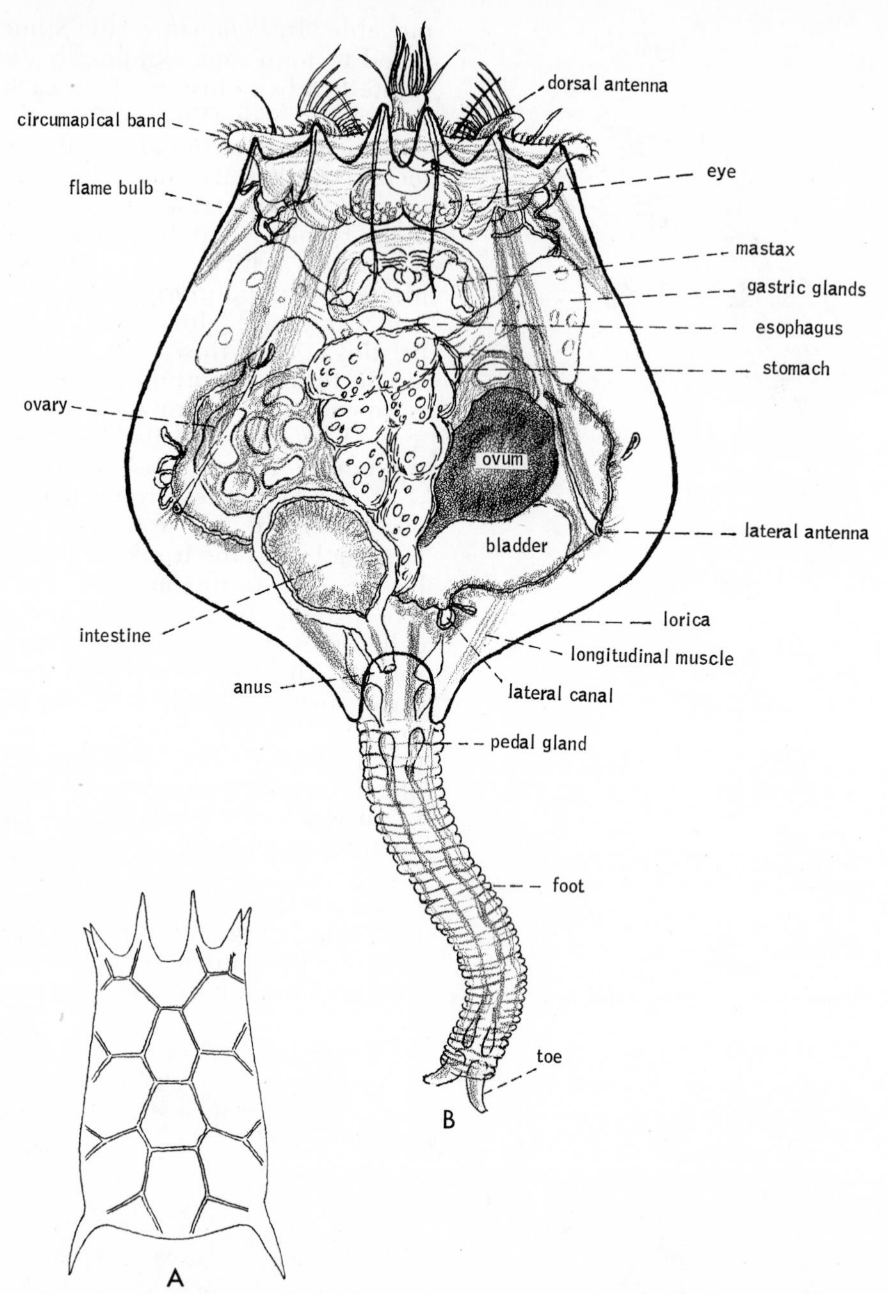

Figure 9–11. *A*. Lorica of *Keratella quadrata* (dorsal view). (Modified after Ahlstrom from Pennak.) *B*. A brachionid rotifer. (After Hudson.)

the end of the foot. Functionally the spurs replace the toes, which are very small in these genera. In some sessile rotifers, the secretion from the pedal glands is used in the construction of vaselike cases in which such rotifers live (Fig. 9–13*B*). In planktonic rotifers, the foot is usually reduced or has disappeared altogether. Often it is turned ventrally.

Body Wall and Pseudocoel. The cuticle is secreted by the underlying epidermis, which is thin and syncytial and always possesses a constant number of nuclei. Beneath the epidermis are the body muscles (Fig. 9–12*B*). Although some of the muscle fibers are circular (ring muscles) and some are longitudinal (retractor muscles), the body-wall musculature is not organized into distinct circular and longitudinal sheaths as in flatworms. The pseudocoel lies beneath the body wall and surrounds the gut and other internal organs. The pseudocoel is filled with fluid and a syncytial network of branching ameboid cells.

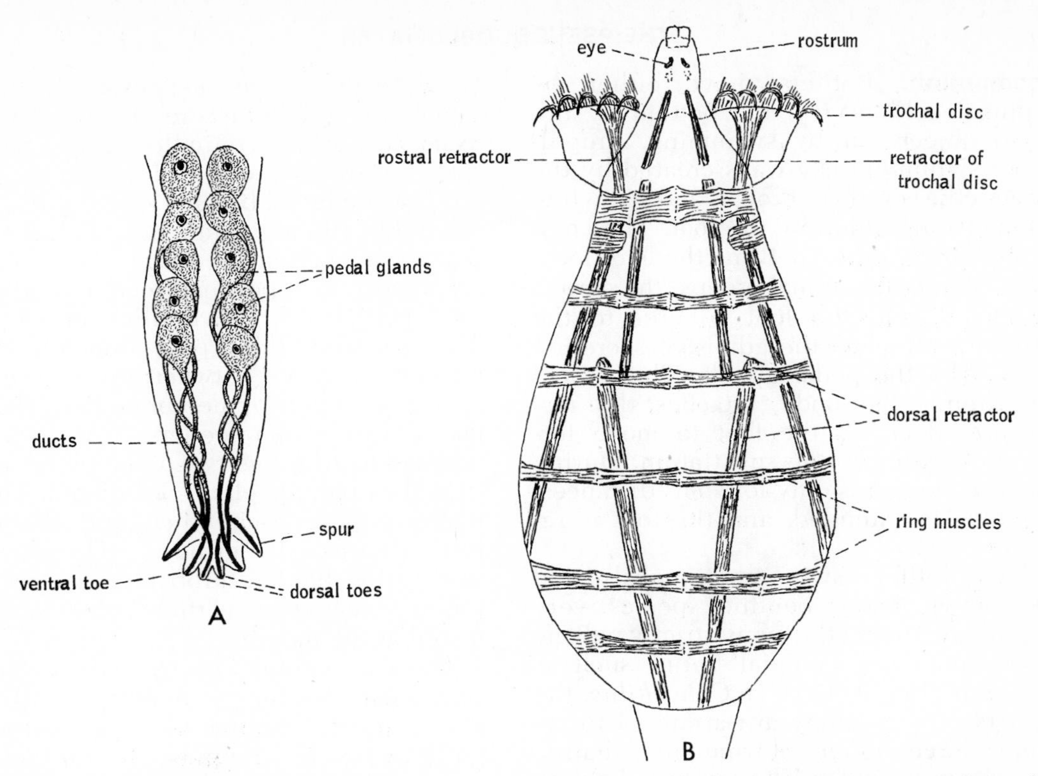

Figure 9–12. *A.* Foot of *Embata parasitica*, showing pedal glands (Dorsal view). *B. Rotaria* musculature (Dorsal view). (Both after Brakenhoff from Hyman.)

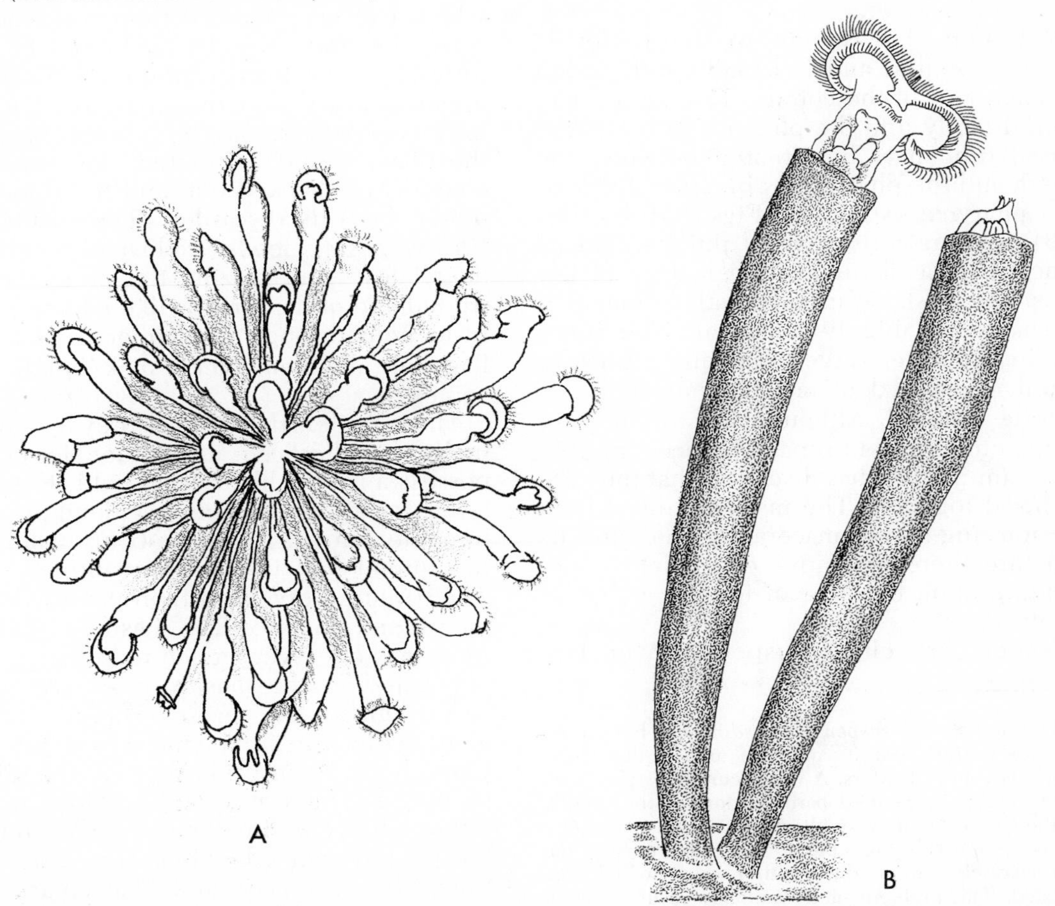

Figure 9–13. *A. Conochilus volvox*, a colonial pelagic rotifer. *B. Limnias ceratophylli*, a tubiculous rotifer in which the tube is secreted. (Both after Hudson.)

Locomotion. Rotifers move either by creeping leechlike over the bottom of submerged objects, or by swimming. Propulsion for swimming is always created by the beating cilia of the corona, and crawling movements are aided by the foot. The common bdelloids utilize both methods of locomotion. When the animal creeps, the corona is retracted, and the foot adheres to the substratum by using the adhesive secretion produced by the pedal glands. The animal then extends the body, attaches the rostrum, and detaches the foot to move forward and again grip the substratum. During swimming, which is only for short distances, the corona is extended, and the foot is retracted.

Pelagic rotifers swim continuously and never crawl; many benthic species very rarely swim. Among the many strictly pelagic species are a few colonial forms, such as *Conochilus* (Fig. 9–13A). In *Conochilus,* the members of the colony are arranged to resemble a large number of trumpets radiating from a common center. The combined ciliary action of the coronae propels the colony through the water.

Nutrition. The mouth of the rotifer is typically ventral and is usually surrounded by some part of the corona. The mouth may open directly into the pharynx, or a ciliated buccal tube may be situated between the mouth and the pharynx, as in ciliary feeders. The pharynx, or mastax (Figs. 9–6 and 9–11*B*), is characteristic of all rotifers, and its structure is a distinguishing feature of the class. The mastax is usually oval or elongated in shape and highly muscular. The inner walls carry seven large projecting pieces, or trophi, composed of an acid mucopolysaccharide material. Although the trophi were thought to represent separate pieces (Fig. 9–14), scanning EM has disclosed that they are all fused together. The mastax is used both in capturing and in macerating food, and its structure therefore varies considerably, depending upon the type of feeding behavior involved.

Rotifers are either suspension* or raptorial feeders. The suspension feeders, of which the bdelloids are the most notable examples, feed on minute organic particles that are brought to the mouth in the water current produced by the coronal cilia. In the bdelloids, the membranelles of each trochal disc create a circular current. These two currents move in opposing directions and sweep food particles to the midline of the head. The particles then pass into the mouth through a groove between the anterior and posterior circlets of membranelles. The mastax of suspension feeders (Fig. 9–14*A*) is adapted to grinding; two of the pieces are extremely large, platelike, and ridged. The two plates oppose each other, and the ridges form a surface for grinding. The mastax of suspension feeders probably also acts as a pump, sucking in particles that have collected at the mouth.

The carnivorous species, which feed on protozoans, rotifers, and other small metazoan animals, capture their prey either by grasping or by trapping. In the grasping species, the pieces of the mastax are constructed like a pair of forceps, the end of which can be projected from the mouth to seize the prey (Fig. 9–14*B* and *C*). These forms have no buccal tube, and the mouth opens directly into the pharynx. After the prey is seized, it may be brought back into the pharynx and macerated; commonly the contents are sucked out, and the remainder of the body is discarded. The carnivorous *Chromogaster* feeds exclusively on dinoflagellates. After seizing the prey, the protruded end of the forceps-like mastax drills a little hole through the armor of the dinoflagellate and then quickly sucks out the body contents.

Those rotifers that capture prey by trapping include *Collotheca* and other forms that possess a funnel-like buccal field (Fig. 9–9). When small protozoans accidentally swim into the funnel, the setae-bearing lobes of the funnel fold inward, preventing escape. The captured organism then passes into the mouth and pharynx. The mastax of trapping rotifers is often very much reduced.

Located in the mastax walls are two to seven glandular masses called salivary glands (Fig. 9–6*A*). The salivary glands open through ducts just in front of the mastax proper. A tubular esophagus, which may be either ciliated or lined with cuticle, connects the pharynx with the stomach. At the junction of the esophagus and stomach are a pair of gastric glands (Figs. 9–6 and 9–11*B*), each of which opens separately by a pore into each

---

*The designation *suspension feeding* has been used here for rotifers, but it could be argued that these animals are filter feeders. A water current is produced which directs suspended particles toward the mouth, but there is no straining, or filtering, mechanism present. Yet these animals are able to use cilia to sort out large particles from the smaller ones which are ingested. The problem of terminology results merely because the feeding mechanism of these animals does not neatly fit the definitions.

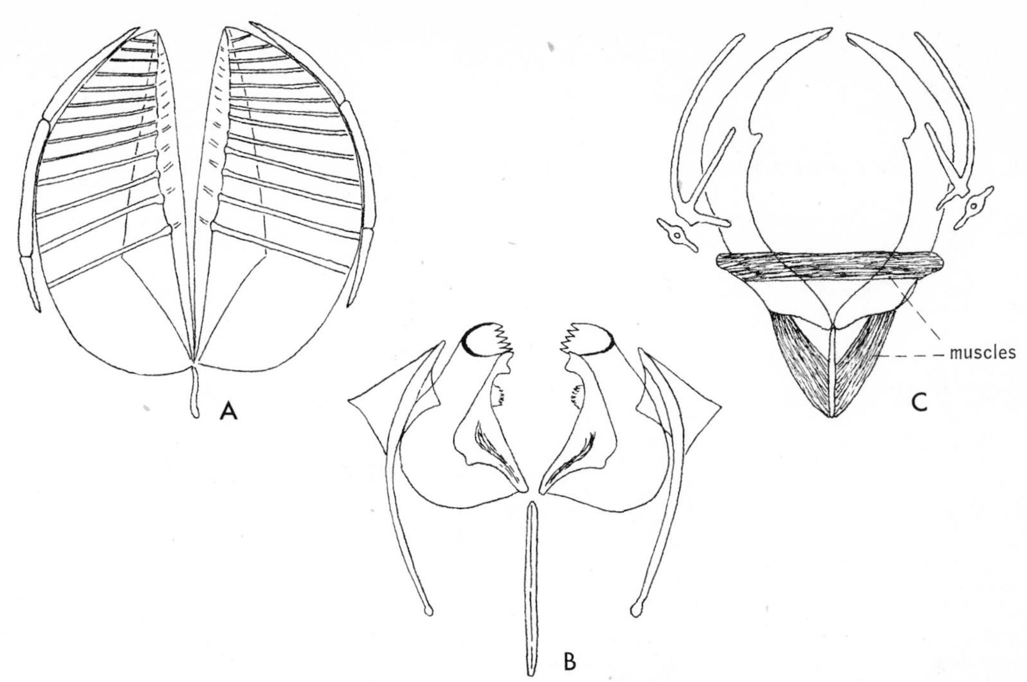

Figure 9–14. Three types of mastax trophi. *A.* Mastax with trophi adapted for grinding. Note the two large ridged plates that provide grinding surfaces. (After Beauchamp from Hyman.) *B.* Mastax with trophi adapted for grasping. *C.* Mastax of *Synchaeta* with grasping, forceps-like trophi. (*B* and *C* after Hyman.)

side of the digestive tract. The stomach is a large sac or tube composed of a constant number of distinct granular cells, except in the bdelloids and some others. The stomach passes into a short intestine. The excretory organs and the oviduct open into the terminal end of the intestine, which is sometimes called the cloaca. The anus opens onto the dorsal surface near the posterior of the trunk.

The gastric glands, and perhaps the salivary glands, are believed to produce enzymes; digestion is probably extracellular except in some pelagic forms that have lost the intestine and anus. Absorption takes place in the stomach.

Water Balance. Typically, two protonephridia are present, one on either side of the body. The two protonephridial tubules empty into a bladder (Fig. 9–11*B*), which opens into the ventral side of the cloaca; in the bdelloids, there is a constriction between the stomach and the intestine, and the somewhat bulbous cloaca acts as a bladder (Fig. 9–6*A*). The contents of the bladder or cloaca are emptied through the anus by constriction; often the pulsation rate is between one and four times per minute. Such a high rate of discharge is evidence of the osmoregulatory function of the protonephridia; water enters through the mouth by swallowing rather than by penetration of the cuticle.

The Nervous System. The brain consists of a dorsal ganglionic mass lying over the mastax (Figs. 9–8*B* and 9–15), and gives rise to a varying number of nerves that extend to the anterior sense organs and to other parts of the body. The sense organs consist of sensory bristles located in various parts of the ciliated crown, two ciliated pits (Fig. 9–8*A*), and one to five eyes (Fig. 9–7*A*). The eyes are simple ocelli composed of a number of pigmented photoreceptor cells.

Reproduction. Reproduction is entirely sexual, and like most other aschelminths, rotifers are dioecious. The males (Fig. 9–16) are always smaller than the females, and certain structures, such as the cloaca, are degenerate or absent in the male. The difference in size between sexes varies from species in which the males are only slightly smaller to forms in which the male is only one-tenth the size of the female. Parthenogenesis is common, and in the bdelloids no males have ever been reported.

The female reproductive system in the majority of species consists of a single ovary and a yolk-producing vitellarium, both of which form a single body (germovitellarium) located anteriorly in the pseudocoel (Figs. 9–6 and 9–11*B*). The vitellarium supplies yolk to the developing eggs by direct flow

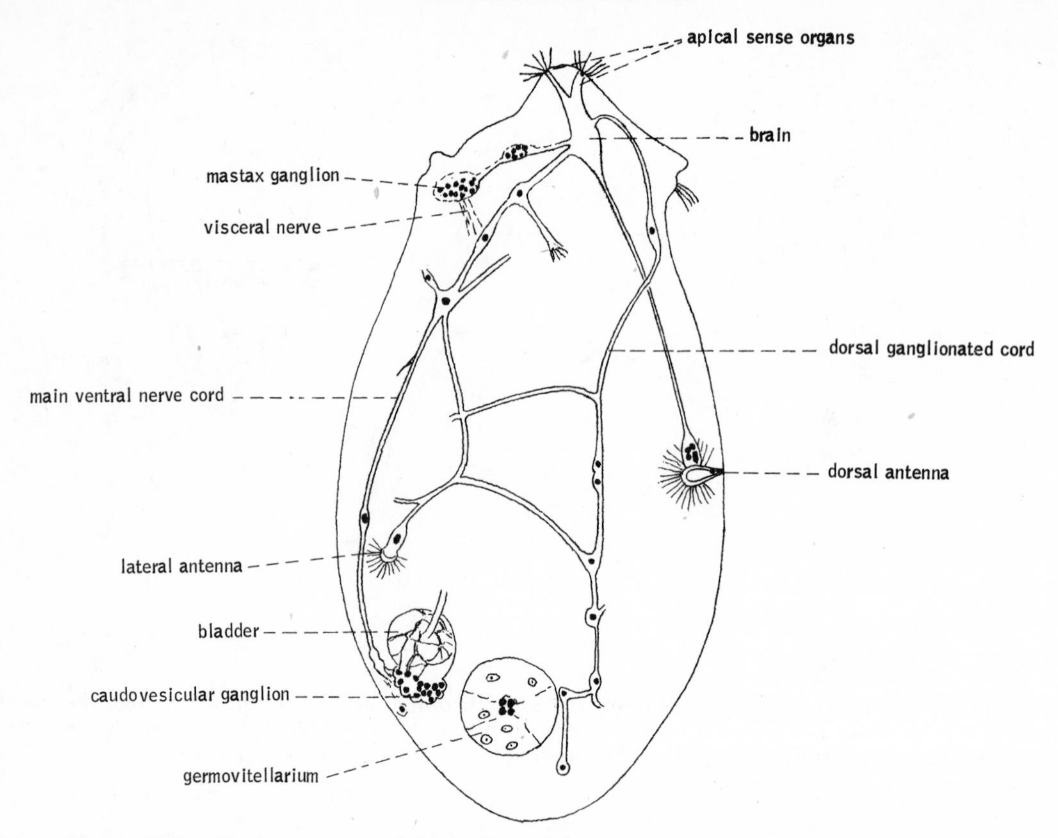

Figure 9–15. Nervous system of *Asplanchna* (lateral view). (After Nachtwey from Hyman.)

through cytoplasmic bridges. The eggs then pass through an oviduct into the cloaca.

In the male a single saclike testis and a ciliated sperm duct (Fig. 9–16) are present. Because a cloaca is usually absent in males, the sperm duct runs directly to a gonopore that is homologous to the anus in females and has the same position. Two or more glandular masses called prostate glands are associated with the sperm duct, and the end of the sperm duct is usually modified to form a copulatory organ. For example, the entire gonopore wall may protrude to act as a penis; or the end of the sperm duct may be modified into a tube with a heavy cuticular lining so that the whole tube can be everted.

Copulation is by hypodermic impregnation. In the planktonic *Asplanchna,* penetration is preceded by adhesion of the penis to the female cuticle, and the male may be pulled about by the female. In *Brachionis,* penetration occurs only in the softer coronal region. In some species, copulation occurs within a few hours of hatching, when the female's cuticle is still soft.

Each ovarian nucleus forms one egg, and since only some ten to twenty such nuclei exist in most species, there is a corresponding limit to the number of eggs produced in the lifetime of a particular female. Each egg is surrounded by a shell and a number of egg membranes, all of which are secreted by the egg itself. The eggs are attached to objects on the substratum or may be attached to the body of the female. A few rotifers,

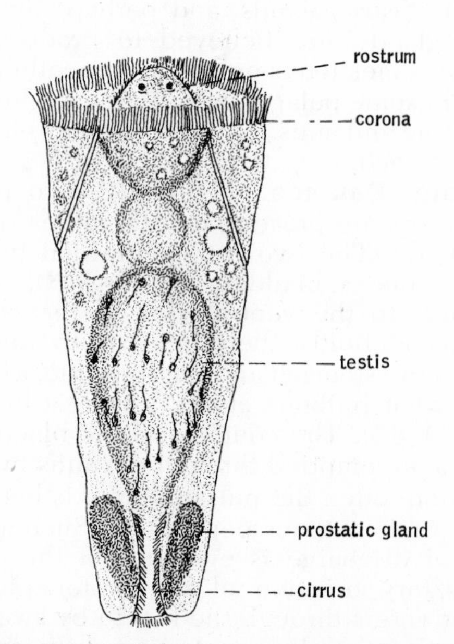

Figure 9–16. Male of *Chonochilus* (female shown in Fig. 9–13A). (After Wesenberg-Lund from Hyman.)

such as *Asplanchna* and *Rotaria*, brood their eggs internally.

Bdelloids, in which there are no known males and development is parthenogenetic, produce eggs that hatch into females only. In other rotifers (monogonont), however, several different types of eggs are produced. One type, called an amictic egg, is thin-shelled, cannot be fertilized, and develops by parthenogenesis into amictic females. Typical meiosis does not take place in maturation, and the eggs are diploid (Fig. 9–17).

A second type of egg, called a mictic egg, is also thin-shelled but is haploid. If these eggs are not fertilized, they produce males parthenogenetically. If they are fertilized, they accumulate a larger amount of yolk material and secrete a heavy resistant shell. Such eggs are now called dormant eggs; in contrast to the thin-shelled unfertilized amictic and mictic eggs, which hatch in several days, these dormant eggs are capable of withstanding desiccation and other adverse environmental conditions and may not hatch for several months. Dormant eggs hatch into females.

Parthenogenesis and the production of thin-shelled eggs that hatch in a short time and of dormant eggs are particularly characteristic of rotifers that live in temporary ponds and streams, which are subject to environmental stress. Dormant eggs are able to withstand adverse environmental conditions, and parthenogenesis facilitates a rapid increase in population levels. The reproductive pattern tends to be cyclic. After spring rains with the advent of warmer temperatures, dormant eggs that have passed through the winter hatch into females. These females produce a number of generations of parthenogenetic females, each having a life span of one to two weeks. In the late spring or early summer when this population reaches a peak, mictic eggs are produced and males appear (Fig. 9–17). Dormant eggs carry the species through until the next season. Rotifers inhabiting large permanent bodies of fresh water may display a number of cycles or population peaks during the warmer months or may be present during the whole year. The production of mictic eggs is induced by specific environmental factors, such as high population density, dietary substances, and increased photoperiod, but the importance of different factors varies from species to species.

Embryogeny. Cleavage in rotifers is spiral and determinate but modified from the pattern described in the previous chapter. Nuclear division is completed early in development and never occurs again. In free-moving species, no larval development takes place. When the females hatch, they have all the adult features and attain sexual maturity after a growth period of a few days. The smaller males do not undergo a growth period but are sexually mature when they leave the egg. The sessile rotifers hatch as free-swimming "larvae" that are structurally very similar to free-swimming species. After a short period they settle down, attach, and assume the characteristics of the sessile adults.

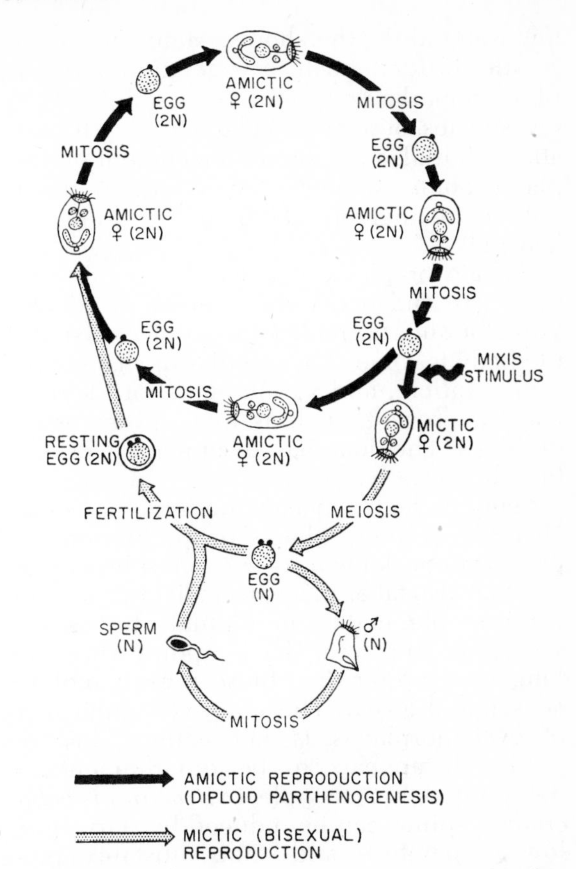

Figure 9–17. Life cycle of monogonont rotifers. (From Birky, C. W., 1964: J. Exp. Zool., *155*:273–292.)

Ecology and Distribution. Rotifers are largely cosmopolitan in distribution, most species living throughout the world, although some are restricted to particular regions. There are only approximately 50 strictly marine forms, most of which are littoral. However, to these must be added a considerable number of littoral species that also inhabit fresh water. The majority of freshwater rotifers inhabit the substratum or live on submerged vegetation and other objects. In large bodies of fresh water most of these bottom-dwelling species are limited to shallow water along the shores. Many species of sessile rotifers attach to vegetation and display a remarkable restrictiveness, not only to the species of algae or plant to which they attach, but also the site of attachment on the plant. Some species, for example, attach only in the angle formed by the stem and the leaf stalk.

Pelagic or planktonic species swim continuously and display a number of adaptations for such a mode of existence. Usually the body is somewhat saccular in shape with a thin cuticle, and various flotation devices may be present, such as long spines or oil droplets. The foot has disappeared or has turned ventrally.

Many pelagic species undergo seasonal changes in body shape or proportions, a phenomenon known as cyclomorphosis and one that also takes place in small crustaceans. For example, certain individuals during one season of the year have spines that are longer or shorter than those during another season of the year. The adaptive significance of cyclomorphosis is not certain, but in rotifers it appears to be environmentally determined. For example, in *Brachionis calyciflorus* spines can be induced by starvation, low temperature, and some substance produced by a predator rotifer, and the factors have an additive effect.

Most terrestrial rotifers are associated with mosses and lichens and are active only during the short periods when these plants are filled with water. During this time, these terrestrial rotifers swim about in the water films on the leaves and stems of the plants. These species are capable of undergoing desiccation, usually without the formation of a cyst, and can remain in a dormant state for as long as three to four years. Their resistance to both low temperatures and lack of moisture in such a dormant state is remarkable. Some species have been placed in liquid helium (−272°C) and in desiccators and vacuums without damaging effects.

There are quite a number of epizoic and parasitic rotifers which live primarily on small crustaceans, particularly on the gills. Endoparasitic species inhabit snail eggs, heliozoans, the interior of *Volvox*, the intestine and coelom of earthworms, freshwater oligochaetes, and slugs. One genus, *Proales*, is parasitic within the filaments of the freshwater alga *Vaucheria* and produces gall-like swellings. In parasitic rotifers either the foot or the mastax becomes modified as an attachment organ, and the corona is reduced.

SYSTEMATIC RÉSUMÉ OF PHYLUM ROTIFERA

Class Seisonacea. A small class of marine rotifers epizoic on certain crustaceans (*Nebalia*). Elongate body with reduced corona. Gonads paired. *Seison*.

Class Bdelloidea. Anterior end retractile and usually bearing two trochal discs. Mastax adapted for grinding. Swimming and creeping rotifers. Males absent; female with two germovitellaria. *Philodina, Embata, Rotaria, Adineta*.

Class Monogonta. Swimming and sessile rotifers. Mastax most often of the grasping type. With only one germovitellarium. *Notommata, Chromogaster, Synchaeta, Polyarthra, Brachionis, Keratella, Euchlanis, Proales, Asplanchna, Testudinella, Pedalia, Trochosphaera, Limnias, Floscularia, Chonochilus, Collotheca*.

PHYLUM KINORHYNCHA

The phylum Kinorhyncha, also known as Echinodera, consists of a small group of marine pseudocoelomates living in the muddy bottoms of coastal waters. The members of this phylum are somewhat larger than rotifers and gastrotrichs but are always less than 1 mm. in length. The general body shape is similar to that of the gastrotrichs, there being a head and trunk separated by a neck region (Fig. 9–18A). However, kinorhynchs differ from both rotifers and gastrotrichs in their lack of external cilia.

A distinguishing feature of the class is the division of the cuticle into clearly defined segments (zonites). The head composes the first segment, the neck makes up the second segment, and the trunk in all adults constitutes eleven segments. The cuticle of each segment is subdivided into one dorsal and two to three ventral plates, and the cuticle between the plates is very thin and flexible, permitting articulation (Figs. 9–18A and 9–19). The dorsal plate of each segment bears large median and lateral, recurved spines, from which the name Echinodera—spiny skin—is derived. The spines are movable and hollow, being filled with epidermal tissue. Usually a single pair of adhesive tubes is located on the ventral surface of the anterior of the trunk, but there are species which possess numerous lateral adhesive tubes.

The anteriorly located mouth is situated at the end of a protrusible cone, which is

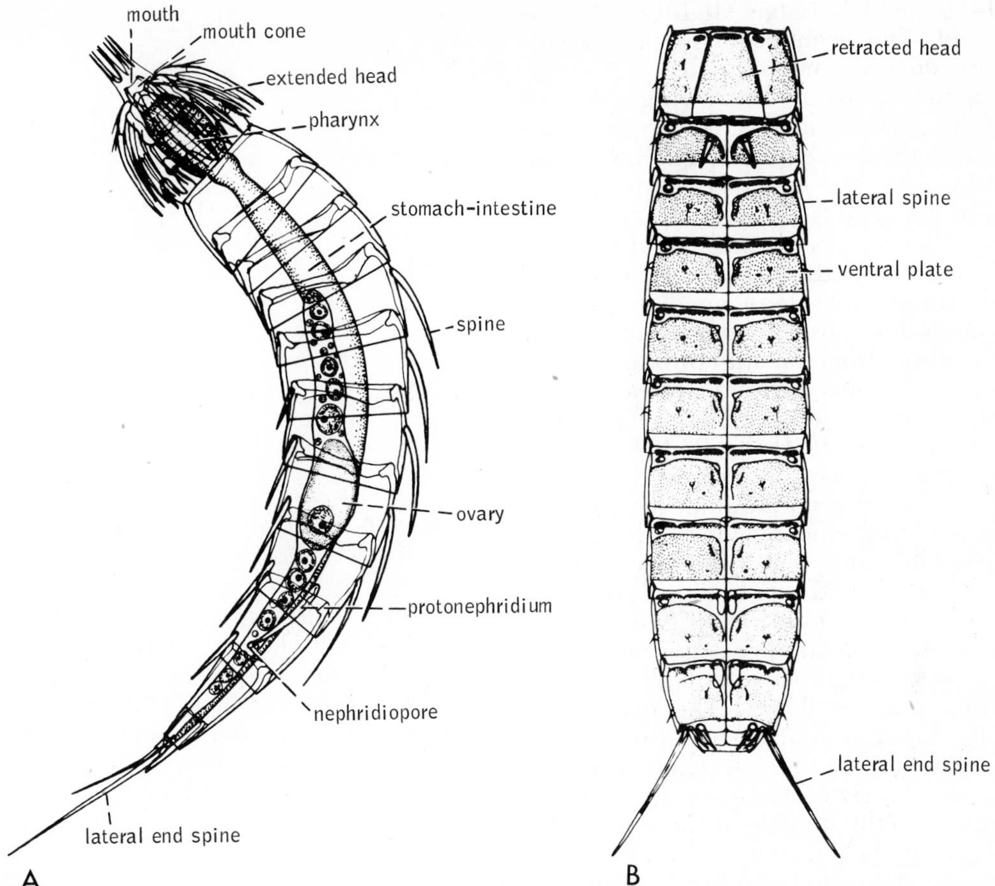

Figure 9–18. *A. Echinoderes* (lateral view). *B. Pycnophyes frequens* (ventral view). (Both figures courtesy of R. P. Higgins.)

surrounded at the tip and the base by circlets of spines (scalids) (Fig. 9–18*A*). The entire head can be withdrawn either into the neck or into the first trunk segments; hence the name Kinorhyncha—movable snout. In the former case, the cuticular plates of the neck placids are adapted for closing over the retracted head. A similar closing apparatus is present on the first trunk segments in those species in which both head and neck are retracted into the trunk.

The body wall of the kinorhynch is essentially like those of gastrotrichs and rotifers, although the epidermis is thickened along the middorsal line and along each lateral angle to form longitudinal epidermal cords

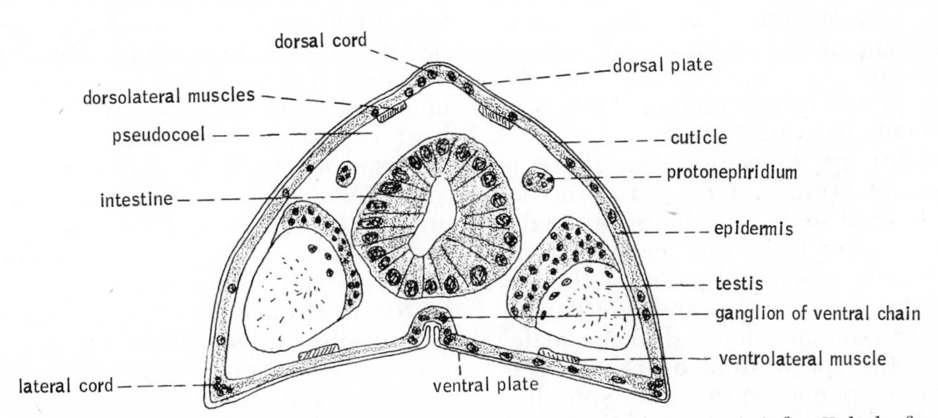

Figure 9–19. The kinorhynch, *Pycnophyes* (transverse section at level of intestine). (After Zelinka from Hyman.)

(Fig. 9–19), which bulge slightly into the pseudocoel. The pseudocoel is well developed and contains ameboid cells.

Kinorhynchs move by burrowing, because most of them live in mud and silt. The head extends into the mud and is anchored by the recurved spines. The body is then drawn forward until the head is retracted into the neck or trunk. The head is again pushed forward into the mud, and the process is repeated.

A few kinorhynchs feed on diatoms; the others are all deposit feeders and obtain organic material from the mud through which they burrow. A buccal cavity in the mouth cone leads into a barrel-shaped pharynx lined with cuticle (Fig. 9–18*A*). Radial muscle fibers are present, as in gastrotrichs, but are located outside the epithelium. The entire pharynx is surrounded by longitudinal protractor muscle bands. The esophagus is also lined with cuticle and anteriorly receives the secretions of two dorsal and two ventral salivary glands. Posteriorly, two or more similar glands, called pancreatic glands, also empty into the esophagus. The esophagus leads into a tapered stomach-intestine, which lacks a cuticular lining. A short hindgut connects the stomach-intestine with the anus, which opens to the outside at the end of the trunk. The details of digestion are unknown.

There are two protonephridia, one on either side of the intestine, and each contains a modified flame bulb. The two nephridiopores are located on the dorsolateral surface of the eleventh segment.

The nervous system is closely associated with the epidermis. The brain consists of a nerve ring around the anterior of the pharynx, and from this ring arises a single, midventral nerve cord containing one cluster of ganglion cells in each segment. Other clusters of ganglion cells are located dorsally and laterally in each segment but are not interconnected to form a distinct cord. The sense organs consist of pigment-cup ocelli at the anterior of the body of a few species, and sensory bristles located over the general body surface, especially the trunk.

Kinorhynchs are dioecious, but there is little sexual dimorphism. One pair of saclike ovaries is located in the middle of the body, and a short oviduct leads from each ovary to a distinct female gonopore located on the last segment. One pair of testes and one pair of gonopores are similarly located in the male. The end of the sperm duct usually carries two to three penial spines or spicules. Copulation has not been observed, but eggs numbering three to six have been found in the shed cuticle of a Baltic species. The young are not clearly segmented and do not possess spines or a head (Fig. 9–20). The digestive tract is also incompletely developed. Periodic molts of the cuticle occur in the attainment of adulthood, after which molting does not occur.

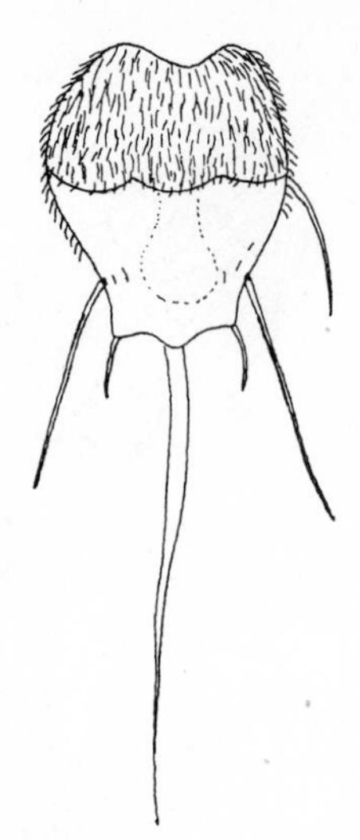

Figure 9–20. Early juvenile of the kinorhynch, *Echinoderes*. (After Nyholm from Hyman.)

PHYLUM NEMATODA

The Nematoda, called roundworms, is the largest aschelminth phylum (10,000 described species) and contains some of the most widespread and numerous of all multicellular animals. Free-living nematodes are found in the sea, in fresh water, and in the soil. They occur from the polar regions to the tropics in all types of environments, including deserts, hot springs, high mountain elevations, and great ocean depths. They are often present in enormous numbers. One square meter of bottom mud off the Dutch coast has been reported to contain as many as 4,420,000 nematodes. An acre of good farm soil has been estimated to contain from several hundred million to billions of terrestrial nematodes. A single decomposing apple on the ground of an orchard has yielded 90,000 roundworms belonging to a number of different species, and a single fig has been reported to contain as many as 50,000 nematodes belonging to at least eight species.

In addition to free-living species, there are many parasitic nematodes. The parasitic forms display all degrees of parasitism and attack virtually all groups of plants and animals. The numerous species that infest food crops, domesticated animals, and man himself make this phylum one of the most important of the parasitic animal groups.

In order to accommodate the many courses in invertebrate zoology which do not include the parasitic groups, the initial discussion of this phylum will emphasize the free-living members and the parasitic forms will be dealt with at the end of the section.

External Structure. Nematodes have a slender, elongated body with both ends gradually tapered in most species (Fig. 9–21). The majority of free-living nematodes, especially freshwater and terrestrial forms, are less than 1 mm. in length and are often microscopic. However, some marine species attain a length of 5 cm.

Two distinctive characteristics of nematodes are the relatively perfect cylindrical shape of the body and the rather striking radial or biradial arrangement of structures around the mouth. Some nematode specialists believe this symmetry is evidence that the ancestral members of the phylum at first assumed a sessile existence, in which the posterior of the animal was attached to the substratum with the body extended vertically in the water. As the result of such a sessile existence, a radial or biradial symmetry became superimposed upon the original bilateral plan of structure. The nematode body cannot be divided into distinct regions as can be done in other aschelminths. The nematode mouth is located at the tapered anterior of the animal and is surrounded by lips and sensory papillae or bristles.

In many marine nematodes (Fig. 9–22), which are considered the most primitive members of the phylum, six liplike lobes border the mouth, three on each side, but as a result of fusion, there are often only three lips in terrestrial and parasitic species. The lips and the anterior surface outside the lips bear papillae and may also carry a variety of cuticular projections, ranging from simple rounded eminences to complex branched or even feather-like projections (Figs. 9–22 and 9–23). In some species the cuticle at the base of the lips has developed into large overlapping plates called head shields; the lips of many carnivorous nematodes carry crowns of teeth. As in other aschelminths, the cuticle of the general

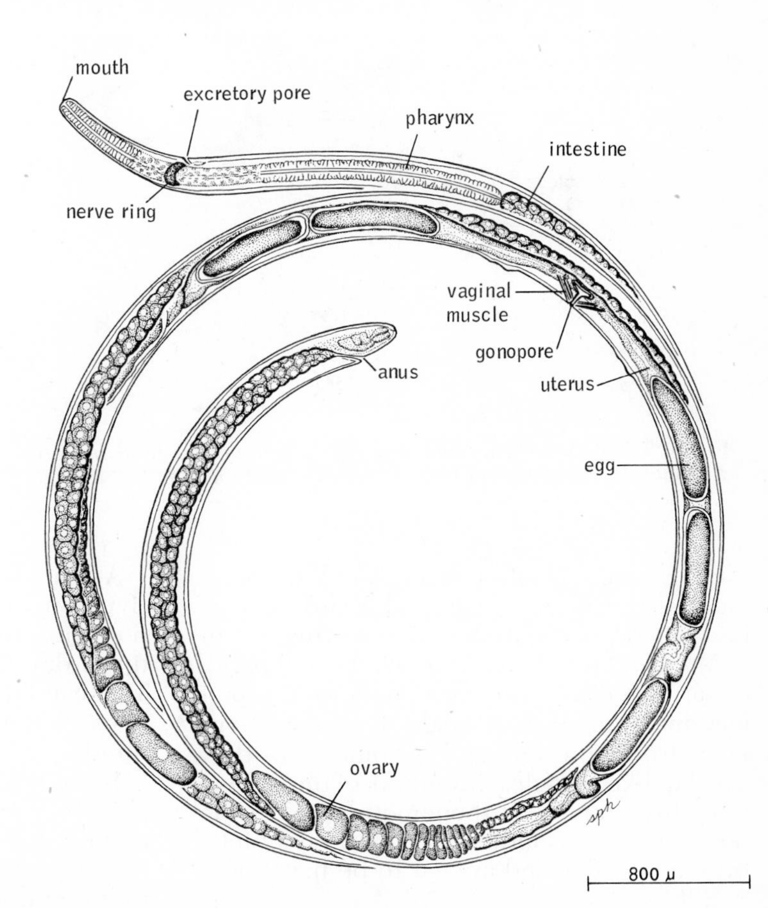

Figure 9–21. Female of the marine nematode *Pseudocella*. (After Hope.)

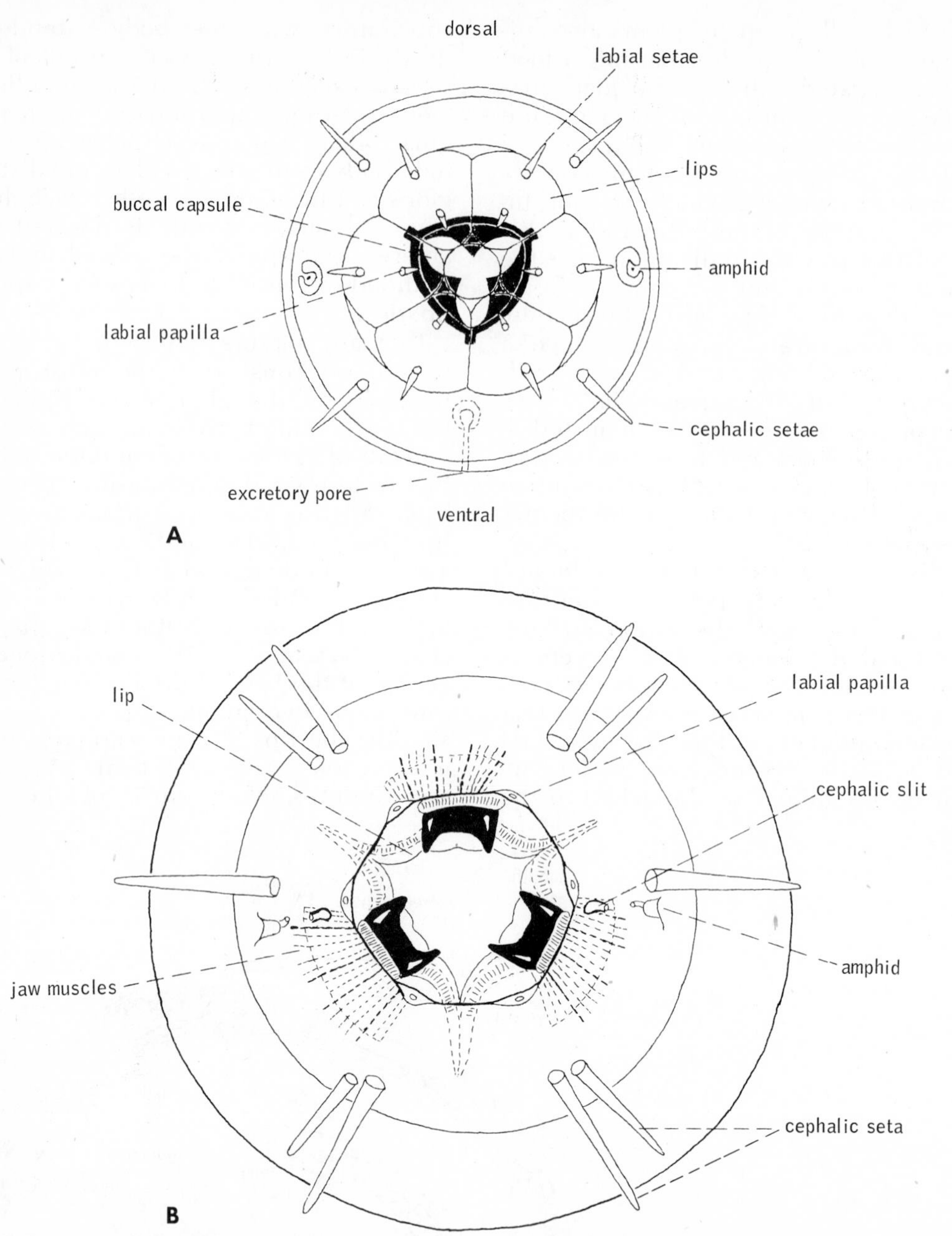

Figure 9–22. *A.* Anterior end of a generalized nematode. *B.* Anterior end of the marine nematode *Enoplus.* (*A* and *B* after de Coninck from Hyman.) (*Figure continued on facing page.*)

body surface is often sculptured or ornamented in different ways. The surface may be pitted, annulated, ridged, ringed, or striated and may bear setae, warts, and papillae arranged in a variety of pattern-like projections (Fig. 9–23). Scale-like projections may also be present, and in some cases they overlap like shingles. Members of the marine interstitial family Stilbonemotidae have the body surface clothed with a symbiotic blue-green alga and appear to be hairy (Fig. 9–23*F*).

A caudal gland is typical of most free-living nematodes. The gland opens at the posterior tip of the body, which is sometimes drawn out to resemble a tubelike tail. The caudal gland in nematodes is homologous to the adhesive tubes of other aschelminths.

Body Wall. The nematode cuticle is considerably more complex than that of other aschelminths. It consists of secreted collagen, which appears to be unique to nematodes, and it is organized within three main

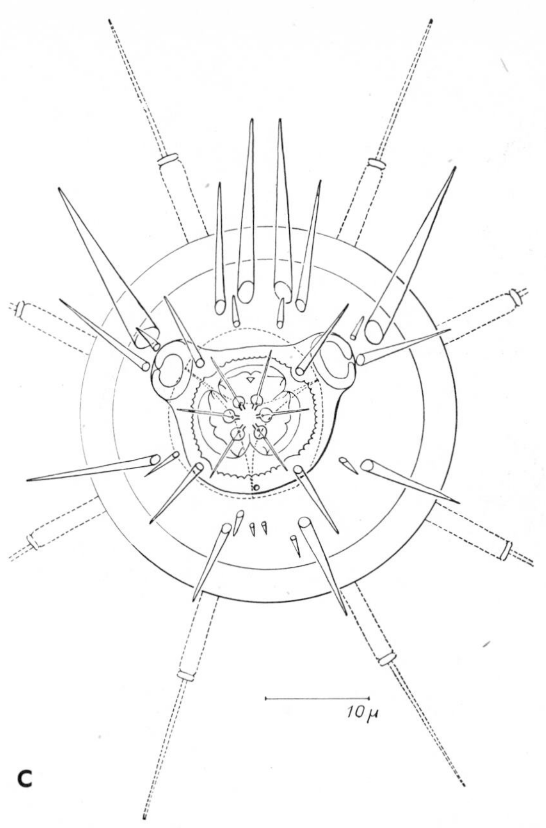

Figure 9–22. *Continued. C.* Anterior end of *Tristicochaeta*, showing the asymmetrically arranged setae. The subcephalic ones are jointed. The two tubular ovals are amphids. (From de Coninck, L., 1965: *In* Grassé, P. (Ed.), Traité de Zoologie, vol. IV, 2nd Fascicule. Masson et Cie, Paris.)

layers (Fig. 9–24). The outer keratinized cortex may exhibit quinone tanning. The median layer varies from no apparent structure in some species to the occurrence of struts, skeletal rods, fibrils, or canals in others. The basal layer commonly contains columellae, rods, or canals.

The cuticle of nematodes differs from the external covering of arthropods in a number of respects. Although preadult molts take place in which the old cuticle is shed whole or in fragments, the cuticle continues to grow during adulthood. Molting does not occur following attainment of adulthood. The physical structure provides the cuticle with flexibility and elasticity. Finally, the presence of enzymes and other chemical substances reflects a metabolic activity that may be important in the physiology of the worm. There is little doubt that additional studies will show that some of these features are shared with other aschelminths.

The epidermis is usually cellular but may be syncytial in some species. A striking feature of the nematode epidermis is the expansion of the cytoplasm into the pseudocoel along the middorsal, midventral, and midlateral lines of the body (Fig. 9–25). The bulging epidermis thus forms longitudinal cords that extend the length of the body. All epidermal nuclei are restricted to these cords and are typically arranged in rows. Secondary or subsidiary cords are present in some species and are located between the four longitudinal cords.

The muscle layer of the body wall is composed entirely of longitudinal, obliquely striated fibers arranged in bands, each strip occupying the space between two longitudinal cords. The fibers may be relatively broad and flat, with the contractile filaments limited to the base of the fiber, or they may be relatively tall and narrow, with filaments at the base and sides. In both types the base of the cell containing the contractile fiber is located against the hypodermis, and the side of the cell with the nucleus is directed toward the pseudocoel (Fig. 9–26). Nematode muscle fibers are unique in that each possesses a slender arm that extends from

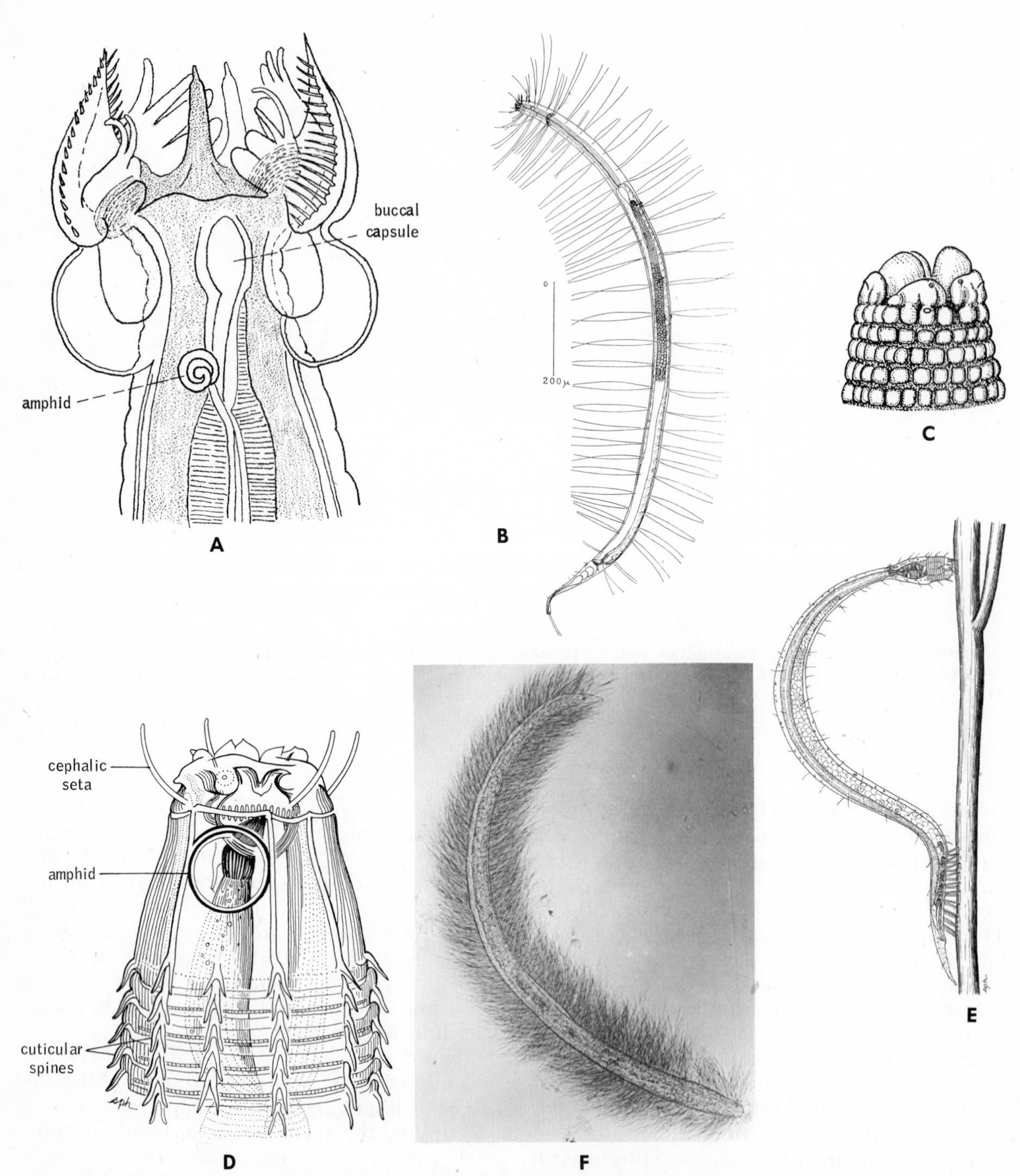

Figure 9–23. Nematode diversity. *A.* Anterior end of *Wilsonema,* showing ornate featherlike projections carried by the lips. (After Steiner from Hyman.) *B. Trichotheristus,* possessing long body setae. *C.* Anterior end of *Placodira,* in which cuticle is ornamented with plaques. (*B* after Schuurmans, Stekhoven, and de Coninck; *C* after Thorne; both from de Coninck, L., 1965: *In* Grassé, P. (Ed.), Traité de Zoologie, vol. IV, 2nd Fascicule. Masson et Cie, Paris.) *D.* Anterior end of *Monopisthia,* in which the cuticle is annulated and bears spines. (After de Coninck.) *E.* Marine draconematid, with preanal ambulatory setae. (Adapted from several sources.) *F.* A stilbonemotid nematode *(Eubostrichus dianae),* in which the body is clothed with algae. (Courtesy of Bruce E. Hopper.)

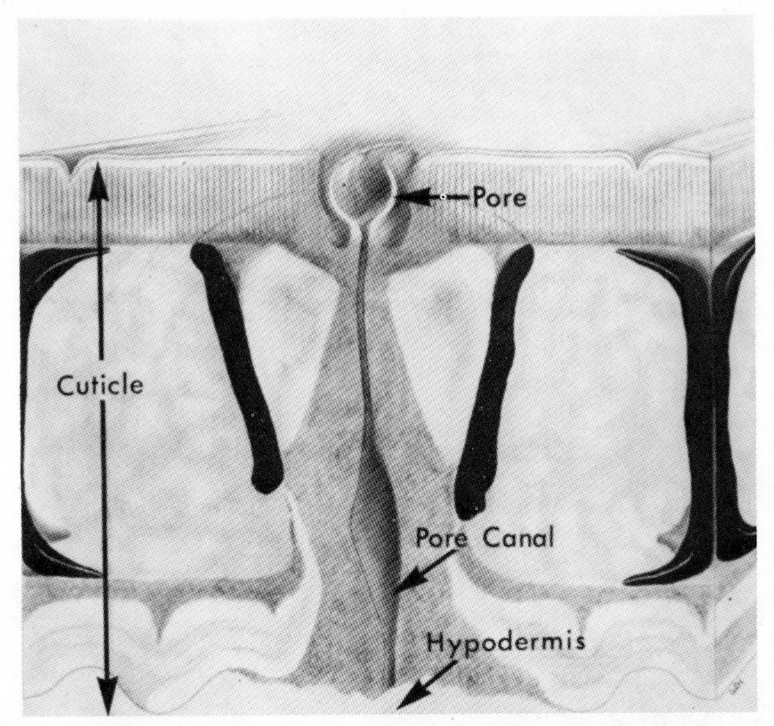

Figure 9–24. Section through the cuticle of the nematode *Acanthonchus*. (From Wright, K., and Hope, D. Reproduced by permission of the National Research Council of Canada from the Canadian Journal of Zoology, *46*, pp. 1005–1011, 1968.)

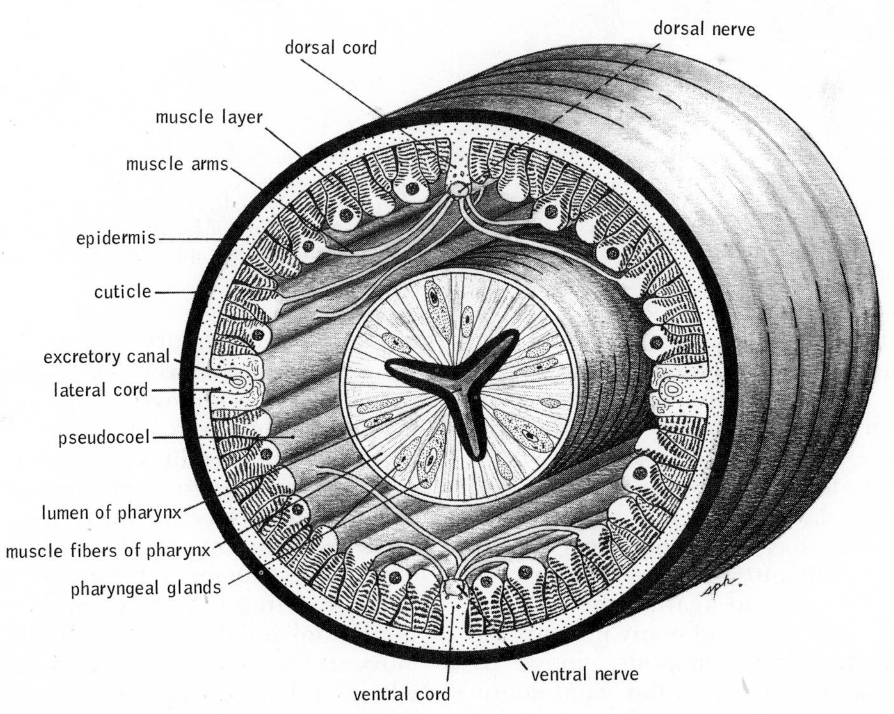

Figure 9–25. Diagrammatic cross section through the body of a nematode at the level of the pharynx. Note that the muscles send extensions, or arms, to the nerve cord, instead of the more usual reverse arrangement. Only a few of the many muscle arms are shown. (Adapted from several sources.)

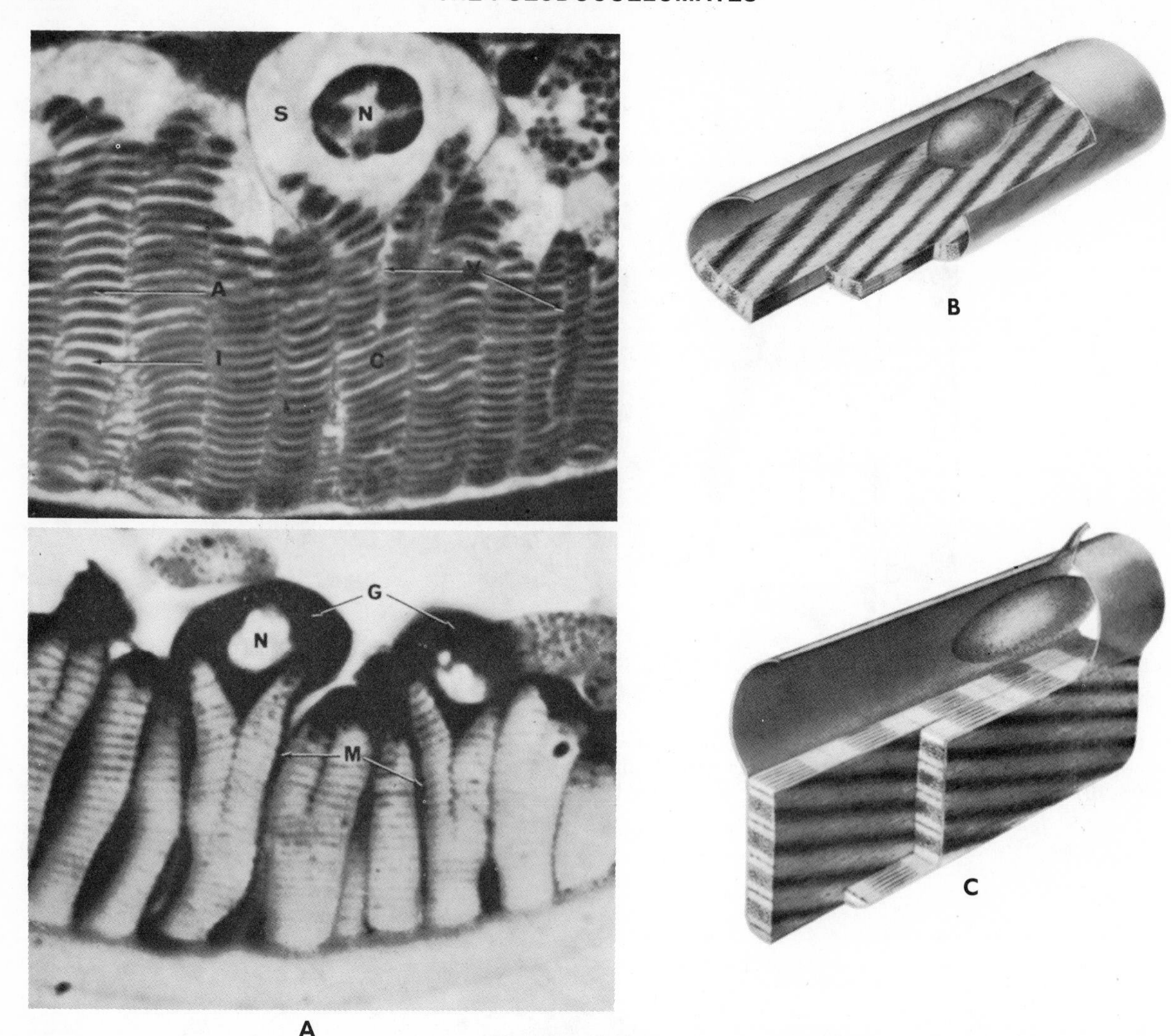

Figure 9–26. Nematode musculature. *A*. Photomicrograph of a cross section through the body wall musculature. Contractile part of cell (C) is directed toward hypodermis, which appears black at bottom of photograph. Nucleus (N) and non-contractile part of cell (S) are located on pseudocoel side of body wall. *B* and *C*. Diagrams of a section of two nematode muscle cells. The oval body is the nucleus, and the non-contractile part of the cell appears hollow. Note the obliquely striated fibrils. *C* is the same type of cell shown in the photograph in *A*. (All from Hope, D., 1969: Proc. Helminthological Soc. Washington, *36*(1):10–29.)

the fiber to either the dorsal or ventral longitudinal nerve cord where innervation occurs, rather than having nerve processes extending from the nerve cord to the main body of the muscle (Fig. 9–25).

The nematode pseudocoel is spacious and filled with fluid. No free cells are present, but fixed cells, located either against the inner side of the muscle layers or against the wall of the gut and against the internal organs, are characteristic of many nematodes.

Locomotion. Most nematodes move by undulatory waves of muscular contraction passing along the longitudinal muscle fibers of the body wall. The hydrostatic skeleton of the pseudocoel fluid and the elasticity of the cuticle are important contributors to the nematode undulatory ability. The undulations produce snake-like gliding. When nematodes are removed from sediment, as is usual in laboratory observations, their movement becomes undirected and of a whipping and thrashing nature. An aqueous environment is necessary for movement, and water films are thus important in the locomotion of terrestrial species. It should be kept in mind that free-living nematodes are largely interstitial inhabitants, and their undulatory movements are effective for progression only when applied against substratum particles or the surface tension of water films. Nematodes move through soil pores of 15 to 45 microns in diameter, and the pore size for optimum movement is about 1.5 times the worm's diameter. A worm also achieves maximum mobility when its length is approximately three times the diameter of the soil particles through which it is moving (Wallace, 1970).

A few species can crawl. The cuticle is sculptured, which aids in gripping the surface. In one species, which possesses a ringed cuticle, crawling is similar to that of earthworms. In some other species crawling involves a caterpillar-like or looping movement, in which the adhesive gland and special elongated ventral bristles provide a means of attachment (Fig. 9–23*E*).

Nutrition. Many free-living nematodes are carnivorous and feed on small metazoan animals, including other nematodes. Other species are phytophagous. Many marine and freshwater species feed on diatoms, algae, and fungi. Algae and fungi are also important food sources for many terrestrial species, but there is one group of fungi which trap nematodes. The worms are caught when they pass through special hyphal loops, which close on stimulation. A large number of terrestrial nematodes pierce the cells of plant roots and suck out the contents. Such nematodes, which could be called parasites, can be responsible for serious damage to commercial plants. There are also many deposit feeding marine, freshwater, and terrestrial species. They consume mud, from which they digest bacteria and organic materials. Other nematodes live on dead organic matter such as dung or on the decomposing bodies of plants and animals. However, some species that live in dead organic material apparently feed only on associated bacteria and fungi. This is probably true of the common vinegar eel, *Turbatrix aceti,* which lives in the sediment of nonpasteurized vinegar.

The mouth of the nematode opens into a buccal capsule (Fig. 9–21, 9–22, and 9–27A), which is somewhat tubular and lined with cuticle. The cuticular surface is often strengthened with ridges, rods, or plates; or it may bear a large number of teeth. The teeth are especially typical of carnivorous nematodes; they may be small and numerous or limited to a few large jawlike processes (Fig. 9–22*B*). The feeding habits of *Mononchus papillatus*, which is a toothed nematode, have been described by Steiner and Heinly (1922). This terrestrial carnivore, which has a large dorsal tooth opposed by a buccal ridge, consumes as many as 1000 other nematodes during its life span of approximately 18 weeks. In feeding, the lips of this nematode are attached to the prey and an incision is made in it by the large tooth. The contents of the prey are then pumped out by the pharynx.

In some carnivores, as well as in many

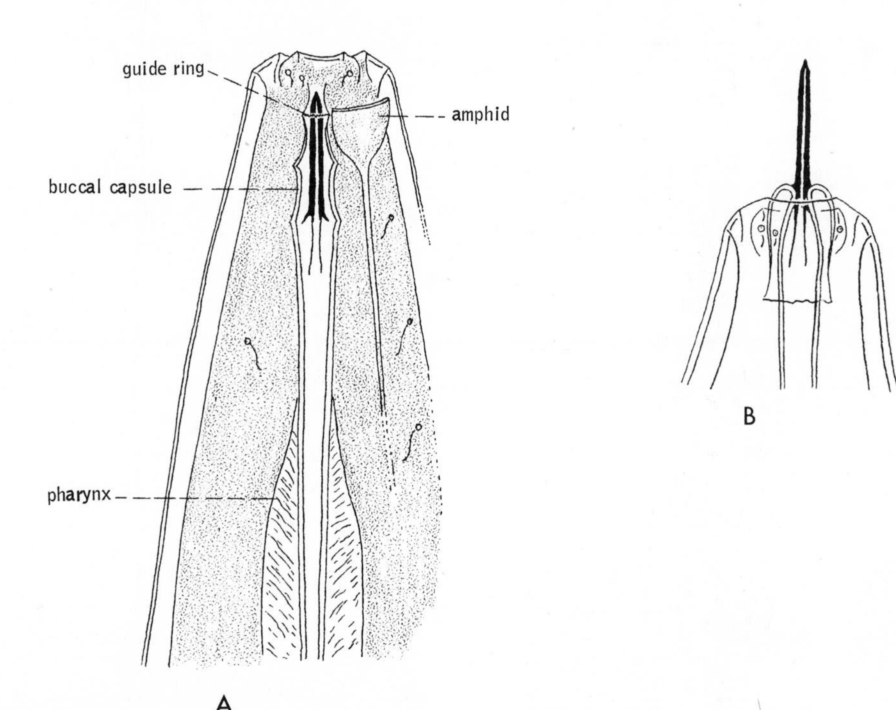

Figure 9–27. *A. Nygolaimus*, a dorylaim nematode; anterior end with spear retracted. *B*. With spear projected. (*A* and *B* after Thorne from Hyman.)

species that feed on the contents of plant cells, the buccal capsule carries a long hollow or solid spear (stylet) (Fig. 9–27), which can be protruded from the mouth. The stylet is used to puncture the prey; and when the stylet is hollow, it may act as a tube through which the contents of the victim are pumped out. In the stylet-bearing herbivores, it is used to penetrate the root cell walls, being thrust rapidly forward and backward. In both groups, pharyngeal enzymes are secreted which initiate digestion of the prey or the plant cell contents and may even aid in the penetration of the plant cell wall.

The buccal capsule leads into a tubular pharynx, sometimes referred to as the esophagus (Figs. 9–21 and 9–27*A*). The pharyngeal lumen is triradiate in cross section, with one angle always located toward the midventral line (a characteristic feature of nematodes), and the pharyngeal lumen is lined with cuticle (Fig. 9–25). The wall is composed of a syncytial epithelium, gland cells, and radial muscle fibers. The anterior walls of the pharynx usually contain special pharyngeal glands, which open into the anterior lumen of the pharynx or into the buccal cavity.

Frequently the pharynx contains one or more glandular and/or muscular swellings or bulbs (Figs. 9–28*A* and 9–30*B*). Muscular bulbs pump food into the intestine and may be provided with a valve. From the pharynx, a long tubular intestine of columnar epithelium extends the length of the body. A valve is located at each end of the intestine and can be opened only by muscular contractions. These valves prevent food from being forced out of the intestine by the fluid pressure of the pseudocoel. A short cuticle-lined rectum (cloaca in males) connects the intestine with the anus, which is located on the midventral line just in front of the posterior tip of the body (Fig. 9–21). There has recently been discovered a deep sea nematode that has no gut.

Digestive enzymes are produced by the pharyngeal glands and the intestinal epithe-

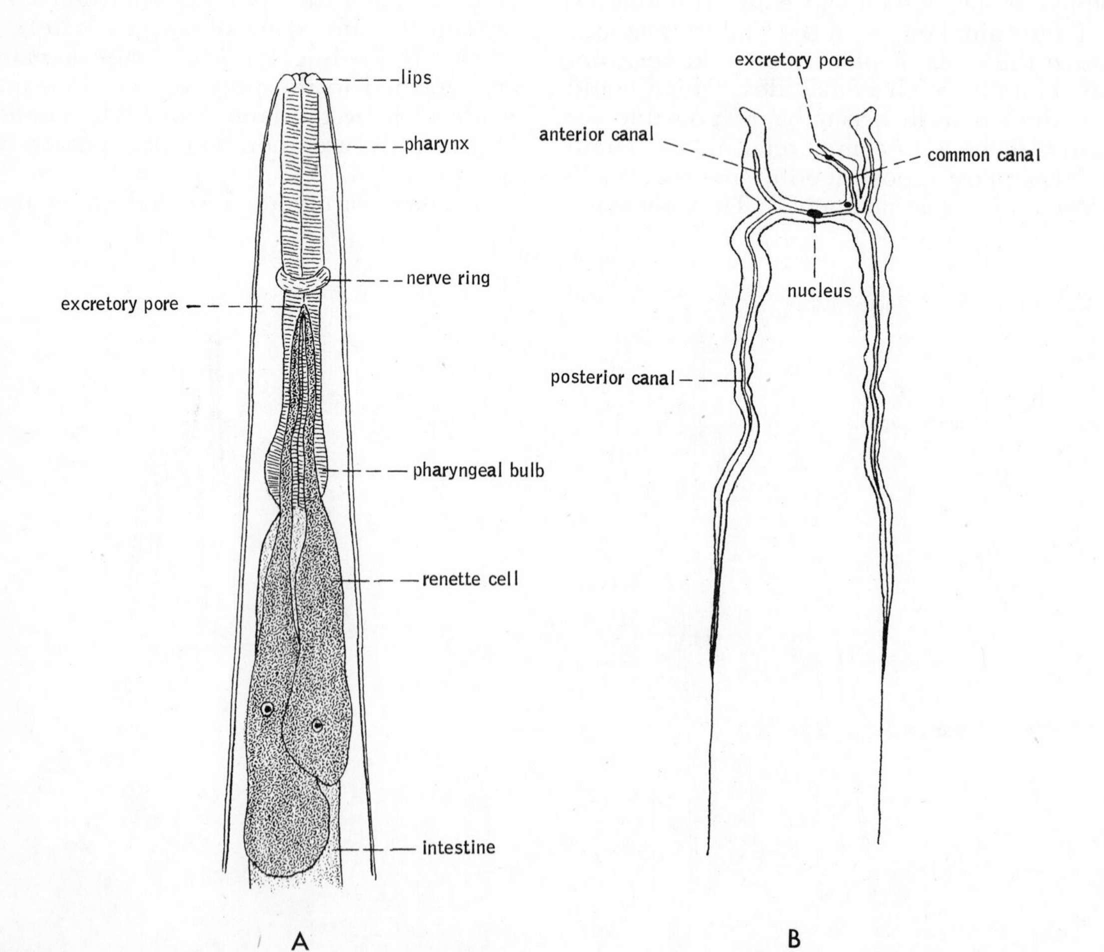

Figure 9–28. *A*. Two-celled renette of *Rhabdias*. (After Chitwood from Hyman.) *B*. H-type tubular excretory system. (After Törnquist from Hyman.)

lium, and digestion is completed extracellularly in the intestine. The intestine is also the site of absorption.

Excretion and Osmoregulation. Protonephridia are absent in all nematodes and apparently disappeared with the ancestral members of the class. However, nematodes do possess a peculiar system of gland cells, with or without tubules, that may have an osmoregulatory function, although conclusive evidence is still lacking. Primitively, as is true in marine nematodes, there is usually one but sometimes two large gland cells. This gland cell (or cells) is called a renette gland (Fig. 9–28*A*) and is located ventrally in the pseudocoel near the pharynx. The gland cell is provided with a necklike duct that opens ventrally on the midline as an excretory pore.

From such a primitive glandular structure a more specialized tubular system is believed to have evolved. In the tubular system, three long canals are arranged to form an H (Fig. 9–28*B*). Two are lateral and extend inside the lateral longitudinal cords. The two lateral canals are connected by a single transverse canal, from which a short common excretory canal leads to the excretory pore, located ventrally on the midline. In many nematodes, that part of the two lateral canals anterior to the transverse canal has disappeared, so that the system is shaped like a horseshoe.

Ammonia is the principal nitrogenous waste of nematodes and is removed through the body wall and eliminated from the digestive system along with indigestible residues. Some species, such as the parasite *Ascaris,* can increase the production of urea from 7 to 52 per cent under experimental osmotic stress. How widespread the ability to make such a shift exists in other nematodes and whether it is a normal occurrence is unknown.

The primary function of the renette system is probably osmoregulation, but the importance of the system varies with the environment. Water passes readily through the cuticle in marine and freshwater species, and in some nematodes the cuticle is known to serve as a pathway for water and ion regulation. For example, the marine nematode *Deontostoma californicum* can osmoregulate in hypertonic NaCl but not in hypotonic solutions. In this species, uptake of ions and entry of water occurs over the entire body surface (Croll and Viglierchio, 1969). Some nematodes which live in soil, in mats of moss, and in certain plants are able to tolerate periods of desiccation. During such times the worm passes into a state of anabiosis (i.e., inactivity accompanied by some water loss and a very low metabolic rate).

Nervous System. The nervous system does not differ greatly from that of other aschelminths (Fig. 9–29*A*). The brain is represented by a circumenteric nerve ring with ganglia attached dorsally, laterally, and ventrally. From the brain, ring nerves extend anteriorly, innervating the labial and cephalic papillae or bristles and the amphids (sense organs that are described later). Dorsal, lateral, and ventral nerves extend posteriorly from the brain and run within the longitudinal cords. The largest of these is the ventral nerve, which arises as a double cord from the ventral side of the nerve ring. In the region of the excretory pore, the double cord fuses and continues posteriorly as a ganglionated chain. The dorsal nerve is motor, the lateral nerves are at least partially sensory and ganglionated, and the ventral nerve is composed of both motor and sensory fibers and is also ganglionated.

A visceral nervous system is present and is composed of pharyngeal nerves from the brain and rectal nerves from the anal ganglion at the end of the ventral cord.

The principal sense organs in nematodes are papillae, setae, and amphids. The labial and cephalic papillae are projections of the cuticle (Fig. 9–22), and each contains a nerve fiber from the papillary nerves. Sensory setae may be especially prevalent on the head, but they are also present elsewhere on the body surface. The amphids, which reach their highest form of development in the aquatic nematodes, and especially in the marine species, are blind invaginations of the cuticle. One amphid is situated on each side of the head; they are commonly located just posterior to the cephalic setae (Figs. 9–22, 9–23, and 9–29). The nature of the invagination varies considerably. It may resemble a simple pouch, opening to the exterior through a slitlike pore; or the amphids may be looped or spiral troughs on the surface of the cuticle, one end of which then passes into and through the cuticle as a tube. Some evidence also exists that the amphids may have gland cells associated with them, and these may be the source of the gelatinous substance sometimes observed in the amphid. The amphids are believed to be chemoreceptors. Electron microscopy of certain nematodes has revealed that the sensory processes of the amphids are actually modified cilia. Cilia had previously been thought to be absent in the Nematoda.

In the tail region of the nematode are a

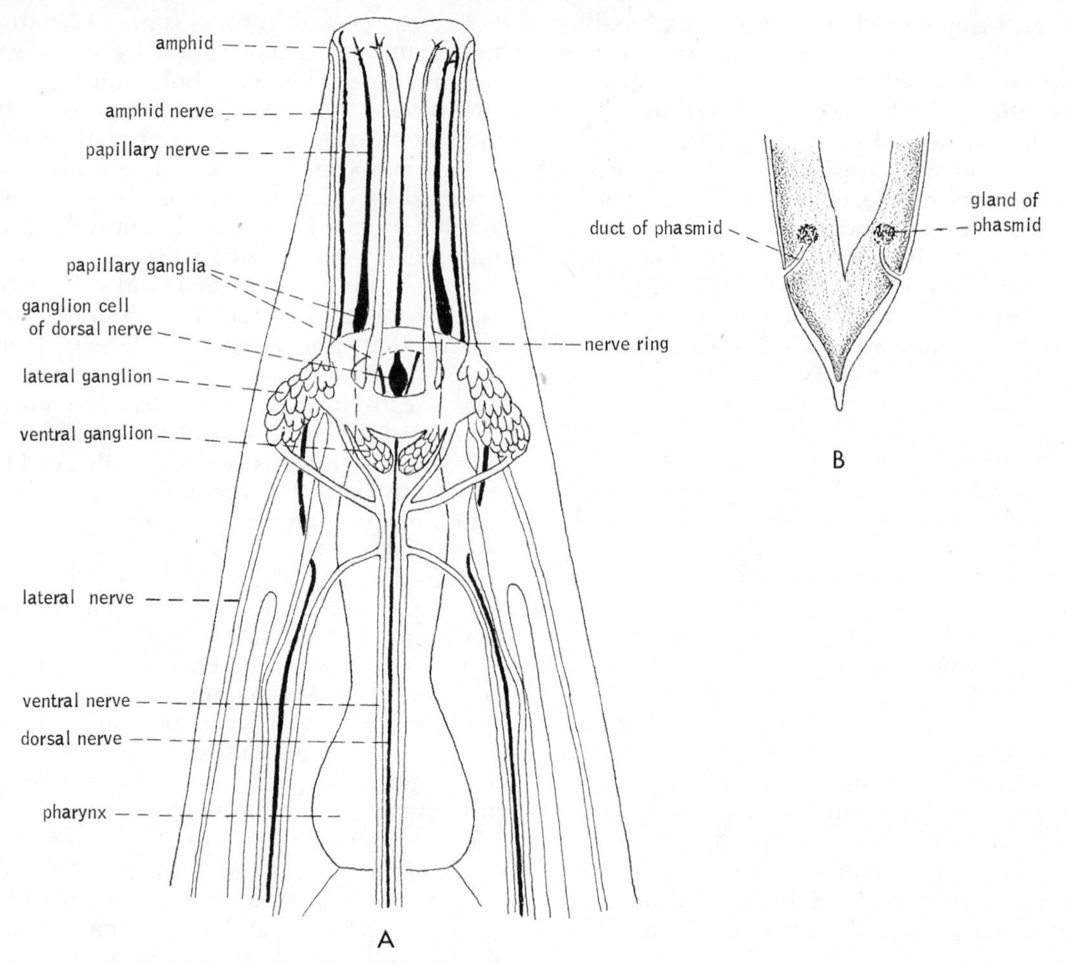

Figure 9–29. *A*. Anterior part of *Cephalobellus* nervous system. (After the Chitwoods from Hyman.) *B*. Phasmids at posterior of *Rhabditis*. (After Chitwood from Hyman.)

pair of unicellular glands, called phasmids (Fig. 9–29*B*), which open separately on either side of the tail. The phasmids are perhaps glandulo-sensory structures that function in chemoreception. They reach their best development in parasitic nematodes. A pair of simple ocelli are located one on each side of the pharynx of a few marine and freshwater nematodes. In the marine nematode *Deontostoma* each ocellus consists of one pigment cell filled with stacked concentric lamellae.

Reproduction. Most nematodes are dioecious. Males are typically smaller than females, and the posterior of the male is curled like a hook (Fig. 9–30*A*). The gonads are tubular and may be especially long and coiled in parasitic species. In most nematode orders, the germ cells arise from a large terminal cell located at the distal end of the tube; the germ cells gradually pass through the length of the gonad, during which time growth and maturation take place. In males there may be one or two tubular testes, which pass more or less imperceptibly into a long sperm duct. The sperm duct eventually widens to form a long seminal vesicle.

A muscular ejaculatory duct, containing a varying number of prostatic glands, connects the seminal vesicle with the rectum. The prostatic secretions are adhesive and supposedly aid in copulation. The wall of the rectum, or cloaca (since it functions as a part of the reproductive system), is evaginated to form two pouches, which join before they open into the cloacal chamber. Each pouch contains a spicule of cuticle having a cytoplasmic core (Fig. 9–30*A*). The spicules vary considerably in length and shape but are usually short and shaped somewhat like a pointed curved blade. Special muscles cause the spicules to protrude through the cloaca and out of the anus or vent. In many nematodes the dorsal, and sometimes even the ventral and lateral, walls of the pouch bear special, cuticular pieces that act as guides in the movement of the spicules through the cloacal chamber. Where more than one piece exists, they are usually fused to form one structure, the gubernaculum (Fig. 9–30*A*).

Usually, two straight or coiled ovaries, which are oriented in opposite directions, are present (Fig. 9–21). The ovary gradually extends into a tubular oviduct and then in turn into a much-widened, elongated uterus. Each of the two uteri opens into a short, common, muscular tube called the vagina, which leads to the outside through the gonopore. The female gonopore is located on the midventral line, usually in the midregion of the body. The upper end of the uterus usually functions as a seminal receptacle; in some nematodes the seminal receptacles are separate pouches that open laterally from the upper end of the uterus.

The females of some nematodes are known to produce a pheromone which attracts males. In copulation, the curved posterior of the male nematode is usually coiled around the body of the female in the region of the genital pores (Fig. 9–30*B*). The copulatory spicules of the male are extended through the cloaca and anus and are used to hold open the female gonopore during the transmission of the sperm into the vagina. A nematode sperm is peculiar in that it lacks a flagellum, and some may move in a more or less ameboid manner. An exception to the usual copulatory mechanism occurs in *Trichosmoides crassicauda,* a nematode parasitic in the urinary tract of rats. The male of this species is minute and lives inside the vagina of the female.

After copulation, the sperm migrate to the upper end of the uterus, where fertilization takes place. The fertilized egg throws off a thick fertilization membrane, which hardens to form the inner part of the shell. To this inner shell is added an outer layer, which is secreted by the uterine wall. The outer surface of nematode eggs is sculptured in various ways and is often important in the detection of parasitic infections from fecal smears. Nematode eggs may be stored in the uterus prior to deposition, and not infrequently embryonic development begins while the eggs are still in the female.

Some terrestrial nematodes, particularly

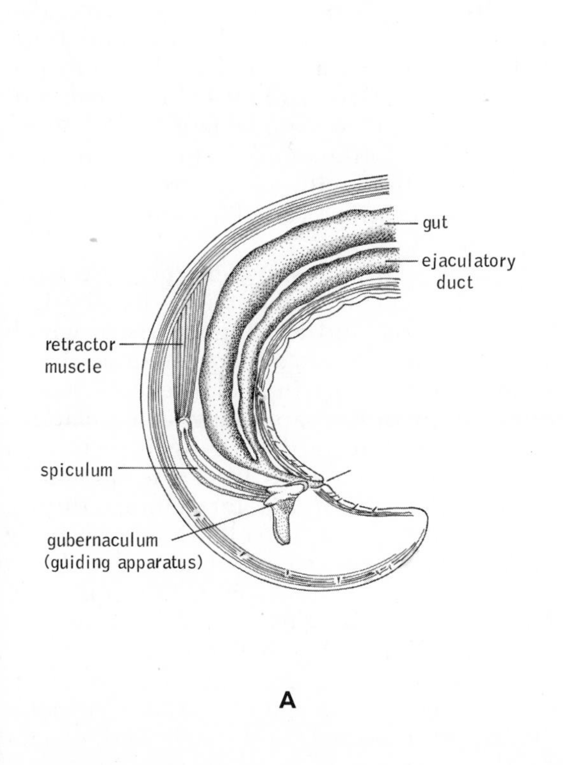

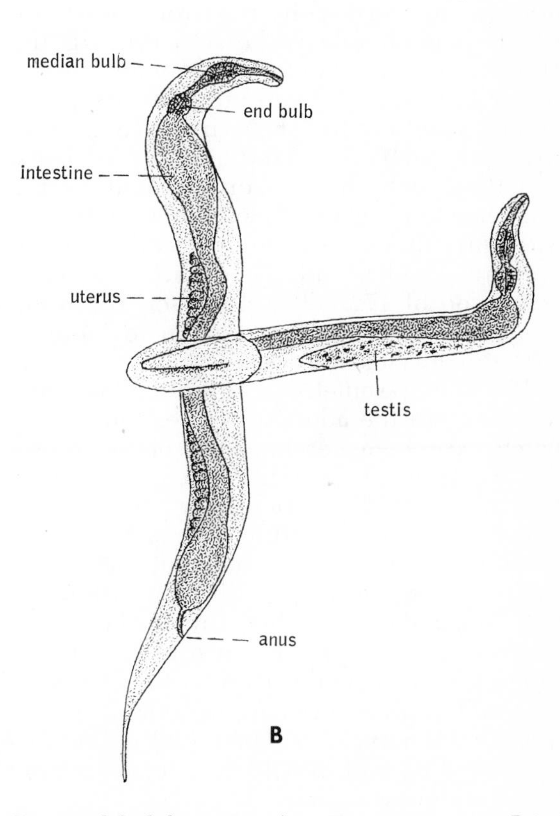

Figure 9–30. *A.* Posterior end of the male of *Pseudocella.* (Modified from Hope.) *B.* Termite parasite, *Pristionchus aerivora,* in copulation. (After Merrill and Ford from Hyman.)

the rhabditoids, are hermaphroditic. Sperm develop before the eggs do and are stored. Self-fertilization takes place after the formation of the eggs. Parthenogenesis also occurs in some terrestrial nematodes. A most remarkable type of parthenogenesis is reported to take place in *Mesorhabditis belari*. In this species, males are present, sperm are produced, and copulation occurs. However, the sperm only initiate cleavage and do not fuse with the nucleus of the eggs that are destined to develop into females.

The deposition of the eggs of free-living nematodes is still not well known. Marine species rarely produce more than 50 eggs, which are often deposited in clusters. Terrestrial species may produce up to several hundred eggs, which are deposited in the soil. Many parasitic nematodes and some free-living forms, such as the vinegar eel, are ovoviviparous.

Embryogeny. Cleavage is determinate, but the spiral pattern has disappeared. A coeloblastula is formed, and gastrulation is by epiboly. The blastocoel persists as the pseudocoel. A remarkable feature of the embryogeny of at least some nematodes is the very early separation of the future germ cells from the somatic cells. In such cases, only in the germ cells is chromosomal integrity maintained. In the somatic cells the chromosomes tend to become somewhat disintegrated through breakage and loss. As in other aschelminths, the various organs of the body contain a fixed number of cells, and these cells have been attained by the time hatching takes place. For example, in *Rhabditis* there are 200 nerve cells and 120 epidermal cells, and the digestive tract is composed of 172 cells. However, the number of intestinal cells is known to increase with growth in some species.

The young, sometimes called larvae, have almost all of the adult structures when they hatch except for certain parts of the reproductive system. Growth is accompanied by four molts of the cuticle, the first two of which may occur within the shell before hatching. Adults do not molt but many continue to increase in size. The few studies to date indicate that molting is under hormonal control. An important constituent of the molting fluid is synthesized by the excretory cell under stimulation from neurosecretory cells in the nerve ring. The molting fluid dissolves the base of the old cuticle, separating it from the new cuticle.

Ecology. Nematodes exist in all types of habitats. With few exceptions, aquatic species are limited to the substratum, and marine forms are found everywhere from the shore to great depths. Marine nematodes occur in enormous numbers wherever the bottom mud is rich in organic matter, regardless of the latitude; they are probably the most abundant metazoan animals of ocean bottoms. Freshwater nematodes exist in all types of freshwater habitats. Some species are even adapted for existence in fast flowing mountain streams. Anchorage is maintained in such forms by the adhesive caudal glands at the end of the body.

Other unusual aquatic habitats include hot springs, in which the water temperature may reach 53° C., and the water in tropical, epiphytic bromeliads. In large lakes, there is often a distinct zonation of the nematode species from the shore line to deeper water.

Terrestrial species live in the film of water that surrounds each soil particle, and they are therefore actually aquatic. In fact, there are species which are found in both soil and freshwater. Although nematodes exist in enormous numbers in the upper few inches of soil, the population decreases rapidly at greater depths. Moreover, the numbers are greater in the vicinity of plant roots. In addition to the more typical terrestrial habitats, nematodes have also been reported to exist in the accumulations of detritus in leaf axils and in the angles of tree branches. Mosses and lichens maintain a characteristic nematode fauna; and these species, like rotifers and other moss inhabitants, are able to withstand extreme and prolonged desiccation, as well as great extremes of temperature.

A great many freshwater and terrestrial nematodes have a cosmopolitan distribution. Birds, animals, and floating debris to which small amounts of mud adhere are undoubtedly important agents in the spread of nematodes. Many of the saprophagous nematodes that inhabit dung utilize dung insects as a means of transportation from one habitat to another. *Rhabditis dubia*, for example, during a special stage, wraps itself around the body of the dung-breeding psychodid fly. When the fly alights on new dung, the nematode detaches. Dung beetles are similarly utilized by other species of nematodes.

Parasitism. The great numbers of parasitic species of nematodes attack virtually all groups of animals and plants and, except for the absence of ectoparasitic forms on animals, display all degrees and types of parasitism. Moreover, some of the major groups of nematodes contain *both* free-living and parasitic species. All of these facts suggest that parasitism has evolved many times within the phylum and that the radiation of the group is a fairly recent event. Schaefer (1971) suggests that nematode radiation was coupled with the evolution of flowering plants, insects, and higher vertebrates. The types of nematode host-parasite relationships have been outlined by Hyman (1951). A modification of this outline is followed here:

1. Completely free-living. Life cycle is direct, and all stages are free-living.

2. Ectoparasites of plants. Worms feed on the external cells of plants by puncturing the cell wall with stylets and sucking out the cell contents.

3. Endoparasites of plants. Worms in juvenile stage enter the plant body and feed on the living cells, producing death of tissue or gall-like structures. Reproduction takes place within the host, and the new generation of juveniles migrates to other plants.

4. Saprophagous type of zooparasitism. Adults and juveniles are free-living in soil, but worm in late juvenile stage enters an invertebrate. The host is not injured, and the worm feeds on the dead tissues when the host dies.

5. Zooparasitic in juvenile stages only. The juveniles are parasitic in an animal host, usually an invertebrate, for at least part of the early life cycle. Adults are free-living. Perhaps the best known nematodes illustrating this type of life cycle are members of the family Mermithidae. The mermithid nematodes range in length from less than 1 mm. to an extreme of 50 cm. The adults live in soil or mud and do not feed. The intestine is blind and the cells are packed with stored food material. The adult life span varies from a few days to months, depending upon the stored food supply. Eggs are deposited in the soil and following hatching, the larvae, which have an initial free-living period, penetrate crustaceans, arachnids, mollusks, or most commonly an insect. A period of development occurs within the host (Fig. 9–31); the worm then leaves to become an adult in the original soil or aquatic mud environment.

6. Phytoparasitic juveniles and zooparasitic adults. The female worm produces juveniles within a plant-feeding insect host. When the insect punctures the plant tissue, the juvenile worms enter the plant and remain as endoparasites. When mature and after copulation, the female enters the larva of the host, which lives on the same plant. The larva metamorphoses into an adult and deposits a new generation of juvenile worms. *Heterotylenchus aberrans,* a parasite of onion flies, illustrates this type of life cycle.

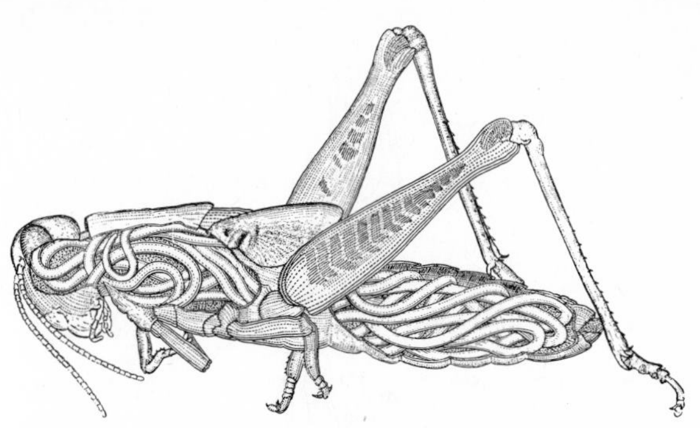

Figure 9–31. Juvenile mermithid nematode within a grasshopper host. (After Christie.)

7. Zooparasitic juveniles and phytoparasitic adults. Early stages of development take place within an invertebrate host. Later, worms in juvenile stages leave the host and enter the plant on which the host feeds. Worms complete development and reproduce as phytoparasites. The new generation of young then enter the host.

8. Zooparasitism in adult females only. The young become adults in the soil. After copulation, the male dies, and the female enters an invertebrate host to produce the next generation.

9. Adult zooparasites with one host. Adult worms of both sexes are parasitic within a vertebrate or invertebrate host. Transmission from one host to another is by eggs or newly hatched young, which may be free-living for a varying part of their development. Many nematode parasites possess this type of life history, of which the following are a few of the more interesting or important examples.

The ascarid nematodes are intestinal parasites of man, dogs, cats, pigs, cattle, horses, chickens, and other vertebrates. They are large nematodes; the female of the horse nematode, *Parascaris equorum,* may attain a length of 35 cm. The life cycle typically involves transmission by the ingestion of eggs passed in another host's feces. The larval developmental stages are not confined to the intestinal lumen and frequently involve temporary invasion of other tissues of the host (Fig. 9–32).

The dog ascarid, *Toxocara canis,* for

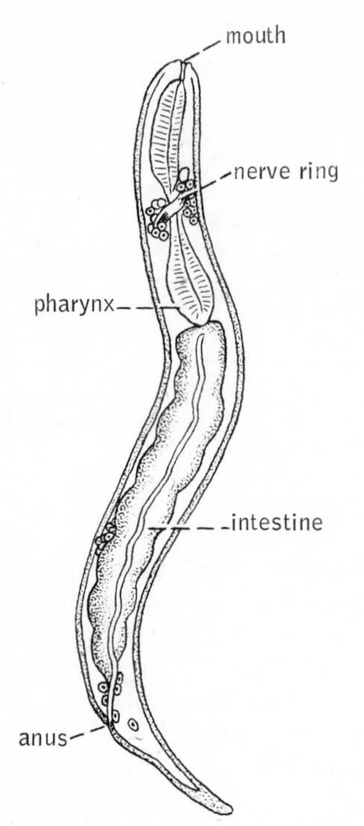

Figure 9–32. Juvenile (rhabditiform larva) ascarid nematode. (After Noble and Noble.)

which puppies are usually wormed, illustrates the life cycle. The adult male worm of this species reaches 10 cm. in length and the female 18 cm. Eggs begin development in the host's feces and become infective in about five days. If ingested by another host, the eggs hatch in the small intestine. The young worms penetrate the intestinal walls to enter the hepatic portal system, where they are carried by the blood to the lungs. Here they break into the alveoli and migrate up the air passageways. They are swallowed in the esophagus and eventually reach the intestine. Variations in this history may occur in older dogs or alternate hosts.

The human ascarid, *Ascaris lumbricoides*, which is usually a typical example in introductory courses, is certainly one of the best known parasitic nematodes. Physiological studies suggest that *Ascaris* produces enzyme inhibitors that protect the worm from the host's digestive enzymes. If this is true, such a mechanism is probably utilized by most other gut-inhabiting nematodes. The ascarids feed on the host's intestinal contents.

The strongylids are another group of parasites of the digestive tract of vertebrates and include the hookworms. Most members of this group feed on the host's blood. The mouth region is usually provided with cutting plates, hooks, teeth, or combinations of these structures for attaching and lacerating the gut wall (Fig. 9–33). A heavy infestation of hookworms can produce serious danger to the host through loss of blood and tissue damage. Hookworms are one of the parasites causing serious infections in humans. It is estimated that some 59 million Asians are infected with the Old World species, *Ancylostoma duodenale*.

The life cycle of the strongylids involves an indirect migratory pathway by the larva, as in ascarids. In hookworms, for example, the eggs leave in the host's feces and hatch outside the host's body. The larva gains reentry by penetrating the host's skin and is carried to the lungs in the blood. From the lungs the larva migrates to the esophagus where it is swallowed and passes to the intestine. Skin penetration is not true of all species of this order. Some species have a pattern similar to the dog ascarid.

Oxyurid nematodes, known as pinworms, have a simpler life cycle. These small nematodes are parasitic in the gut of vertebrates and invertebrates. Infection usually occurs through the ingestion of eggs passed in feces. The eggs hatch and larvae develop within the gut of the new host. The human pinworm, *Enterobius vermicularis*, is a scourge, though not dangerous, of children throughout the world. The female worm deposits eggs at night in the perianal region. Itching is caused by the migration of the female depositing her eggs, and scratching by the child contaminates the fingernails and hands with eggs. The eggs are thus easily spread to other children or reinfect the same child.

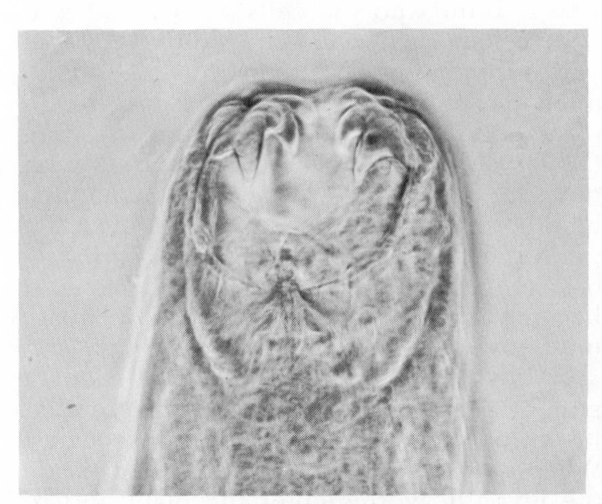

Figure 9–33. Anterior end of the dog hookworm, *Ancylostoma caninum*, showing buccal region and teeth.

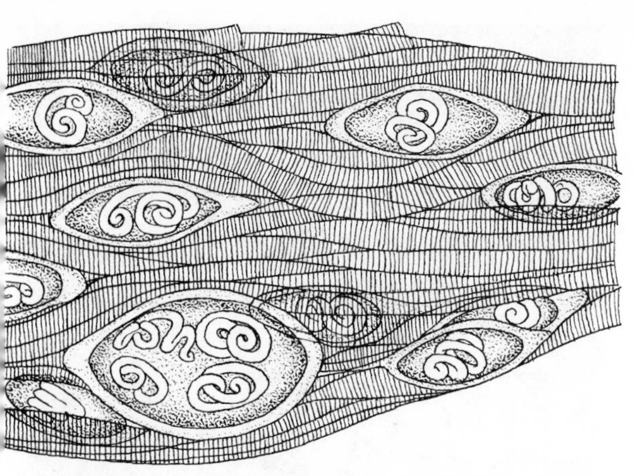

Figure 9–34. Larvae of *Trichinella spiralis* within calcareous cyst in striated muscle tissue of host. (After Chandler and Read.)

Trichurids are also parasites of the alimentary tract of vertebrates, especially birds and mammals. The whipworms, *Trichuris,* which infect man, dogs, cats, cattle and other mammals, are relatively small worms (the human whipworm, *Trichuris trichiura,* is about 4 cm.) and have a life cycle similar to that of the pinworms. Other trichuroid genera, such as *Capillaria,* follow a pattern similar to that of the ascarids. Certainly the most familiar of the trichurids is *Trichinella spiralis* of mammals, the cause of the disease trichinosis. The minute worm, which lives in the intestinal wall, is ovoviviparous, and the larvae following birth are carried to the striated muscles in the blood. The larvae form calcified cysts in the striated muscles and, if infection is high, can produce pain and stiffness (Fig. 9–34). Transmission to another host can occur only if the flesh containing encysted larvae is ingested by another host. It is obvious that in some animals, such as the rat, this can be a one-host parasite; in others, such as man and the pig, it would normally require two hosts.

10. Zooparasites with one intermediate host. Varying degrees of juvenile development take place within an intermediate host, after which there is a reinfection of the primary host, where reproduction occurs.

This type of life cycle is illustrated by a number of nematode parasites, including the familiar filariids and dracunculids. The filariids are threadlike worms which inhabit the lymphatic glands, coelom, and some other sites of the vertebrate host, especially birds and mammals. The female is ovoviviparous and the larvae are called microfilariae. Bloodsucking insects, such as fleas, certain flies, and especially mosquitoes, are the intermediate hosts. A number of species parasitize man, producing filariasis.

The chiefly Asian *Wucheria bancrofti* illustrates the life cycle. The male is 40 mm. × 0.1 mm. and the female is about 90 mm. × 0.24 mm. The adults live in the lymph glands. The microfilariae are found in the blood and are present in the peripheral blood stream. A distinct periodicity of larval migration to the peripheral circulation, coinciding with the activity of the mosquito, has been demonstrated by a number of investigators. When certain species of mosquitoes bite the host, the microfilariae enter with the host's blood. Development within the intermediate host involves a migration through the gut to the thoracic muscles and after a certain period into the proboscis. From the proboscis the microfilariae are introduced back into the primary host when the mosquito feeds. In severe filariasis the block-

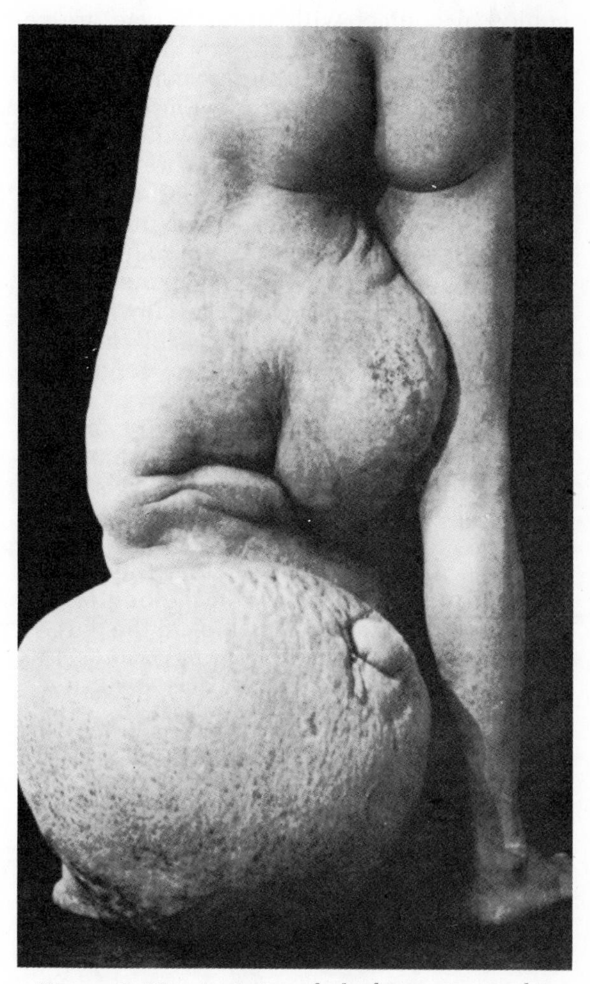

Figure 9–35. A victim of elephantiasis, resulting from severe filariasis. (Courtesy of Mayo Clinic.)

ing of the lymph vessels by large numbers of worms results in extreme growth and swelling of host tissue, particularly in the legs, breast, and scrotum. Such enlargement is called elephantiasis (Fig. 9–35).

Loa loa, the African eye worm, lives in the subcutaneous tissues of man and baboons. The worm migrates about in the tissue and sometimes passes across the eyeball, accounting for the name eye worm (Fig. 9–36).

The dracunculids are also rather thread-like worms. They live in the connective tissue and body cavities of vertebrate hosts. The most notable example is the guinea worm, *Dracunculus medinensis,* which parasitizes man and many other mammals, especially in Asia and Africa. The female is about 1 mm. in diameter and up to 120 cm. in length. After a period of development in the body cavity and connective tissue of the host, the gravid female migrates to the subcutaneous tissue and produces an ulcerated opening to the exterior. If the ulcerated area of the host comes in contact with water, larvae are released. After a short free-living stage, the larvae are ingested by species of copepod crustaceans (*Cyclops*) where development continues in the host's hemocoel. When the primary host swallows copepods while drinking, the nematode larvae are released and penetrate the intestinal wall to reach the coelom or subcutaneous tissue. The worms can be removed surgically, but the ancient method, still practiced, is to slowly wind them out on a small stick (Fig. 9–37). Breaking of the worm will cause severe inflammation.

There are other nematode groups besides the filariids and dracunuculids that possess one intermediate host.

11. Zooparasitic with two intermediate hosts. Although life cycles with two intermediate hosts are not common, there are a number of nematodes which belong to this category. An example is one of the largest known nematodes, *Dioctophyma renale* (Fig. 9–38), which parasitizes the kidneys and coelom of mammals, chiefly carnivores. The female may attain a length of 103 mm. and a diameter of 12 mm. and can destroy the kidney, resulting in the host's death. The eggs leave the body of the host in the urine and must reach water. If the eggs are ingested by certain freshwater annelids (branchiobdellids) that are epizooic on crayfish gills, a larva hatches. The larva penetrates the worm's intestine and enters the coelom, where it encysts. If the crayfish containing the attached larva is eaten by a certain fish, the nematode encysts in the fish's coelom. The parasite reaches the primary host when the latter eats the fish.

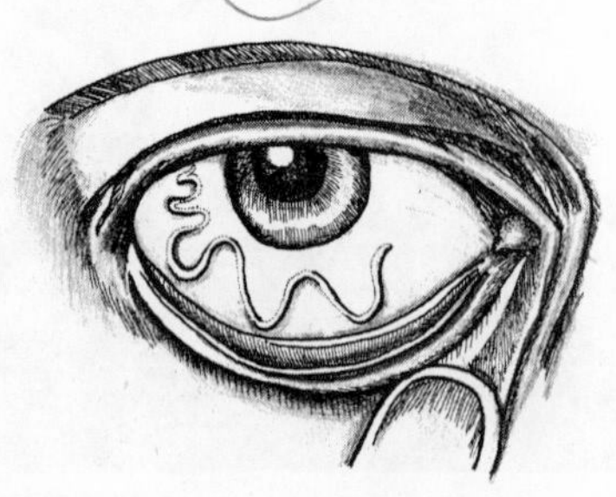

Figure 9–36. The eye worm, *Loa loa,* in the cornea. (From Cheng.)

Figure 9–37. A guinea worm, *Dracunculus medinensis,* being removed from ulcerated opening of arm by slow winding on a match stick. (Courtesy, Institute of Public Health Research, Teheran University School of Public Health.)

12. Zooparasitic adults alternating with free-living adults. This strange life-cycle pattern occurs in a few nematodes, such as *Rhabdias bufonis,* which parasitizes the lungs of frogs, including species of frogs used for laboratory experiments. The parasitic phase is also unusual in that the worms are hermaphroditic, although sperm are produced before the eggs. Eggs are coughed up into the pharynx and then swallowed by the frog. They hatch in the intestine and the larvae leave the host in the feces. The

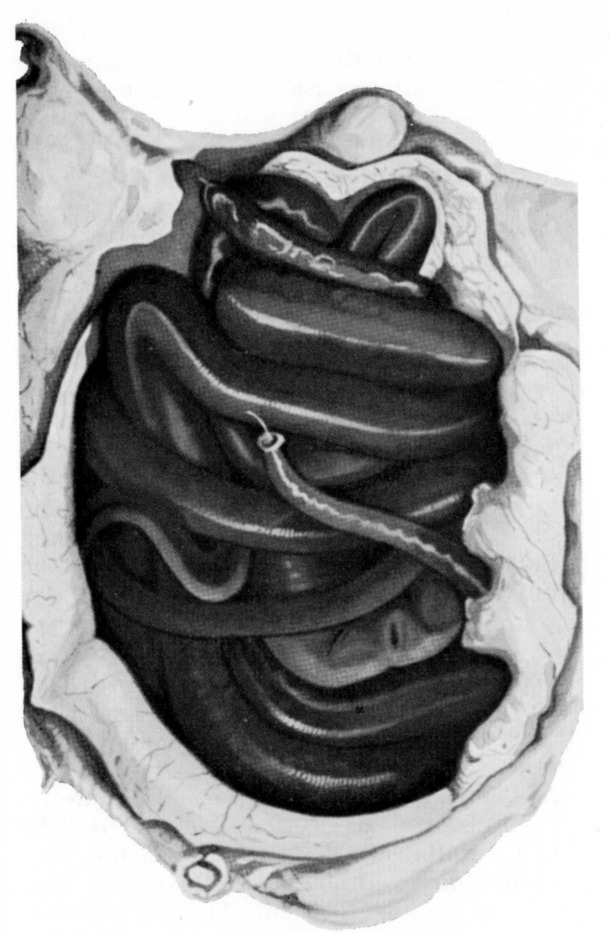

Figure 9–38. The kidney of a dog which has been opened to show a specimen of the large nematode, *Dictophyme renale.* (After Lukasiak.)

liberated larvae develop into dioecious soil-inhabiting adults. Larvae of the free-living worms penetrate the frog's skin and are carried in the lymphatic system back to the lungs.

Classification. The phylum Nematoda is divided into two classes and fourteen orders. The class Adenophorea, or Aphasmida, contains nematodes that lack phasmids. Although there are parasitic members, the majority of free-living forms are marine and freshwater. To this class belong the orders Enoplida, Chromadorida, Desmodorida, Desmocolecida, Monohysterida, Araeolaimida, Dorylaimida, Dioctophymatida, and Trichinellida.

The class Secernentea, or Phasmida, contains nematodes that possess phasmids. The majority of parasitic forms are members of this class, and the free-living species are largely soil inhabitants. To this class belong the orders Rhabditida, Tylenchida, Strongylida, Ascarida, and Spirurida.

PHYLUM NEMATOMORPHA

The remaining aschelminths are a small group (230 species) of extremely long worms, commonly known as hairworms, which comprise the phylum Nematomorpha. The adults are free-living, but the juveniles are all parasitic in insects and crustaceans. In most hairworms, which comprise the class Gordioidea, the adults live in fresh water and damp soil. The single genus *Nectonema,* which makes up the class Nectonematoidea, is pelagic in marine waters. The body of nematomorphs (Fig. 9–39) is extremely long and threadlike, and lengths of 36 cm. or more are typical. The diameter, however, is usually not much more than 1 mm. The hairlike nature of these worms is so striking that it was formerly thought that they arose spontaneously from the hairs of a horse's tail. There is no distinct head, and the color is usually dark brown to black.

Like nematodes, the nematomorph body wall is composed of a thick outer cuticle covering the body surface; a cellular epidermis with a ventral, or with a dorsal and a ventral, longitudinal cord; and a muscle layer of longitudinal fibers only. Except in *Nectonema,* the pseudocoel is somewhat reduced by mesenchymal tissue. The digestive tract is vestigial, and the adults do not feed. The nervous system is very similar to that of kinorhynchs in that it is composed of an anterior nerve ring and a ventral cord. The

Figure 9–39. Three female gordioid nematomorphs, or horsehair worms. (Courtesy of R. W. Pennak.)

sexes are separate, and in gordioids the two long cylindrical gonads extend the length of the body. As in nematodes, the sperm ducts empty into the rectum, or cloaca, but there are no copulatory spicules present. The ovaries are peculiar in that they transform into uteri because of the development of lateral diverticula after the production of eggs. The oviducts also empty into the cloaca.

Hairworms live in all types of freshwater habitats in temperate and tropical regions of the world. The females are very inactive, although the males commonly swim or crawl about by whiplike motions of the body. Copulation is similar to that in nematodes, and the eggs are deposited in the water. The young on hatching have a protrusible proboscis armed with spines (Fig. 9–40).

After hatching, the young enter an arthropod host living in or along the water's edge (commonly beetles, cockroaches, crickets, or grasshoppers). Recently leeches have been reported as hosts. The larvae of *Nectonema*, the only marine hairworms, parasitize hermit crabs and true crabs. On penetration of the host, the young enter the hemocoel where development is completed. Their nutrition as parasites is apparently accomplished by direct absorption of food materials through the body wall, perhaps followed by the production of enzymes that break down the host tissues in the vicinity of the worm. After several weeks to several months of development, during which a number of molts occur, the worms leave the host as almost completely formed adults. Emergence only occurs when the host is near water. Sexual maturity is shortly attained during the free-living adult phase of the life cycle.

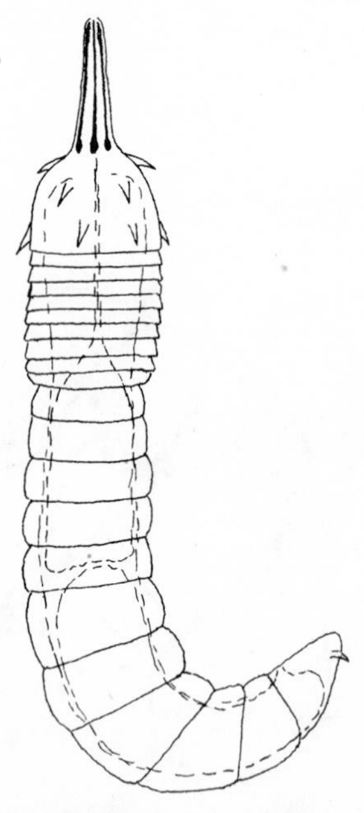

Figure 9–40. Gordioid worm larva. Proboscis protrudes. (After Pennak.)

The remaining pseudocoelomate phylum, the Acanthocephala, is not included among the aschelminths.

PHYLUM ACANTHOCEPHALA

The acanthocephalans are a phylum of some 500 species of parasitic, wormlike pseudocoelomates. The ultrastructure of the body wall, muscles, and sperm would seem to indicate a close relationship to the aschelminths (Whitfield, 1971). However, the embryogeny, particularly the origin of the pseudocoel, is peculiar.

All are endoparasites requiring two hosts to complete the life cycle. The juveniles are parasitic within crustaceans and insects, while the adults live in the digestive tract of vertebrates. The body of the adults is elongated and composed of a trunk and a short anterior proboscis and neck region (Fig. 9–41*A*). Most acanthocephalans are only a few centimeters long, although some species may attain a length of 50 cm. The proboscis is covered with recurved spines (Fig. 9–41*B*), hence the name Acanthocephala—spiny head.

The acanthocephalan proboscis and neck can be retracted into the anterior of the trunk. The trunk is typically covered with spines, and not infrequently is divided into superficial segments. The retractable proboscis and anchoring spines provide the means by which acanthors (larvae) move within the host. To either side of the neck an evagination of the body wall projects posteriorly into the trunk pseudocoel. These two evaginations, called lemnisci, are permeated with lacunae. The body wall is peculiar in that it is composed of a surface cuticle, a syncytial, fibrous epidermis that contains channels or canals. The canals may form either a network or distinct longitudinal channels. The canals are not connected with the exterior or the interior of the animal. They are filled with fluid and are believed to represent a means of circulation of nutritive material absorbed from the host. These worms possess no digestive system, and food is absorbed directly through the body wall from the host. Two protonephridia, which are

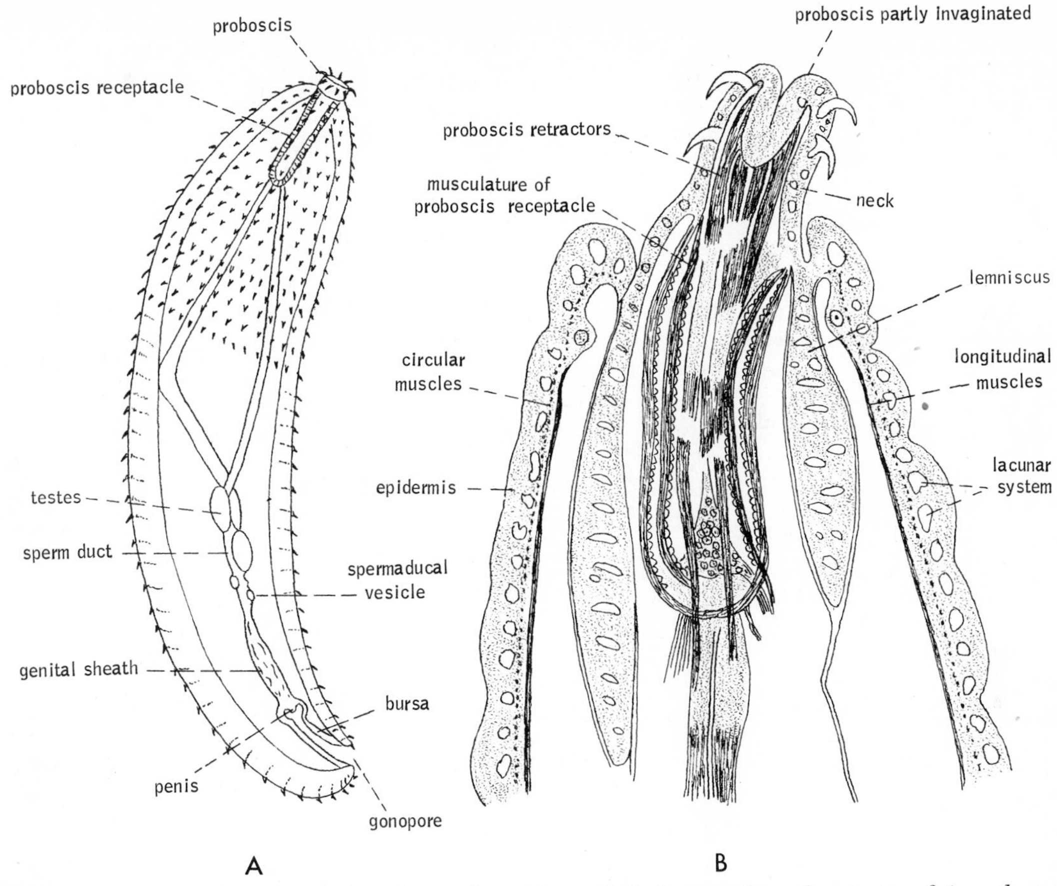

Figure 9–41. *A. Acanthogyrus.* (After Thapur from Hyman.) *B.* Section through anterior of *Acanthocephalus.* (Modified after Hamann from Hyman.)

associated with the reproductive system, are present in one order. The nervous system is composed of a ventral, anterior ganglionic mass, from which arises a varying number of single and paired longitudinal cords.

The sexes are separate, and the gonopore is located at the end of the trunk. The male system includes a protrusible penis, and the eggs are fertilized within the body of the female. Development of the egg takes place within the female pseudocoel; and when a larval stage having an anterior crown or rostellum with hooks is attained, each larva is enclosed within a shell. The developed eggs (i.e., the encased larvae) then pass out of the vertebrate host with the feces. If the eggs are eaten by certain insects, such as roaches or grubs, or by aquatic crustaceans, such as amphipods, isopods, or ostracods, the larvae emerge from the eggs, bore through the gut wall of the host, and become lodged in the hemocoel. The larva, called an acanthor, possesses a rostellum with hooks that is used in penetrating the host's tissues.

Development in acanthocephalans ceases when the worms have almost become adults. When the intermediate host is eaten by a fish, a bird, or a carnivorous mammal, the worms attach to the intestinal wall of the vertebrate host, by using the spiny proboscis. Acanthocephalans often exist in great numbers within the vertebrate host and can do considerable damage to the intestinal wall. As many as 1000 acanthocephalans have been reported in the intestine of a duck, and 1154 in the intestine of a seal.

PHYLUM GNATHOSTOMULIDA

The Gnathostomulida is a small phylum of minute acoelomate worms that live in the interstitial spaces of marine sands in shallow water. This group of worms represents the latest phylum of animals to be discovered. Although observed earlier, the first description of a gnathostomulid was not published until 1956 by Ax. Since that time over 80 species and 18 genera have been described from many parts of the world, es-

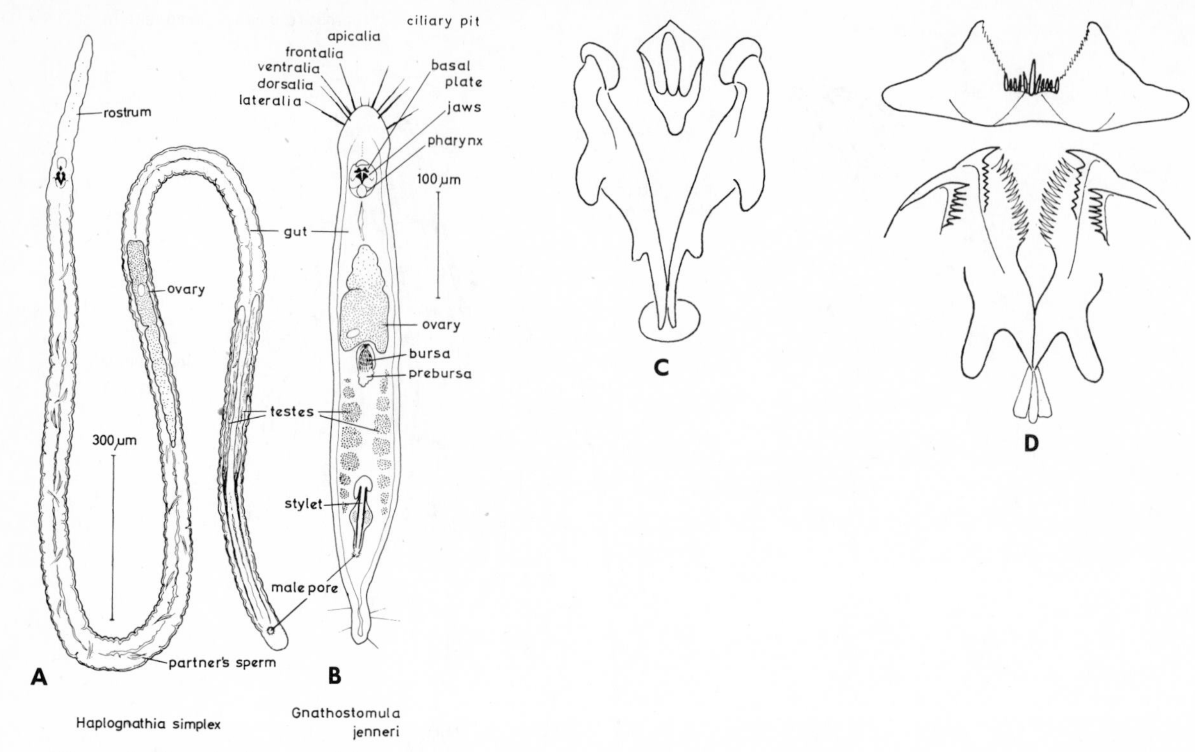

Figure 9–42. Gnathostomulida. *A. Haplognathia simplex. B. Gnathostomula jenneri. C.* Basal plate and jaws of *Haplognathia simplex. D.* Basal plate and jaws of *Gnathostomula mediterranea.* (All from Sterrer, W., 1972: Syst. Zool., *21*(2):151–173.)

pecially from the east coast of North America. They have been found to be widespread, and often to occur in enormous numbers. The tardiness of their discovery can be attributed to their minute size, the necessity of studying them alive (they become greatly distorted when preserved), the general lack of attention received by the interstitial fauna prior to the past twenty years, and their ability to tolerate anaerobic conditions, which results in their being among the last interstitial animals to come to the surface of a sample of stagnating bottom material.

Gnathostomulids are mostly between 0.5 and 1.0 mm. in length. All are elongate, and some are even threadlike. The more or less cylindrical transparent body consists of an anterior head separated from a trunk by a slightly constricted neck region (Fig. 9–42). The body tapers posteriorly into a tail.

Like flatworms, the gnathostomulids are ciliated, but each epithelial cell bears but a single cilium, a distinctive feature of gnathostomulids. They move by a slow gliding type of swimming and the body can be contracted and twisted by the three or four paired groups of longitudinal fibers that constitute the body wall musculature. Unlike flatworms, ciliary propulsion in gnathostomulids can be reversed. Mesenchyme is poorly developed. The nervous system is epidermal and the sense organs are ciliary pits and sensory cilia, which are especially well developed at the anterior end.

Gnathostomulids are like flatworms in having a gut that lacks an anal opening, but the mouth, which is located ventrally behind the head, bears a comblike plate on the ventral lip and a pair of toothed lateral jaws within the mouth cavity (Fig. 9–42*C* and *D*). The jaws are operated by muscles. The remainder of the gut is a tubular intestinal sac. The principal food appears to be bacteria and fungi, which are scraped up by the comb on the ventral lip and passed into the intestinal sac by snapping movements of the jaws.

The majority of gnathostomulida must be capable of anaerobic respiration, for most occur within black marine sands where no oxygen is present.

Gnathostomulids are hermaphroditic, although individuals with only male or female systems occur. The female reproductive system usually consists of a single ovary and associated bursa, or sperm storage sac. A vagina is present in some species. The male system consists of a pair of posterior testes and a copulatory organ, which in some species has the form of a stylet. The male gonopore is at the posterior end. In the few species observed to copulate, sperm are injected by the stylet into the body of another individual and then make their way to the bursa. A single large egg is laid at each oviposition. The egg ruptures through the body wall and briefly adheres to the bottom. The worm regenerates very quickly. Cleavage is spiral, and development is direct. It appears that at least some gnathostomulids may exhibit feeding non-sexual stages that alternate with non-feeding sexual stages, the reproductive system degenerating and regenerating from one stage to another.

The Gnathostomulida are placed at the end of this chapter only for the benefit of comparison. They are like flatworms in their ciliated body covering, acoelomate structure, absence of an anus, and hermaphroditic reproductive system. But they lack the flatworms' 1-9 flagellar ultrastructure, the mesenchyme is greatly reduced, and the ciliary beat can be reversed. Moreover, they display a number of features, such as a head, jaws, and sensory bristle-like cilia, that are found in rotifers and gastrotrichs. Gnathostomulids are thus clearly related to the lower bilateral animals, but their unique combination of characters demands their classification within a separate phylum.

Bibliography

Baer, C. J., 1961: Acanthocéphales. *In* Grassé, P. (Ed.), Traité de Zoologie, Vol. 14, pt. 1. Masson et Cie, Paris, pp. 733–782.

Bird, A. F., 1971: The Structure of Nematodes. Academic Press, N. Y. 318 pp.

Bird, A. F., and Bird, J., 1969: Skeletal structures and integument of Acanthocephala and Nematoda. *In* Florkin, M., and Scheer, B. J. (Eds.), Chemical Zoology, Vol. 3, Academic Press, N. Y., pp. 253–288.

Cheng, T. C., 1964: The Biology of Animal Parasites. W. B. Saunders, Philadelphia. (A parasitology text with a good account of the parasitic pseudocoelomates.)

Chitwood, M. B., 1969: The systematics and biology of some parasitic nematodes. *In* Florkin, M., and Scheer, B. J. (Eds.), Chemical Zoology, Vol. 2, Academic Press, N. Y., pp. 223–244.

Crofton, H. D., 1966: Nematodes. Hutchinson University Library, London. 160 pp.

Croll, N. A., 1970: The Behaviour of Nematodes. St. Martin's Press, N. Y.

Croll, N. A., and Viglierchio, D. R., 1969: Osmoregulation and uptake of ions in a marine nematode. Proc. Helminthol. Soc. Wash., *36*(1):1–9.

Crompton, D. W. T., 1970: An Ecological Approach to Acanthocephalan Physiology. Cambridge University Press, N. Y. 136 pp.

D'Hondt, J. L., 1971: Gastrotricha. Oceanogr. Mar. Biol., Ann. Rev., *9*:141–192.

Edmondson, W. T., Ward, H. B., and Whipple, G. C. (Eds.), 1959: Freshwater Biology, 2nd edition. John Wiley and Sons, N. Y. [Keys and figures to the freshwater pseudocoelomates. Nemata (B. G. Chitwood and M. W. Allen), pp. 368–401; Gordiida (B. G. Chitwood), pp. 402–405; Gastrotricha (R. B. Brunson), pp. 406–419; Rotifers (W. T. Edmondson), pp. 420–494.]

Ferris, V. A., 1971: Taxonomy of the Dorylaimida. *In* Zuckerman, B. M., Mai, W. F., and Rhode, R. A. (Eds.), Plant Parasitic Nematodes, Vol. 1, pp. 163–189. Academic Press, N. Y.

Golden, A. M., 1971: Classification of the genera and higher categories of the order Tylenchida (Nematoda). *In* Zuckerman, B. M., Mai, W. F., and Rhode, R. A. (Eds.), Plant Parasitic Nematodes, Vol. 1, pp. 191–232. Academic Press, N. Y.

Goodey, T., 1951: Soil and Freshwater Nematodes. John Wiley and Sons, N. Y.

Gosner, K. L., 1971: Guide to Identification of Marine and Estuarine Invertebrates; Cape Hatteras to the Bay of Fundy. Wiley-Interscience, N. Y. Aschelminthes, pp. 184–210.

Grassé, P. (Ed.), 1965: Némathelminthes, Rotifères, Gastrotriches, et Kinorhynques. *In* Traité de Zoologie, Vol. 4, pts. 2 and 3. Masson et Cie, Paris. (A detailed general account of the pseudocoelomate invertebrates.)

Hope, W. D., and Murphy, D. G., 1972: A Taxonomic Hierarchy and Checklist of the Genera and Higher Taxa of Marine Nematodes. Smithsonian Contribution to Zoology, *137*:1–101.

Hummon, W., 1971: Biogeography of sand beach Gastrotricha from the northeastern United States. Biol. Bull., *141*(2):390.

Hyman, L. H., 1951: The Invertebrates: Acanthocephala, Aschelminthes, and Entoprocta. Vol. 3. McGraw-Hill, N. Y.

Hyman, L. H., 1959: The Invertebrates: Smaller Coelomate Groups. Vol. 5. McGraw-Hill, N. Y., pp. 750–754.

Jennings, J. B., and Colam, J. B., 1970: Gut structure, digestive physiology and food storage in *Pontonemi vulgaris* (Nematoda: Enoplida). J. Zool. London, *161*:211–221.

Kaestner, A., 1967: Invertebrate Zoology, Vol. 1. Wiley-Interscience, N. Y., pp. 217–263.

Lee, D. L., 1965: The Physiology of Nematodes. W. H. Freeman, San Francisco.

Levine, N. D., 1968: Nematode Parasites of Domestic Animals and of Man. Burgess Publishing Co., Minneapolis. 600 pp.

Light, S. F., Smith, R. I., Pitelka, F. A., Abbott, D. P., and Weesner, F. M., 1967: Intertidal Invertebrates of the Central California Coast. University of California Press, Berkeley. Aschelminths, pp. 63–108.

Noble, E. R., and Noble, G. A., 1972: Parasitology, 3rd edition. Lea & Febiger, Philadelphia. 617 pp.

Pennak, R. W., 1953: Freshwater invertebrates of the

United States. Ronald Press, N. Y. (Chapters on the rotifers, gastrotrichs, nematodes, hairworms, and entoprocts, each with a brief introduction to the group and keys and illustrations for the identification of common freshwater forms.)

Pratt, I., 1969: The biology of the Acanthocephala. *In* Florkin, M., and Scheer, B. J. (Eds.), Chemical Zoology, Vol. 3. Academic Press, N. Y., pp. 245–252.

Read, C. P., 1970: Parasitism and Symbiology. Ronald Press, N. Y. 316 pp. (A text which approaches parasitology from the standpoint of host-parasite relationships rather than from a survey of parasitic groups.)

Rogers, W. P., 1962: The Nature of Parasitism. Academic Press, N. Y. (A study of host-parasite relationships with particular emphasis on nematodes.)

Roggen, D. R., Raski, D. J., and Jones, N. O., 1966: Cilia in nematode sensory organs. Science, *152*:515–516.

Schaefer, C., 1971: Nematode radiation. Syst. Zool., *20* (1):77–78.

Steiner, G., and Heinly, H., 1922: Possibility of control of *Heterodera* by means of predatory nemas. J. Washington Acad. Sci., *12*:367–386.

Sterrer, W., 1972: Systematics and evolution within the Gnathostomulida. Syst. Zool., *21*(2):151–173.

Thorne, G., 1961: Principles of Nematology. McGraw-Hill, N. Y. 553 pp.

Tietjen, J. H., 1969: The ecology of shallow water meiofauna in two New England estuaries. Oecologia, *2*:251–291.

Wallace, H. R., 1970: The movement of nematodes. *In* Fallis, A. M. (Ed.), Ecology and Physiology of Parasites. University of Toronto Press, Toronto, pp. 201–212.

Whitfield, P. J., 1971: Phylogenetic affinities of Acanthocephala: An assessment of ultrastructural evidence. Parasitology, *63*(1):49–58.

Yeats, G. W., 1971: Feeding types and feeding groups in plant and soil nematodes. Pedobiologia, *11*(2): 173–179.

Zuckerman, B. M., Mai, W. F., and Rhode, R. A., 1971: Plant Parasitic Nematodes: Cytogenetics, Host-Parasite Interactions and Physiology. Vol. 2. Academic Press, N. Y. 347 pp.

Chapter 10

THE ANNELIDS

The phylum Annelida comprises the segmented worms and includes the familiar earthworms and leeches, plus a great number of marine and freshwater species of which most persons are completely unaware. A shovelful of muddy sand taken from the shore along a coastal sound at low tide usually brings to light a much richer and far more spectacular collection of "worms" than could be found in a backyard garden.

In general, the annelids attain the largest size of any of the wormlike invertebrates and display the greatest structural differentiation. The most distinguishing characteristic of the phylum is metamerism, the division of the body into similar parts, or segments, which are arranged in a linear series along the anteroposterior axis. The youngest segments occur at the posterior end of the series. The segmented part of the body is always limited to the trunk. The head, represented by the prostomium and containing the brain, is not a segment, nor is the pygidium, the terminal part of the body which carries the anus. However, in all metameric animals there has been a tendency for anterior trunk segments to fuse, in varying degrees, with the unsegmented head, prostomium, or acron. This fusion gives rise to a secondary compound "head." It is thus important to understand what the term *head* refers to. The formation of new segments in a metameric animal always takes place just in front of the pygidium. The oldest body segments are therefore anterior, and the youngest are posterior.

Metamerism appears to have evolved twice within the Animal Kingdom, in the chordates as an adaptation for undulatory swimming and in the annelids as an adaptation for burrowing. In both groups the primary metameric structures are the body wall muscles. Also important in annelid metamerism is the segmentation of the coelom by transverse septa. Each septum is composed of two layers of peritoneum, one derived from the segment in front and one from the segment behind. As an accommodation to serve the primary segmentation of the coelom and body wall musculature, the lateral nerves, blood vessels, and excretory organs are also segmentally arranged.

Coelomic fluid functions as a hydraulic skeleton against which the muscles act to change the body shape. Contraction of the longitudinal muscles causes the coelomic fluid to exert a laterally directed force, and the body widens. Contraction of the circular muscles causes the coelomic fluid to exert an anteriorly and posteriorly directed force, and the body elongates (Fig. 10–1). The significance of the compartmentation of the coelom by transverse septa now becomes apparent. The coelomic fluid skeleton is localized, and the body shape changes of widening and elongation can be restricted to certain segments. Waves of peristaltic contraction pass down the length of the body, bringing about elongation and then shortening of a number of segments encompassed within the wave. Those segments in which the longitudinal muscles are contracted have an increased diameter and are anchored against the burrow wall. Segments in which the circular muscles are contracted are elongated and move forward (Fig. 10–1). Powerful force can be generated, enabling the animal to thrust the anterior end through the substratum or to move rapidly through a previously excavated burrow.

Annelids possess a more or less straight digestive tract running from the anterior mouth to the posterior anus. The gut is located in the coelom by longitudinal mesenteries and by the septa, through which the gut penetrates (Fig. 10–2). Digestion is extracellular. Excretion takes place by means of nephridia, primarily metanephridia, and characteristically there is one pair per segment. There is usually a well-developed blood-vascular system, in which the blood is usually confined to vessels—that is, the system is closed. The nervous system consists

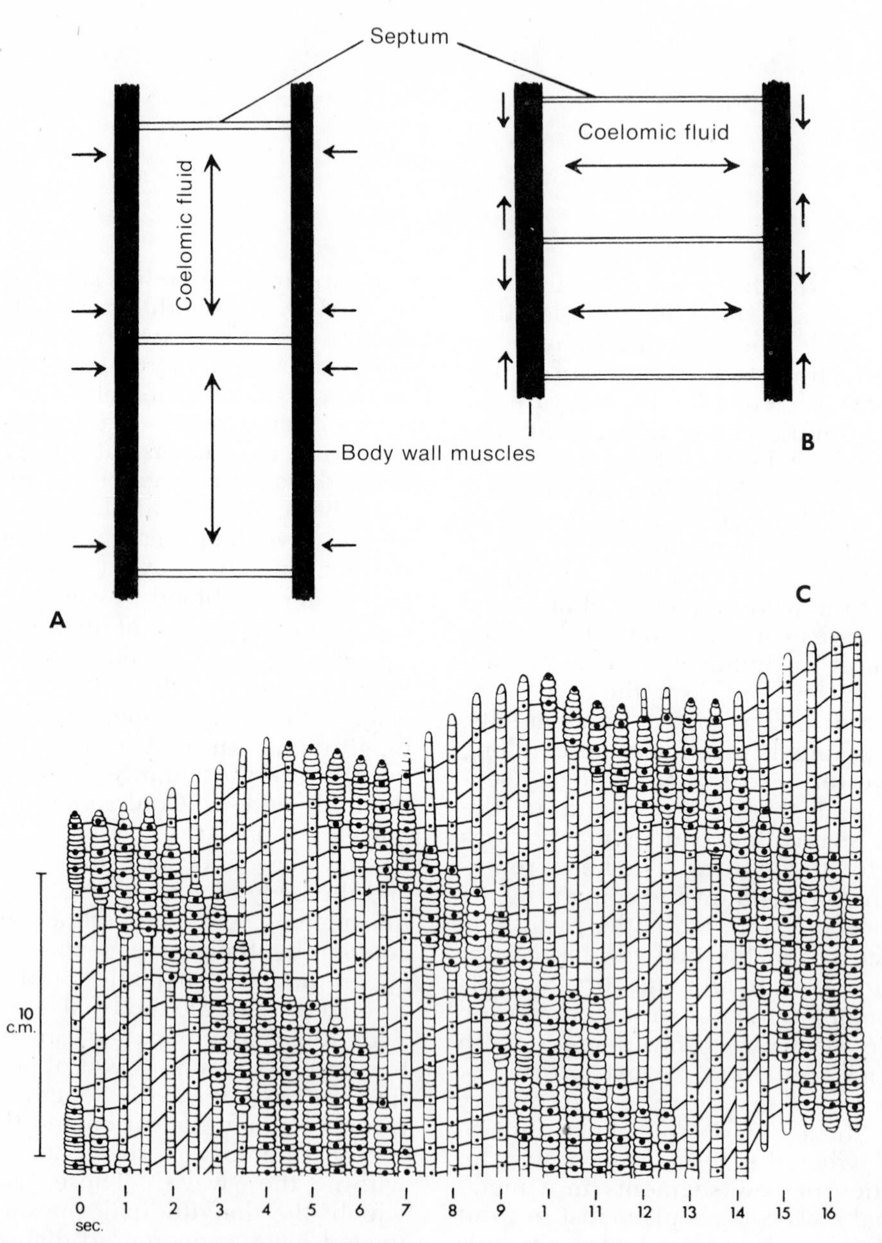

Figure 10–1. *A* and *B*. Diagrammatic frontal section through two annelid segments. External arrows indicate direction of force exerted by body wall muscles. Internal arrows indicate direction of force exerted by coelomic fluid pressure: (*A*) during circular muscle contraction, and (*B*) during longitudinal muscle contraction. *C*. Diagram showing mode of locomotion of an earthworm. Segments undergoing longitudinal muscle contraction are marked with larger dot and drawn twice as wide as those undergoing circular muscle contraction. The forward progression of a segment during the course of several waves of circular muscle contraction is indicated by the horizontal lines connecting the same segments. (After Gray and Lissmann.)

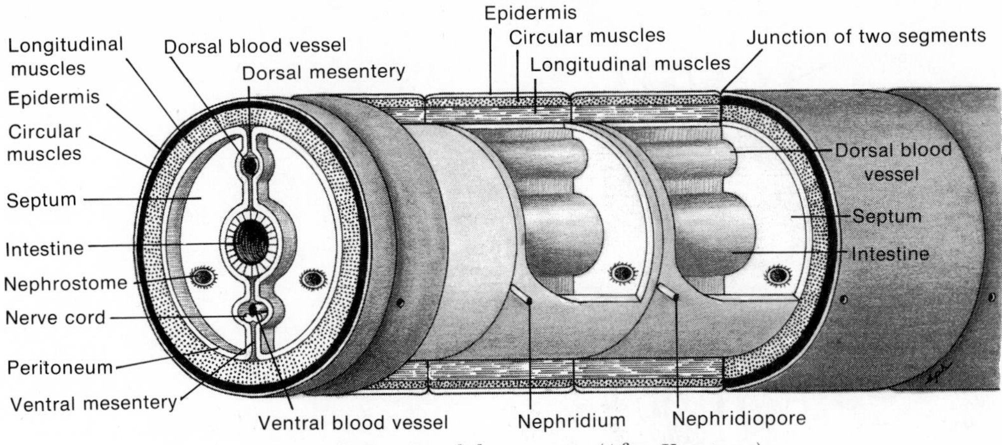

Figure 10-2. Annelid segments. (After Kaestner.)

of an anterior, dorsal ganglionic mass, or brain; a pair of anterior connectives surrounding the gut; and a long double or single ventral nerve cord with ganglionic swellings and lateral nerves in each segment.

The phylum contains over 8700 described species, which are placed into three classes: Polychaeta, Oligochaeta, and Hirudinea. The ancestral annelids were probably marine animals burrowing in the bottom sand and mud of shallow coastal waters. The class Polychaeta contains most of the living marine species. The class Oligochaeta, which includes the freshwater annelids and the terrestrial earthworms, may have stemmed from some early polychaetes, but more likely they evolved independently from the ancestral annelids. The class Hirudinea, the leeches, clearly arose from some stock of freshwater oligochaetes.

CLASS POLYCHAETA

Polychaete worms are very common marine animals, but their secretive habits result in their being overlooked by casual observers. Over 5300 species have been described. The majority are 5 to 10 cm. long with a diameter ranging from 2 to 10 mm.; but some syllids are no longer than 2 mm., whereas one species of *Eunice* may attain a length of 3 m. Many polychaetes are strikingly beautiful and are colored red, pink, or green or possess a combination of colors; some are iridescent.

The "typical" polychaete is perfectly metameric, with each of the cylindrical body segments being identical and each bearing a pair of lateral, fleshy, paddle-like appendages called parapodia (Fig. 10-3). At the anterior of the worm is a well-developed head, called the prostomium, which bears eyes, antennae, and a pair of palps. The mouth is located on the ventral side of the body between the prostomium and the first trunk segment, which is called the peristomium. The terminal segment, the pygidium, carries the anus. Few polychaetes possess such a typical structure. The different modes of existence displayed by the worms of this class have led to varying degrees of modification in this basic plan of structure.

Polychaetes can be divided into two groups—errant (free-moving) forms and sedentary forms, although the distinction between the two groups is not always sharp. The errant polychaetes, or Errantia, include some species that are strictly pelagic, some that crawl about beneath rocks and shells, some that are active burrowers in sand and mud, and many species that construct and live in tubes.

The sedentary polychaetes, or Sedentaria, are largely tubicolous—that is, are tube-dwellers, or inhabit permanent burrows. Usually only the head of the worm ever emerges from the opening of the tube or burrow.

Polychaete Structure. The "typical" polychaete body plan is most nearly attained in the errant families. The dorsal preoral prostomium is well developed and bears numerous sensory structures. The prostomial sense organs usually consist of eyes, antennae, ventrolateral palps, and ciliated pits or grooves, called nuchal organs (Fig. 10-3). The prostomium projects forward like a shelf over the mouth and behind it is the peristomium, which forms the lateral and ventral

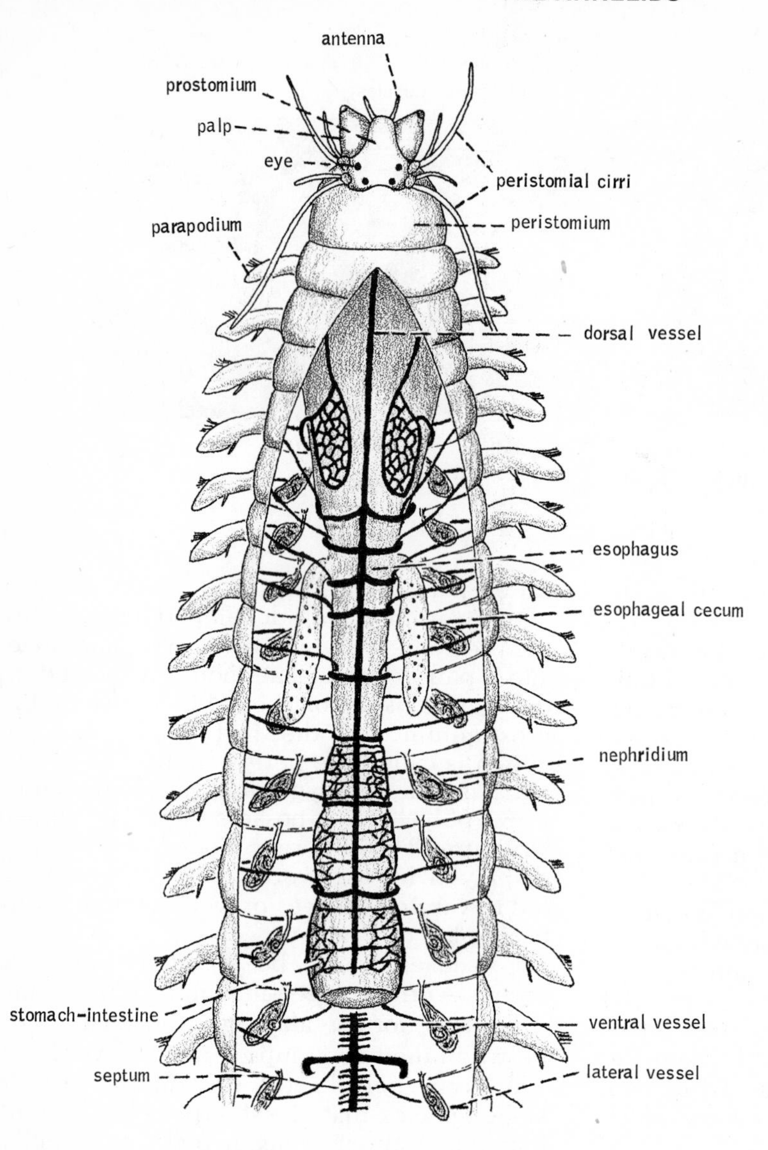

Figure 10–3. *Nereis* (dorsal view). (After Brown.)

margins of the mouth. With some exceptions the peristomium represents in part the first trunk segment or in some instances two or more fused trunk segments; however, these segments are usually modified with sensory structures such as peristomial cirri, and the entire region can be considered part of the polychaete "head." In most errant species the peristomial parapodia are reduced or absent.

The most distinguishing feature of polychaetes is the presence of parapodia, the paired lateral appendages extending from the body segments. A typical parapodium is a fleshy projection extending from the body wall and is more or less laterally compressed (Fig. 10–4*A*). The parapodium is basically biramous, consisting of an upper division, the notopodium, and a ventral division, the neuropodium. Each division is supported internally by one or more chitinous rods, or acicula. A cirriform process projects from the dorsal base of the notopodium and from the ventral base of the neuropodium.

The notopodia and neuropodia assume various shapes in different families and may be subdivided into several lobes. Usually the two main divisions (the notopodium and the neuropodium) are differently shaped, or one may be somewhat reduced. In fact, reduction in some polychaetes has resulted in uniramous parapodia. This is true, for example, in most phyllodocids, in which all of the notopodium has disappeared except the broad flattened dorsal cirrus (Figs. 10–5*C* and 10–7*A*).

The distal ends of the parapodial rami

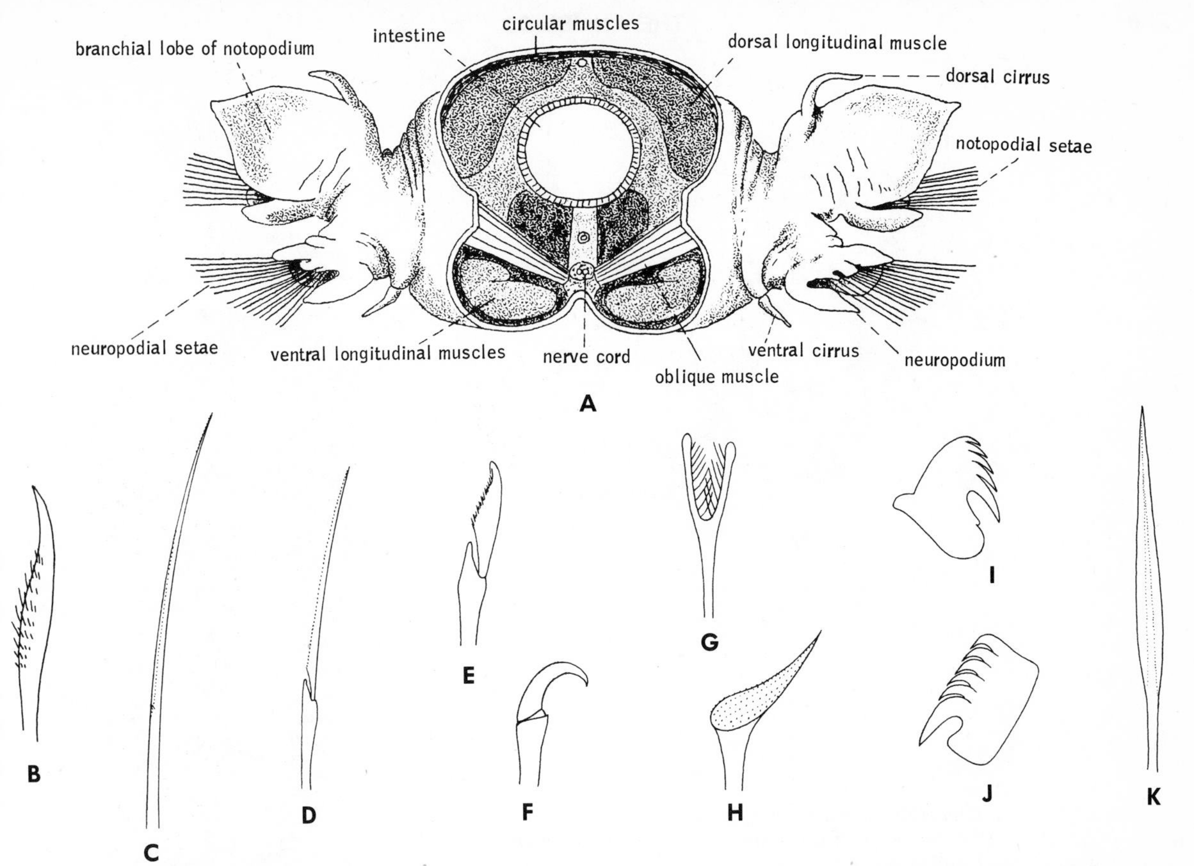

Figure 10–4. *A*. Body segment of *Nereis* (posterior view). (Modified from Snodgrass and Clark.) *B* to *K*. Various types of polychaete setae: *B*. Aphroditidae. *C*. Glyceridae. *D* and *E*. Nereidae. *F*. Spintheridae. *G*. Orbiniidae. *H*. Sabellariidae. *I* and *J*. Uncinate setae of Terebellidae and Serpulidae. *K*. Sabellidae. (*B-K* after Day.)

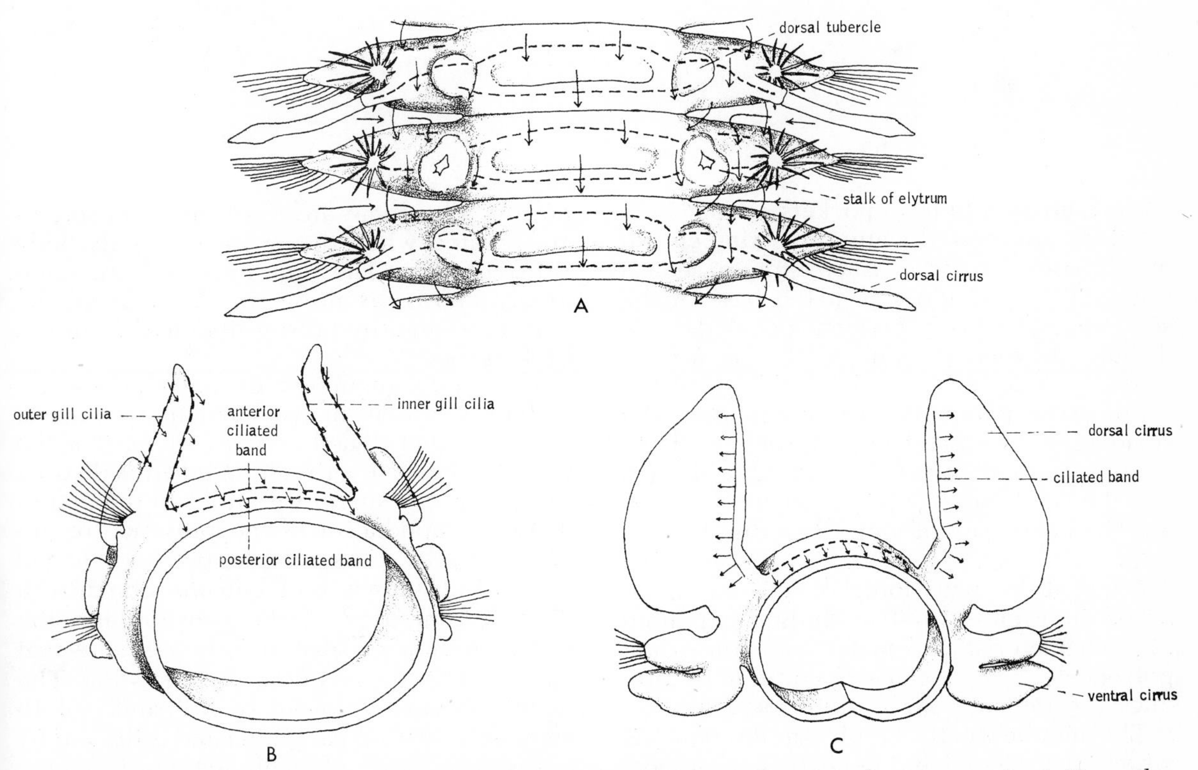

Figure 10–5. Surface ciliation in three polychaetes. Arrows indicate direction of water currents. A. *Harmothoe imbricata*. B. *Scolelepis squamata*. C. *Phyllodoce laminosa*. (After Segrove.)

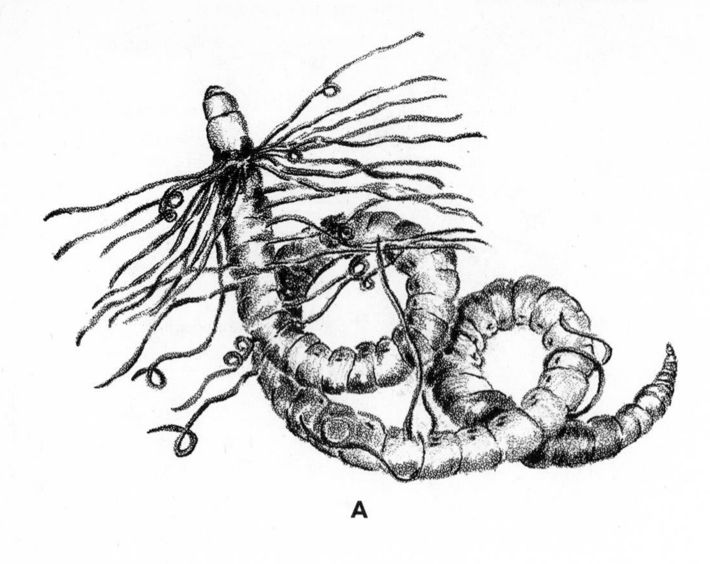

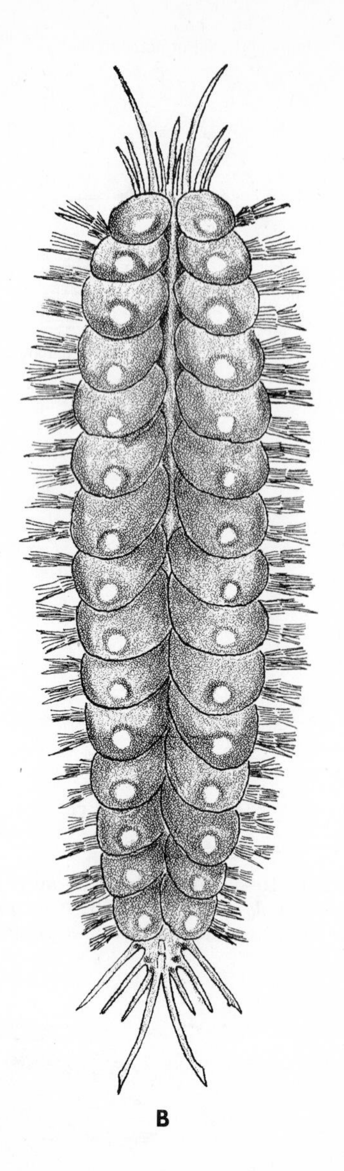

Figure 10–6. *A. Cirratulus cirratus*, a polychaete with long thread-like dorsal cirri. (After McIntosh from Fauvel.) *B. Lagisca flocculosa*, a scale worm. (After McIntosh.)

are invaginated to form pockets or setal sacs in which are located many projecting chitinous bristles, or setae. Each simple seta is secreted by a single cell at the base of the setal sac and usually projects a considerable distance beyond the end of the parapodial lobe.

Since the parapodia are compressed, the setae tend to spread out like a fan. They assume a great variety of different shapes, and the setal bundles of a particular species may be composed of more than one type of seta.

A few of the many forms of setae are illustrated in Figure 10–4, including a jointed type characteristic of many common errant polychaetes. New setae are continually produced by the setal sac as older setae are lost.

The members of the family Amphinomidae, which live in coral and beneath stones, have brittle, tubular, calcareous setae containing poison. The setae are used for defense and are erected when the worm bends its body (Fig. 10–7*C*). The amphinomids are commonly known as fireworms because of the pain produced by poison liberated from the broken setae.

The body segments of polychaetes are generally similar; in many sedentary species, however, there has been a tendency for the trunk to become differentiated into distinct regions, as a result of variations in the nature of the parapodia, or of the presence or absence of gills. For example, the body of the burrowing lugworm *Arenicola* contains a "head" composed of the prostomium, the peristomium, and several body segments that have lost their parapodia (Fig. 10–8*A*). The head region is followed by a "trunk" of 19 segments bearing parapodia and gills, and by a posterior "tail" of small segments lacking appendages.

The polychaete integument is composed of a single layer of columnar epithelium, which is covered by a thin layer of cuticle. Sometimes the cuticle is striated, as in *Diopatra*, producing iridescent purple and green colors. A number of polychaetes, such as syllids, scale worms, and some chaetopterids, are luminescent. The luminescent material is secreted by certain of the epithelial gland cells. Luminescent mucus of *Chaetopterus* often disperses into the water, producing a cloudlike luminescence. The significance of luminescence in most species is still not understood.

Beneath the epithelium lie in order a thin layer of connective tissue, a layer of circular muscle fibers, a much thicker layer of longitudinal muscle fibers and a thin layer of peritoneum (Figs. 10–2; 10–4*A*; and 10–8*B*). Although the muscles of the body wall essentially comprise two sheaths, the longitudinal fibers typically are broken up into four bundles—two dorsolateral and two ventrolateral. Oblique muscles are commonly present (Fig. 10–8*B*); they consist of strands of muscle fibers that extend from the midventral line to the midlateral line on each side of the worm where they join the circular layer and are concentrated primarily at the level of the parapodia. The body wall muscles of annelids are obliquely striated, as was described for nematodes.

The spacious coelom is compartmented by transverse septa, each of which is composed of a double layer of peritoneum (Fig. 10–2). The septa are penetrated by the gut, suspended dorsally and ventrally by longitudinal mesenteries. Thus each coelomic compartment is divided into right and left halves. Actually, however, such perfect compartmentation is rarely encountered, since the mesenteries are almost never complete and the septa have partially or completely disappeared in many polychaetes.

Adaptive Diversity in Polychaetes. If the first annelids were burrowers in sand and

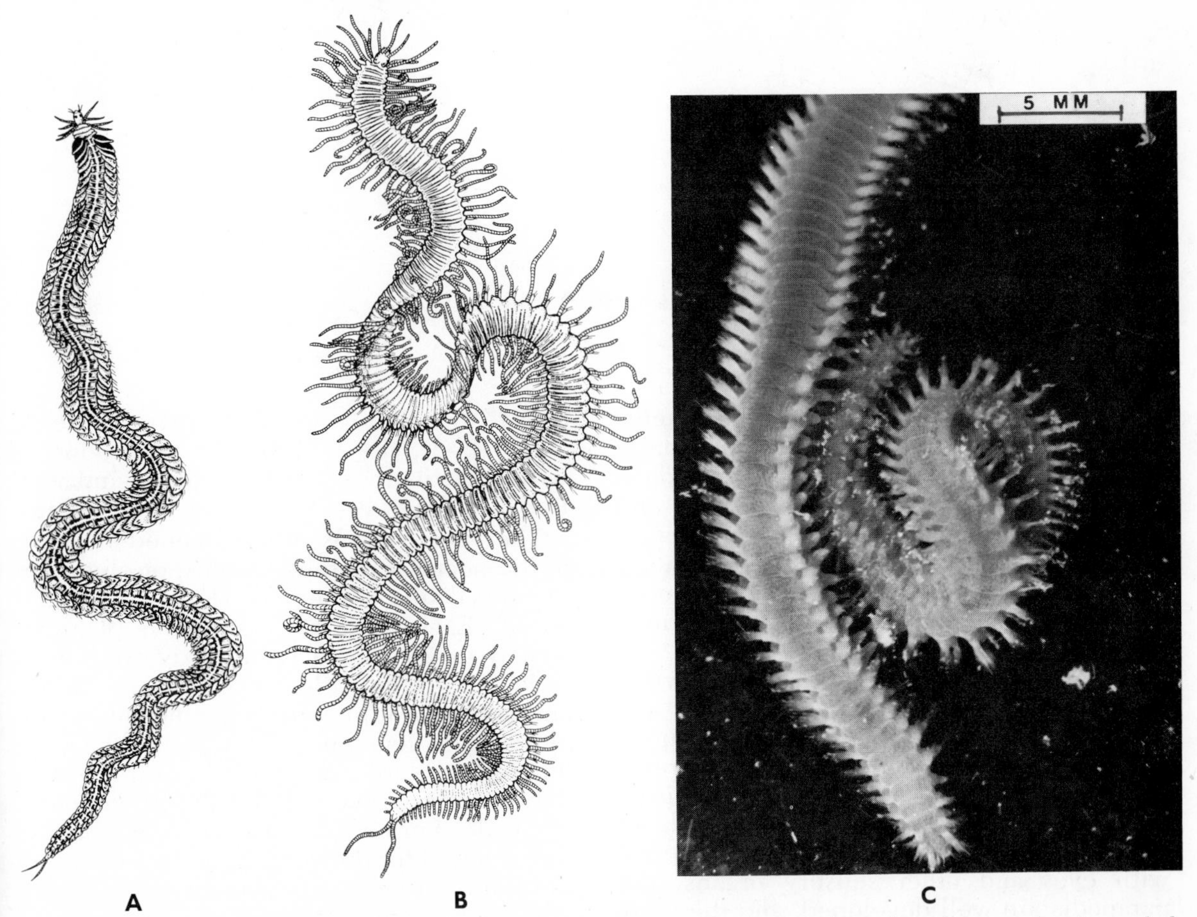

Figure 10–7. Three surface-dwelling polychaetes. All have well-developed heads and parapodia. *A*. A phyllodocid, *Phyllodoce maculata*, with large flattened dorsal cirri. *B*. A syllid, *Trypanosyllis zebra*, with long tentacular dorsal cirri. (*A* and *B* after MacIntosh.) *C*. An amphinomid, a fireworm. Note the bundles of hollow calcareous setae, which contain an irritant. This West Indian species is very common beneath stones. (By Betty M. Barnes.)

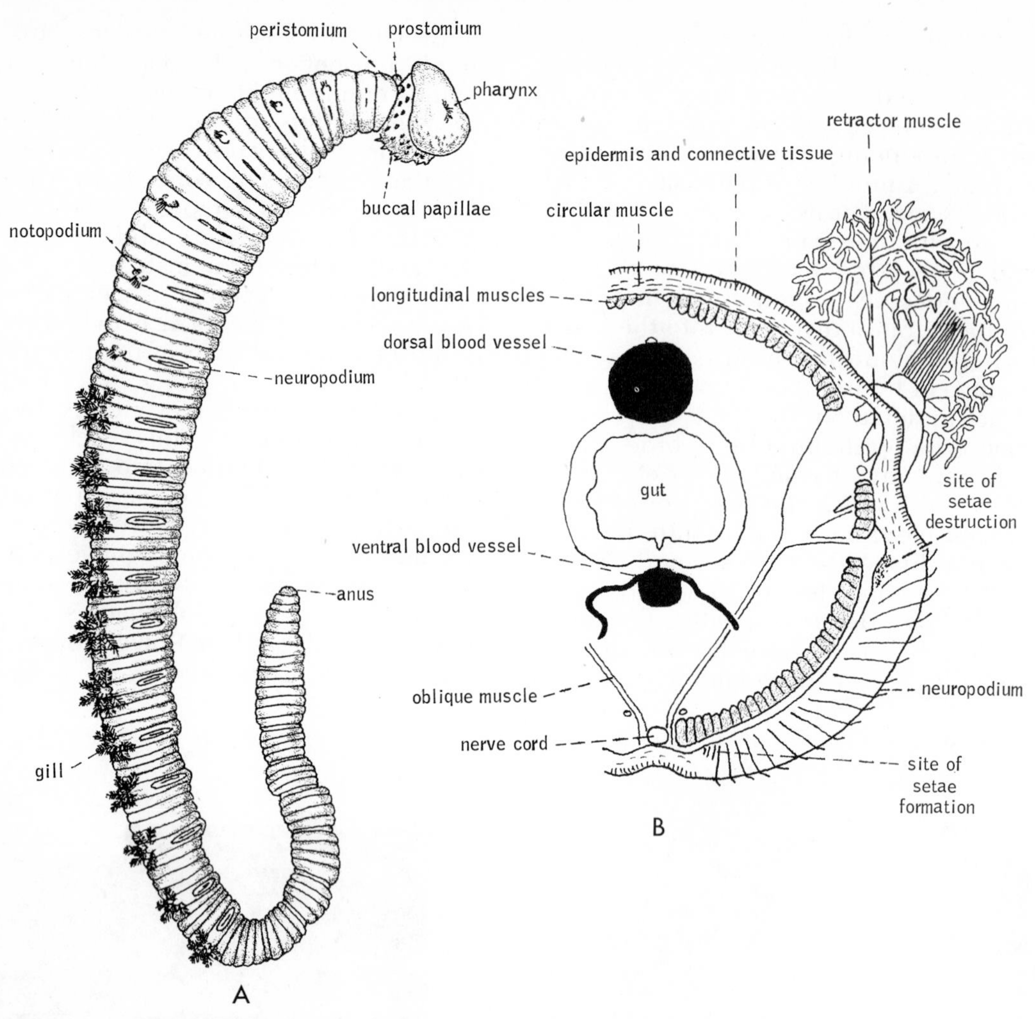

Figure 10–8. *A. Arenicola* (lateral view). (After Brown.) *B.* A setigerous segment of *Arenicola marina* (transverse section). (Modified from Wells.)

mud of shallow water, then the polychaetes constitute a line which may have originally diverged in the exploitation of a surface existence. The well-developed heads and parapodia of errant groups would therefore constitute primitive features. The sedentary burrowers and tube dwellers would be more specialized and reflect secondary adaptations for later modes of existence.

Crawling Polychaetes. The crawling polychaetes, which live beneath stones and shells, in coral crevices, in algae, and among hydroids, bryozoans, and other sessile organisms, most nearly approach the "typical" generalized body form just described. In these worms the prostomium is equipped with eyes and other sensory organs, the parapodia are well developed, and the body segments are generally similar. In this category belong members of a number of polychaete families, including many nereids (Fig. 10–3), the syllids, the phyllodocids with their flattened leaflike dorsal and ventral cirri (Fig. 10–7), and the scale worms.

The scale worms are so named because of the peculiar platelike scales, or elytra, on the dorsal side of the body (Figs. 10–6*B* and 10–9). Each scale is a flattened plate attached to the dorsolateral body wall by a little stalk (Fig. 10–10). They are paired and may appear on most segments or only on alternating segments; they may be overlapping and large enough to cover the entire dorsal surface. As will be described later, the scales perhaps serve to provide a protective channel for the ventilating current when the worms are wedged beneath stones and in other tight-fitting places.

In *Polynoe* and other genera of scale worms, the elytra can be shed when the worm is irritated and later new elytra can be regenerated. On the other hand, the scales of *Iphione* are firmly attached and so sclerotized that they serve as a protective coat of armor over the back of the worm. Some scaleworms possess luminescent glands on the surface of the elytra. The scales may luminesce one after another or even be detached, radiating blue or green light. In the scale worm *Aphrodita* (Fig. 10–11*A*) (called a sea mouse) the entire dorsal surface, including the elytra, is covered by hairlike "felt," composed of threadlike setae that arise from the notopodia and trail back over the dorsal surface of the animal (Fig. 10–9*B*).

In crawling species, the longitudinal muscle layer is better developed than the circular layer, which may even be absent, and the septa tend to be incomplete.

Movement is brought about by the combined action of the parapodia, the body wall musculature, and to some extent the coelomic

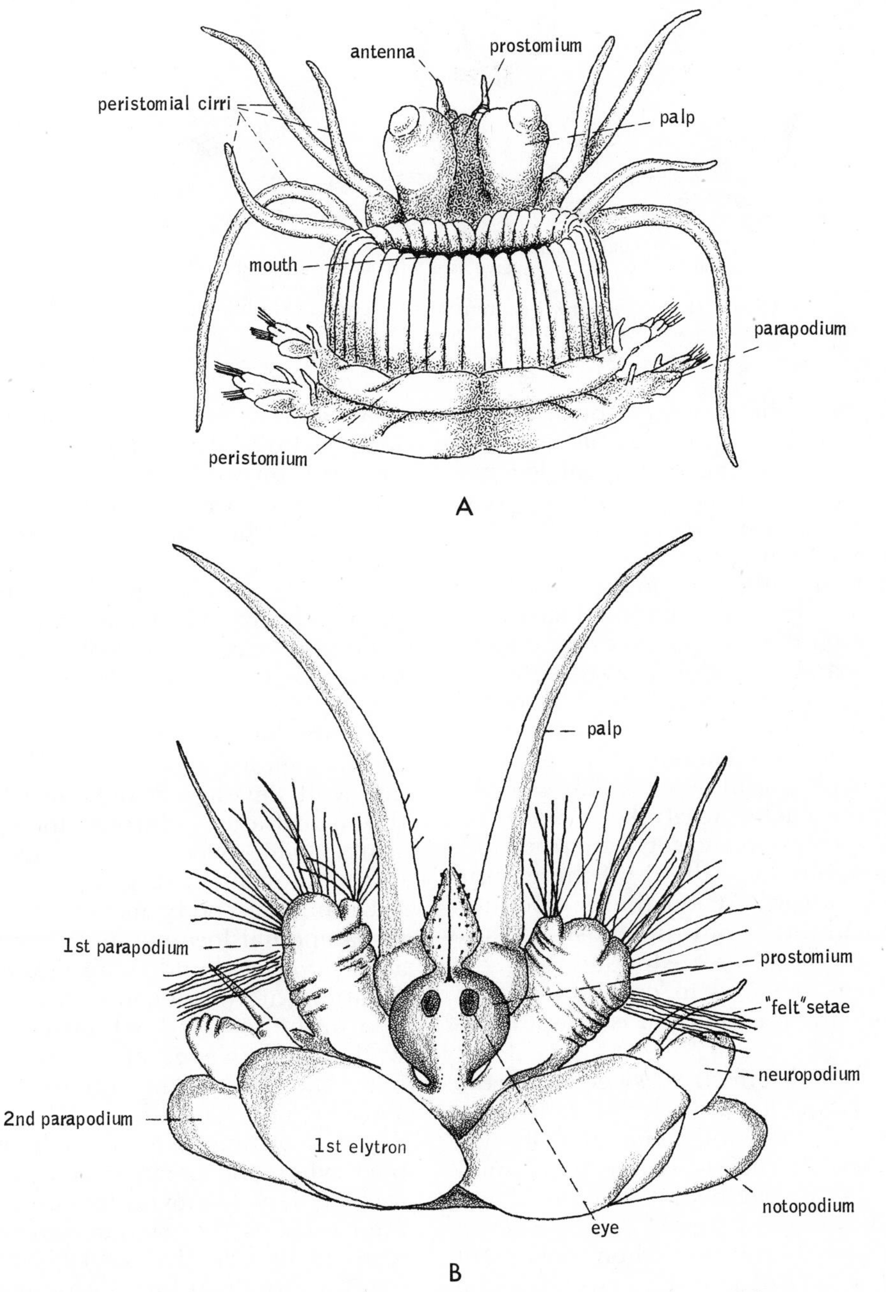

Figure 10–9. *A*. Anterior of *Nereis* (ventral view). (After Snodgrass.) *B*. Anterior of the sea mouse, *Aphrodita aculeata* (dorsal view). (After Fordham.)

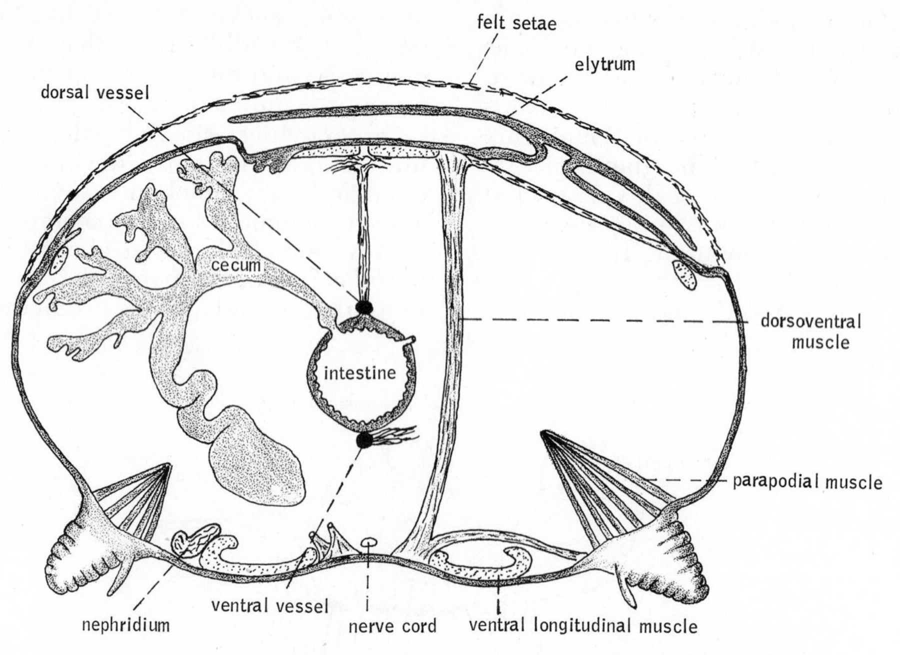

Figure 10–10. *Aphrodita* (transverse section). (Modified after Fordham.)

fluid. The parapodia and setae push against the substratum but rarely act as paddles. Gray (1939) studied the pattern of locomotion in *Nereis,* and his observations apply to many other errant polychaetes. The slow crawling movement of *Nereis* results entirely from the action of the parapodia. Parapodial movement involves an effective backward stroke in which the parapodia are in contact with the substratum and a forward stroke in which the parapodia are lifted from the ground. Each parapodium describes an ellipse each time it completes one of these two-stroke cycles. Following a pause in the cycle after the backward stroke of the parapodium, the aciculum and setae are retracted, and the parapodium is lifted off the ground and moved forward. When the parapodium reaches an anterior oblique position, the aciculum and setae are extended, and the parapodium is lowered to come in contact with the substratum and then is swept backward.

The combined effective sweeps of the numerous parapodia propel the worm forward. The right and left parapodia of each segment work alternately rather than simultaneously, and parapodial action takes place in successive waves along the side of the worm. Gray found that when movement is initiated, a wave of activation sweeps along the worm, affecting every fourth to eighth parapodium.

As soon as one parapodium begins its effective stroke, the next preceding parapodium sweeps forward and starts its effective stroke. The waves of activation thus move forward rather than backward. Activation takes place on both sides of the body, but the waves alternate with each other; when a particular group of parapodia on one side is executing an effective stroke, the parapodia on the opposite side are executing the recovery stroke.

Parapodial stepping is highly developed in the sea mouse *Aphrodita.* The neuropodia are well developed and raise the body off the substratum during their propulsive stroke. The body shape does not change during movement (Fig. 10–12).

Undulatory body movements in addition to parapodial locomotion give the worm the ability to crawl and swim rapidly. Body undulations are produced by waves of contraction in the longitudinal muscles of the body wall. These waves of contraction coincide with the alternating waves of parapodial activity just described. The longitudinal muscles on one side of each segment contract when the parapodium on that side of the segment is moved forward; the muscles then relax as the parapodium sweeps backward in its effective stroke. As Gray points out, the principal force of propulsion in this type of movement comes, not from parapodial movement, but rather from body contractions pulling against points of contact made by the parapodia.

Pelagic Polychaetes. Six families of polychaetes contain exclusively planktonic or pelagic species (Dales and Peter, 1972). They are rather like crawling forms but tend to be transparent, as are many other planktonic animals. Their swimming movements are similar to those just described for crawling species. The Alciopidae, which have enormous eyes (Fig. 10–37*B*), and the Tomopteridae, which have lost the setae and possess membranous parapodial pinnules, are the most highly specialized of the pelagic families (Fig. 10–13).

Burrowers. Many polychaetes have become adapted for burrowing. The more active burrowers include species of glycerids,

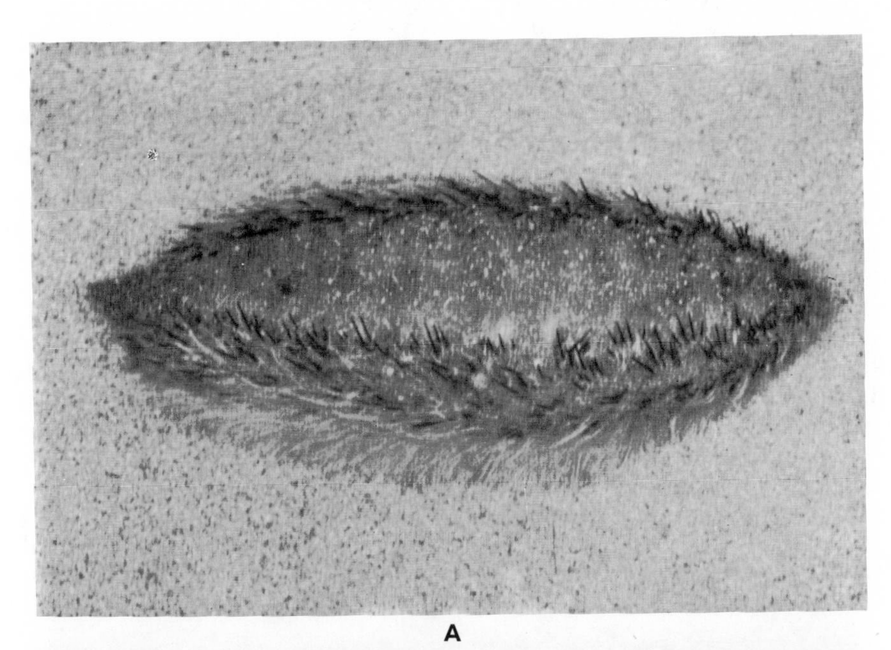

A

B

Figure 10–11. *A*. The sea mouse, *Aphrodita aculeata* (dorsal view). The wide uniform dorsal strip is covered by felt setae. *B*. The fan or peacock worm, *Sabella pavonina*, showing the expanded radioles projecting from the aperture of the tubes. (Both courtesy of D. P. Wilson.)

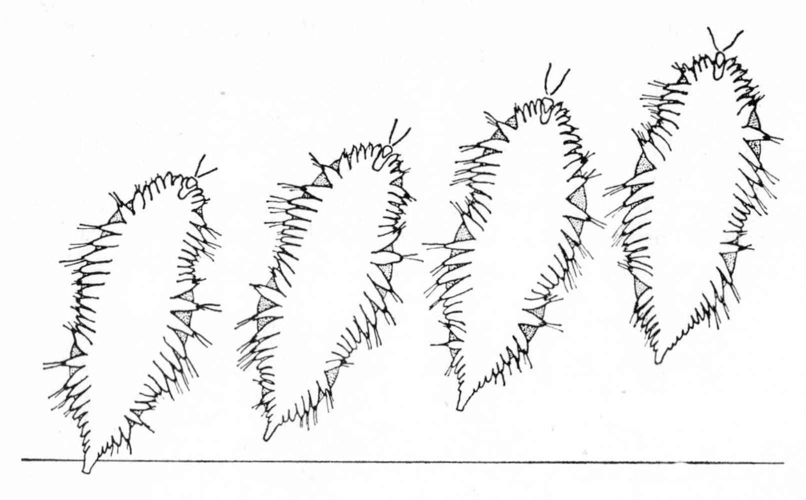

Figure 10–12. Parapodial movement in polychaete *Aphrodita* (sea mouse) crawling. Waves of movement on one side of body alternate with the opposite side. (After Mettam.)

arabellids, lumbrinerids, opheliids, and capitellids. Most of them construct a system of mucus-lined galleries, within which they move about (Figs. 10–14 and 10–15). The adaptations of many of these burrowers remarkably parallel those of the earthworms among the oligochaetes. The prostomium is reduced and pointed, and eyes, palps, and antennae are usually absent (Figs. 10–14*A* and *C*, and 10–15). Parapodia tend to be smaller than those of crawling surface dwellers.

Like earthworms, active burrowers such as the gallery-dwelling lumbrinerids and capitellids move through the substratum by peristaltic contractions. The circular muscle layer of the body wall is well developed, and the septa effectively compartment the coelomic fluid and localize its skeletal function.

Many polychaetes occupy more or less fixed simple burrows excavated in the substratum. These sedentary burrowers include members of such families as the Arenicolidae (the lugworms, Fig. 10–16), the Terebellidae (Fig. 10–17), and the Cirratulidae (Fig. 10–6*A*). The prostomial sensory appendages

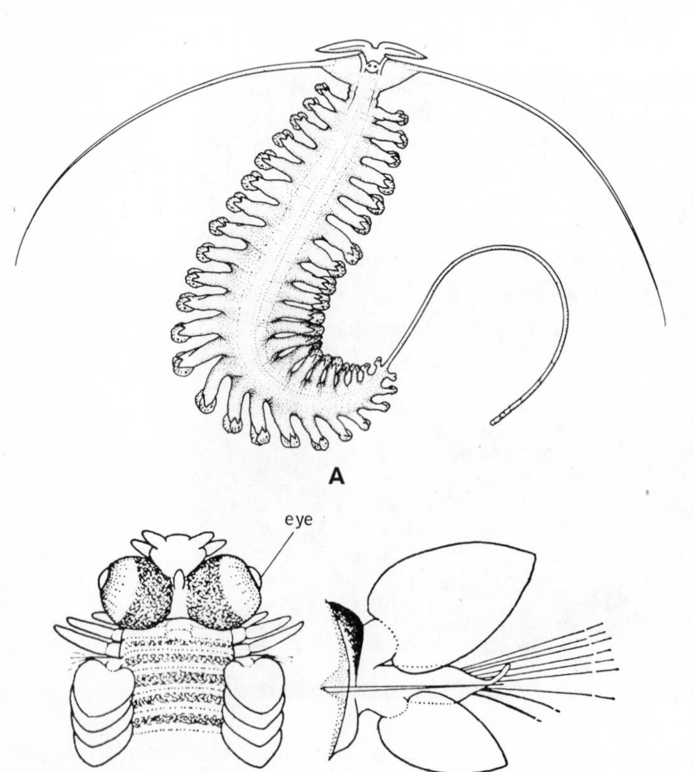

Figure 10–13. Pelagic polychaetes. *A*. *Tomopteris renata*, a tomopterid. *B*. Dorsal view of head of *Rhynchonerella angelina*, an alciopid. *C*. Parapodium of *Rhynchonerella*. (From Day, J., 1967: Polychaeta of South Africa. British Museum, London.)

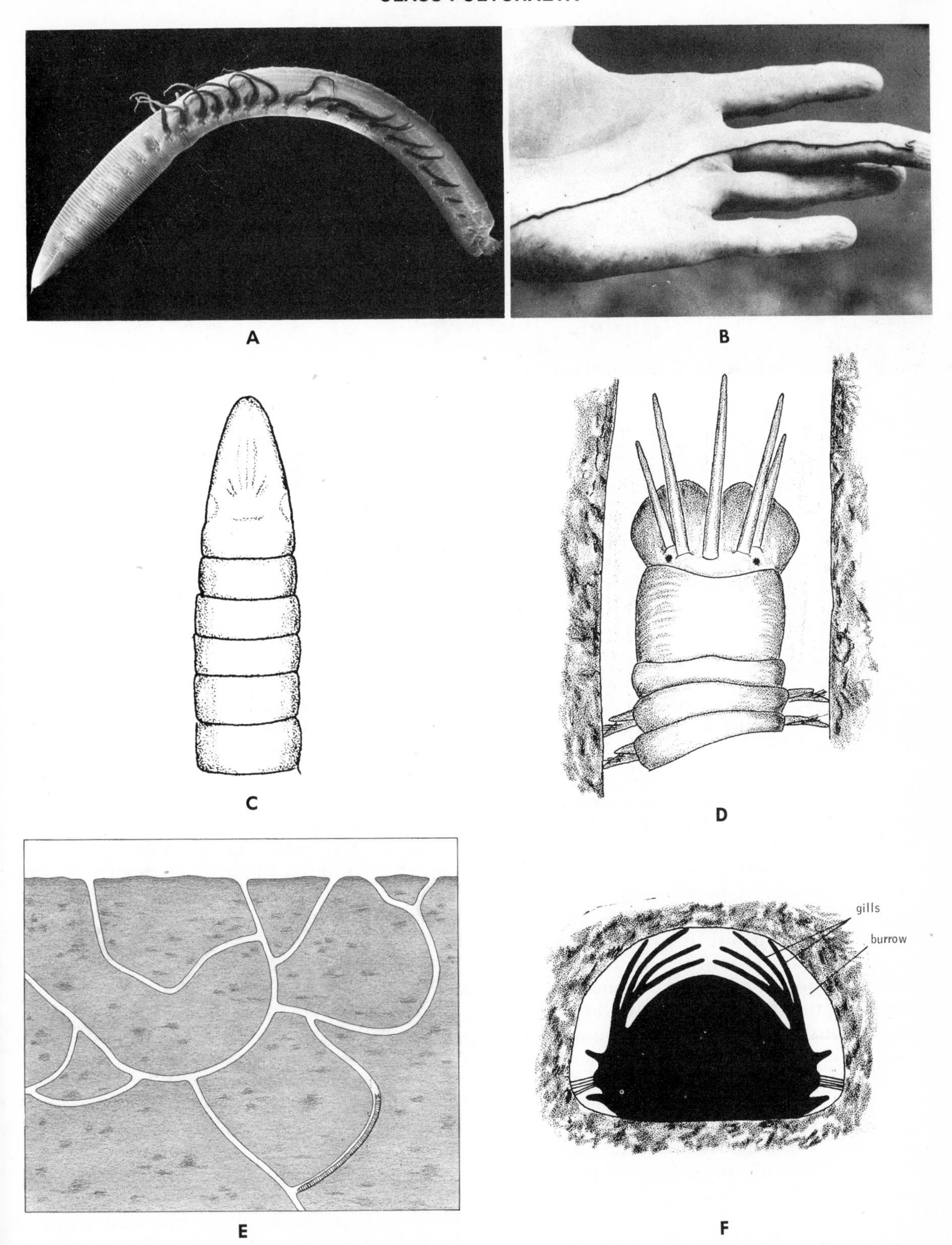

Figure 10–14. Burrowing polychaetes. *A.* Lateral view of *Ophelia denticulata.* Note the small pointed prostomium. The long projections are gills. (Courtesy of C. R. Gilmore.) *B.* The threadworm *Drilonereis,* the burrows of which are similar to those of *Marphysa* shown in *E.* (By Betty M. Barnes.) *C.* Dorsal view of anterior end of *Drilonereis. D.* Dorsal view of anterior end of *Marphysa sanguinea,* a common gallery dweller that has retained eyes and antennae. The palps are large and shovel-like. *E.* Burrow system of *Marphysa sanguinea,* with worm in one of the galleries. The burrow system is very irregular and possesses multiple openings to surface. *F.* Cross section of *Marphysa* in burrow. Dorsal projections are gills.

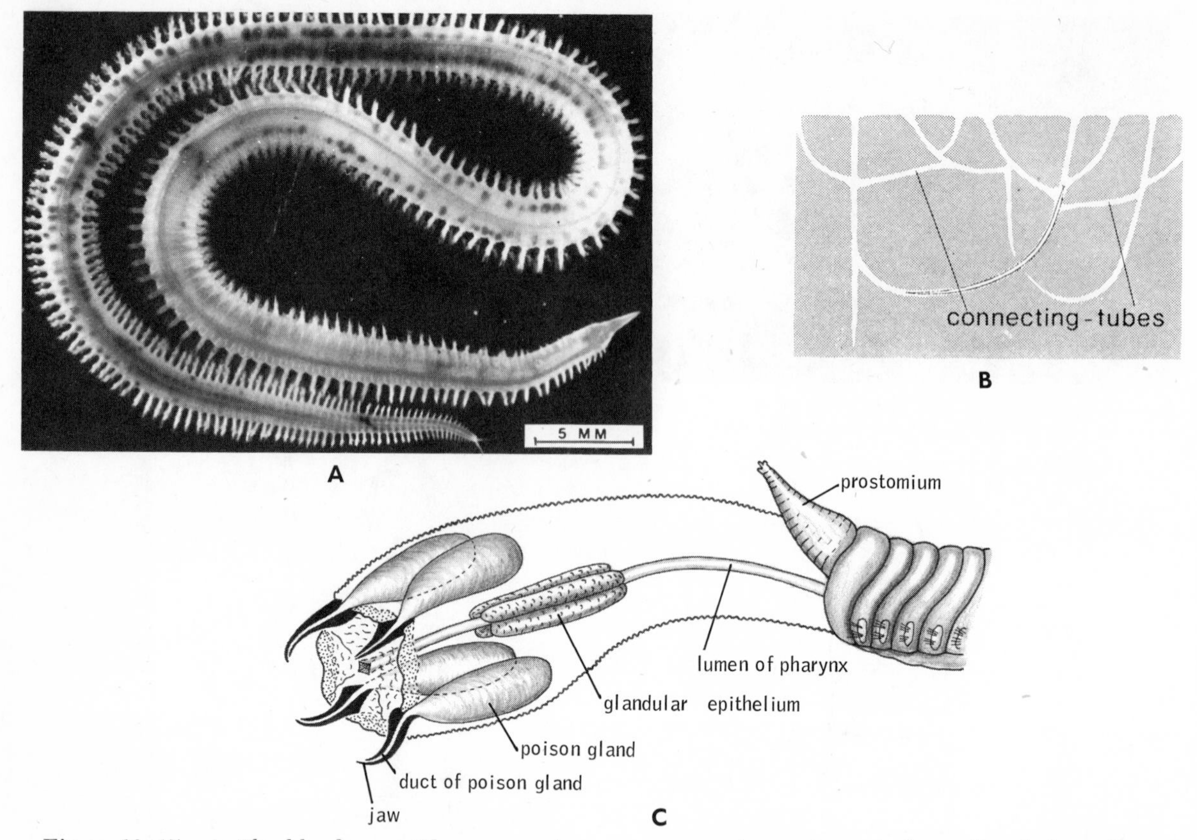

Figure 10–15. *A.* The bloodworm, *Glycera americana*, a common burrowing polychaete found along the east coast of the United States. Note the pointed prostomium. (Courtesy of G. M. More.) *B.* Burrow system of *Glycera alba*, showing worm lying in wait for prey. (After Ockelmann and Vahl.) *C.* Structure of the proboscis of *Glycera*. (After Michel.)

are generally absent, but the head region may carry specialized feeding structures (Fig. 10–17). Movement through the burrow is usually by peristaltic contractions; the parapodia are greatly reduced, and are in part represented by transverse ridges provided with setae modified into hooks called uncini (Figs. 10–4*I* and *J*; and 10–8*B*). The trunk of sedentary burrowers may be differentiated into regions (Fig. 10–8*A*).

Tubicolous Polychaetes. A tube-dwelling habit has evolved in many families of polychaetes. The tube may serve the worm as a protective retreat or as a lair for catching passing prey. It may provide access to clean oxygenated water above a muddy or sandy bottom. It may permit the worm to inhabit hard, bare surfaces such as rock, shell, or coral.

One group of tubicolous worms, including some members of the families Eunicidae and Onuphidae, are typically carnivorous and extend from the opening of the tube to seize passing prey. They are not greatly different from the surface polychaetes. Prostomial sensory appendages are well developed and the parapodia, which are used in crawling through the tube, are not markedly reduced (Fig. 10–18*A*). The tube may be straight, planted vertically in the sand or mud, and composed of sand grains cemented together, or a secreted organic material, or a combination of secreted and foreign material. Tube secretions are produced by glands located on the ventral surfaces of the segments.

Species of the onuphid genus *Hyalinoecia* construct secreted hyaline quill-like tubes, which lie horizontally on the bottom and may be moved about by the worm. The onuphids *Diopatra* and *Onuphis* build heavy, conspicuous, membranous tubes that may occur in great numbers in intertidal areas. The projecting chimney of the tube is bent over and flares at the end like a ship's funnel (Fig. 10–18*B*). The chimneys are covered with bits of shell, seaweed, and other debris that the worm collects and places in position with its jaws.

The majority of tubicolous polychaetes belong to the Sedentaria and are highly modified for a tube-dwelling existence, with

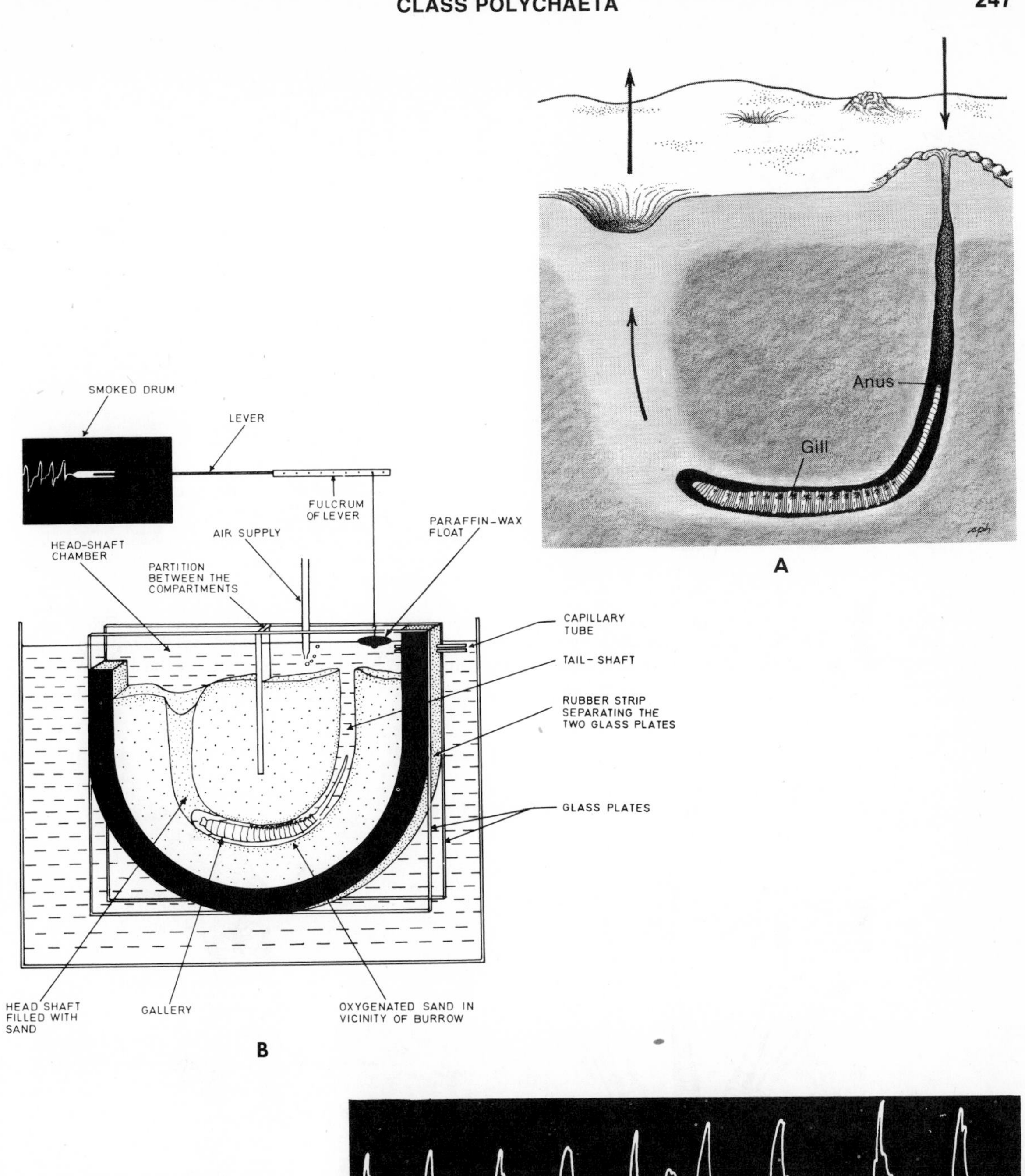

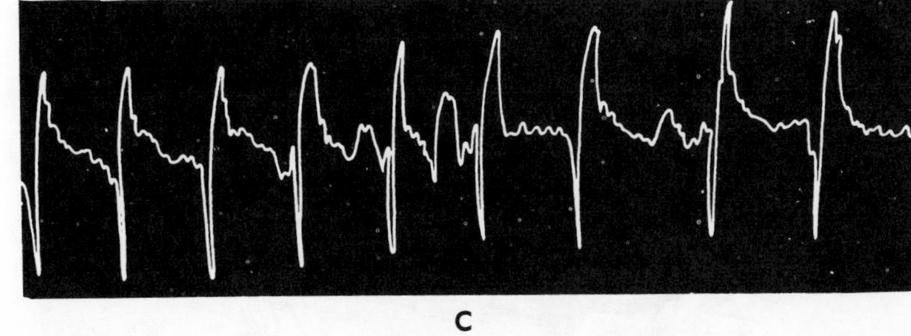

Figure 10–16. *A*. The lugworm, *Arenicola,* in burrow. Arrows indicate the direction of water flow produced by the worm. The worm ingests the column of sand on the left, through which water is filtered. Pile of sand at burrow opening is defecated castings. *B*. Apparatus devised for recording activity cycles of *Arenicola marina*. *C*. Kymograph tracing of activity cycles of *Arenicola* over a period of six hours. The downstroke reflects the worm backing up to the burrow opening to defecate; the sharp upstroke reflects the worm moving back down to the head of the burrow and vigorously resuming ventilation contractions and deposit feeding. Intervals between defecations are about 40 minutes. (*B* and *C* after Wells, 1959, from Newell, R. C., 1970: Biology of Intertidal Animals. American Elsevier Co., N.Y.)

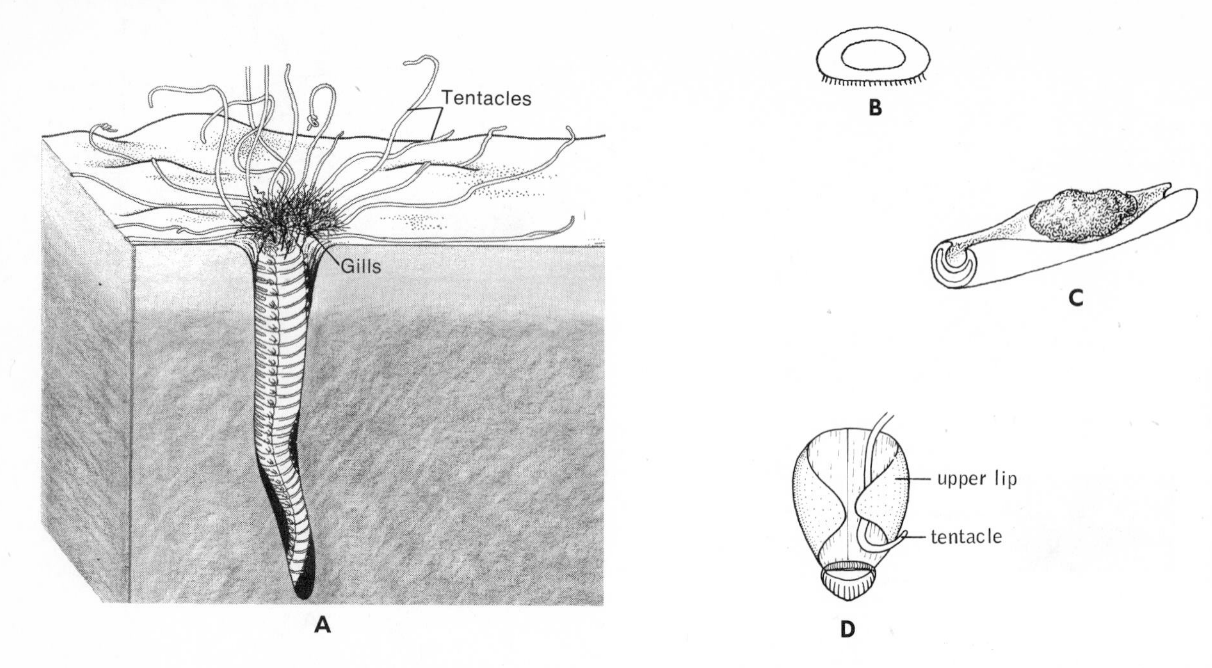

Figure 10–17. *A. Amphitrite* at aperture of burrow with tentacles outstretched over substratum. *B.* Cross section through tentacle of *Terebella lapidaria*, creeping over substratum. *C.* Section of tentacle of *Terebella lapidaria* rolled up to form a ciliary gutter, transporting deposit material. *D.* A tentacle being wiped by one of the lips. (*B, C,* and *D* after Dales.)

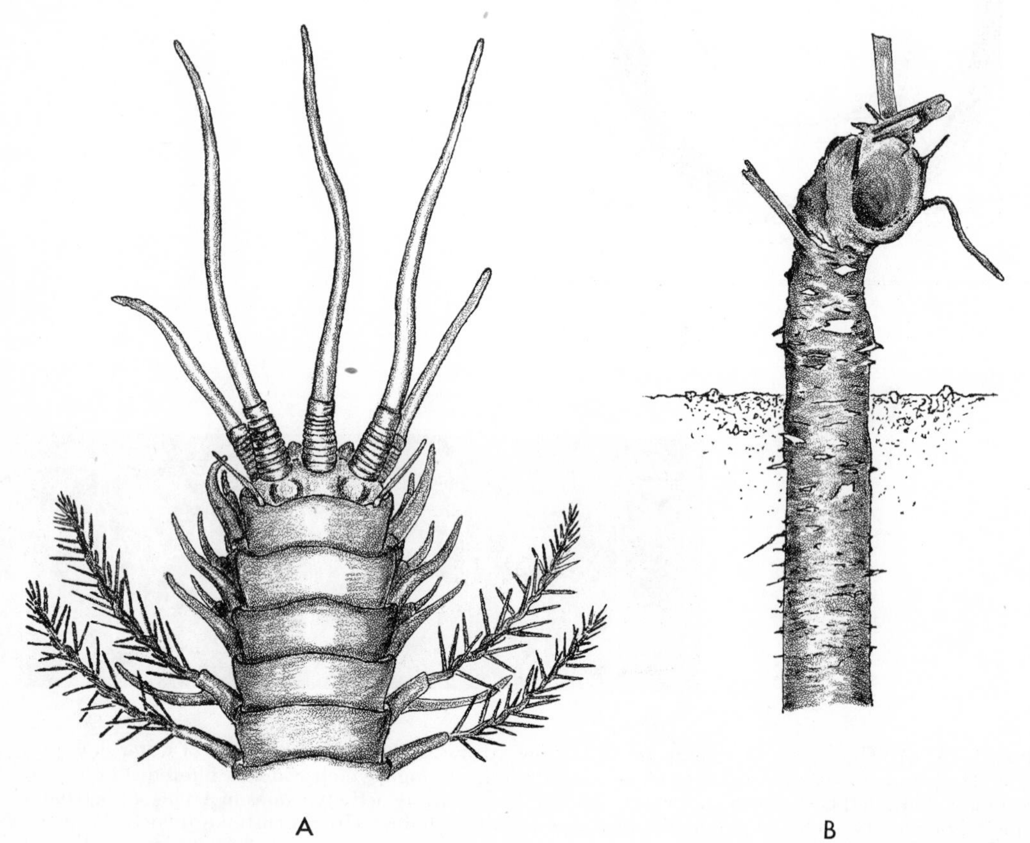

Figure 10–18. *A.* Head and first two gill-bearing segments of *Diopatra* (dorsal view). *B.* Funnel-shaped parchment tube of *Diopatra.*

adaptations similar to the more sedentary burrowers. Prostomial sensory appendages are reduced or absent and there are commonly special anterior feeding structures. The worms usually move within the tube by peristaltic contractions, and the parapodia are reduced and provided in part with uncinate setae for gripping the tube wall. The body segments are commonly differentiated into regions.

The sand-grain tubes of some members of the Maldanidae, called bamboo worms, are common in the intertidal zone. These worms, which are oriented anterior end downward in the tube, have a truncate head and the parapodia are reduced to ridges that have the appearance of cane joints (Fig. 10–19*A-C*).

Both the Sabellariidae and the Pectinariidae construct sand-grain tubes and have

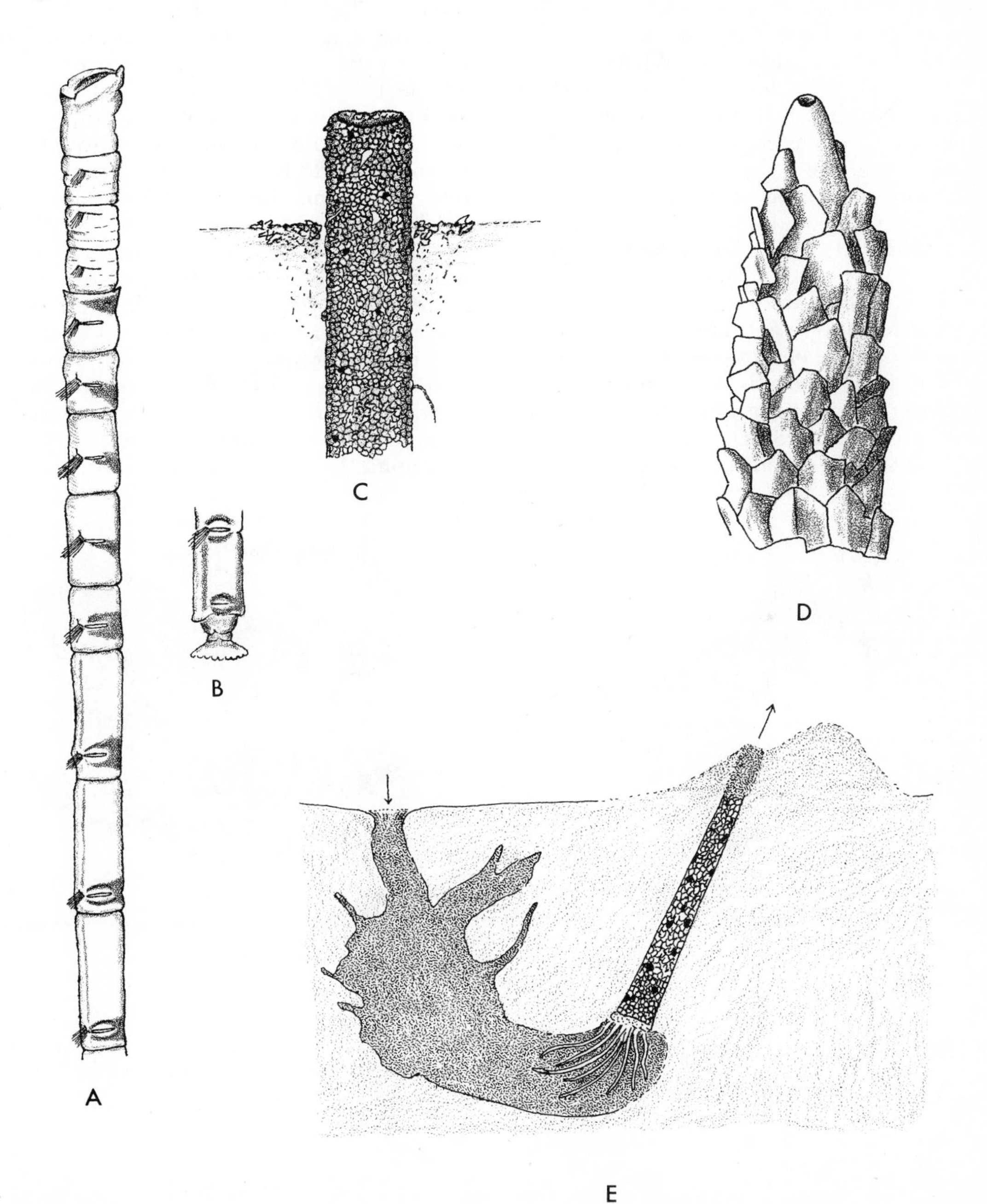

Figure 10–19. *A.* Anterior half of the bamboo worm, *Clymenella*. *B.* Posterior end of *Clymenella*. *C.* Exposed end of sand-grain tube of *Clymenella*. *D.* Anterior of *Owenia* sand-grain tube, with overlapping sand grains. The membranous secreted part of the tube projects at the tip. (After Watson.) *E. Pectinaria* and sand-grain tube in normal, buried position. Arrows indicate path of water current. (After Wilcke from Kaestner.)

highly modified heads bearing heavy conspicuous setae. In the Sabellariidae two fused segments have grown forward, and dorsally form an operculum for blocking the tube entrance (Fig. 10–20*A*). Species of *Sabellaria* and related genera build their tubes on top of each other, creating honeycomb-like aggregations (Fig. 10–20). Such colonies may be composed of millions of individuals and assume reef-like proportions. An enormous amount of shell fragments and sand grains may be locked up and stabilized by these worms. Studies by Wilson (1971) on *Sabellaria alveolata* on the coast of England have shown that a colony is continually adding new members through settlement of larvae.

The sand-grain tube of the pectinariids is conical, with the smaller end opening at the surface (Fig. 10–19*E*). The heads of these worms bear rows of large conspicuous setae that are used in digging.

Owenia of the Oweniidae carries its tube about and uses the chimney end like a screw. The tube is composed of an inner membranous secreted lining and an outer layer of sand grains. Flexibility is attained by the use of flat sand grains, attached at one edge and overlapping adjacent grains (Figs. 10–19*D* and 10–21*B*). The outer surface sand resembles tile roofing, with the free edges of the grains projecting upward. Each of the first seven segments of the body contains a pair of long glands that lie in the lateral portions of the coelom. A duct extends from each gland through the body wall and opens in the vicinity of the setal bundles (distinct parapodial lobes are absent). The glands produce a viscous material that forms the membranous lining of the tube, shaped by the worm revolving in the tube; the setae or the gland-bearing segments act as brushes for applying the secretion to the wall as it flows from the duct openings. Additions to the outer sand-grain and inner membranous layers are accomplished by separate operations. Sand grains are collected during feeding, and those grains suitable for tube construction are stored in a ventral pouch located beneath the mouth. During construction of the tube, the pouch projects outward and downward and fastens a grain to the margin of the tube (Fig. 10–21*B*). The pouch, apparently, also secretes the cementing material.

Among the most beautiful of the sedentary

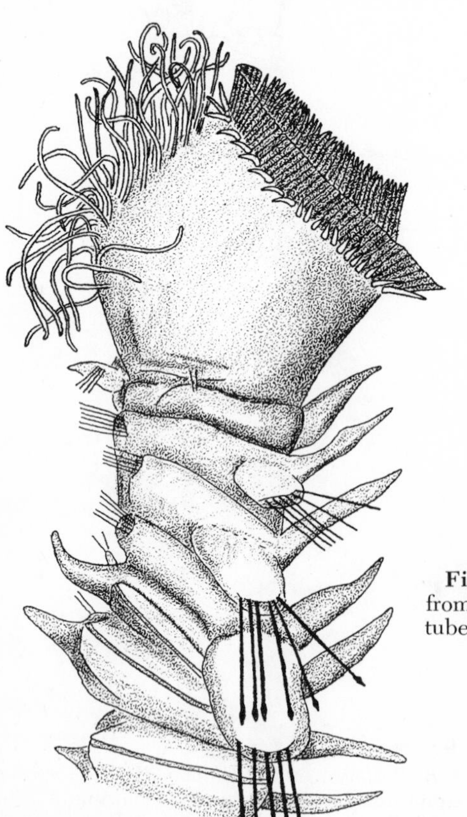

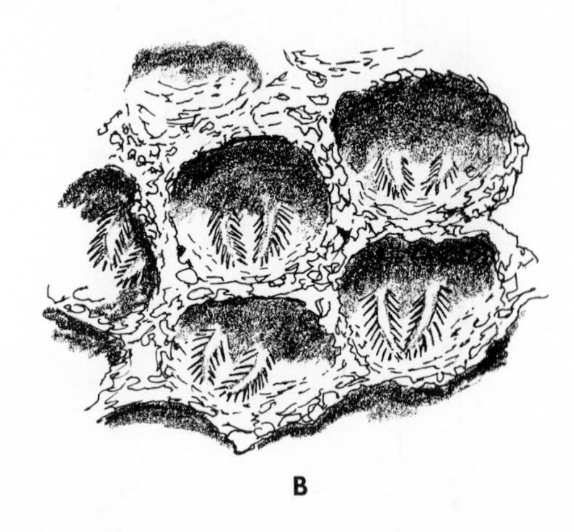

Figure 10–20. *A*. Anterior of *Sabellaria* (lateral view). (After Ebling from Kaestner.) *B*. Anterior ends of *Sabellaria* projecting from sand-grain tubes.

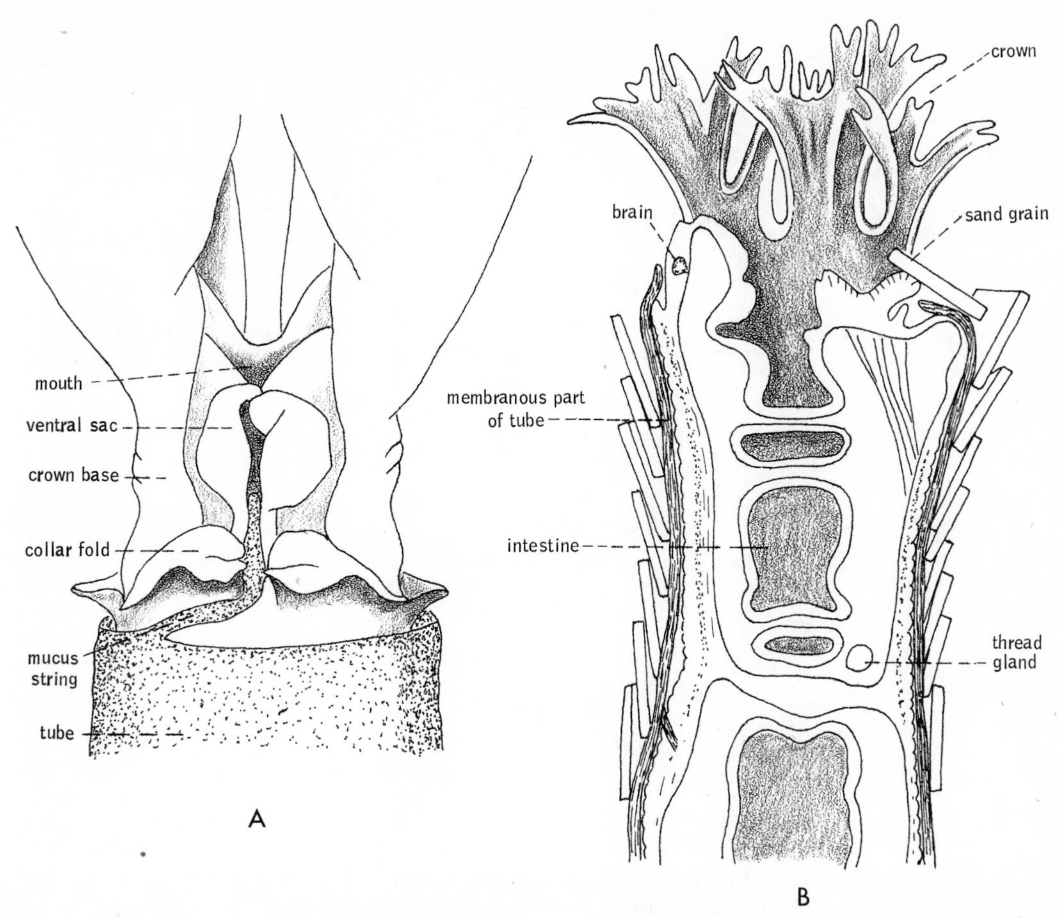

Figure 10–21. *A*. The fanworm, *Sabella pavonina*, in process of tube construction (ventral view). (After Nicol.) *B*. *Owenia*, adding a sand grain to tube (longitudinal section). (After Watson.)

polychaetes are the fanworms, or feather dusters, which comprise the families Sabellidae and Serpulidae. In all of the many common members of this group—*Sabella* (Fig. 10–11*B*), *Potamilla, Hydroides, Serpula*, and *Spirorbis* (Fig. 10–22)—the prostomium has developed to form a funnel-shaped or spiral crown consisting of few to many pinnate processes called radioles. The radioles are rolled up or closed together when the worm withdraws its anterior into the free end of the tube. Sabellids build membranous or sand-grain tubes. Serpulids secrete calcareous tubes which are attached to rocks, shells, or algae, enabling these worms to live on an otherwise inhospitable hard substratum. The most dorsal radiole on one or both sides of a serpulid is modified into a long stalked knob called an operculum (Fig. 10–22), which acts as a protective plug at the end of the tube when the crown is withdrawn.

In both sabellids and serpulids, the peristomium is folded back to form a very distinct collar, which fits over the tube opening and is the principal structure used to mold additions on the end of the tube. In the serpulids, two large calcium carbonate secreting glands open beneath the collar folds. When additions are made to the tube, secretions flow out between the collar and the body wall. This space then acts as a mold in which the secretion hardens and is simultaneously fused as a new ring on the end of the tube.

Nicol (1930) has given an excellent account of the process in *Sabella*, which constructs a tube of sand grains imbedded in mucus. The worms sort detritus collected by the ciliated radioles, and sand grains of suitable size for tube construction are stored in a pair of opposing ventral sacs located below the mouth. The walls of the sacs produce mucus, which is mixed with the sand grains. When additions are to be made at the end of the tube (Fig. 10–21*A*), the ventral sacs deliver a ropelike string of mucus and sand grains to the collar folds below, which is

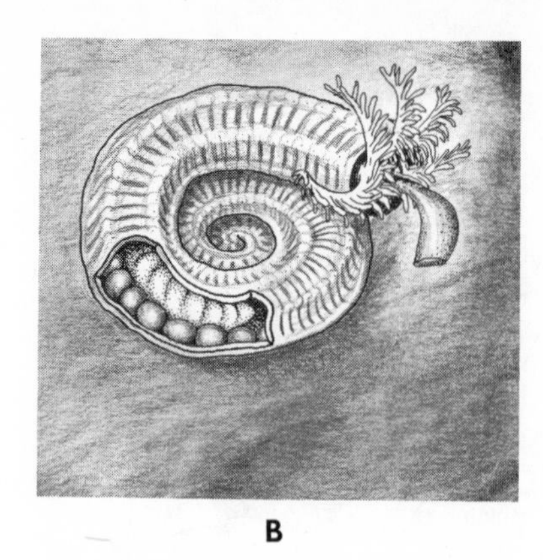

Figure 10–22. *A.* The serpulid, *Hydroides,* with radioles and operculum extended from end of calcareous tube attached to a rock. *B. Spirorbis,* a common serpulid with a snail-like tube found attached to a variety of substrates, including algae. Cutaway of shell shows eggs being brooded within tube.

divided midventrally, like the front of a shirt collar. The string of building materials is received at the collar folds. The worm rotates slowly in the tube, and the collar folds act like a pair of hands, molding and attaching the rope to the end of the tube. The operation is quite similar to an Indian method of making pottery.

In both sabellids and serpulids, the ventral surface of each segment bears a pair of large mucus-secreting pads, or glandular shields. When the worm rotates, these glands lay down a mucous coating on the inner surface of the tube. Ventral glandular shields are also found in the terebellids and are used to line the burrow, which is excavated in mud or sand (Fig. 10–23). Glandular shields are found in many other tube-building polychaetes and may be located on the dorsal surface instead of the ventral. In other polychaetes, the mucus used in tube construction is secreted by glands around the parapodia.

Boring Polychaetes. Representatives of a number of different polychaete families, including the Eunicidae, Spionidae, Terebellidae, and Sabellidae, excavate protective retreats within dead or living calcareous shells. Boring is begun by a newly settled young worm, but the mechanism is still unknown. Species of the spionid *Polydora* attack live oyster shells. The worm partially fills the excavation with debris, leaving room for a **U**-shaped burrow. When boring takes place between the mantle edge and shell, the oyster attempts to wall out the worm with new shell, creating unsightly "mud blisters" and reducing the market value of the mollusk.

Interstitial Polychaetes. The interstitial includes polychaetes of many different families (Fig. 10–24). All are minute and some, such as the archiannelids (see p. 284), are very aberrant.

Nutrition. The feeding methods of polychaetes are closely correlated with the various modes of existence the members of the class display.

Raptorial Feeders. Raptorial feeders include members of many families of crawling species, many pelagic groups, tubicolous eunicids and onuphids, and the glycerids and other gallery dwellers. The prey consists of various small invertebrates, including other polychaetes, which are captured by means of an eversible pharynx, or proboscis. The proboscis commonly bears two or more chitinous jaws (Fig. 10–26). The mouth lies beneath the prostomium and is bordered laterally and ventrally by the periostomium (Fig. 10–26A). The mouth opens into the proboscis, which is derived from the stomodeum and is thus lined with a cuticle similar to the integument.

The polychaete proboscis is rapidly everted; this places the jaws at the anterior of the body and causes them to open. The food is seized by the jaws and the pharynx is retracted. Although protractor muscles may be present, an increase in coelomic pressure resulting from the contraction of body-wall muscles is an important factor in the eversion of the proboscis. When pressure on the coelomic fluid is reduced, the proboscis is withdrawn by the retractor muscles, which extend from the body wall to the pharynx (Fig. 10–27A).

Raptorial tube-dwellers may leave the tube partially or completely when feeding, depending on the species. *Diopatra* uses its hood-shaped tube as a lair (Fig. 10–18*B*). Chemoreceptors monitor the ventilating current of water passing into the tube, and when approaching prey is detected, the

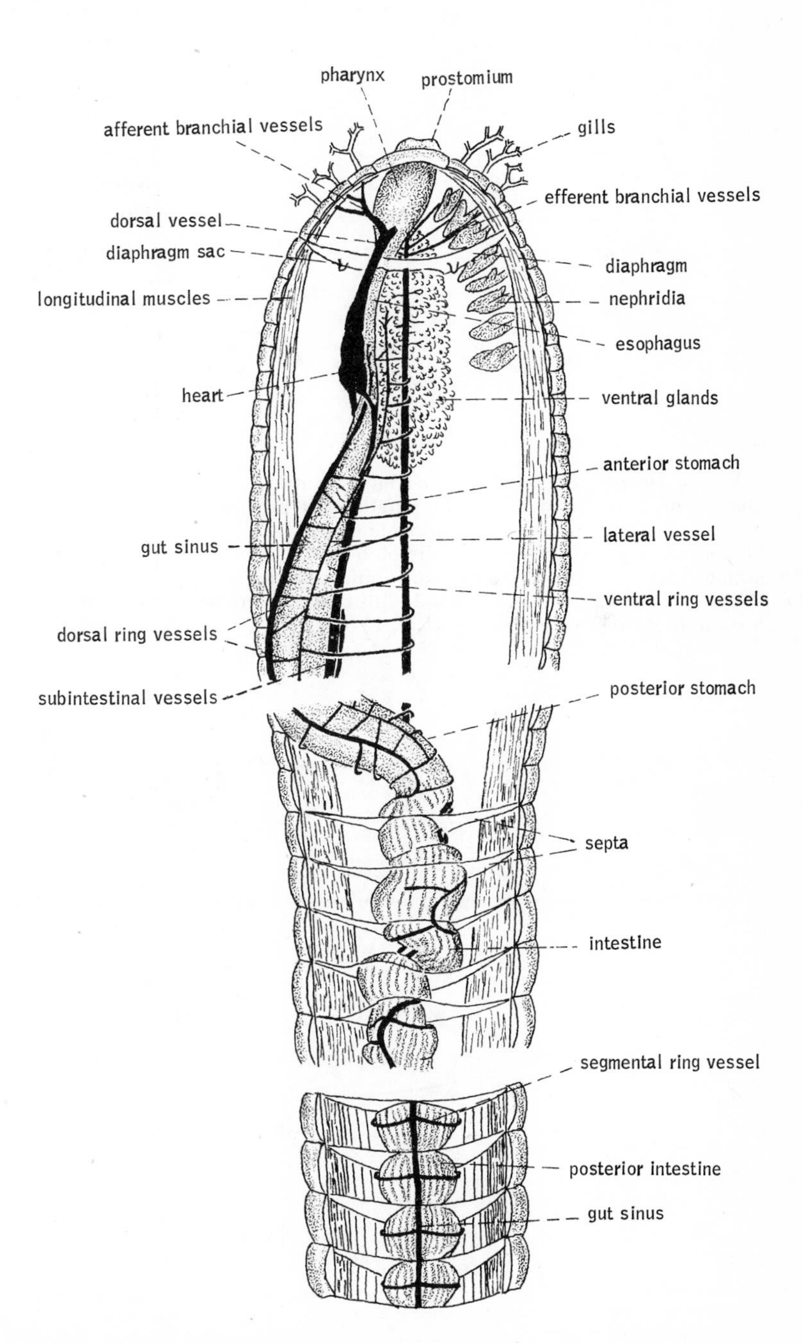

Figure 10–23. *Amphitrite* (dorsal dissection). (After Brown.)

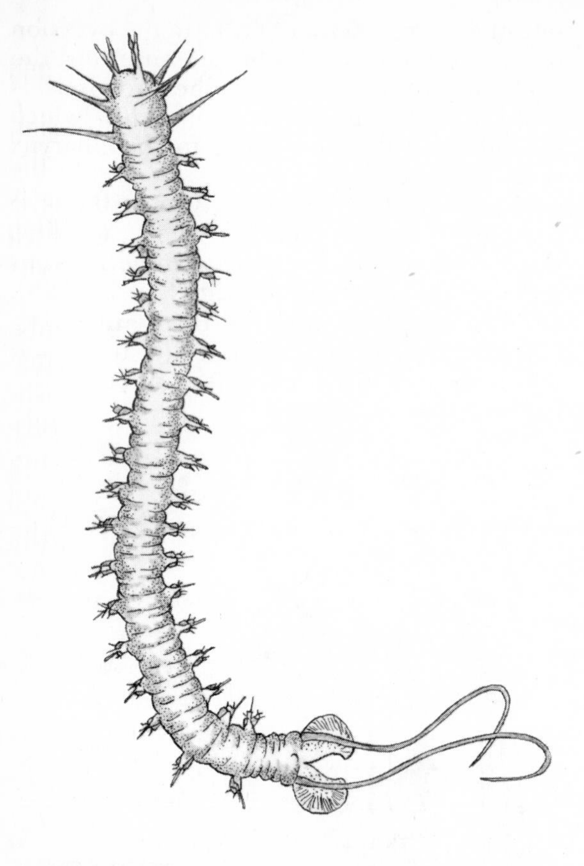

Figure 10–24. *Hesionides arenaria*, a minute polychaete, no more than 2 mm. in length, that inhabits interstitial spaces in the intertidal zone of surf beaches of north Europe. The ventrally directed parapodia are adopted for crawling between sand grains. (From Ax, P., 1966: Veroeffentlungen des Institut für Meeresforschung Bremerhaven. Suppl. II, p. 15–66.

worm partially emerges from the tube opening and seizes the victim with a complex pharyngeal armature of teeth. During feeding the prey may be clasped with the enlarged anterior parapodia.

Some raptorial feeders have a long tubular proboscis. *Glycera* lives within a gallery system constructed in muddy bottoms. The system contains numerous loops which open to the surface (Fig. 10–15). Lying in wait at the bottom of a loop, the worm can detect the surface movements of prey by changes in water pressure. It slowly moves to the burrow opening and then seizes the prey, such as small crustaceans and other invertebrates, with the proboscis.

The proboscis can be protruded to about one-fifth the length of the body (Fig. 10–15*C*); when it is retracted, it occupies approximately the first twenty body segments (Fig. 10–27*A*). At the back of the proboscis are located four jaws arranged equidistantly around the wall. The proboscis is attached to an **S**-shaped esophagus. No septa are present in these anterior segments, and the proboscis apparatus lies free in the coelom. Just prior to eversion of the proboscis the longitudinal muscles contract violently, sliding the proboscis forward and straightening out the esophagus. The proboscis is then everted with explosive force, and the four jaws emerge open at the tip (Fig. 10–15*C*). Each jaw contains a canal which delivers poison from a gland at the jaw base.

Autolytus, another polychaete with a tubu-

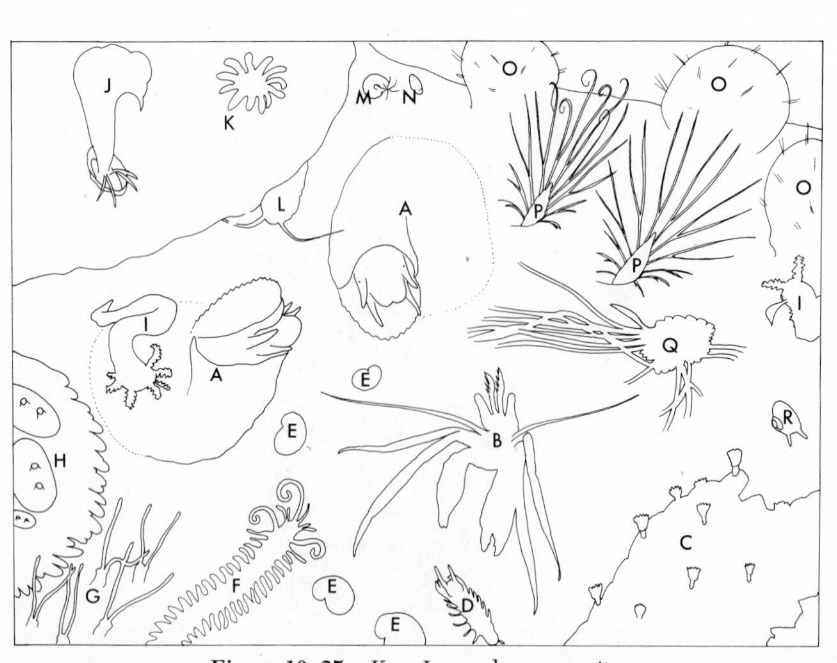

Figure 10–25. Key. Legend on opposite page.

A. Sessile vermetid gastropod, *Dendropoma*
B. Hermit crab, *Clibanarius tricolor*
C. Encrusting bryozoan
D. Tanaid crustacean
E. Young vermetid gastropods, *Dendropoma*
F. Syllid polychaete
G. Spionid polychaetes
H. Colonial tunicate which forms encrusting sheet
I. Serpulid polychaete
J. Boring acrothoracican barnacle, *Berndtia*
K. Scleractinian coral, *Porites astreoides* (Scale of drawing is smaller than other animals; polyps are at least 2 mm. across.)
L. Isopod crustacean
M. Serpulid polychaete
N. Ostracod crustacean
O. Sessile foraminiferan, *Homotrema*
P. Boring acrothoracican barnacle, *Berndtia*
Q. Tentacles of terebellid polychaete extending from burrow
R. Gastropod

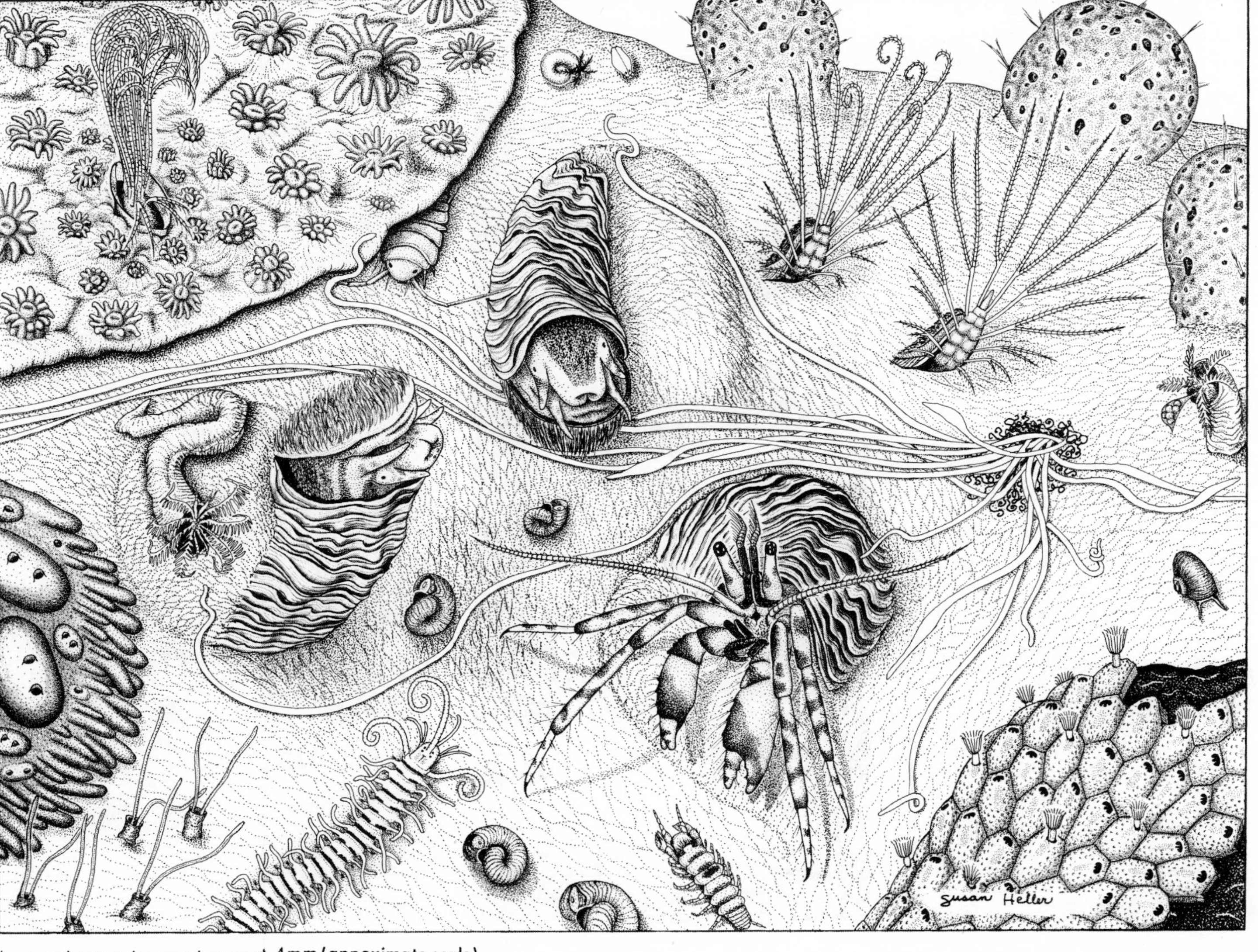

Figure 10–25. Marine invertebrates inhabiting a coralline stone. In tropical waters the surfaces of subtidal coralline rocks and old coral heads, especially the undersurfaces, are inhabited by a great diversity of small invertebrates. Some are sessile, some are borers, and some live in crevices and old excavations. This figure, drawn from living animals, illustrates such a fauna from stones in Bermuda. Although the figure is a composite of many sketches, the diversity, numbers, and proximity of animals is not exaggerated.

lar proboscis, measures about 2 cm. in length and lives free or in tubular retreats attached to the stalks of hydroids (Fig. 10–27*B*). They crawl about over the hydroid and feed upon the polyps. The proboscis of *Autolytus*, as well as those of some other syllids, is a tube that is free at its anterior and lies within a buccal cavity. The proboscis is lined by a thick layer of chitin. Posteriorly the proboscis opens into a bulb, the walls of which are composed of radial muscles that produce a sucking action. The bulb in turn opens into the intestine. During feeding, the proboscis is extended from the mouth, and the projecting crown of teeth is used to cut off tentacles and other parts of the polyps. Food (parts of the polyp) is then sucked in by the pumping action of the bulb. A closely related genus, *Procerastea*, is reported to insert its proboscis into the gastrovascular cavity of the polyps and suck out the contents.

Herbivores, Omnivores, Scavengers, and Browsers. Not all errant polychaetes that possess jaws are necessarily carnivores. A scavenger or omnivorous habit has evolved in many polychaetes. The jaws may, for example, be used to tear off pieces of algae. These polychaetes generally belong to the same families as do the carnivores, and they are similarly adapted.

Studies on *Nereis* by Goerke (1971) demonstrate the diversity of feeding habits that may exist even within a genus. The species of *Nereis* possess a muscular, eversible pharynx with a pair of heavy jaws (Fig. 10–26*B*). Some species of *Nereis*, such as *N. pelagica*, *N. virens*, and *N. diversicolor*,

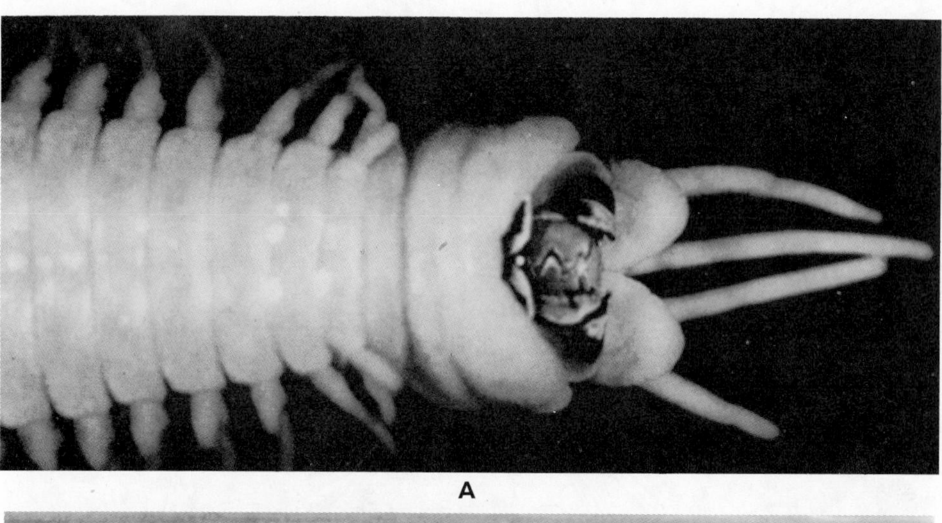

A

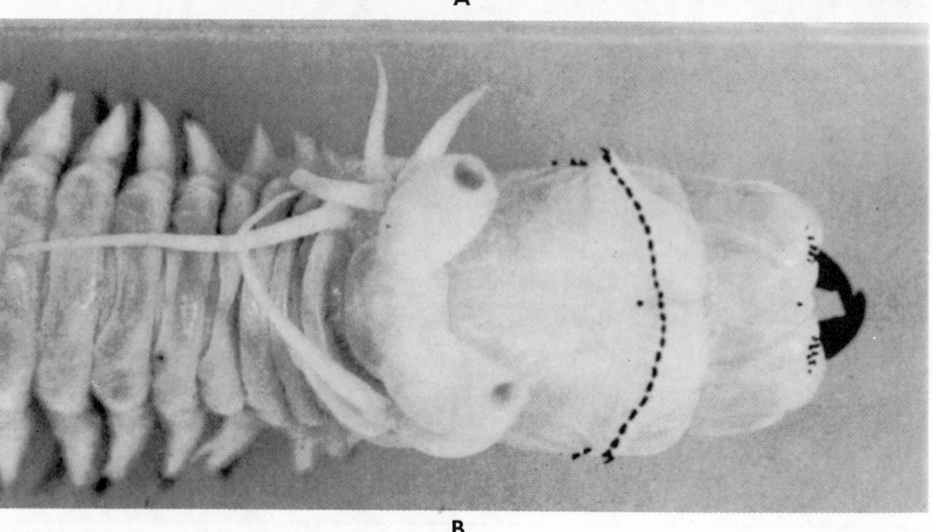

B

Figure 10–26. Polychaete jaws. *A*. Ventral view of anterior end of *Eunice*, showing protruded jaws. The complex buccal armature is composed of a number of pieces, of which two are especially large and opposing. *B*. Dorsal view of anterior end of *Perinereis nuntia*, with protruded everted pharynx. The pharynx is composed of two sections, one of which bears a circle of denticles and the other a pair of heavy chitinous jaws. (From Bennett, I., 1966: Fringe of the Sea. Rigby Ltd., Adelaide, Australia.)

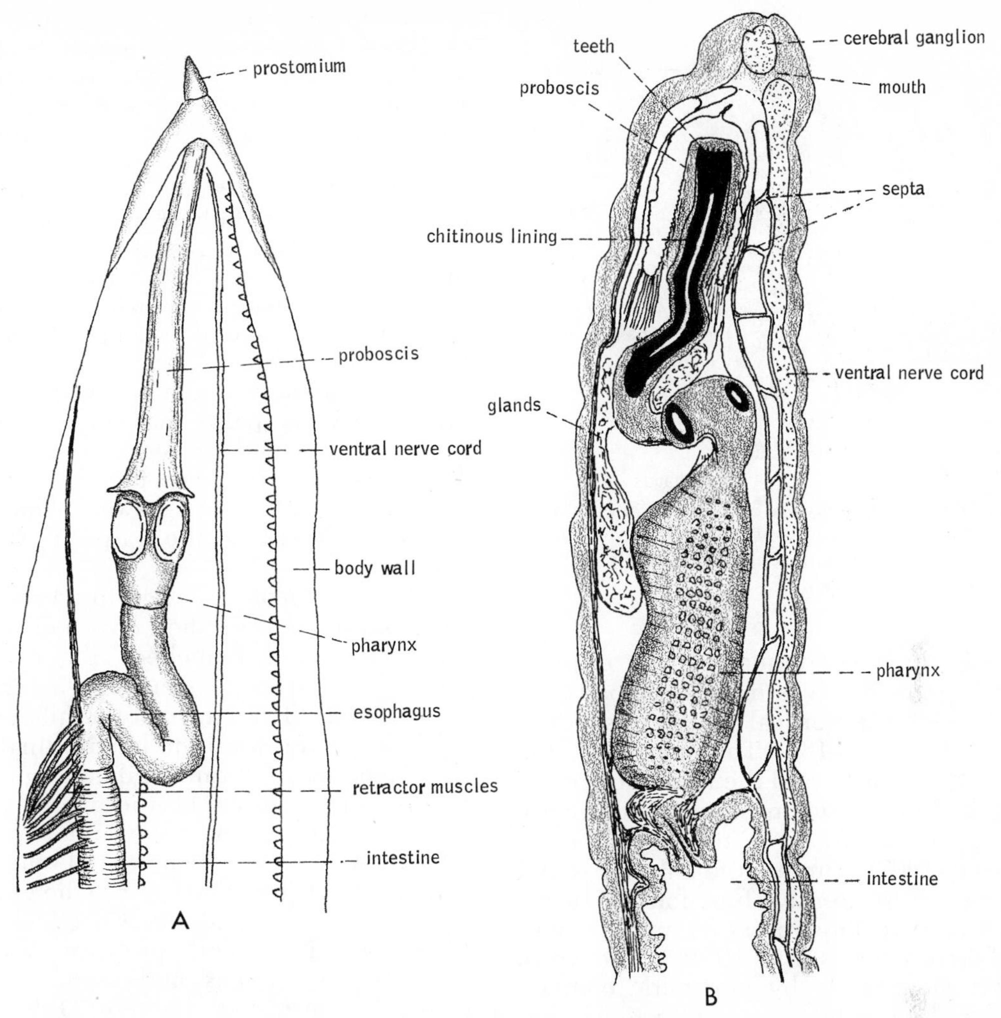

Figure 10–27. *A.* Anterior of *Glycera* digestive tract (dorsal view). (After Wells.) *B.* Anterior of *Autolytus* (sagittal section). (After Okada.)

are omnivorous and feed on algae, other invertebrates, and even detritus. *Nereis succinea* and *N. longissima* feed primarily on detritus material in the substratum. However, *N. fucata,* a commensal in hermit crab shells, is carnivorous.

Direct Deposit Feeders. Organic material which settles to the bottom accumulates around and between mineral particles. This deposit material is an important source of food for many invertebrates. The organic material is digested and the mineral portion egested as castings. The deposit-feeding polychaetes can be differentiated into direct and indirect deposit feeders. The direct deposit feeders consume sand or mud directly when the mouth is applied against the substratum. Ingestion is generally facilitated by means of a simple nonmuscular proboscis, which is everted by elevated coelomic fluid pressure (Fig. 10–28). Among polychaetes, direct deposit feeders include burrowers or tube dwellers. The less stationary burrowers include capitellids and opheliids, both of which ingest the substratum through which they burrow.

McConnaughey and Fox (1949) took population counts of the little burrowing opheliid *Euzonus (Thoracophelia) mucronatus*, which measures about 25 mm. long and not more than 2 mm. in diameter. These worms inhabit the intertidal zone on the Pacific coast of the United States and form colonies that occupy extensive stretches of protected beaches. In such colonies, the number of worms averages 2500 to 3000 per square foot. McConnaughey and Fox estimated that worms occupying a typical strip of beach 1 mi. long, 10 ft. wide, and 1 ft. thick would ingest approximately 14,600 tons of sand each year.

The lugworms, members of the family

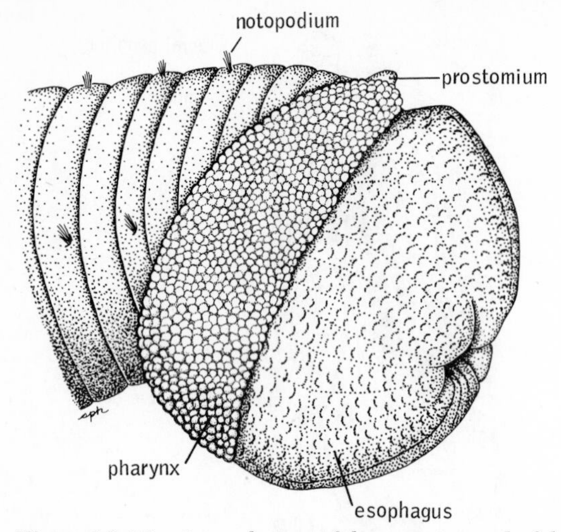

Figure 10–28. Lateral view of the anterior end of the capitellid *Notomastus*, a burrowing direct deposit feeder. The everted pharynx obscures the greatly reduced prostomium. (After Michel.)

Arenicolidae, are common direct deposit feeders. *Arenicola* lives in an **L**-shaped burrow in which the vertical part opens to the surface (Fig. 10–16). The head of the worm is directed toward the blind horizontal part of the burrow, where sand is continually ingested by means of a simple proboscis (Jacobsen, 1967). Three large anterior septal diaphragms apparently function in regulating coelomic fluid pressure in the eversion of the proboscis (Fig. 10–29). The worm irrigates the burrow by peristaltic contractions that drive water into the burrow opening. Water leaves the burrow by percolating up through the sand.

Bamboo worms, the Maldanidae, are examples of direct deposit feeding tube dwellers. These worms live upside down and ingest the substratum at the bottom of the sand-grain tube. Following a distinct rhythm, the feeding halts and the worm backs up to the top of the tube to defecate the mineral particles passed through the gut.

Indirect Deposit Feeders. Indirect deposit feeders lack a proboscis. Special head structures extend out over or into the substratum. Deposit material adheres to mucous secretions on the surface of these feeding structures and is then conveyed to the mouth along ciliated tracts or grooves.

The prostomial ingestive organs of the terebellids, e.g., *Amphitrite* and *Terebella*, are formed of large clusters of contractile tentacles, which stretch out over the surface of the substratum by ciliary creeping (Figs. 10–17 and 10–25). Surface detritus adheres to the mucus secreted by the tentacular epithelium. The tentacles are nonselective, and all food particles that can be carried are moved down a ciliated groove on the aboral side of the tentacles. Food accumulates at the base of the tentacles, which are individually wiped at different times over the upper lip bordering the mouth (Fig. 10–17*D*). Cilia on the lip then drive the food into the mouth. The ampharetids also trap detritus on a cluster of ciliated tentacles, but the tentacles in these worms can be retracted into the mouth.

The tubicolous spionids also possess two long tentacular palps that project from the tube opening and extend over the bottom (Fig. 10–30*D*). Particles that adhere to the surface are propelled toward the mouth down the length of the palps in a ciliated channel.

In the tubicolous oweniid *Owenia*, the peristomium bears a short crown of two clusters of flattened, branched, ribbon-like filaments, each of which ends in a bifid lobe (Fig. 10–31). The edges of the filaments are rolled inward toward the longitudinal axis of the filament to form a ciliated gutter. In feeding, the crown of the worm projects from the end of the tube and expands to about twice the diameter of the body. Adhering detritus is swept into the gutters by long cilia located along the edges of the filaments. The food particles are then entangled in mucus and carried by cilia down the gutter to the mouth. Only the oral surfaces of the filaments are ciliated, although mucous glands are present on both sides. A

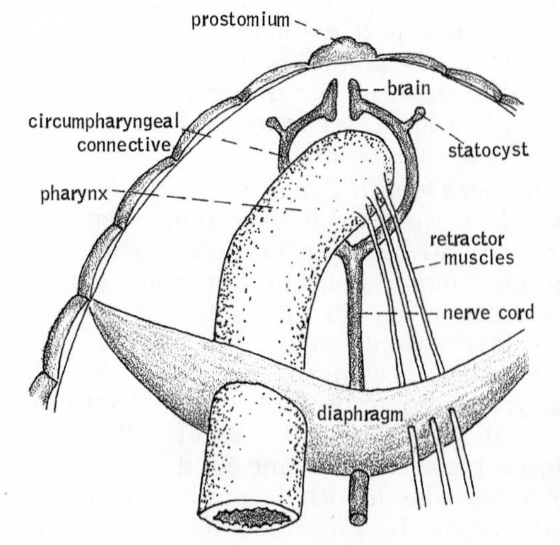

Figure 10–29. Anterior of *Arenicola* (dorsal dissection). (After Ashworth from Brown.)

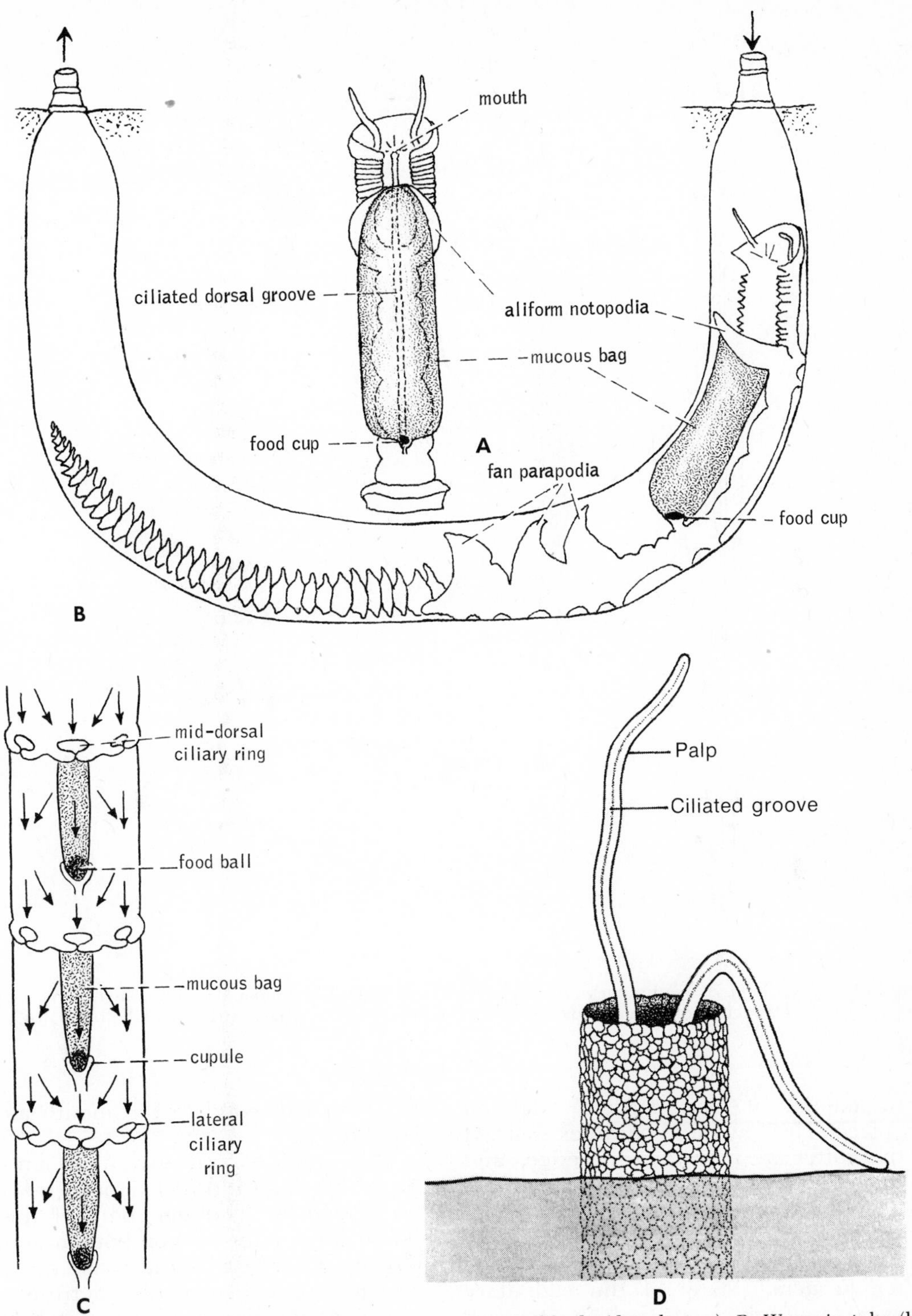

Figure 10–30. *Chaetopterus* during feeding. *A*. Anterior part of body (dorsal view). *B*. Worm in tube (lateral view). Arrows indicate direction of water current through tube. (After MacGinitie.) *C*. Dorsal view of three segments of middle body region of *Spiochaetopterus*, showing the position of mucous bags and the formation of food balls. Arrows indicate the direction of water currents. (After Barnes.) *D*. Palps of a spionid polychaete projecting from its sand-grain tube. Ciliated groove of palp conveys back to mouth detritus material picked up from substratum or suspended in water.

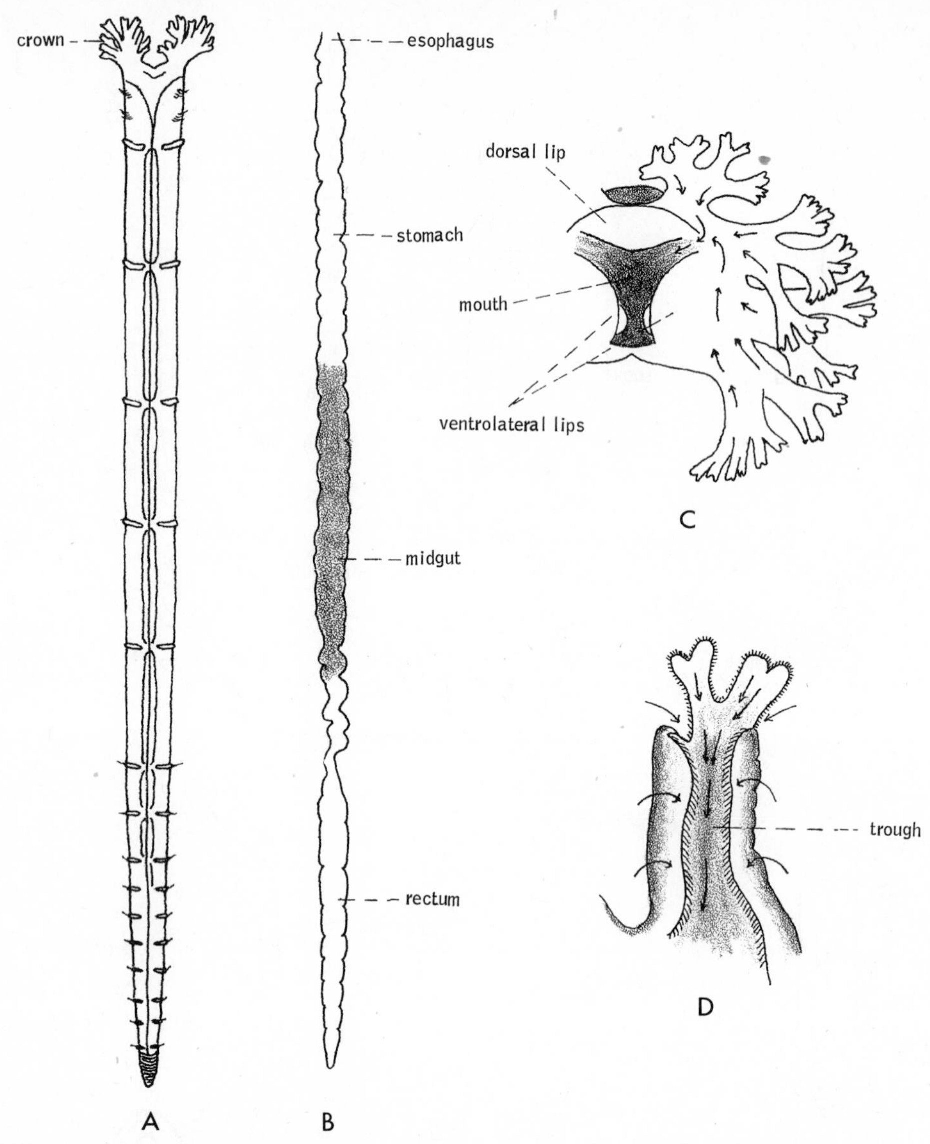

Figure 10–31. *Owenia. A.* Ventral view. *B.* Digestive tract. *C.* Mouth region and half of crown, showing direction of ciliary currents. *D.* Section of crown, showing trough and direction of ciliary currents. (All after Dales.)

certain amount of food particle selection takes place. The few large particles that get into the gutter are moved to the edges and then dislodged by periodic muscular contractions of the filaments. Ciliary tracts move the food particles from the base of the crown into the mouth.

Filter Feeders. Many of the sedentary burrowers and tubicolous polychaetes are filter feeders. Like indirect deposit feeders, there is no proboscis and the head is usually equipped with special feeding processes, used to collect detritus and plankton from the surrounding water. The particles adhere to the surface of the feeding structures and are then conveyed to the mouth along ciliated tracts.

The spionids and oweniids are transitional between indirect deposit feeders and filter feeders. The members of these two families not only feed on bottom deposits as already described, but also use the palps or crown to collect suspended detritus.

The crown of serpulid and sabellid fanworms is composed of two opposing half circles of bipinnate radioles, forming a funnel when they are expanded outside the end of the tube (Fig. 10–32). Beating of the cilia located on the pinnules produces a current of water that flows through the radioles into the funnel, and then flows upward and out. Particles are trapped on the pinnules and are driven by cilia into a groove running the length of each radiole. The particles are carried along the groove down to the base of the radiole, where a rather complex sorting mechanism takes place. The largest particles are rejected, and fine material is carried by ciliated tracts into the mouth. In many sabellids, particles are sorted into three grades, with the medium size grade being stored for use in tube construction.

The feeding mechanisms of the chaetopterids is quite different from that of other filter feeders (Figs. 10–30 and 10–33). *Chaetopterus*, which lives in **U**-shaped parchment tubes, has a highly modified body structure. The notopodia on the twelfth segment are extremely long and aliform (i.e., winglike), and the epithelium is ciliated and richly supplied with mucous glands. The notopodia on segments 14 to 16 are modified and fused, forming semicircular fans that project like piston rings fit against the cylindrical wall of the tube.

The beating of the fans produces a current of water that enters the chimney of the **U**-shaped tube near the anterior of the worm, flows through the tube, and then flows out of the opposite chimney. Each fan beats

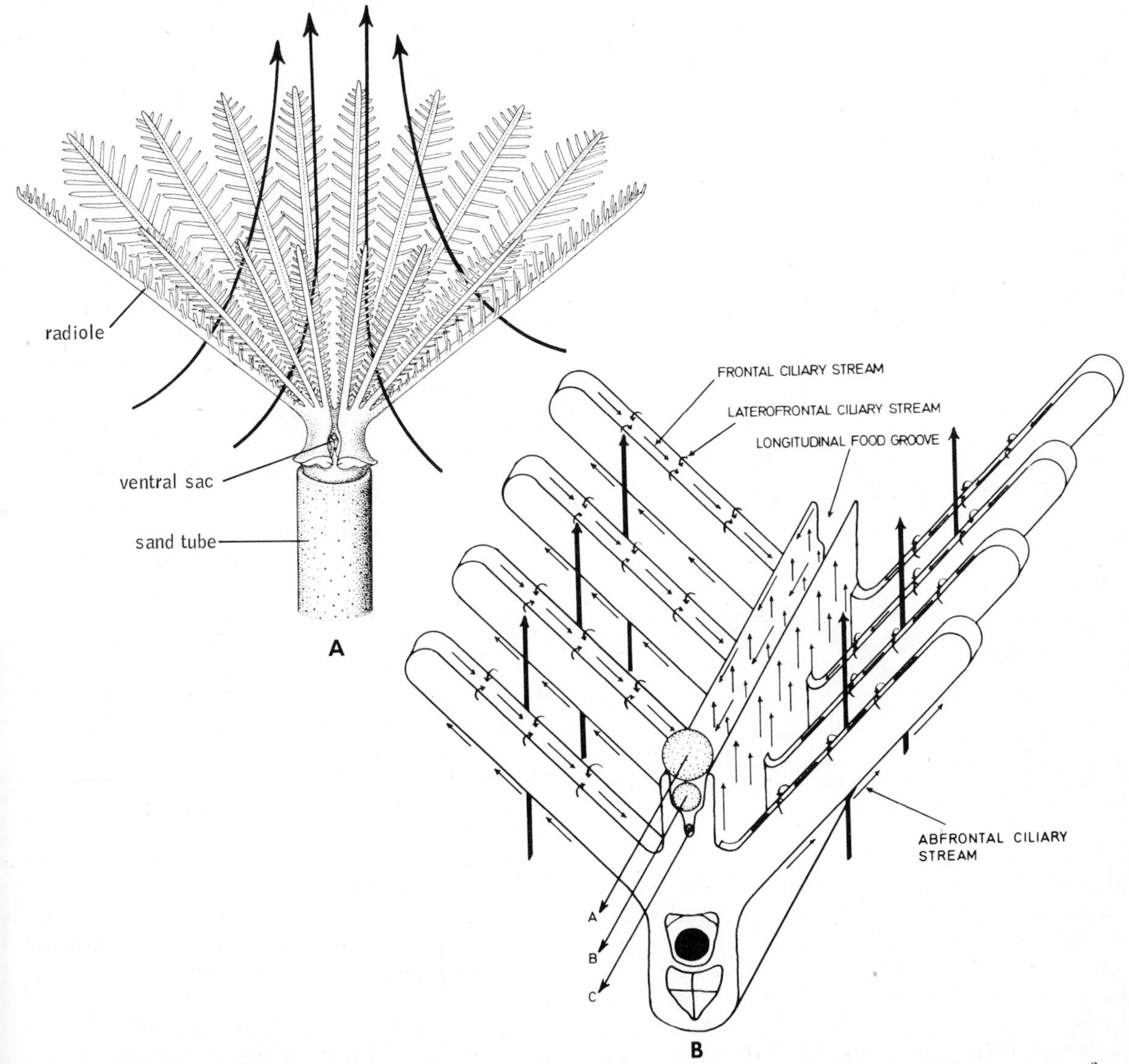

Figure 10–32. Filter feeding in the fanworm *Sabella*. *A*. Water current passing through radioles projecting from opening of tube. *B*. Water current (large arrows) and ciliary tracts (small arrows) over a section of one radiole. The letters A, B, and C indicate different size particles sorted. (After Nicol from Newell, R. C., 1970: Biology of Intertidal Animals. American Elsevier Co.)

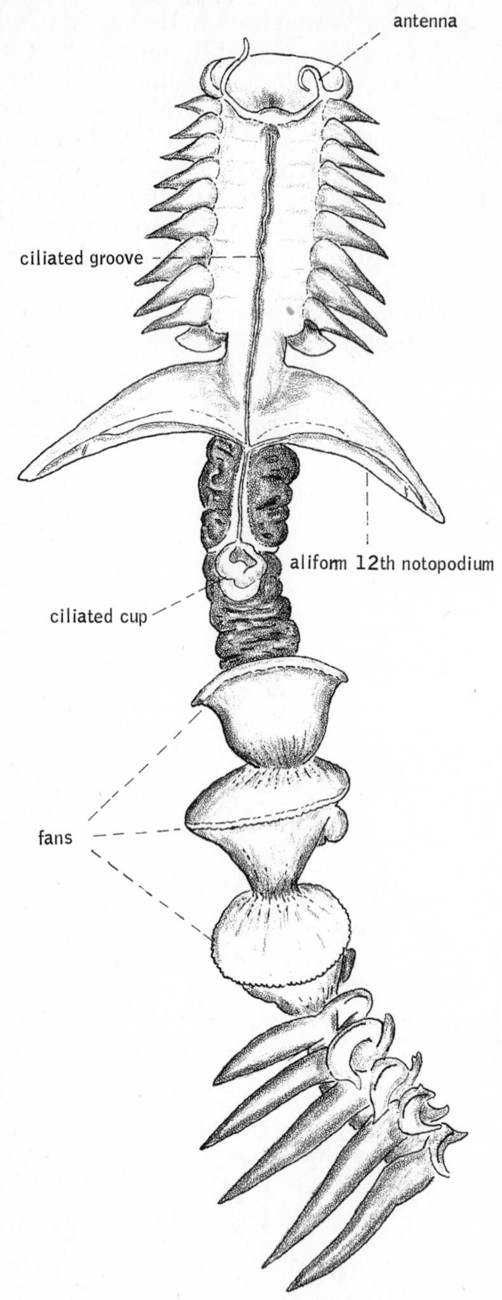

Figure 10-33. *Chaetopterus variopedatus.* (From life.)

about 60 times a minute to produce the water currents. The paired aliform notopodia are stretched out around the walls of the tube, and a sheet of mucus is secreted between them, resembling a net from a basketball hoop. The mucous film is continuously secreted from each notopodium where the film is attached to it. As a result, the sheet assumes the shape of a bag. The posterior end of the bag is grasped by a ciliated, cup-like structure located on the middorsal side of the worm a short distance behind the aliform notopodia. Water brought into the tube by the rhythmic beating of the fan parapodia passes through the mucous bag which strains out all suspended detritus and plankton.

Large objects brought into the tube by the water current are detected by peristomial cilia and shunted to either side; the aliform notopodia then are raised to let the large objects pass by. The food-laden mucous bag is continuously being rolled up into a ball by the dorsal cupule, or food cup. When the ball reaches a certain size, the bag is cut loose from the notopodia and rolled up with the ball. The cupule then projects forward and deposits the mucous food ball onto a ciliated middorsal groove, which extends to the anterior of the worm; and the ball is carried to the mouth. An 18 to 24 cm. long specimen of *Chaetopterus* may produce mucous film for the bag at the rate of approximately 1 mm. per second, with food balls averaging 3 mm. in diameter.

The other members of the Chaetopteridae build straight, vertically oriented tubes, but utilize mucous bags for filter feeding. The number of mucous bags and the site of their formation vary—as many as 13 are formed at one time in *Spiochaetopterus*. In a number of genera the water current is activated by cilia rather than by pumping (Fig. 10-30*C*).

The Alimentary Canal. Typically, the alimentary canal of polychaetes is a straight tube extending from the mouth at the anterior of the worm to the anus located in the pygidium. The canal is differentiated into a number of regions. Although some variation exists, the regions most often encountered are the proboscis, or pharynx (or buccal cavity when the proboscis is absent), which is derived from the stomodeum, a short esophagus, a stomach, an intestine, and a rectum. Often, as in *Nereis*, no stomach is present and the esophagus opens directly into the intestine (Fig. 10-3).

In many species, such as the bamboo worms (*Clymenella*), *Owenia* (Fig. 10-31*B*), and *Euzonus* (*Thoracophelia*), these regions can only be detected histologically, and the gross appearance of the digestive tract is that of a simple uniform tube. As a general rule, no specialized digestive glands are present; the epithelial lining of the stomach, or the intestine when the stomach is absent, elaborates enzymes for extracellular digestion. Cellulose has been reported in the omnivorous *Nereis virens*. The intestine is the site of absorption; not infrequently the walls are folded to increase the intestinal surface area.

In some worms, such as *Amphitrite*, increased intestinal surface area for digestion or absorption is attained by intestinal coiling; in other species, the increased area is attained by evaginations of the gut, or cecum. In *Nereis* two large glandular ceca open into the esophagus (Fig. 10–3). The function of these ceca is still uncertain, but they probably secrete digestive enzymes. A similar pair of esophageal ceca are present in *Arenicola* and are known to produce lipase and protease.

A more peculiar structure exists in the sea mouse. Here, pairs of lateral ceca empty into the intestine throughout its entire length (Fig. 10–10). Digestion begins in the intestine, and digested and partially digested food is forced into the ceca by fluid pressure from the surrounding coelom. The entrance into each cecum is guarded by projecting epithelial cells, which form a sieve and strain out all but the finest particles. Digestion is completed and absorption takes place within the cecum.

Movement of food through the digestive tract is brought about by the contraction of the muscle cells in the gut wall, by the general ciliary action of the gut lining, by a ventral ciliated groove, or by a combination of these factors.

The egested wastes from worms living in a tube with double openings, such as *Chaetopterus*, are readily removed by water currents. Such flushing, however, is accomplished less efficiently when the tube is closed or, more commonly, when the tube is deeply buried in mud and sand. Many polychaetes turn around in the tube or burrow and thrust the pygidium out of the opening during defecation. The tubicolous bamboo worms and the sedentary burrow-inhabiting arenicolids, both of which are direct deposit feeders and live oriented head downward in the tube, periodically back up to the surface to release castings (Fig. 10–16). Defecation appears to follow a rhythmic cycle controlled by an internal pacemaker system (Fig. 10–16).

The fanworms have a ciliated midventral groove, which extends from the anus anteriorly. In the thoracic region, the groove curves around the side of the body to the dorsal surface and extends anteriorly to the head. Fecal material is molded with mucus into distinct pellets in the rectum and is then carried out of the tube by the ciliated groove.

The situation in the chaetopterids with straight tubes is very similar to that in fanworms. Molded fecal pellets, when released from the anus, are carried forward along a middorsal groove. Anteriorly this groove is the same as the one that carries the food balls, but the fecal pellets, rather than entering the mouth, are transferred to a special ejection tract that runs the length of each palp. The tip of the palp is projected from the mouth of the tube and the fecal pellet falls away from the tube opening.

Gas Exchange. Gills are common among the polychaetes, but they vary greatly in both structure and location. This lack of consistency would seem to indicate that gas exchange organs have arisen independently within the class a number of times. The ancestral polychaetes were probably devoid of gills, and diffusion of gases took place through the general body surface. This mode of gas exchange is still retained in polychaetes which are very small or which possess long, thread like bodies, as in the burrowing Lumbrineridae, Arabellidae, and Capitellidae. General diffusion of gases through the body surface accounts for some gas exchange (percentages vary according to species) even when gills are present.

In the polynoid scale worms, gas exchange is largely restricted to the dorsal surface, which is roofed over by the elytra (Fig. 10–5A). Cilia on the dorsal surface create a current of water flowing posteriorly beneath the elytra. Some water enters over the head, but the principal inflow is lateral. Cilia on the parapodia propel the water upward over the parapodia and then inward between the bases of the elytra, where it becomes confluent with the main stream of water flowing posteriorly.

The felt-covered sea mice (*Aphrodita*) lack cilia (Figs. 10–10 and 10–11A), but a similar dorsal water current is produced by the animal, which tilts the elytra upward and then rapidly brings them down. The movement begins at the anterior of the body and sweeps posteriorly over successive elytra, thus driving water through the underlying channel.

Most commonly the gills are associated with the parapodia and in most cases are modified parts of the parapodium. The notopodium may possess a flattened branchial lobe, which acts as a gill (Fig. 10–4A).

More commonly, the dorsal cirrus of the parapodium is modified to serve as a gill. The cirrus may be formed like a long inverted cone as in *Sabellaria* and *Scolelepis* (Fig. 10–5*B*); a large flattened lobe, as in the phyllodocids (Fig. 10–5*C*); or irregularly branched filaments as in *Arenicola* (Fig. 10–8*B*). In the Eunicidae and Onuphidae, the gills arise from the base of the dorsal cirrus and may be cirriform, as in *Nothria;* pectinate, as in *Marphysa* and *Eunice* (Fig. 10–14*F*); or spirally branched, as in *Diopatra* (Fig. 10–18*A*).

The gills are not always associated with the parapodia. The gills of the terebellids, such as *Amphitrite,* are arborescent and are located on the dorsal surface of the anterior segments and can thus be projected from the opening of the burrow or tube (Fig. 10–17*A*). The cirratulids have long threadlike gills arising from many segments (Fig. 10–6*A*).

The sabellid and serpulid fanworms do not have gills associated with the parapodia, but the bipinnate radioles composing the fans serve as sites of gas exchange, along with the general body surface.

Ciliary action commonly produces the ventilating water currents flowing over the gills. Usually, bands of cilia on the dorsal surface of each segment create a water current moving anteriorly or posteriorly. Ciliary tracts on the gills produce lateral currents either entering or leaving the main dorsal current (Figs. 10–5*B* and *C*).

The Nereidae, Eunicidae, and Glyceridae are devoid of any surface ciliation; these worms, which are largely burrowers and tube-dwellers, keep water flowing over the gills and through the tubes or burrows by their undulatory body movements. Some tube-dwellers, such as the sabellid fanworms, pump water through the tube by means of peristaltic contractions of the body wall. Others, such as many chaetopterids, drive water through the tube by means of cilia.

In most tubicolous polychaetes, the circulating water passes through the tube and out through the sand or mud in which the tube is buried. However, in serpulid fanworms, which live in a blind calcareous tube, inhalant currents enter and exhalant currents leave the opening of the tube. These currents are produced by special ciliary tracts.

Circulation. In the majority of polychaetes there exists a well-developed blood-vascular system, in which the blood is enclosed within vessels. The basic plan of circulation is relatively simple. Blood flows anteriorly in a dorsal vessel situated over the digestive tract; at the anterior of the body, the dorsal vessel is connected to a ventral vessel by one to several vessels or by a network of vessels passing around the gut. (Fig. 10–34*A*.) The ventral vessel carries blood posteriorly beneath the alimentary tract.

In each segment, the ventral vessel gives rise to one pair of ventral parapodial vessels, which supply the parapodia, the body wall, and the nephridia, and to a single ventral intestinal vessel that supplies the gut (Fig. 10–35*A*). The dorsal vessel, in turn, receives a corresponding pair of dorsal parapodial vessels and a dorsal intestinal vessel. The dorsal and ventral parapodial vessels and the dorsal and ventral intestinal vessels are interconnected by a network of smaller vessels.

There are many variations of this basic circulatory pattern. In worms with differentiated trunk regions and gills that are limited in number and position, the circulatory pathways are considerably modified and more complicated (Fig. 10–34). Instead of a network of vessels, many species, such as *Arenicola* and the fanworms, possess a large sinus lying between the epithelium and the muscle layers in the walls of the intestine and stomach. The sinus partially or completely surrounds the gut and replaces the dorsal vessel in this region.

In general, the blood vessels are contractile, particularly the dorsal vessel. The contraction of these vessels, rather than a distinct pumping center or heart, causes propulsion of the blood. When blood fills a section of a vessel after a peristaltic wave, the contraction is again initiated.

A peculiar feature of the circulatory system of many polychaetes is the presence of blind contractile vessels. Such vessels are found in the nereids and the opheliids, as well as in other families, but they reach the height of development in the fanworms, where an ebb and flow within the contractile vessels virtually replaces a continuous-flow capillary system.

A number of structures called hearts or heart bodies appear in polychaetes, but they certainly are not all homologous. The heart in *Chaetopterus* is a bulbous swelling in the

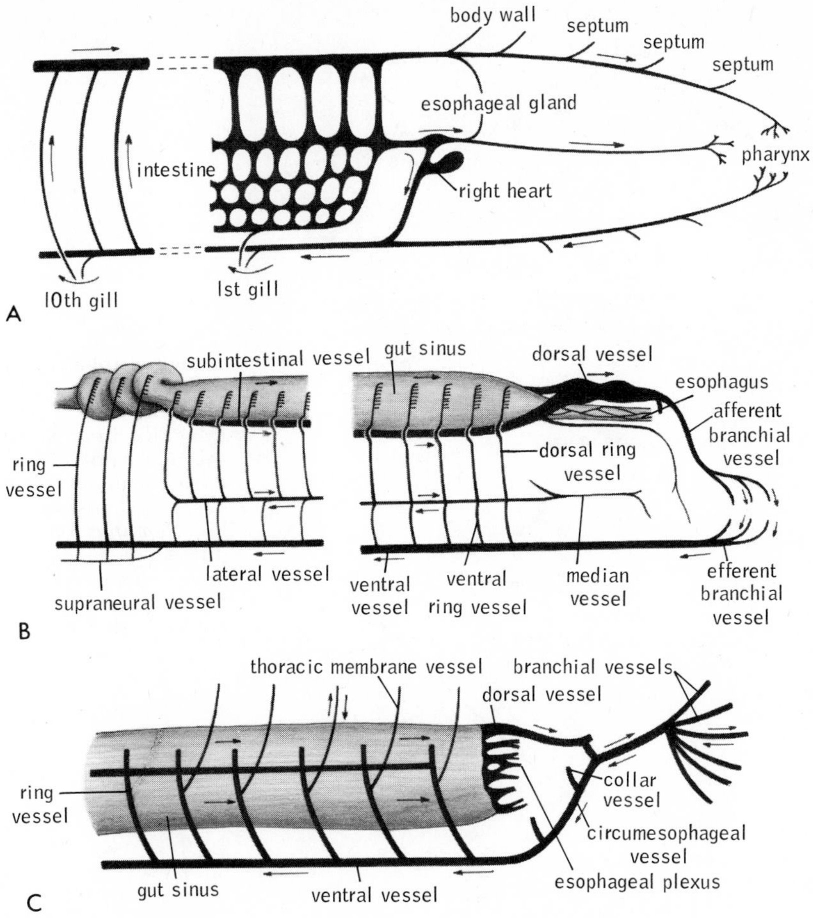

Figure 10–34. Circulatory systems of three polychaetes. *A. Arenicola.* (After Bullough.) *B. Amphitrite. C. Pomatoceros.* (*B* and *C* after Thomas.)

dorsal vessel and is reported to be a pumping center. There is a pair of lateral hearts in the arenicolids associated with the gut sinus (Fig. 10–34A) and there is a heart behind the gills in the terebellids and ampharetids. Such hearts have probably evolved to counteract backflow.

Primitively, as in scale worms, phyllodocids, and syllids, the blood is a colorless fluid containing a few amebocytes; but the blood of most polychaetes, especially the larger species, contains respiratory pigments dissolved in the plasma. Hemoglobin is the most common of these pigments, but chlorocruorin is characteristic of the blood of the serpulid and sabellid fanworms and also of the Flabelligeridae and Ampharetidae. Chlorocruorin is an iron porphyrin like hemoglobin, but the slight difference in side chains gives it a green rather than a red color. *Magelona* has a blood-vascular system with corpuscles containing a third iron-bearing protein pigment (not a porphyrin) that is called hemerythrin.

The molecules of plasma hemoglobin and chlorocruorin are always very large. The plasma hemoglobin of *Arenicola,* for example, contains 96 heme units. Each heme is attached to two peptide chains, in contrast to the one heme to one polypeptide of mammalian hemoglobin (Waxman, 1971). There are numerous polychaetes, including *Glycera, Capitella,* and some terebellids, in which the blood-vascular system is reduced or absent and the coelomic fluid functions in internal transport. The coelom of these worms contains hemoglobin located in coelomic corpuscles. Such coelomic hemoglobin, like the corpuscular hemoglobin of vertebrates, is always a small molecule. In *Glycera,* for example, it consists of simple monoheme units.

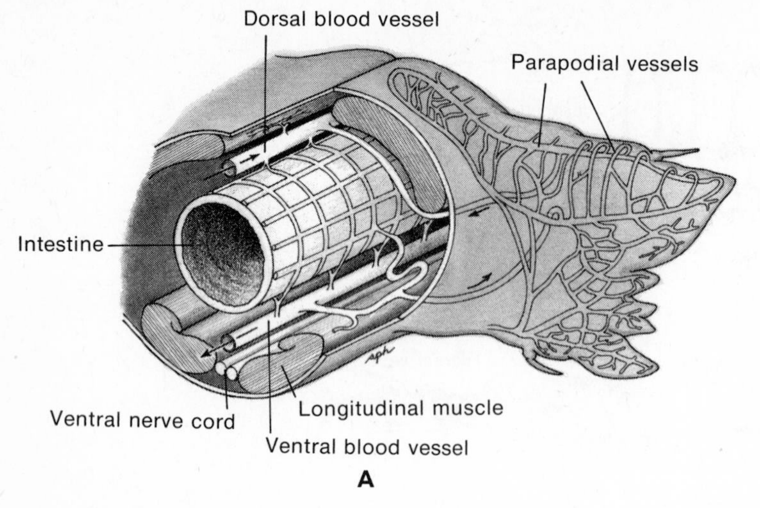

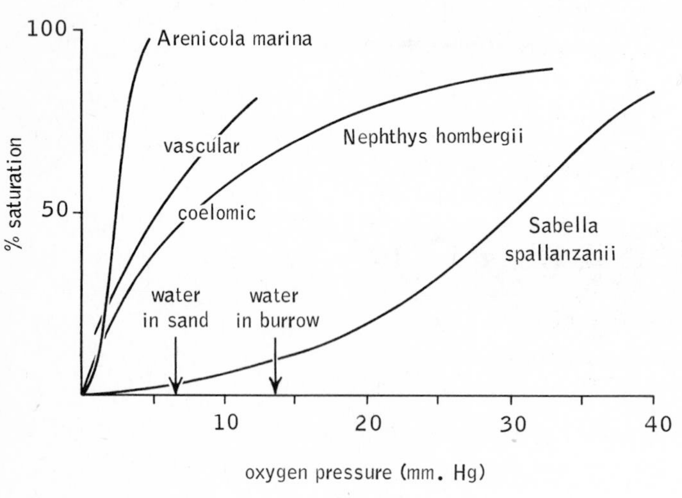

Figure 10–35. *A*. Vascular system within a segment of *Nereis virens*. Arrows indicate direction of blood flow. (Modified from Nicoll.) *B*. Oxygen dissociation curves of the respiratory pigments of three species of polychaetes. *Arenicola* possesses vascular hemoglobin. *Nephthys hombergii* possesses two different hemoglobins, one vascular and one coelomic. The respiratory pigment of *Spirographis spallanzanii* is chlorocruorin. (After Jones from Dales.)

The significance of the difference between polychaete blood-vascular hemoglobin and coelomic hemoglobin is not yet clear. The large size of the molecule in plasma may perhaps prevent these proteins from producing too high a colloid osmotic pressure of the blood. If the large amount of small hemoglobin molecules of vertebrates was dissolved in plasma, rather than in corpuscles, the osmotic pull exerted by the blood would be far too great. The significance may also be related to the energy requirements of synthesis. If the oxygen transport needs of the animal can be met by large molecules dissolved in plasma, the energy cost of synthesizing such molecules is likely to be less than the cost of construction of special cells to house small molecules. The question then would be why the coelomic hemoglobin is in corpuscles. The answer is not certain; it has been suggested that corpuscles prevent the loss of coelomic hemoglobin from the excretory or reproductive tubules.

There are a number of interesting exceptions to the usual disposition of the respiratory pigments just described. The blood of *Serpula* contains both hemoglobin and chlorocruorin. *Terebella lapidaria* possesses not only coelomic red corpuscles, but also a blood-vascular system with a different hemoglobin. The same is true of *Nephtys hombergii*. The two hemoglobins are not alike, as indicated by their oxygen dissociation curves (Fig. 10–35*B*).

In the majority of polychaetes, the respiratory pigments certainly function in oxygen transport. The plasma hemoglobin of *Nephtys hombergii* and the chlorocruorin of *Spirographis spallanzanii*, for example, produce dissociation curves essentially like those of most vertebrates. But this is not true of all polychaetes. The dissociation curve of the plasma hemoglobin of *Arenicola* is far to the left of the graph, being saturated with oxygen at very low oxygen tensions (Fig. 10–35*B*). It has been suggested that the hemoglobin of such species might therefore function as an oxygen reserve during low tide, when the oxygen tensions of the burrow and surrounding sand would be considerably decreased. But the amount of hemoglobin present would seem to permit an oxygen reserve lasting only a few minutes. However, like many invertebrates, polychaetes are oxyconformers; i.e., oxygen consumption is regulated in part by the amount of oxygen available in the surrounding environment. Mangum (1970) has suggested that the decreasing metabolic demands for oxygen during stagnation might enable the worm to stretch out over a significant period of time what would otherwise be an inadequate store of oxygen provided by the hemoglobin.

The Nervous System. The general plan of the nervous system in polychaetes is relatively simple and more or less uniform throughout the class, as well as in the other classes of annelids. The polychaete nervous system is characterized by a dorsal ganglionic mass, or brain, in the prostomium (or the brain may extend back through the anterior body segments); a pair of connectives surrounding the anterior gut; and a ventral nerve cord extending through the length of the body. The ventral cord typically bears segmental swellings or ganglia, from which lateral nerves extend into each segment.

Brain. The brain is usually bilobed, lying in the prostomium beneath the dorsal epithelium (Figs. 10–29 and 10–36). The greatest degree of differentiation appears in the errant polychaetes, in contrast to the sedentary polychaetes in which the brain has a relatively simple structure. In many errant species the brain can be divided into a forebrain with nerves to the palps, a midbrain with nerves to the eyes and antennae, and a hindbrain with nerves to the nuchal organs; but there is great variation, depending upon the degree of development of sense organs.

Typically, a single pair of circumpharyngeal or circumesophageal connectives surround the anterior gut and interconnect the brain and the ventral nerve cord. However, in some species, such as *Sabella*, the connectives are doubled; and in others, such as *Arenicola* (Fig. 10–37*G*) and *Amphitrite*, both of which have very reduced brains very closely associated with the overlying integument, the single pair of connectives is located in the body wall just beneath the epidermis. The connectives are often ganglionated, with nerves extending to the tentacular cirri of the peristomium or into the anterior trunk segments.

Many polychaetes, particularly those with an eversible proboscis, have a system of stomatogastric nerves, arising from the

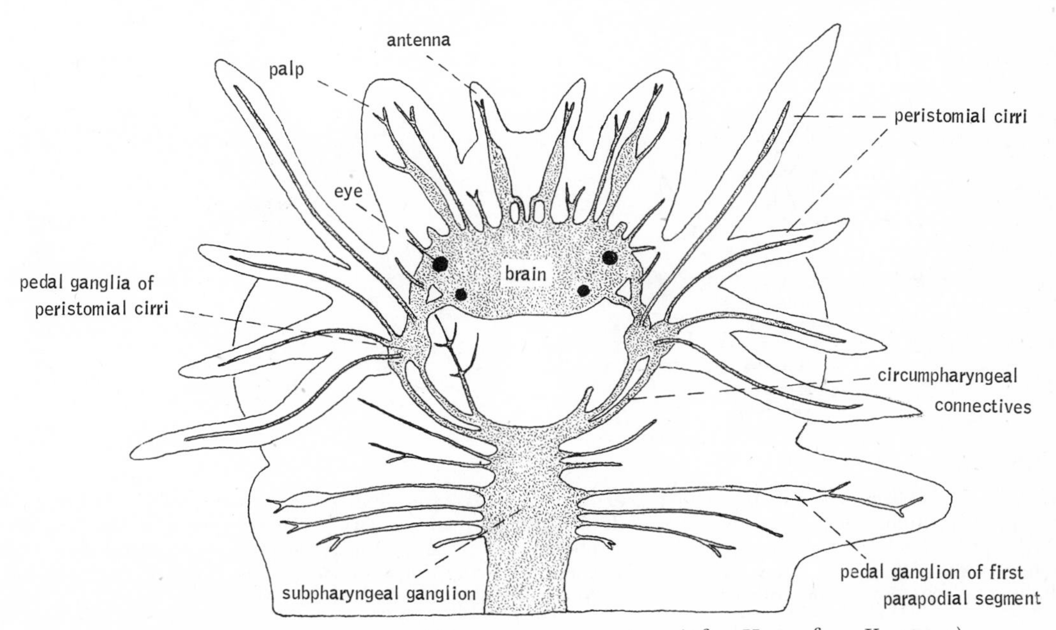

Figure 10–36. Anterior part of *Nereis* nervous system. (After Henry from Kaestner.)

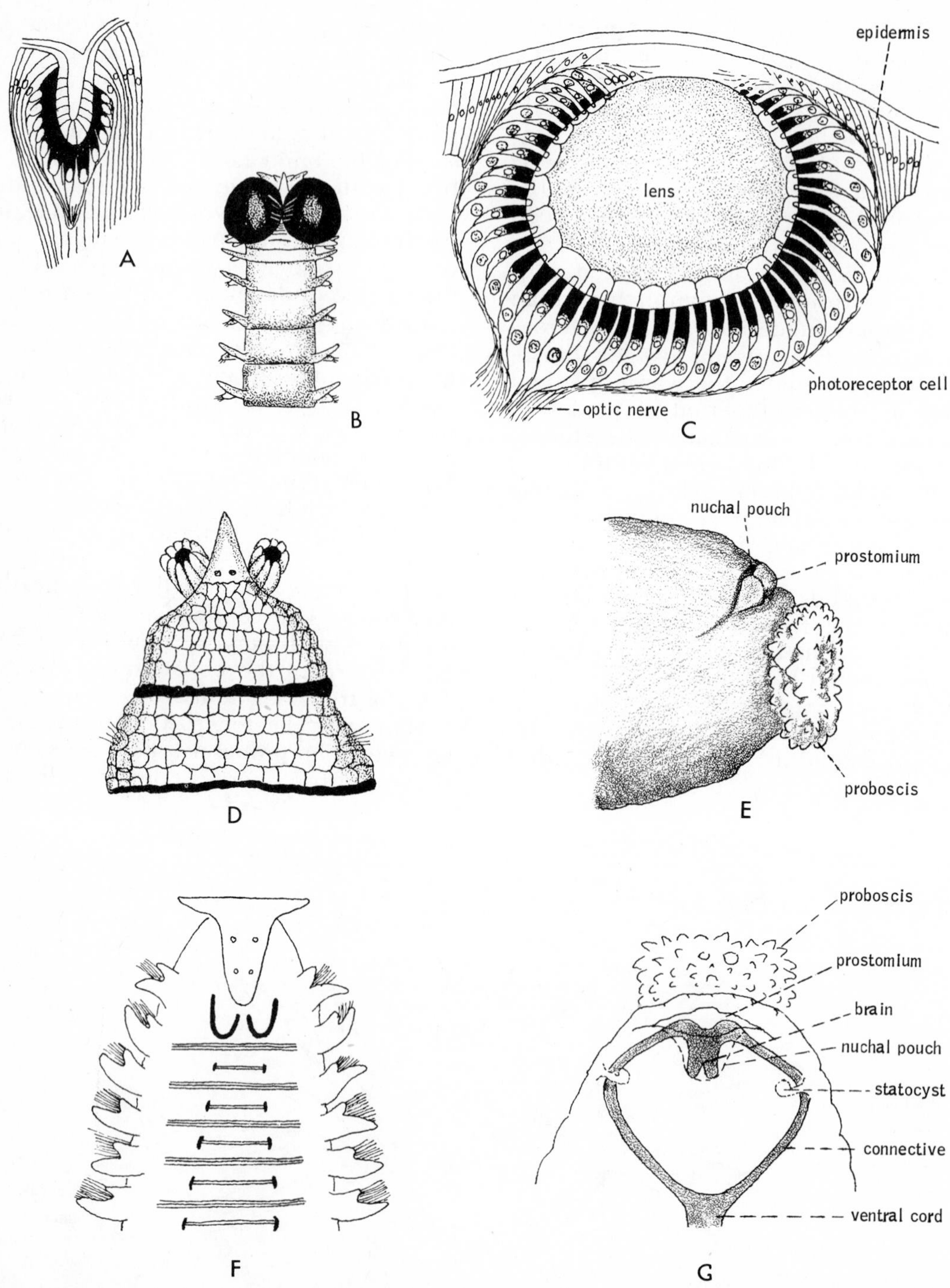

Figure 10–37. *A.* Simple eye of *Mesochaetopterus.* (After Hesse from Fauvel.) *B.* Anterior of the alciopid, *Vanadis grandis,* showing highly developed eyes. (After Izuka from Fauvel.) *C.* Eye of *Nereis.* (After Hesse from Fauvel.) *D.* Anterior of *Notomastus latericeus* with everted nuchal organs. (After Rullier from Fauvel.) *E.* Head of *Arenicola* (side view). (After Wells.) *F.* The spionid, *Malacoceros,* showing dorsal sense organs in black. (After Söderström from Fauvel.) *G.* Statocysts and anterior part of nervous system of *Arenicola.* (After Wells.)

posterior part of the brain or connectives, which are involved in the motor control of the proboscis or pharynx.

Ventral Nerve Cord. As a primitive structure, the ventral nerve cord is completely double throughout, with transverse connectives between the separate ganglia (in fanworms, for example); but in most polychaetes the two cords are fused in varying degrees. In most polychaetes the ventral nerve cord lies within the muscle layers; but in a few forms, such as *Clymenella*, the ventral nerve cord is associated with the ventral integument.

There is typically one ganglionic swelling per segment, although there are two per segment in *Sabella*. From each ganglion emerge two to five pairs of lateral nerves that innervate the body wall of that segment. The lateral nerves contain both sensory and motor components; that is, the lateral nerves are mixed. The sensory nerve components form a nerve plexus beneath the surface epithelium; it is here that the sensory cell bodies are located. The motor neurons innervate the parapodial muscles and the muscles of the body wall; the motor cell bodies are located in the segmental ganglia along with the cell bodies of interneurons. The cell bodies of motor neurons innervating the parapodia are not always located in the ventral ganglia. Some polychaetes have a separate system of pedal ganglia located peripherally near the parapodia (Figs. 10–36A, 10–38A and B). Annelid ganglia have the same structure as those of mollusks and crustaceans, which have been studied much more extensively. A description is given in connection with these two groups (see page 359).

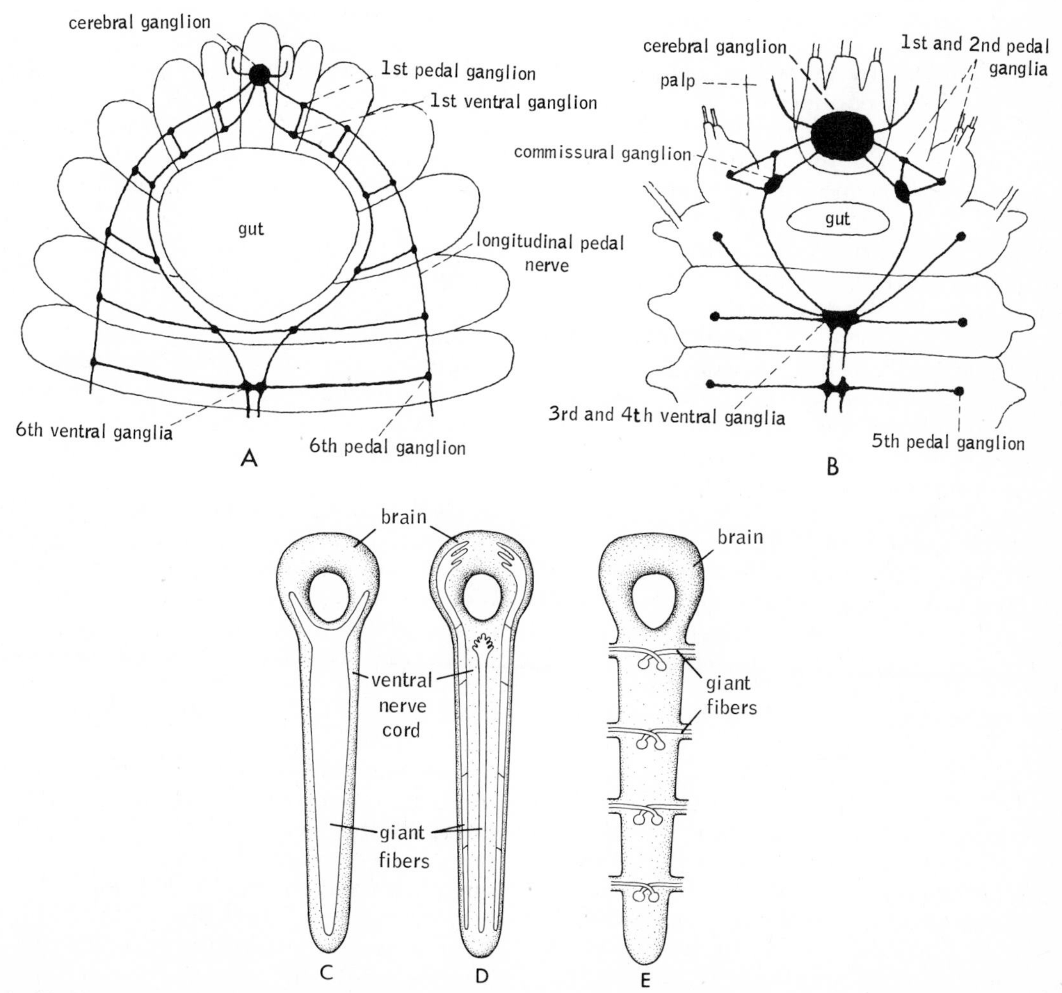

Figure 10–38. Nervous systems of scale worms: *A. Hermodice; B. Lepidasthenia.* (Both after Storch from Fauvel.) Nervous systems of three polychaetes, showing types of giant axons: *C. Eunice*, with a single giant axon; *D. Nereis*, with medial and lateral giant fibers; *E. Thalenessa*, with intrasegmental giant axons. (*C* and *D* after Nicol; *E* after Rhodes from Nicol.)

The remainder of the cord contains longitudinal fiber tracts. These tracts are composed of fine fibers and frequently giant axons, and the interconnections between the peripheral neurons and the longitudinal fibers are often complex. In *Nereis*, whose nervous system is the best known of any polychaete, the sensory fibers of the four pairs of segmental nerves display a distinct distribution pattern with regard to the specific longitudinal tracts with which they are associated and to their innervation areas. Sensory neurons of the first and fourth segmental nerves innervate the general segmental integument; those of the second nerve innervate the parapodia; and those of the third nerve are principally proprioceptors of the segmental muscles. The motor innervation of the segmental muscles is effected through a multiple linkage system involving primary, secondary, and sometimes tertiary motor neurons. Connections between these neurons may be through typical dendritic synapses or by means of axon-to-axon contact. The latter connection also occurs in the longitudinal fiber tracts.

Giant Axons. An important defense of most polychaetes against their many predators is the ability to contract very rapidly. The rapid end-to-end contraction reflex is particularly well developed in the tube-dwellers, which emerge from their protective housing to feed. Correlated with this ability to contract rapidly is the presence of giant axons in the ventral nerve cords. The enlarged diameter of the axon increases the rapidity of conduction and therefore makes possible simultaneous contractions of the segmental muscles.

Although giant axons are present in the majority of polychaetes (Nicol, 1948) and reach their greatest development among tube-dwellers, the arrangement and structure are far from uniform; the axons have probably evolved independently within a number of groups. The number of giant fibers ranges from one, as in *Myxicola* and *Eunice*, to many, as in *Oenone (Halla)*. *Nereis* possesses five giant fibers, three of which are very large and two, small. A fiber may consist of a single giant neuron or it may be syncytial in nature, being composed of a large number of anastomosing neurons. Both types may be present in the same worm, as in *Nereis*. A few examples of different giant axon systems are illustrated in Figure 10–38*C-E*.

Nicol's studies of the sabellid fanworm *Myxicola*, which has a single large giant fiber, have contributed much to the knowledge of how these neurons function. The giant fiber of *Myxicola* obeys the all-or-none law, can be fired at any level along the length of the body, and will conduct an impulse in either direction. Conduction along the giant fiber is 12 m./sec. as compared to about 0.5 m./sec. along ordinary longitudinal tracts. Efferent branches from the giant axon run to the longitudinal muscles. The giant fiber is not involved in the conduction of ordinary locomotor impulses, for if the fiber is severed locomotion is not inhibited; only rapid contraction is blocked.

Most of the characteristics of the giant fiber system of *Myxicola* are true of other polychaetes, with some exceptions. In *Nereis*, for example, the giant fibers will only conduct posteriorly, and conduction to the segmental muscles is provided by special motor neurons which are in touch contact with the giant axons.

Neuroendocrine System. Increasing evidence indicates that the glandular epithelium (infra-cerebral gland) associated with the underside of the posterior part of the brain constitutes a neuroendocrine complex. The majority of studies have demonstrated a regulating role for this system in reproduction, which will be described later.

Sense Organs. The principal specialized sense organs of polychaetes are eyes, nuchal organs, and statocysts.

Eyes. Eyes are most conspicuous in errant polychaetes, but even here they are sometimes absent. The eyes are located on the surface of the prostomium and number two, three, or four pairs (Figs. 10–3 and 10–9*B*). They may be very simple or highly developed, depending on the worm's mode of existence. In general, the polychaete eye is of the retinal-cup variety, the wall of which is composed of rodlike photoreceptors, pigment, and supporting cells (Figs. 10–37*A* and *C*). The photoreceptors are directed toward the lumen of the cup; these eyes are therefore of the direct type. An optic nerve issues from the back of the cup and extends the short distance to the brain. The interior of the eye is filled with a refractive body, which may or may not be connected with the surface cuticle. The surface cuticle acts as a cornea. Such eyes function only in light detection; most polychaetes are negatively phototactic.

The annelid eye reaches its highest development among the pelagic raptorial Alciopidae (Fig. 10–37*B*). The two large eyes protrude greatly from the surface of the prostomium. Each eye consists of an inner retinal cup, a vitreous body, and an outer bulging lens and cornea. This eye possesses a remarkable mechanism for accommodation which indicates that the eye is probably capable of object detection. Focusing is effected by two methods. An ampulla to the side of the eye forces fluid into the optic vesicle, located between the lens and vitreous body; this pressure forces the lens outward. Also, the lens can be flattened by the action of muscle fibers attached to its side. The lens surface is even striped to control the amount of light entering the eye.

A few species, such as the opheliid *Polyophthalmus,* have a simple eyespot located on the body wall between each pair of parapodia in addition to the eyes on the prostomium.

The sabellid fanworms, like many other sedentary polychaetes, have lost the prostomial eyes. Some members of this family, however, possess eyespots on the branchial filaments. These branchial eyes are located on both sides of the main axis of the filament and in *Branchiomma,* for example, number 10 to 15 pairs per filament. Each branchial eye is composed of a cluster of pigment and photoreceptor cells. Each photoreceptor cell contains an outer large refractive body, so that the eye is essentially compound. The nerve fibrils of the receptors pass into the general epidermal nerve network.

Although the serpulid fanworms lack branchial eyes, the epidermis of the crown contains photoreceptor cells. These worms are very sensitive to sudden light reductions and will immediately withdraw into the tube. This behavior probably represents a protective adaptation against passing predators. The branchial eyes of sabellids probably serve the same purpose.

Nuchal Organs. Nuchal organs consist of a pair of ciliated sensory pits, or slits, often eversible, that are located in the head region of most polychaetes (Fig. 10–37*D* and *E*). Although of varied structure, all nuchal organs are innervated by the posterior portion of the brain, and are composed of ciliated columnar and sensory cells. Gland cells are sometimes also present.

The nuchal organs appear to be chemoreceptors, which are important in the detection of food. When the organs are destroyed experimentally, the worms no longer feed. Nuchal organs attain their greatest development in the predatory species. They are completely absent or rarely appear in filter feeders.

Ciliated Sense Organs. Specialized ciliated sense organs on the trunk segments are characteristic of some polychaetes. Orbiniids and spionids possess ciliated sensory tubercles, ridges, or bands, called dorsal organs, located on the dorsal surface of the segments (Fig. 10–37*F*). Some families, such as the Orbiniidae and the Opheliidae, possess ciliated sensory structures called lateral organs, located between the notopodium and neuropodium of each parapodium.

Statocysts. Statocysts are found in some polychaete families, including the Orbiniidae, the Arenicolidae, the Terebellidae, and the Sabellidae—all are sedentary burrowers or tube-dwellers. The statocysts of *Arenicola* are located within the body wall of the head and are bulbous, with a canal that opens to the outer lateral body surface (Fig. 10–37*G*). The statocysts are associated with the circumpharyngeal connectives, which in *Arenicola* run within the body wall. Wells (1950) reports that the statocysts of *Arenicola* contain spicules, diatom shells, and quartz grains, all covered with a chitinoid material. *Arenicola* always burrows head downward, and if an aquarium containing a worm is tilted 90 degrees, the worm makes a compensating 90-degree turn in burrowing. If the statocysts are destroyed, this compensating ability is lost. The Sabellid *Megalomma* also has statocysts, but they orient the worm so that it burrows backward, since its crown must be directed upward.

Tactile Cells. In addition to specialized sense organs, the general body surface of polychaetes is supplied with tactile cells, which are particularly concentrated on the parapodial and head appendages. *Nereis,* for example, responds so positively to the sides of an artificial burrow that it can be killed by other environmental factors to which it would normally give a negative response.

Excretion. Polychaete excretory organs are either protonephridia or metanephridia. There may be one pair of excretory organs per segment or one pair for the entire animal. The anterior end (nephrostome) of the nephridial tubule is located in the coelom of the segment immediately anterior to that from which the nephridiopore opens (Fig. 10–2). The tubule penetrates the posterior septum of the segment, extends into the next segment, where it may be coiled, and then opens to the exterior in the region of the neuropodium. Both the preseptal portion of the nephridium and the postseptal tubule are covered by a reflected layer of peritoneum from the septum.

Protonephridia, which are more primitive than metanephridia, are found in phyllodocids, alciopids, tomopterids, glycerids, nephtyids and a few others. The solenocytes are always located at the short preseptal end of the nephridium and are bathed by coelomic fluid. The solenocyte tubules are very slender and delicate and arise from the nephridial wall in bunches. In *Phyllodoce* the preseptal tube is branched, and the solenocyte tubules originate at the tips of the branches (Fig. 10–39). Fluid apparently enters through the walls of the solenocyte tubules and is then driven by a long flame flagellum into the lumen of the nephridium. The nephridial tubule is ciliated, and fluid is forced to the exterior through the nephridiopore.

All other polychaetes possess metanephridia, in which the preseptal end of the nephridium possesses an open ciliated funnel, the nephrostome, instead of solenocytes. The metanephridium is thus open at both ends. Goodrich (1945) has divided such open nephridia into several categories depending upon their supposed evolutionary origin and their employment as gonoducts. This auxiliary function is discussed in the subsequent section on reproduction. The metanephridium is treated here as a single morphological type.

The typical metanephridium is found in the nereids (Fig. 10–40*C*). The nephrostome possesses an outer investment of peritoneum and the interior is heavily ciliated. The postseptal canal, which extends into the next successive segment, becomes greatly coiled to form a mass of tubules, which are enclosed in a thin saclike covering of peritoneal cells. The nephridiopore opens at the base of the neuropodium on the ventral side. The entire lining of the tubules is ciliated.

The metanephridia of most other polychaetes differ only in minor details. The nephrostome may be developed to various

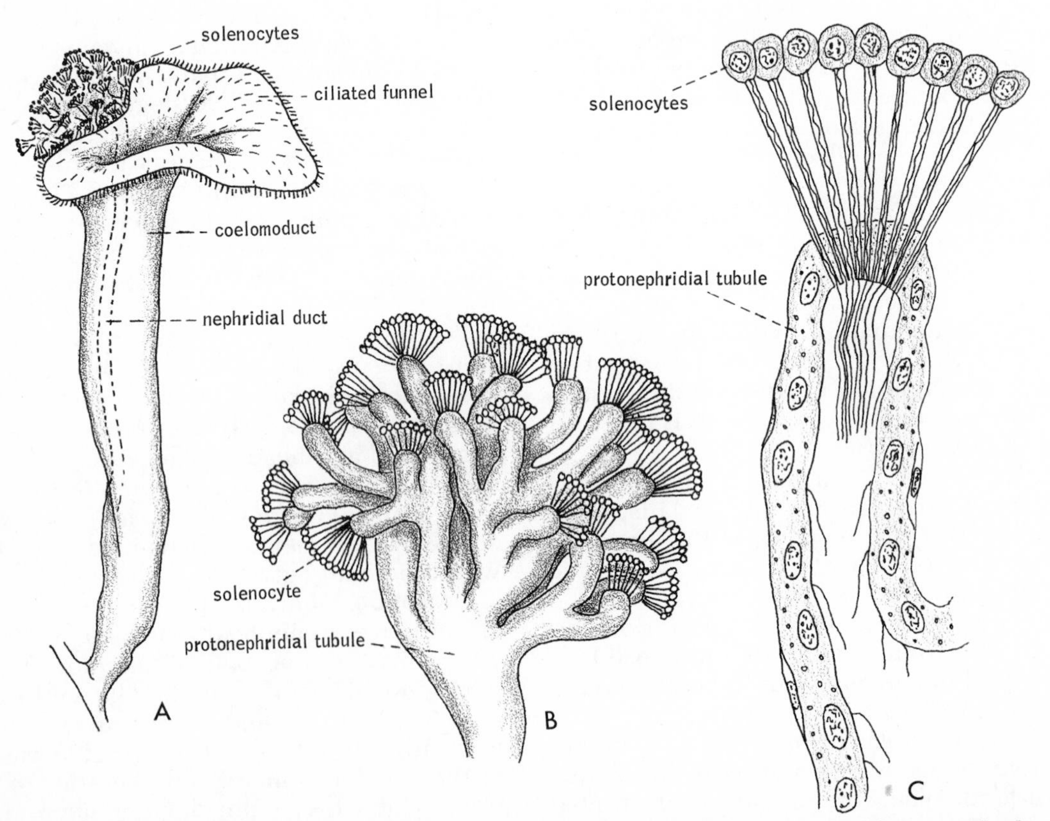

Figure 10–39. Protonephridium and coelomoduct of *Phyllodoce paretti*. *A*. Relation of protonephridium and coelomoduct. *B*. Branched end of protonephridium. *C*. Solenocytes of one protonephridial branch. (All after Goodrich.)

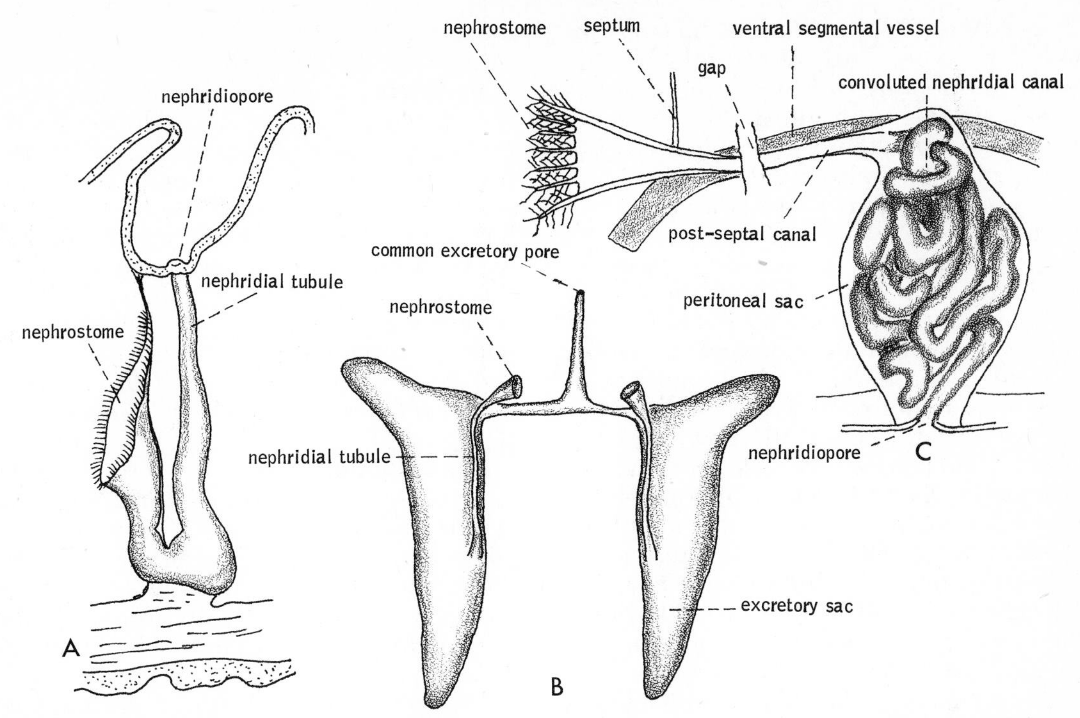

Figure 10–40. Metanephridia of three polychaetes. *A. Scolelepis.* (After Goodrich.) *B.* The serpulid fanworm, *Pomatoceros.* (After Thomas.) *C. Nereis vexillosa.* (After Jones.)

degrees. The post-septal tubule may be **U**-shaped, as in *Scolelepis* (Fig. 10–40*A*), *Odontosyllis,* and the scale worms, rather than coiled.

The absence of septa does not necessarily limit the number of nephridia present in polychaetes. The nephridia, like other segmental organs, may display various degrees of regional restriction in the more specialized families, reaching its zenith in the sedentary tubicolous flabelligerids and fanworms, where only one pair of functional anterior nephridia are to be found. In the fanworms the two nephridia join at the midline to form a single median canal, which extends forward to open through a single nephridiopore on the head (Fig. 10–40*B*). Excretory waste is thus deposited directly outside, and fouling of the tube is avoided.

In *Owenia* the problem of tube-fouling has been solved in a different manner. The nephridiopores open into an epidermal canal that extends forward on both sides of the body to the sixth segment, from which each canal opens to the outside. The canal represents an epidermal groove that has closed over.

The principal nitrogenous waste in polychaetes is ammonia. Wastes diffuse from the coelomic fluid or from the blood to the nephridial tubule and then are transferred to the lumen for removal to the outside. The greatly coiled tubule mass in the nereids appears to be a modification that increases surface area for tubular secretion.

Chloragogen tissue, coelomocytes, and the intestinal wall may play accessory roles in excretion. Chloragogen tissue is composed of brown or greenish cells, located on the wall of the intestine, on various blood vessels, or within the heart body of certain forms. Chloragogen tissue, which has been studied much more extensively in earthworms (see page 292), is an important center of intermediary metabolism and hemoglobin synthesis. Not infrequently, the intestinal wall is green in color owing to the presence of coproporphyrin and pheophorbide pigments. These same green pigments also are present in the epidermis of *Owenia* and other polychaetes. Although the physiological significance of these pigments has not been determined, they may well be excretory products.

In polychaetes the association of the blood vessels with the nephridia is variable. The fanworms and the arenicolids lack a well-developed nephridial blood supply, and the coelomic fluid must be the principal route for waste removal. In other polychaetes, such as *Aphrodita*, the nephridia are surrounded by a network of vessels. In the nereids, the nephridial blood supply varies from one species to the next. For example, in *Nereis vexillosa* there is no distinct association of blood vessels with the nephridium, while in other nereids there is a very close relationship.

Many polychaetes, particularly nereids, can tolerate low salinities, and have become adapted to life in brackish sounds and estuaries. A small number of species live in fresh water, often in lakes which geologically have been rather recently cut off from the sea. This is not true of all, however; the sabellid *Manayunkia speciosa* occurs in enormous numbers in certain regions of the Great Lakes, such as around the mouth of the Detroit River. There are a few terrestrial polychaetes, all tropical Indo-Pacific nereids, which burrow in soil or live in moist litter.

The osmoregulatory ability of polychaetes has been studied most in the European estuarine *Nereis diversicolor*. The salinity tolerance of this species ranges from nearly that of fresh water to salinities greater than sea water. The worms are most abundant at points along the estuary where there are maximum salinity fluctuations. The blood and tissue fluids of *Nereis diversicolor* are isotonic with the surrounding environment over a wide range of salt concentrations, but below a certain point the animal maintains a salt concentration greater than that of the environment. Under these conditions the nephridia produce a urine which is hyposmotic to the coelomic fluid.

Regeneration. Polychaetes have relatively great powers of regeneration. Tentacles, palps, and other attenuated parts that are lost are soon replaced. Some worms even display self-amputation, or autotomy. For example, the elytra of certain scale worms and the posterior trunk segments of some eunicids and onuphids, such as *Marphysa* and *Diopatra*, are detached when the worms are disturbed. The lost parts are later regenerated. In general, new trunk segments are more readily regenerated in worms with undifferentiated body regions, and both a new head and a new tail can be re-formed. In worms with a thorax and abdomen or with a tail, caudal regeneration may take place, but the formation of a new head is less common. Cells for regeneration are supplied by the remains of whatever tissues have been lost; i.e., new epidermal cells are derived from epidermis, and mesoderm from dedifferentiated muscle cells, coelomocytes, and similar tissues. Experimental studies, especially with oligochaetes, indicate that the nervous system plays an important inductive role in regeneration and in some way there is an involvement of the neuroendocrine system. Severing the nerve cord alone can cause the formation of a new head where the cut is made. A lateral secondary head will form if the severed end of a cord, cut just behind the subesophageal ganglion, is pulled through a hole in the lateral body wall.

A good review of some of the studies on polychaete regeneration is presented by Rose (1970).

Reproduction. Asexual reproduction is known in some polychaetes, including syllids, sabellid fanworms, and spionids; it takes place by budding or division of the body. The chaetopterid *Phyllochaetopterus* normally reproduces by division of the body into a number of sections, each of which can then form a new individual.

Most polychaetes reproduce only sexually, and the majority of species are dioecious. Polychaete gonads are not distinct organs but are masses of developing gametes, which develop as projections or swellings of the peritoneum in different parts of some segments. For example, in *Nereis* the ventral septal peritoneum is the site of gamete formation.

Primitively, most of the segments produce gametes; this is true in many scale worms, many nereids, and the eunicids. In general, restriction of segments for gamete formation coincides with specialization of the trunk into different regions. When there is a distinct thoracic and abdominal region, the gonads are usually limited to the abdomen. In some fanworms, all of the abdominal segments form gametes, which arise from the anterior peritoneum of each segment. However, in *Amphitrite* and *Arenicola*, gamete-forming segments are much more limited in number. In *Arenicola* there are only six reproductive segments, which contain relatively distinct gonads, located one pair to a segment on gonadial vessels associated with the nephridia.

The gametes are usually shed into the coelom as gametogonia or primary gametocytes, and maturation takes place largely in the coelomic fluid. When the worm is mature, the coelom is packed with eggs or sperm; in species where the body wall is thin or not heavily pigmented, the gravid condition is easily apparent. For example, the abdomen of a ripe male *Pomatoceros* appears white, while the abdomen of a female is bright pink or orange owing to the color of the sperm or the eggs.

Some polychaetes, such as the capitellids, possess special gonoducts, but the gametes in most species leave the body through the metanephridia or in some cases by the dehiscence, or breakdown, of the body wall.

When production of gametes takes place in most segments, the metanephridia function both as excretory and genital ducts; but when the gonads are restricted to specific segments, the metanephridia serve only in the excretory capacity in the sterile segments and serve in the reproductive capacity only or in both capacities in the fertile segments. Much of our knowledge of nephridial structure in annelids results from the valuable work of Goodrich (1945), who made extensive studies on the role of nephridia in sperm and egg transport.

In fanworms, only the large anterior nephridia are excretory, and the nephridia of the fertile abdominal segments function only as exits for the genital products. As the gametes of the fanworm pass through the nephridiopore, they are carried to the ciliated median ventral groove, where they are moved anteriorly and out of the tube.

Gametes probably are moved through the nephridia primarily by cilia, although, in *Arenicola,* body contractions and hydrostatic pressure are important. Nephrostomial cilia supposedly act as a sorting mechanism and prevent the passage of either unripe eggs or coelomocytes.

The escape of gametes through the rupture in the body wall is perhaps a specialized condition. Rupturing is found in some nereids, syllids, and eunicids, all of which become pelagic at sexual maturity. After the rupture of the body wall, the adults die.

Although in most polychaetes fertilization of eggs takes place in the surrounding sea water, copulation does occur in some groups. The genital nephridia of *Pisione remota* and some of the parapodial cirri of the planktonic alciopids have been modified as seminal receptacles. Also, copulatory setae or parapodia have evolved in a few species. *Platynereis megalops,* which occurs in Woods Hole, Massachusetts, displays a very unusual method of internal fertilization. During swarming the male is supposed to insert his anus into the mouth of the female while entwined about her. The sperm are into the egg-filled coelom of the female, the gut of the female having undergone regression. Almost immediately the fertilized eggs are shed from the posterior end of the female.

Epitoky. Epitoky is a reproductive phenomenon characteristic of nereids, syllids, and eunicids. Epitoky is the formation of a reproductive individual, or epitoke, that differs in a varying number of secondary sexual characteristics from a nonsexual form, or atoke. Epitokal modifications include changes in the formation of the head, the structure of the parapodia, the size of the segments, and the segmental musculature, among others.

Often the gamete-bearing segments are the most strikingly modified, so that the body of the worm appears to be divided into two markedly different regions. For example, the epitoke of *Nereis irrorata* has large eyes and reduced prostomial palps and tentacles (Fig. 10–41). The anterior 15 to 20 trunk segments are not greatly modified, but the remaining segments, comprising the epitokal region and packed with gametes, are much enlarged; their parapodia contain fans of long spatulate swimming setae. In *Palola* (=*Eu-*

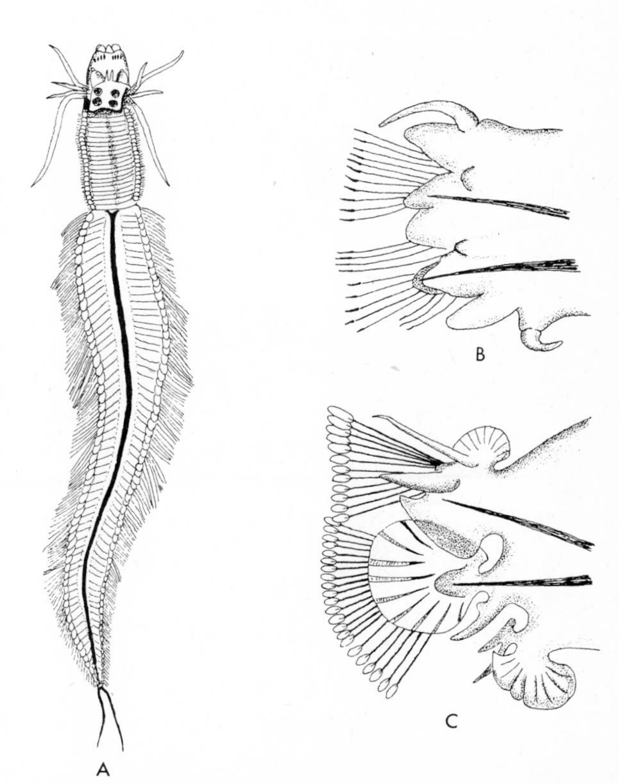

Figure 10–41. A. Epitokous male of *Nereis irrorata.* (After Rullier from Fauvel.) *B* and *C.* Parapodia of atoke (*B*) and epitoke (*C*) of *Nereis irrorata* male. (After Fauvel.)

nice) viridis, the Samoan palolo worm, the anterior of the worm is unmodified, and the epitokal region consists of a very long chain of narrower but longer posterior segments, each with an eye spot on the ventral side (Fig. 10–42*A*).

Epitokous individuals arise from an atokous form either by direct transformation or by budding. Transformation is characteristic of the nereids and the eunicids, while asexual budding of epitokes is common in some syllids. Syllid epitokes, often called stolons, usually form at the caudal end, either as a single body or as a chain or cluster of individuals (Fig. 10–42*C*); but in *Syllis ramosus* numerous epitokes bud from the sides of the body and may in turn form secondary buds (Fig. 10–42*B*).

Swarming. Usually, epitokous polychaetes swim to the surface during the shedding of the eggs and sperm. This behavior, known as swarming, congregates sexually

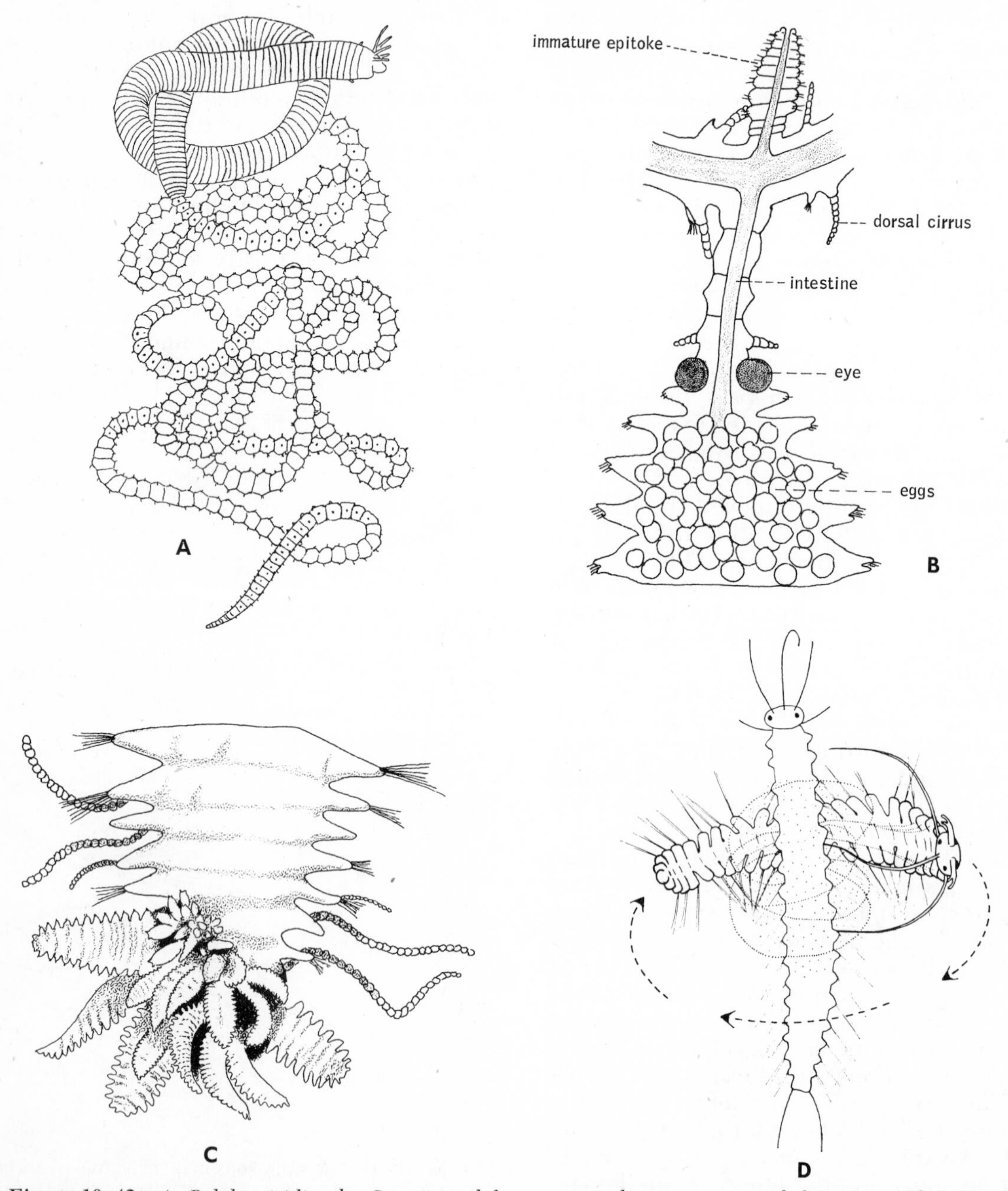

Figure 10–42. *A. Palola viridis*, the Samoan palolo worm, with posterior epitokal region. (After Woodworth from Fauvel.) *B. Syllis ramosus*, showing epitokes developing from sides of atoke. (After MacIntosh from Potts.) *C.* Posterior of *Trypanosyllis*, showing cluster of budding epitokes. (After Potts from Fauvel.) *D.* Syllid polychaetes during swarming. Male is swimming around female and releasing sperm. (After Gridholm.)

mature individuals and increases the likelihood of fertilization. Experimental evidence indicates that the female produces a pheromone that attracts the male and stimulates shedding of the sperm. The sperm in turn stimulate the shedding of the eggs. The male syllid *Autolytus*, for example, swims in circles around the female, touching her with the antennae and releasing sperm (Fig. 10–42*D*). A West Indian species of *Odontosyllis* luminesces when the worms come to the surface at dusk during swarming.

To ensure fertilization, swarming in the male and female must coincide and take place over a relatively short period of time. Thus the swarming phenomenon is usually marked by a very distinct periodicity. Swarming is induced in *Autolytus edwardsi* by changes in light intensity, and the worms leave the bottom and swim to the surface at dawn and dusk.

Swarming often coincides with lunar periods, the most striking examples being displayed by the so-called palolo worms. The name palolo originally referred to a Samoan species of eunicid, *Palolo (=Eunice)*, but it is now applied to other species as well. The Samoan palolo worm occupies rock and coral crevices below the low-tide mark, and releases epitokes in October or November at the beginning of the last lunar quarter. The natives, who consider the epitokes a great delicacy, eagerly await the predicted night of swarming in boats and scoop up great numbers of the worms from the ocean surface. Needless to say, swarming also provides a day of feasting for fish, birds, and other predators.

Clark and Hess (1940) have made a rather complete study of the Atlantic palolo, *Eunice schemacephala*. This West Indian worm lives in habitats similar to those of the Samoan palolo. The worms are negatively phototactic and emerge from the burrows to feed only at night. Swarming takes place in July during the first and third quarters of a lunar cycle. At three or four o'clock in the morning during such a period, the worm backs out of its burrow, and the caudal sexual epitokal region breaks free. The epitoke swims to the surface, where it undergoes spiral swimming movements. By dawn the ocean surface is covered with sexual bodies, and at the rising of the sun the epitokes rupture. Fertilization immediately follows dehiscence, and acres of eggs may cover the sea. A ciliated larval stage is attained by the next day, and in three days the larvae sink to the bottom.

The physiological causes of swarming and swarming periodicity are still poorly understood. Certainly, a complex of factors is involved. The inherent swarming reflex increases with sexual maturity and is very probably under hormonal control. Studies on nereids and syllids have shown that neurosecretory cells in the brain or in the proventriculus (syllids) elaborate some substance that inhibits the formation of gametes and epitokal transformation or budding. For example, if the brain of a young nereid is removed, epitokal characters appear; or if the brain of a young worm is transplanted into the coelom of a mature individual, the epitokal characters in the host are suppressed.

Environmental factors may greatly influence the time of swarming. Clark and Hess (1940) have shown that wave action and turbidity suppress swarming in the Atlantic palolo. Light also suppresses swarming, even moonlight. But the actual role of the moon in swarming is difficult to understand. The relation of lunar phases to swarming periods is different in different species. This variation makes any hypothesis based merely on light intensity difficult to prove.

Hermaphroditism. There are some hermaphroditic polychaetes, including a number of fanworms. In most hermaphroditic fanworms, the anterior abdominal segments produce eggs and the posterior segments produce sperm. However, in *Branchiomma*, a sabellid, sperm and eggs are formed in the same segments.

Egg Deposition. In many polychaetes, the eggs are shed freely into the sea water, where they become planktonic and separate. In some polychaetes, however, the eggs are laid in mucous masses that are attached to the tubes or to other objects. *Clymenella* (Fig. 10–43*A*), a bamboo worm, produces a large ovoid egg mass that is attached to the chimney of the tube.

Arenicola cristata also produces mucous egg masses that are anchored to the surface of the sand bottom. *Scoloplos armiger* displays the same characteristic, with each egg mass being approximately 1 × 2 cm. and containing 600 to 1800 eggs. The jelly is believed to be produced by gland cells lining the female nephridia.

Many polychaetes brood their eggs. There are tubicolous species, such as some spionids and serpulids, which brood their eggs attached within the tube. Some scale worms use the respiratory water channel beneath the elytra as a brood chamber. Some species of

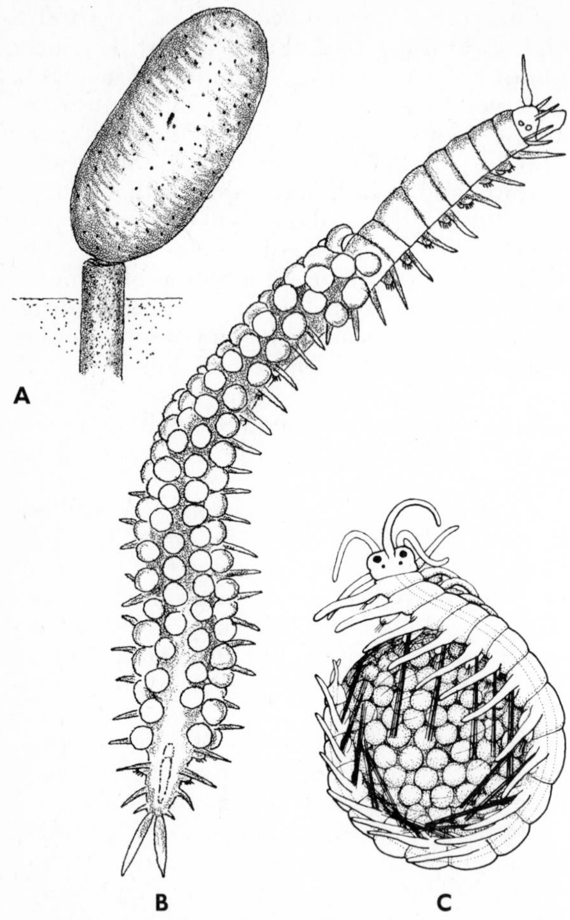

Figure 10–43. *A. Clymenella mucosa* egg mass attached to end of tube. *B. Brania* with eggs attached to dorsal surface. (After Viguier and Pierantoni from Potts.) *C. Autolytus* carrying egg mass beneath body. (After Thorson.)

Spirorbis brood their eggs in the cavity of the operculum, and *Autolytus* broods its eggs within a secreted sac attached to the ventral surface of the body. The parapodia may be used to retain the eggs. In *Pionosyllis*, an egg is attached to the base of each dorsal cirrus. Another syllid, *Brania*, attaches three or four large eggs directly to the dorsal surface of each segment (Fig. 10–43*B*), and the phyllodocid polychaete, *Notophyllum*, uses the same site to attach agglutinated bands of eggs.

There are a few polychaetes which display internal brooding. The freshwater *Nereis limnicola*, for example, develops its larvae within the coelom. The serpulid *Filograna implexa* and some nereids release developing larvae, but this is relatively rare. *Ctenodrilus* is reported to be truly viviparous, nutrition for the embryo being supplied by maternal blood vessels.

Embryogeny. The polychaete egg is telolecithal with a variable amount of yolk, and cleavage is spiral. A displaced blastocoel is usually present, but a stereoblastula develops in *Nereis*, *Capitella*, and others. Gastrulation takes place by means of invagination, epiboly, or both.

The Trochophore. After gastrulation, the embryo rapidly develops into a larval stage known as a trochophore (Figs. 10–44*B* and 10–45). The trochophore larva is not characteristic only of polychaetes, but also occurs in mollusks and other phyla. In polychaetes, the trochophore structure usually begins to appear on about the second day of development. The embryo assumes a top-shaped form, the cells of the first quartet being located at the apex, or aboral end. A characteristic feature of the trochophore is the presence of a girdle of ciliated cells, called the prototroch; it rings the larva just above the equator.

The prototroch cells are derived in part from the first quartet—specifically the $1a^2$ to $1d^2$ cells, called the primary trochoblasts. The primary trochoblasts form only the group of ciliated cells in each quadrant. The completion of the girdle comes from the secondary trochoblasts, which fill in between the primary groups. In many polychaetes, cells derived from the first quartet form a conspicuous crosslike pattern. The $1a^{112}$ to $1d^{112}$ cells form the rays of the cross, which extend from the apical sensory plate. Cells of the apical sensory plate usually bear a tuft of cilia as in the larvae of turbellarians and rhynchocoels.

At the oral end of the embryo, the blastopore becomes elongated, extending along the lower side of the embryo. The upper end of the blastopore forms the mouth and stomodeum, which, because of differential growth, develops so that it lies just beneath the prototroch. The remainder of the blastopore closes, and a new opening, the anus, forms at the site of the lower blastoporal slit and connects with the gut.

Two additional girdles of cilia commonly appear after the formation of the prototroch and after further differentiation of the oral end of the embryo. The first girdle of cilia is called the telotroch and rings the embryo just above the anus (Fig. 10–44*A* and *C*); the second girdle of cilia, which appears still later and is called the metatroch, forms just below the mouth (Fig. 10–44*A*).

Internally the two teloblast cells, derived from the *4d* cell, each give rise to a band of mesoderm, which extends dorsally toward

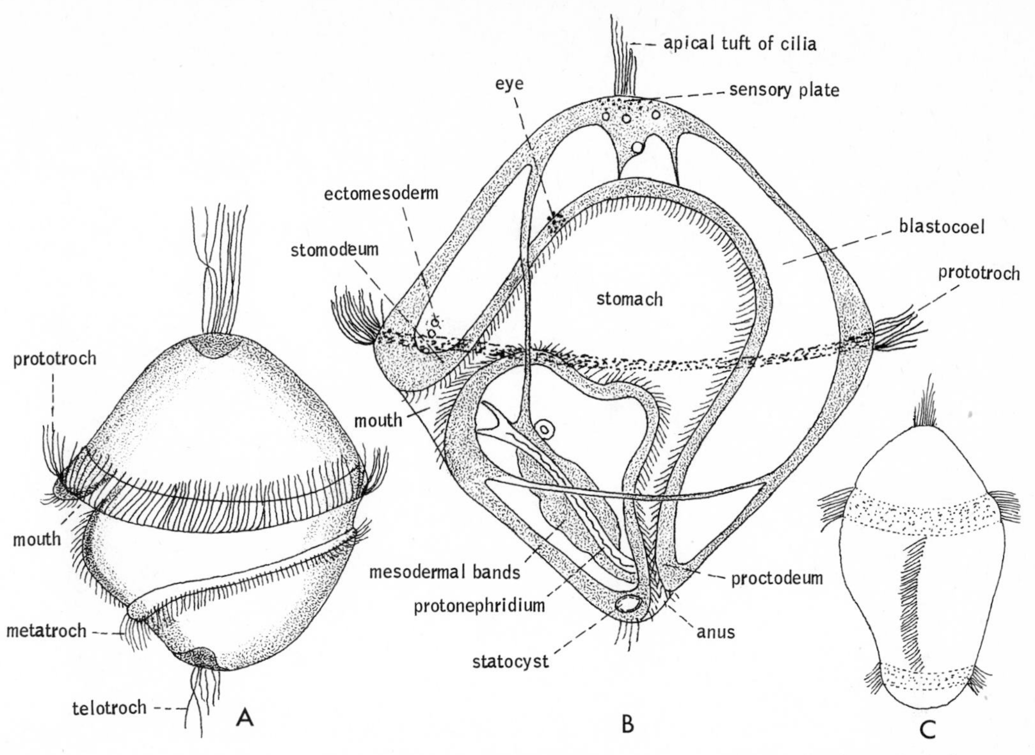

Figure 10–44. *A*. Trochophore of *Polygordius*, showing prototroch, metatroch, and telotroch. (After Dawydoff.) *B*. Structure of an annelid trochophore. (After Shearer from Hyman.) *C*. Trochophore of *Marphysa*. (After Dawydoff.)

the mouth from both sides of the anus (Fig. 10–44*B*). The mesodermal bands form all of the entomesoderm. The larval alimentary tract consists of the stomodeum, formed from second and third quartet ectoderm, and the stomach and intestine, derived from the endoderm. Blastomeres *4A*, *4B*, *4C*, and *4D*, plus a few offspring from the *4d* cells, form all of the entoderm.

The old blastocoel remains in the form of a large cavity lying between the gut and the outer ectoderm; into this cavity is proliferated ectomesoderm from the second and third quartets. This ectomesoderm forms larval muscle bands that become attached to the gut and to the ciliary girdles. Third quartet ectoderm also forms two larval protonephridia that lie in the blastocoel cavity on both sides of the intestine. Finally, in a highly developed trochophore, radial nerves emerge from a ganglionic mass lying beneath the apical plate. The radial nerves connect with one to several longitudinal nerves encircling the larva.

The fully developed trochophore larva can be divided into three regions; the prototrochal region, consisting of the apical plate, the prototroch, and the mouth region; the pygidium, consisting of the telotroch and the anal region behind it; and the growth zone, which includes all of the larva between the mouth region and the telotroch. The growth region eventually forms all of the trunk segments of the body.

Primitively, the trochophore is a planktonic, free-swimming larva. Planktonic trochophores appear in the development of *Podarke*, *Hydroides*, certain nereids, *Spirorbis*, *Sabellaria*, and *Scoloplos*. The greatest development of larval structures is attained in these trochophores, which not only are free-swimming but are often also feeding larvae.

Metamorphosis. Additional development takes place as the trochophore metamorphoses into the adult body form (Fig. 10–45). In polychaetes with planktonic feeding trochophores, metamorphosis is somewhat indirect and results in the loss of many of the larval structures, such as nephridia, ectomesodermal muscle bands, and ciliated girdles.

The most conspicuous feature of metamorphosis is the gradual lengthening of the growth zone—the region between the mouth and the telotroch—owing to the

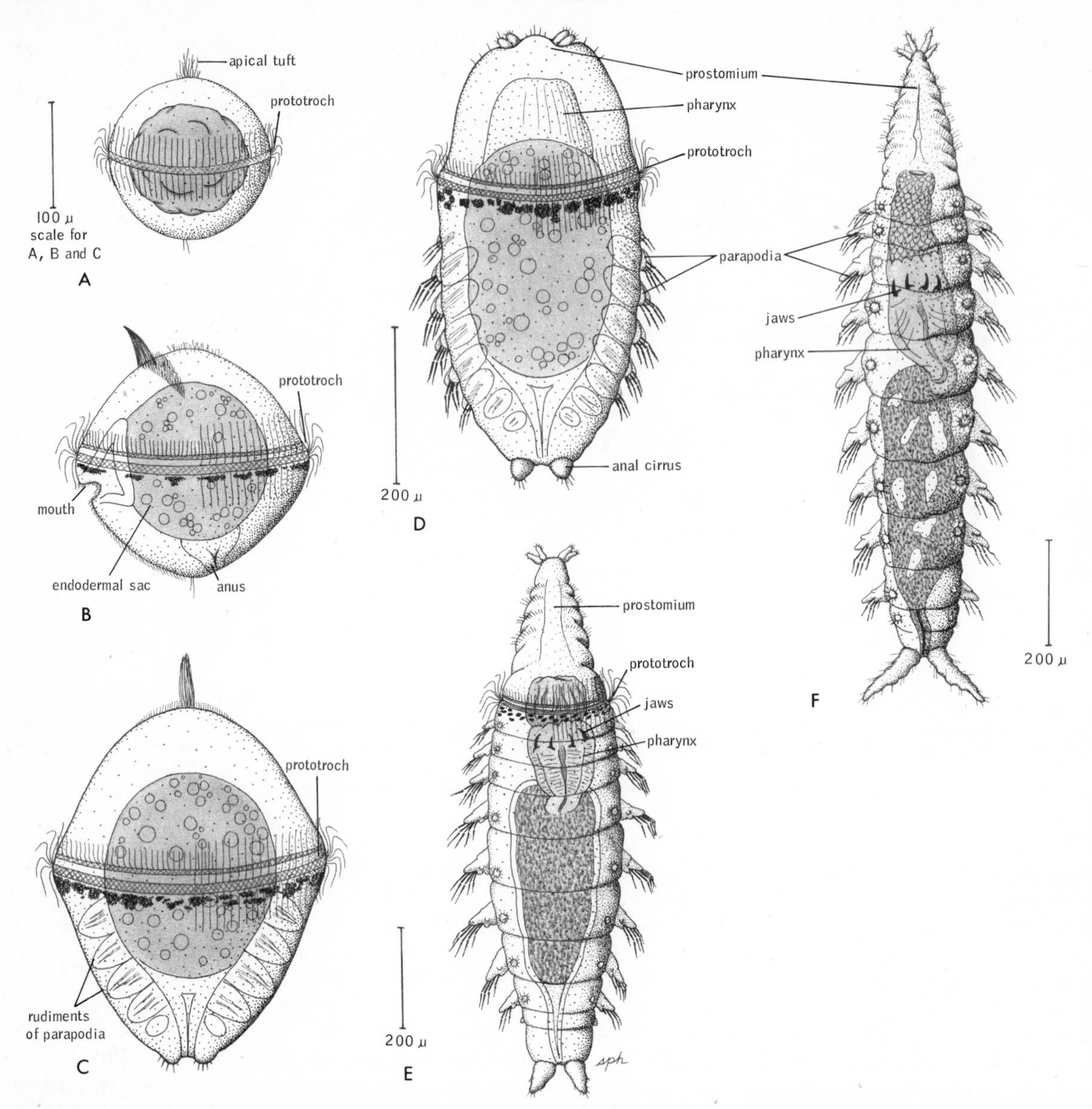

Figure 10–45. Larval stages of *Glycera convoluta*. *A*. Early trochophore (15 hours). *B*. Later trochophore (ten days). *C*. Young metatrochophore (four weeks). *D*. Metatrochophore at seven weeks. Although still a swimming stage, it frequently comes to rest on the bottom. *E*. Post-larva at eight weeks. Metamorphosis follows metatrochophore illustrated in *D*. Larva becomes benthic and raptorial. *F*. Young worm at two months. Compare to Figure 10–15. (All after Cazaux, C., 1967: Vie et Milieu, *18*:559–571.)

formation and development of trunk segments (Fig. 10–45). The segments develop from anterior to posterior, and the germinal region always remains just in front of the terminal pygidium. The mesodermal bands undergo active proliferation and anteriorly lay down a pair of somites for each newly formed segment.

Each somite becomes hollow, forming half the coelomic compartment; the mesodermal cells form the peritoneum, the body wall and gut musculature, the blood vessels, and the nephridia for each segment. Since somite development takes place in the blastocoel, this cavity is gradually obliterated. In almost all polychaetes studied so far, all trunk ectoderm originates from blastomere *2d*. This ectoderm forms not only the integument but also the ganglia of each segment. The formation of body segments is marked externally by the development of setal sacs and setae. Also, each of the early segments is often ringed by a girdle of cilia.

In the prototrochal region, which originally formed the major part of the body of the trochophore, the cells of the apical plate form the prostomium and the brain.

The brain gives rise to the circumpharyngeal connectives, which grow ventrally around the gut and connect with the subesophageal ganglion. Mesodermal tissue grows forward and invades the prostomium.

The mouth region may give rise to the peristomium, in which case the peristomium is without setae (achetous) and does not contain the subesophageal ganglion. Commonly, as in *Nereis, Capitella,* and the fanworms, the mouth region fuses with the first trunk segments to form the peristomium, in which case the peristomium contains the subesophageal ganglion—the ganglion of the first trunk segment. The peristomium, however, may still be achetous, as in *Nereis,* owing to resorption of the setal sacs. Thus, this difference in peristomial origins indicates that the peristomia of different polychaetes are not necessarily homologous.

Metamorphosis may result in the immediate termination of a planktonic existence, but more often the elongated, metamorphosing larvae remain planktonic for varying lengths of time. The metamorphosing stages of spionids, sabellariids, and oweniids even possess greatly enlarged anterior setae that serve as flotation or protective devices. *Nereis fucata* settles to the bottom 22 days after the trochophore stage has begun; at the time of settling, there are six or seven setae-bearing segments. The larvae of *Ophelia bicornis* are free-swimming until three pairs of setae-bearing segments have developed. In both *Arenicola marina* and *Nereis diversicolor,* the trochophore larvae are not active swimmers and remain on the bottom.

In many polychaetes there is no free trochophore; the stage is passed in the egg prior to hatching. In such species metamorphosis is more direct, since larval structures are never greatly developed to begin with. When the trochophore stage is suppressed, hatching occurs at various times during advanced development; and there may yet be a free-swimming, post-trochophoral, larval stage. For example, in *Autolytus* an elongated achetous larva breaks free from the brood sac of the mother. On the other hand, *Clymenella mucosa* and *Scoloplos armiger* have no planktonic stages and assume the adult mode of existence on emerging from the jelly egg case. In any event, the larvae eventually sink to the bottom, where they complete postlarval development and assume the habits of the adult.

Symbiosis

Commensalism. In their commensal relationship with other animals, polychaetes may be hosts or guests. As might be expected, the role of host is played primarily by the tube-dwellers and the burrowing polychaetes. Their guests include scale worms and crustaceans, particularly species of little crabs. Perhaps the most notable example is a species of crab that lives exclusively in *Chaetopterus* tubes and belongs to the genus *Pinnixa.* Other species of *Pinnixa* live in maldanid tubes. The crabs enter the tubes as juveniles and feed on food material brought in by the water currents. A male and a female crab are almost always found together; the larval stages are swept out by water circulating through the tube. A few nontube-dwellers also harbor commensals; for example, *Pseudopythina rugifera,* a small clam, lives attached to the ventral surface of the sea mouse.

Polychaetes known to be commensal with other animals are distributed in numerous families throughout the class. They live in tubes and burrows of other polychaetes and crustaceans, with hermit crabs, on echiuroid worms, in the ambulacral grooves of sea stars, and on sea cucumbers and other animals. Several species of *Arctonoe,* Pacific scale worms, live associated with sea cucumbers, sea stars, and mollusks; they display colors similar to those of the host and have the setae modified for clinging.

Parasitism. Parasitism is not common among polychaetes. *Labrorostratus,* an arabellid, lives in the coelom of other polychaetes and may be almost as big as its host. Clark (1956) has noted that endoparasitic polychaetes are not markedly modified, as is usually the case in endoparasitism. The life histories of these parasites are unknown; perhaps they leave the host and assume a free-swimming existence at sexual maturity. Polychaete ectoparasites include the Histriobdellidae (Fig. 10–46*A*), which are parasites in the gill chambers of crustaceans, and the Ichthyotomidae (Fig. 10–46*B*) which attach to the fins of marine eels. The Ichthyotomidae suck blood by means of piercing stylets and a pumping pharynx.

A most interesting group of commensal and parasitic polychaetes are the myzostomes, sometimes placed in a separate phylum or class, the Myzostomida. These little worms are rarely more than a few millimeters long and resemble flatworms (Fig. 10–46*C*). The body is greatly flattened and the five pairs of parapodia are carried on the undersurface. The most peculiar feature of the myzostomes is their constant association with echinoderms. Moreover, except for a

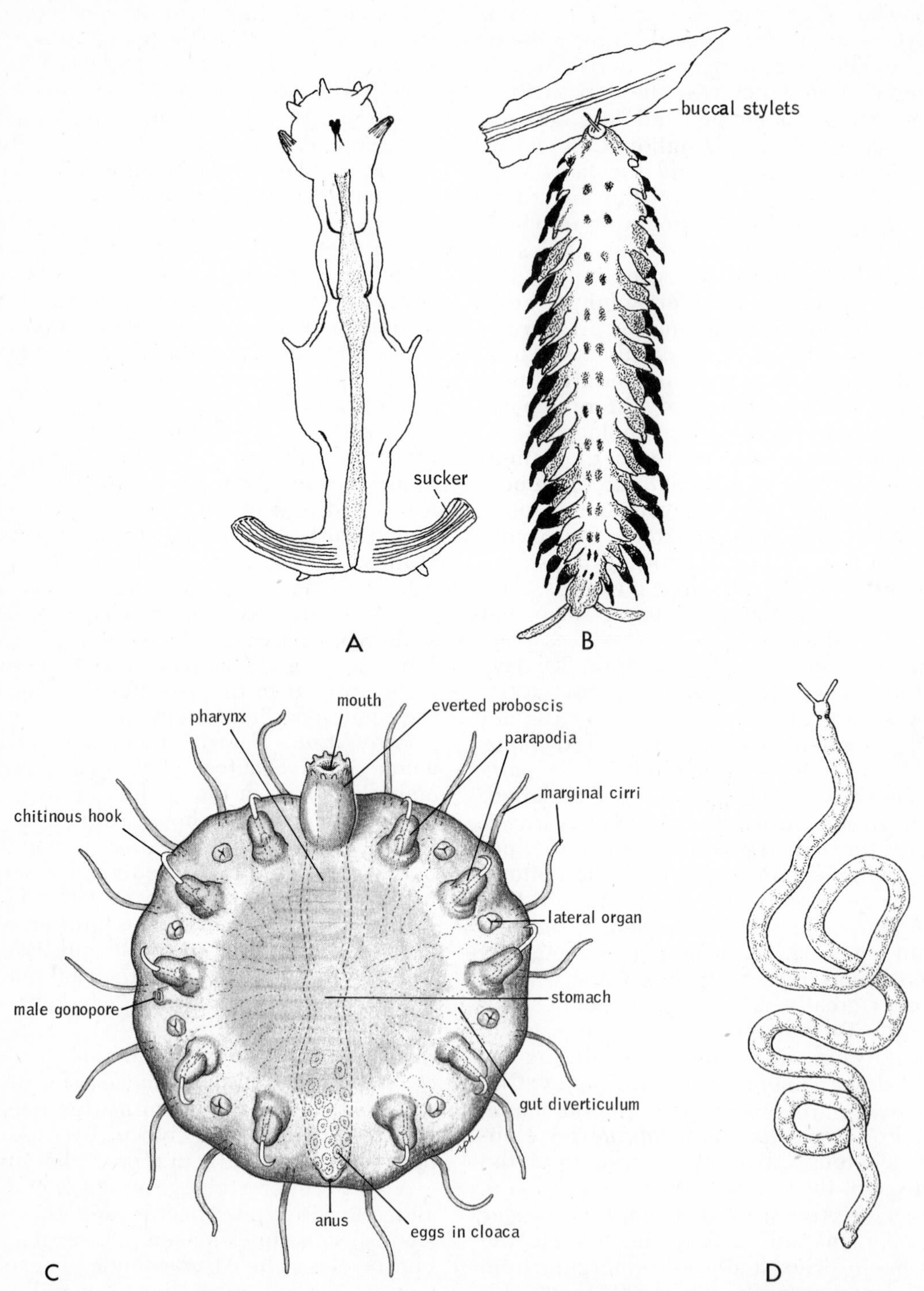

Figure 10–46. *A. Histriobdella*, an ectoparasite in crustacean gill chambers. (After Haswell from Fauvel.) *B. Ichthyotomus*, an ectoparasite on fins of marine eels. (After Eisig from Fauvel.) *C. Myzostoma*, a commensal on crinoids. *D. Polygordius neapolitanus*, an archiannelid. Segmentation is poorly indicated externally. (After Fraipont.)

relatively small number of species that live on brittle stars and sea stars, the majority of the myzostomes are restricted to crinoids, or sea lilies. Most myzostomes are commensals, living on plankton gathered by the arms of the host; and either they are sedentary or crawl about over the surface of the host's body. Parasitic myzostomes live in the body wall, the coelom, or the digestive tract of the host. Development includes a typical free-swimming polychaete larva.

SYSTEMATIC RÉSUMÉ OF CLASS POLYCHAETA

This résumé has been adapted from those of Fauvel (1959) and Day (1967), who, along with many modern authorities, consider the polychaete families too heterogeneous to be grouped into orders. However, a system of orders based largely on the structure of the pharynx has been proposed by Dales (1963), and one based largely on larval structure has been proposed by Mileikovsky (1968). It is doubtful that the Errantia and Sedentaria are natural divisions, although these two taxa are used below. This list includes the more common families to which references have been made in the preceding discussion. Lack of space permits mention of only the distinguishing characteristics.

Subclass Errantia. Polychaetes with most of the following characteristics: Segments numerous and similar. Parapodia well developed, with acicula and setae. Head with sensory structures. Pharynx with jaws or teeth. Swimming, crawling, burrowing, and tube-dwelling worms.

Family Aphroditidae. Sea mice. Long setae forming a feltlike covering of the dorsal surface. *Aphrodita, Laetmonice.*

Family Polynoidae and Sigalionidae. Scale worms. Dorsal surface bears elytra. Chiefly crawling polychaetes. *Lepidonotus, Harmothoe, Polynoe, Lepidasthenia, Lagisca, Iphinoe, Arctonoe, Euthalenessa, Sigalion, Thalanessa.*

Family Phyllodocidae. Uniramous parapodia with flattened, leaflike cirri. Crawling polychaetes. *Notophyllum, Phyllodoce.*

Family Amphinomidae. Crawling polychaetes with brittle, poisonous setae. Fireworms. *Hermodice, Amphinome.*

Family Pisionidae. Crawling polychaetes without prostomial antennae, but with two pairs of peristomial cirri. *Pisione.*

Family Alciopidae. Planktonic worms; possess transparent bodies and two large eyes. *Alciopa, Vanadis.*

Family Tomopteridae. Planktonic worms with membranous pinnules in place of setae. *Tomopteris.*

Family Hesionidae. Crawling polychaetes with well-developed prostomial sense organs and long dorsal parapodial cirri. *Podarke.*

Family Syllidae. Small crawling worms with long, delicate bodies and uniramous parapodia. *Syllis, Odontosyllis, Pionosyllis, Autolytus, Procerastea, Trypanosyllis, Brania (=Grubia).*

Family Nephtyidae. Rapid-crawling worms with well-developed prostomial sense organs located between parapodial rami. *Nephtys.*

Family Glyceridae. Errant burrowing worms; possess conical prostomium and long proboscis with four jaws. *Glycera.*

Superfamily Eunicida. Elongated worms with proboscis armature of at least two pieces. This large family contains a number of ecologically diverse but closely related families. The Eunicidae and Onuphidae, with eyes and antennae, contain many tubicolous species. *Eunice, Marphysa, Palola, Lysidice; Onuphis, Diopatra, Hyalinoecia, Nothria.* The Lysaretidae are crawling polychaetes. *Oenone, Halla.* The Arabellidae and Lumbrineridae are threadlike errant burrowers with reduced prostomial sensory structures. *Arabella, Drilonereis, Labrorostratus; Lumbrinereis.*

Family Histriobdellidae. Ectoparasitic. *Histriobdella.*

Family Ichthyotomidae. Ectoparasitic. *Ichthyotomus.*

Family Myzostomidae. Greatly flattened commensals and parasites of echinoderms, particularly crinoids. *Myzostoma.*

Subclass Sedentaria. Polychaetes with most of the following characteristics: Body commonly displays regional differentiation. Parapodia reduced, without acicula or compound setae. Prostomium without sensory appendages, but head commonly provided with palps, tentacles, and other structures for feeding. No teeth or jaws present.

Family Orbiniidae. Sedentary burrowers with conical or globular prostomium, without any appendages. *Orbinia, Scoloplos.*

Family Spionidae. Tubicolous polychaetes with two long prostomial palps. *Spio, Scolelepis (=Nerine), Polydora, Mala.*

Family Magelonidae. Sedentary burrowers with one pair of long papillate palps. *Magelona.*

Family Chaetopteridae. Tubicolous polychaetes with one pair of long palps. *Chaetopterus, Phyllochaetopterus, Spiochaetopterus, Mesochaetopterus (=Ranzania).*

Family Cirratulidae. Segments bear long, threadlike gills. *Cirratulus, Cirriformia (=Audouinia), Ctenodrilus.*

Family Flabelligeridae. Sedentary burrowers with cephalic gills that can be retracted into the mouth. *Flabelligera.*

Family Opheliidae. Errant burrowers with conical prostomium. *Ophelia, Polyophthalmus, Euzonus (=Thoracophelia).*

Family Capitellidae. Errant burrowers with conical prostomium and long body. *Dasybranchus, Capitella, Notomastus.*

Family Arenicolidae. Sedentary burrowers without head appendages. Lugworms. *Arenicola.*

Family Maldanidae. Tubicolous polychaetes with small prostomium fused to peristomium, and without head appendages. The bamboo worms. *Clymenella, Maldane.*

Family Oweniidae. Tubicolous species without prostomial appendages, or with foliaceous prostomial crown. *Owenia.*

Family Sabellariidae. Tubicolous polychaetes in which head is modified to form operculum for closing tube. *Sabellaria.*

Family Pectinariidae, or Amphictenidae. Tubicolous polychaetes constructing conical tubes. Head with heavy, forward-directed, golden setae used as an operculum and for digging. *Pectinaria (=Cistenides).*

Family Ampharetidae. Tubicolous polychaetes with retractile buccal tentacles. *Amphicteis, Ampharete.*

Family Terebellidae. Prostomium bears numerous long, filiform tentacles. Tubicolous or sedentary burrowers. *Terebella, Amphitrite, Polymnia.*

Family Sabellidae. The fanworms or feather-duster worms. Noncalcareous tubes. *Branchiomma (=Dasychone), Sabella, Myxicola, Potamilla, Manayunkia, Megalomma, Spirographis.*

Family Serpulidae. The fanworms or feather-duster worms. Tubes calcareous. *Serpula, Apomatus, Spirorbis, Pomatoceros, Filograna (=Salmacina), Hydroides (=Eupomatus).*

Archiannelida. Most modern authorities on the annelids agree that the old phylum or class Archiannelida represents an assortment of enigmatic annelids. At the present time the archiannelid group has been reduced to five small families. The great majority are marine, and most are members of the interstitial fauna. Setae and parapodia are commonly absent, and the epidermis may be smooth or ciliated (Fig. 10–46*D*). Internal structures are reduced to various degrees. The name Archiannelida was created in the belief that these worms were primitive, but Hermans (1969), who has recently reviewed their systematic position, has concluded that their structural peculiarities are largely adaptations for an interstitial existence and that they should be considered a separate order of polychaetes. *Polygordius, Protodrilus, Nerilla, Dinophilus.*

CLASS OLIGOCHAETA

The class Oligochaeta contains some 3100 species of annelids, including the familiar earthworms and many species that live in fresh water. Some freshwater forms burrow in the bottom mud and silt; others live among submerged vegetation. In general, oligochaetes approximate the polychaetes in size, but reach greater extremes. Some species of the freshwater genus *Chaetogaster* and members of the family Aeolosomatidae are less than 0.5 mm. long, whereas the giant earthworm of Australia and other parts of the world may exceed 3 m. in length (Fig. 10–47).

External Anatomy. Oligochaetes perhaps evolved directly from the ancestral marine annelids, retaining adaptations for burrowing and for invading freshwater sediments and soil. Metamerism is highly developed, parapodia are absent, and the prostomium is usually a small rounded lobe or a small cone without sensory appendages (Figs. 10–48*A* and 10–50). In a few genera, such as *Stylaria* (Fig. 10–48), the prostomium may be drawn out into a tentacle. The prostomium is usually distinct from the peristomium, but the two regions may be indistinguishably fused.

Although oligochaetes have no parapodia, with few exceptions, they have setae. Although not as diverse as those of polychaetes, oligochaete setae may be long or short; heavy or needle-like; straight or curved; and blunt, pointed, forked, pectinate, or even plumose. Commonly the shaft of the oligochaete seta is **S**-shaped with a swelling or nodulus, in the middle (Fig. 10–48*B*).

Figure 10–47. An Australian giant earthworm. (Courtesy of Globe Photos.)

In general, the longer setae are characteristic of aquatic species (Fig. 10–48*E*), while the setae of earthworms project only a short distance beyond the integument. On each side of a segment there are setal sacs, in which the setae are secreted and from which they emerge as groups or bundles. Two of the groups are ventral and two are ventrolateral or dorsolateral. The number of setae per bundle varies from one to as many as twenty-five (Fig. 10–48*E*). In any case, they are less numerous in these worms than in polychaetes, hence the origin of the name Oligochaeta—"few setae."

In most earthworms, such as *Lumbricus*, and in some aquatic families, the setal number is limited to eight, with two setae forming each group (Fig. 10–48*A*). But in several groups of earthworms, such as the Megascolecidae, a more specialized condition exists, in which 50 to more than 100 setae form a ring or girdle around each segment. Attached to the base of each seta are protractor and retractor muscles that allow the seta to be extended or withdrawn (Fig. 10–48*B*).

In mature oligochaetes, certain adjacent segments are thickened and swollen by glands that secrete mucus for copulation and also secrete the cocoon. The glandular area of these segments, collectively called the clitellum, partially or completely covers the segments and often forms a conspicuous girdle around the body (Fig. 10–48*A*). The position of the clitellum along the trunk is variable but is always located on the anterior half of the worm. The clitellum is discussed in more detail in the section dealing with reproduction.

Body Wall and Coelom. The structure and histology of the oligochaete body wall is essentially like that of burrowing polychaetes. Circular muscles are well developed, and the septa partitioning the coelom are relatively complete except at the anterior and posterior extremities. There is usually a large perforation in the septum through which the ventral nerve cord extends. In *Lumbricus*, as well as oligochaetes, this perforation is provided with a sphincter to control the flow of coelomic fluid from one segment to another.

Muscle fibers are present between the peritoneal layers of each septum and apparently aid in regulating coelomic fluid pressure during contractions of the body wall muscles. The septa of certain cephalic segments of the larger earthworms extend posteriorly, resembling pockets, and they are very muscular. The contraction of these pouches increases the coelomic pressure and can cause the eversion of the pharynx or, if the mouth is kept closed, can cause elongation of the first segment for burrowing.

In most earthworms each coelomic compartment, except at the extremities, is connected to the outside by a sphinctered middorsal pore, located in the intersegmental furrows. These pores exude coelomic fluid, which aids in keeping the integument moist. Some giant earthworms squirt fluid several centimeters when disturbed.

Locomotion. Movement in oligochaetes is by peristaltic contractions, as described for

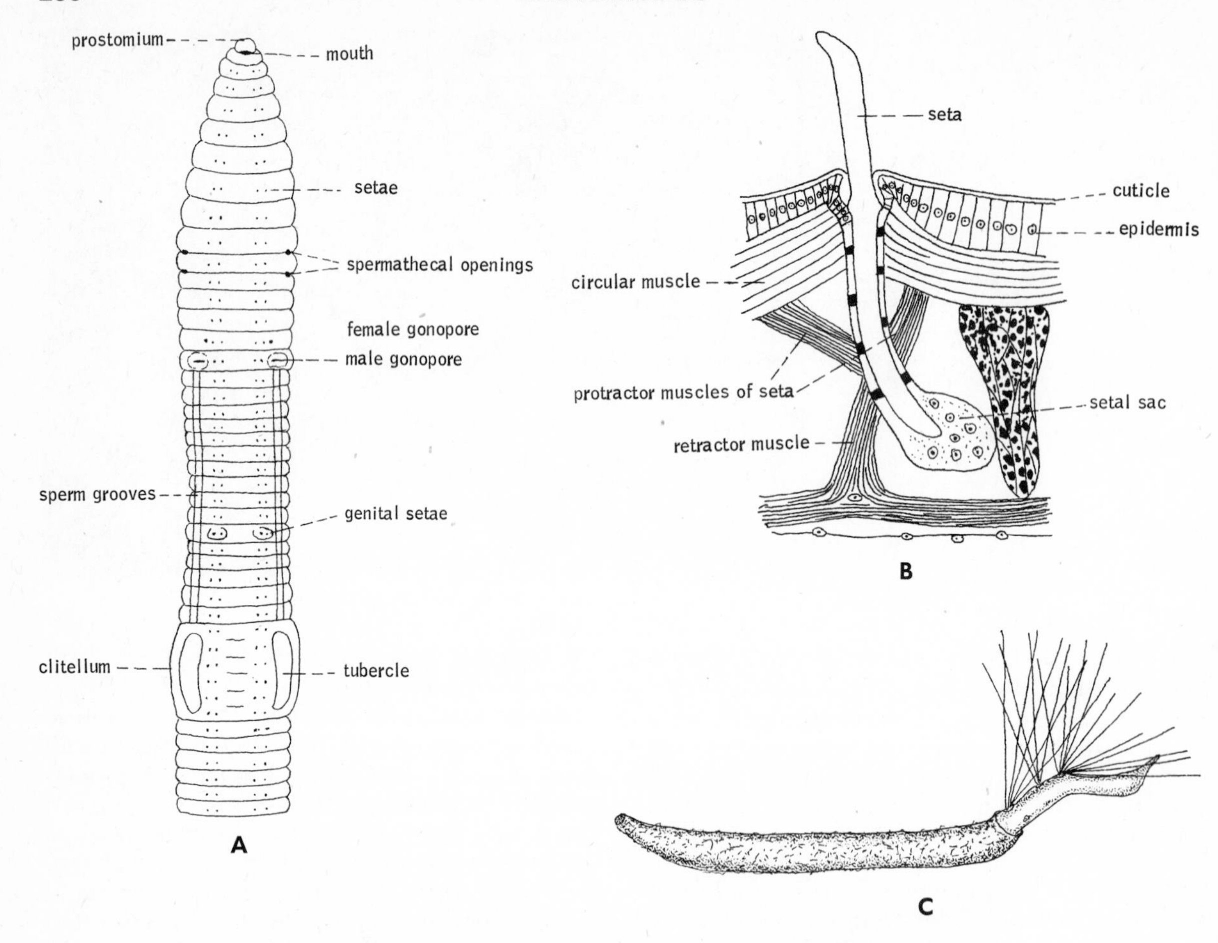

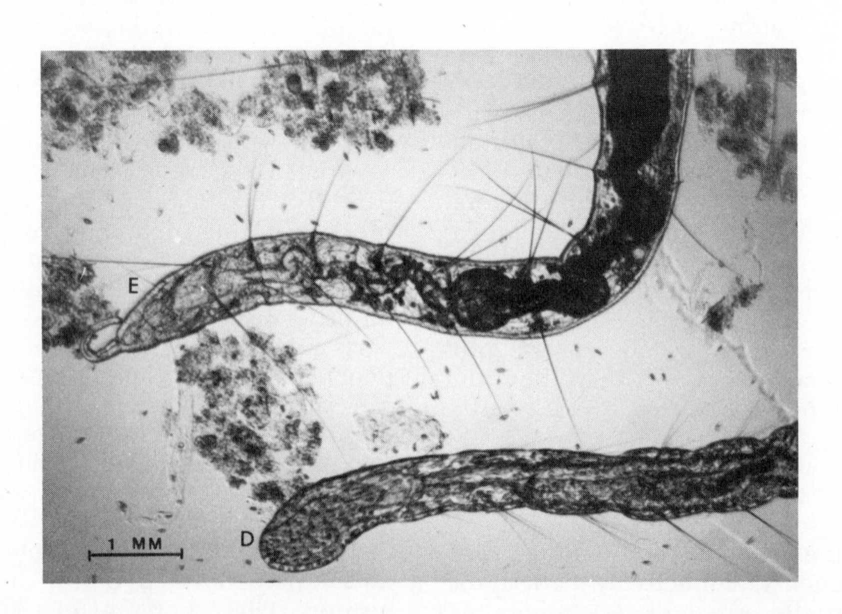

Figure 10–48. *A*. Anteroventral surface of the earthworm, *Lumbricus terrestris*. (After Stephenson from Avel.) *B*. Body wall of the earthworm, *Pheretima* (transverse section). (After Bahl from Avel.) *C*. The tubicolous freshwater naidid, *Ripistes parasita*. (After Cori from Avel.) *D* and *E*. Anterior of two freshwater oligochaetes, *Aeolosoma* (*D*) and *Stylaria* (*E*). *Stylaria* possesses a long tentacular prostomium. (Both by Betty M. Barnes.)

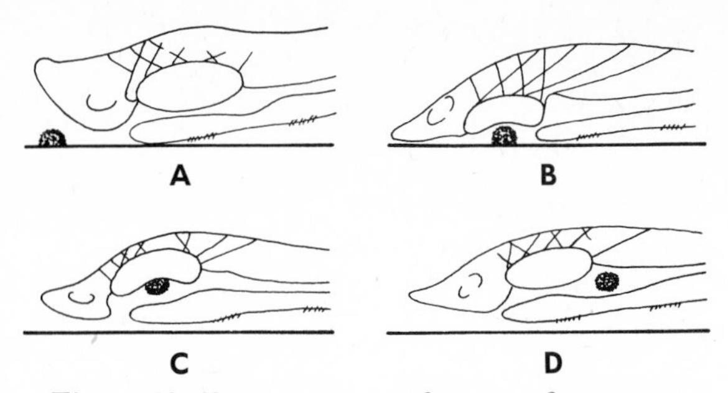

Figure 10–49. *A* to *D*. Mechanism of ingestion in *Aulophorus carteri*, which possesses a dorsal pad-like pharynx. (After Marcus and Avel.)

burrowing polychaetes (p. 233). Earthworm locomotion has been the subject of extensive studies, of which Seymour's (1969) is one of the most recent. Circular muscle contraction and the consequent elongation of segments is most important in crawling, and always generates a coelomic fluid pressure pulse. Longitudinal muscle contraction is more important in burrowing, dilating the burrow or anchoring the segments against the burrow wall. When the worm is crawling, longitudinal contraction may not be strong enough to generate a pressure pulse, but it does so in burrowing.

Setae are extended during longitudinal contraction and retracted during circular contraction. Each segment moves forward in steps of 2 to 3 cm., at the rate of seven to ten steps per minute (Fig. 10–1*C*). The direction of contraction waves can be reversed, thus enabling the worm to crawl backward.

Although freshwater species move in the same manner as do the earthworms, an undulatory mode of progression is characteristic of the few worms that are able to swim. The microscopic aeolosomatids swim by means of a ciliated prostomium.

Nutrition. The majority of oligochaete species, both aquatic and terrestrial, are scavengers and feed on dead organic matter, particularly vegetation. Earthworms feed on decomposing matter at the surface and may pull leaves into the burrow. They also utilize organic material obtained from mud or soil that is ingested in the course of burrowing.

Algae and other microorganisms are an important food source for many tiny freshwater species. The little tube-dwelling naidid, *Ripistes*, is an indirect deposit feeder. This worm has long setae on the sixth, seventh, and eighth segments (Fig. 10–48*C*). In feeding, the anterior of the body is extended from the tube and the long setae are moved about in the water. Detritus collected on the setae is then periodically wiped off in the worm's mouth. Members of the genus *Chaetogaster*, little oligochaetes that are commensals on freshwater snails, are raptorial and catch amebas, ciliates, rotifers, and trematode larvae by the sucking action of their pharynx.

The Pharynx. The digestive tract is straight and relatively simple (Fig. 10–50). The mouth, located beneath the prostomium, opens into a small buccal cavity, which in turn opens to a more spacious pharynx. The dorsal wall of the pharyngeal chamber is both very muscular and glandular, and forms sort of a bulb or pad, which is the principal ingestive organ.

In aquatic forms, in order to ingest food the pharynx is everted, and the muscular disc is applied to the surface of food particles like a suction cup (Fig. 10–49*A-D*). The mucous covering of the disc causes particles to adhere to its surface, and the pharynx is then retracted.

In earthworms the pharynx acts as a pump. The anterior of the worm extends from the burrow, and the mouth is pressed against bits of humus lying on the ground. The pharynx then undergoes a series of contractions that pump the food into the mouth. In both aquatic and terrestrial species the action of the pharynx is augmented by the action of strands of muscle fibers extending from the pharynx to the body wall.

The pharyngeal glands produce a salivary secretion containing mucus and, at least in the lumbricids, a protease. The glands are often highly developed and include more than one type. Those of the aquatic naidids, *Dero* and *Aulophorus*, project through the outer surface of the pharynx. In some species, such as *Eisenia*, certain of the glands, called septal glands, are located entirely outside the pharynx on the surfaces of adjacent septa. A long duct extends from the glandular mass through the dorsal pharyngeal wall and opens into the lumen of the pharynx.

The Esophagus. The pharynx opens into a narrow tubular esophagus, which may be modified at different levels to form a stomach, a crop, or a gizzard. Both a crop and a gizzard are commonly found in terrestrial oligochaetes, although their positions vary. In *Lumbricus*, as well as in other members of the same family, both a crop and a gizzard are present at the end of the esophagus (Fig. 10–50). On the other hand, the crop and gizzard of earthworms belonging to the family Megascolecidae are located at the anterior of

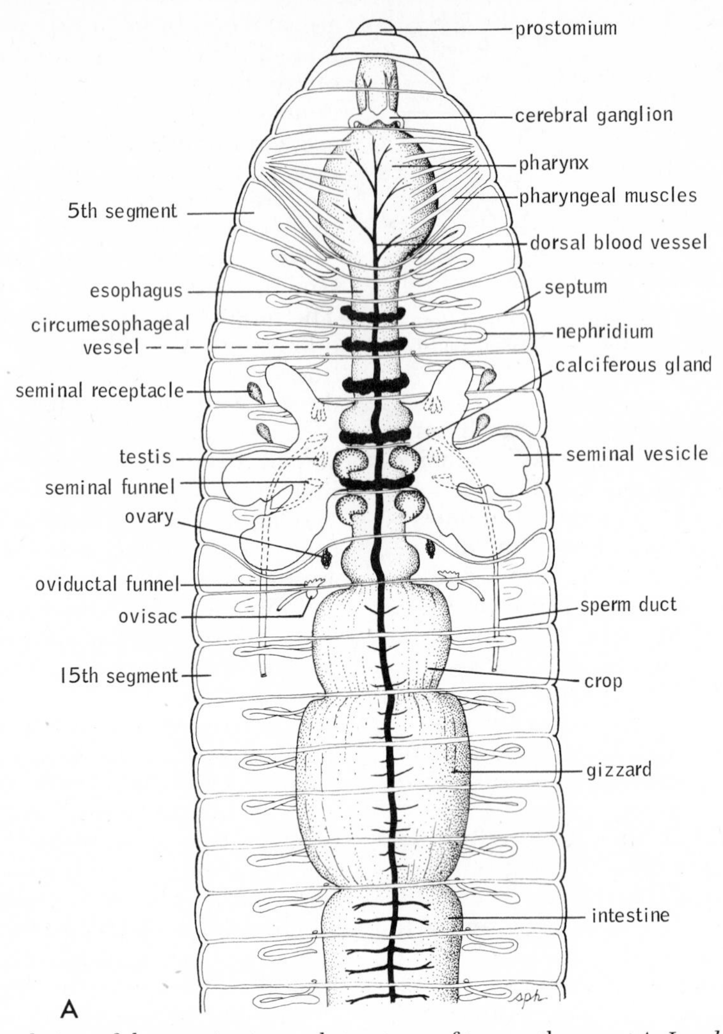

Figure 10–50. Dorsal view of the anterior internal structures of two earthworms. *A. Lumbricus.*
Illustration continued on opposite page.

the esophagus shortly behind the pharynx. In some forms, there is a series of gizzards, numbering two to ten, each occupying a separate segment. Where a single gizzard is the rule, as in *Lumbricus*, it occupies several segments.

The crop is thin-walled and acts as a storage chamber. The gizzard, which is used for grinding food particles, is lined with cuticle and is very muscular. That the gizzard represents an adaptation to terrestrial feeding is indicated by the very great size reduction of the gizzard in those earthworms that have returned to an aquatic environment or at least to wet boggy soils.

In the aquatic naidids, the esophagus is dilated posteriorly to form a stomach. However, the function of the stomach is not certain, since histologically it appears to be very similar to the intestine.

Calciferous Glands. A characteristic feature of the oligochaete esophagus is the presence of calciferous glands. The gland cells are located in special evaginations of the esophageal wall at different levels, depending on the family. Calciferous glands attain their greatest development in the lumbricids, in which the posterior esophageal wall has become greatly folded, and the inner tips of the folds have fused to form a new interior wall (Figs. 10–50 and 10–51). Thus, the glandular region becomes completely separated from the esophageal lumen and may appear externally as lateral or dorsal swellings. However, outlets to the lumen of the esophagus are retained to allow passage of the glandular products.

The calciferous glands are excretory rather than digestive organs; they function in ridding the body of excess calcium taken up

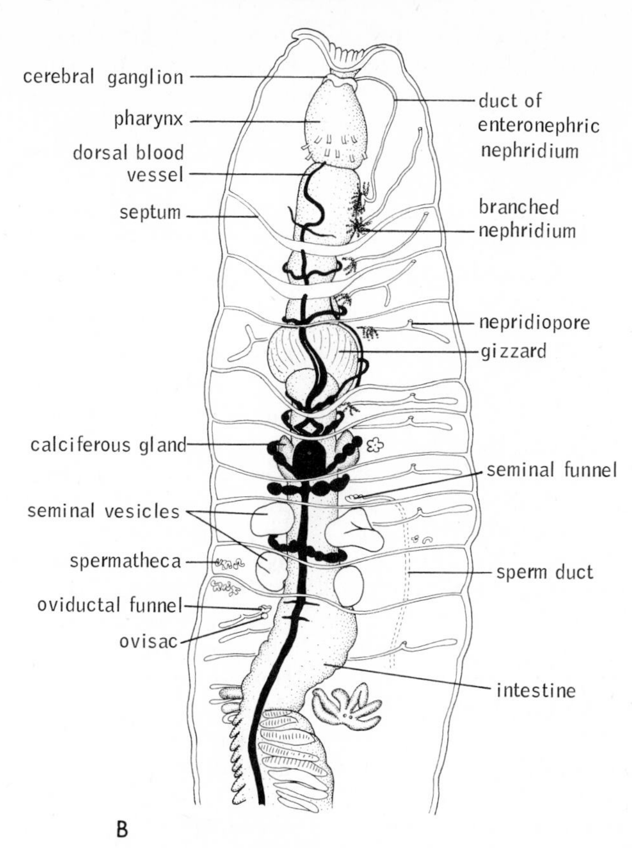

Figure 10–50. *Continued. B.* The tropical glossoscolecid *Microchaetus.* (After Brinkhurst and Jamieson.)

from food, and in maintaining a constant pH in the blood and coelomic fluid by controlling the levels of calcium and carbonate ions. When the level of CO_2 in the blood becomes excessive (with a consequent lowering of the pH), the carbonate ion becomes bound to calcium in the calciferous glands and is excreted into the esophagus as calcite, which is not absorbed in transit through the intestine.

The Intestine. The intestine forms the remainder of the digestive tract and extends as a straight tube through all but the anterior quarter of the body. The intestine is the principal site of digestion and absorption, and the intestinal epithelium consists of secretory cells and absorptive cells. In addition to the usual classes of digestive enzymes, the earthworms, at least, secrete cellulase. The absorbed food materials are passed to blood sinuses that lie between the mucosal epithelium and the intestinal muscles.

A ridge or fold, called a typhlosole, projects internally from the middorsal wall in most terrestrial oligochaetes. The size of the typhlosole is variable; it may consist of a simple fold or of multiple folds. The typhlosole attains its greatest degree of development in the lumbricids. A typhlosole is lacking in freshwater oligochaetes.

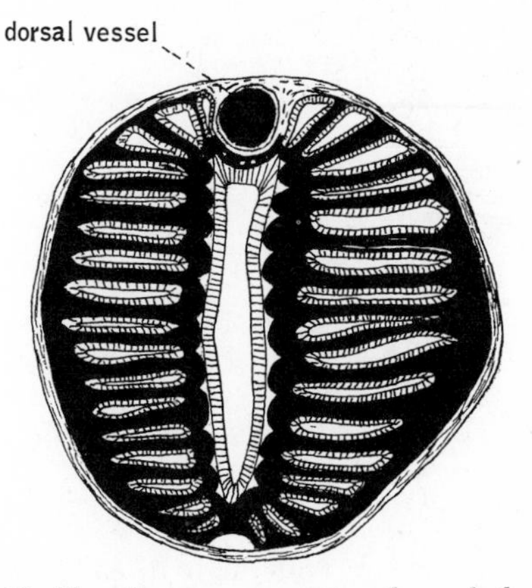

Figure 10–51. Transverse section through the esophagus of the earthworm, *Allobophora,* showing surrounding calciferous gland. (After Combault from Avel.)

Gas Exchange. Gas exchange in almost all oligochaetes, both aquatic and terrestrial, takes place by the diffusion of gases through the general body integument, which in the larger species contains a capillary network within the outer epidermal layer. In terrestrial species, the film of moisture necessary for the diffusion of gases into the body surface is supplied by mucous glands, coelomic fluid, and nephridial excretions.

True gills rarely occur in oligochaetes. Species of the aquatic genera *Dero* (Fig. 10–52*B*) and *Aulophorus* have a circle of finger-like gills at the posterior of the body. A tropical genus, *Branchiodrilus,* like the cirratulid polychaetes, has a long filament extending from the lateral surfaces of each segment. A tubificid, *Branchiura* (Fig. 10–52*A*), also has filamentous gills, but they are located dorsally and ventrally rather than laterally and are limited to segments in the posterior quarter of the body. A South American earthworm that lives in mud has tuft-like gills on the lateral surface of each segment.

Respiratory pigments are lacking in the blood of the Aeolosomatidae and in many Naididae and Enchytraeidae, all of which are small. Other oligochaetes usually have hemoglobin dissolved in the plasma. There is a significant reduction in oxygen consumption when oligochaetes with hemoglobin are exposed to carbon monoxide, indicating the oxygen transporting function of the respiratory pigment. The hemoglobin of *Lumbricus* transports 40 per cent of the oxygen utilized.

Many aquatic oligochaetes tolerate relatively low oxygen levels and, for a short period of time, even a complete lack of oxygen. Members of the family Tubificidae, which live in stagnant mud and lake bottoms, are notable examples. There are members of this family, such as *Tubifex tubifex,* that die on long exposure to ordinary oxygen tensions. *Tubifex* supposedly facilitates ventilation in stagnant water by exposing its posterior end out of the mud and waving it about (Fig. 10–52*C*). Like that of the polychaete *Arenicola,* the hemoglobin of *Tubifex* is reported to be saturated with oxygen at very low oxygen tensions.

Circulation. The circulatory system of oligochaetes is basically like that of polychaetes. In *Aeolosoma,* the dorsal vessel begins over the anterior of the intestine and drains an intestinal sinus. Two pairs of commissural vessels surround the anterior of the gut and connect the dorsal and ventral vessels. These commissural vessels provide the only connection between the two longitudinal vessels, aside from the drainage to and from the intestinal sinus.

In most oligochaetes, the dorsal and ventral vessels are not associated with the sinuses, and they are connected to each

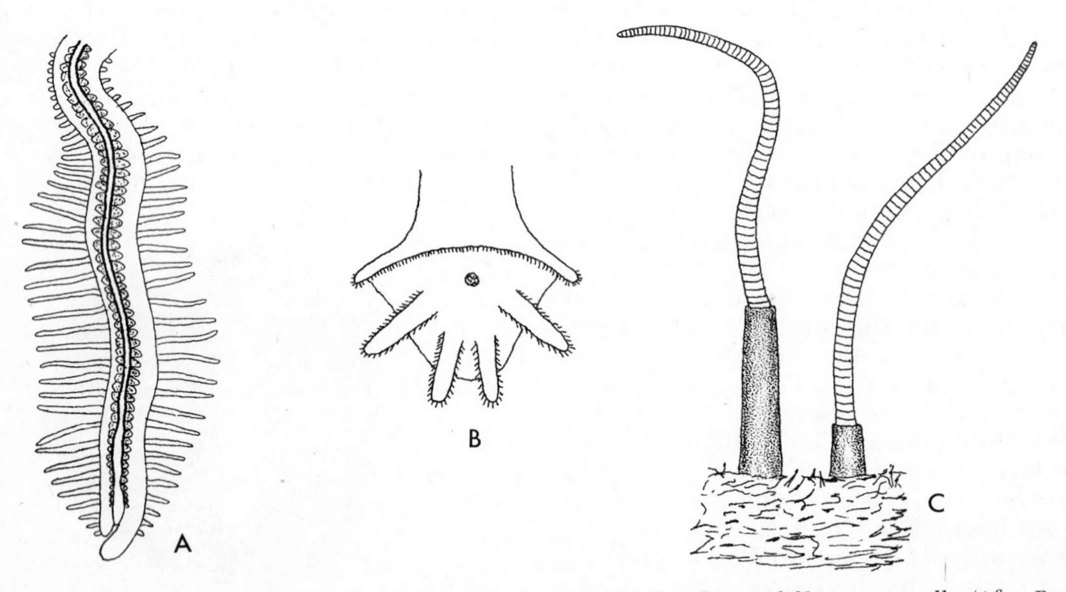

Figure 10–52. *A.* Posterior of *Branchiura sowerbyi,* showing dorsal and ventral filamentous gills. (After Beddar from Avel.) *B.* Posterior of *Dero,* showing circlet of gills around anus. (After Pennak.) *C.* Posterior of body of *Tubife* projected from tube and waved about in water facilitating gas exchange. (After Pennak.)

other by lateral vessels in every segment (Fig. 10–50). Branches from the segmental vessels irrigate the integument and supply the various segmental organs. As in polychaetes, the dorsal vessel of oligochaetes receives blood from the intestinal sinus and the segmental vessels and carries it anteriorly. The ventral vessel carries blood posteriorly and distributes it to the intestinal sinus and to the segmental vessels.

The dorsal vessel is contractile and is the principal means by which the blood is propelled. The vessels in oligochaetes commonly referred to as hearts are certain anterior commissural vessels that are conspicuously contractile. The number of such hearts varies. Five are present in *Lumbricus* and surround the esophagus (Fig. 10–50). Only one pair of hearts is present in *Tubifex*, and this pair is circumintestinal. A single pair is also characteristic of the naidids. These segmental hearts function as accessory organs for blood propulsion.

Valves in the form of endothelial folds are present in the hearts and may also be found in the dorsal vessel at junctions containing segmental vessels.

Excretion. The oligochaete excretory organs are always metanephridia, although in some cases a secondary loss of the nephrostome has occurred. Typically, there is one pair of metanephridia per segment except at the extreme anterior and posterior ends; the structure of the metanephridia is basically like that of polychaetes. In the segment following the nephrostome, the tubule is greatly coiled, and in some species, such as *Lumbricus*, there are several separate groups of loops or coils (Fig. 10–53). Before the nephridial tubule opens to the outside, it is sometimes dilated to form a bladder. The nephridiopores are located on the ventrolateral surfaces of each segment.

In contrast to the majority of oligochaetes, which possess in each segment a single typical pair of nephridia called holonephridia, many earthworms of the families Megascolecidae and Glossoscolecidae are peculiar in possessing several additional types of modified nephridia. One type, called meronephridia, are multiple nephridia and arise from a division of the original embryonic cord of cells that in other annelids forms a single nephridium. Another type of nephridium in these two families is branched, having as many as 100 united divisions. The holonephridia, the meronephridia, and the branched nephridia may open to the outside

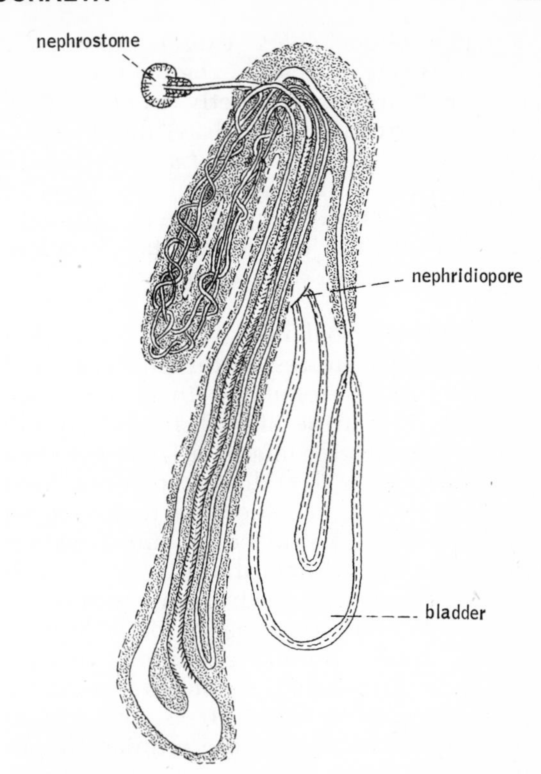

Figure 10–53. Nephridium of *Lumbricus*. (After Maziarski from Avel.)

through nephridiopores; or they may open into various parts of the digestive tract, in which case they are termed enteronephric (Fig. 10–50*B*).

A single worm may possess a number of different types of these nephridia, each being restricted to certain parts of the body. *Pheretima posthuma*, a megascolecid earthworm of India, exemplifies the complexity that can be found in some oligochaete excretory systems. The nephridia of the fourth, fifth, and sixth segments are branched. Each nephridium consists of hundreds of branches in which the nephrostomes have degenerated. The common tubule of each nephridial cluster opens into the pharynx; these branched nephridia are thus enteronephric. The seventh through the fifteenth segments each contain 200 to 250 very small separate meronephridia, which lack nephrostomes and open onto the body surface. Each of the remaining posterior segments contains a single pair of large holonephridia, which are enteronephric and empty into the intestine by way of a pair of longitudinal excretory canals located above the intestine, one on each side of the typhlosole. These canals open into the lumen of the intestine in each segment.

Aquatic oligochaetes excrete ammonia, and terrestrial species excrete urea; but earthworms are less perfectly ureotelic than are other terrestrial animals. Although urea is present in the urine of *Lumbricus* and other earthworms and although the level of urea varies depending upon the condition of the worm and the environmental situation, ammonia remains an important excretory product.

Salt and water balance, which is of particular importance in freshwater and terrestrial environments, is regulated in part by the nephridia. The urine of both terrestrial and freshwater species contains a far lower concentration of salts than do the coelomic fluid and blood. Thus, considerable reabsorption of salts must take place as fluid passes through the nephridial tubule.

In the terrestrial earthworms water uptake and loss occur in large part through the skin. Under normal conditions of adequate water supply, the nephridia excrete a copious hypotonic urine. It is not certain whether reabsorption by the ordinary nephridia is of importance in water conservation, but the enteronephric nephridia do appear to represent an adaptation for the retention of water. By passing the urine into the digestive tract, most of the remaining water can be reabsorbed as it goes through the intestine. Worms with enteronephric systems can tolerate much drier soils; or they do not have to burrow as deeply during dry periods. The nephridia are also employed in removal of the products of hemoglobin destruction. The walls of certain sections of the tubule where this function is performed are filled with hematochrome pigments.

Chloragogen Tissue. Surrounding the intestine and investing the dorsal vessel of oligochaetes is a layer of yellowish cells, called chloragogen cells, which play a vital role in intermediary metabolism, similar to the role of the liver in vertebrates. Chloragogen tissue is the chief center of glycogen and fat synthesis and storage. Deamination of proteins, the formation of ammonia, and the synthesis of urea also take place in these cells. In terrestrial species, silicates obtained from food material and the soil are removed from the body and deposited in the chloragogen cells as waste concretions.

Histologically, the chloragogen tissue is derived from the peritoneum. Its color is due to green-yellow lipid inclusions. Chloragogen cells are released into the coelom as free cells called eleocytes.

Nervous System. In most oligochaetes the two ventral nerve cords are fused and located within the muscle layers of the body wall. The oligochaete brain has shifted posteriorly and in lumbricids lies in the third segment, above the anterior margin of the pharynx (Fig. 10–54*A*).

Most aquatic oligochaetes have four pairs of lateral nerves per segment; usually only three pairs are present in terrestrial families. Studies on the lumbricids indicate that the segmental nerves contain both sensory and motor components. Moreover, the sensory fibers of each nerve innervate epidermal sections of both the adjacent preceding and following segments, in addition to epidermal sections of the segments in which the nerve itself is located (Fig. 10–54*B*).

As in polychaetes, giant nerve fibers are present in oligochaetes except for the aeolosomatids. The earthworms possess five giant nerve fibers. Three are quite large and are grouped at the middorsal side of the ventral nerve cord. The other two giant nerve fibers are less conspicuous; they are situated midventrally and are rather widely separated. The middorsal fiber is fired by anterior stimulation, and the two dorsolateral fibers are fired by posterior stimulation.

The relation of the nervous system to locomotion has been studied extensively in earthworms. If the ventral nerve cord is cut or destroyed, the muscles of the body wall posterior to the point of severance do not contract, demonstrating that such contraction is not intrinsic within the muscles themselves. Although the basic peristaltic rhythm appears to arise from the ventral nerve cord, contraction can be initiated indirectly by way of reflex arcs involving sensory neurons in the body wall. A wave of longitudinal muscular contraction exerts a pull on the following segments. This pull apparently stimulates the sensory neurons of those following segments, and a reflex action is initiated that causes the contraction of the circular muscle layer and the elongation of the segments. This traction-stimulated reflex was beautifully illustrated by Friedlander (1888), who severed a worm and loosely connected the two parts by a thread. Although the nerve cord was cut, a peristaltic wave continued from one part of the worm to the other, resulting from the pull of the thread on the severed posterior half and from the initiation of the traction reflex.

The anterior end of the ventral nerve cord exhibits a distinct dominance over the more

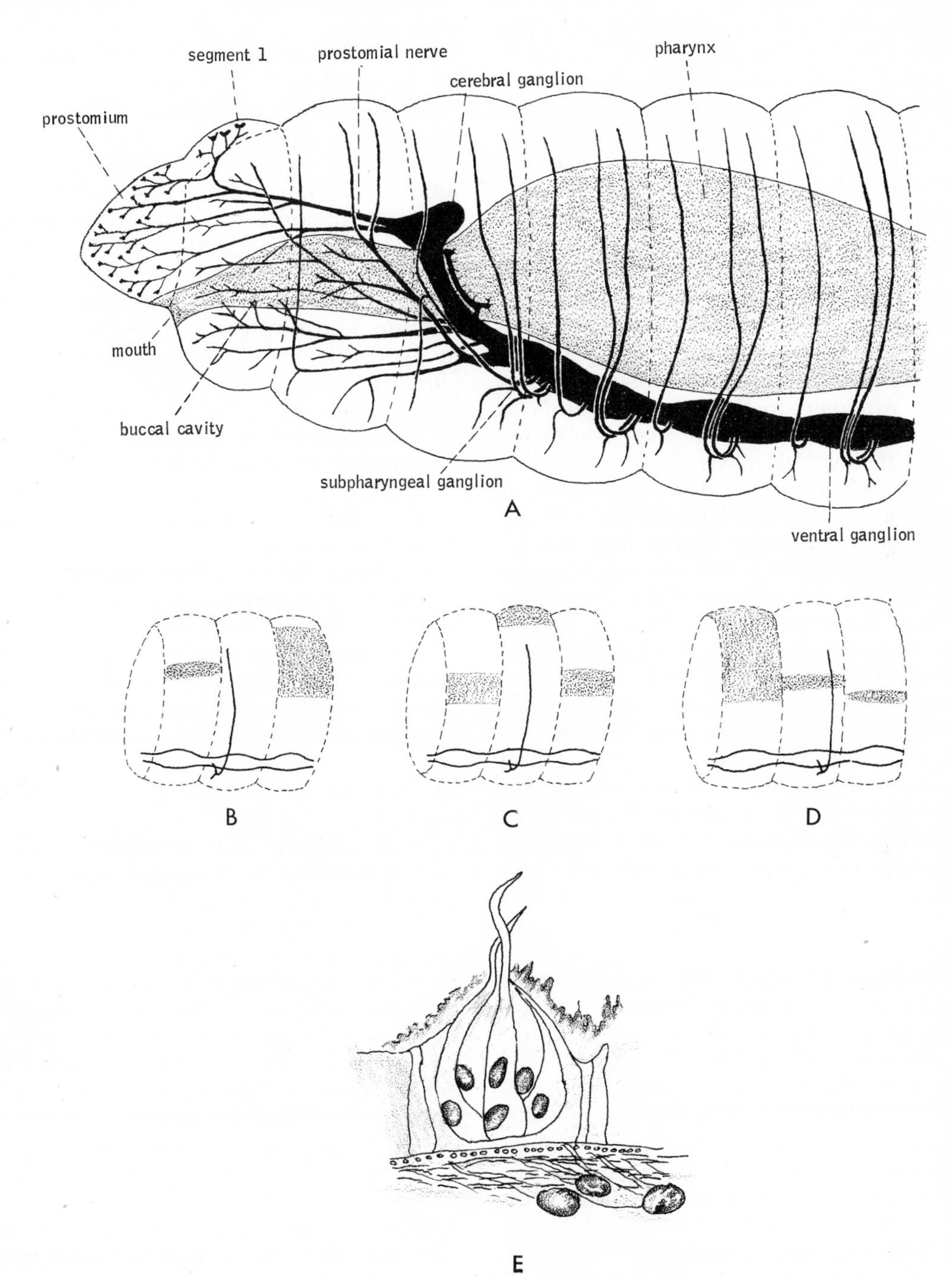

Figure 10–54. *A. Lumbricus* nervous system (lateral view). (After Hess from Avel.) *B–D.* Sensory innervation of anterior, middle, and posterior lateral nerves of each segment. Shaded areas are *not* innervated by the particular nerve depicted. Each nerve supplies not only its own segment, but also anterior and posterior adjacent segments. (After Prosser from Avel.) *E.* Sensory tubercle on prostomium of *Slavina.* (After Marcus from Avel.)

posterior sections. The subpharyngeal ganglion is the principal center of motor control and vital reflexes, and dominates the succeeding ganglia in the chain. All movement ceases when the subpharyngeal ganglion is destroyed. Motor control continues normally following removal of the brain, but the worm loses its ability to correlate movement with external environmental conditions. The relation of the subpharyngeal ganglion to the brain is thus somewhat analogous to the relation of the medulla to the higher brain centers in vertebrates.

Studies on *Lumbricus* have demonstrated the presence of a well-developed visceral nerve system, which in many respects parallels that of vertebrates (Millot, 1943). A similar system undoubtedly exists in other annelids. Attached to the posterior ends of the circumpharyngeal connectives is a series of pharyngeal ganglia, which form a relay center for motor impulses from the brain and subpharyngeal ganglia to the muscle fibers of the gut wall. Sensory fibers from the digestive tract also enter the central nervous system by way of these ganglia. Impulses from this relay center appear to effect some control in the total tone of the gut musculature. Within each segment there also exists a dual system of motor fibers arising from the ventral cord and innervating the muscles of the gut wall. One set of fibers, which leaves the ventral nerve cord by way of the two posterior segmental nerves, is excitatory and has been shown to secrete acetylcholine; the other fibers are associated with all three segmental nerves, have an inhibitory effect on nerve impulse transmission, and secrete norepinephrine.

Sense Organs. Oligochaetes lack eyes, except for a few aquatic forms that have simple pigment-cup ocelli. However, they have a dermal light sense, for the integument is well supplied with isolated photoreceptors located in the inner part of the epidermis. Each photoreceptor is lens-shaped, with a basal process that enters the subepidermal nerve plexus. Although they exist in every segment, photoreceptors do not occur on the ventral surface of the body, and are more abundant at both the anterior (particularly on the prostomium) and the terminal posterior segment of the worm.

In *Lumbricus*, many of the prostomial photoreceptors occur in clusters well beneath the epidermal surface, and are directly connected to branches of the cerebral nerves. Most oligochaetes are negatively phototactic to strong light and positively phototactic to weak light. There is some evidence that these two opposing responses are associated with two different types of photoreceptors.

The other sensory structures have a more or less general distribution in the integument. Clusters of sensory cells that form a projecting tubercle, with sensory processes extending above the cuticle, appear to be chemoreceptors (Fig. 10–54*E*). The basal processes from each cell in the cluster unite to form a nerve, which joins a segmental nerve. The tubercles form three rings around each segment and are particularly numerous on the more anterior segments, and especially on the prostomium, where (in *Lumbricus*) there may be as many as 700 per square millimeter.

The integument is richly supplied with free nerve endings, which may be tactile in function. Some other epidermal cells are undoubtedly sensory, but their exact function is unknown.

Reproduction and Development

The Reproductive System. The reproductive system of oligochaetes differs from that of polychaetes in a number of striking respects. Oligochaetes are all hermaphroditic, they possess distinct gonads, and the number of reproductive segments present is very much limited. The oligochaete arrangement is undoubtedly a specialized one, and the polychaete condition is probably primitive. Brinkhurst and Jamieson (1972) have suggested that the ancestral annelids might have looked like oligochaetes, but have had polychaete reproductive systems.

The only exception to the oligochaete pattern appears in the Aeolosomatidae, which, although hermaphroditic, are similar to polychaetes in lacking distinct gonads and in the large number of segments capable of producing gametes. Also, these worms possess no gonoduct and the sperm, at least, utilize the nephridia to exit from the body. However, these little worms may not be oligochaetes.

In most aquatic groups there is usually one ovarian segment and one testicular segment; but in terrestrial families, two male segments may be present. The reproductive segments are located in the anterior half of the worm, and the female segment or segments are located behind the male segments (Fig. 10–55). The exact location of fertile segments along the trunk varies in different families.

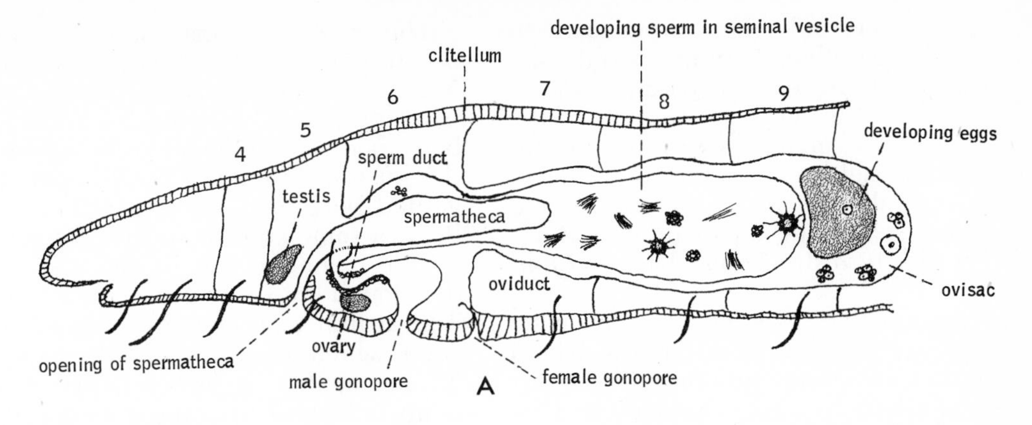

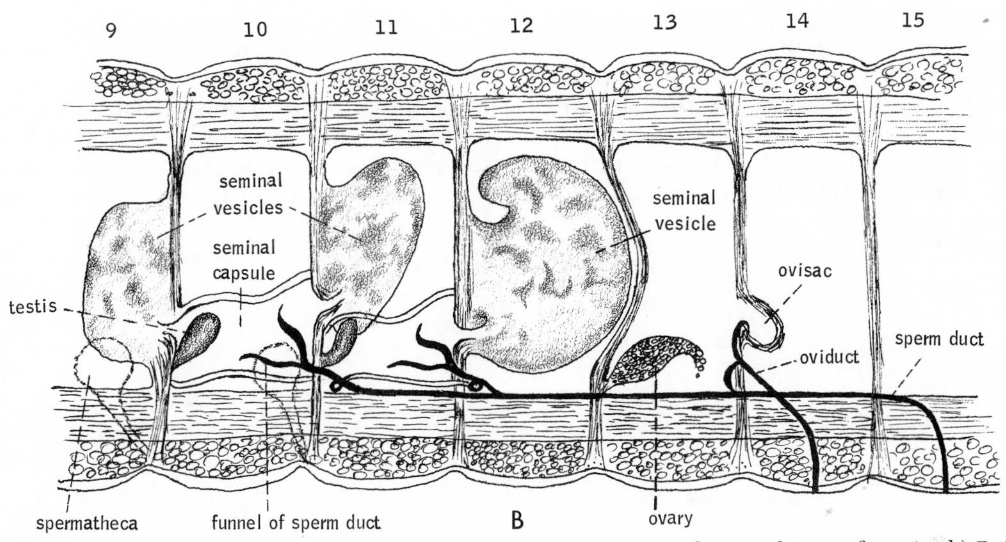

Figure 10–55. *A.* Reproductive system of a Naididae (lateral view). (After Stephenson from Avel.) *B.* Reproductive segments of the earthworm, *Lumbricus* (lateral view). (After Hesse from Avel.)

The ovaries and testes, both of which are typically paired, are situated in the fertile segments on the lower part of the anterior septum and project into the coelom. Both are small and often pear-shaped, particularly the ovaries; the testes may be lobed or digitate. As in polychaetes, the gametes are released from the gonads as gametogonia, and maturation takes place mostly in the coelom, where the gametes are unattached. However, in earthworms the eggs are detached as oocytes.

Although maturation of the gametes is completed in the coelom, it is typically restricted to special coelomic pouches called seminal vesicles and ovisacs. Both arise as outpocketings of the septa of the reproductive segments, but the number, size, and position vary. For example, in the naidids, in which only one pair of testes is present, there is also a single pair of very large seminal vesicles, which project backward from the posterior septum of the male segment and penetrate the four succeeding segments (Fig. 10–55*A*). The tubificids possess, in addition, a smaller second pair of seminal vesicles that arise from the anterior septum of the male segment and project forward into the preceding segment.

In *Lumbricus*, there are two male segments with three pairs of seminal vesicles (Fig. 10–55*B*). One pair projects backward from the posterior septum of each of the two male segments, and one pair projects forward from the anterior septum of the first male segment. The aeolosomatids and most species of the Enchytraeidae and the Moniligastridae lack seminal vesicles.

The ovisacs closely resemble the seminal

vesicles, but since only one ovarian segment usually is present, there is typically only one pair of ovisacs. The pair of ovisacs projects backward from the posterior septum of the single ovarian segment. Ovisacs are absent in the enchytraeids; are large and project through several segments in the naidids and the tubificids; and are small or sometimes absent in the terrestrial oligochaetes (*Lumbricus*).

The reproductive segments, whether male or female, are each provided with a pair of ducts, either vasa deferentia or oviducts, for the exit of sperm or eggs. The male and female ducts are quite similar, and their basic plan is very much like that of a nephridium. The internal opening is through a ciliated funnel located on the posterior septum of the fertile segment. From the funnel a tubule extends backward, penetrating the septum and passing through one or more segments before opening on the ventral surface of the body. The tubule is at least partially ciliated.

The sperm duct may be coiled, or it may be straight, as in the lumbricids (Fig. 10–55*B*). Furthermore, in earthworms the two pairs of sperm ducts on each side of the body become confluent prior to opening to the outside through a single common male genital pore, which has a raised border or lips. In some aquatic species the atrium is a simple chamber, but in many species it is developed in various ways for copulation. The atrium may contain a penis (Fig. 10–56*C*), which often has a chitinous sheath, or it may contain a pseudopenis. A pseudopenis is formed from the enlarged tip of the atrium or from an inversion of the body wall. In the latter case the pseudopenis is eversible. Glandular tissues, called prostate glands, are commonly associated with the male gonoducts. In most aquatic groups, the posterior portion of the sperm duct wall is glandular and forms the prostate. Often, however, the prostate is a separate glandular mass that is drained by a duct into the vas

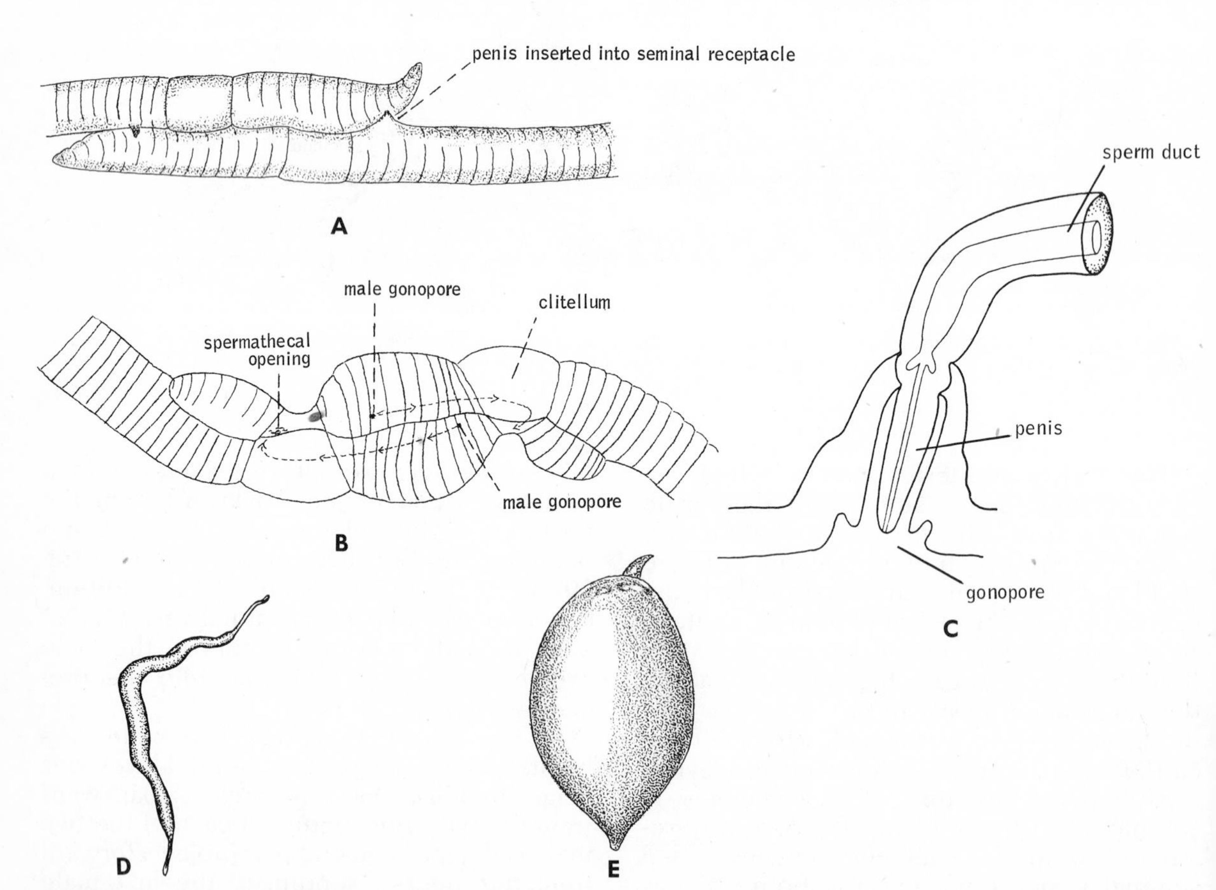

Figure 10–56. *A*. Copulation by direct sperm transmission in the earthworm, *Pheretima communissima*. (After Oishi from Avel.) *B*. Copulation by indirect sperm transmission in the lumbricid, *Eisenia foetida*. Arrows indicate path of sperms from male gonopores to openings of spermathecae. (After Grove and Cowley from Avel.) *C*. Penis of the lumbriculid *Rhynchelmis*. *D*. Cocoon of the glossoscolecid *Alma nilotica*. (*C* and *D* from Brinkhurst, R., and Jamieson, B. G., 1972: Aquatic Oligochaeta of the World. Toronto Univ. Press.) *E*. Cocoon of the lumbricid, *Allobophora terrestris*. (After Avel.)

deferens at a point near the gonopore. In some megascolecid earthworms, the prostates are not connected to the vas deferens, and they open separately onto the ventral surface of segments adjacent to those bearing the male gonopore.

Although constructed on the same plan as the sperm ducts, the oviducts are smaller, simpler, and open through two female genital pores located on the segment following the ovarian segment (Fig. 10–55*B*). Forming a part of the female reproductive system, but completely separated from the female gonoducts, are the seminal receptacles. These storage chambers are simple pairs of sacs, opening onto the ventral surface of the segment containing them. The number of seminal receptacles ranges from one to five pairs, each pair in a separate segment. Although they are usually located in certain segments anterior to the ovarian segment, the exact position is variable.

In some terrestrial species, including *Lumbricus,* the male segment is partitioned so that the testes, the sperm duct funnel, and the opening to the seminal vesicles are enclosed in a special ventral compartment called a seminal capsule (Fig. 10–55*B*). The seminal capsule is completely separated from the larger, remaining portion of the coelomic cavity. For this reason, the testes are not visible in the usual dorsal dissection. A somewhat similar seminal capsule, as well as an ovarian chamber, are present in the Moniligastridae.

The general plan of the oligochaete reproductive system is relatively uniform, but the numbers of various structures, the segments in which they are located, and the segments from which the genital pores open are extremely variable. This variation is of considerable importance in the taxonomy of oligochaetes.

The Clitellum. The clitellum is a reproductive structure characteristic of oligochaetes. It consists of certain adjacent segments in which the epidermis is greatly swollen with unicellular glands that form a girdle, partially or almost completely encircling the body from the dorsal side downward (Figs. 10–48*A* and 10–56*B*). The clitellum is always poorly developed ventrally. The number of segments composing the clitellum varies considerably; there are two clitellar segments in most aquatic forms, six or seven in *Lumbricus,* and as many as sixty in certain Glossoscolecidae. In aquatic species, the clitellum is often located in the same region as the genital pores; while in the lumbricids, the clitellum is considerably posterior to the gonopores. The degree of development of the clitellum varies not only from group to group but seasonally as well. In aquatic species the clitellum may be only one cell thick, whereas in many earthworms it forms a heavy girdle. The development of the clitellum generally coincides with sexual maturity, but there are some worms in which the clitellum becomes conspicuous only during the breeding season.

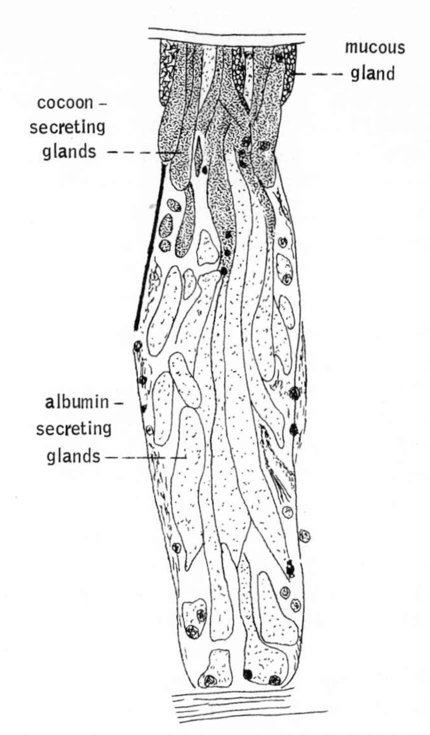

Figure 10–57. Section through clitellum of *Lumbricus terrestris.* (After Grove from Avel.)

The glands of the clitellum produce mucus for copulation, secrete the wall of the cocoon, and secrete albumin in which the eggs are deposited within the cocoon. In the earthworms, the glands performing each of these three different functions form three distinct layers (Fig. 10–57). The large albumin-secreting glands comprise the deepest and thickest layer, while the mucous glands make up the surface layer. In immature individuals the clitellum may be absent or inconspicuous.

Copulation. Copulation is the rule in oligochaetes, and mutual dissemination of sperm occurs. The ventral anterior surfaces of a pair of copulating worms are in contact, with the anterior of one worm directed toward the posterior of the other. In all oligochaetes except the lumbricids, the male genital pores of one worm directly appose the seminal receptacles of the other. The two worms are held in position by a common, enveloping mucous coat secreted by the clitellum; they may also be hooked together by genital setae. The genital setae are modified ventral setae generally located in the region of the male gonopore or the seminal receptacles.

Additional anchorage is obtained when the raised rim or lips of the male pore, or the everted penis or atrium, are inserted into the opening of the seminal receptacle. Transmission of sperm into one pair of seminal receptacles in the earthworm *Pheretima communissima* takes over 1½ hrs. and then is repeated for each of the other two pairs of receptacles (Fig. 10–56A).

In the lumbricids, the male genital pores do not appose the seminal receptacles during copulation, so that sperm must travel a considerable distance from one opening to the other (Fig. 10–56*B*). During copulation, the posteriorly placed clitellum of one worm attaches to the segments containing the seminal receptacles of the opposite worm. Attachment is accomplished by means of an adhesive slime tube and by the genital setae. The genital setae on the clitellum pierce the body wall of the apposing worm; those setae near the seminal receptacles act as claspers. The intervening region between the two clitella is less rigidly attached, although each worm is covered by a slime tube. At the emission of the sperm, certain muscles in the body wall of the segments posterior to the male gonopores contract and form a pair of ventral sperm grooves, which extend posteriorly to the clitellum. Because the grooves are roofed over by the enveloping slime tube, the sperms are actually passed through an enclosed channel.

The movement of sperm down the sperm groove is effected by a greater contraction of the muscles producing the groove. A pit is thus formed that travels the length of the groove, carrying small amounts of semen. When the semen reaches the region of the clitellum, the semen then passes over to the other worm and enters the seminal receptacles. The mucous tube obviously must be incomplete in this region. The emission of semen may or may not be accomplished simultaneously in both members of the copulating pair. Copulation in *Lumbricus* continues from two to three hours.

The Cocoon. A few days after copulation, a cocoon is secreted for the deposition of the eggs. First a mucous tube is secreted around the anterior segments, including the clitellum. Then the clitellum secretes a tough chitin-like material that encircles the clitellum like a cigar band; this material forms the cocoon. The deeper layer of clitellar glands secretes albumin into the space between the wall of the cocoon and the clitellum.

When the cocoon is completely formed, it slips forward over the anterior end of the worm in the same manner as removing the band from a cigar. In *Eisenia,* the eggs are discharged from the female gonopores, and they somehow enter the cocoon before it leaves the clitellum. Sperm are deposited in the cocoon as it passes over the seminal receptacles. As the cocoon slips over the head of the worm and is freed from the body, the mucous tube quickly disintegrates, and the ends of the cocoon constrict and seal themselves. The cocoons of terrestrial species are left in the soil, and the cocoons of aquatic species are left in the bottom debris or in mud, or are attached to vegetation.

The cocoons are yellowish in color and ovoid in shape (Fig. 10–56*D* and *E*). Cocoons of *Tubifex* are 1.60 mm. × 0.85 mm., while those of *Lumbricus terrestris* are approximately 7 mm. × 5 mm. The largest cocoons, 75 mm. × 22 mm., are produced by the giant Australian earthworm, *Megascolides australis.* A cocoon contains anywhere from one to twenty eggs, depending on the species. A succession of cocoons may be produced; *Lumbricus terrestris* mates continually during the spring and fall, and cocoons are formed every three or four days. Tubificids and lumbriculids generally reproduce only once a year. The reproductive system is then resorbed, and is reformed the following year.

Self-fertilization (which takes place inside the cocoon) is known to occur in a few isolated cases. The seminal receptacles of isolated individuals of the tubificid *Limnodrilus udekemianus* have been found to contain sperm, and such isolated specimens produce cocoons and young. Parthenogenesis also occurs normally in a few species.

Embryogeny. In general, the eggs of aquatic groups, particularly the primitive families, contain a relatively large amount of yolk. On the other hand, terrestrial species have much smaller eggs with much less yolk; the abundant albumin in the cocoon supplies most of the nutritive needs of the embryo. In both aquatic and terrestrial groups, development is direct with no larval stages, and all development takes place within the cocoon.

The cleavage pattern, while retaining some traces of the spiral character of the cleavage pattern of the polychaetes, is considerably modified in oligochaetes, especially in earthworms. Segmentation is complete and unequal in the oligochaete cleavage pattern but much less regular than in typical spiral cleavage. The mesoderm, however, still arises from special mesoblast cells that form mesodermal cords as in the polychaetes.

The oligochaete young hatch from cocoons after eight days to ten weeks of development, depending not only on the species, but also on environmental conditions. *Lumbricus* hatches in two to three weeks. Usually only some of the original number of eggs deposited in the cocoon hatch. In *Lumbricus terrestris* only one egg develops. The young worms emerge from the ends of the cocoon after hatching.

Asexual Reproduction. Asexual reproduction is very common among many species of aquatic oligochaetes, particularly the aeolosomatids and the naidids. In fact, there are many asexually reproducing naidids in which sexual individuals are rare or have never been observed; others reproduce asexually in the summer and sexually in the fall. Asexual reproduction always involves a transverse division of the parent worm into two or more new individuals. When regeneration and formation of new segments take place only after separation, such division is called fragmentation. Fragmentation is not common but is known to occur in a few species. Usually the parent divides into more than two pieces. In *Allonais paraguayensis*, six to eight fragments are formed, each of which regenerates into a new individual; in two months a single worm can produce 15,000 descendents.

Division by fission zones is much more frequently encountered. In this type of division, regeneration precedes the separation of the daughter individuals to a certain extent. Before fission, a zone of new cells appears at some point along the body of the parent, giving rise to the precursors of a new head and a tail for the future daughter individual. Division always takes place through the fission zone, some of the bud cells going to one individual and some to the other.

Not infrequently, as in species of *Nais*, a new fission zone forms before division has occurred in an old one. Such delayed divisions produce chains of individuals or zooids, similar to those in certain turbellarian flatworms. Chains containing as many as eight zooids have been reported in some species. A clone of *Aulophorus furcatus* was traced through 150 generations for a three-year period with no appearance of sexual individuals and with no diminishing of the asexual fission rate.

Needless to say, oligochaetes have considerable capability for regeneration. The results from regeneration experiments are essentially the same as those obtained from experiments with polychaetes.

Many oligochaetes live several years. Earthworms in aquaria have been known to live for six years, but their life span is probably much shorter in more unprotected conditions.

Ecology and Distribution. Aquatic species live in all types of freshwater habitats, but are most abundant where the water is shallow. At depths of more than 1 m. the oligochaete fauna decrease greatly. The Tubificidae are an exception and inhabit the bottoms of deep lakes in concentrations as great as 8000 per square meter. Abundance of different species of aquatic oligochaetes can serve as a good indication of water pollution (Goodnight, 1973).

Some species crawl about submerged vegetation and other objects, where they feed on surface debris. Most aquatic forms burrow in mud and debris on the bottom. Members of the naidid genus *Aulophorus* construct tubes, and a tropical species, *Aulophorus carteri*, builds its tube with spores from aquatic ferns.

A few oligochaete species are capable of encystment during unfavorable environmental conditions. The worm secretes a tough mucous covering that forms the cyst wall. Some species form summer cysts for protection against desiccation; others form winter cysts when the water temperature becomes low.

The littoral zone, particularly the intertidal regions of sounds and estuaries, contains a surprisingly large number of oligochaetes. These marine species belong to a number of families, including the tubificids and enchytraeids. Although the majority of these marine species are members of the interstitial fauna, are shallow burrowers, or live beneath intertidal rocks or in algal drift, some live below the low-tide mark, even in the bathyal and abyssal zones of the sea.

Many species of the Haplotaxidae and the Enchytraeidae are amphibious or transitional between a strictly aquatic and a strictly terrestrial environment. These worms live in marshy or boggy land and around the margins of ponds and streams.

The terrestrial oligochaetes include the five families of earthworms—Glossoscolecidae, Lumbricidae, Megascolecidae, Eudrilidae, Moniligastridae—and many enchytraeids, and representatives from other families. All are burrowers and are found everywhere except in deserts. They often abound in tremendous numbers. As many as 8000 enchytraeids and 700 lumbricids have been reported from one square meter of meadow soil.

Soils containing considerable organic matter, or at least a layer of humus on the surface, maintain the largest fauna of worms, but other edaphic factors are important to the distribution of terrestrial species. The amount of moisture, the degree of acidity, the amount of oxygen (a factor in wet soils), and the soil texture may all impose limitations on distribution. Acid soils are particularly unfavorable habitats owing to the lack of free calcium ions necessary for the worm to maintain a higher pH in the blood (see p. 288, calciferous glands).

A cross section through soil reveals a distinct vertical stratification of worms. The tunnels of larger species, such as *Lumbricus terrestris*, range from the surface to several meters in depth, depending on the nature of the soil. Young worms and small species are restricted to the few inches of upper humus, while others have a wider distribution but are still limited to the upper level of the soil that contains some organic matter.

The terrestrial oligochaete constructs its burrow by forcing its anterior end through crevices and by swallowing soil. The egested material and mucus are plastered against the burrow wall, forming a distinct lining. Some egested material is removed from the burrow as castings. The burrows may be complex, with two openings and horizontal and vertical ramifications. During dry seasons or during the winter, earthworms migrate to deeper levels of the soil, down to 10 ft. in the case of certain Indian species. Since the amount of oxygen in the soil decreases with depth, the availability of oxygen can limit the extent of such migrations.

After moving to deeper levels, during dry periods an earthworm often undergoes a period of quiescence, losing as much as 70 per cent of its water. Balance is quickly restored and activity resumed as soon as water is again available. Species of the lumbricid genus *Allobophora* undergo a more profound period of inactivity that is essentially a type of diapause. The worm rolls up into a ball inside a deep mucus-lined chamber and remains this way for as long as two months. The starting and terminating of diapause is apparently intrinsic and not a direct result of environmental conditions, as is quiescence.

The activities of earthworms have a beneficial effect on the soil. The extensive burrows increase soil drainage and aeration, but more important is the mixing and churning of the soil resulting from burrowing. Deeper soil is brought to the surface as castings and organic material is moved to lower levels. Soil churning can be demonstrated by placing earthworms in a container that has the bottom half filled with sand and the upper half filled with potting soil. Five worms will thoroughly mix 500 cc. of sand with 500 cc. of soil in several months. Darwin's *The Formation of Vegetable Mould through the Action of Worms* is a classic treatment of this subject.

In the tropics, a number of arboreal species have been reported. The terrestrial forms live in accumulated humus and detritus in leaf axils and branches of trees. Aquatic species inhabit the water reservoirs of bromeliad epiphytes, living on the trunks and branches of tropical trees.

There are no parasitic oligochaetes, but some naidids are commensal on aquatic snails, and certain commensal enchytraeids live attached to the anterior of lumbricids.

Members of the major aquatic families occur throughout the world wherever suitable habitats exist. Many terrestrial species, however, have very limited geographical distribution. Among the four families of earthworms, which have been most studied in this respect, many species are endemic, that is, have very restricted distributions; others are widely distributed throughout the world.

The distribution of endemic species of the family Lumbricidae collectively forms a belt encircling the temperate regions of Europe, Asia, and eastern North America. In Europe, no endemic lumbricids live north of a line running through the Low Countries and middle Germany. This line represents the southernmost extent of glaciation during the Pleistocene epoch. On the other hand, other lumbricids, mostly species of European origin, have been reported from many localities throughout the world, particularly in the Southern Hemisphere. Man has certainly been an important agent in the immigrations of these species, and in many instances these species have displaced native endemic forms. For example, the earthworm fauna of the larger Chilian cities is comprised of European species only.

SYSTEMATIC RÉSUMÉ OF CLASS OLIGOCHAETA

The following classification is based on that of Brinkhurst and Jamieson (1972).

Order Lumbriculida. Four setae per segment. At least one pair of male funnels in same segment as male gonopores. Clitellum one cell thick and including male and female gonopores. A single family of freshwater oligochaetes, the Lumbriculidae. *Lumbriculus.*

Order Moniligastrida. Four setae per segment. Testes and funnels in paired dorsal suspended testis sacs. Clitellum one cell thick and including male and female gonopores. Single family of tropical Asian terrestrial species, the Moniligastridae. Includes some giant forms.

Order Haplotaxida. Male funnel always at least one segment in front of its associated gonopore.

Suborder Haplotaxina. Four setae per segment or two setae in each of four bundles. Male gonopores open immediately behind the two testicular segments. Clitellum one cell thick. A single family of freshwater and semiterrestrial species, the Haplotaxidae.

Suborder Tubificina. Setae bundles usually with numerous setae. One pair of testes and ovaries, and male gonopores in segment immediately behind testicular segment. Clitellum one cell thick.

Family Enchytraeidae. Aquatic and terrestrial worms, some white in color. *Enchytraeus.*

Family Tubificidae. Marine and freshwater oligochaetes, many species widely distributed in stagnant water. *Tubifex, Branchiura, Limnodrilus.*

Family Naididae. Very small aquatic worms, some with elongate proboscis. *Nais, Ripistes, Aulophorus, Branchiodrilus, Slavina, Stylaria, Dero,* the carnivorous *Chaetogaster.*

Family Phreodrilidae. Small family of freshwater, marine, and commensal species.

Family Opistocystidae and Family Dorydrilidae.

Suborder Lumbricina. Eight setae per segment. Male pores at least two segments behind those containing the two pairs of testes. This suborder contains the small family Alluroididae and three families of earthworms. The latter also contain some inhabitants of fresh water and boggy soils.

Family Glossoscolecidae. *Pentoscolex, Alma.*

Family Lumbricidae. Many common earthworms of temperate regions. *Lumbricus, Eisenia, Dendrobaena, Allophora.*

Family Megascolecidae. Earthworms, chiefly of the Southern Hemisphere. To this family belong the Australian giants of the genus *Megascolides.* Other genera include *Pheretima* and *Megascolex.*

Family Eudrilidae. African earthworms.

Members of the family Aeolosomatidae, which are minute freshwater species, have in the past been considered to be primitive oligochaetes. On the basis of their reproductive system (see p. 294), and other features, Brinkhurst and Jamieson (1972) believe that they are unrelated to other oligochaetes and should be removed from the class.

CLASS HIRUDINEA

The Class Hirudinea contains over 500 species of marine, freshwater, and terrestrial worms, commonly known as leeches. Although they are all popularly considered to be bloodsuckers, a large number of leeches are not parasitic.

As a group, the leeches are certainly the most specialized annelids, and most of the distinguishing characterisitcs of the class have no counterpart in the other two annelid groups. But they do display many oligochaete features. Both groups lack parapodia and head appendages; both are hermaphroditic, with gonads and gonoducts restricted to a few segments; and both possess direct development within cocoons secreted by a clitellum. These similarities clearly suggest a common ancestry, and the two groups are commonly considered subclasses within the class Clitellata. Like oligochaetes, leeches are basically a freshwater group, with some invasions of land and secondary invasions of the sea.

Leeches are never as small as many poly-

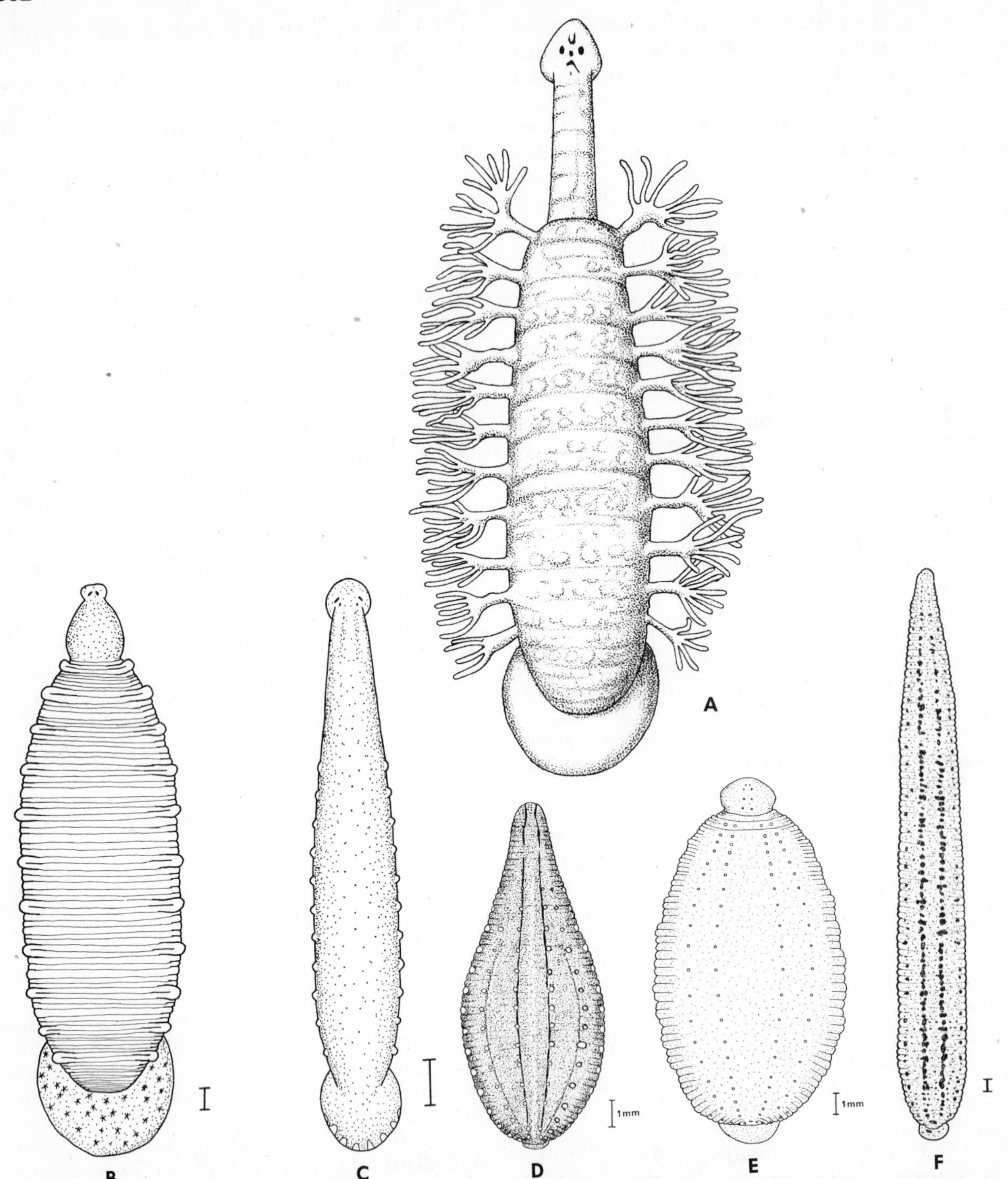

Figure 10–58. External dorsal views of different species of leeches. *A* to *C*. Fish leeches (Piscicolidae). *A*. *Ozobranchus*, showing lateral gills. (After Oka from Mann.) *B*. *Cystobranchus*. *C*. *Piscicola*. *D* and *E*. Glossiphoniid leeches: *D*, *Glossiphonia complanata*, a very common European and North American leech which feeds on snails; *E*, *Theromyzon*, a cosmopolitan genus of leeches which attack birds. *F*. *Erpobdella punctata* (Erpobdellidae), a common North American scavenger and predatory leech. *Illustration continued on opposite page.*

chaetes and oligochaetes. The smallest leeches are 1 cm. in length, and most species are 2 to 5 cm. long. Some species, including the medicinal leech (*Hirudo medicinalis*), may attain a length of 20 cm., but the giant of the class is the Amazonian *Haementeria ghiliani*, which reaches 30 cm. The coloration is often striking. Black, brown, olive-green, and red are common colors, and striped and spotted patterns are not unusual.

External Anatomy. The anatomy of leeches is remarkably uniform. The body is typically dorsoventrally flattened and frequently tapered at the anterior (Fig. 10–58).

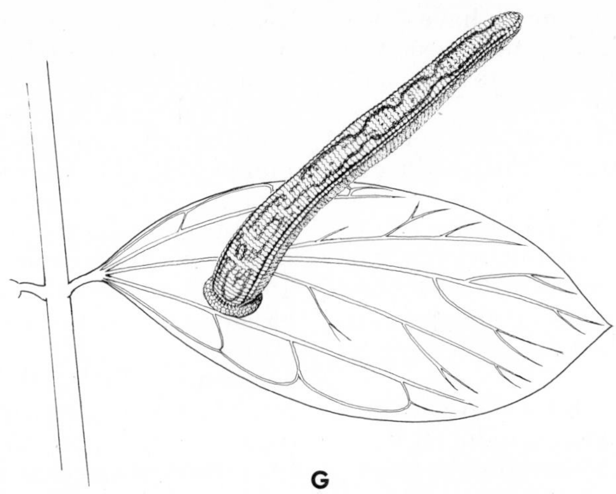

Figure 10–58. *Continued. G. Haemadipsa* (Haemadipsidae), a blood sucking terrestrial leech of south Asia poised on a leaf. (*B–F* from Sawyer, R. J., 1972: North American Freshwater Leeches, Univ. Illinois Press; *G* adapted from Keegan et al.)

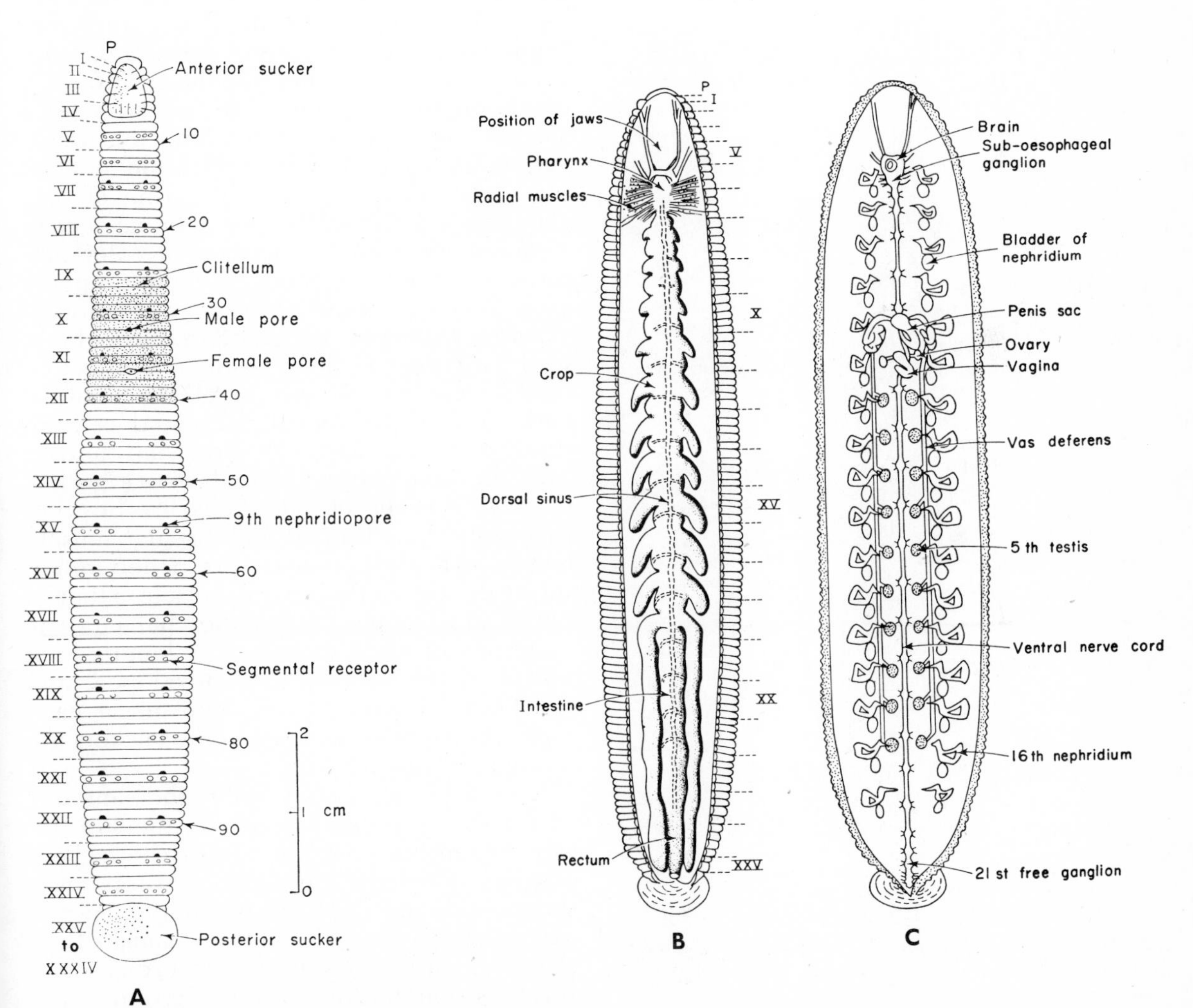

Figure 10–59. The medicinal leech, *Hirudo medicinalis*. *A*. External ventral surface. *B* and *C*. Dorsal view of internal structure. (From Mann, K. H., 1962: Leeches. Pergamon Press.)

The segments at both extremities have been modified to form suckers. The anterior sucker is usually smaller than the posterior one, and frequently surrounds the mouth. The posterior sucker is disc-shaped and turned ventrally. Metamerism is very much reduced. The number of segments, unlike other annelids, is fixed at 34, but secondary external annulation has obscured the original external segmentation. There are no setae present. As in oligochaetes, a clitellum is present and is always formed by segments IX, X, and XI.

The body of a leech can be divided into five regions (Fig. 10–59*A*). A reduced prostomium and six segments compose the head, or cephalic region. Dorsally, the head bears a number of eyes; ventrally, it bears the anterior sucker surrounding the mouth (Figs. 10–60 and 10–61*A*).

The head is followed by a preclitellar region of four segments and a clitellar region of three segments (IX, X, and XI). Midventrally, the clitellar segments bear a single male gonopore and a single female gonopore, usually separated by one complete segment. The clitellum is rarely conspicuous, except during reproductive periods; in the Glossiphoniidae it is rarely conspicuous.

The middle region of the body, containing 15 segments (XII to XXVI), comprises the greater part of the trunk. The terminal, or posterior, region is composed of eight fused segments, which are modified to form the large ventral posterior sucker. The anus opens dorsally in front of the sucker, a necessary adaptation.

The number of annuli per segment varies not only in different regions of the body, but also in different species. For example, in *Glossiphonia*, segments I and II are composed of one annulus; segment III of two annuli; and segments IV to XXII of three annuli each. However, the midregion of the body of *Hirudo* contains five annuli per segment. The best means of determining the segmentation of leeches and the relationship of annuli to the primary segments is by a study of the nervous system and the innervation of the annuli by segmental nerves. However, the occurrence of a ring of sensory papillae around the first annulus of each segment, the serial repetition of color patterns, and the placement of nephridiopores give good indications of the segmentation.

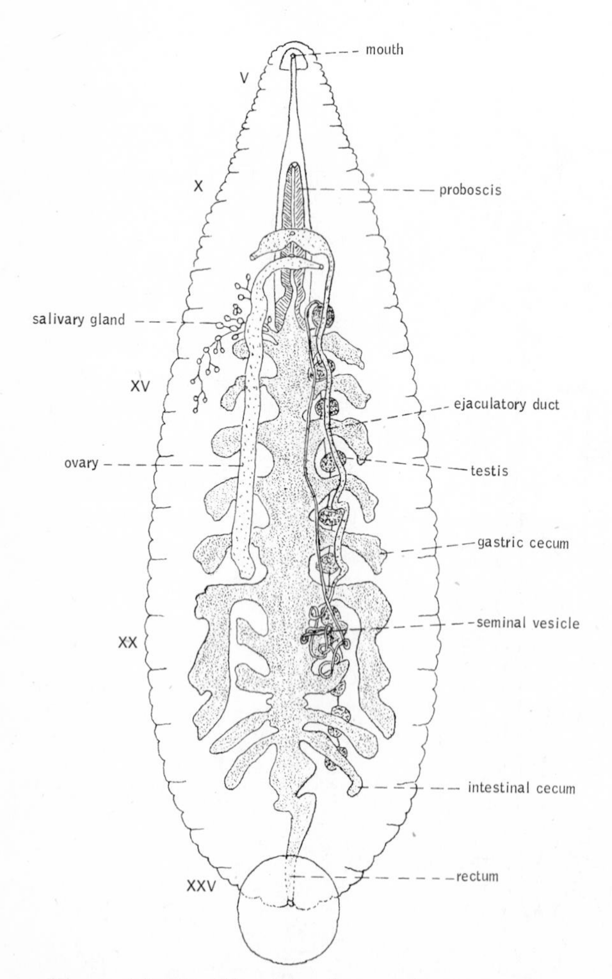

Figure 10–60. *Glossiphonia complanata*, a glossiphoniid leech (ventral view). (After Harding and Moore from Pennak.)

Body Wall and Locomotion. The body wall contains a more distinct connective dermis than is present in other annelids, and some of the unicellular gland cells of the integument are very large and sunken into the connective tissue layer (Fig. 10–62). A layer of oblique muscles lies between the longitudinal and circular muscle layers, and all three layers are penetrated at intervals by dorsoventral muscle strands, which are responsible for the characteristic flattening of some species. The longitudinal muscle layer is powerfully developed, and the fibers converge in the suckers at each end. The circular muscles in the suckers are arranged concentrically.

Leeches crawl, but only the anterior and posterior suckers anchor to the substratum. When the posterior sucker is attached, a wave of circular contraction sweeps over the animal, and the body is lengthened and extended forward. The anterior sucker then attaches, and the posterior sucker releases. A wave of longitudinal contraction then oc-

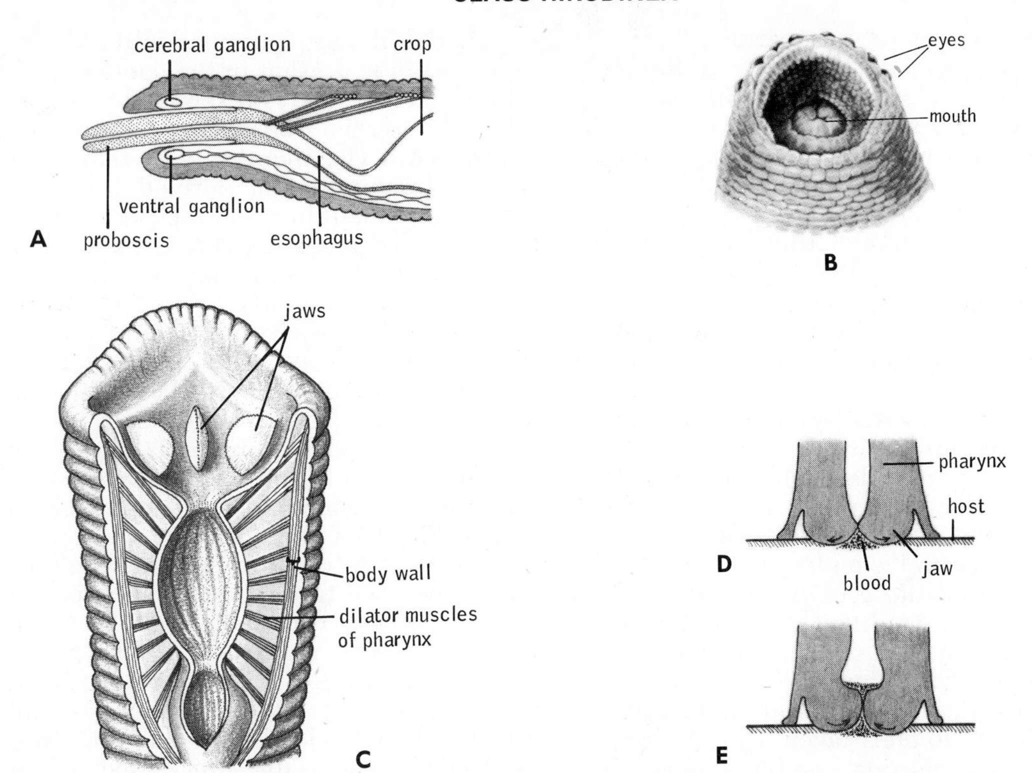

Figure 10–61. *A*. Sagittal section through anterior end of *Glossiphonia*, showing protruded tubular pharynx. (Modified after Scribin.) (Figure 10–60 shows non-protruded state.) *B*. Ventral view of oral region of a terrestrial blood sucking haemadipsid leech. Jaws are not exposed. (From Keegan, H. L., et al., 1968: 406th Medical Laboratory Special Report. U.S. Army Med. Command, Japan.) *C*. Ventral dissection of anterior end of *Hirudo*, showing three jaws. (Modified after Pfurtscheller.) *D* and *E*. Ingestion by *Hirudo*. Outward movement of teeth (*D*), followed by medial movement of teeth and dilation of pharynx (*E*). (After Herter.)

curs, shortening the animal and moving the posterior sucker forward. In the swimming leeches, primarily hirudinids and erpobdellids, the dorsoventral muscle bands are maintained in a constant state of contraction, flattening the body. Waves of contraction along the longitudinal muscles then produce vertical undulations.

Coelom, Circulation, and Respiration. A striking difference between leeches and other annelids is loss of the distinct coelom by the leeches. Only in the five anterior segments of the primitive *Acanthobdella* are there coelomic compartments and separating septa. Interestingly, in this same leech there is no anterior sucker, and the anterior seg-

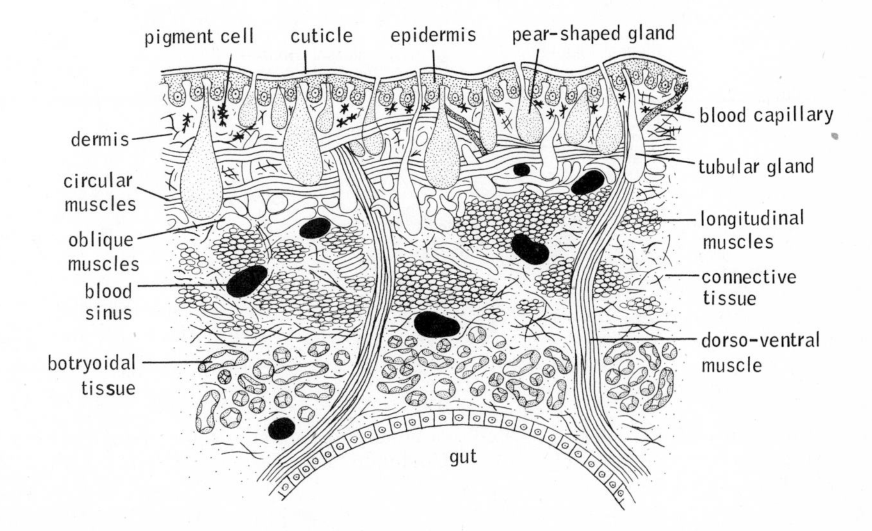

Figure 10–62. Section through the body wall of *Hirudo*. (After Mann.)

ments with coelomic compartments bear setae. *Acanthobdella* provides additional evidence linking leeches with oligochaetes. In all other leeches, septa have disappeared, and the coelomic cavity has been invaded by connective tissue. This connective-tissue invasion has reduced the coelom to a system of interconnected coelomic sinuses and channels (Figs. 10–62 and 10–63). In the hirudinid leeches, the coelom is further restricted by the proliferation of chlorogogen and loose mesenchymal (botryoidal) tissue from the sinus walls.

The coelomic sinuses act as an auxiliary circulatory system in the glossiphoniids and piscicolids (rhynchobdellids), which have retained the blood-vascular system of oligochaetes (Fig. 10–63); but in the other leech orders the ancestral circulatory system has disappeared, and the coelomic sinuses and fluid have been modified to form a true blood-vascular system. In all leech orders, the coelomic sinuses form a regular system of channels (Fig. 10–63), and in those in which it has completely taken over the function of internal transport, the blood (actually the equivalent of coelomic fluid) is propelled by the contractions of the lateral longitudinal channels.

Gills are found only in the Piscicolidae, the general body surface providing for gas exchange in other leeches. The piscicolid gills are lateral leaflike or branching outgrowths of the body wall (Fig. 10–58*A*). The gills are vesicular and filled with coelomic fluid, and ventilation takes place when they pulsate. Leeches without gills ventilate by undulating the body with only the posterior sucker attached.

Respiratory pigment is found only in the gnathobdellid and pharyngobdellid leeches. Hemoglobin is present in the coelomic fluid of these leeches and is responsible for about half of the oxygen transport.

Nutrition. The mouth opens into an extensive foregut region, derived from ectoderm and containing the ingestive organs. Two types of ingestive organs exist in the leeches. A proboscis is characteristic of the order Rhynchobdellida (families Glossiphoniidae and Piscicolidae) (Fig. 10–61). The members of the order Gnathobdellida (families Hirudinidae and Haemadipsidae) and the order Pharyngobdellida (family Erpobdellidae) lack a proboscis but possess instead a muscular, sucking pharynx, which is provided with toothed jaws in the gnathobdellids (Fig. 10–61).

In leeches with a proboscis, the proboscis is an unattached tube lying within a proboscis cavity, which is connected to the ventral mouth by a short, narrow canal. The proboscis is highly muscular, has a triangular lumen, and is lined internally and externally with cuticle. Cuticle also lines the proboscis cavity. Ducts from large unicellular salivary glands open into the proboscis. When feeding, the animal extends the proboscis anteriorly out of the mouth, forcing it into the tissue of the host.

In the gnathobdellids, which lack a proboscis, the mouth is located in the anterior sucker and is flanked by an upper and lower lip (Fig. 10–61*B* and *C*). Just within the mouth cavity, three large, oval, bladelike jaws each bear along the edge a large number of small teeth. The three jaws are arranged in a triangle, one dorsally and two laterally. When the animal feeds, the anterior sucker is attached to the surface of the

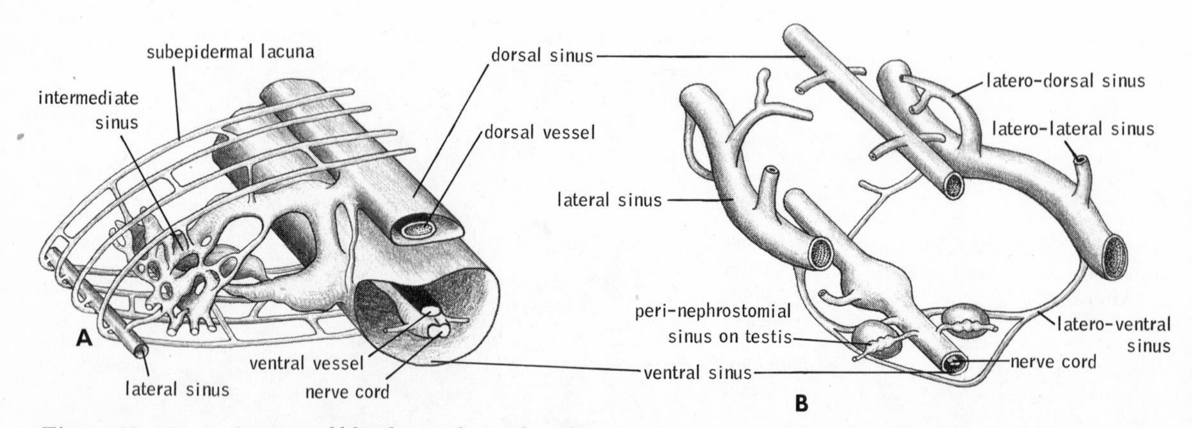

Figure 10–63. *A*. Section of blood vascular and coelomic sinus system of the glossiphoniid *Placobdella costata*. (After Oka from Harant and Grassé.) *B*. Coelomic sinus system of *Hirudo*, in which the original blood vascular system has disappeared. (After Mann.)

prey or host, and the edges of the jaws slice through the integument (Fig. 10–61*D* and *E*). The jaws swing toward and away from each other, activated by muscles attached to their bases.

Immediately behind the teeth, the buccal cavity opens into a muscular, pumping pharynx. Between the pharyngeal wall and the body wall are masses of unicellular salivary glands that open between the teeth on the jaws. The salivary glands secrete an anticoagulant called hirudin. The erpobdellids also have a pharynx, but the jaws are replaced by muscular folds.

The remainder of the digestive tract is relatively uniform throughout the class. A short esophagus opens into a relatively long stomach, or crop. The stomach may be a straight tube, as in the erpobdellids, but more commonly it is provided with one to eleven pairs of lateral ceca (Figs. 10–59 and 10–60). The posterior pair of ceca are usually greatly elongated, and in some Piscicolidae the last two pairs have fused together.

The posterior third of the alimentary canal is composed of the intestine, which is often separated from the stomach by a pyloric sphincter. The intestine may be a simple tube or, as in the rhynchobdellids, may be provided with four pairs of slender lateral ceca. The intestine opens into a short rectum, which empties to the outside through the dorsal anus, located in front of the posterior sucker.

Many leeches are predacious, but about three-fourths of the known species are bloodsucking ectoparasites. However, in many cases the difference lies only in the size of the host. The Hirudinidae especially demonstrate a gradation from predation to parasitism. The Erpobdellidae contain the greatest number of predacious leeches, but this type of feeding habit is found in other families as well. Predatory leeches always feed on invertebrates. Prey includes worms, snails, and insect larvae. Feeding is relatively frequent and the prey is usually swallowed whole. Many glossiphoniids suck all the soft parts from their hosts and are best regarded as specialized predators.

The bloodsucking parasitic leeches attack a variety of hosts. Some, primarily species of *Glossiphonia* and *Helobdella,* feed only on invertebrates, such as snails, oligochaetes, crustaceans, and insects, but vertebrates are hosts for most species. Piscicolidae are parasites of both freshwater and marine fish, sharks, and rays (Figs. 10–58*A* and *B*). The glossiphoniids feed on amphibians, turtles, snakes, alligators, and crocodiles. Species of the cosmopolitan glossiphoniid genus *Theromyzon* attach to the nasal membranes of shore and water birds. The aquatic Hirudinidae and the terrestrial Haemadipsidae feed primarily on mammals, including man (Fig. 10–58*G*).

Parasitic leeches are rarely restricted to one host, but they are usually confined to one class of vertebrates. For example, *Placobdella* will feed on almost any species of turtles and even alligators, but they rarely attack amphibians or mammals. On the other hand, mammals are the preferred hosts of *Hirudo,* but this leech will also attack snakes, turtles, and amphibians. Furthermore, some species of leeches that are exclusively bloodsuckers as adults are predacious during juvenile stages.

The notorious mammalian bloodsuckers, such as *Hirudo,* select a thin area of the host's integument, attach the anterior sucker very tightly to this area, and then slit the skin. The jaws of *Hirudo* make about two slices per second. The incision is anesthetized by a substance of unknown origin. The pharynx provides continual suction, and the secretion of hirudin prevents coagulation of the blood. Penetration of the host's tissues is not well understood in the many jawless proboscis-bearing species that are bloodsuckers. The proboscis becomes rigid when extended, and it is possible that penetration is aided by enzymatic action.

Jennings and van der Lande's (1967) comparative studies have contributed considerably toward understanding of many of the peculiarities of leech digestion. The gut secretes no amylases, lipase, or endopeptidases. The last are enzymes which cleave a protein molecule in the middle, breaking it into two smaller fragments. Endopeptidases are characteristic of most animals, but the gut of leeches produces only exopeptidases, which cleave away small external groups from the larger molecule. This type of digestion perhaps explains the fact that digestion in leeches is so slow.

Also characteristic of the leech gut is a symbiotic bacterial flora that is apparently important in nutrition. In both the bloodsucking medicinal leech *Hirudo medicinalis* and the predacious *Erpobdella octoculata,* the gut bacteria are responsible for a considerable part of digestion. Although their contribution is less in other species, digestion by bacterial action may be significant in all leeches. The bacteria may also produce vitamins and other compounds that are used by the leech host.

Bloodsuckers feed infrequently, but when they do, they can consume an enormous quantity of blood. *Haemadipsa* may ingest ten times its own weight, and *Hirudo* two to five times its own weight. Following ingestion, water is removed from the blood and excreted through the nephridia. The digestion of the remaining blood cells then takes place very slowly. These leeches can then tolerate long periods of fasting. Medicinal leeches have been reported to have gone without food for one and half years without fatal results, and since they may require 200 days to digest a meal, they need not feed more than twice a year in order to grow.

Excretion. Leeches contain from ten to seventeen pairs of nephridia, located in the middle third of the body, one pair per segment (Fig. 10–59). The leech nephridium differs from that of the oligochaetes in a number of respects (Fig. 10–64). As a result of the coelom reduction and the loss of septa in the leech body, the nephridial tubules are imbedded in connective tissue, and the nephrostomes project into the coelomic channels. Each nephrostome opens into a nonciliated capsule.

The nephridial tubule in leeches is especially peculiar. It is in part or completely composed of a cord of cells through the interior of which runs a nonciliated, intracellular canal (Fig. 10–64). In primitive leeches the canal opens into the capsule, but in most leeches the cavities of the capsule and the canal do not connect and the two parts of the nephridium may even have lost all structural connection. The nephridial canal drains many finer branching canals, which are also intracellular (Fig. 10–64*A*). A short terminal ciliated section, which is derived from ectoderm, opens to the outside through the nephridiopore. The nephridiopores are ventrolaterally located except in the terrestrial Haemadipsidae, where the moisture necessary to permit proper functioning of the suckers is provided by the urine of the displaced first and last pairs of nephridiopores.

In the piscicolids, the nephridial tubules are interconnected, but no reduction in the number of nephridiopores or other parts has taken place. The nephridia of the hirudinids are further specialized because of the conversion of the coelomic channels into a blood-vascular system. The capsules, which are enclosed within a blood-filled sinus, contain multiple funnels and are sometimes called ciliated organs (Fig. 10–64*B*). In hirudinids the terminal part of the tubule is enlarged, forming a distinct bladder.

Ammonia is the principal waste and is removed from the blood by cells of the nephridial tubule. There is evidence that in *Hirudo*, at least, some of the ammonia results from the degradation of more complex nitrogenous substances by the action of symbiotic bacteria.

The function of the nephridial capsules is believed to be the production of coelomic fluid—the blood of many leeches—by the

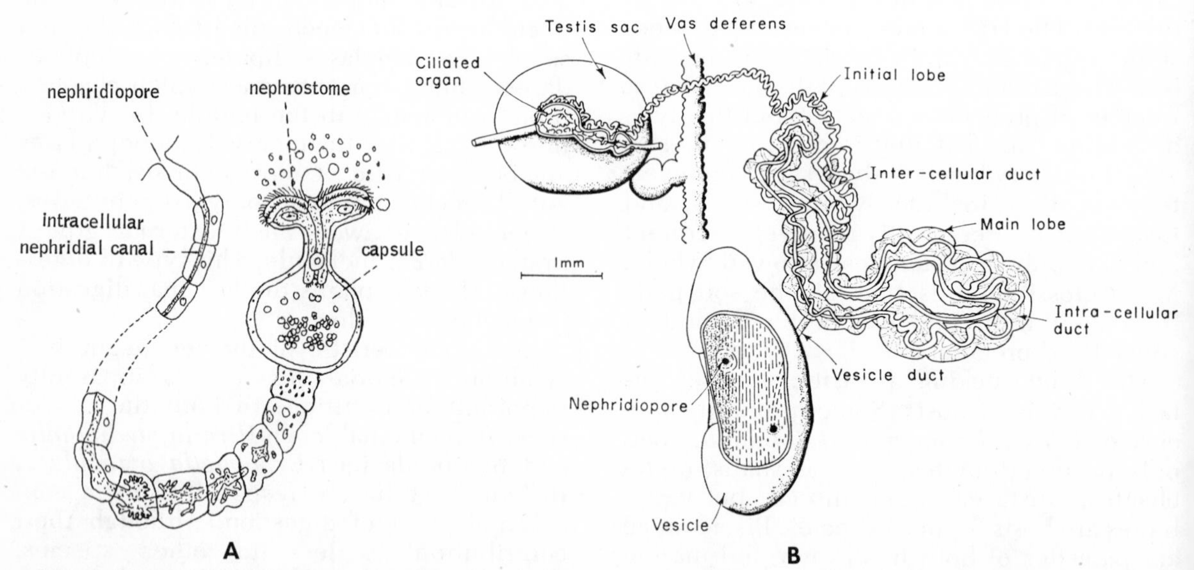

Figure 10–64. *A.* Nephridium of a glossiphoniid. (After Graf from Harant and Grassé.) *B.* Nephridium of *Hirudo.* The capsule, or ciliated organ, lies on top of a testis. (After Bhatia from Mann.)

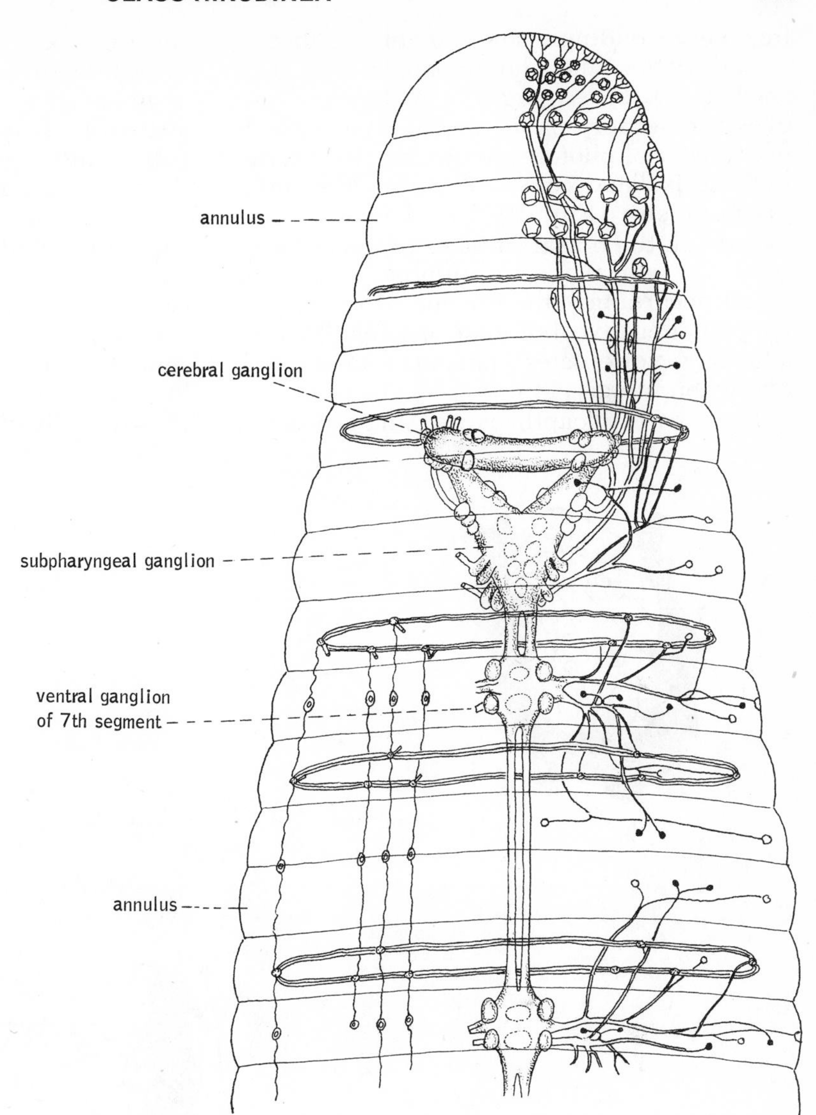

Figure 10–65. Nervous system of *Erpobdella punctata.* (After Bristol from Harant and Grassé.)

outward beating of the nephrostome cilia. The coelomocytes are phagocytic and engulf particulate matter, but the eventual fate of the waste-laden cells is not certain. They may migrate to the epidermis or to the epithelium of the digestive tract. Particulate waste is also picked up by the chlorogogen and mesenchymal tissue of hirudinid leeches and by coelomic epithelial cells and pigment cells (adipose cells) of the coelomic connective tissue in glossiphoniids and piscicolids.

The Nervous System. The nervous system of leeches reflects their specializations of body structure. The cell bodies of the ganglia are grouped into distinct masses, or follicles (Fig. 10–65). Each ganglion is composed of six such follicles, arranged in two transverse triads. In the fifth and sixth segments, a large ganglionic nerve ring surrounds the pharynx or proboscis. This collar represents the brain, the circumpharyngeal connectives, and the subpharyngeal ganglion of other annelids, and the ganglia of the first three or four segments that have migrated posteriorly. There are two ventral nerve cords, although each pair of segmental ganglia are fused (Fig. 10–65).

Some rhynchobdellid leeches can dramatically change color as a result of pigment movements in large chromatophores, which are under neurohumoral control. The significance of the color change is not certain, for these leeches do not adapt to background coloration.

Sense Organs. As in oligochaetes, the general epidermis in leeches contains dispersed sensory cells of a number of types—

free nerve endings, sensory cells with terminal bristles, and photoreceptor cells. The specialized sense organs in leeches are eyes and sensory papillae. The eyes consist of a cluster of photoreceptor cells essentially like the isolated cells and surrounded by a pigment cup (Fig. 10–66). The eyes are located on the dorsal surfaces of the anterior segments and vary in number from two to ten, according to the species. Most leeches are photonegative, but some, especially those attacking vertebrates, becomes photopositive when hungry.

The sensory papillae are small projecting discs arranged in a dorsal row or in a complete ring around one annulus of each segment (Fig. 10–66*B*). Each papilla consists of a cluster of many sensory cells and supporting epithelium. The sensory cells each bear a terminal bristle, which projects above the cuticle surface. The function of the papillae is still uncertain. The number and arrangement of both eyes and sensory papillae have considerable taxonomic value.

Despite the lack of highly organized concentrated sense organs, leeches can detect low levels of many types of stimuli, and the sensitivity is often an adaptation for finding

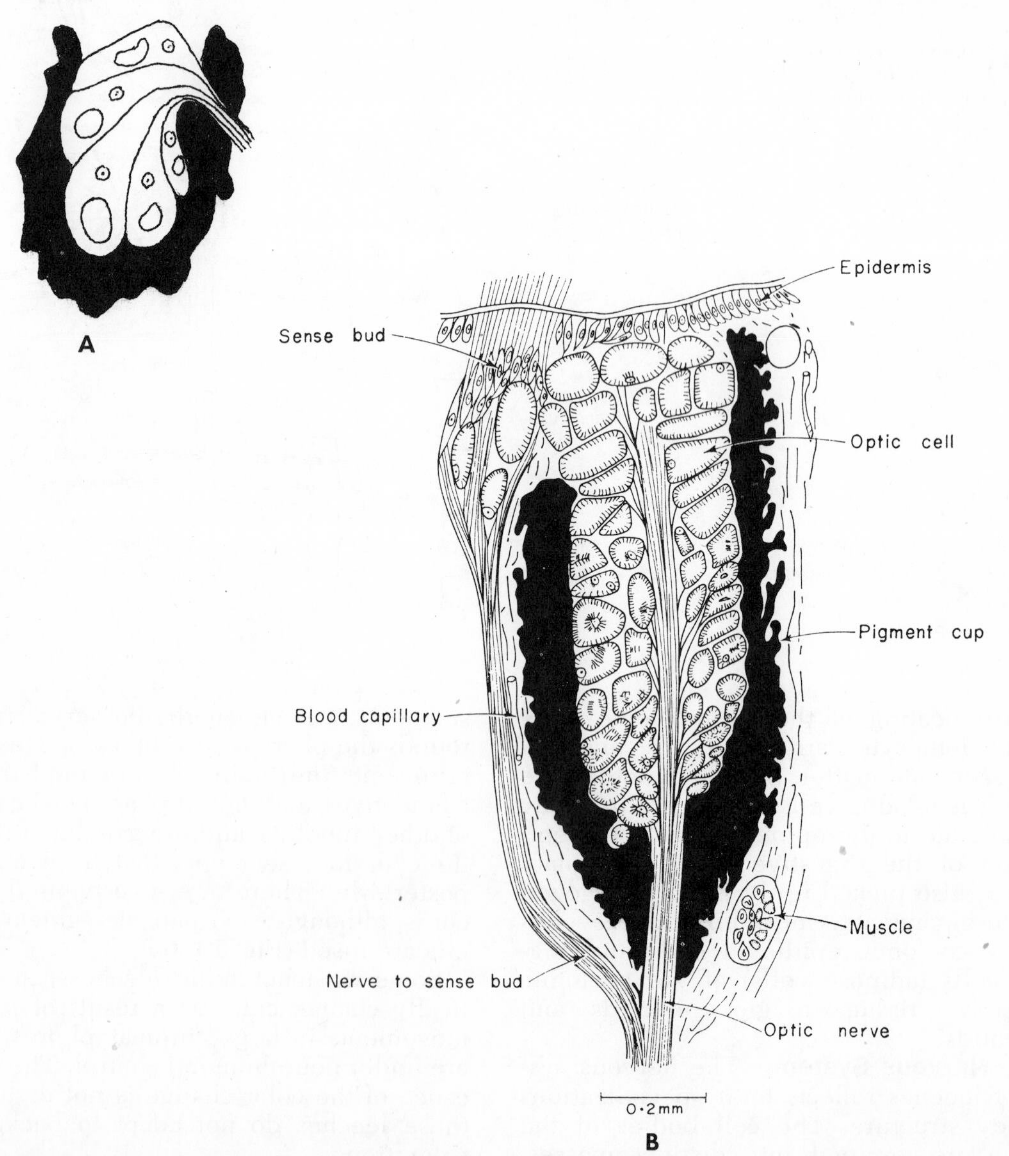

Figure 10–66. *A*. Section through eye of *Erpobdella*. (After Hesse from Harant and Grassé.) *B*. Section through the more complex eye of *Hirudo*. A cluster of epidermal sensory cells (sensillae) are shown to one side of the eye. (After Hesse from Mann.)

prey or a host. Fish leeches respond to moving shadows and water pressure vibrations. Both predatory and bloodsucking leeches will attempt to attach to an object smeared with various host or prey substances—fish scales, tissue juices, oil gland secretions, sweat, and other substances. *Hirudo,* which is a bloodsucker of warm-blooded animals, will be attracted to the host by the elevated temperature. The same leech is also attracted by body secretions and has been reported to swim toward a man standing in water. Supposedly, the terrestrial bloodsucking Haemadipsidae of the tropics, which are attracted by passing warm-blooded mammals, will move over vegetation and converge on a man standing in one place.

Reproduction. Unlike many other annelids, leeches do not reproduce asexually, nor can they regenerate lost parts. Like oligochaetes, all leeches are hermaphroditic, with distinct gonads and ducts having relatively fixed positions. Although the ovaries lie in front of the testes, the position of the gonopores is reversed. The single median male gonopore is located on the ventral side of segment X just in front of the single female gonopore on segment XI. The cavity of each ovary and of each testis represents a specialized part of the coelom.

Four to many pairs of spherical testes are arranged segmentally, beginning in either segment XII or segment XIII (Fig. 10–67). A short vas efferens connects each testis to a vas deferens, which runs anteriorly on each side of the body. Anterior to the first pair of testes, each vas deferens enlarges or becomes greatly coiled to form an epididymis (Fig. 10–67). An ejaculatory duct issues from each vesicle to empty into a single median atrium, which opens to the outside through the male gonopore.

The atrium is highly muscular and is surrounded by gland cells that open through the atrium walls into the lumen. The atrium of erpobdellids and most glossiphoniids and piscicolids is not eversible and does not act as a penis. Instead, it is a chamber for the formation and expulsion of a spermatophore. The spermatophore varies in shape, depending on the shape of the atrium, but it always consists of two bundles of sperm, one bundle from each ejaculatory duct, fused together to varying degrees (Fig. 10–68*E*).

In the hirudinids, the atrium is more complex, and its distal wall can be everted through the gonopore to form a penis (Fig. 10–67*B*). The proximal enveloping glands are called a prostate. Spermatophores are not formed.

Spermatogenesis in leeches takes place primarily in the lumen of the testis, just as it does in the coelom of other annelids.

The ovaries are always limited to a single pair, located between the most anterior pair of testes and the male atrium. Each ovary is a mass of germinal tissue, or sometimes a number of masses, enclosed within an ovisac (Fig. 10–67*A*). The cavity of the ovisac is derived from the coelom. Each ovisac is usually elongated and sometimes folded forward on itself.

A short oviduct extends anteriorly from each ovisac and joins its opposite member to form a vagina, which opens through the female gonopore on the ventral surface of segment XI. The vagina is sometimes preceded by a common oviduct, as in *Haemopis.* The eggs leave the ovaries as oocytes, and maturation is completed in the fluid of the ovisac. Gametogenesis in leeches has been shown to be under neurosecretory control.

Sperm transfer in the hirudinids, most of which possess a penis, is similar to direct sperm transmission in earthworms (Fig. 10–68*A*). The ventral surface of the clitellar regions of a copulating pair are brought together, with the anterior of each worm directed toward the posterior of the other. Thus the male gonopore of one worm apposes the female gonopore of the other. The penis is everted into the female gonopore, and sperm are introduced into the vagina, which probably also acts as a storage center.

Sperm transfer in most glossiphoniids, piscicolids, and erpobdellids, all of which lack a penis, is by hypodermic impregnation. The two copulating worms commonly intertwine and grasp each other with the anterior sucker. The ventral clitellar regions are in apposition, and by muscular contraction of the atrium, a spermatophore is expelled from one worm and penetrates the integument of the other. The site of penetration is usually in the clitellar region, but spermatophores may be inserted some distance away. As soon as the head of the spermatophore has penetrated the integument, perhaps by a combination of expulsion pressure and a cytolytic action by the spermatophore itself, the sperm are discharged into the tissues (Fig. 10–68*D*). Following liberation from the spermatophore, the sperm are carried to the ovaries in the coelomic channels. There

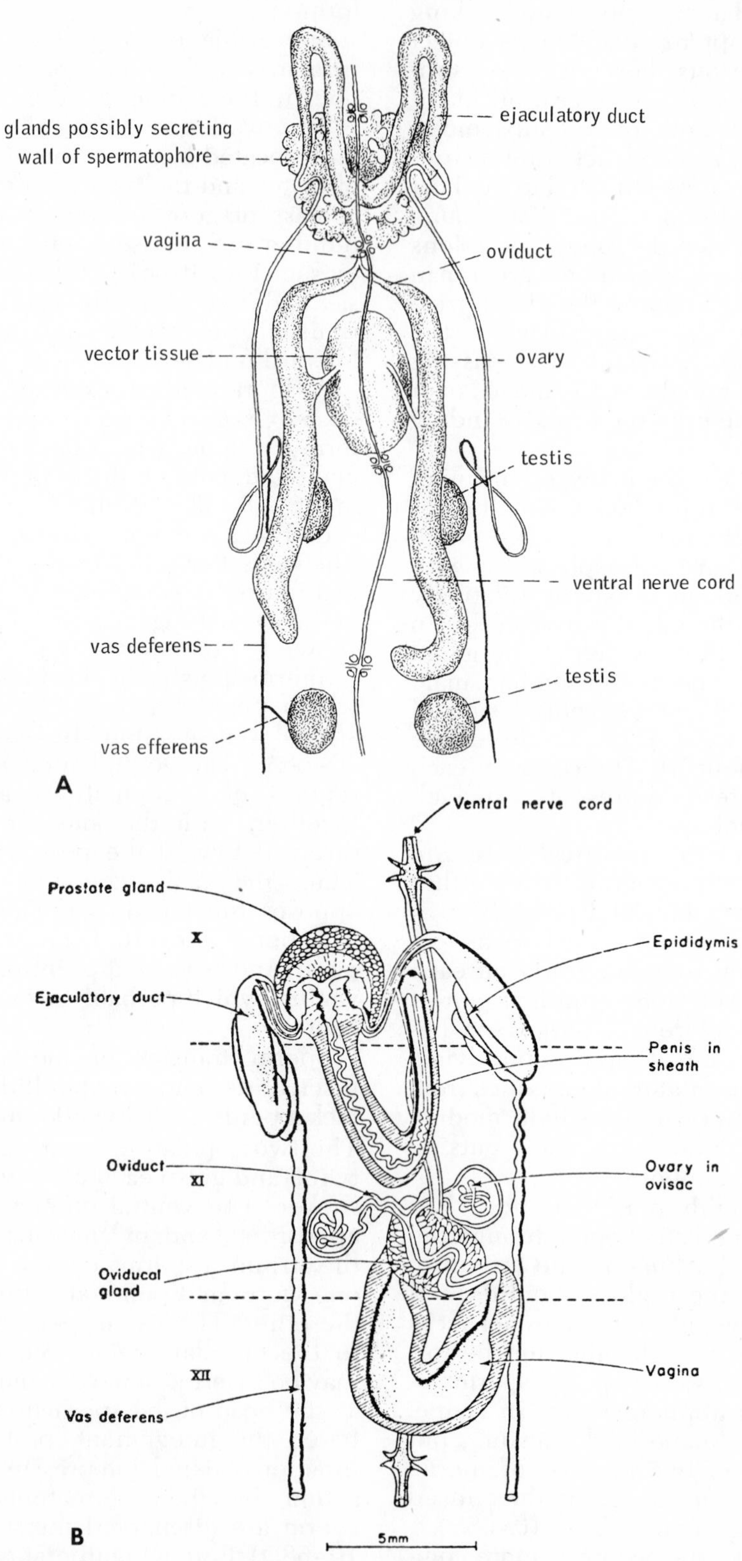

Figure 10–67. *A*. Reproductive system of *Piscicola geometra*. (After Brumpt from Harant and Grassé.) *B*. Reproductive organs of *Hirudo*. Testes are associated with vas deferens in more posterior segments. (After Leuckart and Brandes from Mann.)

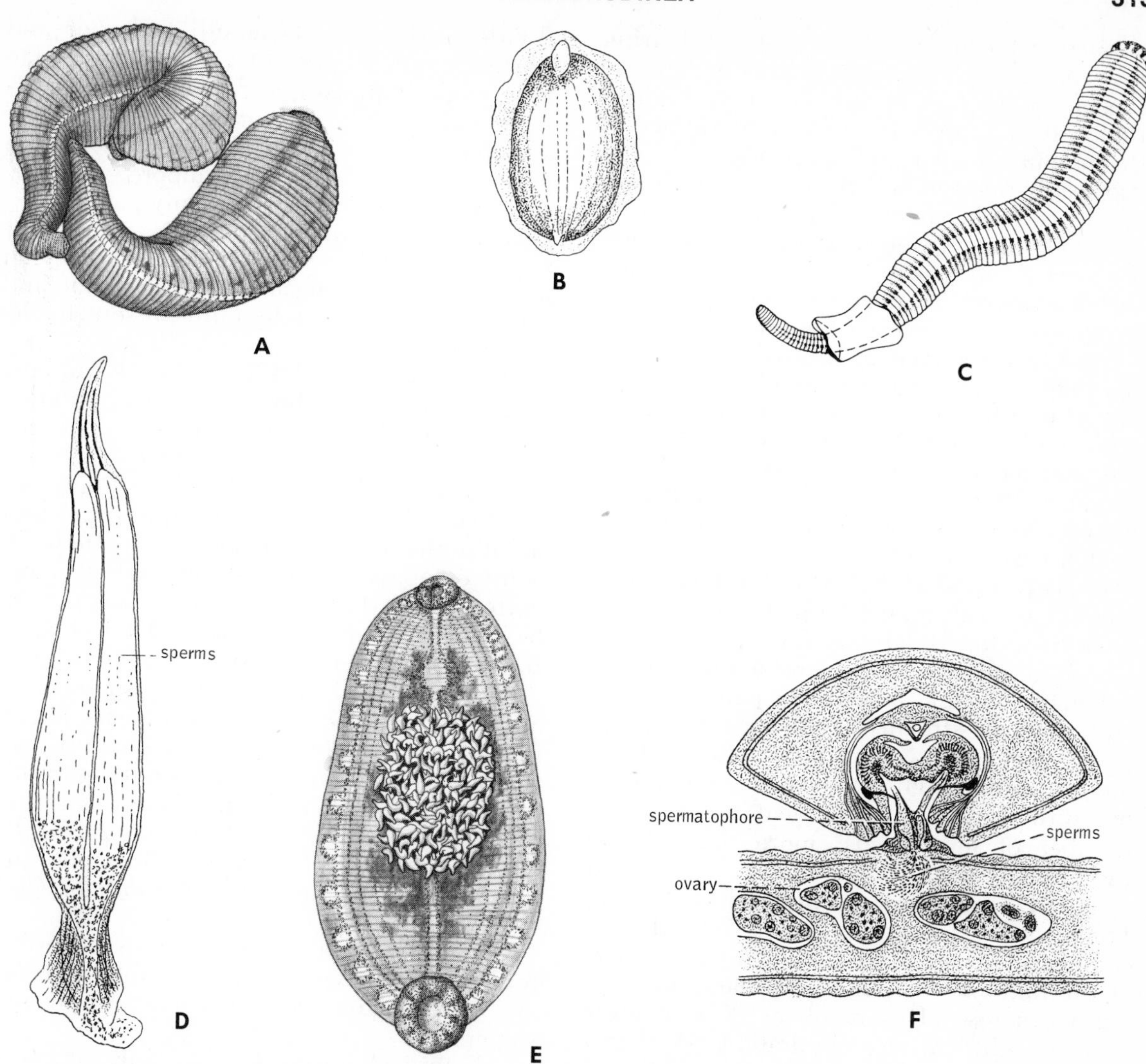

Figure 10–68. *A. Hirudinaria* copulating. (After a photograph by Keegan et al.) *B.* Cocoon of *Erpobdella octoculata.* (After Pavlovsky from Harant and Grassé.) *C. Erpobdella* withdrawing body from cocoon. (From Nagao, Z., 1957: J. Fac. Sci. Hokkaido Univ. Ser. VI, Zool., *13*:192–196.) *D.* Spermatophore of the rhynchobdellid, *Haementeria.* (After Pavlovsky from Harant and Grassé.) *E.* Glossiphoniid brooding young. *F.* Section through two copulating *Erpobdella.* Upper individual is injecting spermatophores into lower individual. (After Brumpt.)

are some fish leeches, however, in which the spermatophore is stabbed into a special copulatory area. The special tissue, called vector tissue, lies beneath the copulatory area and connects with the ovisacs, (Fig. 10–67*A*). This vector tissue acts as a pathway for sperms moving to the ovisacs where fertilization takes place.

Eggs are laid from two days to many months after copulation. At this time, the clitellum becomes conspicuous and in most families secretes a cocoon as is done in the oligochaetes (Fig. 10–68*C*). The cocoon is filled with a nutritive albumin produced by certain of the clitellar glands. The cocoon then receives the fertilized eggs as it passes over the female gonopore. A cocoon contains only one egg in the Piscicolidae, but in other leeches the number varies from a few to many. Beginning in May, *Erpobdella punctata* in Michigan produces some ten cocoons, each with five eggs, which hatch three to four weeks later. The cocoons are affixed to submerged objects or vegetation (Fig. 10–68*B*). Some piscicolids attach their cocoons to their host. Terrestrial species place them in damp soil beneath stones and other objects, and the hirudinids, such as *Hirudo* and *Haemopis*, leave the water to deposit their cocoons in damp soil.

The glossiphoniids brood their eggs. In some the cocoons are attached to the bottom and covered and ventilated by the ventral surface of the worm. In others, the atypical cocoons are membranous and transparent and are attached to the ventral surface of the parent (Fig. 10–68*F*). During the course of development the embryonic leeches break free of the cocoon and attach themselves directly to the ventral surface of the parent by a special anterior ectodermal process. When the posterior sucker is developed, the sucker is used for attachment.

Development in leeches follows essentially the same pattern as in oligochaetes, although the mesoderm is derived from the third instead of the fourth blastomere.

Most leeches have an annual cycle, breeding in the spring and maturing by the following year. Life cycles are correlated in part with feeding habits. Some, such as *Hirudo*, are associated with the host only during actual feeding; others, such as *Hemibdella*, never leave the host. But most leave the host at least to breed.

Ecology and Distribution. Although some leeches are marine, most aquatic species live in fresh water. Relatively few species tolerate rapid currents; most prefer the shallow vegetated water bordering ponds, lakes, and sluggish streams. Acid waters support poor leech faunas. In favorable environments, often high in organic pollutants, overturned rocks may reveal an amazing number of individuals; more than 10,000 individuals per square meter have been reported from Illinois (Richardson, 1925). Some species are able to estivate during periods of drought by burrowing into mud at the bottom of a pond or stream, and can survive a loss of as much as 90% of their body weight during dehydration. Mass upstream migrations of swimming forms during breeding periods have been reported.

There is a tendency toward amphibious habits in the hirudinid and erpobdellid leeches. For example, there is a terrestrial hirudinid, *Haemopis terrestris*, which is occasionally plowed up in fields in midwestern United States. Complete terrestriality has been attained in the Haemadipsidae, which inhabit humid jungles of southern Asian and Australian regions.

Although leeches are found throughout the world, they are most abundant in north temperate freshwater lakes and ponds. Much of the North American leech fauna is shared with Europe.

SYSTEMATIC RÉSUMÉ OF CLASS HIRUDINEA

Annelids composed of 34 segments, which lack parapodia and setae, and which are subdivided externally as annuli. Anterior and posterior suckers present. Coelom reduced to a network of sinuses, and internal metamerism evident only in nervous and excretory systems. All are hermaphroditic, with one male and one female gonopore. Marine, freshwater, and terrestrial; many parasitic.

Order Acanthobdellida. A primitive order, contains a single North European species parasitic on salmonid fish. Setae and a compartmented coelom are present in the five anterior segments.

Order Rhynchobdellida. Strictly aquatic leeches with a proboscis and a circulatory system that is separate from the coelomic sinuses.

Family Glossiphoniidae. Flattened leeches, typically with three annuli per segment in the mid-region of the body. Includes many ectoparasites of both invertebrates and vertebrates. *Marsupiobdella, Glossiphonia, Helobdella, Placobdella, Theromyzon, Haementeria, Hemiclepsis.*

Family Piscicolidae. The fish leeches. Subcylindrical body that often bears lateral gills. Usually more than three annuli per segment. Most marine leeches belong to this family. Parasites of marine and freshwater fish and, rarely, crustaceans. *Piscicola, Pontobdella, Trachelobdella, Branchellion, Ozobranchus, Illinobdella, Myzobdella, Cystobranchus.*

Order Gnathobdellida. Aquatic or terrestrial leeches having a noneversible pharynx and three pairs of jaws. Five annuli per segment.

Family Hirudinidae. Chiefly amphibious or aquatic bloodsucking leeches. *Haemopis, Hirudo, Macrobdella, Philobdella.*

Family Haemadipsidae. Terrestrial tropical leeches of Australasian region, attacking chiefly warm-blooded vertebrates. *Haemadipsa, Phytobdella.*

Order Pharyngobdellida. Contains five families, all having a nonprotrusible pharynx; teeth are lacking, although one or two stylets may be present. Primarily aquatic, with some semiterrestrial forms.

Family Erpobdellidae. Predaceous leeches. *Erpobdella, Dina.*

BRANCHIOBDELLID ANNELIDS

The Branchiobdellidae is an enigmatic group of annelids which are parasitic or commensal on freshwater crayfish. They

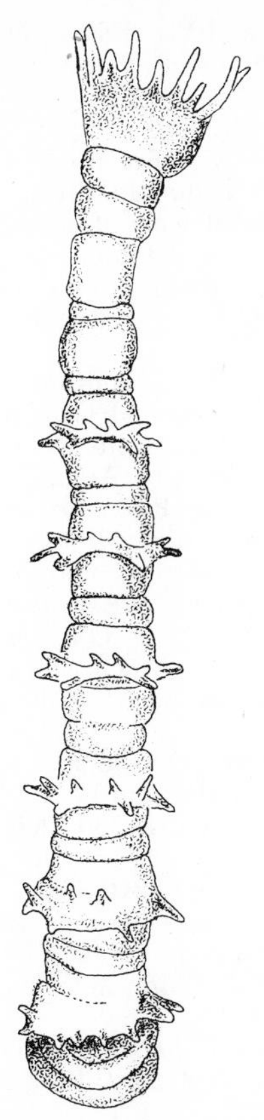

Figure 10–69. *Stephanodrilus,* a member of the leechlike family Branchiobdellidae; they are parasitic and commensal with freshwater crayfish. (After Yamaguchi from Avel.)

show similarities to both oligochaetes and leeches. They are usually included with the oligochaetes, but it is now believed that they diverged early from the base of the common oligochaete-hirudinean stock (Holt, 1968; Brinkhurst and Jamieson, 1972).

All branchiobdellids are very small, and are composed of only 14 to 15 segments (Fig. 10–69). The head is modified into a sucker with a circle of finger-like projections. The buccal cavity contains two teeth. The posterior segments are also modified to form a sucker, and all of the segments lack setae. Some species are ectoparasitic on the gills of crayfish; others live on the outer surface of the exoskeleton and probably are merely commensal.

Bibliography

Anderson, D. T., 1966: The comparative early embryology of the Oligochaeta, Hirudinae, and Onychophora. Proc. Linn. Soc. N. S. W., *91*:10–43.

Brinkhurst, R. O., and Jamieson, B. G., 1972: Aquatic Oligochaeta of the World. Toronto University Press, Toronto. 860 pp.

Cather, J., 1971: Cellular interactions in the regulation of development in annelids and molluscs. Advances in Morphogenesis, *9*:67–124.

Clark, L. B., and Hess, W. N., 1940: Swarming of the Atlantic palolo worm, *Leodice fucato.* Tortugas Lab. Papers, *33*(2):21–70.

Clark, R. B., 1956: *Capitella capitata* as a commensal, with a bibliography of parasitism and commensalism in polychaetes. Ann. Mag. Nat. Hist., *9*(102):433–448.

Clark, R. B., 1969: Systematics and phylogeny: Annelida, Echiura, Sipuncula. *In* Florkin, M., and Scheer, B. J. (Eds.), Chemical Zoology, Vol. 4. Academic Press, N. Y., pp. 1–68.

Dales, R. P., 1957: The feeding mechanism and structure of the gut of *Owenia fusiformis.* J. Mar. Biol. Assoc. U. K., *36*(1):81–89.

Dales, R. P., 1963: Annelids. Hutchinson University Library, London.

Dales, R. P., and Peter, G., 1972: A synopsis of the pelagic polychaeta. J. Nat. Hist., *6*(1):55–92.

Day, J., 1967: Polychaeta of Southern Africa. Pt. I, Errantia; Pt. II, Sedentaria. British Museum of Natural History.

Fauvel, P., Avel, M., Harant, H., Grassé, P., and Dawydoff, C., 1959: Embranchement des Annelides. *In* Grassé, P. (Ed.), Traité de Zoologie, Vol. 5, pt. 1. Masson et Cie, Paris, pp. 3–686.

Friedlander, B., 1888: Über das Kriechen der Regenwürmer. Biol. Zbl., *8*:363–366.

Goerke, H., 1971: Die Ernahrungweise der Nereis-Arten der deutschen Kusten. Veroff. Inst. Meeresforsch. Bremerh. *13*:1–50.

Goodnight, C. J., Hartman, O., and Moore, J. P., 1959: Oligochaeta, Polychaeta, and Hirudinea. *In* Edmondson, W. T., Ward, H. B., and Whipple, G. C. (Eds.), Freshwater Biology. 2nd Edition. John Wiley and Sons, N. Y., pp. 522–557. (A guide for the identification of freshwater annelids. Keys and figures.)

Goodnight, C. J., 1973: The use of aquatic macroinvertebrates as indicators of stream pollution. Trans. Amer. Micr. Soc., *92*(1):1–13.

Goodrich, E. S., 1945: The study of nephridia and genital ducts since 1895. Quart. J. Micr. Sci., *86*:113–392.

Gosner, K. L., 1971: Guide to Identification of Marine and Estuarine Invertebrates: Cape Hatteras to the Bay of Fundy. Wiley-Interscience, N. Y. Annelids, pp. 326–387.

Gray, J., 1939: Studies in animal locomotion, VIII. The kinetics of locomotion of *Nereis diversicolor.* J. Exp. Biol., *16*:9–17.

Gray, J., and Lissmann, H. W., 1938: Studies in animal locomotion, VII. Locomotory reflexes in the earthworm. J. Exp. Biol., *15*:506–517.

Gray, J., Lissman, H. W., and Pumphrey, R. J., 1938: The mechanism of locomotion in the leech. J. Exp. Biol., *15*:408–430.

Hanson, J., 1949: The histology of the blood system in Oligochaeta and Polychaeta. Biol. Rev., *24*:127–173.

Hartman, O., 1951: Literature of the Polychaetous Annelids. Vol. 1, Bibliography. Privately published. (A complete listing of the literature on polychaetes from Linnaeus, 1758 through 1959. A required reference for any serious student of this group. 1959: Vol. 2, Catalogue of the polychaetous annelids of the world. Pts. 1 and 2. Allan Hancock Found. Publ., Occas. Paper no. 23.)

Hermans, C. O., 1969: The systematic position of the Archiannelida. Syst. Zool., *18*:85–102.

Holt, T. C., 1968: The Branchiobdellida: Epizootic annelids. The Biologist, Vol. L, Nos. 3–4, pp. 79–94.

Jacobsen, V. H., 1967: The feeding of the lugworm, *Arenicola marina*. Ophelia, *4*:91–109.

Jennings, J. B., and van der Lande, V. M., 1967: Histochemical and bacteriological studies on digestion in nine species of leeches. Biol. Bull., *133*: 166–183.

Kaestner, A., 1967: Invertebrate Zoology. Vol. 1. Wiley-Interscience, N. Y., pp. 454–566.

Light, S. F., Smith, R. I., Pitelka, F. A., Abbott, D. P., and Weesner, F. M., 1967: Intertidal Invertebrates of the Central California Coast. University of California Press, Berkeley. Annelida, pp. 63–108.

McConnaughey, B., and Fox, D. L., 1949: The anatomy and biology of the marine polychaete *Thoracophelia mucronata*. Univ. Calif. Publ. Zool., *47*(12):319–339.

MacGinitie, G. E., 1939: The method of feeding of *Chaetopterus*. Biol. Bull., *77*:115–118.

Mangum, C., 1970: Respiratory physiology in annelids. Amer. Sci., *58*(6):641–647.

Mann, K. H., 1962: Leeches (Hirudinea), Their Structure, Physiology, Ecology, and Embryology. Vol. 2. Pergamon Press, N. Y.

Mileikovskii, S. A., 1968: Morphology of larvae systematics of Polychaeta. Zool. Zh., *47*:49–59.

Millott, N., 1943: The visceral nervous system of the earthworm. II. Evidence of chemical transmission. Proc. Roy. Soc. Lond., ser. B, *131*:362–373.

Nicol, E. A. T., 1930: The feeding mechanism, formation of the tube, and physiology of digestion in *Sabella pavonina*. Trans. Roy. Soc. Edinburgh, *56*(3):537–598.

Nicholls, J. G., and Van Essen, D., 1974: The nervous system of the leech. Sci. Amer., *230*(1):38–48.

Nicol, J. A. C., 1948: The giant axons of annelids. Quart. Rev. Biol., *23*(4):291–323.

Pennak, R. W., 1953: Freshwater Invertebrates of the United States. Ronald Press, N. Y., pp. 278–320. (A brief general account of the freshwater annelids, figures and keys to families and genera.)

Richardson, R. E., 1925: Illinois River bottom fauna in 1923. Illinois Natural History Survey Bulletin, *15*: 391–423.

Rose, S. M., 1970: Regeneration: Key to Understanding Normal and Abnormal Growth and Development. Appleton-Century-Crofts, N. Y.

Sawyer, R. T., 1972: North American freshwater leeches, exclusive of the Piscicolodae, with a key to all species. Illinois Biol. Monogr. no. 46. 154 pp.

Seymour, M. K., 1969: Locomotion and coelomic pressure in *Lumbricus*. J. Exp. Biol., *51*:47.

Stephenson, J., 1930: The Oligochaeta. Oxford University Press, N. Y. (A classic account of the oligochaetes. Out of date in some areas but still of great value.)

Watson, A. T., 1901: On the structure and habits of the Polychaeta of the family Ammocharidae. J. Linn. Soc. London, *28*(181):230–260.

Waxman, L., 1971: The hemoglobin of *Arenicola cristata*. J. Biol. Chem. *246*(23):7318–7327.

Wells, G. P., 1950: Spontaneous activity cycles in polychaete worms. Symp. Soc. Exp. Biol., *4*:127–142.

Wells, G. P., 1959: Worm autobiographies. Sci. Amer., *200*(6):132–141.

Wilson, D. P., 1971: *Sabellaria* colonies at Duckpool, North Cornwall. J. Mar. Biol. Assoc. U. K., *51*:509.

Chapter 11

THE MOLLUSKS

Members of the phylum Mollusca are among the most conspicuous invertebrate animals and include such familiar forms as clams, oysters, squids, octopods, and snails. This phylum is one of the few invertebrate groups that has attained any popularity with laymen and amateur collectors. During the late eighteenth and nineteenth centuries, when natural history occupied the time of many well-to-do gentlemen, shell collections were as popular as they are today. Courses in how to make a shell collection were even included in the curriculum of some finishing schools for young ladies. Such collections, often containing species gathered from various parts of the world, have contributed considerably to our knowledge of the phylum. As a result, mollusks rank at present next to birds and mammals, as the best known taxonomically of all the larger groups of animals.

In abundance of species, mollusks comprise the largest invertebrate phylum aside from the arthropods. Over 80,000 living species have been described. In addition, some 35,000 fossil species are known, for the phylum has had a long geological history; and the possession of a mineral shell by the animal, which increases the chances of preservation, has resulted in a rich fossil record that dates back to the Cambrian.

A superficial survey of the classes of living mollusks seems to indicate a heterogeneous assemblage. Clams, for instance, appear to have little structural similarity to squids, and snails seem quite different from either group. Yet all mollusks are built on the same fundamental plan.

As a means of understanding the basic design of molluscan structure, we will begin by examining a hypothetical ancestral mollusk. Since mollusks are pre-Cambrian in origin, evidence for the ancestral condition can be drawn only from a comparative study of living forms. This description is therefore entirely hypothetical.* However, it does represent the basic plan of the phylum. Many features of this hypothetical ancestor, such as the structure and function of the digestive tract and gills, are encountered among primitive living species. Therefore, the following description can for the most part be considered that of a generalized mollusk.

The ancestral mollusk was an inhabitant of the pre-Cambrian oceans, where it probably crawled about over rocks and other types of hard bottom in shallow water (Fig. 11–1A). It was bilaterally symmetrical, probably not much over 1 cm. in length, and had a somewhat ovoid shape. The ventral surface was flattened and muscular to form a creeping sole, or foot. The dorsal surface was covered by an oval, convex, shieldlike shell which protected the underlying internal organs or visceral mass. At first the shell of this hypothetical ancestor was probably little more than a tough cuticle composed of a horny organic material called conchiolin; but later the shell became reinforced with calcium carbonate. The underlying epidermis, called the mantle (or pallium), secreted the animal's shell. As in living mollusks, secretion would have been most active around the edge of the mantle, although some new material was added to the older portions of the shell. Thus

*The description of the ancestral mollusk in this chapter is based on the work of a number of authors but largely embodies the views of Yonge and Stasek.

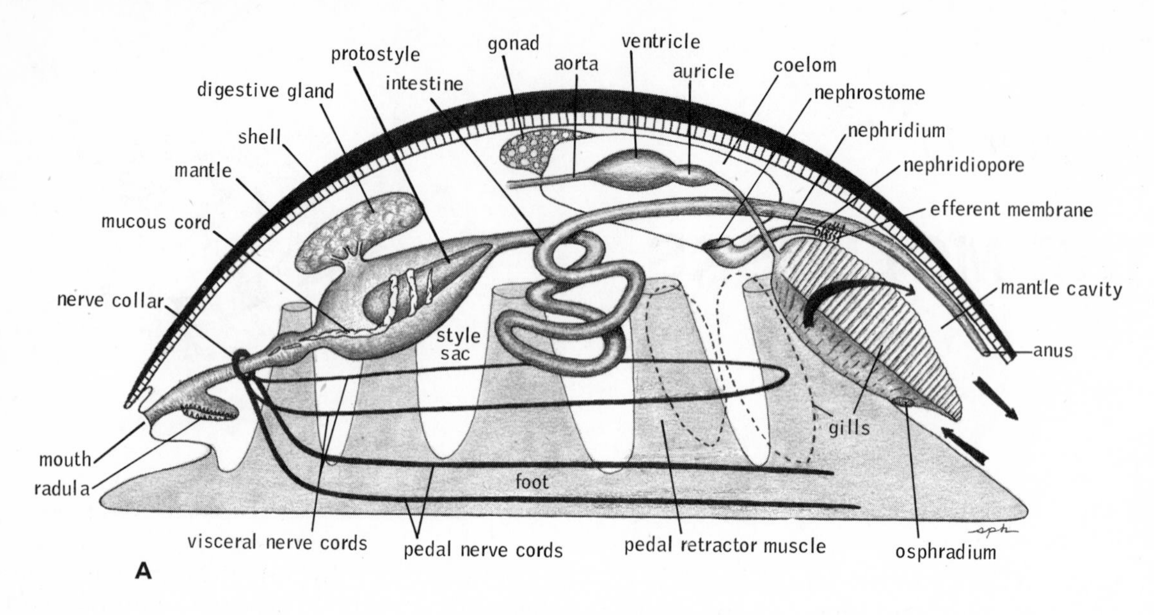

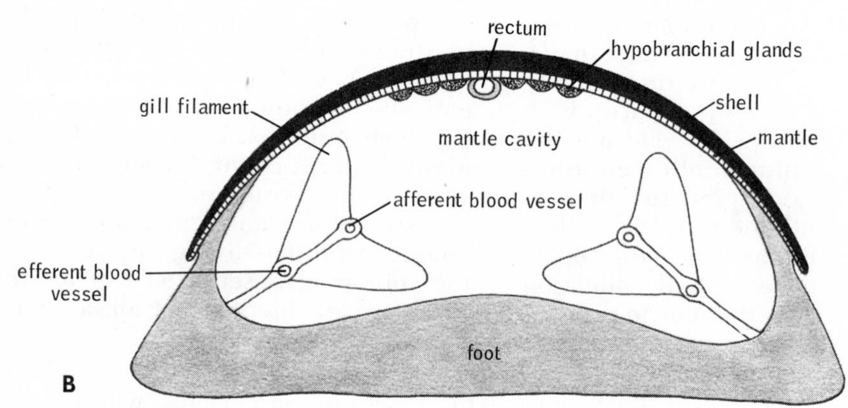

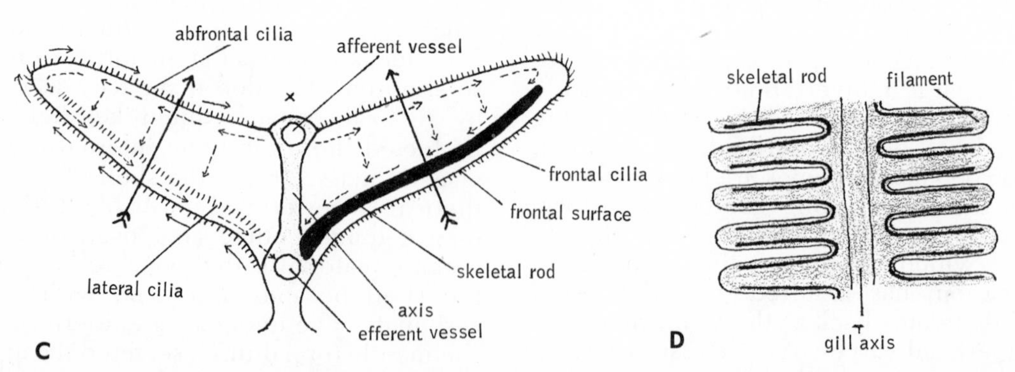

Figure 11–1. *A*. Hypothetical ancestral mollusk (lateral view). Arrows indicate path of water current through mantle cavity. (Adapted from various authors.) *B*. Transverse section through body of ancestral mollusk at level of mantle cavity. *C*. Transverse section through gill of the primitive gastropod, *Haliotis*. Large arrows indicate direction of water current over gill filaments; small, solid arrows indicate direction of cleansing ciliary currents; small, broken arrows indicate direction of blood flow within gill filaments. *D*. Frontal section through primitive gill, showing alternating filaments and supporting chitinous rods. (*B* to *D* after Yonge.)

the shell increased in diameter and thickness at the same time.

A series of pairs of retractor muscles enabled this ancestral form to pull its protective shell down tightly against the hard bottom on which it lived. If dislodged, the animal would have been very vulnerable to predators, for the shield-shaped shell was too shallow to provide a retreat into which the broad, exposed ventral surface could be withdrawn. Each retractor muscle was attached to the inner surface of the shell and was inserted into each side of the foot.

The periphery of the shell, as well as its underlying mantle, overhung the body only slightly, except toward the posterior, where the overhang was so great that it created a chamber called the mantle cavity. Within this protective chamber were located a number of gills, as well as openings from a pair of nephridia.

The gills of the ancestral mollusk were very probably like those of primitive living gastropods and bivalves. Such a gill consists of a long flattened axis projecting from the anterior wall of the mantle cavity and contains blood vessels, muscles, and nerves (Fig. 11–1A). To each side of the broad surface of the axis are attached flattened triangular or wedge-shaped filaments, which alternate in position with those filaments on the opposite side of the axis (Fig. 11–1*C* and *D*). The gills are located on opposite sides of the mantle cavity and are held in position by a ventral membrane and a dorsal membrane. The ventral membrane connects the ventral margin of the gill axis with the lateral floor of the mantle cavity. The dorsal, or efferent, membrane suspends the gills from the midline of the roof but is limited to the more anterior part of the gill axis. Functionally, the position of the gills divides the mantle cavity into an upper and lower chamber (Fig. 11–1A). Water enters the lower, or inhalant, chamber from the posterior, passes upward through the gills into the dorsal, or exhalant, chamber, and then moves posteriorly out of the cavity again.

The margin of the gill filament first encountering the inhalant current is called the frontal surface (roughly equivalent to the ventral margin); the opposite margin is called the abfrontal surface (Fig. 11–1*C*). To prevent the filaments from collapsing under the flow of water, the frontal margin is reinforced by a tiny chitinous skeletal rod. Propulsion of water through the mantle cavity is largely effected by the beating of a powerful band of lateral cilia located on the gills just behind the frontal margin. Sediment brought in by water currents and trapped on the gills is carried upward first by frontal cilia and then by abfrontal cilia toward the axis, where it is swept out by the exhalant current.

Two blood vessels run through the gill axis. The afferent vessel, which carries blood into the gill, runs just within the abfrontal margin. The efferent vessel, which drains the gill, runs along the frontal margin. Blood diffuses through the filaments from the afferent to the efferent vessel (Fig. 11–1*C*), and thus constitutes a countercurrent to the external water stream flowing from the frontal to the abfrontal margin.

In the ancestral mollusk, as in most living mollusks, not only the mantle epidermis, but also the epidermis of the remainder of the exposed body parts, including the foot, were covered by cilia and contained mucous gland cells. Mucous glands were especially prevalent on the foot, from which they lubricated the substratum for locomotion. As in the case of flatworms, the ancestral mollusk undoubtedly moved by means of a gliding motion over the substratum, employing a combination of ciliary action and muscular contractions that will be described in the next section (Gastropoda).

In all probability, the ancestral mollusk was a microphagous herbivore that fed on fine algae growing on rocks in shallow water, and possessed the same unique feeding apparatus that is found in the majority of living mollusks. In many of these forms, the anterior mouth opens into a chitin-lined buccal cavity (Fig. 11–2A). The posterior wall of the buccal cavity is evaginated to form the pocket-like radula sac, which contains on its floor the feeding organ, called the radula. The radula apparatus consists of an elongated cartilaginous base, the odontophore. Over the odontophore and around its anterior is stretched a membranous belt, the radula proper, which bears a number of longitudinal rows of recurved chitinous teeth.

Protractor and retractor muscles are attached to the odontophore, as well as to the radula. Thus, not only can the odontophore be projected out of the mouth, but the radula can move to some extent over the odontophore. Within the sac the lateral margins of the radula tend to roll up; this action protects the walls of the sac from the teeth. As the odontophore is projected out of the mouth, the changing tension causes the radula belt to flatten over the odontophore tip. The

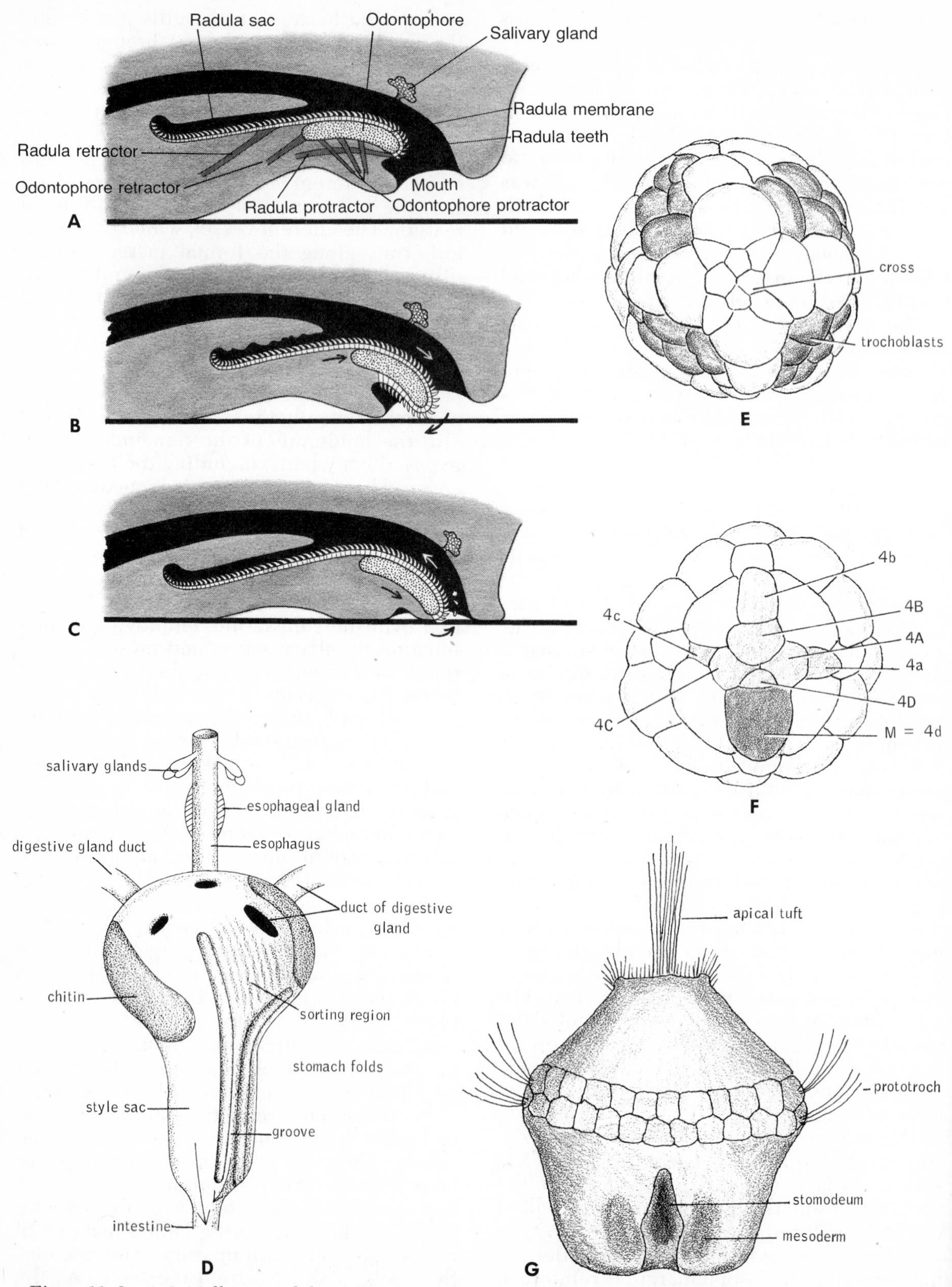

Figure 11–2. *A-C.* Molluscan radula. *A.* Mouth cavity, showing radula apparatus (lateral view). *B.* Protraction of the radula against the substratum. *C.* Forward retracting movement, during which substratum is scraped by radula teeth. *D.* Diagram of a primitive molluscan stomach. (Modified from Owen.) *E-F.* Blastula of the chiton *Ischnochiton*. *E.* Animal hemisphere. *F.* Vegetal hemisphere. *G.* Trochophore larva of the gastropod *Patella*. (After Patten.)

flattening in turn brings about the erection of the teeth. In living forms which are grazers, the radula functions as a scraper (Fig. 11–2*B* and *C*). The odontophore is extended from the mouth against the substratum and then retracted. Since the radula teeth recurve posteriorly, the effective scraping stroke is forward and upward when the odontophore is retracted (like licking). In this way, algae and other particles are scraped away from the surface of the rock and carried up and back into the mouth (Fig. 11–2*C*).

To compensate for the continual wear and loss of radular teeth at the anterior end of the ribbon, new ones are continuously secreted at the posterior end. Over a period of time the radula slowly moves forward over the odontophore.

A pair of salivary glands open onto the anterior dorsal wall of the buccal cavity. These glands secrete mucus, which lubricates the radula and entangles the ingested food particles. Food in mucous strings passes from the buccal cavity into a tubular esophagus, from which it is moved posteriorly toward the stomach (Fig. 11–2*D*). In primitive living mollusks the stomach is shaped like an ice cream cone with a broad hemispherical anterior, into which the esophagus opens, and a tapered posterior, which leads into the intestine. The anterior region of the stomach is lined with chitin except for a ciliated, ridged sorting region and the entrance point for two ducts from a pair of lateral digestive glands (liver), or diverticula. The posterior conical region of the stomach, called the style sac, is lined with cilia but devoid of chitin. Two folds extend the length of the stomach floor, bounding a median ciliated groove.

The contents of the stomach are rotated by the style sac cilia, which beat at right angles to the long axis of the sac. The rotation winds up the mucous food strings, drawing them along the esophagus and into the stomach (Fig. 11–1*A*). The rotating mucus mass is called a protostyle. The size and consistency of particles within the string vary greatly, and the chitinous lining of the anterior part of the stomach protects the wall from damage by sharp surfaces (Fig. 11–2*D*). The acidity of the stomach fluid (pH of 5 to 6 in living mollusks) decreases the viscosity of the mucus and aids in freeing the contained particles. Such particles are eventually swept against the sorting region, in which they are graded by size. Lighter and finer particles are driven by the cilia of the ridges to the duct openings of the digestive diverticula. Heavier and larger particles are carried in the grooves between the ridges to the large groove running along the floor of the stomach to the intestine. Posterior bits of the protostyle are also periodically pinched off and passed into the intestine.

Particles utilized as food pass into the ducts of the digestive glands, and digestion occurs intracellularly within the cells of the distal tubules. Although intracellular digestion appears to be primitive in mollusks, at least some extracellular digestion occurs within the stomach cavity of most living species.

The long coiled intestine functions largely in the formation of fecal pellets. The anus opens middorsally at the posterior margin of the mantle cavity and the wastes are swept away by the exhalant current (Fig. 11–1*A*). The formation of compact fecal pellets decreases the risk of contaminating the mantle cavity and gills with wastes.

In mollusks, the coelomic cavity is located in the middorsal region of the body (Fig. 11–1*A*). The cavity surrounds the heart dorsally and a portion of the intestine ventrally, and thus represents both a pericardial and perivisceral coelom. The heart consists of a pair of posterior auricles and a single anterior ventricle. The auricles drain blood from each gill and then pass it into the muscular ventricle, which pumps it anteriorly through a single aorta. The aorta branches into smaller blood vessels that deliver the blood into sinuses, where the tissues are bathed directly. This, then, is an open circulatory system. Return drainage through the sinuses causes the blood to travel by way of the gills back to the auricles. The blood contains amebocytes as well as the respiratory pigment called hemocyanin (see page 357).

The excretory organs of the ancestral mollusk probably consisted of a pair of tubular metanephridia, commonly called kidneys (Fig. 11–1*A*) in living species. At one end the nephridium connects with the coelom through a nephrostome; and at the other end, the nephridium opens into the posterior of the mantle cavity through a nephridiopore that is located laterally to the intestine. The pericardial coelom receives waste from two sources. The heart wall delivers a filtrate from the blood, and glands in the pericardium secrete waste into the coelom. The pericardial fluid then passes through the nephrostome into the nephridium. Here further secretion, as well as some selective reabsorption, occurs from the tubule wall, and the final urine is emptied into the mantle cavity.

The ground plan of the molluscan nervous system is composed of a nerve ring around the esophagus, from the underside of which two pairs of nerve cords extend posteriorly. The ventral pair, called the pedal cords, innervates the muscles of the foot; the dorsal pair, called the visceral cords, innervates the mantle and visceral organs.

The sense organs of the ancestral mollusk are highly conjectural, but many living species possess eyes, a pair of statocysts in the foot, and osphradia. The osphradia are patches of sensory epithelium located on the posterior margin of each of the afferent gill membranes; they function as chemoreceptors and also determine the amount of sediment in the inhalant current (Fig. 11–1*A*).

Primitively in mollusks, a pair of anterior dorsolateral gonads are present in each side of the coelom. When ripe, the eggs or sperm break into the coelomic cavity and are transported to the outside through the nephridia. Fertilization takes place externally in the surrounding sea water.

The embryogeny of the ancestral mollusk must certainly have been very similar to the early embryogeny of most presently existing species. In existing species, the zygote undergoes typical spiral cleavage (Fig. 11–2*E* and *F*), and gastrulation is by epiboly or invagination, or both. The resulting gastrula quickly develops into a free-swimming trochophore larva (Fig. 11–2*G*). The structure of the trochophore and the fate of the cleavage quartets is virtually identical to that of the annelids. In most of the molluscan classes the trochophore passes into a more highly developed veliger larva, in which the foot, shell, and other structures make their appearance (Fig. 11–36). But in the ancestral mollusks, a veliger stage may have been absent. In any event, metamorphosis occurs at the end of larval life, the prototroch degenerates, and the larva sinks to the bottom to assume the benthic habit of the adult.

Of the seven classes of mollusks, the monoplacophorans, the chitons, and the gastropods each display features that might most closely ally them with the ancestral mollusks. The classic presentation typically begins with the monoplacophorans or chitons. We will break with tradition and begin with the gastropods. The biology of this large class is better known and can serve to some extent to introduce the other classes. Moreover, the gastropods will provide a basis for comparison in a later discussion of the interesting and controversial monoplacophorans.

CLASS GASTROPODA

The class Gastropoda is the largest class of mollusks. Over 35,000 existing species have been described, and to this total should be added some 15,000 fossil forms. The class has had an unbroken fossil record beginning with the early Cambrian period, and has undergone the most extensive adaptive radiation of all the major molluscan groups. Considering the wide variety of habitats the gastropods have invaded, they are certainly the most successful of the molluscan classes. Marine species have become adapted to life on all types of bottoms, as well as to a pelagic existence. They have invaded fresh water, and the pulmonate snails and several other groups have conquered land with the elimination of the gills and conversion of the mantle cavity into a lung.

Origin and Evolution. In many respects, gastropods exhibit the least change from the ancestral molluscan plan, for they possess a well-developed head, a shell of one piece, and a broad, flat creeping foot. The most significant modification, and the one to which most of the other changes are related, is the twisting, or torsion, that the body has undergone. Torsion is not the coiling of the shell; all evidence indicates that coiling of the shell evolved *before* torsion. The pre-torsion shell was planospiral; that is, the shell was not a conical spiral, but rather each spiral was located completely outside of the one preceding it and in the same plane, like a hose that is coiled flat on the ground (Fig. 11–3*E*). The shell, though spiral, was bilaterally symmetrical. Torsion and the spiraling of the shell were therefore separate evolutionary events.

Torsion was a much more drastic change than the spiraling of the shell. Viewing the animal dorsally, most of the body behind the head, including the visceral mass, mantle, and mantle cavity, was twisted 180 degrees counterclockwise (Fig. 11–3*A* and *B*). The mantle cavity, gills, anus, and the two nephridiopores were now located in the anterior part of the body behind the head. Internally the digestive tract and nervous system were twisted into a **U**-shape. The shell remained a symmetrical spiral.

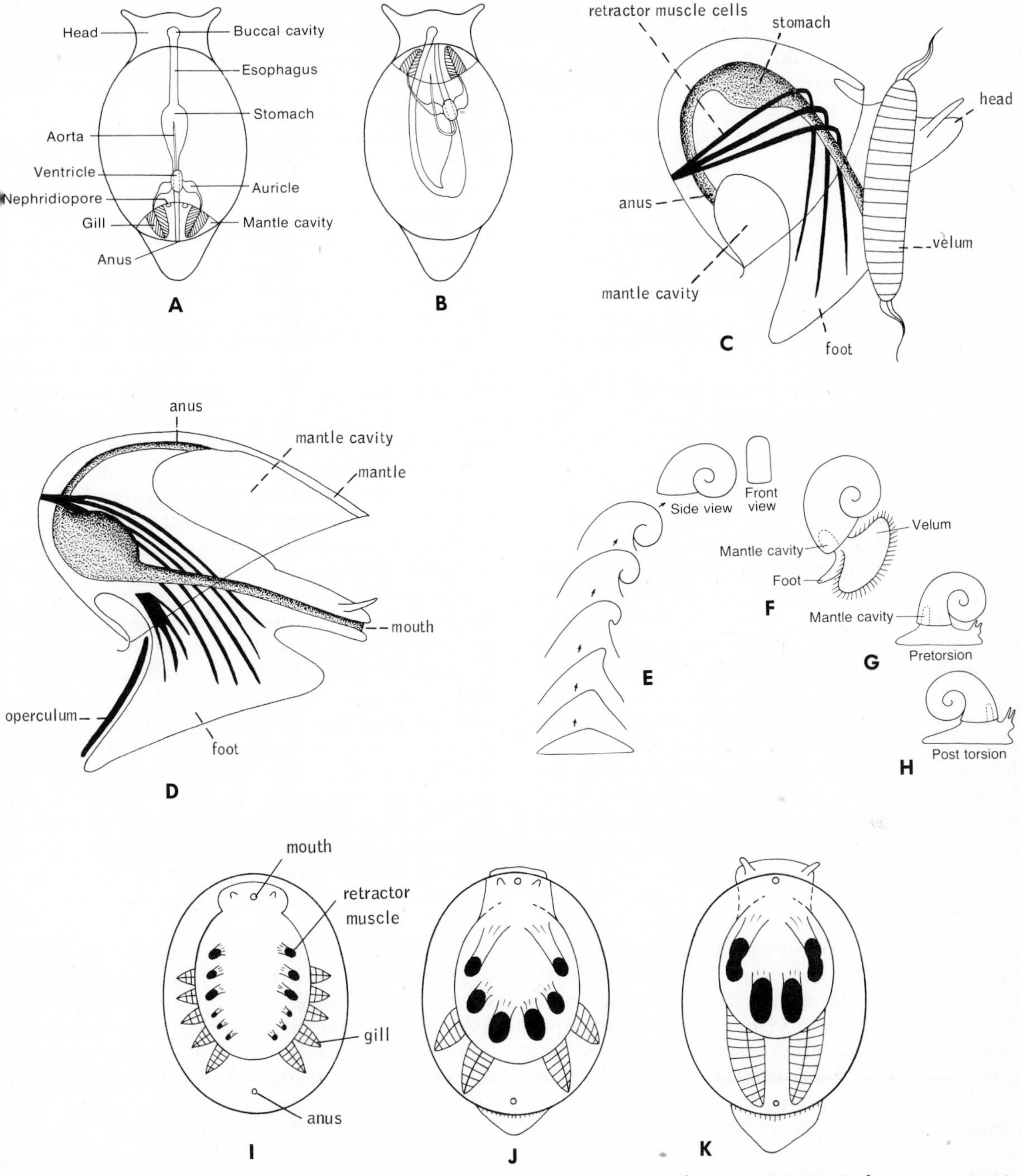

Figure 11–3. Dorsal view of ancestral gastropod. *A*. Prior to torsion. *B*. After torsion. *C-D*. Embryonic torsion in the gastropod *Patella* (a limpet). *E*. Evolution of a planospiral shell. Height of the shield-like shell of hypothetical ancestral mollusk increases, and peak forms. Peak is pulled forward and coiled under. Aperture is reduced and animal can withdraw into spiral shell, which is more compact and less awkward to carry than would be a straight conical shell. Note that shell is bilaterally symmetrical. *F*. Diagram of a gastropod larva before torsion. Velum is the ciliated larval locomotor organ. *G*. Hypothetical pretorsion "gastropod" with a planospiral shell. *H*. Post-torsion gastropod. Torsion does not affect planospiral shell except to place coils of shell posteriorly. Mantle cavity is now anterior. *I-K*. Reduction in number of gill and retractor muscles, increase in lateral extent of mantle cavity, and postulated increase in size of head of three monoplacophorans. Similar forms may have been ancestral to the gastropods. (*A* and *B* modified from Graham; *C* and *D* after Boutan; *F* after Yonge; *I-K* after Stasek.)

Torsion is not merely an evolutionary hypothesis, for it appears in the embryogeny of living gastropods. The larva is at first bilaterally symmetrical and then quite suddenly undergoes twisting.

No really convincing explanation of the evolutionary significance of torsion has yet been advanced. It has been suggested that torsion represents an adaptation to facilitate flushing the mantle cavity with clean water. Since animals frequently face into water currents, an anterior mantle cavity would thus be advantageous. However, this idea would seem to be contradicted by the general tendency in many aquatic gastropods to restrict the opening into the mantle cavity.

A number of authors have postulated that torsion represents a larval adaptation. Reexamining the ideas of Garstang (1928) and others, Ghiselin (1966) has suggested the following sequence of evolutionary changes to account for torsion. The shield-like shell of the ancestral mollusk was of little protective value to the planktonic larva. There was thus selection for a conical shell with a more restricted aperture. Such a larval shell would expose less of the body and could serve as a retreat into which the larva could further withdraw. A straight conical shell being cumbersome to carry in swimming, the shell became spiraled in the course of transformation. Spiraling was forward over the head because of developmental complications; i.e., for some unknown reason more shell could be deposited along the posterior than the anterior margin (Fig. 11–3*F*). This compact planospiral shell was satisfactory for larval life, but on settling its forward projection would have been disadvantageous for a crawling adult existence. Torsion is postulated as an adaptation to correct this difficulty. Twisting the visceral mass 180° displaced the coils of the shell to a trailing position behind the animal (Fig. 11–3*H*). The occurrence of torsion in later larval life would thus have prepared the animal for settling.

Stasek (1972) postulates that torsion was an adult adaptation to permit withdrawal of the larger and more developed head that probably evolved in the line of mollusks (monoplacophorans) leading to the gastropods. In these forms the planospiral shell was higher and possessed a more reduced aperture than the shell described for the hypothetical ancestral mollusk. The number of retractor muscles had become reduced to a single pair, and they inserted on the shell well back from the edge (Fig. 11–3*I* to *K*). Thus, the body region ("neck") between the head-foot and the visceral mass was more constricted than in other mollusks. Twisting of the body to bring the mantle cavity to a lateral position, and later to an anterior position, provided additional room for the withdrawal of the head when the animal retracted.

Regardless of the evolutionary significance of torsion, it seems to have posed a number of functional problems, and many of the modifications exhibited by gastropods represent attempts to adjust to these problems. The chief problem was fouling. After the development of torsion, if the water circuit through the mantle cavity had remained as it had been in the ancestral mollusk, waste from the anus and nephridia would have been dumped on top of the head. This problem seems to have been solved in the primitive gastropods by a cleft or split in the shell and mantle over the mantle cavity (Fig. 11–4*C*). At the same time, the anus was withdrawn from the edge of the mantle cavity to a position beneath the inner margin of the cleft. The inhalant current continued to come in over the head and pass over the gills, but now instead of making a **U**-turn and passing out in the same direction, the water current flowed up and out through the cleft in the shell, taking waste from the anus and nephridia with it. Such a slit first appeared in the fossil coiled monoplacophorans and may also have been present in the gastropod ancestors as a preadaptation to torsion (Rollins and Batten, 1968).

Up to this point we have been considering a gastropod with a symmetrical planospiral shell. Such a form is not entirely hypothetical, for there are early fossil species (the Bellerophontacea, Fig. 11–4*A* and *B*), with symmetrical planospiral shells bearing a cleft along the anterior middorsal edge. The cleft indicates the midline of the mantle cavity.

Eventually, a final change took place that resulted in the typical gastropod structure. This change involved the shell. Although there are fossil species with planospiral shells, all existing gastropods possess asymmetrical shells; or if the shells are symmetrical, this symmetry has been secondarily derived. The planospiral shell had the disadvantage of not being very compact; since each coil lay completely outside of the

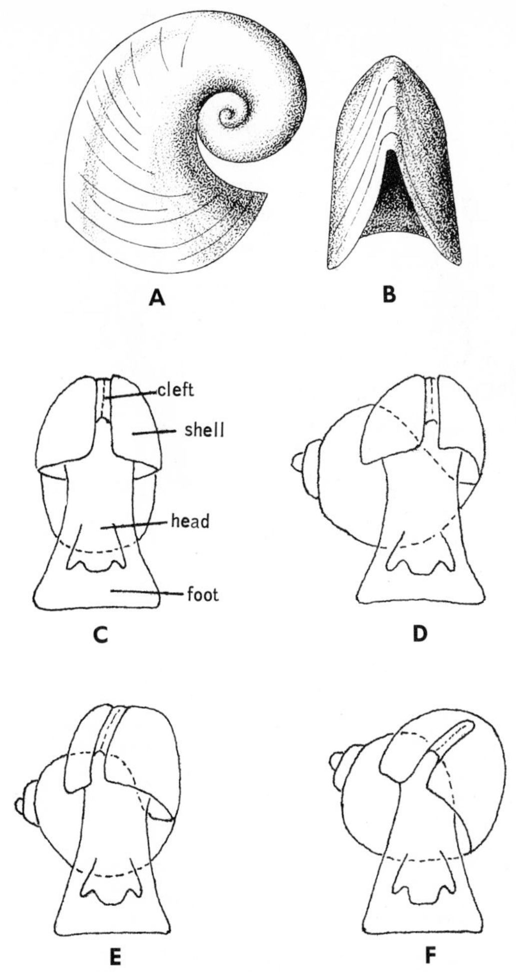

Figure 11–4. Side view (*A*), and front view (*B*), of the planospiral shell of *Strepsodiscus*, a genus of the fossil Bellerophontacea, the earliest known Archeogastropoda. *C* to *F*. Evolution of the asymmetrical gastropod shell. Slot in shell for exhalant water current marks location of mantle cavity. *C*. Ancestral post-torsion gastropod with planospiral shell. *D*. Apex of spiral is drawn out, producing a more compact shell. *E*. Position of shell over body shifted, providing more equal distribution of weight. *F*. Final position of shell over body, typical of most living gastropods. Axis of shell is oblique to long axis of body, and mantle cavity is located on left side of body. Right side of body is compressed by shell. (*C* to *F* after Yonge.)

preceding one, the diameter of the shell could become relatively great (Fig. 11–4*C*). The problem was solved with the evolution of asymmetrical coiling, in which the coils were laid down around a central axis called the columella and each coil lay beneath the preceding coil (Fig. 11–4*D*). Such a shell is much more compact and does not require the rapid increase in diameter characteristic of the ancestral planospiral shell.

The new spiral cone shell obviously could not be carried as was the old planospiral shell, because all the weight would hang on one side of the body (Fig. 11–4*D*). In order to obtain the proper distribution of weight, the shell position became shifted, so that the axis of the spiral slanted upward and somewhat posteriorly. The net result was that the shell was eventually carried obliquely to the long axis of the body, as in living gastropods.

The new position of the shell restricted the mantle cavity to the left side of the body, for the mantle cavity on the right side was occluded by the bulging whorl of the visceral mass (Fig. 11–4*F*). This occlusion has had profound effects; it has resulted in the decrease in size, or the complete loss, of the gill, auricle, and nephridium on the right side of the body.

With this background of possible gastropod evolutionary origins, we must now pause to consider the manner in which existing gastropods are classified. Gastropods are divided into three subclasses. The first, known as the Prosobranchia, includes all gastropods that respire by gills and in which the mantle cavity, gills, and anus are located at the anterior of the body—in other words, those gastropods in which torsion is evident. The majority of gastropods are prosobranchs.

From the Prosobranchia evolved the two other subclasses, the Pulmonata and Opisthobranchia. In the subclass Pulmonata, which includes the land snails, the gills have disappeared, and the mantle cavity has been modified into a lung. The Opisthobranchia display detorsion. The shell and mantle cavity are usually either reduced in size or absent, and many species have become secondarily bilaterally symmetrical. The sea hares and the sea slugs (nudibranchs) are perhaps the most familiar members of this subclass.

Shell and Mantle. The typical gastropod shell is a conical spire composed of tubular whorls and containing the visceral mass of the animal (Fig. 11–5*A*). Starting at the apex, which contains the smallest and oldest whorls, successively larger whorls are coiled about a central axis, called the columella; the largest whorl eventually terminates at the opening, or aperture, from which the head and foot of the living animal protrude. A shell may be spiraled clockwise or counterclockwise, or, as it is more frequently stated, displays a right-handed (dextral) or left-handed (sinistral) spiral. A shell possesses a right-handed spiral when the aperture opens

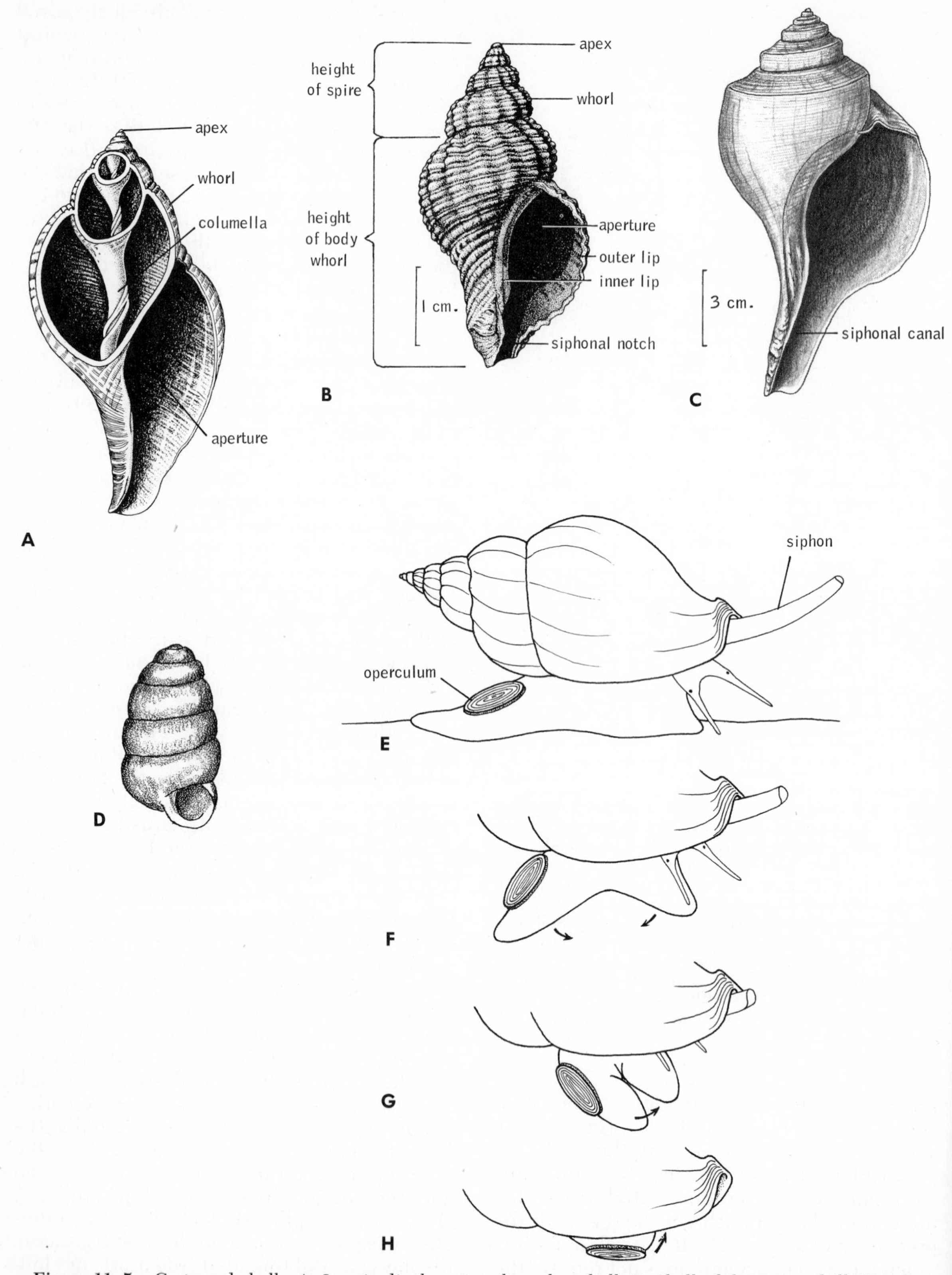

Figure 11–5. Gastropod shells. *A.* Longitudinal section through a shell. *B.* Shell of the oyster drill *Urosalpinx cinerea,* showing commonly designated features. (After Turner.) *C. Busycon canaliculatum,* the common Atlantic Coast whelk. *D. Pupilla,* a land snail only slightly more than 1 mm. in height. (After Baker.) *E.* Gastropod with an operculum. *F–H.* Withdrawal into shell and closure by operculum.

to the right of the columella (if the shell is held with spire up and aperture facing observer) and left-handed when it opens to the left. Most gastropods are right-handed, a few are left-handed, and some species have both right-handed and left-handed individuals.

The first shell is laid down by the larva and is called the protoconch; it is represented by the smallest whorl at the apex of the shell. The protoconch can usually be differentiated from the newer parts of the shell, because in most gastropods it is smooth and lacks the sculpturing characteristic of the later parts of the shell. Except for the protoconch, the rest of the shell has uniform structure and is laid down by the edges of the mantle.

A gastropod shell typically consists of four layers. The outer periostracum is thin and composed of a horny protein material called conchiolin or conchin. The inner layers consist of calcium carbonate. The outermost calcareous layer is generally prismatic; i.e., the mineral is deposited as vertical crystals of calcite, each surrounded by a thin protein matrix. The inner calcareous layers, usually two but sometimes more, are laid down as aragonite sheets, or lamellae, over a thin organic matrix.

The gastropod head and foot are withdrawn into the shell by a retractor muscle. This muscle, called the columella muscle, actually arises in the foot, but appears to emerge from the mantle on the right side. The muscle is inserted onto the columella of the shell. Primitively, two retractor muscles are present (Fig. 11–6*B*); this paired arrangement is found in a few living species, although the left muscle is usually very small. In most gastropods the left muscle has disappeared, and only the right one remains.

The foot of a great many shelled gastropods bears a horny disc called the operculum on its posterior dorsal surface (Fig. 11–5*E*). The operculum neatly fills the shell aperture and thus acts as a protective door or lid (Fig. 11–5*F* and *G*). Although the operculum usually is composed of the same material as the periostracum, in some species it may have calcareous deposits on a periostracal base.

Gastropod shells display an infinite variety of colors, patterns, shapes, and sculpturing, but at this point only two of the more radical modifications in shell form will be mentioned. In a considerable number of gastropods, the shell is conspicuously spiraled only in the juvenile stages. The coiled nature disappears with growth, and the adult shell represents a single large expanded body whorl. In the abalone, *Haliotis* (Fig. 11–6*A*), and in the slipper shells, *Crepidula*, the shell remains asymmetrical; but in the limpets, of which there are a number of unrelated groups, the shell has become secondarily symmetrical and looks like a Chinese hat (Fig. 11–7*A*).

The second modification that must be mentioned here is shell reduction and shell loss, a condition that has occurred many times in the history of gastropods. When the shell is greatly reduced, it often becomes buried within the mantle tissues.

Other shell modifications will be described later in connection with ventilation, movement, and habitation.

Water Circulation and Gas Exchange. The great diversity of gastropods reflects adaptive radiation at various points in the evolutionary history of the group. The main lines of this evolution are perhaps best reflected in the modifications for water circulation and gas exchange. We will therefore begin our examination of the biology of gastropods with these two problems, using them as a way of gaining an initial overview of the class. Such an overview has been diagrammatically depicted in Figure 11–8, which can be used in following the discussion below.

Prosobranchs. The most primitive type of gill structure and water circulation occurs in those prosobranchs with cleft or perforated shells, which, as stated earlier, may have been the primitive gastropod solution to the sanitation problems caused by torsion. These prosobranchs belong to the primitive suborder Archaeogastropoda and include *Scissurella, Haliotis,* and the keyhole limpets. In all three groups, two gills are present and display the primitive bipectinate condition of the ancestral mollusks. The rectum and anus are removed from the edge of the mantle cavity and open beneath the shell perforation or cleft.

Scissurella and *Pleurotomaria* have typical spiral shells, but the anterior margin of the shell and the underlying mantle are deeply cleft (Fig. 11–6*D* and *E*). The ventilating current produced by the action of the lateral cilia of the gills enters the mantle cavity at the anterior of the body. It passes between the gill filaments and then continues upward and out through the shell cleft. Although only the notched planospiral shell remains of the extinct bellero-

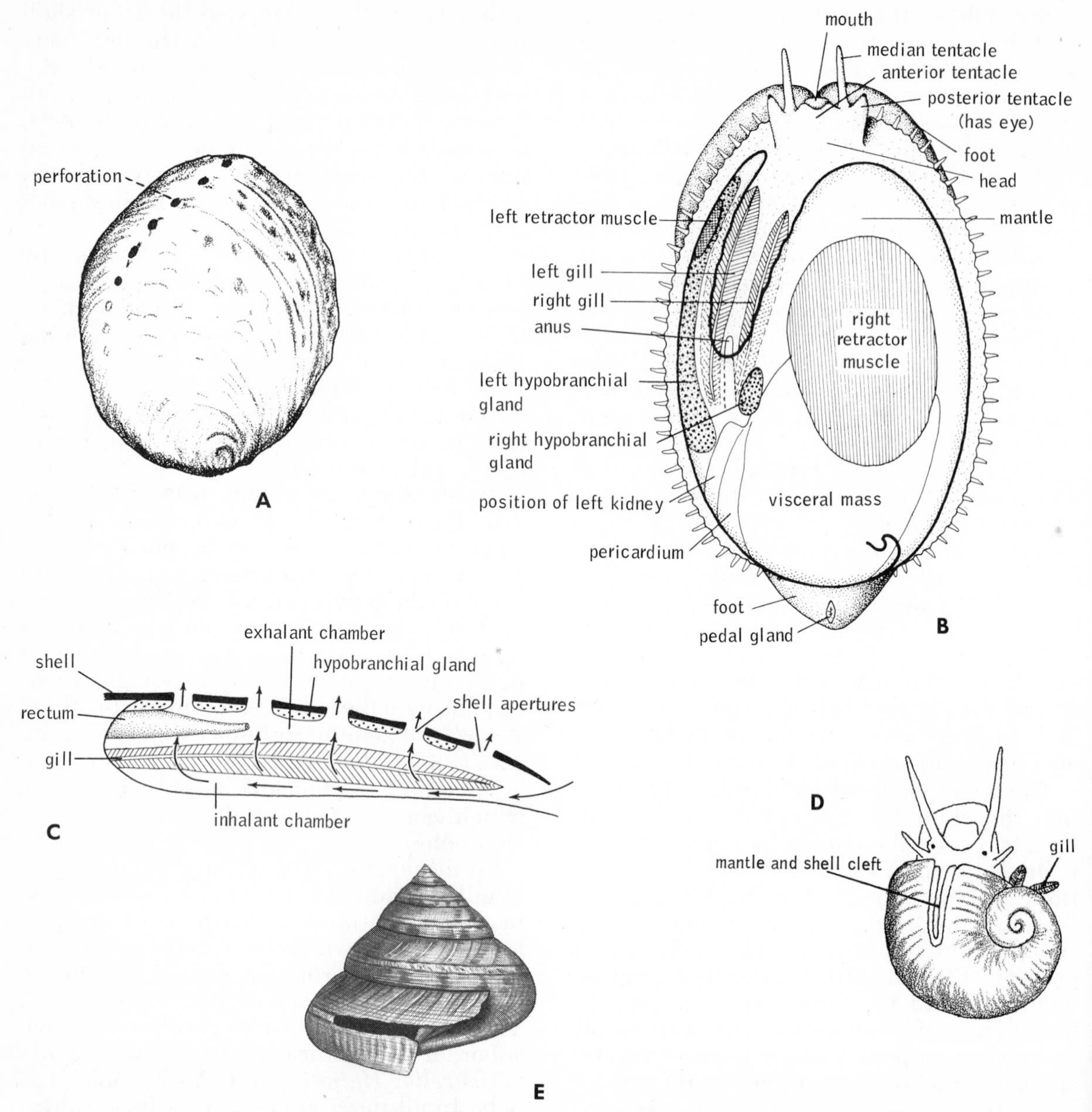

Figure 11–6. *A-C.* The abalone *Haliotis. A.* Dorsal view of shell. *B.* Dorsal view of animal with shell removed, showing two bipectinate gills. *C.* Diagrammatic lateral view of mantle cavity, showing path of ventilating current. *D-E. Scissurella* (*D*) and *Pleurotomaria*(*E*), gastropods with slotted shells for exhalant current. (*B* after Bullough; *C* after Yonge; *E* drawn from a photograph by Abbott.)

phontacean (Fig. 11–4*B* and 11–8), these earliest known archaeogastropods most certainly possessed two bipectinate gills and a ventilating current similar to those of *Scissurella.*

The abalones (*Haliotis*) and the keyhole limpets are intertidal and shallow-water inhabitants of wave-swept rocks. The broad shells of both groups are designed for minimum water resistance and as protective shields. They are perforated instead of slotted, making possible a ventilating current like that of *Scissurella* but avoiding the structural weakness of a deep notch.

The low, shield-like shell of the abalone is asymmetrical and constitutes in large part a single expanded whorl (Fig. 11–6*A*). As in most other prosobranch gastropods, the mantle cavity is displaced to the left side of the body. In *Haliotis tuberculata* the shell above the cavity contains a line of five holes (Fig. 11–6*A*). The mantle is cleft along the line of shell perforations, and the edges of the mantle fit together and project into the shell openings to form a lining for each hole. The anus and nephridial openings lie beneath one of the perforations. The ventilating current is similar to that of *Scissurella,* and the exhalant current flows dorsally out through all of the five shell perforations. The current flowing through the fourth opening also carries out waste from the anus and nephridia as well as the genital products (Fig. 11–6*C*).

The keyhole limpets have a conical, secondarily symmetrical shell, which has either a cleft at the anterior margin or a hole at the apex. The latter arrangement appears in *Fissurella* and *Diodora* (Fig. 11–7*A* to *C*). The opening arises as a notch along the shell margin during early stages of development. The notch then becomes enclosed, and through differential growth it gradually assumes a position at the apex of the shell. The mantle projects through the opening to form a siphon. Water enters the mantle cavity anteriorly, flows over the gills, and issues as a powerful stream from the opening at the shell apex. The anus and urogenital openings are located just beneath the posterior margin of the shell opening.

The Patellacea is another group of archaeogastropod limpets that are very common intertidal and shallow-water rock inhabitants in many parts of the world (the west coast of the United States, Britain, and northern Europe). The limpet shell probably evolved independently from that of the keyhole limpets, and shell openings or clefts are lacking. This group displays reduction and loss of the original gills. *Acmaea* possesses only the left gill, which projects to the right side of the body. As in all limpets, the mantle and shell overhang produces a distinct groove on each side between the foot and the mantle edge (Fig. 11–7*D*). The inhalant ventilating

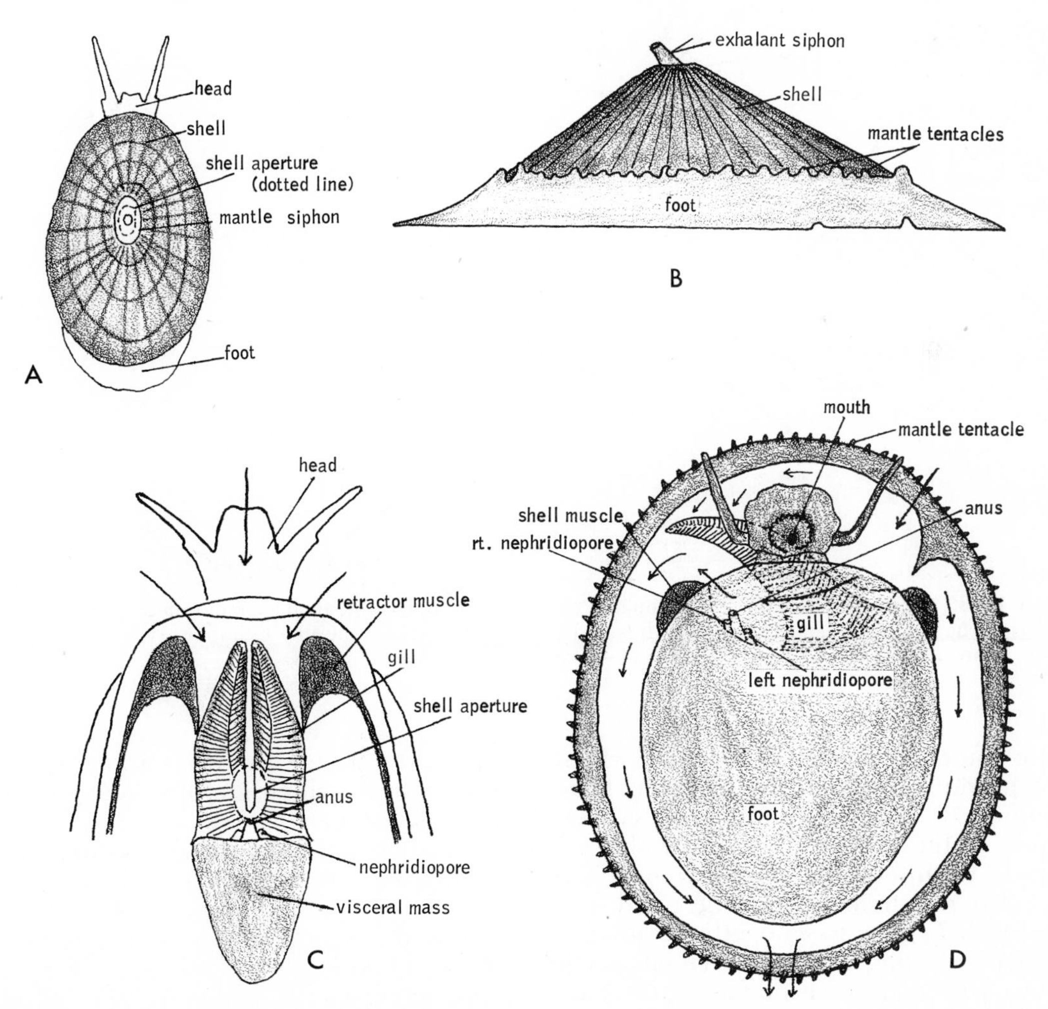

Figure 11–7. *A-C.* The keyhole limpet, *Diodora. A.* Dorsal view. *B.* Lateral view. *C.* Exposed mantle cavity (dorsal view). *D.* The marine limpet, *Acmaea* (ventral view). Arrows indicate path of water circulation. (All after Yonge.)

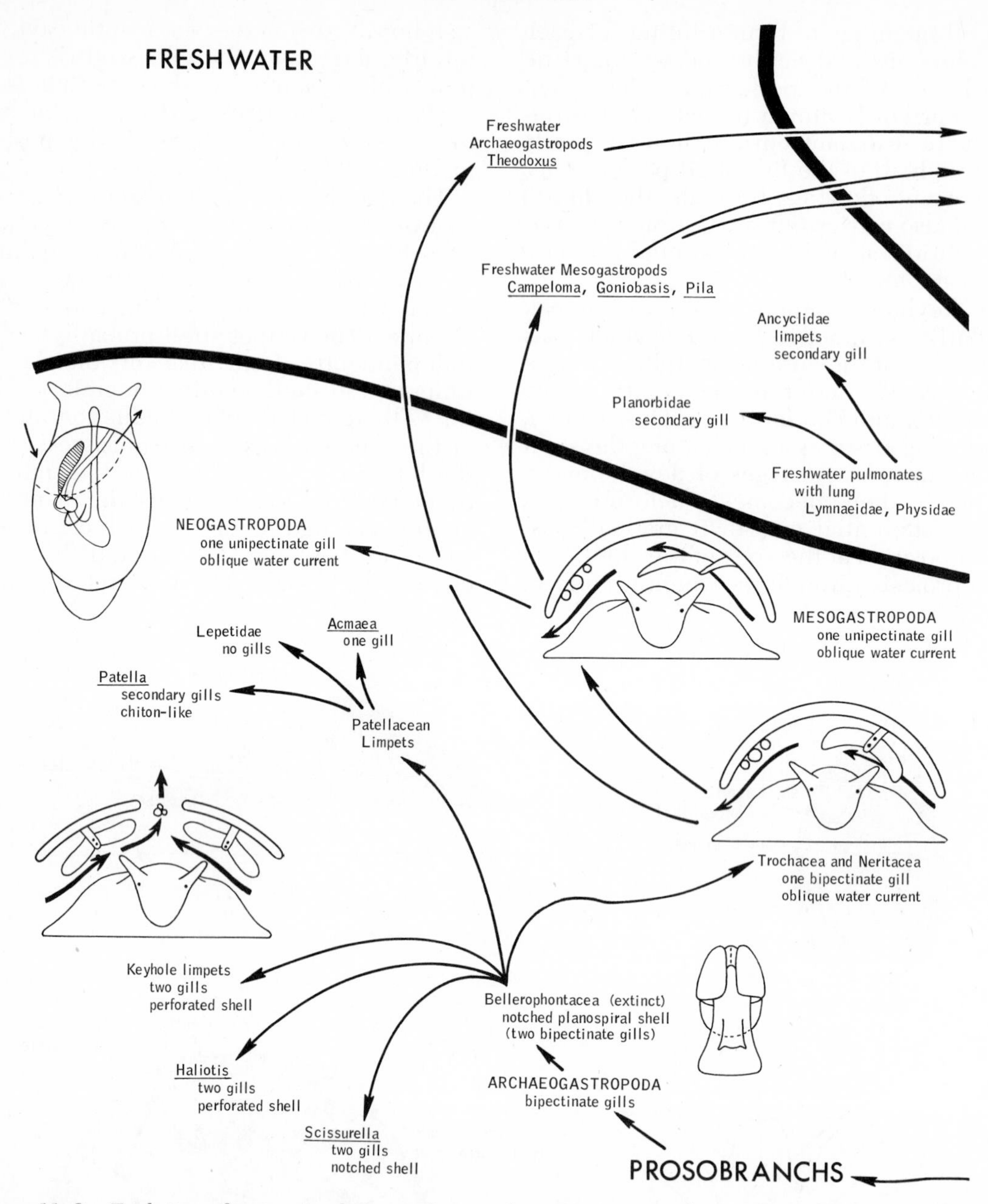

Figure 11–8. Evolution of water circulation and gas exchange in gastropods. Diagram reflects phylogenetic relationships of the subclasses and orders. Families and genera listed represent only examples. (Figures adapted from Graham, Hyman, Morton, and Yonge.) *Figure 11–8 continued on opposite page.*

current enters the mantle cavity anteriorly on the left side. Part of the current flows posteriorly in the left lateral mantle groove; the rest of the current flows over the gill and then down the right mantle groove. The two exhalant streams converge and exit posteriorly. *Lepeta* has no gills at all, gas exchange occurring across the general mantle surface. In *Patella,* secondary gills have formed as folds of the mantle, and they project into the pallial groove along the side of the body. These limpets strikingly parallel the chitons, which are also adapted for life on rocky shores.

The remaining archaeogastropods—the Neritacea and Trochacea—possess only the left gill, and their ventilating current enters the mantle cavity on the left side of the head and exits on the right (Fig. 11–8). The anus opens at the right edge of the mantle cavity, and wastes are carried away in the exhalant water stream. Such an

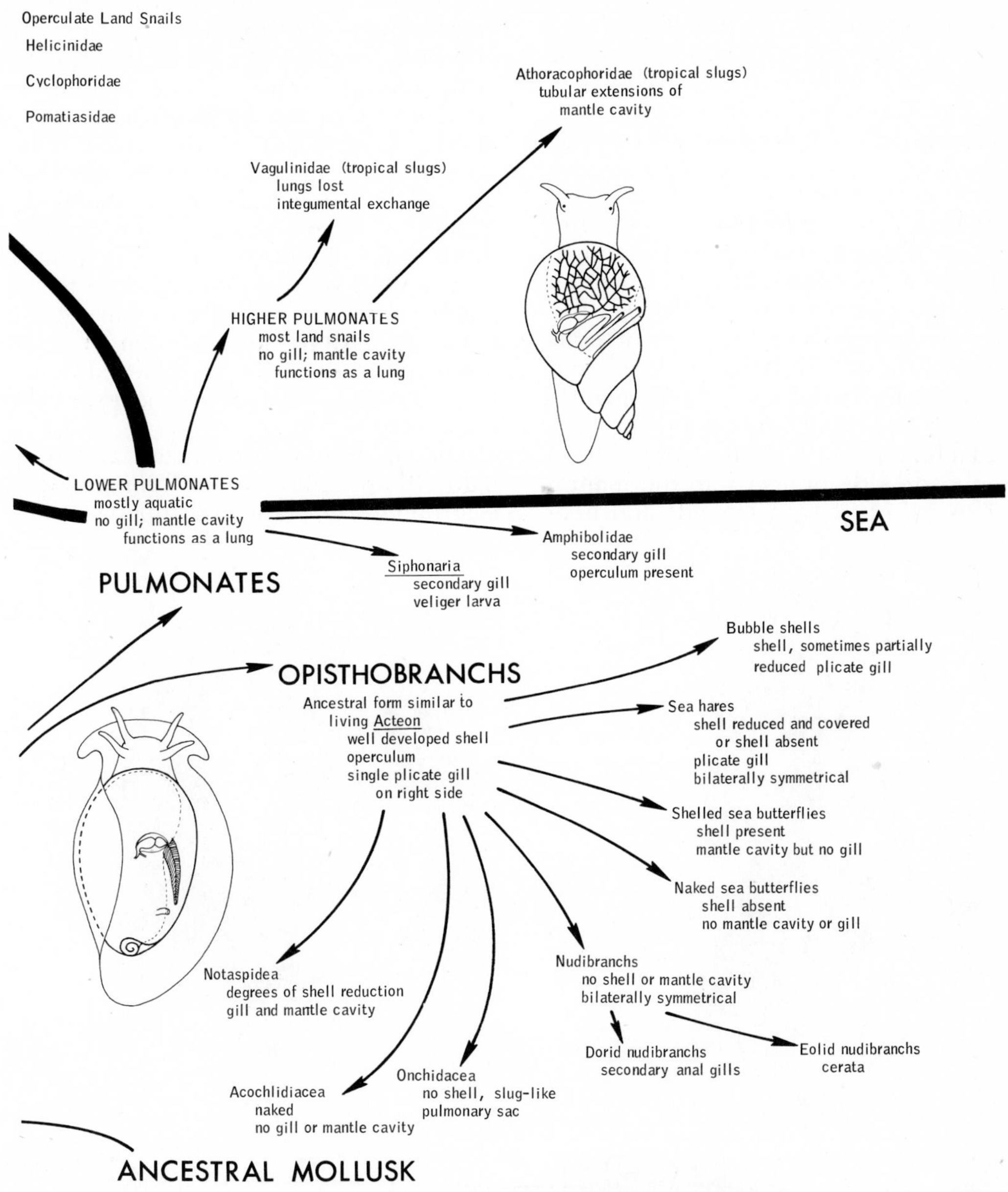

Figure 11–8. *Continued.*

oblique water current is apparently an efficient solution to the fouling problems imposed by torsion, for it is found in most prosobranchs.

The neritaceans include many common rocky intertidal species, such as the semitropical and tropical species of *Nerita* (Fig. 11–16*D*). Members of this group have also invaded fresh water (*Theodoxus*), and perhaps from some freshwater stock evolved a family of tropical land snails, the Helicinidae (Fig. 11–8). These land snails possess an operculum and are called operculate land snails, as opposed to the more familiar pulmonate land snails, which lack an operculum. The gill of the Helicinidae has been lost, and gas exchange occurs across the roof of the mantle cavity.

Of the some 20,000 species of prosobranchs, the majority are not archaeogastropods but mesogastropods and neogastropods. The archaeogastropods are largely restricted to rocky surfaces. The great adaptive diversity of the mesogastropods and neogastropods is in part correlated with their ability to exploit other types of habitats, and especially soft bottoms. This ability may be related to a major change in the structure of the gills. The dorsal and ventral membranes which suspend the bipectinate gills of the archaeogastropods present considerable surface areas that could be fouled with sediment carried within the ventilating current, and this perhaps accounts for the restriction of these primitive prosobranchs to the cleaner water over rocky bottoms. In the mesogastropods and neogastropods, the membranous suspension of the gills has disappeared, and the gill axis is attached directly to the mantle wall (Fig. 11–8). The filaments on the side of the attachment have disappeared; those on the opposite side project into the mantle cavity. The gill of mesogastropods and neogastropods is thus unipectinate, or monopectinate.

A further modification associated with the ventilating current of many species of these two higher orders of prosobranchs, especially the mesogastropods, is the development of an inhalant siphon by the extension and inward rolling of the mantle margin (Fig. 11–9*B* and *C*). In many species the anterior margin of the shell aperture is notched (Fig. 11–5*B*) or drawn out as a siphonal canal to house the siphon (Figs. 11–5*C* and 11–25*B*). The siphon may provide access to surface water in some species that burrow; or, by making possible the selection of restricted areas of water, the mobile siphon may function as a sense organ, especially in carnivores.

The oblique water current and unipectinate gill are relatively uniform among mesogastropods and neogastropods, and the

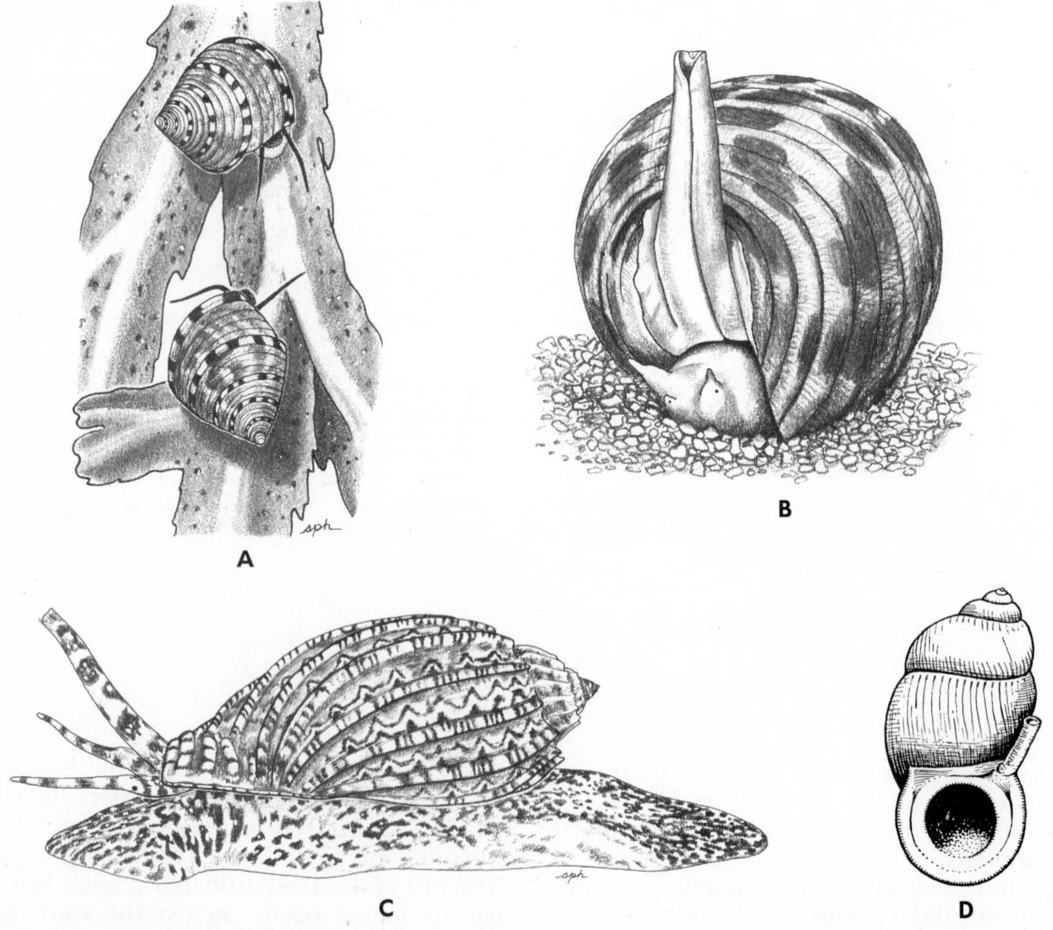

Figure 11–9. Prosobranchs. *A. Calliostoma,* an Australian trochid (an archaeogastropod). *B.* Anterior end of the neogastropod *Mitra,* showing the folded origin of the siphon. *C.* Lateral view of a harp shell, showing siphon, tentacles, and eyes. *D.* Shell of a cyclophorid land snail, showing the breathing tube, which permits gas exchange when the animal is withdrawn and the aperture is sealed by the operculum. (*B* based on a photograph by Paul Zahl; *C* drawn from a photograph by Abbott; *D* after Rees from Purchon, R. D., 1968: Biology of the Mollusca, Pergamon Press, N.Y.)

diversity of these higher prosobranchs results from variations in adaptations for locomotion, habitation, feeding, and other functions. Only their immigrations from the sea need be mentioned here. Mesogastropods are well represented in fresh water, as a result of a number of independent invasions (Fig. 11–8). The majority are tropical, but there are many genera, such as *Goniobasis*, *Pleurocera*, *Viviparus*, *Campeloma*, and *Valvata*, that contain temperate species. *Viviparus* and *Campeloma* have lost the gill, and *Valvata* is peculiar among mesogastropods in possessing a bipectinate gill. The tropical Ampullariidae (*Ampullarius* and *Pila*) are amphibious and have a divided mantle cavity, with a lung on one side and a gill on the other.

There are two large families of mesogastropod land snails, the Cyclophoridae and the Pomatiasidae (Fig. 11–8). Like the archaeogastropod Helicinidae, they are largely tropical and operculate, they have no gill, and gas exchange occurs across a vascularized mantle wall within the mantle cavity. A notch or a breathing tube in the shell aperture permits the entrance of air when the operculum is closed (Fig. 11–9*D*).

Opisthobranchs. The remaining two subclasses of gastropods, the Opisthobranchia and the Pulmonata, are probably both derived from prosobranchs that possessed only the left gill. The some 1100 highly diverse species of opisthobranchs are largely marine and are characterized by detorsion, in which the mantle cavity and the structures it contains have shifted around to the right side (Fig. 11–8). The reasons for detorsion are unknown. Complicating the study of the origin of opisthobranchs is the fact that the gill is plicate, or folded, rather than filamentous, and does not appear to be homologous with the prosobranch gill.

Primitive opisthobranchs are asymmetrical and possess the more or less typical coiled gastropod shell; a few species even have an operculum. But throughout the subclass, apparently correlated with detorsion, there has been a tendency toward shell reduction and loss, reduction of the mantle cavity and associated loss of the original gill, and attainment of a secondary bilateral symmetry. The some 12 orders and 70 families of opisthobranchs appear to have evolved along a number of radiating lines that probably stemmed from a primitive shelled ancestral group (Fig. 11–8). To varying degrees within each of these lines, shell and mantle cavity reduction has occurred; and some lines culminate in groups which are secondarily bilaterally symmetrical.

The Cephalaspidea, containing the bubble shells (*Acteon*, *Scaphander*, *Hydatina*, *Bulla*), is the largest order of opisthobranchs, and the one to which belong the more primitive members of the subclass (Fig. 11–10*A*). A shell closed by an operculum is found in members of the family Acteonidae, and in *Acteon* (the most primitive known opisthobranch) the nervous system is still twisted. However, there are many bubble shells in which the reduced shell is partially or completely (*Philine*) enclosed by the mantle, and the strange *Gasteropteron* is slug-like (Fig. 11–10*B*).

The order Anaspidea contains the sea hares, which reach the largest size (40 cm.) of any opisthobranch. The reduced shell is buried in the mantle or completely lost, and the body is bilaterally symmetrical. The mantle cavity and gills are still present, and the posterior edge of the mantle can be rolled to form an exhalant siphon (Fig. 11–10*C*).

The pteropods comprise two orders of small, swimming, pelagic opisthobranchs. The shelled pteropods (order Thecosomata) possess a typical shell with an operculum (Fig. 11–10*D*). The naked pteropods (order Gymnosomata) lack a shell (Fig. 11–15*F*). A gill is absent in most pteropods, and the naked forms have no mantle cavity. Gas exchange occurs across the general body surface.

Members of the order Acochlidiacea are tiny naked forms that are peculiar in having an elongate visceral mass that is distinctly separate from the foot (Fig. 11–10*E*). There is no gill or mantle cavity. The small number of freshwater opisthobranchs are members of this group.

Opisthobranchs of the order Notaspidea possess a mantle cavity and gill, but display all degrees of shell reduction. Some are even limpet-like or slug-like (Fig. 11–10*F*).

The sea slugs, members of the order Nudibranchia, certainly rank among the most spectacular and beautiful mollusks. Shell, mantle cavity, and original gill have disappeared, and the body is secondarily bilaterally symmetrical. The dorsal body surface is greatly increased in many nudibranchs by numerous projections called cerata, which are usually arranged in rows (Figs. 11–10G and 18–6). The cerata may be club-shaped, as in *Eolidia*, or branched, as in *Dendronotus*.

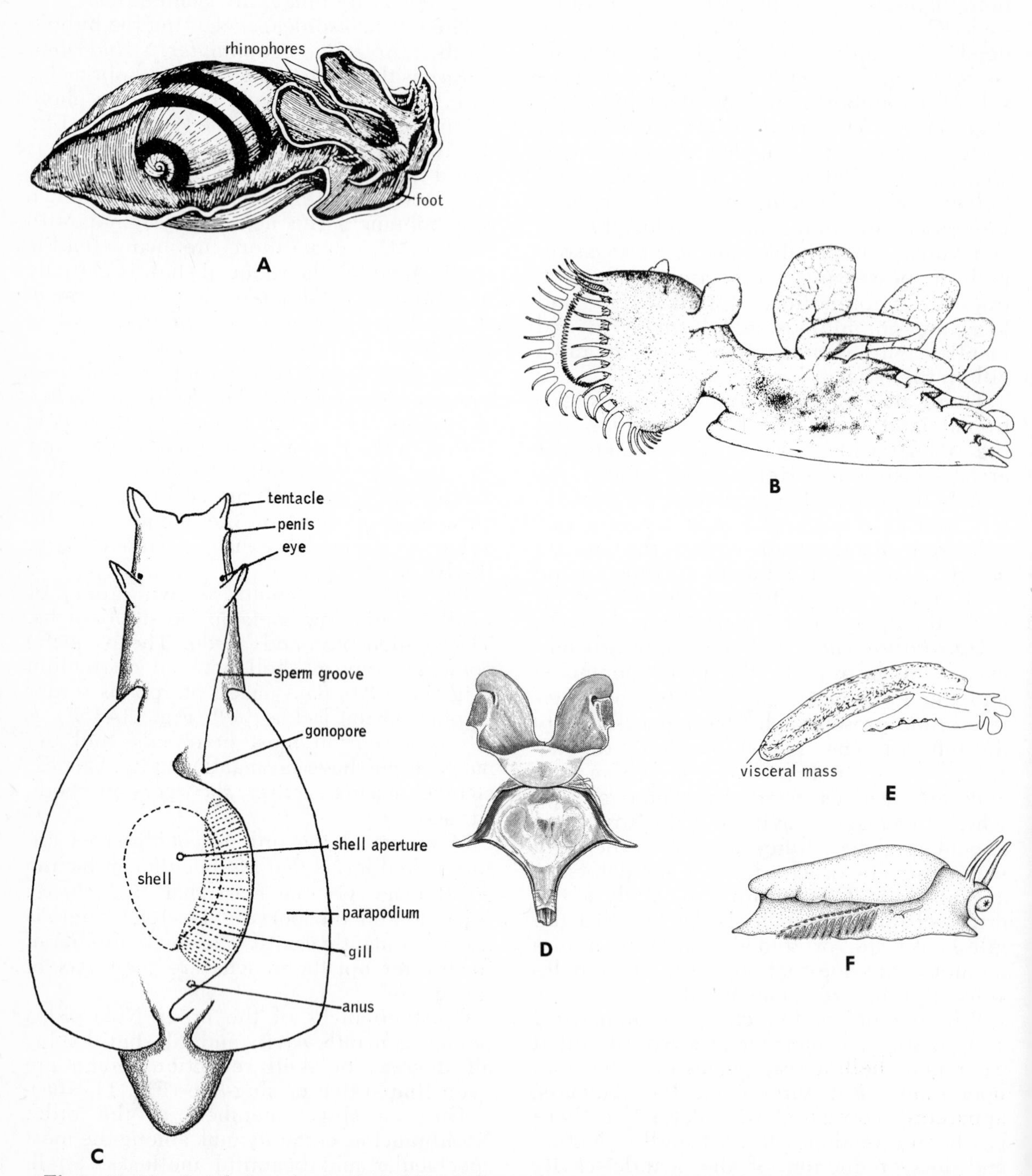

Figure 11–10. Opisthobranchs. *A.* The bubble shell *Hydatina*, a cephalaspidean. (Modified from several authors.) *B. Gasteropteron*, a slug-like cephalaspidean. (After Hornell from Hyman.) *C.* Diagrammatic dorsal view of a sea hare, an anaspidean. (After Bullough.) *D. Cavolina*, a shelled sea butterfly (pteropod; Thecosomata). (After a photograph by Abbott.) *E. Microhedyle* (Acochlidiacea). (After Odhner from Hyman.) *F.* Slug-like *Pleurobranchus* (Notaspidea). (After Vayassière and Hyman.) *Figure 11–10 continued on opposite page.*

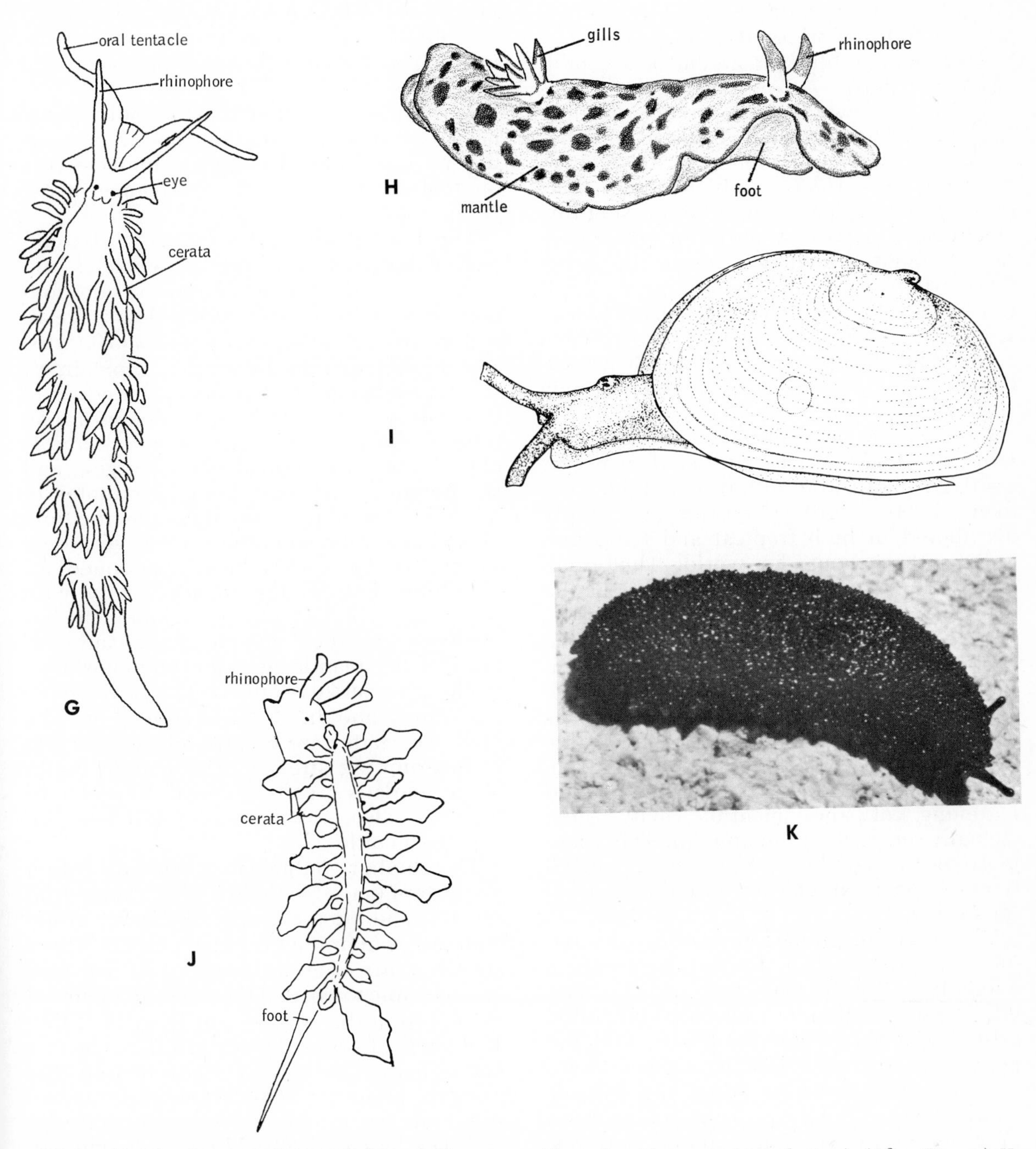

Figure 11–10. *Continued. G. Eolidia papillosa,* a nudibranch with cerata (dorsal view). (After Pierce.) *H. Glossodoris,* a nudibranch with secondary anal gills. (After a photograph by P. Zahl.) *I. Berthelinia,* an opisthobranch with a bivalve shell (Sacoglossa). (After Kawaguti from Hyman.) *J. Hermaea,* a sacoglossan with cerata. (After Marcus from Hyman.) *K.* The onchidacean *Onchidella floridana,* a slug-like intertidal opisthobranch with a posterior pulmonary sac. (By Betty M. Barnes.) (*B, F, I,* and *J* from Hyman, L. H., 1967: The Invertebrates, Vol. 6, Mollusca, Pt. I. McGraw-Hill Book Co., N. Y.)

Cerata are lacking in some nudibranchs, such as *Doris*, but these sea slugs have secondary gills arranged in a circle around the posterior anus (Fig. 11–10*H*). The cerata, as well as other parts of the nudibranch body, are usually brilliantly colored and commonly are red, yellow, orange, blue, or green, or a combination of colors.

There are several other groups of slug-like forms. The Sacoglossa contain a number of species, such as members of the genera *Hermaea* and *Alderia*, which have cerata (Fig. 11–10*J*). The amphibious slug-like Onchidacea possess a posterior pulmonary sac, through which gas exchange takes place when the animal is emergent at low tide (Fig. 11–10*K*). The sac communicates with the exterior by means of a pore. When submerged, the pore is closed and gas exchange takes place across the general body surface.

Pulmonates. The order Pulmonata contains the highly successful land snails as well as many freshwater forms, and the more than 14,000 described species are widely distributed in both tropical and temperate regions throughout the world. The group takes its name from the conversion of the mantle cavity into a lung. The edges of the mantle cavity have become sealed to the back of the animal except for a small opening on the right side called the pneumostome (Figs. 11–8 and 11–11*A* and *C*). The gill has disappeared and the roof of the mantle cavity has become highly vascularized. Ventilation is facilitated by the arching and flattening of the mantle cavity floor (actually the back of the animal). The pneumostome generally remains open at all times, or opens and closes with the ventilating cycle.

The origin of pulmonates is obscure, but they probably evolved from some group of operculate prosobranchs that had a single gill. These ancestral forms perhaps inhabited estuarine marshes and mud flats, and the pulmonate condition could have evolved as a means of gas exchange when the animals were confined to small stagnant puddles of water or to wet but exposed surfaces. These conditions are similar to those postulated for the origin of amphibians.

There are a few marine species, all of which live on intertidal rocks or in estuarine habitats. The limpet *Siphonaria* is believed to be the most primitive known pulmonate (Figs. 11–11*E* and 11–16*E*). It is one of the few pulmonates that possess a veliger larva, indicating that the marine habit does not represent a secondary return to the sea. There is a secondary gill within the mantle cavity, but some malacologists believe that part of the gill is homologous to the gill of prosobranchs. Water enters and leaves the mantle cavity through two siphons on the right side (Fig. 11–11*E*).

Amphibola, another marine pulmonate, has a typical shell but is unusual in possessing an operculum. In other pulmonates the operculum is lost during the course of development.

The lower pulmonates (suborder Basommatophora), with one pair of tentacles and with eyes at the tentacle base, include both marine and freshwater forms. Many freshwater species, such as *Lymnaea*, *Bulinus*, and *Physa*, come to the surface to obtain air for gas exchange. In *Lymnaea*, the edges of the mantle cavity can be extended as a long tube for this purpose. The pneumostome is closed when the animal is submerged, and submergence may last from 15 minutes to more than an hour, depending upon the time of year and other conditions. However, some deep lake lymnaeids have abandoned air breathing and fill the mantle cavity with water. A secondary gill (pseudobranch) has evolved in other aquatic pulmonates as folds of the mantle near the pneumostome. Such a secondary gill is found in some planorbids and in the ancyclids (Fig. 11–11*B*). The latter are limpets adapted for life in fast-running streams. The mantle cavity of ancyclids is greatly reduced. The planorbids may also have secondary gill folds inside the mantle cavity.

The higher pulmonates (suborder Stylommatophora) include the terrestrial species, and are a considerably larger group than that containing the aquatic forms. There are two pairs of tentacles (the first pair may be inconspicuous), and the eyes are mounted at the top of the second pair (Fig. 11–11*D*). The shell of these terrestrial pulmonates is not as heavy as those of many marine gastropods, although some variation may result from the availability of calcium in the soil. The largest shells, 27 cm. in height, are possessed by members of the African species *Achatina fulica*.

Shell reduction or loss has occurred independently a number of times within the higher pulmonates, and such naked species are called slugs (*Arion*, *Phylomycus*, *Limax*). The shell is generally absent or reduced and buried within the mantle, but in *Testacella*

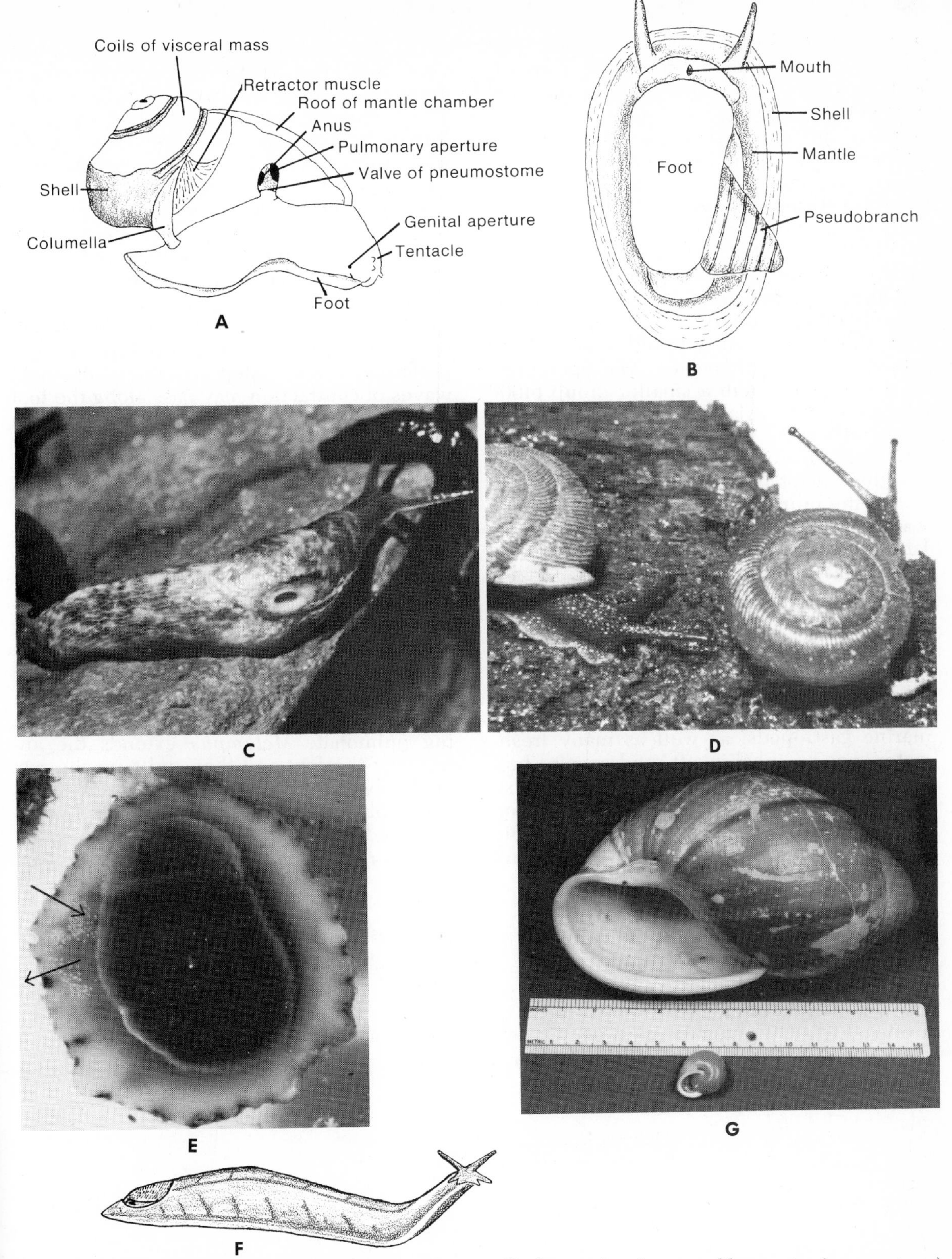

Figure 11–11. ***Pulmonates.*** *A.* The land snail *Helix,* partially dissected to show pore-like opening (pneumostome) into lung, or mantle cavity. *B.* The freshwater pulmonate *Ferrissia,* a limpet, with secondary gill (ventral view). *C.* A terrestrial slug. Opening into lung seen at lower edge of saddle-like mantle. *D.* A land pulmonate (Stylommatophora), showing the eyes mounted at the ends of the tentacles. *E.* Ventral view of *Siphonaria,* showing openings into mantle cavity. *F. Testacella,* a land slug with a reduced shell. *G.* Shell of specimen of the large South American pulmonate *Strophocheilus,* compared to two common species of temperate American land snails, *Polygyra* and *Retinella* (smallest, on ruler). (*A* after Rowett, H. G., 1957: Histology; *B* after Pennak; *F* after Baker; photographs by Betty M. Barnes.)

a little shell is perched on the back (Fig. 11–11*F*). The pneumostome of slugs is usually a conspicuous opening on the right side of the body (Fig. 11–11*C*).

There are only a few exceptions to the pulmonate mode of gas exchange in these terrestrial forms. The mantle cavity of one family of tropical slugs, the Athoracophoridae, is extended by blind tubules which ramify the surrounding wall. In another group of tropical slugs, the Veronicellidae, the mantle cavity is lost and gas exchange is entirely integumentary.

It should be remembered that not all land snails are pulmonates. The operculate land snails, although a smaller group (4000 species of terrestrial prosobranchs as compared with some 15,000 species of terrestrial pulmonates), are very common in the tropics, and their adaptation for life on land parallels that of the pulmonates in many ways. Freshwater snails may be pulmonates or operculate mesogastropods, or even operculate archaeogastropods.

Locomotion and Habitation. The typical gastropod foot is a flat, creeping sole similar to that of the ancestral mollusk, but it has become adapted for locomotion over a variety of substrata. Some species have remained on the ancestral hard bottom, but many marine gastropods, as well as many freshwater species, inhabit soft sand or mud bottoms. Others live on seaweed or, in the case of the pulmonates, on terrestrial vegetation, rotten leaves, and logs. Typically, the sole is ciliated and provided with numerous gland cells. The glands of the foot elaborate a mucous trail over which the animal moves. Some very small snails, as well as species which live on sand and mud bottoms (e.g., *Nassarius* and *Polinices*), move by ciliary propulsion. Most hard-bottom gastropods, and even large soft-bottom species when moving rapidly, are propelled by waves of fine muscular contraction that sweep from the posterior to the anterior or sometimes in the reverse direction. In the trough of the wave the foot is lifted, and then is returned to the substratum a little ahead of the previous point of contact (Fig. 11–12*B*). It thus performs a small step. As many as eight waves of contraction may pass along the foot simultaneously. In some species a wave extends across the entire width of the foot (monotaxic) (Fig. 11–12*A*), but in many forms a wave involves only half of the width of the foot and the waves on the right side move independently of those on the left (ditaxic) (Fig. 11–12*C*). Snails with ditaxic pedal waves appear to have a sort of shuffling gait. Also important in the overall function of the foot are the shifting and changing of blood volumes within the hemocoelic spaces of the foot.

A few gastropods employ a different manner of pedal progression. The marsh-dwelling pulmonate *Melampus* extends the anterior of the foot and then pulls up the rest of the body behind it. Some prosobranchs and cephalaspid opisthobranchs that live on soft sand bottoms have become adapted for burrowing. In *Polinices*, the front of the foot,

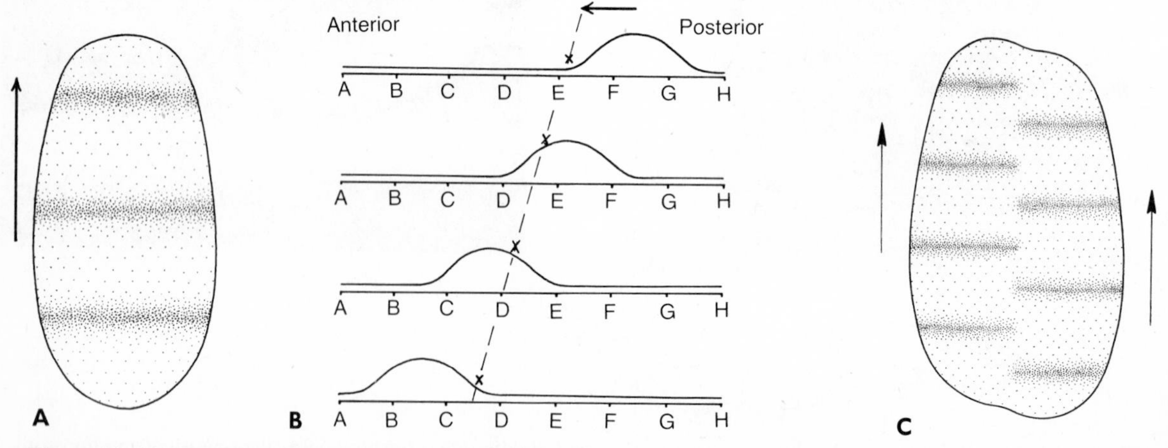

Figure 11–12. ***Pedal creeping in gastropods.*** *A*, Ventral view of foot of a gastropod showing three transverse waves of contractions moving from back to front of foot. Waves are monotaxic. *B*, Diagrammatic sections of a region of the foot during the course of movement of a wave of contraction (arrow). In region of contraction the foot is lifted from substratum. In the course of contraction a region of the foot (x) is placed back upon the substratum in advance of the point from which it was lifted. Letters indicate fixed positions of substratum. Diagonal dashed line shows forward movement of contracted region of foot. *C*, Ventral view of foot in which locomotion is by forward ditaxic waves.

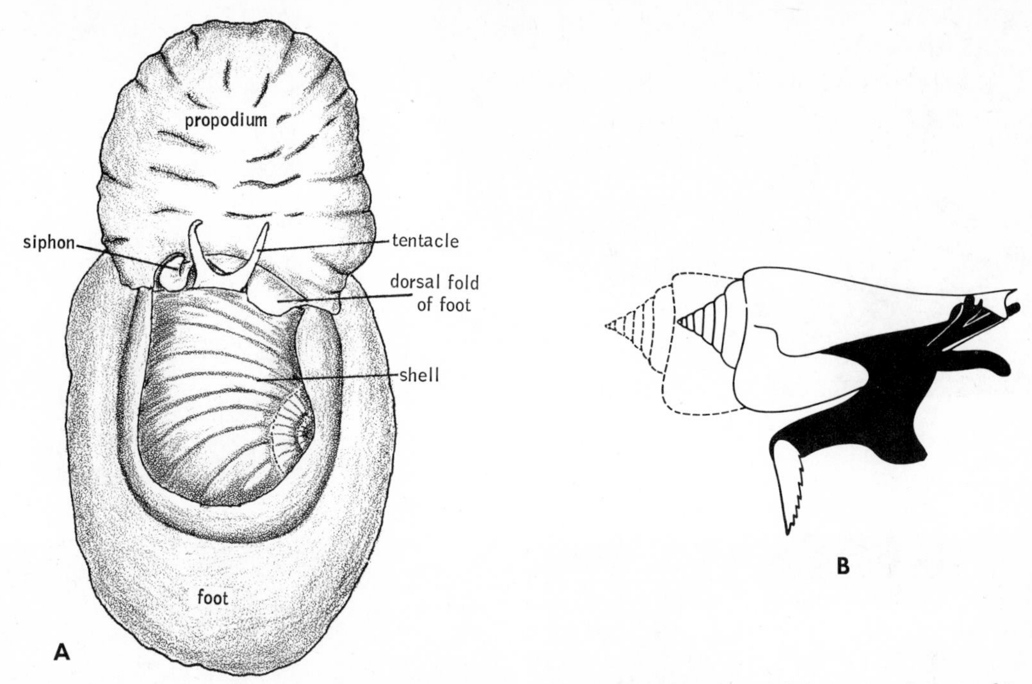

Figure 11–13. *A. Polinices*, a burrowing gastropod (dorsal view). (Drawn from life.) *B.* Lateral view of *Strombus*. Black is foot, to which the toothed, bladelike operculum is attached. Dotted outline indicates length of one "leap." (From Morton, J. E., 1964: *In* Wilbur, K. M., and Yonge, C. M. (Eds.), Physiology of Mollusca, Vol. 1, Academic Press, N. Y.)

called the propodium, acts like a plough, and a dorsal flap-like fold of the foot covers the head as a protective shield (Fig. 11–13*A*). *Sinum*, another burrower, which has a reduced shell almost covered by the reflexed mantle, resembles a lump of lard.

The conch *Strombus* crawls over sand in a very different fashion from other gastropods. The foot is somewhat reduced in size and the operculum is formed into a large claw (Fig. 11–13*B*). The claw digs into the sand and the animal then "poles" forward by rapid contraction of the columella muscle.

Limpets, abalones, and slipper shells are especially adapted for clinging to rocks and shells. All have low broad shells that can be pulled down tightly. The large foot functions as an adhesive organ as well as for movement, and the surrounding shirt-like mantle margin, which may bear tentacles, serves as an important sense organ (Fig. 11–14*B*). The homing ability of many intertidal limpets—*Acmaea, Patella, Fissurella*—has been the subject of numerous studies. A slight depression in the rock surface, over which the edges of the shell have come to fit, constitute the limpet's "home" (Fig. 11–14*A*). At high tide the animal may wander from 10 to 150 cm. away in feeding, depending upon the species, but then returns to its original home. The home is recognized by its surface configuration, and abrading the rock surface or the edge of the shell with a rasp greatly reduces the limpet's ability to identify the site. Not all individuals, even in species which exhibit homing, return to the original home.

A small number of gastropods are adapted for a sessile existence. The worm shells, members of three unrelated mesogastropod families (Vermetidae, Turitellidae, and Siliquariidae), have typical larval and juvenile shells; but as the animal grows older, the whorls become completely separated, and the adult shell looks like a corkscrew or is completely irregular (Fig. 11–14*C*). Worm shells live attached to sponges, to other shells, or to rocks, and the separated whorls provide greater surface area for attachment. The foot is reduced, but an operculum is present (Figs. 11–14*D* and *E;* and 10–25).

Hipponix, a relative of the near-sessile slipper shells, has a cap-like shell and attaches to rocks and other shells by a calcareous piece secreted by the foot.

A swimming pelagic existence has been adopted by the heteropods and by some opisthobranchs, notably the pteropods or sea butterflies. In most of these groups the foot has become modified as an effective swimming organ. The Heteropoda are laterally compressed, and the foot is trans-

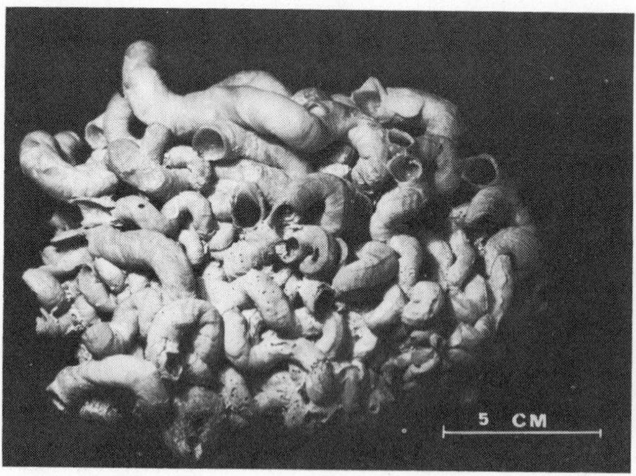

Figure 11–14. *A. Acmaea*, a common intertidal patellacean limpet on temperate rocky shores. *B.* Ventral view of the keyhole limpet *Fissurella*, showing the sensory tentacles on the mantle margin. *C.* A mass of vermetid worm shell. *D.* A minute West Indian worm shell, *Dendropoma irregularis* (Vermetidae), which may cover the surface of intertidal rocks. Each black circle is the aperture of a shell, closed by the operculum. The tube-like shell itself is difficult to recognize. (All photographs by Betty M. Barnes.) *E.* The worm shell *Serpulorbis* laying down mucous threads. Shell has been removed. (After Morton from Hyman, L. H., 1967: The Invertebrates, Vol. 6, Mollusca, Pt. I. McGraw-Hill Book Co., N. Y.)

formed into a ventral fin, even though these animals swim upside down (Fig. 11–15*I*). In *Carinaria*, the shell is much reduced; but in *Atlanta* it is well developed and symmetrically compressed, and it displays keeled whorls. *Carinaria japonica* occupies surface waters at night and lower depths during the day; it may migrate as much as 100 meters between the two levels.

The swimming foot of opisthobranchs is modified differently. Two fins, called parapodia, arise as lateral projections from the side of the foot. In the sea hares the fins arise from the middle of the body and are very broad (Fig. 11–10*C*). The shelled pteropods, or sea butterflies, have long, anteriorly located parapodia, which function as oars (Fig. 11–15*A* to *E*), and the animals swim upside down. The naked pteropods (Gymnosomata) are fast swimmers. The parapodia are ventrally located and perform a rapid sculling motion, each stroke describing a figure eight (Fig. 11–15*F* and *H*).

Of still other modes of existence that might be described, space permits only brief mention of the prosobranch violet shells (*Janthina*), which float beneath a raft of bubbles secreted by the foot (Fig. 11–17*A*); the Coralliophilidae, whose often strangely shaped shells are buried in coral (Fig. 11–17*B* and *C*); the carrier shell, *Xenophora*, which attaches foreign objects, including other gastropod and bivalve shells, to its own shell with a foot secretion (Fig. 11–18); a few minute naked interstitial opisthobranchs; and a group of algae-inhabiting sacoglossan opisthobranchs that have secondarily derived bivalve shells (Fig. 11–10*I*).

Nutrition. Virtually every type of feeding habit is exhibited by gastropods. There are herbivores, carnivores, scavengers, deposit feeders, suspension feeders, and parasites. Despite great differences in feeding habits, it is possible to make a few generalizations. (1) A radula is usually employed in feeding. (2) Digestion is always at least partly extracellular. (3) With few exceptions, the enzymes for extracellular digestion are produced by the salivary glands, esophageal pouches, the digestive diverticula, or a combination of these structures. (4) The stomach is the site of extracellular digestion, and the digestive diverticula are the sites of absorption and of intracellular digestion, if such digestion takes place. (5) As a result of torsion, the stomach has been rotated 180°, so that the esophagus enters the stomach posteriorly and the intestine leaves anteriorly (Fig. 11–3*B*). In the higher gastropods, there has been a tendency for the esophageal opening to migrate forward again toward the more usual anterior position. (6) Food is moved by ciliary tracts rather than by muscular contraction.

In most gastropods, the radula has become a highly developed feeding organ acting as a grater, rasp, brush, cutter, grasper, or conveyor. The teeth vary in number from 16 to thousands, and are always arranged in rows. Usually there is a median longitudinal row of teeth, on each side of which is a varying number of rows of marginal and lateral teeth (Fig. 11–19*C*). The median, marginal, and lateral teeth usually differ from one another in shape and structure. The character and form of the radula teeth are relatively constant at the family level and above, and they have systematic importance.

Primitive Gastropods. The most primitive feeding habit and digestive tract is found in the archaeogastropods, and it is from these forms that we have drawn many of our speculations about the ancestral mollusks. The marine archaeogastropods, which are mostly rock inhabitants, are microphagous, grazing on fine algae. The radula bears many teeth, at least 12 in each transverse row (Fig. 11–19*C* and *D*). When the radula is retracted, the lateral teeth direct the ingested particles into the center of the gutter produced by the lateral folding of the radula ribbon. A food-laden mucous string is formed within the gutter, and is pulled into the esophagus and then the stomach by the rotating mucous mass within the style sac.

The esophagus enters the right side of the posterior end of the stomach, and the intestine leaves anteriorly at the end of a style sac (Fig. 11–19*B*). The openings of the ducts of the digestive diverticula lie near the opening of the esophagus. Part of the posterior stomach wall is covered with a chitinous lining, called the gastric shield, and part is occupied by a large sorting area. The sorting area projects into a deep pouch, the caecum. Also extending into the caecum is the major typhlosole, a fold bordering the intestinal groove. As particles are dislodged from the mucus, they are swept against the sorting area and into the caecum, where sorting occurs. Fine particles are carried by cilia along the crests of the ridges, to emerge from the caecum as a stream going to the ducts of the digestive diverticula. Heavy rejected particles are carried in the grooves toward the large in-

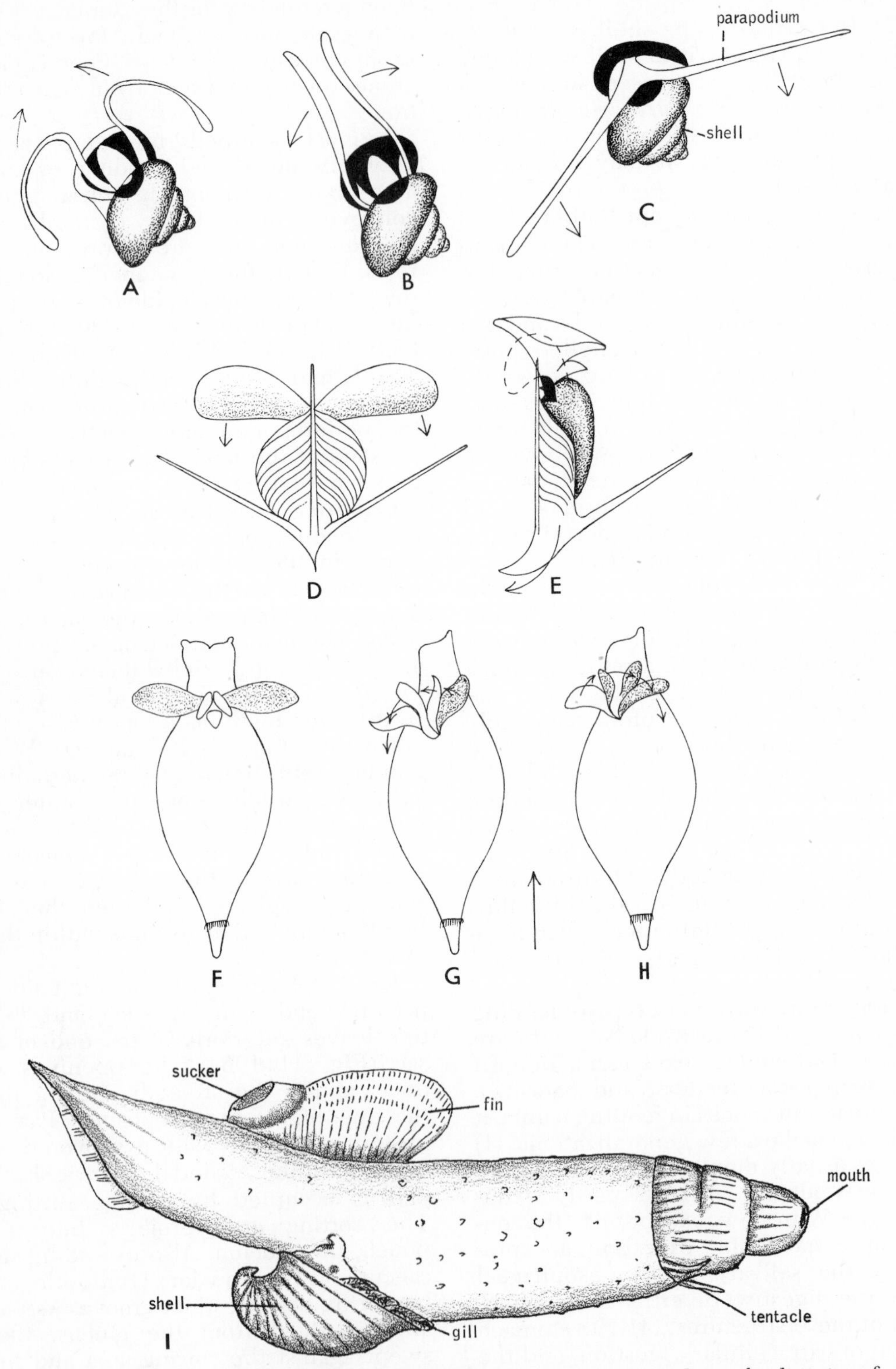

Figure 11–15. *A-C*. Swimming in the shelled sea butterfly, *Limacina*. Arrows indicate the direction of movement of the parapodia. *A* shows the recovery stroke; *B*, the beginning of the effective stroke; and *C*, the middle of the effective stroke. *D* and *E*. Ventral and side view of the shelled sea butterfly, *Euclio pyramidata*, showing parapodial movements in swimming. *F-H*. Parapodial movement in the swimming of the naked sea butterfly, *Clione limacina*. *F*, ventral view; *G* and *H*, side views. (All After Morton.) *I*. *Carinaria*, a pelagic prosobranch (Heteropoda). (After Abbott.)

Figure 11–16. Distribution of intertidal mollusks on a rocky tropical shore. *A*. South shore at Devonshire Bay, Bermuda, at low tide. Number 3 indicates the lower intertidal zone, which is covered by a heavy mat of algae. Number 2 marks the upper intertidal zone, which is covered with a fine algal felt and appears pale in the photograph. Number 1 indicates the spray zone, which is black in color and supports a microscopic blue-green alga. *B. Littorina ziczac,* found in the upper intertidal zone. The members of this mesograstropod family (Littorinidae) are common intertidal snails throughout the world. *C. Tectarius muricatus*. This common West Indian littorinid occurs by the thousands in the supratidal splash zone, where it occupies small pit-like depressions in the coralline rock. *D. Nerita versicolor* is an archaeogastropod inhabiting the upper intertidal zone. Most of the other members of this widely distributed tropical family are intertidal animals. *E. Siphonaria pectinata*. This limpet, which inhabits the upper intertidal zone, is one of the few marine pulmonates. *F. Chiton tuberculatus* and the keyhole limpet, *Fissurella angusta*. These two mollusks are found in the upper intertidal zone at this locality, but in the absence of a heavy algal mat, they may be found at lower levels. The rocks on which these specimens are located, as well as much of the upper intertidal zone, are covered by the worm shell *Dendropoma* (see Fig. 11–14*D* taken at the same locality). (All by Betty M. Barnes.)

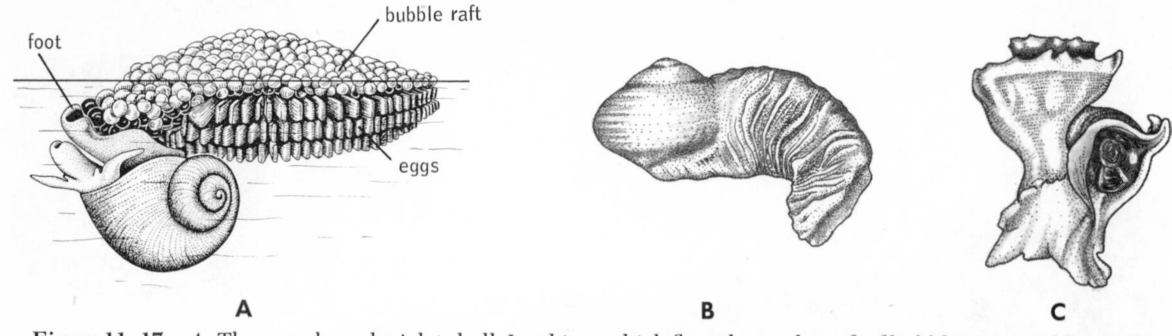

Figure 11–17. *A*. The prosobranch violet shell, *Janthina*, which floats beneath a raft of bubbles secreted by the foot. (After Fraenkel.) *B* and *C*. Two members of the Indo-Pacific Coralliophilidae, which live buried in coral. *Magilus* (*B*) and *Latiaxis* (*C*). (*B* and *C* after photographs by Abbott.)

testinal groove that runs down the style sac into the intestine.

Digestion is partly extracellular, by enzymes that are elaborated by glands in the esophageal region, and partly intracellular within the digestive diverticula.

Higher Gastropods. Correlated perhaps with the invasion of soft bottoms and other habitats, the diets and feeding habits of higher gastropods became extremely diverse, especially among the mesogastropods and opisthobranchs (Fig. 11–20). Most higher gastropods are macrophagous. Digestion has become entirely extracellular and takes place in the stomach, which has lost most of its primitive features—chitinous lining, sorting area, style sac—and has become more or less a simple sac (Fig. 11–21*C*). Enzymes are supplied by the digestive diverticula or by glands associated with the esophagus or buccal region.

The highly adaptable mesogastropod radula (taenioglossan) bears seven teeth in a transverse row (Fig. 11–21*A*), of which the lateral ones are hook-shaped. The neogastropods, which are mostly carnivores, have radulas with only three teeth (or, in some, only one tooth) per transverse row, but the teeth are heavy and usually bear several cusps (Fig. 11–21*B*). The largely herbivorous pulmonates have radulas with the largest number of teeth of any gastropods (up to 150 small teeth per transverse row). The opisthobranch radula is highly variable. In a number of gastropod groups, the radula has been lost in the adult.

In addition to the radula, feeding is facilitated in many gastropods, especially pulmonates, by jaws, which are thickened cuticular pieces in the front of the buccal cavity.

Herbivores. The many herbivorous gastropods include some marine prosobranchs, the freshwater prosobranchs, the operculate land snails, a variety of opisthobranchs, and the majority of the pulmonates. Marine forms feed on algae, but freshwater and land forms

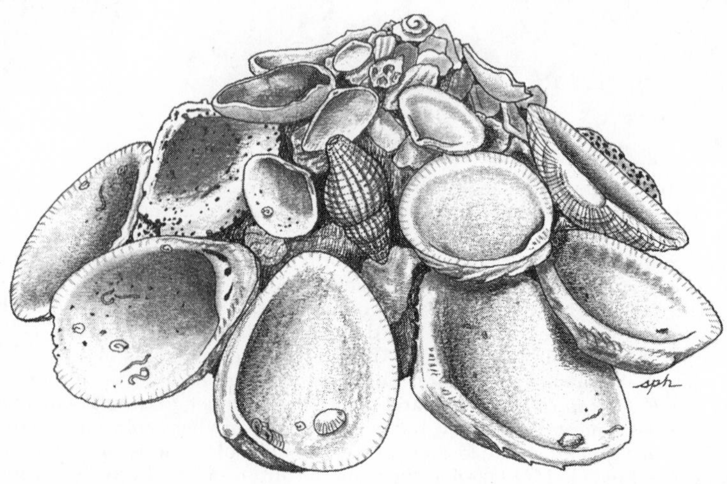

Figure 11–18. The carrier shell *Xenophora*, which attaches shells and other foreign objects to its own shell. (Based on a photograph by T. Abbott.)

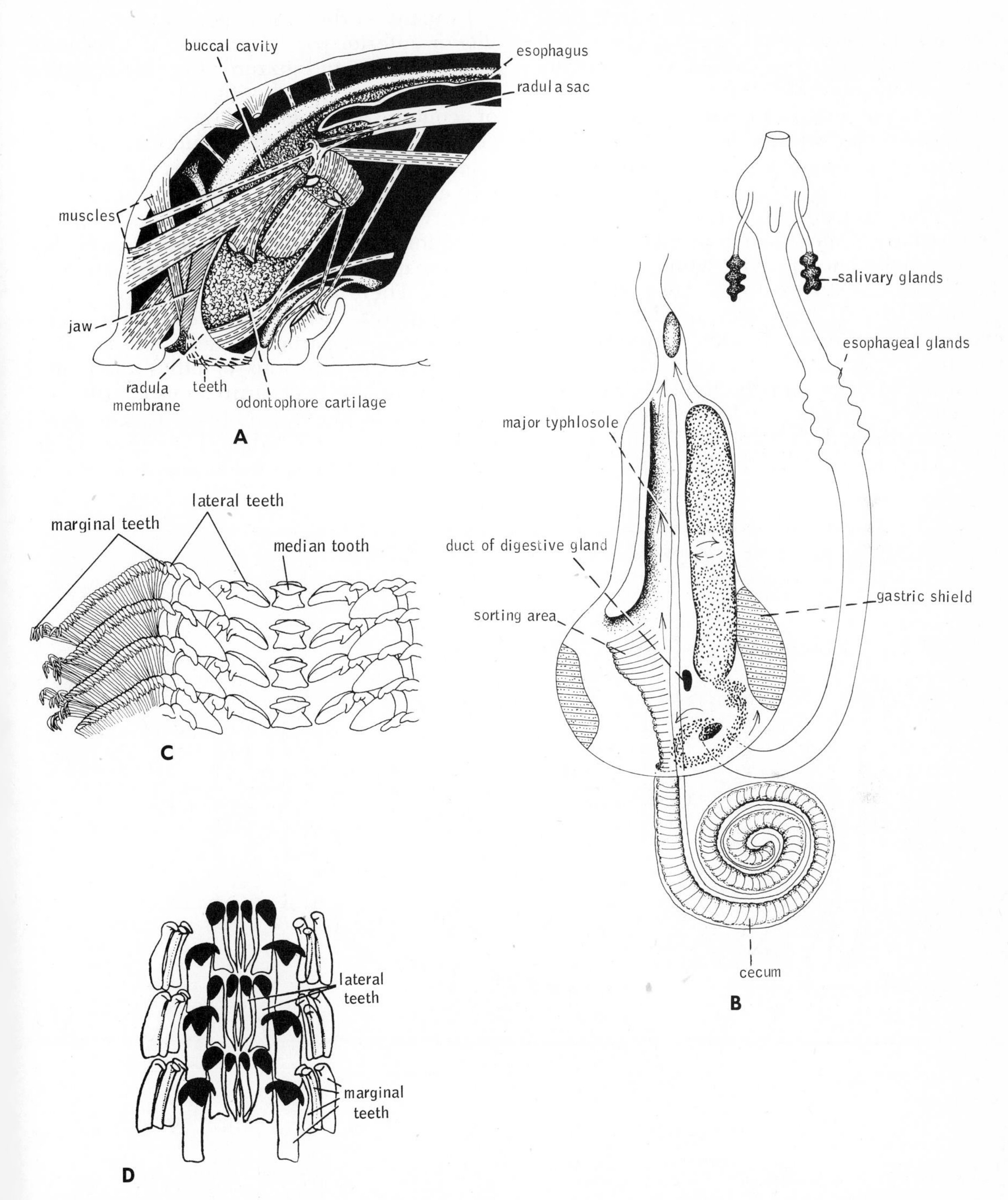

Figure 11–19. *A*. Radula and odontophore of *Monodonta*, at mouth opening, pressed against substratum in feeding. (After Nisbet.) *B*. Diagram of digestive tract of a primitive prosobranch (Trochidae). Arrows show ciliary currents and rotation of mucous mass (protostyle) within style sac. (After Owen.) *C*. Part of radula of the abalone *Haliotis* (archaeogastropod), showing four transverse rows of teeth. Marginal teeth on one side have been omitted. *D*. Part of the radula of the archaeogastropod *Patella*, showing three transverse rows of 12 teeth each. (*C* and *D* from Fretter, V., and Graham, A., 1962: British Prosobranch Molluscs. Ray Society, London, p. 171.)

consume the tender parts of aquatic and terrestrial vascular plants, decaying vegetation, or fungi. A few terrestrial snails and slugs are serious agricultural pests. The giant African snail *Achatina* can be very destructive, and considerable effort has been expanded to prevent its spread. Although such intertidal rock-dwelling mesogastropods as *Littorina* and *Tectarius* rasp up fine algae, (Fig. 11–16*B* and *C*), many marine herbivores feed on larger algae and are capable of cutting. The sea hare, *Aplysia*, can pull a 2 cm. strip of algae into the buccal cavity with its radula; it then holds the strip with the jaws while the radula severs it. Jaws are an adaptation of many herbivorous gastropods, including pulmonates.

The sacoglossan opisthobranchs are quite specialized herbivores. These tiny gastropods have the radula reduced to a single longitudinal row of teeth, which are used to slit open algal cells. The contents of the cells are then sucked out.

In many herbivorous species, the esophagus or anterior part of the stomach is modified as a crop and gizzard (Fig. 11–22). The gizzard may be lined with cuticle (sea hare) or contain sand grains (many freshwater snails). Amylases and cellulases are produced by the esophageal glands or the digestive glands. Terrestrial pulmonates possess a powerful array of digestive enzymes, and digestion is initiated in the large crop that to a great extent replaces the stomach (Fig. 11–23). There is no gizzard.

Carnivores. Although a few of the many carnivorous gastropods are pulmonates, feeding on earthworms or other snails and slugs, the majority of carnivorous families are prosobranchs and opisthobranchs. The radula of these marine families is variously modified for cutting, grasping, tearing, scraping, or

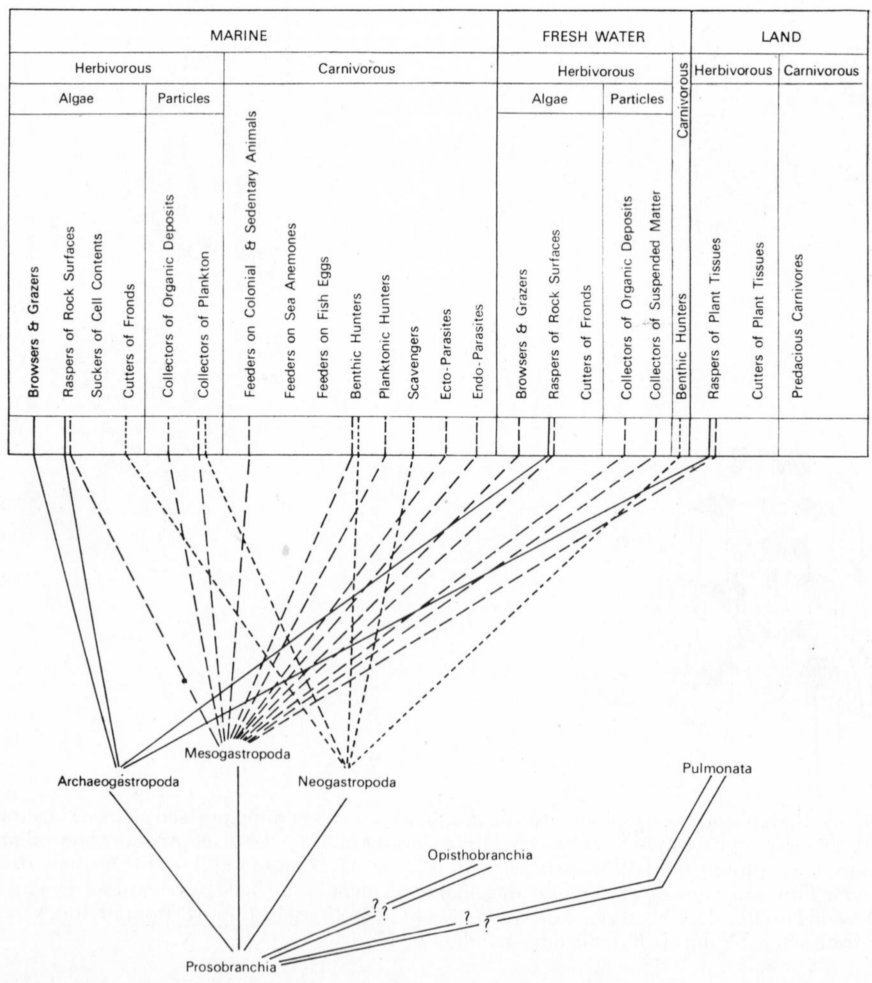

Figure 11–20. Diagram illustrating the adaptive radiation in prosobranch feeding habits. (From Purchon, R. D., 1968: Biology of the Mollusca. Pergamon Press, London, p. 73.)

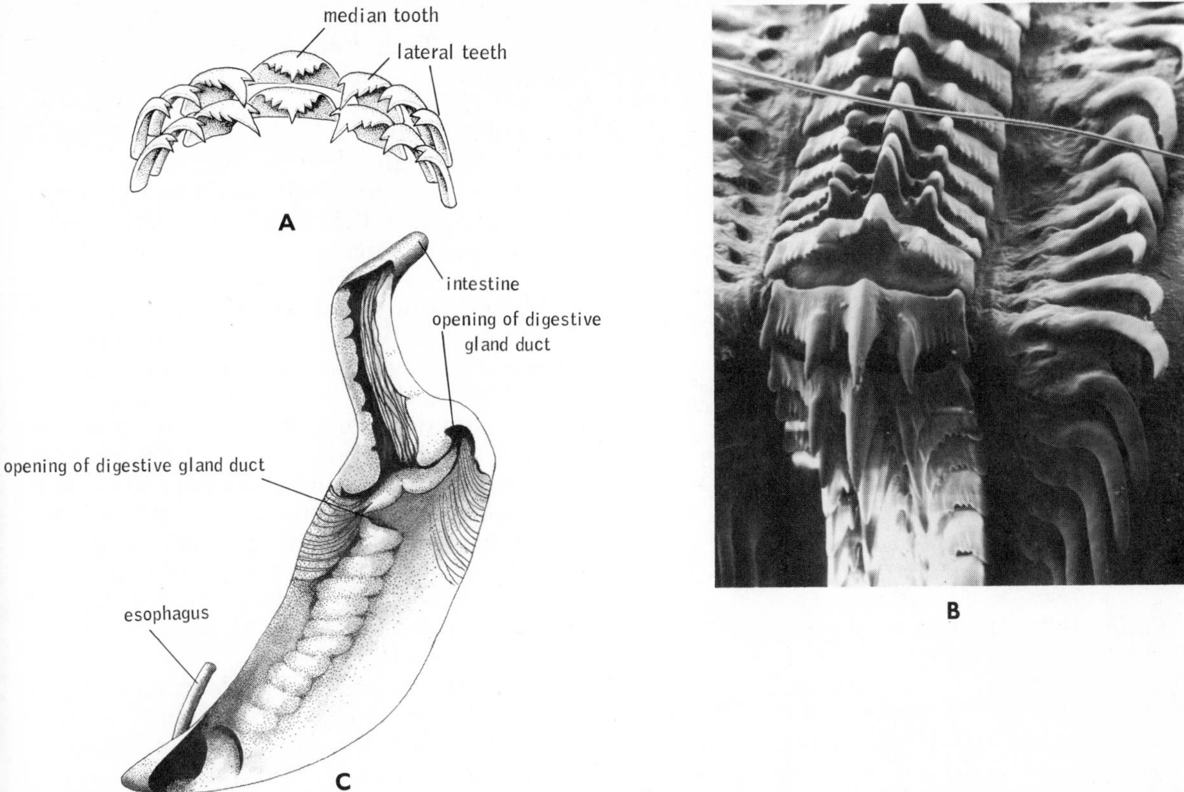

Figure 11–21. *A*. Two transverse rows of radula teeth of the mesogastropod *Lanistes*. (Modified from Turner.) *B*. Scanning electron micrograph of the radula of *Urosalpinx*, a carnivorous drilling neogastropod. There are only three teeth to a transverse row, but middle tooth bears several cusps. (From Carriker, M. R., 1969: Excavation of boreholes by the gastropod *Urosalpinx*. Amer. Zool., 9:917–933.) *C*. Sac-like stomach of a carnivorous prosobranch (*Natica*). Stomach has been opened dorsally. (From Fretter, V., and Graham, A., 1962: British Prosobranch Molluscs. Ray Society, London, p. 226.)

conveying. Jaws are sometimes present. The most common adaptation of carnivorous prosobranchs is a highly extensible proboscis, which enables the snail to reach and penetrate vulnerable areas of the prey. The proboscis (which contains the esophagus, buccal cavity, and radula and, at the tip, the true mouth) lies within a proboscis sac, or sheath, which has a mouthlike opening to the outside (Fig. 11–24). In feeding, the proboscis is projected out of the opening of the proboscis sac (Fig. 11–25). When the proboscis is fully extended, the proboscis sac is obliterated, and the buccal region is then equivalent to that of a prosobranch which lacks a proboscis. The proboscis is retracted within the proboscis cavity when feeding is terminated.

The larger bottom-dwelling carnivores, the neogastropods and some mesogastropods, commonly feed upon bivalve mollusks, other gastropods, and echinoderms. Such carnivores, many of which burrow into the sand to reach their prey, include volutes, bonnets, helmets, olive shells, harp shells, and whelks. Some species in these groups smother the victim with their foot. The whelks, *Buccinum*, *Busycon*, *Fasciolaria*, and *Murex*, may grip the bivalve with the foot, pulling or wedging the two valves apart with the edge of the shell (Fig. 11–26*A*). The bivalve's gape permits entrance of the proboscis. To accomplish the wedging, the gastropod may first partially pull open the valves or erode the valve margin with the lip of its own shell.

A number of prosobranchs are adapted for drilling holes in the shells of prey, particularly those of bivalves (Fig. 11–26*C* and *D*). The two best known such families are the neogastropod Muricidae (*Urosalpinx*, *Murex*, *Eupleura*) and the mesogastropod Naticidae (moon shells—*Natica*, *Polinices*). The mechanism has been most extensively studied in the Muricidae, and particularly in *Urosalpinx*, which causes great damage

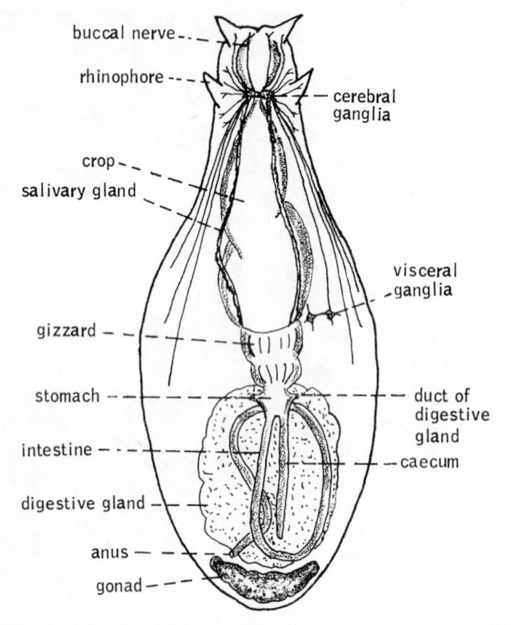

Figure 11–22. Digestive and nervous systems of the sea hare, *Aplysia*. (After Bullough.)

in oyster beds. The anterior sole of the foot contains an eversible gland, which is applied to the area to be drilled. The exact nature of the secretion produced by this gland is still uncertain, but it is known to be acid; it both reduces the organic framework and demineralizes the shell. Penetration is primarily a result of glandular activity rather than that of the radula. The snail drills with the radula for about a minute, and then applies the eversible gland to the site of the operation for about 30 to 40 minutes, repeating the cycle until the bivalve shell is penetrated. Approximately eight hours would be required to penetrate a shell 2 mm. thick, and penetration to a depth of 5 mm. has been recorded. The beveled sides readily identify the hole as having been drilled by a gastropod. When drilling is completed, the proboscis is extended through the hole, and the soft tissues of the prey are torn by the radula and ingested. In the naticids, the shell softening gland is located at the proboscis tip rather than on the foot. The adaptation for drilling apparently evolved early in geological history, for perforated bivalve and brachiopod shells have been found from the Devonian period.

One of the most remarkable groups of carnivores is the neogastropod genus *Conus*. Cone shells are tropical and subtropical in distribution and are found mainly in the western Atlantic and Indo-Pacific areas. They feed primarily on polychaete worms, other

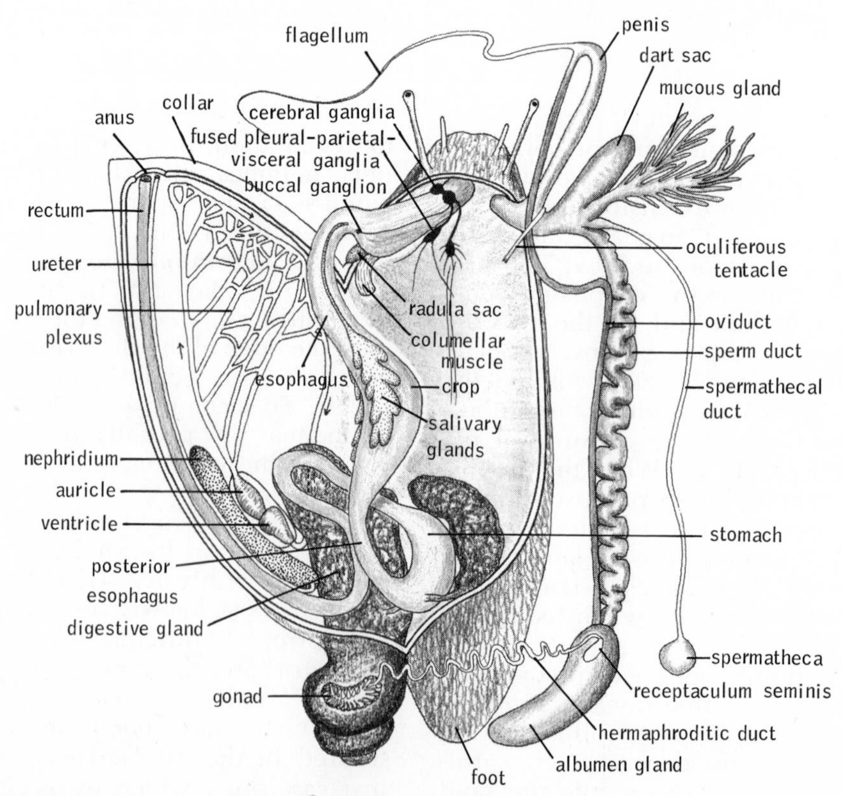

Figure 11–23. Dissection of the land snail, *Helix*. (Adapted from several sources.)

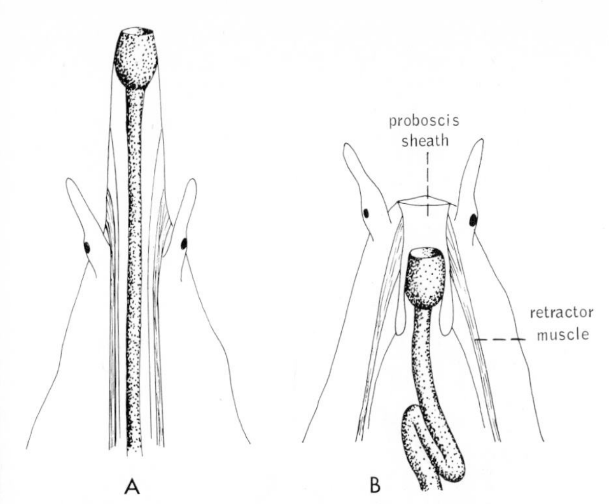

Figure 11–24. Diagram showing the proboscis of a prosobranch (*A*), protracted and (*B*), retracted. (After Fretter and Graham.)

gastropods, or fish, which they stab and poison with the radular teeth. In these marine gastropods, the odontophore has disappeared, and the radula is greatly modified (Fig. 11–27*A*). The teeth, which function singly, are long, grooved, and barbed at the end (Fig. 11–27*B*), and they are attached to the radula membrane by a slender cord of tissue. A large muscular bulb functions as an injector and is connected to the buccal cavity by a duct, which secretes the poison.

The cone shells have a long, highly maneuverable proboscis. When the proboscis is projected, the barbed end of a single radula tooth slips out of the radula sac into the buccal cavity, and is then hurled into the prey. In those species that feed upon polychaetes and snails, the tooth is injected free of the proboscis, but in forms which feed on fish (Fig. 11–27*C*), the cone does not strike until the fish is very close and retains a grip on the end of the tooth. The victim is very quickly immobilized by the neurotoxic poison, which enters the wound through the hollow cavity of the tooth.

The bite, or sting, of a number of South Pacific species, such as *Conus marmoreus*, *C. aulicus*, *C. tulipa*, *C. textile*, and *C. geographicus*, is highly toxic to humans; a few deaths have been reported, in one case within four hours.

There are many carnivorous opisthobranchs. The principal raptorial groups are the naked sea butterflies and the bubble shells. The former prey upon shelled sea butterflies, and the latter upon bivalves and gastropods, which are seized with the hooked teeth of the radula and swallowed whole.

The nudibranchs are grazing carnivores that feed upon sessile animals, such as hydroids, sea anemones, bryozoans, sponges, ascidians, barnacles, and fish eggs. There usually is no proboscis, but jaws are commonly present. In the Aeolidiidae, most of which feed on hydroids and sea anemones, the buccal cavity contains a pair of bladelike jaws, which are used to cut small pieces of tissue from the prey. Neither a proboscis nor esophageal pouches are present; the salivary glands secrete mucus and perhaps enzymes that pass with food into a simple bilaterally symmetrical stomach (Fig. 11–28*A*). The stomach floor contains cilia that cause currents to move posteriorly toward the three to five ducts of the digestive diverticula. The digestive diverticula in these carnivorous nudibranchs is peculiar in being confined entirely within the numerous club-shaped cerata, which cover the dorsal surface of the body (Fig. 11–10*G*).

Each ceras contains a tubule of the digestive diverticula, and the tubules are joined together by a system of branching ducts leaving the stomach (Fig. 11–28*A*). Thus the digestive diverticula of these nudibranchs is a diffuse rather than a compact organ, as it is in many mollusks. The secretory cells of the digestive tubules in the cerata produce a protease and diastase, which pass into the stomach and act extracellularly. Absorption takes place in the stomach walls and in the digestive diverticula. According to Graham (1938), some intracellular digestion of small particles, largely glycogen, takes place within the digestive diverticula.

The most remarkable feature of these nudibranchs is their utilization of the prey's nematocysts. Ciliary tracts in the stomach and in the ducts from the digestive diverticula carry the undischarged nematocysts to the cerata, in which they are engulfed but not digested. Instead, these undischarged nematocysts are moved to the distal tips of the cerata, called cnidosacs, which open to the exterior. There the nematocysts are used by the nudibranchs for defense. The discharge mechanism is not thoroughly understood. Graham (1938) suggests that discharge of the nematocyst is caused by pressure exerted either by circular muscle fibers around the cnidosac or else by the enemy.

The prosobranch counterparts of the predaceous sea butterflies are the pelagic heteropods, which feed upon small fish; the cowries (mesogastropods), like the nudibranchs, are grazers that feed upon ascidians (Fig. 11–26*B*).

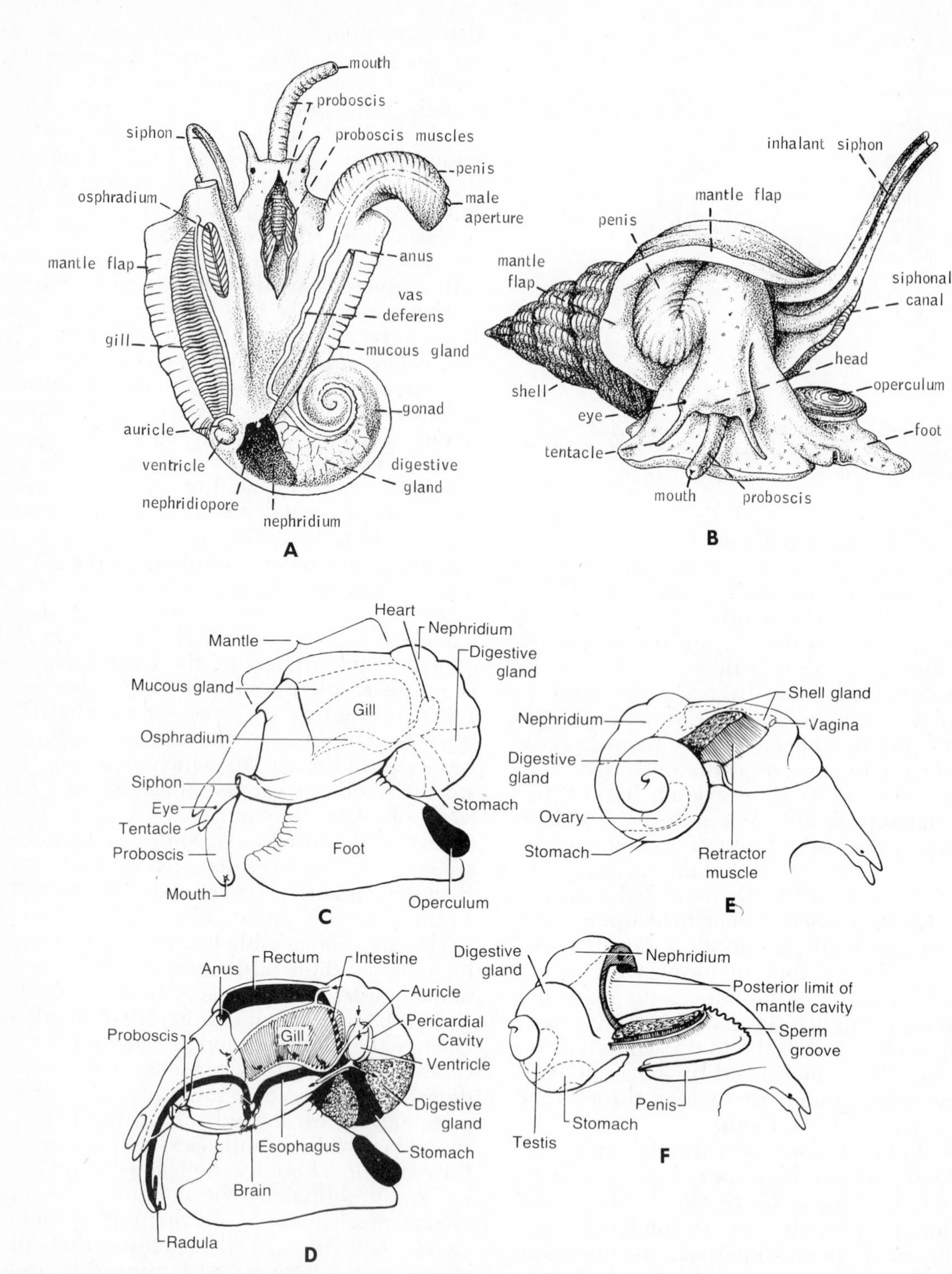

Figure 11–25. Anatomy of two marine prosobranchs (neogastropods). *A* and *B*. *Buccinum undatum*. *A*. Dorsal view, with shell removed and wall of mantle cavity cut and reflected. *B*. Animal crawling with proboscis protruded. (After Cox.) *C-F*. *Busycon canaliculatum*, with shell removed. Shell, which is shown in Figure 11–5*C*, is similar to that of *Buccinum* and is carried in the same position. *C*. Left side, showing external organs and internal organs visible through the integument. *D*. Same view with digestive, respiratory, circulatory, and nervous systems indicated. *E*. Female, showing portion of the right side. *F*. Male, portion of right side with mantle and retractor muscle cut short. In *E* and *F* the proboscis is withdrawn.

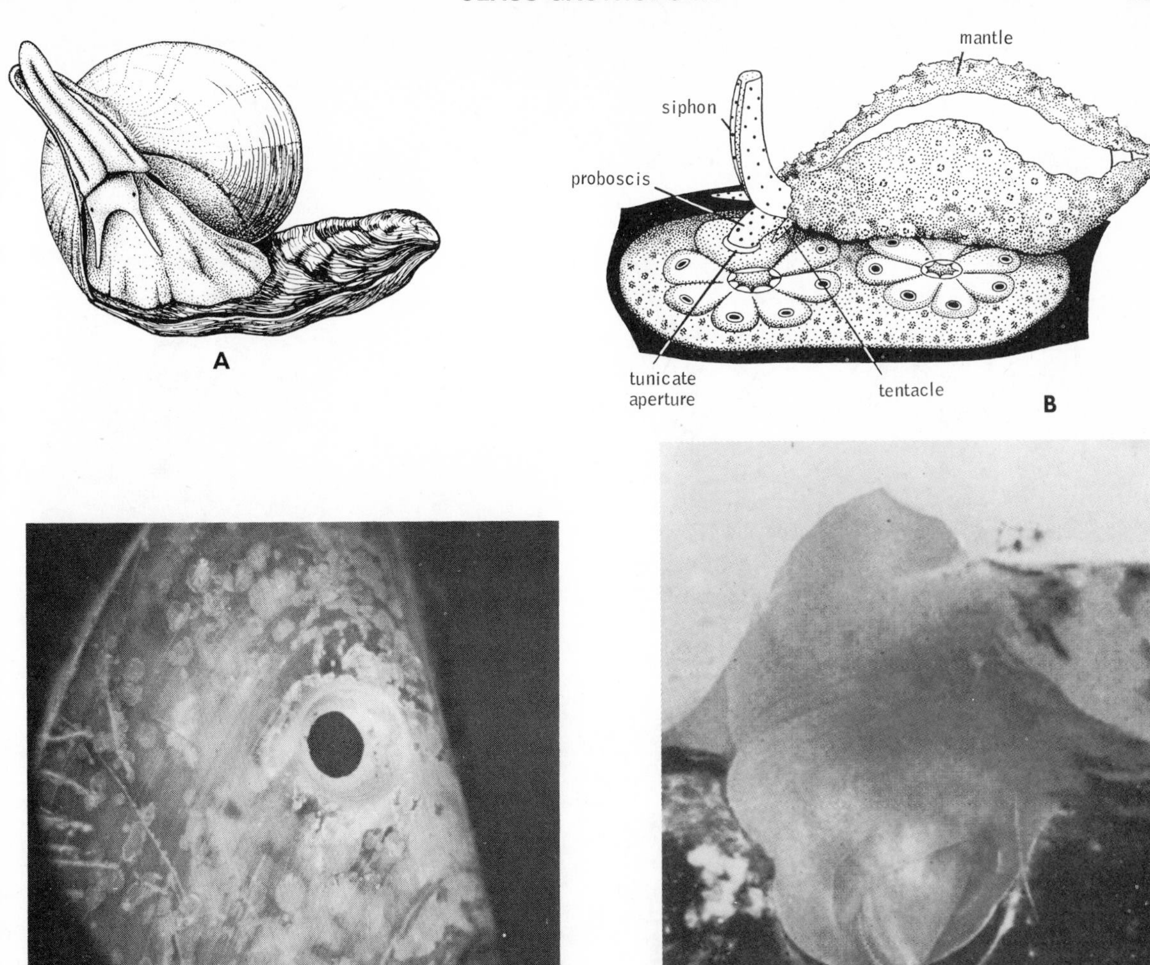

Figure 11–26. *A. Busycon* using the edge of its shell to pry open an oyster. *B.* The cowrie *Erato voluta* feeding on a colonial tunicate. The proboscis of the snail is thrust into the buccal opening of the tunicate. The shell of the cowrie is partially covered by the reflexed mantle; the erect structure is the siphon. (From Fretter, V., 1951: Proc. Malac. Soc. London, 29:15.) *C.* Photograph of a perforation through a bivalve shell produced by the radula of a drilling gastropod. Note the beveled edge. *D.* Radula of the oyster drill *Urosalpinx* rasping across bottom of a partially excavated borehole. (From Carriker, M. R., 1969: Amer. Zool., 9:920.)

Scavengers and Deposit Feeders. A scavenging habit has been adopted by numerous gastropods, of which *Nassarius* is a notable example. The feeding habits of different species of this genus of little neogastropods range all the way from a carnivorous habit to deposit feeding. On quiet, protected beaches along the Atlantic coast of the southeastern United States, *Nassarius obsoletus* may occur in enormous numbers at low tide, feeding on organic material deposited in the intertidal zone (Fig. 11–28*C*). This same species, however, is also a facultative carrion feeder and will consume the flesh of fresh fish. Specific proteins from oysters or crabs have been demonstrated to elicit a search response with the protruded proboscis.

The conch *Strombus* is also a deposit feeder, and the large, mobile proboscis sweeps across the bottom like a vacuum cleaner.

Suspension Feeders. Suspension feeding has evolved a number of times within the Gastropoda. In *Crepidula,* a filtering suspension feeder, the gill filaments have been tremendously lengthened to provide an increased surface area for trapping plankton (Fig. 11–29*A* and *B*). The shell and mantle edges are held tightly against the substratum except for a slight gap on each side of the anterior. Water enters the left side and

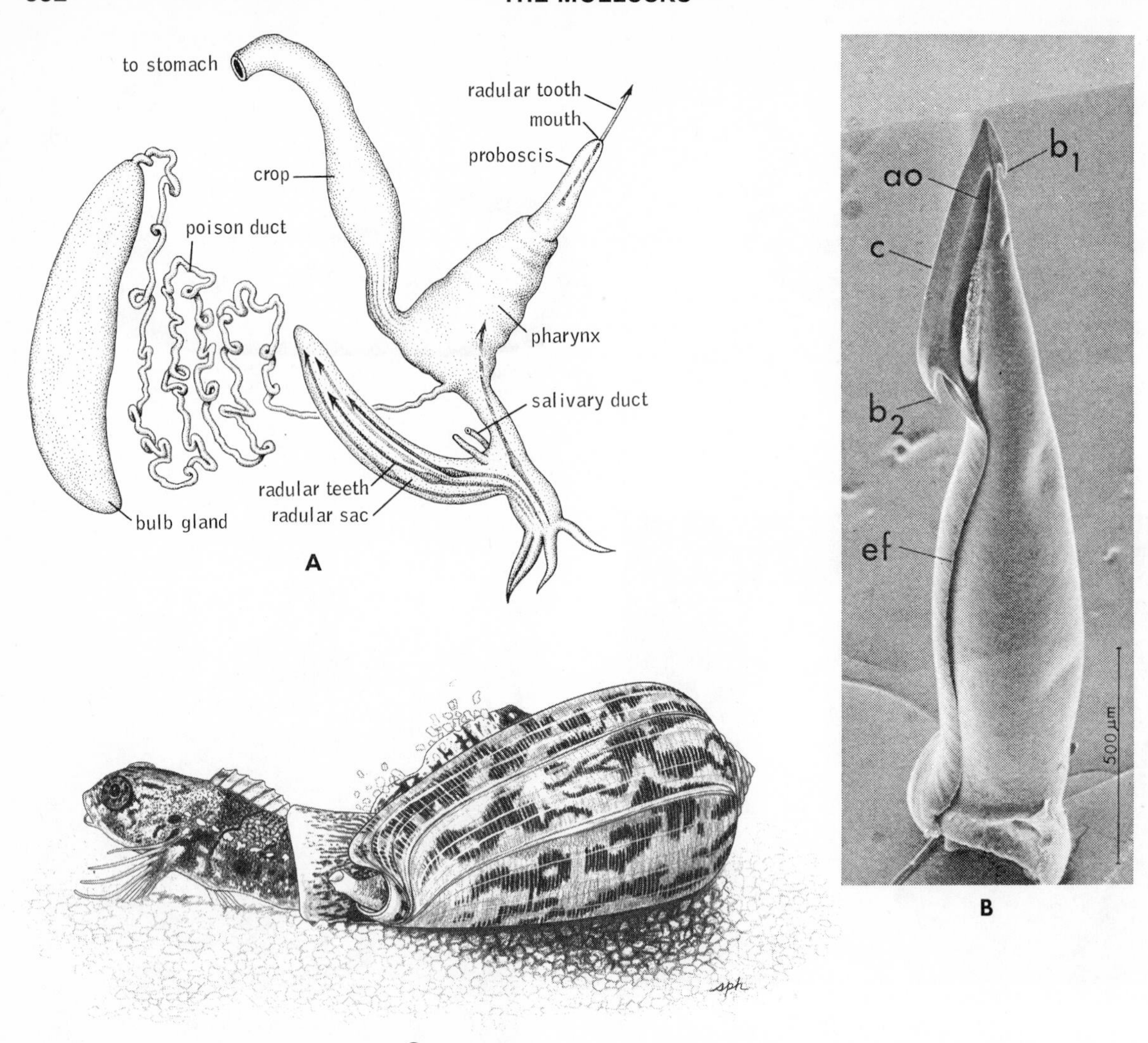

Figure 11–27. Feeding mechanism of the cone shells (*Conus*). *A.* Buccal structures of *Conus striatus.* (Modified after Clench.) *B.* Scanning electron photomicrograph of the harpoon-like radula tooth of *Conus imperialis.* Note the folded structure and barbed end. (From Kuhn, A. J., et al., 1972: Science, *176*:49–51. Copyright 1972 by the American Association for the Advancement of Science.) *C.* A cone shell swallowing a fish. (Based on a photograph by Robert F. Sisson and Paul Zahl.)

leaves from the right; and as plankton passes through the mantle cavity, it is trapped in a mucous sheet lying across the frontal surface of the long gill filament. The mucus is secreted by glands at the base of the gill.

Gill cilia carry the sheet containing entangled particles to the right toward the tips of the filaments, which curl back against the sides of the body. The particles then pass into a longitudinal groove, which lies along the side of the body and which is in contact with gill filaments. Here the particles collect and are compacted into a mucous string. Another mucous sheet—actually a web of mucous threads—screens the entrance into the mantle cavity (Fig. 11–29*B*). This sheet filters out large particles, which are either carried directly to a pocket in front of the mouth or expelled anteriorly as pseudofeces.

The anterior end of the longitudinal groove is also located close to the mouth, and at intervals mucus from the pocket or the anterior end of the mucous string in the groove is seized by the radula and pulled into the buccal cavity. The food particles are then carried to the stomach along an esophageal food groove.

Struthiolaria feeds in much the same way as *Crepidula,* but the animal lies buried in the bottom and utilizes surface deposit material thrown into suspension. Many of the sessile worm shells use the gills as a food trapping surface. Some also secrete a trap of mucous threads in the vicinity of the shell opening. The threads are produced by the pedal gland and laid down by pedal tentacles. They entangle small particles and are pulled into the mouth by the radula.

The shell-bearing sea butterflies are suspension feeders. Most of them apparently trap food particles in the mucus covering the parapodia, but *Limacina* feeds primarily on diatoms and dinoflagellates that are removed from water circulating through the mantle cavity. This food is trapped in mucus on the walls of the mantle cavity and then compacted to form a mucous string. A ciliated tract carries the string out of the mantle cavity to the mouth, where portions of it are detached by the radula.

Some sea butterflies, such as *Gleba* and *Corolla*, secrete an enormous floating mucous net measuring as much as two meters in diameter (Fig. 11–28*D*). The animal hangs beneath the net by its extended proboscis. There is no radula, and food particles trapped in the net are pulled in by proboscis cilia.

Crystalline styles are found in many suspension and deposit feeders—*Crepidula, Struthiolaria*, many worm shells, *Strombus*,

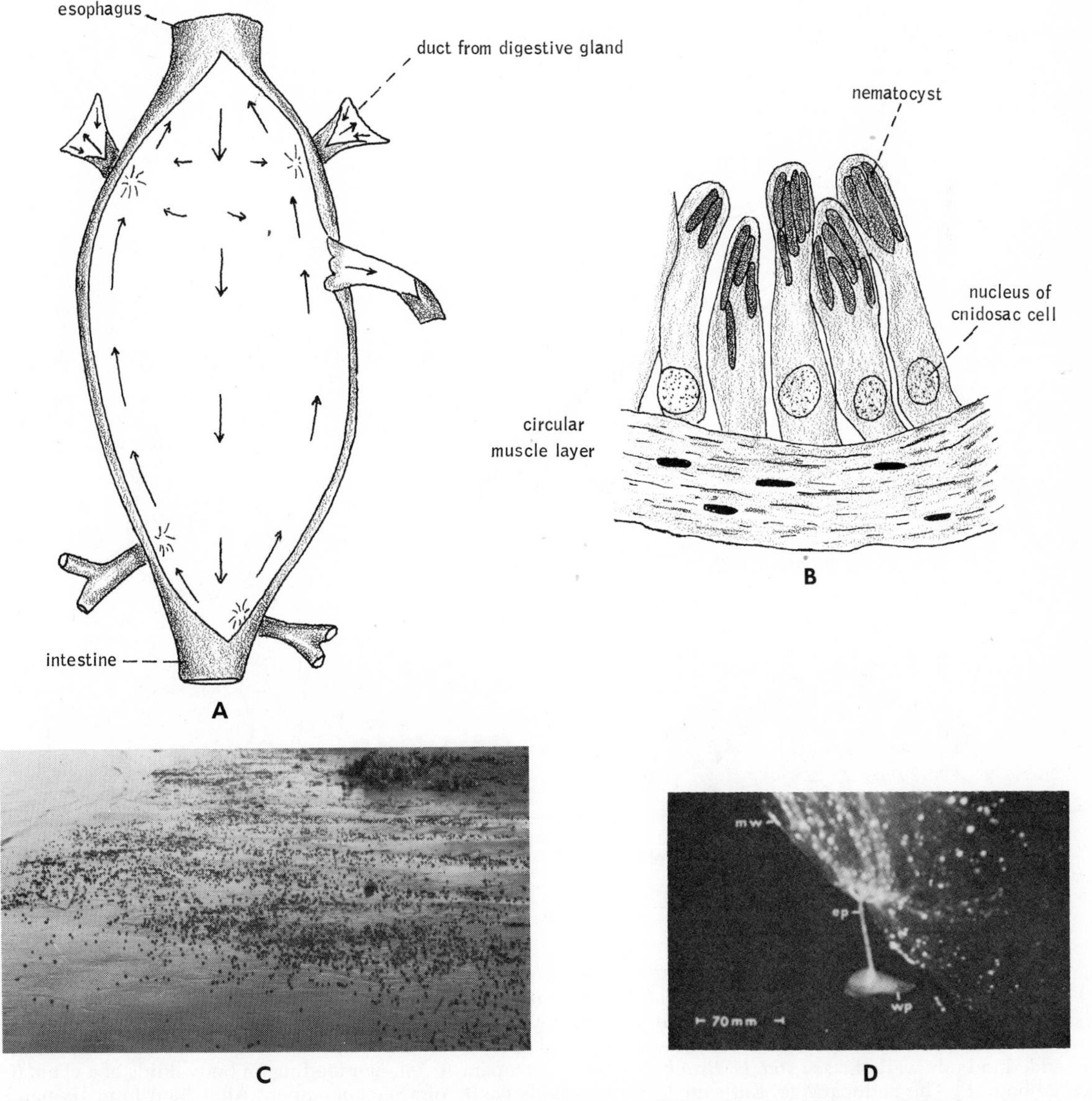

Figure 11–28. *A*. Stomach of the sea slug, *Eolidia papillosa*, opened dorsally. *B*. Cnidosac of *Eolidina* (transverse section). (Both after Graham.) *C*. *Nassarius obsoletus* feeding at low tide. This intertidal prosobranch scavenger and deposit feeder occurs in enormous numbers on protected beaches on the east coast of the United States. (By Betty M. Barnes.) *D*. The sea butterfly *Gleba cordata* feeding from its large, delicate mucous web. The long process labeled *ep* is the proboscis. The label *wp* indicates one of the wing-like extensions of the foot. (From Gilmer, R. W., 1972: Science, *176*:1240. Copyright 1972 by the American Association for the Advancement of Science.)

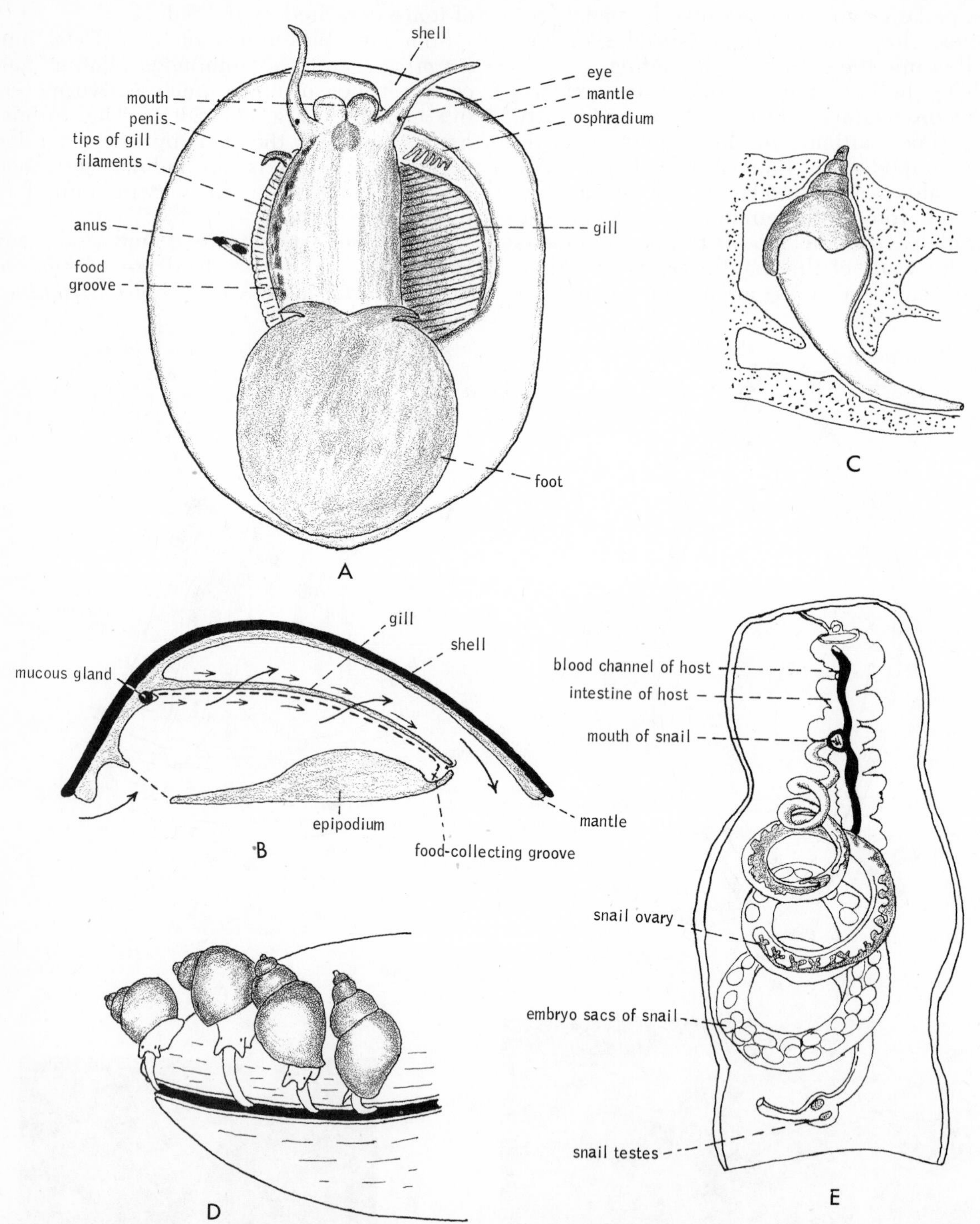

Figure 11–29. *A.* The slipper shell, *Crepidula fornicata*, a ciliary feeder (ventral view). (Drawn from life.) *B. Crepidula*, showing direction of water current (large arrows) through mantle cavity. Ciliary currents (small arrows) carry food particles. Dashed lines indicate filtering sheet of mucus. (Modified from Morton.) *C.* The parasite, *Stylifer*, embedded in body wall of a sea star. *D. Brachystomia*, an ectoparasite, shown feeding on body fluids of a clam. (*C-D* after Abbott.) *E.* The endoparasite, *Entoconcha*, within body cavity of a sea cucumber. (After Baur from Hyman.)

species of *Nassarius*, and some sea butterflies among the opisthobranchs. This structure is typical of bivalve mollusks and is described in detail on page 400. The presence of a crystalline style in gastropods is associated with the more or less continuous feeding of the animal on a diet of phytoplankton or organic detritus.

Parasites. Parasitism has evolved in a number of gastropods, which along with certain free-living forms present an interesting adaptive series leading from an epizoic to an ectoparasitic existence and thence to endoparasitism. The ectoparasites are modified chiefly with respect to the nature of the buccal region and digestive system. Members of the opisthobranch family Pyramidellidae possess chitinous jaws, stylets, and a pumping pharynx for sucking blood from bivalve mollusks and polychaetes (Fig. 11–29*D*). The mesogastropod Eulimidae, which are ectoparasites on echinoderms, also have a sucking proboscis but lack a buccal armature.

The endoparasites belong to two related families of mesogastropods. Members of the Styliferidae, such as *Stylifer*, live embedded in the body wall of echinoderms, on which they feed (Fig. 11–29*C*). The shell is still present in these snails, but the foot is vestigial and the proboscis is developed for parasitic feeding. The most modified parasitic gastropods are members of the Entoconchidae, which live within sea cucumbers. The body is wormlike and a shell is present only during larval development. *Entoconcha* feeds on the blood of the host by attaching the mouth region to one of the host's blood vessels (Fig. 11–29*E*). *Enderoxenos* has lost the digestive tract and absorbs food through the body wall.

Excretion and Water Balance. Members of the Archaeogastropoda possess two nephridia, but in all other gastropods the right nephridium has disappeared, except for a small section that contributes to the reproductive duct. Also, as a result of torsion, the nephridium is located anteriorly in the visceral mass (Fig. 11–25). The structure of the nephridium is sac-like and the walls are greatly folded to increase the surface area for secretion. Wastes are eliminated through a short ureter. Primitively, the distal end of the nephridium connects with the pericardial cavity, but in most gastropods the connection between nephridium and pericardial cavity is a small renopericardial canal (Fig. 11–34*A*). This canal opens into the more proximal part of the nephridium, in some cases at the level of the ureter. The pericardial glands of gastropods are located either above the auricle or along the sides of the pericardial cavity.

The nephridiopore of both the prosobranchs and the lower opisthobranchs opens at the back of the mantle cavity, and wastes are removed by the circulating water current (Fig. 11–8). Such an arrangement is not possible in many pulmonates, because the mantle cavity functions as a lung. As a result, the ureter has lengthened along the right wall of the mantle in higher pulmonates and opens to the outside near the anus and the pneumostome.

Excretion in gastropods, as is true of other mollusks, involves filtration into the coelom and reabsorption and secretion in the nephridium. The degree to which these various processes occur in different gastropods depends in large part upon the environment in which they live.

Aquatic gastropods, like most other aquatic invertebrates, excrete ammonia or ammonium compounds. The blood of the great majority of marine gastropods is isosmotic with sea water, and when such animals are placed in brackish water the blood concentration merely falls to equal that of the environment. Freshwater species maintain a rather low blood salt concentration level (although higher than that of the water in which they live), and the kidneys are capable of excreting a hyposmotic urine by reabsorption of salts. Large amounts of water are expelled by way of the nephridia in freshwater species.

Terrestrial pulmonates and operculate land snails, in order to conserve water, convert ammonia to relatively insoluble uric acid. This adaptation to conserve water is particularly striking in the Indian apple snail, which is seasonally amphibious. During its aquatic phase, this snail excretes ammonium compounds; and during its terrestrial phase, it excretes uric acid. Perhaps as an adaptation for water conservation, the renopericardial canal of pulmonates has a small orifice opening into the pericardial cavity, and there are no pericardial glands. Wastes are thus largely a result of nephridial secretion.

Terrestrial pulmonates have not been very successful in controlling desiccation through the body surface, and considerable water is lost in the production of the slime trail in crawling. However, many pulmonates can survive excessive desiccation; *Helix* can survive a water loss equal to 50% of the body weight, and the slug *Limax* can survive an 80% loss.

As would be expected, the majority of pulmonates require a humid environment. They either are nocturnal or live in damp environments, such as beneath logs or in leaf mold on forest floors. Snails of the genus *Partula* on the mountainous Society Islands (Tahiti) of the South Pacific illustrate some of the diverse habits of land snails living in humid tropical areas. The members of this genus feed on decaying vegetation and fungi. Some species live only on the ground. Other species are arboreal during the day, when they are inactive, attaching themselves to the undersides of leaves. At night they descend to the ground to feed. Still other species spend their entire lives in the trees.

There are pulmonates that inhabit dry rocky areas, dunes, and even deserts. Such species are active only at night or following rains. During dry periods in the tropics, in deserts, and during the winter in temperate regions, snails become inactive, either estivating or hibernating. They may first burrow into humus or soil, or climb up onto vegetation and attach the aperture edge of the shell with dried mucus. The edges of the mantle are drawn together in front of the shell aperture, and a protective mucous membrane that hardens when dry (or a thin calcareous membrane) is secreted over the opening. Several such epiphragms may be secreted as the snail withdraws further into the shell. Freshwater snails also estivate when ponds dry up, and hibernate when the water is frozen.

In most gastropods, the digestive gland appears to play some role in excretion. A number of histological studies have demonstrated the presence of excretory cells in this organ; waste from the gland would be eliminated from the body by way of the stomach and intestine.

Internal Transport. As a result of torsion, the heart of gastropods is located anteriorly in the visceral mass (Fig. 11–3*B*). Aside from this change in the position of the heart, the primitive archaeogastropods have retained the circulatory plan of the ancestral mollusk, and they possess two auricles. However, in all other gastropods, the right auricle either has become vestigial or, in most instances, has disappeared as a result of the loss of the right gill, which supplied it with blood (Figs. 11–8 and 11–25). Contraction of the heart is myogenic, as in all mollusks. The ventricle gives rise to a posterior aorta supplying the visceral mass and an anterior aorta supplying the head and foot; or there may be a single short aorta, which then divides into an anterior and a posterior artery. An enlargement of the anterior vessel, or second heart, may be present in the head region, as in *Busycon*, and it probably functions in controlling blood pressure.

From the arterial sinuses, blood eventually collects in a large venous cephalopedal sinus (Fig. 11–30). Much of this blood passes through the nephridium before entering the branchial circulation, but some may return directly to the heart. In the return flow to the heart in pulmonates, all blood from the venous sinuses passes through the pulmonary capillary network in the roof of the lung (Fig. 11–23).

Prosobranchs and pulmonates possess the

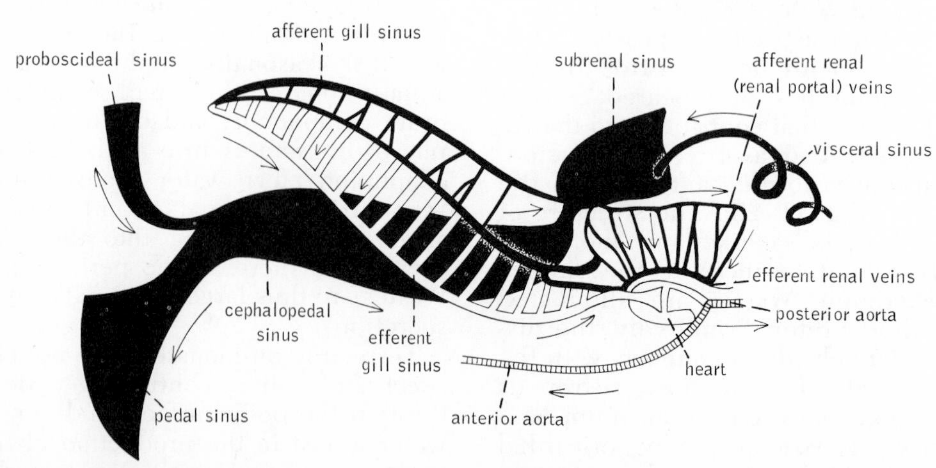

Figure 11–30. Diagram of the circulatory system of *Struthiolaria*, a mesogastropod prosobranch. (After Morton.)

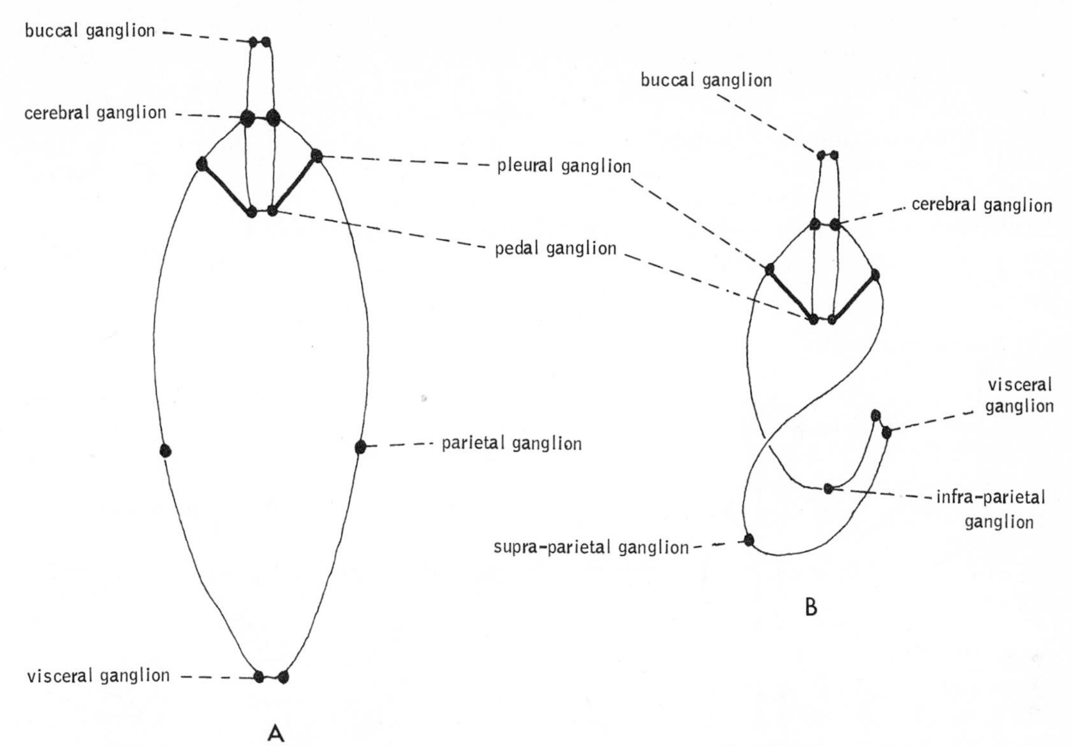

Figure 11–31. *A.* Hypothetical pre-torsion nervous system. *B.* Post-torsion nervous system.

respiratory pigment hemocyanin, which is dissolved in the plasma. Hemocyanin is a protein in which a molecule of oxygen is carried between two copper atoms. The functional units, which have a molecular weight of at least 50,000, are combined to form large molecules attaining molecular weights of several million. Oxyhemocyanin is pale blue; deoxyhemocyanin is colorless. The freshwater Planorbidae (pulmonates) possess hemoglobin instead of hemocyanin in the plasma. The opisthobranch gas transport provisions are poorly known, but the blood of the sea hare *Aplysia*, which has been extensively studied, is reported to lack respiratory pigments.

Nervous System. The nervous system of gastropods may be more easily understood if the ground plan is first described as if torsion had not taken place (Fig. 11–31A). A pair of adjacent cerebral ganglia lie over the posterior of the esophagus and give rise to nerves that connect anteriorly to the eyes, tentacles, statocysts, and a pair of buccal ganglia, which are located in the back wall of the buccal cavity. The buccal ganglia innervate the muscles of the radula and other structures in this vicinity. A nerve cord issues ventrally from each cerebral ganglion on each side of the esophagus. These are the two pedal connectives, which extend ventrally to a pair of ganglia located in the midline of the foot and which innervate the foot muscles.

A pair of connectives arise from the cerebral ganglia and extend back to a pair of pleural ganglia, which supply the mantle and the columella muscle. One pair of connectives join the pleural and pedal connectives. A second pair of cords leave the pleural ganglion and extend posteriorly until they terminate in a pair of visceral ganglia which are located in the visceral mass and which supply organs in this region. A pair of parietal, or intestinal, ganglia innervate the gills and osphradium; they are located along the length of the visceral nerves.

The Effect of Torsion. Although the visceral cords have just been described as bilateral, they are actually asymmetrical and twisted into a figure eight, as a result of torsion (Fig. 11–31*B*). This twisting reverses the original positions, so that the former right intestinal ganglion is now on the left, and the former left intestinal ganglion is now on the right. Furthermore, the present left visceral nerve and intestinal ganglion (the pre-torsional right ganglion) are located higher in the visceral mass than their opposite counterparts; the two ganglia are therefore sometimes called supra- and infra-intestinal ganglia. The pleural ganglia are not affected by torsion because of their anterior position.

The twisted condition is a primitive feature of the gastropod nervous system, since torsion has developed during the evolutionary history of all gastropods. Another primitive feature of the gastropod nervous system is the separation of the ganglia by nerve cords, as previously described. Such a primitive nervous system is found, with some modifications, in many prosobranchs, such as *Patella* and *Haliotis* (Fig. 11–32A). However, in most gastropods the original arrangement is obscured because of two evolutionary tendencies. First, there has been a tendency toward both concentration

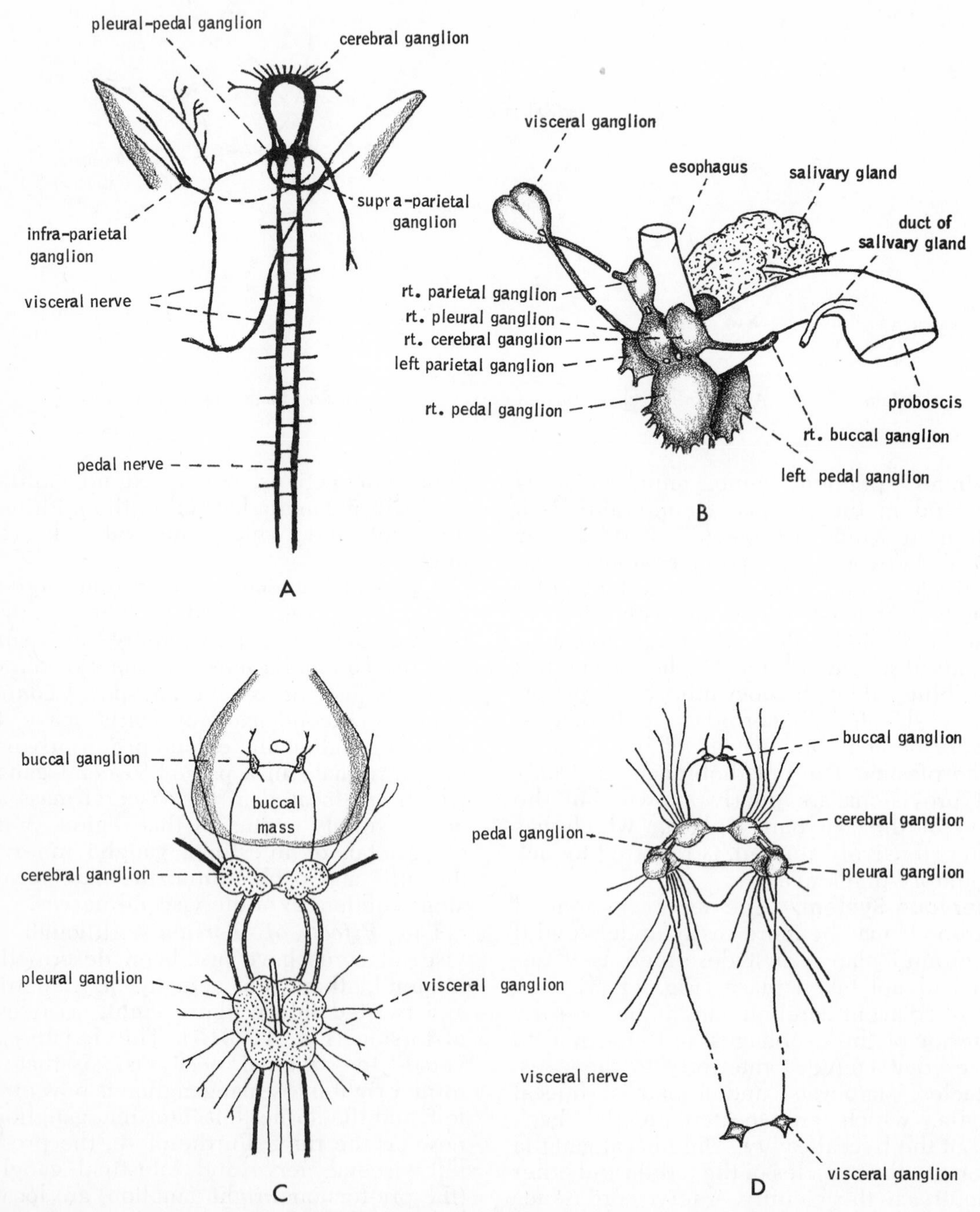

Figure 11–32. *A*. Primitive nervous system in *Haliotis*. (After Lacaze-Duthiers from Parker and Haswell.) *B*. Concentrated nervous system in *Busycon*. (After Pierce.) *C* and *D*. Secondarily symmetrical nervous systems. *C*. In the pulmonate, *Helix*. *D*. In the opisthobranch, *Aplysia*. (Both after Bullough.)

and fusion of ganglia with a consequent shortening of the connectives between ganglia. Second, there has been a tendency for the ganglia and cords to adopt a secondary bilateral symmetry.

Virtually every gastropod displays some degree of concentration of ganglia. The pleural ganglia are always located adjacent to the cerebral ganglia. Not infrequently, as in *Patella* and *Busycon,* the two visceral ganglia have fused to form a single center (Fig. 11–32*B*). Considerable concentration is present in higher prosobranchs, such as *Busycon,* where all ganglia except the visceral ones have migrated forward and are located around the esophagus and below the cerebral ganglia (Fig. 11–32*B*). All connectives between ganglia, except those between the parietal and visceral ganglia, have thus disappeared.

In the pulmonates the visceral ganglia have migrated forward (Fig. 11–32*C*). This shortening of all the connectives has resulted in a secondary bilateral symmetry of the nervous system.

Primitive cephalaspid opisthobranchs (*Acteon*) still display a twisted nervous system, but other opisthobranchs have an untwisted system in which the ganglia tend to be concentrated around the anterior end of the gut. In *Aplysia,* for example, the pleural, pedal, and cerebral ganglia are arranged around the gut, and the intestinal ganglia have fused and lie beside the fused visceral ganglia (Fig. 11–32*D*).

The ganglia of mollusks, annelids, and arthropods are organized differently from those of vertebrates. The periphery of a ganglion is occupied by cell bodies of unipolar motor neurons or interneurons, forming a rind or cortical layer. The interior of the ganglion, the neuropile, is filled with intertwined processes. A considerable number of the processes arise from the cell bodies of the cortical layer, but there are also present the processes of interneurons whose cell bodies are in other ganglia, and processes of sensory neurons whose cell bodies lie near receptor terminals. The parallel bundles of fibers that run between ganglia compose the nerve cords or connectives.

It is within the neuropile that synaptic connections occur, and not at the cell body as is often true in the vertebrates. The branching processes make possible multiple connections and contribute to the complexity of the circuitry. A single interneuron may have connections in the neuropiles of several ganglia, and it may receive and distribute impulses at many synaptic junctions. Some single interneurons, called "command" neurons, have such an array of connections that impulses from them generate relatively complex motor responses, such as swimming and turning. Sea hares and some other opisthobranchs have been the principal molluscan subjects of neurophysiological investigations of specific neuronal connections and control of behavioral reflexes. Interested students should consult D. Kennedy et al. (1969).

Sense Organs. The sense organs of gastropods include eyes, tentacles, osphradia, and statocysts. Eyes are characteristic of most gastropods, and one is typically located at the base of each cephalic tentacle. The conch *Strombus,* however, has large eyes located at the ends of long stalks, or peduncles, and the eyes are also carried at the ends of special optic tentacles in the higher pulmonates. Primitively, as in *Patella,* the eye is a simple pit containing photoreceptor and pigment cells (Fig. 11–33*A*), but in most higher gastropods the pit has become closed over and differentiated into a cornea and a lens (Fig. 11–33*B*). The tentacular eyes of gastropods are always direct; that is, the photoreceptor cells are directed toward the source of light. The most highly developed eyes are found among the pelagic heteropod prosobranchs. The large eyes of these fast-moving hunters are telescopic and are supposedly superior to the eyes of many fish. The eyes of most gastropods would appear to detect only changes in general light intensity.

A single pair of cephalic tentacles are displayed by prosobranchs, and two pairs occur in higher pulmonates and many opisthobranchs. In addition to bearing the eyes, the cephalic tentacles contain tactile and chemoreceptor cells. In opisthobranchs, especially nudibranchs, the upper half of the second pair bears plate-like folds that apparently increase the surface area for chemoreception. These modified opisthobranch tentacles are called rhinophores (Fig. 11–10*G*). Tentacles are not always restricted to the head; they may be found on the foot or the mantle margin. Tentacles are especially well developed around the margin of the skirt-like mantle of many limpets (Fig. 11–14*B*).

A pair of closed statocysts are located in the foot near the pedal ganglia of gastropods, although they may be innervated by the cerebral ganglia. Statocysts are absent from sessile forms.

The evolution of the osphradium of

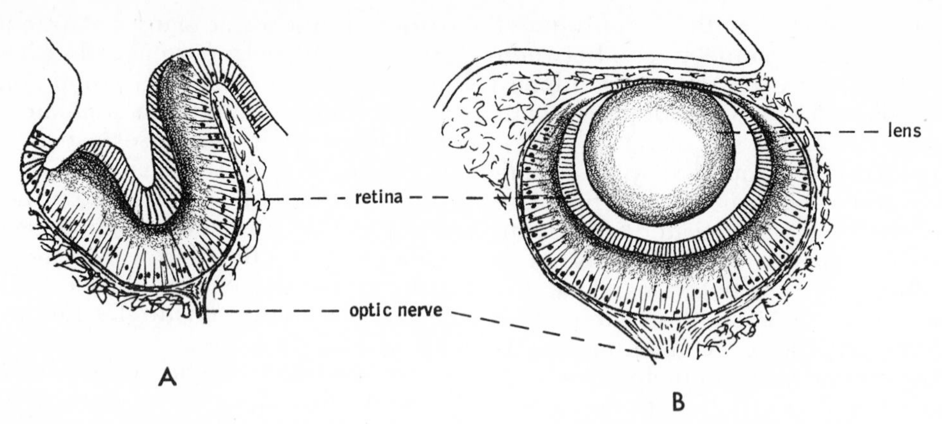

Figure 11–33. Eyes of two marine prosobranchs. *A. Patella*, a limpet. *B. Murex*. (Both after Helger from Parker and Haswell.)

gastropods closely parallels that of the gills. In the primitive Archaeogastropoda, an osphradium is present for each gill and is essentially like that described for the ancestral mollusk. In the other prosobranchs, which possess but one gill, there is also only one osphradium, which is located on the mantle cavity wall anterior and superior to the attachment of the gill (Figs. 11–25 and 11–29*A*). In most cases, the osphradium has become either filamentous or folded, thereby increasing the surface area. The osphradium is greatly reduced or has disappeared in those gastropods that have lost the gill, that possess a reduced mantle cavity, or that have taken up a strictly pelagic existence. This variation in the presence of the osphradium in gastropods tends to support Yonge's (1947) view that an important function of the osphradium is to detect sediment in the water passing over the gills. Yet the osphradium probably also functions in chemoreception. This organ is highly developed in prosobranch carnivores and scavengers, which can locate carrion, animal juices, or prey from a considerable distance, as much as 2 m. in the scavenger *Cominella*. *Nassarius* and other gastropods continually wave their siphon about as they move over the bottom, selecting and monitoring water from various parts of the environment.

Sea stars, snails, leeches, and other predators elicit marked escape reactions from some gastropod prey. The escape response is usually a rapid retraction into the shell and is apparently mediated through chemoreceptors, for related but nonpredatory species do not produce the response. In some instances contact is required; in other cases mere proximity of the predator or even a part of the predator, such as the isolated tube foot of a sea star, is sufficient to bring about the escape response.

Reproductive System. Many gastropods are dioecious. The single gonad, either ovary or testis, is located in the spirals of the visceral mass next to the digestive gland (Fig. 11–25). The gonoduct ranges from a very simple to a highly complex structure, but in all cases it has developed in close association with the right nephridium. In the primitive Archaeogastropoda, both nephridia are functional, and the right nephridium provides outlet for either the sperm or eggs (Fig. 11–34*A*). The gametes pass through a short duct that extends from the gonad and opens into the nephridium at various points; the gametes are then conducted by the nephridium into the mantle cavity through the nephridiopore. The genital duct is thus formed from two elements—the gonoduct proper and the right nephridium. In this type of reproductive system, the eggs are provided, at most, with gelatinous envelopes produced by the ovary, since the gonoduct is not sufficiently specialized to produce tertiary membranes. There is thus no need for copulatory organs, and fertilization takes place in the sea water after the eggs are swept out of the mantle cavity.

In all the other gastropods, the right nephridium has degenerated except for a portion that functions solely as part of the genital duct. Furthermore, the genital duct has become considerably lengthened by a third addition, derived from the mantle; as a result, the genital pore is located at the opening of the mantle cavity. According to Fretter (1946), this third section, which might be called the pallial duct, probably first arose as a ciliated groove extending from the nephridiopore, but in at least the females of existing gastropods the groove has become closed over to form a distinct tube (Fig. 11–34*B*). In many species, the pallial duct has become separated from the mantle surface. It is the pallial portion of the genital duct that has undergone elaboration or differentiation to provide for sperm storage and for egg membrane formation.

In reproductive systems where the production of tertiary egg membranes by the pallial section of the oviduct takes place, a penis has evolved in the male so that fertilization takes place before the formation of the egg capsules or cases. This penis is a long cylindrical extension or fold of the body wall arising just behind the right cephalic tentacle. In some gastropods, including the neritids, littorinids, and cerithids, the vas deferens continues to open at the back of the mantle cavity, and the pallial section appears in part as a glandular ciliated groove, which extends to the tip of the penis (Fig. 11–25). In others, the groove has closed over, and the vas deferens opens at the penile tip. At least a portion of the pallial section forms a prostate for producing seminal secretion. The duct from the testis is usually coiled and provides a storage place for the sperm before copulation. The entire male duct thus consists of: a coiled duct from the testis, representing the original gonoduct; a short renal contribution to the vas deferens; and the pallial vas deferens, containing the prostate, which runs in the floor of the mantle cavity.

In the female, the pallial section of the oviduct is modified to form both an albumin gland and a large jelly gland or capsule gland (Fig. 11–34*B*). Species such as *Cerithium* and *Littorina* imbed the eggs in jelly masses produced by the jelly gland; but in most of the higher prosobranchs, the eggs are enclosed in a capsule. At the end of the pallial oviduct proximal to the gonad is a seminal receptacle for the reception and storage of sperm from the penis. Also, the eggs are fertilized at this point before they enter the secretory portion of the oviduct. In a great many groups, such as *Littorina* and neogastropods, the sperm are not directly received by the seminal receptacle; instead, during copulation, they enter a copulatory bursa at the distal end of the oviduct (Fig.

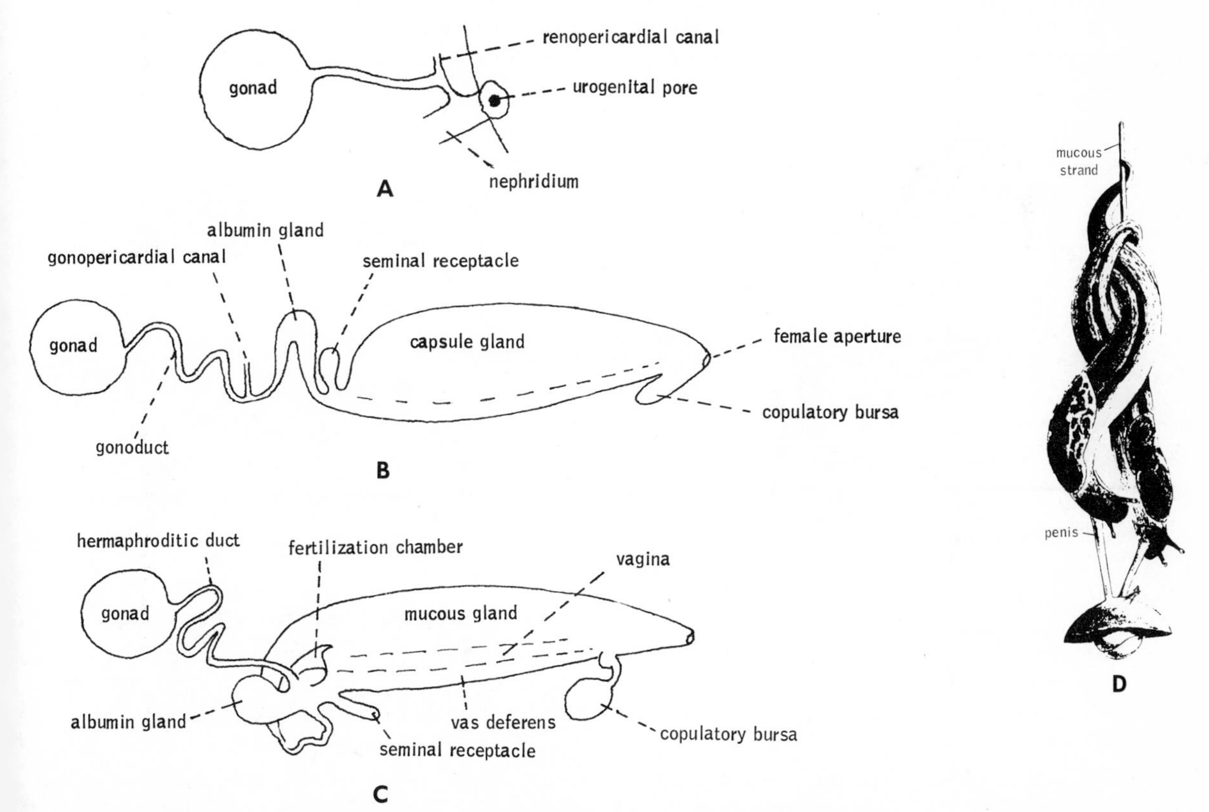

Figure 11–34. Reproductive tracts of three gastropods. *A*. A trochid, a primitive dioecious prosobranch. *B*. Female system of *Nucella*, a prosobranch. *C*. The sea hare, *Aplysia*, a monoecious opisthobranch. (All after Fretter.) *D*. Copulating slugs (*Limax maximus*), suspended on a mucous strand. Penes are unrolled. (After Adams from Hyman, L. H., 1967: The Invertebrates, Vol. 6, p. 602.)

11–34*B*). From the bursa the sperm later pass posteriorly along a ventral channel in the oviduct and enter the seminal receptacle, or the bursa becomes the definitive seminal receptacles and the original seminal vesicle functions to digest excess sperm. Not uncommonly, the opening to the bursa is adjacent to the opening of the gonopore, so that there are two female gonopores.

The outgoing eggs pass through the dorsal portion of the oviduct, where membranes are applied. Commonly, there is a groove that conducts eggs from the female gonopore down to the foot for attachment to the substratum.

A small number of prosobranchs are hermaphroditic. Protandric hermaphrodites include some limpets and the slipper shells (*Crepidula*). In the more or less sessile slipper shells, individuals tend to live stacked up on one another (Fig. 11–35*A*). The right shell margins are adjacent, thereby permitting the penis of the upper individual to reach the female gonopore of the individual below. Young specimens are always males. This initial male phase is followed by a period of transition in which the male reproductive tract degenerates; the animal then develops into a female or another male. The sex of each individual is influenced at least partly by the sex ratio of the association. An older male will remain male longer if it is attached to a female. If such a male is removed or isolated, it will develop into a female. The scarcity of females influences certain of the males to become females. Once the individual becomes female, it remains in that state.

Pulmonates are simultaneous hermaphrodites, and opisthobranchs are simultaneous or protandric hermaphrodites. The reproductive systems are very complicated and display endless variations. Only a few generalizations can be made here. In opisthobranchs the single gonad, or ovitestis, of simultaneous hermaphrodites contains egg-producing and sperm-producing follicles, and a common hermaphroditic duct leads from the ovitestis to a seminal vesicle (ampulla). From the seminal vesicle in some species there is a common gonoduct and gonopore opening into the mantle cavity. Beyond the gonopore a sperm groove extends to the penis behind the right tentacle. In other opisthobranchs there is a separate sperm duct and oviduct distal to the seminal receptacle, and the sperm duct terminates in the penis. In either case, the system is provided with a bursa, glands (mucous, albumin, and prostate), and tracts for the

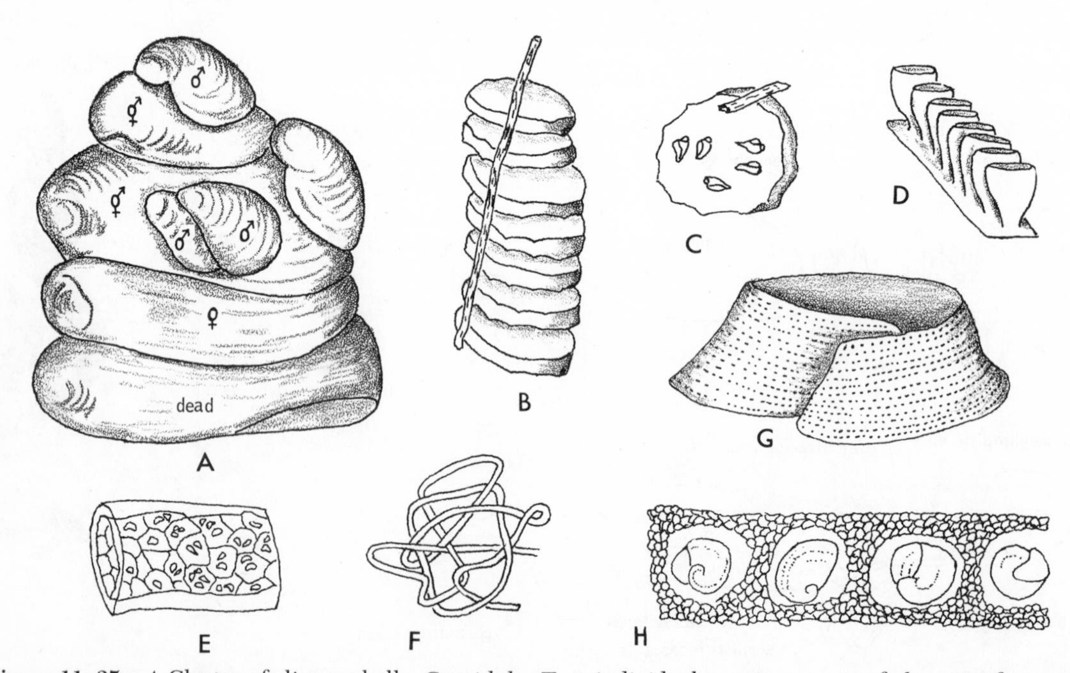

Figure 11–35. *A*. Cluster of slipper shells, *Crepidula*. Two individuals are in process of changing from male to female sex. Male at top will fertilize individual below when female phase is reached. (After Coe.) *B* and *D*. Parchment egg cases. *B* and *C*. *Busycon*. *D*. *Conus*. *E* and *F*. Gelatinous egg string of sea hare, *Aplysia*. *E*. Enlarged section. *F*. Entire string. *G* and *H*. Sand-grain egg case of the moon shell, *Lunatia heros*. *G*. Entire case. *H*. Section of case. (*B* to *H* after Abbott.)

separation of gametes; and the penis may be complicated in various ways, including the presence of a stylet.

Copulation with mutual sperm transfer is the rule for most opisthobranchs. The sea hare *Aplysia* forms chains of copulating animals in which a given member of the chain receives sperm from the animal behind and provides sperm for the one in front. Insemination by hypodermic impregnation occurs in some opisthobranchs.

The hermaphroditic system of pulmonates is built on the same general plan as that of many opisthobranchs. There is an ovotestis, a common hermaphroditic duct containing the seminal vesicle, and a separate sperm duct and oviduct (Fig. 11–23). Although many species have separate male and female gonopores, in some, including many land pulmonates, the distal end of the male system rejoins the female system to empty through a common gonopore. The penis of pulmonates may be a complicated structure. Details of the reproductive system are important in the systematics of the group, and occasionally the system must be dissected out for species identification.

Copulation in freshwater snails involves one member acting as a male and one as a female. In land pulmonates there is mutual exchange of sperm, usually in the form of spermatophores, and copulation is commonly preceded by a "courtship" involving circling, oral and tentacular contact, and intertwining of the bodies. Bizarre sexual behavior occurs in some species, particularly slugs. In the shelled Helicidae (*Helix*) the vagina contains an oval dart sac, which secretes a calcareous spicule. When a pair of snails are intertwined, one member drives its spicule dart into the body wall of the other. Copulation follows this rather drastic form of stimulation. A new dart is secreted later. Copulating limacid slugs hang intertwined from a cord of mucus attached to a tree trunk or branch, and in the process of exchanging sperm the penes are unrolled to a length of 10 to 25 cm. (and sometimes as much as 85 cm.) and are twisted together at the tips (Fig. 11–34*D*).

Although sperm are received in the distal seminal receptacle, fertilization takes place at the end of a common hermaphroditic duct. Self-fertilization has been recorded for a number of species.

Embryogeny. The eggs of primitive prosobranchs (archaeogastropods) are fertilized externally and are either shed singly or aggregated in gelatinous masses. However, all other gastropods possess internal fertilization, typically following copulation involving the insertion of a penis into the female gonopore. Egg deposition in gelatinous strings, ribbons, or masses is characteristic of most mesogastropods and opisthobranchs, but neogastropods and some mesogastropods embed their eggs in an albumin mass surrounded by a capsule or case, which is usually attached to the substratum. The size and shape of the case, the nature of the wall (which may be leathery or gelatinous), and the number of cases attached together are extremely variable and are characteristic of the species (Fig. 11–35).

Aquatic pulmonates deposit their eggs in gelatinous capsules. Terrestrial species produce a small number of large, separate, yolky eggs, which are often provided with thin calcareous shells. The eggs are laid in a heap in leaf mold or in other damp, sheltered locations. A number of land snails are reported to be ovoviviparous, the eggs being brooded in the oviduct.

The free-swimming trochophore larva is found only in primitive Archaeogastropoda that shed their eggs directly into the sea water; in all the other gastropods, the trochophore stage is suppressed and is passed before hatching.

More characteristic of marine gastropods is a free-swimming larval type called a veliger. The veliger larva is derived from a trochophore but represents a more complex stage of development (Fig. 11–36). The characteristic feature of the veliger is the swimming organ, called a velum, which consists of two large semicircular lobes bearing long cilia. The velum forms as an outward extension of the prototroch of the trochophore. Other structures of the veliger larva show a greater degree of development than those in the trochophore larva. The foot, eyes, and tentacles appear; the shell, which first formed in the trochophore, develops spirally in the veliger because of unequal growth; the stomodeum develops into the buccal structures and esophagus, and connects with the stomach; and larval retractor muscles are formed.

The veliger larva is a suspension feeder, and the long cilia of the velum function not only in locomotion but also in feeding. The beating of the long velar cilia brings fine plankton in contact with the shorter cilia of a subvelar food groove. Within the food groove particles become entangled in

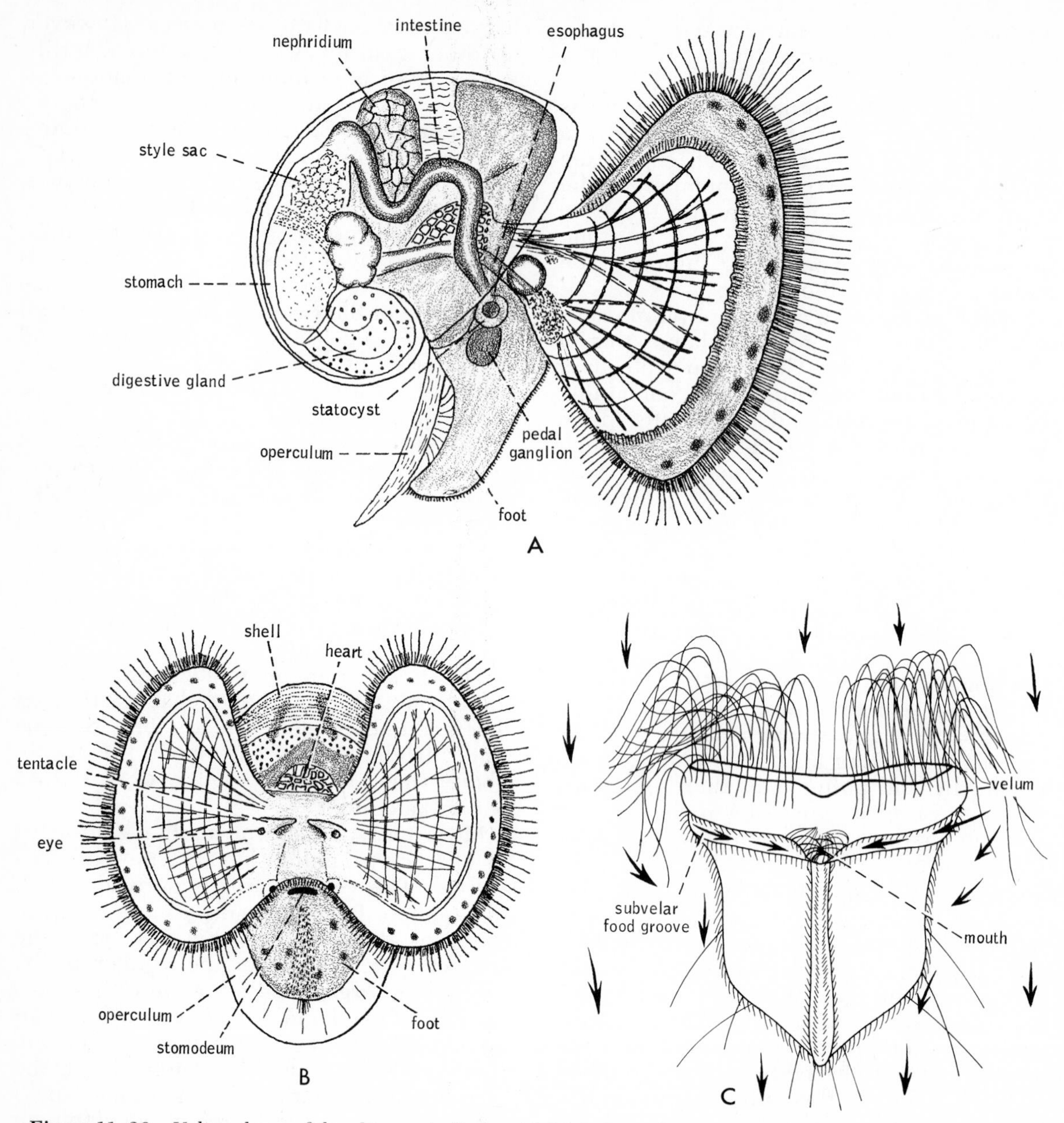

Figure 11–36. Veliger larva of the slipper shell, *Crepidula*. *A*. Lateral view. *B*. Frontal view. (Both after Werner from Raven.) *C*. Suspension feeding in the early veliger of *Archidoris*. Arrows indicate direction of movement of food particles. (After Thompson.)

mucus and are conducted to the mouth (Fig. 11–36*C*).

During the course of the veliger stage, torsion occurs and the shell and visceral mass twist 180 degrees in relation to the head and foot. Torsion may be very rapid (only about three minutes in the marine limpet *Acmaea*), or it may be a gradual process (ten days in the terrestrial prosobranch *Pomatias*). In many archaeogastropods, such as *Haliotis*, *Patella*, and *Calliostoma*, torsion takes place in two 90-degree phases, in which the first is relatively rapid and the second more gradual.

When torsion is rapid, the development of the larval retractor muscle appears to be the primary cause of the torsion (Fig. 11–3*C* and *D*). This muscle is composed of six large spindle-shaped muscle cells, which connect as a bundle to the pre-torsional right side of the shell apex. The opposite end of the bundle fans out into the foot rudiment, the mantle, and the stomodeal region. Four of these cells are sickle-shaped when they first develop, but they straighten out just before they reach a functional state. Torsion coincides with, and is probably largely caused by, the asymmetrical growth or straightening of these cells, which forces the visceral mass and foot to twist.

As development proceeds, the veliger reaches a point at which not only can it swim by means of the velum, but the foot is also sufficiently formed to allow creeping. Settling and metamorphosis occur. The velum is lost and the final features of the adult form are attained. Settling sites are of critical importance in the survival of the larvae, and many species can delay metamorphosis until specific types of substrates can be reached. Some nudibranchs, for example, must contact certain species of hydroids, bryozoans, or ascidians before metamorphosis will occur. Metamorphosis appears to be induced by chemical rather than physical characteristics of the substratum.

There is little recapitulation of torsion and detorsion in opisthobranch development. However, in nudibranchs and other forms that are shell-less as adults, a shell appears in the veliger and then is cast off during metamorphosis.

Some marine prosobranchs, especially the Neogastropoda, nearly all freshwater prosobranchs, and almost all pulmonates have no free-swimming larvae; but both trochophore and veliger stages can be recognized within the egg. At hatching, a tiny snail emerges from the protective shell or case.

Many gastropods reach adult size and sexual maturity at six months to two years, but slow growth continues and larger species may not reach maximum size for many years. The life span is highly variable: 5 to 16 years for the limpet *Patella vulgata;* 4 to 10 years for the periwinkle *Littorina littorea;* 3 years for the intertidal deposit feeder *Nassarius obsoletus;* 1 to 2 years for many freshwater pulmonates; 5 to 6 years for the land snail *Helix aspersa;* and only 1 year for many nudibranchs.

Economic Importance. Gastropods serve as food for numerous other animals, particularly as veliger larvae and newly settled young; but the principal predators of aquatic gastropods are fish, aquatic birds, and mammals, many of which are adapted to a molluscan diet. Land snails serve as food for birds, salamanders, and small mammals.

Various species of gastropods are utilized as food by man in different parts of the world. The west coast abalone, *Haliotis,* is consumed in California, but state law prevents its exportation from the state. Conch is eaten in southern Florida. North Americans, except for patrons of French restaurants, have never acquired a taste for snails, but in southern Europe the large land snail, *Helix pomatia,* is considered a delicacy.

Of greater importance to human nutrition are the destructive activities of some gastropods. The predation of muricids on oysters has already been mentioned. Certain species of land snails and slugs do extensive damage to cultivated plants. Slugs attack both the underground and aerial portions of plants and are a common pest in greenhouses and in truck gardens. The introduction of carnivorous snails has been used as a means of controlling destructive species in some truck farming areas.

From a medical standpoint, snails are of considerable importance as intermediate hosts for many trematode parasites of man and domesticated animals. Developmental stages of blood flukes, liver flukes, and lung flukes take place in freshwater snails in various parts of the world where these diseases are epidemic (p. 173). Marine snails also serve as hosts for trematode parasites of shore birds and fish.

SYSTEMATIC RÉSUMÉ OF CLASS GASTROPODA

Subclass Prosobranchia. Marine, freshwater, and terrestrial forms in which the mantle cavity and contained organs are located anteriorly. Aquatic species possess one or two gills within mantle cavity. Shell and usually an operculum present. Mostly dioecious.

Order Archaeogastropoda (Aspidobranchia). Primitive forms in which there are commonly two gills, two auricles, and two nephridia. The right gill may be reduced or absent, but even if only one gill is present, it is bipectinate. Shell coiled or secondarily symmetrical in limpets. Largely marine. Forms with slit or perforated shells—*Pleurotomaria, Scissurella, Haliotis;* the keyhole limpets—*Emarginula, Fissurella, Diodora;* patellacean limpets—*Acmaea, Patella, Lepeta, Lottia;* Trochacea (top shells)—*Trochus, Calliostoma, Monodonta, Turbo, Margarites.* The Neritacea includes marine species (*Nerita*) as well as tropical forms which have invaded fresh water (*Theodoxus*) and land (Helicinidae).

Order Mesogastropoda (Pectinibranchia). Possess a single unipectinate gill, one auricle, and one nephridium; these organs on the right side have disappeared. Chiefly marine, but many freshwater and terrestrial genera. This is the largest order of gastropods and contains many common species: periwinkles—*Littorina;* worm shells—*Vermetus, Vermicularia, Siliquaria; Cerithium, Bittium, Batillaria;* violet shells—*Janthina;* slipper shells and allies—*Crepidula, Capulus, Hipponix;* carrier shells—*Xenophora; Struthiolaria;* conches—*Strombus, Lambis;* cowries—*Cypraea;* pelagic heteropods—*Atlanta, Carinaria;* moon shells—*Polinices, Natica, Lunatia;* tun and helmet shells—*Tonna, Cassis;* freshwater operculate snails—*Valvata, Viviparus, Campeloma, Ampullarius, Pila;* tropical operculate land snails—Cyclophoridae, Pomatiasidae.

Order Neogastropoda (Stenoglossa). Members of this order also have a single unipectinate gill, but differ from mesogastropods in having a bipectinate osphradium, a concentrated nervous system, and a shell with a siphonal notch or canal. Mostly carnivorous species having a proboscis with a radula containing three large teeth in a transverse row. Marine. Boring muricids—*Murex, Urosalpinx* (oyster drill), *Eupleura, Purpura, Thais; Anachis;* whelks—*Buccinum, Neptunea, Busycon, Nassarius, Bullia;* Olives, volutes, and harp shells—*Oliva, Voluta, Harpa;* the poisonous toxoglossans—*Conus, Turris, Terebra.*

Subclass Opisthobranchia. Have one gill, one auricle, and one nephridium, but display detorsion. Reduction and loss of shell and mantle cavity common. Many are secondarily bilaterally symmetrical. Head commonly bears two pairs of cephalic tentacles. Hermaphroditic. Mostly marine.

Order Cephalaspidea. Bubble shells. Dorsal surface of head shield-like. Shell generally present but reduced, absent in some. *Acteon, Hydatina, Philine, Scaphander, Bulla.*

Order Pyramidellacea. Ectoparasites of bivalve mollusks and polychaetes. Shell and operculum present. Proboscis contains a stylet instead of a radula. *Pyramidella, Odostomia, Brachystomia.*

Order Acochlidiacea. Small naked species with no gills, and having the visceral mass sharply set off from the rest of the body. Some members of this order found in fresh water. *Acochlidium.*

Order Philinoglossacea. Similar to acochlidiaceans except that the visceral mass is not sharply set off from rest of body.

Order Anaspidea, or Aplysiacea. Sea hares. Large opisthobranchs with more or less bilaterally symmetrical external form. Reduced shell buried in mantle; gill and mantle cavity present; foot with lateral parapodia. *Aplysia, Bursatella, Akera.*

Order Notaspidea. Shelled or naked opisthobranchs. Gill present. *Pleurobranchus.*

Order Sacoglossa. Shelled and slug-like opisthobranchs with a radula bearing a single row of teeth adapted for suctorial feeding on algae. *Elysia, Alderia;* the bivalved *Berthelinia.*

Order Thecosomata. Shelled pteropods, or sea butterflies. Shelled pelagic species with large parapodia. *Limacina, Cavolina, Spiratella, Clio, Cymbulia, Gleba.*

Order Gymnosomata. Naked pteropods. Pelagic species with no shell or mantle cavity. Parapodial fins present. *Pneumoderma, Cliopsis.*

Order Nudibranchia. Nudibranchs, or sea slugs. Shell and mantle cavity absent, and body secondarily bilaterally symmetrical. Doridaceans, with secondary gills around anus—*Doris, Chromodoris, Glossodoris, Jorunna, Onchidoris.* Dendronotaceans, with simple to branched cerata—*Tritonia, Dendronotus.* Eolidaceans, with simple cerata—*Eolidia, Glaucus.*

Order Ochidiacea. Slug-like intertidal opisthobranchs having a pulmonary sac for aerial gas exchange. *Onchidium.*

Order Parasita. Worm-like endoparasites of sea cucumbers. *Entoconcha, Enteroxenos.*

Subclass Pulmonata. One auricle and one kidney; gills absent, and mantle cavity, which is on right side, is converted into a vascularized chamber for gas exchange in air or secondarily in water. Nervous system concentrated and symmetrical. Shell usually present, but operculum lacking. Hermaphroditic.

Order Basommatophora. Pulmonates with one pair of tentacles; eyes located near tentacle base. Primarily freshwater forms, a few marine. Marine limpets—*Siphonaria, Otina;* marine *Amphibola,* the only operculate pulmonate; freshwater snails—*Lymnaea, Planorbis, Helisoma, Bulinus, Physa;* freshwater limpets—*Ancylus, Ferrissia.*

Order Stylommatophora. Pulmonates with two pairs of tentacles, the second pair bearing eyes at the tip. Terrestrial. *Partula, Achatina, Retinella, Polygyra, Helix, Bulimus.* Land slugs—*Arion, Limax, Deroceras, Phylomycus, Testacella.*

CLASS MONOPLACOPHORA

The members of the following two classes, the Monoplacophora and the Polyplacophora, are similar to gastropods in possessing a flat creeping foot; both are considered to be primitive groups of mollusks.

In 1952, ten living specimens of *Neopilina*, a group of mollusks previously known only from Cambrian and Devonian fossils, were dredged from a deep ocean trench off the Pacific coast of Costa Rica. Since this initial discovery, specimens belonging to seven different species have been collected from deep water (2000 to 7000 m.) in various parts of the world—South Atlantic, Gulf of Aden, and a number of localities in the eastern Pacific.

Neopilina belongs to a group of mollusks known as the Monoplacophora. As the name implies, the monoplacophorans possess a single symmetrical shell, which varies in shape from a flattened shield-like plate to a short cone. The monoplacophorans had been classified with either the chitons or the gastropods, but all shared a peculiar and distinctive feature. The undersurface of the shell displayed three to eight muscle scars.

The living specimens of *Neopilina* are a little more than 3 cm. long and externally resemble a combination of gastropod and chiton. There is a single, large, bilaterally symmetric shell, in which the apex is directed anteriorly (Fig. 11–37*B*). The undersurface of *Neopilina* very much resembles that of a chiton (Fig. 11–37A). The head is reduced in size, and a broad flat foot is present. A pallial groove (the mantle cavity) separates the edge of the foot from the mantle on each side. The mouth is located in front of the foot, and the anus is located in the pallial groove at the posterior of the body. In front of the mouth is a preoral fold, or velum, which extends laterally on each side as a rather large ciliated palp-like structure. Another fold lies behind the mouth and projects to either side as a mass of postoral tentacles.

The pallial groove contains five or six pairs of unipectinate gills (Fig. 11–37*D*). It seems very probable that the ventilating current is similar to that of chitons (p. 372).

Internally, eight pairs of pedal retractor muscles are present. The digestive system includes a radula and a subradula organ within the buccal cavity. The stomach contains a style sac and style; the intestine is greatly coiled. Stomach contents consist of diatoms, forams, and sponge spicules.

Blood from the gills passes through the nephridia and then into two pairs of auricles (Fig. 11–37*E*). The first pair receives blood from the first four pairs of gills and the second from the last pair of gills. The auricles open into two ventricles, one on each side of the rectum. The ventricles surround the rectum but do not communicate with it. Thus the heart is completely paired. Each ventricle gives rise to an anterior aorta, which shortly fuses with its mate to form a single typical anterior vessel. The heart is surrounded by a paired pericardial coelom, the median partition following the line of division of the ventricles.

Unlike other mollusks, there is a pair of large flattened sacs of unknown derivation and function lying in front of the pericardial coelom. Six pairs of nephridia are located on each side of the body. Each pair contains a central sac from which extend numerous secretory lobules and diverticula. Except for the first pair, each nephridium opens into the coelom; three pairs open into the dorsal sacs, and two pairs into the pericardial coelom. The nephridiopores open into the pallial groove. The nervous system is essentially similar to that of chitons. The only ganglia are located at the anterior end, and the pedal and lateral cords are connected by ten pairs of transverse commissures.

The sexes are separate, and two pairs of gonads are located in the middle of the body. Each gonad connects through a gonoduct to one of the two pairs of kidneys in the middle of the body. Fertilization obviously occurs externally, but whether it takes place within the pallial groove is unknown.

The survival of species of *Neopilina* is undoubtedly correlated with their adaptation for life at great depths, and they are perhaps considerably more specialized than were other members of the class. Fossil species appear to have evolved along two lines. In one group (subclass Cyclomya) there was an increase in the dorsoventral axis of the body, leading to a planospiral shell and a reduction of gills and retractor muscles (Fig. 11–37*F*). Although they disappeared from the fossil record in the Devonian, this group may have been ancestral to the gastropods. The other line (subclass Tergomya) retained

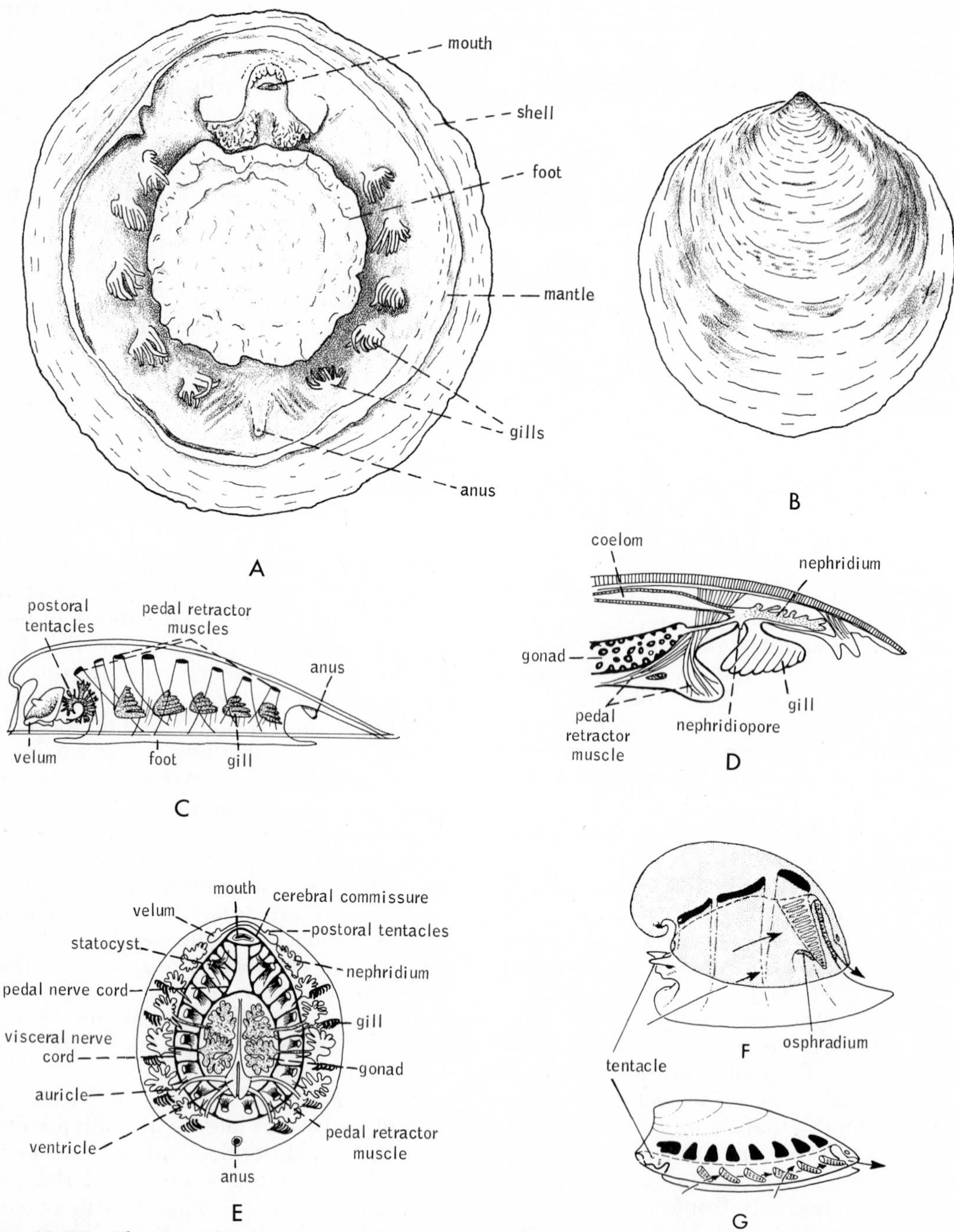

Figure 11–37. The monoplacophoran, *Neopilina*. *A*. Ventral view. *B*. Dorsal view of shell. *C*. Side view. *D*. Transverse section through one half of body. *E*. Internal anatomy. (All adapted from Lemche and Wingstrand.) *F*. Reconstruction of a cyclomyan monoplacophoran. Black bars indicate point of attachment of retractor muscles to shell, and arrows indicate possible ventilating current. *G*. Reconstruction of a tergomyan monoplacophora. (*F* and *G* from Stasek, C. R., 1972: *In* Florkin, M., and Scheer, B. J. (Eds.), Chemical Zoology, Vol. 7. Academic Press, N.Y., p. 21.)

a flattened shell with five to eight retractor muscles (Fig. 11–37*G*). Although this group also disappeared from the fossil record in the Devonian, it is believed to have survived in the genus *Neopilina*.

CLASS POLYPLACOPHORA

The class Polyplacophora contains the chitons. Although some features of their structure and embryogeny are primitive, chitons have become highly adapted for adhering to rocks and shells. They are bilaterally symmetrical, with an ovoid body that is greatly flattened dorsoventrally (Fig. 11–38*A*). There are no cephalic eyes or tentacles, and the head is very indistinct. The mantle, usually called the girdle in chitons, is very heavy, and the foot is broad and flat to facilitate adhesion to hard substrata.

Chiton species range in size from a few centimeters or less, as in the case of the little Atlantic coast chiton *Chaetopleura apiculata* (Fig. 11–38*A*), to over 30 cm. in the giant Pacific species *Cryptochiton stelleri* (Fig. 11–39*A*). However, most species are 3 to 12 cm. in length. Chitons are usually not brightly colored; drab shades of red, brown, yellow, and green are common.

Shell and Mantle. The most distinctive characteristic of chitons is the shell, which is divided into eight overlapping transverse plates. From the nature of the shell is derived the name of the class, Polyplacophora—bearer of many plates. Each plate is similar to the others except for the first and last, the cephalic and anal plates (Fig. 11–38). Except for the overlapping posterior edge, the margins of each plate are covered by mantle tissue. Thus, the shell of chitons is partially imbedded in the mantle.

The degree of lateral mantle reflexion varies. In *Lepidochitona, Placiphorella,* and *Chaetopleura* (Fig. 11–38*A*), most of the plate width is exposed. In *Katharina,* only the midsection of each plate is uncovered; thus the visible part of the shell appears as a narrow median strip extending over the dorsal surface of the animal (Fig. 11–39*A*). In *Symmetrogephyrus,* only the apex of each plate is visible. Finally, in *Cryptochiton,* the shell is completely covered by the mantle (Fig. 11–39*A*).

The shell plates are composed of several layers (Fig. 11–40). The upper layer, called the tegmentum, consists of an organic conchiolin matrix impregnated with calcium carbonate and is sculptured on its exposed surface. The tegmentum is covered by a thin periostracum. Beneath the tegmentum lies a thicker, denser layer, called the hypostracum, which is composed entirely of calcium carbonate. The anterior articulating projections of each plate are composed only of the hypostracum, which in these specialized areas of the plate is called the articulamentum. In chitons whose mantle partially or completely covers the shell plates, there is a corresponding reduction of the tegmentum.

The peripheral area of the mantle, called the girdle, is very heavy and extends a considerable distance beyond the lateral margins of the plates. It is believed to be homologous to the middle mantle lobe of other mollusks (p. 377). The girdle surface displays a variety of ornamentation. It may be naked and smooth, or covered with scales, bristles, or calcareous spicules (Fig. 11–38*C*). The bristles or spicules may be so long and dense

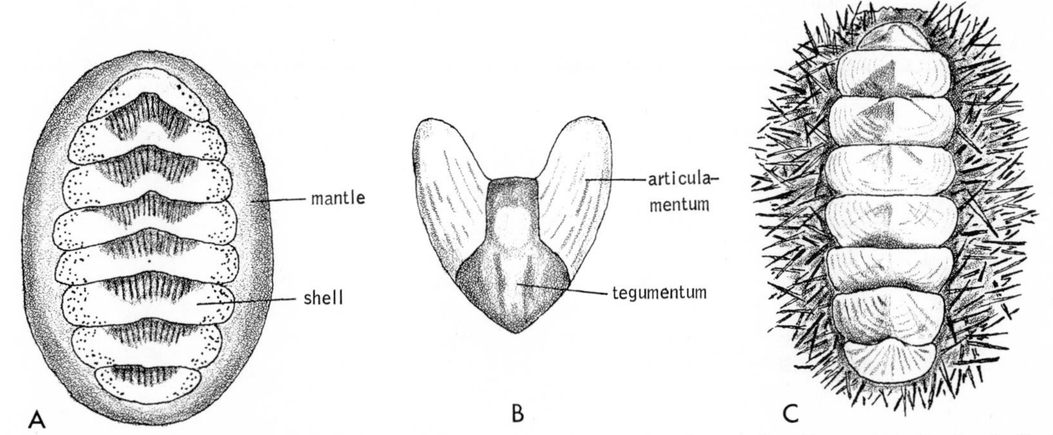

Figure 11–38. *A.* Common Atlantic coast chiton, *Chaetopleura apiculata.* (After Pierce.) *B.* Single shell plate of *Katharina. C. Chiton.* (After Borradaile and others.)

Figure 11–39. *A.* Two species of chitons from the northwest Pacific. The shell plates of the larger species (*Cryptochiton*) are completely covered by the mantle; those of the smaller species (*Katharina*) are partially covered. *B.* The West Indian chiton, *Chiton tuberculatus*, exposed at low tide. (Both by Betty M. Barnes.)

that the animal has a mossy or shaggy appearance.

Foot and Locomotion. The broad flat foot occupies most of the ventral surface and functions in adhesion as well as in locomotion (Fig. 11–41*A*). Chitons creep slowly in the same manner as snails. The division of the shell into transverse plates and their articulation with one another enable them to move over and adhere to a sharply curved surface. Chitons roll up into a ball if dislodged, and although this may serve as a defense mechanism, it also enables the animal to right itself.

Adhesion is effected by both the foot and the girdle. The foot is responsible for ordinary adhesion, but when a chiton is disturbed, the girdle is also employed. The girdle is clamped down especially tightly against the substratum, and the inner margin is then raised. This creates a vacuum that enables the animal to grip the substratum with great tenacity.

Chitons are sluggish animals and may remain in one area for a long period. They are commonly rocky intertidal inhabitants and, like limpets, are motionless for hours when exposed at low tide (Fig. 11–39*B*). When the rock surface is submerged, they move about to feed. They are negatively phototactic and thus tend to locate themselves under rocks and ledges. They are most active at

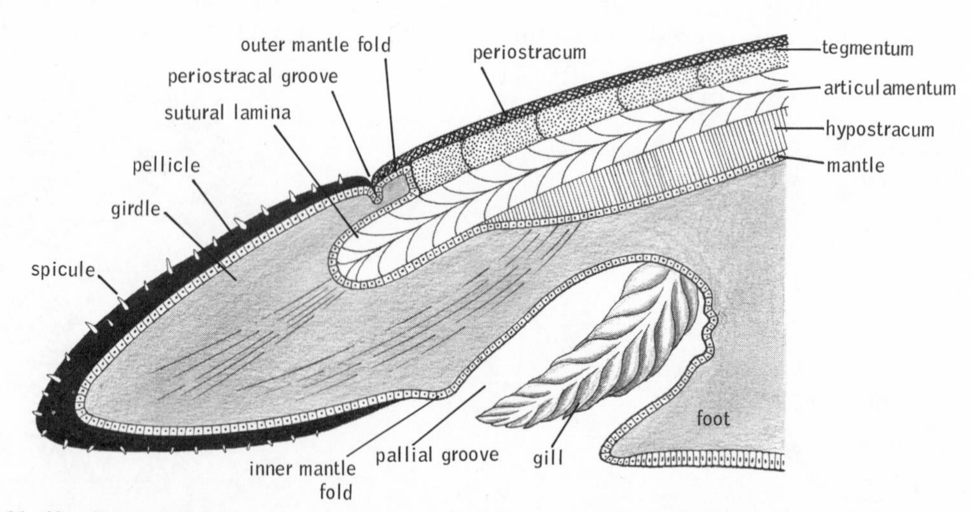

Figure 11–40. Diagrammatic transverse section through girdle, mantle (pallial) groove, and foot on one side of the body of a chiton. (Modified from Mutvei.)

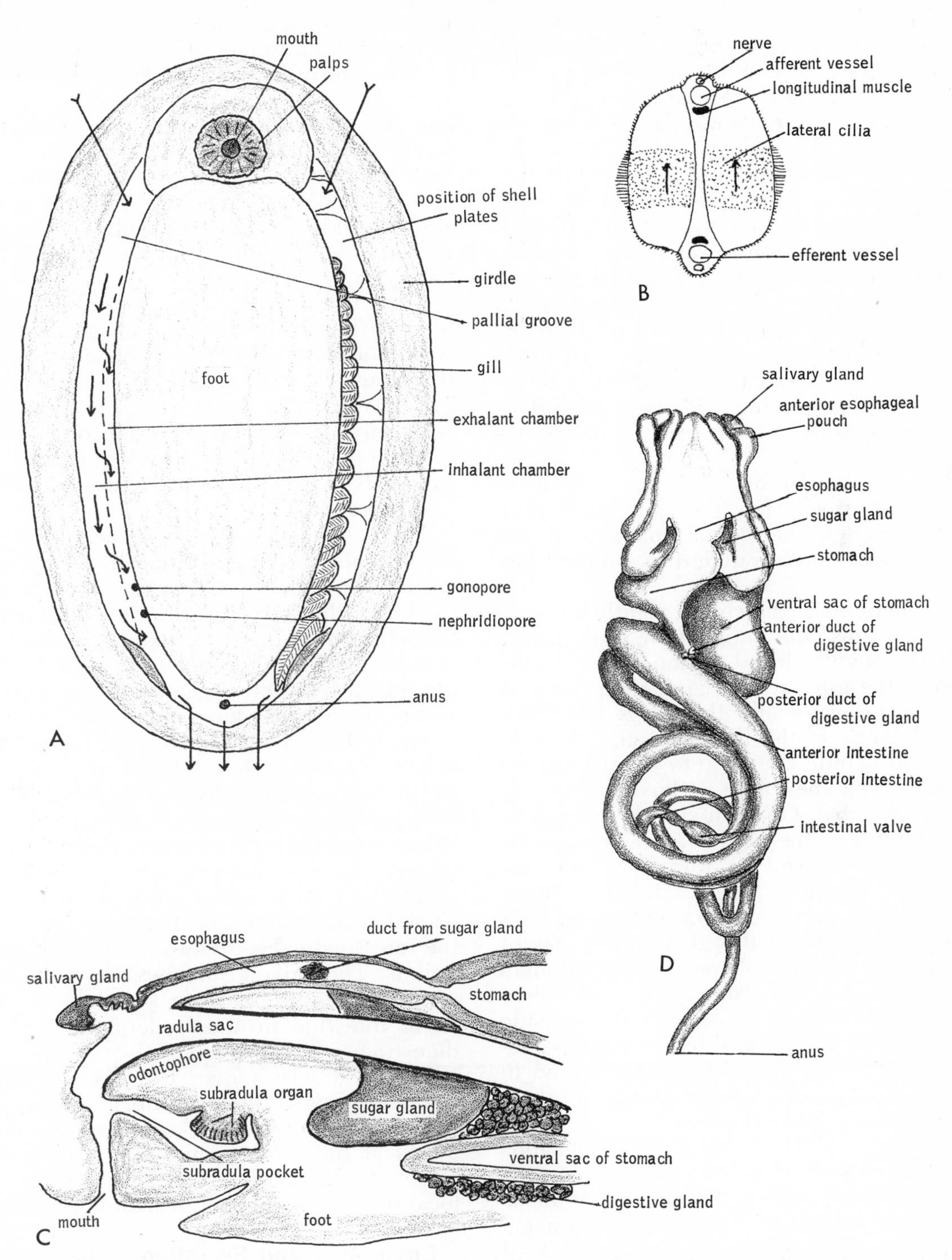

Figure 11–41. The chiton, *Lepidochitona cinerea*. *A*. Mantle groove, showing direction of water currents (ventral view). *B*. Transverse section through gill axis. Arrows indicate direction of water current over filament surface. (*A* and *B* after Yonge.) *C*. Buccal region (lateral view). *D*. Digestive tract (dorsal view). (*C* and *D* after Fretter.)

night if the tide is right. Some species have been reported to exhibit homing.

Water Circulation and Gas Exchange. The mantle cavity of chitons consists of a trough, or groove, on each side of the body between the foot and the mantle edge (Fig. 11–41*A*). This condition is correlated with the extreme dorsoventral flattening of the body, one of the adaptations of chitons for life on rocky surfaces. A large number of paired gills are arranged in a linear series within the two mantle troughs (Fig. 11–41*A*). The number of pairs varies in different species and even within a single species. For example, the number of gills in *Lepidopleurus asellus* varies from 11 to 13 pairs, and in *Tonicella marmorea* from 19 to 26 pairs. The size and number of gills is apparently related to the gas exchange surface area required by animals of a particular volume.

Structurally, chiton gills are similar to those described for the ancestral mollusk, although the lateral margins of the filaments are more rounded in chitons (Fig. 11–41*B*).

The margins of the chiton mantle are held tightly against the substratum, making the pallial groove a closed chamber. Within the groove, the gills, which hang from the roof, curve downward, and their tips touch the lower margin of the foot. The gills thus divide the mantle trough into a ventrolateral inhalant chamber and a dorsomedian exhalant chamber. On each side of the mantle trough toward the anterior end, the mantle margins are raised to form two inhalant openings. Water enters each inhalant chamber through these anterior openings, runs along the course of the groove, and passes up through the gills into each dorsal exhalant chamber. The two exhalant currents converge posteriorly and pass to the outside through one or two exhalant openings created by the locally raised mantle. Arrows in Figure 11–41*A* indicate the course of water currents in *Lepidochitona*.

Nutrition. Most chitons are microphagous and feed on fine algae and other organisms that they scrape from the surface of rocks and shells with the radula. The radula bears 17 teeth in each transverse row, and the lateral teeth are capped with magnetite (iron). The mouth is located on the ventral surface in front of the foot and opens into the chitin-lined buccal cavity (Fig. 11–41*A*). A long radula sac opens into the back of the buccal cavity, as does a smaller more ventral evagination called the subradula sac (Fig. 11–41*C*). The latter contains a cushion-shaped sensory structure, the subradula organ, hanging from the roof. A pair of salivary glands open through the anterior dorsal wall of the buccal cavity.

When a chiton feeds, the subradula organ is first protruded and applied against the rock. If food is present, the odontophore and its radula project from the mouth and scrape. After each stroke, the subradula organ is protruded and tests the substratum again. The salivary secretion contains no enzymes and acts merely as a lubricant for the radula and as a medium for transporting food particles.

From the buccal cavity, mucus-entangled food particles enter the esophagus and are carried along a ciliated food channel toward the stomach (Fig. 11–41*C* and *D*). During this passage, the food particles are mixed with amylase secretions produced by a pair of large pharyngeal glands (or sugar glands), the ducts from which open at the beginning of the esophagus.

The esophagus opens into the anterior of an irregularly shaped stomach, which contains a large ventral sac. In the stomach, the food is further mixed with proteolytic secretions from the digestive gland. Digestion is almost entirely extracellular and takes place in the digestive gland, in the stomach, and in the anterior intestine.

The anterior intestine loops and then joins a large coiled posterior intestine (Fig. 11–41*D*). Between these two intestinal divisions is a sphincter, forming an intestinal valve. Waste leaving the intestine is compacted in mucus that is secreted along the entire length of the digestive tract. As the compacted waste enters the posterior intestine, the intestinal valve divides it into short fecal pellets. The valve also slows down the passage of food through the stomach and anterior intestine to allow adequate time for digestion and absorption.

Ciliary tracts move the fecal pellets through the posterior intestine, where further consolidation of waste takes place. The anus opens at the midline just behind the posterior margin of the foot, and the egested fecal pellets are swept out with the exhalant current.

Circulation and Excretion. The pericardial cavity is large and located beneath the last two shell plates (Fig. 11–42). A single pair of auricles collects blood from all of the gills; only one aorta is present, and it issues anteriorly from the ventricle.

The two nephridia are large and extend anteriorly on each side of the body as long,

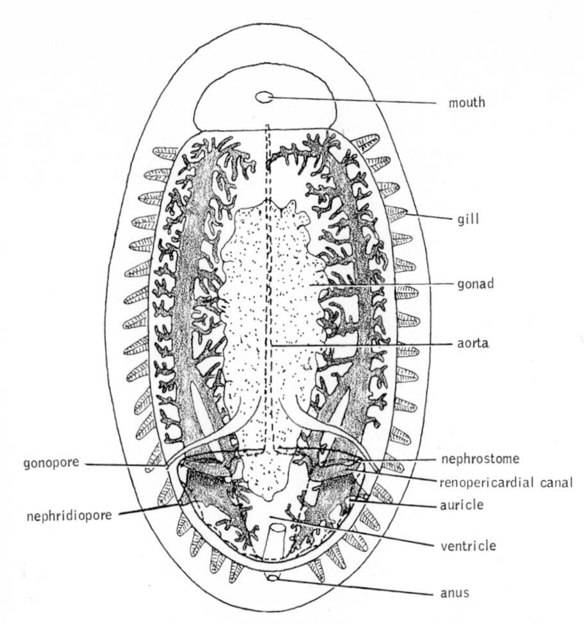

Figure 11–42. Internal structure of a chiton. (After Lang and Haller.)

somewhat **U**-shaped tubes. Their walls are greatly evaginated to form many diverticula that ramify in the hemocoel. Each tube connects with the pericardial cavity, and the nephridiopore opens into the pallial groove on each side between two of the more posterior pairs of gills (Fig. 11–41*A*).

Nervous System and Sense Organs. The nervous system is primitive (Fig. 11–43*A*). Ganglia are lacking or at least poorly developed. A nerve ring surrounds the esophagus, and nerves issue from it anteriorly to innervate the buccal cavity and subradula organ. Posteriorly the nerve ring gives rise to a large pair of pedal nerve cords, innervating the muscles of the foot, and a large pair of lateral, or palliovisceral, nerve cords, innervating the mantle, visceral organs, and shell aesthetes. The palliovisceral nerve cords are interconnected posteriorly to form a complete nerve ring, and both pairs of nerve cords give off a large number of lateral interconnecting nerves.

Statocysts, cephalic eyes, and tentacles are absent. The chief sense organs are the subradula organ and the aesthetes. Aesthetes, which are unique to chitons, are mantle sensory cells that are lodged within vertical canals in the upper tegmentum (Fig. 11–43*B*). The canals and sensory endings terminate beneath a chitinous cap on the surface of the shell plates. The aesthetes provide the exposed dorsal surface of the animal with sensory reception, despite the covering shell plates and the reduction of the head.

The complexity of the aesthetes varies considerably. Some aesthetes are simple receptors of uncertain function. But there are chitons that also possess more complex

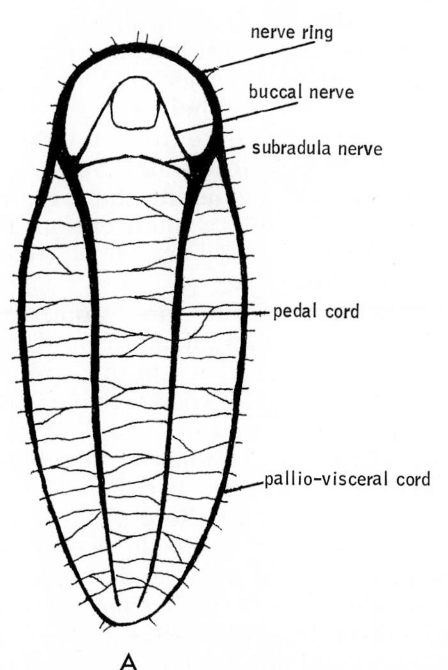

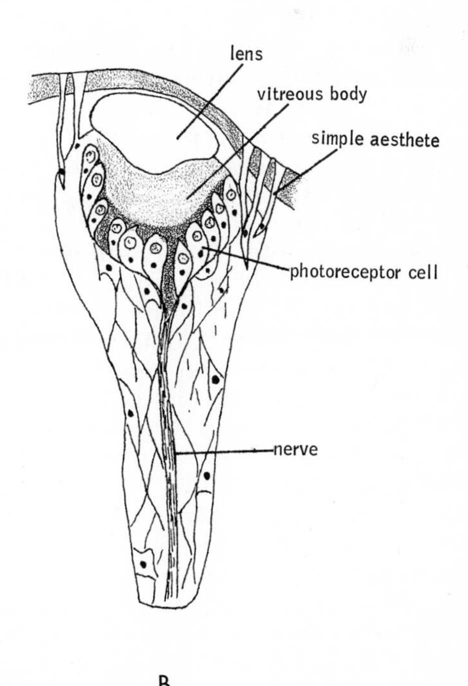

Figure 11–43. *A*. Nervous system of a chiton. (After Thiele from Parker and Haswell.) *B*. Aesthete eye of *Acanthopleura*. (After Nowikoff from Parker and Haswell.)

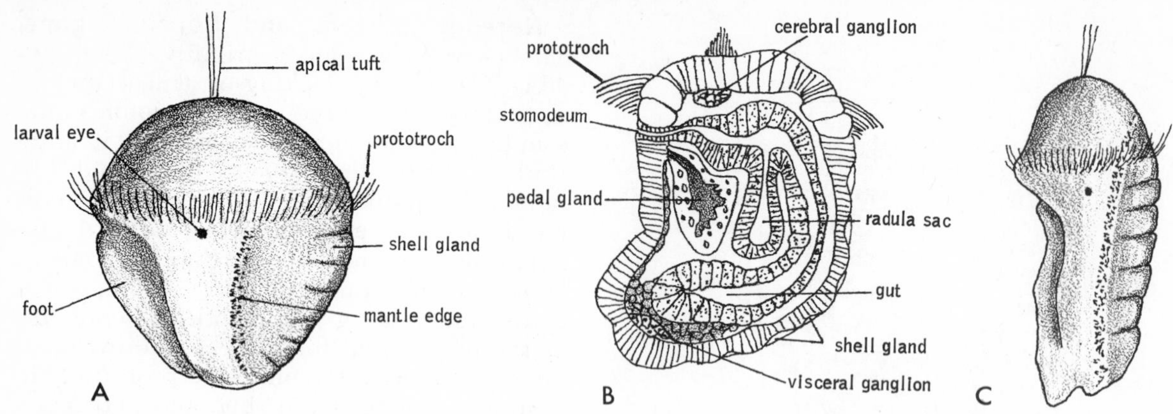

Figure 11–44. Development of chitons. *A*. Trophophore of *Ischnochiton*. *B*. Section through trochophore of *Chiton*. *C*. Metamorphosis of *Ischnochiton*. (*A* and *C* after Heath from Dawydoff; *B* after Kowalevsky from Dawydoff.)

aesthetes, consisting of bundles of sensory cells that form an ocellus. These shell eyes may be composed of merely a cluster of photoreceptor cells and pigment, or they may be modified to contain a cornea, lens, and retina, as in *Acanthopleura*. These chiton ocelli may number in the thousands per individual and are especially concentrated on the anterior shell plates. It has been reported that young chitons are negatively phototactic and that this response may disappear in older animals as the aesthetes become covered by organisms encrusting the shells.

The mantle surface that extends beyond the margins of the shell plates is liberally supplied with tactile and photoreceptor cells even when this surface is ornamented with scales and spines.

Patches of sensory epithelium are located at various places on the walls of the mantle trough in chitons and may function like the osphradia of other mollusks.

Reproduction and Embryogeny. Most chitons are dioecious. A single median gonadial organ is located in front of the pericardial cavity beneath the middle shell plates (Fig. 11–42). The gametes are transported to the outside by two special gonoducts, instead of by the nephridia. A gonopore is located in each pallial groove in front of the nephridiopore (Fig.11–41A).

Copulation does not take place. Sperm leave the male in the exhalant currents and fertilization occurs in the sea or within the mantle cavity of the female. The usual gregariousness of chitons facilitates fertilization. The eggs, which are enclosed within a spiny envelope, are shed into the sea either singly or in strings. In some species the eggs are brooded within the mantle cavity, and *Callistochiton viviparus* gives birth to its young, which have undergone development in the ovary.

Typical spiral cleavage leads to a free-swimming trochophore, except in some of those forms that brood their eggs (Fig. 11–44*A*). A shell gland develops on the dorsal surface of the post-trochal region and is very soon divided by transverse grooves (Fig. 11–44*B*). The foot develops on the side opposite the shell gland, and a pair of larval eyes form just behind the prototroch.

A primitive feature of chiton embryogeny is the direct metamorphosis of the trochophore into the adult form. There is no veliger stage. In the metamorphosis of the chiton trochophore, the post-trochal region elongates to form the major part of the body (Fig. 11–44*C*). The shell gland extends up into the pretrochal region, and the shell plates are formed in the transverse grooves. The prototroch degenerates, and the animal sinks to the bottom as a young chiton. The larval eyes are retained for some time after metamorphosis.

Distribution. There are approximately 600 existing species of chitons. The fossil record, which dates back to the upper Cambrian period, is rather sparse. Some 350 fossil chitons are known.

The majority of chitons are inhabitants of shallow water and are common throughout the world wherever there is a hard substratum. The rocky Pacific coast of the United States, Canada, and Alaska has a rich fauna of over 100 species. Chitons are abundant in the West Indies, but only a few species are found further north along the Atlantic coast of the United States. Some chitons live in deep water; *Lepidochitona bentha* has been dredged from a depth of 4200 m.

CLASS APLACOPHORA

The class Aplacophora comprises some 130 species of strange worm-shaped mollusks, called solenogasters. They are found throughout the oceans of the world to depths of 9000 m. Some burrow in mud bottoms and others creep on hydroids and corals. Most specimens have been collected by dredging, and the biology of the group is poorly known.

Solenogasters are usually less than 5 cm. in length. The head is poorly developed and the typical molluscan features of shell, mantle, and foot are absent (Fig. 11–45). However, the integument contains layers of embedded calcareous spicules. The worm-shaped body is derived by the in-rolling of the mantle margins ventrally. In those species that live on hydroids and corals or on the mud surface, there is a midventral longitudinal groove containing one or more ridges (Fig. 11–45*A* and *B*), which are probably homologous to the foot of other mollusks. The groove is absent from the cylindrical bodies of burrowing forms. The posterior end of the body contains a cavity into which the anus and a pair of coelomoducts empty. The posterior cavity is believed to represent a mantle cavity, and in burrowing species it houses a pair of gills (Fig. 11–45*C*).

The gut is a more or less straight tube, but the buccal cavity contains a radula. The burrowers appear to be selective carnivores and scavengers. Other solenogasters feed upon the coelenterates on which they live.

The aplacophorans are hermaphroditic or dioecious; copulation probably occurs in those with the former condition, and spawning in the latter.

To what extent aplacophorans are specialized and to what extent they are primitive mollusks is uncertain. Some features, such as the similar nervous systems, suggest an affinity with the chitons, and in the past the two groups were placed together in the class Amphineura. Their position within the phylum will be discussed further at the end of the chapter.

CLASS BIVALVIA

The class Bivalvia, also called Pelecypoda or Lamellibranchia, is comprised of mollusks known as bivalves and includes such common forms as clams, oysters, and mussels. Bivalves are all laterally compressed and possess a shell with two valves, hinged dorsally, that completely enclose the body. The foot, like the remainder of the body, is also laterally compressed, hence the origin of the name Pelecypoda—hatchet foot. The head is greatly reduced in size. The mantle cavity is the most capacious of any class of mollusks, and the gills are usually very

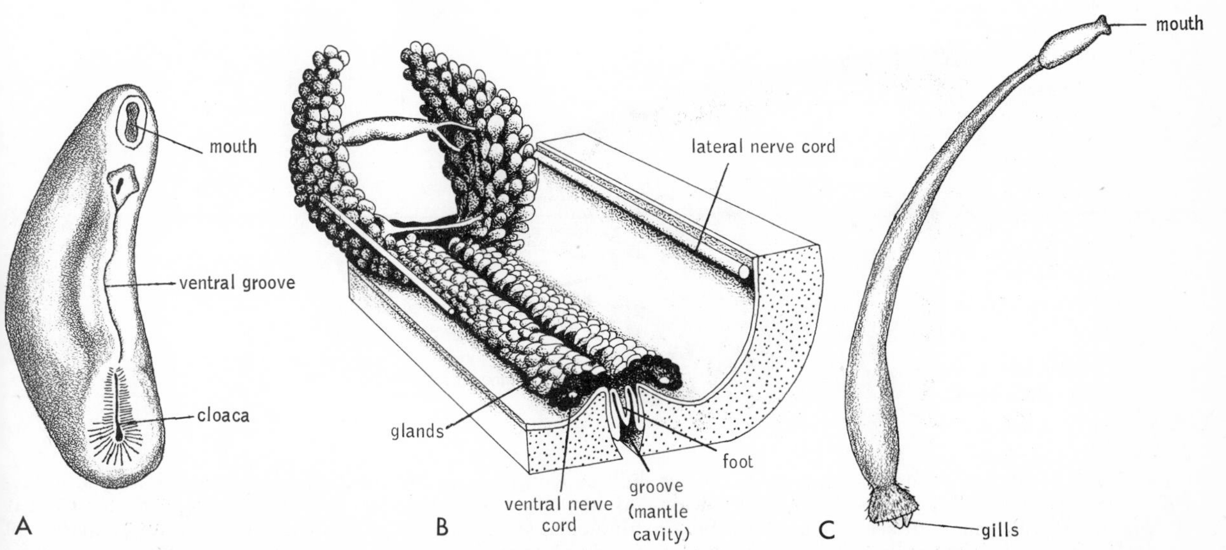

Figure 11–45. Aplacophorans. *A. Neomenia.* (After Hansen from Parker and Haswell.) *B.* Cross section of ventral body portion of *Proneomenia antarctica.* (After Hoffman.) *C. Crystallophrisson.* (After Simroth from Parker and Haswell.)

large, having assumed in most species a food-collecting function in addition to that of gas exchange. Most of these characteristics represent modifications that enabled bivalves to become soft bottom burrowers, for which the lateral compression of the body is better suited. Although modern bivalves have invaded other habitats, the original adaptations for burrowing in mud and sand have taken bivalves so far down the road of specialization that they have become largely chained to a relatively sendentary existence.

Bivalvia consists of three subclasses: the Protobranchia, the Lamellibranchia, and the Septibranchia. The protobranchs are generally believed to be the most primitive of existing bivalves. The septibranchs are highly specialized and have lost their gills. The lamellibranchs contain the majority of the bivalve species.

Shell, Mantle, and Foot. A typical bivalve shell consists of two similar, more or less oval, usually convex valves, which are attached and articulate dorsally with each other (Fig. 11–46*A* and *B*). Each valve bears a dorsal protuberance called the umbo,

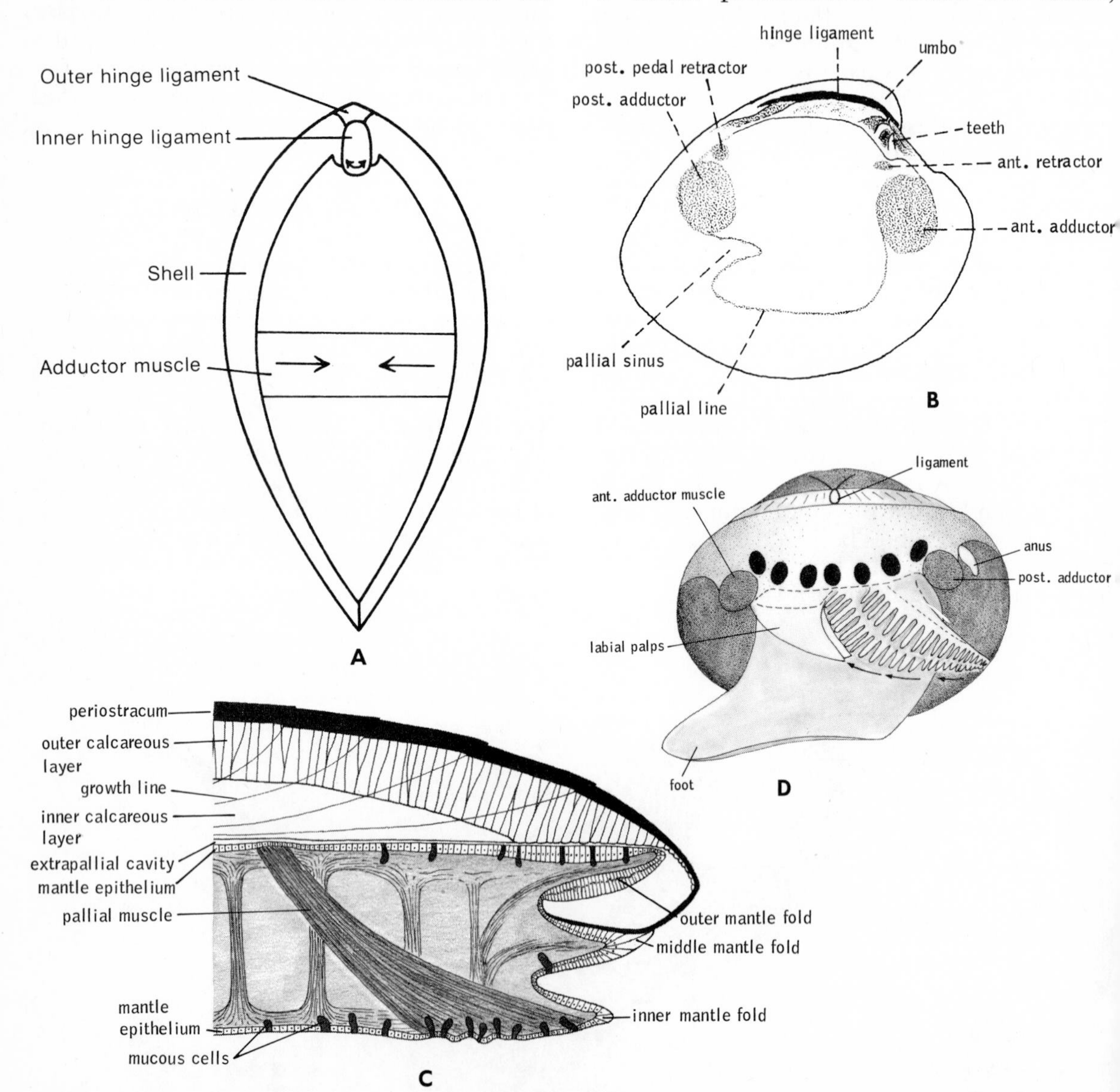

Figure 11–46. *A*. Transverse section of a bivalve shell, showing antagonistic functions of hinge ligaments and adductor muscles. When valves are closed by adductor muscles, outer hinge ligament is stretched and inner ligament is compressed. *B*. Inner surface of left valve of the marine clam *Mercenaria*. *C*. Transverse section through margin of shell and mantle of a bivalve, showing mantle lobes and points of shell secretion. Although periostracum is shown pressed against middle mantle lobe, this layer is actually secreted by inner side of outer mantle lobe. (After Kennedy.) *D*. Reconstruction of the Ordovician bivalve *Babinka*, which possessed multiple pedal retractors. (From Stasek, C. R., 1972: *In* Florkin, M., and Scheer, B. J. (Eds.), Chemical Zoology, Vol. 7. Academic Press, N. Y., p. 33.)

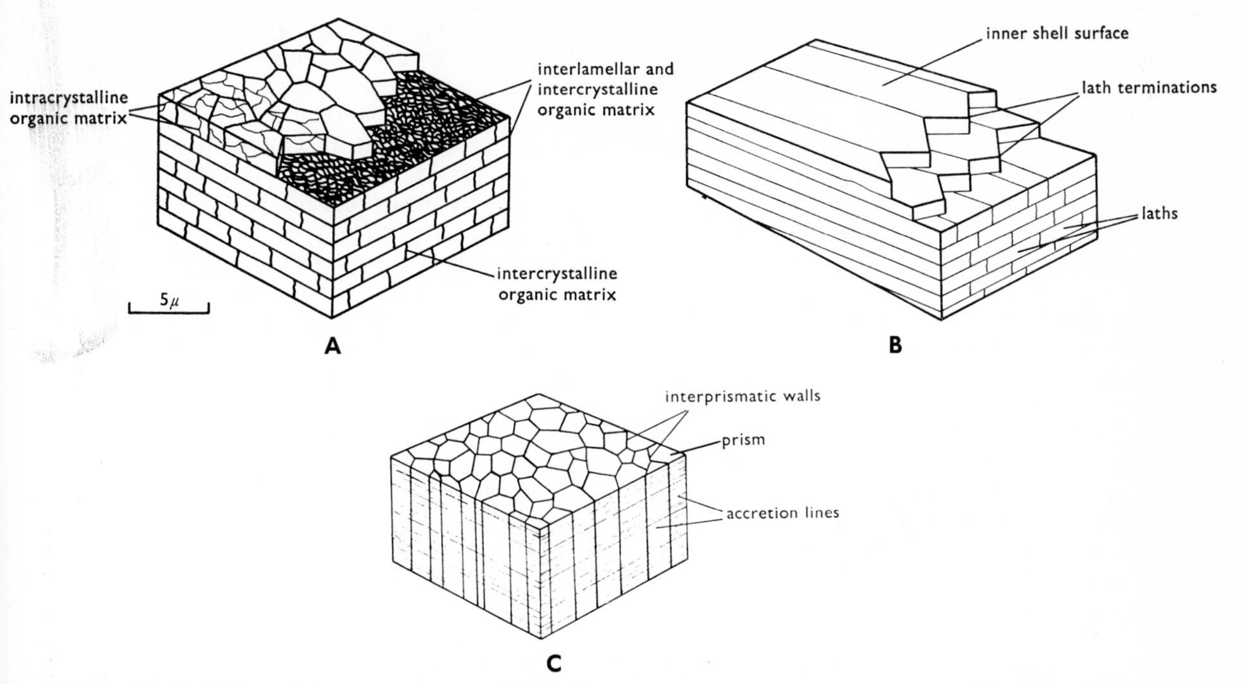

Figure 11–47. Three of the various types of shell ultrastructure. *A*. Nacreous structure. *B*. Foliated structure built of laths. *C*. Prismatic structure. (From Kennedy, W. J., et al., 1969: Biol. Rev., *44*:499–530.)

which rises above the line of articulation. The umbo is the oldest part of the shell, and the concentric lines around it are lines of shell growth. The two valves are attached by an elastic band called the hinge ligament, which is composed of the same horny organic material (conchiolin) as the periostracum and which is continuous with the periostracum. The hinge ligament is so constructed that when the valves are closed, the dorsal or outer part is stretched and the ventral or inner part is compressed (Fig. 11–46*A*). Thus, when the adductor muscles relax, the ligament causes the valves to open. To prevent lateral slipping, the two valves in most species are locked together by teeth and ridges, and opposing sockets and grooves, located on the hinge line of the shell beneath the ligament (Fig. 11–46*B*).

The valves of the shell are pulled together by two large dorsal muscles, called adductors, which act antagonistically to the hinge ligament (Fig. 11–46*A* and *B*). The adductors extend transversely between the valves anteriorly and posteriorly; there are scars on the inner surface of the valves where these muscles are attached.

The bivalve shell is composed of an outer periostracum covering two to four calcareous layers. The latter may be entirely aragonite or a mixture of aragonite and calcite, and they may be deposited as prisms or as minute laths or tablets arranged in sheets (nacre), lenses, or more complex forms (Fig. 11–47). The basic elements—prisms, tablets, and so forth—are always deposited within an organic framework. Although shell structure is not uniform for the class, it is constant and characteristic for different groups of bivalves (Fig. 11–46*C*).

The shells of bivalves exhibit an infinite variety of sizes, shapes, surface sculpturing, and colors. The shells range in size from the tiny seed shells of the fresh-water family Sphaeriidae, which usually do not exceed 2 mm. in length, to the giant clam *Tridacna* (Fig. 11–67*A*) of the South Pacific, which attains a length of over a meter and may weigh over 1100 kg.

Like the shell, the mantle greatly overhangs the body, and it forms a large sheet of tissue lying beneath the valves (Fig. 11–65). The edge of the mantle bears three folds—an inner, middle, and outer fold (Fig. 11–46*C*). The innermost fold is the largest and contains radial and circular muscles. The middle fold is sensory in function. The outer fold is related to the secretion of the shell. Shell secretion has been most studied in bivalves. The inner surface of the outer fold lays down the periostracum, and the outer surface secretes the first calcareous layer. The entire mantle surface secretes the remaining calcareous portion.

The mantle epithelium is in actual contact with the shell surface only in the periostracal groove and at points of muscle attachment (Fig. 11–46*C*). Elsewhere, there is a minute extrapallial space between mantle and shell. It is into this space that shell materials are first secreted; from the extrapallial fluid of this space are then deposited both the calcareous elements and the surrounding organic framework (Fig. 11–48).

The mantle is attached to the shell by the muscle fibers of the inner lobe along a semicircular line a short distance from the shell edge. The line of mantle attachment is impressed on the inner surface of the shell as a scar, called the pallial line (Fig. 11–46*B*).

Despite the attachment of the mantle, occasionally some foreign object, such as a sand grain or a parasite, lodges between the mantle and the shell. The object then becomes a nucleus around which are laid concentric layers of nacreous shell. In this manner a pearl is formed. If the nucleus is infolded within the mantle and moved about during secretion, the pearl becomes spherical or ovoid. Commonly, however, the developing pearl adheres to or even becomes completely imbedded in the shell.

Pearls can be produced by all shell-bearing mollusks, but only those with shells having an inner nacreous layer produce pearls of commercial value. The finest natural pearls are produced by the pearl oysters, *Pinctada margaritifera* and *Pinctada mertensi,* which inhabit most of the warmer Pacific areas. Bead pearls are produced by purposely introducing a shell bead between the mantle and shell of the pearl oyster. The oyster then lays down a calcareous coating approximately one millimeter thick around the bead. Most cultured pearls are started by placing a microscopic globule of liquid or ground "seeds" of unionid (freshwater clam) shells into an oyster. The resulting year-old seed pearl is then transplanted in another oyster. A pearl of marketable size is obtained three years after transplantation.

The foot of most bivalves has become compressed, bladelike, and directed anteriorly as adaptations for burrowing. Foot movement, which will be described in more detail later in connection with burrowing, is effected by a combination of blood pressure and muscle action. Engorgement of the foot with blood, coupled with the action of a pair of pedal protractor muscles, produces extension. The protractors extend from each side of the foot to the opposite valve, where they are attached to the shell just below and behind the anterior adductor muscle (Fig. 11–46*B*). Withdrawal of the foot is effected by the contraction of an anterior pair and a posterior pair of retractors, which are also attached to the foot and the shell. The scars of the pedal retractors are situated just above those of the adductors (Fig. 11–46*B*). Bivalve pedal retractors and protractors are homologous to the pedal retractors of other mollusks. The shells of some early bivalves show multiple pedal muscle scars (Fig. 11–46*D*).

Other specializations of the bivalve shell, mantle, and foot will be described later in connection with different adaptive groups.

Evolution of Bivalve Feeding. It is generally agreed that the early bivalves were shallow burrowers in soft substrata. They belonged to the subclass Protobranchia, which

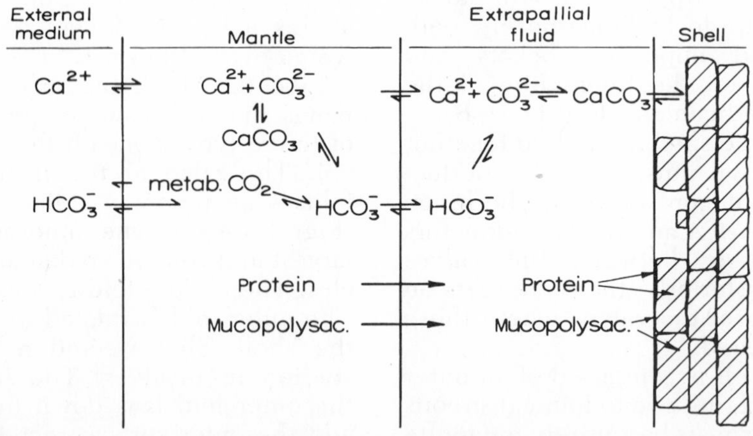

Figure 11–48. Processes of shell secretion in mollusks. (After Wada and Greenaway from Wilbur, K. M., 1972: *In* Florkin, M., and Scheer, B. J. (Eds.), Chemical Zoology, Vol. 7, Academic Press, N. Y., p. 105.)

is represented by some of the oldest fossil forms (Ordovician, perhaps Cambrian) as well as by some living species—*Nucula, Nuculana, Yoldia, Solemya,* and *Malletia.* Most extant species of protobranchs live oriented in the substratum with the anterior end directed downward and the posterior end directed toward the surface. They possess a single pair of posterolateral bipectinate gills, from which the name Protobranchia—first gills—is derived. The ventilation current usually enters the mantle cavity through the shell gape posteriorly and ventrally, passes up through the gills, and exits posteriorly and dorsally (Fig. 11–49). Lateral gill cilia create the water current, and frontal cilia remove sediment trapped on the gill surface, as described for the ancestral mollusks.

Most living protobranchs are indirect deposit feeders, and this is believed by many malacologists to have been the mode of feeding of the early and extinct members of the group. In the ancestral mollusks, the mouth rested against the hard bottom over which the animal crawled. However, when bivalves became adapted for burrowing in sand or mud, the mouth was lifted above the substratum as a result of both the lateral compression of the body and the greatly increased height of the dorsoventral axis. The radula has disappeared. Contact with the substratum is maintained by a pair of tentacles (proboscides), elongations of the margins of the mouth. Each tentacle is associated with two large flap-like folds, called labial palps, located to either side of the mouth (Fig. 11–49). During feeding the tentacles are extended into the bottom sediments. Deposit material adheres to the mucus-covered surface of the tentacle and then is transported by cilia back to the palps, which function as sorting devices. The inner opposing surfaces of each pair are ridged and ciliated (similar to Fig. 11–49*E*). Light particles are carried by certain cilia to the mouth; heavy particles are carried by other cilia to the palp margins, where they are ejected into the mantle cavity.

In some group of early protobranch bivalves, filter feeding evolved. An explosive evolution followed this development, and the filter feeders, called lamellibranchs, came to dominate the bivalve fauna. The gills and ventilating current of protobranchs preadapted them for filter feeding. As the lamellibranchs evolved, plankton in the ventilating current came to be utilized as a source of food, the gills became filters, and the gill cilia which originally served for the transport of sediment became adapted for the transport of plankton trapped in mucus from the filter to the mouth.

The principal modification of the gills for filtering was the lengthening and folding of the gill filaments, which greatly increased their surface area. Many filaments were added to the gills so that they extended far anteriorly, reaching the palps. Perhaps during the evolution of the lamellibranch gills, these attenuated filaments at first projected somewhat laterally and then became bent or flexed downward in the middle (Fig. 11–50). At the angle of flexure, the frontal surface of each filament developed an indentation, or notch, which, when lined up with the notches of adjacent filaments, formed a food groove that extended the length of the underside of the gill. Yonge (1941) suggests that the appearance of the food groove probably preceded folding, since the groove was necessary for nutrition. In any case, each gill filament on each side of the axis became folded, or **U**-shaped. The arm of the **U** that is attached to the axis of the gill is called the descending limb, and the free arm is called the ascending limb. Since the filaments on both sides of the axis have become folded, the net result has been to transform the original single gill into a pair of gills, or demibranchs; the original outer filaments form one member of the pair, and the original inner filaments form the other (Fig. 11–50). The lengthened folded filaments and their attachment to one another give the gill a sheet-like form, hence the name of the subclass Lamellibranchia—sheet gill. Four large, broad filtering surfaces (lamellae) are present, two on each gill (demibranch).

These modifications in gill structure have necessitated a change in both ciliation and circulation. The frontal cilia carry food particles trapped on the gill surface vertically to the food groove (Figs. 11–51*B* and 11–52*B*). Lateral cilia still produce the water current through the gills. Along the angles of the limbs, between the lateral and frontal cilia, is present a new ciliary tract composed of laterofrontal cirri. Each cirrus is a bundle of many fused cilia, with the free ends bent at right angles to the bundle. Opposing cirri form a fine mesh that filters out particles from the water entering the gill and move it onto the frontal cilia. The respiratory water current enters the inhalant, or infrabranchial, chamber at the posterior end of the animal, flows between the filaments, and then moves

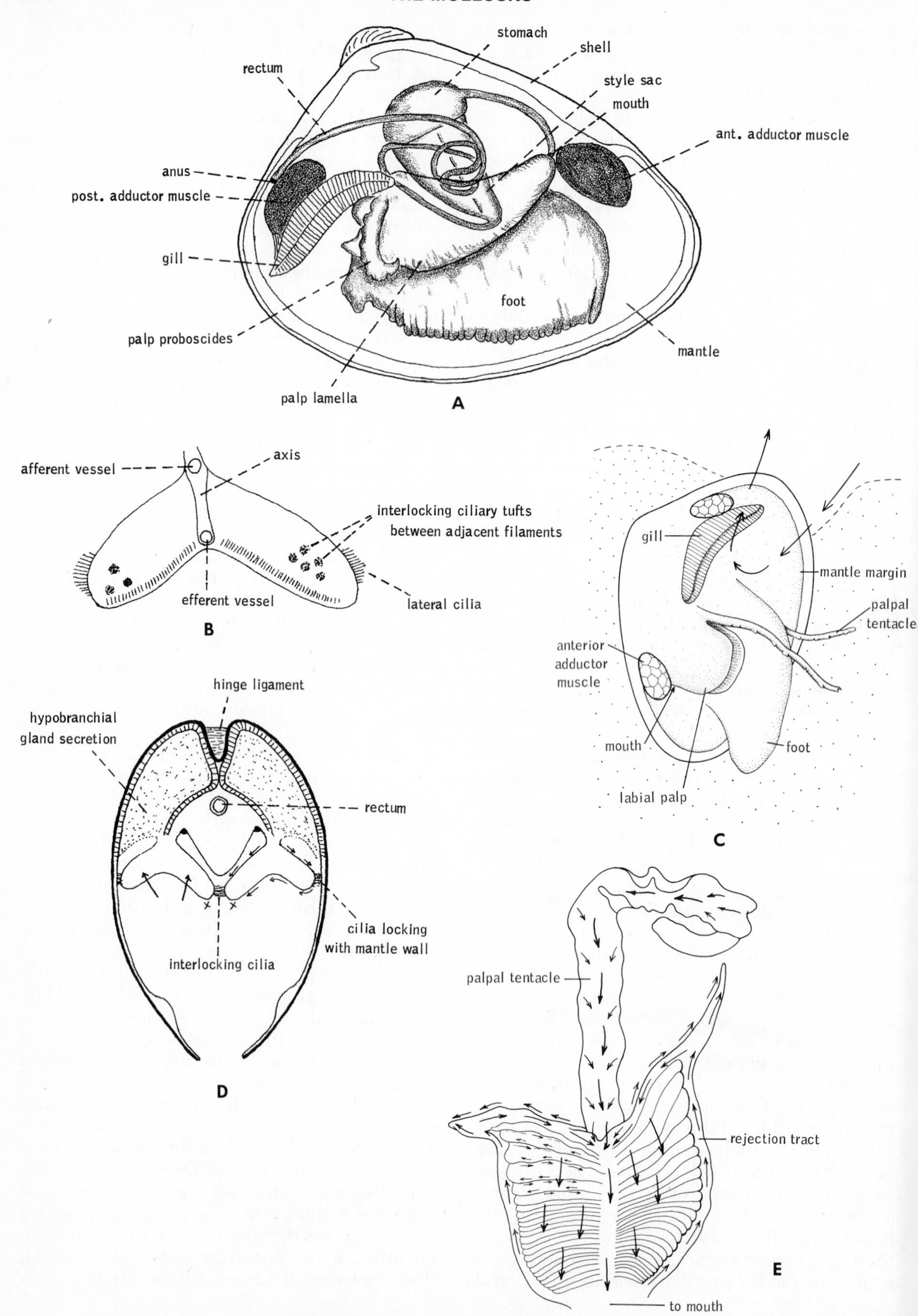

Figure 11–49. *See opposite page for legend.*

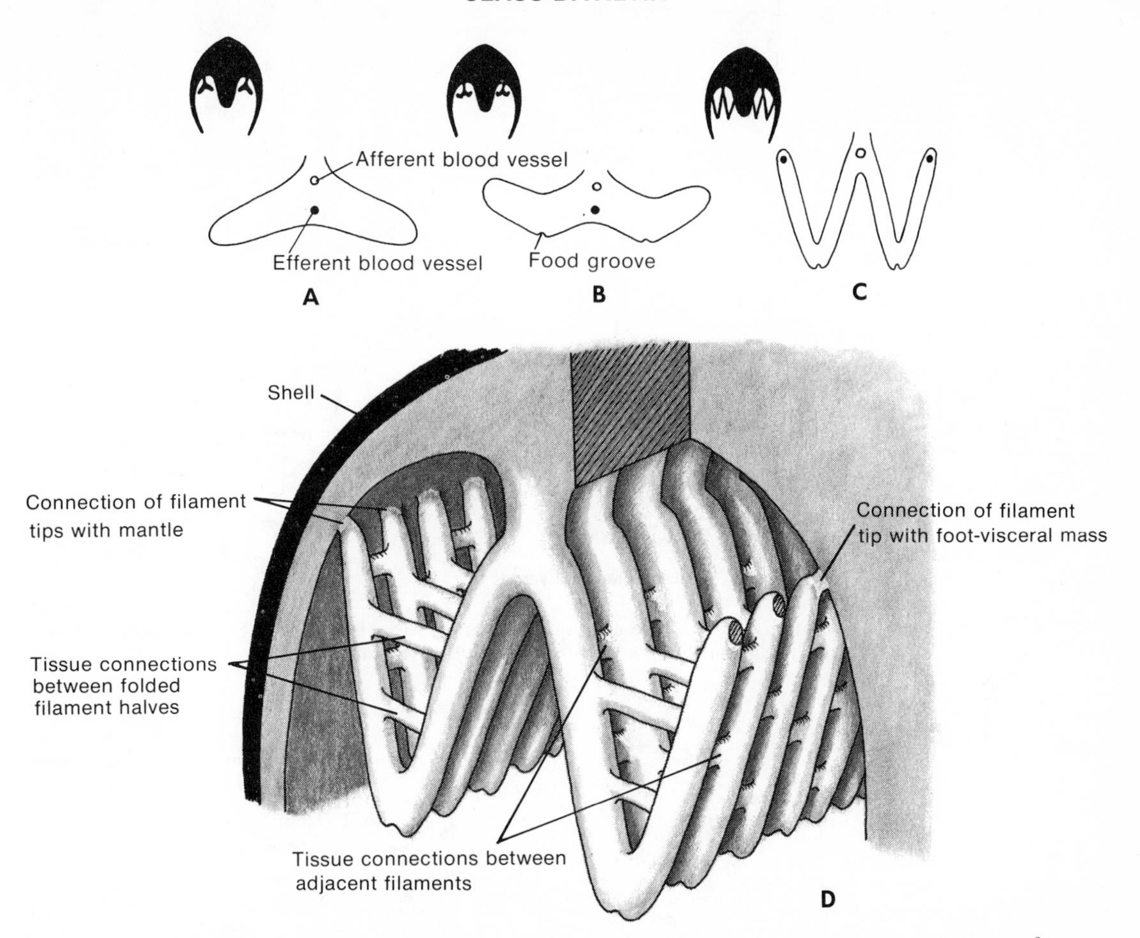

Figure 11–50. Evolution of lamellibranch gills. *A.* Primitive protobranch gill (position relative to foot-visceral mass and mantle indicated in cross section. *B.* Development of food groove in hypothetical intermediate condition. *C.* Folding of filaments at food groove to produce the lamellibranch condition. *D.* Tissue connections which provide support for the folded lamellibranch filaments.

up between the two lamellae. From the interlamellar spaces, water passes into the exhalant, or suprabranchial, chamber and finally flows out through the posterior exhalant opening. Hypobranchial glands have disappeared as a result of the more tightly meshed gill structure, which prevents all but the finest sediment from entering the suprabranchial cavity with the water current.

Support for the long folded filaments is provided by three kinds of new tissue connections at various points within the gill: (1) cross connections, called interlamellar junctions, between the folded filament halves, or lamellae; (2) connections, called interfilamentous junctions, between adjacent filaments; (3) connections between the tips of the filaments and the mantle or foot (Fig. 11–50*D*). The extent of these connections varies in different groups of lamellibranchs and accounts for several different types of lamellibranch gills.

In cases where the individual filaments are still more or less separate, the gill is known as a filibranch gill (Fig. 11–51*A*). Bars of tissue, the interlamellar junctions, have grown between to two limbs of each **U** at intervals, and there may be tissue union of

---

Figure 11–49. The protobranch *Nucula*. *A.* Body with right valve and mantle removed (lateral view). *B.* Gill, showing lateral filaments (transverse section). Note similarity to gill of the primitive gastropod *Haliotis*, shown in Fig. 11–1*C*. *C.* A generalized protobranch, illustrating shallow-burrowing and deposit-feeding habit. Arrows indicate direction of water current. *D.* Position of gills in mantle cavity (transverse section). *E.* Labial palps and palpal tentacle of *Nuculana minuta*. Opposing surfaces of palps are pulled back, and arrows indicate direction of ciliary currents. (*A* to *D* after Yonge; *E* after Atkins.)

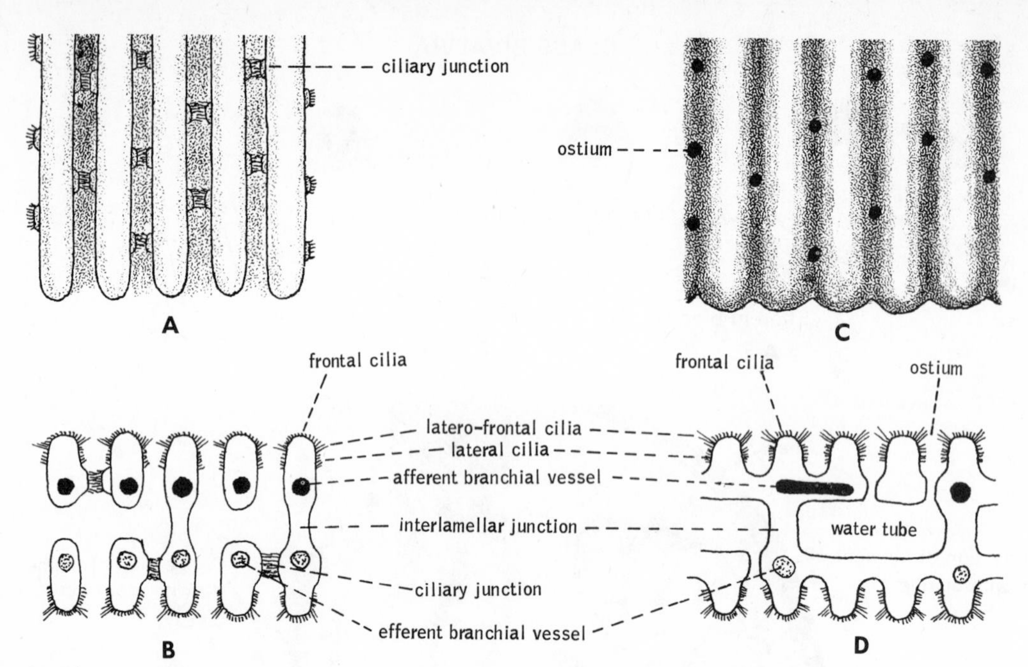

Figure 11–51. *A* and *B*. Filibranch gill. *A*. Five adjacent filaments (surface view). *B*. Frontal section. *C* and *D*. Eulamellibranch gill. *C*. Five fused, adjacent filaments (surface view). *D*. Frontal section.

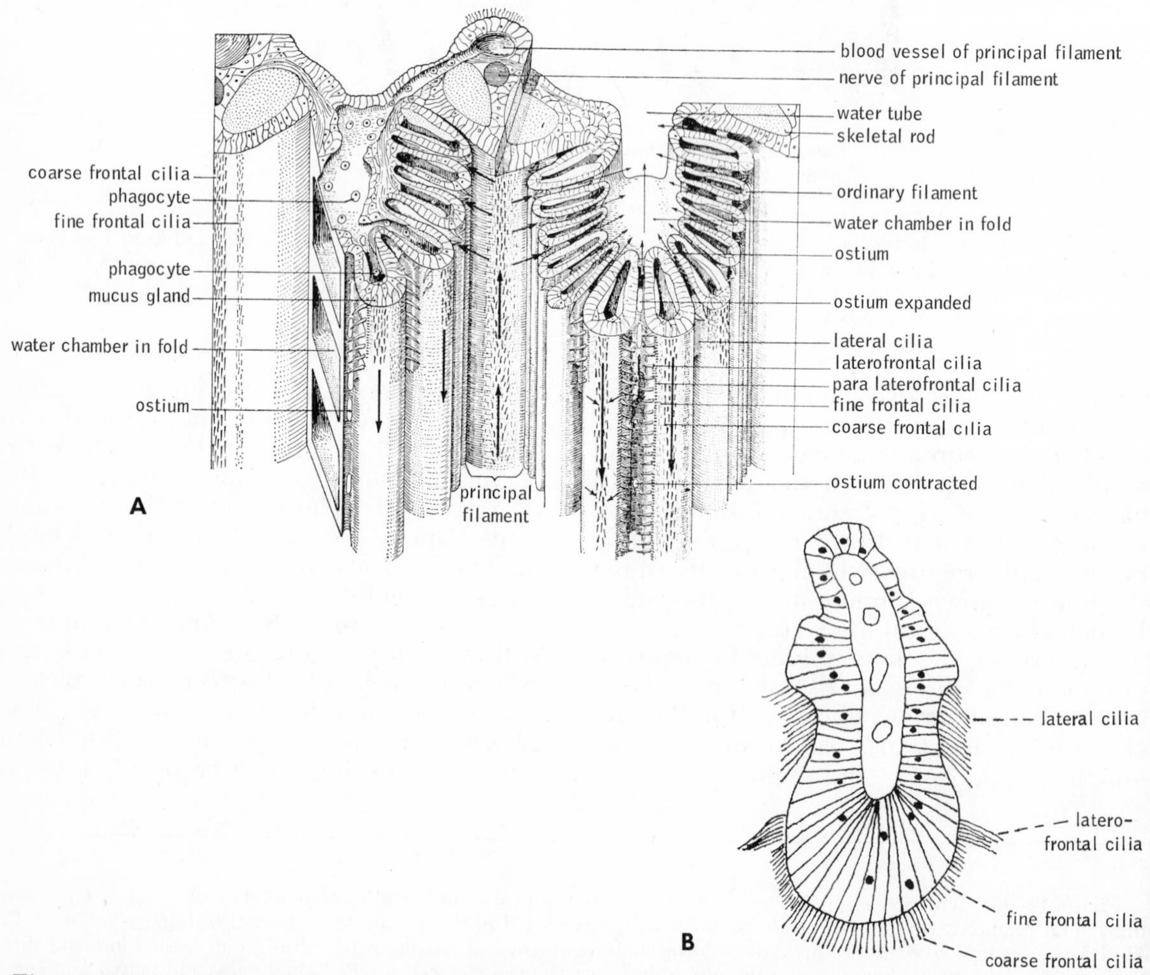

Figure 11–52. *A*. Stereodiagram of part of a plicate gill (*Crassostrea*). (After Nelson from Jorgensen, C. B., 1966: Biology of Suspension Feeding. Pergamon Press, London, p. 71.) *B*. Ordinary filament of *Solen marginatus*. (After Atkins.)

adjacent filaments at the bottoms or tops of the lamellae. But throughout most of their length, filaments are attached to adjacent filaments only by ciliary tufts. Filibranch gills are found in such bivalves as the ark shells (*Arca*), the scallops (*Pecten*), and mussels (Mytilidae).

In the oysters and pen shells, which are said to have a pseudolamellibranch type of gill, the filaments are bound together with some (though not extensive) interfilamentous tissue junctions. In the primitive gills of protobranchs, the efferent, or drainage, vessel ran within the axis of the filament beneath the afferent vessel, as in the ancestral molluscan gill. With the elongation and folding of the filament and the fusion of the ascending limbs, the drainage vessel of the pseudolamellibranch gill and some filibranch gills migrated to the ends of the ascending limbs of each filament (Fig. 11–50*C*).

The most specialized lamellibranch gill is known as a eulamellibranch gill. In gills of this type, the union of filaments has developed further, so that the lamellae actually consist of solid sheets of tissue (Fig. 11–51*C* and *D*). Furthermore, interlamellar junctions have increased in number and extend the length of the lamellae (dorsoventrally). Thus the interlamellar space is partitioned into vertical water tubes. Instead of diffusing through the lamellae, the blood is carried in vertical vessels that course within the tissue junctions. The tips of the ascending limbs have become fused with the upper surface of the mantle on the outside and the foot on the inside, morphologically separating the inhalant chamber from the suprabranchial chamber. Ciliation remains the same, for the frontal edges of the filaments are not involved in the interfilamentous fusion. Thus, a frontal section of the eulamellibranch gill exhibits a lamella with a ridged outer surface, each ridge representing one of the original filaments (Fig. 11–51*C* and *D*).

Water in the inhalant chamber circulates between the ridges and enters the water tubes through numerous pores (ostia) in the lamella. Oxygenation takes place as the water moves dorsally in the water tubes. The water then flows into the suprabranchial cavity and out the exhalant opening.

In some lamellibranchs, such as cockles, razor clams, scallops, and oysters, the surface area of the lamellae has been increased further by folding along the length of the gill, and the plicate gill surface of these bivalves presents an undulated appearance (Fig. 11–52*A*).

Most lamellibranchs feed on fine plankton, largely phytoplankton. Food particles, in some cases as small as 1 μ, are filtered by the long laterofrontal cirri from the water currents passing between filaments or entering the ostia. The particles are then passed onto the frontal cilia, where they are entangled in mucus and moved up or down the margin of the filament to a food groove.

Primitively there are five food grooves transporting particles anteriorly to the palps. Three of the grooves are located at the tip of the gills between or outside the demibranchs; the other two are located ventrally, one along the margin of each gill (Fig. 11–53*G*). The frontal cilia are divided into two tracts of coarse and fine cilia, one carrying particles upward and one downward. Such a two-way vertical tract system with five food grooves is found in oysters and scallops. From such a primitive condition, the great variation in number and location of food grooves and direction of vertical tracts encountered in other lamellibranchs is believed to be derived by deletion. Many of these variations are illustrated in Figure 11–53, in which the forward-moving food grooves are indicated by black circles. Note that the dual tract system has disappeared in most lamellibranchs.

Various mechanisms provide for some prepalpal sorting by the gills and for coping with sediment-laden water short of halting ventilation. The gills of the families Arcidae and Anomiidae display a two-way tract system of frontal cilia, but the heavy particles that are carried to the ventral margin of the gills are transported *posteriorly* and rejected onto the mantle (Fig. 11–53*F*). In a number of families that have plicate gills, such as the Ostreidae (oysters), Pectinidae (scallops), and Solenidae, the frontal cilia of most filaments are arranged in the antagonistic tract system (Fig. 11–53*G*). However, the filament that lies between the folds, called the principal filament (Fig. 11–52*A*), only carries light particles upward. When the water is relatively clean, the gills are expanded and the upward-moving tracts are largely in operation. When there is a lot of sediment in the water, the gills are stimulated to contract, placing the principal filament deep within the folds (Fig. 11–54*A* and *B*). Coarse cilia are stimulated to activity, driving much of the trapped material to the ventral grooves, which if too heavily laden

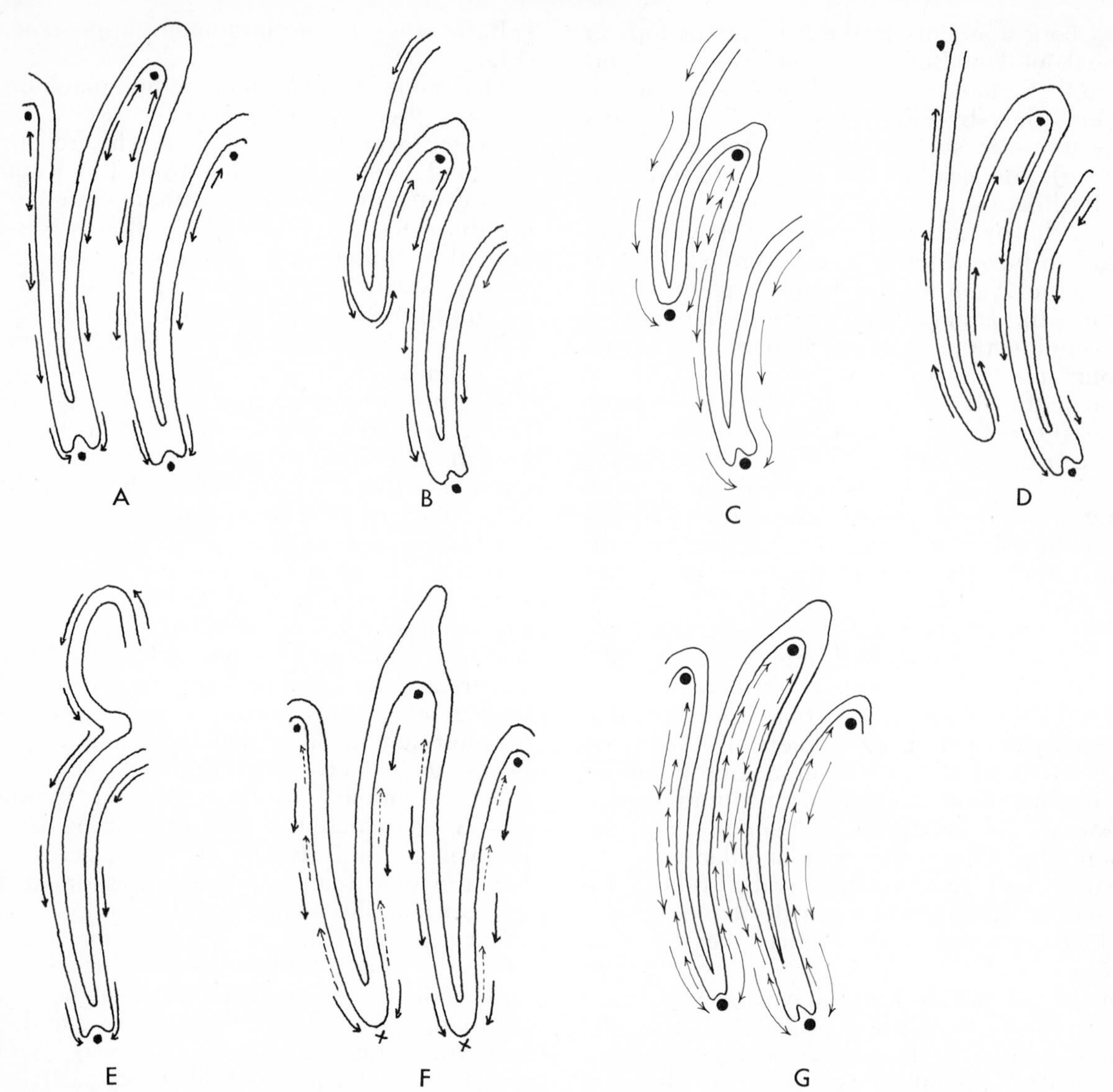

Figure 11–53. Transverse sections of different lamellibranch gills, showing direction of frontal cilia beat and position of anteriorly moving food tracts (black dots). In *F* and *G*, broken arrows indicate fine frontal cilia carrying food particles upward; solid arrows indicate coarse frontal cilia carrying particles ventrally. Crosses represent posteriorly moving channels. Inner demibranch, or gill, is on right in all cases. *A.* Mytilidae and Pinnidae. *B* and *C.* Many eulamellibranchs. *D.* Unionidae (fresh-water). *E.* Most Tellinidae and Semelidae. *F.* Arcidae and Anomiidae. *G.* Ostreidae and Pectinidae. (All after Atkins.)

will drop their loads into the bottom of the mantle cavity.

The ability of the ventral grooves to handle only a limited volume of material or to receive particles no greater than a certain size provides for some sorting in many lamellibranchs (Fig. 11–54*A* and *B*).

The lamellibranch palpal lamellae have the same sorting and conveying function as in the Protobranchia (Fig. 11–54*C* and *D*). Particles are sorted by size and by their weight. Small, light particles are retained for ingestion and are carried up the palpal surface across the crest of the ciliated ridges (Fig. 11–54*D*). Large particles, destined to be rejected, are carried to the edge of the lamellae in the grooves between ridges and fall to the mantle or foot. There are also special resorting tracts running dorsally and ventrally along the ridges. Rejected material, called pseudofeces, from both the palps and the gills leaves the mantle cavity through the pedal gape or, most commonly, by way of the inhalant aperture. In the latter case the particles are carried posteriorly along a ventral ciliated tract of the mantle to accumulate behind the inhalant aperture. When the valves are closed periodically, water is forced out of the inhalant opening, taking these accumulated wastes with it.

From some group of protobranchs evolved still another subclass of bivalves, the Septibranchia (*Poromya* and *Cuspidaria*). The members of this small group have become carnivores or scavengers. The gills are degenerate and have been modified to form a pair of perforated muscular septa, which separate the suprabranchial chambers from the inhalant chambers (Fig. 11–55*C*). By muscular contractions, the septum moves up and down, forcing water into the inhalant chambers and out of the suprabranchial chambers. The force of the pumping septa is sufficient to bring small animals, such as crustaceans and worms, into the mantle cavity. These prey are then seized by the reduced but muscular labial palps and carried to the mouth.

There are some malacologists who believe that filter feeding, rather than deposit feeding, is primitive for bivalves (see Stasek, 1972), but most would agree that it was not until the evolution of the lamellibranch form that bivalves were able to exploit plankton fully as a food source and to spread into new habitats.

Adaptive Radiation of Bivalves. The evolution of filter feeding freed lamellibranchs from dependence upon deposit material and made possible the colonization of many habitats that were uninhabitable for their protobranch ancestors. The success of this adaptive radiation is reflected in the fact that of the some 20,000 described species and 75 families of bivalves, most are lamellibranchs.

Soft Bottom Burrowers (Infauna). It should not be thought that all lamellibranchs departed from the ancestral habitat. The majority of species are inhabitants of soft bottoms, exploiting the protection offered by a subterranean life in marine sand and mud while utilizing food suspended in water brought in from above the surface. Some live just beneath the surface; some are adapted for deep burrowing; some move between the surface and lower levels; and some are especially adapted for shallow, rapid burrowing in a shifting environment. In addition to the many marine forms, soft bottom lamellibranchs also include most of the freshwater clams and mussels.

The actual mechanism of burrowing involves the coordinated action of a number of forces (Trueman, 1966). The protraction of the foot is initiated by contraction of the

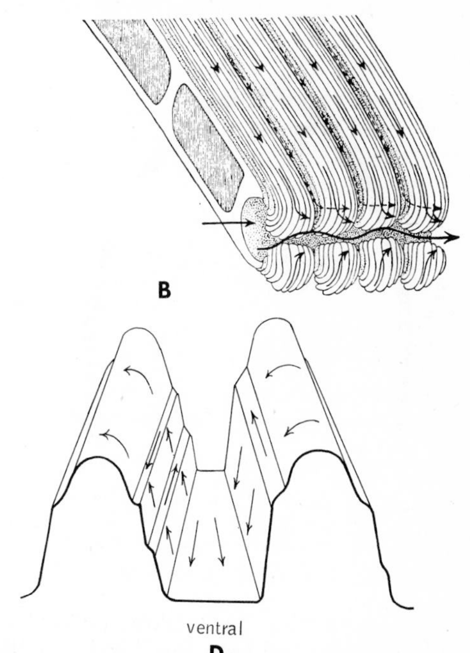

Figure 11–54. *A.* Ventral margin of a section of the plicate gill of *Pinna fragilis* when relaxed and expanded. *B.* Same gill as *A*, when contracted. Note that contraction buries the principal filaments that are located between the folds (plicae) and closes the marginal food groove. In members of this family, most frontal ciliary tracts carry particles downward. (See Fig. 11–53*A*.) (From Atkins, D., 1937: Quart. J. Micr. Sci., *79*:348.) *C.* One pair of palps spread apart and anterior section of gills of *Ostrea edulis*. Arrows indicate ciliary tracts. (After Yonge.) *D.* Section of two ridges of a lamellibranch palp. Mouth lies to left, and ventral edge of palp is toward viewer. Arrows indicate ciliary tracts. (After Stasek.)

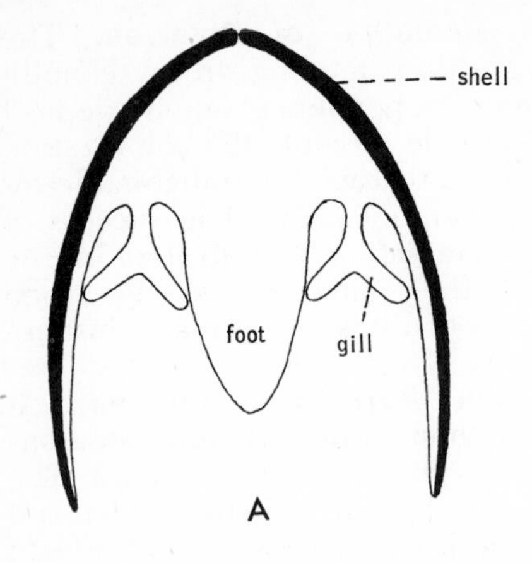

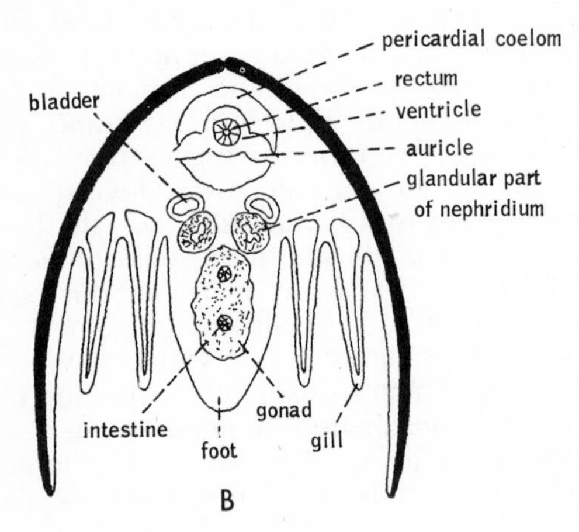

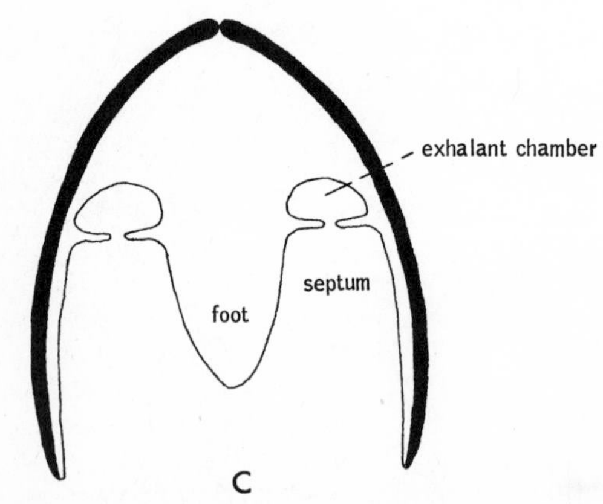

Figure 11–55. *A.* Protobranch bivalve (transverse section). *B.* Lamellibranch bivalve (transverse section). *C.* Septibranch bivalve (transverse section).

pair of pedal protractor muscles. These muscles extend from each side of the foot to the adjacent valve, to become attached just beneath the anterior adductor muscle. The projecting foot probes and pushes into the surrounding sand. As the foot is protruded, the valves begin closing by contraction of the adductor mucles. Some water from the mantle cavity is expelled, which loosens the sand or mud and facilitates the movement of the foot. The water remaining in the mantle cavity and the blood act as a hydrostatic skeleton. The pressure of these two fluids is elevated by the adducted valves. Blood from the visceral mass is forced down into the pedal hemocoel, causing the foot to dilate and anchor into the substratum (Fig. 11–56*D*). With the foot anchored, an anterior pair and a posterior pair of pedal retractor muscles contract, pulling the shell downward. The pedal retractors also originate on the shell in the region of the adductor muscles (Fig. 11–46*B*). In many species, retraction by the anterior pedal muscle occurs before that of the posterior muscle. The effect is to rock the shell which facilitates its movement through the substratum. Ridges or other sculpturing of the shell surface may increase traction. Following pedal retraction, relaxation occurs and the valves gape. To return to the surface or to a higher level within the substratum, some soft-bottom bivalves actually turn around and burrow upward, but most back out, *pushing* against the anchored end of the foot.

In the primitive protobranchs, such as *Nucula*, *Solemya*, and *Yoldia*, the foot bears a little flattened sole (Fig. 11–56*B*). However, the two sides of the sole can be folded together producing a bladelike edge; in this condition the foot is thrust into the mud or sand. The sole then opens and serves as an anchor and the remainder of the body is pulled downward after the sole. *Solemya*, a rapid mud burrower, can completely disappear under the mud surface with two thrusts of the foot.

A major problem arising from burrowing in soft bottoms is that of sediment brought in by water currents. Blood engorgement within the mantle margin enables the mantle edges to be appressed together, even when the valves gape slightly (the muscles of the inner lobe provide for retraction); but since circulation of water through the mantle cavity is necessary for gas exchange and, in most species, for feeding, at least the posterior part of the mantle must be opened for entrance of water, with which sediment enters.

As a means of reducing sediment fouling, there has been a tendency to seal the mantle edges morphologically where openings are not necessary (Fig. 11–57). The most frequent point of fusion of the mantle edges is at the posterior between the inhalant and

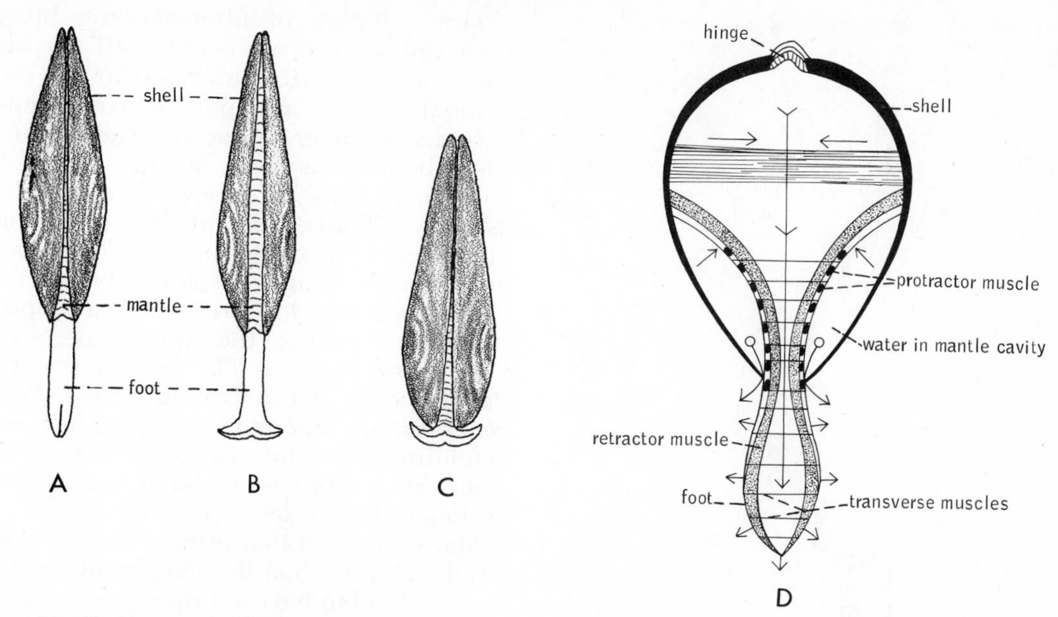

Figure 11–56. Operation of foot of *Yoldia limatula*, a protobranch. *A*. End of foot folded and extended into substratum. *B*. Foot opened and anchored. *C*. Body drawn down into substratum. (All after Yonge.) *D*. Diagrammatic cross-section of a bivalve, showing hydrostatic forces that produce dilation of foot. Central vertical arrow indicates flow of blood into foot. (Modified after Trueman.)

exhalant openings. This fusion forms a distinct exhalant aperture. A functional inhalant opening is present since the mantle edges are pressed together below the opening, although the mantle edges are not morphologically fused. This type of structure is found in the protobranch *Nuculana* and in the freshwater lamellibranch clams *Anodonta* and *Unio*. A second point of fusion just below the inhalant opening, forming a distinct inhalant aperture, has evolved in many bivalves.

Still further mantle fusion has developed in some species, and most of the ventral margin anterior to the inhalant aperture has become sealed. Thus, three apertures remain—the inhalant and exhalant apertures, and an anterior pedal aperture through which the foot protrudes. *Mya* and the razor clams *Ensis* and *Tagelus* are examples of bivalves with pedal apertures. Extensive mantle fusion not only contributes to the reduction of fouling but is probably equally important in maintaining the hydraulic pressure within the mantle cavity necessary for burrowing.

Commonly, when structural inhalant and exhalant apertures are present as a result of fusion, the mantle edges surrounding the apertures have become elongated to form tubular siphons of varying lengths (Fig. 11–58). The advantages of such siphons are obvious; the animals can be completely buried in the mud and only the siphon tips need project above the bottom. The siphons are extended by blood pressure or by water pressure within the mantle cavity when the

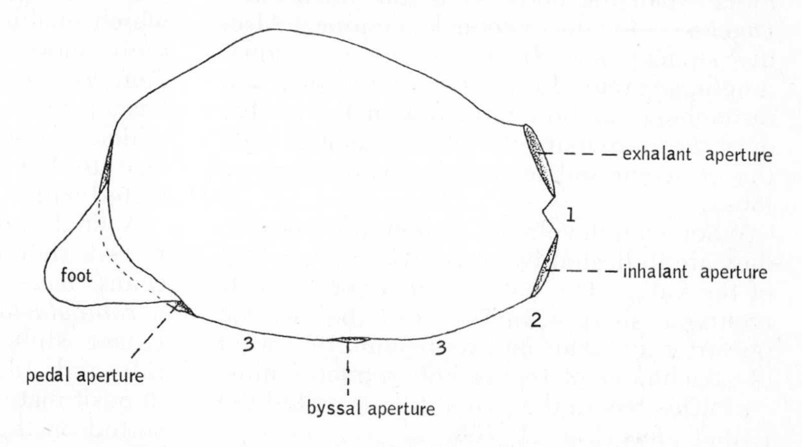

Figure 11–57. Areas of mantle fusion in bivalves. *1*. Between inhalant and exhalant apertures or siphons, the most common point of fusion. *2*. Fusion below inhalant aperture or siphon. *3*. Fusion between inhalant aperture and foot, leaving only a pedal aperture for extension of foot. A byssal aperture is present in only a very few bivalves.

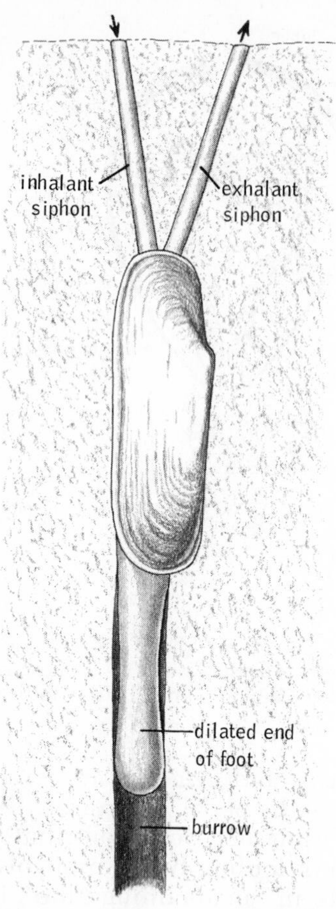

Figure 11–58. *Tagelus*, a common marine bivalve, occurring in intertidal and coastal sands of the Atlantic Ocean. (Drawn from life.)

valves are closed, and are withdrawn by siphon retractor muscles derived from the muscle tissue of the innermost mantle fold. Considerable variation exists in siphon structure. They may be short and only slightly developed; or they may, when extended, be longer than the body, as in the razor clam *Tagelus* and the geoduck *Panope*. Also, the siphons may be of equal or unequal length, separate, fused at the base, or fused throughout. Siphon formation may involve only the muscular lobe of the mantle, only the muscular and sensory lobes, or all three lobes.

When well developed siphons are present, they are indicated by scars on the inner face of the valve. The pallial line impression is recurved sharply inward just below the posterior adductor and represents the point of attachment of the siphon retractor muscles. This bay in the pallial line is called the pallial sinus (Fig. 11–46*B*).

The radiation of filter feeding bivalves with siphons did not occur until the Mesozoic and Cenozoic, but it led to the development of many groups which invaded not only deeper layers of the substratum but offshore bottoms as well, habitats from which Paleozoic bivalves were largely absent (Stanley, 1968). Most modern soft bottom filter feeders possess siphons. Those that do not are rather sluggish shallow burrowers.

In *Mya*, which burrows to a depth of some seven inches, the siphons are covered with periostracum. The geoducks of the Pacific coast of the United States, *Panope generosa*, are among the deepest burrowers, extending down to a depth of over a meter. They have siphons so large that they can no longer be retracted (Fig. 11–59*B*).

Many bivalves that burrow deeply (i.e., to depths greater than the lengths of their own bodies) tend to have semipermanent or even permanent burrows. The walls of the burrows become coated with mucus, which also reduces sediment fouling.

The primitive form of lamellibranch soft bottom burrowers is considered to be one in which the adductor muscles are more or less equal (isomyarian), the mantle is unfused, and the valves are equal and circular in outline. We have already described modifications involving mantle fusion. In many groups the valves are also modified, becoming more streamlined for burrowing by being flattened or elongated. Species of *Donax*, which are inhabitants of surf beaches, have shells that are pointed anteriorly and blunt posteriorly. They back out on incoming waves and reburrow with great rapidity as the wave recedes. The margins of the inhalant siphon are fringed with infolded tentacles, which keep out swirling sand grains (Fig. 11–59*A*).

The razor clams *Ensis* and *Tagelus* have greatly elongated valves and an elongate foot, which enable them to move rapidly within their more or less permanent burrows. *Tagelus* has long siphons, each of which has a separate opening to the surface. *Ensis*, which has short siphons, comes to the surface to feed from the deeper, more protected part of its burrow.

A final specialization of soft bottom burrowers should be mentioned. Some tellinaceans have reverted to deposit feeding. *Scrobicularia*, for example, extends its inhalant siphon above the surface at low tide and, like a vacuum cleaner, sucks in deposit material (Fig. 11–60), which is then sorted on the gills. Deposit feeding is gen-

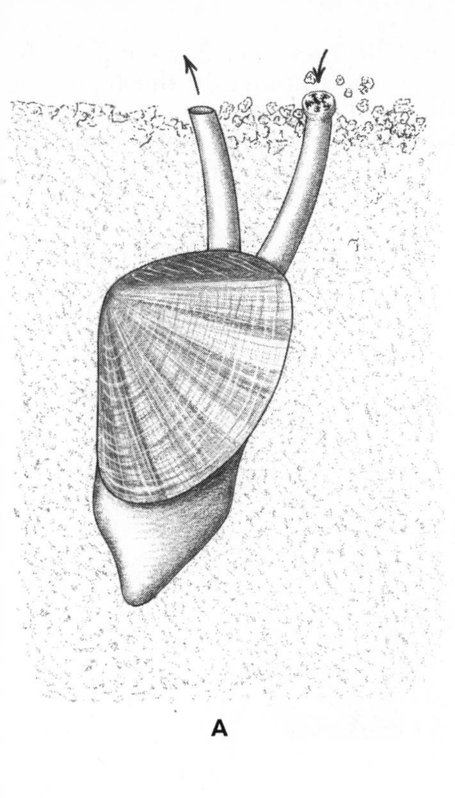

Figure 11–59. A. *Donax variabilis,* a common inhabitant of surf beaches. Rapid burrowing is facilitated by thin pointed foot. Opening of inhalant siphon is frilled, preventing entrance of sand grains. *B.* The geoduck, *Panope generosa,* a giant Californian bivalve in which body and siphon cannot be enclosed within valves. (From Milne & Milne, 1959: Animal Life. Prentice-Hall, Englewood Cliffs, N.J.)

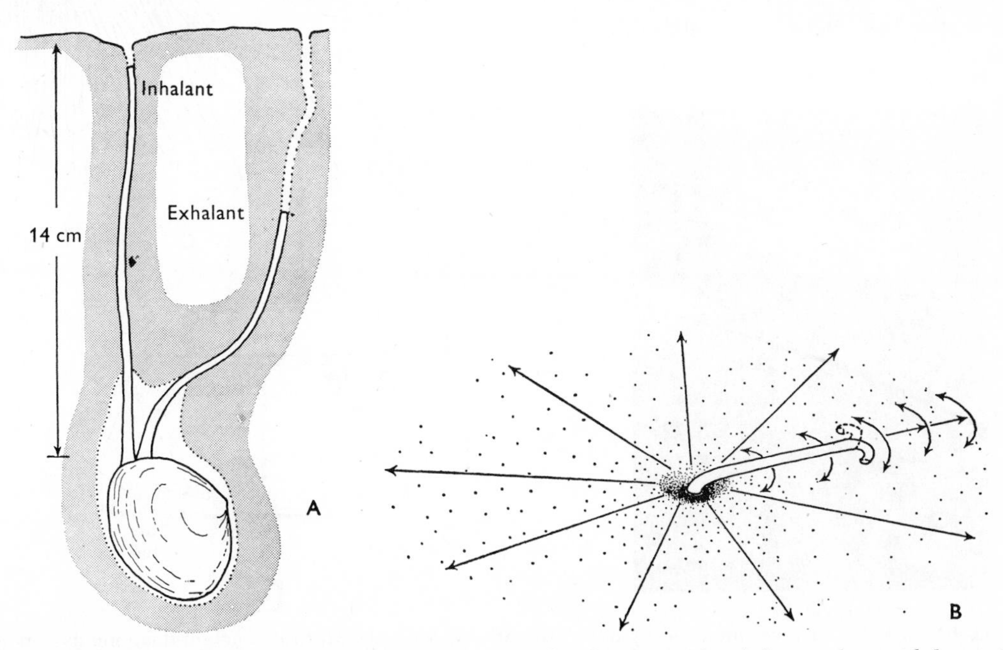

Figure 11–60. Deposit feeding in *Scrobicularia. A.* Animal in burrow with inhalant siphon withdrawn. *B.* Feeding movements of inhalant siphon at low tide. (From Hughes, R. N., 1969: J. Mar. Biol. Assoc. U.K., *49*:807.)

erally utilized as an addition to, rather than as a substitute for, filter feeding.

Attached Surface Dwellers (Epifauna). A number of evolutionary lines of lamellibranchs have invaded firm substrates—peat, wood, shell, coral, rock, or man-made sea walls, jetties, and pilings. Attachment is provided either by byssal threads or by one valve fused to the substratum. Byssal threads are tough, horny threads secreted by a gland in the foot. The secretion flows out to the substratum along a groove in the foot (Fig. 11–61*B* and *C*). Then, after the fiber hardens on exposure to sea water, the foot is removed, leaving the thread anchored to the substratum at one end and to the byssus opening of the foot at the other (Fig. 11–61*A*). A byssus retractor muscle may enable the animal to pull against its anchorage.

Among living surface dwellers attached by

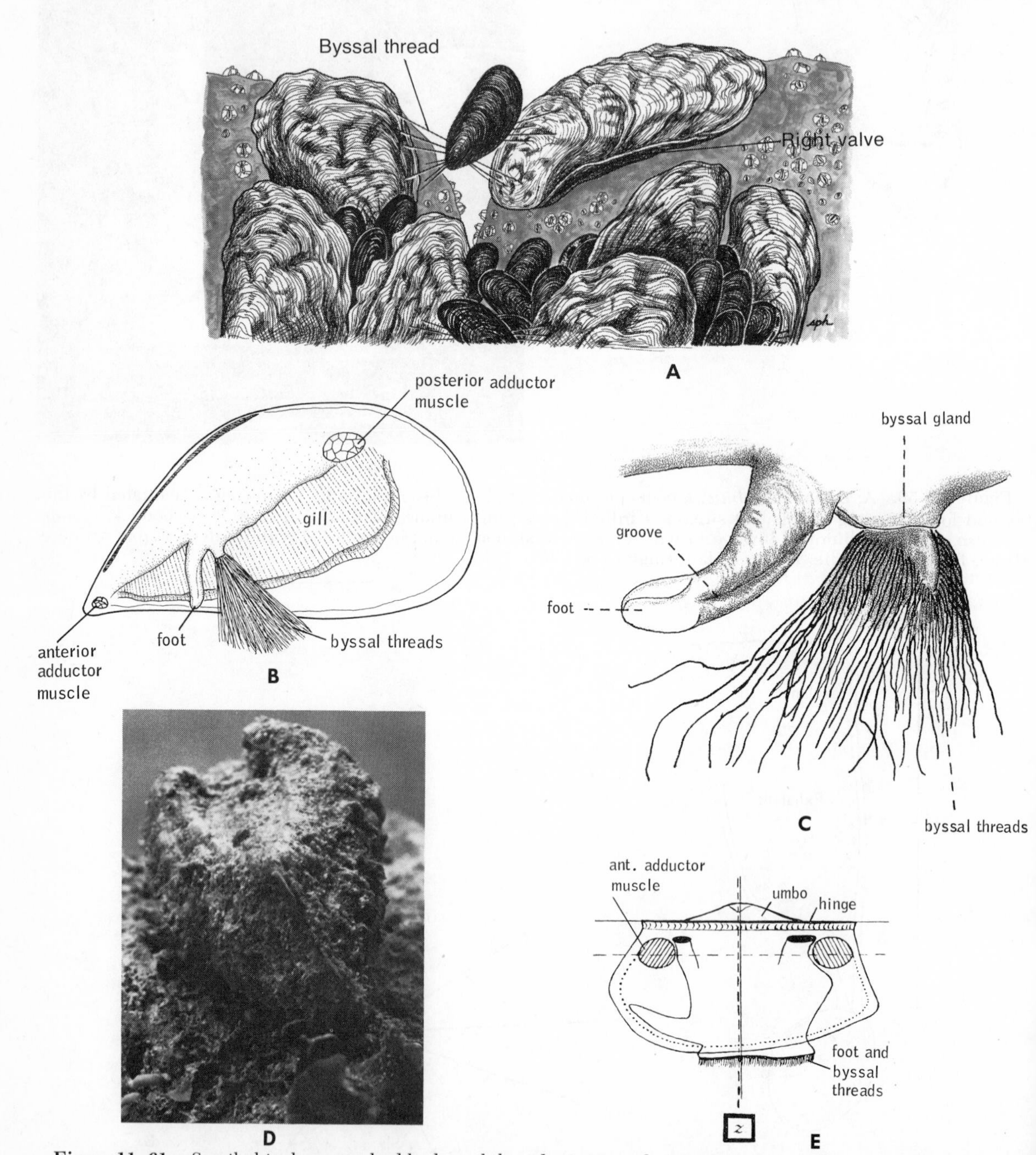

Figure 11–61. Sessile bivalves attached by byssal threads. *A*. Mussels (*Mytilus*) anchored among oysters (*Crassostrea*). *B*. Lateral view of a mytilid with left valve removed. *C*. Foot and byssal gland of a mytilid. *D*. The West Indian *Arca zebra* attached to a coralline rock. *E*. Diagrammatic lateral view of *Arca*. (From Yonge, C. M., 1953: Phil. Trans. Roy. Soc. London B, 237:365). *Figure 11–61 continued on opposite page.*

byssal thread, the widely distributed mussels, members of the family Mytilidae, are perhaps the most familiar. They live attached to wharf pilings, sea walls, and rocks or among oysters, often in great numbers. The threads, laid down by the little, finger-like foot, often radiate outward like guy wires (Fig. 11–61*A*).

Other surface inhabitants attached by byssal threads include many of the heavy-bodied ark shells (Arcidae) (Fig. 11–61*D* and *E*), which are very common on tropical coralline substrates; mangrove oysters (*Isognomon*) (Fig. 11–61*G*), which hang in clusters from mangrove roots; and winged oysters (pteriids), which live attached to sea fans and other gorgonian corals (Fig. 11–61*F*).

Utilization of byssal threads by adults of some 18 different bivalve superfamilies is

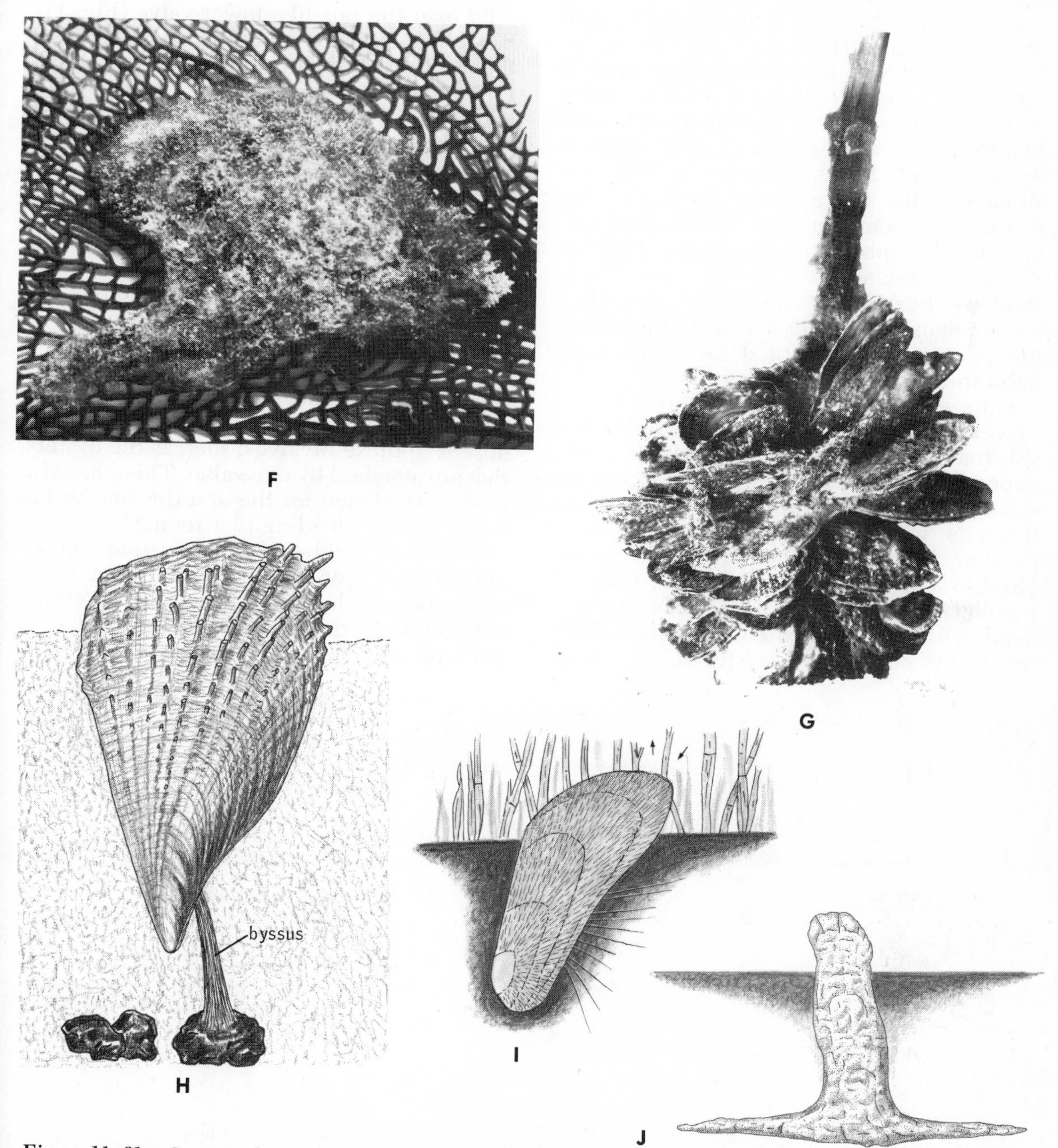

Figure 11–61. *Continued. F. Pteria*, a winged oyster, attached to a sea fan. *G. Isognomon*, a mangrove oyster (*D, F*, and *G* by Betty M. Barnes.) *H.* A pen shell, *Atrina rigida. I. Modiolus* partially buried among intertidal marsh grass. *J.* The hammer oyster, *Malleus malleus.* (*I* and *J* after Yonge from Stanley, S. M., 1972: J. Paleontol., *46*(2):165–212.)

believed to represent a persistent larval adaptation, for the larvae of many unattached burrowing forms produce a byssus for initial temporary anchorage on settling (Fig. 11–62*A*). Adults of a few living species use byssal threads to anchor root-like in soft bottoms. The mytilid genus *Modiolus* live partially buried in peat or coarse substrata (Fig. 11–61*I*), and the hammer oyster *Malleus* (Fig. 11–61*J*) and the pen shells *Pinna* and *Atrina* (Fig. 11–61*H*) occupy a similar position in sand, attaching the byssal threads to small stones. However, such an endobyssate habit was a common adaptation of Paleozoic bivalves (40 per cent of Ordovician bivalves were endobyssate), and Stanley (1972) believes that adult retention of byssal threads for stabilization in soft substrata was the evolutionary route that led to the more common byssally attached surface habit (epibyssate) encountered today. Although most burrowing forms evolved from shallow burrowing ancestors described earlier, Stanley believes that some burrowing groups may have returned to a burrowing habit from endobyssate ancestors.

Surface dwelling bivalves that are attached by cementation lie on one side, fixed to the substratum by either the right or left valve, depending on the species. Such sessile bivalves include four different evolutionary lines, of which the oysters are the most familiar. However, the name oyster is applied to a wide variety of species, some of which attach by byssal threads. In the family Ostreidae, which contains the edible American east coast oyster *Crassostrea virginica* and the European *Ostrea edulis*, the metamorphosing veliger is initially anchored with a drop of adhesive fluid from the byssus gland upon settling. Then the mantle margin attaches the valve to the substratum in the process of shell secretion (Fig. 11–61*A*). Shell attachment has led to varying degrees of inequality in the size of the two valves, the lower being larger or smaller than the upper one. In *Chama*, the tropical jewel boxes, the upper valve forms a lid over the box-like lower valve (Fig. 11–63*A*). The extreme condition was reached in the extinct Mesozoic rudists, in which the lower valve was shaped like a tube or horn (Fig. 11–64). The rudists often occurred in reef-like aggregations.

The common jingle shells or toenail shells, members of the family Anomiidae, possess features of both the byssally attached and the cemented bivalves. They lie on one side but are actually anchored by a large calcified byssal thread which passes through the attached valve (Fig. 11–63*B*).

Attached bivalves share a number of features. As would be expected, the foot is reduced to varying degrees, although not in the primitive ark shells; and it is completely absent in those bivalves, such as the oysters, that are attached by one valve. There has also been a tendency for the anterior end to become smaller, leading to a reduction (anisomyarian) (Fig. 11–61*B*) or loss (monomyarian) (Figs. 11–65 and 11–66) of the anterior adductor muscle. It has been suggested that anterior reduction in mussels (Mytilidae) is perhaps an adaptation to elevate the pos-

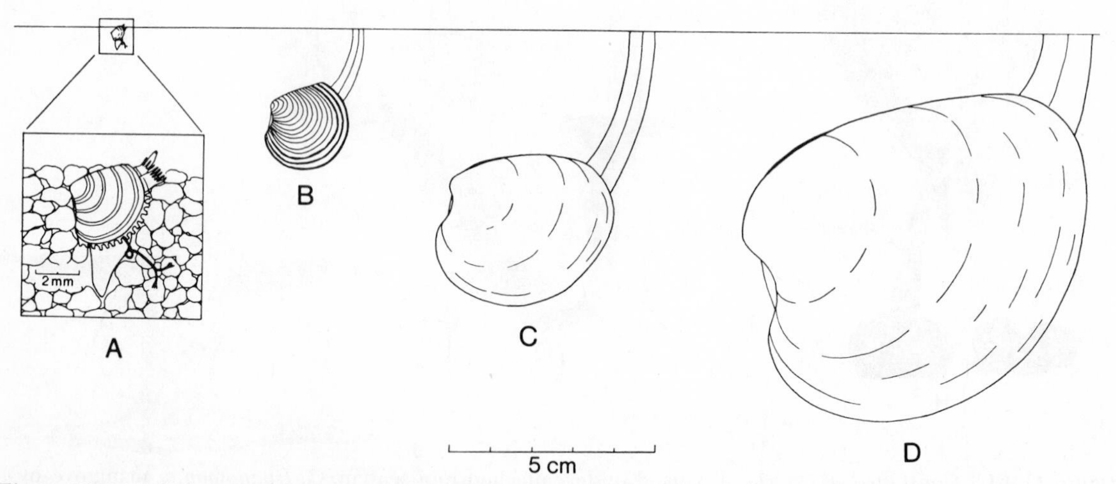

Figure 11–62. Habitation of *Mercenaria mercenaria* at different ages. *A*. Newly settled clam anchored in sand by byssal threads. *B*. First year individual provided with heavy stabilizing concentric ridge on shell. *C* and *D*. Older individuals, showing changes in siphon length. (From Stanley, S. M., 1972: J. Paleontol., *46*(2):165–212.)

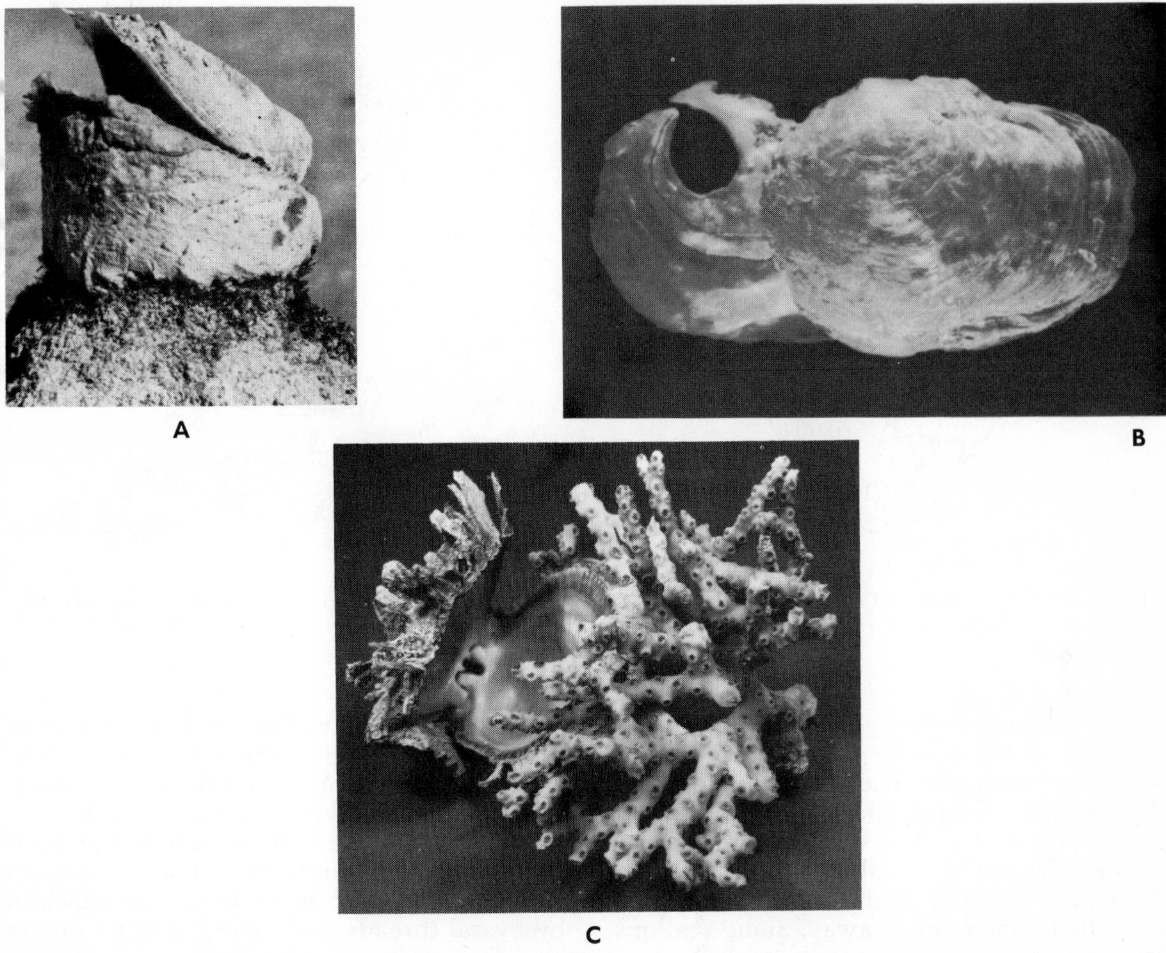

Figure 11–63. *A*. The jewel box, *Chama*, a common tropical sessile bivalve, in which the upper valve forms a lid over the larger lower valve. *B*. A jingle shell, or toenail shell (Anomiidae). Attached valve contains large hole for byssus thread. (By Gates, J. B., from Andrews, J., 1971: Sea Shells of the Texas Coast. Univ. Texas Press, Austin. p. 168.) *C*. The thorny oyster, *Spondylus*, attached to a piece of coral (*Oculina*). (*A* and *C* by Betty M. Barnes.)

terior end, thereby reducing the likelihood of obstruction that the dense aggregations of these bivalves might create.

Mantle fusion and siphon formation have not occurred in sessile bivalves, since they are above the surface and generally on hard substrata, where sedimentation is less of a problem. However, oysters and mussles, which occur in dense beds, are dependent upon the cleansing action of tidal currents to prevent the animals from becoming completely buried in their own feces and pseudofeces.

Unattached Surface Dwellers. Among the small number of bivalves which live free on the surface, the scallops (Pectinidae) and the file shells (Limidae) are the most familiar (Fig. 11–66). There are members of both families which live anchored by byssal threads, but others are unattached, or only weakly attached. They lie on their sides, the foot is reduced, and the anterior adductor muscle has disappeared. Free-living file shells and scallops have evolved the ability to swim by rapid ejection of water from the mantle cavity with the clapping of the valves. The solitary posterior adductor muscle has shifted to a more central position and is divided into smooth and striated sections (Fig. 11–66). The rapid contraction of the striated fibers provides for swimming, and the sustained contraction of the smooth fibers provides for prolonged closure of the valves. The muscular lobe of the mantle margin, when appressed against the lobe on the opposite mantle surface, controls the direction of the water jet, permitting it (in scallops) to exit on either side of the hinge line or opposite the hinge line. The mantle sensory lobe is also highly developed, bearing tentacles and, in scallops, many small blue eyes.

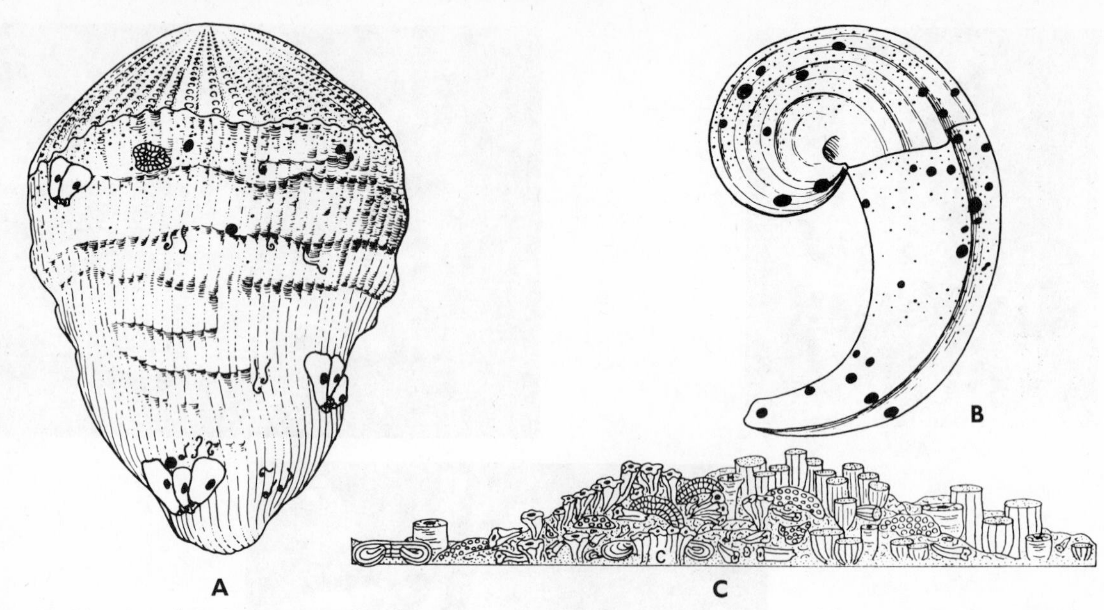

Figure 11–64. *A* and *B*. Two types of rudists, Mesozoic bivalves in which a cap-like upper valve covered a horn-like or tube-like lower valve. *C*. An aggregation of rudists. (From Kauffman, E. G., and Sohl, N. F., 1973: Verh. Nat. Ges.)

The swimming ability of scallops and file shells is used primarily to escape predators or other sudden disturbing conditions. For example, if a predatory starfish, or even one tubefoot of such a starfish, touches the mantle margins of a scallop, it will evoke a swimming response and the scallop will be propelled a meter or so away. Some scallops use the water jets to blow out a depression in the sand surface into which they settle. File shells typically nest in crevices beneath stones and only swim when disturbed.

The Tridacnidae, which include the giant clams (*Tridacna*) of the Indo-Pacific, are also surface dwellers. Some species are attached by byssal threads and some bore, but others

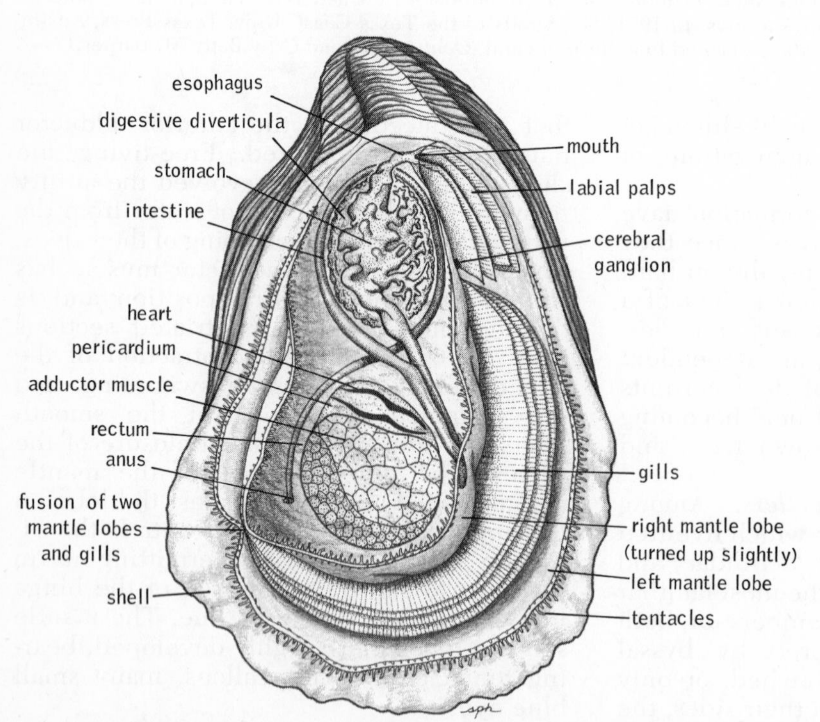

Figure 11–65. Anatomy of the American oyster *Crassostrea virginica*. (Modified after Galtsoff.)

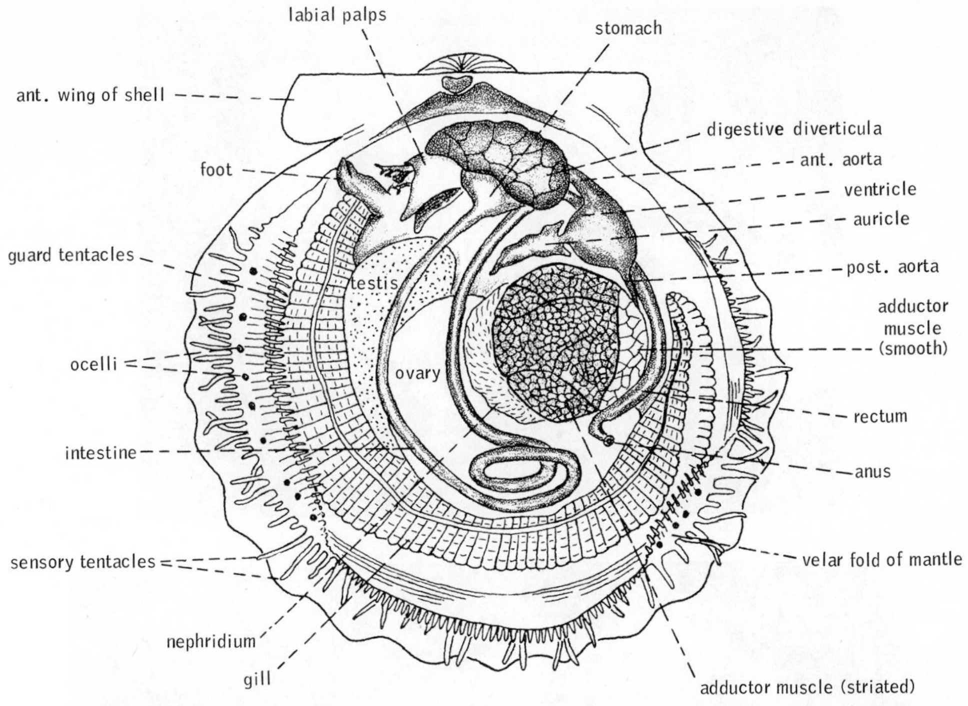

Figure 11–66. Structure of the scallop, *Pecten* (viewed from the left). (After Pierce.)

are unattached. They rest on the bottom with the gape directed upward and with the large mantle surface (actually the siphons) protruded across the greatly fluted shell for maximum exposure to light (Fig. 11–67). The mantle tissue contains symbiotic algae that provide the clam with an accessory source of nutrition.

Other free surface dwellers include some odd jingle shells, *Enignomia aenigmatica* (Anomiidae), which live on mangrove trees in Australia. Unlike other jingle shells, which are highly adapted for an attached existence (p. 392), these species can crawl up and down the mangrove trunks throughout life by means of a gastropod-like foot, which protrudes through a hole in the shell.

Boring Bivalves. The ability to penetrate and live beneath the surface of firm substrate—peat, clay, sandstone, shell, coral, coralline rock, and wood—has evolved in nine superfamilies of lamellibranchs. The ancestors of some of these boring bivalves were soft bottom inhabitants that evolved the ability to burrow in successively firmer substrates. The ancestors of others were surface inhabitants attached to hard substrates by byssal threads.

All boring bivalves begin excavation following settling of the larva, and slowly enlarge and deepen the burrow with growth. They are forever locked within their burrows, and only the siphons project to the small surface opening. If a boring bivalve is removed and placed on the surface, it cannot excavate a new chamber.

Boring bivalves can hold to the side of the burrow by byssal threads or by the ventral surface of the foot, which has developed a sucker-like surface (Fig. 11–68*B*). In the great majority of species, drilling is a mechanical process and the anterior ends of the valves, which are frequently serrated, are the abrading surfaces (Fig. 11–68*A*). The drilling movements are simply adaptations of the movements found in their non-boring ancestors. In those bivalves derived from surface dwelling forms, abrasion is produced by contraction of the byssal or pedal retractor muscle driving the anterior ends of the valves against the head of the burrow with an up thrust or a down thrust. In those derived from soft bottom ancestors, the abrasive force comes primarily from rocking movements. Coupled with rocking there may be an abrasive opening thrust of the anterior valve margin produced by the contraction of the posterior adductor. Modifications to increase valve articulation and

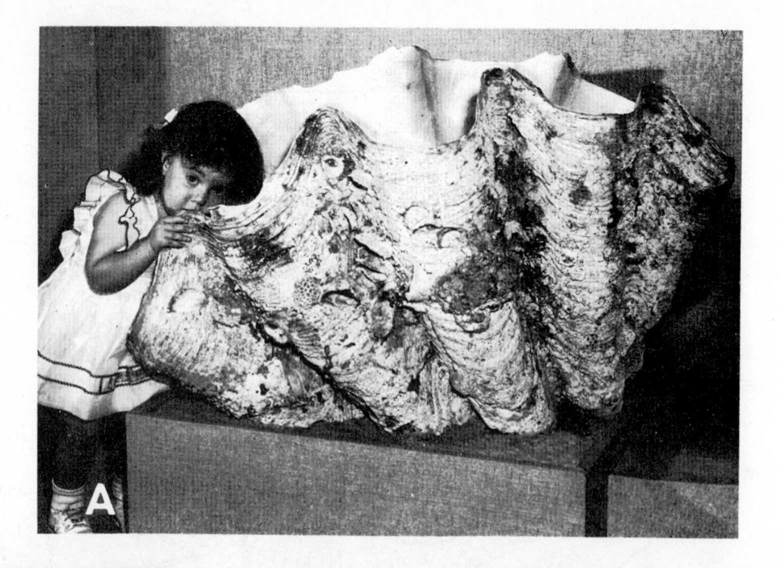

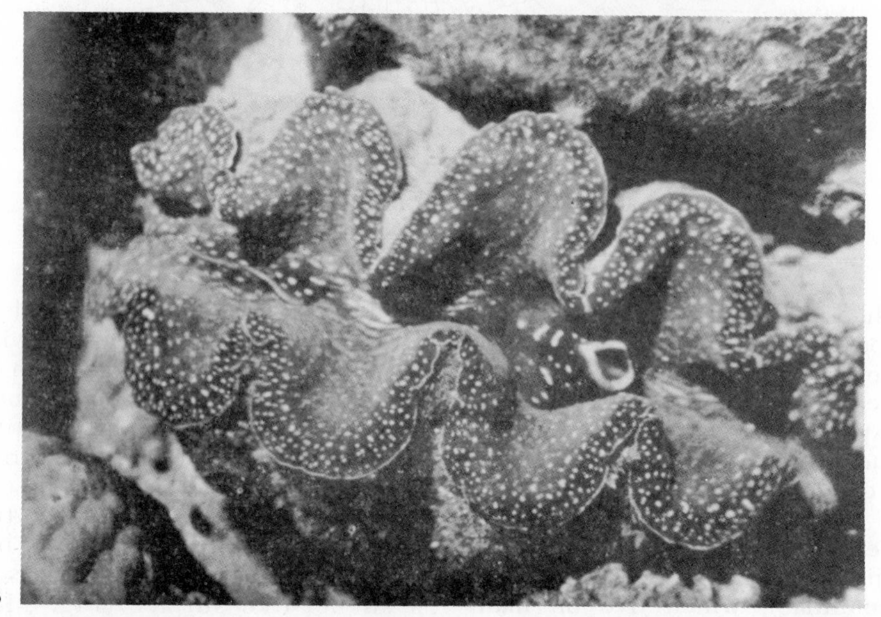

Figure 11–67. The giant clam, *Tridacna*. *A*. Shell. (Courtesy of Cranbrook Institute of Science.) *B*. *Tridacna derasa*, looking down at expanded specimen. Mantle extends over shell fluting. Conical aperture is exhalant siphon. (Courtesy of the British Museum.)

cutting movements include reduction of the hinge ligament and a shift of the anterior adductor muscle to a position *above* the hinge line. This region may then be covered by accessory shell plates (Pholadidae) (Fig. 11–68*A*).

Some boring bivalves rotate within the burrow, i.e., change position, and as a result the burrow cross-section is round. Others remain attached in one place, and the burrow tube takes the shape of the shell. Such forms may even have two openings to the surface if the ends of the siphons secrete some material between them (*Gastrochaena*) (Fig. 11–68*E*). The sediment produced in drilling is eliminated by taking it into the mantle cavity and then ejecting it with the pseudofeces or through the gut.

Lithophaga, a very common cigar-shaped borer in shell and coral, excavates chemically. An acid mucus secreted by the mantle margin softens the calcareous substratum, which is then scraped away with the valves. The shell of *Lithophaga* is protected from its own mucus by the periostracum.

Many boring bivalves inhabit wood. The wood-boring Pholadidae, such as *Martesia* and *Xylophaga*, are adapted in much the same way as the many rock- and shell-boring members of the same family. The most specialized wood borers are the shipworms, members of the family Teredinidae. The natural habitats of the some 60 species of this widely distributed family are mangrove roots and timber swept into the sea by rivers, and they play an important ecological role in the reduction of sea-borne wood. They are a serious pest of piers, pilings, and other wooden structures placed by man in the sea, and much expense and research has been devoted to anti-fouling measures.

The body of the shipworm is greatly elongated and cylindrical (Fig. 11–69*A*). The shell is reduced to two small anterior valves. Cutting of the wood is effected by an opening thrust of the valves while the anterior end of the body is attached to the burrow by the small foot. The mantle, enclosing the greater part of the body behind the valves, produces a calcareous lining within the tunnel. The long, delicate siphons open at the surface of the wood, and the burrow entrance is plugged by the pallets when the siphons are retracted. The burrow increases with the growth of the shipworm that fills it, and may reach a length of 18 cm. to 2 m., depending on the species. The life span is one to several

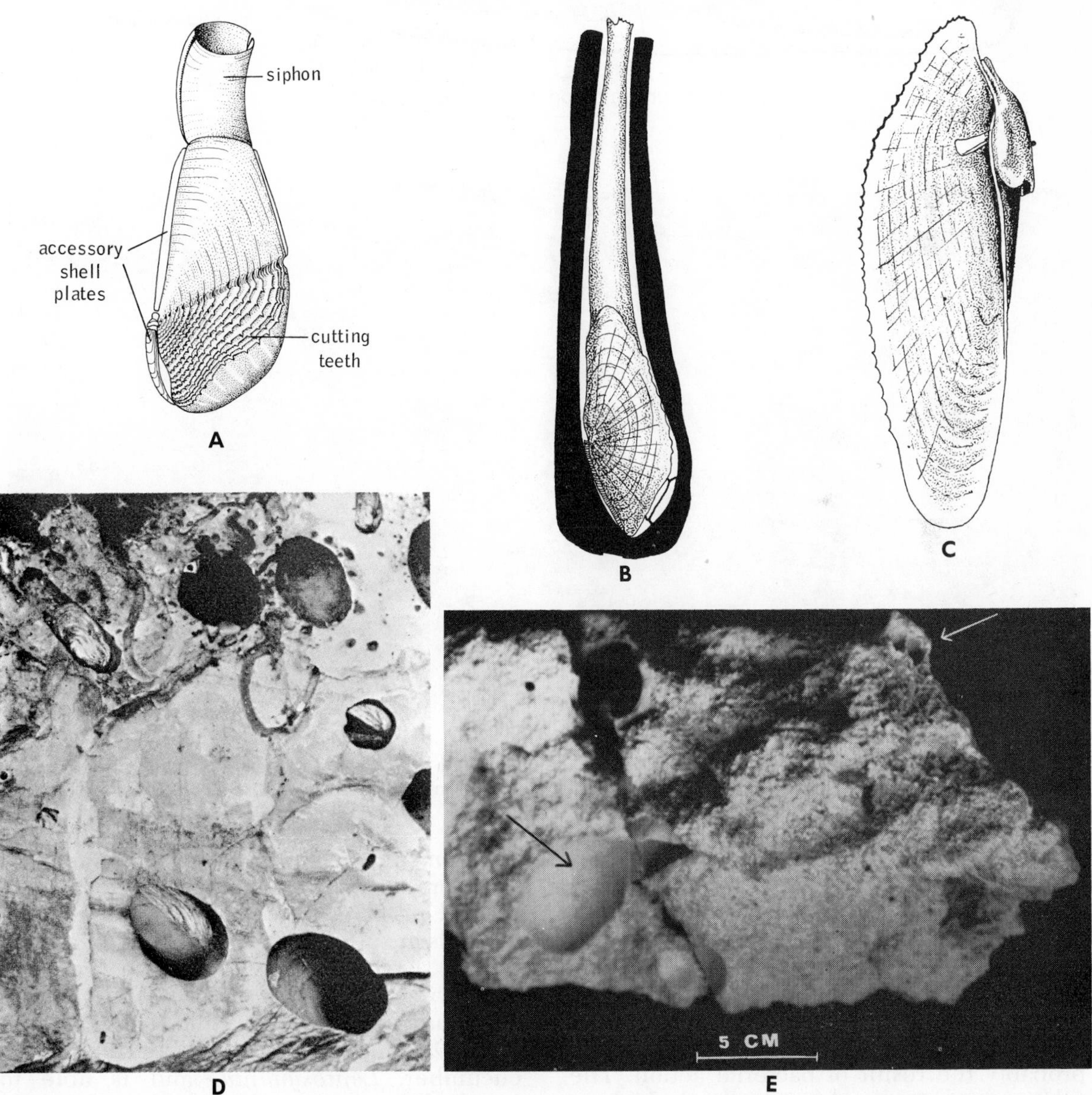

Figure 11–68. A. Structure of a pholadid. *B. Pholas* within burrow. Note attachment by foot and extension of siphons. *C.* The rock borer, *Barnea parva.* (*A-C* after Yonge.) *D.* Burrows of rock-boring bivalves; excavated in limestone. (By E. B. Wilson.) *E.* Burrow of *Gastrochaena hians,* a common West Indian borer in coralline rock. This species does not turn in burrow (black arrow), and the deposition of calcareous material around the siphon tips leads to each having a separate opening to the exterior (white arrow). (By Betty M. Barnes.)

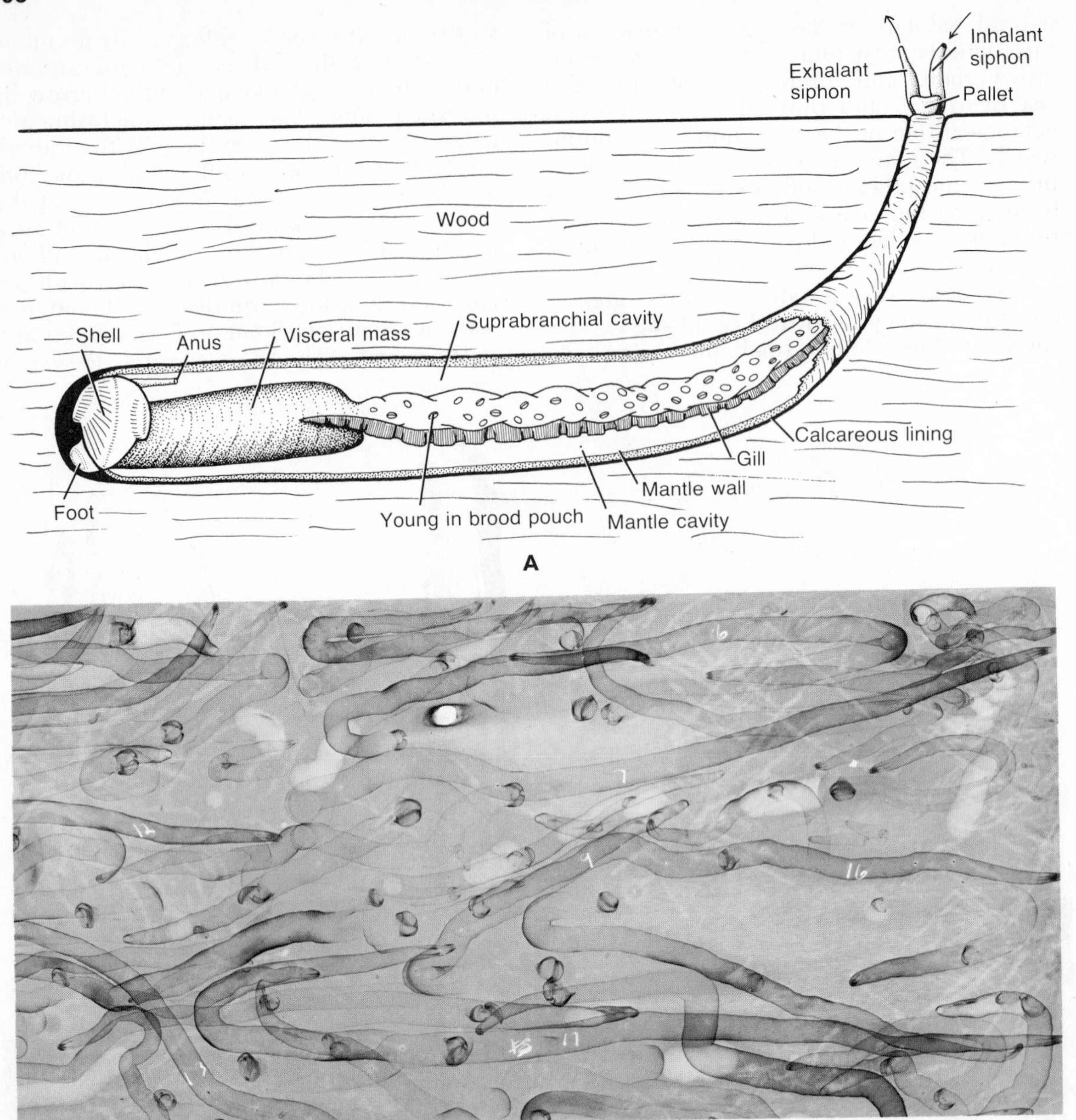

Figure 11–69. *A.* A shipworm, a wood-boring bivalve. (Modified from several sources.) *B.* X-ray photograph of a marine timber section showing shipworms. (By C. E. Lane in Scientific American, February, 1961.)

years, again depending on the species. Timbers can become completely riddled with tunnels (Fig. 11–69*B*).

Shipworms use the excavated sawdust for food. The stomach is provided with a caecum for sawdust storage, and a section of the digestive gland is specialized for handling wood particles. Cellulose digestion is probably the result of bacterial action. The relative importance of sawdust and plankton as food is reflected in the size of the caecum and the gills in different groups of shipworms.

Commensals and Parasites. A small number of bivalves have evolved a commensal or parasitic relationship. The commensals are almost all members of the superfamily Erycinacea, which also contains free-living forms such as *Lasaea*. The hosts are usually burrowing echinoderms. The bivalve *Devonia* lives on a small burrowing sea cucumber, *Leptosynapta*, and is able to crawl about over its host with a flattened foot. Others (*Montacuta, Phlyctaenachlamys*) live in the burrows of certain brittle stars, heart urchins, and mantis shrimp. The nature of the commensal dependence is not yet understood.

Entovalva, which lives in the gut of sea cucumbers, is the only known parasitic bivalve.

Digestion. The structure and physiology of the digestive tract of some of the deposit feeding protobranchs have retained a number of primitive features and are similar to those of archaeogastropods (Fig. 11–70*B*).

Food enters the mouth and is conducted posteriorly to the stomach through a short ciliated tubular esophagus. The stomach, which is surrounded by the large digestive gland, is divided into two regions: a dorsal portion, into which the esophagus and ducts of the digestive gland enter, and a ventral style sac. The dorsal portion of the stomach is lined by chitin except for a large folded and ciliated sorting region, into which the ducts of the digestive gland open. At the apex of the stomach is a toothlike projection arising from the chitinous girdle and called the gastric shield.

Food enters the dorsal part of the stomach and becomes wound into the large mu-

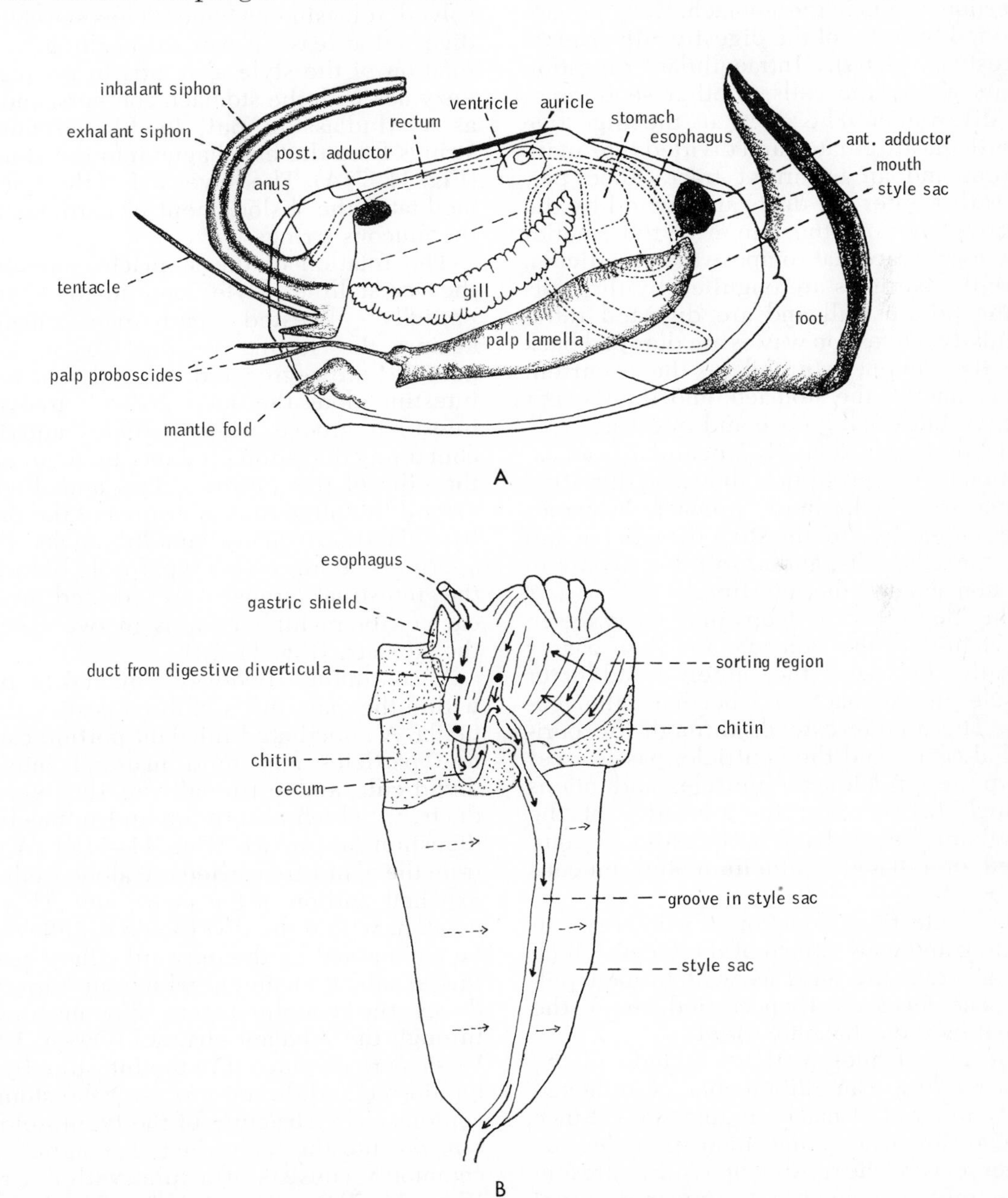

Figure 11–70. *A.* The protobranch, *Malletia,* with right mantle and valve removed (lateral view). *B.* Stomach of *Malletia,* opened. Solid arrows indicate food tracts; broken arrows indicate direction of beat of style-rotating cilia. (Both after Yonge.)

cous mass being rotated by the cilia of the style sac. As particles are freed from the mucus, they are thrown against the large sorting area. Here fine particles are conveyed by ciliated ridges to the duct openings of the digestive glands. Larger rejected particles are carried in the grooves of the sorting region down to the intestinal groove, which runs along the anterior wall of the style sac. The intestinal groove carries the rejected material directly into the intestine.

Digestion in the protobranchs is both extracellular and intracellular. Extracellular digestion occurs in the stomach. Enzymes are secreted by areas of the digestive diverticula (digestive gland). Intracellular digestion occurs within the cells of other sections of the diverticula. The ducts of the digestive diverticula are ciliated and divided into an incurrent and an excurrent tract. Food particles that enter the ducts are carried by the incurrent tract to the lumen of the smaller tubules and sacs that compose the diverticula. Here the particles are engulfed by the cells of the tubule wall and are digested intracellularly. Digestion wastes are dumped back into the lumen, returned by the excurrent ciliary tract to the stomach, and then swept into the intestinal groove and intestine.

The protobranch family Nuculidae is an exception to the description above, all digestion being extracellular in the stomach. Enzymes are secreted by the digestive diverticula and the stomach walls, and absorption occurs in the stomach and the intestine.

The long intestine loops once or twice in the vicinity of the stomach, and then passes dorsally between the anterior adductor muscle and stomach and becomes the rectum. The rectum extends through the pericardial cavity and the ventricle, passes over the posterior adductor muscle, and opens through the anus at the posterior of the suprabranchial cavity. The intestine is composed of ciliated epithelium and mucous glands.

The intestine functions primarily in molding and conveyance of the feces, which emerge from the anus as well formed pellets. The feces are then carried out of the animal by the exhalant current.

The use of finer particles as food in the filter feeding Lamellibranchia is reflected in a number of stomach modifications. Since a need for macerating food particles no longer exists, the girdle of chitin, present in the protobranchs, has disappeared except for a small plate that represent a true gastric shield (Fig. 11–71). A style sac is present, but the mucus has become consolidated into a very compact and often long rod, the crystalline style. The structure and function of the crystalline style are very similar to those in the gastropods. In addition to the protein matrix of the style itself, the style sac secretes amylase, cellulase, and probably lipase, which are adsorbed onto the style in its formation. The projecting anterior end of the style is rotated against the platelike gastric shield by the style sac cilia, and in the process the style end is abraded and dissolved, releasing enzymes. Thus starches are digested at least in part extracellularly. The rotation of the style also aids in mixing the enzymes with the stomach contents, and acts as a windlass to pull food-laden mucous strings from the esophagus into the stomach (Fig. 11–*71A*). The lower pH of the stomach facilitates the dislodgment of particles from the mucous strings.

The rotation of the stomach contents by the crystalline style continually throws partially digested particulate material against the sorting region. Coarse heavy particles are segregated out and sent to the intestine along the deep stomach groove, or intestinal groove. Fine particles and fluid containing digestion products are retained by the cilia of the sorting ridges and directed toward the numerous apertures of the digestive glands. In many lamellibranchs, these apertures, along with a typhlosole bordering the intestinal groove, are located in one caecum or in higher forms in two caeca of the stomach (Fig. 11–71).

A continuous two-way flow takes place within the main ducts of the digestive diverticula. A nonciliated inhalant portion carries fine particles and fluid material into the diverticula and eventually to the terminal ducts, in which absorption and intracellular digestion take place (Fig. 11–71*E*). Wastes from the gland are carried out along a ciliated exhalant portion of the main duct. This circulation within the diverticula is believed to be maintained by the outward ciliary beat in the exhalant channel, which in turn produces the countercurrent flowing inward through the inhalant channel (Owen, 1955). Wastes are conducted to the intestine by the typhlosole and do not mix with the stomach contents. The structure of the typhlosole extension into the main ducts is complex and commonly consists of a tube within a tube (Fig. 11–71*B*). The typhlosole plays an important role in regulating the flow of

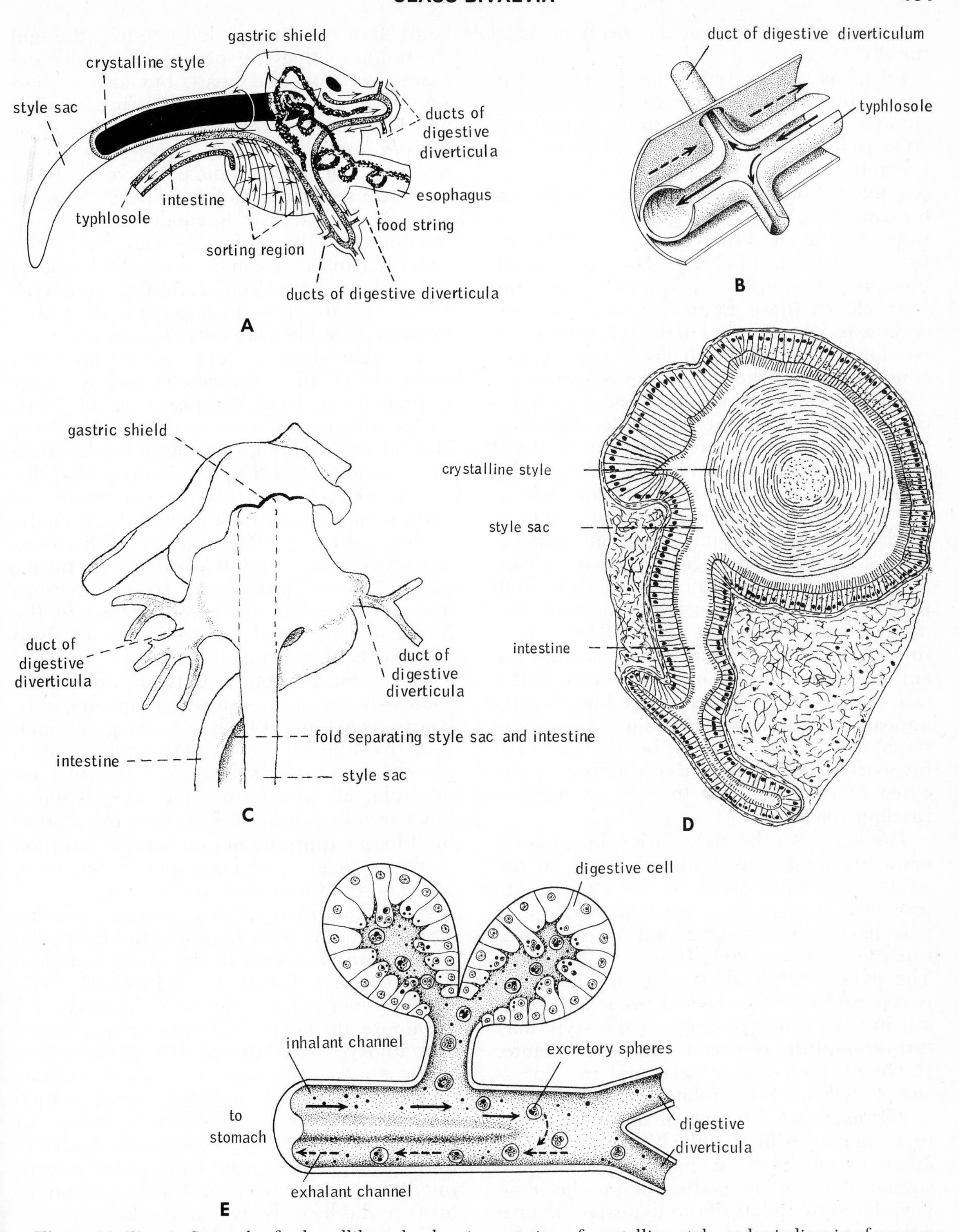

Figure 11–71. *A.* Stomach of a lamellibranch, showing rotation of crystalline style and winding in of mucous food string. Arrows indicate ciliary pathways. (After Morton.) *B.* Typhlosole within cecum of stomach, showing extensions into ducts of digestive diverticula. Solid arrows indicate inhalant ciliary currents; broken arrows indicate exhalant currents. (After Owen.) *C.* Stomach of the oyster *Ostrea edulis. D.* Style sac and intestine of the freshwater clam, *Lampsilis anodontoides* (transverse section). (*C* after Yonge; *D* after Nelson from Yonge.) *E.* Diagram of a section of digestive diverticulum, showing absorption and intracellular digestion of material passed inward from stomach (solid arrows) and outward passage of wastes (broken arrows). (After Owen.)

material into and out of the main ducts of the diverticula.

In many forms, such as *Donax*, *Mya*, *Ensis*, and *Meretrix*, the style sac has become separated from the stomach and the style is located in a sac that projects from the anterior of the intestine, or is located in the anterior of the intestine itself. In the latter condition, the style and secretory epithelium are separated from the intestinal lumen by a fold (Fig. 11–71*D*). The style itself, however, continues to project into the stomach. In these forms, rejected particles and waste are conveyed to the intestine along the fold on the stomach floor. This fold is continuous with the fold in the intestine.

The presence of a crystalline style precludes much, if any, extracellular digestion of proteins, since a protease would digest the style itself. The style sac continually secretes additions to the base of the style as the projecting anterior end of the style is eroded. In at least some intertidal bivalves, such as *Crassostrea virginica*, *Lasaea rubra*, and *Cardium edule*, which feed only at high tide, the different digestive processes display a tidal rhythm (Fig. 11–72). The crystalline style is dissolved at low tide when the animal is not feeding, and is reformed as the tide comes in. A circadian feeding rhythm appears to be present in some species of freshwater clams. It may be that in most bivalves, feeding and the different processes of digestion are to varying degrees rhythmic or phasic.

The length of the style varies, but it is remarkably long considering the size of the animal. In many bivalves, the style is approximately 3 cm. long; but a 12 cm. *Tagelus* may have a 5 cm. style, and Yonge (1932) reported a 36 cm. style in a 1 m. *Tridacna*. The style of the freshwater clam, *Anodonta*, is reported to rotate eleven times a minute, but in very young oyster spat the style may turn as rapidly as seventy times a minute. However, temperature, pH, and food pressure all influence the rotation rate.

The muscular stomach of the carnivorous septibranchs is lined with chitin and acts as a crushing gizzard. Supposedly, material squeezed from the bodies of prey is conveyed into the ducts of the digestive diverticula, where digestion occurs intracellularly. The style is reduced to a small rod, which barely projects into the lumen of the stomach.

Circulation and Gas Exchange. In the majority of bivalves, the ventricle of the heart has become folded around the gut (rectum), so that the pericardial cavity encloses not only the heart, but also a short section of the digestive tract (Figs. 11–55*B* and 11–73*B*). In a few Protobranchia such as *Nucula* and in *Ostrea*, shipworms, some mytilids, and some limids, the rectum and heart are separated, but even here the separation is apparently secondary. The contractions of the ventricle can be easily observed in large clams from which one of the valves has been carefully removed. Pulsations are slow; in *Anodonta* the pulsations average about twenty per minute.

A single anterior aorta issues from the ventricle in all protobranchs and in many filibranchs such as *Mytilus* (Fig. 11–74*A*). In the eulamellibranchs, a posterior aorta is also present, although it is often smaller than the anterior vessel (Figs. 11–66 and 11–73*B*). In *Mercenaria* and others, the posterior aorta is enlarged just before it issues from the pericardial cavity. The function of this muscular swelling, called the bulbus arteriosus, has not been determined. Its contractions may provide additional pumping force for the propulsion of blood, or it may be involved only in regulating blood pressure.

The typical open molluscan circulatory route—heart, tissue sinuses, nephridia, gills, heart—is exhibited in the bivalves, although some modifications of this circuit have taken place in different species. In *Mytilus*, for example, all blood from the body sinuses flows into the kidneys (Fig. 11–74*A*). Part of the blood within the nephridium is returned to the auricles; part circulates through the gills and then returns to the kidney.

In *Anodonta*, only a small amount of blood is sent from the nephridium to the heart; the major portion is sent to the gills, and from there it returns directly to the heart (Fig. 11–74*B*). Nevertheless, in both *Mytilus* and *Anodonta*, the heart receives both oxygenated and unoxygenated blood. In all bivalves, there is a more or less well-developed circulatory pathway through the mantle, which is an additional site of oxygenation. In *Pecten* and *Anodonta*, the mantle pathway forms a separate circuit with blood returning directly to the heart; in *Mytilus*, blood returns to the heart by way of the kidney.

Gas exchange takes place as water moves dorsally within the gills. The amount of oxygen removed from the water current is quite low compared to that in other mollusks (2.5 to 6.8 per cent for the scallop compared to 48 to 70 per cent for the abalone).

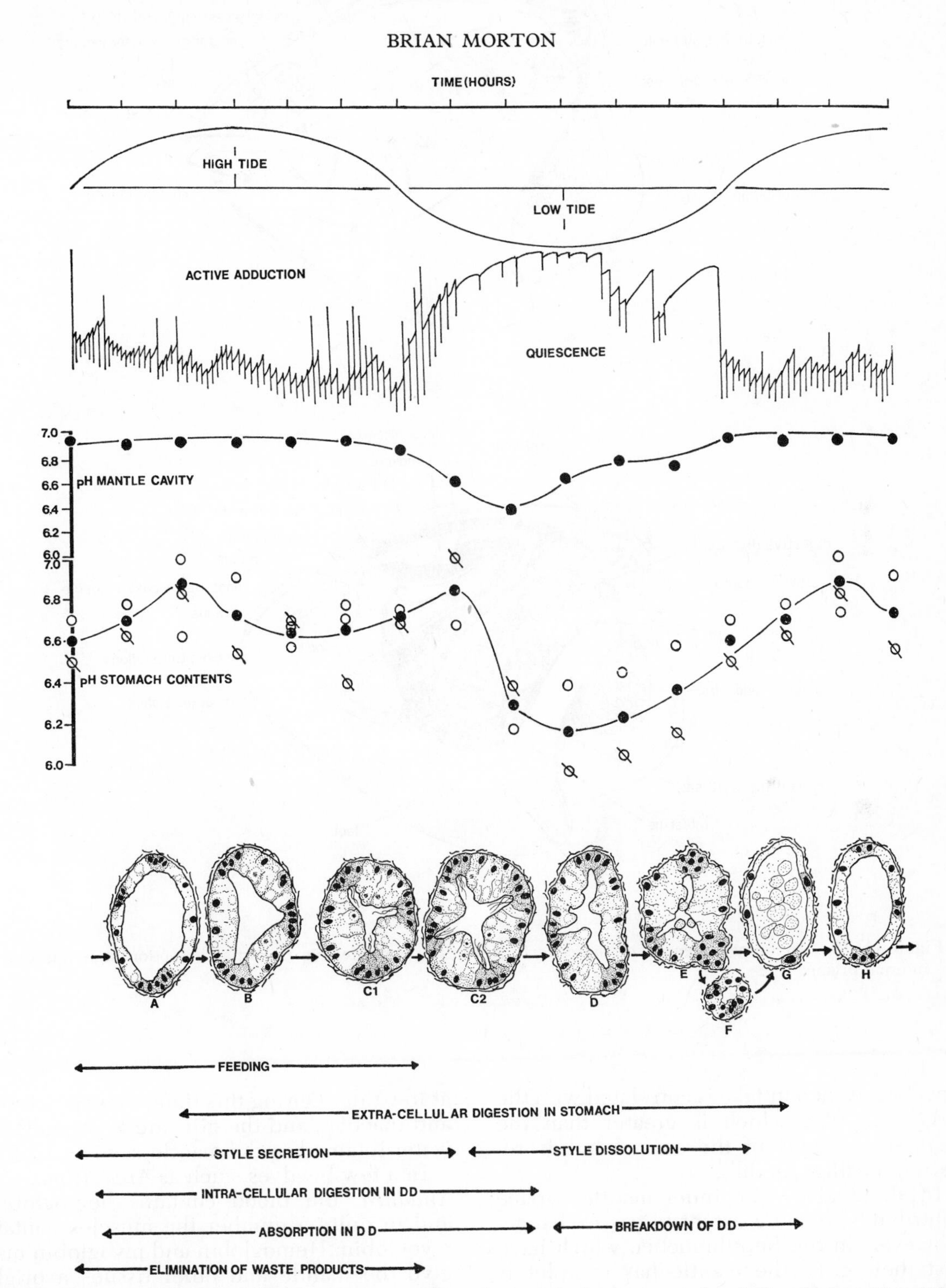

Figure 11–72. Feeding and digestive processes of *Cardium edule* during a period including high and low tides. (From Morton, B., 1970: J. Mar. Biol. Assoc. U. K., *50*:506.)

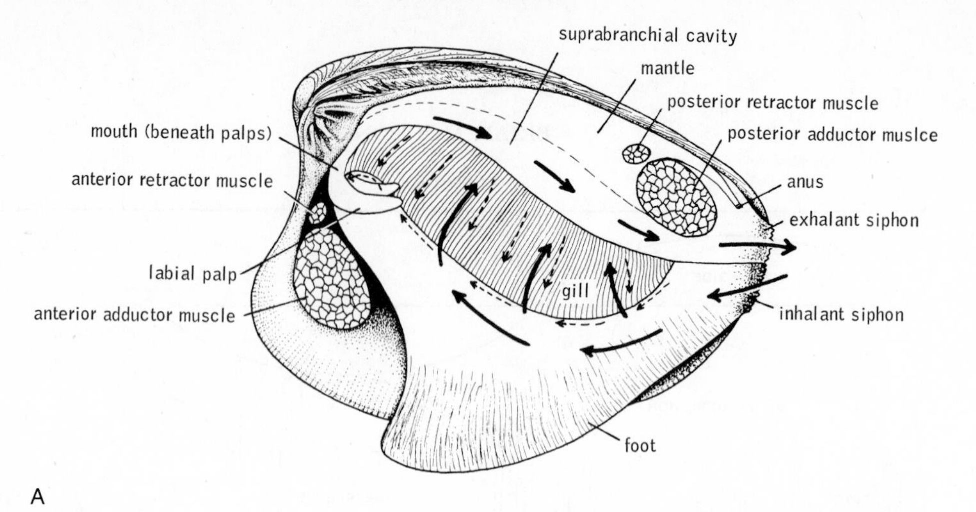

Figure 11–73. Anatomy of *Mercenaria mercenaria*. *A*. Interior of right valve. *B*. Partial dissection, showing some of the internal organs.

This low oxygen uptake is correlated with the large gill size, which is greater than the respiratory needs of the animal but is required for filter feeding.

In all bivalves, the inner mantle surface contributes to some extent to gas exchange. However, in the Septibranchia, which have lost their gills, the mantle has completely taken over this function.

Although bivalves are strictly aquatic, many groups, such as mussels and oysters, live in the intertidal zone and are exposed to air at low tide. During this time, they are closed and inactive, and the gills are kept moist by water retained in the mantle cavity.

In a few bivalves, such as *Arca*, *Lima*, and *Anadara*, the blood contains hemoglobin, and in quite a number the muscles contain myoglobin. Hemoglobin and myoglobin may give the mantle and other tissues a bright red color. The blood of most bivalves, however, lacks any respiratory pigment.

Excretion. The two nephridia of bivalves are located beneath the pericardial cavity or slightly posterior to it, and are folded to form a long **U** (Figs. 11–55*B*, 11–66, and 11–73*B*). One arm of the **U** is glandular and opens into the anterior of the pericardial cavity; the other arm forms a bladder and opens through a nephridiopore at the anterior of the suprabranchial cavity. In the primitive Protobranchia, there are no distinct bladder and glandular portions; instead the unfolded walls of the tube are glandular throughout.

Except for random representatives of different marine families such as the oysters, which can tolerate low salinities and have invaded brackish estuaries and marshes, the freshwater bivalves are members of six families—the Sphaeriidae, the Corbiculidae, the Margaritiferidae, the Unionidae, the Etheriidae (tropical freshwater "oysters" of the Southern Hemisphere), and the Mutelidae (mussels of the Southern Hemisphere). North American freshwater bivalves are primarily members of the Unionidae and the Sphaeriidae, although one Japanese *Corbicula* has been introduced into the North American drainage system.

As in the gastropods, freshwater bivalves excrete large amounts of water through the nephridia. Most salts are reabsorbed and the urine is very hypotonic.

Nervous System and Sense Organs. The nervous system is bilateral and relatively simple. There are three pairs of ganglia and two pairs of long nerve cords. On each side of the esophagus is located a cerebropleural ganglion, which is connected to its opposite member by a short commissure dorsal to the esophagus. From each cerebropleural ganglion arise two more or less posteriorly directed nerve cords. The upper pair of nerve cords (one from each ganglion) extend directly back through the viscera and terminate in a pair of closely adjacent visceral ganglia located on the anteroventral surface of the posterior adductor muscle. The second pair of cords arising from the cerebropleural ganglia extend posteriorly and ventrally into the foot and connect with a pair of pedal ganglia. Foot movement and the anterior adductor muscle are under the control of the pedal and cerebral ganglia, but the visceral ganglia control the posterior adductor muscles and the siphons. Coordination of pedal and valve movements is a function of the cerebral ganglia.

The margin of the mantle, particularly the middle fold, is the principal location of most of the bivalve sense organs. In many

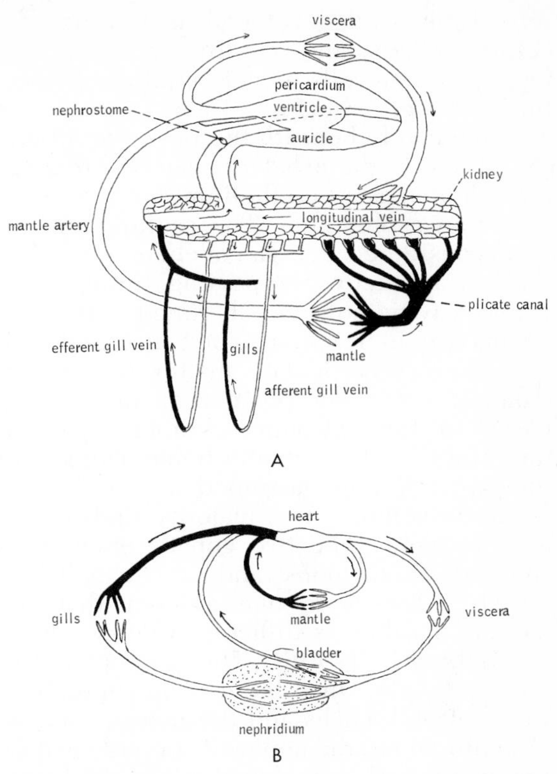

Figure 11–74. *A*. Circulatory system of *Mytilus*. (After Field from Borradaile and others.) *B*. Circulatory system of the fresh-water clam, *Anodonta*. (After Borradaile and others.)

species, the mantle edge bears pallial tentacles, which contain tactile and chemoreceptor cells. The entire margin may bear tentacles, as in the swimming *Pecten* and *Lima* (Fig. 11–66); more commonly tentacles are restricted to the inhalant or exhalant apertures or siphons, or they may even fringe the pedal aperture. *Malletia* and other Nuculanidae are peculiar in that they possess only one long tentacle attached to the underside of the base of the fused siphons (Fig. 11–70A).

A pair of statocysts is usually found in the foot of bivalves and is either located near or embedded in the pedal ganglia. In the primitive *Nucula*, the statocyst remains a pitlike form and is connected to the outer surface by means of a canal. The statocysts of some attached forms, such as oysters, are reduced, and in the scallop the left statocyst is much larger than the right.

Ocelli, which can detect sudden changes in light intensity, may be present along the mantle edge or even on the siphons in some bivalves; in most cases, they are simple pigment-spot ocelli. However, in *Spondylus* and the swimming *Pecten*, the ocelli are remarkably well developed, consisting of a cornea, a lens, and a retina in which the photoreceptor cells are directed away from the source of light (Fig. 11–66).

Immediately beneath the posterior adductor muscle in the exhalant chamber, there is a patch of sensory epithelium that is usually called an osphradium. The osphradium has been considered an organ of chemoreception for testing the water passing through the mantle cavity; however, its function is by no means certain. It may also be concerned with detection of particulate matter in the circulating water. The position of this sensory tissue in the exhalant chamber makes it doubtful that it is actually homologous with the osphradium of gastropods.

Reproduction. The majority of bivalves are dioecious. The two gonads encompass the intestinal loops and are usually so closely adjacent to one another that the paired condition is difficult to detect (Figs. 11–55*B* and 11–73*B*). The gonoducts are always simple, since there is no copulation. In the protobranchs and the more primitive lamellibranchs, the nephridia provide exit for the sperm and eggs, but the gonads do not open into the pericardial cavity; instead they connect to the nephridia directly. In general, the short gonoduct opens into the upper part of the nephridium in the protobranchs and into the lower part nearer the nephridiopore in other bivalves. In most Lamellibranchia, the gonoducts are no longer associated with the nephridia and open separately into the mantle cavity, but still very close to the nephridiopore. In *Mercenaria* and many other species, the gonopore and nephridiopore are located on a common papilla.

The hermaphroditic bivalve species include *Poromya,* shipworms, some species of cockles, oysters, and scallops, the freshwater Sphaeriidae, and a few Unionidae. In hermaphroditic scallops, the gonad is divided into a ventral ovary and a dorsal testis, both of which lie on the anterior side of the adductor muscle (Fig. 11–66). The European and North American oysters, *Ostrea edulis* and *Crassostrea virginica,* are protandric hermaphrodites, although populations of *C. virginica* along the east coast of the United States are reported to be unisexual. *O. edulis* not only shifts from male to female, as is true of many protandric hermaphrodites, but also changes back from female to male. An individual may exhibit both active male and female phases each year.

Embryogeny. In most bivalves, fertilization occurs in the surrounding water; the gametes are shed into the suprabranchial cavity and then swept out with the exhalant current. Factors reported to induce spawning include water temperature, tides, and substances produced by the gametes of the opposite sex.

In bivalves which brood their eggs, fertilization occurs within the mantle cavity by sperm brought in with the ventilating current. Brooding may take place within the suprabranchial cavity, as in some shipworms, or within the gills, as in *Ostrea edulis* and the freshwater Unionidae and Sphaeriidae.

The development of a free-swimming trochophore, succeeded by a veliger larva, is typical in marine bivalves. The structure of these larvae is essentially like that in the gastropods, but the veliger is always symmetrical, since torsion does not take place (Fig. 11–75). The shell and shell gland are present at first in the form of a single dorsal plate, which then grows laterally and ventrally. As a result, the dorsal plate becomes folded and forms the two valves characteristic of bivalves.

The veliger of some marine lamellibranchs is long-lived, larval life lasting for as long as several months. In others it is brief. Some bivalves, such as *Ostrea edulis* and species of shipworms, are larviparous, releasing the veligers following an initial period of brooding (eight days in *O. edulis*).

Metamorphosis is characterized by a sudden shedding of the velum. Settling may be random, with only a few individuals locating on suitable substrates, or it may involve considerable testing of the substratum and delayed metamorphosis. *Ostrea edulis* swims upward and attaches to the shaded underside of objects. Shipworms will settle only on wood.

With the exception of *Dreissena* and *Nausitoria,* which have a free-swimming veliger larva, freshwater bivalves exhibit modified development. Direct development is characteristic of the freshwater Sphaeriidae, which brood the eggs in the water tubes between the gill lamellae. Although direct, embryonic development shows stages similar to those in the larvae of marine bivalves. Both the nonmotile trochophore and a veliger are present; but the prototroch is very much reduced, and the velum is absent. At the completion of development, the young clams are shed from the gills.

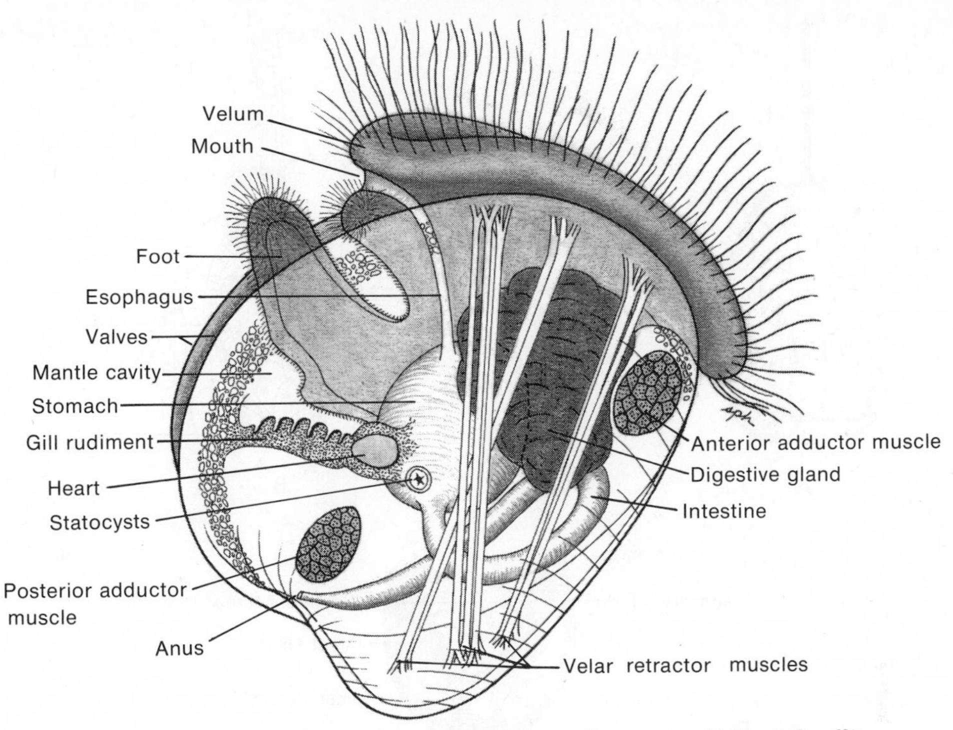

Figure 11–75. A fully developed veliger larva of an oyster. (After Galtsoff.)

The freshwater Unionidae and Mutelidae display an indirect but very specialized development. As in the sphaeriids, the eggs are brooded between the gill lamellae, where they develop through the veliger stage. However, in the Unionidae and Mutelidae the veliger, called a glochidium or a haustorius, has become highly modified for a parasitic existence. The glochidium larva is enclosed by two valves, each edge of which bears a hook in *Anodonta* (Fig. 11–76*C*). The shell valves cover a larval mantle, which bears four clusters of sensory bristles. A rudimentary foot is present, to which is attached a long adhesive thread. There is neither mouth nor anus, and the digestive tract is rather poorly developed.

When mature, the glochidia range in size from .05 mm. to .5 mm., depending upon the species. In *Unio* and *Anodonta,* the glochidia leave the gills through the suprabranchial cavity and exhalant aperture. In *Lampsilis,* the glochidia emerge directly from gills through temporary openings. In either case, the freed larvae sink to the bottom. Here they remain until they come in contact with a fish swimming over the bottom.

The glochidia of *Anodonta,* which bear hooks, immediately clamp onto the fins and other parts of the body surface of the fish. Hookless glochidia are picked up by the respiratory currents of the fish and attach to the gills. In either case, the tissue of the fish in the vicinity of the attached glochidium is stimulated to grow around the parasite and form a cyst. The larval mantle contains phagocytic cells that feed on the tissues of the host and obtain nutrition for the developing clam. During this parasitic period, which lasts from ten to thirty days, many of the larval structures (sensory bristles on the mantle, the adhesive thread, the larval adductor muscle, the larval mantle) disappear, and the adult organs begin developing. Eventually the immature clam breaks out of the cyst, falls to the bottom, and burrows in the mud. Here the remainder of development is completed, and the adult habit is gradually assumed.

Some of the larger freshwater mussels may produce as many as 3,000,000 glochidia, and a single fish has been reported to contain 3000 glochidia. Adult fish are apparently not harmed by the parasitic glochidia, but young fry may die from secondary infection.

Some bivalve glochidia require certain species of fish as hosts; others can tolerate a wide range of host species. A genus of Unionidae, *Simpsoniconcha,* parasitizes the mud puppy, *Necturus,* but is reported also to be able to undergo direct development within the parent gills.

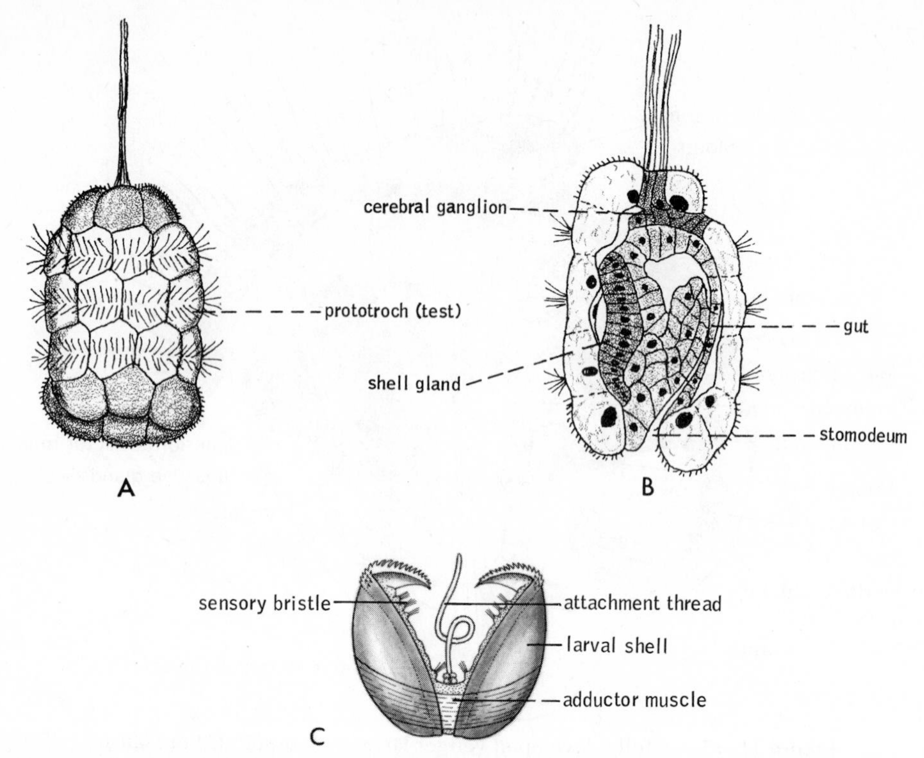

Figure 11–76. *A* and *B*. Trochophore of *Yoldia limatula*. *A*. External view. *B*. Longitudinal section. (Both after Drew from Dawydoff.) *C*. Glochidium of the fresh-water clam, *Anodonta*. (Redrawn from Harms.)

SYSTEMATIC RÉSUMÉ OF CLASS BIVALVIA

The following older system of classification has the advantage of simplicity and is widely encountered. An alternative and perhaps better system, though technically more difficult, is that proposed by Newell (1969) in the Treatise on Invertebrate Paleontology.

Subclass Protobranchia. Primitive. Gill filaments not folded; foot with flattened ventral surface; palpal proboscides frequently present. *Nucula, Nuculana, Yoldia, Solemya, Malletia.*

Subclass Lamellibranchia, or Polysyringia. Gill filaments folded and adjacent filaments attached by ciliary or tissue junctions.

Order Taxodonta. Hinge teeth numerous and similar; anterior and posterior adductor muscles approximately equal in size. Mantle margins unfused; gill filaments unattached, with no interlamellar junctions. *Arca, Anadara, Barbatia.*

Order Anisomyaria. Gills of the filibranch type with interlamellar junctions. Usually sessile; foot small; anterior adductor reduced or absent. No siphons. Mussels—*Mytilus, Modiolus;* the oysters—*Ostrea, Crassostrea, Spondylus, Pinctada; Anomia; Lima;* the scallops—*Pecten; Pinna;* the rock-boring *Lithophaga.*

Order Heterodonta. Hinge teeth consisting of a few large cardinal teeth with or without elongated lateral teeth. Gills of the eulamellibranch type; adductor muscles of equal size. Mantle fused, to varying degrees; siphons commonly present. *Cardium;* the freshwater Sphaeriidae and Corbiculidae; the common edible clam, *Mercenaria; Lasaea;* the commensals *Montacuta* and *Phlyctaenachlamys;* the boring clams—*Petricola, Tagelus;* the commensal clams—*Lepton; Dosinia; Donax; Tridacna; Congeria* and *Dreissena* of fresh and brackish water.

Order Schizodonta. Hinge teeth variable in size, shape. Gills of the eulamellibranch type. Freshwater Unionidae—*Unio, Elliptio, Lampsilis, Anodonta,* and *Simpsoniconcha;* the freshwater Margaritiferidae, Mutelidae, and Etheriidae.

Order Adapedonta. Hinge teeth reduced or absent; ligament reduced or absent, and shell gapes. Gills of the eulamellibranch type; mantle margins sealed except for pedal gape; long siphons. Deep burrowing and boring clams. Razor clams—*Solen, Ensis; Panope; Mya;* boring clams—*Hiatella, Barnea, Martesia, Xylophaga, Pholas, Gastrochaena;* the shipworms—*Teredo.*

Order Anomalodesmata. No hinge teeth. Gills of eulamellibranch type but outer demibranch reduced and turned up dorsally. Hermaphroditic. *Lyonsia; Pandora;* watering-pot shells—*Clavagella.*

Subclass Septibranchia. Gills modified as a pumping septum between inhalant chamber and suprabranchia cavity. *Poromya; Cuspidaria.*

CLASS SCAPHOPODA

The class Scaphopoda contains about 200 species of burrowing marine mollusks that are popularly known as tusk or tooth shells. These names are derived from the shape of the shell, which is an elongated cylindrical tube usually shaped like a trumpet or elephant's tusk. Both ends of the tube are open, and when tusk- or tooth-shaped, as in *Dentalium,* the diameter of the tube increases from one end to the other, and the whole tube is slightly curved (Fig. 11–77*B*). In the genus *Cadulus,* the shell is somewhat cucumber-shaped (Fig. 11–77A).

The shells of most scaphopods average 3 to 6 cm. in length, but *Cadulus mayori,* found off the Florida coast, does not exceed

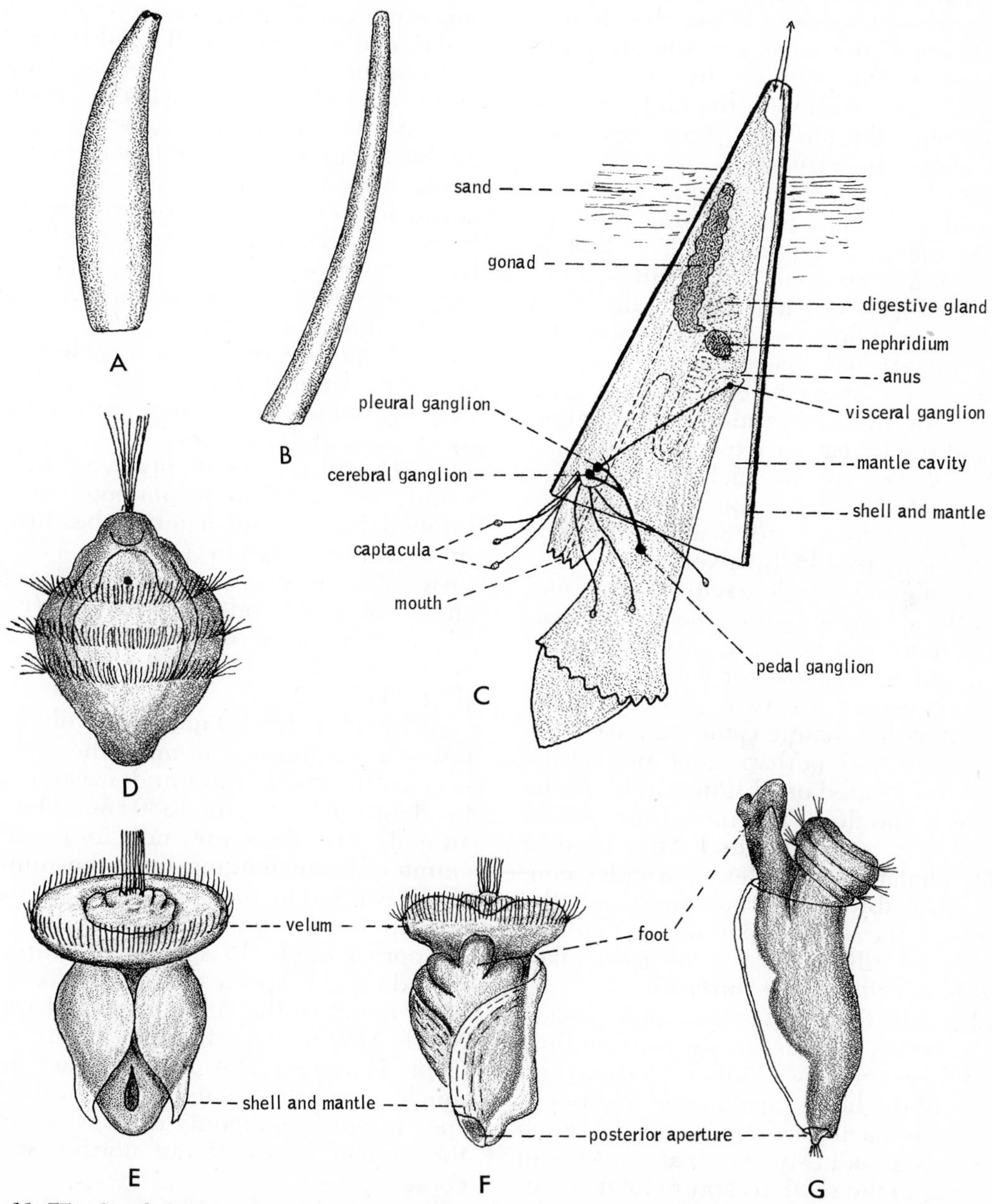

Figure 11–77. Scaphopoda. *A.* Shell of *Cadulus. B.* Shell of *Dentalium.* (Both after Abbott.) *C.* Structure of *Dentalium.* Arrows indicate water current direction through mantle cavity. (After Naef from Borradaile and others.) *D* to *G.* Development in *Dentalium. D.* Trochophore larva. *E.* Veliger. *F* and *G.* Metamorphosis. (*D* and *G* after Lacaze-Dutheirs from Dawydoff.)

4 mm. in length. *Dentalium vernedei*, found off Japan, is the largest living species, reaching a maximum length of 15 cm. However, a fossil species of *Dentalium* has a shell 30 cm. long with a maximum diameter of well over 3 cm. The shells of most scaphopods are white or yellowish, but an East Indian species of *Dentalium* is a brilliant jade green.

The body of the scaphopod is greatly elongated along the anterior-posterior axis (Fig. 11–77*C*). The head and foot project from the larger and anterior aperture of the shell, and the small shell aperture marks the posterior of the body. When the shell is slightly curved, the concavity lies over the dorsal surface. Scaphopods live buried in sand, head downward with the body steeply inclined; only the small posterior aperture projects above the surface.

Adapted to a burrowing habit, the head is reduced to a short conical projection, or proboscis, bearing the mouth. The foot is shaped like a short cone in *Dentalium*, and can be projected downward into the sand when the animal burrows. An encircling lobe that can be erected like a flange increases anchorage. In species of *Cadulus* and allied genera, the foot is more or less elongated, and the tip can be expanded to form a disc that serves as an anchor. Following anchorage of the foot, retraction pulls the shell and body downward. Scaphopods thus burrow essentially like bivalves.

The scaphopod mantle cavity is large and extends the entire length of the ventral surface. The posterior aperture serves for both inhalant and exhalant water currents. Water circulation occurs in two phases. Water slowly enters the mantle cavity as a result of ciliary action and perhaps foot protection. The cilia are located on oblique ridges of the mantle wall and floor. The remaining mantle surface is very weakly ciliated. After 10 to 12 min. of inhalation, a violent muscular contraction (probably foot retraction) expels the water from the same opening it entered. There are no gills; exchange of gases takes place through the mantle surface.

Scaphopods feed on microscopic organisms, especially forams, in the surrounding sand and water. Two lobes, located on each side of the head, bear a large number of threadlike tentacles called captacula. Each tentacle has an adhesive knob at the tip and extends out in the sand to capture food. Small food particles are conveyed back to the mouth by tentacular cilia; larger particles are grasped by the tentacle tips and brought to the mouth by retraction of the tentacle (Gainey, 1972). The buccal cavity contains a median jaw and a well-developed radula with large flattened teeth. The radula functions in the ingestion of the prey seized by the captacula. The stomach and the digestive gland are located in the middle of the body, and digestion is extracellular. The intestine extends anteriorly from the stomach and then bends back posteriorly to open through the anus into the mantle cavity.

The circulatory system is reduced to a system of blood sinuses, and there is no heart. The nervous system is ganglionated and not concentrated. The typical cerebral, pleural, pedal, and visceral ganglia and their corresponding nerve cords are present. However, there are no eyes, tentacles, or osphradia. A pair of nephridia is present, and the nephridiopores are located near the anus.

Scaphopods are dioecious. The unpaired gonad fills most of the posterior part of the body, and the sperm or eggs reach the outside by way of the right nephridium. The eggs are shed singly and are planktonic. Fertilization is external.

Scaphopod development very closely parallels that of the marine bivalves. There is a free-swimming trochophore larva, succeeded by a bilaterally symmetrical veliger (Fig. 11–77*D* and *E*). As in bivalves, the larval mantle and shell in scaphopods are at first bilobed, but then the mantle lobes fuse along their ventral margin (Fig. 11–77*E*). This fusion thus results in a cylindrical mantle and shell that remain open at each end (Fig. 11–77*F*). Metamorphosis is gradual and is accompanied by elongation of the body (Fig. 11–77*G*).

Although a few scaphopods inhabit shallow water or burrow in mud, the majority of species burrow in sand under water ranging in depth from 6 to 1830 m. The living animals are therefore not frequently encountered, but judging from the number of shells washed up on beaches, scaphopods are not rare benthic animals.

Approximately 14 species of scaphopods, including 11 species of *Dentalium*, have been found in the Atlantic coastal waters of North America; the majority occur south of Cape Hatteras. The Pacific coast fauna is much sparser, although the shells of one species were commonly used as currency by the Indian tribes of the northwest Pacific coast.

Several lines of evidence seem to indicate that scaphopods are an offshoot from the ancestral bivalve stock. The symmetry of the body and its orientation within the shell, the reduction of the head, the burrowing habit, the symmetrical veliger, and the embryonic bilobed mantle and shell are strikingly similar to the respective bivalve characteristics. The captacula may well be homologous to the proboscides of protobranchs.

CLASS CEPHALOPODA

The class Cephalopoda contains the nautili, cuttlefish, squids, and octopods. Although some cephalopods such as the octopus have secondarily assumed a less active bottom-dwelling habit, the class as a whole is adapted for a swimming existence and contains the most specialized and highly organized of all mollusks. The head projects into a circle, or crown, of large prehensile tentacles, or arms, which are homologous to the anterior of the foot of other mollusks. In the evolution of the cephalopods, the body became greatly lengthened along the dorso-ventral axis; and, as a result of a change in the manner of locomotion, this axis became the functional anterior-posterior axis (Fig. 11–78). The circle of tentacles is thus located at the anterior of the body, and the visceral hump is posterior. The original posterior mantle cavity is now ventral.

The cephalopods have attained the largest size of any invertebrates. Although the majority range between about six and seventy cm. in length, including the tentacles, some species reach giant proportions. The largest cephalopods are the giant squids, *Architeuthis* (Fig. 11–83), of the North Atlantic, one specimen of which has been reported to have attained a total body length of 16 m. including the tentacles. The tentacles alone were 6 m. long, and the circumference of the body was 4 m. Giant octopods exist only in stories. The body of *Octopus punctatus* (Pacific coast), one of the larger species of this group, does not exceed 36 cm., although the rather slender arms may reach 5 m. in length.

From an evolutionary standpoint, cephalopods appear to be a waning group, for only about 650 species now exist, as compared to the more than 7500 different fossil forms. The class first appeared in the Cambrian period and then, once during the Paleozoic era and once in the Mesozoic era, underwent two great periods of evolutionary development with the formation of many species. The abundance and number of fossil species make the cephalopods an extremely important group of index fossils.

The Shell. A completely developed shell is found only in *Nautilus* and the fossil representatives of the class. In squids and cuttlefish the shell is reduced and internal, and in the octopods it is completely lacking. The shell of *Nautilus* is coiled over the head in a bilaterally symmetrical planospiral (Fig. 11–79). Only the last two whorls

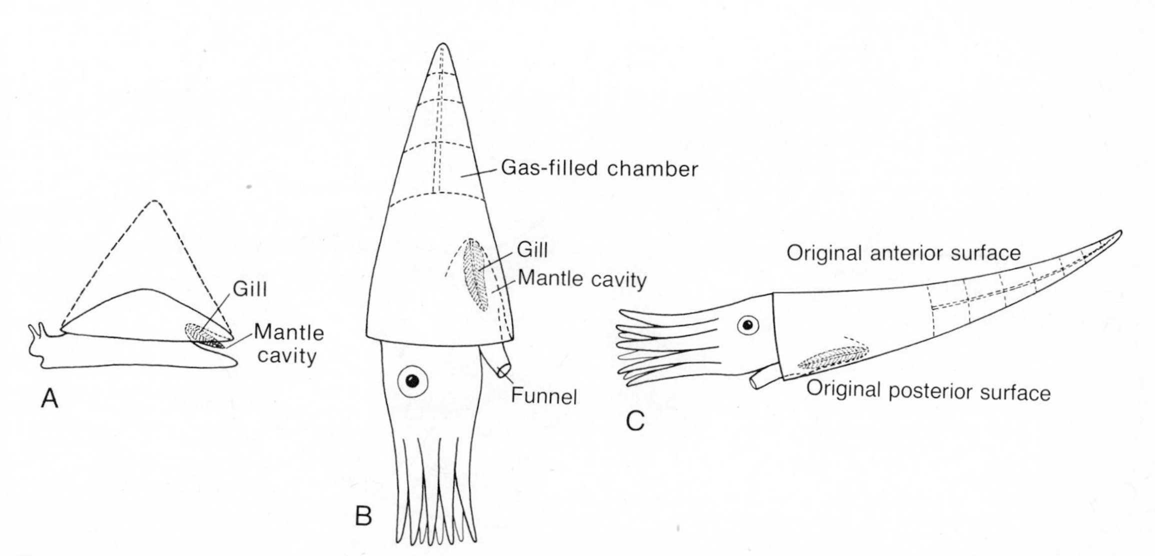

Figure 11–78. Evolution of the cephalopod form. *A.* Lateral view of hypothetical ancestral mollusk. Dashed line indicates increase in length of dorso-ventral axis and a change in the shape of the shell from a shield to a cone. *B.* An early cephalopod oriented in a position that is comparable to the ancestral mollusk. *C.* Actual swimming position of an early cephalopod.

are visible, since they cover the inner whorls. Although coiled, the shell of *Nautilus* is radically different from the gastropod shell. The shell of *Nautilus*, like all cephalopod shells, is divided by transverse septa into internal chambers and only the last chamber is occupied by the living animal. As the animal grows, it periodically moves forward, and the posterior part of the mantle secretes a new septum.

Viewed from the anterior, each septum is concave—that is, is bowed backward. Where the septum joins the wall of the shell, there is a scar, or suture. The septa are perforated in the middle, and through the opening a cord of body tissue, called the siphuncle, extends from the visceral mass. The siphuncle secretes gas into the empty chambers, making the shell buoyant and allowing the animal to swim. The shell is composed of an outer porcelaneous layer, containing prisms of calcium carbonate in an organic matrix, and an inner nacreous layer. The outer surface is often tinted with alternating bands of orange and white, or it may be pearly white.

Not all fossil cephalopods had coiled shells, nor were the shells of all coiled

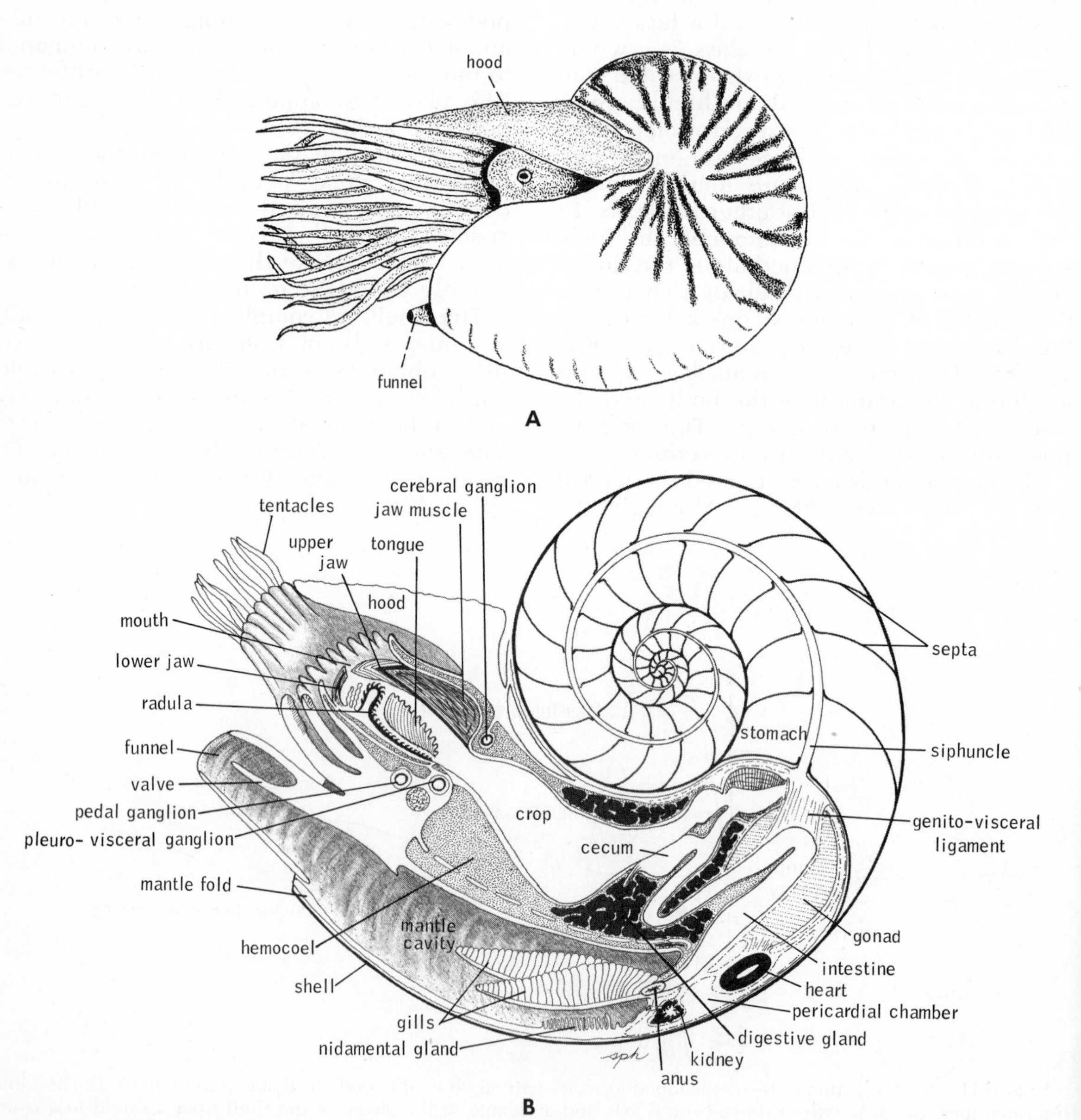

Figure 11–79. *A. Nautilus*, the only living shelled cephalopod, shown swimming. (After Moore and others.) *B*, Sagittal section. (After Stenzel.)

species similar to *Nautilus*. The ancestral cephalopods probably possessed a cone-shaped, curved shell (Donovan, 1964). The apical chambers perhaps were filled with fluid, and only the more recently formed chambers were filled with gas. The animal would thus have swum in a horizontal position, with the convex side of the shell directed upward (Fig. 11–80*I*). Straight cones may have evolved from curved cones in conjunction with increased swimming efficiency. Fluid in the apical chambers or a heavy apical calcareous deposit acted as a counterweight. The earliest fossil cephalopod shells have either curved or straight cones. The straight shells of some species from the Ordovician period exceeded 5 m. in length and the aperture was 36 cm. in diameter. However, a straight or curved shell is not always an indication of primitiveness, for at different periods in the course of cephalopod evolution, the shells of various groups became secondarily uncoiled and straight again (Fig. 19–62).

Coiling in the fossil cephalopods typically resulted in a planospiral shell (Fig. 11–80*C*), but the apex in a few groups was drawn out out as in the snails (Fig. 11–80*D*). In some cases the shell was so loosely coiled that the whorls were unconnected (Fig. 11–80*B*); others displayed fused coils, but these coils were not always impressed into one another, as in *Nautilus*, and each whorl was visible (Fig. 11–80*C*). Many were similar to *Nautilus*. The coils of some fossil forms were exogastric as in *Nautilus*—that is, the ventral surface was located along the outer perimeter of the whorl (Fig. 11–80*G* and *H*). Others were endogastric—the shell was coiled beneath the animal, and the dorsal surface was on the outside of the whorl (Fig. 11–80*I* and *J*). The largest fossil species with a coiled shell was *Pachydiscus seppenradensis* from the Cretaceous period, which had a shell diameter of 3 m. However, many of these fossil cephalopods were small species with shells only 3 cm. in diameter.

One of the most important characteristics for the classification of fossil cephalopods is the nature of the internal sutures—the point of junction between the septum and the wall of the shell. Although the sutures were internal, the chambers became filled with sediment in the course of preservation, and the details of the suture pattern have been beautifully preserved on the outer surface of the filling. The simplest suture lines were straight or slightly waved, as in *Nautilus* (Fig. 11–80*K*); but one large group, the ammonoids, developed elaborate sutures that were zigzagged (Fig. 11–80*L*), or more frequently minutely crinkled (Fig. 11–80*M*). Such sutures indicate a corresponding complexity in the nature of the septum and the part of the mantle that secreted it.

The current system of cephalopod classification separates those forms with complete shells (the old subclass Tetrabranchia) into two subclasses—the Nautiloidea and the Ammonoidea. The Nautiloidea is characterized by straight or coiled shells and simple sutures. The Nautiloidea first appeared in the Cambrian period and is represented today by *Nautilus*. All members of the Ammonoidea were coiled and displayed complex septa and sutures. They appeared in the Silurian period after the nautiloids, and disappeared at the end of the Cretaceous period.

We know nothing concerning the soft parts of fossil ammonoids and nautiloids and can only assume that they were somewhat similar to those of *Nautilus*.

Cephalopods with internal shells or without shells are placed in the subclass Coleoidea (the old subclass Dibranchia). These cephalopods are believed to have evolved from some early straight-shelled nautiloid, and during this evolutionary stage the shell became completely enclosed by the mantle. The basic structure of the shell is illustrated by the fossil belemnoids, the oldest known coleoids. Without introducing any new nomenclature, the belemnoid shell differs from that of the nautiloids only in that the apical and lateral walls were greatly thickened and that the anterior dorsal wall projected as a shield, to which the mantle muscles were attached (Fig. 11–81).

From some primitive belemnoid form the coleoid shell evolved in four directions, all leading to modern forms and all involving a reduction in the weight of the shell. This evolutionary development is illustrated in Figure 11–81. In *Spirula*, a common worldwide deep-water cuttlefish, the thickened wall and shelf have disappeared, and the shell has become coiled. In the evolutionary line leading to *Loligo*—the long bodied squids and the genus to which the common squid of the North American Atlantic coast belongs—all of the shell, including the septa, has disappeared except the shelf and a strip of the dorsal wall. The shell is horny and is called a pen.

A third evolutionary line, represented by the European cuttlefish *Sepia*, has retained the septa, but the shelf and thickened parts of the wall have disappeared. Finally, in

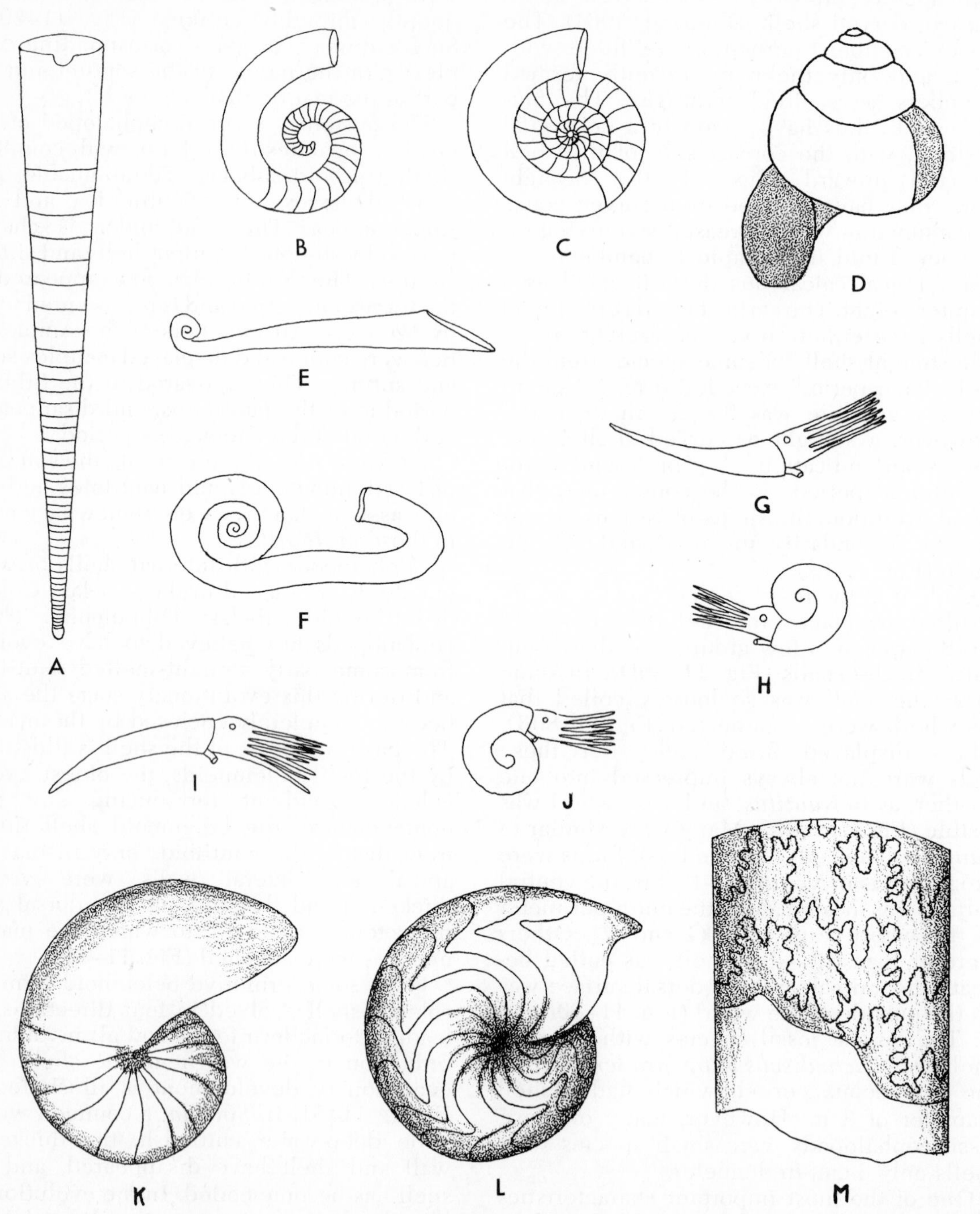

Figure 11–80. Fossil cephalopod shells. *A.* An early straight shell. *B.* Loosely coiled shell. *C.* Planospiraled shell (*A* to *C* after Flower from Shrock and Twenhofel.) *D.* Asymmetrically coiled shell. Shading indicates area occupied by animal. (After Moore and others.) *E* and *F.* Modified coiled shells. *E. Baculites*, from the Cretaceous period. *F. Scaphites* from the Cretaceous period. *G* and *H.* Exogastric coiling, in which ventral surface is located along outer perimeter of shell. *I* and *J.* Endogastric coiling, in which dorsal surface is located along outer perimeter of shell. (*E* to *J* after Shrock and Twenhofel.) *K. Nautilus undulatus*, a Cretaceous cephalopod with simple sutures. (After Sharp from Shrock and Twenhofel.) *L.* The Mississippian *Muensteroceras rotarius*, an ammonoid with angular sutures. (After Shrock and Twenhofel.) *M. Buculites*, a Cretaceous ammonoid with complex sutures. (After Meek from Shrock and Twenhofel.)

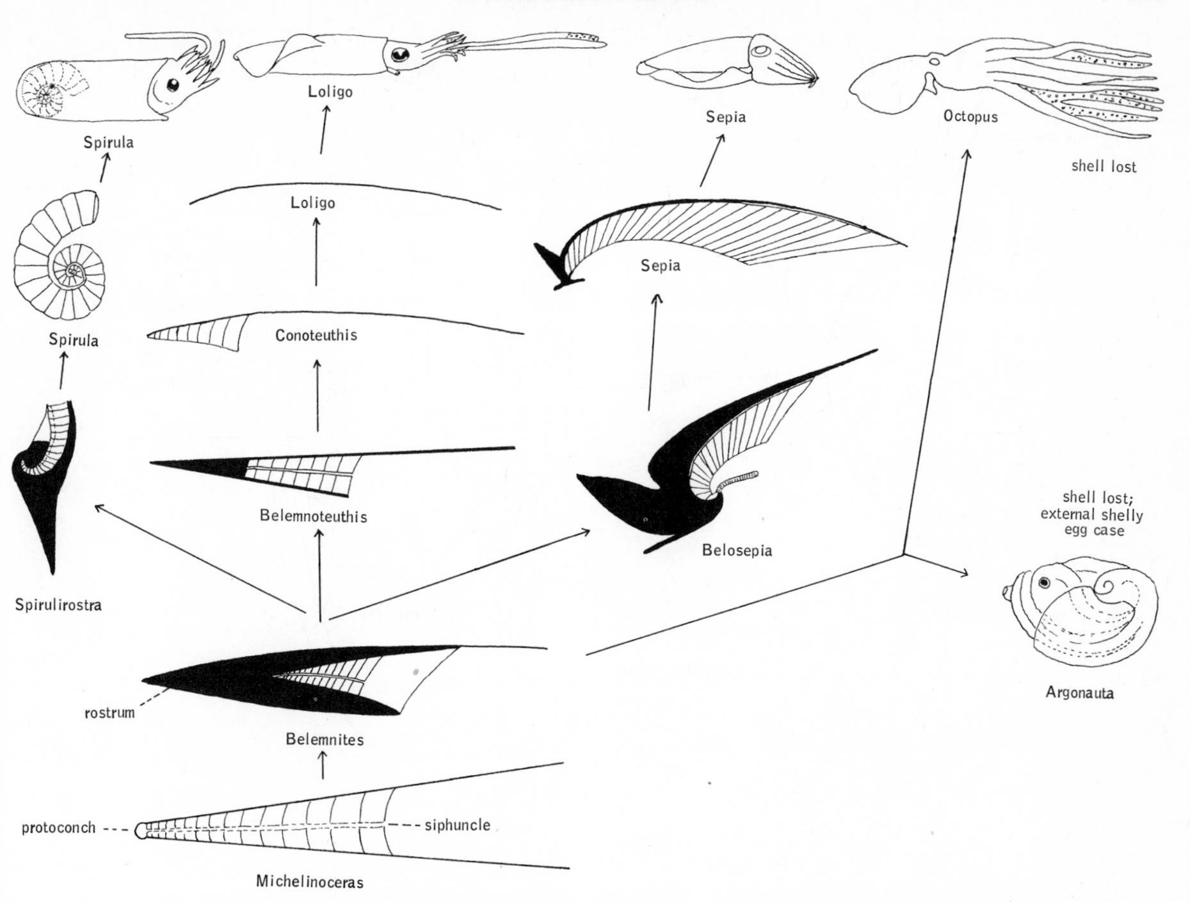

Figure 11–81. Evolution of shells of the Coleoidea. (After Shrock and Twenhofel.)

Octopus the shell has disappeared completely. Not only has the shell of squids undergone great reduction to facilitate swimming, but the body has become more rigid through the development of an internal cartilaginous skeleton. Cartilage plates are located in the mantle wall, in the "neck" region, and surrounding the brain. The cartilage plates that surround the brain form a protective encasement.

Locomotion and Adaptive Diversity in Cephalopods. Most cephalopods swim by rapidly expelling water from the mantle cavity. The mantle contains both radial and circular muscle fibers. During the inhalant phase of water circulation, the circular fibers relax and the radial muscles contract. This action increases the volume of the mantle cavity, and water rushes in dorsally, laterally, and ventrally between the anterior margin of the mantle and the head (Fig. 11–82*B*). When the mantle cavity is filled, the action of the mantle muscles is reversed.

The contraction of the circular muscles not only increases the water pressure within the cavity but also locks the edges of the mantle tightly around the head. Thus water is forced to leave through the ventral tubular funnel. The force of water leaving the funnel propels the animal in the opposite direction. The funnel is highly mobile and can be directed anteriorly or posteriorly, resulting in either forward or backward movement. The rapidity of movement depends largely on the force with which water is expelled from the funnel, but the movements are beautifully controlled. The fastest movement is achieved in backward escape swimming, when powerful contractions of the mantle eject water from the anteriorly directed funnel.

The squids and cuttlefish are best adapted for swimming by water jet. They can hover, perform subtle swimming movements, slowly cruise, or dart rapidly. Squids attain the greatest swimming speeds of any aquatic invertebrate. In general, the squid body is long and tapered posteriorly, and has a pair of posterior lateral fins, which function as stabilizers. The "flying squids" (Onycoteuthidae), which have long tapered bodies and

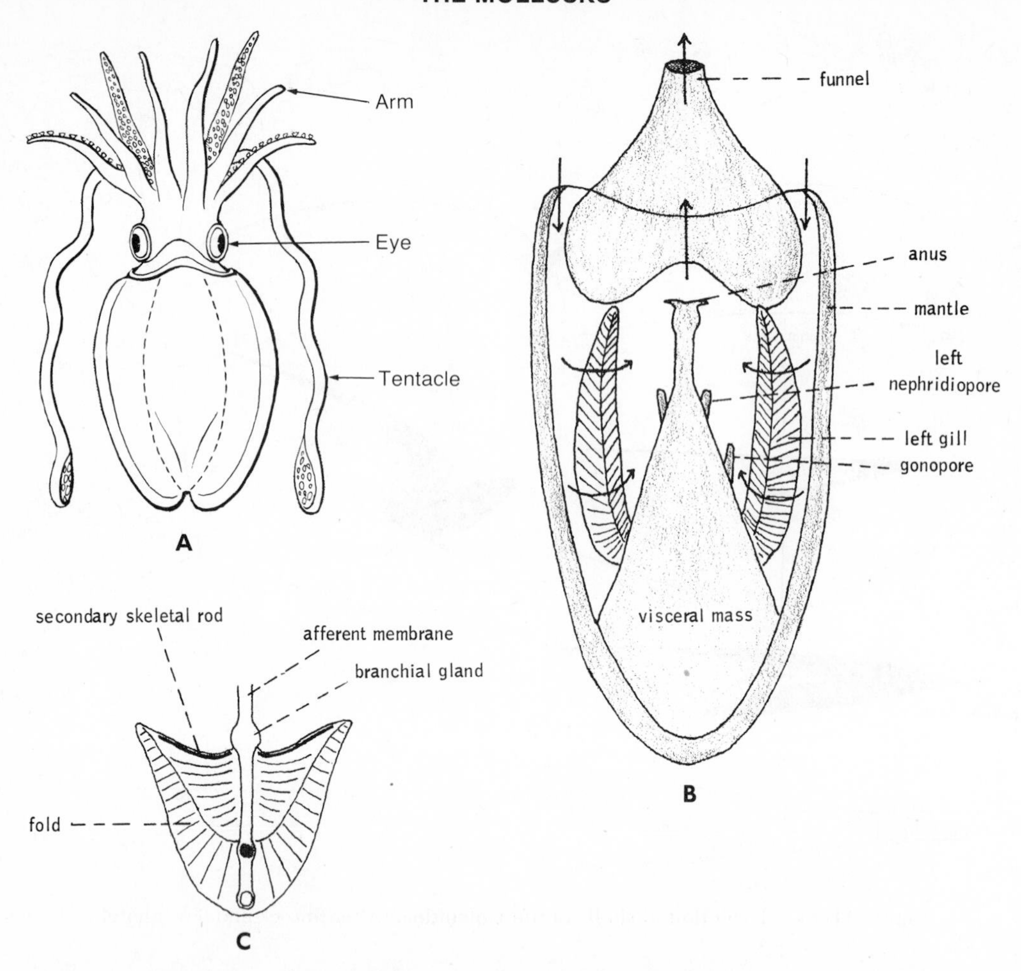

Figure 11–82. *A.* Dorsal view of a cuttlefish. *B.* Diagrammatic frontal section through a cuttlefish, showing direction of water current through mantle cavity (ventral view). *C.* Gill filament of a squid. (*B* and *C* after Yonge.)

highly developed fin vanes and funnel, shoot out of the water and glide for some distance. There have been reports of squids leaping onto the decks of ships 12 ft. above the water's surface. *Dosidicus gigas* reaches a speed of 16 m.p.h. in the air.

The giant architeuthid squids inhabit depths of 300 to 600 m. over the continental slopes (Fig. 11–83). They are not rapid swimmers. Two other interesting groups of deep-water squids are the bathypelagic chiroteuthids and the cranchiids. The chiroteuthids are rather small squids, which have very long slender bodies—the thickness of a pencil in *Dorotopsis*—and long whiplike tentacles (Fig. 11–84*A*). The cranchiids are planktonic and small and are often strangely shaped (Fig. 11–84*C*). Two thirds of the entire body volume of cranchiids is provided by the enormous fluid-filled coelom, which functions as a buoyancy chamber. Most of the seawater cations are replaced in the coelomic fluid with ammonium ions derived from the metabolic wastes. Although isotonic with sea water, the coelomic fluid has a specific gravity (1.010) below that of sea water (1.022 at 50 m. depth and 28°C) and counteracts the heavier body tissues of the animal (specific gravity 1.046). A similar buoyancy mechanism exists in certain other families of oceanic squids (Histioteuthidae, Oetopoteuthidae, and Chiroteuthidae), but many small vacuoles distributed throughout the body, rather than the coelom, constitutes the buoyancy chamber (Gilpin-Brown, 1972).

The body of cuttlefish, such as *Sepia* and *Spirula*, tends to be short, broad, and flattened (Fig. 11–82*A*). Cuttlefish are slower swimmers than the more streamlined squids. As in squids, the fins in cuttlefish function as stabilizers, but they undulate and are also used for steering and propulsion.

The shell of cuttlefish, despite its reduction, still functions in providing buoyancy for the animal. In *Sepia* and *Spirula*, spaces between the thin septa contain fluid and gas—largely nitrogen. By regulating the relative amounts of fluid and gas, the degree of bouyancy can be varied. Light is an important factor controlling the regulating mechanism. During the day the cuttlefish lies buried in the bottom; at night the animal becomes active, swimming and hunting for food. Buoyancy decreases when the animal is exposed to light and increases in the dark.

The smallest cuttlefish are species of the genus *Idiosepius*, which are about 15 mm. long. They live in tide pools and possess a dorsal sucker on the mantle for attaching to algae.

Nautilus is active at night and rests on the bottom during the day, attached with its tentacles to rubble or the walls of crevices. Whether it is swimming or resting, the gas-filled chambers keep the shell upright and provide buoyancy which counteracts the shell weight. Since *Nautilus* ranges from depths near the surface to greater than 600 m., the amount of gas must be regulated to accommodate the animal to the pressures of different depths. The work of Denton and Gilpin-Brown (1966) suggests that the regulatory mechanism is essentially like that

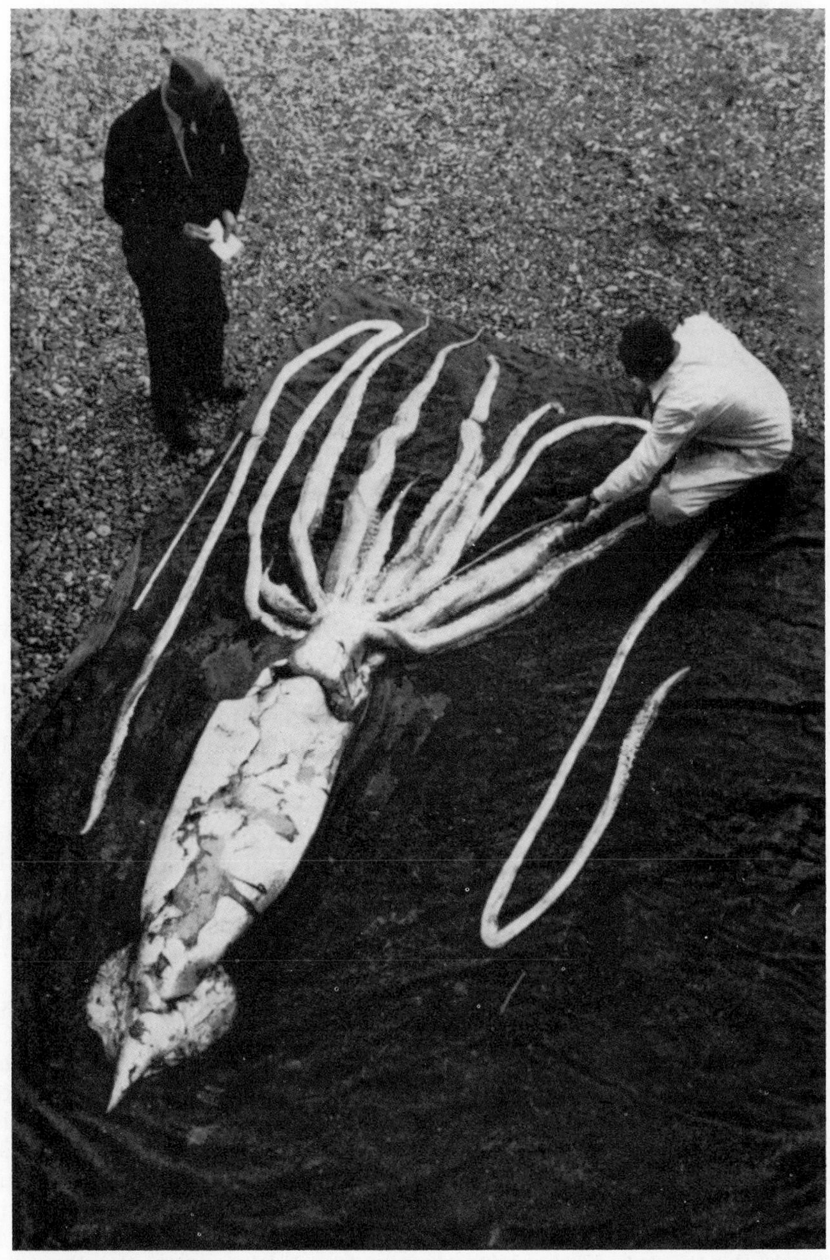

Figure 11–83. A giant squid of the genus *Architeuthis* stranded at Rahneim, Norway in 1954. (By Clarke, M. R., 1966: Adv. Marine Biol., *4*:103.)

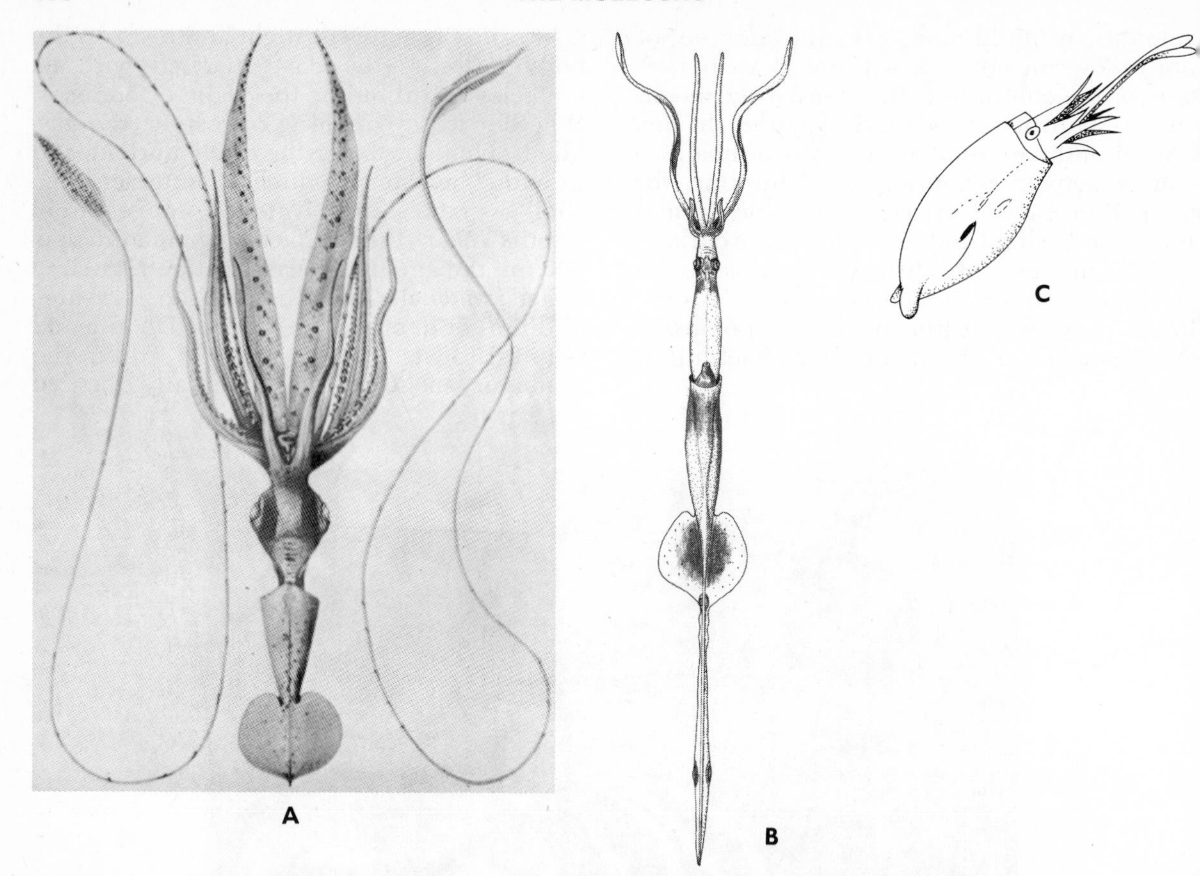

Figure 11–84. Bathypelagic cephalopods. *A* and *B*. The squid *Chiroteuthis veranyi*. Mature (*A*) and immature (*B*) individuals. (After Pfeffer from Lane.) *C*. The squid *Cranchia*. (After Morton.) *Figure 11–84 continued on opposite page.*

of *Sepia*. Gas and fluid are exchanged through the siphuncle to the chambers. Salts are actively pumped in by the living tissue of the siphuncle. Water then passively diffuses out of the chamber and is replaced by gas—chiefly nitrogen—which diffuses into the vacated space. Fluid exchange is restricted to the more anterior chambers. In fact, fluid fills the space between the last septum and the posterior end of the animal when a new septum is being secreted. Fluid is retained in the chamber until the septum is sufficiently strong to withstand pressure changes.

Except when feeding, *Nautilus* swims backward, with about the same speed as man doing a slow breast stroke (Haven, 1972). The locomotor mechanism is the same as that of squids, although the ejection of water through the funnel results from the retraction of the body and contraction of the muscles of the funnel rather than by the mantle.

The octopods have reverted to more sedentary habits. The body is globular and bag-like, and there are no fins (Fig. 11–85). The mantle edges are fused dorsally and laterally to the body wall, resulting in a much more restricted aperture into the mantle cavity. Although octopods are capable of swimming by water jets, with tentacles trailing (Fig. 11–85A), they more frequently crawl about over the rocks around which they live. The arms, which are provided with adhesive suction discs, are used to pull the animal along or anchor to the substratum. Species of *Octopus* usually occupy a den or retreat, from which they make feeding excursions and which they defend.

Two different modes of existence than that usually associated with *Octopus* have evolved in octopods. *Argonautus* and related genera are pelagic surface dwellers. On the other hand, a number of families of deep-water octopods, some bathypelagic and some abyssalbenthic, have abandoned the conventional method of locomotion by water propulsion. Instead, the arms have become webbed like an umbrella and the animal

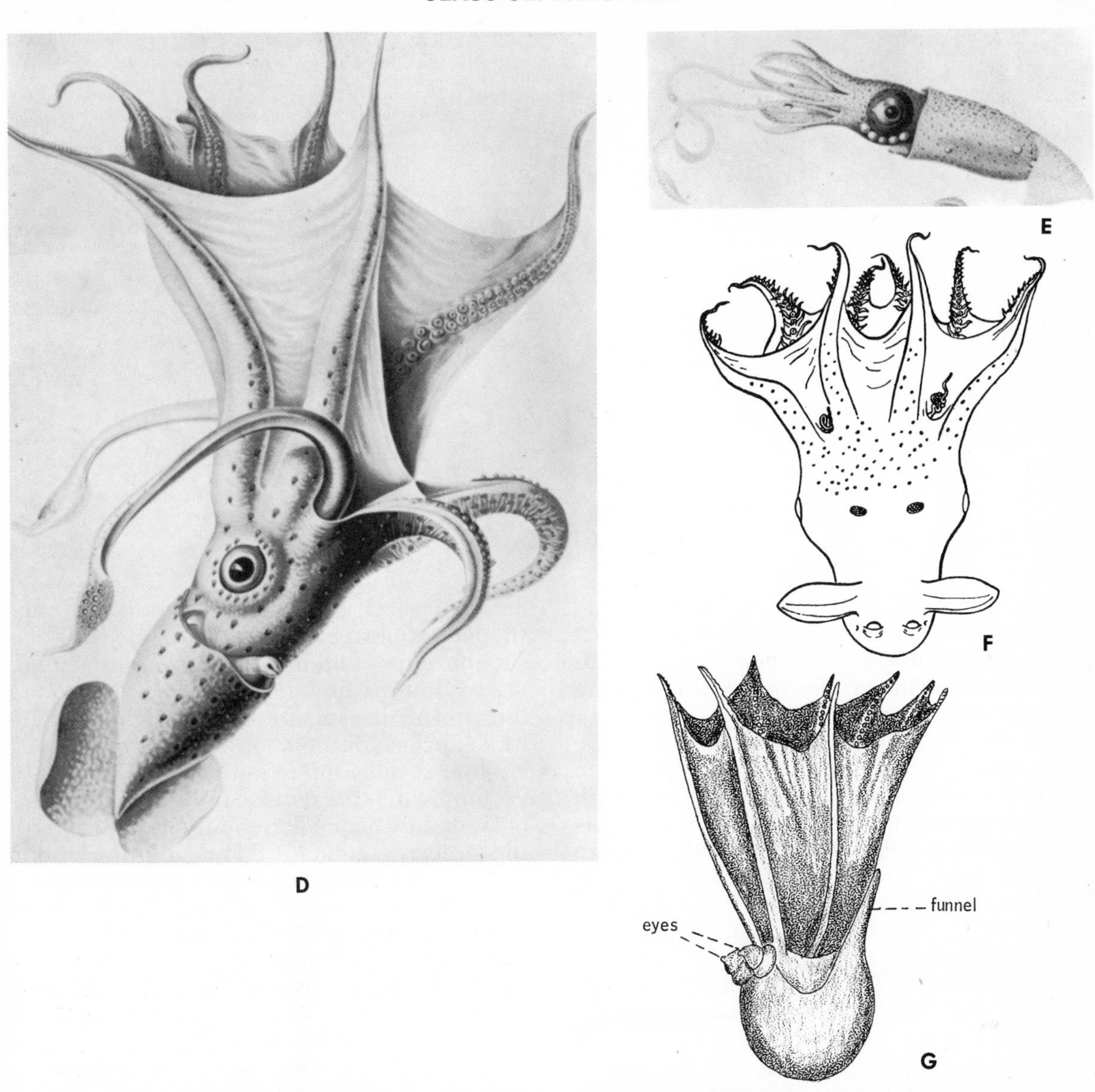

Figure 11–84. *Continued.* *D. Histioteuthis bonellii,* a squid with webbed arms. Body is covered with photophores. (After Chun from Lane.) *E. Lycoteuthis.* Bead-like bodies are photophores. (After Chun from Lane. *A, B, D,* and *E* from Lane, F. W.: Kingdom of the Octopus. Sheridan House, N.Y.) *F. Vampyroteuthis infernalis.* Small coiled tentacles can be seen between webbed arms. (From Pickford, G. E., 1950: Zoologica, 35:87–95. Reprinted with the permission of the New York Zoological Society.) *G. Amphitretus,* a vitreledonellid octopod with webbed arms. (After Hoyle from Parker, T. J., and Haswell, W. A., 1962: Textbook of Zoology, 7th edition. St. Martin's Press, N.Y.)

swims somewhat like a jellyfish. The strange *Cirrothauma* is eyeless, and the body, which contains a thick, jelly-like subcutaneous connective tissue, is transparent. It swims slowly by opening and closing the arm web. Swimming by means of webbed arms has also evolved in two groups of squids—the Cirroteuthacea and the vampire squids, both of which are deep-water cephalopods (Fig. 11–84).

Gas Exchange. The circulation of water through the mantle not only produces the power for locomotion but, as in other mollusks, also provides oxygen for the gills. The nautiloids (*Nautilus*) have four gills, but the coleoids have the usual two. The surface area of the gill filament has been increased by a type of folding similar to that of a fan, and all cilia have disappeared (Fig. 11–82*C*). The loss of cilia is not surprising. Removal of sediment is not a problem for pelagic animals, and the mantle contractions create the water current passing through the mantle cavity. Supporting chitinous rods are present in the filaments, but on the abfrontal margin instead of the frontal margin as in gastropods. Another modification in coleoid gills is the conduction of blood through the

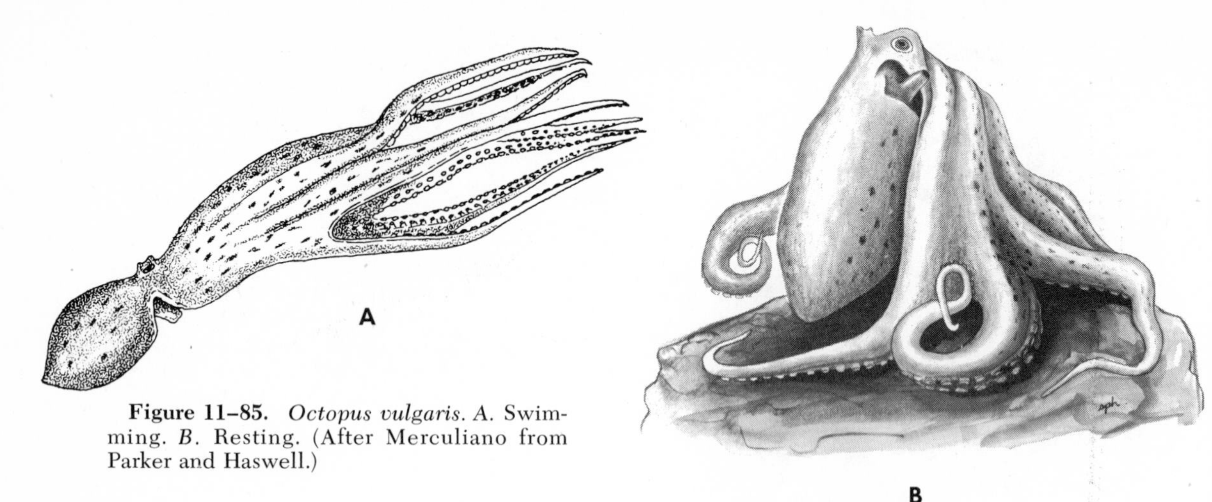

Figure 11–85. *Octopus vulgaris. A.* Swimming. *B.* Resting. (After Merculiano from Parker and Haswell.)

filament in capillaries, increasing the speed of circulation. In many cephalopods that swim with webbed arms, the gills are vestigial and gas exchange takes place through the general body surface.

The two pairs of gills in *Nautilus* may be the more primitive cephalopod state. The coleoid condition perhaps arose through loss of one of the original pairs, but gas exchange was improved as a result of more efficient water circulation, capillaries in the filaments, and the addition of a pair of branchial hearts, which caused an increase in blood pressure.

Nutrition. Cephalopods are highly adapted for raptorial feeding and a carnivorous diet. Prey is located with the highly developed eyes, and capture is effected by the tentacles or arms.. *Nautilus* displays approximately 90 tentacles, arising from lobes that are arranged in both an inner and an outer circle around the head (Fig. 11–79). The tentacles lack adhesive suckers or discs. Each tentacle can be withdrawn into a sheath. Located above the head and tentacles is a large leathery protective hood that represents a fold of one of the tentacular lobes. When the animal withdraws into the shell, the hood acts like an operculum and covers the aperture. A hood is not found in the other living cephalopods.

Squids and cuttlefish possess only ten arms arranged in five pairs around the head (Figs. 11–82A and 11–86). Eight are short and heavy and are called arms; the fourth pair down from the dorsal side are twice the length of the arms and are called tentacles. The inner surface of each arm is flattened and covered with stalked, cup-shaped adhesive discs that function like suction cups. Often a horny reinforcement, or in some cases hooks, are located around the rim of the suction cup. Attached to the inner surface of the floor of the cup are muscle fibers, the contraction of which creates a vacuum when the cup is applied. Suckers are present only on the flattened spatulate ends of the longer tentacles. The highly mobile tentacles are shot out with great rapidity to seize the prey, which is then drawn toward the mouth; the arms aid in holding the prey.

Octopods have only eight tentacles, or arms. The arms are all of equal length and are similar to the arms of squids except that the suckers are sessile and lack horny rings and hooks. The numerical difference in tentacles in dibranch cephalopods provides the basis for dividing the subclass into two orders, the Decapoda (an unfortunate name since it is also used for an order of crustaceans), which contain the squids and cuttlefish with ten tentacles, and the eight-armed Octopoda.

A radula is present, but more important is the pair of powerful beak-like jaws located in the buccal cavity. The beak can bite and tear off large pieces of tissue, which are then pulled into the buccal cavity by the radula, using a tonguelike action, and then swallowed (Fig. 11–86*B*). The location of the buccal mass within a blood sinus permits turning of the entire buccal apparatus and enables the animal to use the jaws with great dexterity. The radula is rudimentary or absent in some deep-sea octopods that have apparently adopted a bottom detritus

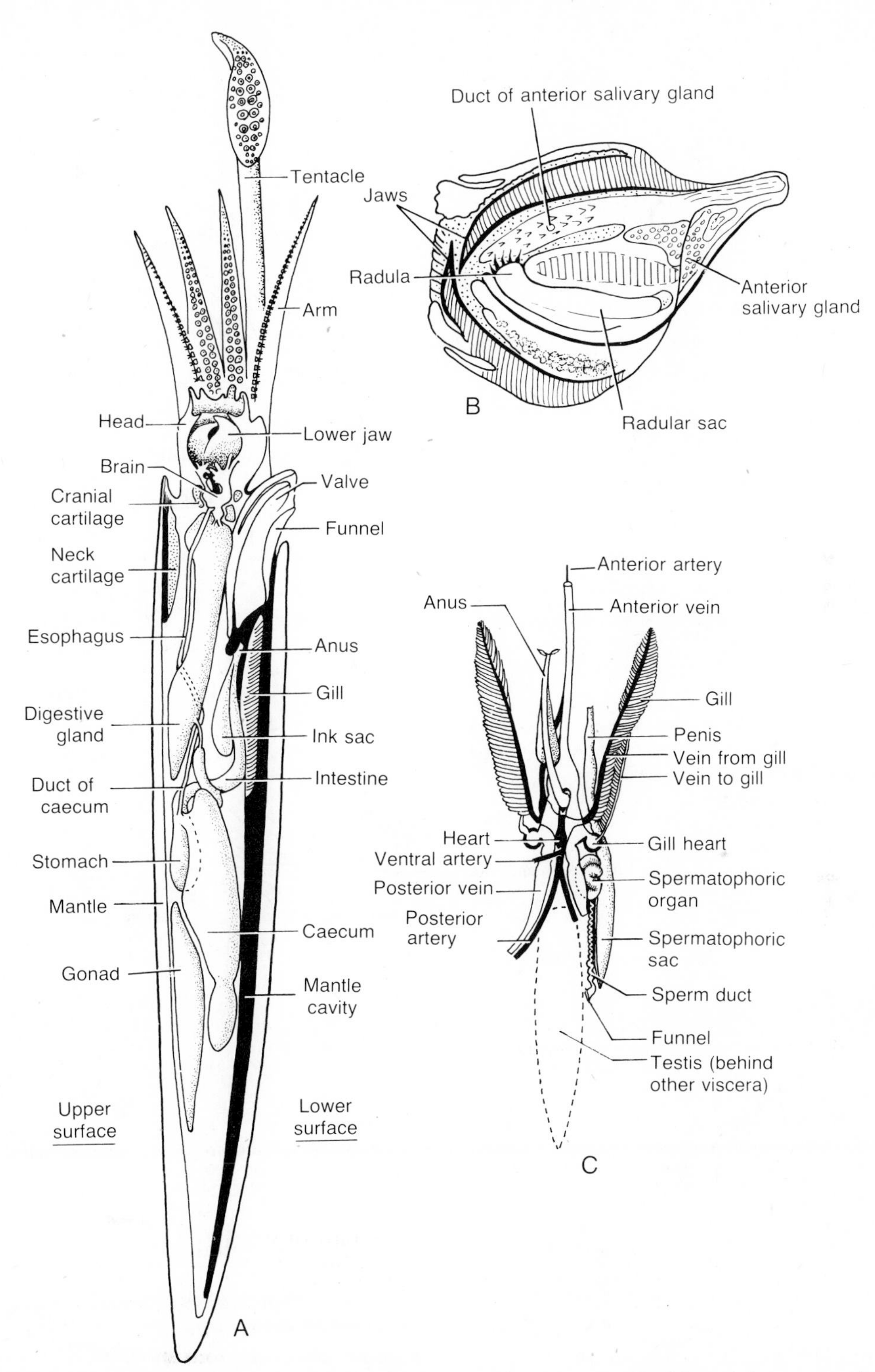

Figure 11–86. Anatomy of *Loligo*. *A*. Lateral view with body wall removed, showing digestive and nervous systems. *B*. Sagittal section through buccal mass of *Loligo*. *C*. Ventral view of part of the circulatory system and of the male reproductive system (drawn to the same scale as *A*, so that it can be turned on edge and fitted into *A*). (*A* and *C* after Williams; *B* after Owen.)

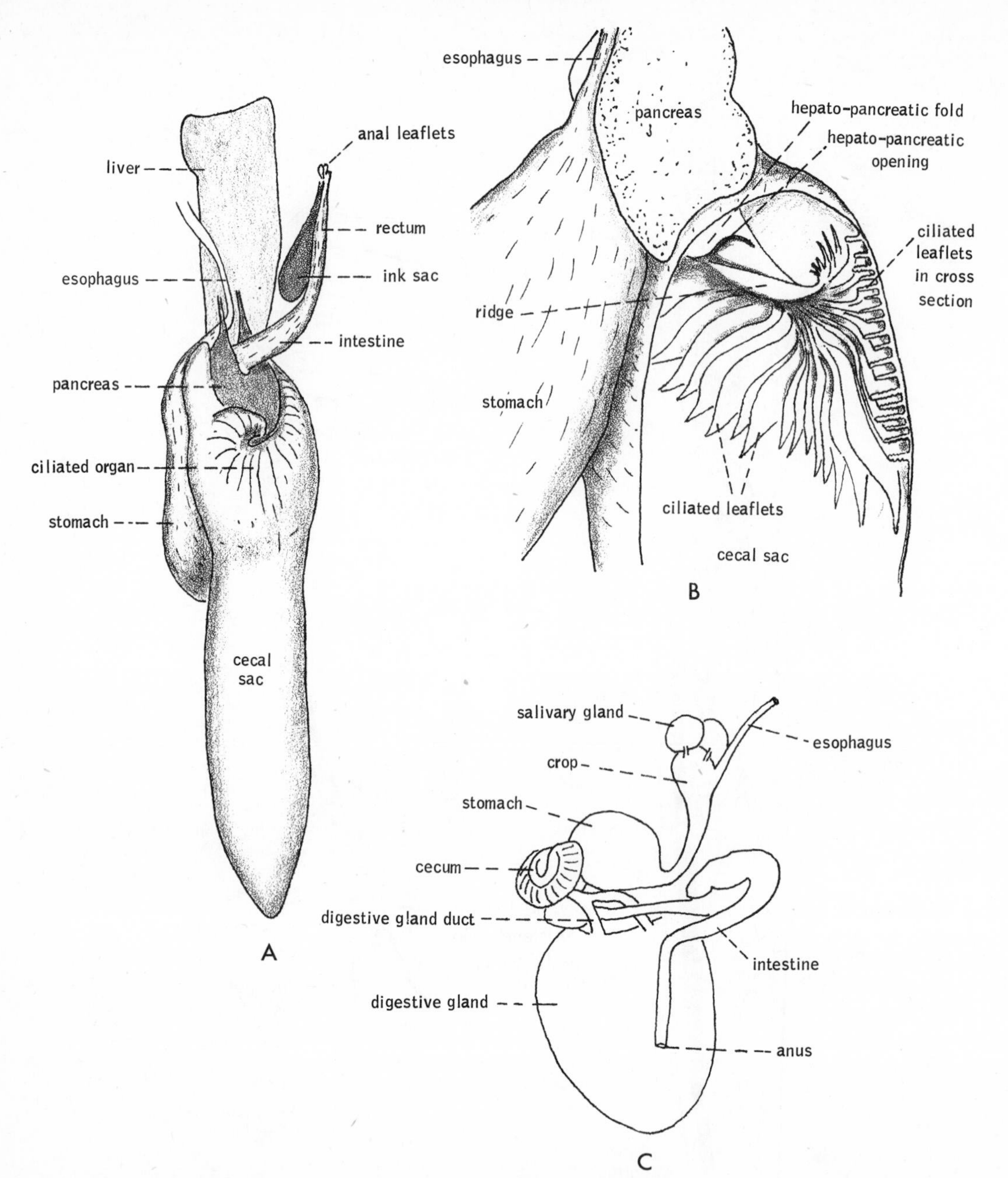

Figure 11–87. *A*. Digestive tract of the squid, *Loligo* (ventral view). *B*. Ciliated folds in cecum of *Loligo* (ventral view). (Both after Bidder.) *C*. Digestive tract of *Octopus vulgaris*. (After Masao.)

feeding habit. In coleoids two pairs of salivary glands empty into the buccal cavity. The nature of the secretions is still known only for relatively few cephalopods. The anterior pair secretes mucus, and in *Octopus vulgaris* a dipeptidase has also been found. The posterior salivary glands, which are much larger in octopods than in decapods, secrete poison and, at least in *Octopus*, proteolytic enzymes. The poison (tyramine) enters the tissues of the prey through the wound inflicted by the jaws.

The diet of cephalopods depends upon the habitats in which they live. Pelagic squids, such as *Loligo* and *Alloteuthis*, feed on fish and pelagic shrimp. *Loligo* will dart rapidly into a school of young mackerel, seize a fish with the tentacles, and quickly bite out a chunk behind the head or bite off the head. The fish is devoured with small bites of the beak until only the gut and tail remain. These parts are then dropped to the bottom.

Cuttlefish swim about over the bottom and feed upon surface-inhabiting invertebrates, especially shrimp and crabs. *Sepia* may rest on the bottom, lying in wait for passing prey. *Spirula* will stir up prey with water jets from the funnel. The victim is seized with the tentacles as in squids.

Octopods live in dens located in crevices and holes. They make nocturnal excursions in search of food, or lie in wait near the entrances of their lairs. Passing snails, fish, and especially crustaceans are seized and dragged into the lair, where they are eaten. To remove gastropods from their shells, the octopus will drill a hole through the shell with the radula and inject poison directly into the occupant.

Nautilus appears to be a bottom feeder, catching small crustaceans and scavenging. When feeding, it swims forward, searching with extended tentacles. Many of the highly modified deep-water octopods and squids with webbed arms have become suspension or detritus feeders.

The esophagus is muscular and conducts food by peristaltic action into the stomach, or, as in *Nautilus* and *Octopus*, into the crop, which is an expansion of one end of the esophagus.

The stomach is very muscular, and attached to its anterior is a large cecum (Figs. 11–86 and 11–87). The cecum is straight in *Loligo* and spiraled in *Sepia*, *Octopus*, and *Nautilus*. The digestive gland in cephalopods is divided into a small spongy diffuse portion, sometimes called the pancreas, and a large solid "liver" (paired in *Sepia*), which is probably homologous to the digestive gland of the other mollusks. In squids the two divisions of the digestive gland are morphologically separated from each other, and the pancreas empties into the liver duct. Digestion is entirely extracellular. Enzymes from both digestive gland divisions empty through a common duct into the junction between the stomach and cecum. Secretions from the two glands, which may be released in separate phases, can be sent to the stomach or cecum through regulation of a groove.

In those species studied, there are some variations in the details of the digestive process, but in all forms digestion begins in the stomach. Here food is kneaded and mixed with glandular secretions (pancreatic in *Loligo;* hepatic in *Sepia*). The contents of the stomach are periodically released into the cecum along with additional enzymatic secretions (hepatic in *Loligo;* pancreatic in *Sepia*) and digestion is completed. The anterior walls of the cecum contain an elaborate series of spiral ciliated folds that separate nondigestible particles from the cecal contents and conduct the particles by way of a ciliated groove back into the stomach and thence into the intestine (Fig. 11–87*B*).

In *Loligo* absorption takes place in the cecal walls and not in the digestive gland as in the other mollusks. Large, indigestible residues in the stomach can be passed directly into the intestine. Some intestinal absorption takes place. The straight or coiled intestines carry waste to the anus, which opens into the mantle cavity.

In *Octopus* and the cuttlefish *Sepia* the liver is the site of absorption. The absorptive and secretory functions alternate, and digestion takes considerably longer than in squids. Digestion has been reported to take as short a time as 4 hours in the squid *Loligo*, but up to 18 hours in *Octopus*. As has been pointed out by Bidder (1950), the digestive modifications of cephalopods, particularly squids, probably represent an adaptation for rapid digestion correlated with an active pelagic life and a carnivorous diet.

Excretion. The conspicuous part of the cephalopod nephridium is a large renal sac (Fig. 11–88). Each sac opens to the mantle cavity through a nephridiopore and communicates with the pericardium by way of a renopericardial canal. Before reaching the branchial heart, the afferent branchial vein crosses the renal sac. The vein contains large evaginations, called renal appendages, which project into the renal sacs. Coinciding with the beating of the branchial heart, blood is forced into and out of the renal appendages of the branchial veins. During this passage the secretion of waste into the renal sacs occurs. Attached to each branchial heart is an appendage, which contacts the pericardial coelom and is thought to correspond to the pericardial glands of other mollusks. In decapods the entire branchial heart lies in the coelom. A filtrate is presumed to pass into the pericardial cavity from the branchial hearts and their appendages. The pericardial fluid is conducted to the renal sacs by the renopericardial canals, which open near the nephridiopores. Some reabsorption by the renopericardial canals has been demonstrated.

The four nephridia of *Nautilus* are basically similar, but they have lost any connection with the pericardial cavity.

Circulation. The circulatory system of cephalopods is closed and therefore consists of the most extensive system of vessels of any mollusks. In coleoids, blood returns from the head region in the large vena cava,

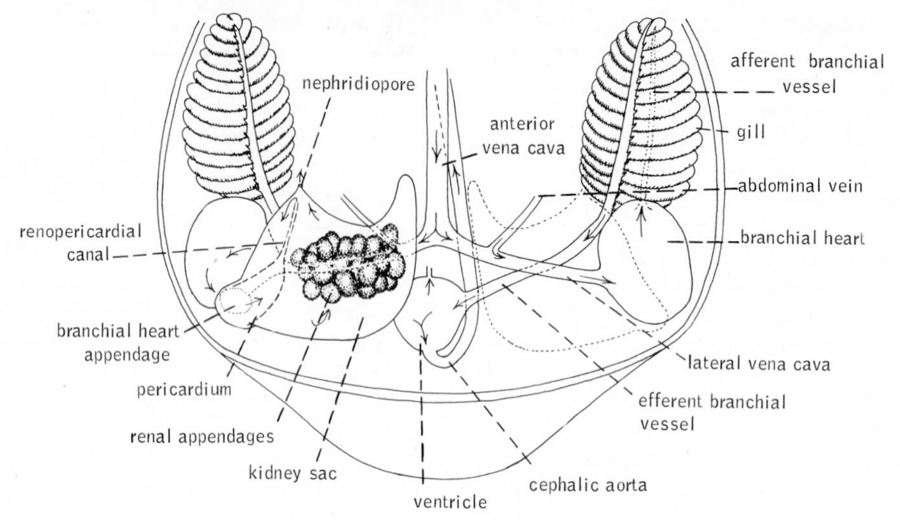

Figure 11–88. Excretory and circulatory systems of *Octopus dolfleini*. (After Potts.)

which divides into two branches near the nephridia (Fig. 11–88). Each branch penetrates a nephridial sac and then passes through a muscular branchial heart before entering the gills. The right branch of the vena cava also receives blood from a vein draining the gonads. Finally an anterior pair and a posterior pair of mantle, or abdominal, veins return blood from the mantle and viscera.

The structure and physiology of the cephalopod circulatory system are closely correlated with the higher metabolic rate of these animals as compared to other mollusks. The existence of capillaries, some contractile arteries, and the branchial hearts increases the blood pressure and the speed of blood flow. The contraction of the branchial hearts, which receive unoxygenated blood from all parts of the body, boosts the pressure of the blood, sending it through the capillaries of the gills. The two auricles of the heart drain blood from the gills and then pass it into the median ventricle. The ventricle pumps blood out to the body through both an anterior and a posterior aorta and eventually through smaller vessels into tissue capillaries. The blood of cephalopods contains a hemocyanin, which unloads at relatively high oxygen pressures.

Nervous System and Sense Organs. The high degree of development of the cephalopod nervous system is unequaled among other invertebrates, and is correlated with the locomotor dexterity and carnivorous habit of these animals. There is great cephalization. All of the typical molluscan ganglia are concentrated and more or less fused to form a brain that encircles the esophagus (Fig. 11–89*A*). Moreover, the ganglia themselves have become so well differentiated that it has been possible to detect experimentally certain areas, or centers, within particular ganglia that control particular regions or functions of the body. For example, the brain centers for both forward swimming and the closing of the suckers have been located in the cerebral ganglia.

The large cerebral ganglia, or supraesophageal region, comprise that part of the brain lying above the esophagus. They give rise to a pair of buccal nerves that run anteriorly to a pair of superior buccal ganglia and then, by way of a commissure around the esophagus, to a pair of inferior buccal ganglia. Both pairs of buccal ganglia lie just behind the buccal mass. Also, a large optic nerve extends from each side of the cerebral ganglia.

In the subesophageal regions of the brain, the pedal ganglia supply nerves to the funnel, and anterior divisions of the pedal ganglia, called brachial ganglia, send nerves to each of the tentacles. The innervations by the pedal and brachial ganglia are evidence that the tentacles and the funnel of cephalopods are homologous to the foot of other mollusks.

The visceral ganglia (perhaps representing both the pleural and visceral ganglia of other mollusks) give rise to three pairs of posteriorly directed nerves: (1) One pair of visceral nerves innervates the various internal organs (Fig. 11–89*B*). One branch from each of these nerves innervates the gills

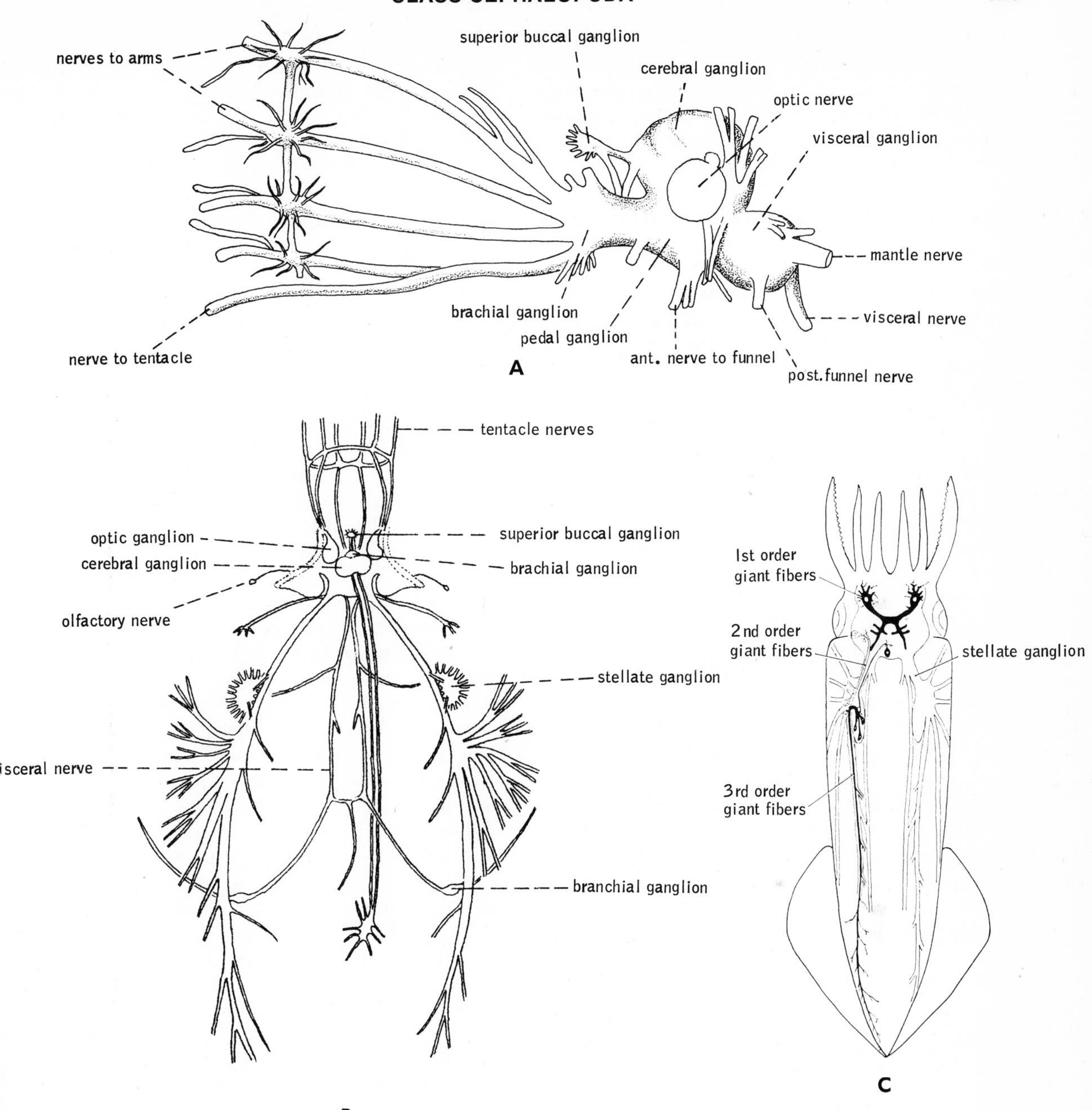

Figure 11–89. *A*. Brain of *Sepia* (lateral view). (After Hillig from Kaestner.) *B*. Nervous system of *Sepia*. (After Hillig from Borradaile and others.) *C*. Giant fiber system of the squid *Loligo*. (From Young, J. Z., 1936: Cold Spring Harbor Symposia on Quantitative Biology, *4*:4.)

and contains a branchial ganglion. (2) A pair of sympathetic nerves, which join together in a gastric ganglion between the stomach and the cecum, innervates the stomach region. (3) The third pair of nerves is large and supplies the mantle.

Ordinary swimming and ventilating contractions of the mantle musculature result from impulses conveyed through a system of many small motor neurons radiating from two stellate ganglia, one in each half of the mantle wall. The rapid escape movements of swimming cephalopods, such as squids, result from a highly organized system of giant motor fibers which bring about powerful and synchronous contractions of the circular muscles of the mantle.

The command center of the system is a pair of very large first order giant neurons which lie in the median ventral lobe of the fused visceral ganglia (Fig. 11–89*C*). These neurons are probably fired by a barrage of impulses from the sense organs. They run to another center within the visceral ganglia, where each makes a connection to a second order giant neuron, which traverses the mantle nerve to the stellate ganglion. Here connections are made with third order giant neurons supplying the circular muscle fibers. The diameters of the third order fibers are not uniform. The greater the distance between the stellate ganglion and the muscle terminal, the greater the diameter of the fiber, thus ensuring that all impulses arrive at the muscle fibers simultaneously and produce a powerful synchronous contraction of the mantle musculature.

Like the nervous system, the sense organs of cephalopods are highly developed. Particularly well developed are the eyes, which in squids, cuttlefish, and octopods are strikingly similar in structure to those of vertebrates (Fig. 11–90). A spherical housing containing cartilaginous plates fits onto a sort of orbit or socket of cartilages associated with those surrounding the brain. The lens, which is suspended by a ciliary muscle, is a rigid sphere with a fixed focal length, and in front of the lens is an iris diaphragm, which can control the amount of light entering the eye. The pupil is slit-shaped, and the eyes are always so aligned that the slit is kept horizontal. The retina contains closely packed, long, rod-like photoreceptors that are directed toward the source of light. The eye is thus of the direct type, instead of indirect as in vertebrates. The rods are connected to retinal cells that send fibers back to

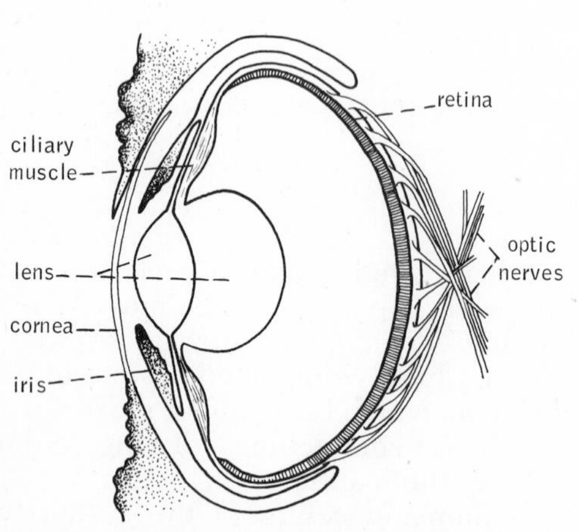

Figure 11–90. Eye of *Octopus*. (After Wells.)

an optic ganglion. These ganglia in turn are located at the outer ends of the optic nerves.

The cephalopod eye undoubtedly forms an image, but the animal's visual perception is certainly quite different from that of a man, which is greatly dependent upon interpretation by the brain. The cephalopod optic connections appear to be especially adapted for analyzing vertical and horizontal projections of objects in the visual field. Within the optic ganglia, the dendritic endings of the neurons receiving impulses from the retinal fibers are arranged along vertical and horizontal axes.

Functioning in an aquatic environment, the cornea of the cephalopod eye contributes little to focusing, because there is almost no light refraction at the corneal surface, as there is in an air-corneal surface. Also, accommodation by the lens is brought about by forward or backward movement as in some vertebrates. The backward movement is effected by the ciliary muscles, and the forward movement by the muscles of the eyeball. Studies of *Octopus* indicate that at least this cephalopod is quite nearsighted. The usual poor visibility in water, the myopic nature of the eyes, and the important sensory role of the tentacles have caused some investigators to question whether true focusing actually occurs in cephalopods. The mechanism for lens movement may function only to maintain normal optical conditions when the eye is subjected to pressure resulting from sudden movement of the body. All of this, however, is debatable.

The cephalopod eye can accommodate itself to light changes both by modifications in the pupil's size and by the migration of pigment in the retina. On the basis of behavioral studies, it seems likely that cephalopods can detect color.

The eyes of *Nautilus* are large and are carried at the end of short stalks, but they are much less complex than those of coleoids. The eye lacks a lens and is open to the external sea water through a small aperture, the pupil. Supposedly, the eye functions like a pinhole camera.

Statocysts are found in both nautiloids and coleoids, but are particularly well developed in the latter, in which they are large and are embedded in the cartilages located on each side of the brain. They not only provide information about static spatial orientation (i.e., body position in relation to gravitational pull), but are so constructed that, like the semicircular canals of vertebrates, they inform the animal of changing positions in motion, such as turning. Without the statocysts a cephalopod can neither keep the pupil slits of the eyes horizontal nor discriminate between horizontal and vertical surfaces.

Osphradia are present only in *Nautilus.*

The arms, and especially the sucker epithelium, are liberally supplied with tactile cells and chemoreceptor cells, especially in the benthic hunting octopods. Textural and chemical discrimination of a surface by the arms has been demonstrated in Octopus, but these animals cannot determine shape with the tentacles, perhaps because the arms contain no divisions or structures that could function as a unit of measurement, as for example the distance between thumb and forefinger in primates or leg joints in arthropods.

Cephalopod behavior has been, and continues to be, an area of great interest. Learning, memory, tactile and visual discrimination, and localization of motor and sensory functions within the brain have been studied with rewarding results. Among the many investigators of cephalopod behavior, the British zoologists J. Z. Young, M. J. Wells, J. Wells, and B. B. Boycott are especially prominent. For a more detailed account of the cephalopod nervous system and cephalopod behavior, the reader should refer to the excellent résumés by M. J. Wells (1962, 1966) and J. Z. Young (1972).

Chromatophores, Ink Gland, and Luminescence. The unusual coloration of cephalopods, other than *Nautilus,* is caused by the presence of chromatophores in the integument. The expansion and contraction of these cells result from the action of tiny muscles that are attached to the periphery of the cell. When the muscles contract, the chromatophore is drawn out to form a large flat plate; when the muscles relax, the pigment is concentrated and less apparent. Particular species possess chromatophores of several colors—yellow, orange, red, blue, and black—and the chromatophores of a particular color may occur in groups. The chromatophore effect is enhanced in *Sepia* by a deeper layer of iridocytes, which reflect the color changes of the overlying pigment cells. The chromatophores are controlled by the nervous system and probably by hormones, with vision being the principal initial stimulus. The degree to which these animals can change color and the stimulus for color change vary considerably. The cuttlefish *Sepia officinalis* displays complex color changes and may simulate the background hues of sand, rock, and so forth, but most changes appear to be correlated with behavior. Most species exhibit color change when alarmed. For example, the littoral squid *Loligo vulgaris* is generally very pale and only darkens when disturbed; others lighten. An octopus when alarmed may flatten its body and present an elaborate "defensive" color display, including color changes flowing over the body and large dark spots around the eyes. Color displays also are associated with courtship in many cephalopods and are described in the next section.

In cephalopods other than *Nautilus* and some deep-water species, a large ink sac is located in the region of the intestine (Figs. 11–86*A* and 11–87*A*). The apparatus consists of an ink-producing gland located in the wall of a large reservoir, and a duct with a terminal ejecting ampulla that opens into the rectum just behind the anus. The gland secretes a brown or black fluid containing a high concentration of melanin pigment, which is stored in the reservoir. When an animal is alarmed, the ink is released through the anus and the cloud of inky water forms a "dummy," confusing the predator. It is also believed that the alkaloid nature of the ink may be objectionable to predators, particularly fish, in which it may anesthetize the chemoreceptive senses.

Many of the deep-sea squids are bioluminescent, with luminescent photophores of different colors arranged in various patterns over the body, even on the eyeball (Fig. 11–84*D* and *E*). The photophores of some species develop to remarkable complexity, including a reflector and lens, and sometimes an overlying chromatophore shutter. *Heteroteuthis* secretes a luminescent ink. The complex light organ of the sepiolids is associated with the accessory nidamental gland and, unlike the photophores of many squids, contains luminescent symbiotic bacteria. Luminescent bacteria also occur in the Loliginidae and the Sepiidae. The bacteria are located in the skin and ducts of the accessory nidamental glands.

Reproduction. Except in a few species, cephalopods are dioecious, and the gonad is located at the posterior of the body (Fig. 11–86). The testis is a saccular structure with an internal cavity, considered to be derived from the coelom. Sperm are formed in the wall of the sac, pass into the lumen, and then issue through a ciliated opening into the vas deferens. The highly coiled vas deferens conducts the sperm anteriorly to a seminal vesicle, which has ciliated grooved walls. Here in the seminal vesicle the sperm are rolled together and encased in a very elaborate spermatophore. The spermatophore is completed in a hardening gland and then in a finishing gland. From the seminal vesicle spermatophores pass into a large storage sac or reservoir called Needham's sac, which opens into the left side of the mantle cavity.

In females the ovary is also saccular and occupies a position similar to that of the testis. The oviduct is looped and terminates in an oviducal gland. In octopods two oviducts are present.

Fertilization may take place within the mantle cavity or outside, but in either case involves copulation. Because of the restricted opening into the mantle cavity, one of the arms of the male—the right or left fourth in squids and cuttlefish and the right third in octopods—has become modified as an intromittent organ, a phenomenon called hectocotyly. The modified arm is called a hectocotylus. The degree of modification varies. In *Sepia* and *Loligo* several rows of suckers are smaller and form an adhesion area for the transport of spermatophores. In octopods the modification is much greater. The tip of the arm carries a spoonlike depression in *Octopus;* in *Argonauta* and others, there is actually a cavity or chamber where the sperm are stored.

Before copulation, a male cephalopod performs various displays that serve to identify it to the female. The cuttlefish *Sepia* presents a striped color pattern and establishes a temporary bond with a female, swimming about above her. The display is also directed toward intruding males, and the weaker male will depart. A male octopod, which is smaller than the female, may present a color display or, as in *Octopus vulgaris,* lift up its arms and display its large seventh or eighth suckers on the second and third arms.

In pelagic cephalopods, copulation occurs while the animals are swimming, the male seizing the female head-on (Fig. 11–91). During copulation the hectocotylus receives spermatophores from the male's funnel or plucks a mass of spermatophores from the Needham's sac. The male hectocotylus is then inserted into the mantle cavity of the female and deposits the spermatophores on the mantle cavity wall near the openings from the oviducts, or into the genital duct itself in *Octopus.* In octopods with storage chambers at the end of the hectocotylus, the arm may break off and remain in the mantle cavity of the female, as in *Argonauta* and *Philonexis.* The detached hectocotylus was originally thought to have been a parasitic worm and was described as *Hectocotylus;* hence the derivation of the term.

In *Loligo* and other genera of the same family, the spermatophores may be received by the mantle cavity as just described; or the hectocotylus may be inserted into a horseshoe-shaped seminal receptacle, located in a fold beneath the mouth. The buccal membrane alone receives the spermatophores in *Sepia.*

The spermatophore is shaped like a baseball bat and consists of an elongated sperm mass, a cement body, a coiled springlike ejaculatory organ, and a cap (Fig. 11–92A). The spermatophore is discharged by the removal of the cap. This results, in squids, from traction on the filament when the spermatophores are pulled from Needham's sac or, in ocotopods, by the uptake of water when removed from the sac. With the

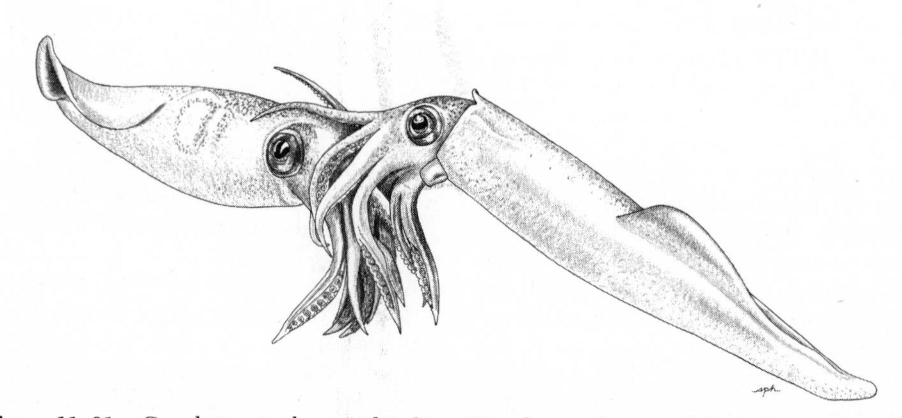

Figure 11–91. Copulation in the squid *Loligo.* (Based on a photograph by Robert F. Sisson.)

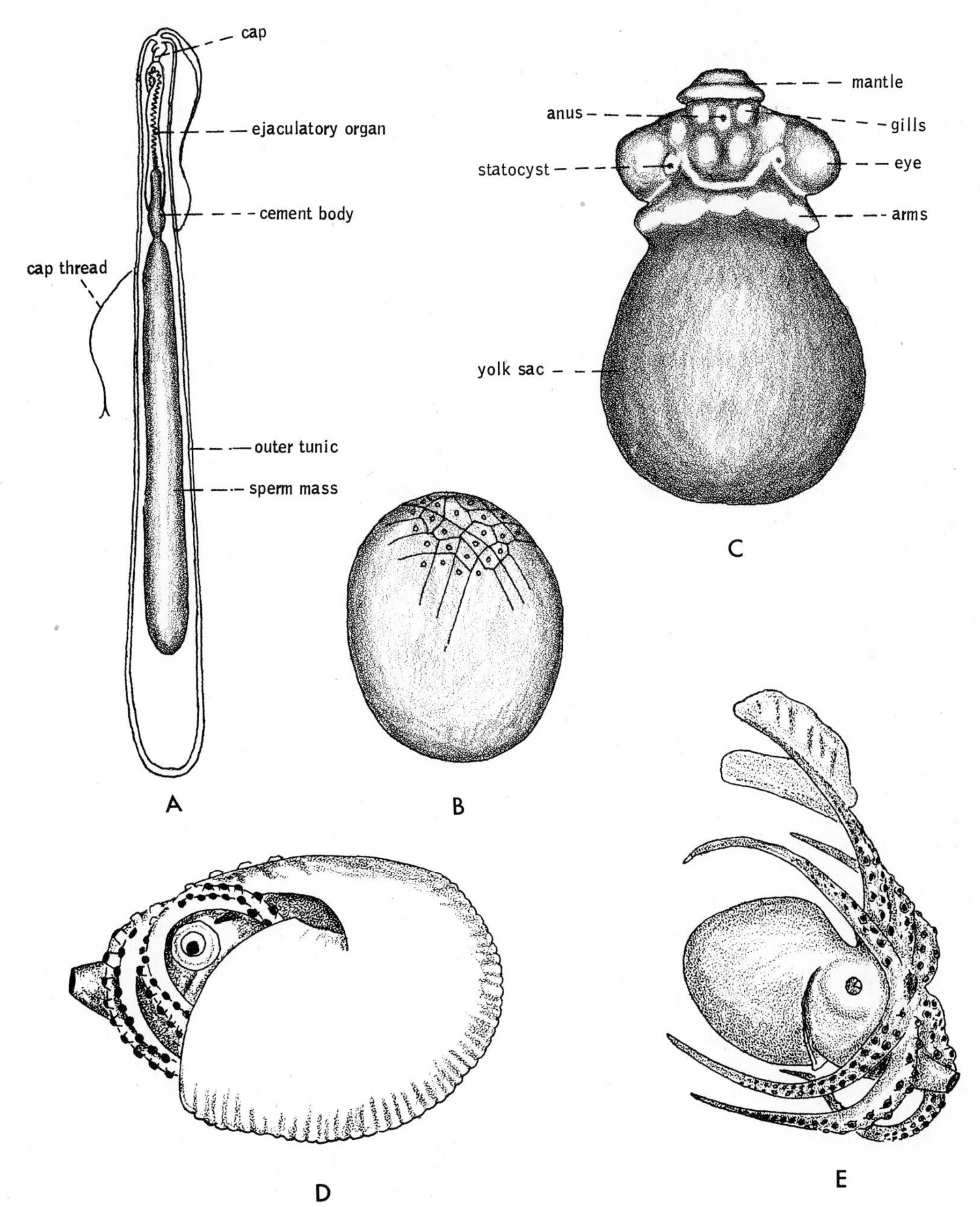

Figure 11–92. *A*. Spermatophore of *Loligo*. *B*. Discoidal meroblastic cleavage in *Loligo*. (After Watase from Dawydoff.) *C*. Embryo of *Loligo*. (After Naef from Dawydoff.) *D*. *Argonauta argo* female in shelly egg case. (After Naef from Kaestner.) *E*. *Argonauta hians*. (After Verrill from Abbott.)

cap removed, the ejaculatory organ is everted, pulling out the sperm mass. The cement body adheres to the seminal receptacle or mantle wall and the sperm mass disintegrates, liberating sperm for as long as two days.

As the eggs are being discharged from the oviduct, each is enveloped by a paired membrane or capsule in the oviducal gland. Additional protective covering is produced by a nidimental gland, which is located in the mantle wall near the oviducal gland and opens independently into the mantle cavity. In *Loligo*, secretions from the nidimental gland surround the eggs in a gelatinous mass. After leaving the mantle cavity, the egg string is held by the arms, and the eggs may be fertilized by sperm from the seminal receptacle under the mouth. The female then attaches the fertilized eggs to the substratum in a cluster of ten to fifty elongated strings, each containing as many as one hundred eggs. The gelatinous covering of each mass hardens on exposure to sea water, and the individual egg capsules swell to several times the original diameter, leaving a distinct space around the egg. Large numbers of *Loligo* come together to copulate and spawn at the same time, and a "community pile" of egg strings may be formed on the bottom. Death of the adult soon follows spawning.

Sepia deposits its eggs singly but attaches them by a stalk to seaweed or other objects. *Octopus* forms egg clusters that resemble a bunch of grapes and are attached within rocky recesses.

Deep-water and strictly pelagic squids, such as the oegopsids, deposit their eggs singly, and the eggs are free-floating rather than attached. Octopods remain to care for the eggs after they are deposited. Debris and sediment are continually removed from the deposited eggs by the tentacles and water is periodically ejected over the eggs from the funnel. The females of at least *Octopus vulgaris* die after brooding their eggs.

Perhaps the most remarkable adaptation for egg deposition is found in a genus of pelagic octopods, *Argonauta*, commonly known as the paper nautilus (Fig. 11–92*D* and *E*). The two dorsal arms of the female are greatly expanded at the tip to form a veil or membrane. The expanded portion of each arm secretes one half of a beautiful calcareous bivalved shell into which the eggs are deposited. The shell is carried about and serves as both a brood chamber and a retreat for the female. The posterior of the female usually remains in the shell; when disturbed, she withdraws completely into the retreat. The much smaller male does not secrete a shell and frequently is a cohabitant in the shell of the female.

The reproductive system of *Nautilus* is somewhat similar to that of coleoids, although the ducts are paired. Little is known about copulation and fertilization. The male copulatory organ consists of four tentacles. The eggs are deposited and attached to the substratum singly within rather elaborate capsules.

Cephalopod eggs are much more heavily yolk-laden than other mollusk eggs. The eggs of *Sepia* and *Eledone* are very large and may reach 15 mm. in diameter. The eggs of *Loligo* and *Octopus* contain less yolk.

Embryogeny. Embryonic development is direct in all cephalopods. As in the telolecithal eggs of reptiles and birds, cleavage in cephalopods is meroblastic. Cleavage results in the formation of a germinal disc, or cap, of cells at the animal pole (Fig. 11–92*B*). Here the embryo will form (Fig. 11–92*C*). The margin of the disc grows down and around the yolk mass and forms a yolk sac; the yolk is gradually absorbed during development. The eyes arise as ectodermal invaginations. The arms form posteriorly and then migrate forward to locations surrounding the mouth. The funnel, which represents a part of the foot, ultimately is located within the mantle cavity owing to overgrowth and enclosure by the developing mantle. Nothing is known concerning development in *Nautilus*.

Economic Importance. The principal enemies of cephalopods are fish and whales. The same species (such as mackerel) that when young served as food for the cephalopod, may in adult life prey on the cephalopod. Squids are important to the diet of sperm whales. These whales sometimes bear circular sucker scars on the skin from a battle with a giant squid.

Squids used for bait are caught commercially in the Grand Banks region. Cephalopods are an important human seafood eaten in many parts of the world, including Europe.

SYSTEMATIC RÉSUMÉ OF CLASS CEPHALOPODA

Subclass Nautiloidea. Possess external shells, which may be coiled or straight; sutures not complex. Existing forms possess many slender, suckerless tentacles. Two pairs of gills and two pairs of nephridia are present. All members are extinct except *Nautilus;* the class has been in existence from the Cambrian period to the present time. Genera include *Endoceras* and *Nautilus*.

Subclass Ammonoidea. Fossil forms with coiled external shells having complex septa and sutures. Existed from the Silurian period to Cretaceous period. Includes *Ceratites*, *Scaphites*, and *Pachydiscus*.

Subclass Coleoidea. Shells are internal or absent. A few tentacles bear suckers. One pair of gills and one pair of nephridia are present. Members have existed from the Mississippian period to the present.

Order Decapoda. The cuttlefish and squids. Two long tentacles and eight shorter arms present; body with lateral fins.

Suborder Belemnoidea. Extinct species. Shell internal, chambered but with a posterior solid rostrum and a dorsal shield-like extension. *Belemnites, Belemnoteuthis.*

Suborder Sepioidea. Cuttlefish and sepiolas. Shell with septa, or shell greatly reduced or lost. Body mostly short and broad or saclike. *Sepia, Idiosepius, Spirula, Sepiola.*

Suborder Teuthoidea. Squids. Shell or pen a flattened plate. Body mostly elongate with long tentacles. *Loligo, Architeuthis, Chiroteuthis, Onycoteuthis, Cranchia.*

Order Vampyromorpha. Vampire squids. Small deep-water, octopod-like forms with eight arms united by a web, but two small retractile tentacles also present. *Vampyroteuthis.*

Order Octopoda. The octopods. Possess eight arms; body globular with no fins. *Octopus, Eledone, Eledonella, Vitreledonella, Amphitretus, Cirroteuthis, Argonauta.*

THE ORIGIN OF THE MOLLUSKS

The striking similarity between the embryogeny of the mollusks and that of the polychaete annelids has long been recognized. Both kinds of animals exhibit spiral cleavage and virtually identical trochophore larvae. This similarity has been the principal evidence to support the view that the annelids and mollusks arose from some common ancestral stock.

The discovery of living monoplacophorans contributed fresh views not only to a molluscan-annelidan relationship but also to the nature of the ancestral mollusks. The striking feature of *Neopilina,* of course, is the apparent metameric plan of structure. *Neopilina* displays eight pairs of retractor muscles, five pairs of gills, and six pairs of nephridia. A number of zoologists (Fretter and Graham, 1962; and Lemche and Wingstrand, 1959) believe that the metamerism displayed by *Neopilina* is evidence that the mollusks comprise a fundamentally metameric phylum. This belief has in turn led to the view that mollusks evolved from the annelids. The molluscan trochophore is thus believed to be additional evidence for an annelidan origin of mollusks.

However, many zoologists do not find the evidence for metamerism very convincing. First, the replicated structures differ in numbers and do not display any particularly striking coincidence in their sequence. Therefore, it is difficult to determine either functionally or structurally what would represent a segment.

The affinity between mollusks and annelids seems probable even without *Neopilina,* but it appears more likely that the two groups diverged from some common ancestor prior to the appearance of either molluscan or annelidan characters, especially metamerism. The nature of such an ancestor is debatable, but some replication of structures along the length of the body probably was present. In annelids this replication developed into metamerism as an adaptation for burrowing. In mollusks the replication tended to become reduced with the assumption of new modes of existence. Vagvolgyi (1967) and Stasek (1972) have suggested the flatworms as being near the common ancestral stock of annelids and mollusks.

Current ideas about molluscan phylogeny have recently been reviewed by Stasek (1972). He suggests that the gastropods, bivalves, scaphopods, and cephalopods, all of which appear to be derived from forms possessing a shell of a single piece, evolved from monoplacophorans. The aplacophorans and polyplacophorans are believed to have diverged early in molluscan history, before the formation of the single shell. Thus, the absence of a shell in aplacophorans would be a primitive feature rather than a secondary one.

Bibliography

Abbott, R. T., 1954: American Sea Shells. D. Van Nostrand, Princeton. (A semitechnical guide to the marine mollusks of the Atlantic and Pacific coasts of North America.)

Andrews, J., 1971: Sea Shells of the Texas Coast. University of Texas Press, Austin. 298 pp. (An excellent illustrated systematic account of the mollusks of the Texas coast; useful for much of the Gulf Coast and West Indies.)

Beklemishev, V. N., 1963: On the relationship of the Turbellaria to other groups of the animal kingdom. *In* Dougherty, E. C. (Ed.), The Lower Metazoa. University of California Press, Berkeley, pp. 234–244.

Bidder, A. M., 1950: Digestive mechanisms of European squids. Quart. J. Micr. Sci., new series, *91*(1):1–43.

Burch, J. B., 1962: How to Know the Eastern Land Snails. W. C. Brown Co., Dubuque, Iowa. 214 pp.

Carriker, M. R., and D. V. Zandt, 1972: Predatory behavior of a shell-boring muricid gastropod. *In* Olla, B. L., and Winn, H. E. (Eds.), Behavior of Marine Animals: Invertebrates, Vol. 1. Plenum Publishing Corp., N. Y., pp. 157–244.

Cheng, T. C., 1967: Marine mollusks as hosts for symbioses. Adv. Mar. Biol., 5:1–424.

Clarke, M. A., 1966: A review of the systematics and ecology of oceanic squids. Adv. Mar. Biol., *4*:91–300.

Clench, W. J., 1959: Mollusca. *In* Edmondson, W. T., Ward, H. B., and Whipple, G. C. (Eds.), Freshwater Biology, 2nd Edition. John Wiley and Sons, N. Y., pp. 1117–1160. (Key to the common freshwater mollusks of the United States.)

Coe, W. R., 1936: Sexual phases in *Crepidula*. J. Exp. Zool., *72*(3):455–477. (A classic paper on protandry in *Crepidula*.)

Crofts, D. R., 1955: Muscle morphogenesis in primitive mollusks and its relation to torsion. Proc. Zool. Soc. London, *125*:711–750.

Denton, E. J., and Gilpin-Brown, J. B., 1966: On the buoyancy of the pearly nautilus. J. Mar. Biol. Assoc. U. K., *46*:723–759.

Denton, E. J., Gilpin-Brown, J. B., and Howarth, J. V., 1961: The osmotic mechanism of the cuttlebone. J. Mar. Biol. Assoc. U. K., *41*:351–364.

Donovan, D. T., 1964: Cephalopod phylogeny and classification. Biol. Rev., *39*(3):259–287.

Emerson, W. K., 1962: A classification of the scaphopod mollusks. J. Paleontol., *36*(3):461–482.

Fretter, V., and Graham, A., 1962: British Prosobranch Molluscs. Ray Society, London.

Gainey, L. F., 1972: The use of the foot and captacula in the feeding of *Dentalium*. Veliger, *15*(1):29–34.

Galtsoff, P. S., 1964: The American oyster. U. S. Fisheries Bull., *64*:1–480.

Garstang, W., 1928: Origin and evolution of larval forms. Report British Assoc., section D, p. 77.

Ghiselin, M. T., 1966: The adaptive significance of gastropod torsion. Evolution, *20*:337–348.

Gilpin-Brown, J. B., 1972: Buoyancy mechanisms of cephalopods in relation to pressure. Symp. Soc. Exp. Biol. *26*:251–259. (Good review, with bibliography.)

Goreau, T. F., Goreau, N. I., and Yonge, C. M., 1973: On the utilization of photosynthetic products from zooxanthellae and of a dissolved amino acid in *Tridacna maxima*. J. Zool., *169*:417–454.

Gosner, K. L., 1971: Guide to Identification of Marine and Estuarine Invertebrates: .Cape Hatteras to the Bay of Fundy. Wiley-Interscience, N. Y. Mollusca, pp. 249–325.

Graham, A., 1938: The structure and function of the alimentary canal of aeolid mollusks, with a discussion of their nematocysts. Trans. Roy. Soc. Edinburgh, pt. 2, *59*(9):267–305.

Graham, A., 1949: The molluscan stomach. Trans. Roy. Soc. Edinburgh, pt. 3, *61*(27):737–778. (A comparative treatment of the molluscan stomach; it deals primarily with gastropods and bivalves.)

Graham, A., 1955: Molluscan diets. Proc. Malacol. Soc. London, *31*:144–159. (A list of molluscan groups and their diets.)

Graham, A., 1971: British Prosobranch and Other Operculate Gastropod Molluscs. Academic Press, N. Y. 120 pp.

Haven, N., 1972: The ecology and behavior of *Nautilus pompilius* in the Philippines. Veliger, *15*(2):75–80.

Hyman, L. H., 1967: The Invertebrates. Vol. 6, Mollusca. I. McGraw-Hill, N. Y. (This volume, the last of Dr. Hyman's contributions to the Invertebrate Series, covers a part of the phylum Mollusca. In it, four classes are treated: Aplacophora, Polyplacophora, Monoplacophora, and Gastropoda.)

Jørgensen, C. B., 1966: Biology of Suspension Feeding. Pergamon Press, N. Y.

Kaestner, A., 1964: Invertebrate Zoology, Vol. 1. Wiley-Interscience, N. Y., pp. 269–424.

Keen, A. M., and McLean, J. H., 1971: Sea Shells of Tropical West America. Stanford University Press, Stanford. 1080 pp.

Kennedy, D., Selverton, A. I., and Remler, M. P., 1969: Analysis of restricted neural networks. Science, *164*: 1488–1496.

Kennedy, W. J., Taylor, J. D., and Hall, A., 1969: Environmental and biological controls on bivalve shell minerology. Biol. Rev., *44*:499–530.

Lebour, M. V., 1937: The eggs and larva of the British prosobranchs. J. Mar. Biol. Assoc. U. K., *22*:105–166.

Lemche, H., 1959: Molluscan phylogeny in light of *Neopilina*. Proc. 15th Int. Cong. Zool., London.

Lemche, H., and Wingstrand, K. G., 1959: The anatomy of *Neopilina galatheae*. Galathea Rep., *3*:9–71.

Light, S. F., Smith, R. I., Pitelka, F. A., Abbott, D. P., and Weesner, F. M., 1967: Intertidal Invertebrates of the Central California Coast. University of California Press, Berkeley. Mollusca, pp. 211–270.

Marcus, E., and Marcus, E., 1968: American Opisthobranch Mollusks. University of Miami Press, Coral Gables, Fla. 256 pp.

Mason, J., 1972: Cultivation of the European mussel, *Mytilus edulis*. Oceanogr. Mar. Biol. Ann. Rev., *10*:437–460.

Moore, R. C. (Ed.), 1960–1971: Treatise on Invertebrate Paleontology. Mollusca, Vols. L-N. Geological Society of America and University of Kansas Press. (In addition to descriptions of fossil species, the introductory sections cover various aspects of the biology of mollusks.)

Morton, B., 1970: The tidal rhythm and rhythm of feeding and digestion in *Cardium edule*. J. Mar. Biol. Assoc. U. K., *50*:499–512.

Morton, J. E., 1963: The molluscan pattern: evolutionary trends in a modern classification. Proc. Linn. Soc. London, *174*:53–72.

Morton, J. E., 1967: Molluscs, 4th Edition. Hutchinson University Library, London.

Nair, N. B., and Saraswathy, M., 1971: The biology of wood-boring teredinid molluscs. Adv. Mar. Biol., *9*:336–509.

Nakazima, M., 1956: On the structure and function of the midgut gland of Mollusca, with a general consideration of the feeding habits and systematic relations. Jap. J. Zool., *2*(4):469–566. (A good comparative account of the digestive diverticula of mollusks.)

Newell, N. D., 1969: Classification of Bivalvia. *In* Moore, R. C. (Ed.), Treatise on Invertebrate Paleontology, Vol. N, pt. 1, Mollusca 6, Bivalvia. pp. N205–N224.

Orton, J. H., 1913: The mode of feeding of *Crepidula*. J. Mar. Biol. Assoc. U. K., 9:444–478.

Owen, G., 1955: Observations on the stomach and digestive diverticula of the Lamellibranchia. I. The Anisomyaria and Eulamellibranchia. Quart. J. Micr. Sci., *96*:517–537.

Owen, G., 1956: Observations on the stomach and digestive diverticula of the Lamellibranchia. II. The Nuculidae. Quart. J. Micr. Sci., *97*:541–567.

Pennak, R. W., 1953: Freshwater Invertebrates of the United States. Ronald Press, N. Y., pp. 667–726. (A general account of the freshwater gastropods and bivalves with figures and keys to the North American genera.)

Pilsbry, H. A., 1939–1946: Land Mollusca of North America. Vol. 1, pts. 1 and 2; vol. 2, pts. 1 and 2.

Acad. Nat. Sci., Monogr. 3, Philadelphia. (This is the most authoritative taxonomic work on the North American land snails.)
Portmann, A., Fischer-Piette, E., Franc, A., Lemche, H., Wingstrand, K. G., and Manigault, P., 1960: Introduction to the Mollusks, the Chitons, the Monoplacophorans, and the Pelecypods. *In* Grassé, P. (Ed.), Traité de Zoologie, Vol. 5. Masson et Cie, Paris, pp. 1625–2164.
Potts, W. T. W., 1967: Excretion in molluscs. Biol. Rev., *42*(1):1–41.
Purchon, R. D., 1968: The Biology of the Mollusca. Pergamon Press, N. Y. 596 pp. (Good general account of many aspects of molluscan biology, especially gastropod and bivalve habitation and feeding.)
Raven, C. P., 1958: Morphogenesis: The Analysis of Molluscan Development. Pergamon Press, N. Y. (A very complete account of molluscan embryogeny with extensive bibliography.)
Rollins, H. B., and Batten, R. L., 1968: A sinus-bearing monoplacophoran and its role in the classification of primitive molluscs. Paleontology, *11*(1):132–140.
Roper, C. F., Young, R. E., and Voss, G. L., 1969: An illustrated key to the families of the order Teuthoidea (Cephalopoda). Smithsonian Contributions to Zoology, *13*:1–32.
Runham, W. W., and Hunter, P. J., 1971: Terrestrial Slugs. Hutchinson University Library, London. 184 pp. (A general biology of pulmonate slugs.)
Salvini-Plawen, L. V., 1972: Zur Morphologie und Phylogenie der Mollusken. Z. wiss. Zool., *184*:205–394.
Shrock, R. R., and Twenhofel, W. H., 1953: Principles of Invertebrate Paleontology. McGraw-Hill, N. Y., pp. 350–502. (A good general treatment of the fossil mollusks, particularly the cephalopods.)
Stanley, S. M., 1968: Post-Paleozoic adaptive radiation of infaunal bivalve molluscs—a consequence of mantle fusion and siphon formation. J. Paleontol., *42*(1):214–229.
Stanley, S. M., 1970: Relation of Shell Form to Life Habits of the Bivalvia (Mollusca). Geol. Soc. Amer., Mem. 125. 296 pp.
Stanley, S. M., 1972: Functional morphology and evolution of byssally attached bivalve mollusks. J. Paleontol., *46*(2):165–212.
Stasek, C. R., 1961: The ciliation and function of the labial palps of *Acila castrensis* (Protobranchia, Nuculidae) and an evaluation of the role of the protobranch organs of feeding in the evolution of Bivalvia. Proc. Zool. Soc. London, *137*:511–538.
Stasek, C. R., 1965: Feeding and particle-sorting in *Yoldia ensifera* (Bivalvia, Protobranchia), with notes on other nuculanids. Malacologia, *2*:349–366.
Stasek, C. R., 1972: The molluscan framework. *In* Florkin, M., and Scheer, B. J. (Eds.): Chemical Zoology, Vol. 3. Academic Press, N. Y., pp. 1–44.
Trueman, E. R., 1966: Bivalve mollusks: fluid dynamics of burrowing. Science, *152*:523–525.
Trueman, E. R., 1968: The burrowing process of *Dentalium*. J. Zool. Soc. London, *154*:19–27.
Turner, R. D., 1966: A Survey and Illustrated Catalogue of the Teredinidae. Museum of Comparative Zoology, Harvard University, Cambridge.
Vagvolgyi, J., 1967: On the origin of molluscs, the coelom, and coelomic segmentation. Syst. Zool., *16*: 153–168.
Wells, M. J., 1962: Brain and Behavior in Cephalopods. Stanford University Press, Stanford.
Wells, M. J., 1966: Cephalopod sense organs. *In* Wilbur, K. M., and Yonge, C. M. (Eds.), Physiology of Mollusca, Vol. 2. Academic Press, N. Y., pp. 523–545.
Wilbur, K. M., and Yonge, C. M. (Eds.), 1964: Physiology of Mollusca, Vol. 1; 1966, Vol. 2. Academic Press, N. Y.
Yonge, C. M., 1932: The crystalline style of the Mollusca. Sci. Progr., *26*:643–653.
Yonge, C. M., 1939: On the mantle cavity and its contained organs in the Loricata. Quart. J. Micr. Sci., *323*:367–390.
Yonge, C. M., 1941: The protobranchiate Mollusca: a functional interpretation of their structure and evolution. Phil. Trans. Roy. Soc. London, series B, *230*:79–147.
Yonge, C. M., 1947: The pallial organs in the aspidobranch Gastropoda and their evolution throughout the Mollusca. Phil. Trans. Roy. Soc. London, series B, *232*:443–518.
Yonge, C. M., 1957: Mantle fusion in the Lamellibranchia. Staz. Zool. Napoli, *29*:151–171.
Young, J. Z., 1972: The Anatomy of the Nervous System of *Octopus vulgaris*. Oxford University Press, N. Y. 690 pp.

There are some scientific journals devoted exclusively to mollusks: Proceedings of the Malacological Society of London, the Journal of Conchology, Nautilus, Veliger, and Malacologia.

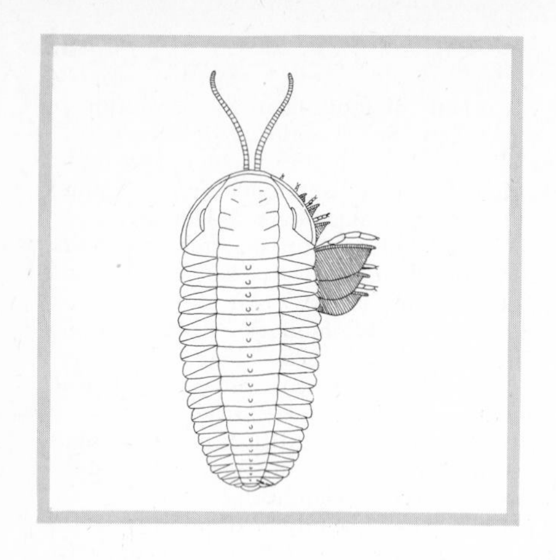

Chapter 12

INTRODUCTION TO THE ARTHROPODS; THE TRILOBITES

THE ARTHROPODS

The phylum Arthropoda is a vast assemblage of animals. At least three quarters of a million species have been described; this is more than three times the number of all other animals species combined. The tremendous adaptive diversity of arthropods has enabled them to survive in virtually every habitat; they are perhaps the most successful of all the invaders of the terrestrial habitat.

Arthropods represent the culmination of evolutionary development in the protostomes. They arose either from a primitive stock of polychaetes or from an ancestor common to both; and the relationship between arthropods and annelids is displayed in several ways:

1. Arthropods, like annelids, are metameric (Fig. 12–1*A*). Metamerism is evident in the embryonic development of all arthropods and is a conspicuous feature of many adults, especially the more primitive species. Within a number of the arthropodan classes, notably the arachnids and crustaceans, there has been a tendency for metamerism to become reduced. In such forms as mites and crabs, it has almost disappeared. Loss of metamerism has occurred in three ways. Segments have become lost; segments have become fused together; and segmental structures, such as appendages, have become structurally and functionally differentiated from their counterparts on other segments. Different structures having the same segmental origin are said to be serially homologous. Thus, the second antennae of a crab are serially homologous to the chelipeds (claws), for both evolved from originally similar segmental appendages.

2. Primitively, each arthropod segment bears a pair of appendages (Fig. 12–1*B*). This same condition is displayed by the polychaetes, in which each metamere bears a pair of parapodia. However, the exact homology between parapodia and arthropodan appendages is uncertain.

3. The nervous systems in both groups are constructed on the same basic plan. In both, a dorsal anterior brain is followed by a ventral nerve cord containing ganglionic swellings in each segment (Fig. 12–1*A*).

4. The embryonic development of a few arthropods still displays holoblastic determinate cleavage, with the mesoderm in these forms arising from the *4d* blastomere.

EXOSKELETON

Although arthropods display these annelidan characteristics, they have undergone a great many profound and distinctive changes in the course of their evolution. The distinguishing feature of arthropods, and one which has been an important factor in the evolutionary success of the group, is the chitinous exoskeleton or cuticle, which covers the entire body. Movement is made possible by the division of the cuticle into separate plates. Primitively, these plates are confined to segments, and the plate of one segment is connected to the plate of the adjoining segment by means of an articular membrane, a region in which the cuticle is very thin and flexible (Fig. 12–1*C*). Basically the cuticle of each segment is divided into four primary plates—a dorsal tergum, two lateral pleura, and a ventral sternum (Fig. 12–1*B*). This pattern has frequently

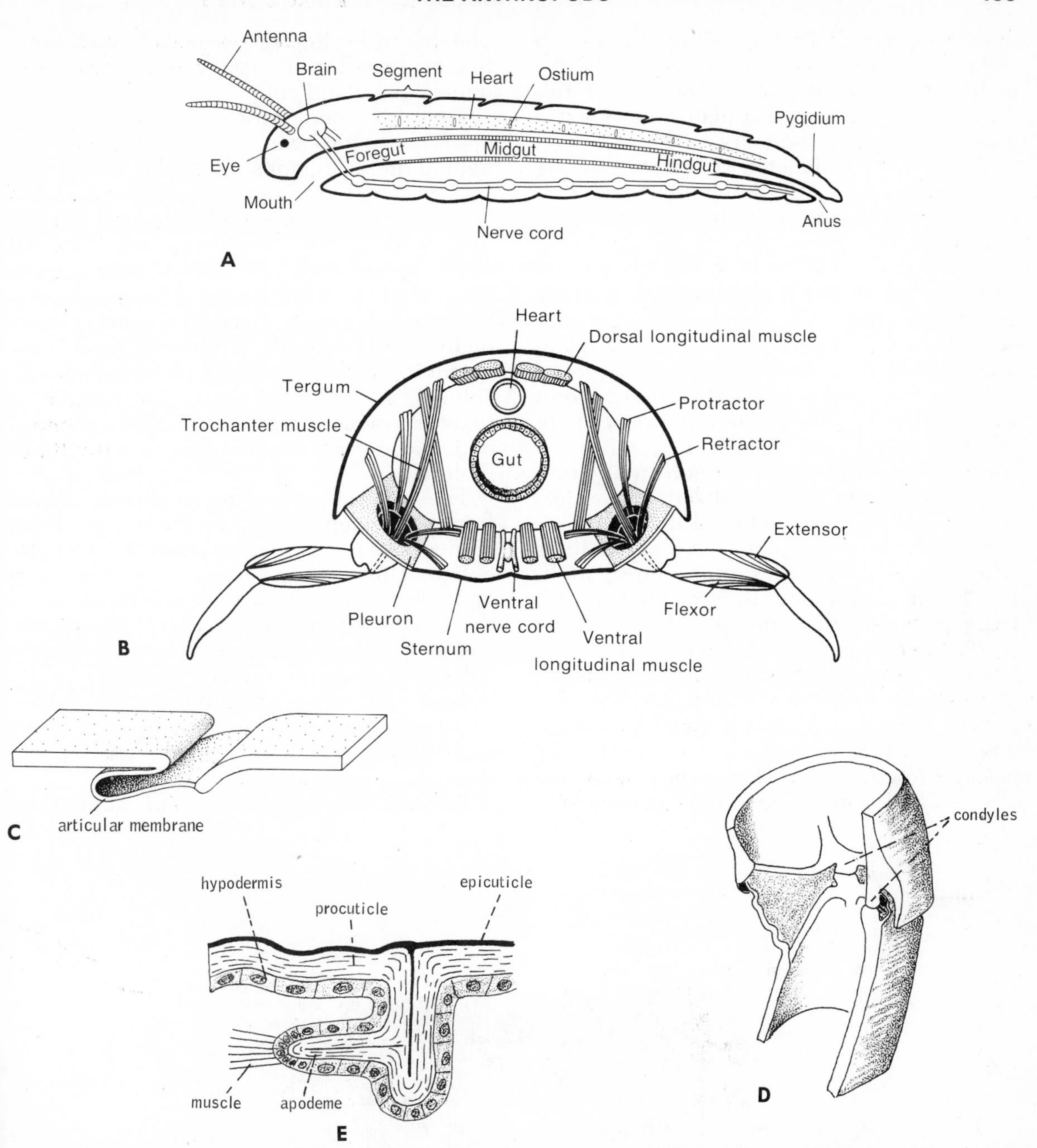

Figure 12–1. *A* and *B*. Structure of a generalized arthropod. *A*, sagittal section; *B*, cross section. *C*. Intersegmental articulation. Note articular membrane folded beneath segmental plane. (After Weber from Vandel.) *D*. Appendicular articulation, showing condyles. (After Weber from Vandel.) *E*. An apodeme. (After Janet from Vandel.)

disappeared because of either secondary fusion or subdivision.

The cuticular skeleton of the appendages like that of the body, has been divided into tubelike segments, or sections, connected to one another by articular membranes, thus creating a joint at each junction. Such joints enable the segments of the appendages, as well as those of the body, to move (hence the name of the phylum Arthropoda—jointed feet). In most arthropods, the articular membrane between body segments is folded beneath the anterior segment (Fig. 12–1*C*). In some arthropods, the additional development of articular condyles and sockets is suggestive of vertebrate skeletal structures (Fig. 12–1*D*).

In addition to the external skeleton, there has also been the development of what is sometimes called the endoskeleton. This may be an infolding of the procuticle that produces inner projections, or apodemes, on which the muscles are inserted (Fig. 12–1*E*); or it may involve the sclerotization of internal tissue, forming free plates for muscle attachment within the body.

The arthropod skeleton is secreted by the underlying layer of integumentary epithelial cells known as the hypodermis. It is composed of a thin outer epicuticle and a much thicker procuticle (Fig. 12–2). The epicuticle is composed of proteins and, in many arthropods, wax. The fully developed procuticle consists of an outer exocuticle and an inner endocuticle. Both layers are composed of chitin and protein bound together to form a complex glycoprotein, but the exocuticle in addition has been tanned; i.e., with the participation of phenols, its molecular structure has been further stabilized by the formation of additional cross linkages. Exocuticle is absent at joints and along lines where the skeleton will rupture during molting. In many arthropods the procuticle is also impregnated with mineral salts. This is particularly true for the Crustacea, in which calcium carbonate and calcium phosphate deposition takes place in the procuticle. Where the exoskeleton lacks a waxy epicuticle and is thin, it is a relatively permeable covering and allows the passage of gases and water. The cuticle is generally penetrated by fine pore canals, which function as ducts for the passage of secretions of underlying gland cells.

The arthropod cuticle is not restricted entirely to the exterior of the body. The hypodermis develops from the embryonic surface ectoderm, and all infoldings of this original layer, such as those parts of the gut that develop from the stomodeum and the proctodeum, thus are lined with cuticle (Fig. 12–1*A*). Other such ectodermal derivatives include: the tracheal (respiratory) tubes of insects, chilopods, diplopods, and some arachnids; the book lungs of scorpions and spiders; and parts of the reproductive systems of some groups. All of these internal cuticular linings are also shed at the time of molting.

The color of arthropods commonly results from the deposition of brown, yellow, orange, and red melanin pigments within the cuticle. However, iridescent greens, purples, and other colors result from fine striations of the epicuticle, which cause light refraction and give the appearance of color. Often body coloration does not directly originate in the cuticle, but instead is produced by subcuticular chromatophores or is caused by blood and tissue pigments, which are visible through a thin transparent cuticle.

Despite the locomotor and supporting

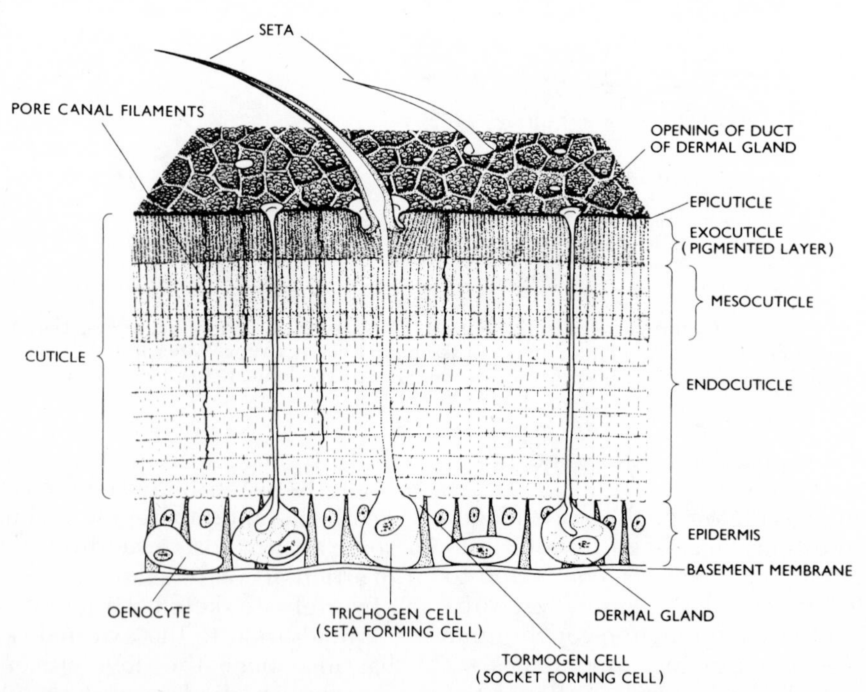

Figure 12–2. Diagrammatic section through the arthropod integument. (From Hackman, R. H., 1971: *In* Florkin, M., and Scheer, B. J. (Eds.), Chemical Zoology, Vol. 6. Academic Press, N.Y.)

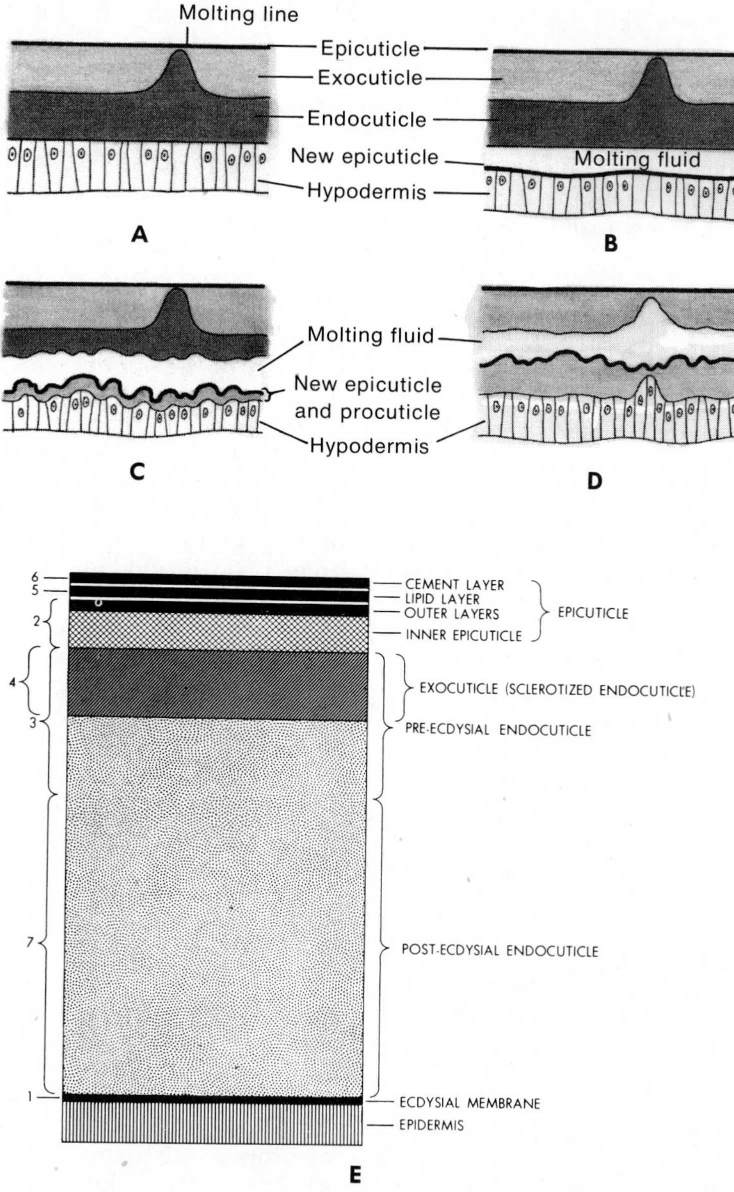

Figure 12–3. Molting in an arthropod. *A*. The fully formed exoskeleton and underlying epidermis between molts. *B*. Separation of the epidermis and secretion of molting fluid and the new epicuticle. *C*. Digestion of the old endocuticle and secretion of the new procuticle. *D*. The animal just before molting, encased within both new and old skeleton. *E*. Sequence in which different parts of the cuticle are deposited. (From Hackman, R. H., 1971: *In* Florkin, M., and Scheer, B. J. (Eds.), Chemical Zoology, Vol. 6. Academic Press, N.Y.)

advantages of an external skeleton, it poses problems for a growing animal. The solution to this problem evolved by the arthropods has been the periodic shedding of the skeleton, a process called *molting* or *ecdysis*.

Before the old skeleton is shed, the epidermal layer (hypodermis) becomes detached from the skeleton and secretes a new epicuticle. The hypodermis now secretes enzymes—chitinase and protease—which pass through the new epicuticle and erode away the untanned endocuticle (Fig. 12–3). Muscle attachments and nerve connections are unaffected, and the animal can continue to move. Following the digestion of the endocuticle, the epidermis secretes a new procuticle. At this point the animal is encased within both an old and a new skeleton (Fig. 12–3*D*). The old skeleton now splits along certain predetermined lines and the animal pulls out of the old encasement. Immediately after shedding, the new skeleton is soft and pliable and is stretched to accommodate the increased size of the animal. Some stretching is brought about by pressure resulting from tissue growth during the intermolt period, but it is also facilitated by the uptake of water or air by the animal. Hardening of the cuticle results from tanning of the protein and from stretching.

Additional procuticle may be added following actual ecdysis, and in some arthropods, such as insects, additions are made to the epicuticle by secretions through the pore canals. The final surface of the epicuticle is often formed by a cement layer (Fig. 12–3*E*).

The stages between molts are known as *instars*, and the length of the instars becomes longer as the animal becomes older. Some arthropods, such as lobsters, continue to molt throughout their life, although the event becomes less and less frequent. Other arthropods, such as insects and spiders, have more or less fixed numbers of instars, the last being attained with sexual maturity.

Molting is under hormonal control. Ecdysone, secreted by certain endocrine glands (for example, the prothoracic glands in insects), is circulated by the blood stream and acts directly on the epidermal cells. The production of ecdysone is in turn regulated by other hormones. Although first studied and best understood in insects, recent investigations indicate that ecdysone controls molting in all arthropods (Krishnakumaran and Schneiderman, 1968).

MOVEMENT AND MUSCULATURE

As movement in arthropods has become restricted to flexion between plates and cylinders of the cuticle, a profound change has taken place in the nature of the body musculature. In annelids the muscles take the form of a longitudinal and a circular sheath-like layer of fibers lying beneath the epidermis. Contraction of the two layers exerts force on the coelomic fluid, which then functions as a hydrostatic skeleton. In arthropods, on the other hand, these muscular cylinders have become broken up into striated muscle bundles, which are attached to the inner surface of the skeletal system (Fig. 12–1*B*).

The muscles are not attached to the hypodermis, but are attached to the inner surface of the procuticle by fibrils (tonofibrils) derived from specialized cells. Flexion and extension between plates are effected by the contraction of these muscles, with muscles and cuticle acting together as a lever system. This cofunctioning of the muscular system and skeletal system to bring about locomotion is essentially the same as in vertebrates. The principal difference is that the muscles in the arthropods are attached to the inner surface of an external skeleton, whereas in the vertebrates the muscles are attached to the outer side of an internal skeleton. Extension, particularly of the appendages, is accomplished, in part or entirely, by an increase in blood pressure.

Arthropods employ as their chief means of locomotion jointed appendages, which act either as paddles in aquatic species or as legs in terrestrial groups. Our knowledge of arthropodan locomotion, especially locomotion on land, results largely from the extensive studies of S. M. Manton (1951). In contrast to the parapodia of polychaetes, the locomotor appendages of arthropods tend to be more slender, longer, and located more ventrally. Despite the more ventral position of the legs, the body usually sags between the limbs (Fig. 12–4). In the cycle of movement of a particular leg, the effective step, or stroke, during which the end of the leg is in contact with the substratum, is closer to the body than is the recovery stroke, when the leg is lifted and swung forward (Fig. 12–4). Among the several factors determining speed of movement, the length of stride is of obvious importance, and stride length increases as the length of the leg increases. Field overlap tends to be considerably reduced in those arthropods utilizing five, four, or three pairs of legs. Stride length is greater and the problems of mechanical interference are further decreased because of differences in leg length and the relative placement of the leg tip. In arthropods which have retained a large number of legs, such as centipedes, the fields of movement of individual legs overlap those of other legs (Fig. 12–4*C*). Thus, the difference in proximity of the legs to the body during the effective and recovery strokes prevents mechanical interference.

The arthropodan gait involves a wave of leg movement, in which a posterior leg is put down just before or a little after the anterior leg is lifted. The movement of the legs on the opposite sides of the body alternate with one another. That is, one limb of a given pair is moving through its effective

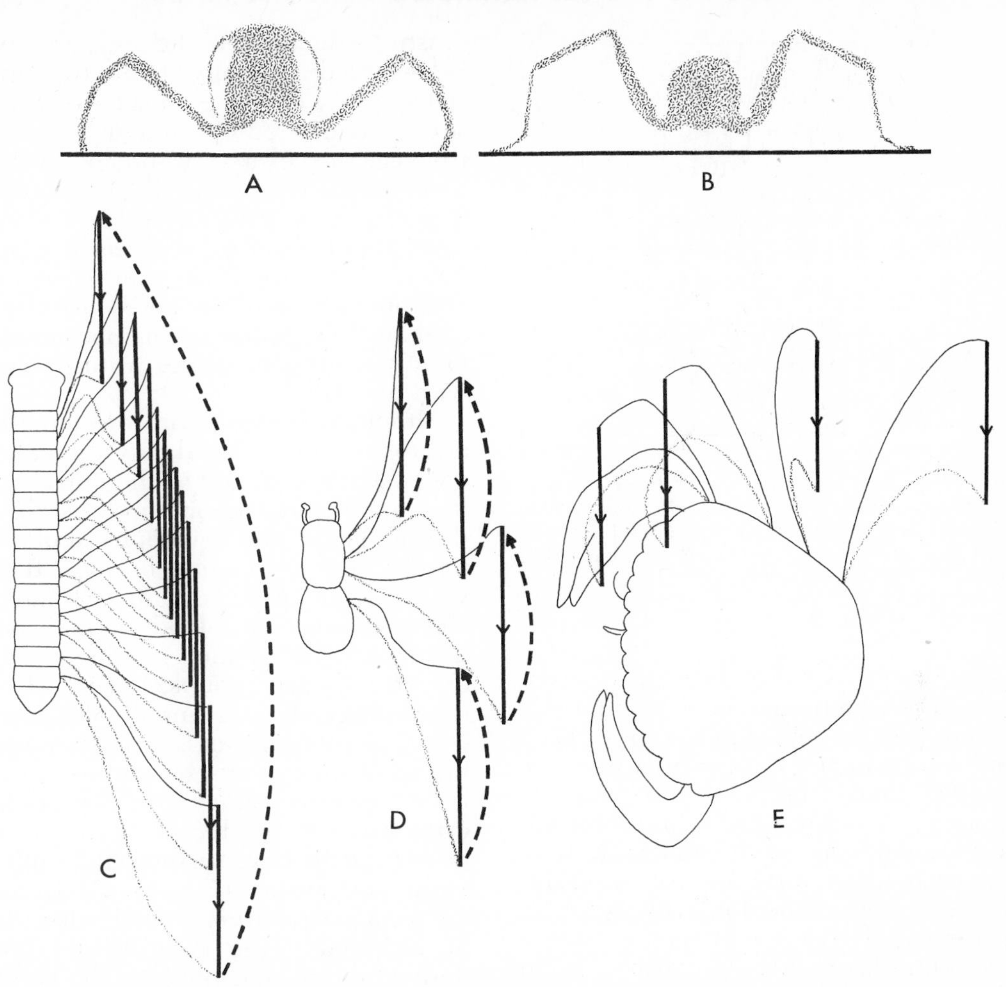

Figure 12–4. Arthropodan locomotion. Stance of a crayfish (*A*) and a spider (*B*). Leg movement in the gait of different arthropods. The effective stroke and the stride length are indicated by the heavy bars. Each leg is therefore drawn twice; at the beginning and at the end of the effective stroke. The recovery stroke is shown as a dashed line (*C* and *D*), and its position is only approximate. *C*. Strokes of the centipede *Scutigera*. *D*. Strokes of a spider. *E*. Strokes of a crab. (All after Manton.)

stroke while its mate is making a recovery stroke. Alternate leg movement tends to induce body undulation. This tendency is counteracted by increased body rigidity, such as the fused leg-bearing segments that form the thorax and cephalothorax of insects, some crustaceans, and arachnids.

An exoskeleton is a highly efficient supporting and locomotor framework for animals, such as arthropods, which are no more than a few centimeters long. For large animals, such as vertebrates, an internal skeleton is more efficient; an external skeleton would be very heavy and unwieldy.

In contrast to the condition in vertebrates, each arthropod muscle contains relatively few fibers and is innervated by only a small number of neurons. Many axon terminals are provided to one muscle fiber (Fig. 12–5), and one neuron may supply more than one muscle. Moreover, several types of motor neurons—phasic (fast) neurons, tonic (slow) neurons, and inhibitory neurons—may supply a single muscle (Fig. 12–5). The terms phasic and tonic, or fast and slow, do not refer to the speed of transmission but to the nature of the muscle response. The impulses of phasic motor neurons produce rapid but brief contractions, which are often involved in rapid movements. The impulses of tonic motor fibers produce slow, powerful, prolonged contractions, which are involved in postural activities and slow movements. The impulses of inhibitory neurons block contractions.

The neuromuscular system may be fur-

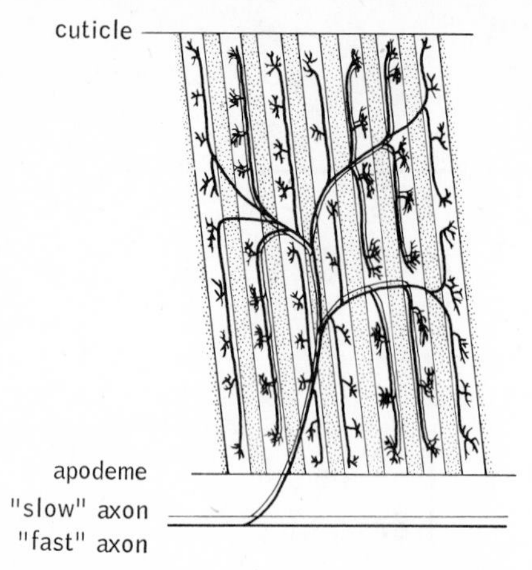

Figure 12–5. Dual innervation of an insect muscle. Fast, or phasic, neuron is depicted with a heavy line. Slow, or tonic, neuron is depicted with a fine line. Note that slow axon does not innervate every fiber. (After Hoyle.)

ther complicated, as in at least the crustaceans, by the differentiation of the muscle fibers into phasic and tonic types, each having a distinctive ultrastructure and physiology. Some muscles are entirely phasic, some are entirely tonic, and some are mixed. Phasic motor neurons innervate only phasic muscle fibers; tonic motor neurons innervate both phasic and tonic fibers or, in some instances, only tonic fibers.

In vertebrates, graded responses are in large part dependent upon the number of motor units contracting. As can be seen, arthropod muscles are not organized as motor units, and graded responses are dependent upon the type of muscle fibers contracting, the type of neuron fired, and the interaction of different types of neurons. For example, two different extensor muscles are innervated by the same motor fiber in a crayfish claw, but the two muscles function independently because each is innervated by separate inhibitory neurons.

The organization of arthropod ganglia is like that of annelids and mollusks described on page 359. Giant fiber systems are frequently well developed, and "command" systems have been identified. Arthropod neural networks and neuromuscular systems have been best studied in crustaceans. Interested students should consult Kennedy et al. (1969) and Atwood (1973).

COELOM AND BLOOD-VASCULAR SYSTEM

The well-developed, metameric coelom characteristic of the annelids has undergone drastic reduction in the arthropods and is represented by only two structures: the space or cavity of the gonads—the gonocoel and its associated ducts; and in certain arthropods, the excretory organs, which are homologous to the annelidan coelomoducts. The change is probably related to the shift from a fluid internal skeleton to a solid external skeleton. The other spaces of the arthropodan body do not constitute a true coelom but rather a hemocoel—that is, merely sinuses or spaces in the tissue filled with blood.

Although derived from the annelids, the arthropod blood vascular system is an open one. What functioned as the dorsal vessel in the annelids has become in the arthropods a distinct heart. The development of a distinct heart is not surprising, because the dorsal vessel in annelids is contractile and is the chief center for blood propulsion. The heart varies in position and length in different arthropodan groups, but in all of them the heart consists essentially of one or more chambers with muscular walls arranged in a linear fashion and perforated by pairs of lateral openings called ostia (Fig. 12–1A). The ostia enable the blood to flow into the heart from the large surrounding sinus known as the pericardium (Fig. 13–25). In this case, the term *pericardium* is misleading, because in arthropods the pericardium does not derive from the coelom as in the vertebrates, but instead is a part of the hemocoel. After entering the heart, blood is pumped out to the body tissues through vessels frequently called arteries, and is eventually dumped into sinuses (collectively the hemocoel) in which it bathes the tissues directly. The blood then returns by various routes to the pericardial sinus.

The blood of arthropods is essentially the same as the blood of annelids. The respiratory pigments hemoglobin and hemocyanin are sometimes present, dissolved in the plasma. Several forms of amebocytes are present in the blood.

The development of an open blood vascular system in arthropods is perhaps correlated with the reduction of the coelom. Certainly correlated with coelom reduction has been the disappearance of the nephridia and the development of new excretory organs.

DIGESTIVE TRACT

The arthropod gut differs from that of most other animals in having large stomodeal and proctodeal regions (Fig. 12–1A). The derivatives of these ectodermal portions are lined with chitin and constitute the foregut and hindgut. The intervening region, derived from endoderm, forms the midgut. The foregut is chiefly concerned with ingestion, trituration, and storage of food; its parts are variously modified for these functions depending upon the diet and mode of feeding. The midgut is the site of enzyme production, digestion, and absorption; however, in some arthropods enzymes are passed forward and digestion begins in the foregut. Very commonly the surface area of the midgut is increased by outpocketings, forming pouches or large digestive glands. The hindgut functions in the absorption of water and the formation of feces.

BRAIN

There is a high degree of cephalization in arthropods. The increase in brain size is correlated with well-developed sense organs, such as eyes and antennae, and many arthropodan groups display rather complex behavioral patterns. The arthropodan brain consists of three major regions—an anterior protocerebrum, a median deutocerebrum, and a posterior tritocerebrum (Fig. 12–6*B*). The nerves from the eyes enter the protocerebrum, which contains one to three pairs of optic centers (neuropiles). The optic and other neuropiles of the protocerebrum function in integrating photoreception and movement and are probably the centers for the initiation of complex behavior.

The deutocerebrum receives the antennal nerves (first antennae in crustaceans) and contains their association centers. Antennae are lacking in the chelicerates (scorpions, spiders, mites), and in these arthropods there is a corresponding absence of the deutocerebrum (Fig. 12–6*C*).

The third brain region, the tritocerebrum, gives rise to nerves that innervate the labium (lower lip), the digestive tract (stomatogastric nerves), the chelicerae of chelicerates, and the second antennae of crustaceans. The commissure of the tritocerebrum is postoral, i.e., is located behind the foregut.

There is considerable controversy regard-

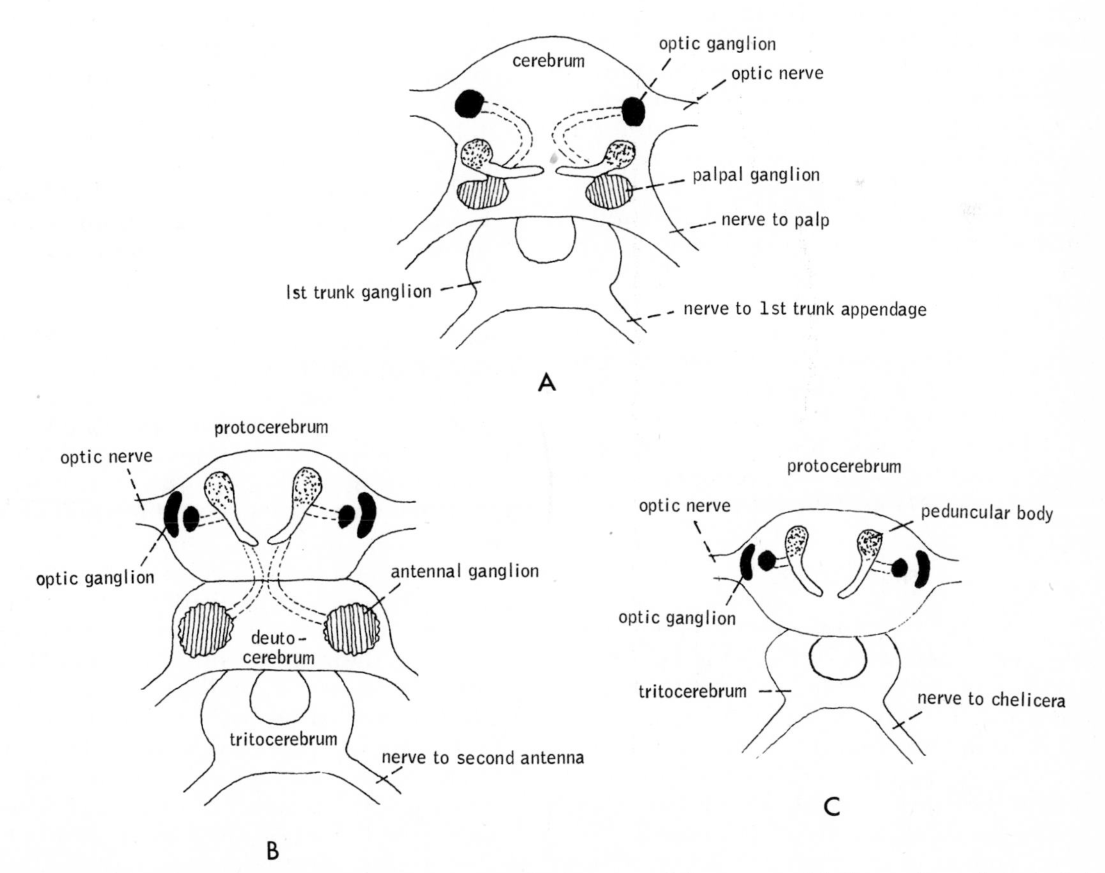

Figure 12–6. Annelid and arthropod brains. *A*. A polychaete annelid. *B*. A mandibulate arthropod. *C*. A chelicerate arthropod. (All after Hanstrom from Vandel.)

ing the evolutionary origin of the arthropodan head and brain. More specifically, the debate centers on the extent to which the arthropodan head is a segmented structure. Most zoologists agree that the tritocerebrum is a segmental ganglion that has shifted anteriorly. Its postoral commissure alone is good evidence of such an origin. With regard to the origins of the protocerebrum and the deutocerebrum, there are two general schools of thought. One view holds that the two anterior brain regions represent the nonsegmented acron, or archicerebrum, which is homologous to the cerebral ganglia of most annelids. The other view is that the protocerebrum and deutocerebrum contain segmental ganglia that have fused with the original acron. Among zoologists who argue for the segmental nature of the anterior brain regions, there are differing opinions about the number of segments involved. A résumé of various ideas on the subject is outlined by Bullock and Horridge (1965).

The embryogeny of the arthropodan head unfortunately does not provide a clear-cut answer to the problem; in fact, embryogeny has been the major source of conflicting opinions. The protocerebrum and deutocerebrum, as well as other head structures, arise from a number of separate embryonic rudiments, which later fuse. It is these separate rudiments that have been considered segmental derivatives by many authors.

The view that the protocerebrum and deutocerebrum are nonsegmental in nature and that the tritocerebrum is the ganglion of the *first* trunk segment encounters the least number of conflicts when the comparative anatomy and the embryogeny of the entire phylum are considered. This view is also held by Bullock and Horridge (1965). The division of the original archicerebrum, or acron, into a protocerebrum and a deutocerebrum is believed to be associated with the development of the sense organs.

SENSE ORGANS

The sensory receptors of arthropods are usually associated with some modification of the chitinous exoskeleton, which otherwise would act as a barrier to the detection of external stimuli. An important and very common type of receptor is one connected with hairs, bristles, or setae. The bristle may be so designed that when it is moved, the receptor ending in the shaft or base is stimulated (Fig. 13–7*D*), or the bristle may carry chemoreceptor endings (Fig. 13–7*E*). Other common modifications for receptors are canals, slits, pits, or other openings in the exoskeleton (Fig. 13–7*F*). They may house chemoreceptors, or the opening may be covered by a thin membrane, to the underside of which is attached a nerve ending (Fig. 13–7*F*). Such sense organs detect vibrations or other forces which change the tension of the skeleton. All of the receptors described so far may be scattered over the body surface as well as concentrated in joints or on certain appendages, such as the antennae or legs. There are also proprioreceptors attached to the inside of the integument or to tendons and muscles.

Most arthropods have eyes, but the eyes can vary greatly in complexity. Some are simple and have only a few photoreceptors. Others are large, with thousands of retinal cells, and can form a crude image. In all arthropods the skeleton contributes the transparent lens-cornea to the eye, and the focus is always fixed because the immovable lens is continuous with the surrounding skeleton.

Insects and many crustaceans, such as crabs and shrimp, have a type of eye called compound, because it is composed of many long, cylindrical units possessing all of the elements for light reception. Each unit, or ommatidium, is covered at its outer end by a translucent cornea derived from the skeletal cuticle (Fig. 12–7). The cornea functions as a lens. The external surface of the cornea, called a facet, is usually square or hexagonal. Behind the cornea, the ommatidium contains a long cylindrical or tapered element called the crystalline cone, which functions as a second lens. The crystalline cone is usually produced by four cells and often has a somewhat four-part structure.

The basal end of the ommatidium is formed by the receptor element (the retinula). The center of the retinula is occupied by a translucent cylinder (the rhabdome), around which are arranged elongated photoreceptor, or retinular, cells (Fig. 12–7). There are commonly seven similar retinular cells plus one eccentric cell, but other numbers and arrangements occur. The retinular cells have their inner surfaces thrown into tubular folds, or microvilli, oriented at right angles to the axis of the ommatidium (Fig. 12–7*E* and *F*). The area of tubules constitutes a rhabdomere, and it projects centrally to make up the major part of the rhabdome. Commonly there are a smaller number or rhabdomeres (4) than there are retinular cells. In some compound eyes, the projecting rhabdomeres touch one another, and

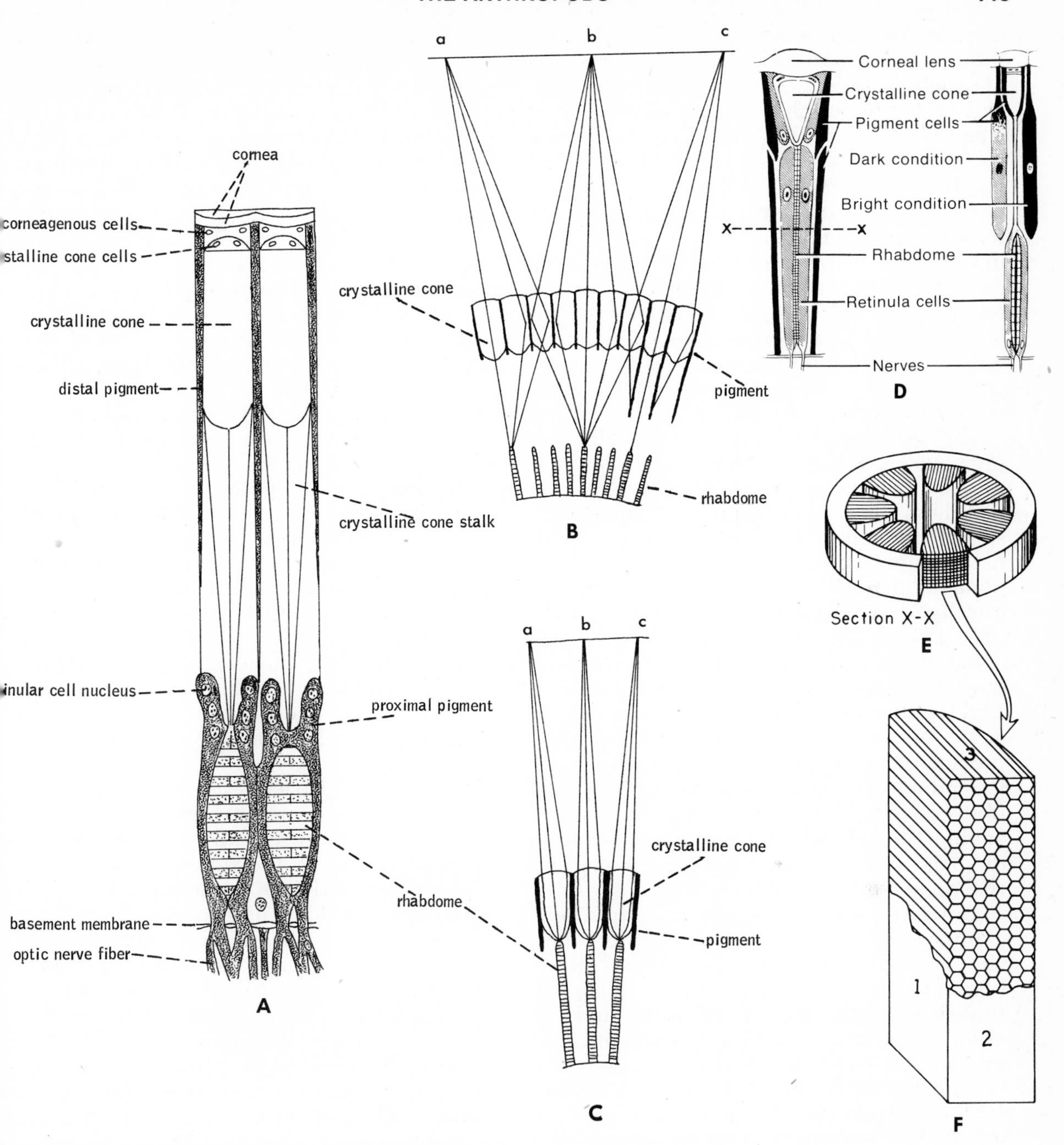

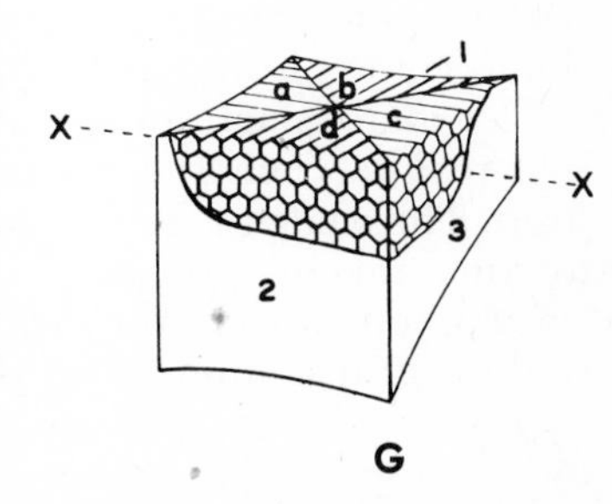

Figure 12–7. *A.* Two ommatidia from compound eye of the crayfish, *Astacus* (longitudinal view). (After Bernhards from Waterman.) *B.* Compound eye specially adapted for superposition image formation. Light rays from points *a* and *b* are being received as a superposition image. Pigment is retracted and light rays, initially received by a number of ommatidia, are concentrated on a single ommatidium. Point of light *c* is being received as an apposition image. Pigment is extended, preventing light rays from crossing from one ommatidium to another. *C.* Compound eye especially adapted for apposition image formation. (*B* and *C* after Kühn from Prosser and others.) *D.* Insect ommatidia, showing a diurnal type (left) and a nocturnal type (right). In the nocturnal type, pigment is shown in two positions, adapted for very dark conditions on the left side, and for relatively bright conditions on the right. *E.* Section through an open type of rhabdome of an insect (level X-X in part *D*). *F.* Microtubules of one of the rhabdomeres in part *E*. *G.* Section through a closed type of rhabdome. (*E-G* from Wolken, J. J., 1971: Invertebrate Photoreceptors. Academic Press, N.Y.)

the resulting rosette virtually fills the exterior of the retinula (Fig. 12–7G). In others the rhabdomeres are not so closely packed, and there is a clear, fluid-filled central area (Fig. 12–7E). The rhabdome is formed by all of the rhabdomeres and the enclosed central space. The rhabdomeres of an ommatidium function as a single photoreceptor unit and transmit a signal that represents a single light point. From each retinula cell extends an axon, and thus a bundle of seven or eight axons leaves each ommatidium. The axons make connections with second order neurons in an optic ganglion within the brain or, in some crustaceans, within an eye stalk.

The retinular cells contain black or brown pigment granules, which comprise the proximal retinal pigment. Distally the ommatidium is surrounded by a number of special pigment cells, forming the distal retinal pigment. Either the proximal or distal pigment or both can migrate centrally or distally, depending on the intensity of light (Fig. 12–7D).

In bright light, a compound eye produces an apposition image (Fig. 12–7C). Light enters an ommatidium either at an angle or perpendicular to the facet. The proximal and distal retinal pigments are extended and act as a screen to prevent light from passing from one ommatidium to another, and thus light rays are restricted to the axial region of the crystalline cone and rhabdome.

The crystalline cone is essentially a lens cylinder and is formed of concentric lamellae. The outer lamellae have lower refractive indices than do the central lamellae and probably function by refraction to eliminate oblique rays and to direct the more axial light rays to the rhabdome. There is no mechanism for accommodation, because the crystalline cone provides for a fixed focus. On emerging from the crystalline cone, light rays pass to the rhabdome. Studies by Shaw (1969) have demonstrated that in the apposition eyes of honeybees and locusts, only 0.1 to 1 per cent of the light reaching a rhabdome comes from facets other than its own.

The total image formed by the compound eye results from the number of ommatidia fired. The image is thus somewhat analogous to the image produced on a television screen, on which the "picture" is essentially a grid composed of dots of light. The apposition image formed by a compound eye is often called a mosaic image, since the total image results from small pieces put together like a mosaic. Unfortunately, the term *mosaic image* implies a special or different sort of image. Actually, the human eye receives and transmits light stimuli in a somewhat similar manner, a single point of light stimulating a functional unit of approximately seven cones; the only difference is that the mosaic in the human eye is of a finer grain (composed of smaller and more numerous spots).

An advantage of the compound eye found in many arthropods is its facility to detect movement. A slight shift in a point of light results in a corresponding shift in the ommatidia being stimulated. The overlapping feature should be of value here, because several ommatidia would then be responding to the same point of light. Another advantage of the compound eyes found in many arthropods is that the total corneal surface is greatly convex, resulting in a wide visual field. This is particularly true for the stalked compound eyes of crustaceans, in which the cornea may cover an arc of 180 degrees or more (Fig. 14–70A).

In weak light, the compound eyes function in a somewhat different manner and form what has been called a "superposition image" (Fig. 12–7B and D). The pigment is retracted, so no screening effect is present. Thus, light can pass from one ommatidium to another, and one rhabdome responds to light rays that originally entered several adjacent facets. In the superposition eyes of the crayfish, at least 50 per cent of the light striking a rhabdome originally entered facets other than its own (Shaw, 1969). The term *superposition image* is also unfortunate because there is probably no image formed. Rather, this condition appears to be an adaptation for light gathering in weak light, making it more likely that the rhabdome will be activated than if it was dependent upon light received only from its own facet.

Most current investigators of compound eyes believe that the terms apposition and superposition are inappropriate and should be abandoned. But as yet there are no generally accepted substitutes.

The compound eyes of most crustaceans are able to adapt, to at least some extent, to both bright and weak light, but in general, they tend to be specially modified for functioning under one of these two conditions. Thus, compound eyes can be classified as either apposition or superposition eyes, although there are many intergradations between the two types.

Arthropods that are diurnal and live in

well lighted habitats, such as terrestrial and littoral species, usually possess apposition eyes (Fig. 12–7*C* and *D*). The screening pigment is well developed. The length of the crystalline cone is approximately equal to its focal length, and the lower end of the cone and the upper end of the rhabdome are contiguous, or nearly so. The retinular cells are quite long, extending from the crystalline cone to the basal membrane of the retina. All of these modifications tend to confine light entering a single ommatidium and to funnel the light down the axis of the ommatidium to the rhabdome.

Superposition eyes are found in nocturnal species or those that live in poorly lighted habitats. However, there are many exceptions, and it is now recognized that not all superposition eyes can be considered nocturnal eyes, since they are also found in diurnal arthropods. The superposition eye is especially modified for collecting and concentrating onto one ommatidium light originally striking a large patch of facets (Fig. 12–7*B* and *D*). Screening pigment is usually present but may be reduced or absent in cave-dwelling and bathypelagic species. The crystalline cone tends to be twice as long as its focal length, and there is considerable space between the end of the crystalline cone and the rhabdome. The retinular cells are much shorter than those in apposition eyes, and they are restricted to the base of the ommatidium. A special reflecting pigment is often present around the retinula and may be movable.

Of the several factors that affect the degree of visual acuity in compound eyes (i.e., how well various parts of the field can be resolved), the number of ommatidia composing the eye in relation to eye size is of greatest importance. The larger the number of ommatidia, the finer is the retinal grain and the less crude is the image. The number of ommatidia in crustacean lateral eyes varies enormously. The eye of certain ants possess only a few ommatidia, while the eye of a dragonfly may possess 10,000 ommatidia. Studies on crustaceans with well developed compound eyes indicate some ability to distinguish in regard to form and size, but the degree of visual acuity is certainly slight and the total image crude as compared to that formed by the human eye.

Color discrimination has been demonstrated in a number of arthropods. For example, the hermit crab, *Pagurus*, can discriminate between painted yellow and blue snail shells and shells colored different shades of grey. The chromatophores of the shrimp, *Crangon*, adapt to backgrounds of yellow, orange, or red, but not to any shade of grey (the chromatophore changes are mediated through the eyes). Cockroaches have been shown to have receptors for ultraviolet and green wavelengths.

REPRODUCTION AND DEVELOPMENT

With few exceptions, arthropods are dioecious. Furthermore, many employ modified appendages during copulation. Fertilization is always internal in terrestrial forms but may be external in aquatic species. The eggs of most arthropods are extremely rich in yolk and are centrolecithal. In the centrolecithal egg the nucleus is surrounded by a small island of non-yolky cytoplasm in the middle of a large mass of yolk (Fig. 12–8*A*). There is also a peripheral sphere of non-yolky cytoplasm. The presence of such a large amount of yolk and the change in its distribution have effected considerable modification in the development of arthropods when compared with development in the typical polychaetes. Although some arthropods have holoblastic cleavage and a few display a determinate pattern, most have a modified type of cleavage that is associated with centrolecithal eggs and is called superficial cleavage (Fig. 12–8*B* to *E*).

After fertilization, the centrally located nucleus undergoes mitotic divisions, but without the formation of cell membranes and without any cleavage of the yolk. The result, after several such divisions, is an uncleaved egg containing a large number of syncytial nuclei within the center. As division continues, the nuclei gradually migrate to the periphery, where cell membranes form but do not extend into the yolky interior. This stage of development represents a stereoblastula. Development continues with the formation of a primordial germinal disc located on one side of the embryo (Fig. 12–8*F*). This germinal center proliferates entoderm and mesoderm and eventually forms the embryonic body, which appears to be wrapped around the egg (Fig. 12–8*G*).

In many arthropods, there is a type of cleavage that is somewhat intermediate between total cleavage and superficial cleavage. After a few divisions of the nucleus, the yolk undergoes segmentation or breaks up into pyramids, each of which contains one of the nuclei (Fig. 12–8*H*). Gradually the yolk pyramids disappear, and development shifts to the superficial type.

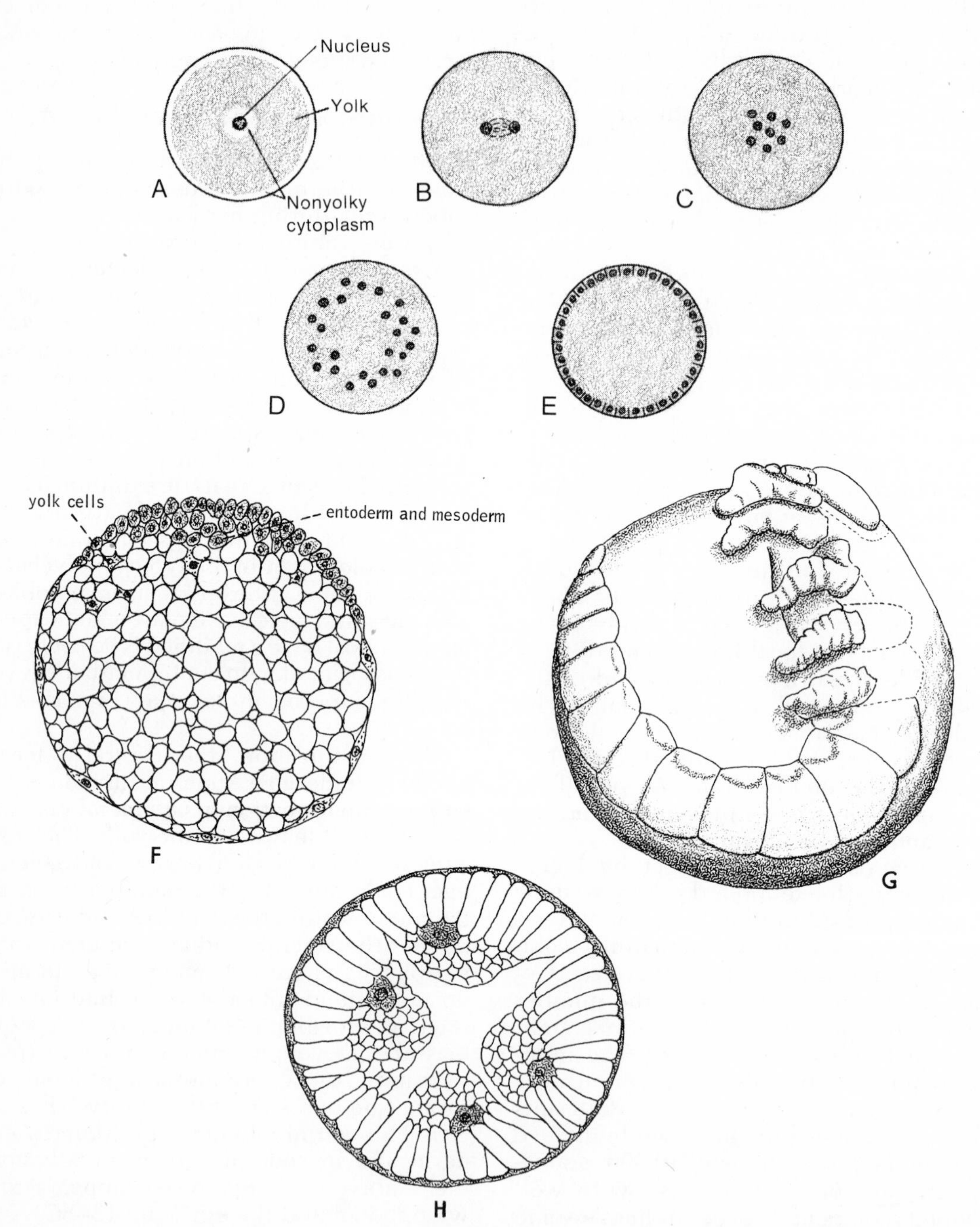

Figure 12–8. *A* to *E*. Superficial cleavage. *A*. Centrolecithal egg. *B* and *C*. Nuclear division. *D*. Migration of nuclei toward periphery of egg. *E*. Blastula. Cell membranes have developed, separating adjacent nuclei. *F*. Gastrula of the spider, *Theridion maculatum*, showing development of germinal center. (After Dawydoff.) *G*. Embryo of the arachnid, *Thelyphonus*. (After Kaestner.) *H*. Initial complete cleavage in the spider, *Theridion maculatum*. (After Morin from Dawydoff.)

SYNOPSIS OF THE ARTHROPOD CLASSES

Subphylum Trilobitomorpha. The fossil trilobites.
Subphylum Chelicerata.
Class Merostomata. The living horseshoe crabs and fossil eurypterids.
Class Arachnida. The largest of the chelicerate classes; includes the scorpions, harvestmen, spiders, ticks, and mites.
Class Pycnogonida. The sea spiders.
Subphylum Mandibulata.
Class Crustacea. The crustaceans.
Class Insecta. The insects.
Class Chilopoda. The centipedes.
Class Diplopoda. The millipedes.
Class Symphyla. The symphylans.
Class Pauropoda. The pauropodans.

THE TRILOBITES

The subphylum Trilobitomorpha is the most primitive of all known arthropodan groups and from an evolutionary standpoint represents a good starting point in discussing the arthropodan classes. Trilobites are an extinct group of marine arthropods, which were once abundant and widely distributed in Paleozoic seas. They reached their height of distribution and abundance during the Cambrian and Ordovician periods and disappeared at the end of the Paleozoic era. Over 3900 species have been described from fossil specimens.

The trilobite body was in general somewhat oval and flattened and displayed a dorsal cuticle, or exoskeleton, that was much heavier and thicker than the ventral surface, which carried the appendages (Fig. 12–9). This difference is the reason that in most cases only the dorsal skeleton has been preserved to form the fossil record. Most trilobite species ranged from 3 to 10 cm. in length, although some planktonic forms were only 0.5 mm. in length; some species attained a length of little less than a meter, about the same size as living horseshoe crabs.

The trilobite body was divided into three more or less equal sections—a solid anterior cephalon; an intermediate thorax or trunk region, consisting of a varying number of separate segments; and a posterior pygidium. Each of these body divisions was in turn divided into three regions by a pair of furrows running from anterior to posterior and forming a median axial lobe flanked on each side by a lateral lobe. The name Trilobita refers to this transverse trilobation of the dorsal body surface.

The anterior body section, the cephalon, was composed of four fused segments in addition to the acron, and was covered by a shieldlike carapace with the posterior lateral margins projecting backward. The original head segmentation was frequently indicated by short transverse grooves on the carapace. The cephalic shield not only covered all the dorsal surface but was folded under at its margins so that the original ventral surface was quite narrow and restricted. The pleural skeleton was reduced to thin membranes to which the appendages were attached. This condition was true not only for the cephalon, but also for the other two divisions of the body. On each side of the middle of the carapace was a pair of eyes that varied considerably in size, depending upon the species.

The mouth (Fig. 12–10*B*) was located in the middle of the underside of the cephalon just behind and beneath a liplike prominence called the labrum (Fig. 12–9*B*). On each side of the labrum was a long sensory antenna, believed to be homologous to the first pair of antennae of the crustaceans and the antennae of insects.

Behind the mouth were located four pairs of similar appendages identical to the appendages of the trunk and pygidium. Each appendage was biramous and consisted of an inner walking leg (telopodite) and an outer gill-bearing branch (pre-epipodite) (Fig. 12–10*A*). The pre-epipodite was situated between the lateral extension of the pleural lobe and the walking leg and consisted of numerous segments, each of which bore (or at least the terminal segment bore) a fringe of filaments extending posteriorly. The fringe of one appendage overlapped the fringe of the next. These filaments were perhaps gills, with gases diffusing between blood within the filament and the external water.

The trilobite trunk (thorax) consisted of a varying number of separately articulating segments; some species were able to roll up into a ball as do sow bugs. Like the cephalon, the thorax possessed segments with posterior lateral margins that were frequently prolonged and extended backward. The lateral margins of the trunk tergites were folded ventrally. Each segment bore ventrally a pair of appendages similar to the appendages of the cephalon.

The pygidium was constructed on the same plan as the thorax except that the seg-

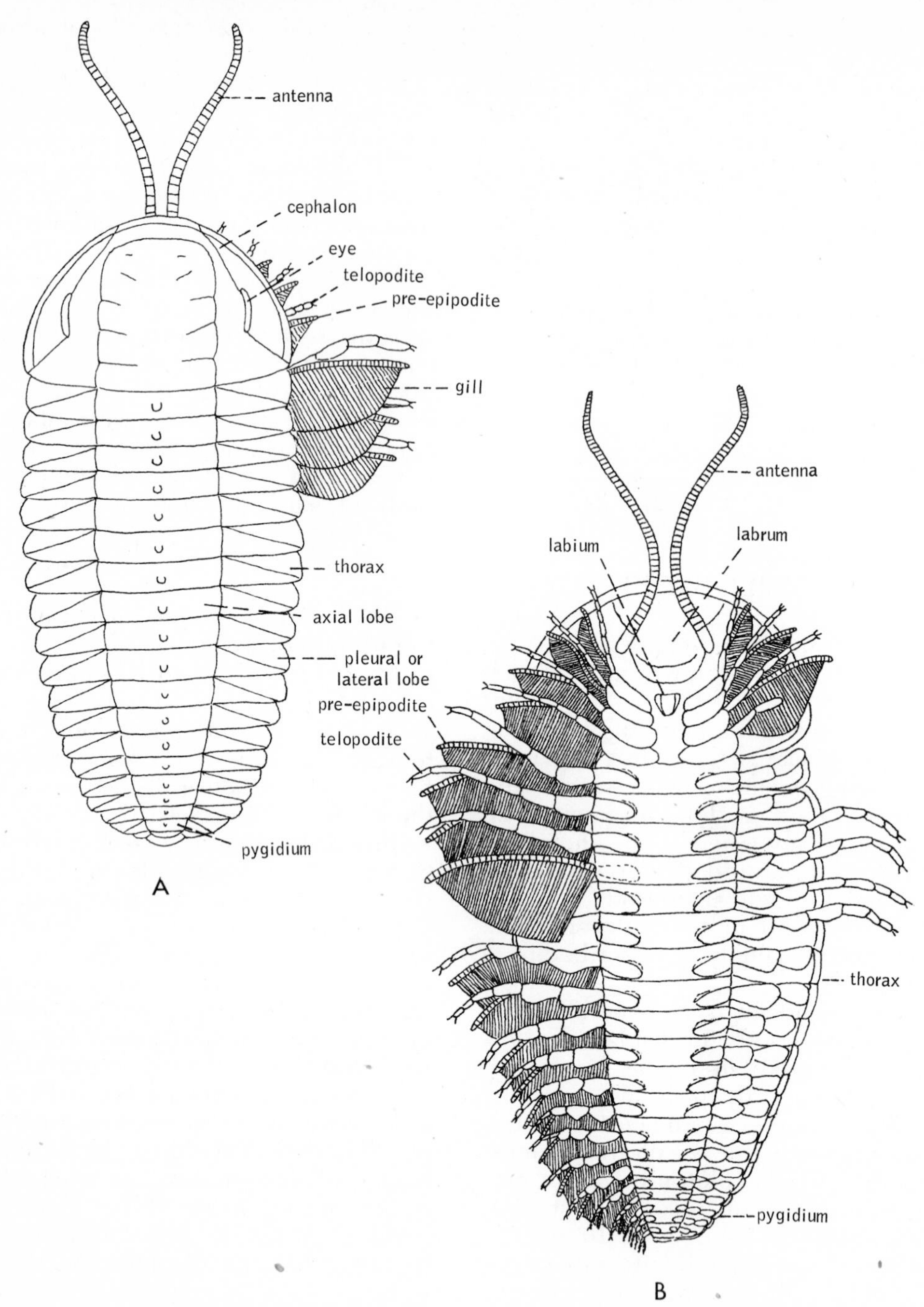

Figure 12–9. The Ordovician trilobite, *Triarthrus eatoni*. *A*. Dorsal view. *B*. Ventral view. (Both after Walcott and Raymond from Störmer.)

ments were fused and formed a solid shield. The appendages of the pygidium usually were successively smaller toward the posterior end.

The majority of trilobites were bottom dwellers, and crawled over sand and mud using the walking legs. The flattened body and dorsal eyes were adaptations for this type of existence. Some trilobite groups had a shovel-shaped or plow-shaped cephalon adapted for burrowing (Fig. 12–11A). Such bottom-dwelling forms were probably scavengers or consumed mud and silt and then digested the organic materials from it, as do many annelids. In addition, the branchial filaments possibly could have sifted food materials from the surrounding water and then passed the food anteriorly to the

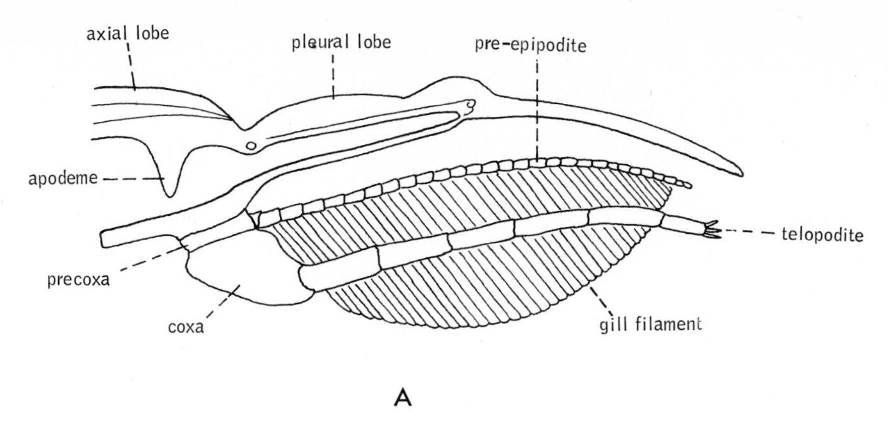

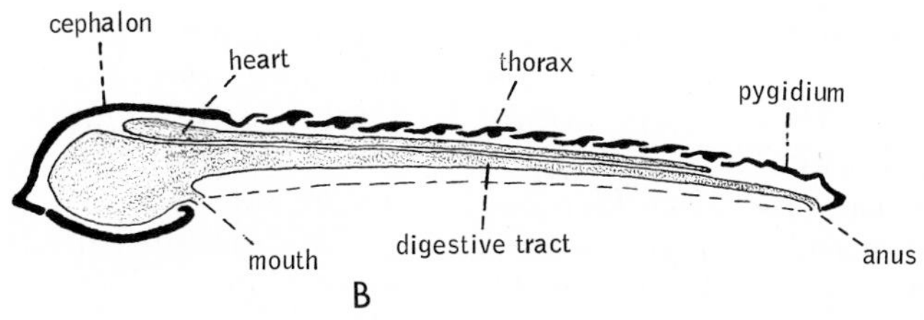

igure 12–10. *A*. Trilobite, showing a trunk appendage (transverse section through left half). (Modified after Walcott Raymond from Störmer.) *B*. Trilobite (sagittal section). (After Moore, Lalicker, and Fischer.)

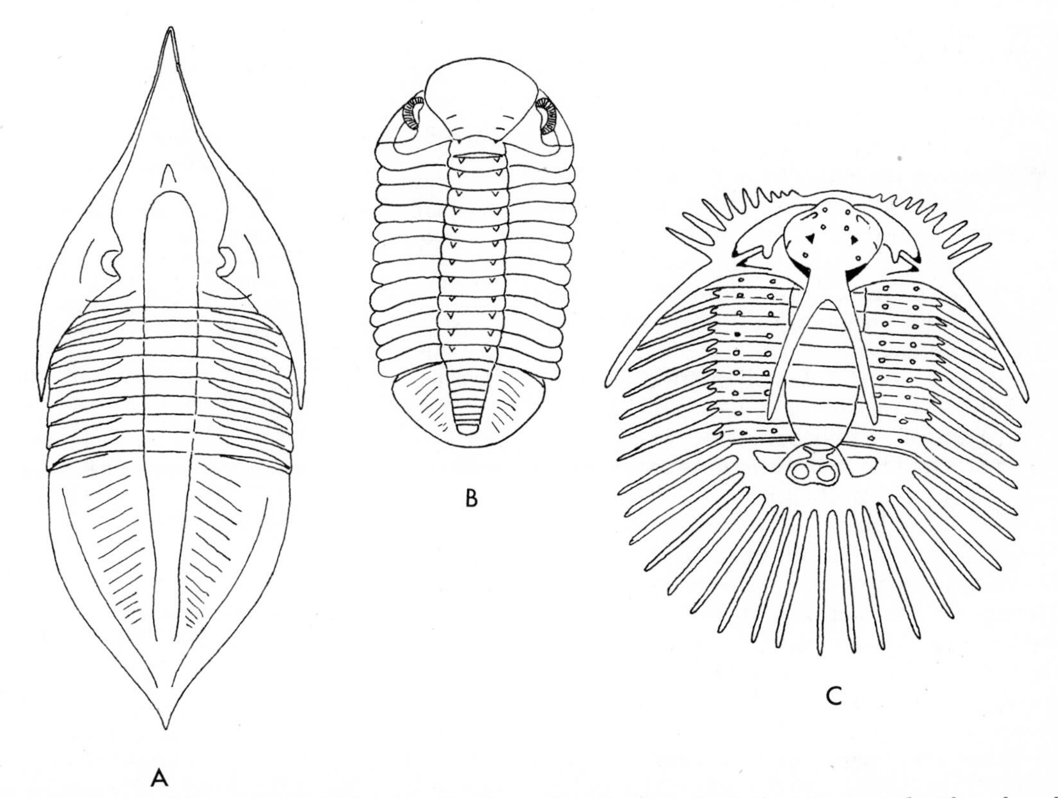

Figure 12–11. *A*. A burrowing trilobite, *Megalaspis acuticauda,* from the Ordovician period with a plow-shaped cephalon. (After Störmer.) *B*. A pelagic trilobite, *Phacops steenbergi,* from the Silurian period. (After Barrande from Störmer.) *C*. A planktonic trilobite, *Radiaspis radiata,* from the Devonian period. The long spines may have been flotation devices. (After Störmer.)

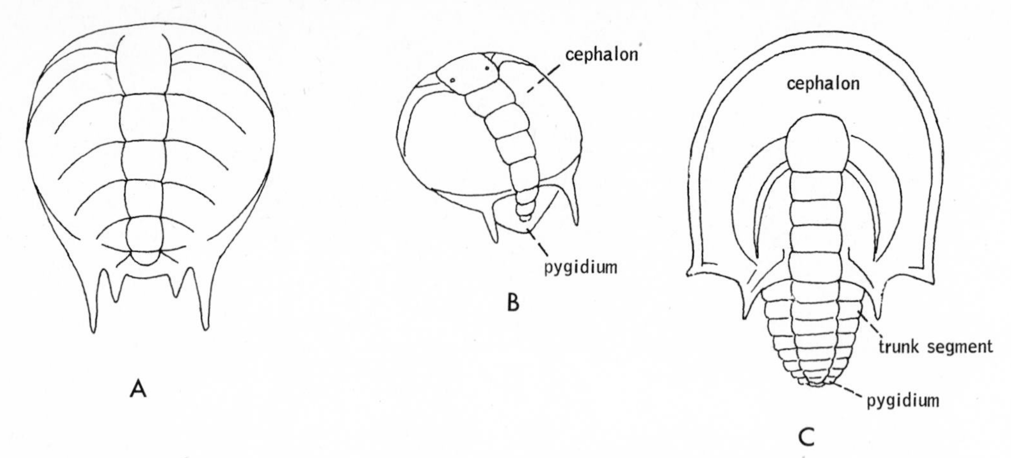

Figure 12–12. Trilobite larvae. *A*. Protaspis of *Olenus*, with four segments. *B*. Late protaspis of *Olenus*. Pygidium just formed. *C*. Meraspis of *Paedeumias*. (All after Moore, Lalicker, and Fischer.)

mouth from one overlapping fringe to the next.

Some groups of trilobites were apparently not confined to the bottom but took up a swimming existence (Fig. 12–11*B*). In these forms the body was narrower, and the eyes were located on the sides of the head. Although nothing is known about their appendages, the walking legs may have been flattened and adapted for swimming. The smallest species of trilobites were planktonic. The dorsal surface displayed long radiating spines that have been considered flotation devices (Fig. 12–11*C*).

Fossil material has not only yielded much information on the structure of adult trilobites, but has also made possible a remarkably complete understanding of the developmental stages. For some forms, such as species of *Sao* and *Olenus*, almost the entire larval series has been determined from fossil specimens.

During their postembryological development, trilobites passed through three larval periods, each of which consisted of a number of instars. The trilobite emerged from the egg as a tiny protaspis larva. The protaspis was undoubtedly planktonic, measuring from 0.5 mm. to 1 mm. in length, and the body was covered by a single dorsal carapace consisting of the acron and four postoral segments (Fig. 12–12*A* and *B*). After passing through several instars, in which additional segments were added to the carapace, the protaspis passed into a meraspis larva. In this larval stage the pygidium was located behind the cephalon (Fig. 12–12*C*). The thoracic region appeared during succeeding molts as segments were gradually freed from the anterior border of the pygidium.

The final larval stage is known as a holaspis larva, which, though still very small, displayed the general adult structure. Additional molts merely involved minor changes and an increase in size.

The relationship of the trilobites to the other groups of arthropods will be discussed in the concluding section of Chapter 16.

Bibliography

Atwood, H. L., 1973: An attempt to account for the diversity of crustacean muscles. Amer. Zool., *13*: 357–378.

Bernhard, C. G. (Ed.), 1966: The Functional Organization of the Compound Eye. Wenner-Gren International Symposium 7, Pergamon Press.

Bullock, T. H., and Horridge, G. A., 1965: Structure and Function in the Nervous Systems of Invertebrates, Vol. 2. W. H. Freeman, San Francisco, pp. 801–1719.

Hackman, R. H., 1971: The integument of arthropods. *In* Florkin, M., and Scheer, B. J. (Eds.), Chemical Zoology, Vol. 6, pt. B. Academic Press, N. Y., pp. 1–62.

Kennedy, D. A., Selverston, A. I., and Remler, M. P., 1969: Analysis of restricted neural networks. Science, *164*:1488–1496.

Krishnakumaran, A., and Schneiderman, H. A., 1968: Chemical control of molting in arthropods. Nature, *220*:601–603.

Manton, S. M., 1951: The evolution of arthropodan locomotory mechanisms. Pt. 2. General introduction to the locomotory mechanisms of the Arthropoda. J. Linnean Soc. (Zoology), *42*:93–117.

Manton, S. M., 1970: Arthropods: Introduction. *In* Florkin, M., and Scheer, B. J. (Eds.), Chemical Zoology, Vol. 5. Academic Press, N. Y., pp. 1–34.

Moore, R. C. (Ed.), 1959: Treatise on Invertebrate Paleontology. Part O, Arthropoda 1. Geological Society of America and University of Kansas Press. (This volume covers trilobites.)

Moore, R. C., Lalicker, C. G., and Fischer, A. G., 1952:

Invertebrate Fossils. McGraw-Hill, N. Y., pp. 475–520. (A good morphological account of the trilobites.)

Redmond, J. R., 1971: Blood respiratory pigments—Arthropoda. *In* Florkin, M., and Scheer, B. J. (Eds.), Chemical Zoology, Vol. 6. Academic Press, N. Y., pp. 119–144.

Sehnal, F., 1971: Endocrines of arthropods. *In* Florkin, M., and Scheer, B. J. (Eds.), Chemical Zoology, Vol. 6, pt. B. Academic Press, N. Y., pp. 307–345.

Shaw, S. R., 1969: Optics of arthropod compound eyes. Science, *165*:88–90.

Shrock, R. R., and Twenhofel, W. H., 1953: Principles of Invertebrate Paleontology. McGraw-Hill, N. Y., pp. 536–641. (A general account of the trilobites; good bibliography.)

Störmer, L., 1949: Sous-embranchement des Trilobitomorphes. *In* Grassé, P. (Ed.), Traité de Zoologie, Vol. 6. Masson et Cie, Paris, pp. 159–216.

Vandel, A., 1949: Généralités sur les Arthropodes. *In* Grassé, P. (Ed.), Traité de Zoologie, Vol. 6. Masson et Cie, Paris, pp. 79–158. (A general comparative treatment of the anatomy, embryogeny, and evolution of the arthropods.)

Wolken, J. J., 1971: Invertebrate Photoreceptors. Academic Press, N. Y., 179 pp. (A review of research in some invertebrate photoreceptor systems.)

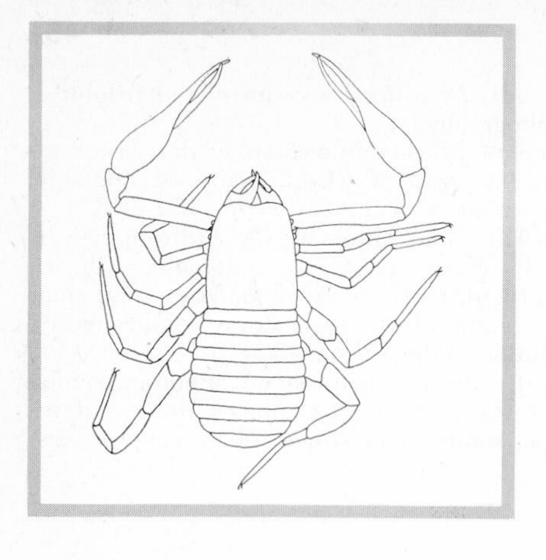

Chapter 13 THE CHELICERATES

All of the animals described in this chapter belong to the subphylum Chelicerata, one of the principal evolutionary lines within the Arthropoda. The body of chelicerates is divided into a cephalothorax (or prosoma) and an abdomen (or opisthosoma). There are no antennae, and the first preoral appendages (embryonically postoral) are feeding structures called chelicerae. The first postoral appendages are called pedipalps and are modified to perform various functions in the different classes.

CLASS MEROSTOMATA

The Merostomata are aquatic chelicerates characterized by five or six pairs of abdominal appendages modified as gills and by a spikelike telson at the end of the body. The group can be divided into two distinct subclasses—the Xiphosura (horseshoe crabs) and the extinct Eurypterida.

Subclass Xiphosura

Although the fossil record of the Xiphosura extends back to the Ordovician period, three genera and four species compose the only living representatives today. One of these is the horseshoe crab, *Limulus polyphemus*, common to the northwestern Atlantic coast and Gulf of Mexico. All the other members of this group are found along Asian coasts from Japan and Korea south through the East Indies and Philippines.

Horseshoe crabs live in shallow water on soft bottoms, plowing through the upper surface of the sand. They reach a length of 60 cm. and are dark brown in color. The carapace is smooth, horseshoe-shaped, convex above and concave below, with the posterior lateral angles prolonged backward to about half the length of the abdomen (Fig. 13–1). Its shape not only facilitates pushing through the sand and mud but provides a protective covering for the ventral appendages. To the outside of each of two dorsal lateral ridges is located a large eye, and to each side of a median ridge at the anterior end is a small median eye.

The anterior dorsal surface is reflected ventrally, and in the front forms a large triangular surface that tapers back toward the mouth (Fig. 13–2*A*). A frontal organ and a pair of degenerate eyes are located on the ridge formed by this triangle. Behind the apex of this ridge is a narrow labrum or upper lip. A pair of trisegmented chelicerae are attached to each side of the labrum; the last two segments form a pair of pincers. The mouth is located behind the labrum and is followed by a short narrow sternum.

Five pairs of walking legs are located posterior to the chelicerae on the underside of the cephalothorax (Fig. 13–2*A*). The first four pairs of walking legs are all similar and have the coxae (most proximal segment) heavily armed with spines located on the median side. These gnathobases macerate and move food anteriorly. The two tarsal (distal) segments form pincers. The coxae of the last pair of legs are not modified to form gnathobases but bear on the median side a short spatulate process known as a flabellum, which is used for cleaning the gills. Furthermore, the fifth or last pair of walking legs is not chelate and possesses four leaflike processes attached to the end of the first tarsal

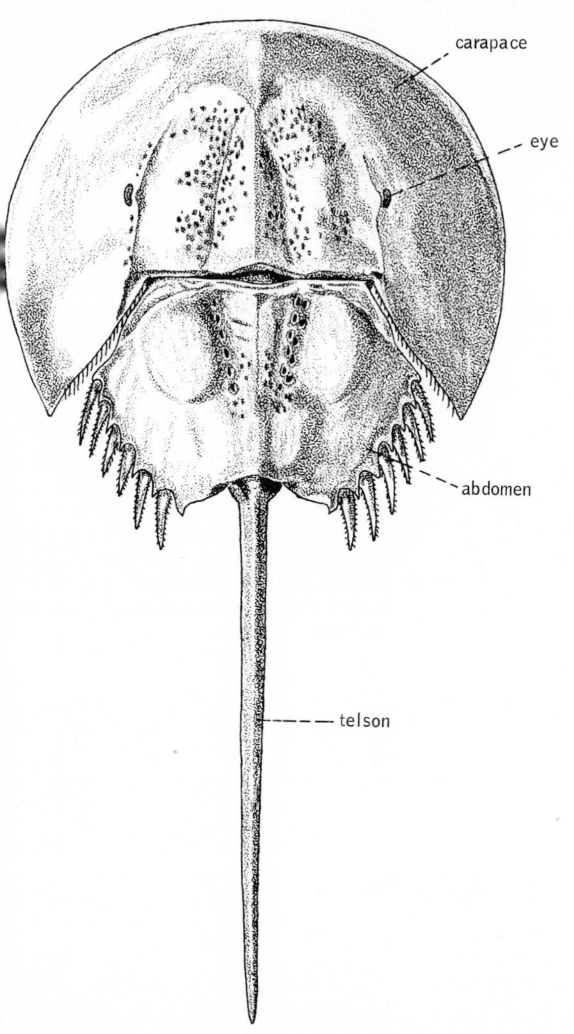

Figure 13–1. A horseshoe crab (dorsal view). (After Van der Hoeven from Fage.)

segment. This last pair of appendages is used for pushing and for clearing and sweeping away the mud and silt during burrowing.

The last pair of cephalothoracic appendages are known as chilaria and are located behind the sternum between the last pair of walking legs. These appendages are believed to represent the coxae of a pair of degenerate appendages associated with the seventh, or pregenital, segment. Each chilarium consists of a single article armed with hairs and spines, as are the gnathobases, and probably functions similarly.

The abdomen (opisthosoma) is unsegmented and fits into the concavity formed by the posterior border of the cephalothorax and its lateral extensions (Figs. 13–1 and 13–2A). Along the two median furrows are six small pits marking the locations of internal muscle attachments. Six short mobile spines border the posterior edge of the abdomen.

A long, triangular, spike-like tail or caudal spine (telson) articulates with the posterior of the abdomen. The telson of horseshoe crabs is not actually a true telson, since it does not bear the anal opening. Rather, the tail of the horseshoe crab represents a series of fused tergites (dorsal plates) of the more anterior abdominal segments. The tail of these animals is highly mobile and may be used for pushing and for righting the body when it is accidentally turned over. It is never used for defense, and the horseshoe crab can be picked up and carried by it.

The abdomen bears six pairs of appendages (Fig. 13–2A). The first pair forms the genital operculum, a large membranous flaplike structure resulting from the median fusion of the paired appendages of the eighth, or genital, segment. Two genital pores are located on the underside of this flap.

Posterior to the genital operculum are located five pairs of appendages modified as gills. Like the genital operculum, they are flaplike and membranous, and each pair is fused along the midline. The undersurface of each flap is formed into many leaflike folds called lamellae, which provide the actual surface for gaseous exchange (Fig. 13–2). Each gill contains approximately 150 lamellae. This arrangement of leaflike lamellae has caused the appendages to be called book gills. The movement of the gills maintains a constant circulation of water over the lamellae, and in addition the gills function as paddles during the upside down swimming of small individuals.

Horseshoe crabs are scavengers and feed on mollusks, worms, and other organisms, including bottom-dwelling algae. Food material is picked up by the chelate appendages, passed to the gnathobases where it is macerated, and then moved anteriorly to the mouth.

The mouth, located just behind the chelicerae, opens into an esophagus that is sclerotized and forms longitudinal folds; it extends anteriorly through a dilated portion, the crop, and into a grinding chamber, the gizzard (Fig. 13–3A). The longitudinal folds of cuticle in the gizzard possess denticles, and the whole structure is provided with strong muscles. After the food is ground in the gizzard, the large undigestible particles are regurgitated through the esophagus, while the usable food material passes posteriorly through

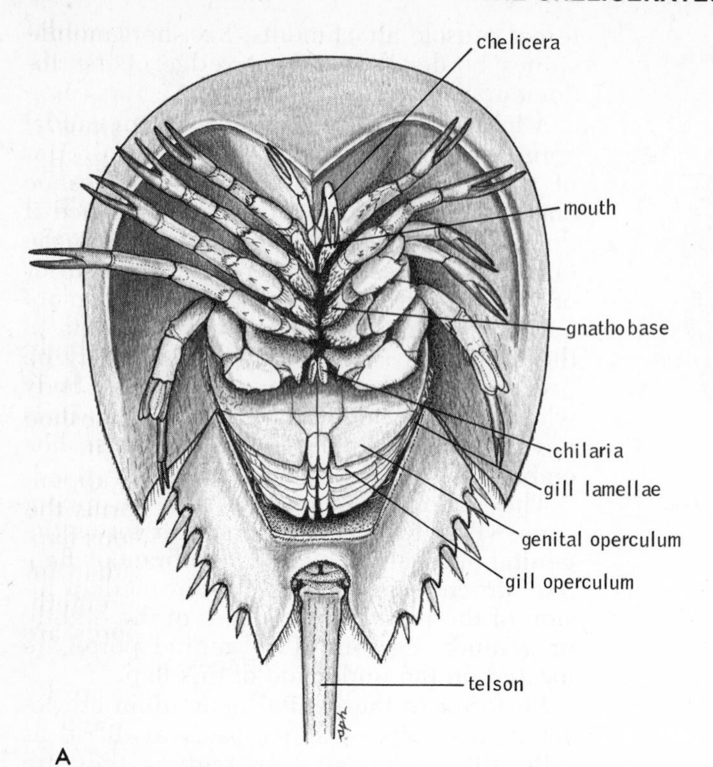

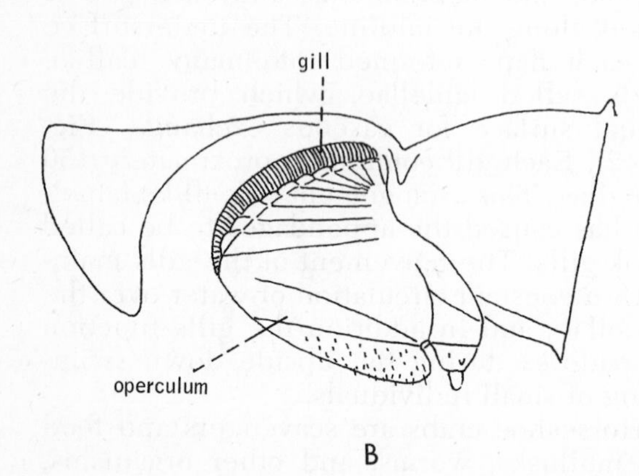

Figure 13–2. *A. Limulus polyphemus*, showing appendages on one side (ventral view). (From life.) *B.* Posterior surface of a gill operculum of *Limulus* in position beneath the body. (After Störmer from Fage.)

a valve into the enlarged anterior part of the nonsclerotized midgut known as the stomach. The remainder of the midgut, called the intestine, extends posteriorly into the abdomen. Opening into each side of the stomach is one of two pairs of ducts from two large glandular hepatic ceca that ramify throughout the cephalothorax and abdomen. Enzyme production and digestion take place within the midgut region and the hepatic ceca. The hepatic ceca are also the principal areas for absorption of digested food materials. Digestion is not entirely extracellular; dipeptides are digested intracellularly within the hepatic ceca.

Wastes are egested through a short sclerotized rectum and out of the anus, which is located on the ventral side of the abdomen just in front of the telson.

The circulatory system is well developed. A dorsal tubular heart with eight ostia is located throughout most of the length of the intestine (Fig. 13–3*A*). From the heart, blood is then pumped out through three large anterior arteries and four pairs of lateral arteries, which then branch into a well-developed arterial system supplying the body. The arteries eventually terminate in tissue sinuses, and the blood collects ventrally in two large longitudinal sinuses.

From the ventral sinuses the blood flows into the five book gills located on each side of the body. Within these gills, the blood is oxygenated. The exchange of gases takes

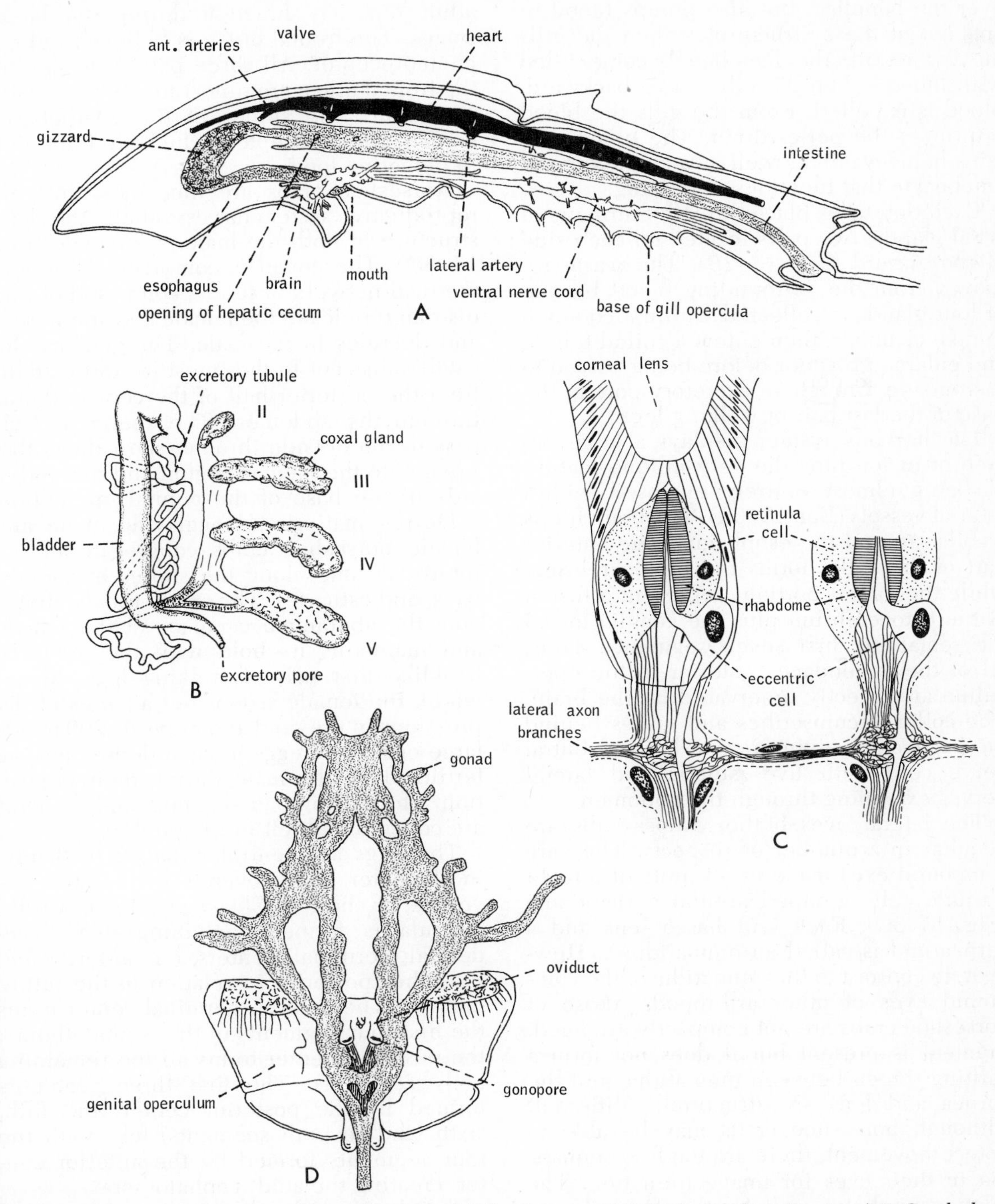

Figure 13-3. *A.* Sagittal section through *Limulus.* (After Patten and Redenbaugh from Kaestner.) *B.* Coxal glands and ducts of *Limulus.* (After Patten and Hazen from Fage.) *C.* Longitudinal section through two ommatidia of the eye of *Limulus.* (After MacNichol.) *D.* Female reproductive system of *Limulus.* (After Owen from Fage.)

place between the blood in the lamellae and the surrounding water. The movement of the gills not only causes the water to circulate over the lamellae, but also pumps blood in and out of these structures. When the gills move forward, the lamellae become filled with blood; when the gills move backward, blood is expelled. From the gills the blood returns to the pericardium. The blood contains hemocyanin as well as a single type of amebocyte that functions in clotting.

Excretion takes place through four pairs of coxal glands, two pairs located on each side of the gizzard (Fig. 13–3*B*). The waste removed from the surrounding blood by one of four glands is collected within a common saclike chamber, then enters a coiled tubule and enlarged bladder before being passed to the outside through an excretory pore at the base of the last pair of walking legs.

The nervous system displays a large degree of fusion plus the very unique feature of being almost entirely enclosed within arterial vessels (Fig. 13–3*A*). The brain forms a collar around the esophagus. The anterior part of the collar forms the protocerebrum, while the lateral portions represent a fusion of the tritocerebrum plus the ganglia for all the remaining first seven segments. Thus, all of the appendages anterior to the operculum are directly innervated by the brain. The collar circumscribes and unites behind the esophagus, giving rise to both a ventral nerve cord with five ganglia and lateral nerves extending through the abdomen.

The lateral eyes of horseshoe crabs are peculiar in a number of respects. They are compound eyes made up of units of 8 to 14 retinula cells grouped around a rhabdome (Fig. 13–3*C*). Each unit has a lens and a cornea and is called an ommatidium. However, in contrast to the ommatidia of the compound eyes of other arthropods, those of horseshoe crabs are not compactly arranged. Pigment is present but it does not form a shifting screen between ommatidia, and the cornea and lens are structurally different. Although horseshoe crabs may be able to detect movement, there are too few ommatidia in their eyes for image formation. The relationship between light stimulus and axon firing can be studied in a relatively simple state in horseshoe crab eyes, and this has resulted in their being used extensively in neurophysiological research.

The median eyes are invaginated cups. The interior of the cup forms a lens that is continuous with an exterior cornea and is surrounded by retinal cells. On each side of the ventral frontal organ is located a third pair of eyes. These are degenerate in the adult but may function during the larval stages. The frontal organ is believed to be a chemoreceptor. All three pairs of eyes and the frontal organ are innervated by the protocerebrum. The spines of the gnathobases contain chemoreceptors which are important in detecting food.

Horseshoe crabs are dioecious, and the reproductive system has essentially the same structure in both the male and female (Fig. 13–3*D*). The gonad is comprised of a symmetrical network of tissue, composed of tiny ovarian tubules in the female and sperm sacs and ductules in the male. The gonad is located subjacent to the intestine and extends from the posterior half of the cephalothorax through the abdomen. The sperm or eggs pass to the outside through short ducts that open onto the median region and the underside of the base of the genital operculum.

During mating and egglaying, male and female horseshoe crabs congregate in the intertidal zone along the shores of sounds, bays, and estuaries. The smaller male climbs onto the abdominal carapace of the female and maintains its hold with the modified hooklike first pair of walking legs. Meanwhile, the female scoops out a series of depressions in the sand and deposits 200 to 300 large eggs. The eggs in each depression are fertilized by the male during their deposition. The mating pair separate, and the eggs are covered and left in the sand.

The eggs are centrolecithal, 2 to 3 mm. in diameter, and covered by a thick envelope, or chorion. Cleavage is total. A solid gastrula is formed containing two mesodermal (germinal) centers, one anterior and the other posterior in relation to the future embryo. The anterior germinal center forms the first four segments of the cephalothorax; the posterior center forms all the remaining body segments. The first three segments formed by the posterior center (the fifth, sixth, and seventh segments) fuse with the four segments formed by the anterior center, creating the adult cephalothorax.

As it hatches, the trilobite larva, so named because of its superficial similarity to trilobites, emerges from the egg. This larva is approximately 1 cm. long, and actively swims about and burrows in sand. The cephalothorax of the horseshoe crab is very similar to the cephalon of the trilobite. The caudal spine is very small and does not project beyond the abdomen. Only two of the five pairs of book gills are present, although all an-erior appendages are present. As successive molts take place, the remaining book gills appear, the caudal spine increases in length, and the young crab assumes the adult form. Sexual maturity is not reached until the third year. In Japanese species, it has been calculated that attainment of sexual maturity requires 13 instars in the male and 14 in the female.

Subclass Eurypterida

The second group composing the class Merostomata are the extinct giant arthropods, which belong to the subclass Eurypterida (or Gigantostraca). The eurypterids were aquatic and existed from the Cambrian to the Permian period. The eurypterids probably attained the largest size of any of the arthropods. One species of the genus *Pterygotus* was almost 3 m. long.

Eurypterids were quite similar to horseshoe crabs in their general body plan (Fig. 13–4). They had the same prosomal appendages and a telson, but the cephalothorax was smaller and lacked posterior lateral extensions. Lateral and median eyes were present. The chief differences between the two groups are that the abdomen of the eurypterids was composed of separate segments and consisted both of a seven-segmented preabdomen (mesosoma) bearing appendages and a postabdomen (metasoma) of five narrower segments lacking appendages. The telson was attached to the last abdominal segment.

The appendages of the cephalothorax varied in size within different genera. They consisted of one pair of chelicerae, usually small; four pairs of walking legs; and one pair of large, elongated, oarlike appendages. The fourth pair of walking legs was also often elongated and paddle-like. Most eurypterids, judging from the appendages, not only were able to crawl over the bottom but were active swimmers. Two modes of swimming have been suggested. Either the paddle-like legs were used as flippers or the animal swam on its back, using the abdominal ap-

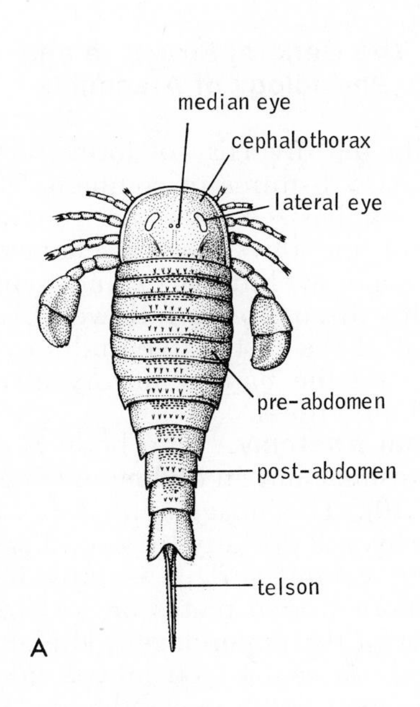

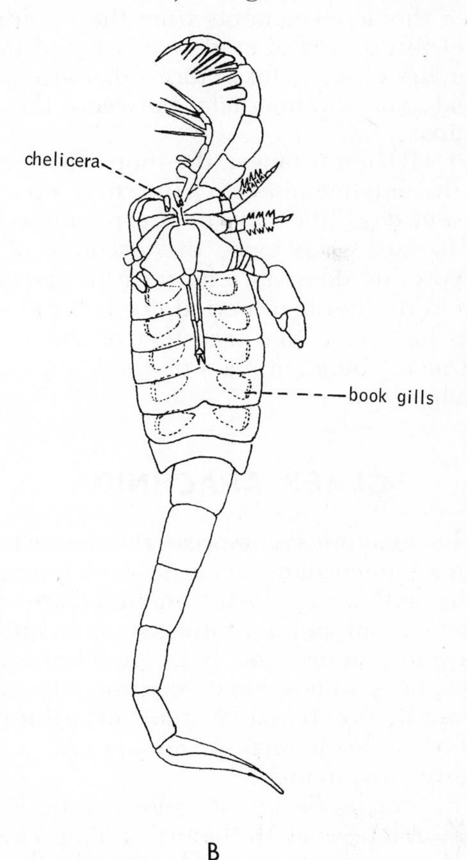

Figure 13–4. Eurypterida. *A. Eurypterus remipes* (dorsal view). (After Nieszkowski.) *B. Mixopterus kiaeri*, showing appendages of one side (ventral view). (After Störmer from Fage.)

pendages for propulsion and the last two elongated legs as balancers.

The abdominal appendages consisted of six pairs of gills, the first pair also forming the operculum. The exact nature of the gills, however, is still undetermined.

Fossil larval stages indicate a postembryonic development in the eurypterids similar to that of the horseshoe crabs.

The location of fossils indicates that the eurypterids, after a marine origin, gradually invaded both brackish water and fresh water and perhaps even assumed a terrestrial existence.

The merostomes display an apparently close relationship to the trilobites, and the trilobites may have been ancestral to the merostomes. Evidence for this affinity is illustrated in the presence, during the embryonic development of the horseshoe crabs, of two germinal centers equivalent to the cephalon and the larval pygidium of the trilobites. The abdomen of the adult horseshoe crab thus corresponds to the trilobite pygidium, while the cephalothorax corresponds to the cephalon, to which have fused the three thoracic segments from the pygidium. A second source of evidence is found in the structure of many fossil forms that are transitional and intermediate between the two groups.

Of all merostomes, the horseshoe crabs are the only members to have survived to the present day. But they too are members of a dying race. However, the account of the eurypterids does not end with their extinction in the Permian period, for it is believed that they were the ancestors of the largest and most abundant class of chelicerates, the Arachnida.

CLASS ARACHNIDA

The arachnids comprise the largest and from a human standpoint the most important of the chelicerate classes; included are many common and familiar forms, such as spiders, scorpions, mites, and ticks. Arachnids also have the dubious honor of probably being the most objectionable group of arthropods as far as the layman is concerned, a bias largely unwarranted.

The arachnids are an old group. Fossil representatives of all the orders date back to the Carboniferous period, and fossil scorpions have been found dating from the Silurian period. The early arachnids were undoubtedly aquatic and were contemporaries of the eurypterids, from which they are believed to have evolved. The Silurian scorpions were aquatic; the first terrestrial arachnids, both scorpions and some nonscorpion forms, appear in the Devonian. Arachnids may well have been the first terrestrial arthropods.

Except for the few groups that have adopted a secondary aquatic existence, living arachnids are terrestrial chelicerates. Like other evolutionary conquests of land, this migration from an aquatic to a terrestrial environment required certain fundamental morphological and physiological changes. The epicuticle became waxy, reducing water loss. The book gills became modified to use air, resulting in the development of the arachnid book lungs and tracheae. In addition, the appendages became better adapted for terrestrial locomotion. Furthermore, once a terrestrial existence was established, a great many unique innovations evolved independently along different lines. The development of silk in spiders, pseudoscorpions, and some mites and of poison glands in scorpions, spiders, and pseudoscorpions are but two examples.

The General Structure and Physiology of Arachnids

Despite the diversity of forms, arachnids exhibit many features in common. This discussion emphasizes only the general morphological and physiological characteristics and the major evolutionary trends within the class. This discussion is followed by a brief treatment of the distinctive and specialized features and the natural history of each of the arachnid orders.

External Anatomy. The body is divided into a prosoma and an abdomen (Figs. 13–9 and 13–10). The unsegmented prosoma is usually covered dorsally by a solid carapace, while the ventral surface is provided with one or more sternal plates or is covered by the coxae of the appendages. The abdomen primitively is segmented and divided into a preabdomen and a postabdomen. In most arachnids other than scorpions these two subdivisions are no longer conspicuous (Fig. 13–13), and a tendency for segmentation to disappear because of fusion has developed. In mites, primary segmentation has become lost, and the abdomen has fused with the prosoma to form a single body region (Fig. 13–44).

The appendages common to all arachnids are those arising from the prosoma and consist of a pair of chelicerae, a pair of pedipalps, and four pairs of legs (Fig. 13–13). The chelicerae are used in feeding, but the pedipalps serve a number of functions and are variously modified.

Nutrition. The majority of arachnids are carnivorous, and digestion partly takes place outside the body. Prey, usually small arthropods, are captured and killed by the pedipalps and chelicerae. While the prey is held by the chelicerae, enzymes secreted by the midgut are poured out over the torn tissues of the prey. Digestion proceeds rapidly and a partially digested broth is produced. This fluid is then taken into the prebuccal cavity, located in front of the mouth. The roof of this chamber is formed by the anterior wall of the prosoma and is known as the labrum. An anterior sternal piece frequently forms the floor, and the sides are usually formed by the coxae of the pedipalps.

The liquid food passes through the mouth and into the sclerotized pharynx and esophagus of the foregut (Figs. 13–29 and 13–45*A*). The tubular pharynx is the chief pumping organ, driving the liquid food into the foregut. Its walls are composed of longitudinal cuticular strips connected by membranes. The diameter of the tube can be increased or decreased by the action of externally attached muscles, thus effecting a sucking action. In some arachnids the esophagus is enlarged, forming an additional pump.

The esophagus conveys the food to the midgut or mesenteron, which consists of a central tube with lateral diverticula (Fig. 13–29). The diverticula are located in both the prosoma and the abdomen, and become filled with the partially digested liquid pumped into the foregut. The midgut wall is composed of secretory cells and absorptive cells. The secretory cells produce the enzymes for external digestion and other secretions that complete digestion after the food reaches the mesenteron.

When digestion is completed, food products are taken up by absorptive cells. Since not all of the absorbed food is immediately needed for metabolism, much of it is stored in the interstitial cells surrounding the diverticula. The mesenteron extends to the posterior part of the abdomen, where it is connected to the anus by a short sclerotized intestine, forming the hindgut.

Many arachnids are capable of surviving for long periods without feeding. Periods of starvation exceeding two years for certain species of spiders and more than one year for a scorpion have been recorded.

Excretion. Guanine is the most important excretory product of arachnids. The excretory organs are coxal glands and Malpighian tubules. Some groups possess both; some one or the other. Coxal glands are so named because they open onto the posterior of the coxae of the appendages. The coxal gland itself is located at the side of the prosoma, and is a thin-walled spherical sac immersed in blood (Fig. 13–5). Waste materials are absorbed from the blood by the cells of the gland, and passed into the lumen. A long convoluted tubule connects the gland to the excretory pore. Arachnids never have more than four pairs of coxal glands, and the excretory pore, or pores, are located on different coxae in different orders. This variation has arisen because the glands are derived from coelomic sacs, which in the embryo are located one pair per segment. Different pairs of coelomic sacs are retained as the coxal glands in the adults by the different orders.

Malpighian tubules, the second type of arachnid excretory organ, consist of one or two pairs of slender tubes that arise from the posterior of the mesenteron at its junction with the intestine (Fig. 13–25). The tubules are directed anteriorly, and ramify between the abdominal diverticula of the hindgut. The thin syncytial walls of the tubule absorb waste from the blood in the hemocoel. The waste is then excreted into the lumen of the tubules as guanine crystals, and passed into the intestine.

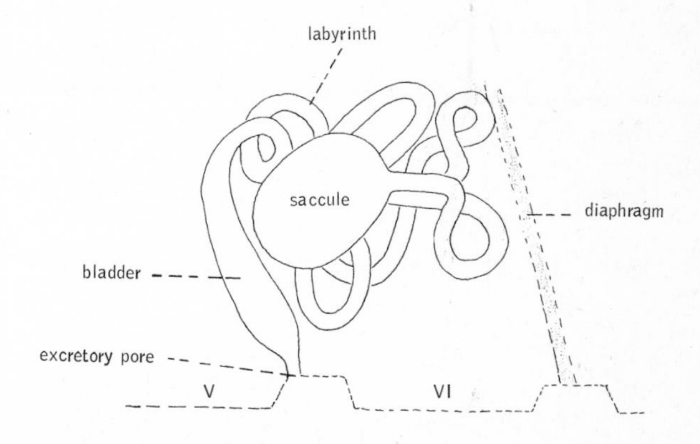

Figure 13–5. Coxal gland of a scorpion. (After Millot and Vachon.)

In addition to Malpighian tubules and coxal glands, arachnids possess certain large cells, called nephrocytes, that are localized in clusters in certain parts of the prosoma and abdomen. These cells have been found to pick up dyes injected experimentally in the animal, but their exact function has not been determined.

Nervous System. The nervous system is greatly concentrated except in the relatively primitive scorpions, resulting from ganglionic fusion (Fig. 13–6*A*). The brain, composed of the protocerebrum and tritocerebrum, is an anterior ganglionic mass lying above the esophagus. The protocerebrum contains the optic centers and optic nerves; the tritocerebrum contains the nerves supplying the chelicerae. The remainder of the central nervous system is subjacent to the esophagus. In many orders, most or all of the ganglia originally located in the thorax and abdomen have migrated anteriorly and fused with the subesophageal ganglion—the ganglion of the pedipalpal segment (Fig. 13–6*B*). Thus the arachnid nervous system commonly resembles a collar or ring surrounding the esophagus. The posterior ventral half of this collar gives rise on each side to nerves innervating the appendages, and a single posterior nerve bundle extends back into the abdomen.

Three types of sense organs are common to most arachnids—sensory hairs, eyes, and slit sense organs. The sensory hairs can be either simple innervated setae (Fig. 13–7*C* and *E*), movable setae, or fine hairs of great length, called trichobothria (Fig. 13–7*D*). The base of a trichobothrium is expanded to form a small ball that fits into a large, structurally complicated socket in the integument. The hair base contains a process from a sensory nerve cell of the hypodermis and is stimulated by very slight vibrations or air currents. Sensory hairs are scattered over the

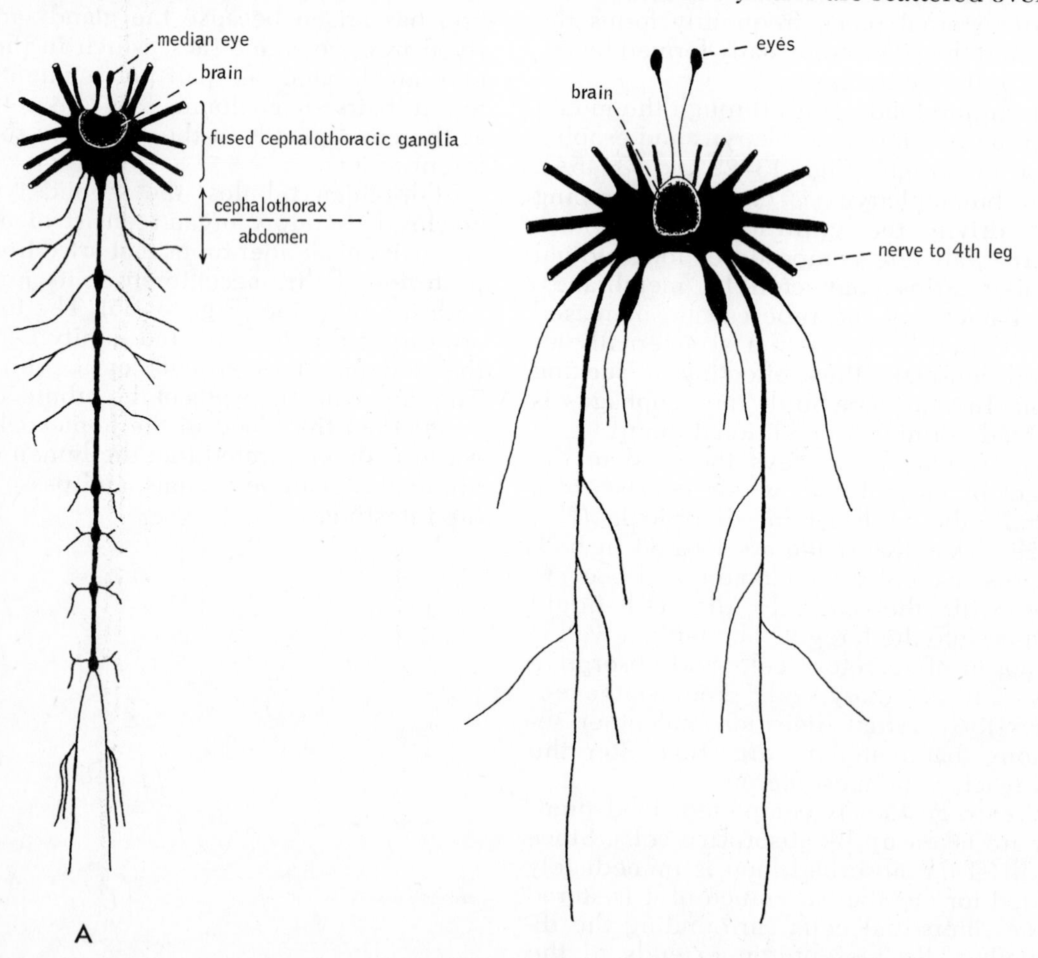

Figure 13–6. *A*. Scorpion nervous system, in which abdominal ganglia are distinct. *B*. An opilionid nervous system, in which all ventral ganglia have migrated forward and fused. (Both after Millot.)

body surface but are particularly prevalent on the appendages. For most arachnids, they are believed to be the most important sense organs.

The eyes of all arachnids are similar (Fig. 13–7*A* and *B*). They are always composed of a combined cornea and lens, which is continuous with the cuticle but much thicker. Beneath the lens is a layer of hypodermal cells known as the vitreous body. The retinal layer, containing the photoreceptor cells, lies behind the vitreous body. Finally, the retina is backed by a postretinal membrane.

The photoreceptors are oriented either toward the light source (a direct eye) or toward the postretinal membrane (an indirect eye). In the indirect eyes, the postretinal membrane may function as a reflector, called the tapetum, that reflects the light toward the receptors. Some arachnids possess only direct or only indirect eyes; many, such as spiders, have both. The number of receptors is directly correlated with the ability of the eye to form an image, a phenomenon achieved by relatively few arachnids. This problem is more fully discussed in the section on spiders.

A slit sense organ consists of a slitlike pit in the cuticle, covered by a very thin membrane which bulges inward (Fig. 13–7*F*). The undersurface of the membrane is in contact with a hairlike process, which projects upward from a sensory cell. Slit sense organs may occur in great numbers, either singly or in groups (lyriform organs) on the appendages and body of most arachnids. For example, the spider *Cupiennius salei* has 3000 slit sense organs, about half of which are grouped (Barth and Libera, 1970).

Slit sense organs respond to slight changes in the tension of the exoskeleton and function in proprioception and in detecting vibrations in the frequency range of sound.

Gas Exchange. Arachnids possess either book lungs or tracheae, or both. Book lungs are more primitive and probably are a modification of book gills, an adaptation associated with the migration of the arachnids to a terrestrial environment. The two are very similar in structure, but in contrast to book gills the book lungs of arachnids are internal. Book lungs occur in pairs and are located on the ventral side of the abdomen. Scorpions have as many as four pairs, each pair occupying a separate abdominal segment.

Each book lung consists of a sclerotized pocket that represents an invagination of the ventral abdominal wall (Fig. 13–8). The wall on one side of the pocket is folded into leaflike lamellae. The lamellae are held apart by bars that enable the air to circulate freely. Diffusion of gases takes place between blood circulating within the lamellae and the air in the interlamellar spaces (Fig. 13–8). The nonfolded side of the pocket forms an air chamber (atrium) that is continuous with the interlamellar spaces, and that opens to the outside through a slitlike opening (spiracle). Some ventilation results from the contraction of a muscle attached to the dorsal side of the air chamber. This contraction dilates the chamber and opens the spiracle, but most gas movement is by diffusion.

The tracheal system of arachnids is similar to that of insects but has evolved independently. In fact, the tracheal system appears, in some cases, to be derived from the book lungs. This certainly seems to be true in spiders and is discussed in more detail in the section dealing with spiders. Tracheae tend to be most highly developed in small arachnids, which would otherwise be subject to greater water loss with book lungs.

Ricinuleids, pseudoscorpions, and some spiders possess sieve tracheae derived from book lungs. In tracheal systems of this type, the spiracle opens into an atrial or tubelike chamber from which arises a great bundle of tracheae. Mites, harvestmen, solifugids, and most spiders possess tube tracheae, in which the tracheae do not arise as a bundle but are simple branched or unbranched tubes (Fig. 13–31*B*). The spiracles may open into an initial atrium or directly into a tracheal tube. Whether the tracheal system is of the sieve or tube type, the chitin-lined tracheae eventually terminate in small fluid-filled tubules supplying various tissues directly with oxygen. As would be expected, in those arachnids that obtain oxygen exclusively by tracheae, the circulatory system plays little part in oxygen transport and the vessels may be reduced. On the other hand, the blood of some arachnids, such as scorpions and many spiders with book lungs, contains the respiratory pigment hemocyanin.

Internal Transport. The arachnid heart, as it is in most arthropods, is a dorsal tube within a pericardial chamber. In arachnids the heart is almost always located in the anterior half of the abdomen. In its primitive condition, the heart is segmented, corresponding to the metamerism of the abdomen, and is expanded between each segment because of the attachment of suspensory ligaments (Fig. 13–29). Located one on each

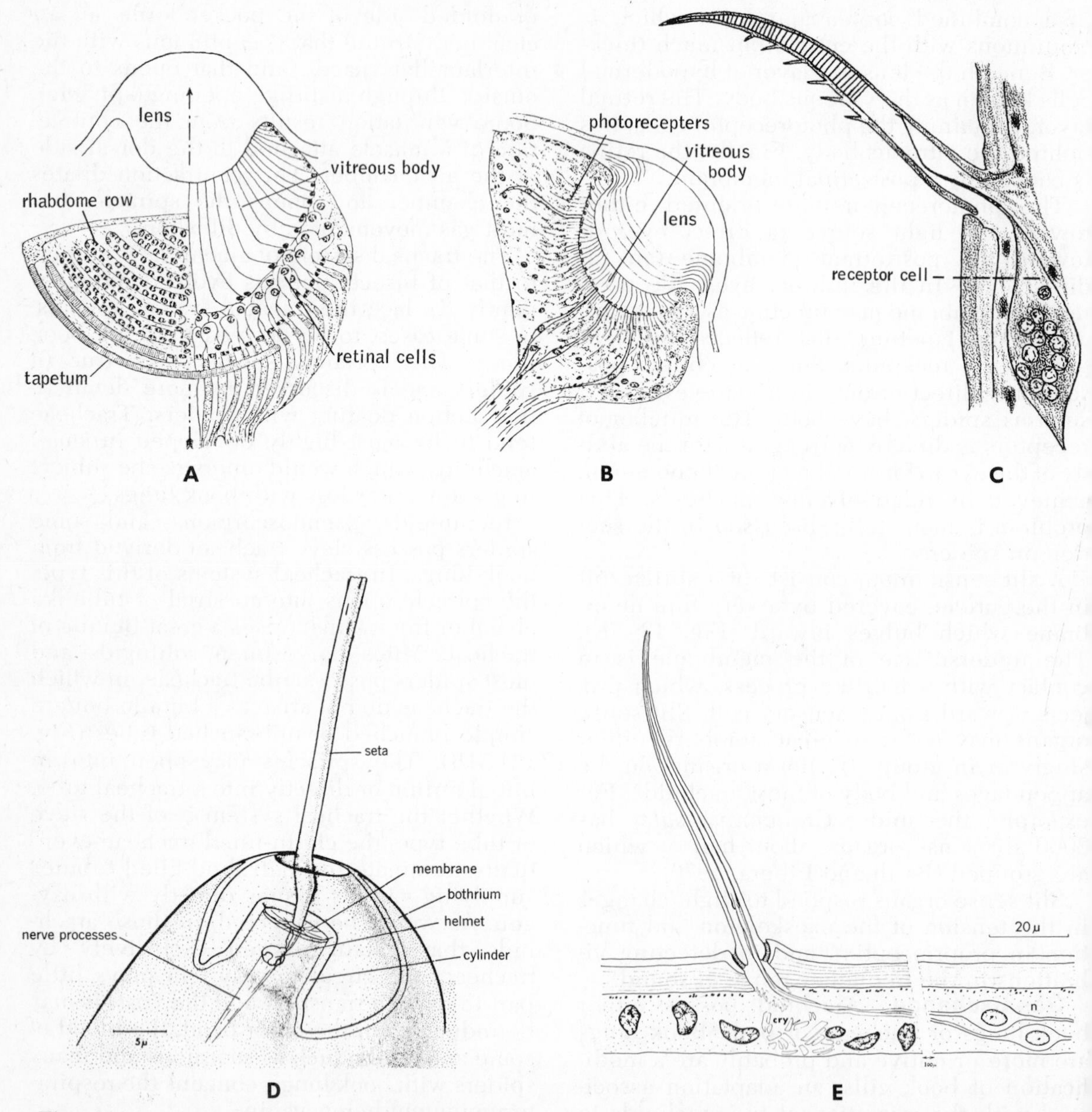

Figure 13–7. Arachnid sensory structures. *A*. Diagram of an indirect eye of a spider having a tapetum. Sagittal view to right of arrow; three-dimensional view to left of arrow. *B*. Sagittal section of a direct eye of a spider, *Aname*. (*A* and *B* from Homann, H., 1971: Z. Morphol. Tiere, *69*(3):201–273.) *C*. Sensory seta of a mite. (After Gossel from Millot.) *D*. Trichobothrium of a spider. Only lower part of hair is shown. (From Gorner, P., 1966: Cold Spring Harbor Symposium Quant. Biol., *30*:69–73.) *E*. Chemosensory hair of a spider (*n*, nerve). (From Foelix, R. F., 1970: J. Morphol., *132*:313–334.)

Figure 13–7 continued on opposite page.

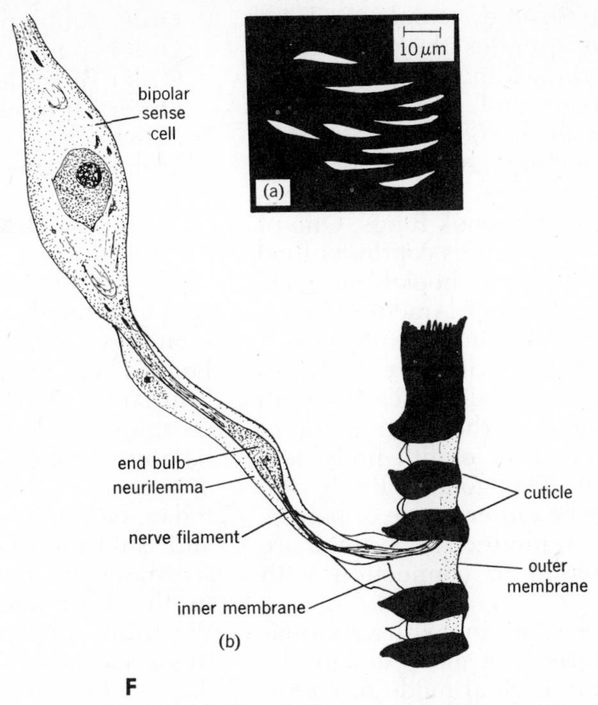

Figure 13–7. *Continued. F.* Slit sense organ of a spider. Surface view is shown in (a) and sectional view in (b). (From Salpeter, M. M., and Walcott, C., 1960: Exp. Neurol., *2*:232–250.)

side are a pair of slits, or ostia, which allow blood to enter the heart from the pericardial chamber. The greatest number of heart segments, totaling seven, appear in scorpions. However, reduction in all degrees of the number of heart segments has occurred. The number of segments in some forms is reduced even to a single segment or, in certain mites, is absent.

Histologically, the heart wall is composed of an outer epithelial layer, encompassing two muscle layers and an inner lining of endothelium. The muscles are responsible for heart contraction, while the pull exerted by the ligamentous suspension causes dilation. During diastole, blood flows through the ostia from the pericardial chamber to fill the heart. On contraction the valves guarding the segmentally arranged ostia close to prevent backflow, and blood is forced out of the

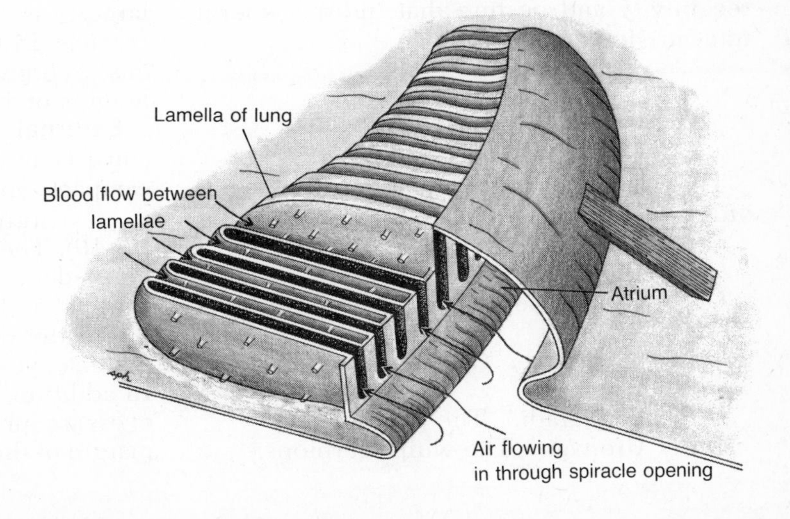

Figure 13–8. Diagrammatic section through a book lung.

heart through the several vessels—a large anterior aorta that supplies the prosoma, a small posterior aorta leading to the posterior half of the abdomen, and from each heart segment a pair of small abdominal arteries. Small arteries eventually empty the blood into tissue spaces and then into a large ventral sinus that bathes the book lungs. One or more pairs of venous channels conduct blood from the ventral sinus or the book lungs back into the pericardial chamber. Arachnid blood is a colorless fluid containing amebocytes.

Reproduction. Arachnids are always dioecious, and the genital orifice in both sexes is usually found on the ventral side of the second abdominal, or eighth body segment (Fig. 13–10). The gonads lie in the abdomen, and may be either single or paired. The ducts of the reproductive system are more easily discussed in connection with each of the orders.

Indirect sperm transmission with a spermatophore is characteristic of many arachnids. It appears to be the original mode of sperm transfer for the class and an adaptation for reproduction on land. The behavior of some living arachnids suggests that, primitively, the spermatophores were probably randomly deposited by the male. If a female encountered a spermatophore, she would be attracted chemotactically and would take it up within her gonopore. However, in most arachnids with a spermatophore, the male himself has become involved in attracting the female to the spermatophore. There is often a complex "courtship" or behavioral pattern preceding mating. The sexes, especially the female, of different groups respond to certain chemical, tactile, or visual cues. Such cues in an aggressive raptorial group such as the arachnids provide for recognition and elicit the necessary responses of receptivity and posture that indirect sperm transmission demands.

SUMMARY OF CLASS ARACHNIDA

The class Arachnida includes these ten orders with living representatives:

Order Scorpiones. The scorpions.
Order Pseudoscorpiones. The false scorpions.
Order Solifugae. The sun spiders, or wind scorpions.
Order Palpigradi. The palpigrades.
Order Uropygi. The whip scorpions.
Order Amblypygi. The amblypygids.
Order Araneae. The spiders.
Order Ricinulei. The ricinuleids.
Order Opiliones. Harvestmen, or daddy longlegs.
Order Acarina. The mites and ticks.

Order Scorpiones

The scorpions are the oldest known terrestrial arthropods and may have been the first members of this phylum to have conquered land. Their fossil record dates back to the Silurian period. As stated earlier, Silurian scorpions were aquatic, possessing gills and having no tarsal claws. Terrestrial scorpions appear in the Devonian. While abundant today, scorpions are most common in tropical and subtropical areas. In North America scorpions are found as far north as Virginia in the East and British Columbia in the West, but they are not common at these extreme latitudes. They are much more abundant in the Gulf states and the Southwest. About 800 species have been described.

Scorpions are generally secretive and nocturnal, hiding by day under wood and stones and in burrows in the ground. But there are also species associated with vegetation. They are often found near dwellings, and the custom in the deserts of shaking out shoes in the morning is a wise precaution. Scorpions are popularly believed to inhabit desert regions, but although many desert species exist, they are by no means restricted to arid situations. Many scorpion species require a humid environment and live in tropical rain forests and similar jungle habitats.

Scorpions are large arachnids, most ranging from 3 to 9 cm. in length. The smallest species is the Middle Eastern *Microbuthus pusillus*, which is only 13 mm. long, and the largest is the African *Pandinus*, which reaches 18 cm. However, some Carboniferous scorpions are known that attained lengths of 44 and 86 cm.

External Anatomy. The scorpion body consists of a prosoma covered by a single carapace, and a rather long abdomen, ending in a stinging apparatus (Figs. 13–9 and 13–10). The prosoma is relatively short and four-sided, and the anterior border is narrower than the posterior border. In the middle of the dorsal carapace is a pair of large median eyes, each raised on a small tubercle. In addition, two to five pairs of small lateral eyes are present along the anterior lateral margin of the carapace. The coxae of the legs

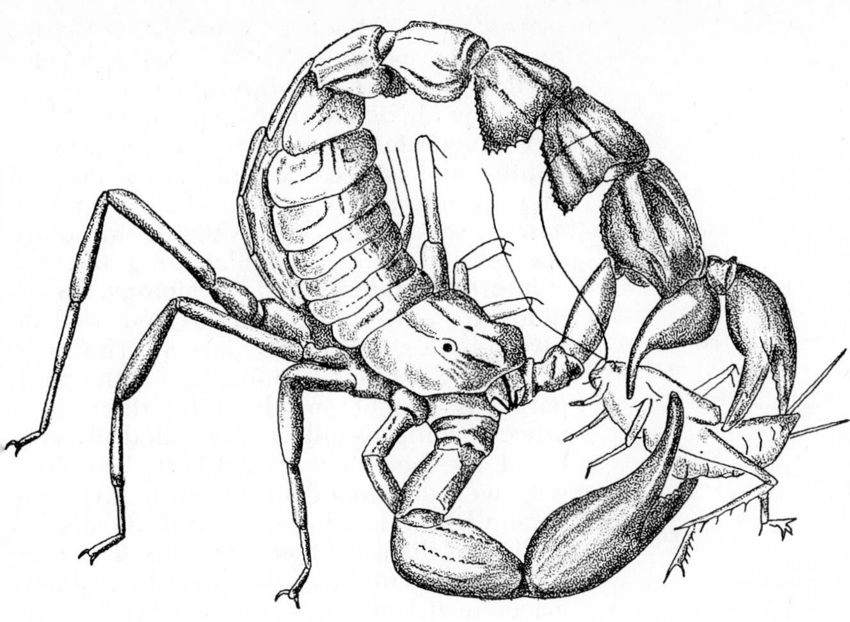

Figure 13–9. The North African scorpion, *Androctonus australis*, capturing a grasshopper. (After Vachon from Kaestner.)

occupy most of the ventral surface, and the original sternites are represented by a fused median plate, the sternum.

The chelicerae are small, triarticulate, and chelate; and they project anteriorly from the front of the carapace (Fig. 13–9). The pedipalps are greatly enlarged and form a pair of pincers for capturing prey. Only the pseudoscorpions have similarly enlarged, clawlike pedipalps. The legs each terminate in two pairs of claws.

The scorpion abdomen is believed to be very primitive. It is composed of a seven-segmented preabdomen and a postabdomen of five narrow segments, so the two regions are clearly differentiated. The genital opercula are located behind the sternum on the ventral side of the first abdominal segment (which is the eighth body segment; the seventh body segment, or first abdominal segment, of most arachnids has been lost in scorpions). The opercula consist of two small plates, contiguous along the midline of the ventral side, and covering the genital opening.

Posterior to the genital plates and attached to the second abdominal segment is a pair of sensory appendages known as the pectines, which are peculiar to scorpions. Each pectine is made up of three rows of chitinous plates forming an elongated axis that projects to each side from the point of attachment near the ventral midline. Suspended from the body of the pectine is a series of tooth-like processes that give the whole appendage the appearance of a comb.

The second through the fifth abdominal segments each bear on the ventral side a pair of transverse slits (spiracles), opening into the book lungs.

The segments of the postabdomen, sometimes called the tail, resemble narrow rings. The last segment bears the anal opening on the posterior ventral side and also bears the stinging apparatus characteristic of scorpions.

Stinging Apparatus and Venom. The sting is attached to the posterior of the last segment and consists of a bulbous base and a sharp curved barb that injects the venom (Fig. 13–10). The venom is produced by a pair of oval glands, each enveloped by a layer of smooth muscle fibers within the base of the apparatus. By a violent contraction of the muscular envelope, the liquid venom is ejected from the lumen of the glands into a sclerotized common duct that leads to the outside through a subterminal opening of the stinging barb. The scorpion raises the postabdomen over the body so it is curved forward, and a stabbing motion is used in stinging.

The venom of most scorpions, while sufficiently toxic to kill many invertebrates, is not harmful to man. At most, the sting is equivalent to that of a hornet or a yellow jacket. The scorpions of the southeastern

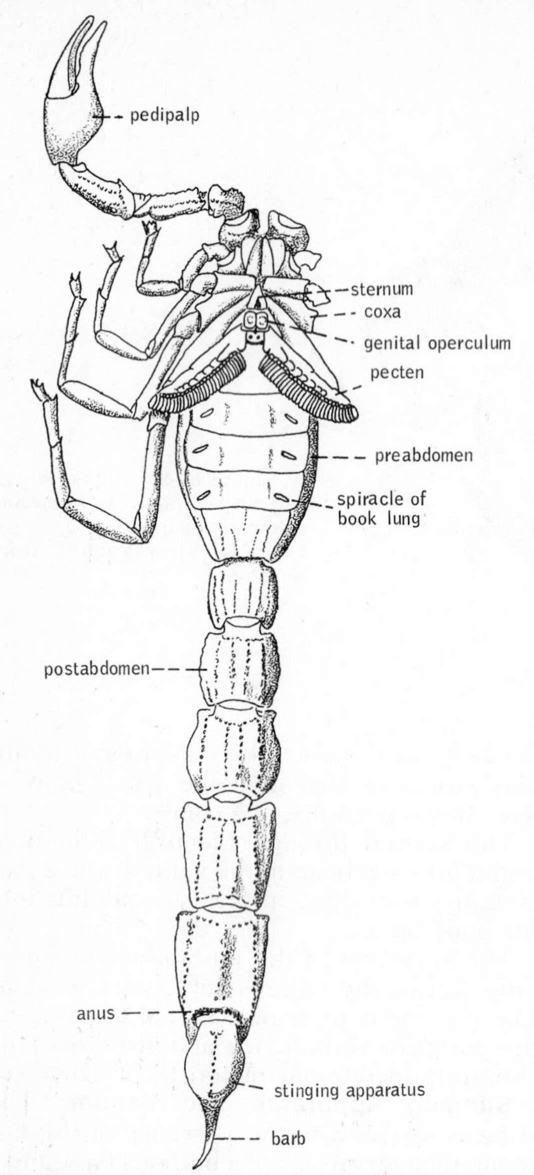

Figure 13–10. *Androctonus australis* (ventral view). (After Lankester from Millot and Vachon.)

United States and Gulf Coast region would fall in this category, as well as many of our midwestern and western forms.

However, certain species exist that possess a highly toxic venom that can be fatal to man. The most notorious of these are *Androctonus* of North Africa and various species of *Centruroides* in Mexico, Arizona, and New Mexico. Vachon (1949) states that *Androctonus australis* of the Sahara Desert has venom equivalent in toxicity to cobra venom, and this venom can kill a dog in seven minutes. With human victims, death usually occurs in six to seven hours. Species of *Centruroides* have been responsible in Mexico for deaths, mostly in children. One species of this genus, *C. sculpturatus* of Arizona and New Mexico, can be dangerous.

The neurotoxic venom of scorpions is very painful and may cause paralysis of the respiratory muscles, or cardiac failure in fatal cases. Antivenoms are available for those species appearing in populated regions.

Internal Anatomy and Physiology. Scorpions are entirely carnivorous and feed on invertebrates, particularly insects. The prey, located by the long trichobothria of the pedipalps, is caught and held by their large pincers while usually being killed or paralyzed by the sting (Fig. 13–9). Scorpions with well developed stings and highly toxic venom (Buthidae) have less well developed pedipalps than do other scorpions. The prey is transferred to the chelicerae, which slowly macerate it, and digestion begins. The preoral chamber is framed at the top by the chelicerae, on the sides by the coxae of the pedipalps, and at the bottom by the maxillary processes of the first pair of walking legs.

The scorpion heart is seven-segmented and extends the length of the preabdomen. Gas exchange is accomplished solely by book lungs. Excretion takes place through the action of two pairs of Malpighian tubules and a single pair of coxal glands, which open on the coxae of the third pair of walking legs.

Some species of scorpions at times raise their bodies well off the ground. This behavior, called stilting, permits the circulation of air beneath the animal, which tends to prevent excessive elevation of body temperature and desiccation.

The scorpion nervous system, unlike that of other arachnids, retains a distinct nerve cord with seven unfused ganglia (Fig. 13–6*A*). Giant fibers integrate the pedipalpal grasp and the stinging thrust. The sense organs of scorpions consist of the eyes, trichobothria, slit sense organs, and pectines. The eyes are of the direct type, but are not highly developed. The pectines are undoubtedly sensory, for the ventral side of each tooth of the comb is liberally provided with sensory cells (Fig. 13–11*A*). The distal ends of these cells are contained within little spigot-like projections. During movement of the scorpion, the pectines are held out from the sides of the body in a horizontal position so that the teeth touch the ground. The functions of these organs are still uncertain. They

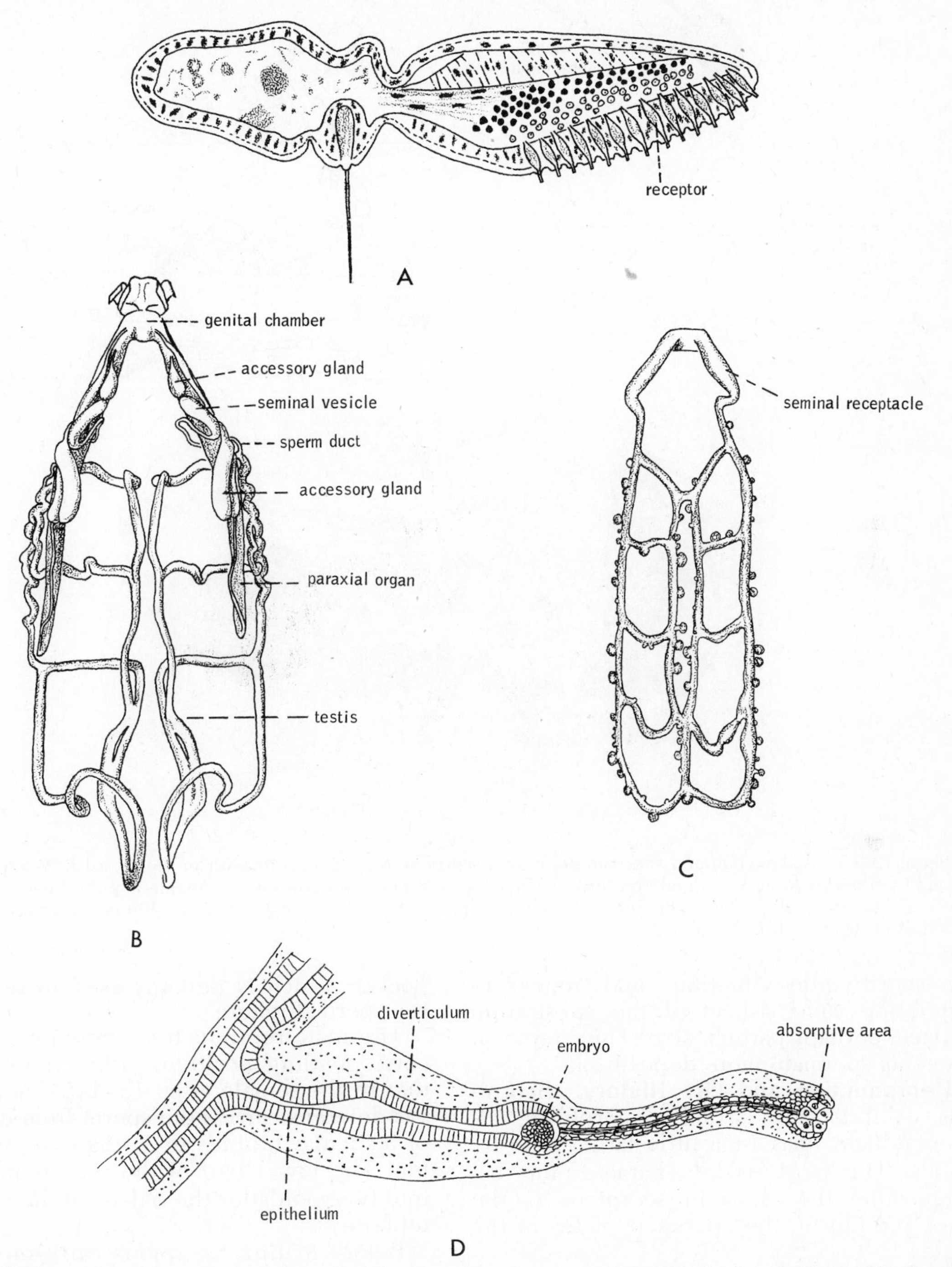

Figure 13–11. *A*. Transverse section through the pectine of a scorpion. (After Schröder from Millot and Vachon.) *B*. Male reproductive system of the scorpion, *Buthus* (dorsal view). (After Millot and Vachon.) *C*. Female reproductive system of the scorpion, *Parabuthus* (dorsal view). (After Pavlovsky from Millot and Vachon.) *D*. Ovarian diverticulum of the tropical Asian scorpion, *Hormurus australasiae*, a viviparous species. (After Pflugfelder from Dawydoff.)

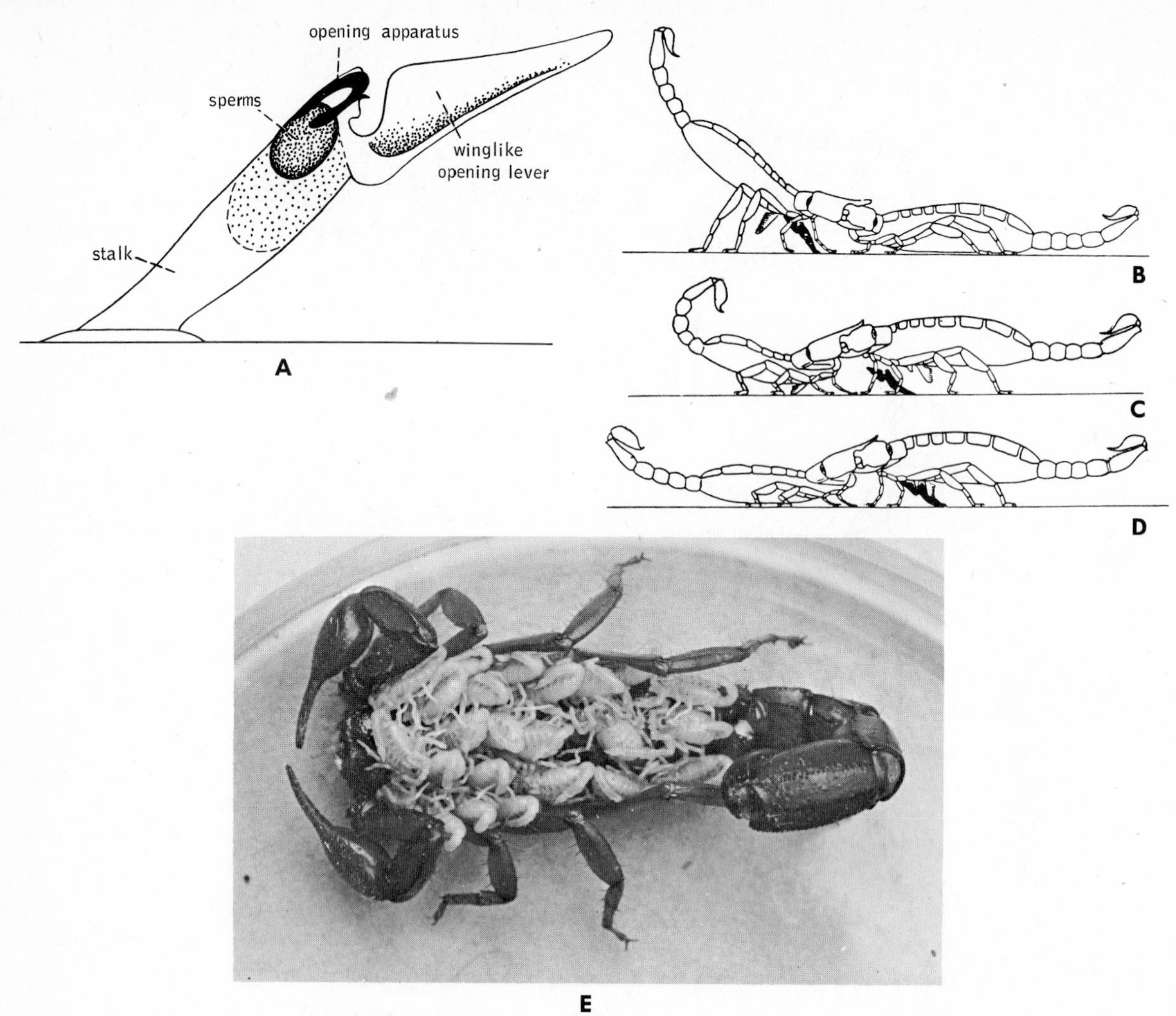

Figure 13–12. *A.* Diagram of a spermatophore of a scorpion. *B-D.* Sperm transfer in scorpions: *B.* While holding female's pedipalps in his own, male on left deposits spermatophore on ground. *C.* Female is pulled over spermatophore. *D.* Spermatophore taken up into female's gonopore. (*A-D* after Angermann.) *E.* Female scorpion carrying young. (Courtesy of H. L. Stahnke.)

are sensitive to vibrations and appear to determine some aspect of the substratum surface, perhaps particle size. Their removal prevents spermatophore deposition.

Reproduction and Life History. Scorpions exhibit little sexual dimorphism, although males may be a little larger than females. The most useful character for distinguishing the sexes in scorpions is the hook present on the opercular plates of the male.

The ovarian tubules are located between the midgut diverticula in the preabdomen (Fig. 13–11*C*). Prior to emptying into a single genital atrium, each oviduct dilates to form a small seminal receptacle. The genital atrium opens to the outside between the genital opercula. In some forms the genital atrium projects posteriorly into a medium pocket, which is perhaps used in receiving the spermatophore.

The male reproductive system occupies a corresponding position within the body to that in the female (Fig. 13–11*B*). A slender vas deferens carries the sperm from each set of testicular tubules to a single genital atrium. The lower part of the tract is modified into two molds for the halves of the spermatophore.

Before mating, scorpions carry on an extended courtship. In some species the male and female face each other; each extends its abdomen high into the air and moves about in circles (Fig. 13–12*B–D*). The male then seizes the female with his pedipalps, and together they walk backward and forward. This behavior may last hours or even days. Eventually the male deposits a spermatophore, which is attached to the ground. A winglike lever extends from the spermatophore. The male then maneuvers the female so that her genital area is over the spermatophore. Pressure on the spermatophore lever releases the sperm mass, which is taken up into the female orifice (Fig. 13–12*A*).

Development in scorpions is particularly interesting because the entire order is either ovoviviparous or truly viviparous; that is, they brood their eggs within the female reproductive tract. The ovoviviparous species have very large telolecithal eggs exhibiting meroblastic cleavage; development takes place in the lumen of the ovarian tubules. The eggs of viviparous species possess little yolk and display total and equal cleavage. This type of development is found in the tropical Asian species, *Hormurus australasiae*. Its eggs develop within the diverticula of the ovary. Each diverticulum in turn develops a tubular appendage distally that contains a cluster of absorbing cells at the end (Fig. 13–11*D*). These cells rest against the maternal digestive ceca, from which nutritive material is absorbed. The nutritive material passes through the tubule to the embryo at the base, an arrangement that is a little like the mammalian umbilical cord.

Development takes several months or even a year or more with anywhere from six to ninety young produced, depending on the species. At birth the young are only a few millimeters long, and they immediately crawl upon the mother's back (Fig. 13–12*E*). The young remain there through the first molt, which occurs in about one week. The young scorpions then gradually leave the mother and become independent. They reach the adult stage in about one year. About seven molts take place during this period in *Heterometrus longimanus*, a Philippine species.

Order Pseudoscorpiones

Pseudoscorpions are tiny arachnids, rarely longer than 8 mm. They live in leaf mold, in soil, beneath bark and stones, in moss and similar types of vegetation, and in the nests of some mammals. A few species inhabit caves, and some species of several genera are common inhabitants of algae and drift in the intertidal zone. A cosmopolitan species, *Chelifer cancroides*, is found in houses.

Because of their small size and the nature of their habitat these animals are rarely seen, although they are actually quite common. A few handfuls of leaf mold sifted through a Berlese funnel usually will yield several individuals. Pseudoscorpions are found throughout the world, and about two thousand species have been described.

Pseudoscorpions superficially resemble the true scorpions but lack the long abdomen and sting (Fig. 13–13). Moreover, no adult

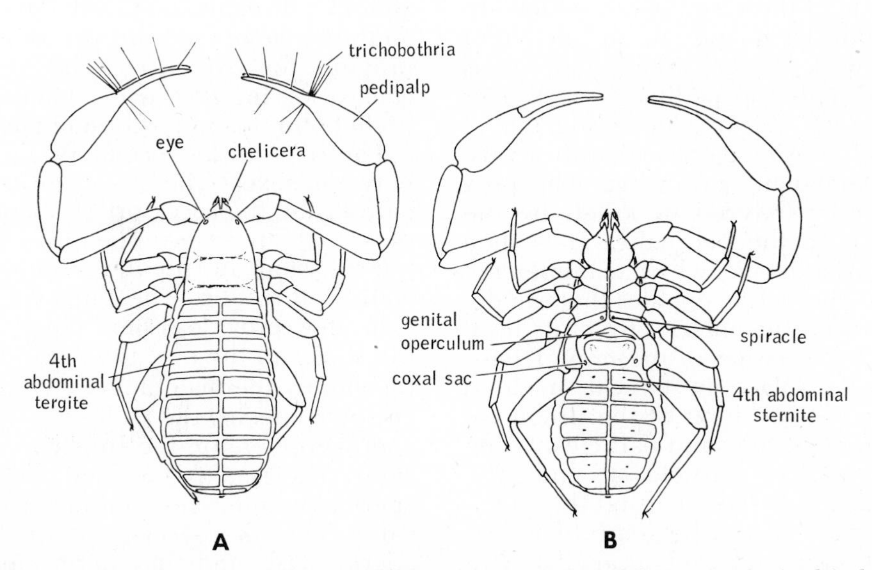

Figure 13–13. Dorsal (*A*) and ventral (*B*) views of the pseudoscorpion *Chelifer cancroides* (male), showing external structure. (After Beier from Weygoldt, P., 1969: Biology of Pseudoscorpions. Harvard Univ. Press.)

scorpions are as small as pseudoscorpions. The rectangular carapace bears one or two eyes at each anterior lateral corner, or the eyes may be absent. There are never median eyes. The ventral surface of the prosoma is entirely occupied by the coxae of the walking legs and pedipalps. The relatively wide abdomen forms a broad juncture with the prosoma, and is rounded posteriorly.

Pseudoscorpion appendages are much more highly modified than those of the scorpions. The chelicerae are made up of two articles forming a pair of pincers and bearing several accessory structures (Fig. 13–14). Within the movable finger of the chelicera are located several ducts leading from a pair of silk glands in the prosoma to a group of pores on the end of the upper side of the finger. Frequently the ducts open not at the surface but at the end of hornlike processes collectively known as the galea. The movable finger and the fixed finger are each provided with a comblike structure along their inner borders. These structures are known as the serrula externa and serrula interna, respectively. On the internal face of each chelicera at the base of the fixed finger is a cluster of short heavy stiff hairs called flagella. The function of these structures is discussed later.

Pseudoscorpion pedipalps are similar to those of the scorpions. However, these pedipalps are peculiar in that usually each has a poison gland located in one or both fingers or in the hand. A duct issues from the poison gland and opens at the end of a tooth at the tip of the finger. Pseudoscorpions are slow-moving and hold their pedipalps, which are supplied with trichobothria, to the front when they walk. If the pseudoscorpion is disturbed, it pulls the pedipalps back over the carapace, and becomes immobile.

Pseudoscorpions feed on small arthropods, such as collembolans and mites. The prey is caught, and paralyzed or killed by the poison glands in the pedipalps. It is then passed to the chelicerae, which tear open the exoskeleton. This action enables the pseudoscorpion to insert the anterior tip of its head (labrum) into the tissues of the prey. Digestion then takes place in typical arachnid fashion. Hairs at the front of the prebuccal chamber strain out the solid particles. When a large mass of solid particles accumulates, the chelicerae are withdrawn, and the flagellar hairs thereon catch the mass of debris and eject it from the prebuccal chamber. This process is repeated whenever the front

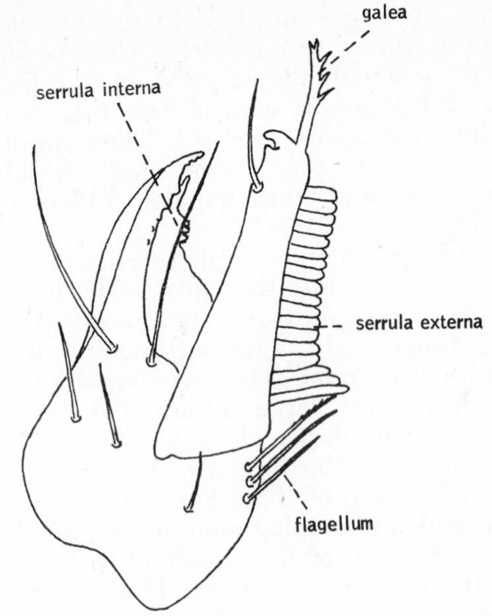

Figure 13–14. Chelicera of *Lamprochernes oblongus* (dorsal view). (After Hoff.)

of the prebuccal chamber becomes clogged. Eventually all the internal tissues of the prey are dissolved, and the meal is terminated.

After feeding, the buccal pieces are cleaned by the comblike internal and external serrulae on the fingers of the chelicerae (Fig. 13–14).

Gas exchange in pseudoscorpions is accomplished by means of a tracheal system that opens through two pairs of spiracles on the ventral side of the third and fourth abdominal segments. Coxal glands that open on the coxae of the third pair of walking legs provide for excretion. The sense organs consist of indirect eyes, tactile hairs and trichobothria, and lyriform organs.

There is little secondary differentiation between sexes. The ovary and testis are unpaired, and the accessory structures are complex in both sexes. The process of sperm transmission in pseudoscorpions is especially interesting. The range of complexity exhibited provides some clues as to possible stages in the evolution of indirect sperm transfer by spermatophores (Weygoldt, 1969). In some species the male deposits a stalked spermatophore on the substratum in the absence of a female. If a female encounters the spermatophore, she is attracted chemotactically. She takes a stance over the terminal sperm mass, and the sperm are taken into the female orifice. Uptake is facilitated by swelling of the sperm mass, triggered by some substance produced in the female atrium. This is probably the most primitive method of sperm transfer in pseudoscorpions. The course of subsequent evolution appears to have increased the chance of the female's obtaining the spermatophore.

There are certain other species in which the male, on encountering a female, lays down silk signal threads after depositing the spermatophore. When the female comes across the thread, she is guided to the spermatophore.

With pairing behavior, the male himself directs the female to the spermatophore in the more evolved patterns. The male grasps the female with his chelipeds and, following a dancelike courting behavior, maneuvers the female over the spermatophore until it is in the proper position to be taken up. Finally, in the most specialized pattern, the male maneuvers the female over the spermatophore without touching her. Two long tubelike organs are evaginated from the sexual region of the male abdomen. The female is attracted, and the two sexes dance without touching each other. The male then attaches a spermatophore to the substratum and backs away. The female follows, and the male then seizes her pedipalpal femurs. When she is in the proper position, he aids in the uptake of the spermatophore with his forelegs and by making pushing movements (Fig. 13–15*A*–*C*).

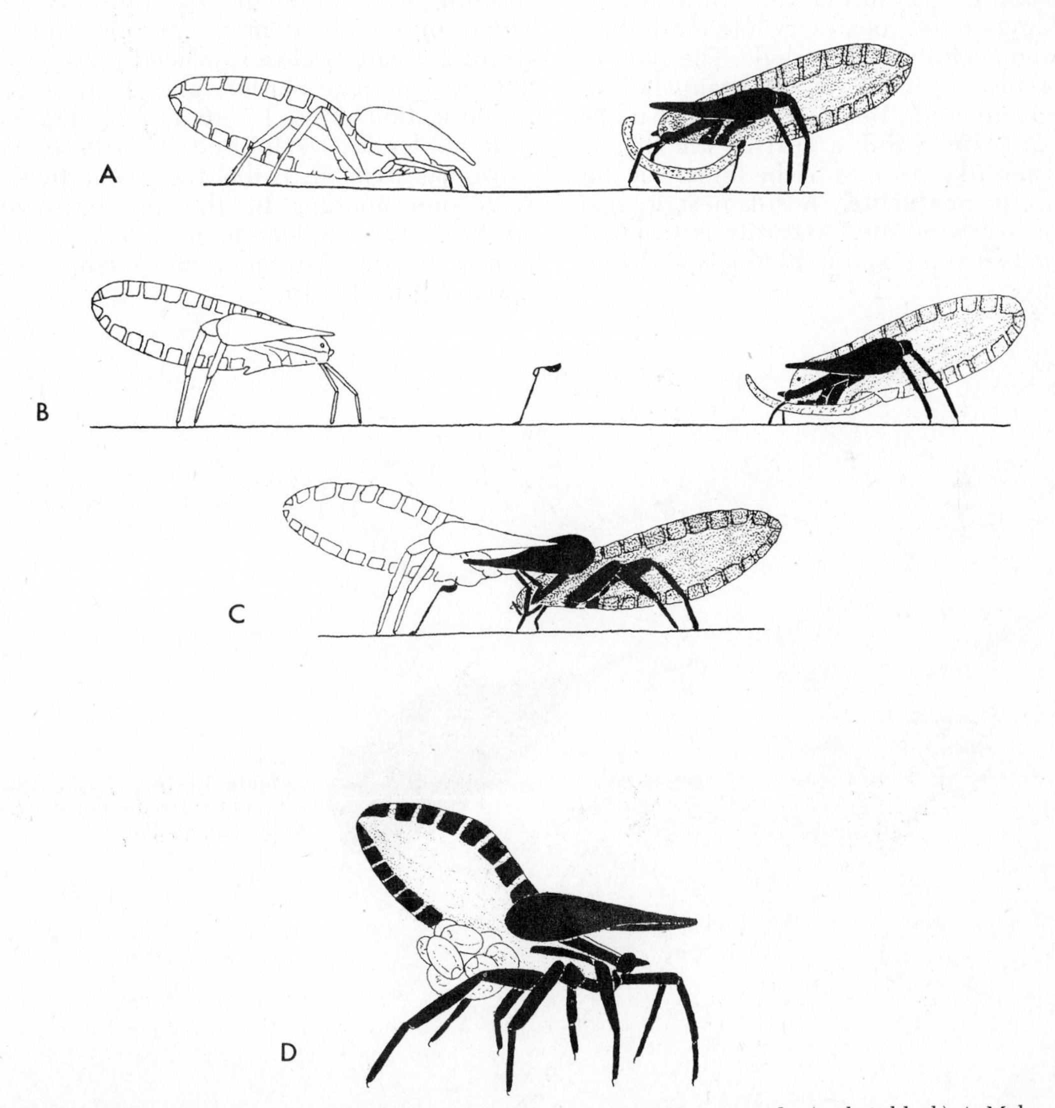

Figure 13–15. *A* to *C*. Courtship and sperm transmission in *Chelifer cancroides* (male is black). *A*. Male, producing spermatophore. *B*. Spermatophore, attached to substratum. *C*. Male pressing female down onto spermatophore. *D*. Female of *Chelifer cancroides*, carrying embryos. (All after Vachon.)

The eggs appear after sperm transmission (a month in *Chelifer cancroides*). Prior to egglaying, the female uses small bits of dead leaves and debris to build a nest and lines it with silk emitted from the galeae of the chelicerae. First the liquid silk is attached, and then the thread is drawn out as the chelicerae are moved to another position.

After the eggs are laid, they remain in a membranous sac that is attached to the genital opening on the ventral side of the body. Development takes place within the sac (Fig. 13–15*D*). In a later stage of development, the embryo is supplied with a nutritive material secreted by the mother's ovaries. The young undergo one molt before hatching and one during hatching, and emerge from the brooding sac during the third instar. Depending on the species, two to more than fifty young may be brooded. The young pseudoscorpion molts twice again before becoming an adult. Before each instar, molting takes place within a nest of silk that is constructed like the nest of the female at the time of egg production. A silk nest is also used for overwintering. Maturity is reached in one to two years, and individuals may live two to five years.

Order Solifugae

The solifugids, or solpugids, are an interesting group of about 800 tropical and semitropical arachnids, sometimes called sun spiders because of the diurnal habits displayed by many species, or wind scorpions because of the great speed with which they can run. In the United States a few species have been found in Florida, and over a hundred have been found in the Southwest, some as far north as Colorado. Many solifugids perfer an arid environment and for this reason are common in the hot desert regions of the world. They hide under stones and in crevices, and many species burrow.

Solifugids are large arachnids, sometimes reaching 7 cm. in length. The prosoma is divided into both a large anterior carapace bearing a pair of closely placed eyes on the anterior median border, and a short posterior section (Figs. 13–16 and 13–17). The unique feature exhibited by these two prosomal divisions is that they can articulate with one another in the non-burrowing species. The abdomen is large, broadly joined to the prosoma, and visibly segmented (Fig. 13–16).

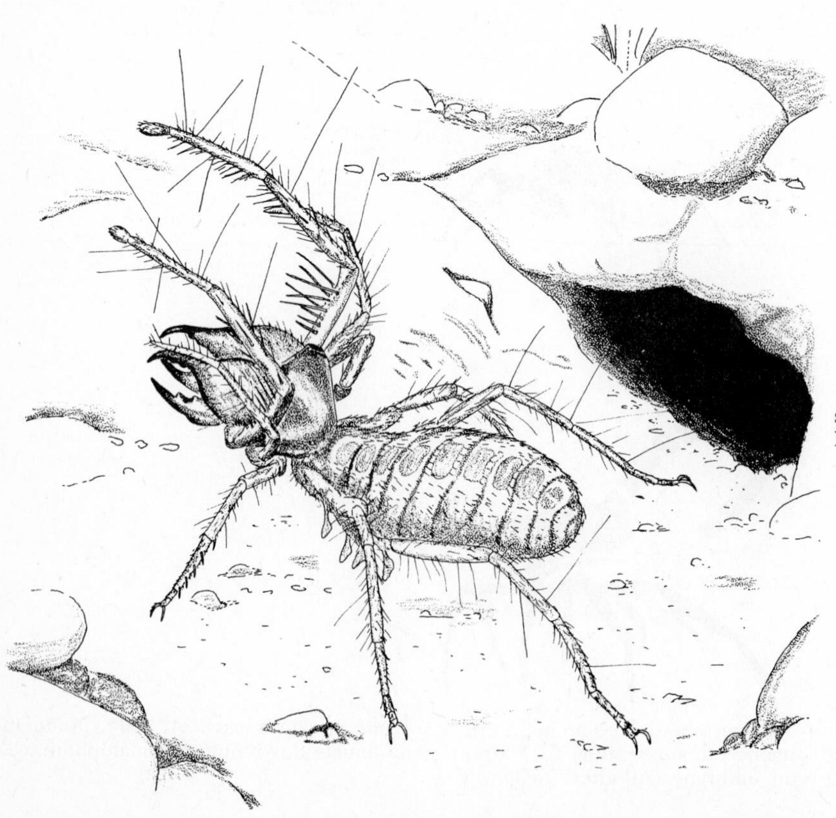

Figure 13–16. North African solifugid, *Galeodes arabs*. (After Millot and Vachon.)

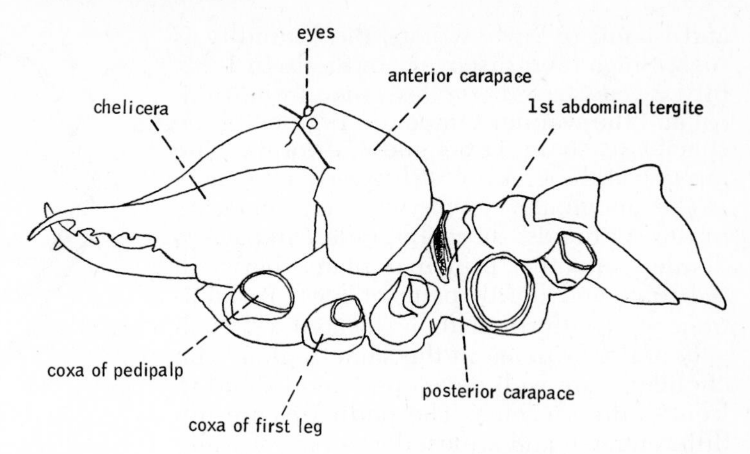

Figure 13–17. Cephalothorax of the solifugid, *Galeodes graecus* (lateral view). After Kaestner from Millot and Vachon.)

The most striking characteristic of the solifugids is the enormous size exhibited by the chelicerae, which project in front of the prosoma and can be directed upward by the flexing of the subdivided prosoma. These chelicerae are extremely heavy, and the length of each exceeds the length of the prosoma. Each chelicera is composed of two pieces forming a pair of pincers that articulate vertically. The pedipalps are leglike but terminate in a specialized adhesive organ used in the capture of prey. The first pair of legs are somewhat reduced in size and are used as tactile organs; the remaining three pairs of legs are used for running.

Solifugids possess voracious appetites and feed on all types of small animals, including vertebrates. The pedipalps seize the prey and pass it to the chelicerae, which kill the captured animal and tear apart the tissues. For gas exchange the animal uses a highly developed tracheal system that opens to the outside through three pairs of ventral slit-like spiracles on the prosoma and abdomen. Excretion is accomplished both by a pair of coxal glands and by a pair of Malpighian tubules.

Males generally seize the females preparatory to copulation. In some solifugids there is then a brief period of stroking and palpation by the male, which throws the female into a passive state. The male turns the female over and opens her genital orifice with his chelicerae. He then emits a globule of semen on the ground, picks it up with his chelicerae, and deposits it into the genital orifice of the female. The entire act takes only a few minutes, and the male then leaps away. In American solifugids sperm transmission is direct, although there is precopulatory behavior and the male inserts the chelicerae into the female orifice before and after sperm transfer. The female deposits 50 to 200 eggs in burrows in the ground.

Order Palpigradi

The palpigrades are a small order of little arachnids, not exceeding 3 mm. in length (Fig. 13–18). They are sometimes called microwhip-scorpions because they possess a terminal whip-like flagellum similar to the Uropygi. Palpigrades live beneath stones

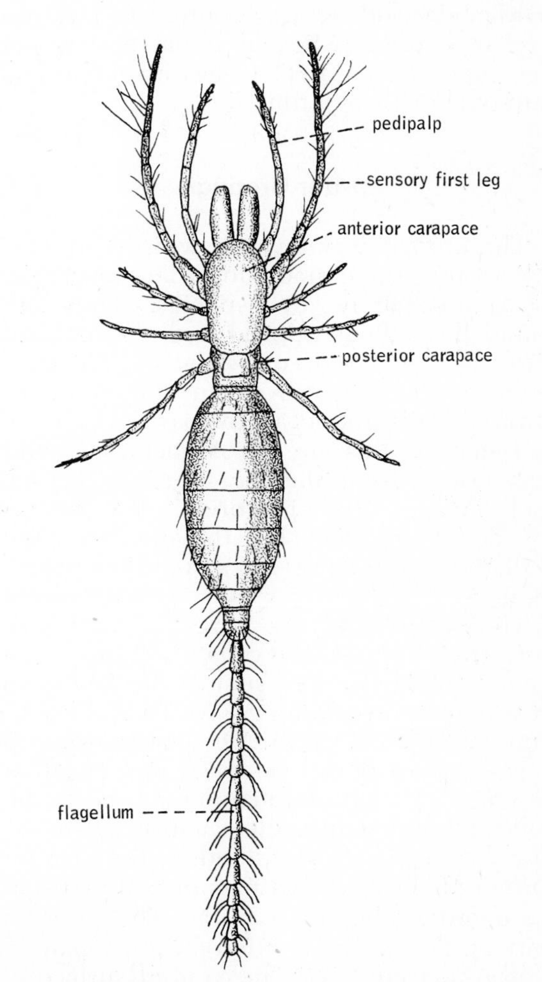

Figure 13–18. *Koenenia*, a palpigrade. (After Kraepelin and Hansen from Millot.)

and in soil or caves where the humidity is high. Since their discovery in Sicily in 1885, fifty species have been described from tropical and the warmer temperate regions; three species occur in Texas and California. The biology of the group is still poorly known.

The integument is very thin and pale. As in the solifugids, the palpigrade carapace is divided into two principal plates between the third and fourth pairs of legs. In addition, a tiny third plate is located on each side of the prosoma in the same region. The chelicerae are well developed and extend in front of the prosoma. The pedipalps are undifferentiated and are used as a pair of walking legs. However, the first pair of legs has a sensory function and is held off the ground during locomotion. Eyes are lacking.

The segmented abdomen is broadly joined to the prosoma and terminates in a long, narrow, and highly mobile flagellum composed of many articles.

Feeding habits are unknown. Three pairs of book lungs have been described in some palpigrades; others are reported to lack gas exchange organs. Sexual behavior has not yet been observed. The eggs are large and laid only a few at a time.

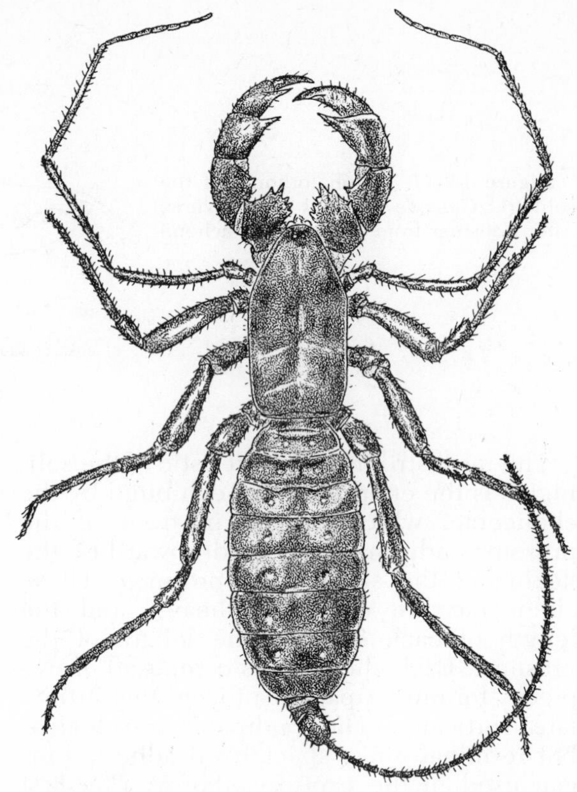

Figure 13–19. American whip scorpion, *Mastigoproctus giganteus.* (After Millot.)

Order Uropygi

The uropygids, known as whip scorpions, are another small arachnid order, consisting of approximately 130 species. They are found throughout tropical and semitropical parts of the world; one species, *Mastigoproctus giganteus,* lives in the southern United States from coast to coast.

The uropygids are all of nocturnal habit and hide during the day beneath leaves, rocks, logs, and other debris. A few species are desert dwellers, but the majority have a distinct preference for a humid environment.

The whip scorpions range in size from very small forms approximately 2 mm. long up to the large species, such as the American *Mastigoproctus giganteus*, which reaches 65 mm. in length (Fig. 13–19). The prosoma is covered by a dorsal carapace that is either single or divided in a manner similar to the solifugid and palpigrade carapace. The divided carapace is restricted to the smaller forms, which are sometimes placed in a separate order, the Schizomida. The anterior part of the carapace often bears a pair of closely placed eyes. The ventral surface of the prosoma is largely covered by the coxae of the appendages.

Uropygid chelicerae are only moderate in size and are two-segmented. The distal piece of each chelicera is fashioned to form a hook or fang that folds against the large basal piece. The pedipalps are stout, heavy, and relatively short; the last two articles of the pedipalps are frequently modified to form a pincer used in seizing prey. The first pair of legs is very long, and the distal articles are often secondarily segmented. The first pair of legs is not used in locomotion but has a sensory-tactile function. The last three pairs of legs are unmodified. Except when excited, the whip scorpions move rather slowly, using only the last three pairs of legs. The pedipalps and tactile first pair of legs are held in front of the animal as it moves forward, with the tactile legs frequently touching the ground.

The abdomen is segmented and the last, or twelfth, segment bears the anus and a long or short flagellum similar to that of palpigrades.

The posterior half of the abdomen contains a pair of large anal glands, which open one on each side of the anus. When irritated, the animal elevates the end of the abdomen and sprays the attacker with fluid secreted from these glands. The secretion in *Mastigoproctus* is 84% acetic acid and 5% caprylic acid, the latter permitting penetration of the acetic acid into the arthropod predator. The fluid can burn human skin, and the repellent odor has resulted in these animals sometimes being called vinegaroons.

During feeding, the prey is seized and torn apart by the pedipalps and then passed to the chelicerae, which complete maceration of the tissues. Two pairs of book lungs are located on the ventral side of the second and third abdominal segments. Excretory organs are coxal glands and Malpighian tubules.

Sperm transfer is indirect and involves the deposition of a spermatophore. A curious courtship dance is performed in some species, which is similar to that exhibited by scorpions (Fig. 13–20*A*). During this dance, the female is held by her long modified sensory legs, which the male crosses in front of his chelicerae. In *Thelyphonus* the female picks up the spermatophore with her genital area, but in *Mastigoproctus* the male inserts the spermatophore with his chelicerae (Fig. 13–20*B*).

When the time for egg laying is at hand, the female secludes herself in some sort of shelter or retreat and lays from seven to thirty-five large eggs. She remains in the shelter with the eggs attached to her body until they have hatched and undergone sev-

A

B

Figure 13–20. Mating in the uropygid *Mastigoproctus giganteus*. *A*. Male grasping the antenniform legs of the female with his chelicerae. *B*. Male inserting the spermatophore into the female gonopore with his chelicerae. (By Weygoldt, P., 1972: Zeitschrift des Kölner Zoo, *15*(3):98 and 99.)

Figure 13–21. Female of *Mastigoproctus giganteus* within her brood chamber, carrying newly hatched young. (By Weygoldt, P., 1972: Zeitschrift des Kölner Zoo, *15* (3):100.)

eral molts (Fig. 13–21). The female dies shortly after the young disperse. Some forms require up to three years to mature sexually.

Order Amblypygi

The amblypygids are a tropical and semitropical group of approximately 60 species, which were formerly grouped taxonomically with the uropygids. The order is represented in the United States by several forms in Florida. The amblypygids, like the uropygids, are a nocturnal and secretive group, hiding during the day beneath logs, bark, stones, leaves, and similar objects. In addition, several species inhabit caves. In general, the members of this group prefer a humid environment.

Amblypygids range in length from 4 to 45 mm. and have a somewhat flattened body, which resembles that of spiders (Fig. 13–22). The carapace bears a pair of median eyes anteriorly and two groups of three eyes each laterally. The chelicerae are similar to those of spiders. The pedipalps are heavy and raptorial. The first pair of legs are modified as sensory-tactile appendages; they are extremely long and whiplike and are composed of many articles. The last three pairs of legs are unmodified and are used for locomotion. The gait of the amblypygid is quite crablike

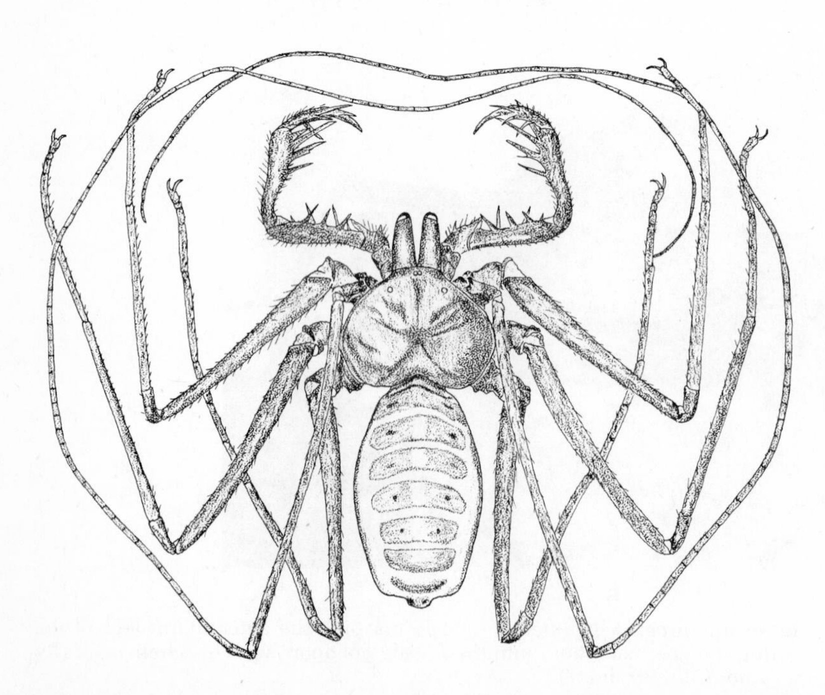

Figure 13–22. The African amblypygid, *Charinus milloti.* (After Millot.)

because of its flattened body and its ability to move laterally. One of the long tactile legs is always pointed toward the direction of movement; the other may explore the areas to either side of the animal. The segmented abdomen lacks the flagellum characteristic of the uropygids.

The insect prey is located with the long first legs and is then quickly captured and immobilized with the spiny pedipalps. The chelicerae tear out pieces of the prey, from which juices are sucked into the digestive system in typical arachnid fashion. Two pairs of book lungs are located on the ventral side of the second and third abdominal segments. The excretory organs consist of coxal glands and Malpighian tubules.

In the few species in which reproductive habits have been observed, a courting pair face each other, each palpating the body of its mate with the long, delicate, whiplike first legs (Fig. 13–23). The male deposits a spermatophore stalk, followed slightly later by a separate deposition of two sperm masses. Using his pedipalps or first legs, the male guides the female over the spermatophore, and she takes up the sperm masses. From six to sixty large eggs are laid. When egg laying is at hand, the reproductive glands secrete a parchment-like membrane that holds the eggs to the underside of the female abdomen. The mother carries the eggs in this manner until hatching and the first molt of the young have taken place. The young climb onto the abdomen of the mother until the next molt, following which one after another descends and departs via her posterior end. Any young that fall off the mother's back she eats.

Figure 13–23. Male of the amblypygid *Heterophrynus longicornis* tapping the body of a female with his antenniform legs. (By Weygoldt, P., 1972: Zeitschrift des Kölner Zoo, *15*(3):100.)

Order Araneae

Except for perhaps the Acarina, which comprise the mites and ticks, the spiders constitute the largest order of arachnids. Approximately 32,000 species have been described, and this probably represents only a portion of the actual number. Also, spider populations are very large. Bristowe (1958) has calculated that an acre of undisturbed, grassy meadow in Great Britain contained 2,265,000 spiders.

A number of adaptations of spiders make them especially interesting animals—the great variety of uses to which their silk has been put in different families; their feeding habits and their utilization of venom; the well developed vision of some hunting spiders; and the modification of the pedipalps in the male to form a copulatory organ.

External Structure. Spiders range in size from tiny species less than 0.5 mm. in length to large tropical mygalomorphs (called tarantulas, bird spiders, or monkey spiders in different parts of the world) with a body length of 9 cm.; leg span can be much greater. The convex carapace usually bears eight eyes anteriorly (Figs. 13–24 and 13–32A). A large sternum is present on the ventral surface, and a small median plate known as the labium is attached directly in front of the sternum (Fig. 13–25).

Each chelicera is of moderate size and, like uropygid and amblypygid chelicerae, consists of a distal and a basal piece (Fig. 13–25). The distal piece is a fang, which, when the animal is at rest, fits into a groove at the end of the large basal piece.

The female pedipalps are short and leglike, but in the male they have become modified, forming copulatory organs. The last segment is greatly enlarged and knoblike and resembles a boxing glove. The structure of this segment is discussed later. The legs are variable in length and usually consist of eight

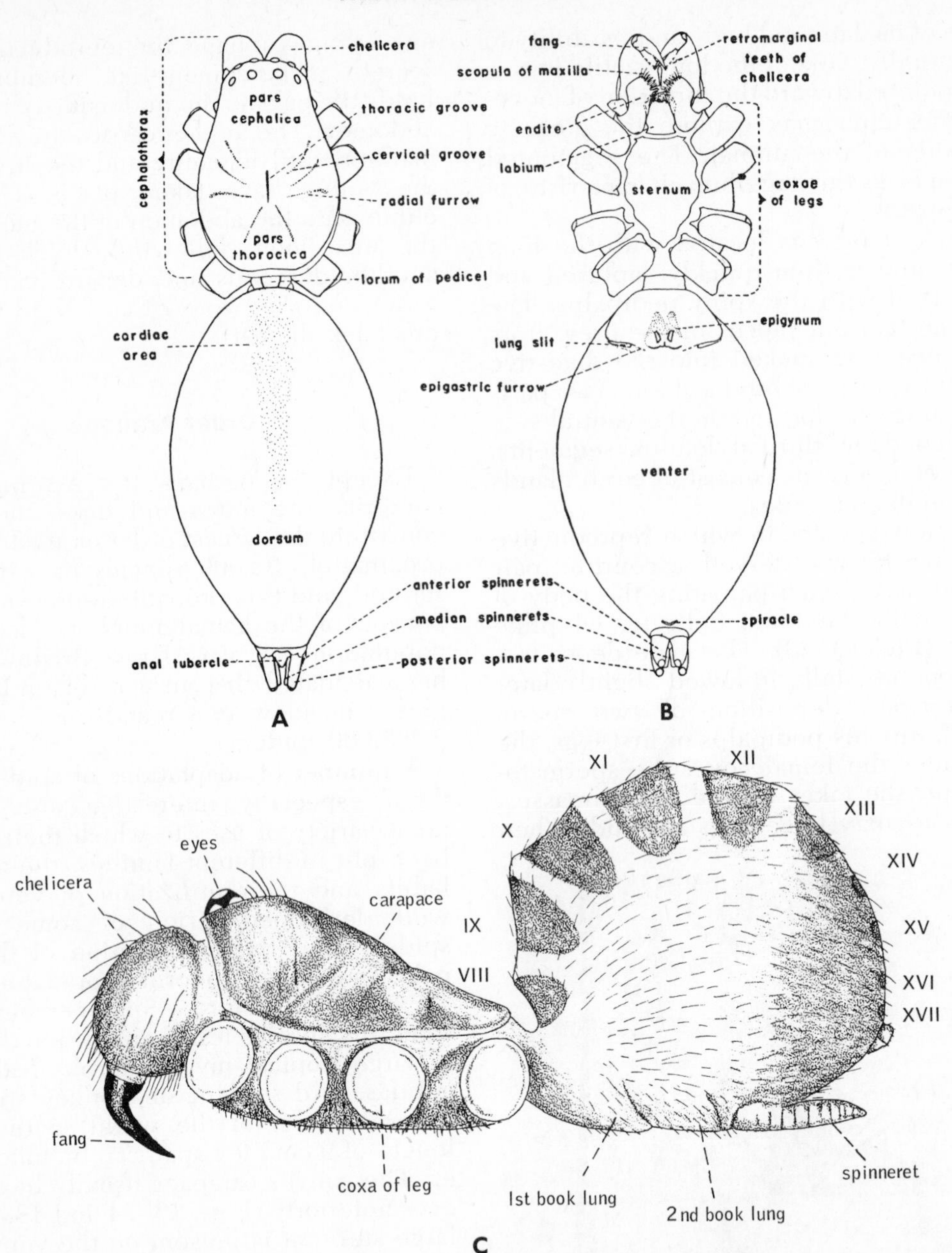

Figure 13–24. Dorsal (*A*) and ventral (*B*) views of a spider, showing external structure. (From Kaston, B. J., 1948: Spiders of Connecticut. Bull. 70, State Geol. Nat. Hist. Survey, 13.) *C*. The primitive Asian mygalomorph spider, *Liphistius malayanus*, with legs removed (side view). (After Millot.)

segments—a basal coxa, a small trochanter, a long femur, a short patella, a long tibia and metatarsus, a tarsus, and a distal minute pretarsus bearing two or three claws.

The globe-shaped or elongated abdomen is unsegmented except in a few groups of primitive spiders. However, segmentation is often reflected in the color pattern. The abdomen is connected to the prosoma by a short and narrow portion called the pedicel (Fig. 13–24). Anteriorly, the ventral side of the body bears a transverse groove known as the epigastric furrow (Fig. 13–24). The reproductive openings are located in the middle of this furrow, and the spiracles of the book lungs on each side of it. The end of the abdomen bears a group of modified appendages known as spinnerets (Fig. 13–24). These are spinning organs and are located on the ventral surface just anterior to the terminal anus. In primitive spiders, eight spinnerets are located more anteriorly on the abdomen; but most spiders have only six, including a vestigial or modified pair, and they are located at the posterior of the abdomen.

Silk. Each spinneret is a short, conical

structure bearing many spigots, the openings from the silk glands (Figs. 13–24 and 13–26). The glands themselves are large and are located within the posterior half of the abdomen. The ducts from the silk glands extend posteriorly and terminate at the end of the spinnerets.

The silk of spiders is a protein similar to the silk of caterpillars. It is emitted as a liquid, and hardening results not from exposure to air but probably from the actual drawing-out process itself. A single thread is composed of several fibers, each drawn out from liquid silk supplied at a separate spigot. Most spiders produce more than one type of silk, and the various types are secreted from two to six different kinds of silk glands.

Silk plays an important role in the life of the spider and is put to a variety of uses, but not all families of spiders build webs to catch prey. One function of silk that is common to most spiders is its use as a dragline (Fig. 13–27). Spiders continually lay out a line of dry silk behind them as they move about. At intervals it is fastened to the substratum with adhesive silk. The dragline acts as a safety line similar to that used by mountain climbers. The common sight of a spider suspended in mid-air, after being brushed off some object, results from its continual retention of its dragline. Many spiders build silken nests beneath bark and stones, which they may use as retreats or in which they may overwinter. All spiders wrap up their eggs within silken cocoons, and the young on hatching are carried on air currents by means of a silken strand. These functions of silk will be more fully described later.

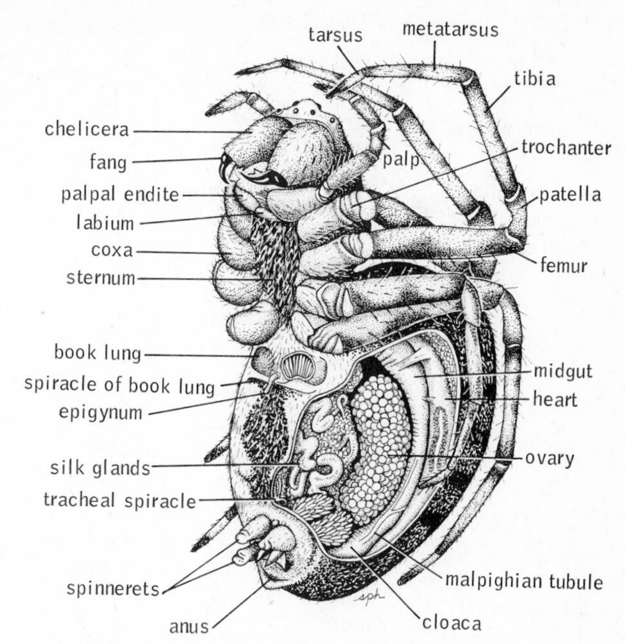

Figure 13–25. The orb weaver, *Araneus diadematus* (ventral view). (After Pfurtscheller from Kaestner.)

Feeding. Spiders, like most other arachnids, are predatory and feed largely on insects, although small vertebrates may be captured by large species of spiders. The prey either is pounced upon by members of the more active hunting groups or is caught in a silken snare, as in the case of members of web-building families.

Hunting forms include wolf spiders, fisher spiders, crab spiders, jumping spiders, and many mygalomorph spiders; the latter is a more primitive group that contains the so-called tarantulas and the trap-door spiders. Species that stalk their prey typically have heavier legs than do web-builders. Prey is detected by tactile and visual stimuli, and in some families such as the wolf spiders and the colorful jumping spiders, the eyes are highly developed (Fig. 13–32*B* and *C*). Jumping spiders leap by sudden extension of the legs, resulting from a very rapid elevation of blood pressure. Hunting spiders lay down a dragline, and some ground spiders tie up their prey by running around them. Trap-door spiders construct silk-lined burrows that are closed by a lid covered with moss, soil, and other material (Fig. 13–37*A*). The spider lies in wait beneath the lid, and some species may detect passing prey by means of silk lines that radiate out over the ground (Buchli, 1969).

Prey varies with habitat and distribution of the spider. Studies on the little wolf spider *Pardosa amentata*, which lives on sunny moist ground in Europe, indicate that 70 per cent of the diet is composed of dipterans (flies and gnats). About one insect per day was caught, and feeding activity was greater in the morning than in the afternoon (Edgar, 1970).

The dragline may have been the origin of the snare of web-building families. The first web was perhaps the crisscrossing threads of the dragline as the spider moved about within a limited space. The trapping of insects would have endowed this web with a selective advantage, and from such a prototype evolved the highly developed triangles, orbs, funnels, sheets, and meshes characteristic of the webs of different spider families and species. Many webs are composed of both dry and adhesive strands. The latter are usually lines with an outer unpolymerized

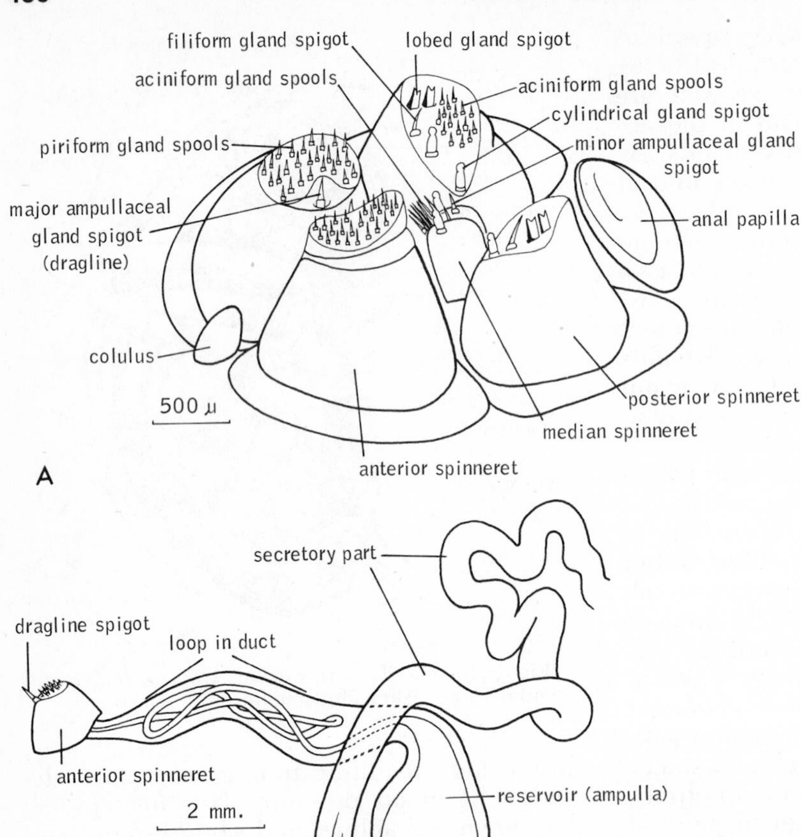

Figure 13–26. Spinnerets of the orb-weaving spider *Araneus diadematus*. A. Distribution of spigots for different silk glands on spinnerets. *B*. Anterior spinneret and gland supplying the dragline spigot. (From Wilson, R. S., 1969: Amer. Zool., 9(1):103–111.)

viscous layer of silk that greatly increases the trapping quality of the web; in some spiders it is a very fine entangling loose "wool" attached to the line.

The construction of an orb web by a spider is a remarkable feat. Web building is dependent upon morphological and physiological factors such as weight, leg length, silk supply, appetite, and an instinctive behavioral pattern involving the integration of sensory information and locomotor activity. No aspect is learned. The most complex web can be constructed by members of that species immediately upon hatching. Considerable knowledge regarding web building has been provided by pharmacologists, who have found the alterations in a spider's web-building behavior to reflect different drug effects.

Most of the familiar orb webs are produced by members of the family Araneidae. Web construction is begun with a horizontal line, which may be quite long. To place this line, the spider may let out a strand of silk which is carried by air currents. When the sticky end contacts another surface, it is pulled

Figure 13–27. Jumping spider, showing silk dragline (white diagonal line in upper right corner). (By Betty M. Barnes.)

tight and anchored. Alternatively, the spider may first attach the free end and then drop or be blown to the opposite point of attachment. After laying down the horizontal line, which is often strengthened with additional threads, the spider drops from the center, laying out a vertical line. The vertical line is pulled tight, converting the **T** to a **Y** frame, which provides the basic scaffolding of the web. Additional frame lines are placed, and then the radii, or spokes, are attached from outside to center, first on one side and then the other, using a pre-existing spoke as a guide. With the radii in place, a temporary dry spiral of thread is laid down from the center to the outside. Then the permanent adhesive spiral is placed from the outside inward, and the spider removes the dry temporary spiral in the process. Since spacing of lines is determined by leg span, the size of the web is partly dependent on the size of the spider and increases with the spider's growth. Some spider species have a wide mesh, and others have a fine mesh, depending on the potential prey size. The entire web, or at least the adhesive spiral, is replaced periodically, for the wet silk loses its stickiness within a few days. The old silk may be eaten, and very quickly much of the protein finds its way back to the silk glands. A web may be replaced every night.

Web-building spiders are aerialists, and they usually have more slender legs than do hunting spiders (Fig. 13–38*B*). In climbing about the web, the lines are hooked by a small middle claw which lies between the two large claws found in all spiders, and they are held against two outer setae called accessory claws. Why the spider does not stick to the web and how it can cut silk with its chelicerae is not known. Eyesight is not well developed in web-building spiders, but they are highly sensitive to vibrations. A web builder can determine from thread vibrations the size and location of the trapped prey, and many species will respond to different stimuli with different attack patterns (Fig. 13–28) (Robinson, 1969). Orb weavers, tangle-web spiders, and others commonly swath prey in silk before or after biting it. Some of the many interesting variations in prey capture are described by Kaestner (1968).

Spiders bite their prey with the chelicerae, which may also hold and macerate the tissues during digestion. The chelicerae are unique among arachnids in being provided with poison glands that open near the tip of the fang (Fig. 13–29). The glands themselves are located within the basal segments of the chelicerae and usually extend backward into the head. A duct runs from the poison gland out through the fang. When the spider bites, the fangs are raised out of the groove in which they lie, and are rammed into the prey. Simultaneously, muscles around the poison gland contract, and fluid from the gland is discharged from the fang into the body of the prey.

The venom of most spiders is not toxic to man, but a few species have dangerous bites. Among these are several species of black widows (*Lactrodectus*), which are found in most parts of the world, including the United States and southern Canada (Figs. 13–30 and 13–36*A*). The venom is neurotoxic. Although the bite may be unobserved, the symptoms are severe and very painful; they include pain in abdomen and legs, high cerebrospinal fluid pressure, nausea, muscular spasms, and respiratory paralysis. Fatal cases are rare and are more apt to result from incorrect diagnosis and mistreatment (such as removal of the appendix) than from the spider venom. Antivenin and other treatments are available.

The recluse spiders, members of the genus *Loxosceles*, have a hemolytic venom and produce a local necrosis, or ulceration, of tissue that spreads from the bite. The ulcer is slow to heal. The bite of *Loxosceles reclusa*, the brown recluse spider, which is found in the southern and midwestern United States, can be dangerous, as can that of the large South American *Loxosceles laeta*. In southeastern Brazil, a ctenid, *Phoneutria*, and a wolf spider of the genus *Lycosa* are venomous.

North American mygalomorph "tarantulas," despite their size and reputation, are not venomous; but there are mygalomorphs, such as the Australian *Atrax* and the South American *Trechona*, both funnel-web builders, that are dangerous. However, many "tarantulas" have urticating setae on the abdomen, which are brushed off with a leg when the spider is disturbed. These setae are barbed and irritating, and readily penetrate the skin of a potential predator. In man they can cause a rash.

Spiders having chelicerae with teeth (Fig. 13–24*B*), such as many hunting forms, chew their prey, aiding the digestion of tissues by the enzymes poured out from the mouth. The undigestible skeletal remains are discarded as a wad. Spiders with toothless chelicerae do not chew, but introduce en-

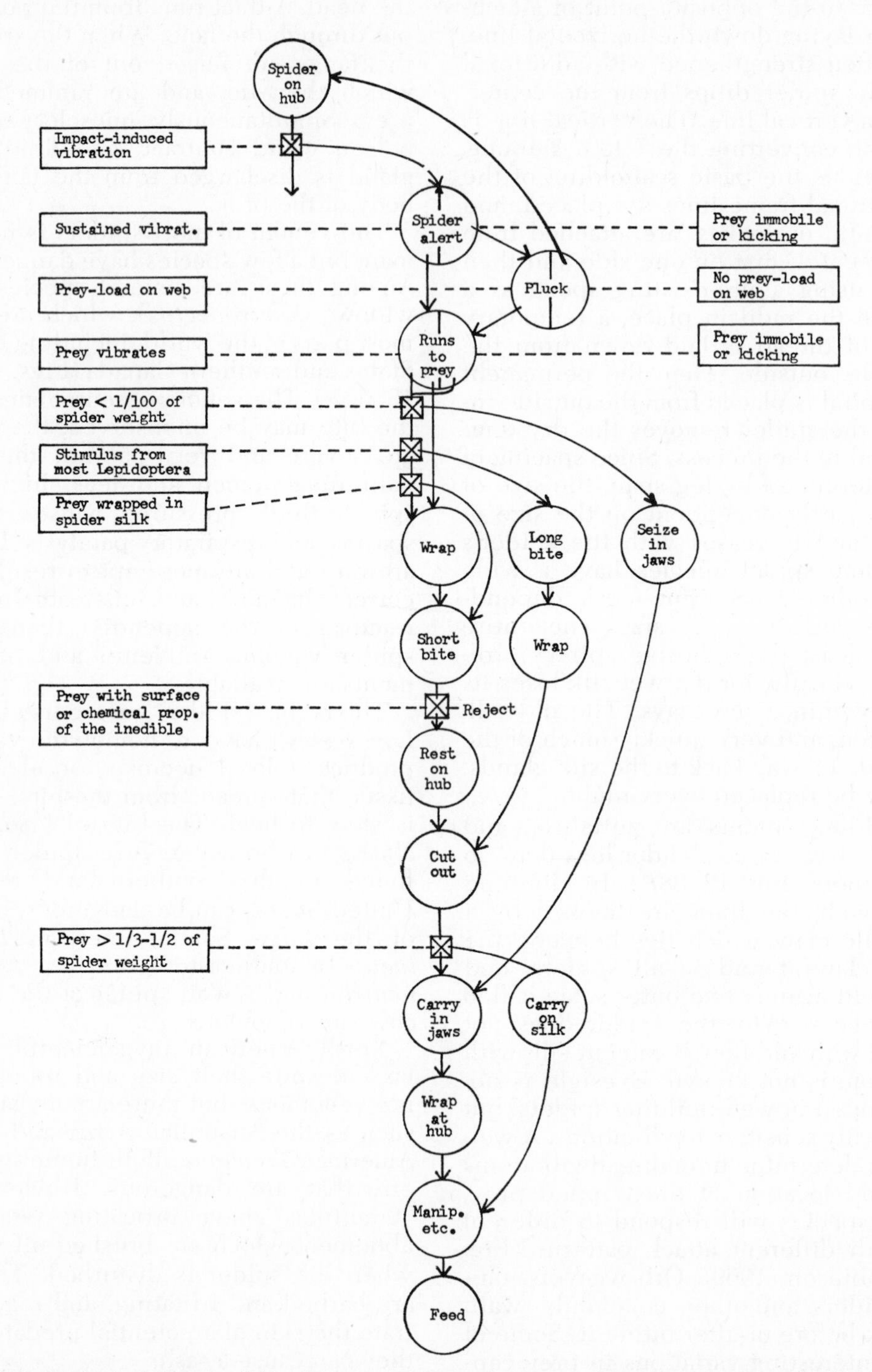

Figure 13–28. Predatory behavior of the orb-weaver *Argiope argentata*, showing sequence and relationship to prey size. Circle depicts action of spider; square indicates prey. Sequence begins with spider waiting in center of web (on hub). When insect strikes web, spider becomes alert, following which it may run to prey or pluck web. Vibrations from web-plucking provide information on nature of prey. Large prey are given a long bite, injecting a larger amount of poison than in a short bite. They are then wrapped with silk. Small prey are first wrapped and then given a short bite. (From Robinson, M. H., 1969: Amer. Zool., 9:161–173.)

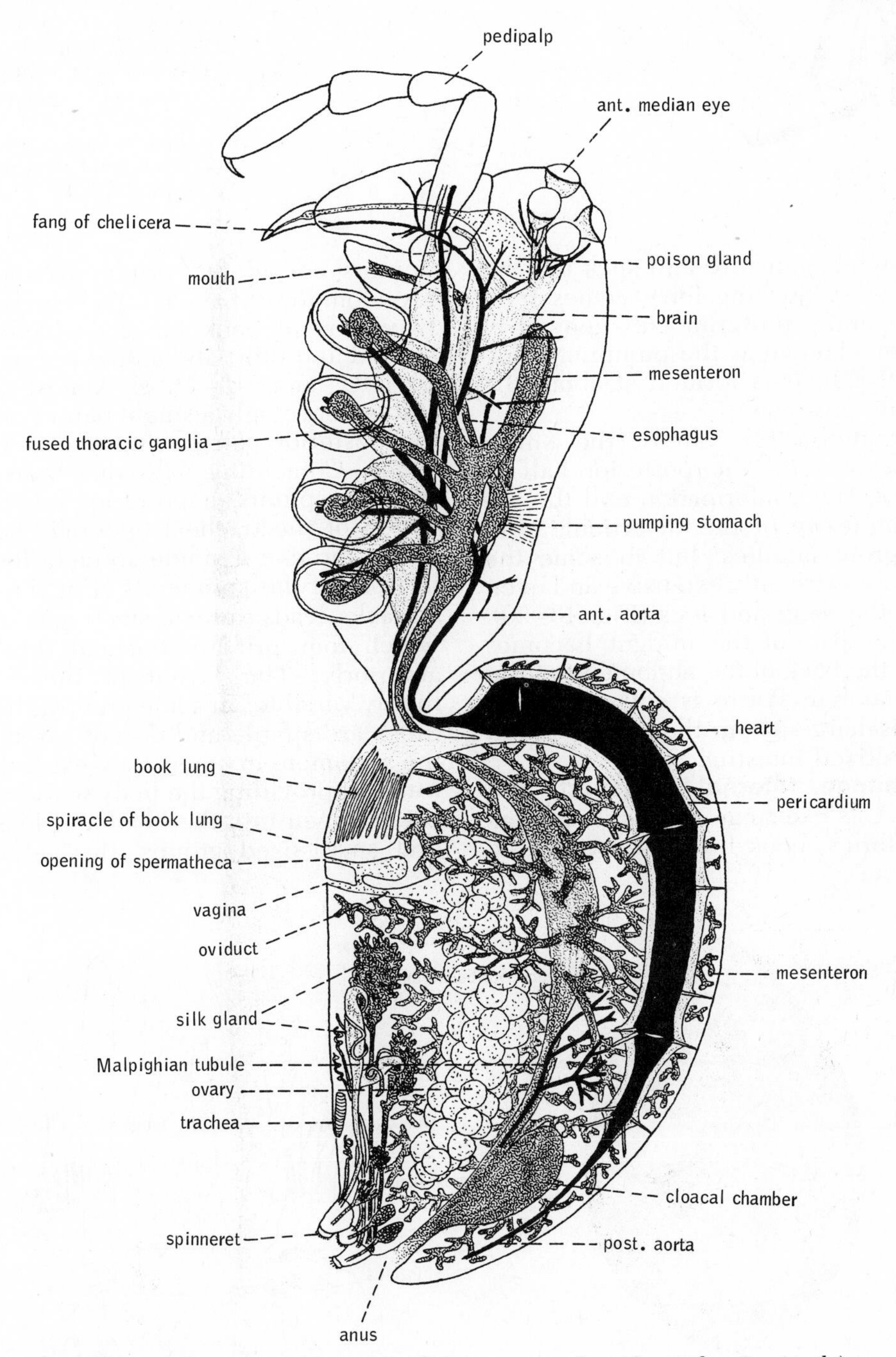

Figure 13–29. Internal anatomy of an arancomorph spider. (After Comstock.)

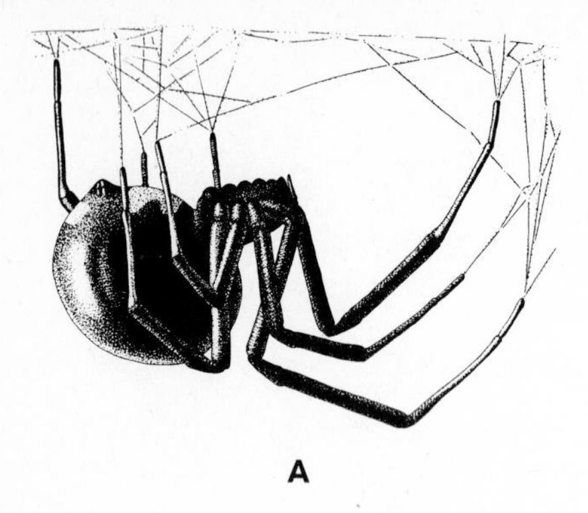

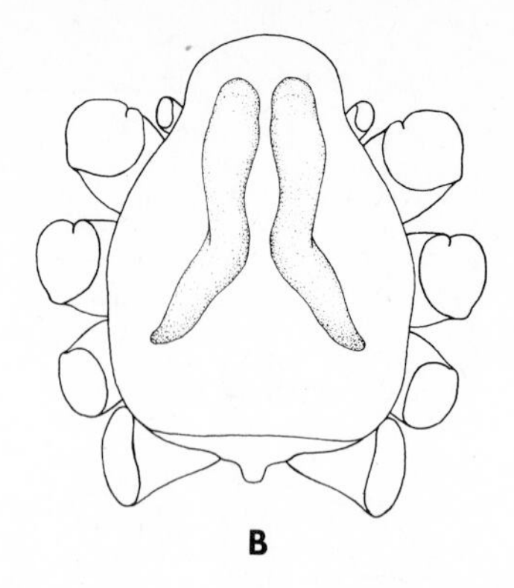

Figure 13–30. Venomous spiders. *A.* Female black widow, *Latrodectus mactans,* hanging in web. Color is shiny black with red hour glass marking on ventral surface of abdomen. This species of the cosmopolitan genus is found in eastern and midwestern United States. *B.* Location of poison glands in cephalothorax of *Latrodectus.* (Both from Kaston, B. J., 1970: Trans. San Diego Soc. Nat. Hist., *16*(3):33–82.)

zymes through a puncture and suck out the digested tissues. Sucking force comes from the pharynx and a posterior enlargement of the esophagus known as the pumping stomach (Fig. 13–29). It is located at about the middle of the prosoma.

The mesenteron fills almost the entire abdomen, as well as the posterior half of the prosoma. The conformation and the degree of complexity of the diverticula vary within different families, but in some the diverticula are extremely extensive and even ramify into the head and legs (Fig. 13–29). The posterior part of the midgut becomes enlarged in the back of the abdomen to form a cloacal chamber. Waste is collected here and then discharged from the anus through a short sclerotized intestine.

Gas Exchange, Internal Transport, and Excretion. Gas exchange organs in spiders are of two forms, book lungs and tracheae. Spiders considered primitive, such as the mygalomorphs, have no tracheae but display two pairs of book lungs derived from the second and third abdominal segments (Figs. 13–24*C* and 13–31*A*). Almost all other spiders have only a single pair of book lungs, the posterior pair having evolved into tracheae. Coinciding with this transformation in most spiders, a posterior migration and fusion of the tracheal openings have taken place, forming a single spiracle located just in front of the spinnerets (Fig. 13–29). The spiracle leads into a small chamber from which four primary tracheal tubes extend anteriorly. The layout of these tubes is highly variable. In some groups, the tracheal tubes are simple and do not extend beyond the abdomen; in others they exceed the book lungs in providing the body with oxygen and extend even into the head and legs. In several small-sized groups, the anterior book

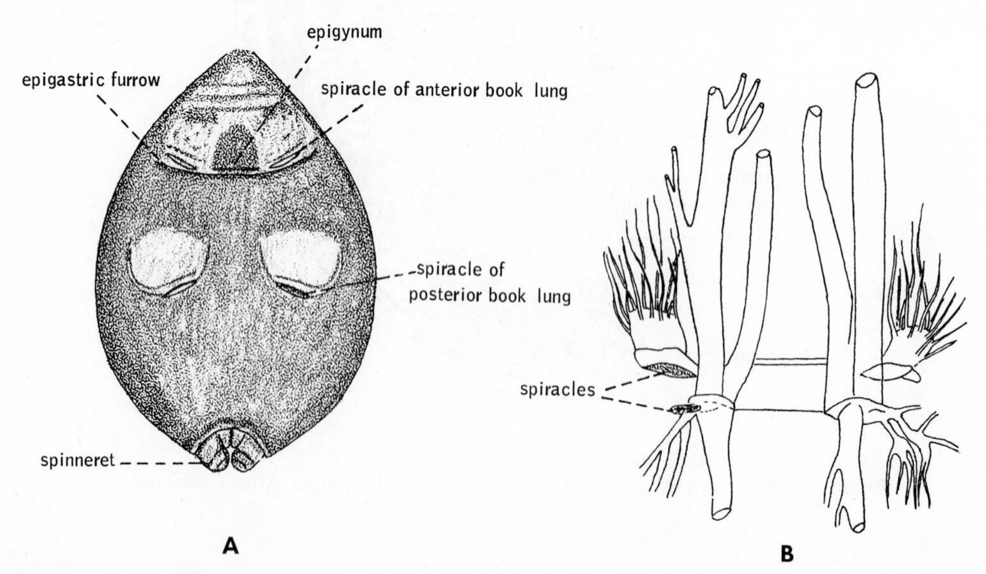

Figure 13–31. *A.* Abdomen of a mygalomorph spider, showing the two pairs of book lungs (ventral view). (After Comstock.) *B.* Central part of tracheal system of the caponiid spider, *Nops coccineus,* in which both pairs of book lungs have been replaced by tracheae. (After Bertkan from Kaestner.)

ungs have also been transformed into tracheae, so that gas exchange is accomplished entirely by this means (Fig. 13–31*B*). Although the tracheal system of spiders appears to have developed in most cases from the book lungs, the modification probably evolved independently within a number of different spider groups (Levi, 1967).

The spider circulatory system is similar to that of scorpions, although the heart varies in length and in the number of ostia (Fig. 13–29). These variations are apparently related to the nature of the gas exchange organs. On one extreme, the heart is five-segmented with a corresponding number of ostia. This is the more primitive structure and is found only in forms with four book lungs. As the book lungs become reduced in size and the tracheae become increased in complexity, there is a corresponding loss in the number of heart chambers, of ostia, and of arteries, until in the exclusively tracheate spiders, most of which are small, the number of ostia is often reduced to two pairs.

Another peculiarity of the spider circulatory system is that the anterior aorta is considerably larger than the posterior aorta; thus, the anterior part of the abdomen and the cephalothorax receive greater irrigation with oxygenated blood. The tracheae, when present, compensate for this since they almost always supply at least the posterior part of the abdomen.

Blood pressure in a resting spider is equal to that of man and may double in an active spider or one ready to molt. Blood pressure also causes the legs to extend in opposition to flexor muscles. A valve in the anterior aorta is believed to maintain the high pressure.

Coxal glands are not as well developed in spiders as in other arachnids. Groups considered primitive have two pairs of coxal glands opening onto the coxae of the first and third pair of walking legs. The coxal glands are highly developed in these forms. In all others, only the anterior pair of glands remain and they display various stages of regression.

Although coxal glands in many spiders have become reduced functionally as excretory organs, it is possible that the silk glands and perhaps even the poison glands evolved from them.

More important functionally as excretory organs are the two Malpighian tubules, which are connected to the cloacal chamber in the posterior part of the abdomen (Fig. 13–29). Each tubule extends anteriorly to branch over the abdominal diverticula. Guanine wastes are excreted not only by the cells of the Malpighian tubules but also by the cloacal wall. Certain cells located beneath the integument and over the surface of the intestinal diverticula also produce guanine crystals. These waste products eventually pass into the lumen of the diverticula, where they dissolve and are passed posteriorly to be excreted from the cloacal pocket. The white color of some spiders is caused by guanine stored in the body wall.

Sense Organs. The eyes of some spiders surpass those of all other arachnids in degree of development. Usually, there are eight eyes arranged in two rows of four each along the anterior dorsal margin of the carapace (Fig. 13–32*A*). Frequently, the arrangement of the eyes is characteristic of certain families. In the wolf spiders and jumping spiders, the eyes are situated over the surface of the anterior half of the carapace, resulting in perception over a wide field (Fig. 13–32*B* and *C*). Spiders in a few families have the eyes reduced in number to six, four, or even two, and some completely lack eyes.

Of the two rows of four eyes usually found in spiders, the anterior median eyes are of the direct type, and all the remaining eyes are indirect. Since the indirect eyes are provided with a tapetum, which reflects light rays, the anterior median eyes appear dark and the others often appear pearly white. In some families, notably the hunting wolf spiders, the tapetum has developed to such an extent that these spiders can easily be located at night and captured by using a flashlight to look for the reflection of the eyes.

In most spiders, the eyes are unable to form an image owing to an insufficient number of receptors. The number of receptors is much greater in cursorial (hunting) species than in the sedentary web builders. In the hunting spiders, eyes are important for detecting movement and locating objects. The jumping spiders are capable of perceiving a sharp image of considerable size. In this family, the number of receptors is extremely large, particularly in the anterior median eyes, in which the tapered ends of the retinal cells are greatly narrowed and compact. Short focal length supposedly compensates for lack of accomodation, but in many species of jumping spiders there has been an increase in the depth of the anterior median eyes, resulting in a somewhat tubular struc-

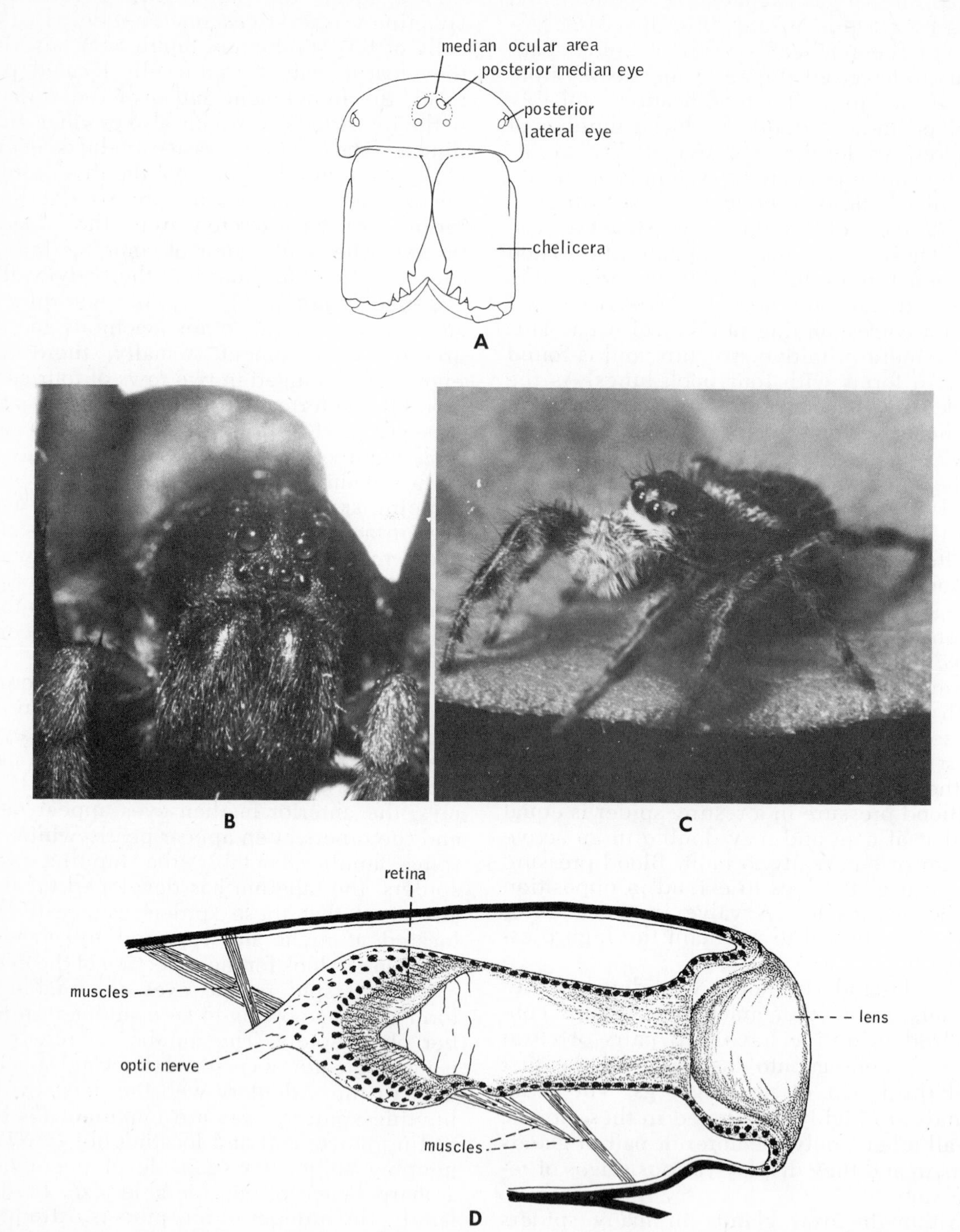

Figure 13–32. *A.* Front view of a spider, showing eight eyes arranged in anterior and posterior rows of four each, a pattern exhibited by many families. (After Kaston.) *B.* Face of a wolf spider. Posterior median eyes are very large; the posterior lateral eyes are located back over the carapace and can just be seen. *C.* A jumping spider. In this family, the anterior median eyes are very large and the posterior eyes are located back over the carapace. (*B* and *C* by Betty M. Barnes.) *D:* Sagittal section of the anterior median eye of the jumping spider, *Salticus scenicus*, showing the greatly elongated structure. (After Scheuring from Kaestner.)

ture (Fig. 13–32*D*). Some members of this family, such as species of the genus *Phidippus,* can focus on an object more than 36 cm. away.

Like certain insects and crustaceans, some spiders (the funnel-web spider *Agelena* and the wolf spider *Arctosa*) utilize the sun and polarized sky light to orient themselves to their surroundings. *Arctosa,* which has been studied extensively, uses the angle of the sun in addition to sky polarization, and posseses an internal clock, which corrects for the changing angle during the course of the day.

Chemosensitive tubular hairs on tips of appendages, trichobothria, and slit sense organs are very important sense organs in all spiders; in most spiders, especially the sedentary web-building forms, they are more important than the eyes. The funnel-web builder *Agelena* can determine the position of prey when it is one centimeter distant by means of the trichobothria. Isolated or grouped slit sense organs are located over the entire surface of the body and are highly developed in spiders. Studies of certain web-building species have indicated that the slit sense organs in the joint between the tarsus and the metatarsus (the last two long leg segments) are especially sensitive and enable the spider to discriminate vibration frequencies transmitted through the silk strands of the web or even through the air. Spiderlings, in species in which the spiderlings remain in the parental web after hatching, can be recognized by the parent. The spider can also discern the size of the entrapped prey or even the kind of insect, when the insect is a species that produces buzzing vibrations. The spiders themselves may produce vibration signals. The male may tweak the strands of the female's web, or the mother may produce vibration signals which are detected by the spiderlings.

Reproduction and Life History. The ovaries of the female consist of two elongated parallel sacs located in the ventral part of the abdomen (Figs. 13–25, 13–29, and 13–33). The germinal epithelium gives rise to the oocytes, which in the process of development and enlargement bulge outward from the ovarian lumen. During egg formation, this gives the ovary the appearance of a cluster of small grapes. In a gravid female, the ovaries may occupy two thirds or more of the abdomen. The epithelial cells not involved in egg formation secrete an adhesive material used to agglutinate the eggs in the cocoon. At maturity, the eggs rupture into the lumen of the ovary and pass into the curved oviduct leading from each ovary. Each oviduct extends forward, downward, and toward the midline and converges to form a median tube (Fig. 13–33). This tube extends ventrally and backward to join with a short chitinous vagina, which opens at the middle of the epigastric furrow.

Associated with the vagina and uterus are two or more seminal receptacles and glands (Fig. 13–33). In most spiders, these have separate openings to the outside and are connected internally to the vagina. These external openings are for the reception of the male copulatory organ during mating and are located just in front of the epigastric furrow on a special sclerotized plate called the epigynum (Fig. 13–24*B*). Fertilization occurs at the time of egg laying, mating having taken place some time earlier. Sperm are stored in the seminal receptacles, or spermathecae.

The male reproductive system is relatively simple. The testes consist of two large tubes lying ventrally along each side of the abdomen. When the sperm leave the testes, they pass anteriorly into two convoluted sperm ducts, which extend downward and open to the outside through a single genital pore in the middle of the epigastric furrow.

The copulatory organs of the male are not

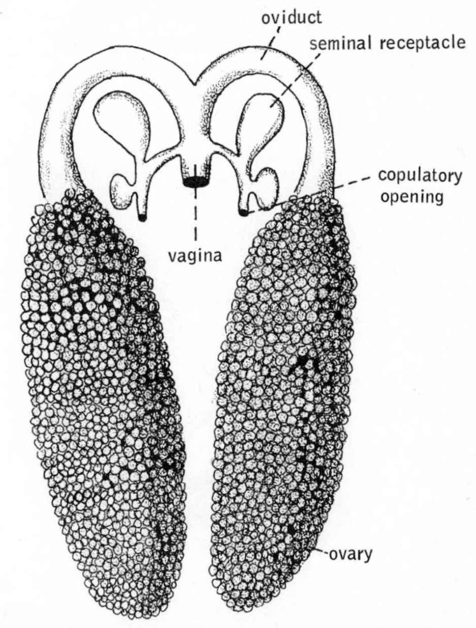

Figure 13–33. Reproductive system of female spider. (After Millot.)

connected to the sperm duct opening but are located at the ends of the pedipalps. The tarsal segment of these appendages has become modified to form a truly remarkable structure for the transmission of sperm. Basically, each palp consists of a bulblike reservoir from which extends an ejaculatory duct (Fig. 13–34*B* and *C*). This leads to a penis-like projection called the embolus. At rest, the bulb and embolus fit into a concavity, the alveolus, on one side of the tarsal segment.

During mating (Fig. 13–35*A*), the tarsal segment becomes engorged with blood, causing the bulb and embolus to be extended and project out of the alveolus (Fig. 13–

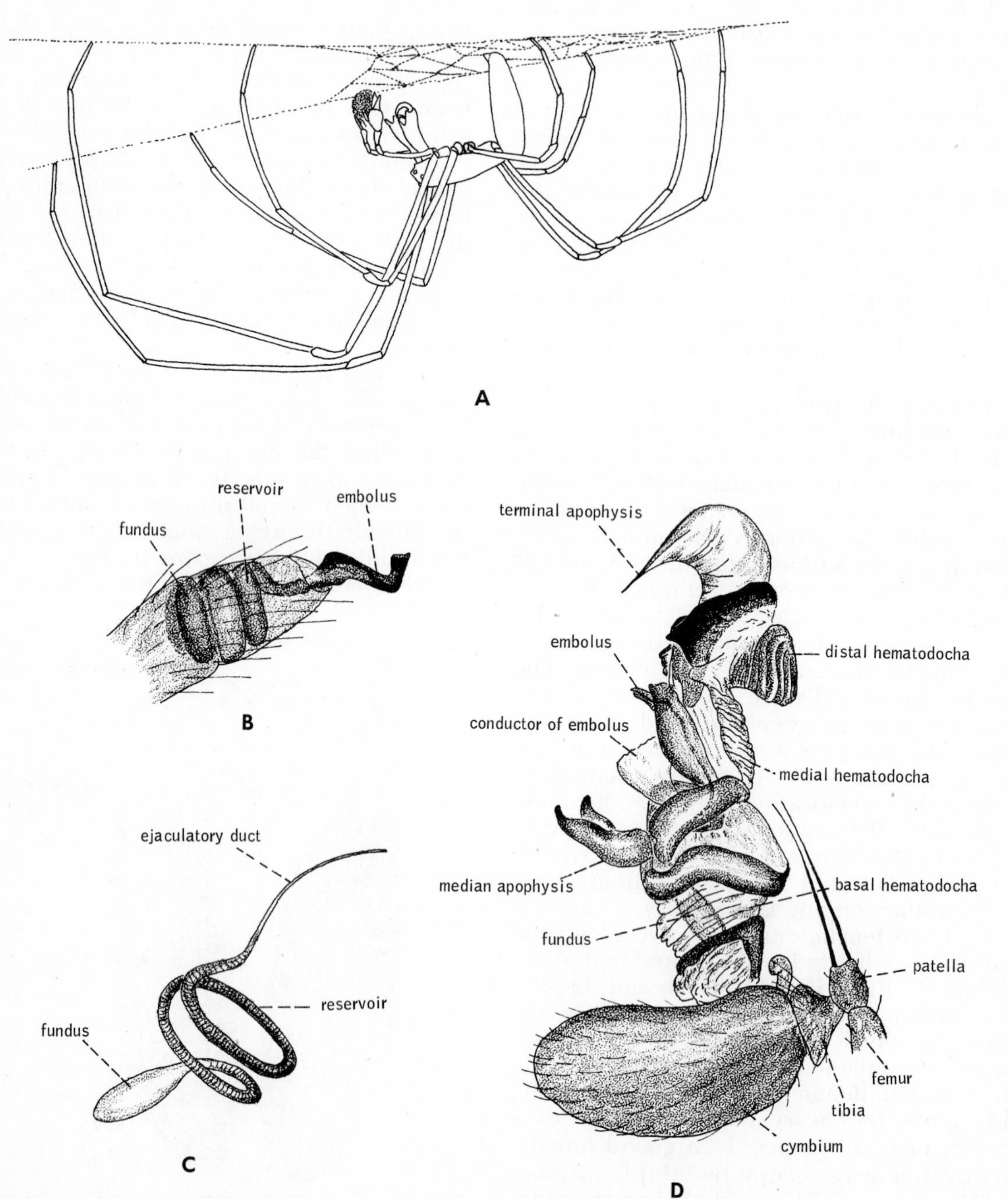

Figure 13–34. *A*. Male tetragnath spider in sperm web, filling palps from globule of semen. (After Gerhardt from Millot.) *B*. Simple palp of *Filistata hibernalis*. *C*. Receptaculum seminis, the semen-containing part of the male palp. *D*. Complex palp of the orb weaver, *Aranea frondosa*, in expanded state. (*B* to *D* after Comstock.)

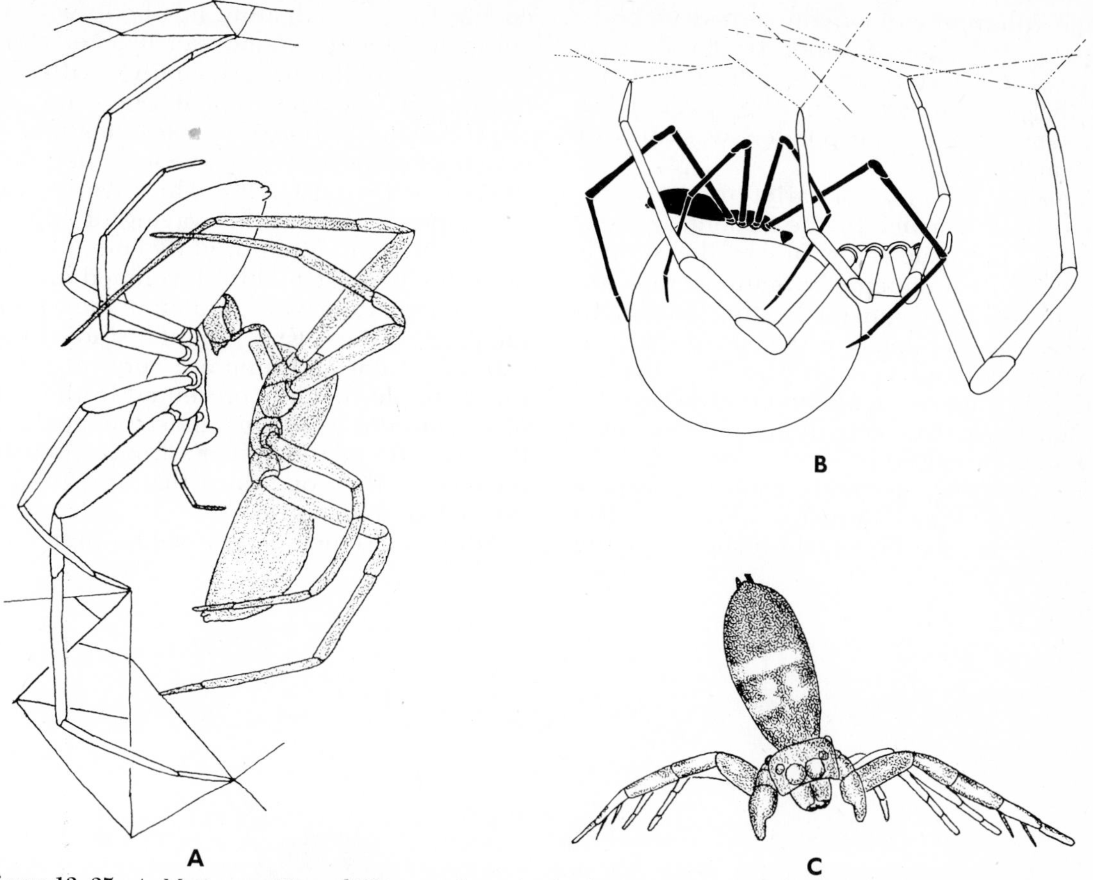

Figure 13–35. *A*. Mating position of *Chiracanthium* (male shaded). (After Gerhardt from Kaston.) *B*. Mating position of a black widow, *Latrodectus* (male black). (From Kaston, B. J., 1970: Trans. San Diego Soc. Nat. Hist., *16*(3): 33–82.) *C*. Courting posture of the male jumping spider, *Gertschia noxiosa*. (After Kaston.)

34*D*). The embolus is then inserted into one of the female reproductive openings leading to the seminal receptacles.

In groups of spiders believed to be primitive, the palpal organ has remained simple and consists only of the basic parts previously described (Fig. 13–34*B*). In the majority of families, however, the palp has become much more complicated with the addition of a great many accessory parts, such as the conductor, bulbal apophyses, and various other processes (Fig. 13–34*D*). These structures, in combination with the sclerotized configurations of the female epigynum, aid in the orientation of the palp and in the insertion of the embolus into the female openings.

The structure of the palp and the epigynum remains relatively constant within a species, while differing from those of other species. As a result, these organs are the primary structures used by araneologists for classifying and identifying species.

The female and the male reproductive systems and the male palpal organ are not completely formed until the last molt has taken place. The male then fills, or charges, the pedipalps with sperms in the following manner. He first spins a tiny sperm web, or at least a strand of silk, with special silk glands that open onto the anterior ventral surface of the abdomen. Upon this web is ejaculated a globule of semen (Fig. 13–34*A*). Next the palps are dipped into the globule until all of the semen is taken up into the reservoirs. With the palps filled, the male spider then seeks a female with which to mate.

The predatory habits of spiders, as in other arachnids, make recognition of the sexual partner especially important, and highly complex precopulatory behavior patterns have evolved in many species. In all spiders, chemical and tactile cues are of primary importance. On encountering a dragline, a male spider can determine whether it was produced by a mature female of the same species. In certain wolf spiders, a substance (pheromone) on the female's body initiates the male courting response. Evidence of such a substance is indicated by the fact that a male will court the severed leg of a female or the evaporated washings of the female's body.

Females respond to a variety of cues produced by the male. In the sedentary web-builders, the male often plucks the strands of webbing, producing vibrations that are detected and recognized by the female. In some cases the male is so small that he clambers ignored over the body of the female (for example, in *Mastophora bisaccata*, the female averages 12 mm. and the male 1.85 mm. in length).

The cursorial, or hunting, forms display the most unique courtship behavior. The approach may be direct, the male pouncing on the female, palpating her body with his pedipalps and legs, and causing her to fall into an immobile state. In families with well-developed eyesight, visual cues are also important and courtship takes the form of dancing and posturing by the male in front of the female (Fig. 13–35*C*). This involves various movements and the waving of appendages, which are often brightly colored. Such behavior is most highly developed and has been most extensively studied in the colorful jumping spiders. Experimental studies indicate that color and form are very important. For example, models presented to the male of *Corythalia xanthopa* must be no larger than five times the spider's frontal area and no smaller than one third that area to elicit the courting response.

Adults of some species mate a number of

Figure 13–36. *A*. Black widow, *Latrodectus mactans*, with egg case. *B*. Female wolf spider carrying young. (Both by W. Van Riper, Colorado Museum of Natural History.)

times during their lifetime; others mate only once. The embolus of the male palp breaks off within the female duct in some spiders. There are also many species in which a plug forms, filling the openings into the seminal receptacles and preventing a second mating.

Some time after copulation, the female lays her eggs. Several to 3000 eggs are laid in one to several batches, depending on the species. Just before the deposition of eggs, the female spins a small sheet and the eggs are fertilized with stored sperm as they are deposited. A second sheet is then spun over the egg mass. The two sheets are sealed at the edges and usually are given an additional covering of silk so that they assume a spherical shape, the egg sac or cocoon (Fig. 13–36*A*). The cocoon is either attached to webbing (in web-building families), hidden in the spider's retreat, or attached to the spinnerets and dragged about after the mother (as in the wolf spiders and a few others). In many spiders the female dies after completing the cocoon; in others she may remain with the young for some time after hatching.

The spiderlings hatch inside the cocoon and remain there until they undergo the first molt. They usually emerge from the cocoon either in late summer and fall or in the following spring. The spiderlings of many species, as well as the adults of some small forms, use a form of transportation known as ballooning, which aids in the dispersal of the species. The little spider climbs to the top of nearby twigs or grass and releases a strand of silk; as soon as air currents are sufficient to produce a tug on the strand, the spider releases its hold on the plant and is carried away by the currents. Ballooning spiders have landed on ships several hundred miles off shore and have been collected from airplanes flying at 5000 feet. The female wolf spider carries her young on her back after they hatch (Fig. 13–36*B*). The spiderlings gradually disperse over a period of time.

Although some of the "tarantulas" (mygalomorph spiders) have been kept in captivity for as long as twenty-five years, most spiders live only one to two years. The numbers of molts required to reach sexual maturity varies, depending upon the size of the species. Large species undergo up to 15 molts before the final instar, while tiny species molt only a few times. Males molt fewer times than do females of the same species, and within a species there is always some variation in the number of instars before the final molt.

SYSTEMATIC RÉSUMÉ OF THE ORDER ARANEAE

Suborder Orthognatha (mygalomorphs). This suborder includes the "tarantulas" (also called "bird spiders" or "monkey spiders" in various parts of the world) and trapdoor spiders (Fig. 13–37). Although most stalk or lie in wait for their prey, there are some web builders. There are numerous representatives in North America. They are rather large and possess two pairs of book lungs and two pairs of coxal glands; the fang of the chelicera articulates in the same plane as the long axis of the body.

Suborder Labidognatha (araneomorphs). This is the largest suborder and includes most of the spiders. With few exceptions, only a single pair of book lungs is present, and in all cases the plane of articulation of the chelicerae is at right angles to the long axis of the body.

The araneomorph families can be roughly divided into web-building and cursorial forms. A description of a few of the larger and more common families follows.

Family Pholcidae. Small and often long-legged, resembling phalangids or daddy longlegs (Fig. 13–38*A*). Members spin small webs of tangled threads in sheltered recesses. Several species commonly live in houses and, with other web-building house dwellers, are responsible for cobwebs.

Family Theridiidae. A large family of tangled web (cobweb) builders known as comb-footed spiders because of the presence of a series of serrated spines on the fourth tarsus. These spines comb out a band of silk used for trussing up the prey. To this family belongs the black widow, *Latrodectus*, and also one of the most common house spiders, *Achaearanea tepidariorum* (Fig. 13–38*B*).

Family Araneidae. The orb-weaving spiders (Fig. 13–38*D*). Members of this family spin circular webs. Many species are of considerable size and brightly colored, such as the black and yellow *Argiope*, or "writing" spider.

Family Linyphiidae. The sheet-web spiders. Webs are horizontal silken sheets or bowls (Fig. 13–38*C*). The larger species construct their webs in vegetation. The great number of very tiny species that this family contains live in fallen leaves and humus.

Family Agelenidae. Funnel-web spiders. Although web builders, they are more closely related to certain hunting spider

A

B

Figure 13–37. Mygalomorph spiders. *A*. Trap-door spider capturing a beetle. (By W. Van Riper, Colorado Museum of Natural History.) *B*. "Tarantulas" from the southwestern United States. (From Buchsbaum, R., and Milne, L., 1960: Lower Animals: Living Invertebrates of the World. Doubleday and Co.)

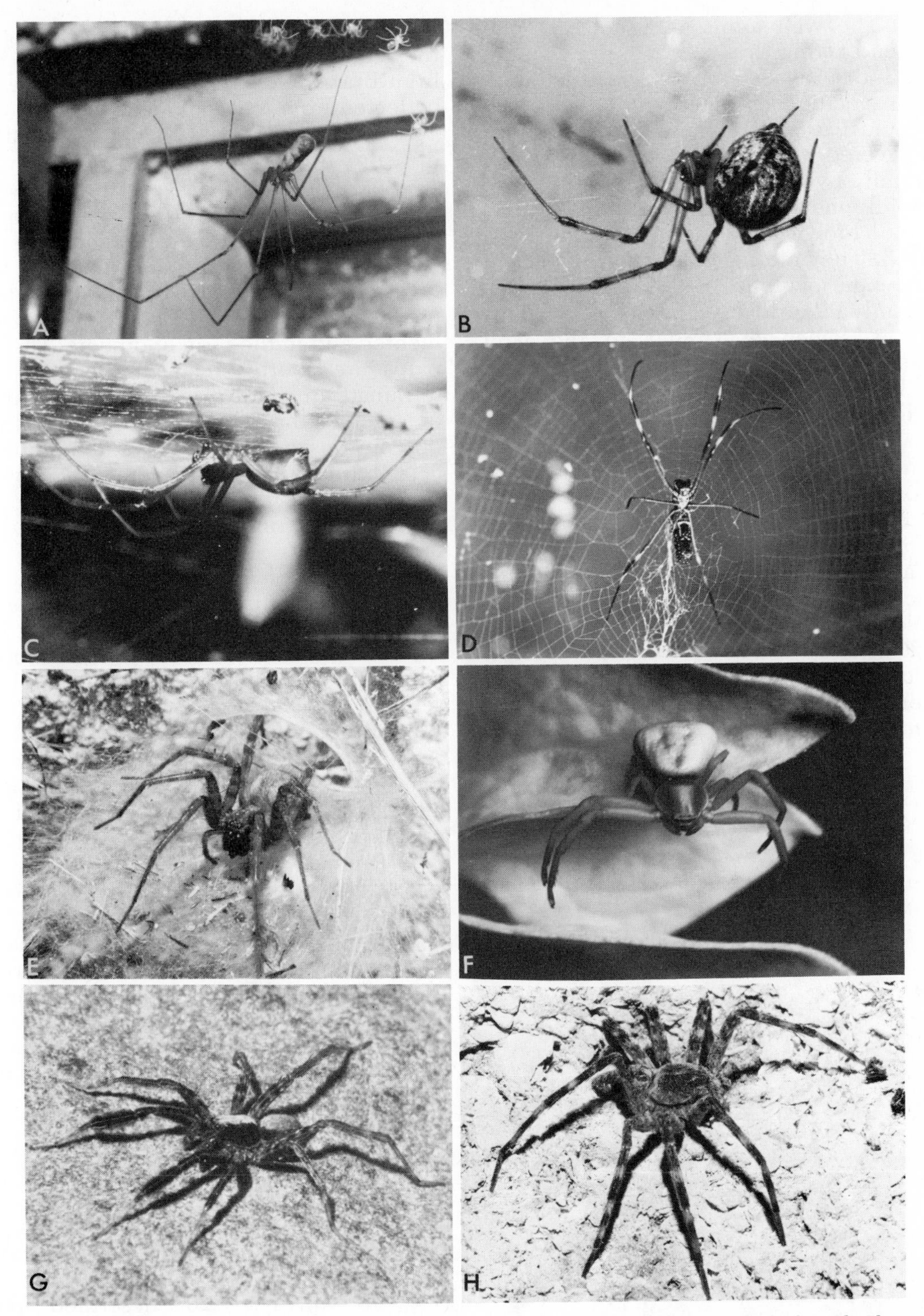

Figure 13–38. Representatives of some common families of araneomorph spiders. *A*. Pholcidae. This long-legged species, a member of the genus *Pholcus*, is a common house spider, building a web of irregularly placed threads (cobweb). Newly hatched young can be seen in the web at the top of the photograph. *B*. Theridiidae (comb-footed spiders). *Achaearanea*, a very common house spider. Web is an irregular network of threads. *C*. Linyphiidae (sheet-web spiders). The members of this family hang beneath horizontal sheetlike webs. *D*. Araneidae (orb-weavers). Some species of this tropical genus, *Nephila*, reach a very large size. *E*. Agelenidae (funnel-web spiders). Spider sits in narrow end of funnel. *F*. Thomisidae (crab spiders). This family of hunting spiders is common on flowers. *G*. Lycosidae (wolf spiders). *H*. Pisauridae (fisher spiders). Members of this family are commonly found along the margins of streams and ponds. (All by Betty M. Barnes.) (Members of the Salticidae, jumping spiders, are shown in Figs. 13–27 and 13–32*C*.)

groups. The web forms a funnel, the narrowed end acting as a retreat for the spider (Fig. 13–38*E*). The web is constructed in dense vegetation or in crevices of logs or rocks, and it is easily visible, especially in grass covered with dew.

Family Lycosidae. Wolf spiders. These are rapidly moving and rather hairy spiders with dull brown and black coloration (Fig. 13–38*G*). They are most active at night and are common members of the ground fauna.

Family Pisauridae. Fisher spiders. This family is somewhat similar to the wolf spiders in appearance but with longer legs (Fig. 13–38*H*). They are common around the edges of ponds, lakes, and streams.

Family Thomisidae. Crab spiders. This and the following family commonly lie in wait on vegetation. Crab spiders get their name from their crablike movements and the position of their legs (Fig. 13–38*F*).

Family Salticidae. Jumping spiders. Species of this family are heavy-bodied and capable of jumping short distances (Fig. 13–32*C*). They are often brightly colored and possess the best eyesight of all spiders. Numerous species live in the temperate and tropical regions of the world.

Order Ricinulei

The ricinuleids are a small order of arachnids containing 25 described species.

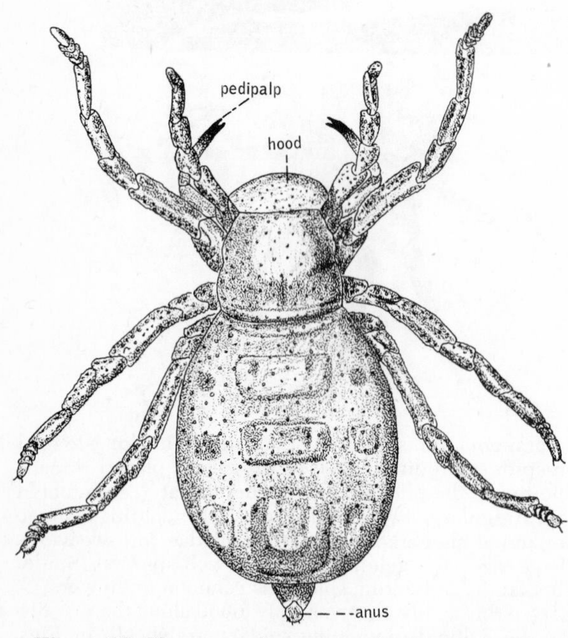

Figure 13–39. *Ricinoides*, a ricinuleid. (After Millot.)

They are found in Africa (*Ricinoides*) and in the American hemisphere (*Cryptocellus*) from Brazil to Texas, where they have been collected from leaf mold and caves.

Ricinuleids are heavy-bodied animals that measure from 5 to 10 mm. in length (Fig. 13–39). The cuticle is very thick and often sculptured. Attached to the anterior margin of the carapace is a curious hoodlike structure that can be raised and lowered. When lowered, the hood covers the mouth and chelicerae. The chelicerae are two-segmented and in the form of pincers. The pedipalps are shorter than the legs and also terminate in small pincers, but they are not enlarged as those in the scorpions and pseudoscorpions are. The legs are unmodified except in the male, and movement is slow and sluggish. The third pair of legs in the male have become modified to form copulatory organs.

The abdomen is narrowed in front, forming a pedicel attached to the prosoma, and behind, forming a short tubercle that bears the anus at the end. The nine segments composing the abdomen are fused and highly sclerotized.

Ricinuleids have been observed to feed on other small arthropods, and the digestive tract is more or less like that of other arachnids. The excretory organs consist of Malpighian tubules and a pair of coxal glands. The circulatory system is degenerate, and gas exchange takes place through tracheae. There are no eyes or trichobothria.

The third walking leg of the male has become modified for transferring a spermatophore. Little is known about egg laying, but a species of *Ricinoides* carries a single egg under the hood. The young, on hatching, is a six-legged "larva," similar to that of mites and ticks.

Order Opiliones

The order Opiliones, or Phalangida, contains the familiar long-legged arachnids known as daddy longlegs, or harvestmen (Fig. 13–40). The 3200 described species live in both temperate and tropical climates, and most prefer humid habitats. They are abundant in vegetation, on the forest floor, on tree trunks and fallen logs, in humus, and in caves. Many species are nocturnal, while others display diurnal activity.

The average body length is from 5 to 10 mm. exclusive of the legs, but some of the

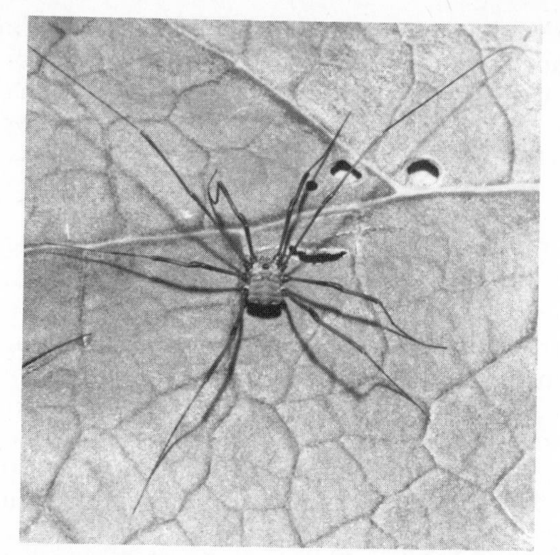

Figure 13–40. Dorsal view of a harvestman on a leaf. (By Betty M. Barnes.)

tropical giants reach 20 mm. and have a leg length of 160 mm. In contrast to these are certain tiny short-legged mitelike species that are never larger than 1 mm. in length.

The prosoma of opilionids is broadly joined to the short abdomen with no constriction between the two divisions (Fig. 13–41*A*). As a result, the body is rather elliptical in shape. In the center of the simple prosomal carapace is a tubercle of varying shape and size, with an eye located on each side of it. In some groups such as the genus *Caddo*, the tubercle extends almost the entire width of the carapace, so that the eyes appear along the sides of the carapace rather than in the middle.

Along the anterior lateral margins of the carapace are located the openings to a pair of scent, or repugnatorial, glands (Fig. 13–41*A*), which produce secretions, often quinones, having an acrid odor. Some harvestmen spray an intruder; certain species pick up a droplet of secretion and thrust it at a would-be predator.

The chelicerae are small, rather slender and triarticulate, the last two segments forming a pair of pincers. The pedipalps are usually somewhat leglike and similar to those of spiders (Fig. 13–41*A*). The legs in many opilionids are extremely long and slender, and exceed the body length many times (Fig. 13–40). The tarsus is always multisegmented with many articles, and as a result is very flexible. The normal gait of opilionids is often slow, but when disturbed they can run very rapidly. Self-amputation of a leg is an important means of defense against predators.

The abdomen is externally segmented. The sternites of the first two abdominal segments are fused together to form an anterior plate known as the operculum, which bears the genital openings (Fig. 13–41*B*).

In contrast to other arachnids, many har-

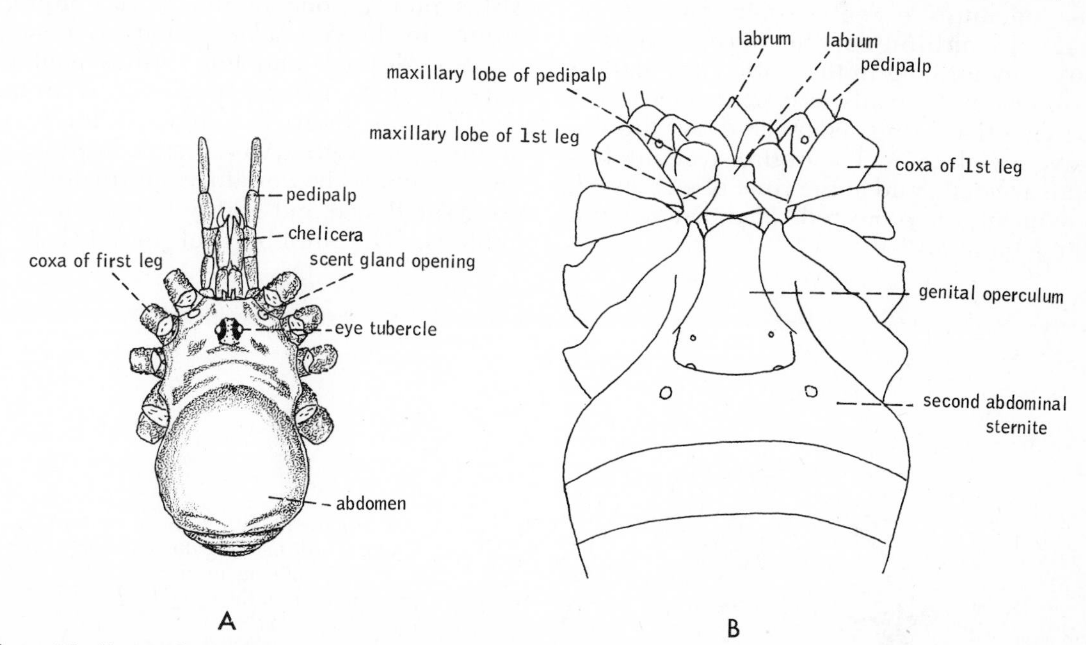

Figure 13–41. *A*. The opilionid, *Leiobunum flavum*, with only bases of legs shown (dorsal view). The abdomen of this species is not as conspicuously segmented as are many other opilionids. (After Bishop.) *B*. Ventral surface of *Phalangium*. (After Hansen and Sörensen from Berland.)

vestmen are omnivorous. North American and European opilionids have been observed feeding on small invertebrates, dead animal matter, and pieces of fruits and vegetables. Predatory species feed on other small arthropods, and some species feed on snails.

The prey or food is seized by the pedipalps and passed to the chelicerae, which hold and crush it. The coxal endites of the appendages are then used to pass food into the mouth. Suction is applied by the pharynx and, unlike other arachnids, the ingested food is not limited to liquid material but includes small particles. Thus, a greater part of digestion must take place in the midgut. These arachnids cannot tolerate prolonged periods of food and water deprivation, as can many spiders.

A pair of coxal glands, opening on each side between the third and fourth coxae, provide for excretion. Respiration takes place through tracheae, which are probably not homologous with the tracheae of other arachnids. The spiracles are located on each side of the second sternite, and in the active, long-legged harvestmen, there are secondary spiracles on the tibia of the legs. Sense organs consist of two direct eyes and also tactile hairs and lyriform organs, located on the prosoma and the appendages, especially the long second pair of legs.

The female reproductive system includes an ovipositor, a structure that is absent in other arachnids except certain mites (Fig. 13–42). In opilionids, the ovipositor is a tubular organ lying in the midventral part of the abdomen. Its walls in many species contain sclerotized rings connected by thin extensive membranes. The entire structure lies within a sheath, and at the time of egg laying the ovipositor is projected some distance out of the genital orifice.

Mating is not preceded by any distinct courtship, although males may fight other males for possession of a female. The male in many species faces the female and projects a tubular penis, which commonly passes between the female's chelicerae before entering the female orifice.

Shortly after mating, the female seeks a damp location for depositing her eggs. The long ovipositor is extended, and when the proper location is found, such as in humus, moss, rotten wood, and snail shells, the ovipositor penetrates the substratum to deposit the eggs (Fig. 13–42). The number of eggs laid at one time ranges in the hundreds, and several batches are laid during the life of a female. In temperate regions the life span is only one year. The individual may winter over in the egg or as an immature form, depending only upon the time of hatching. Parthenogenesis has been reported to occur in a number of species.

Order Acarina

The order Acarina, which contains the mites and ticks, is without question the most important of the arachnid orders from the standpoint of human economics. Numerous species are parasitic on man, his domesticated animals, and his crops, while others are destructive to food and other products. Mites rank as one of the most ubiquitous groups in the Animal Kingdom. Even polar regions, deserts, and hot springs contain a mite fauna. Terrestrial species are extremely abundant, particularly in moss, fallen leaves, humus, soil, rotten wood, and detritus, and their success is certainly correlated with their small size and ability to exploit microhabitats. The numbers of individuals are

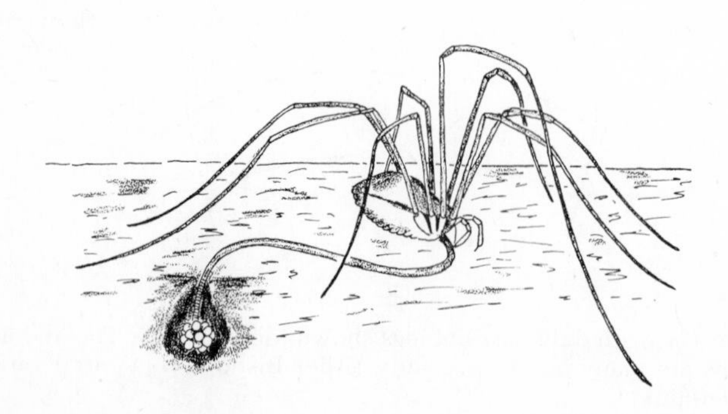

Figure 13–42. Female of the opilionid, *Leiobunum rotundum,* depositing eggs in ground. (After Henkin from Kaestner.)

enormous, certainly surpassing all other arachnid orders. A small sample of leaf mold from a forest floor often contains hundreds of individuals belonging to numerous species. Despite their abundance, the taxonomy and biology of mites are still not well known as compared to the other major arachnid orders.

Twenty-five thousand species have been described to date (compared to about 32,000 described species of spiders), but some acarologists believe that this number is only a fraction (20 per cent?) of the total and that most species of mites will become extinct through disappearance of rain forests and other habitats before they are ever recognized by man.

Because of the economic importance of mites, many zoologists direct their attention entirely to this group of arachnids; as a result, a special field has developed known as acarology, the study of mites.

Much of acarology belongs in the realm of parasitology, but the mites should not be considered as an entirely parasitic group. Many species are free-living, and others are parasitic only for a brief period during their life cycle.

Unfortunately, acarologists have developed a special nomenclature, largely borrowed from entomology, that tends to isolate the order rather than reflect its relationship to other arachnids. To avoid confusion, much of this terminology is avoided here and the group is treated in the same manner as the previous orders. This treatment affords a much better comparison with the other arachnids.

The Acarina are so morphologically diverse and display so few unifying features that acarologists consider the order to be polyphyletic. This makes a general discussion rather difficult, and it should be realized that exceptions exist for almost every statement.

External Morphology. The majority of adult mite species average 1 mm. or less in length, although many are larger. The ticks are the largest members of the order, some species reaching 3 cm. in length (Fig. 13–43).

The most striking characteristic is the apparent lack of body divisions (Fig. 13–44*A* and *B*). Abdominal segmentation has disappeared, and the abdomen has fused imperceptibly with the prosoma. Except in a few primitive forms, segmentation is secondary when present. Thus the positions of the appendages, the eyes, and the genital orifice

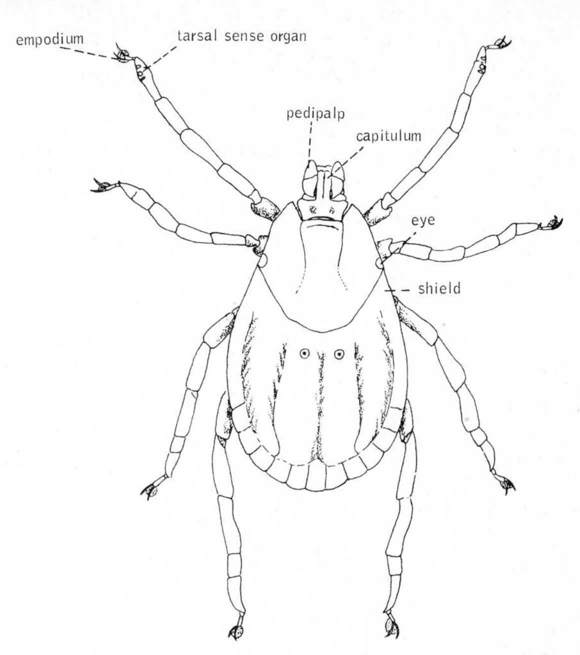

Figure 13–43. A tick, *Dermacentor variabilis.* (After Snodgrass.)

are the only landmarks that differentiate the original body regions. Coinciding with this fusion, the entire body has become covered with a single sclerotized shield, or carapace, in many forms.

Another general feature of the group is the change that has taken place in the head region carrying the mouth parts, this region being called the capitulum or gnathosoma (Fig. 13–45*A* and *B*). The dorsal body wall projects forward to form a rostrum, or tectum. Ventrally the large pedipalpal coxae extend forward to form the floor and sides of the prebuccal chamber. The roof of the chamber is formed by a labrum. These processes, which house the prebuccal chamber, together form a buccal cone, which fits into a sort of socket at the anterior of the mite body and which is roofed over by the rostrum. The chelicerae are attached to the back wall of the socket above the buccal cone, while the pedipalps are attached to both sides of the cone. The attachment of the buccal cone into the anterior socket is such that extension and retraction of the cone is possible in some species.

The chelicerae and pedipalps are variable in structure, depending on their function. They are usually composed of two or three segments and may terminate in pincers, or the distal piece may fold back on the base to form a crushing and grasping appendage (Fig. 13–45*B* and *E*). In some species of

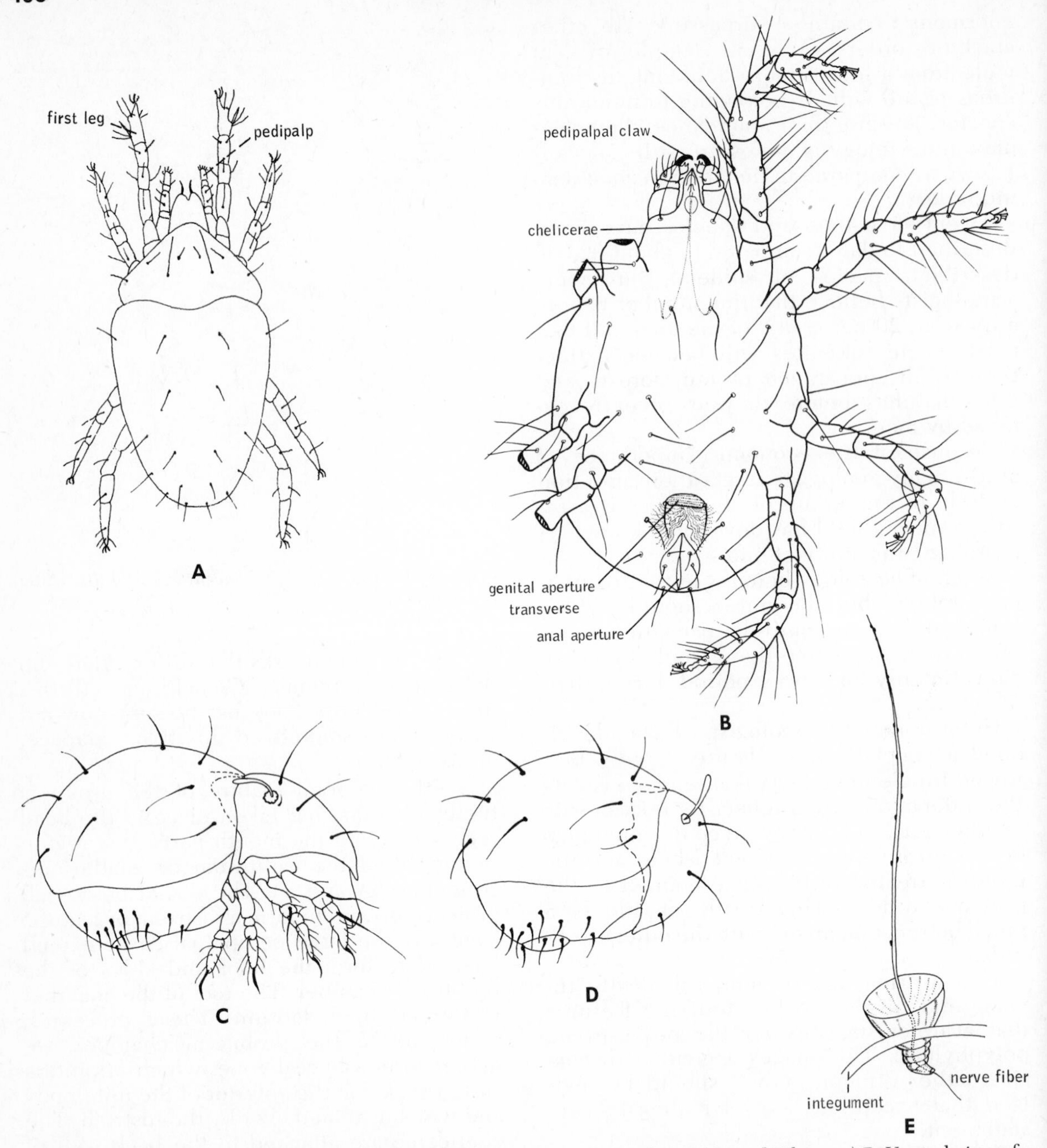

Figure 13–44. *A*. A trombidiform mite, *Tydeus starri* (dorsal view.) (After Baker and Wharton.) *B*. Ventral view of a spider mite, *Tetranychus*. (After G. W. Krantz.) *C* and *D*. Side view of the oribatid mite *Mesoplophora*, (*C*) before and (*D*) after "closing up" the body. The antenna-like structure on the cephalothorax is the pseudostigmatic organ. (After Schaller.) *E*. Pseudostigmatic organs of the oribatid mite *Belba*. (After Beck.)

mites the chelicerae have become needle-like for piercing (Fig. 13–45*C*), and in ticks the fingers of the pincers are provided with large teeth for anchoring into the integument of the host (Fig. 13–45*D*). The pedipalps may be relatively unmodified and leglike; they may be heavy and chelate like an additional pair of chelicerae; or in some parasitic forms they may be vestigial.

In many forms the four pairs of legs are each six-segmented and are composed of a coxa, a trochanter, a femur, a genu, a tibia, and a tarsus, the last bearing a pair of claws. These appendages have become modified in some groups for purposes other than walking. In the swimming mites these appendages have long setae with which the mite paddles through water (Fig. 13–46*D*), while in some other mites certain legs have become adapted for jumping or clinging (Fig. 13–46*B*).

In mites and ticks, the ventral side of the body is covered by plates that vary in form and number, depending on the family (Fig. 13–44*B*). The genital plate, located between the last two pairs of legs, bears the genital orifice (Figs. 13–44*B* and 13–46*B*). This location indicates a forward migration of abdominal segments, as in the harvestmen.

Hairs, or setae, cover the mite body; they assume every conceivable shape and form, from simple hairs to pilose, club-shaped, and flattened types. Many are sensory. The nature and position of the setae in mites are extremely important diagnostic characteristics in the classification and identification of species. The symmetrical arrangement of the variously shaped setae and the ornate sculpturing of their cuticle make many mites beautiful and spectacular animals (Fig. 13–46*A*).

Mites compete with spiders in the variety of coloration. Most mites are varying shades of brown, but many display a wide range of hues, such as black, red, orange, green, or combinations of these colors.

Two groups of mites deserve special mention. The Oribatei (beetle mites) are a particularly striking group of free-living, terrestrial mites, which are very abundant in humus and moss. These little mites are usually globe-shaped, with the dorsal surface covered by a convex, highly sclerotized shield, so that they often resemble tiny beetles. The ornamentation of the cuticle and the symmetrical retention of shed skins

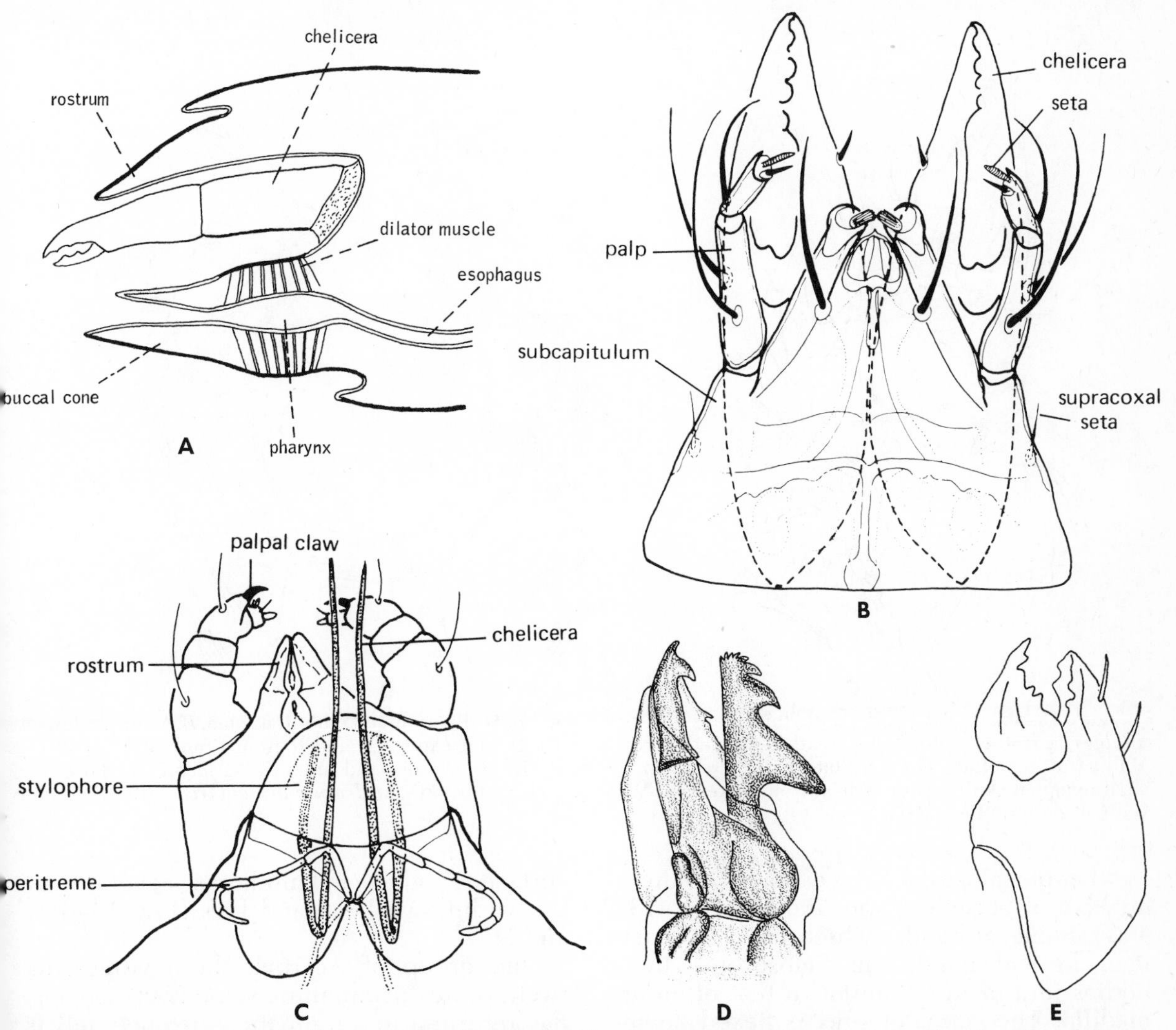

Figure 13–45. *A*. Head of a mite (sagittal section). (After Snodgrass.) *B*. Ventral view of head region of *Glycyphagus*. *C*. Stylet-like chelicerae of a spider mite, *Tetranychus*. (*B* and *C* from Krantz, G. W., 1971: Manual of Acarology. Oregon State Univ. Press, pp. 13 and 209.) *D*. Hooked chelicera of the tick, *Ixodes reduvius*. (After Neumann from André.) *E*. Chelate chelicera of *Belba verticillipes*.

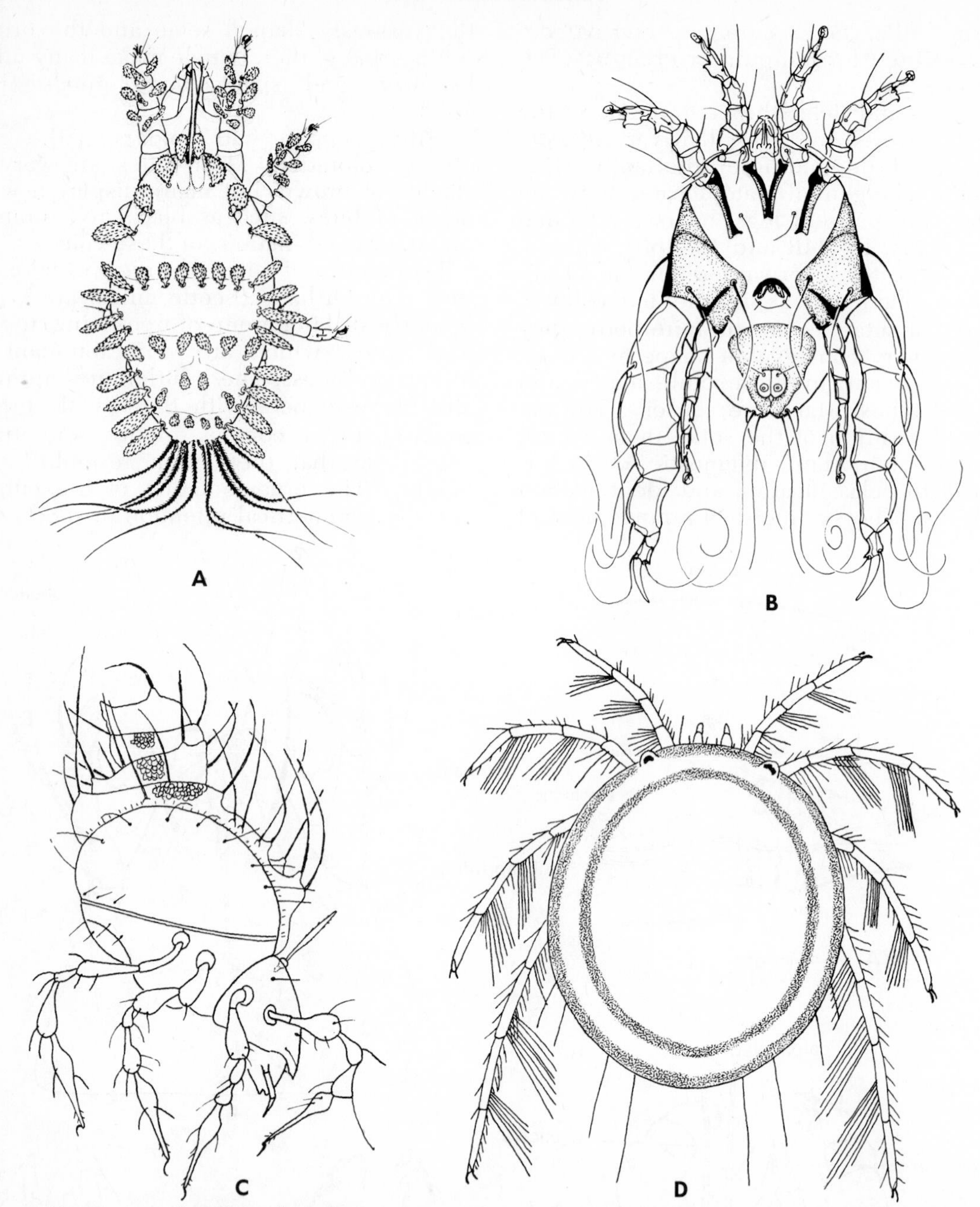

Figure 13–46. *A. Tuckerella*, a phytophagous mite, with symmetrically arranged club-like setae. *B*. Ventral view of *Analges*, a feather mite. Note greatly enlarged third pair of legs and claws. (*A* and *B* from Krantz, G. W., 1971: Manual of Acarology. Oregon State Univ. Press, pp. 211 and 275.) *C*. An oribatid mite, *Belba jacoti*, carrying five shed nymphal skins. (After Wilson from Baker and Wharton.) *D*. Water mite, *Mideopsis orbicularis*. (After Soar and Williamson from Pennak.)

on the dorsal surface give some of these mites a spectacular appearance (Fig. 13–46*C*). Some oribatids, which are very common in leaf mold, can "close up" their bodies in a manner similar to that of an armadillo. The cephalothorax is flexed downward and backward so that the legs and head fit into a concavity on the ventral surface of the abdomen. A plate carried on the dorsal surface of the cephalothorax covers over the withdrawn head and legs (Fig. 13–44*C* and *D*).

One group of Acarina, the Hydracarina (water mites), containing some 2800 species, has returned to an aquatic existence and is found in both fresh and salt water (Fig. 13–46*D*). The marine forms live largely in shallow water and are frequently encountered. They do not swim but crawl about over algae, bryozoans, hydroids, and sponges. Most water mites, however, are not marine but live in fresh water. They are often bright red, and many are active swimmers with long hairs on their legs. The larval stages may be parasitic on the gills of freshwater clams or aquatic insects.

Internal Anatomy and Physiology. Mites exhibit a tremendous diversity and specialization of diets and feeding habits. Carnivores which live in soil and humus feed on nematodes and small arthropods, including eggs, insect larvae, and other mites. Small crustaceans are the principal prey of water mites. The chelicerae of carnivorous mites are variously modified depending upon the prey. Some mites tear off pieces of prey; others suck out the tissues.

Many herbivorous species, such as spider mites (Tetranychidae), have chelicerae modified as needle-like stylets (Figs. 13–44*B* and 13–45*C*). The mites pierce plant cells and suck out the contents. A number of spider mites are serious agricultural pests of fruit trees, clover, alfalfa, cotton, and other crops. Spider mites construct protective webs from silk glands which open near the base of the chelicerae. The minute gall mites (Tetrapodili), which also feed on plant cells, have stylet-shaped chelicerae and include some forms which are agricultural pests (Fig. 13–47*D*). Both the spider mites and the gall mites could be considered plant parasites. Other herbivores include some fungus-feeding mites and some marine water mites that feed upon algae.

Many mites are carrion feeders or scaven-

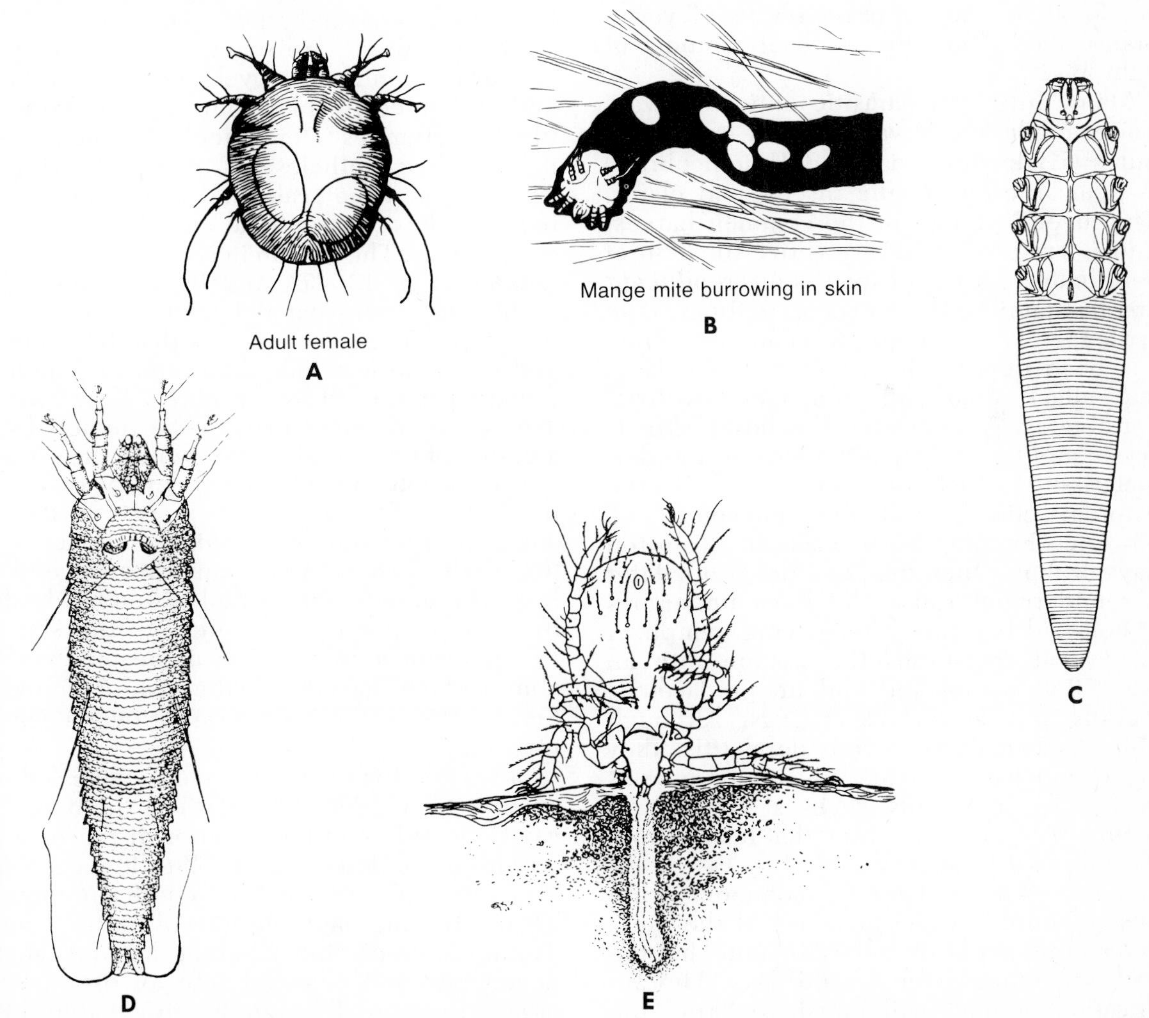

Figure 13–47. *A* and *B*. The mange mite, *Sarcoptes scabiei*. *A*. Adult female. *B*. Mite burrowing in skin and depositing eggs in tunnel. (Both after Craig and Faust.) *C*. Ventral view of a sheep hair follicle mite, *Demodex ovis* (0.23 mm.). (After Hirst from Kaestner.) *D*. Ventral view of a gall mite, *Eriophyes*, a minute plant parasite (0.2 mm.). (After Nalepa from Kaestner.) *E*. A chigger, larva of *Neotrombicula*, feeding on skin (0.25 mm.). (After Vitzthum from Kaestner. *C* to *E* from Kaestner, A., 1968: Invertebrate Zoology, Vol. 2, Wiley-Interscience, N.Y.)

gers. The soil-inhabiting oribatid mites feed on decomposing plant and animal material. A large number of "scavengers" have highly specialized diets. For example, different species of storage mites (Acaridae) and allied families feed on flour, dried fruit, mattress and upholstery stuffing, hay, and cheese. *Dermatophagoides* is commonly associated with house dust. The classification of scavengers should probably also include the feather mites (Fig. 13–46*B*) and some species which live on the fur of animals. They feed on oil, dead skin, and feather fragments, and are not actually parasites.

The majority of parasitic mites are ectoparasites of animals, both vertebrates and invertebrates, but other forms of parasitism exist. Some mites have become internal parasites through an invasion of the respiratory tracts of their hosts. There are species that are found in the air passageways of vertebrates and within the tracheal systems of arthropods.

Many mites are parasitic only as larvae. For example, the larval stages of freshwater mites (Hydrachnellidae and Unionicolidae) are parasitic on aquatic insects and clams. The juvenile stages of the common harvest mites (Trombiculinae) parasitize the skin of vertebrates. Larvae of species of *Trombicula* are the familiar chiggers, or redbugs. The six-legged larva emerges from an egg, which has been deposited on the soil. The larva may attack almost all groups of terrestrial vertebrates. In attacking, the host's skin is bitten and the young mite feeds on the dermal tissue, which is broken down by the external action of proteolytic enzymes (Fig. 13–47*E*). Feeding takes place for up to ten days or more; then the larva drops off. After a semidormant stage, the larva undergoes a molt and becomes a free-living nymph. A later molt transforms the nymph into an adult. Both nymph and adult are predaceous, feeding largely on insect eggs. Although chiggers can cause severe dermatitis, they are of much greater medical importance as vectors for pathogens, such as Asian scrub typhus. The intense itching that results from the bite of a chigger is caused by the mite's oral secretions and not, as commonly supposed, simply by the presence of the mite. Scratching quickly removes the mite, but the irritation remains for several days. The application of fingernail polish to "kill" the mite is therefore of little use as a cure.

Many acarines are parasitic during their entire life cycle but are attached to the host only during periods of feeding. The dermanyssid mites of birds and mammals (red chicken and other fowl mites) and the ticks illustrate this type of life cycle. Ticks penetrate the skin of the host by means of the highly specialized hooked mouth parts and feed on blood (Fig. 13–45*D*). The body is not highly sclerotized and is capable of great expansion when engorged with blood. This is especially true of the female. With a few exceptions, the tick drops off the host after each feeding and undergoes a molt. Many species can live for long periods, well over a year, between feedings. Copulation occurs while the adults are on the host feeding, and the female after feeding drops to the ground and deposits an egg mass. A six-legged "seed" tick hatches from the egg.

Ticks attack all groups of terrestrial vertebrates. In man they are responsible for the transmission of Rocky Mountain spotted fever, tularemia, and other pathogens.

Finally, there are parasitic mites that spend their entire life cycle attached to the host. Included in this group are the wormlike follicle mites (Demodicidae), which live in the hair follicles of mammals (Fig. 13–47*C*), and the scab- and mange-producing fur mites (Psoroptidae and Sarcoptidae) of mammals. The human itch mite (*Sarcoptes scabiei*) (Fig. 13–47*A* and *B*), the cause of scabies or seven-year itch, tunnels into the epidermis. The female is less than 0.50 mm. and the male less than 0.25 mm. in length. Irritation is caused by the mite's secretions. The female deposits eggs in the tunnels for a period of two months, after which she dies. Up to 25 eggs are deposited every two or three days. The eggs hatch in several days and the larvae follow the same existence as the adult. The infection can thus be endless. The mite is transmitted to another host by contact with infected areas of the skin. Itch mites have in recent years become very abundant in some cities in the eastern United States, after having been rare for over a century.

Most free-living mites have a typical digestive tract (Fig. 13–48). The prebuccal cavity, mouth, and the anterior of the pharynx lie in the buccal cone. With few exceptions, the pharynx is the chief pumping organ. In one large group of mites, the Trombidiformes, the hindgut has become degenerate and changed into an excretory organ. But the different feeding habits of others make it probable that there is considerable variation in the digestive physiology of the group.

Excretory organs of mites consist of one to

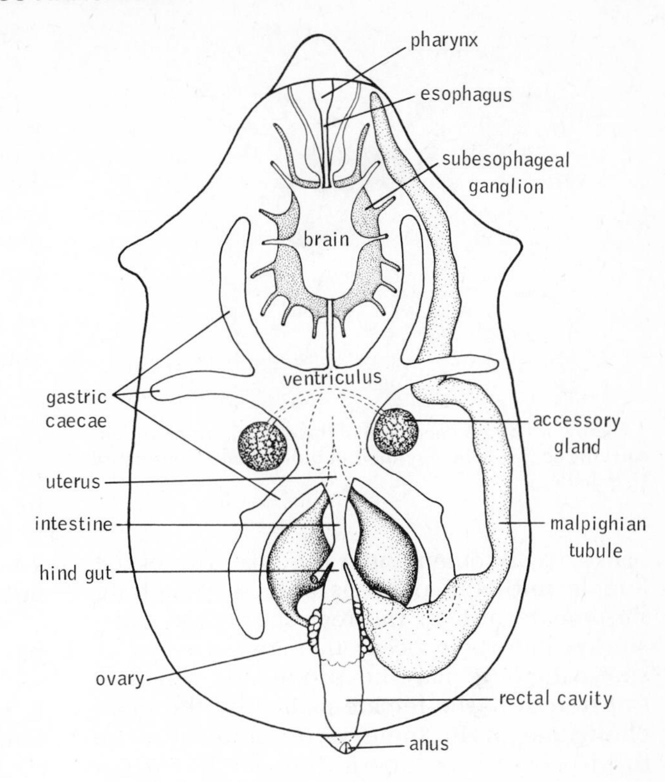

Figure 13–48. Internal anatomy of a mesostigmatid mite, *Caminella*. (After Ainscough from Krantz, G. W., 1971: Manual of Acarology. Oregon State Univ. Press.)

four pairs of coxal glands, or a pair of Malpighian tubules, or both (Fig. 13–48). In the trombidiform mites, these typical arachnid excretory organs are lacking and have been replaced by special excretory organs modified from the hindgut. The midgut of mites, like that of spiders, has taken on an excretory function.

The circulatory system is reduced and, except in a few groups, consists only of a network of sinuses. Circulation probably results from contraction of body muscles.

Although in some mites the gas exchange organs have completely disappeared, most mites have tracheae. The spiracles vary in number from one to four pairs, located on the anterior half of the body.

Sensory setae are probably the most important of the sense organs. In certain mites, the first pair of legs are especially well supplied with setae. These legs have lost their locomotor function and have become entirely sensory in nature. The oribatid mites possess a peculiar form of sensory seta called a pseudostigmatic organ, which is probably similar to a trichobothrium. The seta itself, which has various shapes, arises out of a cupule or pit. Two such pseudostigmatic organs are located on the cephalothorax (Fig. 13–44*C–E*). These setae are thought to detect air currents to which the mite responds by moving down deeper into the leaf mold. These organs would thus represent an adaptation against desiccation.

Although many mites are blind, some trombidiforms and certain other groups possess eyes. There may be one or two pairs or, in the oribatids, a single median eye. Some water mites have five eyes. Only rarely are the eyes well developed. Innervated pits and slits are common in mites and are perhaps similar to the slit sense organs of other arachnids.

Reproduction and Development. The male reproductive system consists of a pair of lobate testes, located in the midregion of the body, and these are often quite extensive. A vas deferens, sometimes supplied with accessory glands, extends from each testis. These may join ventrally to open through a median gonopore or through a chitinous penis that can project through the genital orifice.

In the female there is usually a single ovary of varying size, which is connected to the genital orifice by an oviduct (Fig. 13–48). A seminal receptacle and accessory glands are also present.

There is great variation in the mechanisms of sperm transfer. Sperm are transmitted directly in most mites. For example, in one group of mesostigmatid mites, the male

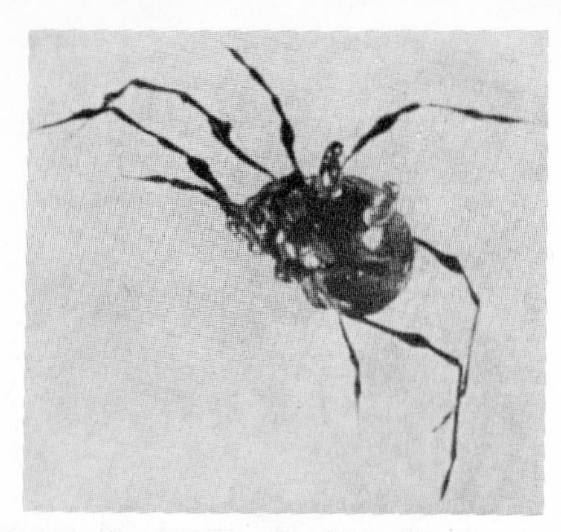

Figure 13–49. A specimen of the oribatid mite, *Belba*, carrying eggs attached by another individual. (Courtesy of F. Schaller.)

backs his abdomen under the posterior of the female until the penis is able to penetrate the female orifice. Indirect sperm transmission is known to occur in many species. A spermatophore may be produced, which is transferred to the female orifice by the male chelicerae or in some water mites by the third pair of legs. Sperm transfer in certain oribatids is remarkably similar to that in some pseudoscorpions. The male produces a stalked cupule containing a drop of semen. Interestingly, the spermatophores are deposited by means of a penis which is homologous to the female ovipositor. The sperm droplets are deposited on the substratum in large numbers in the absence of any females. When a female encounters a globule of sperm, she apparently is attracted chemotactically and with her open genital orifice, she removes the sperm droplet from its stalked cup. The number of eggs laid varies in different forms; they are deposited in soil and humus. The oribatid *Belba* attaches its large eggs to the bodies of other individuals, which carry them about until hatching (Fig. 13–49). A few species are ovoviviparous, and the oribatid mites possess an ovipositor.

After an incubation period of two to six weeks, a six-legged "larva" hatches from the egg. The newly hatched young differ from the adult in that they lack the fourth pair of legs and in certain other features. The fourth pair of legs is acquired after a molt, and the "larva" changes into a protonymph. Successive molts transform the protonymph into a deutonymph, a tritonymph, and finally an adult. During these stages, adult structures are gradually attained. Modifications of this typical life cycle are frequently encountered.

SYSTEMATIC RÉSUMÉ OF THE ORDER ACARINA

Suborder Onychopalpida. Primitive mites with typical claws on the pedipalps and more than a single pair of lateral spiracles. Two groups comprise the suborder, the notostigmatid and the holothyroid mites. The notostigmatids are brightly colored mites resembling harvestmen. They have segmented abdomens and chelate chelicerae. The holothyroids are represented by a single genus of large Indo-Pacific mites.

Suborder Mesostigmata. A large group of mites with a pair of tracheal spiracles located beside the coxae of the third pair of legs. Body covered with brown plates, which vary in number and position. Parasitic and free-living species. The Dermanyssidae are economically important parasites of birds (red mites) and mammals.

Suborder Ixodides. Large species of parasitic acarines known as ticks. Mouth with recurved teeth modified for piercing. A tracheal spiracle located behind third or fourth pair of coxae. Hard ticks or wood ticks (Ixodidae)—*Ixodes, Dermacentor;* soft ticks (Argasidae)—*Argas.*

Suborder Trombidiformes. Mites with a single pair of spiracles located anteriorly near the mouth parts. Chelicerae commonly adapted for piercing. This large suborder contains many plant parasites, including the so-called spider mites (Tetranychidae) and parasites on invertebrates and vertebrates. Among the latter are the harvest mites (Trombiculinae). There are free-living species, many of which are predaceous. Also included in the suborder are the follicle mites (Demodicidae), marine water mites (Halacaridae), and freshwater mites (Hydrachnellidae and Unionicolidae).

Suborder Sarcoptiformes. Tracheal system absent or, if present, without typical spiracular openings to the exterior. Chelicerae typically chelate. The free-living oribatid mites (Oribatei), called beetle or moss mites, are the largest group comprising the Sarcoptiformes, but also members of this suborder are the storage mites (Acaridae), the scabies, mange, or itch mites (Sarcoptidae and Psoroptidae), and the feather mites.

Suborder Tetrapodili. This suborder contains the gall mites, minute mites (less than 0.25 mm.) with elongate annulated bodies lacking the last two pairs of legs. They tunnel through plant tissues and feed on the contents of plant cells with their stylet-like chelicerae.

CLASS PYCNOGONIDA

The Pycnogonida, or Pantopoda, is a small group of some 600 species of marine animals known as sea spiders. The name sea spider is derived from the somewhat spider-like appearance of these animals, which crawl about slowly on long legs. Although largely unknown to the layman, sea spiders are actually common animals. Careful examination of bryozoans and hydroids scraped from a wharf piling or rocks will always yield a few specimens. They live in all oceans from the Arctic and Antarctic to the tropics, and there are many littoral forms, as well as species that live at great depths. The majority of species are found in cold waters.

The structure of the brain, the nature of the sense organs, and the presence of chelicerae have been considered by many zoologists as justification for placing the pycnogonids among the chelicerates. However, their exact relationship to arachnids and merostomes is by no means clear, for pycnogonids are aberrant in many respects. The presence of multiple gonopores, ovigerous legs, the segmented trunk, and the additional pairs of walking legs in many species have no counterpart in the other chelicerate classes. Hedgpeth believes that the peculiarities of pycnogonids are sufficiently great to warrant their being given equal rank with the chelicerates. Tiegs and Manton (1958) also believe that pycnogonids are rather distantly related to the other chelicerates or even that they may be an aberrant offshoot from some ancestral arthropodan group. Sharov (1966) goes so far as to suggest that the pycnogonids are the most primitive living arthropods, but it is unlikely that this view will receive very wide acceptance. Although the evolutionary affinities of pycno-

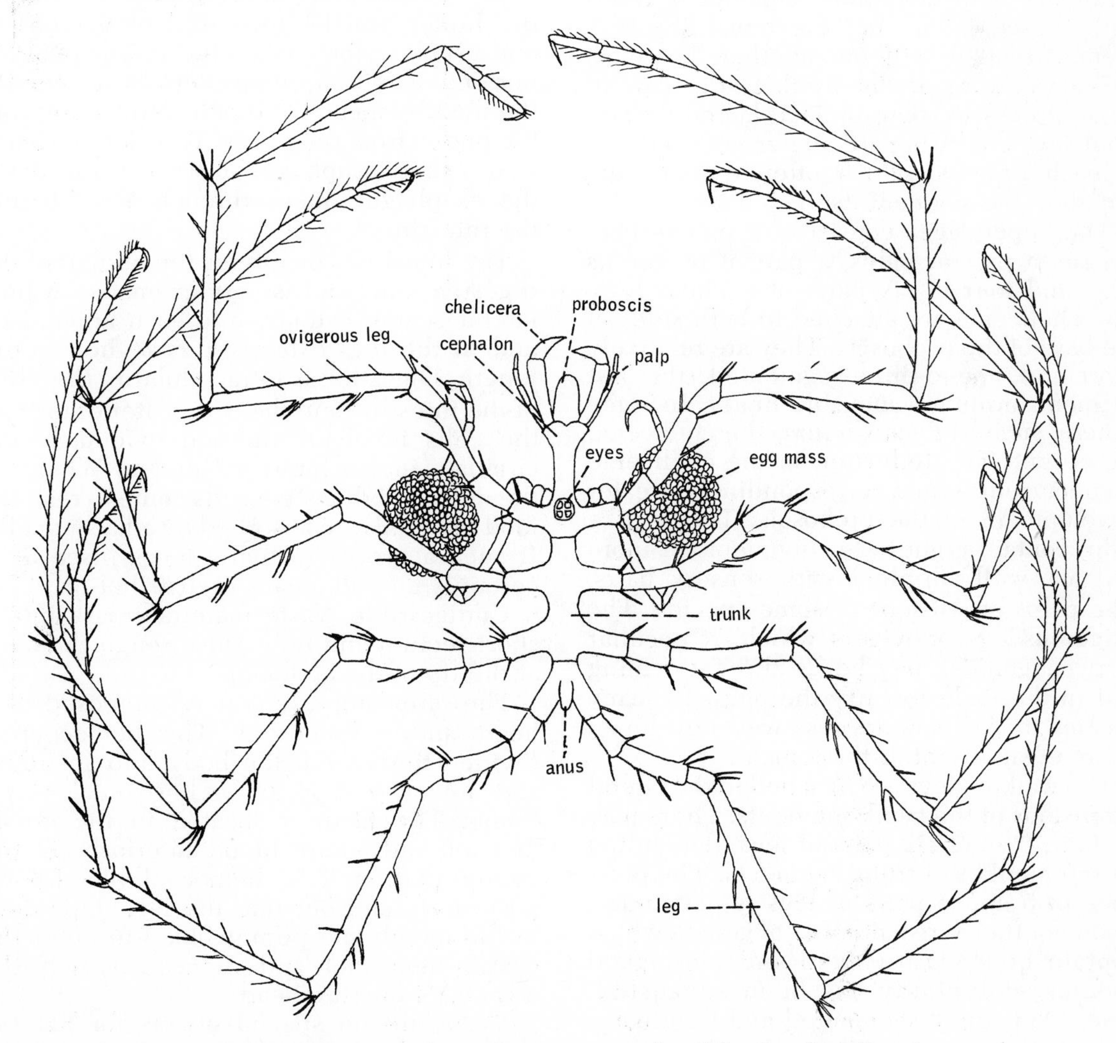

Figure 13–50. *Nymphon rubrum*, a sea spider. (After Sars from Fage.)

gonids are uncertain, there seems little doubt that they represent a very aberrant group of arthropods.

Pycnogonids are mostly small animals with a body length ranging from 1 to 10 mm., but specimens of the deep sea reach more than 6 cm. in body length and have a leg span of almost 75 cm. Although most pycnogonids are drab in color, there are some which are green, and some deep sea forms which are red.

The body is often long and narrow and is composed of a number of distinct segments (Fig. 13–50). The head, or cephalon, bears at the anterior end a cylindrical proboscis that is directed forward or, in some species, is curved ventrally. The posterior part of the cephalon is narrowed to form a short neck that bears on its dorsal surface four eyes mounted on a central tubercle. Posterior to the head is a trunk of four to six cylindrical segments. The first trunk segment is fused with the cephalon, but the remaining segments articulate with one another. The most striking feature of the trunk is the pair of large processes that project laterally from each segment. A leg articulates with the end of each process, which often exceeds the length of the segment itself.

The appendages consist of a pair of chelicerae, a pair of palps, a pair of ovigerous legs, and four to six pairs of walking legs. The chelicerae are attached to both sides of the base of the proboscis. They are relatively short and one-to-three-segmented, the last segment forming a movable finger. In some genera, such as *Pycnogonum*, the chelicerae are absent. The reduction in the chelicerae is concurrent with a correspondingly greater development of the proboscis. The leglike palpi contain eight to ten segments or more and are well supplied with sensory hairs. The palps are absent in some species. The ovigerous legs, or ovigers, which are peculiar to pycnogonids, may be used for grooming and in the male to carry the eggs. In many species the ovigers are less well developed or are even absent in the female.

The walking legs are attached to the lateral extensions of the trunk segments. There may be four, five, or six pairs of legs, depending on the number of trunk segments. The presence of 5 and 6 pairs of legs is not understood, for there are a number of genera which contain both eight-legged and ten-legged species. Polyploidy has been suggested. Each leg is eight-segmented and terminates in a claw. The legs are very long in pycnogonids with slender bodies but are shorter in the more compact species. There are no abdominal appendages.

Most pycnogonids are bottom dwellers and crawl about over hydroids and bryozoans (Fig. 18–6); some are able to swim, using vertical flapping movements of the legs, and have been picked up in plankton samples.

Pycnogonids are carnivorous and feed on hydroids, soft corals, anemones, bryozoans, and sponges. Some apply the proboscis directly to the prey and suck up the tissues; others wrench off the polyps with the chelicerae and pass them to the mouth at the tip of the proboscis. The mouth contains three liplike teeth that not only regulate the size of the opening but also act as a rasp. From the mouth a tubular pharynx extends through the proboscis (Fig. 13–51*A*). The wall of the pharynx is composed of three articulating sclerotized plates, resulting in a triangular lumen. By means of exterior muscles attached to the pharyngeal plates, the size of the lumen can be increased or decreased, and a sucking force is produced. The pharynx not only acts as a pump, but also macerates the food by means of bristles projecting into the lumen from the plates. The pharynx leads into a short esophagus located in the neck; the esophagus opens through a valve into the intestine.

The intestine constitutes the midgut of the digestive tract and is very extensive. A long lateral cecum extends into each appendage and in the legs extends almost their entire length. Digestion is intracellular in the walls of the intestine and the ceca. After digestion, the cells involved are said to detach and circulate in the intestinal lumen. Wherever the detached digestive cells come to rest, the food materials are absorbed by adjacent cells. If such a process actually takes place, it is certainly an odd means of distribution.

Undigestible waste materials pass into a short rectum and then out through the anus at the tip of the abdomen.

The circulatory system is composed of a heart and a hemocoel. The hemocoel is divided throughout the body into an upper and lower portion by a horizontal membrane. The heart is located in the dorsal division and pumps blood anteriorly. In the ventral portion of the hemocoel, blood flows posteriorly and out into the legs. Openings in the membrane permit blood to enter the dorsal hemocoel and thence through the ostia back into the heart.

There are no special organs for gas exchange and excretion.

The brain is located beneath the ocular

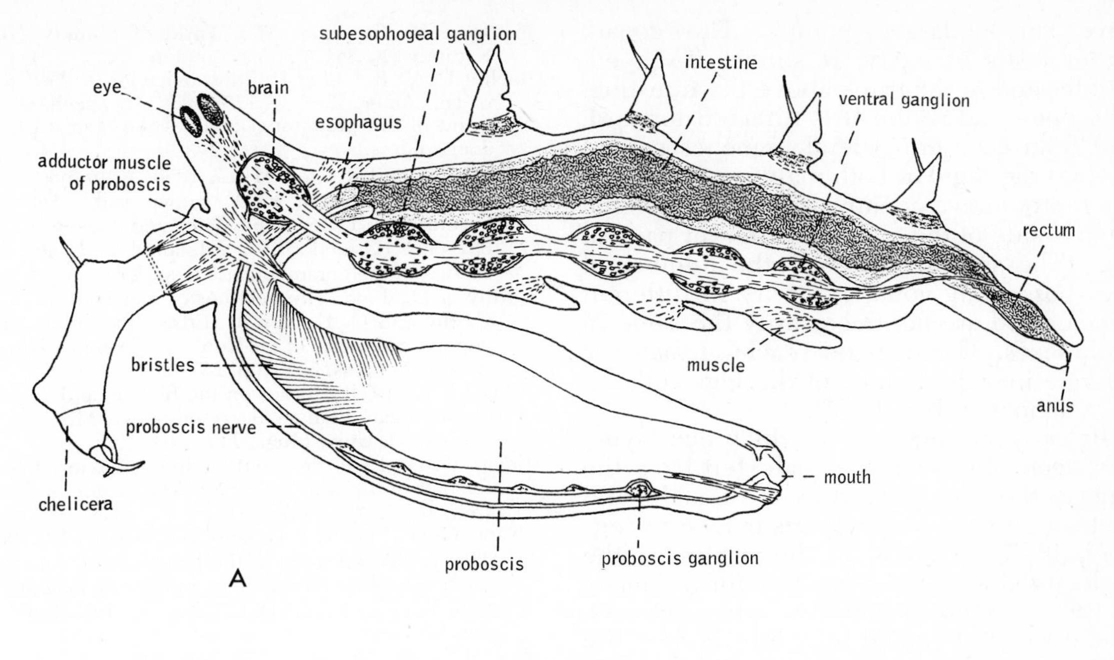

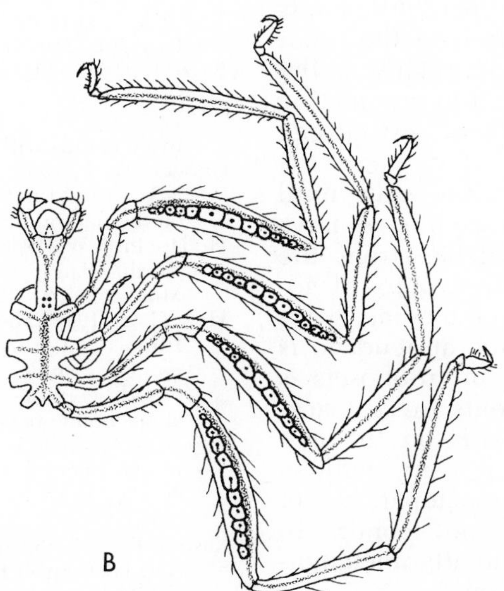

Figure 13–51. *A*. The pycnogonid, *Ascorhynchus castelli* (sagittal section). (After Dohrn from Fage.) *B*. Female of *Pallene brevirostris*, with eggs in femurs. (After Sars from Fage.)

tubercle and consists of a protocerebrum and tritocerebrum. A pair of ocular nerves from the protocerebrum and a pair of cheliceral nerves from the tritocerebrum provide for the innervation of the eyes and chelicerae. The lack of a deutocerebrum and the innervation of the chelicerae by the tritocerebrum has been considered evidence that pycnogonids are chelicerates.

The brain connects with a subesophageal ganglion, from which extend nerves for the palps and ovigerous legs. A pair of ventral nerve cords extend posteriorly from the subesophageal ganglion and bear a pair of fused ganglia for each trunk segment. A terminal nerve or pair of nerves from the last pair of ganglia innervates the abdomen.

The sense organs consist of sensory hairs and four eyes mounted on the dorsal tubercle.

Pycnogonids are dioecious. Females can be distinguished from males by the poorly developed condition of ovigerous legs or by their complete absence. Also, gravid females have very enlarged femurs. The gonad, either testis or ovary, is single, **U**-shaped, and located in the trunk above the intestine. The open end of the **U** is directed forward, and from each arm lateral branches extend far into the legs. In both males and females, the reproductive openings are located on the ventral side of the coxae of different pairs of legs. The specific legs and the number of legs possessing gonopores vary in different species and are not necessarily the same in both sexes. The eggs on reaching maturity migrate into the femurs of the legs containing gonopores (Fig. 13–51*B*).

In most pycnogonids in which egg laying has been observed, the male fertilizes the eggs as they are emitted by the female, and he then gathers them into his ovigerous legs (Fig. 13–50). Glands on the femurs of the male provide cement for forming as many as 1000 eggs into an adhesive spherical mass. The eggs are brooded by the male, and the egg masses are held around the middle joints of the ovigerous legs. Bristles on the inner side of the segments aid in retaining the eggs. The male is not limited to one mating during a season and may carry several egg clusters.

The eggs are brooded by the male until they hatch. In most pycnogonids, a larva called a protonymphon hatches from the egg. The larva has only three pairs of appendages, representing the chelicerae, palps and ovigerous legs, and each appendage is only three-segmented. A short proboscis is present, but the trunk segments are still lacking. The larva either remains on the ovigerous legs of the male or, in many shallow-water species, becomes an ectoparasite or endoparasite on hydroids and corals. In either case the larva eventually metamorphoses into a young pycnogonid.

Bibliography

Baker, E. W., and Wharton, G. W., 1952: An Introduction to Acarology. Macmillan, N.Y.

Balogh, J., 1972: The Oribatid Genera of the World. Akademiai Kiado, Budapest, Hungary. (Keys and descriptions of the 700 described genera of the soil-inhabiting beetle mites.)

Barth, F. G., and Libera, W., 1970: Ein Atlas der Spaltsinnesorgan von *Cupiennius salei.* Z. Morph. Tiere, *68*(4):343–369.

Bishop, S. C., 1949: The Phalangida of New York. Proc. Rochester Acad. Sci., *9*(3): 159–235. (A taxonomic study of New York harvestmen. Keys, figures, and bibliography of taxonomic papers included.)

Bonnet, P., 1945–1961: Bibliographia Araneorum. Toulouse.

Bristowe, W. S., 1958: The World of Spiders. New Naturalist Series, Collins, London.

Buchli, H. H. R., 1969: Hunting behavior in the Ctenizidae. Amer. Zool., *9*:175–193. (An excellent account of the predatory habits of the ctenizid trapdoor spiders.)

Buck, J. B., and Keister, M. L., 1950: Arthropoda: *Argiope aurantia. In* F. A. Brown (Ed.), Selected Invertebrate Types. John Wiley and Sons, N.Y., pp. 382–394. (Directions for a detailed study and dissection of the common garden spider.)

Carthy, J. D., 1968: The pectines of scorpions. *In* J. D. Carthy and G. E. Newell (Eds.), Invertebrate Receptors. Symposia of the Zool. Soc. London. No. 23. Academic Press, N.Y., pp. 251–261.

Edgar, A. L., 1971: Studies on the biology and ecology of Michigan Phalangida (Opiliones). Misc. Publ. Mus. Zool. Univ. Mich., *144*:1–64.

Edgar, W. D., 1970: Prey and feeding behavior of adult females of the wolf spider *Pardosa amentata.* Neth. Jour. Zool., *20*(4):487–491.

Evans, G. O., Sheals, J. G., and MacTarlane, D., 1961: The Terrestrial Acari of the British Isles. An introduction to their morphology, biology and classification. Vo. I. Introduction and biology. British Museum, London.

Fage, L., 1949: Classe des Merostomaces. *In* P. Grassé (Ed.), Traité de Zoologie, Vol. 6. Masson et Cie, Paris, pp. 219–262.

Fage, L., 1949: Classe des Pycnogonides. *In* P. Grassé (Ed.), Traité de Zoologie, Vol. 6. Masson et Cie, Paris, pp. 906–941. (A good general account of the pycnogonids with an extensive bibliography.)

Gertsch, W. J., 1949: American Spiders. Van Nostrand, N.Y. (An old but still useful semitechnical volume on the natural history of spiders.)

Hedgpeth, J. W., 1948: The Pycnogonida of the Western North Atlantic and the Caribbean. Proc. U.S. Nat. Mus., 97:157–342.

Hoff, C. C., 1949: The pseudoscorpions of Illinois. Bull. Illinois Nat. Surv., vol. 24, art. 4. (A good starting point for students interested in the taxonomy of pseudoscorpions. Keys, figures, and bibliography of taxonomic papers included.)

Kaestner, A., 1968: Invertebrate Zoology, Vol. 2. Wiley-Interscience, N.Y. 472 pp. (This volume is largely devoted to a very good general account of the arachnids.)

Kaston, B. J., 1948: Spiders of Connecticut. State Geol. Nat. Hist. Surv. Bull., *70*:874. (This work is essential for any American student interested in the taxonomy of spiders. Keys for most families and genera are provided; almost every species is illustrated. The major part of the Connecticut spider fauna is found throughout the eastern United States.)

Kaston, B. J., 1964: The evolution of spider webs. Amer. Zool., *4*:191–207.

Kaston, B. J., 1970: Comparative biology of American black widow spiders. Trans. San Diego Soc. Nat. Hist., *16*(3):33–82.

Kaston, B. J., 1972: How to Know the Spiders. 2nd edition. W. C. Brown, Dubuque, Iowa.

Krantz, G. W., 1970: A Manual of Acarology. Oregon State University Bookstores, Inc. 335 pp. (A very useful introduction to the techniques, use of keys and the literature necessary for the identification of ticks and mites.)

Levi, H. W., 1967: Adaptations of respiratory systems of spiders. Evolution, *21*(3):571–583.

Levi, H. W., and Levi, L. R., 1968: A Guide to Spiders

and Their Kin. A Golden Nature Guide. Golden Press, N.Y. (An excellent small semipopular guide for the identification of common spiders. Many colored figures and much information.)

Lockhead, J. H., 1950: Arthropoda: *Xiphosura polyphemus*. *In* F. A. Brown (Ed.), Selected Invertebrate Types. John Wiley and Sons, N.Y., pp. 360–381. (Directions for a detailed laboratory study and dissection of the Atlantic horseshoe crab.)

McCreve, J. D., 1969: Spider venoms: biochemical aspects. Amer. Zool., *9*:153–156.

Millot, J., Dawydoff, C., Vachon, M., Berland, L., André, M., and Waterlot, G., 1949: Classe des Arachnides. *In* P. Grassé (Ed.), Traité de Zoologie, Vol. 6. Masson et Cie, Paris, pp. 263–905.

Moore, R. C. (Ed.), 1955: Treatise on Invertebrate Paleontology. Part P, Arthropoda 2. Geological Society of America and University of Kansas Press. (This volume covers the chelicerates.)

Muma, M. J., 1970: A synoptic review of North American, Central American, and West Indian Solipugida. Arthropods of Florida, vol. 5. Contribution No. 154, Bureau of Entomology, Florida Dept. Agriculture and Consumer Services. 62 pp.

Newell, I. M., 1959: Acari. *In* W. T. Edmondson, H. B. Ward, and G. C. Whipple (Eds.), Freshwater Biology, 2nd edition. John Wiley and Sons, N.Y., pp. 1080–1116. (Keys and figures to the common water mites of the United States.)

Pennak, R. W., 1953: Freshwater Invertebrates of the United States. Ronald Press, N.Y., pp. 470–487. (A valuable introductory work on the taxonomy of freshwater Hydracarina. Keys, figures, and bibliography are included, as well as an introductory section on the biology of the group.)

Robinson, M. H., 1969: Predatory behavior of *Argiope argentata*. Amer. Zool., *9*:161–173.

Savory, T., 1964: Arachnida. Academic Press, N.Y. (A general treatment of the biology of arachnids.)

Schaller, F., 1968: Soil Animals. Univ. Michigan Press, Ann Arbor. 144 pp. (An excellent little volume on the biology of animals living in soil and humus.)

Sharov, A. G., 1966: Basic Arthropodan Stock. Pergamon Press, N.Y.

Snodgrass, R. E., 1948: The feeding organs of Arachnida, including mites and ticks. Smithsonian Misc. Collect., *110*(10):1–93.

Tiegs, O. W., and Manton, S. M., 1958: The evolution of the Arthropoda. Biol. Rev., *33*:255–337.

Turnbull, A. L., 1973: Ecology of true spiders. Ann. Rev. Entomol., *18*:305–848.

Vachon, M., 1949: See under Millot.

Weygoldt, P., 1969: The Biology of Pseudoscorpions. Harvard University Press, Cambridge. 145 pp.

Weygoldt, P., 1972: Geisselskorpione und Geisselspinnen (Uropygi und Amblypygi). Z. des. Kölner Zoo, *15*(3):95–107. (A good account of the biology of these two arachnid orders.)

Wilson, R. S., 1969: Control of drag-line spinning in certain spiders. Amer. Zool., *9*(1):103–111.

Witt, P. N., Reed, C. F., and Peakall, D. B., 1968: A Spider's Web. Springer-Verlag, N.Y. 107 pp. (An analysis of the regulatory mechanisms in webbuilding.)

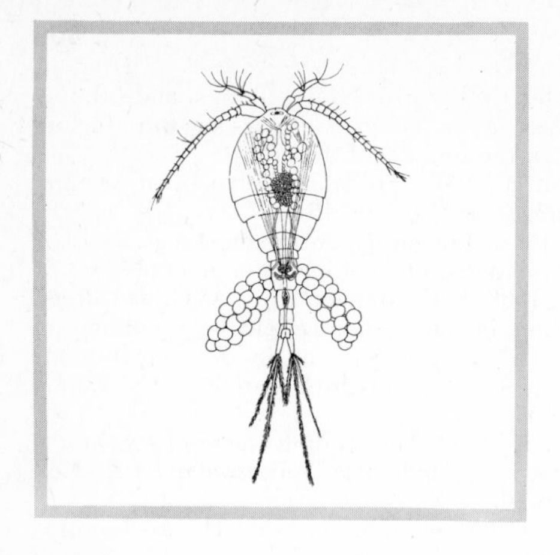

Chapter 14

THE CRUSTACEANS

The subphylum Mandibulata has commonly been used to contain all of those arthropods which possess mandibles as feeding appendages and antennae as sensory structures. It includes three major groups: the largely aquatic crustaceans and the terrestrial insects and myriapods (centipedes and millipedes). Most zoologists agree that the myriapods and insects evolved from some common ancestral group. Crustaceans, on the other hand, display many fundamental differences from the terrestrial mandibulates. As will be elaborated at the end of Chapter 16, there is growing acceptance of the view that the crustacean mandibles and the insect-myriapod mandibles reflect convergent evolution; i.e., these appendages have arisen independently in each group. The Mandibulata therefore probably does not constitute a natural assemblage, and the term *mandibulates* should be used as a grouping of convenience rather than in a taxonomic sense.

The 26,000 species of the class Crustacea include some of the most familiar arthropods, such as crabs, shrimp, lobsters, crayfish, and wood lice. In addition, there are myriads of tiny crustaceans that live in the seas, lakes, and ponds of the world and occupy a basic position in aquatic food chains.

The Crustacea is the only large class of arthropods that is primarily aquatic. Most crustaceans are marine, but there are many freshwater species. In addition, there are some semiterrestrial and terrestrial groups, but in general, the terrestrial crustaceans have never undergone any great adaptive evolution for life on land.

It is convenient to divide crustaceans into two groups, the entomostracans and the malacostracans. The entomostracans include the smaller species, such as fairy shrimp, water fleas, copepods, and barnacles. The malacostraceans include the larger, more familiar crustaceans, such as crabs, shrimp, and lobsters. The Malacostraca is an actual subclass of crustaceans, but the Entomostraca, containing four principal natural groups—the Cephalocarida, the Branchiopoda, the Ostracoda, and the Branchiura-Cirripedia-Copepoda—has no formal taxonomic significance.

External Anatomy. The head is more or less uniform throughout the class and bears five pairs of appendages (Figs. 14–1A and 14–31A). Anteriorly, there is a first pair of antennae, or antennules, which have been generally considered to be homologous to the antennae of the other mandibulate classes. They are innervated by the deutocerebrum and are probably not true segmental appendages. The first antennae are followed by the second antennae, often referred to simply as the antennae. The presence of two pairs of antennae constitutes a distinguishing feature of crustaceans. In no other group of antennate arthropods is there a second pair. Innervated by the tritocerebrum, the second antennae are postoral in origin and probably represent the appendages of the most anterior body segment, which has become incorporated in the head. As such, the antennae would be homologous to the arachnid chelicerae.

Flanking and often covering the ventral mouth are the third pair of appendages, the mandibles (Fig. 14–1A). These are usually short and heavy with opposing grinding and biting surfaces. Behind the mandibles are two pairs of accessory feeding appendages, the first and second maxillae. In front of and behind the mouth there are the variously developed upper and lower processes, or the labrum and the labium respectively. The labrum is developed from the body wall, and the labium is developed from that part of the foregut forming the posterior and ventral margin of the mouth.

The trunk is much less uniform than the head. Primitively, the trunk is composed of a series of many distinct and similar segments and a terminal telson bearing the anus at its base. However, only the fairy shrimp and the Cephalocarida approximate this condition (Fig. 14–9). In most crustaceans, the trunk segments are characterized by varying degrees of regional specialization, of reduction or restriction in number, of fusion, and of other modifications (Fig. 14–31*A*). Usually a thorax and an abdomen are present, but the number of segments they contain varies in a characteristic manner from group to group.

In many common crustaceans, the thorax, or anterior trunk segments, is covered by a dorsal carapace (Fig. 14–31*A*). The carapace generally arises as a posterior fold of the head and may be fused with a varying number of the tergites located behind it. Usually the lateral margins of the carapace overhang the sides of the body at least to some extent, and in extreme cases the carapace may completely enclose the entire body like the valves of a clam (Fig. 14–16).

Crustacean appendages are typically biramous (Fig. 14–2). There is a basal protopodite composed of two pieces—a coxopodite (coxa) and a basopodite (basis). To the basopodite is attached an inner branch (the endopodite) and an outer branch (the exopodite), each of which may be composed of one to many pieces. There are innumerable variations of this basic plan. Sometimes an appendage has lost one of the branches and has become secondarily uniramous. Often different parts of the appendage bear highly developed processes or extensions that have been given special names. Most frequently encountered are the terms *epipodite, exite,* and *endite,* referring to processes respectively borne by the coxopodite, exopodite, and endopodite.

Primitively, the appendages are numerous and similar and together perform a number of different functions. But, as in other arthropods, the evolutionary tendency in crustaceans has been toward a reduction in the number of appendages and a specialization of different appendages for particular functions. Thus, in a family of swimming crabs, whereas one pair of appendages is modified for swimming, others are modified for crawling, prehension, sperm transmission, eggbrooding, and food handling.

Integument. The cuticle of most large crustaceans, in contrast to that of most other arthropods, is usually calcified. Both the epicuticle and the procuticle contain depositions of calcium salts. The procuticle consists of three layers. The outer, thin, pigmented layer contains tanned proteins and is calcified. Beneath the pigmented layer is a calcified layer of untanned chitin that composes most of the procuticle. It does not contain the pigment typical of the layer above. The innermost layer of the procuticle is thin, untanned, and uncalcified.

Associated with the integument, but located beneath the hypodermis, are tegumental glands and chromatophores. The tegumental glands are composed of a cluster of secretory cells and a long duct that penetrates the exoskeleton and opens onto the surface of the epicuticle. The function of the tegumental glands is still uncertain. Chromatophores are especially characteristic of malacostracans and are discussed later in the chapter.

Locomotion. The ancestral crustaceans were probably small swimming epibenthic suspension feeding animals, and some modern forms have retained this primitive mode of existence. Propulsion in swimming is produced by the oarlike or propeller-like beating of certain appendages. The swimming appendages are usually supplied with fringed swimming setae that increase the water-resisting surface.

The majority of crustaceans have taken up a crawling habit. Although some swimming ability is often retained, certain appendages have usually become heavier and adapted to crawling and burrowing.

Nutrition. A great range of diets and feeding mechanisms are used by crustaceans. At least some representatives of almost every order are filtering suspension feeders, eating plankton and detritus. Filter feeding in the arthropods involves setae instead of cilia. Fine setae on certain appendages function as a filter for the collection of food particles (Fig. 14–3). The necessary water current is produced by the beating of the filtering appendages or, more commonly, by special appendages modified for this purpose. The collected particles are removed from the filter setae by special combing or brushing setae, and these particles are transported to the mouthparts by other appendages or sometimes in a ventral food groove.

There is usually little if any qualitative selection of food particles; instead, the size of the filtering net determines the nature of the

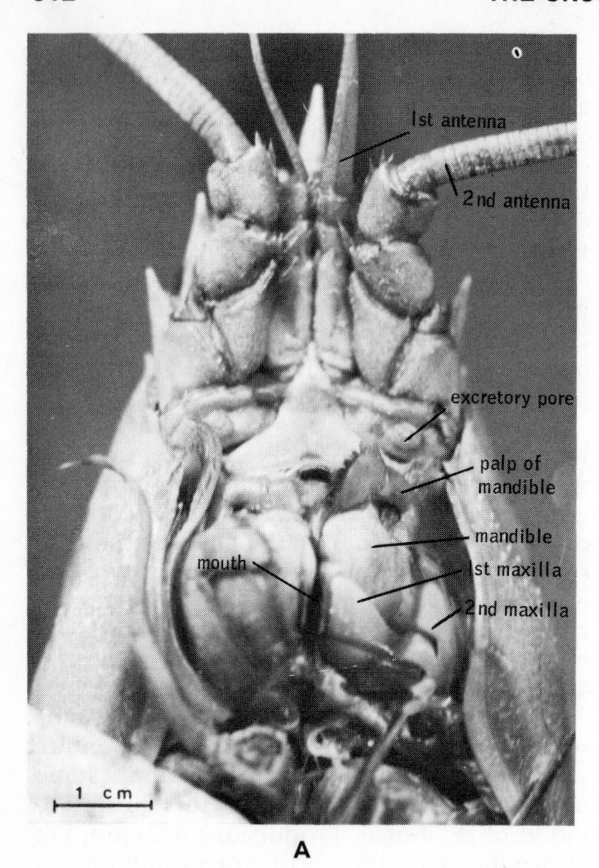

Figure 14–1. Crustacean structure. *A.* Head appendages of a lobster. (By Betty M. Barnes.) *B.* Internal structure of a crayfish (lateral view). *C.* Cross section of a crayfish just behind third pair of legs. (*B* and *C* after Howes.)

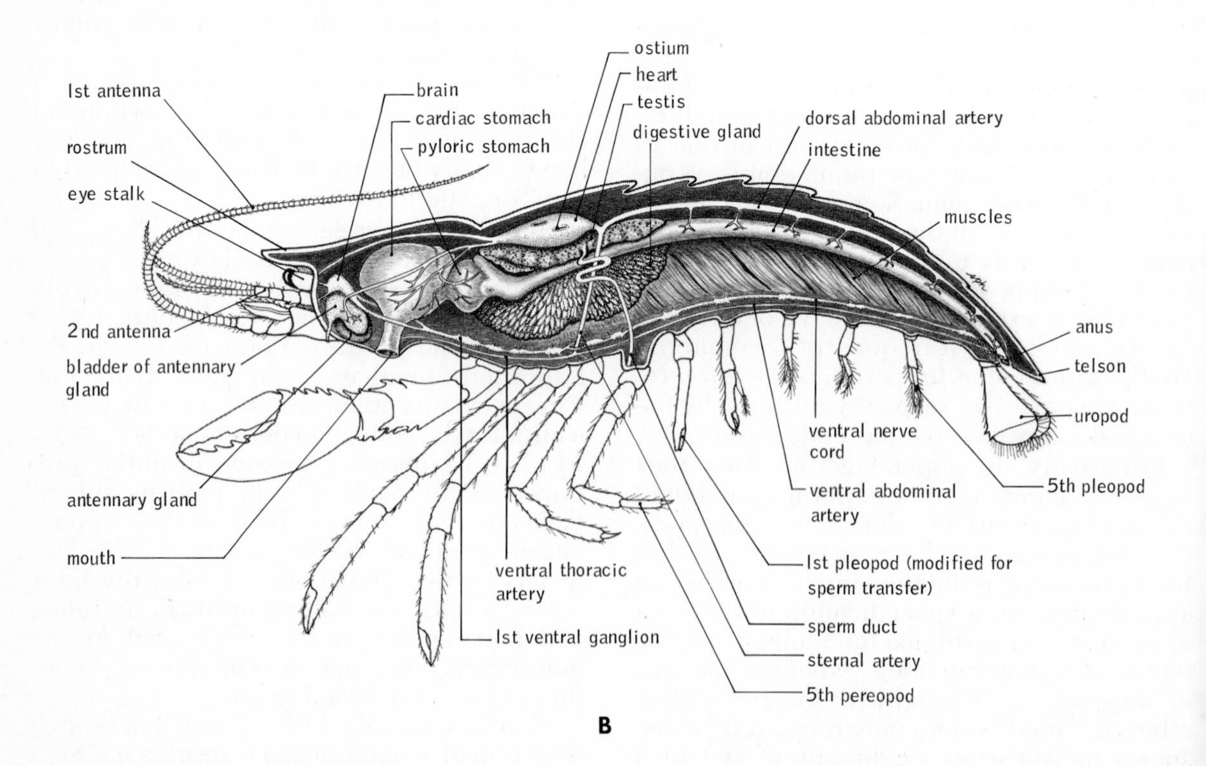

Illustration continued on opposite page.

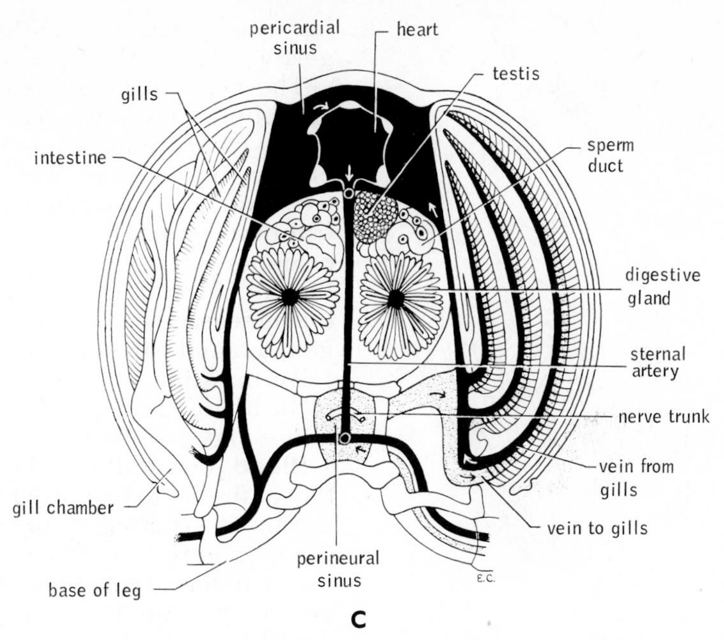

Figure 14–1. *Continued.*

material that is trapped (Fig. 14–3). Commonly, each filtering seta is provided with lateral rows of smaller bristles (setules), and the overlapping of adjacent setae can provide a very fine mesh. In smaller crustaceans, the setules may be as close together as four microns, permitting the collection of very small diatoms and other phytoplankton. Some filtering species have been cultured solely on a diet of bacteria. Some idea of the amount of food trapped and ingested by filter-feeding crustaceans is indicated by the studies of Marshall and Orr (1955) on the marine copepod, *Calanus finmarchicus.* Using radioactive diatom cultures, these zoologists reported that this little crustacean, which is about five millimeters long, may filter out and ingest from 11,000 to 373,000 diatoms every 24 hours, depending upon the size of the diatoms. Detritus is primarily used by benthic crustaceans, and some forms stir up the bottom sediment to increase the concentration of particles for filtering.

Filter feeding undoubtedly evolved independently a number of times within the Crustacea, and virtually every pair of appendages, even the antennae and mandibles, may be modified in one group or another for filter feeding. Filter feeding probably first arose in connection with swimming and is therefore primitively associated with the trunk, the same limbs creating both the swimming and feeding currents. The tendency in most groups has been for the filtering apparatus to develop forward, nearer the mouth, and to use only the anterior trunk appendages or the head appendages.

Other crustaceans are scavengers, herbivores, or carnivores. Either the anterior trunk or the thoracic appendages are adapted for grasping or for picking up food, and the maxillae and mandibles function in holding, biting, and maceration. Raptorial feeding is most highly developed in the larger crustaceans, and frequently certain appendages are modified for seizing or stunning prey.

Many crustacean species utilize more than

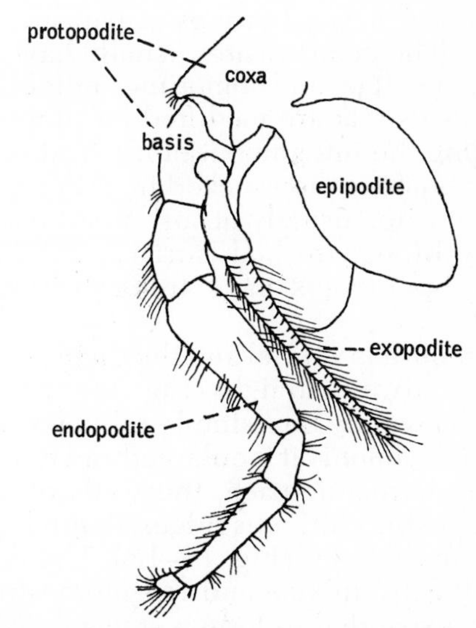

Figure 14–2. Right 5th pereopod, or leg, of the syncarid, *Anaspides tasmaniae,* showing basic structure of a crustacean appendage. (After Waterman and Chace.)

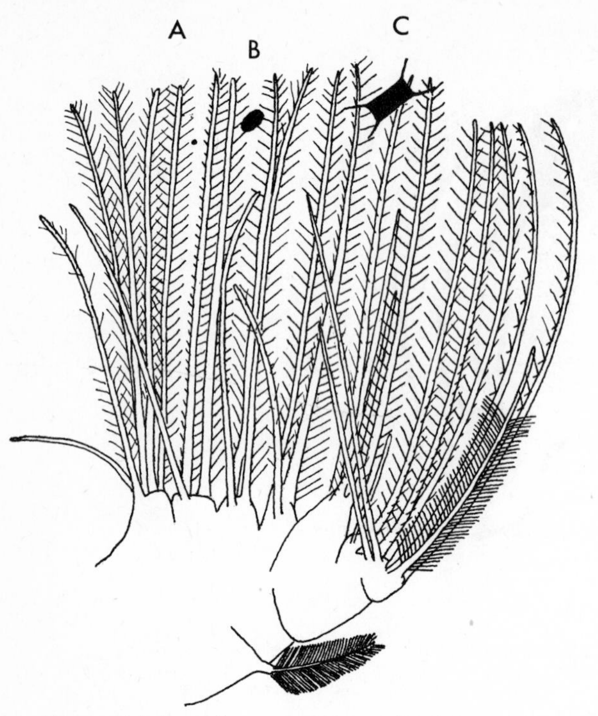

Figure 14–3. Filtering setae on the left maxilla of the copepod, *Calanus*. Three diatoms, representing species of different sizes, are drawn to scale to indicate the filtering ability of setae. Diatom *B* is approximately 25 microns long. The diatoms are: *A*, *Nannochloris oculata; B*, *Syracosphaera elongata;* and *C*, *Chaetoceros decipiens.* (After Dennell from Marshall and Orr.)

one method of feeding. For example, some scavengers supplement their diet with filter feeding or also feed upon living plants and animals; some filter feeders occasionally capture and consume small animals.

There are several groups of parasitic crustaceans. The ectoparasites usually have mandibles, maxillae, and sometimes neighboring appendages that are modified for piercing or lacerating the integument of the host. Tissue fluid or blood is then sucked from the wound. Endoparasites usually absorb food materials directly through the body surface, and in these forms appendages have frequently disappeared.

The mouth is ventral and the digestive tract is almost always straight (Fig. 14–12*A*). The foregut may only be a simple tubular esophagus, but commonly it is enlarged and functions as a triturating stomach, the walls of which bear opposing chitinous ridges, denticles, and calcareous ossicles (Fig. 14–1*B*). The midgut varies greatly in size and in entomostracans may be expanded to form a stomach. One to several pairs of ceca are almost always present. One pair of ceca, especially in malacostracans, has usually become modified to form large digestive glands (the hepatopancreas).

The hepatopancreas is composed of ducts and blind secretory tubules and is relatively solid. Its secretions are the primary source of digestive enzymes. The action of the digestive fluid takes place in the midgut and in the triturating stomach of the foregut when this chamber is present. Absorption is confined to the midgut walls and the tubules of the hepatopancreas, which also contains cells for glycogen, fat, and calcium storage.

An intestine, composed in part or entirely of the hindgut, comprises the posterior portion of the digestive tract.

Blood Circulation. The circulatory system is similar to that of chelicerates, although the heart, or arteries, or both are absent in some groups of entomostracans. The heart varies in form from a long tube to a spherical vesicle (Fig. 14–4). It is usually located in the dorsal part of the thorax, but when it is tubular, it may extend through the entire length of the trunk.

The number of ostia is greatest in the more primitive tubular heart; when the heart is a small vesicle, the ostia may be reduced to a single pair. The arteries are elastic but usually lack muscle fibers. Except in those groups that lack an arterial system, the heart always gives rise to a median anterior aorta. In addition, there may also be a median posterior abdominal artery, lateral arteries, and a ventral artery all leaving the heart.

In some small crustaceans, the arterial system is restricted to a tiny unbranched anterior artery; in other forms the principal arteries are large and branch extensively to supply various areas of the body (Fig. 14–1*B*). They may even branch into capillary networks, especially around the nerve cord and brain. Regardless of the development of the arterial system, the terminal vessels almost always empty into tissue spaces (lacunae) and thence into intercommunicating sinuses. The sinuses are bounded by membranes derived from adjacent connective tissue. Return blood flow to the pericardium and heart is generally through distinct sinus channels, which in many larger forms may assume the character of vessels.

Membranous partitions usually cut through the sinuses, resulting in more definite pathways and facilitating blood flow. For example, in appendages there is commonly a longitudinal septum dividing the appendage sinus into two channels. One channel receives the arterial flow, and the other drains into the pericardium.

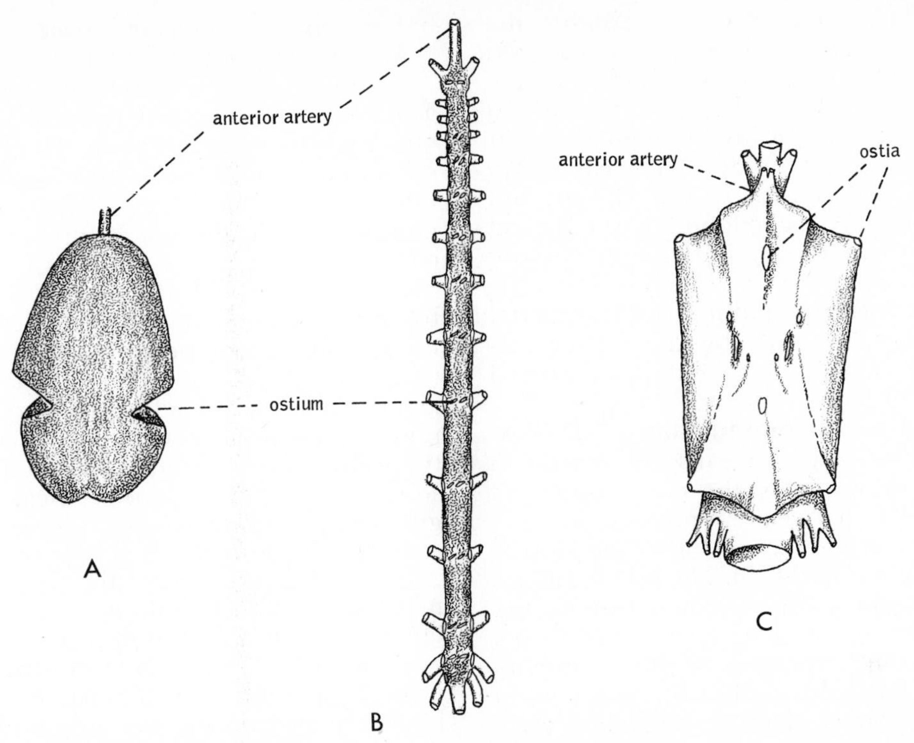

Figure 14–4. Crustacean hearts. All dorsal views. *A.* Copepod, *Calanus finmarchicus.* (After Lowe from Maynard.) *B.* Stomatopod, *Squilla mantis.* (After Alexandrowicz from Maynard.) *C.* Crayfish, *Astacus.* (After Baumann from Maynard.)

Accessory hearts, or blood pumps, that increase pressure are not uncommon and usually consist of a local arterial enlargement onto which muscles are inserted. Such an accessory heart, called the cor frontale, is found in most malacostracans and represents an enlargement of the anterior median artery in front of the triturating stomach. In small crustaceans, the beating of the appendages and the movement of internal organs, such as gut peristalsis, play an important role in augmenting blood flow. Blood flow is further regulated by valves that, aside from the afferent ostia, are most commonly present at the junction of heart and arteries; but arterial valves sometimes are also present, and crustaceans that lack hearts often have valves in the sinus channels.

The blood contains small hyaline and larger granular amebocytes, but the numerous variations and intergradations may indicate that these differences are all phases of a single type. The number of cells range from 1000 to over 50,000 per cubic centimeter of plasma, depending on the species, the age of the individual, the nutritional state, the proximity to ecdysis, and other physiological factors.

In addition to phagocytosis, the blood cells are also involved in clotting. Under certain irritating conditions such as amputation, certain amebocytes called explosive cells disintegrate and liberate a substance that converts plasma fibrinogen to fibrin. Islands of coagulated plasma then appear, to which other islands connect and in which blood cells are trapped, thus forming a clot. Other clotting mechanisms that do not involve explosive cells also may be present, but their presence has not as yet been definitely confirmed.

Gas Exchange. Gills are the usual gas exchange organs of crustaceans and are typically associated with the appendages. In their simplest and most primitive form, each gill is a platelike process (lamella) on the coxa of most trunk appendages (Fig. 14–10*B*). In malacostracans the gills are usually restricted to the thoracic appendages, and the surface area is often greatly increased by the development of filaments or lamellae arranged along a central axis (Fig. 14–65).

In many groups, particularly those that are very small, gills are absent, and gas exchange takes place over the general body surface. Integumentary gas exchange is probably of varying importance in all crustaceans.

The water current for ventilation is generally provided by the beating of the appendages. These appendages may be gill-bearing, or certain appendages may be modified for producing a ventilating current. Within the circulatory medium, oxygen is transported either in simple solution or bound to hemoglobin or hemocyanin. Hemocyanin is found only in the malacostracans, but both pigments have a sporadic distribution. The respiratory pigments are dissolved in the plasma, but hemoglobin may also be found in muscle and nervous tissue and even in the eggs of some species.

Modifications for amphibious and terrestrial gas exchange are considered in later discussions of groups that contain terrestrial species.

Excretion and Osmoregulation. The excretory organs of crustaceans are similar in structure and origin to the coxal glands of chelicerates. Crustacean excretory organs are paired and composed of an end sac and an excretory tubule (Fig. 14–5). The end sac arises from an anterior coelomic compartment, and the duct may represent a persisting coelomoduct. In adult crustaceans, the excretory organs are located in either the antennal or second maxillary segments, and are therefore called antennal or maxillary glands. The excretory pores of the antennal glands open on the underside of the bases of the second antennae, and those of the maxillary glands open onto or near the bases of the second maxillae. Both antennal and maxillary glands are commonly present in crustacean larvae, but usually only one pair persists in the adult. A few groups possess both types.

Although small amounts of urea and uric acid have been detected, the principal nitrogenous waste product is ammonia. This is true even in terrestrial forms, an indication of the rather limited terrestrial adaptation of these crustaceans. In addition to ammonia, there is considerable excretion of amines.

Excretion through the end sac takes place by filtration, but this organ in most crustaceans is not highly developed for waste elimination (Fig. 14–5*B*). If, for example, the antennal glands are removed or the excretory pores are plugged, the removal of ammonia is only partially hindered, and in some cases only slightly so. Experimental evidence indicates that the gills are important accessory organs for the excretion of ammonia.

In most crustaceans, the antennal and maxillary glands also do not play an important role in osmoregulation. Most crustaceans, even many freshwater and terrestrial species, produce a urine that is isosmotic with the blood. In such forms, the only ionic regulations effected by the excretory glands appear to be the conservation of potassium and calcium and the elimination of excess magnesium and sulfates. This is not true, however, of the freshwater crayfish, in which the antennal glands do elaborate a hypotonic urine. For most crustaceans, the gills are the chief

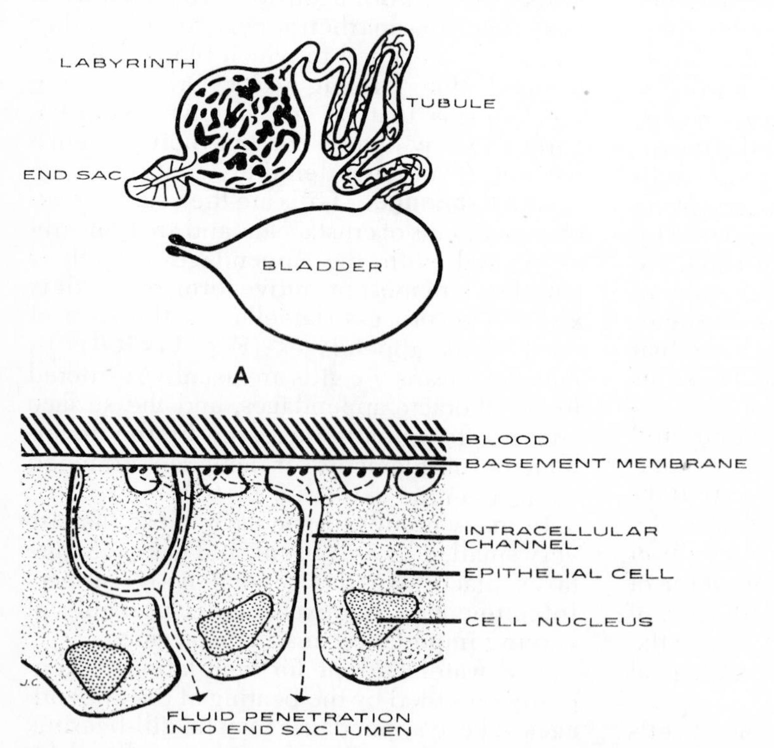

Figure 14–5. Antennal gland (excretory organ) of a decapod crustacean. *A*. Entire excretory organ. Nephridiopore opens onto base of second antenna. *B*. Diagrammatic section through end sac wall, showing site of filtration. (Redrawn from Kummel by A. P. M. Lockwood, 1967: Physiology of Crustacea. W. H. Freeman and Co., San Francisco.)

organs for maintaining salt balance. In fresh-water and brackish-water forms, the gills actively absorb salts.

Crustaceans, like most other arthropods, possess nephrocytes—cells capable of picking up and accumulating waste particles. Nephrocytes are most commonly located in the gill axes and in the bases of the legs.

Nervous System. As in most other arthropods, the nervous system of crustaceans displays the usual tendency toward concentration and fusion of ganglia. In the most primitive condition, the ventral cord displays a ladder-like configuration (Fig 14–6). Varying degrees of medial and longitudinal fusion have developed independently within most crustacean groups. In crayfish, for example, medial fusion has taken place, resulting in a single cord with single, rather than obviously paired, ganglia. In most crabs, all ventral ganglia have fused with the subesophageal ganglia to form a single mass.

Four pairs of major nerves arise from the brain —a pair to the antennules, a pair to the antennae, a pair to the compound eyes, and a pair of circumesophageal connectives to the ventral cord. In addition, there may be a pair of dorsal nerves to the head integument and a nerve to the median nauplius eye.

As in annelids, giant fibers are found in the central nervous system of many crustaceans. The giant fibers are particularly well developed in forms such as shrimp and crayfish, which can dart rapidly backward by a sudden flexion of the tail. Decapods, in which these fibers have been most studied, usually possess a pair of dorsomedian giant fibers arising from cell bodies in the brain and a pair of dorsolateral giant fibers.

The dorsolateral giant fibers are composed of a series of neurons. The cell body of each component nerve cell is located in a ventral ganglion, from which an axon runs dorsally and anteriorly to the next ganglion. Here the axon terminates in contact with the axon arising from the ganglion of that segment. Impulses for the median giant fibers arise in the brain, but impulses for the lateral fibers can arise from any ganglion, and transmission between members of the series can occur in any direction. A single impulse along either the medial or lateral fibers is sufficient to evoke the complete escape reflex.

Crustacean neuromuscular systems were described in Chapter 13 (p. 440). Lack of space does not permit discussion of the numerous studies on crustacean reflexes (see Wiersma, 1961, and Kennedy et al., 1969). It is interesting, however, that some of the more complex reflexes such as feeding and copulation occur even when the brain is removed. The brain apparently exercises an inhibiting and regulating control over these functions rather than being the center of initiation, and in this respect is rather like the annelid brain.

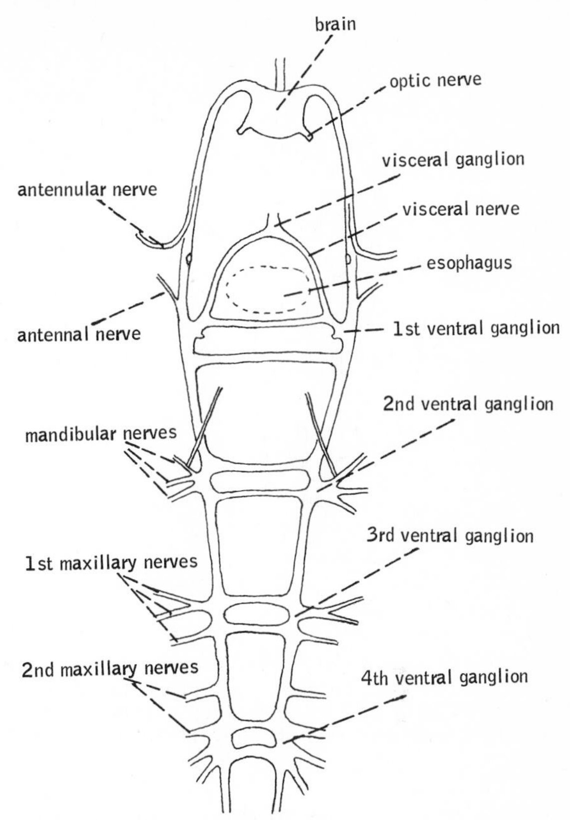

Figure 14–6. Anterior of nervous system of *Triops cancriformis*, a tadpole shrimp. (After Lankester and Pelseneer from Parker and Haswell.)

Sense Organs. The sense organs of crustaceans consist of eyes, statocysts, proprioceptors, general tactile receptors, and chemoreceptors. The eyes are of two types, median eyes and compound eyes. A median eye is a characteristic feature of the nauplius larva of crustaceans and is therefore often referred to as the nauplius eye (Figs. 14–8 and 14–17A). It may degenerate or persist in the adult. The median eye is composed of three or sometimes four small pigment-cup ocelli of the inverse type (Fig. 14–7*B*). The cups are located directly over the protocerebrum, where they either form a compact mass or are somewhat separated.

Each ocellus is composed of only a few photoreceptor cells, which possess rhabdome microvilli on their opposing surfaces; in most ocelli, no lens is present. A median nerve

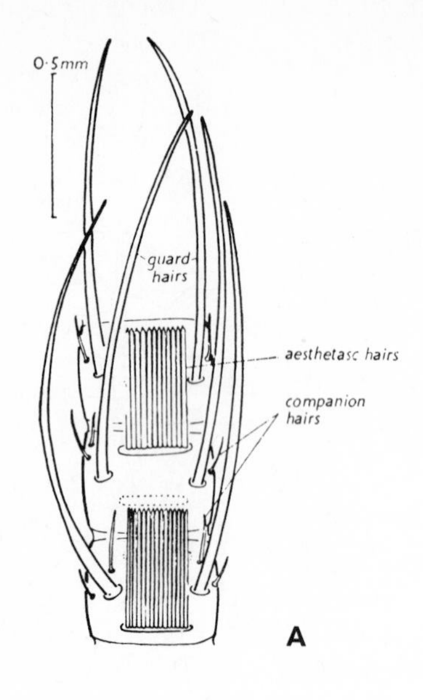

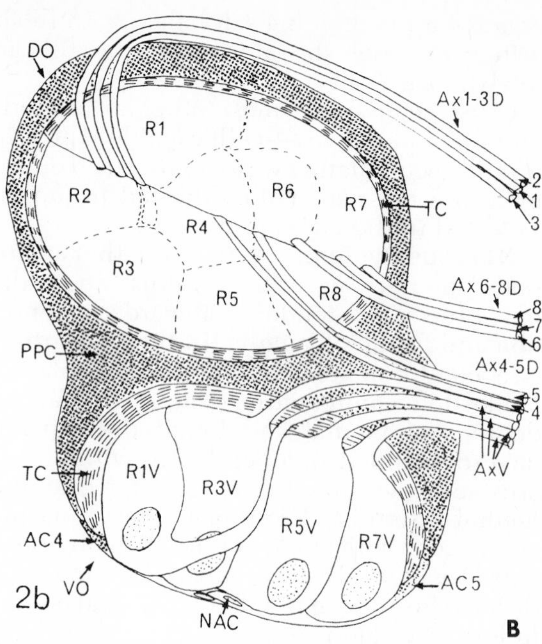

Figure 14–7. *A*. Rows of chemosensory hairs, called esthetascs, on the antenna of the lobster *Panulirus*. Only one of the two rows on each segment has been shown. (From Laverack, M. S., 1968: Oceanogr. Mar. Biol. Ann. Rev., 6:249–324.) *B*. Section through one ocellus of the nauplius eye of the copepod *Doropygus*. (See Figure 14–18*A* for position.) (From Dudley, P., 1969: La Cellule, 68(1):12.)

extends from the eye to the protocerebrum. Both the function and physiology of the median eye are obscure. The median eye is probably largely an organ of orientation, at least in planktonic larvae and in such groups as the copepods, which lack compound eyes. As such, the eye would enable the animal to determine the direction of the light source and thus enable the animal to locate the upper surface of the water or, in burrowing forms, to locate the surface of the substratum.

Two compound (lateral) eyes are found in the adults of most species. These compound eyes typically are well separated on each side of the head, but median fusion has taken place in some groups. The eyes may be located at the end of a usually movable stalk (peduncle) (Fig. 14–63*A*), or they may be sessile (Fig. 14–48). The total corneal surface is often greatly convex, resulting in a wide visual field. This is particularly true for stalked compound eyes, in which the cornea may cover an arc of 180 degrees or more (Fig. 14–69*A*).

To at least some extent, the compound eyes of many crustaceans are able to adapt by screening pigments to both bright and dim light. In general, however, they tend to be especially modified for functioning under one of these two conditions. Crustaceans that are diurnal and live in well lighted habitats, such as terrestrial and shallow-water species, usually possess apposition eyes with well developed screening pigment. Superposition eyes are found in nocturnal and deep-water crustaceans. Screening pigment is usually present but may be reduced or absent in cave-dwelling and bathypelagic species.

The number of ommatidia in crustacean compound eyes varies enormously. A single eye in the wood louse, *Armadillidium*, is composed of not more than 25 ommatidia, while the eye of the lobster, *Homarus*, may possess as many as 14,000 ommatidia. Studies on crustaceans with well developed compound eyes indicate some ability to distinguish in regard to form and size, but the degree of visual acuity is certainly slight and the total image crude compared to that formed by the human eye.

Figure 14–8. First nauplius larva of *Cyclops fuscus*, a copepod. (After Green.)

Color discrimination has been demonstrated in a number of crustaceans, and may be of general occurrence throughout the class. For example, the hermit crab, *Pagurus*, can discriminate between painted yellow and blue snail shells and shells colored different shades of grey. The chromatophores of the shrimp, *Crangon*, adapt to a background of yellow, orange, or red, but not to any shade of grey (the chromatophore changes are mediated through the eyes). The mechanism of color vision is still unknown, and in some entomostracans such as *Daphnia* it has not been definitely proved that color responses are actually mediated through the compound eye.

Statocysts are restricted to a few groups of Malacostraca. Only a single pair are present and are located in the base of the antennules or in the base of the abdomen, uropods, or telson. Each statocyst arises as an ectodermal invagination and usually retains an opening to the exterior. The statolith may be secreted, but commonly it is composed of a mass of agglutinated sand grains. The physiology of crustacean statocysts is best known in the decapods (shrimp, crayfish, lobsters, and crabs) and is discussed with these groups.

Special proprioceptive structures called muscle receptor organs are found in malacostracans, especially the decapods. Each organ consists of a specially modified muscle cell, two of which are located in the dorsal musculature of each abdominal segment and the last two thoracic segments, at least in the lobster, *Homarus*. The muscle cell displays a complex dual sensory innervation, which is stimulated by the stretching of the dendritic processes within the muscle cell. The processes may be stretched by the alternate contraction and extension of the muscle receptor organ itself or by the action of adjacent muscle cells.

Proprioceptors also are present in the appendages. Although these sensory receptors at present have been found only in the decapods, at least similar receptors will probably be found in other crustaceans as well. One type of sensory receptor, called an elastic receptor, is present in the joints of the walking legs and consists of a specially innervated strand of connective tisue that extends across the joint. On the basal side of the joint, the strand is connected to the tendon of the flexor muscle, and on the distal side it is attached to the wall of the segment. The sensory neurons are stimulated by both the stretching and the movement of the connective tissue strand.

Another type of sensory receptor, called a myochordotonal organ, is found in the basal end of the third segment of each walking leg in decapods; it is thought to be proprioceptive from its structure. The organ is a thin membrane containing the cell bodies of a cluster of bipolar sensory neurons. The dendritic processes are anchored in the ventral wall of the segment.

The remaining sensory structures are represented by isolated receptors located over the surface of the body. The most important of these are the sensory (tactile) hairs, which are essentially like those of other arthropods. Each hair consists of a hollow chitinous shaft, which often possesses lateral setules. The shaft is swollen at the base and articulates with the surface skeleton. A canal in the integument enables processes from two bipolar neurons to enter the base of the hair. The hair must be moved rather than touched for the receptor to be stimulated.

Although tactile hairs can appear anywhere on the body surface, they are especially prevalent on the appendages. Among other functions, the tactile hairs are probably important in orienting the animal with regard to the substratum. If the statocysts of the spiny lobster, *Panulirus*, are removed, the loss of postural orientation is quickly regained once the legs establish contact with the substratum. The long, delicate tactile hairs or hair fans that are found on many crustaceans appear to function in the detection of water currents. Such hairs are present on the first antennae of the crab, *Carcinus*, which can detect the movement of objects in water even when blinded. Tactile hairs are also present on the first and second antennae of many raptorial tube-dwelling crustaceans, which project these appendages from the entrance of the tube or burrow to detect passing prey.

Chemoreception is the function of a variety of sensory structures found on the appendages, especially the two pairs of antennae, and feeding appendages. Of these, esthetascs are the most common and occur in the majority of crustaceans. Esthetascs are long, delicate sensory hairs that are usually present in rows (Fig. 14–7A). The dendritic processes of a group of bipolar sensory neurons extend through the entire length of the hair. In terrestrial crustaceans, the esthetascs have the form of tiny plates instead of hairs.

Other supposed chemoreceptors are funnel canals and pores, which are found in decapods. Both are canals in the cuticle containing dendritic processes of sensory neurons.

Although chemoreception has been demonstrated in many crustaceans, evidence that the receptors described above are involved is only indirect. The wood louse, *Oniscus*, prefers to congregate on filter paper soaked in distilled water rather than on one soaked in a dilute sugar solution, but the ability to discriminate disappears with the loss of the ends of the antennae containing the esthetascs. The amphibious marine isopod, *Ligia*, can distinguish fresh water from salt water either with the antennae or with the tips of the legs, depending on the species. Response to food stimuli on contact with the mouthparts in blinded decapods has been repeatedly demonstrated.

Thermoreception has been experimentally demonstrated in a number of different crustaceans, but the mechanism is still unknown. Sensitivity to water pressure has been demonstrated in a number of crustaceans, including some crab larvae and other planktonic crustaceans. Increased pressure stimulates the crab zoea larva to swim upward toward light, and thus to maintain itself in the plankton. When development reaches the megalops stage and the statocysts are acquired, the larva responds to gravity and settles to the bottom.

Reproduction and Development. Most crustaceans are dioecious, but barnacles and some scattered members of other groups are hermaphroditic. The gonads of crustaceans are typically elongated paired organs, lying in the dorsal portion of the thorax or abdomen, or in both (Fig. 14–12*A*). The oviducts and sperm ducts are usually paired simple tubules that open either at the base of a pair of trunk appendages or on a sternite. However, the segments that bear the gonopores vary from one group to another.

Copulation is the general rule in crustaceans, the male usually having certain appendages modified for clasping the female. In many crustaceans, the sperm lack a flagellum and are nonmobile, and in some crustaceans the sperm are transmitted as spermatophores. A seminal receptacle is sometimes present in the female. The seminal receptacle may be located near the base of the oviduct, but frequently it is a separate pouchlike ectodermal invagination of the genital segment or a neighboring segment. In a few groups, the sperm ducts open at the end of a penis, or certain appendages may be modified for the transmission of sperm.

Most crustaceans brood their eggs for different lengths of time. The eggs may be attached to certain appendages, may be contained within a brood chamber located in various parts of the body, or may be retained within a sac formed when the eggs are expelled.

Although the eggs are typically centrolecithal and cleavage is superficial, holoblastic cleavage is not uncommon and in some it is even determinate. Often only the initial cleavage divisions are complete, as in some arachnids, but there are a considerable number of species, both entomostracans and malacostracans, in which a hollow blastula is formed. A blastopore even develops in some cladocerans and shrimp.

A free-swimming planktonic larva is characteristic of most marine species and even some freshwater forms. The earliest and basic type of crustacean larva is a nauplius larva (Fig. 14–8). Only three pairs of appendages are present—uniramous first antennae, the biramous second antennae, and the mandibles. The second antennae and mandibles bear swimming setae. No trunk segmentation is evident, and a single median, or nauplius, eye is borne on the front of the head.

In the course of successive molts, trunk segments and additional appendages are usually gradually acquired, development proceeding from anterior to posterior. When the first eight pairs of trunk appendages are free of the carapace, the larva in higher malacostracans is called a zoea (Fig. 14–76*A-D*). In contrast to the nauplius larva, more posterior appendages provide for propulsion in swimming. Since the corresponding stage of development in other crustaceans is not as clearly defined from the earlier stages, the term *zoea* is generally not applied. With the acquisition of a full complement of functional appendages, the young crustacean is called a postlarva.

The postlarva may be quite similar to the adult in general appearance, or may still be strikingly different in some respects. For example, the postlarva of crabs has a crablike cephalothorax, but the abdomen is still large and extended (Fig. 14–76*E*). After additional molts, all of the adult features are attained except for size and sexual maturity.

The basic developmental pattern of nauplius, zoea (or its equivalent), and postlarva is very frequently modified. The malacostracan nauplius does not feed, and in many groups there has been a tendency for some or all of the larval stages to be suppressed. In most decapods, for example, the nauplius stage is passed in the egg, and the zoea larva is the hatching stage. Both the nauplius and zoea may be suppressed, and the young hatch as postlarvae. All larval development is suppressed in some crustaceans, such as the crayfish.

When development is direct or when there is a graded series of larval stages without radical changes from one stage to the next, development is said to be anamorphic. When the young hatch as postlarvae, development is said to be epimorphic. However, in most crustaceans development tends to be metamorphic. Metamorphosis is especially striking in the sessile barnacles and in many parasites. In barnacles the larva, after a free-swimming existence, attaches and undergoes radical structural changes to assume the adult structure.

Since the postlarval stages of different groups are usually distinctive, these stages have often received special names, such as the megalops of crabs, and the acanthosoma of sergestid prawns. Moreover, the zoea, if particularly distinctive, may be given a special designation such as the mysis of lobsters. The intermediate stages have also received different names. For example, the later nauplius instars are called metanauplii, and the prezoeal instar is called a protozoea.

Other aspects of crustacean behavior and physiology are discussed at the end of this chapter, following the survey of crustacean groups.

SYSTEMATIC RÉSUMÉ OF CLASS CRUSTACEA

The diversity of crustaceans requires the use of a greater hierarchy of taxa in their classification than is usually necessary for other animal groups. The system of classification presented in R. C. Moore's *Treatise on Invertebrate Paleontology* (1969) is preferred by many specialists on the Crustacea. However, that system makes the Crustacea a superclass. In order to retain the rank of class for each of the major arthropod groups in this text, the system used here follows that of Waterman and Chace (1960). In respects other than relative ranking, the two classifications are similar. A synopsis is given below. The number after each name refers to the approximate number of described species that the group contains.

- Class Crustacea (26,000+)
  - Subclass Cephalocarida (4)
  - Subclass Branchiopoda (800)
    - Order Anostraca
    - Order Notostraca
    - Order Diplostraca
      - Suborder Conchostraca
      - Suborder Cladocera
  - Subclass Ostracoda (2000)
    - Order Myodocopa
    - Order Cladocopa
    - Order Podocopa
    - Order Platycopa
  - Subclass Mystacocarida (3)
  - Subclass Copepoda (4500)
    - Order Calanoida
    - Order Harpacticoida
    - Order Cyclopoida
    - Order Notodelphyoida
    - Order Monstrilloida
    - Order Caligoida
    - Order Lernaeopodoida
  - Subclass Branchiura (75)
  - Subclass Cirripedia (900)
    - Order Thoracica
      - Suborder Lepadomorpha
      - Suborder Verrucomorpha
      - Suborder Balanomorpha
    - Order Acrothoracica
    - Order Ascothoracica
    - Order Rhizocephala
  - Subclass Malacostraca (18,000)
    - Series Phyllocarida
      - Order Leptostraca
    - Series Eumalocostraca
      - Superorder Syncarida (29)
        - Order Anaspidacea (4)
        - Order Bathynellacea (25)
      - Superorder Hoplocarida (180)
        - Order Stomatopoda (180)
      - Superorder Peracarida (9000)
        - Order Thermosbaenacea (6)
        - Order Spelaeogriphacea (1)
        - Order Mysidacea (450)
          - Suborder Lophogastrida
          - Suborder Mysida
        - Order Cumacea (770)
        - Order Tanaidacea (350)
        - Order Isopoda (4000)
          - Suborder Gnathiidea
          - Suborder Anthuridea
          - Suborder Flabellifera
          - Suborder Valvifera
          - Suborder Asellota
          - Suborder Phreatoicidea
          - Suborder Epicaridea
          - Suborder Oniscoidea
        - Order Amphipoda (4600)
          - Suborder Gammaridea
          - Suborder Hyperiida
          - Suborder Caprellidea
          - Suborder Ingolfiellidea
      - Superorder Eucarida
        - Order Euphausiacea (90)
        - Order Decapoda (8500)

Suborder Natantia (1930)
Section Penaeidea
Section Caridea
Section Stenopodidea
Suborder Reptantia (6391)
Section Macrura (693)
Superfamily Eryonidea
Superfamily Scyllaridea
Superfamily Nephropsidea
Superfamily Thalassinidea
Section Anomura (1270)
Superfamily Coenobitoidea
Superfamily Paguroidea
Superfamily Galatheoidea
Superfamily Hippoidea
Section Brachyura (4428)
Subsection Gymnopleura
Subsection Dromiacea
Subsection Oxystomata
Subsection Brachygnatha
Superfamily Brachyrhyncha
Superfamily Oxyrhyncha

SUBCLASS CEPHALOCARIDA

The Cephalocarida is the most recently discovered and perhaps the most primitive of the existing groups of crustaceans. Only four species, belonging to three genera, comprise the subclass, which was first described in 1955 from specimens collected from the bottom sand and mud of Long Island Sound. At the present time, records of cephalocarids also include findings in the West Indies and the coasts of California and Japan.

Cephalocarids are tiny shrimplike crustaceans that measure less than 4 mm. in

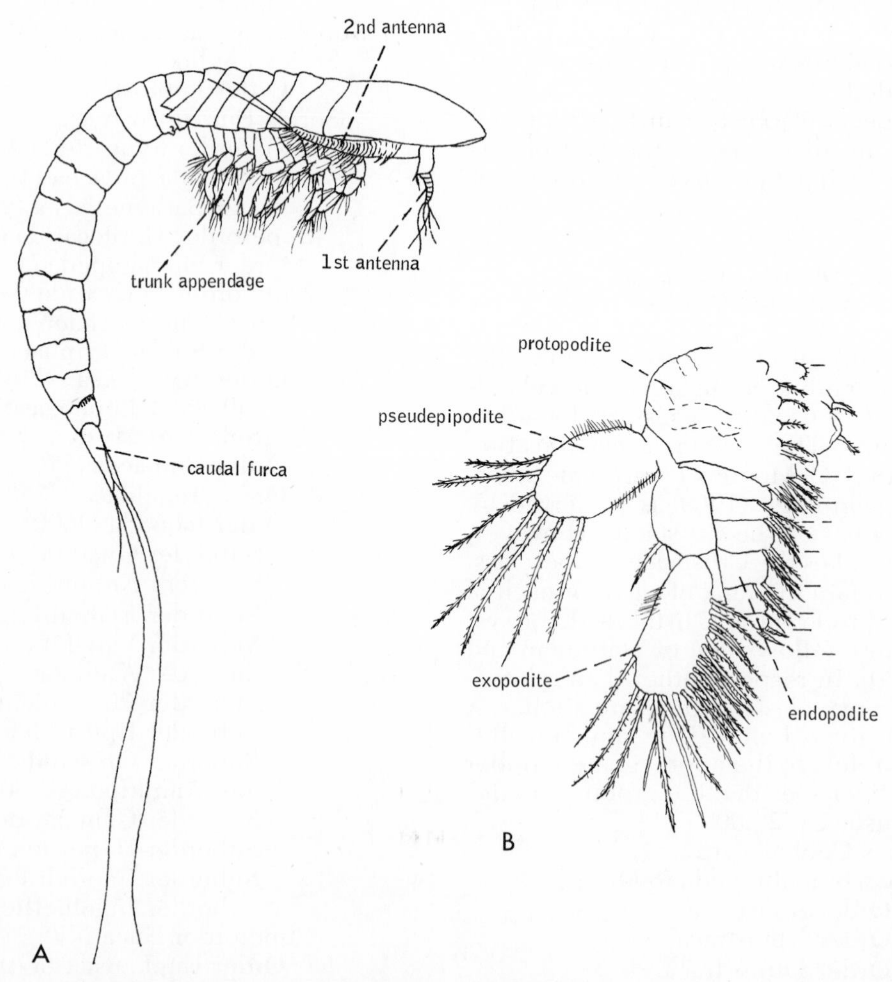

Figure 14–9. *A.* The cephalocaridan, *Hutchinsoniella macracantha.* (After Waterman and Chace.) *B.* A cephalocaridan trunk appendage. (After Sanders.)

length (Fig. 14–9). The horseshoe-shaped head is followed by an elongated trunk of nineteen segments, of which only the first nine bear appendages. Both pairs of antennae are short, and the eyes have disappeared, probably as a result of life in the mud-water interface. The trunk appendages are interesting in several respects. Not only are all the trunk appendages similar, but they are not markedly different from the second pair of maxillae. Also, the basal section of each appendage bears a large flattened outer piece (pseudepipodite) that gives the limb a triramous rather than the usual biramous structure (Fig. 14–9*B*).

The internal anatomy and the physiology and habits of cephalocarids are still not well known. They are indirect deposit feeders. The sweeping action of the second antennae and trunk appendages throws deposit material into suspension. The suspended detritus is then sucked into the median space between the limbs and carried forward to the mouth by limb spines in a midventral food groove. Unlike most free-living crustaceans, they are hermaphroditic and possess a common gonoduct which opens on the sixth segment. In *Hutchinsoniella macracantha*, a single egg develops within an ovisac carried on the first trunk segment that lacks appendages, and the hatching stage is a metanauplius.

SUBCLASS BRANCHIOPODA

Branchiopods are small crustaceans that are largely restricted to fresh water. Although several structurally diverse groups compose the subclass, all are characterized by having trunk appendages that are of a flattened leaflike structure (Fig. 14–10*B*). The exopodite and endopodite each consist of a single flattened lobe, bearing dense setae along the margin. The coxa is provided with a flattened epipodite that serves as a gill; hence the name Branchiopoda—"gill feet." In addition to gas exchange, the trunk appendages are usually adapted for filter feeding and commonly for locomotion. The first antennae and second maxillae are vestigial. The last abdominal, or anal, segment bears a pair of large terminal processes called cercopods.

The Branchiopoda is composed of four distinct groups. The order Anostraca, called the fairy shrimps, is characterized by an elongated trunk containing 20 or more segments, of which the anterior 11 to 19 segments bear appendages (Fig. 14–10*A*). There is no carapace (hence the name Anostraca—"without carapace"), and the compound eyes are stalked.

In the order Notostraca, or tadpole shrimp, the head and anterior half of the trunk are covered by a large shieldlike carapace (Fig. 14–10*C*). The anterior half of the trunk bears 35 to 70 pairs of appendages. Since there are only 25 to 44 "segments," some of which carry ten or more pairs of appendages, the original external segmentation has probably been replaced by a secondary annulation. Only vestiges of the second antennae remain. The compound eyes have become sessile and are located close together beneath the dorsal shield.

The order Diplostraca contains two suborders of laterally compressed branchiopods, in which the body is at least partially enclosed within a bivalve carapace. In the suborder Conchostraca, called the clam shrimps, the entire body is nearly or completely enclosed within the carapace, and externally the animal looks strikingly like a little clam (Fig. 14–11). Dorsally the carapace is folded rather than hinged, and the two valves are closed by a transverse adductor muscle. Ten to thirty-two trunk segments are present, each bearing a pair of appendages. The second antennae are well developed, biramous, and setose, and the compound eyes are sessile.

The second suborder of diplostracans is the Cladocera, or water fleas, of which the genus *Daphnia* is a common representative (Fig. 14–12*A*). The carapace of most cladocerans encloses the trunk but not the head, and often terminates posteriorly in an apical spine. The head projects ventrally and somewhat posteriorly as a short beak, so that the body has the appearance of a plump bird. External segmentation has been lost, and the number of trunk appendages is reduced to five or six pairs. The tip end of the trunk, commonly called the postabdomen, is turned ventrally and forward and bears special claws and spines for cleaning the carapace.

The majority of branchiopods are only a few millimeters in length, and some are as small as 0.25 mm. The largest are the fairy shrimps, which are usually more than 1 cm. long and may attain a length of 10 cm. Most branchiopods are pale and transparent, but rose or red colors sometimes are found, caused by the presence of hemoglobin within the body; and some cladocerans and clam shrimp are brown or black.

Branchiopods live almost exclusively in fresh water; only some cladocerans are

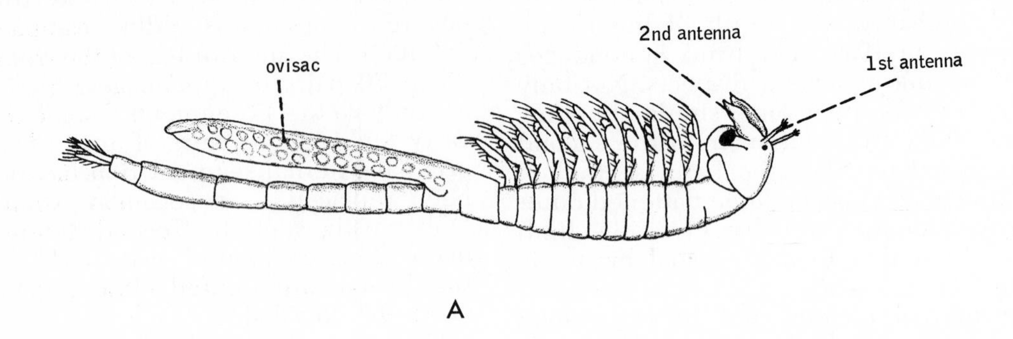

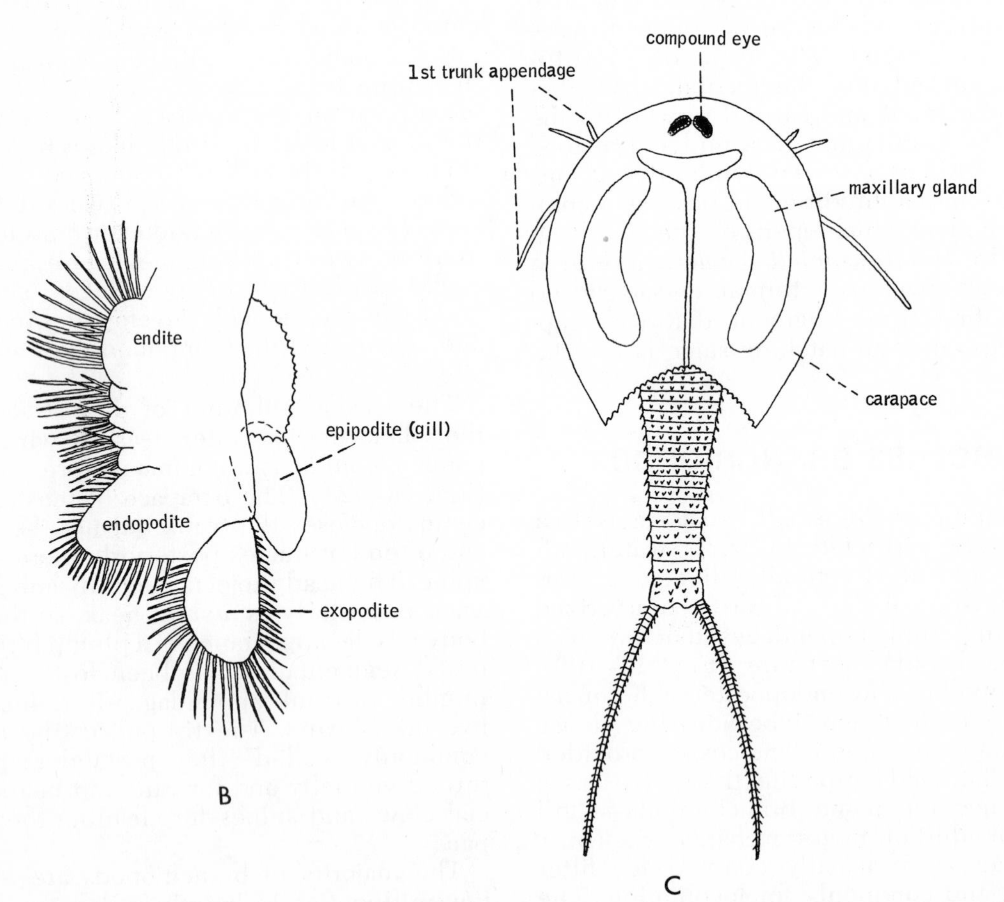

Figure 14–10. Branchiopoda. *A.* Fairy shrimp, *Branchinecta*, an anostracan (lateral view). (After Calman.) *B.* Trunk appendage of *Branchinecta paludosa*, showing foliaceous structure. (After Sars from Pennak.) *C.* Tadpole shrimp, *Triops*, a notostracan (dorsal view). (After Pennak.)

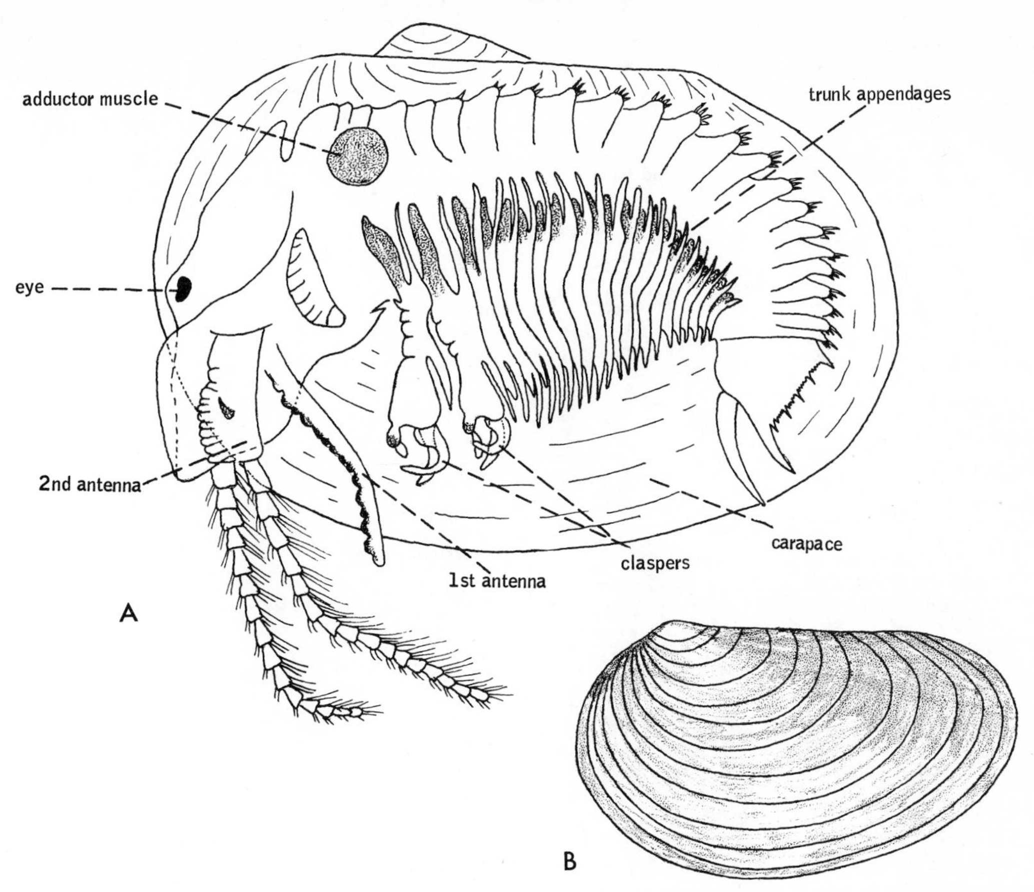

Figure 14–11. *A.* Male of the conchostracan, *Cyzicus mexicanus*, with left valve removed. (After Mattox from Edmondson, Ward, and Whipple.) *B.* Left shell valve of the conchostracan, *Cyzicus*. (After Sars from Calman.)

marine. Moreover, only cladocerans inhabit streams, large ponds, and lakes. The other groups are all confined to temporary pools, springs, and small ponds. The only exception is the brine shrimp, *Artemia*, an anostracan that is found in salt lakes and ponds throughout the world. The noncladoceran branchiopods are also peculiar in that they are frequently absent from certain ponds but present in adjacent ones; or they may inexplicably disappear from a certain pond for a few years, then reappear. Moreover, many species are highly ephemeral, appearing only briefly during the short existence of pools formed by melting snows or spring rains. The restriction of the larger branchiopods to small bodies of fresh water may be correlated with the absence of fish in such habitats.

Locomotion. The water fleas swim by means of powerful second antennae (Fig. 14–12A). Movement is largely vertical and usually jerky. The downstroke of the antennae propels the animal upward; then it slowly sinks, using the antennae in the manner of a parachute. A few species move in a smoother manner, beating the antennae continuously, as in the aberrant predaceous cladoceran, *Leptodora*, which swims with long oarlike antennal strokes (Fig. 14–12*C*). Although cladocerans ordinarily maintain the head upright, a few swim upside down. Also unusual are certain species that hang from the surface film by supposedly non-wettable antennal setae.

Other branchiopods use the trunk appendages in swimming, although the clam shrimp use the second antennae as well. Fairy shrimps usually swim upside down; however, when lighted from below in an aquarium, they roll over and swim right side up. Members of the other groups normally swim dorsal side up. Many clam shrimp and tadpole shrimp swim and crawl over the bottom, and some clam shrimp plow through the bottom sediment. Some cladocerans have also taken up a bottom existence. *Graptoleberis* is said to plow through mud, and some forms, such as species of *Camptocercus*, have become adapted for jumping. The postabdomen is used as a spring or

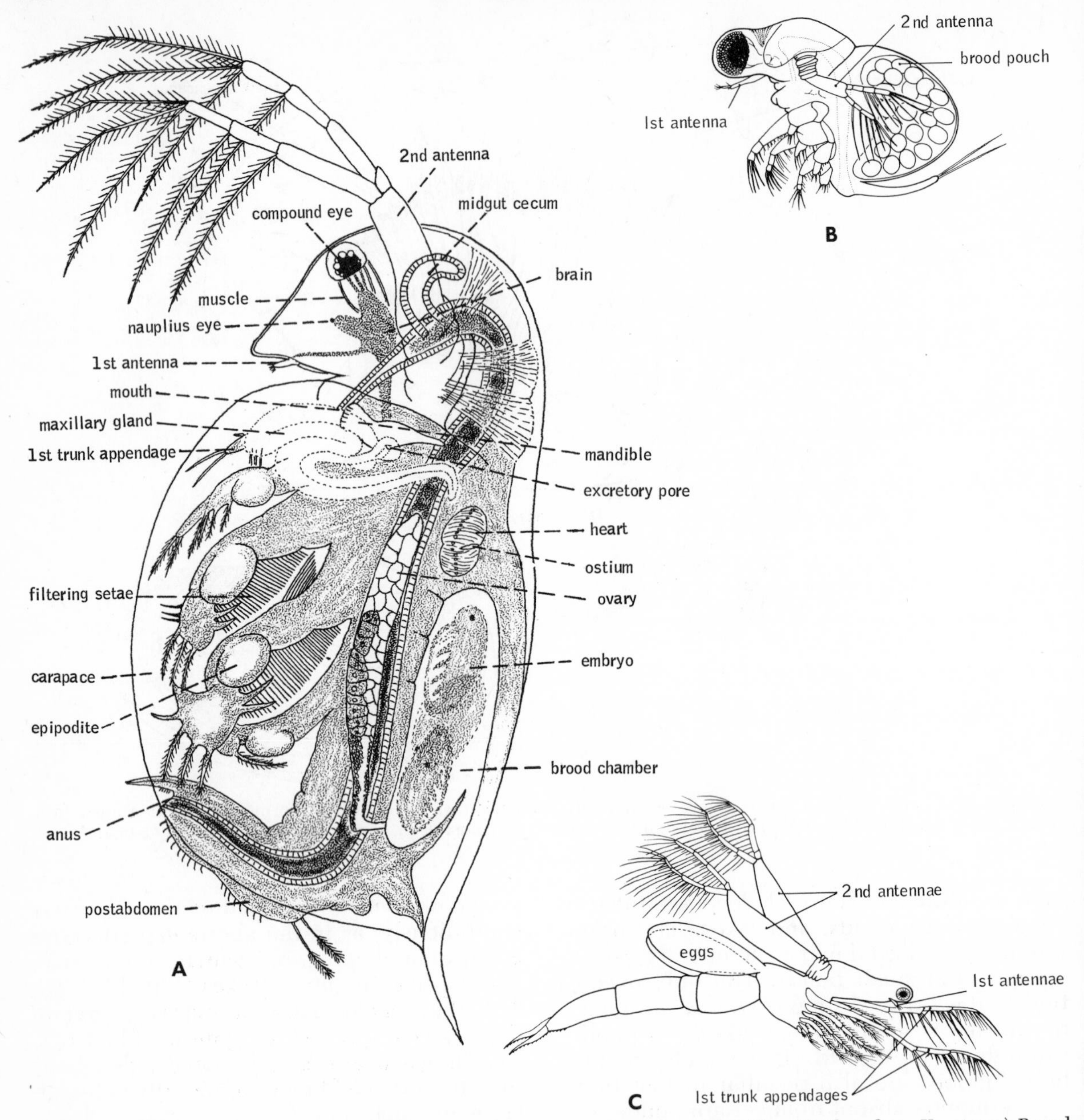

Figure 14–12. *A.* Female of the cladoceran, *Daphnia pulex* (lateral view). (After Matthes from Kaestner.) *B* and *C.* Two aberrant cladocerans: *B*, *Polyphemus pediculus*; *C*, *Leptodora*. (Both from Pennak; *B* after Lilljeborg, *C* after Sebestyén.)

lever, tossing the animal forward several times its body length.

The nauplius eye is persistent in almost all branchiopods (Fig. 14–12*A*). The sessile cladoceran compound eyes are unusual not only in being fused into a single median eye, but also in that this single median eye can be rotated by special muscles (Fig. 14–12*A*). The compound eye, at least in many cladocerans, is used to orient the body in swimming.

Feeding. With few exceptions, branchiopods are filter feeders, and the margins of the trunk appendages are supplied with fine filtering setae. In the fairy shrimp, *Chirocephalus*, the space between limbs increases as the limbs move forward (Fig. 14–13*B*). Water is sucked into this space from the midventral line, and the filtering setae collect particles from the incoming stream. On the backstroke, water is forced out of the interlimb space posteriolaterally and distally (Fig. 14–13*A*).

By several complex mechanisms, the collected food particles are transferred to a mid-ventral food groove extending forward to the mouth and bordered by curved setules. The forward movement of particles in the groove is accomplished by slight anteriomedian spurts of water from the interlimb spaces whenever a limb changes from backstroke to forestroke. Glands in the walls of the food groove, as well as in the labrum, secrete an adhesive material (perhaps mucus), and the entangled particles are pushed into the mouth by the adjacent appendages, particularly the first maxillae.

The feeding mechanism of other branchiopods works on the same principle but with various modifications. In clam shrimp, only the anterior trunk appendages are used for filtering, and the posterior appendages are modified as jaws for grinding large particles. The cladocerans, which possess only four to six pairs of trunk appendages, have the second to fifth pairs or the third and fourth pairs (in Daphniidae) adapted for filtering, and the filter setae are usually arranged on the appendage to form a distinct comb (Fig. 14–12*A*). The water current passes from anterior to posterior, and collected particles are transferred into the food groove by special setae (gnathobases) at the basal part of the appendages. In tadpole shrimp such gnathobases also move the food forward in the food groove.

Food particles for the branchiopods are either plankton (in the case of planktonic forms) or detritus (in the case of benthic forms). The efficiency of the filtering setae of *Daphnia* is indicated by the fact that some members of this genus can be grown using a diet of bacteria.

Some branchiopods scrape up food material from plant and other surfaces, and some are predacious. The planktonic cladocerans, *Leptodora*, *Bythotrephes*, and *Polyphemus*, have their anterior appendages modified for grasping. During feeding, these anterior appendages form a trapping basket, in which prey, particularly other small crustaceans, are captured. The prey is then chewed with the sickle-like blades of the mandibles.

The branchiopod foregut forms a short esophagus, and the midgut is often enlarged to form a stomach. In cladocerans, however, the midgut is more or less tubular and not easily distinguished from other parts of the digestive tract. There are often two small digestive ceca (Fig. 14–12*A*). The intestine in some cladocerans is coiled one to several times.

Gas Exchange, Internal Transport, Osmoregulation. The thin vesicular or lamellar epipodite on each of the trunk appendages of branchiopods is considered to function as a gill, but it is probable that the entire appendage and even the general integumentary surface are also important in gas exchange (Fig. 14–10*B*).

Hemoglobin has been found in the blood, muscles, nervous tissue, and eggs of representatives of all branchiopod orders, although its distribution is sporadic. Moreover, the presence of hemoglobin is frequently dependent upon the amount of oxygen in the water, so that both colored and colorless individuals of the same species exist. This is true for *Daphnia*, in which the animals are colorless in well-aerated water and pink in stagnant water. Hemoglobin is lost when the aquatic medium becomes well-aerated. Under this condition, either the hemoglobin may be transferred to the ovary and eggs, or the iron, after hemoglobin destruction, may be excreted through the maxillary glands.

The heart of cladocerans is a small glob-

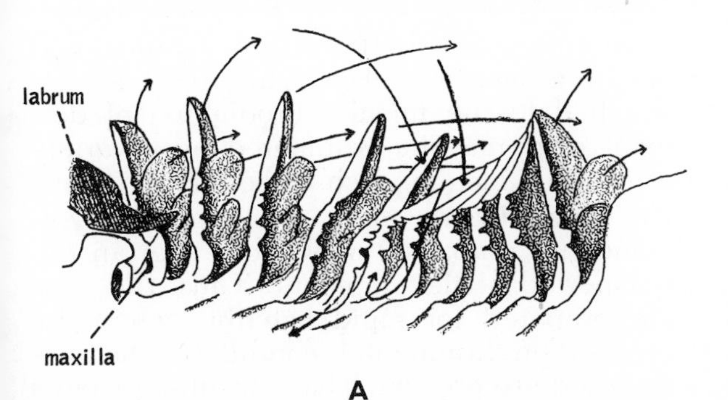

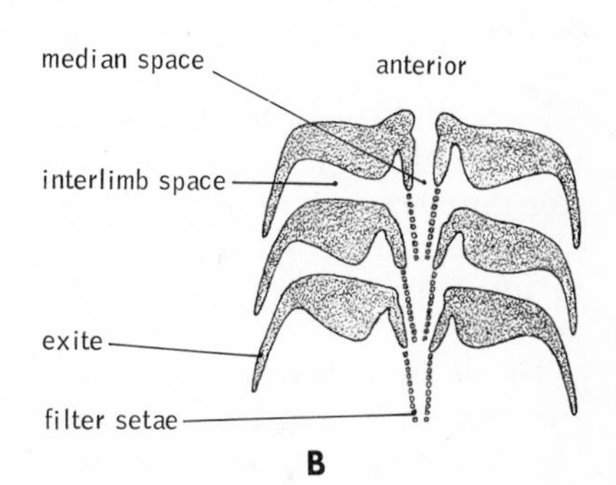

Figure 14–13. A. Trunk appendages of the fairy shrimp, *Branchinella*, showing swimming and feeding currents (lateral view). Setae have been omitted. (After Cannon.) *B*. *Branchinella*, showing three pairs of trunk appendages (frontal section). (After Cannon from Green.)

ular sac with only two ostia, and it lies at the anterior of the trunk (Fig. 14–12A). In all other branchiopods, the heart is tubular with many ostia and may extend into the abdomen, as in the fairy shrimp. The arterial system is restricted to a short unbranched anterior aorta, but blood circulates through regular pathways.

The excretory organs are maxillary glands, usually called shell glands when the duct can be seen coiled within the carapace wall. Little is known regarding water balance mechanisms except in the brine shrimp, *Artemia,* which has been studied extensively. This crustacean can tolerate salinities ranging from 10 per cent sea water to the saturation point for sodium chloride. The internal osmotic pressure varies only slightly with external conditions. This pressure never falls below the equivalent of approximately 1 per cent NaCl nor exceeds 2.8 per cent, even when in a 30 per cent solution of NaCl. Ionic regulation is maintained by the absorption or excretion of salts through the gills. Also, *Artemia* is known to be capable of excreting a urine hyperosmotic to its blood. In brine, the osmotic pressure of the urine is four times that of the blood.

Reproduction and Development. The united or separate male gonopores of cladocerans open near the anus or the postabdomen, which may be modified in the form of a copulatory organ; and the oviducts open into a dorsal brood chamber located beneath the carapace. In other branchiopod species, the gonopores open ventrally on varying segments located more toward the middle of the trunk.

During copulation, the male fairy shrimp clasps the dorsal side of the female abdomen with the large, modified clasping second antennae, and the male inserts the paired eversible copulatory processes containing the openings of the vas deferens into the single median female gonopore. The female gonopore leads into a uterine chamber where the two oviducts also open. Similar copulatory behavior is displayed by the tadpole shrimp, but in clam shrimp the mating pair are oriented at right angles to each other, the male using modified, hooked, anterior trunk appendages for clasping (Fig. 14–11A).

In all branchiopods, the eggs are brooded for varying lengths of time. In the fairy shrimp, a special sac is formed on extrusion of the eggs from the glandular uterine chamber (Fig. 14–10A). Both cladocerans and clam shrimp brood their eggs dorsally beneath the carapace (Fig. 14–12A). In the tadpole shrimp, *Triops,* eggs are carried in a "box" formed on the eleventh limb, the epipodite of which acts as a lid.

The eggs are produced in clutches of two to several hundred, and a single female may lay several clutches. Development in most cladocerans is direct, and the young are released from the brood chamber by the ventral flexion of the postabdomen of the female. In the other branchiopods, embryonated eggs are released by the female and fall to the bottom after only a brief brooding period; or they may remain attached to the female and reach the bottom when she dies. In any case, the eggs typically hatch as nauplius larvae.

Parthenogenesis is common in branchiopods, and in some species males are uncommon or unknown. In cladocerans, parthenogenetic eggs hatch into females for several generations until certain factors, such as change in water temperature or the decrease of the food supply as a result of population increase, induce the appearance of males, and fertilized eggs are produced. The fertilized eggs are large, and only a few are produced in a single clutch. The walls of the brood chamber are now transformed into a protective, saddle-like capsule (ephippium) that is cast off at the next molt, either separating from or remaining with the rest of the detached exoskeleton (Fig. 14–14A). The ephippia float or sink to the bottom and can withstand drying and freezing. By means of such protected resting eggs, cladocerans may be dispersed by wind or animals for some distances and can overwinter or survive during summer droughts.

Thin-shelled summer eggs and thick-shelled resting (dormant) eggs are also produced by many species of the other branchiopod groups, but both types of eggs may be either parthenogenetic or fertilized. Development in thin-shelled eggs is rapid, and hatching may take place while the eggs are attached to the female. Production of dormant eggs may be stimulated by a variety of external factors, such as population density, temperature, and photoperiod. Dormancy (or diapause) may occur after an initial period of development. Thus, the eggs are prepared for rapid hatching when the precise environmental conditions that are required are present. The controlling factors that break dormancy—oxygen, salinity, temperature, illumination—vary and are related

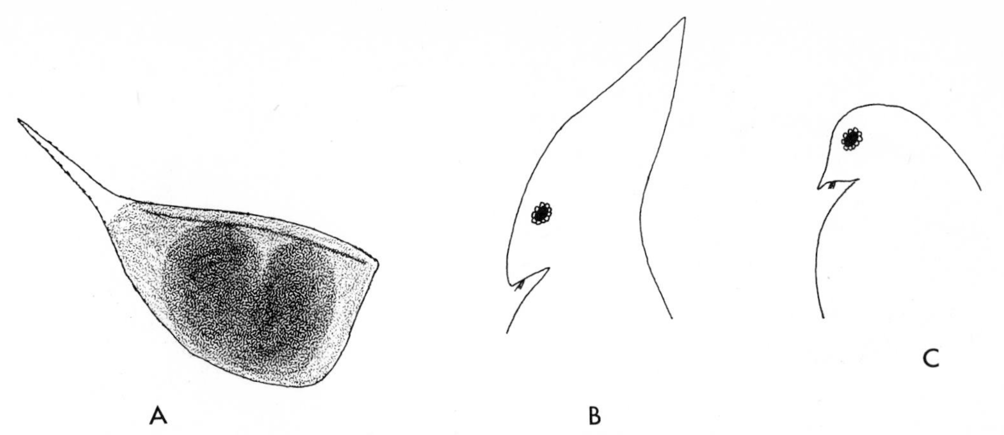

Figure 14–14. *A*. Ephippium of *Daphnia pulex*. (From a photograph by Pennak.) *B* and *C*. Cyclomorphosis in *Daphnia*. *B*. Summer long-head form. *C*. Spring and fall round-head form. (*B* and *C* after Pennak.)

to the type of habitat for which the species is adapted.

Population pulses, or cycles, are common in branchiopods. This is particularly true of many cladoceran species. Some cladocerans, such as species of *Diaphanosoma*, *Moina*, and *Chydorus*, exhibit a single population rise and fall during the warmer months in some lakes. Others are dicyclic and exhibit both spring and fall population peaks. The phenomenon is by no means always predictable. For example, some species of *Daphnia* may be monocyclic, dicyclic or even acyclic, depending on the lake or pond in which they live.

Many of these reproductive phenomena, such as two types of eggs, parthenogenesis, and population peaks, also are found in freshwater rotifers. A further parallel is seen in cyclomorphosis. In some cladoceran species, such as the lake-inhabiting *Daphnia dubia* and *Daphnia retrocurva*, the head progressively changes from a round to helmet-like shape between spring and midsummer (Fig. 14–14*B* and *C*). From midsummer to fall, the head changes back to a normal round shape. Cyclomorphosis is still poorly understood. It may, as in rotifers, result from internal factors; or it may result from an interaction between external conditions (temperature) and internal factors (perhaps genetic).

SUBCLASS OSTRACODA

Ostracods, called mussel or seed shrimps, are small crustaceans that are widely distributed in the sea and in all types of freshwater habitats. Over 2000 living species have been described. They superficially resemble clam shrimp in having a body completely enclosed in a bivalve carapace. Perhaps this group, rather than the Conchostraca, should have the name clam shrimp, for the ostracod carapace has evolved along lines that are even more strikingly like those of bivalves. In ostracods, the valves are rounded or elliptical and the outer wall of each valve is impregnated with calcium carbonate. Dorsally there is a distinct hinge line formed by a noncalcified strip of cuticle.

The valves are closed by a cluster of transverse adductor muscle fibers that are inserted near the center of each valve. In some ostracods, such as the Cytheridae, the valves may be locked together by hinge teeth and ridges at the hinge line. The surface of the valves may be covered with setae and sculptured with pits, tubercles or irregular projections (Fig. 14–15*B*). In the order Myodocopa there is a notch in the anterior margin of the valves permitting the protrusion of the antennae when the valves are closed (Fig. 14–15*A*).

The ostracod fossil record is continuous from the Cambrian period and is the most extensive of any group of crustaceans. The small size and the calcification of the valves have undoubtedly been primary factors in preservation. The valves are virtually the sole remains of the approximately 10,000 fossil species, and the classification of fossil forms is based entirely on valve morphology.

The head region comprises much of the ostracod body, for the trunk is very much reduced in size (Fig. 14–16). The head appendages, particularly the antennules and antennae, are well developed. All external

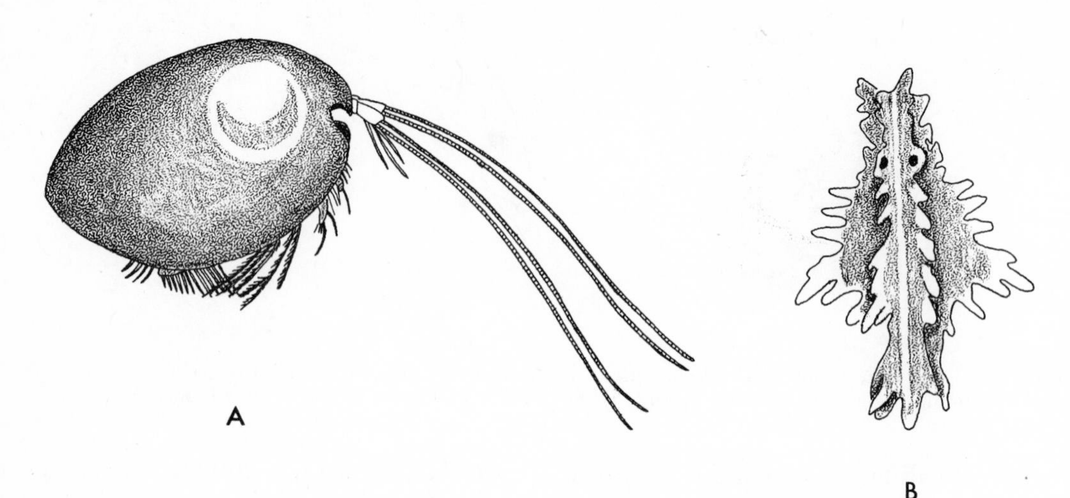

Figure 14–15. *A.* A myodocopid ostracod with antennal notches in valves (lateral view). *B.* The ostracod, *Cythereis*, with greatly sculptured valves (dorsal view). (Both after Müller from Schmitt.)

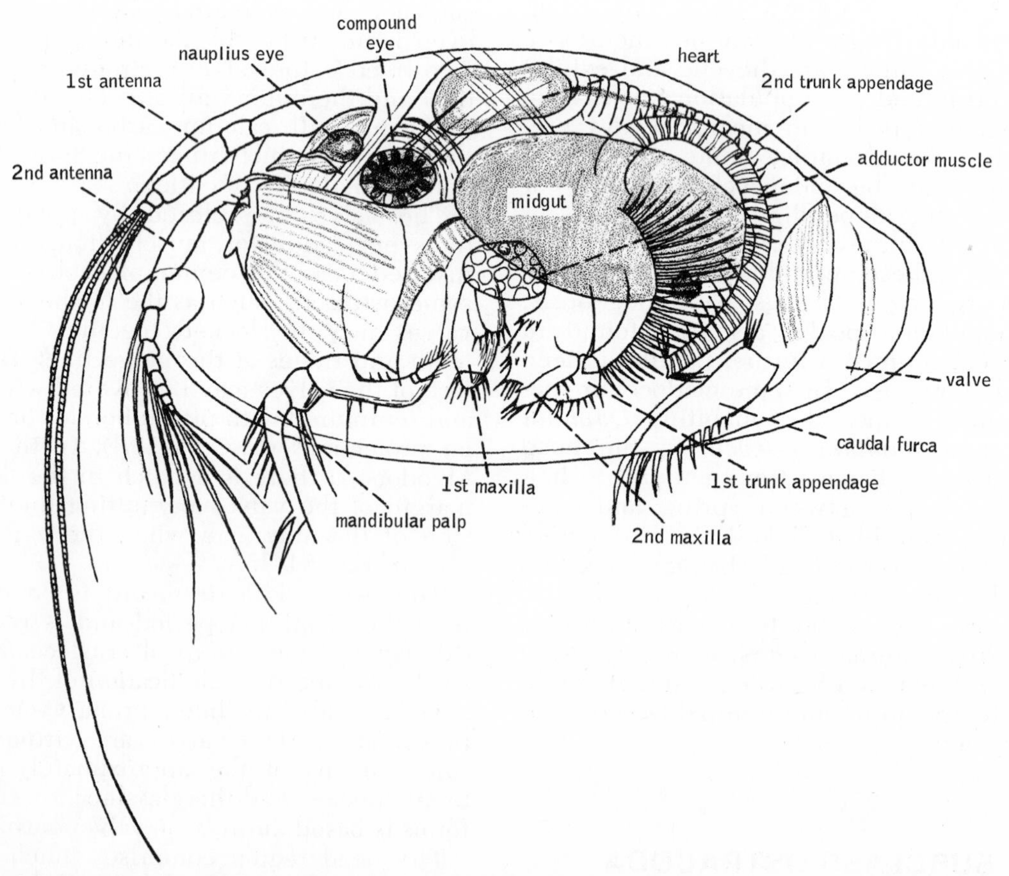

Figure 14–16. Female marine myodocopid ostracod, *Skogsbergia*, with left valve and left appendages removed (lateral view). (After Claus from Calman.)

trunk segmentation has disappeared, and the trunk appendages are usually reduced to two pairs. A few species have only one pair of trunk appendages (*Cytherella*), and in the small suborder Cladocopa there are no trunk appendages. The appendages, including the maxillae, may be more or less leglike, or modified for swimming, feeding, clasping, or other functions. Moreover, the function of an appendage may change during the course of development. The trunk terminates in two long caudal rami that project ventrally and forward.

Most ostracods are minute, ranging from less than 1 mm. to several millimeters in length. The giant of the group is the pelagic deep-sea *Gigantocypris mulleri,* which reaches a length as great as 3 cm. Hues of gray, brown, or green are the most common colors, but some species may be red or yellow.

Locomotion. Although there are some planktonic ostracods, the majority live near the bottom, where they swim intermittently or scurry over or plough through the upper layer of mud and detritus. Many species rest on algae, water plants, or other submerged objects. The second antennae or both pairs of antennae are the principal locomotor appendages and are variously modified, depending on the habits of the animal. In swimming ostracods, the two pairs of antennae are supplied with long setae, while in burrowing species the antennules are stout with short spines and are used for digging.

The fresh-water *Darwinula* uses the first antennae for clearing away debris in front of the body, and *Candona* uses its first antennae to balance upright on the bottom. Some species have first antennae supplied with hooked setae that aid in climbing aquatic vegetation. Hopping is displayed by *Cypridopsis vidua;* the second antennae and the second pair of legs move simultaneously, kicking the animal forward on the backstroke. Perhaps the most remarkable ostracods are the New Zealand and South African terrestrial species of *Mesocypris,* which can plow through forest humus.

Feeding. Ostracods display diverse feeding habits. There are carnivores, herbivores, scavengers, and filter feeders. Algae are a common plant food, and the prey of carnivorous species include other crustaceans, small snails, and annelids. Detritus particles, often stirred up by the antennae or mandibles, are also a common source of food. In the filter feeding cylindroleberids, a water current is produced by the beating of the second maxillae and passes backward over the mouth region. Particles are caught on the anteriorly pointing filter setae of the first maxillae and first trunk appendages. In other forms, the filtering setae are located on both the mandibles and the first maxillae.

The last pair of legs of cypridinaceans are long and wormlike and are used for clearing the interior of the valves of debris (Fig. 14–16).

Scavengers include species of *Vargula* and *Cypridopsis,* which hold large food particles with the mandibles or antennae while they are torn by the maxillae. *Gigantocypris* is predacious and is reported to capture other crustaceans and even small fish with its antennae.

Gas Exchange, Internal Transport, Excretion, and Sense Organs. Although gill-like structures are present in the Cylindroleberididae (Myodocopa), gills are lacking in other ostracods and gas exchange is integumentary, the locomotor and feeding currents providing for ventilation. Hemoglobin has been reported in a few species. A heart and blood vessels are present only in the marine order Myodocopa. Blood circulates between the valve walls in all ostracods.

Some ostracods possess antennal glands, some have maxillary glands, and some are among the few crustaceans that possess both types of excretory organs in the adult. The maxillary glands are large and coiled, and lie between the inner and outer walls of the valves.

A nauplius eye is present in all ostracods, but compound eyes appear only in the Myodocopa and are sessile (Fig. 14–16A). The most important sense organs are probably the sensory hairs found not only on the appendages but also on the valves.

Luminescence. Ostracods were the first crustaceans in which luminescence was observed. Three marine genera, *Cypridina, Vargula,* and *Conchoecia,* contain the most common luminescent species. A cloud of bluish light is produced externally by secretions from a gland in the labrum; it occurs as spontaneous flashes lasting only one to two seconds. The gland contains, at least in *Vargula,* two to four kinds of cells that empty to the outside by separate adjacent pores. The light is produced only on stimulation by a reflex of muscle contraction, and a beam of light can initiate the reflex.

Reproduction and Development. The ovaries are paired tubes, which in the

Cypridae and Cytheridae lie at least partly in the space between the valve walls. The oviducts are usually paired, and each terminates in a seminal receptacle. The female gonopores are located ventrally between the last pair of appendages at the caudal end of the trunk. In the Conchoeciidae, the oviducts are united, and they empty through a single gonopore on the left side.

The testis usually consists of four long and often coiled tubes that may lie within the shell valves. The paired sperm ducts each terminate in an ejaculatory duct, which opens through one of the two large sclerotized penes projecting ventrally in front of the caudal furca. In the marine family, Cypridinidae, the sperm ducts do not open through the penes but through a single median gonopore between the two penes. Corresponding to the unilateral female system, male conchoeciids have a single penis, located on the right side.

The sperm are motile and in some cyprids are of a remarkably large size. *Pontocypris monstrosa,* which is less than 1 mm. in length, has sperm as long as 6 mm.

During copulation in some species, the female is clasped dorsally and posteriorly by the second antennae or the first pair of legs of the male, and the penes are inserted between the valves of the female into the gonopores. Some species of marine cypridinid *Philomedes* possess benthic females and short-lived planktonic males. There is a period of swarming, during which the females swim to join the males. Following copulation, the female sinks to the bottom and cuts off her swimming setae.

Most commonly the eggs are shed freely in the water or are attached singly or in groups to vegetation and other objects on the bottom; but the eggs are brooded in the dorsal part of the shell cavity in some ostracods, including marine cypridinids and the freshwater *Darwinula.* The eggs hatch as nauplius larvae, but each nauplius is enclosed in a bivalve carapace like the adult.

Parthenogenesis is common in the freshwater Cypridae, and there are species in which males are unknown.

SYSTEMATIC RÉSUMÉ OF SUBCLASS OSTRACODA

There are approximately 2000 known species of living ostracods, and over 10,000 fossil species have been described. The living ostracods are contained within four orders.

Order Myodocopa. Marine ostracods having a shell with antennal notches. The usually biramous second antennae have a very large basal segment. Two pairs of trunk appendages are present.

Family Cylindroleberididae. *Asterope.*

Family Cypridinidae. *Vargula, Gigantocypris, Cypridina.*

Family Conchoeciidae. *Conchoecia.*

Order Cladocopa. Marine ostracods with no trunk appendages. Second antennae biramous. *Polycope.*

Order Podocopa. Second antennae uniramous. Two pairs of trunk appendages are present. This large order includes marine as well as freshwater species.

Family Cypridae. *Cypris, Pontocypris, Candona, Cypridopsis, Mesocypris.*

Family Darwinulidae. *Darwinula.*

Family Cytheridae. *Cythere.*

Order Platycopa. Marine ostracods having biramous second antennae and a single pair of trunk appendages. *Cytherella* is the only genus.

SUBCLASS COPEPODA

The Copepoda is the largest subclass of entomostracans, over 7500 species having been described. Most copepods are marine, but there are many freshwater species and a few that live in moss and humus. Also, there are many that are parasitic on various marine and freshwater animals, particularly fish. Marine copepods exist in enormous numbers, and since most planktonic species feed on phytoplankton, they are an important link in many marine food chains. A major part of the diet of many marine animals, such as whales and fish, is composed of copepods.

Copepods are very small crustaceans, most ranging from less than 1 mm. to several millimeters in length. Parasitic forms are generally larger. For example, species of the genus *Penella,* which is parasitic on fish and whales, may be over 32 cm. in length (Fig. 14–20). Although most copepods are rather pale and transparent, some species may be colored a brilliant red, orange, purple, blue, or black.

The body of free-living copepods is usually short and cylindrical (Fig. 14–17). The trunk is composed of ten segments and consists of a thorax and an abdomen. The head

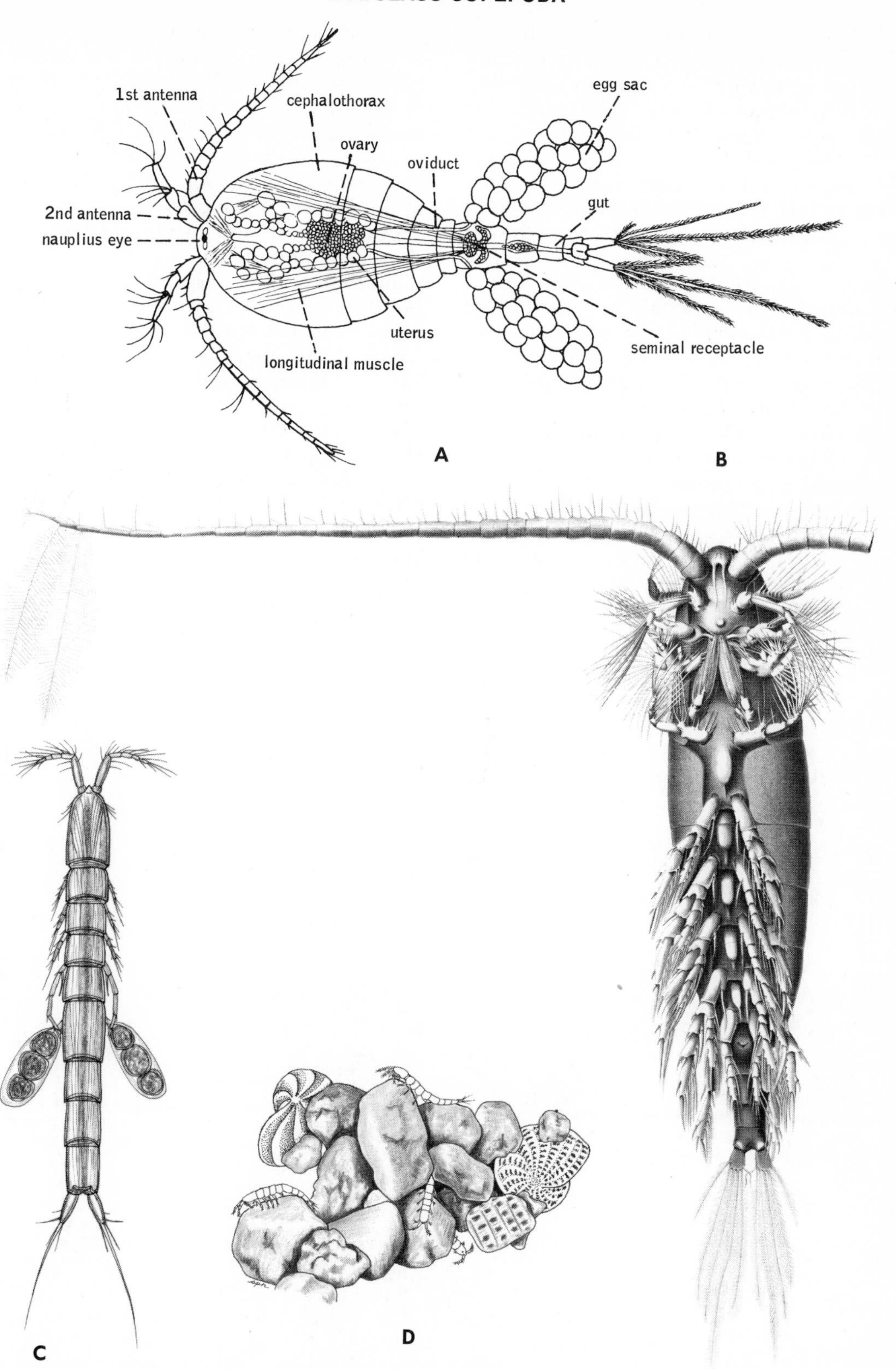

Figure 14–17. *A.* A cyclopoid copepod, *Macrocyclops albidus* (dorsal view). (After Matthes from Kaestner.) *B.* Ventral view of *Calanus*, a typical calanoid copepod, with appendages shown. (After W. Giesbrecht. Fauna and Flora Golfes Neapel. Monogr. *19*:1–831; 1892.) *C.* A cylindropsyllid harpacticoid copepod (dorsal view). (After Sars). *D.* Interstitial harpacticoid copepods crawling among sand grains. Two foram shells are among sand grains. (Drawn from life.)

is either rounded anteriorly or may bear a rostrum. Compound eyes are absent, but the median naupliar eye is typical of all copepods and consists of two lateral and one median ocellus. Posteriorly the head is fused with the first thoracic segment and sometimes the second segment, as well. The thoracic region is usually tapered and is composed of three, four, or five unfused segments with appendages. Sometimes the last two thoracic segments are fused. The abdomen is narrow, cylindrical, and devoid of appendages. The anal segment bears two caudal rami, which in some planktonic marine species are spectacularly developed. For example, in *Calocalanus pavo*, each ramus is turned laterally and bears four long setae with bilaterally arranged setules, each seta resembling a feather.

The uniramous first antennae are long and conspicuous (Fig. 14–17). The second antennae are smaller, and the exopodite has disappeared in most cyclopoids. The first pair of thoracic appendages have become modified to form maxillipeds for feeding. The remaining thoracic appendages, except the last one or two pair, are all more or less similar and rather symmetrically biramous (Fig. 14–18A). Each ramus usually consists of three segments beset with setae. The last one or two pairs of thoracic appendages are either variously modified or vestigial, depending upon the species and the sex.

Locomotion. Of the three free-living orders of copepods, the calanoids are largely planktonic; the harpacticoids are largely benthic; and the cyclopoids contain both planktonic and benthic species. The thoracic limbs are the principal locomotor organs for rapid swimming, and such movement is usually jerky, particularly in swimming forms. The jerky sudden motions are produced by the extremely rapid oarlike beating of the legs, during which the first antennae lie back against the body. When the beating of the legs ceases, the first antennae are extended laterally to prevent sinking. After a short interval, the action of the thoracic limbs is repeated.

Planktonic copepods have the very long

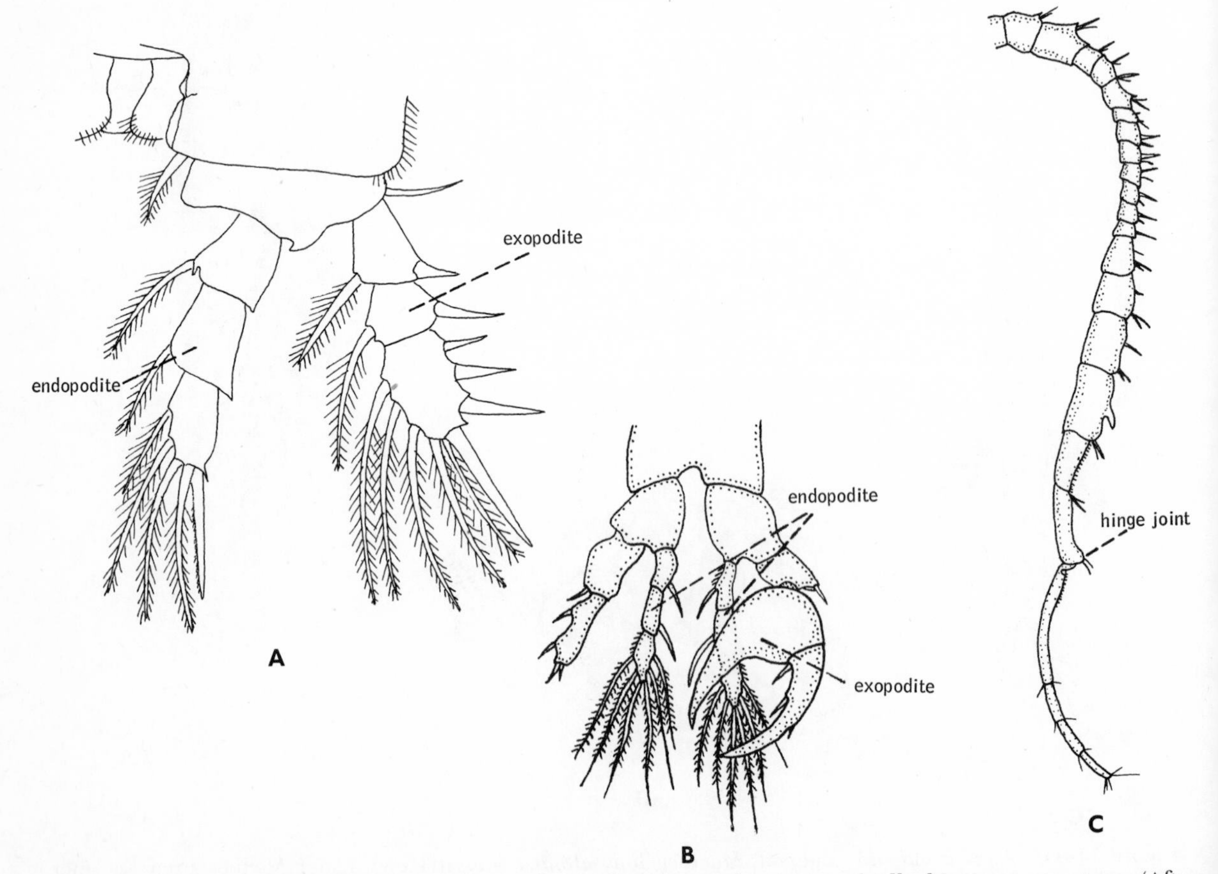

Figure 14–18. *A*. Right second trunk appendage of *Cyclops*, showing markedly biramous structure. (After Pennak.) *B*. Last pair of thoracic appendages of a male *Centropages*, in which the right exopodite is modified as a claw for clasping female abdomen. *C*. The clasping right first antenna of a male *Centropages*. (*B* and *C* after Sars from Calman.)

antennae and the appendages clothed with long delicate setae, which increase flotation. They sink very slowly. Many calanoids display a slow gliding movement, resulting from the feeding current produced by the appendages around the mouth, especially the second antennae. The gliding is punctuated by the jerky propulsions caused by the thoracic limbs. Some gliders, such as *Calanus* and *Diaptomus*, can also swim with the ventral side upward.

Although most planktonic copepods live in the upper 50 meters, there are many species that are found at different levels down to 2000 m. and a few at even greater depths. Vertical swimming is oriented by light. Many species exhibit daily vertical migration, a phenomenon described at the end of this chapter (page 609). Great swarms of copepods sometimes occur when favorable environmental conditions coincide with reproductive periods. A slowly moving swarm of *Calanus finmarchicus* described from the North Sea covered an area 100 by 40 km. and at one period reached a density of 100,000 individuals of various ages beneath a square meter of water surface.

The bottom-dwelling harpacticoids and some cyclopoids crawl over or burrow through the substratum, and many harpacticoids live between sand grains. The thoracic limbs are used in crawling, and this is accompanied in harpacticoids by lateral undulations of the body. Interstitial harpacticoids have slender worm-like bodies and short antennae (Fig. 14–17*D*).

Feeding. Planktonic copepods are chiefly filter feeders in which the second maxillae are modified as filtering appendages. Phytoplankton constitutes the principal part of their diet. In *Calanus* and *Diaptomus* the second antennae, the mandibular palps, and the first pair of maxillae are clothed with setae, and these appendages vibrate regularly between 600 and 2640 times a minute (Fig. 14–19*B*). These vibrations create two swirls of water passing along both sides of the body, producing the gliding movement described earlier (Fig. 14–19*A*). That part of the swirl flowing toward the midline of the body spins anteriorly and is sucked forward into a median filter chamber (Fig. 14–19*B*).

The feeding current is sucked out of the chamber anteriorly and laterally by the vibrations of the first maxillae; during the passage of the feeding current, it is filtered by the dense screen of setae on the second maxillae. The collected particles are cleaned off by the endites of the first maxillae and by special long setae on the base of the maxillipeds, and are then passed to the mandibles and mouth.

The maxillae of *Diaptomus* and *Calanus* are passive filters, but in some copepods, such as species of *Acartia*, they act as a scoop net. Not all planktonic copepods are filter feeders; some capture larger organisms in addition to filtering for phytoplankton, and some are strictly predacious. Species of *Anomalocera* and *Pareuchaeta* even capture young fish. Some freshwater Cyclopidae are herbivorous; others are carnivorous. Members of the common freshwater genus, *Cyclops*, are predatory, as are some of the other cyclopoid genera. Most bottom-dwelling harpacticoids feed on bacteria and detritus.

Gas Exchange and Internal Transport. There are no gills in free-living copepods, and except in the calanoids and some parasitic species, there is neither heart nor blood vessels. The excretory organs are maxillary glands.

Many luminescent copepods, such as *Metridia*, *Pleuromamma*, and *Oncaea*, have been reported. The luminescent secretion is produced by groups of gland cells in the integumentary epidermis. The number of light organs varies from 10 to 70, and they are distributed at various points over the surface of the body.

Reproduction and Development. In copepods, the ovary is paired or single, and the one or two glandular oviducts are provided with lateral diverticula (Fig. 14–17*A*). The oviducts open onto the ventral surface of the first abdominal segment, which also contains a pair of seminal receptacles with separate openings to the exterior. There is usually an internal connection between the seminal receptacles and the oviducts.

The copepod testis is usually single in free-living forms. The sperm duct may be paired, but in the calanoids and some harpacticoids it is present only on one side of the body. Copepods are one of the few entomostracan groups that form spermatophores, and the lower end of the sperm duct is modified for this purpose. The male opening is also located on the first abdominal segment.

During copulation, the male clasps the female with his first antennae, which may be so modified that the distal half of the appendage folds over the basal half (Fig. 14–18*C*). In all male calanoids only one first

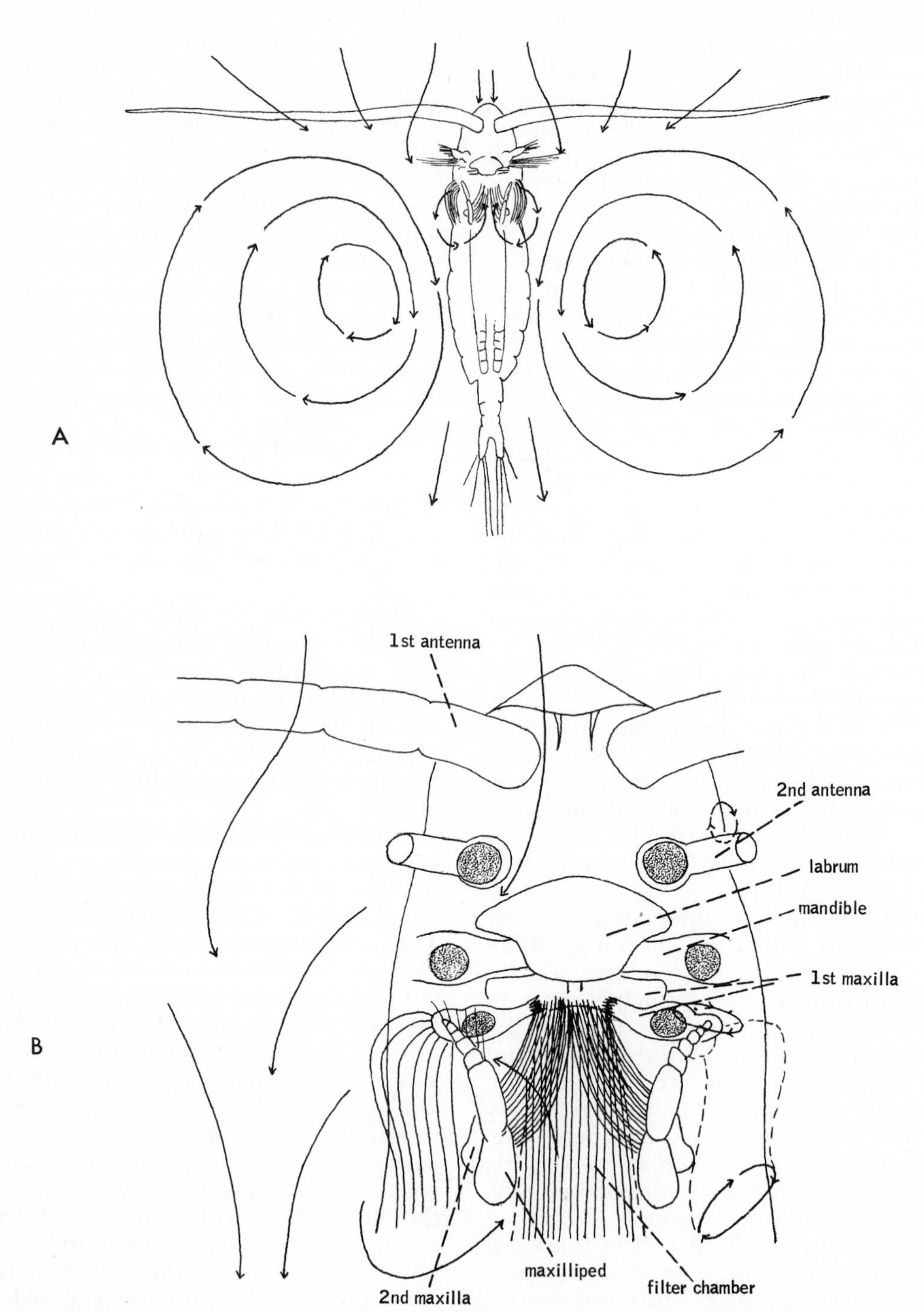

Figure 14–19. *A.* Swimming and feeding currents in *Calanus finmarchicus*. *B.* Anterior of *Calanus finmarchicus*, showing filter currents and filter apparatus (ventral view). (Both after Cannon.)

antenna is specialized for copulation, and in *Diaptomus* the right sixth or fifth swimming leg has become modified to form a large claw for grasping the female abdomen (Fig. 14–18*B*). However, many, perhaps a majority, of the calanoids do not display any specialization of either of the male antennae for grasping the female. The spermatophores are transferred to the female by the thoracic appendages of the male and adhere to the receptacle openings by means of a special adhesive cement. The copulatory behavior of the calanoid, *Diaptomus*, involves pursuit by the male. After the female is caught with the first antenna of the male, her abdomen is clasped with the hooked right sixth thoracic appendage. The opposite sixth appendage transfers the spermatophore.

Copepod eggs may be stored for some time before fertilization and egg laying. Most calanoids shed their eggs singly into the water. However, the eggs of other copepods are usually enclosed within an ovisac. The ovisac is produced by oviduct secretions when the eggs are emitted from the oviduct and remain attached to the female genital segment, in which they function as brood chambers (Fig. 14–17*A*). One or two sacs are formed, depending upon the number of oviducts. Each sac contains from a few to 50 or more eggs, and clutches may be produced at frequent intervals. For example, in the fresh-water *Cyclops*, the eggs hatch in from 12 hours to five days, and a new clutch is then immediately produced, all from a single mating. In some copepods, such as certain harpacticoids, the last pair of legs in the female have become modified, forming platelike shields to protect the ovisacs. The ovisacs are set free in a few species.

The eggs typically hatch as nauplius larvae. After five or six naupliar instars, the larva passes into the first copepodid stage. The first copepodid larva displays the general adult features; but the abdomen is usually still unsegmented, and there may be only three pairs of thoracic limbs. The adult structure is attained typically after six naupliar and five copepodid instars. Molting does not occur after the adult instar is reached. The entire course of development may take as little as one week or nearly as long as one year. Six months to a little over a year is the maximum life span of most free-living species.

Some freshwater calanoids and harpacticoids produce both thin-shelled eggs and thick-shelled dormant eggs, but this adaptation has not been as widely developed as in other entomostracans. In many freshwater copepods, the copepodid instars (or even adults) secrete an organic cystlike covering and become inactive under unfavorable conditions. Such cysts, buried in mud, are particularly adapted to withstand desiccation and enable the copepod to estivate during a temporary drying of the pool or pond in which it lives. They also provide a means of dispersal on the muddy feet or bodies of birds and other animals.

Parasitic Copepods. The Copepoda contain most of the parasitic crustaceans. There are a few parasitic cyclopoids and harpacticoids, and the orders Notodelphyoida, Monstrilloida, Caligoida, and Lernaeopodoida are exclusively parasitic. These four orders contain over one thousand species and display a great diversity of structural modifications and variations in life cycles.

Marine and freshwater fish are the most common hosts of caligoid and lernaeopodoid copepods, which are called fish lice. Fish lice are ectoparasitic and attach to the gill filaments, the fins, or the general integument (Fig. 14–20). Other copepods are commensal or endoparasitic within polychaete worms, in the intestine of echinoderms (particularly crinoids), and in the digestive tract of tunicates and bivalves. Some copepods are parasitic on other crustaceans.

All degrees of modification from the free-living copepod form are exhibited by these parasites. Species of notodelphyoids that are commensal within the pharynx of tunicates are only very slightly modified and move about freely within the host. Many cyclopoid fish lice are also only slightly modified (Fig. 14–21*A*). On the other hand, other ectoparasitic and endoparasitic copepods are so highly modified and bizarre that they no longer have any resemblance to the free-living species (Fig. 14–21*C*, *D*, and *E*). External segmentation is reduced or altogether absent; most of the appendages are reduced or have disappeared; and the body tends to become vermiform.

Certain appendages have usually become specialized as holdfast organs in the ectoparasites. In the Notodelphyoidea, which are commensal in the digestive tract of tunicates, the exopodite of the thoracic limbs bears a curved claw for attachment. More commonly, the second antennae and maxillae terminate in hooks used for holding onto the host (Fig. 14–21*A*). In the common lernaeopodoid fish lice, a special attachment button and thread

Figure 14–20. Parasitic copepods, *Penella exocoeti*, on a flying fish. The copepods are in turn carrying the barnacle, *Conchoderma virgatum* (striped body). (Modified after Schmitt.)

are produced by a frontal gland on the head. The button attaches to the gill filament of the host, and the thread provides for temporary attachment. Later the frontal gland produces another button (the bulla) into which the second maxillae are inserted for permanent attachment (Fig. 14–21*B*).

The mouth parts of ectoparasites are adapted for piercing and sucking. Commonly the labrum and labium together form a tube containing the sickle-like or stylet-like mandibles. In the fish louse, *Lepeophtheirus pectoralis*, as well as in others, the antennae, maxillae, and maxillipeds are also used to lacerate the surface of the host.

Internal parasites have often lost their mouthparts, and food is absorbed directly from the host. For example, the body of the

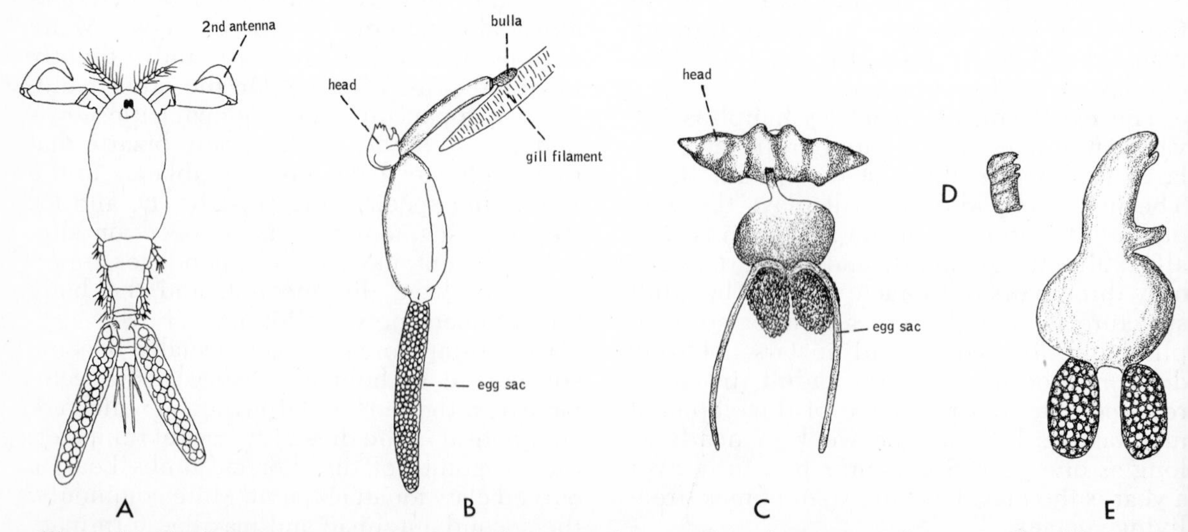

Figure 14–21. Parasitic copepods. *A. Ergasilus versicolor,* a cyclopoid parasite. It lives on gills of freshwater fish. Only adult female is parasitic, hooking to fish with clasping antennae. (After Wilson from Pennak.) *B. Salmincola salmonea,* mature female attached to gill of European salmon. (After Friend.) *C. Spyrion.* Head is imbedded in skin of fish, remainder of body hangs free. (From Parker and Haswell.) *D* and *E.* Male (*D*) and female (*E*) of *Brachiella obesa* live on gills of red gurnard. (After Green.)

larva of *Monstrilla anglica,* which is endoparasitic within the blood vessels of polychaetes, possesses two or four large absorptive processes (arms). Some endoparasites display a very intimate association with the tissues of the host. The saclike *Xenocoeloma brumpti* lives in the body wall of polychaetes, with one end of its body opening into the coelom and the other end opening to the exterior. The body is completely covered by epithelium from the host.

Generally, the female copepod is more highly modified for a parasitic existence than the male, and sometimes the male is entirely free-living and displays the typical copepod appearance.

In most parasitic copepods, the adults are adapted for parasitism, and the larval stages are usually typical and free-swimming. Contact with the host occurs at various times during the life cycle of the copepod, and modifications appear with each molt. The salmon gill maggot, *Salmincola salmonea,* which is parasitic on the gills of the European salmon, has a typical life cycle. When the salmon enters the coastal estuaries on its migration to fresh water, the copepod, in the form of a first copepodid larva, attaches to the gills. The larva attaches by a structure resembling a "button and thread" that is held by the second maxillae; the larva then undergoes a series of molts. The male matures first, and copulation takes place before the female is mature. The male then dies. The female undergoes a final molt and becomes permanently attached to the host by the second maxillae fused with the bulla, which is embedded in the tissue of the host (Fig. 14–21*B*). Egg sacs are then formed and may be as long as 11 mm. Several clutches are produced by a single female.

Larval parasitism is exhibited by the monstrilloids and some other copepods. In the Monstrilloida, the first nauplius larva and the adults are free-swimming, but the intervening stages are endoparasitic in polychaetes. Such larval parasites are often highly modified.

SYSTEMATIC RÉSUMÉ OF SUBCLASS COPEPODA

Order Calanoida. Free-living, largely planktonic copepods with the prosome-urosome articulation (principal point of trunk articulation) located between the fifth and sixth postcephalic segments (thorax and abdomen). First antennae are long, at least half the length of the body. Second antennae are biramous. *Calanus, Calocalanus, Diaptomus, Metridia, Pleuromamma, Centropages* and *Lucicutia.*

Order Harpacticoida. Prosome–urosome articulation located between the fourth and fifth postcephalic segments. First antennae are very short, and second antennae are biramous. Abdomen is almost as wide as thorax, and the body is often linear in shape. Largely benthic in fresh and salt water; a few parasitic forms. *Harpacticus, Canthocamptus.*

Order Cyclopoida. Prosome–urosome articulation located between the fourth and fifth postcephalic segments. First antennae are usually shorter than the length of the head and thorax. Second antennae are uniramous. Includes planktonic and benthic copepods living in fresh and salt water; many are parasitic. *Cyclops, Sapphirina, Oncaea, Lernaea.*

Order Notodelphyoida. Prosome–urosome articulation located between fourth and fifth postcephalic segments in males and between the first and second abdominal segments in females. Commensals in tunicates but are not highly modified. *Doropygus.*

Order Monstrilloida. Second antennae and mouthparts of adults are absent. Includes marine copepods with larval stages parasitic in polychaetes. *Monstrilla, Xenocoeloma.*

Order Caligoida. Metasome–urosome articulation located between the third and fourth thoracic segments, but this articulation may be absent in the more modified females. Second antennae are modified for attachment to host. Are largely ectoparasitic on freshwater and marine fish. *Caligus* and *Lepeophtheirus.*

Order Lernaeopodoida. Segmentation is reduced or absent. Thoracic appendages are reduced or absent, especially in females. Second maxillae are modified for attachment to host. Are ectoparasites on freshwater and marine fish. *Salmincola, Penella, Brachiella,* and *Lesteira.*

SUBCLASS MYSTACOCARIDA AND SUBCLASS BRANCHIURA

The Mystacocarida and the Branchiura are two small subclasses that are related to the copepods and barnacles. The Mystacocarida was first described in 1943 by Pennak and Zinn from specimens collected off

Massachusetts. The only known genus, *Derocheilocaris*, has since been reported from the coasts of South America, southern Africa, and the western Mediterranean. These little crustaceans are only approximately 0.5 mm. or less in length and are adapted for living between intertidal sand grains (Fig. 14–22*A*).

The elongated cylindrical body is divided as is the copepod body, but the head is divided transversely by a constriction, and the thoracic segment bearing the maxilliped is not completely fused with the head. The thorax proper is composed of five segments, four of which each bear a pair of appendages. Unlike the thoracic appendages of copepods, those of the mystacocarids are reduced to a single simple lamella. Both antennae are long and prominent, and the mouth appendages are more elongate than those in copepods and are provided with setae, probably for collecting detritus and other particulate matter. Only the nauplius eye is present. The sexes are separate, and nauplius larvae are the hatching stage.

The subclass Branchiura contains 75 species of common blood-sucking ectoparasites on the skin or in the gill cavity of freshwater and marine fish and on some amphibians. The most striking differences between branchiurans and copepods are the presence of a pair of sessile compound eyes and a large shieldlike carapace covering the head and thorax (Fig. 14–22*B*).

The abdomen is small, bilobed, and unsegmented. Both pairs of antennae are very small, and the first pair is provided with a large claw for attachment to the host. Also important in attachment (except in *Dolops*)

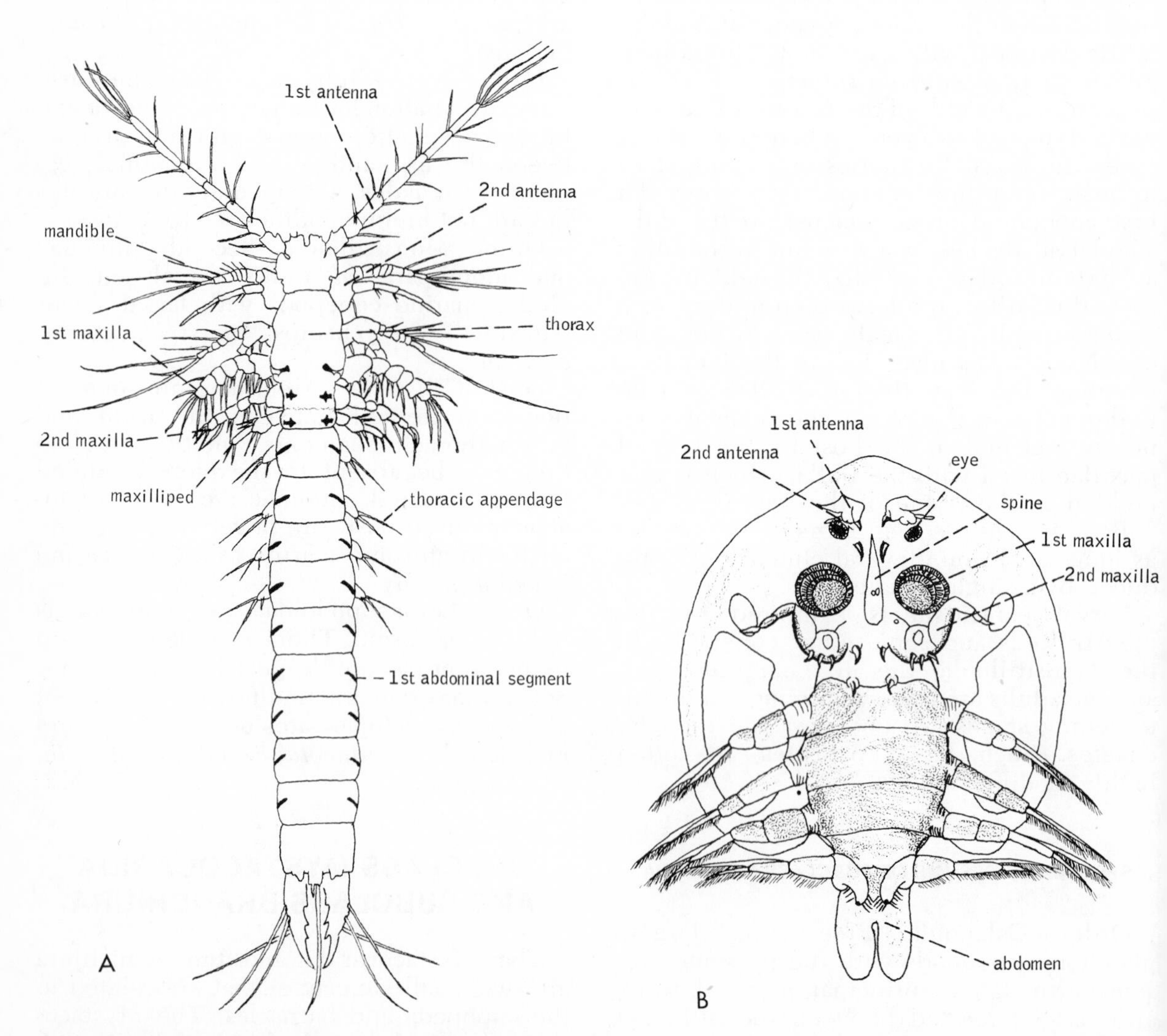

Figure 14–22. *A*. *Derocheilocaris*, a mystacocaridan. (After Delamare and Chappuis.) *B*. *Argulus foliaceus*, a branchiuran fish parasite. (After Wagler from Kaestner.)

are two large suckers modified from the bases of the first maxillae, the rest of the appendage being vestigial. As in many copepods, the labrum and labium form a sucking mouth cone. In *Argulus*, there is also a large sheathed hollow spine located in front of the mouth cone (Fig. 14–22*B*). The spine is used for puncturing the skin of the host and is supplied by a gland that is supposedly poisonous. The second maxillae are heavy and uniramous, and terminate in claws. There are no maxillipeds. The four thoracic appendages are large and biramous with swimming setae, for branchiurans can detach and swim from one host to another.

Copulation occurs while the parasites are on the host, but the eggs are deposited on the bottom. A post-nauplius stage hatches from the egg and is parasitic like the adult. In contrast to copepods, branchiurans continue to molt after reaching sexual maturity.

SUBCLASS CIRRIPEDIA

The Cirripedia includes the familiar marine animals known as barnacles. Cirripedes are the only sessile group of crustaceans, aside from the parasitic forms, and as a result they are one of the most aberrant groups of Crustacea. In fact, it was not until 1830, when the larval stages were first discovered, that the relationship between the barnacles and other crustaceans was fully recognized, and the barnacles were not classified as members of the Mollusca.

Barnacles are exclusively marine. Approximately two thirds of the nearly 900 described species are free-living, attaching to rocks, shells, coral, floating timber, and other objects. Some barnacles are commensal on whales, turtles, fish, and other animals, and a large number of barnacles are parasitic. The wholly parasitic Rhizocephala is so highly specialized that all traces of arthropod structure have disappeared in the adult.

External Structure. Louis Agassiz described a barnacle as "nothing more than a little shrimplike animal, standing on its head in a limestone house and kicking food into its mouth." A further analogy might be drawn if one can imagine an ostracod turned upside down and attached to the substratum by the anterior end.

The cypris larva of barnacles, which indeed looks very much like an ostracod, settles to the bottom and attaches to the substratum by means of cement glands located in the base of the first antenna (Fig. 14–30*B* to *D*). The larval carapace, which encloses the entire body as it does in ostracods, persists as the enveloping carapace, or mantle, of adult barnacles. It becomes covered externally with calcareous plates in the ordinary barnacles (Thoracica). The carapace opening is therefore directed upward and enables the animal to project its long thoracic appendages for scooping plankton.

The free-living, or thoracican, barnacles can be divided into stalked and nonstalked (sessile) types. In stalked barnacles, sometimes called goose barnacles, there is a long stalk (peduncle) that is attached to the substratum at one end and bears the major part of the body (capitulum) at the other (Fig. 14–23*A*). The peduncle represents the preoral end of the animal and contains both the vestiges of the larval first antennae and the cement glands that opened on them. The peduncle is provided with longitudinal and oblique muscles and is highly movable. The capitulum contains all but the preoral part of the body and is surrounded by the carapace (mantle). The mantle surface bears at least five basic plates (Fig. 14–24). The dorsal margin of the mantle is covered by a keel-like carina, and the two lateral surfaces are covered posteriorly by a pair of terga and anteriorly by a pair of scuta. The unattached upper and opposing margins of the terga and scuta can be pulled together for protection or opened for the extension of the appendages. A large adductor muscle runs transversely between the two scuta (Fig. 14–23).

There is no peduncle in sessile barnacles (Fig. 14–25). The attached undersurface of the barnacle is called the basis and is either membranous or calcareous. This is the preoral region of the animal and contains the cement glands. Certain plates located on the capitular base of many stalked barnacles are present in the sessile barnacles as mural plates, which form with the carina and rostrum a vertical wall ringing the animal (Fig. 14–25*B* and *C*). The top of the animal is covered by an operculum, formed by the paired movable terga and scuta. Usually, however, the vertical wall extends upward considerably beyond the base of the terga and scuta, so that they appear to be at the bottom of a sort of vestibule. The plates composing the wall overlap one another, and may be simply held together by living tissue or by interlocking teeth, or actually may be fused to some extent.

Within the mantle the body is flexed back-

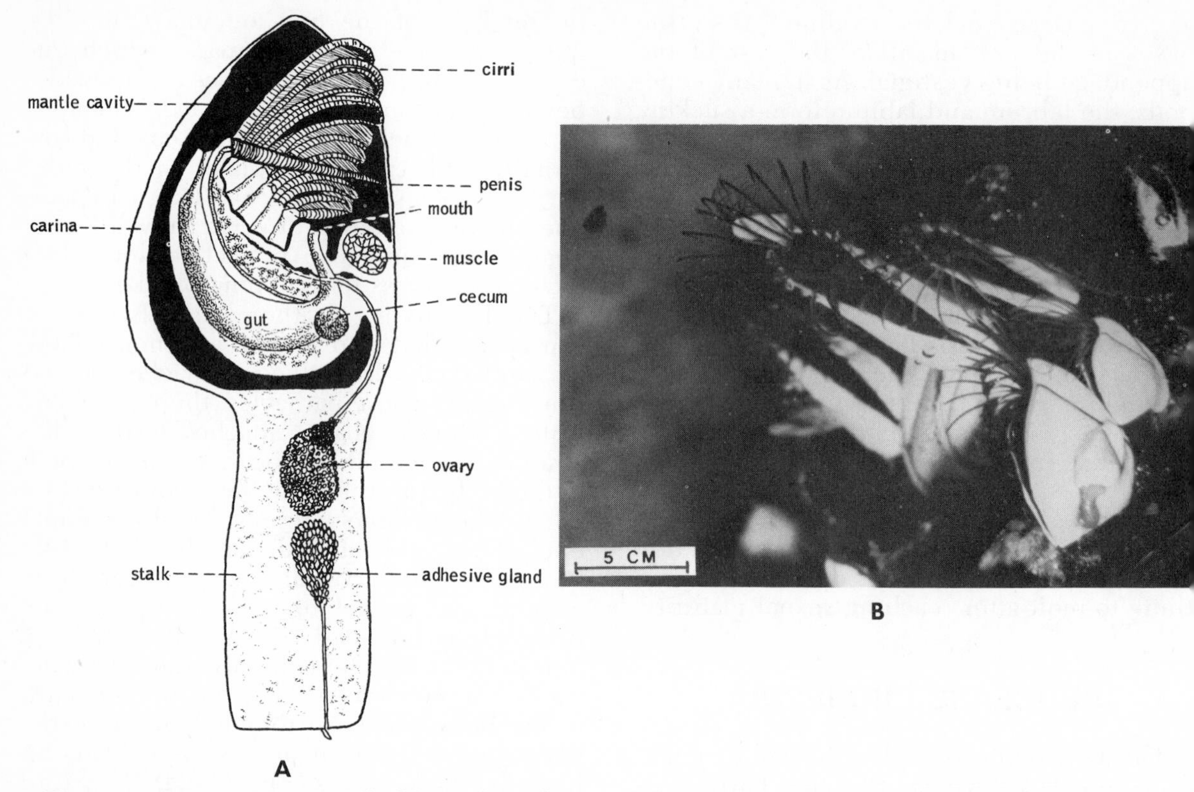

Figure 14–23. *A. Lepas,* a stalked barnacle. (After Broch from Kaestner.) *B. Lepas* attached to a fishing float. (By Betty M. Barnes.)

ward at an angle of 90 degrees from the point of attachment of the head to the body, so that the long axis of the body is at right angles to the vertical axis of the mantle (Figs. 14–23*A* and 14–25*A*). Thus, the appendages are directed upward toward the mantle aperture rather than toward the side. The major part of the body consists of a cephalic region and an anterior trunk (thoracic) region. External segmentation is very indistinct.

The first antennae are vestigial except for the cement glands, and the second antennae are present only in the larva. The second pair of maxillae are united. There are typically six pairs of thoracic feeding appendages (cirri), from which the name Cirripedia is derived. The exopodite and endopodite of each limb are very long, segmented, and provided with long setae.

When not encrusted with other sessile organisms, barnacles are often colorful animals. White, yellow, pink, red, orange, and purple are common colors, and striped patterns are encountered. Stalked barnacles, including the peduncle, range from a few millimeters to 75 cm. in length. The majority of sessile species are a few centimeters in diameter, but some are considerably larger. *Balanus psittacus* from the west coast of South America reaches a height of 23 cm. and a diameter of 8 cm. The smallest barnacles, which are usually only a few millimeters in length, are species of the little order Acrothoracica, the members of which burrow into mollusk shells and coral.

Evolution and Ecological Distribution of Barnacles. Of the two types of common free-living (thoracican) barnacles, the stalked barnacles are considered to be the most primitive. On the basis of fossil forms, such as *Cyprilepas,* the cypris larva common to all cirripeds (Fig. 14–30*B*), and the cyprid-like morphology of the ascothoracica (Fig. 14–28*C*), the ancestral barnacle was probably a cyprid-like, bivalved crustacean attached to the substratum by its first antennae (Fig. 14–24). Initially each valve was covered by a single plate, but to facilitate the opening of the valves and extension of the

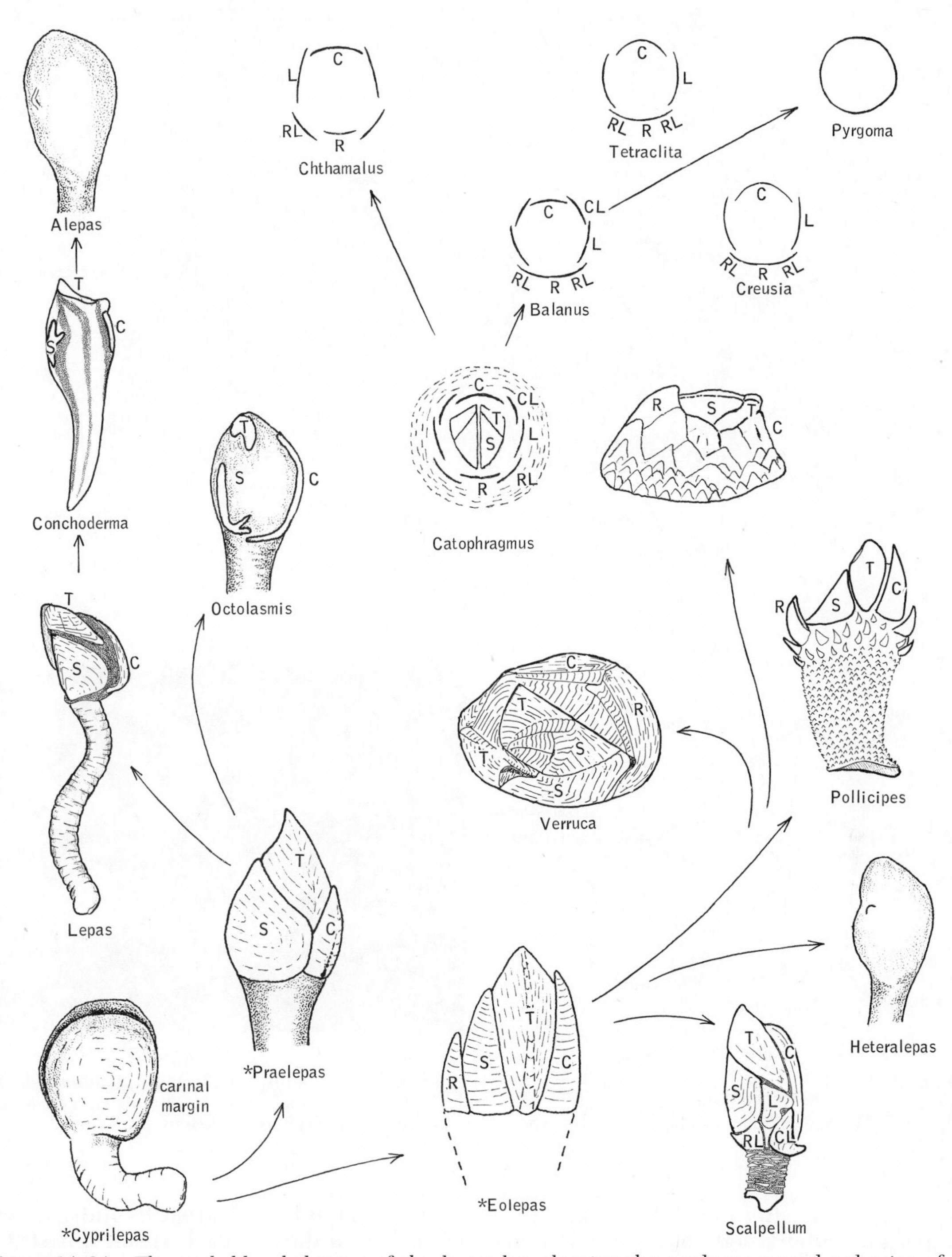

Figure 14–24. The probable phylogeny of the barnacles, showing the tendency toward reduction of shell plates. The genera illustrated represent types within the principal phylogenetic lines. *Cyprilepas*, *Praelepas* and *Eolepas* are extinct genera. The tergum and scutum are omitted from plans of the sessile barnacles except for *Verruca* (Verrucomorpha) and *Catophragmus*. Plates designated include: carina (C), carinolateral (CL), lateral (L), rostrum (R), rostrolateral (RL), scutum (S), tergum (T).

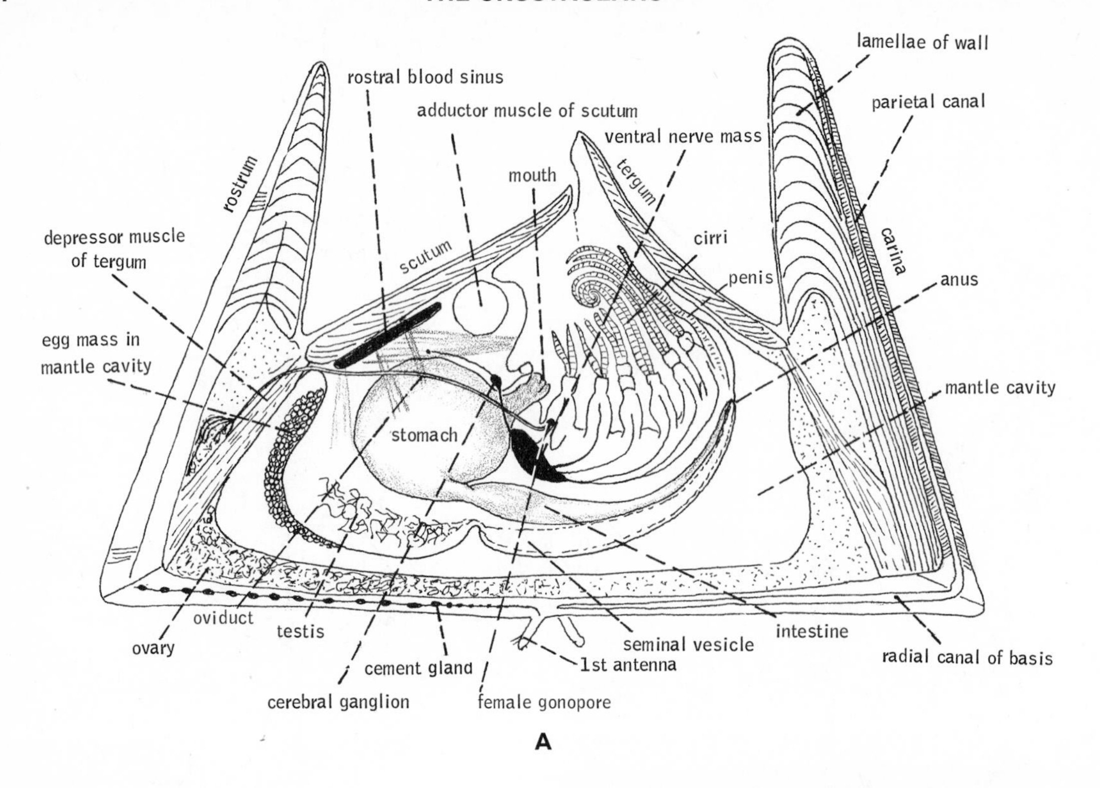

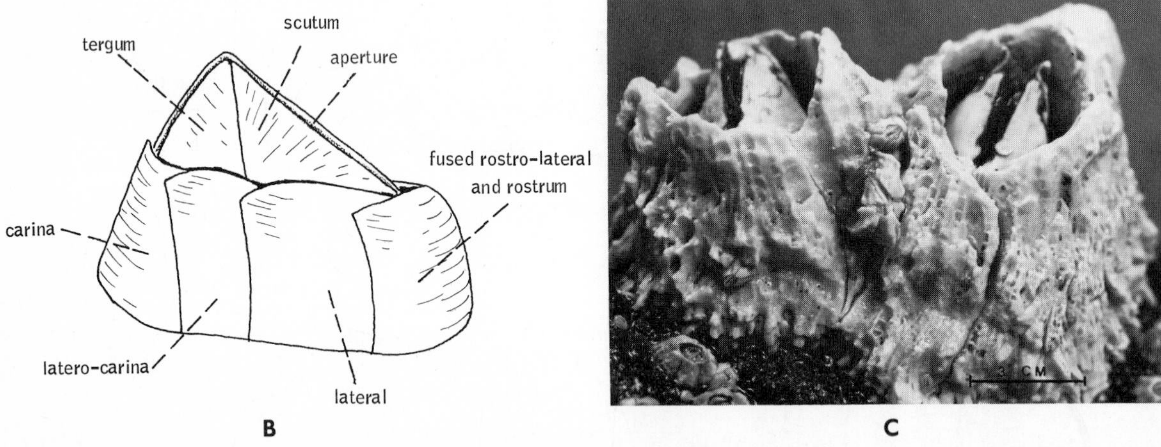

Figure 14–25. *A. Balanus*, a sessile barnacle (vertical section). (After Gruvel from Calman.) *B. Balanus*, showing number and position of shell plates (diagrammatic lateral view). (After Broch from Kaestner.) *C. Tetraclita*, a sessile barnacle, showing the circular wall and the projecting movable tergal plates. (By Betty M. Barnes.)

cirri without exposing the entire body, the plate became subdivided. The ventral aperture became guarded by two pairs of plates, the terga and scuta. These were supported by a posterior dorsal plate, the carina. From such an ancestral stalked barnacle, two principal lines are believed to have evolved: one leading to the existing stalked barnacles known as the lepadids, and another leading to a group of living stalked barnacles known as the scalpellids. The scalpellids, in turn gave rise to the sessile barnacles. A striking feature in the evolution of each of these three groups has been the parallel tendency toward a reduction in the calcareous plates, either in size or number, or both.

In the lepadids, which attach to floating objects or other animals, the peduncle, or stalk, has remained naked and the capitulum has never become covered with more than the ive original basic plates—one carina, two erga, and two scuta (Fig. 14–24). This form s well illustrated by the common *Lepas*. The tendency toward plate reduction is evident in commensal genera. For example, in *Conchoderma*, which lives on whales, turtles, ships' bottoms, and other floating or moving objects, the five plates are reduced o a vestigial state and are widely separated, or all but two of the plates have disappeared Fig. 14–26*C*). Further reduction is seen n *Alepas*, which occurs only on jellyfish.

In the scalpellids, the peduncle became overed with calcareous plates or scales, which generally increase in size toward the apitulum. In many genera, the base of the apitulum is surrounded by several whorls of accessory plates, one of which, a rostral plate, covers the anteroventral end of the apitulum (Fig. 14–24). In some species of *Pollicipes*, some of these lateral plates are onsidered to be homologous with the wall plates of the sessile barnacle *Catophragmus*. *Scalpellum*, on the other hand, possesses accessory lateral plates, but these are arranged differently and are fewer than in *Pollicipes*. The extreme in plate reduction is seen in the ommensal naked *Heterolepas*. One group of calpellids has become adapted for boring in coralline rock (Fig. 14–27*E*).

The sessile barnacles (balanomorphs) are hought to have arisen from the stalked scalpellids by a shortening and disappearance of the peduncle. The large attachment surface and the low, heavy, circular wall make sessile barnacles highly adapted for life on current-swept and wave-pounded intertidal rocks. The wall (mural plates) is believed to have formed from the lateral plates covering the capitulum of stalked barnacles. In the primitive genus *Catophragmus*, the wall is composed of many whorls of imbricated plates, with the interior and largest whorl consisting of eight plates—three paired mural plates, the carina, and rostrum (Fig. 14–24). In most other sessile barnacles, only this inner whorl of eight plates remains and composes the wall; these plates are usually reduced in number through loss or fusion (Fig. 14–24). The remaining plates are large and provide a heavy protective ring. *Octomeris* has eight wall plates; the common *Chthamalus* and *Balanus* have six; *Tetraclita* has four, which are deeply creased, resulting in the name "thatch barnacle"; and in *Pyrgoma*, fusion has taken place to such an extent that the wall is continuous, with no evidence of the original separate plates.

Although there are some deep-water balanomorphs, the group as a whole is intertidal or just subtidal, and species are typically restricted to particular zones. Some are limited to the low tide mark, and some live in the mid-intertidal zone. A few are

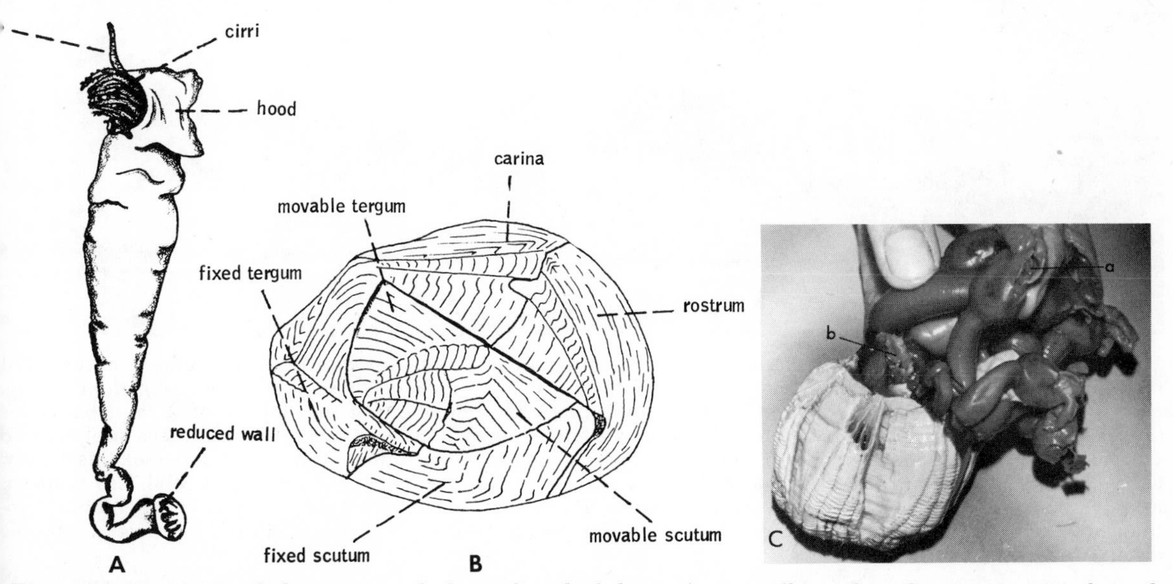

Figure 14–26. *A*, *Xenobalanus*, a sessile barnacle, which lives commensally on fins of cetaceans. Note loss of opercular plates. (After Gruvel from Calman.) *B*. *Verruca* (apical view). (After Pilsbry.) *C*. A cluster of the stalked barnacle, *Conchoderma*, attached to *Coronula*, a sessile whale barnacle. Opening into mantle cavity of *Conchoderma* (a) and *Coronula* (b). (By V. B. Scheffer, U.S. Fish and Wildlife Service.)

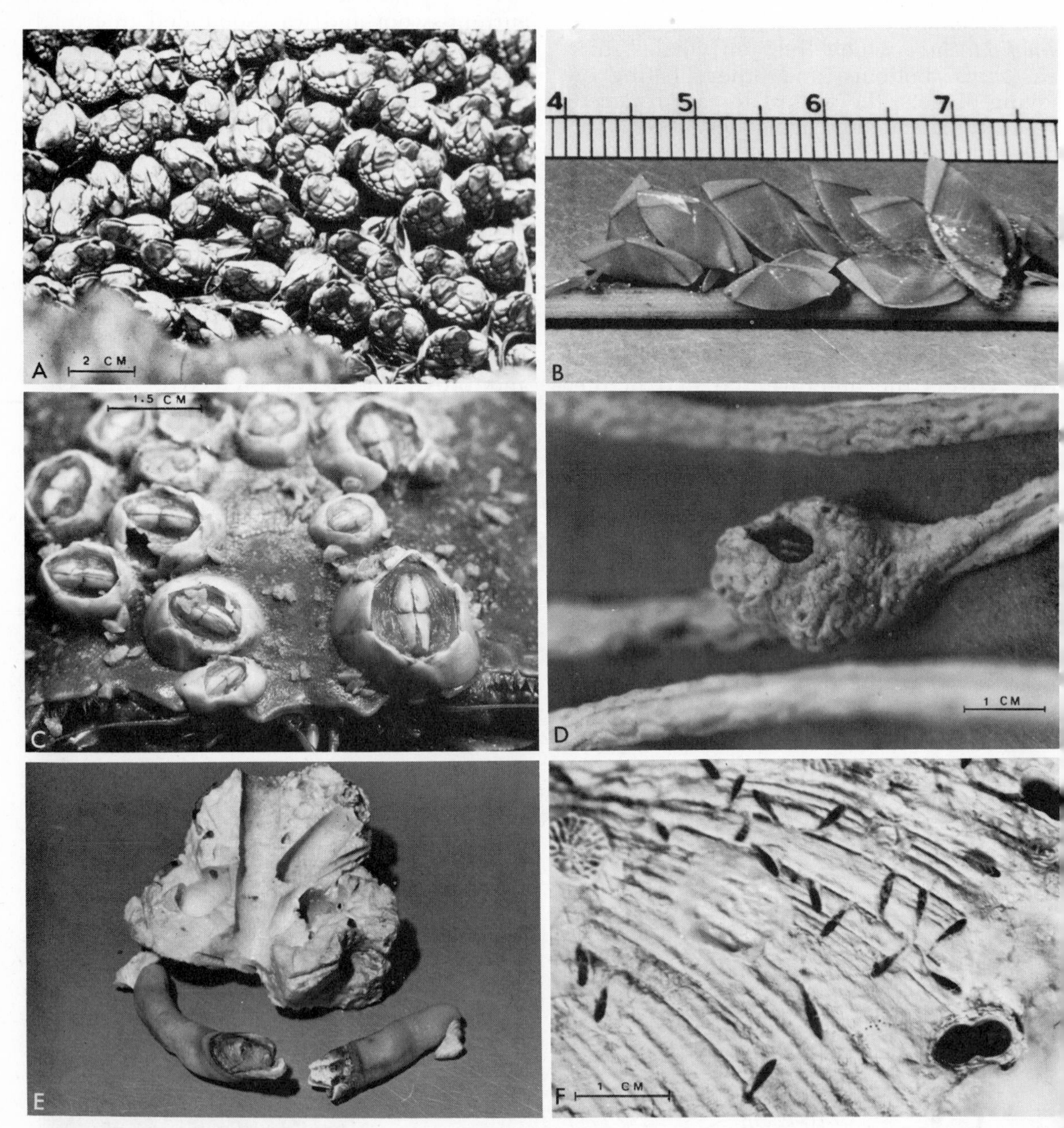

Figure 14–27. *A. Pollicipes polymerus*, a stalked scalpellid barnacle that occurs in large numbers on intertidal rocks along the west coast of the United States. Plate details are similar to those shown for *Scalpellum* in Fig 14–24. *B.* A species of *Poecilasma* on a sea urchin spine. The members of this family of stalked barnacles (Poecilasmatidae) are related to the lepadids but are attached to bottom objects, including sea urchins and crustaceans. *Octolasmis* (shown in Fig. 14–24), which is found on crab gills, is also a member of this family *C. Chelonebia* on the carapace of a blue crab. *D.* A species of *Balanus* which lives on whip coral. *E. Lithotrya* a scalpellid barnacle which bores into coralline rock. *F.* Slit-like openings of the burrows of the boring acrothoracican barnacle *Kochlorine* in an old clam shell. The dumbbell-shaped openings are the burrows of a boring clam. (All by Betty M. Barnes.)

adapted for life in the spray zone at the high tide mark on wave-lashed rocks. Intertidal barnacles commonly occur in enormous numbers, as indicated by the figures in Table 14–1, which give the location and aggregation of two species of barnacles on the English coast.

Some sessile barnacles have become adapted for life on other surfaces besides rocks. There are chthamalids which attach to the stems of intertidal grasses. *Chelonebia* lives on the shells of sea turtles and swimming crabs (Fig. 14–27*C*). Certain species of the usually rock-inhabiting *Balanus* live on gorgonian corals (Fig. 14–27*D*), and *Pyrgoma* lives on madreporarian corals, where it eventually becomes covered by calcareous deposition. Plate reduction has occurred in some commensal balanomorphs. In the whale barnacle *Coronula,* the terga and scuta are reduced and the mantle integument along the upper margins of the wall has extended to form a slight hood (Fig. 14–26*C*). *Coronula* comes to be attached deep into the host's skin. Folding has taken place on the outer surface of the plates, and the canals or chambers so formed contain extensions of the epidermis of the host. Extreme reduction of plates has been attained in *Xenobalanus,* which is commensal on the fins of cetaceans (Fig. 14–26*A*). Here the terga and scuta are gone, and the wall is reduced to a vestige. The body is greatly lengthened, producing a superficial resemblance to a stalked barnacle, and the distal end is protected by a highly developed hood.

Most peculiar are the boxlike species of the suborder Verrucomorpha, in which lateral plates have disappeared, and the wall is formed from the tergum and scutum of one side and the carina and rostrum (Fig. 14–26*B*). The remaining tergum and scutum form a movable lid. The Verrucomorpha are believed to have evolved independently from the pedunculate scalpellids.

The barnacles described thus far and illustrated in Figure 14–24 are all members of the order Thoracica. Probably in the early evolution of barnacles there appeared another line represented today by some thirty species of the order Acrothoracica. Acrothoracicans are all little cirripeds, usually only a few millimeters in length, that bore into calcareous material (Fig. 10–24). The preoral end of the body bears an attachment disc; the mantle is not covered with calcareous plates; and one pair of cirri are located near the mouth, while the remainder are at the end of the trunk.

Trypetesa lives inside snail shells, but only when the shell is occupied by a hermit crab. Other members of the order Acrothoracica, bore into coral and into the exterior of mollusk shells. An old clam shell may have its surface covered with tiny slits approximately 1 mm. in length marking the site of a boring barnacle (Fig. 14–27*F*). When feeding, the naked sac-shaped animal projects its cirri through the slit, sometimes holding the appendages open like a fan and turning them several times before retracting them into the shell (Fig. 10–25).

Parasitic Barnacles. Considering the common occurrence of commensal barnacles, it is not surprising that parasitism has evolved within the class. There are a few parasitic Thoracica, and two other orders are exclusively parasitic. The most important of these is the order Rhizocephala, which is largely parasitic on decapod crustaceans. The body is saccular, and the mantle is devoid of calcareous plates. There is also a complete absence of appendages and segmentation.

In *Peltogasterella,* one of the well known rhizocephalans, a cyprid, destined to be a

TABLE 14–1. VERTICAL DISTRIBUTION OF TWO SPECIES OF SESSILE BARNACLES ON THE PLYMOUTH COAST OF ENGLAND*

	Numbers of Individuals Per Square Meter					
	Balanus balanoides			*Chthamalus stellatus*		
Height Above Mean Tide	Adults	Young	Total	Adults	Young	Total
+3.4 m.	0	0	0	0	0	0
+2.7m.	0	0	0	15,200	9,200	24,400
+1.8 m.	0	400	400	54,000	38,000	92,000
+0.8 m.	4,000	12,400	16,400	55,600	35,200	90,800
−0.2 m.	40,400	20,400	60,800	400	4,800	5,200
−2.1 m.	0	0	0	0	0	0

*(Moore, 1936.)

female, attaches itself by its first antennae to the bases of a crab's setae. A perforation in the host's integument is made, permitting the entrance of a mass of dedifferentiated cells of the parasite (Fig. 14–28A). Within the host, growth of the parasite takes place through the ramification of nutrient rhizoids. Sexual development involves the formation of an external brood chamber, which projects mushroom-like through a new opening produced in the crab's integument near the base of the abdomen. A male cyprid can then attach itself to the opening of the brood chamber and extrude dedifferentiated cells, which migrate into a special chamber in the female, in which they redifferentiate into a testis. The female is now converted into an hermaphrodite. The effects of this parasitism on the crab are numerous and severe. Two of the most striking effects are the inhibition of molting and parasitic castration. The development of the gonads is retarded, or the

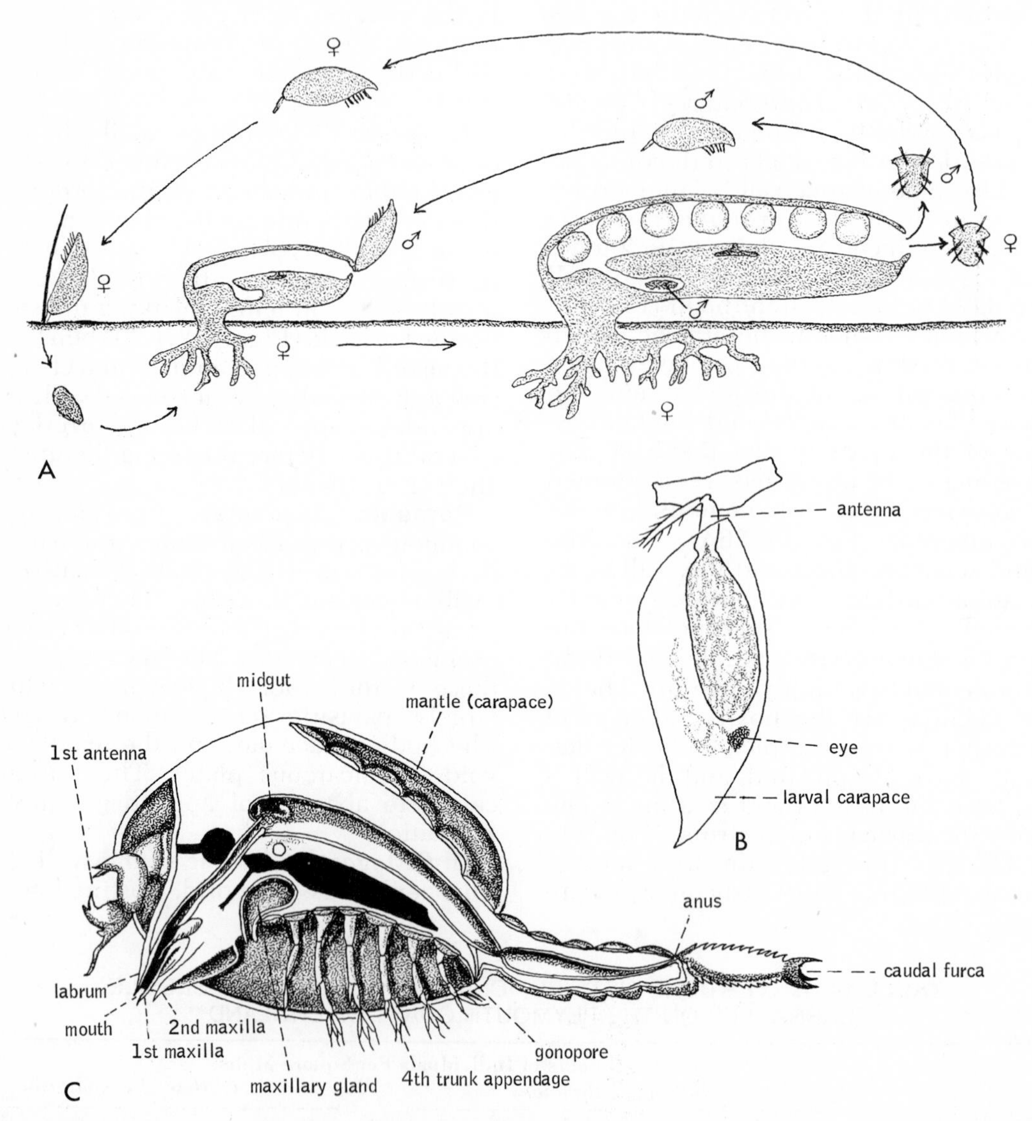

Figure 14–28. *A*. Life cycle of a rhizocephalian barnacle. The horizontal shaded line represents the integument of the crab host. At the left the cypris larva of a female rhizocephalian attaches itself to a seta and penetrates the host's integument. Toward the middle a male cypris larva attaches itself to the protruding brood chamber of a young female parasite. Cells of the male larva pass into the female brood chamber and then into a special chamber (seminal receptacle) in which they develop into sperm. At the right is the brood chamber of a mature female rhizocephalian parasite. Sperm are shown in the small chamber under the main brood chamber. At the extreme right are two nauplii, which have hatched from the eggs contained in the brood chamber of the female. (Modified from Yanagimachi). *B*. Cypris larva of a rhizocephalian shortly after attaching itself to host. (After Delage.) *C*. *Ascothorax ophioctenis*, a parasite in the bursae of brittle stars. (After Wagin.)

gonads atrophy. Parasitic castration is in turn accompanied by varying degrees of change in the secondary sexual characteristics.

Members of the Order Ascothoracica are ectoparasites on sea lilies and serpent stars, or endoparasites in octocorallian corals and in the coelom of sea stars and echinoids (Fig. 14–28*C*). This order was thought to differ so greatly from other Cirripedia that it was once considered a separate subclass. However, the acrothoracicans are now believed to be the most primitive of the Cirripedia, despite their adaptations for parasitism. A chitinous, basically bivalved carapace encloses the body. There are six pairs of thoracic appendages, although these may be reduced or lost. There are no second antennae, but all species possess prehensile first antennae and usually a limbless abdomen. In the ectoparasites, the mouth appendages are modified for sucking; in the endoparasites, the mantle bears absorptive lobes or papillae.

Internal Anatomy and Physiology. During feeding, the paired scuta and terga open, and the cirri unroll and extend through the aperture (Fig. 14–23*B*). When outstretched, a number of cirri on each side form one side of a basket. The two sides sweep toward each other and downward, each half acting as a scoop net. The action is somewhat analogous to the opening and closing of your two fists simultaneously when the bases of the palms are placed together. The rate of beating (one cycle of opening and closing) averages approximately 40 times a minute, but temperature and other environmental conditions influence the scooping rate. In currents some barnacles may hold the cirri outstretched for a period of time.

On the closing stroke, food particles suspended in the water are trapped by the setae, and the first one to three pairs of cirri are used to scrape these particles off and transfer them to the mouth parts. The size of plankton used for food varies. Some species of *Lepas* and *Tetraclita* capture copepods, isopods, and amphipods and could therefore be classified as predacious rather than filter feeders. When such large food particles are consumed, a cirrus that has captured some animal bends down to the mouth field independently of the others.

Many barnacles feed on microplankton, but in at least some of these barnacles, such as *Balanus improvisus*, the setae are too widely separated to act as filters. In these cases the last four pair of cirri perform the usual scooping motion but produce a water current rather than functioning as a filter. The water current is directed over the first two pairs of cirri, which remain within the mantle cavity close to the mouth. These anterior cirri have the setae closely placed, and the cirri filter phytoplankton from the water current.

In most barnacles, an esophagus leads into an enlarged, somewhat **U**-shaped midgut containing a pair of anterior ceca (Figs. 14–23*A* and 14–25*A*).

A heart and arteries are absent, but circulation of the blood is facilitated by a blood pump, consisting of a large sinus located between the esophagus and the adductor muscle (Fig. 14–25*A*). Muscle bands are inserted onto the lower side of the membrane enclosing the sinus, and they act as dilators. Blood circulates through the mantle walls and also into the peduncle.

Gills are lacking, and the mantle and cirri are probably the principal sites of gas exchange. In sessile barnacles and also in *Lepas*, the surface area of the mantle around the upper margin has been increased by folding to produce leaflike projections. Whether these are concerned with gas exchange or with egg brooding is not known.

The excretory organs are maxillary glands.

The cerebral ganglia lie in front of the rather vertical esophagus. The ganglia of the ventral cord are concentrated within a single mass in the sessile barnacles (Fig. 14–25*A*). In pedunculate species, four ganglionic masses can be distinguished behind the subesophageal ganglion. Sensory hairs on the thoracic appendages are the most important sense organs. A nauplius eye is retained, but it is buried in the tissues or under the shell.

Reproduction and Development. Most parasitic barnacles and the boring acrothoracicans are dioecious. Thoracican barnacles are mostly hermaphroditic and are the only large group of hermaphroditic crustaceans. However, cross fertilization is generally the rule, for a suitable substratum almost always contains a large number of adjacent individuals. The ovaries of sessile barnacles (Fig. 14–25*A*) lies in the basis and in the walls of the mantle, and in stalked forms they are located in the peduncle (Fig. 14–23*A*). The paired oviducts open at or near the bases of the first pair of cirri. The testes are located in the cephalic region but may extend into the thorax (Fig. 14–25*A*). The two sperm ducts unite within a long penis, which is attached in front of the anus (Fig. 14–23*A*). The penis can be protruded out of the body and into the mantle cavity of another individual for the deposition of sperm.

Tiny dwarf males that attach themselves to female or hermaphroditic individuals are found in some of the pedunculate genera, such as *Scalpellum* and *Ibla*, in a few species of *Balanus* (Fig. 14–29*C*) and in the boring Acrothoracica (Fig. 14–29*D*). In addition to being very small, these males show all degrees of modification through degen-

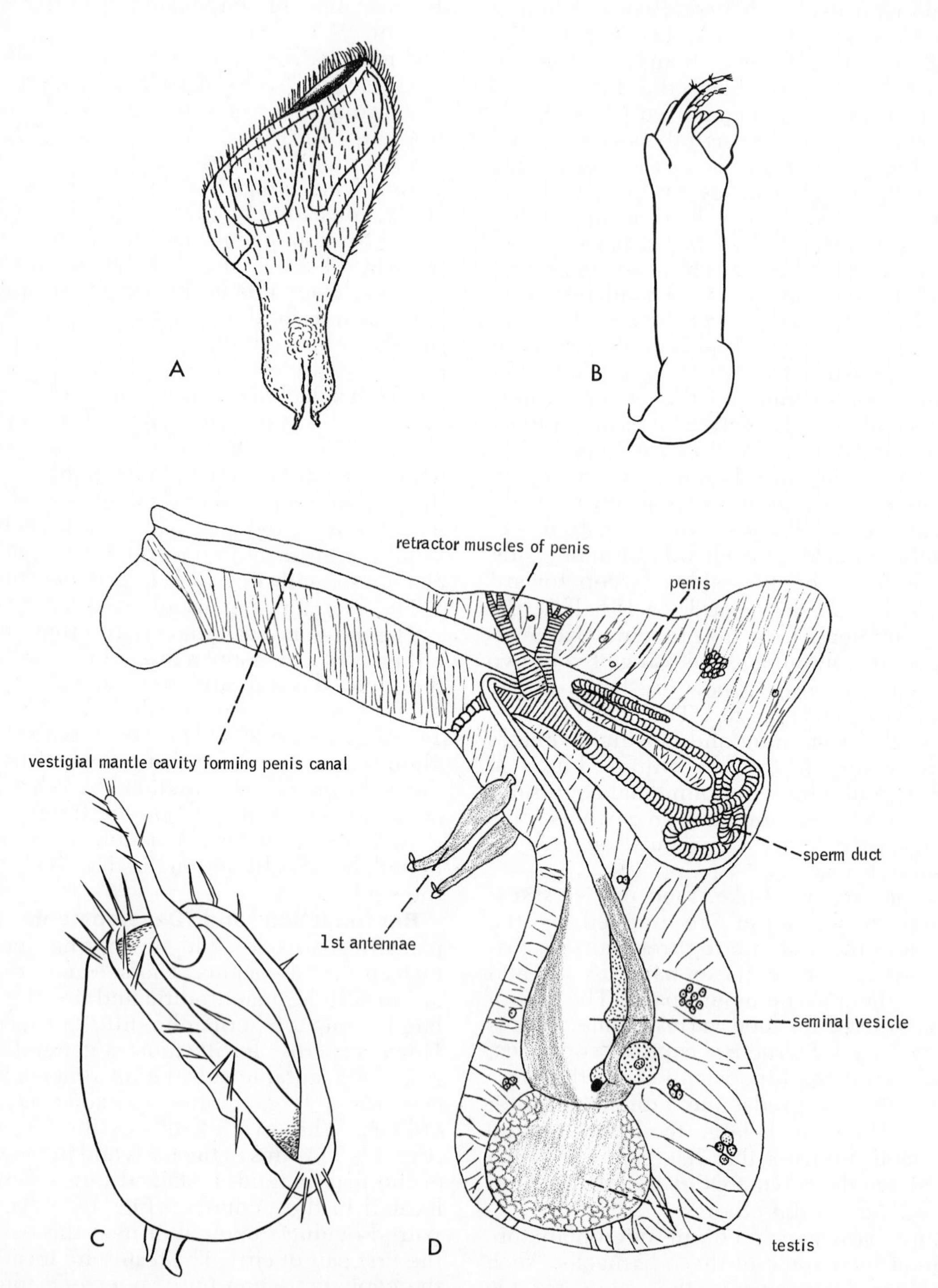

Figure 14–29. Dwarf male barnacles. *A. Scalpellum* with shell plates still present. (After Gruuel from Calman.) *B. Ibla.* (After Newman.) *C.* Complemental male of a species of *Balanus*. (After Henry and McLaughlin.) *D. Trypetesa*, in which male is greatly modified. (After Bernak from Calman.)

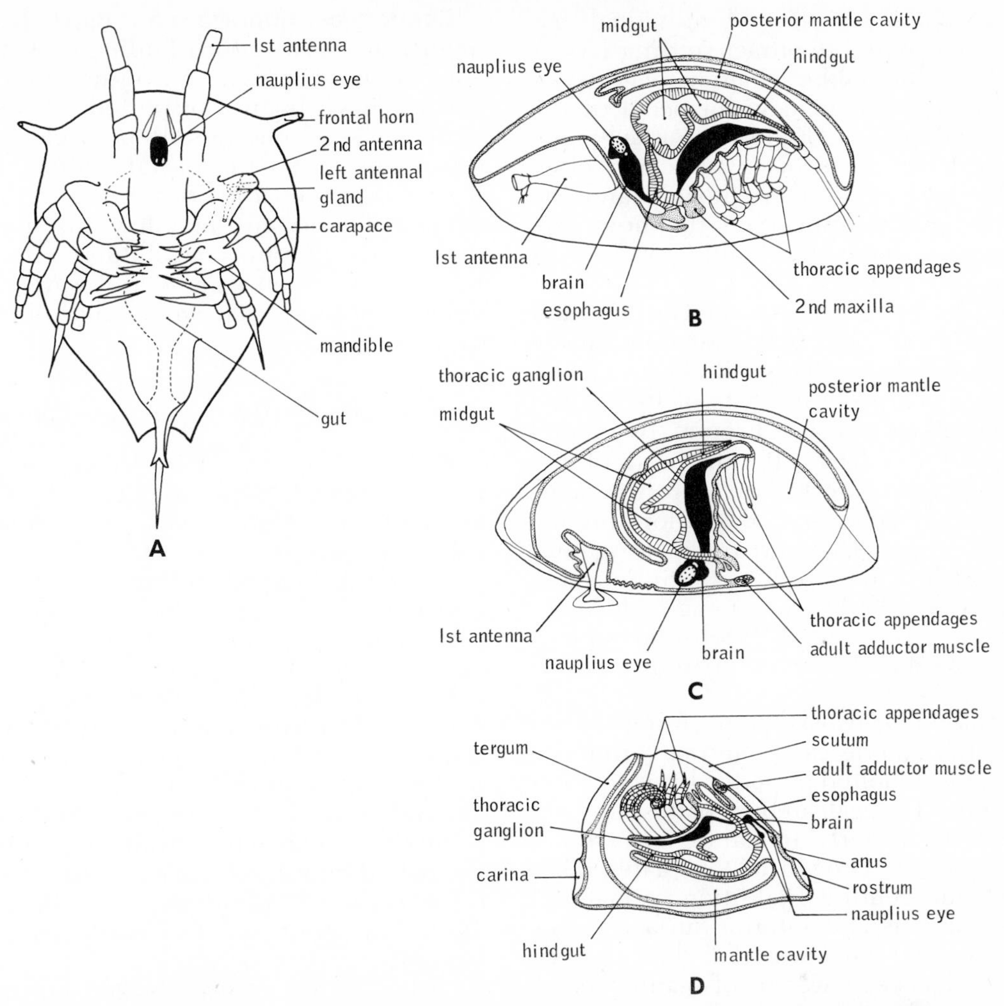

Figure 14–30. *A.* Nauplius larva of *Balanus.* Setae omitted. *B* to *D.* Metamorphosis of cypris larva of *Balanus.* *B.* Free-swimming cypris larva. *C.* Cypris shortly after attachment to substratum. Note rotation of body. *D.* Young barnacle. (All from Walley, L. J., 1970: Phil. Trans. Roy. Soc. London, B, Biol. Sci., *256*(807):237–280).

eracy or loss of structures. In some species of *Scalpellum*, the males are only miniatures of the female or hermaphroditic individual. When attached to a hermaphrodite, the males are called complemental males (Fig. 14–29A). When the males are greatly modified, the species is usually dioecious, and the larger host individuals are females.

An extreme modification is found in the genus *Trypetesa*, which bores in snail shells. The dwarf males of this genus have lost the digestive tract and appendages except for the first antennae (Fig. 14–29*C*). The body consists of little more than a sac containing the testes, sperm ducts, and an enormous penis, all housed within the mantle. A single female of *Trypetesa* may contain as many as fourteen males attached to the outside of her mantle cavity.

The eggs are brooded within an ovisac in the mantle cavity in all barnacles. A nauplius larva represents the hatching stage in most species and can be easily recognized by the triangular shield-shaped carapace (Fig. 14–30*A*). A single gravid individual may release over 13,000 larvae. Six nauplius instars are succeeded by a non-feeding cypris larva, which is so named because of its resemblance to an ostracod. The entire body is enclosed within a bivalve carapace and possesses a pair of sessile compound eyes and six pairs of thoracic appendages. The cypris larva is the settling stage and attaches to a suitable substratum by using the cement glands in the first antennae (Fig. 14–30*B*).

A number of factors appear to operate to increase the likelihood of dense settling, upon which reproduction in the adult depends. Simultaneous shedding of the nauplius by aggregations of individuals has been observed in some species. A protein in the exoskeleton of older attached individuals has been demonstrated to attract settling larvae. There is considerable selection of the surface, and attachment may not occur at the first surface contact. Following attachment, metamorphosis takes place; the cirri elongate, the body undergoes flexion, and the primordial plates appear on the new exoskeleton lying beneath the old cypris valves (Fig. 14–30*C* and *D*).

According to the studies of Costlow and Bookhout (1953, 1956) of the sessile barnacle *Balanus improvisus,* the shell of an attached one-day-old barnacle consists of the paired opercular plates (terga and scuta), two mural plates, and the basis. By the end of the fourth day, one plate has divided into three parts and two have been added, forming the six mural plates characteristic of the genus. Shell growth is more or less continuous and independent of body growth and ecdysis. In this species, the first 20 molts take place between two and three days apart on an average.

The cuticle, or exoskeleton, lining the interior of the mantle cavity and covering the appendages is molted periodically as in other arthropods. The calcareous plates are secreted by the underlying mantle and are not shed at ecdysis. Growth of the plates takes place by the continual addition of material to their margins and interior surfaces, thus increasing their thickness and diameter. In sessile barnacles, a wedge of mantle tissue lying between the junction of the basis and the mural plates adds material to the periphery of the basis. This mantle tissue also permits a continual outward and upward growth of the mural plates to accommodate the increase in the diameter and height. The young barnacle is about 3 mm. in diameter at the end of a month.

Cement is elaborated throughout the life of an individual, and repair of partial detachment is possible. However, detachment is most likely rare, for the cement provides a very strong bond that can withstand enormous pressures. Barnacle cement has received much attention from investigators in recent years because of possible applications to dentistry.

Most barnacles that survive the heavy mortality in the period immediately following settling probably live for two to six years. However, longevity and growth rates are greatly influenced by environmental conditions. Smothering by other individuals and other sessile organisms is a common cause of death.

Economic Importance. Barnacles are among the most serious fouling problems on ship bottoms, buoys, and pilings, and many species have been transported all over the world by shipping. Sessile barnacles are acquired by ships in coastal waters; larvae from floating lepadids settle on ships at sea. A badly fouled ship may have its speed reduced by 30 per cent. Much effort and money have been expended toward the development of special paints and other antifouling measures.

SYSTEMATIC RÉSUMÉ OF SUBCLASS CIRRIPEDIA

Order Thoracica. Free-living and commensal barnacle with six pairs of well developed cirri. Mantle usually covered with calcareous plates.

Suborder Lepadomorpha. Pedunculate barnacles. *Lepas, Scalpellum, Pollicipes, Conchoderma,* and *Heteralepas.*

Suborder Verrucomorpha. Asymmetrical sessile barnacles. Wall composed of carina and rostrum, and one tergum and one scutum; the other tergum and other scutum form the operculum. *Verruca.*

Suborder Balanomorpha. Symmetrical sessile barnacles with a wall surmounted by the paired movable terga and scuta. *Balanus, Chthamalus, Catophragmus, Octomeris, Tetraclita, Pyrgoma, Coronula,* and *Xenobalanus.*

Order Acrothoracica. Naked boring barnacles with a chitinous attachment disc and six to four pairs of cirri (one at mouth, remainder terminal). They live in any calcareous substratum, especially shells and corals. *Trypetesa, Kochlorine, Berndtia.*

Order Ascothoracica. Naked barnacles that parasitize echinoderms and corals. Bivalve or saccular mantle. Prehensile first antennae and abdomen are usually present. *Dendrogaster, Ascothorax.*

Order Rhizocephala. Naked barnacles, parasitic primarily on decapod crustaceans; a few parasitic on tunicates. Appendages and digestive tract are absent; peduncle forms footlike absorptive processes. *Peltogasterella, Sacculina.*

SUBCLASS MALACOSTRACA

The subclass Malacostraca contains almost three quarters of all the known species of crustaceans, as well as all the larger forms, such as crabs, lobsters, shrimp, and wood lice.

The trunk of malacostracans is typically composed of fourteen segments and the telson, of which the first eight segments form the thorax and the last six the abdomen (Fig. 14–31*A*). All the segments bear appendages. The first antennae are often biramous. The exopodite is often in the form of a flattened scale. The mandible usually bears a palp, and the triturating surface of the mandible is divided into a grinding molar process and a cutting incisor. Each of the two maxillae bears a palp.

Primitively, the thoracic appendages, or legs, are similar, and the endopodite is the most highly developed of these appendages, being used for crawling or prehension (Fig. 14–31*B*). In most malacostracans, the first one, two, or three pairs of thoracic appendages have turned forward and become modified to form maxillipeds. Primitively, the thorax was probably covered by a carapace, but this has been lost in some orders. With a few exceptions, the gills are modified thoracic epipodites.

The first five pairs of abdominal appendages, called pleopods, are similar and biramous. The two rami of each abdominal appendage can often be hooked together by special spines. The pleopods may be used for swimming, for carrying eggs in the female, and sometimes for gas exchange. In the males, the first one or two pairs of pleopods are often modified, forming copulatory organs. Usually each ramus of the sixth abdominal appendages (uropods) is composed of single large flattened piece, and together with the usually flattened telson, forms a tail fin, which is most frequently used in swimming.

Most malacostracans have the foregut modified as a two-chambered stomach bearing triturating teeth and comb-like filtering setae. Within the stomach, food is chewed and digestion begins. The fine particulate products of this action are filtered out and passed to the midgut outpocketings, called the digestive gland or hepatopancreas.

The female gonopores are always located on the sixth thoracic segment, and the male gonopores are on the eighth. The nauplius

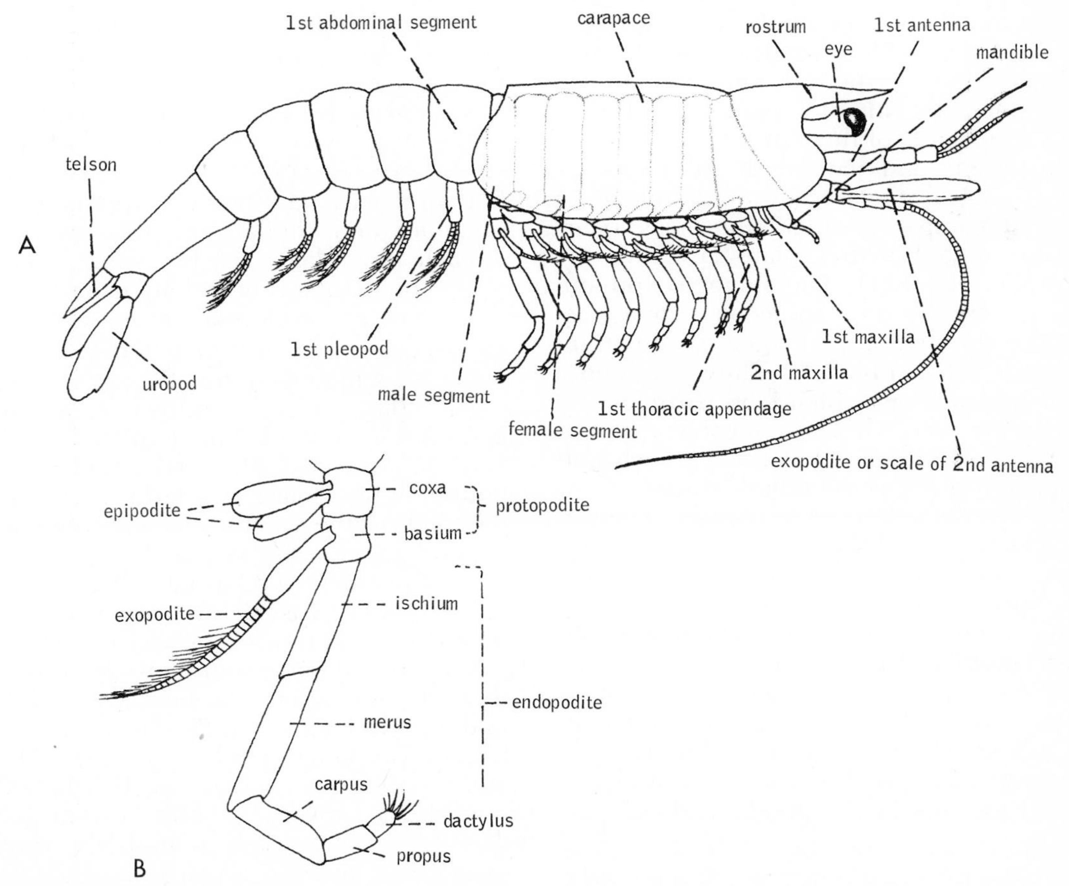

Figure 14–31. A generalized malacostracan. *A*. Lateral view of body. *B*. A thoracic appendage. (Both after Calman.)

larva is usually passed within the egg, but when present, it doesn't feed.

Because of the size and diversity of the Malacostraca, each of the orders is considered separately.

Series Phyllocarida; Order Leptostraca*

This cosmopolitan order is composed of about 10 species of small marine crustaceans that differ from the basic malacostracan plan in that they have eight abdominal segments instead of six (sometimes the leptostracans are stated to have seven segments, but the so-called telson is actually an eighth segment). For this reason, the Leptostraca are placed in a separate series, the Phyllocarida, instead of with the Eumalacostraca, which includes the remaining malacostracans. Morphologically, the phyllocaridans are believed to represent the most primitive existing malacostracans. The fossil record bears this out, for the earliest known malacostracans were phyllocaridans, which appeared in the Silurian period.

Although *Nebaliopsis* is bathypelagic, most species are littoral, living in shallow water. *Nebalia bipes*, which reaches about 12 mm. in length, lives in bottom mud and in seaweed along the Atlantic coast, as well as in many other parts of the world.

The thorax and part of the abdomen of leptostracans is enclosed by an unhinged but bivalved carapace, which anteriorly bears a little hinged rostral plate covering the head (Fig. 14–32*A*). The thoracic appendages are similar and foliaceous, somewhat like the thoracic appendages of the branchiopods (Fig. 14–32*B*). They produce a water current from which food particles and detritus are filtered. The collected material is then transferred to a ventral groove and moved anteriorly to the mouth parts.

The first four pairs of pleopods are used for swimming, and the last two pairs are reduced. The seventh abdominal segment lacks appendages.

The thoracic epipodites, which are flattened lamellae, form the gills. However, the inner surface of the carapace is also important to gas exchange. The compound eyes are stalked. The eggs are carried on the thoracic appendages of the female, and each egg hatches as a postlarva, called a manca stage, which still bears an incompletely developed carapace.

*Also classified as series Leptostraca; order Nebaliacea.

Series Eumalacostraca; Superorder Syncarida; Orders Anaspidacea, Stygocaridacea, and Bathynellacea

The Syncarida is represented by three small orders of primitive fresh-water crustaceans, the Anaspidacea, the Stygocaridacea, and the Bathynellacea. The five species of Anaspidacea occur in pools and lakes in Tasmania and South Australia, and probably represent relicts of a once more widely distributed order. The Tasmanian species *Anaspides* attains a length of 5 cm. Members of the Stygocaridacea and Bathynellacea are blind interstitial species and include some of the smallest malacostracans. *Bathynella natans* of European underground waters is about 2 mm. long. *Parabathynella neotropica* is only 1.2 mm. in length (Fig. 14–33). No species have yet been described from North America, but bathynellaceans have been reported from most other parts of the world.

Although living syncaridans are found only in fresh water, there are many extinct fossil marine species.

A carapace is lacking, and the first thoracic segment is fused with the head in the Anaspidacea (Fig. 14–34), but is free in the Bathynellacea. The thoracic appendages are biramous and similar, and the coxopodites of each appendage bear two simple lamellar gills. The pleopods have long fringed exopodites, but the endopodites are reduced or absent except for the first two pairs in the male, on which they function as copulatory organs. The uropods and telson are broad and form a well developed tail fan.

Anaspidaceans swim and crawl over the bottom or on aquatic vegetation. The fringed pleopods serve as the principal swimming appendages for the syncarids, and for very rapid movement the tail fan is used in some species. The Tasmanian *Anaspides* is reported to sometimes leap out of the water.

Syncarids feed chiefly on detritus, particularly the bathynellaceans, which inhabit underground water. In *Paranaspides*, which live in Tasmanian pools and lakes, the second maxillae have become modified for filtering, and a forward feeding current is produced by the beating of both the thoracic exopodites and the second maxillae. Filter feeding may well be secondary, for fossil

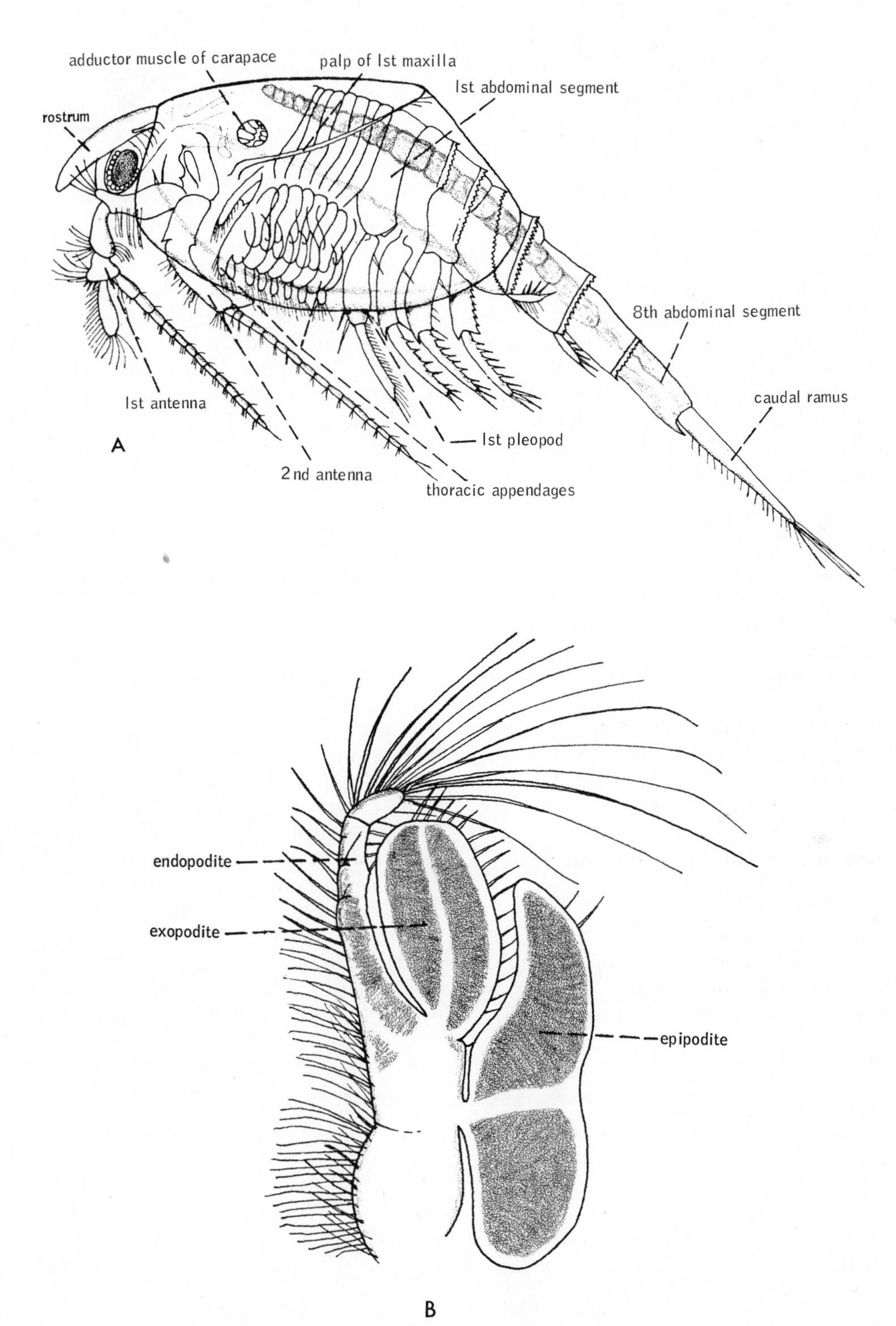

Figure 14–32. *A.* Female *Nebalia bipes,* a leptostracan. *B.* First thoracic appendage of *Nebalia.* (Both after Claus from Calman.)

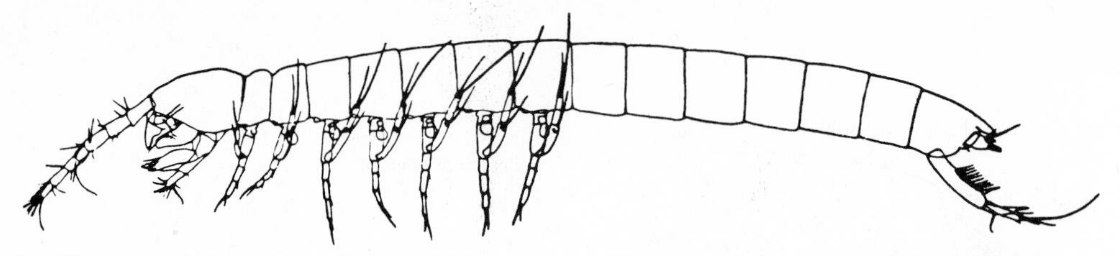

Figure 14–33. Lateral view of *Parabathynella neotropica*, a bathynellacean. Total length is only 1.2 mm. (From Noodt, W., 1965: Natürliches System und Biogeographie der Syncarida. Gewässer und Abwässer, 37–38:77–186.)

syncarids are thought to have been carnivorous (Brooks, 1962). *Anaspides* feeds on algae, detritus, tadpoles, and worms scraped up or captured by the anterior thoracic appendages. The food is macerated by the mandibles, and the maxillary filter collects the small particles, which are then transferred to the mouth.

The heart is tubular, extending through both thorax and abdomen. The excretory organs are maxillary glands. A statocyst is located in the base of each second antenna, and the eyes are either stalked or sessile.

The anterior pleopods of the male function as copulatory organs, and a median pocket-like seminal receptacle is located on the sternum of the female between the last pair of thoracic appendages. The eggs are not brooded and there are no free-swimming larval stages.

Superorder Hoplocarida and Order Stomatopoda

The marine crustaceans called mantis shrimps comprise the Stomatopoda, which is the only order of hoplocaridans. The body is dorsoventrally flattened and elongate (Fig. 14–35*A*). There is a rather shieldlike carapace, which is fused with the first two thoracic segments only (Fig. 14–35*B*). Anteriorly the carapace terminates short of the front of the head, so that the eyes and first antennae are attached beneath a small free rostrum. The third and fourth thoracic segments are vestigial, but the last four are well developed and unfused. The abdomen is large, broad, and distinctly segmented. The surface of the carapace, the free thoracic and abdominal tergites, and the broad telson are usually ornamented with keel-like ridges and spines.

The well developed compound eyes are large and stalked. Between them is a nauplius eye. The first antennae are large and triramous. The most conspicuous feature of the smaller second antennae is the large fringed scale that is almost as long as, or longer than, the endopodite. The most distinctive feature of the stomatopods is the structure of the thoracic appendages. The first five pairs are uniramous and subchelate. The second pair is enormously developed for raptorial feeding. The inner edge of the movable finger is provided with long spines or is shaped like the blade of a knife. The finger folds back into a deep groove in the

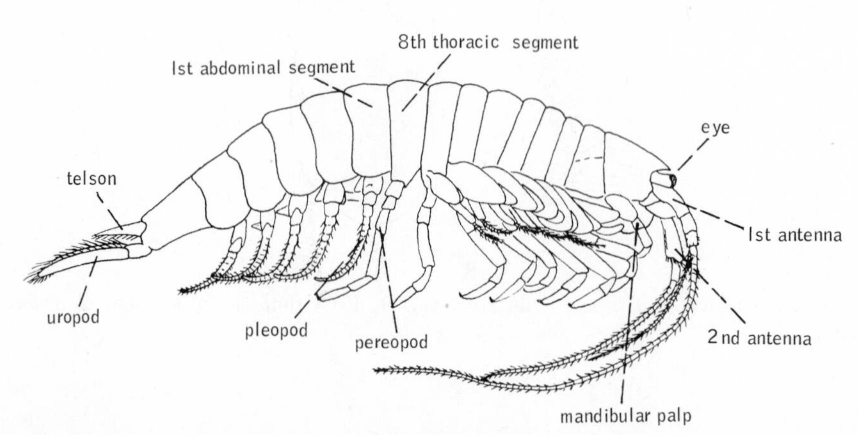

Figure 14–34. *Anaspides tasmaniae*, a syncaridan. (After Calman from Waterman and Chace.)

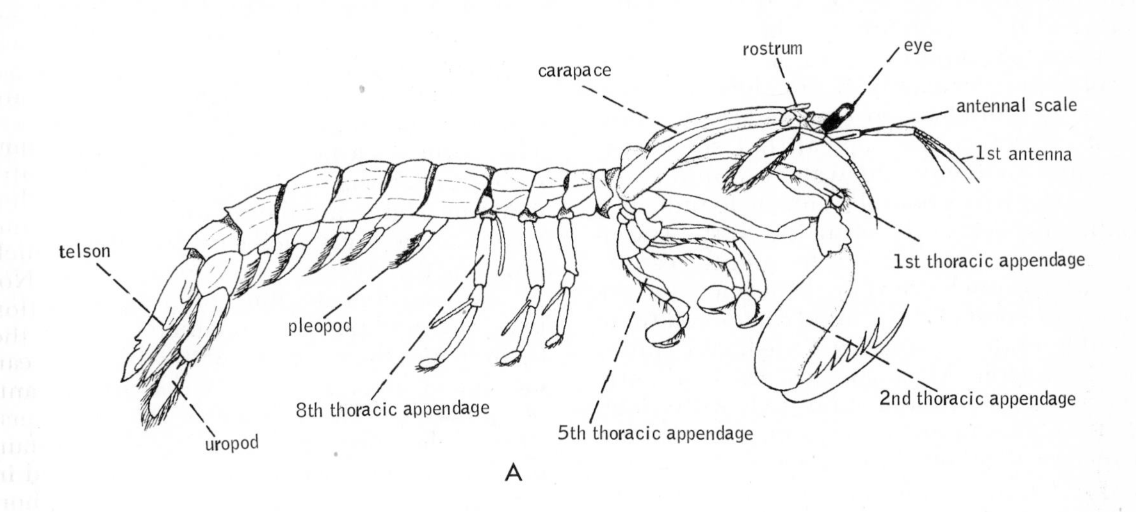

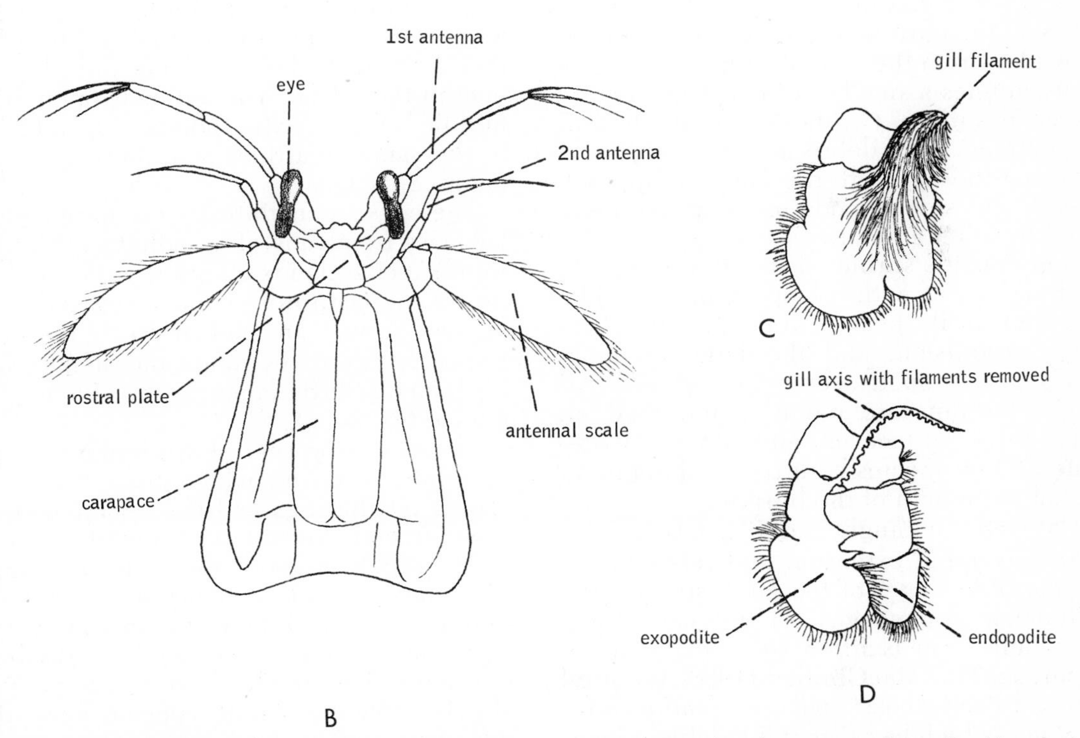

Figure 14–35. Stomatopoda. *A. Squilla mantis* (lateral view). *B.* Head and carapace of *Squilla* (dorsal view). *C.* Second pleopod and gill of *Squilla*. *D.* Second pleopod with gill filaments removed. (All after Calman.)

heavy penult segment, which has a sharp and finely toothed margin. The last three thoracic appendages are slender and not chelate. The pleopods are well developed and bear the gills (Fig. 14–35*C* and *D*). The uropods and the telson are very large.

Mantis shrimps range in size from small species approximately 5 cm. long to giant forms greater than 36 cm. in length. Most mantis shrimps are tropical, but some species live in temperate waters. *Squilla empusa*, which is about 18 cm. in length, is a common species inhabiting the North American Atlantic coast and is frequently caught in shrimp trawls. Shrimp fishermen on the southeast coast of the United States call them "double-enders" because of the heavy spines on the telson. Most species are brilliantly colored. Green, blue, and red with deep mottling are common, and some species are striped or display other patterns.

Most stomatopods live in burrows excavated in the bottom, or live in rock or coral crevices. One coral-inhabiting species, *Gonodactylus guerini*, has the entire surface of its telson ornamented with radiating spines. This armored telson is used to plug the entrance to the burrow when the mantis shrimp is inside. From the exterior, the telson accurately mimics a small sea urchin attached to the coral surface. *G. bredini* closes its burrow entrance with debris at night and blocks it during the day with the raptorial appendages. Some stomatopods use burrows excavated by other animals.

Many mantis shrimps leave the burrow to feed and swim with a looping motion. The powerful oarlike beating of the pleopods provides propulsion, and the large antennal scales and uropods serve as rudders. All mantis shrimps are raptorial and feed on small fish, crustaceans, and other invertebrates. Some species hunt prey; others lie in wait at the mouth of the burrow.

The victim is caught and dispatched by an extremely rapid extension and retraction of the movable finger of the large second pair of thoracic appendages. At the same time, the whole limb is lifted somewhat like an uppercut. The MacGinities (1968) reported seeing a Pacific coast species, *Pseudosquilla bigelowi*, which has a sharp bladelike finger, cut a shrimp neatly in two with one slice. Sometimes the prey is struck with the folded end joint of the appendage.

Mantis shrimps are the only malacostracans, aside from isopods, that have abdominal gills. These gills are filamentous, and the slender axis arises on the front sides of the exopodite near the basal suture. The heart is tubular and extends through most of the thorax and abdomen. It is strikingly segmental, displaying 13 pairs of ostia. Hemocyanin is present in the blood of at least some stomatopods. The excretory organs are maxillary glands.

The highly developed stalked compound eyes play an important role in prey capture. The corneal surface forms a large, swollen oval disc at the end of the peduncle and contains a large number of ommatidia, which have only a slight angle of divergence. Not only do the eyes provide for the detection of moving objects but, judging from the proximity and position in which the eyes can be held and the accuracy with which the animal can swim to prey, depth perception must be possible. The antennae are important sites of chemoreception and are also used in prey detection when the range is short enough.

The testes are delicate convoluted tubules located in the abdomen and joined together within the telson. The sperm duct opens at the end of a pair of sclerotized penes attached to the coxae of the last pair of legs. The ovaries extend from the middle of the thorax through the abdomen and also connect in the telson. The oviducts open on the middle of the sixth thoracic sternite, and in the same area there is a pouchlike seminal receptacle.

The eggs are agglutinated to form a globular mass by means of an adhesive secretion from special glands opening onto the sternites of the last three thoracic segments. The eggs in *Squilla* are picked up by the maxillipeds as they are extruded and then kneaded for hours with the anterior thoracic appendages (Fig. 14–36).

The entire process of spawning may take four hours, during which time the female stands on the legs and the telson with her body arched upward. The agglutinated egg mass, which is about the size of a walnut and may contain as many as 50,000 eggs (*Squilla*), is carried by the smaller chelate appendages and is constantly turned and cleaned. The female does not feed while she is brooding. Some species keep their egg mass inside the burrow. For example, a Bahamian species of *Gonodactylus*, which lies curled up in a coral crevice, holds the egg mass over the back of the carapace and abdomen like a cap and does not pick it up unless disturbed (Fig. 14–36).

A zoeal larva is the hatching stage and bears a much larger carapace than the adult

Figure 14–36. Two species of stomatopods caring for eggs. *Squilla mantis* (*left*) carrying egg mass; *Gonodactylus oerstedi* (*right*) with egg mass in burrow. (Modified after Schmitt.)

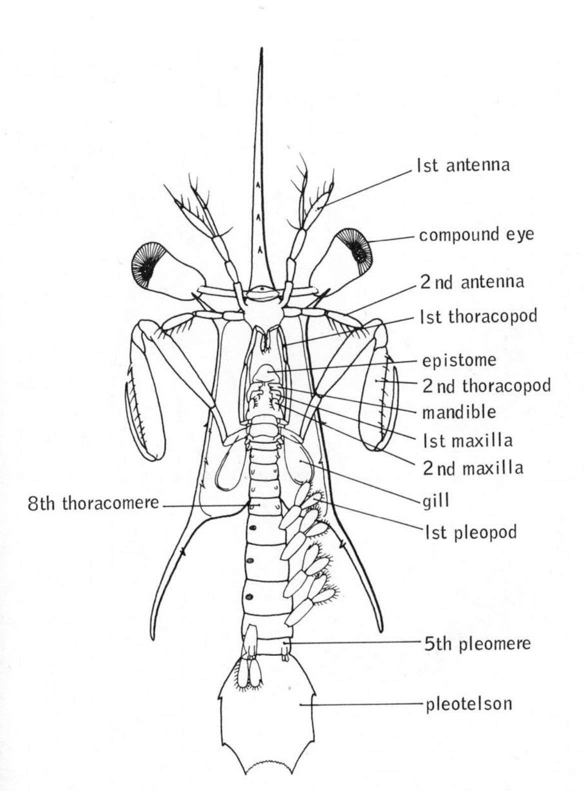

Figure 14–37. First pelagic antizoeal larva of *Squilla*, a mantis shrimp. (After Brooks). Vol. III. Brooks, W. K. 1878. "The larval stages of *Squilla empusa*." in Chesapeake Zoological Laboratory: Scientific Reports, pp 143–170.

(Fig. 14–37). Planktonic larval life may last for three months.

Superorder Peracarida

The subsequent five orders comprise the superorder Peracarida.* Primitively, the thorax is covered by a carapace, but in some orders, such as the amphipods and isopods, the carapace is greatly reduced. At least the first thoracic segment is always fused with the head, and the last four thoracic segments are almost always unfused, even when a carapace is present. The nauplius eye never persists in the adult.

A distinctive characteristic of the group is the presence of a ventral brood pouch, or marsupium, in the female. The marsupium is formed by large platelike processes (oostegites) on certain thoracic coxae. The oostegites project inward horizontally and overlap with one another to form the floor of the marsupium (Fig. 14–49). The thoracic sternites form the roof. The marsupial oostegites may make their appearance as small projections during juvenile instars; but their development is under hormonal control, and

*Two other orders, the Thermosbaenacea containing six species (which are exceptional among the peracaridans in that they carry eggs in a dorsal brood pouch) and the Spelaeogriphacea containing a single species, are omitted.

they are not completely formed until the reproductive instar. Development is direct, and on release from the brood chamber the young, called post-larvae, have most of the adult features.

Primitively, peracaridans are maxillary filter feeders, as in many mysids, cumaceans, and tanaidaceans, but the tendency in most higher peracaridans has been toward a direct mode of feeding.

Order Mysidacea

Mysidaceans look much like little shrimps. The majority are from 1.5 to 3 cm. in length, but some, such as the bathypelagic *Gnathophausia*, may attain a length of 35 cm. (Fig. 14–38*A*). A few species live in fresh water, but most are marine.

The thorax is covered by a carapace, but as in all peracaridans, the carapace is not

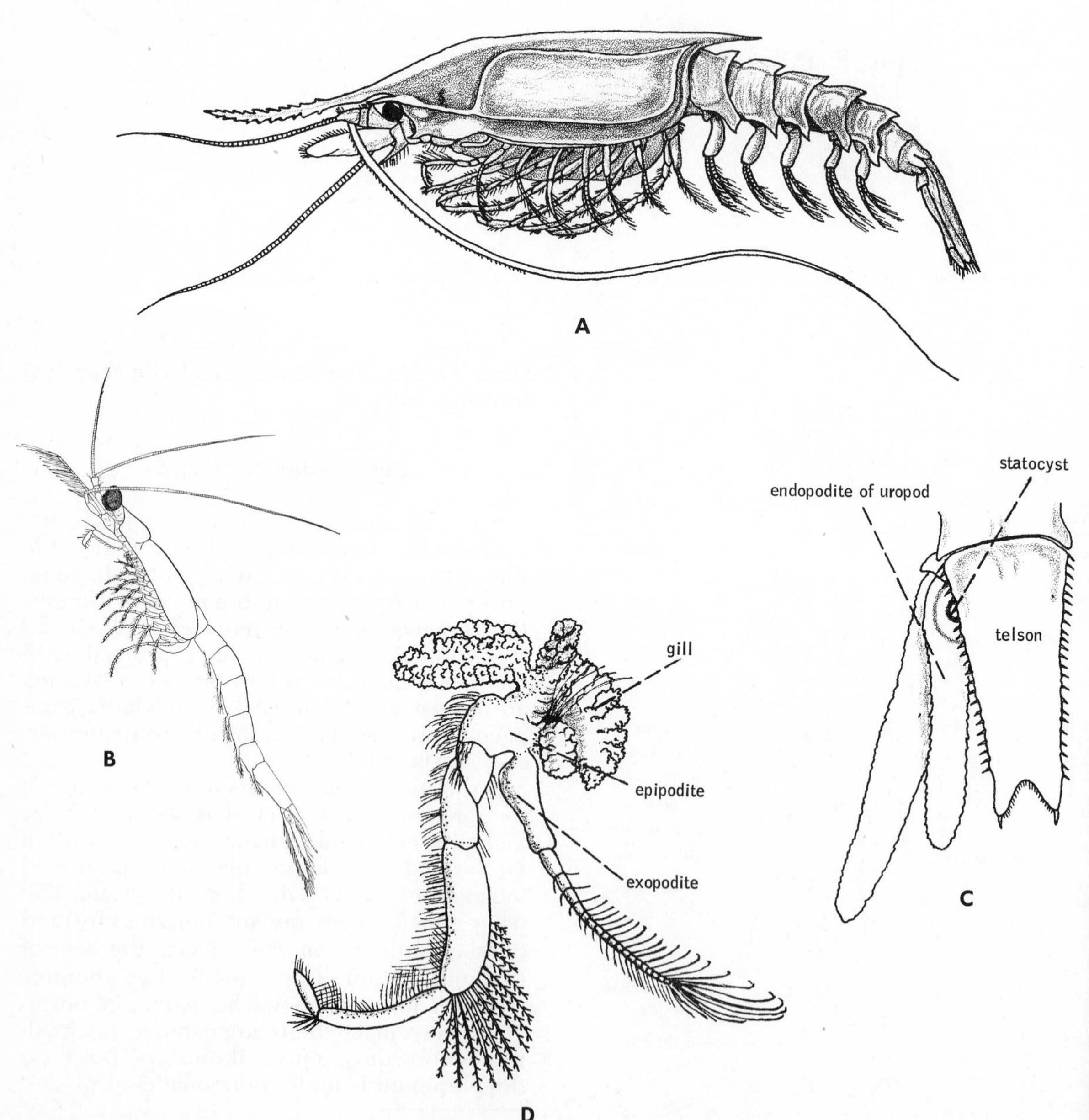

Figure 14–38. Mysidacea. *A. Gnathophausia* (lateral view). *B. Mysis oculata*, a freshwater mysid 3 cm. in length. (From Pennak, R. W., 1953: Freshwater Invertebrates of the United States. Copyright © 1953, The Ronald Press, N.Y.) *C.* Telson and one uropod of *Mysis*. *D.* Second thoracic appendage of *Gnathophausia*. (*A, C,* and *D* after Sars from Calman.)

united with the last four thoracic segments. Anteriorly, the carapace often extends forward as a rostrum, below which project stalked compound eyes. The first thoracic appendages and sometimes the second pair as well (in the Mysidae) are modified as maxillipeds. The remaining six or seven thoracic appendages are more or less similar, although some may be subchelate. The thoracic exopodites are filamentous and sometimes bear swimming setae. The pleopods in some groups, such as the Mysidae, are reduced in the female; in males the fourth pleopod is usually modified and very long.

Some genera, such as *Gnathophausia*, swim using the pleopods; others, such as *Mysis*, use the thoracic exopodites for swimming, particularly when the pleopods are reduced. Benthic forms crawl over the bottom or plow through the surface mud and silt.

Thoracic gills are present in the lophogastrids and are branched and foliaceous (Fig. 14–38*D*). In the Mysidae, the inner surface of the carapace acts as a gill. The ventilating current is produced by the thoracic exopodites and the maxillipeds and flows from posterior to anterior.

Most mysidaceans are filter feeders and feed while swimming. In *Gnathophausia*, the food current is produced by the beating of a large feathered exite on the second maxilla, and food particles are collected by filtering setae on the same appendage. Thus, in this genus, there is a swimming current produced by the pleopods, a ventilating current produced by the thoracic exopodites and maxillipeds, and a feeding current created by the maxillary exite; and all three currents are more or less separate and independent. This probably represents the primitive peracaridan condition.

In those genera, such as *Hemimysis*, in which pleopods are reduced and the exopodites have secondarily assumed a natatory function, the current produced by the thoracic appendages provides to some extent for all three functions—that is, for swimming, gas exchange, and feeding. In swimming, the exopodite performs a slightly rotatory movement, pulling water up along the axes of the leg to the midline of the body. From here the water is then moved forward by the beating of the maxillipeds and the exite of the maxillae, and food particles are filtered by setae on the maxillary endites. *Hemimysis* also feeds on large food masses, in addition to filter feeding. Many bathypelagic forms, such as the lophogastrids, have completely lost the filtering mechanism and are scavengers.

Excretory organs are antennal glands, but the primitive lophogastrids possess maxillary glands in addition.

About 450 mysidaceans have been described. Marine species often live in large swarms and form an important part of the diet of such fish as shad and flounder. Many marine forms are found in algae and tidal grass, and can tolerate relatively foul conditions. The few freshwater species appear to be rather recent immigrants from the sea. Of the several North American species, one has been reported from fresh and brackish water of Louisiana and Texas. Another, *Neomysis mercedis*, lives in estuaries, rivers, and lakes on the west coast, and a third, *Mysis relicta*, is confined to cold deep-water lakes of the northern United States, Canada, and Europe (Fig. 14–38*B*). Lake trout feed extensively on this species.

The stalked compound eyes, well developed carapace, thoracic gills, and long tubular heart, and the presence of both antennal and maxillary glands are believed to be primitive features and suggest that the mysidaceans stem from near the base of the peracaridan line.

Order Cumacea

Cumaceans are marine peracaridans of approximately the same size as or a little smaller than mysids. Most cumaceans inhabit the bottom, where they live buried in sand and mud. The body shape is very distinctive (Fig. 14–39). The head and thorax are greatly enlarged, while the abdomen is very narrow and terminates in slender elongated uropods. A carapace is present and is fused with the first three or four thoracic segments. The carapace is peculiar in having two anterior-lateral extensions. These extensions swing together in front of the animal to form a ventral false rostrum. When eyes are present, they are located on a common median prominence situated above the rostrum base.

The antennae are vestigial in the female; in the male they are extremely long and are borne folded back along the sides of the body, sometimes in a groove. The first three thoracic appendages are modified as maxillipeds, and the fourth pair is long and prehensile. Exopodites are present on some of

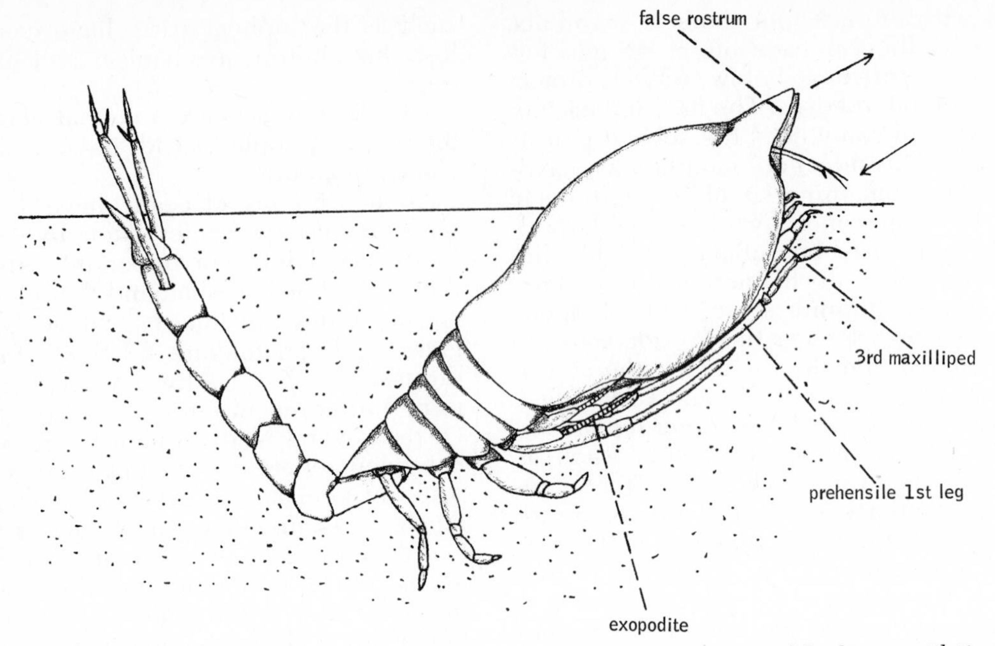

Figure 14–39. *Diastylis*, a cumacean, buried in sand. Arrows indicate direction of feeding–ventilating current. (Modified from several authors.)

the legs. Females lack pleopods, but they are usually present in males.

The thoracic exopodites are used in swimming and may be assisted by lateral movements of the abdomen. The more posterior legs are used for burrowing. The animal moves backward into the burrow as the sand is excavated and ends up in an inclined position with its head projecting above the surface (Fig. 14–39).

A series of filamentous gills are located on the epipodite of each first maxilliped. This epipodite also forms the floor of a branchial chamber that is walled and roofed in by the sides of the body and the carapace. As a modification for the partially buried existence, the inhalant ventilating current has become shifted anteriorly. The movement of the epipodite acts as a ventilating pump, and water is drawn in anteriorly over the mouth parts and passes upward through the branchial chamber. The water is then sent anteriorly again by the long, concave, platelike exopodite of the first maxilliped, which fits against the pseudorostrum and forms an exhalant siphon. The exhalant current thus emerges anteriorly on both sides of the head as two parallel streams.

In *Diastylis*, the inhalant ventilating current, while passing over the mouthparts, is filtered for food particles by setae on the second maxillae. Not all cumaceans are filter feeders. Some, such as *Cumopsis*, *Lamprops*, and *Iphinoe*, scrape organic matter from sand grains. While the animal is half buried in sand, the grains are swept up by the first pairs of legs, passed to the third maxillipeds, and then passed to the first and second. The latter two pairs of maxillipeds are modified to form a cup in which the sand grain is rotated and the food material is rasped from its surface. The cleaned grain is either thrown into the exhalant ventilating current or thrown over the eye onto the side of the carapace, where it rolls off. When the animal becomes too deeply buried or when too large a hole has been formed by the sweeping, it moves and reburies itself.

Excretory organs are maxillary glands.

Swarming at the time of mating is characteristic of many cumaceans. Large numbers, particularly males, leave their burrows and swim to the surface.

More than 770 species of cumaceans have been described. Most live at depths of less than 500 meters; where bottom conditions are favorable, some species may attain population densities of thousands per square meter.

Order Tanaidacea

The Tanaidacea display similarities to both the Cumacea and to the Isopoda, with which they were formerly classified. These peracaridans, of which some 350 species have been described, are almost exclusively marine and generally small, most of them being only 1 to 2 mm. in length (Fig. 14–40). The majority are bottom inhabitants of the littoral zone, where they live buried in mud, construct tubes, or live in small holes and crevices in rocks.

A small carapace covers the anterior part of the body and is fused to the first and second thoracic segments. Many species lack eyes, but when eyes are present, they are located laterally on immovable processes. The first pair of thoracic appendages are maxillipeds and the second pair (gnathopods) are large and chelate, a distinctive feature of the tanaidaceans. The third pair of thoracic appendages are adapted for burrowing. The exopodites have disappeared in all thoracic appendages except the second and third, probably as a result of the burrowing habit. Pleopods are sometimes reduced or absent in females. The last abdominal segment is fused with the telson.

The inner surface of the carapace functions as a gill. A posterior to anterior feeding current is produced by the beating of the epipodites of the maxillipeds and, to a lesser extent, by the exopodites of the second and third thoracic appendages. The second maxillae filter some food particles, but in *Apseudes* and other members of the same family, large particles are retained on the brushing setae of the maxillipeds. The diet is also supplemented by raptorial feeding. This tendency toward the loss of filter feeding in the Apseudidae is culminated in the Tanaidae, which are entirely raptorial. Associated with this change in feeding habits, the branchial chambers are less developed, the maxillae are greatly reduced, and the remaining thoracic exopodites have disappeared.

Some tanaidaceans are hermaphroditic, but the eggs are brooded as in other peracaridans.

Order Isopoda

The order Isopoda is the largest order of crustaceans aside from the Decapoda. Most of the 4000 described species are marine, but a considerable number live in fresh water; and the pill bugs, or wood lice, are the only large group of truly terrestrial crustaceans. The order also includes many parasitic forms. The isopods represent the culmination of the cumacean-tanaidacean line that probably originated in some ancestral form near the lophogastrid mysidaceans.

The most striking characteristic of the isopods is the dorsoventrally flattened body (Figs. 14–41 and 14–42*B*). The head is usually shield-shaped, and the terga of the thoracic and the abdominal segments tend to project laterally. A carapace is absent (remember the progressive reduction through the Cumacea and Tanaidacea), although the first one or two thoracic segments are fused with the head. The abdominal segments may be distinct or fused to varying degrees. The last abdominal segment is almost always fused with the telson; and in the Asellota, which includes the greatest number of freshwater species, all but the first or second abdominal segments are fused with the telson to form a large abdominal plate (Fig. 14–41*A*). The abdomen is usually the same width as the thorax, so that the two regions may not be clearly demarcated dorsally.

There are exceptions to most of these characteristics. Members of the mostly marine suborder Anthuridea have an elongated cylindrical body (Fig. 14–42*A*); and the fresh-

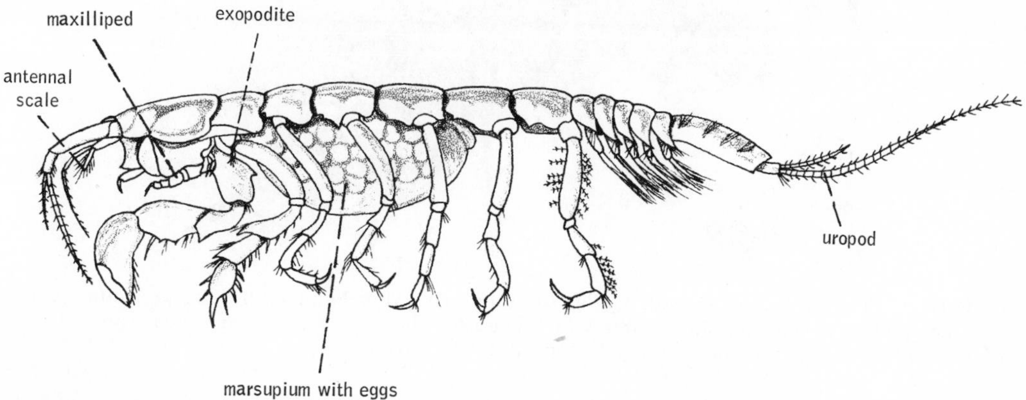

Figure 14–40. A tanaidacean, *Apseudes*, with eggs in marsupium (After Sars from Calman.)

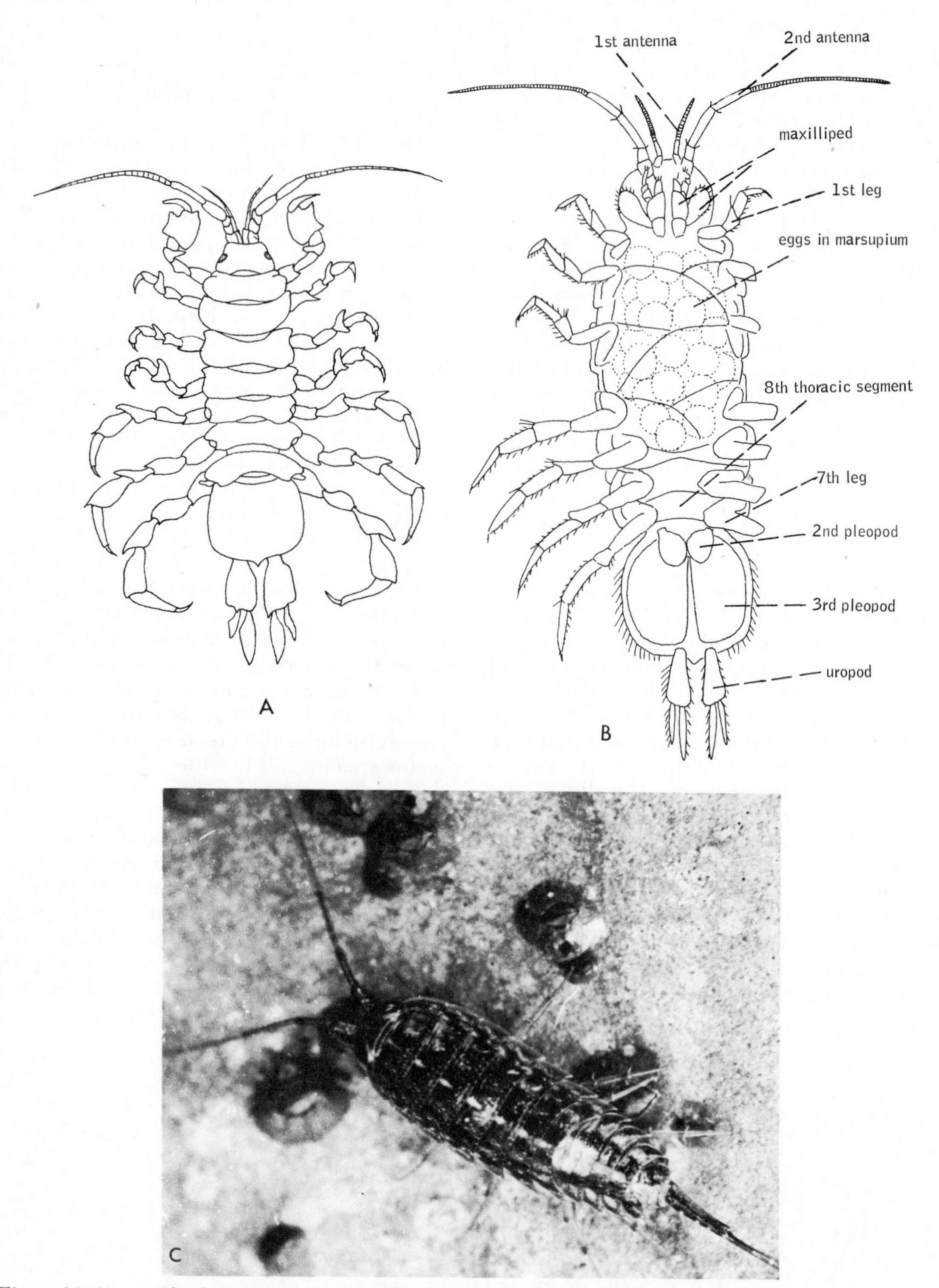

Figure 14–41. *A.* The freshwater isopod, *Asellus* (dorsal view). (After Pennak.) *B. Female Asellus* (ventral view). Appendages complete only on one side. (After Van Name.) *C.* The isopod *Lygia* on a sea wall at low tide. The barnacles are a little less than one cm. in diameter. (By Betty M. Barnes.)

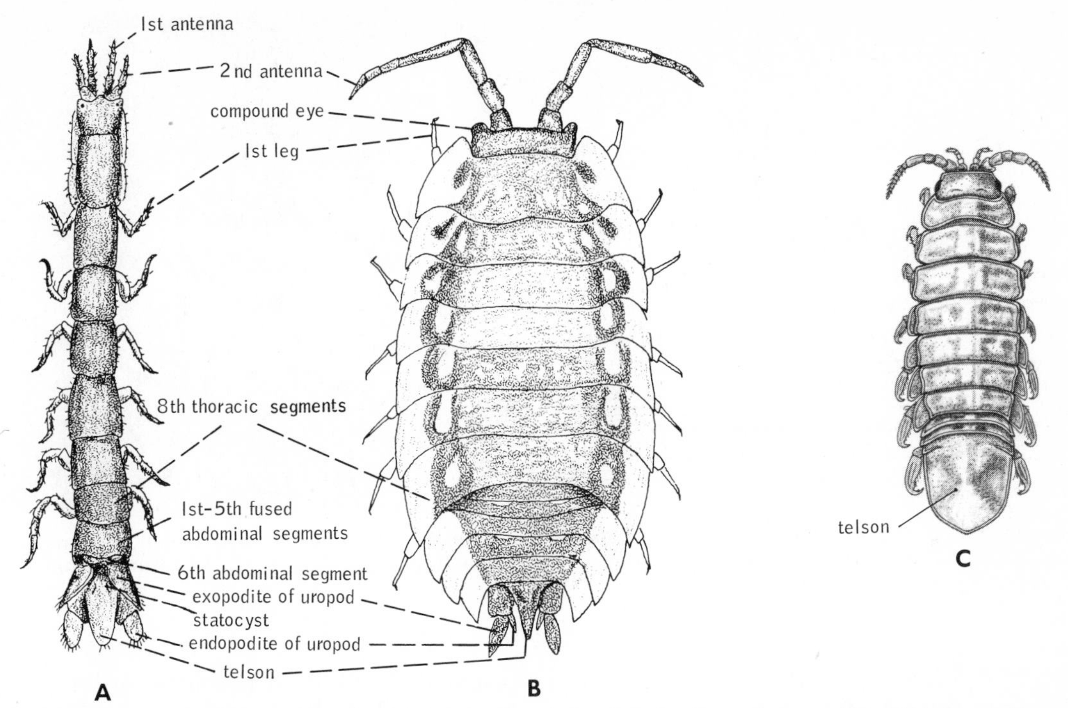

Figure 14-42. *A.* Marine isopod, *Cyathura carinata.* (After Gruner from Kaestner.) *B. Oniscus asellus,* a terrestrial isopod. (After Paulmier from Van Name.) *C. Idothea pelagica,* a pelagic marine isopod. (After Sars.)

water *Phreatoicus* is even laterally compressed. The abdomen of the Gnathiidae, which look like insects, is narrow and taillike (Fig. 14-47*B* and *C*).

The first antennae are short and uniramous, and in terrestrial isopods they are vestigial. The second antennae, however, are well developed except in some parasitic forms, but the exopodite is typically absent (Fig. 14-41). The compound eyes are always sessile.

The first pair of thoracic appendages are modified to form maxillipeds; the other thoracic appendages are usually adapted for crawling. In some groups the more anterior pairs are modified as prehensile gnathopods (Fig. 14-41A). The thoracic appendages are uniramous, and the coxopodite is commonly expanded, forming a coxal plate, which may be fused with the tergum and the sternum. Unlike those in most other crustaceans, some of the isopod pleopods are modified for gas exchange. The uropods are usually fan-shaped or styliform.

Most isopods are 5 to 15 mm. in length. The giant of the group is the deep sea *Bathynomus giganteus,* which reaches a length of 42 cm. and a width of 15 cm. The coloration in isopods is usually drab, shades of gray being the most common.

Locomotion. Isopods are benthic animals, and most are adapted for crawling. *Lygia,* one of the most common isopods along coastlines, can run very rapidly over exposed wharf pilings and rocks. Manton (1952) states that each leg in *Lygia* may take 16 steps per second, a frequency that is surpassed by few other arthropods. Some isopods, such as *Cruregens,* can run backward as rapidly as forward. Some terrestrial wood lice can climb vegetation, and *Lygia* commonly runs about upside down beneath stones or dock planking. *Astacilla,* which climbs over seaweed and hydroids, has posterior thoracic limbs that are adapted for clinging.

Many aquatic isopods burrow. Commonly the anterior legs are pressed down into the mud and then moved to the side like a breast stroke in swimming. Some burrowing species construct tunnels through the substratum, packing excavated material against the walls. *Limnoria* tunnels through wood and can cause extensive damage to docks and pilings (Fig. 14-43). *Sphaeroma tenebrans* bores into the prop roots of mangroves. Particularly remarkable is the tropical American isopod, *Cirolana salvadorensis,* which plows through dry beach sand.

Aquatic isopods can usually swim, as well as crawl. In some forms, such as *Munnopsis,*

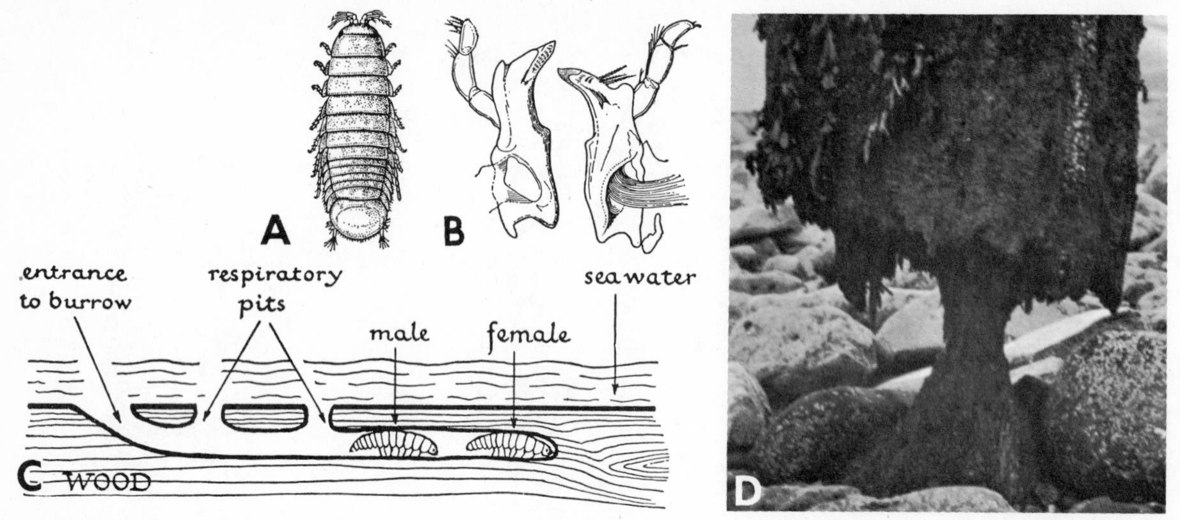

Figure 14-43. *A*. The wood-boring isopod, *Limnoria lignorum*. *B*. Mandibles of *L. lignorum*. *C*. Diagrammatic section of burrow of *L. lignorum*. *D*. Jetty piling nearly eaten through at base by *L. lignorum*. (From Yonge, C. M., 1949: The Seashore. Collins, London.)

certain of the legs are flattened and paddle-shaped, and they beat in an oarlike or rotary manner. However, most commonly the pleopods are used for swimming, and in the families Sphaeromidae and Serolidae, the first three pairs of pleopods are especially adapted for swimming; gas exchange is restricted to the more posterior pleopods. The tail fin is rarely employed in swimming, in contrast to its use in many other malacostracans.

The ability to roll up in a ball has evolved in many terrestrial Oniscoidea, affording protection and aiding in the reduction of water loss from evaporation. *Oniscus* and *Armadillidium* assume a perfectly spherical shape when they roll up, from which characteristic the name *pill bug* is derived. Many marine sphaeromids also roll up, with the sharp and spiny tips of the uropods and telson exposed.

Feeding. Isopods have completely dispensed with filter feeding, and the majority are scavengers and omnivorous, although some tend toward a herbivorous diet. Wood lice feed on algae, fungi, moss, bark, and any decaying vegetable or animal matter. One species that is commensal in ant nests feeds on the feces of the host. A few wood lice are carnivorous.

The mouth parts tend to be compact, forming a prognathus mound that is protected by a small labrum in front and by maxillipeds behind. The maxillipeds form a protective operculum, and often the maxilliped of one side is hooked to its opposite member by special coupling setae. During feeding, food is usually held up by the anterior legs, which may be subchelate, while chewed upon by the mouth parts.

The triturating, or cardiac, stomach is well developed, particularly so in the terrestrial wood lice. The midgut contains one to three pairs of ceca. The digestive secretions of the wood borer, *Limnoria*, and species of *Sphaeroma* are reported to contain cellulase; in the wood louse, *Porcellio*, cellulose digestion results from symbiotic bacteria.

There are several groups of parasitic isopods. The Gnathiidae in the larval stages and the adult Cymothoidae are ectoparasitic on the skin of fish and have mandibles adapted for piercing (Fig. 14-44). Piercing mouth parts are also present in the parasites composing the suborder Epicaridea, all of which are bloodsuckers, parasitic on many different crustacean groups. Many epicaridans are highly modified and show little resemblance to free-living forms (Fig. 14-45*A*). This is especially true of members of the family Entoniscidae, which live in the gill chambers of crabs. They occupy a deep invagination of the body wall, through which water from the branchial chamber of the host circulates (Fig. 14-45*B*).

Gas Exchange and Excretion. The shift of the gills from the thorax to the abdomen in isopods is perhaps correlated with the reduction of the protective branchial chamber provided by the carapace. Primitively, each pleopod ramus is modified as a large flat lamella, and both rami of each pleopod function both in gas exchange and in swimming (Fig. 14-41*B*). However, there is usually some modification of this arrangement. In some isopods, gas exchange and swimming are divided among the pleopods, the anterior ones being fringed and coupled for swimming and the posterior pleopods being for gas exchange.

The pleopods typically lie flat against the underside of the abdomen and are often protected by a covering (the operculum) derived in a variety of ways. In some forms, the first pair of pleopods is modified, forming an operculum that covers the more posterior pairs; the second or third exopodites may also form opercula. In the marine Valvifera, the uropods are greatly elongated and meet at the midline ventrally to form a gill covering resembling two doors. Sometimes the gill surface is increased by folding or by filamentous or villous projections, particularly in parasitic isopods.

In the amphibious and terrestrial Oniscoidea, the exopodite of each pleopod forms an operculum covering the gill. The undersurface of the abdomen is thus covered by two rows of overlapping opercula. If gas exchange through the gill is blocked in *Lygia* and *Oniscus*, oxygen consumption is reduced by only about 50 per cent, indicating that the general integumentary surface is equally important as a site of gas exchange.

As in cumaceans and tanaidaceans, the excretory organs are maxillary glands.

Sense Organs. The compound eyes are always sessile and are located on each side of the head. Eyes are absent in some parasitic, cave-dwelling, and deep-sea species; in the fresh-water *Asellus*, the eyes contain only a few ommatidia. On the other hand, in *Bathynomus*, each eye is composed of approximately 3000 visual units.

A pair of statocysts are located in the telson of some isopods, such as the burrowing *Cyathura*, and experiments on *Cyathura carinata* demonstrate a loss of ability for vertical burrowing when the statocysts are removed or destroyed (Fig. 14–42A).

Adaptations for Life on Land. Almost all of the terrestrial isopods, the wood lice or pill bugs, are members of the suborder Oniscoidea. They are believed to have invaded land directly from the sea, rather than by way of fresh water, and have come to occupy a wide range of habitats and to display varying degrees of toleration to desiccating conditions. Some live at the edge of the sea and some in marshes. Many species live beneath stones, in bark, and in leaf mold in both temperate and tropical regions, but there are also species which live beneath stones in deserts.

Shore-inhabiting forms include the widespread *Lygia*, which lives on pilings, jetties, and rocks at the water's edge, and *Tylos*, which lives beneath drift or sand at the high tide mark on protected beaches. *Tylos* emerges at low tide to feed and orients to the high tide mark by means of beach slope and sand moisture. Some species can also use the sun's angle for orientation.

Most terrestrial isopods possess some

Figure 14–44. An isopod fish louse, *Renocila heterozota*, on an anemone fish. Finger-like projections in photograph are tentacles of sea anemone. (By Mariscal, R. N., 1969: Crustaceana, *14*:7–104.)

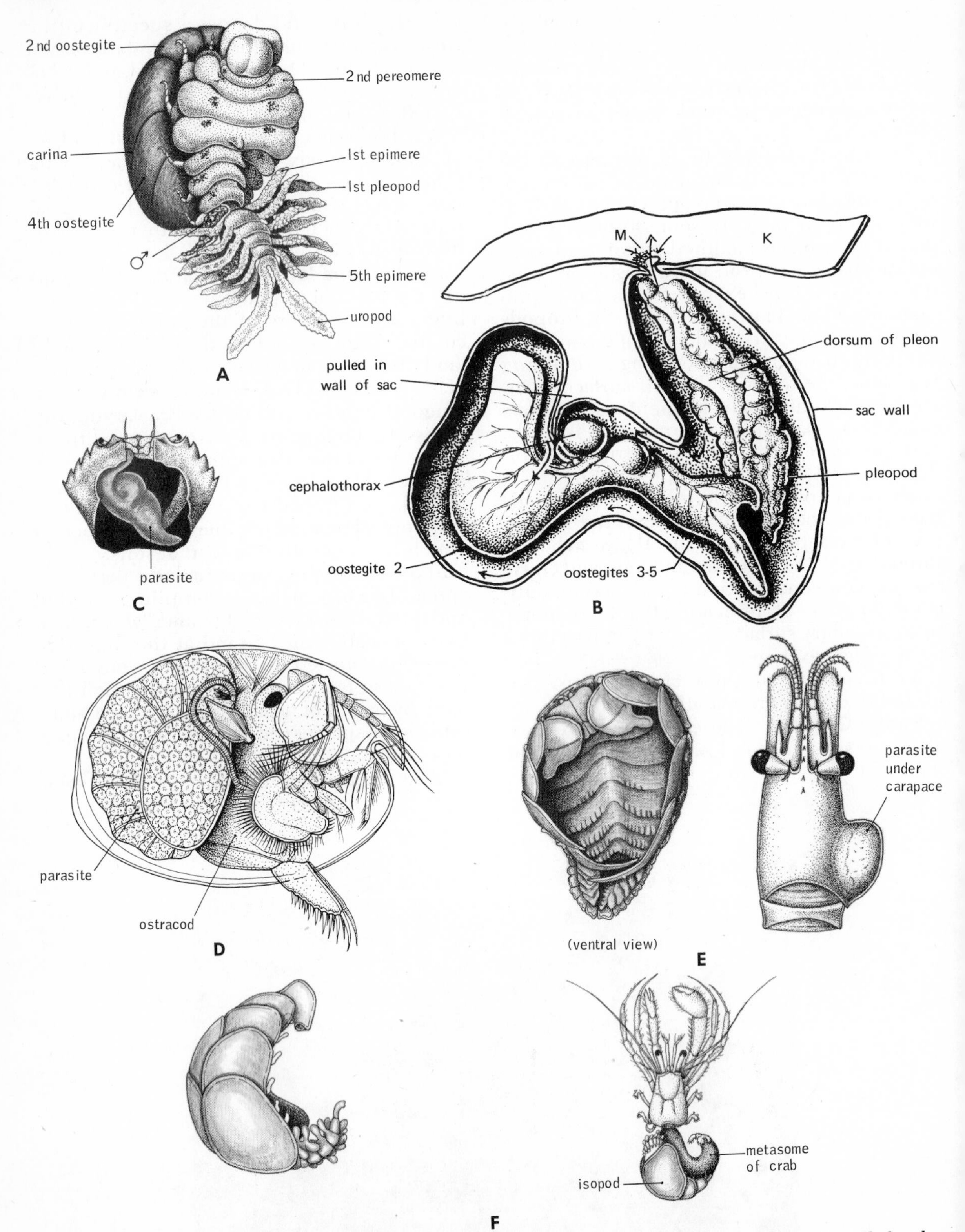

Figure 14–45. Parasitic epicaridean isopods. *A*. The bopyrid *Cancricepon elegans*, parasitic in the gill chambers of certain crabs. *B*. The entoniscid *Portunion maenadis* lying within a sac-like invagination of the body wall of a crab gill chamber. Arrows indicate flow of water through sac. *C*. Area within the crab carapace occupied by *Portunion*. (*A* and *C* after Giard and Bonnier; *B* from Kaestner, A., 1970: Invertebrate Zoology, Vol. 3, Wiley-Interscience, N. Y.) *D*. Isopod *Cyproniscus* in ostracod *Cypridina*. *E*. Ventral view of *Bopyrus squillarum* and location in branchial cavity of shrimp. *F*. *Athelges tenuicaudis* on abdomen of hermit crab. (*D* to *F* after Sars.)

adaptations to reduce water loss, but the group as a whole is considerably less well adapted in this regard than are other terrestrial arthropods. Wood lice tend to be nocturnal and live beneath stones and in other places where the environment is humid. They have never evolved a waxy epicuticle of the type responsible for reducing integumental evaporation in insects and spiders. The thin ventral exoskeleton is a primary site of evaporation, and the ability to roll up into a ball is probably an adaptation to cut down on water loss. In general, wood lice are photonegative and strongly thigmotactic, and can discriminate between relatively slight differences in humidity, all of which tend to keep them beneath protective retreats during the day. The commonly observed aggregation of wood lice under certain stones or wood results in part from their being attracted by the body odor of other individuals of their own species.

The eyes of wood lice are poorly developed, probably correlated with their nocturnal, secretive behavior and their diet of decaying vegetation. Repugnatorial glands are used in defense against such predators as spiders and ants.

Wood lice continue to use the gills for gas exchange, but in the more terrestrial species the operculum contains a lung-like cavity (in the Oniscidae) or tube-like invaginations or pseudotracheae (as in the Porcellionidae and Armadillidiidae) (Fig. 14–46*B* and *C*). Wood lice with pseudotracheae can tolerate much drier air than can wood lice without them.

The replacement of water lost from integumental evaporation comes from moist food and drinking. The gills must retain a covering film of moisture. The fact that they lie within a depression of the covering exopodite facilitates water retention, but water must be replaced periodically (Fig. 14–46*A*). In some wood lice, the two endopodites of the uropods are held together like a tube and dipped into a droplet of dew or rain. Water taken up is distributed to the gills. Other wood lice possess a system of surface channels which carry any water that comes in contact with the animal's back to the ventral surface and then back to the gills.

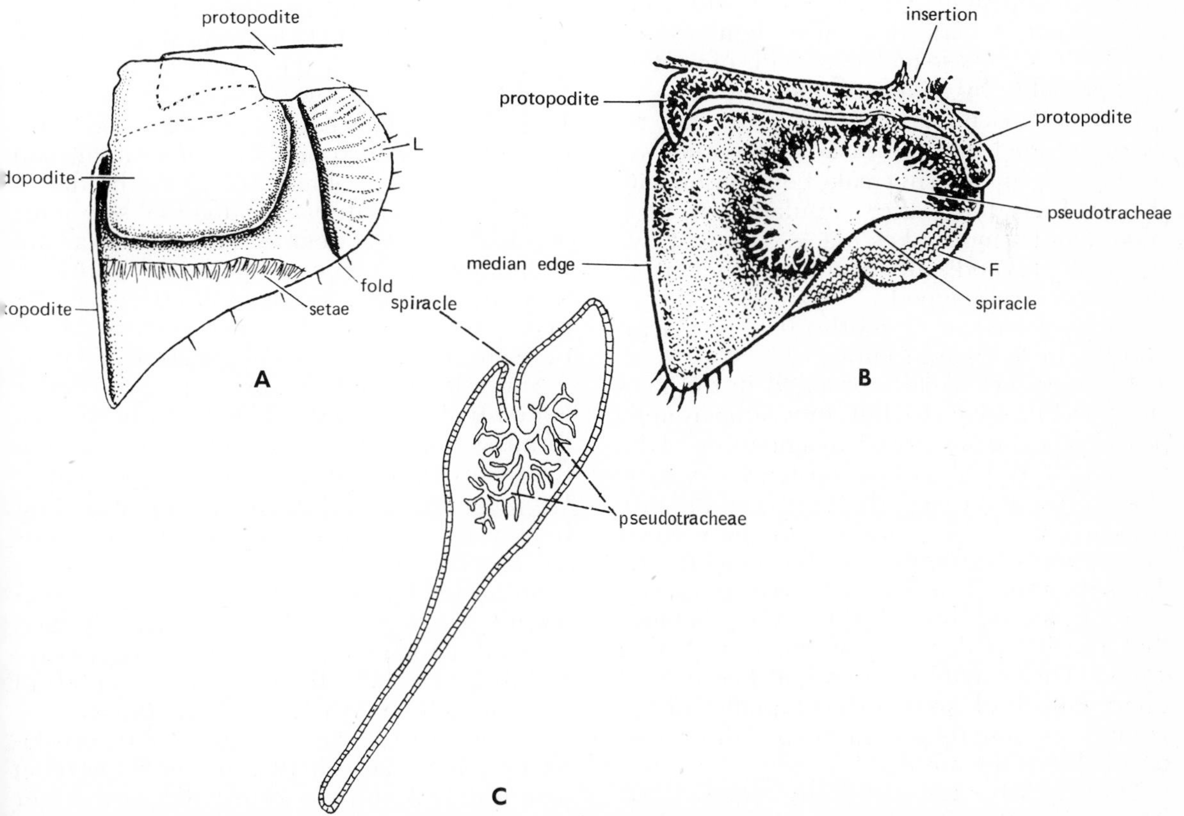

Figure 14–46. *A*. Fifth pleopod of the wood louse *Oniscus asellus*, showing depression in exopodite in which respiratory endopodite lies. Setae keep out foreign material in water entering depression. *B*. Posterior or dorsal view of the exopodite of the wood louse *Porcellio scaber*, showing spiracle and location of pseudotracheae. *C*. Section through exopodite figured in *B*. (All after Unwin from Kaestner, A., 1970: Invertebrate Zoology, Vol. 3. Wiley-Interscience, N.Y.)

The maxillary glands of wood lice are poorly developed and, despite their terrestrial habit, these isopods release nitrogenous wastes as ammonia, at least in part as gas.

Reproduction and Development. The gonads are paired and separate. The male sperm ducts open onto the sternum of the genital segment by way of either separate or united papillae or penes. The first pleopod of the male bends the penis back to the endopodites of the second pleopods, and the cavity in the second pleopods is filled with sperm. The female gonopores are also paired sternal openings. In the Asellidae, each oviduct is enlarged basally, forming a seminal receptacle. During copulation, the male injects sperm into one of the gonopores, first on one side and then on the other. In the American freshwater asellid, *Lirceus*, the male seizes a preadult female that is about to, or has just, shed the posterior half of its preadult cuticle, and he presses his ventral surface against the side of her body. The endopodites of the second pair of pleopods vibrate during sperm transmission. The male then moves to the other side of the female's body, where the process is repeated. The eggs are fertilized in the oviduct. A male recognizes females and determines their sexual state with his antennae, probably by means of pheromones.

In the Oniscoidea, the gonopores are reported to lead into blind saclike seminal receptacles, which do not make connection with the oviducts until after copulation is completed and a molt takes place. The first two pairs of pleopods are copulatory organs in most terrestrial isopods.

Many parasitic Cymothoidae and Epicaridea are hermaphroditic.

The eggs are usually brooded in the marsupium (Fig. 14–41*B*), but some sphaeromids possess special pouchlike invaginations of the thoracic sternites, and the oostegites may be absent. In aquatic isopods, there is often a process from the basal portion of the maxilliped that projects backward into the marsupium. The vibrations of this process create a water current passing through the brood chamber that facilitates aeration of the developing young. The marsupium of wood lice is kept filled with fluid, so that development of the young is essentially aquatic despite the terrestrial habit of the adults.

Thirty to fifty eggs are usually brooded, but the number of young that complete development is significantly smaller. As in cumaceans and tanaidaceans, the hatching stage is a postlarva (manca stage), with the last pair of legs incompletely developed. The young usually do not remain with the female after they leave the marsupium; but in *Arcturus*, the female carries the young about attached to her long antennae (Fig. 14–47*A*).

SYSTEMATIC RÉSUMÉ OF ORDER ISOPODA

Suborder Gnathiidea. First and seventh thoracic segments reduced, so that only five large segments are visible dorsally. Eighth thoracic appendages are absent. Abdomen small and much narrower than thorax. Manca stage is ectoparasitic on marine fish. *Gnathia.*

Suborder Anthuridea. Body long and cylindrical; last abdominal segment not fused with telson. First pair of legs are heavy and subchelate; first pair of pleopods form an operculum covering other pairs. All are marine except a few freshwater species. *Anthura, Cyathura, Cruregens.*

Suborder Flabellifera. Body is more or less flattened; some abdominal segments may be fused together; the last segment is fused with the telson. The uropods are fan-shaped and form a tail fan together with the telson. Coxae are expanded to form coxal plates; they may be fused with the body. Members are largely marine; a few found in fresh water. Included are the wood borer, *Limnoria; Cirolana; Bathynomus; Serolis; Sphaeroma;* and the ectoparasites of fish, the Cymothoidae.

Suborder Valvifera. Abdominal segments are fused to some degree. Thoracic coxae are expanded to form coxal plates. Exopodite of uropods are either vestigial or absent; endopodite forms an operculum over gills. Members are marine. *Astacilla, Arcturus, Idotea.*

Suborder Asellota. Three to five posterior abdominal segments fused with telson, forming a large plate. Coxae of thoracic appendages are never expanded to form plates. Uropods are styliform. Includes marine and freshwater species. *Asellus, Lirceus, Munnopsis, Jaera.*

Suborder Phreatoicidea. Abdominal segments are not coalesced. Body is laterally compressed. Thoracic coxae are small; uropods are styliform. Includes the freshwater isopods of Australia and South Africa. *Phreatoicus.*

Suborder Epicaridea. Parasites on crustaceans. Suctorial mouthparts and piercing mandibles; both pairs of maxillae are either vestigial or absent. Females are often greatly modified; some without segmentation or appendages. *Bopyrus, Entoniscus, Portunion, Liriopsis.*

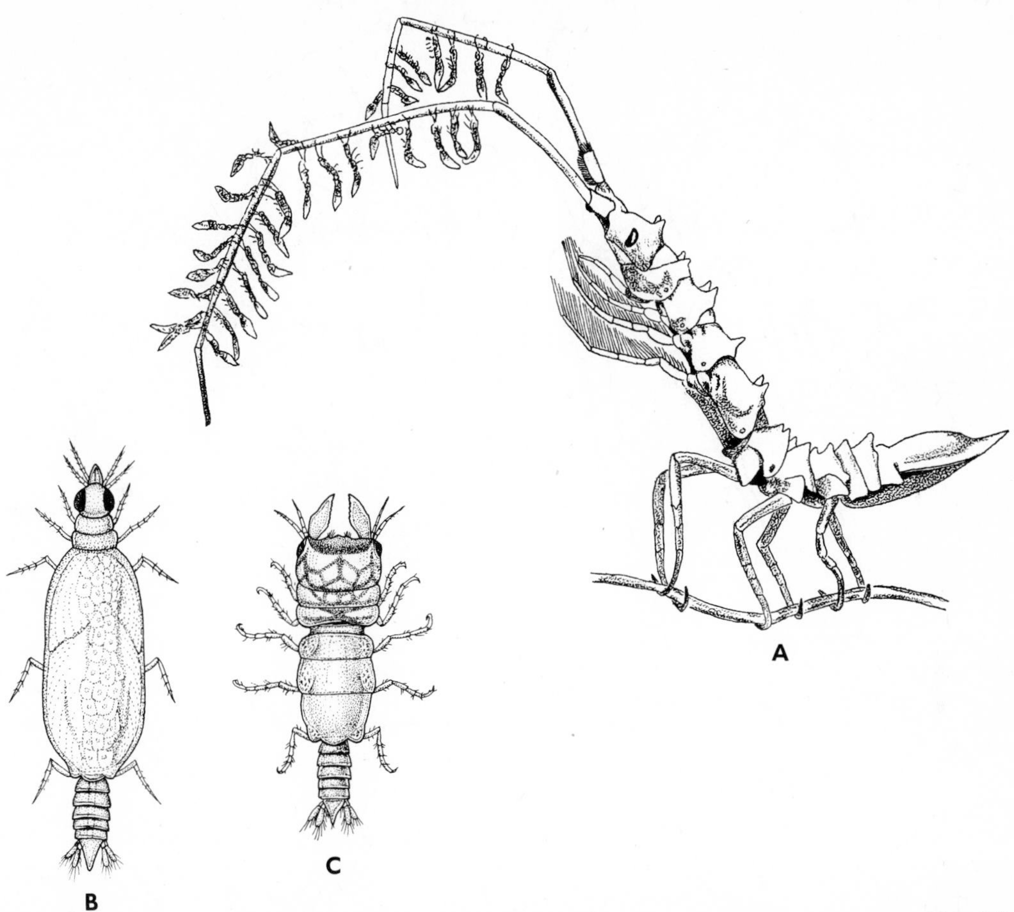

Figure 14–47. Female of *Arcturus baffini*, a marine isopod, carrying young on antennae. (Modified after Schmitt.) *B* and *C*. An aberrant gnathidean isopod, *Gnathia maxillaris*, which looks like an insect. The larval stage (*B*) is parasitic on fish and possesses crowded sucking mouthparts. Both larva and adult (*C*) are less than 3 mm. long. (Based on living specimens and figures by Sars.)

Suborder Oniscoidea. Abdominal segments are usually distinct. First antennae are vestigial. Coxae are expanded to form coxal plates and are fused with body. Amphibious and terrestrial members. *Ligia*, *Oniscus*, *Porcellio*, *Armadillidium*, *Tylos*.

Order Amphipoda

The 4600 species of amphipods comprise the second of the two major groups of peracaridans and probably represent a separate line of evolution from some ancestral mysidacean. Most are marine, but there are many freshwater species; there are also some semiterrestrial and terrestrial forms. The structure of amphipods displays some convergence with that of isopods (Fig. 14–48). The eyes are sessile. There is no carapace, although the first thoracic segment and sometimes the second (as in the Caprellidea) are fused with the head. Thoracic exopodites have disappeared. Also, the abdomen is usually not distinctly demarcated from the thorax in either size or shape. In contrast to isopods, however, the amphipod body tends to be laterally compressed, giving the animal a rather shrimplike appearance (Fig. 14–48). Also, the gills and heart are thoracic and not abdominal.

There are a few blind species, but sessile compound eyes are typical in most amphipods. In the Hyperiidea, the eyes are usually enormous, covering most of the sides of the head (Fig. 14–50*C* and *D*). Many bathypelagic and other species in the Hyperiidea are peculiar in having each eye divided into an upper and lower portion, so that these animals have essentially double eyes. On the other hand, the eyes of the Oedicerotidae are fused together at the end of the rostrum. Except in the family Ampeliscidae and the

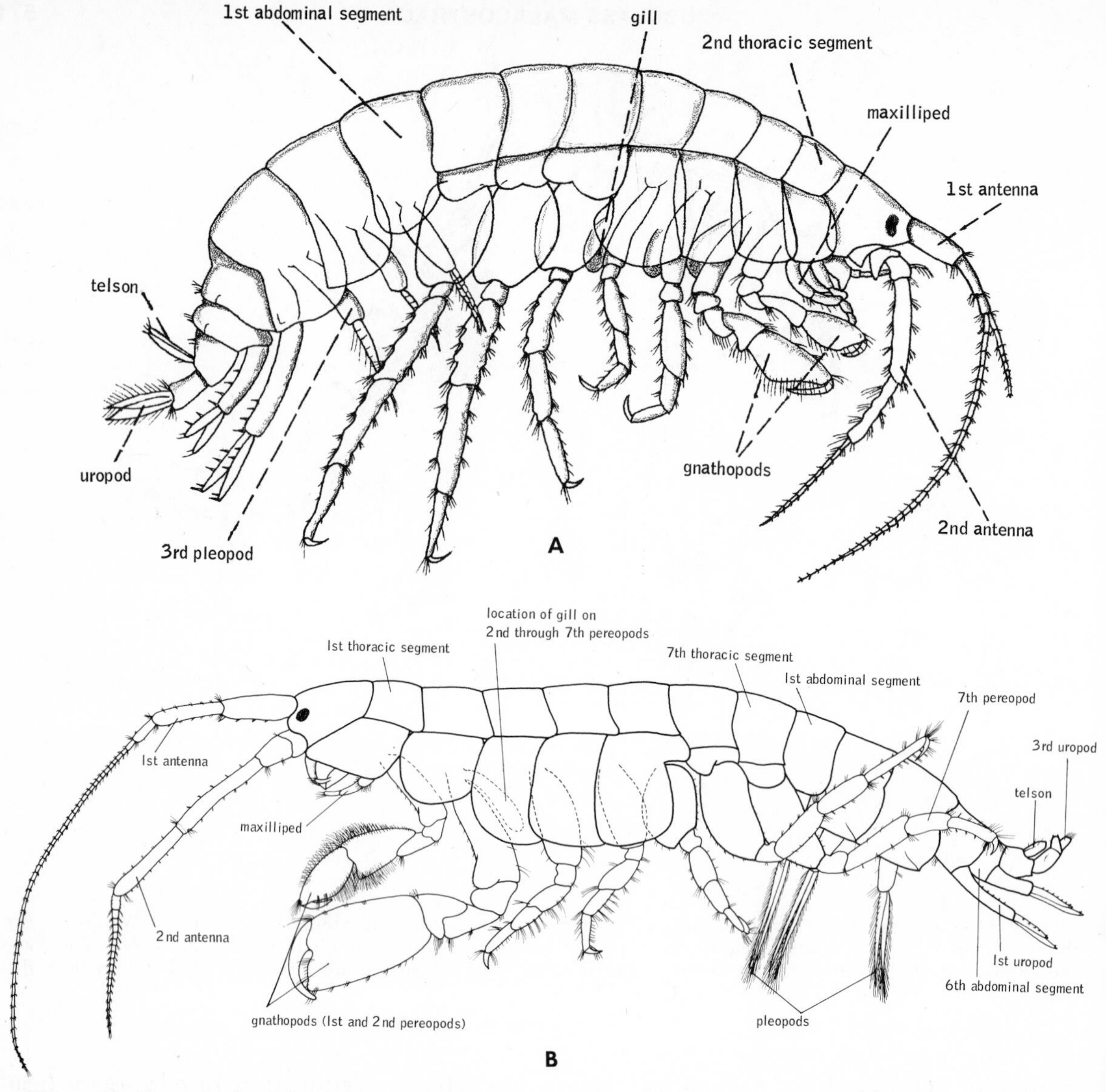

Figure 14–48. *A*. Male of the amphipod, *Gammarus*. (After Sars from Calman.) *B*. Male of the marine amphipod *Ampithoe lacertosa* (drawn from life).

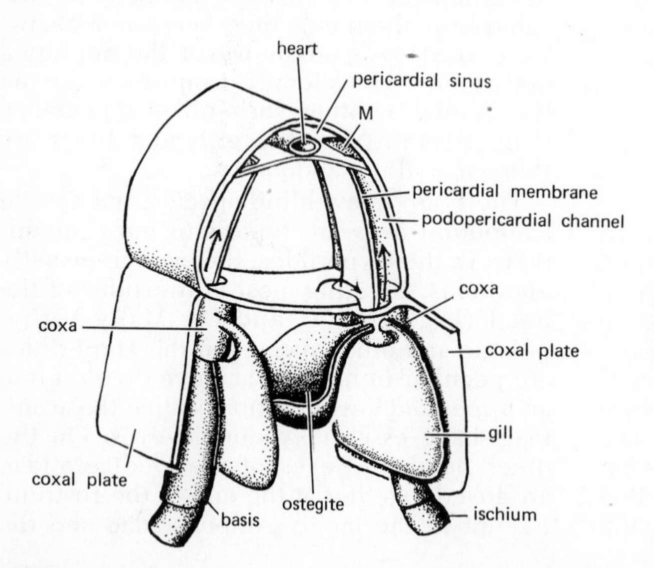

Figure 14–49. Diagrammatic cross section through the body of an amphipod. Only one marsupial plate, or oostegite, is shown. (After Klövekorn and others, from Kaestner, A., 1970: Invertebrate Zoology, Vol. 3, Wiley-Interscience, N. Y.)

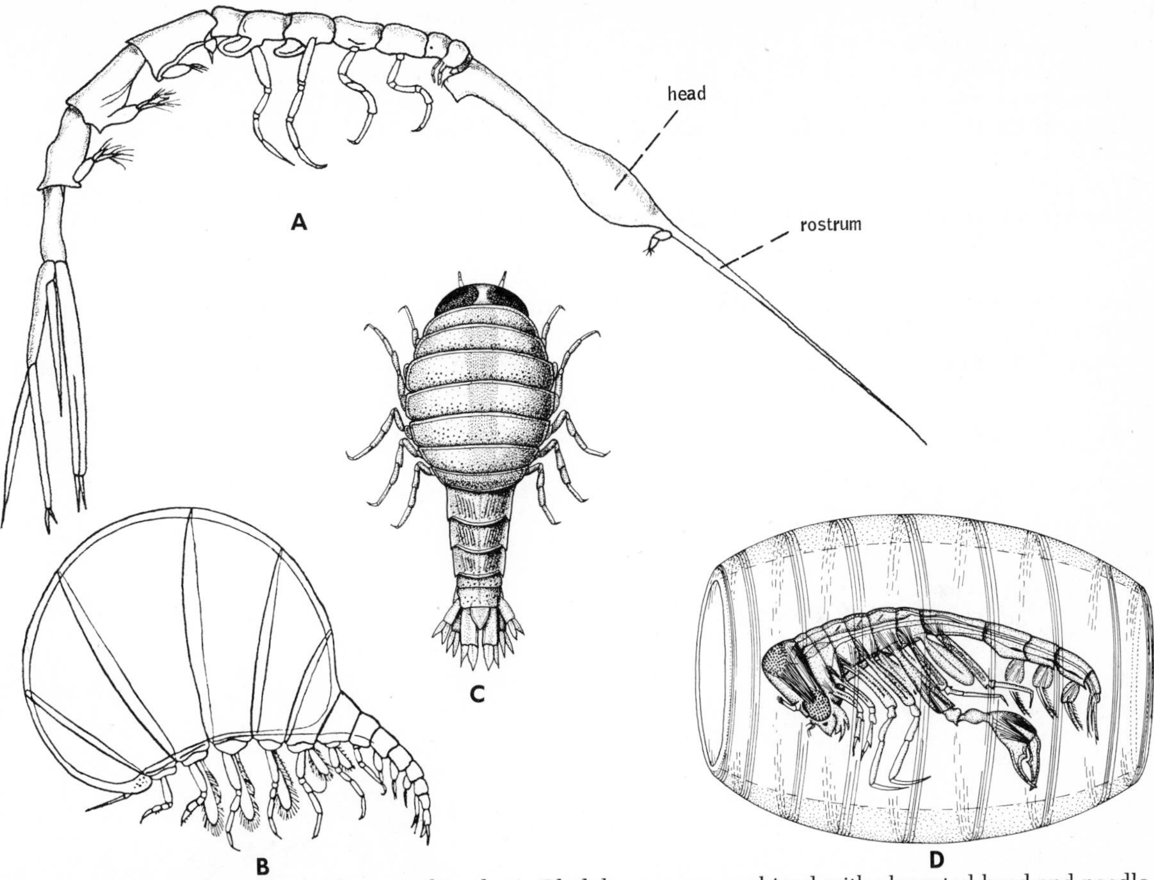

Figure 14–50. Pelagic hyperiidean amphipods. *A. Rhabdosoma*, an amphipod with elongated head and needle-like rostrum. (After Stebbing from Calman.) *B. Mimonectes*, 2.5 cm. long. (After Schellenberg. From Kaestner, A. 1970: Invertebrate Zoology, Vol. 3, Wiley-Interscience N. Y.) *C. Hyperia*. Species of this genus are commonly found attached to pelagic coelenterates and ctenophores. 2 cm. long. (After Sars.) *D. Phronima sedentaria*, which lives within the tunic of urochordates. 3 cm. long. (After Claus.)

genus *Phronima*, the eyes of amphipods are peculiar in lacking corneal facets, i.e., the exoskeleton forms a continuous sheet over the eyes.

The first and second antennae are usually well developed, although they lack exopodites. The first pair of thoracic appendages are modified to form maxillipeds, and their coxae are fused together. The coxae of the other thoracic appendages are usually long flattened plates that increase the appearance of lateral body compression. The second and third thoracic appendages are usually modified for prehension and are called gnathopods. Pleopods are present and are usually divided into two groups. The anterior three pairs are fringed and are used for swimming and for ventilation. The others are directed backward and are similar to the uropods.

Exceptions to all of these typical amphipod features can be found, and some groups have diverged widely from the general plan. In the caprellids, which are common marine amphipods found on seaweed, hydroids, and bryozoans, the abdomen is vestigial, and the thorax is long and slender (Fig. 14–51*D*). The whale lice (Cyamidae) also have a vestigial abdomen, and the thorax is dorsoventrally flattened as in the isopods (Fig. 14–52). A dorsoventrally flattened body also appears in some other amphipods, such as *Platyischnopus*. Other aberrant amphipods include the pelagic genera *Rhabdosoma* (Fig. 14–50*A*), which has an elongated head with an extremely long needle-like rostrum, and *Mimonectes* (Fig. 14–50*B*), in which the body is enormously swollen and spherical, perhaps for mimicking jelly fish, on which *Mimonectes* is commensal or parasitic.

Amphipods are about the same size as isopods. The giant of the order was thought to be the marine *Alicella gigantea*, which may reach 14 cm. in length; but in 1968 an undescribed 28 cm. benthic lysianannid amphipod was photographed from 5300 meters in the Pacific. The smallest forms are only a few millimeters long. Most amphipods are translucent, brown or gray in color, but some species are red, green, or blue-green.

Locomotion and Habitation. All hyperiideans are pelagic, and the strange swollen shape of the head and thorax of some species (*Hyperia, Mimonectes, Rhabdosoma*) is probably a flotation device. Gammarideans are bottom dwellers, but most members can swim, even if infrequently. Propulsion for swimming is provided by the anterior pleopods and in some gammarids by the thoracic appendages. Usually swimming takes plac intermittently between crawling and bur rowing; in leaving the substratum, initia thrust is commonly gained by a backwar flip of the abdomen. Walking is effected b the legs; but in rapid movement over th bottom, both legs and pleopods are used and the animal often leans far over to on side.

Some amphipods, especially the caprel lids, are adapted for climbing. The tips o the legs are provided with grasping claws fo clinging to hydroids, bryzoans, and alga

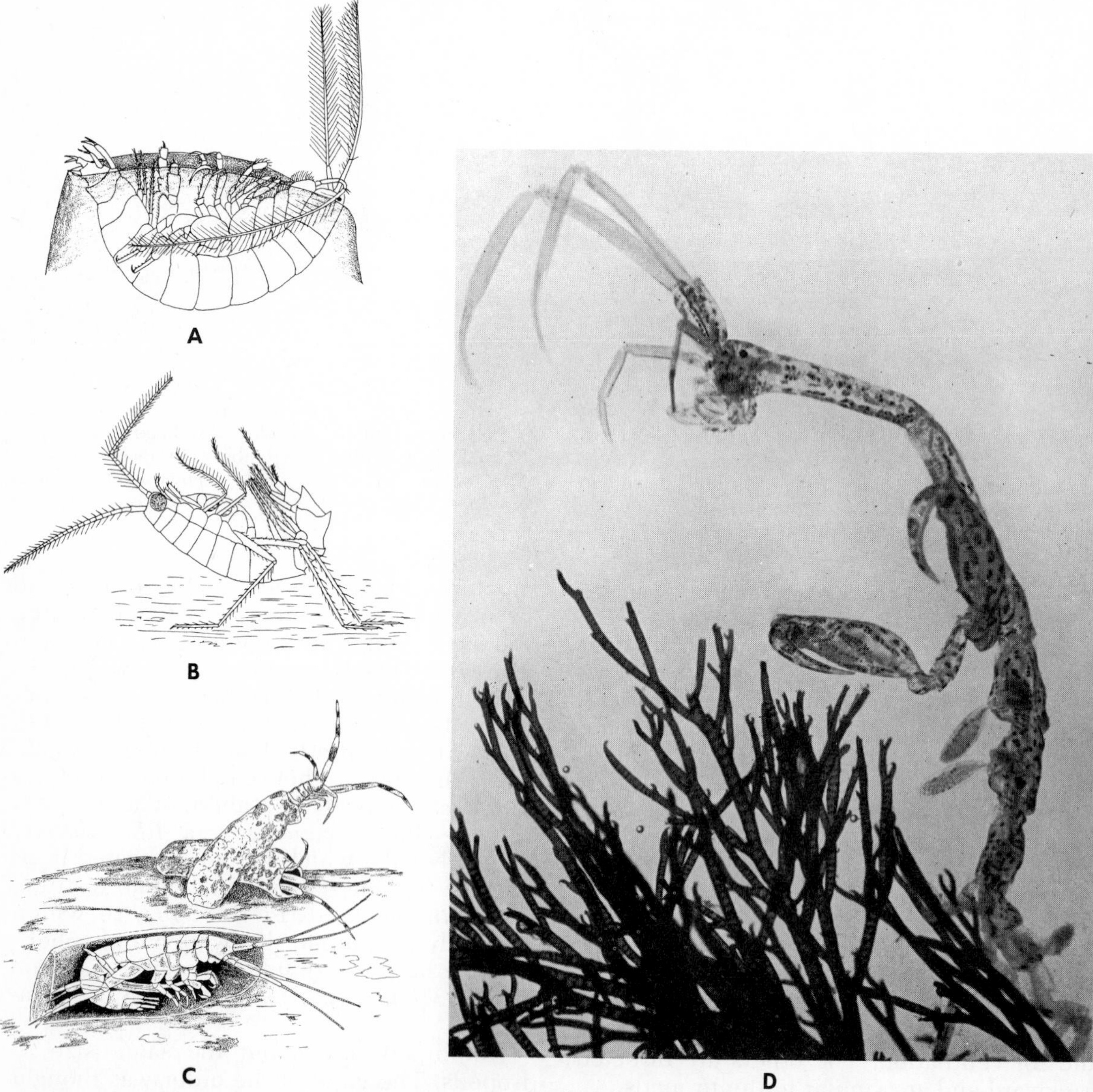

Figure 14–51. *A. Haploops tubicola*, hanging on edge of tube in feeding position. Margin of tube is shaded but shown as transparent. *B. Melphidippella macra* in feeding stance. (*A* and *B* after Enequist.) *C. Cerapus*, a tubicolous amphipod. The tube, which the animal can carry about, is composed of an inner secreted layer covered with fragments of foreign material. (Modified after Schmitt.) *D. Caprella equilibra*, clinging to seaweed. (By D. P. Wilson.)

Fig. 14–51*D*), over which they crawl like an inchworm. Some caprellids live on sea stars or even on spider crabs.

Many amphipods are accomplished burrowers, and some construct tubes of mud or of secreted material. The methods of burrowing vary, but the anterior legs (gnathopods) are most commonly the appendages employed and are often specially adapted for the purpose.

The excavated material may be swept backward (as done by the gammarids) and removed by the uropods and telson; or it may be moved forward. In *Neohela,* the hands of the gnathopods are placed together and function as a scoop. In some amphipods, such as *Haustorius,* the antennae may aid in digging; and in *Tryphosites,* as well as others, the pleopods are used, the animal burrowing straight down from the swimming position. Usually the posterior ventilating current produced by the beating pleopods is important in sucking away the loosened material.

A great many burrowing species construct temporary or permanent burrows, tunnels, or tubes. The burrows may be horizontal, vertical, or **U**-shaped with two openings. Sometimes the walls are simply packed, but often they are reinforced with secreted material.

Tube construction is particularly common among amphipods, and a variety of materials, such as mud, clay, sand grains, and shell and plant fragments may be used. The material is usually bound together with a cementing secretion produced from glands opening at the bases of the fourth and fifth thoracic appendages. *Haploops* mixes the secretion with clay, and the resulting stiff mass is then drawn out between the end of the abdomen and gnathopods and applied to the walls. Cement glands are not always confined to the third and fourth thoracic appendages. They may appear on other appendages; and in the beach fleas, which line their burrows with sand, the glands are located all over the body.

Several tube dwellers, such as species of *Siphonoecoetes* and *Cerapus,* build unattached tubes of shell fragments and sand grains and carry the tubes about with them (Fig. 14–51*C*). *Siphonoecoetes* drags itself about with its second antenna. *Cyrtophium* makes its tubes out of a section of a hollow plant stem, which it lines with secreted materials; the animals can even swim with the tube.

A few amphipods have rather unusual retreats. Some species of *Siphonoecoetes* live in old tooth-shells, and *Polycheria* constructs a burrow in the tunic of ascidians. Like the isopod, *Limnoria,* the amphipod, *Chelura terebrans* bores in wood. Particularly remarkable are the species of the pelagic genus, *Phronima,* which fashion a home out of what are reported to be the swimming bells of siphonophorans or old pelagic tunicate tests (Fig. 14–50*D*). In *Polycheria sedentaria,* only the adult female occupies the case; the males and immatures are found free-swimming and closer to the surface.

Members of the genus *Ingolfiella,* some only 1 to 2 mm. long, are adapted for an interstitial existence.

Freshwater amphipods, of which about 50 species are known in the United States, are common benthic animals in algae and other vegetation of streams, ponds, and lakes and sometimes are found in great numbers. Pennak (1953) reports population densities for *Gammarus* of 10,000 per square meter in certain springs.

Lake Baikal in Siberia contains a remarkable endemic gammarid fauna of many species, some brightly colored red and blue and some reaching a size of over 6 cm. There are both pelagic and benthic species.

A considerable number of freshwater amphipods have been reported from cave pools and streams, are usually pale, blind, and provided with an abundance of sensory hairs.

In contrast to isopods, amphipods have not been very successful at adapting to life on land. The relatively small number of terrestrial and semi-terrestrial species are beach fleas, members of the family Talitridae. Most live beneath drift or stones near the high-tide mark, but a few are found in moist humus and soil away from the shore.

Beach fleas scull rapidly over the sand, gaining additional power with pushing strokes from the abdomen. They can jump, using a sudden backward extension of the abdomen and telson. *Talorchestia,* 2 cm. in length, can leap forward a distance of 1 m., a feat unequaled by any other animal of comparable size. Many beach fleas also burrow. In *Talorchestia,* the body is braced with the second and third pairs of legs, while the gnathopods sweep the material back to the uropods and telson, which flip it away. When burrowing, a beach flea looks very much like a dog digging.

Like the isopod *Tylos,* the beach flea

Talitrus has been shown to use the eyes to obtain astronomical clues for locating the high-tide zone, in which these amphipods normally live. If displaced either above or below the high-tide mark, the animals migrate accurately back to their normal zone. According to the studies of Pardi and others (1952, 1954, 1955), the angle of the sun is used as a compass in conjunction with a map sense of the east-west orientation of the particular beach they inhabit. Moreover, the young supposedly inherit the parents' map sense of a particular locality!

That the angle of the sun is a primary clue for orientation is proved by the fact that the direction of the movement of the animal can be controlled by reflecting the rays of the sun experimentally from different angles. An internal clock mechanism provides interpolation for the changing angle of the sun during the course of the day. This aspect of the mechanisms was proved by transporting *Talitrus* in the dark to a beach at different longitude. On liberation, the immigrants operated on the same time as that of the original location. In the absence of direct sunlight, these amphipods are reported to use sky polarization in the same manner.

Feeding. Most amphipods are detritus feeders or scavengers. Mud or dead animal and plant remains are picked up with the gnathopods, or detritus is raked from the bottom with the antennae, particularly the second pair. Sometimes the mouth-parts attack the food directly. Some burrowing forms scrape detritus and diatoms from sand grains. The beach fleas feed on dead animal remains and seaweed washed up by the tide.

Filter feeding is utilized by a number of amphipods. The calliopiids *Aora, Corophium,* and others strain fine detritus through filter setae on the gnathopods, the feeding current being provided by the pleopods. The tubicolous *Corophium* may first sweep up the mud with the antennae; *Maera* and *Eriopisa* dig it up with the gnathopods.

Other filter-feeding amphipods use the first or second antennae as filters. Many gammarids, including the tube-dwelling *Haploops,* extend the antenna into the natural water current as a net. *Haploops* hangs upside down in the mouth of the tube, clinging to the rim with the legs (Fig. 14–51*A*). The first antennae are extended into the current as a trapping net and are periodically cleaned by the gnathopods. *Melphidippella,* which sits upside down on the bottom, uses not only the second antennae but also the third and fourth pairs of thoracic appendages as a filter. All three limbs are held outstretched and catch falling detritus (Fig. 14–51*B*). At intervals the limbs are scraped by the gnathopods.

A maxillary filter (second maxillae) is present in a few forms, such as *Haustorius arenarius.* In this species, the feeding current is produced by the maxillae and moves in the opposite direction (anteriorly) from the ventilating current.

Although many amphipods supplement their diet by catching small animals, strictly predacious feeding is not common. The most notable examples of predacious feeders are the pelagic hyperiids and some caprellids (skeleton shrimp). Raptorial caprellids attach themselves to hydroids and to bryozoan stems with the last pair of thoracic appendages and project themselves motionless and outstretched, waiting to seize passing prey with the gnathopods. However, many caprellids feed on detritus that collects on the surfaces of the organisms on which they live (Keith, 1969). Species of the gammaridean *Dulichia,* which live on sea urchin spines have a similar feeding habit.

Parasitism is much less prevalent among amphipods than among isopods. There are a few ectoparasites of fish, such as *Lafystius* and *Trischizostoma.* The mouthparts are suctorial. The cyamids, called whale lice, have the legs adapted for clinging to the host. Although often stated to be parasitic, they probably feed on diatoms and debris

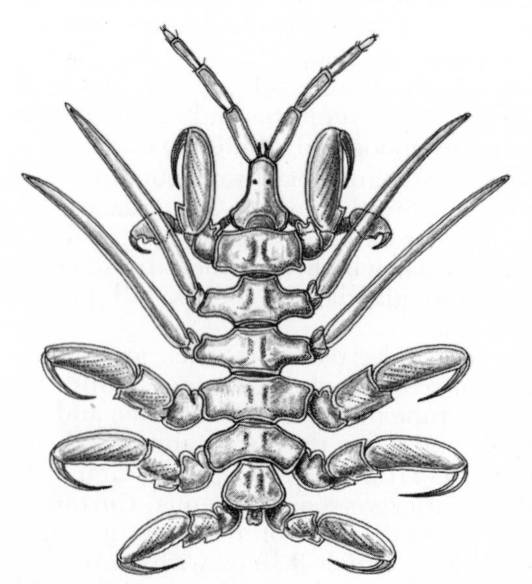

Figure 14–52. *Cyamus boopis,* an amphipod commensal on whales. (After Sars.)

at accumulates on the whale's skin (Fig. 4–52).

Gas Exchange, Internal Transport, and xcretion. The gills, which are thoracic in ontrast to those of isopods, are usually simle lamellae or vesicles (Fig. 14–48). Ocasionally, as in some gammarids and the hale lice, the surface area of the gills is inreased by ridges, folds, or ramifications. he number of gills ranges from two to six airs, but gills are never present on the first nathopods. The talitrid beach fleas are unsual in having accessory gills on the first abominal segments; in some gammarids and yamids, the thoracic appendages bear acessory gills, and some freshwater and cave ammarids have sternal gills. Typically, the osteriorly flowing ventilating current is roduced by the anterior pleopods, and the longate coxae and ventral extensions of the erga provide a protected channel through hich the water flows. A ventilating current s particularly important in those amphipods hat dwell in burrows or tubes; in a few tube wellers, the antennae, rather than the pleoods, provide for ventilation.

Although they live out of water, the beach leas have retained gills, but they are reuced in size. In order for beach fleas to repire, the air must be humid, and they are hus restricted to living in moist sand beeath drift or in other damp areas away from he sea. They may come out to feed at night. alitrids are common in forest litter in Indoacific areas. Integumentary gas exchange is ndoubtedly of considerable importance in hese amphipods, for they can live for some ime with the gills blocked.

The heart is tubular, with one to three airs of ostia, and lies in the thorax. The rterial system is not as greatly developed as n the isopods.

Reproduction. The gonads are paired nd tubular. The male gonopores open at the nd of a pair of long penis papillae on the ternum of the last thoracic segment, and the emale oviducts open on the sixth thoracic oxae. Pleopods do not function as copulaory organs. Males are attracted by female heromones. In species of *Gammarus* and ome other genera, the males carry the feales around beneath them for days, sepaating briefly to permit the final preadult olt of the female. Actual sperm transfer is ccomplished quickly. The male twists his bdomen around so that his uropods touch he female marsupium; when sperm are mitted, they are swept into the marsupium by the ventilating current of the female. The pair separate, and the eggs are soon released into the brood chamber where fertilization takes place. The ventilating current also provides for the ventilation of the eggs in the marsupium. The marsupium of most gammarideans bears plumose marginal setae, which aid in preventing the eggs from falling out. One brood per year of 15 to 50 eggs is common in most temperate freshwater species. In marine forms, there may be 2 to 750 eggs in a clutch, and more than one brood per year is common. Development is direct in almost all amphipods.

SYSTEMATIC RÉSUMÉ OF ORDER AMPHIPODA

Suborder Gammaridea. Head not fused with second thoracic segment. Maxilliped with a palp, thoracic legs with well-developed coxal plates. Last two abdominal segments separate. This largest suborder of amphipods includes marine and all of the freshwater species. *Gammarus, Maera, Eriopisa, Calliopius, Alicella,* Melphidippidae, *Polycheria.* The semiterrestrial Talitridae (beach fleas) include *Talitrus, Orchestia, Talorchestia;* the widespread North American freshwater *Hyallela azteca.* The Oedicerotidae; *Ampelisca, Haustorius, Corophium, Cyrtophium, Siphonoecoetes, Chelura, Aora, Ampelisca, Haploops;* the parasites, *Lafystius* and *Trischizostoma.*

Suborder Hyperiidea. Head not fused with second thoracic segment. Maxillipeds have no palp; coxae often small or fused with body. Eyes usually very large, covering greater part of head. Last two abdominal segments fused. Body more or less transparent. Entirely marine and pelagic. *Scina, Hyperia, Phronima, Rhabdosoma, Mimonectes.*

Suborder Caprellidea. Head fused with second thoracic segment. Abdominal segments fused, and reduced; has vestigial appendages. Includes the skeleton shrimp, Caprellidae; and the whale lice, Cyamidae.

Suborder Ingolfiellidea. Head not fused with second thoracic segment. Elongate body. Coxae are small. Abdominal segments are distinct, all but fourth and fifth pairs of abdominal appendages vestigial. Interstitial. *Ingolfiella.*

Superorder Eucarida

The Eucarida contains most of the larger malacostracans. The carapace is highly de-

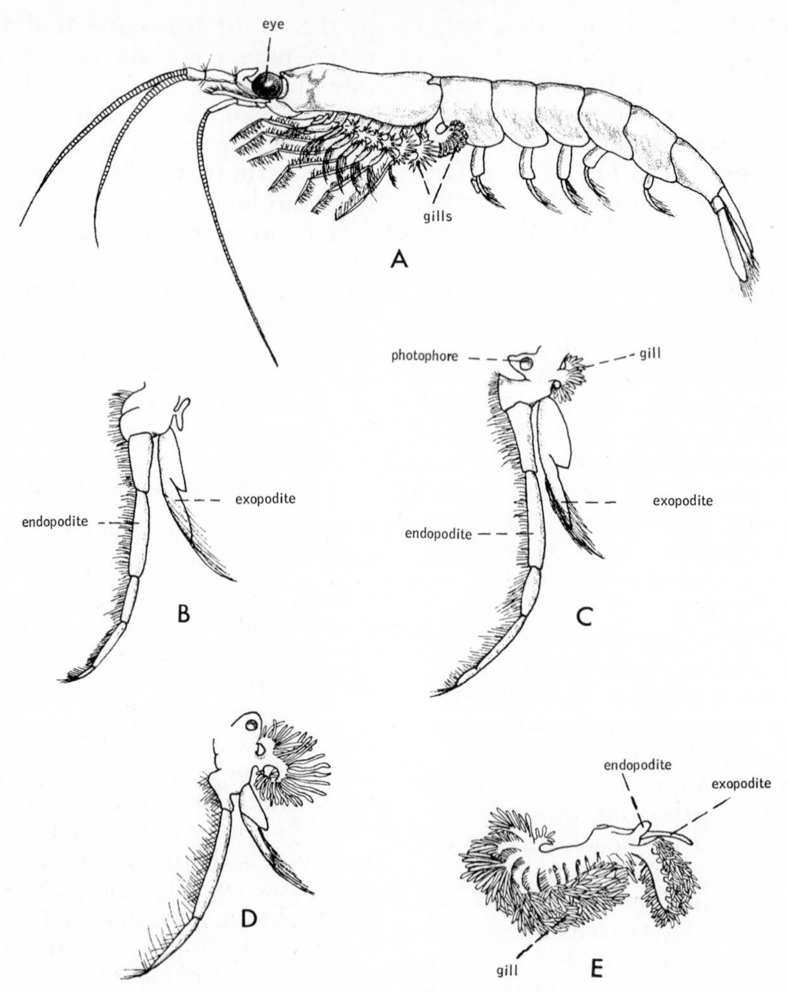

Figure 14–53. *A. Meganyctiphanes*, a euphausiacean. *B* to *E*. Thoracic appendages of *Meganyctiphanes*. *B*. First. *C*. Second. *D*. Seventh. *E*. Eighth. (All after Calman.)

veloped and fused with all of the thoracic segments. The eyes are stalked, and the basal section of the second antenna is composed of only two segments. There is no thoracic brood chamber as in the peracaridans, and development is usually indirect, with a zoea larva. The Eucarida is composed of only two orders, the Euphausiacea and the Decapoda. The basic form is shrimplike (caridoid) (Fig. 14–31).

Order Euphausiacea

Euphausiaceans, known as krill, are pelagic shrimplike crustaceans approximately 3 cm. in length and are the most primitive eucaridans. They are all marine. The sides of the carapace do not tightly enclose the gills, as is true of many decapods (Fig. 14–53*A*). None of the thoracic appendages are specialized as maxillipeds, and each bears an exopodite (Fig. 14–53*C*). The pleopods are well developed, have long setae, and are used for swimming. The telson bears a pair of movable spines, which have been interpreted as being the vestigial caudal furca. The similarity of euphausiaceans to mysidaceans was the basis for formerly combining the two groups and probably does indicate that they are both near the common base of the eucaridan and peracaridan lines respectively. However, as already pointed out, euphausiaceans lack the peracaridan marsupium and the carapace is united with all of the thoracic segments.

Most euphausiaceans are filter feeders. The first six thoracic appendages compose the filtering apparatus. Each of the leglike endopodites bears a long fringe of setae on one side, and together with the other limbs of that side they form one half of a funnel-shaped net or basket (Fig. 14–53*A*). The feeding current passing through the net probably results from the forward swimming movement of the body. Many species, such as members of the genus *Euphausia*, trap microplankton; others consume zooplankton. *Meganyctiphanes*, which feeds on zooplankton, possesses a maxillary filter in addition to the filtering net formed by the thoracic appendages. A few euphausiaceans, such as *Stylocheiron*, are predaceous, capturing arrowworms and other larger pelagic animals; the filtering apparatus is not developed, and the third pair of thoracic appendages are very long and chelate.

Large exposed filamentous gills, called podobranchs, are present on all of the thoracic appendages, although poorly developed on the first (Fig. 14–53*D*). The ventilating current is produced by the thoracic exopodites.

Most euphausiaceans are luminescent. The light-producing material is not secreted, as in ostracods and other entomostracans, but is intracellular, located within special light-producing organs called photophores. Each photophore is composed of a cluster of light-producing cells, a reflector, and a lens. A photophore is usually located on the upper end of the ocular peduncle, on the coxae of the seventh thoracic appendages, and one in the middle of each of the sternites of the first four abdominal segments (Fig. 14–53*D*). The luminescence is probably an adaptation for swarming and reproduction.

The sperm are transferred to the female as spermatophores, and the eggs may be liberated into the sea water when they are laid or they may be retained briefly within the filtering basket or attached to the undersurface of the sternum of the posterior thoracic segments. A nonfeeding, nauplius larva represents the hatching stage.

The approximately 90 species composing the order Euphausiacea are strictly pelagic crustaceans. Such species as the Antarctic *Euphausia superba*, which is about 6 cm. in length and transparent, are surface forms; others live at deeper levels or undergo vertical migrations. The red bathypelagic genus, *Bentheuphausia*, lives at depths of 1000 to 1700 m.

Many members, including *Euphausia superba*, live in great swarms and constitute the chief food of many commercially important species of whales. A blue whale may consume two to three tons of euphausiaceans at one feeding. A swarm may cover an area equivalent to several city blocks, and one seen from the air looks like a giant amoeba slowly moving and changing shape. Although vertically the swarm may occupy a layer five or more meters thick, the several meters of surface water contain the greatest concentration and may reach densities of 63,000 individuals per $m.^3$, all of which are adults or near adults. Early stages of development are restricted to depths of 700 to 1500 m., where the water is warmer.

Order Decapoda

The order Decapoda contains the familiar shrimp, crayfish, lobsters, and crabs and includes the largest and some of the most highly specialized crustaceans. It is also the largest order of crustaceans; the 8500 described decapods represent almost one third of the known species of crustaceans. Most of the decapods are marine, but the crayfish, some shrimps, a few anomurans, and a number of crabs have invaded fresh water. There are also some amphibious and terrestrial crabs.

Decapods are distinguished from the euphausiaceans, as well as from other malacostracans, in that their first three pairs of thoracic appendages are modified as maxillipeds. The remaining five pairs of thoracic appendages are legs, from which the name Decapoda is derived (Fig. 14–54*B*). The first pair is frequently much heavier than the remaining pairs, and when so constructed, the limb is called a cheliped (Fig. 14–55). The legs usually lack exopodites. The head and thoracic segments are fused together dorsally, and the sides of the overhanging carapace enclose the gills within well-defined, lateral branchial chambers (Figs. 14–66*A* and *E*).

All colors and combinations of colors produced by pigments in the exoskeleton and by chromatophores are found in decapods. Many shrimps can adapt body coloration to the background. Also, the body surface and coloration of some decapods are especially adapted for certain habitats. The shrimps and crabs that live in *Sargassum* are colored olive yellow with mottling and spots, as is true of most other animals in this floating menagerie. The crab, *Parthenope investigatoris*, from the Indian Ocean has the body surface sculptured like a piece of old coral. Particularly remarkable are the brachyuran decorator crabs (Fig. 14–56*A* and *B*). These

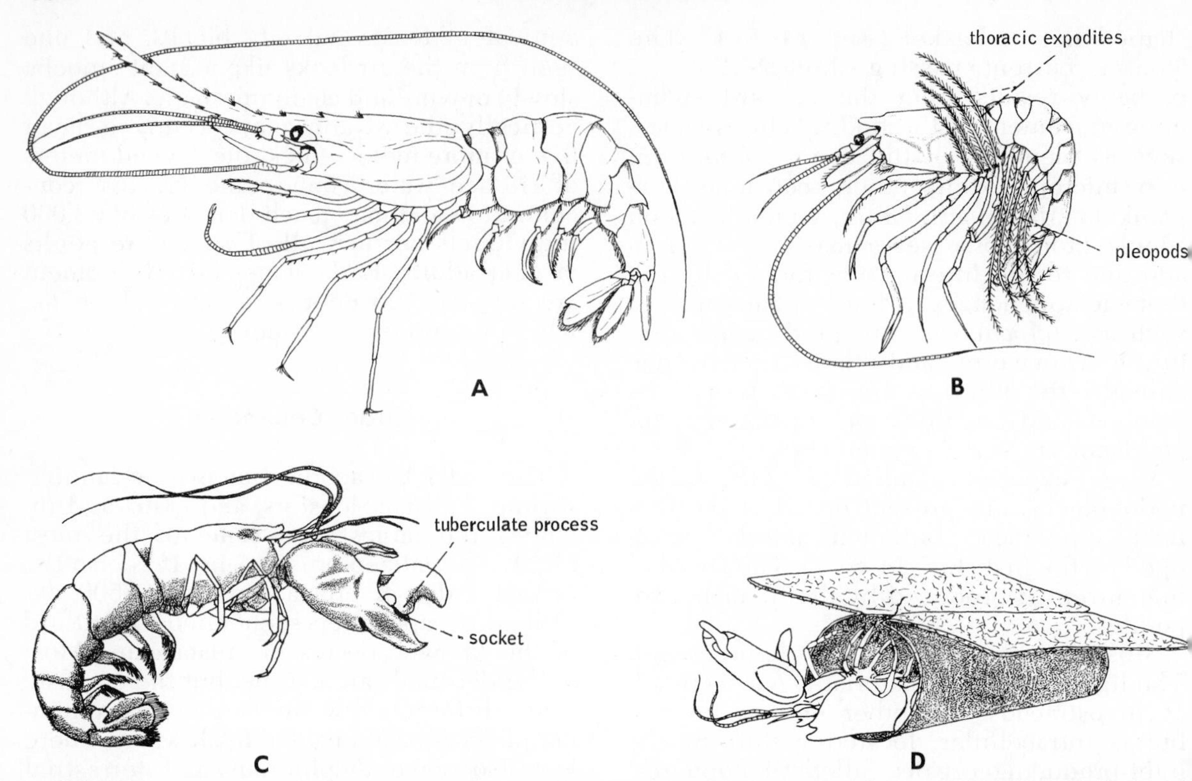

Figure 14–54. Natantian decapods (shrimp). *A. Heterocarpus* (Caridea). *B. Psathyrocaris fragilis* (Caridea) with well-developed thoracic exopodites. (Both after Alcock from Calman.) *C. Alpheus*, a pistol or snapping shrimp (After Schmitt.) *D.* The snapping shrimp, *Alpheus pachychirus*, sewing together an algal mat. (After Cowles from Schmitt.)

are spider crabs that attach pieces of seaweed, hydroids, bryozoans, and sponge to the carapace and legs.

The smallest decapods are species of the brachyuran crab, *Dissodactylus*, which are commensal on sand dollars (Fig. 14–57*B*). The cephalothorax of these species is only a few millimeters wide. *Macrocheira kaempferi,* a Japanese spider crab, has the greatest leg span of any living arthropod (Fig. 14–57A). The cephalothorax may attain a length of 45 cm. and the chelipeds, a span of 4 m. However, some lobsters and crayfish have longer bodies.

Decapod Diversity—Locomotion and Habitation. The diversity of forms encountered among decapods can more easily be appreciated if examined from the standpoint of adaptations for locomotion and habitation.

The Decapoda is divided into two suborders—the Natantia, containing the shrimp or prawns; and the Reptantia, containing the lobsters, crayfish, and crabs. In the Natantia, the body is generally adapted for swimming (natant—"swimming"), and it tends to be laterally compressed with a well-developed abdomen (Fig. 14–54). The cephalothorax often bears a keel-shaped serrated rostrum and generally slender legs. The first two or three pairs of legs are chelate, and one of the chelate pairs may be longer and heavier than the others. The legs of some primitive shrimps bear exopodites (Fig. 14–54*B*). The exoskeleton is relatively thin and flexible.

There are some pelagic and bathypelagic shrimp (in fact, they are the only pelagic decapods), but most shrimp are bottom dwellers that swim intermittently. The pleopods, which are large and commonly fringed are the principal swimming organs, although rapid ventral flexion of the abdomen with the tail fan is used for quick backward darts. Abdominal flexion also provides for vertical steering.

Benthic shrimp are found in all sorts of bottom habitats. Species of *Penaeus,* which are the most important shrimp of commercial fisheries throughout the world, live on sand bottoms in 5 to 30 meters of water Some shrimp are shallow burrowers in soft bottoms, and the sand shrimp *Crangon* uses the beating pleopods for excavating. Many small species live within algae, beneath stones and shells, and within holes and crevices in coral and rock.

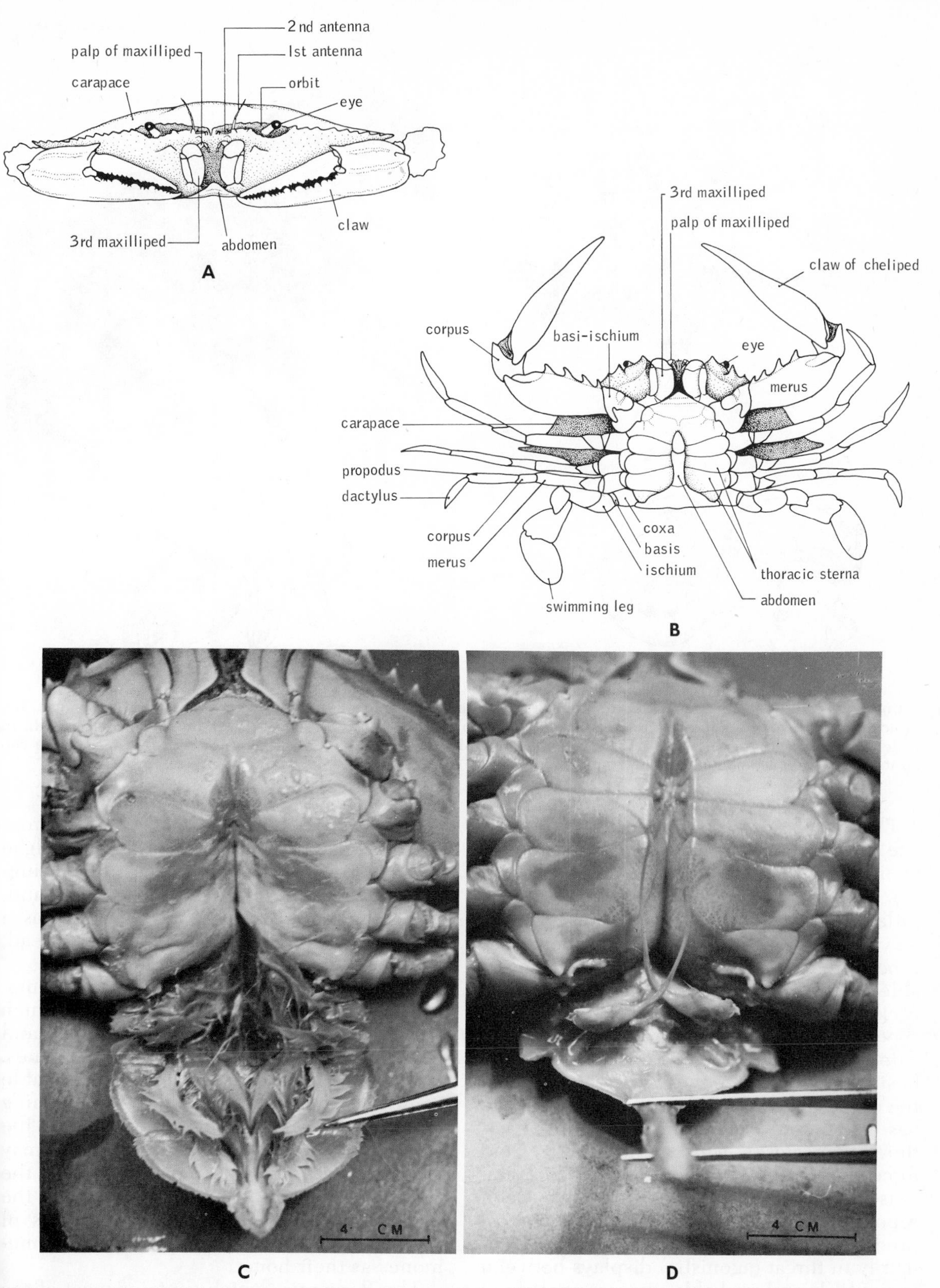

Figure 14–55. A brachyuran crab of the family Portunidae. The fifth pair of legs is adapted for swimming. *A.* Frontal view. *B.* Ventral view. (After Schmitt from Rathbun.) *C.* Abdomen of a female. Pleopods are used for carrying eggs. *D.* Abdomen of a male. Note that it is much narrower than that of the female. Only the anterior two pairs of pleopods, which are used as copulatory organs, are present. (Both by Betty M. Barnes.)

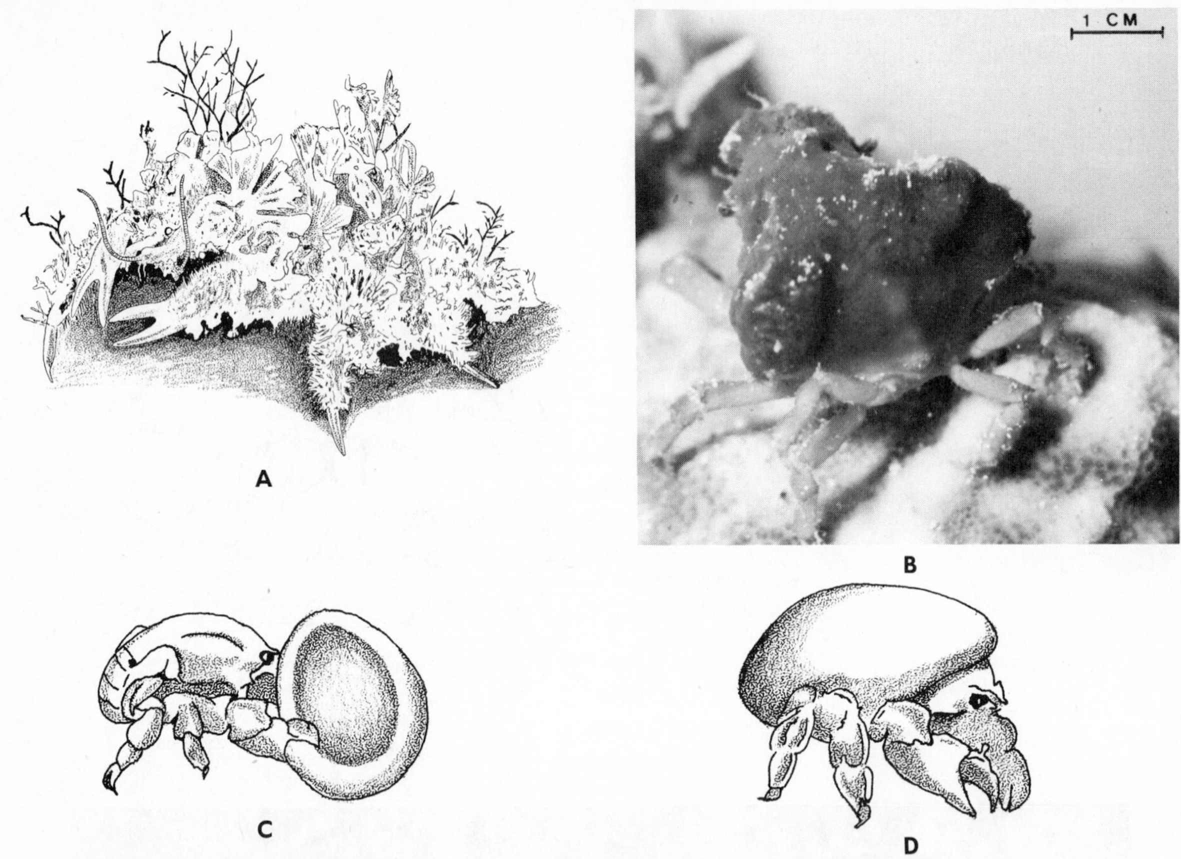

Figure 14–56. *A.* A decorator crab (spider crab), *Oregonia gracilis*, camouflaged with a covering of algae, hydroids, and bryozoans. (Modified after Schmitt.) *B. Macrocoeloma trispinosum*, a West Indian decorator crab, in which the carapace is covered with sponge. (By Betty M. Barnes.) *C.* and *D.* A dromiid crab putting sponge cap on its back. (After Dembowska from Schmitt.)

The pistol or snapping shrimp (Alphaeidae) are a common and widely distributed family. These little shrimp, which are 3 to 6 cm. long, have one of the chelipeds greatly enlarged (Fig. 14–54*C*). The base of the movable finger contains a large tuberculate process that fits into a socket on the immovable finger. The movable finger is locked, or cocked, when contact is made between two specialized discs, one at the base of the elevated finger and one on the hand. The adhesive force between the discs prevents closing of the finger until the contracting muscle has generated a counteracting pull. Then the finger closes with great rapidity and force, producing a snapping or popping noise. Although some species have been reported to use the snapping mechanism in predation and defense, it functions primarily in threat (agonistic) displays between individuals and probably ensures spacing of a population (Nolan and Salmon, 1970). Snapping shrimp live in natural or constructed retreats or burrows.

Alpheus pachychirus constructs tubes from filamentous mats formed from certain algae (Fig. 14–54*D*). Lying on its back, the shrimp pulls the mat around itself like a cloak and, using one of the slender pointed legs as a needle and the algal filaments as thread, stitches the edges of the mat together.

There are a number of snapping shrimp, as well as other species of shrimp, which live in sponges, tunicates, bivalves, and corals or with sea urchins and sea anemones. A large sponge may become a veritable apartment house of shrimp. The cleaning shrimp *Periclimenes* and *Stenopus* clean the surfaces of certain reef fish; the fish may swim between the shrimp's chelae, or the shrimp may even insert its chelae into the fish's gill region (Fig. 14–58). Species of *Periclimenes* use the tentacles of sea anemones as their home.

The Reptantia, which contains most of the decapods, consists of benthic animals that have become more highly adapted for crawling than have most shrimp (reptant—"crawl-

Figure 14–57. *A.* The Japanese spider crab, *Macrocheira kaempferi,* the largest living arthropod. (Courtesy of the American Museum of Natural History.) *B. Dissodactylus,* a commensal pea crab which lives on sand dollars. This is one of the smallest decapods. *C. Stenorhynchus,* a West Indian spider crab with a long spine-like rostrum that lives on sea anemones. *D.* Three coral galls formed by a crab of the family Hapalocarcinidae. The ventilating current of a young female causes the formation of the coral enclosure, in which the crab spends the remainder of her life. The tiny male and larvae can enter and leave through the openings visible in the photograph. (*B-D* by Betty M. Barnes.)

Figure 14–58. Peppermint shrimp *Hippolysmata grabhami* cleaning a fairy bass. Several other fish are waiting their turn to be cleaned. (By Faulkner, D., 1970: Hidden Sea. Viking Press. p. 74.)

ing"). The body tends to be dorso-ventrally flattened at least to some degree. The legs are usually heavier than those of natantians and never bear exopodites. The first pair of legs usually are powerful chelipeds. When pleopods are present, they are never adapted for swimming.

In many reptantians, the more primitive long-bodied form of the natantians has been greatly shortened through folding of the abdomen ventrally beneath the anterior part of the body. This evolution of a "crab" form occurred independently a number of times in reptantian evolution as an adaptation for locomotion or habitation.

Lobsters, crayfish, and burrowing shrimp, which comprise the section Macrura ("large tail"), still possess a large extended abdomen bearing the full complement of appendages. The carapace is always longer than it is broad, and it may bear a depressed rostrum. Macrurans crawl with the legs but can swim rapidly backward to escape by flexing the abdomen ventrally.

Lobsters are heavy-bodied decapods and are generally inhabitants of holes and crevices of rocky and coralline bottoms. The Nephropidae have large chelipeds and are similar in form to crayfish. The American lobster *Homarus americanus,* which may reach a length of 60 cm. and a weight of 22 kg. (48 lbs.), are caught commercially in pots, or traps, which the animal enters as a home or for bait. The spiny lobsters (*Panulirus* and *Jasus*) of the Mediterranean, West Indies, coast of southern California, and other parts of the world lack chelipeds, but the very large spiny antennae are used for defense (Fig. 14–59A). The Spanish, or shovel nose, lobsters (*Scyllarus*) have lateral shelf-like projections from the flattened carapace. Crayfish are similar to nephropsid lobsters and will be discussed on page 597.

Burrowing shrimp, or mud shrimp (*Callianassa, Upogebia, Thalassina*) are shallow-water or intertidal macrurans that live in long deep burrows excavated in sand or mud. Like true shrimp, the exoskeleton is soft and flexible and the coloration is typically pale. *Thalassina* is found in mangroves. Species of *Upogebia* are common in shelly intertidal mud on both the Atlantic and Pacific coasts of the United States (Fig. 14–59*B*). The burrows of both *Callianassa* and *Upogebia* have two entrances and enlarged chambers for turning around (Fig. 14–59*C*).

Members of the section Brachyura, the true crabs, have the most highly specialized short body form and, judging from the number of species (4500), are probably the most successful decapods. The abdomen is greatly reduced and fits tightly beneath the cephalothorax (Fig. 14–55). Uropods have disappeared. In the female, pleopods are retained for brooding eggs (Fig. 14–55*C*); in the male, only the anterior copulatory pleopods are present (Fig. 14–55*D*). The carapace is very broad, at least as wide as it is long and commonly much wider, increasing the flattened appearance of the body.

The evolution of abdominal reduction and flexion in brachyurans was probably a locomotor adaptation, shifting the center of gravity forward to a point beneath the locomotor appendages. Crabs can crawl forward slowly, but they commonly move sideways, especially when crawling rapidly. In such a gait, the leading legs pull by flexing, and the trailing legs push by extending. Chelipeds are not used in crawling.

The ghost crabs, *Ocypode,* which live in burrows above the high-tide mark on sandy

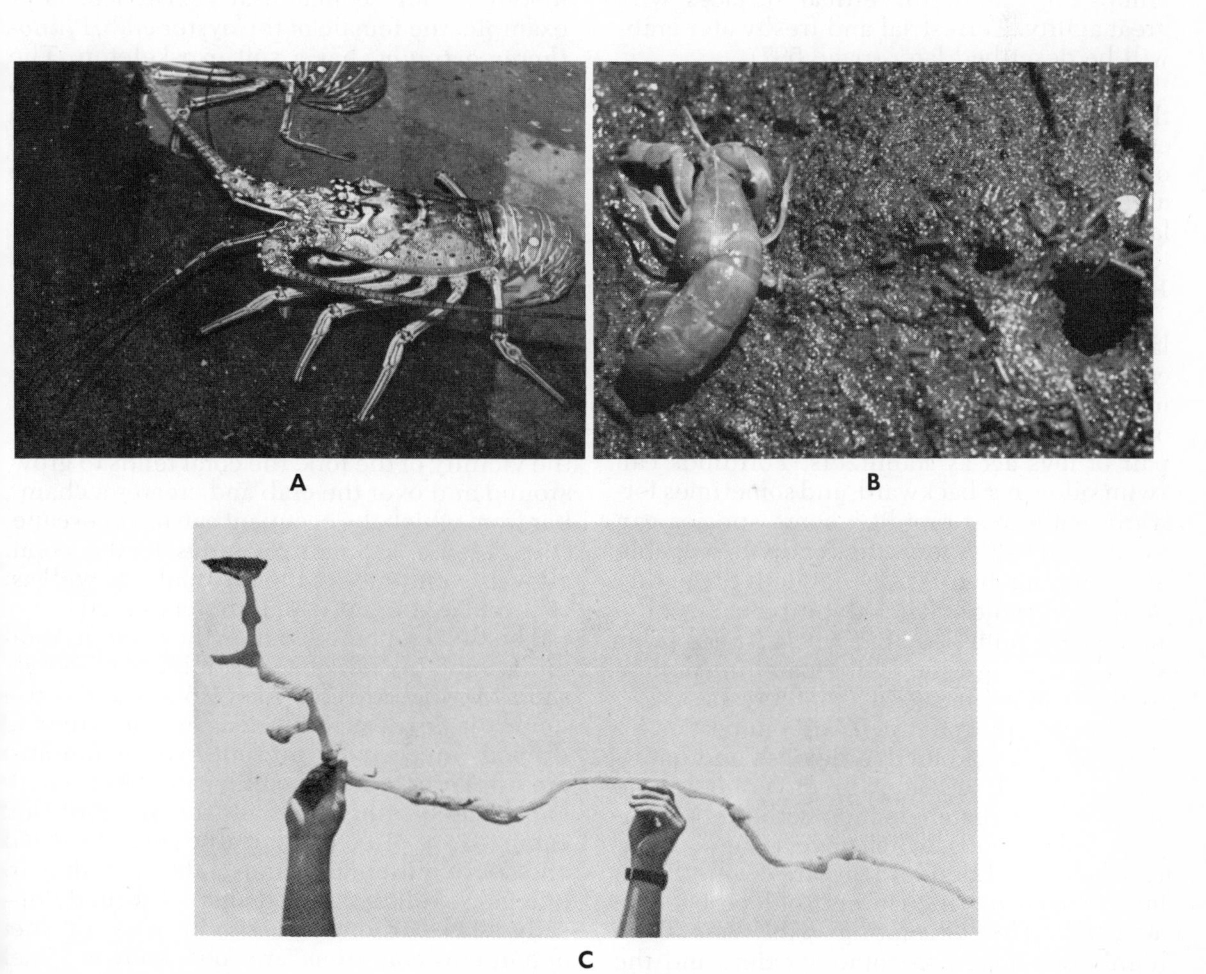

Figure 14–59. Macruran decapods. *A. Panulirus,* a spiny lobster. Note the heavy thorny second antennae, which have a defensive as well as a sensory function. *B.* A species of burrowing shrimp (*Upogebia*) from the West Coast of the United States. (*A* and *B* by Betty M. Barnes.) *C.* Mold of burrow of *Ubogebia affinis* from the coast of Georgia. (By R. W. Frey and J. D. Howard, 1969: Trans. Gulf Coast Assoc. Geol. Soc., *19*:427–444.)

beaches and among dunes, are among the fastest running crabs, some species reaching speeds of better than 1.6 meters per second (Fig. 14–69A). When running, the animal moves sideways; and at top speed, the body is raised well off the ground, and only two or three pairs of legs are used. In order to distribute the work load between the flexors and extensors on each side of the body when running rapidly, ghost crabs frequently stop abruptly and turn the body 180 degrees. Such a turn reverses the leading and trailing sides of the body.

Brachyuran crabs are found in all types of habitats and to great depths. Many abyssal crabs have long slender legs for crawling about over soft bottoms. The abyssal Homolodromiidae are among the most primitive brachyurans. At the other extreme the common tropical *Grapsus grapsus,* which lives just above the water on wave-washed rocks, climbs and clings to vertical surfaces with great agility. Terrestrial and freshwater crabs will be described later (page 597).

Most crabs cannot swim, but members of the family Portunidae, which includes the common edible blue crab *Callinectes sapidus* of the Atlantic coast, are the most powerful and agile swimmers of all crustaceans. The last pair of legs in members of this family terminate in broad flattened paddles (Fig. 14–55*B*).

During swimming, the limb is extended laterally and somewhat above the level of the carapace, and describes a figure eight in its movement. The action is essentially like that of a propeller, and the counterbeating fourth pair of legs act as stabilizers. Portunids can swim sideways, backward, and sometimes forward, with great rapidity; some species can even catch fish. A few other crabs are capable of swimming, but usually not with the agility of the portunids. The amphibious woolly-handed, or mitten, crab, *Eriocheir,* can swim sideways, using a sort of dog-paddling motion by the walking legs (Fig. 14–68A).

Although the chelipeds are important in defense, other protective devices and habits have evolved in many brachyurans. Some species carry sea anemones with their chelipeds. Spider crabs, which have triangular convex bodies and slender legs, are covered with hooked setae to which foreign objects become attached. The decorating habit has been highly developed in some species, and the body becomes completely overgrown with sponge and other sessile organisms (Fig. 14–56A and *B*).

The Dromiidae and Dorippidae use the small dorsally directed fourth and fifth pairs of legs to hold objects over the body. *Hypoconcha* and *Dorippe* usually cover themselves with one half of a bivalve shell, although *Dorippe* may use other objects, including an old fish head! *Dromia* cuts out a cap of sponge with its chelipeds and fits the cap upon its back like an oversized beret (Fig. 14–56*C* and *D*).

Most of the commensal crabs belong to the family Pinnotheridae, called pea crabs because of the small size of many of its species. In addition to inhabitants of polychaete tubes and burrows, there are species that live in the mantle cavity of bivalves and snails, in the cloaca of sea cucumbers, on sand dollars (Fig. 14–57*B*), in the tunnels of the burrowing shrimps, *Callianassa* and *Upogebia,* in the pharynx of tunicates, and in other animals. Often the body has become considerably modified for commensal existence. For example, the female of the oyster crab, *Pinnotheres ostreum,* has a soft exoskeleton. The male of this species, which leaves its host to find a female, has normal chitinization. Pea crabs respond positively to substances produced by the host, and can detect and move up a water stream that has passed over a host.

The Hapalocarcinidae contains the coral gall crabs. These crustaceans inhabit certain types of corals that have branching growth forms. A gall is initiated by a little immature female, which after her larval stages, comes to rest in a fork of the coral. The water currents created by the crab ventilating its gills affect the growth pattern of the coral. In the vicinity of the fork, the coral tends to grow around and over the crab and creates a chamber from which the occupant can never escape (Fig. 14–57*D*). Small openings in the coral allow the entrance of the tiny male, as well as of the plankton on which the crab feeds.

The third reptant division, the section Anomura, contains a diverse assemblage of forms, some of which are crab-like. However, the abdomen is never as reduced as in brachyurans; uropods are usually present. All anomurans are similar in having the fifth pair of legs small and located either beneath the sides of the carapace or directed dorsally (Fig. 14–61*C* and *D*). In most brachyuran crabs the fifth pair of legs is well developed and not turned dorsally. The origin and significance of the anomuran condition are not known. The Anomura may be a polyphyletic grouping.

Two superfamilies of anomurans, the Coenobitoidea and the Paguroidea, have evolved the habit of housing the abdomen within gastropod shells. The hermit crab condition, which evolved independently in the two groups, probably had its origin in forms which utilized crevices and holes as protective retreats. Indeed, among the primitive symmetrical Pylochelidae and even the more specialized Clibanariidae, there are species which still utilize such retreats; others live in bamboo or hollow mangrove roots (Figs. 10–24 and 14–60*A*).

In most hermit crabs, the abdomen is not flexed beneath the cephalothorax but is modified to fit within the spiral chamber of gastropod shells (Fig. 14–60*C* and *D*). The abdomen is asymmetrically developed, with a thin, soft, nonsegmented cuticle, and at least the pleopods on the short side have been lost. Those on the long side are retained in the females to carry eggs. The twist of the abdomen is adapted for right-handed spirals, although left-handed shells may be used.

Hermit crabs always use empty shells and never kill the original occupant. Most species inhabit the shells of a number of different gastropods, depending on what is available in a particular locality. When the crab becomes too large for its shell, it seeks another, but does not leave the old shell until a suitable new one has been found.

Hermit crabs locate a prospective shell with the eyes, and then inspect it with the chelae, inserting one into the interior. If the shell appears suitable, the crab will leave its old shell and try out the new one. If the fit, weight, or movability is bad, the crab returns to the old shell. Two hermit crabs will fight for the possession of an empty shell or the shell inhabited by one of the combatants. Crabs carrying visually larger or heavier shells more frequently win shell fights than do those carrying small shells (Hazlett, 1970).

The shell is held in several ways. The uropods are greatly modified, and the left one is used for hooking to the columella of the shell. Contraction of the longitudinal abdominal muscles presses the surface of the abdomen against the inner walls of the shell, and the last two pairs of legs are also pushed against the wall of the shell opening. One or both chelipeds may be adapted for blocking the aperture of the shell when the crab is withdrawn.

Not all hermit crabs use gastropod shells for retreats. A Pacific Coast species, *Pylopagurus minimus*, lives in *Dentalium* shells. *Xylopagurus* lives in wood. Another species of *Pylopagurus* lives within a mass of encrusting sponges or bryozoans. It is reported that this crab first inhabits a snail shell, which is supposedly dissolved after becoming covered with bryozoans. The hermit crab is also supposed to gradually dissolve the inner skeletal material of the bryozoan colony, and thus the crab is never forced to find a larger retreat.

From hermit ancestors evolved two groups of anomurans, the coconut crabs (Fig. 14–62*A*) and the lithode crabs, which have short body forms and which no longer house the abdomen within shells. The abdomen, although folded crab-like beneath the cephalothorax, is asymmetrical. Other anatomical features also provide evidence of a hermit ancestry.

The lithode crabs, which have very heavy and sometimes sculptured carapaces, are inhabitants of cold oceans (Fig. 14–61*A*, *B*, and *C*). To this group belongs the Alaskan king crab, *Paralithodes camtschatica* which is trapped commercially in the north Pacific.

Two other groups of anomurans have flexed crab-like abdomens, but the abdomen is symmetrical and not greatly reduced as in brachyurans. The galatheids use holes and crevices for retreats. The little porcelain crabs are common shallow-water decapods in many parts of the world. They live beneath stones and, in their general appearance and sideways gait, they look very much like brachyuran crabs (Fig. 14–61*D*). Perhaps the abdominal flexion is an adaptation for motility.

The superfamily Hippoidea contains the sand crabs, or mole crabs. Abdominal flexion in this group appears to be an adaptation for burrowing backward in sand. In the mole crabs, the body is somewhat cylindrical and there are no chelipeds (Fig. 14–63). Species of *Emerita* live on open beaches and dig with the uropods and fourth pair of legs. In surf, mole crabs are usually washed out with each wave; while the wave is receding, they rapidly burrow backward in the soft sand until only the antennae are visible.

Nutrition. The mouth of decapods is more or less ventral and flanked by the feeding appendages, which lie on top of one another (Fig. 14–1*A*). The third maxillipeds are the outermost appendages and in higher forms cover the other appendages. The mouth parts are framed above and anteriorly by a transverse plate (the epistome) and laterally by the sides of the carapace (Fig. 14–55*A*). The arrangement of the mouth parts is particularly striking in higher crabs, in which the ends of the epistome have fused with the carapace and the third maxilliped is represented

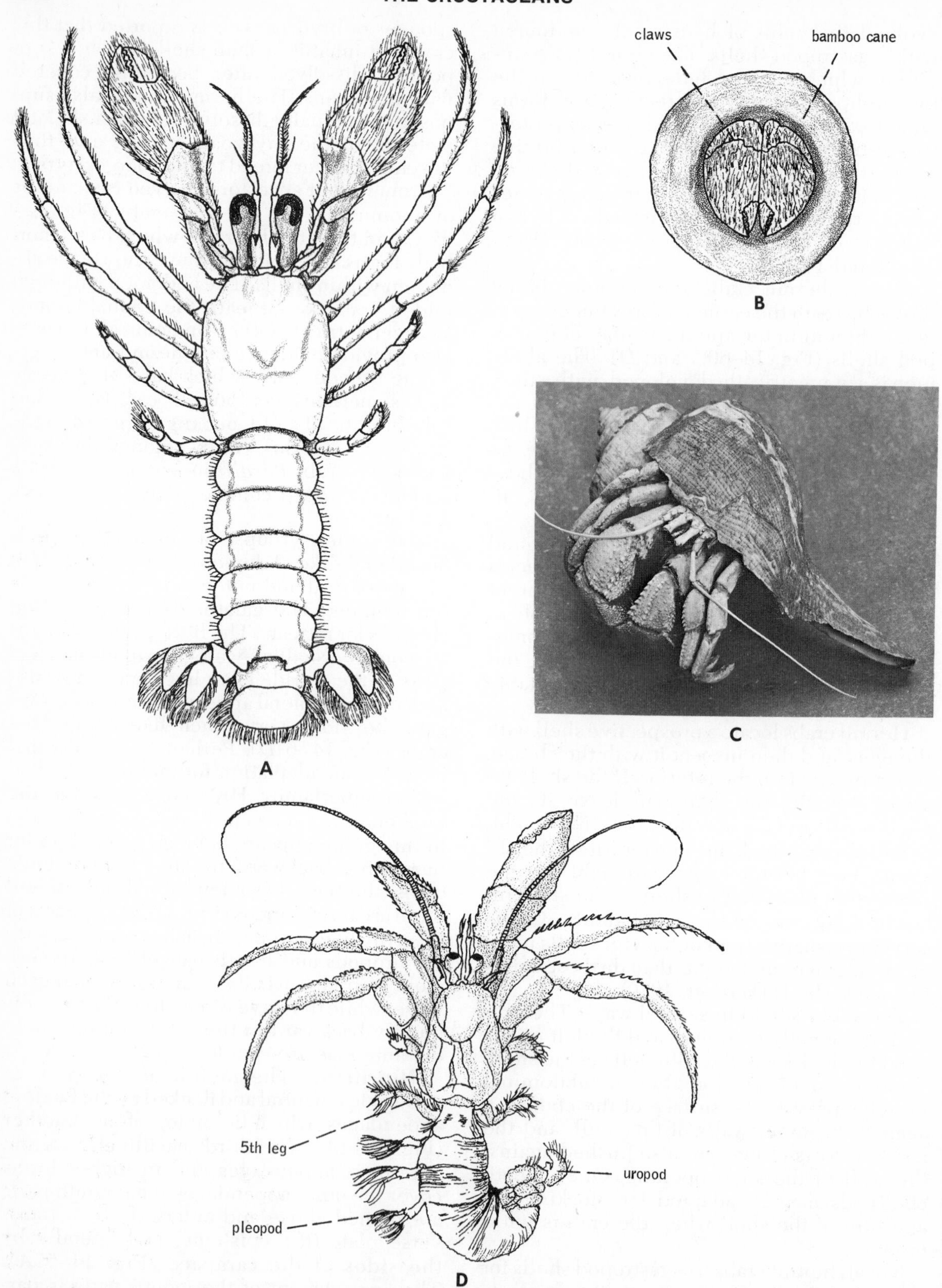

Figure 14–60. *A. Pylocheles miersi*, an anomuran that inhabits sections of old bamboo canes (dorsal view). Note how little modified the abdomen of this anomuran is compared to that of the pagurid illustrated in part *D* of this figure. *B. Pylocheles* in cane with claws blocking opening. (Both after Alcock from Calman.) *C.* A pagurid hermit crab. (Courtesy of American Museum of Natural History.) *D.* The hermit crab *Eupagurus bernhardus*, out of shell (dorsal view). (After Calman from Kaestner.)

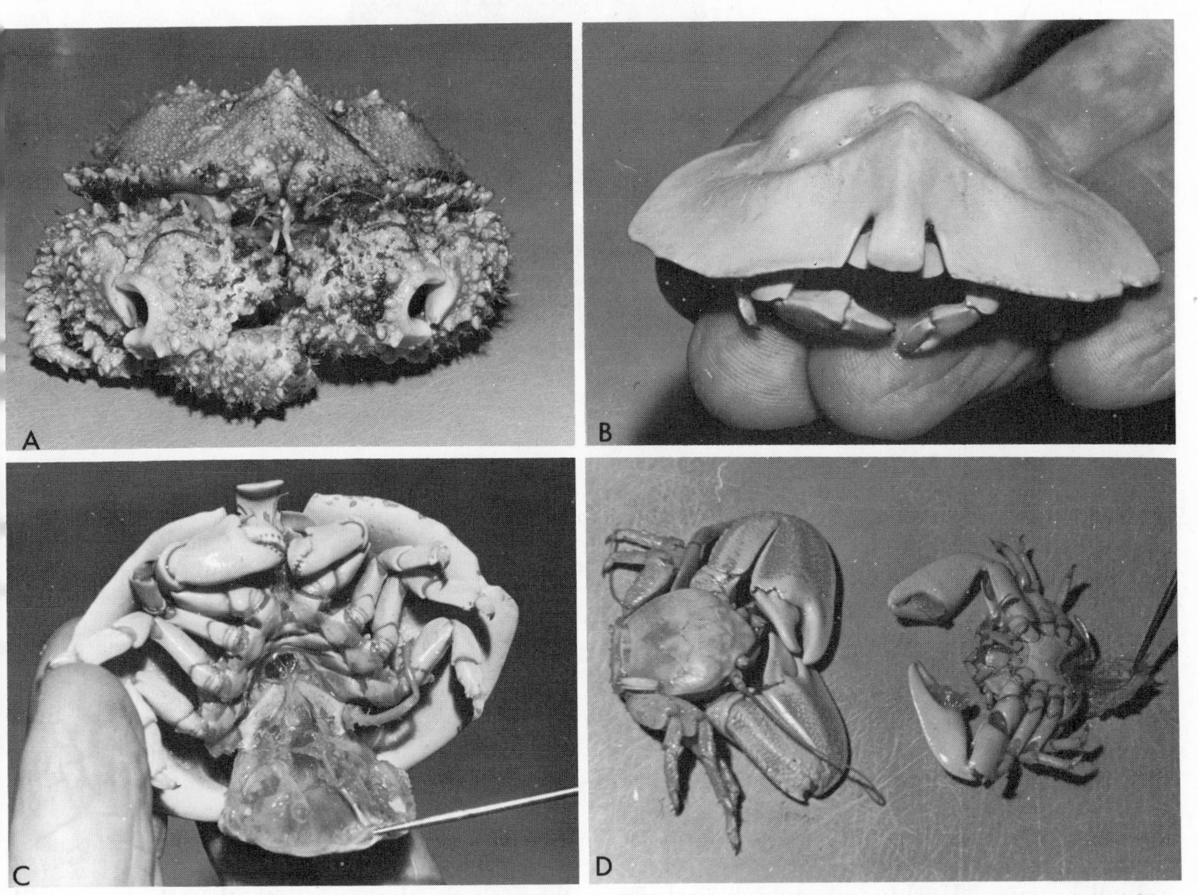

Figure 14–61. Anomuran crabs. *A.* Front view of the lithode crab *Lopholithodes foraminatus.* The siphon-like openings are located at a point of articulation in the chelipeds. *B. Cryptolithodes sitehansis,* a little lithode crab with a helmet-like carapace. *C.* Undersurface of *Cryptolithodes,* showing the reduced fifth pair of legs and the asymmetrical abdomen pulled back. *D.* The porcelain crab, *Petrolisthes eriomerus.* Dorsal view of one specimen; ventral view of another with abdomen pulled back. (All by Betty M. Barnes.)

Figure 14–62. *A.* The South Pacific robber, or coconut, crab, *Birgus latro.* (By Betty M. Barnes.) *B.* Juvenile individual of *Birgus,* in which the abdomen is housed within a gastropod shell in typical hermit crab manner. (From Reese, E. S., 1968: Science, *161*:385–386. Copyright 1968 by the Association for the Advancement of Science.)

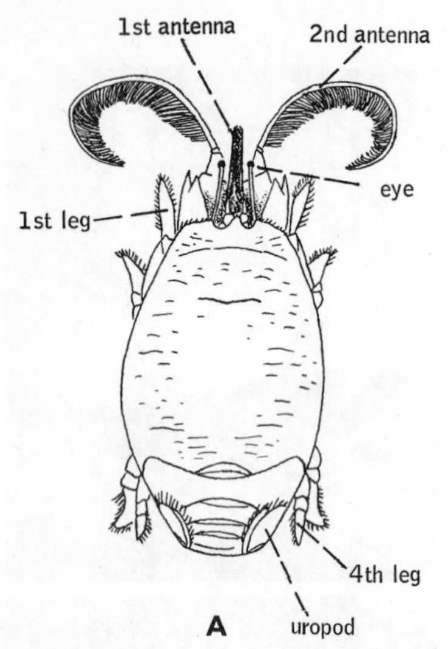

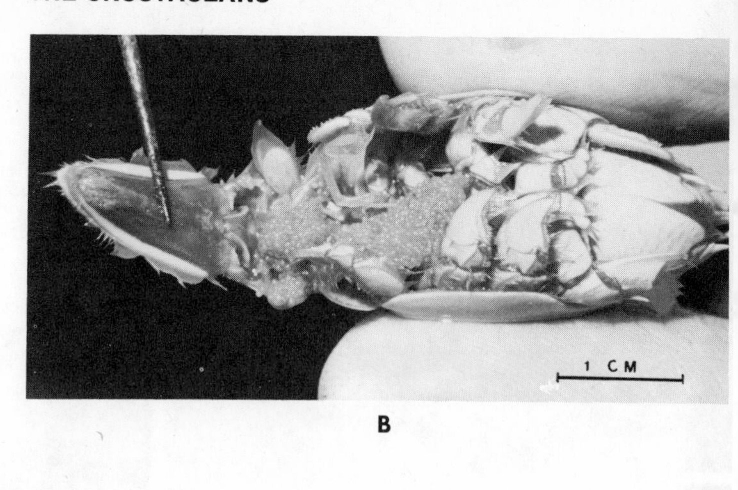

Figure 14–63. Anomuran crabs. *A.* The mole crab, *Emerita talpoida* (dorsal view). (After Verrill, Smith, and Harger from Waterman and Chace.) *B.* Undersurface of a brooding female mole crab with abdomen pulled back, showing greatly reduced fifth pair of legs. (By Betty M. Barnes.)

by two flattened rectangular plates (Fig. 14–55*B*). These rectangular plates completely fill the usually square buccal frame, covering the inner mouth appendages like double doors.

Most decapods are predacious, being either scavengers or omnivores. There is no sharp line between predatory forms and scavengers, and most predacious decapods feed on dead animal remains. Herbivores, which include most freshwater and terrestrial species, will usually also feed on carrion.

Prey or other food is caught or picked up with the chelipeds and then passed to the third maxillipeds, which push it between the other mouth parts. While a portion is bitten by the mandible, the remainder is torn away by the maxillae and maxillipeds. The severed piece is then pushed into the pharynx and another bite is taken.

A scavenger–detritus-feeding habit is characteristic of porcelain crabs, many hermit crabs, and others. Detritus feeding grades into filter feeding, for in many species material is collected with the chelipeds or third maxillipeds and then sifted by other mouth parts. For example, fiddler crabs scoop up mud and detritus with the cheliped. The doorlike third maxillipeds open, and the collected material is dumped into the buccal frame. Here the material is worked over and sifted by brush and straining setae on the second and first maxillipeds. An important aid to this process is the utilization of water pumped into the buccal cavity from the branchial chamber. The water brings about flotation of light material in a direction away from heavier particles. After organic material has been removed, the residue is spit out as round pellets, which may eventually surround the burrows, or cover the surface of the beach in forms that feed on detritus washed up by the receding tide (Fig. 14–70*B*).

Some species of the burrowing *Callianassa* and *Upogebia* create a current of water through the burrow with the pleopods and strain out plankton and detritus with fringed setae on the chelipeds and first legs. *Callianassa californiensis* digs up mud within its system of tunnels and strains it through the first two pairs of legs. The maxillipeds are used as combs for transferring the collected material to the other mouth appendages.

The maxillipeds are used as filtering appendages in many anomurans and in the commensal pea crabs. In the pea crab, *Pinnixa,* which lives in polychaete tubes, the endopodites of the maxillipeds are provided with filtering setae and are extended laterally or above the head like a net. The water current through the tube is provided by the host. A similar method of feeding is probably employed by many other members of this family, although some oyster crabs are known to feed on the mucous food strings collected by the host.

Among the most interesting filter feeders are the mole crabs, *Emerita* (Fig. 14–63*A*). The long, densely fringed second antennae project above the sand surface after the crabs are buried. In species that inhabit open beaches, such as *Emerita talpoida,* the second antennae filter plankton and detritus from the receding wave current. The crab buries itself seaward and its outstretched second antennae for a characteristic V on the sand surface, during the backwash of each wave; in species that live in quieter water, the projecting antennae are moved about.

In the typical decapod digestive tract, a short esophagus leads into a large capacious cardiac stomach (Figs. 14–1 and 14–64). This stomach is separated by a constriction from a much smaller and more ventral pyloric stomach, which lies in the posterior half of the thoracic region. A long intestine extends from the pyloric stomach through the abdomen and opens to the exterior on the ventral side of the telson. The esophagus, cardiac stomach, and the anterior half of the pyloric stomach are derived from the foregut and are lined with chitin. The midgut may be extremely short, as in crayfish and crabs, when it is limited to the posterior half of the pyloric stomach. On the other hand, in the lobster, *Homarus*, and the snapping shrimp, *Alpheus*, all but the terminal portion of the intestine, lying in the last abdominal segment, is a part of the midgut.

One to several ceca (hepatopancreas) arise from the midgut in most decapods. In the crayfish there is a very small single dorsal cecum pressed closely against the pyloric stomach. In lobsters there is also a dorsal midgut cecum, but it is located in the sixth abdominal segment near the junction with the hindgut. In most crabs there are two very long ceca that run anteriorly and laterally from their origin just behind the pyloric stomach.

The cardiac and pyloric stomachs contain a triturating gastric mill in all decapods, but the degree of development varies. The mill is least developed in the Natantia, in which the mandibles and other mouthparts perform a relatively efficient job of chewing before the food reaches the cardiac stomach. The greatest development of the gastric mill is found in macrurans and crabs, in which large pieces of food are quickly consumed. In crayfish, for example, food reaches the cardiac stomach in the form of long kneaded strands.

In such a stomach, the walls of the cardiac stomach and the anterior half of the pyloric stomach have been greatly strengthened with the development of a large number of sclerites. Three large opposing teeth, one dorsal and two lateral, project into the lumen (Fig. 14–64*B*). Also, there has developed a complex muscular system, which controls the movement of the stomach walls. Some of these muscles lie within the stomach wall; others run from the body wall to the stomach.

The triturating action of the gastric armature, plus the enzymatic action of the digestive secretion passed forward from the hepatopancreas, reduces food in the cardiac stomach to a very fine consistency.

The lumen of the pyloric stomach is greatly reduced by the complex folded walls. Setae on the folds form a filtering system, with definite channels leading to the two ducts of the hepatopancreas and the intestine. The channels conducting material to the duct openings begin behind the esophagus as a single median ventral groove on the cardiac stomach floor. Prior to reaching the narrow cardiac-pyloric junction, the channel divides into a right and a left pathway.

Within the pyloric stomach each channel division passes through a gland filter provided with extremely fine setae. The gland filter appears externally as a small swelling on each side of the pyloric stomach. Only the smallest particles and fluid are permitted to enter the ducts, larger particles being directed to the hindgut. The walls of the pyloric stomach are also strengthened with plates, and a system of muscles controls the size of the channels and aids in squeezing material through the filters.

The hepatopancreas is the site of enzyme production, absorption, and food storage. The digestive secretions, which are initially passed forward to the cardiac stomach, contain the usual array of enzymes and, in some species, cellulase and chitinase as well.

The hepatopancreas of decapods is usually a very large and compact paired mass that fills much of the cephalothoracic region. The pagurid hermit crabs are peculiar in having the digestive gland located in the abdomen. With few exceptions, each gland opens by a single duct into the sides of the midgut end of the pyloric stomach.

Gas Exchange. Primitively, there are four gills on each side of every thoracic segment. One gill (pleurobranch) arises from the body wall above the point of attachment of the appendage. Two gills (arthrobranchs) are attached to the articulating membrane between the appendage and the body wall; and one gill (podobranch) arises from the coxa of the appendage and represents a modification of part of the epipodite (Fig. 14–66). The remainder of the epipodite extends between the gills as a supporting lamella (mastigobranch, an unfortunate term for it is not actually a gill).

If all the gill series were present on all seg-

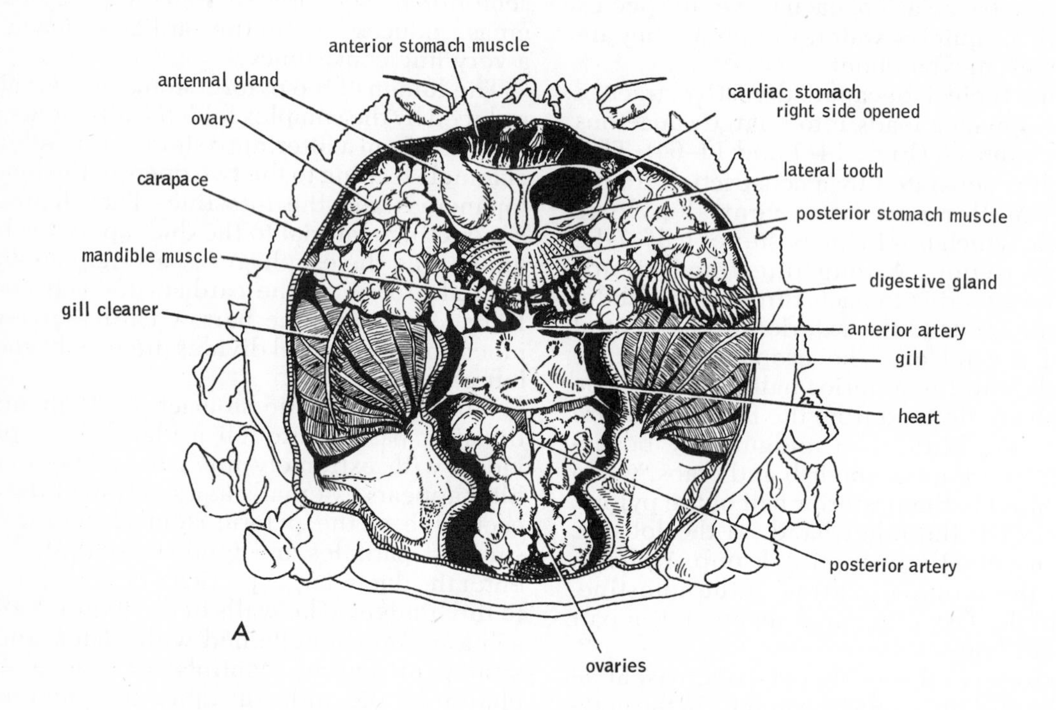

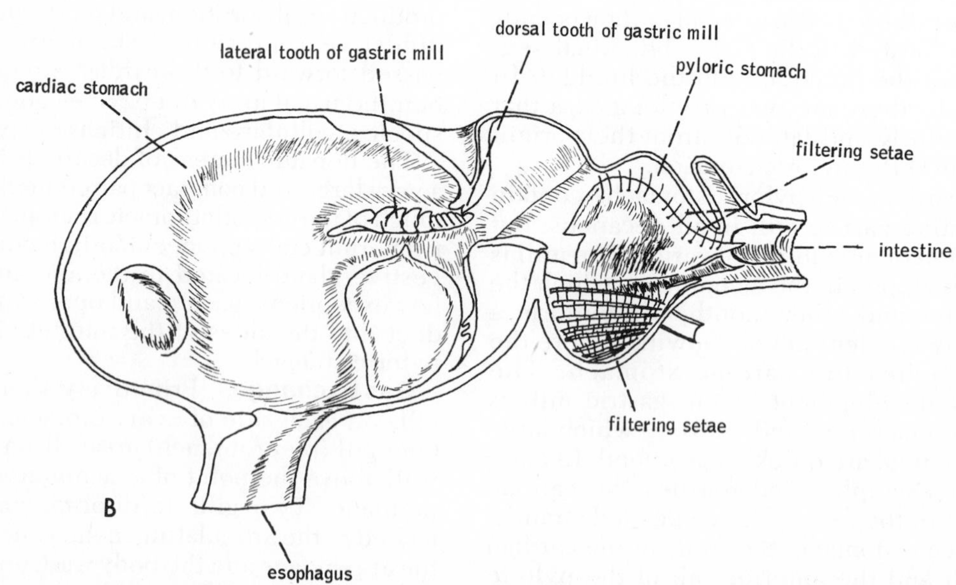

Figure 14-64. *A.* The brachyuran crab, *Eriocheir sinensis* (dorsal dissection). (After Panning from Kaestner.) *B.* Stomach of the crayfish, *Astracus* (lateral view). (After Kaestner.)

ments, there would be a total of 32 gills on each side of the body, but no decapod has retained this maximum number. The peneid shrimp, *Benthesicymus*, with 24 gills, has the greatest number, but reduction to far less than this is the general rule. A few examples serve to illustrate the great variation in branchial formulae. The snapping shrimp, *Alpheus*, with a total of six gills on each side, has one arthrobranch on the third segment (maxilliped), and one pleurobranch on each of the leg segments. The lobster, *Homarus*, has a complement of 20 gills on each side of the body, distributed as follows: a podobranch on the second and third maxillipeds and the first four pairs of legs; two arthrobranchs on the third maxilliped and the first four pairs of legs; and a pleurobranch on the last four pairs of legs. The branchial formula of many crayfish is similar to *Homarus*, except that most of the pleurobranchs tend to be vestigial or absent. Nine pairs of gills are common in marine crabs, but the swimming crab, *Callinectes*, has eight gills to one side—a podobranch on the second maxilliped; one arthrobranch on the second maxilliped and two on the third maxilliped and cheliped; and a pleurobranch on the second and third pairs of legs. Finally, the little pea crab, *Pinnotheres*, has only three gills to a side—one arthrobranch on the third maxilliped and two arthrobranchs on the chelipeds. In all decapods, the gills of the first maxillipeds are vestigial or absent.

Three types of gill structure are found in decapods, the particular type being characteristic of all gills of a species regardless of the gill position. The gill is composed of a central axis along which are arranged lateral extensions or branches. In the peneid shrimps, the axis bears biserially arranged main branches that are in turn subbranched. This type of gill is said to be dendrobranchiate or dendritic (Fig. 14–65*A* and *B*). The caridean shrimps and most anomurans and brachyurans have phyllobranchiate (lamellar) gills. In this type the gill axis bears flattened branches, which are usually arranged in two series along the axis (Fig. 14–65*E* and *F*). The third type of gill is called trichobranchiate (filamentous) and is typical of most macrurans, a few genera of anomurans, and certain primitive dromiid crabs (Fig. 14–65*C* and *D*). The branches are filamentous but not subbranched, and there are several series along the axis.

In the axis of each gill runs an afferent and an efferent branchial channel (Fig. 14–66*A*). From the afferent channel, blood flows into each filament or lamella and then back into the efferent channel. A longitudinal partition in each filament divides it into two channels. One channel conducts blood to the tip of the filament. The other channel conducts it back and thence into the outflowing channel of another filament at the same level, until the blood eventually reaches the efferent axial channel. In the brachyurans, blood flows through the lamellae within a fine sinus network.

The blood of decapods contains hemocyanin for oxygen transport. The pigment is always dissolved in the blood plasma and never is found in the tissues, as is frequently the case with hemoglobin in many entomostracans.

The ventilating current is produced by the beating of paddle-like scaphognathite, or gill bailer, a projection of the second maxilla (Fig. 14–66*D*). Water is pulled forward and the exhalant current flows out anteriorly in front of the head. In the Natantia, the ventral margins of the carapace fit loosely against the sides of the body, and water can enter the branchial chamber at any point along the posterior and ventral edges of the carapace (Fig. 14–66*A*). The macruran carapace is applied somewhat more tightly, and the entrance of water is limited to the posterior carapace margins and around the bases of the legs (Fig. 14–67*B*). The point of entrance of the ventilating stream is even more restricted in the brachyurans, in which the margins of the carapace fit very tightly against the ventrolateral body wall and the inhalant opening is located around the bases of the chelipeds (Figs. 14–66*C* and 14–67*A*). The forward position of the inhalant openings in brachyurans results in water taking a **U**-shaped course through the gill chambers. On entering the inhalant opening, the water passes posteriorly into the hypobranchial part of the chamber, and then moves dorsally, passing between the gill lamellae. The exhalant current flows anteriorly in the upper part of the gill chamber and issues from paired openings in the upper lateral corners of the buccal frame (Fig. 14–67*A*).

Since the majority of reptantians are bottom dwellers and include many burrowers, a variety of different mechanisms have evolved to prevent clogging of the gills with silt and debris. The bases of the chelipeds in crabs and the coxae of the legs of crayfish and lobsters bear setae that filter the incurrent stream. The gills are cleaned in crabs by the fringed epipodites of the three pairs of maxillipeds. These processes are elongated, especially that of the first maxilliped, and they sweep up and down the surface of the gills, removing

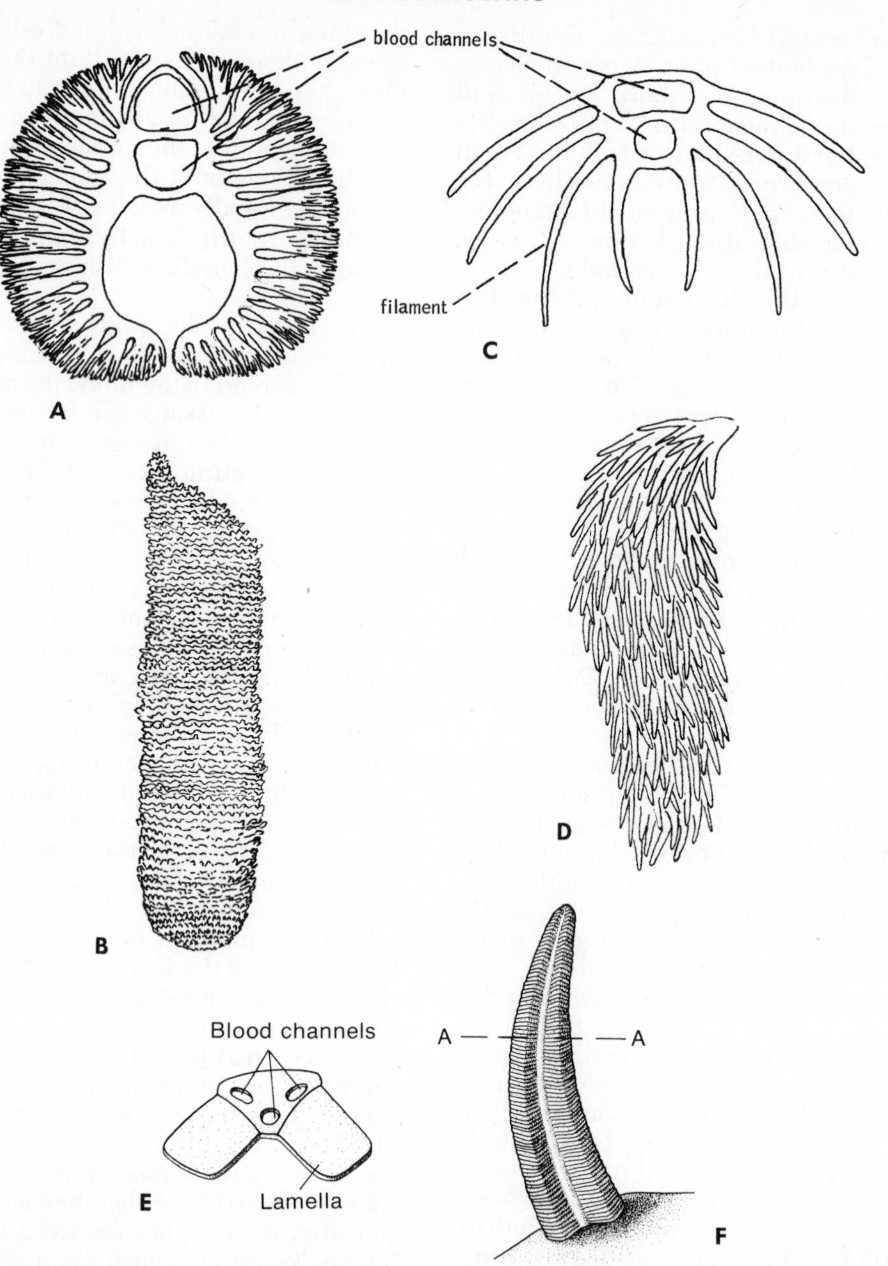

Figure 14–65. Decapod gills. *A* and *B*. Dendrobranchiate gill of the shrimp, *Penaeus*. *A*. Transverse section. *B*. Entire gill. *C* and *D*. Trichobranchiate gill of the crayfish, *Astacus*. *C*. Transverse section. *D*. Entire gill. *E* and *F*. Phyllobranchiate gill of the shrimp, *Palaemon*. This gill type is also characteristic of most anomurans and brachyurans. *E*. Transverse section. *F*. Entire gill. (All after Calman.)

detritus (Fig. 14–66*E*). As a further aid to the cleaning of the gills and branchial chamber, the gill bailer may periodically reverse its beat and thus reverse the direction of flow through the chamber.

In burrowing species, when the ventral parts of the body are covered by mud and sand, a reversed current is used for ventilating purposes; when the body is free, a forward current is used. A further ventilating adaptation in many burrowing decapods is the development of inhalant siphons. For example, in some species of mole crabs and in certain shrimps, the first antennae are held together, forming an inhalant passageway between them. The same thing is done with antennal scales in the shrimp, *Metapenaeus*, and with the second antennae in the burrowing crab *Corystes cassivelaunus* (Fig. 14–67*C*). The large box crabs of the genera *Calappa* and

Hepatus, which are sluggish bottom inhabitants, have greatly flattened chelipeds, which when folded fit tightly against the face but leave an inhalant channel between the inner side of the cheliped and the carapace (Fig. 14–67*B*). As in other oxystomatous crabs, the maxillipeds of *Calappa* roof a pair of exhalant channels.

Internal Transport. The heart is a coffin-shaped vesicle located in the thorax and is provided with three pairs of ostia, one ostium at each lateral angle of the heart and one dorsal pair (Fig. 14–4*C*); or in crabs, there are one lateral pair and two dorsal pairs.

Five arteries leave the heart anteriorly (Fig. 14–1*B*)—an anterior median (opthalmic) artery, a pair of lateral cephalic (antennary) arteries, and a pair of hepatic arteries. A median abdominal artery leaves the heart posteriorly, and a sternal artery arises either from the underside of the heart or from the base of the abdominal artery. The sternal artery descends to the sternum, passing to one side of the gut and piercing the nerve cords.

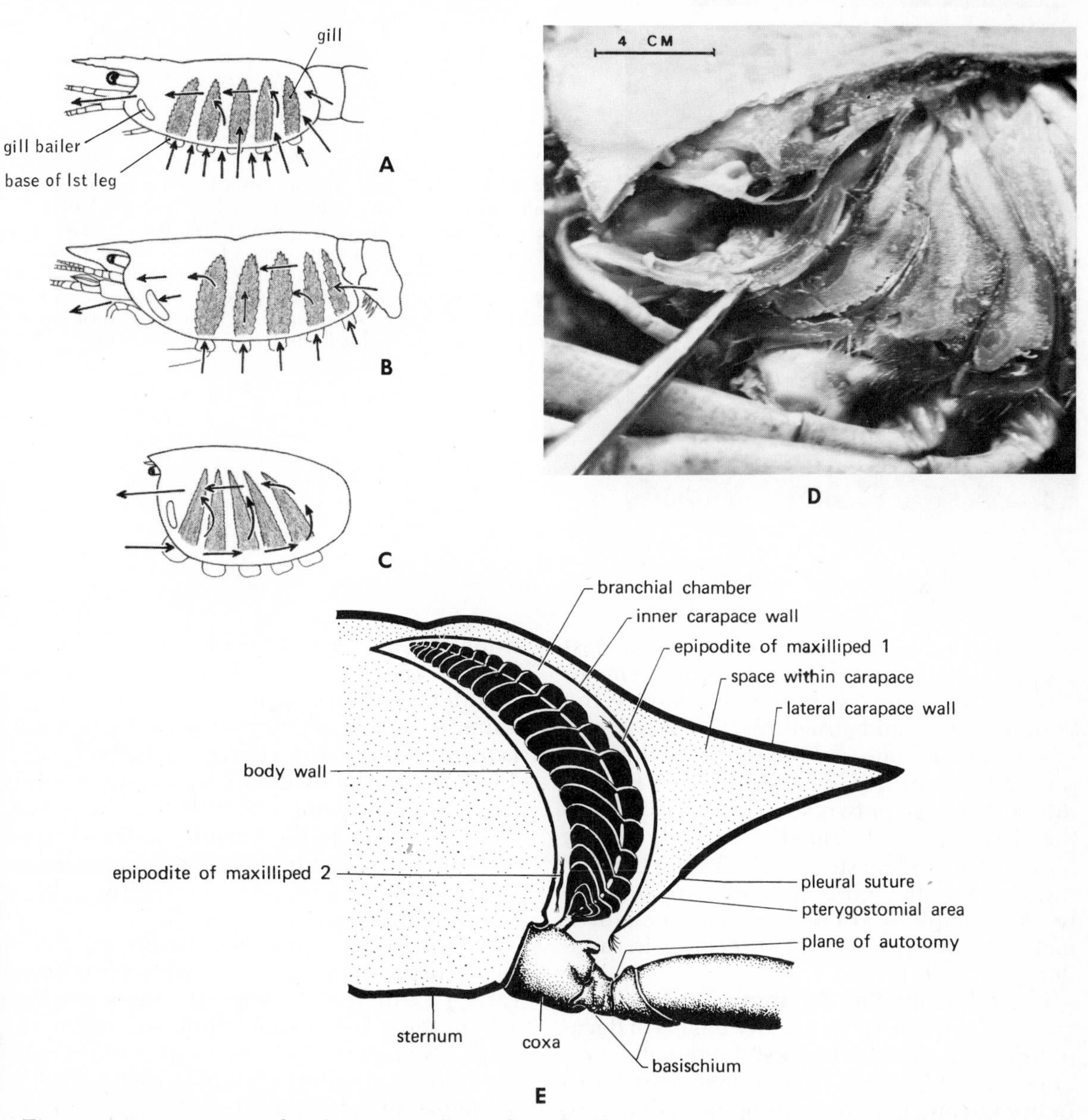

Figure 14–66. *A* to *C*. Paths of water circulation through gill chamber of three decapods, showing progressive restriction of openings into chamber. *A*. Shrimp. Water enters along entire ventral and posterior margin of carapace. *B*. Crayfish. Water enters at bases of legs and at posterior carapace margin. *C*. Crab. Water enters only at base of cheliped. *D*. Gill bailer (held by forceps) of a crayfish, showing its position within branchial chamber. (By Betty M. Barnes.) *E*. Cross section through the gill chamber of a crab. (From Kaestner, A., 1970: Invertebrate Zoology, Vol. 3. Wiley-Interscience, N. Y.)

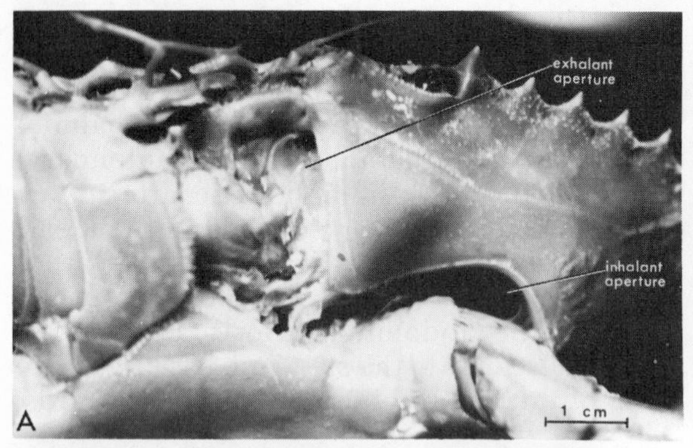

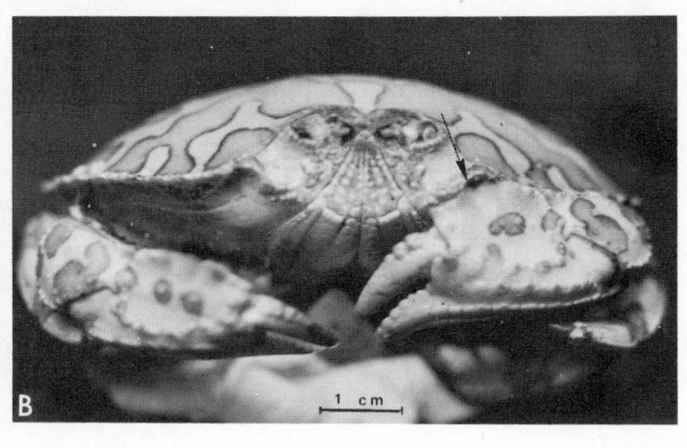

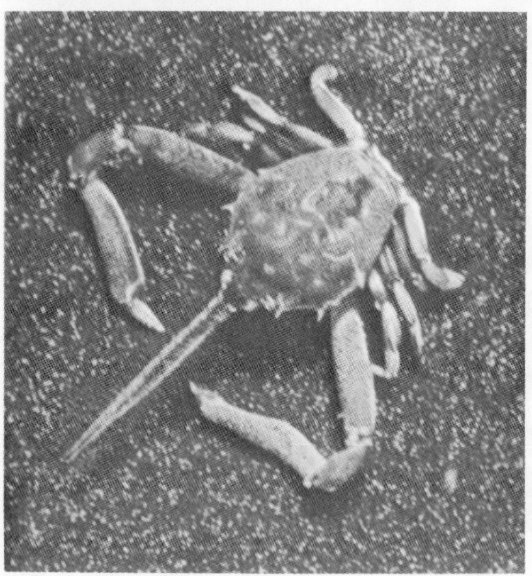

C

Figure 14–67. *A.* Anterior view of a blue crab, *Callinectes.* The left maxillipeds have been removed and the gill bailer can be seen within the exhalant aperture. *B.* The box crab *Hepatus.* Arrow indicates opening of left siphon formed by the opposing carapace and cheliped. The siphonal groove on the carapace can be seen on the right side. (Both by Betty M. Barnes.) *C.* The masked crab, *Corystes cassivelaunus.* Antennae form an inhalant tube when animal is buried. (By. D. P. Wilson.)

It then divides into an anterior and a posterior subneural artery. Each of these major arteries branches extensively, supplying various organs and structures.

Around the brain and the region of the ventral nerve cord, the smaller arteries branch into a capillary network before passing into the tissue sinuses. Blood eventually drains into a large median sternal sinus prior to passing through the gills and returning to the heart. Complete circulation has been estimated to take from 40 to 60 seconds in large decapods.

Excretion and Salt Balance. Antennal (or green) glands are the excretory organs of decapods and reach their highest degree of development in this group. The end sac, which lies in front of and to both sides of the esophagus, is typically divided into a saccule and a labyrinth. The labyrinth walls are greatly folded and glandular, converting the original vesicle into a spongy mass. The saccule may be a single vesicle, or its interior may also be partitioned. The labyrinth leads into a bladder by way of an excretory tubule. The decapod bladder varies in complexity. It may either be a simple vesicle or give rise to many diverticula that extend to various parts of the body (Fig. 14–5). From the bladder a short duct extends into the basal segment of the second antenna, where it opens to the outside on the summit of a little papilla (Fig. 14–1A). In brachyurans the excretory pore is covered by a little movable operculum.

Although the antennal glands are called excretory organs, most nitrogenous waste (NH^3) is diffused from the body surface where the exoskeleton is thin, such as on the gills. The antennal glands appear to function in controlling internal fluid pressure and, to some extent, ion content. Uptake of water by the blood increases the filtration pressure within the antennal gland and the passage of fluid into the saccule (Fig. 14–5). A copious amount of urine may be excreted, depending upon internal and external conditions. In normal seawater (3.4% NaCl), the European swimming crab, *Carcinus maenas,* produces a daily amount of urine equivalent to 3.6% of the body weight; in brackish water (1.44% NaCl), daily urine production is equal to one third of the body weight. The maintenance of a constant body volume is indicated by the fact that the size increase of a newly molted blue crab (*Callinectes*) is the same regardless of the salinity of the environment. The antennal glands apparently function to maintain this constancy.

Except in crayfish and certain shrimp, the antennal glands do not play a major role in osmotic regulation, for in most decapods the urine is isosmotic with the blood even when the animal is in brackish water. Within the higher salinity ranges of sea water, most decapods act as osmoconformers; i.e., the blood salt content is the same as that of the external medium. Those species that can tolerate lower salinities become osmoregulators, maintaining a minimum blood salt concentration that is hyperosmotic to the surrounding medium. As in fish, the gills pick up salts from the ventilating current, replacing ions lost in the urine and elsewhere.

Freshwater and Terrestrial Decapods. There have been numerous invasions of freshwater and land by decapods, but except for the crayfish, the record has not been especially successful. It is nonetheless interesting.

There are five families of freshwater shrimp. The Atyidae are inhabitants of streams, pools, and lakes in tropical and subtropical parts of the world, including southern Europe, the West Indies, and California. They are detritus feeders, living among small stones and water plants. The European *Atyaephyra desmarestii,* which is about 2 cm. long, occurs in schools.

The Palaemonidae contains many marine and brackish shallow water species, such as *Palaemonetes,* in both tropical and temperate regions. Freshwater species include some stream and pool inhabitants (*Macrobrachium*), some that live in cave waters, and some that are found in both brackish and fresh water (*Palaemon*). The Palaemonidae osmoregulate by means of the gills, and this is probably also the case with Atyidae.

Merguia rhizophorae lives on mangrove roots along the coast of Central and South America and is the only known semi-terrestrial shrimp. It is capable of leaping.

The crayfish are the only freshwater macrurans, but they are the most successful freshwater decapods. The more than 500 species are found throughout the world in streams, ponds, lakes, and caves. Some live beneath stones or within debris. Many species excavate burrows, which they use as retreats and for overwintering. Burrowing crayfish commonly inhabit bottom lands where the water table is not too far below the surface, and they may be semi-terrestrial. Most crayfish are about 10 cm. in length, but the Australian *Euastacus* reaches the size of lobsters. Crayfish are omnivores. Unlike those of other decapods, the antennal glands of crayfish can excrete a hyposmotic urine and play an important role in salt balance. The initial filtrate is isotonic with the blood; but in its passage to the bladder, and perhaps even within the bladder, there is reabsorption of ions until the urine is very dilute, containing as little as one tenth the amount of salts in the blood.

The only freshwater anomurans are the crab-like members of the family *Aeglidae,* which are found in the more temperate parts of South America. The anomurans also contain one group of terrestrial decapods, the tropical land hermit crab *Coenobita* and the closely related coconut crab *Birgus* (Fig. 14–62*A*). Both are found in Indo-Pacific regions, and *Coenobita* also occurs in the West Indies.

The land hermits live close to the shore and periodically enter the water to wet the body and interior of the shell. The juveniles of the coconut crab house the abdomen within a shell when they first emerge from the sea following larval development (Fig. 14–62*B*), but the adults have abandoned the hermit crab habit and have acquired a crab-like form with a flexed abdomen. The adults live in burrows further back from the sea, but are still a coastal species. *Birgus* can climb trees and has been reported to climb to the tops of palms and mangroves 60 feet tall, using the sharp heavy ends of the legs like spikes. It descends backwards. Coconut crabs feed on carrion and both decaying and fresh vegetation. They obtain water by drinking. Neither *Coenobita* nor *Birgus* can excrete a concentrated (hypertonic) urine. They both have reduced gills and have converted the moist branchial chamber into lungs with highly vascularized areas for gas exchange. The vascular surface is on the floor of the chamber in *Coenobita;* in *Birgus* highly vascularized epithelial folds, or appendages, hang from the roof of the branchial chamber. *Coenobita* is also provided with an accessory vascularized gas exchange area on the anterodorsal surface of the abdomen.

There are many brachyuran crabs that can tolerate either very brackish or fresh water, but must return to salt water to breed.

In this category belongs the Chinese mitten crab, *Eriocheir* (Fig. 14–68*A*), which is found in the rivers and rice paddies of Asia and has been introduced into the rivers of Europe, probably by ship ballast. Another is the brachyuran, *Rhithropanopeus,* of the American coast, which is found in salt water as well as in freshwater ditches. *Rhithropanopeus* has been introduced into northern Europe and San Francisco Bay, possibly in the same way as *Eriocheir.* Many species of *Uca* live in very brackish estuaries and marshes. *Callinectes sapidus* is quite euryhaline and ranges far up into brackish estuaries and bays.

Strictly freshwater crabs—forms that complete the entire life cycle in fresh water—include the Potamonidae (river crabs), which are found throughout the semitropical and tropical parts of the world (*Potamon edulis* is found in the rivers of Italy, Greece, and North Africa), and a few grapsids. The little Jamaican grapsid, *Metropaulias,* is found only in water contained within bromeliads.

All of these crabs, both temporary and permanent residents of brackish and fresh water, excrete urine which is isosmotic with the blood and regulate their salts by means of ion absorption through the gills.

Except for the land hermits and coconut crabs, most of the decapods that have invaded land are brachyuran crabs. This is not surprising, considering the motility of crabs and their tightly enclosed branchial chambers. Although the living terrestrial crabs are derived from a number of different land invasions, most display similar adaptations.

Gills continue to be utilized for gas exchange, but there is a tendency for the gill number or the gill surface area to be reduced. The reduction is possible because of the greater amount of oxygen per unit volume of air compared to water. The reduction is advantageous in decreasing water loss from the gills through evaporation. A further correlation with increased dependency upon air as a medium for gas exchange has been a tendency for the branchial chamber to become more capacious. The gill bailer continues to provide for ventilation, but moves air rather than water.

Water loss by evaporation is much greater in land crabs than in other terrestrial arthropods (arachnids, insects), and most of this is through the carapace. Water is replaced, for the most part, by drinking. No land crab can excrete a hypertonic urine. Such a capability would aid in conservation of water. However, land crabs save water by excreting very little urine.

All land crabs live in burrows, or at least beneath stones, and are usually active only at night. They are primarily vegetarians and scavengers. All can run very rapidly. Eyes are generally well developed.

The Ocypodidae contains some of the most familiar amphibious crabs. This family includes *Dotilla* and the fiddler crabs, *Uca,* both of which are intertidal, and the more terrestrial ghost crab, *Ocypode,* which burrows in dunes above the high-tide mark (Fig. 14–69*A*). The amphibious fiddler crabs are found along the North American coast and in many other parts of the world (Fig. 14–70). Some species inhabit protected sand beaches of bays and sounds; others can tolerate brackish water and burrow in the mud of marshes and estuaries. The burrows are located in the inter-

A

B

Figure 14–68. *A.* The Chinese mitten crab, *Eriocheir sinensis,* in a rice paddy. This amphibious freshwater brachyuran is native to southern Asia but has been introduced into the Rhine and Elbe rivers of Europe. *B.* A river crab, *Potamon anomalus* (Potamonidae). (By Betty M. Barnes).

Figure 14–69. *A.* Ghost crab, *Ocypode ceratophthalma. B.* The Indo-Pacific soldier crab, *Mictyris longicarpus.* (Based on a photograph by Healy and Yaldwyn.)

tidal zone. At low tide, the crabs come out to feed, and enormous numbers may cover the beach (Fig. 14–70*D*).

The burrows of most species do not extend much deeper than 36 cm., and the excavated sand is usually carried to the surface as small pellets or balls. The common Atlantic fiddler, *Uca pugilator,* carries the sand balls cradled in the second, third, and fourth legs of one side (Fig. 14–70*B*). The west coast fiddler, *Uca crenulata,* carries its sand balls away from the entrance. As the incoming tide approaches the burrows, the crabs return and plug the entrance with sand or mud.

The Indo-Pacific soldier crabs of the family Mictyridae are similar in habit to *Uca* (Fig. 14–69*B*). When they are disturbed, they quickly burrow sideways like a corkscrew.

The family Grapsidae contains perhaps the most ecologically diverse assemblage of crabs. There are marine, brackish-water, freshwater, amphibious, and terrestrial species. Amphibious grapsids include *Sesarma,* which lives beneath drift and stones, the agile rock-inhabiting *Grapsus,* and numerous mangrove inhabitants, such as *Aratus.*

Most amphibious species are periodically immersed in water, and the branchial chamber contains some water when the crab is on land. The water remaining within the chamber is aerated by the vibrations of the gill bailer. In *Ocypode* and *Uca,* air enters the branchial chambers through special posterior openings between the third and fourth or fourth and fifth legs, and the usual anterior inhalant openings in aquatic crabs are used for exhaling air in these genera. These same openings may also be used for replacing water lost from the gill chambers.

Some amphibious genera, such as *Sesarma,* aerate the water remaining in the branchial cavity externally. The scaphognathite drives the water out of the exhalant openings, and it flows back over the surface of the body in special sculptured channels, being aerated in the process. The water then reenters the branchial chambers through the openings between the third and fourth legs or near the first legs.

The more truly terrestrial crabs include the ghost crabs *Ocypode,* some grapsids, some potamonids, and the gecarcinids. The Gecarcinidae, all of which are terrestrial and to which the name *land crab* usually refers, are found in tropical and subtropical America, West Africa, and the Indo-Pacific area; it includes *Cardisoma,* which lives in fields and woods in southern Florida and the West Indies, the *Gecarcinus,* which inhabits grasslands and forests along the coasts of Africa and Central America and in the West Indies (Fig. 14–71).

Although terrestrial, the gecarcinids are nevertheless coastal in distribution, for females must return to the sea to release their spawn. The terrestrial potamonids are found the greatest distance from the coast, often hundreds of miles. However, these crabs are derived from freshwater species and need only return to fresh water to breed.

The lamellae of the gecarcinid and grapsid land crabs are rigid and are held apart by various structures to permit the circulation of air between them. Usually there is also a

Figure 14–70. *A.* Male fiddler crab, an amphibious brachyuran. (Courtesy of American Museum of Natural History.) *B.* Fiddler crab by burrow. The large balls of sand were excavated from burrow; the small pellets are composed of sand that has been filtered of organic material and ejected by the mouth parts. *C.* Mold of burrow of *Uca pugilator.* (By R. W. Frey and J. D. Howard, 1969: Trans. Gulf Coast Assoc. Geol. Societies, *19*:427–444.) *D.* Large numbers of *Uca pugilator* on a beach at low tide. (By Betty M. Barnes.)

Figure 14–71. West Indian gecarcinid land crabs. *A. Gecarcinus lateralis. B. Cardisoma guanhumi.* (Both by Betty M. Barnes.)

supplementary gas exchange surface formed by the development of the capillary network in the median wall of the branchial chamber.

The maintenance of gill moisture in certain gecarcinid land crabs has been elucidated by Bliss (1963) and Mason (1970). In the Bahamas, the burrows of *Cardisoma guanhumi* extend down to the level of ground water. This is not true of the burrows of *Gecarcinus lateralis*. Most crabs possess a pair of pericardial sacs in the posterior end of the cephalothorax. The sacs are filled with vacuolated connective tissue and some muscle tissue, and are permeated with blood sinuses. One part of the sac is in contact with the pericardial cavity and part of the surface is exposed to the back of the branchial chamber. These pericardial sacs are more extensive and bulge much further into the branchial chamber in terrestrial crabs than they do in the marine species. In fact, the gills lie across the bulge of the sac.

The function of the pericardial sacs is to regulate the hemostatic pressure of the blood as water is absorbed prior to or during molting. In *Gecarcinus lateralis* droplets of dew or rain water are conducted via a tiny channel from the base of the fifth leg to the posterior extension of the pericardial sac (Fig. 14–72). Water is rapidly carried by capillary action up the surface of the sac to the gills, where it serves in part to maintain the humidity of the branchial chamber and gills. The water source may differ in *Cardiosoma*, but the pericardial sacs may function in the same way.

It should be noted that the conversion of the gill chamber into a lung in land crabs has taken place in a manner strikingly parallel to that in the gastropod mollusks.

The ghost crabs, *Ocypode*, are in some respects transitory between an amphibious and a truly terrestrial habit. As in other amphibious crabs, some species may wet the body in the sea; others do not. There are also species which have vascularized areas in the branchial walls for supplementary gas exchange.

Nervous System and Sense Organs. Natantians and macrurans possess the largest number of unfused ventral ganglia; only the first two thoracic ganglia are fused with the subesophageal, leaving six unfused thoracic and six abdominal ganglia. The ventral ganglia, in different groups, show varying degrees of additional fusion, culminating in the brachyurans, in which all of the abdominal ganglia have migrated anteriorly to fuse with the thoracic ganglia, forming a single ventral mass.

Although there are some blind decapods, particularly among deep-sea forms and cave-dwelling crayfish, compound eyes are usually highly developed. The ocular peduncle is composed of two or rarely three segments and is much more mobile than in most other crustaceans. In some crabs, one of the peduncular segments is greatly lengthened, placing the eyes at the ends of extremely long stalks. The corneal surface is usually terminal, but this is not always the case. In some crabs, such as some species of *Ocypode*, the pedun-

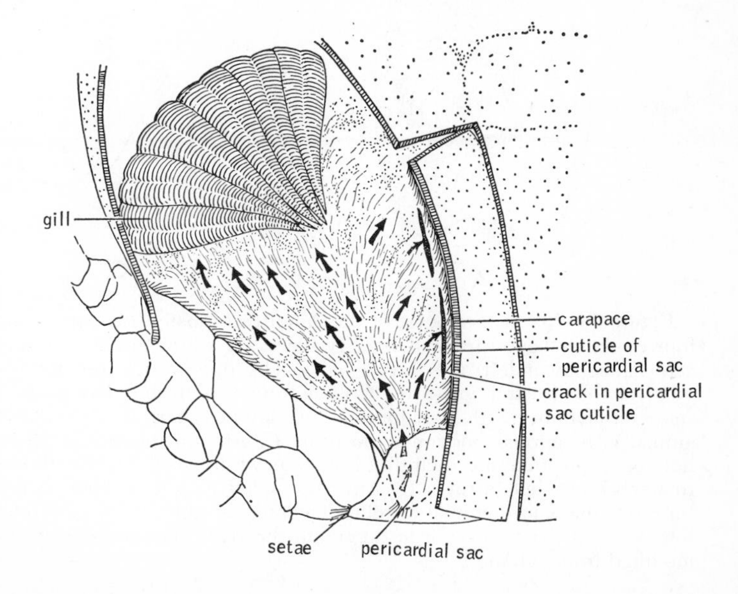

Figure 14–72. Uptake of water by land crab *Gecarcinus* via pericardial sacs. Water enters at base of fifth pair of legs and is conducted over surface of sacs (arrows) to gills. Some water enters sacs through cracks in old cuticle. (From Mason, C. A., 1970: J. Exp. Zool., *174*(4):381–390.)

cle extends beyond the corneal surface, which is wrapped around the stalk (Fig. 14–69*A*).

The carapace is commonly modified in the region of the eyes. In shrimps and macrurans, the carapace is frequently notched above the eye. The snapping shrimps are peculiar in that the anterior margin of the carapace has grown down and completely enclosed the eyes; however, the covering is probably thin enough to permit light perception. In brachyurans, there are transverse orbits in which the eyes rest when not erected.

A pair of statocysts is present in nearly all decapods and is located in the basal segment of the first antennae. The sac is always open to the exterior, although in crabs and some others the opening is reduced to a slit and is functionally closed. The statolith may be composed of fine sand grains bound together by secretions from the statocyst wall (Fig. 14–73). Since the statocyst is an ectodermal invagination, its lining along with the statolith is shed at each molt. The sand grain statoliths are replaced when the head is buried in sand, or the animal actually inserts sand grains into the sac. Along the floor of the sac are a number of rows of sensory hairs with which the statolith is directly connected or is in intermittent contact. The hairs arise from receptor cells in the sac wall, and the receptor cells are innervated by a branch of the antennular nerve.

Continual impulses from the receptor cells of one statocyst, independent of the statolith pull, tend to cause the animal to rotate or list to the opposite side. However, since the tendency initiated by one statocyst counteracts the other, rotation does not occur except when one statocyst is injured or removed.

When the animal is in a normal horizontal position, the floor of the statocyst is inclined about 30 degrees, which results in a medial gravitational pull on the statolith. This pull stimulates the receptor cells to initiate additional impulses causing the animal to rotate, but again one statocyst counterbalances the other (Fig. 14–73*B*). If the animal is rotated or lists 30 degrees to the right side, this places the floor of the right statocyst in a horizontal position, and the statolith exerts no pull on the statolith hairs. However, the floor of the left statocyst is now inclined 60 degrees and the impulses initiated by its receptor cells cause the animal to rotate in the opposite direction—that is, back to a horizontal position (Fig. 14–73*C*). Stimulation of the receptor cells by the statolith is maxi-

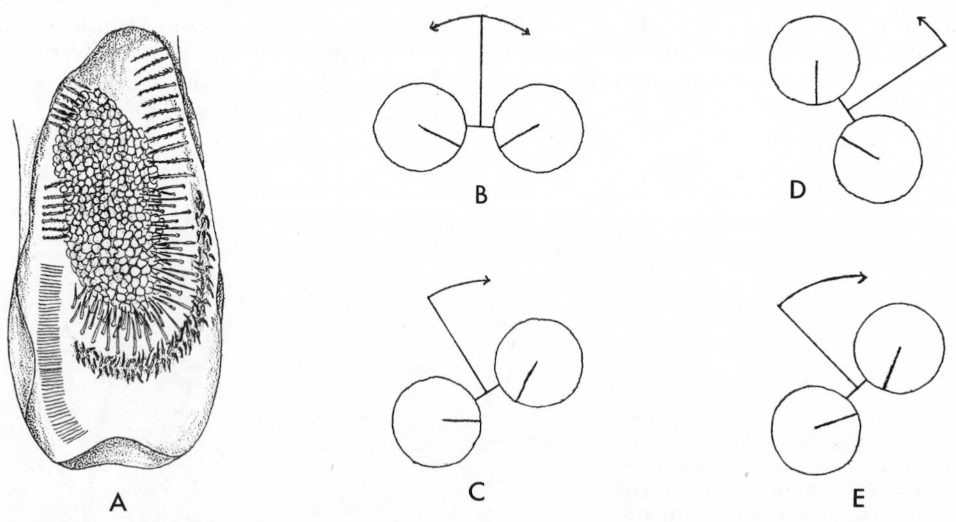

Figure 14–73. *A*. Statocyst of the American lobster, *Homarus americanus* (with dorsal wall removed). Note the four crescent-shaped rows of sensory hairs; the inner three are in contact with the statolith. The row of long, delicate hairs (lower left) are thread hairs. The opening of the statocyst to the exterior is at upper left beyond the edge of the illustration. (After Cohen from Cohen and Dijkgraaf.) *B* to *E*. Statocyst function. Circle represents statocyst chamber with floor inclined 30 degrees (bar). *B*. Counterbalancing pull exhibited by two statocysts when animal is in normal horizontal position. *C*. Animal rotated 30 degrees to right, placing floor of right statocyst in horizontal position and inclining left statocyst floor 60 degrees. Impulses initiated by left statocyst cause animal to rotate back to left (arrow). *D*. Animal rotated to left so floor of right statocyst is inclined 90 degrees, initiating impulses that cause animal to rotate back to the right. *E*. Animal rotated to right. The more inclined position of the left statocyst floor causes it to dominate the right statocyst, and the animal rotates back to the left. (*B* to *E* greatly modified from Schöne.)

mum when the floor of the statocyst is vertical, or tilted 90 degrees (Fig. 14–73*D*). When the floors of both statocysts are tilted, the one in which the receptor cells are most stimulated dominates the other (Fig. 14–73*E*).

The statocysts of most shrimp, as well as many Reptantia, are simple sacs. More complex statocysts are found in crabs and in some macrurans and anomurans. The statocyst invagination in crabs is deep and irregular and may contain a labyrinth compartment, as well as a statocyst chamber. In forms with such complex statocysts, these organs indicate not only the gravitational position of the animal, but also its position in regard to movement. The rotation receptors give rise to special free sensory hairs, which are not associated with the statolith and are stimulated by the motion of the fluid within the statocyst chamber when the animal moves (Fig. 14–73*A*). The mechanism is thus like that of the semicircular canals in vertebrates.

Quite a few decapods possess special structures for producing sound, but none are known to be able to hear in the sense of possessing special receptors for the detection of air- or water-borne pressure waves. Sounds, especially those produced on land, are largely or entirely detected as substrate vibrations (Horch and Salmon, 1969; Salmon, 1971). A few species apparently can detect pressure waves through the general tactile receptors, and probably all can detect vibrations through the substratum.

Stridulation, in which one part of the body is rubbed against another, is a common mechanism for sound production. Rapping with the chela is another. The ghost crabs, *Ocypode*, produce a filing, creaking, or buzzing sound by rubbing a ridge of tubercles on the inner side of one hand against a process on the second segment of the same appendage. The noise apparently serves to warn other individuals against entering its burrow. At least some species of *Uca* also stridulate when another crab enters its burrow. In *Uca*, the sound is produced by the crab's rubbing the cheliped file against the front of the carapace. Species of spiny lobsters, *Panulirus*, use the second antennae to make a sound that has been compared to that of the filing of a saw or the creaking of leather.

The popping sound produced by snapping shrimp is very audible and easily detected when underwater or on a rocky breakwater at low tide.

Luminescence. Seventeen genera of shrimp contain species known to be luminescent. The luminescent organs are most commonly photophores with lenses and reflectors, and may arise anywhere on the body, even the roof of the branchial chamber (as in *Sergestes*). Among the best known luminescent species are those belonging to the genus *Sergestes*, one of which, *Sergestes challengeri*, possesses 150 light organs. Many species of this genus have a peculiar type of internal photophore formed from modified cells of the tubules of the digestive glands. Foxton's (1970) study of the vertical distribution of this genus near the Canary Islands disclosed that species of the upper mesopelagic zone were semitransparent and possessed internal ventral photophores. On the other hand, species of the deep mesopelagic zone were red and possessed dermal photophores, when photophores were present. Some deep-sea shrimp emit a luminous cloud-like secretion into the water.

Reproduction and Development. The paired but connected testes usually lie in the thorax but may extend posteriorly into the anterior portion of the abdomen. Like those of most other malacostracans, decapod sperm have no tails and are sometimes star-shaped. The sperm may be transmitted in the form of spermatophores, in which case the more distal portion of the sperm duct is glandular and modified for spermatophore formation. The terminal end of the sperm duct is a muscular ejaculatory duct, which usually opens onto the coxa or the articulating membrane between the coxa and sternum. There is a single tubular penis in some hermit crabs and a pair in all brachyurans. The first two pairs of pleopods in male decapods are usually modified to aid in sperm transfer. For example, in brachyurans, the first pelopod is in the form of a cylinder into which fits the piston-like, second pleopod. Sperm emitted from the penis are pumped through the conducting anterior pleopod (Fig. 14–74).

The ovaries are similar in structure and location to the testis (Fig. 14–64*A*). The oviducts are usually unmodified and open onto the coxae. In brachyurans (except the primitive superfamilies Gymnopleura and Dromiacea), however, the terminal end of each oviduct is modified as a glandular seminal receptacle and vagina for the reception of the male penis. A median seminal receptacle, opening separately to the outside, occurs in most other decapods and is in the form of a ventral pouch created by processes of the last one or two thoracic sternites.

Precopulatory "courtship" is characteristic

A

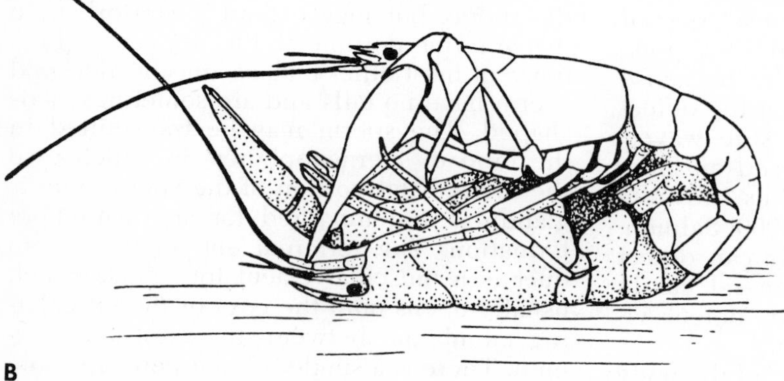

B

Figure 14–74. Copulatory structures of the brachyuran crab, *Ranina*. The nipple-like structure in front of the end of the dissecting needle is the penis. Just above the penis is the base of the second copulatory pleopod, which projects into the somewhat cylindrical first pleopod. The two pleopods are directed forward and lie in the sternal depression covered by the abdomen. *B*. Copulating crayfish. Female is stippled. (From Andrews in Kaestner, A., 1970: Invertebrate Zoology, Vol. 3, Wiley-Interscience, N.Y.)

of many decapods, especially anomurans and brachyurans. Precopulatory behavior in hermit crabs involves the male holding the female with one cheliped and tapping and stroking her with the other, or pulling her back and forth. Mating takes place as soon as the female completes her pre-adult molt, and both animals partially come out of their shells.

Considerable sexual dimorphism is present in many species of decapods. Some male mole crabs are small and neotenous, and possess an adhesive pad or some other means of clinging to the female carapace. Dimorphism is particularly striking in fiddler crabs, for example, in which one claw of the male is very large, in contrast to the two small claws of the female.

Pheromones are important in sexual attraction of some aquatic crabs; visual and acoustical signals are of especial importance for terrestrial forms. The semi-terrestrial fiddler crabs go through an elaborate courtship behavior, during which the male waves or signals with his large claw like a semaphore. Each species follows a distinct pattern of movement, and some rituals may also involve curtsies and rapping on the ground with the large claw. The addition of acoustical signals appears to be an adaptation of temperate species which court at night as well as during the day. At the end of the courtship ritual, the male attempts to seize and copulate with the female above ground; or, as in *Uca pugilator* and *U. annulipes*, the male entices the female into his burrow, or in some species carries her into his burrow.

In the brachyuran families Cancridae (*Cancer*) and Portunidae (*Callinectes*, *Portunus*, and *Carcinus*), there is a premolt attendance of the females by the male, in which the male carries the female about beneath him, her carapace beneath his sternum (Fig. 14–75). He releases her so that she can molt. Copulation occurs shortly afterwards.

In natantians the copulating pair are usually oriented at right angles to each other with the genital regions appressed together. Mating may be preceded by caressing behavior on the part of the male. The male crayfish turns the female over and pins back her chelipeds with his own (Fig. 14–74*B*). The first two pairs of pleopods are then lowered to an angle of 45 degrees and held in this position by one of the last pair of legs (folded beneath the pleopods).

In crayfish possessing a seminal receptacle, the tips of the first pleopods are inserted into the chamber, and sperm flow along grooves in the pleopods. Copulation in lobsters is very similar to that in crayfish. When the seminal receptacle is absent, the sperm are transmitted as spermatophores, which are attached to the body of the female, particularly at the bases of the last two pairs of legs. These are frequently seen as "tar spots" on the females of the spiny lobsters, *Panulirus*.

During copulation in crabs, the female lies beneath the male, or in the reverse position, with ventral surfaces opposing each other. The first pair of pleopods, which conducts the sperm, is inserted into the openings of the female. Terrestrial crabs mate on land.

Egg laying takes place shortly after copulation in forms with no seminal receptacles, but when sperm are deposited in seminal receptacles, which are commonly sealed and plugged by the semen (e.g., brachyuran crabs), egg laying may not take place until some time later.

Peneid shrimp shed their eggs directly into the sea water. In all other decapods the eggs on extrusion are typically attached to the pleopods with a cementing material. The cementing material is associated with the egg membrane.

Fertilization is internal in brachyurans, but in most decapods the eggs are probably fertilized at the moment of egg laying. In crayfish the female lies on her back and curls the abdomen far forward, creating a chamber into which the eggs are driven by a water current produced with the beating pleopods. In crabs the usually tightly flexed abdomen is lifted to a considerable degree to permit brooding, and the egg mass, which

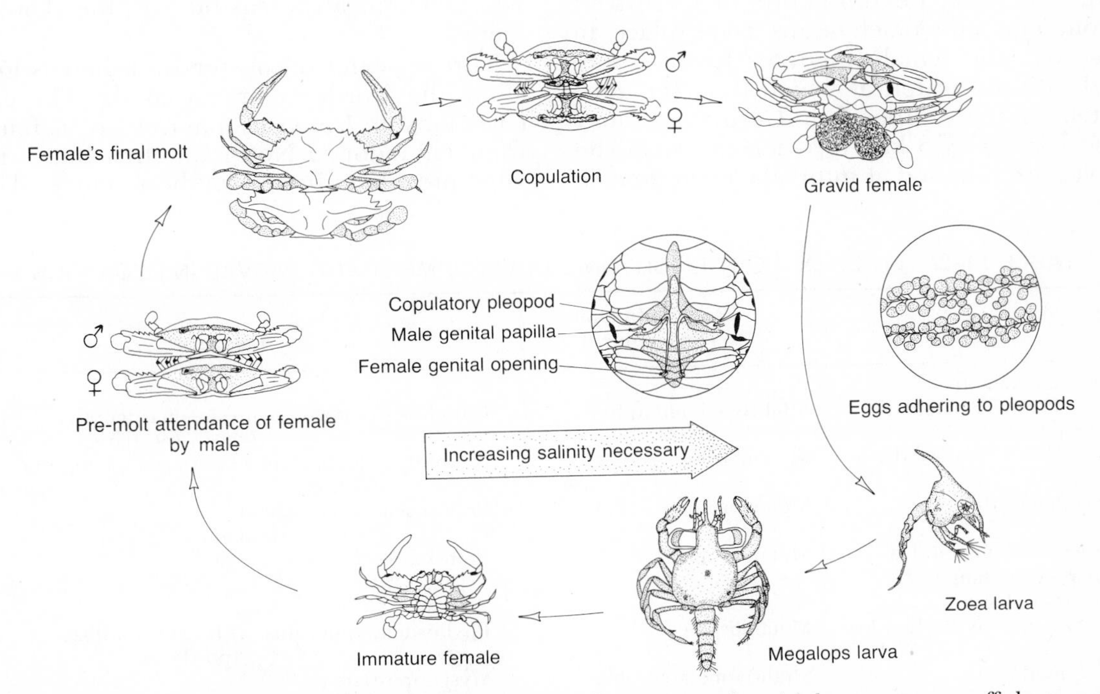

Figure 14–75. Life cycle of the commercial blue crab, *Callinectes sapidus*. Adults are common off shore as well as in very brackish water. Females living in brackish water must return to water of high salinity to permit hatching of the eggs. The hatching stage is a zoea larva, which lives in the plankton for a period of time. Successive molts transform the zoea to a megalops larva, possessing many crab features but with a still unfolded abdomen. Further development leads to the adult body form. Prior to the female's last molt she is carried about beneath a male, and copulation occurs after molting. Eggs are laid some time later and are attached to pleopod setae beneath the folded abdomen. (Courtesy of C. Piling; adapted from various sources.)

often becomes orange in color, is sometimes called a sponge (Fig. 14–75).

The hatching stage varies greatly. In some nonbrooding peneid and sergestid shrimp, the eggs hatch as nauplius, metanauplius, or protozoea larvae, but in almost all other marine decapods hatching takes place at the protozoeal or zoeal stage. The special names given the zoeal and postlarval stage of the different decapod groups are listed in Table 14–2.

The larval stages of crabs are particularly interesting. The zoea is easily recognized by the very long rostral spine and sometimes by a pair of lateral spines from the posterior margin of the carapace (Fig. 14–76*A*). The postlarval stage, called a megalops, recapitulates the macruran ancestry of this group, for the abdomen is large, unflexed, and bears the full complement of appendages (Fig. 14–76*E*).

As is true of many other invertebrates, there is a tendency for larval life to be shortened in decapods which inhabit cold oceans or abyssal depths. Larval stages are usually absent in the strictly freshwater decapods (see page 597). Brackish-water forms and river immigrants, such as *Eriocheir*, return to more saline water for breeding (Fig. 14–75). Likewise, development in terrestrial anomurans and brachyurans takes place in the sea. The female migrates to the shore and into the water, at which time the zoea hatch and are liberated. *Cardisoma* may travel as far as 8 km. to reach the sea. The larvae are released at intervals as the female periodically immerses herself. She remains at the sea about one night. The potamonid crabs, which are terrestrial or amphibious, return to fresh water to breed, although there are no free larval stages.

Economic Significance. The decapods equal or surpass the bivalve mollusks in importance as a source of human food, and the two groups together comprise the so-called shellfish industry. In the United States the commercial shrimpers are centered primarily along the southeastern Atlantic coast and the Gulf of Mexico. Species of *Penaeus* comprise most of the catch.

Shrimp are caught by trawling, in which a very long **V**-shaped net is towed behind the shrimp boat. Each arm of the **V** is connected by ropes to the boat, and at the apex the net forms a bag. The nets are towed along the bottom at slow speeds for half an hour to an hour, depending on the catch. In addition to shrimp, a wealth of other marine animals, called "trash" by shrimpers, is also collected. In the Gulf of Mexico, trawling takes place further off shore and may continue for days or weeks before the trawler returns to port. The Japanese farm shrimp, rearing them from eggs to marketable size (Webber, 1968), but shrimp farming is not yet economically feasible in the United States.

Three species of lobster are fished as food along the North American coasts. The east coast lobster, *Homarus americanus*, is found from Labrador to North Carolina but is not very prevalent in its southern range. This

TABLE 14–2. TYPES OF POST-EMBRYONIC DEVELOPMENT AND LARVAE IN DECAPODS

Group	Post-Embryonic Development	Larvae
Suborder Natantia		
Family Penaeidae	Slightly metamorphic	Nauplius→protozoea→mysis→mastigopus (zoea) (postlarva)
Family Sergestidae	Metamorphic	Nauplius→elaphocaris→acanthosoma→mastigopus (protozoea) (zoea) (postlarva)
Section Caridea	Metamorphic	Protozoaea→zoea→parva (postlarva)
Section Stenopodidea	Metamorphic	Protozoea→zoea→postlarva
Suborder Reptantia		
Section Macrura		
Superfamily Scyllaridea	Metamorphic	Phyllosoma→puerulus, nisto, or pseudibaccus (zoea) (postlarva)
Superfamily Nephropsidea	Slightly metamorphic	Mysis→postlarva (zoea)
Section Anomura	Metamorphic	Zoea→glaucothöe in pagurids, grimothea (postlarva)
Section Brachyura	Metamorphic	Zoea→megalopa (postlarva)

(After Waterman and Chace. *In* T. H. Waterman (Ed.): Physiology of Crustacea. Vol. 1, 1960.)

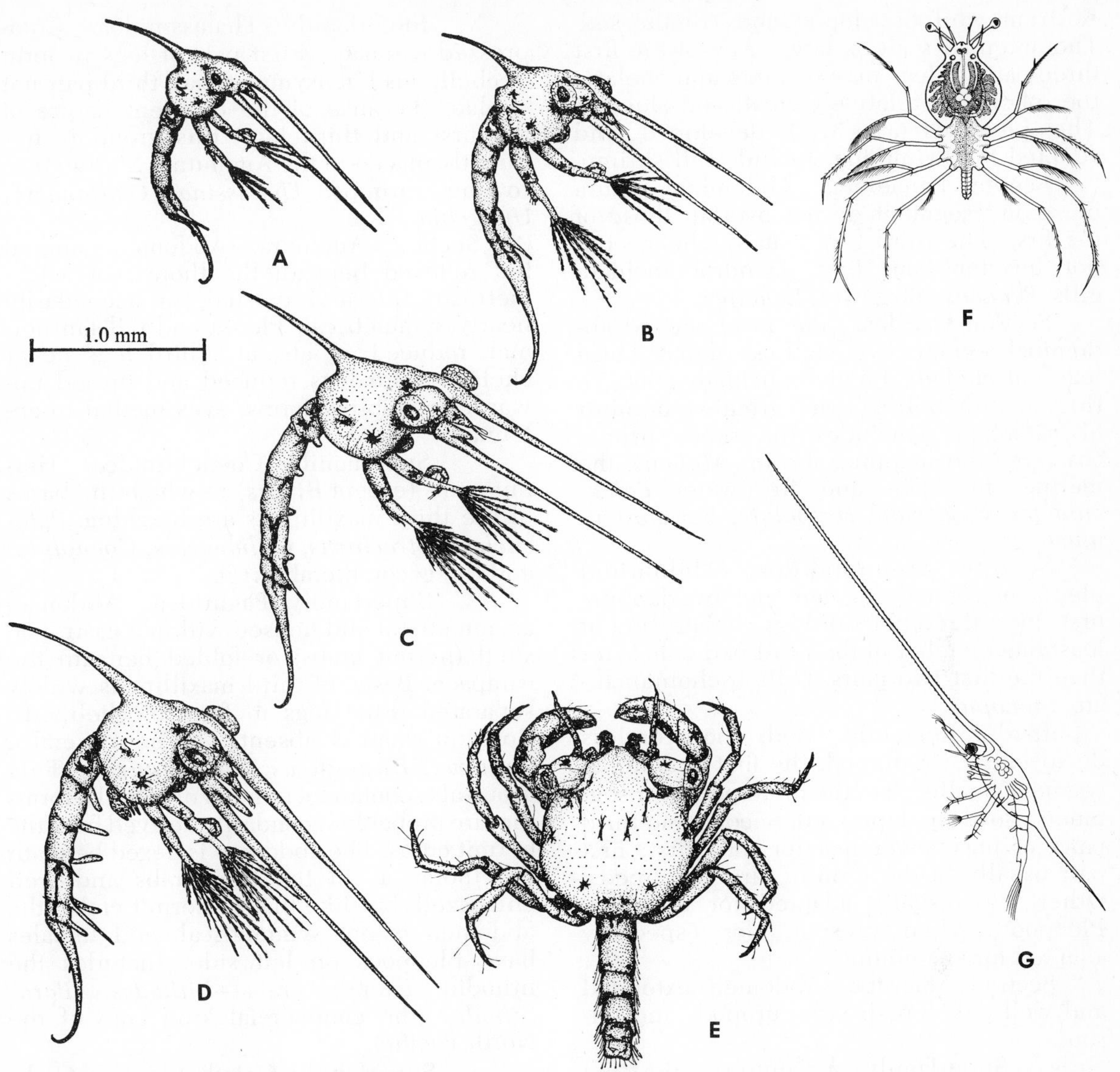

Figure 14–76. Decapod larvae. Four zoeal stages (*A, B, C, D*) and megalops (*E*) of the brachyuran crab *Rhithropanopeus harrisii.* (From Costlow, J. D., and Bookhout, C. G., 1971: Fourth European Marine Biology Symposium, Cambridge Univ. Press, p. 214.) *F.* Phyllosoma (zoea) larva of the spiny lobster, *Palinurus.* (Modified after Claus.) *G.* Metazoea of the anomuran crab *Porcellana.* (Adapted from various sources.)

species has enormous chelipeds, which may contain nearly as much meat as the abdomen, or "tail." Off the Gulf and Pacific coasts, species of the spiny lobster, *Panulirus,* are used as food. The body is covered with spines, and in contrast to *Homarus* the first legs are similar to the others rather than clawed. The lobster tails sold in markets are usually from spiny lobsters caught and shipped from various parts of the world. The animals are commonly caught in traps, or pots, that the lobster enters as a retreat.

The portunid swimming crab, *Callinectes sapidus,* or blue crab, is the commercially important crab, occurring along the east and Gulf coasts of the United States. It is caught in shallow water with a trap or line, but commerical fishermen also catch them in large numbers by trawling when fishing for shrimp, or in winter by means of a crab dredge, which dislodges crabs which have buried themselves in the sea bottom. On the west coast and in Europe, species of *Cancer,* a nonswimming crab, are used as food and are caught by trapping.

SYSTEMATIC RÉSUMÉ OF ORDER DECAPODA

Suborder Natantia. Body more or less compressed; the first abdominal segment no smaller than the more posterior segments.

Rostrum well developed and compressed. The antennal scale is large. Any of the first three pairs of legs may be large and chelate; the others are relatively small and slender. The pleopods are well developed and adapted for swimming. Includes all shrimp.

Section Penaeidea. Abdominal pleura of second segment do not overlap those of the first. The third legs usually chelate but not heavier than first. Dendrobranchiate gills. *Penaeus, Sergestes, Leucifer.*

Section Caridea. Pleura of second abdominal segment overlap those of first. Third legs not chelate. Phyllobranchiate gills. To this section belongs the greatest number of natantians. Includes the sand shrimp, *Crangon;* the snapping shrimp, *Alpheus;* the marine, brackish- and freshwater *Palaemonetes; Palaemon; Hippolyte; Benthesicymus.*

Section Stenopodidea. Abdominal pleura of second segment not overlapping first. First three pairs of legs chelate, and at least one member of the third pair is heavier than the first two pairs. Gills trichobranchiate. *Stenopus.*

Suborder Reptantia. Body more or less dorsoventrally flattened; the first abdominal segment smaller than the more posterior segments. Rostrum depressed. Second and third pairs of legs never heavier than first; first pair usually in the form of large chelipeds. Other legs usually adapted for crawling. Pleopods, when present, not especially adapted for swimming.

Section Macrura. Abdomen extended and well developed; large uropods and telson.

Superfamily Eryonidea. Rostrum small or absent. First segment of second antennae not fused with epistome. First four pairs of legs chelate. Broad depressed carapace. Marine, usually in deep water. *Polycheles.*

Superfamily Scyllaridea. Rostrum small or absent. First segment of second antennae fused with epistome. First four pairs of legs not chelate. Marine. The spiny lobsters and Spanish lobsters. *Panulirus, Jasus, Scyllarus, Ibacus.*

Superfamily Nephropidea. Cylindrical carapace with a well-developed rostrum. First three pairs of legs chelate; first pair typically in form of heavy chelipeds. Includes the lobsters, Homaridae–*Homarus, Nephrops;* the freshwater crayfish—*Astacus, Cambarus.*

Superfamily Thalassinidea. Compressed carapace. First pair of legs in form of chelipeds but asymmetrical; third pair not chelate (because of the different nature of the first and third legs, this group is frequently placed in the Anomura). Marine burrowing shrimps. *Thalassina, Callianassa, Upogebia.*

Section Anomura. Abdomen normal but reflexed beneath the thorax; or asymmetrical, soft and twisted; or secondarily nearly symmetrical. Pleura and tail fan normal, reduced, or absent. Third legs never chelate. Fifth legs reduced and turned upward. In crablike forms, eyes medial to antennae.

Superfamily Coenobitoidea. Hermit crabs (except *Birgus*) in which the bases of the third maxillipeds are touching. *Pylocheles, Petrochirus, Clibanarius, Coenobita,* and the coconut crab *Birgus.*

Superfamily Paguridea. Abdomen asymmetrical and housed within a gastropod shell (hermit crabs) or folded beneath the carapace. Bases of third maxillipeds widely separated. First legs in form of chelipeds. Rostrum small or absent. The hermit crabs, *Pagurus, Eupagurus, Pylopagurus.* This group also contains a number of crablike forms that are probably secondarily derived from the hermit crabs. The abdomen is flexed beneath the thorax as in the true crabs and well chitinized, but like that in hermit crabs, the abdomen is not symmetrical, and females have pleopods on left side. Includes the lithodid (stone) crabs—*Lithodes, Paralithodes* (the commercial king crab of the North Pacific).

Superfamily Galatheidea. Crablike forms with symmetrical abdomen flexed beneath thorax. Well developed tail fan. Rostrum often well developed; carapace not fused with epistome. First legs in form of chelipeds; fifth legs within gill chambers. *Galathea, Munida;* the porcelain crabs, *Petrolisthes, Pachycheles, Porcellana, Polyonyx;* the South American freshwater *Aegla.*

Superfamily Hippidea. Symmetrical abdomen flexed beneath thorax. Rostrum reduced or absent. Cephalothorax more or less cylindrical. First legs chelate or subchelate at the most. Commonly in sand in surf zone. The sand, or mole, crabs—*Hippa, Emerita, Blepharipoda, Lepidopa, Albunea.*

Section Brachyura. Abdomen reduced and tightly flexed beneath thorax. Carapace fused with epistome. First legs in form of heavy chelipeds; third legs never chelate. Eyes usually lateral to second antennae. The true crabs.

Subsection Gymnopleura. Primitive burrowing crabs with more or less trapezoidal or elongate carapace, subchelate first legs, and with some or all remaining legs flattened and expanded for burrowing. *Ranina*.

Subsection Dromiacea. Primitive brachyurans; marine. Last pair of legs often dorsal in position and modified for holding objects over the crab. Carapace usually not broader than long. Uropods present but much reduced. In females, sternum deeply and complexly grooved; genital apertures on coxae. *Dromia, Hypoconcha*.

Subsection Oxystomata. Last pair of legs normal or modified as in Dromiacea. Mouth frame triangular rather than quadrate. Uropods absent. Female reproductive openings on sternum. Marine. *Dorippe;* the box crab, *Calappa; Lithadia*.

Subsection Brachygnatha. Mouth frame quadrate, and last pair of legs unmodified in size and position. Female genital openings on sternum.

Superfamily Brachyrhyncha. Carapace not narrowed anteriorly. Body shape round, transversely oval, or square. Pseudorostrum very small or absent. Orbits well developed. This superfamily contains the majority of crabs: Corystidae—*Corystes*. The cancer crabs, Cancridae—*Cancer, Pirimela*. The swimming crabs, Portunidae—*Portunus, Callinectes, Carcinus, Arenaeus, Ovalipes*. The freshwater crabs, Potamonidae (actually three families according to modern carcinologists)—*Potamon, Pseudothelphusa*. The mud crabs, Xanthidae—*Xantho, Menippe* (stone crab), *Pilumnus, Rhithropanopeus, Panopeus, Neopanopeus, Eurypanopeus*. The commensal pea crabs, Pinnotheridae—*Pinnotheres, Pinnixa, Dissodactylus*. The amphibious crabs, Oxypodidae—*Oxypode* (ghost crabs), *Uca* (fiddler crabs), *Dotilla;* Mictyridae (soldier crabs). Marine, freshwater, amphibious, and terrestrial Grapsidae—*Geograpsus, Grapsus, Pachygrapsus, Planes, Sesarma, Eriocheir* (Chinese mitten crab), *Aratus, Metaplax, Metopaulias*. The land crabs, Gecarcinidae—*Cardisoma, Epigrapsus, Gecarcoidea, Gecarcinus, Ucides*. The coral gall crabs, Hapalocarcinidae—*Hapalocarcinus, Cryptochirus, Troglocarcinus*.

Superfamily Oxyrhyncha. Carapace narrowed anteriorly into a pseudorostrum. Body shape roughly triangular. Incomplete orbits. Marine. The decorator and spider crabs—*Maja, Inachus, Macrocheira, Hyas, Libinia, Pelia, Parthenope, Pugettia, Loxorhynchus, Stenorhynchus*.

Vertical Migration in Crustaceans

In no other aquatic animals is the phenomenon of vertical migration of such wide occurrence nor has it been more extensively studied than in the Crustacea. Such movement is exhibited by some representatives of almost every order, and is particularly prevalent among cladocerans, calanoid copepods, and euphausiaceans. The adults, larval stages, or both may migrate. For example, the Antarctic *Euphausia superba* lives at the surface in the adult stages and the nauplii remain in deep water; the postnaupliar larval stages migrate.

Most commonly, vertical migratory movement follows a diurnal pattern, as in the copepod, *Calanus finmarchicus*. A few hours before sunset the animals begin to move upward from the depth of water occupied during the day to accumulate at the upper level by late evening. Around midnight the descent begins. This is interrupted briefly by a slight upward movement at dawn, after which descent continues to the day depth.

There are many variations on the typical pattern just described. Ascent and descent may occur earlier or later, or they may even be reversed, as in the copepod, *Acartia clausii*, and the euphausiacean, *Nyctiphanes couchii*, with upward movement occurring during the day and downward movement at night. Seasonal variation is also common.

In larger crustaceans, both the ascending and descending movement is probably kinetic—that is, through the locomotor activity of the animal. In smaller crustaceans the descent is, in part at least, a passive sinking, especially in such groups as copepods and cladocerans, in which the animal tends to be oriented vertically.

The distance traveled varies considerably, depending primarily upon the swimming ability of the animal. The larger crustaceans, such as shrimp and euphausiaceans, may travel a distance as great as 600 to 1000 m. A study on pelagic caridean shrimp in the region of the Canary Islands (Foxton, 1970) showed that six species are restricted to depths below 600 meters during the day, but that at night the populations migrate in part or as a whole to levels as high as 100 m. The smaller groups, such as copepods and cladocerans, have a migratory range of 30 to 150 m. Although the upper position is commonly at the surface, it may be at lower levels. For example, bathypelagic species rarely rise to a level higher than 200 m. below the surface.

For the majority of crustaceans, light is the primary controlling factor in vertical migration. The animals are photopositive to certain light intensities, most commonly dim light, and tend to remain at that level where light conditions are optimum. Gravity is probably involved chiefly in orientation.

The value of diurnal vertical migration to the animal is still far from certain. It is commonly held that such movement brings the animal from the nonproductive twilight zone up into the rich photosynthetic stratum for feeding, at a time (night) when the animal would not be as exposed to predation or be injured by high light intensities.

Molting, Chromatophores, Hormones, Physiological Rhythms, and Autotomy

Molting and Growth. In many crustaceans, including barnacles, crayfish, and the lobster, *Homarus*, molting and growth continue throughout the life of the individual, although molts become spaced further and further apart. Such crustaceans may live to be quite old and some may become very large. In others, such as some crabs, molting and growth cease with the attainment of sexual maturity or of a certain size, or after a certain number of instars.

The process of molting has probably been investigated in more detail in crustaceans, especially decapods, than in any other arthropods. It was formerly thought that all physiological processes involved in molting ceased during most of the intermolt period (i.e., the time between two periods of actual shedding, or ecdysis). It is now realized that molting is virtually a continuous process in the life of a crustacean and that as much as 90 per cent or more of the intermolt period may be involved with concluding and preparatory processes associated with the preceding and the future molt. This is especially true in species that molt all year round. In species that molt seasonally, such as species of the crayfish, *Cambarus*, and the fiddler crab, *Uca*, there is a rather definite intervening rest period during the intermolt. But even during the rest period, food reserves are accumulated for the next molt.

The preparatory phase, or proecdysis, is marked by a continuing accumulation of food reserves and a rise in blood calcium, probably resulting from the activity of the hepatopancreas and resorption of calcium from the cuticle. In some crustaceans, such as crayfish and land crabs, the stomach epithelium secretes calcareous concretions, called gastroliths, which function as calcium storage centers. Eventually the membranous layers and part of the calcified layers of the old exoskeleton are digested away. Resorption of calcium and digestion of the calcified layer are especially great where splitting later occurs or where the old skeleton must be stretched or broken to permit extraction of a large terminal part of an appendage, such as a claw. After the separation of the old cuticle from the epidermis and the secretion of the new epicuticle, the animal is prepared for the actual process of ecdysis and usually seeks some protected retreat or remains in its burrow. The body swells from the uptake of water through the gills and quickly emerges from the old skeleton, which is commonly eaten to gain calcium salts. Water uptake appears to result from a rise in the osmotic pressure of the blood, which may be related in turn to the increase in blood calcium. A large part of the water eventually enters the cells, which at this time display an increase in free amino acid levels.

During the concluding phase of ecdysis (postecdysis), the endocuticle is secreted and calcification and hardening of the skeleton take place. The animal remains in its retreat and does not feed during the first part of this phase. This is the longest part of the molt cycle and may occupy a considerable part of the intermolt period.

Chromatophores. A characteristic and striking feature of the integument of many malacostracans is the presence of chromatophores. The crustacean chromatophore is a cell with branched, radiating, noncontractile processes (Fig. 14–77A). Pigment granules flow into the processes in the dispersed, or stellate, state and are confined to the center of the cell in the contracted, or punctate, state.

White, red, yellow, blue, brown, and black pigments may be present. The red, yellow, and blue pigments are carotene derivatives obtained from the diet. The red compound so conspicuous in boiled crabs, lobsters, and shrimp is astaxanthin (carotenalbumin). In the exoskeleton of the living animal, this

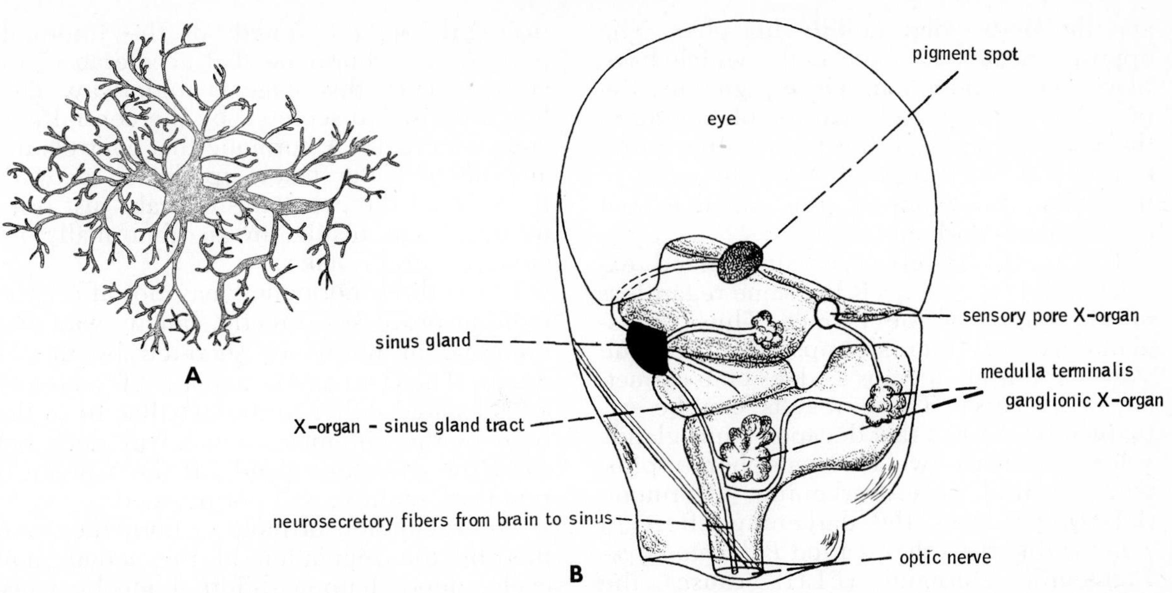

Figure 14–77. *A.* Crustacean chromatophore in expanded state. (After Green.) *B.* Eye and eyestalk of the shrimp, *Leander,* showing neurosecretory centers and tracts. (After Carlisle and Knowles.)

pigment is conjugated with a protein and is blue or some other color characteristic of the conjugated state. Curiously, a single chromatophore may possess one, two, three, or even four color pigments, any one of which can move independently of another. Crustacean chromatophores are thus usually classified as monochromatic, bi- or dichromatic, and polychromatic. In general, polychromatic chromatophores are found only in shrimp.

Two types of color changes occur. One type, called morphological color change, involves the loss or formation of pigments within the chromatophores or change in the number of chromatophores when the animal is subjected to a constant background and constant light intensity for a long time. The other type, called physiological color change, is a rapid color adaptation to background and results from the dispersal and concentration of pigments within the chromatophores.

The most common type of physiological color change is a simple blanching (or lightening) and darkening. This response is typical of many crabs, such as the fiddler crab, *Uca.* Many crustaceans, however, especially shrimp, can adapt to a wide range of colors. The little shrimp, *Palaemonetes,* for example, possesses trichromatic chromatophores with red, yellow, and blue pigments; through the independent movement of these three primary colors it can adapt to any background color, even black. Other species have similar abilities.

Hormones. The hormones of crustaceans and insects are the best known of any invertebrates. Crustacean hormones are neurosecretions or secretions of three endocrine tissues—the Y-organ, the androgenic gland, and the ovary. The movement of chromatophore pigments is controlled by hormones elaborated by a neurosecretory system in the eye stalk (or below the eye, when the eyes are sessile) and in other parts of the central nervous system (Fig. 14–77*B*). Located between the two basal optic ganglia is a body called the sinus "gland," which is a center for hormone release. The sinus gland is composed of the swollen endings of nerve fibers which have their origin in neurosecretory cell bodies located in the eye stalk ganglia. The neurosecretory material synthesized within the cell bodies migrates along the axon to the swollen endings composing the sinus gland. Here it is released into the hemolymph. In shrimp there are two masses of neurosecretory cell bodies. One is called the medulla terminalis X-organ, and the other is called the sensory pore X-organ. In a crab, the X-organs are situated close together and collectively are termed the X-organ.

In shrimp (aside from *Crangon*) which have red, yellow, blue, and white pigments, removal of the eye stalks results in a darkening of the body through dispersion of the red and yellow pigments. If these eye-stalkless animals are then injected with sinus-gland extract, the white pigment disperses, and the body color rapidly blanches. The opposite occurs in some crabs which have black, red, yellow, and white pigments. Removal of the eye stalks causes blanching of the body color, resulting from a concentration of the black pigments and dispersal of the white. Injections of sinus-gland extract cause a body darkening.

The particular effect of sinus-gland extract in the two groups is the same regardless of the origin of the extract. Thus, sinus-gland extract from shrimps produces the same darkening in crabs as does crab extract, and vice versa. These results can be explained by the fact that decapod sinus glands release at least two chromatophorotropins. One, called *Uca*-Darkening Hormone (UDH), produces the darkening effect in some crabs; the other, called *Palaemonetes*-Lightening Hormone (PLH), causes the blanching of shrimp.

Other parts of the central nervous system also elaborate three or four chromatophorotropins. Some of these produce effects similar to the sinus-gland factors; others apparently antagonize the sinus-gland chromatophorotropins.

Eye-stalk hormones control not only the functioning of the chromatophores but other processes as well. In different crustaceans any one or a combination of the three retinal pigments—distal, proximal, and reflecting pigments—may migrate distally or proximally in adapting the eye to bright or weak light. Experimental evidence indicates that at least the movement of the distal retinal pigment and the reflecting pigment is under the control of sinus-gland hormones. Thus, if specimens of *Palaemonetes*, in which the eyes are dark-adapted, are injected with sinus-gland extract, the distal and retinal pigments move into the light-adapted position. Similar results for the distal pigment have been found for the crayfish, *Cambarus*, and certain other decapods. The cause of proximal pigment migration is still unknown. It does not seem to be under the control of the sinus gland, and there is some question as to whether it is under hormonal control at all.

The physiological processes involved in molting are regulated by complex hormonal interactions. The X-organ–sinus-gland complex may produce three hormones: (1) A hormone that inhibits the various physiological processes involved in preparation for proecdysis, such as the storage of food reserves and calcium. This hormone is thus important in regulating the length of the intermolt period. (2) A hormone that accelerates proecdysis, once this stage is underway. This hormone is also secreted by the central nervous system. (3) A hormone that controls the amount of water taken up during ecdysis. Removal of the eye stalks of crabs just prior to this stage results in an abnormally increased water uptake.

These three hormones may not affect the molting processes directly, but by way of a hormone or hormones secreted by the Y-organ. The Y-organ is a mass of secretory cells located either in the maxillae or in the base of the antennae, whichever does not carry the excretory glands. If the Y-organ is removed, molting will not proceed.

The regulation of molting hormones, and thereby the regulation of the actual molt cycle, depend upon different mechanisms. In crayfish, which molt seasonally, day length is the controlling factor (Aiken, 1969); in the crab *Carcinus*, tissue growth is the controlling factor (Adelung, 1971). Although sex in crustaceans is determined genetically, the development and function of the gonads and the development of secondary sexual characteristics is under hormonal control (Fig. 14–78). If the Y-organ is removed prior to sexual maturity, gonadal development is seriously impaired; if the animal is adult, the gonads are unaffected. As yet, it is uncertain whether these results are caused by the loss of a specific hormone produced by the Y-organ or by a general metabolic malfunction from the absence of the Y-organ.

In female malacostracans there exists a hormonal interrelationship between the ovary and the X-organ–sinus-gland complex that is somewhat similar to the pituitary-ovary relationship of vertebrates. The sinus gland produces a hormone that inhibits the development of eggs during the nonbreeding periods of the year. During the breeding season, a gonad stimulatory hormone is secreted, probably by the central nervous system (Fig. 14–78). The blood level of gonad inhibiting hormone declines, egg development begins, and the ovary elaborates a hormone, initiating structural changes preparatory for egg brooding, such as the development of ovigerous setae on the pleopods or the development of oostegites (marsupium) in peracaridans. These characteristics appear at the next molt. See Adiyodi and Adiyodi (1970) for a detailed review of these hormonal relationships.

The development of the testes and male

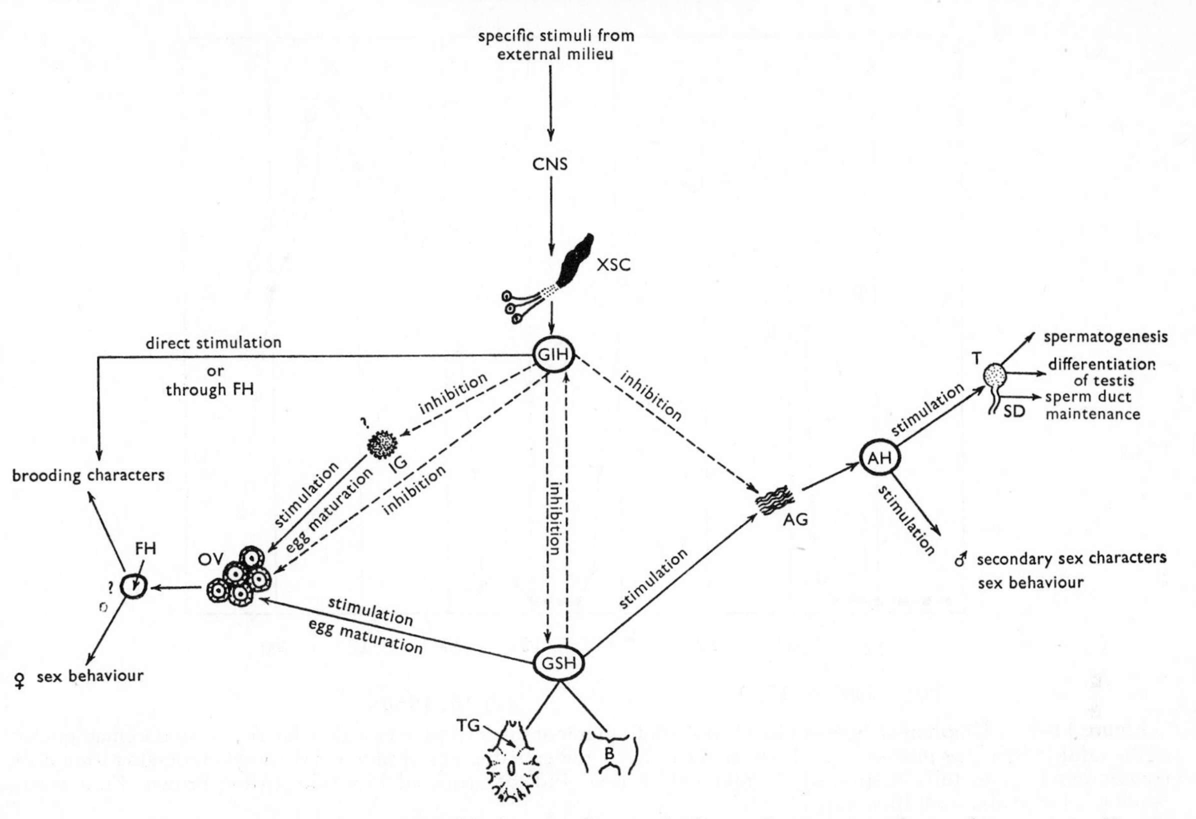

Figure 14–78. Diagram illustrating possible hormonal control of reproduction in a decapod crustacean. Key to lettering: AG, androgenic gland; AH, androgenic hormone; B, brain; CNS, central nervous system; FH, female hormone; GIH, gonad inhibiting hormone; GSH, gonad stimulatory hormone; IG, hypothetical intermediate gland; OV, ovaries; SD, sperm duct; T, testes; TG, thoracic ganglion; XSC, X-organ–sinus-gland complex. (From Adiyodi, K. G., and Adiyodi, R. G., 1970: Biol. Rev., *45*:121–165.)

sexual characteristics in malacostracans is controlled by hormones produced by a small mass of secretory tissue, the androgenic gland. This gland is located at the end of the vas deferens—except in isopods, in which it appears to be located in the testis itself. Removal of the androgenic gland is followed by a loss of male characteristics and conversion of the testes into ovarian tissue. If an androgenic gland is transplanted into a female, the ovaries become testes and male characteristics appear.

Physiological Rhythms. Pigment movement and other physiological processes often display a rhythmic activity in crustaceans. Such physiological rhythms, or physiological "clock mechanisms," have been studied extensively in the fiddler crabs, *Uca*, by F. A. Brown and fellow workers. Through chromatophore changes, the body color of *Uca* is dark during the day and pale at night. That such a periodicity is not mediated through the eyes can be demonstrated by the fact that the rhythm persists in *Uca* even when the animal is kept in continual darkness. Superimposed on this diurnal rhythm is a tidal rhythm.

Fiddler crabs remain inactive within the burrows at high tide, and emerge with the receding tide. Within the burrows, the body color tends toward the nocturnal paleness and the motor activity and metabolic rate are at a minimum, regardless of the time of day. Thus, maximum dispersion of black pigment does not consistently occur at noon, but shifts across the daylight hours, following a 12.4 hour tidal cycle (Fig. 14–79). In addition to changes in body coloration, the spontaneous locomotor activity and metabolic rate also display distinct tidal rhythms.

Similar rhythms have been recorded for many other crustaceans. The blue crab, *Callinectes sapidus*, and the amphibious isopod, *Lygia exotica*, possess a color-change rhythm similar to that of *Uca*, although it is only diurnal in *Lygia*. Species of crayfish display a diurnal rhythm of locomotor activity, but such a rhythm is never present in cave species. Diurnal rhythms have also been reported for the retinal-pigment move-

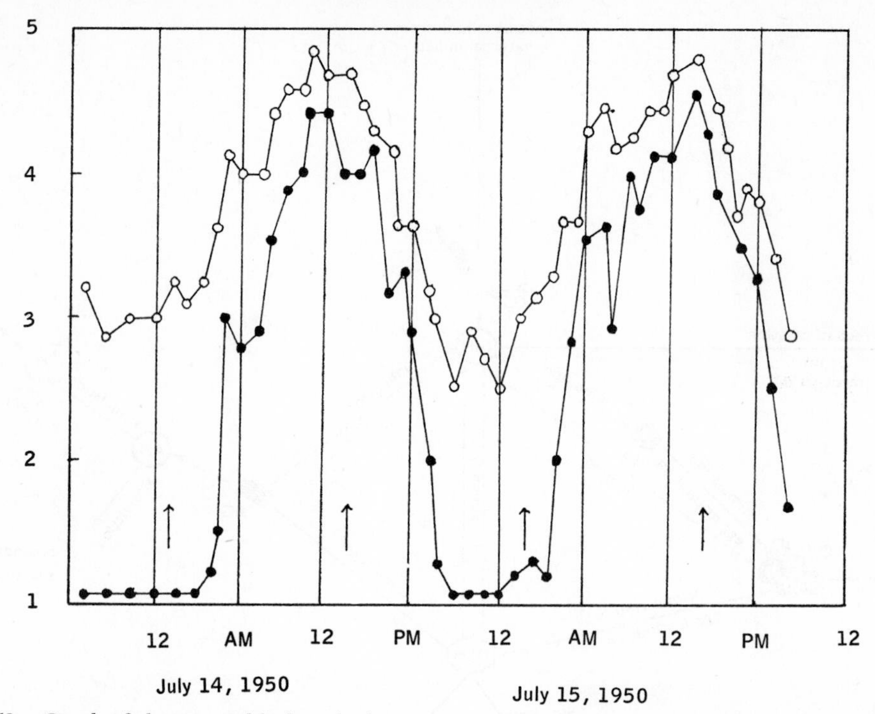

Figure 14–79. Graph of changes in black and white pigment dispersion over a two-day period in chromatophores of the fiddler crab, *Uca pugnax*. Numbers on vertical axis indicate degrees of pigment dispersion, ranging from fully concentrated (1) to fully dispersed (5) pigment. Arrows indicate time of low tide. (After Brown, Fingerman, Sandeen, and Webb from Brown.)

ment in certain crayfish, shrimp, and *Uca*, for luminescence in certain copepods, and for vertical migration in the copepod, *Acartia*.

Such physiological rhythms are typically adjusted to the solar and tidal conditions of the particular geographical location of the animal. For example, the *Uca* populations of two geographical areas with a four-hour difference in the high tide display a similar four-hour difference in their physiological rhythms. Moreover, the original geographical influence persists for a while even when a population is moved to a new locality. This was demonstrated by Brown and others (1955) by flying a population of *Uca* from Woods Hole, Massachusetts to Berkeley, California. The physiological rhythms of the transplanted population coincided exactly with those of a control population simultaneously observed at Woods Hole.

Physiological rhythms in Crustacea are not completely rigid, and some degree of phase shifting can be induced. For example, very low temperatures can inhibit the rhythms, which when restored may either stabilize in a new phase or shift back to the old. Similar shifts have been induced by manipulating light at critical periods during the daily cycle.

The causal mechanism is still far from understood. The best hypothesis at the present time is that external geophysical forces, such as daily and lunar cycles, create corresponding metabolic cycles in the animal. These in turn act as pacesetters for basic continuous internal rhythms. The metabolic

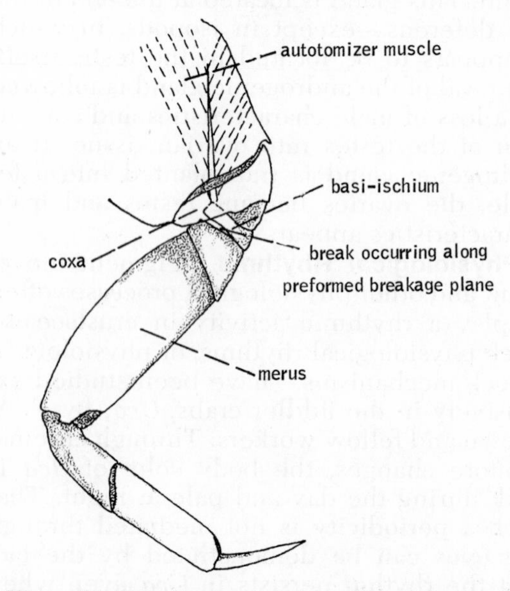

Figure 14–80. Autotomy in leg of the crab, *Carcinus maenas*. (After Wood and Wood from Bliss.)

cycle would be environmentally induced; the internal rhythms, genetically induced.

Autotomy. Many malacostracans are capable of limb autotomy. It is especially well developed in brachyurans and some anomurans and is apparently absent in stomatopods and most isopods.

Severance always takes place at a preformed breakage plane, which runs across the basi-ischium (Fig. 14–80). Internally there is a corresponding double membranous fold that divides the segment into basal and proximal halves. The membrane is perforated by a nerve and blood vessels. When the limb is cast off, the plane of severance passes between the two membranes, leaving one membrane attached to the basal stub. The membrane constricts around the perforations, so that there is very little bleeding.

In some species, severance can take place only if the limb is pulled either by the animal itself or by an outside force; but in its most highly developed state, as in most Reptantia, autotomy is a unisegmental reflex. An autotomizer muscle, which also functions in locomotion under normal conditions, originates from the thoracic wall and inserts on the basal half of the basi-ischium. If a leg is caught or damaged by a predator, a reflex is set up, and the autotomizer muscle is

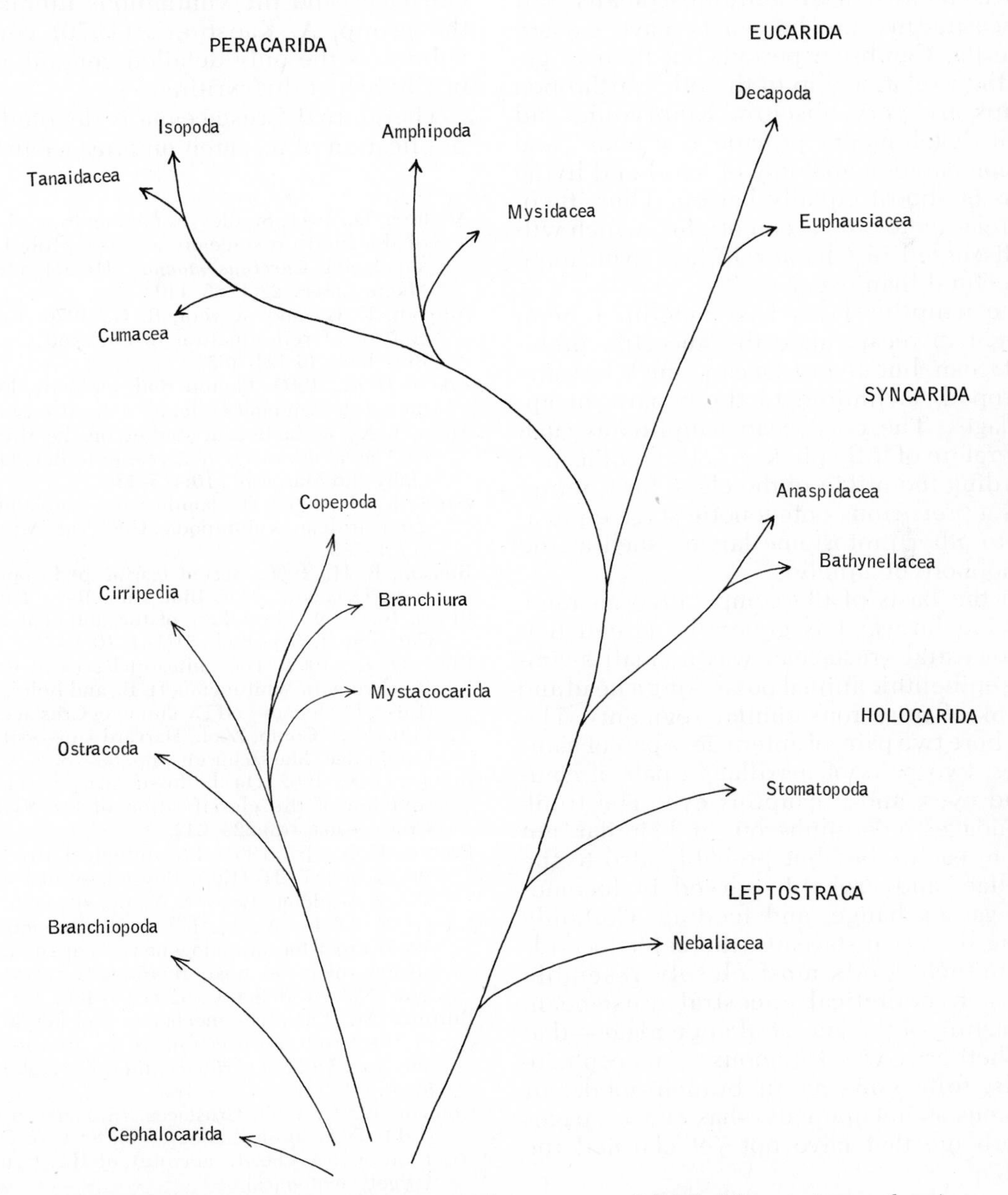

Figure 14–81. A suggested phylogeny of the Crustacea. (Modified from various authors.)

stimulated to undergo extreme contraction. The basi-ischium is pulled partially beneath the coxa, and fracturing takes place along the breakage plane.

In brachyuran crabs, an autotomy reflex exists in all five pairs of legs, but in the pagurid hermit crabs the reflex is absent in the last two pairs of legs; and in some macrurans, such as the lobster, *Homarus*, only the chelipeds exhibit the reflex. The other legs must be pulled off.

Crustacean Phylogeny

Crustaceans, both entomostracans and malacostracans, are known to have existed since the Cambrian period, but their origin and their relationship to the other arthropod groups are very obscure. Embryonic and larval development provide few clues, and the comparative anatomy of fossil and living forms is almost equally barren. Thus, theories relating to crustacean origins, which will be discussed in Chapter 16, are even more conjectural than usual.

The nauplius larva has sometimes been thought to recapitulate the ancestral protocrustacean, but no crustacean, much less any arthropod, is limited to three pairs of appendages. The crustacean nauplius as such is therefore of little phylogenetic significance regarding the origin of the class, for it represents a precocious ontogenetic stage equivalent to other protostome larvae, such as the trochophore of annelids.

On the basis of the comparative anatomy of living forms, it is generally agreed that the ancestral crustacean was a small swimming epibenthic animal possessing a head and a trunk of numerous similar segments. The head bore two pairs of antennae, a pair of mandibles, two pairs of maxillae, a pair of compound eyes, and a nauplius eye. The trunk appendages were numerous and similar, not only to each other but probably also to the maxillae, and probably served in locomotion, gas exchange, and feeding. Certainly among living crustaceans, the cephalocarids and branchiopods most closely resemble such a hypothetical ancestral crustacean. The nature of the ancestral appendage—that is, whether it was triramous as in cephalocarids, foliaceous an in branchiopods, or biramous as in copepods—has evoked pages of verbiage that have not yet clarified the problem.

During the evolution of the existing crustacean groups, the ancestral stock probably divided early into three principal lines—one leading to the Branchiopoda, one to the Branchiura-Cirripedia-Copepoda, and one to the malacostracans. The cephalocarids are considered by many to be sufficiently generalized to stand near the stem line of all three (Fig. 14–81). The malacostracan line culminates in two major groups, the peracaridans and eucaridans.

Bibliography

Despite the size and diversity of the Crustacea and the voluminous literature on the group, A. Kaestner's (1970) very fine volume is the only detailed general account in English at this writing.

The journal Crustaceana is devoted to the publication of research on crustaceans.

Adelung, D., 1971: Studies on the moulting physiology of decapod crustaceans as exemplified by the shore crab *Carcinus maenas*. Helgolander Wiss. Meeresunters. *22*(1):66–119.

Adiyodi, K. G., and Adiyodi, R. G., 1970: Endocrine control of reproduction in decapod Crustacea. Biol. Rev., *45*:121–165.

Aiken, D. E., 1969: Photoperiod, endocrinology, and the crustacean molt cycle. Science, *164*:149–155.

Allen, J. A., 1972: Recent studies on the rhythms of post-larval decapod Crustacea. Ann. Rev. Oceanography and Mar. Biol., *10*:415–436.

Barnard, J. L., 1969: The families and genera of marine gammaridean Amphipoda. U.S. Nat. Mus. Bull., *271*:1–535.

Benson, R. H., 1966: Recent marine podocopid ostracods. Oceanogr. Mar. Biol. Ann. Rev., *4*:213–232.

Binns, R., 1969: Physiology of the antennal gland of *Carcinus*. J. Exp. Biol., *51*(1):1–10.

Bliss, D. D., 1963: The pericardial sacs of terrestrial Brachyura. *In* Whittington, H. B., and Rolfe, W. O. I. (Eds.), Phylogeny and Evolution of Crustacea. Spec. Pub., Mus. Comp. Zool., Harvard University Press, Cambridge, Massachusetts, pp. 59–78.

Brooks, H. K., 1962: On the fossil Anaspidacea, with a revision of the classification of the Syncarida. Crustaceana, *4*(3):229–242.

Brown, F. A., Jr., 1961: Physiological rhythms. *In* Waterman, T. H. (Ed.), Physiology of Crustacea, Vol. 2, Academic Press, N. Y., pp. 401–430.

Brown, F. A., Jr., Webb, H. M., and Bennett, M. F., 1955: Proof for an endogenous component in persistent solar and lunar rhythmicity in organisms. Proc. Nat. Acad. Sci. U. S., *41*:93–100.

Burrows, M., 1969: The mechanics and neural control of the prey capture and strike in the mantid shrimps *Squilla* and *Hemisquilla*. Z. vergl. Physiol., *62*(4):361–381.

Calman, W. T., 1909: Crustacea. *In* Lankester, E. R. (Ed.), Treatise on Zoology, Vol. 8, A. & C. Black, London. (A classic account of the Crustacea. Largely anatomical.)

Carlisle, D. B., and Knowles, F., 1959: Endocrine Control in Crustaceans. Cambridge University Press, N.Y.

Chace, F. A., and Hobbs, H. H., 1969: The freshwater and terrestrial decapod crustaceans of the West Indies with special reference to Dominica. Smithsonian Institution, United States National Museum, Bull. *292*:1–258.

Cloudsley-Thompson, J. L., 1958: Spiders, Scorpions, Centipedes and Mites. Pergamon Press, N. Y. (A good discussion of the biology of wood lice is presented in Chapter 1.)

Costlow, J. D., Jr., 1956: Shell development in *Balanus improvisus*. J. Morph., *99*:359–416.

Costlow, J. D., Jr., and Bookhout, C. G., 1953: Moulting and growth in *Balanus improvisus*. Biol. Bull., *105*:420–433.

Crane, J., 1957: Basic patterns of display in fiddler crabs. Zoölogica, *42*:69–82.

Dahl, E., 1956: On the differentiation of the topography of the crustacean head. Acta Zool., *37*:123–192. (This paper includes considerable discussion of the feeding mechanisms of crustaceans.)

Darwin, C., 1851–1854: A Monograph on the Subclass Cirripedia. 2 vols. Ray Society, London. (A classic and still valuable account of the barnacles.)

Edmondson, W. T., Ward, H. B., and Whipple, G. C. (Eds.): Freshwater Biology. 2nd Edition. John Wiley and Sons, N. Y., pp. 558–901. (Keys and figures to the common genera and species of North American freshwater crustaceans.)

Enequist, P., 1949: Studies on the soft-bottom amphipods of the Skagerrak. Zool. Bidrag. frän Uppsala, *18*:297–492.

Foster, B. A., 1971: Desiccation as a factor in the intertidal zonation of barnacles. Mar. Biol., *8*:12.

Foxton, P., 1970: The vertical distribution of pelagic decapods collected on the Sond Cruise 1965: I. J. Mar. Biol. Assoc. U. K., *50*(4):939–1000.

Friend, G. F., 1941: The life history and ecology of the salmon gill-maggot *Salmincola salmonea*. Trans. Roy. Soc. Edinburgh, *60*:503–541.

Gosner, K. L., 1971: Guide to Identification of Marine and Estuarine Invertebrates: Cape Hatteras to the Bay of Fundy. Wiley-Interscience, N. Y. Crustacea, pp. 423–555.

Green, J., 1961: A Biology of Crustacea. Quadrangle Book, Chicago.

Gregoire, C., 1971: Hemolymph coagulation in arthropods. *In* Florkin, M., and Scheer, B. J. (Eds.), Chemical Zoology, Vol. 6. Academic Press, N. Y., pp. 145–189.

Hammer, W. M., Smyth, M., and Mulford, E. D., 1969: The behavior and life history of a sand-beach isopod, *Tylos punctatus*. Ecology, *50*(3):442–453.

Harris, J. E., 1953: Physical factors involved in the vertical migration of plankton. Quart. J. Micr. Sci., *94*:537–550.

Hartnoll, R. G., 1969: Mating in Brachyura. Crustaceana, *16*:161–181.

Hazlett, B. A., 1968: The sexual behavior of some European hermit crabs. Pub. Staz. Zool. Napoli, *36*(2): 238–252.

Hazlett, B. A., 1970: The effect of shell size and weight on the agonistic behavior of a hermit crab. Z. Tierpsychol., *27*(3):369–374.

Herreid, C., 1963: Observations on the feeding behavior of *Cardisoma guanhumi* in southern Florida. Crustaceana, *5*:176–180.

Highnam, K. G., and Hill, L., 1969: The Comparative Endocrinology of the Invertebrates. American Elsevier, N. Y.

Horch, K. W., and Salmon, M., 1969: Production, perception, and reception of acoustic stimuli by semiterrestrial crabs. Forma Functio, *1*:1–25.

Ivanov, B. G., 1970: On the biology of the Antarctic krill *Euphausia superba*. Mar. Biol., *7*:340.

Kaestner, A., 1970: Invertebrate Zoology, Vol. 3, Crustacea. Wiley-Interscience, N. Y., 523 pp. (An excellent general account of the crustaceans; systematically arranged and covering all aspects of the biology.)

Kennedy, D., Selverston, A. I., and Remler, M. P., 1969: Analysis of restricted neural networks. Science, *164*:1488–1496.

Kieth, D., 1969: Aspects of feeding in *Caprella californica* and *Caprella equalibra*. Crustaceana, *16*(2): 119–124.

Light, S. F., Smith, R. I., Pitelka, F. A., Abbott, D. P., and Weesner, F. M., 1967: Intertidal Invertebrates of the Central California Coast. University of California Press, Berkeley. Crustacea, pp. 113–186.

Lockwood, A. P. M., 1967: Physiology of Crustacea. W. H. Freeman and Co., San Francisco, 328 pp. (A good, concise summary of crustacean physiology.)

MacGinitie, G. E., and MacGinitie, N., 1968: Natural History of Marine Animals, 2nd ed. McGraw-Hill, N.Y.

Manning, R. B., 1969: Stomatopod Crustacea of the Western Atlantic. University of Miami Press, 460 pp.

Manton, S. M., 1952: The evolution of arthropodan locomotory mechanisms. Pt. 2. J. Linnean Soc. London (Zool.), *42*:93–117. (General introduction to the locomotory mechanisms of the Arthropoda.)

Marshall, S. M., and Orr, A. P., 1955: On the biology of *Calanus finmarchicus*. VIII. Food uptake, assimilation, and excretion in adult and stage V *Calanus*. J. Mar. Biol. Assoc. U. K., *35*:495–529.

Mason, C. A., 1970: Function of the pericardial sacs during the molt cycle in the land crab *Gecarcinus lateralis*. J. Exp. Zool., *174*(4):381–390.

Mau Chline, J., and Fischer, L. R., 1969: The biology of euphausiids. Adv. Mar. Biol., vol. 7. Academic Press, N. Y.

McCain, J. C., 1968: The Caprellidae (Crustacea: Amphipoda) of the western North Atlantic. Bull. U. S. Nat. Mus., No. 278, pp. 1–147.

Miller, D. C., 1961: The feeding mechanisms of fiddler crabs, with ecological considerations of feeding adaptations. Zoologica, *46*(8):89–101.

Moore, H. B., 1936: The biology of *Balanus balanoides*, V. Distribution in the Plymouth area. J. Mar. Biol. Assoc. U.K., *20*:701–716.

Moore, R. C. (Ed.), 1969: Treatise on Invertebrate Paleontology, Pt. R, Arthropoda 4, vols. 1 and 2. Geological Society of America and University of Kansas Press. (These two volumes cover the Crustacea except for ostracods, which are treated in Part Q, Arthropoda 3, 1961.)

Naylor, E., 1972: British Marine Isopods. Academic Press, N. Y.

Nolan, B. A., and Salmon, M., 1970: The behavior and ecology of snapping shrimp. Forma Functio, *4*:289–335.

Pardi, L., 1954: Über die Orientierung von *Tylos latreillii*. Z. Tierpsychol., *2*:175–181.

Pardi, L., and Grassi, M., 1955: Experimental modification of direction finding in *Talitrus saltator*. Experimentia, *11*:202–203.

Pardi, L., and Papi, F., 1952: Die Sonne als Kompass bei *Talitrus saltator*. Naturwissenschaften, *39*: 262–263.

Pennak, R. W., 1953: Freshwater Invertebrates of the United States. Ronald Press, N. Y., pp. 321–469. (A brief general account of the biology of the different groups of freshwater crustaceans, with keys and figures for the identification of common genera and species.)

Pennak, R. W., and Zinn, D. J., 1943: Mystacocarida, a new order of Crustacea from intertidal beaches in Massachusetts and Connecticut. Smithsonian Misc. Collect., *103*:1–11.

Pilsbry, H. A., 1907: The barnacles contained in the collections of the U. S. Natural Museum. Smithsonian Inst. Bull., *60*:1–114. (An old but still valuable taxonomic treatment of the American stalked barnacles. The following paper covers the sessile barnacles.)

Pilsbry, H. A., 1916: The sessile barnacles contained in the collections of the U. S. National Museum; including a monograph of the American species. Smithsonian Inst. Bull., *93*:1–357.

Riegel, J. A., 1971: Excretion—Arthropods. *In* Florkin, M., and Scheer, B. J. (Eds.), Chemical Zoology, Vol. 6. Academic Press, N. Y., pp. 249–277.

Rudjakov, J. A., 1970: The possible causes of diel vertical migrations of planktonic animals. Mar. Biol., *6*:98.

Salmon, M., 1965: Waving display and sound production in the courtship behavior of *Uca pugilator*, with comparisons to *U. pugnax*. Zoologica, *50*:123–150.

Salmon, M., 1967: Waving display, sound production, and coastal distribution of Florida fiddler crabs. Animal Behavior, *15*:449–459.

Salmon, M., 1971: Signal characteristics and acoustic detection by fiddler crabs, *Uca rapax* and *Uca pugilator*. Physiol. Zool., *44*:210.

Sanders, H. L., 1957: The Cephalocarida and crustacean phylogeny. Syst. Zool., *6*:112–128.

Sanders, H. L., 1959: The significance of the Cephalocarida in crustacean phylogeny. XV Internat. Congr. Zool., London (1958), Proc., pp. 337–340.

Sanders, H. L., 1963: The Cephalocarida. Memoirs Conn. Acad. Arts and Sci., *15*:1–80.

Schmitt, W. L., 1965: Crustaceans. University of Michigan Press, Ann Arbor. (An interesting popular treatment of crustaceans.)

Schmitz, E. H., and Schultz, T. W., 1969: Digestive anatomy of terrestrial Isopoda: *Armadillidium vulgare* and *Armadillidium nasatum*. Amer. Midl. Nat., *82*(1):163–181.

Schultz, G. A., 1969: How to Know the Marine Isopod Crustaceans. Brown, Dubuque, Iowa.

Siewing, R., 1960: Neuere Ergebnisse der Verwandtschaftsforchung bei den Crustacean. Wiss. Zeitschr. Universität. Rostock, Math.-Nat. Reihe, *9*:343–358.

Southward, A. J., 1955: Feeding of barnacles. Nature, *175*:1124–1125.

Stevcic, Z., 1971: The main features of brachyuran evolution. Syst. Zool., *20*(3):331–340.

Tiegs, O. W., and Manton, S. M., 1958: The evolution of the Arthropoda. Biol. Rev. Cambr. Phil. Soc., *33*:255–337.

Tombes, A. S., 1970: An Introduction to Invertebrate Endocrinology. Academic Press, N. Y., 217 pp.

Tomlinson, J. T., 1969: The burrowing barnacles (Cirripedia: Order Acrothoracica). Bull. U. S. Nat. Mus., No. 296, pp. 1–162.

Van Name, W. G., 1936: The American land and freshwater isopod crustaceans. Amer. Mus. Bull., *71*:1–535. (A very complete taxonomic treatment of the American wood lice and freshwater isopods.)

Van Weel, P. B., 1970: Digestion in Crustacea. *In* Florkin, M., and Scheer, B. J. (Eds.), Chemical Zoology, Vol. 5, Pt. A. Academic Press, N. Y., pp. 97–115.

Waterman, T. H. (Ed.), 1960: The Physiology of Crustacea. Vol. 1, Metabolism and Growth. *Also*: Vol. 2, Sense Organs, Integration, and Behavior, 1961. Academic Press, N. Y.

Waterman, T. H., and Chace, F. A., 1960: General crustacean biology. *In* Physiology of Crustacea (see above), Vol. 1, pp. 1–33.

Webber, H. H., 1968: Mariculture. Bio. Sci., *18*(10): 940–945.

Whittington, H. B., and Rolfe, D. W. I. (Eds.), 1963: Phylogeny and Evolution of Crustacea. Spec. Pub. Mus. Comp. Zool., Harvard University Press, Cambridge, Massachusetts. (A collection of papers devoted to various aspects of crustacean evolution.)

Wiersma, C. A. G., 1961: Reflexes and the Central Nervous System. *In* Waterman, T. H. (Ed.), Physiology of Crustacea. Vol. 2. Academic Press, N. Y., pp. 241–279.

Chapter 15

THE INSECTS

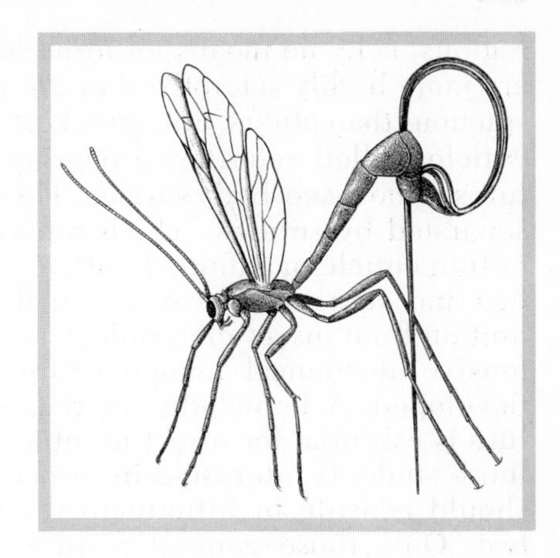

The class Insecta, or Hexapoda, containing more than 750,000 described species, is the largest group of animals; in fact, it is larger than all the other animal groups combined. Obviously, only a brief, rather superficial treatment of insects is possible here. For more extensive accounts, especially those dealing with the details of morphology and the insect orders, the student must refer to textbooks of entomology.*

The success of insects is evidenced by the tremendous number of species and individuals and by their great adaptive radiation. Although they are essentially terrestrial animals and have occupied virtually every environmental niche on land, insects have also invaded aquatic habitats and are absent only from the deeper waters of the sea. This success of insects can be attributed to a number of factors, but certainly the evolution of flight endowed these animals with a distinct advantage over other terrestrial invertebrates. Dispersal, escape from predators, and access to food or more optimum environmental conditions were all greatly enhanced. The powers of flight also evolved in reptiles, birds, and mammals, but the first flying animals were insects.

Insects are of great ecological significance in the terrestrial environment. Two thirds of all flowering plants are dependent upon insects for pollination. The principal pollinators are bees, wasps, butterflies, moths, and flies, and the three orders represented by these insects have an evolutionary history that is closely tied to that of the flowering plants, which underwent an explosive evolution in the Cretaceous.

Insects are of enormous importance for man. Mosquitoes, lice, fleas, bedbugs, and a host of flies can contribute directly to human misery. More importantly, these and others contribute indirectly as vectors of human diseases or of diseases of man's domesticated animals: mosquitoes (malaria, elephantiasis, yellow fever); tsetse fly (sleeping sickness); lice (typhus and relapsing fever); fleas (bubonic plague); and the housefly (typhoid fever and dysentery). Our domesticated plants are dependent upon some insects for pollination but are destroyed by others. Vast sums are expended to control insect pests, which can greatly reduce the high agricultural yields necessary to support large human populations. But the overzealous use of pesticides can in turn be hazardous to the environment.

Insects are distinguished from other arthropods by their possession of three pairs of legs and usually two pairs of wings carried on the middle, or thoracic, region of the body. In addition, the head typically bears a single pair of antennae and a pair of compound eyes. A tracheal system provides for gas exchange and the gonoducts open at the posterior end of the abdomen.

EXTERNAL MORPHOLOGY

Although the exoskeleton of an arthropod segment is composed of a tergum, a sternum, and two pleura, the thickness of the cuticle (i.e., the degree of sclerotization within these

*This chapter has been designed to meet the requirements of those invertebrate zoology courses that include a brief coverage of insects. This account is not intended to be equivalent to the extensive treatment accorded to the other invertebrate groups.

regions) is by no means uniform. Some parts are more highly sclerotized or are more conspicuous than others. Such thickened areas of cuticle, called sclerites, are prominent features of the insect body surface. They are often separated by sutures, which represent lines of thin cuticle or infolded cuticle. The sclerites and sutures have received detailed attention from insect morphologists and an extensive anatomical nomenclature has been developed. A knowledge of this nomenclature is essential for insect identification, and those students interested in insect taxonomy should consult an introductory entomology text. Only those general features of insect anatomy that provide a basis for comparison with other arthropods will be discussed in the following paragraphs.

Body Regions. The head of most insects is oriented so that the mouth parts are directed downward (hypognathus; Fig. 15–1*B*), although in some predaceous species, such as the carabid and tiger beetles, the conspicuous mouth parts are directed forward (prognathus). In a generalized hypognathus insect, the more lateral and dorsal surfaces of the head bear one pair of compound eyes and one pair of antennae. Between the eyes and antennae are usually located three ocelli (Fig. 15–1*A*). Three pairs of appendages contribute to the mouth parts (Fig. 15–1*A* and *B*). One pair of mandibles is located anteriorly, followed by a pair of maxillae and then by the labium. Although single, the labium actually represents a fused pair of second maxillae. Anteriorly, the mandible is covered by a shelflike extension of the head, forming an upper lip, or labrum. From the floor of the prebuccal cavity a median lobelike process, called the hypopharynx, projects. The hypopharynx arises behind the maxillae near the base of the labium. The modifications of the mouth parts associated with different feeding habits will be discussed later.

The thorax, which forms the middle region of the insect body, is composed of three segments—a prothorax, a mesothorax, and a metathorax (Fig. 15–1*C*). On each of the three segments a pair of legs articulates with the pleura. The thoracic terga of insects are called nota, and it is with the notal and pleural processes of the mesothorax and the metathorax that the two pairs of wings articulate. The basal section of the leg articulating with sclerites in the pleural area is the coxa, which is followed by a short trochanter (Fig. 15–1*D*). The remaining sections consist of a femur, a tibia, a tarsus, and a pretarsus. The tarsus is composed of one to five segments. The pretarsus is represented chiefly by a pair of claws. In other arthropods, such as arachnids, the pretarsus is not usually considered separate from the tarsus. The legs of insects are generally adapted for walking or running, but one or more pairs may be modified for such functions as grasping prey, jumping, swimming, and digging.

The abdomen is composed of nine to eleven segments, but the eleventh segment is complete only in the primitive proturans and in embryos (Fig. 15–1*E*). The telson is vestigial. The only abdominal appendages in the adult are a terminal pair of sensory cerci borne on the eleventh segment. The reproductive structures are thought by some entomologists to represent segmental appendages, but this is by no means certain. A variety of abdominal appendages serving different functions are present in many insect larvae.

Insect Wings and Flight. Wings are characteristic features of insects, but a wingless condition occurs in a number of different groups. In some, the absence of wings is obviously secondary. For example, ants and termites have wings only in the males and females at certain periods of the life cycle; workers always lack wings. Some parasitic insect orders, such as the lice and the fleas, have lost the wings completely. On the other hand, it is fairly certain that the insects comprising the orders Protura, Thysanura, Collembola, and a few others arose from ancestral wingless insects independently from the stock in which wings evolved. Those groups in which the wingless condition is considered primary are classified as the subclass Apterygota and are believed to represent the most primitive members of the class. Winged insects and those that are secondarily wingless form the subclass Pterygota.

The sequence of events in the evolutionary development of insect wings is unknown. The earliest known fossil insects are from the Devonian period and include both wingless and a primitive winged species. No intermediate types have yet been discovered. The most widely accepted theory of wing origin is that wings were originally flat lateral flanges of the notum, which enabled the insect to alight right side up when jumping. These flanges then gradually enlarged into winglike structures, making gliding possible. The last step was the development of hinges, enabling wings to move. The earliest functional wing was thought to be a fan-shaped membrane with trusslike supporting veins.

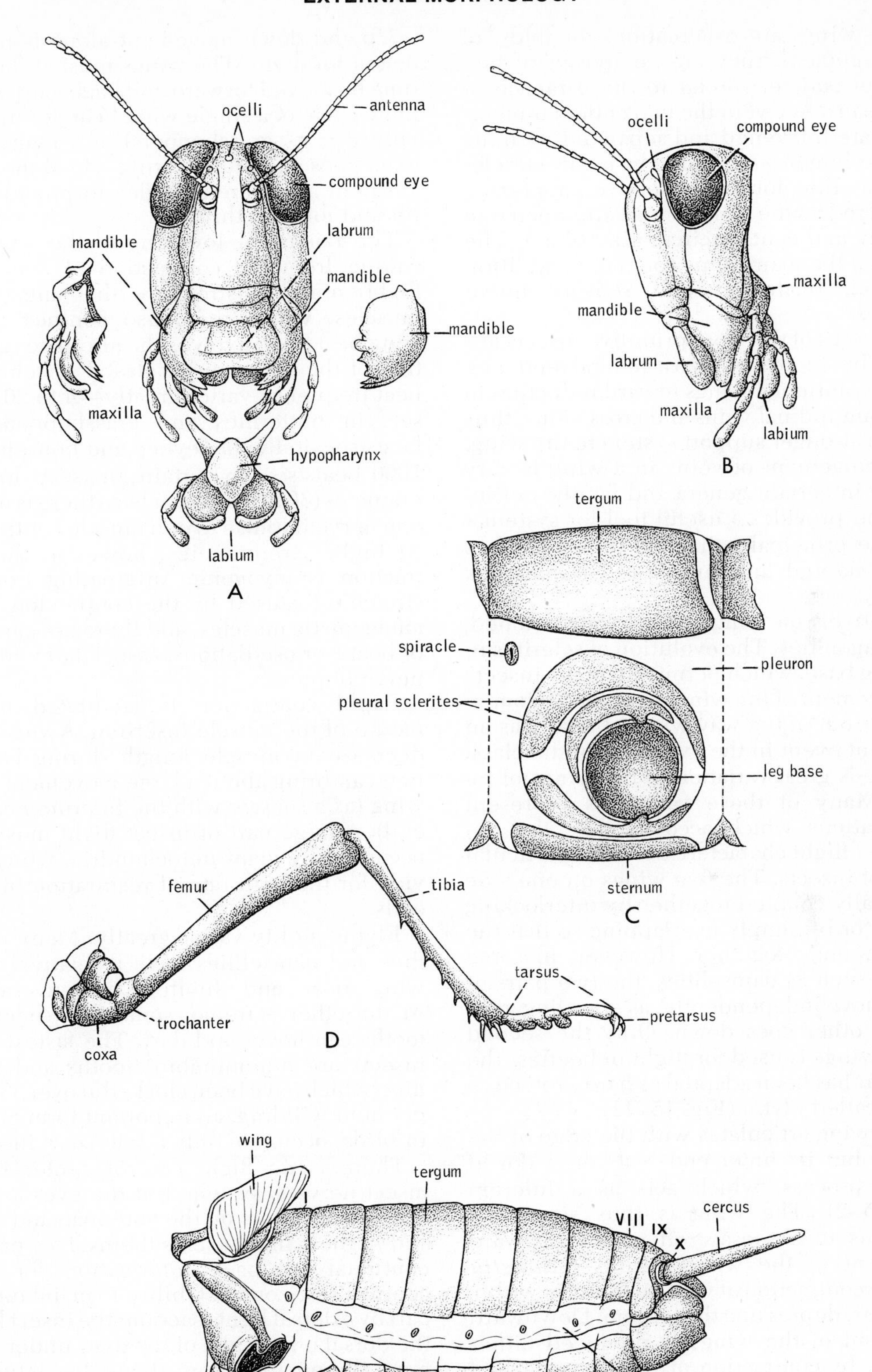

Figure 15–1. External morphology. *A.* Anterior surface of the head of a grasshopper. *B.* Lateral view of the head of a grasshopper. (After Snodgrass from Ross.) *C.* Lateral view of a wingless thoracic segment. *D.* Leg of a grasshopper. *E.* Lateral view of the abdomen of a male cricket. (*C–E* after Snodgrass.)

Since wings are evaginations, or folds, of the integument, they are composed of two sheets of cuticle, similar to the carapace of crustaceans. At a vein the two cuticular membranes are thickened and separated, forming a tubular lumen surrounded by heavy cuticle; the veins thus form an effective supporting skeletal rod for the wing. Wing veins open into the body and contain circulating blood. The lumina of the main veins contain, in addition to blood, tracheoles and sensory nerve branches.

The wings of the more primitive insects are netlike, but there has been a general tendency in the evolution of wings toward reduction to a few longitudinal veins and cross veins, thus giving a stronger support system to the wing. The arrangement of veins in a wing is very specific in certain genera and families of insects and provides a useful tool for systematists; the principal veins and their branches are all named and numbered (Comstock system).

Primitively the wings are held outstretched, as in dragonflies. The evolution of sclerites in the wing base, which permits in many insects the placement of the wings over the abdomen and thus out of the way when at rest, was an important event in the evolution of the class.

There is great variation in the wings of insects. Many of these variations represent modifications which accommodate the demands of flight characteristic of the particular group of insects. The two wings on one side are usually coupled together by interlocking devices or by simply overlapping so that the wings operate together. However, in some insects, such as damselflies, the two pairs of wings move independently; as one wing goes up, the other goes down. Only the second pair of wings is used for flight in beetles; the front pair has been adapted as hard protective plates, called elytra (Fig. 15–11).

Each wing articulates with the edge of the tergum, but its inner end rests on a dorsal pleural process, which acts as a fulcrum (Fig. 15–2). The wing is thus somewhat analogous to a seesaw off center. Upward movement of the wings results *indirectly* from the contraction of vertical muscles within the thorax, depressing the tergum. Downward movement of the wings is produced either *directly*, by contraction of muscles attached to the wing base (e.g., dragonflies and roaches); or *indirectly*, by the contraction of transverse horizontal muscles raising the tergum (e.g., bees, wasps, and flies), or by both direct and indirect muscles (grasshoppers and beetles).

Up and down movement alone is not sufficient for flight. The wings must at the same time be moved forward and backward. A complete cycle of a single wing beat describes an ellipse (e.g., grasshoppers) or a figure eight (e.g., bees and flies), during which the wings are held at different angles to provide both lift and forward thrust.

The raising or lowering of the wings resulting from the contraction of one set of flight muscles stretches the antagonistic muscles, which then also contract. Insect wing beat thus involves the alternate contraction of the antagonistic elastic systems. The beat frequency varies greatly—4 to 20 beats/sec. in butterflies and grasshoppers; 190 beats/sec. in the honeybee and housefly; and 1000 beats/sec. in certain gnats. At low frequencies (30 beats/sec. or less) there is usually one nerve impulse to one muscle contraction. At higher frequencies, however, the contraction is myogenic, originating from the stretching caused by the contraction of the antagonistic muscles, and there are a number of beats, or oscillations, associated with each nerve impulse.

Rapid contraction is facilitated by the nature of the muscle insertion. A very slight decrease in muscle length during contraction can bring about a large movement of the wing (as a seesaw with the fulcrum near one end). A large part of insect flight muscles is occupied by giant mitochondria, which provide for the high rate of respiration in these cells.

Flying ability varies greatly. Many butterflies and damselflies have a relatively slow wing beat and limited maneuverability. At the other extreme, some flies, bees, and moths can hover and dart. The fastest flying insects are hummingbird moths and horseflies, which have been clocked at over 33 miles per hour. Gliding, an important form of flight in birds, occurs in only a few large insects.

There is no flight control center in the insect nervous system, but the eyes and the sensory receptors on the antennae and wings and in the wing muscles themselves provide continual feedback information for flight control. Horizontal stability is maintained in part by a dorsal light reaction: the insect keeps the dorsal ommatidia of the eyes under maximum illumination from above. Deviation because of rolling is corrected by slight changes in wing position to bring the dorsal part of the eyes back to maximum illumination.

Members of the order Diptera (flies, gnats, and mosquitoes) have the second pair of

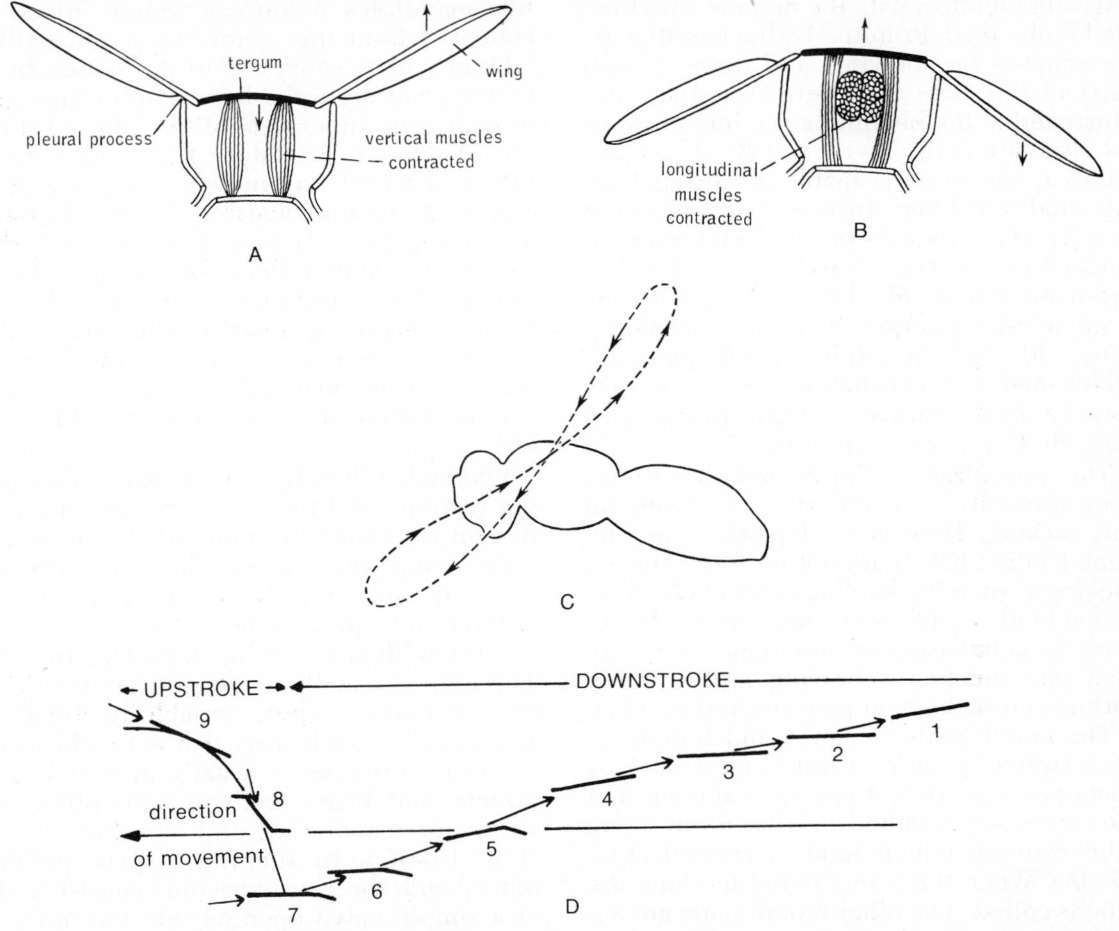

Figure 15–2. Diagrams showing relationship of wings to tergum and pleura, and the mechanism of the basic wing strokes in an insect. *A*. Upstroke resulting from the depression of the tergum through the contraction of vertical muscles. *B*. Downstroke resulting from the arching of the tergum through the contraction of longitudinal muscles. *C*. An insect in flight, showing the figure 8 described by the wing during an upstroke and a downstroke. *D*. Changes in the position of the forewing of a grasshopper during the course of a single beat. Short arrows indicate direction of wind flowing over wing and numbers indicate consecutive wing positions. (*A* and *B* after Ross; *C* after Magnan from Fox and Fox; *D* after Jensen from Chapman: The Insects: Structure and Function. London, University of London Press Ltd., 1971. Courtesy of the American Elsevier Publishing Co., Inc.)

wings reduced to knobs, called halteres (Fig. 15–15*R*). The halteres beat with the same frequency as the forewings and function as gyroscopes for the control of flight instability (pitching, rolling, and yawing).

Flight speed is probably determined by air flow over receptors on the antennae and movement of objects from front to back across the eyes. Flight is inhibited by contact of the tarsi with the ground.

Insect flight muscles are very powerful. The fibrils are relatively large and the mitochondria are huge (about half the size of a human red blood cell). Insects are the only poikilothermic fliers, and a low body temperature and a corresponding low metabolic rate impose limitations on mobility. On a cold day some butterflies are known to literally "warm up" before flight. They remain stationary on a tree trunk or some other location, and move the wings up and down until sufficient internal heat is generated to permit the stroke rate necessary for flight.

INTERNAL ANATOMY AND PHYSIOLOGY

Nutrition. Insects have adapted to all types of diets. The mouth parts may be highly modified, but the modifications are associated less with diet than with the method by which food is obtained. Primitively, the mouth parts are adapted for chewing, and it was mouth parts of this type that were described and illustrated in the beginning section on external structure (Fig. 15–1*A* and *B*). The mandibles are heavy and capable of cutting, tearing, and crushing. Insects with chewing mouth parts include the primitive apterygotes, dragonflies, crickets, grasshoppers, beetles, and many others. The larvae of such insects as moths and butterflies have chewing mouth parts, although the adult's mouth parts are highly modified. The diets of chewing insects may be herbivorous or carnivorous, and some diets are very restrictive.

The specialization of insect mouth parts has been primarily in modifications for piercing and sucking. However, adaptations for the same feeding habits are not uniform, since a sucking or piercing feeding habit evolved independently in different insect orders. Moreover, the mouth parts may be adapted for more than one function—chewing and sucking, cutting and sucking, or piercing and sucking.

The mouth parts of moths and butterflies are adapted for sucking liquid food, such as nectar, from flowers. A part (galea) of each of the two greatly modified maxillae forms a long tube through which food is sucked (Fig. 15–3*A*). When the insect is not feeding, the tube is coiled. The other mouth parts are absent or are vestigial.

Piercing mouth parts are characteristic of herbivorous insects, such as aphids and leafhoppers, which feed on plant juices. Predaceous insects, such as assassin bugs and mosquitoes, which utilize the body fluids of other animals as food also have piercing mouth parts. In all these insects the mouth parts are elongated and are organized in various ways to form a beak. For example, the beak of the plant feeding and predaceous bugs (Hemiptera and Homoptera) consists of a stylet composed of the mandibles and the maxillae that lie in a groove on the heavier labium. The stylet contains a lumen for the outward passage of salivary secretions and another for sucking in fluids (Fig. 15–3*B–F*). Other parts of the beak do not penetrate.

Bees and wasps have mouth parts adapted for both chewing and sucking. In bees, for example, nectar is gathered by the elongated maxillae and the labium. Pollen and wax are handled by the labrum and mandibles, which have retained the chewing form.

In biting flies, such as horseflies, the knifelike mandibles produce a wound. Blood is collected from the wound by a spongelike labium and is conveyed to the mouth by a tube formed from the hypopharynx and epipharynx (the inner side of the labrum) (Fig. 15–4*A*). Some predatory flies and hemipterans inject salivary secretions into the prey and suck up the digested tissues. Certain non-biting flies, such as houseflies, use the spongelike labium alone for obtaining food, the mandibles and maxillae being reduced. Such insects are not restricted to liquid foods. Saliva can be exuded through the labium onto the solid material, and the fluid then can be sucked back into the mouth (Fig. 15–4*B*).

Food taken into the mouth passes through the pharynx and then into the foregut. The foregut is commonly subdivided into an anterior esophagus, a crop, and a narrower proventriculus (Fig. 15–5). The anterior end of the esophagus may be called the pharynx and is modified as a pump in sucking insects. The crop is usually a storage chamber. The proventriculus is quite variable in structure and function. In insects that eat solid food, the proventriculus is usually modified as a gizzard and bears teeth or hard protuberances for macerating and shredding food (Fig. 15–5*C*). In sucking insects, on the other hand, the proventriculus consists only of a simple valve opening into the midgut. Between these two extremes are some beetles and honeybees, in which the proventriculus acts as a regulatory valve permitting fluids but not solid food to enter the midgut. This function is particularly important in the separation of pollen from nectar in bees.

Most insects possess a pair of salivary, or labial, glands which lie below the midgut and have a common duct opening into the buccal cavity (Fig. 15–5*B*). Mandibular glands, in addition to salivary glands, are functional in the apterygotes and in a few groups of pterygotes. The function of salivary glands varies and has not been determined in all insects. The glands usually secrete saliva, which moistens the mouth parts and may be a solvent for the food. The salivary glands may also produce digestive enzymes, which are mixed into the food mass before it is swallowed. In some of the hymenopterans, the glands secrete silk used to make the pupal cells. Other special secretions of the salivary glands in various insects include mucoid materials, a pectinase that hydrolyzes the pectin of cell walls, venomous spreading

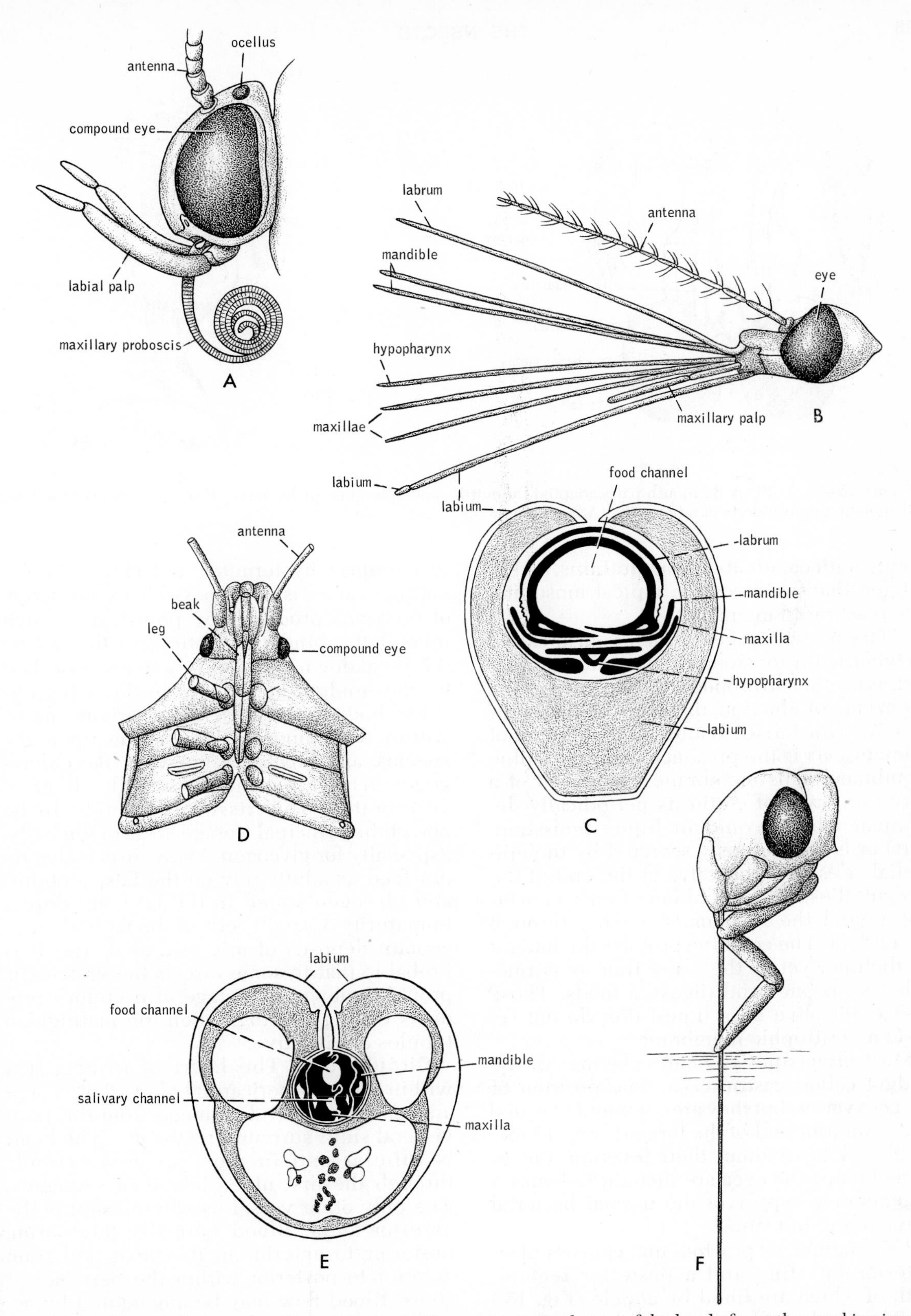

Figure 15–3. Mouth parts of sucking and piercing insects. *A*. Lateral view of the head of a moth, a sucking insect. (After Snodgrass.) *B*. Lateral view of the head of a mosquito, showing separated mouth parts. *C*. Cross section of mouth parts of a mosquito in their normal functional position. (*B* and *C* after Waldbauer from Ross.) *D*. Ventral view of anterior half of a hemipteran, showing beak. (After Hickmann.) *E*. Cross section through a hemipteran beak, showing the food and the salivary channels enclosed within the stylet-like maxillae and the mandibles. (After Poisson.) *F*. A hemipteran penetrating plant tissue with its stylets. (After Kullenberg.)

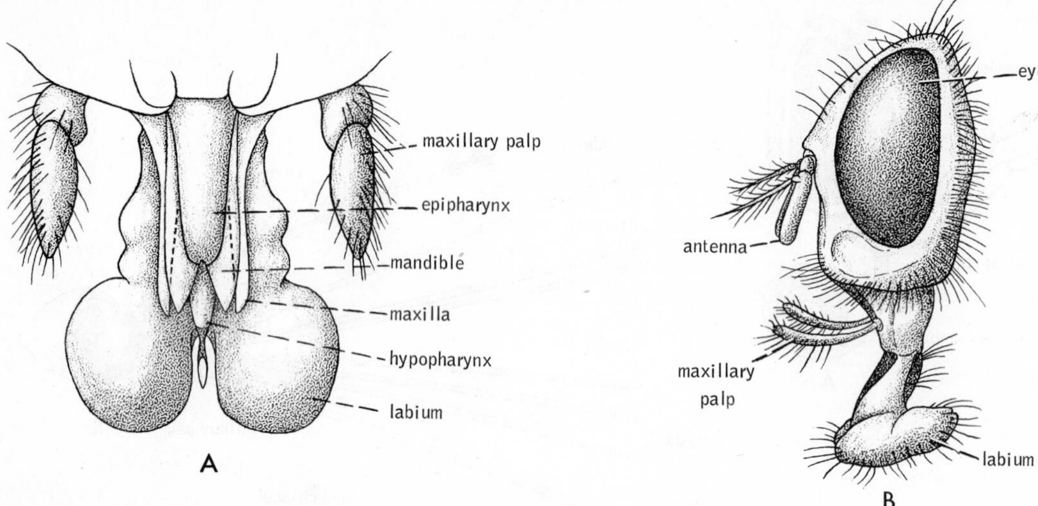

Figure 15–4. *A.* Black fly mouth parts adapted for cutting and sponging. (After Ross.) *B.* Lateral view of the head and sponging mouth parts of a housefly. (After Snodgrass.)

agents, anticoagulants and agglutinins, and an antigen that produces the typical mosquito-bite reaction in man.

The insect midgut, which is also called the ventriculus or stomach, is usually tubular and, as in other arthropods, is the principal site of enzyme production, digestion, and absorption. A characteristic feature of the midgut of many insects is the presence of a peritrophic membrane. This membrane, composed of a very thin layer of chitin, is periodically delaminated by the midgut lining (grasshoppers) or is continuously secreted by the epithelial cells near the valve at the end of the foregut (flies). The membrane forms a covering around the food mass moving through the midgut. The covering protects the midgut epithelium, yet at the same time is permeable to enzymes and digested foods. Those insects that live on a liquid diet do not secrete a peritrophic membrane.

Most insects possess outpocketings of the midgut called gastric ceca. The position of the ceca varies, but they are commonly located at the anterior end of the foregut (Fig. 15–5). Little is known about their function, but in some groups the ceca are thought to house a regenerative supply of the normal bacterial fauna of the intestine.

The hindgut, or proctodeum, consists of an anterior intestine and a posterior rectum, both of which are lined by cuticle (Fig. 15–5). The hindgut functions in the egestion of waste and in water and salt balance. In most insects, rectal pads, or glands, occur in the epithelium. These organs are thought to be the principal sites of reabsorption. Digestion of cellulose by termites and certain wood-eating roaches is made possible by the action of enzymes produced by protozoans which inhabit the hindgut. Acetic acid formed by the breakdown of wood is actively absorbed by the hindgut epithelium in these insects.

Fat bodies are present in various places within the hemocoel, depending upon the species, and function somewhat like chlorogogen tissue in annelids and the liver of vertebrates. This tissue is thought to be one of the principal storage areas of the body, especially for glycogen. Many insects that do not feed as adults rely on the fats, proteins, and glycogen stored in the fat body during immaturity. Certain cells of the fat body may contain deposits of uric acid or urates. It is probable that in some insects the urate cells provide temporary storage of excretory products during molting, when the malpighian tubules are inactive.

Circulation. The heart of insects lies within a dorsal pericardial sinus that is separated by a dorsal diaphragm from the perivisceral sinus surrounding the gut. The heart is tubular and in most species extends through the first nine abdominal segments. The only other vessel usually present is the anterior aorta. Blood normally flows from posterior to anterior in the heart and from anterior to posterior within the perivisceral sinus. Blood flow may be augmented by accessory pulsating structures in the head, thorax, legs, or wings and by contractions of the dorsal diaphragm. In many rapid-flying insects, there is an additional thoracic "heart," which draws blood through the

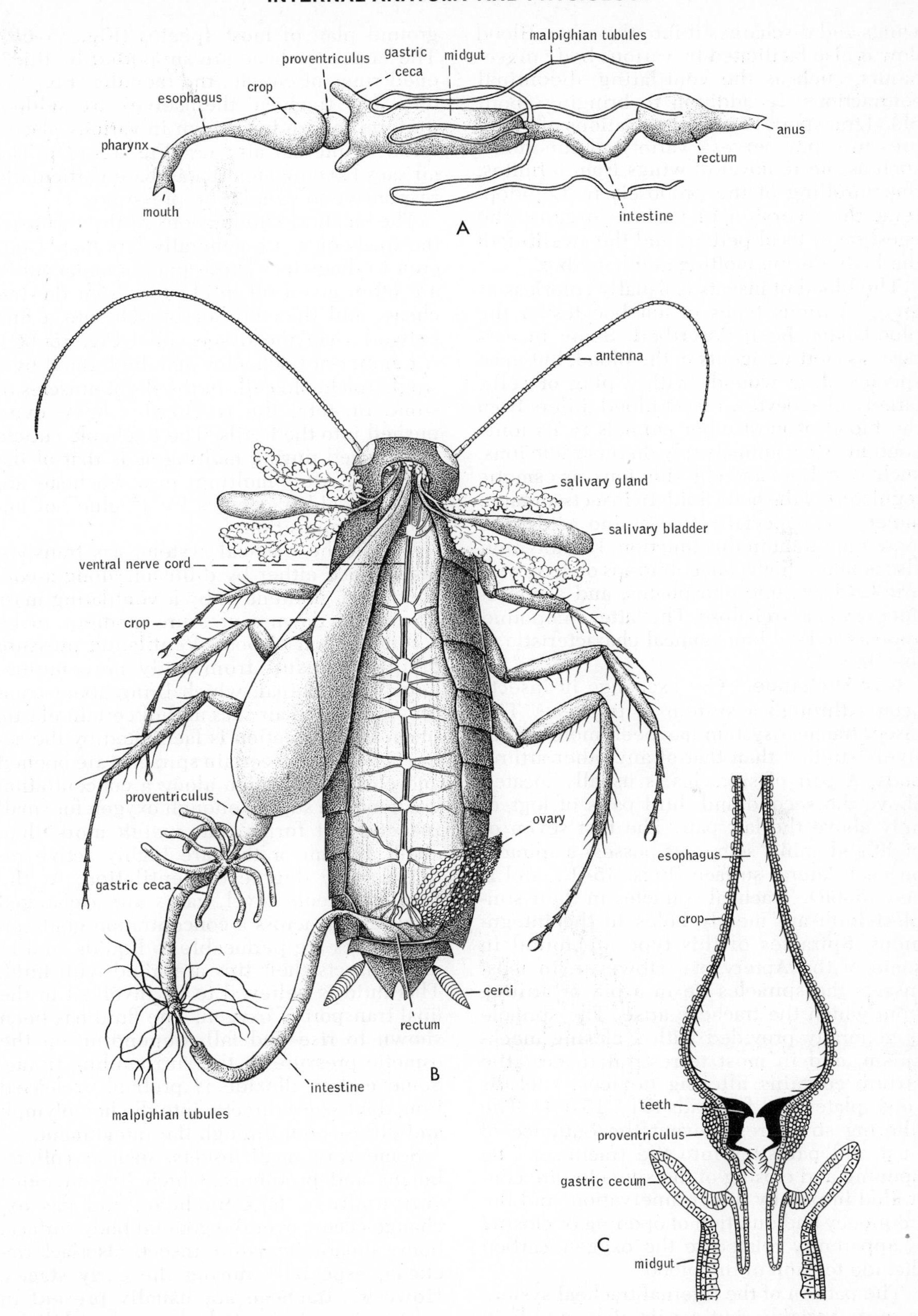

Figure 15–5. *A*. Schematic diagram of the digestive tract in the insect. *B*. Internal structure of a cockroach. *C*. Longitudinal section through the foregut and anterior part of the midgut of a cockroach. (*A* and *C* after Snodgrass; *B* after Rolleston.)

wings and discharges it into the aorta. Blood flow is also facilitated by various body movements, such as the ventilating abdominal contractions. In addition to bringing about blood transport, localized elevations of blood pressure may serve a variety of functions, such as the removal of wings from termites, the unrolling of the proboscis in Lepidoptera, the eversion of various organs, the egestion of fecal pellets, and the swelling of the body during molting and hatching.

The blood of insects is usually colorless or green. Various types of amebocytes in the blood have been described. Some insects possess clotting agents in the blood, but most species close wounds with a plug of cells, largely phagocytes. Insect blood differs from the blood of most other animals in its ionic content. Most animals rely on inorganic ions, such as sodium and chloride ions, as osmotic regulators of the body fluid. In insects, organic molecules, especially free amino acids, are more important in this function. Hemolymph also contains high concentrations of dissolved uric acid, organic phosphates, and a nonreducing sugar, trehalose. The latter compound appears to be a biochemical characteristic of the class.

Gas Exchange. Gas exchange in insects occurs through a system of tracheae. The insect tracheal system has been more extensively studied than that of any other arthropods. A pair of spiracles is usually located above the second and third pairs of legs or only above the last pair. The first seven or eight abdominal segments possess a spiracle on each lateral surface (Figs. 15–1*C* and *E* and 15–6*B*). Tracheal spiracles in their simplest form are merely holes in the integument. Spiracles of this type are found in some of the Apterygota. However, in most insects the spiracles lie in a pit, or atrium, from which the tracheae arise. The spiracle is generally provided with a closing mechanism, and in most terrestrial insects the atrium contains filtering devices, such as sieve plates and felt pads (Fig. 15–6*A*). The filtering structures prevent the entrance of dust and parasites into the tracheae. The opening and closing of the spiracles are controlled in part by direct innervation, and the frequency and duration of opening or closing is apparently related to the oxygen–carbon dioxide tension of the blood.

The pattern of the internal tracheal system is quite variable, but a pair of longitudinal trunks with cross connections form the ground plan of most species (Fig. 15–6*B*). The larger tracheae are supported by thickened rings of cuticle, the taenidia (Fig. 15–6*A*). The tracheae themselves are seldom uniform in size but widen in various places, forming internal air sacs (Fig. 15–6*D*). The air sacs have no taenidia and are particularly responsive to ventilation pressures.

The smallest subdivisions of the tracheae, the tracheoles, are generally less than 1 micron in diameter. These fine cuticular tubes are often given off in clusters from the tracheae, and then further branch into a fine network over the tissue cells (Fig. 15–6*C*). A number of tracheoles may be formed by a single tracheole cell. In the flight muscles of some insects, the tracheoles have even pushed into the fibrils. The tracheole cuticle is not shed during molting as is that of the tracheae. After molting, new tracheae are joined to old tracheoles by a "glue" of unknown composition.

Within the tracheal system, gas transport takes place either by diffusion along a concentration gradient or by a ventilating mass flow of air down a pressure gradient, or by a combination of both. Ventilating pressure gradients result from body movements, largely abdominal, which bring about compression of the air sacs and of certain elastic tracheae. Ventilation is facilitated by the sequence in which certain spiracles are opened and closed. Diffusion along a concentration gradient can supply enough oxygen for small insects, but forms that weigh more than about 1 gram or that are highly active require some degree of ventilation. At the tissue-tracheole level, gases are exchanged by diffusion across a concentration gradient. Tracheoles are permeable to liquids, and in most insects their tips are filled with fluid. This fluid is believed to be involved in the final transport of oxygen. The fluid has been shown to rise and fall, depending on the osmotic pressure in the surrounding tissue. Some carbon dioxide is probably released from the tissues directly into the hemolymph and diffuses out through the integument.

Some very small insects, such as collembolans and proturans, which live in moist surroundings, lack tracheae, and gas exchange occurs over the general body surface. Some aquatic immature insects also lack tracheae, especially during the early stages. However, tracheae are usually present in aquatic immatures and always in adult insects that live in water. The adults merely

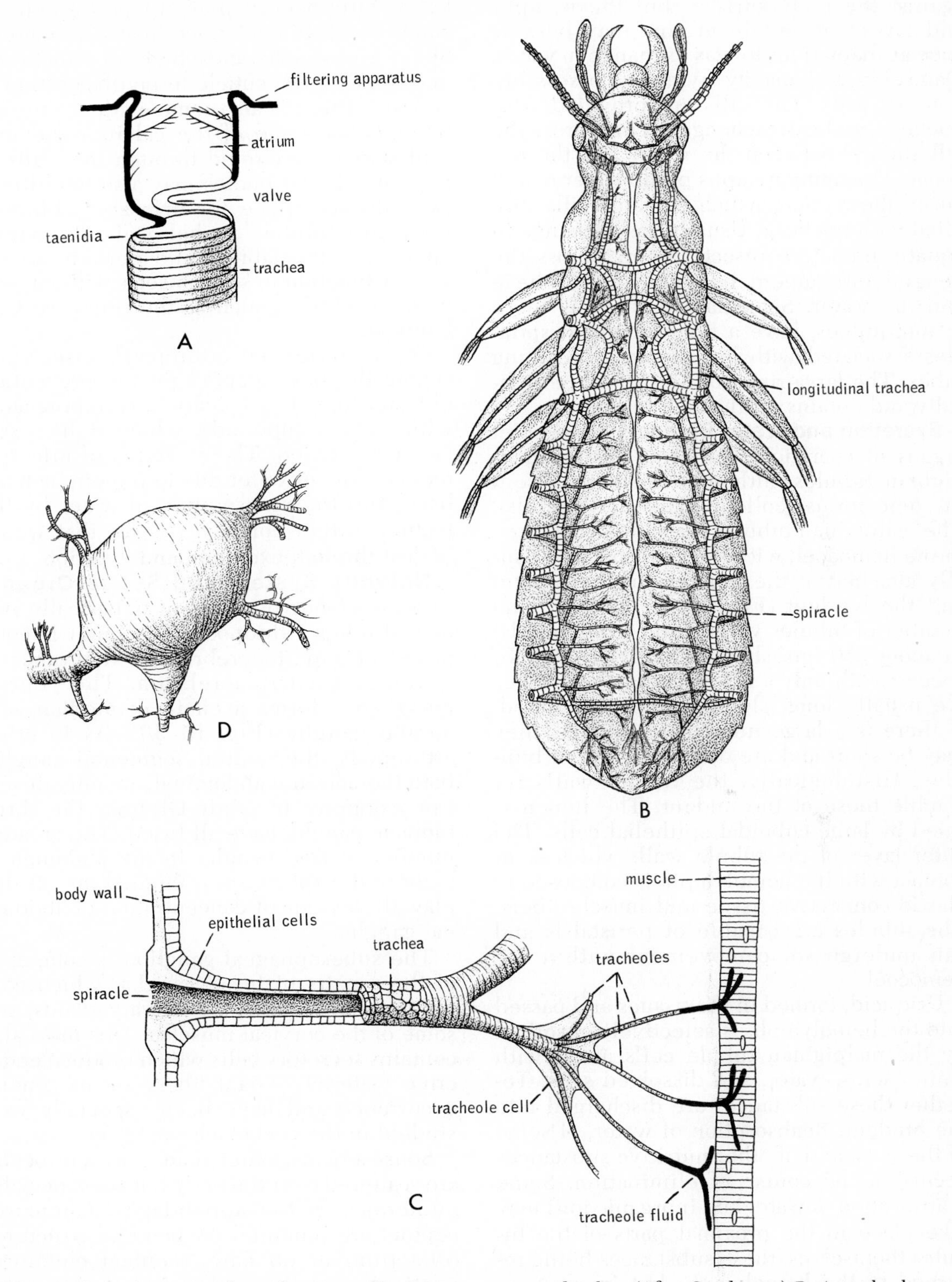

Figure 15–6. *A.* A spiracle with atrium, filtering apparatus, and valve. (After Snodgrass.) *B.* A tracheal system of an insect. (After Ross.) *C.* Diagram showing relationship of spiracle and tracheoles to tracheae. (Modified from Ross.) *D.* An air sac. (After Snodgrass.)

utilize air from air bubbles or films trapped against the body surface, but the nymphs and larvae of certain groups may possess special adaptations for gas exchange in water. Damselfly and mayfly nymphs possess abdominal gills. The gills are provided with tracheae, and gas exchange occurs across the gill surface between the water and the tracheae. Dragonfly nymphs pump water in and out of the rectum, which contains gills supplied with tracheae. Usually gas exchange in aquatic immature insects occurs across the general integument between the tracheae and the water. Some larvae, such as those of mosquitoes, have a few functional spiracles associated with one or more breathing tubes. The larva rises to the surface periodically and obtains air through the tube.

Excretion and Water Balance. The chief organs of excretion in insects are the malpighian tubules, although they are absent in some forms (collembolans and aphids). The malpighian tubules lie more or less free in the hemocoel, with the proximal end usually attached at the junction of the midgut and the hindgut (Fig. 15–5*A* and *B*). The number of tubules varies from two (coccids) to about 250 (grasshopper *Schistocerca*). In species with only a few tubules, the tubules are usually long, slender, and convoluted. If there is a large number of tubules, they may be short and are often grouped in bundles. Histologically, the tubule walls resemble those of the midgut. The lumen is lined by large cuboidal epithelial cells. The outer layer of the tubule wall, which is in contact with the hemolymph, is composed of elastic connective tissue and muscle fibers. The tubules are capable of peristalsis and can undergo some movement within the hemocoel.

Uric acid, formed in the tissues and passed into the hemolymph, is selectively absorbed by the malpighian tubule cells along with amino acids, water, and dissolved salts. Together these substances are discharged into the hindgut. Reabsorption of water, of some of the salts, and of other nutritive substances occurs in the course of elimination. Some reabsorption of water and inorganic ions may take place in the proximal parts of the tubules themselves, these substances being returned to the hemolymph; but in most insects, the rectal epithelium transfers these substances back into the hemolymph.

Not all the waste products are removed by the malpighian tubules. Pericardial cells, or nephrocytes, typically located on or near the heart, supposedly pick up particulate or complex waste for intracellular degradation. Some excess salts and other substances are deposited in the cuticle to be disposed of at ecdysis. There is also some evidence for believing that some of the calcium and uric acid salts are excreted through the walls of the gut. The fat bodies may also be utilized as a storage place for uric acid. Finally, there are tubular cephalic glands, which open onto the labium and perform an excretory function in some insects without malpighian tubules, such as the primitive Collembola.

Of the terrestrial arthropods, insects are among the best adapted for the prevention of water loss. The epicuticle is impregnated with waxy compounds, which reduce surface evaporation. The excretion of uric acid reduces loss of water due to protein metabolism. And the reabsorption of water by the rectum further conserves water that would be lost through excretion and egestion.

Nervous System and Sense Organs. The insect nervous system is basically like that of other arthropods. The brain is composed of a protocerebrum, a deuterocerebrum, and a tritocerebrum. The ventral nerve cord forms a chain of median segmental ganglia (Fig. 15–5*B*). As in other arthropods, the ventral segmental ganglia, both thoracic and abdominal, are often fused. For example, in adult Diptera, the three thoracic ganglia have all fused. The greatest number of free ganglia in the abdomen is eight in the apterygotes. Winged insects display all degrees of concentration of abdominal ganglia.

The subesophageal ganglion is composed of three pairs of fused ganglia, which control the mouth parts, the salivary glands, and some of the cervical muscles. This mass also contains secretory cells which produce endocrine materials. Giant fibers are of general occurrence and have been especially well studied in the cockroach.

Sense organs, other than eyes and ocelli, are scattered over the body but are especially numerous on the appendages. Chemoreceptors are mounted on peg-like structures (olfaction) or on hairs (contact chemoreception), and the endings of the receptor cells pass to the surface of the cuticle through fine canals (Fig. 15–7*A*). Chemoreceptors are especially abundant on the antennae, legs, and mouth parts. Contact

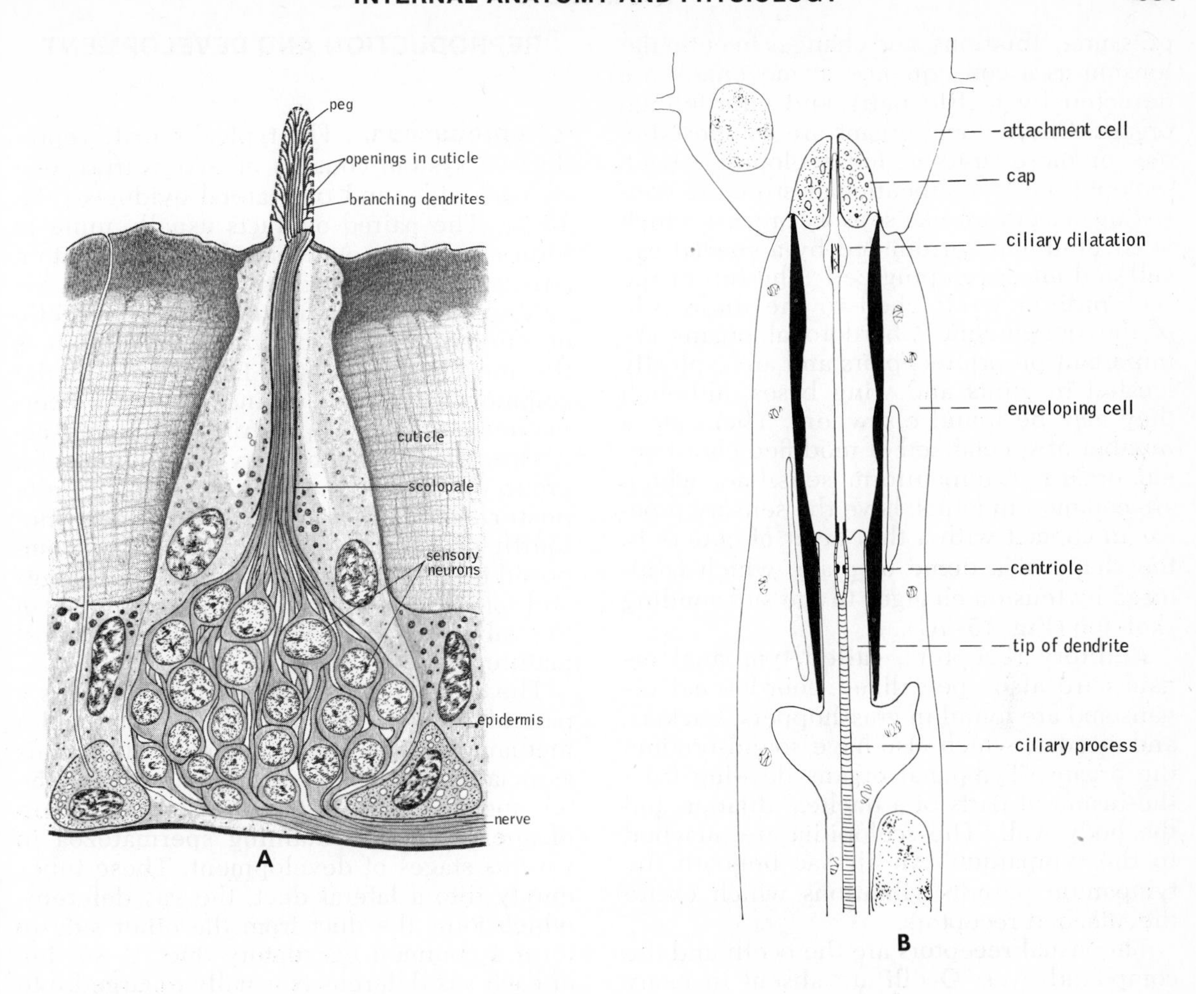

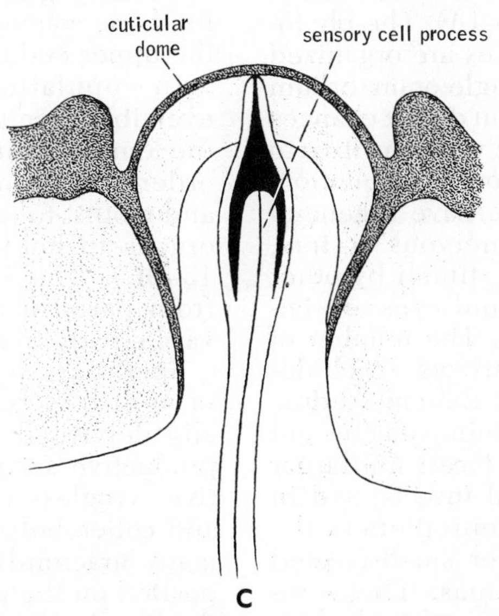

Figure 15–7. *A.* A chemosensory peg organ from the antenna of a grasshopper. (After Slifer et al.) *B.* Structure of a scolopidium from the tympanal organ of a locust. (From Howse, P. E., 1968: *In* Invertebrate Receptors, Zoological Society of London, 23rd Symposium, Academic Press, N.Y., p. 181.) *C.* A campaniform sense organ. (After Snodgrass.)

pressure, vibrations, and changes in cuticular tension as a consequence of movement are detected by tactile hairs and chordotonal organs. Chordotonal organs are composed of one or more units called scolopidia. Each scolopidium is a subcuticular structure consisting of a cilium-like sensory process which is covered and surrounded by a special cap cell and an enveloping cell. The top of the scolopidium is attached to the underside of the integument. Chordotonal organs are important proprioreceptors and are typically located in joints and wing bases, although they may be found elsewhere. There are a number of specialized or modified chordotonal organs. Campaniform sensillae, which are common in joints, have the sensory process in contact with a thin layer of cuticle in the shape of a dome or plate, which is altered by tension changes in the surrounding skeleton (Fig. 15–7*C*).

Auditory receptors, called tympanal organs, are also specialized chordotonal organs and are found in grasshoppers, crickets, and cicadas, which also have sound-producing organs. Tympanal organs develop from the fusion of parts of a tracheal dilation and the body wall. The scolopidia are attached to the tympanum. An air sac beneath the tympanum permits vibrations which excite the attached receptors.

The visual receptors are the ocelli and the compound eyes. Ocelli are absent in many adult insects, but when present, there are usually three, found on the anterior dorsal surface of the head (Fig. 15–1*A*). The photoreceptor cells on each ocellus are organized somewhat like those of a single ommatidium of a compound eye. Ocelli can detect changes in light intensity and may be very sensitive to low intensities. They function in orientation, and in some way appear to have a general stimulatory effect on the nervous system, enhancing the reception of stimuli by other sensory structures. Compound eyes are laterally situated on the head. The number of facets is greatest in flying insects which depend on vision for feeding. Extreme reduction of facets is found in certain parasitic and cave-dwelling insects. The facets are larger in nocturnal than in diurnal insects, and in some crepuscular flying neuropterans the eyes are divided into upper small-faceted and lower large-faceted areas. There are also some insects, such as the gyrinid water beetles, in which the compound eyes are divided into upper and lower halves.

REPRODUCTION AND DEVELOPMENT

Reproduction. The typical female reproductive system consists of two ovaries, one on each side, and two lateral oviducts (Fig. 15–8). The paired oviducts usually unite to form a common oviduct, which leads into a vagina. The vagina in turn opens onto the ventral surface behind the seventh, eighth, or ninth segments, of which the eighth is the most common site. Diverticula of the common oviduct or vagina include a spermatheca, a bursa copulatrix, and paired accessory glands. Each ovary is made up of a group of tubules, the ovarioles, which unite posteriorly into a calyx. The upper one-fourth or one-third of an ovariole is composed of germinal epithelium in which eggs and follicle cells develop. The remainder of the tubule is filled with mature, or almost mature, eggs.

The male reproductive system includes a pair of testes, a pair of lateral ducts, and a median duct opening through a ventral penis associated with the eighth segment (Fig. 15–9*A* and *B*). Each testis consists of a group of sperm tubes containing spermatozoa in various stages of development. These tubes empty into a lateral duct, the vas deferens, which joins the duct from the other side to form a common ejaculatory duct. A section of each vas deferens is usually enlarged into a seminal vesicle where sperm are stored. Accessory glands, which secrete seminal fluid, are commonly present as pouches from the upper end of the ejaculatory duct.

In copulation the often extensible, or eversible, penis of the male is inserted into the female genital orifice. The males of many orders—dragonflies, true flies, butterflies, and moths, to name a few—possess clasping organs to hold the female abdomen (Fig. 15–9*C*). The clasping organs are derived from parts of the terminal segments and vary greatly in structure.

Sperm are transferred in spermatophores in most insects. The spermatophore is usually deposited directly into the female reproductive system, but among a few primitive wingless insects, such as thysanurans and collembolans, transfer is indirect. As in many arachnids, the spermatophore is deposited on the ground and then taken up by the female.

The Odonata (dragonflies and damselflies) also exhibit a form of indirect sperm

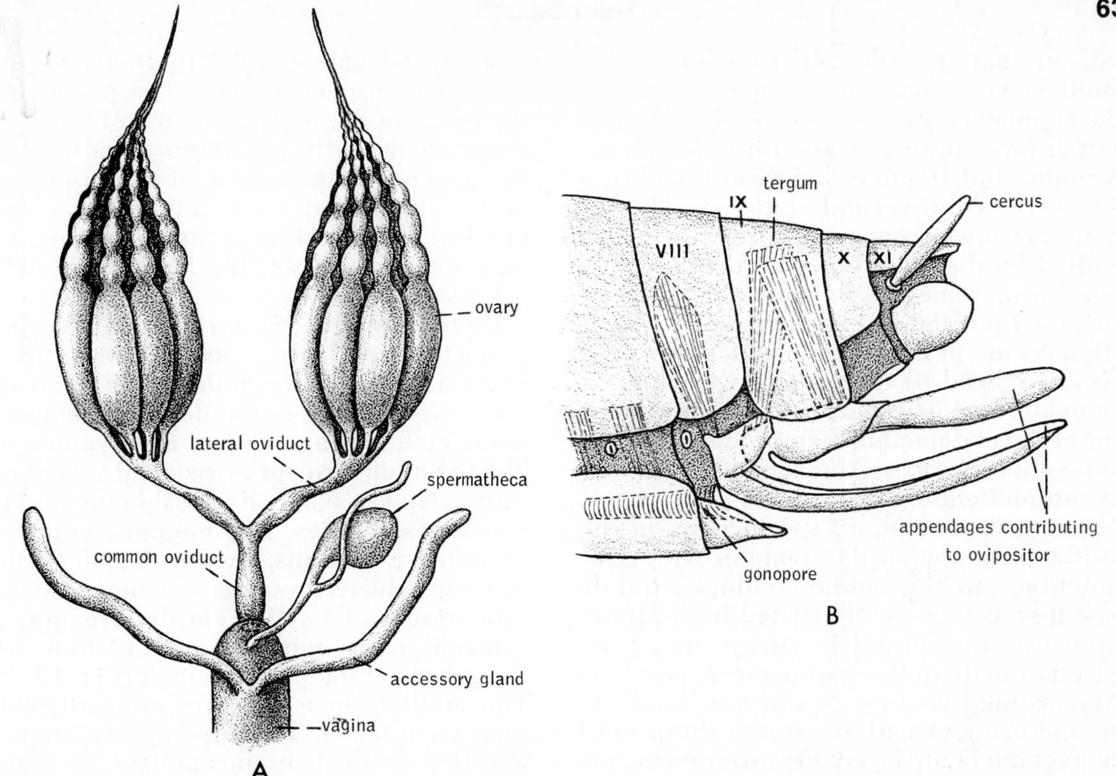

Figure 15–8. *A.* Reproductive system in a female insect. *B.* Lateral view of the posterior end of the abdomen, showing reproductive opening and appendages, forming ovipositor. (Both after Snodgrass.)

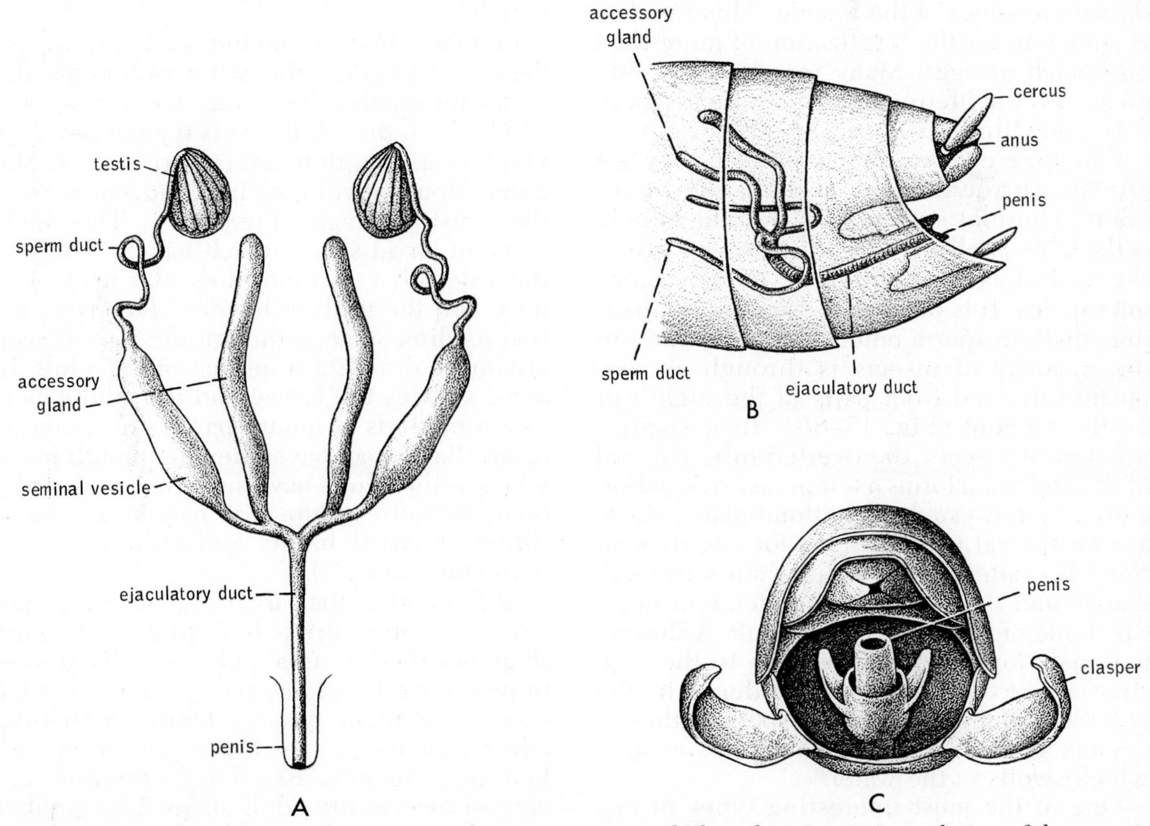

Figure 15–9. Reproductive system in a male insect. *A.* General plan of system. *B.* Lateral view of the posterior end of the abdomen, showing reproductive opening and other structures. *C.* Posterior view of the abdomen, showing penis and claspers. (All after Snodgrass.)

transfer that is strikingly parallel to that found in some arachnids, such as spiders. The copulatory organ of the male is a median ventral tube located on the third abdominal sternum, and it projects forward within a groove onto the second segment. The intromittent organ contains a reservoir, which is filled by the male prior to copulation. In transferring semen to the reservoir, the male curls the tip of the abdomen ventrally. Copulation occurs in flight, the male clasping the thorax or head of the female with his abdominal cerci. The female gonopore at the end of the abdomen is then brought forward to the undersurface of the male to unite with his intromittent organ.

During insemination, sperm may be deposited in the vagina, the common oviduct, or sometimes in the lateral oviducts, but in most insects they are deposited in the bursa copulatrix. In only a few groups are sperm placed directly in the spermatheca. Not long after mating, however, sperm are found in the spermatheca and are stored there until the eggs are being layed. Fertilization occurs as the eggs pass through the oviduct at the time of egg deposition. At each mating a large number of sperm are transferred to the spermatheca of the female. This number is sufficient for the fertilization of more than one batch of eggs. Many insects mate only once in their lifetime, and none mate more than a few times.

When the eggs reach the oviduct, they are already surrounded by a shell-like membrane (chorion) secreted by ovarian follicle cells. This shell may be up to seven layers thick, and is perforated by one or more micropyles. It is through these minute openings that the sperm enter. Egg deposition in the majority of insects is through an ovipositor, derived from parts of the eighth or ninth segment (Fig. 15–8*B*). In a smaller number of insects, the everted tubelike end of the abdomen forms a temporary ovipositor, and in a few, such as collombolans, there are no special modifications for egg deposition. The site of egg laying varies tremendously and is in large part dependent upon the mode of existence of the adult. Adhesive materials for attaching the eggs to the substratum or to each other are produced by the accessory glands. In aquatic species, the accessory glands provide a gelatinous coating which swells in the water.

One of the most interesting types of egg deposition is that associated with gall formation. The females of gall wasps (hymenopterans) and gall gnats (dipterans) deposit their eggs in plant tissues. The plant tissue surrounding the eggs is induced to undergo abnormal growth and forms a gall, which has a shape characteristic of the insect producing it. The gall forms a protective chamber for the developing eggs, larvae, and often even the pupae. The larvae feed on the gall tissues.

Development. Superficial cleavage is characteristic of insects; collembolans are an exception in having holoblastic cleavage. Insect young vary in the degree of development at hatching. Young apterygotes are like the adults except in size and sexual maturity. Newly hatched grasshoppers, cockroaches, stoneflies, leaf hoppers, and bugs resemble the adults, except that the wings and reproductive organs are undeveloped. The wings of early nymphs are merely external pads, which begin to look like wings only at the preadult molt (Fig. 15–10). The adult form is reached gradually with successive molts. This type of development is called gradual, or incomplete, metamorphosis (hemimetabolous development); all the immature stages from hatching to the adult are termed nymphs, or when aquatic, naiads.

In many insects, including bees, wasps, flies, and beetles, the wing rudiments develop internally; the wings seem to appear suddenly in the adults. This type of development is a complete metamorphosis (holometabolous development), and consists of three distinct stages (Fig. 15–11). The newly hatched larval stage, which has no wings, is the caterpillar of butterflies, the maggot of flies, and the grub of beetles. This is an active feeding stage, although the food is usually quite different from that of the adult. In some species the larvae and the adults have different kinds of mouth parts. For example, caterpillar larvae have chewing mouth parts, whereas the adults have sucking mouth parts. Some parasitic groups may have two or more different larval habits and structures (hypermetamorphosis).

At the end of the larval period, the young become nonfeeding and quiescent. This stage is called a pupa and is usually passed in protective locations, such as the ground, a cocoon, or plant tissues. During pupation, adult structures are developed from embryonic rudiments. Few larval structures are carried over to the adult stage. The number of molts required to reach the adult stage ranges from as few as three or four to over

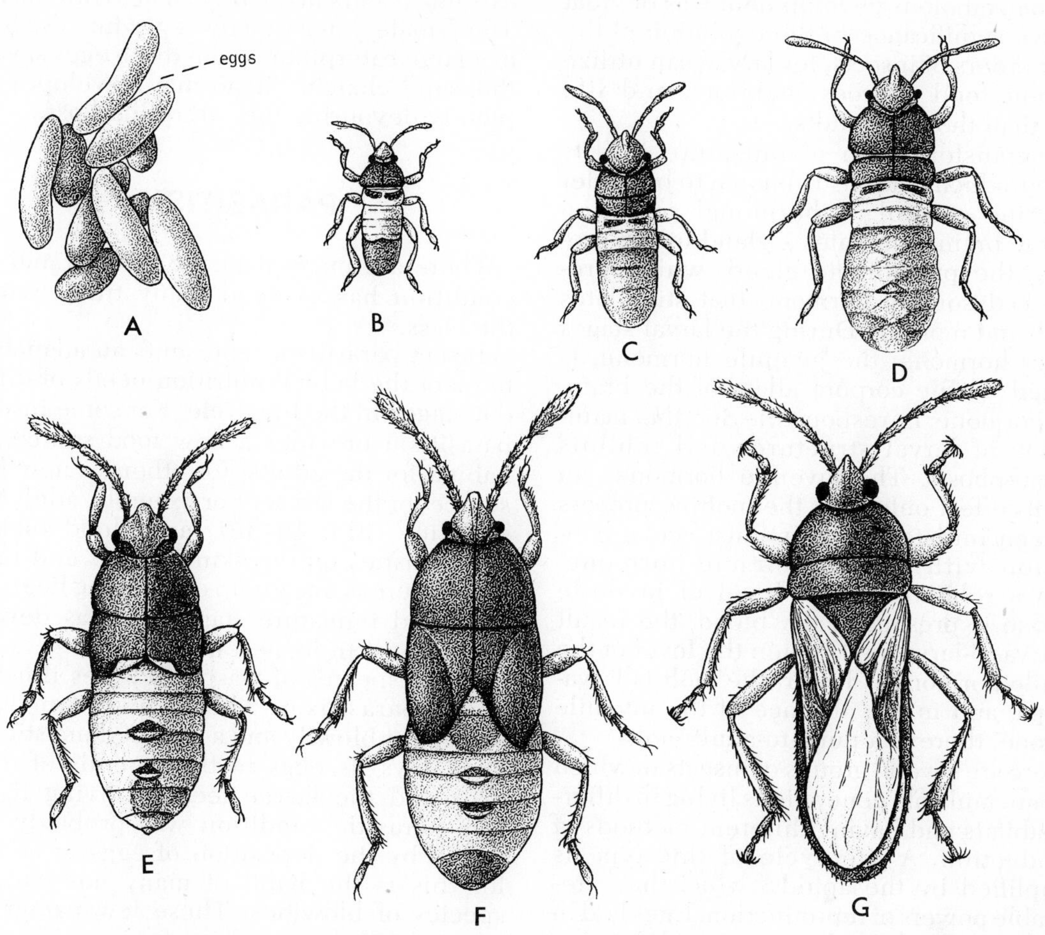

Figure 15–10. Stages in the gradual metamorphosis of a chinch bug. (After Ross.)

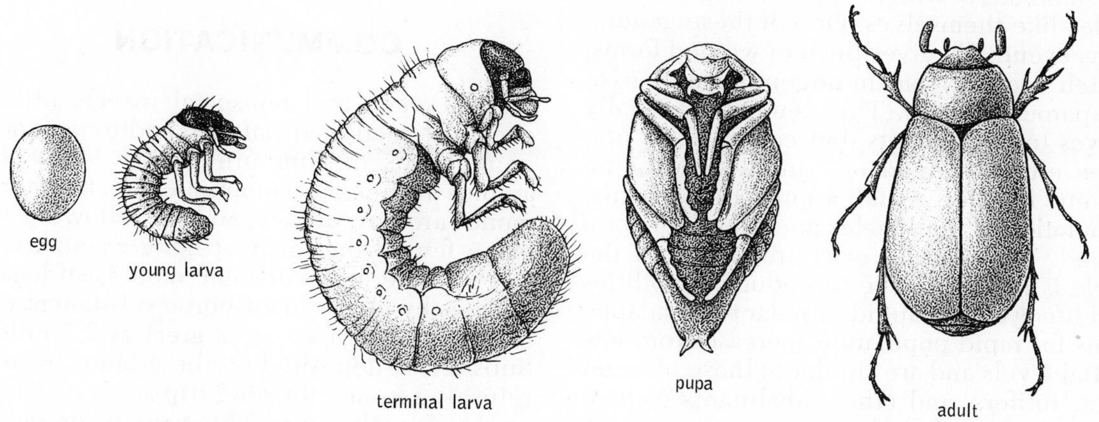

Figure 15–11. Stages in the complete metamorphosis of a beetle. (After Ross.)

30 and is dependent in large part on the type of development.

Holometabolous development was of great adaptive significance in the evolution of the higher orders of insects, for larvae can utilize different food sources, habitats, and life styles than those of adults.

The transformation of immature insects into reproducing adults is known to be under endocrine control. A hormonal secretion from the brain stimulates a gland in the prothorax, the prothoracic gland, which produces ecdysone, a hormone that stimulates growth and molting. During the larval stages another hormone, the juvenile hormone, is secreted by the corpora allata of the brain. This hormone is responsible for the maintenance of larval structures and inhibits metamorphosis. The juvenile hormone can exert its effect only after the molting process has been initiated. It thus must act in conjunction with the prothoracic hormone. When a relatively high level of juvenile hormone is present in the blood, the result is a larva-to-larva molt. When the level of the juvenile hormone is lower, the molt is larva-to-pupa, and in the absence of the juvenile hormone, there is a pupa-to-adult molt.

There are several groups of insects in which there are multiple generations living in different habitats and having different methods of reproduction. A life cycle of this type is exemplified by the aphids, which have remarkable powers of reproduction. Eggs laid in the autumn hatch in the spring and develop into wingless parthenogenetic females. The females, sometimes called stem mothers, are ovoviviparous and may give birth to any number of broods of wingless, parthenogenetic females like themselves. One of these generations eventually gives birth to winged forms, which are still parthenogenetic and ovoviviparous females. This generation usually moves to other plants, but continues to produce either winged or wingless parthenogenetic females. As fall approaches, a sexual generation of both males and females is produced. These mate, eggs are laid, and the cycle is repeated. The reproductive abilities and life cycles of aphids appear to be adaptations for rapid population increase from low initial levels and are similar to those of water fleas, rotifers, and other inhabitants of temporary bodies of freshwater.

Polyembryony, in which the initial mass of embryonic cells gives rise to more than one embryo, is highly developed in some parasitic hymenopterans and is a classic example of this developmental phenomenon. Two to many larvae are formed from a simple egg deposited in the body of another insect host. The extreme occurs in the tiny chalcid *Litomastix*. The female deposits a few eggs into the body of a large caterpillar. From these eggs several thousand chalcid larvae may develop, completely devouring the caterpillar host.

PARASITISM

There are many parasitic insects, and the condition has evolved many times within the class.

Insect parasitism represents an adaptation to meet the habitat-nutrition needs of different stages in the life cycle. For some insects, parasitism provides a new food source and habitat for the adults; for others, a new food source for the larvae. For example, adult fleas and lice (Fig. 15–15*J*) are blood-sucking ectoparasites on the skin of birds and mammals (there is one group of chewing lice). The eggs and immature stages of fleas develop off the host in its nest or den.

Many species of wasps and flies illustrate larval parasitism. The screw-worm fly, a species of blowfly and a pest of domestic animals, lays its eggs in the wounds of mammals and the larvae feed on living tissue. The parasitic condition was probably preceded by the deposition of eggs in carrion, for this is the habit of many non-parasitic species of blowflies. These few paragraphs cannot begin to provide an appreciation of our present knowledge of the many parasitic insect species. The interested reader should consult Askew (1971).

COMMUNICATION

Both social and nonsocial insects utilize chemical, tactile, visual, and auditory signals as methods of communication. Many examples of chemical communication by pheromones are now known, and the following are just a few of the many striking examples in insects. The males of some moths can locate females by means of air-borne substances detected from a distance as great as 2.7 miles. Substances deposited on the ground by ants returning from a foraging trip serve as a trail marker for other ants. This type of communication is especially important in the complex movements of tropical army ants. Substances produced by the death and decomposition of the body of an ant within the colony stimulate other workers to remove the body. If a live ant is painted with an extract from a decomposing body, the painted ant will be carried live and struggling from the nest.

Among the more unusual visual signals are the luminescent flashings of fireflies, which function in sexual attraction.

Sound production is especially notable in grasshoppers, crickets, and cicadas. The chirping sounds of the first two are produced by rasping. The front margin of the forewing (crickets) or the hind leg (katydids) acts as a scraper and is rubbed over a file formed by a vein of the forewing. Where scraper and file are both located on the forewings, as in crickets, the wings cross over, and one forewing functions as a scraper and the other as a file. Each species of cricket produces a number of songs which differ from the songs of the other species. Cricket songs function in sexual attraction and aggression. The static-like sounds of cicadas, which serve to aggregate individuals, are produced by vibrations of special chitinous abdominal membranes.

The remarkable mechanisms of communication in bees are described in the next section.

SOCIAL INSECTS

Colonial organization has evolved in a number of animal phyla, but only among some insects and vertebrates are individuals functionally interdependent, yet morphologically separate. The condition is therefore usually described as being a social organization.

Social organizations have evolved in two orders of insects, the Isoptera, which contains the termites, and the Hymenoptera, which includes the ants, bees, and wasps. In all social insects, no individual can exist outside of the colony nor can it be a member of any colony but the one in which it developed. There is cooperative brood care and an overlap of generations. All social insects exhibit some degree of polymorphism, and the different types of individuals of a colony are termed castes. The principal castes are male, female (or queen), and worker. Males function for the insemination of the queen, which produces new individuals for the colony. The workers provide for the support and maintenance of the colony. Caste determination is a developmental phenomenon regulated by the presence or absence of certain substances provided in the immature stages by other members of the colony.

Termites live in galleries constructed in wood or soil, and in some species the colony may be huge and structurally very complex. Termites differ from social hymenopterans in that workers are sterile individuals of both sexes and the reproductive male is a permanent member of the colony. The colony is built and maintained by workers, some of which may be soldiers (Fig. 15–12). The soldiers have large heads and mandibles and serve for the defense of the colony. Workers and soldiers are wingless; wings are present in the males and queens only during a brief nuptial flight, during which pairing and dispersion occur.

The great variety of termite nests are remarkable in their construction and are often symmetrical. Some sort of ventilating system that regulates the oxygen supply and the temperature may be present. One species of termite that lives on the Ivory Coast of Africa constructs a huge nest (termitarium), which is covered by an insulating layer. The center of the nest contains fungus gardens, which the termites maintain and utilize for food, and a royal chamber occupied by the queen. The center receives fresh air by way of channels from a cellar cavity. Other channels conduct stale air from the center to the surface of the nest.

Except for the fungus growing species, termites depend upon cellulose as a food source and upon symbiotic flagellates for cellulose digestion. The symbiosis was probably an important factor in the evolution of social behavior in termites (Wilson, 1971).

Ant colonies resemble those of termites and are housed within a gallery system in soil or wood, or beneath stones. Like bees and wasps, workers are always sterile females. There may be a soldier caste of workers. Wings are present only during the nuptial period of reproductive males and females. Copulation occurs at this time and the male never becomes a functional part of the colony.

Polymorphism is less highly developed in wasps and bees. There is no soldier caste and workers are winged, but many of these insects exhibit remarkable adaptations for a social organization.

The honeybee, *Apis mellifera,* is the best known social insect. This species is believed to have originated in Africa and to be a recent invader of temperate regions, for unlike other social bees and wasps of temperate regions, the honeybee colony survives the winter and multiplication occurs by the division of the colony, a process called swarming. Stimulated at least in part by the crowding of workers (20,000 to 80,000 in a single colony), the mother queen leaves the hive along with part of the workers (a swarm) to found a new colony. The old colony is left with developing queens. On hatching, a new queen takes several nuptial flights during which copulation with males (drones) occurs, and she accumulates enough sperm to last her lifetime. The male dies following copulation, when his reproductive organs are literally exploded into the female. A new queen may also depart with some of the workers as an afterswarm, leaving the remaining workers to yet another developing queen. Eventually the old colony will consist of about one-third of the original number of workers and their new queen.

Honeybee colonies are large. The workers' life span is not long, and a queen may lay one thousand eggs per day. The diet provided these larvae by the nursing workers results in their developing into sterile females, i.e., additional workers. The nursing behavior of the workers is a response to a pheromone ("queen substance") produced by the queen's mandibular glands. At the advent of the swarming or when the vitality of the queen diminishes, the production of this pheromone declines. In the absence of the inhibiting effect of the pheromone, the nursing workers construct royal cells, into which eggs and royal jelly are placed. The exact composition of this complex food is still unknown, but those larvae fed upon it develop into queens in about 16 days. At the same time that queens are being produced, unfertilized eggs are deposited into cells similar to those for workers. These haploid eggs develop into drones.

A remarkable feature of honeybee social organization is the temporal division of the workers. As can be seen from the graph in Figure 15–13*A*, the first activities of the worker are maintenance tasks within the hive. During this period there is secretion by wax, mandibular, and other glands involved in comb construction, food storage, and larval care. After about three weeks, such glandular activity declines, and the bee begins a period of foraging outside of the hive, its final service to the colony. The many functions performed in the lifetime of a worker are not strictly sequential, but a worker shifts from one task to another (Fig. 15–13*A*). A large amount of time is spent by the older worker bees in resting and patrolling. Patrolling, or "determination" of hive needs, plus the ability of workers to change tasks, enables a colony to adjust to changing environmental conditions and is believed to have contributed to the success of the species.

Communication between members of a honeybee colony is highly evolved; some aspects, such as the tail-wagging dance, set the honeybees apart from all other social insects. A successful foraging scout returns to the hive and communicates to other workers

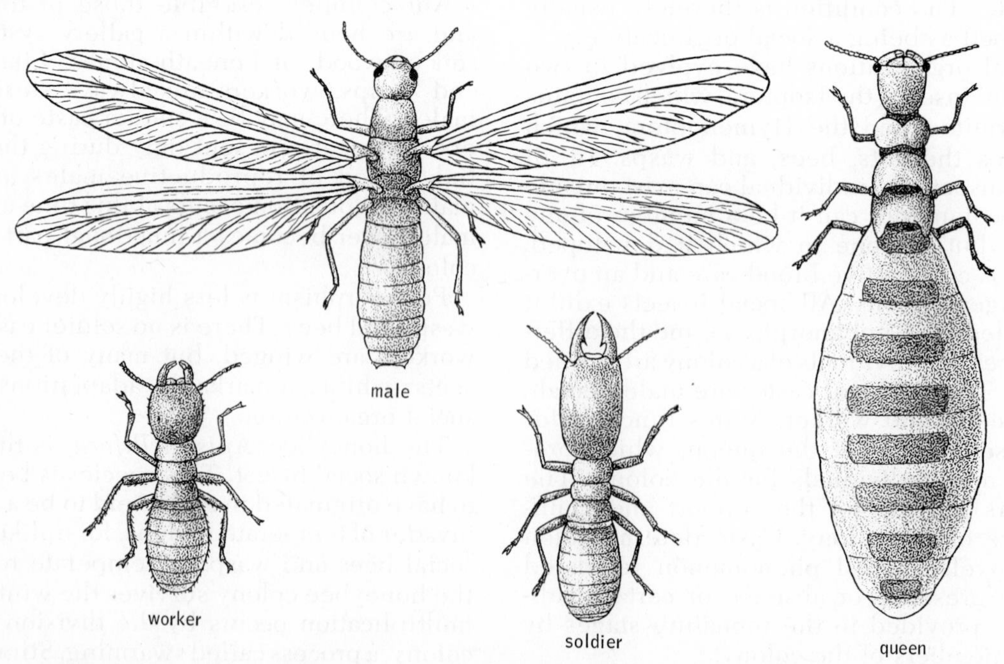

Figure 15–12. Caste of the common North American termite, *Termes flavipes*. (After Lutz.)

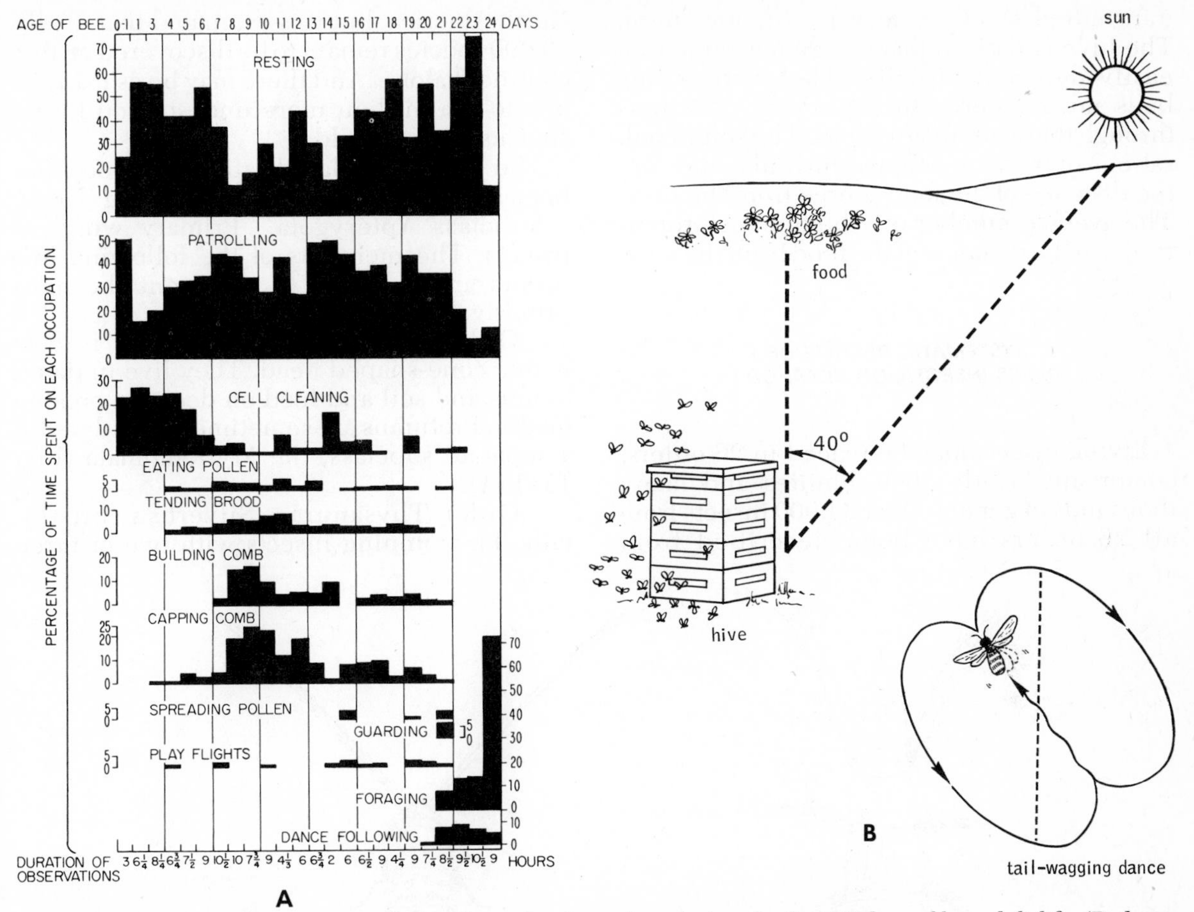

Figure 15–13. *A.* The activities of a single worker honeybee during the first 24 days of her adult life. (Redrawn from Ribbands, 1953; based on data of Lindauer, 1952.) *B.* Diagram illustrating the inclination of the straight tail-wagging run by a scout bee to indicate the location of a food source by reference to the sun. The food source is located at an angle 40 degrees to the left of the sun. The tail-wagging run of the scout bee is therefore upward (indicating that food is toward sun) and inclined 40 degrees to the left (indicating the angle of the food source to the sun). (After Von Frisch.)

the nature, direction and distance of a food source. The nectar and pollen on the scout's body provide the information about the kind of food that has been found. The scout bee also executes an excited dance which is a ritualization of the flight path. The dancing bee circles to the right and to the left, with a straight-line run between the two semicircles (Fig. 15–13*B*). During the straight-line run, the bee wags her abdomen and emits audible pulsations. Von Frisch, a pioneer in the study of communication in bees, discovered that the orientation of the circular movements shows the direction of the food, and that the frequency of the tail-wagging runs indicates the distance. The closer the food source is to the hive, the greater the frequency of tail-wagging runs. Bees use the angle of the sun and light polarization as a means of orientation, and the dance of the scout bee indicates the location of the food in reference to the sun's position. If the tail-wagging run is directed upward, the food is located toward the sun; if the tail-wagging run is directed downward, the food is located away from the sun. The inclination of the run to the right or to the left of vertical indicates the angle of the sun to the right or to the left of the food source. An internal "clock" compensates for the passage of time between discovery of the food and the start of the dance, so that the information is correct even though the sun has moved during the interval. On cloudy days, the polarization of the light rays and ultraviolet light act as indirect references in the absence of the sun. If the food source is closer than 80 meters, the clues provided by chemoreception are sufficient for finding the food, and the tail-wagging dance is not performed by the scout bee.

Although the tail-wagging dance has been decoded, the sensory modality by which it is transmitted to other bees is still uncertain. The hive is dark so that the dance cannot be easily detected visually. The surrounding bees must receive the dancer's vibrations through their antennae or legs. The sound pulsations of the dancer apparently also indicate the distance of the food source from the hive. The average number of vibrations is proportional to the distance of the food from the hive.

SYSTEMATIC RÉSUMÉ OF CLASS INSECTA, OR HEXAPODA

Living insects may be placed in 26 orders, comprising nearly 1000 families and many thousands of genera. Over 84,000 insects from all 26 orders have been described from North America, but it is estimated that nearly 25,000 species remain to be discovered on that continent alone. And there may be as many as one to ten million more undescribed insect species in the world.

The following classification of insects has been adopted from Borror and Delong (1964):

Subclass Apterygota. Primary wingless insects. The members of the following five orders are believed to represent the most primitive living insects.

Order Protura. Eyeless, small insects with a cone-shaped head. They live in damp humus and soil and feed on decayed organic matter. Proturans are sometimes considered as a separate subclass, the Myrientomata (Fig. 15–14*A*).

Order Thysanura. Silverfish, bristletails. Fast-running insects with two or three

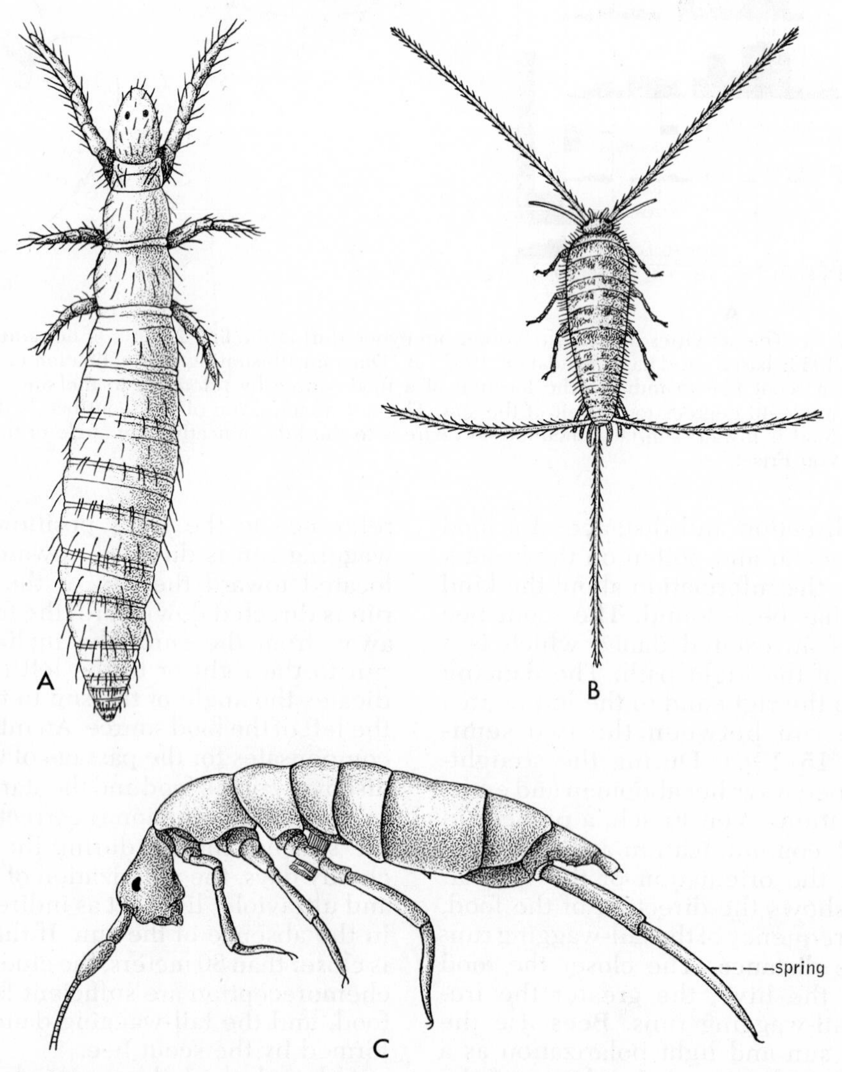

Figure 15–14. Subclass Apterygota. *A*. Order Protura: a proturan. (After Ewing from Ross.) *B*. Order Thysanura: a silver fish. (After Lutz.) *C*. Order Collembola: a springtail. (After Willem.)

styliform appendages on the abdomen. They live in dead leaves and wood and around stones. Some species are found in houses, where they eat books and clothing (Fig. 15–14*B*).

Order Collembola. Springtails. Small insects with an abdominal jumping organ, well-developed legs and antennae; eyes either absent or represented by isolated ommatidia. Abundant in moist leaf mold, soil, rotten wood. They are sometimes considered as a separate subclass, the Oligoentomata (Fig. 15–14*C*).

Subclass Pterygota. Winged insects, or if lacking wings, the wingless condition is secondary.

Order Ephemeroptera. Mayflies. Elongate insects with net-veined wings, of which the first is larger than the second. Two or three caudal filiform appendages. Antennae small and mouth parts of adults vestigial. Gradual metamorphosis (Fig. 15–15*A*).

Order Odonata. Dragonflies and damselflies. Predaceous insects with long, narrow, net-veined wings, large eyes, and chewing mouth parts. Gradual metamorphosis. Nymphs aquatic. Dragonflies are stout-bodied, damselflies are slender and delicate (Fig. 15–15*B*).

Order Orthoptera. Grasshoppers, katydids, crickets, roaches, mantids, and walking sticks. Large-headed insects with strong chewing mouth parts and compound eyes. Many species have femur of hind leg enlarged for jumping. Winged and wingless species. Most winged forms have membranous hindwings protected by leather-like forewings. Largely herbivorous, sometimes causing vast crop damage. Gradual metamorphosis (Fig. 15–15*C*).

Order Isoptera. Termites. Social insects. Winged and wingless individuals composing the colony. Soft-bodied, with abdomen broadly joined to thorax. Gradual metamorphosis (Fig. 15–12).

Order Plecoptera. Stoneflies. Adults have long antennae, chewing mouth parts, and two pairs of well-developed, membranous wings. Gradual metamorphosis. Nymphs aquatic. Adults emerge during winter in certain groups (Fig. 15–15*D*).

Order Dermaptera. Earwigs. Elongate insects resembling beetles, with chewing mouth parts, compound eyes, and large forceps-like cerci. Most species have fan-shaped wings and elytra. Nocturnal, with omnivorous food habits. Gradual metamorphosis (Fig. 15–15*E*).

Order Embioptera. Webspinners. Small, slender, soft-bodied insects with large heads and eyes. They feed on plants and live in silken tunnels, which they weave ahead of themselves to create routes. Silk glands and spinning hairs located on front tarsi. They are gregarious and live in large colonies. Tropical and semitropical. Gradual metamorphosis (Fig. 15–15*F*).

Order Psocoptera. Booklice, barklice, psocids. Small, fragile, pale insects, with chewing mouth parts. Winged or wingless. When winged, wings membranous and front pair a little larger than hind pair. Gradual metamorphosis. They live in a wide variety of habitats—bark, foliage, under stones. Some species infest buildings and are found in books (Fig. 15–15*G*).

Order Zoraptera. Small, pale, soft-bodied insects resembling tiny termites with chewing mouth parts. Both winged and wingless forms. Gradual metamorphosis. These are rare insects, living in colonies under dead wood in warm climates. Only some twenty known species and one genus (Fig. 15–15*H*).

Order Mallophaga. Chewing lice, bird lice. Wingless, flattened insects that live as ectoparasites on birds and mammals. The eyes are reduced or absent; legs are short and thorax is small. Gradual metamorphosis. Feed on scales, feathers, hair, skin, and sometimes on dried blood around wounds. Many species infest domestic birds and livestock and cause considerable damage by skin irritation (Fig. 15–15*I*).

Order Anoplura. Sucking lice. Similar to the chewing lice but mouth parts adapted for sucking. Ectoparasites of birds and mammals. A number of species are parasitic on domestic animals, and the head louse and crab louse are parasites of man. More serious than the irritation produced by these parasites is their role as vectors of disease, such as typhus fever (Fig. 15–15*J*).

Order Thysanoptera. Thrips. Small slender insects with mouth parts adapted for rasping and sucking. Winged and wingless. Wings, when present, are narrow, with few veins, and fringed with hairs. Development peculiar in that there are nymphlike early instars but a preadult pupa stage. A large number of species feed on flowers, but some thysanopterans are predaceous and feed on mites and smaller insects (Fig. 15–15*K*).

Order Hemiptera. True bugs. Piercing-sucking mouth parts, in which the beak arises from the front of the head and extends ventrally and posteriorly. Forewings with a

(*Text continued on page 648.*)

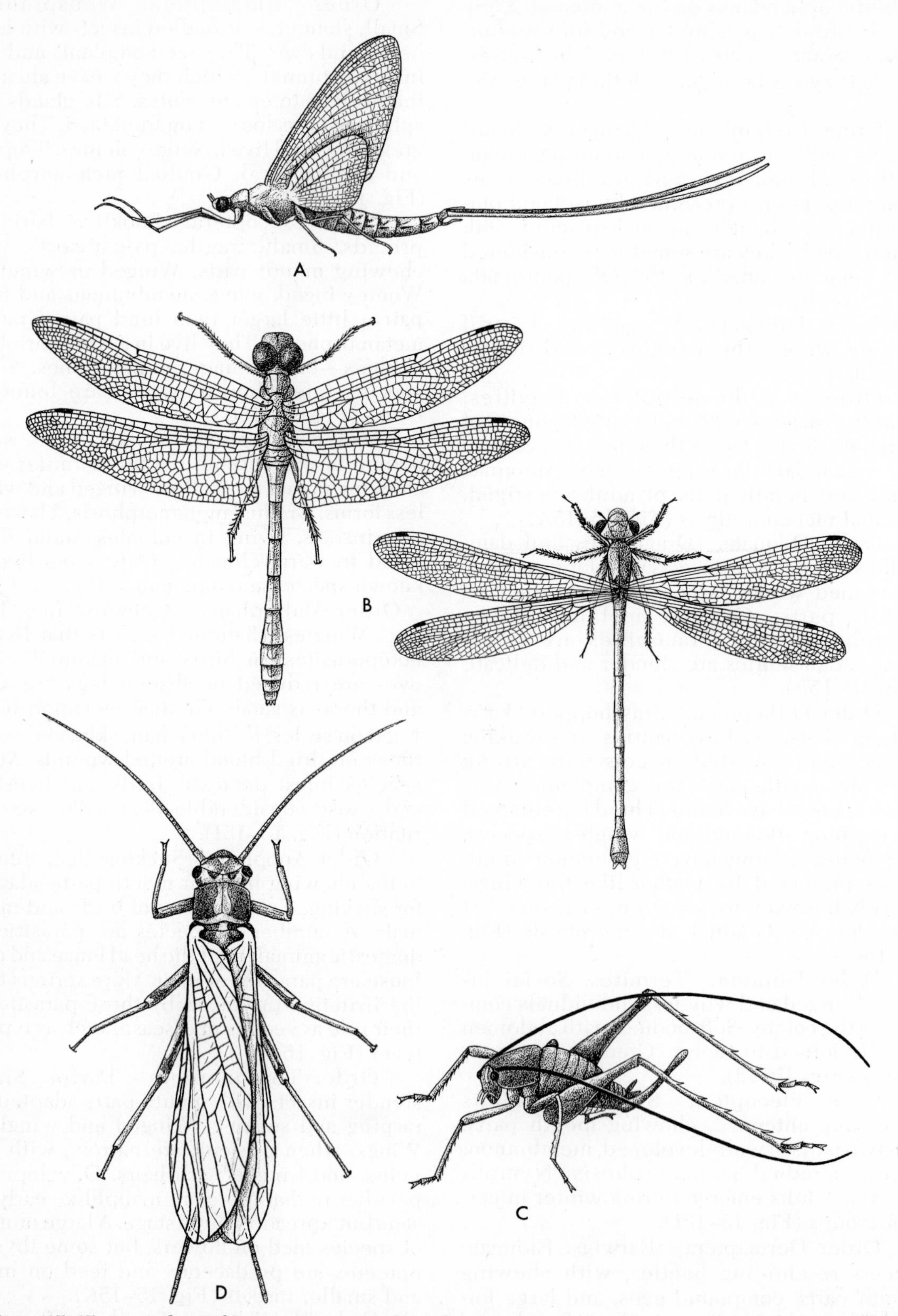

Figure 15–15. *A.* Order Ephemeroptera: a mayfly. (After Ross.) *B.* Order Odonata: a dragonfly (left) and (right) a damselfly. (After Kennedy from Ross.) *C.* Order Orthoptera: a camel cricket. (After Lutz.) *D.* Order Plecoptera: a stonefly. (After Ross from Illinois Nat. Hist. Survey.)

Illustration continued on opposite page.

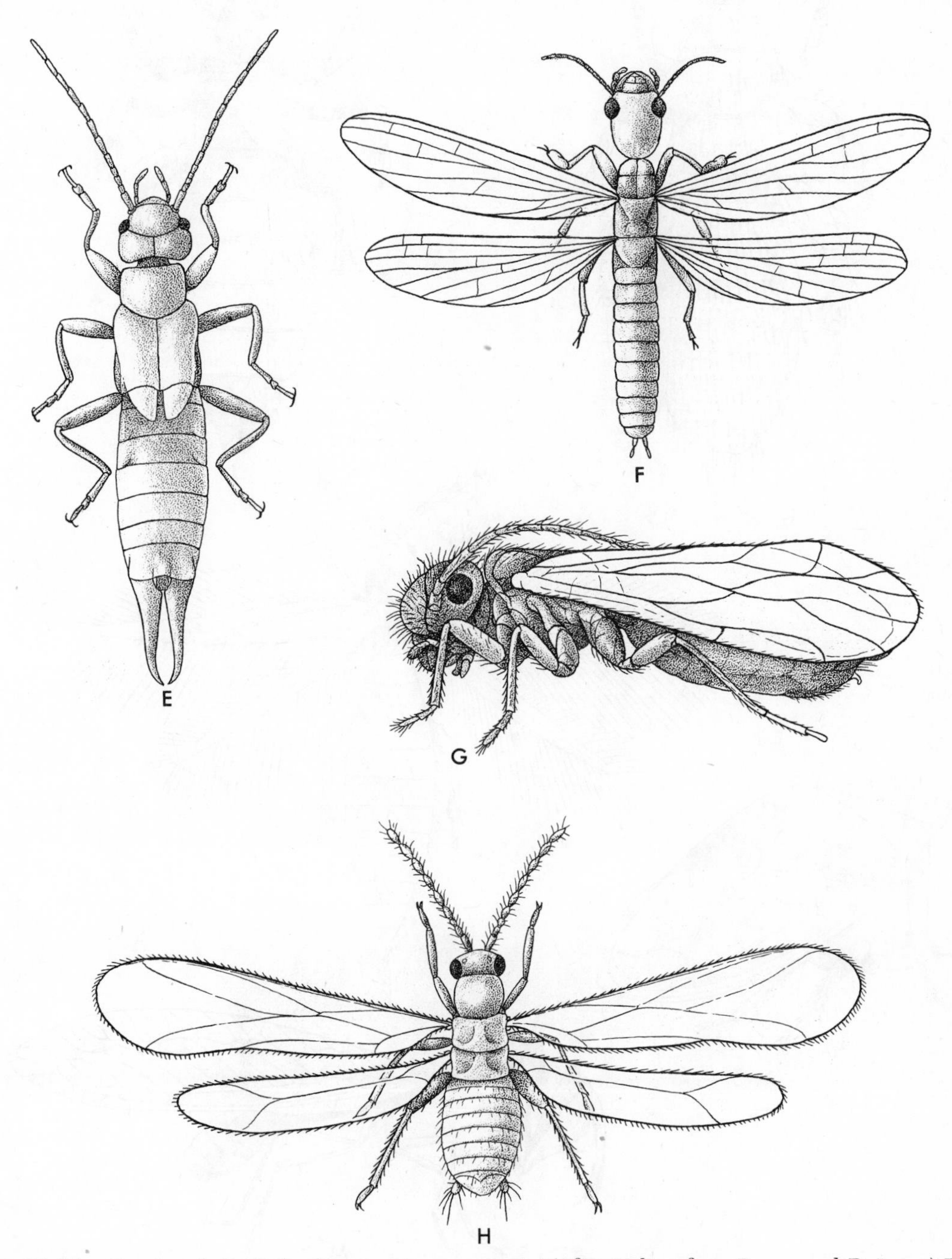

Figure 15–15. *Continued. E.* Order Dermaptera: an earwig. (After Fulton from Borror and DeLong.) *F.* Order Embioptera: a webspinner. (After Enderlein from Comstock.) *G.* Order Psocoptera: a psocid. (After Sommerman from Ross.) *H.* Order Zoraptera: *Zorotypus.* (After Caudell from Comstock.)

Illustration continued on following page.

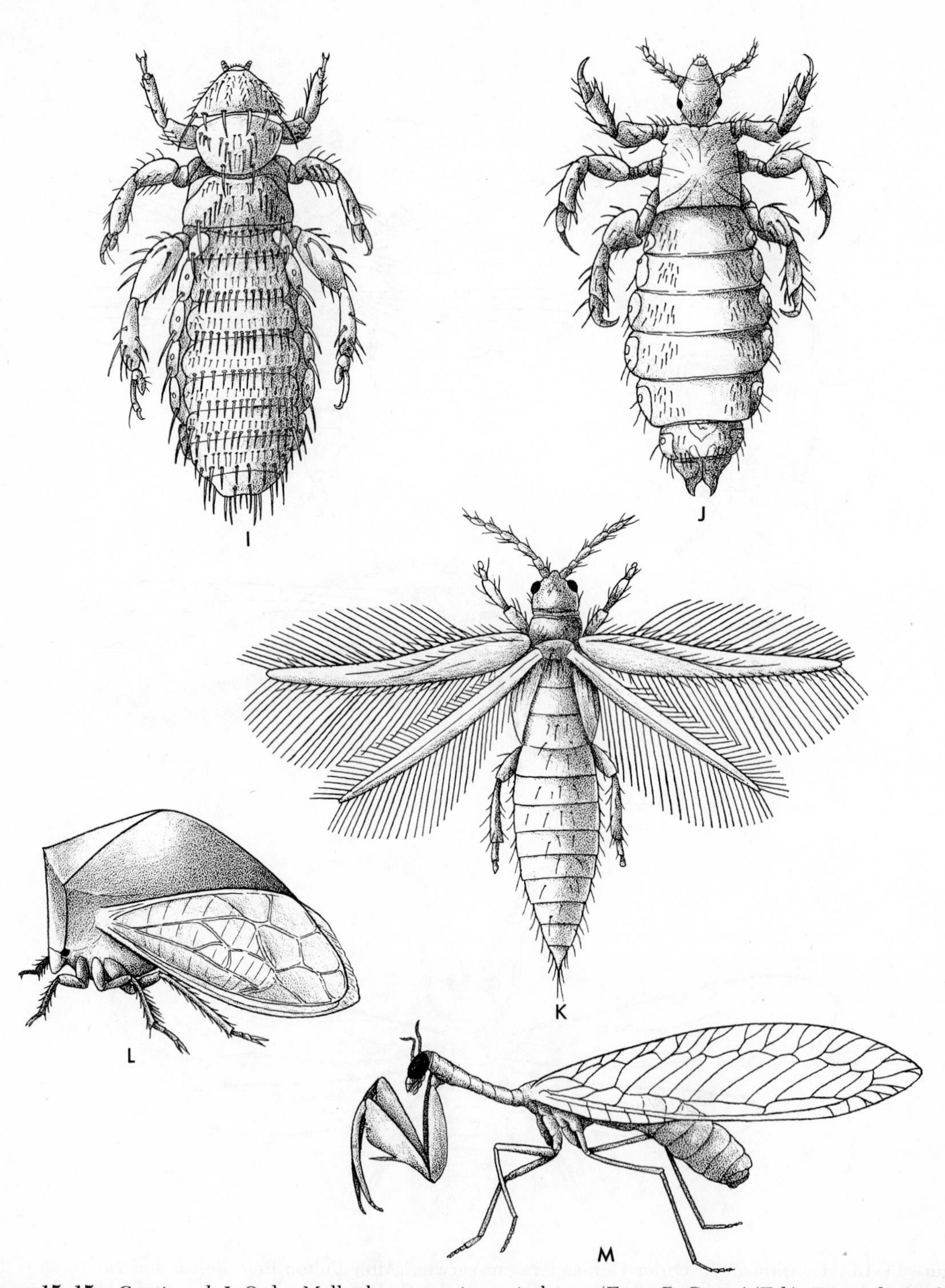

Figure 15–15. *Continued. I.* Order Mallophaga: a guinea pig louse. (From P. Grassé (Ed.): Traité de Zoologie.) *J.* Order Anoplura: a body louse of man. (From P. Grassé (Ed.): Traité de Zoologie.) *K.* Order Thysanoptera: a thrip. (After Moulton from P. Grassé (Ed.): Traité de Zoologie.) *L.* Order Homoptera: buffalo treehopper. (After Irving from H. Curran, 1954: Golden Playbook of Insect Stamps. Simon & Schuster, New York.) *M.* Order Neuroptera: mantispid. (After Banks from Borror & DeLong.) *Illustration continued on opposite page.*

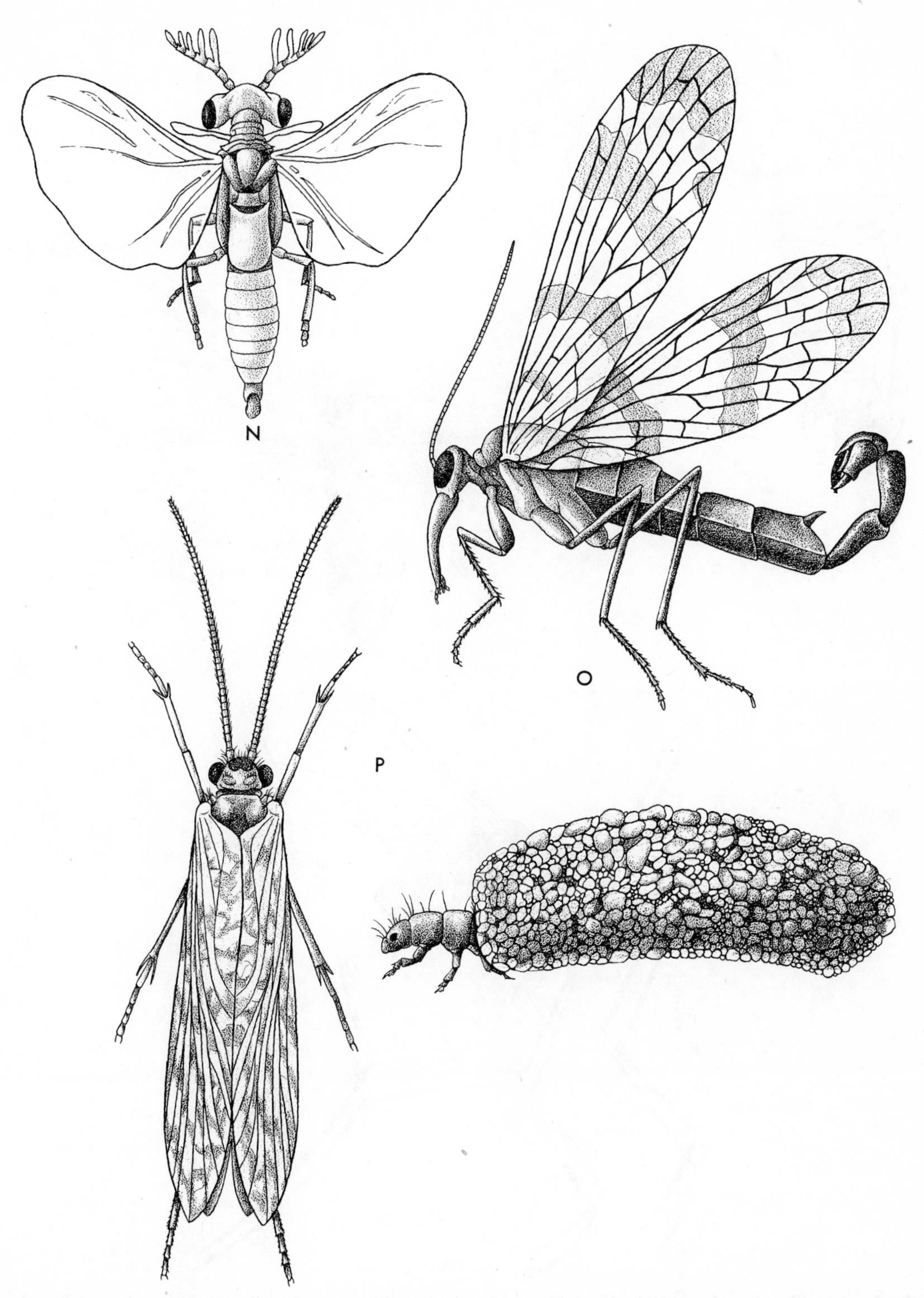

Figure 15–15. *Continued. N.* Order Strepsiptera: twisted wing parasite. (After Bohart; Entomological Society of America, from Borror & DeLong.) *O.* Order Mecoptera: a scorpionfly. (After Taft from Borror & DeLong.) *P.* Order Trichoptera: an adult caddis fly and larva in case. (After Mohr; Illinois Nat. Hist. Survey, from Ross.)

Illustration continued on following page.

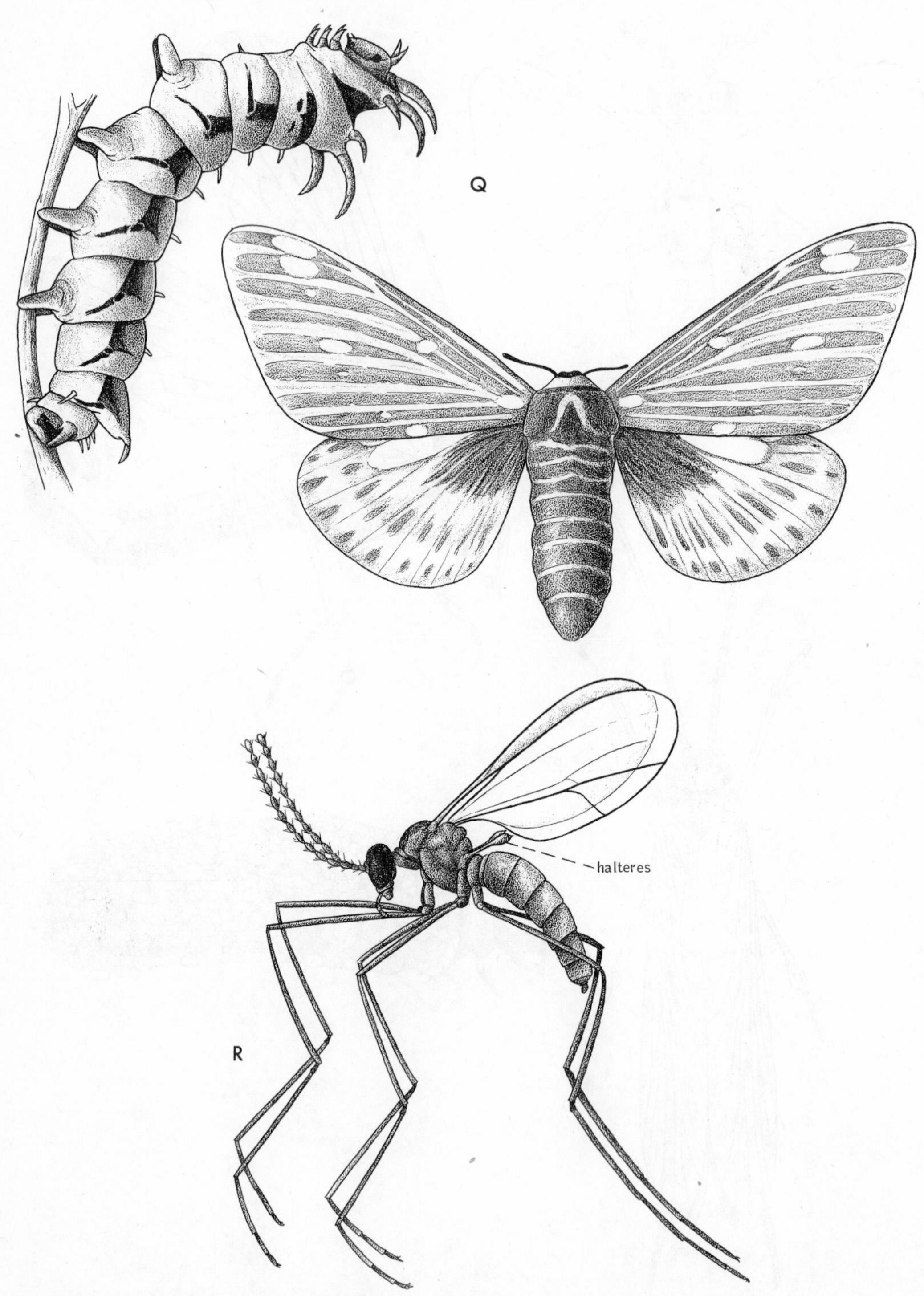

Figure 15–15. *Continued. Q.* Order Lepidoptera: the royal walnut moth and its caterpillar, the hickory-horned devil. (After Lutz.) *R.* Order Diptera: a gall gnat. (After Usda from Borror & DeLong.)

Illustration continued on opposite page.

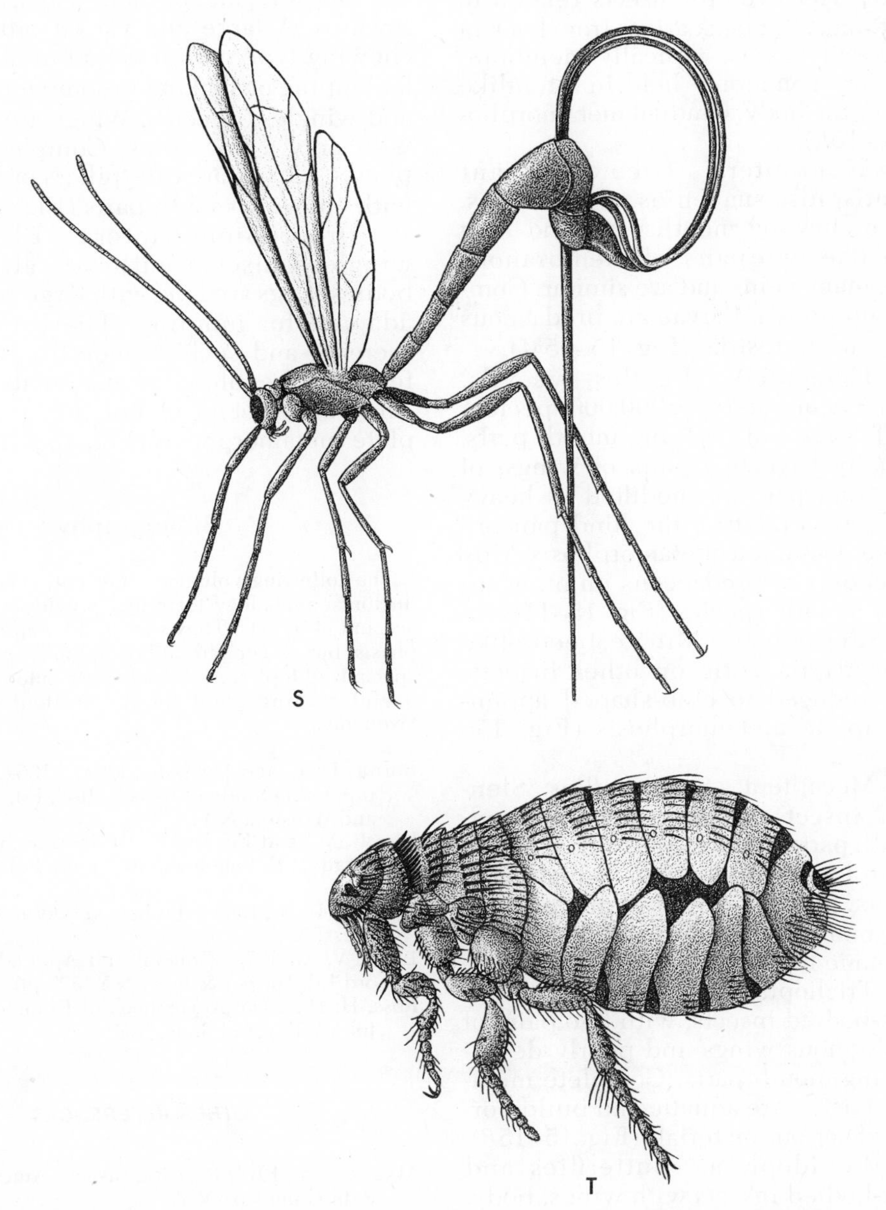

Figure 15–15. *Continued. S.* Order Hymenoptera: a parasitic female ichneumon fly. (After Lutz.) *T.* Order Siphonaptera: a flea. (After Bouché from Borror & DeLong.)

thickened basal and distal membranous section. The membranous portion of the forewings overlaps when at rest over the abdomen. Hindwings entirely membranous. Gradual metamorphosis. Herbivorous and predaceous (Fig. 15–10*G*).

Order Homoptera. Cicadas, leaf hoppers, aphids. Herbivorous insects related to the hemipterans, but beak arises from back of head and forewings are typically membranous. Wings are commonly held in a tentlike position over the body. Gradual metamorphosis (Fig. 15–15*L*).

Order Neuroptera. Lacewings, ant lions, mantispids, snakeflies, dobsonflies. Adults have chewing mouth parts and long antennae. The two pairs of membranous wings have many veins and are similar. Complete metamorphosis. Larvae are predaceous and usually are terrestrial (Fig. 15–15*M*).

Order Coleoptera. Beetles, weevils. Largest order of insects (over 300,000 species), with hard bodies and chewing mouth parts. Adults usually have two pairs of wings, of which the front pair are modified as heavy protective covers (elytra); the hind pair are membranous. Complete metamorphosis. Most are plant feeders or predaceous on other insects. Some aquatic species (Fig. 15–11).

Order Strepsiptera. Minute, beetle-like insects, mostly parasitic on other insects. Forewings reduced to club-shaped appendages. Complete metamorphosis (Fig. 15–15*N*).

Order Mecoptera. Scorpionflies. Slender-bodied insects, often vividly colored. Biting mouth parts prolonged as a beak. Most species with long narrow wings, which have many veins. Complete metamorphosis. Adults are omnivorous, and grublike larvae feed on organic matter (Fig. 15–15*O*).

Order Trichoptera. Caddisflies, water moths. Soft-bodied insects, with two pairs of hairy membranous wings and poorly developed chewing mouth parts. Complete metamorphosis. Larvae are aquatic and build portable cases of various materials (Fig. 15–15*P*).

Order Lepidoptera. Butterflies and moths. Soft-bodied insects with wings, body, and appendages covered with pigmented scales. Mouth parts are modified as a coiled proboscis used for sucking flower nectar. Compound eyes large. Complete metamorphosis. Larvae are caterpillars and usually are plant feeders; adults feed little or not at all (Fig. 15–15*Q*).

Order Diptera. True flies. Large order, all of which have functional front wings and reduced, knoblike hind wings (halteres). Mouth parts are variable, as is body form. Complete metamorphosis. Group includes mosquitoes, horseflies, midges, gnats. Adults often vectors of diseases; larvae frequently damaging to vegetables and domestic animals (Fig. 15–15*R*).

Order Hymenoptera. Ants, bees, wasps, sawflies. A large and varied order, all with chewing type mouth parts but also modified for lapping or sucking in some forms. Winged and wingless species. Wings are transparent with only a few veins. Complete metamorphosis. Larvae are caterpillars or are grublike with chewing mouth parts (Fig. 15–15*S*).

Order Siphonaptera. Fleas. Small, wingless insects with laterally flattened bodies. Legs are long with large coxae and are adapted for jumping. These insects have piercing and sucking mouth parts and they feed on the blood of mammals and birds. They are vectors of bubonic plague. Complete metamorphosis (Fig. 15–15*T*).

Bibliography

The following volumes represent some of the introductory entomology texts that provide a good general account of insects. They vary in their approach and emphasis, but all contain bibliographic references to more specialized topics. An introductory entomology text is a useful starting point for the student interested in taxonomy.

Borror, D. J., and De Long, D. M., 1964: An Introduction to the Study of Insects. Rev. Ed. Holt, Rinehart and Winston, N.Y.

Fox, R. M., and Fox, J. W., 1964: Introduction to Comparative Entomology. Reinhold Publishing Corp., N.Y.

Lanham, U. N., 1964: The Insects. Columbia University Press, N.Y.

Little, V. A., 1972: General and Applied Entomology. 3rd Ed. Harper & Row, N.Y. 527 pp.

Ross, H. H., 1965: A Textbook of Entomology. 3rd Ed. John Wiley and Sons, N.Y.

OTHER REFERENCES

Askew, R. R., 1971: Parasitic Insects. American Elsevier Publishing Co., N.Y.

Barbosa, P., and Peters, T. M., 1972: Readings in Entomology. W. B. Saunders Co., Philadelphia. 450 pp. (A selection of a wide range of articles reflecting the directions of interest in contemporary entomological research.)

Beck, S. D., 1968: Insect Photoperiodism. Academic Press, N.Y. 288 pp.

Brian, M. V., 1965: Social Insect Populations. Academic Press, N.Y.

Carthy, J. D., 1965: The Behavior of Arthropods. Oliver and Boyd, Edinburgh and London.

Chapman, R. F., 1969: The Insects: Structure and Function. American Elsevier Publishing Co., N.Y. 819 pp.
Dadd, R. H., 1970: Digestion in insects. *In* Florkin, M., and Scheer, B. J. (Eds.), Chemical Zoology, Vol. 5, Pt. A. Academic Press, N.Y., pp. 117–145.
Danilevskii, A. S., 1965: Photoperiodism and Seasonal Development of Insects. Oliver and Boyd, Edinburgh and London.
Davey, K. G., 1965: Reproduction in Insects. W. H. Freeman, San Francisco.
Dethier, V. G., 1963: The Physiology of Insect Senses. John Wiley and Sons, N.Y.
Gilmour, D., 1965: The Metabolism of Insects. W. H. Freeman, San Francisco.
Metcalf, C. L., and Flint, W. P., 1962: Destructive and Useful Insects. McGraw-Hill Book Co., N.Y.
Muirhead-Thompson, R. C., 1968: Ecology of Insect Vector Populations. Academic Press, N.Y. 174 pp.
Patton, R. L., 1963: Introductory Insect Physiology. W. B. Saunders Co., Philadelphia.
Riegel, J. A., 1971: Excretion—Arthropods. *In* Florkin, M., and Scheer, B. J. (Eds.), Chemical Zoology, Vol. 6, Pt. B. Academic Press, N.Y., pp. 249–277.
Rockstein, M. (Ed.), 1965–1973: The Physiology of Insecta. Vol. 1 (revised, 1973); Vol. 2 (1965); Vol. 3 (revised, 1973). (An extensive treatment of the physiology and behavior of insects. Well organized by topics contributed by many physiologists and entomologists.)
Smith, D. S., 1968: Insect Cells: Their Structure and Function. Oliver and Boyd, Edinburgh and London. (Electron micrographs topically arranged with accompanying explanatory text.)
Wenner, A. M., 1964: Sound communication in honeybees. Scientific American, *210*(4):116–124.
Wigglesworth, V. B., 1966: Insect Physiology. 6th Ed. John Wiley and Sons, N.Y.
Wilson, E. O., 1971: The Insect Societies. Harvard University Press, Cambridge. 548 pp.

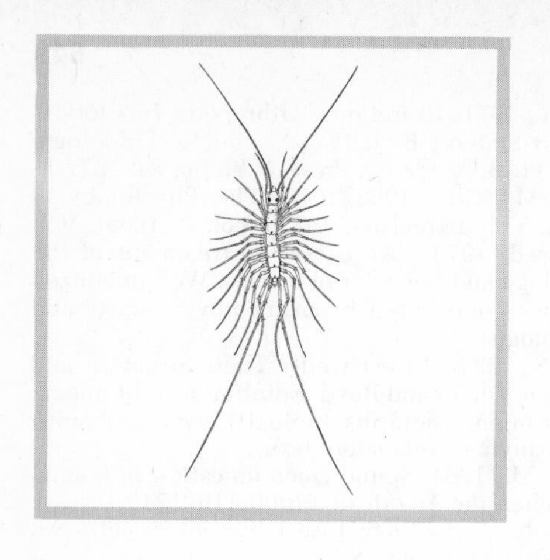

Chapter 16

THE MYRIAPODOUS ARTHROPODS: THE ONYCHOPHORANS; ARTHROPOD PHYLOGENY

Four groups of terrestrial mandibulates comprising some 10,500 species—the centipedes, the millipedes, the pauropods, and the symphylans—have a body composed of a head and an elongated trunk with many leg-bearing segments. This common feature was formerly considered sufficient reason for uniting all four groups within a single class, the Myriapoda. Although these arthropods are perhaps more closely related to each other than to the Insecta and the Crustacea, they exhibit marked differences. Most zoologists have therefore abandoned the Myriapoda except as a convenient collective name, and each of the four groups is now considered as a separate class. Manton (1970) believes that the Myriapoda should be revived as a higher taxon to indicate the affinity between the four classes.

Most myriapods require a relatively humid environment and live beneath stones and wood and in soil and humus. They are widely distributed in both temperate and tropical regions.

The head bears a pair of antennae and sometimes ocelli, but, except in certain centipedes, true compound eyes are never present. The mouth parts lie on the ventral side of the head and are directed forward. An epistome and labrum form the upper lip and the roof of a preoral cavity. The lower lip is formed by either a first or a second pair of maxillae, and enclosed within the oral cavity are the pair of mandibles and a hypopharynx. The mandibles have similar mechanics for movement in all myriapods.

Gas exchange is typically by a tracheal system in which the spiracles cannot be closed. Excretion takes place through malpighian tubules. The heart is a dorsal tube extending through the length of the trunk, with a pair of ostia in each segment. There is usually little if any development of an arterial system. The nervous system is not concentrated, and the ventral nerve cord bears a ganglion in each segment. Indirect sperm transfer by spermatophore is highly developed and the myriapods parallel the arachnids in many aspects of this process.

Class Chilopoda

The members of the class Chilopoda, known as centipedes, are perhaps the most familiar of the myriapodous arthropods. They are distributed throughout the world in both temperate and tropical regions, where they live in soil and humus and beneath stones, bark, and logs. The approximately 3000 described species are distributed within four principal orders. The order Geophilomorpha is composed of long wormlike centipedes that are adapted for living in soil (Fig. 16–1*B*). The orders Scolopendromorpha (Fig. 16–1*A*) and Lithobiomorpha (Fig. 16–1*D*) both contain heavy-bodied centipedes that live beneath stones, bark, and logs. The

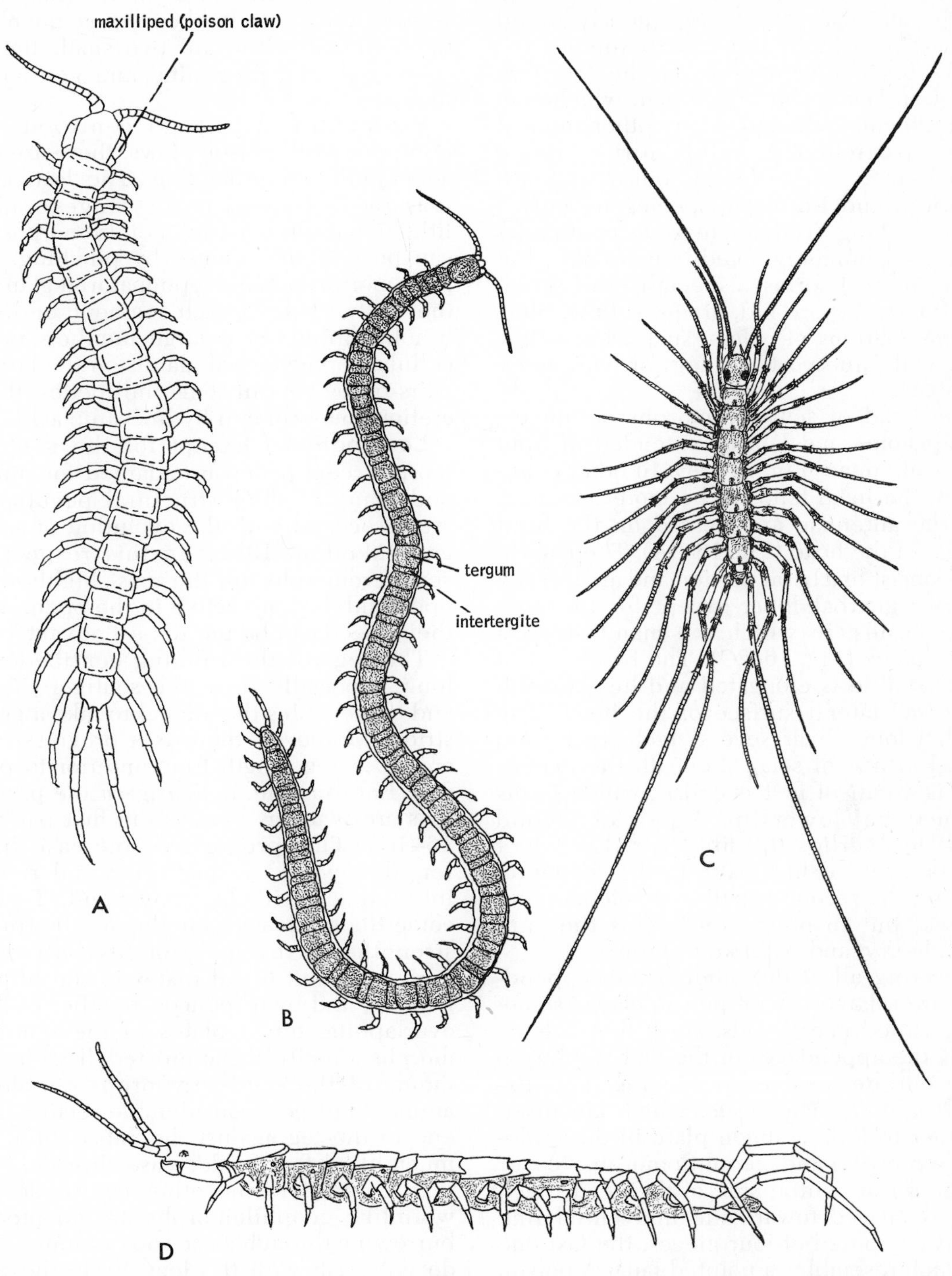

Figure 16–1. Chilopoda. *A*, *Otocryptops sexspinnosa*, a scolopendromorph centipede. *B*. A geophilomorph centipede. *C*. *Scutigera coleoptrata*, the common house centipede, a scutigeromorph. *D*. *Lithobius*, a lithobiomorph centipede. (All after Snodgrass.)

Scutigeromorpha are long-legged forms, some of which live in and around human habitations (Fig. 16–1*C*). *Scutigera coleoptrata*, which is found in both Europe and North America, is not infrequently found trapped in bathtubs and wash basins.

The largest centipede is the tropical American *Scolopendra gigantea*, which may reach 26 cm. in length. Many other tropical forms, particularly scolopendrids, range from 18 to 24 cm. in length, but most North American and European species are only 3 to 6 cm. long. Temperate zone centipedes are most commonly a red-brown color, but many tropical forms, especially the scolopendromorphs, are red, green, yellow, blue, or combinations of colors, such as a yellow body with blue posterior legs or with green crossbars.

The head of scutigeromorphs is convex, the epistome and labrum extended in front of the antennae (Fig. 16–2*A*). In other centipedes, the head tends to be more flattened, and the antennae are located on the front margin of the head (Fig. 16–1*A*). The mandibles consist of a basal portion and an unfused anterior gnathal lobe, except for the geophilomorphs, in which the mandible is a single piece (Fig. 16–2*C*). The basal part of the mandible is elongated and lies beneath the ventrolateral surface of the head. The gnathal lobes bear several large teeth and a thick fringe of setae. Beneath the mandibles is a pair of first maxillae, which forms a functional lower lip. A pair of second maxillae overlies the first pair. Each first maxilla bears a short palp. In the scutigeromorphs, the second maxillae are slender and leglike, but in other centipedes they are short, heavy, and palplike in form.

Covering all of the other mouth appendages are a large pair of poison claws, sometimes called maxillipeds, since they are actually the appendages of the first trunk segment but are involved in feeding (Fig. 16–2*A*, *B* and *D*). The coxae, which are fused together to form a single plate in the scolopendromorphs and geophilomorphs, cover the posterior ventral side of the head. Each claw is curved toward the midventral line and is composed of four pieces, the last one of which resembles a pointed fang. A poison gland is usually lodged within each claw and is drained by a duct that opens near the tip of the last segment.

Posterior to the first trunk segment, which carries the maxillipeds, are 15 or more leg-bearing segments. The tergal plates vary considerably in size and number, and the differences are correlated with locomotor habits. The coxae of the legs are attached laterally to each side of the sternal plates. Between the last leg-bearing segment and the terminal telson are two small limbless segments—the pregenital and genital segments.

Protection. Although centipedes are equipped with poison claws, there are other adaptations for protection. The last pair of legs in centipedes is the longest, and in lithobiomorphs and scolopendromorphs they can be used in defense by pinching. Geophilomorphs possess repugnatorial glands on the ventral side of each segment, and some lithobiomorphs bear large numbers of unicellular repugnatorial glands on the last four pairs of legs. As in some millipedes, the secretions may contain hydrocyanic acid.

Locomotion. Except for the geophilomorphs, centipedes are adapted for running and many of their structural peculiarities are associated with the evolution of a rapid gait (Manton, 1952). In this respect, the scutigeromorphs are the most highly developed and they are active in the open, where their speed can be put to best advantage.

The legs of the running centipedes are long, especially so in the scutigeromorphs, and this enables the animal to take a greater stride. Moreover, there is a progressive increase in leg length from anterior to posterior. For example, in *Scutigera* the posterior legs are twice as long as the first pair (Fig. 16–1*C*). This progressive increase in leg length toward the posterior end reduces interference with leg movement. To overcome the tendency to undulate, the trunk is strengthened by more or less alternately long and short tergal plates in the lithobiomorphs, and by a reduced number of large, overlapping tergal plates in the scutigeromorphs. Finally, the annulated distal leg segments of the scutigeromorphs enable the animal to place a considerable section of the end of the leg against the substratum, very much like a foot, to decrease slippage.

In contrast to the other centipedes, the wormlike geophilomorphs are adapted for burrowing through loose soil or humus. They do not push with the legs as do the millipedes, but rather the pushing force is provided by extension and contraction of the trunk, as in earthworms. A British species of *Stigmogaster*, for example, can increase its body length by as much as 68 per cent. Powerfully developed longitudinal muscles

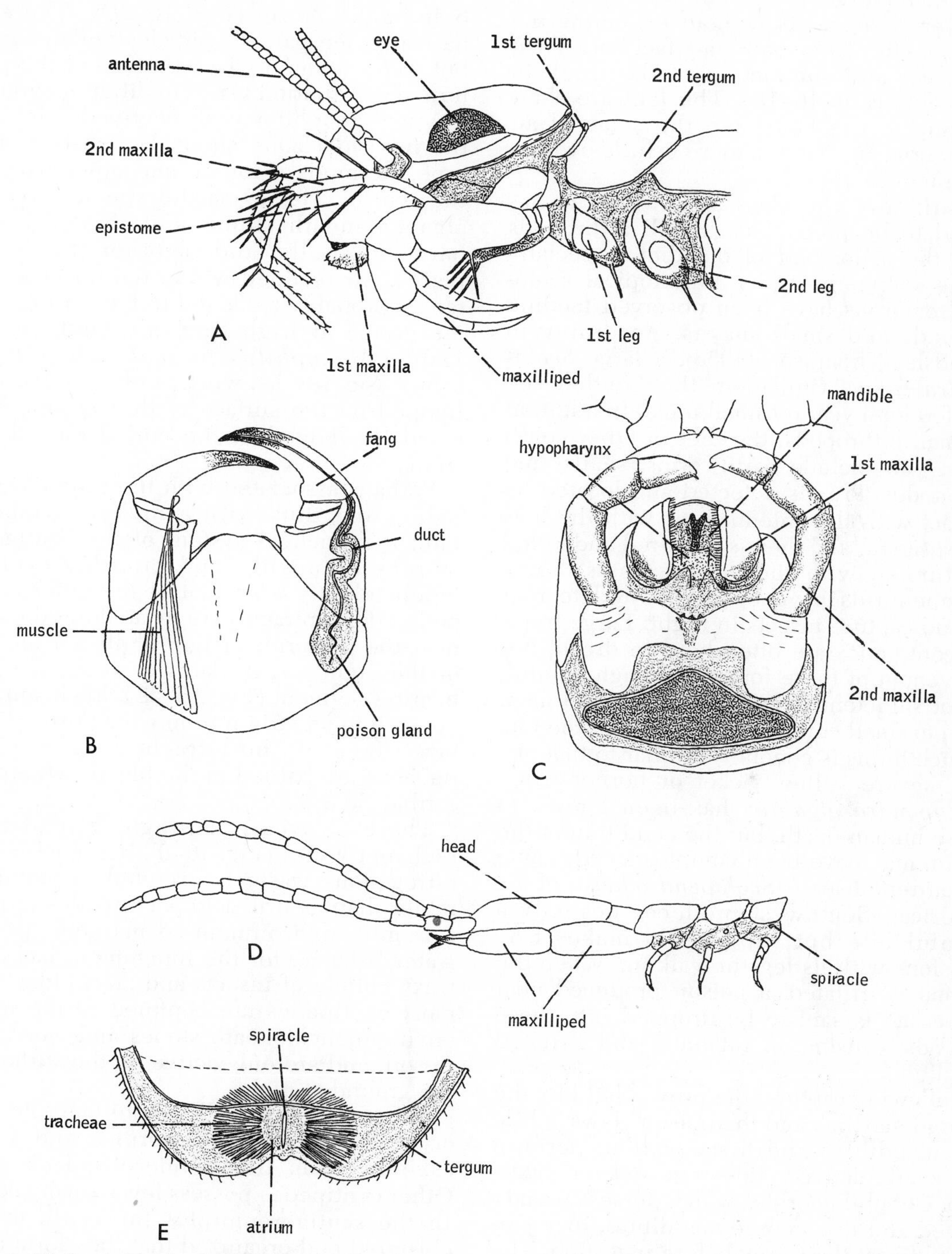

Figure 16–2. *A.* The head of *Scutigera coleoptrata* (lateral view). *B.* Maxillipeds, or poison claws, of *Otocryptops* (ventral view). *C.* The head of *Otocryptops* (ventral view). *D.* The head and anterior trunk segments of a geophilid (lateral view). *E.* Posterior of *Scutigera* tergum showing pair of tracheal lungs. (All after Snodgrass.)

of the body wall, an elastic pleural wall, the presence of a small intersternite and an intertergite between each of the larger main sternal and tergal plates, and the increase in number of segments without extending the total length of the body, all facilitate great extension and contraction of the trunk in burrowing (Fig. 16–1*B*). The legs are short and anchor the body like the setae of an earthworm. In walking, there is little overlap of leg movement.

Nutrition. The class as a whole is believed to be predacious. Small arthropods form the major part of the diet, especially of the scutigeromorphs, but tropical scolopendromorphs have been observed feeding on toads and small snakes. According to Cloudsley-Thompson (1958), a large *Scolopendra* from Trinidad in the London zoo was fed for a year on small mice. In addition to small arthropods, the diet of other centipedes can include earthworms, snails, and nematodes. Prey is detected and located by contact with the antennae, or with the legs in *Scutigera,* and then is captured and killed or stunned with the poison claws. Some scolopendrids have been reported to rear up and capture insects in flight. Large tropical centipedes are often held in dread, but the venom of these forms, although painful, is not sufficiently toxic to be lethal to man, even to small children. The effect of the bite of such forms is generally similar to that of a very severe yellow jacket or hornet sting. *Scolopendra gigantea* has been known to cause human death, but the condition of the victim may have been complicated by some concurrent disease. *Scolopendra heros* of the American Southwest produces not only a painful bite but supposedly makes tiny incisions with its legs in walking. When the animal is irritated, a poison produced near the coxae is said to be dropped into these wounds, causing an inflamed and irritated condition.

Following capture, the prey is held by the second maxillae and the poison claws, while the mandibles and first maxillae perform the manipulative action required for ingestion. Geophilomorphs, which possess weakly armed and less mobile mandibles, may partially digest their prey before ingestion. The digestive tract is a straight tube, with the foregut occupying from one-seventh to two-thirds of the length, depending upon the species (Fig. 16–3). The hindgut is short. Salivary secretions are provided by glands associated with each of the feeding appendages.

Gas Exchange, Circulation, and Excretion. Except in the scutigeromorphs, the spiracles of the tracheal system lie in the membranous pleural region above and just behind the coxae (Fig. 16–2*D*). There is basically one pair of spiracles per segment; but some segments lack them, and the pattern of distribution varies in different groups.

A few geophilomorph centipedes inhabit the intertidal zone along the coasts of Europe, Bermuda, Florida, and other parts of the world. These marine species live in algae, beneath stones and shells, in old barnacle shells, and in other protective places. Air retained within the tracheal system is probably sufficient to last during submergence at high tide, although in the Indian *Myxophilus indicus* and perhaps other species as well, additional air is trapped on the surface of the coxae and as a bubble lodged in the curled end of the trunk.

Perhaps associated with their more active habits and thus with a higher metabolic rate, the tracheal system of the scutigeromorphs is lunglike and probably evolved independently from that of the other centipedes. The spiracles are located middorsally near the posterior margin of all but the last of the eight tergal plates covering the leg-bearing segment (Fig. 16–2*E*). Each spiracle opens into an atrium from which extend two large fans of short tracheal tubes. The tracheae are bathed in the blood of the pericardial cavity.

There is usually a single pair of malpighian tubules (Fig. 16–3), but much of the nitrogenous waste is excreted as ammonia rather than as uric acid. Centipedes require a humid environment to maintain proper water balance, for the integument lacks the waxy cuticle of insects and arachnids. Thus most centipedes are confined to the moist environment beneath stones and wood and in soil, and are only active on the surface of the ground at night.

Sense Organs. All geophilomorphs and cryptopid scolopendromorphs, and a few cave-dwelling lithobiomorphs, lack eyes. Other centipedes possess few to many ocelli. In the scutigeromorphs, the ocelli are so clustered and organized that they form compound eyes. The optical units, of which there are up to two hundred, form a compact group on each side of the head and tend to be elongated with converging optic rods. In *Scutigera,* the combined corneal surface is greatly convex, as are the compound eyes of insects and crustaceans, and each unit is

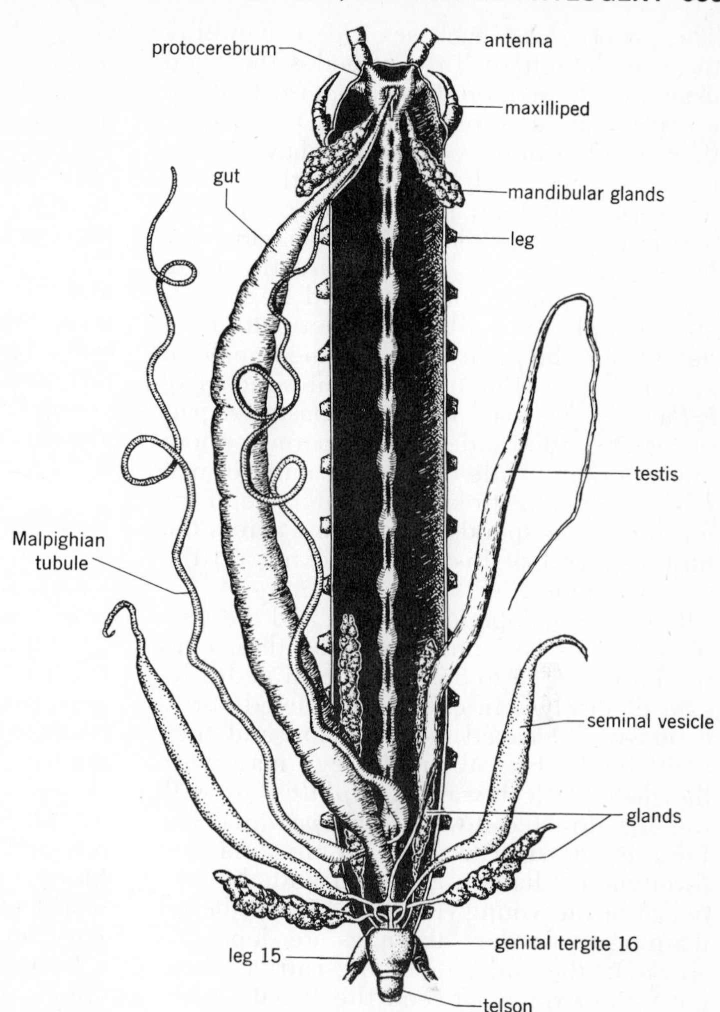

Figure 16–3. Internal structure of the centipede *Lithobius*. (From Kaestner, A., 1968: Invertebrate Zoology, Vol. II. Wiley-Interscience, N. Y.)

remarkably similar to an ommatidium (Fig. 16–2A). However, there is no evidence that the compound eyes of *Scutigera*, nor the eyes of any other centipedes, function in more than the simple detection of light and dark. Many centipedes are negatively phototactic.

A pair of organs of Tömösvary are present on the head at the base of the antennae in all lithobiomorphs and all scutigeromorphs. Each sense organ consists of a disc with a central pore into which the endings of subcuticular sensory cells converge. The few studies on the organs of Tömösvary suggest that they detect vibrations, perhaps auditory ones.

The long last pair of legs of many centipedes have a sensory function, especially in lithobiomorphs and scutigeromorphs; they are modified to form a pair of posteriorly directed, antennae-like appendages.

Reproduction and Development. The ovary is a single tubular organ located above the gut, and the oviduct opens through a median aperture onto the ventral surface of the posterior, legless genital segment. The female aperture is flanked by a tiny pair of appendages, sometimes called gonopods. In the male, one to twenty-four testes are located above the midgut. The testes are connected to a single pair of sperm ducts, which open through a median gonopore on the ventral side of the genital segment (Fig. 16–3). In all centipedes, the male aperture is located on a median penis, which may be variously developed. The genital segment carries small gonopods.

Sperm transmission is indirect in centipedes as in other myriapods. Except in scutigeromorphs, the male constructs a little web of silk strands secreted by glands located at the posterior end of the body. A spermatophore as large as several millimeters is emitted and placed on the webbing. The female picks up the spermatophore and takes it into her reproductive opening.

The gonopods of each sex aid in handling the spermatophore. In centipedes the male usually does not produce a spermatophore until a female is encountered. Moreover, there is often initial courtship behavior. The two sexes may palpate one another's posterior end with their antennae, while moving in a circle (Fig. 16–4*A*). This behavior may last as long as one hour before the male spins a spermatophore web and deposits a spermatophore. Following spermatophore deposition, the male "signals" the female in various ways. For example, in species of *Lithobius*, the male keeps his posterior pair of legs to either side of the spermatophore and webbing, while turning the anterior part of his body and stroking the antennae of the female. She responds by crawling across the posterior end of his body, picking up the spermatophore.

Both the scolopendromorphs and the geophilomorphs lay and then brood their eggs in clusters of 15 to 35. These centipedes locate themselves in cavities hollowed out in a piece of decayed wood or soil and then wind themselves about the egg mass with the dorsal side inward (*Geophilus*) or with the legs directed inward (*Scolopendra*). The female guards the eggs in this manner through the hatching period and the dispersal of the young (Fig. 16–4*B*). In the remaining two orders, the eggs are deposited singly in the soil after being carried about for a short time between the female gonopods.

In the orders Scolopendromorpha and Geophilomorpha, development is epimorphic; i.e., the young display the full complement of segments when they hatch. Development in the other two orders is anamorphic; i.e., the young on hatching have only a part of the adult complement of segments. Young *Scutigera* on hatching have four pairs of legs and in the subsequent six molts pass through stages with 5, 7, 9, 11, and 13 pairs of legs. There are then four epimorphic stages with fifteen pairs of legs before sexual maturity is reached. Development in *Lithobius* is similar, although newly hatched young have seven pairs of legs; several years are required to reach sexual maturity.

SYSTEMATIC RÉSUMÉ OF CLASS CHILOPODA

Subclass Epimorpha. Eggs brooded; young possess all segments on hatching.

Order Geophilomorpha. Slender burrowing centipedes, with 31 to 170 pairs of legs. Eyes absent. Widely distributed in both temperate and tropical regions throughout the world. *Geophilus, Strigamia, Mecistocephalus.*

Order Scolopendromorpha. Twenty-one or 23 pairs of legs. With or without eyes. Many species distributed throughout the world, especially in the tropics. *Scolopendra, Theatops, Otocryptops.*

Subclass Anamorpha. Brooding absent; young do not possess full complement of segments on hatching. Adult has 15 pairs of legs.

A

B

Figure 16–4. *A.* Courting scolopendromorph centipedes. *B.* Female with brood. (By H. Klingel.)

Order Lithobiomorpha. Spiracles paired and lateral. Worldwide in distribution, but most genera and species are found in temperate and subtropical zones. *Lithobius*, *Bothropolys*.

Order Scutigeromorpha. Legs and antennae very long. Eyes large and compound. Spiracles unpaired and located middorsally on tergal plates. Distributed throughout the world, especially in the tropics. *Scutigera*.

Class Symphyla

The Symphyla is a small class containing approximately 120 known species, which live in soil and leaf mold. They have evoked considerable interest among some zoologists as being myriapods which display a number of characteristics like those of insects. Symphylans are found in most parts of the world, although the northern limit in distribution appears to be determined by their inability to tolerate temperatures below −15 degrees F.

Symphylans are between 2 and 10 mm. long and superficially resemble centipedes (Fig. 16–5*A*). The trunk contains 12 leg-bearing segments, which are covered by 15 to 22 tergal plates. The last (thirteenth) segment carries a pair of spinnerets, or cerci, and a pair of long sensory hairs (trichobothria). The trunk terminates in a tiny oval telson.

A well-developed epistome and labrum project in front of the laterally placed antennae (Fig. 16–5*B*). Each mandible bears a toothed and independently movable gnathal lobe and lies beneath the epistome and labrum. The mandibles are covered ventrally by a pair of long first maxillae. The second pair of maxillae are fused, forming a labium (Fig. 16–5*C*). The apparent similarity of symphylan mouth parts to those of insects has often been cited as evidence for the supposed affinity of the two groups. However, Manton (1970) and others claim that the mouth parts are only superficially similar and are functionally quite dissimilar. They reject the idea that insects are most closely related to symphylans.

The trunk structure, especially the presence of the additional tergal plates, is undoubtedly correlated with the locomotor habits of these animals. Most symphylans can run very rapidly and can twist, turn, and loop their bodies when crawling through the crevices within humus. This ability is probably an adaptation to escape predators, for symphylans feed on decayed vegetation. *Scutigerella immaculata* may attack living plants and can be a serious pest of vegetable and flower crops, especially in greenhouses. The tracheal system is limited to the anterior part of the body. There is a single pair of spiracles that open onto the sides of the head, and the tracheae supply only the first three trunk segments. Attached to the body wall beneath the base of each leg are an eversible coxal sac and a small appendage (the stylus), structures also present in primitive insects. The coxal sacs take up moisture. The function of the stylus is unknown, although it is probably of a sensory nature. There are no eyes, but two organs of Tömösvary are well developed (Fig. 16–5*B*).

Like those of the diplopods and pauropods, the genital openings are located on the ventral side of the third trunk segment. The copulatory behavior of *Scutigerella* is known and is most unusual. The male deposits a spermatophore at the end of a stalk. The female, on encountering the spermatophore, eats it, but instead of swallowing the sperm, stores them in special buccal pouches. She then removes the eggs with her mouth from the single gonopore and attaches them to the substratum (Fig. 16–5*D* and *E*). In handling the eggs with her mouth, the female smears each egg with a little semen and fertilizes it.

The eggs are laid in clusters of about 35 and are attached to the walls of crevices or to moss or lichen. Parthenogenesis is common. The role of the spinning organs in reproduction is unknown. Development is anamorphic; the young on hatching have six or seven pairs of legs. *Scutigerella immaculata* lives for as long as four years and molts throughout its lifetime.

Class Diplopoda

The Diplopoda are commonly known as millipedes or thousand-leggers. They are secretive and largely shun light, living beneath leaves, stones, bark, and logs and in soil. Some inhabit the old burrows of other animals, such as earthworms; a few are commensal in ant nests. Quite a number of millipedes are cave inhabitants. The more than 7500 described species comprise the greatest number of myriapodous arthropods. They live throughout the world, especially in the tropics; but the best known faunas are those of North America and Europe.

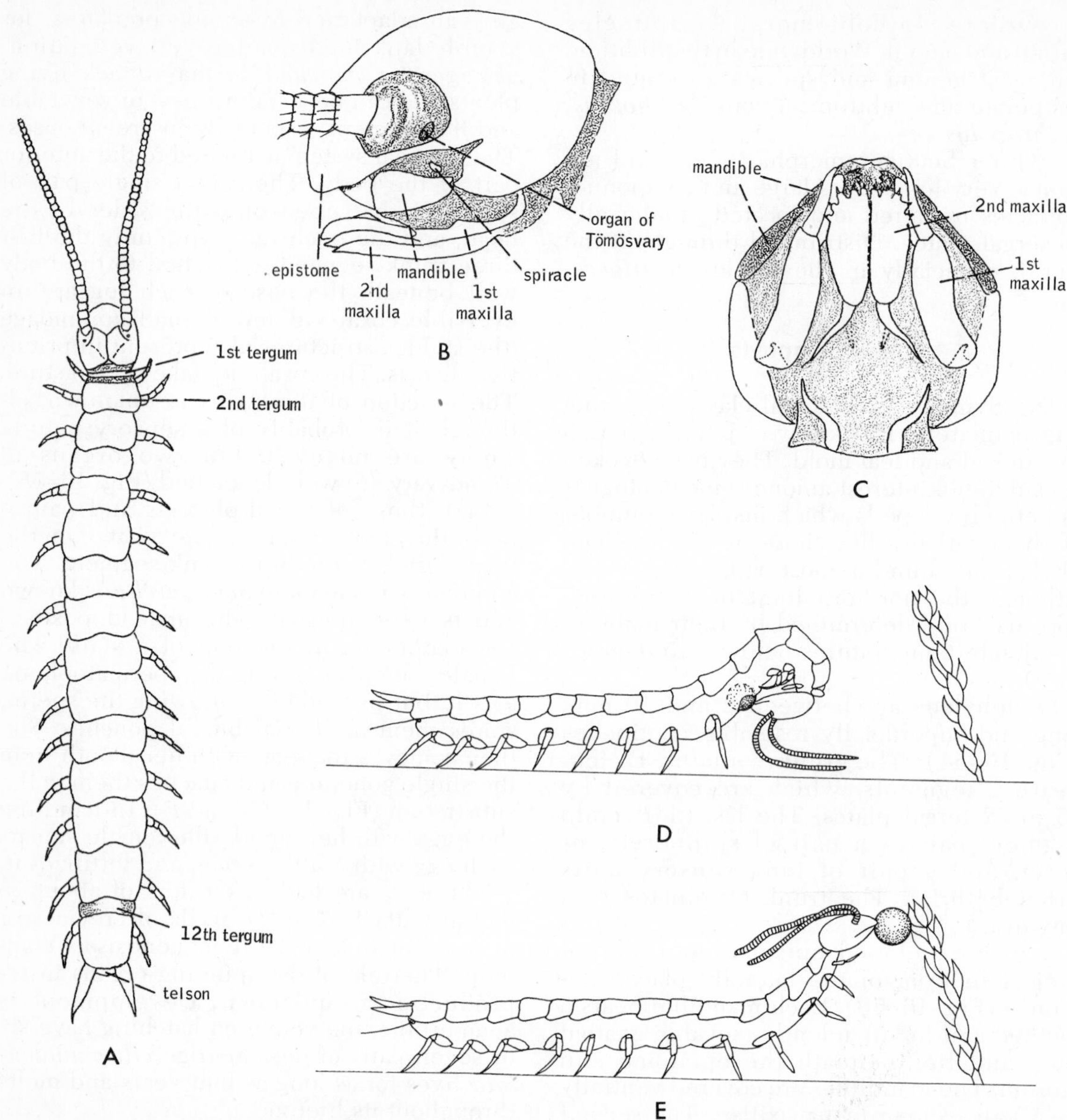

Figure 16–5. Symphyla. *A. Scutigerella immaculata* (dorsal view). *B.* The head of *Hanseniella* (lateral view). *C.* The head of *Scutigerella immaculata* (ventral view). (All after Snodgrass.) *D* and *E.* Female of *Scutigerella* removing egg from gonopore with her mouth parts and attaching it to moss. The egg when carried by the mouth parts is smeared with semen stored in buccal pouches. (After Juberthie-Jupeau.)

A distinguishing feature of the class is the presence of doubled trunk segments, or diplosegments, derived from the fusion of two originally separate somites. Each diplosegment bears two pairs of legs, from which the name of the class is derived (Fig. 16–6 and 16–7*B*). The diplosegmented condition is also evident internally, for there are two pairs of ventral ganglia and two pairs of heart ostia within each segment.

The diplopod head tends to be convex dorsally and flattened ventrally, with the epistome and labrum extending anteriorly in front of the antennae (Fig. 16–8). The sides of the head are covered by the convex bases of the very large mandibles. Distally the mandible bears a movable gnathal lobe, which has teeth and a rasping surface. The floor of the preoral chamber is formed by the maxilla, often called the gnathochilarium (Figs. 16–7A and 16–8). It is a broad, flattened plate attached to the posterior ventral surface of the head and bearing distally a few small lobes. The head of diplopods

Figure 16–6. Diplopoda. *A*. A juliform millipede. (By Betty M. Barnes.) *B*. Pselaphognath millipedes of the genus *Polyxenus*. These tiny millipedes are only 4 mm. in length. The legs are obscured by the large scalelike spines. (By K. H. Schömann.) *C*. A flat-backed millipede. *D*. The pill millipede *Glomeris*. *E*. *Glomeris* rolled up. (*D* and *E* by F. Schaller.)

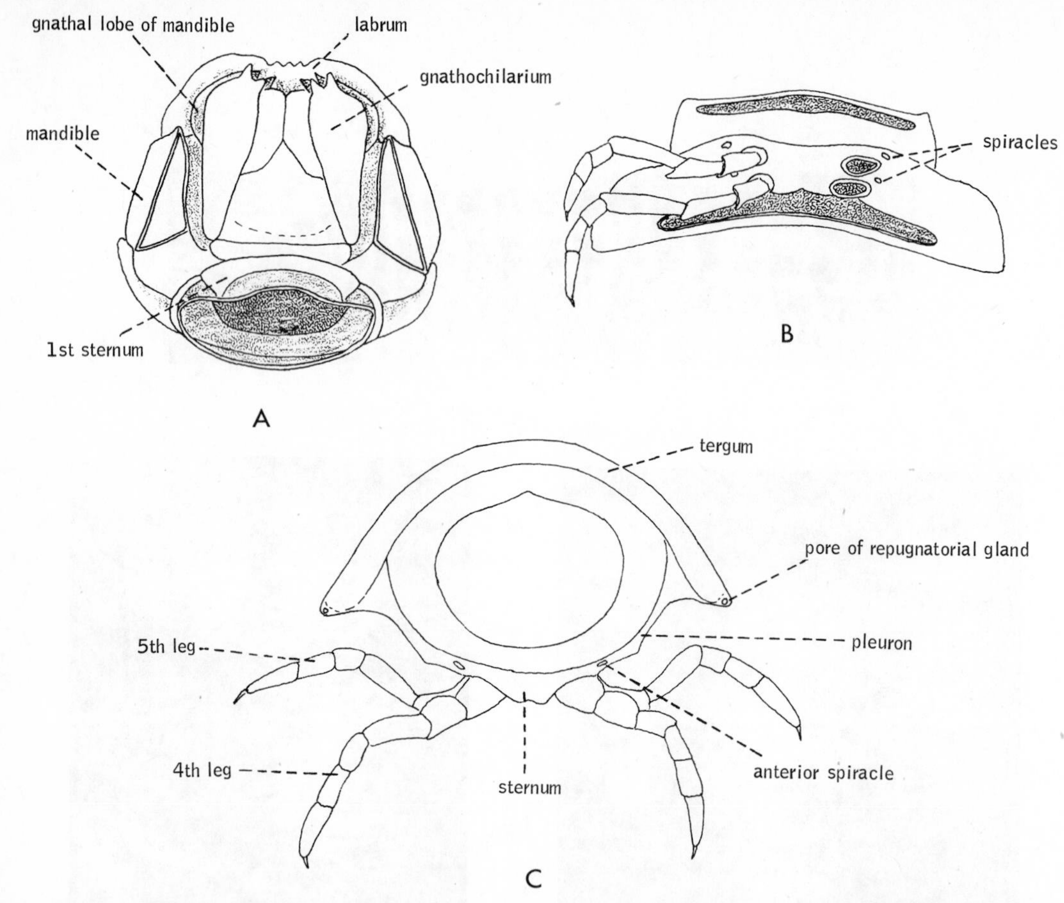

Figure 16–7. *A.* Head of *Habrostrepus*, a juliform millipede (ventral view). *B.* A diplosegment of *Apheloria*, a flat-backed millipede (ventral view). A diplosegment of *Apheloria* (transverse section). (All after Snodgrass.)

does not contain a second maxillary segment. The posterior floor of the preoral chamber bears a median and two lateral lobes, which are variously developed in different diplopod groups and represent the hypopharynx.

The trunk may be dorsoventrally flattened, as in the order Polydesmoidea, the flat-backed millipedes (Fig. 16–6*C*), and some colobognaths; or it may be essentially cylindrical, as in the familiar millipedes of the order Juliformia (Fig. 16–6*A*). A typical segment (diplosegment) is covered by a convex dorsal tergum that, in the flat-backed forms and some others, extends laterally as a shelf, called a carina or paranotum (Fig. 16–7*C*). Ventrolaterally, there are two pleural plates, and ventrally, two sternal plates. A median sternal plate is also commonly present. The sternal plates bear the legs. Primitively, the plates composing a segment may be separate and distinct, but coalescence of varying degrees has usually taken place. In flat-backed and juliform millipedes, all of the plates are fused together, and in the latter group they form a nearly cylindrical ring.

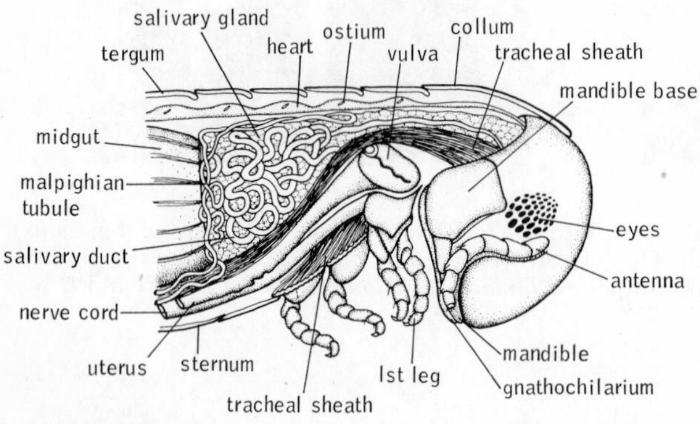

Figure 16–8. Head and anterior trunk segments of the juliform millipede, *Narceus* (lateral view). (After Buck and Keister.)

The extreme anterior segments differ considerably from the others and are probably not diplosegments. The first (the collum) is legless and forms a large collar behind the head (Fig. 16–8). The second, third, and fourth segments carry only a single pair of legs. In some millipedes (polydesmoids), the last one to five segments are also legless.

The body terminates in the telson, on which the anus opens ventrally.

The integument is hard, particularly the tergites, and, like the crustacean integument, is impregnated with calcium salts. The surface is often smooth, but in some groups the terga bear ridges, tubercles, spines, or isolated bristles.

The external anatomy just described applies to the subclass Chilognatha, which contains the vast majority of millipedes. The members of the small subclass Pselephognatha are quite different. The integument is soft and covered with tufts and rows of hollow scalelike spines (Fig. 16–6*B*). The gnathochilarium differs from that of other diplopods. The trunk is composed of 11 to 13 segments, of which the first four bear a single pair of legs each; the last two are limbless.

Diplopods vary greatly in size. The pselaphognaths are minute, some species of *Polyxenus* being only 2 mm. long. There are also some minute chilognaths, such as the European *Adenomeris* and *Chamaesoma*, which are less than 4 mm. in length; but most members of this subclass are several centimeters or more in length. The largest millipedes are tropical species of the family Spirostreptidae, which may be 28 cm. long. The number of segments is also extremely variable, ranging from 11 in the pselaphognaths to more than 100 in the juliform groups. Moreover, in the juliform groups, the number varies within certain limits even within a particular species.

Most diplopods are black and different shades of brown; some species are red or orange, and mottled or spotted patterns are not uncommon. Some southern California millipedes are luminescent.

Locomotion. In general, diplopods are not very agile animals, most species crawling slowly about over the ground. According to the extensive studies of Manton (1954) on arthropod locomotion, the gaits of diplopods, although slow, are adapted for exerting a powerful pushing force, enabling the animal to push its way through humus, leaves, and loose soil. The force is exerted entirely by the legs, and it is with the evolution of such a gait that the diplosegmented structure is probably associated. The backward pushing stroke is activated in waves along the length of the body and is of longer duration than the forestroke. The number of legs involved in a single wave is proportional to the amount of force required for pushing. Thus, while the animal is running, as few as 12 legs or less may compose a wave, but when pushing, a single wave may involve as many as 52 legs in some juliform millipedes.

The head-on pushing habit has been most highly developed in the juliform species, which burrow into relatively compact leaf mold and soil. This habit is reflected in the smooth, fused, rigid cylindrical segments, the rounded head, and the placement of the legs close to the midline of the body. The flat-backed millipedes, which are the most powerful millipedes, open up cracks and crevices by pushing with the whole dorsal surface of their bodies. The lateral keels in these millipedes provide a protected working space for the more laterally placed legs. Ability to climb is particularly striking in some colobognaths and lysiopetalids, which inhabit rocky situations. These millipedes can climb up smooth surfaces by gripping with opposite legs. These rock dwellers also include the swiftest of the millipedes. For example, the lysiopetalid *Callipus longobardius* can move at a maximum speed of 8.0 cm./sec., compared with 0.4 cm./sec. for the juliform *Diploiulus ceruleocinctus*. Speed is correlated with their predatory and scavenging feeding habits and the need to cover greater distances in finding food.

Protective Devices. To compensate for the lack of speed in fleeing from predators, a number of different protective mechanisms have evolved in millipedes. The long, many-segmented millipedes, such as the colobognaths and the juliform groups, coil the trunk into protective spirals when at rest or disturbed. Members of the order Glomerida (Oniscomorpha), called pill millipedes, as well as some others, can roll up into a ball (Fig. 16–6*D* and *E*). The body of glomerids is greatly convex dorsally and flattened ventrally, and contains only 15 to 17 trunk segments (actual rings). The last tergite is expanded laterally and covers the head when the animal is rolled up. Some tropical species, when rolled up, are larger than a golf ball.

Repugnatorial glands are present in many millipedes, including the polydesmid and juliform groups. There is one pair of glands per segment, although they are entirely absent from some segments. The openings are located on the sides of the tergal plates, or (in the flat-backed millipedes) on the margins of the tergal lobes (Fig. 16–7*C*). Each gland consists of a large secretory sac, which empties into a duct and out through an external pore. The principal component of the secretion varies in different species. Aldehydes, quinones, phenols, and hydrogen cyanide have been identified. The hydrogen cyanide is liberated when a precursor and an enzyme are mixed from a double-chambered gland. The secretion is toxic or repellent to other small animals; the secretion of some large tropical species is reportedly caustic to the human skin. The fluid is usually exuded slowly, but lysiopetalids and the juliform spirobolid millipedes can discharge it as a spray or jet for considerable distances. Cloudsley-Thompson (1958) states that a Haitian species, *Rhinocricus lethifer*, can emit a spray of fluid for a distance as great as 70 cm. or more to either side of the body. Ejection probably is caused by the contraction of the trunk muscles adjacent to the secretory sac.

Nutrition. Millipedes are primarily herbivorous, feeding mostly on decomposing vegetation. Food is usually moistened by secretions and chewed or scraped by the mandibles. However, the families of the order Colobognatha exhibit a progressive development of suctorial mouth parts with a corresponding degeneration of the mandibles. These changes culminate in the tropical Siphonophoridae, in which the labrum and gnathochilarium are modified to form a long piercing beak for feeding on plant juices. The beak contains the very much reduced mandibles. In the Polyzoniidae, which is represented in temperate regions (*Polyzonium*), the mouth parts are semisuctorial.

A carnivorous or omnivorous diet has been adopted by the rock-inhabiting lysiopetalids and some other millipedes. It has been reported that prey includes phalangids, insects, centipedes, and earthworms.

The digestive tract is typically a straight tube with a long midgut. Salivary glands open into the preoral cavity. The anterior pair are located in the head; the posterior pair surround the foregut and open into the inner surface of the gnathochilarium (Fig. 16–8). A constriction demarcates the midgut from the long hindgut (rectum).

Gas Exchange, Circulation, and Excretion. There are four spiracles per diplosegment, located just anterior and lateral to each of the four coxae (Fig. 16–7*B* and *C*). Each spiracle opens into an internal tracheal pouch from which arise numerous tracheae.

The heart ends blindly at the posterior end of the trunk, but anteriorly a short aorta continues into the head (Fig. 16–8). There are two pairs of lateral ostia for each segment except for the anterior segments, in which there is only a single pair. Two malpighian tubules arise from each side of the midgut-hindgut junction and are often long and looped.

Like centipedes, millipedes lack a waxy epicuticle and most species cannot tolerate desiccating conditions. However, some colobognaths and lysiopetalids are found in arid habitats. These millipedes possess coxal sacs, which supposedly function to take up water, such as dew drops.

Sense Organs. Eyes may be totally lacking, as in the flat-backed millipedes, or there may be two to eighty ocelli. These are arranged about the antennae in one or several transverse rows or in two lateral clusters (Fig. 16–8). Most millipedes are negatively phototactic, and even those without eyes have integumental photoreceptors. The antennae contain tactile hairs and peg- and conelike projections richly supplied with what are probably chemoreceptors. The animal continually taps the substratum with the antennae as it moves along.

As in centipedes, organs of Tömösvary are present in many millipedes; they are absent in colobognaths and in the juliform groups.

Reproduction and Development. A pair of long, fused, tubular ovaries lie between the midgut and the ventral nerve cord. Two oviducts extend anteriorly to the third, or genital, segment, where each opens into an atrium, or vulva (Fig. 16–8). The vulvae are protractable pouches that open onto the ventral surface near the coxae (Fig. 16–9*C* and *D*). When retracted, a vulva is covered externally by a sclerotized hoodlike piece, and internally a small operculum covers the opening into the uterus. At the bottom of the vulva, a groove leads into a seminal receptacle.

The testes occupy positions corresponding to those of the ovary but are paired tubes with transverse connections. Anteriorly, near the region of the genital segment (the third), each testis passes into a sperm duct, which either opens through a pair of penes on or near the coxae of the second pair of legs, or opens through a single median penis, or into a median groovelike depression behind the coxal bases.

Sperm transfer in millipedes is indirect. The actual copulatory organs are usually modified trunk appendages (gonopods), and these are critical structures in identification of species. In most millipedes one or both

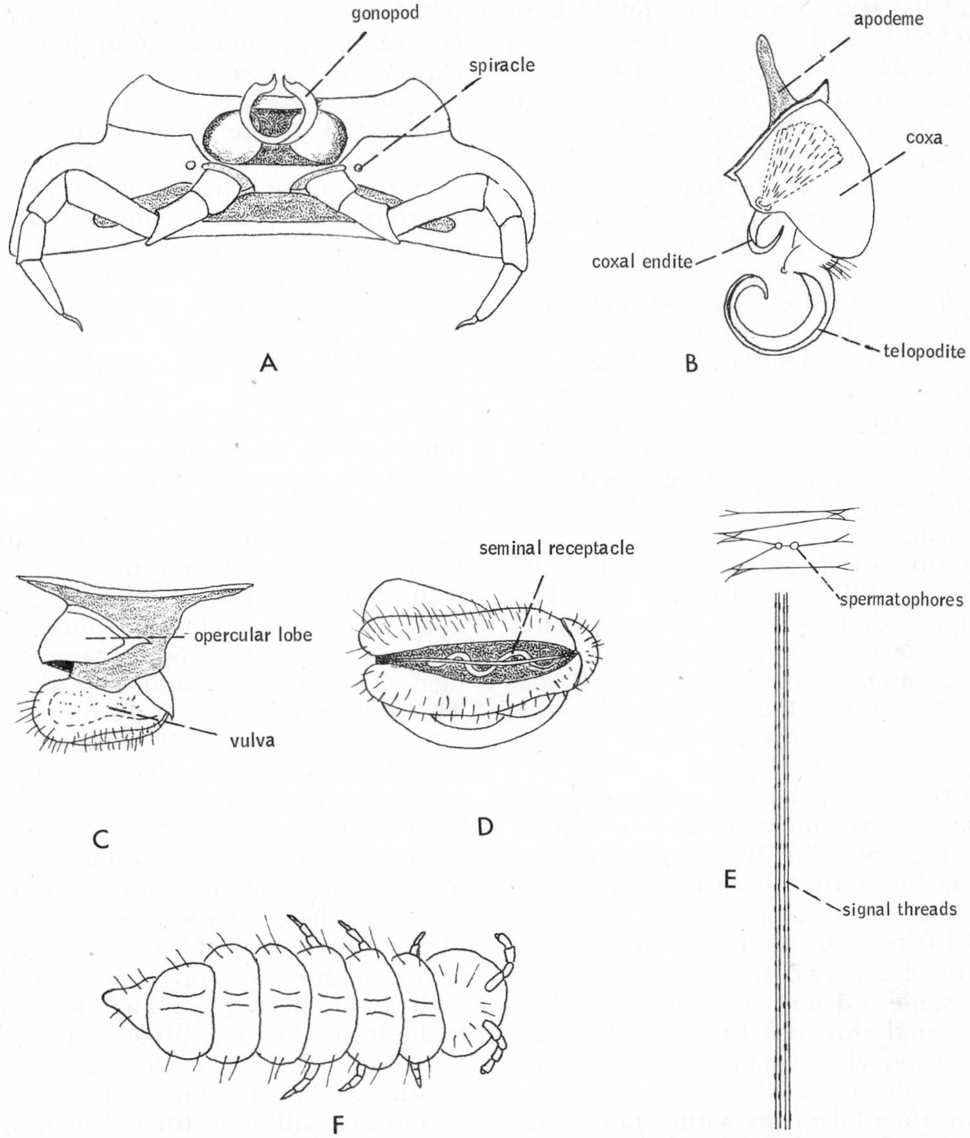

Figure 16–9. *A.* Seventh diplosegment of a male *Apheloria,* showing gonopods and legs (ventral view). *B.* Left gonopod of *Apheloria. C.* Right vulva of third segment of *Apheloria* (lateral view). *D.* Vulva (ventral view). (*A* to *D* after Snodgrass.) *E.* Signal threads leading to spermatophore web of *Polyxenus.* (After Schömann.) *F.* A newly hatched millipede. (After Cloudsley-Thompson.)

pairs of legs of the seventh segment serve as gonopods. In the flat-backed millipedes, *Apheloria,* for example, where the first pair of legs on the seventh segment are gonopods, each gonopod consists of a coxal segment to which are attached a small hooklike process and a larger sickle-shaped piece (the telopodite) (Fig. 16–9*B*). The telopodite contains a depression on its base from which a canal extends out to the tip. The two gonopods lie close together in a common ventral depression, and the sickle-shaped telopodites are so oriented that together they form a median ringlike opening (Fig. 16–9*A*). When the male charges the gonopods with sperm, he bends the anterior part of the body ventrally and posteriorly, inserting the two coxal penes on the third segment through the ring formed by the telopodites. In this position sperm are deposited into the basal depression (reservoir) of the telopodites. In the juliform millipede, *Spirobolus,* both pairs of legs of the seventh segment are gonopods, but only the second pair contains sperm reservoirs and canals. The telopodite resembles a short arm, and the hooklike process is absent. In the Colobognatha, the gonopods are the second legs of the seventh segment and the first legs of the eighth. Both pairs are leglike and little modified.

During copulation, the body of the male is twisted about or stretched out against that of the female so that the gonopods are opposite the vulvae, and the body of the female is held by the legs of the male. The gonopods are protracted, and sperm are transferred through the tip of the telopodite into the vulva. In the European flat-backed millipede, *Polydesmus angustus*, the male runs up the back of the female; he then crawls around to her ventral surface and seizes her gnathochilarium with his mandibles. A fertilized female will run away as soon as the male touches her anal segment, and unfertilized females can be distinguished from fertilized ones by stroking the anal segment with a brush.

Copulation in the Oniscomorpha (pill millipedes) is still not well understood. The legs of the seventh segment are not adapted for sperm transfer. The last one or two pairs of legs, commonly called gonopods, are modified for clasping the female, and sperm are transferred with the aid of the mandibles.

Sperm transfer in the tiny pselaphognath *Polyxenus* is very similar to that in certain pseudoscorpions. The two gonopores are located on the ends of two papillae at the bases of the second pair of legs. The male spins a little web of zigzag threads on which are placed two spermatophores. Extending some 1.5 cm. from the web is a pair of signal threads (Fig. 16–9*E*). The signal threads, when contacted by a female, serve to guide her to the spermatophores, which are then taken up within her reproductive openings. If another male comes across the signal threads, he too will be led to the spermatophores, which he eats. This male then deposits another set of spermatophores. Such behavior by males has been interpreted as an adaptation to insure the presence of fresh spermatophores for females.

The diplopod eggs are fertilized at the moment of laying, and anywhere from ten to three hundred eggs are produced at one time, depending on the species. Some deposit their eggs in clusters in soil or humus; others, such as *Spirobolus*, regurgitate a material that is molded into a cup with the head and anterior legs. A single egg is laid in the cup, which is then sealed and polished. The capsule is deposited in humus and crevices, and it is eaten by the young millipede on hatching. The European pill millipede, *Glomeris*, has similar habits but forms the capsule with excrement.

Many millipedes construct a nest for the deposition of the eggs. Some flat-backed species and colobognaths construct the nest from excrement. The rectum of the female is everted and rapidly drying excrement is deposited as she moves in a circular path. A thin-walled domed chamber is gradually built up and topped by a chimney. The vulvae are applied against the chimney opening, and the eggs fall into the chamber as they are laid. The opening is then sealed, and the chamber is covered with grass and other debris. The female, and in some species the male, may remain coiled about the nest for several weeks.

Some flat-backed and juliform millipedes construct the nest in soil, reinforcing the walls from the inside with excrement. The flat-backed *Strongylosoma pallipes* builds several such nests, each containing 40 to 50 eggs. The nest is closed from the outside. The nests of members of the order Chordeumida are made with silk secreted by glands mounted on two papillae on the terminal segment. *Polyxenus* covers its egg clusters with shed tail setae.

Development is anamorphic. The eggs of most species hatch in several weeks, and the newly hatched young usually have only the first three pairs of legs and not more than seven segments (Fig. 16–9*F*). With each molt, additional segments and legs are added to the trunk. Many millipedes, including the Polydesmoidea, the Juliformia, and the silk-spinning Chordeumida, undergo ecdysis within specially constructed molting chambers similar to the egg nests. The shed exoskeleton is generally eaten, perhaps to aid in calcium replacement. Most millipedes live for several years.

Parthenogenesis is common in the pselaphognaths, and males are rare.

SYSTEMATIC RÉSUMÉ OF CLASS DIPLOPODA

Subclass Pselaphognatha. Minute millipedes with a soft integument bearing tufts and rows of serrated setae. Trunk bears 13 to 17 pairs of legs. No gonopods. No repugnatorial glands. Widespread. *Polyxenus*, *Lophoproctus*.

Subclass Chilognatha. Integument hard. Setae few and not serrated. Repugnatorial glands often present. Certain legs of male are modified as gonopods.

Superorder Pentazonia (Oniscomorpha, Opisthandria). Last one or two pairs of legs of male are modified as clasping gonopods.

Order Oniscomorpha. Pill millipedes. Trunk is covered with 11 to 14 arched tergites, of which second and last are large, enabling body to roll into a tight ball that conceals head and legs. No repugnatorial glands. Largely tropical. Rare in North America; *Glomeris* is common in Europe.

Order Glomeridesmida (Limacomorpha). Small, eyeless, tropical millipedes. Trunk is composed of 22 arched segments. Cannot roll into a ball. No repugnatorial glands. *Glomeridesmus.*

Superorder Colobognatha. Trunk composed of 30 to 192 depressed rings; often with wide lateral tergal keels. Sternites not fused with pleurotergal arch. Head and mouth parts small; often formed into a snout. Repugnatorial glands present. *Polyzonium, Brachycybe, Siphonophora.*

Superorder Helminthomorpha (Proterandria). One or two pairs of legs of seventh segment are modified as gonopods.

Order Chordeumida (Nematophora). Usually with eyes. Trunk composed of 28 to 60 rings, which are either cylindrical or with lateral tergal keels. Sternites not fused with pleurotergal arch. One or three pairs of spinnerets on last tergal plate. Repugnatorial glands sometimes present. Widely distributed. *Cleidogono, Chordeuma,* the lysiopetalids—*Abacion, Lysiopetalum, Callipus.*

Order Polydesmida. Flat-backed millipedes. No eyes. Trunk usually composed of 20 rings with prominent, lateral, tergal keels. Sternites fused with pleurotergal arch. Many segments have repugnatorial glands. Widely distributed; many species. *Polydesmus, Oxidus, Apheloria.*

Order Juliformia. Usually with eyes. Trunk composed of 40 or more cylindrical segments. Sternites fused with pleurotergal arch. Repugnatorial glands on most trunk segments. Widely distributed. This large group of some 3000 species is now divided into four orders: Order Julida, with many temperate species—*Nemasoma, Blaniulus, Diploiulus, Julus;* Order Cambalida—*Cambala;* Order Spirobolida—*Narceus* (the large common millipede of the eastern United States), *Spirobolus, Rinocricus;* Order Spirostreptida, with mostly tropical species—*Graphidostreptus, Orthoporus.*

Class Pauropoda

The pauropods constitute a small class of soft-bodied, rather grublike animals that inhabit leaf mold and soil (Fig. 16–10). All are minute, ranging from 0.5 to 2 mm. in length. Although once considered rare, pauropods have now been found to be frequently abundant in forest litter. There are approximately 360 described species, which are widespread in both temperate and tropical regions.

Pauropods are similar to millipedes in a number of ways. The trunk contains eleven segments, nine of which each bear a pair of legs. The first segment (collum) and the eleventh segment and telson are legless. Six of the dorsal tergal plates are very large and overlap adjacent segments. As in certain centipedes, such larger terga perhaps represent an adaptation to prevent lateral undulations during rapid movement. All except the first terga carry a pair of long, laterally placed setae. Unlike the collum of diplopods, that of the pauropods is very inconspicuous dorsally and expanded ventrally.

On each side of the head there is a peculiar disclike sensory organ which is perhaps homologous to the organ of Tömösvary of other myriapods. The antennae are two-branched. One division terminates in a single flagellum; the other, in two flagella and a peculiar club-shaped sensory structure. Each mandible is a single elongated piece

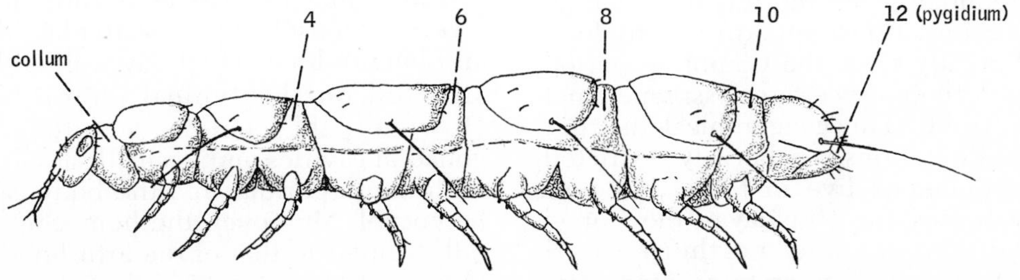

Figure 16–10. The pauropod, *Pauropus silvaticus* (lateral view). (After Tiegs from Snodgrass.)

carrying a comb of curved teeth. The lower lip is probably homologous to the gnathochilarium of diplopods, for it apparently represents the first maxillae, still distinct distally, fused to a triangular first maxillary sternite.

The diet and feeding habits are very poorly known. Different species have been reported to feed on fungal hyphae, humus, and the bodies of dead animals. There is neither heart nor (except in some primitive species) trachea, their absence probably being associated with the small size of these animals.

As in diplopods, the third trunk segment is the genital segment. A single ovary lies below the gut, and a single oviduct opens into a depression between the legs. A separate seminal receptacle also opens into the depression. The testes are located above the gut, and the sperm ducts open between the coxae of the third legs. Sperm are transferred via a spermatophore, which is deposited by the male along with two threads. As in the little pselaphognath millipedes, spermatophore deposition is not dependent upon the presence of a female. The eggs are laid in humus, either singly or in clusters. Development is anamorphic, and as in diplopods, the young hatch with only three pairs of legs. In *Pauropus sylvaticus*, development to sexual maturity takes about 14 weeks.

PHYLUM ONYCHOPHORA

The onychophorans have been described as the "missing link" between annelids and arthropods. The validity of such a statement may be questioned, but this little group of animals does have many interesting similarities to both annelids and arthropods. The phylum is introduced at this point for the purpose of comparison and also for the benefit of the final discussion of the evolution of arthropods.

There are only about 65 existing species of onychophorans, but the phylum is an ancient one and does not appear to have changed greatly since the Cambrian period, from which the only certain fossil specimen was discovered. The geographical distribution of existing forms is relatively restricted. All onychophorans live in tropical regions (the East Indies, the Himalayas, the Congo, the West Indies, and northern South America) or south temperate regions (Australia, New Zealand, South Africa, and the Andes). No species are found north of the Tropic of Cancer.

Most onychophorans are confined to humid habitats, such as in tropical rain forests, beneath logs, stones, and leaves, or along stream banks. During winter snows and low temperatures, or during dry periods, they become inactive and remain in protective retreats.

External Structure. Onychophorans look very much like slugs with legs; in fact, they were thought to be mollusks when first discovered by Guilding in 1825 (Fig. 16–11*A*). The body is more or less cylindrical and ranges from 1.4 cm. to 15 cm. in length. The anterior carries a pair of large annulated antennae and a ventral mouth, which is flanked on each side by short, conical oral papillae and a pair of clawlike mandibles. The mandibles represent modified segmental appendages, as in arthropods. The legs vary in number from 14 to 43 pairs, depending on the species and the sex. Each leg is a large, conical, unjointed protuberance bearing a pair of terminal claws (Fig. 16–11*B*). At the distal end of the leg on the ventral side are three to six transverse pads, on which the leg rests when walking. The entire surface of the body is covered by large and small tubercles, which are arranged in rings or bands encircling the legs and trunk. The larger tubercles each terminate in a sensory bristle, and the tubercles are covered by minute scales. Onychophorans are colored blue, green, orange, or black, and the papillae and scales give the body surface a velvety and iridescent appearance.

Internal Structure and Physiology. The body surface is covered by a chitinous cuticle, but unlike the exoskeleton of arthropods, the cuticle of onychophorans is only one micron thick, flexible, and very permeable, and it is not divided into articulating plates. The absence of a rigid exoskeleton enables onychophorans to squeeze their bodies into very confining places. Beneath the cuticle is a single layer of epidermis, which overlies a thin connective-tissue dermis and three layers of muscle fibers—circular, diagonal, and longitudinal. The body wall is thus constructed on the typical annelidan plan. However, the coelom is reduced to the gonadial cavities and to small sacs associated with the nephridia, and the body cavity is a hemocoel. Moreover, the hemocoel is markedly similar to that of the arthropods in that it is partitioned into a dorsal pericardial

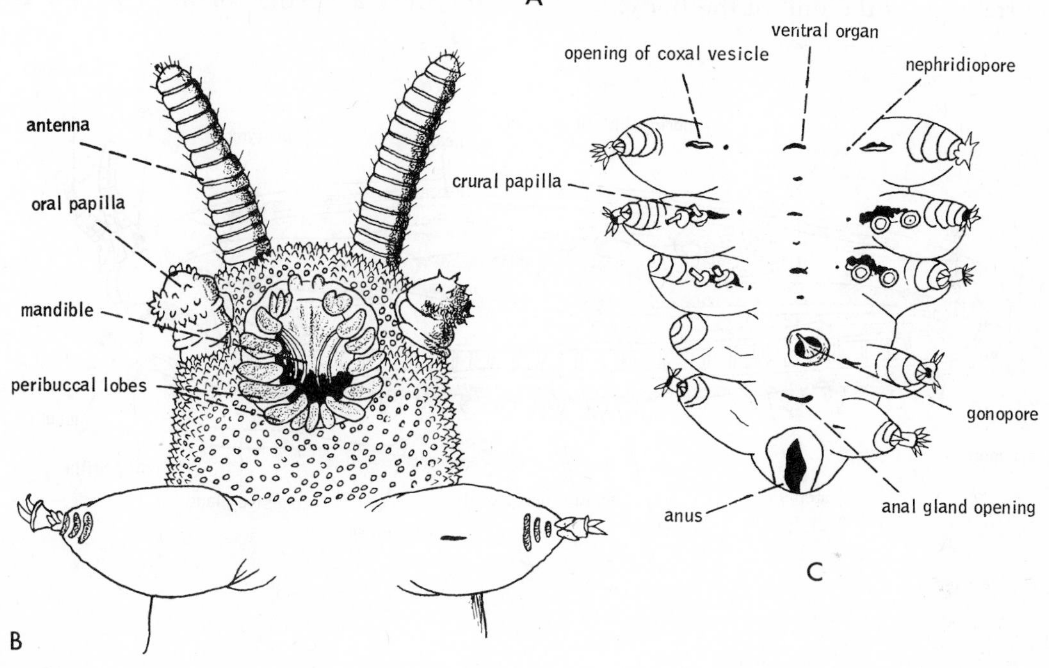

Figure 16–11. Onychophora. *A. Peripatus.* (By H. Sturm.) *B.* Anterior of *Peripatopsis capensis* (ventral view). (After Cuénot.) *C.* Posterior of a male *Peripatus corradoi* (ventral view). (After Bouvier from Cuénot.)

sinus, a median periintestinal sinus, a mid-ventral perineural sinus, and two ventrolateral sinuses.

Onychophorans crawl by means of the legs and by extension and contraction of the body, which is held off the ground. Waves of contraction progress from the anterior to the posterior. When a segment is extended, the legs are lifted from the ground and moved forward. A pushing force is exerted in the effective stroke. As in arthropods, the legs are located more ventrally than are the parapodia of annelids. Locomotion is slow.

Most species are predaceous and feed on small invertebrates, such as snails, insects, and worms. A number of species display a particular preference for termites.

When disturbed, onychophorans use for defense a special pair of glands that secrete an adhesive material. The glands open at the ends of the oral papillae (Fig. 16–11*B*), and the secretion is discharged as two streams for a distance as great as 50 cm.; it hardens very rapidly.

The mouth is located at the base of the prebuccal depression, which is surrounded by cutaneous lobes (Fig. 16–11*B*). Within the depression lie the lateral clawlike mandibles, which are used for grasping and cutting prey. A pair of salivary glands, representing modified nephridia, open into a median dorsal groove that extends backward behind the dorsal tooth. Salivary secretions are passed into the body of the prey, and the digested tissues are then sucked into the mouth. The prebuccal depression opens into the chitin-lined foregut, composed of a pharynx and an esophagus (Fig. 16–12A). A large, straight intestine is immediately posterior to the esophagus and is the site of the remaining digestion and of absorption. The hindgut (rectum) is tubular and loops forward over the intestine before passing posteriorly to the anus. The anus opens on the ventral side at the end of the body.

The circulatory system is similar to that of the arthropods (Fig. 16–12A). A tubular heart, open at each end and provided with a pair of lateral ostia in each segment, lies within the pericardial sinus and propels blood forward into the general hemocoel. The partitions between sinuses are perforated by openings that facilitate blood circulation. The blood is colorless and contains phagocytic amebocytes.

Each segment contains a single pair of nephridia located in the ventrolateral sinuses (Fig. 16–12A). The ciliated funnel and nephrostome lie within an end sac, which represents a vestige of the coelom. Before

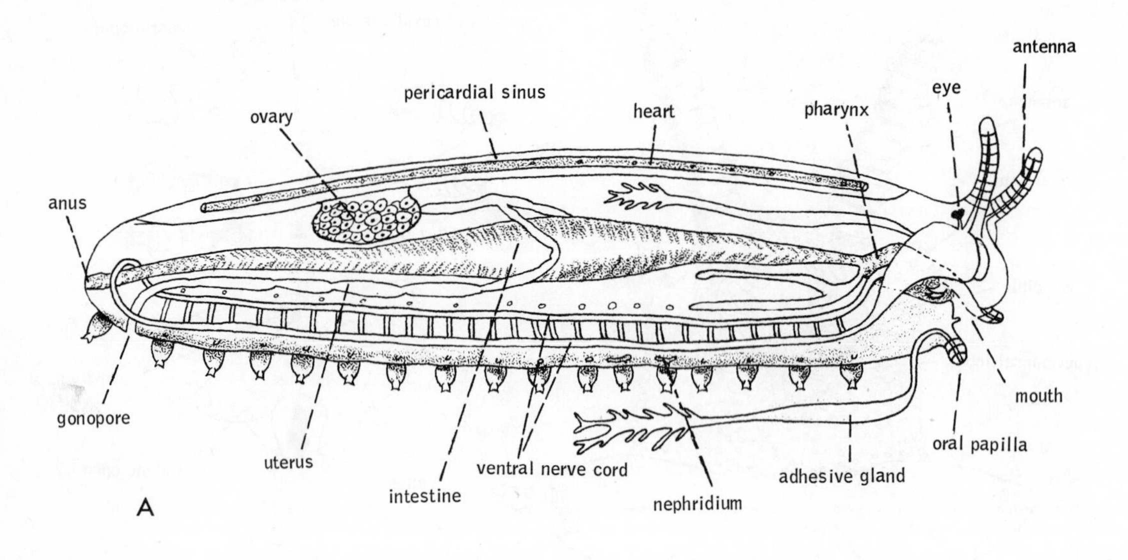

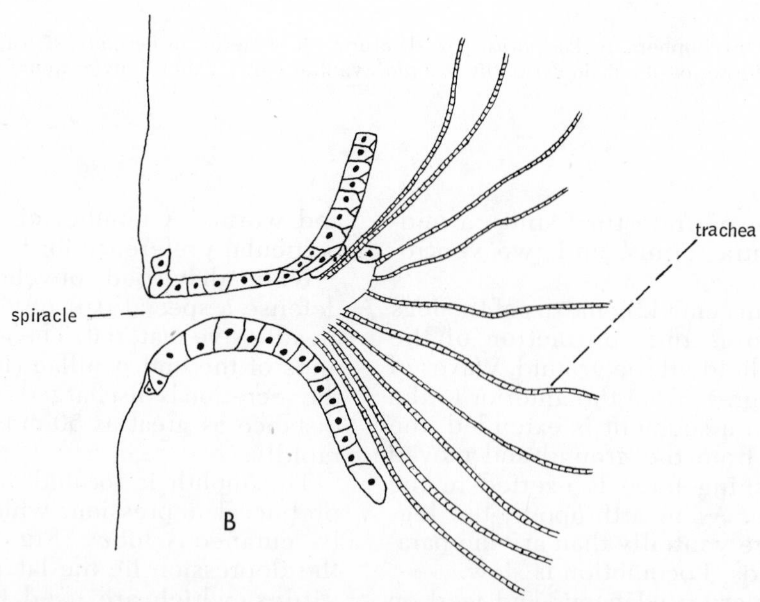

Figure 16–12. *A*. Internal structure of an onychophoran (lateral view). (After Cuénot.) *B*. Section through spiracle and associated tracheae of *Peripatopsis capensis*. (After Balfour from Borradaile and others.)

opening to the outside, the tubule becomes enlarged to form a contractile bladder. The nephridiopore is located on the inner base of each leg, except for the fourth and fifth segments in which it is mounted on a more distal tubercle (Fig. 16–11*C*). The nature of the excretory waste is not known. The anterior nephridia are modified as salivary glands, and the posterior nephridia are modified as gonoducts in the female.

The legs of some onychophorans, such as *Peripatus*, possess a thin-walled vesicle opening to the exterior near the nephridiopore (Fig. 16–11*C*). The vesicle perhaps functions to take up moisture as do the coxal glands of some myriapods.

The gas exchange organs are tracheae. The spiracles are minute openings and are present in large numbers all over the surface of the body between bands of tubercles. The spiracle opens into a very short atrium, at the end of which arises a tuft of minute tracheae (Fig. 16–12*B*). Each trachea is a simple straight tube and extends directly to the tissue that it is supplying.

The nervous system is composed of a large bilobed brain lying over the pharynx and a pair of ventral nerve cords connected together by commissures (Fig. 16–12*A*). The brain supplies nerves to the tentacles, the eyes, and the mouth region. In each segment, the ventral nerve cords contain a ganglionic swelling and give rise to a number of paired nerves supplying the legs and body wall.

There is a small eye at the base of each antenna. The eyes are of the direct type with a large chitinous lens and a relatively well-developed retinal layer. Onychophorans avoid light and are largely nocturnal. The large tubercles and other areas of the integument are supplied with sensory cells.

Reproduction. The sexes are always separate; the males are a little smaller than the females and often have fewer legs. Both sexes are usually supplied with special glands called crural glands, which are thought to have some sort of sexual function. The glands are housed in some, or all, of the legs and open near the nephridiopore at the end of a prominent papilla (Fig. 16–11*C*).

The ovaries are a pair of fused, elongated organs, located in the posterior part of the body (Fig. 16–12*A*). Each ovary is connected to an elaborate genital tract containing a seminal receptacle and a uterus. The ends of each uterus join together and open to the exterior through a common genital pore, situated ventrally near the posterior of the body.

The male system contains two elongated nonfused testes and relatively complex paired genital tracts. Prior to reaching the exterior, the two tracts join to form a single tube in which the sperm are formed into spermatophores. The spermatophores are as long as 1 mm. and are enclosed in a chitinous envelope. The male gonopore is ventral and posterior, like that of the female (Fig. 16–11*C*).

In the mating of the South African *Peripatopsis*, which lacks seminal receptacles, the male crawls over the body of the female and deposits a spermatophore at random on her sides or back, and over a period of time a female may accumulate many spermatophores.

The spermatophore stimulates blood amoebocytes to bring about dissolution of the underlying integument. Sperm then pass from the spermatophore in to the female hemolymph. They eventually reach the ovaries, where fertilization of the eggs takes place.

Sperm transfer in onychophorans with seminal receptacles is not understood.

Onychophorans are oviparous, ovoviviparous, or viviparous, and in most species reproduction appears to be continuous. Oviparous forms are limited to the Australian genera, *Ooperipatus* and *Symperipatus*. The females of these genera are provided with an ovipositor, and the large yolky eggs are laid in moist situations and enclosed in a chitinous shell. As in most arthropods, cleavage is superficial.

All other onychophorans are oviviviparous or viviparous, and the eggs develop within the uterus. The eggs of viviparous onychophorans are small and have little yolk, and cleavage either is superficial or has become holoblastic. Uterine secretions provide for the nutrition of the embryo, and the nutritive material is obtained by the embryo through a special embryonic membrane or through a "placental" connection to the uterine wall.

Geographical Distribution. The geographical distribution of onychophorans is peculiar in a number of respects. The phylum consists of two families. Each has a wide, discontinuous distribution around the world, but neither is found in the same area with species of the other family. The family Peripatidae is more or less equatorial in distribution, while the Peripatopsidae is limited to the Southern Hemisphere. Species within the same genus may be widely separated. For example, the genus *Opisthopatus* is found in Chile and South Africa, and *Mesoperipatus* of the Congo has its nearest allies in the Caribbean Islands.

Considering the geological antiquity of the onychophorans and the improbability of their being spread by other animals, the distribution of the phylum has been used as evidence to support the belief that the land masses of the equatorial regions and the southern hemisphere were connected during past geological ages.

The structure of the body wall, the nephridia, the thin flexible cuticle, and the non-jointed appendages are certainly annelidan in character. But in other respects onychophorans are more like arthropods. The coelom is reduced, and the cuticle is chitinous. A pair of appendages is modified for feeding, and there is an open circulatory system with a dorsal tubular heart containing ostia. Thus, onychophorans appear to be more closely allied to arthropods than to annelids. They may represent an offshoot from near the base of the arthropod line, or, as will be discussed in the following section, they may be on the main line leading to the arthropods.

Manton (1970) includes onychophorans within the Arthropoda. Such a classification is based largely on phylogenetic viewpoints described in the next section. The soft body and non-jointed appendages, among other annelidan characters, which are certainly primary, present serious problems, and to consider onychophorans as arthropods demands a fundamental redefinition of the phylum Arthropoda. It is questionable whether such a redefinition is justified or will be widely accepted.

ARTHROPOD PHYLOGENY

Although the arthropods constitute a well-defined assemblage, the phylogenetic interrelationships within the phylum are by no means clear. There is a lack of fossil forms to bridge many important gaps, and comparative anatomy and embryology more commonly provide fuel for argument than for the resolution of problems. Yet the dearth of evidence has been no hindrance to speculations regarding the origin and evolution of arthropods. Much has been written on the subject and, needless to say, there is a diversity of answers to a number of major questions which can be posed: Do the arthropods constitute a monophyletic or a polyphyletic group? Do the Mandibulata represent a natural grouping? What are the affinities of the Crustacea?

Before attempting to examine some of the answers to these questions, it might be useful first to indicate certain relationships which are rather generally recognized as being probable (even here there are dissenters). Most zoologists agree that the arthropods evolved from some primitive polychaete stock, or at least from some common ancestral form in which metamerism had already appeared. There is general agreement that the chelicerates (i.e., merostomes and arachnids) constitute a natural group. As indicated earlier, there are many zoologists who would prefer to omit the pycnogonids from the chelicerate grouping. Finally, most zoologists would concur that insects are related to the myriapods.

Accepting these affinities, there remains unanswered the nature of the relationships between the myriapod–insects, the crustaceans, the trilobites, and the chelicerates. Contemporary views concerning these more difficult relationships are largely dependent upon a fundamental question posed earlier: Have these groups evolved from a common arthropod ancestor or have they evolved independently? In other words, do the arthropods constitute a monophyletic or a polyphyletic phylum?

The concept of a monophyletic origin of arthropods is old and has been described with variations by a number of authors. During this century, it has been elaborated by such zoologists as Handlirsch (1937), Snodgrass (1938), and, more recently, the Russian Sharov (1966). All of these authors would place the trilobites or a prototrilobite near the base of the arthropod stem. From the trilobites or their immediate ancestors evolved one line leading to the mandibulate arthropods—crustaceans, myriapods, and insects—and another line culminating in the chelicerates. The mandibulates are thus considered to represent a natural group, and such common structures as antennae, mandibles, and compound eyes are thought to reflect a common evolutionary origin.

There are two principal difficulties with this phylogeny. First, there is little evidence to link or bridge the major groups. Second, the mandibulates do not appear to be a natural group. Crustaceans possess a second pair of antennae, biramous appendages, a mandibular palp, and other features that strongly suggest that they are not related to the myriapod-insect assemblage.

A polyphyletic origin for the arthropods has recently been most strongly argued by Manton (Tiegs and Manton, 1958; Manton, 1964, 1970), and her views have received relatively wide acceptance. According to this theory, a metameric animal evolved from some group of primitive polychaete annelids or a group phylogenetically near the annelids. From such a "protonychophoran" stock, some group invaded land to give rise to the onychophorans, which are still represented today, reflecting the beginnings of arthropodization. The onychophorans are considered to be the ancestors of the myriapod-insect assemblage of arthropods. Arthropodization in this line was associated with life on land.

Also from the protonychophorans stemmed a second line leading to the other arthropods, which were marine or at least initially so. The evolution of arthropodization perhaps resulted from the selection pressures associated with a marine-bottom existence. At the base of this line was some pretrilobite, which gave rise to one branch leading to the crustaceans and to another leading to the trilobites and the chelicerates.

A polyphyletic origin for the arthropods would explain the uniqueness of the crustaceans and the absence of any evidence linking the myriapod-insect group with the other arthropods. The implications of Manton's phylogeny are numerous. The mandibulates constitute an artificial assemblage in contrast to the chelicerates. The crustaceans are in no way related to the other mandibulates—the myriapods and the insects. In fact, the onychophorans are more closely related to the insects than are the insects to the crustaceans. Not only has arthropodization evolved more than once, but such complex structures as compound eyes have arisen independently in crustaceans and in insects.

Not only does paleontology shed little light on the controversies outlined here, but there are several small fossil groups of unknown relationships. Such a group is the Pseudocrustacea, known from the early Paleozoic era. The pseudocrustaceans were similar to both crustaceans and trilobites. They ranged in length from 10 to 105 mm. The limbs were similar to those of trilobites, but the stalked eyes and certain other features were similar to those of crustaceans (Fig. 16–13). The pseudocrustaceans have been regarded as perhaps being an offshoot of the line from which the trilobites and crustaceans were derived.

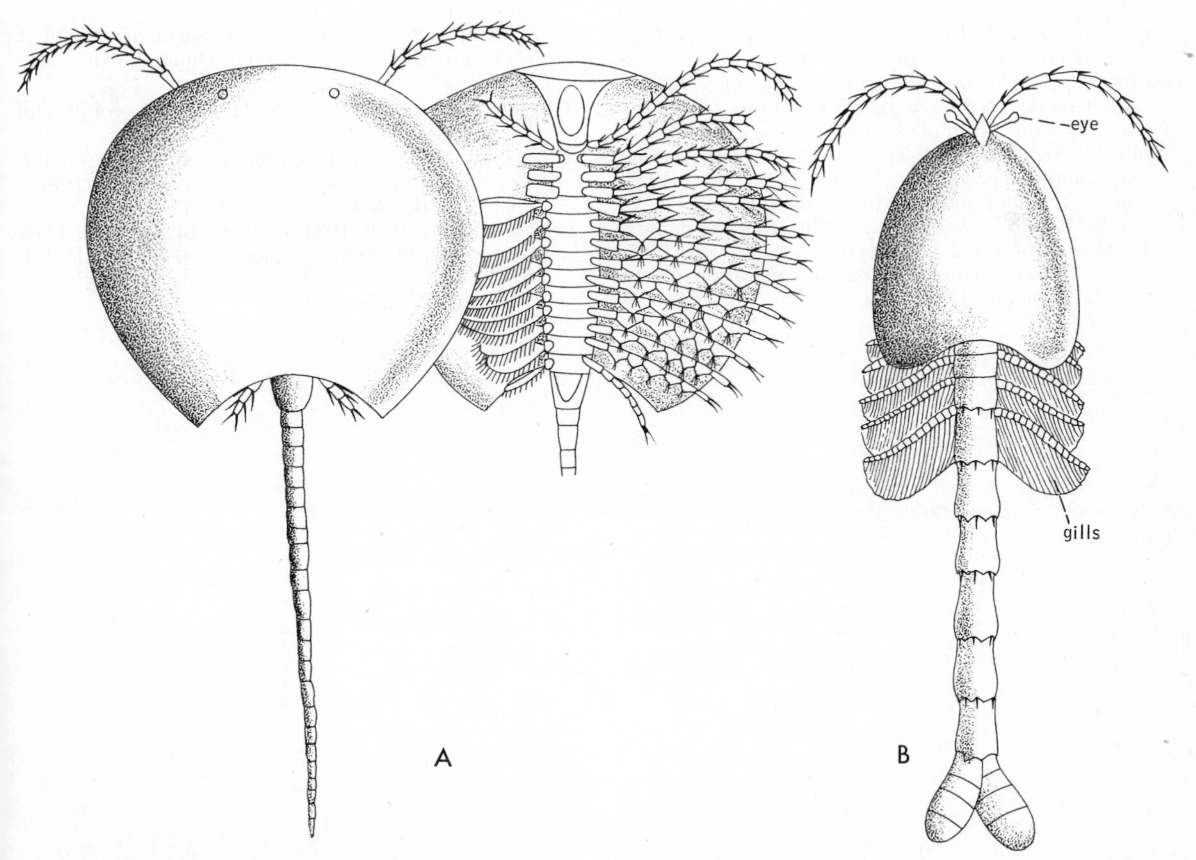

Figure 16–13. Two Pseudocrustacea of the Cambrian period. *A*. Dorsal and ventral views of *Burgessia*. *B*. Dorsal view of *Waptia*. (Both after Walcott.)

Bibliography

Attems, G., 1926–1940: *In* W. Kukenthal and T. Krumbach (Eds.), Handbuch der Zoologie. Vol. 4, Progoneata, Chilopoda, 1926; Vol. 52, Myriapoda, Geophilomorpha, 1929; Vol. 54, Chilopoda, Scolopendromorpha, 1930; Vols. 68–70, Diplopoda, Polydesmoidea, 1937–40. W. de Gruyter, Berlin and Leipzig. (This and the works of Verhoeff (see below) contain the most extensive and detailed accounts of the myriapodous classes.)

Brolemann, H. W., 1932: Chilopodes. Faune de France, Vol. 25.

Brolemann, H. W., 1935: Myriapodes. Diplopodes: Chilognathes 1. Faune de France, Vol. 29.

Buck, J. B., and Keister, M. L., 1950: *Spirobolus marginatus*. *In* F. A. Brown (Ed.), Selected Invertebrate Types. John Wiley, N.Y., pp. 462–475.

Cloudsley-Thompson, J. L., 1948: *Hydroschendyla submarina* in Yorkshire: with an historical review of the marine Myriapoda. Naturalist, pp. 149–152.

Cloudsley-Thompson, J. L., 1958: Spiders, Scorpions, Centipedes, and Mites. Pergamon Press, N.Y. (A discussion of the natural history and ecology of the myriapodous arthropods is presented in Chapters 2, 3 and 4.)

Cuénot, L., 1949: Les Onychophores. *In* P. Grassé (Ed.), Traité de Zoologie, Vol. 6. Masson et Cie, Paris, pp. 3–75.

Handlirsch, A., 1937: Neue Untersuchungen über die fossilen Insekten. Ann. Naturh. (Mus.) Hofmus., Wien, *48*:1.

Hoffmann, R. L., and Payne, J. A., 1969: Diplopods as carnivores. Ecology, *50*(6):1096–1098.

Kaestner, A., 1968: Invertebrate Zoology, Vol. 2. Wiley-Interscience, N.Y., 472 pp. (This volume covers the myriapodous arthropods and the onychophorans.)

Manton, S. M., 1952–1961: The evolution of arthropodan locomotory mechanisms. J. Linn. Soc. (Zool.), 1952, pt. 3. The locomotion of the Chilopoda and Pauropoda. *42*:118–167; 1954, pt. 4. The structure, habits, and evolution of the Diplopoda. *42*:229–368; 1956, pt. 5. The structure, habits, and evolution of the Pselaphognatha (Diplopoda). *43*:153–187; 1958, pt. 6. Habits and evolution of the Lysiopetaloidea (Diplopoda), some principles of the leg design in Diplopoda and Chilopoda, and limb structure in Diplopoda. *43*:487–556; 1961, pt. 8. Functional requirements and body design in Colobognatha (Diplopoda), together with a comparative account of diplopod burrowing techniques, trunk musculature, and segmentation, *44*:483–561.

Manton, S. M., 1964: Mandibular mechanisms and the evolution of Arthropods. Phil. Trans. Roy. Soc. London, B, *247*:5–183.

Manton, S. M., 1970: Arthropods: Introduction. *In* Florkin, M., and Scheer, B. J. (Eds.), Chemical Zoology, Vol. V(A). Academic Press, N.Y., pp. 1–34.

Moore, R. C. (Ed.), 1969: Treatise on Invertebrate Paleontology. Pt. R, Arthropoda 4, Vol. 2. Geological Society of America and University of Kansas Press. (This volume covers the myriapods. The Pseudocrustacea is covered in Pt. O, Arthropoda 1, 1959.)

Schaller, F., 1968: Soil Animals. University of Michigan Press. 144 pp.

Sharov, A. G., 1966: Basic Arthropodan Stock. Pergamon Press, N.Y.

Silvestri, F., 1950: Segmentazione del capo dei Colobognati. Proc. Eighth Internat. Congr. Entomol., Stockholm, pp. 371–576.

Snodgrass, R. E., 1938: Evolution of the Annelida, Onychophora, and Arthropoda. Smithsonian Misc. Coll., *97*:1–159.

Snodgrass, R. E., 1952: A Textbook of Arthropod Anatomy. Cornell University Press, Ithaca, N.Y. (An old but still useful description of the external structure of the various myriapodous classes.)

Tiegs, O. W., 1945: The postembryonic development of *Hanseniella agilis* (Symphyla). Quart. J. Micr. Sci., *85*:191–328.

Tiegs, O. W., and Manton, S. M., 1958: The evolution of the Arthropoda. Biol. Rev., *33*:255–337.

Tiegs, O. W., 1947: The development and affinities of the Pauropoda, based on a study of *Pauropus silvaticus*. Quart. J. Micr. Sci., *88*:165–267, 275–336.

Verhoeff, K. W., 1926–1934: *In* H. G. Bronn (Ed.), Klassen und Ordnungen des Tierreichs. Chilopoda, Bd. 5, II (1); Diplopoda, Bd. 5, II (2); Symphyla and Pauropoda, Bd. 5, II (3).

Chapter 17

SOME LESSER PROTOSTOMES

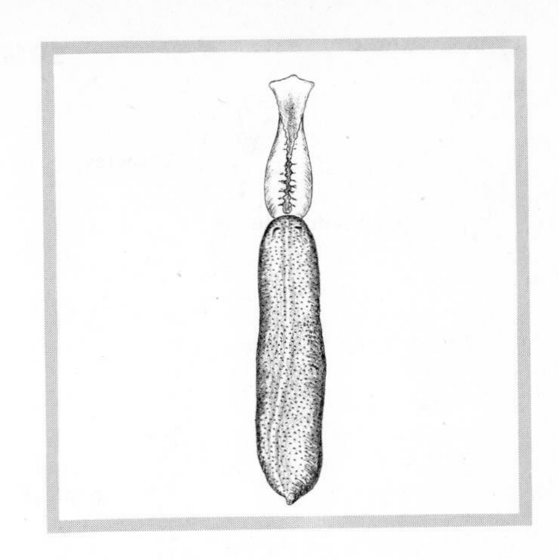

This discussion treats six small phyla of coelomates, which probably stemmed from various points along the protostome line. Three of these groups, the sipunculans, the echiurans, and the priapulids, were once united as members of the phylum Gephyrea. The artificiality of this arrangement was soon recognized, and the three groups have since been separated. Possible phylogenetic relationships will be indicated in the discussion of each phylum.

PHYLUM POGONOPHORA

By the middle of the 19th century, almost all phyla of animals had been at least observed by zoologists. The Pogonophora, however, were completely unknown prior to the 20th century. The first specimen was dredged from Indonesian waters in 1900. Since that time more than 80 species have been described and more are being discovered. The northwest Pacific has yielded the richest pogonophoran fauna, but this discovery perhaps reflects the concentrated work of Russian oceanographic vessels in this area. It is now becoming apparent that the phylum is fairly widespread in the world's seas, especially along the continental slopes. In the western Atlantic, pogonophorans have been collected from Nova Scotia to Florida and from the Gulf of Mexico.

Pogonophorans are almost exclusively deep-water animals (more than 100 m.), a fact which accounts for their delayed discovery. They are sessile, living in secreted, stiff, chitinous tubes that are usually fixed upright in bottom ooze. However, there are some species that construct their tubes in decaying wood or other debris. The long, wormlike body ranges in length from 10 to 85 cm., and is composed of a short forepart, a long trunk, and a short opisthosoma (Fig. 17–1).

The forepart consists of an anterior cephalic lobe and a posterior glandular region which provides secretions for the formation of the tube. Beneath the cephalic lobe, the forepart bears long tentacles, which are the distinguishing feature of the phylum (Pogonophora—"beard bearer"). Depending upon the species and also upon the age of the individual, the tentacles range in number from a single spiral tentacle to more than 250 (Fig. 17–1*A*). The tentacles are especially peculiar in usually bearing minute pinnules that are extensions of single epithelial cells. Tracts of cilia run the length of each tentacle (Fig. 17–1*F*).

The long trunk typically bears papillae and, in the midregion, two girdles of short toothed setae. The terminal opisthosoma (Fig. 17–1*A* and *B*) is composed of 5 to 23 short segments with setae, which are essentially like those of annelids in structure and origin. The opisthosoma and setae probably function in anchoring the body within the tube and aid in burrowing in the bottom ooze below the open lower end of the tube. The opisthosoma is fragile and easily breaks away from the rest of the body when the animal is pulled. As a result, most specimens dredged from the bottom lack this region, and it has been only in recent years that the opisthosoma has been known to exist.

Internally, there is a coelomic compartment in each of the body divisions, and there are extensions of the coelom into the tentacles. In the opisthosoma there are septa between the segmental coelomic compartments. A well-developed, closed, blood-vascular sys-

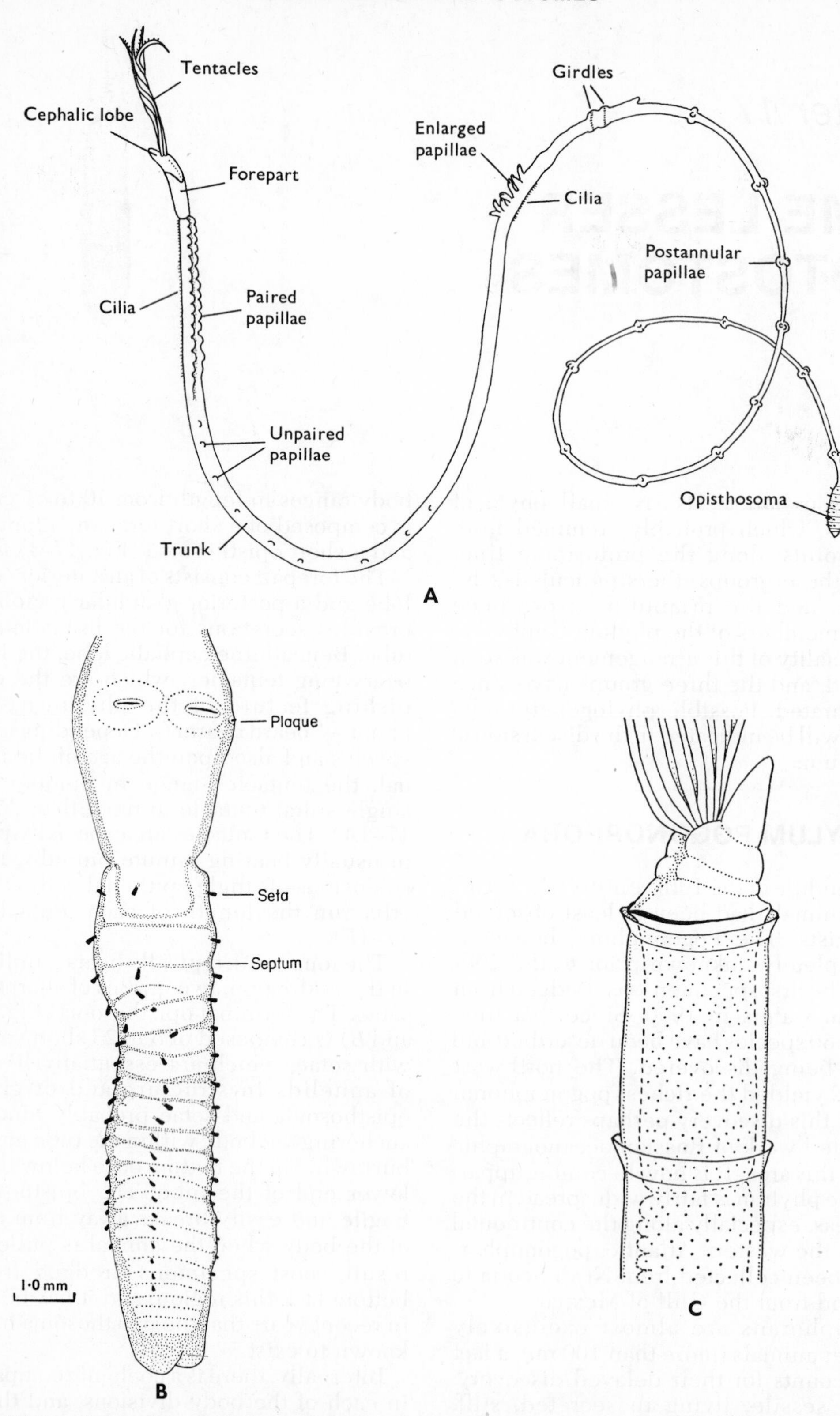

Figure 17–1. Phylum Pogonophora. *A.* Diagram of a typical pogonophoran. Length has been greatly reduced. (From George, J. D., and Southward, E. C., 1973: J. Mar. Biol. Assoc. U. K., *53*(2):403–424.) *B.* Opisthosoma of *Polybrachia canadensis*, showing segments and setae. *C.* Anterior of *Polybrachia canadensis*, showing postulated position of animal during tube secretion. The glandular region of the forepart occupies the entire upper section of the tube. (*B* and *C* from Southward, E. C., 1969: J. Zool., *157*:449–467.) (*Illustration continued on opposite page.*)

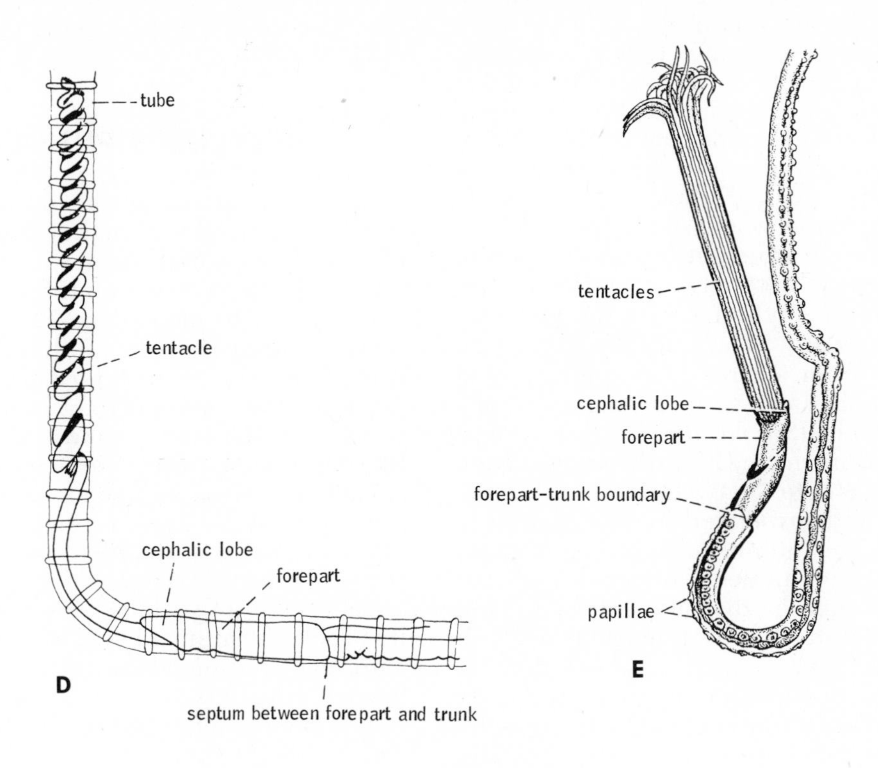

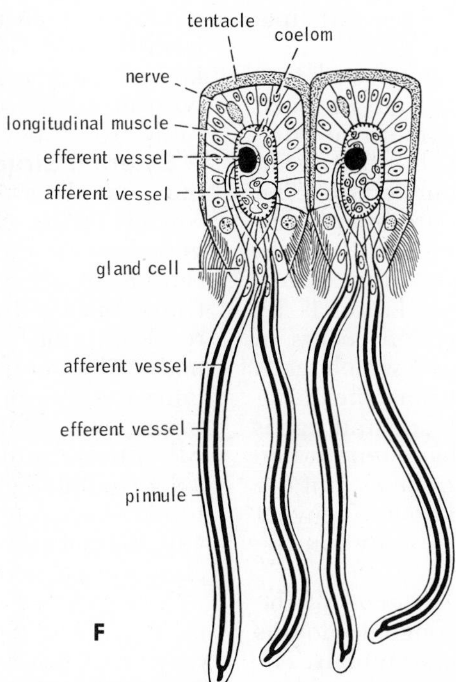

Figure 17–1. *Continued. D.* Anterior of *Siboglinum*, a pogonophoran with a single tentacle, in its tube. (After Caullery from Hyman.) *E.* Anterior of *Lamellisabella johanssoni*, a pogonophoran with 18 tentacles fused together to form a hollow cylinder. *F.* Cross section of two adjacent tentacles of *Lamellisabella*. (*E* and *F* after Ivanov.)

tem is present, and each tentacle is supplied with two vessels (Fig. 17–1*F*).

The most remarkable feature of pogonophorans is the complete absence of a mouth and a digestive tract. In fact, zoologists who examined the first specimen thought that part of the body was missing. In the absence of a digestive tract, the mode of nutrition in these animals is puzzling to say the least. Ivanov (1963) suggests that the tentacles are probably extended from the mouth of the tube and are used to gather organic detritus suspended in the surrounding water. The tentacles are arranged to form a cylinder with the pinnules directed inwardly. The cylinder formation may result from a circular arrangement of straight tentacles or a spiral arrangement of a single tentacle (*Siboglinum*) or many tentacles (*Spirobrachia*). Cilia drive water from the anterior end to the posterior end through the funnel and suspended food particles are trapped on the pinnules. Enzymatic gland cells flank the pinnules. Collected food particles might then be digested externally and absorbed directly by the tentacular epithelium. Uptake of amino acids has been demonstrated to be possible anywhere on the body surface, and the fibrous cuticle that covers the body is everywhere provided with microvilli. The tentacles are probably also the principal sites of gas exchange.

Pogonophorans are dioecious, and two cylindrical gonads are located one on each side in the trunk coelom. The sperm are disseminated as spermatophores, but the mechanism of fertilization is still uncertain. The eggs of at least some species are brooded in the tube, and embryos collected with tubes have yielded information concerning development. In *Siboglinum*, gastrulation is by epiboly, and mesoderm formation occurs in two phases. In the second phase, which leads to the formation of the opisthosoma, the mesoderm arises from teloblast cells and undergoes segmentation. Development has been studied through rather late embryonic stages, but the point of hatching is as yet unknown. To what extent the hatching stage is a swimming larva is also unknown.

Prior to 1970, when the existence of the opisthosoma was unknown, pogonophorans were thought to be deuterostome animals. The body appeared to be tripartite, like that of deuterostomes, and the general form of these animals appeared somewhat similar to that of pterobranch hemichordates. However, the segmented setiferous terminal part of the body, the similarity of the setae to those of annelids, and the origin and segmentation of the mesoderm clearly indicate the protostome position of the Pogonophora and suggest that they are closely related to the Annelida.

PHYLUM SIPUNCULA

The sipunculans are a group of approximately 330 species of marine animals, sometimes called peanut worms. Sipunculans are rather drab-colored worms that range in length from 2 mm. to more than 72 cm., although most are from 18 to 36 cm. long. All are bottom dwellers, the majority occurring in shallow water, and they are rather sedentary in habit. Some live in sand and mud where, like *Sipunculus*, they are active burrowers; some live in mucus-lined excavations. Others live in coral crevices, in empty snail shells (*Phascolion*) or annelid tubes, and in other sorts of protective retreats (Fig. 17–2*B*). There are a number of species which bore in coralline rock, the animal being oriented with the anterior end directed toward the opening of its burrow (Fig. 17–3). Boring sipunculans are very common in tropical reef limestone. Densities as great as 700 individuals per square meter of coralline rock have been reported from Hawaii. The mechanism of boring is not known. Although the burrow may be lined with mucous secretions, sipunculans do not build true tubes.

External Structure. The cylindrical body of sipunculans is divided into an anterior narrowed section, called the introvert, and a larger posterior trunk (Fig. 17–2*A*). The introvert can be retracted into the anterior end of the trunk, but the introvert is not a proboscis. It represents the head and the anterior part of the body. The anterior end of the introvert contains the mouth surrounded by a scalloped fringe, lobes, tentacles, or tentaculate lobes. All of these projections are ciliated and bear a deep ciliated groove on their inner side. Behind the anterior end, the surface of the introvert is typically covered with spines, tubercles, and other ornamentations.

The trunk is a simple cylinder, and its surface is not strikingly ornamented except in some rock-boring forms, such as species of *Paraspidosiphon* and *Cloeosiphon*. In these crevice-inhabiting sipunculans, the surface of the trunk is thickened at the anterior end to form a dorsal or a collar-like shield, which is used to block the opening of the retreat when the introvert is invaginated (Fig. 17–3).

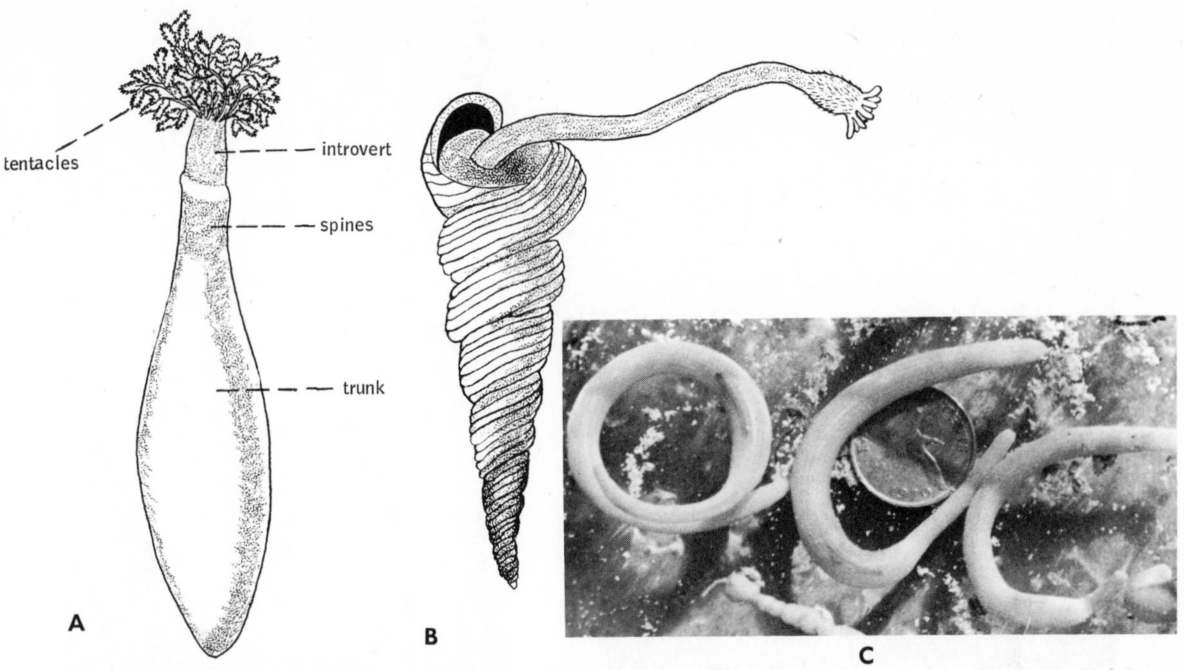

Figure 17–2. Sipuncula. *A. Dendrostomum pyroides*, a burrowing sipunculan (3/5 natural size). (After Fisher from Tétry.) *B. Phascolion* in a snail shell. Introvert is extended out of aperture. (After Cuénot from Tétry.) *C.* Specimens of West Indian *Sipunculus* dug from sand in shallow water. (By Betty M. Barnes.)

Internal Structure and Physiology. The body wall is constructed like that of the annelids. There is an external cuticle secreted by a simple layer of epidermis, a thin connective-tissue dermis, and a circular muscle layer and a longitudinal muscle layer, which is covered on the inner side by peritoneum.

A large coelom extends the length of the body; elevated coelomic fluid pressure brings about the protrusion of the introvert. The pressure is controlled by the contraction of the body wall. Retraction of the introvert is brought about by special retractor muscles that originate on the body wall of the trunk and run anteriorly. They are inserted on the end of the esophagus (Fig. 17–4A).

The tentacles of the introvert are hollow but are not connected to the coelom. Rather, the tentacular lumen connects by a system of canals with one or two blind tubular sacs that are connected to the esophagus (Fig. 17–4A). These sacs are contractile and receive fluid from, or supply fluid to, the tentacles when they are contracted or expanded.

Sipunculans are mostly deposit feeders. The introvert and the tentacles are extended in feeding, but the exact mode of feeding varies in different species. In burrowing forms, such as *Sipunculus*, the worm ingests sand and silt through which it burrows. Some rock borers, such as *Phascolosoma antillarum*, hold their tentacular crown open at the mouth of the burrow and perhaps depend upon ciliary currents for obtaining food. Other hard-bottom forms extend the introvert out over the rock surface. The expanded tentacles are placed against the substratum, and surface detritus is trapped in mucus and driven into the mouth by the beating cilia. After the introvert collects material, it then invaginates for the purpose of ingestion. *Golfingia procera* from the North Sea is reported to prey upon the sea mouse, *Aphrodite*. The sipunculan penetrates the body wall of the annelid with its introvert and then sucks out the contents.

The digestive tract is **U**-shaped and coiled (Fig. 17–4A). The mouth opens into an esophagus, which leads in turn into a long intestine. The intestine descends to the end of the trunk, turns, and ascends anteriorly. The entire intestinal loop is twisted in a single spiral coil. Anteriorly the intestine passes into a short rectum, which opens to the outside through the anus. The anus is located middorsally at the anterior end of the trunk except in a single genus, *Onchnesoma*, in which it opens on the introvert.

There is no blood-vascular system and no gas exchange organs, but the coelomic fluid functions in circulation and contains abundant corpuscles bearing hemerythrin. Hemerythrin is a respiratory pigment similar in structure

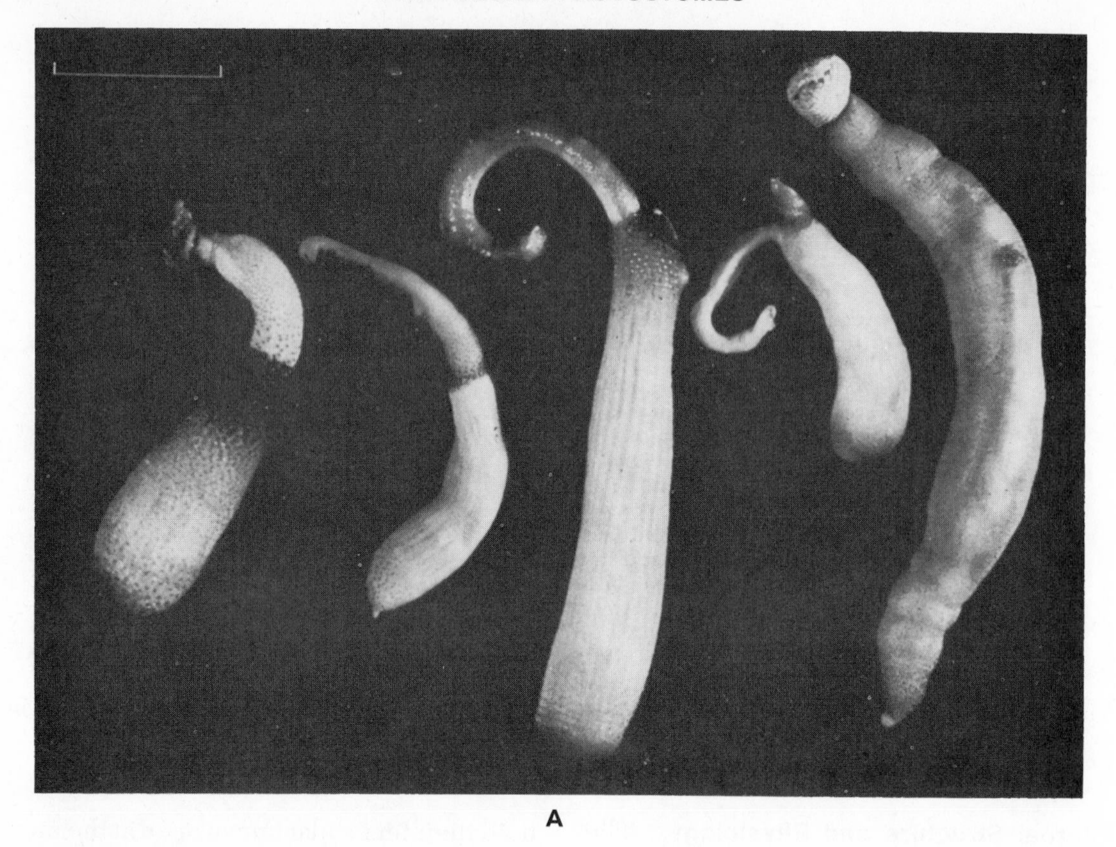

A

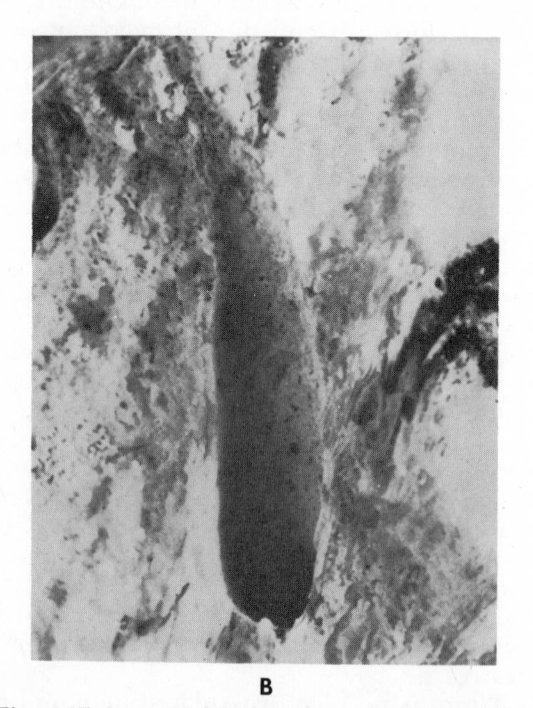

B

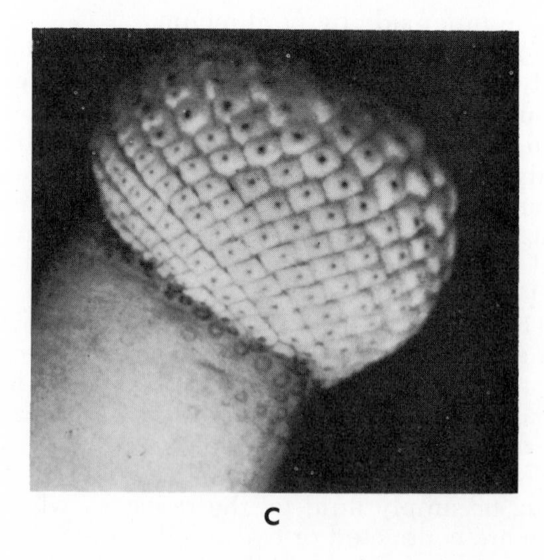

C

Figure 17–3. *A*. Rock-boring sipunculans. From left to right: *Phascolosoma antillarum, Phascolosoma dentigerum, Paraspidosiphon steenstrupi, Lithacrosiphon gurjanovae, Cloeosiphon aspergillum.* All from the Caribbean except for *C. aspergillum,* which is from the Maldive Islands in the Indian Ocean. The introvert of *P. steenstrupi* and *L. gurjanovae* is displaced ventrally by the anterior dorsal shield. Scale is 5 mm. *B*. Burrow of *P. antillarum* in coralline rock. Scale is 5 mm. *C*. Anterior calcareous shield of *C. aspergillum.* Introvert is retracted in center. (All from Rice, M. E., 1969: Amer. Zool., 9:803–812.)

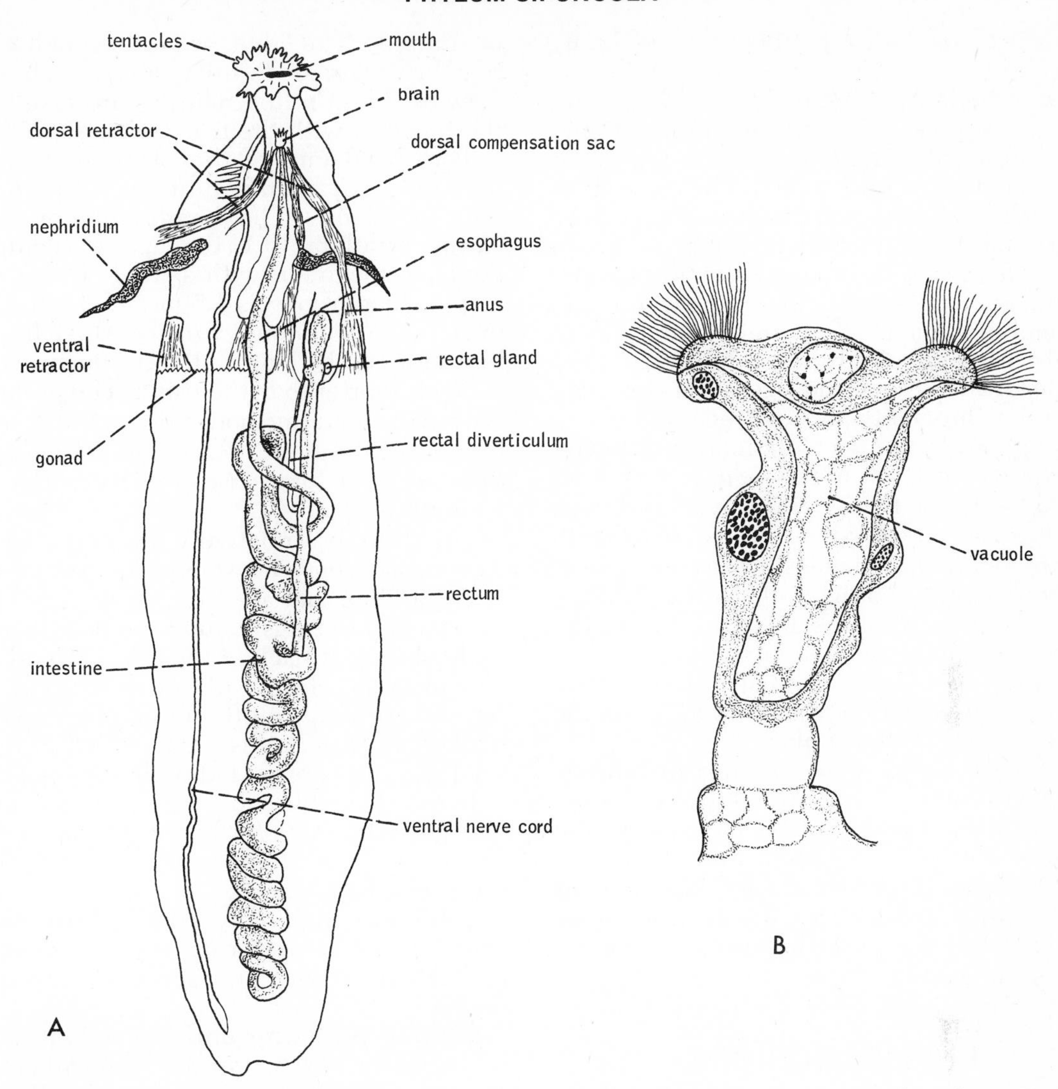

Figure 17–4. *A*. Dissection of *Sipunculus nudus*. (After Metalnikoff from Hyman.) *B*. Fixed urn from *Sipunculus nudus*. (After Selensky from Hyman.)

to hemocyanin, but the metallic atoms to which the oxygen molecules are bound are iron rather than copper.

Sipunculans possess a single pair of large sac-like metanephridia, and the nephridiopores open anteriorly and ventrally at about the same level as the anus (Fig. 17–4*A*). Associated with excretion in sipunculans are peculiar cell clusters called fixed and free urns. Fixed urns are clusters of peritoneal cells, each cluster elevated like a vase and capped by a ciliated cell (Fig. 17–4*B*). The location on the peritoneum varies in different species. Free urns are fixed urns that have become detached from the peritoneum and move about in the coelomic fluid. The free urns moving in the coelomic fluid gather particulate waste material, forming a trailing aggregate that is eventually dumped in various places within the coelom or removed by the nephridia.

The nervous system is essentially similar to that of the annelids but is in no way metameric. The brain lies over the anterior end of the esophagus, and a pair of circumesophageal connectives join ventrally with the single ventral nerve cord, which runs the length of the body and lacks conspicuous ganglionic swellings (Fig. 17–4*A*).

Sensory cells are particularly abundant on the end of the introvert that is used to probe the surrounding environment. In *Golfingia*, *Phascolosoma*, and a few other genera, the dorsal end of the introvert bears a pair of ciliated pits, called nuchal organs, which may be chemoreceptors. Also, a pair of pigment-cup ocelli are imbedded in the brain of many species.

Reproduction. Sipunculans are dioecious, and the sex cells arise from the coelomic epithelium covering the origin of the retractor muscles on the body wall. The immature gametes are shed into the coelom, where maturation is completed; ripe eggs or sperm leave the body by way of the nephridia. Fertilization takes place in the sea water, and the emission of sperm by the males induces neighboring females to shed their eggs.

The fertilized eggs undergo spiral cleavage. Later development may be direct or lead to a trochophore larvae. In *Golfingia,* the trochophore is typical, but in *Sipunculus* it is elongated. After a free-swimming existence ranging from one day (in *Golfingia*) to a month (in *Sipunculus*), metamorphosis takes place, and the young worms sink to the bottom. There is no trace of metamerism during the course of embryonic development.

A few sipunculans are known to reproduce asexually by constriction and separation of the posterior end of the trunk.

Although sipunculans are not metameric animals, they are probably related to the annelids. The construction of the body wall, the nature of the nervous system, and the embryology are annelidan in character. Sipunculans perhaps diverged from the line leading to the annelids at some point before the development of metamerism.

PHYLUM ECHIURA

Echiurans are marine worms that are somewhat similar to sipunculans in size and general habit. Many species, such as *Echiurus, Urechis,* and *Ikeda,* live in burrows in sand and mud; others live in rock and coral crevices (Fig. 17–5*D*). *Thalassema mellita,* which lives off the southeastern coast of the United States, inhabits the test of dead sand dollars. When the worm is very small, it enters the test and later becomes too large to leave. The majority of echiurans live in shallow water, but there are also deep-sea forms. About 100 species have been described.

External Structure. The body of echiurans is composed of a sausage-shaped cylindrical trunk and an anterior proboscis (Fig. 17–5*A*). The proboscis is a large flattened projection of the head and cannot be retracted into the trunk. The proboscis is actually a cephalic lobe and, since it contains the brain, is probably homologous with the prostomium of annelids. The edges are rolled ventrally so that the underside forms a gutter, and in some species the in-rolled margins are fused near the junction with the trunk (Fig. 17–5).

The distal end of the proboscis is often flared and may even be bifurcated, as in *Bonellia* (Fig. 17–5*C*). The length of the proboscis varies considerably. In the Japanese echiuran, *Ikeda,* a specimen with a trunk 40 cm. long may have a proboscis 1.5 m. in length. However, in the common *Echiurus,* which lives on both sides of the North Atlantic, the proboscis is much shorter than the trunk. The proboscis is used in obtaining food and is very extensible in some echiurans. For example, a specimen of *Bonellia* 8 cm. long can extend its proboscis to a length of 1 m.

The trunk is a relatively uniform cylinder. The surface may be smooth or ornamented with papillae that are irregularly distributed or arranged in rings around the body. On the underside, just back of the anterior end, is a pair of large, closely placed setae (Fig. 17–5*A*). The setae are chitinous and are curved or hooked. As in annelids, each seta arises from a setal sac and is provided with a special musculature to enable movement. In addition to anterior setae, *Urechis* and *Echiurus* possess one or two circlets of setae around the posterior end of the trunk.

Echiurans are largely a drab gray or brown color; but some are green, such as *Bonellia,* and others are red or rose. A few are transparent.

Internal Structure and Physiology. The body wall is like that of the sipunculans and the annelids. The proboscis is especially glandular, and its undersurface is ciliated. There is no indication of metamerism in the adult, and the coelom is continuous throughout the trunk and is noncompartmented.

Most echiurans, such as *Echiurus,* are detritus feeders. The proboscis is projected from the burrow or retreat, and the ventral face is stretched out over the substratum (Fig. 17–5*D*). Detritus adheres to the mucus on the proboscis surface, from which it is then driven into a median ciliated groove. Here the particles are conducted back into the mouth.

Urechis, which lives along the California coast in **U**-shaped burrows, has a mode of feeding that is somewhat similar to the polychaete, *Chaetopterus.* The proboscis is very short, and a circlet of mucous glands girdles the anterior end of the trunk just behind the setae. During feeding, the glandular girdle is brought in contact with the burrow wall, and the glands begin secreting mucus. As the

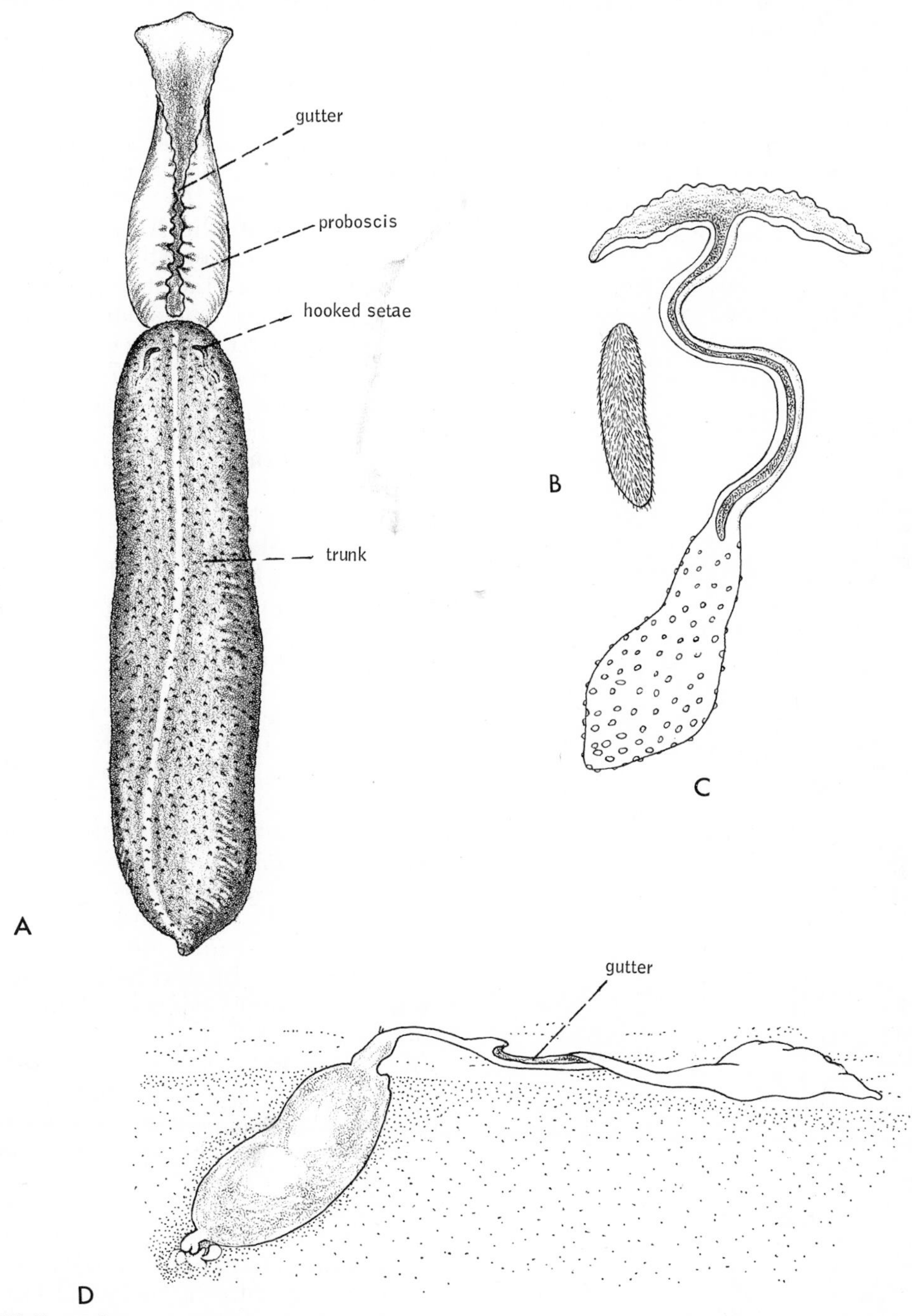

Figure 17-5. Echiura. *A. Echiurus* sp. from life, (ventral view). *B* and *C*. Dwarf male (*B*) and female (*C*) of *Bonellia viridis*, showing the extreme sexual dimorphism in this species. (After MacGinitie and MacGinitie.) *D. Tatjanellia grandis* in feeding position with trunk buried and proboscis extended out over surface of bottom. (After Zenkevitch from Dawydoff.)

mucus is spun out, the worm backs up, and eventually a funnel-shaped mucous collar surrounds the anterior end of the body. After the formation of the mucous collar, water is pumped through the burrow by peristaltic action of the trunk. Because the mucous "net" extends from the wall of the tube to the body of the worm, all sea water must pass through it, and even the finest particles are strained out. When loaded with food material, the net is detached from the body. The worm then moves backward, seizes the accumulated food mass using its short proboscis, and swallows the mass.

The digestive tract is extremely long and greatly coiled (Fig. 17–6). The mouth is located at the base of the proboscis and opens into a short muscular pharynx. A long, coiled esophagus connects the pharynx with a bulbous gizzard, and in some species there is an intervening stomach. A long, greatly coiled intestine comprises the remainder of the digestive tract. The end of the intestine joins a short rectum that opens through the anus at the posterior of the body.

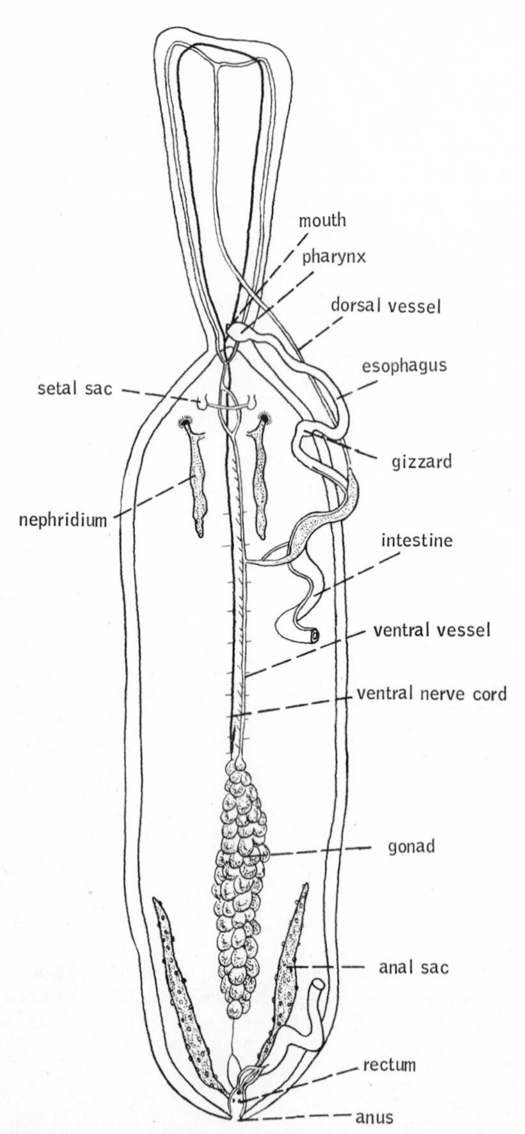

Figure 17–6. Internal structure of an echiuran. (After Delage and Hérourard from Dawydoff.)

A closed blood-vascular system is present, except in *Urechis*, and is constructed on the same plan as that of the annelids. A ventral vessel runs beneath the digestive tract and, at different levels between the pharynx and the anterior part of the intestine, depending on the species, gives rise to a pair of dilated contractile vessels that encircle the intestine and join the dorsal vessel (Fig. 17–6). These periintestinal vessels are involved in the propulsion of blood. The dorsal vessel originates at the junction of the periintestinal vessels and runs forward over the anterior portion of the digestive tract. The blood is colorless and contains some amebocytes.

The coelomic fluid probably aids in circulation because, among the many cells that it contains, there are some with hemoglobin. There are no gas exchange organs.

The excretory organs are metanephridia, which are large sacs similar to those of the sipunculans (Fig. 17–6). However, whereas there is only a single pair of nephridia in the sipunculans, echiurans possess one (in *Bonellia*), two (in *Echiurus*), or up to hundreds of pairs (in *Ikeda*). Moreover, in some genera, such as *Thalassema*, there are many more nephridia in the male than in the female. When there are only one or two pairs of nephridia, they are located anteriorly, and the nephridiopores open just behind the anterior setae.

In addition to the nephridia, echiurans possess a peculiar pair of accessory organs called anal sacs. These are simple, or sometimes branched, contractile diverticula that arise from each side of the rectum. Distributed over the surface of the diverticula and often arranged in tufts are numerous ciliated funnels similar to nephrostomes. The funnels open into the coelom and lead into the lumen of the diverticula. The anal sacs are believed to function in excretion in the same manner as the nephridia, but the waste in this case leaves the body by way of the anus.

Coelomic urns like those of the sipunculans are known to be present in at least one species of echiuran and probably will be found in others.

The nervous system is similar to that of the sipunculans and the annelids. The periesophageal commissures run forward into the proboscis and make their dorsal connection at the anterior of the proboscis margin (Fig. 17–6). The cerebral ganglia are reduced, but the dorsal loop of the periesophageal connectives is their homolog. The single ventral nerve cord lies just beneath the ventral vessel and displays no ganglionic swellings. There are no special sense organs.

Reproduction. The sexes are separate, and the gametes arise from the peritoneum of the ventral mesentery in the posterior of the trunk. As in annelids and sipunculans, the gametes are freed into the coelom, where maturation is completed, and the mature eggs or sperm are released through the nephridia. The anal sacs never function as gonoducts. Fertilization is external in the sea water except in *Bonellia,* in which the eggs are fertilized in the nephridial sacs, which act as uteri.

Species of *Bonellia* are peculiar in displaying an extreme sexual dimorphism. The males of such species are minute. In *Bonellia viridis,* the females may be 1 m. long, whereas the males are only 1 to 3 mm. in length (Fig. 17–5*B*). In addition to the reduction in size, the male is structurally modified in many ways. There is no proboscis or circulatory system, and the digestive tract is reduced. The entire body surface is ciliated, and there is a system of genital ducts. These dwarf males enter the body of the female and live in her uteri or coelom. Of special interest is the mechanism of sex determination in *Bonellia.* Any larva that comes in contact with a female and enters her body is induced by a female hormone to develop into a dwarf male. Larvae that do not come in contact with females develop into females.

Spiral cleavage and free-swimming trochophore larva characterize the embryology of echiurans. During the course of development, the mesodermal bands in some echiurans, such as *Echiurus,* are segmented and develop ten pairs of rudimentary coelomic pouches, which coincide with a similar transitory metamerism of the nerve cord.

The circulatory system, the setae, the presence of multiple nephridia in some species, and the transitory metamerism that appears during the course of embryonic development, in addition to the annelidan character of the body wall and nervous system, all indicate a close phylogenetic relationship between the echiurans and the annelids, closer it would seem than between the sipunculans and the annelids.

It appears probable that the echiurans represent a group of worms that stemmed from the early ancestral polychaetes or at least from the line leading to the polychaetes. Some degree of metamerism had certainly been attained in the ancestral group from which the echiurans arose.

PHYLUM PRIAPULIDA

The phylum Priapulida consists of only eight known species and six genera of cucumber-shaped animals, which live buried in the bottom sand and mud in shallow and deep water. Off the North American coast they range northward from Massachusetts and California. They are also found in the Baltic Ocean and off Siberia; some species have been taken from Antarctic waters. The tiny *Tubiluchus* lives in coralline sediments. Another minute species of a recently discovered genus is tubicolous.

The group has had an unsettled history. After the dismemberment of the old phylum Gephyrea, the priapulids were considered to be pseudocoelomates and were placed by Hyman (1951) within the phylum Aschelminthes. Later studies indicate that the body cavity is a coelom, not a pseudocoel. Thus, the removal of the priapulids from the aschelminth phylum is required.

External Structure. The cylindrical body of the priapulids may attain a length of some 6 cm. and is divided into an anterior proboscis region and a posterior trunk region (Fig. 17–7). The proboscis, which comprises the anterior third of the animal, is somewhat barrel-shaped and ornamented with longitudinal, riblike papillae. One species has the mouth surrounded by a crown of branched tentacles. The mouth invaginates into the anterior of the proboscis. The trunk is covered with small spines and tubercles and is divided superficially into 30 to 100 segments. In the little *Tubiluchus corallicola,* the trunk bears a long terminal tail (Fig. 17–8), and in the genus *Priapulus,* the posterior end of the trunk carries one or two caudal appendages, each consisting of a hollow stalk to which are attached many spherical vesicles. A gas exchange or chemoreceptive function has been suggested for these structures, but neither function has yet been demonstrated.

The body wall is composed of one layer of epithelial cells lying beneath the cuticle, and relatively well-developed circular and

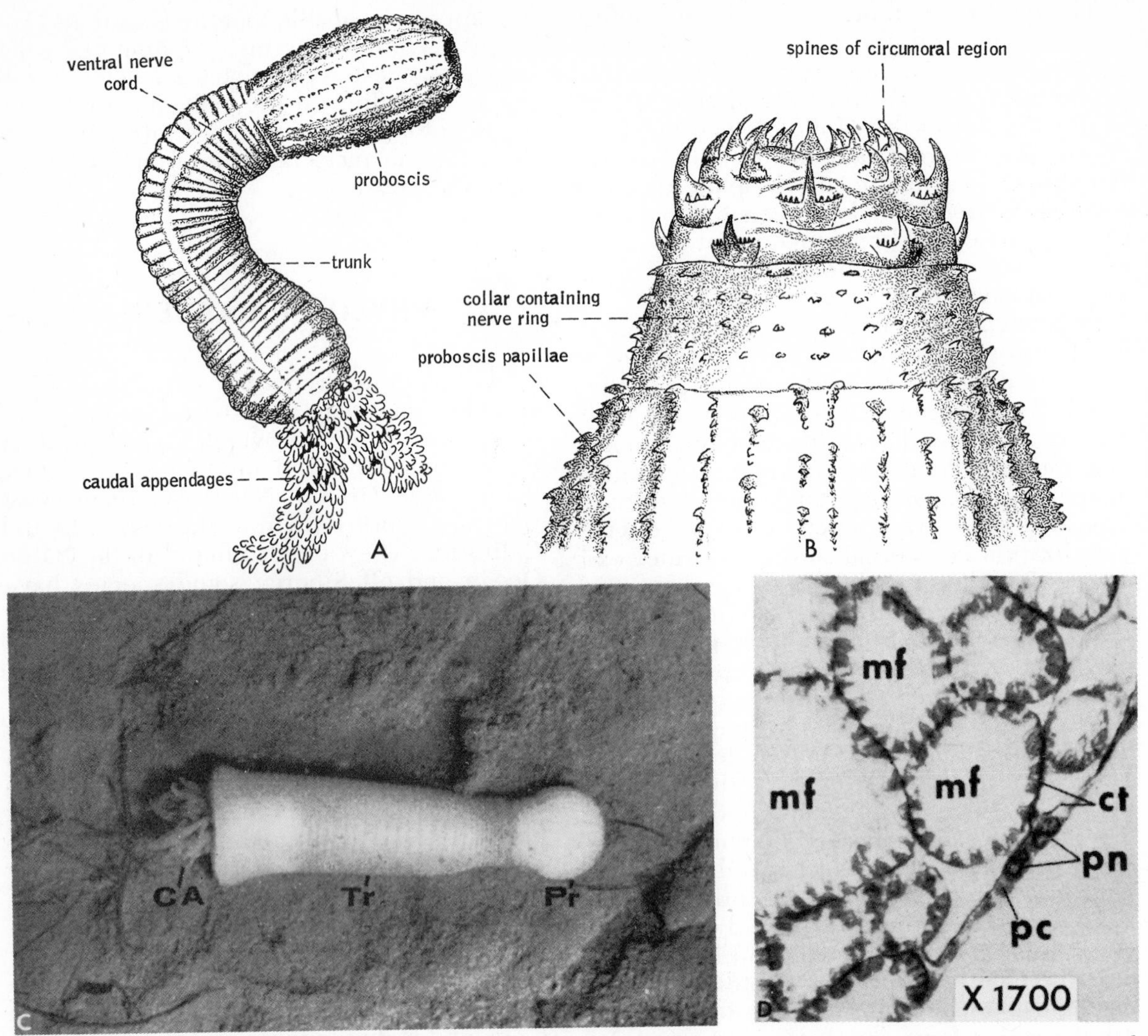

Figure 17–7. *A. Priapulus bicaudatus* (ventral view). *B. Priapulus;* anterior end with mouth region everted. (Both after Theel from Hyman.) *C.* A recently molted and contracted specimen of *Priapulus caudatus.* The ball-like swelling represents the retracted proboscis. The retracted caudal appendage partially projects from the posterior end of the trunk. (From Shapeero, W. L., 1962: The American Midland Naturalist, *68*:237.) *D.* Photograph of section through peritoneum and adjacent muscle fibers; mf, muscle fibers; ct, connective tissue; pn, peritoneal nucleus; pc, peritoneal cytoplasm. (From Shapeero, W. L., Science, *133*:879–880. Copyright 1961 by the American Association for the Advancement of Science.)

longitudinal muscle layers. The cuticle contains chitin and is periodically molted.

Internal Structure and Physiology. The body cavity was known to be bounded by a membrane, which also forms mesenteries for the internal organs. But it was thought that this membrane was acellular and that the body cavity was therefore a pseudocoel. The histological studies of Shapeero (1961) have disclosed that the membrane separating the body wall contains nuclei and is actually cellular in structure (Fig. 17–7*D*). The cellular condition would suggest that the lining of the body cavity is peritoneum and that the body cavity is a coelom. Unfortunately, the embryogeny of priapulids is still not well enough known to provide confirming evidence.

The coelom contains amebocytes and, at least in *Priapulus caudatus,* a large number of corpuscles bearing hemerythrin.

Priapulids are capable of burrowing through the substratum by contractions of the body. Locomotor activity in *Priapulus* is cyclical (Hammond, 1970). During sedentary periods, when it feeds and defecates, the animal lies extended with the proboscis protruded. During movement, contraction of

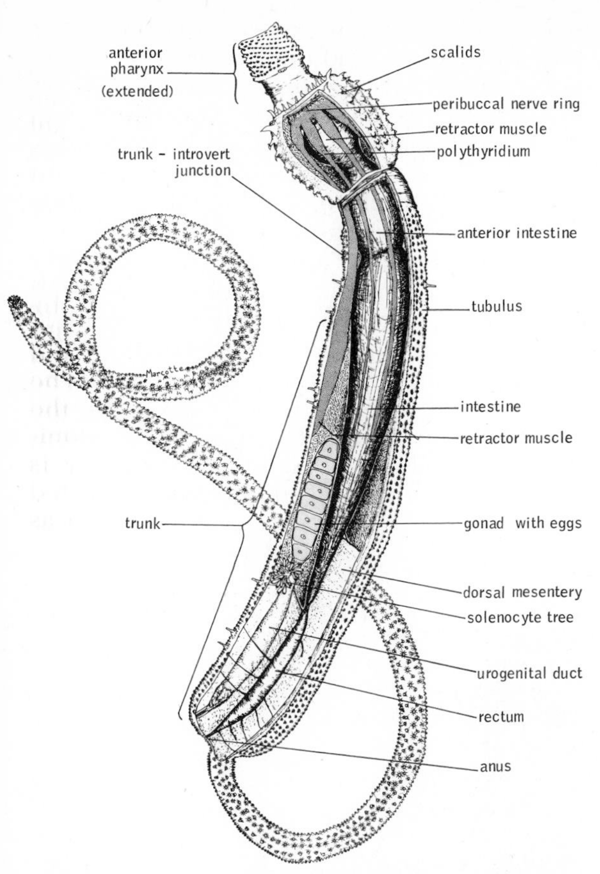

Figure 17–8. Structure of *Tubiluchus corallicola*, a minute priapulid which inhabits coralline sediments. (By Brian Marcotte.)

longitudinal and circular muscles of the body wall first brings about the invagination of the proboscis. This is followed by protrusion of the proboscis, during which it is thrust into the substratum and the animal moves forward.

Priapulids are predacious and feed on soft-bodied, slow-moving invertebrates, particularly polychaete worms. The invaginated mouth region, which is surrounded by several circlets of anteriorly curved spines, is everted during feeding (Fig. 17–7); the spines are used to seize prey. Food is passed to a muscular pharynx, which is lined with cuticle and has teeth. A straight tubular intestine connects the pharynx with the short terminal rectum. The anus is located at the posterior end of the trunk.

The priapulid nervous system is closely associated with the epidermis of the body wall and consists of a nerve ring surrounding the anterior end of the pharynx, and a single midventral ganglionated cord. No eyes are present, but the papillae on the proboscis and trunk probably have a sensory function.

The excretory and genital organs are closely associated. On each side of the intestine, there is an elongated body containing a central protonephridial tubule. Masses of solenocytes are connected to one side of the tubule and a single gonad is attached to the opposite side. The gonad is tubular and connects to the protonephridial tubule, which thus functions as an excretory canal and as a genital duct. Each protonephridial tubule opens separately through a nephridiopore at the end of the trunk.

Embryogeny and Phylogeny. The sexes are separate, but knowledge of the reproduction of these animals is very limited. The egg undergoes radial cleavage and develops into a stereoblastula. Gastrulation is by epiboly and a stereogastrula is the hatching stage. Although the larval stage is known, a description of the intervening period of development between the gastrula and larva is still lacking. Only the appearance of mesodermal strands has been observed in the gastrula. The origin of the coelom remains unknown. The larva is remarkably similar to that of rotifers. The priapulid larva possesses a distinct lorica, as well as a foot and toes with adhesive glands, and inhabits bottom mud as do the adults.

The priapulids display a number of similarities to the pseudocoelomate phyla. There is a cuticle which is molted. The proboscis is similar to that of kinorhynchs and acanthocephalans, and the larval stage is reminiscent of rotifers. However, if the body cavity is a coelom, as the evidence seems to indicate, these pseudocoelomate characters represent merely convergent evolution. Moreover, there is a significant absence of a number of distinctive aschelminth features. The muscle layers and protonephridia are composed of distinct cells, rather than being syncytial as in the pseudocoelomates. There is no cell constancy in priapulids, and the priapulid proboscis only superficially resembles that of kinorhynchs and acanthocephalans.

Granted that the priapulids are coelomates, their phylogenetic relationship to other coelomates is very obscure. Shapeero believes they represent the remnants of an ancient group, which has retained certain primitive features. A better understanding of their proper phylogenetic position may be attained with more complete knowledge of their embryogeny.

Their inclusion here with the pogonophorans, sipunculans and echiurans is largely one of convenience.

PHYLUM TARDIGRADA

The tardigrades are a group of very tiny but highly specialized animals called water bears. Although some reach 1.2 mm., the majority are less than 0.5 mm. in length. Although not uncommon, tardigrades are seldom encountered because of the rather limited habitats in which they live.

There are a few marine interstitial tardigrades that have been collected from both shallow and deep water. Also, there are some freshwater species that live in the bottom detritus or on aquatic algae and mosses. The majority of tardigrades, however, live in the water films surrounding the leaves of terrestrial mosses and lichens. This very restricted habitat is shared primarily with some bdelloid rotifers, and the two groups present many parallel adaptations. About 350 species of tardigrades have been described, but the North American fauna has received less attention than that of Europe.

External Structure. The bodies of tardigrades are short, plump, and cylindrical; and ventrally there are four pairs of stubby legs (Fig. 17–9). The anterior and posterior ends are more or less bluntly rounded, and the head and trunk are not demarcated. The legs are short cylindrical extensions of the ventrolateral body wall, and each leg terminates in four to eight claws. The body is covered by either a smooth or an ornamented cuticle, which in some tardigrades, such as

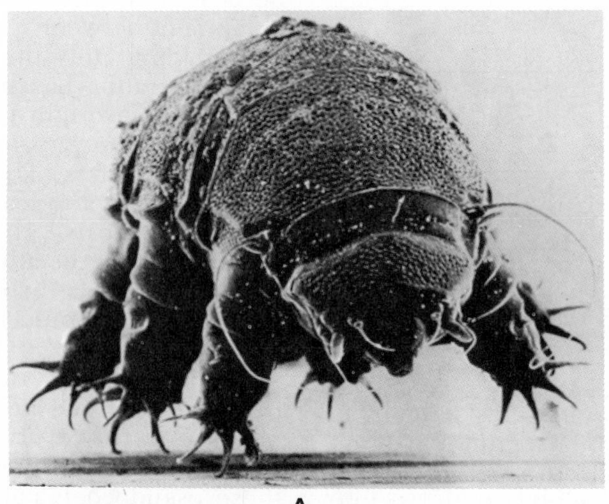

A

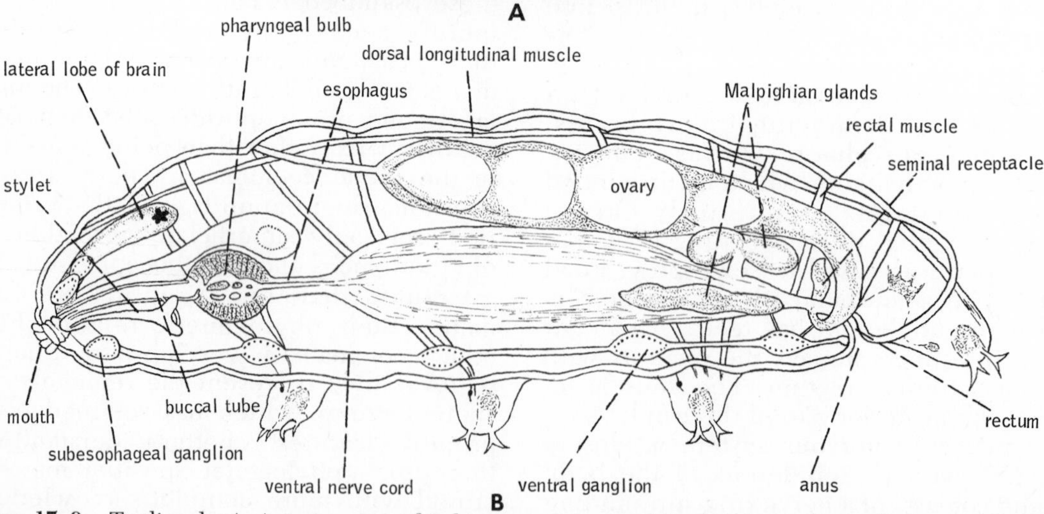

B

Figure 17–9. Tardigrada. *A.* Anterior view of *Echiniscus arctomys.* Mouth is located at end of snout-like anterior projection. (By Crowe, J. H., and Cooper, A. F., 1971: Scientific American, 225:30–36. Copyright ©1971 by Scientific American, Inc. All rights reserved.) *B.* Internal structure of *Macrobiotus hufelandi* (lateral view). (After Cuénot.) *C. Echiniscoides* climbing on an algal filament. (From Marcus, E., 1929: *In* Bronn, H. G. (Ed.), Klassen und Ordnungen des Tierreichs, vol. 5, pt. 4, p. 156. Akademische Verlagsgesellschaft, Frankfurt.)

Illustration continued on opposite page.

Echiniscus, is divided into symmetrically arranged plates (Fig. 17–9A). A limited number of spines or bristles is frequently present.

The cuticle is composed of albuminoids and does not contain chitin. The cuticle is secreted by an underlying epidermis that is composed of a constant number of cells, depending on the species. This constancy in cell number, like that in rotifers, is probably associated with the very small size of these animals. Periodically the old cuticle is shed, and the epidermis secretes a new one. Approximately six days prior to ecdysis, the cuticular pieces forming the buccal apparatus are cast out and re-formed. Also, new claws are formed and lie beneath the old. During molting, the body contracts, pulling away from the old cuticle, which is then slipped off and left behind as a relatively intact casing.

Internal Structure and Physiology. The musculature is not arranged in circular and longitudinal layers but in separate muscle bands, each composed of a single smooth

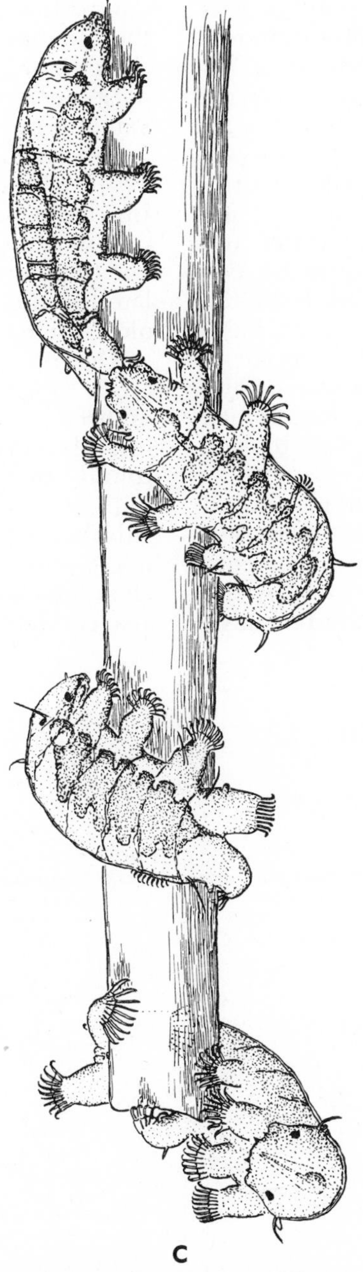

Figure 17–9. *Continued.*

muscle cell, extending from one subcuticular point of attachment to another (Fig. 17–9*B*). Tardigrades move about slowly, crawling on their legs and using the hooks at the ends of the legs for grasping the substratum.

The coelom is confined to the gonadial cavity, but a fluid-filled hemocoel extends between the muscle bands and the other internal organs. There is neither a special gas exchange system nor a circulatory system.

The majority of tardigrades feed on the contents of plant cells, which are pierced with a special feeding apparatus resembling that of herbivorous nematodes and rotifers (Figs. 17–9*B* and 17–10). The anterior mouth opens into the buccal tube, which is followed by the bulbous muscular pharynx. These two divisions comprise the foregut and are lined by cuticle. One stylet lies on each side of the buccal tube. The anterior pointed ends of the two stylets project into the anterior end of the buccal tube, and their posterior ends are supported by two transverse pieces, which extend from the wall of the buccal tube to the stylet base. The buccal tube is also flanked by a pair of large glands, each of which opens into the buccal tube near the projecting stylet points. These glands secrete the stylets prior to each molt.

During feeding, the mouth is placed against the plant cells, and the stylets are projected to puncture the cell wall. The contents of the cell are then sucked out by the pharyngeal bulb. Some tardigrades have been observed to attack rotifers, nematodes, and other tardigrades, but it is not known if such species are strictly carnivorous.

The pharyngeal bulb passes into the tubular esophagus, which opens in turn into a large midgut, or intestine (Fig. 17–9*B*). Here digestion and absorption take place. The end of the intestine leads into a short hindgut (rectum), which opens to the outside through the terminal anus. In some tardigrades, such as *Echiniscus*, egestion is associated with molting, and the feces are left behind in the old cuticle. In these forms, the expulsion of the feces aids in the contraction of the body and separation from the old cuticle.

At the junction of the intestine and rectum of some tardigrades are three large glands, sometimes called malpighian tubules, that are thought to be excretory in function (Fig. 17–9*B*). The buccal glands and intestinal lining have been suggested as playing some role in excretion in one group of tardigrades, and there are indications that wastes accumulated in the epidermis are left behind in the old cuticle at molting.

The nervous system is distinctly metameric (Fig. 17–9*B*). The brain is composed of three median lobes and two large recurved lateral lobes. A pair of commissures surrounds the buccal tube and connects ventrally to a subpharyngeal ganglion. From the subpharyngeal ganglion, a double ventral nerve cord extends posteriorly, connecting a chain of four ganglia, the first of which is also connected dorsally to the lateral lobes of

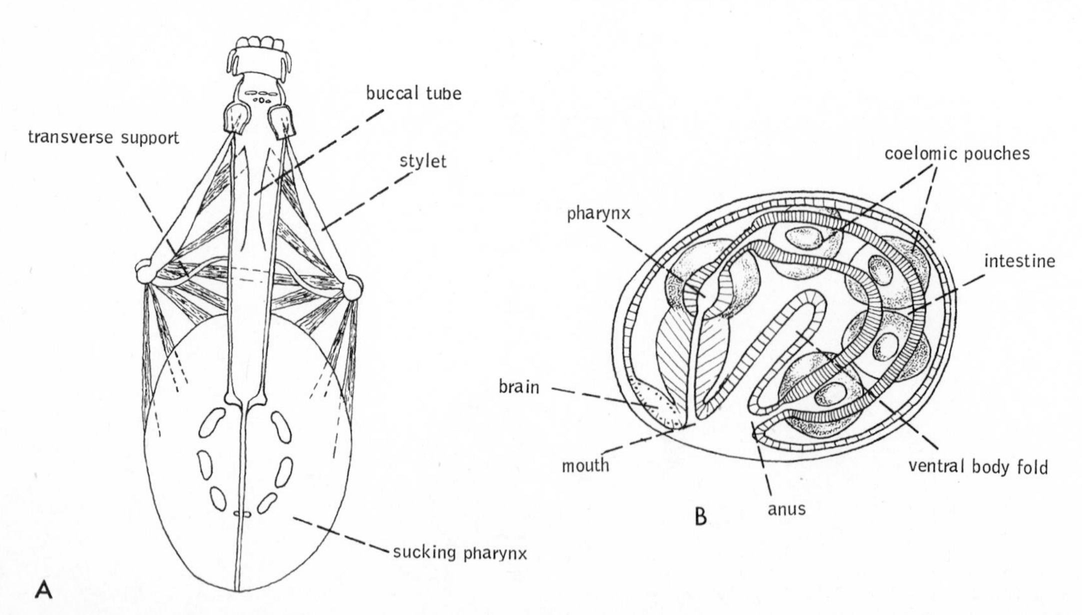

Figure 17–10. *A*. Pharynx and buccal apparatus of a tardigrade (dorsal view). (After Pennak.) *B*. Embryo of *Hypsibius*. (After Marcus from Cuénot.)

the brain by a pair of connectives. From each ganglion issue several pairs of lateral nerves, one of which terminates in pedal ganglia in each pair of legs. Some of the bristles and spines are sensory, particularly those in the head region. Most tardigrades also possess a pair of simple eye spots, each containing a single red or black pigmented cell.

Reproduction. Tardigrades are all dioecious and have a single saccular gonad, testis or ovary, located above the intestine. In the male, two sperm ducts leave the testis posteriorly, encircle the rectum, and open through a single median gonopore just in front of the anus. In the female, there is only a single oviduct, which passes to one side or the other of the rectum and either opens through the female gonopore above the anus or opens into the ventral side of the rectum (Fig. 17–9*B*). In the latter case a small adjacent seminal receptacle also opens into the rectum.

Females are often more numerous than males, and in some genera, such as *Echiniscus*, males are unknown. In many tardigrades, particularly terrestrial species, fertilization takes place within the ovary. In such cases, sperm are either deposited in the cloaca, where they are temporarily stored in the seminal receptacle, or the sperm are deposited in the hemocoel by impregnation through the cuticle. In many aquatic species, sperm are deposited beneath the old cuticle of the female just prior to molting. When the female molts, the eggs are also deposited in the old cuticle, where they are fertilized.

One to thirty eggs are laid at a time, depending on the species. In aquatic tardigrades, they may be either deposited in the old cuticle, or attached singly or in groups to various objects. Like rotifers and gastrotrichs, some aquatic tardigrades are reported to produce thin-shelled eggs when environmental conditions are favorable and thick-shelled eggs when environmental conditions are adverse. The eggs of terrestrial species typically possess a very thick sculptured shell that can resist the frequent periods of desiccation to which mosses are subjected. Parthenogenesis is common.

Development is always direct. Cleavage is holoblastic but does not follow any of the patterns that have been previously discussed. After the gut has formed, five pairs of coelomic pouches appear (Fig. 17–10*B*). The coelomic pouches are arranged metamerically, but most peculiarly they arise as outpocketings of the gut. Such a mode of mesoderm formation (enterocoelous) is characteristic of the deuterostomes but not the protostomes. The last pair of coelomic sacs fuse together to form the gonad; the other coelomic pouches disintegrate, and their cells form the body musculature. Development is completed within 14 days or less, and the little tardigrades hatch by breaking the shell with their stylets. Further growth is attained by the increase in the size of cells rather than by the addition of cells.

Anabiosis. Tardigrades, like moss-inhabiting rotifers, exhibit remarkable ability to withstand extreme desiccation and exposure to low temperatures. Thus during periods of dry weather when the moss becomes desiccated, the tardigrade inhabitants pull in their legs, lose water, become contracted and shriveled, and pass into an anabiotic (or cryptobiotic) state. Species of two genera, *Macrobiotus* and *Hypsibius*, can also form cysts by withdrawing from the cuticle, which forms the cyst wall. In this state, metabolism proceeds at a very low rate, and the animal can withstand abnormal environmental conditions. For example, specimens have recovered after immersion in liquid helium (−272 degrees C.), brine, ether, absolute alcohol, and other substances.

When water is again present, the animal swells and becomes active within a few hours. There are records of tardigrades emerging from a state of anabiosis lasting seven years. Under natural conditions, the life of moss-inhabiting tardigrades is undoubtedly frequently interrupted by anabiosis, which may well lengthen the life span of these animals by years.

Phylogeny. The phylogenetic position of tardigrades is uncertain. They show many similarities to the aschelminths, especially rotifers, but their embryology would indicate that they are coelomates. The metameric character of the nervous system and the transitory appearance of metamerically arranged coelomic pouches (despite their enterocoelous origin) could place this group somewhere along the annelid-arthropod line. Some zoologists believe that tardigrades are derived from the mites, the tardigrade cuticle being homologous to the arachnid epicuticle. However, the resemblance between tardigrades and mites is certainly very superficial. Moreover, many of the tardigrade structural peculiarities are undoubtedly specializations for living in their restricted habitats. Although tardigrades are not particularly intermediate in structure between annelids and arthropods, they may have stemmed from the main protostome line at some point between the annelids and the arthropods. At least, this is as good a guess as any on the basis of present knowledge.

PHYLUM PENTASTOMIDA, OR LINGUATULIDA

Pentastomids comprise a little phylum of approximately 70 species, called tongue worms, which are related to the arthropods. All members of the phylum are parasitic and live within the lungs or nasal passageways of vertebrates. The principal hosts are tropical reptiles, such as snakes and crocodiles, but some species parasitize mammals and birds. Pentastomids have also been reported in North America, Europe, Australia, and even in Arctic birds.

The body is from 2 to 13 cm. in length and bears anteriorly five short protuberances, from which the name pentastomid is derived (Fig. 17–11A). Four of these projections are leglike, bearing claws, and are located two to each side of the body. The fifth projection is an anterior, median, snoutlike process bearing the mouth. Not infrequently the legs are reduced to nothing more than the claws, which are used for clinging to the tissues of the host.

The body is covered by a thick chitinous cuticle that is molted periodically during larval development. The body-wall muscles are striated but arranged in layers. The digestive tract is a relatively simple straight tube with the anterior end modified to pump in the blood of the host, on which these parasites subsist (Fig. 17–11B). The nervous system is similar to that of the annelids and the arthropods, with paired, metamerically

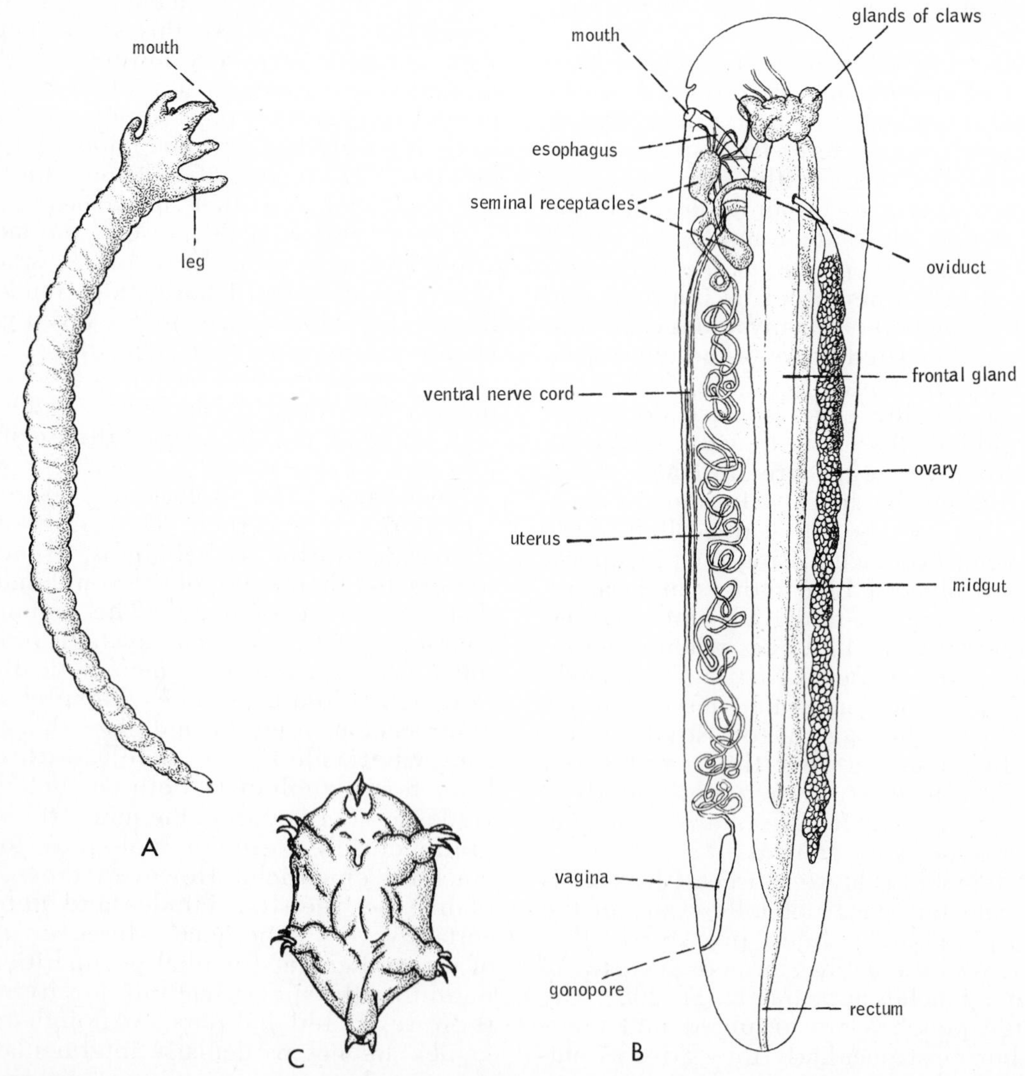

Figure 17–11. Pentastomida. *A. Cephalobaena tetrapoda* from the lung of a snake. (After Heymons from Cuénot.) *B*. Internal structure of a female *Waddycephalus teretiusculus*, parasitic in certain Australian snakes. (After Spencer from Cuénot.) *C*. Larva of *Porocephalus crotali*. (After Penn.)

arranged ganglia located along the ventral nerve cord. There are no gas exchange, circulatory, or excretory organs.

The sexes are separate, with a well-developed genital system (Fig. 17–11*B*). Fertilization is internal, and the embryonated eggs are passed into the digestive system of the host and then to the outside in the feces of the host. In most pentastomids, the life cycle requires an intermediate host, which may be fish (pentastomids in crocodiles), or herbivorous or omnivorous mammals, such as rodents, rabbits (intermediate host for *Linguatula* in dogs), or small ungulates. Actually, larval pentastomids have been reported in almost every class of vertebrates. Larval development takes place within the intermediate host and involves a number of molts. The larvae possess four to six leglike appendages (Fig. 17–11*C*). When the intermediate host is eaten, the parasite is transferred to the stomach of the primary host and reaches the lungs and nasal passageways through the esophagus. Pentastomid larvae have been reported from man but are killed by calcareous encapsulation.

The taxonomic status of the pentastomids is uncertain, and their being ranked here as a phylum is undoubtedly arbitrary. There is little question that the group is related to the arthropods, and perhaps they should be considered a class of that phylum. But there is little agreement among those who consider them arthropods as to where they should be placed within the phylum. As in the case of the tardigrades, some zoologists argue that the pentastomids had an acarine origin, but the evidence appears rather superficial and is probably prejudiced because of the parasitic nature of the pentastomids. A myriapod relationship has also been suggested.

Bibliography

Clark, R. B., 1969: Systematics and phylogeny: Annelida, Echiura, Sipuncula. *In* Florkin, M., and Scheer, B. J. (Eds.), Chemical Zoology, Vol. 4. Academic Press, N. Y., pp. 1–62.

Crowe, J. H., and Cooper, A. F., Jr., 1971: Cryptobiosis. Scientific American, *225*(12):30–36.

Cuénot, L., 1949: Les Onychophores, Les Tardigrades, et Les Pentastomides. *In* Grassé, P. (Ed.), Traité de Zoologie, Vol. 6. Masson et Cie, Paris, pp. 3–75.

Dawydoff, C., 1959: Classe des Echiuriens. *In* Grassé, P. (Ed.), Traité de Zoologie, Vol. 5, pt. 1. Masson et Cie, Paris, pp. 855–907.

George, J. D., and Southward, E. C., 1973: A comparative study of the setae of Pogonophora and polychaetous Annelida. J. Mar. Biol. Assoc. U. K., *53*(2):403–424.

Gosner, K. L., 1971: Guide to Identification of Marine and Estuarine Invertebrates: Cape Hatteras to the Bay of Fundy. Wiley-Interscience, N. Y. 693 pp.

Hammon, R. A., 1970: The burrowing of *Priapulus caudatus*. J. Zool., *162*:469–480.

Heymons, R., 1935: Pentastomida. *In* Bronn, H. G. (Ed.), Klassen und Ordnungen des Tierreichs. Bd. 5, Abt. IV. Akademische Verlagsgesellschaft, Frankfurt.

Hyman, L. H., 1959: The Invertebrates. Vol. 5, The Smaller Coelomate Groups. McGraw-Hill, N. Y., pp. 610–696. (A general treatment of the sipunculans with an extensive bibliography.)

Ivanov, A. V., 1963: Pogonophora. Consultants Bureau, N. Y., 479 pp.

Kaestner, A., 1967–1968: Invertebrate Zoology, Vols. 1 and 2. Wiley-Interscience, N. Y.

Kohn, A. J., and Rice, M. E., 1971: Biology of Sipuncula and Echiura. Bioscience, *21*:583–584. (A brief review of a symposium on the biology of these two phyla.)

Light, S. F., Smith, R. I., Pitelka, F. A., Abbott, D. P., and Weesner, F. M., 1967: Intertidal Invertebrates of the Central California Coast. University of California Press, Berkeley. 446 pp.

Marcus, E., 1929: Tardigrada. *In* Bronn, H. G. (Ed.), Klassen und Ordnungen des Tierreichs, Bd. 5, Abt. IV. Akademische Verlagsgesellschaft, Frankfurt. (At the present time, this is the most authoritative work on tardigrades.)

Marcus, E., 1959: Tardigrada. *In* Edmondson, W. T., Ward, H. B., and Whipple, G. C. (Eds.), Freshwater Biology. 2nd Edition. John Wiley, N. Y., pp. 508–521. (Figures and key to all of the nonmarine genera of the world and 76 species found in fresh water.)

Nørrevang, A., 1970: The position of Pogonophora in the phylogenetic system. Zeitschr. Zool. Syst. Evolutionsforsch., *8*, H. 3, 161–172.

Nørrevang, A., 1970: On the embryology of *Siboglinum* and its implications for the systematic position of the Pogonophora. Sarsia, *42*:7–16.

Pennak, R. W., 1953: Freshwater Invertebrates of the United States. Ronald Press, N. Y., pp. 240–255. (A general discussion of freshwater tardigrades, with key and figures to the common genera in the United States.)

Rice, M. E., 1969: Possible boring structures of sipunculids. Amer. Zool., *9*:803–812.

Rice, M. E., 1970: Asexual reproduction in a sipunculan worm. Science, *167*:1618–1620.

Riggin, G. T., Jr., 1962: Tardigrada of southwest Virginia, with the addition of a description of a new marine species from Florida. Virginia Agr. Exp. Sta. Tech. Bull., *152*:1–145. (This paper is a good source for taxonomic work on North American tardigrades.)

Shapeero, W. L., 1961: Phylogeny of Priapulida. Science, *113*(3456):879–880.

Shapeero, W. L., 1962: The distribution of *Priapulus caudatus* on the Pacific coast of North America. Amer. Midl. Nat., *68*:237–241.

Southward, E. C., 1971: Recent researches on the Pogonophora. Oceanogr. Mar. Biol., Ann. Rev., *9*:193–220.

Southward, E. C., 1971: Pogonophora of the Northwest Atlantic: Nova Scotia to Florida. Smithsonian Contrib. Zool., *88*. 29 pp.

Stephen, A. C., and Edmonds, S. J., 1972: The Phyla Sipuncula and Echiura. British Museum, London. 528 pp.

Van der Land, J., 1970: Systematics, zoogeography, and ecology of the Priapulida. Zool. Verh. Rijksmus. Nat. Hist. Leiden, *112*:1–118.

THE LOPHOPHORATES

Three related phyla of protostomate coelomates remain to be considered—the Phoronida; the Bryozoa, or moss animals; and the Brachiopoda, or lamp shells. All members of these three phyla possess a food-catching organ called a lophophore, and they are usually grouped together as the lophophorate coelomates. Discussion of the lophophorates has been deferred until this point not because they hold a terminal position among the protostomes, but because they possess some characteristics in common with the deuterostomes, the other major evolutionary line of the Animal Kingdom that is discussed in the remaining chapters.

The food-catching organ, or lophophore, that is characteristic of these phyla is a circular or horseshoe-shaped fold of the body wall that encircles the mouth and bears numerous ciliated tentacles (Fig. 18–1*A*). The tentacles are hollow outgrowths of the body wall. Each contains an extension of the coelom. The ciliary tracts on the tentacles drive a current of water through the lophophore, and plankton is collected in the process.

In many deuterostomes, such as echinoderms and hemichordates, the body is divided, at least in the larval stage, into three regions—a protosome, a mesosome, and a metasome; and each region contains a coelomic compartment—a protocoel, a mesocoel, and a metacoel. Lophophorates appear to exhibit such a tripartite body organization. Although there is no well-developed protocoel in lophophorates, perhaps as a result of the reduced head, the existing coelom is clearly divided by a transverse septum into a mesocoel and metacoel. The mesocoel is contained within the anterior end of the body and the lophophore, and the metacoel occupies the trunk region of the body.

In addition to possessing a lophophore and a divided coelom, nearly all members of these phyla are sessile, have a reduced head, secrete a protective covering, and except for some of the brachiopods, possess a **U**-shaped digestive tract (Fig. 18–2*A*). But all of these characteristics are correlated with a sessile existence and, in part at least, represent evolutionary convergence.

From a phylogenetic standpoint, the embryogeny of lophophorates is particularly interesting, for both protostome and deuterostome features are displayed. In phoronids and brachiopods, the mouth is derived from the blastopore, and the larvae of phoronids and bryozoans are modified trochophores. Thus, these phyla are clearly protostomes. But only in some phoronids is there any evidence of spiral cleavage; in bryozoans and especially brachiopods, cleavage is markedly radial. Moreover, in some brachiopods the mesoderm and coelom arise by enterocoelic pouching, as is typical of deuterostomes. For this reason, brachiopods are sometimes separated from the other two lophophorate phyla.

PHYLUM PHORONIDA

The phoronids are a small group consisting of two genera and approximately 15 species of wormlike animals. All members are marine and live within a chitinous tube that is either buried in sand or attached to rocks, shells, and other objects in shallow water (Fig. 18–1*B*). A few species bore into mol-

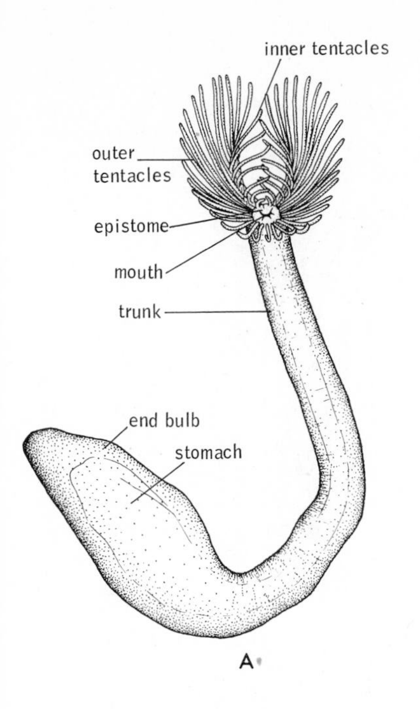

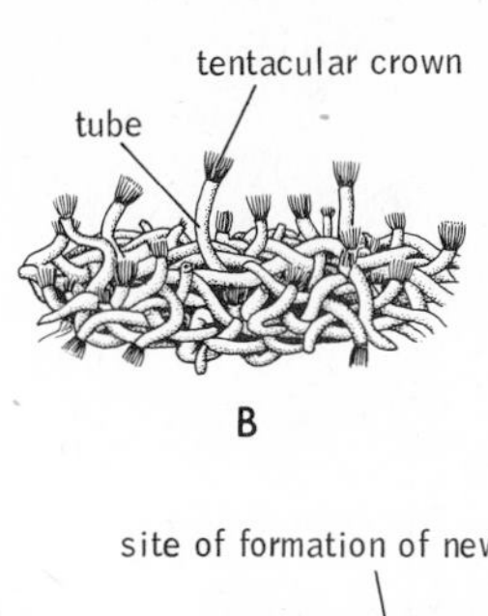

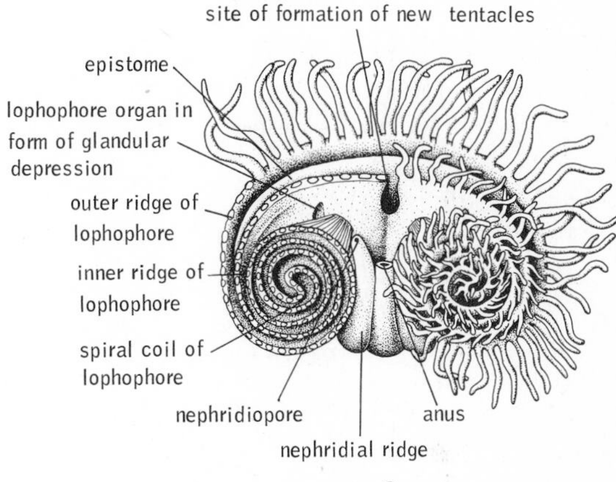

Figure 18–1. *A. Phoronis architecta.* (After Wilson.) *B.* Part of a *Phoronis hippocrepia* colony. (After Shipley.) *C.* Anterior view of *Phoronis australis.* Tentacles on right side have been cut away in order to show spiral arrangement of lophophore. (After Shipley.)

lusk shells. Although the body is free within the tube, only the anterior of the animal ever emerges from the opening of the tube.

The cylindrical body, which in most species is less than 20 cm. long, is somewhat enlarged posteriorly and bears no appendages or regional differentiation except at the anterior end (Fig. 18–1*A*). Here the lophophore is the most conspicuous feature. The lophophore consists of two parallel ridges curved in the shape of a crescent or horseshoe (Fig. 18–1*C*). The bend of the crescent is located ventrally, with one ridge passing above and the other ridge passing below the mouth. The horns of the ridges are directed dorsally and may be rolled up as a spiral.

Each lophophoral ridge bears a large number of hollow tentacles that are slender, ciliated extensions of the body wall (Fig. 18–1*A*). A long, narrow, crescentic fold, called the epistome, overhangs the mouth. The anus opens above the mouth but outside of the upper lophophoral ridge and is flanked on each side by a nephridiopore. The lophophore is separated from the trunk by a groove in *Phoronis* and by a collar-like fold in *Phoronopsis*.

The body wall is composed of an outer epithelium, under which lies a thin layer of circular muscle fibers and a thick longitudinal muscle layer. Gland cells are abundant at the anterior end of the body, particularly on the lophophore.

The coelom is divided into the posterior metacoel, which occupies the trunk region, and the anterior mesocoel, which extends into the lophophore and each of the tentacles. The two divisions are separated by a peritoneal septum. The coelomic fluid contains red corpuscles and coelomocytes.

Like all lophophorates, phoronids are filter feeders. The tentacular cilia beat downward, creating a water current directed toward a groove between the two lophophoral ridges. The current then follows the grooves inward on each side toward the mouth and exits dorsally over the anus and nephridiopores. Plankton and suspended detritus that are caught in the lophophoral current are entangled in mucus on contact with the tentacles and the basal groove. Cilia on the basal groove convey the food particles toward the mouth, the lateral angles of which are continuous with the lophophoral groove.

The digestive tract is **U**-shaped (Fig. 18–2*A*). The mouth opens into a buccal region and then into a long tubular esophagus. The esophagus passes gradually into the enlarged

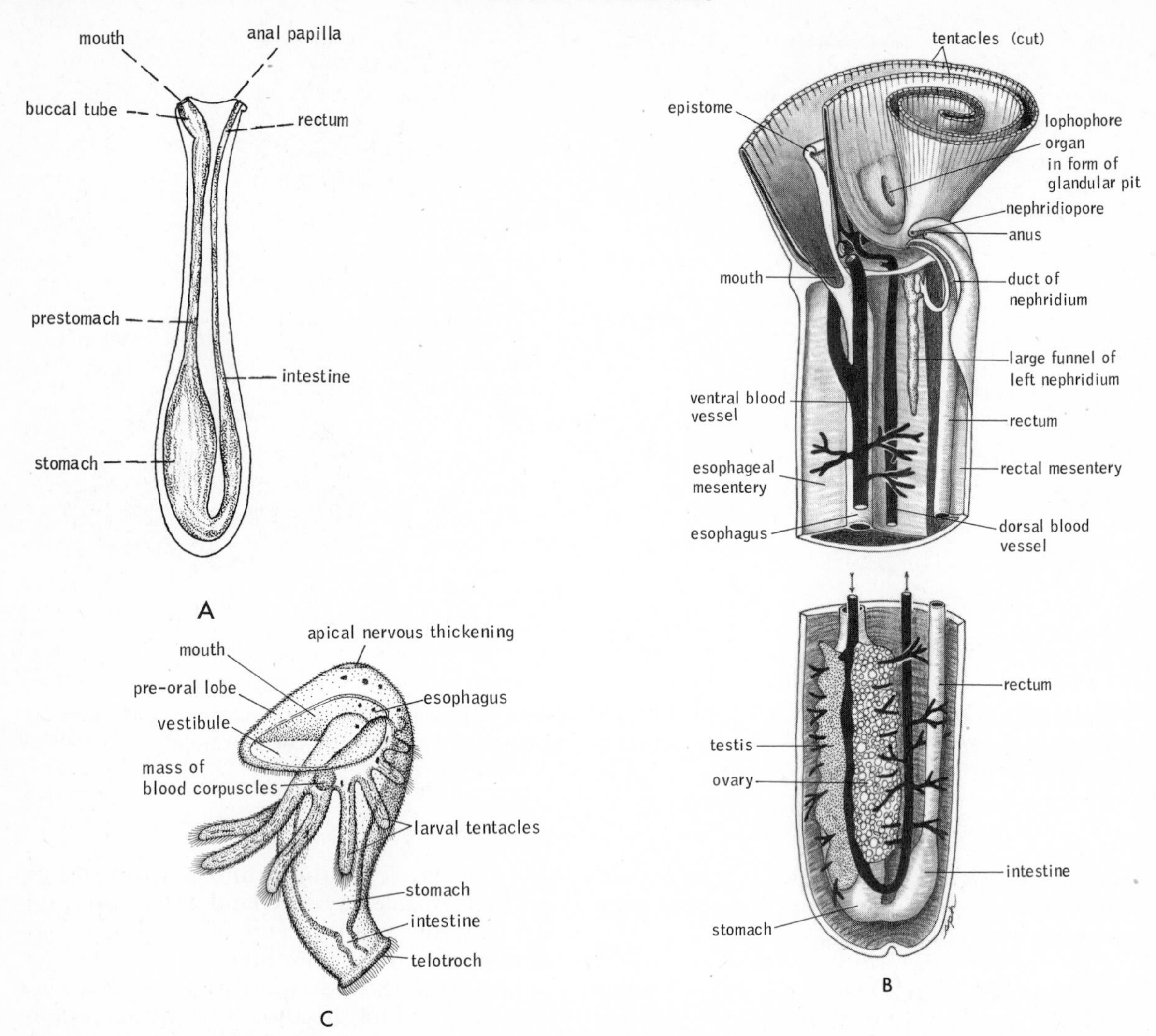

Figure 18–2. *A.* Digestive tract of a phoronid. (After Hyman.) *B.* Internal structure of anterior and posterior of *Phoronis australis.* (After Benham.) *C.* Actinotroch larva of *Phoronis.* (After Wilson.)

stomach in the posterior end of the animal. The stomach joins a long tubular intestine, and at this point of junction, the digestive tract makes a 180-degree turn so that the intestine extends forward and dorsally to the anterior half of the alimentary tract. A terminal rectum opens to the outside through the dorsal anus located in the midline above the lophophore. Digestion probably takes place in the stomach and is believed to be intracellular.

A definite blood-vascular system is present in phoronids. There are two main vessels—a dorsal vessel that lies between the two limbs of the gut and carries blood anteriorly, and a ventral vessel that lies to the left of the esophagus and delivers blood to the posterior of the animal (Fig. 18–2*B*). Anteriorly, the dorsal vessel supplies each tentacle with a small vessel. The lophophore is drained by the ventral vessel. Posteriorly, the dorsal and ventral vessels are connected by a network of sinuses within the stomach wall. Along its course, the ventral vessel gives off a large number of small blind contractile vessels called capillary ceca. The blood possesses red corpuscles that contain hemoglobin. The blood is driven through the system by waves of contraction that sweep over the dorsal and ventral vessels.

Excretion is accomplished by a pair of ciliated **U**-shaped metanephridia located at the anterior of the trunk coelom (Fig. 18–2*B*).

The nervous system consists of a nerve ring located in the epidermis at the outer base of the lophophore. The nerve ring gives rise to nerves supplying the tentacles, to motor fibers innervating the longitudinal muscle layer of the body wall, and to a single giant motor fiber that courses through the epidermis on the left side of the body as the lateral nerve. A small fiber is situated on the right side in a few species. The nerve ring is also continuous with a nervous layer located in the base of the epidermis over the entire body wall. The epidermal nerve plexus supplies motor fibers to the body wall musculature and also supplies the epidermis with numerous sensory cells. The sensory cells comprise the principal part of the sensory system of phoronids.

At least one species, *Phoronis ovalis,* that is found in aggregations, is known to reproduce asexually, both by budding and by transverse fission. The majority of phoronids are hermaphroditic. The sex cells arise from the peritoneum covering the blind capillary ceca of the ventral vessel, and the gonads are merely masses of cells surrounding these vessels, the ovaries located on one side and the testes on the other. The gametes are shed into the coelom of the trunk and escape to the outside by way of the nephridia. The eggs of most species are fertilized externally and either are planktonic or are brooded in the concavity formed by the two arms of the lophophore. In brooding species, the eggs are held in place by the inner tentacles and probably by secretions from the lophophoral organs, a glandular area located on each side of the lophophoral concavity.

The early embryogeny of some species shows traces of spiral cleavage. The gastrula develops into an elongated ciliated larva called an actinotroch, which is a modified trochophore (Fig. 18–2*C*). A large preoral hood covers the mouth, and behind the mouth is a slanting collar bearing ciliated tentacles. A heavily ciliated telotroch, which is probably the principal locomotor organ, rings the posterior of the trunk. There is a complete and more or less straight digestive tube, and like the trochophore of mollusk and annelids, the larval excretory organs are protonephridia. After several weeks of a free-swimming and feeding planktonic existence, the actinotroch undergoes a rapid metamorphosis and sinks to the bottom, where it secretes a tube and takes up an adult existence.

PHYLUM BRYOZOA

The phylum Bryozoa, or Polyzoa, or Ectoprocta, is the largest and the most common of the lophophorate phyla and contains approximately 4000 living species. The group constitutes one of the major animal phyla, despite the fact that it has largely been left to the interest of specialists. The Bryozoa formerly included the Entoprocta; however, the two groups are probably only distantly related. The tentacular crown of entoprocts is not a lophophore, since the tentacles enclose the anus. Moreover, the entoproct body cavity appears to be a pseudocoel. With the separation of the Entoprocta, some zoologists, notably Hyman (1959), encouraged the use of the name Ectoprocta for the remaining bryozoans and the dropping of the name Bryozoa. These authors suggested that such a change would avoid confusion and would emphasize the distinction between the two groups. Although the name Ectoprocta has been adopted in many texts, including the first edition of the present one, the majority of European and many American zoologists and specialists prefer to retain the name Bryozoa. They argue with some justification that the removal of the small entoproct group from the Bryozoa does not warrant changing a well-established name for the remaining predominant members of the phylum.

With very few exceptions, bryozoans are all colonial and sessile animals, and the individuals composing the colonies are usually less than 0.5 mm. in length. Individuals of most species are encased in a nonliving envelopment that contains an opening for the protrusion of the lophophore. The interior of the body is occupied largely by the rather spacious coelom and the **U**-shaped digestive tract. There are no gas exchange, circulatory, or excretory organs, probably owing to the small size of these animals.

Farmer et al. (1973) have suggested that the bryozoans evolved from some line of phoronids which took up an epibenthic existence, and that most of the peculiarities of bryozoans are associated with miniaturization and with the evolution of colonial organization and a skeletal covering.

The phylum is divided into three classes —the Phylactolaemata, the Gymnolaemata, and the Stenolaemata. The class Stenolaemata contains some living marine species and over 500 genera of fossil species. The class Gymnolaemata is almost entirely marine and includes the great majority of living bryozoans, as well as many fossil species. The class Phylactolaemata, on the other hand, is restricted to fresh water and, although it is widely distributed, contains only about 50 species.

The Structure of the Zooid. The bryo-

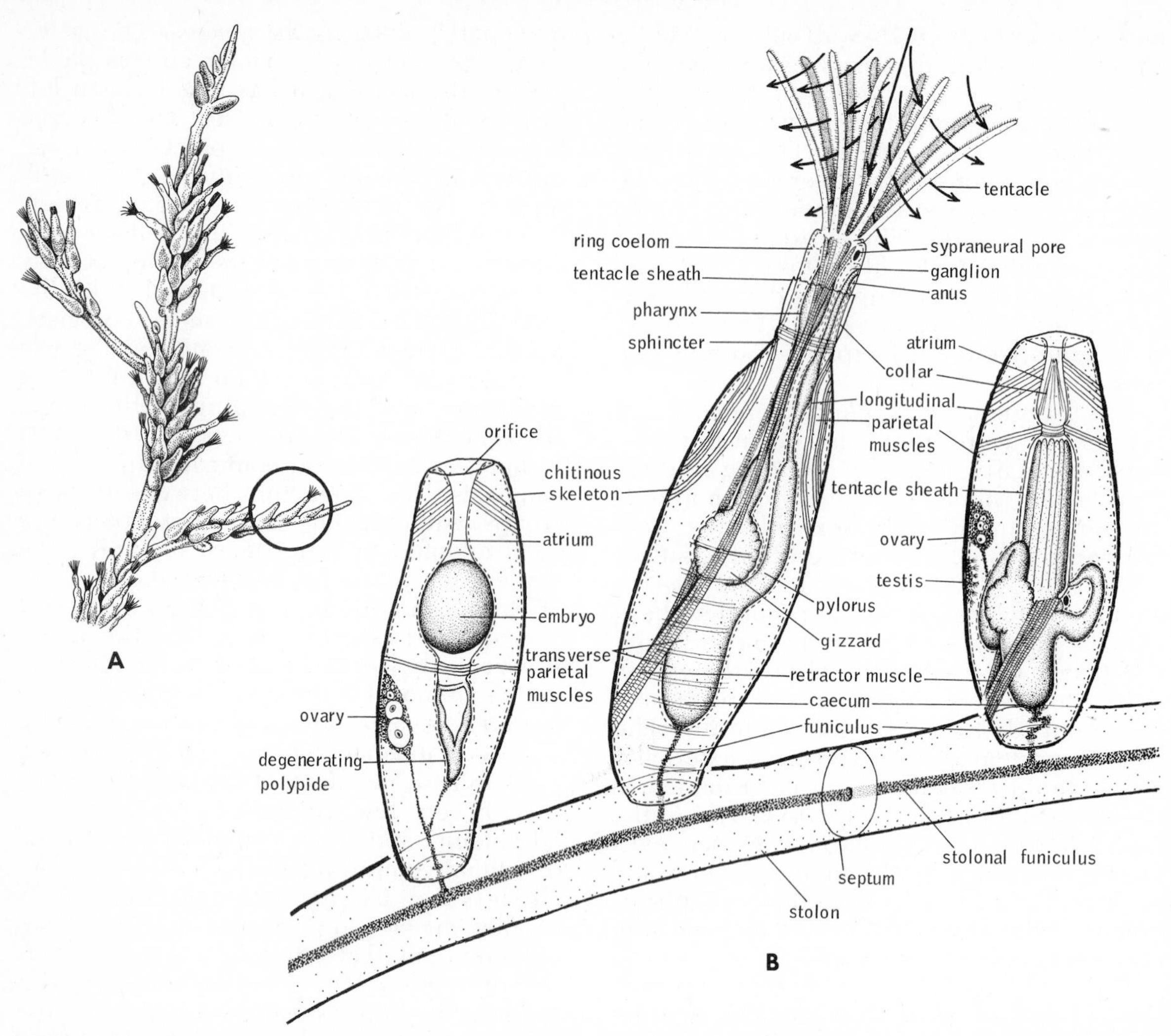

Figure 18–3. *Bowerbankia*, a stoloniferous gymnolaemate bryozoan. *A*. Section of colony. (Modified from Gay in Prenant and Bobin.) *B*. Stolon, brooding individual, and two autozooids of *Bowerbankia*, from circled portion of *A*. One autozooid has lophophore protruded and the other has it retracted. (Modified from Ryland and others.)

zoan individual is boxlike, oval, vaselike, or tubular in shape (Figs. 18–3 and 18–4). In gymnolaemates, on which the following description is based, the nonliving outer covering (zoecium) is composed of chitin or of chitin overlying a thick layer of calcium carbonate. The latter forms an especially rigid external skeleton. Some impregnation of the chitinous layer with calcium carbonate may be present, even when a calcareous layer is absent. This is true of many ctenostome-gymnolaemates, including the common *Bugula*.

An opening (the orifice) enables the lophophore to protrude, and in a considerable number of marine bryozoans the orifice is provided with a lid (operculum), which closes when the lophophore withdraws (Fig. 18–7*A*). At the orifice, the exoskeleton is turned inward for some distance to form a chamber called the atrium. The bottom of the atrium may be provided with a sphincter, or in a few forms with a collar (Figs. 18–3 and 18–8*C*).

The outer covering, or exoskeleton, is secreted by the epidermis of the body wall (Figs. 18–3 and 18–7*A*). As would be expected, there are no muscle layers composing the body wall underlying the rigid exoskeleton of gymnolaemates, and the epidermis overlies a thin, delicate peritoneum.

Within the body wall, there is a large coelom surrounding the digestive tract. From the underside of the diaphragm a sheath of body wall, called the tentacular sheath, extends downward and is attached to the

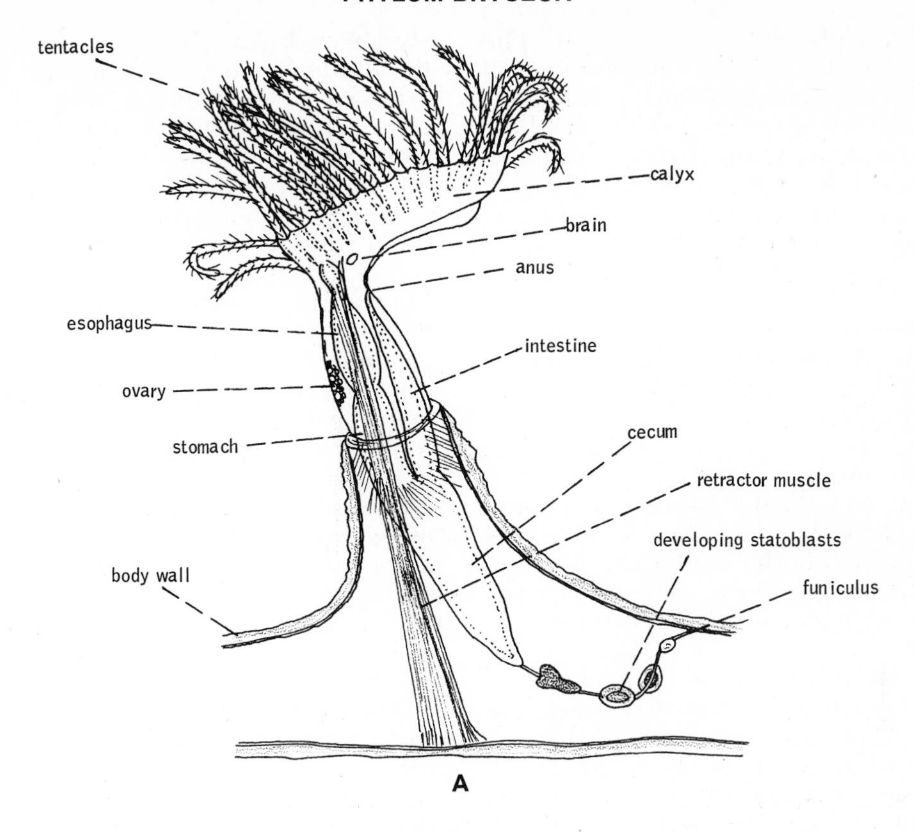

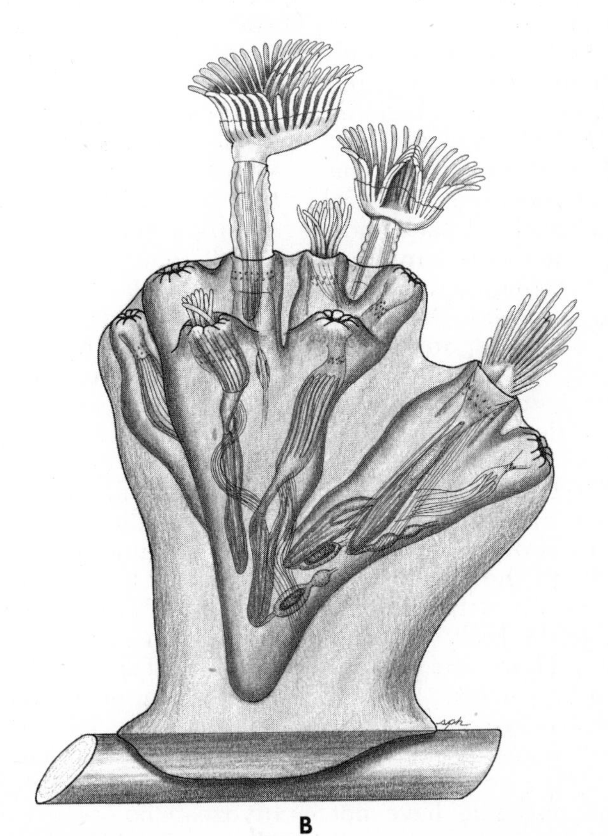

Figure 18–4. Phylactolaemate (freshwater) bryozoans. *A.* Structure of *Plumatella.* (After Pennak.) *B.* Small colony of *Lophopus crystallinus* attached to a water plant. External covering is gelatinous. (After Allman.)

base of the lophophore (Fig. 18–3). The tentacular sheath encloses the tentacles of the lophophore when the animal is retracted. When the animal is feeding, the lophophore protrudes through the diaphragm, vestibule, and orifice, and the tentacular sheath everts, forming the covering of the extended necklike anterior of the trunk.

In gymnolaemates, the lophophore is circular and consists of a simple ridge bearing 8 to 34 tentacles; there is no epistome (Fig. 18–3). When retracted, the tentacles are bunched together within the sheath; when protruded, the tentacles fan out, forming a funnel with the mouth at the base.

The outer surface of each tentacle is unciliated, but on both the lateral surface and the inner median surface is a longitudinal tract of cilia. Within the epidermal layer are longitudinal muscle fibers, and through the center of each tentacle runs an extension of the coelom.

The mouth at the center of the lophophore opens into a **U**-shaped digestive tract, the details of which are considered in the discussion on nutrition (Figs. 18–3 and 18–4). The anus opens through the dorsal side of the tentacular sheath. The coelom is divided by a septum into an anterior portion occupying the lophophore with extensions into the tentacles, and a larger posterior trunk coelom. The trunk coelom is crossed by muscle fibers and cords of tissue, one of the most conspicuous of which is a cord called the funiculus, which extends between the posterior end of the stomach and the back body wall. The most conspicuous of the internal muscles are two sets of lophophore retractors, which extend on each side from the base of the lophophore or the septum to the back body wall.

The nervous system is composed of a ganglionic mass on the dorsal side of the pharynx, from which a nerve issues on each side to encircle the pharynx as a nerve ring (Fig. 18–3). The ganglion and ring give rise to nerves extending to each of the tentacles, to the tentacular sheath, to the diaphragm, and to the digestive tract. A subepithelial nerve plexus is located in the body wall and in the tentacular sheath. There are no specialized sense organs in bryozoans.

As stated earlier, there are no special blood vascular, gas exchange, or excretory systems.

The terms cystid and polypide have not been used in the preceding description of the zooid, but they will be encountered by students who may have occasion to deal with the literature on bryozoans. The term *cystid* refers to the exoskeleton and body wall. The term *polypide* refers to the contents of the zooid within the body wall, i.e., the lophophore, gut, muscles, and other structures.

Zooids of the freshwater Phylactolaemata differ from those of gymnolaemates in many ways. The body wall contains circular and longitudinal muscle layers, and the epidermis is covered by a cuticle or, as in the common *Pectinatella,* by a gelatinous layer. The lophophore of freshwater phylactolaemates, with the exception of the circular lophophore of *Fredericella,* is essentially like that of phoronids—that is, the lophophore is horseshoe-shaped and is composed of two ridges bearing a total of 16 to 106 tentacles (Fig. 18–4). As in phoronids, one ridge passes above and one passed below the mouth at the bend of the horseshoe. A dorsal hollow lip (epistome) overhangs the mouth.

Colony Organization. With this brief background of the structure of a zooid, the organization of bryozoan colonies can now be examined. Gymnolaemates are very common and abundant marine animals. Although some species have been recorded from depths as great as 5500 m., most species are found in coastal waters and attach to rocks, pilings, shells, algae, and other animals. Some species form stoloniferous colonies, which probably represent the more primitive condition (Fig. 18–3). The erect or creeping stolons are composed of modified zooids that give the stolon a jointed appearance. Those zooids that are unmodified are attached by the posterior end to the stolon and are completely separate from one another. The exoskeleton of stoloniferous bryozoans usually lacks calcium carbonate. A Caribbean species of *Amathia* forms colonies over mangrove or other objects that look and feel like great masses of branching spaghetti. However, the vast majority of marine bryozoans are not stoloniferous, and the colony is formed by the more direct attachment and fusion of adjacent zooids. Moreover, the orientation of the body to the substratum is different. The dorsal surface is attached to the substratum, or to other zooids, and the ventral surface becomes the exposed surface, now called the frontal surface (Fig. 18–7A).

The growth patterns of nonstoloniferous bryozoan colonies vary greatly (Fig. 18–5). Many noncalcified species, such as the common Atlantic *Bugula,* form erect branching colonies that look like seaweed. In *Bugula*, for example, such a plantlike growth form is attained commonly through a biserial attachment of zooids (Fig. 18–5*D*). The most common type of colony is the encrusting form, in which the zooids are organized as a sheet attached to rocks and shells. The exoskeleton is usually calcareous, and since the lateral and end walls of the zooids are fused, the orifice has typically migrated toward the exposed ventral, or frontal, surface. *Membranipora, Microporella,* and *Schizoporella* are very common encrusting genera (Figs. 18–5*A* and 18–6), and a large colony may be composed of as many as two million members.

There are also colonies composed of two sheets of zooids attached back to back. Other colonies are tuft-like, and some have the zooids radially arranged. The colony of the Paleozoic *Archimedes* looks like an old-fashioned wood bit or screw (Fig. 18–5*B*).

The zooids of a colony communicate by pores through the transverse end walls or the lateral walls, or both, depending on the growth pattern of the colony (Figs. 18–3, 18–7*A*, and 18–8). The pores permit not only exchange via coelomic fluid, but also interconnections of strands of the funiculus, the cord of mesenchymal tissue attached to the underside of the stomach. In all gymnolaemates, the pores are plugged with epidermal cells, and only a slow diffusion through the pores is permitted; in phylactolaemates, on the other hand, the pores are open or transverse septa may be incomplete, so direct flow of coelomic fluid between members of the colony is possible.

The colonies of most gymnolaemates are polymorphic. The zooid is a typical feeding individual, called an autozooid, and makes up the bulk of the colony. Reduced or modified zooids that serve other functions are known as heterozooids. A common type of heterozooid, and one that is found in many gymnolaemates, is that which is modified to form stolons, attachment discs, rootlike structures, and other such vegetative parts of the colony. These individuals are so reduced that they consist of little more than the body wall and the strands of funiculus tissue passing through their interior (Fig. 18–3).

Other common types of heterozooids, called avicularia and vibracula, are found in many cheilostomes. An avicularium is usually smaller than an autozooid, and the internal structure is greatly reduced (Fig. 18–7*C*). However, the operculum is typically highly developed and modified, forming a movable jaw. Avicularia may be sessile or stalked; when stalked, they are capable of rapid bending or "pecking" movements. Such stalked avicularia are found in *Bugula* (but not in the common *Bugula neritina*) and resemble little bird heads attached to the colony. Avicularia defend the colony against small organisms, particularly settling larvae of other animals.

In a vibraculum, the operculum is modified to form a long bristle, sometimes called a seta, which can be moved in one plane and is apparently used to sweep away detritus and settling larvae (Fig. 18–7*D*). Genera with vibracula include the common and cosmopolitan *Scrupocellaria.* Modification of zooids for reproductive purposes is very common and will be discussed later.

Although the zooids are microscopic, the colonies themselves are one to several centimeters in diameter or height and may attain a much greater size (Fig. 18–5*C*). Some erect species reach a height of over one foot, and some encrusting colonies may attain a diameter of several feet. White or pale tints are typical of most colonies, but darker colors may be found. The taxonomy of marine bryozoans is based almost exclusively on the structure of the exoskeleton and the colonial organization.

The colonies of freshwater phylactolaemates are of two types, the lophopodid and the plumatellid. In the lophopodid type, of which *Lophopus, Cristatella,* and *Pectinatella* are examples, the zooids project from one side of the soft colony sac, resembling the fingers of a glove (Fig. 18–4*B*). *Pectinatella* colonies secrete a gelatinous base, sometimes several feet in diameter, to which surface the zooids adhere. The plumatellid type of colony, of which *Plumatella, Fredericella,* and *Stolella* are examples, has a more or less plantlike growth form, in which there are either erect or creeping branches composed of a succession of zooids. The colonies of freshwater bryozoans are attached to vegetation, submerged wood, rocks, and other objects. *Cristatella,* in which the colony is a flattened gelatinous ribbon, is not fixed and creeps over the substratum at a rate of not more than one inch a day.

Physiology. During feeding, the lophophore is pushed outward through the diaphragm and vestibule, causing the tentacular sheath to evert. The tentacles then ex-

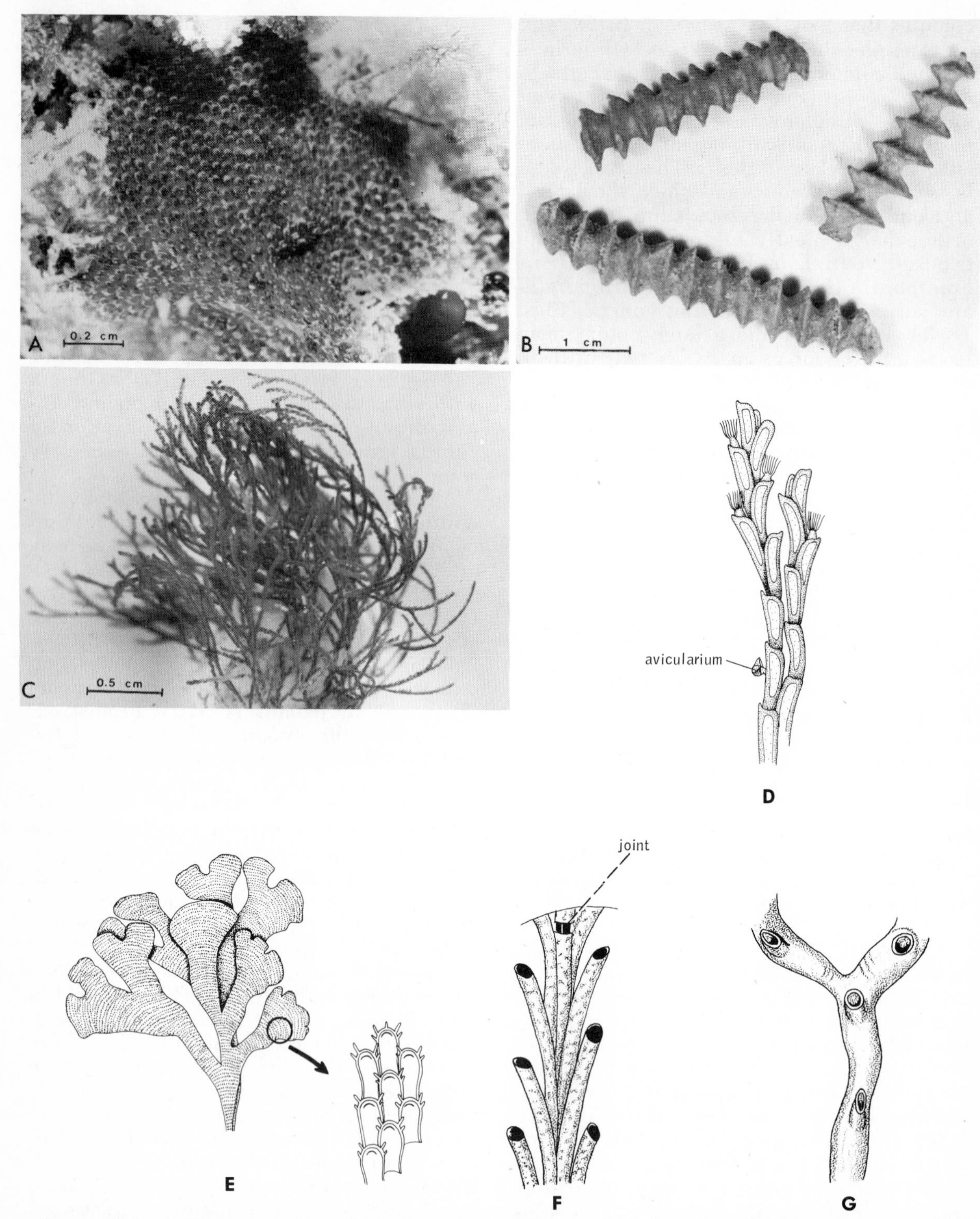

Figure 18–5. *A.* Encrusting colony of a cheilostome gymnolaemate. *B. Archimedes,* a Paleozoic species, in which the colony has the form of a screw. *C* and *D. Bugula,* a common cheilostome gymnolaemate, in which biserially arranged zooids form erect arborescent colonies. (*A, B,* and *C* by Betty M. Barnes.) *E.* Foliaceous colony of *Flustra.* (After Hincks.) *F.* Section of the erect colony of the cyclostome *Crisia.* (After Rogick and Croasdale.) *G.* Uniserial arrangement of zooids in a colony of the cyclostome *Stomatopora.* (After Hincks from Hyman.)

Illustration continued on opposite page.

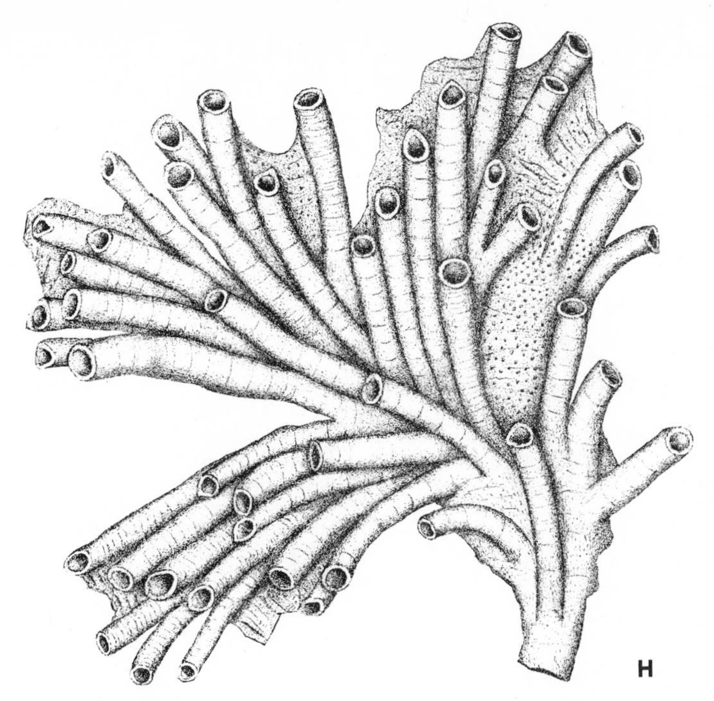

Figure 18–5. *Continued. H.* Cyclostome *Tubulipora*. (From Hincks, T., 1880: A History of the British Marine Polyzoa.)

pand, forming a funnel. Protraction of the lophophore is effected in all cases by the elevation of the coelomic fluid pressure, although the mechanism accomplishing this varies. In the freshwater phylactolaemates, the coelomic fluid pressure is elevated by contraction of the body wall musculature. In marine species with flexible chitinous body coverings, the contraction of transverse and longitudinal parietal muscle bands, which are attached to the walls (Fig. 18–3), causes the flexible but inelastic wall to bow inward, elevating the coelomic fluid pressure.

In calcareous species and even some forms with rigid chitinous exoskeletons, the exposed frontal surface contains a window covered by a thin chitinous membrane (the frontal membrane) (Figs. 18–7 and 18–9). The parietal muscles are inserted on the inner side of the membrane. On contraction, the flexible frontal membrane is bowed inward, increasing the coelomic pressure. Other modifications for lophophore protraction will be described in the section on evolutionary tendencies.

Retraction of the lophophore back into the orifice is effected by the contraction of the lophophoral retractor muscles.

When the lophophore is protracted, the lateral ciliated tracts on the tentacles create a current that sweeps downward into the funnel and passes outward between the bases of the tentacles (Fig. 18–3). Phytoplankton, which is probably the principal food, is driven downward into the open mouth. Ingestion is facilitated by the ciliation of the pharyngeal lining, as well as by rapid dilations of the lower part of the pharynx. These dilations produce suction in the pharynx and aid particularly in the ingestion of large particles. The mouth can be closed, and the ciliary beat of the tentacles can be slowed or reversed, causing rejection of the food particles.

From the mouth, food particles pass through the pharynx and collect in the esophagus. Only the pharynx and the posterior part of the stomach are ciliated, but the entire digestive tract is provided with muscle fibers. Periodically, waves of contraction pass down the esophagus, forcing the collected food particles into the stomach, which includes a caecum. In marine bryozoans, food particles are entangled in mucous cords within the stomach and then are rotated very rapidly by the posterior cilia. Protein and starches appear to be digested extracellularly and fats intracellularly. The stomach is also the site of absorption. Nondigested particles are passed into the intestine and rectum, in which they are molded into fecal pellets and egested when the lophophore is protracted. In phylactolaemates, the food

Text continued on page 705.

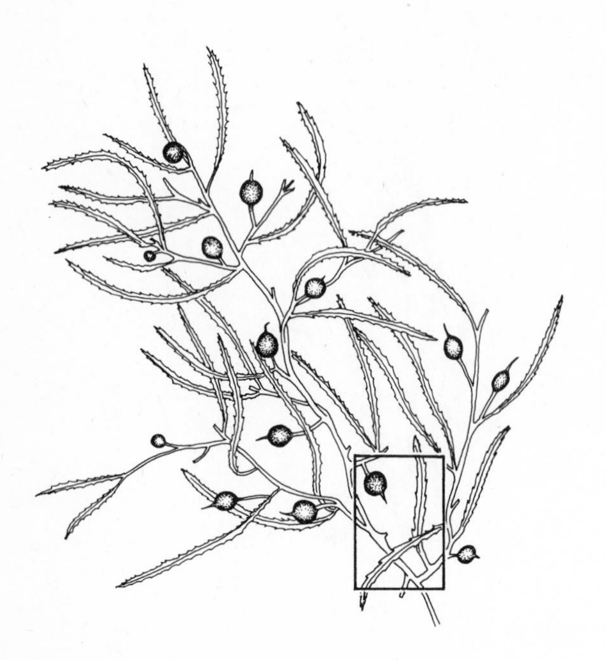

Figure 18–6. Invertebrate animals which live on floating *Sargassum* (Gulf weed), a brown algae, in the Sargasso Sea. Although probably derived from West Indian ancestors, many of these species have become especially adapted for an epiphytic life on floating *Sargassum*. The figure at left shows a large piece of *Sargassum* from which the figure on the opposite page has been taken.

A. *Litiopa* (gastropod mollusk)
B. *Amphithoë* (amphipod crustacean)
C. *Luconacia* (caprellid amphipod)
D. *Anemonia sargassensis* (sea anemone)
E. *Platynereis* (nereid polychaete)
F. *Sertularia* (hydroid cnidarian)
G. *Clytia* (hydroid cnidarian)
H. *Scyllaea* (nudibranch gastropod)
I. *Spirorbis* (serpulid polychaete)
J. Copepod crustacean
K. *Zanclea* (hydroid cnidarian)
L. *Ceramium* (an epiphytic red alga)
M. *Membranipora* (bryozoan)
N. *Doto* (nudibranch gastropod)
O. *Gnescioceros* (polyclad flatworm)
P. *Obelia* (hydroid cnidarian)
Q. *Anoplodactylus* (pycnogonid) feeding on a nudibranch
R. *Fiona* (nudibranch gastropod)

Illustration continued on opposite page.

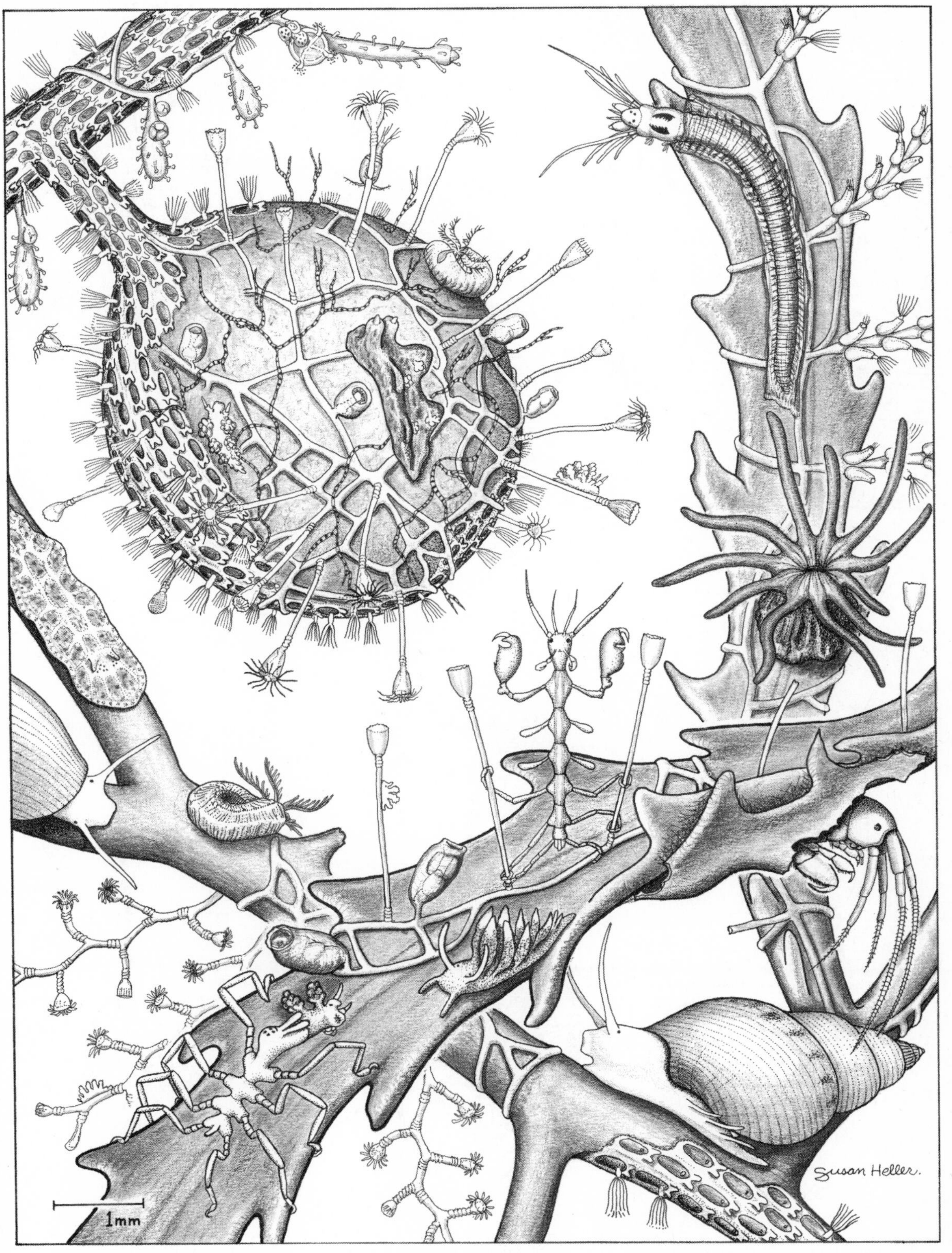

Figure 18–6. *Continued.*

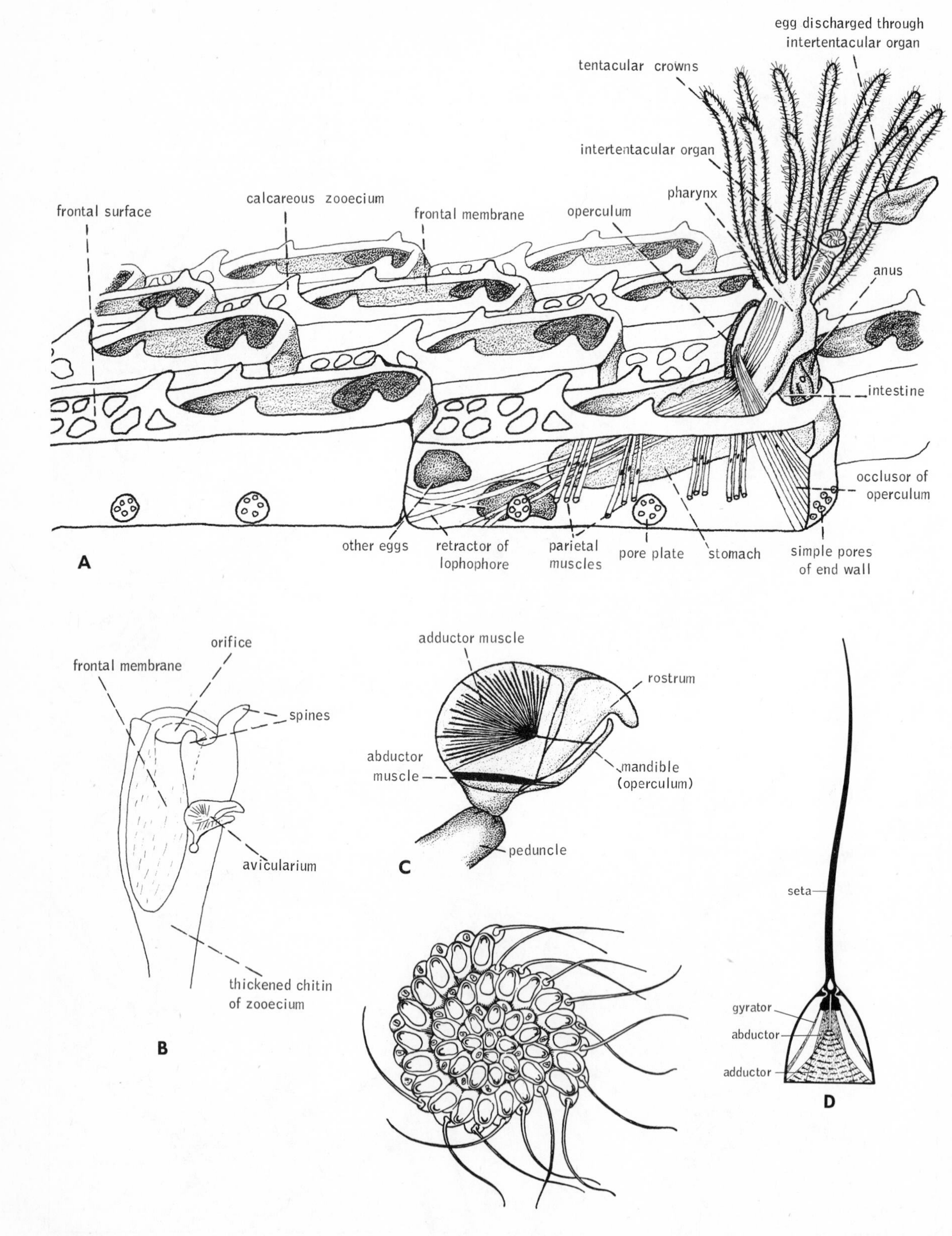

Figure 18–7. *A.* Part of a colony of the encrusting cheilostome, *Electra.* (Modified from Marcus.) *B.* Exoskeleton of *Bugula.* (After Hyman.) *C.* Avicularium of *Bugula.* (After Maturo.) *D.* Vibraculum. (Based on Marcus and Ryland.) *E.* Part of colony of *Heliodoma,* showing vibracula. (From Moore, R. C. (Ed.), Treatise on Invertebrate Paelontology, Geol. Soc. Amer. and Univ. Kansas Press.)

particles are shuttled back and forth within the stomach, which has no ciliated portion, and digestion is apparently entirely extracellular. The intestine may play some role in absorption. In all bryozoans, the epithelial lining of the digestive tract is the primary site for storage of food reserves.

Gas exchange occurs across the exposed body surface, and internal transport of food, gases, and wastes is provided by the coelomic fluid. The coelomic fluid contains coelomocytes, which engulf and store waste materials.

Evolution and Distribution of Bryozoans. Of the two classes of living bryozoans, the freshwater phylactolaemates are considered to be the more primitive. The cylindrical zooids, the anterior orifice, the horseshoe-shaped lophophore, the presence of an epistome, and the non-polymorphic colonies are all considered primitive features. In many other ways, however, phylactolaemates are specialized.

Although not usually common, freshwater bryozoans are widely distributed in lakes and streams that do not contain excessive mud or silt. Many species, such as *Fredericella sultana* and *Plumatella repens,* are cosmopolitan. Although phylactolaemates were formerly a problem in water reservoirs and purification plants, these animals are of little economic significance today.

Unfortunately, there are no fossil phylactolaemates, and thus we know little about their relationship with the marine classes. The first known marine bryozoan is a questionable fossil species from the late Cambrian. But beginning with the Ordovician there is a rich fossil record, and thousands of fossil species have been described. Stenolaemates, of which there are three distinct orders, dominate the Paleozoic fauna, although there were Paleozoic ctenostome gymnolaemates. Cheilostomes, the dominant marine forms today, made their appearance in the Cretaceous.

Marine bryozoans are highly successful animals, exploiting all types of hard surfaces—rock, shells, coral, and wood (e.g., mangrove roots)—and capable of utilizing very restricted spaces. A few species even bore in calcareous substrates. The stolons of boring species are located in tunnels within the substratum, and the autozooids open to the surface. The mechanism of boring is unknown.

From an economic standpoint, marine bryozoans are one of the most important groups of fouling organisms. Over 130 species, of which different species of *Bugula* are among the most abundant, have been taken from ship bottoms.

Along with hydroids, bryozoans rank among the most abundant marine epiphytic animals. Large brown algae are colonized by many species, which display distinct preferences for certain types of algae. Evidence indicates that at the time of settling, the larvae of epiphytic species are attracted to the algal substrate, perhaps by some substance produced by the algae. The widely distributed encrusting *Membranipora,* species of which are abundant on floating *Sargassum* (Fig. 18–6), displays a number of adaptations for an epiphytic life. The frontal surface is uncalcified and the lateral walls are jointed (i.e., broken), permitting some flexibility of the sheet of zooids attached to a bending algal thallus. Although the colony is encrusting, growth in one direction predominates.

Stoloniferous colonies with cylindrical, upright, more or less separate autozooids probably represent the more primitive organization, but as we have seen, the great majority of bryozoans are adnate, with contiguous autozooids having the frontal surface exposed. Of various trends that are evident in the evolution of marine bryozoans, adaptations associated with protection of the vulnerable frontal surface are especially striking.

Primitively, the frontal surface behind the orifice is covered only by the frontal membrane, which must be thin and flexible in order to bow inward in the process of elevating the coelomic fluid pressure and protruding the lophophore. Such a frontal surface is found in *Membranipora,* in which the contrasting calcified lateral walls are especially conspicuous (Figs. 18–6 and 18–9*A* and *B*).

A number of groups have reduced the vulnerability of the frontal membrane by reducing its area to an oval. Protection has also been afforded by the development of spines on the frontal margin of the lateral walls (Fig. 18–9*C*), or in a few species by a single long spine which curves back over the frontal membrane.

Elaboration of the protective spines led to the cribrimorphs, which include many fossil and some living species. In these forms, overaching spines form a protective cover but permit access of water to the frontal membrane. In some, the spines merely touch at

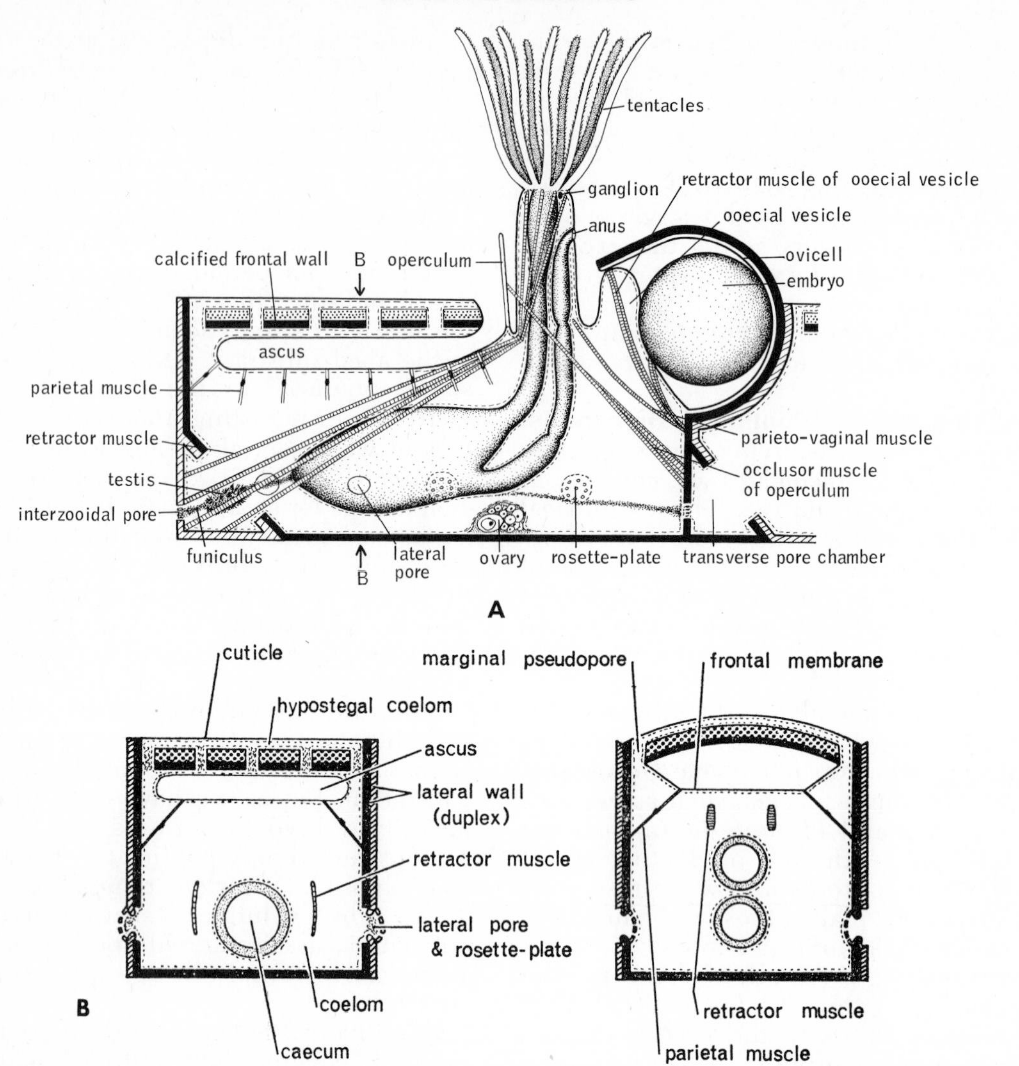

Figure 18–8. *A.* Diagrammatic sagittal section through an ascophoran cheilostome. (Modified from Ryland.) *B.* Transverse sections through same individual (proximal at left, distal at right.)

Illustration continued on opposite page.

their tips, but in others the spines also fuse laterally except for pore-like openings (Fig. 18–9*D* to *F*).

A similar modification to that of the cribrimorphs occurred independently in another group, in which the frontal membrane became covered by an overarching vault that was solid but open at the front. Another adaptation, illustrated by the genera *Cellaria, Micropora,* and others, was the protection of the zooid by the development of a wall *beneath* the frontal membrane, resulting in a double-chambered (cryptocyst) zooid. The wall was perforated to permit passage of the parietal muscle bands responsible for inbowing of the frontal membrane (Fig. 18–9*G*).

On the basis of numbers of species, one of the most successful adaptations for the protection of the frontal membrane is that found in the Ascophora. Here the frontal surface has become solid, and an invagination of the body wall near the orifice projects backward into the coelom as a compensation sac, called an ascus (Fig. 18–8*A*). The parietal muscles are attached to the underside of the sac; when they retract, the sac dilates, water enters, coelomic fluid pressure rises, and lophophore is protruded. The operculum, which covers not only the orifice but also the entrance to the ascus, is pivotally hinged so that when the back of the operculum goes down, opening the ascus, the front goes up and opens the orifice.

Modifications described thus far developed in the Gymnolaemata. The Stenolaemata, of which the cyclostomes are the only living representatives, evolved a method of lophophore protrusion that, while it was dependent upon elevation of coelomic fluid pressure, did not involve deformation of the body wall. Moreover, there is no evidence that stenolae-

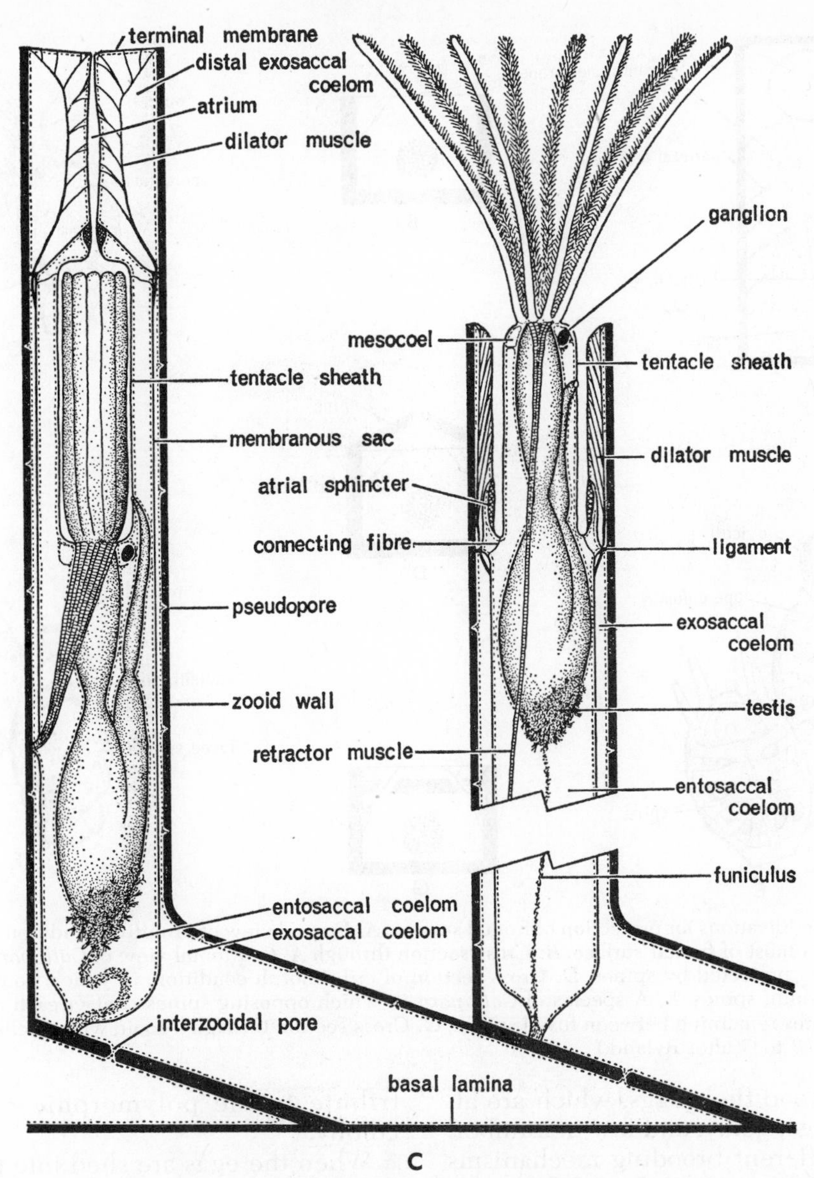

Figure 18–8. *Continued. C.* Two zooids of a cyclostome stenolaemate. (*B* and *C* from Ryland, J. S., 1970: Bryozoans. Hutchinson and Co., London.)

mates, which are represented by an enormous number of extinct Paleozoic forms, evolved from forms that possessed a frontal membrane. The zooids are tubular and calcified, and the orifice, which lacks an operculum, is located at the anterior, or distal, end. An atrium occupies the distal end of the zooid, and between the atrial wall and the outer body wall is a distal extension of the coelom (Fig. 18–8*C*). Dilator muscles are attached to the coelomic sides of the atrial wall. When these muscles contract, opening the atrium, fluid in the distal extension of the coelom is forced backward. The coelomic fluid pressure is thus elevated and the lophophore is protruded.

Reproduction. All freshwater bryozoans and most marine species are hermaphroditic. Simultaneous production of eggs and sperm may take place, but a tendency toward protandry is not uncommon. The one or two ovaries and the one to many testes are masses of developing gametes covered by peritoneum; the masses bulge into the coelom. The ovaries are located in the distal end of the animal and the testes in the basal end (Fig. 18–3). In dioecious species, the entire colony may be composed of zooids of the same sex, or both male and female individuals may be present. There are no genital ducts, and the eggs and sperm rupture into the coelom.

Some marine species shed small eggs directly into the sea water. The vast majority

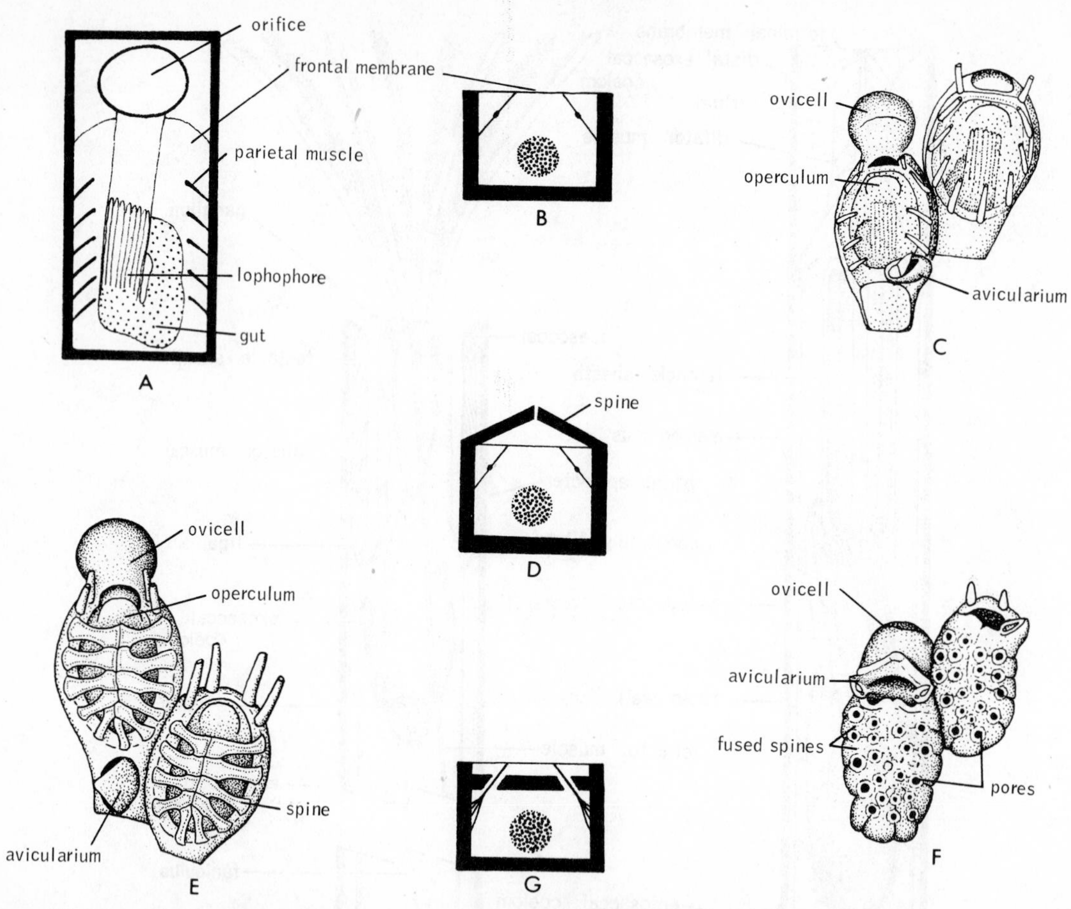

Figure 18–9. Modifications for protection of frontal surface. *A.* Frontal view of primitive condition, in which frontal membrane occupies most of frontal surface. *B.* Cross section through *A. C.* Frontal view of *Callopora,* in which oval frontal membrane is protected by spines. *D.* Cross section of cribrimorph condition, in which frontal membrane is covered by overarching spines. *E.* A species of *Callopora* in which opposing spines contact each other distally. *F. Cribrilina,* with pores remaining between fused spines. *G.* Cross section through a zooid with calcified wall beneath frontal membrane. (*B* to *G* after Ryland.)

of bryozoans brood their eggs, which are almost always large, yolky, and few in number. A variety of different brooding mechanisms are employed. A few species brood their eggs within the coelom, but most brood them externally. The digestive tract and lophophore often degenerate to provide space for the egg. The cavity of the tentacular sheath or invaginations of the atrial wall are common sites for brooding (Fig. 18–3). Many cheilostomes, including the common *Bugula,* brood their eggs in a special external chamber called an ovicell (Figs. 18–8*A* and 18–10). The body wall at the distal end of the zooid grows outward, forming a large hood (the ovicell). A second, smaller evagination, which is directly connected to the coelom, bulges into the space formed by the ovicell. A single egg is brooded in the space between the two evaginations. Zooids that become highly modified during the course of egg production and brooding are considered reproductive individuals and contribute to the polymorphic nature of the colony.

When the eggs are shed into the sea water or are brooded externally, they escape from the coelom by way of a special opening in the region of the lophophore. This coelomopore may be a simple opening or may be mounted at the end of a projection called an intertentacular organ (Fig. 18–7*A*). In those species with ovicells, the egg is extruded through the coelomopore as a "stream" and then rounds up within the ovicell cavity.

The simultaneous presence of sperm and eggs and the lack of knowledge concerning sperm liberation led many zoologists to believe that self-fertilization was widespread among bryozoans. Silén, on genetic grounds, doubted the general occurrence of self-fertilization, and his studies of *Electra* (1966) revealed the mechanism of cross fertilization in this genus, which has since been found to hold true for others as well. In the species studied by Silén, the sperm develop before the complete maturation of the eggs. The sperm are shed through two tentacles of the lophophore and are disseminated into the sea water. The liberated sperm, when caught in the feeding currents of other individuals, adhere to the tentacular surfaces of the lophophore or enter the intertentacular organ. In the former case, the eggs are fertilized as they leave the intertentacular organ. In those species in which the sperm enter the intertentacular organ, this structure is the site of fertilization. Silén suggests that the coelomophore may be the means by which sperm enter the coelom in species that brood their eggs internally.

Cleavage in marine species is radial but leads to a modified trochophore larva, which escapes from the brood chamber. The larvae vary considerably in form (Fig. 18–11), but all possess a locomotor ciliated girdle or corona, an anterior tuft of long cilia, and a posterior adhesive sac. The larvae of some species of nonbrooding gymnolaemates, such as *Electra* and *Membranipora*, called cyphonautes larvae, are triangular and greatly compressed, and each lateral surface is covered by a chitinous valve. Only the larvae of nonbrooding bryozoans possess a functional digestive tract and feed during larval existence. Feeding larvae may have a larval life of several months, but the larvae of brooding species, which are nonfeeding, have a very brief larval existence prior to settling. Larvae are at first positively phototactic, which may facilitate escape from the brood chamber and dispersal, but as the time for settling approaches, the larva becomes negatively phototactic and is attracted to shadow surfaces.

During settling the adhesive sac everts and fastens by means of secretions to the substratum. The larval structures of the attached larva undergo retraction and histolysis, followed by development into an adult. The first zooid is called an ancestrula. By budding, the ancestrula gives rise to several other zooids. These zooids, in turn, bud off new individuals, and thus by subsequent asexual reproduction a new colony arises and gradually increases in size. Budding involves the cutting off of a part of the parent zooid by the formation of a body wall partition. The new chamber evaginates either before or after its formation and the skeleton is "stretched" by the addition of new material. New internal structures develop from the ectoderm and peritoneum of the body wall. The exact pattern of budding (i.e., the number and location of buds) determines the growth pattern of the colony.

In the erect dendritic *Bugula*, growth occurs at the tips of each branch. Encrusting species have a peripheral growing edge.

The life span of bryozoan colonies varies greatly. Some live only a single year, with growth taking place with increase of water temperatures in temperate regions; libera-

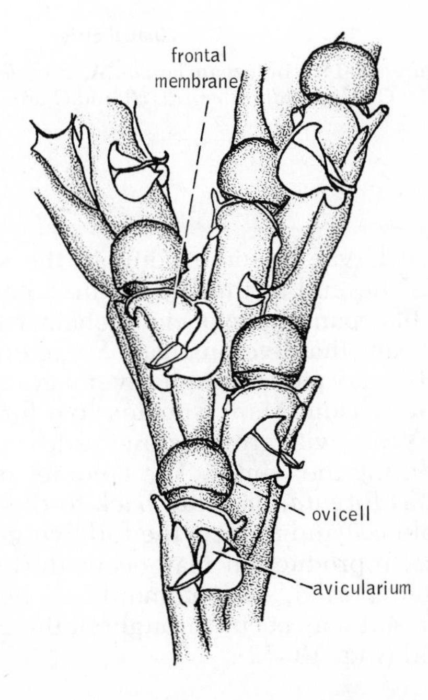

Figure 18–10. Part of a *Bugula* colony with ovicells. (After Maturo.)

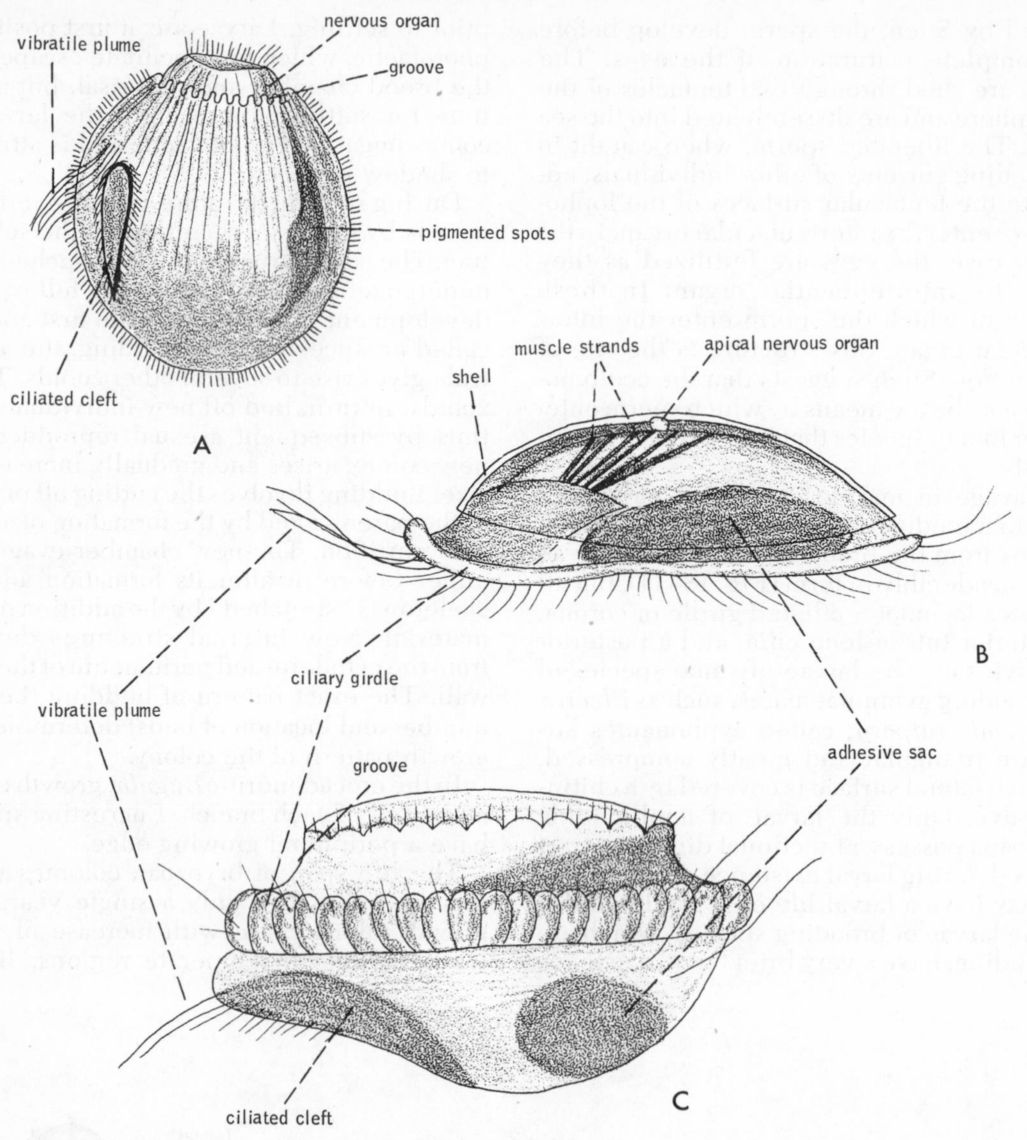

Figure 18–11. Bryozoan larvae. *A. Bugula.* (After Cavel from Hyman.) *B.* Cyphonautes larva of *Flustrellidra hispida. C. Alcyonidium mytili.* (*B* and *C* after Barrois from Hyman.)

tion of larvae at some point in the summer marks the end of the life of the colony. Annual life spans are especially characteristic of bryozoans that live on algae. Some epiphytic species may pass through several generations in one season. Many species live for two or more years, with growth slowing down or halting during the winter. The colonies of some, such as *Bugula*, may die back to the stolons or hold fast and re-form the following season. Sexual reproduction may occur during a restricted period; or, commonly in perennial species, it may occur throughout the growing period (Fig. 18–12).

As pointed out earlier, brooding is commonly accompanied by degeneration of the internal organs of the adult. The gut and lophophore are phagocytized, and the remains of these organs are lodged in the coelom as dark balls, called brown bodies, that are often conspicuous. In many species, degeneration does not indicate the end of the life span of the individual. Following brooding, new internal organs may be regenerated. Alternating phases of degeneration and regeneration are especially common in bryozoan colonies that live a number of years, and even in some annual species. A large colony may

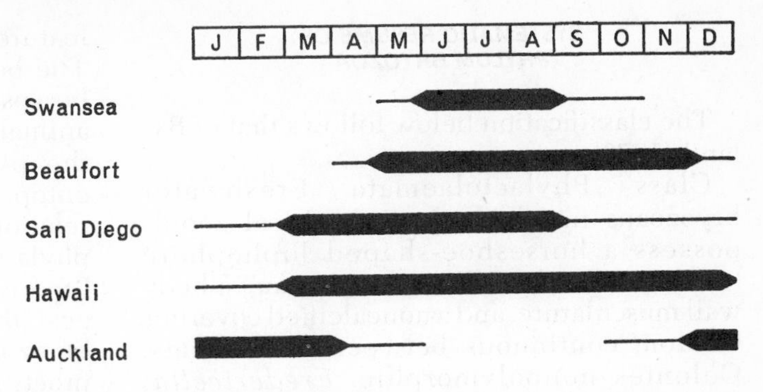

Figure 18-12. Period of egg production and larval settling of *Bugula neritina* in different parts of the world. (From Ryland, J. S., 1970: Bryozoans. Hutchinson Co., London.)

exhibit zones of individuals in various stages of development. For example, at the outer perimeter of an encrusting colony, or at the tips of erect branches in a form such as *Bugula,* budding and development of new individuals occur. Further inward is a zone of fully developed individuals, which are feeding and reproducing. Still further inward is a zone of degenerating members containing dark brown bodies. Within the degenerate zone may be a zone of regenerated feeding individuals.

Development in freshwater phylactolaemates takes place within an embryo sac that bulges into the coelom. Development leads to the formation of a cystid sac, which then proceeds to bud off one to several zooids. This young ciliated colony or "larva" is released from the parent colony. It swims about for a short time prior to settling. After attachment, retraction and degeneration of the once ciliated outer cystid wall ensues. Meanwhile, the young colony that had been developing inside the confines of the cystid wall continues to bud successive zooids until an adult colony is formed. The parent zooid dies after

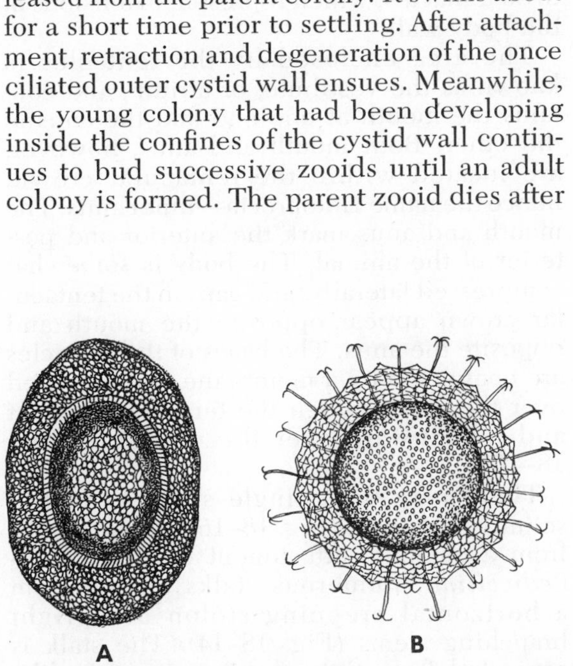

Figure 18-13. Statoblasts of freshwater bryozoans. *A.* A floating statoblast of *Hyalinella punctata.* (After Rogick from Pennak.) *B.* Statoblast with hooks from *Cristatella mucedo.* (After Allman.)

producing a number of daughter individuals. Thus, in branching colonies, only the tips of branches contain living zooids; and in the flattened gelatinous colonies, living zooids are restricted to the periphery.

In addition to sexual reproduction and to budding, freshwater bryozoans also reproduce asexually by means of special resistant bodies called statoblasts (Fig. 18-13). One to several statoblasts develop on the funiculus and bulge into the coelom as masses of peritoneal cells that contain stored food material, and epidermal cells that have migrated to the site of statoblast formation. After organizing cellularly, each mass secretes both an upper and a lower chitinous valve that form a protective covering for the internal cells. Since the rims of the valves often project peripherally to a considerable extent, the statoblasts are usually somewhat disc-shaped. Statoblasts are continuously formed during the summer and fall.

The structure and shape of statoblasts are important in the taxonomy of the freshwater bryozoans. Some types of statoblasts adhere to the parent colony or fall to the bottom; others contain air spaces and float. These floating statoblasts are sometimes armed with hooks around the margins.

Statoblasts remain dormant for a variable length of time. During this period they may be spread considerable distances by animals, floating vegetation, or other agents and are able to withstand desiccation and freezing. When environmental conditions become favorable, as in the spring, germination takes place, the two valves separate, and a zooid develops from the internal mass of cells. The number of statoblasts produced by a freshwater bryozoan is enormous. Brown (1933) reported drifts of statoblast valves one to four feet wide along the shores of Douglas Lake in Michigan and estimated that *Plumatella repens* colonies in one square meter of littoral lake vegetation produce 800,000 statoblasts.

SYSTEMATIC RÉSUMÉ OF PHYLUM BRYOZOA

The classification below follows that of Ryland (1970).

Class Phylactolaemata. Freshwater bryozoans in which the cylindrical zooids possess a horseshoe-shaped lophophore (except in *Fredericella*), an epistome, a body wall musculature, and a noncalcified covering. Coelom continuous between individuals. Colonies nonpolymorphic. *Fredericella, Plumatella, Pectinella, Lophopus, Cristatella.*

Class Stenolaemata. Marine bryozoans. Zooids tubular, with calcified walls that are fused with adjacent zooids; orifices circular and terminal. Lophophore protrusion not dependent on deformation of body wall. Order Cyclostomata contains some living and many fossil species: *Crisia, Lichenopora, Stomatopora, Tubulipora.* Orders Cystoporata, Trepostomata, and Cryptostomata all became extinct at the end of the Paleozoic.

Class Gymnolaemata. Primarily marine bryozoans with polymorphic colonies. Zooids cylindrical or flattened; lophophore circular; an epistome and body wall musculature are lacking. Protrusion of the circular lophophore depends on body wall deformation.

Order Ctenostomata. Stoloniferous or compact colonies in which the uncalcified exoskeleton is membranous, chitinous, or gelatinous. The usually terminal orifices lack an operculum. *Amathia, Alcyonidium, Aeverrillia, Bowerbankia,* the freshwater *Paludicella.*

Order Cheilostomata. Colonies composed of boxlike zooids that are adjacent but have separate walls. Exoskeleton is usually calcified. Orifice is provided with an operculum (except in *Bugula*). Avicularia, vibracula, or both may be present. Eggs are commonly brooded in ovicells. *Aeta, Callopora, Electra, Flustra, Membranipora, Tendra, Bugula, Micropora, Cellaria, Cribulina, Microporella, Schizoporella.*

PHYLUM ENTOPROCTA

The Entoprocta is a small phylum of some 60 species of mostly sessile animals which are very similar to bryozoans. They were formerly included in the phylum Bryozoa, but were removed because of a number of differing features, including the absence of a coelom. The body cavity of entoprocts is believed to be a pseudocoel by some zoologists, and these animals were placed by Hyman (1951) with the pseudocoelomate phyla. However, if entoprocts are indeed pseudocoelomates, the relationship with the other pseudocoelomate phyla is very obscure. Recent studies on the life cycle of entoprocts (Nielsen, 1971) suggest that they are not as remote from the bryozoans as was formerly thought. The entoprocts are therefore discussed at this point in order that they may be more easily compared to bryozoans.

Except for the single freshwater genus *Urnatella* (Fig. 18–14*A*), all entoprocts are marine and live attached to rocks, shells, or pilings, or are commensal on sponges, polychaetes, bryozoans, and other marine animals. Members of the commensal family Loxosomatidae are solitary; the other two families are colonial. All are very small and never exceed 5 mm. in length. The body (Figs. 18–15 and 18–16) consists of a somewhat ovoid or boat-shaped structure called the calyx, which contains the internal organs, and a stalk by which the calyx is attached to the substratum (Fig. 18–14*B*). The attached underside of the calyx was originally the dorsal surface, and the upper or free side was originally the ventral surface. The upper margin of the calyx bears an encircling crown of 8 to 30 tentacles, which represent extensions of the body wall.

The area enclosed by the tentacles is known as the vestibule, or atrium, and contains the mouth at one end and the anus at the other. Both mouth and anus, however, are located within the tentacular crown; hence the name Entoprocta—inner anus. The mouth and anus mark the anterior and posterior of the animal. The body is somewhat compressed laterally, and gaps in the tentacular crown appear opposite the mouth and opposite the anus. The bases of the tentacles are connected by a membrane that is pulled over the crown when the tentacles contract and fold inward over the vestibule (Fig. 18–16*B*).

There may be a single stalk, as in the solitary *Loxosoma* (Fig. 18–16); several stalks from a common attachment disc; or, as in *Pedicellina*, numerous stalks arising from a horizontal creeping stolon or upright branching stems (Fig. 18–14). The stalk is separated from the calyx by a septum-like fold of the body wall and is commonly partitioned into short cylinders, or segments. In some species, certain segments are swol-

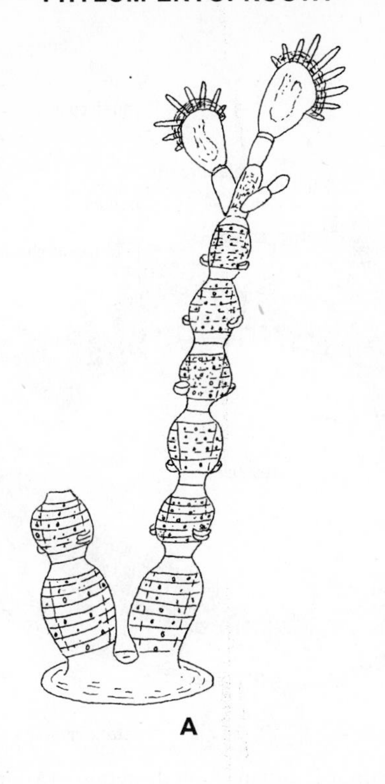

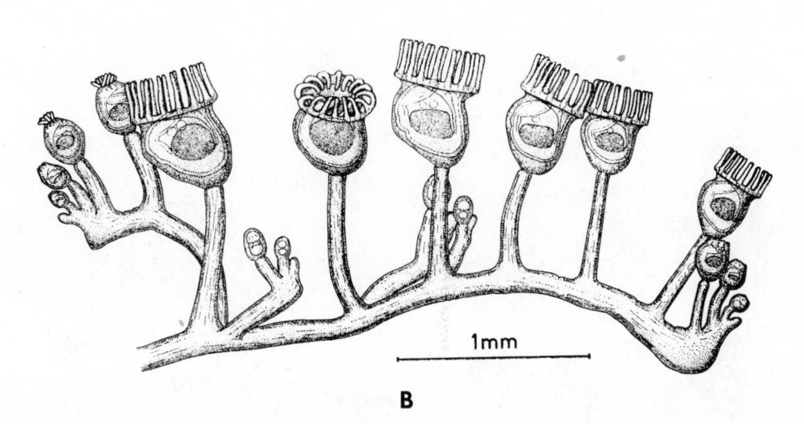

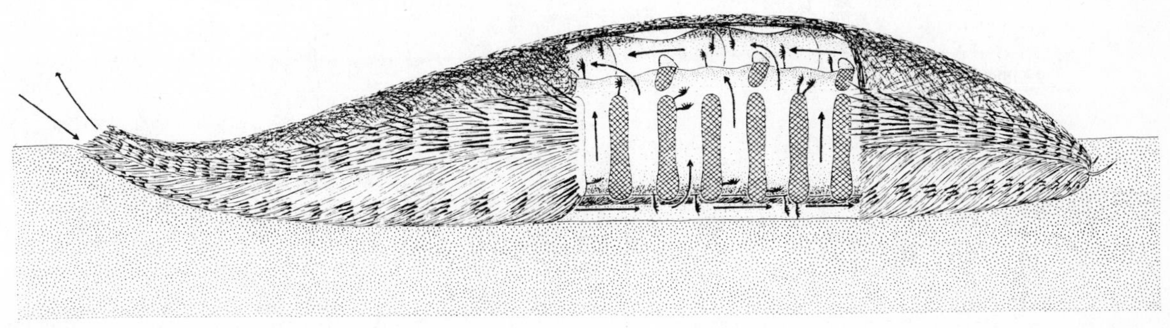

Figure 18–14. A. *Urnatella gracilis*, a freshwater entoproct. (Modified after Leidy from Pennak.) *B*. Part of a colony of the marine entoproct, *Pedicellina*. (After Ehlers from Hyman.) *C*. Semidiagrammatic figure of a sea mouse (polychaete) showing individuals of the solitary commensal species *Loxosomella fauveli* attached to the parapodia and other parts of the body. Arrows indicate ventilating current of the worm. (*B* and *C* from Nielsen, C., 1964: Ophelia, *1*(1):1–76.)

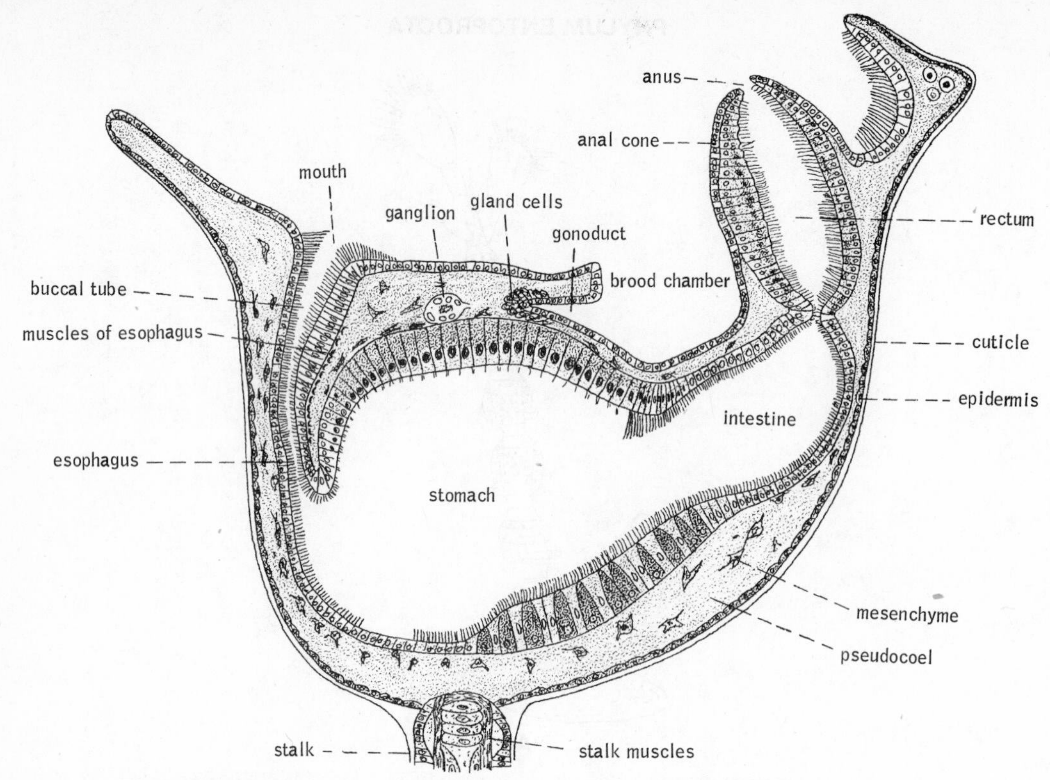

Figure 18–15. *Pedicellina* (median sagittal section). (After Becker from Hyman.)

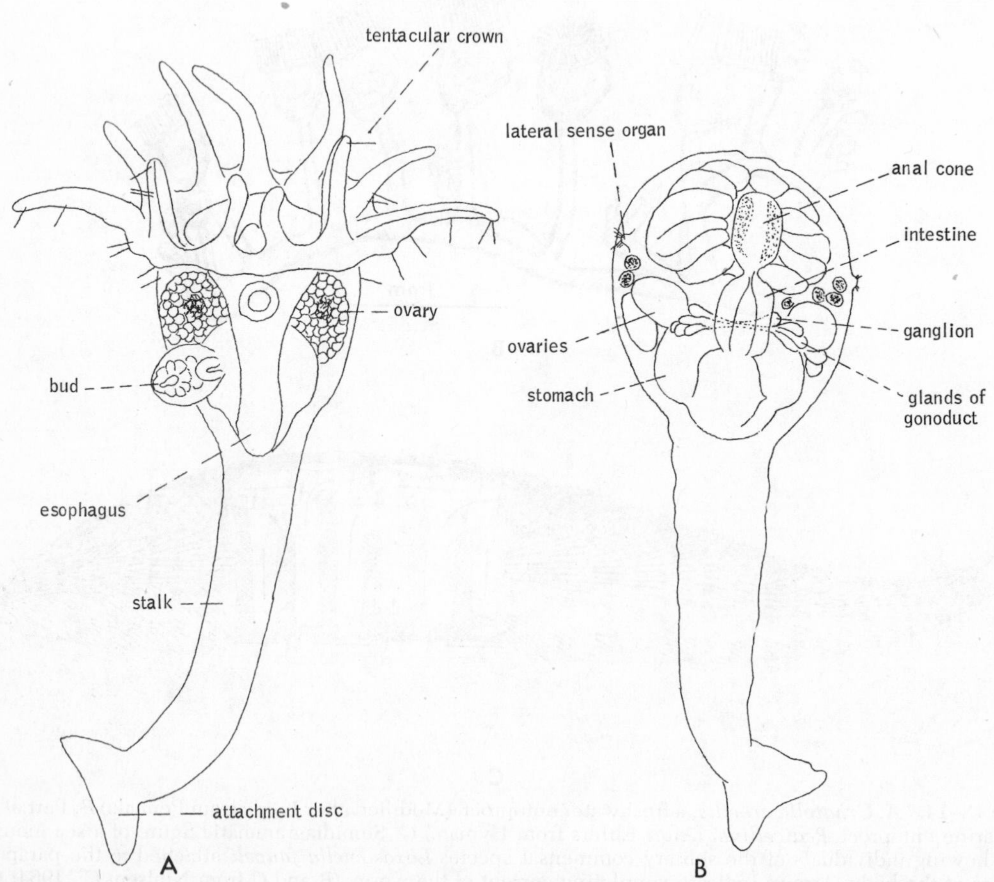

Figure 18–16. Solitary marine entoproct, *Loxosoma*. *A*. Expanded. *B*. Retracted. (Both after Atkins from Hyman.)

len with longitudinal muscle fibers that, on sudden contraction, produce a curious flicking motion in members of the colony.

The body wall consists of a cuticle and an underlying cuboidal epithelium. The muscle layer is limited to longitudinal fibers along the inner wall of the tentacles, in the tentacular membrane, and in certain areas of the calyx. The extensive pseudocoel, which also extends into the tentacles, is filled with a gelatinous "mesenchyme" containing both free and fixed cells.

Entoprocts are filter feeders, consuming organic particles and small plankton, such as diatoms and small protozoans. The beating of cilia on the sides of the tentacles causes a water current to pass into the vestibule between the tentacles and then to pass upward and out (Fig. 18–17). When suspended food particles pass between the tentacles, they become trapped on frontal cilia located on the inner surface of the tentacles. These frontal cilia beat downward, carrying the food particles to the base of the tentacle. Here the food particles are carried in ciliated vestibular grooves that run along the inner base of the tentacular crown on both sides toward the mouth. If disturbed or if encumbered with excessive detritus, an entoproct ceases feeding and folds its tentacles over the vestibule (Fig. 18–16*B*). Commensal species appear to be dependent upon ventilating and other water currents of the host for maintaining a continual supply of unfiltered water (Nielsen, 1964).

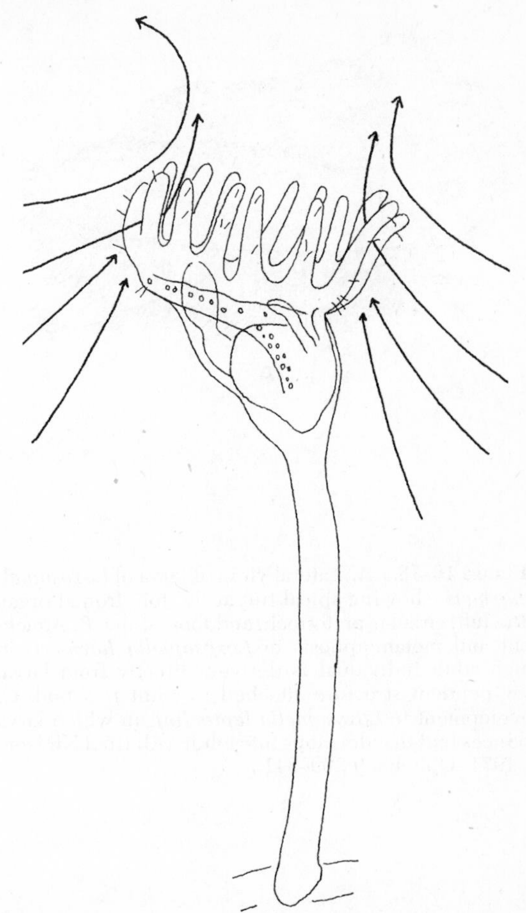

Figure 18–17. *Loxosoma,* showing water currents during feeding. (After Atkins from Hyman.)

The digestive tract is **U**-shaped and completely ciliated (Fig. 18–15). Food particles entering the mouth pass into a short buccal cavity and then into a large bulbous stomach, which composes the major part of the alimentary canal. Certain areas of the stomach wall are glandular and are believed to produce enzymes for extracellular digestion. Absorption is believed to take place in the stomach and in the short ascending intestine located at the posterior of the stomach. The intestine opens into a rectum, and wastes are egested through the anus, located at the posterior of the vestibule. Not infrequently the anus is mounted on a projection called the anal cone.

Two flame bulbs are located between the stomach and the vestibule. The two nephridial ducts from the flame bulbs unite prior to opening through a single median nephridiopore located just behind the mouth. The nervous system contains a single, large, median ganglion situated between the stomach and the vestibule (Fig. 18–15). The ganglion gives rise to three pairs of nerves to the tentacles and three pairs of nerves to the stalk and to certain parts of the calyx. Sensory cells with projecting bristles are located over the body surface, especially on the outer side of the tentacles and along the calyx margin.

Asexual reproduction by budding is common in all entoprocts, and it is by this means that extensive colonies are formed. In most species, the buds arise from segments of the stolon or from the upright branches. In the solitary entoprocts, the buds develop from the calyx (Fig. 18–16*A*), separate from the parent, and then attach as new individuals. Many entoprocts are capable of shedding the calyx when environmental conditions are unfavorable. New calyces are regenerated when environmental conditions are again suitable.

The phylum includes both dioecious and hermaphroditic species. In dioecious species, two rounded gonads are located between

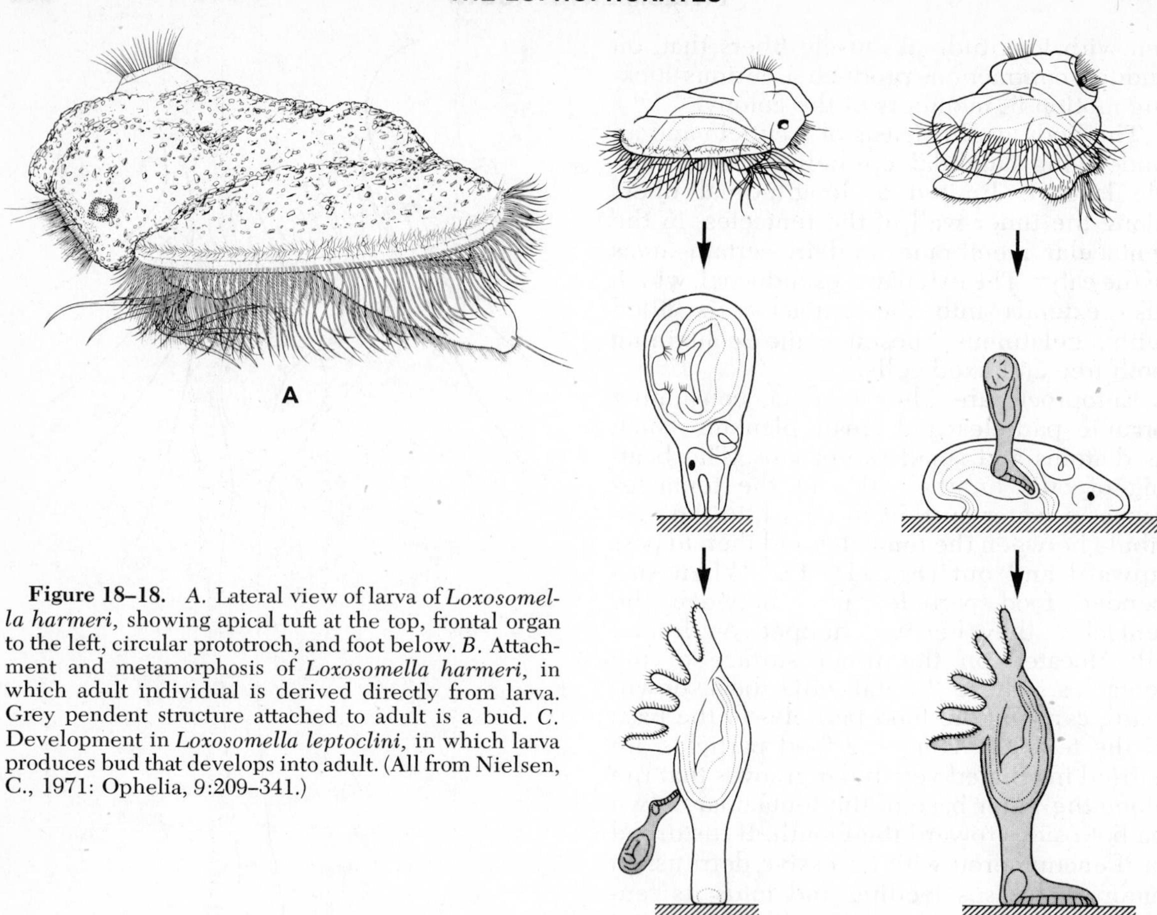

Figure 18–18. *A.* Lateral view of larva of *Loxosomella harmeri,* showing apical tuft at the top, frontal organ to the left, circular prototroch, and foot below. *B.* Attachment and metamorphosis of *Loxosomella harmeri,* in which adult individual is derived directly from larva. Grey pendent structure attached to adult is a bud. *C.* Development in *Loxosomella leptoclini,* in which larva produces bud that develops into adult. (All from Nielsen, C., 1971: Ophelia, *9*:209–341.)

the vestibule and the stomach (Fig. 18–16*A*). The simple gonoducts become confluent and empty through a single, median gonopore located just posterior to the nephridiopore. In hermaphroditic species, there are two testes and two ovaries, and all four gonoducts become confluent prior to reaching the gonopore.

The eggs are believed to be fertilized in the ovaries, although how sperm reach the ovaries is uncertain. After ovulation, each egg is covered with a membranous envelope secreted by the wall of the gonoduct. When the egg leaves the gonopore, the egg membrane becomes attached to the vestibule wall between the gonopore and the anus. The region of the vestibule acts as a brood chamber (Fig. 18–15); it is here that embryonic development with spiral cleavage takes place. A ciliated, free-swimming larva hatches from the egg. The larva, which is a trochophore more or less like that of annelids and mollusks, possesses an apical tuft of cilia at the anterior end, a frontal organ, a ciliated girdle around the ventral margin of the body, and a ciliated foot (Fig. 18–18). After a short free-swimming existence, the larvae settle to the bottom, where they creep over the surface with the ciliated foot and eventually attach with the frontal organ. Some undergo a complex metamorphosis, in which the future calyx rotates 180 degrees to attain the inverted condition of the adult. In some, the larva does not develop into an adult, but produces buds from which the adults are derived (Fig. 18–18).

On the basis of their development, Nielsen (1971) believes that entoprocts are related to other groups with trochophore larvae and probably had a common ancestry with the coelomate ectoprocts. He would reunite them within the phylum Bryozoa.

PHYLUM BRACHIOPODA

The last of the three lophophorate phyla is the phylum Brachiopoda, commonly known as lamp shells. These animals resemble bivalve mollusks in possessing a mantle and a calcareous shell of two valves that approximates that of mollusks in size. In fact the phylum was not separated from the mollusks until the middle of the 19th century. However, the resemblance to mollusks is superficial, for in brachiopods the two valves enclose the body dorsally and ventrally instead of laterally, and the ventral valve is typically larger than the dorsal (Fig. 18–19*A* and *B*). Moreover, the ventral valve is usually attached to the substratum directly or by means of a cordlike stalk (Fig. 18–21*C*). The mantle lobes secrete the shell and enclose the lophophore.

All brachiopods are marine, and very few species are found at depths beyond the edge of the continental shelf. Most species live attached to rocks or other firm substrata, but some forms, such as *Lingula*, live in vertical burrows in sand and mud bottoms (Fig. 18–21*A*). Although fossil species were widely distributed and abundant on reefs, modern forms are largely inhabitants of cold waters and are of rather spotty occurrence.

The approximately 280 species of living

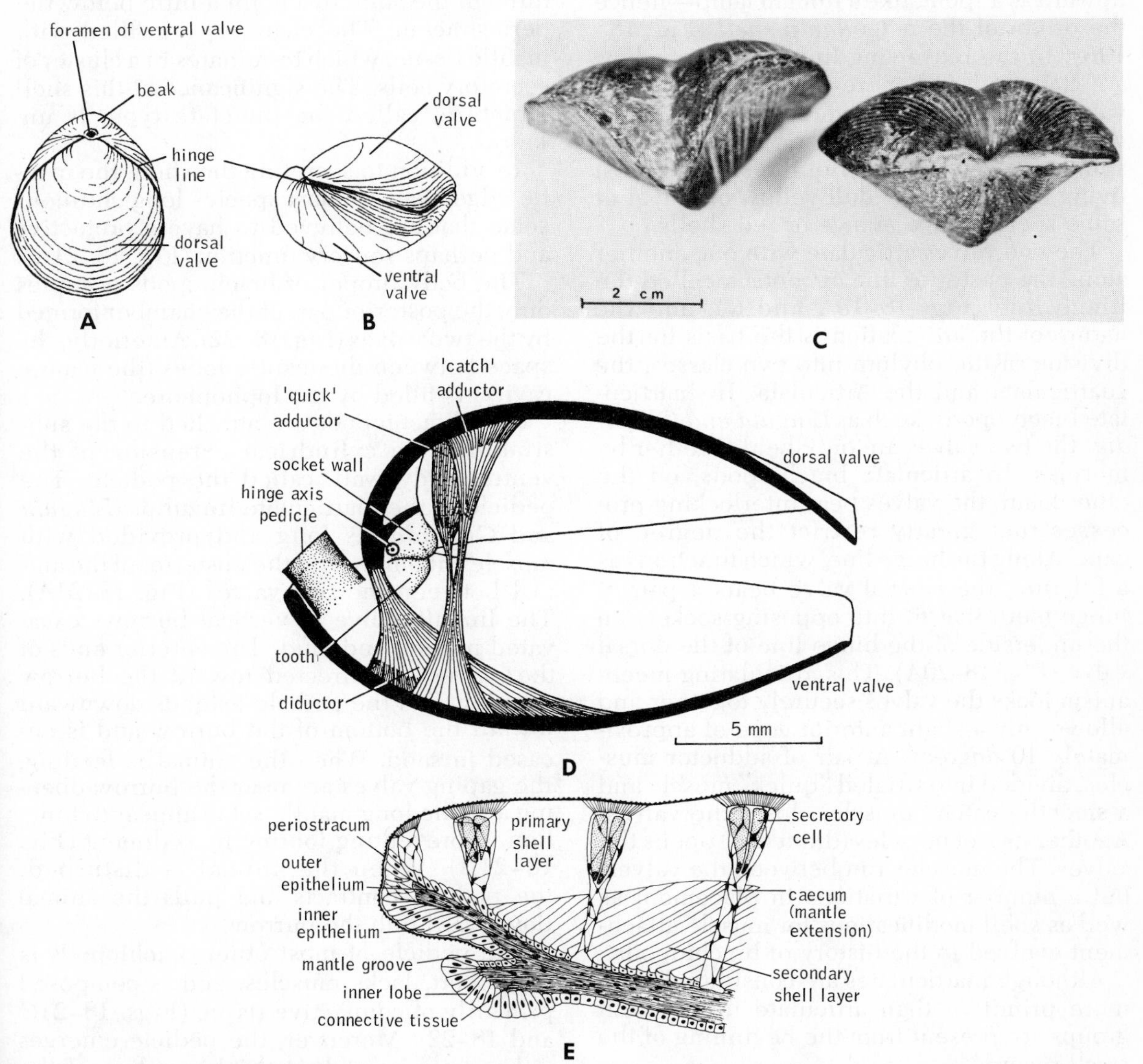

Figure 18–19. *A.* The articulate brachiopod, *Terebratula* (dorsal view). *B. Hemithyris* (lateral view). (*A* and *B* after Woods.) *C. Paraspirifer*, a fossil articulate brachiopod. (By Betty M. Barnes.) *D.* Sagittal section through the terebratulid *Waltonia*, showing relationship of valves, muscles, and pedicle. (From Rudwick, M. J. S., 1970: Living and Fossil Brachiopods. Hutchinson Co., London.) *E.* Section through mantle edge of an articulate brachiopod. (Modified from Williams and Rowell.)

brachiopods are but a fraction of the 30,000 described fossil species that flourished in the seas of the Paleozoic and Mesozoic eras. The phylum made its appearance in the Cambrian period and reached its peak of evolutionary development during the Ordovician period. After the Mesozoic era, the group rapidly declined to its present number and rather restricted occurrence. The genus *Lingula* (but none of the present living species) dates back to the Ordovician period.

External Anatomy. Each of the two valves is bilaterally symmetrical and is usually convex. The smaller dorsal shell fits over the larger ventral shell, the apex of which in some groups is drawn out posteriorly and upward as a spout, like a Roman lamp—hence the origin of the name *lamp shell* (Fig. 18–19*B*). In the burrowing lingulids, the valves are flattened and more equal in size. The valves may be ornamented with concentric growth lines and a fluted, ridged, or even a spiny surface. The color of the shell in most living brachiopods is dull yellow or gray, but some species have orange or red shells.

The two valves articulate with one another along the posterior line of contact, called the hinge line (Fig. 18–19*A* and *C*), and the nature of the articulation is the basis for the division of the phylum into two classes, the Inarticulata and the Articulata. In inarticulate brachiopods, such as *Lingula and Glottidia,* the two valves are only held together by muscles. In articulate brachiopods, on the other hand, the valves bear interlocking processes that greatly restrict the degree of gape. Along the hinge line, which functions as a fulcrum, the ventral valve bears a pair of hinge teeth that fit into opposing sockets on the underside of the hinge line of the dorsal valve (Fig. 18–20*A*). This articulating mechanism locks the valves securely together and allows only a slight anterior gape of approximately 10 degrees. A pair of adductor muscles, divided into striated "quick" muscle and a smooth "catch" muscle, closes the valves. Another pair of muscles (diductors) opens the valves. The muscles run between the valves, but a number of variations in placement as well as shell modifications for muscle attachment evolved in the history of brachiopods.

Although inarticulates are considered to be more primitive than articulate forms, both groups are present from the beginning of the fossil record.

The shell is secreted by the underlying dorsal and ventral mantle lobes (Fig. 18–19*E*). The shell of inarticulates is usually composed of a mixture of calcium phosphate and chitin (chitinophosphate), and the shell of articulates is composed of calcium carbonate in the form of calcite. The chitinophosphate shell appears to be the more primitive, for the Cambrian shells were of this type; calcite shells did not appear until the Ordovician. The outer surface of the shell is covered by a thin organic periostracum. As in most mollusks, the periostracum and the outer layers of mineral deposition are secreted by the mantle edge, and the inner layer of shell is secreted by the entire outer mantle surface.

The shells of many fossil and living articulate brachiopods contain uniformly distributed vertical channels that extend upward through the shell to a point a little below the periostracum. The channels are filled with mantle tissue, which terminates in a cluster of secretory cells. The significance of this shell structure, called the punctate type, is unknown.

In addition to secreting the shell, the mantle edge bears in most species long chitinous setae that are believed to have a protective and perhaps sensory function (Fig. 18–21*C*).

The body proper of brachiopods occupies only the posterior part of the chamber formed by the two valves (Fig. 18–22). Anteriorly, the space between the mantle lobes (the mantle cavity) is filled by the lophophore.

Most brachiopods are attached to the substratum by a cylindrical extension of the ventral body wall, called the pedicle. The pedicle of the inarticulate lingulids (*Lingula* and *Glottida*) is long and provided with muscles; it emerges at the posterior of the animal between the two valves (Fig. 18–21*A*). The lingulids live in vertical burrows excavated in sand and mud. The anterior ends of the valves are directed toward the burrow opening, and the pedicle extends downward toward the bottom of the burrow and is encased in sand. When the animal is feeding, the gaping valves are near the burrow opening and the long mantle setae appear to function in preventing fouling by sediment (Fig. 18–21*B*). When the animal is disturbed, the pedicle contracts and pulls the animal downward into the burrow.

The pedicle of most other brachiopods is very short, lacks muscles, and is composed primarily of connective tissue (Figs. 18–21*C* and 18–22). Moreover, the pedicle emerges either from a notch at the hinge line of the ventral valve or through a hole at the upturned apex (Fig. 18–19*A*). This means, of course, that the pedicle emerges from the dorsal side of

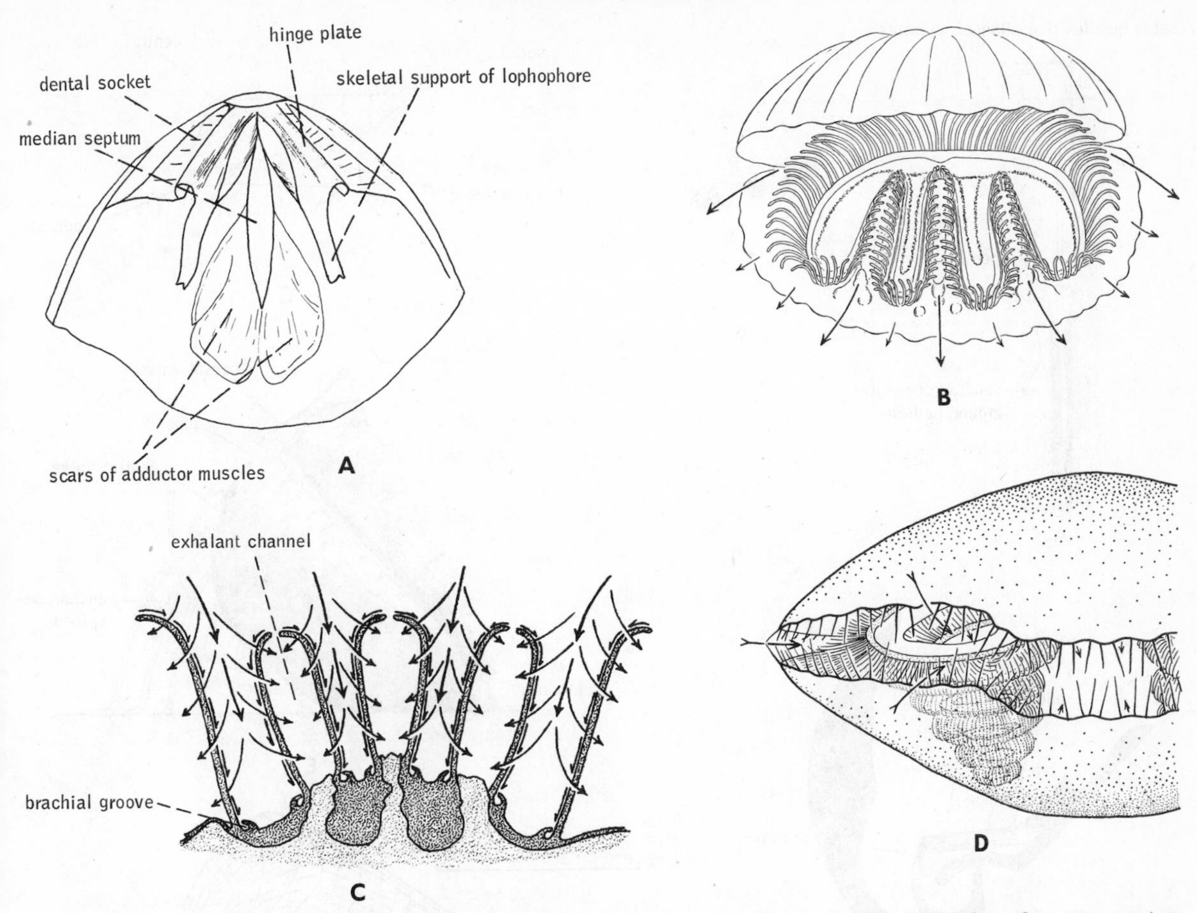

Figure 18–20. *A.* Internal surface of dorsal valve of an articulate brachiopod. (After Davidson from Hyman.) *B.* View into anterior gape of *Megathyris*, showing lophophore and exhalant water currents. *C.* Diagrammatic cross section through lophophore of *Megathyris*, showing water currents (large arrows) passing between filaments into exhalant channels, and the direction of ciliary tracts (small arrows) carrying food particles to the brachial grooves. (*B* and *C* after Atkins.) *D.* Front view of gape of *Notosaria*, showing setal grille and spiral lophophore. (Modified from Rudwick.)

the ventral valve, which extends posteriorly considerably beyond the dorsal valve. These brachiopods are attached to the substratum upside down with the valves held in a horizontal position (Fig. 18–21*C*). Muscles within the valves, which are inserted on the pedicle base, permit erection, flexion, and even rotation of the animal on the pedicle. The end of the pedicle adheres to the substratum by means of rootlike extensions or short papillae.

The pedicle has been lost completely in a few brachiopods of both classes, such as *Crania* (Inarticulata) and *Lacazella* (Articulata). Such species are cemented directly to the substratum by the more posterior surface of the ventral valve and are thus oriented in a normal manner with the dorsal side up (Fig. 18–21*D* and *E*). The more posterior part of the ventral valve being the point of attachment, the anterior margin is directed somewhat upward and clear of the substratum. Some species of fossil brachiopods that attached by cementation were important contributors to Paleozoic reefs.

The shell form of a number of fossil groups suggests that they were adapted for living free on the surface of soft bottoms. Spines, long shell "wings," and flattened ventral surfaces appear to have been devices to prevent sinking (Fig. 18–21*F* and *G*).

Lophophore and Feeding. As in other lophophorates, the brachiopod lophophore is basically a crown of tentacles surrounding the mouth; but here, as a means of increasing the surface area, the lophophore projects anteriorly as two arms, or brachia, from which the name brachiopod is derived. In its simplest form, this is like the horseshoe-shaped lophophore of freshwater bryozoans. The arms are further looped or spiraled in complicated ways (Fig. 18–23). Each brachium bears a row of tentacles, and the tentacle-bearing ridge is

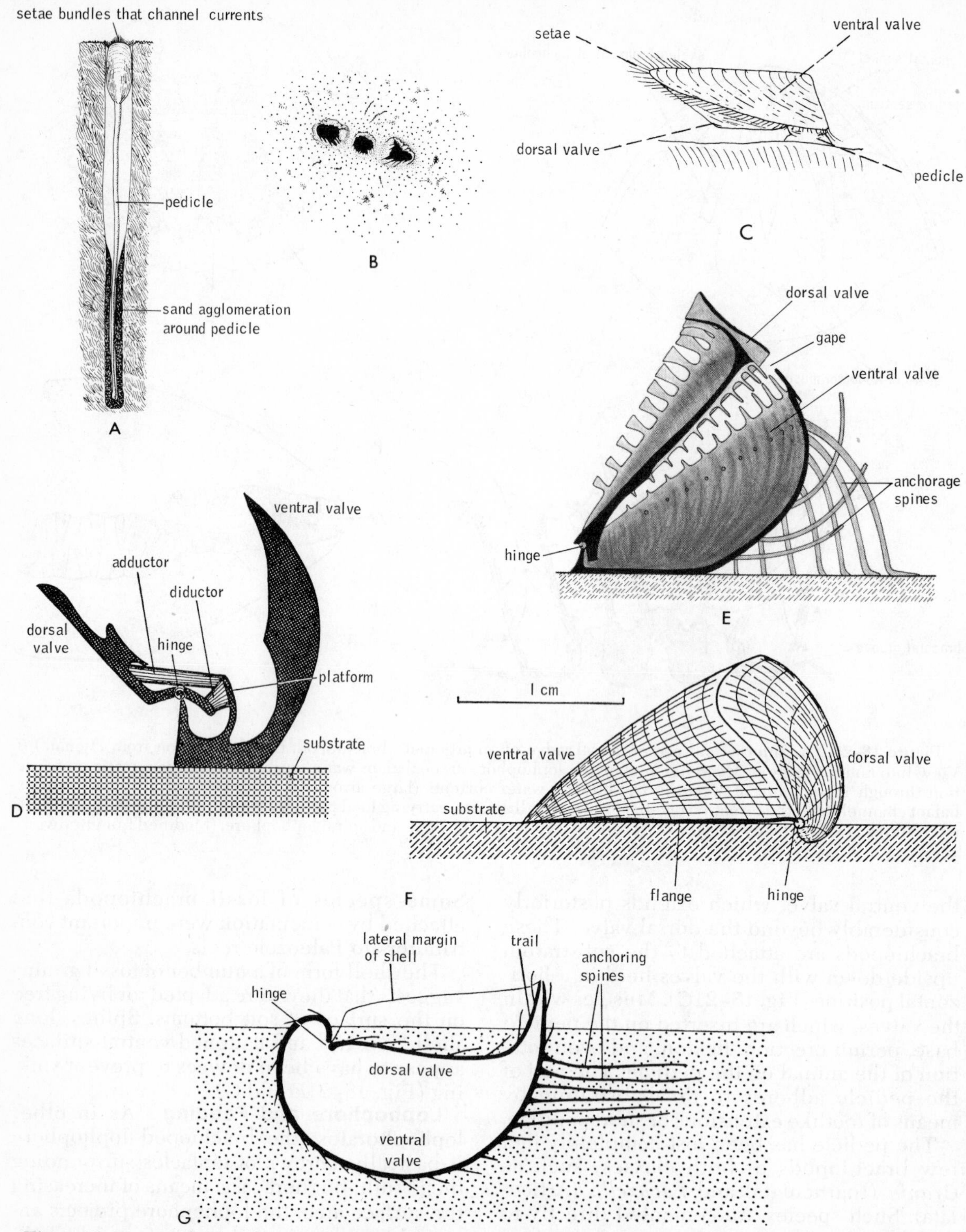

Figure 18–21. Brachiopod diversity and relationship to substratum. *A.* The inarticulate *Lingula* in feeding position within burrow. (Modified from Francois.) *B.* Burrow opening of *Lingula* when feeding. Setae surround middle exhalant and lateral inhalant apertures. (Modified from Rudwick.) *C.* Living articulate *Discinisca,* attached by pedicle. (After Morse from Hyman.) *D.* Diagrammatic lateral view of living articulate *Lacazella,* which lives attached directly to substratum by ventral valve. *E.* Sagittal section through Permian articulate *Chonosteges,* which cemented ventral valve directly to substratum. (Modified from Rudwick.) *F.* Devonian articulate *Spyringospira,* which is believed to have lived unattached on soft bottoms. *G.* Permian articulate *Waagenoconchia,* which is believed to have lived partially buried in soft bottoms. (*D, F,* and *G* from Rudwick, M. J. S., 1970: Living and Fossil Brachiopods. Hutchinson Co., London.)

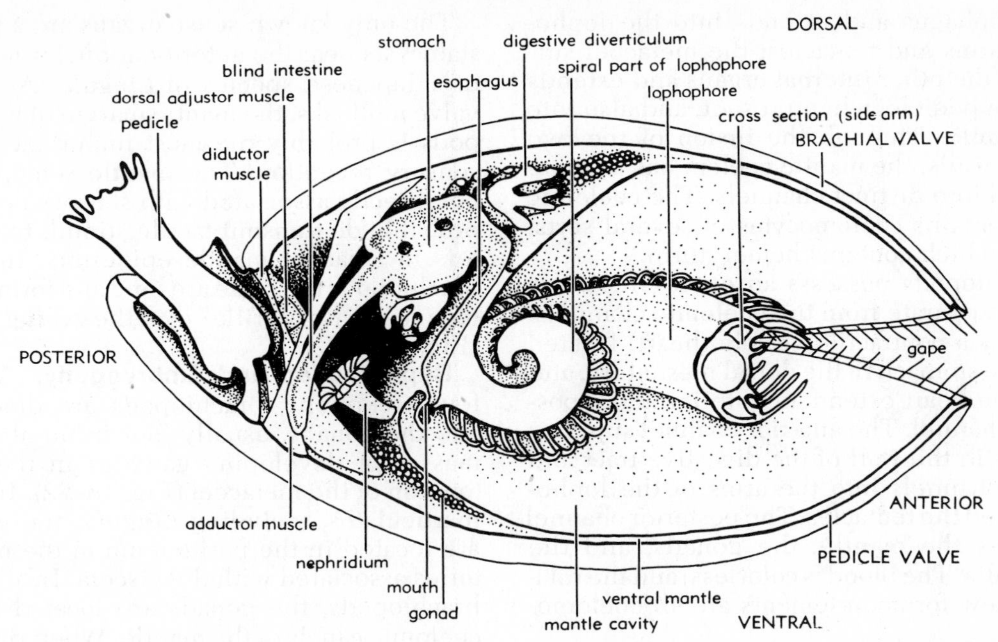

Figure 18–22. Section through an articulate brachiopod. (From Williams, A., and Rowell, A. J., 1965: *In* Moore, R. C. (Ed.), Treatise on Invertebrate Paleontology, Geol. Soc. Amer. and Univ. Kansas.)

flanked by a brachial groove at its base. The lophophore is supported by a cartilaginous axis and a fluid-filled canal within each brachium. Also, in many brachiopods, the dorsal valve bears complicated supporting processes, and the processes and inner valve surface may be grooved and ridged for the reception of the lophophore (Fig. 18–20A).

When the brachiopod is feeding, water enters and leaves the valve gape through distinct inhalant and exhalant apertures and chambers created by the lophophore (Fig. 18–20). In its circuit over the lophophore, water passes between the tentacles, driven by lateral tentacular cilia. Particles, especially fine phytoplankton, are screened by the lateral cilia of two adjacent filaments, trapped in mucus on the frontal cilia, and then transported down the tentacles to the brachial groove. The brachial groove conducts food to the mouth (Fig. 18–20*C*). Rejected particles are carried away in the median outward-flowing current. Some species can reverse the water current, and even the direction of the beat of the frontal cilia, when there is too heavy an accumulation of particles within the lophophore.

McCammon (1969) believes that plankton is only of secondary importance as food for articulate brachiopods, and that dissolved and colloidal particles are of much greater importance. Gut contents, the absence of an anus, the complex filtering apparatus and the distribution of articulate brachiopods in cold water, where dissolved nutrient concentrations are higher, are cited as evidence.

The mouth leads into an esophagus that extends dorsally and a short distance forward prior to turning posteriorly and joining a dilated stomach (Fig. 18–22). The stomach is surrounded by a digestive gland that opens through the stomach wall by means of one to three ducts on each side. In articulates, a blind intestine extends posteriorly from the stomach, but in inarticulates the intestine opens to the outside through a rectum and anus. The anus is a posterior median opening between the valves in *Crania;* in *Lingula* and others, it opens into the mantle cavity on the right side. According to Chuang (1959), digestion in *Lingula* is chiefly intracellular within the digestive gland.

Internal Anatomy and Physiology. The outer epithelium of the body wall is attached dorsally and ventrally to the inner surface of the valves. Beneath the epithelium is a layer of connective tissue, and where the body is free of the shell, there is a layer of longitudinal muscle fibers. Internally the body wall is lined with peritoneum. Since the mantle lobes are extensions of the anterior body wall, each is composed of a double wall with coelom in between. The epidermis of the exposed inner surface of each mantle lobe is ciliated.

The coelom is similar to that of other lophophorates. The anterior mesocoel surrounds the esophagus and extends into the lophophore arms and tentacles; the metacoel surrounds the other internal organs and extends into the pedicle, when present, and also into the mantle. Through the fusion of the two mantle walls, the mantle coelom has become divided into distinct channels. The coelomic fluid contains coelomocytes of several sorts, one of which contains hemerythrin.

Brachiopods possesss an open circulatory system separate from the coelomic channels. There is a contractile vesicle (heart) located over the stomach in the dorsal mesentery, and from the heart extend an anterior and a posterior channel. The anterior channel supplies sinuses in the wall of the digestive tract and sends channels into the arms of the lophophore and the tentacles. The posterior channel supplies the mantle, the gonads, and the nephridia. The blood is colorless, and the relatively few formed elements are all coelomocytes.

Although contraction of the heart has been observed, the exact role of the blood-vascular system in the physiology of brachiopods is not definitely known. The circulation of food materials is perhaps its primary function. There are no specialized gas exchange organs, but the lophophore and mantle lobes are probably the principal sites of gaseous exchange. Oxygen transport is probably provided by the coelomic fluid, for there is a definite circulation of coelomic fluid through the mantle channels and it is carried, at least in part, by the hemerythrin in the coelomocytes.

One or two pairs of metanephridia are present in brachiopods. The nephrostomes open into the metacoel on each side of the posterior end of the stomach, and the tubules then extend anteriorly to open into the mantle cavity through a nephridiopore situated posteriorly and to each side of the mouth (Fig. 18–22). The lining of the nephridia is glandular and ciliated. Coelomocytes are active in the ingestion of particulate waste, and such waste-laden cells, as well as free waste particles, are picked up by the nephrostome from the coelomic fluid and then passed to the outside.

An esophageal nerve ring with a small ganglion on its dorsal side and a larger ganglion on the ventral side forms the nerve center of brachiopods. From the ganglia and their connectives, nerves extend anteriorly and posteriorly to innervate the lophophore, the mantle lobes, and the valve muscles. In inarticulates, the nervous system is closely associated with the epidermis, and in all brachiopods there is probably a subepidermal nerve plexus.

The only known sense organs are a pair of statocysts near the anterior adductor muscles in a Japanese species of *Lingula.* As in bivalve mollusks, the mantle margin of brachiopods is probably the most important site of sensory reception. The mantle setae, while not directly associated with sensory neurons, probably do transmit tactile stimuli to receptors in adjacent mantle epidermis. In some brachiopods the setae are long and form a protective "sensory grille" over the gaping valves (Fig. 18–20*D*).

Reproduction and Embryogeny. With a few exceptions, brachiopods are dioecious and the gonads, usually four in number, are masses of developing gametes in the peritoneum of the metacoel (Fig. 18–22). In most inarticulates, including *Lingula,* the gonads are located in the peritoneum of the mesenteries associated with the viscera. In all other brachiopods, the gonads are located in the coelomic canals of the mantle. When ripe, the gametes pass into the coelom and are discharged to the exterior by way of the nephridia.

Except for a few brooding species, the eggs are shed into the sea water and fertilized at the time of spawning. The embryogeny of brachiopods has evoked considerable interest because of the similarities to development in deuterostomes. Cleavage is radial and nearly equal, and leads to a coeloblastula that undergoes gastrulation by invagination, except in *Lacazella,* in which entoderm develops by delamination or ingression. In contrast to the typical method of mesoderm formation in protostomes, the mesoderm in brachiopods appears to be enterocoelic—that is, it arises from the archenteron. Masses of mesodermal cells may be proliferated laterally from the archenteron, as in *Lingula,* and later form an internal coelomic cavity; or the mesoderm may arise as pouches to form the archenteron. In *Terebratulina,* the anterior end of the archenteron separates as a primordial coelomic sac, which subsequently divides. This method of coelom formation occurs in many of the echinoderms.

The embryo eventually develops into a free-swimming and feeding larva. The larva of inarticulates resembles a minute brachiopod (Fig. 18–23*A*). The pair of mantle lobes and the larval valves enclose the body and the ciliated lophophore, which acts as the larval locomotor organ. The pedicle, which in this group is derived from the mantle, is coiled in the back of the mantle cavity. As additional shell is laid down, the larva becomes heavier and sinks to the bottom. There is no metamor-

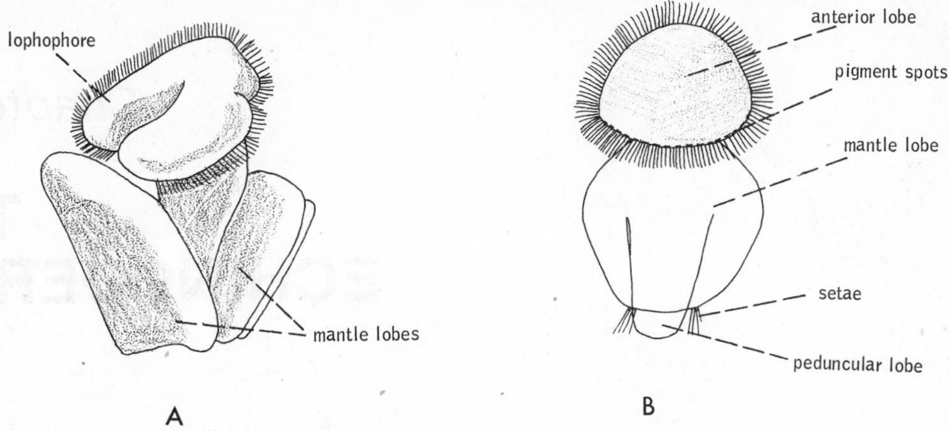

Figure 18–23. *A.* Larva of the inarticulate, *Lingula.* (After Yatsu from Hyman.) *B.* Larva of the articulate, *Terebratella inconspicua.* (After Percival from Hyman.)

phosis in *Lingula;* the pedicle attaches to the substratum, and the young brachiopod takes up an adult existence. Articulate larvae differ in having a ciliated anterior lobe representing the body and lophophore, a posterior lobe that forms the pedicle, and a mantle lobe that is directed backward (Fig. 18–23*B*). In *Terebratulina,* the larva settles after a short free-swimming existence of approximately 24 to 30 hours, and then it undergoes metamorphosis. The mantle lobe reverses position and begins the secretion of the valves, and the adult structures develop from their larval precursors.

Bibliography

Brown, C. J. D., 1933: A limnological study of certain freshwater Polyzoa with special reference to their statoblasts. Trans. Amer. Microsc. Soc., *52*:271–316.

Chuang, S. H., 1959: Structure and function of the alimentary canal in *Lingula unguis.* Proc. Zool. Soc. London, *132*:293–311.

Farmer, J. D., Valentine, J. W., and Cowen, R., 1973: Adaptive strategies leading to the ectoproct groundplan. Syst. Zool., *22*(3):233–239.

Gosner, K. L., 1971: Guide to Identification of Marine and Estuarine Invertebrates: Cape Hatteras to the Bay of Fundy. Wiley-Interscience, N.Y. Bryozoa, Phoronida, Brachiopoda, pp. 222–248.

Hyman, L. H., 1951: The Invertebrates. Vol. 3. Acanthocephala, Aschelminthes and Entoprocta. McGraw-Hill, N.Y.

Hyman, L. H., 1959: The Invertebrates. Vol. 5. Smaller Coelomate Groups. McGraw-Hill, N.Y., pp. 228–609.

Larwood, G. P., 1973: Living and Fossil Bryozoa: Recent Advances in Research. Academic Press, N.Y., 652 pp.

Light, S. F., Smith, R. I., Pitelka, F. A., Abbott, D. P., and Weesner, F. M., 1967: Intertidal Invertebrates of the Central California Coast. University of California Press, Berkeley. Bryozoa, Phoronidea, Brachiopoda, pp. 271–284.

MacGinitie, G. E., and MacGinitie, N., 1968: Natural History of Marine Animals. McGraw-Hill, N.Y.

McCammon, H. M., 1969: The food of articulate brachiopods. J. Paleontol., *43*(4):976–985.

Moore, R. C. (Ed.), 1953: Treatise on Invertebrate Paleontology. Bryozoa, Pt. G, 253 pp.; 1965: Brachiopoda, Pt. H, Vols. 1 and 2, 926 pp. Geological Society of America and University of Kansas Press.

Nielsen, C., 1964: Studies on Danish Entoprocta. Ophelia, *1*(1):1–76.

Nielsen, C., 1971: Entoproct life cycles and the entoproct/ectoproct relationship. Ophelia, *9*(2):209–341.

Pennak, R. W., 1953: Freshwater Invertebrates of the United States. Ronald Press, N.Y., pp. 256–277. (A general discussion of freshwater bryozoans and key to the species of the United States.)

Rogick, M. D., 1959: Bryozoa. *In* Edmondson, W. T., Ward, H. B., and Whipple, G. C. (Eds.), Freshwater Biology. 2nd Edition. John Wiley and Sons, N.Y., pp. 495–507.

Rudwick, M. J. S., 1970: Living and Fossil Brachiopods. Hutchinson University Library, Hutchinson and Co., London, 199 pp. (An excellent general account of living and fossil brachiopods.)

Ryland, J. S., 1970: Bryozoans. Hutchinson University Library, Hutchinson and Co., London. (An excellent general account of living and fossil bryozoans.)

Schneider, D., 1963: Normal and phototropic growth reactions in the marine bryozoan *Bugula avicularia. In* Dougherty, E. C. (Ed.), The Lower Metazoa. University of California Press, Berkeley, pp. 357–371.

Silen, L., 1966: On the fertilization problem in the gymnolaematous Bryozoa. Ophelia, *3*:113–140.

Chapter 19

THE ECHINODERMS

Members of the phylum Echinodermata are among the most familiar marine invertebrates, and such forms as the sea stars have become virtually a symbol of sea life. The phylum contains some 5300 known species and constitutes the only major group of deuterostome invertebrates.

Echinoderms are exclusively marine and are largely bottom dwellers. All are relatively large animals, most being at least several centimeters in diameter. The most striking characteristic of the group is their pentamerous radial symmetry—that is, the body can usually be divided into five parts arranged around a central axis. This radial symmetry, however, has been secondarily derived from a bilateral ancestral form, and the echinoderms are in no way related to the other radiate phyla—the sponges, the cnidarians, and the ctenophores. Furthermore, echinoderms are true coelomate animals and have a higher level of structure than do the other radiate groups.

Characteristic of all echinoderms is the presence of an internal skeleton. The skeleton is composed of calcareous ossicles that may articulate with one another, as in sea stars and brittle stars, or may be fused together to form a rigid skeletal test, as in sea urchins and sand dollars. Typically the skeleton bears projecting spines or tubercles that give the body surface a warty or spiny appearance, hence the name echinoderm—spiny skin.

The most distinctive feature of echinoderms is the presence of a unique system of coelomic canals and surface appendages composing the water-vascular, or ambulacral, system. Primitively, the water-vascular system probably functioned in collecting and transporting food, but in many echinoderms it has assumed a locomotor function.

Echinoderms possess a spacious coelom in which is suspended a well-developed digestive tract. There is no excretory system. Gas exchange structures vary in nature from one group to another and appear to have arisen independently within the different classes. Most members of the phylum are dioecious. The reproductive tracts are very simple, for there is no copulation and fertilization is external in sea water.

Echinoderm Development. The eggs are typically homolecithal, and the early embryogeny is relatively uniform throughout the phylum and displays the basic features of deuterostome development. In contrast to that of protostomes, cleavage in echinoderms is radial and indeterminate instead of spiral and determinate (Fig. 3–3).

As cleavage progresses, the blastomeres decrease in size, a segmentation cavity appears in the middle of the mass of cells, and eventually a blastula displaying a large blastocoel is formed. Gastrulation takes place primarily by invagination, forming a narrow tubular archenteron (primitive gut). The archenteron grows forward and eventually connects with the anterior stomodeum, which will form the mouth. The blastopore remains as the larval anus. Thus, the origin of the mouth in relation to the blastopore is different from that in protostomes. In deuterostomes, the anus either is derived from the blastopore or is secondarily formed near the site of the closed blastopore. The mouth forms as a secondary opening (deuterostome—second mouth) at the opposite end from the blastopore.

Prior to the formation of the mouth, the advancing distal end of the archenteron proliferates some mesenchyme into the blastocoel and then, primitively at least, gives rise to two lateral pockets or pouches that eventually separate from the archenteron (Fig. 19–34A). The cavities of the pouches represent the future coelomic cavity, and the cells composing the pouch wall become the mesoderm. The remaining archenteron becomes the entoderm of the gut. Thus, the mesoderm of deuterostomes is enterocoelous in origin—that is, the mesoderm arises by pouching (evaginations) of the archenteron. Furthermore, the coelom appears more or less simultaneously with the formation of mesoderm. The number of such pouches formed varies in different groups of deuterostomes.

Primitively, in echinoderms, the two original pouches, one on each side, each give rise by evagination or subdivision to coelomic vesicles arranged one behind the other and called respectively the axocoel, the hydrocoel, and the somatocoel (Fig. 19–12A). These coelomic vesicles correspond to the paired protocoel, mesocoel, and metacoel of lophophorates and other oligomerous coelomates. The two somatocoels meet above and below the gut to form the gut mesenteries. The left axocoel opens dorsally through a pore called the hydropore. This primitive and somewhat hypothetical plan of the coelomic vesicles is considerably modified in the development of existing echinoderms. This plan does, however, provide a basis for understanding the origin and arrangement of the coelomic vesicles in the different classes that are discussed later.

The gastrula rapidly develops into a free-swimming larva (Fig. 19–11). The most striking feature of the echinoderm larva is its bilateral symmetry, which is in marked contrast to the radial symmetry of the adult. Wound over the surface of the larva are a varying number of ciliated locomotor bands. There is a complete functional digestive tract with a large ciliated stomodeum, an esophagus, a stomach, an intestine, and an anus. Food is obtained from the current of water produced by the stomodeal cilia. Later larval development in most echinoderms involves the formation of short or long slender projections (arms) from the body wall, and the nature and position of these arms, or the lack of them, distinguishes the larvae of the different echinoderm classes. The arms disappear in later development and are not equivalent to the arms of certain adult echinoderms, such as sea stars and brittle stars.

After a free-swimming planktonic existence, the bilateral larva undergoes a metamorphosis in which the radial symmetry of the adult is developed.

Ecology and Distribution. A large number of echinoderms are adapted for life on rock and other hard substrata. Indeed, a hard bottom appears to have been the ancestral environment for at least some of the living classes. But within each class, there are also many species which are specialized for life on or within sand or mud. These soft-bottom echinoderms display many interesting adaptations, which differ from class to class.

Echinoderms are unique in being virtually the only major phylum in the Animal Kingdom in which there are no parasitic species. Moreover, there are only a very few echinoderms that are commensal on other animals. Although there are no parasitic echinoderms and only a few commensal species, echinoderms are favorite hosts for an enormous number of commensals and parasites from other animal groups. Parasitic myzostomes, snails, clams, and copepods and commensal scale worms, shrimp, and crabs are but some animals for which echinoderms supply food and shelter.

Echinoderms are found in all oceans from littoral waters to great depths. Many tend to live in distinct aggregations and in enormous numbers. Sea urchins may be found in such densities in certain favorable spots that the entire bottom is covered. Similarly, sand dollars can be collected by the bushels on some sand flats. There are many reports of high population densities of feather stars.

Echinoderms are one of the most abundant groups of deep-sea invertebrates. There are representatives from every class, but the greatest number of abyssal echinoderms are holothuroids (sea cucumbers).

The center of the greatest species density of echinoderms, excluding sea stars, is the Indo-Pacific, particularly the area encompassed by the Philippines, Borneo, and New Guinea. From this center a relatively rich fauna spreads out in all directions as far as the Red Sea and southern Japan. For example, there are 91 species of crinoids (sea lilies and feather stars) known from southern Japan in contrast to some 85 species reported from the entire Atlantic basin.

Although the crinoids (sea lilies) are the most primitive class of living echinoderms, the more familiar asteroids (sea stars) are treated first in the discussion of the echinoderm classes and serve to introduce the basic features of echinoderm structure. The classification used throughout this chapter follows that used in the Treatise on Invertebrate Paleontology (Moore, 1966, 1967). Characterization of the orders in the Systematic Résumés has frequently been limited by the necessity of omitting the highly technical descriptions of skeletal features on which these taxa are largely based.

CLASS STELLEROIDEA; SUBCLASS ASTEROIDEA

The class Stelleroidea contains those star-shaped free-moving echinoderms in which the body is composed of rays, or arms, projecting from a central disc. All living species are members of one of two subclasses, the Asteroidea (containing the sea stars, or starfish) and the Ophiuroidea (containing the brittle stars).

The 1600 described species of sea stars are common and familiar animals that crawl about over rocks and shells or live on sandy or muddy bottoms. They are found throughout the world, largely in coastal waters; but the northwest Pacific, particularly from Puget Sound to the Aleutians, possesses the greatest concentration of asteroid species of any place in the world. Seventy species are endemic to the Vancouver Island area alone.

Sea stars are commonly a drab yellow, but many species are more brightly colored. Red, orange, blue, purple, green, and darker shades are not infrequent. Also, combinations of colors exist. For example, in a common North Atlantic *Astropecten*, the upper surface of the disc and rays is bluish-purple, margined with a pale band (Fig. 19–1*A*).

External Structure. The sea stars are typically pentamerous, with most species possessing five arms that grade into the disc (Fig. 19–2*A* and *B*). However, a greater number of arms are characteristic of many asteroids. For example, there are seven to fourteen arms in the European sun star, *Crossaster papposus* (Fig. 19–1*B*), and as many as 40 arms or more in *Heliaster* of the American West coast. Most asteroids range from 12 to 24 cm. in diameter, but there are some sea stars that are less then 2 cm. in diameter, while the 20-rayed star (*Pycnopodia*) of Puget Sound may measure almost one meter across.

Unlike those of brittle stars, the arms of asteroids are not sharply set off from the central disc—that is, the width of the arm usually increases toward the base and grades into the disc. In most species the arm length ranges from one to three times the diameter of the central disc. From this average, there deviate some forms that have extremely long slender arms (Fig. 19–2*E*) and others that have very short arms. In the cushion stars, *Plinthaster* (Fig. 19–2*C*) and *Goniaster* (Fig. 19–1*D*), each arm has the shape of an isosceles triangle, and in *Culcita* (Fig. 19–2*D*), the arms are so short that the body appears pentagonal.

The mouth is located in the center of the underside of the disc, and the entire undersurface of the disc and arms is called the oral surface (Fig. 19–2*B*). The opposite, or upper, side of the body is the aboral surface. From the mouth extends radially a wide furrow into each arm (Figs. 19–2*B* and 19–6*B*). Each furrow (ambulacral groove) contains two or four rows of small tubular projections, called tube feet or podia. These tube feet are the locomotor organs and form part of the water-vascular system.

The margins of the ambulacral grooves are guarded by movable spines that are capable of closing over the groove. The tip of each arm bears a small tentacle and a red pigment spot. The aboral surface bears both the inconspicuous anus, when present, in the center of the disc and a large button-like structure (the madreporite) toward one side of the disc between two of the arms (Fig. 19–2*A*). The general body surface is typically rough and beset with spines, tubercles, or ridges. The reef-inhabiting *Acanthaster* has spines 3 cm. in length projecting from the aboral surface (Fig. 19–2*F*). At the other extreme, there are many phanerozone sea stars in which the surface appears relatively smooth because of the flattened nature of the spines, or paxillae, which will be described later. Many members of this same order have arms and a disc, which are bordered by large conspicuous plates (Fig. 19–1*A*).

Body Wall. The outer surface of the body is covered by an epidermis composed of ciliated (flagellated?) columnar epithelium (Fig. 19–3*A*). The epithelial cells are attenuated, at least at the base, and covered on the outer surface by a thin cuticle. The epidermis also contains slender neurosensory cells and mucous gland cells that provide the body surface with a protective mucous coating. Detritus that falls on the body is trapped in the mucus and then swept away by the epidermal flagella.

Beneath the epidermis is a layer of nerve cells forming a subepidermal plexus (Fig. 19–3*A*). Beneath the nervous layer is a thick layer of connective tissue (the dermis) that houses the skeletal system.

The asteroid skeleton is composed of separate ossicles in the shape of rods, crosses, or plates. The ossicles are so arranged that they

Figure 19–1. *A. Astropecten irregularis*, a burrowing sea star (aboral view). *B.* The sun star, *Crossaster papposus* (aboral view). (Both by D. P. Wilson.) *C. Oreaster*, a large sea star with an elevated aboral surface. *D. Goniaster* (dried specimen), a greatly flattened sea star. (*C* and *D* by Betty M. Barnes.)

form a lattice network and are bound together by connective tissue (Fig. 19–4*A* and *B*). The amount of space between ossicles varies, depending on the form of the ossicles, and the space is particularly reduced when the ossicles are in the form of plates.

The calcareous ossicles that compose the echinoderm skeleton display many interesting features. As in vertebrate bone, there is a considerable turnover of the mineral constituents. The ossicles are irregularly perforated (fenestrated), perhaps representing an adaptation for reduction of weight and increase in strength. The most remarkable characteristic is that each ossicle of echinoderm skeletons is believed to represent a single crystal of magnesium-rich calcite, 6(Ca,Mg)CO_3 (Donnay and Pawson, 1969). The crystal is formed within a cell of the dermis. As the crystal increases in size, it becomes surrounded by a large number of cells, all of which are daughter cells of the original cell which initiated the formation of the crystal. The shape and size of the ossicles of different

Figure 19–2. *A* and *B*. Aboral (*A*) and oral (*B*) views of *Echinaster spinulosus* from the west coast of Florida. (By Betty M. Barnes.) *C*. Aboral view of *Plinthaster dentatus*. *D*. Oral view of *Culcita*, a sea star with very short arms. *E*. *Linckia*, a common and widespread genus with long arms. (By Betty M. Barnes.) *F*. The Pacific coral-eating *Acanthaster planci*. (*C*, *D*, and *F* courtesy of the National Museum of Natural History.)

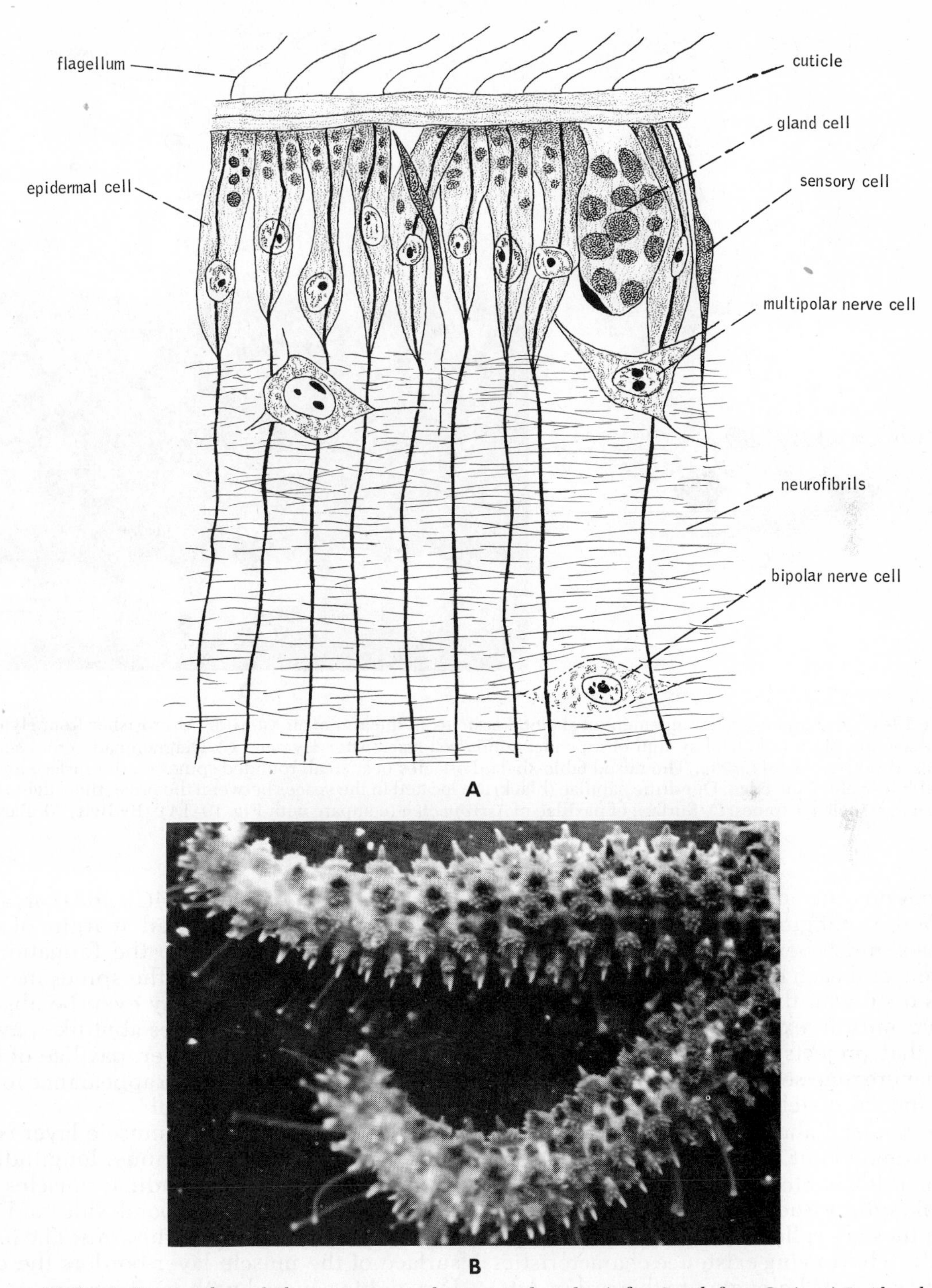

Figure 19-3. *A.* Section through the integument of *Asterias glacialis.* (After Smith from Cuénot.) *B.* Aboral surface of two arms of *Coscinasterias.* The crown-like projections are aboral spines bearing a wreath of pedicellariae. The darker small clusters between the spines are papulae. Tube feet project from the oral surface. (By Betty M. Barnes.)

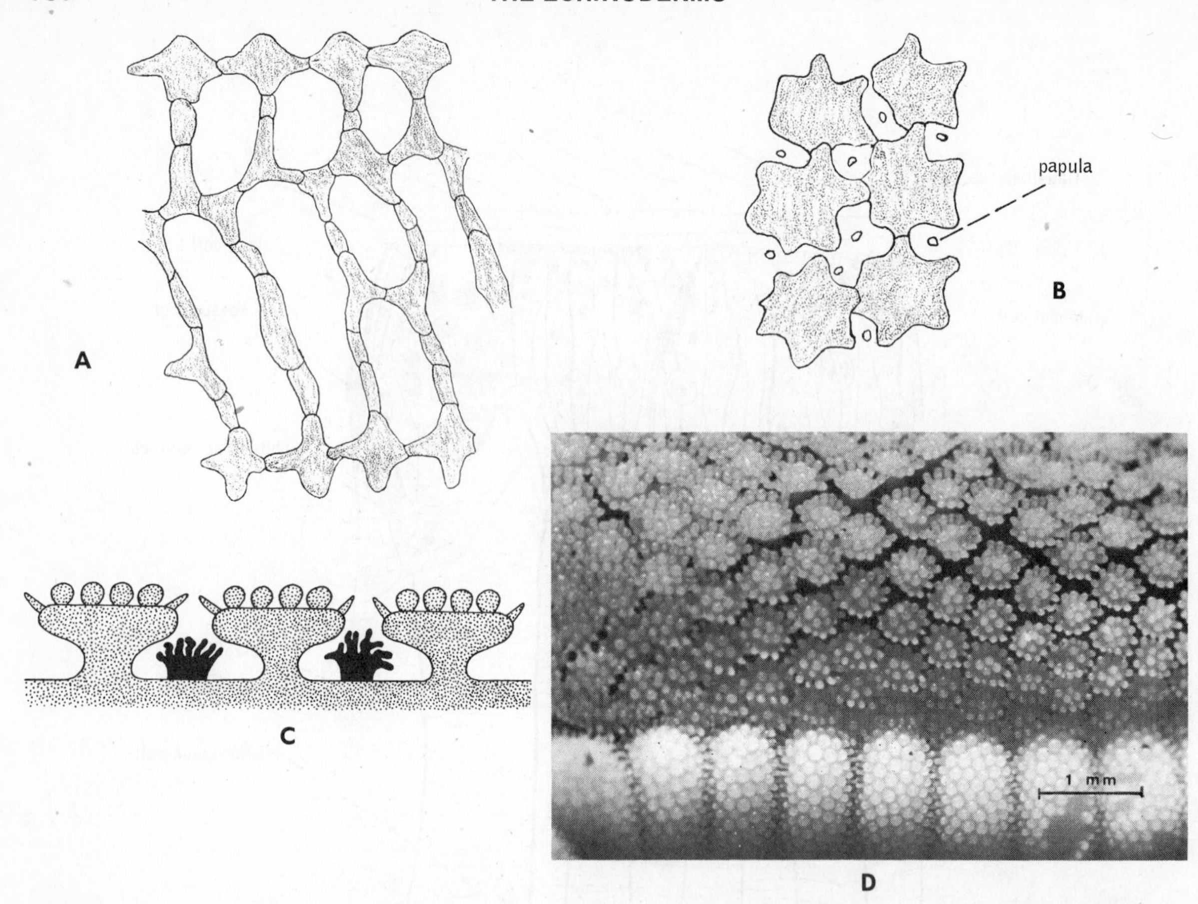

Figure 19–4. *A.* Lattice-like arrangement of skeletal ossicles in the arm of an asteriid. (After Fisher from Hyman.) *B.* Small section of endoskeletal system of a phanerozonic sea star. (After Hyman.) *C.* Diagrammatic cross section through several paxillae of *Luidia.* The raised table-shaped ossicles bear small rounded spines on the surface and flat movable spines along the edge. Dendritic papulae (black) are located in the spaces between the projecting edges of the paxillae and associated spines. *D.* Surface of paxillae of *Astropecten* (compare with Fig. 19–1*A*). (By Betty M. Barnes.)

echinoderms are apparently determined in part by the configuration of the dermis.

Spines and tubercles are also part of the skeleton, and each either consists of separate pieces resting on the deeper dermal ossicles or represents an extension of the dermal ossicles that projects to the outer surface. In the phanerozone sea stars, the aboral surface bears special ossicles, the central portion of which is raised above the body surface and even extended out like a parasol. The raised part of the ossicle is crowned with small movable spines. Such an ossicle and its associated spines are called a paxilla; it is an adaptation for a burrowing existence characteristic of many sea stars belonging to this order. Adjacent paxillae create a protective space above the aboral integument. Through this space flows respiratory and feeding currents. The surrounding sediment is held back by the paxillae. The movable spines on the paxillae can fold down as a roof over the space between the spines (Fig. 19–4*C* and *D*) or, as in *Astropecten,* the extended margin of the paxillae may contribute to the formation of the roof. In the latter case the spines may be reduced to granules or may even be absent, whereas the adjacent spines abut like paving stones. As mentioned earlier, paxillae of this type (tabular) lend a smooth appearance to the aboral surface of the animal.

Beneath the dermis is a muscle layer composed of outer circular and inner longitudinal smooth fibers. The longitudinal muscles are best developed on the aboral side and are involved in the bending of the arms. The inner surface of the muscle layer borders the coelom and is covered with peritoneum.

In two orders of sea stars, the body surface bears small specialized jawlike appendages (pedicellariae) that are used for protection, especially against small animals or larvae which might settle on the body surface of the sea star. The pedicellariae are of two types—

stalked and sessile. The stalked pedicellariae are characteristic of the order Forcipulata, which includes *Asterias*, *Pycnopodia*, and *Pisaster*, and are more specialized than the sessile type. These pedicellariae each consist of a short fleshy stalk surmounted by a jaw-like apparatus. This apparatus is composed of three small ossicles that are arranged to form forceps or scissors. In the forceps type, two opposing ossicles rest upon and articulate with a basal ossicle (Fig. 19–5*C*). In the scissors type, the two opposing ossicles have curved bases that overlap and lie on each side of the basal ossicle (Fig. 19–5*D*). By means of special adductor and abductor muscles, the jaws can be opened and shut.

The distribution of stalked pedicellariae is variable. They may be scattered over the body surface, situated on the spines, or commonly, as in *Asterias*, they may form a wreath around the base of the spines (Fig. 19–3*B*).

Sessile pedicellariae are limited to the order Phanerozonia, which includes *Astropecten*. Sessile pedicellariae are composed of two or more short movable spines located on the same or adjacent ossicles. The spines oppose each other and articulate against one another to act as pincers (Fig. 19–5*A*). Primitively, the spines composing the pedicellariae are not markedly different from other spines. In many phanerozones, however, the pedicellariae are distinctly differentiated. For example, in one type of sessile pedicellariae, called alveolar pedicellariae, the spines are very short, broad, and shaped like the valves of a clam (Fig. 19–5*B*). The two valves are imbedded in the body surface.

The papulae and podia, which are also appendages of the body wall, are discussed in connection with gas exchange and the water-vascular system.

Water-Vascular System. The water-vascular system is a system of canals and appendages of the body wall that are unique to echinoderms. Since the entire system is derived from the coelom, the canals are lined with a ciliated epithelium and filled with fluid. The water-vascular system is well developed in asteroids and functions as a means of locomotion. The internal canals of the water-vas-

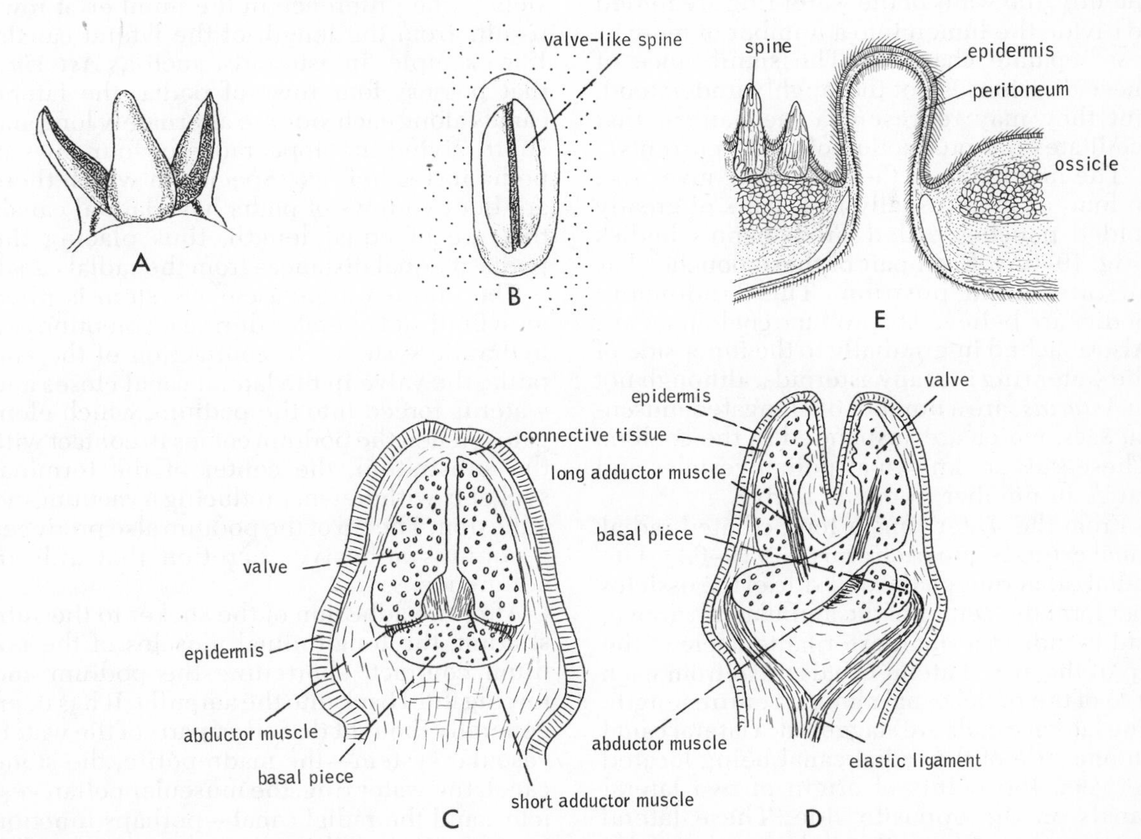

Figure 19–5. *A*. Sessile pedicellaria with three valves, from *Luidia*. *B*. A sessile pedicellaria with two clamlike valves, from *Hippasteria* (frontal view). (*A* and *B* after Cuénot.) *C* and *D*. Distal ends of two-stalked pedicellariae from *Asterias*. *C*. Forceps type. *D*. Scissors type. (*C* and *D* after Hyman.) *E*. Section through an asteroid papula. (After Cuénot.)

cular system connect to the outside through the button-shaped madreporite located on the aboral surface (Figs. 19–2*A* and 19–6*A*).

The surface of the madreporite is creased with many fine furrows covered by the flagellated epithelium of the body surface. The bottom of each furrow contains pores that open into pore canals passing downward through the madreporite. A madreporite may be perforated by as many as 250 pores and canals and is thus constructed essentially like a sieve. The pore canals join collecting canals that open into a small space (the ampulla) beneath the madreporite.

The ampulla leads into a vertical stone canal that descends to the oral side of the disc (Fig. 19–6*A*). The stone canal is so named because of the calcareous deposits located in its walls. A rolled projection of the stone canal wall divides the lumen of the canal into two passageways so that water may simultaneously circulate orally and aborally. On reaching the oral side of the disc, the stone canal joins a circular canal (the water ring) that is located just to the inner side of the ossicles that ring the mouth (Fig. 19–6*A*). Not infrequently, the walls of the water ring are folded to divide the lumen into a number of more or less separate channels. The significance of these divisions is not thoroughly understood, but they may represent a mechanism that facilitates the production of water currents.

The inner side of the water ring gives rise to four, or more usually five, pairs of greatly folded pouches called Tiedemann's bodies (Fig. 19–6*A*). Each pair of these pouches has an interradial position. The Tiedemann bodies are believed to produce coelomocytes. Also attached interradially to the inner side of the water ring in many asteroids, although not in *Asterias*, are a number of elongated muscular sacs, which are suspended in the coelom. These sacs are known as polian vesicles and range in number from one to five.

From the water ring, a long ciliated radial canal extends into each arm (Fig. 19–6*A*). The radial canal runs on the oral side of the ossicles that form the center of the ambulacral groove, and it ends in a small external tentacle at the tip of the arm. Lateral canals arise from each side of the radial canal along its entire length. The lateral canals are staggered, a lateral canal on one side of the radial canal being located between the points of origin of two lateral canals on the opposite side. These lateral canals pass between the ambulacral ossicles on each side of the groove and enter the coelom (Fig. 19–6).

Each lateral canal is provided with a valve and terminates in a bulb and tube foot (Fig. 19–6). The bulb (ampulla) is a small muscular sac that bulges into the aboral side of the coelom. The wall of the ampulla contains circular and longitudinal fibers and connective tissue, and its inner and outer surface is clothed with peritoneum. The ampulla opens directly into a canal that passes downward between the ambulacral ossicles and leads into a tube foot (podium).

The podium is a short tubular external projection of the body wall located in the ambulacral groove. Commonly, the tip of the podium is flattened, forming a sucker. Like the body wall, the podium is covered on the outside with a flagellated epithelium and internally with peritoneum. Between these two layers lie connective tissue and longitudinal muscle fibers. Contraction of the muscles on one side of the podium brings about bending of the appendage.

The podia are arranged in two or four rows along the length of the ambulacral groove, and the ampullae occupy a corresponding position on the coelomic side of the ambulacral ossicles. The difference in the number of rows results from the length of the lateral canals. For example, in asteroids, such as *Asterias*, that possess four rows of podia, the lateral canals along each side are alternately long and short, giving the appearance of four rows of podia instead of two. Species in which there are but two rows of podia have lateral canals that are of equal length, thus placing the podia at equal distances from the radial canal.

The entire water-vascular system is filled with fluid and operates during locomotion as a hydraulic system. On contraction of the ampulla, the valve in the lateral canal closes and water is forced into the podium, which elongates. When the podium comes in contact with the substratum, the center of the terminal sucker is withdrawn, producing a vacuum and adhesion. The tip of the podium also produces a copious adhesive secretion that aids in adherence.

After the adhesion of the sucker to the substratum, the longitudinal muscles of the podium contract, shortening the podium and forcing fluid back into the ampulla. It has been generally thought that other parts of the water-vascular system—the madreporite, the stone canal, the water ring, the muscular polian vesicles, and the radial canal—perhaps function in maintaining the proper water pressure necessary for the operation of the ampullae and podia. But Binyon (1964), in his study of

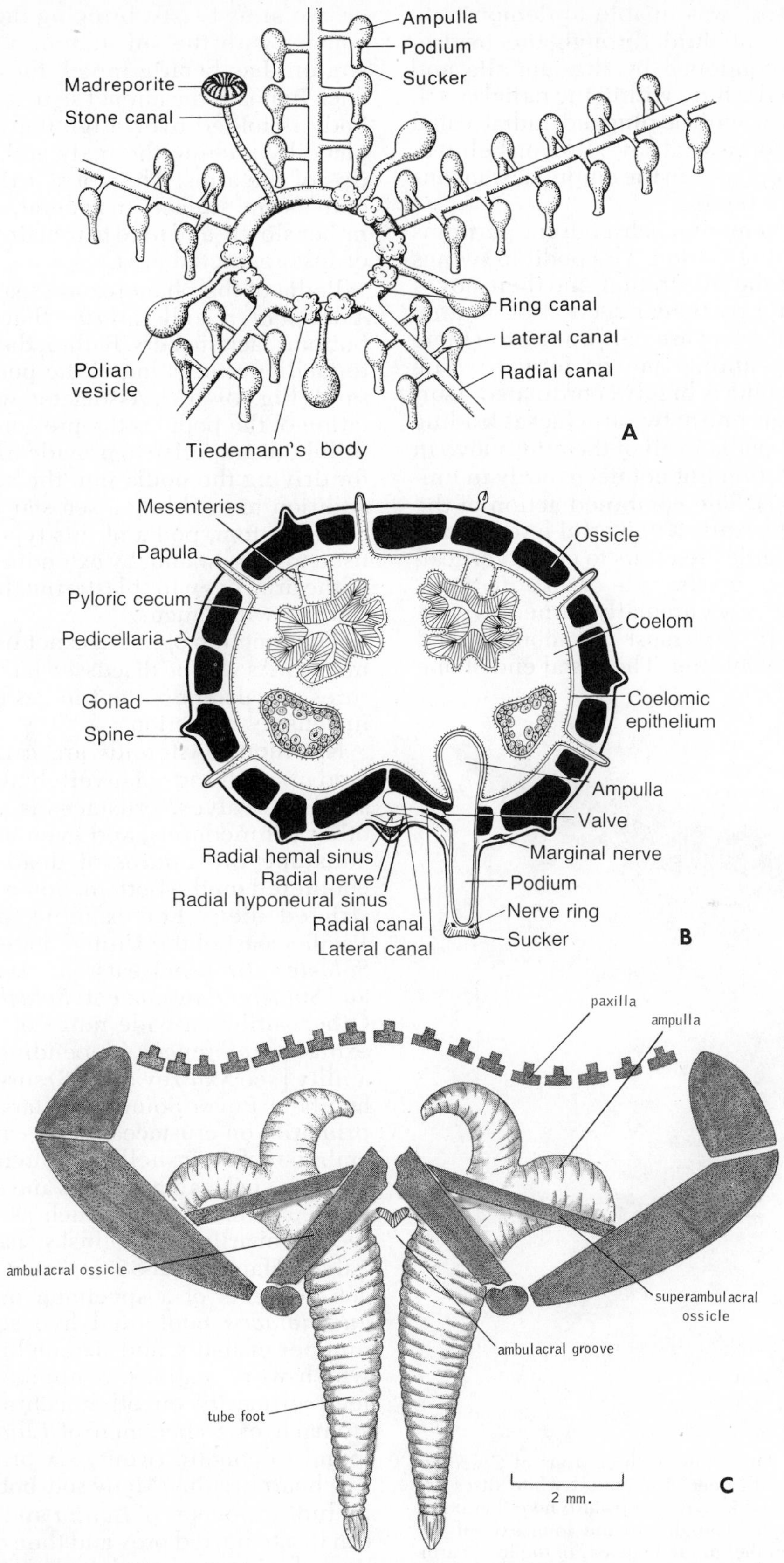

Figure 19–6. *A.* Diagram of the asteroid water-vascular system. *B.* Diagrammatic cross section through the arm of a sea star. *C.* Transverse section through arm of *Astropecten,* a soft-bottom sea star. (From Heddle, D., 1967: Echinoderm Biology. Academic Press, N.Y.)

Asterias rubens, was unable to demonstrate any exchange of fluid through the madreporite, or dependence by the ampulla and podium upon the fluid within the radial canal. A severed arm with a plugged radial canal moved about for several days. Binyon believes that water is replaced in the ampulla by means of a potassium pump.

During movement, each podium performs a sort of stepping motion. The podium swings forward, grips the substratum, and then moves backward. In a particular section of an arm, most of the tube feet are performing the same step, and the animal moves forward. The action of the podia is highly coordinated. During progression, one or two arms act as leading arms, and the podia in all of the arms move in the same direction but not necessarily in unison (Fig. 19–7). The combined action of the podial suckers exerts a powerful force for adhesion and enables sea stars to climb vertically over rocks or up the side of an aquarium.

If a sea star is accidentally turned over, it can right itself. The most common method employed is by folding. The distal end of one or two arms twists, bringing the tube feet in contact with the substratum. Once the substratum has been gripped, these arms move back beneath the animal so that the rest of the body is folded over. Righting may also take place by arching the body and rising on the tips of the arms. The sea star then rolls over onto its oral surface. In general, sea stars move rather slowly and tend to remain within a more or less restricted area.

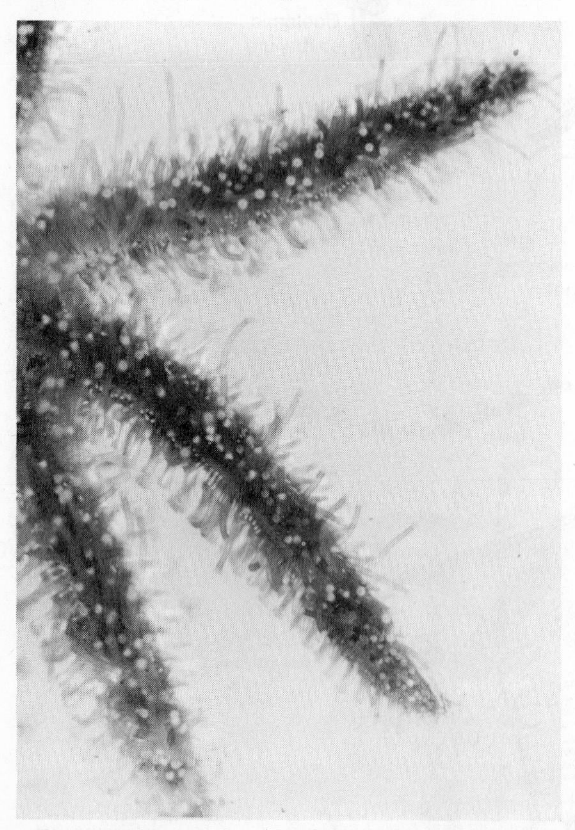

Figure 19–7. Oral view of three arms of *Coscinasterias*, showing tube feet. The round white discs are the suckers. Sea star is moving upward; note that podia are swinging at right angles to the long axis of the upper arm but parallel to the long axis of the lower arm. (By Betty M. Barnes.)

Podia of the phanerozone sea stars, such as *Astropecten* and *Luidia*, that live on soft bottoms lack suckers. Rather, the tip is pointed to facilitate thrusting of the podium into the sand (Fig. 19–6*C*). Associated with this adaptation of the podia is the presence of doubled (bilobed) ampullae to provide increased force for driving the podia into the substratum. In addition to enabling a sea star to move over a soft bottom, podia of this type may also be used to burrow and, by extending up the sides of the arm, even for plastering the walls of the burrow with mucus.

The function of podia is not limited to locomotion. As will be discussed later, these structures are also important in gas exchange and in sensory reception.

Nutrition. Asteroids are carnivorous and feed upon all sorts of invertebrates, especially snails, bivalves, crustaceans, polychaetes, other echinoderms, and even fish. They also consume the bodies of dead animals encountered on the bottom. Some have very restricted diets. For example, on the North Pacific coast of the United States, the sunstar *Solaster stimpsoni* eats only sea cucumbers, and *Solaster dawsoni* eats *Solaster stimpsoni*. Others utilize a wide range of prey but will exhibit preferences, depending upon availability [see Mauzey's (1968) survey of feeding habits of Puget Sound sea stars]. Some feed primarily on crustaceans and can pull hermit crabs out of their shells, although smaller crustaceans, such as amphipods and copepods, are also consumed. Some, such as the asteriids, feed primarily on mollusks, particularly bivalves. Hamann (1885) reported that the stomach contents of a specimen of *Astropecten auranciacus* contained five scaphopods, a number of snails, and sixteen bivalves, ten of which were scallops. A number of asteroids feed primarily on other echinoderms. The stomach of a specimen of *Luidia sarsi* was found to contain twenty-six brittle stars and five heart urchins. Many soft-bottom sea stars, including species of *Luidia* and *Astropecten*, can locate buried prey and then dig down into the substratum to reach it. *Stylasterias forreri* is remarkable in being able to add to its usual molluscan diet small fish that are caught by the pedicellariae when the fish come to rest against the aboral surface of the sea star.

There are some asteroids which feed on sponges and the polyps of hydroids and corals. The tropical Pacific *Acanthaster planci* (crown of thorns sea star) has attained considerable notoriety in recent years as a result of its consumption of coral polyps. High population levels of this sea star have destroyed large numbers of reef corals in some areas. Attempts have been made to control predation by *Acanthaster*, and there has been considerable debate as to whether or not the high population levels of this sea star have resulted from man's modification of the environment (Chesher, 1969; Newman, 1970).

Some sea stars are suspension feeders; this may be the primitive mode of feeding in the class. Plankton and detritus (*Porania, Henricia*), or mud (*Ctenodiscus*), that come in contact with the body surface are trapped in mucus and then swept toward the oral surface by the epidermal cilia. On reaching the ambulacral grooves, the food-laden mucous strands are carried by ciliary currents to the mouth. Some sea stars, such as *Astropecten* and *Luidia*, which are largely carnivorous in habit, utilize ciliary feeding as an auxiliary method of obtaining food.

The digestive system is short, straight, and radial, and it extends between the oral and aboral sides of the disc (Fig. 19–8). The mouth is located in the middle of a tough circular peristomial membrane that is muscular and provided with a sphincter. The mouth leads into a short esophagus, which in some genera, such as *Echinaster*, gives rise to ten esophageal pouches of uncertain significance. The esophagus opens into a large stomach that fills most of the interior of the disc and is divided by a horizontal constriction into a large oral chamber (the cardiac stomach) and a smaller flattened aboral chamber (the pyloric stomach).

The walls of the cardiac stomach are pouched and connected to the body wall by mesenteries. The most important of these attachments are ten pairs of triangular mesenteries called gastric ligaments, which contain a considerable amount of connective tissue. There are two such ligaments connecting the stomach to the ambulacral ossicles of each arm.

The smaller aboral pyloric stomach is often star-shaped because of the entrance of the ducts from the pyloric ceca (Fig. 19–9). There are two pyloric ceca in each arm, each of which is composed of an elongated mass of glandular cells suspended in the coelom of the arm by a dorsal mesentery (Fig. 19–6*B*). A longitudinal duct extends the length of each cecum and gives rise to many lateral ducts, into which the secretory lobules open. The secretory lobules are composed of secretory, absorptive, and storage cells. The longitudinal duct may open separately into the pyloric stomach, or may join with the longitudinal duct of the other cecum in the same arm, before emptying into the pyloric stomach.

A short tubular intestine extends from the aboral side of the pyloric stomach to open through the anus in the middle of the aboral surface of the disc. The intestine typically bears a number of small out-pocketings called rectal ceca (Fig. 19–8). In some species of asteroids, the anus or both anus and intestine

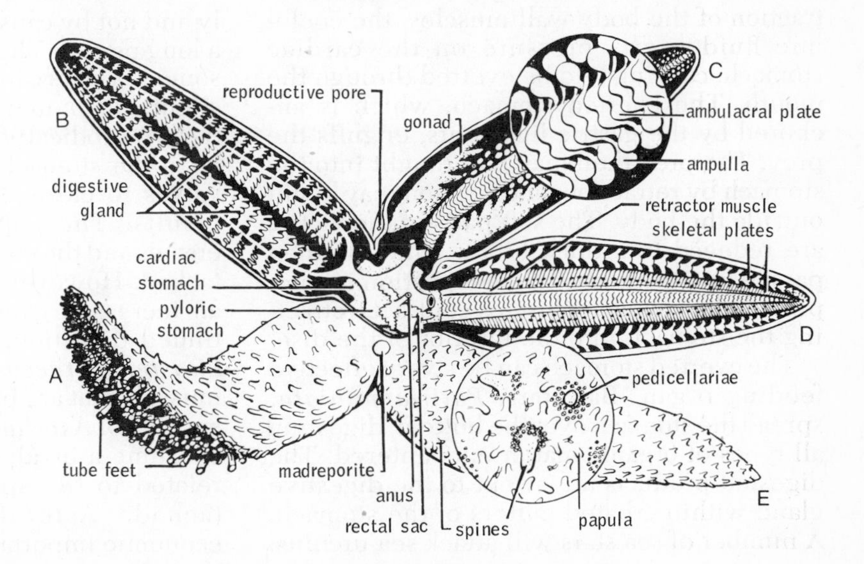

Figure 19–8. *Asterias* viewed from above with the arms in various stages of dissection. *A.* Arm turned to show lower side. *B.* Upper body wall removed. *C.* Upper body wall and digestive glands removed, with a magnified detail of the ampullae and ambulacral plates. *D.* All internal organs removed except the retractor muscles, showing the inner surface of the lower body wall. *E.* Upper surface, with a magnified detail showing surface features.

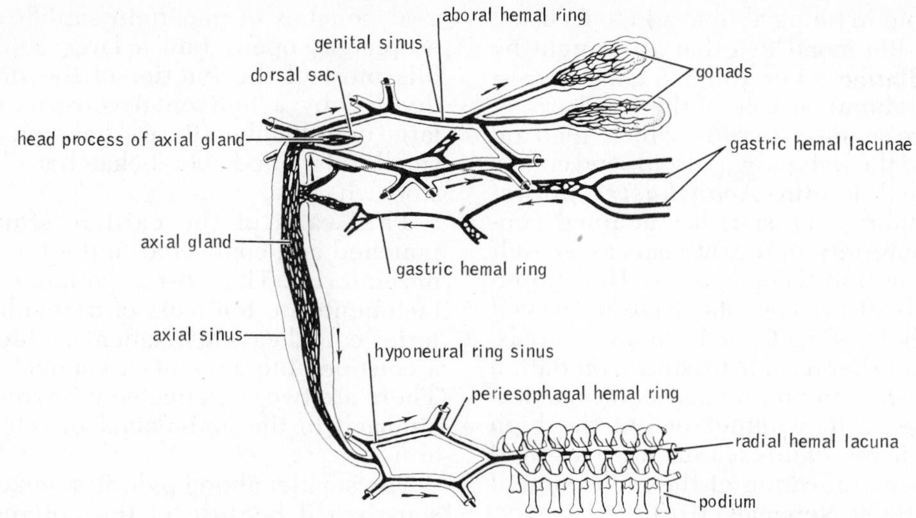

Figure 19–9. Asteroid axial complex and hyponeural and hemal systems. (From Ubaghs, G., 1967: *In* Moore, R. C. (Ed.): Treatise on Invertebrate Paleontology, Pt. S, Vol. 1. Courtesy of The Geological Society of America and The University of Kansas.)

are lacking. The entire digestive tract is lined with a ciliated epithelium, and in the ducts of the pyloric ceca the cilia are so arranged as to create fluid currents, both incoming and outgoing. Gland cells are particularly abundant in the cardiac stomach lining.

In those sea stars that have short, or more or less inflexible, arms and in forms in which the podia lack suckers, such as *Astropecten* and *Luidia*, the prey is swallowed whole and digested within the stomach, although the stomach wall must be in contact with the tissues being digested. Shells and other indigestible material are then cast out of the mouth. Asteroids that have relatively long flexible arms, such as *Asterias*, display a different method of feeding. Through the contraction of the body-wall muscles, the coelomic fluid exerts pressure on the cardiac stomach, causing it to be everted through the mouth. The everted stomach, which is anchored by the gastric ligaments, engulfs the prey. The prey then may be brought into the stomach by retraction, or digestion may begin outside the body. The soft parts of the victim are reduced to a thick broth, which is then passed into the body. When digestion is completed, the stomach muscles contract, retracting the stomach into the interior of the disc.

The everted stomach can be a very effective feeding organ. Bat stars, *Patiria miniata*, spread the stomach over the bottom, digesting all types of organic matter encountered. The digested products are swept to the digestive gland within ciliated gutters of the stomach. A number of sea stars will attack sea urchins. The spines are forced down as the sea star crawls over the aboral surface of the urchin, and then the everted stomach digests away the epithelium and muscles at the spine base.

A considerable number of sea stars, particularly the asteriids, feed almost exclusively on bivalves and are notorious predators of oyster beds. During feeding, such a sea star extends itself over a clam, holding the gape of the clam upward against the mouth of the sea star and applying the arms against the sides of the clam valves. The sea star inserts the everted stomach through minute openings between the imperfectly sealed edges of the valves, or the pull exerted by the sea star is sufficiently great to produce a very slight gape in the clam. The gape is produced quite rapidly and not by causing the clam to fatigue over a long period. The everted cardiac stomach of some sea stars can squeeze through a space as slight as 0.1 mm. Furthermore, the echinoderm periodically relaxes its pull on the valves once the stomach has entered, allowing the valves to close. No damage to the stomach results. The gape increases as digestion ensues and the clam's adductor muscles are attacked. However, according to Feder (1955), an increased gape is not necessary for continued digestion, at least not in the Pacific *Pisaster ochraceus*. This species can consume mussels that are bound with wire. A Japanese species of *Asterias* requires 2½ to 8 hours to consume a bivalve, the time variation being related to the species of bivalve being attacked. Asteroids are of considerable economic importance as predators of oysters, and they are sometimes removed from commercial oyster beds by dragging a large moplike apparatus over the bottom. The sea stars grasp, or become entangled in, the mop threads with their pedicellariae and are brought to the surface and destroyed.

In suspension feeders such as *Porania* and *Henricia,* the pyloric stomach gives rise to five pairs of pouches, called Tiedemann's pouches, that extend back beneath the pyloric ceca. According to Anderson (1960), these pouches are ciliary pumping organs for drawing food particles from the stomach into the pyloric ceca.

Digestion appears to be primarily extracellular, and enzymes—protease, amylase, and lipase—are produced by the pyloric ceca. Some particles are swept along the ciliated ducts into the pyloric ceca, where some intracellular digestion is believed to take place. The pyloric ceca are probably also the primary sites of absorption. Products of digestion may be stored in the cells of the pyloric ceca or may be passed through the ceca into the coelom for distribution.

Some sea stars have been shown to augment their regular diets through direct uptake by surface epithelium of dissolved organic materials in the surrounding sea water. Similar reports for other groups of animals may indicate that this source of food substances is widely utilized by marine invertebrates.

Internal Transport, Gas Exchange, and Excretion. The large fluid-filled coelom surrounding the internal organs within the disc and the arms provides the principal means for internal transport. The coelomic peritoneum is composed of ciliated cuboidal cells, and the beating of the cilia causes a continual circulation of the coelomic fluid. The body fluids of all asteroids, as well as those of other echinoderms, are isosmotic with sea water. Their inability to osmoregulate prevents them from inhabiting estuarine waters. The coelomic fluid contains phagocytic coelomocytes which are produced by the coelomic peritoneum and perhaps by the Tiedemann bodies. In addition to the major part of the coelom, there are three tubular coelomic divisions: They are the water-vascular system already described, the hemal system, and the perihemal system.

A blood-vascular (hemal) system is found in asteroids but is very much reduced and probably plays little role in circulation. The system consists of small fluid-filled sinus channels that lack a distinct lining. The channels are surrounded by special separate extensions of the coelom called perihemal spaces or sinuses. The principal hemal channels consist of an oral hemal ring located just beyond the periphery of the peristome (Fig. 19–9). From this ring a radial hemal sinus extends into each arm and lies beneath the oral surface along the midline of the ambulacral groove (Figs. 19–6*B* and 19–9). From the oral hemal ring a channel ascends through a dark elongated mass of spongy tissue (the axial gland). This gland extends along the length of the stone canal, and its lumen consists of hemal channels (Fig. 19–9). The channel along the stone canal communicates with a small dorsal sac located in the madreporite.

In addition to the ascending channel from the oral hemal ring, the axial gland also receives small channels from the pyloric ceca and the walls of the cardiac stomach. After passing through the axial gland, the channel from the oral ring connects with an aboral hemal ring lying beneath the aboral surface of the disc. Branches to the gonads issue from the aboral hemal ring. The function of the hemal system and axial gland is poorly known. Coelomocytes are present, and the system may represent a pathway for the distribution of food material carried by these cells. Circulation through the system is thought to be effected by pulsations of the dorsal sac.

The asteroid perihemal system is composed of spaces, which surround and parallel much of the radial part of the hemal system.

As is true of other echinoderms, ammonia is the principal nitrogenous waste, and removal is accomplished by general diffusion through thin areas of the body surface, such as the tube feet and the papulae. Studies involving the injection of dyes into the coelom indicate that coelomocytes engulf waste and, when laden, some migrate to the papulae where they collect at the distal end. The tip of the papula then constricts and pinches off, discharging the coelomocytes to the outside. Other coelomocytes may pass to the outside through the epithelium of the suckers of the podia or at other sites. Some wastes may be excreted by the cells of the pyloric ceca and are removed with the egested waste.

The papulae and tube feet provide the principal gas exchange surfaces for asteroids. The ciliated peritoneum that forms the internal lining of the papulae produces an internal current of coelomic fluid; the outer ciliated epidermal investment produces a current of sea water flowing over the papulae (Fig. 19–5*E*).

In the burrowing phanerozones the branched papulae are protected by the paxillae, and the ventilating current flows through the channel-like spaces located beneath these spines. A few phanerozone sea stars have papulae concentrated between the marginal plates to form what are called cribriform organs. The tube feet play an equally important role as gas exchange surfaces and are perhaps the primary gas exchange structures in those asteroids with greatly reduced numbers of papulae. If the ambulacral grooves of *Asterias rubens* are covered, oxygen consumption decreases by as much as 60 per cent.

The Nervous System. The asteroid nervous system, like that of some other echinoderms, is not conspicuously ganglionated, and the greater part of the system is intimately associated with the epidermis. The nervous center is a somewhat pentagonal circumoral nerve ring that lies just beneath the peristomial epidermis. The circumoral ring supplies fibers to the esophagus and to the inner parts of the peristome. From each angle of the ring a large radial nerve extends into each of the arms (Fig. 19–6*B*). The radial nerves are continuous with both the epidermis and the subepidermal nerve net, and these nerves form a **V**-shaped mass along the midline of the oral surface of the ambulacral groove. The radial nerve supplies fibers to the podia and ampullae.

At the margins of the ambulacral groove, the subepidermal nervous layer is thickened to form a pair of marginal nerve cords that extend the length of the arm (Fig. 19–6*B*). The radial and superficial system thus far described is mainly sensory, and the radial cords provide through-conducting pathways. The muscles of the podia, ampullae, and the body wall are supplied by a deeper predominantly motor system. This deep system consists of fibers running beneath (aboral to) the superficial radial nerves and motor centers located in the vicinity of the podia and the podia-ampullae junctions.

There are many experimental studies on the role of the nervous system in the movement of sea stars. The integrity of both the radial nerves and the circumoral nerve ring is essential in the coordination of the podia. Although the podia may not all "step" in unison, they are coordinated to the extent of their involvement in stepping—that is, to step or not to step—and they are coordinated in that they step in the proper direction, depending upon which arm is leading.

If the radial nerve of a particular arm is severed, the podia, although capable of movement, are not coordinated with podia of other arms. Moreover, if the circumoral nerve ring is cut between two radial nerves, all movement is inhibited, for coordination of the podia in all of the arms is lost. It is believed that a nerve center exists at the junction of each radial nerve with the circumoral nerve ring and that this center in a leading arm exerts a temporary dominance over the nerve centers of the other arms. In the majority of sea stars including *Asterias*, any arm can act as a dominant arm, and such dominance is determined by reaction to external stimuli. In a few species, one arm is permanently dominant.

With the exception of the eye spots at the tips of the arms, there are no specialized sense organs in the asteroids. The sensory cells contained within the epidermis are the primary sensory receptors and probably function for the reception of light, contact, and chemical stimuli. This is true of other echinoderms as well. Each sensory cell consists of a distal hairlike process that runs to the epidermal surface and a proximal process that joins the subepidermal network. These epidermal sensory cells are present in enormous numbers and are particularly prevalent on the suckers of the tube feet, on the tentacles, and along the margins of the ambulacral groove. In the latter area, totals of 70,000 per square millimeter have been reported.

The eye spot at the end of each arm lies beneath the tentacle on the oral side of the arm tip and is composed of a mass of 80 to 200 pigment-cup ocelli that form an optic cushion. Each ocellus consists of a cup of epidermal cells containing red pigment granules that are said to be photolabile—that is, they change chemically in the presence of light. The photoreceptors are elongated sensory cells that bulge into the lumen of the cup. The thickened cuticle, which acts as a lens, overlies the cup.

The importance of the optic cushions in reactions to light stimuli apparently varies in different species. In some species, the tip of the arm turns upward to expose the ocelli to light. The photopositive *Asterias rubens* and *Astropecten irregularis* can be led around by shining a light on the optic cushions of a particular ray. Others are negatively phototactic. However, all asteroids react to light, even when the optic cushions are covered, apparently because of photoreceptor cells in the general body epidermis.

Of all the reactions to external stimuli, that of contact of the podia with the substratum appears to be dominant and probably accounts for the righting reaction.

Regeneration and Reproduction. Asteroids exhibit considerable powers of regeneration. Any part of the arm can be regenerated, and destroyed sections of the central disc are replaced. Studies on *Asterias vulgaris* have shown that if there is at least one-fifth of the central disc attached to an arm, an entire starfish will be regenerated. If the remaining section of the disc contains the madreporite, even less of the disc is required. Regeneration is typically slow and may require as long as one year for complete re-formation to take place.

A number of asteroids reproduce normally by asexual reproduction. Commonly this involves a division of the central disc so that the animal breaks into two parts, a process termed fissiparity. Each half then regenerates the missing portion of the disc and arms, although extra arms are typically produced (Fig. 19–10*A*). *Linckia,* a genus of common sea stars in the Pacific and other parts of the world, is remarkable in being able to cast off its arms near the base of the disc. Unlike those of other asteroids, the severed arm regenerates a new disc and rays. Such regenerating specimens with small regenerating arms at the base of the original arm are popularly called comets (Fig. 19–10*B*).

With a few exceptions, asteroids are dioecious, and there are ten gonads, two in each arm. Each gonad is unattached within the coelomic cavity except for a point near the junction of the arm and the disc (Fig. 19–8). The gonads are tuftlike or resemble a cluster of grapes, and they vary greatly in size depending on proximity to the time of spawning. When filled with eggs or sperm, the gonads almost completely fill each of the arms. At other times, the gonads are quite small and occupy only a small area at the base of the arm. There is a gonopore for each gonad, usually located between the bases of the arms. In a number of astropectinids, as well as in some other groups, each arm contains many gonads, which are arranged in rows along the length of the arm. In such species, the gonopores open on the oral surface.

There are a few hermaphroditic asteroid species, such as the common European sea star, *Asterina gibbosa.* This species is protandric; small or young individuals are males, but as they become older and increase in size, they develop into females.

In the majority of asteroids, the eggs and sperm are shed freely into the sea water where fertilization takes place. As in other echinoderms, the presence of eggs or sperm in the sea water acts as a stimulant for the shedding of

A

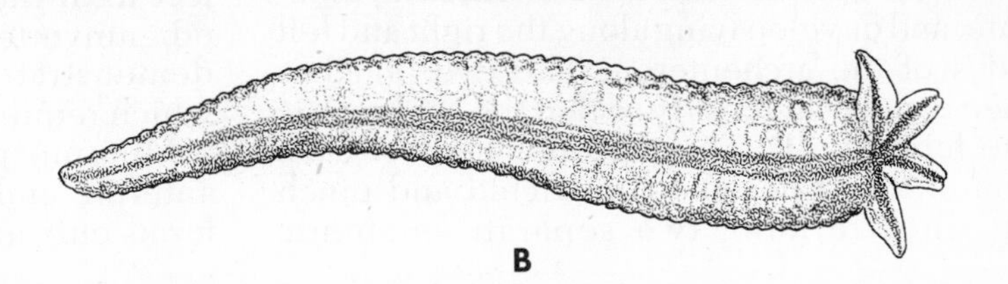

B

Figure 19–10. *A.* Regenerating arms in a specimen of the fissiparous *Coscinasterias.* (By Betty M. Barnes.) *B.* Comet of *Linckia.* Regeneration of body at base of detached arm. (After Richters from Hyman.)

the sex cells of an individual of the opposite sex. Binyon (1964) has suggested that the axial organ may function as a chemoreceptor for substances received through the madreporite, thereby triggering release of gametes. It has been shown for several sea stars that both egg maturation and spawning are stimulated by a substance released from the nervous system. The substance causes contraction of the ovary wall in bringing about expulsion of the eggs. There is usually only one breeding season per year, which occurs in the spring in sea stars of temperate waters; during spawning, a single female may shed as many as 2,500,000 eggs.

In most asteroids, the liberated eggs and individuals in the later developmental stages are planktonic. However, a number of Arctic and Antarctic sea stars brood their eggs. The eggs of such brooding species are usually large and have considerable yolk material. A relatively small number are produced in these brooding species and, unlike the majority of asteroids, development is direct with no larval stage. A variety of brooding methods may be exhibited. In some species, the female rises up on the tips of her arms and broods her eggs in the space between the substratum and the oral side of the disc. In other species, the eggs are brooded in depressions on the aboral surface of the disc or in brooding baskets formed by spines between the bases of the arms. The pterasterids have the entire aboral surface modified to serve as a chamber for the brooding of their eggs. *Leptasterias groenlandica*, a circumpolar Arctic species, broods its eggs in the cardiac stomach. While not a brooding species, *Asterina gibbosa* is unusual in that it attaches its eggs to stones and other objects.

Embryogeny. Except in brooding species, development in the asteroids is indirect and involves the formation of a free-swimming larval stage. The eggs of many sea stars are small, homolecithal, and contain relatively little yolk material, but there are some species with planktonic development that have more yolky eggs. The early stages of development conform to the pattern described in the introduction of this chapter. In most species, the coelom arises from the tip of the advancing archenteron as two lateral pouches. The two pouches separate from the archenteron, elongate, and develop lying along the right and left sides of the archenteron. The two pouches then connect anteriorly so that the coelom has the form of a **U**. The posterior ends of each arm of the **U** become constricted and pinch off, thus forming two separate coelomic vesicles, the right and the left somatocoel (Fig. 19–11*A* and *H*). The remaining anterior coelomic vesicles represent the hydrocoel and axocoel, but they never separate. The left hydrocoel connects with the dorsal surface to form the hydropore.

In the European *Asterina gibbosa*, the formation of the coelom takes place somewhat differently. The anterior part of the archenteron is cut off to form the coelom, and the remaining posterior portion of the archenteron then forms the gut.

The asteroid embryo becomes free-swimming at some point between the blastula and gastrula stages. Feeding commences as soon as the digestive tract is completely formed. Diatoms that are collected from the water current produced by the stomodeal cilia are the principal food. At first the entire larval surface is covered with cilia, but as development proceeds, the surface ciliation becomes confined to a definite locomotor band (Fig. 19–11). This locomotor band consists of two lateral longitudinal bands that extend anteriorly and then ventrally to connect in front of the mouth as a preoral loop.

Posteriorly, the two lateral bands turn forward on the ventral surface and connect in front of the anus, forming a preanal loop. The preoral loop later separates or in some cases arises separately from the rest of the locomotor band to form an anterior ciliated ring around the body. After the formation of the locomotor bands, projections, called arms, arise from the body surface. The locomotor bands extend along the arms, which in some species may be quite long. This larval stage is then known as a bipinnaria larva; several weeks may elapse in the attainment of this stage.

The bipinnaria larva becomes a brachiolaria larva with the appearance of three additional arms at the anterior end (Fig. 19–11*B*). These arms are short, ventral in position, and covered with adhesive cells at the tip. Between the bases of the three arms, there is a glandular adhesive area that forms a sucker. The three arms and the sucker represent an attachment device, and the brachiolaria then settles to the bottom. The tips of the arms provide a temporary attachment to some object until the sucker itself is attached. As in other invertebrates, delayed settling has been demonstrated in larvae of certain species which require a particular type of substratum.

Metamorphosis then takes place. The anterior end of the larva degenerates and forms only an attachment stalk, and the adult

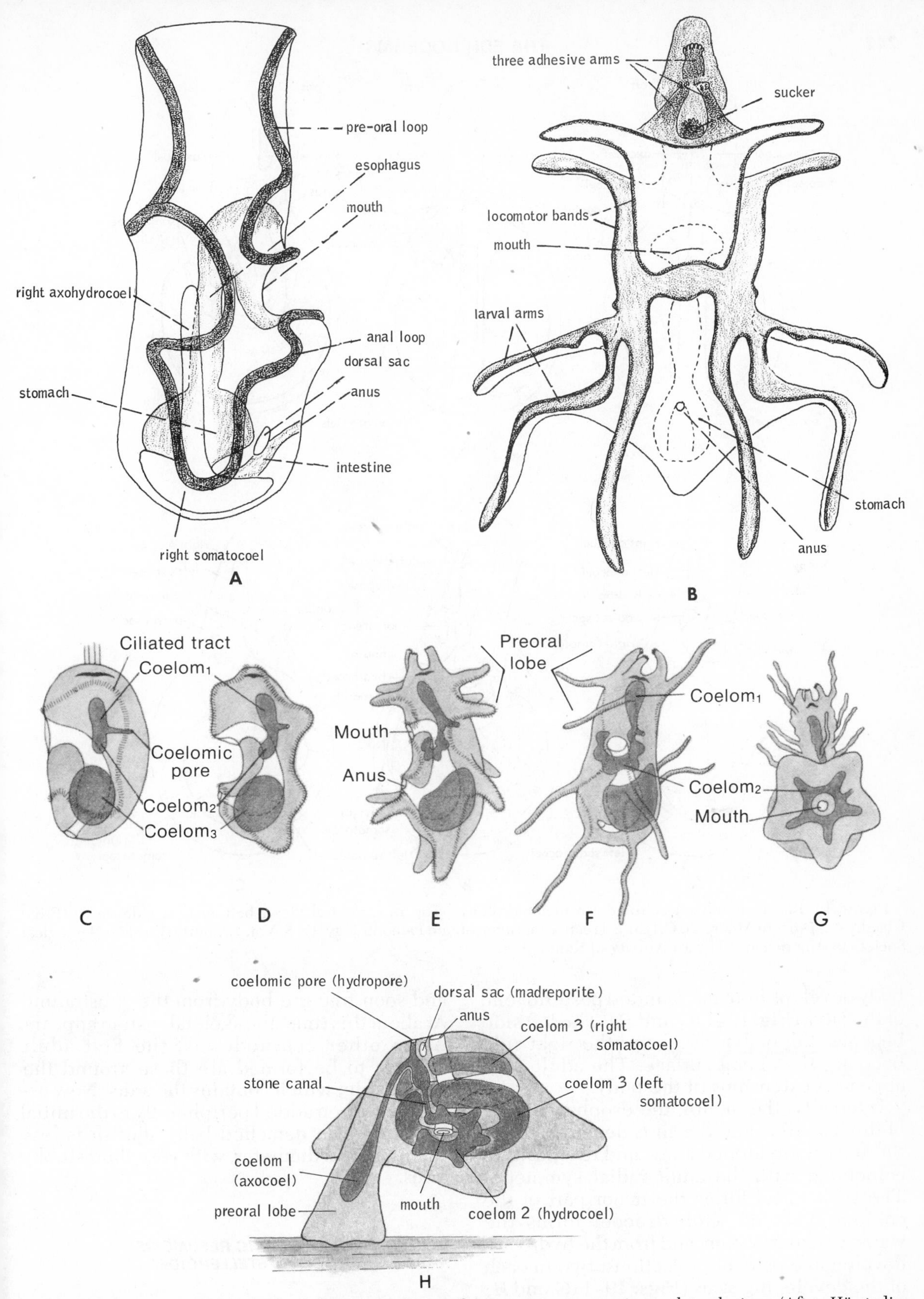

Figure 19–11. *A.* Early (14 days) bipinnaria larva of *Astropecten auranciacus,* lateral view. (After Hörstadius from Hyman.) *B.* Brachiolaria larva of *Asterias,* ventral view. (After Agassiz from Cuénot.) *C* to *H.* Diagrammatic lateral views of larval development and metamorphosis of a sea star, showing development of coelom and water-vascular system. *C* and *D,* early bipinnaria larva; *E,* brachiolaria larva; *F,* attached metamorphosing larva; *G* and *H,* young starfish developing from posterior part of old larva. (Adapted from various sources.)

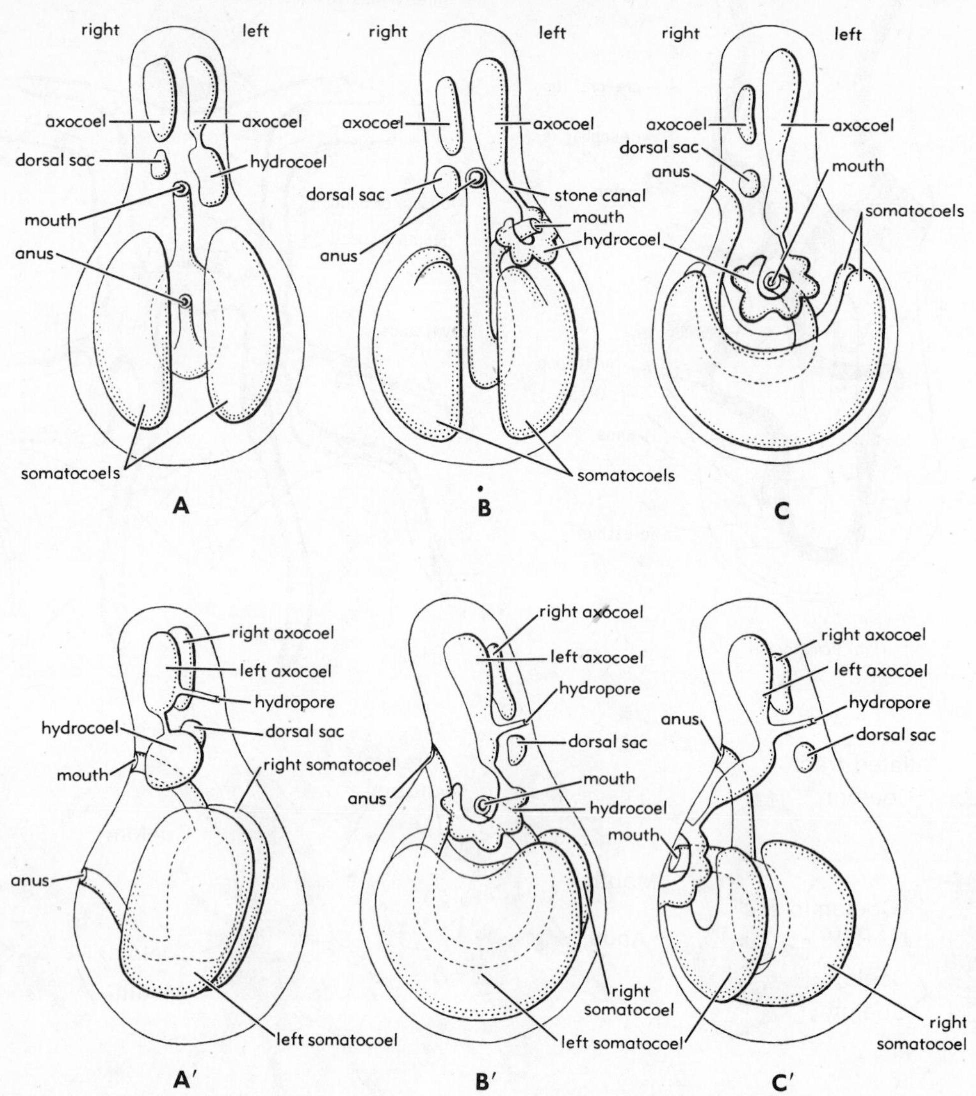

Figure 19-12. Generalized echinoderm metamorphosis. Top row, ventral view; bottom row, side view. (From Ubaghs, G., 1967: *In* Moore, R. C. (Ed.): Treatise on Invertebrate Paleontology, Pt. S, Vol. 1. Courtesy of The Geological Society of America and The University of Kansas.)

body develops from the rounded posterior end of the larva (Fig. 19–11*G* and *H*). The left side becomes the oral surface, and the right side becomes the aboral surface. The adult arms appear as extensions of the body.

Internally, the mouth, the esophagus, part of the intestine, and the anus degenerate. All these parts are formed anew and in a position coinciding with the adult radial symmetry. The somatocoel forms the major part of the coelom. The left axohydrocoel forms the water-vascular system, and from the hydrocoel develop five pairs of projections, two in each of the developing arms (Figs. 19–11*G* and *H*; and 19–12). These projections represent the cavity and coelomic lining of the first pair of podia in each arm. As soon as additional podia are formed, they begin to grip the substratum and soon free the body from the substratum. At about this time, the skeletal system appears. As in other echinoderms, the first adult ossicles to be formed are those around the aboral pole, which contains the anus. New ossicles are then added peripherally to the initial skeleton. The detached baby starfish is less than 1 mm. in diameter, with very short stubby arms.

SYSTEMATIC RÉSUMÉ OF CLASS STELLEROIDEA

Unattached echinoderms having a body composed of flattened central disc and radially arranged arms, or rays.

Subclass Stomasteroidea. Fossil Paleozoic sea stars in which the skeletal structure of the arms displays a number of primitive and extinct features. A single living species, *Platasterias latiradiata*, is known to inhabit deep water off the Pacific coast of Mexico.

Subclass Asteroidea. Stelleroid echinoderms in which there are open ambulacral grooves and a large coelomic cavity in the relatively wide arms.

Order Platyasterida. Primitive, mostly extinct sea stars. The common soft-bottom *Luidia*.

Orders Paxillosida and Valvatida. Sea stars with marginal plates, and usually with paxillae on the aboral surface. Pedicellariae of the sessile type. Many members of these orders are burrowers in soft bottoms, and the podia of these species lack suckers. *Astropecten, Ctenodiscus, Culcita, Goniaster, Oreaster, Linckia, Porania.*

Order Spinulosida. The members of this order are not always distinct from phanerozone species, but, in general, conspicuous marginal plates are absent and the tube feet are suckered. The aboral surface is covered with low spines which give the order its name. There are no pedicellariae. *Asterina, Patiria, Echinaster, Henricia, Acanthaster, Crossaster, Pteraster.*

Order Forcipulatida. Sea stars with pedicellariae composed of a short stalk and three skeletal ossicles. *Freyella, Heliaster, Pycnopodia, Asterias, Leptasterias, Pisaster.*

The systematic résumé of the class Stelleroidea is continued on page 751, following the discussion of the other subclass, the Ophiuroidea.

CLASS STELLEROIDEA; SUBCLASS OPHIUROIDEA

The subclass Ophiuroidea contains those echinoderms known as basket stars and serpent stars, or brittle stars (Fig. 19–13). The 2000 described species make this the largest of the major groups of echinoderms. They are found in all types of marine habitats, and they are often abundant on soft bottoms in the deep sea.

Ophiuroids resemble asteroids in that they both possess arms. However, in other respects the two classes are quite different. The extremely long arms of ophiuroids are more sharply set off from the central disc. There is no ambulacral groove, and the podia play little role in locomotion. Moreover, the arms have a relatively solid construction compared with those of the asteroids.

External Structure. Ophiuroids are relatively small echinoderms. The disc in most species ranges from 1 to 3 cm. in diameter, although the arms may be quite long. The basket stars are the largest members of the class and the disc in some species of this group may attain a diameter of almost 12 cm. A great variety of colors are found in the ophiuroids, but in most cases the coloration is not particularly conspicuous because of their small size and the mottled and banded patterns.

The central disc is flattened, displaying a rounded, pentagonal, or somewhat star-shaped circumference (Fig. 19–14). The aboral surface varies from smooth to granular with small tubercles or spines, and sometimes small calcareous plates, called shields, are conspicuous. The largest of the plates are a pair of radial shields located opposite the base of each arm (Fig. 19–14*B*).

There are typically only five arms; however, in basket stars the arms branch either at the base or more distally, and the subdivisions repeatedly branch to produce a great mass of coils that resemble tentacles (Fig. 19–13*C*). The arms characteristically appear jointed, because of the presence of four longitudinal rows of shields (Figs. 19–14*A* and *B*; and 19–15*A*). There are two rows of lateral shields, one row of aboral shields, and one row of oral shields. A single set—that is, two lateral, one aboral and one oral shield—completely surround the arm and correspond in position to an internal skeletal ossicle, which is described later. Not infrequently, the oral and aboral shields are reduced by the large size of the lateral shields, which may even meet on the oral and aboral surfaces (Fig. 19–14*A*).

Each lateral shield usually bears two to fifteen large spines arranged in a vertical row (Fig. 19–15*A*). These spines vary considerably in size and shape, depending upon the species. For example, in *Ophiothrix* and others, the most oral spine in each row is hook-shaped; in *Ophiohelus* the spines toward the end of each arm are shaped like little umbrellas; and in *Ophiopteron* some of the spines are interconnected by webbing. Some serpent stars have what are probably poisonous spines, for the end of such a spine is provided with a thick glandular covering.

There is no ambulacral groove on the oral surface of the arms. The podia are small, tentacle-like papillate appendages that extend between the oral and the lateral shields, and there is typically one pair of podia per joint

Text continued on page 747.

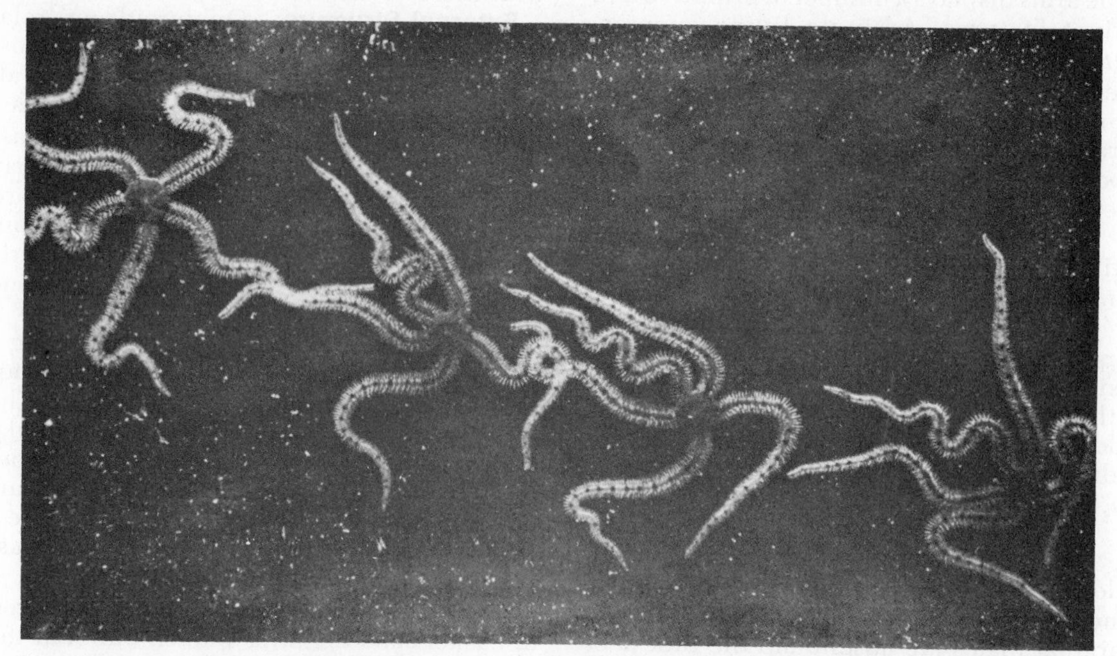

A

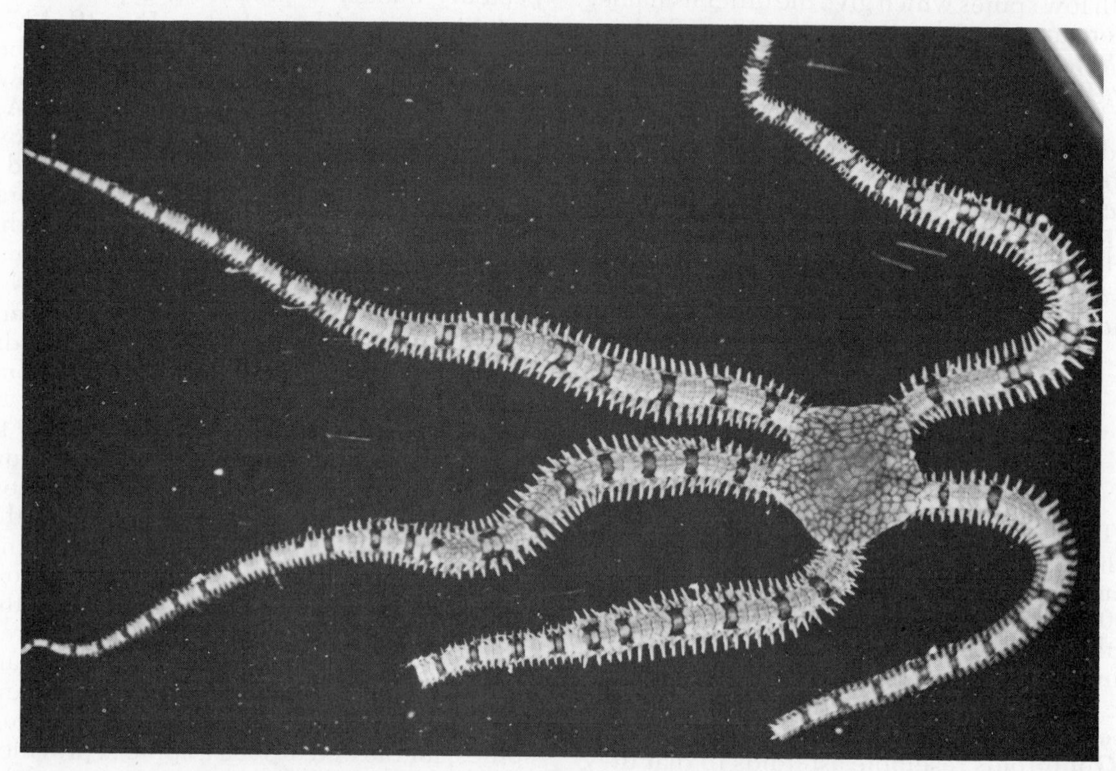

B

Figure 19–13. *A.* A Caribbean brittle star, shown in repetitive flash photographs, pulling itself along with its two anterior arms and shoving with the other three. Ophiuroids are far more agile and flexible than are sea stars. (By Fritz Goro, courtesy of LIFE Magazine, ©1955, Time, Inc.) *B.* Aboral view of the brittle star *Ophionereis.* (By Betty M. Barnes.) *C.* The basket star *Gorgonocephalus.* (Courtesy of the American Museum of Natural History.)

Illustration continued on opposite page.

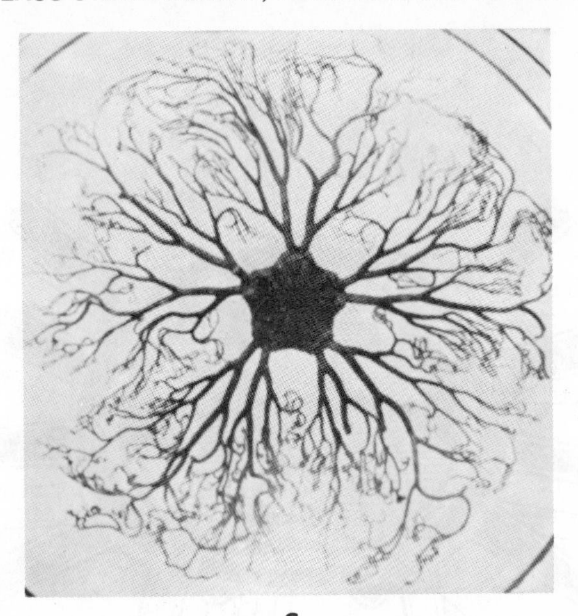

Figure 19–13. *Continued.*

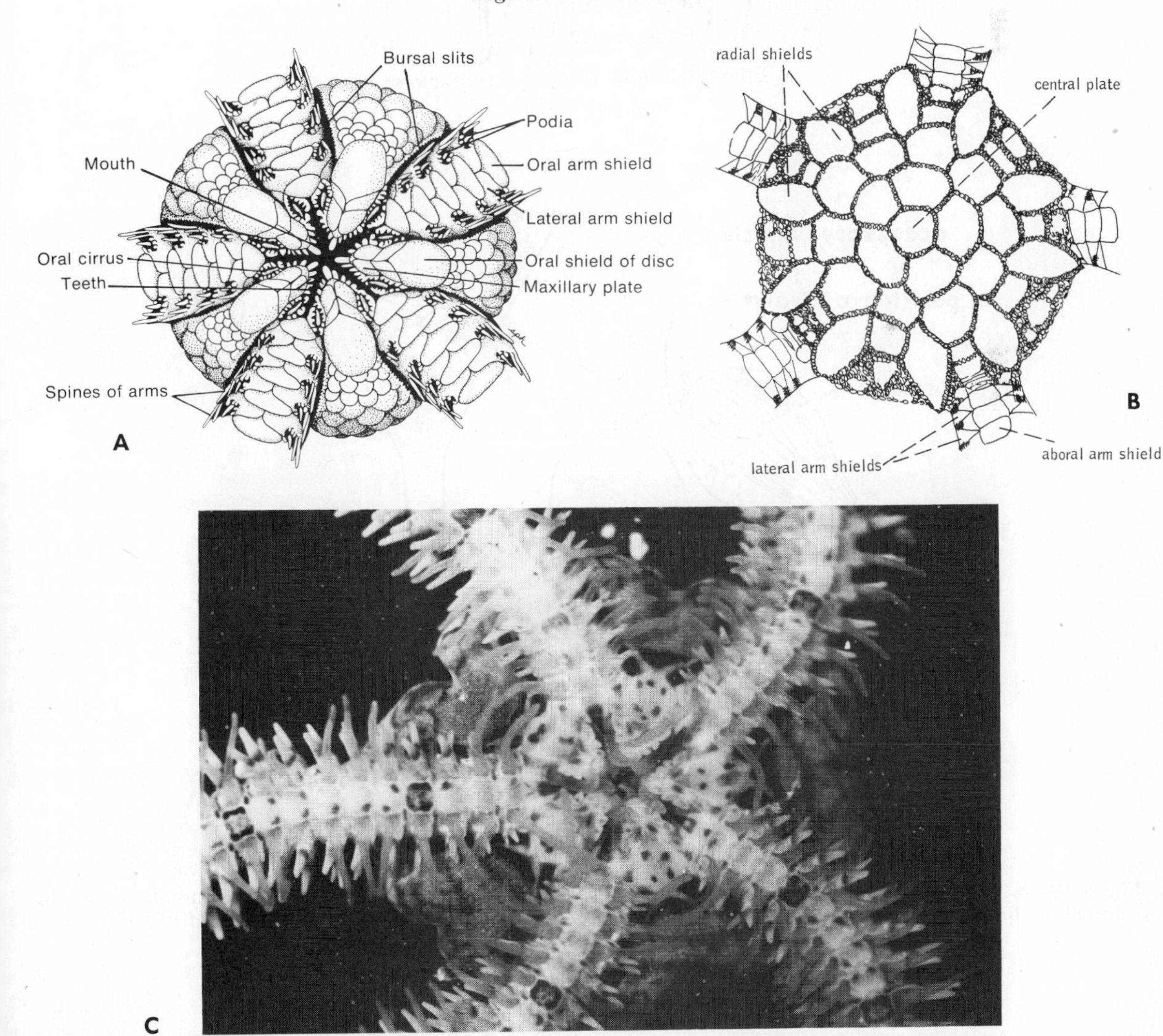

Figure 19–14. A. The disc of *Ophiura sarsi* (oral view). (After Strelkov.) *B*. The disc of *Ophiolepis* (aboral view). (After Hyman.) *C*. Oral view of a West Indian species of *Ophionereis*. (By Betty M. Barnes.)

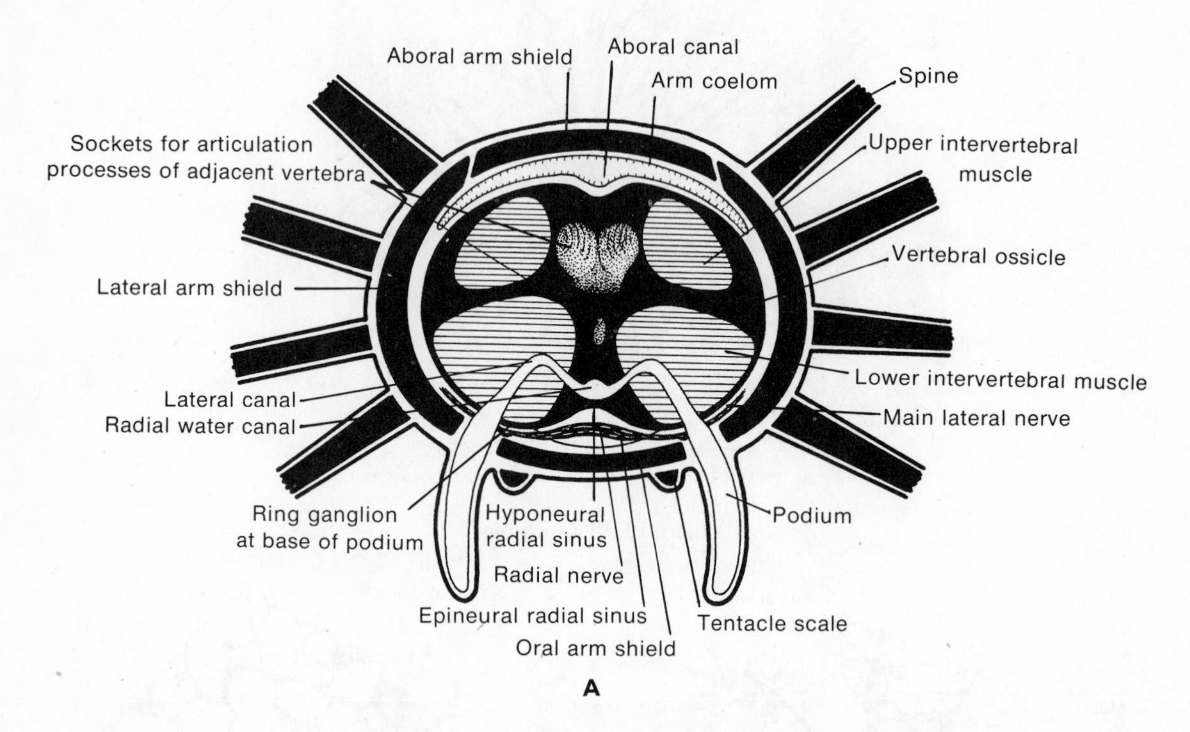

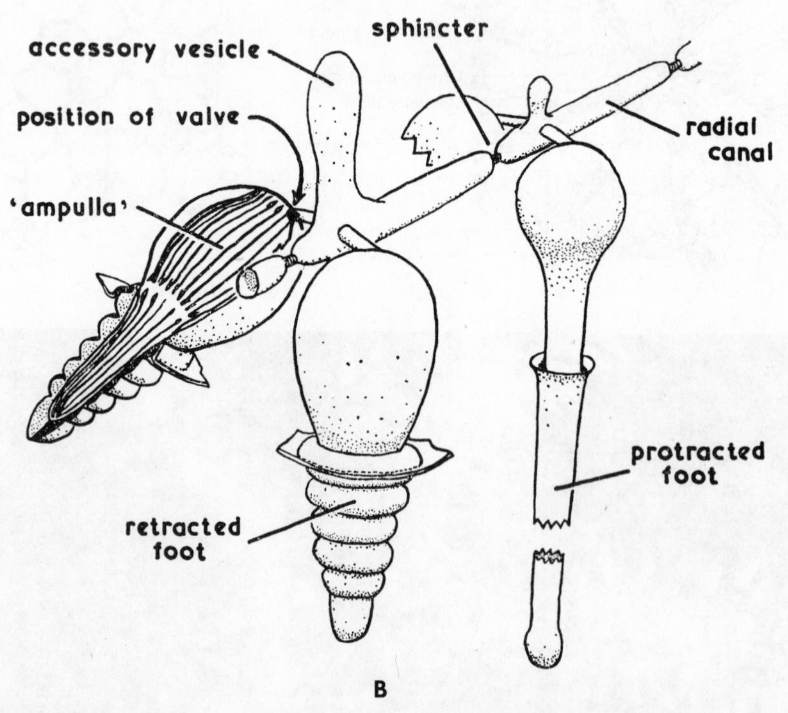

Figure 19–15. *A.* Diagrammatic section through the arm of a brittle star. *B.* Diagram of the water-vascular system in the arm of the brittle star *Amphiura.* (From Buchanan, J. B., and Woodley, J. D., 1963: Nature, *197*:616–617.)

(Fig. 19–15*A*). Neither papulae nor pedicellariae are present in ophiuroids.

The center of the oral surface of the disc is occupied by a complex series of large plates that frame the mouth area and also form a chewing apparatus (Fig. 19–14*A*). The plates are organized into five triangular jaws that are oriented longitudinally with the apex directed toward the center of the disc. Each jaw is composed of several plates, of which the most conspicuous are a pair of half jaws and a large basal oral shield. The jaws bear many small toothlike plates along their margin. At the apex, the teeth also extend aborally in a vertical row (Fig. 19–16).

In most ophiuroids, one oral shield is modified, forming a madreporite. Thus, the madreporite is located on the oral surface, in contrast to its aboral position in echinoids and asteroids. The arms extend inward to the jaws on the oral surface of the disc, leaving five large, somewhat triangular interradial areas, having essentially the same surface structure as the aboral side of the disc.

Body Wall and Skeleton. With the exception of the basket stars, the epidermis of the ophiuroids is a reduced syncytium, continuous with the underlying dermis. Moreover, aside from certain specialized areas, there are no surface cilia. The dermis contains the more superficial skeletal shields, as well as the large deeper ossicles of the arms. These ossicles are highly specialized, and it is believed that each represents two fused ambulacral ossicles. Each ossicle is a large disc-shaped skeletal piece (a vertebra) that almost fills the entire interior of the arm (Fig. 19–15*A*).

The vertebral ossicles are arranged linearly from one end of the arm to the other, and each ossicle is covered by four superficial arm shields. The two end surfaces of a vertebral ossicle are structurally adapted for articulation with adjacent vertebrae. In serpent stars, this articulation allows great lateral mobility of the arm, but no vertical movement. However, the basket stars, which have a somewhat different vertebral articulation, the arms can bend and coil in any direction.

In basket stars, there is a muscle layer beneath the dermis, but in serpent stars the muscle layer is absent, and the coelomic peritoneum covers the inner surface of the dermis. There are, however, well-developed intervertebral muscles for the movement of the arms (Fig. 19–15*A*).

A considerable number of ophiuroids, such as the cosmopolitan *Amphipholis squamata*, are luminescent. A yellow or yellow-green light emanates from the sides of the arms, particularly in the region of the spines. Large gland cells in these areas are believed to be the source of the luminescence.

Locomotion. The ophiuroids are the most highly mobile of all echinoderms. During movement the disc is held above the substratum, with one or two arms extended forward and one or two arms trailing behind. The remaining two lateral arms perform a rapid rowing movement against the substratum that propels the animal forward in leaps or jerks. Serpent stars show no arm preference and can move in any direction.

Righting is accomplished by extending two adjacent arms away from each other so that they form a straight line tangent to the disc. Using the extended arms as a pivot, the disc is raised and pushed over by the other arms. In clambering over rocks or in seaweed and hydroid colonies, the ends of the supple arms often coil about objects (Fig. 19–17*C*). A few ophiuroids reportedly are able to swim, using

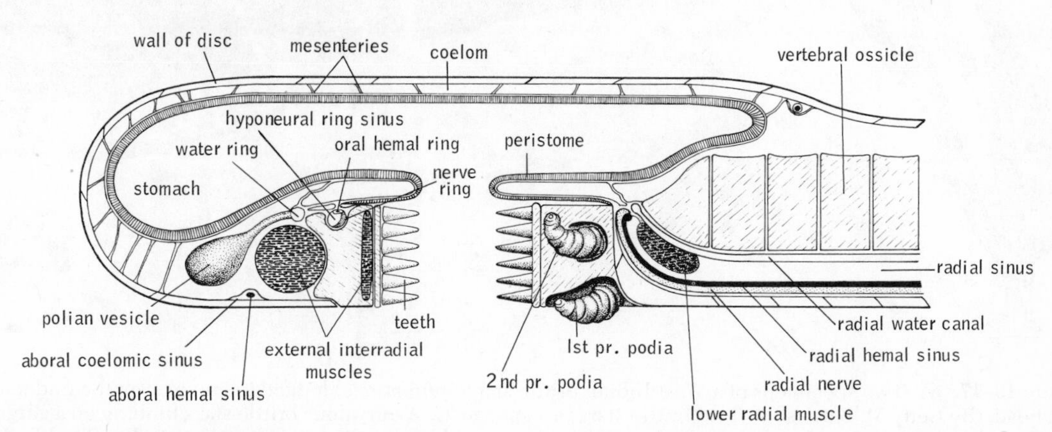

Figure 19–16. The disc and base of an arm of a brittle star (vertical section). (After Ludwig.)

the same method of progression as in crawling. Some forms also burrow; *Amphiura* covers itself completely in sand except for the ends of the arms. The podia are used in digging. *Amphioplus* lives in sand burrows, which are connected to the surface by tube-like channels (Fig. 19–17*D*).

Some echinologists consider the ophiuroids to be the most successful group of living echinoderms; they attribute this success in part to their motility, small size, and ability to utilize the protective cover of crevices, holes, spaces beneath stones, and other natural retreats (Fig. 19–17*A*). Although they are inconspicuous, the numbers of ophiuroids, both species and individuals, on many tropical reefs greatly exceeds those of other echinoderms. The only commensal echinoderms are ophiuroids. Large sponges may contain great numbers of ophiuroids living in the water

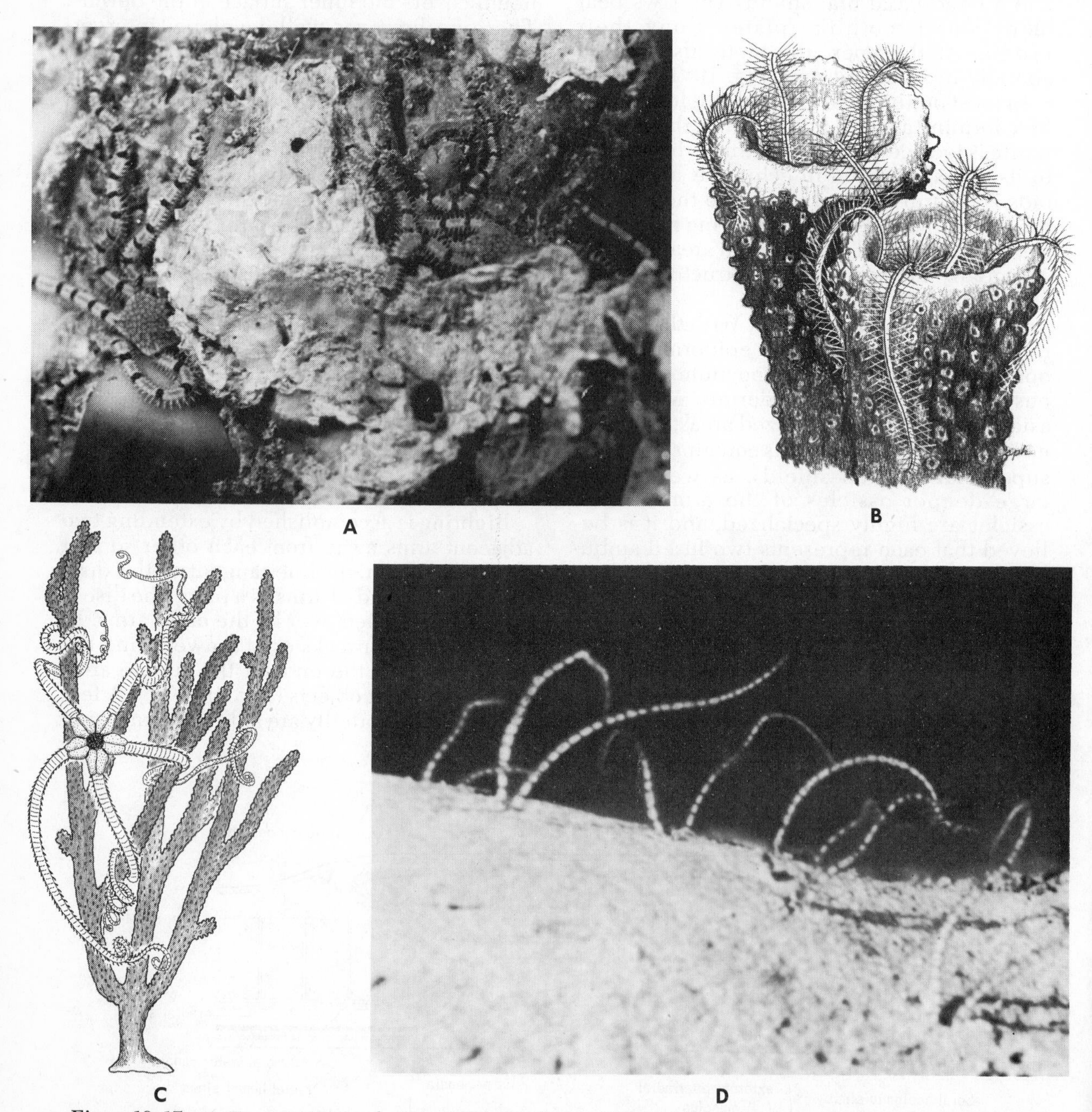

Figure 19–17. *A.* Two specimens of a West Indian brittle star (*Ophionereis*) lodged in crevices on the underside of a coral head. (By Betty M. Barnes.) *B.* Two brittle stars in a sponge. *C.* A euryalous brittle star climbing on a gorgonian coral. These brittle stars are related to the basket stars and are capable of coiling their arms vertically. (Modified from Hyman.) *D.* Specimens of *Amphioplus* projecting two arms from the tube-like burrows and trapping suspended particles from the passing water current. (From Fricke, H. W., 1970: Helgolander wiss. Meeresunters., *21*:124–133.)

canals (Fig. 19–17*B*). Species of *Ophiomaza*, an Indo-Pacific brittle star, live on the oral surfaces of feather stars, clutching the calyx with the arms. There are also several tiny Indo-Pacific brittle stars that live on the undersurfaces of sand dollars.

Water-Vascular System. The oral shield that forms the madreporite in ophiuroids usually bears but a single pore and canal. In addition to the oral position of the madreporite, the stone canal ascends to the water ring, which is located in a groove on the aboral surface of the jaws. The water ring bears four polian vesicles and also gives rise to the radial canals, which penetrate through the lower side of the vertebral ossicles of the arms (Fig. 19–15*A*). In each ossicle, the radial canal gives rise to a pair of lateral canals, which may extend directly to the pair of ventrolateral podia or may first loop upward and then downward to the podia (Fig. 19–15). The paired lateral canals of ophiuroids contrast with the staggered arrangement of other echinoderms. Ampullae are absent, but a valve is present between the podium and the lateral canals. Fluid pressure for protraction is generated by a dilated ampulla-like section of the podial canal and in some forms by localized contraction of the radial water canal. As in asteroids, the radial canal terminates in a small external tentacle at the tip of the arm. The entire water-vascular system is lined with a ciliated peritoneum.

Nutrition. Some ophiuroids feed on bottom detritus and small dead or living animals. Large food particles are raked into the mouth by the arms, while smaller particles are passed to the mouth by means of the podia or perhaps in some cases by special flagella on the oral surface of the arms. The Pacific brittle star, *Ophiocomina nigra*, utilizes a variety of mechanisms, which is probably true of other species of ophiuroids as well. *Ophiocomina* is a filter feeder, deposit feeder, browser, and scavenger. In filter feeding, the arms are lifted from the bottom and waved about in the water. Plankton and detritus adhere to mucous strands strung between the adjacent arm spines. The trapped particles may be swept downward toward the tentacular scale by ciliary currents or collected from the spines by the tube feet, which extend upward for this purpose. The tube feet are then scraped across the tentacular scales, depositing collected particles in front of the scale (Fig. 19–18). This is also where the ciliary tracts deposit their material. From this point on each side, the food particles are picked up by adjacent podia, compacted into a bolus, and passed along the midoral line of the arm toward the mouth. The food balls are moved by the podia until they reach the proximal parts of the arm, in which movement toward the mouth is facilitated by cilia.

Deposit feeding on intermediate size particles is performed by the podia. The podia collect the particles from the substratum, compact them into food balls, and move them toward the mouth as previously described.

Large food material, such as dead animal matter, is swept into the mouth by the looping motion of an arm. Browsing over algae or carrion, the animal utilizes its teeth or oral tube feet.

Similar types of suspension and deposit feeding mechanisms involving mucus-laden spines and podia have been described by Pentreath (1970) for ophuroids from New Zealand reefs. Such indirect mechanisms have the advantage of permitting the animal to extend only two or three feeding arms out from its protective retreat, as well as to utilize a variety of food sources (Fig. 19–17).

Basket stars are capable of capturing relatively large prey, especially crustaceans.

The digestive tract is extremely simple (Fig. 19–16). The jaws frame a shallow prebuccal cavity, which is roofed aborally by the peristomial membrane containing the mouth.

The esophagus connects the mouth with a

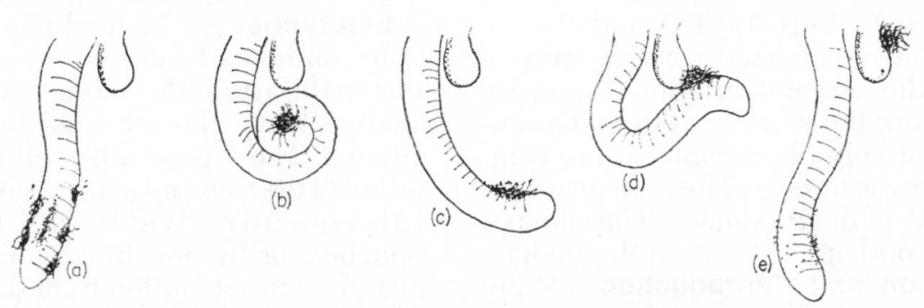

Figure 19–18. Feeding activity of a podium of the suspension-feeding brittle star *Ophionereis fasciata*. *A*. Particles collected by podium from spines. *B*. Particles consolidated by podium into one mass. *C* to *E*. Mass transferred from podium to tentacle scale. (From Pentreath, R. J., 1970: J. Zool., *161*:395–429.)

large saclike stomach. The stomach fills most of the interior of the disc, and in most ophiuroids the margins are infolded to form ten pouches. There is no intestine or anus, and with the exception of a single species, no part of the digestive tract extends into the arms. Although the stomach must obviously be the site of digestion and absorption, little is known about the details of the digestive process.

Circulation, Gas Exchange, and Excretion. The coelom in ophiuroids is much reduced when compared with that of other echinoderms. The vertebral ossicles restrict the coelom to the aboral part of the arms (Fig. 19–15A); the stomach, bursae (see below), and gonads leave only small coelomic spaces in the disc (Fig. 19–16). The coelomocytes are active ameboid cells. The hemal system is essentially like that of asteroids (Fig. 19–16).

Gas exchange in ophiuroids takes place by means of ten internal sacs (bursae) that represent invaginations of the oral surface of the disc. Internally each bursa is lodged between two stomach pouches. The bursae are connected to the outside by way of a slit that runs along the margins of the arms on the oral surface of the disc (Fig. 19–14A). Commonly the two sides of the slit connect in the middle to form two openings. The bursae may be lined with ciliated epithelium, especially the slits. The beating cilia create a current of water that enters the peripheral end of the slit, passes through the bursa, and flows out the oral end of the slit. Many species also pump water into and out of the bursae by raising and lowering the oral or aboral disc wall, or by contraction of certain disc muscles associated with the bursae.

The thin-walled respiratory bursae may well be the principal center for waste removal.

Nervous System. The nervous system is doubled, and is composed of the outer sensory and motor ectoneural system and the inner motor hyponeural system. The two systems parallel each other and are separated only by a thin layer of connective tissue. As in asteroids, they are composed of a circumoral nerve ring and radial nerves (Figs. 19–15A and 19–16).

There are no specialized sense organs in ophiuroids, the general epithelial sensory cells composing the sensory system. Ophiuroids are negatively phototropic, as are echinoids and many asteroids. They are also able to detect food without contact. The chemoreceptors are perhaps located on the podia.

Regeneration and Reproduction. Many ophiuroids can cast off, or autotomize, one or more arms if disturbed or seized by a predator. A break can occur at any point beyond the disc; the lost portion is then regenerated. In most ophiuroids, the disc does not have the regenerative powers of the arms, and in at least some species the entire disc and at least one arm must be present for the remaining arms to regenerate.

Asexual reproduction by fissiparity takes place in some ophiuroids, notably the six-armed serpent star, *Ophiactis*. The disc divides into two pieces, each piece with three arms. Fission can take place along any plane; the missing half is then regenerated.

The majority of ophiuroids are dioecious. They usually exhibit no sexual dimorphism; however, the females of three species are known to carry dwarf males about, the two sexes being attached mouth to mouth. The gonads are small sacs attached to the coelomic side of the bursae near the bursal slit. There may be one, two, or numerous gonads per bursa with various positions of attachment. Hermaphroditic species are not uncommon. Many are protandric; the gonad first produces sperm and then eggs. In other ophiuroids, the bursae bear separate ovaries and testes.

When the gonads are ripe, they discharge into the bursae, probably by rupture, and the sex cells are carried out of the body in the respiratory water current. Fertilization and development take place in the sea water in many species, but brooding is common. The bursae are used as brood chambers, and development takes place within the mother until the juvenile stage is reached. In most species, only a few young are brooded in each bursae, but as many as 200 embryos reportedly have been contained within each bursa in an Antarctic form. In some brooding serpent stars, such as *Amphipholis squamata*, development is viviparous, for the embryo is attached to the bursal wall and receives nourishment from the mother. In a few species, such as the Antarctic *Ophionotus hexactis*, a single egg is formed and passes into an ovarian lumen, in which both fertilization and development take place; the young serpent star ruptures into the bursa.

Embryogeny. In nonbrooding oviparous ophiuroids, early development is similar to that in the asteroids. A free-swimming stage is attained at the blastula stage that is completely ciliated. The coelom arises either by the separation of the free end of the archenteron, which subsequently divides into right and left pouches, or by pouching, which takes place just prior to separation from the archenteron. The larva of many species, called an ophiopluteus, is very similar to the echinopluteus of the

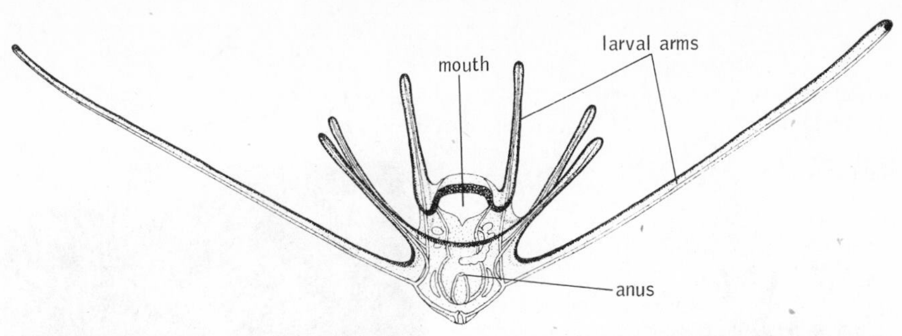

Figure 19–19. Ophiopluteus larva of *Ophiomaza* (oral view). (After Mortensen.)

echinoids (Fig. 19–19). Metamorphosis takes place while the larva is still free-swimming, and there is no attachment stage. The tiny serpent star sinks to the bottom and takes up an adult existence.

SYSTEMATIC RÉSUMÉ OF CLASS STELLEROIDEA

Subclass Ophiuroidea.

Order Stenurida. Paleozoic ophiuroids.

Order Oegophiurida. Mostly extinct Paleozoic ophiuroids, except for a single deep-water species, *Ophiocanops*.

Order Phrynophiurida. This order contains ophiuroids in which the disc and arms are covered by skin. Dorsal arm shields absent.

Suborder Ophiomyxina. Primitive brittle stars in which disc and arm plates are covered by a thick soft skin. *Ophiomyxa*.

Suborder Euryalina. Arms simple or branched (basket stars) but capable of coiling vertically. *Asteronyx*, *Gorgonocephalus*.

Order Ophiurida. Mostly small ophiuroids, usually with five arms. Arms capable of transverse movement only. Dorsal arm shields present. This order contains most of the brittle stars and serpent stars. *Amphiura*, *Amphipholis*, *Ophiopholis*, *Ophiactis*, *Ophiothrix*, *Ophioderma*, *Ophiocoma*, *Ophiolepis*, *Ophiomusium*, *Ophiomaza*, *Ophionereis*.

CLASS ECHINOIDEA

The echinoids are free-moving echinoderms commonly known as sea urchins, heart urchins, and sand dollars. About 800 species have been described. The name Echinoidea, which means "like a hedgehog," is derived from the fact that the bodies of these animals are covered with spines. The echinoid body does not possess arms like that of asteroids. Rather, the shape is circular or oval, and the body is spherical or greatly flattened along the oral-aboral axis. The class is particularly interesting from the standpoint of symmetry, for while the sea urchins are radially symmetrical, many soft-bottom members of the class display various stages in the attainment of a secondary bilateral symmetry. A third distinctive feature of echinoid structure is the fusion of the skeletal ossicles into a solid case (the test).

External Structure. ***Regular Echinoids.*** The radial, or regular, members of the class are known as sea urchins. In these forms, the body is more or less spherical in shape and armed with relatively long movable spines (Fig. 19–20). Sea urchins are brown, black, purple, green, white, and red, and some are multicolored. Most sea urchins are 6 to 12 cm. in diameter, but some Indo-Pacific species may attain a diameter of nearly 36 cm.

The sea urchin body can be divided into an aboral and an oral hemisphere, with the parts arranged radially around the polar axis. The oral pole bears the mouth and is directed against the substratum. The mouth is surrounded by a peristomial membrane that is thickened along the inner edge to form a lip (Fig. 19–21*A*). The peristome of sea urchins contains small imbedded supporting ossicles, and the surface bears a number of different structures arranged in a radial manner. There are five pairs of short, heavy modified podia, called buccal podia, and five pairs of bushy projections, called gills.

In addition to the buccal podia and the gills, the area around the peristome bears small spines and pedicellariae. The aboral pole contains the anal region, known as the periproct (Fig. 19–21*B*). The periproct is a small circular membrane containing the anus, usually in the center, and a varying number of imbedded plates. In the Arbaciidae, there are four large

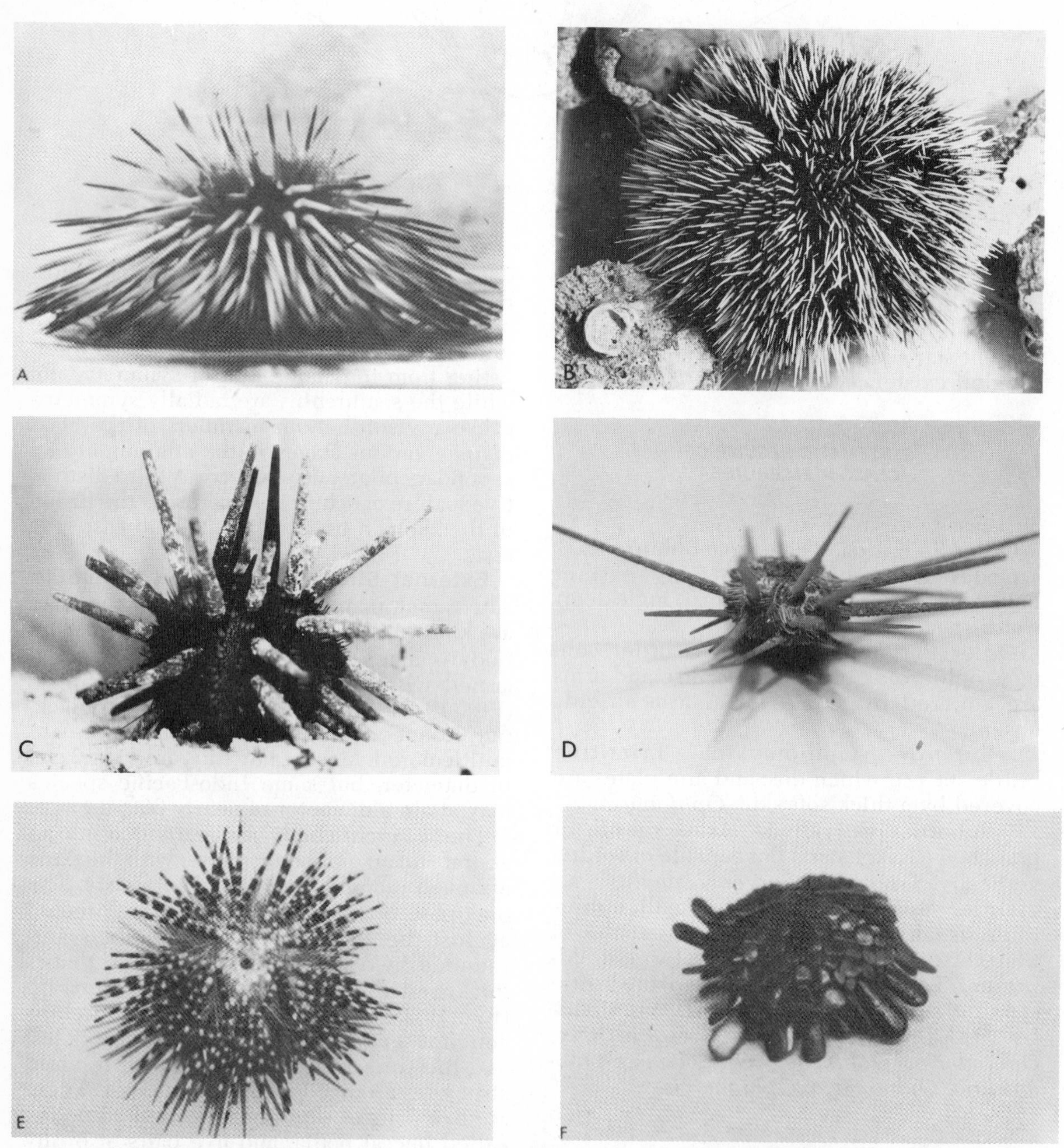

Figure 19–20. Regular urchins. *A*. Side view of the common Atlantic sea urchin, *Arbacia punctulata* showing long spines and podia. *B*. A West Indian species of *Tripneustes*, viewed from above. *C*. A species of *Eucidaris* with very small secondary spines around the base of the heavy primary spines. *D*. *Stylocidaris*, with long tapered primary spines. *E*. *Echinothrix* from Hawaii. The small slender secondary spines lie in distinct bands between the larger striped primary spines. Note the conspicuous anus. *F*. *Colobocentrotus*, a pacific sea urchin with blunt aboral spines which fit together to form a smooth surface. Such spines are perhaps an adaptation for living on wave-washed rocks. (All by Betty M. Barnes.)

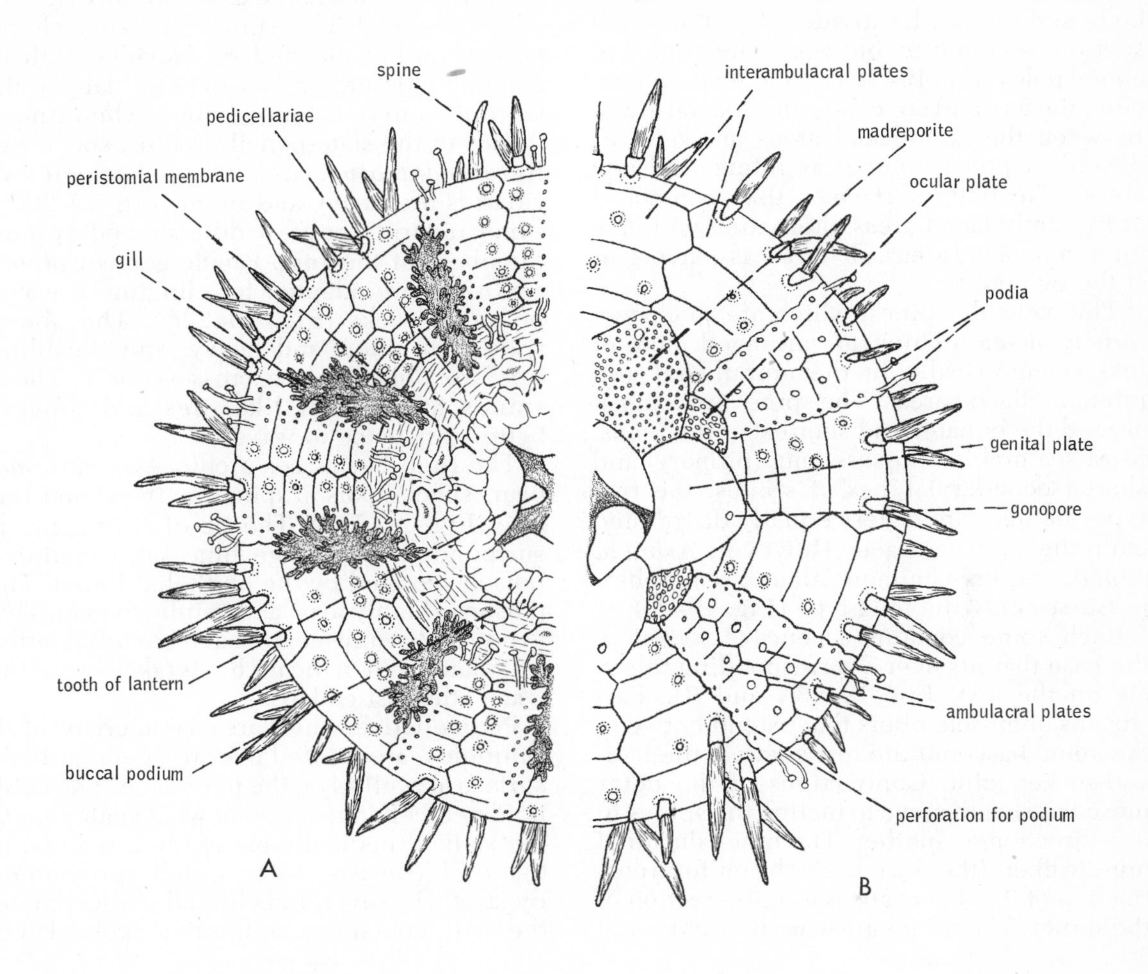

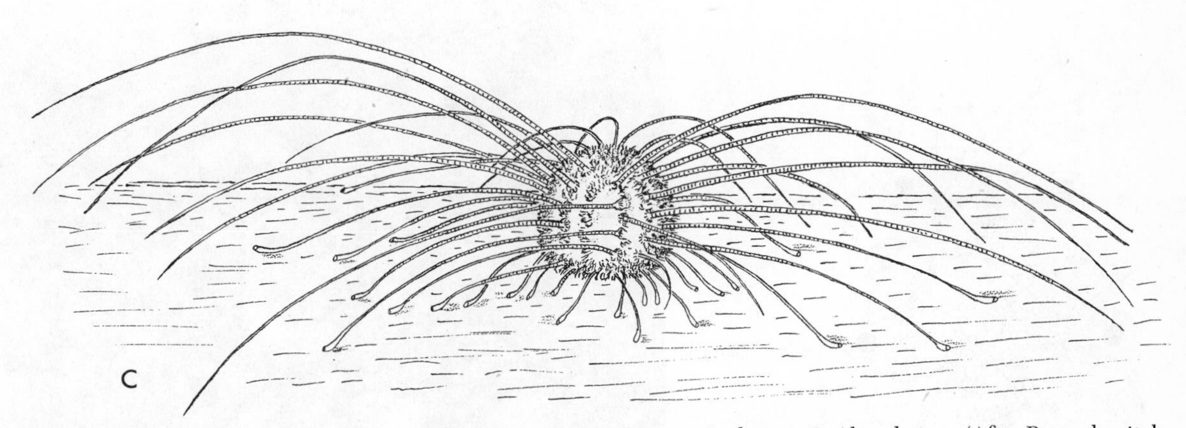

Figure 19–21. *A* and *B*. The regular urchin, *Arbacia punctulata*. *A*. Oral view. *B*. Aboral view. (After Petrunkevitch from Reid.) *C*. *Plesiodiadema indicum*, a deep-water, regular urchin from the Indo-Pacific. Note long, curved spines with flattened ends. (After Mortensen from Cuénot.)

periproct plates (Fig. 19–21*B*), but in other families of sea urchins the plates are smaller and more numerous (Fig. 19–22). The globose body surface can be divided into ten radial sections extending between the oral and aboral poles (Fig. 19–21*A*). Five sections contain tube feet and are called ambulacral areas. Between the ambulacral areas are sections devoid of podia, known as interambulacral areas. The body surface is thus composed of five ambulacral areas alternating with five interambulacral areas, all ten areas converging at the two poles.

The movable spines, which are so characteristic of sea urchins, are arranged more or less symmetrically in the ambulacral and interambulacral areas. The spines are longest around the equator and shortest at the poles. Most sea urchins possess long (primary) and short (secondary) types of spines, the two types being more or less equally distributed over the body surface. However, *Arbacia punctulata*, the common Atlantic sea urchin, possesses only the long type (Fig. 19–20).

Each spine contains a concave socket at the base that fits over a corresponding tubercle on the test (Fig. 19–23*A* and *B*). Two sheaths of muscle fibers that extend between the spine base and the test encircle the ball-and-socket joint. Contractions of the outer muscular sheath serve to incline the spines in one direction or another. The inner sheath of muscle fibers (the cog muscle) by uniform contraction of its fibers causes a rigid erection of the spines. The spines are usually circular and taper to a point, but many species depart from this generalization. *Diadema* of tropical reefs has spines, which may exceed 36 cm. in length (Fig. 19–24*A*). The spines are very sharp, hollow, and brittle, and are provided with an irritant. This urchin can inflict a dangerous, painful wound if stepped upon. The primary spines of the slate-pencil urchins, species of *Heterocentrotus*, and of some species of cidarids are heavy and blunt (Fig. 19–20*C*); some cidarids have paddle-shaped spines. Members of the Indo-Pacific genus *Colobocentrotus* are adapted for clinging to wave-pounded rocks (Fig. 19–20*F*). The aboral spines are short and heavy, and the blunt tips are polygonal in cross section. These spines fit together like tiles and provide better resistance to waves.

The Indo-Pacific sea urchin, *Asthenosoma*, bears special poison spines on the aboral surface (Fig. 19–23*C*). The tip of each spine is surrounded by a large blue sac containing poison secreted by the epithelial lining. The poison is highly toxic and painful to man. This genus is also unusual in that it has articulating test plates. When the urchin is taken out of the water, the test collapses.

Pedicellariae, which are characteristic of all echinoids, are located over the general body surface as well as on the peristome. The echinoid pedicellariae are somewhat analogous to the stalked pedicellariae of the asteroids, in that each consists of a long stalk surmounted by jaws. However, in echinoid pedicellariae, the stalk contains a supporting skeletal rod,

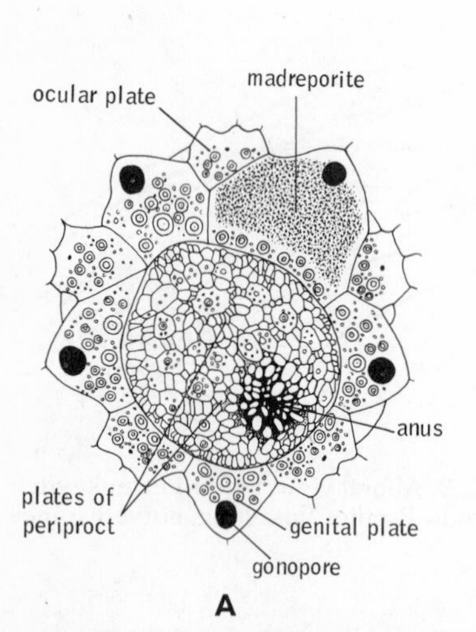

Figure 19–22. *A*. Periproct and surrounding plates of the regular urchin, *Strongylocentrotus*. (After Lovén from Ludwig.) *B*. Surface view of a section of a sea urchin test, showing tubercles on which spines are located, paired perforations for tube feet, and junction line (groove) of fused ossicles. Compare with Fig. 19–31. (By Betty M. Barnes.)

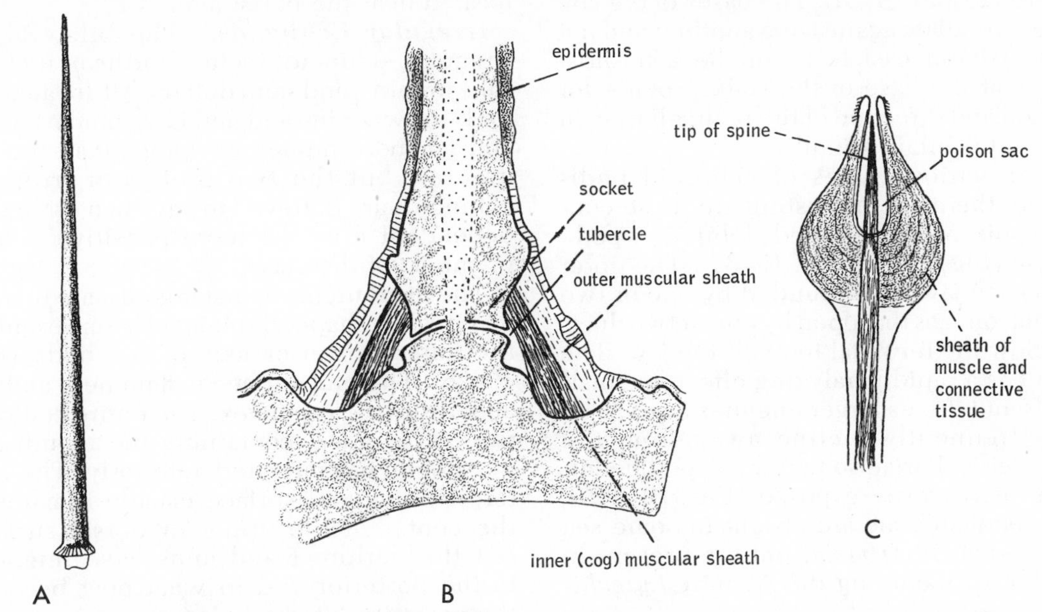

Figure 19–23. Spines of regular urchins. *A.* Primary spine of an urchin such as *Arbacia.* (After Hyman.) *B.* Section through the base of a *Cidaris* spine, showing muscular sheaths. (After Cuénot.) *C.* Poison spine of *Asthenosoma varium,* an Indo-Pacific species. (After Sarasins from Hyman.)

Figure 19–24. *A. Diadema.* Species of this genus, which occurs in both the Caribbean and Indo-Pacific, possess long hollow needle-like spines, which can inflict painful punctures when handled or stepped upon. This West Indian species is common on reefs, where it lives in sheltered or protected recesses. (By C. Gebelein.) *B. Lytechinus.* The members of this genus, along with certain others, cover the aboral surface with shells, algae, and other objects. The author has a golf ball he found on a specimen of *Lytechinus* in Bermuda. (By Betty M. Barnes.)

and there are usually three opposing jaws (Figs. 19–25 and 19–31). The bases of the jaw ossicles articulate against one another, and not against a basal ossicle as in the asteroids. Muscles at the base of the stalk provide for elevation and direction of the pedicellariae in response to certain stimuli.

Of the various types of echinoid pedicellariae, the most interesting are those containing poison glands, called globiferous pedicellariae (Figs. 19–25 and 19–31). The outer side of each jaw is surrounded by one or two large poison sacs that open by one or two ducts just below the terminal tooth of the jaw. The poison has a rapid paralyzing effect on small animals and drives larger enemies away. The spines frequently incline away from the poison pedicellariae so that these pedicellariae are more easily exposed (Fig. 19–25*B*). Poison pedicellariae are absent in some sea urchins, such as *Arbacia,* but are present in many forms, including the Atlantic *Lytechinus*.

Other types of pedicellariae are used for defense or for cleaning the body surface, wherein they bite and break up small particles of debris that are then removed by the surface cilia. When the pedicellariae are touched on the outside, they snap open; when touched on the inside, they snap shut. Pedicellariae also respond to chemical stimuli.

In most sea urchins, the ambulacral areas bear hard stalked spherical or ovoid bodies (spheridia) that contain a statocyst vesicle with statoliths. The spheridia may be limited in number and located only on the oral side, or there may be many throughout the length of the ambulacrum. In *Arbacia*, there is only one spheridium per ambulacrum, and each is located near the peristome.

Irregular Echinoids. The bilateral, or irregular, echinoids include the heart urchins, cake urchins, and sand dollars. All are adapted for burrowing in sand and have much smaller and far more numerous spines than the sea urchins, but the two groups of irregular echinoids are believed to have had origins independent from the more primitive regular members of the class.

The heart urchins (spatangoids) are more or less oval in shape, the long axis representing the anteroposterior axis of the body (Fig. 19–26). The oral surface is flattened, and the aboral surface is convex. The entire center of the oral surface, containing the mouth and peristome, has migrated anteriorly. The center of the aboral surface usually remains in the center of the upper or dorsal surface, but the periproct and anus have migrated to the posterior end in what now becomes the posterior interambulacrum.

Podia are degenerate or absent around the circumference of the body, so that functional podia are confined to the oral and aboral surfaces. The conspicuous aboral ambulacral areas are each shaped like a petal radiating from the aboral center and are known as petaloids. The podia of the petaloids are modified for gas exchange. The oral ambulacral areas (phyllodes) also have a flower-like arrangement and contain specialized podia for obtaining food particles. The small spines form a dense covering over the body surface but have the same basic structure as in the sea urchins.

Certain specialized spines (clavules) are ex-

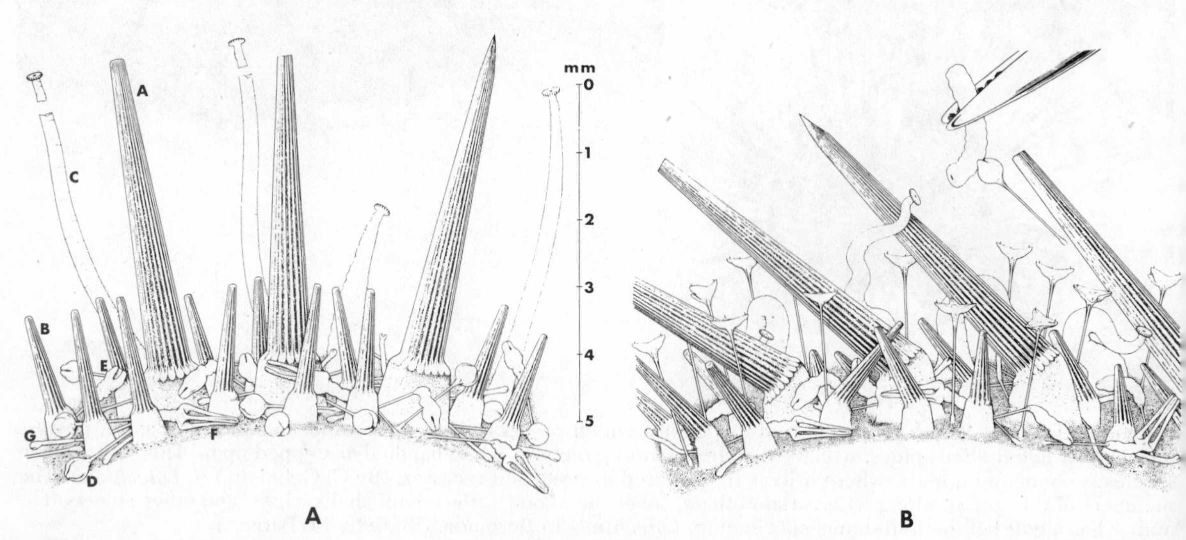

Figure 19–25. Responses of spines and pedicellariae of the sea urchin *Psammechinus miliaris* to a tube foot of a predatory sea star. *A*. Undisturbed surface. *B*. Stimulated surface. (From Jensen, M., 1966: Ophelia, 3:214–215.)

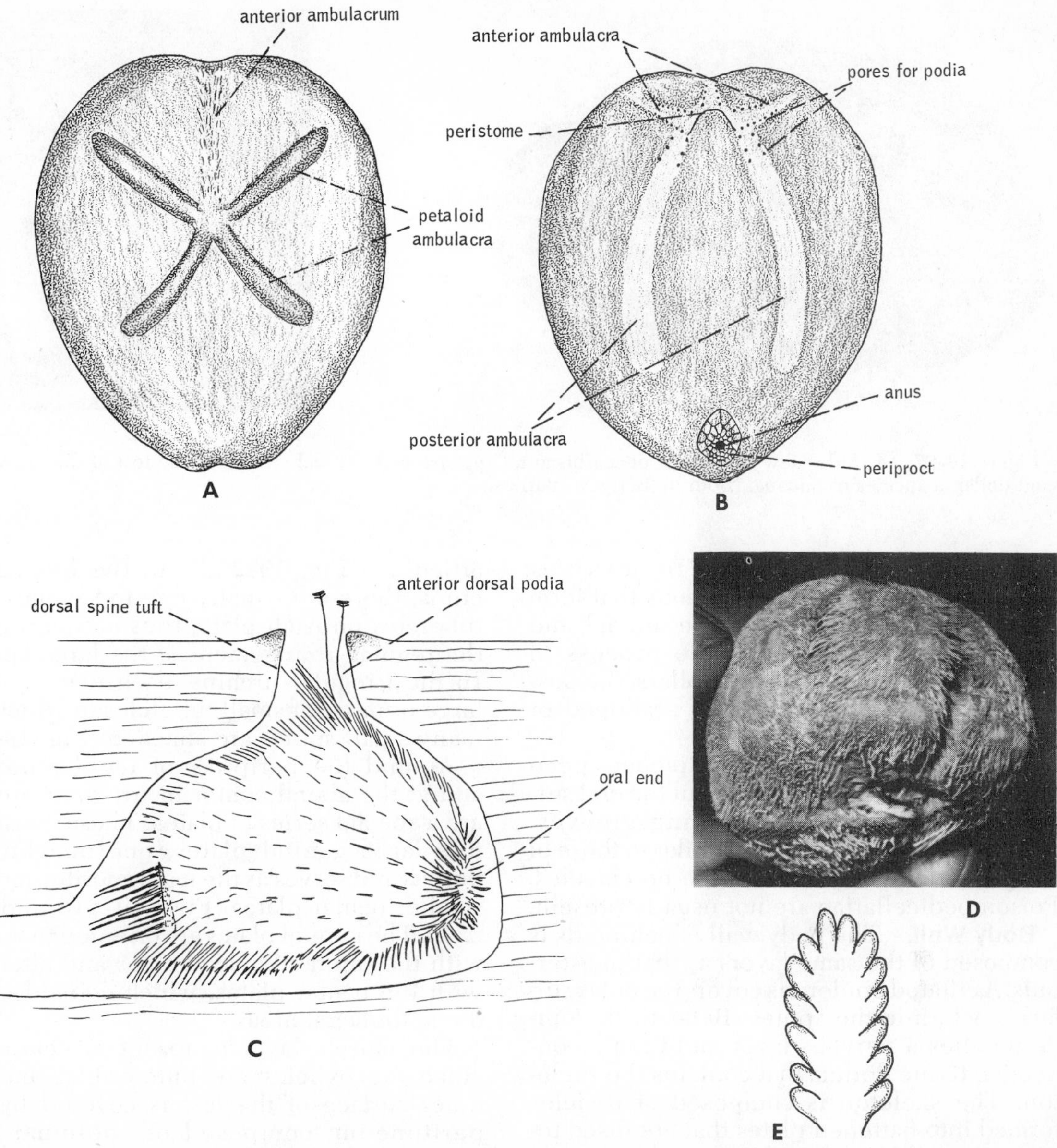

Figure 19–26. Irregular echinoids, spatangoids (heart urchins). *A. Meoma ventricosa* from the West Indies (aboral view). *B. Meoma* (oral view). (Both after Hyman.) *C. Echinocardium flavescens* in its sand burrow (lateral view). (Modified after Gandolfi-Hornyold.) *D.* Anterior end of *Moira atropos,* a heart urchin from the Atlantic Coast of the southeastern United States. (By Betty M. Barnes.) *E.* Branchial podium from the petaloid of *Spatangus.* (After Hoffman from Hyman.)

tremely minute, ciliated at the base, and flattened at the distal end. The clavules appear together in tracts located over certain parts of the surface, such as around each of the petaloids. It is believed that the clavules produce water currents and secrete large amounts of mucus for burrow maintenance. Pedicellariae are present. The spheridia are located in the phyllodes.

The cake urchins or sand dollars (clypeasteroids) differ from the heart urchins in a number of respects. A few species, such as the sea biscuits (Fig. 19–27A), are shaped somewhat like heart urchins, but in typical sand dollars the body is greatly flattened, displaying a circular circumference (Fig. 19–28). The aboral center and the oral center, which contains the mouth, are both centrally located. The periproct, however, is ventral and, like that of the heart urchins, is located in the posterior interambulacrum.

The body of some common sand dollars, called keyhole sand dollars (*Mellita*), contains large elongated notches or openings known as lunules (Fig. 19–28*A* and *C*). Lunules vary in number from two to many

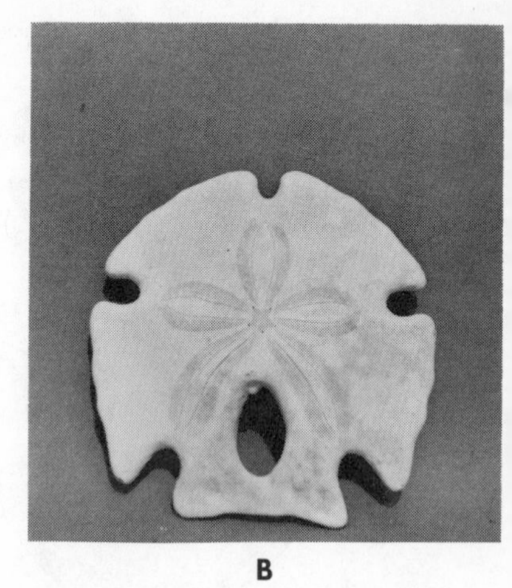

Figure 19–27. *A.* Side view of the test of sea biscuit, *Clypeaster. B.* Aboral surface of the test of the arrowhead sand dollar, a species of *Encope.* (Both by Betty M. Barnes.)

and are symmetrically arranged. In most cases, the lunules arise from indentations that form along the circumference of the animal and then become enclosed in the process of growth. In the scutellid sand dollars, the posterior circumference is deeply scalloped or notched (Fig. 19–28*B*).

The aboral surface bears conspicuous petaloids. There are no phyllodes, but the oral surface contains distinct radiating grooves. Spheridia and spination are similar to those of the heart urchins, but there are no clavules. Poison pedicellariae are not usually present.

Body Wall. The body wall of echinoids is composed of the same layers as that in asteroids. A ciliated epidermis covers the outer surface, including the spines. Beneath the epidermis lies a nervous layer and then a connective tissue dermis that contains the skeleton. The skeleton is composed of ossicles formed into flattened plates that are fused together to produce a solid immovable test.

In all echinoids, the plates are arranged in rows running from the oral pole to the aboral pole. Each ambulacral area is composed of two rows of ambulacral plates, and each interambulacral area is composed of two rows of interambulacral plates. There are thus twenty rows of plates—ten ambulacral and ten interambulacral (Fig. 19–21*B*). The ambulacral plates are pierced by holes enabling the canals to connect with the ampullae and podia (Fig. 10–22*B*).

Echinoids are the only group of *living* echinoderms in which the canals for the podia penetrate the ossicles; in all others, the canals are located between the ossicles. Each of the plates, both ambulacral and interambulacral, bears rounded tubercles on which the spines articulate (Fig. 19–22*B*). In the regular urchins, there are usually one to several large tubercles on each plate, thus accounting for the regular arrangement of the large spines. In the irregular urchins, each plate bears a large number of small tubercles to which the many small spines are attached (Fig. 19–33).

Around the periproct of regular urchins and at the aboral center in irregular urchins are a special series of plates. These consist of five large genital plates, one of which is porous and serves as the madreporite, and five smaller ocular plates (Figs. 19–21*B* and 19–22*A*). The genital plates are oriented to line up with the interambulacral areas and alternate with the ocular plates, which coincide with the ambulacral areas.

The muscle layer is absent in echinoids, since the ossicles are immovable, and the inner surface of the test is covered by the peritoneum, composed of columnar epithelium.

Locomotion. Sea urchins are adapted for an existence on rocks and other types of hard bottoms, and spines and podia are used in movement. The tube feet function in the same manner as those of the sea stars, and spines may be used for pushing. Sea urchins can move in any direction, and any one of the ambulacral areas can act as the leading section. If overturned, these animals right themselves by attaching the aboral podia of one of the ambulacral areas. Attachment of the podia progresses in an oral direction, gradually turning the animal over onto the oral side. In forms with long spines, such as *Arbacia,* righting also involves specialized movements of the spines.

Some sea urchins tend to seek rocky de-

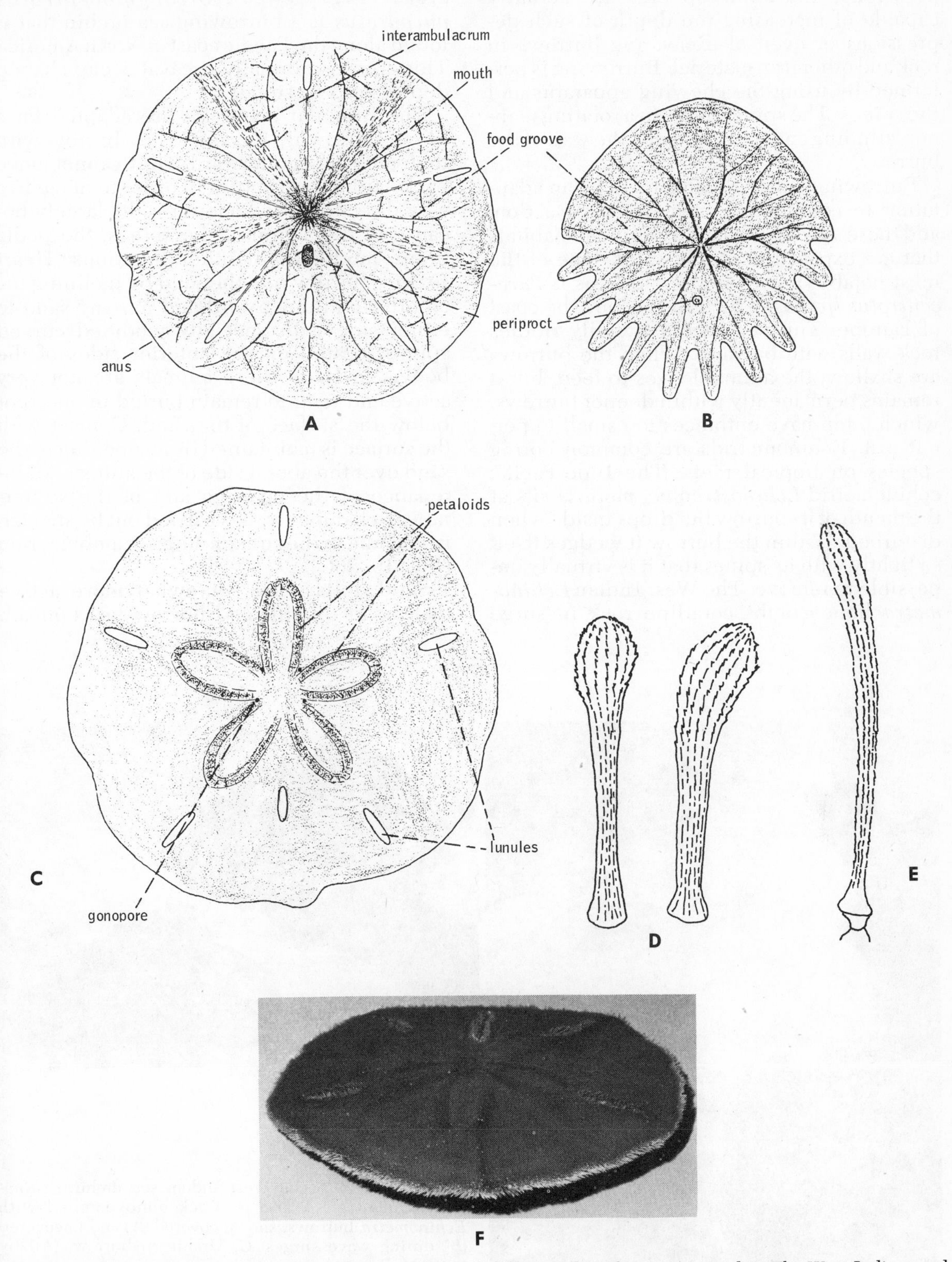

Figure 19–28. Irregular echinoids, clypeasteroids (sand dollars and sea biscuits). *A* and *C*, The West Indian sand dollar, *Leodia sexiesperforata*. *A*. Oral view. *C*. Aboral view. *B*. The African scutellid sand dollar, *Rotula orbiculus* (oral view). *D*. Aboral spines of *Mellita*. *E*. Oral spines of *Mellita*. (All after Hyman.) *F*. The Atlantic Coast five slotted sand dollar, *Mellita quinquiesperforata*. (By Betty M. Barnes.)

pressions, and some species are actually capable of increasing the depth of such depressions or even of excavating burrows in rock and other firm material. Burrowing is performed by using the chewing apparatus and the spines. The spines move in a rotating manner, grinding and wearing away the wall of the burrow.

Burrowing behavior appears to be an adaptation to counteract excessive wave action, and these species are largely found in habitats that are exposed to rough water. One of the most notable burrowing sea urchins is *Paracentrotus lividus*, which lives along the coast of Europe. This sea urchin literally riddles rock walls with burrows. When the burrows are shallow, the animal leaves to feed, but it remains permanently within deeper burrows, which often have entrances too small to permit exit. Echinometrids are common boring species on tropical reefs. The Indo-Pacific echinometrid *Echinostrephus molaris* sits at the mouth of its burrow but drops inside when disturbed. Within the burrow it wedges itself so tightly with its spines that it is virtually impossible to remove. The West Indian *Echinometra* honeycombs coralline rock in surge areas (Fig. 19–29). *Strongylocentrotus purpuratus* is a burrowing sea urchin that is found along the Pacific coast of North America. This species normally excavates cup-shaped depressions in stone.

The irregular echinoids are adapted for a life of burrowing in sand. They burrow with their anterior end forward. They cannot move backward but must turn in order to move in a new direction. Movement results largely because of the action of the spines, the podia being modified for other functions. Heart urchins burrow into the sand by inclining the anterior end downward and moving sand to either side with specially modified curved spines located on the anterior sides of the body. Normally these animals are not very active and tend to remain buried in one spot below the surface of the sand. Contact with the surface is maintained by an opening in the sand over the aboral side of the animal. Maintenance of the opening and of the subterranean chamber wall is carried out by specialized podia in the sunken anterior ambulacrum (Fig. 19–26*C*).

Sand dollars are somewhat more active than heart urchins and burrow just beneath

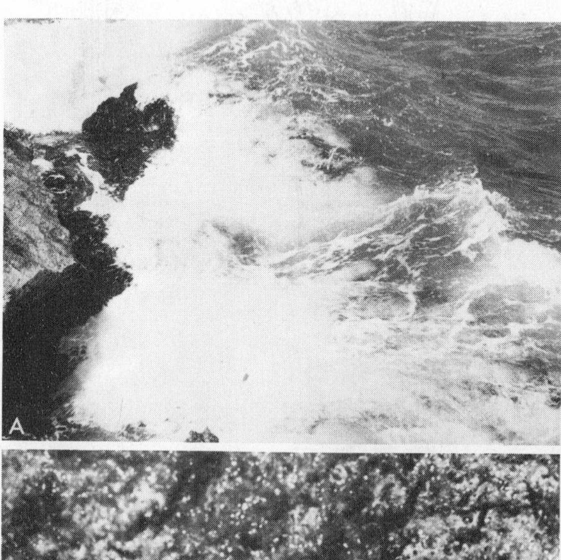

Figure 19–29. The West Indian sea urchin *Echinometra lacunter*. *A* and *B*. Rock honeycombed with *Echinometra* burrows, shown covered (*A*) and uncovered (*B*) during wave surges. *C*. Urchin in burrow. (All by Betty M. Barnes.)

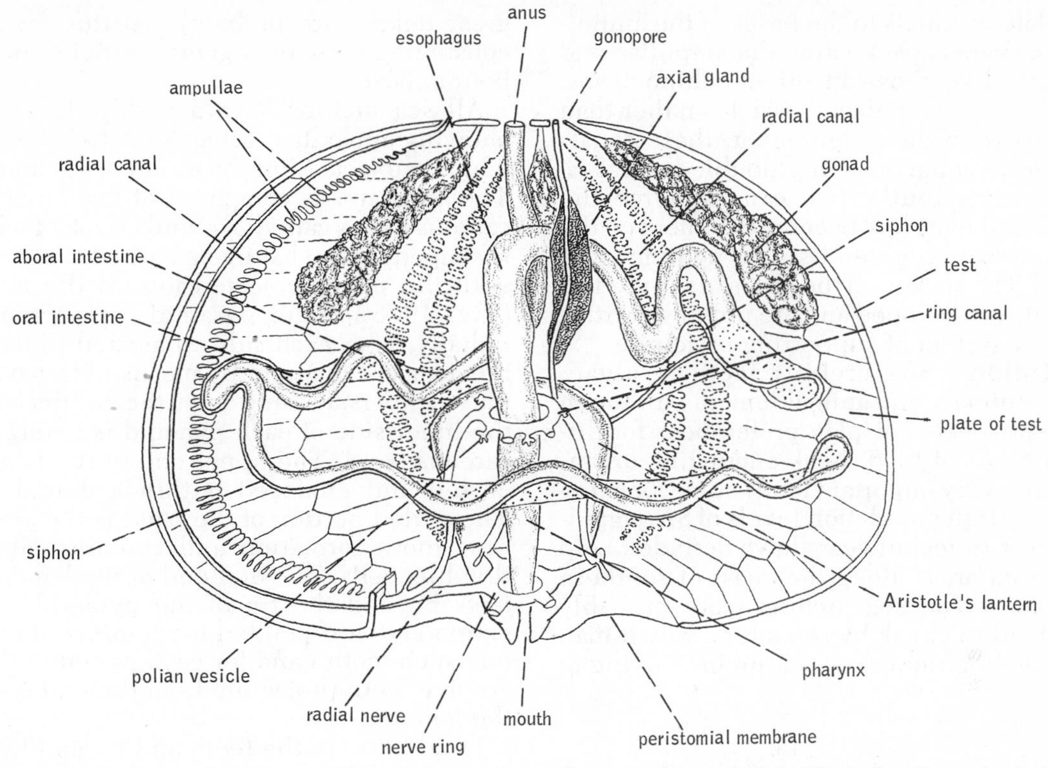

Figure 19–30. Internal structure of the regular urchin, *Arbacia* (side view). (Modified after Petrunkevitch from Reid.)

he sand surface. Some species, such as the common sand dollars of the east coast of the United States, *Mellita quinquiesperforata*, cover themselves completely with sand; in others, the posterior end projects obliquely above the sand surface.

The method of burrowing differs in different species, but in all cases the spines are of primary importance in moving sand grains.

In many species, including the common Atlantic five-lunuled sand dollar (*Mellita quinquiesperforata*), the spines on the oral surface push the animal forward into the sand. The entire operation takes up to 20 minutes. The six-lunuled *Leodia* of the Caribbean burrows by rotating its anterior edge back and forth, slicing its way through the sand. *Leodia* can completely bury itself in one to four minutes. Some sand dollars can right themselves if turned over. In righting itself, the animal burrows its anterior end into the sand, gradually elevates its posterior end, and eventually flips its body over. The Atlantic five-lunuled species partially elevates its body and is then apparently dependent upon water currents to be turned back over onto the oral surface.

Ikeda (1941) studied the Japanese keyhole sand dollar, *Astriclypeus*, and found that the anterior lunules are essential in burrowing and becoming upright. The spines that border these lunules also drive sand through them. If the lunule is plugged with paraffin, burrowing is considerably prolonged. Observations of the Atlantic *Mellita quinquiesperforata* suggest that perhaps the downward movement of sand through the posterior lunule controls the inclination of the animal in burrowing.

Water-Vascular System. The water-vascular system of echinoids is essentially like that of the sea stars (Fig. 19–30). One of the genital plates around the periproct contains pores and pore canals and functions as the madreporite (Fig. 19–22A). A stone canal descends orally to the water ring, which lies above the peristome in heart urchins or just above the chewing apparatus in regular urchins and sand dollars. The radial canals extend from the water ring and run along the underside of the ambulacral areas of the test.

Each radial canal terminates in a small protrusion called a terminal tentacle, which penetrates the most apical ambulacral plate. The radial canal gives off alternately on both sides lateral canals to the bases of the ampullae. The canals connecting the ampullae and podia, unlike those in other echinoderms, penetrate the ambulacral ossicles rather than pass between them. Moreover, these canals are also peculiar in being doubled—that is, from each ampulla, two canals pierce the ambulacral plate and become confluent on the outer surface to enter a single podium (Fig. 19–31). The suckers of the podia of sea urchins are highly developed and are provided with a complex system of supporting ossicles.

Nutrition. Sea urchins feed on algae, sessile animals, and animal remains, although different species display preferences for different types of food. Rock-encrusting organisms are a very important food source for many species. High population levels of *Strongylocentrotus* in recent years have denuded kelp beds from areas along the coast of southern California. The sea urchins were probably once held in check by sea otters, which man has largely eliminated. Sea urchins living at great depths are probably detritus feeders consuming minute organic particles in the bottom ooze.

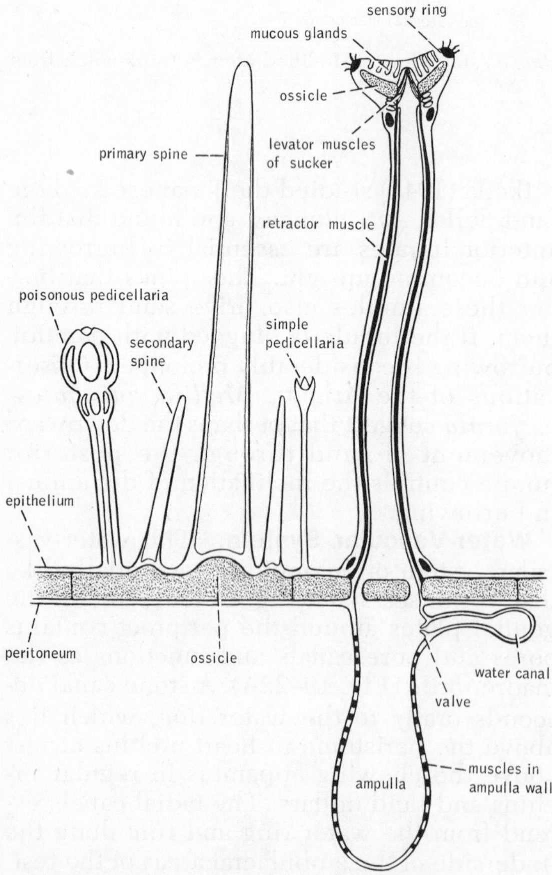

Figure 19–31. Diagrammatic section through the body wall of a sea urchin, showing one ambulacral and one interambulacral ossicle and associated structures. (After Nichols, in part.)

All sea urchins have a highly developed chewing apparatus called Aristotle's lantern which projects slightly through the mouth. The apparatus is composed of five large calcareous plates called pyramids, each of which is shaped somewhat like a barbed arrowhead with the point projected toward the mouth (Fig. 19–32). The pyramids are arranged radially, with each side connected to that of the adjacent pyramid by means of transverse muscle fibers. Passing down the midline along the inner side of each pyramid is a long calcareous band. The upper end of the band is curled and enclosed within a dental sac. The curled portion of the band is the area of new tooth formation and contains rapidly dividing cells. The oral end of the band projects beyond the tip of the pyramid as an extremely hard pointed tooth. Since there is one such tooth band for each pyramid, there are five teeth projecting from the oral end of the lantern.

In addition to the teeth and pyramids, the Aristotle's lantern is composed of a number of smaller rodlike pieces located at the aboral end. By means of special muscles, the lantern can be partially protracted and retracted through the mouth. Other muscles control the teeth, which can be opened and closed. The lantern is an efficient chewing structure; but feeding is not rapid, and the consumption of a bunch of seaweed may take weeks.

In *Psammechinus miliaris* it has been found that there is some external digestion by coelomocytes which pass from the gills to the narrow space between the oral surface and substratum. The products of digestion are sucked in by the pharynx. A similar digestive process may occur in other sea urchins.

The interior of Aristotle's lantern contains both a buccal cavity and a pharynx that ascends through the apparatus and passes into an esophagus (Fig. 19–32*B*). The esophagus descends along the outer side of the lantern and joins an intestine (Fig. 19–30). At the junction of the esophagus and intestine, a blind pouch or cecum is usually present. The intestine is very long and can be divided into a proximal small intestine and a distal large intestine. The small intestine makes a complete turn around the inner side of the test wall, to which it is suspended. It then passes into the larger aboral intestine, which makes a complete turn in the opposite direction. The large intestine then ascends to join the rectum,

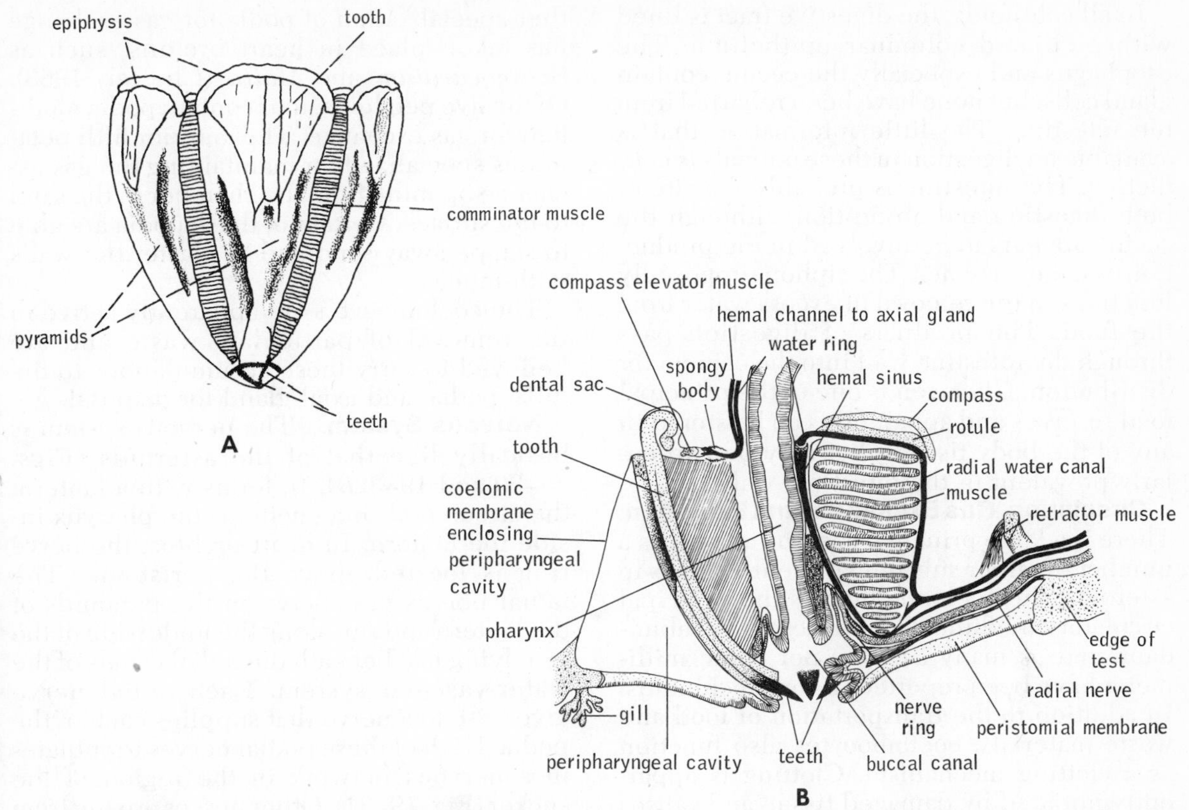

Figure 19–32. *A.* Aristotle's lantern of *Tripneustes esculentus*, a regular urchin (lateral view). (After Hyman.) *B.* Vertical section through Aristotle's lantern and peristomial region of the regular urchin, *Paracentrotus lividus.* (After Cuénot.)

which empties through the anus within the periproct.

In most echinoids a narrow tube, called a siphon, parallels the intestine for about one half of its length. The ends of the siphon open into the lumen of the intestine.

Irregular urchins all feed on minute organic particles in the sand in which they burrow. In the heart urchins, food is obtained by means of modified podia that occupy the phyllodes on the oral surface. During feeding, these podia grope about the sand surface of the chamber, picking up food particles. The relatively complex burrowing and feeding behavior of *Moira* has been described by Chesher (1963).

The chewing lantern is absent in heart urchins, but otherwise the alimentary tract is essentially like that of the sea urchins. Since the anus has shifted away from the center of the aboral surface, the rectum passes posteriorly from the intestine.

The sand dollars possess a modified lantern. It consists primarily of large pyramids and cannot be protracted from the mouth. These echinoids feed on minute organic particles like the heart urchins. Sand and debris through which the sand dollar burrows are passed backward over the aboral surface by special club-shaped spines. Fine particles, constituting the food materials, drop down between the spines, become entangled in mucus, and are carried by ciliary currents to the margins of the body or the lunules and thence to the oral surface. Here branched food grooves conduct masses of particles to the mouth. Transport is provided by the pushing action of tube feet which border the grooves. In *Dendraster exocentricus* on the west coast of the United States suspended particles, particularly diatoms, are a major source of food, for this sand dollar lives with the posterior half of the body projecting above the sand surface. The alimentary canal of sand dollars differs from that of other echinoids only in that the rather large intestine makes but a single turn around the circumference of the test. The rectum descends to the posterior ventral anus. Some spatangoids, including *Echinocardium,* build a drain at the back of the burrow to collect feces (Fig. 19–26*C*). The drain is constructed by specialized posterior podia.

In all echinoids, the digestive tract is lined with a ciliated columnar epithelium. The esophagus and especially the cecum contain gland cells, but none have been reported from the intestine. The little information that is available on digestion in these animals is conflicting. The intestine is probably the site of both digestion and absorption, although the cecum appears to be involved in the production of some enzymes. The siphon supposedly functions in the removal of excess water from the food. The products of digestion pass through the intestine wall into the coelom for distribution. Glycogen is one of the principal food reserves and apparently can be stored in any of the body tissues, but may be particularly prevalent in the intestinal wall.

Circulation, Gas Exchange, and Excretion. There is a large principal coelom, as well as a number of minor subcompartments, and as in asteroids, the coelomic fluid is the principal circulatory medium. Coelomocytes are abundant, and as many as 7000 per cubic millimeter have been reported in some echinoids. In addition to the transportation of food and waste materials, coelomocytes also function as a clotting mechanism. Clotting is apparently initiated by damaged tissue and causes the formation of a meshwork of coelomocytes at the place of injury.

A hemal system is present and has the same basic plan of structure as in the asteroids.

In regular echinoids, the five pairs of peristomial gills are probably the chief centers of gas exchange (Figs. 19–21*A* and 19–32*B*). Each gill is a highly branched outpocketing of the body wall and is therefore lined within and without by a ciliated epithelium. The internal cavities of the gills are filled with fluid and are continuous with a separate division of the coelom, called the peripharyngeal coelom, that surrounds the lantern (Fig. 19–32*B*). Special ossicles and their associated muscles, composing the aboral part of the lantern, act as a pumping mechanism, changing the coelomic pressure within the peripharyngeal coelom and thereby forcing fluid into and out of the gills. Pumping is not continuous but depends on the need for oxygen, and the apparatus is under the control of the nerve ring. The nerve ring is stimulated to emit impulses to the lantern by an increase in acidity resulting from an accumulation of carbon dioxide.

There are no peristomial gills in heart urchins and sand dollars. In these animals, the modified podia of the petaloids, which are thin-walled and lobulate, act as gas exchange structures (Figs. 19–26*E* and 19–33). A further specialization of podia for gas exchange has taken place in heart urchins, such as *Echinocardium* and *Moira* (Chester, 1963). Of the five petaloids, four contain podia modified for gas exchange. The anterior fifth petaloid is specialized for maintaining the gas exchange opening from the chamber in the sand to the surface. Certain of these podia are able to scrape away sand and to plaster the walls with mucus.

The coelomocytes are apparently active in the removal of particulate waste and are believed to carry these accumulations to the gills, podia, and axial gland for disposal.

Nervous System. The nervous system is basically like that of the asteroids (Figs. 19–30 and 19–32*B*). In forms with a lantern, the circumoral ring encircles the pharynx inside the lantern. In heart urchins, the nerve ring is located above the peristome. The radial nerves pass between the pyramids of the lantern and run along the underside of the test, lying just beneath the radial canals of the water-vascular system. Each radial nerve gives rise to a nerve that supplies each of the podia. Each of these podial nerves terminates in a nervous network in the region of the sucker (Fig. 19–31). Other nerves arising from the radial nerve also penetrate the test and supply both the pedicellariae and the spines with a nervous network that is continuous with the subepidermal plexus.

Studies of muscle innervation in pedicellariae and ampullae have shown that an extension or "tail" of the muscle fiber is the junction point with the motor axon terminal. A somewhat similar arrangement occurs in nematodes.

The numerous sensory cells in the epithelium, particularly on the spines, pedicellariae, and podia, compose the major part of the echinoid sensory system. The buccal podia of sea urchins, the podia around the circumference of heart urchins, and the podia of the oral surface of sand dollars are all important in sensory reception. As already described, the spheridia are statocysts which function in orienting the animal to gravitational pull. It has been found, for example, that if the spheridia of a sand dollar are removed, the righting reaction is greatly delayed. Although ocelli are absent in almost all species, photoreceptor cells are located on the tube feet. Echinoids are in general negatively phototropic and tend to seek the shade of crevices in rocks and shells. Some species of sea urchins, such as *Lytechinus*, cover themselves with shell fragments and other objects, using the tube feet

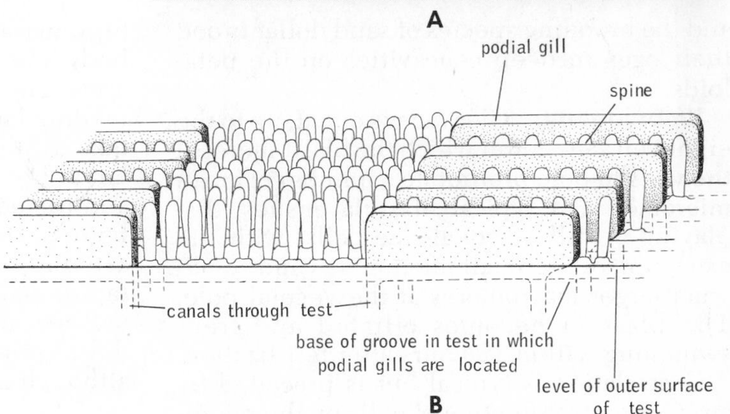

Figure 19–33. *A*. Surface of the aboral center of the test of the sand dollar *Mellita*. Holes opposite the points of the central star are the gonopores. There is no gonad in the posterior interambulacrum. In the petaloid included in the lower half of the photograph, the grooves in which the modified branchial podia are located are conspicuous (compare with diagram in *B*). At the ends of the grooves can be seen the perforations for the podial canals penetrating the test. The inner of the two canals is very conspicuous; in this photograph, the outer canals can be seen only in the areas opposite the gonopores. (By Betty M. Barnes.) *B*. Diagram of branchial podia across one petaloid of a sand dollar.

(Fig. 19–24*B*). The significance is uncertain but clearly seems to be a response to light.

Reproduction. All echinoids are dioecious, and the majority display no sexual dimorphism. In regular echinoids, there are five gonads suspended along the interambulacra on the inner side of the test (Fig. 19–30). In most irregular echinoids, the gonad of the posterior interambulacrum has disappeared. Each gonad is covered on the outside with coelomic peritoneum and lined internally with a germinal epithelium. Between these two layers are found muscle cells and connective tissue. A short gonoduct extends aborally from each gonad and opens through a gonopore located on one of the five genital plates (Figs. 19–22*A*, 19–30, and 19–33*A*).

Sperm and eggs are shed into the sea water by the contraction of the muscle layers of the gonads, and fertilization takes place in the sea water. As in asteroids, spawning in temperate species takes place in the spring and early summer. Brooding is displayed by some cold-water sea urchins and heart urchins, and there is a brooding species of sand dollar. Brooding sea urchins retain their eggs on the peristome or around the periproct and use the spines in holding the eggs in position. The heart urchins

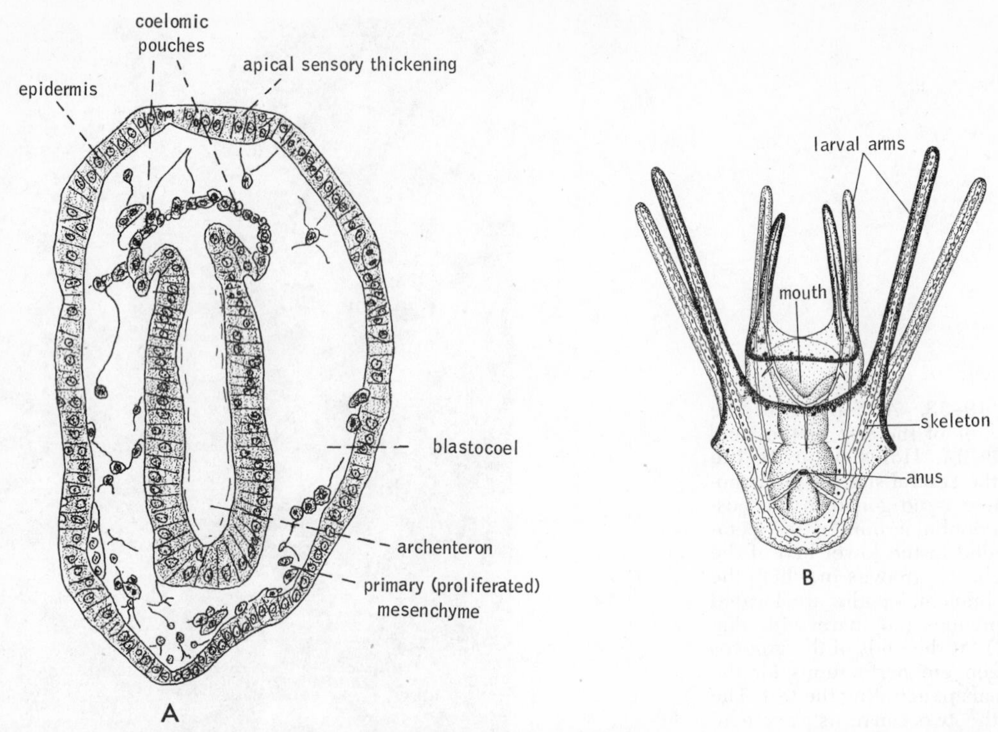

Figure 19–34. *A*. Gastrula of the sea urchin, *Echinus esculentus*. (After McBride from Hyman.) *B*. Echinopluteus larva of the sand dollar, *Fibularia craniola*. (After Mortensen.)

and the brooding species of sand dollar brood their eggs in deep concavities on the petaloids.

Embryogeny. Cleavage is equal, up to the eight-cell stage, after which the blastomeres at the vegetal pole proliferate a number of small micromeres. A typical blastula ensues, displaying a wall of equal-size cells, with the exception of the small micromeres and somewhat larger macromeres at the vegetal pole. The blastula becomes ciliated and free-swimming within 12 hours after fertilization.

Gastrulation is typical but is preceded by an interior proliferation of cells by the micromeres, which forms the mesenchyme. Additional mesenchymal cells are proliferated from the tip of the archenteron. The coelom is formed by the separation of the free end of the archenteron. This separated portion then divides into right and left pouches, or lateral divisions may appear prior to the separation from the main portion of the archenteron (Fig. 19–34*A*). The gastrula becomes somewhat cone-shaped and gradually develops into a larval stage (the echinopluteus), which displays five or six pairs of elongated arms supported by calcareous rods and bearing ciliated bands (Fig. 19–34*B*).

The echinopluteus differs from the asteroid bipinnaria in the general shape of the larval body and in the number and position of the arms. The echinopluteus is a planktonic and a feeding larva; its complete development may take as long as several months. During later larval life, the adult skeleton begins to form, and the echinopluteus gradually sinks to the bottom. However, there is no attachment as in asteroids, and metamorphosis is extremely rapid, taking place in about an hour. Young urchins are no larger than 1 mm. Internal development is basically like that of asteroids, although differing in detail.

SYSTEMATIC RÉSUMÉ OF CLASS ECHINOIDEA

Subclass Perischoechinoidea. Largely primitive fossil urchins of the Paleozoic seas, which made their first appearance in the Ordovician period with *Bothriocidaris*.

Order Cidaroida. Of the four orders of the subclass Perischoechinoidea, this order is the only one that evolved the condition of having two rows of plates for each ambulacrum and interambulacrum. It is also the only one that survived the Mesozoic era and became the ancestor of the remaining echinoids, most of which belong to the next subclass. The existing members of the Cidaroida are characterized by widely separated primary spines and small secondary spines. Gills are absent. *Cidaris, Notocidaris.*

Subclass Euechinoidea. This subclass contains the majority of living species of echinoids.

Superorder Diadematacea. Sea urchins with perforated tubercles. Gills usually present.

Order Pedinoida. Rigid test with solid spines. Ten buccal plates on peristomial membrane. *Caenopedina* is the only living genus.

Order Diadematoida. Rigid or flexible test with hollow spines. Ten buccal plates on peristomial membrane. *Diadema, Plesiodiadema.*

Order Echinothuroida. Flexible test with hollow spines. Simple ambulacral plates on peristomial membrane. Gills inconspicuous or lost. *Asthenosoma.*

Superorder Echinacea. Sea urchins with rigid test and solid spines. Gills present. Peristomial membrane with ten buccal plates.

Order Arbacioida. Periproct with four or five plates. *Arbacia.*

Order Salenioida. Anus located eccentrically within periproct because of the presence of a large plate (suranal plate). *Acrosalenia.*

Order Temnopleuroida. Test usually sculptured. Camerodont lantern (large epiphyses are fused across top of each pyramid). *Toxopneustes, Lytechinus.*

Order Phymosomatoida. Like the previous order but primary tubercles nonperforate. *Glyptocidaris.*

Order Echinoida. Camerodont lantern and nonsculptured test with imperforate tubercles. *Echinus, Paracentrotus, Colobocentrotus, Heterocentrotus, Strongylocentrotus.*

Superorder Gnathostomata. Irregular urchins in which mouth is in center of oral surface but anus has shifted out of apical center. Lantern present.

Order Holectypoida. No petaloids. Many fossil members were essentially regular in shape. The two living genera, *Echinoneus* and *Micropetalon,* are oval.

Order Clypeasteroida. True sand dollars. Petaloids present. Test greatly flattened. No phyllodes present. *Clypeaster, Fibularia, Mellita, Encope, Rotula.*

Superorder Atelostomata. Irregular urchins with lantern absent.

Order Holasteroida. Oval or bottle-shaped echinoids with thin delicate test. Petaloids and phyllodes not developed. Deep-water species. *Pourtalesia.*

Order Spatangoida. Heart urchins. Oval and elongated echinoids. Oral center shifted anteriorly and anus shifted out of aboral apical center. Petaloids present but may be sunk into grooves. Phyllodes present. *Spatangus, Echinocardium, Moira, Meoma.*

Order Cassiduloida. Mostly extinct echinoids with round to oval test and a central or slightly anterior apical center. Phyllodes with intervening smaller areas (bourrelets) and poorly developed petaloids present. The few existing species are somewhat similar to sand dollars. *Echinolampas.*

Order Neolampidoida. Like the Cassiduloida but without petaloids or phyllodes. *Neolampas.*

CLASS HOLOTHUROIDEA

The holothuroids are a class of some 500 echinoderms known as sea cucumbers. Like echinoids, the body of the holothuroids is not drawn out into arms, and the mouth and anus are located at opposite poles. Also, there are ambulacral and interambulacral areas arranged meridionally around the polar axis. However, holothuroids are distinguished from other echinoderms in having the polar axis greatly lengthened, which results in the body having an elongated cucumber shape (Fig. 19–35). This shape forces the animal to lie with the side of the body, rather than the oral pole, against the substratum. The class is further distinguished from other echinoderms by the reduction of the skeleton to microscopic ossicles and by the modification of the buccal podia into a circle of tentacles around the mouth.

External Anatomy. Most sea cucumbers are colored black, brown, or olive green; but there are some rose, orange, or violet species, and striped patterns are not infrequent. There is considerable range in size. The smallest species are less than 3 cm. in length (oral to aboral end), while *Stichopus* from the Philippines may attain a length of a meter and a diameter of 24 cm. Most of the common North American and European species, such as *Cucumaria, Holothuria, Thyone,* and *Leptosynapta,* range from 10 to 30 cm. in length.

The body shape varies from almost spherical, as in *Sphaerothuria,* to long and worm-

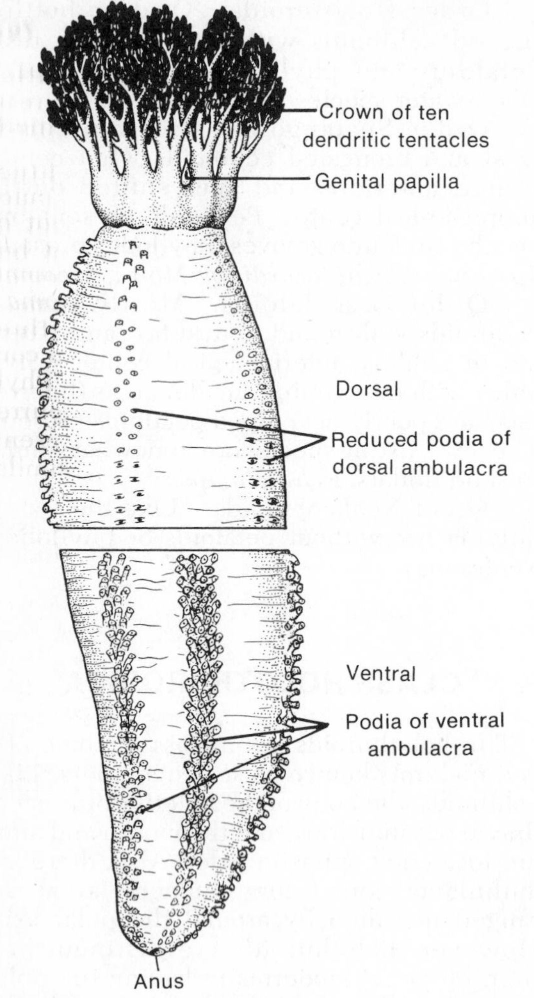

Figure 19–35. The North Atlantic sea cucumber, *Cucumaria frondosa.*

like, as in *Synapta, Leptosynapta,* and *Euapta* (Figs. 19–36*A* and *B;* and 19–37*A*). Not infrequently, the mouth and anus are displaced from the ends of the long axis of the body. Although there are a few genera, such as *Psolus* and *Theelia*, that are provided with a protective armor of calcareous plates (modified surface ossicles), the body surface in the majority of sea cucumbers is leathery as the result of reduction of the skeletal ossicles to microscopic size.

Holothuroids lie with one side of the body against the substratum, and this ventral surface is composed of three ambulacral areas (the trivium), commonly called the sole (Fig. 19–35). The dorsal surface consists of two ambulacral areas. As might be expected, there has been a tendency for the dorsal and ventral surfaces to become differentiated, thus producing a secondary bilateral symmetry. Note that the bilateral symmetry of holothuroids has evolved in an entirely different manner from that of the irregular echinoids.

The degree of differentiation between the dorsal and ventral surfaces varies considerably and is most easily seen in the nature of the regular podia. In the common genera, *Thyone* and *Cucumaria*, podia are present on both surfaces of the body, but suckers are best developed on the podia of the sole (Fig. 19–35). In other species, such as *Holothuria*, the dorsal and lateral podia are reduced to warts, tubercles, or papillae. In still others, such as *Psolus*, dorsal and lateral podia have completely disappeared. The functional podia are thus restricted to the creeping sole.

There is a general tendency for the podia of sea cucumbers, whether reduced or not, to lose their radial distribution and to become more or less randomly scattered over the body surface. The primitive radial configuration is seen in *Cucumaria*, on which the podia are more or less restricted to the five ambulacral areas; in *Thyone*, on the other hand, podia are scattered over the entire body surface. Members of the order Apodida (*Synapta, Leptosynapta, Euapta*), which are elongated and wormlike, and the order Molpadiida, completely lack podia on any part of the body.

The mouth is always surrounded by 10 to 30 tentacles, which represent modified buccal podia and are thus part of the water-vascular system. The tentacles are highly retractile, and both mouth and tentacles can be completely retracted by pulling the adjacent body wall over them. The form of the tentacles varies considerably. They may be long and irregularly branched, as in *Cucumaria, Thyone*, and *Psolus*, or the branches may be horizontal and restricted to the end of a basal stalk, like a mop (Fig. 19–35). In *Synapta* and *Leptosynapta*, the tentacles consist of a long central axis with regular side branches (Fig. 19–36*A*), and in the Molpadiida they are short and finger-like.

Body Wall. The epidermis is composed of nonciliated columnar epithelium that is covered externally by a thin cuticle. The basal ends of the epidermal cells merge with the dermis, which consists of a thick layer of connective tissue surrounding the microscopic calcareous ossicles. The ossicles display a great variety of shapes (Fig. 19–36*E*). These different shapes are important in the taxonomy of holothuroids. Beneath the epidermis is a layer of circular muscle that overlies five single or double bands of longitudinal fibers located in the ambulacral areas.

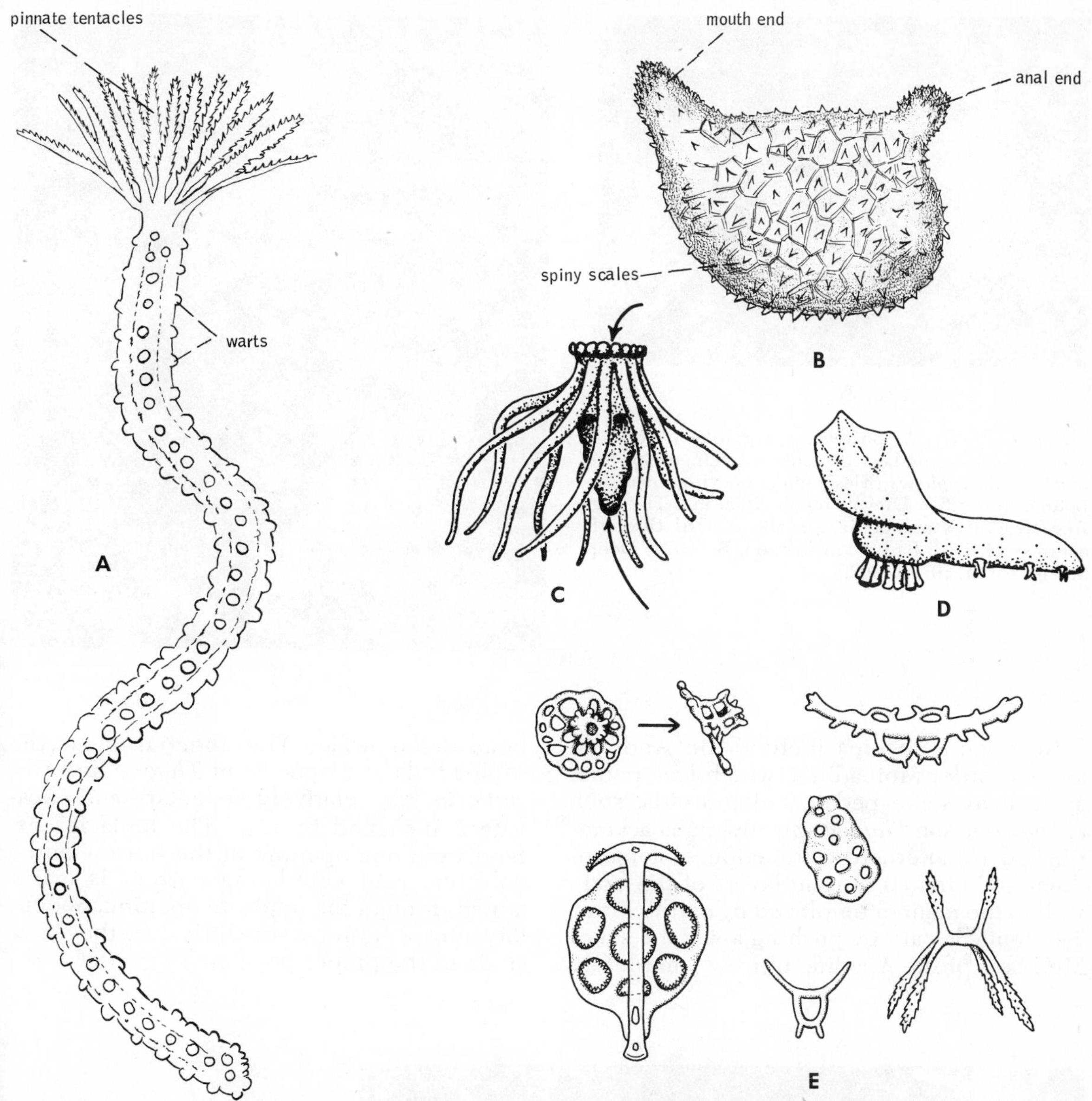

Figure 19–36. *A. Euapta lappa*, a member of the order Apodida from the Bahamas. (After Hyman.) *B. Sphaerothuria*, a short-bodied holothuroid with scaly armor. (After Ludwig from Hyman.) *C.* The pelagic elasipod sea cucumber, *Pelagothuria*, showing circlet of long webbed papillae behind mouth and tentacles. *D.* Another elasipod sea cucumber, *Peniagone*, possessing a dorsal sail. (*C* and *D* after Nichols.) *E.* Microscopic ossicles from sea cucumbers. (After Bell.)

The body wall of sea cucumbers (called trepang) is a culinary delicacy in the Orient. Large species of sea cucumbers are collected and boiled, which causes their bodies to contract and thicken and also brings about evisceration of the internal organs. The body wall is then dried and sold, mostly to Chinese, as trepang or bêche-de-mer. Trepang, when cooked in certain dishes, imparts a distinctive flavor to the food.

Locomotion. Sea cucumbers are relatively sluggish animals and live on the bottom surface or burrow in sand and mud. Forms with podia may creep along on the sole, with the podia functioning as those in asteroids; such epibenthic forms include species of *Cucumaria*, *Holothuria*, *Stichopus*, and *Psolus*. Righting is accomplished by twisting the oral end around until the podia contact the substratum. Many hard-bottom forms live beneath stones; others, such as species of the large, tough *Stichopus*, crawl exposed on the surface (Fig. 19–38). There are also species which live on algae.

A

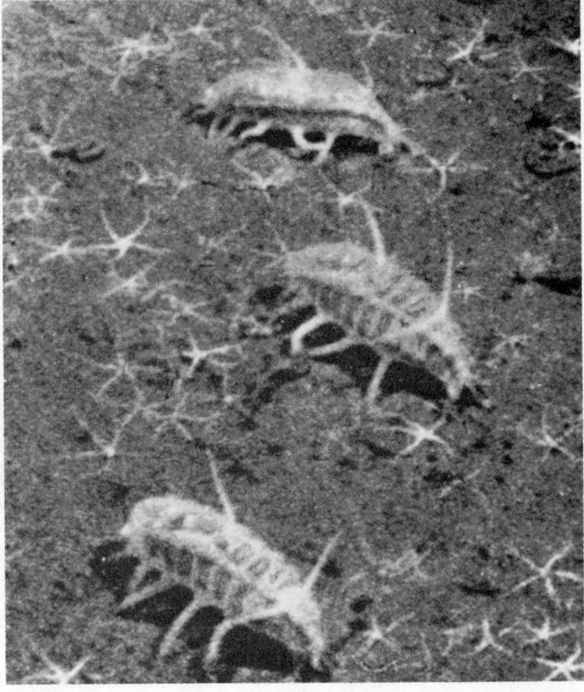

B

Figure 19–37. *A.* Burrowing wormlike (Apodida) synaptid sea cucumbers. *B.* Three specimens of the deep-sea *Scotoplanes* (Elasipodida) crawling over the bottom in the San Diego Trough (1060 m.). The leglike structures are podia. (By R. F. Dill through courtesy of E. G. Barham in Hansen, B.: 1972: Deep Sea Research, 19:461–462.)

Burrowing species include the Apodida and the order Molpadiida, which lack podia, as well as some pedate holothuroids, such as the common *Thyone*. Burrowing is accomplished by alternate contraction of longitudinal and circular muscle layers of the body wall in the manner employed by earthworms. The tentacles aid by pushing away the sand. Members of the Apodida burrow completely beneath the surface. The other burrowers, the Molpadiida and species of *Thyone* and *Cucumaria*, are relatively sedentary and excavate a **U**-shaped burrow. The tentacles extend from one opening of the burrow and a pulsating anal ventilating current is maintained through the opposite opening. Sedentary burrowers move very little once they have attained the proper position.

Figure 19–38. The north Pacific *Stichopus californicus* in a tank with bat stars, *Patiria miniata,* both from the west coast of the United States. Species of the widely distributed genus *Stichopus* are large sea cucumbers (over 36 cm.) with tough body walls. (By Betty M. Barnes.)

The Elasipodida are a curious group of largely deep-sea holothuroids, some of which are benthic and some pelagic. The small number of podia are greatly enlarged and may be used by benthic species for walking (Fig. 19–37*B*), or may be webbed together in various ways to form fins or sails. One genus, *Pelagothuria*, that has been collected at the surface, has a circlet of long, webbed papillae just behind the tentacles and probably swims like a jellyfish (Fig. 19–36*C*). *Peniagone* possesses a dorsal sail (Fig. 19–36*D*).

Nutrition. Sea cucumbers are chiefly deposit or suspension feeders. More or less stationary burrowers, such as *Thyone* and *Cucumaria*, or forms which live beneath stones, stretch out their branched tentacles and either sweep them over the bottom or hold them out in the sea water. In either case, particulate material is trapped in mucus on the tentacular surfaces. One at a time, the tentacles are then stuffed into the pharynx, and the adhering food particles are wiped off as the tentacles are pulled out of the mouth. Studies on the suspension-feeding European *Cucumaria elongata* indicate that mucus is obtained by the tentacles when they are placed in the pharynx (Fig. 19–39) (Fish, 1967).

More mobile epibenthic forms, such as the large *Stichopus* and the deep-sea elasipods, are deposit feeders, grazing on the bottom with their tentacles. The sand or mud castings of some species are very conspicuous. The subsurface wormlike Apodida are also deposit feeders; they consume the bottom material through which they burrow. The column of ingested sand in some species is easily visible through the transparent body wall.

The mouth is located in the middle of a buccal membrane at the base of the tentacular crown. The mouth opens into a pharynx, which is surrounded anteriorly by a calcareous ring (Figs. 19–40 to 19–42). The calcareous ring, which is perhaps homologous to the Aristotle's lantern of echinoids, is usually composed of five radial and five interradial rectangular plates, arranged somewhat like the staves in a barrel. The calcareous ring provides support for the pharynx and the water ring, and it serves as the site for the anterior insertion of the longitudinal muscles of the body wall and the retractor muscles of the tentacles and mouth region. The tentacles and mouth can be pulled completely within the anterior end of the body when the animal is disturbed. Protraction is brought about through elevation of the coelomic fluid pressure.

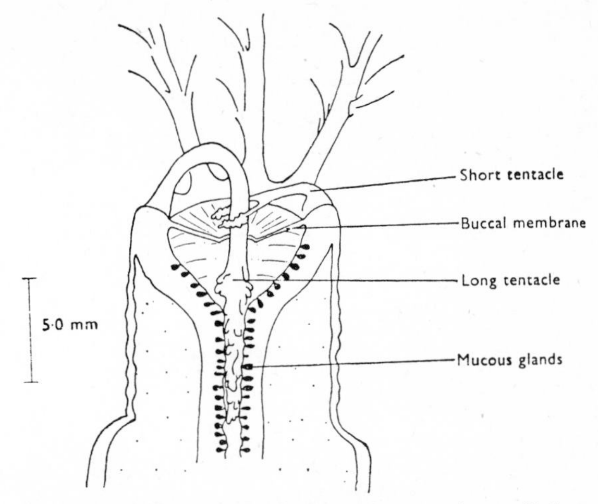

Figure 19–39. *Cucumaria* with tentacle stuffed in pharynx, removing food particles. (From Fish, J. D., 1967: J. Mar. Biol. Assoc. U.K., *47*:129–143.)

In some holothuroids, such as *Thyone*, the pharynx leads into a slender esophagus, which in turn opens into a short muscular stomach (Fig. 19–40). Frequently the esophagus is absent, as in *Cucumaria*, and if the stomach is poorly developed, which is true of *Cucumaria* and *Holothuria*, the pharynx appears to open directly into the intestine (Fig. 19–41). The long intestine is looped three times through the length of the body and is supported by mesenteries. In many holothuroids, the intestine terminates in a cloaca prior to opening to the outside through the anus.

The digestive tract is lined with columnar epithelium that is sometimes ciliated. Gland cells are often prevalent in the pharynx and the stomach but may be absent in the intestine. Beneath the lining epithelium, there is well developed connective tissue and circular and longitudinal muscle fibers.

Coelomocytes supposedly play an active and unusual role in digestion. These cells traverse the intestinal wall carrying digestive enzymes into the intestinal lumen. The products of digestion are then picked up by coelomocytes and transported out of the intestine for general distribution. However, recent studies of several species of sea cucumbers have demonstrated ordinary absorption of sugars by the gut wall in *in vitro* preparations. Protein- and carbohydrate-splitting enzymes are present, but fat digestion is negligible.

Circulation, Gas Exchange, and Excretion. The coelom is large and lined with a thin ciliated epithelium, and the usual divisions are present. The peritoneal cilia produce a current or coelomic fluid that contributes to

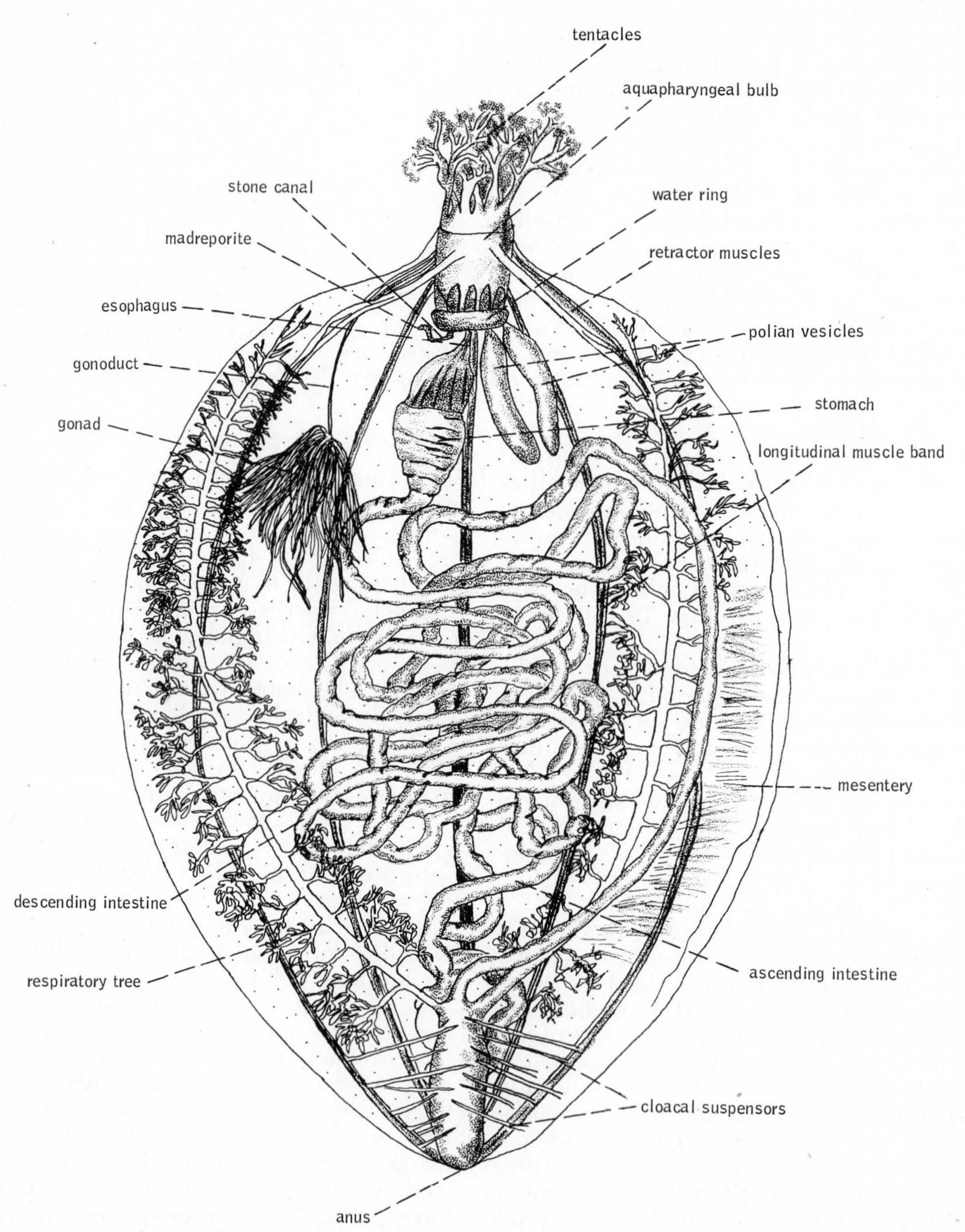

Figure 19–40. Internal structure of *Thyone briaereus*, a common sea cucumber that inhabits North Atlantic coastal waters. (After Coe from Hyman.)

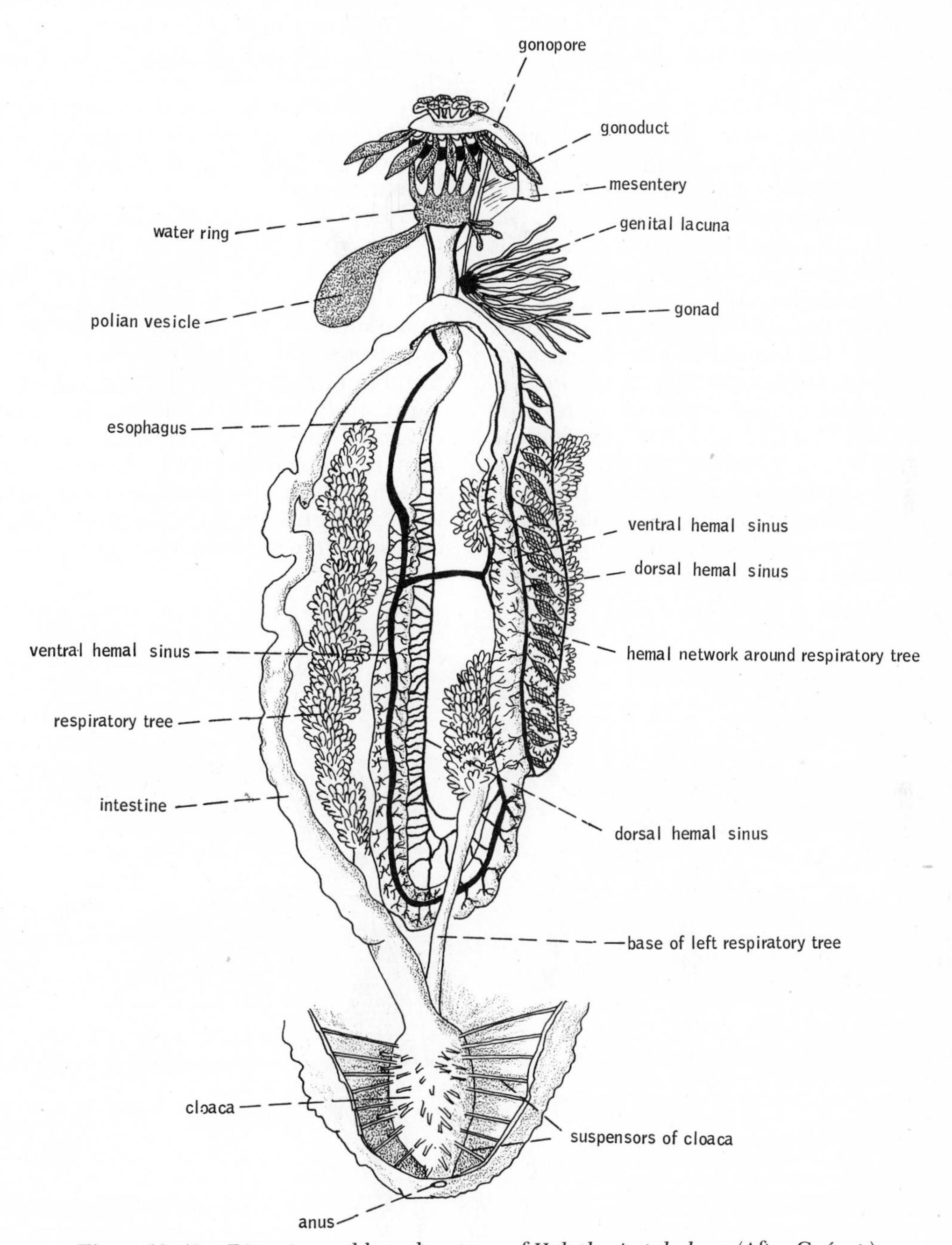

Figure 19–41. Digestive and hemal systems of *Holothuria tubulosa*. (After Cuénot.)

the general circulation of materials within the body.

Of the several types of coelomocytes, one flattened discoidal type, called a hemocyte, contains hemoglobin. When they are present in large numbers, hemocytes give a red color to the coelomic fluid and hemal fluid.

Holothuroids also possess the most highly developed hemal system of any of the echinoderm classes. The general organization of the system is essentially like that of other echinoderms. A hemal ring and radial hemal sinuses parallel the water ring and the radial canals of the water-vascular system (Figs. 19–41 to 19–43). The most conspicuous features of the system, at least in larger species, are a dorsal and a ventral sinus that accompany the intestine (Fig. 19–41). These main intestinal sinuses supply the intestinal wall with a large number of smaller channels. Between the arms of the first great loop of the intestine, formed by what may be called the descending and the ascending intestine, the dorsal sinus gives rise to a complex interconnecting network of channels containing many lacunae.

As in other echinoderms, the hemal channels do not have distinct linings, although muscle fibers and connective tissue are present. In the intestine, the sinuses are merely spaces in the connective-tissue layer. The hemal fluid is essentially the same as that in the coelomocytes; in fact, the coelomocytes are formed in the walls of the hemal channels. The primary role of the hemal system is apparently the distribution of food materials that are brought into the system by the amebocytes. The dorsal sinus is contractile, and fluid is pumped through the intestinal sinuses into the ventral sinus and anteriorly into the hemal ring.

Except for the pelagic Elasipodida and burrowing Apodida, which obtain oxygen through the general body surface, gas exchange in holothuroids is accomplished by means of a remarkable system of tubules called respiratory trees. The respiratory trees are two in number and are located in the coelom on the right and left sides of the digestive tract (Figs. 19–40 and 19–41). Each tree consists of a main trunk with many branches, each of which ends in a tiny vesicle. The trunks of the two trees emerge from the upper end of the cloaca either separately or by way of a common trunk.

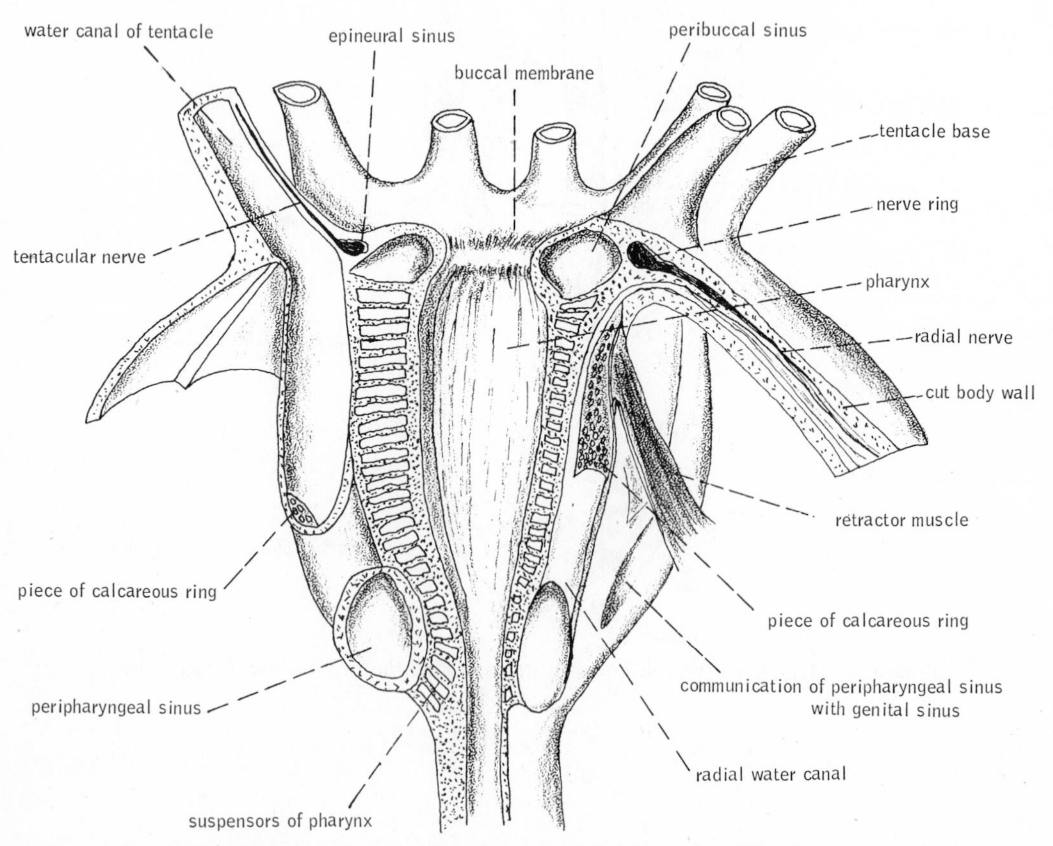

Figure 19–42. Anterior of *Cucumaria*. (After Hérouard from Hyman.)

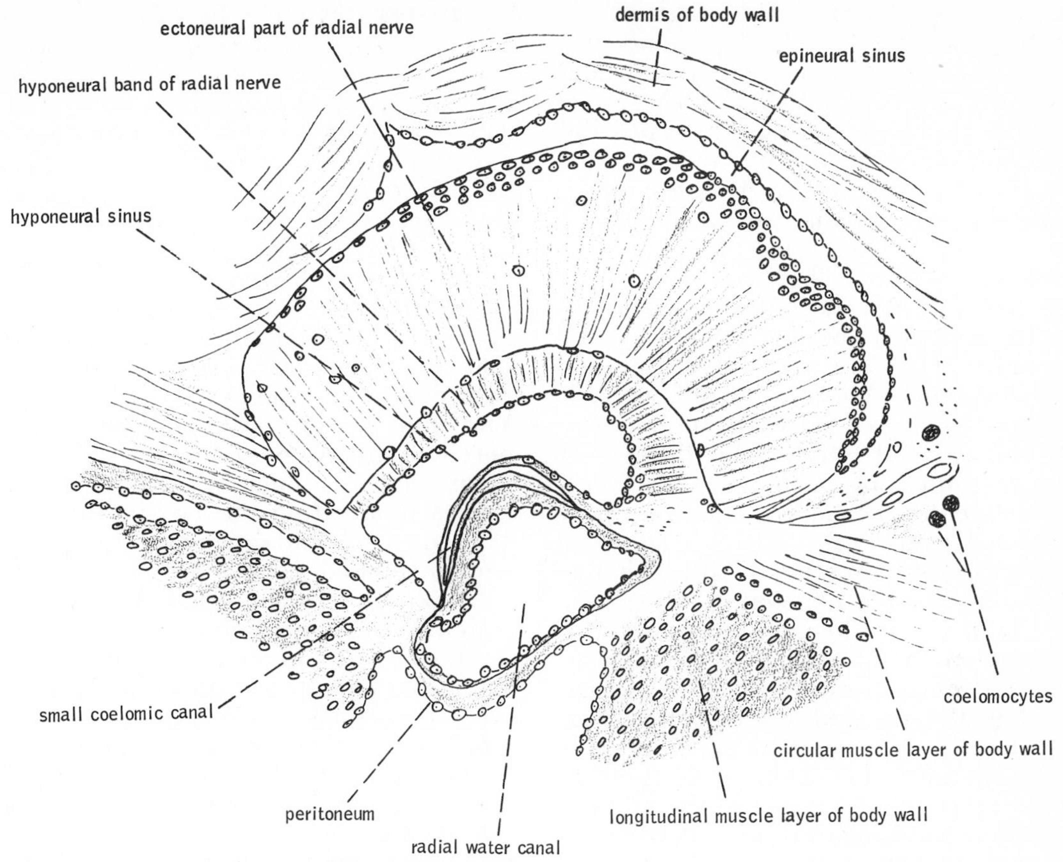

Figure 19–43. An ambulacrum of *Caudina* (transverse section). (After Danielsson and Koren from Hyman.)

Water circulates through the tubules by means of the pumping action of the cloaca and the respiratory trees. The cloaca dilates, filling with sea water (Fig. 19–44). The anal sphincter then closes, the cloaca contracts, and water is forced into the respiratory trees. Water leaves the system because of the contraction of the tubules and the reverse action of the cloaca. Pumping is slow; *Holothuria* requires six to ten cloacal dilations and contractions to fill the trees, each contraction taking one minute or more. All the water is expelled in one action.

The branches of the left tree are intermingled in the hemal network between the ascending and descending intestine, and experimental evidence indicates that oxygen passes from the terminal vesicles of the left tree into the coelomic fluid and then into the hemal network (Fig. 19–41).

An interesting commensal relationship is that between the slender tropical pearlfish and sea cucumbers. These little fish, which are about 15 cm. long, make their home in the trunk of one of the respiratory trees of certain sea cucumbers. The fish leaves the host at night while it searches for food; after such excursions, the fish forces its way into the anus and cloaca and back to the shelter of the respiratory tree.

Most ammonia probably exits by diffusion through the respiratory trees. Particulate

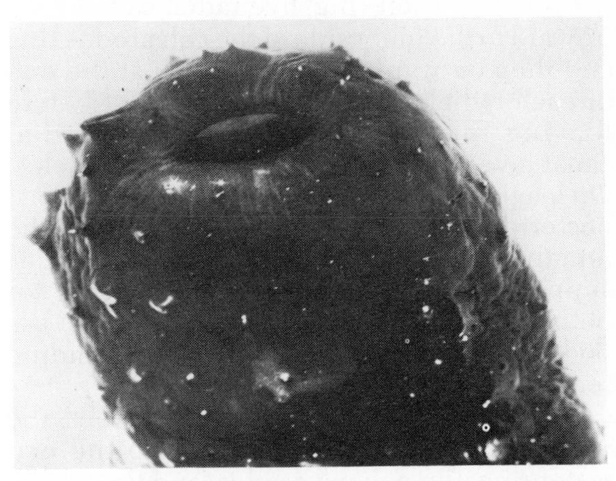

Figure 19–44. Posterior end of *Stichopus*, showing dilated anus during ventilation. (By Betty M. Barnes.)

waste, as well as nitrogenous material in crystalline form, is carried by coelomocytes from various parts of the body to the gonadal tubules, the respiratory tree, and the intestine. Waste then leaves the body by way of the lumens of these organs.

The Apodida, which lack respiratory trees, possess peculiar coelomic organs (ciliated funnels or ciliated urns) that probably have an excretory function. The urns are attached to the bases of the mesenteries or to the coelomic wall near the attachments of the mesenteries. They are relatively numerous and are sometimes clustered together. Each ciliated urn is covered externally with peritoneum and is lined internally with ciliated columnar epithelium. Coelomocytes laden with waste enter the funnels and then pass into the body wall.

Water-Vascular System. Although the water-vascular system of holothuroids is basically like that of other echinoderms, the madreporite in most species is peculiar in having lost connection with the body surface and in being unattached in the coelom (Fig. 19–40). Pores and pore canals are still present in the madreporite, but coelomic fluid rather than sea water enters and leaves the system. The madreporite hangs just beneath the base of the pharynx and is connected to the water ring by a short stone canal.

The water ring encircles the base of the pharynx and gives rise to the elongated or rounded polian vesicles, which hang into the coelom (Figs. 19–40 and 19–41). The number of polian vesicles varies considerably. There may be one (*Cucumaria*), three, four (*Thyone*), or as many as ten or even fifty in certain Apodida. The vesicles are believed to function as expansion chambers in maintaining pressure within the water-vascular system.

From the water ring, five radial canals pass upward to the inner side of the calcareous ring and then outward through a notch at the end of each radial plate (Fig. 19–42). Just before leaving the calcareous ring, each radial canal gives off smaller canals to the tentacles. On leaving the ring, the radial canals then pass posteriorly within the body wall along the length of the ambulacra. Here lateral canals supply the podia. Ampullae are present for both podia and tentacles, although when the podia are reduced, there is a corresponding reduction in the ampullae.

In the Apodida, which lack tube feet, the water-vascular system is limited to the oral water ring, the polian vesicles, and the buccal podia (tentacles).

Nervous System. The circumoral nerve ring lies in the buccal membrane near the base of the tentacles (Fig. 19–42). The ring supplies nerves to the tentacles and also to the pharynx. The five radial nerves, on leaving the ring, pass through the notch in the radial plates of the calcareous ring and run the length of the ambulacra in the coelomic side of the dermis. Like the radial nerves of the ophiuroids, the radial and tentacular nerves of sea cucumbers are ganglionated. Each radial nerve consists of both an ectoneural and hyponeural portion, but the hyponeural portion dwindles away before it reaches the nerve ring (Fig. 19–43). The outer sensory ectoneural band receives nerves from the podia and body wall; the inner motor hyponeural band gives rise to nerves innervating the body-wall muscles. There is a subepidermal nerve plexus and a well-developed nerve plexus in the dermis.

Unlike most other echinoderms, the circumoral nerve ring does not play a dominating role in nervous integration. Early studies on *Thyone* indicated that most of the normal reactions—such as movement, righting, and reactions to light—take place normally in an animal deprived of the anterior part of the body, which contains the nerve ring. Motor control of respiratory pumping is centered in the cloaca and is thus also independent of the nerve ring.

Sensory cells in the epidermis, which are most numerous at the two ends of the animal, compose the sensory system in most of the holothuroids. In the burrowing Apodida, such as *Synapta* and *Leptosynapta*, the warts or tubercles of the body surface bear a cluster of sensory cells surrounded by gland cells. Fibers from the bundle of nerve cells descend as a nerve into a ganglion in the dermal plexus. The exact function of these sensory tubercles is unknown.

The Apodida also possess statocysts. There is one statocyst adjacent to each radial nerve, located near the point at which the nerve leaves the calcareous ring. Each statocyst is a hollow epidermal sphere containing one to many specialized vacuolated cells acting as statoliths (lithocytes) and containing an inorganic material. Although all holothuroids react to light, only a few species possess specialized photoreceptor organs. Photoreceptor organs are found in a few synaptids and consist of a simple cluster of photoreceptor cells with pigment located at the base of the tentacles.

The response of most holothuroids to adverse stimuli is contraction of the entire body or of parts of the body, such as the tentacles. Many sea cucumbers avoid light, and most species are nocturnal, being relatively inactive during the day. The burrowing species, which possess statocysts, exhibit positive geotropism and tend to always maintain the oral end in a downward position.

Evisceration and Regeneration. The expulsion of sticky tubules from the anal region is commonly associated with sea cucumbers, but this defensive phenomenon is actually limited to some species of the genera *Holothuria* and *Actinopyga*. Such sea cucumbers possess from a few to a large mass of white, pink, or red blind tubules (tubules of Cuvier) attached to the base of one (frequently the left) or both of the respiratory trees, or to the common trunk of the two trees (Fig. 19–45). When these sea cucumbers are irritated or attacked by some predator, the anus is directed toward the intruder, the body wall contracts, and by rupture of the cloaca the tubules are shot out of the anus.

During the process of expulsion, each tubule is greatly elongated by water forced into its lumen, and the tubules break free from their attachment to the respiratory tree. In *Holothuria*, the detached tubules are sticky and entangle the intruder in a mesh of adhesive threads. Small crabs and lobsters may be rendered completely helpless and left to die slowly, while the sea cucumber crawls away.

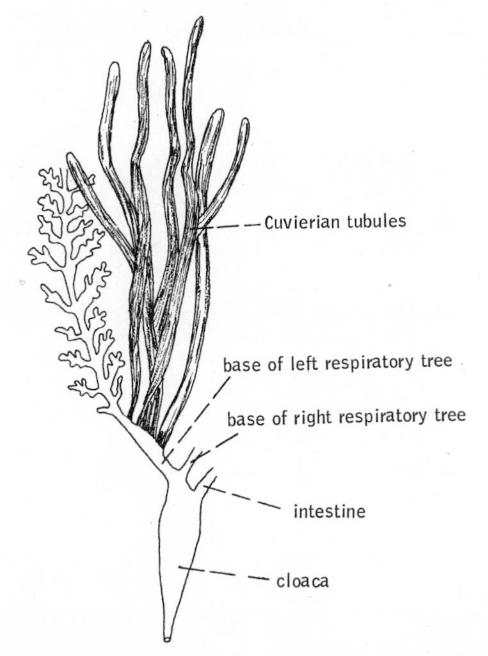

Figure 19–45. Base of a respiratory tree of *Holothuria impatiens*, showing Cuvierian tubules. (After Russo from Hyman.)

After discharge the tubules of Cuvier are regenerated.

Often confused with the discharge of the tubules of Cuvier is a more common phenomenon (called evisceration) that occurs in many holothuroids. Evisceration in the case of some genera, such as *Holothuria*, *Stichopus*, and *Actinopyga*, involves the rupture of the cloaca and the expulsion of one or both respiratory trees, the digestive tract, and the gonads. In *Thyone* and other sea cucumbers, the anterior end ruptures, and the tentacles, pharynx and associated organs, and at least part of the intestine are expelled.

The phenomenon has largely been observed in laboratory specimens that were subjected to crowded conditions, to foul water, to the injection of chemicals into the coelom, or to other abnormal conditions. But eviscerated specimens or individuals in the process of regeneration have been reported from natural habitats during certain times of the year, and it is probable that evisceration is a normal seasonal phenomenon in some species (Swan, 1961). Evisceration is followed by regeneration of the lost part; the remaining stubs of the eviscerated organs or the associated mesenteries are the sites of the initial regenerative growth.

In the case of more radical loss of body parts, holothuroids differ considerably in their regenerative powers. In many forms, particularly species of *Cucumaria*, *Thyone*, and *Holothuria*, the cloacal region is the center of regeneration. In transversely bisected animals, each half regenerates, but if the animal is cut into numerous small sections, only the terminal piece, containing the cloaca, regenerates. In the burrowing synaptids, on the other hand, only the section of the body containing the anterior end is capable of regeneration, and all posterior pieces eventually die.

Reproduction. Holothuroids differ from all other living echinoderms in possessing a single gonad, a condition that is believed to be primitive. Most cucumbers are dioecious, and the gonad is located anteriorly in the coelom beneath the middorsal interambulacrum (Figs. 19–40 and 19–41). There are a few protandric hermaphrodites. Typically the gonad is composed of a large cluster of simple or branched tubules joining together at the end of the gonoduct and attached to the left side of the dorsal mesentery suspending the anterior region of the gut. The gonoduct plus the tubules thus resemble a mop. Not infrequently, the tubules are divided into two clusters, one on each side of the dorsal mesentery; in the synaptids, the gonad is composed of only two branched tubules. The tubule walls are lined with germinal epithelium, contain muscle fibers, and are covered externally with peritoneum. The gonoduct has a ciliated lumen and runs anteriorly in the mesentery to the gonopore, which is located middorsally between the bases of two tentacles or just behind the tentacular collar (Fig. 19–35).

Some thirty brooding species are known, over half of which are cold-water forms, largely Antarctic. During spawning, the eggs are caught by the tentacles and transferred to the sole or to the dorsal body surface for incubation. Frequently these two areas contain special brooding pockets for retaining the eggs; in a few species, such pockets have become so invaginated that they are actually internal. Even more remarkable is coelomic incubation, which takes place in the Californian *Thyone rubra*, in *Leptosynapta* from the North Sea, and in a few species from other parts of the world. The eggs pass from the gonads into the coelom and are fertilized in an undiscovered manner. Development takes place within the coelom, and the young leave the body of the mother by way of a rupture in the anal region. A deep-sea elasipodan is known to brood its young in the ovary.

Embryogeny. Except in brooding species, development takes place externally in the sea water, and the embryo is planktonic. Development through gastrulation is like that of the asteroids. The anterior half of the archenteron separates to develop as the coelom, leaving a shorter posterior portion to become the gut.

By the third day of development, a larval stage called an auricularia has been reached (Fig. 19–46*A*). The auricularia is very similar to the bipinnaria of the asteroids, and possesses a ciliated locomotor band that conforms to the same development as the locomotor band of the bipinnaria. The auricularia of most species is 1 mm. or less in length, but giant 15 mm. larvae of unknown adults have been collected in plankton off Japan, Bermuda, and the Canary Islands.

Further development leads to a barrel-shaped larva, called a doliolaria, in which the original ciliated band has become broken up into three to five flagellated girdles (Fig. 19–46*B*).

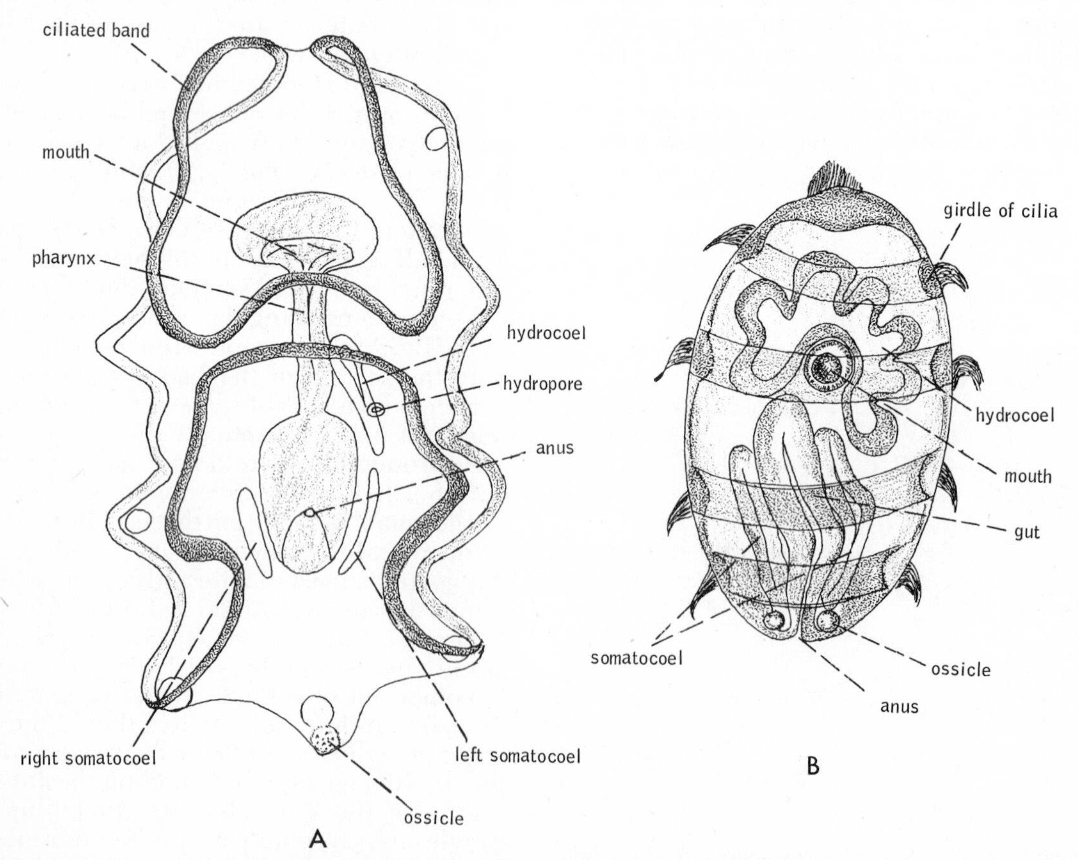

Figure 19–46. *A*. An auricularia larva (oral view). (After Mortensen from Hyman.) *B*. The doliolaria larva of *Leptosynapta inhaerens*, a common North Atlantic holothuroid (oral view). (After Runnström from Cuénot.)

There are some species of holothuroids (Dendrochirotida) that possess a non-feeding, barrel-shaped vitellaria. This type of larva, which is found in crinoids and a few ophuiroids, possesses ciliated bands but no arms. Fell (1966) believes the vitellaria to be the most generalized type of echinoderm larva. Gradual metamorphosis, forming a young sea cucumber, takes place during the latter part of planktonic existence. The tentacles, which are equivalent to buccal podia, appear prior to the appearance of the functional podia. At this stage, the metamorphosing animal is sometimes called a pentactula larva. Eventually the young sea cucumber settles to the bottom and assumes the adult mode of existence.

SYSTEMATIC RÉSUMÉ OF CLASS HOLOTHUROIDEA

Order Dactylochirotida. Primitive sea cucumbers, in which the tentacles are simple and the body is enclosed within a flexible test. *Sphaerothuria, Echinocucumis.*

Order Dendrochirotida. Buccal podia, or tentacles, dendritic and not provided with ampullae. Podia occurring on the sole, on all the ambulacra, or over the entire surface. *Cucumaria, Thyone, Psolus.*

Order Aspidochirotida. Tentacles peltate, or leaflike. Podia present, sometimes forming a well-developed sole. *Holothuria, Actinopyga, Stichopus.*

Order Elasipodida. Aberrant sea cucumbers with large conical papillae and other appendages. Tentacles leaflike. Almost all are deep-sea species. *Pelagothuria, Peniagone.*

Order Molpadiida. Posterior end of body narrowed to a tail. Fifteen digitate tentacles, but regular podia absent. *Molpadia, Caudina.*

Order Apodida. Wormlike sea cucumbers with only buccal podia, or tentacles, present. *Leptosynapta, Synapta,* and *Euapta.*

CLASS CRINOIDEA

The crinoids are the most ancient and in some respects the most primitive of the living classes of echinoderms. Attached stalked crinoids, called sea lilies, flourished during the Paleozoic era, and some 80 species still exist today. The majority of living crinoids, however, belong to a more modern branch of the class, the suborder Comatulida. The comatulids, or feather stars, are nonsessile, free-swimming crinoids. There are approximately 550 species centered primarily in Indo-Pacific waters.

External Structure. The body of existing crinoids is composed of a basal attachment stalk and a pentamerous body proper, called the crown (Fig. 19–47). A well-developed stalk is present in sea lilies but has largely disappeared in the free-swimming feather stars (Fig. 19–47*B*). In the sessile sea lilies, the stalk may reach 72 cm. in length but is usually much shorter. However, there are fossil species with 25 m. stalks. The basal end bears a flattened disc or, in some cases, rootlike extensions by which the animal is fixed to the substratum.

The internal skeletal ossicles give the stalk a characteristic jointed appearance. The stalk of many crinoids bears small, slender, jointed appendages (cirri) that are displayed in whorls around the stalk (Fig. 19–48*A*). Although the stalk is vestigial in comatulids, the most distal cirri of the stalk remain, and spring as one or more circles from around the base of the crown (Fig. 19–47*B*). The cirri of comatulids are used for grasping the substratum when the animal comes to rest. They are long and slender in forms that rest on soft bottoms, and stout and curved in species that grasp rocks, seaweed, and other objects.

The pentamerous body (the crown) is equivalent to the body of other echinoderms and, like those of the asteroids and the ophiuroids, is drawn out into arms. The crown is attached to the stalk by its aboral side; thus, in contrast to other living echinoderms, the oral surface is directed upward. The skeletal ossicles are best developed in the aboral body wall, usually called the calyx, which is thus somewhat cuplike. The oral wall (tegmen) forms a more or less membranous covering for the calyx cup, although it may contain imbedded calcareous plates (Figs. 19–47A and 19–48*B*). The mouth is located in the center of the oral surface; five ambulacral grooves extend peripherally from the mouth to the arms. The tegmen is thus divided into five ambulacral and five interambulacral areas. The anus opens onto the oral surface and is usually located in one of the interambulacral areas at the top of a prominence called the anal cone (Fig. 19–48*B*).

The arms issue from the periphery of the crown and have a jointed appearance like the stalk. Although there are some primitive species that possess five arms, in most crinoids each arm forks immediately upon leaving the crown, forming a total of ten arms (Fig. 19–

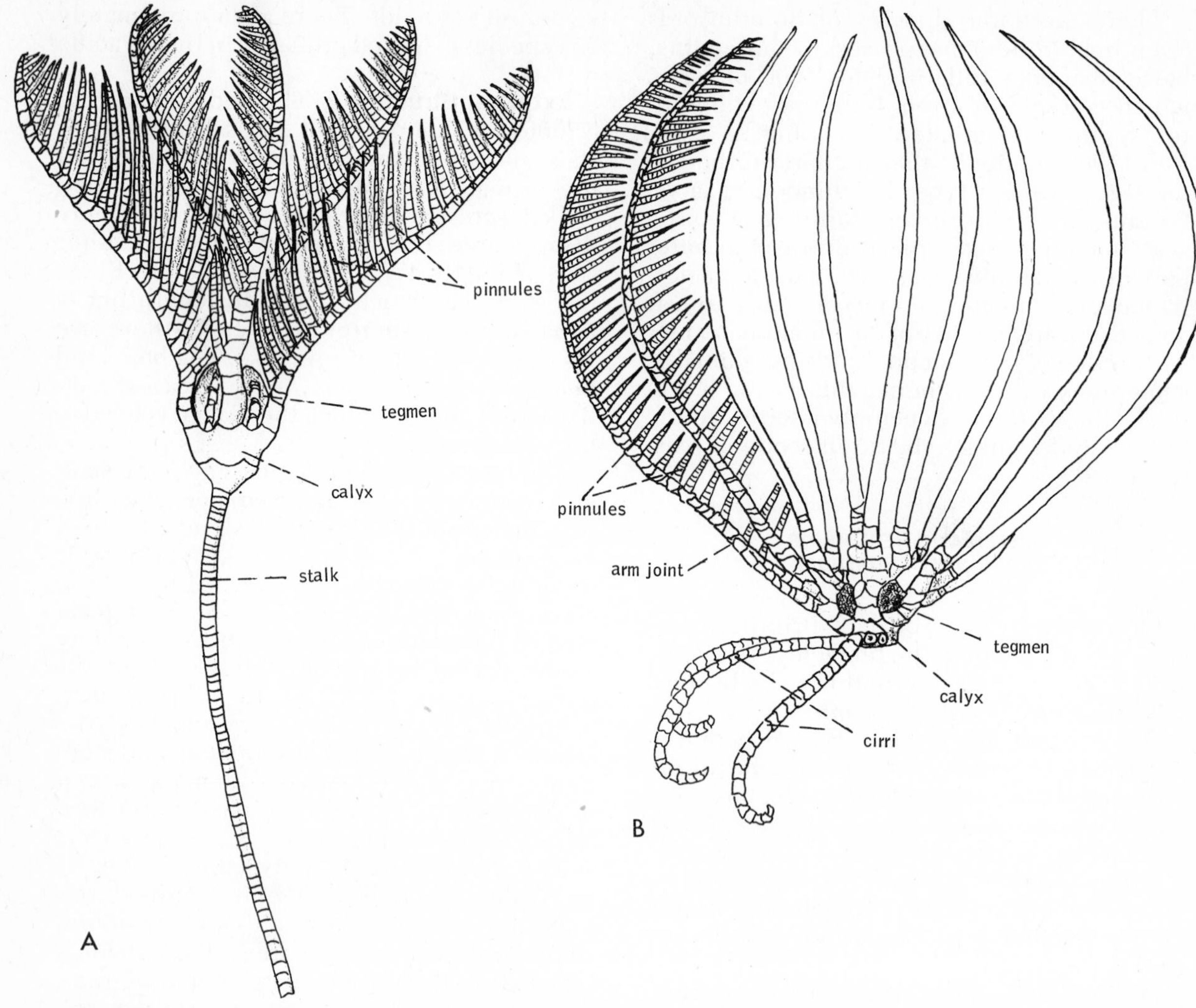

Figure 19–47. *A. Ptilocrinus pinnatus*, a stalked crinoid (or sea lily) with five arms. *B.* A Philippine thirty-armed comatulid (or feather star), *Neometra acanthaster.* (Both after Clark from Hyman.)

47*A*). However, some comatulids possess 80 to 200 arms resulting from the repeated forking of the original arm. The arms are usually only a few centimeters in length but may reach almost 35 cm. in some species.

On each side of the arm is a row of jointed appendages called pinnules, from which the name feather star is derived (Figs. 19–47*A* and 19–48*C*). The ambulacral grooves on the oral surface extend along the length of both the arms and the pinnules. The margins of the grooves are bordered by movable plates, called lappets, which can expose or cover the groove. On the inner side of each lappet are three podia united at the base (Figs. 19–48*C* and 19–49*A*). Both podia and lappets also extend onto the pinnules. In the mouth region the podia, called oral podia, appear singly. Also paralleling the ambulacral grooves, but located to the outer side of the lappets, are small spherical surface bodies called saccules (Fig. 19–49*A*).

Cold-water crinoids and those of the eastern Pacific are usually straw-colored, but littoral species from tropical waters, especially comatulids, display a variety of colors in both solid and variegated patterns.

Body Wall. In the region of the ambulacral grooves, the epidermis is composed of ciliated columnar epithelium, but elsewhere the epidermis is for the most part a thin, non-ciliated syncytium that is poorly demarcated from the underlying dermis and may be lacking altogether (Fig. 19–49*B*). Most of the dermis is occupied by the skeletal ossicles. The stalk, cirri, arm, and pinnules are of a solid construction, being composed almost entirely of a series of thick disc-shaped ossicles; this accounts for the jointed appearance of these appendages (Figs. 19–47*B* and 19–49*A*).

The oral and peripheral surfaces of the ossicles articulate to permit at least some movement, similar to the ossicles composing the ophiuroid arms. The ossicles of the stalk are more securely interlocked than those of the arms, but even here some bending is possible. In more or less the center of the appendicular ossicles, there is a small canal through which runs a nerve and also, in the case of the stalk and cirri, an extension of the coelom. Also, the arm and the pinnule ossicles each bear a deep groove on the oral surface in which is situated the ciliated ambulacral groove (Figs. 19–48*C* and 19–49*B*).

The ossicles are bound together by distinct connective tissue bands called ligaments. The ligaments must possess some contractile powers, since the stalk and cirri are capable of bending movements. Muscles are present in the arms and pinnules and act

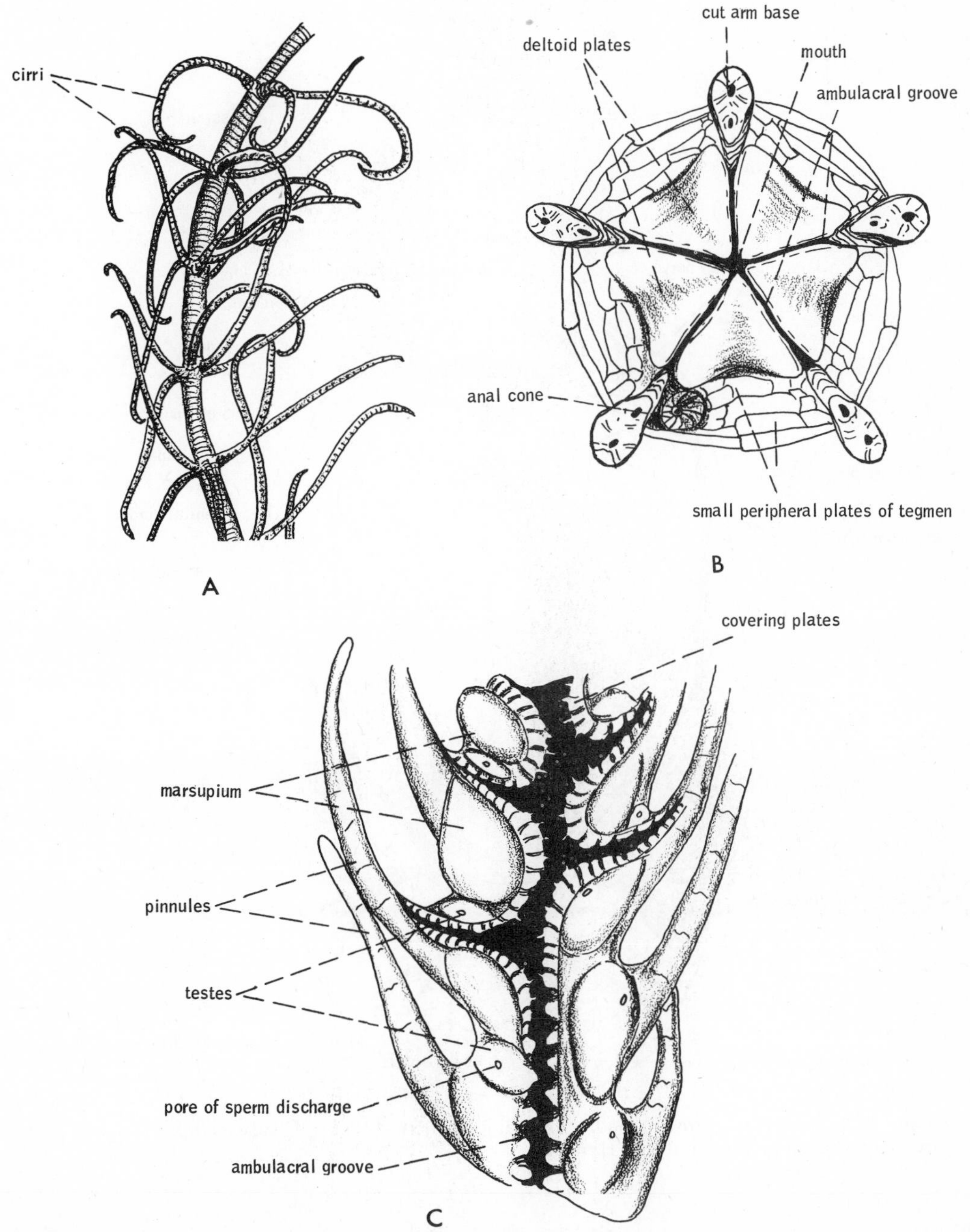

Figure 19–48. *A*. Part of stalk of the West Indian crinoid, *Cenocrinus asteria*, showing whorls of cirri. *B*. Tegmen of *Hyocrinus*, a stalked crinoid (oral view). (*A* and *B* after Carpenter from Hyman.) *C*. An arm section from *Notocrinus virile*, a comatulid (oral view). (After Hyman.)

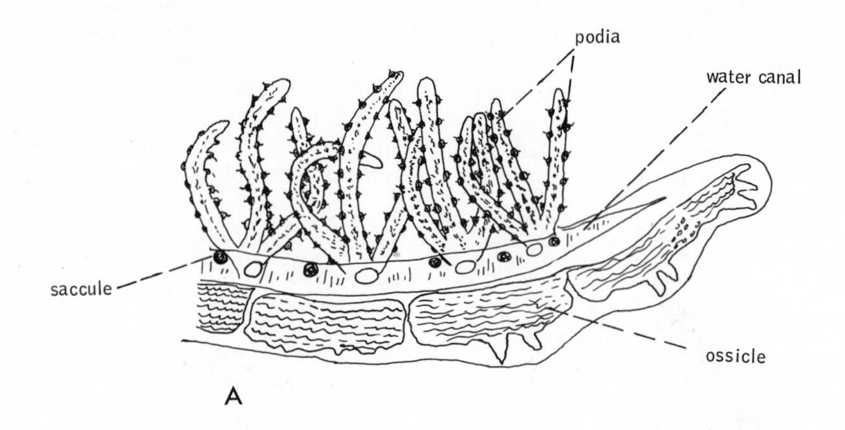

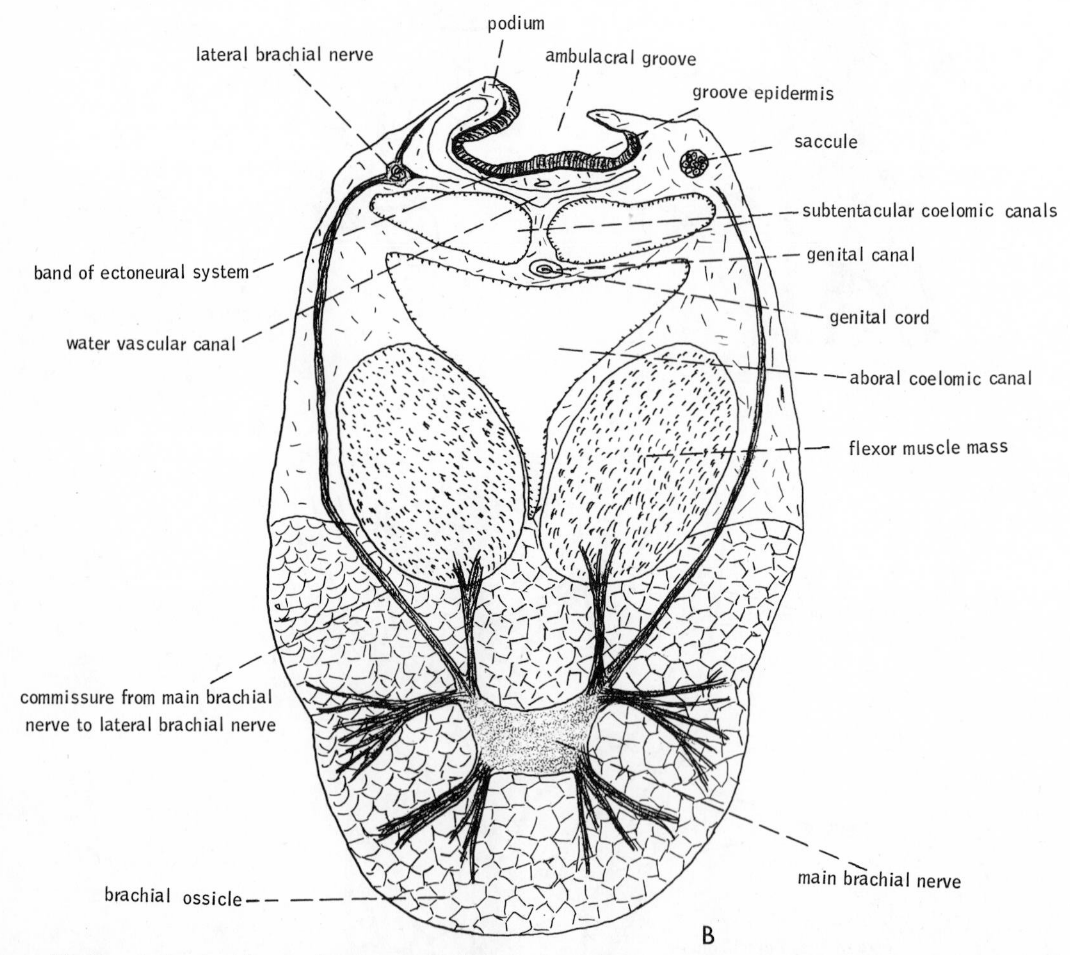

Figure 19–49. *A*. A crinoid arm, showing podia (longitudinal section). (After Chadwick from Hyman.) *B*. A crinoid arm (transverse section). (After Hamann from Hyman.)

antagonistically to the ligaments. The position and elasticity of the ligaments is such that they cause the arms to extend. The pair of large muscles that extend between the ossicles and are located toward the oral side bring about flexion of the arms toward the oral side or the crown (Fig. 19–49*B*).

Locomotion. The sessile sea lilies are limited to bending movements of the stalk and flexion and extension of the arms. The stalkless comatulids, however, are free-moving and are capable of both swimming and crawling. The oral surface is always directed upward, and these animals right themselves if turned over. The cirri are strongly thigmotactic and appear to control the righting reflex. If a comatulid is inverted in such a way that the cirri maintain contact with the substratum, righting does not take place. As soon as the cirri are released, the animal turns over.

Swimming is performed by raising and lowering one set of arms alternately with certain others. In the ten-armed species, every other arm sweeps downward while the alternate set moves upward. In species with more than ten arms, the arms still move in sets of five, but sequentially. For example, in a forty-armed comatulid, there are eight sets of arms acting in sequence. Crawling movements are accomplished by the lifting of the body from the substratum and moving about on the tips of the arms. The arms are often used to grasp and pull the animal over irregular and vertical surfaces. Feather stars swim and crawl only for short distances and sit perched on the bottom for long periods by means of the grasping cirri (Fig. 19–50).

Figure 19–50. Perching feather stars.

Nutrition. Crinoids are suspension feeders. During feeding, the arms and pinnules are held outstretched and the podia are erect. The podia are shaped like small tentacles and bear long, slender, mucus-secreting papillae along their length (Fig. 19–49*A*). Planktonic organisms or detritus that come in contact with the podia are trapped in mucus and tossed into the ambulacral grooves by a sudden whiplike action of the podia. The cilia of the ambulacral grooves beat in an oral direction, carrying food down the arms, across the disc, and into the mouth. Studies on some Red Sea crinoids disclosed that the food was largely zooplankton in the size range of several hundred microns (Rutman and Fishelson, 1969).

There is some correlation between the number and length of the arms and the food supply of the habitat. Crinoids living at great depths or in cold water, where detritus or plankton is rich, usually have a small number of arms (ten or less); the reverse is true of littoral warm-water species. The total length of food-trapping ambulacral surface may be enormous. The Japanese stalked crinoid *Metacrinus rotundus*, with 56 arms 24 cm. in length, possesses a total ambulacral groove length of 80 meters.

Crinoids are believed to display the primitive method of feeding used by the echinoderms and also to illustrate the original function of the water-vascular system—that is, the water-vascular system originally evolved as a means of capturing food and secondarily assumed a locomotor function in those groups that have become free-moving and inverted.

The mouth leads into the short esophagus, which then opens into an intestine (Fig. 19–51). The intestine descends aborally and laterally; when it reaches the calyx, it makes a complete turn around the inner side of the calyx wall. The terminal portion then passes upward into the short rectum, which opens through the anus at the tip of the anal cone. In most crinoids, the intestine bears a number of outpocketings along its inner side. The entire digestive tract, except the terminal portion of the rectum, is lined with cilia, and gland cells are abundant in both the esophagus and the rectum.

The details of digestion are still unknown. Wastes are egested as large compact mucus-

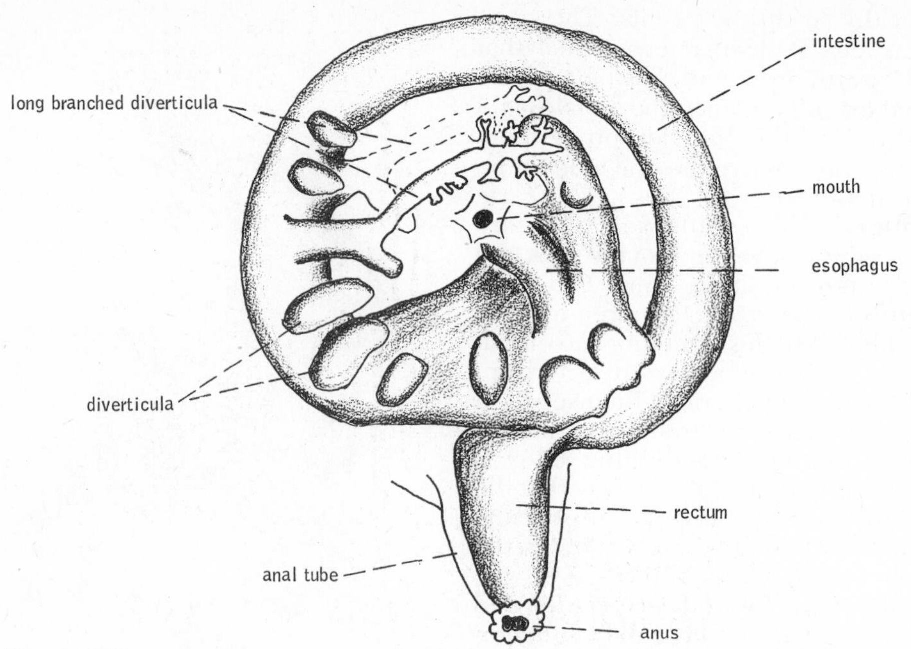

Figure 19–51. Digestive tract of the comatulid, *Antedon*. (After Chadwick from Hyman.)

cemented balls that fall from the anal cone into the surface of the disc and then drop off the body. Water is reportedly pumped into and out of the rectum through the anus, but whether this water circulation is of respiratory significance or is an aid to fecal removal is uncertain.

Circulation, Gas Exchange, and Excretion. The coelom is reduced to a network of communicating spaces as a result of invasion by connective tissue. In the oral side of the arms, the coelom extends as five parallel canals, and in the stalk five coelomic canals pass through the central perforation in the ossicles and give off one canal into each cirrus (Fig. 19–49). Several types of coelomocytes are present and move throughout the tissues of the body, as well as through the fluid of the coelomic, hemal, and water-vascular systems.

The hemal system is a network of spaces and sinuses within the connective tissue strands invading the coelom. There is a definite plexus that surrounds the esophagus, and from this plexus branches extend downward (aborally) through the center of the crown into a spongy mass of cells. This spongy mass is closely associated with the axial gland, which in crinoids is an elongated tubular mesh of glandular tissue occupying the polar axis of the crown. Branches of the hemal system also supply the intestine. There are no hemal spaces in the stalk, but one hemal sinus extends through the arms, along with the coelomic canals, and is involved in the transport of sex cells.

Gas exchange takes place through any thin part of the body surface exposed to sea water. The podia are undoubtedly the primary center of gaseous exchange, and the great surface area presented by the branching arms makes unnecessary any special respiratory surfaces such as are found in some other echinoderms.

Wastes gathered by coelomocytes are believed to be deposited in little saccules located in rows along the sides of the ambulacral grooves. Supposedly, the saccules periodically discharge to the exterior.

Water-Vascular System. There is no madreporite in crinoids. The water ring encircles the mouth and gives off at each interradius a large number of short stone canals, which open into the coelom (the feather star *Antedon* possesses about 50 canals at each inter-radius). At each radius of the ring canal, a radial canal extends into each arm just beneath the ambulacral groove and forks into all of the branches and into the pinnules (Fig. 19–49*B*). From the radial canals extend lateral canals supplying the podia. There are no ampullae, and one lateral canal supplies the cluster of three podia except for the single buccal podia. Hydraulic pressure for extension of the podia is generated by contraction of the radial water canal, which is provided with cross-muscle fibers. Peculiar to crinoids are 500 to 1500 minute ciliated canals (called ciliated funnels), which perforate the wall of the tegmen and open into the underlying coelomic spaces. These openings perhaps compensate for the absence of a madreporite, permitting maintenance of proper fluid pressure within the body and therefore indirectly within the water-vascular system.

Nervous System. In contrast to the nervous system of other echinoderm classes, the chief system in crinoids is an aboral (or entoneural) system located as a cup-shaped mass in the apex of the calyx. From the aboral system, a nervous sheath surrounding the five coelomic canals passes downward through the stalk ossicles and gives off nerves to the cirri along its course. Also arising from the aboral center are five brachial nerves, which after passing through an outer concentric pentagonal nerve ring, proceed through a canal in the arm ossicles and supply each pinnule with a pair of nerves (Fig. 19–49*B*). The brachial nerve also gives off along its course a pair of nerves that innervate the muscles of the ossicles and adjacent epidermis, a pair of nerves innervating the aboral side of the arm, and a pair of commissures that connect to the lateral brachial nerves.

The lateral brachial nerves belong to the hyponeural system and lie along the sides of the arm (Fig. 19–49*B*). These nerves innervate the pinnules, the podia, and the water-vascular canals of the arm. Orally, the five primary pairs of nerves from all of the arms connect to a pentagonal nerve ring that lies just peripheral to the water ring. This ring also supplies some nerves to internal structures in the crown.

Finally, there is an oral (ectoneural) nerve system that is homologous to the principal system of other echinoderms. The crinoid ectoneural system consists of a subepidermal band of nervous tissue that runs just beneath the epidermis of the ambulacral groove in the arms (Fig. 19–49*B*). After crossing the tegmen, the nerve bands descend in the region of the mouth as a nervous sheath encircling the esophagus and the intestine.

Regeneration and Reproduction. Crinoids possess considerable powers of regeneration, and in this respect are quite similar to the asteroids and the ophiuroids. Part or all of an arm can be cast off if seized or if subjected to unfavorable environmental conditions. The lost arm is then regenerated. In like manner, pinnules and cirri are easily regenerated.

In *Antedon*, an animal can replace the loss of one fifth of the disc and the corresponding arm, and can sustain the simultaneous loss of four of the five pairs of arms without death. The aboral nerve center is essential in regeneration, and, provided this tissue remains intact, all of the visceral mass can be regenerated. The coelomocytes play an active role in regeneration. They migrate to the site of the wound in large numbers, carrying food materials and phagocytizing tissue debris.

Crinoids are all dioecious but not sexually dimorphic. Moreover, there are no distinct gonads. The gametes develop from germinal epithelium within an expanded extension of the coelom (the genital canal) located within the pinnules, as in *Antedon*, or within the arms (Fig. 19–49*B*). Not all of the pinnules are involved in the formation of sex cells, but only those along the proximal part of the arm length.

When the eggs or sperm are mature, spawning takes place by rupture of the pinnule walls. In *Antedon* and others, the eggs are cemented onto the outer surface of the pinnules by means of the secretion of epidermal gland cells. Hatching takes place at the larval stage. In other crinoids, the eggs are shed into the sea water. Brooding by cold-water species (all antarctic comatulids) is displayed by the crinoids as in the other echinoderms. The brood chambers are saclike invaginations of the arm or the pinnule walls adjacent to the genital canals, and the eggs probably enter the brood chamber by rupture (Fig. 19–48*C*).

Embryogeny. Development through the early gastrula stage is essentially like that in the asteroids and the holothuroids. The archenteron then divides transversely in the middle to form both an anterior enterohydrocoel and a posterior somatocoel. As in holothuroids, the right hydrocoel and the right axocoel never form, although the somatocoel divides into a right and left portion.

During the formation of the coelomic sacs, the embryo elongates and development proceeds toward a free-swimming larval stage. The crinoid larva, a non-feeding vitellaria, is essentially like the vitellaria of holothuroids, being somewhat barrel-shaped with an anterior apical tuft and a number of transverse ciliated bands (Fig. 19–52*A*). In crinoids, the vitellaria is attained directly without the formation of the intervening bipinnaria-like larva of sea cucumbers. At least in *Antedon*, in which the eggs at spawning adhere to the pinnule surface, hatching takes place at the formation of the vitellaria, which is free-swimming for only a few days at the most.

After a free-swimming existence, the vitel-

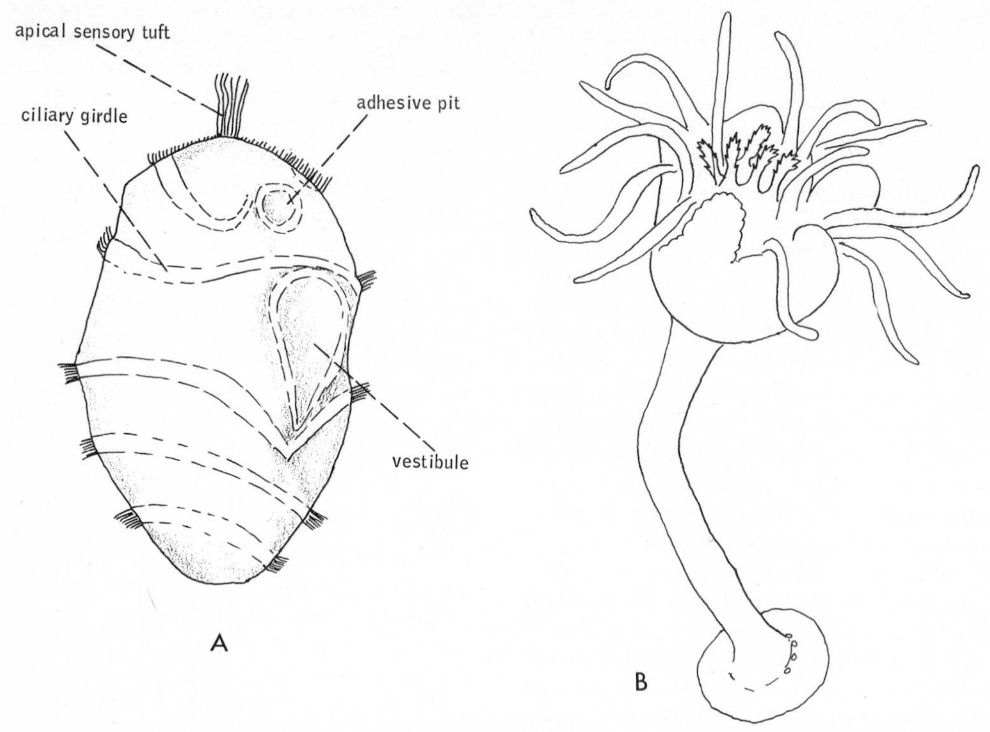

Figure 19–52. *A.* Vitellaria larva of the comatulid, *Antedon mediterranea,* with five ciliary girdles (lateral view). (After Bury from Hyman.) *B.* Pentacrinoid larva of *Antedon.* (After Thomson from Hyman.)

laria settles to the bottom and attaches, employing a glandular midventral depression (the adhesive pit) located near the apical tuft. Then there ensues an extended metamorphosis resulting in the formation of a minute stalked sessile crinoid. In the comatulid, *Antedon,* actually the only crinoid for which development is well known, metamorphosis also results in a stalked sessile stage (the pentacrinoid larva) that resembles a minute sea lily (Fig. 19–52*B*). The pentacrinoid of *Antedon* is a little over 3 mm. long when the arms appear, and it requires about six weeks from the time of attachment of the vitellaria to attain this stage. After several months as a pentacrinoid, during which time the cirri are formed, the crown breaks free from the stalk, and the young animal assumes the adult free-swimming existence.

SYSTEMATIC RÉSUMÉ OF CLASS CRINOIDEA

There are three subclasses of crinoids. They are the extinct subclasses Eocrinoidea and Paracrinoidea, and the subclass Eucrinoidea. The eucrinoids are placed within four orders. Three of these orders, the Inadunata, the Flexibilia, and the Camerata, are composed solely of fossil species. A fourth order, the Articulata, contains all the living stalked crinoids as well as the feather stars. The skeletal structure of the calyx is the principal character on which these divisions are based. A more extensive systematic résumé has therefore not been attempted here.

FOSSIL ECHINODERMS AND ECHINODERM PHYLOGENY

The echinoderms rank with mollusks, brachiopods, and arthropods in having one of the richest and oldest fossil records of any group in the Animal Kingdom. Echinoderms first appeared in the lower Cambrian period and were extremely abundant during the later periods of the Paleozoic era, when a number of fossil classes reached the peak of their evolutionary development.

The crinoids are the only living echinoderms which are attached, but an attached condition was characteristic of the majority of fossil forms, including a number of extinct classes. These groups are particularly interesting in that they display many primitive features of the phylum.

Like living crinoids, most fossil forms were probably suspension feeders in which the podia and the ambulacral grooves were food-catching and food-conducting structures. In most attached fossil species, the peripheral end of the groove was extended upon a slender arm (the brachiole) (Fig. 19–53*B*). The term *brachiole* is used to distinguish these body extensions from the heavier arms of crinoids and stelleroids. Brachioles varied in number from a few to hundreds and undoubtedly evolved as a means of increasing the surface area for food collection. Since these fossil forms were sessile, and since the

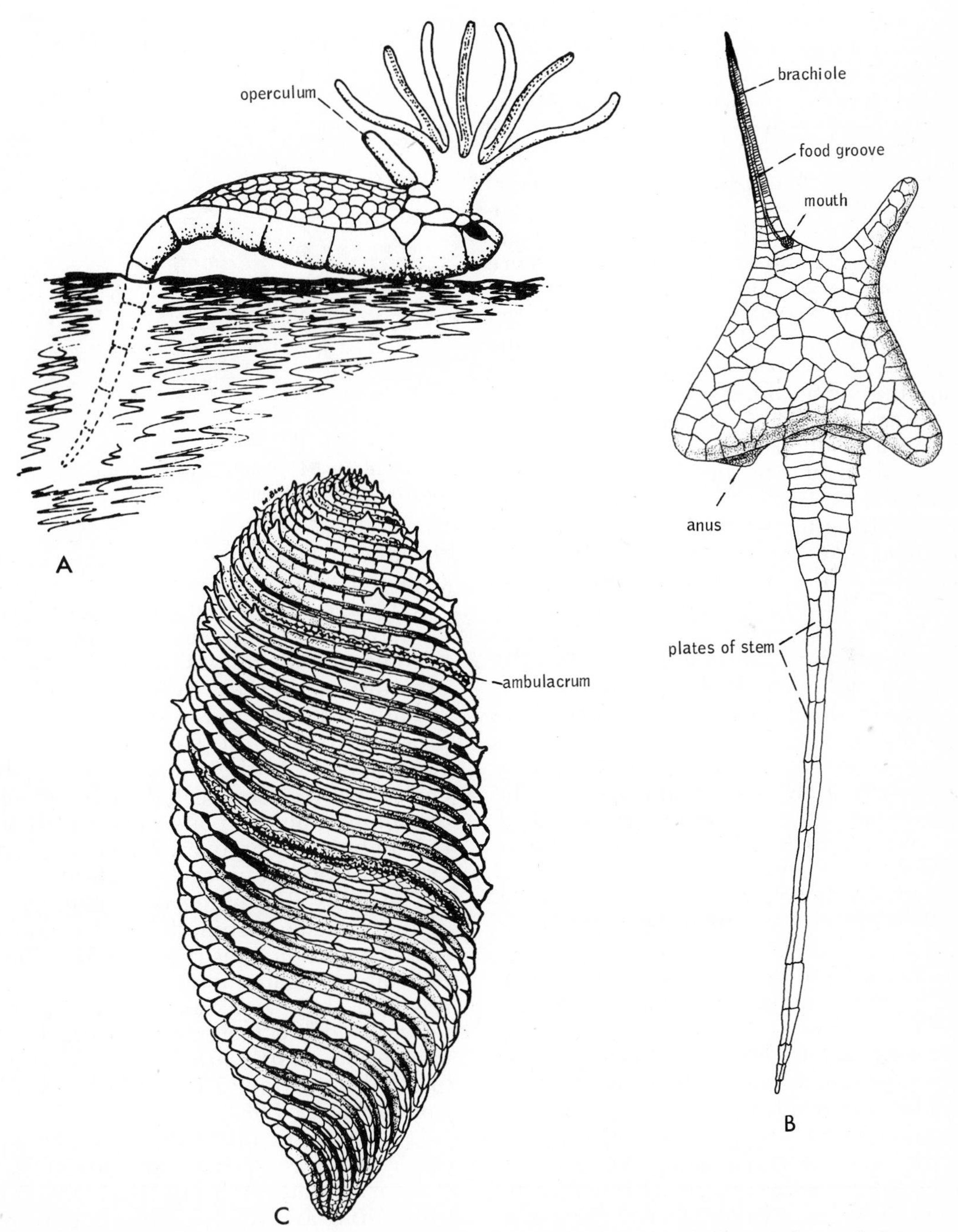

Figure 19–53. *A*. Reconstruction of the carpoid, *Gyrocystis* in the process of feeding. (After Nichols.) *B*. *Dendrocystites*, a later carpoid possessing one brachiole. (Modified from Bather.) *C*. A hypothetical restoration of a partially contracted helicoplacoid, *Helicoplacus*. (After Durham and Caster.)

oral surface was always directed upwards, the original function of the podia could not have been locomotor.

The skeletal system of the extinct attached echinoderms was essentially like that of the crinoids, with the exception that the ossicles (plates) of the crown were not limited to the aboral surface (calyx), but extended orally to the mouth, thus enclosing the internal organs within a test (or theca). The anus was usually located eccentrically in one of the interradii, as in the crinoids.

An early small group of extinct echinoderms were the carpoids, known from the Cambrian to the Devonian. Formerly placed in the class Carpoidea, they are now separated into four classes (p. 792). These animals are of especial interest because they were bilaterally symmetrical, with no indication of radial symmetry (Fig. 19–53*A* and *B*). Some were stalked and sessile, but the body was apparently bent over so that the crown was oriented horizontally to the substratum—that is, one side of the oral-aboral axis faced toward the substratum and the other side faced away from the substratum. The oral surface carried a large and a small opening. The small opening has been thought to be the anus and the larger opening the site of the mouth. It has been suggested that the mouth was surrounded by a lophophore-like crown of tentacles (Fig. 19–53*A*). The tentacular crown could be retracted and the oral opening covered by an operculum. Later carpoids possessed one or two brachioles near the mouth (Fig. 19–53*B*). A groove ran the length of the brachiole.

Another bilateral class of fossil echinoderms is the recently discovered Helicoplacoidea (Durham and Caster, 1963). The helicoplacoids were spindle-shaped animals with the mouth at one end of the body (Fig. 19–53*C*). There was a single branched ambulacrum. The body wall was covered with pleated plates, which allowed expansion and contraction. This flexibility perhaps permitted the animal to retract its body, which may have been situated vertically in the sand. The expanded anterior end projected above the surface of the sand when the animal was feeding.

The class Cystoidea was a more typical group of attached species that ranged from the middle Ordovician period to the Permian. The theca of cystoids was more or less oval, with the oral end directed upward and the aboral end attached to the substratum directly or sometimes by a stem (Figs. 19–54 and 19–62). A characteristic feature of the cystoids was a system of pores that perforated

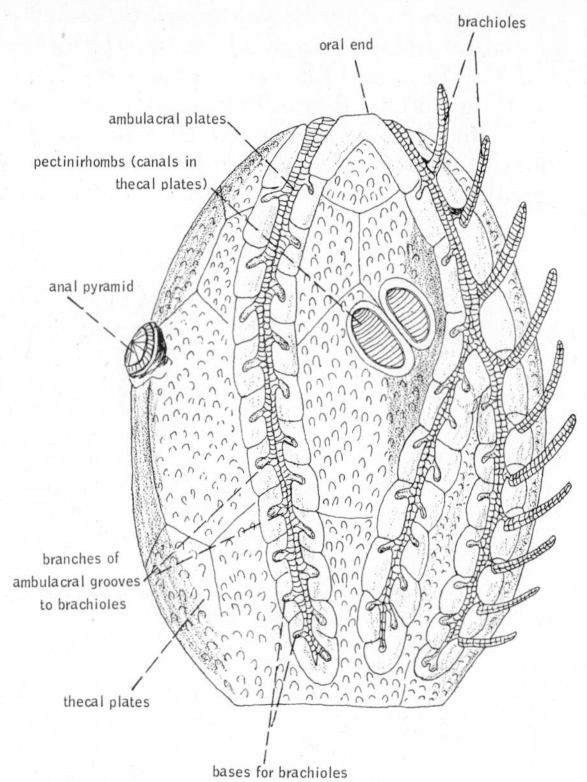

Figure 19–54. Lateral view of the cystoid, *Callocystites.* (After Hyman.)

the theca. The pores were either dispersed over the body surface (diplopores), or confined to special areas (rhombopores). In either case, the pores were perhaps part of a system of internal folds for gas exchange. The three, or more commonly five, ambulacra were radially arranged and extended outward and downward over the sides of the theca, in some species to the aboral pole. Small brachioles were either located around the mouth or mounted on the plates along each side of the ambulacral grooves. Branches from the groove extended up each brachiole.

The class Blastoidea is another group of extinct attached echinoderms that flourished in the Paleozoic seas (Fig. 19–55). The blastoids first appeared in the Ordovician period, reached their peak in the Mississippian period, and became extinct during the Permian period. *Pentremites* is a very common blastoid fossil found in Mississippian limestone (Fig. 19–55*B*).

Like cystoids, blastoids possessed an oval theca that was attached aborally by a short stalk or was attached directly to the substratum. Five ambulacra extended from the mouth at the oral pole down the sides of the theca. Characteristically, however, the am-

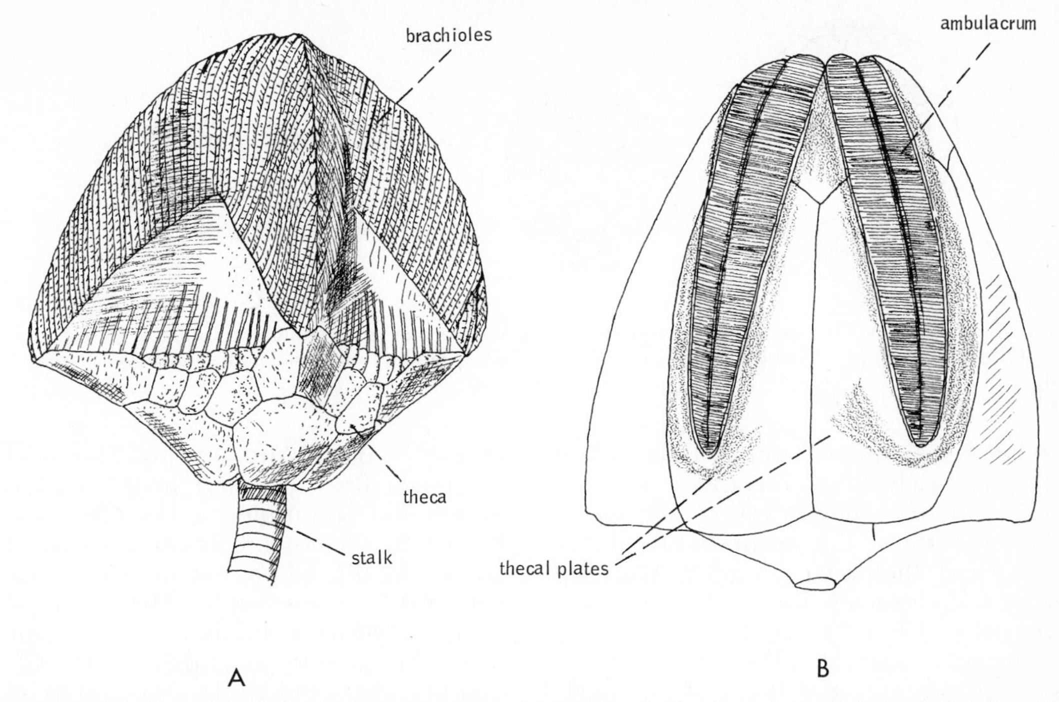

Figure 19–55. Class Blastoidea. *A. Blastoidocrinus,* showing brachioles (lateral view). (After Jaekel from Hyman.) *B. Pentremites* (lateral view, brachioles not shown). (After Hyman.)

bulacral areas were elevated as broad ridges that alternated with the depressed interambulacral areas. The margins of the ambulacra were bordered by a single row of slender closely placed brachioles, onto which extended branches from the ambulacral groove. In oral view, the brachioles gave the appearance of a thick fringe along the border of a five-pointed star.

Peculiar to blastoids was a system of folds, or folds and pores, called hydrospires, located at each side of the ambulacra. The hydrospires are thought to represent a gas exchange mechanism by which water circulated inwardly through the thecal plates, allowing greater surface area for the exchange of gases between the coelom and the sea water. The hydrospires were thus somewhat like reversed asteroid papulae, the folds projecting into the coelom instead of externally from the coelom (Fig. 19–56).

The Edrioasteroidea is a class of extinct attached echinoderms that first appeared in the lower Cambrian period and ranged into the Pennsylvanian period. Although there are only a relatively small number of fossil species, the group is of phylogenetic interest in perhaps being on the evolutionary line leading to some of the unattached echinoderms. The theca was oval and spherical and composed of imbricated, abutting, or fused plates. The presence of tubercles on the plates indicates that these animals possessed spines. There is little to indicate that other extinct attached echinoderms had spines. Some species had stalks (Fig. 19–57*A*) but others were apparently unattached and could move about on the aboral surface and temporarily attach themselves by means of an aboral sucker (Fig. 19–57*B*). There were no brachioles present, and the five ambulacra, which extended over the theca, were straight

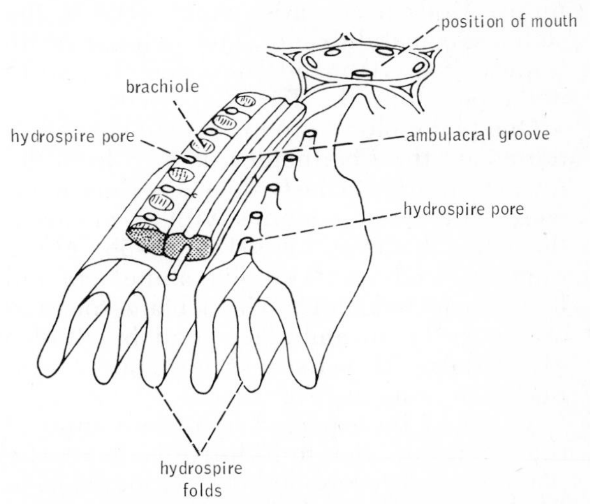

Figure 19–56. Hypothetical gas exchange system of a blastoid. (After Nichols.)

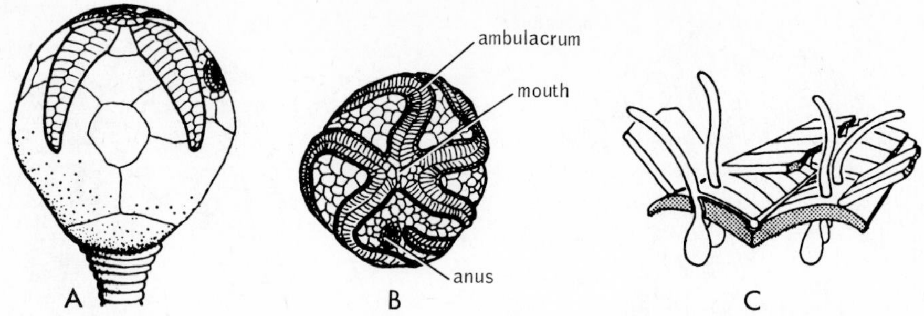

Figure 19–57. Edrioasteroids. *A. Steganoblastus*, a stalked species with straight ambulacra. *B. Edrioaster*, an unattached species with curved ambulacra. *C.* Ambulacral groove, showing ampullae, podia, and cover plates. (All after Nichols.)

or curved. Some of these animals thus looked very much like a brittle star wrapped around a ball. Of particular interest was the presence of pores between the ambulacral plates through which the podia extended. Movable cover plates protected the podia and food grooves (Fig. 19–57*C*). In this respect the edrioasteroids anticipated the echinoids.

The crinoids possess the richest fossil record of all the echinoderm classes and are the only attached species that still exist today. Typical crinoids first appeared in the Ordovician period, and they left a fossil record in every succeeding period. The class reached its climax in the Mississippian period, although there was a second somewhat lesser evolutionary development during the Permian period. During these periods, certain shallow seas supported enormous faunas. During the Permian period, such a sea covered the island of Timor, and the rocks composing this island today are one of the richest sources of fossil specimens. Paleozoic crinoids were all stalked, and the earlier species lacked pinnules. Modern crinoids, which contain the stalkless comatulids and which belong to a different subclass than the Paleozoic species, did not appear until the Triassic period.

One of the oldest groups of extinct echinoderms are the Eocrinoidea, known from the lower Cambrian to the Ordovician. Eocrinoids were stalked echinoderms with an enclosed theca like that of cystoids (Fig. 19–58). At the upper, or oral, end were five ambulacra and five to many brachioles. The pentamerous and oral position of ambulacra and brachioles gives these animals a superficial resemblance to crinoids (Fig. 19–62).

As would be expected from the nature of the skeleton, the holothuroids have the poorest fossil record and the echinoids have the best fossil record of the living classes of unattached echinoderms; but compared with attached forms the fossil record of unattached species is sparse. Although fossil asteroids and echinoids first appeared in the Ordovician period and ophiuroids in the Mississippian, Paleozoic species are relatively rare; these classes are much better represented in Mesozoic and Cenozoic rocks. The earlier echinoids, including a number of extinct orders, were all regular forms, and the first fossil heart urchins and sand dollars appeared in the Triassic period. Rich deposits of fossil echinoids are known, but the fossil distribution of this class is too localized for them to be of any value as index fossils.

A class of extinct unattached echinoderms, the Ophiocistioidea, lived from the Ordovician period to the Devonian. Only a few dozen

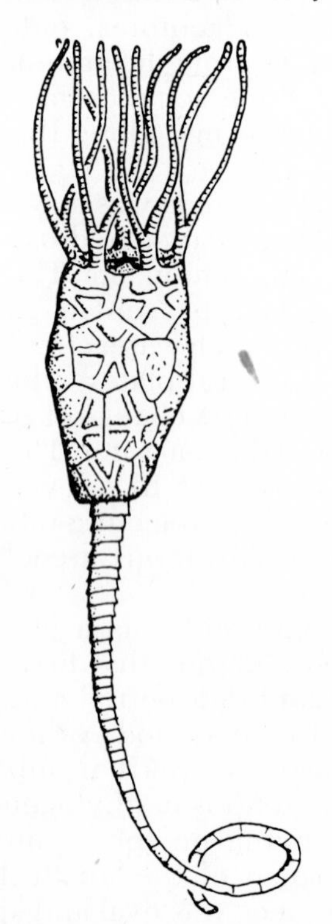

Figure 19–58. *Macrocystella*, an eocrinoid. (After Nichols.)

fossil species are known. The body was encased in a test that was flattened orally and was dome-shaped aborally (Fig. 19–59). Except in *Volchovia,* there was no anus and the madreporite was located eccentrically on the oral surface (Fig. 19–59*A*). The mouth, which was surrounded by a peristome, was located in the center of the oral surface. A chewing apparatus composed of five jawlike plates was also present. There was apparently only one gonad, but multiple gonopores opened in the same interambulacrum as the madreporite. Five radially arranged ambulacra extended peripherally from the peristome but were limited to the oral surface. Most remarkable were the three pairs of giant podia that occupied the ambulacra (Fig. 19–59*B*). Each podium was covered by small overlapping

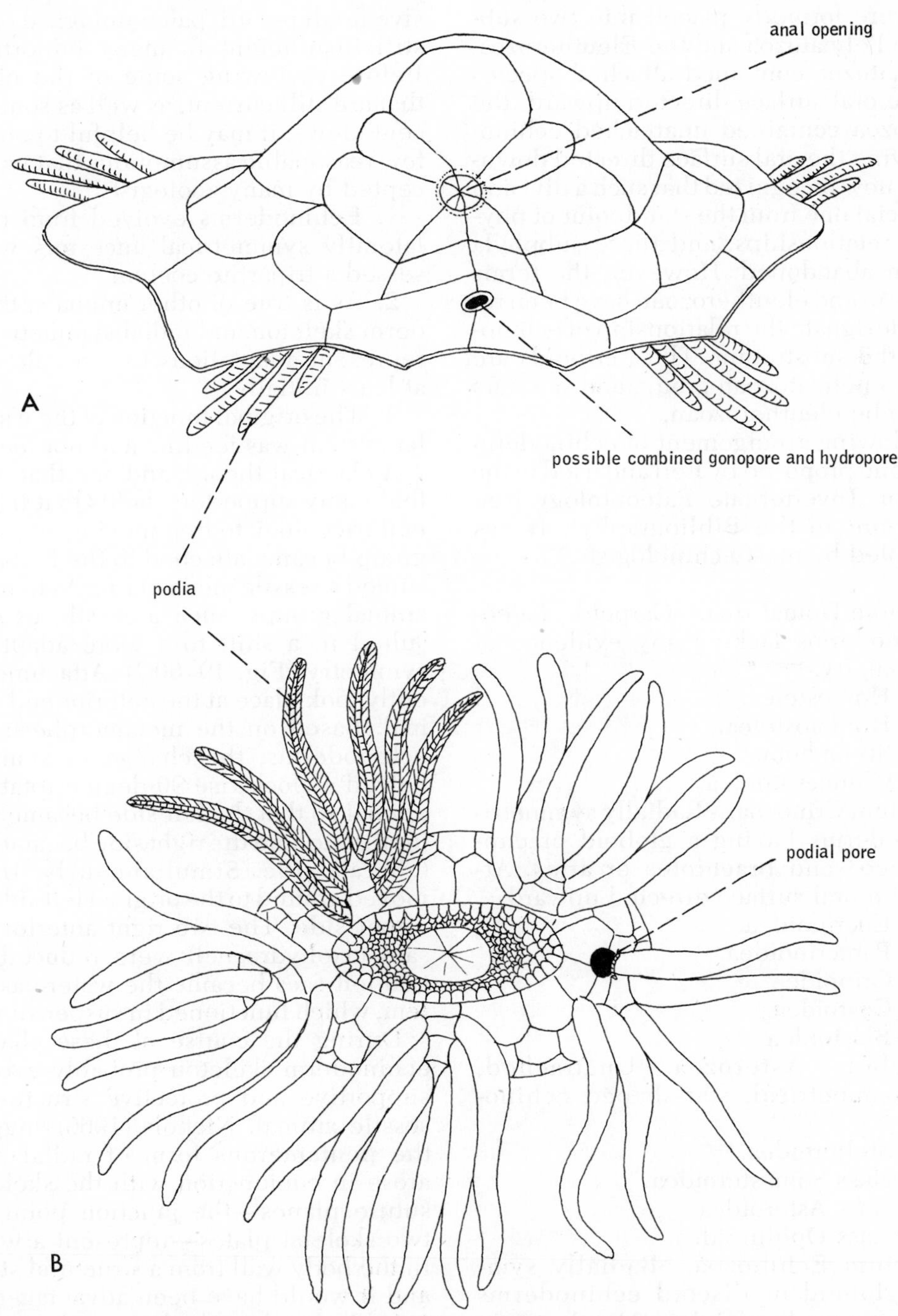

Figure 19–59. The extinct unattached echinoderms, class Ophiocistioidea. *A. Volchovia* (aboral view). (After Regnell from Hyman.) *B. Sollasina* (oral view). Podial scales only shown on podia of one ambulacrum; one podium removed to show podial pore. (Modified after Sollas from Hyman.)

scales and was thus preserved with the test. There seem to have been pores present between the podial plates, from which papula-like structures could have emerged.

The Machaeridia, Cycloidea, and Cyamoidea are three extinct fossil groups that are commonly considered to be echinoderms, but the nature and the position of these fossils are still highly uncertain.

The classes of echinoderms, both living and extinct, were formerly placed into two subphyla, the Pelmatozoa and the Eleutherozoa. The Pelmatozoa contained attached species having the oral surface directed upward; the Eleutherozoa contained unattached echinoderms having the oral surface directed downward. It is now recognized that such a division is an artificial one from the standpoint of phylogenetic relationships, and these subphyla have been abandoned. However, the terms pelmatozoan and eleutherozoan have been retained to designate the relationship of echinoderms to the substratum. Thus, crinoids are said to be a pelmatozoan group, and sea stars are said to be eleutherozoan.

The following arrangement of echinoderm classes is one proposed by Fell and used in the Treatise on Invertebrate Paleontology (see under Moore in the Bibliography). It has been adopted by most echinologists.

Subphylum Homalozoa. Carpoids. Paleozoic echinoderms lacking any evidence of radial symmetry.
- Class Homostelea.
- Class Homoiostelea.
- Class Stylophora.
- Class Ctenocystoidea.

Subphylum Crinozoa. Radially symmetrical echinoderms having a globoid or cup-shaped theca and brachioles or arms. Attached, with oral surface directed upward.
- Class Eocrinoidea.
- Class Paracrinoidea.
- Class Crinoidea.
- Class Cystoidea.
- Class Blastoidea.

Subphylum Asterozoa. Unattached, radially symmetrical, star-shaped echinoderms.
- Class Stelleroidea.
  - Subclass Somasteroidea.
  - Subclass Asteroidea.
  - Subclass Ophiuroidea.

Subphylum Echinozoa. Radially symmetrical globoid or discoid echinoderms without arms or brachioles. Mostly unattached.
- Class Helicoplacoidea.
- Class Edrioasteroidea.
- Class Holothuroidea.
- Class Echinoidea.
- Class Ophiocistioidea.

The origin of the echinoderms and the phylogenetic relationships of the subphyla continue to be unsolved questions and subjects of much speculation. Despite the extensive fossil record, paleontological evidence is still insufficient in many important areas. Before reviewing some of the older ideas that are still current, as well as some more recent views, it may be helpful to enumerate a few reasonable assumptions that would be accepted by many zoologists.

1. Echinoderms evolved from motile, bilaterally symmetrical ancestors which possessed a tripartite coelom.
2. As is true of other animals, the echinoderm skeleton and radial symmetry probably represent adaptations to a sessile existence, at least initially.
3. The original function of the water-vascular system was feeding and not locomotion.

A classical theory, and one that would still find many supporters, holds that from a bilateral tricoelomate free-moving ancestor, some group became attached to the bottom and assumed a sessile mode of life. As in many other animal groups, such a sessile existence resulted in a shift to a more adaptive radial symmetry (Fig. 19–60*C*). Attachment apparently took place at the anterior end of the animal. Based on the metamorphosis of living echinoderms, the change in symmetry involved a clockwise 90-degree rotation of the animal so that the left side became the upper (oral) side, and the right side became the lower (aboral) side. Simultaneously, the mouth moved around to the original left side, now the upper side. The two right anterior coelomic sacs (axohydrocoel) were reduced, and the two left sacs became the water-vascular system, which functioned in suspension feeding.

During the course of these changes, the echinoderm skeleton probably evolved as a supportive and protective structure for the sessile animal. Nichols (1966) suggests that the pentamerous form of radial symmetry arose in conjunction with the skeleton. The suture planes—the junction point between two skeletal plates—represent a weak point in the body wall from a structural standpoint, and it would have been advantageous to the animal not to have had two such suture planes opposite one another. This advantage could be

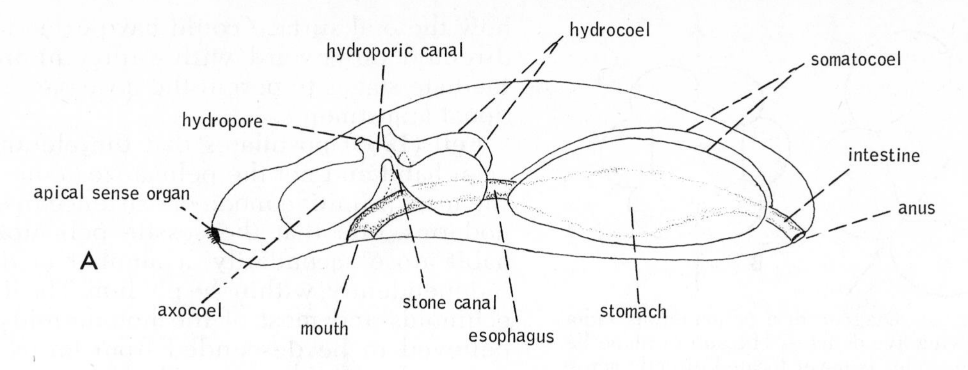

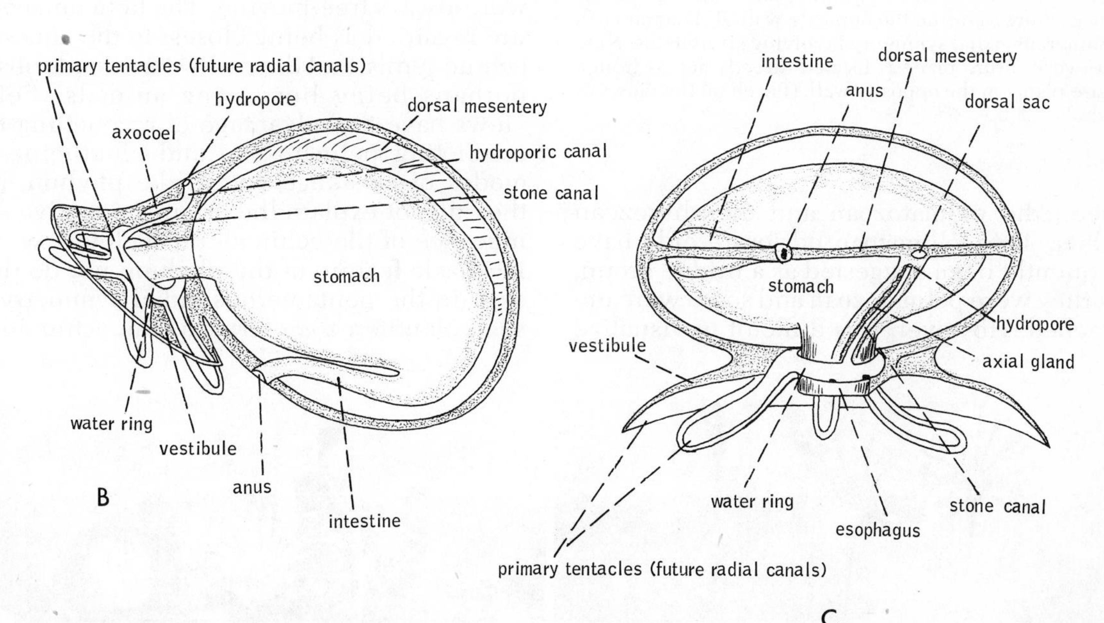

Figure 19–60. Hypothetical ancestors of echinoderms. A. Dipleurula ancestor. (After Bather from Hyman.) *B.* Bilateral pentactula ancestor. *C.* Pentactula ancestor after torsion. (*B* and *C* after Bury from Hyman.)

attained only by an odd number of ossicles forming the circumference of the body wall. The smallest number would be five if the animal were to be truly radial (Fig. 19–61). At this point, we have a suspension-feeding echinoderm that was both sessile and radial. This stage in the evolution of echinoderm symmetry is illustrated by the extinct and living Crinozoa.

After attaining a radial symmetry, some of these sessile echinoderms became detached and reassumed a free-moving existence. The radial symmetry was retained, but the oral surface, which was directed upward in sessile forms, was placed against the substratum; the aboral surface became the functional upper side of the animal, and the water-vascular system was utilized in locomotion. Sea stars, brittle stars, and sea urchins all illustrate such a free-moving, radial existence.

This theory is supported by the extinct and living crinozoan fauna and by embryological evidence, such as the attached metamorphosis of asteroids. However, there are also some objections to this theory. The oldest echinoderms, all from the early Cambrian, are three very diverse groups: the eocrinoids, the edrioasteroids, and the carpoids. The pelmatozoan habit of the eocrinoids and the edrioasteroids is compatible with the idea that the first echinoderms were radially symmetrical and attached animals; but the irregular carpoids are not. Either the carpoid symmetry is secondary, in which case the earliest echinoderms are Precambrian, or the first echinoderms were not pentamerous, radially symmetrical animals.

A second principal problem with this classical theory of echinoderm evolution is the lack of any forms that bridge the great gap be-

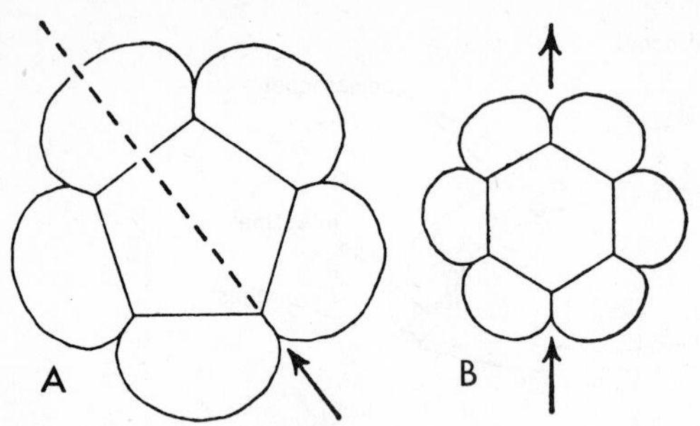

Figure 19–61. *A.* Diagram of a pentamerous radial symmetry involving five ossicles. The suture plane between any two ossicles is never located directly across from a suture plane on the opposite wall. *B.* Diagram of a hexamerous radial symmetry involving six ossicles. Note that every suture plane is located directly across from a suture plane on the opposite wall. (Based on the views of Nichols, 1966.)

tween the pelmatozoan and eleutherozoan habits. The echinozoan edrioasteroids have frequently been suggested as a bridge group, for they were pelmatozoan and some were unattached. However, it is difficult to visualize how the oral surface could have come to be directed downward with sufficient intermediate stages to permit the necessary functional adjustments.

Fell (1965) postulates that the eleutherozoan habit and not the pelmatozoan one was the most primitive mode of existence in echinoderms, and that the sessile pelmatozoan habit arose secondarily, a number of times independently, within the phylum. The living echinoids and most of the holothuroids are believed to be descended from forms that were always free-moving. The helicoplacoids are regarded as being closest to the ancestral echinoderms, and the earliest echinoderms as perhaps being burrowing animals. Fell's views have the advantage of reconciling the conflicting pelmatozoan and eleutherozoan modes of existence within the phylum, but they do not explain the original adaptive significance of the echinoderm skeleton, which is a basic feature of the phylum; nor do they explain the pentamerous radial symmetry of such eleutherozoan forms as the echinoids.

Figure 19–62. Reconstructions of the marine fauna of Paleozoic seas. *A.* Eocrinoids; a coiled cephalopod is on right. *B.* A long stalked crinoid in front of a cluster of rugose corals. To the left of the rugose corals is a brachiopod, and in the front left is a gastropod. *C.* Straight cone nautiloid cephalopods, with rugose corals in lower right corner. *D.* A large coiled cephalopod resting on bottom. Brachiopods in lower left corner, and rugose corals in background. (Photographs by Betty M. Barnes. Models are in the National Museum of Natural History, Smithsonian Institution.)

Embryology sheds little light on the interrelationships of living echinoderm classes. Similarity of the ophiuroid and echinoid pluteus larvae was formerly thought to indicate an evolutionary relationship. However, many echinologists now believe that there is considerable convergence in the larval forms of different classes.

There has also been much speculation about the nature of the ancestral pre-echinoderms. A classical view has been that of Bather (1900), who postulated that echinoderms evolved from a creeping, soft-bodied, bilateral animal (called a dipleurula ancestor), which possessed three pairs of coelomic sacs (Fig. 19–60*A*). The water-vascular system was a new development of the coelom in the evolution of the phylum.

Another theory, originally proposed by Semon (1888) and Bury (1895), has been supported by Hyman (1955). According to these zoologists, the common ancestor of echinoderms, which they call a pentactula, was a bilateral animal with hollow tentacles around the mouth, which were used for capturing food (Fig. 19–60*B* and *C*). The tentacles, each containing an extension of the coelom, a nerve, and a blood channel, were thus very much like a lophophore. The coelom into which the tentacular coelom opened had separated from the main coelom forming the water ring, and it had a special opening to the outside, the hydropore and the stone canal. According to this Pentactula Theory, each tentacle became a radial canal of the water-vascular system. With the assumption of a sessile existence and suspension feeding, the podia arose as side branches of the original tentacles. The arms (brachioles) developed with the addition of protective plates in the tentacle walls, and the ambulacral groove evolved as a means for conducting captured food particles to the mouth.

The Pentactula Theory is attractive in that it provides a functional precursor of the water-vascular system and also links the echinoderms and the lophophorates. Most recently, Nichols (1967) has suggested the Sipuncula, with their circle of hollow oral tentacles, as ancestors of both the lophophorates and the echinoderms.

Bibliography

Anderson, J. M., 1960: Histological studies on the digestive system of a starfish, *Henricia*, with notes on Tiedemann's pouches in starfishes. Biol. Bull., *119*(3):371–398.

Bather, F. A., 1900: The echinoderms. *In* Lankester, R. (Ed.), A Treatise on Zoology, Vol. 3, A. & C. Black, London.

Binyon, J., 1964: On the mode of functioning of the water vascular system of *Asterias rubens*. J. Mar. Biol. Assoc. U. K., *44*:577–588.

Boolootian, R. A. (Ed.), 1966: Physiology of Echinodermata. John Wiley, N. Y. (An excellent review of echinoderm ecology, behavior, and physiology.)

Burnett, A. L., 1960: The mechanism employed by the Starfish *Asterias forbesi* to gain access to the interior of the bivalve *Venus mercenaria*. Ecology, *41*: 583–584.

Bury, H., 1895: The metamorphosis of echinoderms. Quart, J. Micr. Sci., *38*:44–135.

Chesher, R. H., 1963: The morphology and function of the frontal ambulacrum of *Moira atropos*. Bull. Mar. Sci. Gulf and Caribbean, *13*:549–573.

Chesher, R. H., 1969: Destruction of Pacific reefs by sea star *Acanthaster planci*. Science, *165*:280.

Chia, F. S., 1969: Some observations of the locomotion and feeding of the sand dollar, *Dendraster excentricus*. J. Exp. Mar. Biol. Ecol., *3*(2):162–170.

Christensen, A. M., 1970: Feeding biology of the sea star *Astropecten irregularis*. Ophelia, *8*(1):1–134.

Clark, A. H., 1915, 1921, 1931, 1941, 1947, 1950: A monograph of the existing crinoids, Vol. 1, pts. 1, 2, 3, 4a, 4b, 4c. Bull. U. S. Nat. Mus. 82. (A monumental work on living crinoids.)

Clark, A. M., 1962: Starfishes and Their Relations. British Museum, London.

Cuénot, L., 1948: Anatomie, Éthologie, et Systématique des Échinodermes. *In* Grassé, P. (Ed.), Traité de Zoologie, Vol. II, Échinodermes, Stomocordes, Procordes. Masson et Cie, Paris, pp. 1–363.

Donnay, G., and Pawson, D. L., 1969: X-ray diffraction studies of echinoderm plates. Science, *166*:1147–1150.

Durham, J. W., and Caster, K. E., 1963: Helicoplacoidea, a new class of echinoderms. Science, *140*:820–822.

Feder, H. M., 1955: On the methods used by the starfish *Pisaster ochraceus* in opening three types of bivalved mollusks. Ecology, *36*:764–767.

Fell, H. B., 1963: The phylogeny of sea stars. Phil. Trans. Roy. Soc. London B, *246*:381–485.

Fell, H. B., 1965: The early evolution of the Echinozoa. Breviora, No. 219.

Fell, H. B., 1966: Ancient echinoderms in modern seas. Oceanogr. Mar. Biol. Ann. Rev., *4*:233–245.

Ferguson, J. C., 1969: Feeding, digestion, and nutrition in Echinodermata. *In* Florkin, M., and Scheer, B. J. (Eds.), Chemical Zoology, Vol. 3, Echinodermata. Academic Press, N. Y., pp. 71–100.

Fish, J. D., 1967: Biology of *Cucumaria elongata*. J. Mar. Biol. Assoc. U. K., *47*:129–143.

Fontaine, A. R., 1965: Feeding mechanisms of the ophiuroid *Ophiocomina nigra*. J. Mar. Biol. Assoc. U. K., *45*:373.

Gosner, K. L., 1971: Guide to Identification of Marine and Estuarine Invertebrates: Cape Hatteras to the Bay of Fundy. Wiley-Interscience, N. Y. Echinodermata, pp. 556–590.

Hamann, A., 1885: Beiträge zur Histologie der Echinodermen. Heft 2, Die Asteriden.

Hyman, L. H., 1955: The Invertebrates. Vol. 4, Echinodermata. McGraw-Hill, N. Y. (The chapter *Retrospect* in Vol. 5 (1959) of this series summarizes the literature on echinoderms from 1955 to 1959.)

Ikeda, H., 1941: Function of the lunules of *Astriclypeus*. Annot. Zool. Japon., *20*:79–83.

Kanatani, H., and Shirai, H., 1970: Mechanism of starfish spawning. Develop., Growth. Differ., *12*(2): 119–140.

Langeloh, H., 1937: Uber die Bewegungen von *Antedon*. Zool. Jahrb. Abt. Allg. Zool., *57*(3):235–279.

Lewis, J. B., 1968: The function of the sphaeridia of sea urchins. Canad. J. Zool., *46*:1135–1138.

Light, S. F., Smith, R. I., Pitelka, F. A., Abbott, D. P., and Weesner, F. M., 1967: Intertidal Invertebrates of the Central California Coast. University of California Press, Berkeley. Echinodermata, pp. 285–294.

Mauzey, K. P., Birkeland, C., and Dayton, P. K., 1968: Feeding behavior of asteroids and escape responses of their prey in the Puget Sound region. Ecology, *49*(4):603–619.

Millot, N. (ed.), 1967: Echinoderm Biology. Academic Press, N. Y. 240 pp. (A collection of papers presented at a symposium in 1966.)

Moore, R. C. (Ed.), 1966 and 1967: Treatise on Invertebrate Paleontology. Echinodermata, Pts. U and S. Geological Society of America and University of Kansas Press.

Newman, W. A., 1970: *Acanthaster:* a disaster? Science, *167*:1274–1275.

Nichols, D., 1959: Changes in the chalk heart-urchin, *Micraster*, interpreted in relation to living forms. Phil. Trans. Roy. Soc. London B, *242*:347–437.

Nichols, D., 1960: The histology and activities of the tube-feet of *Antedon bifida*. Quart. J. Micr. Sci., *101*(2):105–117.

Nichols, D., 1966: Echinoderms. Hutchinson University Library, London.

Nichols, D., 1967: The origin of echinoderms. *In* Millot, N. (Ed.), Echinoderm Biology. Academic Press, N. Y., pp. 209–229.

Pantreath, R. J., 1970: Feeding mechanisms and the functional morphology of podia and spines in some New Zealand ophiuroids. J. Zool., *161*:395–429.

Rutman, J., and Fishelson, L., 1969: Food composition and feeding behavior of shallow water crinoids at Eilat (Red Sea). Mar. Biol., *3*(1):46–57.

Semon, R., 1888: Die Entwicklung der *Synapta*. Jena. Ztschr. Wiss., 20.

Swan, E. F., 1961: Seasonal evisceration in the sea cucumber *Parastichopus californicus*. Science, *133*(3485):1078–1079.

Ubaghs, G., 1969: General characteristics of the Echinodermata. *In* Florkin, M., and Scheer, B. J. (Eds.), Chemical Zoology, Vol. 3, Echinodermata. Academic Press, N. Y., pp. 3–46.

Chapter 20

THE LESSER DEUTEROSTOMES

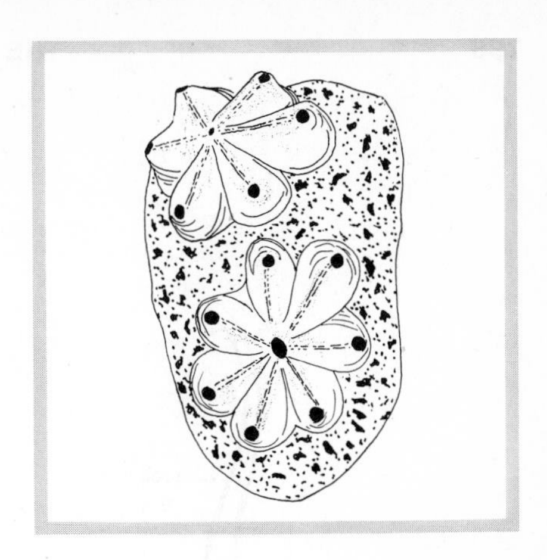

In addition to the echinoderms, there are four other groups of invertebrate deuterostomes—the Hemichordata, the Chaetognatha, the Urochordata, and the Cephalochordata. The last two groups are subphyla of the phylum Chordata. Although all are deuterostomes, they do not represent a close phylogenetic unit but stem from different points along the deuterostome line. The term *lesser* used in the chapter title refers only to the relatively small numbers of species comprising each group. Many are widely distributed and common invertebrate animals, and all are highly specialized.

PHYLUM HEMICHORDATA

The hemichordates are a small group of wormlike marine animals that until recently had been considered a subphylum of the chordates. As a consequence of this position, the common North Atlantic hemichordates, *Saccoglossus kowalevskii* and *Balanoglossus aurantiacus* of the Carolina coast, have acquired some fame in comparative anatomy classrooms. Alliance with the chordates was based on the presence of gill slits and what was thought to be a notochord in hemichordates. It is now generally agreed that the hemichordate "notochord" is neither analogous nor homologous with the chordate notochord, and that other than the common possession of pharyngeal clefts the two groups are dissimilar. The hemichordates have thus been removed from the Chordata in all modern treatments and have been given the rank of a separate phylum.

The hemichordates are composed of two classes—the Enteropneusta (acorn worms) and the Pterobranchia. The acorn worms are the most common and best known hemichordates. The pterobranchs consist of three genera of small tube-dwelling animals and for the most part are not found in European and North American waters. The salient features of the Pterobranchia are described after discussion of the Enteropneusta.

CLASS ENTEROPNEUSTA

The Enteropneusta (acorn worms) are inhabitants of shallow water. Some live under stones and shells, but many common species, including *Saccoglossus*, burrow in mud and sand. Exposed tidal flats are frequently dotted with the coiled ropelike casting of these animals.

Acorn worms are relatively large animals, the majority ranging from 9 to 45 cm. in length. The Brazilian species, *Balanoglossus gigas*, may exceed 1.5 m. in length and constructs burrows 3 m. long. The cylindrical and rather flaccid body is composed of an anterior proboscis, a collar, and a long trunk (Fig. 20–1A). These regions correspond to the typical deuterostome body divisions—protosome, mesosome, and metasome.

The proboscis is usually short and conical, from which the name acorn worm is derived, and is connected to the collar by a narrow stalk. The collar is a short cylinder that anteriorly overlaps the proboscis stalk and ventrally contains the mouth. The trunk comprises the major part of the body. Behind the collar, the trunk bears a longitudinal row of gill pores at each side of a middorsal ridge. More laterally, the anterior half of the trunk contains the gonads, and in some hemichor-

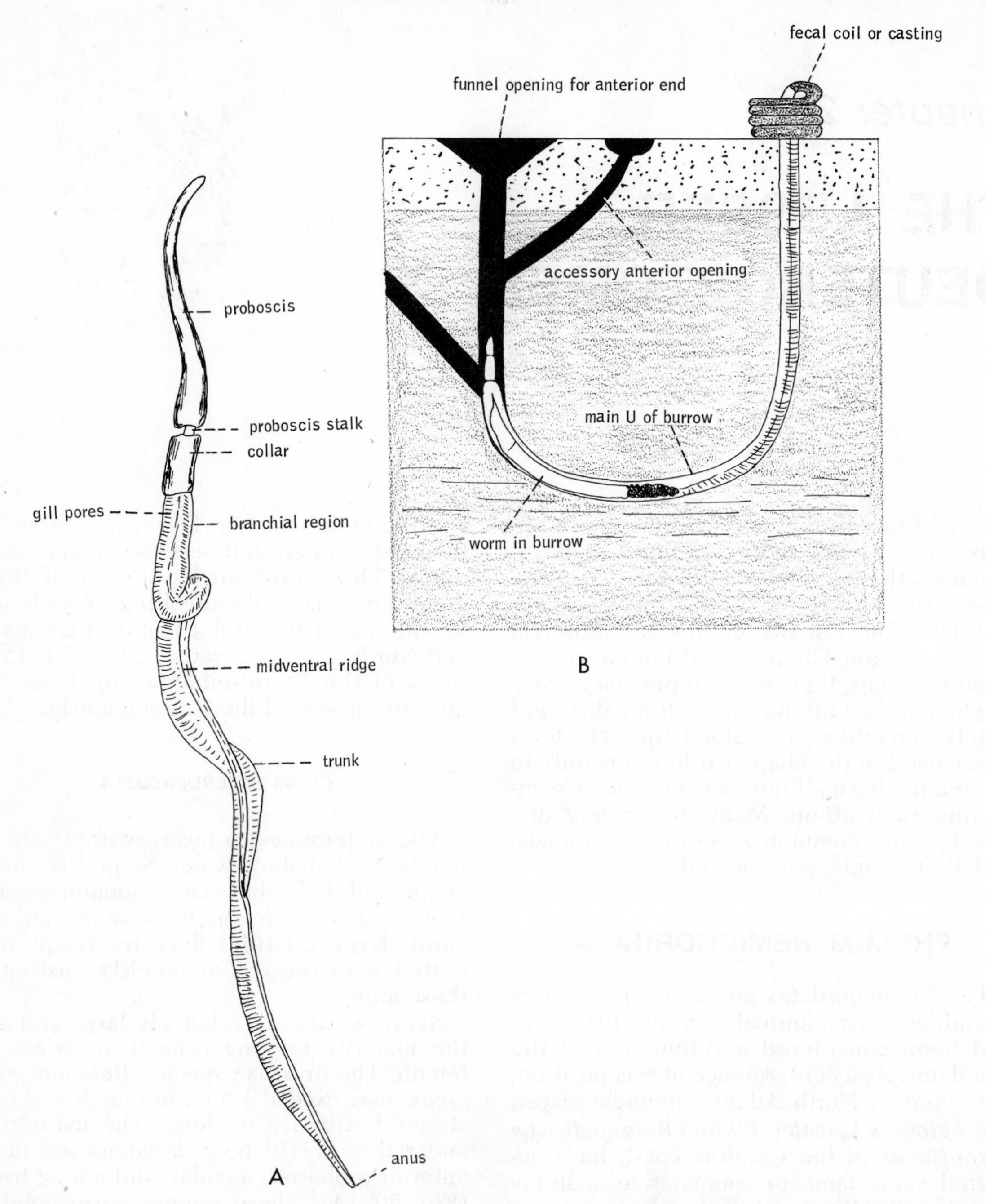

Figure 20–1. A. *Saccoglossus kowalevskii*, an acorn worm common to the Atlantic Coast of North America and Europe. (After Hyman.) *B.* Burrow system of the Mediterranean *Balanoglossus clavigerus*. (After Stiasny from Hyman.)

dates, such as *Balanoglossus*, the lateral body wall in this region is drawn out on each side, forming winglike plates (Fig. 20–4*B*). The remaining postbranchial region of the trunk may be undifferentiated, or the darkened wall of the intestine may be visible through the body wall, forming the hepatic region. The hepatic region may also be marked externally by sacculations of the body wall. When a hepatic region is present, the trunk is terminated by a caudal region.

The body is covered by ciliated columnar epithelium that is well provided with gland cells, especially in the collar and trunk regions. A well-developed nervous layer is present in the lower part of the epidermis.

Hemichordates display the usual tricoelomate structure of deuterostomes. A single coelomic cavity occupies the proboscis, a pair of cavities is present in the collar, and a pair of cavities is present in the trunk. However, enteropneusts are peculiar in that the coelomic epithelium has formed connective tissue and muscle fibers that fill much of the original coelomic cavity, and a distinct peritoneal lining has disappeared. Moreover, this coelomic musculature in large part replaces the typical body-wall musculature.

The protocoel is restricted to the posterior end of the proboscis and opens to the exterior through a middorsal pore (Fig. 20–2A). The anterior part of the old coelom is filled with a mass of longitudinal muscle fibers and connective tissue. The paired coelomic cavities

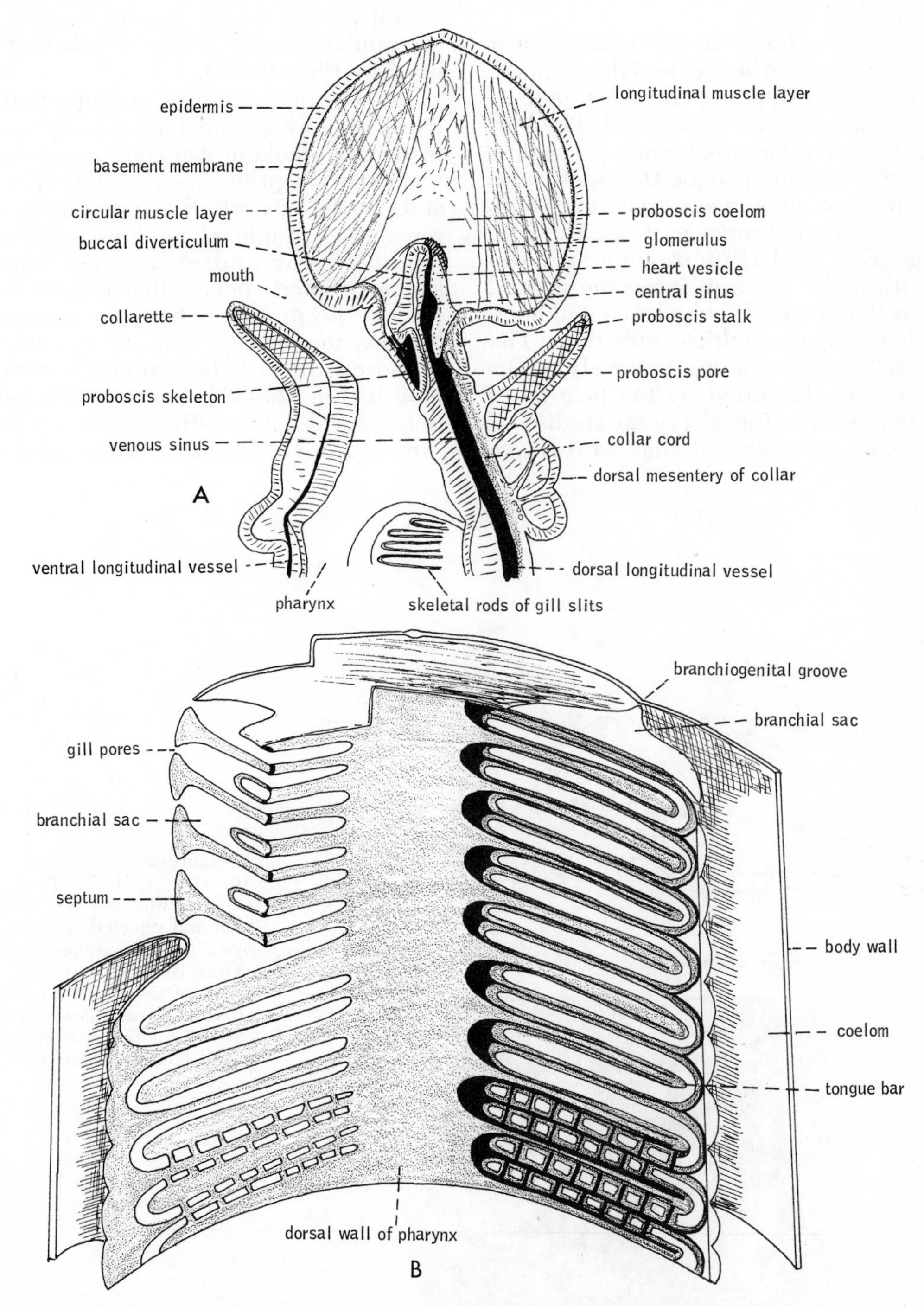

Figure 20–2. *A*. Anterior of *Glossobalanus minutus* (sagittal section). (After Spengel from Hyman.) *B*. Stereodiagram of gill region of an enteropneust. (After Delage and Hérouard from Hyman.)

of the collar also open to the exterior through a pair of canals and pores located on each side of the middorsal line. Muscles are poorly developed in the collar. The trunk coelom is also paired, the dorsal and ventral mesenteries of the gut separating the two cavities on each side. There is no opening to the exterior. Radial fibers traverse the trunk coelom, and longitudinal fibers are well developed ventrally.

Acorn worms have limited locomotor powers and are rather sluggish animals. Many burrowing species construct mucus-lined excavations in mud and sand. The burrows of species of *Balanoglossus*, *Saccoglossus*, and other genera may be **U**-shaped, with two openings to the surface, and one or both ends of the worms at times protrude from the openings (Fig. 20–1*B*). Burrowing or movement within the burrow is accomplished largely by the proboscis, which is lengthened and anchored by peristaltic contractions. The trunk may be pulled along passively, or its movement may be aided by the beating of trunk cilia. Knight-Jones' (1952) studies of *Saccoglossus* have shown that worms can move backward by reverse peristalsis of the proboscis and reverse beating of the trunk cilia. Other acorn worms live in masses of seaweed, under rocks and stones, or buried in sand and mud, and they move about relatively little.

Many burrowing enteropneusts consume sand and mud, from which organic matter is digested. The quantity of substrate ingested is indicated by the great piles of castings that accumulate at the posterior opening of the burrow (Fig. 20–3*B*).

Suspension feeding is an important method of obtaining food for some species. Detritus and plankton that come in contact with surface of the proboscis are trapped in mucus and carried posteriorly by strong ciliary currents. This mode of feeding is utilized by nonburrowing and even many burrowing species. Some species that inhabit burrows project the proboscis from the mouth of the burrow, moving it about. At the base of the proboscis the cilia beat ventrally toward the mouth (Fig. 20–3*A*). Some of the food particles carried ventrally by these cilia pass into a groove, forming the preoral ciliary

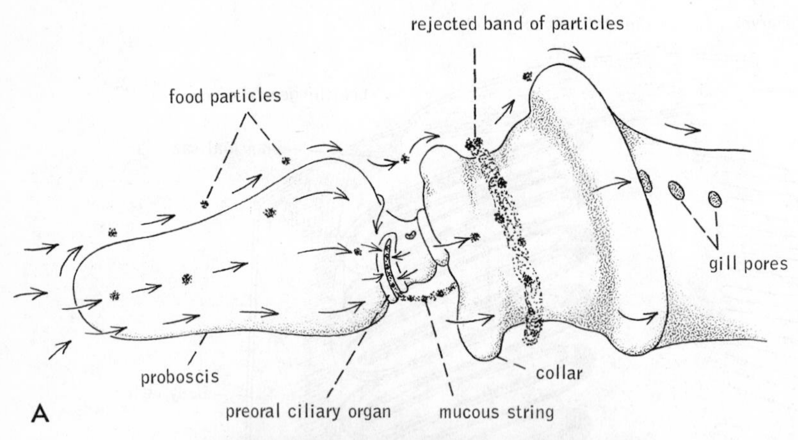

Figure 20–3. *A.* Side view of anterior end of *Protoglossus köhleri*, showing the direction of food and rejected particles carried by cilia of the proboscis and collar. (After Burdon-Jones.) *B.* Photograph of the anal end of *Balanoglossus aurantiacus* depositing casting on the surface of an exposed sand flat at low tide.

organ. Within the groove the particles are conducted ventrally. The function of the preoral ciliary organ is not certain, but the presence of receptor cells and a concentration of underlying nerve cells suggest a sensory function, perhaps that of testing water and food particles passing into the mouth.

Although particles are carried into the mouth by cilia, their passage is probably facilitated by the water current flowing into the mouth and out of the pharyngeal clefts. Feeding may be halted or large particles rejected by covering the mouth with the edge of the collar. Particles then pass over the collar instead of into the mouth.

The digestive tract is a straight tube that is histologically differentiated into a number of regions. The large mouth, located between the ventral anterior margin of the collar and the dorsal proboscis stalk, leads into a buccal tube within the collar (Fig. 20–2*A*). Dorsally, a long narrow diverticulum extends from the buccal tube and projects forward into the proboscis. It is this proboscis diverticulum that was for so long thought to be a "notochord" and accounted for the placement of the acorn worms among the chordates.

Histologically, the wall of the diverticulum is identical to the wall of the buccal tube into which it opens; Hyman (1959) and current investigators of the hemichordates agree that the buccal diverticulum is apparently nothing more than a preoral extension of the gut, and is certainly not a notochord. The buccal tube passes posteriorly into the pharynx, which occupies the branchial region of the trunk and is laterally perforated by the gill slits. The gill slits are usually limited to the dorsal half of the pharynx, and the alimentary portion of the tract occupies the ventral half of the pharynx.

Behind the pharynx, the gut continues as an esophagus, and in two families, containing the genera *Saccoglossus, Spengelia,* and *Glandiceps,* it opens externally to the dorsal surface through a number of canals and pores. The intestine comprises the remainder of the gut. Anteriorly the intestinal epithelium of the hepatic region is colored green or brown because of inclusions, and in some species, such as *Balanoglossus,* it is sacculated. Behind the hepatic region, the intestine exhibits no further histological differentiation. In *Saccoglossus,* the intestine opens directly to the exterior through the terminal anus, but commonly the anus is preceded by a rectum.

The middle region of the esophagus is constricted, and food is driven through the constriction by peristaltic contractions of the anterior section of the esophagus. The constricted region molds the food particles and mucus into a food-laden mucous cord. Water squeezed out in the formation of the cord is believed to escape through the esophageal pores.

Digestion and absorption occur in the intestine, although knowledge of the digestive process is still very incomplete. In *Glossobalanus minutes,* the mucus covering the proboscis surface, where food is initially trapped, has been found by Barrington (1940) to contain amylase. Muscles in the gut wall are weakly developed, and food is apparently moved posteriorly in ciliated grooves.

Enteropneusts possess an open blood-vascular system composed of two main contractile vessels and a system of sinus channels. The blood, which is colorless and is largely lacking in cellular elements, is carried anteriorly in a dorsal vessel located in the mesentery suspending the digestive tract (Fig. 20–2*A*). At the level of the collar, the dorsal vessel passes into a venous sinus and then into a central sinus located at the base of the proboscis.

The central sinus is situated beneath a closed fluid-filled sac (the heart vesicle) that contains muscle fibers in its ventral wall (Fig. 20–2*A*). The pulsations of this wall aid in driving blood through the central sinus chamber. From the central sinus, all of the blood is delivered anteriorly into a special system called the glomerulus. The heart vesicle, buccal diverticulum, and central sinus together bulge into the coelom at the base of the proboscis. The peritoneal covering of the faces of these bulging structures is greatly evaginated into the proboscis coelom. These evaginations contain blood sinuses and collectively make up the glomerulus, and all blood from the central sinus is driven anteriorly through the glomerular sinus system. Although the glomerulus is thought to have an excretory function, there is as yet no experimental proof.

From the glomerulus, blood is delivered by a system of channels to the ventral longitudinal vessel, which runs posteriorly beneath the digestive tract. Along the length of its course, the ventral vessel supplies both the body wall and the gut wall with a rich network of sinuses that eventually drain back into the dorsal vessel.

The pharyngeal gill slits in the anterior trunk region are assumed to be the gas exchange organs of enteropneusts. The number of slits can range from a few to one hundred or more pairs, since new slits are continually being formed during the life of the worm. Each slit opens through the side of the pharyngeal wall as a **U**-shaped cleft with the arms of the **U** directed upward (Fig. 20–2*B*). The pharyngeal wall between clefts (the septum) and that part of the wall projecting downward between the arms of the **U** (the tongue bar) are supported by skeletal pieces that have arisen as thickenings of the basement membrane of the pharyngeal epithelium.

The **U**-shaped pharyngeal clefts perforate the pharyngeal wall and then open into the branchial sacs. There is one sac per gill slit, and each sac opens to the exterior by a dorsolateral gill pore, which along with the other pores on that side is often located in a longitudinal groove.

The septa and tongue bars are ciliated on both the pharyngeal and the lateral faces and internally contain a plexus of blood sinuses from the ventral longitudinal vessels. The beating cilia produce a stream of water passing into the mouth and out through the gill slits.

The gill slits of the hemichordates and the chordates probably evolved originally as a feeding mechanism, in which small particles were strained out of the water current passing through the pharyngeal clefts. This is still the method of feeding in the tunicates and the cephalochordates. A gas exchange function has been secondarily assumed by the gill slits, and in hemichordates it is believed that gas exchange takes place between the water current and the hemal sinuses in the septa and tongue bars.

The nervous system is relatively primitive (Fig. 20–4*A*). In different regions of the body, the nerve plexus at the base of the surface epithelium has become thickened to form nerve cords, in which the nerve fibers are arranged longitudinally. For the most part, these nerve cords retain their epidermal location and their connection with the rest of the epidermal nerve plexus.

The principal nerve cords are the midventral and the middorsal proboscis cords connected at the proboscis base by an anterior nerve ring. Similarly, there is a midventral and middorsal trunk cord connected at the anterior end of the trunk by a circumenteric nerve ring. The ventral trunk cord terminates at the collar, but the dorsal cord continues into the collar as the collar cord.

The collar cord actually becomes internal —that is, it is separated from the epidermis above and is continuous with the general epidermal nerve plexus only at the two ends. In some acorn worms, the collar cord is hollow and may even open to the outside through an anterior and a posterior neuropore. The collar cord possesses giant nerve cells and is apparently a conduction path. The giant nerve cells are involved in rapid transmission, as is true of giant cells in other animals. Digging and peristaltic waves are initiated in the proboscis, and the dorsal nerve cord conducts these impulses posteriorly. Cutting the dorsal nerve cord anterior to the trunk therefore greatly slows down conduction. There seems to be little coordination and integration of response. In the trunk, conduction can take place through the general epidermal plexus, as well as through the ventral and dorsal cord.

Except for the preoral ciliary organ already described (Fig. 20–3*A*), neurosensory cells scattered throughout the surface epithelium comprise the sensory system of hemichordates.

Acorn worms are fragile animals, and the larger species are especially difficult to collect intact. Most species probably can regenerate at least missing parts of the trunk. Asexual reproduction has been reported for several species, including members of the genera *Glossobalanus* and *Balanoglossus*.

Enteropneusts are all dioecious. The saclike gonads are located in the coelom along each side of the trunk, beginning in the branchial region and extending to or through the hepatic region. In some species, the gonads are located behind the gill slits. There are a number of genera, such as *Balanoglossus*, *Stereobalanus*, and *Ptychodera*, in which the sides of the body containing the gonads bulge to the outside as genital wings or ridges. Usually one such wing is present on each side and below the gill pores, but in *Stereobalanus* there is both a dorsal and a ventral wing with the gill located in between (Fig. 20–4*B*). Each gonad opens to the exterior through a pore that is often located in the same groove as the gill pores.

Masses of eggs imbedded in mucus are shed from the burrow and are fertilized externally by sperm emitted from nearby males that are apparently stimulated by the presence of the released eggs. The mucous masses are soon broken up by tidal currents and the eggs are dispersed.

Early development is strikingly like that of

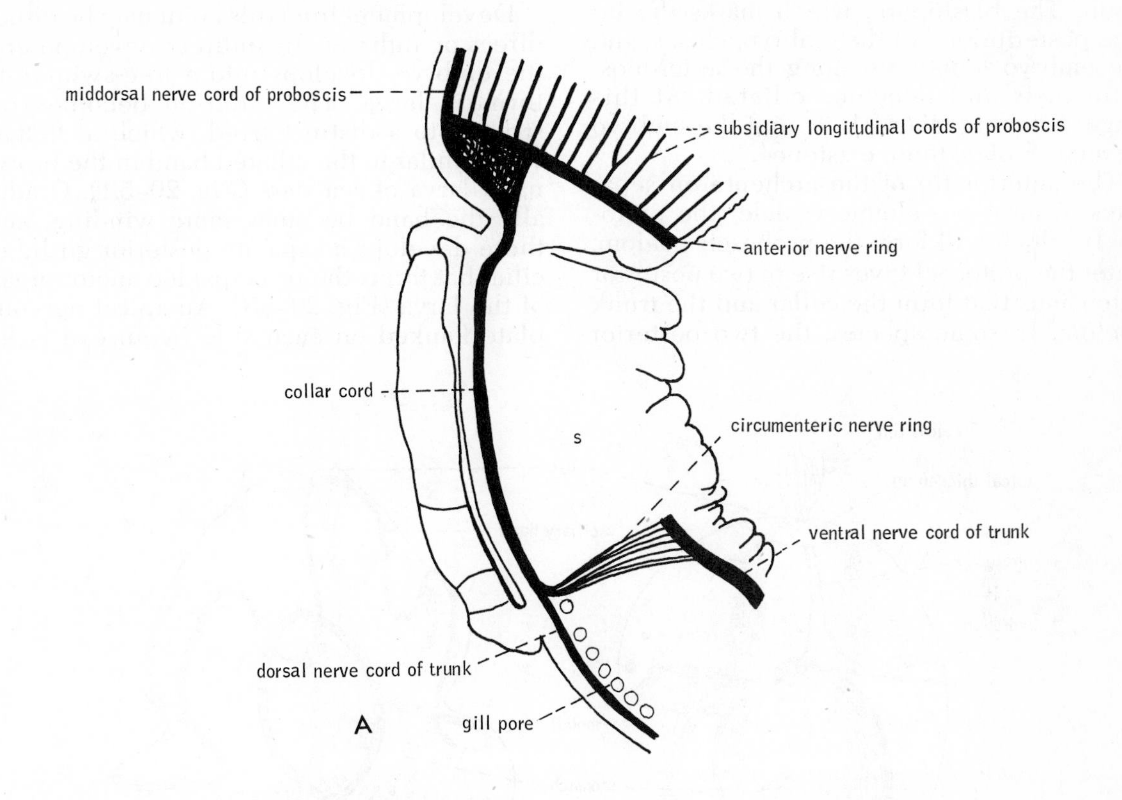

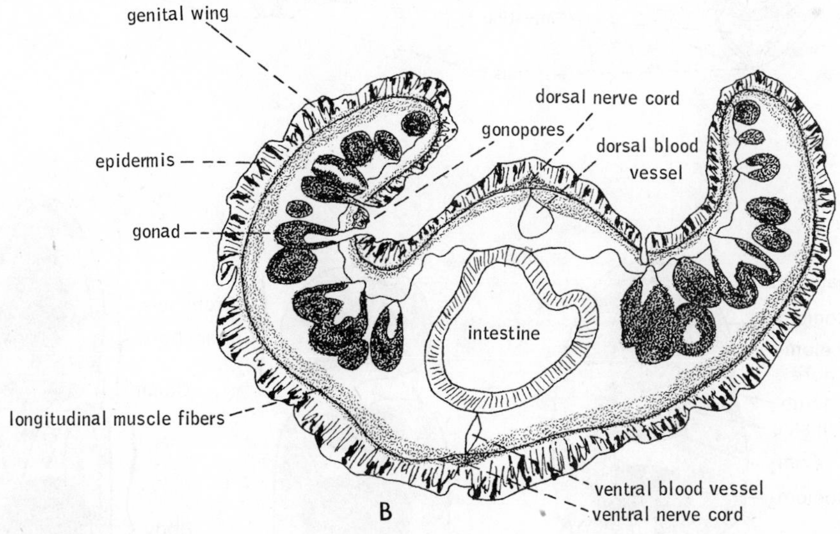

Figure 20–4. *A.* Anterior part of the *Saccoglossus cambrensis* nervous system (longitudinal view). Nerve networks not shown. (After Knight-Jones from Hyman.) *B.* Trunk of *Ptychodera bahamensis*, showing genital wings (transverse section). (After Van der Horst from Hyman.)

the echinoderms. Equal holoblastic cleavage leads to a coeloblastula, which then undergoes invagination to form a narrow archenteron. The blastopore, which marks the future posterior end of the embryo, closes, and the embryo lengthens along the anteroposterior axis and becomes ciliated. At this stage, hatching takes place, and the embryo assumes a planktonic existence.

The anterior tip of the archenteron separates to form a coelomic vesicle (the protocoel), which will form the proboscis coelom. Later the protocoel gives rise to two posterior extensions that form the collar and the trunk coelom. In some species, the two posterior coelomic divisions arise as evaginations of the archenteron independent of the formation of the protocoel.

Development from this point may be either direct or indirect. In indirect development, the embryo develops into a free-swimming tornaria larva. The ciliation becomes restricted to a distinct band, which at first is very similar to the ciliated band in the bipinnaria larva of sea stars (Fig. 20–5*A*). Gradually the band becomes more winding, and there develops a separate posterior girdle of cilia that forms the principal locomotor organ of the larva (Fig. 20–5*B*). An apical nervous plate flanked on each side by an eye is lo-

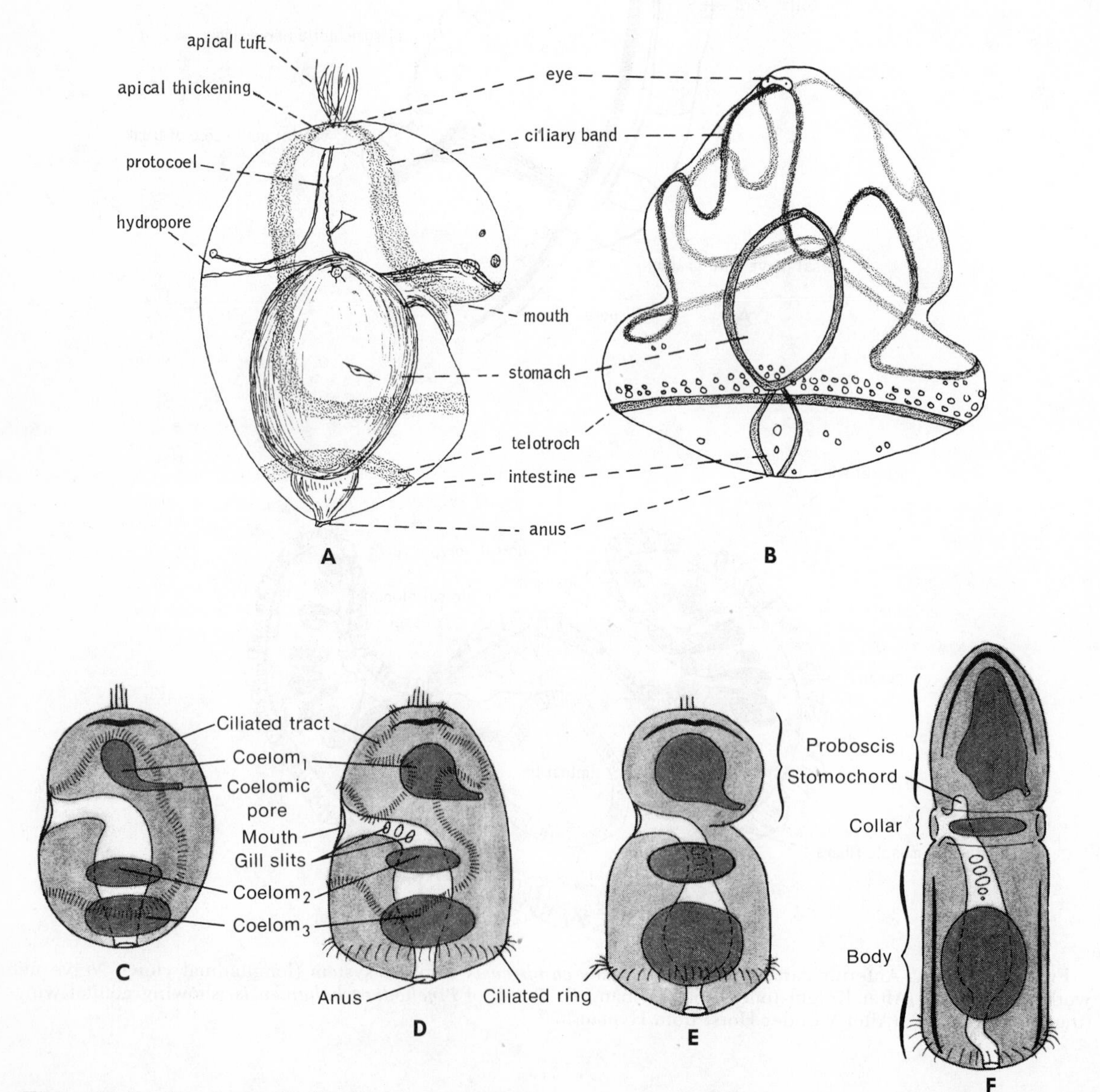

Figure 20–5. Development of *Balanoglossus clavigerus*. *A*. Early tornaria larva (lateral view). *B*. Fully developed tornaria (lateral view). (*A* and *B* after Stiasny from Hyman.) *C* to *E*. Diagrammatic side views of the larval development of a hemichordate. Compare with Fig. 19–11. *C*. Early larva. *D*. Later larva. *E* and *F*. Metamorphosis.

cated at the anterior end. After a planktonic feeding existence of several days to several weeks, the larva becomes girdled by a constriction initiating the division between proboscis and collar (Fig. 20–5*C–F*). The larva elongates, sinks to the bottom, and assumes an adult existence.

Development is direct in a number of enteropneusts, including the Atlantic acorn worm, *Saccoglossus kowalevskii.* A ciliated gastrula may hatch from the egg, or hatching may take place at a later stage; but in any case a tornaria larva never forms, and development proceeds directly, terminating in the young worm. In *Saccoglossus kowalevskii* the eggs hatch as young worms.

Burdon-Jones' (1952) studies of *Saccoglossus horsti* have disclosed that the young worms possess a postanal tail, which is used for anchoring the body in the burrow. It is suggested that this larval tail region is homologous to the stalk of the pterobranchs, members of a second class of hemichordates described below.

CLASS PTEROBRANCHIA

The Pterobranchia consists of a small number of species belonging to three genera, and these are rarely encountered. All are bottom dwellers in relatively deep water, and except for *Rhabdopleura,* which has been dredged up close to the European coast, most species are found in the Southern Hemisphere. With the exception of *Atubaria,* pterobranchs live in secreted tubes, which are organized in aggregations *(Cephalodiscus)* or colonies *(Rhabdopleura)* on the bottom (Fig. 20–6). Species of both genera are attached by a stalk, but individuals forming the colonies of *Rhabdopleura* are connected together by a stolon.

The proboscis is shield-shaped, but the most striking features of these worms are the arms and tentacles carried on the dorsal side of the collar. In *Rhabdopleura,* there are two recurved arms, and in *Cephalodiscus* there are five to nine pairs of arms (Fig. 20–6). The arms bear numerous small tentacles that are heavily ciliated. The tentacles supposedly capture minute organisms, which are then driven by the cilia to the mouth; however, the entire ciliated body surface of *Cephalodiscus* may collect suspended particles. Both the arms and the tentacles are hollow, each containing an extension of the mesocoel, and in this respect they are thus somewhat similar to a lophophore. There are no gill clefts in the genus *Rhabdopleura,* and there is only one pair in *Cephalodiscus.* And as is true of many sessile tube-dwelling animals, the gut is **U**-shaped, the anus opening anteriorly on the dorsal side of the collar. The sexes are separate, and a ciliated larva, unlike a tornaria, is known for *Cephalodiscus.* From the initial individual that develops from the larva, the colony, or aggregation, is formed by budding of the stalk or stolon.

Hemichordate Phylogeny

The evidence of a phylogenetic relationship between hemichordates and both the echinoderms and the chordates is very convincing. Although the adults are quite different, the early embryogeny of the hemichordates is remarkably like that of the echinoderms. The formation of the gastrula and the coelom is very similar to these stages in echinoderms, and the early tornaria larva is virtually identical to the bipinnaria of the asteroids. Of the two major classes of hemichordates, the pterobranchs are considered the more primitive, and the lophophore-like arms and tentacles are thought to represent a primitive feature of the phylum that has been lost in the enteropneusts.

It is believed by Hyman (1959) and others that the pterobranchs may be similar to the common ancestor of both the echinoderms and the hemichordates; and it may have been from such arms and tentacles that the echinoderm water-vascular system, which is believed to have been originally a food-catching device, arose. Certainly a close relationship between the hemichordates and the echinoderms is difficult to deny, and Hyman places the hemichordates close to the base of the echinoderm line.

An affinity with the chordates is also indicated, although not as close a one as that with the echinoderms. Only in the hemichordates and the chordates are pharyngeal clefts found. Also, the dorsal collar nerve cord of hemichordates, which is sometimes hollow, is somewhat similar to the dorsal hollow nerve cord of chordates, and perhaps the two structures are homologous. However, the lack of a notochord and the differences in the general body structure exclude hemichordates from the phylum Chordata. As suggested by Hyman, the same line from which the hemichordates stemmed probably culminated in the chordates.

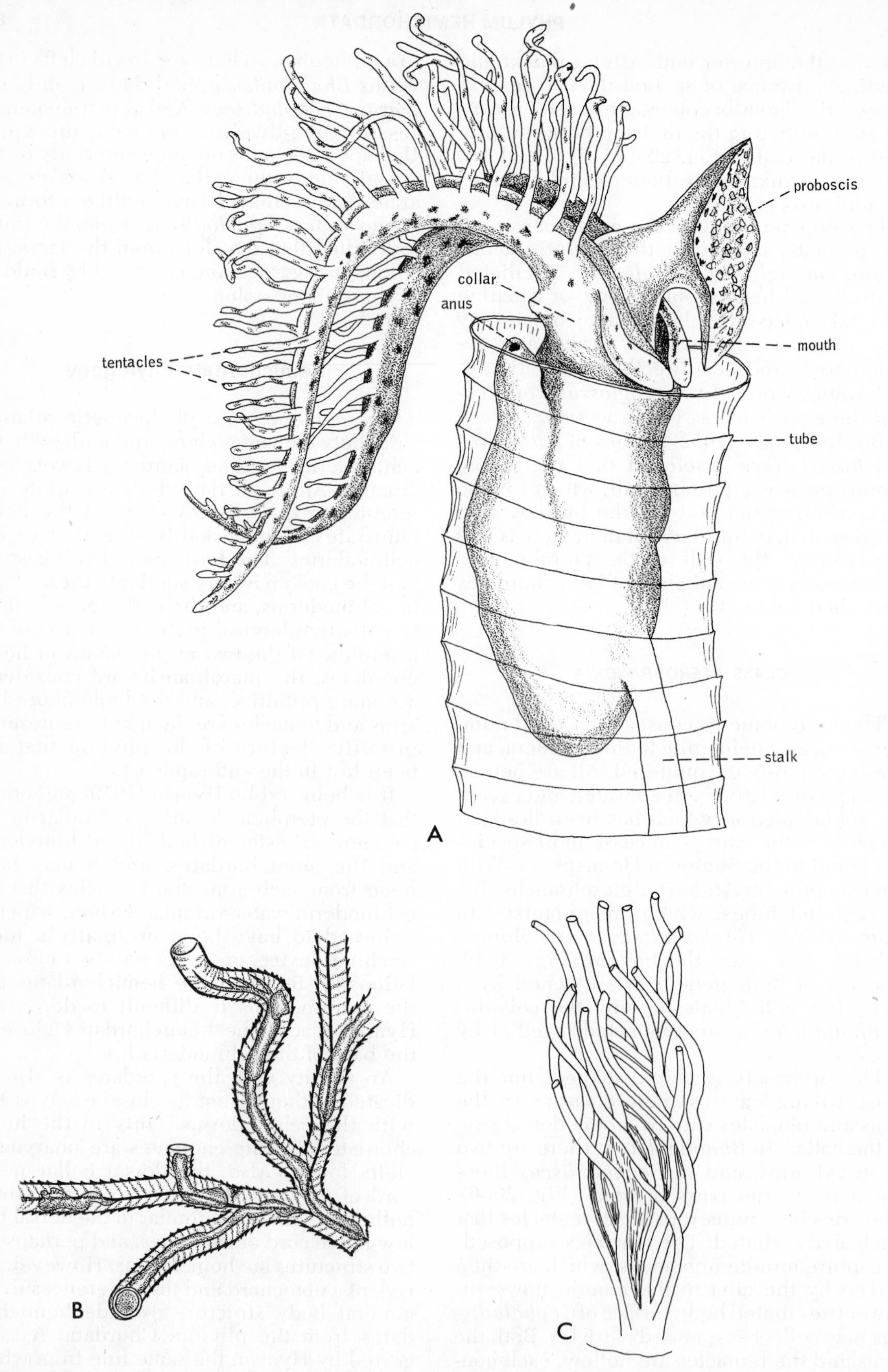

Figure 20–6. *A*. The pterobranch hemichordate, *Rhabdopleura*, in its tube (lateral view). (After Delage and Hérouard from Dawydoff.) *B*. Small portion of colony of *Rhabdopleura*, showing tube and individuals within connected by stolon. (After Lankaster.) *C*. Tubes of a colony of *Cephalodiscus densus*. (After Andersson from Hyman.)

PHYLUM CHORDATA

The chordates are the largest phylum of deuterostomes, but most chordates are vertebrates and fall outside the scope of this book. Two subphyla, the Urochordata and Cephalochordata, lack a backbone but possess the three distinguishing chordate characteristics —at some time in the life cycle, there can be found a notochord, a dorsal hollow nerve cord, and pharyngeal clefts. Since cephalochordates (*Amphioxus*) initiate most courses in comparative vertebrate anatomy, this group is not treated here. The urochordates, however, deserve some attention, for members of this group are less familiar, although they are very common marine animals.

Subphylum Urochordata

Adult urochordates, commonly known as tunicates, little resemble other chordates. Most are somewhat barrel-shaped animals attached by one end to the substratum. Only the larval stage, which looks like a microscopic tadpole, possesses distinct chordate characteristics. The tunicates consist of three classes—the Ascidiacea, the Thaliacea, and the Larvacea. The ascidians contain the majority of species and the most common and typical tunicates. The other two classes are specialized for planktonic existence. About 1300 species of urochordates have been described.

CLASS ASCIDIACEA

External Structure. Ascidians, often called sea squirts, are sessile tunicates and are common marine invertebrates throughout the world. The majority are found in littoral waters; they attach to rocks, shells, pilings, and ship bottoms or are sometimes fixed in mud and sand by filaments or a stalk (Fig. 20–7*B*). There are even a few interstitial ascidians. Some species have been dredged from considerable depths.

Ascidians often occur in great numbers. Piling surfaces in temperate waters are often covered with large species. On tropical reefs there is a great diversity of species, with many minute colonial forms living in crevices and beneath old coral heads. Others form large conspicuous clusters on gorgonian corals.

The bodies of solitary species range from spherical to cylindrical in shape (Figs. 20–8*A* and 20–9*A*). One end is attached to the substratum, and the opposite end contains two openings that may be extended as two separate siphons. Although gray and green colors are common, all shades of coloration are found in ascidians, and some colonial species are very beautiful. The body ranges in size from that of a pea to that of a large potato, which some species closely resemble. *Halocynthia pyriformis*, which is found on the Atlantic coast north of Cape Cod, is called the sea peach because of its similarity to the fruit in size, shape, and coloration. Some large irregular ascidians are commonly covered by other smaller sessile organisms, which make the ascidian even more inconspicuous.

The two openings at the free end of the body are the buccal siphon and the cloacal (atrial) siphon. These provide for the passage of a current of water through the animal. On the basis of the position of various organs within the body of the free-swimming larva, it is known that the buccal siphon in the metamorphosed adult marks the anterior end of the body. The atrial siphon indicates the dorsal side, and the visceral organs at the attached end of the body represent the morphological posterior end of the animal.

Body Wall, Atrium, and Pharynx. The body of ascidians is covered by a single layer of epithelial cells, but this epidermal covering does not form the external surface. Instead, the entire body is invested with a special mantle (the tunic) that is characteristic of most members of the subphylum and from which the name *tunicate* is derived (Figs. 20–7*A* and 20–8*A*). The tunic is usually quite thick but varies from a soft delicate consistency to one that is tough and similar to cartilage. The tunic of *Amaroucium stellatum*, called sea pork, has both the appearance and texture of salted pork. The tunic may be colored and commonly looks and feels like marble or glass. Not infrequently the tunic is translucent, and the colors of the internal organs account for the coloration of the animal.

A fibrous matrix composes the greater part of the tunic. Curiously, a principal constituent is a type of cellulose, called tunicine, but there are some species in which it occurs in relatively small amounts or may even be absent. In addition to cellulose, the tunic also contains proteins and inorganic compounds, and in some species calcium salts are precipitated in the form of distinct spi-

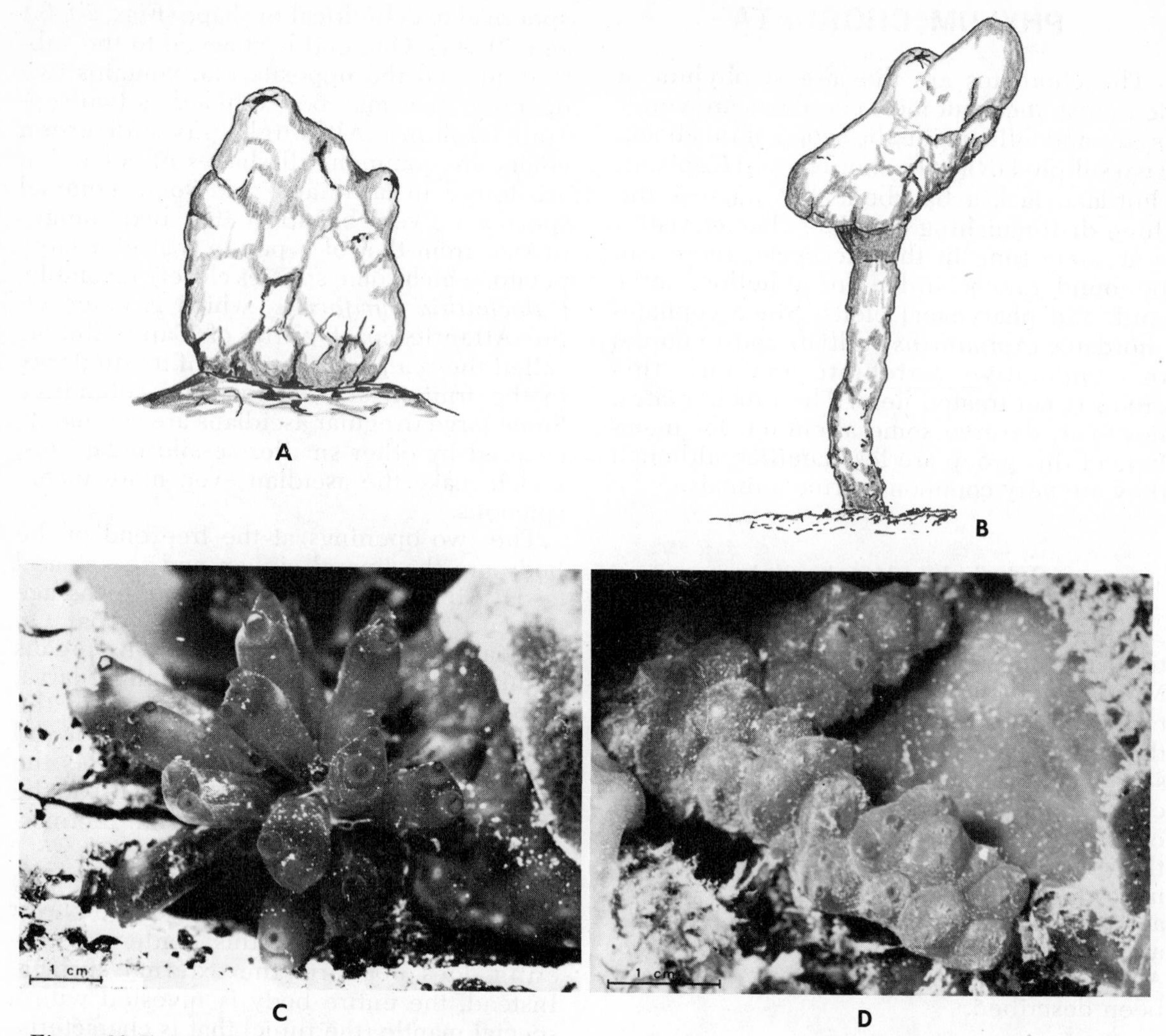

Figure 20–7. *A. Polycarpa pomaria,* a European shallow-water solitary ascidian with an irregular tough tunic. *B.* An Indo-Pacific deep-water ascidian which attaches in soft bottoms by means of a stalk. *C. Ecteinascidia turbinata,* a colonial West Indian ascidian which is orange in color. *D.* A small tropical ascidian which forms sheet-like colonies beneath stones. (*C* and *D* by Betty M. Barnes.)

cules. Another unusual feature of the tunic is the presence of ameboid cells and blood cells that have migrated from the body mesenchyme. Moreover, in some species such as *Ciona*, the tunic is supplied with blood vessels. It is by means of the tunic that ascidians adhere to the substratum, and the tunic is often roughened or papillose in this region. Often rootlike extensions called stolons ramify from the base of the body, and these too are covered by the tunic (Fig. 20–9*A*).

Within the tunic, the body of ascidians can be conveniently divided into three regions —an anterior or distal pharyngeal region containing the pharynx, an abdominal region containing the digestive tract and other internal organs, and a postabdomen (Fig. 20–9*A*). The postabdomen is the most basal part of the body and is commonly indistinct from the abdominal region (Fig. 20–8*A*). In some species, however, the postabdomen is as long as the thoracic and abdominal regions combined and contains the heart and reproductive organs (Fig. 20–9*B*).

The anterior buccal siphon opens internally into a large pharyngeal chamber. The walls of the pharynx are perforated with small slits, permitting water to pass from the pharyngeal cavity into the surrounding atrium and then out by way of the atrial siphon.

Within the buccal siphon is a circlet of projecting tentacles that prevent large objects from coming in with the water current (Fig. 20–8*A*). Exterior to the tentacles, the wall of the siphon is lined with an infolded layer of the epidermis and the tunic. Below the tentacles, the pharyngeal wall begins, and its lining is derived from entoderm.

The pharynx may be cylindrical or somewhat laterally compressed. On the ventral side—the side opposite the atrial siphon—a deep groove (the endostyle) extends the length of the pharyngeal wall (Fig. 20–8*A* and *B*). The lateral walls of the groove are ciliated and project inward toward each other, so that the endostyle is shaped somewhat like a keyhole. The bottom of the groove is lined with both mucus-secreting cells and a median row of cells with long flagella. Posteriorly, the endostyle terminates in a little pit at the base of the pharynx, but its lateral walls continue across the pharynx floor to the esophagus as two parallel adjacent ciliated ridges called the retropharyngeal band. On reaching the anterior end, the lateral ridges of the endostyle separate and encircle the top of the pharynx, forming the right and left peripharyngeal ridges.

Above and adjacent to each peripharyngeal ridge is another ridge that may or may not be ciliated. The two ridges thus form a groove between them. On reaching the dorsal side, each peripharyngeal ridge passes downward a short distance and then to each side of a large projecting ridge (the dorsal lamina), or to a row of finger-like processes (languets) that run posteriorly to the esophagus along the dorsal side of the pharynx. The margin of the ridge curves to the right and forms a gutter along that side of the ridge (Fig. 20–8*C*).

Between the dorsal lamina and the endostyle, the side walls of the pharynx are perforated by vertical slits called stigmata (Figs. 20–8*A* and 20–9*A*). The stigmata are arranged in horizontal rows with horizontal

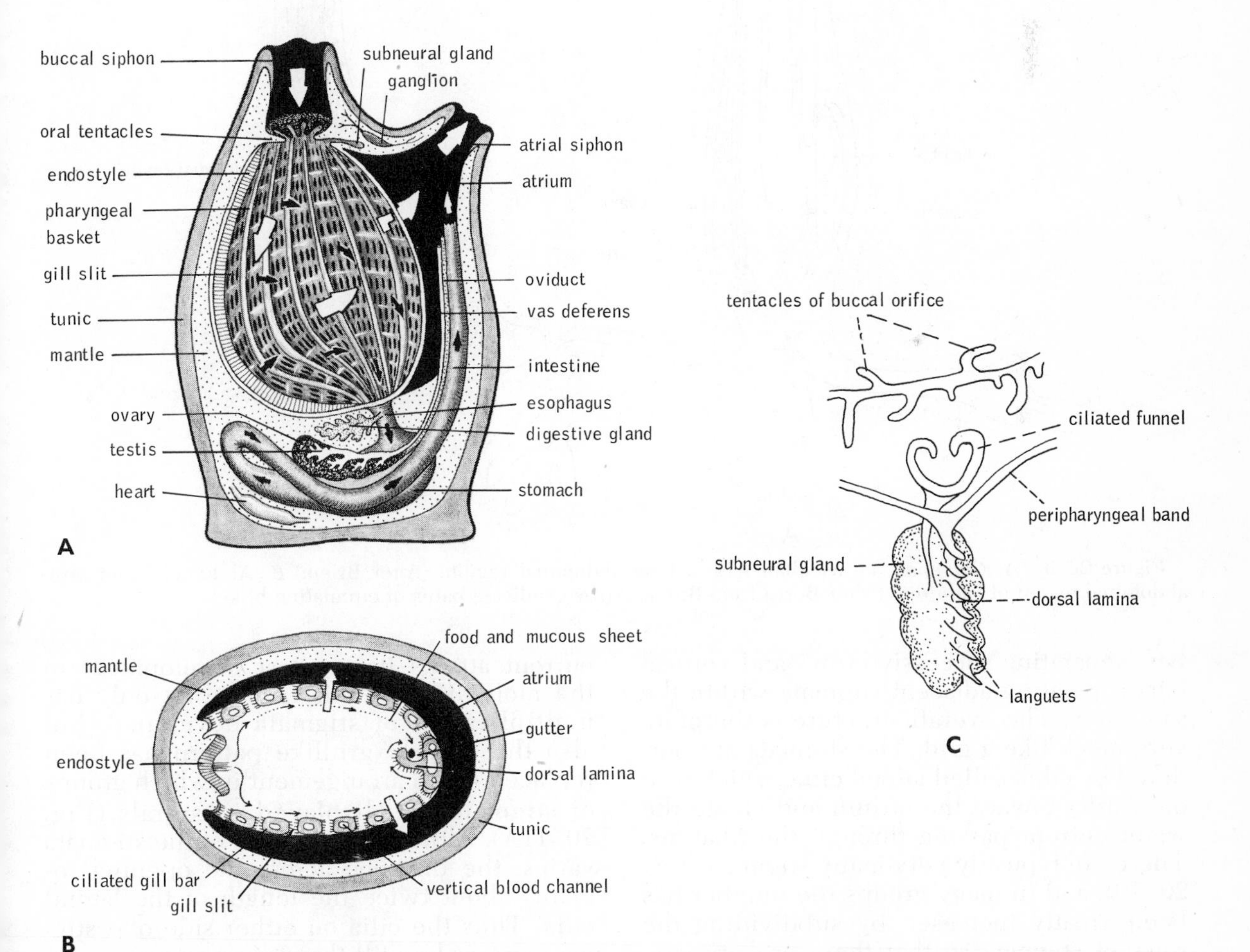

Figure 20–8. *A* and *B*. Diagrammatic lateral view (*A*) and cross section (*B*) of a tunicate, showing major internal organs. Large arrows represent the course of the current of water; small arrows, that of food and mucous sheet. Stomach, intestine, and other visceral organs are embedded in the mantle. *C*. Subneural gland and adjacent pharyngeal area of *Ciona*. (After Bullough.)

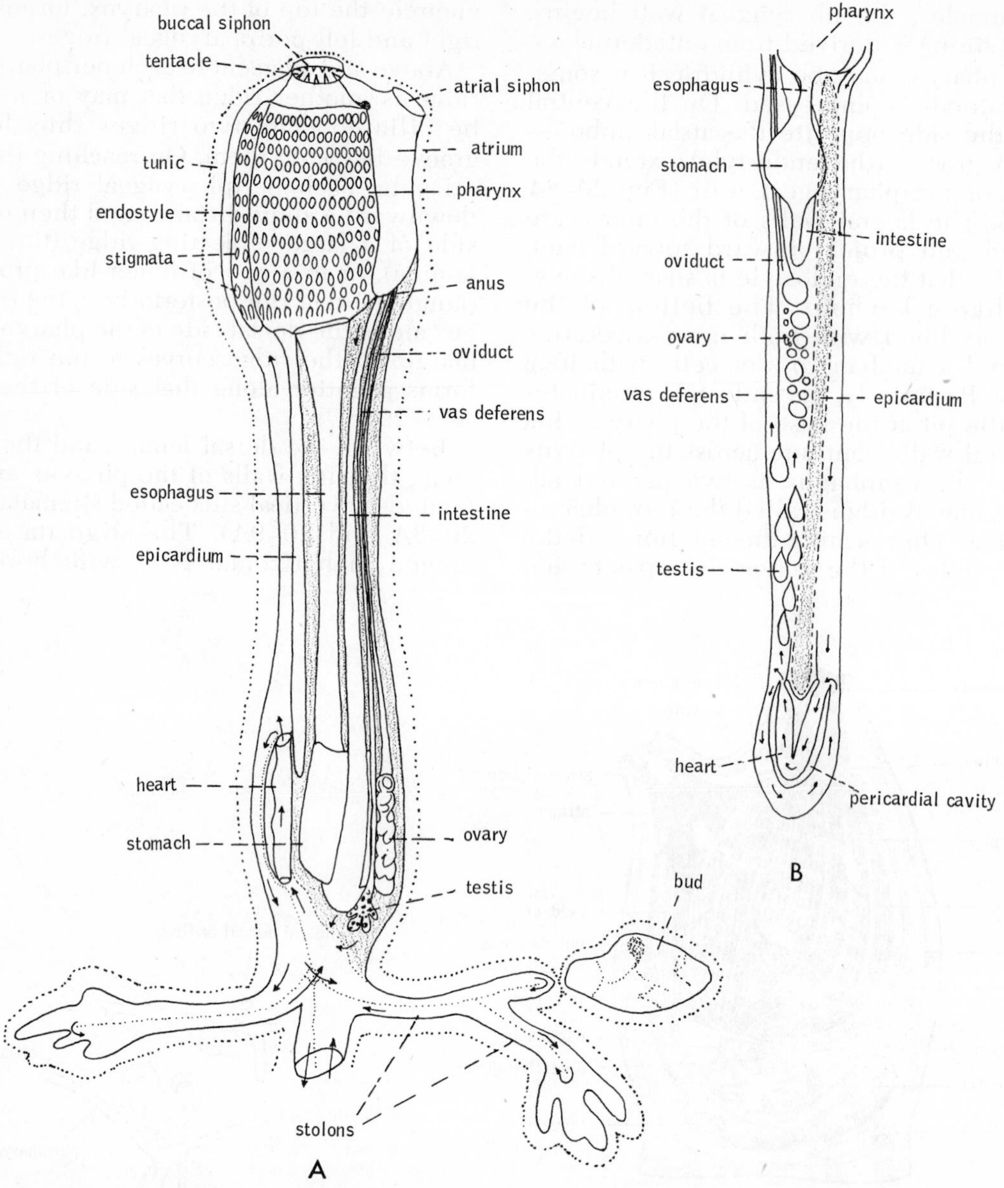

Figure 20–9. *A. Clavelina*, an ascidian with a long abdominal region. (After Brien.) *B.* Abdominal and post-abdominal region of *Sydnium.* (After Berril from Brien.) Arrows indicate paths of circulating blood.

bars separating successive rows and vertical bars separating adjacent stigmata within the same row. The overall structure is therefore very much like a grid. The stigmata are bordered by cilia, called lateral cilia, which beat outwardly toward the atrium and create the water current passing through the pharynx. There are typically very many stigmata (Fig. 20–10), and in many groups the number has been greatly increased by subdividing the rows of stigmata so that there are primary, secondary, and tertiary rows. This tendency to increase the surface area of the pharynx for filtering food from the passing water current attains its greatest development in the molgulids. In this family, not only has multiplication of stigmata developed, but also the orderly gridlike pattern has been replaced by an arrangement in which groups of stigmata are displayed in spirals (Fig. 20–11A). Although the shape of the stigmata varies, the distance across the opening remains about twice the length of the lateral cilia. Thus the cilia on either side of a stigma more or less fill the opening.

In some ascidians, water flow through the body is augmented by rhythmic muscular contractions of the body wall.

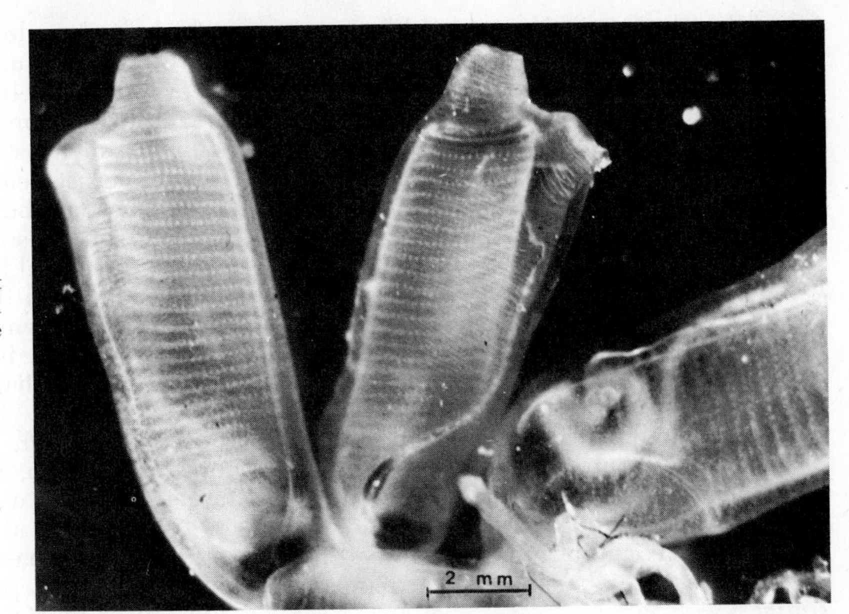

Figure 20–10. A transparent ascidian in which the pharyngeal basket can be seen through the tunic. (By Betty M. Barnes.)

The pharynx is completely surrounded by the atrium except along the midventral line, where the pharynx is attached to the body wall. In addition, the atrium is crossed by cordlike strands of tissue that apparently function to limit the expansion of the cavity during the flow of water through the body. Dorsally the atrium opens to the exterior through the atrial siphon. The atrial region just in front of the siphon is sometimes called the cloaca, because the anus and the gonoducts empty here. All of the atrium, both the pharyngeal and outer sides, is lined with an epithelium derived from ectoderm and is continuous through the atrial siphon with the external epidermis. The inner lin-

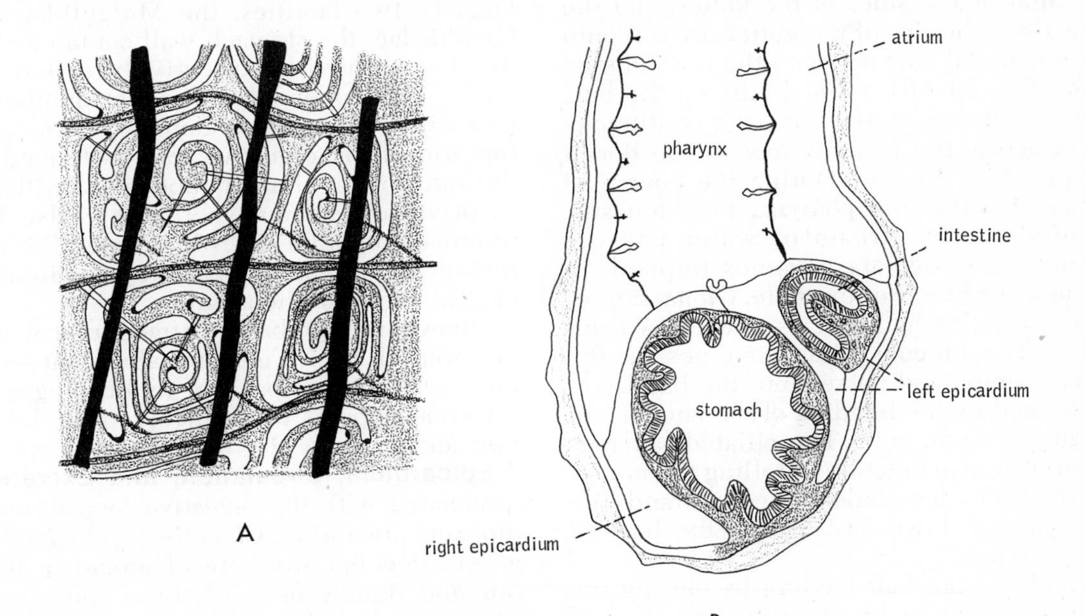

Figure 20–11. A. Spiral stigmata of *Corella parallelogramma.* (After de Selys Longchamps from Brien.) *B.* Abdominal region of *Ciona intestinalis,* showing epicardia (longitudinal section). (After Damas from Brien.)

ing of the pharynx is thus derived from entoderm and the outer covering from ectoderm; between these two layers lies mesenchymal tissue. The body wall in the atrial region consists of an inner and an outer layer of ectodermal epithelium with mesenchyme in between.

The body-wall mesenchyme contains striated muscle bands extending longitudinally toward the siphons. Circular bands are also present and are particularly well developed in the siphon walls, where they act as sphincters. The body-wall muscles can cause a limited degree of general body contraction, depending upon the thickness and rigidity of the tunic. When an animal is exposed at low tide or taken out of the water, contractions of the body and siphons cause the water in the pharynx and atrium to be forced from the siphons as jets—hence the name *sea squirt.*

Nutrition. Tunicates are filter feeders and remove plankton from the current of water that passes through the pharynx. The water current is produced by the beating of long cilia located on the margins of the stigmata, and an enormous quantity of water is strained for food. A specimen of *Phallusia* only a few centimeters long can pass 173 liters of water through the body in 24 hours.

The endostyle is the principal center for the elaboration of mucus. The apparently rather immobile long flagella at the bottom of the gutter probably deflect the mucus to the right or left sides of the gutter, and the cilia along the lips of the gutter drive it onto the horizontal bars between the rows of stigmata (Fig. 20–8*B*). Frontal cilia on the bars carry the mucus in the form of a continuous sheet across the pharynx toward the dorsal lamina (Fig. 20–8*B*). During the course of transport across the pharynx, plankton suspended in the stream of water passing through the stigmata becomes trapped on the mucous film. For example, *Ciona* can remove particles as small as one to two microns. The mucous film converges on the dorsal lamina and passes into the basal gutter formed by the bending of the lamina and languets to one side. The ciliated languets apparently also function in rolling the mucus in cords. The food-laden mucous strands are now carried downward toward the base of the pharynx.

Ascidians can halt feeding by closing the buccal siphon or by stopping the flow of mucus from the endostyle.

There are a few soft-bottom ascidians which feed on deposit material from the surrounding sediment. There are also some deep-water species that have lost the pharyngeal perforations and that feed on small animals such as nematodes and small epibenthic crustaceans.

The base of the pharynx on the dorsal side contains the esophageal opening, toward which the dorsal lamina and the retropharyngeal band (from the endostyle) are directed. The post-pharyngeal part of the digestive tract is located in the abdomen and is arranged in a **U**-shaped loop (Figs. 20–8*A* and 20–9*A* and *B*). The esophagus forms the descending arm, the stomach occupies the turn of the loop, and the intestine forms the ascending arm, which opens into the cloacal region of the atrium beneath the atrial siphon. This basic plan may be variously modified; not infrequently, the loop is twisted or the intestine is coiled to various degrees, and the loop may be greatly abbreviated or quite long.

The esophagus is a narrow tube of varying length. Its walls lack muscles, but the internal lining is ciliated for the movement of food particles. The stomach is an enlargement of the digestive loop at the turn of the **U**, and there is usually a relatively well-developed sphincter at each end. Internally, the stomach is lined with secretory and absorptive cells, and in many species, such as *Clavelina* and *Styela*, there is a ciliated gutter extending the length of the dorsal surface. In two families, the Molgulidae and Cynthiadae, the stomach wall contains glandular outgrowths, collectively called the "liver," that open by a variable number of pores into the stomach lumen. The ascending arm of the digestive tract is formed by the intestine, a slender tube lined with absorptive and mucus-secreting cells. The terminal end of the intestine is modified as a rectum and opens through the anus into the cloacal region of the atrium.

Digestion is probably extracellular within the stomach, which contains abundant secretory cells. On the basis of its histology, the intestine appears to be involved in absorption and to be a site for glycogen storage.

Epicardium, Circulation, and Excretion. Associated with the digestive loop of many tunicates, including *Clavelina* and *Ciona*, is a very peculiar structure of unknown function and significance called the epicardium. The epicardium is usually a simple tube that parallels the digestive loop (Fig. 20–9*A* and *B*). It arises as a double evagination of the base of the pharynx on each side of the retropharyngeal band and is thus entodermal. Distal to their origin from the pharynx, the two evaginations fuse and extend downward to one side of the digestive loop. In *Ciona*, the distal unpaired portion has disappeared, but the basal right and left tubes are present and are located one on each side of the digestive loop. Moreover, these tubes are greatly enlarged, and they surround the stomach, intestine, and other organs in the same manner as a coelom (Fig. 20–11*B*).

The tunicate blood-vascular system is remarkable in many respects. The entire system is open, even the heart. The heart is a short tube lying in a pericardial cavity at the base of the digestive loop. The heart is peculiar in that it is a fold of the pericardial wall, it is supplied with muscle fibers, and it bulges into the pericardial cavity (Figs. 20–8*A* and 20–9*A* and *B*). Thus, the heart is really only a specialized part of the pericardium and has no true wall or lining. The heart is somewhat curved or **U**-shaped, and one end is directed dorsally and the other ventrally. Each end opens into a large vessel or channel, but all of the circulatory pathways lack true walls and are merely sinus channels in the mesenchyme.

From the ventral end of the heart, blood passes beneath (outside) the endostyle by way of a large subendostylar vessel that runs along the ventral side of the pharynx. Along its course, the subendostylar vessel gives off transverse channels to the bars between the rows of stigmata. These channels also connect with vertical channels. Thus, blood circulating through the pharyngeal grid is provided with a great surface area for gas exchange with the water current passing through the pharynx.

Dorsally, the network of pharyngeal channels drains into a longitudinal vessel (median dorsal sinus) that runs beneath the dorsal lamina of the pharynx. On reaching the abdomen, the median dorsal sinus breaks up into many smaller channels supplying the digestive loop and other visceral organs. These channels eventually drain into a dorsal abdominal sinus, which leads to the dorsal end of the heart. In those ascidians, such as *Ciona*, in which the tunic is provided with blood vessels, a vessel to the test usually arises from each of the two main channels leading into the heart, and both of these vessels penetrate the test at the posterior ventral side of the animal.

One of the most interesting features of the tunicate circulatory system is the periodic reversal of blood flow direction. For a time, the heart pumps blood out of its dorsal end into the dorsal abdominal sinus, and blood enters the heart from the subendostylar sinus. After a short period of rest, the direction of contraction is reversed, and blood is pumped in the opposite direction. The heart beat appears to be neurogenic but not under direct control of the dorsal ganglion. There is an excitation center at each end of the heart, which is responsible for initiating contractile waves over the heart from that point, and each alternates in dominance over the other.

The blood of tunicates is no less interesting than other parts of the vascular system. The plasma, which is slightly hypertonic to sea water, contains numerous amebocytes (or lymphocytes). A number of different types have been described. Some actively migrate between the blood and the tissues; some are phagocytic.

In certain ascidians, such as the Ascidiidae and the Perophoridae, the blood contains a type of green cell called a vanadocyte. These cells are unique in containing vanadium that is bound up with hydrosulfuric acid and perhaps protein. Considering the low concentration of vanadium in sea water, the ability of these ascidians to take up and concentrate vanadium to high levels is remarkable. The element is removed from the sea water by the pharyngeal mucus, on which it is absorbed. In two families, the Cionidae and the Diazonidae, the vanadium chromogens are found in the blood plasma and are concentrated in certain regions of the body rather than in the cells. Vanadium chromogens in ascidians do not function in oxygen transport, but the exact function is unknown. This vanadium compound is kept in a reduced state by the presence of a high concentration of sulfuric acid within the same cell.

Some amebocytes are filled with crystals or with yellow, pink, red, white, black, or blue pigment. These cells constitute the excretory system of ascidians. After accumulating waste in the course of circulation, the laden cells in most tunicates tend to become fixed in certain regions of the body, especially the digestive loop and gonads, where they form a diffuse covering on the outer walls of the organs. Within this network of fixed cells, vesicles develop that contain accumulations of uric acid. The deposition of excretory cells begins in the larval stage, accelerates during metamorphosis, and during the life of the animal steadily increases in thickness and density. The production of uric acid results from purine metabolism. Ammonia is produced, as in other marine animals, from protein metabolism and is excreted into the sea water passing through the pharynx.

In all tunicates is another network of vesicles and tubules (pyloric glands) on the outer walls of the intestine. By way of one or many collecting canals, this network drains into the anterior end of the stomach. Although they are found in the same region as the excretory vesicles, the pyloric glands appear to be distinct from the excretory vesicles. The products of the pyloric glands are reported to be calcareous and probably are of an excretory nature.

Nervous System. The nervous system is relatively simple and consists of a cylindrical to spherical cerebral ganglion, or "brain," located in the mesenchyme of the body wall between the two siphons (Figs. 20–8*A*). When the ganglion is cylindrical or ovoid, as is usually the case, one end is directed toward one siphon and the other end toward the opposite siphon. From each end of the ganglion there arises a variable number of mixed nerves, and in some species there are also lateral nerves. The nerves arising from the anterior end of the ganglion supply the buccal siphon; those issuing from the posterior end innervate the greater part of the body—the atrial siphon, the gills, and the visceral organs.

Beneath the cerebral ganglion lies a glandular body called the neural or subneural gland (Fig. 20–8*A* and *C*). The lumen of the gland extends anteriorly as a duct opening into the pharynx by way of a large ciliated funnel. Posteriorly the gland continues as a cord of tissue that represents the larval nerve cord and is therefore homologous to the neural tube of other chordates.

Experiments involving the removal of the cerebral ganglion indicate that the cerebral ganglion is not an especially important center for the control of vital reflexes in the physiology of ascidians. *Ciona* is at first quite flaccid following decerebration, but after a 24-hour period of recovery, all of the body functions are carried out normally. Body contractions, however, are somewhat slower.

There are no special sense organs in ascidians, but sensory cells are very abundant on the internal and external surfaces of the siphons, on the buccal tentacles, and in the atrium. Tactile cells and probably chemoreceptors are included among these sensory cells and very likely play a role in controlling the current of water passing through the pharynx. Pigmented cups containing cylindrical cells with distal tufts of cilia reportedly have been found in a number of species, but whether these structures warrant the name *eyes* is not certain.

Reproduction. With few exceptions, tunicates are hermaphroditic. There are usually a single testis and a single ovary that lie in close association with the digestive loop (Figs. 20–8*A* and 20–9*A* and *B*). The ovary is typically located above the stomach and is a saccular body with two bandlike ventrolateral thickenings. These bands are the germinal areas of the ovaries, and the developing eggs deflect the ovarian wall outward to form follicles. When mature, the eggs are carried in an oviduct that runs parallel to the intestine and opens into the cloaca in front of the anus.

The testis lies below the ovary and is composed of a cluster of small sacs that open into a sperm duct. The sperm duct parallels the course of the oviduct and opens into the cloaca.

In a few families, which include *Molgula* and *Styela,* the gonads are doubled, and an ovary and a testis are located in the right body wall of the branchial region; a left set is located within the digestive loop. The gonoducts are also doubled and empty separately into the cloaca.

Colonial Organization. Most of the larger ascidians, such as *Styela, Ascidia,* and species of *Clavelina* and *Molgula,* are solitary forms and are often called simple ascidians. There are, however, many colonial or compound species. Colonial organization has arisen independently a number of times within the class, and a number of different types of colonies occur. In general, the individuals composing a colony are very small, although the colony itself may reach a considerable size.

In the simplest colonies, the individuals themselves are discrete, but are united by stolons. For example, in *Perophora* the colony is like a vine with a long, trailing, branching stolon to which are attached the globular individuals (Fig. 20–12*A*). In others of these compound forms, such as some species of *Clavelina* and *Molgula,* the stolons are short, and the individuals are joined, forming tuftlike groups (Fig. 20–7*C*) or a sheet of adjacent individuals.

A more intimate association is seen in some ascidians, in which not only the stolons

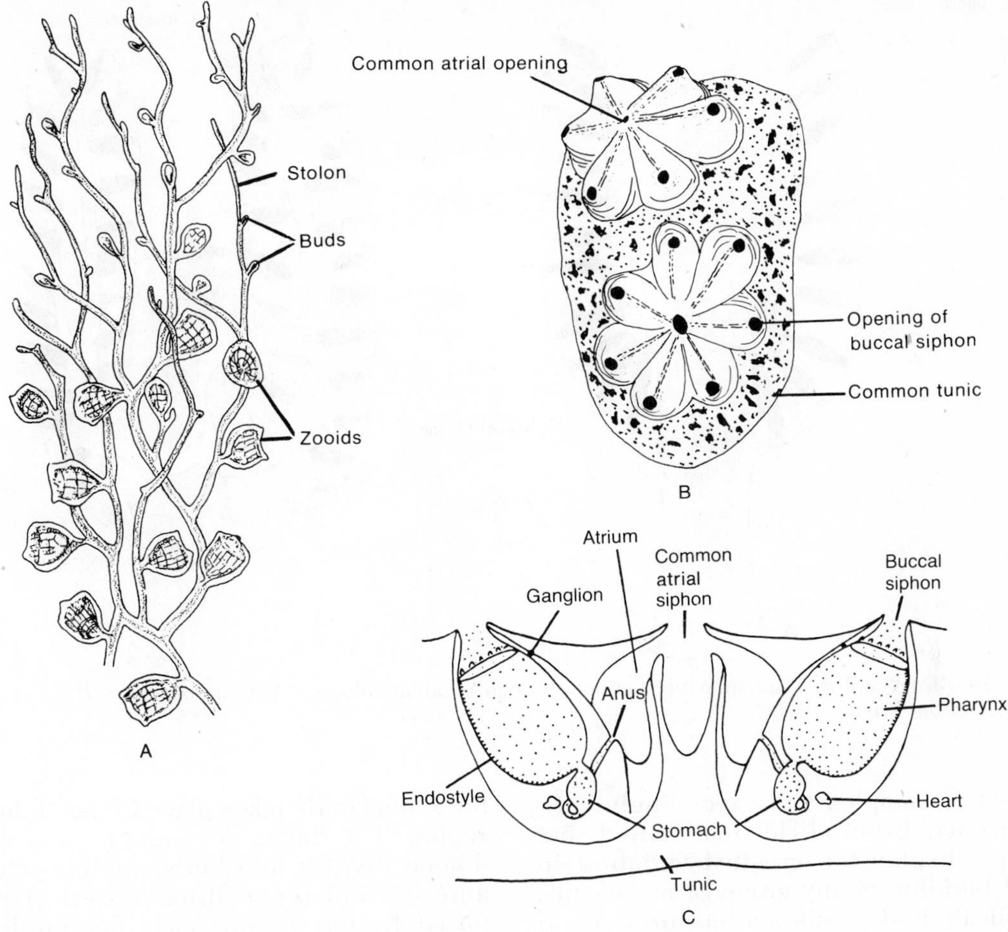

Figure 20–12. *A*. Colony of *Perophora viridis*. (After Miner.) *B*. *Botryllus schlosseri*, a compound ascidian. (After Milne-Edwards from Yonge.) *C*. Vertical section through *Botryllus*. (After Delage and Hérouard.)

but also the basal parts of the bodies are joined to other individuals, forming a common tunic (Fig. 20–7*D*).

In the most specialized colonial families, all of the individuals composing the colony are completely imbedded in a common tunic. Usually there is a very regular arrangement of individuals within the test. *Botryllus* forms flat encrusting colonies, in which the members are organized in a star-shaped pattern (Fig. 20–12*B* and *C*). The buccal siphons of each member open separately to the exterior, but the atria open into a common cloacal chamber, which has one central aperture in the middle of the colony. The individuals of *Botryllus* are only a few millimeters in diameter, but since a single tunic may contain a number of star-shaped clusters, the entire colony may be 12 to 15 cm. in diameter.

Similar configurations are found in other compound ascidians. In *Cyathocormus*, the colony is shaped like a goblet with a stalk-like attachment to the substratum and a large common cloaca (Fig. 20–13*A*). The individuals are oriented horizontally with the buccal siphons located on the external surface and the atrial siphons opening into the common cloaca. In *Coelocormus*, the wall of the cup is doubled by folding so that each wall of the cup contains two layers of individuals (Fig. 20–13*B*). The buccal siphons of one layer open to the exterior and those of the other open to the interior. Between the two walls, the atria open into an internal cloacal canal that exits through a single aperture at the bottom of the cup.

Asexual Reproduction. Regeneration and asexual reproduction are highly developed in colonial ascidians but absent in the families Cionidae and Ascidiidae. Asexual reproduction takes place by means of bud-

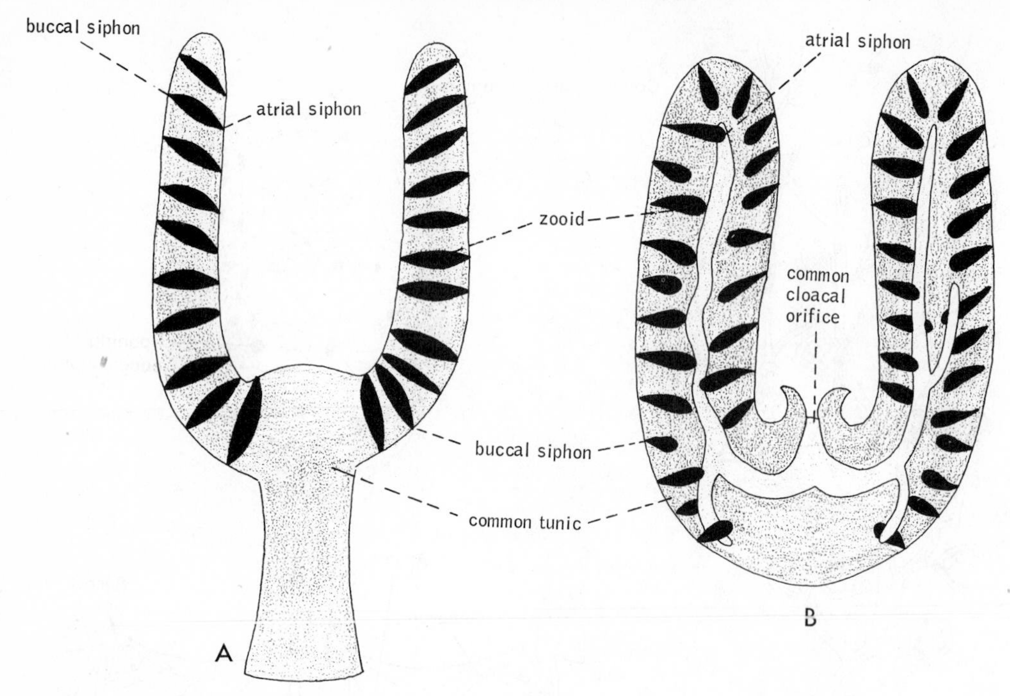

Figure 20–13. *A* and *B*. Diagrammatic sections through ascidian colonies. *A*. *Cyathocormus*. *B*. *Coelocormus*. (After Okada from Brien.)

ding, but is complex and exceedingly variable. In fact, Brien (1948) maintained that ascidians display the greatest variation in asexual budding of any group of metazoans.

A tunicate bud is called a blastozooid and originates in different parts of the body in different groups of ascidians. With the variation of the site of bud formation, there is a corresponding variation in the germinal tissues included within the bud. The most primitive type of budding appears in *Clavelina* and other forms, in which the bud arises from the stolon (Fig. 20–9A). The bud first appears as a swelling of the stolon filled with reserve food material, and may not develop until the stolon breaks away from the parent or the parent degenerates. In *Clavelina* and other members of this family, epidermis and mesenchyme comprise the germinal tissues of such stoloniferous buds.

In the colonial Polyclinidae, which have a long postabdomen bearing an extension of the epicardia and a cord of tissue from the gonads, the postabdomen breaks up within the tunic like a string of sausages (Fig. 20–14A). Each section becomes a bud, with the epidermis, genital cord, and epicardial tissue acting as germinal tissues. The blastozooids reorganize in the base of the old tunic to form a new colony.

Budding in *Diazona* and other members of the same family takes place in the abdominal region. The thorax degenerates, and the abdomen divides into buds, starting with the anterior and proceeding posteriorly. Germinal tissues in this case are epidermis, epicardium, and parts of the digestive loop.

In a number of families, budding is precocious and begins at the larval stage. For example, in the Didemnidae a few hours after the settling of the larva, two buds form in the abdominal region (Fig. 20–14*B* and *C*). The two buds fuse, one forming the thoracic half and one the abdominal half of the new individual. In the Botryllidae, which forms the star-shaped colonies described earlier, a single bud develops on the larva. After fixation, the larva dies, and the bud develops as the initial member of the colony. The star-shaped pattern of the adult colony results from geometrical secondary budding.

Embryogeny. It is in the embryogeny and the larval stage of tunicates that the chordate affinities of the class become strikingly apparent. Solitary species generally have small eggs with little yolk. The eggs are shed from the atrial siphon, and development takes place in the sea water. The eggs of such oviparous species are frequently surrounded by special membranes that act as flotation devices. The eggs of colonial species are typically richer in yolk material and are usually brooded in the atrium, which sometimes contains special incubating pockets. Hatching may take place at the larval stage, and the larva then leaves the parent; or the entire course of development may take place within the atrial cavity. In general, development in brooding species with considerable yolk material is more rapid and condensed than in nonbrooding forms.

Cleavage is complete and slightly unequal, and leads to a flattened coeloblastula. Gastrulation is accomplished by epiboly and invagination, and the large archenteron completely obliterates the old blastocoel. The blastopore marks the posterior end of the embryo and gradually closes while the embryo elongates along the anteroposterior axis. Along the middorsal line, the archenteron gives rise to a supporting rod—the notochord. Laterally, the archenteron proliferates mesodermal cells that form a cord of cells along each side of the body. In this respect, development departs from that shown in *Amphioxus* and other deuterostomes, because there is no pouching of the archenteron. The mesoderm forms the body mesenchyme, and a coelomic cavity never appears nor is there any segmentation. The ectoderm along the middorsal line differentiates as a neural plate, sinks inward, and rolls up as an internal neural tube.

Continued development leads to an elongated microscopic larva, called an appendicularia or more commonly a "tadpole" larva (Fig. 20–15*A*). A distinct tail represents the posterior half of the larva and contains the notochord and the neural tube. Dorsally, the anterior half of the larva contains a pigmented cup, a statocyst, and the dilated end of the neural tube, which becomes the cerebral ganglion. The mouth, which later becomes the buccal siphon, is located anteriorly and may not be open during the larval stage. The mouth leads into the pharynx, which in turn is followed by a twisted digestive loop with a dorsally directed intestine. The pharynx contains a ventral endostyle, but at first the pharynx possesses only two stigmata. The stigmata empty into a small pocket, the future atrium. At the extreme anterior of the larva are three ectodermal projections—the fixation papillae. The entire larva is covered by a tunic secreted by the surface ectoderm, and in the tail region the tunic is extended dorsally and ventrally to form a fin. The tail is the larval locomotor organ, and movement is brought about by special muscle bands.

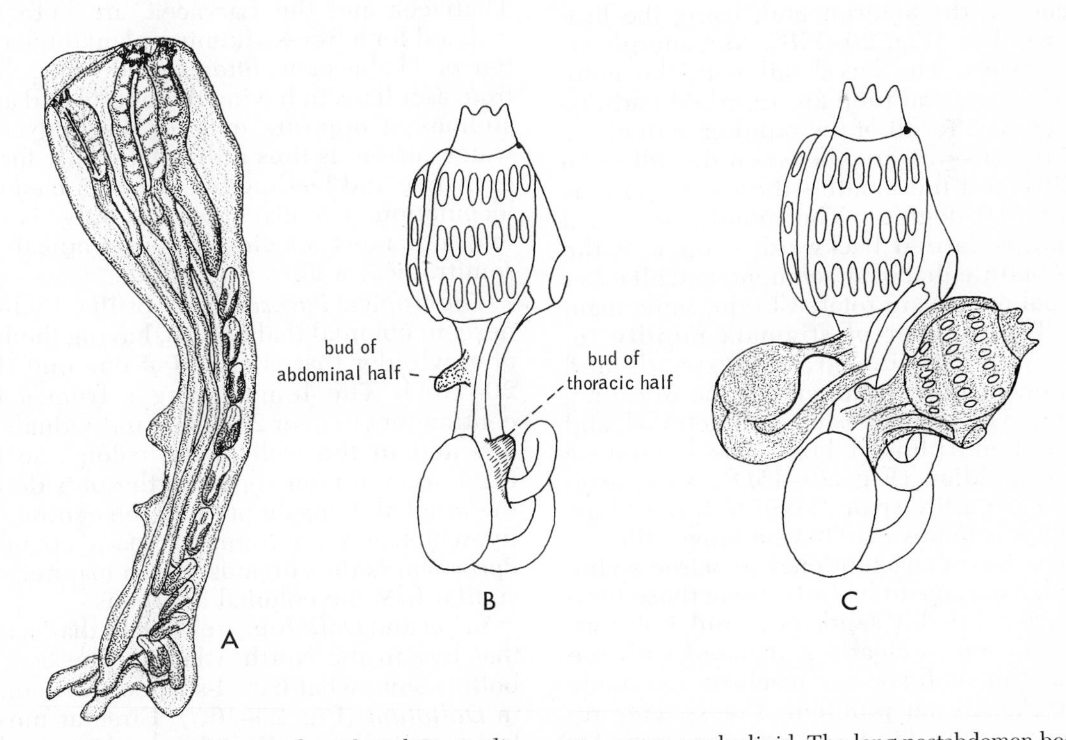

Figure 20–14. A. Budding in the colonial *Circinalium concrescens*, a polyclinid. The long postabdomen becomes sectioned into buds within the common tunic of the colony. (After Brien.) *B* and *C*. Budding in the Didemnidae. *B*. Two buds develop in abdominal region of parent body. *C*. Buds fuse, one forming the abdominal half of the new individual and one forming the thoracic half. (After Salfi from Brien.)

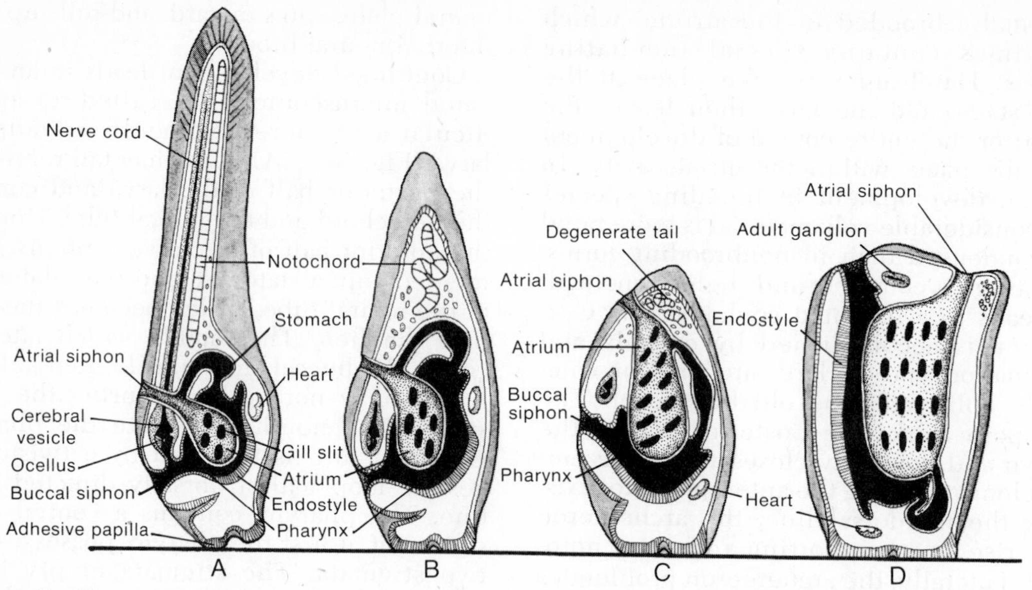

Figure 20–15. *A*. Diagrammatic lateral view of a urochordate tadpole larva, which has just attached to the substratum by the anterior end. *B* and *C*. Metamorphosis. *D*. A young individual just after metamorphosis. (*B* modified after Seeliger from Brien.)

After a free-swimming, but nonfeeding, planktonic existence of a few minutes to a few days, the larva settles to the bottom and attaches at the anterior end, using the fixation papillae (Fig. 20–15*B*). Metamorphosis now ensues. The larval tail with the notochord and neural tube are resorbed and disappear. As a result of the rapid growth of the chin region—the area between the adhesive papillae and the mouth—the entire body is turned 180 degrees. The mouth, or buccal siphon, is carried backward to open at the end opposite that of attachment, and all other internal organs are rotated in the same manner. The number of stigmata rapidly increases; the atrium greatly expands and makes contact with the end of the intestine, or anus. The siphons become functional, and the metamorphosed larva has become a young ascidian (Fig. 20–15*D*). Most ascidians have a life span of one to three years, although colonies may have a longer life.

There has been a tendency for a free-swimming larval stage to be lost among those tunicates that inhabit sand and mud buttoms. This loss seems clearly associated with the fact that in such species reaching a suitable substratum is no problem. For species requiring certain kinds of firm substrata for attachment, a free-swimming larva is indispensable.

CLASS THALIACEA

The other two classes of tunicates, the Thaliacea and the Larvacea, are both specialized for a free-swimming planktonic existence. Thaliaceans, often called salps, differ from ascidians in having the buccal and atrial siphons at opposite ends of the body. The water current is thus utilized not only for gas exchange and feeding but also as a means of locomotion. The class contains only six genera, and most species live in tropical and semitropical waters.

The tropical *Pyrosoma* are brilliantly luminescent colonial thaliaceans, having the form of a cylinder that is closed at one end (Fig. 20–16*A*). The length ranges from a few centimeters to over 2 m. The individuals are oriented in the wall of the colony, so that the buccal siphons open to the outside and the atrial siphons empty into the central cavity, which acts as a common cloacal chamber. *Pyrosoma* is thus organized in a manner very similar to some colonial ascidians.

Salpa and *Doliolum* are solitary thaliaceans that live in the North Atlantic. The body in both is somewhat barrel-shaped, particularly in *Doliolum* (Fig. 20–16*C*). Circular muscle bands, complete in *Doliolum* and incomplete in *Salpa*, produce contractions of the body wall that drive water through the atrial cavity. *Salpa* is especially peculiar in having only two gill clefts, which are so enormous that there are virtually no side walls of the pharynx remaining.

The life cycles of all thaliaceans involve asexual budding. Since the sexually reproducing stages are derived as buds (blastozooids) from the individual formed from the egg (oozooid), thaliaceans are often considered to exhibit metagenesis, but there seems no more justification for considering these animals metagenetic than there is for the hydroid coelenterates.

There is no larva in *Pyrosoma*. Each member of the colony produces a single yolky egg that undergoes meroblastic cleavage in the parent atrium. The later embryo (oozooid) produces four buds along the length of a stolon, which is coiled about the oozooid (Fig. 20–16*B*). Rupture of the parental atrium liberates the oozooid and buds, which sink to the bottom. The oozooid degenerates,

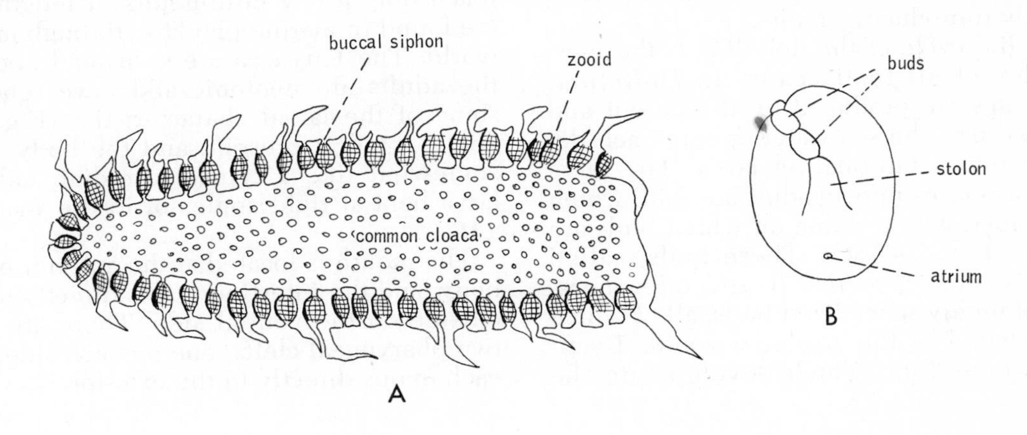

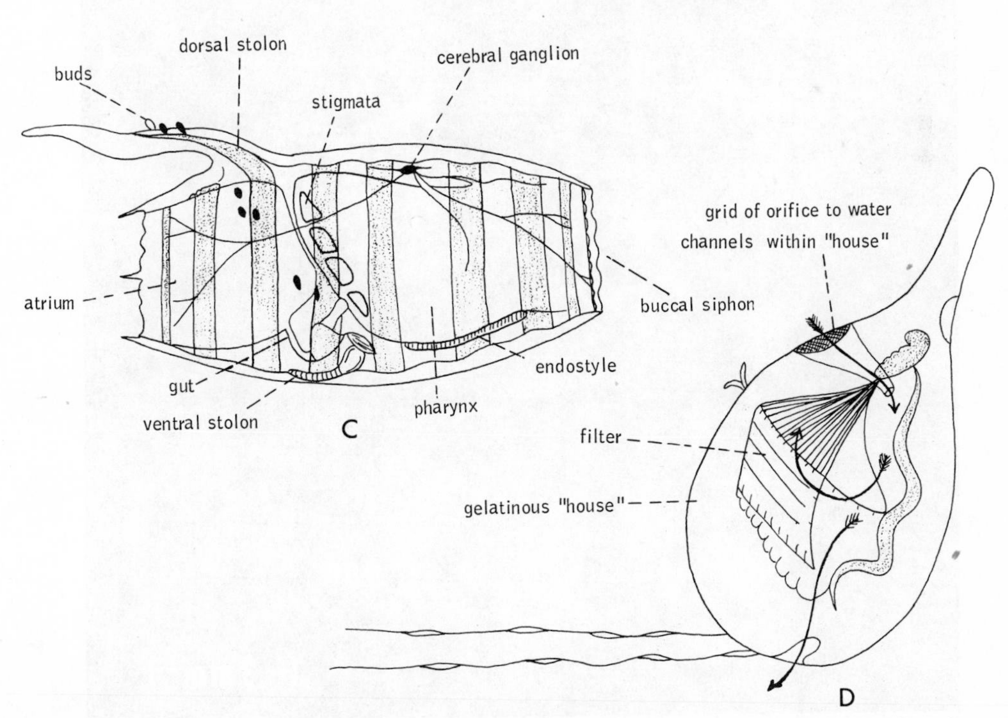

Figure 20–16. *A.* A colony of the thaliacean, *Pyrosoma* (longitudinal section). (After Grobben from Brien.) *B.* Oozooid of *Pyrosoma* with stolon and buds. (After Borradaile and others.) *C.* Oozooid of the solitary thaliacean, *Doliolum.* (Modified after Uljanin and Barrois from Borradaile and others.) *D. Oikopleura albicans*, a larvacean. (After Borradaile.)

and the four primordial buds, already organized in a ring, form a new colony by secondary budding. The new colony then assumes a pelagic existence.

Like *Pyrosoma*, the solitary salpids produce a single egg that develops in the parent, attached to the atrial wall, from which it obtains nutrition. There is no larval stage, and when development is completed, the oozooid breaks free of the parent and is free-swimming. A trailing stolon develops a chain of buds that eventually separate to form sexually reproducing adults.

The life cycle of the doliolids is the most complex of all thaliaceans. In *Doliolum*, three eggs are produced by the parent and are shed into the sea water, where each develops into a tail-bearing larva. The larva metamorphoses into an adult oozooid, which is provided with a stolon on which buds are formed (Fig. 20–16*C*). These buds, called prebuds, then give rise to several generations of highly specialized buds, all of which are attached to the parent oozooid. Eventually, buds form which develop into the sexually reproducing adults. These detach and swim away. The adult oozooid and sexual blastozooid differ in only minor details of structure.

CLASS LARVACEA

The last class, the Larvacea, or Appendicularia, is the most specialized of all tunicates. These are tiny transparent animals that reach only a few millimeters in length and are found in marine plankton throughout the world. The Larvacea are so named because the adults are neotenic and have retained some of the larval characteristics (Fig. 20–16*D*). A tail is present, and the body looks somewhat like a typical ascidian tadpole larva bent at right angles or in the shape of a **U**.

The mouth is located at the anterior of the body, and the intestine opens directly to the outside on the ventral side. There are only two pharyngeal clefts, one on each side, and each opens directly to the exterior.

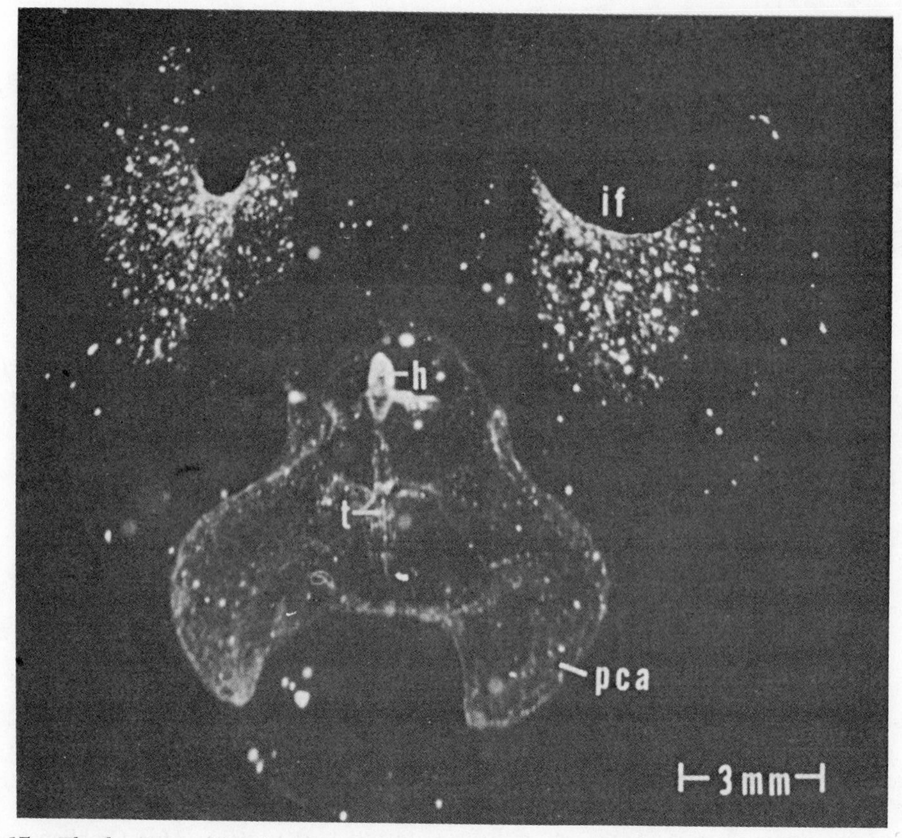

Figure 20–17. The larvacean *Megalocercus* within its mucous house. The letters **h** and **t** indicate the head and tail of the larvacean; **pca** is the inner part of the house; **if** is one of the two incurrent filters of the delicate outer part of the house, which is only partially visible. (From Alldredge, A. L., 1972: Science, *177*:885–887.)

A remarkable feature of the Larvacea is the "house" in which the body is enclosed or to which it is attached (Fig. 20–17). There is no cellulose tunic in larvaceans, but the surface epithelium secretes a delicate gelatinous material that completely encloses the body in several genera. In *Fritillaria,* the animal lies outside of the "house" but is attached beneath it. In *Oikopleura,* the gelatinous enclosure is somewhat egg-shaped with a projecting peak and is much larger than the body of the animal (Fig. 20–16*D*).

The interior of the "house," in which the animal is suspended, contains a number of interconnecting cavities and both an incurrent and an excurrent orifice (Fig. 20–17). The beating of the tail of the animal creates a water current that passes through the "house." The orifice through which water enters is covered by a grid or screen of fine fibers that keep out all but the finest plankton. During its passage through the "house," the water is strained a second time through a finer net. This second straining delivers plankton to the anterior of the body, where it enters the mouth with water and is strained a final time in passing through the pharynx. The "house" is continually shed and replaced, and a single "house" is kept no longer than three hours in *Oikopleura,* probably as a result of a clogging of the filters.

Only sexual reproduction occurs in the Larvacea. Development leads to a free-swimming tadpole larva that undergoes metamorphosis without settling.

Urochordate Phylogeny

The tunicates undoubtedly departed early from the chordate line of evolution. Most of their adult peculiarities are associated with a sessile mode of existence, but they differ in two basic respects from other chordates. There is no evidence of segmentation, and there is no coelom in the adult nor does one appear in the course of embryological development. The coelom has undoubtedly been lost, since it is characteristic of other deuterostomes, and the coelom was certainly present in the ancestral chordate stock. Regarding metamerism, the complete lack of any segmentation, even in embryonic development, seems to indicate that the ancestral chordate was not a metameric animal and that metamerism evolved in the line leading to the cephalochordates and vertebrates *after* the departure of the tunicates.

PHYLUM CHAETOGNATHA

The chaetognaths, known as arrowworms, are common animals found in marine plankton. The entire phylum of some 65 species is marine, and except for the benthic genus *Spadella,* all arrowworms are adapted for a planktonic existence. The adults possess none of the features common to the other deuterostome phyla, and they are like aschelminths in many respects. Only the embryogeny of arrowworms would suggest a deuterostome position for these animals.

External Structure. The body of arrowworms is shaped like a torpedo or feathered dart and is commonly about 3 cm. long, although a length of 10 cm. is attained in some forms (Fig. 20–18). The body is divided into a head, a trunk, and a postanal tail region, and a distinct narrowed neck separates the head and trunk. On the underside of the rounded head is a large chamber (the vestibule) that leads into the mouth (Fig. 20–19).

Hanging down from each side of the head and flanking the vestibule are four to fourteen large curved spines that are used in seizing prey. Several rows of much shorter spines (anterior and posterior teeth) that are curved around the front of the head also assist in capturing prey. All the teeth and spines are composed of chitin, a compound that is relatively uncommon in deuterostomes. A pair of eyes is located posteriorly on the dorsal surface. In the neck region is a peculiar fold of the body wall (the hood) that can be pulled forward to enclose the entire head. The hood is thought to perhaps protect the spines when they are not in use and reduce water resistance during swimming.

The remainder of the body is composed of the elongated trunk and the tail. A characteristic feature of the chaetognaths is the horizontal fins that border these regions of the body. In some arrowworms such as *Sagitta,* there are two pairs of lateral fins, but in most species a single pair of lateral fins projects from the sides of the body along the posterior half of the trunk and overlaps the tail region (Figs. 20–18*A* and *B*). Posteriorly, a large spatula-like caudal fin encompasses the end of the tail. Both the caudal and the lateral fins contain raylike supports.

Internal Structure and Physiology. The construction of the body wall is particularly reminiscent of aschelminths. The epidermis, which is covered on the outer surface with a thin cuticle, is multilayered and contains

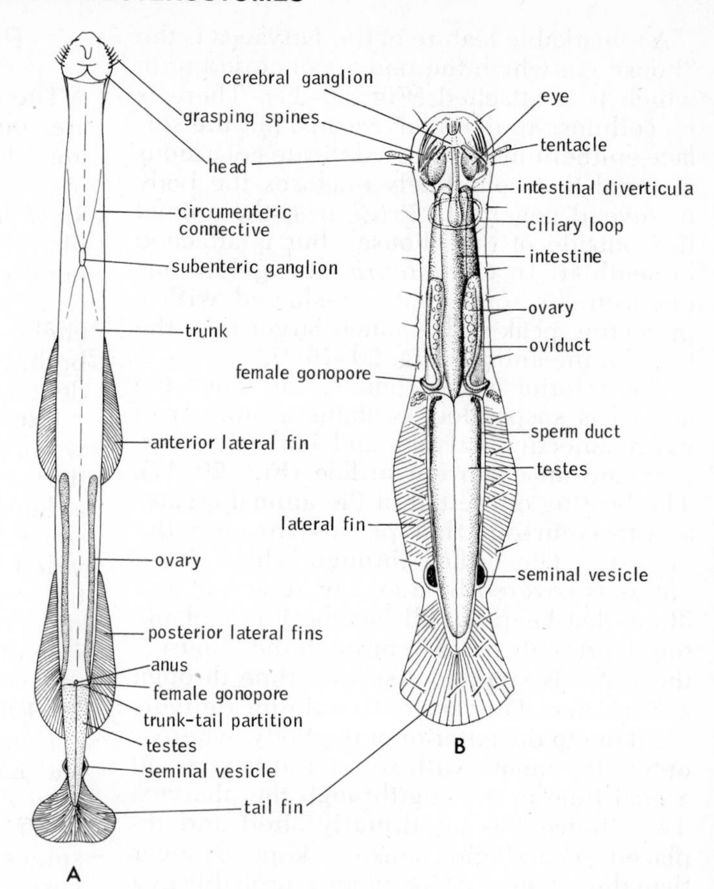

Figure 20–18. Phylum Chaetognatha. *A. Sagitta elegans* (ventral view). (After Ritter-Zahony). *B. Spadella* (dorsal view). (After Hertwig).

large vacuolated cells. A basement membrane lies beneath the epithelium and is thickened to form the supporting rays between the two epithelial layers of the fins; this membrane also forms special supporting plates in the head. The muscles of the body walls are all longitudinal and are arranged in two dorsolateral and two ventrolateral bands. In the head are special muscles for operating the hood, the teeth, the grasping spines, and other structures.

The coelom resembles a pseudocoel, because there is no peritoneum. However, the coelom is compartmented. The head contains a single coelomic space that extends into the hood and is separated by a septum from the paired trunk coelomic spaces. One or two coelomic compartments occupy the tail, but these spaces are believed to represent a secondary separation from the trunk coelom.

Chaetognaths alternately swim and float. The fins play no role in propulsion and are only flotation devices. When the body begins to sink, the longitudinal trunk muscles contract rapidly, and the animal darts swiftly forward. This forward motion is then followed by an interval of gliding and floating. The benthic *Spadella* adheres to bottom objects by means of special adhesive papillae, but it can swim short distances.

Arrowworms are all carnivorous and feed on other planktonic animals, particularly copepods. *Sagitta* has reportedly consumed young fish and other arrowworms as large as itself. In capturing prey, chaetognaths dart forward, the hood is withdrawn, and the grasping spines are spread. The teeth are used in holding, while the spines close down over the victim. The entire operation is accomplished with great rapidity.

The disgestive tract is simple (Figs. 20–

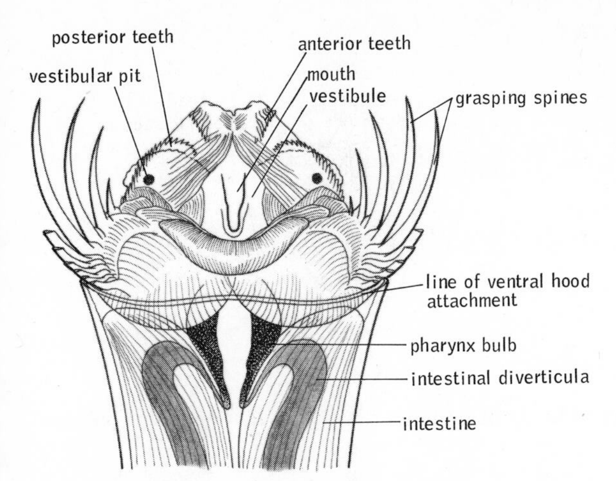

Figure 20–19. Head of *Sagitta elegans* (ventral view.) (After Ritter-Zahony).

18*B* and 20–19). The mouth leads into the bulbous pharynx that penetrates the head-trunk septum to join a straight intestine. The intestine extends through the length of the trunk, and at its anterior gives rise to a pair of lateral diverticula. Muscles are best developed in the pharyngeal wall, but a thin circular layer is present in the intestine. Although the intestine is suspended by a dorsal and a ventral mesentery, these mesenteries appear to be extensions of the basement membrane of the body wall and are not cellular.

After capture, the prey is pushed into the mouth, in which it is lubricated by pharyngeal secretions and then passed to the posterior of the intestine. Here the food is rotated and moved back and forth until it is broken down. The intestine is lined with secretory and absorptive cells, and digestion is probably extracellular.

There are no gas exchange or excretory organs, and the coelomic fluid acts as a circulatory medium.

The nervous center of chaetognaths is a nerve collar surrounding the pharynx. The ring contains dorsally a large cerebral ganglion, and a number of lateral ganglia (Fig. 20–18*B*). The ganglia give rise to a large number of nerves that innervate different head structures. The sides of the cerebral ganglion give rise to a pair of large circumenteric connectives that extend posteriorly and end in a single midventral subenteric ganglion in the anterior part of the trunk (Fig. 20–18*A*). The subenteric ganglion in turn gives rise to a large number of paired nerves that supply the muscles and sensory receptors of the trunk and tail.

Sense organs include the eyes, the sensory bristles, and possibly a head organ (the ciliary loop) (Fig. 20–18*A* and *B*). Each of the two eyes, at least in *Sagitta*, is composed of five fused pigment-cup ocelli, in which the photoreceptors are directed toward the cup. The sensory bristles are arranged in longitudinal rows along the length of the trunk and are probably tactile in function. The ciliary loop is usually a **U**-shaped tract of cilia that extends from the head back over the neck or the anterior end of the trunk. Its sensory function is uncertain.

Reproduction. Chaetognaths are hermaphroditic, and a pair of elongated ovaries is located in the trunk coelom in front of the trunk-tail septum. A pair of elongated testes is located in the tail coelom behind the septum (Fig. 20–18*A* and *B*). From each testis a sperm duct passes posteriorly and laterally to terminate in a seminal vesicle imbedded in the lateral body wall. Along its course, each sperm duct opens into the tail coelom through a funnel. Sperm leave the testis as spermatogonia, and spermatogenesis is completed in the coelom. When mature, the sperm pass into the ciliated funnel of the sperm duct and from there into a seminal vesicle in which the sperm are formed into a single spermatophore. The seminal vesicle ruptures, enabling the spermatophore to escape.

A canal, the seminal receptacle, runs along the lateral side of each ovary and opens to the exterior through two gonopores, located one on each side of the body just in front of the trunk-tail septum. The eggs do not begin to mature until after spermatogenesis has commenced in the tail coelom.

In the benthic *Spadella*, sperm transfer is reciprocal. Two individuals cross each other, with their heads in opposite directions. During this movement, which is very rapid, each deposits a spermatophore on the middle of the neck region of the other. The spermatophore then breaks down and the sperm stream backward along the midline. The stream eventually divides, and sperm enter the orifice of the seminal receptacle on each side.

Spermatophore deposition has also been reported for some pelagic species, but details of sperm transfer are fragmentary. Some workers believe that *Sagitta* undergoes self-fertilization; others do not.

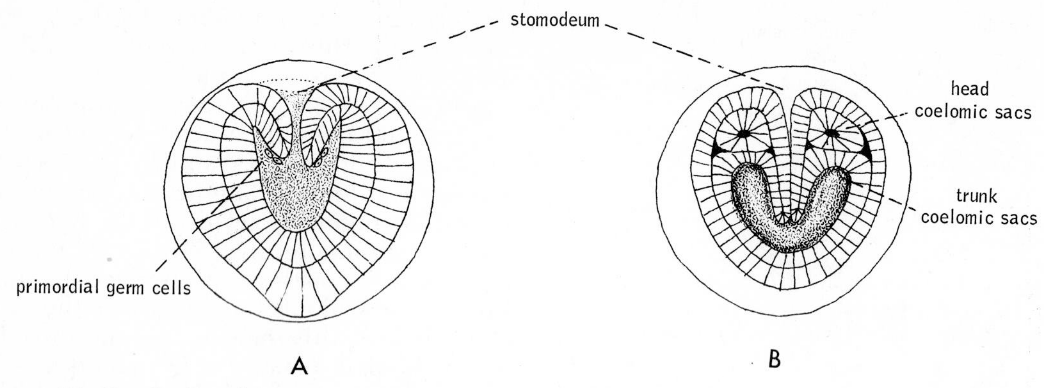

Figure 20–20. Coelom formation in *Sagitta*. *A*. Initial folding of archenteron walls. *B*. Separation of head coelomic sacs. (Both after Burfield from Hyman.)

In *Sagitta*, the eggs are emitted through a temporary oviduct. The eggs are planktonic and are surrounded by a coat of jelly. In other arrowworms, the eggs may be attached to the body surface of the parent and carried about for some time. *Spadella* deposits its eggs in small clusters on algae or other objects.

Embryogeny. Cleavage is radial, complete, and equal and leads to coeloblastula. Gastrulation is accomplished by invagination, and as in the lower chordates, the invaginating mesentoderm fits closely against the outer ectoderm and obliterates the blastocoel. The anterior end wall of the archenteron invaginates, folding backward on each side and cutting off two pairs of lateral coelom sacs (Fig. 20–20*A* and *B*). The coelom is thus enterocoelic in origin. Further development is direct, and although the young are called larvae when they hatch, they are similar to the adult, and no metamorphosis occurs.

There are approximately 50 species of described arrowworms. Most of them are found in tropical waters, but the phylum is represented in the plankton of all oceans; at times enormous numbers appear. A considerable number of species are cosmopolitan. Although all chaetognaths except *Spadella* are planktonic, there is a distinct vertical stratification of species. Many species live only in the upper lighted zone—that is, above approximately 200 m. Of these, some are restricted to coastal waters; others always appear well off shore in open water where there is little fluctuation in salinity. Another group of species occupies a depth ranging from 200 to 900 m.; a few forms live deeper than 900 m. Like many planktonic crustaceans, some arrowworms undergo a diurnal or seasonal migration between the surface and lower depths.

Despite the similarities of adult chaetognaths to aschelminths, the embryogeny of the phylum appears to be deuterostome in nature. There are, however, some peculiarities. For example, the coelom is enterocoelic in origin but does not arise by a direct outpocketing of the archenteron, and only two pairs of coelomic pockets are formed instead of three. Moreover, there is no larval stage comparable to that of the echinoderms and the hemichordates. Thus, the chaetognaths cannot be allied with any specific deuterostome phylum. If chaetognaths are really deuterostomes, the phylum must have departed very early from the base of the deuterstome line and is only remotely related to the other deuterostome groups. Ghirardelli (1968) has reviewed past and present opinions regarding the phylogenetic position of chaetognaths.

Bibliography

Alvarino, A., 1965: Chaetognaths. Oceanogr. Mar. Biol. Ann. Rev., *3*:115–194.

Barrington, E., 1940: Observations on feeding and digestion in *Glossobalanus*. Quart. J. Micr. Sci., *82*:227–260.

Barrington, E., 1965: The Biology of Hemichordata and Protochordata. W. H. Freeman, San Francisco.

Brien, P., 1948: Embranchement des Tuniciers. *In* Grassé, P. (Ed.), Traité de Zoologie, Vol. II, Échinodermes, Stomocordes, Procordes. Masson et Cie, Paris, pp. 553–930.

Bullough, W. S., 1958: Practical Invertebrate Anatomy. Macmillan, N.Y., pp. 446–464. (Descriptions of representative tunicates, including the Thaliacea and the Larvacea.)

Burdon-Jones, C., 1952: Development and biology of the larva of *Saccoglossus horsti*. Phil. Trans. Roy. Soc. London B, *236*:553–590.

Dawydoff, C., 1948: Embranchement des Stomocordes.

In Grassé, P. (Ed.), Traité de Zoologie, Vol. II, Échinodermes, Stomocordes, Procordes. Masson et Cie, Paris, pp. 367–551.

Ghirardelli, E., 1968: Some aspects of the biology of the chaetognaths. Adv. Mar. Biol., *6*:271–375.

Gosner, K. L., 1971: Guide to Identification of Marine and Estuarine Invertebrates: Cape Hatteras to the Bay of Fundy. Wiley-Interscience, N.Y.

Hyman, L. H., 1959: The Invertebrates, Vol. 5, Smaller Coelomate Groups. McGraw-Hill, N.Y., pp. 1–71. (A complete account of the chaetognaths.)

Jones, J. C., 1971: On the heart of the orange tunicate, *Ecteinascidia turbinata*. Biol. Bull., *141*:130–145.

Knight-Jones, E. W., 1952: On the nervous system of *Saccoglossus cambrensis*. Phil. Trans. Roy. Soc. London B, *236*:315–354.

Light, S. F., Smith, R. I., Pitelka, F. A., Abbott, D. P., and Weesner, F. M., 1967: Intertidal Invertebrates of the Central California Coast. University of California Press, Berkeley.

Millar, R. H., 1970: British Ascidians. Academic Press, N.Y. (Keys and notes for the identification of British species.)

Millar, R. H., 1971: The biology of ascidians. Adv. Mar. Biol., *9*:1–100. (A good review of ascidian biology.)

Smith, M. J., 1970: The blood cells and tunic of the ascidian *Halocynthia aurantium*. Biol. Bull., *138*: 354–378.

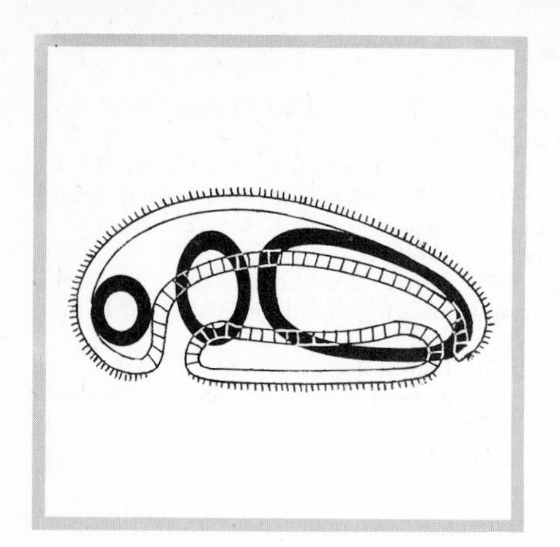

Chapter 21

ANIMAL PHYLOGENY

The evolutionary origin of most of the animal phyla is shrouded in the obscurity of the Archeozoic era. The rocks of this ancient era carry a fossil record which is not only fragmentary but usually difficult to interpret. Moreover, the early metazoans were largely soft-bodied animals and were not likely to have been frequently preserved. Significant fossils may yet be discovered, but current attempts to reconstruct the patterns of major evolutionary relationships within the Animal Kingdom must rely in large part on evidence from embryology and comparative morphology. Some of this evidence is quite convincing. For example, the presence of very similar trochophore larvae in the Mollusca and the Annelida clearly suggests an affinity between these two phyla; and many aspects of arthropod structure indicate an annelidan ancestry or at least a common ancestry for annelids and arthropods. More often, however, morphological and embryological evidence is inconclusive or even lacking. Yet inconclusive evidence has not deterred the construction of phylogenetic schemes. Phylogenetic speculations regarding the relationships between phyla have intrigued zoologists for the past hundred years and still do so today. The purpose of this chapter is to provide an acquaintance with contemporary phylogenetic schemes and an understanding of the rationale underlying them.

A phylogeny of the Animal Kingdom must answer two major questions. First, what was the origin of the Metazoa, and second, what were the principal evolutionary lines leading from the ancestral metazoan? These two problems are clearly interrelated, but for the sake of convenience they will be dealt with separately here.

ORIGIN OF METAZOA

All zoologists agree that metazoans evolved from unicellular organisms; but as to the particular group of unicellular forms involved and the mode of origin, there is anything but unanimity. The current theories can be grouped around three principal viewpoints: that the ancestral metazoan arose from a multinucleate ciliate, which became compartmented or cellularized; that the ancestral metazoan arose by way of a colonial flagellate through increasing cellular specialization and interdependence; or that metazoans have had a polyphyletic origin from different unicellular groups.

Hadži (1953) and Hanson (1958) have been the chief proponents of a ciliate origin for metazoans. Their theory, which may be called the *Syncytial Theory,** holds that multicellular animals arose from a primitive group of multinucleate ciliates. The ancestral metazoan was at first syncytial in structure, but later became compartmented or cellularized by the acquisition of cell membranes, thus producing a typical multicellular condition. Since many ciliates tend toward bilateral symmetry, proponents of the Syncytial Theory maintain that the ancestral metazoan was bilaterally symmetrical and gave rise to the acoel flatworms, which are therefore held to be the most primitive living metazoans. The fact that the acoels are in the same size range as the ciliates, are bilaterally symmetrical, are ciliated, and tend toward a syncytial condition is considered as evidence supporting the primitive position of this

**Syncytial* refers to the histological condition in which cell membranes are absent between adjacent nuclei.

group. The ciliate macronucleus, which is absent in acoels, is assumed to have been absent in the multinucleate protociliate stock from which the metazoans arose and is assumed to have developed later in the evolutionary line leading to the higher ciliates.

There are a number of objections to the Syncytial Theory. For this theory to be acceptable, the embryogeny of the lower metazoans must be considered to have no phylogenetic significance, since nothing comparable to cellularization occurs in the ontogeny of any of these groups.* Actually, the syncytial nature of acoel tissue arises secondarily after typical, cytosomal, embryonic divisions. Furthermore, a ciliate ancestry does not explain the general occurrence of flagellated sperm in metazoans. No comparable cells are produced in ciliates, and it is necessary to assume a *de novo* origin of motile sperm in the metazoan ancestor. The most serious objection to the Syncytial Theory is the necessity for making the acoels the most primitive living metazoans. Bilateral symmetry then becomes the primitive symmetry for metazoans, and the radially symmetrical cnidarians must be derived secondarily from the flatworms. Adherents of the Syncytial Theory consider the biradial anthozoans to be the most primitive and the hydrozoans the most specialized cnidarians. Yet most evidence would seem to indicate that the radial symmetry of cnidarians is primary, and not secondarily evolved from a bilateral ancestor.

The *Colonial Theory*, in which the metazoans are derived by way of a colonial flagellate, is the classic and most frequently encountered theory of the origin of multicellular animals. This idea was first conceived by Haeckel (1874), later modified by Metschnikoff (1887), and revived by Hyman (1940). The Colonial Theory maintains that the flagellates are the ancestors of the metazoans, and in support of such an ancestry the following facts are cited as evidence. Flagellated sperm cells occur throughout the Metazoa. Flagellated body cells commonly occur among lower metazoans, particularly among sponges and cnidarians. True sperm and eggs have evolved in the phytoflagellates. The phytoflagellates display a tendency toward a type of colonial organization that could have led to a multicellular construction; in fact, a differentiation between somatic and reproductive cells has been attained in *Volvox*.

The Colonial Theory holds that the ancestral metazoan probably arose from a spherical, hollow, colonial flagellate. Like *Volvox*, the cells were flagellated on the outer surface; the colony possessed a distinct anterior-posterior axis and swam with the anterior pole forward; and there was a differentiation of somatic and reproductive cells. This stage was called the blastaea in Haeckel's original theory; and the hollow blastula, or coeloblastula, was considered to be a recapitulation of this stage in the embryogeny of living metazoans. According to Haeckel, the blastaea invaginated to produce a double-walled, saclike organism, the gastraea. This gastraea was the hypothetical metazoan ancestor, equivalent to the gastrula stage in the embryonic development of living metazoans. In addition to embryological evidence, Haeckel noted the close structural similarity between the gastraea and some lower metazoans such as the hydrozoan cnidarians and certain sponges. Both of these latter organisms are double-walled, with a single opening leading into a saclike digestive cavity.

Haeckel's gastraea was later modified by Metschnikoff, who noted that the primitive mode of gastrulation in cnidarians is by ingression, in which cells are proliferated from the blastula wall into the interior blastocoel (see p. 104). This produces a solid gastrula. Invagination is apparently a secondary embryonic short cut. Metschnikoff therefore argued that the gastraea was a solid rather than a hollow organism.

In accordance with Metschnikoff's revision, modern elaborators of the Colonial Theory pick up the evolution of the metazoans with Haeckel's blastaea. Through the migration of cells into the interior, the originally hollow sphere became transformed into an organism having a solid structure (Fig. 21–1). The body of this hypothetical ancestral metazoan is believed to have been ovoid and radially symmetrical. The exterior cells were flagellated and, as such, assumed a locomotor sensory function. The solid mass of interior cells functioned in nutrition and reproduction. There was no mouth, and food could be engulfed anywhere on the exterior surface and passed to the interior. Since this hypothetical organism is very similar to the planula larva of cnidarians, it has been called the planuloid ancestor.

From such a free-swimming, radially sym-

*Cellularization does occur in the superficial cleavage of arthropod eggs, but this is a highly specialized condition associated with abundant yolk material.

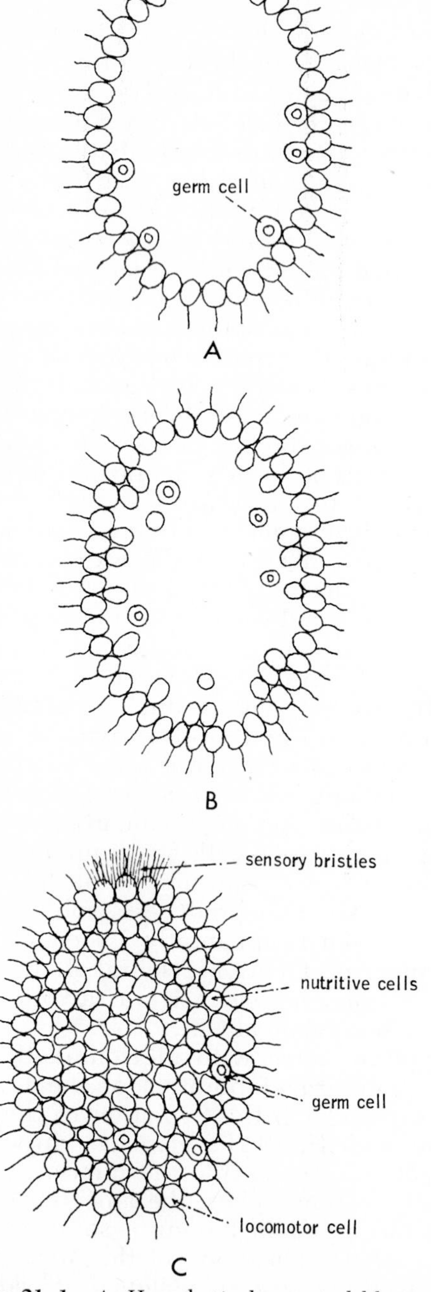

Figure 21–1. *A.* Hypothetical ancestral blastaea. *B.* Multipolar ingression. *C.* Hypothetical planuloid ancestor. (After Hyman.)

metrical, planuloid ancestor the lower metazoans are believed to have arisen. On the basis of this theory, the primary radial symmetry of the cnidarians can thus be accounted for as being derived directly from the planuloid ancestor. The bilateral symmetry of the flatworms would then represent a later modification in symmetry.

The existing flagellates that best fulfill the qualifications for metazoan ancestors are the freshwater volvocid phytoflagellates. These plantlike organisms possess cellulose walls, chlorophyll, and autotrophic nutrition and go through the reduction divisions *following* fertilization. If the phytoflagellates did give rise to the Metazoa, as has indeed been advocated by some authors, then it is necessary to postulate that at some time in the course of the evolutionary events previously described, these plantlike characteristics were lost. A more likely hypothesis is that the metazoans arose from some group of now extinct zooflagellates that possessed a colonial organization similar to that in the volvocid phytoflagellates.

A number of authors, more recently Greenberg (1959), have proposed a polyphyletic origin for the metazoans. Greenberg suggests that the sponges, cnidarians, ctenophores, and flatworms each evolved independently from the protozoans. The sponges and cnidarians derived by way of colonial flagellates, and the flatworms and ctenophores by way of the ciliates or perhaps the mesozoans. It is readily apparent that Greenberg's view is a compromise between the Syncytial Theory and the Colonial Theory and is therefore subject to most of the arguments for and against the other two theories.

Contemporary supporters of the Colonial Theory derive the bilateral phyla from the ancestors of the cnidarians in one of two ways. According to some, bilaterality evolved in conjunction with the evolution of the coelom, as will be described later. According to others, the flatworms are believed to be the most primitive bilateral animals. The anthozoans and the ctenophores were formerly thought by a number of zoologists to provide a bridge from the cnidarians to the flatworms, but this view is not widely held today. A more popular theory places the acoel flatworms at the base of the Bilateria. According to this theory, elaborated by Hyman (1951), certain of the ancestral planuloid stock may have taken up a life on the ocean bottom and as a result developed a creeping mode of movement over rocks and other objects. This could have led to a differentiation between dorsal and ventral surfaces and to the development of a ventral mouth. Such differentiation would have resulted in a bilateral symmetry and would have led to the evolution of the acoeloid ancestor described in Chapter 7.

The acoel flatworms are thus assumed to be the most primitive bilateral animals by adherents of the Syncytial Theory as well as by many supporters of the Colonial Theory.

From the ancestral metazoan ascended the major evolutionary lines leading to the various animal phyla. Within these lines there are generally recognized affinities, and most zoologists would agree to the following phylogenetic groupings:

Cnidarians—Ctenophores
Flatworms—Nemerteans
The Pseudocoelomate Phyla
The Lophophorate Phyla
The Schizocoelous Phyla
(Mollusks—Annelids—Arthropods)
The Enterocoelous Phyla
(Echinoderms — Hemichordates — Chordates)

The problem is how these groups are interrelated. A glance at the phylogenetic schemes diagrammed on the following pages immediately shows the divergent solutions that have been proposed. To understand the rationale behind these schemes, it is necessary to consider first the theories regarding the evolution of the coelom and the evolution of metamerism. Opinion as to how these two conditions evolved provides an important basis for the reconstruction of animal phylogeny.

THE EVOLUTION OF THE COELOM AND METAMERISM

The Gonocoel Theory. One of the most favored theories of coelom origin is known as the Gonocoel Theory. This theory maintains that the coelom represents a persistent gonadial cavity, or gonocoel. At the time of breeding, an enlarged saclike gonad was filled with gametes. When the gametes were shed to the exterior through a short gonoduct, a cavity remained behind, which became the coelom. It is suggested that this cavity may then have provided body suppleness or space for the gut, or may have facilitated nutrition. The annelids provide the principal evidence to support this theory. In polychaetes the coelomic epithelium gives rise to the gametocytes, which then undergo maturation within the coelom. A gravid polychaete's coelom is filled with eggs or sperm.

The Gonocoel Theory was first proposed by Bergh (1885), and subsequently was taken up by Meyer (1890), Lang (1903), and Goodrich (1946). Meyer viewed the coelom as initially arising from a single large pair of gonads, so that the first coelom was paired but unsegmented. However, Bergh and Lang, observing the linearly arranged multiple pairs of gonads of some flatworms

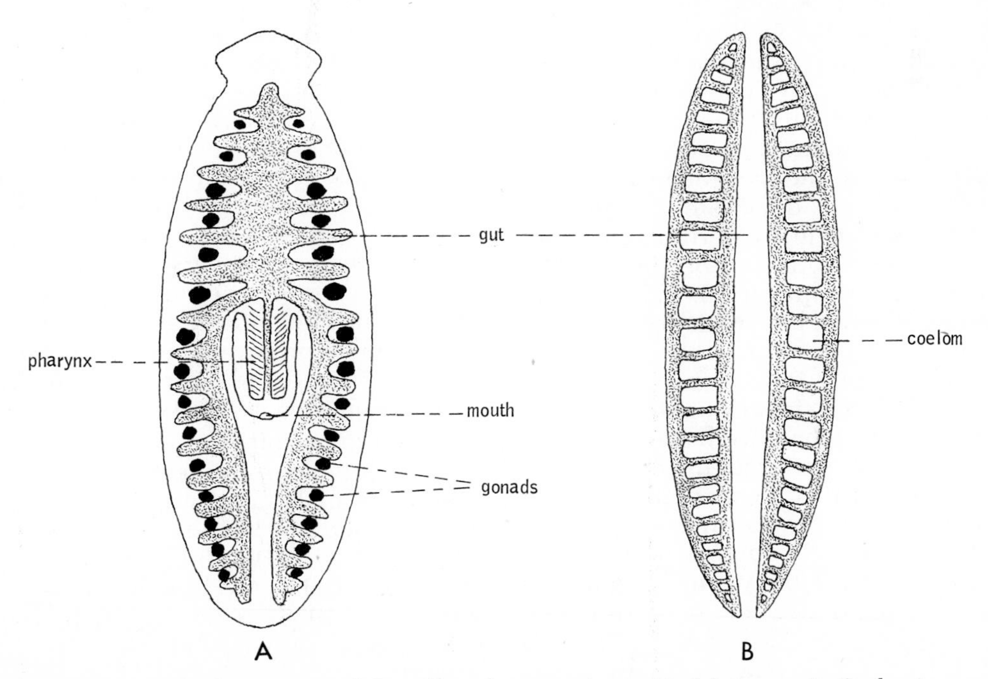

Figure 21–2. *A.* Gonads alternating with branches of enteron in a triclad flatworm. *B.* Coelomic compartments in an annelid worm resulting from fused adjacent expanded gonads. (Both after Hyman.)

(pseudometamerism), believed that the coelom arose in a segmented condition. Each gonad would have given rise to one coelomic cavity. The anterior and posterior walls of adjacent gonads would have become septa separating the coelomic compartments (Fig. 21–2). This modification of the gonocoel explained in the theory most widely supported today thus combines coelom origin with the origin of metamerism.

Adherence to the Gonocoel Theory would lead to a number of phylogenetic conclusions: (1) The acoelomate flatworms would be considered to be primitive and to have given rise to the coelomates. The Gonocoel Theory is thus compatible with either the Syncytial Theory or the Colonial Theory of metazoan origin. (2) Since the annelidan and molluscan coeloms to some degree function as gonocoels, the schizocoelous phyla—annelids, mollusks, and arthropods—would be considered the more primitive coelomates. From the schizocoels would have stemmed the enterocoelous line of evolution. (3) If the coelom arose in a segmented condition, as Lang proposed, segmented coelomates must be more primitive than unsegmented coelomates.

Numerous objections can be raised to the Gonocoel Theory. It is difficult to understand why the gonadial cavity of the ancestral coelomates would not have regressed after the elimination of the gametes. There is little embryological support for this theory. Gonads never arise before the coelom, which should be the case if the coelom evolved from the gonadial cavity. The teloblast cells, which in annelids do give rise to the coelom and germinal peritoneum, produce not only the gonads but also the mesenchyme in flatworms. There is considerable embryological evidence that the coelomic peritoneum is not the primary site of germ cell origin. Finally, the phylogenetic implications of coupling the origin of the coelom with metamerism are difficult to accept.

Although rejecting the Gonocoel Theory, Hyman (1951) considered pseudometamerism to have foreshadowed segmentation. Goodrich (1946) also favored this idea. These authors suggested that segmentation of the body wall musculature was superimposed on a pseudometameric distribution of organs in an elongate wormlike body. Such segmentation was an adaptation for an undulatory mode of swimming. Unfortunately, as pointed out by Clark (1964), longitudinal muscles are not organized on a strictly segmental basis in polychaete annelids, which utilize this method of locomotion. The longitudinal fibers run through several segments.

The Enterocoel Theory. Another popular theory of coelom origin is the Enterocoel Theory. The Enterocoel Theory holds that the coelom evolved as outpocketings of the gut. This event is claimed to have occurred in the metazoan ancestors of the cnidarians. Four gut pouches became separated from the central digestive cavity to give rise to the coelom. Two of these cavities, one anterior and one posterior, subsequently divided so that there were three pairs of coelomic cavities (Fig. 21–3*A–C*). Such an animal would have been the ancestor of the coelomates and would have given rise to the oligomerous coelomates (lophophorates, echinoderms, hemichordates). Support for the Enterocoel Theory is sought in the deuterostome phyla—echinoderms, hemichordates, and chordates—which primitively exhibit coelom formation by outpocketing of the archenteron. The cnidarians are considered to have evolved from the same common ancestor as the coelomates, and the compartmented, or pouched, gut cavity of anthozoans and scyphozoans is believed to have been derived from the original pouched gut of the cnidarian-coelomate ancestors.

The phylogenetic implications of the Enterocoel Theory are far reaching. All metazoan animals above the cnidarians are coelomate. If a coelom is absent, as in the flatworms, it has been secondarily lost. The most primitive coelomates are the oligomerous deuterostomes, which are not only enterocoelous but have a body divided into three regions, each with a pair of coelomic cavities (protocoel, mesocoel, and metacoel). The schizocoelous phyla are secondarily evolved from the enterocoelous line, schizocoely being a secondary slurring over of the primitive enterocoelous mode of coelom formation. The anthozoans are considered to be the most primitive cnidarians because of the compartmented gastrovascular cavity.

The Enterocoel Theory has had a long history and numerous supporters. The idea was first proposed by Lankester (1875), and subsequently elaborated with various modifications by Sedgwick (1884), Masterman (1898), Hubrecht (1904), Söderström (1925), and Lameere (1932). Modern supporters include Jägersten (1955), Remane (1950, 1963), Ulrich (1950), Marcus (1958), and Lemche (1959).

There are formidable difficulties with the

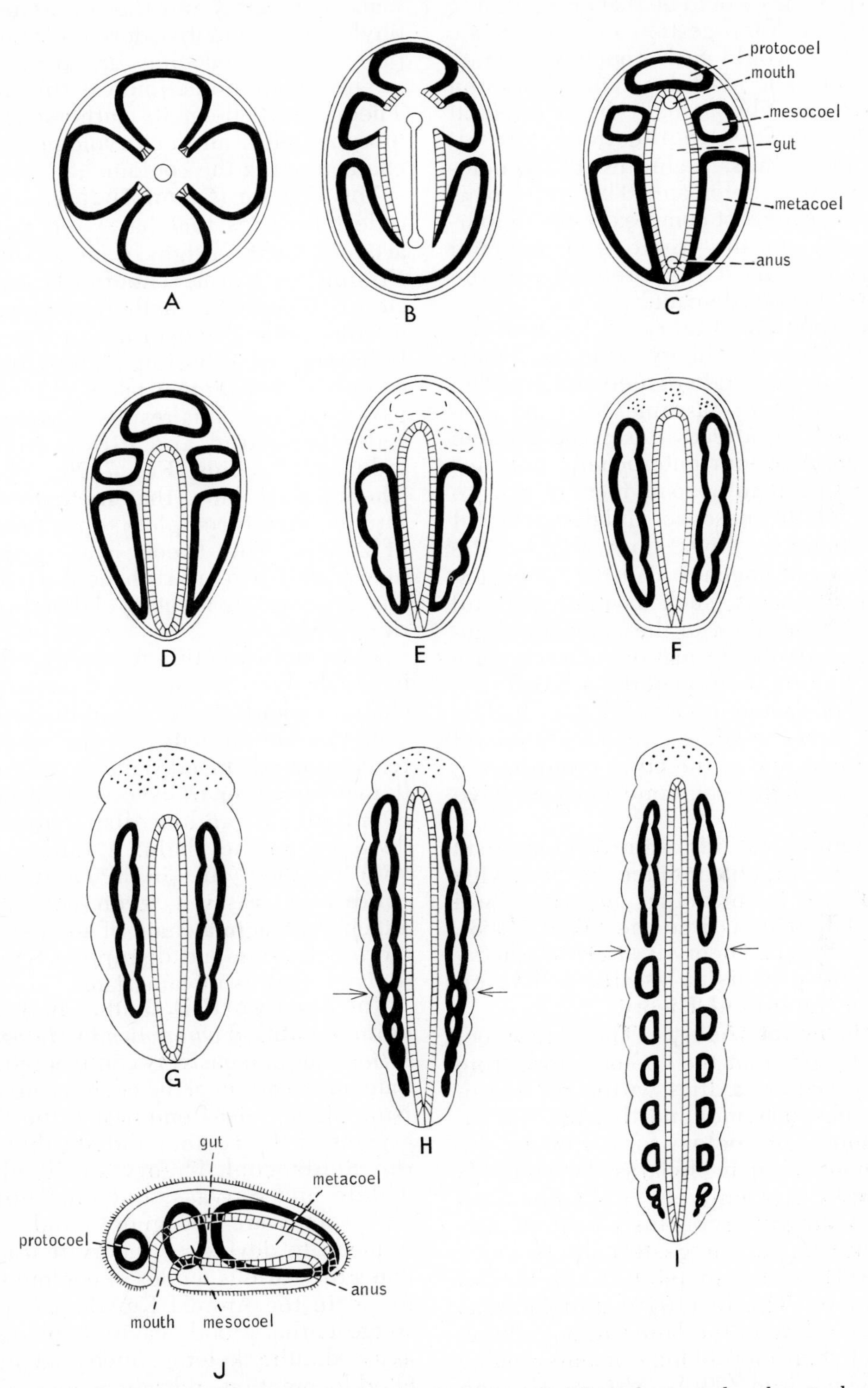

Figure 21–3. Enterocoel-Cyclomerism Theory. *A–C.* Formation of a bilateral ancestral coelomate through enterocoelic pouching of a radiate ancestor. *D–F.* Formation of a single pair of coelomic cavities through loss of protocoel and mesocoel. *G–I.* Posterior budding of metacoel associated with evolution of metamerism. (Based on Remane.) *J.* Lateral view of hypothetical ancestral coelomate based on Enterocoel Theory. (Modified from Jägersten.)

Enterocoel Theory. The obvious advantage of gastric pouches would be that of increasing the surface area for digestion and absorption. But then why would such pouches separate from the gut? The most damaging attacks on the Enterocoel Theory have been directed at the phylogenetic implications. Are all bilateral metazoans, including the acoelomates, basically coelomate? The anthozoans appear to be the most complex of the cnidarians. Can they be the primitive members of the phylum, as most interpretations of the Enterocoel Theory demand?

The Cyclomerism Theory. Closely allied with the Enterocoel Theory is the Cyclomerism Theory of the origin of metamerism. The Cyclomerism Theory assumes that the coelom originated as already described through the separation of four enterocoelic pouches from some protoanthozoan to give rise to the protocoel, the mesocoel and the metacoel of the oligomerous phyla. Loss of all but the posterior pair of pouches leads to the unsegmented coelomates, such as mollusks and sipunculids, and then to the segmented annelids through the formation of secondary segments by proliferation of the original posterior pair of coelomic sacs (Fig. 21–3). The flatworms are considered to have evolved from the same line as the other protostomes, except that all the coelomic cavities have been lost.

The essential features of the Cyclomerism Theory were initially implied by Sedgwick (1884). Today its principal supporter and greatest elaborator is Remane (1950, 1963). The Cyclomerism Theory is subject to many of the same criticisms as those leveled against the Enterocoel Theory.

The Schizocoel Theory. The Schizocoel Theory suggests that the coelom arose from slit-like spaces or cavities within the mesoderm (or mesenchyme) and that the schizocoelous mode of coelom formation in the development of such phyla as the annelids and mollusks is primitive. Various functions have been suggested for this primitive coelom: excretion, as suggested by Faussek (1899, 1911); and a hydraulic skeleton, as suggested by Thiele (1910) and Sarvaas (1933). According to the latter authors, function rather than morphology or embryology defines a coelom. Thus, the molluscan hemocoel, which functions as a hydraulic skeleton in the locomotion of gastropods and bivalves, is considered to be the coelom and not the pericardial cavity.

Phylogenetically, the Schizocoel Theory implies that acoelomates are more primitive than coelomates and that the enterocoelous phyla are secondarily derived from the more primitive schizocoelous groups.

The major objections to the Schizocoel Theory have been its authors' rejection of morphological and embryological criteria in characterizing the coelom.

The Fission (Corm) Theory. One of the oldest theories that has been proposed to account for the origin of metamerism is the Fission, or Corm, Theory. Transverse fission is a common method of asexual reproduction among flatworms, and in some turbellarians new fission planes form before complete detachment takes place at the old fission planes. This results in a chain of partially separated and differentiated individuals, or zooids (Fig. 7–20). The Fission Theory postulates that metamerism arose through such incomplete separation of chains of zooids. This theory was supported by many well-known 19th century zoologists and received its greatest elaboration by Perrier (1882).

The chief objection to the Fission Theory is the lack of gradation of ages in such a chain of zooids, as is true of the segments in a metameric animal. As soon as one zooid reaches an advanced stage of differentiation, it may undergo transverse fission, so that eventually a random distribution of ages develops in the chain. A further difficulty with this theory lies in the nature of the two conditions. Fission, even with persistent attachment as in a chain of zooids, is a much more extensive and disruptive type of body division than is metamerism.

The Theory of R. B. Clark. R. B. Clark, in a volume entitled *Dynamics in Metazoan Evolution*, has persuasively contributed to the debate on the origin of the coelom and metamerism. Clark's ideas emphasize the functional aspects of the coelom and are thus allied to the Schizocoel Theory as developed by Thiele and Sarvaas; but morphology and embryology are not ignored and an attempt is made to develop a theory that accommodates the various modes of coelom formation found in the Animal Kingdom today. Clark suggests that a body cavity evolved initially as a hydraulic skeleton. Such a skeleton facilitated locomotion and represented a response to the increasing body size of metazoan animals. Clark postulates that a body cavity evolved independently a number of times and in various ways—as a persistent blastocoel in the pseudocoelomates, as an enterocoel in the deuterostomes, as a schizocoel, as a gonocoel, and so forth.

According to Clark, metamerism also evolved as an adaptation for locomotion. The ancestral annelids are considered to have been burrowers, and the evolution of a compartmented coelom localized the function of the hydraulic skeleton. This localization greatly facilitated peristaltic contraction of the body wall, the most efficient type of locomotion for wormlike burrowers. The nervous, circulatory, excretory, and muscular systems subsequently became segmentally organized in response to the initial segmentation of the coelom. Metamerism evolved a second time in the chordates. But here segmentation evolved as an adaptation for undulatory swimming movements. The notochord provided support for the body but did little to prevent the alternate contractions of the body wall musculature from producing much more than side-to-side lashings of the trunk and tail. This difficulty was overcome by the segmental organization of the body wall musculature. Each segmental muscle block exerted a local contracting force against the notochord. Alternate waves of contraction sweeping down the body could now effect undulatory body waves and a powerful thrust for swimming. The segmentation of the nervous and circulatory systems evolved in response to the initial muscle segmentation.

From the standpoint of phylogeny, Clark's views on the origin of the coelom and metamerism would place the acoelomates in a primitive position within the Animal Kingdom. From these primitive metazoans a number of radiating lines would have originated, each reflecting a different mode of coelom origin.

Despite the attractiveness of Clark's functional approach, the principal obstacles to his hypothesis are involved in the assumption of multiple modes of coelom origin; for Clark also must cope with certain problems already pointed out concerning some of the theories of coelom formation.

CONTEMPORARY PHYLOGENIES

With some understanding of the various views regarding the origin of the coelom and metamerism, along with a general knowledge of invertebrate zoology, some of the contemporary phylogenetic schemes can be better appreciated.

The phylogenetic scheme of E. Marcus (1958) is diagrammed in Figure 21–4. This scheme is very similar to that of the French zoologist L. Cuénot (1952). The metazoans are assumed to have evolved from a colonial flagellate. This assumption is reflected in the basal position of the radiate phyla. Marcus supports the Enterocoel Theory, as indicated by the primitive position of the anthozoans and the location of the flatworms among the schizocoel coelomates.

The phylogenetic scheme depicted in Figure 21–5 is by Hanson (1958), who is a leading supporter of the Syncytial Theory of metazoan origin. In this scheme the sponges are considered to have evolved independently from the remainder of the Metazoa, which arose from some protociliate. The flatworms are given a primary and central position. From the acoeloid metazoan ancestor arose a number of lines. One of these lines leads to the cnidarians, which are thus considered to be secondarily radiate. The other lines originating from the flatworms lead to the schizocoels, the pseudocoels, and the enterocoels. The position of the mollusks indicates that Hanson believes that primitive mollusks, as exemplified by the monoplacophorans, were segmented.

A third phylogenetic scheme, which reflects the views of Hyman (1940, 1951), is shown in Figure 21–6. A planuloid organism is considered to be the ancestral metazoan, having risen from a colonial flagellate. The planuloid ancestor gave rise to the primarily radiate cnidarians and to the acoel flatworms. The latter form the stem of the bilateral animals. Schizocoely is the primitive mode of coelom formation, and from the schizocoelous line diverged the enterocoelous protostomes. The lophophorates are thought to have stemmed from the schizocoels near the point of divergence from the enterocoels.

The phylogenetic plan of Hadži (1963) departs most radically from the generaly accepted viewpoints (Fig. 21–7). The animal phyla are considered to have evolved along a single straight line. Hadži supports the Syncytial Theory of metazoan origin and suggests that the sponges evolved more or less independently from other metazoans. As expected, the flatworms occupied a basal and central position and from this group have evolved the pseudocoelomates, the cnidarians (via the rhabdocoels), and the ctenophores (via the polyclads). The evolution of metamerism is of fundamental importance in Hadži's scheme. Nonsegmented animals—the Ameria—are considered to be primitive, and to

Text continued on page 827.

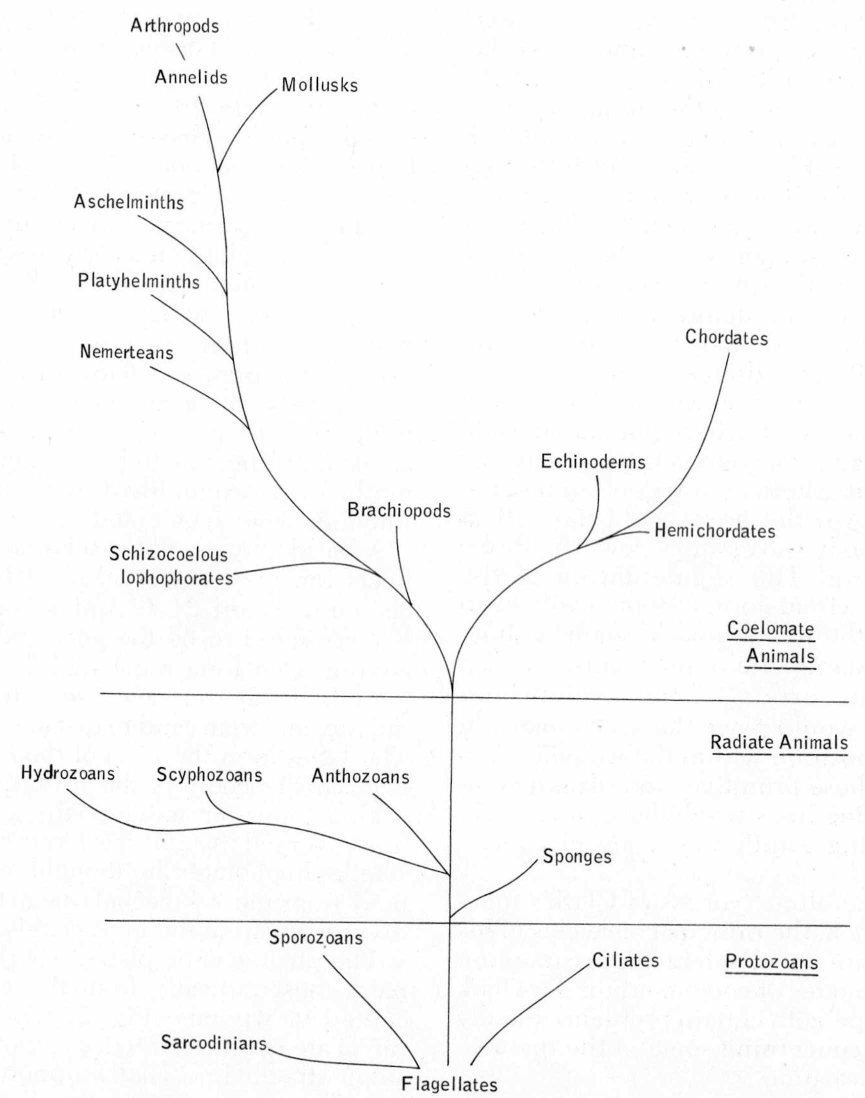

Figure 21–4. Phylogeny of the Animal Kingdom as proposed by Marcus.

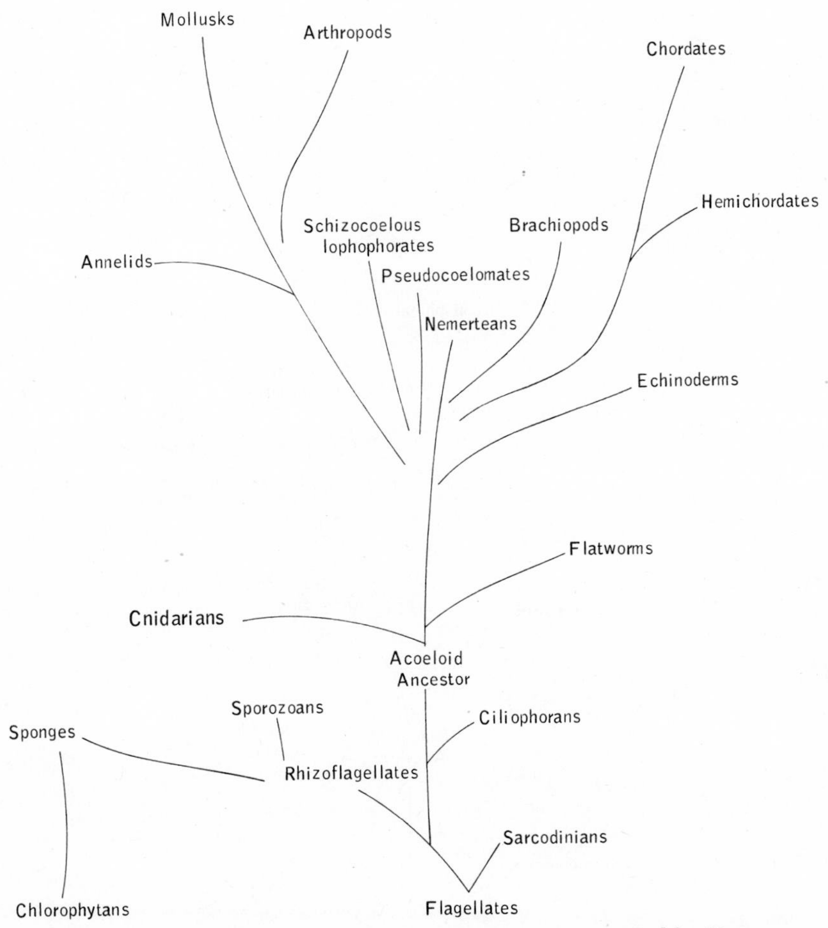

Figure 21–5. Phylogeny of the Animal Kingdom as proposed by Hanson.

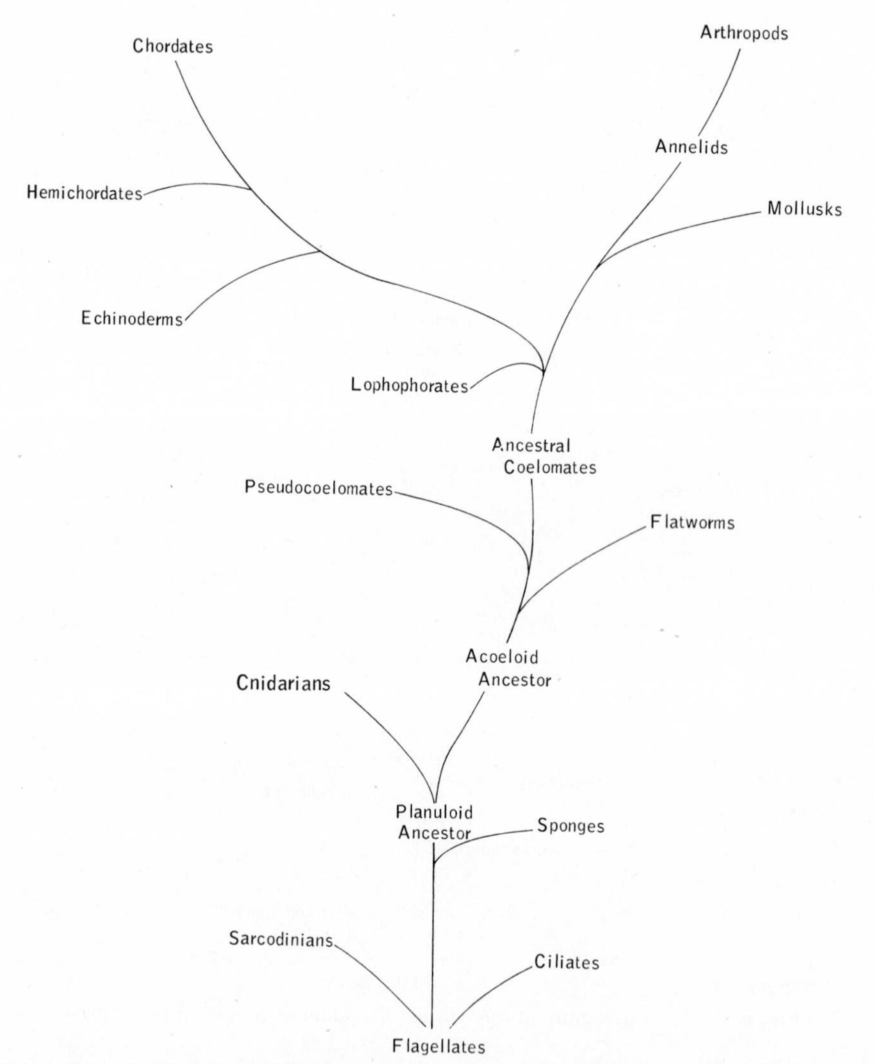

Figure 21-6. Phylogeny of the Animal Kingdom reflecting the views of L. Hyman.

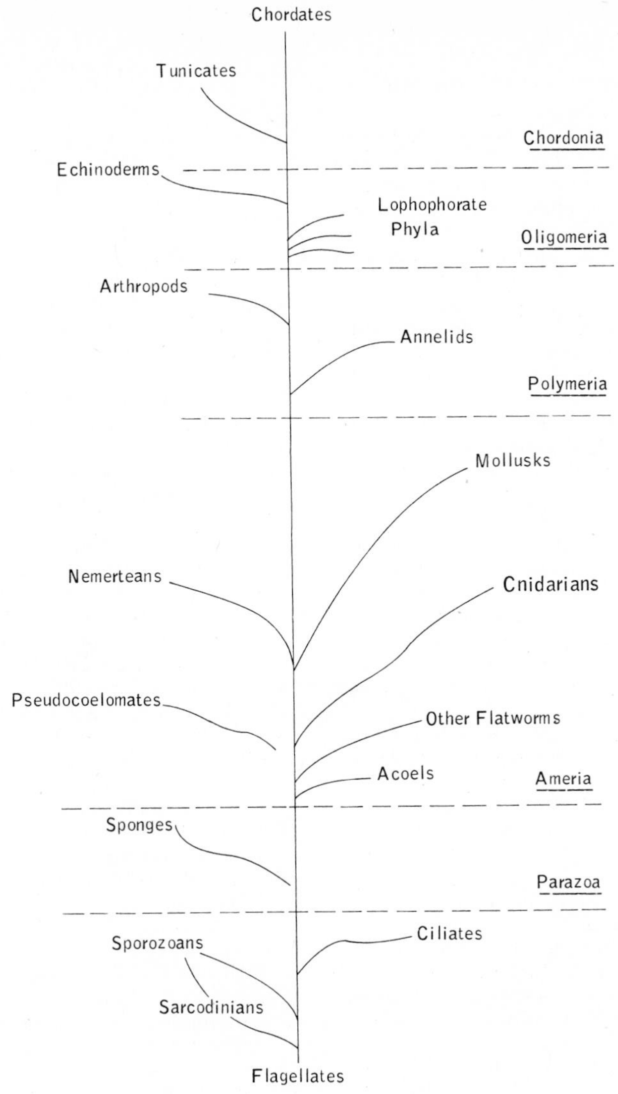

Figure 21–7. Phylogeny of the Animal Kingdom as proposed by Hadži.

have given rise to the Polymeria, which embrace the metameric arthropods and annelids. The Oligomeria, represented by the echinoderms and the lophophorates, evolved through a reduction in segmentation. The chordates, within which metamerism reappears, terminated the line of animal evolution. It should be noted that no affinity is recognized between the annelids and the mollusks, and that the mode of coelom formation is not considered to have phylogenetic significance.

The speculative nature of any attempt to reconstruct an animal phylogeny is obvious. Yet such attempts should not be considered pointless. They have as much value as attempts to fit together any unknown biological pattern. Much of the necessary evidence is still lacking, but hopefully some glimpse of the actual pattern of animal relationships may be obtained in the future.

But beyond the hope of eventual biological truth, there is perhaps another reason for struggling with the riddle of animal phylogeny. Each scheme, no matter how far short of the goal of accuracy it falls, is an attempt to think beyond the isolated evolutionary fragments represented by phyla and classes, and to provide a unified concept and perspective of the entire Animal Kingdom.

Bibliography

Bergh, R. S., 1885: Die Exkretionsorgane der Würmer. Kosmos, Lwow, *17*:97–122.

Clark, R. B., 1964: Dynamics in Metazoan Evolution. Clarendon Press, Oxford. (This volume contains a concise résumé of past and contemporary theories of the evolution of the coelom and metamerism. This résumé is followed by a detailed study of nonappendicular movement in animals, serving as a background for the elaboration of Clark's view on the origin of the coelom and metamerism.)

Clark, R. B., 1969: Systematics and phylogeny: Annelida, Echiura, Sipuncula. *In* Florkin, M., and Scheer, B. J. (Eds.), Zoology, Vol. IV. Academic Press, N.Y., pp. 1–68.

Cuénot, L., 1952: Phylogenese du Règne Animal. *In* Grassé, P. (Ed.), Traité de Zoologie, Vol. 1, Phylogénie. Protozoaires: Géneralites, Flagellés. Masson et Cie, Paris, pp. 1–33.

Dougherty, E. C. (Ed.), 1963: The Lower Metazoa. University of California Press, Berkeley. (A collection of papers from a symposium on the phylogeny and comparative biology of the lower metazoans—sponges, cnidarians, flatworms, nemerteans, pseudocoelomates. The collection contains many papers elaborating or defending contemporary phylogenetic views by the authors or current supporters of these views.)

Faussek, V., 1899: Über die physiologische Bedeutung des Colöms. Trav. Soc. Nat. St. Petersb., *30*:40–57.

Faussek, V., 1911: Vergleichend-embryologische Studien. Z. wiss. Zool., *98*:529–625.

Goodrich, E. S., 1946: The study of nephridia and genital ducts since 1895. Quart. J. Micr. Sci., *86*: 113–392.

Greenberg, M. J., 1959: Ancestors, embryos, and symmetry. Syst. Zool., *8*:212–221.

Hadzi, J., 1953: An attempt to reconstruct the system of animal classification. Syst. Zool., *2*:145–154.

Hadzi, J., 1963: The Evolution of the Metazoa. Macmillan, N.Y. (An exhaustive elaboration and defense of Hadzi's views.)

Haeckel, E., 1874: The gastraea-theory, the phylogenetic classification of the Animal Kingdom and the homology of the germ-lamellae. Quart. J. Micr. Sci., *14*:142–165; 223–247.

Hanson, E. D., 1958: On the origin of the Eumetazoa. Syst. Zool., *7*:16–47.

Hubrecht, A. A. W., 1904: Die Abstammung der Anneliden und Chordaten und die Stellung der Ctenophoren und Plathelminthen im System. Jena. Zool. Naturw., *39*:151–176.

Hyman, L., 1940: The Invertebrates: Protozoa through Ctenophora, Vol. 1. McGraw-Hill, N.Y. (This volume contains a discussion of the Colonial Theory of the origin of the metazoans.)

Hyman, L., 1951: The Invertebrates: Platyhelminthes and Rhynchocoela, Vol. 2. McGraw-Hill, N.Y. (Chapter 1 of this volume discusses the origin of the bilateral animals, the body cavities of animals, and metamerism.)

Jägersten, G., 1955: On the early phylogeny of the Metazoa. The bilaterogastraea theory. Zool. Bidr. Uppsala, *30*:321–354.

Lameere, A., 1932: Origine du coelome. Arch. Zool. Napoli, *16*:197–206.

Lang, A., 1903: Beiträge zu einer Trophocölthoerie. Jena Zool. Naturw., *38*:1–373.

Lankester, E. R., 1875: On the invaginate planula, or diploblastic phase of *Paludina vivipara*. Quart. J. Micr. Sci., *15*:159–166.

Lemche, H., 1959: Protostomian relationships in the light of *Neopilina*. Proc. XV Int. Congr. Zool., pp. 381–389.

Marcus, E., 1958: On the evolution of the animal phyla. Quart. Rev. Biol., *33*:24–58.

Masterman, A. T., 1898: On the theory of archimeric segmentation and its bearing upon the phyletic classification of the Coelomata. Proc. Roy. Soc. Edinburgh, *22*:270–310.

Metschnikoff, E., 1887: Embryologische Studien an Medusen, mit Atlas. A. Holder, Vienna.

Meyer, E., 1890: Die Abstimmung der Anneliden. Der Ursprung der Metamerie und die Bedeutung des Mesoderms. Biol. Cbl., *10*:296–308 (transl. in Amer. Nat., *24*:1143–1165).

Perrier, E., 1882: Les colonies animals et la formation des organisms. Masson et Cie, Paris.

Remane, A., 1950: Die Entstehung der Metamerie der Wirbellosen, Zool. Anz., (Suppl.) *14*:16–23 (See also Dougherty, 1963).

Sarvaas, A. E., 1933: La theorie du coelome. Thesis, University of Utrecht, Netherlands.

Sedgwick, A., 1884: On the nature of metameric segmentation and some other morphological questions. Quart. J. Micr. Sci., *24*:43–82.

Söderström, A., 1925: Die Verwandtschaftbeziehungen der Mollusken. Köhler, Uppsala-Leipzig.

Thiele, J., 1910: Über die Auffassung der Leibeshöhle von Mollusken und Anneliden. Zool. Anz., *35*:682–695.

Ulrich, W., 1950: Vorschläge zu einer Revision der Grosseinteilung des Tierreichs. Zool. Anz., (Suppl.) *15*:244–271.

Valentine, J. W., 1973: Coelomate superphyla. Syst. Zool., *22*(2):97–102.

ACKNOWLEDGMENTS

The following books and articles provided the basis for several illustrations in the text.

Abbott, R. T., 1972: Kingdom of the Seashell. Crown Publishers, Inc., N.Y.

Dörjes, J., 1968: Die Acoela (Turbellaria) der Deutschen Nordsee-Kuste und ein neues System der Ordnung. Z. Zool. Syst. Evolutionsforsch., *6*:56–452.

Galtsoff, P. S., 1964: The American Oyster. Fishery Bull. of the Fish and Wildlife Serv., 64, U.S. Dept. of the Interior.

Gidholm, L., 1965: On the morphology of the sexual stages, mating and egg-laying in *Autolytus* (Polychaeta). Zoologiska Bidrag., Bd. 37, Hft. 1.

Hope, W. D., 1967: A review of the genus *Pseudocella* Filipjev, 1927 (Nematoda: Leptosomatidae) with a description of *Pseudocella trialolaimus* n. sp. Proc. Helminth. Soc. Washington, *34*(1):6–12.

Hoyle, G., 1965: *In* Rockstein, M. (Ed.), Physiology of Insects, Vol. II. Academic Press, N.Y.

Mettam, C., 1971: Functional design and evolution of the polychaete *Aphrodite aculeata*. J. Zool. London, Vol. 163, Pt. 4.

Michel, C., 1970: Rôle physiologique de la trompe chez quatre annelides polychetes: *Eutalia, Phyllodoce, Glycera,* et *Notomastus*. Cahiers de Biologie Marine, *11*:209–228.

Mutvei, H., 1964: Ark. Zool. *16*(2):221–278.

Oekelmann, K. W., and Vahl, O., 1970: On the biology of the polychaete *Glycera alba,* especially its burrowing and feeding. Ophelia, Vol. 8.

Werner, B., Cutress, C. E., and Studebaker, J. P., 1972: Life cycle of *Tripedalia cystophora* Conat (Cubomedusae). Nature (London), Vol. 232.

INDEX

Italic numbers indicate illustrations.